MW01157188

The Safety Professionals Handbook

TECHNICAL APPLICATIONS

Joel M. Haight, Editor
THE PENNSYLVANIA STATE UNIVERSITY

American Society of Safety Engineers
Des Plaines, Illinois, USA

Disclaimer
While the publisher and authors have used their best efforts in preparing this book, they make no representations or warranties with respect to the accuracy or completeness of the contents, and specifically disclaim any implied warranties of fitness for a particular purpose. The information herein is provided with the understanding that the authors are not hereby engaged in rendering professional or legal services. The mention of any specific products herein does not constitute an endorsement or recommendation by the American Society of Safety Engineers, and was done solely at the discretion of the author. Note that any excerpt from a National Fire Protection Association Standard reprinted herein is not the complete and official position of the NFPA on the referenced subject(s), which is represented only by the cited Standard in its entirety.

Library of Congress Cataloging-in-Publication Data

The safety professionals handbook : technical applications / Joel M. Haight, editor.
p. cm.
Includes bibliographical references and index.
ISBN 978-1-885581-53-2 (hardcover : alk. paper)
1. Industrial safety--Management--Handbooks, manuals, etc. I. Haight, Joel M. II. American Society of Safety Engineers.

T55.S215784 2008
658.3'82--dc22

2008005258

Managing Editor: Michael F. Burditt, ASSE
Art Editor: Jeri Ann Stucka, ASSE

Text Design and Composition: Cathy Lombardi
Editors: Nancy E. Kaminski; Sue Knopf; Cathy Lombardi; Publication Services, Inc.
Cover Design: Troy Courson, Image Graphics

Printed in the United States of America on 30% PCW, acid-free paper.

14 13 12 11 10 09 08 7 6 5 4 3 2 1

The Safety Professionals Handbook

Technical Applications

Contents

FOREWORD

THIS BOOK'S 44 chapters are the culmination of several years of work by a very dedicated group of authors, reviewers, editors, and administrators. This was a force of nearly 300 people from all areas of the safety profession. While the writing took over three years to accomplish, the book's historical development also spanned several years. It is the result of the input of over 1100 ASSE members across nearly four years of discussion, planning, strategizing, and topic/subject evaluations. ASSE's Council on Professional Development (CoPD) conceived the idea for such an all-encompassing book in 2000. A task force appointed by the CoPD developed the overall approach and a proposed table of contents. The topic areas are organized such that there is an individual chapter based on each of the following five vantage points: regulatory issues, applied science and engineering principles, benchmarking and performance measurement, cost analysis and budgeting, and best practices. The task force sought and received input through a survey of 7300 ASSE Professional Members from the many areas of specialization in the safety field. ASSE then tabulated and statistically analyzed the survey data to validate changes to the original contents. Once this input was evaluated and the content revised, ASSE's Technical Publications Advisory Committee approved the book's outline and table of contents. The team of authors and technical reviewers was selected from members of the 13 Practice Specialties of the Council on Practices and Standards as well as ASSE members at large.

The Handbook was originally conceived for the purpose of providing a reference for practicing professionals, containing a thorough collection and summary of all that is known in the safety field about each topic area. It is a collection of well-established, well-known, and time-honored information that can help both new and established professionals with the facts and figures needed to complete projects and the day-to-day work in which these professionals (our readers) are involved. An additional purpose was added to the book in mid-2004 with the agreement of the authors present at the 2004 PDC authors' meeting. Since many of the chapters were being written with enough depth and detail to serve as a university textbook, the authors decided that, with some slight changes to the format, the handbook could be made to serve the purposes of both reference handbook for established and new professionals and a textbook for students learning about the safety field for the first time. To support this second purpose, learning objectives were added to each chapter, and ASSE will make available an instructor's guide that will include review questions and answers, exercises, and case studies.

The *Safety Professionals Handbook* represents the diverse, multidisciplinary nature of the safety profession. While each of the 79 chapters in the two volumes captures the vast experience of our authors, each also contains a summary of thoroughly researched, published information from across the profession and pulls it all together in one source. All of the chapters were peer-reviewed, most by three qualified technical reviewers. The authors added their own insights, experiences, opinions, and interpretations of this extensive body of literature for their individual field of expertise. The peer-review process is never easy on authors or reviewers and, quite frankly, most of the time that goes into such a peer-reviewed publication is spent on these reviews and on making the corrections, revisions, additions, and adjustments to the first drafts. This was the case for our book. It was peer-reviewed to ensure that our readers can have confidence in the technical accuracy and the depth and breadth of coverage in the book and are able to gauge their own experience by measuring it against both the summary of the existing, well-known, time-honored, published literature and the validated experience and insight of our authors.

The 79 chapters are grouped into 12 topic areas representing the major areas within the safety profession. These major topic areas are:

- Management of Safety Engineering Work
- Hazard Communication and Worker Right-To-Know
- Environmental Management
- Safety and Health Training
- Workers' Compensation
- Fleet Safety
- Risk Assessment and Hazard Control
- Emergency Preparedness
- Fire Prevention and Protection
- Industrial Hygiene
- Personal Protective Equipment
- Ergonomics and Human Factors Engineering

It was and continues to be the goal of ASSE, in producing the two volumes of the *Safety Professionals Handbook*, to bring the vast body of knowledge available in the safety, health, and environmental arena together in one source that should be found in the office of every safety professional and be easily accessed, read, and understood.

This book also contains two additional sections to enhance its value. One contains the conversions, constants, tables of values, and calculations that safety professionals need when quantifying the hazards they have identified. The other is a foundation chapter covering selected concepts of engineering economics and the financial aspects of safety. This chapter was added to enhance the cost analysis and budgeting chapters throughout the book.

The safety field is a diverse, multidisciplinary, and widely applied field. The only way this type of book could be successfully put together was by inviting input from a large segment of the diverse safety profession. The two volumes bring together the vast number of discipline specialties in the form of 12 topic areas (Industrial Hygiene, Fire Protection, Workers' Compensation, etc.), it also treats each topic area in five different but consistent (regulatory, cost analysis, etc.) ways to explore all the practical aspects of these very different disciplines. The Handbook was written by practitioners, researchers, and academics, and it represents nearly 2000 years of work experience as each author averages roughly 20 years of experience in the field. It is our goal for you, our reader, to both learn something when you read the book and use its content to aid you in performing your very valuable role of protecting people, property, and the environment. We invite your input as you read the book and wish you well in your quest to help keep people safe.

PREFACE

WHETHER YOU NEED to develop and implement a new safety, health, or environmental program or want to improve one of your existing programs, the *Safety Professionals Handbook*, provides SH&E professionals with a complete resource for standards and regulations, applied science and engineering, cost analysis and budgeting, benchmarking and performance criteria, and best practices. The Handbook also explains economic analysis and engineering economy with examples specific to SH&E programs. This peer-reviewed publication developed under the guidance of Dr. Joel Haight is the most comprehensive SH&E resource available. This is an authoritative publication, well referenced and featuring extensive bibliographies throughout.

Attention Instructors: An instructor's guide containing questions and answers for each of the chapters in the *Safety Professionals Handbook* is available at no cost to instructors who have adopted the handbook for a course. You must provide the title and number of the course, the semester/term offered, and the expected enrollment. Please contact ASSE's Customer Service Department for information on how to request a free copy (847-699-2929; email: customerservice@ASSE.org).

Acknowledgments

A BOOK OF this magnitude—ASSE's *Safety Professionals Handbook*—would not be possible if it were not for the extensive input and countless hours of effort by many, many experts and professionals. To our authors, I offer my appreciation for volunteering for this, at times, frustrating but worthwhile experience. Your contribution to the book is recognized and appreciated by ASSE, and its technical value will be recognized and appreciated by our profession as the book makes its way into the hands of professionals in the safety field and students studying to begin a career in the safety field. I emphasize the word "volunteering", so that our readers recognize that you gave of your personal time for the good of the Society and for the good of the profession. Thank you, authors.

Because this handbook content is so widely varied, no one person is expert enough in the field to know and manage all of what is being written in such a book. For this reason, we enlisted the help of 12 topic area coordinators. Each one of these experts managed an individual topic area that most matched their expertise. Without their input and support, this book would not have been possible. Additionally, each one of these topic coordinators authored a chapter themselves. They are recognized in the book with a short bio sketch, but I also would like to thank each one of them here by name. Thank you Adele Abrams, Magdy Akladios, Mike Blayney, Jeff Camplin, Carmen Daecher, Dick Decker, Lon Ferguson, Gayla McCluskey, John Mroszczyk, Jim Ramsay, and Craig Schroll.

Of course, it goes without saying that the authors will be most recognized in the book, as they are the ones who developed the content. However, if it were not for the reviewers and editors providing valuable technical and editorial feedback to our authors, there could be no book. We recognize, in the book, the valuable contribution made by all of our technical and editorial reviewers by name, but I also offer my personal thank you here to this very important group of experts.

To the patient spouses, family, and friends of our authors, I offer my thanks for providing valuable feedback to our authors through informal means as well as for giving up time that you could have been spending with your loved one, but couldn't because they were working on their chapter. Your contribution did not go without notice. Thank you.

I would like to offer a special thanks to our publishing editor Mike Burditt for his professionalism, his patience, his administrative and editorial expertise, and his extensive, knowledgeable input to the writing as well as to the publishing process. Without Mike and the efforts of his very capable and dedicated associate, Jeri Stucka, it would not have been possible to publish this book. Our authors are technical experts, but many are not professional writers, so Mike and Jeri's input was critical in the process of putting together such a significant work where professional writer quality is demanded and necessary. While this book is part of their regular job, they had to do their job while enduring countless delays, many author changes, frequent missed deadlines, as well as sometimes providing a calming influence for frustrated authors who also had multiple other priorities pulling at their available time. They have helped our authors and I make this book one that we can be proud of. Thank you, Mike and Jeri.

Lastly, I would like to thank my wife Janet for her patience in enduring many months of 70- and 80-hour work weeks while I reviewed all of our chapters and for enduring my many frustrations in managing a 4-year endeavor that has finally become this book. I believe that without her support, I would not have been able to continue with this process all the way to successful publication. Thank you, Janet.

I think all of our authors and I have learned much through this process and I hope all of you, our readers, learn just as much through reading and using this important ASSE book.

Joel M. Haight, Ph.D., P.E.

ABOUT THE EDITOR

Joel M. Haight, P.E., CIH, CSP received his M.S. and Ph.D. in Industrial and System Engineering from Auburn University. He is an Associate Professor in Penn State University's Department of Energy and Mineral Engineering. Prior to joining the faculty at Penn State, Dr. Haight worked as a safety engineer for the Chevron Corporation for 18 years. Dr. Haight conducts human error research and intervention effective research in multiple industries. He is a Professional Member of ASSE, AIHA, and the Human Factors and Ergonomics Society, and has authored or co-authored more than 25 peer-reviewed scientific journal articles and book chapters.

ABOUT THE SECTION COORDINATORS*/ CONTRIBUTORS

Magdy Akladios,* P.E., CSHM, CSP, received a Ph.D. in Industrial Engineering from West Virginia University in 1999, an M.S. in Industrial Engineering from West Virginia University in 1996, an M.S. in Occupational Health & Safety from West Virginia University in 1996, and an M.B.A. from West Virginia University in 1995. He is currently an Assistant Professor at the University of Houston–Clear Lake. Prior to that he was a Clinical Associate Professor at Safety and Health Extension, West Virginia University. Dr. Akladios is a Certified Member of the Egyptian Syndicate for Mechanical Engineers and a member in the Industrial Engineers' Honorary Society (Alpha-Pi-Mu).

Susan F. Arnold, MSOH, CIH, is Senior Analyst with The LifeLine Group, Inc., Roswell, Georgia.

Michael B. Blayney,* Ph.D., is the Director of Environmental Health and Safety for Dartmouth College.

Farhad Booeshaghi, Ph.D., P.E., is Adjunct Professor at the Florida State University – Florida A & M University College of Engineering and a managing member and consulting engineer at Global Engineering & Scientific Solutions, LLC.

Craig A. Brown, M.S., P.E., CSP, is an Advisor in Health, Environment and Safety for Chevron.

Brooks Carder, Ph.D., is a Principal in Carder and Associates, LLC, and a Senior Member of the American Society for Quality.

Michael A. Charlton, Ph.D., CHP, CIH, CHMM, CSP, is Assistant Vice President for Risk Management & Safety Environmental Health & Safety Department and Adjunct Associate Professor – Radiological Sciences Program at the University of Texas Health Science Center at San Antonio.

William R. Coffey, CPEA, CSP, is President, WRC Safety and Risk Consultants.

Kathy Jones Council, MPH, MSP, Ph.D. (abd) is completing her dissertation work for her doctorate with a specialization in Safety and Emergency Management at the University of Tennessee. She is a consultant in Health and Safety, Occupational Safety Instructor, and Health Educator with her own business.

Ben D. Cranor, CIH, CSP, is interim Department Head and Assistant Professor in the Department of Industrial Engineering & Technology at Texas A&M University–Commerce.

Jerry Davis, Ph.D., CPE, CSP, is an Assistant Professor in the Department of Industrial and Systems Engineering at Auburn University, Auburn, Alabama.

Dick A. Decker,* P.E., received his Fire Protection Engineering degree in 1967 from the University of Maryland and completed post graduate study in Chemical Engineering at West Virginia University. He is currently Director of Safety at Standard Steel LLC and serves as a consulting Safety Engineer for several companies. Prior to his current position, he served as Safety Director for Cerro Metal Products Company and as a Fire Protection and Safety Design Engineer for Gilbert Associates, Inc. He is a licensed Professional Safety and Fire Protection Engineer in Ohio and Pennsylvania and a member of various professional organizations, including NFPA, SFPE, ASSE, and the Association of Professional Industrial Hygienists. He has presented professional papers and written various professional articles and served on NFPA and ANSI Standards Committees.

David A Dodge, P.E., CSP, is a Senior Consultant with Safety and Forensic Consulting.

Robert L. Edgar, CIH, CET, CHMM, CSP is Health and Safety Manager for BE Services, LLC, in Albuquerque, New Mexico.

David Eherts, Ph.D., CIH, is Vice President of Aviation Safety, EHS and Medical for Sikorsky Aircraft, a subsidiary of United Technologies.

Pamela J. Ferrante, CHMM, CSP, is President of JC Safety & Environmental, Inc.

Jerome F. Fields, M.T., CHMM, is Quality, Health, Safety, Environmental Manager for North America at Atlas Copco Prime Energy, LLC.

Mark A. Friend, Ed.D., CSP, is a Professor and Chair of the Department of Applied Aviation Sciences at Embry-Riddle Aeronautical University, Daytona Beach, Florida.

James G. Gallup, P.E., CSP, is a Senior Fire Protection Engineer for Rolf Jensen & Associates, Inc.

Philip E. Goldsmith, ARM, CSP, is Deputy Chief for Risk Management in the Office of Protection Services, National Gallery of Art, Washington, D.C.

Mark D. Hansen, M.S., P.E., CPE, CPEA, CSP, is Vice President – Environmental & Safety for Range Resources Corporation, Ft. Worth, Texas.

Forrest E. Illing, CIH, is Senior Industrial Hygienist with the Health, Safety and Environmental Services of Bureau Veritas North America, Incorporated.

Carter J. Kerk, Ph.D., P.E., CPE, CSP, is a Professor in the Industrial Engineering Department at the South Dakota School of Mines.

Ken Lewis, M.P.A, CSP, is an adjunct faculty member for Grand Canyon University, and Emergency Management Analyst for Salt River Project in Phoenix, Arizona.

David G. Lilley, Ph.D., D.Sc., P.E., is owner of Lilley and Associates Consulting Engineers, Stillwater, Oklahoma, a company specializing in technical aspects of fire dynamics, fire causation, propagation, and analysis for fire investigation and litigation.

Mohammad A. Malek, Ph.D., P.E., is an adjunct professor of Safety Engineering at the FAMU–FSU College of Engineering, Tallahassee, Florida.

Gayla J. McCluskey* received her M.B.A. in engineering management in 1984 from the University of Dallas and is a Certified Industrial Hygienist, Certified Safety Professional, Qualified Environmental Professional, and Registered Occupational Hygienist. She currently is principal of Global Environmental Health Services and clinical instructor of emergency medicine at Drexel University College of Medicine. She formerly was a managing consultant at Sun Oil Company and director of health, safety, and environmental affairs for North American and Pacific Rim operations at Rhone-Poulenc Rorer. She served for ten years on the Board of the American Industrial Hygiene Association and was its President in 2002 and 2003.

Sheryl A. Milz, Ph.D., CIH, is Assistant Professor at the University of Toledo – Health Science Campus.

John Mroszczyk,* P.E., CSP, received a Master of Science in Mechanical Engineering and a Ph.D. in Applied Mechanics from MIT. Dr. Mroszczyk is a registered Professional Engineer in eight states. He is a member of the American Society of Mechanical Engineers, the American Society of Civil Engineers, the American Society of Safety Engineers, and the National Society of Professional Engineers. Dr. Mroszczyk has been a member of ASSE since 1989. He has served as the newsletter editor, the Assistant Administrator, and Administrator for ASSE's Engineering Practice Specialty. He worked for major corporations such as General Electric before founding his own firm, Northeast Consulting Engineers, Inc., in Danvers, Massachusetts. His work includes design, safety, and consulting services for major corporations, small companies, architects, insurance companies, building contractors, and the legal profession.

Rani A. Muhdi, Ph.D., is Assistant Professor, Department of Engineering Management and Systems Engineering, Old Dominion University, Norfolk, Virginia.

Deborah Imel Nelson, Ph.D., CIH, is the Director of Strategic Initiatives at the Geological Society of America, Boulder, Colorado.

James H. Olds, M.S., ARM, CSP, is the Manager of Safety and Emergency Management for the City of Lakeland, Florida.

Wayne Onyx, P.E., is an Area Senior Vice President of HPR Engineering for Arthur J. Gallagher Risk Management Services.

Steven J. Owen, CET, EIT, is the President of National Code Seminars, Birmingham, Alabama.

William Piispanen, CIH, CEA, CSP, is the Senior Director for Environmental, Safety, and Health for the Washington Division of URS, in Boise, Idaho.

Jon J. Pina, M.S., CSP, is a PA/OSHA Consultant and faculty member in the Safety Science Department at Indiana University of Pennsylvania.

Patrick T. Ragan, M.B.A., CSP, is a Corporate Vice President, Quality, Health, Safety and Environment.

James D. Ramsay, Ph.D., M.A., CSP, is currently the coordinator for the homeland security program at Embry-Riddle Aeronautical University in Daytona Beach, Florida.

James C. Rock, Ph.D., P.E., CIH, is Vice President for Research and Engineering of TUPE, Inc.

Bruce L. Rottner, CSP, is President and Chief Consultant for Scientific Hazard Analysis, Inc.

R. Craig Schroll* has been involved in safety and emergency response since 1972. He began his career as a volunteer and then paid firefighter. He is the president of FIRECON, a consulting and training company with the mission of helping clients prevent, plan for, and control emergencies. Craig serves clients from a wide variety of industries in locations throughout the world. He has a bachelor degree in Fire Science Management from Troy State University and an Associate Degree in Fire Science Technology from the Community College of the Air Force. He is a Certified Safety Professional, Professional Member of both ASSE and SFPE, and belongs to many other organizations, including the NFPA. Craig serves on numerous ANSI and NFPA committees. He is a frequent conference speaker and seminar leader and has been published in numerous professional and trade publications. He is the author of the *Industrial Fire Protection Handbook* published by CRC Press.

Susan M. Smith, Ed.D., MSPH, is Director of the UT Safety Center, and Associate Professor of Safety and Public Health and Safety Programs, at The University of Tennessee.

Kevin E. Stroup, CSP, is a Principal EHS Specialist for Wyeth.

Carol Stuart-Buttle, CPE, is Principal of Stuart-Buttle Ergonomics, Philadelphia, Pennsylvania.

Anthony Veltri, Ed.D., M.S., is an Associate Professor of Environmental Health and Safety at Oregon State University, Corvallis, Oregon.

Stephen J. Wallace, P.E., CSP, is the President of Wallace Consulting Services, LLC, in Washington, D.C.

Montral L. Walker, MSIT, is environmental health and safety representative at Devon Energy in Artesia, New Mexico.

Steven F. Wiker, Ph.D., CPE, is an Associate Professor and Director of the Ergonomics Laboratory, Department of Industrial and Management Systems Engineering, College of Engineering and Mineral Resources at West Virginia University, Morgantown, West Virginia.

TECHNICAL REVIEWERS

The following individuals reviewed one or more chapters in the *Safety Professionals Handbook—Technical Applications*. ASSE and the Editor, Dr. Joel Haight, gratefully acknowledge their contributions.

Mike Behm
Richard Beohm
Earl Blair
Billy Bullock
Srikar V. Chunduri
Sue Cooper
Brian Downie
Terry Eastwood
Rich Fairfax
Joe Fater
Deborah Freeland
Dave Glaser
Greg Green
Ed Grund
Scott Harkins
John Isham
Ashwani K. Gupta
Melina Kinsey
John Lastinger
Steven P. Levine
Bill Marras
Polly McCarty
John Meagher
Gary Mirka
William Moore
Tamara Mueller
Yash Nagpaul
Steve Newell
Michael O'Toole
Judy Reilly
Jim Robbins
Jim Roughton
Roy Sanders
Laurence Smith
Joanne Sullivan
John Taylor
Robert E. Thomas
Pat Thurman
Mary Jane Tremethick
Randy Tucker
Jeff Woldstad

FOUNDATION PRINCIPLES AND APPLICATIONS

Basic Economic Analysis and Engineering Economics

Anthony Veltri and James D. Ramsay

LEARNING OBJECTIVES

- Describe the main motivation for applying economic analysis to occupational safety, health, and environmental affairs.
- Articulate the rationale that supports and the logic that is behind incorporating economic analysis findings into safety, health, and environmental investment proposals.
- Describe the safety, health, and environmental investment strategies available to firms and currently being practiced by firms.
- Characterize what is needed to construct a safety, health, and environmental economic analysis model.
- Describe what will be needed for economic analysis to become an on-going practice within the occupational safety and health profession.

THE ECONOMIC ASPECTS of occupational safety, health, and environmental (SH&E) issues and practices are a timely subject to explore, study, and comprehend. Today, SH&E needs are affecting how business decisions are made, and the needs of business are affecting how SH&E decisions are made. This perspective is expected to dramatically change how proposals for investment in SH&E practices will be put together and presented within an organization's overall investment-allocation process. The primary motivation for applying economic analysis to SH&E investment proposals is to become more competitive when the firm makes decisions about which projects to fund. This indicates that investment allocators will make SH&E investments for the same reasons they make other strategic investments within a firm—because they expect those investments to contribute to economic performance.

Economic analysis was defined by Friedman (1987) as the study of trends, phenomena, and information that are economic in nature. Economic analysis has been used extensively by other internal organizational specialists (i.e., research and development, purchasing, design and process engineering, quality assurance, facility maintenance, operations management, transportation/distribution, and information management). So far, however, SH&E professionals have lagged behind in this effort. The significance of incorporating SH&E in the economic analysis of investment proposals was first recognized over a quarter century ago by Professor C. Everett Marcum, founder and curriculum designer of the West Virginia University graduate degree in SH&E Management Studies. Marcum reasoned in his course lectures that "The design intent (i.e., functionality and form) of a firm's products and technologies, and its operational processes and services, are first expressed by their economic attractiveness; and foremost judged from an economic point of view; and any other features are secondary to the initial economic review."

Bird (1996), in his book entitled *Safety and the Bottom Line*, expressed a similar reasoning in his concept concerning the Axiom of Economic Association. Bird stated that "A manager will usually pay more attention to information when expressed or associated with cost terminology."

These crucial lines of reasoning have generally evaded the practitioners, professors, and students in SH&E management. While they may be well read in the strategic management practices, technical principles, and regulatory aspects that guide decision making and operating actions for the field, practitioners seldom have studied and used the concepts and methods which underlie their economic logic and attractiveness. Most commonly, books, journal articles, and lectures merely mention these in passing.

Concern about analyzing the economic aspects of SH&E issues and practices initially surfaced in the early 1990s and continues today (Henn 1993; Cohan and Gess 1994; Warren and Weitz 1994; Cobas, Hendrickson, Lave, and McMichael 1995; Brouwers and Stevels 1995; Mizuki, Sandborn, and Pitts 1996; Van Mier, Sterke, and Stevels 1996; Lashbrook, O'Hara, Dance, and Veltri 1997; Hart, Hunt, Lidgate and Shankararaman 1998; Timmons 1999; Nagel 2000; Warburg 2001; Adams 2002; Behm, Veltri, and Kleinsorge 2004; and Asche and Aven 2004). During the last fifteen years, there has been a growing need to understand the economic impact that SH&E issues and practices have on competitive performance. Yet the economics of those issues is one of the least-understood subjects in the industry (Tipnis 1994). Increasingly, U.S. firms have taken steps toward better understanding their competitive impact. This trend is evidenced by the development of SH&E sections of national technology roadmaps (Semiconductor Industry Association 1997–1999, The Microelectronics and Computer Technology Industry Environmental Roadmap 1996, and The United States Green Building Council 2003) that incorporate initiatives to reform the way costs linked to SH&E issues and practices are profiled and by the construction and use of various cost-of-ownership (CoO) models (Venkatesh and Phillips 1992, Dance and Jimenez 1995) which have been developed.

Each unit in this volume will have a chapter on cost analysis and budgeting. Therefore, this chapter was developed to advance these efforts by presenting economic analysis as a useful tool for changing how proposals for investment in practices to confront and manage SH&E issues are put together and presented within a firm's overall investment decision-making process. Specifically, this chapter provides (1) a rationale that supports economic analysis and the logic behind incorporating its findings into SH&E affairs and investments, (2) a catalog of SH&E investment strategies available to firms and some currently being used by firms, (3) a blueprint recommended for constructing and using a SH&E economic analysis model, and (4) a summary of elements necessary for economic analysis to become a regular practice in the safety, health, and environmental management profession.

A Rationale for Incorporating Economic Analysis Findings into Safety, Health, and Environmental Investment Proposals

Showing a relationship between investments in SH&E practices and economic performance can be an elusive undertaking (Behm, Veltri, and Kleinsorge 2004). The question that continues to challenge internal organizational stakeholders is "Do investments in practices intended to confront and manage SH&E issues contribute to economic performance?" Many SH&E field practitioners and academics have answered yes (Goetzel 2005, The European Agency for Safety and Health at Work 2004, American Society of Safety Engineers 2002, Jervis and Collins 2001, Smallman and John 2001); however, there is no compelling research that provides a *definitive* financial answer. Many internal stakeholders say no (Asche and Aven 2004, Dorman 2000, Shapiro 1998). They are very skeptical about how SH&E economic analyses are conducted; specifically, they question how cost and potential profitability data are collected, calculated, analyzed, interpreted, and reported. The reality may be that SH&E investments do not routinely set up opportunities to make money. At the same time, the opposite stance that SH&E investments seldom provide a financial payoff is also inaccurate. There should be no denying that investing in practices to confront and manage SH&E issues has always been a complicated proposition with very real methodological issues and economic implications. Even so, most firms invest in SH&E practices despite their economic impact, but they should do so knowingly.

Typically, concern for SH&E performance and economic performance have been viewed as separate lines of attack operating independent of and usually in opposition to one another. However, the actual *interdependence* between these concerns increasingly highlights the need for showing some type of an economic relationship. Generally, SH&E professionals have not incorporated economic analysis as a way of showing how investments in these practices contribute to economic performance (Behm, Veltri, and Kleinsorge 2004). As a result, left out of the firm's competitive business strategy and excused from internal stakeholder expectations that this function justify its internal and external affairs with an economic perspective, SH&E practices tend to be looked at as a necessary cost of doing business, with little economic payback expected (Veltri, Dance, and Nave 2003). To say the least, this is not a viable perception for internal stakeholders to bring and hold onto during the investment-allocation process. Only a focus on the results of economic analysis can provide internal stakeholders with the necessary information to set investment-allocation priorities. The emphasis on the results of economic analysis should not be interpreted to mean there is any intention to deemphasize the importance of ensuring compliance with regulatory mandates. Concern for compliance surely exists, as it rightly should, and employing economic analysis is not intended to replace compliance applications. However, to focus *only* on maintaining compliance with SH&E regulations should not be expected to yield positive financial returns. Alternatively, what we attempt to accomplish with economic analysis is to go beyond compliance in ways that provide pertinent quantitative and qualitative economic information about how a firm's organizational activities (i.e., products, technologies, processes, services) tend to create SH&E issues and how strategic investments in innovative practices to confront and manage these issues might offer financial opportunities and reduce liability.

As a rule, the investment decision-making process hinges on a firm's competitive strategy, its research and development capability, its technology wherewithal, and the human means to productively use and protect organizational resources. The analysis used to reach investment-allocation decisions tends to be heavily slanted toward economic aspects. How well economic analyses are conducted and how well analysis findings support a firm's competitive strategy will usually affect how investments are allocated within a firm. During the last 25 years, existing and emerging SH&E issues (e.g., occupational injuries and illnesses, environmental incidents, natural and man-made hazardous exposures, tough government regulatory requirements, pressure from nongovernment interest groups concerning sustainable resource development and use, and long-term contingent liabilities as a result of past operations) are also increasingly affecting how decisions to fund projects are made within a firm. The real dilemma facing financial decision makers is how can investment choices to confront and manage SH&E issues be made in the absence of sound quantitative economic information. Without economic analysis results that detail the estimated cost and potential profitability of investments, even the most zealous SH&E internal stakeholders are left without a means to objectively make fiscally prudent investment-allocation decisions.

The following are beneficial outcomes that should be expected and leveraged when SH&E economic analyses are effectively conducted (Veltri 1997).

1. A refined understanding of the products, technologies, processes, and services that tend to drive SH&E life-cycle costs.
2. A more complete and objective data set on life-cycle costs and profitability potential of SH&E investments, enabling improvements to product, technology, process, and service designs.

3. An enhanced way of determining which SH&E management strategies and technical tactics to pursue and what level of investment will be required.
4. A new investment analysis structure in which fashioning SH&E issues and practices affects how business decisions are made, and in which business needs affect how SH&E decisions are made.

Despite these leveraging opportunities, there are usually internal organizational barriers to overcome when applying economic analysis to SH&E investments. The following are a sample of internal perceptions that SH&E professionals should be expected to confront:

1. An operations-level perspective that SH&E issues linked to the firm's processes are primarily regulatory-compliance-based and play a very small part in the investment-allocation process of the firm.
2. A design engineering-level perspective that the existing strategy and methodology for performing economic analysis of SH&E issues and practices that affect new product, technology, and process designs are qualitatively and quantitatively immaterial for enhancing design changes.
3. A senior-level executive perspective that proposals for investments in practices to counteract SH&E issues affecting the firm are not financially structured and reported in a manner that allows them to compete with other investment-allocation alternatives.

Such internal organizational barriers can be significant and must be overcome so that SH&E proposals can compete for the firm's investment dollars. The strategy considered most effective in overcoming these barriers is to employ economic analysis in a manner that discloses both the internal and external SH&E-related costs throughout the productive/economic life cycle of a firm's existing, new, and upgraded organizational activities and reveals the financial impact that investments in SH&E practices have on these organizational activity designs.

Available SH&E Investment Strategies

SH&E professionals who wanted to better understand how investment allocation decisions are made have had to satisfy themselves with professional literature that is nonobjective and fragmented with piecemeal approaches, causing them to be disadvantaged during the investment-allocation process. It is imperative that the forward-thinking SH&E specialist, who is interested in making his/her firm more competitive and in advancing his/her own career, understand how the firm makes strategic investment decisions and how it views investment utilization. However, to accomplish this, the SH&E specialist has to first understand the type of investment strategy being used by his/her firm. Figure 1 offers such a framework by providing a catalog of typical SH&E investment strategies that are available to firms or are already being practiced by firms. The framework is a derivative work and borrows heavily from other strategy typologies (Miles and Snow 1978; Porter 1980; Adler, McDonald, and MacDonald 1992; Roome 1992; Schot and Fischer 1993; Welford 1994; Chatterji 1995; Ward and Bickford 1996; Epstein 1995; Day 1998; Brockhoff, Chakrabarti, and Kirchgeorg 1999; Stead and Stead 2000; and Coglianese and Nash 2001).

Each of these levels represents a distinct strategy for how a firm typically makes strategic SH&E investment decisions and how they tend to view investment utilization. Together they represent a way of thinking about SH&E investments that goes beyond existing investment strategies, which are at a distinct disadvantage when competing with the firm's other investment options. Investment allocators are usually reluctant to accept qualitative estimates (i.e., compliance audits performed, behavior-based training provided, perception surveys) when deciding to invest in SH&E investment proposals. They prefer quantitative estimates (i.e., cost and profitability potential).

Internal organizational stakeholders constantly face investment choices among alternatives that are linked specifically to changes in the firm's organizational activities. They may have to decide whether to continue or drop a product or service, acquire certain technologies, or reengineer a process. Generally,

Levels of Investment Strategy

Level 1 Reactive: Posture is to invest only when required, with attention to responding to government directives or insurance carrier mandates

Level 2 Static: Posture is to invest cautiously, with specific attention on preventing occupational injuries, illnesses, and environmental incidents from occurring

Level 3 Active: Posture is to invest assertively, with major emphasis on reducing risk to existing operations, and to lower contingent liability resulting from past operations

Level 4 Dynamic: Posture is to invest strategically, with major emphasis on counteracting the life-cycle risk and cost burdens linked to the firm's organizational activities

Level 1 Reactive

Strategy for financing the firm's SH&E investments at this level can be characterized as a reactive and resistive arrangement. Access to financial resources is based solely on correcting violations cited by government regulatory agencies and mandates from insurance carriers. Additional financial resources needed for providing technical day-to-day SH&E services are provided when it financially suits the company. Tools for performing economic analysis of SH&E issues do not exist because the firm does not want to, does not think it needs to, or is not aware of the potential cost impact of failing to counteract these issues.

Level 2 Static

Strategy for financing the firm's SH&E investments at this level can be characterized as an informal arrangement. A mentality of funding only as much as others in their industry sector is strongly adhered to. An informal pay-as-you-go funding mentality exists; invest to counteract issues only when trying to reduce the outlays associated with injury/illness and environmental incidents. Investments undertaken for preventing occupational injuries, illnesses, and environmental incidents and to meet compliance with regulations generally do not compete for access to financial resources. However, access to financial resources needed to confront and manage more technically discriminating SH&E issues depends upon the capabilities of the firm's SH&E professionals to assemble internal coalitions of support in order to compete for funding. These technically discriminating prevention initiatives tend to have no clear criteria and pattern of funding, thus subjecting them to unpredictable funding outcomes. Tools for performing economic analysis of SH&E investments are considered by internal organizational stakeholders to be qualitatively and quantitatively immaterial for competing with other investment allocation decision alternatives. SH&E cost accounting practices focus on aggregating cost data, causing costs to be hidden in general overhead accounts rather than included throughout the life cycle of the product, service, technology, or process responsible for their generation. As a result, integrated and concurrent design engineering decision-making capabilities required for aggressively controlling SH&E costs are limited and incomplete.

Level 3 Active

Strategy for financing the firm's SH&E investments at this level can be characterized as an applied arrangement. Access to financial resources tends to be allocated when investment requests are intended to reduce risk to products, technologies, processes, and services; enhance compliance with regulatory standards; reduce contingent liability caused by past operations; and minimize outlays associated with accidents, environmental incidents, lawsuits, and boycotts. The funding level tends to be above others in their industry sector and included in the overall budget of the core business units obtaining the services. Tools for performing economic analysis of SH&E investments are chiefly focused on cost-benefit analysis and payback, and sometimes internal rate of return. Costs are accumulated either through the use of cost accounting systems or through the use of cost-finding techniques and are reported on a regular basis for management information purposes. The cost of incidents are charted and charged back to core business units and incorporated into the firm's budget process. However, profiling the cost and profitability of SH&E issues affecting the organizations products, technologies, processes, and services, and integrating cost information into decision-making, does not occur. This condition results in senior-level executives looking at SH&E issues as nonbusiness issues.

Level 4 Dynamic

Strategy for financing the firm's SH&E investments at this level can be characterized as being self-sustaining and a down-to-business arrangement. A strategically opportunistic funding position is taken; this means having sufficient funding for the long-term, while having the financial wherewithal to remain flexible enough to solve new issues and support research and development and other opportunities for innovation that, over time, will lead to significant SH&E performance gains while advancing measurable business goals. Business strategies and SH&E changes are tightly interwoven; changes in products, technologies, processes, and services affect SH&E, and changes in SH&E issues and practices in turn force product, technology, process, and service changes. Access to financial resources and capital is approved for 3 years (typically related to potential business contribution over the long and short term) and is based on factors and circumstances that are causing the firm to fail in its efforts to protect and use resources productively and on conditions/circumstances under which SH&E pays. Senior-level financial executives desire SH&E strategy and activities to become financially self-sustaining and contribute measurably to company competitiveness. Tools for performing economic analysis of SH&E investments provide reliable and timely information on the full cost burdens associated with the firm's products, technologies, processes, and services over their productive and economic life cycle. Major thinking is performed on how to enhance the efficiency and effectiveness of SH&E spending.

FIGURE 1. Available safety, health, and environmental investment strategies

making these decisions requires conducting economic analyses that provide cost and profitability comparisons among mutually exclusive alternatives (i.e., accepting one alternative means not accepting others). Likewise, investment decisions about SH&E practices are also choices between alternatives that are mutually exclusive and linked to the changes in the firm's organizational activities (e.g., products—substituting regulated occupational safety and health resource inputs with unregulated and perhaps less harmful ones, technologies—employing new environmental toxicity monitoring and detection systems, processes—reengineering to eliminate process waste from resource outputs, or services—modifying supply-chain relationships related to SH&E practices). This results in investment-allocation decisions that are usually based on the direct result of the projected economic impact of the mutually exclusive alternatives under analysis. Economic analysis, then, provides a recommended approach to how one might best present proposals for SH&E investments where economic aspects dominate and drive decision making and where economic effectiveness and efficiency are the criteria for choosing which SH&E issues to confront and manage and in which alternative solutions to make selected investments.

An abridged life-cycle costing method, which features net-present-value financial analysis, is the recommended tool for constructing a SH&E economic analysis model. The rationale for this abridged approach is that internal stakeholders have questioned both the relevance of the full life-cycle costing methodology for the actual business decisions they must make and the efficacy of the full methodology for making business decisions in real time. As a result, most firms are encouraging their SH&E professionals to develop and use a more streamlined method that focuses on internal private costs (i.e., costs incurred from organizational activities that result in product-yield quality and process logistical performance problems, injury/illness and environmental incidents, and liability) rather than on external societal costs (i.e., costs incurred as a result of organizational activities that cause pollution of air, water, or soil; natural resource depletion or degradation; chronic or acute health effects; alteration of environmental habitats; and social/economic–welfare effects) to make the economic analysis more relevant and useful for business decision making.

Several abridged life-cycle assessment methods have been described in the literature (Graedel, Allenby, and Comrie 1995), ranging from primarily qualitative approaches to quantitative ones in which expert judgment, a limited scope, and a system boundary keep the life-cycle assessment effort manageable. Experience demonstrates that life-cycle assessment for a complex manufactured product or an industrial manufacturing process works most effectively when it is done semiquantitatively and in modest depth. Unlike the full life-cycle assessment method, an abridged method is less quantifiable and less thorough. It is also quicker and more practical to implement. An abridged assessment will identify approximately 80 percent of the useful SH&E actions that could be taken in connection with corporate activities, and the amounts of time and money consumed will be small enough that the assessment has a good chance of being carried out and its recommendations implemented. The foundation for the abridged architecture was based on the unabridged life-cycle framework developed by the Society of Environmental Toxicology and Chemistry (SETAC 1991).

Present-value financial analysis provides the final link in the architecture. As a dollar today is always preferable to a dollar tomorrow, the sheer nature of SH&E initiatives at the work site often take multiple fiscal years to become fully realized. That is, dollars spent today on SH&E programs and activities may not reap or return benefits to the firm for several years. Hence, present-value financial analysis provides the most reliable means of comparing the financial performance of mutually exclusive alternatives when said alternatives fully mature in subsequent fiscal years (Newman 1983). In this way, present-value financial analysis helps to delineate the long-term financial impact of SH&E investments by presenting the after-tax cash flow and the present-cost value of the investment over a sufficient time horizon. The rationale for using net-present-value financial analysis is that many of the traditional financial analysis techniques employed by SH&E professionals, such as payback and rate of return on investment, fail to take the time value of money into consideration. Although useful tools in the financial analysis of investment decisions, exclusive use of these methods can result in making incorrect decisions, such as accepting SH&E project proposals that lose money, or conversely, rejecting SH&E project proposals that may represent financial opportunities and may reduce contingent liability.

Figure 2 provides an architecture for the SH&E economic analysis model. The architecture is built around three stages: (1) defining and setting the boundaries necessary for managing the economic analysis; (2) conducting an abridged life-cycle inventory analysis and impact assessment of existing SH&E issues and proposed alternatives, from upfront analyses and the acquisition of capital and permits through resource and material use, disposal, and closure; and (3) conducting postimplementation reviews to ensure that the results of implemented solutions are in reasonable agreement with the estimated projections.

Stage I: Defining SH&E Economic Analysis Strategy and Boundaries

Defining the SH&E economic analysis strategy and setting its structural boundaries are key aspects of the economic analysis. As outlined in Figure 2, this initial

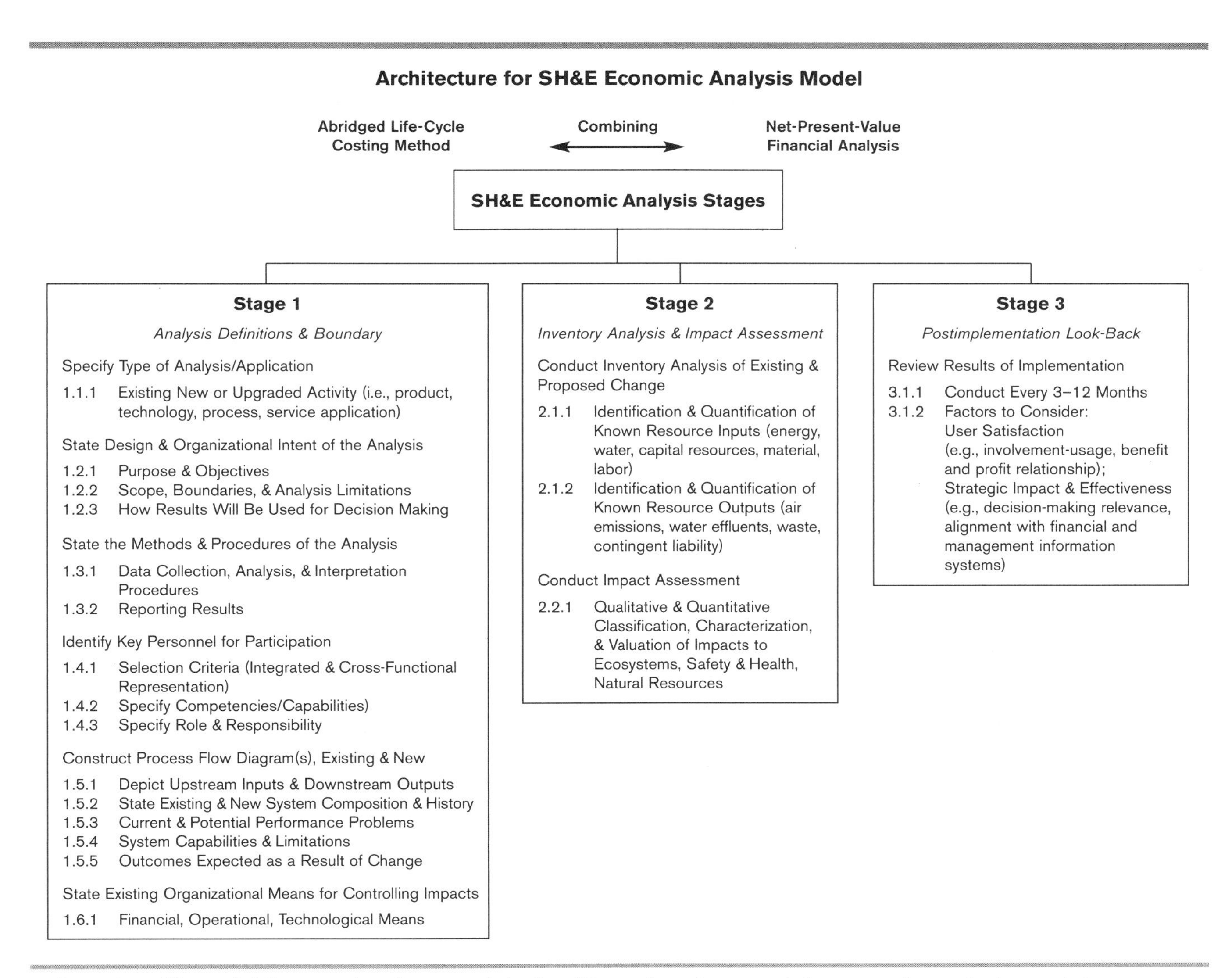

FIGURE 2. Blueprint for constructing a safety, health, and environmental economic analysis model

stage should be accomplished through the following five steps. First, the SH&E professional should consider specifying the type of analysis to be conducted, specifically attending to the following components: (a) a description of the existing, upgraded, or new product, technology, process, or service system; (b) the system's expected economic life (i.e., the equivalent of the estimated amount of time that investments in the system can be expected to have economic value or productive uses and the estimated amount of time recurrent savings and reduced contingent liability can be achieved without having to reinvest as the initial investment ages); (c) the firm's hurdle rate (i.e., the required rate of return in a discounted cash-flow analysis that the firm is using for judging investment proposals); and (d) the existing and potential SH&E issues and impacts (e.g., musculoskeletal disorders resulting in workers' compensation claims, CO_2–NOX emissions resulting in global warming potential acidification) that are linked to the firm's activity under analysis.

The second step is to keep the analysis on course and focused. This requires that the design and organizational intent be stated up front. Possible components include: purpose and objectives, key assumptions and analysis limitations, and how information will be used to drive decision-making capabilities.

The third step is to specify the methodology suggested for performing the analysis (i.e., data collection, analysis and interpretation, and reporting procedures). These should be transparent and stated at the outset.

The fourth step is to identify and empower an integrated SH&E economic analysis project team and assist them in carrying out the study. Individuals on this team must be utilized as supportive personnel in order to carry out the project. Note that it is absolutely essential that their assistance be requested and used. Of course, they must be provided with the advisement and encouragement they will need to perform as expected. The team should be cross-functional in makeup and possess skills in finance, design and process engineering, operations, facility management, procurement, legalities, SH&E affairs, and community relations.

The fifth component of this stage is to construct process flow diagrams of the existing organizational activity and the proposed solution change. The process flow diagram should depict upstream inputs and downstream outputs, the existing and new system composition and history, current and potential performance problems, existing and new system capabilities and limitations, and any beneficial outcomes expected as a result of the change.

Stage II: Inventory Analysis and Impact Assessment

Figure 2 shows that the main function of stage two is to conduct an inventory analysis (i.e., the identification and quantification of known resource inputs such as energy, water, capital, resources, materials, and labor, and known outputs such as air emissions, water effluents, waste, and contingent liability) and an impact assessment (i.e., qualitative and quantitative classification, characterization, and valuation of impacts to ecosystems, human safety and health, and natural resources based on the results of the inventory). It is also sensible to provide investment-allocation decision makers with estimates of the firm's ability and means to control or improve the existing SH&E issue. This will add an additional level of robustness to the analysis. Chief factors to assess should include the firm's (1) financial funding capability (i.e., the existing level of funding available to control or improve the SH&E issue: a high level of funding suggests that the firm has the financial means to effectively control or improve the issue, whereas a low level of funding suggests that the firm has little financial means to affect the issue in the immediate future); (2) human operational capability (i.e., the existing level of human operational wherewithal to control or improve the SH&E issue: a high level of wherewithal suggests that the firm has the human means and capability to control or improve the issue, whereas a low level suggests that the firm has little human operational means to affect the issue in the immediate future); and (3) available technology (i.e., the existing level of technology that is to control or improve the SH&E issue: a high level of available technology suggests that the firm can utilize technology as a way to control or improve the issue, whereas a low level suggests that the firm has little technological means to affect the issue in the immediate future).

The use of impact models (e.g., risk and economic) helps guide the decision making and the operating actions that are necessary for keeping the inventory analysis and impact assessment structured and gives a picture of the life-cycle process-flow inputs and outputs linked to the organizational activity under analysis. In addition, it provides investment-allocation decision makers with an understanding of the extent and magnitude of the issue. A large number of risk assessment and analysis models and documents are available for profiling risk impacts and contingent liability linked to the firm's organizational activities.

Stage III: Postimplementation Review

After investing in SH&E practices, it is very important to determine the degree to which the results of the implemented changes are in reasonable agreement with the estimated projections. For example, if a new technology was purchased because of potential reductions in cost and contingent liability, it is important to see if those benefits are actually being realized. If they are, then the economic analysis projections would seem to be accurate. If the benefits are not being obtained, a review to discover what has been overlooked should be performed. A post-implementation assessment helps uncover the reasons why targets were not met. One possible reason could be that economic projections may have been overly optimistic. Knowing this can help analyzers avoid mistakes in economic cost projections in the future. In order to ensure that economic calculations and cost projections are realistic, everyone involved must know that a review of results will take place. Therefore, three to twelve months after a mutually exclusive alternative has become operational, and regularly afterwards, a postimplementation review should be conducted. Factors to be considered in the look-back should include: user satisfaction (i.e., involvement/usage or cost/profit relationship) and strategic impact and effectiveness (i.e., decision-making relevance, alignment with financial and management information technology systems, and organizational objectives).

An economic analysis of a SH&E investment proposal collects cost information associated with the inventory analysis and impact assessment and uses a financial analysis measure for understanding economic impact. The SH&E professional is cautioned that an economic impact analysis of a SH&E investment proposal is only as accurate as the cost information that it collects—quite literally, the euphemism "garbage in, garbage out" applies! In this sense, SH&E professionals are encouraged to work with their finance and accounting colleagues as estimates for necessary costs are obtained. Therefore, a major component of an economic analysis is gathering data to make reasonable estimates of cost. Appendix A provides an outline of usual as well as potentially hidden SH&E life-cycle cost factors and activity drivers that are typically linked to a firm's organizational activities. Estimated costs are referred to in SH&E economic analysis as incremental costs; they are the difference between the after-tax cash flow of the mutually exclusive alternative(s). Net present value (NPV) analysis is the most applicable financial measure for understanding economic impact because it provides the most reliable method for comparing the financial performance of mutually exclusive alternatives on the basis of their projected after-tax cash flows. Net present value analysis can be thought of as the present value of an investment's future cash flows minus the initial investments required to initiate a particular program (or set of programs). Conventional NPV decision-making rules indicate that projects with profitability indices (PI) of greater than 1.0 should be pursued. When comparing multiple project alternatives, those with higher PIs are understood to be financially more attractive than those with lower, albeit positive, PIs. Alternatively, projects with a PI of 0 will recuperate only the cost of the resources required to make the investment, and conversely, projects with a negative PI represent a financial loss for the investment. Because investment decisions in SH&E are important, proposals should also be supplemented with qualitative information, such as how the investment is expected to maximize sustainable resource development and use practices, enrich the quality of management information, develop human competency and capability, lower contingent liability, maintain regulatory compliance, reduce nongovernment special interest group concerns, and enhance organizational reputation. This type of qualitative information is sufficiently important that it could influence a decision to fund the investment proposal, in spite of the fact that the proposal may not meet the firm's established hurdle rate (i.e., the required return on a discounted investment).

Many firms discontinue their economic analysis after identifying and quantifying resource inputs and outputs. They simply decide to reduce the amount of resource inputs and outputs, taking on a "less is best" strategy rather than investing in the effort necessary for assessing the estimated economic impact. At times, because of the data requirements of impact assessments, it is difficult to relate inventories to an impact analysis and to provide cost and profitability estimates necessary for advancing investment-allocation decision making beyond what has already been collected in the inventory analysis. On the other hand, making an effort to conduct at least a relative impact assessment should provide investment allocators with information that is more meaningful for decision making. For instance, stating the firm's contingent liability (i.e., an estimate of the firm's probability of an accident/incident occurring and the range of cost and economic impact) resulting from increased use and disposal of toxic chemicals is just as easy to understand and assess as providing the change in the reduced level of a chemical input use and/or output waste that was identified and quantified in the inventory analysis. Also, when determining the relative impacts using only inventory analysis, the information provided is limiting when investment-allocation decisions must be made. For example, when the exposure to gases emitted is estimated to be higher for the existing process technology, and the exposure to gases of a different pollutant is also estimated to be higher, which mutually exclusive alternative is preferable for reducing contingent liability and what is the economic impact in terms of cost and profitability potential for making a change? An impact assessment provides investment decision makers with additional information to make such choices. This can be best accomplished by identifying the high risk and cost factors that were linked to the existing situation and performing sensitivity analysis so that the effects of certain changes can be studied and forecast. Using this strategy, benefits and costs are reported in monetary terms and can be estimated over the full life cycle of the product, technology, process, or service under analysis. In addition, a risk and cost impact analysis should be conducted on the countermeasure options to ensure that they do not create additional risk and cost impacts that negate their estimated improvement.

SUMMARY

Questions and uncertainties related to SH&E issues, practices, and investments tend to create business challenges for a firm's internal stakeholders. It is crucial to understand the existing circumstances that drive these issues, their impacts, and their costs. Knowing how to allocate the investment outlays necessary for confronting and managing these issues and how to evaluate the efficacy of those investment outlays is imperative. Most organizations do not understand which products, technologies, processes, or services provide more or less value comparative to their existing SH&E costs. Traditional SH&E costing systems tend to suffer from imprecise cost collection, poor analysis and interpretation procedures, and distorted cost reporting. They offer little transparency of what comprises their costs, fail to consider the financial returns that can be expected later from the investment, and thus lose their decision relevance. It is now time that traditional approaches for justifying SH&E investments yield to a newly fashioned and more economically valued way of thinking.

The last 25 years have shown that changes in a firm's products, technologies, processes, and services are interconnected with its SH&E practices: changes in the firm's products, technologies, processes, and services affect SH&E, and SH&E issues in turn force design changes in the firm's products, technologies, processes, and services. When internal stakeholders are first presented with this connection, many refuse to think about it as an economic opportunity. Viewing it as an additional annoying cost or another regulatory threat, they see it as a move to negotiate a trade-off between operations-related costs and costs related to SH&E practices. This is not the case. In fact, what you will be able to show internal organizational specialists is that you can set acceptable SH&E performance criteria and then compare the life-cycle cost of ownership for mutually exclusive alternatives that meet or exceed those criteria. The comparative approach will provide them with an improved way of deciding between alternative methods for meeting a specific set of criteria. By looking at investments through the SH&E lens, and looking at SH&E issues and practices through a business lens, internal organizational specialists can derive insights that would otherwise go unnoticed.

The use of economic analysis techniques in SH&E is really in its infancy. Economic analysis is not a core accreditation requirement of the Applied Sciences Commission of the Accreditation Board of Engineering and Technology (ABET) which accredits safety programs (ABET 2006). Indeed, the use of economic analysis as applied to SH&E investments currently has reached a point somewhere between the stage of understanding the factors that drive SH&E costs and the stage of using that information for assessing economic impact. Any continued developments in this area will require that SH&E professionals put together investment proposals that are based on sound economic analysis and creatively use the results of the analysis for estimating how countermeasure strategies offer opportunities for reducing costs and enhancing revenues. Like any new concept, economic analysis in SH&E will tend to go through a predictable life cycle. First, the concept will become increasingly appealing to SH&E professionals as a way to enhance the acceptance of investment proposals and make the business case for SH&E issues and practices. Next, firms will tend to hire outside "experts" with SH&E economic analysis backgrounds to help their internal specialists design a SH&E economic analysis model that is congruent with the way they operate their business and to help pave the way for and steer the use of the model. When the model and its use become functional, SH&E specialists will take over. As the SH&E economic model continues to mature, firms will integrate it into their investment-allocation process. At this point, a firm's ability to respond to SH&E issues associated with its products, technologies, processes, and services with appropriate and economically justified SH&E countermeasures may well become a leading indicator of its competitive advantage in the marketplace.

References

Adams, S. "Financial Management Concepts: Making the Bottom Line Case for Safety." *Professional Safety*, August 2002, pp. 23–26.

Adler, P. S., D. W. McDonald, and F. MacDonald. 1992. "Strategic Management of Technical Functions." *Sloan Management Review* 33(2):19–37.

American Society of Safety Engineers (ASSE). 2002. White paper addressing the return of investment for safety, health, and environmental management programs. Des Plaines, IL: ASSE.

Applied Sciences Commission of the Accreditation Board of Engineering and Technology (ABET). 2006 [online]. Accessed 4/03/06. www.abet.org/

Asche, F., and Terje Aven. "On the Economic Value of Safety." *Risk, Decision and Policy* (July–Sept. 2004) 9(3):283–267.

Behm, M., A. Veltri, and I. Kleinsorge. "The Cost of Safety." *Professional Safety*, April 2004, pp. 22–29.

Bird, F. E. 1996. *Safety and the Bottom Line*. Logansville, GA: Febco.

Brockhoff, K., A. K. Chakrabarti, and M. Kirchgeorg. 1999. "Corporate Strategies in Environmental Management." *Research Technology Management* 42(4):26–30.

Brouwers, W., and A. Stevels. 1995. "Cost Model for the End of Life Stage of Electronic Goods for Consumers." Proceedings of the 3rd International Symposium on Electronics and the Environment, Dallas, Texas, pp. 279–284.

Chatterji, D. 1995. "Achieving Leadership in Environmental R&D." *Research Technology Management* 38(2):37–42.

Cobas, E., C. Hendrickson, L. Lave, and F. McMichael. 1995. "Economic Input/Output Analysis to Aid Life Cycle Assessment of Electronics Products." Proceedings of the 3rd International Symposium on Electronics and the Environment, Dallas, Texas, pp. 273–278.

Coglianese, C., and J. Nash. 2001. "Bolstering Private Sector Environmental Management." *Issues in Science and Technology* 17(3):69–74.
Cohan, D., and D. Gess. 1994. "Integrated Life-Cycle Management." Proceedings of the 2nd International Symposium on Electronics and the Environment, San Francisco, California, pp. 149–154.
Dance, D., and D. Jimenez. "Cost of Ownership: A Tool for Environment, Safety and Health Improvements." *Semiconductor International*, September 1995, pp. 6–8.
Day, R. 1998. "The Business Case for Sustainable Development." *Greener Management International* 23:69–92.
Dorman, P. Three Preliminary Papers on the Economics of Occupational Safety and Health, Chapter 3, "Investments in Occupational Safety and Health." April 2000, International Labour Organization, Geneva.
Environmental Protection Agency (EPA). 1995. *An Introduction to Environmental Accounting As A Business Management Tool: Key Concepts and Terms*. Washington D.C.: Office of Pollution Prevention and Toxics.
Epstein, M. J. 1995. *Measuring Corporate Environmental Performance*. New York: McGraw-Hill.
European Agency for Safety and Health at Work. 2004. *Quality of the Working Environment and Productivity – Research Findings and Case Studies*. Belgium: European Agency for Safety and Health at Work.
Friedman, J. 1987. *Dictionary of Business Terms*. Hauppage, NY: Barron's Educational Series, Inc.
Goetzel, R. Z. Policy and Practice Working Group Background Paper. "Examining the Value of Integrating Occupational Health and Safety and Health Promotion Programs in the Workplace." Steps to a Healthier U.S. Workforce Symposium. NIOSH, October 26, 2004, Washington, D.C.
Graedel, T., B. Allenby, and R. Comrie. "Matrix Approaches to Abridged Life-Cycle Assessments." *Environmental Science and Technology* (March 1995) 29(3):134A–139A.
Harrington, J., and A. Knight. 1999. *ISO Implementation*. New York: McGraw-Hill.
Hart, J., I. Hunt, D. Lidgate, and V. Shankararaman. 1998. "Environmental Accounting and Management: A Knowledge-Based Systems Approach." Proceedings of the 6th International Symposium on Electronics and the Environment, Oak Brook, Illinois, pp. 225–230.
Henn, C. L. 1993. "The New Economics of Life-Cycle Thinking." Proceedings of the 1st International Symposium on Electronics and the Environment, Arlington, Virginia, pp. 184–188.
Jervis, S., and T. R. Collins. 2001. "Measuring Safety's Return on Investment." *Professional Safety* 46(9):18–23.
Lashbrook, W., P. O'Hara, D. Dance, and A. Veltri. 1997. "Design for Environment Tools for Management Decision Making: A Selected Case Study." Proceedings of the 5th International Symposium on Electronics and the Environment, San Francisco, California, pp. 99–104.
Microelectronics and Computer Technology Corporation (MCC). 1996. Report MCC-ECESM001-99, *Environmental Roadmap*. 1st ed. Austin, TX: MCC.
Miles, R., and C. Snow. 1978. *Organizational Strategy Structure and Processes*. New York: McGraw-Hill.
Mizuki, C., P. Sandborn, and G. Pitts. 1996. "Design for Environment—A Survey of Current Practices and Tools." Proceedings of the 4th International Symposium on Electronics and the Environment, Dallas, Texas, pp. 66–72.
Nagle, M. "Environmental Supply-Chain Management versus Green Procurement in the Scope of a Business and Leadership Perspective." Proceedings of the 8th International Symposium on Electronics and the Environment, May 2000, San Francisco, California, pp. 219–224.
Newman, Donald. 1983. *Engineering Economic Analysis*. San Jose, California: Engineering Press, Inc.
Porter, M. 1980. *Competitive Strategy*. New York: The Free Press.
Roome, N. 1992. "Developing Environmental Management Strategies." *Business Strategy and the Environment* 1(1):11–24.
Schot, J., and K. Fischer. 1993. "Introduction: The Greening of the Industrial Firm." *Environmental Strategies for Industry*. Washington, D.C.: Island Press.
Semiconductor Industry Association (SIA). 1997 and 1999. *The National Technology Roadmap for Semiconductors Technology Needs*. San Jose, California: SIA.
Shapiro, S. A. 1998. "The Necessity of OSHA." *Kansas Journal of Law and Public Policy* 3(3):22–31.
Smallman, C., and G. John. 2001. "British Directors Perspectives on the Impact of Health and Safety on Corporate Performance." *Safety Science* 38:727–739.
Society of Environmental Toxicology and Chemistry. 1991. *A Technical Framework for Life Cycle Assessments*. Washington D.C.: Society of Environmental Toxicology and SETAC Foundation for Environmental Education, Inc.
Stead, J., and E. Stead. 2000. "Eco-Enterprise Strategy: Standing for Sustainability." *Journal of Business Ethics* 24(4):313–329.
Suhejla, H., M. McAleer, and Laurent Pawels. 2005. "Modeling Environmental Risk." *Environmental Modeling and Software* 20(10):1289–1298.
Timmons, D. M. 1999. "Building an Eco-Design Toolkit at Kodak." Proceedings of the 7th International Symposium on Electronics and the Environment, Danvers, Massachusetts, pp. 122–127.
Tipnis, V. 1994. "Towards a Comprehensive Methodology for Competing on Ecology." Proceedings of the 2nd International Symposium on Electronics and the Environment, San Francisco, California, pp. 139–145.

Van Mier, G., C. Sterke, and A. Stevels. 1996. "Life Cycle Cost Calculations and Green Design Options." Proceedings of the 4th International Symposium on Electronics and the Environment, Dallas, Texas, pp. 191–196.

Veltri, A. 1997. "Environment, Safety and Health Cost Modeling," Technology Transfer Report #97093350A-ENG. Austin, Texas: SEMATECH, Inc.

Veltri, A., D. Dance, and M. Nave. 2003. "Safety, Health and Environmental Cost Model: An Internal Study from the Semiconductor Manufacturing Industry, Part 1." *Professional Safety* 48(7):30–36.

______. 2003. "Safety, Health and Environmental Cost Model: An Internal Study from the Semiconductor Manufacturing Industry, Part 2." *Professional Safety* 48(6):23–32.

Venkatesh, S., and T. Phillips. 1992. "The SEMATECH Cost of Ownership Model: An Analysis and Critique." SEMATECH/SRC Contract No. 91-MC-506 Final Report, Texas SCOE, Texas A&M University.

Warburg, N. 2001. "Accompanying the (re) Design of Products with Environmental Assessment (DfE) On the Example of ADSM." Proceedings of the 9th International Symposium on Electronics and the Environment, Denver, Colorado, pp. 202–207.

Ward, P., and D. Bickford. 1996. "Configurations of Manufacturing Strategy, Business Strategy, Environment, and Structure." *Journal of Management* 22(4):597–626.

Warren, J., and K. Weitz. 1994. "Development of an Integrated Life Cycle Cost Assessment Model." Proceedings of the 2nd International Symposium on Electronics and the Environment, San Francisco, California, pp. 155–163.

Welford, R. 1994. "Barriers to the Implementation of Environmental Performance: The Case of the SME Sector." *Cases in Environmental Management and Business Strategy*. London: Pittman.

Appendix: Life-Cycle Phases, Cost Factors, and Activities

I. UPFRONT. The phase concerned with profiling the SH&E risk and cost burdens associated with an existing, new, or upgraded product, technology, process, or service over its productive/economic life cycle and designing improvement options that maintain a balance between SH&E priorities and other competing business performance factors. The cost of upfront analysis includes all early-stage studies of risk and cost burdens to bring it to a form for decision making.

Note: Activities performed during the upfront phase are considered one-time costs.

Designing for Safety, Health, and Environment. Consideration of SH&E concerns at an early stage in the design engineering of products, technologies, processes, or services to prevent later risk and cost burdens.

Stage I. Concept Development - Specification Setting
Stage II. Detail Design – Design of Components, Parts, Subassemblies, Process Steps
Stage III. Prototype Manufacture and Testing

II. ACQUISITION. The phase concerned with profiling the costs associated with obtaining SH&E permits and procuring capital equipment necessary for controlling hazardous exposures, preventing/controlling pollution, maintaining regulatory compliance, and enhancing business performance. The costs of acquiring a capital asset or permit include both its purchase price and all other costs incurred to bring it to a form and location suitable for its intended use.

Note: Activity performed and capital costs incurred during the acquisition phase are considered one-time costs.

Obtaining Permits. One-time costs associated with obtaining permits (i.e., wastewater discharge; air emissions; handling, storing, and transporting hazardous substances and associated wastes) for the product, technology, process or service. Examples include:

1. **Permit Review/Approval.** Activities performed to study the procedural and performance requirements of the permit, conduct environmental impact studies, make application, lobby for gaining community approval, and sign off on the permit contract.
2. **Permit Fee.** The direct cost associated with the permit.
3. **Process Reengineering.** Activities performed for reengineering and remodeling the process infrastructure to comply with the procedural and performance requirements of the permit, including capital-related equipment and installation and utility hook-up expenses.

Procuring Capital. One-time costs associated with acquiring capital equipment/areas/structures for the product, technology, process, or service (e.g., emission/effluent control equipment for reducing, neutralizing, and minimizing the volume, toxicity, or hazardous properties of process waste; emission/effluent monitoring devices for providing periodic or continuous surveillance, detection, and recording of exposures to process hazards; reclaim equipment for separating process waste for reuse; treatment/storage/disposal facility equipment for the treatment, storage, recycling, or disposal of waste generated by the process, including the consolidation of waste until shipping.

1. **Equipment Review/Signoff.** Activities performed to study capital equipment alternatives; to qualify suppliers; to develop, negotiate, and sign off on equipment contracts; and to make ready the process to receive equipment.
2. **Equipment Cost.** The direct costs associated with capital equipment, including spare parts.
3. **Process Reengineering.** Activities performed for reengineering and remodeling the process infrastructure to accommodate capital, including equipment installation and utility hook-up expenses.

III. USE/DISPOSAL. The phase concerned with profiling the cost burdens associated with protecting and productively using and disposing of process resources in a manner that prevents injury/illness and environmental incidents and that reduces pollution and waste.

Note: Activity costs incurred in the operational phase are considered annual costs.

Operating Capital (CoO). Annual costs associated with operating/owning capital (i.e., equipment, areas, structures). Examples of costs include: utilities, labor, supplies/materials, maintenance, and preventative maintenance.

Resources Consumed. Annual cost of resources consumed by the product, technology, process, or service that has SH&E life-cycle concerns (e.g., effects on natural resource depletion; reduction of raw material; chemical/gas, energy, and water use).

Consumables Used. Annual cost of consumables used by the product, technology, process, or service (e.g., safety, industrial hygiene, ergonomics equipment or supplies for providing employee protection against exposure to process hazards; environmental protection supplies for preventing and controlling environmental incidents; environmental packaging equipment and supplies for consolidating/protecting/improving the handling of waste; hazardous material management equipment and supplies for providing environmental incident response and recovery services; fire-protection equipment and supplies for providing fire prevention and incident-control services; security equipment and supplies for providing process and factory site-monitoring and surveillance; license/certificates for complying with ESH regulations).

Providing Strategic/Technical Support. Annual costs associated with providing strategic and technical support (e.g., strategic management activities such as process strategic planning, reengineering, auditing-process implementation, and managing contracts); technical support activities (e.g., identifying, evaluating, and controlling exposures to hazards; providing training, environmental emission monitoring, and process, safety, and industrial hygiene inspections; advising on regulatory compliance matters; and assisting in manifesting and record-keeping procedures); research/development activities (e.g., testing, conducting studies, and creating innovative ways to protect and use process resources productively).

Training. Annual costs associated with providing training support in areas such as (1) SH&E law required for maintaining compliance with regulatory laws and standards, and (2) SH&E process specific for developing special competencies and capabilities.

Environmental Processing. Annual costs associated with implementing pollution prevention, reuse, and treatment and disposal strategies (e.g., source reduction by process-optimization activities used for limiting pollution before it occurs, including methods for modification of end product to eliminate a waste; revised operating practices; process-modification changes in raw materials, technology, and equipment; reclaim activities used for reusing and recycling a waste based on a closed and open loop system).

Closed Loop: Implies no further processing of a waste material; it is fed directly into the process step.

Open Loop: Implies the material must be processed (e.g., separating a particular component) prior to being reused.

Abatement activities used to control the physical and/or chemical characteristics of a waste; dilution activities used to change the physical and/or chemical characteristics of a waste after its use to reduce the material's volume and toxicity; waste treatment prior to disposal activities used to change the physical and/or chemical characteristics of a waste after its use to reduce the material's volume and toxicity and to improve handling and storage; waste consolidation/packaging activities used to consolidate and store waste before shipping; waste exchange activities used to transfer or sell waste to a brokerage that could use the waste as a raw material; waste shipping and disposal activities for transporting and disposing of a waste.

IV. POSTDISPOSAL. The phase concerned with profiling the cost burdens associated with monitoring the disposal of waste after the waste has left the control of the process and internal factory site and has been transferred to another company for management.

Note: Activity costs incurred during the postdisposal phase are considered to go beyond the productive life of the product, technology, process, or service.

Managing Waste-Site Compliance. Annual costs associated with assuring that waste-site disposal procedures are managed in a manner that maintains compliance with the waste-site disposal contract agreements and federal and state regulations. Examples include:

1. **Waste-Site Review/Selection.** Activities performed to review and select disposal-site alternatives and to develop, negotiate, and sign off on waste disposal contract agreements.
2. **Compliance Monitoring.** Activities performed to assure that the procedural and performance requirements of the contract agreement and federal and state regulations are in compliance.

V. CLOSURE. The phase concerned with profiling the cost burdens associated with retiring the product, technology, process, or service at the end of its useful life and preparing the area for other productive uses.

Note: Activity costs incurred during the closure phase are considered one-time costs.

Decommissioning. One-time costs associated with retiring the product, technology, process, or service following its useful life. Examples include:

1. **Decommissioning Review.** Activities performed for profiling the risk and cost burdens associated with retiring the manufacturing process or factory site.
2. **Dismantling/Cleanup.** Activities required for disassembling components used in the manufacturing process, arranging for disposal, and conducting clean-up procedures.
3. **Component Shipping and Disposal.** Costs incurred for transporting and disposing of dismantled components.

Remediation. One-time costs associated with remediation and preparing the area for other productive uses.

1. **Remediation Plan.** Activities required for developing ways to prepare the area for other productive uses.

VI. INCIDENTS. The area concerned with profiling the cost burdens associated with environmental contamination, pollution, alteration, occupational injury/illness, and noncompliance fines that adversely affect the product, technology, process, or service. Examples include:

Internalities. Incidents that only affect the internal manufacturing process and tend to result in (1) an adversity or disablement to a resource, (2) incurred direct and indirect costs, and (3) production interruption to the process. Examples of costs include:

Direct Costs. Those costs that can be easily identified and calculated or directly assigned to the incident with a high degree of accuracy (e.g., employee financial compensation (both current and reserved), damaged manufacturing property resources, capital replacement expenditures, incident fines, and legal expenses).

Indirect Costs. Those costs that can be intangible and difficult to calculate in the short term (e.g., incident investigation, production delays, loss of training investment, loss of future contribution of employee, replacement of resources, claims management, incident response/recovery/remediation, and business resumption).

Externalities. Internal incidents that affect the outside environment and tend to result in (1) air, water, soil pollution, (2) resource depletion/degradation, (3) chronic/acute health effects, (4) environmental habitat alteration, and (5) social/economic welfare effects.

Direct Costs. Those costs that can be easily identified and calculated or directly assigned to the incident with a high degree of accuracy (e.g., financial compensation for damaged environmental resources, fines, and legal expenses).

Indirect Costs. Those costs that can be intangible and difficult to calculate in the short term (e.g., incident investigation, incident recovery/remediation costs, and claims management).

Noncompliance Fine Facilitation. Citations issued for failing to comply with federal, state, or local environmental, safety, and health agencies.

Direct Costs. Those costs that can be easily identified and calculated or directly assigned to the fine with a high degree of accuracy (e.g., financial payment for the citation; making the facility and the process ready to comply, including any capital expenditures, materials, labor, legal fees, and research).

Indirect Costs. Those costs that can be intangible and difficult to calculate in the short term (e.g., activities needed to study and contest the fine).

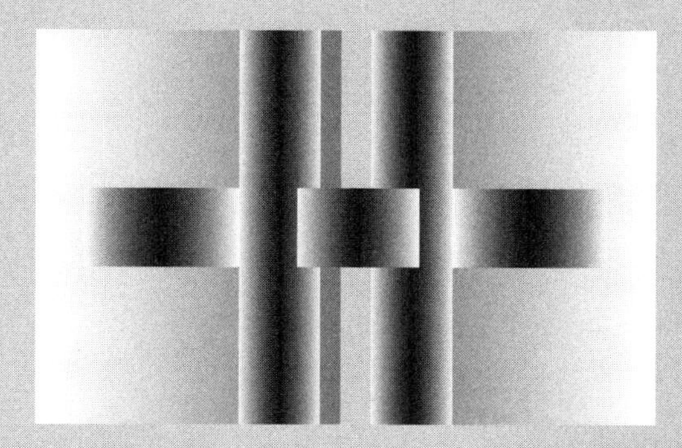

SECTION 1
RISK ASSESSMENT AND HAZARD CONTROL

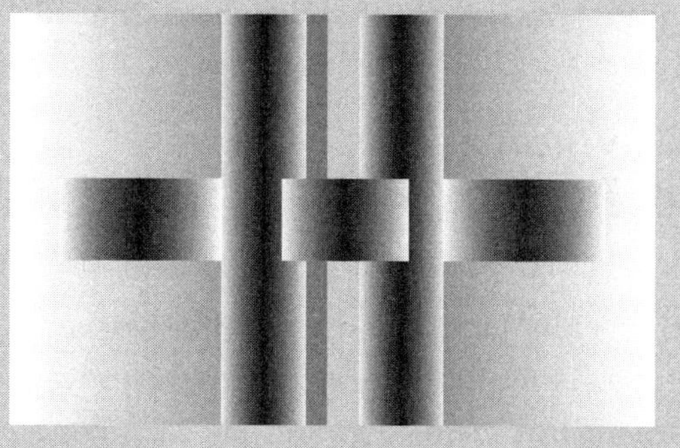

SECTION 1

RISK ASSESSMENT AND HAZARD CONTROL

REGULATORY ISSUES

Jerry Fields

LEARNING OBJECTIVES

- Develop a knowledge of the regulatory requirements to define risk assessment.
- Be able to describe the gap between risk assessment and risk perception and regulatory requirements.
- Gain a basic understanding of the regulatory risk-assessment process.
- Identify three steps in the process of regulatory risk assessment.
- Understand the policies and regulations that embody environmental and occupational health risks.
- Be able to recognize when to use risk assessments for environmental and occupational risks that may have an impact on the environment, workers, and the public.

THIS CHAPTER PROVIDES an introduction to the subject of regulatory issues in risk assessments and hazard controls. The chapter outlines the reasons that risk assessments are completed, and it discusses some regulations that require risk assessments and some reasons why risk assessments are conducted.

DEFINING RISK

Risk is the probability (or likelihood) that a harmful consequence will occur as a result of an action. Risk is a function of hazard and exposure. People take risks every day, in many forms—crossing the street, smoking, drinking alcohol, driving a car: the list is endless.

Somehow we catalog all of these risks, deciding some are acceptable while others are not—and not even considering others. According to Bascietto (1998), for a risk to occur there must be a source of risk (hazard) and an exposure to the hazard. Risk is the possibility of injury, disease, or death. Every risk involves a combination of these two factors: (1) the probability of an undesirable occurrence and (2) the severity of that occurrence (Ropeik 2000).For example, for a person who has measles, the risk of death is one in one million (CDC 2003). A *risk assessment* is an analysis that uses information to estimate a level of risk.

Why Assess Risk?

Assessing risks is done to mitigate indefensible, undesired, or unavoidable risk, to make operating decisions ("go/no go"), and to manage resource distribution for improved control of losses.

Risk assessment is "The process of determining the degree of threat that is posed by one or more hazards to one or more resources, or the product of that process" (Lack 2001). Performing risk assessments provides a systemic approach for ranking risks and making decisions and is a tool to help the organization or individual spend money more intelligently (Alijoyo 2004). Risk assessment is not a substitute for conforming to applicable codes, standards, and regulations (O'Brian 2000). Risk assessment considers information describing an actual or potential opportunity for human contact with chemicals or physical objects, the potential level of exposure, the health effects, and the expected degree of harm (Occupational Safety and Health Regulations 2007a). Risk assessments can be performed for hazards and exposures to chemicals and physical agents in the workplace and many other situations (Occupational Safety and Health Regulations 2007a). Many federal and state regulatory programs use or require risk assessments.

Conducting a Risk Assessment

The four steps of a risk assessment are hazard identification (Can this substance damage health?), dose-response assessment (What dose causes what effect?), exposure assessment (How and how frequently do people contact it?), and risk characterization (combining the other three steps to characterize risk and describe the limitations and uncertainties) (U.S. EPA 2001).

Some problems with risk assessments are that crucial parts of assessment come before (initial assumptions) and after (interpreting results) the actual analysis, models drastically simplify what happens in nature, and accurate accident and equipment-failure data are necessary (O'Brian 2000).

There are several regulatory risk-assessment techniques. More rigorous methods may be used when the regulatory approaches are inappropriate or inadequate. According to Jasanoff (1993), assessments are based on local concerns, guidelines, policies, and assessment review.

The U.S. Environmental Protection Agency (EPA)

History

In July 1970, the White House and Congress worked together to establish the EPA in response to the growing public demand for cleaner water, air, and land (U.S. EPA 1994). Prior to the establishment of the EPA, the federal government was not structured to make a coordinated attack on the pollutants that harm human health and degrade the environment. The EPA was assigned the daunting task of repairing the damage already done to the natural environment and to establish new criteria to guide Americans in making a cleaner environment a reality (U.S. EPA 1994).

Mission Statement

The EPA was established in 1970 to consolidate in one agency a variety of federal research, monitoring, standard-setting, and enforcement activities to ensure environmental protection.

EPA's mission is to protect human health and to safeguard the natural environment—air, water, and land—upon which life depends. For 30 years, the EPA has been working for a cleaner, healthier environment for the American people (U.S. EPA 2001).

The EPA provides leadership in the nation's environmental science, research, education, and assessment efforts. EPA works closely with other federal agencies, state and local governments, and Indian tribes to develop and enforce regulations under existing environmental laws. EPA is responsible for researching and setting national standards for a variety of environmental programs; it delegates to states and tribes the responsibility for issuing permits, and monitoring and enforcing compliance. Where national standards are not met, EPA can issue sanctions and take other steps to assist the states and tribes in reaching the desired levels of environmental quality. EPA also works with industries and all levels of government in a wide variety of voluntary pollution prevention programs and energy conservation efforts (U.S. EPA 1994).

Another federal act to review is the National Environmental Policy Act, which requires risk assessments but is not part of this chapter. (Oliver 2005).

EPA Risk Assessment

After the establishment of EPA, new legislation was developed, implemented, and enforced to regulate chemicals in the environment (U.S. EPA 2001). The EPA uses risk-assessment techniques to assess a chemical's capacity to cause harm (its toxicity), and the potential for humans to be exposed to that chemical in a particular situation; for example, workplace or home (U.S. EPA, 2001).

The most common basic definition of *risk assessment* used within the U.S. Environmental Protection Agency is "a process in which information is analyzed to determine if an environmental hazard might cause harm to exposed persons and ecosystems" (National Research Council 1983).

During the 1970s, risk-assessment procedures for all chemicals were reevaluated and improved. More importantly, they were formalized (Oliver 2005). Standardized tests were developed so consistent evaluations could be performed and the scientific basis of regulations could be more easily applied (U.S. EPA 2001).

According to the EPA, there are two components that make up the definition of risk assessment: *toxicity*, or dose-response assessment, and *exposure assessment* (U.S. EPA 2001). Dose-response assessment is a measure of the extent and types of negative effects associated with a level of exposure. Exposure assessment is a measure of the extent and duration of exposure to an individual. One example of a chemical that was developed based on dose-response assessment data is dioxin (U.S. EPA 2001).

The EPA is responsible for developing and providing state and local government agencies with toxicological and medical information relevant to decisions involving public health (U.S. EPA 2001). State agencies that use such information include all boards and departments within state EPAs, as well as the Department of Health Services, the Department of Food and Agriculture, the Office of Emergency Services, the Department of Fish and Game, and the Department of Justice (U.S. EPA 2001). According to the EPA, risk assessment is the "process by which one attempts to evaluate and predict the likelihood and extent of harm (in qualitative and quantitative terms) that may result from a health or environmental hazard" (U.S. EPA 2001). Information comes from scientific studies, historical data, and actual experiences. A risk assessment provides an estimate of risk to workers who have the potential to be exposed to chemical and physical agents at a work site (McGarity and Shapiro 1993). According to Benner (1983), regulatory risk-assessment analysis has built-in safety margins.

A risk assessment provides essential information about the severity and extent of specific environmental problems for use in risk-management decisions (U.S. EPA 2001). *Risk management* is the process of deciding how and to what extent to reduce or eliminate risk factors by considering the risk assessment, engineering factors (Can procedures or equipment do the job, and for how long and how well?), and social, economic, and political concerns (McGarity and Shapiro 1993).

Why EPA Conducts Risk Assessments

Determining environmental standards, policies, guidelines, regulations, and actions requires making decisions (Ropeik 2000). Environmental decision making is often a controversial process involving the interplay of many forces: science, social, and economic factors; political considerations; technological feasibility; and statutory requirements (Oliver 2005). Environmental decisions are often time-sensitive, for example when public health is known or suspected to be at risk (U.S. EPA 1994).

EPA conducts risk assessment to provide the best possible scientific characterization of risks based on a rigorous analysis of available information and knowledge, a description of the nature and magnitude of the risk, an interpretation of the adversity of the risk, a summary of the confidence or reliability of the information available to describe the risk, areas where information is uncertain or lacking completely, and documentation of all of the evidence supporting the characterization of the risk (U.S. EPA 1994). According to Schoeny, Muller, and Mumford (1998), EPA then incorporates this risk characterization with other relevant information, such as social, economic, political, and regulatory information, in making decisions, in the form of policies and regulations, about how to manage the risk. Risk assessment informs decision makers about the science implications of the risk involved (Schoeny, Muller and Mumford, 1998). Risk assessments that meet their objectives can help guide risk managers to decisions that mitigate environmental risks at the lowest possible cost and which will stand up if challenged in the courts (Schoeny, Muller, and Mumford, 1998).

Two Types of EPA Risk Assessments

Exposure Assessment

Exposure assessment is accomplished in three basic approaches: analysis of the source of exposure in drinking water or workplace air, measurements of the environment, and blood and urine laboratory tests of the people exposed. Sample analyses of air and water often provide the majority of usable data (U.S. EPA 2000). These tests reveal the level of contamination in the air or water to which people are exposed. However, they only reflect concentration at the time of testing and generally cannot be used to quantify either the type or amount of past contamination. Some estimates of past exposures may be gained from understanding how a chemical moves in the environment (U.S. EPA 2003).

Some other types of environmental measurements may be helpful in estimating past exposure levels. Past levels of a persistent chemical can be estimated using the age and size of the fish, and information about how rapidly these organisms accumulate the chemical (CDC 2003).

Analyses of body-fluid levels of possibly exposed people provide the most direct exposure measure. They do not provide good estimates of past exposure levels because the body usually reaches a balanced state, so that there is no longer any change in response to continued exposure (EXTOXNET 1998). Many chemicals are excreted from the body after exposure stops, and the basic understanding of what happens to chemicals in the human body is often lacking for those that do persist. Thus, direct examination of a population may provide information as to whether exposure has occurred but not the extent, duration, or source of the exposure (CDC 2003).

Exposure assessments can be performed reliably for recent events and less reliably for past exposures. The difficulties in exposure assessment often make it the weak link in trying to determine the connection between an environmental contaminant and adverse effects on human health. Exposure-assessment methods will continue to improve, but at the present time there still remains significant uncertainty (U.S. EPA 2001).

Dose-Response Assessment

A distinction must be made between *acute* and *chronic* effects when discussing dose-response assessment. Acute effects occur within minutes, hours, or days, whereas chronic effects appear only after weeks, months, or years. The quality and quantity of scientific evidence gathered is different for each type of effect, and the confidence placed in the conclusions from the test results also differs (U.S. EPA, 2001).

Acute toxicity is easier to deal with. Short-term studies with animals provide evidence as to which effects are linked with which chemicals and the levels at which these adverse effects occur. Often, some human experience is available as a result of accidental exposure. When these two types of evidence are available, it is usually possible to make a good estimate of the levels of a particular toxicant that will lead to a particular acute adverse effect in humans. This approach is the basis for much of the current regulation of toxic substances, especially in occupational situations (OSHA 2006).

Chronic toxicity is much more difficult to assess. There are a variety of specific tests for adverse effects such as reproductive damage, behavioral effects, and cancer (CDC 2003). According to Centers for Disease Control (CDC), cancer assessment will reveal some of the problems inherent in long-term toxicity assessments and also focus on the health effect that seems to be the biggest public concern.

In cancer assessment, it is not only the chronic nature of the disease but also the low incidence that causes difficulty. Society has decided that no more than one additional cancer in 100,000 or one million people is acceptable, so assessment measures must be able to detect this small increase (CDC 2003). Two types of evidence are utilized to determine the dose of chemical that will result in this change. One is based on experiments on animals, and the other is based on experience with humans (CDC 2003).

To detect an increase of one cancer in a million animals, millions of animals would have to be exposed to environmentally relevant amounts of the chemical (CDC 2003). Consequently, investigations are performed on smaller numbers of animals who have been exposed to very large amounts of a chemical. These large amounts are necessary to produce a high enough incidence of cancer to be detectable in this small population (CDC 2003). The results of these CDC types of studies indicate the levels of a chemical that will cause cancer in a high percentage of the population (CDC 2003).

As a result, this information is used to assess and predict the level of a chemical that will cause cancer in a million animals or, more importantly, in a million humans, by utilizing mathematical models (CDC 2003). There are a variety of mathematical models, and the one generally chosen is that which provides the greatest margin of safety (U.S. EPA 2001). In using one of these models, the results are overestimates rather than underestimates of the ability of the chemical to cause cancer (Rhomberg 1997).

The type of evidence utilized in chronic toxicity assessment is known as epidemiological evidence. In this type of study, human populations are carefully observed and possible associations between specific chemical exposures and particular health effects are investigated (CDC 2003). This is not an easy task. It is made even more difficult in cancer assessment by the requirement of detecting very small changes in incidence (one extra cancer in a million people) (CDC 2003).

Epidemiological assessments have been most useful in only certain situations. One is the workplace, where exposure levels to known carcinogens may be elevated, and where the duration of exposure can be determined. According to OSHA, a sizable increase in cancer incidents is needed before a connection can be established. The conclusion that asbestos causes lung cancer is based on this type of situation. An exception to the need for a high cancer incidence rate is where the effect is unique, so that even a few cases are significant. An example of this was the observation that a small number of vinyl chloride workers developed a rare form of liver cancer (CDC 2003). Even with known occupational carcinogens, the question of what happens at low exposures for common environmental chemicals has not been answered (Schoeny, Muller, and Mumford 1998).

The techniques available for assessment of chronic toxicity, especially carcinogenicity, provide rather clear evidence as to whether a particular chemical causes a particular effect in animals (CDC 2003). CDC states there is great uncertainty about the amounts needed to produce small changes in cancer incidence in humans (U.S. EPA 2001). This uncertainty and the difficulties in exposure assessment will continue to make it difficult to draw definite conclusions about the relationship between most environmental exposures and chronic health effects (U.S. EPA 2001).

Examples of EPA Risk Assessments

Health Assessment Document for Diesel Engine Exhaust (U.S. EPA 2002a)

This assessment examined information regarding the possible health hazards associated with exposure to diesel engine exhaust (DE), which is a mixture of gases and particles. The assessment concludes that long-term (i.e., chronic) inhalation exposure is likely to pose a lung cancer hazard to humans, as well as damage the lung in other ways, depending on exposure. Short-term (i.e., acute) exposures can cause irritation and inflammatory symptoms of a transient nature, these being highly variable across the population. The assessment also indicates that evidence for exacerbation of existing allergies and asthma symptoms is emerging. The assessment recognizes that DE emissions, as a mixture of many constituents, also contribute to ambient concentrations of several criteria air pollutants, including nitrogen oxides and fine particles, as well as other air toxics. The assessment's health-hazard conclusions are based on exposure to exhaust from diesel engines built prior to the mid-1990s. The health-hazard conclusions, in general, are applicable to engines currently in use, a category that includes many older engines. As new diesel engines with cleaner exhaust emissions replace existing engines, the applicability of the conclusions in this health-assessment document will need to be reevaluated. To obtain a hard copy or CD of the health-assessment document, contact EPA's National Service Center for Environmental Publications: telephone, 1-800-490-9198 or 513-489-8190; fax, 513-489-8695; mail: NSCEP; P.O. Box 42419; Cincinnati, OH 45242-0419.

Health-Effects Assessment for Mercury (U.S. EPA 1993)

This document represents a brief, quantitatively oriented scientific summary of data on health effects. It was developed by the Environmental Criteria and Assessment Office to assist the Office of Emergency and Remedial Response in establishing chemical-specific health-related goals of remedial actions. If applicable, interim acceptable intakes of chemical-specific subchronic and chronic toxicity are determined for systemic toxicants, or q_1 values are determined for carcinogens for both oral and inhalation routes. A subchronic and chronic interim acceptable intake was determined for mercury based on both oral and inhalation exposure. Estimates for mixed alkyl and inorganic mercury alone are presented.

Risk Management Plan

The Clean Air Act Amendments of 1990 require the EPA to publish regulations and guidance for chemical accident prevention at facilities using extremely hazardous substances (U.S. EPA, 2001). The Risk Management Program Rule (RMP Rule) was written to implement Section 112(r) of these amendments. The rule, which built upon existing industry codes and standards, requires companies of all sizes that use certain flammable and toxic substances to develop a *risk management program*, which must include the following:

- hazard assessment that details the potential effects of an accidental release, an accident history of the last five years, and an evaluation of worst-case and alternative accidental releases
- prevention program that includes safety precautions and maintenance, monitoring, and employee training measures
- emergency response program that spells out emergency health care, employee training measures, and procedures for informing the public and response agencies (e.g. the fire department) should an accident occur
- EPA's risk management plan (RMP) requires an owner or operator of a covered process to prepare and submit a single RMP that includes
- conducting an analysis of a worst-case release scenario
- conducting a hazard assessment
- implementing a streamlined prevention program (hazard control)

For this process analysis EPA has developed three program levels, and each level requires, at a minimum, the three conditions listed above. The three program levels and the requirements of each are as follows:

1. ***Program One***—Processes with no public receptors within the distance to an endpoint from a worst-case release and with no accidents with specific off-site consequences within the past five years. This program imposes limited assessment requirements and minimal prevention and emergency response requirements.
2. ***Program Two***—Processes that are not eligible for program one or subject to program three. This program requires a streamlined prevention (hazard control) program and requires additional hazard assessment, management control systems, and emergency response procedures.
3. ***Program Three***—Processes not eligible for program one and subject to OSHA's Process Safety Management (PSM) Standard 29 CFR 1910.119, Subpart H under federal or state OSHA programs or classified in one of ten

specified North American Industry Classification Systems (NAICS) codes. Program three requires the owner or operator to develop an OSHA PSM program as the prevention program and requires additional hazard assessment, management controls systems, and emergency response procedures (OSHA 1992a).

Summary

Risk assessment is a complex process that depends on the quality of scientific information that is available. Risk assessments are used for assessing acute risks where effects appear soon after exposure occurs. The longer the period of time between exposure and appearance of symptoms, the greater is the uncertainty (U.S. EPA 2001), because of the greatly increased uncertainties in exposure assessment and also the problems involved in using epidemiological or laboratory animal results in such cases (CDC 2003).

The mission of the EPA is to protect human health and to safeguard the natural environment (air, water, and land) upon which life depends. To fulfill this mission, EPA develops and enforces regulations that implement environmental laws enacted by Congress. Implementation of environmental laws includes grants and other financial assistance to state and tribal governments carrying out environmental programs approved, authorized, or delegated by the EPA (U.S. EPA, 1994).

The primary purpose of a risk assessment is to provide data to risk managers' decision-making process (Oliver 2005). The primary purpose of a risk assessment is not to make or recommend any particular decisions; rather, it gives the risk manager information to consider with other pertinent information to make better decisions (Oliver 2005). EPA uses risk assessment as a key source of scientific information for making good, sound decisions about managing risks to human health and the environment (Schoeny, Muller, and Mumford 1998). Examples of such decisions include deciding permissible release levels of toxic chemicals, granting permits for hazardous waste treatment operations, and selecting methods for remediation at Superfund sites (Schoeny, Muller, and Mumford 1998). It is important to consider other factors along with the science when making decisions about risk management. In some regulations, the consideration of other factors is mandated (e.g., costs). Some of these other factors are in the following list, which is adapted from the EPA (2001):

1. ***Economic factors***—the costs and benefits of risks and risk mitigation alternatives
2. ***Laws and legal decisions***—the framework that prohibits or requires some actions
3. ***Social factors***—attributes of individuals or populations that may affect their susceptibility to risks from a particular stressor
4. ***Technological factors***—the feasibility, impact, and range of risk management options
5. ***Political factors***—interactions among and between different branches and levels of government and the citizens they represent
6. ***Public factors***—the attitudes and values of individuals and societies with respect to environmental quality, environmental risk, and risk management

National Center for Environmental Assessment (NCEA)

The National Center for Environmental Assessment conducts risk assessments of national significance for topics such as

- diesel exhaust
- dioxin
- drinking water and disinfection by-products
- ozone
- particulate matter
- PCBs
- perchlorate
- second-hand smoke (ETS)
- watershed assessment
- World Trade Center disaster site

International Organization for Standardization (ISO)

The International Organization for Standardization was established in 1947. ISO is based in Geneva and is the world's largest developer of standards. It provides a single set of voluntary standards that are recognized and respected worldwide (ISO 2004). ISO has developed acceptable standards for all aspects of the workplace.

ISO standards are developed according to three sound principles (ISO 2004):

1. *Consensus*—the views of all interests are taken into account
2. *Conformity*—global solutions to satisfy industries and customers
3. *Voluntarism*—based on voluntary involvement of all interests in the marketplace (market-driven)

ISO has helped to provide a greater acceptance of harmonized International Standards, and to improve competition and eliminate barriers to trade as globalization of industries and markets continues to increase (ISO 2004).

ISO publishes a guide that provides a framework of best practices and procedures that are used by assessment stakeholders, including governmental, private, national, and international systems. This guide for conformity assessment helps to harmonize international practices (ISO 2004). Conformity assessment means checking that products, materials, services, systems, or people measure up to specified requirements.

Environmental Assessment–Management Framework–ISO 14000

Corporate management manages risk, whether consciously or not, and corporations manage risks on an ongoing basis. By using management systems, corporations are defining the organization's environmental policy. These policies will include a commitment to continued improvement, prevention of pollution, and compliance with regulations (ISO 2004). A successful program will include routine auditing and review as key elements to continuous improvement. Other elements for a successful environmental ISO 14000 program will be document control, operational control, control of records, management policies, training, statistical techniques, and corrective and preventive action. One of the major requirements in developing a successful ISO 14000 program is to determine the company's risks through a systematic risk assessment (ISO 2004).

Risk Assessment–ISO 14000 Components

There are three very important requirements for risk assessment and risk management and the development of an Environmental Management System (EMS) (U.S. EPA 2001):

1. Develop a procedure to identify all environmental aspects of all operations. Included are activities, products, and services of a company's operations and of those companies (e.g., suppliers, subcontractors) on which they have influence. The company must determine the environmental aspects that have or can have significant impacts on the environment. These significant impacts must be considered when a company develops its environmental objectives. The procedures developed must be maintained.
2. Develop and work toward environmental objectives.
3. Perform management reviews of EMS and make changes to policy, objectives, and all elements of the EMS when needed.

This program requires a preliminary analysis, a risk analysis, a risk evaluation, and risk control following the initiation. All of these elements are required in the development of a successful program (U.S. EPA 2001).

A preliminary analysis includes the development of a risk information base that will help identify possible exposures to loss using risk scenarios.

- A risk analysis includes the estimated consequences of the risk scenarios identified.
- A risk evaluation includes assessing the acceptability of the risks identified.

Risk control includes the identification of feasible risk-control options in terms of effectiveness and costs, and it assesses stakeholders' acceptance of proposed actions, including regulatory compliance.

Occupational Safety and Health Administration (OSHA)

Requirements for Risk Assessment and Hazard Control

Risk assessment, in occupational safety and health regulations, began when the United States Supreme Court ruled in a benzene decision (*Industrial Union Department v. American Petroleum Institute*, 448 U.S. 607, 1980) that the Occupational Safety and Health Administration could not issue a standard without demonstrating a significant risk of material health impairment. As a result of the Supreme Court ruling, risk assessment became standard practice in OSHA rule making for health standards.

Hazard control is the physical control of various materials, processes, and activities to produce a specific benefit. Safety engineering is the control of system hazards having the potential to cause system damage or system operator injury, or to decrease system benefits. When controlling hazards, OSHA requires the use of engineering controls as a primary method for controlling hazards and reducing the risk of injury to workers (OSHA

According to OSHA, when engineering controls are not feasible, then administrative controls and/or the wearing of personal protective equipment are to be used to reduce the risk of injury (OSHA 1995b). A major goal of OSHA is to prevent workers from being exposed to occupational hazards. Eliminating the hazard by substituting a less hazardous chemical or by engineering the hazards out of the process or equipment is the most effective method of controlling hazards.

OSHA's approach to risk assessment is guided by the Supreme Court (OSHA 2007a). Following the Supreme Court's interpretations of the OSH Act involving benzene and cotton dust, OSHA may not promulgate a standard unless it has determined that there is a significant risk of health impairment at existing permissible exposure levels and that issuance of a new standard is necessary to achieve a significant reduction in that risk (OSHA 2000). For benzene, a risk assessment relating to worker health is not only appropriate but is, in fact, required in order to identify a significant worker health risk and to determine whether a proposed standard will achieve a reduction in that risk (ACGIH 2002). The Court did not require OSHA to perform a quantitative risk assessment in every case; the Court implied, and OSHA as a policy matter agrees, that such assessments should be put in quantitative terms to the extent possible (OSHA 2007a).

It is assumed that mathematical curves are reflective of biological processes that control the biological fate and action of a toxic compound (ACGIH 2002). Many of these factors have not been quantitatively linked to the mathematical models. Biological factors that may play important roles in the risk assessment are (1) dose of the material at the sensitive tissue; (2) the sensitive tissue(s) itself; (3) the nature of the response(s); (4) rates and sites of biotransformation; (5) toxicity of metabolites; (6) chronicity of the compound (cumulative nature of the material or its actions); (7) pharmacokinetic distribution of the material (especially effects of dose on the distribution); (8) the effect of biological variables such as age, sex, species, and strain of test animal; and (9) the manner and method of dosing the test animals (OSHA 2006)

OSHA estimates of risk can result in exposure difficulties in establishing allowable exposure limits related to dermal exposure (OSHA 1995b). There are a number of reasons why this is impractical, among which are the difficulty of quantifying dermal exposures, the inability to select a reliable biological indicator, and the difficulty in correlating the amount absorbed with a precise adverse health affect (OSHA 1995b). OSHA requires adherence to permissible exposure limits that can reduce surface contamination by reducing the opportunity for skin contact through the use of personal protective clothing and equipment that would aid in preventing dermal exposure (OSHA 1995b).

Process Safety Management

Process safety management of highly hazardous chemicals was developed for the prevention of, or for minimizing the consequences of, catastrophic releases of toxic, reactive, flammable, or explosive chemicals into the atmosphere where they may affect the surrounding community. According to OSHA requirements, an initial process hazard analysis is to be completed by the employer on all processes if any of the processes have a highly hazardous chemical that is listed in OSHA's Process Safety Management (PSM) Standard (OSHA 1992a).

There are fourteen key elements of the OSHA PSM Standard. Several of the elements require a risk assessment and methods for hazard control. According to OSHA, process hazard analysis requires the employer to conduct an initial process hazard evaluation on any process covered by this PSM standard (OSHA 1992b). The employer must identify, evaluate, and control the hazards involved in the process. When the PSM took effect, employers had two years to comply with this element of the standard. According to OSHA, employers that plan to start up a new process are to determine if the process will require a PSM program and if it will require a hazard analysis (OSHA 1992b).

Several methodologies can be used by the employer to determine and evaluate the hazards of the process being analyzed. According to OSHA, one very important requirement is that the process hazard analysis is to be performed by a team that is made up of employees with knowledge of the process or processes. Here is a list of acceptable methodologies from OSHA 20 CFR 1910.119:

- what-if scenarios
- checklist
- what-if/checklist combination
- hazard and operability study (HAZOP)
- failure mode and effects analysis (FMEA)
- fault-tree analysis

OSHA requires that hazards of the process are to be addressed when using one of these methodologies. If an employer has had a previous incident that had a likely potential to cause a catastrophic event and potential consequences in the workplace, the employer is required to evaluate the hazards and provide applicable hazard controls. These controls can be engineering and administrative controls that are appropriate to the hazards and the early warning devices for the detection of releases.

The employer must establish a system to address any recommendations or findings the process hazard-analysis team develops and ensure that recommendations are resolved promptly. All activities are to be documented. At least every five years from the initial process hazard analysis, the employer is required to update and revalidate the process hazard analysis for consistency with the current process (OSHA 2000).

Confined Space

An OSHA regulation was promulgated for the purpose of protecting employees in general industry from the hazards associated with permit-required confined spaces. The general requirement of the standard was for employers to evaluate their workplace to determine if any spaces are permit-required confined spaces. For this purpose OSHA has developed a permit-required confined-space decision flow chart (OSHA1998). A confined space and a permit-required confined space had to meet the following criteria:

Confined Space

a. is large enough and so configured that an employee can bodily enter and perform assigned work
b. has limited or restricted means for entry or exit
c. is not designed for continuous employee occupancy

Permit-Required Confined Space

d. contains or has the potential to contain a hazardous atmosphere
e. contains a material that has the potential for engulfing an entrant
f. has an internal configuration such that an entrant could be trapped or asphyxiated by inwardly converging walls or by a floor that slopes downward and tapers to a smaller cross-section
g. contains any other recognized serious safety or health hazards

OSHA requires employers to conduct a risk assessment. OSHA requires an employee to conduct a pre-entry check of the atmosphere with a calibrated direct-read instrument that will measure oxygen content and check for flammable gases and for toxic air contaminants that may cause a health hazard for the employee

(1910.146(c)(5)(ii)(H). According to OSHA 29 CFR 1926.21(b)(6)(i), an employee is required to be instructed in the nature of the hazards involved, precautions to be taken, and the use of personal protective and emergency equipment required (OSHA, 2007b)

Risk Assessments, Risk Characterization, and Regulatory Programs

OSHA believes that risk analysis is a tool necessary for linking sound policy decisions with sound science. The Superfund Amendments and Reauthorization Act, Title III, also referred to as EPCRA, required states and local jurisdictions to develop emergency response plans. Title III created a procedural obstacle course for OSHA's risk-assessment process and prevented a more flexible case-by-case approach. According to OSHA, it often conducts simplified rule-making sessions to streamline or update standards, but that the present OSHA process of risk assessment requires formalized peer review, which is very time-consuming and makes it difficult to lift the burden on industry (OSHA 1992a).

OSHA regulations often address risks that threaten 1 in 100 exposed workers. For example, OSHA completed a risk assessment of workers exposed to ethylene oxide and concluded that the limits prior to regulations estimated that worker lifetime exposure ranged from 63.4 to 109.3 excess cancer deaths per 100 exposed workers (OSHA 2000). According to OSHA, responsible policy makers would not consider these risks marginal. OSHA estimated that if the implementation of the PSM standard would have been delayed one year, the fatalities would have exceeded 130, and the injuries would have exceeded 750 (OSHA 2007a).

OSHA states that risk assessments depend on the nature of the risk; for example, a risk assessment for a safety hazard differs from a risk assessment for toxins (OSHA 2006). Risk assessments for toxic substances, hazardous physical agents, and safety hazards do follow established scientific principles and nationally recognized guidelines established by the National Academy of Sciences. In conducting risk assessments to establish safety standards, all relevant injury and fatality data and any other relevant information will be reviewed. OSHA invites comment on all aspects of the risk-assessment process at the proposal, during public hearings, and at the final rule stage of a standard's development. An obstacle is that OSHA has to follow prescribed procedures, which do not allow for simple risk assessments but only for complex and costly analyses in every case. OSHA is required to demonstrate that the regulation addresses a significant risk, that the regulation will substantially reduce that risk, and that it will do so in a feasible and cost-effective manner (OSHA 2007a). The analyses include detailed examinations of regulatory costs, economic impacts, and benefits (OSHA 2000).

Summary

OSHA has a mandate to protect workers before they are injured or become ill. The statutory framework for OSHA imposes both substantive and procedural obligations. OSHA currently has extensive procedural obligations to assess the impact of proposed regulations and justify the need for a regulation, including mandates to conduct detailed risk assessments. OSHA currently does not publicly reveal the policy preferences it uses in risk assessment; doing so would reveal the extent of the uncertainty that may affect risk estimates.

OSHA risk assessment is subject to divergent, socially conditioned interpretations, which may make it difficult to reach consensus concerning the extent of risk posed by a safety or health hazard (OSHA 2007a). The standards of assessing quality in regulatory science are fluid, controversial, and sensitive to political factors (see OSHA Web site).

According to OSHA, risk analysis is a necessary and appropriate tool for linking sound policy decisions with sound science. The agency has experience in conducting risk analyses to support occupational safety and health regulations. The result of OSHA's approach to risk assessment has been the development of a body of regulations that address and reduce significant risks and are both technologically feasible and economically justifiable. Assessment for a safety hazard, such as fatal falls from roofs, differs substantially from those for toxins, such as cadmium, which can cause many forms of disease. In developing risk assessments for toxic substances, hazardous physical agents, and safety hazards, OSHA follows established scientific principles and nationally recognized guidelines, such as those of the National Academy of Sciences. For health standards, OSHA also carefully explains and justifies its choice of risk-assessment models and discusses the weight of the evidence in a comprehensive manner. For safety standards, the agency describes all relevant injury and fatality data and any other information relevant to the assessment of risk. OSHA takes these steps to ensure that its risk assessments and risk characterizations are as clear and understandable as possible. OSHA invites comment on all aspects of its risk assessments at the proposal, public hearing, and final rule stages of a standard's development (see OSHA Web site).

Before issuing any safety or health regulation, OSHA must demonstrate that the regulation addresses a significant risk, that the regulation will substantially reduce that risk, and will do so in an economically feasible, cost-effective, manner (OSHA 1995b).

Consumer Product Safety Commission (CPSC)

History

The United States Consumer Product Safety Commission (CPSC) is an independent federal regulatory agency created to protect the public from unreasonable risks of injuries and deaths associated with some 15,000 types of consumer products. The CPSC's mission is to keep American families safe by preventing or reducing the risk of injury, illness, or death from primarily household consumer products. The CPSC's empowerment is by way of the older Federal Hazardous Substances Act (FHSA). It regulates consumer products that generate pressure or that are combustible, toxic, corrosive, or radioactive. Its means of product regulation is primarily through label warnings.

CPSC works to reduce the risk of injuries and deaths from consumer products by

- developing voluntary standards with industry
- issuing and enforcing mandatory standards, or by banning hazardous consumer products if no feasible standard would adequately protect the public
- obtaining the recall of products or arranging for their repair
- conducting research on potential product hazards
- informing consumers by using the media (written, radio, television), state and local government agency bulletins, private organizations' publications, and by responding to consumer inquires

According to the CPSC's 2002 annual report (2002a), the staff provided technical support for the development of 64 voluntary safety standards. These were handled by three standards-developing coordinating organizations: The American Society for Testing and Materials International (ASTM), The American National Standards Institute (ANSI), and Underwriters Laboratories Inc. (UL). These standards provide performance safety provisions addressing potential hazards associated with consumer products found in homes, schools, and recreational areas. The CPSC staff continued monitoring conformance to selected voluntary consumer product safety standards.

The Office of Compliance and the regional offices are jointly responsible for identifying consumer products that fail to comply with a specific product safety standard or the CPSC product-related requirements mandated by statute or regulation. CPSC worked cooperatively with the responsible companies to obtain voluntary, corrective action plans monitored by the commission.

CPSC does not deal with the types of products covered by the Department of Transportation (cars, trucks, motorcycles), those covered by the Food and Drug Administration (drugs and cosmetics), or those covered by the Department of the Treasury (alcohol, tobacco, and firearms) (CPSC 2002).

CPSC and Risk Assessments

In mid-2007 the CPSC was actively engaged in more than 60 voluntary standards-development activities on a wide range of consumer products (U.S. CPSC 2007). CPSC provides expert advice, technical assistance, injury and death data and analysis, and supporting research. CPSC submits recommendations concerning new safety standards or modifications of existing standards. Its recommendations are often based on CPSC research, which may include recent injury and death data associated with a product category. CPSC solicits public comment on selected voluntary standards activities by posting its proposed recommendations on the CPSC Web site for a minimum of five days (U.S. CPSC 2007). At the end of the posting period, CPSC will consider any comments received before developing final voluntary recommendations.

The commission was established in 1973 with the task of preventing unreasonable injury to consumers from a wide range of consumer products that are purchased every day. During the 1990s, CPSC was best known for its recalls of a number of consumer products that it thought to be unsafe.

In 1990, CPSC released a study on the potential risk of skin cancer from arsenic on treated-wood playground equipment. The CPSC Environmental Working Group provided new data suggesting a more serious risk of cancers such as bladder and lung cancer associated with chromated-copper-arsenate (CCA) (U.S. CPSC 2002b). CCA is a mix of chromium, copper, and arsenic and is used generally to prevent infestation of wood by insects and fungus (U.S. CPSC 2002b). In 2002, both CPSC and EPA considered issues relating to CCA-treated wood. CPSC studied the amount of CCA released from newly purchased, unused CCA-treated wood used for playground equipment as compared to "used" or "older" wood. The question CPSC raised is how much arsenic children are exposed to when playing on the treated-wood playground equipment (U.S. CPSC 2002b). EPA's Office of Pesticide Programs studied the CCA issue and announced a voluntary decision by industry to phase out the use of CCA-treated wood for consumer use by December 31, 2003 (U.S. EPA 2002b).

Summary

In 1992, the CPSC issued guidelines for assessing health hazards under the Federal Hazardous Substance Act that included risk assessment (U.S. CPSC 2002b). In these guidelines is a series of default assumptions that are used in the absence of evidence but are intended to be flexible to incorporate the latest scientific information. There are provisions to use alternative procedures on a case-by-case basis, provided that the procedures used can be supported by scientific evidence and data. CPSC staff and EPA staff have worked together on several issues related to exposure and potential risk to children and will initiate studies to determine effective methods of reducing the amount of arsenic released from CCA-treated wood (U.S. CPSC 2002b).

National Fire Protection Association (NFPA)

History

Modern fire safety codes and standards were developed in the late 1800s by the National Fire Protection Association to address standards for automatic sprinklers. A total of nine different standards for piping size and sprinkler spacing could be found within 100 miles of the city of Boston (NFPA 1995). The situation was considered a plumber's nightmare, and if left unresolved, it would result in an unacceptably high rate of sprinkler-system failures. After a series of meetings, a committee was chosen to review all past meetings, and in November 1896 the National Fire Protection Association was created. Through technical committees, fire codes were developed over the next 70 years, beginning with a code for automatic sprinklers, which was released in 1897 (NFPA 1995).

Today the NFPA uses its consensus-building process with over 60,000 NFPA members worldwide. Of that number, approximately 24 percent are affiliated with fire departments, the remaining being representatives of the private and public sectors in a wide variety of fields (Watts 2002). The standard-making process requires consensus building, which in turn requires a broad-based representation of interested parties. The public comment period is a mechanism to reach beyond the committee to be even more inclusive. Finally, the general membership of the NFPA must approve before any document becomes a standard.

NFPA's mission is to reduce the worldwide burden of fire and other hazards on the quality of life by providing and advocating scientifically based consensus codes and standards, research, training, and education. (NFPA 2008).

Risk Assessment

The NFPA's use of performance-based fire protection changed significantly during the 1990s. These changes evolved because of global competition, the availability of computer-run analysis tools, unique designs for large public facilities, a rise in new technology, and the occurrence of a few major disasters. Most codes and standards developed were previously based on prescriptive concepts that present design solutions that imply "Build it this way and the design is considered safe." These codes were once adopted and enforced by the local, state, or national jurisdictions having authority. As major losses occurred, the codes were changed or modified. Prescriptive-based codes, however, were grounded in the technology existing at the time they were developed and were not flexible enough as new technologies evolved. In the 1990s, NFPA began developing codes based on performance-based fire protection; these codes focused on design outcome and not simply design solutions (Rose, et al. 2007).

When developing a performance-based code, NFPA looks at the fire risk when conducting a fire risk assessment. With the perception of risk and the acceptance of a risk, this process is always influenced by the values of all stakeholders. For fire safety the hazards are generally fire, explosion, smoke, and toxicity associated with fire. Some of the stakeholders are regulators, employees, community, investors, emergency responders, facility owners and operators, and insurers. All groups have a chance to comment on the proposed code during the development process over a specified period of time during which modifications will be made before a final acceptance of the code is released.

Summary

NFPA codes go through a long process before a new document is published and released for use. First a committee reviews a document that has been placed into a revision cycle. The committee then provides a standardized basis for management and business continuity in private and public sectors by providing common elements, techniques, and processes. A collection of comments is gathered from the public comments process before a draft document is produced. At this point the committee members have to approve the new draft document with a two-thirds majority vote of the committee members. The next step is to have a vote of the NFPA membership followed by an approval from the Standards Council (NFPA 2000).

American National Standards Institute (ANSI)

History

The American National Standards Institute has served as the coordinator of the U.S. voluntary standards system, which includes industry, standards-developing organizations, trade associations, professional and technical societies, government, labor, and consumer groups (ANSI 1999). It has provided a forum where the private and public sectors can work together toward the development of voluntary national consensus standards. ANSI provides the means for the United States to influence global standardization activities and the development of international standards. It is the dues-paying member and sole U.S. representative of the two major nontreaty international standards organizations—the ISO and the International Electrotechnical Commission (IEC)—via the United States National Committee (USNC) (ANSI 1999).

The history of ANSI and the U.S. voluntary standards system is dynamic. Discussions to coordinate national standards development in an effort to avoid duplication, waste, and conflict date back to 1911. In 1916 the American Institute of Electrical Engineers (now IEEE) invited the American Society of Mechanical Engineers (ASME), American Society of Civil Engineers (ASCE), American Institute of Mining and Metallurgical Engineers (AIMME), and the American Society for Testing Materials (ASTM) to join in establishing a national body to coordinate standards development and to serve as a clearinghouse for the work of standards-developing agencies (ANSI 1999).

ANSI, originally founded as the American Engineering Standards Committee (AESC) was formed on October 19, 1918, to serve as the national coordinator in the standards-development process as well as an impartial organization to approve national consensus standards and halt user confusion on acceptability.

ANSI adopted its present name in 1969. Throughout these various reorganizations and name changes, the institute continued to coordinate national and international standards activities and to approve voluntary national standards, now known as American National Standards. Domestic programs were constantly expanded and modified to meet the changing needs of industry, government, and other sectors (ANSI 1999).

ANSI is a private, nonprofit organization that administers and coordinates the U.S. voluntary standardization and conformity assessment system. Its mission is to enhance U.S. global competitiveness and the American quality of life by promoting, facilitating, and safeguarding the integrity of the voluntary standardization system. The institute represents the interests of its company, organizational, governmental, institutional and international members through its headquarters in Washington, D.C., and its operations center in New York City (ANSI 1999).

Conformity Assessment

Conformity assessment is defined as "any activity concerned with determining directly or indirectly that relevant requirements are fulfilled" (ANSI 1999). Many of these conformity-assessment activities are applied in today's marketplace, including accreditation, certification, inspection, registration, supplier's declaration, and testing, but the one dimension that ANSI is directly engaged with is accreditation.

The institute provides accreditation services, specifically in product and personnel areas that recognize the competence of bodies to carry out product or personnel certification in accordance with requirements defined in international standards. ANSI accreditation programs are themselves created in accordance with similar international guidelines as verified by government and peer-review assessments.

Summary

ANSI provides accreditation services, specifically in product and personnel areas that recognize the competence of bodies to carry out product or personnel certification in accordance with requirements defined in international standards. ANSI accreditation programs are themselves created in accordance with similar international guidelines as verified by government and peer-review assessments. Furthermore, in partnership with the Registrar Accreditation Board (RAB), ANSI serves the marketplace in the provision of a National Accreditation Program (NAP) for quality and environmental management systems registrars (ANSI 1999).

Continuing pressures in the global marketplace to preclude redundant and costly barriers to trade drive the need for acknowledgment of equivalency across boundaries. Accordingly, ANSI is involved in several international and regional arrangements for multilateral recognition. These include the International Accreditation Forum (IAF), the Inter-American Accreditation Cooperation (IAAC), and the Pacific Accreditation Cooperation (PAC). ANSI is also recognized by the U.S. Department of Commerce via the National Institute for Standards and Technology (NIST) and their National Voluntary Conformity Assessment System Evaluation (NVCASE) program (ANSI 1999).

Conclusion

Regulatory issues for risk assessment involve a complex process that depends on the quality of scientific information available. Conducting risk assessments is best for acute risks where effects appear soon after exposure occurs. Uncertainty becomes greater as the period of time between exposure and appearance of symptoms or activities grows longer. In many circumstances uncertainties can make it impossible to come to a firm conclusion about risk (Stulz 2003).

Risk assessment is a tool used to facilitate decisions about how to optimize the use of scarce resources. Risk assessment provides the basis for determining the risk involved in certain processes and activities, and it provides justification for actions that have been undertaken.

Properly used, risk assessments often provide an essential ingredient in reaching decisions on the management of hazards. The results of a risk assessment are often used to inform rather than dictate decisions and are only one of many factors taken into account in reaching a decision (Ropeik 2000).

The use of risk assessment is not without controversy. For example, an approach based on the assessment of risk could be seen to underestimate the true impact of a problem and could undermine the adoption of precautionary approaches based on anticipating and averting harm (O'Brian 2000).

A risk assessment is a several-step process that, once completed, will provide a basis for establishing and judging tolerable risk. Information and perspectives are gathered while progressing from one stage to another, often requiring early stages of the process to be revisited. The process is iterative. Stakeholders should be involved at all stages, although final decisions may not always be taken by consensus because the various stakeholders may present and hold different or even opposing views. Risk assessment is a combination of analysis and evaluation that leads to a risk decision. All the regulations require risk assessment and revisions to the risk assessment as necessary.

Important Terms (U.S. EPA 2003)

Hazard: Anything, physical or chemical, that can cause harm

Risk: A chance, high or low, that someone will be harmed by the hazard

Risk assessment: A process by which results of a risk analysis or risk estimates are used to make decisions through ranking or comparison

Risk management: Planning, organizing, leading, and controlling assets and activities through specific means that minimize adverse operational and financial effects of losses

References

Alijoyo, Antonius. 2004. *Focused Enterprise Risk Management*. 1st ed. Jakarta, Indonesia: PT Ray.

American Conference of Governmental Industrial Hygienists (ACGIH). 2002. "Guidelines for Classification of Occupational Carcinogenicity." In *Documentation of the Threshold Limits Values and Biological Exposure Indices*. 7th ed. Cincinnati:ACGIH.

American National Standards Institute (ANSI). 1999, modified July 2005. *ANSI Annual Report: Overview and History* (retrieved June 10, 2005). www.ansi.org/

Baird, S., J. Cohen, J. Graham, A. Shlyakhter, and J. Evans. 1996. "Cancer Risk Assessment: A Probabilistic Alternative to Current Practice." *Human Ecological Risk Assessment* 2(1):79–102.

Barton, A., and A. Sergeant. 1998. "Policy before the Ecological Risk Assessment: What Are We Trying to Protect?" *Human Ecological Risk Assessments* 4(4):787–795.

Bascietto, J. J. 1998. "A Framework for Ecological Risk Assessment: Beyond the Quotient Method." In M. C. Newman and C. L. Strojan, *Risk Assessment: Logic and Measurement*, p. 352. Ann Arbor, MI: Ann Arbor Press.

Benner, Ludwig, Jr. 1983. "What Is This Thing Called a Safety Regulation?" *Journal of Safety Research* 14(4):139–143.

Center for Disease Control (CDC). 2003. *National Report on Human Exposure to Environmental Chemicals*. NCEH Publication No. 02-0716 (retrieved February 10, 2005). www.cdc.gov/exposurereport/

Dear, Joseph A. 1995. *The Hearing of the United States House Science Committee on Risk Assessment and Title III of H.R. 9* (retrieved February 15, 2005). www.osha.gov/pls/oshaweb/owadisp.show_document?p_table=TESTIMONIES&p_id=72

Dorfman, Mark S. 1997. *Introduction to Risk Management and Insurance*. 6th ed. Upper Saddle River, NJ: Prentice Hall.

Extension Toxicology Network (EXTOXNET). 1998. *Toxicology Information Briefs* "Risk Assessment Background." www.extoxnet.orst.edu/tibs/riskasse.htm

Finney, C., and R. E. Polk. 1992. "Developing Stakeholder Understanding, Technical Capability, and Responsibility." *The New Bedford Harbor Superfund Forum Environmental Impact Assessment Review*. 15:517–541.

International Organization of Standardization (ISO). 2004. *Code of Ethics: ISO Action Plan for Developing Countries—2005* (retrieved June 10, 2005). www.iso.com/about_iso/

Jasanoff, Shelia. 1993. "Procedural Choices in Regulatory Science." *Risk: Health, Safety & Environment* 4:143–160.

Kaminski, L., A. Griffiths, M. Buswell, J. Dirsherl, J. Bach, H. Bach, T. Van Dyk, and J. Gibson. 2003. "GICHD—Risk Assessment and Mechanical Application." *EUDEM2-SCOT* 1:335–341.

Lack, Richard W., ed. 2001. *The Dictionary of Terms Used in the Safety Profession*. Des Plaines, IL: American Society of Safety Engineers.

McGarity, Thomas O., and Sidney A. Shapiro. 1993. *Workers at Risk: The Failed Promise of the Occupational Safety and Health Administration*. Westport, CT: Praeger.

McGarity, Thomas O., and Sidney A. Shapiro. 1996. "OSHA's Critics and Regulatory Reform." *Wake Forest Law Review* 31(3):587.

National Research Council (U.S.). 1983. *Risk Assessment in the Federal Government--Managing the Process.* Washington, D.C.: National Academy Press.

National Fire Protection Association (NFPA). 2008. www.nfpa.org

O'Brian, Mary. 2000. *Making Better Environmental Decisions: An Alternative to Risk Assessment.* Cambridge, MA: MIT Press.

Occupational Safety and Health Administration (OSHA). 1992a. 29 CFR 1910.119, Subpart H, *Process Safety Management of Highly Hazardous Chemicals* (retrieved April 1, 2005). www.osha.gov

______. 1992b. CPL 02-02-045 - CPL 2-2.45A - Process Safety Management of Highly Hazardous Chemicals—Compliance Guidelines and Enforcement Procedures.

______. 1993. 29 CFR 1910.146, *Permit-Required Confined Spaces.*

______. 1995a. *All about OSHA*, OSHA pamphlet number 2056. Washington DC: U.S. Department of Labor.

______. 1995b. *Risk Assessment and Title III of H.R. 9.*

______. 1998. 29 CFR 1910.146, Appendix A."Permit-required Confined Space Decision Flow Chart."

______. 2000. *Process Safety Management.* OSHA Document 3132 (retrieved January 29, 2008). www.osha.gov/Publications/osha3132.html

______. 2006. 29 CFR 1910, Subpart Z, *Toxic and Hazardous Substances* (retrieved November 10, 2007). www.osha.gov

______. 2007a. 29 CFR 1910, Section VI, *General Industry Standards* (retrieved November 10, 2007). www.osha.gov

______. 2007b. 29 CFR 1926. *Safety and Health Regulations for Construction.*

Oliver, James. 2005. "The National Environmental Policy Act." Chapter 18 in *Hazardous Materials Management Desk Reference.* 2d ed. Rockville, Md.: Academy of Certified Hazardous Materials Managers.

Rhomberg, L. R., 1997, "A Survey of Methods for Chemical Health Risk Assessment among Federal Regulatory Agencies." Risk Assessment and Risk Management Commission. Washington, D.C.: U.S. EPA.

Ropeik, David. 2000. "Let's Get Real about Risk." *Washington Post*, August 6, 2000, p. B1.

Rose, Susan, Stephanie Flamberg, and Fred Leverenz. 2007. *How to Perform a Risk Assessment According to NFPA 1600 ANNEX A* (retrieved November 10, 2007). Praxiom Research Group Limited. www.praxiom.org/risk-assessment.htm

Russell, M., and M. Gruber. 1987. "Risk Assessment in Environmental Policy-Making," *Science Journal* 236:286–290.

Schoeny, R., P. Muller, and J. Mumford. 1998. "Risk Assessment for Human Health Protection: Applications to Environmental Mixtures." In *Pollution Risk Assessment and Management*, pp. 205–234. P. Douben, ed. New York: John Wiley and Sons.

Stulz, Rene M. 2003. *Risk Management and Derivatives.* 1st ed. Mason, OH: Thomson South-Western.

U.S. Consumer Product Safety Commission (CPSC). 2002a.*U.S. Consumer Product Safety Commission Annual Report to Congress* (retrieved January 29, 2008). www.cpsc.gov/cpscpub/pubs/reports/2002rpt.pdf

______. 2002b, "CPSC & EPA Both Consider Issues Related to CCA-Treated Wood." *CPSC Monitor* (September) 7(9).

______. 2007. "Voluntary Standards Activities FY 2007." *Mid Year Report* (October 2006–March 2007), U.S. Consumer Product Safety Commission staff report. Bethesda, MD: CPSC.

U.S. Environmental Protection Agency (U.S. EPA). 1993. "Health Effects Assessment for Mercury, EPA/540/1-86/042 (NTIS PB86134533)." Washington, D.C.: U.S. Environmental Protection Agency. Prepared in cooperation with Syracuse Research Corp., NY. See also NTIS PB85-123925, PB86-134525, and PB86-134541.

______. 1994. "Model Validation for Predictive Exposure Assessments." Risk Assessment Forum, Washington, DC (retrieved June 10, 2005). www.cfpub.epa.gov/crem/cremlib.cfm

______. 2000. *Risk Characterization Handbook*, EPA 100-B-00-002. Washington, DC: Science Policy Council.

______. 2001. "Developing Management Objectives for Ecological Risk Assessments," Risk Assessment Forum, Washington, DC.

______. 2002a. "Health Assessment Document for Diesel Engine Exhaust, EPA/600/8-90/057F." Washington, D.C.: U.S. Environmental Protection Agency, Office of Research and Development, National Center for Environmental Assessment, Washington Office.

______. 2002b. "Questions & Answers – What You Should Know About Wood Pressure Treated with Chromated-Copper-Arsenate (CCA)." *Pesticides: Topical & Chemical Fact Sheets.* Washington, D.C.: EPA (retrieved January 29, 2008). www.epa.gov/oppad001/reregistration/cca/cca_qa.htm

______. 2003. *Framework for Cumulative Risk Assessment.* EPA/600/P-02/001F. Washington, DC: U.S. Environmental Protection Agency, Office of Research and Development, National Center for Environmental Assessment.

Watts, J. M., Jr. 2002. "Risk Indexing." *SFPE Handbook of Fire Protection Engineering.* 3d ed. Quincy, MA: National Fire Protection Association.

Appendix: Further Reading

Alexander, Carol, and Elizabeth Sheedy, eds. 2004. *The Professional Risk Managers' Handbook: A Comprehensive Guide to Current Theory and Best Practices.* Wilmington, DE: PRMIA Publications.

Almand, Kathleen H. 2007. "Fire Risk Assessment as a Tool." *NFPA Journal* (March/April) (retrieved November 10, 2007). www.nfpa.org/publicColumn.asp

EPA. Information Quality Guidelines Web site. www.epa.gov/oei/qualityguidelines

______. Quality System Web site. www.epa.gov/quality

______. Risk Assessment Guidelines Web site. www.cfpub.epa.gov/ncea/raf/recordisplay.cfm?deid=55907

______. Science Policy Council Web site. www.epa.gov/osp/spc

Health and Safety Executive. *Five Steps to Risk Assessment.* www.hse.gov.uk/risk/fivesteps.htm

______. 1992. *Management of Health and Safety at Work: Approved Code of Practice, L21.* Sudbury, Suffolk, UK: HSE Books.

______. 2006. *Essentials of Health and Safety at Work.* Sudbury, Suffolk, UK: HSE Books.

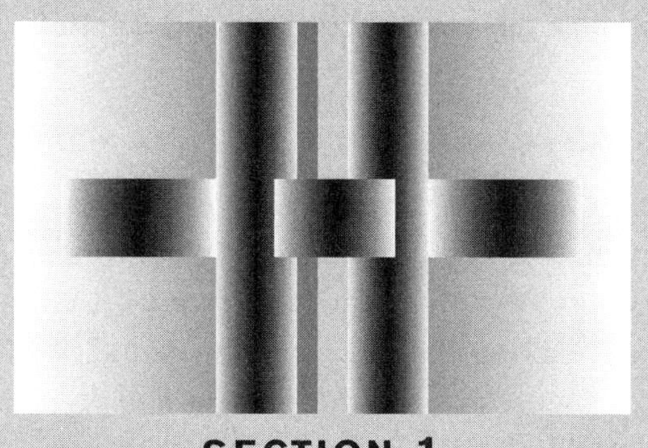

SECTION 1

RISK ASSESSMENT AND HAZARD CONTROL

LEARNING OBJECTIVES

- Become familiar with system safety terminology.
- Be able to recognize various hazardous energy sources and controls.
- Understand the concepts of risk and risk assessment.
- Understand risk mitigation and acceptability criteria.
- Learn about basic design solutions for identified hazards.
- Grasp basic analysis techniques, such as fault tree analysis, event tree analysis, failure modes and effects analysis, hazard and operability studies (HAZOPs), functional and control flow analysis, and sneak circuit analysis.
- Learn the purpose and use of hazard-analysis worksheets and hazard reports.

APPLIED SCIENCE AND ENGINEERING: SYSTEMS AND PROCESS SAFETY

Mark D. Hansen

TO SOME, system safety is an arcane tool, but to others it is a tool used daily in the course of doing their jobs. It has received a bad rap over time because many have acquired mental blocks to this style of applying safety to hazard prevention. Even with all the negative press, system safety is a discipline that has been effectively used to eliminate and control hazards (Roland and Moriarty 1983). System safety has saved many lives in the military, NASA, the Department of Energy, and in many commercial applications.

System safety provides a systematic approach to safety. It begins in the conceptual stages of a project and continues through operation and disposal. It involves the following:

- managing the safety program—planning and execution
- analyzing the system for safety problems at all design stages
- identifying design requirements to make the system safe
- recording results.

SYSTEM SAFETY CONCEPTS

This systematic approach introduced science to the "Fly-Fix-Fly" methods that the Air Force was already using (Malasky 1975). This involved building a system (airplane), operating (fly) it until something went wrong and caused an accident, fixing (fix) the problem, and then operating (fly) it again until something else broke. This was a costly approach in terms of equipment and human loss. By analyzing an aircraft design for hazards and fixing them during design, the Air Force had fewer accidents (Roland and Moriarty 1983). A systematic approach makes safety more of a science than an art. It catches safety problems during the design phase where they can be corrected more economically.

Safety should not be ensured at the work site alone. It must be ensured ahead of time, beginning with the equipment used and the operations performed. This requires advanced planning for safety from the earliest conceptual stages, and continuous safety awareness and analysis throughout design, construction, and operation. The entire system and its interfaces must be considered. The risks must be identified as much as possible, either eliminated or controlled to an acceptable level of risk, and accepted by management.

Hazard Analysis

The basic purpose of system safety is to identify the risk of a system. This involves analyzing the system to identify the following:

- hazards
- outcomes of potential mishaps (hazard effects)
- the potential mishap's occurrence
- the severity and probability of potential mishaps (risk assessment) and how to reduce the likelihood of the potential mishaps occurring (hazard control)

When analyzing a system for hazards, first, make the task manageable. For simple devices such as a toaster or food processor, it may be easy enough to look at the device as a whole and see the hazards involved. With complex systems such as satellites, space vehicles, or chemical plants, the task may initially seem overwhelming. For these systems, the first step is to break the system down into subsystems and the subsystems into components. Next, one looks at the subsystems and components individually for hazards. Then one should analyze the interfaces of these parts for hazards.

During the conceptual design stages, there is little design detail. One may know only what functions the system performs and what types of subsystems may be involved. Still, it is wise to look at the system and ask, what are the basic hazards associated with the system functions and subsystems? A detailed description of the system should be documented to support any analysis (DOD 2000). This at least allows for pinpointing areas of concern for more detailed analysis later on.

Hazard Cause and Effect

As noted earlier, a hazard is the potential for a mishap. Hence, hazards need to be documented in complete statements—generally with a noun and a verb—that accurately describe a condition or event that could result in a mishap. One-word hazards such as *hydrogen* are not adequate. Hydrogen in itself is not a hazard, but merely an element. It is a component of water. It becomes hazardous only when it is released into an area where it could contact an ignition source (creating the potential for fire or explosion) or when it displaces oxygen in an enclosed space (creating the potential for asphyxiation). A better hazard statement would be the following: "Hydrogen inadvertently released into the laboratory."

One must also be careful not to express the hazard as a mishap. Suppose a hazard is labeled "fire." Usually when there is a fire, there is a mishap. The *hazard* is the condition or event that created the potential for fire. Such a hazard might be expressed as the following: "Inadvertent leakage of gasoline onto the garage floor."

The *hazard effect* is the credible worst-case mishap that could result from an uncontrolled hazardous condition. This describes the potential severity of the hazard. One also needs to estimate the probability of the mishap's occurrence. Often, the probability is determined by an educated guess, rather than by exact methods. The combination of severity and probability yields the risk faced from that hazard. Decision makers can then consider all risk factors to determine which hazards get fixed and how, considering available time and money. Knowing the cause of the hazard helps in identifying hazard controls.

For example, suppose a steel tower is being designed. It is next to a wooden building full of people. Assuming the tower is designed with acceptable safety factors (e.g., design codes, consensus standards), one hazard would be weakening of the steel structure by corrosion. The effect would be collapse of the tower onto the building. The severity would be death or severe injury to the building occupants and loss of building and tower. The cause would be exposure of the steel to humid salt air. The mishap is likely to occur in time. Therefore, this poses a high-risk hazard that needs to be controlled. From the cause, it is obvious that the steel needs to be protected—probably with a corrosion-resistant protective coating. The identified hazard information would be as follows:

- Hazard: Weakened structure due to corrosion in the steel
- Cause: Exposure to humid, salt air
- Effect: Collapse of tower onto building, causing death and injury
- Risk: High severity and probability: high risk
- Control: Paint tower with corrosion-resistant protective coating

When describing hazards, keep the following in mind:

- Make a complete hazard statement that describes a hazardous condition or event. Avoid one- or two-word hazard statements.
- Ensure that there is a logical flow from the cause to the hazard to the effect. The cause should cause the hazard, and the hazard should cause the effect.

The following is an example of an incorrect hazard description:

- Hazard: Oxygen pressure
- Cause: Oxygen leakage due to loosening of fittings, chafing, or excess pressure
- Effect: Injury and/or system damage due to fire

Is "oxygen pressure" a hazardous condition or event? Not if it is properly contained. Does "oxygen leakage" cause "oxygen pressure"? Does "oxygen pressure" cause "fire"? Not only does this hazard statement not state a hazardous condition or event, but further, there is no logical flow from cause to hazard to effect.

In this case, a higher oxygen concentration increases the potential for fire. Therefore, a better hazard statement would be "Oxygen-enriched atmosphere due to oxygen leakage and ignition source present." The revised hazard description could read as follows:

- Hazard: Oxygen-enriched atmosphere due to leakage from oxygen lines and ignition source present
- Cause: Loosening of oxygen fittings, chafing, or excess pressure
- Effect: Injury and/or system damage due to fire

There may be other ways to adequately describe this hazard. The important things are the complete hazard statement and logical flow.

Controlling Hazards

There are three factors to consider in controlling hazards: effectiveness, feasibility, and cost. The hazard control should to be effective in reducing the likelihood of the potential mishap and should not introduce any new hazards into the system. Next, the control must be feasible—not degrade system performance to an unacceptable level. Finally, one must consider cost. Can the organization afford the control? Will there be an adequate payback in risk reduction? (e.g., is the organization spending $100,000 to protect a $20,000 piece of equipment?) Are there hazards with a higher risk that need to be fixed first?

Controlling hazards requires use of an appropriate method (engineering design, interlocks, etc.) (Hammer 1989) to reduce the severity and/or probability of occurrence (DOD Directive 3150.2). For example, one could follow what is known as the hazard-reduction precedence sequence (HRPS). The HRPS may also be called by other names, such as risk-reduction sequence, system safety precedence, or safety precedence sequence.

Whatever the name, the sequence is basically the same. In order of most effective to least effective, the HRPS is as follows:

- Design for minimum hazard.
- Install safety devices.
- Install warning devices.
- Use special procedures.

After using this sequence, there may be some residual risk. *Residual risk* is the risk remaining after controls have been put in place to control the hazard. There are few operations that are risk-free. Management must either accept this risk or go through the sequence again.

Designing for minimum hazards is the safest alternative. However, this may render the system useless or be too costly. Eliminating the explosives from a high-tech smart bomb makes it totally safe, but the bomb cannot perform its intended function. On the other side of the coin, using solid fuel instead of liquid fuel may be a safer alternative for a launching system.

If it is not practical or possible to design the hazard out of the system, safety devices must then be installed. These may include pressure-relief valves to prevent overpressure; interlocks that fail to a safe condition to prevent injuries; safe/arm devices that act like safeties on weapons, protective barriers such as laser beams that, once broken, halt machine operation; and so on. Safety devices are part of the hardware, not dependent on human performance, and are effective to the extent that the hardware is used as designed.

If safety devices are not practical or possible, then warning devices may be installed, such as alarms, signs, and so on. These are effective only to the extent that they are used properly.

Special procedures are the least effective control. This includes not only written procedures, but also personal protective equipment such as protective suits, self-contained breathing apparatuses, and so on. Procedures are effective only to the extent that people follow them. Personal protective equipment is only effective to the extent that it is used properly. Procedures are dependent on human factors and human behavior, which is often unpredictable. Training personnel becomes critically important when relying on safety procedures.

Suppose that a robot is taking parts from a bin and putting them on widgets. One hazard to consider would be:

- Hazard: Fast-moving robot arm around people or equipment
- Cause: Excessive speed and lack of barriers to keep out people and equipment
- Effect: Injury or damage due to impact of robot arm on humans or valuable equipment

Now one must develop hazard controls.

First, consider design measures. One could slow down the robot arm so that impact with personnel or equipment will not cause injury or damage. This may not be practical because it would reduce the robot's productivity.

Safety devices may be more practical. There are several options: put a fence around the robot's operating area, install software-hardware interlocks on the gate so that the robot shuts down when the gate is open, put special mats around the robot so that it shuts off when someone steps on them, pad the robot arm.

If management thinks the safety devices are too costly, then one resorts to warning devices and procedures and policies. These may include painting a yellow line around the robot's operating area, installing a red light that flashes when the robot is operating, and instructing employees not to enter the area. One could also put up a warning sign or develop special procedures for people to follow when around the robot.

The important thing to remember is that once a hazard is identified, steps must be taken to control it. Methods such as the HRPS offer a systematic approach or sequence to go through for finding control measures. What control measures are applied will depend on operational effectiveness, cost, and schedule. The high-risk hazards should be controlled first. Management may have to accept some residual risk even after the controls have been applied.

How to Recognize Hazards

There is no one rule for hazard recognition. Hazards come in many different forms. They can result from design deficiencies, human error, environmental factors, and so on. This section gives several general guidelines that help us recognize hazards. Hazard recognition is important early on in the project and in selecting appropriate hazard-recognition techniques to fit the types of hazards or concerns that one desires to control.

Energy Sources

Most hazards are related to an inadvertent or uncontrolled release of energy. An explosion causes shock waves and kinetic energy of shrapnel, which can cause injury or damage. Uncontrolled electrical energy can cause injury or death and property damage. So the first task is to identify all energy sources. Then one needs to determine what targets (people, objects, environment, etc.) could be injured or damaged by any release of that energy. Look for barriers to harmful energy flow between the source and target. Barriers may be physical (design measures) or administrative (policies and procedures). Figure 1 illustrates the relationship of the energy source, barrier, and target and shows how barriers can prevent injuries from energy sources.

Energy may be of one or more of the following types:

- Mechanical—rotating machinery
- Potential—raised crane load, coiled spring; resulting in kinetic energy—falling or flying objects
- Pressure (pneumatic, hydraulic, acoustic)—stored pressure that could be released
- Thermal—high or low temperatures
- Chemical—toxic or corrosive chemicals, reactants
- Electrical—energy that is used to operate equipment
- Radioactive—energy that is used to provide power

Many systems will contain several energy sources. A high-temperature furnace contains both thermal and electrical energy. Also consider interaction among many energy sources. For instance, a tank of gasoline is next to a house in a lightning-prone area. There are two energy sources—electrical energy from the lightning and chemical energy in the gasoline. If the two interact in the right combination, this could cause a fire or explosion.

The targets are the house and its occupants. A hazard statement for this situation should be written in terms of the energy and targets. Barriers that prevent energy transfer to the targets are the hazard controls.

The hazard could be described as follows:

- Hazard: Leaking gasoline tank next to house in lightning-prone area.
- Cause: Careless placement of gasoline tank
- Effect: Death or injury and damage from fire or explosion
- Controls: Move tank a safe distance from the house, install lightning shield around tank, build a blast wall between tank and house, and abandon house before lightning storm

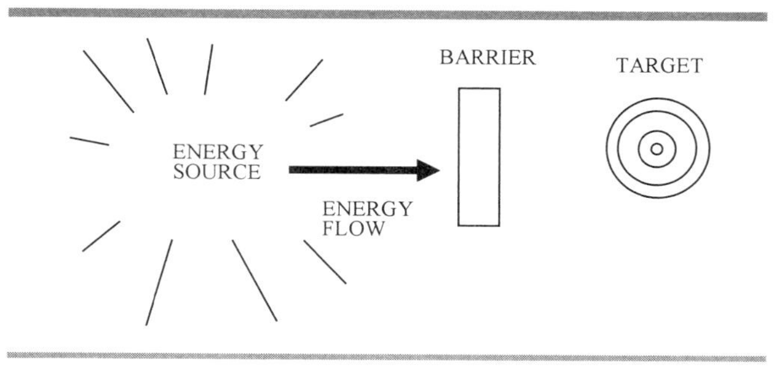

FIGURE 1. Relationships between energy source, barrier, and target (*Source:* Hansen 1993)

Looking for energy sources, targets, and barriers is the best technique to use during the conceptual design stages. One may not see or be able to define all barriers because the design detail is not defined. However, general requirements can be defined. High-energy subsystems are candidates for more detailed analysis later on.

Non-energy-Related Hazards

Examining energy sources enables one to identify most of the serious hazards, but by restricting the process only to energy, one could miss other serious hazards. Such hazards can involve interruptions to normal life functions—for example, asphyxiation, smoke inhalation, or disease—or problems that result from human factors.

- to find such hazards, look for the following:
- presence of inert gas
- potential for loss of breathable atmosphere
- potential for disease
- human interfaces with the system
- software hazards

Software in and of itself is not hazardous. However, it can become hazardous when it interfaces with hardware. Software that controls hardware functions may command an undesired event or condition. Software that monitors hardware may fail to sense or properly process a hazardous condition. There are various methods for analyzing software. It is important that hardware designers identify all software interfaces with the hardware and bring them to the attention of the software designers and analysts. This includes subroutine controls for weapons systems, traffic lights and traffic control, communications systems, cancer irradiation equipment, and so on.

Failures

Many hazards do not result from failures. However, some do. Consider two cases:

1. The failure of safety-critical subsystems—life support, fire alarm or sprinkler, and so on—can cause a hazardous situation. Look at the organization's system to see if it contains any such subsystems. If it does, first, determine the probability that these systems will fail, and then take measures to reduce the probability, if necessary.
2. Component failures can cause hazards or mishaps. For instance, failure of a hydrogen-line coupling can cause hydrogen to leak into a room and create a potential for explosion.

Examine the system to see if any such failures are possible. This can be done in the course of the safety analysis, or one can consult a failure modes and effects analysis, if it is available.

Helpful Hints for Recognizing Hazards

Using the team approach is best. In most cases, no one person can be familiar with all aspects of system design and operation and related hazards. At a minimum, a knowledgeable engineer should analyze the system first and give the results to a safety engineer for review. Consult experts frequently. Experience is invaluable for many reasons, but one should learn from a past history of accidents to avoid repeating them. Inquire about past accidents associated with the organization's system.

Look at all operating modes. Often, a system is scrutinized only while it is in normal operating mode. Problems often do not occur here, but at startup or shutdown. Other examples of operating modes include transport, delivery, installation, checkout, emergency startup and shutdown, and maintenance.

The following is a list of proven methods for finding hazards (Roland and Moriarty 1983):

- Use intuitive engineering sense.
- Examine similar facilities or systems.
- Examine system specifications and expectations.
- Review codes, regulations, and consensus standards.
- Interview current or intended system users or operators.
- Review system safety studies from other similar systems.
- Review historical documents—mishap files, near-miss reports, OSHA-recordable injury rates, National Safety Council data, manufacturers' reliability analyses, and so on.
- Consider external influences, such as local weather, environment, or personnel tendencies.
- Brainstorm—mentally develop credible problems and play "what if?" games: "What could go wrong?" "What is the worst possible thing that could happen?"
- Consider all the energy sources (pressure, motion, chemical, biological, radiation, electrical, gravity, and heat and cold). What's necessary to keep these sources under control? What happens if they get out of control?
- Use checklists such as the hazard category list presented in the following section.

Hazard Categories

The following is a checklist of hazard categories. These are also called *generic hazards*. Note, first, that most are one- or two-word statements and therefore not appropriate for hazard statements. Rather, they are intended as thought joggers. Some of them, such as collision and fire, describe mishaps instead of hazardous conditions or events. In those cases, look for the potential hazard—for example, a drunk driver or bad brakes for a collision. Some are interrelated. For instance, contamination from a corrosive liquid can cause corrosion on a metal structure, which in turn can cause structural failure.

This list is not complete. If one finds hazards that do not fit into any of these categories, add another category to the existing list.

Figure 2 outlines the procedure for using each category as a guide to identifying hazards.

Checklist of Hazard Categories

1. Collision

Collisions are usually thought of as mishaps. Hence, in this category, one looks for the potential for collisions between people and objects or objects and other objects. Collisions often result in injury or death to people and damage to equipment.

Conditions	Causes
Structural failure	Inadequate structure
Inadvertent motion (especially in zero-gravity)	Human error
Falling objects	Horseplay
Flying projectiles	Inadequate handling
Lifting equipment	Equipment failure

2. Contamination

Contamination could be a condition, a hazard or hazard cause, or a mishap. When there is a chemical spill, it is usually considered a mishap because at least environmental damage can be involved. Leakage of corrosive material onto a steel structure is a condition that can cause corrosion, which in turn could cause structural failure. Contamination could involve release of hazardous materials (e.g., toxic, flammable, corrosive), dust and dirt particles (causing degraded performance or health problems), growth of fungus, and so on. It could result in death or illness to people, damage to equipment or environment, or loss of resources (e.g., contaminated fuel or water).

Conditions	Causes
Presence of hazardous materials	Inadequate containment
Leaks	Human error
Dust or dirt in critical systems or components	Improper handling
Presence of fungi	Environmental factors (rust, wind, sand, etc.)
Materials that off-gas (release vapors)	Worn seals, gaskets, joints
Toxic substances	

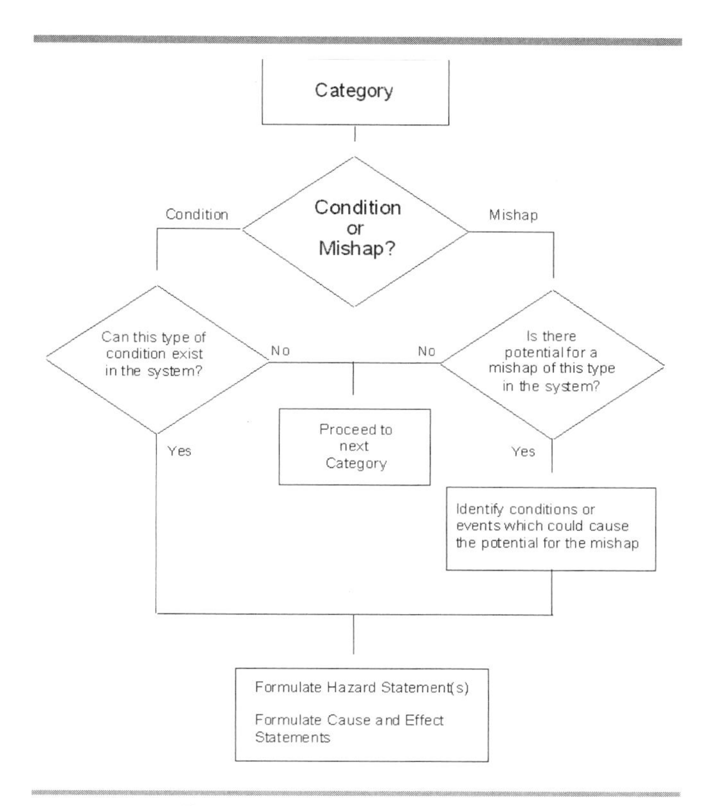

FIGURE 2. Procedure for using each category as a guide to identify hazards (*Source:* Malasky 1974)

3. Corrosion

Corrosion is usually thought of as a condition that could lead to a mishap. It can result from leakage of corrosive or reactive materials (acids, salts, solvents, halogens, etc.) or from other factors, such as galvanic corrosion from joining two dissimilar metals due to a direct current electricity that is chemically generated or exposure to environmental extremes (temperature or humidity). Corrosion can result in death or injury to personnel or in equipment damage (structural or mechanical failure).

Conditions	Causes
Presence of corrosive materials	Contamination
	Inadequate corrosion protection
Joining dissimilar metals	Design errors
Stress corrosion	Environmental factors (vibration, temperature)
Polymerization	Hydrogen embitterment

4. Electrical

In this category one looks for conditions that could cause electrical mishaps—shock, arcing, overcurrent, electrical fire, and so on. Such conditions may include potential for people contacting energized parts, sensitive electronics that need overcurrent protection, short circuits, and potential for environmental electricity (e.g., static electricity, lightning, etc.). Mishaps from this category could include death or injury from electrical shock, equipment damage, overcurrent, overheating and burns, or fire.

Conditions	***Causes***
Exposed electrical conductors	Inadequate insulation
Sensitive electronic parts	Worn insulation
Energized casing or parts	Inadequate grounding
Short circuit	Lightning potential
Lightning/static electricity	Contamination
Flammable materials with electrical parts	Damaged electrical connectors

5. Explosion

An explosion is usually considered a mishap. It is a violent energy release that causes a shock wave and flying fragments. It can result from misuse of explosives, increased temperature or pressure, chemical reactions, and so on. Equipment such as pumps, motors, blowers, generators, and lasers can malfunction and explode. Explosions can result in death or injury to people or damage to equipment both from shock waves or shrapnel and from spread of toxic substances.

Conditions	***Causes***
Presence of explosives and ignitors	Improper handling
Reactive chemical processes	Inadequate pressure relief
High temperature/pressure	

6. Fire

Fire is usually thought of as a mishap. The hazards to look for in this category are conditions that could cause a fire. Fuel (some flammable material), an oxidizer (oxygen, nitric acid, nitrogen tetroxide, etc.), and an ignition source (spark, electrical arc, pilot flame, hot surface, etc.) are necessary to cause a fire. Without any one of these, there is no hazard. Generally, the presence of highly flammable materials is a hazardous condition. Oxygen is present in the air. Potential ignition sources are all around—rotating machinery, heaters, faulty electrical wiring, and so on. Only in highly controlled environments can one claim there are no oxidizers or ignition sources. Fire can result in death, injury, or illness to people or damage to property. Most personnel injury results from smoke inhalation (Hammer 1989). Burning some materials, such as plastic, can spread toxic fumes. Review material safety data sheets because there are some situations where no ignition source is needed to result in fire (concentrated hydrogen peroxide and pyrophorics such as aluminum alkyls are examples). Also, some chemicals may include oxygen sources and burn in low- or no-oxygen atmospheres.

Conditions	***Causes***
Presence of flammable materials	Inadequate containment
Organic substances	Careless handling of flammables or oxidizers
Flammable metals	
Chemicals	
Oxidizers	

7. Physiological

This category deals with conditions that could cause health problems in human beings—physical or mental disability, asphyxiation, disease, and so on.

Conditions

Presence of viruses, bacteria, oxygen-deficient atmospheres, carcinogens, mutagens, teratagens, and so on

Rapid pressure or temperature changes

8. Human Factors

Human error can cause hazards or accidents in any of the categories listed. This category is included to emphasize the importance of looking for ways that human error can cause hazards.

Conditions

Too much physical or psychological stress

Hostile environment

Hardware not designed for humans

Inadequate, confusing, or difficult procedures

Untrained or unqualified people

Recent changes

9. Loss of Capability
This could be a condition or a mishap. Loss of some critical function, such as life support, constitutes a mishap resulting in instant death. If there is back-up life support, then loss of the primary system is a condition that could lead to loss of the backup and a resultant mishap. That may require mission abort or system shutdown. If a spacecraft loses reentry capability, its crew is left stranded in space. Loss of capability can result in death or injury to people. Energy systems and their potential problems were mentioned earlier. Consider, along with energy sources, the loss of the energy source. The loss of electrical power in a chemical plant is a very serious concern and hazard.

Conditions	***Causes***
Loss of primary critical function with redundant systems	Failure
Loss of critical monitoring function	Design deficiency
Component failure	Lack of redundancy
Damage	Operator error
Loss of power or energy source	

10. Radiation
The hazard to look for in this category is exposure of people or sensitive equipment to ionizing and non-ionizing radiation. Ionizing radiation includes alpha, beta, gamma, neutron, and x-ray from radioactive substances. Non-ionizing radiation includes laser, infrared, microwave, ultraviolet, and radio waves. Sources include high-energy antennae, radar, video monitors, welding, infrared and ultraviolet lights, and so on. Exposure to radiation can injure or kill people and damage sensitive equipment.

Conditions	***Causes***
Presence of radioactive materials	Inadequate containment
Radiation sources	Inadequate protection
	Careless handling

11. Temperature Extremes
This category represents conditions of high or low temperatures. Such exposure can injure or kill people or damage sensitive equipment. Increased pressure or flammability from increased temperature can result in fire or explosion.

Conditions	***Causes***
Presence of hot or cold surfaces	Inadequate protection
Hot or cold environments	Design deficiencies
Elevated pressure	Careless handling
Flammability, volatility, reactivity	Environmental factors
Reduced reliability	

12. Mechanical
In this category, look for mechanical conditions that could cause injuries such as cuts, broken bones, and crushed body parts.

Conditions	***Causes***
Exposed sharp edges or points	Lack of guards or failure to use
Moving machinery	Inadequate tie-downs
Pinch points/shear points	Design deficiencies
Elevated objects	
Unstable objects	
Crushing surfaces	
Stored energy release	

13. Pressure
This category deals with conditions related to pressure—high pressure or low pressure. The sudden release of pressure causes a shock wave and possible fragments that could injure people and damage equipment. Slow release of pressure can cause contamination. One can find pressure in gases or liquids (hydraulics).

Conditions	***Causes***
Overpressure	Lack of pressure relief
Vacuum	Inadequate pressure relief
Loose pressure line	Inadequate pressure couplings
Hydraulic ram	Metal fatigue
Inadvertent release—toxic or flammable materials	

Risk Assessment

As discussed earlier, risk is a combination of the severity and probability of a potential mishap. It is not enough to look *only* at the severity. A person can think of many worst-case consequences, such as cutting one's throat while shaving, but the chances of those consequences occurring are so remote that he may not want to worry about it. Hence, in setting priorities for safety activities, one must consider the *total* risk involved, not just the severities.

Severity is the worst-case, yet believable, hazard effect. It could range from minor injury and damage to death and system destruction. Severity is usually easy to determine. There is generally little debate about the worst-case outcome of a potential mishap.

Probability is the chance that a specific outcome will occur. The probability can change with the exposure time. Suppose one predicts that there will be an earthquake somewhere in the world tomorrow. He has a chance of being right. However, if one predicts there will be an earthquake somewhere in the world in the next year, he is almost certain to be right. When stating probabilities, one needs to include a specific time period or exposure interval.

It is usually difficult to determine probability. It is often controversial and subjective. Safety people may tend to estimate a high probability of a system mishap, whereas the designers may tend to estimate a low probability (Roland and Moriarty 1983). The actual probability may be between the two. If the mishap would result from a component failure, and one has significant failure-rate data on that component, then one can just use the failure-rate number. If one has significant experience with a similar system, then one can use that experience to estimate a probability.

In many cases, one is dealing with new systems and technology. However, we can still arrive at some kind of probability estimate. The best approach is to develop prototypes and test them. However, this is expensive and not always practical. Another good approach is to run computer simulations. This is cheaper than testing but is only effective to the extent that the computer model is accurate. The least effective, yet least costly, approach is analysis (Roland and Moriarty 1983). For instance, if a structural analysis is performed and it finds that there is a high safety factor, then one can conclude that there is a low probability of structural failure.

Many analysis situations may not be as straightforward as structural analysis. In these cases, one needs to use "engineering judgment." The following are some factors to consider.

How many causes are there to a particular hazard? For example, a safety person is concerned about gasoline puddles in the garage catching fire. The more gasoline cans and tanks there are in the garage, the greater the likelihood that there will be a leak and resultant fire.

What is the maturity of the system and technology? If a firm is designing a new type of spacecraft, using new technology, then there is probably a good chance of a malfunction or mishap because it has not been tested over time (Stephenson 1991).

How long does it take for a hazardous condition to turn into a mishap, and how can it be detected? For example, a company uses a process that can cause an explosion if it reaches a certain temperature. If its temperature rises slowly enough that it will likely be detected, and the process can be shut down before it reaches the critical temperature, then there is a low probability that the explosion will occur.

Probabilities can be expressed quantitatively or qualitatively. If there is numerical data, it can be used (e.g., 0.0002 per exposure hour, once in one million cycles, etc.). If numerical data is hard to get, it can be expressed qualitatively (e.g., likely to occur soon, likely to occur in time, remote, impossible, etc.).

Risk—putting severity and probability together—is often expressed via a matrix. First, severity levels are set up. The following is an example of a severity categorization:

I — Catastrophic; death or total system destruction
II—Critical; serious injury or major property damage
III—Marginal; minor injury or property damage
IV—Negligible; little or no effect

Then we assign probability levels:

A—Likely to occur immediately
B—Probably will occur in time
C—May occur in time
D—Unlikely to occur

Finally, the risk-assessment matrix is set up using the severity levels:

		Probability level			
		A	B	C	D
	I	1	1	2	3
Severity	II	1	2	3	4
level	III	2	3	4	5
	IV	3	4	5	6

One method of expressing risk is to assign a number to each block of the matrix as shown here. The number is called a RAC (risk assessment code).

Risk Assessment and Decision Making

Making decisions about risks is intrinsic to designing in-process safety. Historically, when designing a new process, engineers examined how the system could break down (Roland and Moriarty 1983). They would then determine the impact of system failures, and estimate their likelihood. Evaluating these issues produced a continuous stream of risk-related design decisions. Unfortunately, these decisions were based more on perceptions of risk than on real measurements of risk (Roland and Moriarty 1983).

The result was an unsystematic and incomplete assessment of the design. This led to an inadequate and sometimes cost-prohibitive, or incompatible, risk-reduction solution. When the process for determining the design basis lacked consistency, it was difficult to know whether the same risk-management philosophy supported all of the company's risk decisions.

By designing in-process safety and decision making using a risk-based approach, safety professionals can improve their ability to understand and reduce risk, control costs, and protect their investments and reputations, as well as protect company employees. Risk-based approaches get integrated with the process and can help safety managers translate the technical complexities of risk analysis into clear messages about risks and options for corporate management.

The Concept of Risk

Before continuing, it will be helpful to look at some terms that are often misused when talking about safety and risk. In everyday conversation, people use words such as risk, hazard, and danger interchangeably. For example, crossing the street without looking both ways might be described as "hazardous," "dangerous," or "risky," all meaning pretty much the same thing.

In chemical process-safety design, risk is defined more precisely in terms of the likelihood and consequences of incidents that could expose people, property, or the environment to the harmful effects of hazards (CCPS 2000).

Hazards are potential sources of harm. Examples of hazards include a reactor vessel under high pressure, a very corrosive chemical, and an improperly sized emergency relief valve (CCPS 2000).

Likelihood is determined in terms of two factors: frequency (how often does this happen?) and probability (what are the odds that it will happen?) (CCPS 2000). Likelihood can be determined by using a risk matrix as shown here:

Criticality or Severity Ratings

1 Catastrophic	Loss of containment of substantial amounts of material that may result in an on-site or off-site death, or damage and production loss greater than $1,000,000
2 Severe	Loss of containment of material that may result in multiple injuries or morbidity, or damage and production loss between $100,000 and $1,000,000
3 Moderate	Loss of containment of small amounts of material that may result in minor injury or morbidity, or damage and production loss between $10,000 and $100,000
4 Slight	Loss of containment of small amounts of material that may result in no injuries or morbidity, or damage and production loss less than $10,000

Frequency Ratings

1 Frequent	Likely to occur more than once per year

2 Probable	May occur several times in 1 to 10 years
3 Occasional	May occur sometime in 10 to 100 years
4 Remote	Unlikely to occur, but possible in 100 to 10,000 years
5 Improbable	So unlikely to occur it can be assumed that occurrence is less than once in 10,000 years

Based on the criticality and frequency definitions presented in the preceding ratings, a risk matrix (see Tables 1 and 2) is used to identify a risk rank (DOD 2000). The risk rankings are defined in Tables 1 and 2 (DOD 2000).

Consequences cover specific outcomes or impacts of an incident, such as a toxic vapor cloud that spreads beyond the plant boundary into a neighborhood or an explosion that causes a fatality.

Risks cannot be completely eliminated from industrial processes any more than they can be eliminated from other activities. Instead, the goal of system safety is to consistently reduce risk to a level that can be accepted by all concerned—facility staff, company management, shareholders, surrounding communities, the public, industry groups, and government agencies. For this reason, defining what constitutes *acceptable* and *unacceptable* risk is a critical part of risk-based design.

TABLE 1

Risk-Ranking Matrix

		Frequency Rating				
		1	2	3	4	5
Criticality Ratings	1	1	1	1	2	4
	2	1	2	3	3	4
	3	2	3	4	4	4
	4	4	4	4	4	4

(*Source:* DOD 2000)

TABLE 2

Risk-Ranking Definitions

Ranking Levels	Description	Action Required
1	Very High	Must be mitigated with engineering and administrative controls before continued operation
2	High	Must be mitigated by engineering controls within six months
3	Medium	Must be mitigated by administrative controls within six months
4	Low	Mitigation is optional depending on cost-benefit

(*Source:* DOD 2000)

Steps in Risk Assessment

A risk-based design approach (see Figure 3) integrates safety where it belongs: at each stage in the design cycle, including laboratory, pilot, production design, and operation. This technique can be incorporated into a company's current design approaches because it derives from process-design engineers' characteristic problem-solving methods. Therefore, it can be applied to simple and complex designs.

Systematic risk-based design helps engineers to include system safety in the design at the earliest development stages, where the most cost-effective solutions to safety challenges tend to be found. Risk-based design supports a disciplined thought process and opens the door to creativity and innovation in risk reduction. This increases the range of possible solutions and focuses attention on risk-reducing options that may be overlooked using historical approaches.

This approach traces the most efficient path possible through the risk-assessment process. In the nine steps that follow (Roland and Moriarty 1983), review and reassessment loops come into play only as needed.

1. ***Identify failure scenarios.*** When engineers have established an initial process design, they can address failure scenarios that might require a safety system. Process hazard-analysis techniques and past experience assist in identifying possible failure scenarios.
2. ***Estimate the consequences.*** In this step, designers establish the consequences of the failure scenarios identified in Step 1. These scenarios typically involve quality, safety, health, and environmental impacts. Consequences of interest include fires, vapor cloud explosions, toxic releases, and major equipment damage. Some potential consequences can be determined through direct observation,

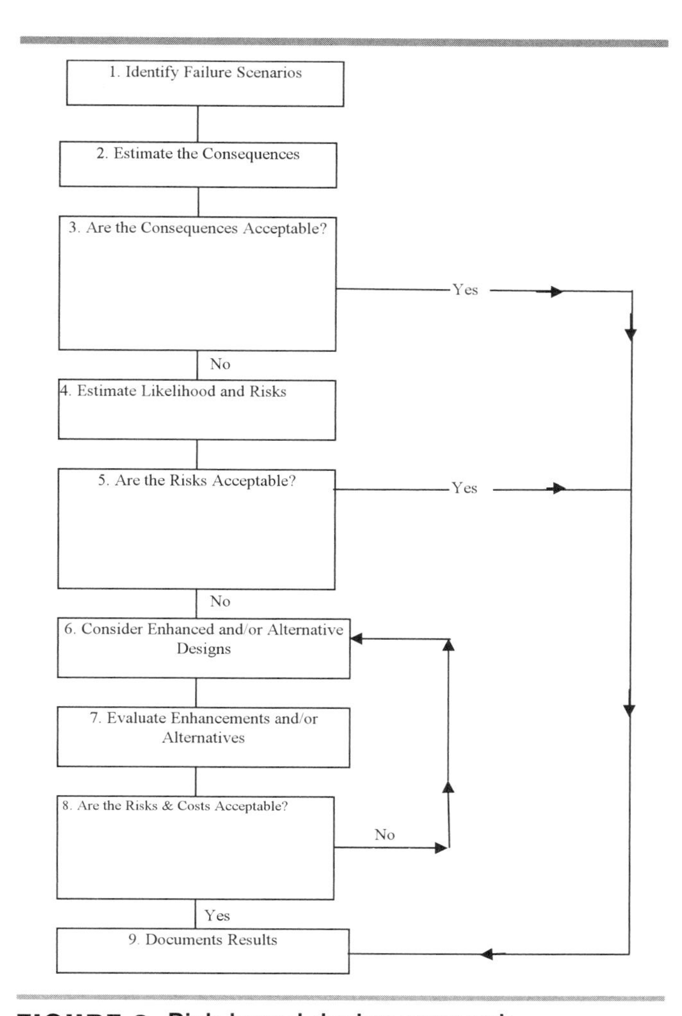

FIGURE 3. Risk-based design approach

engineering judgment, or the use of qualitative consequence criteria. Other cases require experimentation or analytical approaches such as the calculation of maximum hazard distances of vapor cloud dispersion.

3. ***Determine the acceptability of the consequences.*** Accomplishing this requires guidance from established acceptability criteria. These include company-specific criteria, engineering codes and standards, recommended practices, and regulatory requirements.
4. ***Estimate likelihood and risks.*** Estimates of likelihood rest on an understanding of how, and how often, failure scenarios such as those identified in Step 1 might occur. When historical data are available about equipment and processes, these data can be used to estimate failure-scenario frequency. When data are lacking, methods such as fault tree analysis (further described later in this chapter) help in developing quantified estimates. Measures of risk are determined by combining probability and consequence estimates. A detailed review of methods for combining likelihood and consequence estimates to obtain risk measures can be found in system safety literature. Some cases can be resolved through comparisons with similar systems or through the use of qualitative tools such as risk matrices. Others will require quantified approaches such as risk profiles and risk contours.
5. ***Determine risk acceptability.*** Determining risk acceptability means asking, "Can we accept this level of risk?" Guidance on acceptable levels of risk can be gained from established risk criteria. If the criteria indicate an acceptable level of risk, then the design of the process or the emergency-relief system is satisfactory from a risk standpoint. If the criteria indicate unacceptable risk, the next step is to reduce risk through further design refinements.
6. ***Consider enhanced or alternative designs.*** This includes an opportunity to consider the entire process design and define changes that can reduce risk to an acceptable level. The Center for Chemical Process Safety (CCPS) (1992) has classified risk-reduction concepts, in declining order of reliability, as follows: inherently safer (designed in), passive (operates only when a hazard occurs), active (operates whenever the system is operating), and procedural (written procedures).
7. ***Evaluate enhancements or alternatives.*** A design change intended to reduce risk can introduce new failure scenarios and new risks. Therefore, the evaluation of design changes should treat these changes as an integral part of the process. Following Steps 1 through 4, the review should reestimate process risk. The review should also estimate the cost of the proposed changes.

8. ***Determine acceptability of risk and cost.*** As in Steps 3 and 5, established risk criteria can provide guidance on risk acceptability. Cost becomes an issue in this step because, like all designs, system safety designs must meet business criteria. Coupling estimates of cost and risk reduction provides a basis for assessing the cost-benefit trade-off of each alternative design or mitigation solution. The cost-benefit analysis can be qualitative or quantitative. A quantitative approach is especially useful when a large number of competing process-safety systems are being considered. If the analysis yields acceptable risk and cost for a design option, the results should be documented (Step 9). If not, it may be necessary to consider design enhancements and alternatives (Steps 6 through 8).
9. ***Document results.*** The failure scenarios and associated consequence, likelihood, and risk estimates developed during this process document the design basis for process-safety systems and emergency-relief systems. Documentation retains essential information for risk-management situations such as hazard evaluations, management of change, and subsequent design projects. When the findings from Step 3 and Step 5 show that consequences and risk meet acceptability criteria, results still need to be documented. Doing so will cut down on needless repetitions of the analysis and ensure that design or operational changes reflect an understanding of the baseline risk of the design.

Guidelines for Risk Acceptability

Underlying this entire approach is the understanding that risk levels range along a continuum. In most cases, risks cannot be eliminated, only reduced to a level that everyone who has a stake in the activity or process finds acceptable. Because attitudes about the acceptability of risks are not consistent, there are no universal norms for risk acceptability. What your stakeholders view as an acceptable risk will depend on a number of factors, including the following (Post, Hendershot, and Kers 2001):

- ***The nature of the risk.*** Is it a voluntary risk, one that those who are at risk accept, as part of a choice? Or is it involuntary?
- ***Who or what is at risk?*** Does it affect a single person or many people? What about the surrounding environment? Is it an industrial landscape already altered by past uses or a pristine or prized natural setting? Are areas such as schools or residential neighborhoods or resources such as water at risk?
- ***To what degree can the risk be controlled or reduced?*** Designing in process safety focuses on this issue. Making the case for an *acceptable* risk requires that the methods supporting the design basis be technically sound and defensible, clearly documented, and accurate.
- ***What is the risk taker's past experience?*** Uncertainty regarding the impact of the risk influences the risk taker's level of acceptance. For example, the average person understands and accepts the risk of driving an automobile but is uncertain about the risk of nuclear power generation.

Companies that have successfully established risk criteria focus on gaining consistency in their decisions about risk (Post, Hendershot, and Kers 2001). These criteria typically represent levels of risk that the company believes will minimize impacts to continued operations. Risk criteria should also fit with a company's philosophy and culture and match the type of analysis its engineers normally conduct in the design stage. The selection of appropriate risk criteria is a corporate responsibility and requires the involvement and support of senior management, given that it establishes the levels and types of risks the company will accept.

Once a company has established specific risk criteria, these can be used to check outcomes throughout the design process, at Steps 3, 5, and 8 of the approach outlined earlier. This iterative approach builds consistency into the process and increases the likelihood of making risk-based choices early in design—where they are often most cost effective.

Risk-Acceptability Criteria

The following criteria were presented at the Fifth Biennial Process Plant Safety Symposium sponsored by the American Institute of Chemical Engineers (Post, Hendershot, and Kers 2001).

Release limits address the acceptability of potential release consequences by considering the amount of material that could be released. *Acceptable* quantities depend on the physical states and hazardous properties of released materials. A hypothetical release limit for gasoline, for example, might be as much as 5000 pounds, whereas for chlorine, it would be only 200 pounds (Post, Hendershot, and Kers 2001).

Threshold-impact criteria for fence or property line employ standard damage criteria, such as toxicity, thermal radiation, or blast overpressure, together with consequence modeling to determine whether potential impact at the facility's fence or property line exceeds an acceptable threshold.

Single versus multiple component failure criteria provide a qualitative approach to how many component failures will be accepted. For example, a company might choose to accept event scenarios that require three independent component failures, to conduct further analysis of event scenarios triggered by two failures, and not to accept events arising from single failures. *Independent component failures* refers to scenarios where components that fail are not linked to the failure-sequence conditions but fail independent of those conditions.

Critical event frequency addresses event scenarios with a defined high-consequence impact. Examples of such impacts would be a severe injury, a fatality, critical damage to the facility, or impacts on the surrounding community.

Risk-matrix criteria use qualitative and semiquantitative frequency and severity categories to estimate the risk of an event. Events with a low-risk ranking are considered acceptable.

Individual risk criteria consider the frequency of the event or events to which an individual might be exposed, the severity of the exposure, and the amount of time for which the individual is at risk.

Societal risk criteria are risks that can affect the local community. These risks explicitly address both events with a high frequency and minimal consequences and events with a low frequency and serious consequences. This class of criteria can be useful to companies that have recently experienced an adverse event.

Risk matrix and cost thresholds can account for the risk-reduction level provided by a design enhancement and its cost. In cases where the benefit of a risk-reduction step is large, and its cost is small, the way forward is obvious. In more complex situations, a risk matrix and cost threshold with definite "rules" can help clarify decision making.

Cost-benefit criteria are used to determine how much money the company is willing to spend to avert a risk. These criteria help define the amount of risk reduction expected for each dollar expended. They can be developed in conjunction with quantitative estimates of risk. In some cases, companies might use two thresholds—one for the dollars needed to achieve an acceptable risk level and another for any further reduction beyond that level.

Quantitative Analysis

A systematic approach does not necessarily mean a quantitative one. Quantitative analysis is most time- and cost-effective when it is used selectively. In many simple design situations, qualitative approaches are sufficient for selecting the basis of process-safety system design. More complex design cases may occasionally require quantitative risk analysis. But even then, quantitative methods should be used only up to the point where a decision can be made.

For example, consider a company that has toxic-impact criteria limiting off-site vapor cloud concentrations to a specific, quantified level. By performing vapor-cloud dispersion modeling, the company can determine whether specific loss-of-containment scenarios associated with specific failures exceed the toxic-impact acceptability criteria. If the scenario consequences do not exceed the criteria, then there is no need to continue with an analysis of event likelihood or further risk quantification. Specific quantitative analysis methods are discussed later in this text.

Design Solutions: Making the Right Decision

The best decisions about safety and risk reduction in process design bring together technical sophistication and clear business objectives. Decisions about risk should provide definite business value and fit into the business context: What is the business plan for this facility or process? The decisions should reflect consistent thinking and standards for risk-acceptability levels. And they need to be in line with an appropriate cost structure for the safety component of a process.

Understanding the primary types of design choices for safety can help engineers rank their options and introduce modifications where they can do the most good for the least cost. Most design solutions for reducing risk fall into one of these categories (CCPS 2000):

- *Inherently safer*—eliminates or mitigates identified hazards by substituting less hazardous materials and process conditions. Inherently safer solutions tend to require relatively high capital costs, offset by relatively low operating costs. Examples include substituting water for a flammable solvent and reducing large inventories of hazardous "intermediates."
- *Passive*—offers high reliability by operating without active devices that sense or respond to process variables. Like inherently safer design choices, passive systems often require a relatively high initial investment offset by relatively low operating costs. Examples include compatible hose couplings for compatible substances, equipment designed to withstand high-pressure hazards, and dikes that contain hazardous inventories.
- *Active*—uses devices that monitor process variables and trigger mitigation and control systems. Active systems can be less reliable than inherently safer or passive systems because they require more maintenance and more detailed operating procedures. They typically require moderate capital costs, followed by somewhat greater operating costs. Examples include check valves and regulators, pressure safety valves or rupture disks that prevent vessel overpressure, and high-level sensing devices that interlock with inlet valves and pump motors to prevent overfilling.
- *Procedural*—avoids hazards by requiring someone to take action. The capital cost of a procedural system is generally low, but operating costs, including staffing and training, can be high. Reliability, which rests on human variables such as company safety culture and the correct use and handling of mechanical devices, tends to be low. Following procedures also depends on the level of training provided to support procedure knowledge. Without training procedures, how to operate the equipment properly will likely not be well known or understood. Examples of procedural approaches include manually closing a valve after an alarm sounds or carrying out preventive maintenance to reduce the likelihood of equipment failure. Wearing proper personal protective equipment for a particular task would be another example of procedural safeguard.

Incorporating systematic risk assessment in process-safety design is sometimes viewed as an expensive way to achieve greater risk reduction. The reality, however, is that when risk assessment is left out of the

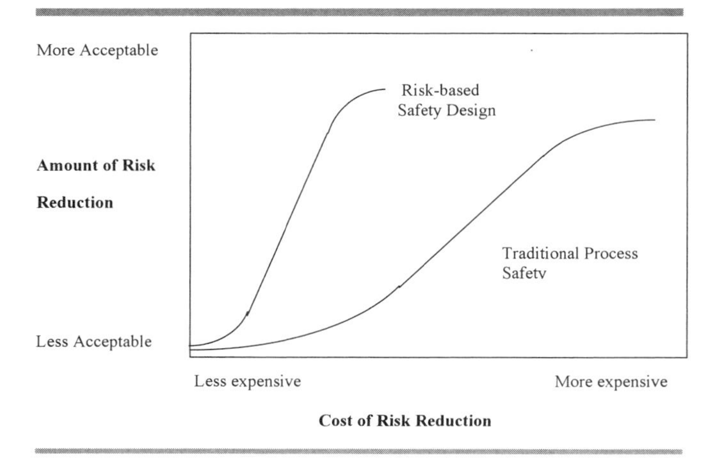

FIGURE 4. General representation of the cost of risk reduction (*Source:* CCPS 2000)

design process, two problems are likely to occur. The system may be overdesigned, with safety protection costing more than it needs to, or the facility may be unprotected from significant, unidentified risks.

Systematic risk-based design helps companies to more fully identify significant risks, rank the risks, and prioritize steps to address them. The result is that capital expenditures, operating expenses, staffing, and other resources are better allocated to risks, enabling companies to buy more risk reduction at a cost that is the same as or less than shown in Figure 4 (CCPS 2000).

When deciding from among the hierarchy of risk-mitigation options, designers should avoid the "project mentality" pitfall, which promotes focusing only on minimizing capital cost. Nor can they simply select the most reliable approach. The key to successful risk-reduction choices is to exploit the detailed technical information about risks and hazards that a systematic design process offers to judge the actual merits and disadvantages of each plausible solution. It is also important to examine the life-cycle costs of each option before making decisions, as shown in Figure 5 (CCPS 2000).

For example, a company was handling a very reactive substance. After several incidents involving the substance, the company reviewed two options for reducing the risk posed by the substance. The first option included total containment of the substance in a vessel rated to withstand a maximum pressure level of 1200 pounds per square inch (psi), which was an inherently safer approach. However, the cost of this vessel was very high, and using it meant having the vessel sit continuously within the facility at a very high pressure, a hazard in and of itself.

The second option was to construct a catch system and allow the reactor to activate an emergency pressure-relief system. This required a reactor vessel with a lower pressure rating and a large vessel to be used as a catch/quench tank. This approach was less expensive, but it required the facility to deal with the potential of a hazardous effluent and to address the reliability of the relief system. At the time, the pressure relief/catch system was found to provide an acceptable risk level overall. The company chose to take advantage of this option's lower cost and to implement mitigation measures that helped attain an acceptable level of risk.

Meeting Stakeholder Needs

Safe design has long been a priority in the energy, chemical, and petrochemical process industries. Today, process-industry companies need to be certain that

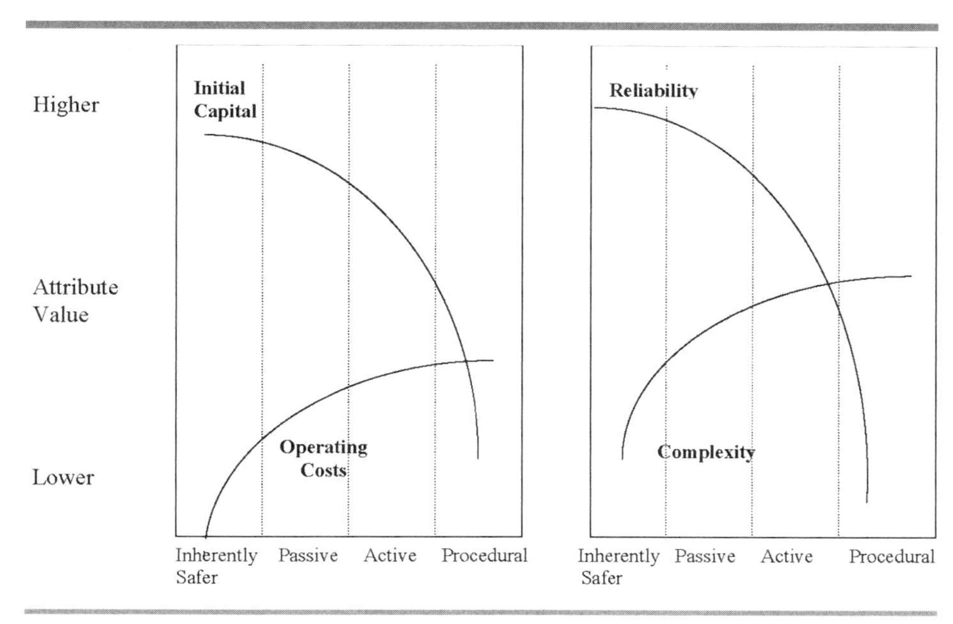

FIGURE 5. Life cycles of each cost option (*Source:* CCPS 2000)

their stakeholders trust how they manage the environmental, health, and safety implications of industrial activities. A safe and documented design basis, along with a formal safety-management system and safety practices, procedures, and training, is critical for providing the level of confidence required for risk management.

In recent years, regulations and industrial standards for acceptable risk have become increasingly stringent. This trend reflects a convergence of public opinion, government regulations, and industry initiatives. The momentum for controlling and reducing risk is likely to continue, with leading companies in process industries setting standards that go well beyond what is required.

At the same time, risk managers and safety environmental managers face unremitting pressure to run their activities "lean and mean" and control and justify costs. Risk-based design can be a key piece in a successful company's toolkit for reaching decisions about incorporating in the design process elements that integrate risk reduction and cost advantages without compromising safety. By communicating options clearly to all concerned stakeholders and by addressing the full life-cycle cost of different options, risk-based design enhances the business value of process-safety activities. Companies that are gaining the benefits of reduced risks and reduced risk-management costs find their competitive position strengthened by lower capital costs and by a more secure franchise to operate. In short, they are more competitive as a result of risk-based design (CCPS 2000).

The Tools and Techniques of System Safety Analysis

Analysis tools and techniques have been referred to earlier and different uses for various applications have been recommended. At this point, a closer look at these options is warranted. Analysis techniques presented in this chapter include the following: checklists, software fault tree analysis, event tree analysis, failure modes and effects analysis (FMEA), SoftTree analysis, hazard and operability (HAZOP) studies, time dependent Petri net analysis, functional flow analysis, information-flow analysis, software sneak analysis, nuclear safety cross-check analysis (NSCCA), real-time logic, and software code analysis. It is not the purpose of this chapter to explain each of these tools and techniques in great detail. However, an explanation is offered to provide a general understanding of each technique and its advantages and disadvantages.

Checklists

Checklists are the simplest and most cost-effective approach and are most often used in conjunction with other analysis techniques. They require the least amount of training to use and are often self-explanatory. Checklists reword the criteria in the form of a question with yes, no, or not applicable being an appropriate response. There is usually a comments section with each checklist item that is used to provide any special explanation regarding the environment in which the checklist was executed. A comments section can also describe why a particular item was marked "no" or "not applicable."

Fault Tree Analysis

A fault tree analysis can simply be described as an analytical tool, where an undesired state of a system is specified and analyzed in the context of its environment and operation. This is done to find all of the possible ways that the undesired event can occur. The fault tree is a graphic model using Boolean algebra symbols (shown in Figure 6) of the various parallel and sequential combinations of faults that will result in the occurrence of the predefined, undesired event. The faults, or top events (see an example of top events in Figure 7) can be events that are associated with component hardware failures, software faults, human errors, or any other pertinent events that can lead to an undesired event. A fault tree (shown in Figure 8) depicts the logical interrelationships of basic events that lead to an undesired event, which is the top event of the fault tree (Hammer 1989).

The fault tree is probably the most used analysis technique throughout system safety because of its inherent advantages (Rausand and Høyland 2004). Fault tree analysis is well understood by most system safety engineers, is well documented, and can be applied at

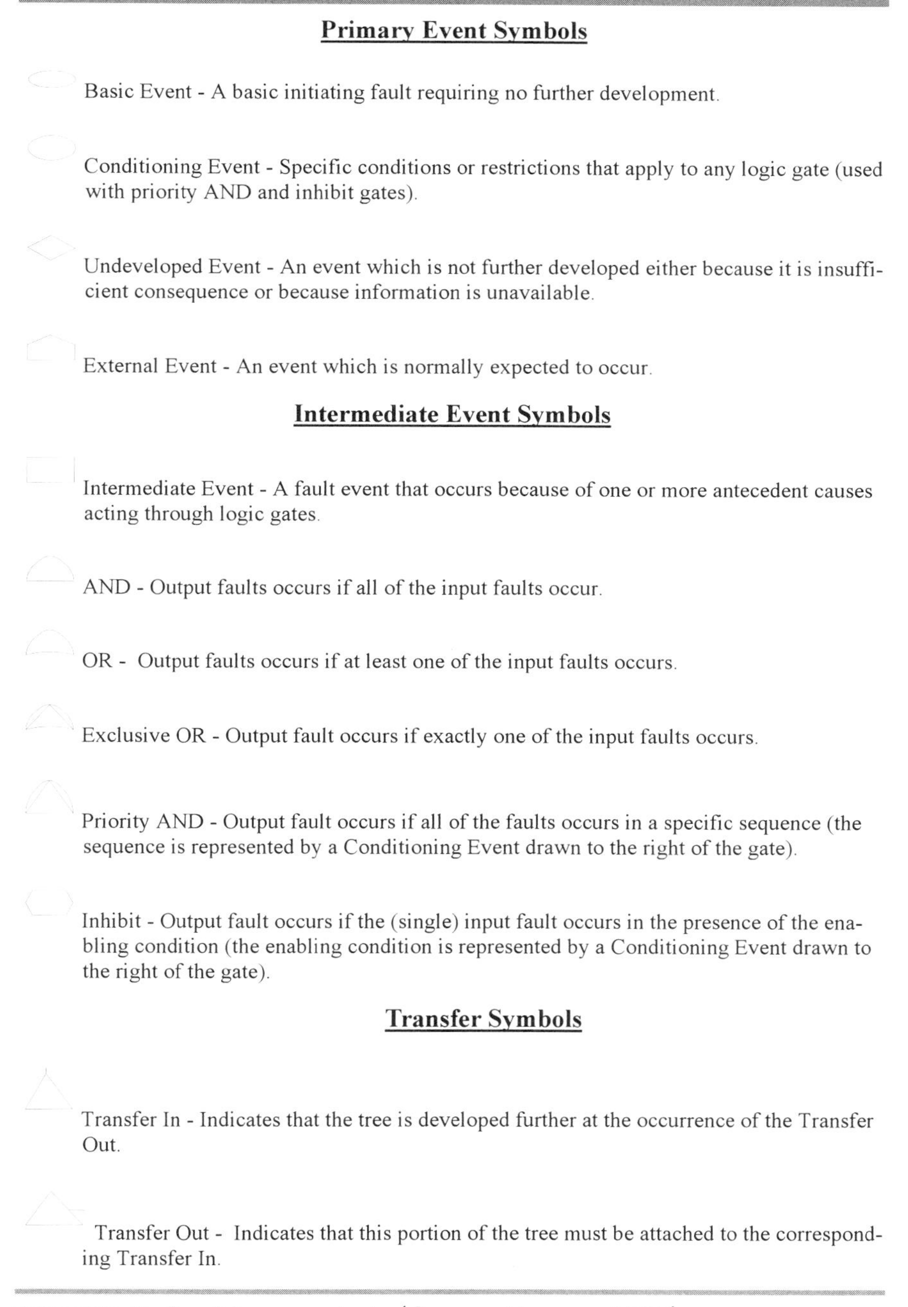

FIGURE 6. Fault tree symbols (*Source:* Hammer 1989)

various levels of the system design (e.g., conceptual, design, and implementation). There are numerous automated tools that perform fault tree analysis. These tools use *standard* trees for logical statements. Boolean algebra is used to quantify probabilities of events based on the tree logic of combined symbols leading up to the event of interest. A few advantages of using fault trees include the following:

- Time is not wasted chasing dead-end events.
- Fault trees readily demonstrate the synergism of failures between related branches.
- Fault trees are easily adaptable to rapidly meet design changes and redesign.
- Fault trees are used to focus on specific safety-critical functions rather than the whole system.

- Fault trees can be used to analyze interfaces between inputs and outputs to and from internal and external systems and subsystems.

However, there are some inherent disadvantages of using fault trees. They are time-consuming and, as a result, expensive to perform. Even though fault trees can be used to determine relevant probabilities, they are not mathematically precise. Fault trees can be used to identify only sequential constructs, and they have no mechanism for handling parallel processes. Because parallel processing is becoming more of a norm in the computer industry, this is a serious shortfall. When using timing constraints, fault trees become extremely cumbersome. There is no formalized way to ensure that human factors are evaluated in a consistent manner. Fault trees are very dependent on the expertise, judgment, and system knowledge of the analyst. Therefore, the results of the fault tree are only as good as the quality of the analyst. Fault trees were developed initially for analyzing electrical and electromechanical systems by Bell Telephone. Events such as undesired missile launches are well suited for this analysis.

1. Injury to ________.
2. Radiation injury ________.
3. Inadvertent start of ________.
4. Equipment ________ activated inadvertently.
5. Accidental explosion of ________.
6. Loss of control of ________.
7. Rupture of ________.
8. Damage to ________.
9. Damage to ________ from ________.
10. Thermal damage to ________.
11. Failure of ________ to operate (stop, close, open).
12. Radiation damage to ________.
13. Loss of pressure in ________.
14. Overpressurization of ________.
15. Unscheduled release of ________.
16. Premature (delayed) release of ________.
17. Collapse of ________.
18. Overheating of ________.
19. Uncontrolled venting of ________ (toxic, flammable, or high-pressure gas).
20. (Operation to be named) inhibited by damage.

FIGURE 7. Example fault tree top events (*Source:* Hammer 1989)

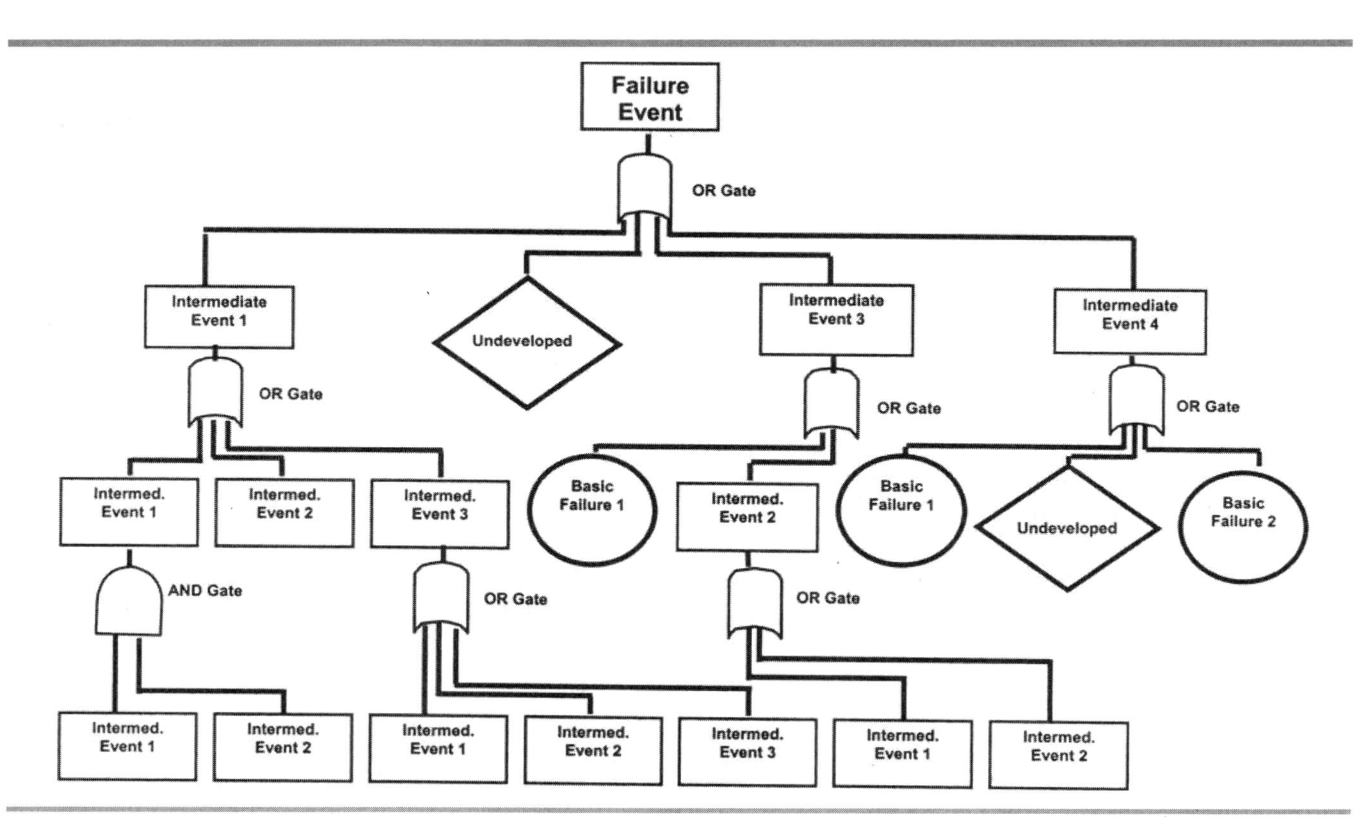

FIGURE 8. Example of fault tree (*Source:* Hammer 1989)

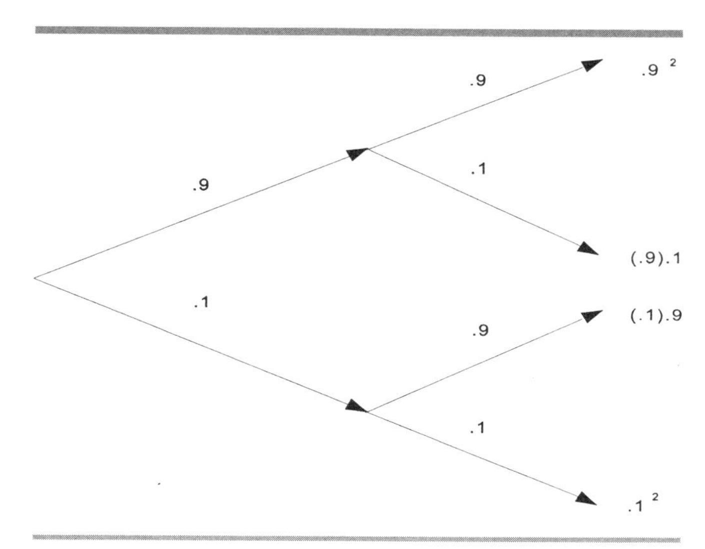

FIGURE 9. Example event tree
(*Source:* Clement 1987)

Event Tree Analysis

Event tree analysis is a bottom-up, deductive technique that explores different responses to challenges. This analysis appears to be a variation of a fault tree analysis. The event tree is developed from left to right, whereas the fault tree is developed top to bottom (as shown in Figure 9). The initiating, or challenge, event of the event tree is similar to the top event of a fault tree, even though gate logic or gates are not used in event trees. A primary difference is that fault trees tend to explore and list only factors that could lead to the failure, and event trees can be converted to fault trees. The procedures and formulae for quantifying each are similar (Clemens 2001).

Failure Modes and Effects Analysis (FMEA)

A FMEA is a reliability analysis technique that examines the effects of all failure modes of systems and subsystems. It is used to determine how long a piece of software or a component will operate satisfactorily and what the effects of any failure of individual components might be on the overall system performance (Hammer 1989). The hardware or software is decomposed all the way down to its lowest component level (e.g., modules). Each module is studied to determine how it could malfunction and cause downstream effects. Effects might result in error or fault propagation into other related software modules, perhaps causing the entire software program to fail. Failure rates for each module are identified and listed. The calculations are used to determine how long a module is expected to operate between failures and the overall probability it will operate for a specific length of time.

An FMEA has inherent advantages. It can be used without first identifying mishaps during any phase of system design. An FMEA helps to reveal unforeseen hazards and is good at helping the analyst identify potentially hazardous single-point failures. However, there are inherent disadvantages to using FMEAs. They are time-consuming and consequently expensive to perform. They are also very dependent on the expertise, judgment, and system knowledge of the analyst. FMEA results are only as good as the analyst (Clement 1987).

Hazard and Operability (HAZOP) Studies

Hazard and operability studies provide a quantitative and qualitative systematic approach to identifying hazards by generating questions ("What if...?") that consider effects of deviations in and from normal parameters of a particular chemical process. HAZOP studies are used to analyze steps during each chemical-process operating phase, operating limits, and safety and health considerations. Analysis of procedures for each operating phase (29 CFR 1910.119 and AIChE 1985) includes initial startup, normal operations, temporary operations, emergency operations, normal shutdown, and startup following a turnaround. Analysis of operating limits includes consequences of deviation, steps to correct or avoid deviation, and safety systems and their functions. Analysis of safety and health considerations includes properties of hazards, precautions necessary to prevent exposures, safety procedures, quality control, and any special or unique hazards (Stephans 1997). HAZOP studies consist of multidisciplinary teams that identify, analyze, and control hazards systematically. The HAZOP team traces the flow of the

Deviation	Meaning	Guide Word
NO	None	Pressure
LESS	Quantitative decrease	Temperature
MORE	Quantitative increase	Flow
PART OF	Qualitative decrease	
AS WELL AS	Qualitative increase	
REVERSE	Opposite of intended direction	
OTHER THAN	Substitution	

FIGURE 10. Example guide words

process, concentrating on specific study nodes. Parameters include normal temperature, pressure, flow, and medium type: gas and chemical. A node may be a vessel or a pipe between vessel 1 and vessel 2. Process parameters will generally define the size of the node. A set of guide words is used to determine types of deviations from prescribed parameters that could occur at each study node. The causes and consequences of the deviations are discussed with the aid of these guide words. Typical guide words and their meanings are shown in Figure 10.

A HAZOP study has inherent advantages. Because the guide words tend to describe failure modes, the HAZOP can be conducted in parallel with an FMEA and complement or perhaps even validate the results. A HAZOP may be used to conduct further analysis of critical items identified in an FMEA. This helps decrease the expense from spending too much time and money on an in-depth FMEA. A HAZOP tends to include human factors and operator errors, whereas other analyses tend to examine system-related failures.

A HAZOP analysis has inherent disadvantages as well. A HAZOP should be performed when the design is relatively firm (Goldwaite 1985; Stephans 1997). A HAZOP should be performed by the 35 percent complete design stage. Because system safety efforts should start as early as possible in the design phase, the HAZOP can be used only as an adjunct tool and not as a stand-alone tool.

Functional and Control Flow Analysis

Functional and control flow analysis is a static analysis technique. It requires the development of functional flow diagrams. Functional flow diagrams illustrate

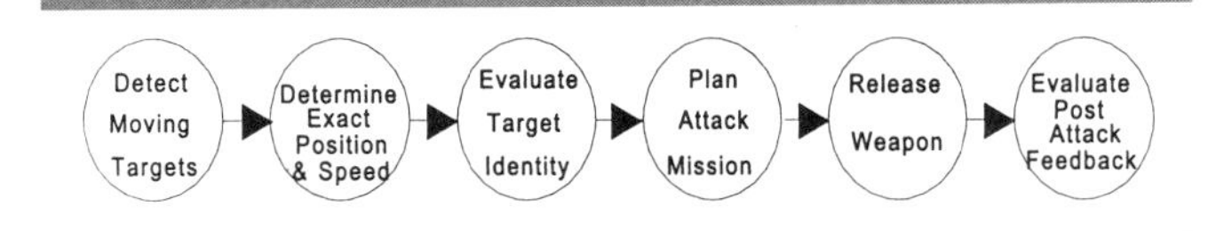

FIGURE 11. Functional and control flow analysis (*Source:* Leveson 2002)

the flow of information through the system based on functions. This analysis is generally applied during the detailed design phase when functional flow diagrams have been completed. An example functional flow analysis is shown in Figure 11 (Leveson 2002).

Functional and control flow analysis has several advantages. It is applicable to all phases of system development. Functional flow diagrams are in various forms of completeness during the design phase and can be examined at any time. Functional flow analysis helps identify potential control problems between functions. As a result, it is a good tool to use for functional grouping and separation of safety-critical functions. It can be used to induce failures in design and to see how they may be propagated to other design elements. It can easily be related to safety-requirement analyses. Functional and control flow analysis has several disadvantages. The software designers usually use a different flow-analysis technique. As a result, the software flow analysis must be translated into intermediate or special language. It is dependent on the expertise of the analyst. Using it for medium-size systems at the detailed design level is complex. The synergistic effect of failures is not easily identified. It provides little benefit in analysis of safety-critical information flow. It requires separate safety analysis to identify safety-critical system states and functions. It is not a complete analysis of a system.

Information and Data Flow Analysis

Information and data flow analysis is also a static analysis of information or data flow through a system. It is generally applied during detailed design phase and requires development of information flow diagrams similar to functional and control flow analysis. An example information flow analysis is shown in Figure 12 (Leveson 2002).

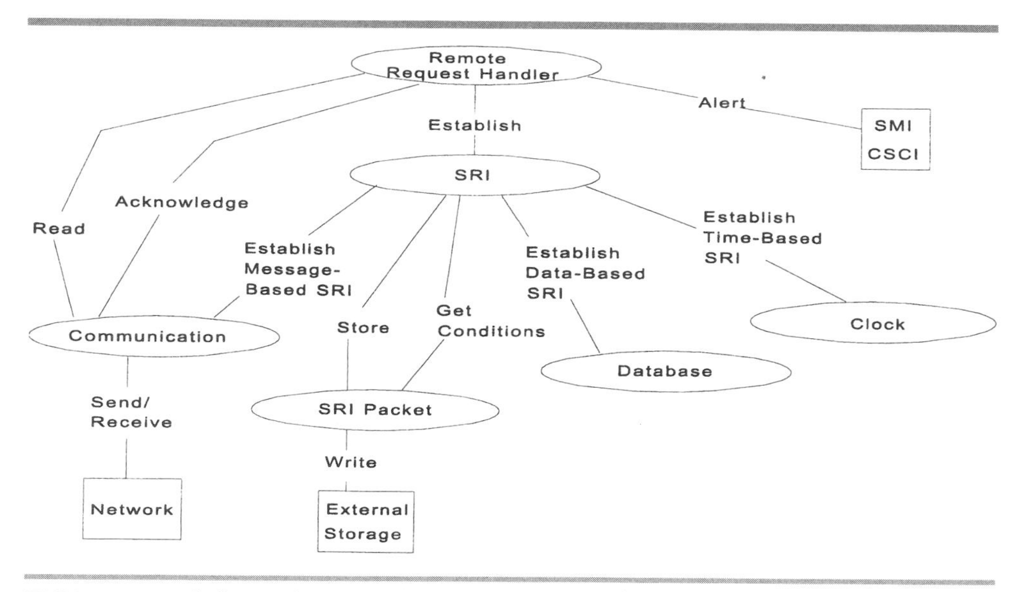

FIGURE 12. Information and data flow analysis (*Source:* Stephans 1997)

Information and data flow analysis has several advantages. It can be used to identify potential bottlenecks in information flow. Safety-critical information is easily tracked through the system. It complements functional flow analysis by coupling functional control with information flow. Information and data flow analysis has several disadvantages as well. It provides no benefit in control-intensive systems. The software designers usually use a different flow-analysis technique. As a result the software flow analysis must be translated into intermediate or special language. Its use is complex for medium to large systems. Again, it is dependent on the expertise of the analyst (Brown 1976).

Sneak Circuit Analysis

Sneak analysis is based on a sneak path, an unintended route, which can allow an undesired error or fault to occur, preventing desired functions from occurring, or which can adversely affect the timing of the functions. Sneak circuit analysis is performed to identify ways in which built-in design characteristics can either allow an undesired event to occur or prevent events from occurring. It is usually applied to electrical or software-hardware systems.

An important feature of sneak analysis is that the sneak paths being investigated are not a result of failures in the system. They are rather the result of the circuit design. A sneak analysis represents program paths as circuit diagrams. The sneak paths may show up only on rare occasions when the software is a unique configuration.

Sneak analysis is usually inductive and can be very difficult to perform without the aid of software tools. Proprietary sneak-analysis tools have been developed by Boeing, General Dynamics, Science Applications International Corporation (SAIC), and several others.

The most common approach to sneak analysis involves visual clues found by comparing circuits with the six basic topographs. The code topology patterns are shown in Figure 13, and the software sneak topology patterns are shown in Figure 14.

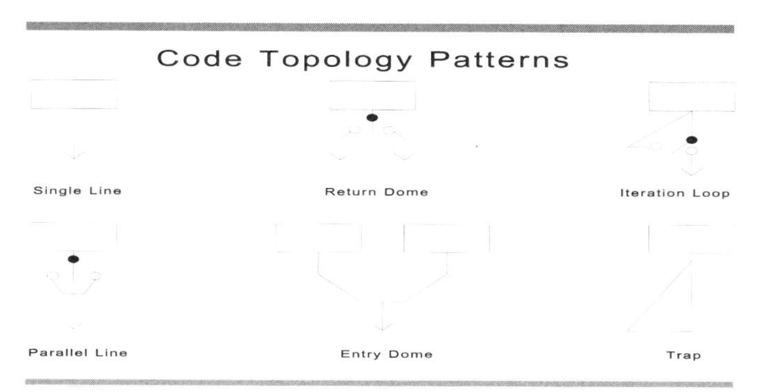

FIGURE 13. Code topology patterns (*Source:* AFISC/SESD)

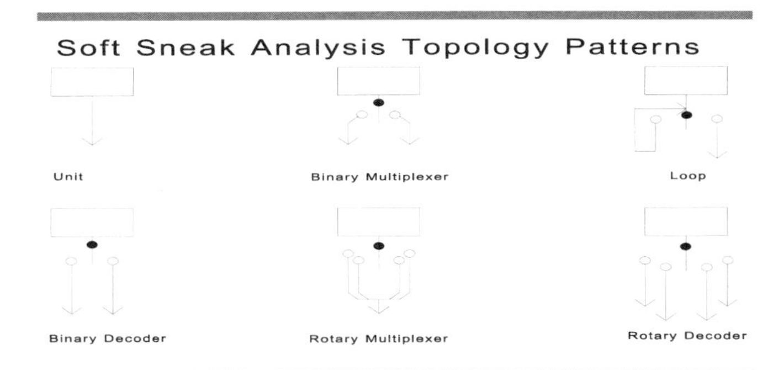

FIGURE 14. Software sneak topology patterns (*Source:* AFISC/SESD)

Nearly every circuit can be broken down into combinations of these topographs. The sneak conditions to be sought are as follows:

1. Sneak paths that can cause error or fault propagation to flow along an unexpected route
2. Incompatible logic sequences that can cause unwanted or inappropriate system responses (sneak timing)
3. Sneak conditions that can cause misleading, ambiguous, or false displays
4. Sneak labels that provide confusing or incorrect nomenclature or instructions on controls and that thus promote operator errors

A sneak analysis has several advantages. It provides an excellent documentation check. It is excellent in isolating alternate (sneak) paths in software. It is a thorough analysis of the source code such that logic errors are easily detected.

A sneak analysis has several disadvantages. It occurs after coding has been substantially completed. Therefore, fixing the problems requires changing code. It is primarily directed at logic errors rather than structural and design errors. As a result it does not validate software algorithms. It is directed at the whole system and does not concentrate on the most costly faults. It also does not consider environmental factors such as temperature, altitude, and radiation issues. Failure modes are handled as special cases. It is very dependent on *clue lists*. It is very labor-intensive and costly. It requires separate safety analysis to identify safety-critical system states and functions. Critical faults are easily lost in the mass of documentation errors (Stephenson 1991).

Hazard Analysis Worksheets and Hazard Reports

Preparation of Hazard Analysis Worksheets (HAWs)

The hazard analysis worksheet (HAW) is a tool used in system safety to analyze and document all hazard/energy sources and accident risk factors identified and evaluated during the various safety analyses performed on a system. The HAW is used to perform two functions: First, it promotes a systematic and thorough analysis of a hazard/energy source or risk factor; and second, it provides a permanent record of analyses performed.

The HAW differs from the hazard report in that it is a record of hazard/energy sources and risk factors evaluated. Typically, the hazard report documents only hazards with a initial risk severity code greater than IV and a frequency code greater than D (described in detail in the "Risk Assessment" section of this chapter). The HAW will generally be kept on file and incorporated in, or appended to, only the appropriate safety analysis documentation. The hazard report will be included in the formal documentation.

The HAW format selected should be formalized to the extent that a standard form and analysis methodology is adopted. The analysis can be handwritten (legibly) to minimize the administrative burden. If a HAW results in a hazard report being prepared, the HAW can be closed as long as reference is made to the hazard report that replaces it. A sample HAW format is shown in Figure 15. The form resembles the hazard-report format because the analysis methodology is similar. This facilitates the transfer of the HAW data to the hazard report. For the most part, the instructions for filling out this particular HAW format are the same as those for the hazard report. Some blocks that may differ from a hazard report are discussed in the paragraphs that follow.

HAZARD ANALYSIS WORKSHEET
SYSTEM ____________ SUBSYSTEM ____________ HAW NO. ____________
EQUIPMENT/COMPONENT ____________ LOCATION ____________
HAZARD/ENERGY FACTOR ____________
ORIGINATOR ____________ DATE ____________ STATUS ____________
HAZARD REPORT SUBMITTED: YES ____ NO ____ HR NUMBER ____________
SINGLE POINT FAILURE: YES ____ NO ____ HR DATE ____________ PHASE(S) OF
OPERATION: ____________
DRAWING/DOCUMENT NUMBER: ____________
REFERENCES:
EXISTING CONTROLS IN THE DESIGN:
ACCIDENT SCENARIO AND TRIGGERING EVENTS:
UNDESIRED EFFECTS: RISK CODES:
INITIAL RISK CODE:
RECOMMENDATIONS:
ORIGINATOR ____________ APPROVED ____________ DATE ____________

FIGURE 15. Typical hazard analysis worksheet (*Source:* Hansen, 1993)

Status: The status block on a HAW will usually be *closed*. The HAW is closed if (1) the initial risk code is IV *or* D, and the reviewing and approving authority concurs with the recommendation that "no additional action is required" as shown, or (2) the HAW is coded higher than IV *and* D, and a hazard report is prepared. In this case the HAW must reference the hazard report number. The HAW will remain open if (1) the recommendations block states that additional analysis, study, research is required, or (2) the present risk of the HAW is coded IV or D, but the design is nevertheless in violation of a safety code or standard. In this case the recommendations block should provide recommended corrective actions required to bring the design into compliance.

Risk Code. The initial risk-code block may often contain a severity code of IV or a frequency code of D (described in detail in the "Risk Assessment" section of this chapter). Many companies use their own tailored version of risk codes to make them meaningful to their business. Examples are a laser printer that has been designed to comply with ANSI Z-136; a cathode ray tube that has been designed to eliminate the possibility of emitting ionizing radiation; lightning hazards properly controlled by a lightning protection system designed, installed, and verified as being in accordance with code; and a floor-loading analysis that shows that floor-loading limits are not exceeded. The risk code is one of the last items to be filled out by the analyst. The reason is that the risk code *cannot* be determined until the accident scenario and triggering events, undesired effects, and existing controls in the design have been determined.

Recommendations. The recommendations block will usually (but not always) contain one of four entries:

1. The hazard is adequately controlled by design features, and no additional actions are required (HAW closed).
2. Recommendations for actions necessary to bring the design into compliance with applicable standards (HAW open).
3. Recommendations for actions necessary to reduce the risk to an acceptable level (HAW closed, hazard report open).
4. Recommendation for studies, analyses, research, or other actions necessary to further qualify or quantify the hazard status (HAW open).

Originator—approved by. At a minimum, the analysis should be reviewed and approved by the safety manager or his supervisor if the system safety manager (SSM) is the HAW originator. This concurrence is the contractor's attestation that the design has an acceptable level of risk or that the corrective actions recommended will reduce the risk to an acceptable level.

Preparation of Hazard Report Forms

A typical hazard report (HR) form is shown in Figure 16. Hazard reports are used to document and guide the complete analysis and risk assessment of a hazard. Each reportable hazard, identified by analyses or other means, should be analyzed and documented on an HR form. The primary purposes of the hazard reports are (1) to document the hazard or accident risk factor, its cases, and the possible accident scenarios and undesired effects that can result from the hazard; (2) to document existing and recommended controls, present risk assessment, and other information pertinent to the assessment; and (3) to provide status tracking, verification, and a record of closure (or risk acceptance). The HR format is continuous to permit as much information as is available to fit in each block. The following paragraphs provide guidance and interpretation for completing each block of this typical hazard report form (Hansen 1993).

Conditions or hazards having an initial risk severity code of IV or a frequency code of D are not to be documented on an HR but are to be discussed and documented in the appropriate analysis and documented on an HAW. All other coded hazards should be documented on an HR.

A hazard report is prepared when approved and released drawings, or the actual physical configuration of the system, contains a hazard with an initial risk code of IIIC or higher. The safety manager is responsible for managing the hazard reports to include revising the HRs, monitoring the status of corrective actions, and reporting the status to program management, when required. The safety manager expeditiously advises program management when a hazard report is originated or revised. The hazard reports sections are described in detail in the following paragraphs.

Short title. A short phrase identifying the particular hazard. Keep it short but specific: not "thermal hazard" but "thermal hazard in [type, model] copier."

System/element. Enter the system or element to which the hazard being analyzed applies.

Rpt No. Hazard reports should be identified numerically by system or subsystem. For example, the first hazard report for a communications system should be COMM-001. This assists in tracking historical events of a particular hazard report.

Operation/Phase. Enter the operations or phases during which the hazard is applicable (e.g., testing,

Hazard Report Form

SHORT TITLE:		
SYSTEM/ELEMENT:		RPT. NO.
OPERATION/PHASE:	REV:	DATE:
EQUIPMENT:	LOCATION:	
ORIGINATOR (NAME):	ORGANIZATION:	DATE:
ACTION (NAME):	ORGANIZATION:	TELE:
SUSPENSE DATE (CLOSURE):		
HAZARD/ACCIDENT RISK FACTOR CODE:		
DESCRIPTION OF HAZARD:		
REQUIREMENTS:		
EXISTING CONTROLS		REFERENCE
ACCIDENT SCENARIO:		
SINGLE POINT FAILURE: YES:	NO:	
UNDESIRED EFFECTS		EFFECT RISK CODE

FIGURE 16A. **Typical hazard report form** (*Source:* Hansen, 1993)

installation, operating, and maintenance (preventive or repair).

Revision and date. Enter the number of this revision (change or update) to the report and the date this revision was prepared. For the initial report enter Rev. 0 and the origination date.

Equipment. Identify the specific equipment or system that contains, or is impacted by, the hazard (e.g., laser printer model XX).

Location. State the location (building, module, room number, etc.) of the system or the equipment involved.

INITIAL RISK CODE CONSIDERING ACCIDENT SCENARIO & EXISTING CONTROLS:		
RECOMMENDED CORRECTIVE CONTROLS	REFERENCE	STATUS
ACTION(S) PLANNED/TAKEN:		
CURRENT RISK CODE BASED ON ACTIONS TAKEN AND VERIFIED:		
VERIFICATION METHODS	REFERENCE	STATUS
RESIDUAL RISK CODE:		
OTHER REFERENCES:		
SERVICE REPORT #:	DATE:	

FIGURE 16B. **Typical hazard report form** (*Source:* Hansen, 1993)

Originator, organization, and date. Enter the name and organization of the person originating the hazard report. Enter the date the HR on this hazard was first originated. This information (name, date) will not change regardless of subsequent revisions.

Action and organization. Enter the individual and agency (or office) responsible for implementing recommended corrective actions. Generally, this is *not* the safety manager or engineer because they usually do not have the responsibility, resources, or authority to order or implement design changes or equipment or facility modifications.

Suspense date. The latest required date for closure. This date should be prior to the implementation unless

ECP #:	DATE:
DRAWING NUMBER(s):	
Design Specification	
OTHER:	
RECOMMENDED CLOSURE:	DATE:
CLOSURE CONCURRENCE:	
SAFETY MANAGER:	DATE:
MANAGER:	DATE:
RISK ACCEPTED:	

RISK ASSESSMENT CODE LEGEND:

		Probability Level			
		A	B	C	D
Severity Level	I	1	1	2	3
	II	1	2	3	4
	III	2	3	4	5
	IV	3	4	5	6

FIGURE 16C. Typical hazard report form (*Source:* Hansen, 1993)

approved workarounds are implemented that adequately control the hazard for the short term. These workarounds or other implemented controls should be explained in the "actions taken" block.

Hazard/risk factor code, description of hazard. Enter the risk factor from the risk-factor list in Table 1 and a description of the hazard. Only a single hazard or risk factor should be documented on each HR to reduce confusion and assure complete analysis of each hazard or risk factor. Each hazard or risk factor may have several applicable hazard reports to differentiate location of the hazard, different accident scenarios, different subcategories of the accident risk factor, and so on. The hazard should be stated as specifically as possible to facilitate complete and accurate analysis and risk assessment. For example, "Building exposed to damage from lightning strikes" is a specific hazard statement, whereas "bad weather" is too broad and vague for either accurate analysis (on one form) or control and closure. State the hazard locations, and list affected equipment, facilities, personnel, and so on. If several locations or items of equipment are impacted, then the accident scenario and the recommended corrective actions should be equally applicable to them all. If this is not the case, one or more separate hazard reports must be prepared.

Requirements. List all applicable codes, standards, specifications, and requirements that relate to the hazard being analyzed. Also include paragraph numbers that are applicable.

Existing controls and reference. List all *existing* controls that actually reduce the risk of the accident or undesired effects. Explain how each existing control specifically controls the hazard, accident scenario, accident risk factors (e.g., triggering events, necessary conditions, etc.), undesired effects, or risk. Explain the effectiveness of each existing control in reducing the risk. An *existing* control is one that is actually in effect and in some way mitigates the potential risk, severity, or frequency. Explain whether the control actually exists (i.e., is physically implemented), is designed but not yet implemented, is specified but not yet designed, or is procedural. In the "reference" block, list the drawing, specifications, procedure title or number, and other specific control documentation.

Accident scenario. Explain the credible accident scenario, including all causes, that can result from the hazard/accident risk factor being analyzed. Explain each step of each accident scenario in the cause-effect or chronological order in which it is likely to occur. The logic relationships (e.g., "and" versus "or") of the steps in the accident scenario must be determined to assure accurate risk coding based on the alternate paths to the undesired event and to assure all "or" paths are controlled. List and explain possible causes or triggering events that can propagate the hazard into an accident. Also explain each intermediate undesired effect and any special conditions that may be present to mitigate or worsen the severity or frequency of the undesired effects. Discuss simultaneous events that may cause other undesired effects. Discuss all accident scenarios that can result in the undesired effects listed in the "undesired effects" block below. A pictorial representation of the hazard/accident scenario flow is shown in Figure 17 (Hansen 1993).

The initial risk code is based on the hazard or accident risk factor being analyzed and the specific accident scenarios, including existing controls. Therefore, the discussion of the accident scenarios must be complete, factual, and thoroughly analyzed to support accurate risk coding and the development of effective recommended controls.

Single-point failure. A single-point failure is one in which the failure of a component or single action by an individual can cause an accident. If a single-point failure can result in an accident, this will be described in the accident scenario.

Undesired effects and effect risk code. List the potential undesired events or effects that can result from the hazard/accident risk factor just described. In the "effect risk code" block, enter the highest present risk code for *each* undesired effect (e.g., personnel II C, equipment III B). Again, these codes *cannot* be assigned until the accident scenario, undesired effects, and existing controls have been determined and analyzed. The highest effect risk code, as determined from Figure 16, is entered in the "initial hazard risk code" block.

Initial risk code. Enter the assessment of the initial risk—that is, the highest risk level arising from the hazard and its potential accident scenarios or effects, considering the effectiveness of any existing (in place) controls. The risk assessment and risk code assigned are based on the specific hazard statement, the possible undesired effects of the hazard and their severity, and the estimated frequency of occurrence of the accident scenarios described. The initial risk code *cannot* be assigned until the accident scenario, undesired effects, and existing controls have been determined and analyzed. The initial risk code is the highest-ranked effect risk code as determined from Figure 16.

Recommended corrective controls, reference, and status. List all recommended corrective controls that will reduce the initial risks. List and explain recommended design changes, specification changes, safety devices, procedures, and other controls that will elim-

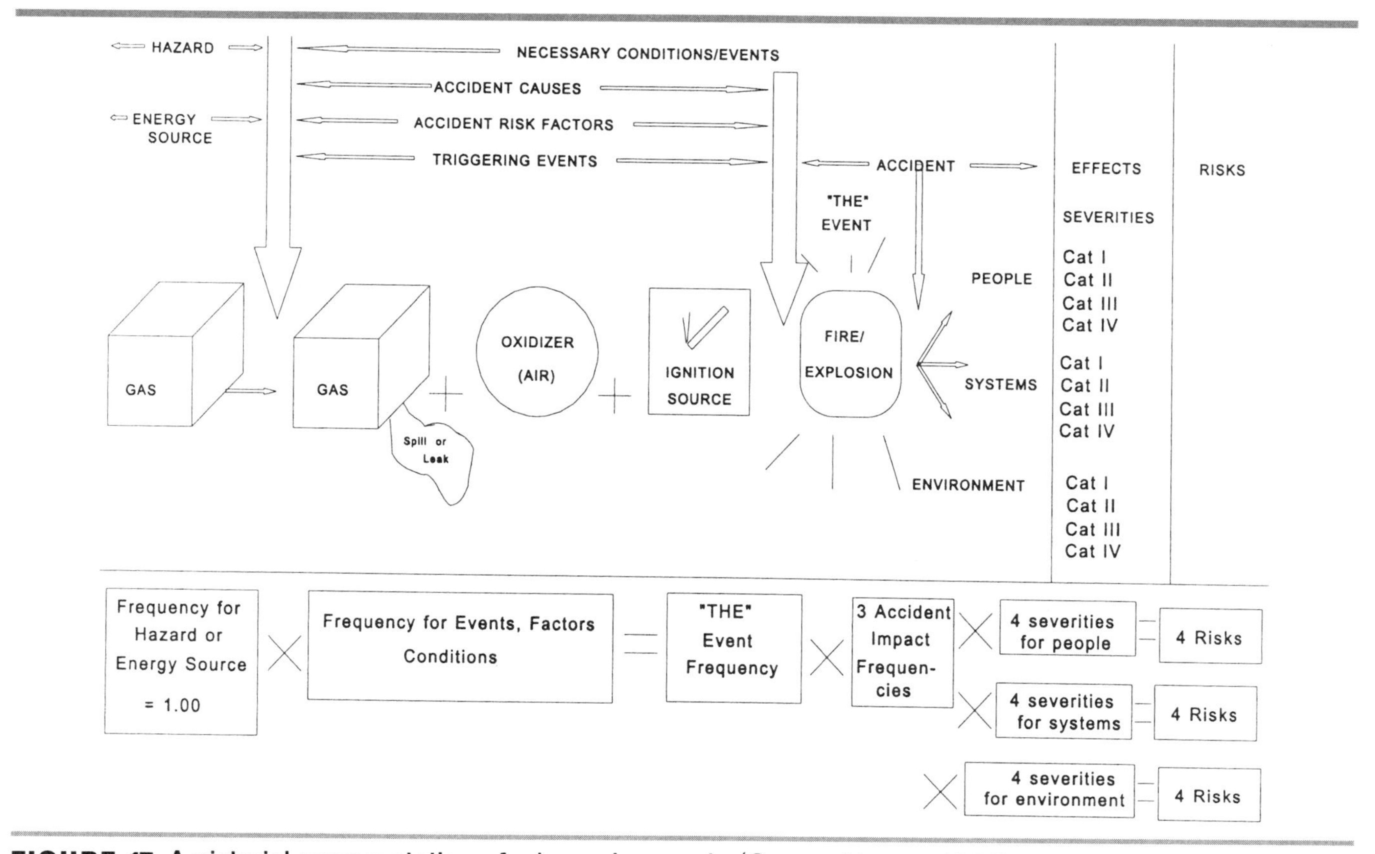

FIGURE 17. A pictorial representation of a hazard scenario (*Source:* Hansen 1993)

inate or mitigate the hazard, control or eliminate accident risk factors, decrease the severity of undesired effects, or reduce the frequency of occurrence of the possible accident scenarios. Recommendations to control undesired effects should be listed and explained. The system-safety precedence specified should be used in prioritizing recommended controls. If any of the recommended controls are alternative to other listed controls, so state. List the status, open or closed, of each recommended control. A control is not considered closed until it has been implemented and verified. For each recommended control, list applicable references (e.g., NFPA 12-1985, Paragraph 2). Recommended corrective actions must be closable and verifiable. As an example, "Personnel will be briefed periodically" cannot be closed and is therefore inappropriate. An appropriate recommendation would be as follows: "(1) A briefing will be prepared and approved by the safety office. (2) A procedure will be established to brief personnel annually and to document all the briefings."

Actions planned/taken. List and explain all actions planned or taken to mitigate the hazard/accident risk and their status (e.g., engineering change proposal, service report, etc.). Include dates of actual or planned completion of each action.

Current risk. The current risk is the initial risk code reassessed based on actions taken and verified as of the date of the HR revision.

Verification methods, reference, and status. List the verification methods (e.g., demonstration, analysis, inspection, test) for each existing and recommended control together with the person or agency that will physically do the verifications. List the status of each verification (i.e., "open" if not verified, "closed" if verified). For each verification listed, also list applicable references (e.g., analysis/test report number, date of inspection). If an item is closed by inspection, include the name of the person who had conducted the inspection as well as the date of the inspection.

Residual risk code. Enter the risk code for the highest remaining risk if all of the recommended controls are implemented. Keep in mind that few operations are risk-free; therefore, it may not be practical or possible to reduce the residual risk to a IV or D category. The residual risk code will be reevaluated, and changed as necessary, based on actions.planned or taken.

Other references. List correspondence and other pertinent documentation relating to this hazard. Also list the service report number, drawing numbers, and the ECP number if they are applicable.

Recommended closure and date. The manager (action person) responsible for correcting the hazard signs and dates the report when, in the manager's professional judgment, the required actions have been implemented and verified and the degree of risk has been reduced to an acceptable level.

Closure concurrence and date. The SSM signs and dates the report when he or she concurs with closure of this hazard report. Closure of any hazard report should not be recommended until the residual risk is considered acceptable, and the controls have been verified as being in place and adequate. This signature is the SSM's attestation that the system can be operated safely.

Manager and date. The manager signs and dates when he or she concurs with closure of this hazard report. This signature constitutes the certification that the hazard is adequately controlled and the systems can be operated safely.

Safety manager and date. The SSM signs and dates the report when the SSM concurs with closure of this hazard report.

Residual risk accepted. This signifies the remaining (residual) risk accepted for this particular hazard. This officially closes the hazard report.

Summary

System safety is a discipline in and of itself within all the other safety disciplines. Its applicability is broad and can be applied to virtually every workplace situation to eliminate and control hazards to an acceptable level. There are many approaches and tools. It provides a systematic framework to identify, analyze, eliminate, and control hazards in all types of systems. Whether the system is simple or complex, system-safety techniques can be used to characterize the system. It is certainly a tool worthy of consideration by a safety professional for addressing workplace hazards.

Important Terms

Safety: Freedom from injury, damage, or loss of resources or system availability. Resources include not only people but also time, equipment, and money.

Reliability: The chance that an item can perform its required function for a specified time under specified conditions.

[Safety and reliability are two different concepts. They do, however, overlap. This happens when a safety-critical system fails to work properly (such as a life-support system or fire-alarm system) or when a component failure could result in injury, damage, or loss (such as the failure of a pipe coupling or allowing toxic gas to escape). Not all safety problems are reliability problems and vice versa.]

System: A collection of "things" that work together to achieve a common goal. The things may be people, equipment, facilities, software, tools, raw materials, procedures, organizations, and so on.

System safety: Involves the application of engineering and management principles and techniques to optimize safety. The system should be as safe as possible, given other constraints such as operational effectiveness, time, and cost. System safety covers all phases of the system life cycle.

System analysis: Separating the system into subsystems and subsystems into components, examining each subsystem and component and their interactions. It also involves documenting the results. Analysis is done to identify hazards and to present options for corrective action to decision makers.

Mishap: An unexpected, unforeseen, or unintended event that causes injury, loss, or damage to personnel, equipment, or mission accomplishment.

Hazard: A condition or event that can result from a mishap if it is uncontrolled. The mishap would be the effect of the hazard.

Hazard cause: A condition that contributes to a hazard. It could be unsafe design, environmental factors, failure, human error, and so on.

Hazard controls: Measures that are taken to eliminate a hazard or reduce the severity or probability of its potential effect.

Risk: The combination of severity and probability of a hazard effect.

References

Air Force Inspection and Safety Center/ Space Electronic Security Division (AFISC/SESD). Nd. *Software Safey Handbook* (Draft). Norton Air Force Base, CA.

American Institute of Chemical Engineers (AIChE). 1985, *Guidelines for Hazard Evaluation Procedures.* New York: AIChE/Battelle Columbus Division.

______. 1992. *Guidelines for Implementing Process Safety Management Systems.* New York: Center for Chemical Process Safety (CCPS).

Brown, D. B. 1976. *Systems Analysis and Design for Safety.* New Jersey: Prentice Hall.

Center for Chemical Process Safety (CCPS). 2000. *Guidelines for Process Safety in Outsourced Manufacturing Operations.* New York: American Institute of Chemical Engineers.

Churchman, C. W. 1981. *The Systems Approach.* New York: Dell.

Clemens, P. L. 2001. *A Charlatan's Guide to Quickly Acquired Quackery: The Trouble with System Safety.* Houston: NASA Training Center.

Goldwaite, William H. 1985. ??

Hammer, W. 1989. *Occupational Safety Management and Engineering.* 4th ed. Englewood Cliffs, N.J: Prentice Hall.

Hansen, M. D. 1993. "CSOC Integrated System Safety Program Plan." United States Air Force Space Command, Loral Command & Control Systems, Colorado Springs Division. December 31, 1993.

Hassl, D. E., N. H. Roberts, W. E. Vesely, and F. F. Goldberg. 1980. *Fault Tree Handbook* (NUREG 0492). Washington, D.C.: U.S. Nuclear Regulatory Commission, Office of Nuclear Research, Division of Systems and Reliability Research.

Leveson, N. G. 2002. "System Safety Engineering: Back to the Future, Aeronautics and Astronautics" (unpublished). Boston: Massachusetts Institute of Technology.

Lowrance, W. W. 1976. *Of Acceptable Risk.* Los Angeles: Kaufman.

Malasky, S. W. 1974. *System Safety: Planning, Engineering, and Management.* Rochelle Park, NJ: Hayden Books.

McGraw Hill. 1982. *Concise Encyclopedia of Science and Technology.* New York: McGraw Hill.

Post, R. L., D. C. Hendershot, and P. Kers. 2001. "Safety and Reliability: A Synergetic Design Approach." 5th Biennial Process Plant Safety Symposium, American Institute of Chemical Engineers, 2001 Spring National Meeting, April 22–26 2001, George R. Brown Convention Center, Houston, TX.

Rausand, M., and A. Høyland. *System Reliability Theory: Models, Statistical Methods and Applications.* 2d ed. New York: Wiley.

Roland, H. E. 1986. "The Fault Tree: Why and How of Quantification, Hazard Prevention." *Journal of the System Safety Society* (May–June). pp. 28–31.

Roland, H. E., and B. Moriarty. 1983. *System Safety Engineering and Management.* New York: Wiley.

Stephans, R. A., and W. W. Talso. 1997. *System Safety Analysis Handbook.* 2d ed. Unionville, VA: System Safety Society.

Stephenson, J. 1991. *System Safety 2000: A Practical Guide for Planning, Managing, and Conducting System Safety Programs.* New York: Van Nostrand Reinhold.

Swain, A. D. 1974. *The Human Element in System Safety: A Guide for Modern Management, Industrial and Commercial Techniques.* London: Swain.

U. S. Department of Defense (DOD). 1990. *DoD Directive 3150.2. Safety Studies and Reviews of Nuclear Weapon Systems.* Washington, D.C.: DOD.

______. 2000. *MIL-STD-882D, Standard Practice for System Safety.* Washington, D.C.: DOD.

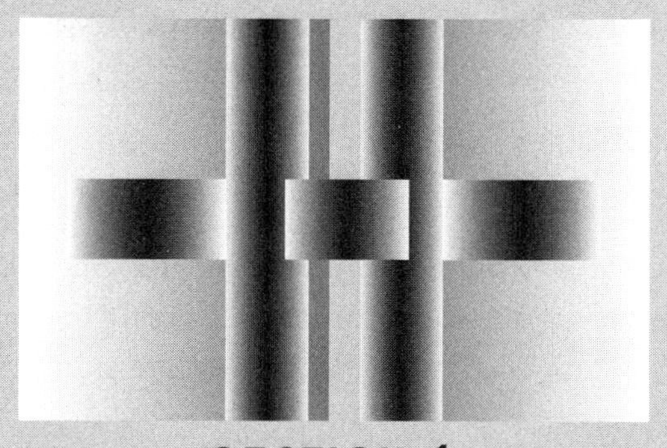

SECTION 1
RISK ASSESSMENT AND HAZARD CONTROL

LEARNING OBJECTIVES

The following objectives apply to working on or near energized electrical conductors and equipment operating at 50 volts or more to ground.

- Understand the hazards involved in such work.
- Know which rules and standards apply to electrical safety for personnel.
- Understand and be able to apply safety-related work practices.
- Be able to properly select and use personal protective equipment, including flame-resistant clothing, for electrical work.
- Be able to ensure proper training and the qualification of personnel performing electrical work.
- Utilize safe switching procedures for personnel safety.
- Understand the need for temporary protective grounding for personnel.

APPLIED SCIENCE AND ENGINEERING: ELECTRICAL SAFETY

Steven J. Owen

ELECTRICITY HAS BECOME an essential part of modern life, both at home and on the job. Some employees, such as electricians, work directly with electricity; other employees work with wiring associated with overhead lines, cable harnesses, or circuit assemblies. Still others, such as office workers and salespeople, work with electricity indirectly. As a source of power, electricity is accepted without much thought as to the hazards encountered. Perhaps because it has become such a familiar part of our daily surroundings, often it is not treated with the respect it deserves.

This chapter has been adapted from the book *Electrical Safety—Procedures and Practices* (Owen 1997). For purposes of discussion, the hazards associated with electricity are broken down into three types: (1) shock, (2) arc flash, and (3) arc blast.

HOW DOES ELECTRICITY ACT?

To handle electricity safely, it is necessary to understand how it acts, how it can be directed, and how parts of the human body react to it. For this purpose, it is helpful to compare the flow of electricity with the flow of water.

Operating an electric switch may be considered analogous to turning on a water faucet. Behind the faucet or switch there must be a source of water or electricity, with something to transport it and with pressure to make it flow. In the case of water, the source is the reservoir or pumping station, the transportation is through pipes, and the force to make it flow is the pressure provided by a pump. With electricity, the source is the power-generating station. Current travels through electric conductors in the form of wires and by pressure, measured in volts, provided by a generator.

Resistance to the flow of electricity is measured in ohms and varies widely. It is determined by three factors: the nature of the substance itself, the length and cross-sectional area of the substance, and the temperature of the substance.

Some substances, such as metals, offer very little resistance to the flow of electric current and are called conductors. Other substances, such as Bakelite, porcelain, and dry wood, offer such high resistance that they can be used to prevent the flow of electric current and are called insulators. The relationship between resistance, current, and voltage is called Ohm's Law and is represented as follows:

Voltage = Current times Resistance

$$E = I \times R \quad (1)$$

Current = Voltage divided by Resistance

$$I = E/R \quad (2)$$

Resistance = Voltage divided by Current

$$R = E/I \quad (3)$$

A *volt* is the unit used to measure the electrical pressure causing the current to flow.

An *ampere* is the unit used to measure current, which is the net transfer of electric charge per unit time (e.g., conduction current in a wire).

An *ohm* is the unit used to measure the resistance of a conductor such that a constant current of one ampere in it produces a voltage of one volt between its ends.

Shock

Electricity travels in closed circuits. The preferred, normal route is through a conductor. Shock occurs when the body becomes a part of the electric circuit. The current enters the body at one point and leave at one or more other points. Shock normally occurs in one of three ways, as the person being shocked comes in contact with

- both wires of the electric circuit,
- one wire of the electric circuit and the ground, or
- a metallic part that has become energized by being in contact with an energized wire while the person is also in contact with the ground or a grounded surface.

To better understand the harm done by electrical shock, we need to understand something about the physiology of certain body parts, including the skin, heart, and muscles.

The Skin

Skin covers the body and is made up of three layers. The most important layer, as far as electric shock is concerned, is the outer layer of dead cells referred to as the horny layer. This layer is composed mostly of a protein called keratin. It is the keratin that provides the largest percentage of the body's electrical resistance. When it is dry, the outer layer of skin may have a resistance of several thousand ohms, but when it is moist, there is a radical drop in resistance, as is also the case if there is a cut or abrasion piercing the horny layer. The amount of resistance provided by the skin will vary widely from individual to individual. The resistance will also vary widely at different parts of the body. For instance, the workman with high-resistance hands may have low resistance on the back of his calf.

The skin, like any insulator, has a breakdown voltage at which it ceases to act as a resistor and is simply punctured, leaving only the lower-resistance body tissue to impede the flow of current in the body. This voltage will vary with the individual, but is in the area of 600 volts. Because most industrial power-distribution systems are 480 volts or higher, people working at these levels need to develop a special awareness of the shock potential.

The Heart

The heart is the pump that sends life-sustaining blood to all parts of the body. The blood flow is caused by the contractions of the heart muscles, which are controlled by electrical impulses. The electrical impulses are delivered by a highly complicated, intricate system of nerve tissue with built-in timing mechanisms. A current, measured in milliamperes, can upset the rhythmic coordinated beating of the heart by disturbing the nerve impulses.

When this happens, the heart is said to be in fibrillation—that is, pumping action stops. Death will occur quickly if the normal coordinated rhythmic beating is not restored. Remarkable as it may seem, what is needed to defibrillate the heart is a shock of an even higher intensity.

Muscle

The other muscles of the body are also controlled by electrical impulses delivered by nerves. Electric shock can cause loss of muscular control, resulting in the inability to let go of an electrical conductor. Electric shock can indirectly cause other injuries because of involuntarily muscle reactions from the electric shock, which can cause bruises, fractures, and even deaths resulting from collisions or falls. The current from the contact with the energized conductor or equipment disrupts the normal electrical impulses, which causes the unexpected action from the muscles.

The Severity of Shock

The severity of shock received when a person becomes a part of an electric circuit is affected by three primary factors:

1. The amount of current flowing through the body, measured in amperes
2. The path of the current through the body
3. The length of time the body is in contact with the circuit

Other factors that may affect the severity of the shock are the frequency of the current, the phase of the heart cycle when shock occurs, and the general health of the person prior to the shock. Effects can range from a barely perceptible tingle to immediate cardiac arrest. There are no absolute limits or known values that show an exact injury at any given amperage range.

A severe shock can cause considerably more damage to the body than is visible. For example, a person may suffer internal hemorrhages and destruction of tissues, nerves, and muscle. In addition, shock is often only the beginning in a chain of events. The final injury may well be from a fall, cuts, burns, or broken bones.

The most common shock-related injury is a burn. Burns suffered in electrical accidents may be of three types—electrical burns, arc burns, and thermal contact burns. Electrical burns are the result of the electric current flowing through the tissues or bones. Heat generated by the current flow through the body causes tissue damage. Electrical burns are one of the most serious injuries you can receive and should be given immediate attention. Because the most severe burning is likely to be on the inside, what may at first appear to be a small surface wound could, in fact, be an indication of severe internal burns. Arc and contact burns are discussed in the following sections.

Arc Flash

Arc burns make up a substantial portion of the injuries from electrical malfunction and may cause serious injury or death to workers. At the initiation of an arc fault, tremendous energy can be released in a very brief time as substantial electrical currents are passing through the air. Metal conductor parts can vaporize, resulting in hot vapors and hot metal being violently spewed. Through direct exposure or by igniting the worker's clothing, the thermal energy can result in severe burns to workers. A large portion (some estimates are as high as 80 percent) of all electrical injuries are a result of an arc flash igniting flammable clothing worn by personnel exposed to an arc flash.

Arc Blast

The third source of possible hazard is the blast associated with an electric arc. This blast comes from the pressure developed by the near-instantaneous heating of the air surrounding the arc and from the expansion of the metal as it is vaporized. The rapid thermal escalation of the air and vaporization of metal can create a very loud explosion and create tremendous pressures. These high sound levels can rupture eardrums. This is why NFPA 70E-2004 requires hearing protection for those working on or near electrical equipment classified as Risk/Hazard Category 2 and higher.

These high pressures associated with the arc blast can cause collapsed lungs and result in forces that can cause workers to lose balance [i.e., the required safe working position required by the Occupational Safety and Health Administration (OSHA) and NFPA 70E]. During an arc blast, copper vapor expands to

67,000 times the volume of solid copper (copper expands by a factor in excess of 67,000 times in boiling). The air in the arc stream expands when heating from the ambient temperature to that of the arc, which is approximately 35,000° F. Examples of this vaporization of metal may involve a noninsulated metallic screwdriver blade, a metal fish tape, or other noninsulated tools or parts coming in contact with energized parts as phase-to-phase or phase-to-ground faults. These pressures can be great enough to hurl people, parts, and equipment considerable distances.

Phase-to-phase faults, also referred as bolted faults, happen when an accidental contact is made between two or more energized ungrounded conductors simultaneously, resulting in a tremendous amount of energy released. An example of a phase-to-phase fault would be if a large tree limb fell across energized overhead lines, contacting two or more lines simultaneously. Another example would be if a noninsulated screwdriver or a metal flashlight was dropped or left lying across two or more exposed energized bus bars simultaneously.

Phase-to-ground faults happen when contact is made between an energized ungrounded conductor and a grounded conductor or grounded surface. An example of a phase-to-ground fault would be if a noninsulated screwdriver was dropped or somehow wedged between an exposed energized bus bar or conductor and a grounded metal surface or grounded conductor.

Other hazards associated with the arc and blast include spewing of molten metal droplets with temperatures as high as 1800° F, which can cause contact burns and associated damage, as well as being inhaled into the person's lungs (as a startled person gasps for air). Another negative effect of an explosion caused by an arc blast is projectiles—for example, parts of the molded case, other metallic or nonmetallic parts of a circuit breaker, or the ferrule of a fuse traveling through air—that impact workers and other equipment. A possible beneficial side effect of the blast is that it can hurl a nearby person away from the arc as well as reduce the time a worker is exposed to the arc flash temperatures.

The total force exerted on a worker's body due to an arcing fault blast is dependent on the body surface exposed to the blast wave. The potential health risk to a worker resulting from the total forces exerted on his or her body depends on the worker's position when the blast contacted the worker. A worker standing on the floor would most likely be able to safely withstand more pressure than a worker on a ladder. A worker on a ladder or working from a scaffold or bucket who is subjected to an arc flash and arc blast has an increased chance of injury from a fall.

Safety-Related Work Practices

General Requirements for Safety-Related Work Practices

The safety-related work practices of OSHA 29 CFR 1910, Sections 1910.331–335, and NFPA 70E-2004, Articles 110, 120, and 130, apply to work by both qualified persons (those who have training in recognition and avoidance of electrical hazards of working on or near exposed energized parts) and unqualified persons (those with little or no training). The rules apply to the following installations:

- premises wiring installations of electrical conductors and equipment within or on buildings or other structures
- wiring for connection to supply installations of conductors that connect to the supply of electricity
- other wiring installations of other outside conductors on the premises
- optical fiber cable installations of optical fiber cable where such installations are made along with electrical conductors

Establishing Safe Work Conditions and Determining Safe Working Distances

Over the years, analytical tools have been developed to better assess the hazards created from arcing faults. Because of injuries and deaths, NFPA 70E-2004 *Electrical Safety Requirements for Employee Workplaces* adopted formulas to define the safe working distance from a potential arc. These formulas are used to determine the type of protective gear a worker needs to wear when working on energized electrical equipment. The formulas for this calculation are based on the work and a technical paper by Ralph Lee (1982).

Lee showed temperature and time thresholds for incurable and just-curable burns. At a distance of 3 feet, 0 inches, from the source of the arc, the arc energy required to produce these temperatures was determined to be 23 MW and 17 MW for incurable and just-curable burns, respectively. Lee also found that the maximum arc energy occurred when it represented 50 percent of the available three-phase bolted fault (bf). For example, the arc from a 46 MVA (million volt-amps) available source for 0.1 second could cause an incurable burn at a distance of 3 feet, 0 inches. The arc from a 34 MVA (34 million volt-amps) available fault for 0.1 second at 3 feet, 0 inches, would result in a just-curable burn. The following formulas were developed by Lee and incorporated into NFPA 70E [3][4][5]:

$$Dc = (2.65 \times \text{MVA bf} \times t)^{1/2} \quad (4)$$

$$Df = (1.96 \times \text{MVA bf} \times t)^{1/2} \quad (5)$$

Dc = distance in feet for a just-curable burn

Df = distance in feet for an incurable burn

MVA bf = bolted three-phases MVA at point of short circuit = $1.73 \times$ voltage L – L $\times$ available short circuit current $\times 10^{-6}$

t = time of exposure in seconds

Example 1: Assume an available 40,896-ampere bolted three-phase fault on a 480-volt system, protected with a noncurrent-limiting fuse with a clearing time of six cycles (0.1 second). Find the distance in feet for a just-curable burn.

$$Dc = (2.65 \times \text{MVA bf} \times t)^{1/2}$$
$$Dc = (2.65 \times 1.732 \times 480 \times 40{,}896 \times 10^{-6} \times 0.1)^{1/2}$$
$$Dc = (9.00)^{1/2}$$
$$Dc = 3 \text{ ft}$$

This means that any exposed skin closer than 3 feet, 0 inches, to this available fault for 0.1 second or longer may not be curable should an arcing fault occur. If employees must work on this equipment where parts of their body would be closer than 3 feet, 0 inches, from the possible arc, suitable protective equipment must be utilized so that the possibility of employee injury is minimized. See OSHA 29 CFR 1910, Section 1910.335(a)(1), and NFPA 70E-2004, Section 130.7(A).

Example 2: Assume an available 40,896-ampere bolted three-phase fault on a 480-volt system, protected by Class J, 200 amp, current-limiting fuses. The opening time is assumed at 1/4 cycle (0.004 seconds), and the equivalent root mean square (RMS) let-through current (as read from the fuse manufacturer's chart) is 6000 amperes.

$$Dc = (2.65 \times \text{MVA bf} \times t)^{1/2}$$
$$Dc = (2.65 \times 1.732 \times 480 \times 6000 \times 10^{-6} \times 0.004)^{1/2}$$
$$Dc = (0.0528)^{1/2}$$
$$Dc = 0.229 \text{ ft (or 2.75 inches)}$$

Thus, the flash-protection boundary was significantly decreased, from 3 feet, 0 inches (36 inches) in Example 1 to 0.23 feet (2.75 inches) in Example 2, by the use of current-limiting fuses, which limited the short circuit current from 40,896 amperes to 6000 amperes let-through and reduced the exposure time from 6 cycles to 1/4 cycle.

Employees must wear, and be trained in the use of, appropriate protective equipment for the possible electrical hazards with which they are faced. Examples of this personal protective equipment include head, face, neck, chin, eye, ear, body, and extremity protection as required. Electrical protective equipment selected for the job task, based on the Hazard/Risk Assessment from Table 130.7(C)(9)(a) and selected from Table 130.7(C)(10), must also meet the minimum arc thermal protective value (ATPV) ratings of Table 130.7(C)(10) of NFPA 70E-2004.

In Example 2, personal protective equipment may be determined by using Table 130.7(C)(9)(a), Hazard/Risk Category 2, based on the type of equipment (e.g., use a 600-volt class switchgear), with a person doing a voltage test on the 600-volt bus in the back side of the switchgear, by opening a hinged door to gain access to the bus. The next step would be to go to Table 130.7(C)(10), Protective Clothing and Personal Protective Equipment Matrix. Select Hazard/Risk Category column 2. Looking vertically from top to bottom, it shows that garments consisting of a t-shirt, long-sleeved shirt, and pants made of nonmelting or untreated natural fiber material should be worn. Over those garments, personnel need to wear a flame-resistant, long-sleeved shirt and flame-resistant pants (or flame-resistant coveralls in lieu of the shirt and pants), along with leather work shoes, hearing protectors, safety glasses, insulated gloves, and a double-layered switching hood. To ensure that the proper selection of flame-resistant clothing has been made, find Table 130.7(C)(11). In this table, for Hazard/Risk Category 2, an APTV rating should be a minimum of at least 8 cal/cm^2. This rating is found on the manufacturer's labels sewn into the clothing.

The hazard/risk category is a determination of the relative danger of a given job task, which is based in part on the incident energy available, the distance from exposed energized parts, and the distance in which an arc flash could reach a person, who may be injured by burns not exceeding second degree or 1.2 cal/cm^2, which are considered to be just curable.

Personal protective equipment, sufficient for protection against an electrical flash, would be required for any part of the body within 3 feet, 0 inches, of the fault in Example 1. Such equipment would include a switching hood (nonconductive hard hat, arc-rated face shield, and flame-resistant head, neck, and face protection), hearing protectors, flame-resistant clothing that meets the minimum requirements of Table 130.7(C)(11), voltage-rated gloves, and leather footwear.

Significantly less personal protective equipment would be required for Example 2 because the flash zone is within 0.23 feet (2.75 inches), and the incident energy available is calculated to be 0.17 cal/cm^2. In this case, the required personal protective equipment might be reduced to a nonconductive hard hat, safety glasses, flame-resistant shirt, flame-resistant pants or blue jeans (minimum 12-ounce fabric weight), voltage-rated gloves, and leather footwear—which is equivalent to Hazard/Risk Category 1—and possibly treated as Risk/Hazard Category 0 (from Table 130.7(C)(11) Protective Clothing Characteristics), with no flame-resistant clothing.

In an actual case where an electrical worker is to work on energized equipment, the safe working distance must be determined by making calculations over the full range of possible currents and estimation of exposure time. The worker is required to wear protective clothing and gear for the worst-case condition. The possible currents would encompass the range of currents up to the maximum available current that could occur if a mishap occurred. The exposure time is dependent on reasonable reaction time and is situational. For example, a worker standing in front of a switchgear might reasonably be expected to get out of the way of a blast in one second. The normal human reaction time (to detect the condition and to move out of the path) has been estimated to be one-fourth of a second, whereas an arc fault may develop into an arc blast in one-sixth of a second. A worker on his or her knees might be exposed for two to three seconds or longer. A worker lying on the ground might be there for three to five seconds or longer. However, a worker in a bucket truck might be exposed to this hazard for many seconds or minutes. These times can be utilized with the time current curves to determine the maximum amount of current at those times where the overcurrent protective device does not open in that reaction or movement time. With that information the hazard can be calculated for the worst-case condition.

Work Permits

A summary of the sources for the topics discussed in this section is presented in the chapter appendix. Readers are advised to consult the full text of the standards for recommended procedures and practices.

Live parts to which an employee might be exposed should be put into an electrically safe work condition before an employee works on or near them, unless the employer can demonstrate that de-energizing introduces additional or increased hazards or is infeasible due to equipment design or operational limitations.

Energized parts that operate at less than 50 volts to ground are not required to be de-energized if there will be no increased exposure to electrical burns or to explosion due to electric arcs.

Where the work to be performed will be on energized parts, the electrical work must be performed by written permit only. The energized electrical work permit should include, but not be limited to, the following items (see Figure 1):

- a description of the circuit and equipment to be worked on and the location
- justification as to need for the work to be performed in an energized condition
- a description of the safe work practices that will be employed
- results of the shock hazard analysis
- determination of shock protection boundaries
- results of the flash hazard analysis
- the flash protection boundary
- the necessary personal protective equipment to safely perform the assigned task
- means employed to restrict the access of unqualified persons from the work area
- evidence of completion of a job briefing, including a discussion of any job-specific hazards
- energized work approval (authorizing or responsible management, safety officer, owner, etc.) signature(s).

Exemptions to Work Permit

Work performed on or near live parts by qualified persons related to tasks such as testing, troubleshooting, voltage measuring, and so on, is permitted without an energized electrical work permit, provided appropriate safe work practices and personal protective equipment are provided and used.

Approach Boundaries

NFPA 70E, Section 130.3(A), provides a basis for determining the flash protection boundary. For systems that operate at 600 volts or less, the flash protection boundary should be 4 feet, 0 inches, based on the product of clearing times of six cycles (0.1 second) and an available bolted fault current of 50 kiloAmperes (kA) or any combination not exceeding 300 kA cycles (5000 ampere seconds). This is the default value. For clearing times and fault currents other than 300 kA cycles, or under engineering supervision, calculating the flash protection boundary is also permitted.

At voltage levels above 600 volts, it is necessary to calculate the flash protection boundary. The flash protection boundary is the distance at which the incident energy equals 5 J/cm^2 (1.2 cal/cm^2).

Flash Hazard Analysis

Flash hazard analysis should be done before a person approaches any exposed electrical conductor circuit part that has not been placed in an electrically safe work condition.

The incident energy exposure determined by the flash hazard analysis should be used to select protective clothing and personal protective equipment for job-specific tasks. The clothing should be flame-resistant. Clothing made from flammable synthetic materials that melt at temperatures below 315° C (600° F), such as acetate, nylon, polyester, polypropylene, and spandex, either alone or in blends with cotton, should not be worn. The suggestion is to avoid wearing materials that may melt against the skin.

Employees working in areas where there are electrical hazards should be provided with and use protective equipment that is designed and constructed for the specific part of the body to be protected and for the work to be performed. This is taken from 29 CFR 1910, Section 1910.335(a)(1), of the OSHA General Industry Standard.

Hazard/Risk Analysis

(Adapted in part from Cooper-Bussman, *Handbook for Electrical Safety*, 2d ed., 2004.)

Hazard/risk analysis is a process that is used to do the following:

- evaluate circuit information drawings—electrical distribution one-line and other appropriate drawings
- determine the degree and extent of hazards
- provide job planning as necessary to safely perform tasks
- determine approach boundary requirements
- evaluate flash protection boundary requirements

ENERGIZED ELECTRICAL WORK PERMIT

Part I: To Be Completed by the Person or Organization Requesting Permit

Name of person/organization requesting permit: ______

______

Today's date: ______

Date for work to be performed: ______

Job number/work order number: ______

Part II: To Be Completed by Qualified Person(s) Performing Work on Energized Circuits and Parts

(1)(a) Description of circuit/equipment to be worked on: ______

______

______

(1)(b) Location of circuit/equipment to be worked on: ______

______

(2) Description of work to be performed: ______

______

______

______

(3) Justification for performing work on energized circuits and parts. Why this work cannot be performed on de-energized circuits and parts:

______

______

______

(4) A description of the safe work practices to be employed: ______

______

______

(5) Results of the shock hazard analysis: ______

______

______

(6) Determination of the shock protection boundaries: ______

______

(7) Results of the flash hazard analysis: ______

______

______

(8) Determination of the flash protection boundary: ______

______

FIGURE 1. Sample energized work permit (Adapted from NFPA 70E – 2004, Annex J)

ENERGIZED ELECTRICAL WORK PERMIT (cont.)

(9) Determination of personal protective equipment necessary to perform job task safely: ______________________

____

____

(10) Evidence of completion of job briefing: ______________________

____

____

____

Part III: Approval(s) To Perform Work on Energized Circuits and Parts

(1) Signatures of authorizing person(s). Sign only if you are in agreement that work can be performed safely. If not in agreement, leave unsigned and return to requester.

(a)(1) Electrical-qualified person to perform work or supervise work:

____

Signature Date: ______________________

(a)(2) Electrical-qualified person to perform work or supervise work:

____

Signature Date: ______________________

(a)(3) Electrical-qualified person to perform work or supervise work

____

Signature Date: ______________________

(b)(1) Responsible management authority and title

____

Signature Date: ______________________

(b)(2) Responsible management authority and title

____

Signature Date: ______________________

(b)(3) Responsible management authority and title

____

Signature Date: ______________________

Add more names as necessary.

FIGURE 1. Sample energized work permit (Adapted from NFPA 70E – 2004, Annex J)

- evaluate personnel qualifications
- determine appropriate personal protective equipment based on the potential hazards present

To engineer additional safety into electrical systems to protect personnel from arc flash and arc blast, engineering options may include the following: replacing existing switchgear with arc-resistant switchgear; installing a secondary main relay that can trip a primary circuit breaker; installing zone interlocking in switchgear (i.e., a signal from a downstream circuit breaker blocks an upstream circuit breaker from tripping); changing overcurrent device type to current-limiting device; installing differential relays; installing provisions for remote racking and remote operation; using a racking wrench or similar device that has a long or extra-long handle to increase working distance; changing the sequence of switching operations to reduce the time when exposure is high (Cooper Bussman 2004).

Training and Qualification of All Personnel

The training requirements that are contained in NFPA 70E-2004, Section 110.6, and OSHA 29 CFR 1910, Section 1910.332(a), standards apply to all employees who face a risk of electrical shock or injury when they are working on or near exposed energized parts or parts that may become energized. If this training has not been completed, then a person cannot be considered qualified.

OSHA defines a qualified person as one who is familiar with the construction and operation of the equipment and the hazards involved. Qualified persons are intended to be only those who are well acquainted with, and thoroughly conversant in, the electrical equipment and electrical hazards involved with the work to be performed.

Whether an employee is considered a qualified person depends on various circumstances in the workplace. It is possible for an individual to be considered qualified with regard to certain equipment in the workplace, but unqualified as to other equipment.

For example, an employee may have received the necessary training to be considered qualified to work on a particular piece of equipment. However, if that same employee were to work on other types of equipment for which he or she had not received the necessary training, he or she would be considered unqualified for that equipment.

In order to be considered qualified (permitted to work on or near exposed energized parts), employees, at a minimum, must be trained in and familiar with the following:

- the skills and techniques necessary to distinguish exposed live parts from other parts of electric equipment
- the skills and techniques necessary to determine the nominal voltage of exposed live parts
- the clearance distances specified in OSHA 29 CFR 1910, Section 1910.333(c), Table 130.2(C), NFPA 70E-2004, and the corresponding voltages to which the qualified person will be exposed
- the safety-related work practices required by OSHA 29 CFR 1910, Section 1910.331-335, Article 130, and NFPA 70E-2004 that pertain to their respective job assignments
- any electrically related safety practices not specifically addressed by OSHA CFR 1910, Section 1910.331-335, Article 130, and NFPA 70E -2004 that are necessary for their safety

The training may be on-the-job, in a classroom setting, or both. All personnel should have training appropriate for their tasks, including safety training. This is required per the new definition of "qualified person" in NFPA 70-2002, National Electrical Code (NFPA 2002). NFPA 70E-2004 is more detailed in the definition of qualified person.

Where does the term "qualified person" fit in? Who uses this term? The definition is found in the OSHA General Industry Standard, 29 CFR 1910, Section 1910.399. The definition is one familiar with the construction and operation of the equipment and the hazards involved. This definition ties into the requirements of Section 1910.332 related to training. The OSHA Construction Standard, 29 CFR 1926, Section 1926.32, defines a qualified person as follows:

"'Qualified' means one who, by possession of a recognized degree, certificate, or professional standing, or who by extensive knowledge, training, and experience, has successfully demonstrated his ability to solve or resolve problems relating to the subject matter, the work, or the project."

The National Electrical Code (2002) revised the definition of qualified person (NFPA 2002). The revised definition reads as follows: "one who has the skills and knowledge related to the construction and operation of the electrical equipment and installations and has received safety training on the hazards involved. "

No matter what the job task is and no matter which trade is involved, personnel need the appropriate level of training to perform a job task safely and efficiently. OSHA standard 29 CFR 1926, Section 1926.20, covering general safety and health provisions, is written as follows: "The employer shall permit only those employees qualified by training or experience to operate equipment and machinery." This is merely the first of a number of requirements dictating training for employees. Why do we keep making references to training? A person cannot be considered competent or qualified (OSHA 2006) without having received the appropriate training that is required by the sections mentioned here.

There are advantages to providing the training to meet the definition of *qualified*. The number one priority is always safety. Only those persons qualified by training are allowed to work on or near exposed, energized electrical circuits, conductors, or parts operating at 50 volts or more to ground. The benefits include a safer workplace and greater efficiency, along with improved employee morale and possibly lower insurance rates.

Use of Equipment

Test Instruments and Equipment

Test instruments, equipment, and their accessories should be rated for circuits and equipment to which they will be connected. Test instruments, equipment, and their accessories should be designed for the environment to which they will be exposed, and for the manner in which they will be used. Test instruments and equipment and all associated test leads, cables, power cords, probes, and connectors should be visually inspected for external defects and damage before the equipment is used on any shift.

If there is a defect or evidence of damage that might expose an employee to injury, the defective or damaged item should be removed from service and no employee should use it until necessary repairs and tests rendering the equipment safe have been made.

Because the use of test instruments can expose employees to live parts of electric circuits, testing on energized electric circuits or equipment should be performed only by qualified persons.

Visual Inspection

To prevent injuries to employees resulting from exposed conductors or defects in the test equipment, a visual inspection of all test equipment is required before it is used. If any defects are found or suspected, the damaged equipment must be taken out of service, and employees must not be allowed to use the defective damaged equipment until it has been repaired.

Using test equipment in improper environments or on circuits with voltages or currents higher than the rating of the equipment can cause equipment failure. Verification of voltage on an incorrect scale or value other than voltage (e.g., ohms or current) has led to meters being damaged and personnel being injured.

Hazards can be exceedingly greater for an incident in a high-energy circuit than one in a low-energy circuit. The same misapplication that results in a blown fuse and a puff of smoke in a low-energy circuit could result in a violent explosion and injury in a high-energy circuit.

Because employees can be injured as a result of this failure, test equipment must be used within its rating and to be suitable for the environment in which it is to be used.

Portable Electric Equipment

Grounded Tools

Portable electrical hand tools are widely used by those in general industry and on construction sites. Because they are so widely used, they are also often abused. Tools that are faulty or damaged can become a source of electrical shock to the user.

It is very important that all tools (unless double-insulated) contain an equipment grounding conductor that is connected to the tool frame and through the supply cord back to the service entrance enclosure. If a ground fault occurs in a defective tool, the grounding conductor must carry enough current to immediately trip the circuit breaker or blow the fuse. This requires that the ground fault path have low impedance. All electrical power tools should be listed by a nationally recognized testing laboratory (NRTL), such as Underwriters Laboratories (UL) or an equivalent. Tools and their cords must be inspected before each and every use. If defects are found, the tool should be either marked and taken out of service until repaired or destroyed. Periodic ground-continuity and insulation-resistance testing should be performed to ensure that the tool can be operated safely.

Handling

Portable electric hand tools should always be handled in a manner that will not damage the tool. Flexible cords that are connected to the tool should not be used as a means of raising and lowering the tool. Flexible cords should never be fastened with staples or otherwise hung in any way that might damage the outer jacket or insulation.

Visual Inspection of Tools and Cords

Each shift, before portable electric tools and extension cords are used, they must be inspected for external defects (such as loose parts, deformed or missing pins, or damage to outer jacket or insulation) and for evidence of internal damage (such as pinched or crushed outer jacket). Cord-and-plug-connected equipment and extension cords that remain connected once they are put into place and are not exposed to damage need not be visually inspected until they are relocated.

If there is a defect or evidence of damage that might expose an employee to injury, the defective tool or cord must be marked and taken out of service and must not be used until the defect or damage has been repaired, and tests have been made to ensure the safety of the tool.

When an attachment plug is to be connected to a receptacle (including a cord set), the relationship of the plug and receptacle contacts should first be checked to ensure they have the proper mating configuration. All extension cords that are used with grounding-type tools should contain an equipment grounding conductor. These attachment plugs and receptacles must not be connected or altered in any way that would prevent continuity of the equipment grounding conductor. Any adapter that will interrupt the continuity of the equipment grounding conductor should not be used.

Quick Checklist for Tools

1. Visually check the tool and its cord for cracks or defects in the cord or the tool.
2. Check all cord-and-plug connected electrical connectors for defects, broken or missing pins, and exposed insulation at connectors on electrical powered tools.
3. Visually check extension cords for cracks, defects, broken or missing pins in the connectors, or exposed insulation at connectors.
4. Check handles to ensure they are in good condition, are of the proper type for the tool or equipment to which they attach, and are properly installed.
5. If any defect or damage is found, immediately remove for service, tag properly, repair, and retest before using or destroy.

Personal Protective Equipment and Other Equipment

Safeguards for Personnel Protection

Under 29 CFR 1910, Section 1910.335, OSHA specifies that employees working in areas where there are potential electrical hazards should be provided with, and use, electrical protective equipment that is appropriate for the specific parts of the body to be protected and for the work to be performed.

All electrical workers, at some time during the performance of their duties, will be exposed to energized circuits or equipment. There are two important considerations. Consideration number one is the use of flame-retardant clothing—that is, a flame-retardant clothing program. NFPA 70E-2004 goes into detail about how to comply with this requirement.

Where an employee is working within the flash-protection boundary, as established by NFPA 70E-2004, Section 130.3, he or she should wear protective clothing and other personal protective equipment in accordance with the tables or the calculation provided in NFPA 70E-2004, Section 130.3.

To protect the body, employees should wear flame-resistant clothing wherever there is possible exposure to an electric arc flash above the threshold incident-energy level for a second-degree burn (1.2 cal/cm^2). Clothing and equipment that provide worker protection from shock and arc flash hazards should be utilized. Clothing and equipment required for the degree of exposure should be worn alone or integrated with flammable nonmelting apparel. Where flame-resistant clothing is required, it should cover associated parts of the body as well as all flammable apparel, while allowing for movement and visibility. NFPA 70E-2004 provides tables that help with the hazard/risk assessment, the selection of flame-resistant clothing and other personal protective equipment, and the proper arc ratings of flame-resistant clothing and equipment.

A second consideration is the use of insulated rubber liners (gloves), with the proper voltage rating for the circuits or equipment that are being worked on. Specifications and in-service care of insulating gloves can be found in the manufacturer's instructions as well as ASTM D120-02 and F496 (ASTM 2002a, 2002b).

Rubber Insulating Liners—Various Classes

ASTM D120-02, Standard Specifications for Insulating Gloves (2002), has designated a specific color coding for the classification of rubber protective equipment. The following is the color, classification, and system voltage rating:

Class 00	Beige	500 volts
Class 0	Red	1000 volts
Class 1	White	7500 volts
Class 2	Yellow	17,000 volts
Class 3	Green	26,500 volts
Class 4	Orange	36,000 volts

Inspection of Protective Equipment

Before rubber protective equipment can be worn by personnel in the field, all equipment must have a current test date stenciled on it and must be inspected by the user. Before insulating rubber liners (gloves) can be worn, they must be visually inspected and air-tested before they are used each day or work shift by the user. They must also be tested during the work shift if their insulating value is ever in question. Because rubber protective equipment is used for personal protection, and serious injury could result from its misuse or failure, it is important than an adequate safety factor be provided between the voltage on which it is being used and the voltage at which it was tested.

Gloves, sleeves, and other rubber protective equipment can be damaged by many different chemicals, especially petroleum-based products such as oils, gasoline, hydraulic fluid, inhibitors, hand creams, pastes, and salves. If contact is made with these or other petroleum-based products, the contaminant should be wiped off immediately. If any signs of physical damage or chemical deterioration are found (e.g., swelling, softness, hardening, stickiness, ozone deterioration, or sun-checking), the protective equipment must not be used. Rings, watches, jewelry, and sharp objects should not be worn on your hands or arms when you are wearing rubber gloves or sleeves.

Inspection Methods

Gloves and sleeves can be inspected by rolling the outside and inside of the protective equipment between the hands. This can be done by squeezing together the inside of the gloves or sleeves to bend the outside area and create enough stress to the inside surface to expose any cracks, cuts, or other defects. When the entire surface has been checked in this manner, the equipment is then turned inside out, and the procedure is repeated.

To check rubber blankets, place the blanket on a flat surface, roll the blanket from one corner, and then roll the blanket toward the opposite corner. If there are any irregularities in the rubber, this method will expose them. After the blanket has been rolled from each corner, it should then be turned over and the procedure repeated.

Holes and other small defects can be detected by inflating the protective equipment. This can be done by rolling the cuff of the glove closed and holding the glove close to the ear and face to detect any air leakage. Gloves and sleeves can also be inspected by using mechanical inflaters.

Storage of Insulating Equipment

Once the protective equipment has been properly cleaned, inspected, and tested, it must be properly stored. It should be stored in a cool, dry, dark place that is free from ozone, chemicals, oils, solvents, or other materials that could damage the equipment. Such storage should not be in the vicinity of hot pipes or direct sunlight. Gloves and sleeves should be stored in their natural shape and should be kept in a bag or box inside of their protectors. They should be stored, undistorted, right side out, and unfolded. Blankets may be stored rolled in containers that are designed for this use with the inside diameter of the roll being at least two inches.

Leather Protectors

Proper-fitting leather protector gloves must be worn over the rubber gloves whenever possible; these protectors must meet the specifications of ASTM F696-02 (ASTM 2002a). This is to prevent damage to the rubber glove from sharp objects and mechanical injury and to reduce ozone cutting. Leather protectors provide only mechanical protection for rubber insulating gloves and when used alone do not provide any protection against serious injury, death, or other potential injuries from electrical shocks and burns. If the leather protector gloves have been used for any other purpose, they must not be used for the protection of rubber gloves.

If work is being performed on small equipment where unusually high finger dexterity is required, the protector gloves for Class 00 (per manufacturer's instructions) and Class 0 insulating gloves may be omitted if the employer can demonstrate that the possibility of physical damage to the gloves is small, and if the class of glove is one class higher than that required for the voltage involved, in accordance with OSHA 29 CFR 1910, Section 1910.137(b). This should only be considered when the likelihood of damage to the glove is low. Rubber gloves being used without a protector must not be used again until they have been properly inspected and electrically tested per the requirements of Section 1910.137(b)(2)(viii) and (ix), 29 CFR 1910.

Protector gloves must not be used if they have holes or tears. Care should be taken to protect the gloves from damage and from grease, oils, and other chemicals that could cause damage to the insulating glove. Whenever the insulating gloves are inspected, the protector gloves should also be inspected inside and out for sharp objects that could damage the insulating gloves. Workers are allowed to wear cloth gloves to keep hands warm in cold weather and to absorb perspiration in hot weather.

Inspection and Testing of Insulating Equipment

Before rubber protective equipment is placed into service, it must be cleaned, inspected, and electrically tested. Gloves and sleeves should be cleaned and washed using a mild detergent and warm water. After they have been cleaned, they should be thoroughly rinsed to remove all the soap and detergent. This cleaning can be done by utilizing a commercial tumble-dry washing machine. After the equipment has been cleaned and properly dried, the equipment must be inspected for any defects (see Table 1).

It is very important that the facility being utilized for the electrical testing be approved and designed to protect the operator. Electrical tests should

TABLE 1

Rubber Insulating Equipment Test Intervals

Type of Equipment	When to Test
Rubber insulating line hose	Upon indication that insulating value is suspect
Rubber insulating covers	Upon indication that insulating value is suspect
Rubber insulating blankets	Before first issue and every twelve months thereafter*
Rubber insulating gloves	Before first issue and every six months thereafter*
Rubber insulating sleeves	Before first issue and every twelve months thereafter*

*If the insulating equipment has been electrically tested, but not issued for service, it may not be placed into service unless it has been electrically tested within the previous twelve months.

(Adapted from OSHA 29 CFR 1910, Section 1910.137, Table I–6)

be performed only by individuals who have been given the proper training and instruction to perform these tests.

Insulating Equipment Failing to Pass Inspections

Insulating equipment failing to pass inspections or electrical test may not be used by employees, except as follows:

- Rubber insulating line hose may be used in shorter lengths with the defective portion cut off.
- Rubber insulating blankets may be repaired using a compatible patch that results in physical and electrical properties equal to those of the blanket.
- Rubber insulating blankets may be salvaged by severing the defective area from the undamaged portion of the blanket. The resulting undamaged area should not be smaller than 22 inches by 22 inches (560 mm by 560 mm) for Class 1, 2, 3, and 4 blankets.
- Rubber insulating gloves and sleeves with minor physical defects, such as small cuts, tears, or punctures, may be repaired by the application of a compatible patch. Also, rubber insulating gloves and sleeves with minor surface blemishes may be repaired with a compatible liquid compound. The patched area should have electrical and physical properties equal to those of the surrounding material. Repairs to gloves are permitted only in the gauntlet area.
- Repaired insulating equipment should be retested before employees can use it.

Rubber Insulating Line Hose and Covers

There are no in-service retest requirements for line hose and covers; however, frequent field inspections should be made. If the line hose and cover being used is type 1, it should not be left in service or on energized lines longer than necessary. Line hose and covers should be inspected inside and out for cuts, corona cutting, scratches, holes, and tears. If damage is found, and it extends beyond one-quarter the wall thickness of the hose, the line hose and cover should be removed from service. Line hose and covers should be wiped clean of any petroleum-based products that could damage the hose. All line hose and covers that appear to be suspect must be taken out of service and electrically tested in accordance with ASTM F 478, Standard Specification for In-Service Care of Insulating Line Hose and Covers (1999).

Head Protection

Whenever employees are working in the vicinity of exposed energized parts, and there is a possibility of head injury due to contact, nonconductive head protection should be worn. Head protection must meet the requirements of ANSI Z89.1, Requirement for Protective Headwear for Industrial Workers, 1997.

Insulated Tools

Whenever employees are working near any exposed energized parts, they must use insulated tools. The tools used must be insulated for a voltage not less than that of the conductors and circuit parts on which they will be used, and they must be suitable for the environment in which they will be used and for conditions of use. Insulated tools provide protection against flashover, shock, and burns. Note that ordinary plastic-dipped tools are not designed for this purpose, and applying electrical tape over the exposed metal parts of tools is not acceptable.

All approved double-insulated tools used on circuits and equipment operating at 1000 volts or less should bear the international 1000-volt symbol. These tools must meet the requirements of ASTM 1505, Standard Specification for Insulated and Insulating Hand Tools (2001). Insulated tools should be inspected prior to each use for cracks, cuts, or other damage. For example, should the inside (red) insulation become visible through the colored (orange) outer layer, the tool should be removed from service immediately, as it is no longer safe to use on energized circuits and parts.

Insulated tools used at other voltages should have a voltage rating appropriate for the voltage on which they are used. These tools are to be used as secondary protection and are not meant to be used in lieu of appropriate personal protective equipment. Rubber liners and leather protectors are still required to protect the hands when voltage testing, troubleshooting, or performing any task where the operating voltage is 50 volts or more to ground.

Protective shields, protective barriers, or insulating materials must be used to prevent injury from shock, burns, or other electrically related injuries while employees are working near exposed energized parts that might be accidentally contacted, or where dangerous electric heating or arcing might occur. When normally enclosed live parts are exposed, for maintenance or repair, they must be guarded to protect unqualified persons from contact with the live parts.

Double-Insulated Tools

Double-insulated tools and equipment generally do not have an equipment grounding conductor. Protection from shock depends on the dielectric properties of the internal protective insulation and the external housing to insulate the user from the electric parts. External metal parts, such as chuck and saw blades, are generally insulated from the electrical system.

All users of double-insulated power tools should be aware of some precautions in their use. These include the following:

- Double-insulated tools are designed so that the inner electrical parts are isolated physically and electrically from the outer housing. The housing is nonconductive. Particles of dirt and other foreign matter from the drilling and grinding operations may enter the housing through the cooling vents and become lodged between the two shells, thereby voiding the required insulation properties.
- Double insulation does not protect against defects in the cord, plug, and receptacle. Continuous inspection and maintenance are required.
- A product with dielectric housing—for example, plastic—protects the user from shock if interior wiring contacts the housing. Immersion in water, however, can allow a leakage path that may be either high- or low-resistance.
- Double-insulated tools and equipment should be inspected and tested as well as all other electrical equipment and should not be used in highly conductive, wet, or damp locations without also using a ground-fault circuit interrupter (GFCI).

Equipment Use in Conductive Work Locations

All portable electric equipment that is used in highly conductive locations such as wet or damp locations must be listed and approved for use in such locations. While working in wet or damp locations, employees' hands often become wet. This can be a hazard to the employee while plugging and unplugging flexible cords and cord-and-plug-connected tools. Energized connections should be handled only with the proper insulating protective equipment in wet or damp conditions. It is also recommended that GFCIs (Class A) be utilized under these circumstances.

Equipment Used to Alert Personnel

Whenever energized parts are exposed, and there is a possibility of contact by other employees, alerting techniques to warn and protect employees and other personnel from hazards that could cause injury due to electric shock, burns, or failure of electric equipment parts must be used. These techniques include use of tags, signs, barriers, and manual signaling.

Safety Signs and Tags

Safety signs, safety symbols, or accident-prevention tags should be used where necessary to warn employees about electrical hazards that may endanger them. Such signs and tags should meet the requirements of ANSI Standard Z535, Series of Standards for Safety Signs and Tags, 1998.

Barricades

Barricades should be used in conjunction with safety signs where it is necessary to prevent or limit employee access to work areas exposing employees to uninsulated energized electric conductors or circuit parts. Conductive barricades should not be used where they might cause an electrical hazard.

Alternate Alerting Techniques

When work areas are such that signs and barricades do not provide adequate warning of and protection from electrical hazards, manual signaling and alerting should be used to warn and protect employees. The primary duty and responsibility of persons providing manual signaling and alerting is to warn exposed employees and provide protection from electric hazards. They should remain in the area as long as employees are exposed to electrical hazards.

Vehicular and Mechanical Equipment

Any vehicle that is capable of contacting overhead lines must be operated so that it at no time comes closer than 10 feet to the overhead lines operating at 50 kilovolts (kV) or less. This would include vehicles capable of being elevated, such as trash trucks, dump trucks, and vehicles transporting high loads.

However, under any of the following conditions, the clearance may be reduced:

- If the vehicle is in transit, with its structure lowered, it may come within 4 feet of energized lines. However, if the voltage is above 50 kV, the distance must be increased 4 inches for each 10 kV over 50 kV.
- If insulating barriers are installed that will prevent contact with the line
- If the equipment is an aerial lift, the boom is insulated for the voltage involved, and the aerial lift is being operated and work is performed by a qualified person

Whenever vehicles or mechanical equipment are being operated near overhead lines, and the lines have not been de-energized and grounded, no person standing on the ground may approach or contact the vehicle or equipment unless

- the person is using protective equipment rated for the voltage, or
- the equipment is located so that no uninsulated part of its structure can come closer to the line than permitted, which is 10 feet maximum.

If any vehicle or mechanical equipment capable of having parts of its structure elevated near energized overhead lines is intentionally grounded, employees working on the ground near the point of grounding may not stand at the grounding location whenever there is a possibility of contact with overhead lines. Additional precautions, such as the use of barricades or insulation, should be taken to protect employees from hazardous ground potentials, depending on earth resistivity and fault currents, which can develop within the first few feet of, or more outward from, the grounding point (step potential).

Step potential is caused by the flow of fault current through the earth. The current flow creates a voltage drop at earth's surface. A person standing with his or her feet apart bridges a portion of this drop, and this places a potential difference from foot to foot. It is for this reason that all personnel and the public should be positioned at a safe distance from the driven ground rod.

Illumination

Whenever employees are working in the vicinity of exposed live parts, it is important to provide enough illumination to ensure that employees can see well enough to avoid contacting exposed live parts.

Where the lack of illumination or an obstruction precludes observation of the work to be performed, employees may not perform tasks near exposed energized parts. Employees should not reach blindly into areas that may contain energized parts. The intent is to eliminate the possibility of shock, burn, or electrocution because the worker was not able to clearly see the energized parts and avoid them.

OSHA provides Table 1926.56 in the construction standard 29 CFR 1926 to assist in the determination of the minimum illumination intensities in foot-candles for specific locations.

Conductive Apparel

Conductive articles of jewelry and clothing (such as watchbands, bracelets, rings, key chains, necklaces, metalized aprons, cloth with conductive thread, or metal headgear) may not be worn where they present an electrical contact hazard with exposed live parts. Because such apparel could provide a conductive path for current flow between adjacent energized parts, which generally results in severe burns to the individual wearing such apparel, all conductive apparel must be removed or rendered nonconductive. This recognizes the use of gloves or wrapping of the conductive apparel with insulating tape as an alternative to removing the items. Be aware that some manufacturers require that no jewelry be worn underneath rubber insulating liners. The best approach is always to require that employees remove such jewelry.

Selection and Use of Work Practices

Whenever work is to be performed on or near exposed energized parts, safety-related work practices should be employed to prevent electric shock or other injuries resulting from either direct or indirect electrical contact. The specific safety-related work practices must be consistent with the nature and extent of the associated electrical hazards. The basic intent is to require that employers use one of three options to protect employees working on electric circuits and equipment:

1. De-energize the equipment involved and lock out its disconnecting means.
2. De-energize the equipment and tag the disconnecting means if the employer can demonstrate that tagging is as safe as locking.
3. Work the equipment energized if the employer can demonstrate that it is not feasible to de-energize it.

Work on De-Energized Parts

A *de-energized* part is obviously safer than an energized part. Because the next best method of protecting an employee working on exposed parts of electrical equipment—the use of personal protective equipment—would continue to expose that employee to a risk of injury from electrical shock, OSHA 29 CFR 1910, Section 1910.333, and NFPA 70E-2004, Section 120.1, suggest that equipment de-energizing should be the primary method of protecting employees.

Obviously there have been no fatalities caused by de-energized circuits, though some fatalities have involved circuits that were thought to be de-energized. For this reason, OSHA has not accepted the argument that a qualified employee can work on energized circuits as safely as he or she can work on de-energized circuits. Therefore, OSHA is not leaving it up to the employers' discretion whether to de-energize electric circuits on the basis of convenience, custom, or expediency.

OSHA requires that all live parts to which an employee may be exposed be de-energized before the employee works on or near them, unless the employer can demonstrate that de-energizing introduces additional or increased hazards or is infeasible due to equipment design or operational limitations. However, live parts that operate at less than 50 volts to ground need not be de-energized if there will be no increased exposure to electrical burns or to explosion due to electric arcs. Examples of increased or additional hazards include the following:

- interruption of life support equipment
- deactivation of emergency alarm systems

- shutdown of hazardous location ventilation equipment
- removal of illumination for an area

Examples of work that may be performed on or near energized circuit parts because of infeasibility of de-energizing due to equipment design or operation limitations include the following:

- testing of electrical circuits that can only be performed with the circuit energized (troubleshooting)
- work on circuits that form an integral part of a continuous industrial process

When the term *additional or increased hazards* is used, it applies to instances where de-energization would be a threat to human life, and not merely to the equipment or process. Removal of illumination from an area is an example of a circumstance that might introduce additional or greater hazards. This would suggest that only in those instances where de-energizing a lighting circuit would present danger to personnel is it permissible to deviate from the basic requirement for de-energizing. If such de-energizing is simply an inconvenience, it should not be interpreted to be an additional or increased hazard.

The determination of what would constitute additional or increased hazards should not be applied without serious consideration. Elimination of the need for de-energizing, prior to working on or near any equipment or circuit, is permitted only where the employer can demonstrate that such de-energization would result in greater hazards or is impractical. This places the burden of proof on the employer. If an accident were to occur where the circuit was not de-energized, the decision not to de-energize would definitely come under scrutiny by an OSHA compliance officer, an insurance representative, or an expert in this type of accident investigation.

There are other instances where de-energizing is not required. Testing, troubleshooting, measurements, and adjustments that can only be performed while the circuit is energized are exempt from OSHA and NFPA requirements of de-energizing if additional steps such as selecting and using the appropriate personal protective equipment or isolating the energized part(s) by use of insulating materials are used. However, after the testing, troubleshooting, measurement, or adjustment is completed, any repairs or additional preventive maintenance would have to be performed with the circuit de-energized.

Work on Energized Parts

If the exposed live parts are not de-energized because of increased or additional hazards or because it is infeasible to de-energize them, other safety-related work practices should be used to protect those who may be exposed to the electrical hazards involved. These work practices, such as selecting and using the appropriate personal protective equipment or isolating the energized part(s) by use of insulating materials, should protect employees against contact between energized parts and any part of their body or indirectly through use of some other conductive object. The work practices developed should be suitable for the work to be performed and for the voltage level of the exposed conductors or circuit parts.

Working on or Near De-energized Parts

Conductors and parts of electric equipment that have been de-energized but have not been locked out or tagged should be considered and treated as energized, if employees are working on or near enough to them to expose them to any electrical hazard they present.

Lockout/Tagout (LOTO) Procedures

All employers who allow employees to perform work under activities requiring LOTO are required to establish a written LOTO procedure and also to establish procedures for enforcement of such a program. The LOTO procedure is intended to protect personnel from injury during servicing and maintenance of machines and equipment. LOTO is not intended to cover normal production operations. In order to have a safe and reliable LOTO program, the procedure must describe the scope, purpose, authorization, rules, responsibilities, and techniques needed to control all hazardous energy sources.

Work on cord-and-plug-connected electrical equipment does not require LOTO where the equipment is unplugged and where the plug is under the exclusive control of the employee who is performing the servicing and maintenance on the equipment.

In general, the kinds of activities covered under LOTO are those such as lubrication, cleaning, unjamming or servicing of machines or equipment, and making adjustments or tool changes. However, activities that are normal production operations—such as minor tool changes and adjustments—are not covered if they are routine, repetitive, and integral to the use of the equipment for production, and the work is performed using methods that provide effective employee protection.

Lockout/Tagout

OSHA has determined (from experience in the field, including accident investigations and record keeping) that lockout is, by far, the most effective means of providing employee protection and is preferred over tagout. However, if the energy-isolating device is not capable of being locked out, a tagout program may be used, provided that the tagout program will provide the same level of safety as a lockout program. Additional means beyond those necessary for lockout are required. These means include the following:

- removal of an isolating circuit element
- blocking of a controlling switch
- opening of an extra disconnecting device
- removal of a valve handle to reduce the likelihood of an inadvertent energization

If the energy-isolating equipment is capable of being locked out, the employer must utilize a lockout program, unless it can be demonstrated that the tagout program will provide the same level of safety.

The energy-isolating devices for all equipment that has been replaced, repaired, renovated, or modified since October 31, 1989, must be designed to accept a lockout device.

Energy-Control Procedure

A written procedure must be developed and enforced for all employees who may be injured by the unexpected startup or re-energization of machines and equipment during service and maintenance. However, if all of the following elements are met for a particular machine or piece of equipment, the employer does not have to document the procedure:

- The machine or equipment has no potential for stored or residual use.
- The machine or equipment has a single source of energy, which can be readily identified and isolated.
- The isolation and locking out of the energy source will completely de-energize and deactivate the machine or equipment.
- The machine or equipment is isolated from the energy source and locked out during servicing or maintenance.
- A single lockout device will achieve a locked-out condition.
- The lockout device is under the exclusive control of the authorized employee performing the servicing and maintenance of the equipment.
- The servicing and maintenance of the equipment does not create a hazard for other employees.
- If this exception is used, the employer must not have had any accidents involving unexpected activation or re-energization of the machine or equipment.

In this written procedure, the employer must clearly and specifically outline the scope, purpose, authorization, rules, and techniques that will be utilized for the control of hazardous energy. Included should be the administrative responsibilities for implementation of the program, training, compliance, and the following:

- intended use of procedure
- steps for shutting down, isolating, blocking, and securing machines or equipment
- steps for placement, removal, and transfer of lockout/tagout devices and the responsibility for them
- requirements for testing a machine or piece of equipment to determine the effectiveness of the lockout/tagout

Interlocks

Where interlocks must be defeated to gain access to enclosures or other areas containing exposed live parts, only qualified persons may disable the interlocks. Because interlocks may be defeated only temporarily, anytime the qualified person working on the equipment leaves the equipment unattended, for whatever reason, the interlocks must be restored. Additionally, when the work is completed, the interlock should be returned to its operable condition.

Electric Power and Lighting Circuits

All disconnecting means that are used for opening, reversing, or closing the circuits under load conditions should be load-rated devices. Pieces of equipment, such as isolating switches, are intended for isolating an electric circuit from the source of power. Cable connectors not of the load-break type, fuses, terminal lugs, and cable splice connections may not be used for these purposes, except for emergencies.

Reclosing Circuits after Protective Device Operation

Anytime a circuit has been de-energized by the operation of a protective device (such as a fuse or circuit breaker), the circuit must be checked by a qualified person to determine if it can be re-energized safely. Under fault conditions, protective devices such as breakers can be damaged. Without such a check, it is possible for an employee to be injured in case of failure of the protective device. The repetitive manual reclosing of circuit breakers or re-energizing through replaced fuses is prohibited per OSHA 29 CFR 1910, Section 1910.334(b)(2), and NFPA 70E-2004, Section 130.6(K).

When it can be determined from the design of a circuit that the automatic operation of a protective device was caused by an overload rather than a fault condition, no examination of the circuit or connected equipment is required before the circuit is re-energized.

Overcurrent Protection Modification

Overcurrent protection of circuits and conductors may not be modified, even on a temporary basis. This provision prevents the use of a fuse or circuit breaker with a rating too high to protect the equipment or conductors involved. This is also intended to prevent the temporary bypassing of protective devices, which could lead to shock and fire hazard.

Lockout/Tagout Devices

The employer is required to furnish locks, tags, chains, and other hardware used for the isolation of hazardous energy sources. These devices must be uniquely identified and used for no other purpose and must meet the following requirements:

- standardized, using one or more of the following:
  - color
  - shape
  - size
  - type
  - format
- clearly visible and distinctive
- designed so that all essential information necessary for the application is provided
- designed in such a manner as to deter accidental or unauthorized removal and substantial enough to prevent removal without the use of excessive force or unusual techniques
- designed for the conditions of the environment in which they are installed
- capable of withstanding the environment to which they will be exposed for the maximum time that exposure is expected

Tagout devices and their attachment means must be substantial and durable enough to prevent accidental or inadvertent removal and must have the following characteristics:

- nonreusable attachment means
- attachable by hand
- self-locking
- unlocking strength of no less than 50 pounds
- at least equivalent, in general design, to a one-piece, all-environment-tolerant nylon cable tie

Identification on LOTO Devices

All lockout and tagout devices must have a means of identifying the person applying the lockout/tagout devices, and the tagout device must warn against hazardous conditions if the equipment is re-energized. The device should include at least one of the following warnings:

- Do not start.
- Do not open.
- Do not close.
- Do not energize.
- Do not operate.

Periodic Inspections

At least annually, an authorized employee, other than the ones utilizing the lockout/tagout procedure, must inspect and verify the effectiveness of the lockout/tagout procedure. These inspections must provide for a demonstration of the procedure and be implemented through random audits and planned visual observations. The inspections are intended to ensure that the lockout/tagout procedures are being properly implemented and to correct any deviations or inadequacies observed.

If the lockout/tagout procedures are used less than once a year, they only need inspecting when used. The periodic inspection must provide for and ensure effective correction of identified deficiencies. These periodic inspections must be documented and include the identity of the machine or equipment on which the lockout/tagout was applied, the date of inspections, the employees included in the inspection, and the person performing the inspection.

Training and Communication

Before lockout/tagout can take place, the employer must ensure that all employees are trained in the purpose and function of the lockout/tagout procedure. The employer must ensure that all employees understand and have the knowledge and skills required for the safe application, usage, and removal of energy controls. OSHA 29 CFR 1910, Section 1910.147, recognizes three types of employees: authorized, affected, and other.

Different levels of training are required for each type of employee based on the roles of each employee in lockout/tagout and the level of knowledge they must have to accomplish tasks and to ensure the safety of their fellow workers.

Authorized Employee: Any employee who is allowed to lock out or implement a tagout procedure on a machine or piece of equipment to perform servicing or maintenance. These employees must receive training in the recognition of hazardous energy sources and the methods and means necessary for the control of the hazardous energy.

Affected Employee: An employee whose job requires him or her to operate or use a machine or piece of equipment that is being serviced or maintained under a lockout/tagout, or who is working in an area in which the lockout/tagout procedure is being performed. These employees must receive training in the purpose and use of the lockout/tagout procedure.

Other Employees: All other employees who work around the area or are in the area where lockout/tagout is utilized must receive training about the procedure and about the serious consequences relating to the attempts to restart or re-energize the equipment.

Minimum Training Requirements

Each training program must cover at a minimum the following three areas:

1. The energy-control program
2. The elements of the energy-control procedures relevant to the employees duties
3. The pertinent requirements of the OSHA lockout/tagout standard, 29 CFR 1910, Section 1910.147, as well as the lockout/tagout requirements in Section 1910.333

Retraining

Retraining for authorized and affected employees must be done under the following conditions:

- whenever there is a change in job assignments
- whenever there is a new hazard introduced due to a change in machines, equipment, or process

- whenever there is a change in the lockout/tagout procedure itself
- whenever the required or other periodic inspections by the employer reveals inadequacies in the company procedures or a lack of knowledge of the employees

Certification of Training

All training that has been completed must be documented, and this documentation must be kept on file and be up to date. The documentation must include the employee's name, the dates of training completion, and the name of the instructor.

Isolation of Electric Circuits and Equipment–Preplanning

Preplanning the safe manner in which equipment or circuits are going to be de-energized must be done. Before the machine, equipment, or circuits are de-energized, the authorized employee must have knowledge of the type of hazardous energy and the methods to control it. For example, preplanning stages would include determining the following:

- types and amount of energy involved
- the individual machine involved
- the processing machine with stored energy involved
- verification of energy-isolating devices

Before machines and equipment are shut down and de-energized, all affected employees must be notified. These notifications must take place before the controls are applied and after they are removed from the equipment. Only an authorized employee can implement the lockout/tagout controls. The methods for disconnecting electric circuits and equipment must include the following procedures:

- The circuits and equipment are required to be disconnected from all energy sources.
- Disconnecting is done only by personnel authorized by the employer.
- The sole disconnecting means must not be a control circuit device such as push buttons, selector switches, or electrical interlocks that de-energize electric power circuits indirectly through contractors or controllers.
- The sole disconnecting means is not allowed to be an electrically operated disconnecting device such as a panic button operating a shunt trip on a large power circuit breaker.
- The authorized employee who turns off a machine or equipment must have knowledge of the type and the magnitude of the energy, the hazards involved, and the methods of means to control the energy.
- An orderly shutdown must be utilized to avoid any additional hazards.

Applying Lockout/Tagout Devices

The lockout/tagout procedure for applying locks or tags, or both, to disconnected circuits and equipment, from all sources of energy, must be included in the written procedure. Where these requirements in the procedure are applied, they will prevent reenergizing of the circuits and equipment. These lockout/tagout devices must be applied only by an authorized employee. Tags are utilized to supplement the locks. A tag is a warning to indicate that the energy-isolated device and the electrical equipment being controlled cannot be operated until the lockout/tagout devices have been removed. The basic rule is that only the authorized employee placing the lockout/tagout device is allowed to remove it. This requirement prohibits an unauthorized person from removing the lockout/tagout device and energizing the circuit or equipment, which could cause serious injury to an employee working on the equipment. Many electrical shocks and other injuries are due to an unauthorized person removing a lockout/tagout device and re-energizing the circuit equipment. The authorized person who places the lockout/tagout device or an energy-isolating device is required to list the following on the tag:

- name of the authorized person applying the lockout/tagout device
- date of application

Applying Locks Only

Where work to be performed involves only a simple circuit and can be completed in a short time, a lock can be used safely without a tag. 29 CFR 1910.333 limits the use of this procedure to one circuit or single piece of equipment that involves a lockout period of no longer that one work shift. Additionally, affected employees are required to be trained in and familiar with this procedure to avoid confusion over the purpose of the lock on the energy-isolating device.

The procedure pertaining to using only a lock must provide a level of safety that is equal to both locks and tags. To use a lock without a tag, the following conditions must be met:

- Only one circuit or piece of equipment is involved in the lockout.
- The lockout period does not extend beyond the work shift.
- Employees exposed to the hazards associated with re-energization of the circuit or equipment are familiar with this procedure.

Applying Tags Only

Tags are allowed to be used without locks where locks cannot be applied to an energy-isolating device because of design limitations, or when the employer demonstrates that tagging alone will provide safety that is equivalent to applying a lock. However, because a person could operate the disconnecting means before reading or seeing the tag, one additional safety measure must be provided. These safety measures include the following:

- removal of a fuse or fuses for a circuit
- removal of a draw-out circuit breaker from a switchboard or switchgear, or a bucket from a motor control center, for example
- placement of a blocking mechanism over the operating handle of a disconnecting means; the handle must be blocked from being placed in the closed position
- the opening of a switch or disconnecting means that opens the circuit between the source of power and the exposed circuits and parts
- the opening of a switch for control circuit that operates a disconnect and disables the system
- grounding the circuit where work is to be done

The additional safety measures are necessary because tagging alone is considered less safe than locking out. A disconnecting means without a lock can be closed by an employer who has failed to recognize the purpose of the tag. The disconnect is also capable of being accidentally closed by an employee who thinks it controls his or her equipment.

Tags should comply with these requirements:

- be marked distinctively
- have a standardized design
- clearly prohibit energizing the energy-isolating device

Stored Energy Release

All stored or mechanical energy that might endanger personnel is required to be released or restrained. Stored energy, such as from capacitors, must be discharged through either the motor windings or other effective means. High-capacitance elements are required to be short-circuited and grounded before the electrical equipment or components are worked on. If the possibility for reaccummulation of this energy to a hazardous level exists, verification of a safe condition must be continued until the work is complete. Mechanical energy, such as springs, must be released or restrained to immobilize the springs. This procedure of containing stored energy of all types will prevent unexpected power or energizing of devices that could cause injury to employees. Methods that can be used to release or restrain mechanical energy are:

- slide gate
- slip blind
- line valve and block
- grounding

Verifying that Equipment Cannot Be Restarted

Only authorized qualified personnel are allowed to operate the equipment operating controls or otherwise verify that the equipment cannot be restarted or re-energized.

Where present, the following devices must be activated to verify that it is impossible to restart the equipment by energizing the circuits and parts:

- push buttons
- selector switches
- electrical interlocks
- opening the switch for a control circuit that operates a disconnect and disables the system
- grounding the circuit on which work will be done

Verifying that Circuits and Equipment Are De-energized

Circuits and equipment must be tested by qualified employees using appropriate test equipment to verify that the circuits and equipment are de-energized. Verification that the circuit elements and equipment parts have been de-energized to ensure that employees will not be subject to live circuits and energized parts is mandatory. The test must also determine whether any inadvertent induced voltage or unrelated voltage backfeed, through specific parts of the circuit, are present on the electrical system. Where the circuit to be tested is over 600 volts, the test equipment is required to be checked for proper operation immediately before and immediately after the circuit to be worked on has been tested. One way to determine whether a circuit has been opened is to operate the controls for the equipment supplied by the circuit to verify that the equipment cannot be restarted. This method of testing has the advantage of not exposing employees to possible live circuits and parts. However, operating the equipment controls is not always reliable in that the circuit has been completely de-energized. It is possible to interrupt a portion of the circuit, while the rest of the circuit to the equipment may still be energized. A suggestion for enhanced safety would be to require further testing, such as a voltage test, to verify that the system is, in fact, de-energized.

Inspection of Machines or Equipment

The work area must be inspected to ensure that all nonessential items have been removed and that all machines or equipment components are operationally intact.

Verifying that Employees Are Clear of Circuits and Equipment

Before disconnects are closed and power restored to circuits and equipment, all affected employees in the work area who are near or working on the circuits or equipment must be warned. A check is to be made to ensure all employees have been safely positioned or removed and are clear of circuits, parts, and equipment. After a visual check to verify all employees are indeed clear, the power can be restored.

Lockout/Tagout Device Removal

A lockout/tagout device is to be removed only by the authorized employee who applied it. However, if the employee who applied the lockout/tagout devices is absent from the workplace, the locks and tags can be removed by another authorized employee. In this case, the following rules should be followed:

- The employer must ensure that the employee who applied the lockout/tagout is absent from the workplace, and all reasonable efforts have been made to contact the employee to inform him/her that the lockout/tagout has been removed.
- The employee must be notified of the removal of the lockout/tagout before he or she resumes work at the facility.
- There must be unique operating conditions involving complex systems present.
- The employer must demonstrate that it is infeasible to do otherwise.

Release for Energizing

When circuits and equipment are deemed ready to be re-energized, employees should be available to assist in any way necessary to ensure that circuits and equipment can be safely energized. Employees who are responsible for operating the equipment or process should be notified that the system is ready to be energized.

Testing or Positioning of Machines and Equipment

Whenever contractors and their employers are utilized, they must follow the lockout/tagout procedure that is in place at the work site. If the contractor's personnel have not been properly trained in the lockout/tagout procedure, the employer must provide an authorized employee to implement lockout/tagout of the energy-isolating devices.

Group Lockout/Tagout

If more than one individual, craft, crew, or department is required to lock out or tag out equipment or processes, the lockout/tagout procedure should afford the same level of protection for the group as that provided by personal locks and tags. The group lockout/tagout should be used in accordance with, but not limited to, the following procedures:

- Primary responsibility for the group lockout/tagout is delegated to one authorized employee for a set number of employees working under the protection of a group lockout/tagout.
- Provisions have been made for the authorized employee to ascertain the exposure status of individual group members with regard to the lockout/tagout.
- When more than one crew, craft, or department is involved, an authorized employee has been assigned overall responsibility for the group lockout/tagout to coordinate affected work forces and ensure continuity of protection.
- Each authorized employee must place his or her own personal lockout/tagout devices to the group lockout/tagout device, group lockbox, or comparable mechanism before they begin work. These devices must be removed by each employee when the employees are done working on the machine or equipment.

Release from the group lockout/tagout must be accomplished by the following steps:

1. The machine or equipment area must be cleared of nonessential items to prevent malfunctions, which would result in employee injuries.
2. All authorized employees must remove their respective locks or tags from the energy-isolating devices or from the group lockbox following procedures established by the company.
3. In all cases the lockout/tagout procedure must provide a system that identifies each authorized employee involved in the servicing and maintenance operation.
4. Before re-energization, all employees in the area must be safely positioned or removed from the area, and all affected employees must be notified that the lockout/tagout devices have been removed.

During all group lockout/tagout operations where the release of hazardous energy is possible, each authorized employee must be protected by his or her personal lockout/tagout device and by the company procedure. A master danger tag used for group lockout/tagout can be used as a personnel tagout device if each employee personally signs on and signs off on it and if the tag clearly identifies each authorized employee who is being protected by it.

Procedures for Shift Changes

The lockout/tagout procedure must ensure continuity of protection if the lockout/tagout is going to extend beyond the shift change or personnel changes. The procedure must provide provisions for the orderly transfer of the lockout/tagout devices between off-going and oncoming employees.

Confined or Enclosed Work Spaces

Some installations of electric equipment provide limited working space for maintenance employees. Such cramped conditions can lead to employees backing or moving into exposed live parts. Areas such as manholes and vaults are examples. To prevent accidental contact from energized parts, precautions must be taken to assure that accidental contact does not occur. For example, protective blanks of the insulating type or the arc-blast type could be used to shield some of the live parts, or portions of the electrical installation could be de-energized. Safe work practices are required when the circuit(s) or part(s) are not de-energized.

OSHA 29 CFR 1910, Section 1910.333(c)(5), also requires that doors and panels be secured if they could swing into employees and cause them to contact exposed energized parts.

Overhead Lines

If work is to be performed in the vicinity of overhead lines, before this work can take place, the lines must be de-energized and grounded, or other protective measures should be used before the work is started.

If power lines are to be de-energized, arrangements must be made with the organization that operates and controls the electric circuits involved, in order to de-energize and properly ground the electric circuits. If protective measures are to be used, the measures should include guarding, isolating, and insulating.

Whichever protective measure is used, it must prevent employees from contacting such lines directly with any part of their body or indirectly through conductive materials, tools, or equipment.

When an *unqualified person* is working near overhead lines, in an elevated position, such as from an aerial device, the person and the longest conductive object with which he or she may be able to contact the line must not be able to come within the following distances:

- for voltages to ground 50 kV or below: 10 feet, 0 inches
- for voltages to ground over 50 kV: 10 feet, 0 inches, plus 4 inches for every 10 kV above 50 kV

When unqualified persons are working on the ground in the vicinity of overhead lines, the above distances still apply.

TABLE 2

Approach Distances for Qualified Employees—Alternating Current

Voltage Range (Phase to Phase)	Minimum Approach Distance
300V and less	Avoid contact
Over 300V, but not over 750V	1 ft 0 in (30.5 cm)
Over 750V, but not over 2kV	1 ft 6 in (46 cm)
Over 2kV, but not over 15kV	2 ft 0 in (61 cm)
Over 15kV, but not over 37kV	3 ft 0 in (91 cm)
Over 37kV, but not over 87.5kV	3 ft 6 in (107cm)
Over 87.5kV, but not over 121kV	4 ft 0 in (122 cm)
Over 121kV, but not over 140kV	4 ft 6 in (137 cm)

(Adapted from OSHA 29 CFR 1910, Section 1910.333(c), Table S-5)

When *qualified persons* are working in the vicinity of overhead lines, whether in an elevated position or from the ground, the qualified persons may not approach or take any conductive object closer to the exposed lines than shown in Table 2.

A qualified person may work closer to exposed energized overhead lines if one of the following protective measures has been taken:

- The qualified person is insulated from the energized parts (e.g., gloves, with sleeves if necessary, rated for the voltage involved are considered to be insulation of the person from the energized part on which work is performed).
- The energized part is insulated both from all other conductive objects at a different potential and from the other persons.
- The person is insulated from all conductive objects and is at a potential different from that of the energized part.

Because the installation of this type of protective equipment would require the person to be closer than 10 feet to the line, the person must be qualified. The persons performing this type of work must have the specialized training (generally provided by the employer) related to overhead line work necessary to perform this work safely (see qualified person, i.e., training).

Electrical Switching Operations

Major causes of personnel injury at industrial plants are the malfunctions that occur during the closing or opening of some types of switches and circuit breakers (Owen 1997). Some of these are due to the inadequacy of the switch; many switches and breakers do not have sufficient capability to withstand possible fault currents, especially with supply systems increasing in fault capacity. In particular, when a switch is to be closed after work has been performed on its load circuit, there is a possibility of switch failure. Such failure frequently initiates a phase-to-phase or ground fault, which either burns through the cover or blasts open the cover or door of the switch, injuring or burning the person who operated the switch. See NFPA 70E-2004, Table 130.7(C)(9)(a), for Hazard /Risk category and Table 130.7(C)(10) for the selection of personal protective equipment, including flame-resistant clothing, required for this task.

Failure of a fuse with insufficient fault-interrupting capability upon switch closure can likewise initiate a fault within the enclosure with similar results.

A well-defined method of closing switches and circuit breakers can, in general, go far toward eliminating personnel hazards involving this operation. Steps in this procedure include:

1. Wear appropriate flame-resistant clothing and other appropriate clothing where necessary.
2. Wear gauntlet-type gloves and other appropriate personal protective equipment.
3. Check NFPA 70E-2004, Tables 130.7(C)(9)(a), 130.7(C)(10), and 130.7(C)(11), or calculate incident energy available and select flame-resistant clothing from Table 130.7(C)(11) and other personal protective equipment from Table 130.7(C)(10).
4. Stand to one side of the switch, not directly in front of it, facing the wall or structure on which the switch is mounted, and as close to it as possible (hug the wall).
5. Use the hand nearest the switch to operate the handle; if standing to the right side of the switch, operate the handle with the left hand.
6. As the switch is operated, turn your face the opposite way (if at the left of the switch, look to the left, and vice versa).
7. Keep other personnel away from the front of the switch and at least 1.2 meters (approximately 47 inches) to either side.
8. Selection of the side to stand at will depend on the proximity of the handle to one side or the other, on ease of operations of the handle, or on the side of the enclosure more remote from the line terminals. The hinges are as likely to rupture from the internal pressure as the latch is to bust.
9. Firm and smart operation is desirable. Never indecisively operate (tease) the switch.

Grounding and Bonding Notes

Electrical systems are solidly grounded to limit the voltage to ground during normal operation and to prevent excessive voltages due to lightning, line surges, or unintentional contact with higher-voltage lines and to stabilize the voltage to ground during normal operation. This protects people and prevents fires.

Electrical connections are made to available grounding electrodes (such as metal water piping systems, effectively grounded structural metal members, concrete-encased electrodes, and ground rings) that are present at buildings or structures. Where necessary, install grounding electrodes (ground rings, "ufer" grounds, ground rods) per National Electrical Code requirements found in Article 250. To enhance this system, equipment grounding conductors are bonded to the system grounded conductor to provide a low-impedance path for fault current that will facilitate the operation of overcurrent devices under ground-fault conditions. The goal of the ground-fault current path is to provide a permanent and adequate path of low impedance so that enough current will flow in the circuit to cause an overcurrent device to operate.

Install equipment grounding conductors for this current to flow on. Install bonding jumpers where required to help make an effective path for the fault current flow on and over. Bonding by definition is the permanent joining of metallic parts to form an electrically conductive path that will ensure electrical continuity and the capacity to safely conduct any current likely to be imposed on it. This is a function of the grounding system. It is a part of the grounding system.

Ground systems for safety. Use personal protective grounding to protect individuals while they perform work on electrical systems. With respect to static electricity, ground and bond equipment in order to prevent a static spark from igniting a hazardous (explosive) atmosphere, which could cause an explosion.

Personal Protective Grounding

The primary use of personal protective grounds is to provide maximum safety for personnel while they are working on or near de-energized lines, buses, or equipment (Owen 1997). This is accomplished by reducing voltage differences at the work site (voltage across the worker) to the lowest practical value in case the equipment or line being worked on is accidentally energized. Another function of personal protective grounds is to protect against induced voltage from adjacent parallel, energized lines.

Proper protective grounding results in a safer working environment. Low-resistance ground cables will limit the voltage drop in the work area to acceptable levels. Personal protective grounds must be designed, assembled, and installed in a manner that satisfies the following basic criteria:

- The personal protective grounding cable to be applied must be capable of conducting the maximum available fault current that could occur if the isolated line or equipment becomes energized from any source.
- The grounding cable must be able to carry the maximum available fault current for a sufficient length of time to permit protective relays and circuit breakers to clear the fault.
- The grounding cable must be terminated with clamps of adequate capacity and strength to withstand all electrical and mechanical forces present under maximum fault conditions.
- The grounding cable must meet the following requirements: it must be easy to apply, it must satisfy the requirements of field conditions, and it must adapt to a wide range of conductor, structural steel, and ground rod/wire sizes.
- Cable length should be as short as possible for the specific task being performed. The greater the length, the longer the time it takes to clear faults.

Grounding Cables and Hardware

The American Society for Testing and Materials (ASTM) Committee F-18, Electrical Protective Equipment for Workers, developed a consensus standard for protective grounds. The ASTM designation is ASTM F-855 *Standard Specification for Temporary Protective Grounds to Be Used on De-energized Electric Power Lines and Equipment* (1997) (see Table 3).

All grounding cables, clamps, and ferrules used to construct grounding cables must meet the requirements of ASTM F-855. Aluminum cables must not be used for personal grounds.

Personal protective grounding cables consist of appropriate lengths of suitable copper grounding cable, with electrically and mechanically compatible ferrules and clamps at each end. In addition, appropriate hot sticks are required for installing and removing the conductor-end clamps to the conductors. Hot sticks are required for attaching ground-end clamps if the grounded system and worker are at different potentials.

NFPA 70E-2004, Table 130.7(c)(9)(a), has established hazard/risk categories for application of safety grounds (after voltage test to ensure that the circuit is de-energized prior to application of safety grounds). The hazard/risk category determined from Table 130.7(c)(9)(a) may then be used to determine the appropriate personal protective equipment and flame-resistant clothing from Table 130.7(c)(10). One important piece of information to take from this section is that even when the circuit has been determined to be de-energized (through voltage-sensing equipment), personal protective equipment and flame-resistant clothing must still be worn when installing the protective safety grounds (Owen 1997).

TABLE 3

Approach Distances for Qualified Employees—Alternating Current

Maximum Fault Current Capability for Grounding Cables

Cable Size (AWG)	Fault Time (cycles)	RMS Amperes (copper)
1/0	15	21,000
	30	15,000
2/0	15	27,000
	30	20,000
3/0	15	36,000
	30	25,000
4/0	15	43,000
	30	30,000

(Adapted from ASTM F-855 *Standard Specification for Temporary Protective Grounds to Be Used on De-energized Electric Power Lines and Equipment*, 1997)

Note: These current values are the "withstand rating" currents for grounding cables taken from Table 5, ASTM F-855, *Standard Specification for Temporary Protective Grounds to Be Used on De-energized Electric Power Lines and Equipment* (ASTM 1997). These values are about 70 percent of the fusing (melting) currents for new copper conductors. They represent a current that a cable should conduct without being damaged sufficiently to prevent reuse.

Size of Grounding Cable

The size of the grounding cable must be selected to handle the maximum calculated fault current of the power system or specific portion. The minimum size of cable allowed by ASTM is #2 American Wire Gage (AWG); however, the maximum available fault current may require larger cables. If larger cables are not available, parallel cables may be used. Most manufacturers and suppliers of grounding cables publish tables to assist the user in selecting the proper cable size for a given fault current (Owen 1997).

Jackets

Cables are normally insulated at 600 volts. When used as grounding cable, the insulation or jacket serves primarily for mechanical protection of the conductor. The flexible elastomer or thermoplastic jackets are manufactured, applied, and tested according to ASTM standards. Black, red, and yellow jackets are usually neoprene rubber compounds, whereas clear jackets are ultraviolet-inhibited polyvinyl chloride (PVC).

All jackets must have the AWG size stamped or printed repeatedly along the length of the cable. The clear jacket allows easy visual inspection of the conductor for strand breakage, but becomes stiff and hard to handle at low temperatures. The clear jacket will split or shatter at low temperatures (Owen 1997).

Ferrules

Ferrules should be of a type specified by ASTM F-855-1997. Ferrules should have the filler-compound vent hole at the bottom of the cable, so that employees can visually check that the cable is fully inserted into the ferrule. Compound should be used with crimped ferrules.

The ferrules should be crimped with the ferrule manufacturer's recommended die. The press must have enough pressure to completely close the die. Heat shrink tubing should be installed over a portion of the ferrule to minimize strand breakage caused by bonding. In all cases, the manufacturer's recommendations should be followed (Owen 1997).

Grounding Cable Length

Excessive cable lengths should be avoided; for example, do not use a 100-foot cable where a 50-foot cable will be sufficient. The greater the length, the longer the time for clearing a fault. Therefore, similar to lightning protection, the shortest possible length is desirable. Slack in the installed cables should be minimal to reduce possible injury to workers. Resistance in the cable increases with cable length, and excessive length will exceed the tolerable voltage drop. Longer-than-necessary cables also tend to twist or coil, which reduces the current-carrying capacity of the cable (Owen 1997).

Grounding Clamps

Grounding clamps are normally made of copper or aluminum alloys. Grounding clamps are sized to meet or exceed the current-carrying capacity of the cable and are designed to provide a strong mechanical connection to the conductor, metal structure, or ground wire/rod. Clamps are furnished in, but not limited to, four types, according to their function and methods of installation:

1. Type I clamps, for installation on de-energized conductors, are equipped with eyes for installation with removable hot sticks.
2. Type II clamps, for installation on de-energized conductors, have permanently mounted hot sticks.
3. Type III clamps, for installation on permanently grounded conductors or metal structures, have T-handles, eyes, and/or square or hexagon-hand screws.
4. Other types of special clamps are designed for specific applications, such as cluster grounds, underground equipment grounding, and so on.

Application of Personal Protective Grounds

Prior to installation, each cable must be visually inspected for mechanical damage. Suspect cables must not be used. Grounds should be placed at the work site.

Before coming within the minimum work distances of high-voltage lines or equipment, workers must isolate, de-energize, test, and properly configure those parts that are required. All conductors and equipment must be treated as energized until tested and properly ground.

The ground-end clamp of each grounding cable should always be the first connection made and the last to be removed. The phase connections should be made on the closest phase of the system first, and then each succeeding phase in order of closeness.

When removing the grounding cable, reverse the order by which the clamps were applied.

Always use appropriate personal protective apparel and equipment when applying and removing personal protective grounds (Owen 1997).

Safety-Related Maintenance Requirements

Qualified persons. Only persons who are considered qualified should be permitted to perform maintenance on electrical equipment.

Single-line diagram. Where single-line diagrams are provided, they should be maintained. It is important to keep drawings accurate and up to date.

Spaces around electrical equipment. All working space and clearances required should be maintained.

Grounding and bonding. Equipment, raceway, cable tray, and enclosure bonding and grounding should be maintained to ensure electrical continuity.

Guarding of live parts. Enclosures should be maintained to guard against accidental contact with live parts and other electrical hazards.

Safety equipment. Locks, interlocks, and other safety equipment should be maintained in proper working condition to accomplish the control purpose.

Clear spaces. Access to the working space and escape passages should be kept clear and unobstructed.

Identification of components. Identification of components, where required, and safety-related instructions (operating or maintenance), if posted, should be securely attached and kept in legible condition.

Warning signs. Warning signs, where required, should be visible, securely attached, and maintained in legible condition.

Identification of circuits. Circuit or voltage identification should be securely affixed and maintained in updated and legible condition.

Single and multiple conductors and cables. Electrical cables and single and multiple conductors should be maintained free of damage, shorts, and ground that would present a hazard to employees.

Flexible cords and cables. Flexible cords and cables should be maintained to avoid strain and damage.

- Damaged cords and cables: cords and cables should not have worn, frayed, or damaged areas that present an electrical hazard to employees.
- Strain relief: strain relief of cords and cables should be maintained to prevent pull from being transmitted directly to joints or terminals.

Testing and visual inspections are required for protective equipment, protective tools, safety grounding equipment, and personal protective equipment. Generally, all equipment needs to be visually checked before each use on any job site.

Employer's Electrical Safety Program

A plan should be designed so that neither workplace conditions nor the actions of people expose personnel unnecessarily to electrical hazards. Objectives of an employer's electrical safety program would include:

- Make personnel aware of rules, responsibilities, and procedures for working safely in an electrical environment.
- Demonstrate compliance with federal law (including appropriate safety standards).
- Provide documentation of general requirements and guidelines for providing electrically safe facilities.
- Provide documentation of general requirements and guidelines that direct the activities of personnel who may be exposed to electrical hazards.
- Encourage and make it easy for each employee to be responsible for his or her own electrical safety self-discipline. Address the safety needs of all employees, as well as contractors and visitors.

The content of the electrical safety program should include the following:

- management commitment
- organizational support
- electrical safety policies
- training and qualification of personnel
- use of protective equipment, tools, and protective methods
- use of electrical equipment
- documentation
- oversight and auditing
- technical support
- emergency preparedness

Responsibilities of the Employer's Electrical Safety Authority

- Assume responsibility of the electrical safety program.
- Develop and revise company electrical-safety standards.
- Provide interpretations of nationally recognized codes and standards.
- Provide guidance for facility management.
- Resolve NFPA 70 (NEC), NFPA 70E, NFPA 70B, NFPA 79, and OSHA issues and questions.
- Establish and document effective safe work practices.
- Provide technical input for OSHA interpretations.
- Provide guidance for electrical training programs.
- Provide guidance for procedure preparation.
- Provide consultation services to management.
- Review electrical safety incidents and participate in investigations.
- Issue summaries and lessons-learned about electrical safety incidents.
- Evaluate nonlisted electrical equipment.
- Develop a program for documentation of standards, practices, procedures, and guidelines; accurate drawings; equipment manuals, inspection records, and histories; findings from audits; and completed training records.

Methods of Achieving Electrically Safe Work Conditions

To provide and maintain electrically safe facilities:

- Design electrical systems with safety in mind.
- Ensure that all electrical installations are code-compliant.
- Inspect new facilities, existing facilities, and renovated or modified facilities.
- Provide for maintenance—predictive and preventive.
- Ensure management and employee familiarity with recognized standards.
- Perform safety audits of workplace conditions.
- Provide appropriate training for personnel.
- Provide a technical authority with the knowledge and experience to respond to questions about design, installation, and maintenance.
- Provide the organizational structure to accomplish electrically safe work conditions.

New Ideas for Safe Work Practices

If at all possible, avoid work on energized electrical equipment. Work on or near exposed live parts should be prohibited, except under approved, controlled, and justified circumstances. If work must be done on energized circuits and parts, require work authorization. NFPA 70E-2004, Section 130.1, refers to this as a "Justification for Work" permit. Make sure that every authorized employee performing work on energized parts is properly trained to understand and apply recognized safe work practice standards. This is a reference to a qualified person. Ensure that the employer has established written safe practices and procedures and that employees understand the importance of following the written practices, as well as the consequences if the written practices are not followed.

Employers are required to provide safety training for all employees who are authorized, affected, and other (NFPA 70E-2004, Section 110.6; OSHA 29 CFR 1910, Section 1910.332[a]). One of the benefits of providing safety training is that employees are recognized as *qualified*. Employers need to perform safety audits and make careful assessments of personnel activities. Best practices would encourage employers to solicit and use oversight groups to ensure the success of their safety programs. Employers need to establish a technical authority to respond to questions regarding safe work practices.

Final Notes and Suggestions

Fatal and survivable electrical accidents suggest that the majority of workplace events are related to work processes and practices. Therefore, it is important that companies have and use effective work practices and that appropriate protective equipment is supplied by the employer and used by the employee. Appropriate personal protective equipment can greatly reduce the chances of receiving flash burns, as well as burns resulting from a direct contact with energized parts. It should be noted that appropriate personal protective equipment provides minimal protection from shrapnel expelled and the explosive pressure exerted. This dictates the need for an effective training program.

The following is a list of suggestions for improved electrical safety, adapted from *Handbook for Electrical Safety* (2d ed.).

- Use finger-safe electrical components when possible. This can reduce the chance that an arc fault will occur.
- Specify the use of insulated bus for equipment such as motor control centers, switchgear, switchboards, and so on. This will reduce the chance that an arc fault will occur. In addition, it has been found that it increases the probability that an arc fault will self-extinguish.
- Use current-limiting overcurrent protective devices such as current-limiting fuses and current-limiting circuit breakers. Obtain verifiable engineering data on the current-limiting ability of the overcurrent protective devices. Specify the most current-limiting devices available where possible; the greater the degree of current-limitation, the less will be the arc fault energy released (when the fault current is in the current-limiting range of the overcurrent protective device).
- Size current-limiting, branch-circuit overcurrent protective devices as low as possible. Typically, the lower the ampere rating, the greater degree of current limitation. Limit the ampere rating size of main and feeders where possible. Where possible, split large feeders into two feeders; for example, rather than one 1600-ampere motor control center, specify two 800-ampere motor control centers. This will reduce short-circuit current and possibly reduce flame-resistant clothing requirements. Also, reduced ampere ratings may allow for lower overcurrent device ratings (which is a result of the ability of the current-limiting overcurrent devices to clear faults more quickly than noncurrent-limiting overcurrent devices) and allow less fault current let-through in the process of clearing the fault. Less let-through means less explosive energy available to potentially develop into an arc blast or arc flash.

- For motor starter (controller) protection, specify Type 2 protection that uses protective device combinations that have been tested for Type 2 protection.
- If noncurrent-limiting overcurrent protective devices are used, utilize high-impedance circuit components to try to limit the arc-fault current potentially available. Do not use circuit breakers with short-time delays. It has been well documented that arc-fault incident energy is directly proportional to the time the fault is permitted to persist. Permitting an arcing fault to intentionally flow for 6, 12, or 30 cycles dramatically increases the hazards to electrical workers. If selective coordination of overcurrent protective is the objective, then use current-limiting fuses that can be selectively coordinated simply by adhering to minimum ampere rating ratios between main feeder fuses or feeder and branch-circuit fuses.

Use NFPA 70E-2004, as a primary reference guide in conjunction with the OSHA standards for compliance with OSHA requirements for electrical safety for employees in the workplace.

References

American Society for Testing Materials. 1997. *Standard Specification for Temporary Protective Grounds to Be Used on De-energized Electric Power Lines and Equipment* (ASTM F 855). West Conshohocken, PA: ASTM.

______. 1998. *Standard Specification for Rubber Covers* (ASTM D 1049). West Conshohocken, PA: ASTM.

______. 1999a. *Standard Performance Specification for Determining the Arc Thermal Performance Value of Materials for Clothing* (ASTM F 1959). West Conshohocken, PA: ASTM.

______. 1999b. *Standard Specification for In-Service Care of Insulating Blankets* (ASTM F 479). West Conshohocken, PA: ASTM.

______. 1999c. *Standard Specification for In-Service Care of Insulating Line Hoses and Covers* (ASTM F 478). West Conshohocken, PA: ASTM.

______. 1999d. *Standard Specification for Rubber Insulating Blankets* (ASTM D 1048). West Conshohocken, PA: ASTM.

______. 1999e. *Standard Test Method for Determining the Ignitability of Non-Flame-Resistant Materials for Clothing by Electric Arc Exposure Method Using Mannequins* (ASTM F 1958). West Conshohocken, PA: ASTM.

______. 2001a. *Standard Guide for Visual Inspection for Electrical Protective Rubber Products* (ASTM D 1236-01). West Conshohocken, PA: ASTM.

______. 2001b. *Standard Specification for Insulated and Insulating Hand Tools* (ASTM F 1505). West Conshohocken, PA: ASTM.

______. 2002a. *Standard Performance Specification for Textile Material for Wearing Apparel for Use by Electrical Workers Exposed to Momentary Electric Arc and Related Thermal Hazards* (ASTM F 1506-02). West Conshohocken, PA: ASTM.

______. 2002b. *Standard Specification for In-Service Care of Insulating Gloves and Sleeves* (ASTM F 496-02). West Conshohocken, PA: ASTM.

______. 2002c. *Standard Specification for Leather Protectors for Rubber Insulating Gloves and Mittens* (ASTM F 696-02). West Conshohocken, PA: ASTM.

______. 2002d. *Standard Specification for Rubber Insulating Gloves* (ASTM D 120-02). West Conshohocken, PA: ASTM.

______. 2002e. *Standard Specification for Rubber Insulating Sleeves* (ASTM D 1051-02). West Conshohocken, PA: ASTM.

______. 2002f. *Standard Test Method for Determining Arc Rating of Face Protective Products* (ASTM F 2178-02). West Conshohocken, PA: ASTM.

Cooper Bussman, Inc. 2004. *Handbook for Electrical Safety*. 2d ed. Saint Louis, MO: Cooper Bussman, Inc.

______. 2004. *Safety BASICs—Bussman Awareness of Safety Issues Campaign*. PowerPoint slides. Distributed by Cooper Bussman, Saint Louis, MO.

Lee, R. 1982. "The Other Electrical Hazard: Electrical Arc Flash Burns." *IEEE Transactions on Industry Applications* 1A-18(3) (May/June):246.

______. 1987. "Pressures Developed by Arcs." *IEEE Transactions on Industry Applications* 1A-23:760–764.

National Fire Protection Association. 2002. *National Electrical Code 2002 edition (NFPA 70)*. Quincy, MA: NFPA.

______. 2004. *Standard for Electrical Safety Requirements for Employee Workplaces, (NFPA 70E)*. Quincy, MA: NFPA.

______. 2005. *National Electrical Code 2005 edition (NFPA 70)*. Quincy, MA: NFPA.

Occupational Safety and Health Administration. 2006. *OSHA General Industry Regulations*. 29 CFR 1910, Subpart S Electrical, 1910.301–399. Washington, D.C.: U.S. Department of Labor.

______. 2006. *OSHA Construction Industry Regulations*. 29 CFR 1926, Subpart K Electrical, 1926.400–449. Washington, D.C.: U.S. Department of Labor.

Owen, S. J. 1997. *Electrical Safety—Procedures and Practices*. Pelham, AL: S. J. Owen.

Appendix: Electrical Safety Topics and Referenced Sources

Chapter Topic	Adapted From
Justification for Work–Work-Related Permits	NFPA 70E-2004, Section 130.1
Exemptions from Work Permit	NFPA 70E-2004, Section 130.1(A)
Approach Boundaries	NFPA 70E-2004, Section 130.3(A)
Flash Hazard Analysis	NFPA 70E-2004, Section 130.3(A)
Training and Qualification of All Personnel	OSHA 29 CFR 1910, Section 1910.332; NFPA 70E-2004, Section 110.6
Visual Inspection	OSHA 29 CFR 1910, Section 1910.334(a)(2)
Portable Electrical Equipment	OSHA 29 CFR 1910, Section 1910.334(a)(2); NFPA 70E-2004, Section 110.6
Handling	OSHA 29 CFR 1910, Section 1910.334(a)(1)
Visual Inspection of Tools and Costs	OSHA 29 CFR 1910, Section 1910.334(a)(2); NFPA 70E-2004, Section 110.9(b)(3)
Safeguards for Personnel Protection	OSHA 29 CFR 1910, Section 1910.335; NFPA 70E-2004, Section 130.6
Rubber Insulating Liners–Various Classes	ASTM D 120-02 Standard Specifications for Insulating Gloves, 2002
Inspection of Protective Equipment	OSHA 29 CFR 1910 Section 1910.137(b)
Inspection Methods	OSHA 29 CFR 1910, Section 1910.137(b)
Storage of Insulating Equipment	OSHA 29 CFR 1910, Section 1910.137(b)
Leather Protectors	OSHA 29 CFR 1910, Section 1910.137(b)
Inspection and Testing of Insulating Equipment	OSHA 29 CFR 1910, Section 1910.137(b)
Insulating Equipment Failing to Pass Inspections	OSHA 29 CFR 1910, Section 1910.137(b)
Rubber Insulating Line Hose and Covers	OSHA 29 CFR 1910, Section 1910.137(b)
Insulated Tools	OSHA 29 CFR 1910, Section 1910.335(a)(2)
Conductive Work Locations	OSHA 29 CFR 1910, Section 1910.334(a)(4)
Alerting Techniques	OSHA 29 CFR 1910, Section 1910.335(b)
Safety Signs and Tags	OSHA 29 CFR 1910, Section 1910.335(b)
Barricades	OSHA 29 CFR 1910, Section 1910.335(b)
Alternate Alerting Techniques	OSHA 29 CFR 1910, Section 1910.335(b)
Vehicular and Mechanical Equipment	OSHA 29 CFR 1910, Section 1910.333(c)
Illumination	OSHA 29 CFR 1910, Section 1910.333(c)(4) and 29 CFR 1926, Section 1926.56
Conductive Apparel	OSHA 29 CFR 1910, Section 1910.333(c)(8)
Selection and Use of Work Practices	OSHA 29 CFR 1910, Section 1910.333
Work on De-energized Parts	OSHA 29 CFR 1910, Section 1910.333(A)(1); NFPA 70E-2004, Section 120.1–2
Energized Parts	OSHA 29 CFR 1910, Section 1910.333(a)
Working on or Near De-energized Parts	OSHA 29 CFR 1910, Section 1910.333(b)
Lockout/Tagout (LOTO) Procedures	OSHA 29 CFR 1910, Sections 1910.147 and 1910.333
Lockout/Tagout	OSHA 29 CFR 1910, Sections 1910.333(b) and 1910.147
Energy-Control Procedure	OSHA 29 CFR 1910, Section 1910.147(c)(4)

Interlocks	OSHA 29 CFR 1910, Section 1910.333(c)(10)
Electric Power and Lighting Circuits	OSHA 29 CFR 1910, Section 1910.334(b)
Reclosing Circuits After Protective Device Operation	OSHA 29 CFR 1910, Section 1910.334(b)(2); NFPA 70E-2004, Section 130.6(K)
Overcurrent Protection Modification	OSHA 29 CFR 1910, Section 1910.334(b)(3)
Lockout/Tagout Devices	OSHA 29 CFR 1910, Section 1910.147(c)(5)
Identification	OSHA 29 CFR 1910, Section 1910.147(c)(5)(D)
Periodic Inspections	OSHA 29 CFR 1910, Section 1910.147(c)(6)
Training and Communication	OSHA 29 CFR 1910, Section 1910.147(c)(7)
Authorized Employee	OSHA 29 CFR 1910, Section 1910.147(c)(7)
Affected Employee	OSHA 29 CFR 1910, Section 1910.147(c)(7)
Other Employees	OSHA 29 CFR 1910, Section 1910.147(c)(7)
Minimum Training Requirements	OSHA 29 CFR 1910, Section 1910.147(c)(7)
Retraining	OSHA 29 CFR 1910, Section 1910.147(c)(7)
Certification of Training	OSHA 29 CFR 1910, Section 1910.147(c)(7)
Isolation of Electric Circuits and Equipment—Preplanning	OSHA 29 CFR 1910, Section 1910.147(c)(8)
Lockout/Tagout Device Application	OSHA 29 CFR 1910, Section 1910.147(d)(4)
Applying Locks Only	OSHA 29 CFR 1910, Section 1910.333(b)(2)(iii)(E)
Applying Tags Only	OSHA 29 CFR 1910, Section 1910.333(b)(2)(iii)(D)
Stored Energy Release	OSHA 29 CFR 1910, Section 1910.147(d)(5)
Verify That Equipment Cannot Be Restarted	OSHA 29 CFR 1910, Section 1910.147(d)(6)
Verify That Circuit and Equipment Are De-energized	OSHA 29 CFR 1910, Section 1910.333(b)(2)(iv)
Inspection of Machines or Equipment	OSHA 29 CFR 1910, Section 1910.147(e)(1)
Verify Employees Are Clear of Circuits and Equipment	OSHA 29 CFR 1910, Section 1910.147(e)(2)
Lockout/Tagout Device Removal	OSHA 29 CFR 1910, Section 1910.147(e)(3)
Release for Energizing	OSHA 29 CFR 1910, Section 1910.147(e)(3)
Testing or Positioning of Machines and Equipment	OSHA 29 CFR 1910, Section 1910.147(f)(1)
Group Lockout/Tagout	OSHA 29 CFR 1910, Section 1910.147(f)(3)
Procedures for Shift Changes	OSHA 29 CFR 1910, Section 1910.147(f)(4)
Confined or Enclosed Work Spaces	OSHA 29 CFR 1910, Section 1910.333(c)(5)]
Overhead Lines	OSHA 29 CFR 1910, Section 1910.333(c)(3)
Unqualified Persons	OSHA 29 CFR 1910, Section 1910.333(c)(3)(i)
Qualified Persons	OSHA 29 CFR 1910, Section 1910.333(c)(3)(ii)
Grounding Clamps	ASTM F-855 Standard Specification for Temporary Protective Grounds to Be Used on De-energized Electric Power Lines and Equipment, 1997
Safety-Related Maintenance Requirements	NFPA 70E-2004, Article 2005
Employers Electrical Safety Program	NFPA 70E-2004, Section 110.7

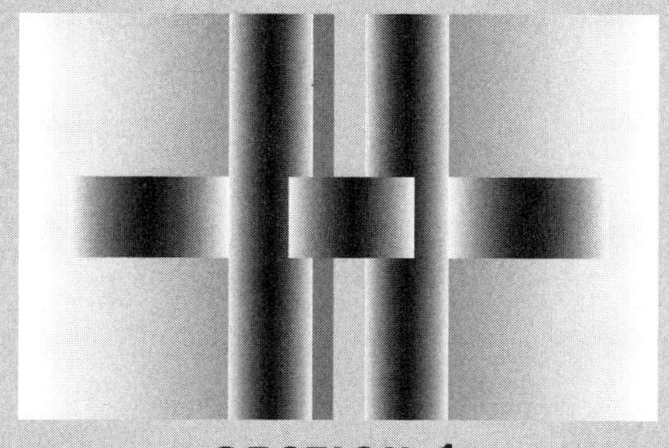

SECTION 1

RISK ASSESSMENT AND HAZARD CONTROL

Applied Science and Engineering: Permit-to-Work Systems

David Dodge

LEARNING OBJECTIVES

- Learn to identify the kind of tasks that need to be controlled by a permit-to-work system.
- Learn what steps must be taken after a task is identified as requiring a work permit so that the task can be performed safely.
- Ensure that only qualified persons issue permits and perform work under a permit.
- Understand the definition of a *qualified person.*
- Be able to take the concepts in this chapter and apply them to unique facilities and operations in order to identify tasks that require permits.

A PERMIT-TO-WORK SYSTEM is a managerial safety technique that enables everyone involved with a particular task to be sure that the procedure associated with the task is safe for its intended purpose. The permit to work is essentially a control document used to apply for permission to do work. Normally the permit-to-work system is used for nonroutine tasks that are performed infrequently and that need specialized equipment and/or training to perform safely. A permit-to-work system provides a formal control system that allows the unusual task to receive special attention by those who are knowledgeable in the techniques necessary to accomplish the task safely. The nonroutine task is not normally familiar to the worker, and therefore may present unique hazards that come from unfamiliarity. The permit, or request for permission, triggers a risk assessment and hazard control analysis that will provide detailed, step-by-step procedures for the job as well as the safeguards and monitoring techniques needed.

As the first step in developing a permit-to-work system, each facility, building, or operation must be reviewed and inspected to determine which tasks performed must be subjected to the permission and review of a permit-to-work system. This chapter discusses seven tasks which, because of their nonroutine and potentially hazardous nature, normally fall under a permit-to-work system. However, you may find that your operation has others. Individual management teams must evaluate their operations by risk assessment means to determine whether they may need permit-to-work systems.

No matter what the other tasks might be, the general approach taken in this chapter can be used to apply the permit-to-work system to them. Remember that a permit is normally required to perform nonroutine, unique tasks that can expose workers to hazards or hazardous conditions if not properly performed.

Once the tasks that require permits to perform them are identified, they should be catalogued. Any changes within the facility or operation that exclude or add similar tasks must be monitored and the catalog updated. For example, a building addition might add a confined space or a laboratory might add a new chemical or storage facility. Each of these changes must be reviewed by a person who is knowledgeable in that particular operation and who can perform a thorough risk assessment of the change. Once a list is made of all tasks to which the permit-to-work system applies, it must be conveyed to workers who might be involved so that they know which jobs will require permission to perform. Many facilities, understanding the importance of the permit system, provide signs in areas where tasks are performed that indicate that a permit is required. Still others merely inform supervisors of the permit-necessary tasks and provide them with a list. However it is accomplished, all those who might be involved must be informed that there is a permit-to-work system and which tasks require a permit.

At this point, it is well to define several terms that will be used frequently in this chapter. A *qualified person* (as defined in the American National Standards, "Safety Regulations for Confined Spaces," ANSI Z117.1) is one who by reason of training, education, and experience is knowledgeable in the operation to be performed and is qualified to judge the hazards involved and specify controls and/or protective measures. An *approving authority* is one who is assigned the responsibility of approving, by signature, the permit to work. The approving authority may or may not be a qualified person, but if he or she is not, they must require the input of a qualified person in order to approve the permit. A *risk assessment* (defined in the American National Standard, "Control of Hazardous Energy Lockout/Tagout and Alternative Methods," ANSI Z244.1) is a comprehensive evaluation of the probability and the degree of the possible injury or damage to health in a hazardous situation in order to select appropriate safeguarding.

Once the tasks that come under the permit-to-work system are identified, a document that serves as the permit should be devised. The permit should be clear and concise, which increases the likelihood that it will be used; however, it should provide enough detail that the approval authority will have enough information to perform a proper risk assessment. When formulating a permit, keep in mind that a worker is asking permission and assistance in performing a duty that he or she does not perform very often. The permit, before approval, details exactly how the job is to be accomplished. Figure 1 is an example of a permit that can be used for several tasks discussed in this chapter.

Normally, the sequence of events within a permit-to-work system are as follows. A worker is assigned a task for which a permit to work is required. The worker then goes to a supervisor or other designated authority to obtain a permit. The worker then fills out the permit and returns it to the supervisor. The permit is assigned a unique number for identification purposes. The permit form must detail the duration for which it is issued, the issue date and time, and the exact location of the job. At this point, a qualified person must do a risk assessment of the task to stipulate the hazard control measures necessary to perform the task safely. American National Standard, Control of Hazardous Energy Lockout/Tagout and Alternative Methods (ANSI/ASSE Z244.1-2003) Annex A is an example of a risk assessment technique. The supervisor can be the approving authority, however, a qualified person might have to be consulted in order to obtain the best possible risk assessment. Depending on the complexity of the task and the number of times it has been performed in the past, the worker might have sufficient information already detailed on the permit form indicating the safeguards necessary. The qualified person must still review the hazards and the proposed safeguards to make sure that the best possible safety measures are used. For well-known permit-required tasks, such as confined space entry or asbestos removal, a preestablished job safety procedure should already have been formulated.

The following list outlines the steps that the permit-to-work system should require:

- type of permit (confined space, hot work, etc.)
- unique permit number
- names of the workers and/or contractors
- location of task

Permit Number ________________________

Location of work (building/Room No.) _______ Contact Name ______________________________

Summary of work to be done

SAFETY PROCEDURES: To be implemented prior to commencement of work

1. The following processes are to be suspended during the course of the work:
2. The following equipment is to be withdrawn from service during the course of the work:
3. All users have been made aware of this suspension withdrawal Y/N
4. Safety Warning Notices have been posted where required Y/N
5. The following steps have been taken to eliminate, control, or contain hazards.

APPROVAL

Signed ______________________________

CONTROL OF RISKS ARISING FROM THE WORK

1. Isolation of services: (please check as appropriate)

   _____ water _____ electrical power _____ fuel lines _____ compressed gases _____ other (specify)
2. Are there safety implications resulting from the isolation? Y/N
3. Lock-out required Y/N Location ______________________________
4. Air monitoring required Y/N
5. Are there hazards associated with the work? Y/N
6. If yes, what safety precautions are required to control the risks?

Authorization

NAME: (Print) ______________________________

Company/Department: ______________________________

Signed: ______________________________ Date: ______________ Time: __________

Permit Validity Period:

From: Date ___/___/___ Time: __________ To: Date ___/___/___ Time: __________

Completion of Work

I confirm that the work has been completed in accordance with this permit. Services have been restored and the work area is ready for inspection.

Signed: ______________________________ Date: ______________ Time: __________

FIGURE 1. Sample permit-to-work permit

- details of the task to be performed
- identification of hazards
- risk assessment of task and hazards
- safeguards required
- personal protective equipment required
- authorization to begin the task
- permit time period
- extension of time period, if necessary
- completion date and time

Once the permit has been reviewed, approved, and signed by the approval authority, a copy of the completed permit should be made available to all involved in the task. This is normally accomplished by posting the permit in the area of the job.

When the work is completed, the permit is returned to the approval authority for cancellation and filing. The approval authority should assure himself or herself that the job site is free of hazards upon completion of the job.

Contractors and subcontractors must be held to the same high safety standards to which host managers, supervisors, and workers are held. Therefore, permits to work must be required of all outside firms who are asked to perform duties for which a permit is normally required. Most of these duties involve operations that affect other workers or operations. The prudent manager will want to be able to assess and monitor the contractor's presence and activities within the facility or operation.A contractor can provide its own qualified person, risk assessment, and hazard control measures, but the facility management will provide the permit approval authority so that oversight authority can be maintained.

The first specific permit-to-work system that is discussed in this chapter is for confined spaces, because it is more detailed due to the complexity of the hazards and safeguards involved, and because its concepts can be used in devising permits to work for any task or operation.

Permit to Work–Confined Spaces

Confined Space Survey

The first step in establishing a confined space permit-to-work system within any facility or operation is to perform a survey to establish a documented inventory of all areas that are to be considered confined spaces (ANSI 2003a, section 3.1). A confined space is defined by the American National Standards Institute standard Z117.1-2003 as "an enclosed area that is large enough and so configured that an employee can bodily enter and has the following characteristics: its primary function is something other than human occupancy; and has restricted entry and exit" (ANSI 2003a). The Occupational Safety and Health Administration (OSHA) standard 29 CFR 1910.146(b) uses a similar definition.

A survey must be performed by a qualified person or persons who have sufficient expertise to define a confined space. The goal, at this point, is simply to establish a list of the spaces that are, by definition, confined spaces. Keep in mind that confined spaces are not only limited to the interior of buildings or obvious external tanks, but can also be hidden in remote locations, be underground, and can be smaller than room size.

A blueprint of the facility is a handy reference tool when conducting the survey, as is a list of all maintenance tasks to be performed on a regularly scheduled basis. Many not-so-obvious confined spaces are those that are accessed infrequently for painting or cleaning and, as a result, might be missed.

In addition, each building, process, or equipment modification or addition must be reviewed by safety personnel or a qualified person to determine whether the changes create a confined space that must then be added to the inventory. The confined space survey must provide the following information:

- the confined space number
- the date of the survey
- the location of the space
- a description of the space
- the number of entry points
- the frequency of entry
- who usually enters the space
- reasons for entering the space
- if the space can be entered by a human
- if there is a hazardous atmosphere
- if there is limited or restricted entry
- if the space is designed for human occupancy
- other safety hazards

- possible atmospheric hazards such as oxygen deficiency, flammable materials, toxic materials
- identification of the specific hazards for flammable and/or toxic materials and their exposure levels
- possible hazardous contents of the space, such as past contents, current contents, and dust
- potential energy sources, such as electrical, hydraulic, pneumatic, and mechanical
- potential hazards of the space itself, such as slippery surfaces, high or low temperatures, and high noise levels
- hazards presented from the configuration of the space

It is not enough to create a list of confined spaces and file the list away where those who need its information most cannot find it. Once a confined space is identified, it must be physically labeled as such at all entry points. The labeling must indicate that the area is a confined space and that it can be entered only after a written confined space permit is obtained from the proper authority listed on the label. Identification labels are available from commercial suppliers.

Hazard Identification

The definition of a confined space does not require that there is a hazard present. The confined space survey might detect confined spaces that are free of hazard and, as a result, can be entered without a permit. These spaces are called "nonpermit confined spaces" by both ANSI Z117.1-2003 and OSHA 1910.146(b) and will not be further addressed in this chapter.

The permit-to-work system is concerned with those confined spaces that contain known or potential hazards and, therefore, require special permission and safeguards to enter. It must be noted that the concern is that KNOWN or POTENTIAL hazards might exist. While a confined space might, at the time of entry, be deemed free of hazards other than those inherent to the space, there can be the POTENTIAL for the creation of a hazard after work has commenced, whether by the nature of the work itself, or by the introduction of a hazardous material through piping or other entry means. Therefore, the hazard identification performed must take into consideration future as well as existing hazards within the confined space. Such future hazards can include the introduction of material through a pipe or other opening into the space, heat generated by the task being performed, or toxic materials generated by the task being performed.

Note that there are entities that do not regard underground sewage systems and their associated manholes, collection basins, and piping as confined spaces. For the purposes of this reference, such spaces fit all the criteria of a confined space and, thus, are considered confined spaces. Every portion of an underground, enclosed area that contains potentially hazardous material and has the potential for oxygen depletion must be considered a confined space and be subject to the permitting process.

The hazard identification process must be accomplished by a qualified person and must take into consideration several criteria during the identification process. First, the configuration of the confined space must be noted to determine means and ease of entry and exit. Many confined spaces require some means of assistance, such as a mechanical lifting device, to get workers in and out. If so, all necessary equipment is a requirement of the permit system. The point of entry and exit must be identified and examined to ensure that it does not interact with other operations in the area. If, for example, a worker within a confined space must exit within a fork truck travelway, appropriate signage and barricades must be provided in order to redirect fork truck travel.

Second, the hazard identification process should consider all uses, past and present, of the confined space. Past uses are important due to the fact that residues can be left behind that might interact with the work progress, such as cleaning chemicals used or hot work performed. Material Safety Data Sheets (MSDS) should be used to determine hazards presented by whatever was in the space before the entry by workers, and to determine how those materials might interact with the process that made the entry necessary.

At this point, a documented list of confined spaces has been established and each space has been surveyed for its physical characteristics and its past and current uses. Next, the confined space evaluation process should determine the EXISTING or POTENTIAL hazards unique to each confined space. This process, too, should be documented so that everyone involved, no matter how far in the future, will have access to the information. It is extremely critical that the hazards be identified by a qualified person who might not necessarily be the same person who identified the confined spaces. The work can require different expertise and it is very important that the present hazards and POTENTIAL hazards are identified. The assistance of an industrial hygienist or someone of similar training and experience might be necessary.

During this phase of the confined space evaluation, the hazards to be assessed include oxygen depletion, oxygen enrichment, flammable and explosive atmosphere, and toxic materials in the confined space (ANSI 2003a, section 3.2). The evaluator must understand that the work that is being performed within the confined space after the permitting process might contribute its own hazard. For example, an internal combustion engine used within the space will, if not properly exhausted, fill the space with carbon monoxide, or a chemical cleaning agent can, if not properly ventilated, fill the space with toxic materials. Each potential hazard must be identified and documented in preparation for a complete permit-to-work system.

Other hazards to be considered are biological hazards, mechanical hazards from any machinery that might be in the space or used in the operation, and physical hazards. Physical hazards include electrical exposures within the space, heat stress possibilities from process sources or in an improperly ventilated space, and engulfment by solid or liquid materials that might inadvertently enter the space by way of unsecured entry pipes or chutes.

Hazard Evaluation

Now that the confined spaces have been identified and the hazards to which each might be subjected have been cataloged, the hazard presented must be evaluated (ANSI 2003a, section 3.3). The evaluation determines the extent and breadth of the hazard as well as the consequences of exposure. At this point in the establishment of the confined space permitting system, the qualified person must determine what is the likelihood of worker exposure to the hazard and, if exposed, what is the likely extent of that exposure. This information will aid in determining what safeguards must be used to make and keep the confined space hazard-free while work is being performed inside. As in every step along the identification and evaluation process, any possibility of changes in the conditions within the confined space while workers could be exposed should be assessed.

At this point, the evaluator must consider how the identified hazards will be safeguarded. The logical risk evaluation must lead you to engineering safeguards that eliminate the hazard before personal protective equipment is considered. In order to best determine the most effective engineering safeguards, the evaluator must have knowledge of and access to updated data such as Threshold Limit Values by the American Conference of Governmental Industrial Hygienists (ACGIH), 29 CFR 1910, subpart Z, and Materials Safety Data Sheets. For safeguards of mechanical and physical hazards, consult the other chapters in this section of the handbook; OSHA 29 CFR 1910 and its referenced publications provide a good resource.

Whatever safeguards are determined to be appropriate must now be documented along with the corresponding confined space and listed hazards. Keep in mind that any one confined space can subject those entering it to more than one hazard, and each must be addressed if the ultimate goal of a safety task performed is to be realized.

Once the method of control for each of the identified hazards is identified, an emergency response plan must be formulated in the event that the established safeguards fail (ANSI 2003a, section 14). Careful consideration must be given to exactly how it will be determined that a safeguard has failed and, next, how to get the workers out of the confined space in the safest, most efficient manner. All of these procedures must be documented, because they will eventually be part of the confined space entry permitting system. Then all involved will know exactly what to do, not only when the system is working well, but also when it is not. The permitting documentation will eventually state that, when emergency evacuation equipment is deemed necessary, it will be present and ready to use at the confined space entrance at all times that workers are inside the space.

The Written Confined Space Entry Permit Program

After performing the identification survey of the confined spaces within the operation or facility, identifying the hazards associated with each space and evaluating those hazards, we are now ready to formulate a written confined space entry program. Remembering that the permit-to-work system provides a systematic, disciplined, and documented approach to assessing the risks of a job and specifies the precautions to be taken when performing work in a confined space, a permit must be established that will lead the user through the proper steps to perform the required task safely.

Once it has been determined that work must be performed in a confined space, the supervisor can then go to the inventory list of confined spaces to determine what hazards are involved and the specified controls for each hazard. Now a permit system must be established that will:

- specify the work to be done and the equipment to be used
- specify the precautions to be taken when performing the task
- give permission for work to start
- provide a check to ensure that all safety considerations have been taken into account
- provide a check of the completed work

To this end, a permit must be established that will communicate all necessary information to all involved and provide for an approval process for the confined space entry. The permit should contain at least the following information:

- the name of the person controlling the confined space, and the date and time of entry
- the name of the person authorizing the entry
- the location of the confined space
- a description of the work to be performed
- the names of all workers required to enter the confined space and the names of all attendants
- the names of any outside contractors who will enter the confined space
- a list of all hazards to be controlled before entry and the controlling method of each
- a list of all safety equipment that will be necessary to control the identified hazards
- the type of atmospheric tests required, the expected results of those tests, and the type of equipment necessary to carry out the tests. In addition, the frequency of atmospheric testing must be stipulated
- the type of emergency rescue equipment necessary, including emergency communication equipment
- the type and number of communication devices present on the job, and to whom they will be assigned
- the duration of the permit

Figure 2 is an example of a confined space entry permit.

Personnel

Now that a permit-to-work permit has been established, the qualifications of those involved in the confined space entry must be established. We have already determined that a qualified person is one who is trained and knowledgeable in the operation and is qualified to judge the hazards involved and specific control measures. Such a person was asked to identify confined spaces and identify and evaluate the hazards. However, there may be an entirely different group of workers who are involved in the actual confined space entry. There are generally five groups of personnel who are involved: the entry supervisor, the entrants, the standby attendants, atmospheric monitoring personnel, and emergency response personnel (ANSI 2003a).

The entry supervisor is the person who is in charge of the permitting process and who should be trained in and have knowledge of all aspects of the confined space entry permit. The entry supervisor is responsible for determining that all requirements of the permitting

Confined Space Entry Permit

Date and Time Issued: ____________________ Date and Time Expired: ____________________

Job site/Space I.D.: ____________________ Job Supervisor: ____________________

Equipment to be worked on: ____________________ Work to be performed: ____________________

Stand-by personnel: ____________________

1. Atmospheric checks: Time: ____________
   - Oxygen ____________%
   - Explosive ____________% L.F.L.
   - Toxic ____________ PPM

2. Tester's signature ____________________

3. Source isolation (No Entry):	N/A	Yes	No
Pumps or lines blinded,	( )	( )	( )
disconnected or blocked	( )	( )	( )

4. Ventilation Modification	N/A	Yes	No
Mechanical	( )	( )	( )
Natural Ventilation only	( )	( )	( )

5. Atmospheric check after Isolation and Ventilation:

Oxygen ____________%	>	19.5	%
Explosive ____________ % L.F.L.	<	10	%
Toxic ____________ PPM	<	10	PPM H(2)S

Time ____________

Tester's signature: ____________________

6. Communication procedures: ____________________

____

7. Rescue procedures: ____________________

____

____

8. Entry, standby, and backup persons:	Yes	No
Successfully completed required training		
Is it current?	( )	( )

9. Equipment:	NA	Yes	No
Direct reading gas monitor–tested	( )	( )	( )
Safety harnesses and lifelines for entry and standby persons	( )	( )	( )
Hoisting equipment	( )	( )	( )
Powered communications	( )	( )	( )
SCBA's for entry and standby persons	( )	( )	( )
Protective clothing	( )	( )	( )
All electric equipment, listed Class I, Division I, Group D and Nonsparking tools	( )	( )	( )

FIGURE 2. Sample confined space entry permit

Confined Space Entry Permit (cont)

10. Periodic atmospheric tests

Oxygen	____%	Time ______	Oxygen	____%	Time ______
Oxygen	____%	Time ______	Oxygen	____%	Time ______
Explosive	____%	Time ______	Explosive	____%	Time ______
Explosive	____%	Time ______	Explosive	____%	Time ______
Toxic	____%	Time ______	Toxic	____%	Time ______
Toxic	____%	Time ______	Toxic	____%	Time ______

We have reviewed the work authorized by this permit and the information contained herein. Written instructions and safety procedures have been received and are understood. Entry cannot be approved if any squares are marked in the "No" column. This permit is not valid unless all appropriate items are completed.

Permit Prepared By:
(Supervisor) ______________________________

Approved by
(Unit Supervisor) ______________________________

Reviewed By:
(CS Operations Personnel): ______________________________

____
(signature)

____
(printed name)

This permit to be kept at the job site. Return job site copy to Safety Office following job completion.

Copies: White Original (Safety Office)
Yellow (Unit Supervisor)
Hard (Job site)

FIGURE 2. Sample confined space entry permit

system have been filled before he or she signs the permit allowing confined space entry. In addition, he or she should ensure that all conditions of the safe work procedure, as detailed in the work permit, are followed for the entire duration of the task. These conditions include atmospheric monitoring and the presence of rescue apparatus and personnel.

An entrant is the one who actually enters the confined space and who should be trained in, and maintain familiarity with, the confined space entry permit system and the individual permit in use at the time of entry. The entrant should use only the equipment specified in the work permit, and should be informed of the symptoms that can be expected if exposure to atmospheric hazards elevates beyond acceptable levels. Further, the entrant should be trained in self-evacuation methods and proper communication techniques between the entrant and the standby attendant. The entrant and the standby attendant should be provided with communication equipment so that they can remain in constant communication throughout time spent within the confined space.

The standby attendant is the one who is stationed outside of the confined space at all times that the space is occupied, and who is responsible for assisting with entry and exit to the confined space. Most importantly, the standby attendant is not, under any circumstances, to enter the confined space so that he or she will be available to communicate evacuation procedures to the entrants. In the event of any deviation from the permit requirements, the standby attendant will be provided with a direct communication link to the rescue team and will not attempt rescue unless he or she is specifically trained in rescue techniques and acts as part of the rescue team. The standby attendant should remain outside of the entrance to the confined space at all times unless relieved by a trained attendant or until all workers have vacated the confined space.

Atmospheric monitoring personnel monitor the atmosphere of a confined space when the risk assessment has stipulated that such testing is necessary. Continuous monitoring/alerting devices, both personal and area, can be used. No person should enter the confined space to conduct monitoring except as provided by written authority. Results of any monitoring should be recorded on the confined space entry permit.

Rescue personnel are those who are trained and qualified in general rescue techniques and who are specifically knowledgeable of the particular techniques involved in the specific confined space. Rescue personnel should also be trained in first aid and cardiopulmonary resuscitation (CPR) techniques. Rescue personnel training should include an actual rehearsal of a rescue using all necessary apparatus and equipment.

Each of these categories of personnel and their duties are vital to the safety of the entrants. Therefore, they should be trained, knowledgeable, and dedicated. A list must be maintained of the names of each qualified person in each category, and the list must contain information concerning training dates and designation.

Work in the Confined Space

Once a permit has been issued for work to be performed within a known and labeled confined space, testing must then take place to assure all of those involved that the space is, indeed, safe to enter. All atmospheric testing within the confined space should be performed from outside of the space, and the atmosphere should be made safe for unprotected work before entry. However, there are times when this is impractical or impossible, and under these circumstances entrants must be protected by personal protective equipment. These events and the proper protective equipment must have been previously identified and evaluated, and therefore are part of the completed confined space work permit.

The atmosphere within a confined space must be tested for (in this order) oxygen levels, flammability, and exposure levels of any hazardous materials that might be present, and all testing must be performed by a qualified person. The atmosphere within the confined space must be tested in a manner that will ensure that all levels and all areas within the space are free from hazard. To accomplish this, testing must take place at all vertical levels within the space and all areas that might trap contaminants. All areas within a confined space that access work areas must be tested, even though work might not be scheduled in all parts of the space. Flammable and toxic vapors migrate from place to place, as do workers. All testing equipment used should be listed for its intended purpose and should be calibrated at the frequency recommended by its manufacturer. A list shall be established and maintained that documents the calibration and service dates of all testing equipment (ANSI 2003a, section 6).

Initial atmospheric testing within the confined space must be accomplished with any temporary or permanent ventilation systems shut down. If it is determined that ventilation is required, the atmosphere must be retested with the ventilation system functioning to determine the quality of the outside air drawn into the confined space. Care should be taken to make sure that ventilation intake equipment is well away from contamination sources, such as engine exhaust or operations that emit toxic vapors. Atmospheric testing with the ventilation system functioning will determine the proper location of the ventilation intake.

Because circumstances surrounding and within a confined space can change rapidly, the risk assessment performed during the confined space evaluation should seriously consider continuous atmospheric monitoring whenever workers are within the confined space. The risk assessment performed during the evaluation of the confined space and its hazards will have already determined the length of time that the entrants can be out of the confined space before retesting is required prior to reentry. Such an assessment must include the hazardous materials involved and their possible reentry mode, the possibility of oxygen depletion while the entrants were out of the space, the type of ventilation, and whether the work performed before departure from the space might have changed the oxygen level or the flammability and/or toxicity of the atmosphere.

The acceptable limits of atmospheric hazards within the confined space will have been established during the hazard identification and evaluation phase of the permit-to-work system establishment and, therefore, will be prominently listed on the permit issued to those on the confined space entry team. The oxygen level will be set at between 19.5 percent and 23.5 percent for all confined spaces.

The flammability and toxicity will change for each space, depending upon the involved hazardous materials. The National Fire Protection Association (NFPA 325M) has established flammability ranges for most chemicals. If a reference source other than the NFPA is used, care must be taken to ensure that the provided data is reliable and accurate. The flammability of the atmosphere within a confined space must be reduced to less than 10 percent of the lower explosive limit (LEL) or below the lower flammable limit (LFL) before an entrant is allowed to enter a confined space. If the flammability of a space is to be lowered using ventilation, great caution must be taken in directing the exhaust of the ventilation system away from any surrounding sources of ignition. The hazard evaluation and risk assessment must also take into consideration the fact that a flammable atmosphere that is above the allowed explosive limit, meaning that it is too rich to ignite, carries another set of hazards that must be addressed, such as the toxicity of the material and the lowering of the flammable levels to within the flammable range during the work process.

The allowable toxicity of the atmosphere within the confined space will have been established during the hazard evaluation, and will be prominently listed on the approved confined space entry permit. The qualified person who evaluates and establishes the acceptable toxic limits has several reference sources available in which predetermined acceptable toxic levels for various materials are listed: The American Conference of Governmental Industrial Hygienists (ACGIH) Threshold Limit Values Document, Material Safety Data Sheets (MSDS), and 29 CFR 1910, subpart Z, are reference sources that are widely used in industry. Any other reference sources used must be reviewed carefully and their accuracy verified.

When either the oxygen level is improper, the flammability of the atmosphere is unsuitable, or the toxicity levels are too high, or any combination of these hazardous conditions has been detected by testing, entry should not take place until the hazardous condition is corrected to acceptable levels or appropriate personal protective equipment is used as detailed in the confined space work permit. The first priority should be to eliminate the identified hazard within the confined space by engineering means. Using personal protective equipment to eliminate the hazard should be the last resort.

The risk evaluation and assessment might have determined that ventilation (either forced or natural) is required to provide an atmosphere in which persons can work and, when it has, the work permit should detail the amount of pre-entry purge ventilation time and the frequency of atmospheric testing once entry has been made. Of consideration as well is whether or not forced ventilation is necessary at all times that the space is occupied. It normally is. If it is determined that natural ventilation is an acceptable means of safeguard, care should be taken not to decrease the size of the air entry and exit points with the placement of tools, equipment, or other apparatus at openings.

Isolation of the Confined Space

Before work can be allowed in the identified confined space, care must be taken to make sure that no hazardous materials enter the space once it has been purged and work commenced (ANSI 2003a, section 8).

The permit to work will have detailed possible modes of entry and methods to isolate the confined space. Because the permit-to-work system has already evaluated the space and its hazards, the physical methods that must be taken to prevent unwanted entry of hazardous materials into the space will be detailed on the work permit. Each potential entry point or hazard source should be evaluated as a separate entity, and the best possible means of isolation or deactivation provided.

One method of entry of toxic or flammable materials into the space is by way of pipelines that lead into the space from other known and previously evaluated sources. Pipelines leading into the confined space must be drained, flushed, and cleaned of hazardous materials before entry, and the pipeline secured in an effective manner to prevent flow. The pipe can be provided with a blank, taken apart and misaligned at a flange, or isolated by valves if so provided. If valves are used, more than one valve should be used with a bleed valve in between. If piping must be left open during work within the confined space, a qualified person must assess the risk of exposure and provide other precautions as needed, such as personal protective equipment (ANSI 2003a, section 8).

Entry of a hazardous material into the confined space is not the only hazard to address. Other sources can include electrical, mechanical, hydraulic, pneumatic, chemical, thermal, and radioactive hazards, and falling objects. Each of these potential hazards must be evaluated for its ability to injure workers just as they would outside of the confined space, and appropriate safeguards used and detailed on the work permit.

If it is determined during the hazard evaluation and risk assessment that a hazard must be controlled using either lockout or tagout methods, a program developed in accordance with ANSI Z244.1-2003 (ANSI 2003b) and OSHA 29 CFR 1910.147 (OSHA 1996a) should be used. Such a requirement will be listed as part of the approved work permit.

Emergency Response in a Confined Space

A documented emergency response procedure should be developed that will detail the exact method by which rescue will take place, the type of personnel and training involved, and the equipment necessary to perform the rescue (ANSI 2003a, section 14). Although some confined spaces are similar and may use the same personnel, equipment, and techniques, many will be dissimilar, and therefore require a different plan and procedure. When assessing and devising the emergency response procedure, you must consider the area and configuration of the confined space and the distance away from an exit point a disabled worker or workers might be. The number of likely disabled workers will have a profound affect on the number of rescue personnel required and the type of equipment necessary. The entrance opening to and from the confined space will affect the type of equipment that rescue workers wear and carry when entering the space. The site of the exit in relation to the size and weight of those to be rescued, as well as the type of entrance, must be considered. If an entrance and exit portal is provided with a ladder, the rescue method will have to provide a harness and mechanical lifting device to lift a disabled worker out of the space. The rescue procedures and the risk assessment from which they are derived must be reviewed before a rescue is required, and the safest rescue operation for all involved, including the rescue personnel, must be provided.

Understanding that a rescue is necessitated by a problem within the confined space, the first priority is to attempt to formulate a rescue plan that makes entry by rescuers into the confined space unnecessary. One way this can be accomplished is by fitting the entrants with body harnesses and lanyards prior to entry.

Establishing a predetermined method by which entrants can perform their own rescue is another rescue method that does not expose emergency personnel to hazards within the confined space. This method normally requires a warning of impending hazard, usually by way of continuous monitoring.

All rescue methods must be predetermined and documented in an emergency response procedure, and all those who are involved with the rescue must be trained in its implementation. Also, they must have practiced the rescue procedure in its entirety.

A rescue procedure is initiated by a standby attendant who, as established by the confined space permit system, has been stationed at the entry point of the confined space with the express duty of monitoring conditions within the confined space and initiating emergency procedures if necessary. The attendant must be warned that his or her first and most important duty in the event of an emergency is to contact rescue personnel as quickly as possible. The attendant must be advised that he or she is never to enter the space to attempt a rescue unless part of a designated, trained rescue team. If the attendant is a member of such a team, no rescue attempt inside of the confined space will be attempted until the entire rescue team is assembled at the confined space entry point. Someone must be left outside of the confined space during the rescue operation.

Outside Contractors

Outside contractors can be hired to perform work in confined spaces over which the plant or facility owner or operator has control. However, the owner or operator might have no direct control over those who actually perform the tasks. In this instance, the owner or operator and the contractor should act as a team to ensure the safety of all involved.

The facility owner or operator should accept no less degree of safety for the contractor's workers than it would for its own. That means the confined space entry must be made under the procedure developed by a well-thought-out permit-to-work system. The contractor should expect information concerning the space necessary to enter the space safely from the facility owner or operator (ANSI 2003a, section 17).

The facility owner or operator must have performed an inventory of its confined spaces and labeled them as such before the confined space can be entered. In this way, both the contractor and the owner or operator know that the space does, in fact, fit the definition of a confined space. Also, the owner or operator will have identified hazards, evaluated them, and stipulated control and monitoring methods.

These duties must be performed before any confined space entry is made, no matter whose employees are entering. Normally, the owner or user of the space will accomplish this task. These duties can be performed by a contracted party, but the owner or operator must stipulate that they are performed in compliance with this chapter and accepted industry standards.

The owner or operator of the confined space must inform the hired contractor of all information concerning the space that has been discovered during the evaluation, such as the size and exit points of the space, the hazards discovered, the hazard safeguards, and rescue methods. The contractor will then have all of the information necessary to perform work safely within the confined space. A work permit must still be issued and approved before entry, even though a contracted party is used. The parties involved must understand that the facility owner will be responsible for the permit to work and its approval. The designation and training of personnel, including rescue personnel, and atmospheric monitoring procedures and equipment can be the responsibility of the contractor. Whoever is designated to perform these duties, responsibilities should be distilled to a document and signed by both parties so that all involved know who is responsible for each phase of the entry.

Permit to Work–Hot Work

Hot work is work that involves temperatures that have the potential to create a source of ignition. These operations include grinding, welding, thermal or oxygen cutting or heating, and other related heat-producing or spark-producing operations. As discussed earlier in this chapter, the risk assessment and hazard control aspects of the confined space permit system can be used to develop a permit-to-work system for other operations where it is determined to be appropriate.

Just as is true for a confined space permit system, the hot work permit must specify:

- the work to be done and equipment to be used
- the precautions to be taken when performing the task
- the clearance zone for flammable materials
- the location, date, and time of the task

- the type of firefighting equipment that must be available while the work is being performed.
- an inspection of the area before beginning the work.
- the assignment of a fire watch at the completion of the job, including the duration of the watch.
- the type of communication devices to be provided to the parties involved.
- an authorization from the approval authority to start the job.
- the area around the operation must be cleared of combustible and flammable materials.

The permit to work for hot work must include space (and prompting text) for the user to completely answer the above questions. Unlike a confined space permit system, the location of the work area is not predetermined. However, most of the safeguards for hot work are well known, and the permit is a mechanism that leads the user to ask the proper questions to assess the risks of the job, and then to provide the proper hazard control measures. Both the assessment and control methods must be reviewed and approved by a qualified person. Figure 3 shows an example of a hot work permit.

The process for a job to commence under a hot work permit system (and for most permitted work systems) is that a worker determines that hot work has to be performed. The worker then goes to the appropriate designated supervisor or other authority to request permission to perform such work and to obtain a blank permit-to-work form. (In order for the

Hot Work Permit

DATE ISSUED: __________ VALID UNTIL: __________

BUILDING: __________ BUILDING # __________ PROJECT # __________

LOCATION OF WORK __________

CONTACT PHONE NO. __________

The following precautions must be taken before work is started:

–Cutting and/or welding equipment must be thoroughly inspected and found to be in good repair, free of damage or defects.

–At least one fire alarm pull station or means of contacting the fire department (i.e., site telephone) must be available and accessible to persons(s) conducting the cutting/welding operation.

–A multi-purpose, dry chemical portable fire extinguisher must be located such that it is immediately available to the work and is fully charged and ready for use.

–Floor areas under and at least 35 feet around the cutting/welding operation must be swept clean of combustible and flammable materials.

–All construction equipment fueling activities and fuel storage must be relocated at least 35 feet away from the cutting/welding operation.

Where applicable, the following precautions will also be taken before the work begins:

–Fire resistant shields (fire retardant plywood, flameproof tarpaulins, metal, etc.) must

–Containers in or on which cutting/welding will take place must be purged of flammable vapors.

The following precautions will be taken during and after the work:

–Person(s) must be assigned to a fire watch during and for at least 30 minutes after all cutting/welding.

–Fire watch person(s) are to be supplied with multipurpose dry chemical, portable fire extinguishers and trained in their use.

FIGURE 3. Sample hot work permit

system to function properly, the worker must have been trained in at least the rudimentary aspects of the permit-to-work system so that he or she understands that a task involving hot work requires permission by way of a permit.)

Once the authority determines that a permit is required, the worker is issued a blank permit and is required to fill it out with all necessary information. Many times the worker requires assistance in determining the appropriate safeguards and assigning the fire watch.

When the permit form is completed, the approval authority determines whether the worker is properly trained to perform the job safely, and whether or not the safeguards are adequate. Upon approval, the worker proceeds to the job site and inspects it for hazards such as combustible or flammable materials in the area, and that the job site is provided with the extinguishing equipment as specified on the permit.

When the hot work is completed and the work area cleaned up, some form of fire watch must be assigned. This can range from requiring the worker or one of the work party to stay at the job site for a period of time (30 minutes) to ensure that there is no possibility of fire, to requiring several inspectors for an extended period of time. The extent of the fire watch is to be determined by the risk assessment performed before the job is started. Whatever fire watch remains after the job is completed (and there should be one no matter how small the job), it should be provided with appropriate fire extinguishing apparatus and knowledge and training in its use.

Once the fire watch has ended, the work permit document is returned to the authorizing person so the end of the job can be logged and the permit filed.

A permit should be issued for no longer than one work shift. If it becomes necessary for new personnel to take over the hot work job, a new permit should be issued to ensure that the workers involved are properly trained.

Permit to Work–Roof Work

Work on roofs normally involves the risk of working at elevations, with the associated hazard of falls from the work area. Therefore, it is well for job rules to be established before the hazardous work is undertaken so that all involved know exactly how to perform the required duties in the safest possible manner.

If the entity for which you work is in the business of doing roof work, no permit-to-work system is required because the work is not out of the ordinary for the workers, and it is expected that all of the proper training and safeguarding will be in place before each day's activities. However, for the organization that only occasionally has to assign its workers job duties on building roofs, it is understood that those workers will not be as well-versed in roof work and the safeguarding necessary. For this reason, occasional roof work must be attempted only as part of a permit-to-work system that will ensure that those undertaking the tasks are trained, and that the job has been subjected to a formal risk assessment and, as a result, hazard controls have been implemented.

The permit document will be one that provides the date and location (building) of the work to be performed, and the exact location of the task upon the roof area. The exact location is necessary for the determination of the personal protection and fall arrest equipment that must be provided. For example, if work is to take place on the edge of a roof without a parapet (within ten feet of the edge), some form of fall protection, such as a lanyard safety system or edge protection, will be required. However, if the roof is large and all work will take place more than ten feet from the unprotected edges, fall protection might not be required. A risk assessment must be performed by a qualified person before each job that requires a roof work permit in order to determine the exact method of hazard (fall) control.

One of the hazards to be considered during the risk assessment is whether or not roof work is to be performed in high winds and/or rain. High winds greatly affect the safety of personnel working with sheet material, and rain affects the coefficient of friction of the roofing material and therefore its slipperiness. Another hazard to be addressed during a roof work risk assessment is the effect the additional weight of workers, repair equipment, and any additional equipment being installed on the roof may have on the roof's structural components. All roofs should be designed to accept additional weight applied during ordinary maintenance, but to the extent that the maintenance becomes extraordinary (such as in combination with a heavy snow load), the risk assessment must take this into consideration. In addition, the affect of the weight of any equipment left on the roof must be considered.

Taking all of this into consideration, the permit document must contain information that will allow the person who must authorize the work to determine the adequacy of the training of those undertaking the job. Roof work can be hazardous and should be performed only by those trained to avoid its risks.

A checklist can be developed that will assist all of those involved in determining the hazards in the roof work and the safeguards necessary to perform the task safely. The checklist should encompass at least the following:

- the location of the work
- expected weather conditions
- roof-loading restrictions
- fall protection equipment necessary
- perimeter or barrier guarding
- communication means
- coordination of crane operation
- adequacy of roof access means (ladders, hoists)
- inspection of ladders
- protection of workers or equipment below
- chimney/fume hood discharge or steam discharge
- overhead electrical wire protection (insulation, distance)

Roofing contractors should be required by contract or written agreement to perform their tasks in compliance with industry safety practices and in a manner that will not adversely affect others within or surrounding the facility on which the work is to be performed. The contractor must apply for permission to work on the facility's roof by way of a permit to work, so that the facility manager will know what is taking place on the roof.

The sequence of events leading up to the authorization of a permit to perform roof work is similar to that for hot work. The permit must be issued for no longer than one day due to the large effect weather has on the safety of roof workers and, therefore, the need to evaluate the job on at least a daily basis. Weather changes during a 24-hour period might necessitate the revocation of a work permit. The permit to work shown in Figure 1 can be used as a guideline.

Not all elevated work takes place on a roof. There can be times within a facility or operation when non-routine work is required at height that demands special knowledge and safeguards. A permit should be required before such work is performed. Many of the hazards involved are similar to those involved in roof work, and therefore the roof work hazard checklist and permit system can be used.

Work performed from a ladder by maintenance workers is usually routine and so does not require a permit. If there is work at height other than roof work to be performed, careful consideration should be given to providing the workers with a simple, protected work platform, such as aerial platforms or scaffolding with proper guarding of their open sides and proper access points (see the section "Permit to Work—Scaffolding" later in this chapter). If such a protected work platform cannot be provided, fall protection safeguards must be provided.

Permit to Work–Excavations

Excavating is a task that is not often undertaken by a facilities maintenance department, and can be an extremely hazardous duty if not properly safeguarded. Therefore, a permit to work must be applied for and approved before any excavation is begun.

The permit itself should be designed in such a manner that it can be used as a tool that leads the person who applies to do the job through the permit, and the person who ultimately authorizes the permit to ensure that all foreseeable hazards are discovered and addressed.

The process of excavating is not a new one, and to those who do it on a daily basis, the hazards associated with it might be commonplace. However, to the occasional excavator, the risks might not be readily apparent. The permit-to-work system provides the vehicle that allows them to recognize potential problems and leads to the proper safeguards. The permit system is meant to serve as a guide to allow all involved to perform a proper risk assessment and to develop adequate safeguards.

A checklist can be developed that will assist those involved in the permitting process to discover hazards and determine safeguards. The checklist should include at least the following:

- the type of soil into which the excavation will be made
- the type of wall protection required (shoring, sloping, trench box)
- type of access (ladder, lift)
- utilities in the areas, such as underground water, sewer, electrical, stream, piping
- the depth of the excavation
- the location of the spoils
- protection of the public
- inspection frequency while work is being performed
- weather, such as rain, which might affect side stability
- positioning of any excavating equipment, such as trucks and backhoes
- stability of adjacent structures and their foundations
- ground water levels

Excavation can be a hazardous operation if not accomplished with the proper safeguards for the soil conditions. Therefore, the risk assessment should be undertaken with the input of a person who is trained, experienced, and competent in the safeguarding of excavations.

The permit-to-work system must require an approved permit for all excavations. For shallow or spot excavations, the permit might only be required to locate underground utilities or other underground installations with no other safeguards required. However, the location of these underground facilities is important enough to mandate the initiation of a permit. The permit, therefore, must lead the applicant to locate all underground installations such as sewer, water, fuel, and electrical lines before ground is broken, and any that are discovered shall be protected from damage or displacement. Utility companies shall be contacted to mark the actual locations of these installations. Contact numbers for "Dig Safe" or other similar programs should be available on the permit.

The permit must require that all excavation work within or near a public way, such as a road or pedestrian walkway, be safeguarded so as to protect both vehicular and pedestrian traffic. Any necessary barricades or diversion devices must be listed on the permit before approval.

The completed permit must also address all necessary shoring materials if the risk assessment of the excavation determines that it is needed. Alternative safeguards, such as sloping of the sides of the excavation or relying upon the consistency of the excavated material to prevent collapse, may be used. However, they must be stipulated on the work permit and ultimately approved by a qualified person. If the excavation process lasts for more than one day, or less because of weather conditions that might affect the collapse safeguarding technique, it must be inspected by a qualified person daily or after the weather incident.

The permit must require that the placement of excavated material be in a safe place. A diagram is often helpful to enable the approval authority to better understand exactly where the material will be placed. The approval authority must require that the permit be specific as to exact safeguarding and material placement. General, vague statements about safety must not be accepted.

If the excavation is to be left unattended or left overnight, the permit applicant must be required to detail what precautions will be taken to prevent inadvertent interaction with the excavation by pedestrians or vehicles. If mobile equipment is to be used adjacent to the excavation, the method of keeping it a safe distance from the sides of the excavation must be stipulated.

The permit must require that the number and means of exit and entry points out of and into the excavation be indicated. Once again, a diagram is extremely helpful. If ladders are required, they should be inspected before use.

If there is any possibility that the excavation might have either insufficient oxygen or hazardous vapors, a confined space permit to work must be applied for as well. The method of job completion, such as backfilling and compacting, must be detailed. Any pedestrian or vehicular pathways disturbed during the excavation must be returned to safe, usable condition.

Lastly, the permit must list all personnel involved in the excavation project so that the approval authority can determine whether or not they are properly trained. As with all work permits, the permit itself must be returned to the approval authority when the job is complete and kept on file.

If a contractor is hired to perform excavation work, the contractor must be required to obtain a permit to work so that the facilities manager can share information concerning the location of underground utilities and equipment, and so that work affecting a public or private travelway can be properly addressed.

Permit to Work–Electrical

Electrical work, on either high- (above 600V to ground) or low-voltage electrical systems that are nonroutine in nature, must be performed only with the permission acquired through the permit-to-work system. While the permit to work must be obtained to start work, and will dictate the location, time, and extent of the electrical work to be performed, all electrical work must be accomplished in conjunction with a sound lockout/tagout system that has been established in compliance with OSHA regulations (1910.147) and the ANSI Standard for the Control of Hazardous Energy – Lockout/Tagout and Alternative Methods (ANSI Z244). The control of hazardous electrical energy through a lockout or tagout system or other alternative method is provided through the establishment of ". . .requirements and performance objectives for procedures, techniques, designs and methods that protect personnel where injury can occur as a result of the unexpected release of hazardous energy." (ANSI 2003b, section 1.2)

Although unexpected energy release can come in many forms, electrical energy control activation is one of the most common hazards within any industrial setting. The primary method of electrical energy control is through a documented lockout/tagout program (ANSI 2003b, section 1.2). The establishment of a lockout/tagout program is addressed in the previous chapter of this text, "Applied Science an Engineering: Electrical Safety." This entails the systematic survey of an electrical control circuit to identify control devices and then demands the application of a means to secure the control device in the off position before any work is attempted on the system. The control device can be secured with a hasp and padlock for which only a designated worker has keys. If several workers are involved with the circuit, there may be several locks on the hasp. In the case of electrical work that requires a permit, the lockout/tagout procedure does not replace a permit to work, but might be required as part of the permitting process (see Figure 1).

Work on high-voltage electrical systems can be extremely hazardous, and requires specialized training and tools in order to be performed safely. Therefore, unless your facility employs workers with such specialized training you might want to limit the activities of your staff to work only on low-voltage electrical equipment and contract others to work on the high-voltage equipment.

Even though a contractor might be hired to work on high-voltage electrical systems, a permit to work should still be required so that the authorizing personnel can understand the scope of the job and how it might affect not only the safety of surrounding employees, but also other facility operations.

If the untrained facilities personnel are required to work within an electrical substation or in the area of high-voltage electrical systems or equipment, a permit to work must still be requested and approved so that the approving authority can have a clear understanding of the extent and location of the work. In this way the possible exposures can be determined by way of a risk assessment and, as a result, the proper safeguards applied. Also, the facility will want to know, through its approving authority, the effect of the work on the remainder of the facility.

The permit document should contain information on the location, date, and time of the job, and the duration of the job. The permit should be limited to one day for high-voltage work unless no changes in the work's influence on the remainder of the facility are expected. The permit should also detail the scope of the work, who will be responsible for the work, and who is performing the work.

Only properly trained workers can be allowed to work on electrical systems or equipment. The permit should detail any necessary safeguards in addition to the requirements of the facility lockout/tagout program. Some electrical work might take place over the period of more than one work shift. Therefore, a means of handing over safety responsibilities for both the permit to work and the lockout/tagout system must be devised and approved.

The permit to work is an application for permission to perform electrical work and will detail the exact safety procedures necessary to accomplish the risk without damaging person or property. In the case of electrical work, the permit to work might well encompass other permits to work, such as a confined space permit or the lockout/tagout system. Figure 1 can be used as a general permit for electrical work.

Permit to Work–Scaffolding

Erecting and working from scaffolding presents workers with risks that are not commonly encountered or well defined, and therefore requires a system that ensures worker safety. A permit-to-work system will accomplish this task by establishing standards for building scaffolding or similar work platforms, and by establishing a system of communication that will allow all involved with the scaffold to determine whether or not it is available for use.

Because a scaffold serves as an elevated work area, its proper erection is essential to the safety of the workers who use it. Thus, the ultimate approval of the permit demands that the person responsible for erecting the scaffold is a qualified person. This can mean that an outside party is hired to oversee this important function or, in the case of a contractor, the credentials of the contractor's qualified person are checked. This check is especially important if your employees will be working from a scaffold erected by someone over whom you have no control.

No matter who erects the scaffolding, the permit to work must stipulate that it be erected in compliance with OSHA regulations (1926, subpart L) and with reference to the ANSI standard Safety Requirements for Scaffolding (ANSI A10.8). Both the OSHA and ANSI standards require that the scaffolding be erected and inspected by a competent person who is ". . . capable of identifying existing and predictable hazards . . ." (ANSI 2001, sections 3.15, 4.44, 4.45; OSHA 1996b)

The competent person will need to maintain an inspection checklist that includes, at the very minimum:

- proper designer credentials
- proper placement of scaffolding tags
- proper placement of guard rails, mid rails and toe boards
- proper footing
- placement of safety netting if required
- secure scaffolding plank
- safe scaffold access (ladder, stairway, etc.)
- proper attachment to the building or other structure, if necessary
- proper distance of scaffolding platform from the building
- the scaffolding is not overloaded (consider weight of workers, supplies, machinery, materials on the scaffolding, and the scaffolding components)
- the use of only scaffold-grade wooden components
- the protection of nearby electrical conductors
- weather conditions, such as high wind, snow, and rain that might affect the safety of the workers
- inspection of any support ropes for abrasion and corrosion
- the stability of freestanding scaffold systems
- protection from nearby vehicles

Scaffolding presents a hazard both during its use and during its erection, modification, and dismantling. However, the permit to work can be used for each phase. The person who undertakes the task of erecting the scaffold must request permission to do so by way of the permit system, which will establish not only the standards to which it will be built, but also the safeguards necessary to build the structure of the scaffold safely. During the erection phase, the proper fall protection must be detailed on the permit. The workers involved must be listed so that the proper training requirements can be established. Therefore, training requirements must be established before the permit to work is requested and a list of authorized workers developed.

The permit-to-work document must list the date of erection, the projected date of erection completion, and the length of time that the scaffold will be available for use. The permit must also establish inspection frequency criteria for the scaffold. Usually, because scaffolding can be subjected to extreme work and weather conditions, a scaffold is inspected for damage or displacement at the beginning of each shift.

The permit-to-work system must also establish a means to notify potential users as to the exact state of the scaffolding, that is, whether it is safe to use, or still in the process of erection or dismantling. Such a communication system will also serve to inform potential scaffolding workers what type of work can be performed from the scaffold and any weight limitations applied. Appendix E of the ANSI A10.8 safety standard details a Scaffolding Tagging Program, in which various colored tags are placed at the entranceways to the scaffold that inform workers of the status (complete or incomplete) of the scaffold. Any system that communicates this vital information to workers can be used and implemented as a necessary portion of the approved scaffold permit to work.

Permit to Work–Laboratory

Educational institutions have had research and testing laboratories for a long time, but these are now appearing more and more in manufacturing facilities. Each laboratory, depending upon the work performed and the materials stored, presents its own, unique hazards. Maintenance workers or others who are asked to perform nonroutine tasks for which safety procedures are not already developed can be subjected to safety hazards that must be subjected to a risk assessment process before the task commences in order to discover and detail the best and safest method to perform the task. This is the purpose for which a permit-to-work system was created. The most likely qualified person to at least assist in implementing the laboratory permit-to-work system will work within the laboratory itself. He or she should be well versed in all aspects of the safe handling of the materials used and stored within the laboratory, and have direct access to up-to-date safety literature.

The permit to work is an application for permission to work within the laboratory. It must stipulate exactly which system within the laboratory will be affected and what, if any, stored material might be in the area. A permit to work should be required for even minor tasks (such as replacing light bulbs, general building maintenance, and window cleaning) that demand general access to the laboratory.

With the permit request, the worker who is asked to enter the laboratory will have access to information concerning laboratory activities that might, if disrupted, create an unsafe condition. This is also true for contractors hired to work within the laboratory. Therefore, they should request permission for entry to the laboratory through the permit-to-work system.

In order to obtain approval, the permit to work should detail all necessary safeguards as determined by a risk assessment. Many permits to work within a laboratory are accompanied by a hot work permit if flammable gases might be released. In addition, biological or radiological hazards might be encountered, and laboratory personnel are most probably the best source of reference for safeguarding measures.

As with all permits to work, location, date, and time of the work to be done should be listed, along with the duration of the permit. Most permits should expire after one day so that any changes within the laboratory can be assessed. The permit should list the qualified person within the laboratory, and provide in detail the safety measures that will be taken. The work permit should not be approved by the laboratory qualified person, but rather by someone who can review the safeguarding methods with an impartial viewpoint.

Any work performed on the waste water or waste disposal systems within the laboratory, up to the first external manhole from the laboratory, should be subjected to the critical review provided by the permit-to-work system so that a risk assessment of the potential hazards of accumulated hazardous materials within the system can be completed. The confined space permit system criteria will serve as a good resource for risk assessment, atmospheric testing, and personal protective equipment requirements, even if the waste system is not determined to be a confined space.

Permit to Work–Asbestos

"Asbestos is a problem because, as a toxic substance and a known carcinogen, it can cause several serious diseases in humans. Symptoms of these diseases typically develop over a period of years following asbestos exposure." (EPA 2005)

Because of the hazards outlined in this statement, all work with asbestos materials shall be performed in compliance with current federal and state regulations. All work involving the removal of asbestos-containing materials must be undertaken by contractors licensed for asbestos removal. However, some maintenance or facilities workers might come into contact with asbestos-containing materials while performing their daily activities. Such a scenario is the type of activity that should be addressed by a permit-to-work system.

The permit system will be of benefit only if all asbestos-containing material within the facility has been identified before the work has begun so that all those involved with the permit-to-work system will be able to adequately assess the risk. Once permission is requested to perform work and the material to be disturbed indicated, the approving authority should confirm, by referring to the previously compiled asbestos inventory, whether or not the material does contain asbestos. If the material is not on the list, and there is any question whatsoever that the material might contain asbestos, the material should be tested before any work is performed. If testing confirms that asbestos is present, it should then be added to the asbestos materials list for future reference.

Once the permit document is received by the approving authority, he or she should then determine whether the task can be performed by in-house staff or a contractor, because the ultimate goal is to prevent atmospheric contamination and the hazard it creates to both workers involved in the task, and to those who are not involved but who might be exposed.

The Occupational Safety and Health Act carries very specific regulations concerning work with asbestos (OSHA 1910.1001) and the permit approving authority will have to assess whether in-house workers are capable of performing the task safely. The facilities management will want to monitor the job carefully through the permit-to-work system, no matter who performs the work. To this end, the permit must accurately locate the work area and note the time and date of the work. The exact nature of the work to be performed should be provided in detail so the approval authority and the qualified person can assess the risk and extent of possible contamination. Safety measures should be listed in detail, including personal protective equipment required, confinement methods, air monitoring procedures, emergency procedures, decontamination, and waste disposal. Because of the risks presented by asbestos contamination, a similarly detailed permit to work must be presented and approved before any work is started by either a licensed contractor or in-house personnel.

A qualified person should be assigned to inspect the work area setup and its safety, control, and monitoring measures after those measures have been put in place, but before any work takes place. A qualified person should inspect the work area after the work is completed, but before any containment structure is removed, to make sure that the area is ready for release. At this time, atmospheric monitoring should indicate that the asbestos fiber count is within the preestablished limits.

References

American National Standards Institute (ANSI). 2001. ANSI/ASSE A10.8-2001. *Scaffolding Safety Requirements*. New York: ANSI.

______. 2003a. ANSI/ASSE Z117.1-2003. *Safety Requirements for Confined Spaces*. New York: ANSI.

______. 2003b. ANSI/ASSE Z244.1-2003. *Control of Hazardous Energy—Lockout/Tagout and Alternative Methods*. New York: ANSI.

Environmental Protection Agency (EPA). 2005. *Asbestos and Vermiculite* (retrieved November 1, 2007). www.epa.gov/asbestos/pubs/asbreg.html

Occupational Safety and Health Administration (OSHA). 1996a. OSHA 29 CFR 1910.147. *The Control of Hazardous Energy (Lockout/Tagout)* (retrieved November 1, 2007). www.osha.gov/pls/oshaweb/owadisp.show_document?p_id=9804&p_table=STANDARDS

______. 1996b. OSHA 29 CFR 1926.451 (a)(3). *General Requirements* (retrieved November 1, 2007). www.osha.gov/pls/oshaweb/owadisp.show_document?p_table=standards&p_id=10752

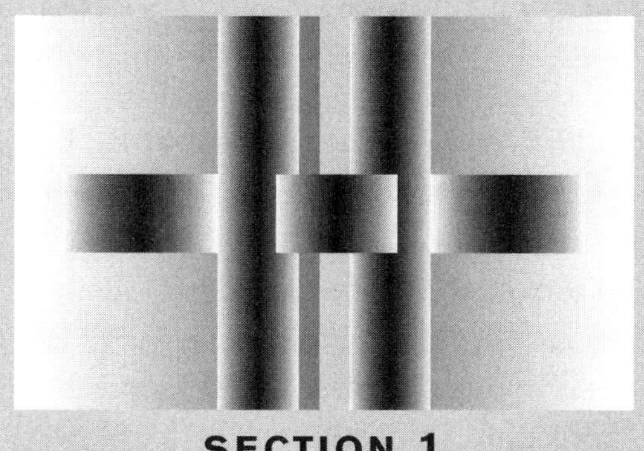

SECTION 1

RISK ASSESSMENT AND HAZARD CONTROL

Applied Science and Engineering: Basic Safety Engineering

John Mroszczyk

LEARNING OBJECTIVES

- Become familiar with *designing for safety*–the thought process involved in identifying hazards and using the design-for-safety hierarchy. From top priority to lowest priority, the hierarchy helps management eliminate the hazard or reduce the risk to an acceptable level by engineering design, providing a safety device, or providing warnings, instructions, and special operating procedures.
- Know the basic areas of safety engineering: noise control, walking-working surfaces, industrial/commercial ventilation, lockout/tagout, power tools, machine safeguarding, and material handling.
- Recognize the hazards involved in each of these basic areas and the role that designing for safety plays in each area.

A *HAZARD* IS the potential to cause injury or damage. *Risk* is the product of the probability of an occurrence and the severity of the injury or damage. A *safe* design is a design in which all hazards have been eliminated, or the associated risk reduced to an acceptable level. A *defective* design is one in which there is little or no attempt to eliminate or control hazards. *Designing for safety* is a formal process that incorporates hazard analysis at the beginning of the design. Safety methodology should be applied to the design of products, equipment, machines, facilities, buildings, and job tasks.

Designing for Safety: The Process

Once a preliminary design concept has been proposed, the first step in designing for safety is to identify the hazards. All available background information should be collected, reviewed, and analyzed. Important background information would include, for example, the foreseeable use and misuse, the environment, and the capabilities and behaviors of the user. The injury history can be useful in assessing what the uses and misuses are. In the workplace, the injury history can be obtained from injury records within the plant or related industries, from Occupational Safety and Health Administration (OSHA) statistics (www.osha.gov), or National Institute of Occupational Safety and Health (NIOSH) statistics (www.cdc.gov/niosh/). The injury history for a product can be obtained by contacting the National Injury Information Clearinghouse.

Information on human capability and behavior can be found in references on ergonomics and human factors, such as Woodson (1992), Hammer (1972), the National Safety Council (NSC 1988), or Plog (2002). Examples of human behavior as they relate to product and equipment design within the corresponding design approach (Woodson 1992) are:

1. Adults often avoid reading instructions. Products and equipment should be designed as nearly as possible to be understandable without reading special instructions.
2. Adults are absent-minded, easily distracted, or may be in a hurry. Products and equipment should have protective features in the event that the user forgets to press a button, turn a knob, follow prescribed steps, or is in a hurry.
3. Adults tend to believe that they are smart enough to take shortcuts. Potential hazards due to taking short cuts should be eliminated during the design.
4. Many adults are willing to take risks. Products and equipment should be designed so that they can only be operated in a safe manner.
5. Children are extremely curious. They investigate, examine, and try to play with many things.
6. Small children investigate by touching things, putting their fingers or hands into openings, and putting things into their mouths. Holes should be small enough in diameter so that a child's finger will not fit. Product finishes should be non-toxic. Products should be designed so that parts that are small enough to be swallowed do not come off.
7. Children like to climb or reach for things on surfaces that they cannot see. Products, such as stoves, should be designed so that children cannot reach the controls or the burners.
8. Children will always use a toy in a different manner. Toys should be strong enough that they do not come apart or break. There should be no sharp edges, corners, or projections.
9. Children are unsteady on their feet. Products, equipment, or other objects that will be used around children should be designed with the consideration that a child may fall on it. There should be no sharp edges, corners, projections, or breakable materials.

Industry standards and regulations also should be reviewed. Some of the organizations that write standards include the American National Standards Institute (ANSI), Underwriter's Laboratories (UL), the American Society for Testing and Materials (ASTM), the Society of Automotive Engineers (SAE), and the National Fire Protection Association (NFPA). However, it should be noted that standards generally have serious limitations that the designer needs to be aware of. There are a number of reasons why deficiencies may exist. A standard cannot address every possible design situation that may be encountered. The standard-writing process (negotiation and compromise) can severely weaken the safety content of a standard. There may be hazards that are not addressed, or a standard for a particular product or piece of equipment may not even exist. Federal regulations (OSHA, Environmental Protection Agency, Department of Transportation, and U.S. Consumer Product Safety Commission) and building codes are other sources of information.

Many hazards can be readily identified. *Kinematic* hazards occur whenever moving parts create cutting, pinching, or crushing points. Any mechanical component that rotates, reciprocates, or moves transversely should be examined for kinematic hazards. Any product or piece of equipment that vibrates or emits noise should be checked for high noise or vibration levels. Hot surfaces and hot substances should be checked to ensure that the temperature will not cause scalding or burn injuries. The uncontrolled release of energy from energy storage devices, such as springs, capacitors, and compressed gas storage cylinders, is hazardous. Electrical circuits, toxic materials, and flammable materials, are some other potential hazards.

There can be hidden hazards. One approach for identifying hidden hazards is a "what if" analysis. Using this method, a series of questions is posed focusing on each component, including manufacturing processes, materials, maintenance, wear and tear, operator error, and operator capabilities. Typical questions might be: "What if component A does not operate as intended?" "What if the user forgets to perform routine maintenance?" "What if the user does not follow the step-by-step instructions?" or "What if part X wears out?"

Once all the hazards have been identified, the first priority is to eliminate or reduce the risk to an acceptable level. Examples of eliminating a hazard by engineering design include using a ramp in place of a single step, placing a guard over critical switches to prevent an inadvertent activation, using an irregular bolt pattern so that a critical bracket cannot be installed upside down, and making components of a child's toy large enough so that they are not a choking hazard. If a design alternative does not eliminate the hazard or provide adequate risk reduction, then a safety device should be considered. Examples of safety devices include *dead man* controls on lawn mowers and snow throwers, guards on table saws, obstruction detection sensors on automatic garage door openers, and chain brakes on chain saws. A dead man-type control is a control that requires continuing operator contact for the machine to operate. The machine stops when the operator lets go of the control.

In some cases it is not possible to achieve adequate risk reduction by a design change or by providing a suitable safety device. Under these circumstances, warnings and/or written instructions should be provided. A warning can be either an audible or visual alarm, such as a backup alarm on a construction vehicle, or a warning label. Examples of residual and latent hazards that would require a warning label are not using a grinding wheel in excess of the rated speed, not using a flammable liquid near an open flame, or exposure to radiation hazards. A warning should never be used in place of an alternative design or safety device.

A warning label or sign should alert the user to the specific hazard, the seriousness of the hazard, the consequences of interaction with the hazard, and the ways to avoid the hazard. The ANSI Z535 series of standards sets forth a simple and straightforward format for designing warning labels and signs (ANSI 2002a-e). A warning sign or label should consist of a signal word panel, a message panel, and an optional symbol panel (see Figure 1). The label can be in a horizontal or a vertical format.

Colors should conform to ANSI Z535.1, *Safety Color Code* (ANSI 2002a). This standard establishes uniformity for colors used in safety signs, safety labels, and other areas where warnings are required. The specifications are in terms of the Munsell Notation System, a system of specifications based on hue, value (lightness), and chroma (IES 2003). The hue scale consists of 100 steps in a circle containing five principal and five intermediate hues. The value scale contains ten steps from black to white (0 to 10). The chroma scale contains steps from neutral gray to highly saturated. For example, standard safety red has a Munsell hue of 7.5R and a Munsell value/chroma of 4.0/14. Standard safety green has a Munsell hue of 7.5G and a Munsell value/chroma of 4.0/9. The Munsell notation for safety white is N9.0 and safety black is N1.5 (ANSI 2002a).

FIGURE 1. Warning label horizontal format (*Source:* ANSI 2002d)

The signal word panel contains the signal word. A safety alert symbol should also be included for personal injury hazards. The three-tiered hierarchy of signal words are DANGER, WARNING, and CAUTION. DANGER with the safety alert symbol indicates an imminently hazardous situation which, if not avoided, will result in death or serious injury. This signal word should be used only in the most extreme situations. The word DANGER shall be safety white on a safety red background. WARNING with the safety alert symbol indicates a potentially hazardous situation which, if not avoided, could result in death or serious injury. The word WARNING shall be safety black on a safety orange background. CAUTION with the safety alert symbol indicates a potentially hazardous situation which, if not avoided, may result in minor or moderate injury. The word CAUTION shall be safety black on a safety yellow background. CAUTION without the safety alert symbol indicates a potentially hazardous situation that may result in property damage.

The message panel should describe the hazard, the consequences of not avoiding the hazard, and ways to avoid the hazard. The lettering should be black on a white background or white on a black background. The message should be written in action sentences rather than passive ones (ANSI 2002d). For example, "Keep hands away from rotating blade" should be used rather than "Your hand must be kept away from rotating blade," or "Lockout power before servicing equipment" should be used instead of "Power must be locked out before servicing equipment." Use wording such as "Turn off power if jam occurs" instead of "Turn off power in the event a jam occurs."

The optional safety symbol is a graphic to convey the hazard without words. ANSI Z535.3 lists criteria for the design of safety symbols (ANSI 2002c). The user population for a proposed symbol should be carefully studied. A symbol that is to be used in a multi-lingual environment should be evaluated with a multilingual audience focus group. Figure 2 shows several safety symbols that have passed the ANSI Z535.3 criteria for acceptance: (a) an entanglement hazard—hand and gears, and (b) a person falling on a surface (ANSI 2002c).

A warning should be durable and placed in an area where it will be visible. The warning label should last as long as the product itself. A label that has faded or fallen off is of no use. Multilanguage labels may be required depending on the users.

In many instances, instructions telling the user how to use the product or machine will be required. It would be preferable to place the instructions directly on the product or machine. When this is not possible, an instruction/operation manual should accompany the product or equipment. The instruction manual should provide directions on the proper use, installation, assembly, and maintenance of the product or piece of equipment. Instructions should be clear, concise, and worded using positive language, which tells the user what to do. Installation/assembly instructions should note the proper installation/assembly sequence and be logical and easily understood. There should be warnings of the consequences of not following the proper installation or assembly procedure. The instruction manual should include a duplication of the warning labels that are on the product or piece of equipment.

FIGURE 2. Examples of acceptable safety symbols (*Source:* ANSI 2002c)

Administrative procedures, such as training and/or special operating procedures, should be implemented when warnings are not suitable. End users should develop and implement training programs based on the proper use, installation, assembly, and maintenance procedures outlined in the manual provided by the manufacturer. For example, forklift drivers need to be trained in the proper use of forklift trucks. Lockout/tagout is an example of a special operating procedure implemented by the end user when machinery or equipment is serviced.

Once a product or piece of equipment becomes operational, its use should be monitored so that the safety can be improved. This requires anticipating, identifying, and evaluating current hazards as well as new hazards, and monitoring field data. The analysis and investigation of incidents in the field can be used to improve current or future designs.

Walking and Working Surfaces

Falls are one of the leading causes of injury and death (Marpet and Sapienza 2003). Many factors may contribute to a fall besides walking-surface hazards. These include mental impairments, physical impairments, age, attentiveness, fatigue, and footwear. Engineers can design safety features into walking surfaces to accommodate humans, such as handrails, skid resistance, proper stairway tread and risers, ramps with an appropriate incline, and color contrast to provide depth of field.

An understanding of the biomechanics of walking provides insight into the causes of falls (Rosen 2000; Templer 1994). The normal gait cycle on a level surface starts with pushing off with the ball and then the big toe of the rear foot. The other leg swings forward while bent at the hip, knee, and ankle. The bending of the leg allows it to clear the ground. The swinging leg straightens at the knee as the heel strikes the ground. The body weight is momentarily supported by the ball/big toe of the rear foot and the heel of the forward foot. The weight then shifts forward as the point of contact on the forward foot shifts from the heel, to the arch, to the ball, then to the big toe. The forward foot then becomes the rear foot, pushing off as the other foot is now swung forward and the cycle repeats.

Walking occurs automatically. We do not consciously think about each foot's placement as we walk. People fall when their normal gait is suddenly interrupted. The body cannot adjust in time, their center of gravity moves outside of the base provided by their feet, and they fall. Typically this occurs when one foot slips or is blocked from moving forward.

Pedestrians generally slip and fall when the force applied to the floor by the feet during walking overcomes the resisting force at the floor/shoe interface. Under dry conditions the resisting force, or traction, is related to the *coefficient of friction* (COF) between heel and floor. The COF is defined as the force required to move one surface over another divided by the force pressing the two surfaces together. The traction provided by a floor surface can be affected by humidity, precipitation, oil, dust, dirt, debris, or any other foreign materials that may be on the floor.

There is some controversy as to whether the static or dynamic COF is most relevant in walking. The difference between the two can be illustrated by the classic high school physics experiment. Consider a block resting on a horizontal surface. The *static* coefficient is the horizontal force that must be applied to the block to start the block moving, divided by the weight of the block. The *dynamic* coefficient is the horizontal force that must be applied to keep the block moving, divided by the weight of the block (Templer 1994). The static coefficient is usually higher than the dynamic coefficient (Sotter 1995). The advocates for static friction argue that no motion is observed between the shoe and the floor during walking. Advocates for dynamic friction argue that the shoe does not stop moving during walking. In the United States, the *static coefficient of friction* (SCOF) is used to rate walking surfaces (Templer 1994; Sotter 1995).

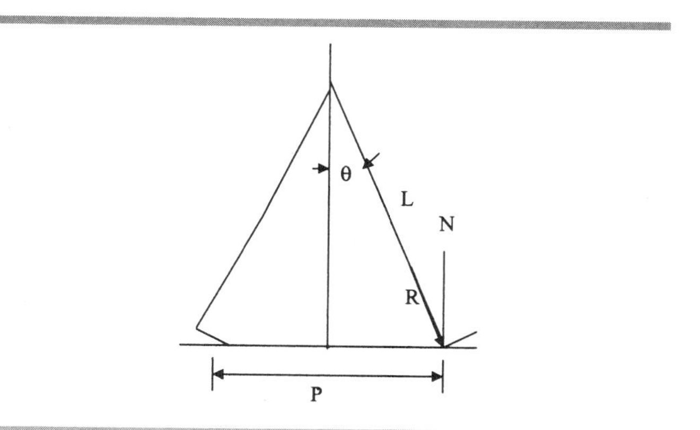

FIGURE 3. Walking on a level surface

A theoretical model (Templer 1994; Meserlian 1995) used for analyzing walking on a level surface is shown in Figure 3. The angle θ is the angle that the forward leg makes with the vertical. The force R is the force applied to the heel. The forces H and N are the components of the resultant force R and at an angle θ to the walking surface. The length L is the length of the leg from hip to heel and P is the pace (step length). The required SCOF, ($\mu = H/N$) is:

$$\mu = \frac{H}{N} = \frac{R \sin \theta}{R \cos \theta} = \tan \theta \text{ and } \sin \theta = \frac{P}{2L} \quad (1)$$

For example, a person with a leg length of 32 inches and a pace of about 21 inches:

$$\sin \theta = \frac{21}{(2)(32)} \text{ or } \theta = 19.16 \quad (2)$$

then

$$\mu = \tan 19.16 = 0.35.$$

The pace and leg length vary in the general population depending on sex and size of the person. The pace can also vary with walking speed. For example, tests done on human subjects (Meserlian 1995) found the range in calculated SCOF to be 0.30 to 0.46 as the walking speed was varied from very slow (69 steps per minute) to very fast (138 steps per minute).

Foreign substances such as liquid, grease, and oil on a walking surface can make an otherwise safe floor slippery. Usually, an unexpected wet, oily, or greasy spot on an otherwise safe floor creates the problem. As the heel touches down, the liquid or substance cannot move out of the way quickly enough. The heel is momentarily supported by the liquid film, not the floor's surface. This phenomenon (hydroplaning) is familiar to anybody who has ever driven too fast on a wet road (Sotter 1995). The presence of the intermediate material makes the SCOF of the bare floor surface irrelevant. Under these circumstances it is more appropriate to use the term *slip resistance* of the surface in place of SCOF. It should be noted that in some of the literature, the term SCOF is used interchangeably with slip resistance.

Slip resistance with foreign substances can be increased by imparting roughness to the surface. Typically this is done by grinding or sanding the surface, applying coatings that contain abrasive particles, or broom finishing in the case of concrete. The roughness should be deep enough and sharp enough to penetrate the film or contaminant. A small representative area should be tested before the entire floor is treated.

Building and fire codes generally require that floor surfaces be *slip resistant* without specifying any particular surface or a numerical value. Historically, a SCOF of 0.50 or higher has been considered adequate for most pedestrian safety (Kohr 1994). A more recent guideline, ANSI A1264.2, reaffirms the 0.50 threshold for safety (ANSI 2001). Since the traction requirement is the same, regardless of what the person is walking on, the 0.50 safety threshold is also used to evaluate surfaces where foreign substances may be present.

There are many factors that must be considered when selecting a surface material or maintaining a walking surface. There is no design standard that can control shoe material. Therefore, any recommendation for flooring material should assume that pedestrians may be wearing slippery shoes. Other factors to be considered include people walking fast, running, or turning, the age and physical capability of the users, and foreign substances. An office area on an upper floor could be assumed to be dry, while a ground floor entrance lobby is probably going to be wet during inclement weather. Other areas where foreign substances will likely be found on the floor are produce areas, cafeterias, repair garages, machine shops, and restaurants. Under some circumstances a slip resistance greater than 0.50 may be required.

TABLE 1

Coefficient of Friction for Selected Dry Materials

Material	Leather Shoe	Neolite Shoe
Brushed concrete, new, against grain	0.75	0.90
Asphalt tile, waxed, heavy use area	0.56	0.47
Smooth steel, rusted slightly	0 .54	0.49
Asphalt, old, in parking lot	0 .53	0.64
Steel checker plate, rusted moderately	0.50	0.64
Quarry tile, unglazed	0 .49	0.60
Thermoplastic, old, on crosswalk	0 .45	0.86
Brick pavers, new, on stair	0 .43	0.73
Exposed aggregate, pea gravel	0 .41	0.57
Granite stairs, old exterior	0.40	0.66
Plywood "A" side, unfinished, with grain	0.39	0.75
Plywood "A" side, unfinished, against grain	0.39	0.75

(*Source:* Templer 1994)

Table 1 lists the SCOF for a representative sample of flooring materials (Templer 1994). Note that many floor materials such as quarry tile, brick pavers, exposed aggregate, and granite, are unsafe with leather soles, using the 0.50 guideline.

Table 2 lists the slip resistance of some flooring materials (Templer 1994). In particular, linoleum, clay tiles, concrete, and terrazzo are unsafe when wet.

The control of slipping hazards may not end with the design and specification of a flooring material. For example, cleaning and maintenance can degrade the slip resistance of many floor materials. Inadequate rinsing of cleaning compounds, inappropriate floor finishes, wax residue, and spilled liquids can create an unsafe condition. Administrative measures (floor safety programs) may therefore be required in some commercial and workplace establishments to deal with these foreseeable events.

A floor safety program should include regular inspections. For some establishments it may be necessary to measure the slip resistance several times during the year. The slip resistance should be checked after the floor is cleaned and/or a new finish is applied. Supermarkets, in particular, should do periodic walk-arounds to check for foreign substances on the floor. The frequency of these inspections may need to be every 15 to 30 minutes, depending on the area. Employees should be trained to clean up foreign substances as soon as they are found. A warning sign or a warning cone should be placed in the area, particularly if an employee cannot remain to warn customers. Extra janitorial service should be employed during inclement weather.

Drag sleds are one class of devices for measuring the SCOF of dry surfaces. A drag-sled test device consists of a known weight, which is drawn across a surface. The force required to overcome the initial friction divided by the weight is the SCOF. ASTM F609-05, *Standard Test Method for Using a Horizontal Pull Slip Meter (HPS)*, standardizes drag sled devices (ASTM 2005). These devices should only be used on dry surfaces. The susceptibility to sticktion, operator-induced variables, and inability to mimic the dynamics of a human foot striking a surface severely limit the usefulness of drag sleds. Sticktion is the temporary bond created from water being squeezed out of the interface between the test foot and the surface. Sticktion produces unrealistically high readings on wet surfaces (DiPilla and Vidal 2002).

Articulated inclinable testers are another class of instruments for measuring surface traction. These devices use the striking angle at the onset of slipping to determine the slip resistance. The English XL uses a footpad that is thrust out onto the surface, just as the heel does during walking, by actuating a force cylinder. The force cylinder is attached to a mast. The angle of the mast is varied with an adjustment wheel. The inclination of the mast at the onset of slipping provides a direct reading of the slip resistance or slip index (the slip index is simply the tangent of the angle). The English XL can be used on dry and wet surfaces. ASTM Standard Test Method (F-1679) for Using a Variable Incident Tribometer (VIT) standardizes the use of this device.

The Brungraber Mark II is another articulated inclinable tester that is approved for dry and wet testing. This device uses a 10-pound weight on an inclinable frame. The angle of the frame is adjusted after each time the weight is released until a slip occurs. ASTM *Standard Test Method (F-1677) for Using a Portable Inclinable Articulated Strut Slip Tester (PIAST)* standardizes the use of this device. The English XL and the Brungraber Mark II produce comparable slip-resistance readings under both dry and wet conditions, provided the Mark II is used with a grooved test foot (Grieser et al. 2002).

A *ramp* is an inclined walking surface. Walking on a ramped surface is similar to walking on a flat surface. However, people tend to lean forward when ascending a ramp and backward when descending. Most building codes, including the Americans with Disability Act (ADA), limit the maximum slope to 1 in 12 (8.3 percent).

Figure 4 shows the same theoretical model as Figure 3 except the person is walking on a ramp with slope α. It is assumed that the pace P is the same as on a level surface. The SCOF ($\mu = H/N$) is:

$$\mu_{\text{ramp}} = \frac{H}{N} = \frac{R\sin(\alpha + \theta)}{R\cos(\alpha + \theta)} = \tan(\alpha + \theta) = \frac{\tan\alpha + \tan\theta}{1 - \tan a \tan\theta} \quad (3)$$

TABLE 2

Slip Resistance of Floor and Tread Finishes

Material	Dry, Unpolished	Wet
Clay tiles (Carborundum finish)	> 0.75	> 0.75
Carpet	> 0.75	0.40 to < 0.75
Clay tiles (textured)	> 0.75	0.40 to < 0.75
Cork tiles	> 0.75	0.40 to < 0.75
PVC (with nonslip granules)	> 0.75	0.40 to < 0.75
PVC	> 0.75	0.20 to < 0.40
Rubber (sheets or tiles)	> 0.75	< 0.20
Mastic asphalt	0.40 to < 0.75	0.40 to < 0.75
Vinyl asbestos tiles	0.40 to < 0.75	< 0.40
Linoleum	0.40 to < 0.75	0.20 to < 0.40
Concrete	0.40 to < 0.75	0.20 to < 0.40
Granolithic	0.40 to < 0.75	0.20 to < 0.40
Cast iron	0.40 to < 0.75	0.20 to < 0.40
Clay tiles	0.40 to < 0.75	0.20 to < 0.40
Terrazzo	0.40 to < 0.75	0.20 to < 0.40

(*Source:* Templer 1994)

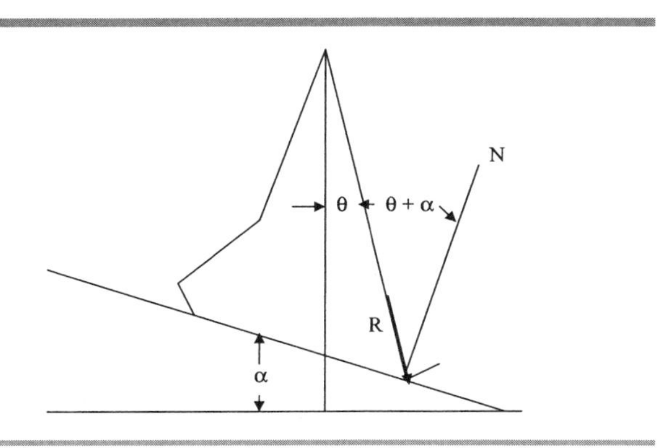

FIGURE 4. Walking on a sloped surface

The tan θ term is the SCOF for a flat walking surface, so that Equation 3 can be rewritten as:

$$\mu_{ramp} = \frac{\tan\alpha + \mu_{flat}}{1 - \mu_{flat}\tan\alpha} \qquad (4)$$

Equation 4 indicates that sloped surfaces require a higher SCOF than flat surfaces. Table 3 shows a tabulation using the mathematical expression for μ_{ramp}. The column on the left lists the SCOF on a level surface. The remaining numbers list the required SCOF necessary to maintain the same level surface SCOF for each slope.

It is interesting to note that for a ramp slope of 1 in 12 (1:12) a SCOF of 0.61 is required to maintain an equivalent 0.50 level surface SCOF.

A trip and fall usually occurs when one foot becomes caught or is blocked from moving forward.

TABLE 3

SCOF for Level and Ramped Walking Surfaces

Level	1:20	1:18	1:16	1:14	1:12	1:10	1:8	1:6	1:4
0.80	0.89	0.90	0.91	0.92	0.95	0.98	1.03	1.12	1.31
0.75	0.83	0.84	0.85	0.87	0.89	0.92	0.97	1.05	1.23
0.70	0.78	0.79	0.80	0.81	0.83	0.86	0.90	0.98	1.15
0.65	0.72	0.73	0.74	0.76	0.78	0.80	0.84	0.92	1.07
0.60	0.67	0.68	0.69	0.70	0.72	0.74	0.78	0.85	1.00
0.55	0.62	0.62	0.63	0.65	0.66	0.69	0.72	0.79	0.93
0.50	0.56	0.57	0.58	0.59	0.61	0.63	0.67	0.73	0.86
0.45	0.51	0.52	0.53	0.54	0.55	0.58	0.61	0.67	0.79
0.40	0.46	0.47	0.47	0.49	0.50	0.52	0.53	0.61	0.72
0.35	0.41	0.41	0.42	0.43	0.45	0.47	0.50	0.55	0.66
0.30	0.36	0.36	0.37	0.38	0.39	0.41	0.44	0.49	0.59

(*Source:* Templer 1994)

Recognized tripping hazards include extension cords, mats and runners, walking surface hardware, torn or loose carpeting, merchandise left on the floor, speed bumps, and wheel stops. Mats and runners should have beveled edges. The underside should have a suction design or be secured so that the mat or runner does not slide. Speed bumps should be clearly marked with a contrasting color according to ANSI Z535.1 (2002a). Wheel stops should also be marked with contrasting color and should not be placed in pedestrian walkways. Tripping hazards in the workplace include tools, pipes, debris, wood, and other objects that are left on walking/working surfaces.

Abrupt changes in elevation are also tripping hazards. Research on the effect of step height on gait dynamics (Gray 1990) established a maximum allowable step height of 6.7 mm (about 0.25 inch). Safety standards, such as ASTM F1637 (2002) limit a vertical change in elevation in a walking surface to less than 0.25 inch. Height changes between 0.25 and 0.50 inch should be beveled with a slope no greater than 1:2. Height changes greater than 0.50 inch should be transitioned by a ramped walking surface.

Low-level objects can also create a tripping hazard. These hazards are usually found in mercantile establishments. Customers may not see a low-level object in their immediate vicinity as they move about the sales floor. Examples of such tripping hazards are merchandise displays, pallets, baskets, boxes, bags, stools, pails, and flat bed trucks. Merchandise displays and any other objects placed on the sales floor should be at least 36 inches high so they will be readily visible.

The gait cycle on stairs is different from walking on level surfaces (Templer 1994). In descent, the lead foot swings forward and stops directly over the nosing. At the same time the heel of the rear foot lifts off the previous tread, the body is supported by the forward foot. The rear foot then begins to swing forward and the cycle is repeated. In ascent, the leading foot moves forward and is lowered on the tread, well forward of the nosing. The rear foot then rises tiptoe, lifting the body upward and forward. The rear leg is then raised and swings forward to the next step. The human mind automatically controls the foot placement on subsequent steps based on the initial step geometry. This results in walking up and down stairs without a lot of conscious thought.

Falls on stairways can be attributed to slippery treads, tripping hazards, poor visual contrast, different tread/riser dimensions, and poor geometry. Slip resistant surfaces are an important factor for safe stairs. The nosing area is most critical because this is the area that the foot contacts during descent. Surfaces with a low SCOF, such as smooth concrete, marble, terrazzo, ceramic tile, and polished surfaces, should be avoided. Abrasive strips, slip resistant nosings, or inlaid abrasive material are some of the measures that can be taken to improve the slip resistance of stair treads.

Anything on the stair treads that can snap or block a foot is a tripping hazard. Tripping hazards on stairways include loose or frayed carpeting and loose or raised leading edges. Raised nails or screws are also common tripping hazards on stairways. Carpeting should be tightly fitted. Stair treads should be flat and planar.

Poor visual contrast can also lead to falls on stairs, especially during descent. This situation usually arises with concrete stairs and carpeted stairs. The lack of visual contrast makes the stair treads appear to blend together. If the leading edge of the tread is not clearly defined, cognitive misinformation can lead to improper foot placement. Overstepping or understepping due to such misleading information can lead to falls. Color contrast provided at the leading edge and step illumination are some of the measures that can be taken to improve the delineation of stair treads. Step illumination can be small lights along the sidewalls or in the risers that illuminate the steps. Illuminated strips embedded in the step nosings can also be used to illuminate steps in dark environments.

Safe stairs are also related to tread-riser geometry (Templer 1994). Risers between 6.3 and 8.9 inches create fewer missteps during ascent. When descending, treads between 11.5 and 14.2 inches and risers between 4.6 and 7.2 inches result in the fewest missteps. OSHA (1910.24) permits tread-riser combinations that result in a rise angle between 30°35′ (approximately 6.5-inch riser/11-inch tread) and 49°54′ (approximately 9.5-inch riser/8-inch tread). Most commercial building codes and the NFPA 101, *Life Safety Code*, require risers between 4 and 7 inches and treads greater than 11 inches (NFPA 2006). The *International Residential Code* limits the maximum riser to 7.75 inches and the minimum tread depth to 10 inches (ICC 2006a). Local and state-adopted building codes should be referred to when designing or building stairs.

An important stairway safety feature is a handrail. A handrail offers support for the user and provides a means to arrest a fall, regardless of the cause of the fall. The required handrail height may vary with local building codes. The *International Residential Code* requires a handrail height of 34 to 38 inches (ICC 2006a). Commercial codes permit a handrail height between 34 and 38 inches. OSHA (1910.23) requires a handrail height of 30 to 34 inches. More than one handrail may be required depending on the width of the stairway. Most building codes require intermediate handrails so that all portions of a stairway width are within 30 inches of a handrail.

Handrail assemblies should be able to resist a single concentrated load of 200 pounds applied in any direction and have attachment devices and supporting structure to transfer this load to the appropriate structural elements of the building. Handrails should also be graspable. Circular cross-sections should have an outside diameter of at least 1.25 inches and not greater than 2 inches. If the handrail is not circular, it should have a perimeter dimension of at least 4 inches and not greater than 6.25 inches with a maximum cross-section of 2.25 inches (ICC 2006b).

Short flights of stairs, one or two risers, are extremely dangerous and should be avoided whenever possible. The reason is that the small change in elevation is so slight that it is not readily apparent. There are several safety measures that can be taken to reduce the risk of a fall for existing short-flight stairs or in cases where they cannot be avoided (NFPA 2006). The safety measures include prominent handrails, contrasting colors, step illumination, contrasting nosings, and warning signs. These measures make the steps more readily apparent to users; however, they do not eliminate the hazard.

Industrial/Commercial Ventilation

Industrial/commercial ventilation systems are mechanical systems designed to supply and exhaust air from an occupied space to reduce or eliminate airborne substances. Safety engineers design ventilation systems to control health hazards (cigarette smoke, dusts, toxic fumes, heat disorders), fire/explosion hazards (flammable vapors, dusts), and other hazards caused by airborne contaminants. Ventilation systems may also be used, for example, to prevent woodworking equipment from clogging by removing the sawdust from the cutting area.

Ventilation systems reduce the risk of a health or explosion hazard by lowering the concentration of the substance below maximum allowable limits. For example, the *threshold limit value* (TLV) of a substance is the concentration at which workers may be exposed on a regular basis without experiencing adverse health effects. OSHA 29 CFR 1910.1000 lists TLVs for many airborne contaminants. In the case of fire/explosion hazards, the lower explosion limit is the important threshold. The *lower explosive limit* (LEL) is the lower limit of flammability or explosibility of a gas or vapor at ambient temperature expressed as a percent of the gas or vapor in air by volume. The *NFPA Fire Protection Handbook* lists the fire hazard properties of many liquids, gases, and volatile solids (NFPA 2005). In the case of heat control, the amount of ventilation should be such that the net effect of body metabolism, external work performed, evaporative heat losses, convective heat exchanges, and radiative heat exchanges do not produce excessive heat gain, potentially resulting in adverse health problems. OSHA regulations, NFPA standards, ANSI standards, and local and state building and fire prevention codes should be referred to in order to assess the amount of ventilation that may be required in a particular situation.

Industrial/commercial ventilation systems consist of fans, blowers, ducts, air-cleaning devices, and hoods. Fans and blowers generate airflow by creating a pressure difference in the system. Air-cleaning devices include electrostatic precipitators, fabric collectors, dust collectors, and filters. Exhaust hoods provide an entry point to capture the air. Ducts transport the air between the inlet, the fan or blower, and the exhaust.

A well-designed ventilation system consists of supply and exhaust. The supply system provides makeup air to replace the air that is exhausted. In most cases the supply air is tempered to suit the space. The amount of supply air should be approximately equal to the amount of air that is exhausted. Air should be discharged outdoors at a point where it will not cause a nuisance and where it cannot be drawn in by a ventilating system. Air should not be discharged into an attic or crawl space. Air-intake openings should be located a minimum of 10 feet from any hazardous or noxious contaminant such as vents, chimneys, plumbing vents, streets, alleys, and loading docks (ICC 2006c).

There are two types of exhaust ventilation systems. *General ventilation* systems dilute the contaminated air with uncontaminated air. These systems are also known as *dilution ventilation* systems. General ventilation systems are most effective when the air contaminants are gases or vapors, are evenly dispersed, and do not pose a high health risk. These systems should be monitored carefully because the level of exposure is controlled by the amount of dilution air.

Local ventilation systems control the contaminants directly at the source. These systems are most effective when there are several large, fixed sources. In many circumstances, a local ventilation system is the only choice, as in the case of a contaminant source located directly in a worker's breathing area.

There are several basic principles of fluid mechanics that are used in the design of a ventilation system. The *static pressure* (SP) in a moving fluid is the pressure that would be exerted upon a surface moving with the velocity of the fluid. In ventilation ducts, the static pressure is the pressure that tends to burst (or collapse) the duct. The *velocity pressure* (VP) or *dynamic pressure* is the pressure that results from the kinetic energy of the moving fluid. The *total pressure* (TP) or *stagnation pressure* is the sum of the static pressure and the velocity pressure. In the design of ventilation systems, it is customary to express the pressure in inches of water (w.g.). The pressure under a 1-inch column of water is .036 psi. The relationship between the total pressure (TP), velocity pressure (VP), and static pressure (SP) is:

$$TP = SP + VP \quad (5)$$

where

$VP = 1/2\ \rho V^2$
ρ = mass density of the air
V = velocity of the air stream

The total pressure (TP), static pressure (SP), and velocity pressure (VP) can be illustrated in the manometer arrangement in Figure 5. In the first manometer, the total fluid pressure consists of the static pressure and the velocity pressure that impinges on the tube. The second manometer only reads the pressure acting on the duct wall, which is the static pressure. The third manometer reads the difference between the total pressure and the static pressure, which is the velocity pressure.

Another fluid mechanics principle is the conservation of mass, which requires that the flow that enters a duct must be the same as the flow that exits the duct. The principle of conversion of energy states that all the energy must be accounted for as the air goes from one point in the system to another. If "1" corresponds to some upstream condition and "2" corresponds to a downstream condition, the conservation of mass and energy can be stated mathematically (ACGIH 2004):

conservation of mass: $\rho_1 V_1 A_1 = \rho_2 V_2 A_2$ (6)

conservation of energy: $TP_1 = TP_2 + H_L$ (7)

or

$SP_1 + VP_1 = SP_2 + VP_2 + H_L$

where

A = duct area
H_L = energy losses (gains)
ρ = fluid mass density

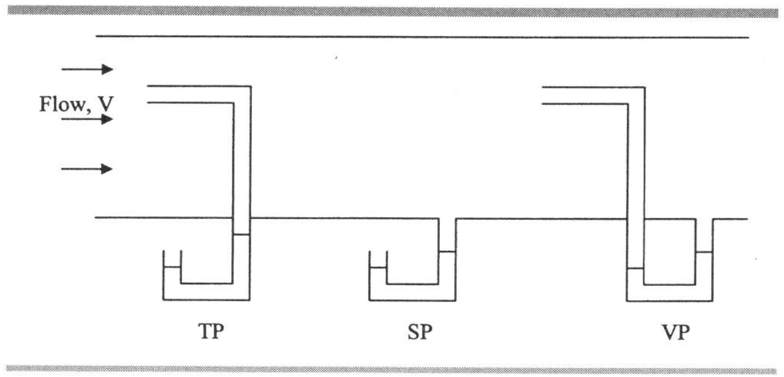

FIGURE 5. Total pressure (TP), static pressure (SP) and velocity pressure (VP) in a pressurized duct

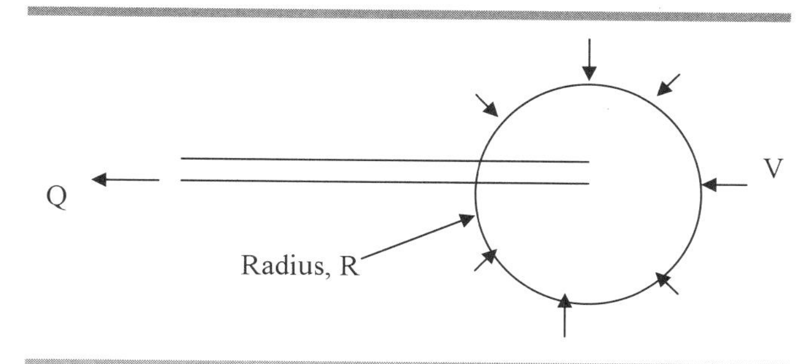

FIGURE 6. Point suction source

The energy losses, H_L, are usually expressed as the product of a loss coefficient and the velocity pressure. Loss coefficients depend on the particular loss mechanism. They can be found in a number of references (ACGIH 2004; Burton 1995) or may be provided by the equipment manufacturer.

The *capture velocity* is the minimum induced air velocity that is required to capture and convey the contaminant. This can be best illustrated by considering a point suction source as shown in Figure 6. Using the principle of conservation of mass, with constant density, the induced velocity over a sphere of radius R is:

$$Q = V(4\pi R^3) \quad (8)$$

where

Q = volume suction flow rate
V = velocity at radius R

In order to entrain a particular contaminant located at a radius R, the suction flow Q (cfm) must be equal to or greater than the capture velocity times the area of capture for that contaminant. Typical capture velocities are listed in Table 4. The velocity contour entering a typical hood is actually quite different from the point suction source depicted in Figure 6. Velocity contours for hoods are generally determined experimentally or provided by the vendor. Some velocity contours for basic hoods can be found in the literature (ACGIH 2004).

Once the contaminant has been drawn into the hood, the velocity in the duct must be such that the contaminant does not settle. The transport velocity is the minimum velocity that will move a particular

TABLE 4

Range of Capture Velocities

Dispersion of Contaminant	Contaminant	Capture Velocity (fpm)
Released with no velocity into quiet air	Evaporation from tanks, degreasing , etc.	50-100
Released at low velocity into moderately still air	Spray booths, intermittent container filling, low-speed conveyor transfers, welding, plating, pickling	100-200
Active generation into zone of rapid motion	Spray painting, barrel filling, conveyor loading, crushers	200-500
Released at high initial velocity into zone at very rapid motion	Grinding, abrasive blasting, tumbling	500-2000

(*Source:* ACGIH 2004)

TABLE 5

Typical Transport Velocities

Contaminant	Example	Transport Velocity (fpm)
Vapors, gases, smoke	Vapors, gases, smoke	1000-2000.
Fumes	Welding	2000-2500
Very fine light dusts	Cotton lint, wood flour	2500-3000
Dry dusts and powders	Rubber dust, leather shavings Bakelite molding power dust	3000-4000
Average industrial dusts	Grinding dust, limestone dust	3500-4000
Heavy dusts	Sawdust, metal turnings, lead dust	4000-4500
Heavy and moist	Moist cement dust, asbestos chunks	4500 and up

(*Source:* ACGIH 2004)

contaminant without settling. Typical values for the transport velocity are listed in Table 5.

The amount of ventilation required to reduce vapor concentrations is also important. The ventilation rate required to reduce steady-state vapor concentrations to acceptable levels can be calculated from Equation 9 (ACGIH 2004).

$$Q = \frac{G}{CK} \tag{9}$$

where

Q = ventilation rate (cfm)
G = vapor generation rate (cfm)
C = acceptable concentration
K = a factor to allow for incomplete mixing

The rate of vapor generation for liquid solvents is expressed as

$$Q = \frac{\text{CONV} \geq \text{SG} \geq \text{ER}}{\text{MV}} \tag{10}$$

where

CONV is the volume in cubic feet that 1 pint of liquid will occupy at STP (70° and 29.2 in. Hg) when vaporized

SG = specific gravity of the liquid
ER = evaporation rate of liquid
MV= molecular weight of liquid

The rate of vapor generation of a certain process can be calculated if the evaporation rate is known. The ventilation rate required to reduce the concentration can then be calculated.

In designing a ventilation system, the required ventilation rate in cfm may be specified by the equipment manufacturer (in the case of woodworking machines), building codes, the required capture velocity, the required transport velocity, or the vapor generation rate. Once the flow enters the duct, there will be losses in static pressure. The fan or blower must supply enough pressure difference to pull the air through the ventilation system.

Losses arise from duct friction, turbulence in elbows, contractions, entries, and expansions. There are also hood entry losses and equipment losses. The losses due to duct friction are a function of velocity, duct diameter, air density, air viscosity, and duct surface roughness. The Reynolds number (Re) and the relative roughness (e/D), both dimensionless quantities, characterize these parameters. The friction factor can be determined from the Moody diagram once the Reynolds number and the relative roughness have been determined. The Moody diagram is a plot of friction factor versus Reynolds number for various relative roughness values. The Moody diagram can be found in any basic text on fluid mechanics, ventilation design, or the *NFPA Fire Protection Handbook* chapter on hydraulics (NFPA 2005).

$$\text{Reynolds number: Re} = \frac{\rho V D}{\mu} \tag{11}$$

$$\text{Relative roughness} = \frac{e}{D} \tag{12}$$

where

e = absolute roughness
D = duct diameter (ft)
V = duct velocity (ft/sec)
ρ = air density (lbm/ft^3)
μ = air viscosity (lb-sec/ft^2)

Typical values for absolute roughness, e, are listed in Table 6. Once the friction factor has been determined from the Moody diagram, the duct friction loss, H_L, can be calculated from the following relationship:

$$H_L = f\left(\frac{L}{D}\right)VP \tag{13}$$

where

f = friction factor
D = duct diameter (ft)
L = duct length (ft)
VP = velocity pressure (psf)

There are many computer solutions, nomographs, and tabulations that make the computation of duct friction loss less tedious.

There are two components of losses associated with hoods. The first component occurs when the fluid is accelerated from still air into the hood. If "1" in the conservation of energy equation is still air, the static pressure SP_1 and the velocity pressure VP_1 are zero (or close to zero), so that $SP_2 = -VP_2$. This says that the static pressure drop due to acceleration of the air into the hood is equal to the velocity pressure.

The other component of losses associated with hoods is due to flow separation in the hood, which causes the flow to contract through a smaller diameter. Entry loss factors have been determined and are tabulated in a number of references (ACGIH or ASHRAE) or can be obtained from the manufacturer. For example, entry loss factors (H_h) can range from 0.93 for a plain inlet to 0.04 for a bellmouth inlet. The total loss associated with a hood, SP_{hood}, is

$$SP_{hood} = VP + (H_h \times VP) \tag{14}$$

where

VP = velocity pressure (psf)
H_h = hood entry loss factor

TABLE 6

Absolute Roughness

Duct Material	Surface Roughness (ft)
Galvanized metal	0.0005
Black iron	0.00015
Aluminum	0.00015
Stainless Steel	0.00015
Flexible duct (wire exposed)	0.01
Flexible duct (wires covered)	0.003

(*Source:* ACGIH 2004)

Most ventilation systems are complex, with many hoods and branches connecting to one main trunk line. Each branch in the system must be balanced to achieve the desired flow. The total flow entering the main trunk line from the branches must be accurate so that the fan or blower can be sized properly.

There are two methods for designing a ventilation system. One method is the blast-gate method, which relies on the adjustment of blast gates after the system is installed to balance the flow in each branch. The main drawback to the blast-gate method is that loss factors can vary greatly depending on the degree to which the blast-gate is closed. Also, in the case of dusts, for example, particles can be trapped by the blast-gate, building up to the point that the duct becomes clogged.

The second method, the static pressure balance method, starts with the entry requirements at the farthest hood. The duct is sized to achieve the required transport velocity. The velocity pressure usually does not change significantly within the ventilation system. Under this condition, the energy equation states that losses will appear as a reduction in static pressure. The static pressure drop is then calculated based on all the loss mechanisms in the system. At each junction, the branches are balanced analytically by redesigning each branch entering a junction so that the static pressure is the same at the point of entry to the junction. Physics states that the air flow distributes itself based on the losses in each branch; the static pressure at a junction can have only one value.

A typical junction is depicted in Figure 7. The static pressure loss for branch "1" is SP_1 entering the

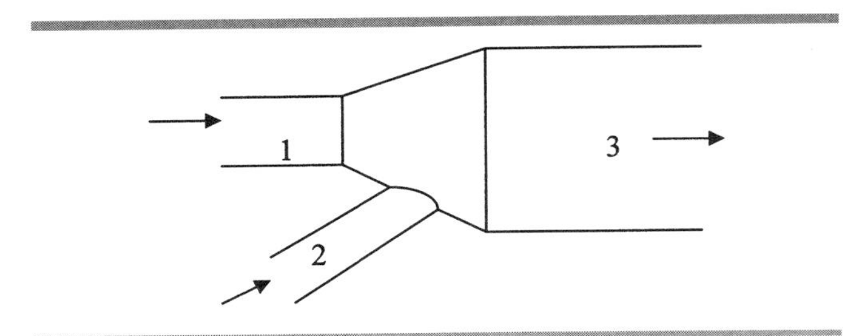

FIGURE 7. Balancing of two branches at a junction, $Q_3 = Q_1 + Q_2$

junction and the static pressure loss for branch "2" is SP_2. Both branches enter the junction and mix together in branch "3", $Q_3 = Q_1 + Q_2$. In the design process, the branch with the lower static pressure must be redesigned by adjusting duct size, elbows, flow, or other parameters until the static pressure drop matches that of the higher branch. This iterative procedure can be systematically done by adjusting the components in that branch using a corrected flow for that branch given by

$$Q_{\text{corrected}} = Q_{\text{design}} = \sqrt{\frac{SP_{\text{governing}}}{SP_{\text{duct}}}} \qquad (15)$$

where

$SP_{\text{governing}}$ = desired SP at the junction

SP_{duct} = original calculated SP for the duct being redesigned

Q_{design} = original calculated flow for the duct

This design method proceeds branch by branch, junction by junction, up to the fan.

Figure 8 depicts a hypothetical junction ventilation system for a welding operation. The BC branch consists of a tapered hood and 25 feet of flexible duct. The exhaust rate into the tapered hood is chosen as 1000 cfm for adequate capture velocity. The AC branch consists of a plain duct end, a 90° elbow, and 25 feet of galvanized duct. The exhaust rate into the plain duct end is chosen as 1335 cfm for adequate capture velocity. The BC and AC branches meet at junction C. At C both branches combine into 40 feet of galvanized duct. The minimum transport velocity for welding fumes is 2500 fpm from Table 5.

An 8-inch diameter duct is chosen for the BC branch. The actual duct velocity for a 1000 cfm exhaust rate is 2865 fpm, which is above the minimum 2500 fpm transport velocity. Using properties for standard air, the velocity pressure is 0.512 inches. The energy loss coefficients for the duct entry and branch entry are obtained from references (ACGIH 2004). The duct friction calculation will use the Moody diagram directly for the purposes of illustration. In practice there are many useful approximations, which make this computation less tedious.

The absolute roughness for flexible duct with wires covered is 0.003 feet from Table 6, so e/D is 0.0045. The Reynolds number is 202,948. The f value of 0.030 is read off the Moody diagram for an e/D of 0.0045 and a Reynolds number of 202,948. The friction loss (per VP), equal to $f(L/D)$, is 1.15. The total loss (per VP) for the BC branch is the summation of 0.25, 1.0, 0.28, and 1.15, or 2.68. The total energy loss is then 2.68 times the velocity pressure, or 1.372 inches (see Table 7). The same calculation procedure is applied to the AC branch.

The BC and AC branches meet at junction C. In reality, the static pressure at C must be the same for both branches. The AC branch in this case governs, since it has a higher static pressure than BC. The flow in the BC branch is corrected as $1000\sqrt{(1.676/1.372)}$, or 1104 cfm. The static pressure is then calculated for the BC branch using a 1104 cfm exhaust rate. The resultant total energy loss, 1.669 inches, is very close to the AC value of 1.676 inches, so no further adjustment is necessary.

Both branches join at C. Using conservation of mass, the total flow in the CD branch is 2439 cfm. The static pressure calculation for the CD branch is as follows. The cumulative pressure loss at the fan is

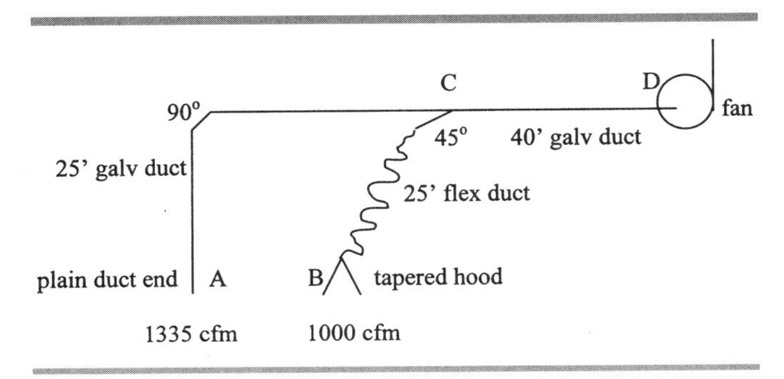

FIGURE 8. Local exhaust junctioned system example

TABLE 7

Static Pressure Calculation, Example Figure 8

	BC	AC	BC redesign	CD
Volumetric flow rate (cfm)	1000	1335	1104	2439
Minimum transport velocity (fpm)	2500	2500	2500	2500
Duct diameter (in.)	8	9	8	12
Duct area (sq ft.)	0.349	0 .441	0.349	0.785
Actual duct velocity (fpm)	2865	3027	3162	3107
Duct velocity pressure (VP) (in. w.g.)	0.512	0.571	0.623	0.601
Energy loss coefficients:				
Duct entry (ACGIH, Fig. 5–12)	0.25	0.93	0.25	--
Acceleration	1.0	1.0	1.0	--
90° elbows (ACGIH, Fig. 5–13)	--	0.33	--	--
Branch entry (ACGIH, Fig. 5–14)	0.28	--	0.28	--
Duct friction				
e/D	(0.0045)	(0.00066)	(0.0045)	(0.0005)
Re = $\rho VD/\mu$	(202,948)	(241,227)	(223,986)	(330,136)
f (Moody Diagram)	(.030)	(.020)	(.030)	(.018)
f(*L*/*D*)	1.15	0.675	1.15	0.754
Total:	2.68	2.94	2.68	0.754
Total Energy Loss (in. w.g.)	1.372	1.676	1.669	0.453
Governing?	No	Yes	--	--
Cumulative Energy Loss (in. w.g.)	--	--	--	2.112

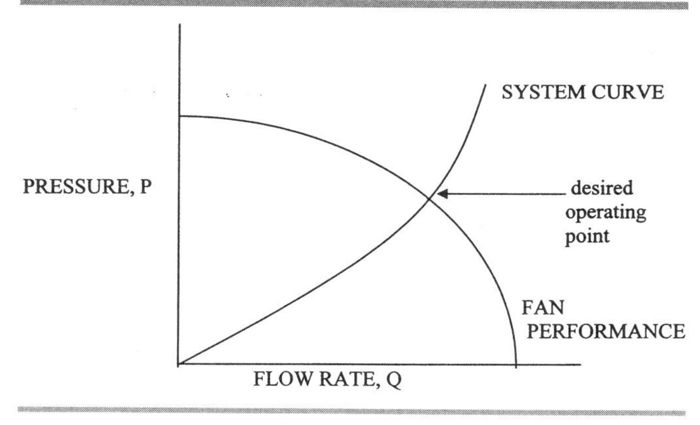

FIGURE 9. Intersection of system curve and fan performance curve

the sum of the static pressures at C (1.669 inches) plus the additional loss in the CD branch (.453 inches), or 2.112 inches.

The design of the ventilation system yields one flow (*Q*) and one static pressure. This point—the desired system operating point—can be plotted on a pressure versus flow graph (see Figure 9). Using the example in Figure 8, the desired operating point would be 2439 cfm at 2.112 inches static pressure. A system curve can be generated by plotting the desired system operating point with other static pressure/flow system values. The curve will be quadratic, since the change in static pressures is proportional to the square of the change in flow.

The fan performance curve, usually obtained from the manufacturer, is then overlaid on the system curve. There may be a number of performance curves corresponding to fan speed, efficiency, horsepower, and diameter. The resulting operating point will be where the system curve and the performance curve intersect. The best fan or blower will be the one that results in the desired operating point. There may be tradeoffs. The most efficient operating points are usually on the flat part of the fan performance curve. The most stable operating points are on the steep part of the curve, where large changes in static pressure result in small changes in flow.

The following fan laws can be used to estimate system changes

$$Q_2/Q_1 = (\text{RPM}_2/\text{RPM}_1) \quad (16)$$

$$Q_2/Q_1 = (\text{SIZE}_2/\text{SIZE}_1)^3(\text{RPM}_2/\text{RPM}_1) \quad (17)$$

$$\text{SP}_2/\text{SP}_1 = (\text{RPM}_2/\text{RPM}_1)^2 \quad (18)$$

$$\text{PWR}_2/\text{PWR}_1 = (\text{RPM}_2/\text{RPM}_1)^3 \quad (19)$$

where

Q = flow
RPM = fan RPM
SP = static pressure
PWR = power
SIZE = size of the fan

The subscripts refer to conditions before and after the change.

Noise Control

Noise is unwanted sound, such as the sound produced by a honking horn, machinery, or a jack-hammer. *Sound* is the pressure wave that travels in air, giving rise to the sensation of hearing. A pressure wave is created when a mechanical disturbance causes air molecules to move one way and then back. This produces areas where the air molecules are spread apart and areas where they are compressed together. A sound wave has wavelength λ, velocity c, and frequency f. The velocity of sound at 70° F is 1128 feet per second. The wavelength, velocity, and frequency are related by $c = f\lambda$.

Noise generally contains many sound waves with different frequencies, amplitudes, and durations that make it difficult to measure. A more useful quantity is the time average or mean squared pressure for approximating the magnitude of the sound. It is customary to report sound pressure levels, Lp, as a quantity that varies linearly as the logarithm of the mean squared pressure. This quantity, having units of *decibels* (dB), is defined as

$$Lp = 10 \log \left(\frac{p^2_{\text{ave}}}{p^2_{\text{ref}}} \right) \tag{20}$$

The p^2_{ave} quantity is the mean squared sound pressure and can be that of the total sound pressure level, one frequency component, or a band of frequencies. The reference pressure p_{ref} is 20 micropascals, the threshold of human hearing. Sound pressure levels for some typical sources (Pierce 1981) are listed in Table 8.

Sound is produced by changes in force, pressure, or velocity. For example, there is no sound if you stand still on a hardwood floor because the force acting on the floor is constant. But if you jump up and down, the force acting on the floor changes and sound (noise) is produced. A balloon full of air does not produce sound because the pressure is constant. If the air is released, then pressure changes and noise are produced. An ordinary house fan produces noise because the air flowing through the fan undergoes rapid changes in velocity.

In the workplace, noise is produced by any number of machines. Any piece of equipment with moving parts such as belt drives, gear boxes, and motors produce noise, especially if they cause a nearby panel to vibrate and radiate sound. Production machinery, including punch presses, saws, grinders, planers, and mills, produces noise. Processes, which release high-pressure fluids or compressed air, will produce noise. The noise exposure can be continuous, intermittent, or from a sudden impact.

TABLE 8

Typical Sound Pressure Levels

Source	Level (dB)
Jet engine	140
Threshold of pain	130
Rock concert	120
Accelerating motorcycle	110
Pneumatic hammer	100
Noisy factory	90
Vacuum cleaner	80
Busy traffic	70
Two-person conversation	60
Quiet restaurant	50
Residential area at night	40
Empty movie house	30
Rustling of leaves	20
Human breathing	10
Hearing threshold	0

(*Source:* Pierce 1981)

The human ear is a complex device that permits us to hear sound. The outer ear acts like a funnel to direct sound pressure waves traveling through the air to the eardrum. The eardrum is simply a membrane that vibrates in response to the pressure waves. The mechanical vibration is transmitted to the middle ear. The middle ear consists of three tiny bones—the hammer, the anvil, and the stirrup. These bones transmit mechanical vibration to the fluid-filled inner ear.

The inner ear includes three semicircular canals and the cochlea. The mechanical vibration input from the stirrup is transmitted to fluid vibrations in the inner ear. The cochlea is lined with thousands of hair cells, which are tuned to different frequencies. As the fluid vibrations excite the appropriate hair cells, nerve endings in the cochlea send electrical impulses along the auditory nerve to the brain.

The human ear can detect sounds from frequencies of 20 Hz to 20,000 Hz. It is most sensitive in the middle and high frequencies. The human ear cannot perceive a change in sound level less than 3 dB (Woodson 1992). A change of 5 dB is noticeable. A change of 10 dB is perceived as twice as loud (or soft). A sound is perceived as much louder (or softer) if the difference is 20 dB.

Long-term exposure to hazardous noise levels will cause permanent changes in the cochlea, resulting in permanent noise-induced hearing loss. This usually starts with a reduction in hearing at 4000 Hz and may extend to the speech frequency range with continued exposure. The onset and extent of hearing loss depends on the level and the exposure duration. It is pervasive and painless, except in extreme situations, such as an explosion.

A hand-held sound-level meter or a noise dosimeter is adequate for assessing most occupational and environmental noise exposures. A frequency analyzer or other instruments may by used when more detailed measurements are needed, such as determining the exact frequency content of the noise emitted by a particular machine. A sound level meter is suitable for a fixed location and when the noise levels are continuous. A noise dosimeter is more appropriate when the noise level varies or when the employee moves around.

A sound-level meter is a hand-held instrument consisting of a microphone, a frequency selective amplifier, and an indicator. Most sound-level meters are equipped to make A-weighted, B-weighted, and C-weighted measurements (see Figure 10). A-weighed measurements (dBA) correct direct noise to match how the human ear hears sound. The human ear is less sensitive at low audio frequencies. Figure 10 shows the A-weighted correction that should be made to an unweighted sound-level measurement as a function of frequency. Other scales, such as B-weighting and C-weighting, emphasize the low frequencies. The main difference between sound-level meters is the accuracy. ANSI S1.4 specifies three accuracy levels for sound-level meters: precision (Type 1), general purpose (Type 2), and survey (Type 3) (ANSI 2006b). OSHA recommends that either Type 1 or Type 2 be used for OSHA noise surveys.

The choice of the meter response depends on the noise being measured, the use of the measurements, and applicable standards. The response of sound-level meters is generally based on either FAST or SLOW averaging. The FAST response corresponds to a 125-millisecond time constant, while the SLOW corresponds to a 1-second time constant. That is, the meter will reach 63 percent of the final steady state value within 1 second.

The employer's legal obligation for occupational noise exposure is specified in 29 CFR 1910.95 (General Industry) and 29 CFR 1926.52 (Construction). There are two action levels that are required by OSHA, corresponding to noise doses of 50 percent and 100 percent. The first action level requires that a hearing conservation program be implemented whenever the 8-hour continuous exposure exceeds 85 dBA (a dose of 50 percent). The second action level requires administrative or engineering controls if the 8-hour continuous exposure exceeds 90 dBA (a dose of 100 percent). If controls are not feasible, then personal protective equipment shall be provided. Appendix B of 29 CFR 1910.95 describes the methodology for determining the adequacy of hearing protection. The Noise Reduction Rating (NRR) is usually written on the hearing protector package. For example, when an employee's A-weighted time-weighted average (TWA) is obtained using a sound level meter, subtract 7 dB from the NRR, and then subtract the remainder from the A-weighted TWA to obtain the estimated A-weighted TWA with the ear protector.

The permissible noise exposures for other sound levels are listed in Table G-16 of 29 CFR 1910.95 and

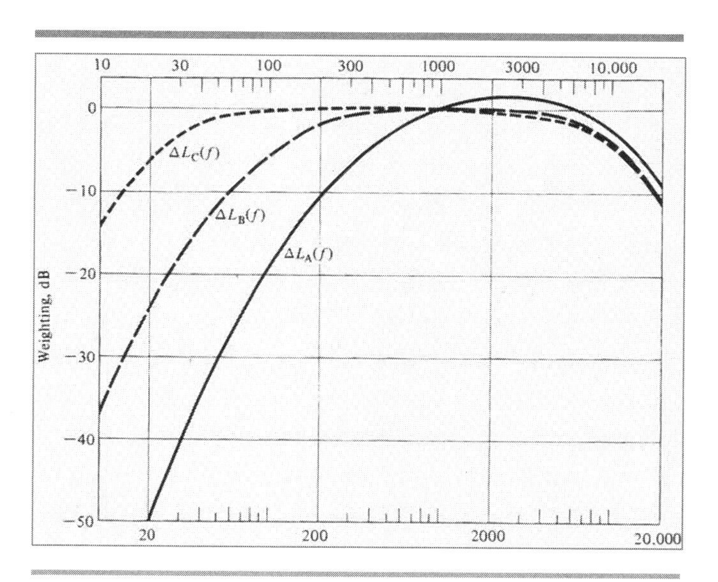

FIGURE 10. A, B, and C sound weighting versus frequency

Table D-2 of 29 CFR 1926.52. These values are replicated in Table 9. When the noise exposure is comprised of more than one period of noise and difference levels, the noise dose is given by

$$D = 100(C_1/T_1 + C_2/T_2 + C_3/T_3 + \ldots C_n/T_n) \quad (21)$$

where C_n is the exposure time at a specific noise level and T_n is the reference permissible duration for that level. The equivalent continuous 8-hour exposure, the 8-hour *time-weighted average* (TWA) noise level, in decibels is

$$\text{TWA} = 16.61 \log10 (D/100) + 90 \quad (22)$$

For example, if a worker operates a machine for 5 hours at 92 dBA, is exposed to 95 dBA for 2 hours, and has 1 hour in an area where the noise level is 75 dBA. The noise dose is

$$D = 100[0 + (5/6) + (2/4)] = 133.33\% \quad (23)$$

$$\text{TWA} = 16.61 \log10 (133.33/100) + 90 = 106.7 \text{ dB} \quad (24)$$

If the worker operates a machine for 5 hours at 85 dBA, is exposed to 92 dBA for 3 hours, and has 1 hour quiet time, the noise dose is

$$D = 100[0 + (3/6) + 0] = 50.0\% \quad (25)$$

$$\text{TWA} = 16.61 \log 10 (50.0/100) + 90 = 85.0 \text{ dB} \quad (26)$$

In the first example, the noise exposure exceeds a TWA of 90 dBA or 100 percent and requires administrative controls, engineering controls, or personal protective equipment. A hearing conservation program would also be required if the engineering controls did not lower the TWA to below 85 dBA. The second example requires a hearing conservation program.

A noise survey should be done to determine the employee 8-hour exposure. The instruments used to measure the noise levels should be regularly calibrated. The survey should begin with a preliminary survey. The purpose of the preliminary survey is to identify areas where hazardous noise levels may exist and to determine if special equipment is needed. For example, measurements might be taken at the center of each work area. If the maximum level does not exceed 80 dBA, then detailed measurements in this area may not be necessary. However, if the maximum noise levels exceed 80 dBA, more detailed measurements should be taken. Areas where it is difficult to communicate with a normal voice level or where workers complain of ringing in their ears should be noted for further investigation.

TABLE 9

Permissible Noise Exposures

Duration per Day (hrs)	Sound Level (dBA) Slow Response
8	90
6	92
4	95
3	97
2	100
1.5	102
1	105
0.5	110
0.25 or less	115

Note: Exposure to impact noise should not exceed 140 dB peak

A detailed survey should then be done to determine the noise levels that exist at each workstation, areas employees may occupy, and any hazardous areas that were identified in the preliminary survey. These results will then be used to define the guidelines for engineering controls, to identify areas where hearing protection is required, and to identify areas where audiometric testing may be required. The survey should be repeated on one or more days to account for potential day-to-day variations. The employer shall notify each employee of the results of the monitoring and the employee (or representative) shall have the opportunity to observe the noise measurements.

For accurate measurements, the sound-level meter should be placed 6 inches to 1 foot from the operator's ear. Care should be taken to ensure that the sound field is not altered. There should be no objects, including the operator, between the noise source and the meter. Objects that could reflect the sound should also be avoided (Thuman and Miller 1986).

If a particular job requires an employee to move in more than one area during the day, or if the noise levels vary above and below 85 dBA, it may be desirable to have the employee wear a dosimeter. A noise dosimeter is basically a sound-level meter with storage and/or computational capability and a clip-on microphone. The instrument stores the readings during the exposure time and computes the percent dose or TWA. The placement location can significantly influence the measurement. For example, a microphone placed on the right side of the body will not accurately measure the noise level from a machine located on the left side of the body. Other factors that affect the accuracy include the type of sound field, the angle of incidence of the sound, the frequency of the sound, and the sound absorption of the person's clothing. For most industrial applications, the microphone should be mounted on the shoulder, collar, hat, or helmet. Dosimeter measurements should be checked with previous measurements to rule out employee sabotage (Harris 1991).

A hearing conservation program should be implemented whenever the employee 8-hour, time-weighted average sound level (TWA) equals or exceeds 85 dBA. The hearing conservation program should include monitoring, employee notification, observation of monitoring, audiometric testing, hearing protection, training, and record keeping. Employees shall be provided with hearing protection when the 8-hour, time-weighted average exceeds 85 dBA.

A baseline audiogram should be taken within 6 months of an employee's first exposure. Audiograms should be done annually thereafter. If there is a 10 dB shift at 2000 Hz, 3000 Hz, or 4000 Hz between the annual audiogram and the baseline, then follow-up steps need to be taken. These steps may include fitting the employee with hearing protection, upgrading the hearing protection, and a clinical audiological evaluation.

The employee training program should include the effects of noise on hearing, the purpose of hearing protection, and the purpose of audiometric testing. The training should be repeated annually. The employer should maintain records of the exposure measurements and audiometric testing. Noise-exposure measurement records should be retained for two years. Audiometric test records should be retained for the duration of the employee's employment.

If the 8-hour continuous exposure exceeds 100 percent, then the employer should implement administrative or engineering controls. If these controls fail to reduce sound levels, then personal protective equipment should be provided for the employees.

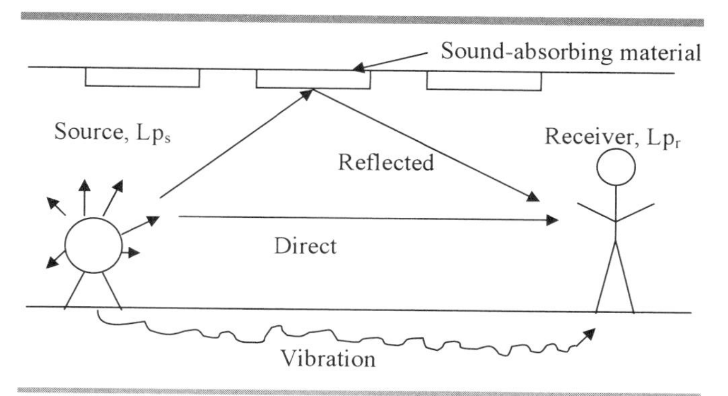

FIGURE 11. Noise source, transmission path, and receiver in a room

In designing for safety, the first priority in controlling hazardous noise is elimination or isolation of the noise by engineering design. There are three areas where noise control can be effective: at the source of the noise, in the transmission path between the source and the receiver, and at the receiver (see Figure 11). The transmission path includes direct sound, reflected sound, and vibration.

Engineering controls that can be taken at the source include reducing the noise output of the source and/or enclosing the source. Reducing the noise output requires an understanding of how the source produces noise (changes in force, pressure, or velocity). Information about the frequency content can also be very useful. For example, any rotating component will produce noise at its rotation rate or multiples of the rotation rate. Noise that has a strong frequency content at 100 Hz could be caused by a bent shaft rotating at 6000 rpm. A 36-tooth gear rotating at 1500 rpm can generate noise at a frequency of $36 \times 1500/60 = 900$ Hz. Corrective measures that can be taken to reduce noise generated by rotating equipment include balancing, replacing bent shafts, replacing worn bearings, lubricating moving parts, or changing the rotation rate.

Machine covers are sometimes made with thin sheet-metal panels that can vibrate and radiate sound. This usually occurs when the driving frequency is close to the natural frequency of the panel. Increasing the thickness of the panel or adding stiffening ribs will change the natural frequency and thereby reduce the noise. Sheet-damping material can be added to dampen the vibration. These sheets are made of visco-elastic material with a self-adhesive backing and can be attached directly to the panel.

The application of engineering controls to reduce the noise output at the source is illustrated in the following examples (Knowles 2003):

1. A control panel mounted to a hydraulic system is determined to be the source of radiated noise. The noise level is reduced by detaching the control panel from the system.
2. In another case, a solid metal cover on a belt drive is found to be radiating noise. A wire mesh cover reduces the vibration so that there is less radiated noise.
3. A urethane rubber coating is clamped to a circular saw blade to reduce noise.
4. Fan noise is reduced by moving the control vanes farther upstream so that the flow entering the fan is less turbulent.
5. A low-frequency resonance in an engine room produces a very loud noise near the walls. The noise is eliminated by changing the revolution speed.
6. Low-frequency noise is produced by a wide belt drive. The wide belt is replaced with several narrow belts.
7. Building vibrations are found to be caused by an elevator. Isolating the elevator drive from the building reduces the noise.

If the noise output of the source cannot be reduced by any appreciable means, then enclosing the source should be considered (see Figure 12). The enclosure should be designed with a dense material, such as sheet metal or plasterboard, on the outside. Sound-absorbing material should be used on the inside. The *transmission loss* (TL) is the logarithmic ratio of the sound power incident on one side of the enclosure wall to the sound power transmitted to the other side. A good rule of thumb (Harris 1991) for enclosures with partial absorption is that the transmission loss of the wall material should be greater than the required noise reduction (NR) plus 15 dB. If the enclosure has complete absorption, then the transmission loss of the wall material should be greater than the required noise reduction plus 10 dB. Install mufflers on cooling air openings. Access panels should also be incorporated for maintenance and service. A well-designed enclosure should reduce the noise by as much as 30 dB.

There are several paths that noise can travel from the source to the receiver. Direct sound is a straight-line propagation of sound waves from the sound source. Moving the receiver farther from the source will reduce the direct sound noise. If the source is nondirectional and is located near the center of a wall or floor in a room, the relationship (Hirschorn 1989) between the source-sound power level (L_w) and sound pressure level at the receiver (Lp_r) located R feet from the source is

$$Lp_r = L_w - 20 \log R + 2.3 \quad (27)$$

In Equation 27, doubling the distance will reduce the sound pressure level at the receiver by approximately 6 dB.

Sound rarely reaches the receiver by a direct path alone. Repeated reflections within the room provide another transmission path from source to receiver. Available methods to reduce reflected noise are to absorb the sound energy and/or to reflect it away from the receiver. Sound-absorbing materials fall into three categories. Sound energy can be converted to heat by internal friction within the sound-absorbing material. Glass fiber is a common absorbing material. Sound energy can also be converted to mechanical energy, such as lead sheets with mass and stiffness that are designed to vibrate. The third option is sound-absorbing cavities in which the sound waves are reflected back

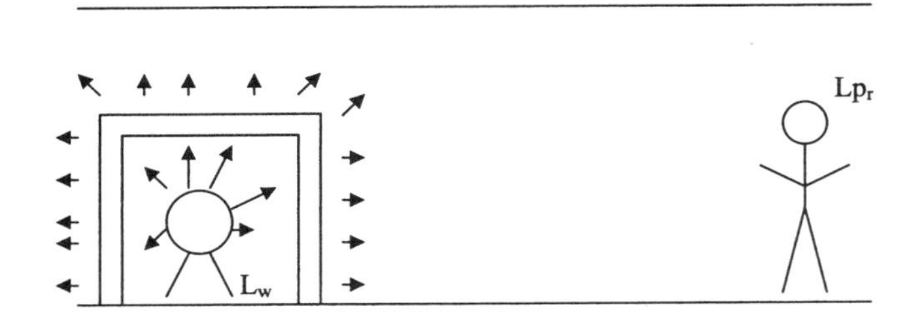

FIGURE 12. Enclosing a noise source to reduce noise at the receiver

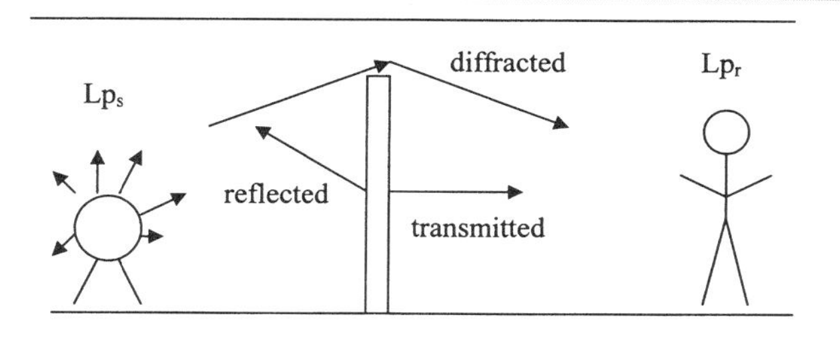

FIGURE 13. Installing a barrier between the source and the receiver

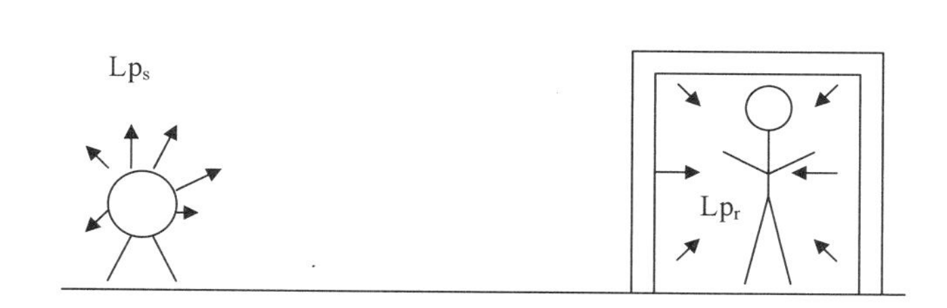

FIGURE 14. Noise reduction by enclosing the receiver

and forth until they are dissipated. Hollow concrete blocks are specially designed for this purpose.

If the reflected sound field is uniform within the room (diffuse), the sound pressure level at the receiver (Lp_r) due to a source with sound power level L_w is (Harris 1991)

$$Lp_r = L_w - 10 \log A + 16.3 \tag{28}$$

where A is the total room absorption in sabins. The total room absorption can be calculated by adding up the product of the area of each absorbing surface times the absorption coefficient for the entire room.

Another noise-control measure is the installation of a barrier between the source and the receiver (Figure 13). The barrier creates an obstacle to the sound waves. Some of the sound waves will be reflected back toward the source, some will be transmitted through the barrier, and some will be diffracted around the barrier. The amount of sound that is reflected depends on the sound absorption of the barrier. The transmitted sound depends on the transmission loss of the barrier. The diffracted sound will vary with the height of the barrier. Rules of thumb (Harris 1991) for the design of barriers include placing the barrier as close to the source as possible without touching the source, extending the barrier beyond the line of sight of the source by one-quarter wavelength of the lowest noise frequency, and using a solid barrier that has a transmission loss at least 10 dB higher than the required attenuation.

Noise can also be transmitted through other paths, such as floor vibration. For example, a machine may cause a floor to vibrate. The vibration may be transmitted through the floor to a worker standing on the floor. Isolating the machine from the floor in this case can reduce noise at the receiver by eliminating or lowering the vibration transmitted through the floor.

A sound-absorbing control room or office can be built to reduce the noise at the receiver (see Figure 14). The room should have adequate ventilation and air conditioning. The control room walls and ceiling should have a high TL and good sealing around the doors and windows. The absorptive treatment on the walls and ceiling are important because the sound transmitted through the walls and ceiling will be reinforced by reflected sound waves within the room. If the source is located in a highly reverberant area and the receiver room walls and ceiling are treated with a material that has a high sound absorption, the noise reduction (NR) will be nearly equal to the transmission loss (TL) of the receiver room. If the receiver room is highly reflective, the noise reduction can be as much as 20 dB. In general (Hirschorn 1989),

$$NR \approx TL + 10 \log (\alpha A/S) \tag{29}$$

where

α = average sound-absorption coefficient of the receiver room

A = total area of the receiver room

S = surface area separating the source room and receiver room

Case studies where engineering controls have been effective (with quantified results) can be found in Jensen et al. (1978). Several of these case studies are summarized below.

1. An 800-ton blanking press was mounted on four footings set on concrete piers. Noise levels

were 120 dB on impact, 105 dB at quasi-peak, and 94.5 dB at the operator position located 4 feet in front of the press. Isolating pads were added. The pads cost about $3000, including labor. The isolator pads reduced the noise level at the operator's station from 94.5 dB to 88 dB.

2. In the folding carton industry, pieces of paper scrap striking the sides of a conveyor caused high noise levels. Sound levels in excess of 90 dB were measured at the operator's platform. A layer of lead damping sheeting was glued to the ducts and other equipment. The sound levels were reduced to the range of 88 to 90 dB.
3. A worker was exposed to high noise levels while operating a metal cut-off saw. The solution was to enclose the saw. The work pieces were transferred through slots in the enclosure. The noise levels were reduced by 13 dB.
4. The sound level at the operating station of a 200-ton punch press was measured as 104 dB. An enclosure was built around the machine. The enclosure included a heat exhaust system and access doors. The sound level for the operator was reduced to 83 dB.

Lockout/Tagout (LOTO)

Lockout/tagout (LOTO) is an administrative procedure for protecting employees from the unexpected start-up of equipment and machines or the release of energy while the equipment is being serviced or repaired. The general industry regulation, 29 CFR 1910.147, *The Control of Hazardous Energy*, sets forth procedures for lockout/tagout. The regulation requires an employer to implement a program to ensure that machines and equipment are rendered inoperable and isolated from energy sources when equipment is serviced. The program shall consist of energy control procedures, employee training, and periodic inspections.

Electrical power to the equipment is the obvious energy source that should be locked out. There are other sources of energy in a typical industrial facility. These include gas, pressurized air, steam, electrical capacitors, hydraulic lines, raised loads, and compressed springs. It is the latent build-up of energy that can be released even when the power is locked out that typically goes unnoticed.

Lockout is a procedure in which locking devices are placed on energy-isolating devices so that the equipment cannot be operated until the locking device is removed. An *energy-isolating device* is a mechanical device that physically prevents the transmission of energy. Examples of energy-isolating devices are manually operated electric circuit breakers, disconnect switches, and line valves. There are a number of different lockout devices. Electrical plug boxes work by inserting the disconnected plug into a small box and locking it. This prevents plug connection during maintenance. Multiple lockout devices can hold up to six padlocks. This is important when several people may be working on a machine. Figure 15 shows a multiple lockout device.

If an energy-isolating device is not capable of being locked out then a tagout system shall be used. *Tagout* is the placement of a tag to indicate that the energy-isolating device and the equipment should not be operated until the tag is removed. Tags warn of specific hazards, such as "Do Not Operate" or "Do Not Start." It should be noted that these tags are warning devices; they do not provide positive restraint as

FIGURE 15. Example of a multiple lockout device and tag

lockout devices do. Figure 15 shows an example of a tag.

There are several pointers to putting together a LOTO program. Simply locking the power does not necessarily mean the equipment is safe to work on. Check for trapped air, gas, springs, raised loads, or other sources of residual energy. For example, bleed air from pneumatic lines so there can be no unexpected release of compressed air or unexpected actuation of an air cylinder. Be careful of multiple workers performing maintenance. Miscommunication can occur, for example, during shift changes. A complete check should be done before the equipment is restarted. Check that all safeguards have been put back in place and that all tools have been removed. The person who applies the lockout device should be the one to remove it. Make sure that all employees are at a safe distance before the equipment is started.

A LOTO program should include the following steps:

1. ***Prepare for shutdown:*** The authorized person should refer to the written company procedure. The authorized person should be trained on the piece of equipment, including the sources of energy, the hazards, and the means to control the energy.
2. ***Shutdown:*** An orderly shutdown should be done. The machine or equipment should be turned off.
3. ***Energy isolation:*** All energy-isolating devices should be located and operated to isolate the machine or equipment from energy sources.
4. ***Lockout or tagout devices applied:*** The authorized employee should apply lockout and/or tagout devices.
5. ***Check for stored energy:*** Residual energy must be restrained, dissipated, or otherwise made safe.
6. The authorized employee must verify the isolation and de-energization of the equipment. This usually involves pressing the push buttons to be sure the equipment will not start.
7. ***Prepare for release from LOTO:*** The equipment should be checked to be sure that all safeguards have been put back in place, the equipment is intact, and all tools have been removed. All workers should be at a safe distance.
8. ***Lockout and/or tagout devices removed:*** Each lockout and/or tagout device should be removed *only* by the person who applied it.

OSHA lockout/tagout requirements for some specific types of equipment are listed below:

1. ***Powered Industrial Trucks:*** Disconnect battery before making repairs to the electrical system (1910.178(q)(4)).
2. ***Overhead and Gantry Cranes:*** The power supply to the runway conductors shall be controlled by a switch or circuit breaker located on a fixed structure, accessible from the floor, and arranged to be locked in an open position (1910.179(g)(5)). The main or emergency switch shall be open and locked in the open position. If other cranes are operating on the same runway, rail stops shall be installed to prevent the operating crane from interfering with the crane being serviced. (1910.179(l)(2)).
3. ***Woodworking Machinery:*** All power-driven woodworking machines should be equipped with a lockable disconnect (1910.213(a)(10)). Controls on machines operated by electric motors should be rendered inoperable (1910.213(b)(5)).
4. ***Mechanical Power Presses:*** The power-press control system should be equipped with a main disconnect that can be locked in the off position (1910.217(b)(8)(i)). The die setter should use safety blocks during die adjustment or repairing the dies in the press (1910.217(d)(9)(iv)).
5. ***Forging Hammers:*** Means shall be provided for disconnecting the power and for locking out or rendering cycling controls inoperable (1910.218(a)(2)(iii)). The ram shall be blocked when the dies are changed or work is being done on the hammer (1910.218(a)(2)(iv)).

Steam hammers should be equipped with a quick-closing emergency valve in the admission piping. The valve should be closed and locked while the equipment is being serviced (1910.218(b)(2)).

6. ***Welding, Cutting, and Brazing:*** Resistance welding machines should be equipped with a safety disconnect switch, circuit breaker, or circuit interrupter so that the power can be turned off during servicing (1910.255(a)(1)).
7. ***Pulp, Paper, and Paperboard Mills:*** The main power disconnect and valves for the equipment should be locked out or blocked off when the equipment is being serviced (1910.261(b)(1)). The valve lines leading to the digester should be locked out and tagged when inspecting or repairing the digester (1910.261(g)(15)(i)). The control valves leading to the pulpers should be locked out and tagged out, or blanked off and tagged before a worker enters the pulper (1910.261(j)(5)(iii). All drives should be equipped with a power switch lockout device (1910.261.(k)(2)(ii)).
8. ***Bakery Equipment:*** The main switch should be locked in the open position prior to servicing an oven or electrical equipment (1910.263(l)(3)(b)).
9. ***Sawmills:*** The hydraulic equipment must be rendered safe prior to servicing (1910.265(c)(13)). The main control switches to the mechanical stackers must be designed so that they can be locked in the open position (1910.265(c)(26)).
10. ***Grain Handling:*** Auger and grain transfer equipment must be de-energized, locked out, and tagged out (1910.272(h)).

Power Tools

Power tools are inherently dangerous for the simple reason that any tool that is powered by electricity, air, powder, or gasoline for the purpose of cutting through wood, metal, concrete, or other materials can easily cut through bone and tissue. The hazards associated with power tools include contact with moving parts, contact with the power source (electricity), losing control of the tool, flying debris, harmful dust and fumes, excessive noise, vibration, and being struck by the work piece. Other factors, such as a less than ideal working environment (uneven construction site versus a clean, level shop) or momentary inattentiveness by the user, can further increase the risk of an injury.

In designing for safety, the first priority is to eliminate the hazard or provide a safety device to reduce the risk to an acceptable level. There are a number of basic safeguards and design features for reducing the risk of an injury from a power tool. All hazardous moving parts of a power tool should be guarded to the greatest extent possible. This would include belts, gears, shafts, pulleys, sprockets, spindles, chains, blades, reciprocating parts, and rotating parts. OSHA 1910.243 addresses the guarding of portable power tools. Refer to the "Machine Guarding" section of this chapter for guarding mechanical hazards.

One of the main hazards with electric-powered tools is electrical shock. An electrical shock can result in injury or death, or cause the user to fall off a ladder or scaffold. To protect the user from electrical shock, the tool should have a three-prong grounding plug that mates to a three-hole receptacle, or be double insulated. In the case of a grounded tool, any current resulting from a defect or short inside the tool will be conducted through the ground wire and not through the operator's body. When an adapter is used to accommodate a two-hole receptacle, the adapter wire must be connected to a known ground. The third prong should never be removed from the plug. Double-insulated tools have an internal protective layer of insulation that isolates the housing. This eliminates the possibility of current flowing through the housing to the operator.

In the case of hand power tools, the handles should be properly designed so that the operator can control the tool. Handles should be located where they allow the operator to balance the tool and apply force in the direction that is needed. Tools with rotating components may need a second handle so that the operator can resist the torque.

Power switches should be conveniently located in such a manner that the operator does not have to relinquish control (grip) of the tool to activate the switch.

A power switch should be protected from inadvertent activation. Lock-on switches, switches that remain engaged without having to keep your finger on them, are used in many hand power tools. However, these switches are generally not recommended, especially for heavy-duty tools, because the power will not shut off if the operator loses control of the tool.

There are a number of general precautions that should be observed by operators of power tools. A power tool should never be carried by the cord or hose. Cords and hoses should be kept away from heat, oil, and sharp edges. Never yank a cord or hose to disconnect the tool. Always disconnect the tool when it is not in use, when changing accessories, or when servicing it. Secure the work piece so that both hands can be used to operate the tool. Do not operate power tools when wearing loose clothing, ties, jewelry, or other apparel that can be caught in moving parts.

Personal protective equipment should be used when operating a power tool. Always wear eye protection with side shields when using power tools. A face shield should be used when large flying particles may be present. Many power tools produce dangerous noise levels. Wear hearing protection with noisy tools. A dust mask or respirator should be worn if there are dusts or harmful fumes present.

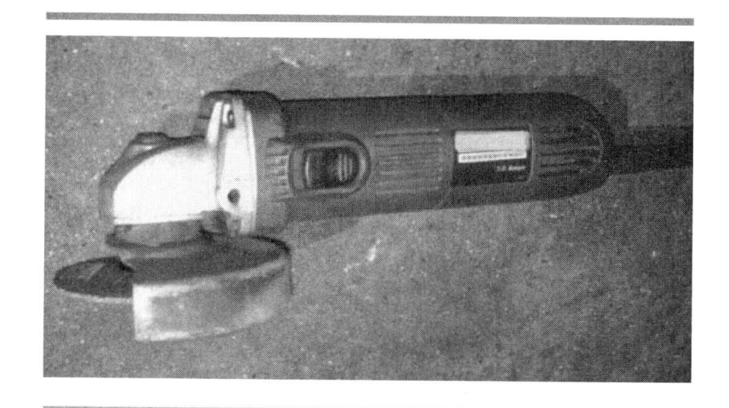

FIGURE 16. A typical hand-held grinder

FIGURE 17. Example of a radial saw

Hand-Held Grinders

Injuries from hand-held grinders occur from particles being thrown into the operator's eye, kickback, and disk fracturing. Although it is impractical to guard the entire disk, the portion between the disk and the operator should be guarded to deflect broken pieces away from the operator. A heavy-duty grinder should have a second handle so the operator can maintain control of the tool. Figure 16 shows a photograph of a typical hand-held grinder.

There are a number of safety precautions that should be followed to prevent fracture of the disk. Only approved flanges should be used to mount the disk. The motor speed must not exceed the safe operating speed on the disk. The disk should be inspected for cracks or chips before using. A disk should not be used if it has been dropped on a hard surface. A grinder should never be used as a cutting tool. This imposes high stresses on the disk.

Radial Saws

Radial saws are powered circular saws. These saws have features that make them more versatile than table saws. For example, the saw arm can be raised, lowered, or swung from side to side to adjust the depth and angle of the cut. The blade can also be replaced with cutters, disks, drum sanders, or other accessories. Figure 17 shows an example of a radial saw.

As with most saws, injuries are the result of contact with the blade (Hagan et al. 2001). To reduce the risk of injury the upper half of the blade, from the top down to the end of the arbor, should be guarded. The lower half of the blade should have a self-adjusting, floating guard that adjusts to the thickness of the stock. The saw should have a limit chain or other positive stop to prevent the blade from traveling beyond the front of the table. There should be a device that will automatically return the saw to the back of the table when it is released at any point in the travel.

Additional precautions should be taken when ripping. A spreader should be used to prevent internal stresses in the wood from causing the wood to bind the blade and cause kickback. Antikickback pawls should also be used. The stock should be fed against the direction of rotation of the blade from the nose of the guard; the side of the guard where the blade rotates upward should be toward the operator. Feeding from the wrong side can cause the operator's hand to be drawn into the blade. To prevent an operator from ripping from the wrong side, the direction of rotation should be marked on the hood. Also, a standard warning label should be affixed to both sides of the rear of the hood stating, "Do not rip from this end."

Hand-Held Circular Power Saws

Hand-held circular power saws are one of the most widely-used power tools. Figure 18 shows a photo of a typical hand-held circular power saw. Most injuries with hand-held circular saws are the result of contact with the blade. To reduce the risk of injury, these saws should have guards above and below the faceplate. The lower guard must retract as the material is cut and maintain contact with the material. The guard should be frequently checked to be sure that it operates properly and covers the blade teeth when not in use. Other safety features include a trigger switch that must be depressed to power the saw.

FIGURE 18. Example of a hand-held circular saw

Many injuries with hand-held circular saws are caused by kickback. Kickback is a sudden reaction that occurs when the blade is pinched. The reaction forces the tool out of the cut and towards the operator. If this occurs quickly enough the lower portion of the blade may contact the operator before the guard retracts. Since this is a hazard that cannot be designed out, the operator should follow some very basic operating procedures to reduce the likelihood of a kickback event and possible injury. These procedures, along with warnings concerning the kickback hazard and the consequences of not following procedures, should be in the manual.

To reduce the likelihood of the blade binding or pinching, never use dull, bent, warped, or broken blades. Avoid cutting wet wood as the wet wood chips can bind the blade and may also bind the guard. Waiting for the blade to stop rotating before removing the saw from the cut will also reduce the chance of kickback. If the blade does bind, releasing the switch immediately can reduce the extent of the kickback. Standing to the side of the cut will put the operator out of the trajectory of the saw if a kickback does occur.

Table Saws

Table saws are used for a variety of tasks, such as ripping, crosscutting, grooving, rabbetting, and dadoing. Most injuries with table saws occur from contact with the saw blade. This can happen if the operator hands are held too close to the blade, if the operator's hands slip off the stock as it is being pushed through the saw, or when removing scrap or finished pieces from the table. Injuries can also occur during ripping operations when the board being cut can be thrown violently backwards toward the operator (kickback). A typical table saw is depicted in Figure 19.

Since table saws are used for a multitude of tasks, it is difficult from an engineering design prospective to design a single guard that covers all foreseeable uses of the saw. All table saws should be designed with a guard that protects the operator for most cutting operations. The top part of the blade, the part that projects above the table, should be guarded with a hood guard. The guard should adjust automatically to the thickness of the work piece, and should remain in contact as the piece is being cut. One chronic problem with hood guards is that users tend to remove them. The exposed part of the saw blade under the table should also be guarded.

The table saw should have a splitter and antikickback devices for ripping. These devices are usually built into the hood guard. Antikickback pawls, along with the spreader, act to reduce kickback. The rip fence should also be parallel to the blade so that the work piece does not bind.

A number of tasks may require special procedures (Hagan et al. 2001). An auxiliary fence and push block should be used for ripping pieces less than 2 inches wide. A push stick should be used when ripping pieces from 2 to 6 inches wide. For other operations (grooving, rabbetting, dadoing) it is impossible to use the spreader and may be impractical to use the hood guard. These tasks will require special jigs to keep the operator's hands away from the blade or cutting head.

FIGURE 19. Example of a typical table saw

FIGURE 20. Example of a typical industrial band saw

Band saws

Band saws are used for cutting curves, miter cuts, and ripping. Injuries from band saws are generally less than from table saws. Figure 20 shows a photo of a typical industrial band saw. Most injuries with band saws occur from contact with the blade. Like many power saws, the point of operation cannot be completely guarded. The wheels of the band saw should be guarded. An adjustable guard should provide protection from contact with the front and right side of the blade above the blade guides.

A band saw should have a tension-control device to indicate the proper blade tension. Other safety devices include an automatic blade tension device to prevent breakage of blades, and a device that prevents the motor from starting if the blade tension is too tight or too loose.

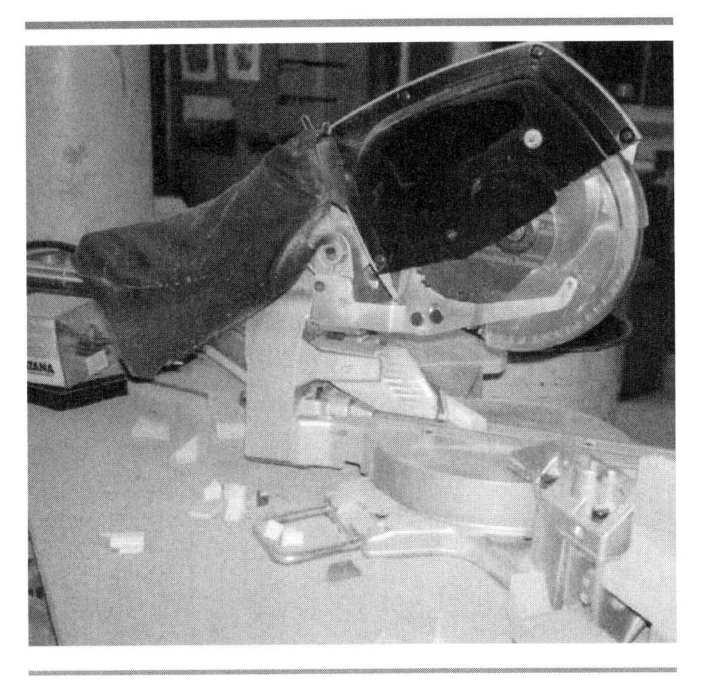

FIGURE 21. Miter saw with a linkage-activated guard

Miter Saws

Miter saws are stationary tools used for crosscutting, mitering, and beveling. These saws consist of a saw head (circular saw blade, arbor, electric motor) mounted to a pivot arm. The pivot arm is supported on a carriage that can be rotated left and right. The work piece is mounted on a horizontal table. The cut is made by swinging the saw head in a downward motion and depressing the trigger. The pivot arm is spring loaded so that the saw head returns to the upright position when it is released. Figure 21 shows a photo of a typical miter saw.

The saw blade should be guarded to protect the operator. Many miter saws use a contact type of guard. This guard covers the sides of the blade, but is open around the periphery. As the saw head is brought down to make the cut, the guard contacts the work piece and retracts. When the saw head is raised back up, the guard returns to its original position. These guards provide some measure of protection, but an operator can still be injured from inadvertent contact with the guard. Body contact with the guard will cause it to retract since the guard cannot distinguish between contact with the work piece and contact with a body part. Many injuries occur when a worker brushes against the guard, exposing the blade, while drawing a board from left to right across the table to make the next cut.

A linkage-driven guard, a guard that retracts and returns to place as the arm is lowered and raised, is a safer design. These guards cover both sides of the blade and the periphery. As the saw head is brought down, a linkage mechanism connected to the pivot arm retracts the guard. The guard cannot be inadvertently displaced. The guard retracts as the pivot arm is drawn down to make the cut and goes back into position as the arm is brought up.

Miter saws should also have a brake that stops the rotation of the blade when the trigger is released. All safety features (guard and brake) should be checked before using the saw to ensure they are operating properly. The operator's hands should be kept away from the cutting plane of the blade when the saw head is lowered.

Chain Saws

Chain saws are another labor-saving tool used for cutting downed trees, branches, large logs, or pieces of wood. However, there is an increased risk of injury because the operator is exposed to a high-speed, powered chain that is sharp and unguarded. Serious injuries or death can occur if an operator falls while carrying the saw, even if it is not running. Injuries can also occur if an operator lowers the saw to his side while the chain is still moving. Figure 22 shows a photo of a typical gas-powered chain saw.

Since it is impractical to guard the chain, the operator should be thoroughly trained in the hazards and

FIGURE 22. A typical gas-powered chain saw

use of a chain saw. There are a number or procedures that should be followed to reduce the likelihood of a chain-saw injury. The chain brake should be engaged when starting the saw so that the chain will not be powered during start-up. Keeping the work area clear, maintaining a firm footing, and turning the engine off when carrying the saw will reduce the likelihood of falling or stumbling onto a powered or stationary chain. Holding the saw with both hands provides optimum control when making a cut.

Aside from being exposed to the sharp chain, the next biggest cause of chain saw injuries is kickback. Kickback is a sudden and violent upward and/or backward movement of the saw toward the operator. It is caused when the motion of the chain is suddenly stopped. This can occur if the chain near the nose or tip of the guide bar contacts an object, such a log or branch. Kickback can also occur when the wood pinches the chain in the cut. Any interference with the moving chain transfers its energy (either gasoline or electric motor) into movement of the entire saw.

There are design safeguards to reduce the likelihood of kickback. Antikickback safety features include a safety nose, a safety chain, and chain brakes. A safety nose is effective is preventing nose-tip kickback. Safety chains use a lower profile, different cutting pitch, and fewer links, thereby reducing the tendency of the chain to get caught. A chain brake stops the chain when kickback occurs. A common method to activate the brake is by a mechanical switch located at the front hand guard. When the saw is propelled backward, the lead hand pushes on the hand guard, activating the brake. It should be noted that these safety devices only reduce the probability of kickback and contact with a moving chain if kickback occurs. They do not prevent kickback.

Pneumatic Nail Guns

Nail guns (pneumatic nailers) are pneumatically powered tools that drive nails into wood (see Figure 23). Small nailers are used for finish work, while larger framing nailers can drive a three-inch framing nail into wood or even concrete. These tools significantly increase work production, but not without an increased risk of injury. Nail guns deserve the same respect as firearms. Nails can be driven into the human body if the gun fires before the operator is ready, if the nail completely penetrates the work piece and comes out the other side, or if the gun is directed at something other than the work piece.

Pneumatic nail guns work with compressed air generated by an air compressor. The *hammer* portion of the gun is a piston that drives a long blade. A trigger mechanism opens a valve that allows compressed air into the chamber at the top of the piston, driving it down and propelling the nail at speeds as high as 1400 feet per second. When the valve closes, the air is directed to the return chamber below the piston, pushing it back up. Safety requirements for compressed air-fastener driving tools are addressed in ANSI SNT-101, *Portable, Compressed-Air-Actuated Fastener Driving Tools* (ANSI 2002f).

Injuries include nails driven into the extremities (hands, feet) to more serious injuries and death from nails driven into heads, eyes, and major organs. For example, a worker shot a nail into his wrist when his wrist brushed against the nail discharge area of the gun. In another case, two workers were positioned on a roof, one above the other. The top worker slipped, the nail gun moved across his body, and the gun

FIGURE 23. A typical pneumatic nail gun

activated. The gun shot a nail into the bottom man's head. Nails have penetrated drywall and struck workers on the other side. Workers have also been injured when a driven nail ricocheted off another nail.

Various engineering designs have been implemented by manufacturers to reduce the risk of injury from unintentional firing. Most nail guns have two separate firing triggers. One trigger is finger-activated and located under the handle, similar to most hand power tools. The other trigger, the nose trigger, is located at the nail discharge (nose). The nose trigger is activated by pressing the nóse of the gun against the work piece. In order to fire the gun, both triggers must be depressed.

There are two types of trigger systems. In the "contact-trip" or "bounce-firing" system, the operator depresses the finger trigger and bounces the nose against the work to fire the gun. This method can be very fast and efficient. However, this system is still dangerous because it allows the operator to fire nails in rapid succession by keeping the finger trigger depressed while quickly bouncing the nose of the gun. Workers can still be injured from double fires that can occur during rapid bouncing of the gun. Injuries can also occur if the nose contacts something other than the work piece while the finger trigger is held.

The sequential-trip firing system requires that the nose be engaged against the work piece before the trigger finger is depressed. This system is safer because only one nail can be fired per finger trigger activation. It eliminates injuries from double firing and from brushing the nose against an object while the finger trigger is depressed. Many nail guns can be converted from sequential mode to bounce mode and vice versa.

The "single-shot" system is a compromise between the bounce-firing system and the sequential-trip system (Arnold 2002). The single shot mode can fire a nail in one of two ways. One way is with the nose guard pressed against the work and the trigger pulled and held. The other way is by depressing the trigger first, then bouncing the nose guard. Either way, the gun will not fire if the trigger is held down and the nose guard is lifted and pressed again. However, if the nose is held against the work piece and dragged to the next nailing spot, and the trigger is released and depressed, the gun will fire again. Nail guns also come equipped with anti-double-firing mechanisms.

There are also nail guns with a *brain* (Arnold 2002). A computer chip senses the prior firing mode, either bounce or sequential. In the sequential-trip mode, the operator has two seconds to pull the trigger before the nose guard has to be reset. In the bounce-fire mode, the operator has one second between shots before the trigger has to be released and pulled again.

MACHINE SAFEGUARDING

Safeguarding is any means of preventing a worker or user from contacting a dangerous part of a machine, product, or other device. There are three areas where mechanical hazards requiring safeguarding are typically found. Any part that moves while the machine is operating should be safeguarded. Power transmission apparatus components, such as pulleys, belts, chains, cranks, shafts, and spindles, should be safeguarded. The *point of operation*, the area where moving parts bend, shape, or cut material, should also be safeguarded. In the workplace, machines that typically require attention include power presses, press brakes, mills, calendars, shears, grinders, and printing machines. Examples of consumer products where safeguarding should be considered include food processors, lawn mowers, power tools, and snow throwers.

Following the designing for safety hierarchy, the first priority is to eliminate the hazard altogether. If this is not possible, there are a number of safeguarding methods that can be considered. A *guard* is a physical barrier that prevents any body part from contacting the hazard. A guard may be fixed or adjustable. A *fixed guard* is one that is permanently attached to the machine, and special tools are required to remove it. *Adjustable guards* are used in cases where openings in the guard, such as for feeding stock, vary from operation to operation. There are a number of design criteria to be considered when designing a guard:

1. The guard must prevent a body part from contacting dangerous parts of a machine.
2. A well-designed guard does not reduce a worker's productivity. A guard that impedes a worker from doing his job might be discarded.

3. The guard must be strong enough so that it does not break under the loads that will be applied to it.
4. The guard must be rigid enough so that it does not vibrate or deflect, causing it to come into contact with other machine parts.
5. The guard must contain flying particles, liquids, or fractured tools.
6. The guard must allow visibility in the working area.
7. The guard must not create additional hazards.

In some cases, it may be necessary to have an opening in the guard to feed material. An opening of 3/8 of an inch or less is generally considered to be safe. However, this opening may be too small to pass material. In these cases, the guard must be designed so that material can be fed in, yet the hand is prevented from reaching the nip point. One formula that is used when the distance from the guard to the danger zone is less than 12 inches is the following (Hagan et al. 2001):

$$\text{Maximum Safe Opening} = 1/4 \text{ in} + 1/8 \text{ in} \times \text{distance from guard to danger zone} \quad (30)$$

ANSI B11.1 (ANSI 1994) provides "opening versus guard distance" data, which is tabulated in Table 10. For example, if a 0.75-inch opening is required in a guard for feeding material, then the guard must be 5.5 to 6.5 inches from the point of operation. This will minimize the chances that an operator's fingertips will reach the danger area.

TABLE 10

Opening Versus Guard Distance

Distance of Opening from Point of Operation (inches)	Maximum Width of Opening (inches)
1/2 to 1 1/2	1/4
1 1/2 to 2 1/2	3/8
2 1/2 to 3 1/2	1/2
3 1/2 to 5 1/2	5/8
5 1/2 to 6 1/2	3/4
6 1/2 to 7 1/2	7/8
7 1/2 to 12 1/2	1 1/4
12 1/2 to 15 1/2	1 1/2
15 1/2 to 17 1/2	1 7/8
17 1/2 to 31 1/2	2 1/8

(*Source:* ANSI B11.1)

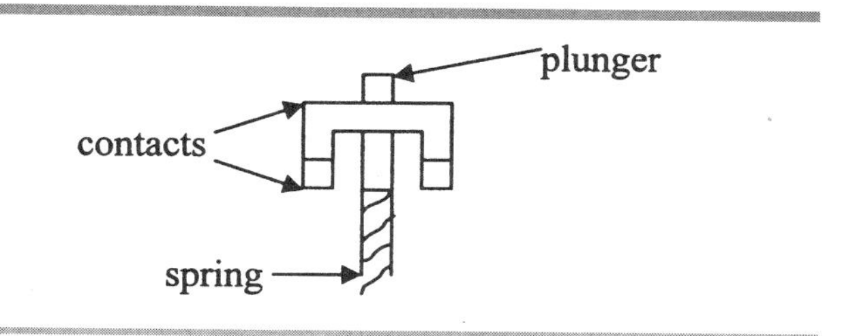

FIGURE 24. Spring-loaded switch

An interlocked guard is required when the operator must enter the point of operation to add material, remove material, or perform maintenance. A safety interlock switch or device prevents the machine from operating or from being started if the guard is not in place. There are many different types of interlocks. Many food processors, for example, have a spring-loaded plunger, which prevents the processor from starting unless the cover is locked into place. In general, an interlock should operate reliably, be designed so that it is not easily defeated, have positively driven contacts, and should not fail in a closed position.

Figure 24 shows a common spring-loaded switch to illustrate the safety issues involved in designing an interlock. It should be noted that the switch shown in Figure 24 should not be used as an *interlock* switch. The switch consists of a spring-loaded plunger and electrical contacts. When the door is opened, the spring pushes the plunger up and the contacts are opened. When the door is closed, the plunger is pushed down and the contacts are closed. The switch is easily defeated since the plunger can be taped down. Also, the operator may inadvertently lean on the plunger while performing maintenance. The contacts are not positively driven. Opening the contacts depends solely on spring pressure. Opening the door does not directly pull the contacts apart. The switch does not fail in the open position. If the contacts become stuck or welded, or if the spring fails, the contacts remain closed. Lastly, the switch has no redundancy.

There are two types of interlocking systems. In a power interlocking system, the removal of the guard directly disrupts the power source. The most common system is a control interlocking system. In this system, the power source is interrupted by the switching of a circuit that controls the power-switching device. Interlocking systems can become very complicated. For example, in most cases not only do you want to prevent machine activation if the guard is removed, but you also want to prevent the operator from entering the area until the machine stops moving. In these cases, it may be necessary to incorporate a guard-locking device so that the guard cannot be opened until the machine has stopped.

In some cases, it is not possible to design a fixed or adjustable guard. For example, in some machines such as metal shears or pizza dough-rolling machines, the operator must hand feed the material into the point of operation. In these cases an *awareness guard* may be the only alternative. An awareness guard alerts an operator that he is entering a dangerous area of a machine rather than preventing the entering as a properly designed barrier guard would. Typically, an awareness guard consists of a series of heavy rollers placed across the feed area. As the operator feeds the material, the weight of the rollers on the operator's hands provides a tactile cue that there is an impending hazard.

Another method to keep the operator's hands out of the point of operation is a two-hand control. A two-hand control usually consists of two palm buttons. To activate the machine, the operator must use both hands, one on each button. The buttons should be spaced far enough apart that the operator cannot activate both controls with, for example, a hand and an elbow, leaving the other hand free. The two-hand control should be located far enough from the point of operation so that the operator cannot reach into the danger area after the machine is activated. A two-hand control only protects the operator; it does not protect other workers who may be in the area.

There are also presence-sensing devices that can be used in place of a physical barrier. Light curtains are one such device. These devices work by sending an array of synchronized, parallel infrared light beams to a receiver. If a hand or body part breaks the beam, the control logic sends a signal to stop the machine. Capacitive sensing devices use an electromagnetic field generated by an antenna. A change in the field caused by a hand or body part sends a signal to the control unit, which stops the machine.

Safety mats are another type of presence-sensing device. A safety mat is basically a large switch. It consists of two metal plates separated by bumpers. When a person steps on the mat, contact is made, sending a control signal to stop the machine. Safety mats are usually used where perimeter guarding is required, such as around robotic workstations and automated equipment. Trip cords, body bars, and trip bars are other types of presence-sensing devices. When a hand or other part of the body contacts these devices, the machine is stopped.

Presence-sensing devices should be installed in such a manner that the worker cannot reach over, under, or around the device. The overall presence-sensing system, including the control circuitry and the braking mechanism, must be able to stop the machine before the worker penetrates the protected area and reaches the point of operation. There are several formulas for calculating the minimum safe distance for the placement of presence-sensing devices (see Figures 25 and 26). The formula presented in ANSI B11.19 (ANSI 2003) is

$$D_s = K(T_s + T_c + T_r + T_{spm}) + D_{pf} \tag{31}$$

where

D_s = the minimum safe distance in inches between the light curtain or the outside edge of the safety mat and the nearest hazard

K = maximum speed at which the operator can approach the hazard in inches per second

T_s = total time it takes for the hazardous motion to stop in seconds

T_c = time it takes for the machine control system to activate the machine's brake in seconds

T_r = response time of the presence sensing device in seconds

T_{spm} = additional stopping time allowed by the stopping performance monitor, seconds

D_{pf} = added distance, in inches, due to the depth penetration factor. For light curtains this is related to how far an object can move through the sensing field before the light curtain reacts. For a safety mat this distance is 48 inches.

Safeguarding of specific classes of machines can be found in the following ANSI standards:

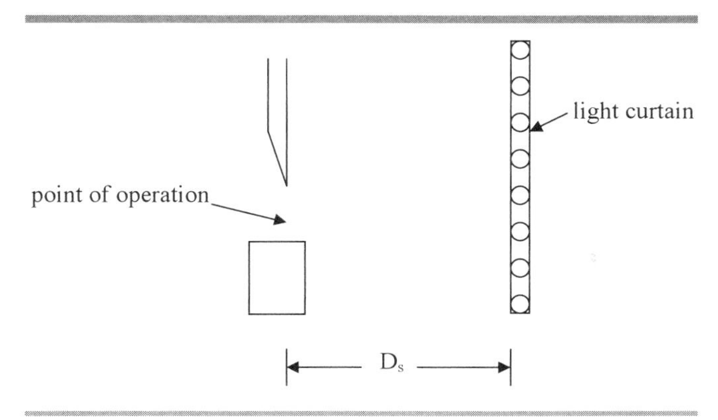

FIGURE 25. Minimum safe distance (D_s) for a light curtain

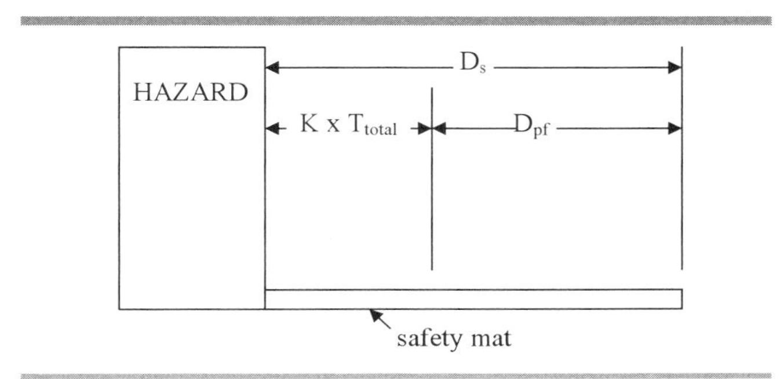

FIGURE 26. Minimum safe distance (D_s) for a safety mat

B7.5	Abrasives
B11	Machine Tools
B15.1	Power Transmission Apparatus
B65	Printing Equipment
B71.1	Garden Equipment
B154.1	Rivet Setting Equipment
B155.1	Packaging Equipment
B173	Hand Tools
B165	Power Tools
B186.1	Portable Air Tools
B209	Hand Tools
B208.1	Pipe Threading Machines
O1	Woodworking Machinery
R15.06	Robots
Z245.5	Baling Equipment

Laser Machining

According to ANSI B11.21, laser machining presents a new set of problems when it comes to machine guarding (ANSI 2006a). Mechanical damage to the protective housing, the laser itself, or the beam delivery system can cause the laser to be aimed at an operator or the viewing window. Electrical hazards include high voltages, stored energy, and high current capability. There can also be thermal hazards. Poorly designed interlock switches, switch assemblies, interlock circuits, gas lines, and gas valves can create hazardous conditions. Radiation hazards can exist in the form of X-rays from plasmas and UV radiation. Potentially hazardous material by-products may be emitted, depending on the material being cut.

ANSI B11.21 (ANSI 2006a) lists safety requirements for laser machine tools. The emergency stop shall deactivate laser-beam generation, automatically position the laser-beam stop, stop all hazardous motion, and eliminate or control all stored energy. There should be a means to conduct exhaust effluent out of the machine tool. Personnel need to be protected from laser radiation. This can be accomplished by floor mats or light curtains that prohibit entrance into a hazard zone.

Material Handling

Conveyors

Conveyors are used in almost every industry. Thousands of miles of conveyors are used to transport materials every day. These devices have many hazardous areas that need to be safeguarded. The number of operating and maintenance injuries from conveyor systems is in the range of 10,000 per year (Schultz 2000, Schultz 2004). Most injuries occur when workers clean or maintain the conveyor, reach into the conveyor to remove objects, tools, or clothing caught there, or fall or reach into a pinch point. Hazardous areas include power transmission, nip points, shear points, pinch points, spill points, exposed edges, areas under the conveyor, and transfer mechanisms.

There are many types of conveyors. Gravity-driven conveyors can have either wheels or rollers that transport the loads. Runaway loads and material falling off the conveyor are some of the hazards associated with gravity conveyors. Powered roller conveyors used powered rollers to move the loads. A belt conveyor uses a moving belt to transport loads. These conveyors consist of the belt, the idlers that support the belt, the pulleys that move the belt, the drive systems, and the supporting structure. Belt conveyor hazards include pinch points, moving parts, attempting repairs to a moving conveyor, and attempting to cross over a moving belt. A chain conveyor uses a series of slats attached to a chain. Other conveyor types include overhead, vertical reciprocating, harpoon, screw, vibrating, and bucket elevator conveyors.

Designing for safety should be applied to the design, installation, and maintenance of conveyor systems. ANSI B20.1, *Safety Standard for Conveyors and Related Equipment*, sets forth minimum safety standards for conveyors (ANSI/ASME 2003). All exposed moving parts that present a hazard should be guarded. When it is not possible to install a guard, such as in cases where the guard would render the conveyor unusable, prominent warnings should be provided. Nip and shear points should be guarded. Antirunaway, brake, or backstop devices should be provided on inclined and vertical conveyors. Openings to hoppers and chutes should be guarded to prevent personnel from falling into them (ANSI/ASME 2003).

There are a number of additional safety issues that should be addressed when selecting a conveyor system. First, on–off controls should be provided near operator stations. The controls should be located where the operator can see as much of the conveyor as possible. Emergency stop controls, such as emergency cables, should be installed in areas where there are exposed moving parts. Warning signs should be provided to alert personnel about standing on conveyors, removing guards, bypassing safety devices, and any hidden hazards.

Railings should be installed in areas where materials can be knocked off the conveyor. Usually this happens when an object protrudes over the edge and gets caught on another object. This can also happen if objects become backed up. Nets or other guards should be installed in areas where a conveyor passes over a pedestrian walkway. Steps or bridges should be installed in areas where workers may need to cross over the conveyor. This would discourage workers from walking on the conveyor. Gates or pull stops should be located in areas where a person could enter a dangerous area after falling onto a conveyor. Select operator stations should be ergonomically designed, including the proper height and reach to avoid worker injuries.

Robots

A *robot* is a programmable materials-handling mechanism that is capable of automatic position control for orienting materials, work pieces, tools, or devices. Figure 27 shows industrial robots in an automotive assembly line. Robots are more hazardous than conventional machines in that robots are programmed, and they do not always follow a continuous, regular cycle of operation. A worker may be tempted to enter the work area thinking the robot is idle or continuing with the same pattern. For example, if a robot happens to be motionless, it should not be assumed that it will remain that way. The program may have a delay where the robot remains idle before taking up the next task. If a robot repeats a particular pattern, it should not be assumed that it will continue to repeat the same pattern.

FIGURE 27. Industrial robots on an automotive assembly line (Courtesy of Kuka Robotics Corporation 2007)

Computers can automatically modify the pattern that the robot was following.

Injuries associated with robotic systems include being struck by the robot, being entrapped between the moving parts of the robot and a fixed object, and being struck by objects that are being handled by the robot. Many injuries involving robots are the result of inadequate safeguarding. A study of 32 robotic injuries (Gainer and Jiang 1987) uncovered the following breakdown:

- Operator in work area while robot in operation
- Operator in work area while robot stopped but put in operation by worker inadvertently hitting start button
- Operator entered work area with robot stopped by compressed air trapped in line
- Operator placed work piece on conveyor during setup and conveyor sensor activated robot
- Adjacent work area within robot work area
- Robot reached over fence
- Worker performed task while robot held the work piece
- Maintenance worker in robot area and colleague started robot
- Maintenance worker in robot area performing test run after maintenance
- Maintenance worker entered robot area to perform maintenance while robot in operation

Sometimes robots are programmed by being physically guided through the desired tasks by the operator using a "teaching pendant." The robot *learns* by storing feedback from its position sensors. Teaching pendants should contain an emergency stop button or deadman switch so the operator can stop the robot. The rate of movement should be limited to 6 inches per second or less when in the teach mode. The robot should be programmed so that the operator cannot put it into automatic cycle using the teaching pendant.

Many of the safeguards that are used for machine safeguarding, such as safety mats and light curtains, can be used to safeguard a robot. Safeguards should prevent entry into the robot's operating area or envelope. The operating envelope is the entire volume swept by all possible motions of the robot. Fences or fixed guards should be placed around the perimeter of the robot operating area. Care should be taken not to make these barriers rigid in areas close to the robot as to avoid entrapment. Fences should not be easy to climb over. Access gates should be interlocked. The interlock should incorporate a blocking device to prevent opening until the robot is shut down. A deliberate manual action should be required to restart the robot. There should be ample clearance between the robot's operational area and any fixed objects such as columns, posts, or poles. Warning signs should be placed in the area and at the access gates. An amber light on the robot will make it more conspicuous and alert nearby workers that the robot is energized.

The following list (Russell 1982) provides additional factors to consider for the design, installation, operation, and maintenance of a robot system:

1. The robot should be positioned to prevent unintentional access. This may require barriers with interlocked gates and/or presence-sensing devices, such as light curtains and safety mats.
2. The control panel for the robot should be located outside the danger area.
3. A "power on" light should be installed so that it is clear to everybody that the robot is energized, even if it isn't moving.
4. The robot should not automatically reactivate if the electricity is interrupted.
5. If the robot stops suddenly, it should not lose its grip on the part it is handling.
6. Consideration should be given to equipping the robot arm with a pressure-sensitive switch that will shut it down if it strikes someone or something.
7. Emergency stop switches and/or trip wires should be installed around the perimeter of the robot's working area.
8. Computer software programs should include preprogrammed reach limits.
9. Fixed mechanical stops should be used to limit the reach of the robot arm.

10. Teaching pendants should have an emergency stop button. The robot should be operated at a reduced speed when in the teaching mode.
11. The robot system should be properly locked out during maintenance.

Powered Industrial Trucks

A powered industrial truck (PIT) is a powered vehicle that is used to transport, stack, and lift material. They travel up to 10 mph. There are many types of powered industrial trucks. Low lift trucks transport loads 4 to 6 inches from the floor and do not raise the load for stacking. Some low lift trucks are battery-powered and allow a rider to stand on the equipment. High lift trucks can transport and lift loads very high. Counterbalance rider trucks are the most common PIT used in industry. These trucks consist of adjustable forks, a tilt mast, and a counterweight. The counterweight acts to balance the truck when the load is lifted. These trucks are powered by electricity, gasoline, liquefied natural gas (LNG), liquefied petroleum gas (LPG), or diesel fuel. Narrow-aisle trucks are designed for material handling in narrow aisles. The operator stands on the operator's platform while working the controls. Order-picking trucks are another form of narrow-aisle truck, except the operator's platform rides up with the forks.

The improper operation of PITs continues to be responsible for worker and pedestrian injuries. Tables 11 and 12 list incident statistics (Swartz 2001). Collisions with fixed objects, such as doors, walls, posts, beams, sprinklers, pipes, racking, electrical boxes, fencing, machinery, and other trucks, can cause operator injuries. Pedestrians are also at risk. Injury to the operator and/or pedestrians can result from a falling load or a truck tipover. Rearward travel (forks trailing) is a common factor with narrow-aisle truck accidents, accounting for over 90 percent of collisions with racking, walls, posts, and other objects.

The risk of injuries can be reduced by a number of PIT design features. PITs that are capable of elevating loads should be equipped with an overhead guard to prevent falling objects from striking the operator. The guard should provide good visibility. PITs should have an audible warning device under the operator's control that can be heard above the ambient noise levels. Flashing lights are another warning device that can alert employees or pedestrians that a truck is approaching. If feasible, a backup alarm should operate when the truck is backing up. The controls should be confined to the operator's station. The operator should not have to reach outside this area where a body part could be pinched between the truck and a fixed object. There should be a deadman control; the truck should stop if the operator leaves the operator's station. ANSI B56.1 lists other specifications for steering, braking, and controls (ANSI 2000).

There are also design measures that can be implemented in the operating area. Barriers, such as horizontal guardrails or vertical posts, should be installed to protect offices, doors, stairs, electrical boxes, gas meters, racking, beams, and pedestrian walkways. Highlighting

TABLE 11

Collision Details of Narrow-Aisle Truck

Incident	Rate
Collisions with racking	33%
Collisions with walls	21%
Collisions with posts	18%
Collisions with other stationary objects	16%
Collision with other forklifts	12%

(*Source:* Swartz 2001, 26)

TABLE 12

Percentage of Forklift Fatalities

Incident	Rate
Tip over	25.3%
Struck by PIT	18.8%
Struck by falling load	14.4%
Elevated employee on truck	12.2%
Ran off dock or other surface	7.0%
Improper maintenance	6.1%
Lost control of truck	4.4%
Employees overcome by CO or propane fuel	4.4%
Faulty PIT	3.1%
Unloading unchocked trailer	3.1%
Employee fell from vehicle	3.1%
Improper use of vehicle	2.6%
Electrocutions	1.0%

(*Source:* Swartz 2001, 26)

posts and beams with yellow paint makes them readily conspicuous compared to the surroundings. Highlighted cushions around fixed objects such as steel columns not only make them conspicuous, but also reduce the force of impact as well. Mirrors should be installed at intersections. Pedestrian walkways should be highlighted. Adequate ventilation should be provided in operating areas to prevent the buildup of carbon monoxide.

Injuries with PITs cannot be reduced by design measures alone. Operators need to be trained. An industrial truck training program involves selecting, training, and testing the operator. The applicant should have a valid motor vehicle license and a good driving record. The operator should be mature, must be able to see well, judge distances, and have good hand and eye coordination. The training program should include classroom instruction, hands-on instruction, a driving test, and a written examination. Each trainee should have the operator's manual for the truck he/she will be operating. The operator training program should be specific to the type of trucks to be used, the company's policies, and the operating conditions. New operators and experienced operators who may be new to the particular facility need to be trained. All operators should receive a periodic refresher course. Any operator who has been involved in an accident while driving a PIT should be retrained before returning to work as an operator. There should be written documentation of the training.

Hands-on training is important, even if the operator has a valid driver's license. There are many differences between driving an automobile and driving a PIT. A PIT has a higher center of gravity and a shorter wheelbase than an automobile. Also, a PIT typically has three points of stability—two front wheels and the center of the rear axle—while an automobile has four. A PIT usually has only two braking wheels instead of four.

Hands-on training should be done in a planned practice area. The practice area should allow sufficient room for maneuvering a truck. The trainer should demonstrate each maneuver; then the trainee should perform the same maneuver. Each operator should successfully pass a written test and a driving skills test before being allowed to drive an industrial truck unsupervised.

Operators should keep their hands, legs, and other body parts inside the operating station of the truck and/or guard. Passengers should never be allowed to ride on the truck. Operators should sound the horn when approaching pedestrians and should never drive near a pedestrian who is standing in front of a fixed object. Operators should stop and sound the horn at blind corners and before going through doorways.

Automated Guided Vehicles (AGVs)

Unlike conventional powered industrial trucks (PITs), automated guided vehicles (AGVs) are driverless vehicles with automatic guidance systems capable of following prescribed paths. In automated factories and facilities, AGVs move pallets and containers. In offices they may be used to deliver and pick up the mail. Figure 28 shows a photo of an AGV.

FIGURE 28. An AGV system (Courtesy of FMC Technologies)

Until about 10 years ago, most AGVs followed electromagnetic wires buried in a floor. Then laser-guided systems came onto the market. These navigation systems allowed the AGV to determine its position in the plant based on the location of reflectors within the area. In-plant global positioning systems may be in the future.

From a safety standpoint, AGVs can reduce injuries due to driver inattention, driving too fast, or personnel not paying attention. However, AGV systems introduce a whole new set of safety concerns. There should be a reliable obstacle detection system so the AGVs do not strike an object or a person in its path. The AGV travel and turning paths need to be clearly marked so that personnel know what areas to stay clear of. Workers need to be trained in the risks associated with AGVs.

The AGV system should include a carefully designed traffic control system to prevent collisions between the AGVs and any towed vehicles. One of the oldest forms of traffic control systems is in-floor zone blocking. With this system, the guide path is divided into zones. Vehicles are allowed to enter a zone only when that zone is completely clear. Sensors in the floor at the zone boundaries detect vehicles as they pass. Vehicles can be released into the zone by having the path energized (Miller 1987)

All areas where there is pedestrian travel or interaction with workers should have a minimum clearance of 18 inches between the AGV and its load and any fixed object. This clearance allows a person to stand while the AGV and its load passes without presenting a safety hazard (Miller 1987).

ANSI B56.5 defines the safety requirements for guided industrial vehicles and automated functions of manned industrial vehicles (ANSI 2005). There are separate responsibilities for the user and for the manufacturer. User responsibilities include load stabilization, marking the travel path on the floor, including turning and maneuvering clearances, and preventative maintenance. The manufacturer is responsible for guidance, travel performance, braking emergency controls and devices, and object detection.

Obstacle detection systems have largely consisted of mechanical bumpers—giant emergency stops (E-stops) that stop the AGV if it contacts a person or obstacle. This low-cost method to prevent contact has a number of drawbacks. A contact-sensitive mechanical bumper has only two states: on or off. Once the contact has been made, additional time is required to stop the vehicle. The stopping time varies with the speed of the vehicle. From a productivity standpoint, the vehicles must operate at lower speeds and the E-stop must be reset after the obstacle is cleared. Mechanical bumpers are also sensitive to vibration and wear, require regular maintenance, and may not be totally effective when an AGV contacts an object when turning.

There are human-factors issues related to contact-sensitive obstacle detection systems. Workers may be uncomfortable with the fact that they have to be hit by the AGV before it will stop. Workers may also be injured trying to beat the AGV, thinking that they can get past the vehicle and a fixed object.

New advanced virtual bumpers using proximity laser scanner and laser scanner interface (PLS/LSI) technology are a significant improvement over contact-sensitive systems. A proximity laser scanner (PLS) creates a sensing field with a pulsed light that is reflected off of a rotating mirror so that it is transmitted in a 180-degree pattern. When an object enters the sensing field, the light is reflected back to the PLS. The distance to the object is computed using the time interval between the transmitted pulse and the reflected pulse, and the angle of the rotating mirror. The sensing field is divided into three areas: safety zone, warning zone, and surveyed zone. The surveyed area is the maximum radius surveyed by the PLS. When the PLS determines that an object is in the safety zone, hazardous motion is stopped. When an object is detected in the warning zone, it initiates a warning or an avoidance maneuver.

A laser scanner interface (LSI) is essentially a computer that interprets information and acts on it. The LSI can interpret data from up to four PLSs. When the PLS and LSI are combined in this way, the PLS acts as the *eyes* and the LSI acts as the *brain*. The LSI can also tell the PLS to change its view depending on the location within the plant. For example, if the LSI gets input from the navigation system that the AGV is near a corner, the PLS view can be changed to go around the corner. The PLS also has a self-checking feature.

The exit window is continuously checked for signal loss by an internal sensing array. Systems approved for AGV use should be able to detect a black vertical object with a diameter of 70 mm and a height of 400 mm (representing a leg in black trousers) anywhere in the AGV route.

There are a number of safety and productivity advantages with virtual bumpers. Speed can be controlled as the vehicle approaches an object and the configuration of the warning and safety zones can be flexible. For example, the vehicle slows down as it approaches an object. Also, a warning can be sounded if an object enters the warning zone. This advanced system permits increased speed and productivity because the size and shape of the stopping/slow-down/warning zones can be adjusted in real time depending on speed and the load. The protective area can be automatically extended on the sides of the vehicle when turning corners. Vehicle speeds can be increased in straight, unobstructed paths for improved productivity. Response times are as low as 60 milliseconds, depending on the system.

Consideration should be given to load size and tuggers when marking travel paths. Although a travel path may be adequate for the AGV, it may not be adequate if the AGV is towing a tugger or carrying a wide load. The travel path should be kept clear of material. Employees should be trained not to ride the AGVs. Training should also include instructing personnel to stay clear of an approaching AGV. The training program should also include contractors that may come into a plant or warehouse to perform work. A safety procedure should be put in place when work has to be done in an area where AGVs are operating. For example, weighted cones or other portable obstacles can be placed where workers may be working on or near an AGV travel path.

References

American Conference of Governmental Industrial Hygienists (ACGIH). 2004. *Industrial Ventilation.* Cincinnati, Ohio: ACGIH.

American National Standards Institute (ANSI). 1994. B11.1, *Safety Requirements for Mechanical Power Presses.* New York: ANSI.

______. 2000. B56.1, *Standard for Low Lift and High Lift Trucks.* New York: ANSI.

______. 2001. A1264.2, *Standard for the Provision of Slip Resistance on Walking/Working Surfaces.* New York: ANSI.

———. 2002a. Z535.1, *Safety Color Code.* New York: ANSI.

———. 2002b. Z535.2, *Environmental and Facility Safety Signs.* New York: ANSI.

———. 2002c. Z535.3, *Criteria for Safety Symbols.* New York: ANSI.

———. 2002d. Z535.4, *Product Safety Signs and Labels.* New York: ANSI.

———. 2002e. Z535.5, *Safety Tags and Barricade Tapes.* New York: ANSI.

______. 2002f. SNT 101, *Portable, Compressed-Air-Actuated, Fastener Driver Tools-Safety Requirements For.* New York: ANSI.

———. 2003. B11.19 2003, *Mechanical Power Presses—Safety Requirements for Construction, Care, and Use.* New York: ANSI.

______. 2005. B56.5 2005, *Safety Standard for Guided Industrial Vehicles and Automated Functions of Manned Industrial Vehicles.* New York: ANSI.

______. 2006a. B11.21, *Safety Requirements for Machine Tools Using a Laser for Processing Materials.* New York: ANSI.

———. 2006b. S1.4, *Specification for Sound Level Meters.* New York: ANSI.

American National Standards Institute and American Society of Mechanical Engineers (ANSI/ASME). 2003. B20.1, *Standard for Conveyors and Related Equipment.* New York: ANSI.

American Society for Testing and Materials (ASTM). 2002. F1637, *Standard Practice for Safe Walking Surfaces.* West Conshohocken, PA: ASTM.

___. 2005. F609-05, *Standard Test Method for Using a Horizontal Pull Meter.* West Conshohocken, PA: ASTM.

Arnold, R. 2002. "Choosing a Framing Nailer." *Fine Homebuilding* (Feb/March), pp. 68–75.

Burton, D. 1995. *Industrial Ventilation Workbook.* Bountiful, Utah: IVE, Inc.

DiPilla, S. and K. Vidal. 2002. "State of the Art in Slip Resistance Measurement." *Professional Safety* (June), pp. 37–42.

Gainer, C. G., and B. Jiang. 1987. *A Cause and Effect Analysis of Industrial Robot Accidents from Four Countries.* Dearborn, MI: Society of Manufacturing Engineers.

Gray, B. 1990. *Slips, Stumbles, and Falls.* Philadelphia: ASTM.

Grieser, B., T. Rhoades, and R. Shah. 2002. "Slip Resistance Field Measurements Using Two Modern Slipmeters." *Professional Safety* (June), pp. 43–48.

Hagan, Philip, et al., eds. 2001. *Accident Prevention Manual for Business and Industry Engineering & Technology.* 12th ed. Itasca, IL: NSC.

Hammer, Willie. 1972. *Handbook of System and Product Safety*. Englewood Cliffs, NJ: Prentice-Hall.
Harris, C. 1991. *Handbook of Acoustical Measurements and Noise Control*. New York: McGraw-Hill.
Hirschorn, Mark. 1989. *Noise Control Reference Handbook*. New York: Industrial Acoustics Company.
International Code Council (ICC). 2006a. *International Residential Code for One- and Two-Family Dwellings*. Country Club Hills, IL: ICC.
______. 2006b. *International Building Code*. Country Club Hills, IL: ICC.
______. 2006c. *International Mechanical Code*. County Club Hills, IL: ICC.
Jensen, P., C. Jokel, and L. Miller. 1978. NIOSH Publication No 79-117, *Industrial Noise Control Manual*. Washington, DC: NIOSH.
Knowles, E., ed. 2003. *Noise Control*. Des Plaines, IL: ASSE.
Kohr, Robert. 1994. "A New Focus on Slip and Fall Accidents." *Professional Safety* (January), pp. 32–36.
Marpet, Mark, and Michael Sapienza, eds. 2003. *Pedestrian Locomotion and Slip Resistance*. West Conshohocken, PA: ASTM.
Meserlian, Donald. 1995. "Effect of Walking Cadence on Static Coefficient of Friction Required by the Elderly." *Professional Safety* (November), pp. 24–29.
Miller, Richard. 1987. *Automated Guided Vehicles and Automated Manufacturing*. Dearborn: Society of Manufacturing Engineers (SME).
National Fire Protection Association (NFPA). 2005. *Fire Protection Handbook*. Quincy, MA: NFPA.
______. 2006. NFPA 101, *Life Safety Code*. Quincy, MA: NFPA.
National Safety Council (NSC). 1988. *Making the Job Easier—An Ergonomics Idea Book*. Itasca, IL: NSC.
Pierce, Alan. 1981. *Acoustics: An Introduction to its Physical Principles and Applications*. New York: McGraw-Hill.
Plog, B., ed. 2002. *Fundamentals of Industrial Hygiene*. 5th ed. Itasca, IL: NSC.
Rosen, Stephen. 2000. *The Slip and Fall Handbook*. 8th ed. Del Mar, CA: Hanrow Press.
Russell, John. 1982. "Robot Safety Considerations . . . a Checklist." *Professional Safety* (December), pp. 36–37.
Schultz, George. 2000. *Conveyor Safety*. Des Plaines, IL: ASSE.
______. 2004. "Conveyor Safety." *Professional Safety* (August), pp. 24–27.
Sotter, George. 1995. "Friction Underfoot." *Occupational Safety and Health* (March), pp. 28–34.
Swartz, George. 2001. "When PITs Strike People & Objects." *Professional Safety* (March), pp. 25–30.
Templer, John. 1994. *The Staircase*. Cambridge, MA: The MIT Press.
Thuman, A and R. Miller.1986. *Fundamentals of Noise Control Engineering*. Englewood Cliffs, NJ: Prentice-Hall.
Woodson, Wesley, et al. 1992. *Human Factors Design Handbook*, 2d ed. New York: McGraw-Hill, Inc.

Appendix: Recommended Reading

American Society of Heating, Refrigerating and Air-Conditioning Engineers (ASHRAE). 1999. *HVAC Applications*. Atlanta: ASHRAE.
______. 1997. *Fundamentals*. Atlanta: ASHRAE.
Baggs, J., et al. 2001. "Pneumatic Nailer Injuries." *Professional Safety* (January), pp. 33–38.
Beohm, Richard. 1998. "Designing Safety into Machines." *Professional Safety* (September), pp. 20–26.
Binder, Raymond. 1973. *Fluid Mechanics*. Englewood Cliffs, NJ: Prentice-Hall.
Constance, John. 1998. "Understanding Industrial Exhaust Systems." *Professional Safety* (April), pp. 26–28.
Etherton, John. 1988. *Safe Maintenance Guide for Robotic Workstations*. Washington, D.C.: NIOSH.
Gallagher, V. 1990. "Guarding by Location: A Dangerous Concept." *Professional Safety* (September), pp. 34–40.
Gierzak, J. 2003. "Duct Liner Materials and Acoustics." *ASHRAE Journal* (December), pp. 46–51.
Guckelberger, D. 2000. "Controlling Noise from Large Rooftop Units." *ASHRAE Journal* (May), pp. 55–62.
Guenther, F. 1998. "Solving Noise Control Problems." *ASHRAE Journal* (February), pp. 34–40.
Hagan, Philip, et al., eds. 2001. *Accident Prevention Manual for Business and Industry Administration & Programs*. 12th ed. Itasca, IL: NSC.
Howard, G. 1982. "Reduce Chain Saw Accidents." *Professional Safety* (December), pp. 15–18.
Hynes, G. 1981. "Occupational Noise Exposure." *Professional Safety* (September), pp. 13–18.
Illuminating Engineering Society (IES). 2003. *IESNA Lighting Ready Reference*. New York: Illuminating Engineering Society of North America.
Jones, Robert. 2003. "Controlling Noise from HVAC Systems." *ASHRAE Journal* (September), pp. 28–34.
Katz, Gary. 2006. "Avoiding Accidents on the Tablesaw." *Fine Homebuilding* (June), pp. 86–92.
Lilly, J. 2000. "Noise in the Classroom." *ASHRAE Journal* (February), pp. 21–29.
Marshall, Gilbert. 2000. *Safety Engineering*. 3d ed. Richard T. Beohm, ed. Des Plaines, IL: ASSE.
McConnell, Steven. 2004. "Machine Safeguarding: Building a Successful Program." *Professional Safety* (January), pp. 18–27.
Moisseev, N. 2003. "Predicting HVAC Noise Outdoors." *ASHRAE Journal* (April), pp. 28–34.

Moore, M., and G. Rennell. 1991. "Kickback Hazard: Do Manufacturer Warnings and Instructions Help Saw Users Understand the Risks." *Professional Safety* (April), pp. 31–34.

Mutawe, A., et al. 2002. "OSHA's Lockout/Tagout Standards: A Review of Key Requirements." *Professional Safety* (February), pp. 20–24.

National Safety Council (NSC). 1993. *Safeguarding Concepts Illustrated*. Itasca, IL: NSC.

______. 1991. *Falls on Floors*. Itasca, IL: NSC.

Occupational Safety and Health Administration (OSHA). 1980. 3027, *Concepts and Techniques of Machine Safeguarding*. Washington, D.C.: OSHA.

Parker, Jerald. 1988. *Heating, Ventilating and Air Conditioning*. New York: John Wiley & Sons.

Parks, R. 1996. "The Basics of Industrial Ventilation Design." *ASHRAE Journal* (November), pp. 29–34.

Roughton, James. 1995. "Lockout/Tagout Standard Revisited." *Professional Safety* (April), pp. 33–37.

Stalnaker, K. 2002. "Making the Transition from Startup to Normal Operation." *Professional Safety* (November), pp. 14–15.

Swartz, George. 1997. *Forklift Safety*. Rockville, MD: Government Institutes.

______. 1998. "Forklift Tipover: A Detailed Analysis." *Professional Safety* (January), pp. 20–24.

Turek, Mark.1991. "Lockout/Tagout Maintaining Machines Safely." *Professional Safety* (November), pp. 33–36.

Winchester, S. 1995. "Safety Fixtures for Table Saws and Shapers." *Fine Homebuilding* (December), pp. 78–80.

SECTION 1

RISK ASSESSMENT AND HAZARD CONTROL

LEARNING OBJECTIVES

- Be able to explain and apply the essential components of a hazard communication program.
- Understand and identify which components of a hazard communication program cost money, and how to assign a dollar value to those components.
- Understand and calculate the value of your time in a company, and how to identify and value the various costs and benefits that accrue to the company for designing and implementing any safety program.
- Demonstrate and describe why basic budgeting practices are insufficient to demonstrate the value added to the company by a hazard communication program.
- Identify and explain the essential differences between cost effectiveness, cost benefit and net present value analyses, and how each may be used to evaluate the impact of the hazard communication program to the company.

APPLIED SCIENCE AND ENGINEERING: PRESSURE VESSEL SAFETY

Mohammad A. Malek

A PRESSURE VESSEL is a closed container designed to withstand pressure, whether internal or external. The pressure may be imposed by an external source, by the application of heat from a direct or indirect source, or by any combination thereof. The pressure vessels are usually subjected to an internal or external operating pressure greater than 15 psig (103 kPa).

Internal pressure in a vessel is developed from the fluid in process application. External pressure on a vessel may be imposed by an internal vacuum or by pressure of the fluid between an outer jacket and the vessel wall. The components of a vessel may fail, causing dangerous accidents, if the vessel cannot withstand the internal or external pressure.

Pressure vessels are designed and constructed in various shapes. They may be cylindrical with heads, spherical, spheroidal, boxed, or lobed. Common types of pressure vessels include boilers, water heaters, expansion tanks, feedwater heaters, columns, towers, drums, reactors, heat exchangers, condensers, air coolers, oil coolers, accumulators, air tanks, gas cylinders, and refrigeration systems.

Pressure vessels contain fluids such as liquids, vapors, and gases at pressure levels greater than atmospheric pressure. Some of the fluids may be corrosive or toxic. All types of workplaces, from workshops and power generators to pulp and paper processors and large petrochemical refiners use pressure vessels. Small workshops use compressed air tanks. Petrochemical industries use hundreds of vessels that include towers, drums, and reactors for processing purposes. Depending upon the application, the vessels are constructed of either carbon steel or alloy steel.

Most pressure vessels are designed in accordance with the codes developed by the American Society of Mechanical Engineers (ASME) and the American Petroleum Institute (API). In addition to these codes, design engineers use engineering practices to make vessels safe. A pressure vessel bears the symbol stamping of the code under which it has been designed and constructed. Because a pressure vessel operates under pressure, safety is the main consideration during its design, construction, installation, operation, maintenance, inspection and repair.

Figure 1 shows a diagram of a typical pressure vessel. The main components are the shell, the head, and the nozzles. This cylindrical vessel is horizontal and may be supported by steel columns, cylindrical plate skirts, or plate lugs attached to the shell. The vessel may be used for any type of industrial process application under internal pressure.

Like any other machine, a pressure vessel is comprised of many components and fitted with controls and safety devices. The major components of a pressure vessel are as follows:

Shell: The main component—the outer boundary metal of the vessel.

Head: The end closure of the shell. Heads may be spherical, conical, elliptical, or hemispherical.

Nozzle: A fitting to inlet and outlet connection pipes.

Types of Pressure Vessels

There are many types of pressure vessels but they are generally classified into two basic categories:

1. Fired pressure vessels burn fuels to produce heat that in turn boils water to generate steam. Boilers and water heaters are fired pressure vessels.
2. Unfired pressure vessels store liquid, gas, or vapor at pressures greater than 15 psig (103 kPa) and include air tanks, heat exchangers, and towers.

This chapter will limit its discussion to unfired pressure vessels, and use of terminology should be assumed to reflect this throughout.

Most pressure vessels are cylindrical in shape. Spherical vessels may be used for extremely high-pressure operation. Vessels may range from a few hundred pounds per square inch (psi) up to 150,000 psi. The operating range of temperature may vary from –100°F to 900°F. The American Society of Mechanical Engineers (ASME), *Boiler and Pressure Vessel Code,* Section VIII – Division I (ASME 2007b) has exempted the following vessels from the definition of pressure vessel:

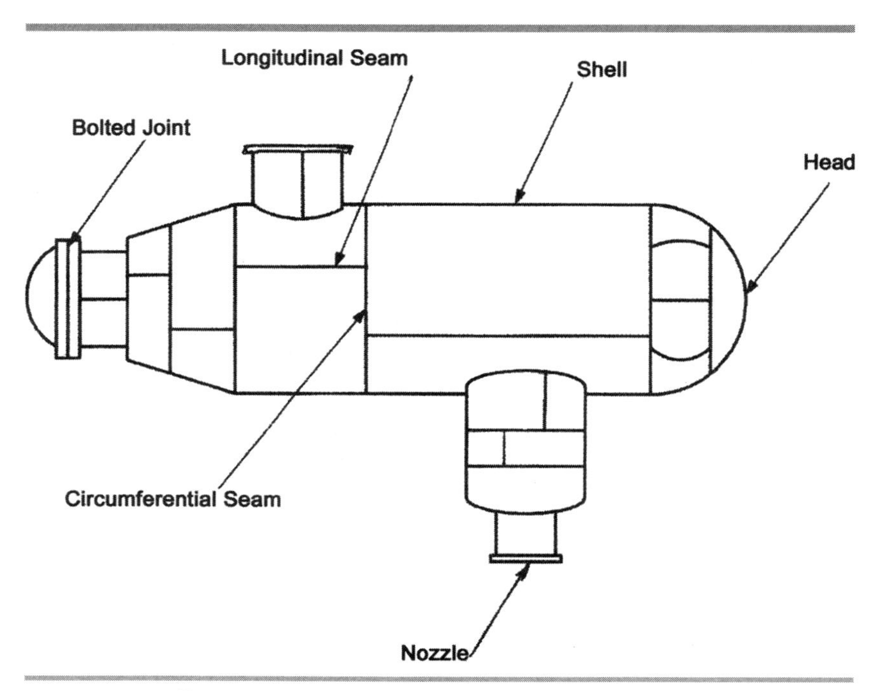

FIGURE 1. Pressure vessel diagram

1. Pressure containers that are integral components of rotating or reciprocating mechanical devices, such as pumps, compressors, turbines, generators, and so on
2. Piping systems, components, flanges, gaskets, valves, expansion joints, and such devices
3. Vessels for containing water under pressure, including those containing air compression to serve merely as a cushion, when neither of the following limitations are exceeded:
   a. design pressure of 300 psi
   b. design temperature of 210°F
4. Hot water supply storage tanks heated by steam or any other indirect means when none of the following limitations are exceeded:
   a. heat input of 200,000 Btu/hr
   b. water temperature of 210°F
   c. nominal water-containing capacity of 120 gal
5. Vessels having an internal or external operating pressure not exceeding 15 psi, regardless of size
6. Vessels having an inside diameter, width, height, or cross section diagonal not exceeding 6 in, regardless of vessel length or pressure
7. Pressure vessels designed for human occupancy

Pressure Vessel Code

Pressure vessels are designed, constructed, inspected, and certified according to the ASME's *Boiler and Pressure Vessel Code* (ASME 2007a), the American Petroleum Institute (API) Code, and the Tubular Exchanger Manufacturers Association (TEMA) Code:

1. ASME Boiler and Pressure Vessel Code (2007a). ASME Code Section VIII is used internationally for construction of pressure vessels. This Code has three separate divisions—Division 1: *Pressure Vessels* (2007b), Division 2: *Alternative Rules* (2007c), and Division 3: *Alternative Rules for Construction of High Pressure Vessels* (2007d).

ASME Section VIII—Division 1: *Rules for Construction of Pressure Vessels* (2007b). These rules contain mandatory requirements, specific prohibitions, and nonmandatory guidance for pressure vessel materials, design, fabrication, examination, inspection, testing, certification, and pressure relief.

ASME Section VIII—Division 2: *Alternative Rules for Construction of Pressure Vessels* (2007c). These rules provide an alternative to the minimum construction requirements for the design, fabrication, inspection, and certification of pressure vessels with maximum allowable working pressure (MAWP) from 3,000 to 10,000 psig.

ASME Section VIII—Division 3: *Alternative Rules for Construction of High Pressure Vessels* (2007d). These rules have to do with the design, construction, inspection, and overpressure protection of metallic pressure vessels with design pressure generally greater than 10,000 psi.

2. American Petroleum Institute (API) Code. API 510, *Pressure Vessel Inspection Code* (2006) is widely used in the petroleum and chemical process industries for maintenance inspection, rating, repair, and alteration of pressure vessels. This inspection code is only applicable to vessels that have been placed in service and that have been inspected by an authorized inspection agency or repaired by a repair organization defined in the code. The code has provisions by which to certify pressure vessel inspectors.

API RP 572, *Inspection of Pressure Vessels* (2001) is a recommended practice (RP) standard for inspection of pressure vessels (towers, drums, reactors, heat exchangers, and condensers). The standard covers reasons for inspection, causes of deterioration, frequency and methods of inspection, methods of repair, and preparation of records and reports.

API 620, *Recommended Rules for Design and Construction of Large, Welded, Low-Pressure Storage Tanks* (2004) provides rules for the design and construction of large, welded, low-pressure carbon steel aboveground storage tanks. The tanks are designed for metal temperature not greater than 250°F and with pressures in their gas or vapor spaces not more than 15 psig. These are low-pressure vessels that are not covered by ASME Section VIII—Division 1 Code (2007b).

API Standard 650, *Welded Steel Tanks for Oil Storage* (2007) covers material, design, fabrication, erection, and testing requirements of aboveground, vertical cylindrical, closed- and open-top, welded steel storage tanks in various sizes and capacities. This standard is applicable to tanks having internal pressures approximately of atmospheric pressure, but higher pressures are permitted when additional requirements are met.

API/ANSI Standard 660, *Shell-and-tube Heat Exchangers for General Refinery Services* (2007) defines the minimum requirements for the mechanical design, material selection, fabrication, inspection, testing, and preparation for shipment of shell-and-tube heat exchangers for general refinery services.

API/ANSI Standard 661, *Air-cooled Heat Exchangers for General Refinery Service, Petroleum and Natural Gas Industries* (2006) covers the minimum requirements for the design, materials, fabrication, inspection, testing, and preparation for shipment of refinery process air-cooled heat exchangers.

3. Tubular Exchanger Manufacturers Association (TEMA) Standards. The *TEMA Standards*, 9th Edition (2007) covers nomenclature, fabrication tolerance, general fabrication and performance information, installation, operation, maintenance, mechanical standard class RCB heat exchangers, flow-induced vibration, thermal relations, physical properties of fluids, and recommended good practices of shell and tube heat exchangers.

Potential Hazards

When a substance is stored under pressure, the potential for hazards exists. Improper vessel design and maintenance increase the risk of pressure vessel failure, posing a serious safety hazard. The risk increases when vessels contain toxic or gaseous substances. A pressure vessel is considered hazardous equipment.

Every year accidents occur to many pressure vessels that are in use in the industry. OSHA statistics indicate that 13 people were injured in 1999, 1 in 1998, 3 in 1997, and 9 in 1996 by pressure vessel accidents. A survey by the National Board of Boiler and Pressure Vessel Inspectors in 2002 shows 1663 accidents involving unfired pressure vessels that caused five fatalities and 22 injuries.

Pressure vessel accidents can be very serious. A serious accident takes human life, damages property, and increases costs for downtime production. Most accidents are caused by one of several things:

Slow rupture. A small crack can allow fluid to escape. The vessel typically remains intact, without fragments. The crack becomes larger over a period of time if not repaired early. Leak hazards are usually determined by the contents of the vessel. Fluid type, such as toxic gases, flammable vapors, and so on, determine the severity of leakage. High pressure inside the vessel may generate high-velocity gases, and with them tremendous cutting or puncturing forces.

Rapid rupture. Sudden increases of internal pressure can cause total structural failure of the vessel. The container is rapidly destroyed, producing fragments and sometimes a shock wave. If the vessel releases toxic or flammable materials, this may increase the likelihood of injury, death, or property damage. Generally, rupture occurs when the internal pressure exceeds the design limits or when structural damage caused by normal wear and tear, corrosion, galvanic action, or acute accident reduces the strength of the vessel. Figure 2 is an example of an air-tank explosion caused by normal wear and tear.

FIGURE 2. Air-tank explosion

Negative pressure. Many vessels collapse under negative pressure. Vacuum breakers can protect structural integrity.

Explosion of reactive chemicals. Rapid chemical reactions in the vessel may produce a large volume of gas in a short period of time. The mixture of gas expands, rapidly producing high temperatures and a shock wave having substantial destructive potential.

Causes of Deterioration

A variety of conditions can cause deterioration in pressure vessels. Common conditions are described below:

Corrosion. Corrosion is the most frequent condition found in pressure vessels. Most common corrosions involve pitting, line corrosion, general corrosion, grooving, and galvanic corrosion.

- In pitting, a vessel is weakened by shallow, isolated, and scattered pitting over a small area. Pitting may eventually cause leakage.
- In line corrosion, pits are nearly connected to each other in a narrow line. Line corrosion frequently near the intersection of the support skirt with the bottom of the vessel.
- General corrosion covers a considerable area of the vessel, reducing material thickness. Safe working pressure should be calculated based on the remaining material thickness.
- Grooving is caused by localized corrosion and may be accelerated by stress corrosion. Grooving is found adjacent to riveted lap joints or welds on flanged surfaces.
- In galvanic corrosion, two dissimilar metals contact each other and with an electrolyte constitute an electrolytic cell. The electric current flowing through the circuit may cause rapid corrosion of the less noble metal (the one having the greater electrode potential). The effects of galvanic corrosion are especially noticeable at rivets, welds, and flanged and bolted connections.

Fatigue. Many vessels are subjected to stress reversals such as cyclic loading. If stresses are high and reversals frequent, failure of components may occur as a result of fatigue. Fatigue failures may also result from cyclic temperature and pressure changes.

Creep. Creep may occur where vessels are subjected to temperatures above those for which they are designed. Because metals become weaker at high temperatures, such distortion may result in failure, especially at points of stress concentration.

Temperature. At sub-freezing temperatures, water and certain chemicals inside a vessel may freeze, causing failure. A number of failures have been attributed to brittle fracture of steels exposed to temperatures below their transition temperature and pressures greater than 20% of the hydrostatic test pressure.

Temper embrittlement. This is a loss of ductility and notch toughness caused by postweld heat treatment or high temperature service above 700°F. Low-alloy steels are prone to temper embrittlement.

Hydrogen embrittlement. This loss of strength and ductility in steels, caused by atomic hydrogen dissolved in steel, occurs at low temperature but is occasionally encountered above 200°F. It is typically caused by hydrogen produced from aqueous corrosion reactions.

Stress corrosion cracking. This cracking of metal is caused by the combined action of stress and a corrosive environment. Stress corrosion can only occur with specific metals in specific environments.

Causes of Accidents

There are many causes of pressure vessel accidents. These are analyzed by experts after accidents, using various methods, such as visual, nondestructive testing, destructive testing, and metallurgical analysis, to judge causes. Causes and points of failure may be categorized as follows:

- Safety valves
- Limit controls
- Improper installation
- Improper repair
- Faulty design
- Faulty fabrication
- Operator error
- Poor maintenance
- Irregular inspection

TABLE 1

Accidents, Injuries, and Deaths Involving Unfired Pressure Vessels

Year	Accidents	Injuries	Deaths
1993	261	24	6
1994	387	19	5
1995	245	65	6
1996	319	22	6
1997	292	41	13
1998	153	12	9
1999	145	73	6
2000	221	3	6
2001	201	18	4
2002	176	22	5

Accident Data

Historical data collected by the National Board shows that more people have died because of accidents involving unfired pressure vessels than because of those associated with fired pressure vessels (such as boilers). Table 1 shows the accidents, injuries, and deaths involving unfired pressure vessels that occurred in a recent ten-year period.

Accident Severity

Any pressure vessel accident can be severe, damaging lives and properties. Such an accident can also cause loss of production time and business, as well as employee layoff and fear. The severity of an accident depends on the pressure, temperature, and type of fluid inside the vessel involved. Vessel size is also important; the bigger the size, the vaster the heating surface exposed to explosive power.

Every accident costs money. Work accidents can for the purpose of cost analysis be classified into two general categories: (1) accidents causing work injuries and (2) accidents causing property damage or interfering with production. Furthermore, there are two types of costs: insured and uninsured.

Each company paying insurance premiums recognizes such expense as part of the cost of accidents. In some cases, medical expenses are covered by insurance. Insurance costs can be obtained directly from the insurance company.

Uninsured costs, sometimes referred to as "hidden costs," include the following:

- Wages paid for time lost by workers who were not injured
- Damage to material and equipment
- Wages paid for time lost by the injured worker, other than compensation payments
- Overtime pay rates necessitated by the accident
- Wages paid supervisors for time required for activities necessitated by the accident
- The effect of the decreased output of the injured worker upon return to work
- The effect of the training period for a new worker
- Uninsured medical cost borne by the company

The following example demonstrates the severity of a pressure vessel accident. At about 3:40 p.m. on April 13, 1994, pulp digester #15 at the Stone Container Corporation in Panama City, Florida, exploded (OSHA Accident: 170670616, Report ID: 0419700). The wood chip pulp digester ruptured and the release of pressure blew the digester through the roof, exposing workers to flying debris and hot wood pulp. The explosion killed three workers and injured five others including one worker who had third-degree burns over 25% of her body. Nearly 600 workers were laid off during the closure. The company's insurance covered property damage and business interruption claims for an undisclosed amount.

Because the State of Florida has no pressure vessel law, the digester was never inspected and certified by the Florida Boiler Safety Program. The condition of the digester before the explosion was unknown. OSHA investigated the accident and proposed an initial penalty of $1 million. After negotiation, the Stone Container Corporation agreed to pay OSHA $690,000 in penalties and to implement improvements at the facility. The corporation agreed to carry out new inspections, maintenance, repairs, and alterations to its digesters in accordance with national codes.

Safety Regulations

Codes and standards have no legal standing; they become mandatory only when adopted by jurisdictions having authority over the locations where the pressure vessels are installed. A jurisdiction is defined as a government authority such as municipality, county, state, or province.

Adoption of codes and standards is accomplished through legislative action requiring that pressure vessels for use within a jurisdiction must comply with the ASME, API, or other Code rules. Designated officials, such as chief boiler and pressure vessel inspectors and their staffs, enforce the legal requirements of the jurisdictions. Legal requirements of pressure vessels vary from jurisdiction to jurisdiction.

Federal Government

The federal requirements of pressure vessels are covered under the Occupational Safety and Health Act of 1970. The federal government has practically handed over pressure vessel safety to the state governments. The U.S. Department of Labor has issued following guidelines for pressure vessel safety:

1. *OSHA Technical Manual* (OTM), Section IV, Chapter 3, "Pressure Vessel Guidelines" (OSHA undated) contains an introduction and descriptions of recent instances of cracking under pressure, deals with nondestructive examination methods, and gives information to aid safety assessment (OSHA undated).
2. OSHA Publication 8-1.5, *Guidelines for Pressure Vessel Safety Assessment* (1989) presents a technical overview of, and information concerning, metallic pressure containment vessels and tanks. The scope is limited to general industrial application vessels and tanks constructed of carbon or low-alloy steels and used at temperatures between –100 and 600°F. Information on design codes, materials, fabrication processes, inspection, and testing that is applicable to these vessels and tanks are presented.

State Governments

Most states have pressure vessel safety laws, rules, and regulations. Pressure vessel safety bills are introduced by legislators, passed by houses and senates, and signed by governors, becoming laws. Rules and regulations are introduced by departments that have adopted pressure vessel safety programs. Codes and standards are adopted under rules and regulations.

Presently the following states enforce pressure vessel laws, rules, and regulations: Alabama, Alaska, Arkansas, California, Colorado, Delaware, Georgia, Hawaii, Illinois, Indiana, Iowa, Kansas, Kentucky, Maine, Maryland, Massachusetts, Minnesota, Mississippi, Missouri, Nebraska, Nevada, New Hampshire, New Jersey, New York, North Carolina, North Dakota, Ohio, Oklahoma, Oregon, Pennsylvania, Rhode Island, Tennessee, Utah, Vermont, Virginia, Washington, Wisconsin, and Wyoming.

The following states, however, are without pressure vessel laws: Arizona, Connecticut, Florida, Louisiana, Michigan, Montana, New Mexico, South Carolina, South Dakota, Texas, and West Virginia.

County Governments

Some counties have pressure vessel laws and rules, including Jefferson Parish (Louisiana) and Miami–Dade County (Florida).

City Governments

Many cities have their own pressure vessel laws and rules. The following cities enforce pressure vessel laws: Buffalo (New York), Chicago (Illinois), Denver (Colorado), Detroit (Michigan), Los Angeles (California), Milwaukee (Wisconsin), New Orleans (Louisiana), New York (New York), Omaha (Nebraska), Seattle (Washington), Spokane (Washington), St. Louis (Missouri), and Washington, D.C.

Certificates

The jurisdictional authorities, abiding by pressure vessel laws and rules, issue certificates to authorized inspectors and to pressure vessels. Certificates issued to inspectors, whether state or insurance company, are called certificates of competency. On the other hand, certificates issued to pressure vessels for operation of those vessels are called certificates of compliance.

Certificate of Competency. An authorized inspector must have a valid certificate of competency for performing inspections in a jurisdiction. The certificate, or commission, is issued by the pressure vessel program of the jurisdiction. This certificate usually expires December 31st yearly, irrespective of date of issue. The certificate indicates the inspector's identification (ID) number, the date of certificate issue, and the certificate expiration date, in addition to the inspector's name and address. A work card usually accompanies the certificate that can be carried in an inspector's wallet.

Certificate of Compliance. Each pressure vessel must have a valid certificate of compliance in order to operate the pressure vessel. The Pressure Vessel Program of the jurisdiction issues this certificate, also called the certificate of operation. This certificate is valid for 2 to 3 years, depending on the type of pressure vessel. A certificate of compliance is shown in Figure 3.

Construction of New Vessels

ASME *Boiler and Pressure Vessel Code,* Section VIII (2007a) establishes design rules for the vessels and their components. The Code delineates the responsibilities of the various organizations involved with the design and fabrication of components. It is the responsibility of the purchaser of a vessel to specify sufficient design information for manufacturers to design and fabricate vessels for the intended service. Users are responsible for providing manufacturers with information, including the intended operating pressure and temperature, as well as any other information relative to any special circumstances (such as lethal or other corrosion problems anticipated, or secondary loadings

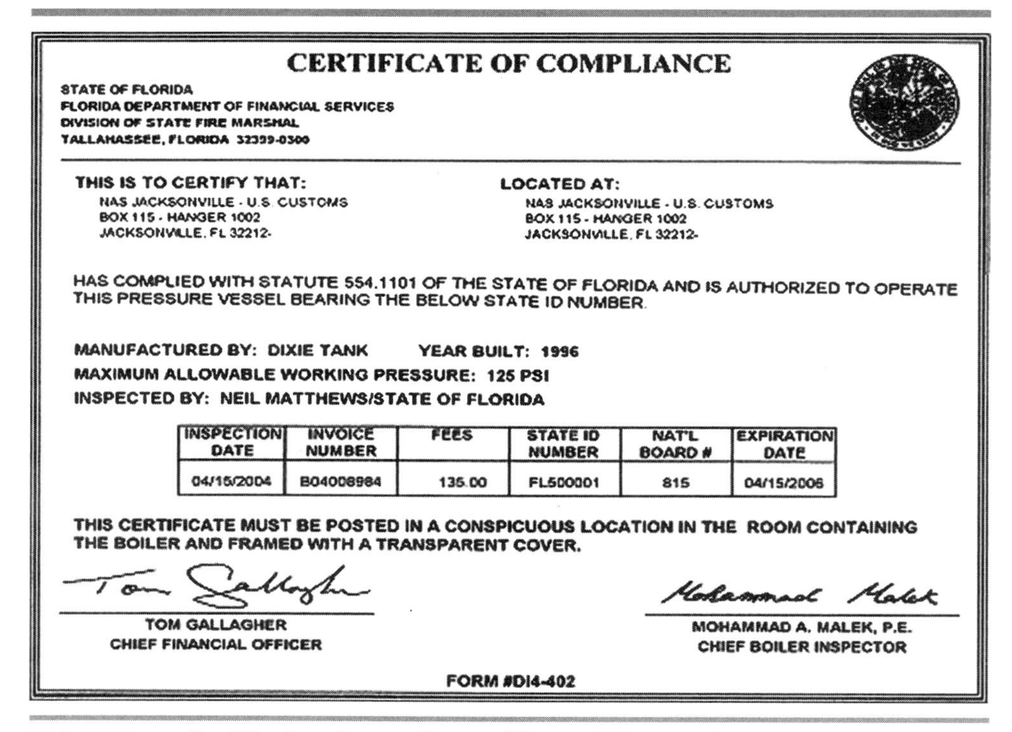

CERTIFICATE OF COMPLIANCE

STATE OF FLORIDA
FLORIDA DEPARTMENT OF FINANCIAL SERVICES
DIVISION OF STATE FIRE MARSHAL
TALLAHASSEE, FLORIDA 32399-0300

THIS IS TO CERTIFY THAT:
NAS JACKSONVILLE - U.S. CUSTOMS
BOX 115 - HANGER 1002
JACKSONVILLE, FL 32212-

LOCATED AT:
NAS JACKSONVILLE - U.S. CUSTOMS
BOX 115 - HANGER 1002
JACKSONVILLE, FL 32212-

HAS COMPLIED WITH STATUTE 554.1101 OF THE STATE OF FLORIDA AND IS AUTHORIZED TO OPERATE THIS PRESSURE VESSEL BEARING THE BELOW STATE ID NUMBER.

MANUFACTURED BY: DIXIE TANK YEAR BUILT: 1996
MAXIMUM ALLOWABLE WORKING PRESSURE: 125 PSI
INSPECTED BY: NEIL MATTHEWS/STATE OF FLORIDA

INSPECTION DATE	INVOICE NUMBER	FEES	STATE ID NUMBER	NAT'L BOARD #	EXPIRATION DATE
04/15/2004	B04008984	135.00	FL500001	815	04/15/2006

THIS CERTIFICATE MUST BE POSTED IN A CONSPICUOUS LOCATION IN THE ROOM CONTAINING THE BOILER AND FRAMED WITH A TRANSPARENT COVER.

TOM GALLAGHER
CHIEF FINANCIAL OFFICER

MOHAMMAD A. MALEK, P.E.
CHIEF BOILER INSPECTOR

FORM #DI4-402

FIGURE 3. Certificate of compliance (*Source:* State of Florida, Florida Department of Financial Services, Division of State Fire Marshal.)

such as earthquakes). In turn, vessel manufacturers are responsible for assuring that designs comply with all applicable requirements of the Code.

A pressure vessel is always designed by taking the safety factor into account. The vessel should withstand the operating conditions and function safely during its service life. Each and every component is designed carefully by using engineering judgment in addition to the code requirements, which are considered minimum requirements. The design engineer must determine all design parameters such as service, design temperature, design pressure, loadings, corrosion, materials, joint category, extent of nondestructive testing, extent of heat treatment, welding process to be used, and maximum allowable stress values.

Calculation of loads is very important in the design of a pressure vessel. The loadings are internal and/or external pressure, dead weight, local loads, cyclic loading, wind and earthquake loads, thermal stresses, and impact and shock loads.

Once the calculations, drawings and specifications are completed, a vessel is ready for construction. The following design for components is based on ASME Section VIII—Division 1 (2007b).

Cylindrical Shells under Internal Pressure

The minimum required thickness or maximum allowable working pressure of cylindrical shells shall be the greater thickness or lower pressure as given by formulas (1) or (2). The following symbols have been used in the formulas:

t = minimum required thickness of shell, in.
P = internal design pressure, psi
R = internal radius of the shell course, in.
S = maximum allowable stress values, psi
E = joint efficiency

(a) Circumferential Stress (Longitudinal Joints). When the thickness does not exceed one-half of the inside radius, or P does not exceed $0.385SE$, the following formula (Section VIII—Division 1) is applicable:

$$t = \frac{PR}{SE - 0.6t} \text{ or } P = \frac{SEt}{R + 0.6t} \tag{1}$$

(b) Longitudinal Stress (Circumferential Joints). When the thickness does not exceed one-half of the inside radius, or P does not exceed $1.25SE$, the following formula (Section VIII—Division 1) is applicable:

$$t = \frac{PR}{2SE + 0.4P} \text{ or } P = \frac{2SEt}{R - 0.4t} \tag{2}$$

Example: What is the MAWP of a cylindrical pressure vessel shell having a thickness of 1.125 inches and a type 2 longitudinal joint with RT 3 examination? The outside diameter of the vessel is 36 inches. The material is SA-515 Gr65 and the vessel design temperature is 800°F.

Solution:

OD = 36 in.
t = 1.125 in.

$$R = \frac{36 - 1.125 \times 2}{2} = 16.875 \text{ in.}$$

S = 11,400 psi (from ASME *Boiler and Pressure Code,* Section II, Part D, Stress Table (2007e))
E = 0.8 [from Table UW-12 of ASME Section VIII—Division 1 (2007b)]

Use formula 6.II.5.1 above:

$$P = \frac{SEt}{R + 0.6t}$$

$$= \frac{11{,}400 \times 0.8 \times 1.125}{16.875 + 0.6 \times 1.125}$$

$$= \frac{10{,}260}{17.55}$$

$$= 584.62$$

$$\approx 585 \text{ psig}$$

Therefore, MAWP of the cylindrical shell is 685 psig.

Formed Heads

Heads are used as enclosures for the shells. There are various types of dished heads used for pressure vessel construction. The most commonly used heads are ellipsoidal, torispherical, hemispherical, conical, or toriconical (cone head with knuckle). Sketches of these types of heads are shown in Figure 4. The following symbols have been used in the formulas of Paragraph UG-32 [Section VIII—Division 1 (2007b)] and Figure 4:

t = minimum thickness of head after forming, in.
P = internal design pressure, psi
D = inside diameter of the head skirt, in.
D_i = inside diameter of the conical portion, in. = $D - 2r\ (1 - \cos^{\alpha})$
r = inside knuckle radius, in.
S = maximum allowable stress values in tension as given in the tables referenced in UG-23, psi
E = lowest efficiency of any joint in the head
L = inside spherical or crown radius, in.

(a) Ellipsoidal Heads. The required thickness of a dished head of semiellipsoidal form, in which half the minor axis equals one-fourth of the inside diameter of the head skirt is calculated by the following formulas:

$$t = \frac{PD}{2SE - 0.2P} \text{ or } P = \frac{2SEt}{D + 0.2t} \quad (3)$$

An acceptable approximation of a 2:00 ellipsoidal head with one with a knuckle radius of $0.17D$ and a spherical radius of $0.90D$.

(b) Torispherical Heads. The required thickness of a torispherical head when the knuckle radius is 6% of the inside crown radius and the inside crown

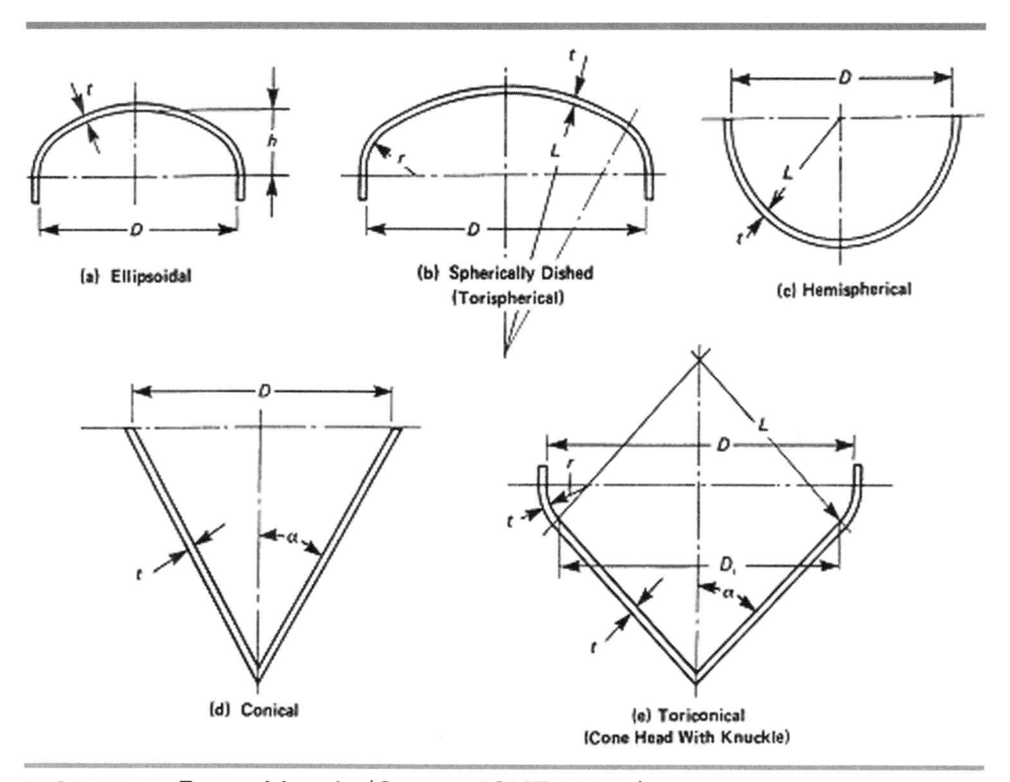

FIGURE 4. Formed heads (*Source:* ASME 2007a.)

radius equals the outside diameter of the skirt is calculated by the following formulas:

$$t = \frac{0.885PL}{SE - 0.1P} \text{ or } P = \frac{SEt}{0.885L + 0.1t} \quad (4)$$

Torispherical heads made of materials having a specified tensile strength exceeding 70,000 psi are designed using a value of S equal to 20,000 psi at room temperature and reduced in portion to the reduction in maximum allowable stress values at temperatures for the material.

(c) Hemispherical Heads. When the thickness of a hemispherical head does not exceed $0.356L$, or P does not exceed $0.665SE$, the following formulas are applicable:

$$t = \frac{PL}{2SE - 0.2P} \text{ or } P = \frac{2SEt}{L + 0.2t} \quad (5)$$

(d) Conical Heads and Sections (Without Transition Knuckle). The required thicknesses of conical heads or conical shell sections that have a half apex angle α not greater than 30 degrees are determined by the following formula:

$$t = \frac{PD}{2\cos\alpha(SE - 0.6P)}$$

or

$$P = \frac{2SEt\cos\alpha}{D + 1.2t\cos\alpha} \quad (6)$$

(e) Toriconical Heads and Sections. The required thickness of the conical portion, in which the knuckle radius is neither less than 6% of the outside diameter of the head skirt nor less than 3 times the knuckle thickness is determined by 6, using D_i in place of D. The required thickness of the knuckle is determined by formula 5 in which

$$L = \frac{D_i}{2\cos\alpha} \quad (7)$$

Toriconical heads or sections may be used when the angle $\alpha \leq 30$ degrees and are mandatory for conical head designs when the angle α exceeds 30 degrees.

Example: Determine the maximum allowable working pressure of a seamless ellipsoidal head with a 96-inch inside diameter and that is 0.25 inches thick. Stress value is 20,000 psi for SA-515 Gr70 at 500°F.

Solution:

D = 96 in.
E = 1.0 (for seamless head)
S = 20,000 psi for SA-515 Gr70 at 500°F
Maximum Allowable Working Pressure formula:

$$P = \frac{2SEt}{D + 0.2t}$$

$$P = \frac{2 \times 20{,}000 \times 1.0 \times 0.250}{96 + 0.2 \times 0.25}$$

$$P = \frac{10{,}000}{96.05}$$

$$P = 104.11$$
$$\approx 104 \text{ psig}$$

Therefore, maximum working pressure for the head is 104 psig.

Pressure Parts

In addition to shells and heads, prefabricated or preformed pressure parts may be used as components for pressure vessels that are subjected to allowable stresses caused by internal or external pressure. Pressure parts such as pipe fittings, flanges, nozzles, welding necks, welding caps, manhole frames, and covers may be formed by casting, forging, rolling, or die forming. No calculations are required for these parts.

Materials

Pressure vessel quality materials are used for the fabrication of pressure vessels. Great care is taken to assure quality of materials. The term *pressure vessel quality* requires that materials withstand the pressure, high temperature, corrosive environment, and other operating conditions for which pressure vessels are designed.

The specifications of pressure vessel quality materials, which are subject to stress caused by pressure, are described in ASME Code, *Materials,* Parts A, B, C, and D (2007e). Most of the materials are covered in Section II; those not so addressed are not permitted for construction uses. Materials for nonpressure parts, such as skirts, supports, baffles, lugs, and clips, need not conform to the specifications to which they are attached. If they are attached by welding, then these parts should be of weldable quality. Material specifications are not limited by production methods or countries of origin. The specifications of materials for major components are given below.

Plate Materials	
SA-285 Grade B	Carbon steel plate
SA-515 Grade 60	Carbon steel plate
SA-516 Grade 70	Carbon steel plate
SA-542 Grade B	2-1/4 Cr-1Mo plate

Pipe and Tubes	
SA-178 Grade A	Carbon steel welded tube
SA-106 Grade A	Carbon steel seamless pipe
SA-335 Grade P2	1/2Cr-1/2Mo seamless pipe
SA-556 Grade C2	Carbon steel seamless tube

Pressure parts	
SA-105	Carbon steel forgings
SA-234 Grade WPB	Seamless and welded fittings
SA-281	Carbon steel forgings
SA-352 Grade LCA	Carbon steel castings

Fabrication

A pressure vessel is fabricated according to the drawings, design, and calculations. The fabrication may be performed partly or wholly in the manufacturer's shop or in the field. Most of the fabrication is done by welding in accordance with ASME Code, Section IX, *Welding and Brazing* (2007f). The fabrication procedures are required to comply with ASME Section VIII.

Shop Inspection

A pressure vessel is inspected and tested in the fabrication shop or field during its construction by an authorized inspector. The manufacturer is responsible for design, construction, and quality control in accordance with ASME Code, Section VIII (2007a). The authorized inspector is responsible for the inspection and testing of a pressure vessel for Code compliance.

Authorized Inspectors. The authorized inspector (AI) is the person responsible for inspection of pressure vessels and their parts during and after construction. An authorized inspector acts as a third party to make sure that all the Code requirements are met.

The authorized inspector is assigned the following responsibilities:

1. Checking the validity of the manufacturer's Certificate of Authorization
2. Ensuring that the manufacturer follows the quality control system accepted by the ASME
3. Reviewing design calculations, drawings, specifications, procedures, records, and test results
4. Checking material to ensure that it complies with Code requirements
5. Checking all welding and brazing procedures to ensure that these are qualified
6. Checking the qualification of welders, welding operators, brazers, and brazing operators
7. Visually inspecting the transfer of material identification
8. Witnessing the proof tests
9. Inspecting each pressure vessel during and after construction
10. Verifying the stamping and nameplate information and attachment
11. Signing Manufacturer's Data Reports.

The authorized inspector ensures that all necessary inspections have been performed to certify that the pressure vessels have been designed and constructed in accordance with Section VIII and that the Code Symbol stamp can be applied (ASME 2007a).

Many pressure vessels, especially large ones, are constructed partly in the shop and partly in the field. The authorized inspector is also responsible for inspection in the field if the manufacturer performs construction there. The field inspection is similar to shop inspection except that inspection procedures are carried out at the site.

Stamping and Data Reports

Each pressure vessel designed according to the rules of Section VIII is required to be stamped with the appropriate Code Symbol stamp. The stamping is done after the hydrostatic test and in the presence of the authorized inspector. Once the stamping has been applied, the manufacturer completes the Data Reports and the authorized inspector signs them (ASME 2007a).

Stamp Holders

The construction of pressure vessels may be performed either in the shop or in the field. In both cases, a manufacturer or stamp holder is required to have a Certificate of Authorization to fabricate and stamp pressure vessels in accordance with the provisions of Section VIII Code (ASME 2007a).

An organization interested in fabricating pressure vessels may apply to ASME for a Certificate of Authorization in accordance with Paragraph UG-105 of Section VIII (ASME 2007a). The prescribed forms are available from the Secretary of the Boiler and Pressure Vessel Committee. The applicant is required to complete the form specifying the ASME stamp required and the scope of activities to be performed. The Secretary issues a Certificate of Authorization for each plant.

If authorization is granted after receiving a recommendation from the Joint Review Team, and the proper administrative fee is paid, an organization is granted a Certificate of Authorization giving permission to use the ASME Symbol stamp for a period of three years. A copy of the Certificate of Authorization for ASME Code symbol "U" is shown in Appendix A.

The Certificates of Authorization and the Symbol stamps are the property of ASME. The stamp holder organization must use them in accordance with the rules and regulations of the Society. The organization must not use the stamp when a certificate expires and shall not allow any other organization to use it. The Certificates of Authorization and stamps are required to be returned to the Society on demand. A stamp holder is required to apply for renewal six months prior to the expiration date of the certificate. The Society has the right to cancel or renew Certificates of Authorization.

Quality Control System

A manufacturer applying for a Certificate of Authorization to use a Code symbol stamp must have a quality control system in place. The quality control system is a written document describing all Code requirements, including material, design, fabrication, and testing by the manufacturer, as well as inspection of pressure vessels and their components by the authorized inspector, that will be met.

A joint review team visits the manufacturer's shop facilities to determine whether the quality control system is adequate to manufacture pressure vessels and their parts. The team consists of a representative of the authorized inspection agency and an ASME designee, who is leader of the team. The ASME designee is selected by the legal jurisdiction. In some cases, a representative of the legal jurisdiction can act as a team leader. The joint review team evaluates the quality control system and its implementation. The manufacturer is required to demonstrate administrative and fabrication functions to show the ability to produce code items according to the description of the quality control system. The demonstration may include a model vessel or part showing fabrication functions.

The joint review team submits its written recommendations to the Society. The Subcommittee on Boiler and Pressure Vessel Accreditation reviews the report, and a certificate of authorization is issued if the recommendation is satisfactory.

The quality control system, which is presented in a document called a "quality control manual," is a very important document for the manufacturer. The manufacturer, the inspector, and the joint review team use this manual to perform all Code activities. The manufacturer must describe the system suitable for the manufacturer's own circumstances. Description of the quality control system is required to meet the requirements of the Mandatory Appendix A-10 of Section VIII—Division 1 (ASME 2007b). The outlines of the features of the quality control system are

1. Authority and responsibility
2. Organization
3. Drawings, design calculations, and specification control
4. Material control
5. Examination and inspection program
6. Correction of nonconformities
7. Welding
8. Nondestructive examination
9. Heat treatment
10. Calibration of measurement and test equipment
11. Records retention
12. Sample forms
13. Inspection of pressure vessels and parts
14. Inspection of safety and safety relief devices

The quality control manual may contain other non-code activities of the manufacturer. The code activities in the shop and the field must be described clearly.

Code Symbol Stamps

The American Society of Mechanical Engineers (ASME) uses six Code Symbol stamps—U, UM, U2, U3, UV, and UD, either individually or in combination—for the design, construction, and assembly of pressure vessels and their parts in accordance with Section VIII (ASME 2007a):

U: Code symbol for vessels designed under Section VIII—Division 1 (2007b)
UM: Code symbol for miniature vessels designed under Section VIII—Division 1 (2007b)
U2: Code symbol for vessels designed under Section VIII—Division 2 (2007c)
U3: Code symbol for vessel designed under Section VIII—Division 3 (2007d)
UV: Code symbol stamp for pressure relief valve
UV3: Code symbol stamp for pressure relief valve, as required by Section VIII—Division 3 (2007d)
UD: Code symbol stamp for rupture disk devices

Required Marking

Data marking. Each pressure vessel, after fabrication, shall be marked with the following data:

1. The official code symbol stamp
2. The name of the manufacturer of the pressure vessel
3. The maximum allowable working pressure: __________ psi at ________ °F
4. The minimum design metal temperature of _____________ °F at ___________ psi
5. The manufacturer's serial number
6. The year built

Type of Construction. The type of construction used for the vessel shall be indicated under the code symbol stamp by applying the appropriate letter(s) as shown in Table 2.

Vessels employing combination types of construction must be marked with all applicable construction letters.

TABLE 2

Marking for Type of Construction

Type of Construction	Letters
Arc or gas welded	W
Pressure welded (except resistance)	P
Brazed	B
Resistance welded	RES

Method of Marking

A manufacturer is required to stamp each pressure vessel after the hydrostatic test. The stamping is done in the presence of the authorized inspector either in the shop or the field. The stamping consists of the appropriate code symbol stamp and data items with letters and figures at least 5/16 in. (8 mm) high. An example of the form of stamping is shown in Figure 5.

If the stamping is so located that it is not visible from the operating floor, the manufacturer is required to provide a metallic nameplate on which all data are stamped. The letters and figures on the nameplate should not be less than 5/32 in. (4 mm) high.

Manufacturer's Data Reports

A Manufacturer's Data Report is similar to a birth certificate. During and after fabrication of a pressure vessel in the shop or field, principal data of the pressure vessel and its components are recorded. The data report provides certain important information regarding manufacturer, purchaser, location, and identification number. The data report also provides summary information about construction details, materials, dimensions, design pressures, hydrostatic test pressures, and maximum designed steaming capacity. This information is used in making decisions about future repairs and alterations of pressure vessels.

There are twelve different Manufacturer's Data Reports for the pressure vessels constructed in accordance with the requirements of Section VIII (ASME 2007a). These data reports are recorded on forms published by the ASME. The Manufacturer's Data Report Forms are

1. Form U-1, Manufacturer's Data Report for Pressure Vessels as required by Section VIII—Division 1 (Appendix B) (ASME 2007b)

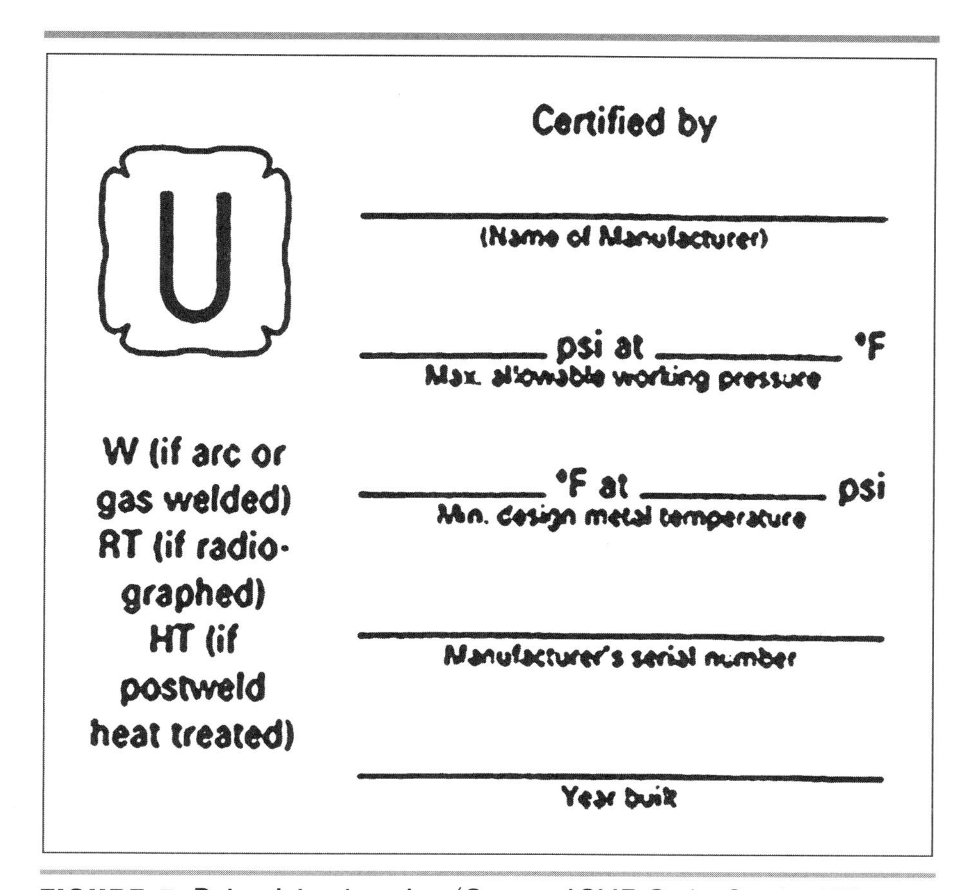

FIGURE 5. Data plate stamping (*Source:* ASME Code, Section VIII–Division 1 2007b)

2. Form U-1A, Manufacturer's Data Report for Pressure Vessels (Alternative Form for Single Chamber, completely shop-fabricated vessel only) as required by Section VIII—Division 1 (ASME 2007b)
3. Form U-2, Manufacturer's Partial Data Report as required by Section VIII—Division 1 (ASME 2007b)
4. Form U-2A, Manufacturer's Partial Data Report (Alternative Form) as required by Section VIII—Division 1 (ASME 2007b)
5. Form U-3, Manufacturer's Certificate of Compliance for Pressure Vessels with UM symbol as required by Section VIII—Division 1 (ASME 2007b)
6. Form U-4, Manufacturer's Data Report Supplementary Sheet as required by Section VIII—Division 1 (ASME 2007b)
7. Form A-1, Manufacturer's Data Report for Pressure Vessels Constructed under Section VIII—Division 2 (ASME 2007c)
8. Form A-2, Manufacturer's Partial Data Report as required under Section VIII—Division 2 (ASME 2007c)
9. Form A–3, Manufacturer's Data Report Supplementary Sheet as required under Section VIII—Division 2 (ASME 2007c)
10. Form K-1, Manufacturer's Data Report for High Pressure Vessels as required by Section VIII—Division 3 (ASME 2007d)
11. Form K-2, Manufacturer's Partial Data Report for High Pressure Vessels as required by Section VIII—Division 3 (ASME 2007d)
12. Form K-3, Manufacturer's Data Report Supplementary Sheet as required by Section VIII—Division 3 (ASME 2007d)

Safety Devices

Pressure sources in a vessel are limited to the maximum allowable working pressure of the lowest system component. Pressure-relief devices are used in case the internal pressure exceeds the maximum allowable working pressure. Common relief devices include pressure relief valves and rupture disks.

A conventional pressure relief valve is a direct spring-loaded pressure relief valve whose operating characteristics are directly affected by changes in the back pressure. This type of valve is used for vessels designed for Section VIII Code (ASME 2007a). A spring-loaded pressure relief valve is shown in Figure 6.

The capacity of a safety relief valve in terms of a gas or vapor other than the medium for which the valve was officially rated may be determined by the following formulas:

For steam

$$W_s = 51.5\,KAP \tag{8}$$

For any gas or vapor

$$W = CKAP\sqrt{\frac{M}{T}} \tag{9}$$

where

W = flow of any gas or vapor, lb/hr
W_s = rated capacity, lb/hr of steam
C = constant for gas or vapor which is function of the ratio of specific heats, $k = C_p/C_v$
K = coefficient of discharge [Para. UG-131(d) and (e) of ASME Section VIII, Div. 1]
A = actual discharge area of the safety relief valve, sq in.
P = (set pressure x 1.10) plus atmospheric pressure, psia
M = molecular weight
T = absolute temperature at inlet (°F + 460)

Example: A safety valve is required to relieve 5000 lb/hr of propane at a temperature of 125°F. The safety valve is rated at 3000 lbs/hr steam at the same pressure setting. Will this valve provide the required relieving capacity in propane service on this vessel?

Solution:

W_p = 5000 lbs/hr
W_s = 3000 lbs/hr
Molecular weight of propane = 44.09
C = 315
T = 125 + 460 = 585

$$W_p = CKAP\sqrt{\frac{M}{T}}$$

Transpose for KAP:

$$KAP = \frac{W_p}{C\sqrt{\frac{M}{T}}}$$

$$KAP = \frac{5000}{315\sqrt{\frac{44.09}{585}}}$$

$$KAP = 57.81857$$

For steam $W_s = 51.5KAP$

$$W_s = 51.5 \times 57.81857$$

$$W_s = 2{,}977.65627 \approx 2978 \text{ lbs/hr}$$

The safety relieving capacity required is 2978 lbs/hr and the capacity provided is 3000 lbs/hr.

Therefore, the valve will provide required capacity in propane on this vessel.

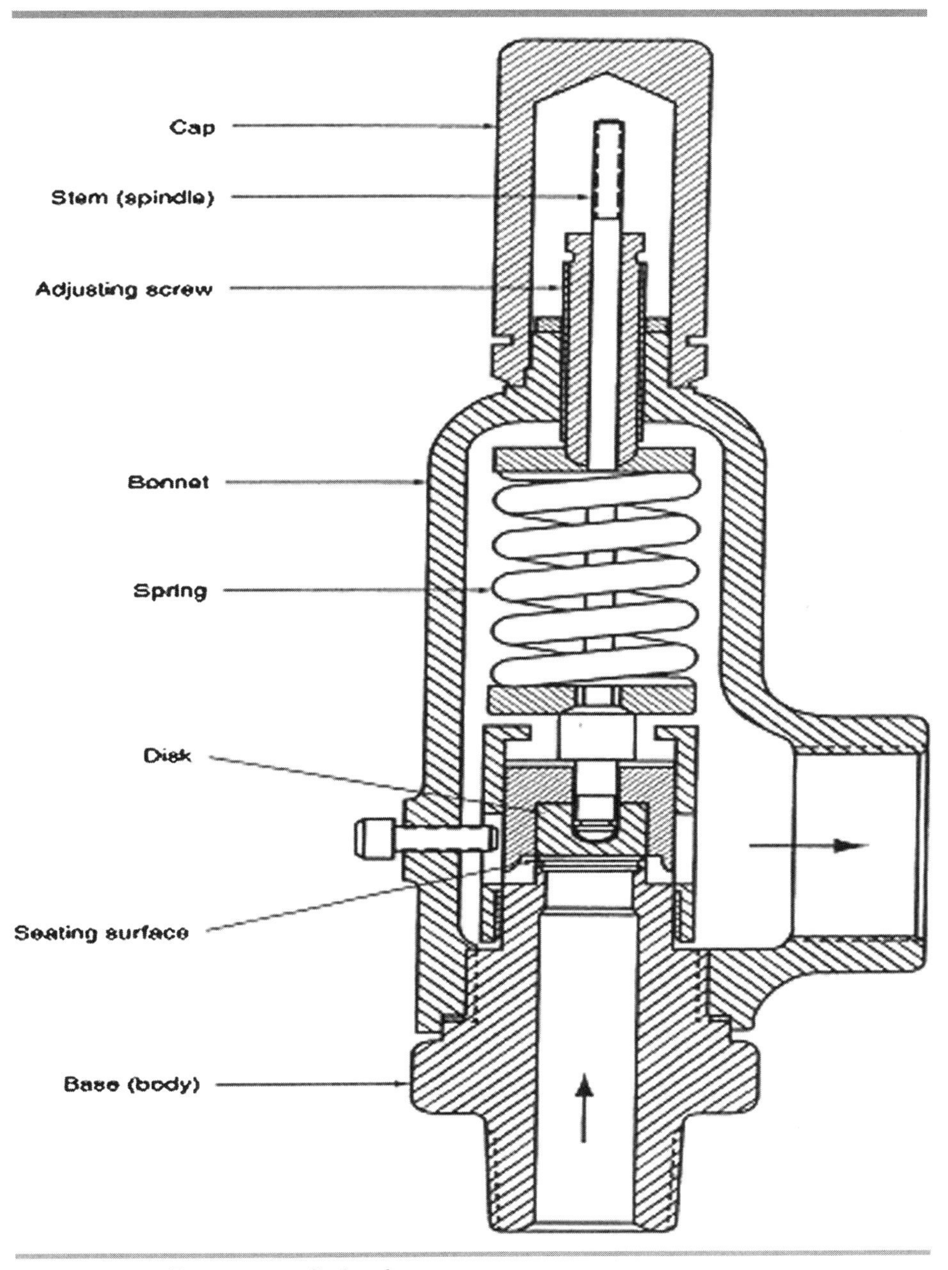

FIGURE 6. Pressure relief valve

A rupture disk is designed to rupture at a predetermined burst rating when installed in a piping system. The rupture disks protect equipment from the effects of overpressurization in static and dynamic pressurized systems. The rupture disk should have a specified bursting pressure at a specified temperature. The manufacturer of rupture disk should guarantee that the disk will burst within ±5% of the specified bursting pressure.

Risk Assessment of Existing Vessels

It is extremely important to assess the safety and hazard implications of pressure vessels in service. The equipment data and current information should be made available to a technical expert for evaluation of safety and risk.

Corrosion Rate Determination

Corrosion rate is defined as loss of metal thickness, in inches (or millimeters) per year. An inspection should determine the vessel's probable corrosion rate.

The corrosion rate of a pressure vessel may be determined by one of the following methods:

1. A corrosion rate may be calculated from data collected by the owner on pressure vessels in the same or similar service.
2. A corrosion rate may be estimated from the owner's experience or from published data on vessels used for comparable service.
3. A corrosion rate may be determined from on-stream after 1000 hours of service by using corrosion-monitoring devices or actual thickness measurements of the vessel.

During inspection, a thorough examination should be performed using ultrasonic thickness measurements or other appropriate means to measure thickness for assessing the integrity of the metal. A representative number of thickness measurements should be taken on all major components such as shells, heads, cone sections, and nozzles. The number and location of thickness measurements should be decided based on previous experience. Once thickness is known, corrosion rate may be calculated by the following formula:

$$Corrosion\ rate = \frac{t_{previous} - t_{actual}}{years\ between\ t_{previous}\ and\ t_{actual}} \quad (10)$$

where corrosion rate is measured in inches (millimeters) per year and equals t_{actual} (the thickness, in inches or millimeters), recorded at the time of inspection for a given location or component minus $t_{previous}$ (the thickness, in inches or millimeters) at the same location as measured during a previous inspection, divided by the lapse in years.

Example: During the in-service inspection of a pressure vessel, an ultrasonic thickness of a cylindrical shell was measured at 2.75 in. The vessel, when installed 5 years prior, was 3 inches thick. The vessel has a MAWP of 2900 psig and an inside diameter of 24 inches. What is the corrosion rate if the vessel is used for ammonia service?

Solution:

t_{actual} = 2.75 in.
$t_{previous}$ = 3 in.
Years between t_{actual} and $t_{previous}$ = 5

$$Corrosion\ rate = \frac{3 - 2.75}{5} = 0.05$$

The corrosion rate of the vessel is 0.05 in. (1.27 mm) per year.

Statistical data may be maintained for corrosion rate calculations for the pressure vessel components. This statistical approach may be used to assess inspection types or to determine inspection intervals.

Remaining Life Determination

Remaining life refers to the number of years a pressure vessel is expected to last in service. This life expectancy is a very important consideration when making decisions regarding pressure vessel replacement, plant production capacity, and appropriate inspection interval.

The period between internal or on-stream inspections shall not exceed half the estimated remaining life of the vessel based on the corrosion rate, or 10 years, whichever is less. In cases where the remaining safe operating life is estimated to be less than 4 years, the inspection interval may be the full remaining safe operating life up to a maximum of 2 years.

For pressure vessels in noncontinuous service that are isolated from process fluids such that they are not exposed to corrosive environments (such as inert gas purged or filled with noncorrosive hydrocarbons), the 10 years shall be the 10 years of actual exposed life.

The remaining life of the pressure vessel can be calculated from the following formula:

$$\text{Remaining life (years)} = \frac{t_{actual} - t_{\min imum}}{Corrosion\ rate} \quad (11)$$

where $t_{minimum}$ equals the minimum allowable thickness, in inches (or millimeters), for a given location or component, divided by the yearly corrosion rate in inches (or millimeters).

Example: During the in-service inspection of a pressure vessel, an ultrasonic thickness of a cylinder shell was measured at 2.8 inches. The vessel, when installed 5 years prior, was 3 inches thick. The vessel has a MAWP of 2500 psig and an inside diameter of 30 in. The vessel is used for propane service and a minimum thickness of 2.5 inches can be allowed for the shell. What is the remaining life of the vessel?

Solution:

t_{actual} = 2.8 in
$t_{previous}$ = 3 in.
$t_{minimum}$ = 2.5 in.
Years between t_{actual} and $t_{previous}$ = 5

$$Corrosion\ \ rate = \frac{3 - 2.80}{5} = 0.04\,\text{in.}\ \ (1.106\,\text{mm})\ \text{yearly}$$

$$\text{Remaining life} = \frac{t_{actual} - t_{previous}}{Corrosion\ rate}$$

$$\text{Remaining life} = \frac{2.80 - 2.50}{0.04} = 7.5\ \text{years}$$

Therefore, the remaining life of the vessel is 7.5 years.

Statistical data should be kept of remaining life calculations for pressure vessel components. This statistical approach may be used to assess inspection types or to determine inspection intervals.

Maximum Allowable Working Pressure Determination

The maximum allowable working pressure for a pressure vessel in service may be determined using the formulas of ASME Code, Section VIII—Division 1 (ASME 2007b). For a pressure vessel in service, it is recommended that the original ASME Code edition or construction code (for non-ASME designed vessels) should be used. The calculated maximum allowable working pressure should not be greater than the original maximum allowable working pressure.

Maximum allowable working pressure calculations require detailed information used in the original code of construction. Details include design data for all the major components, materials specifications, allowable stresses, weld efficiencies, inspection criteria, and service conditions. The wall thickness used in the calculations shall be the actual thickness determined by nondestructive examination at the time of inspection. This actual thickness shall not be greater than the thickness reported in the manufacturer's data report. An example of determining maximum allowable working pressure for an existing pressure vessel in service is given below:

Example: During in-service inspection of a pressure vessel, an ultrasonic measurement taken of a cylindrical shell recorded a thickness of 0.595 inches. The vessel has a MAWP of 1300 psig at 650°F and an outside diameter of 12.75 in. The vessel is fabricated from SA-106 Gr B, 0.688-inch nominal thick pipe with an allowable stress of 15,000 psi. Based on this information, may the vessel be allowed to continue in operation as-is, or will repairs be required?

Solution:

P = 1300 psig at 650°F
D_o = 12.75 in.

$R_o = 6.375$ in.
$t_{measured} = 0.595$ in.
$S = 15{,}000$ psi
Seamless pipe, no radiography, $E = 0.85$

The formula for minimum required thickness is:

$$t = \frac{PR_o}{SE - 0.6P} \quad (12)$$

$$t = \frac{1300 \times 6.375}{15{,}000 \times 0.85 - 0.6 \times 1300}$$

$$t = \frac{8287.5}{11{,}970}$$

$t = 0.692$ in.
(Section VIII—Division 1, 2007b)

Because the measured thickness of 0.595 inches is less than the calculated minimum thickness of 0.692 in., it does not comply with the Code. Therefore, the vessel may not be allowed to continue in operation as-is. A repair is required to build up the minimum required thickness at that location.

In-Service Inspection

In-service inspection plays a vital role in assessment of pressure vessel safety. In selecting the type of inspection, both the condition of the vessel and the environment in which it operates should be taken into consideration. For example, internal inspection is not required for a newly installed pressure vessel, but external inspection is required to verify that installation has taken place according to the code.

An authorized inspector determines the type of inspection to be performed on a particular vessel. The inspection may include various inspection methods such as visual inspection, pressure test, and nondestructive examinations. The appropriate inspection method should provide the information necessary to determine that all components of the vessel are safe to operate until the next scheduled inspection.

Internal inspection. The internal inspection, which may be a visual inspection or a combination of visual and nondestructive examination, is used for assessing the internal condition. Defects such as corrosion, erosion, and cracks are generally revealed as the result of internal inspection. The authorized inspector may replace internal inspection with on-stream inspection. The period between internal or on-stream inspections should not exceed half the remaining life of the vessel (projected by corrosion rate) or 10 years, whichever is less.

On-stream inspection. The on-stream inspection is conducted either while the vessel is out of service, while it is depressurized, or while it is on stream and under pressure. When an on-stream inspection is conducted in lieu of an internal inspection, a thorough examination should be performed using ultrasonic thickness measurements, radiography, or other appropriate means of nondestructive examination (NDE) to measure metal thickness. The authorized inspector should have sufficient access to all parts of the vessels that accurate assessment can be made of vessel condition.

External inspection. Each pressure vessel should be inspected externally when the vessel is under operation. The interval between consecutive external inspections should be 5 years or the same interval as between required internal or on-stream inspections, whichever is less. Such inspection should reveal the condition of insulation and supports, allowance for expansion, evidence of leakage, and general alignment of the vessel on its supports. The inspector may wish to observe conditions under the insulation, but the condition of the insulating system and outer jacketing (such as of the cold box) should also be inspected no less frequently than every 5 years.

Pressure test. A pressure test is necessary when certain repairs and alterations are made to a pressure vessel. The pressure test has also to be done if the authorized inspector believes that a pressure test is necessary to assess the mechanical integrity of the vessel. Pneumatic testing may be applied when hydrostatic testing is impractical because of temperature, foundation, refractory lining, or process reasons.

Risk-Based Inspection

A risk-based inspection (RBI) is a system of combining the assessment of likelihood of failure and the consequence of failure. When an owner chooses to conduct an RBI assessment, it includes a systematic evaluation of both the likelihood of failure and the associated consequence of failure.

The assessment of likelihood of failure must be based on all the possible degradation that could reasonably be expected to affect a vessel. The degradation mechanism in pressure vessels includes metal loss because of corrosion, cracking (including hydrogen and stress corrosion cracking), any further forms of corrosion, erosion, fatigue, embrittlement, creep, and mechanical degradation. The likelihood of failure assessment should be repeated each time the equipment or process changes.

The consequence assessment should consider the potential incidents that may occur because of failures. Such incidents may include fluid release, explosion, fire, toxic exposure, environmental impact, and other health effects.

The American Petroleum Institute (API) has developed an RBI approach. API RP 580, *Risk-Based Inspection* (2002), is used in the petroleum industry to focus inspections and resources on equipment items that have the greatest likelihood or consequence of failure.

Vessel Records

Owners and users shall maintain records throughout the service life of each pressure vessel. The records should be regularly updated to reflect new information regarding operation, inspection, maintenance, alteration, and repair. These records should be evaluated by a technical expert to assess the safety of the pressure vessel.

Pressure vessel records contain five types of information pertinent to mechanical integrity:

1. The vessel identification record indicates general information about the vessel.
2. The design and construction information identifies the design conditions and all codes and standards used for design and construction.
3. The operating history reflects the operating conditions of the vessel.
4. The inspection history provides information about past inspections, verifying that the condition of the vessel is indeed monitored as required.
5. The repair record indicates all repairs or modifications to the vessel.

Hazard Control

The management should employ good engineering principles to reduce the potential hazards from pressure vessels. Plants and facilities should be operated and maintained to protect the environment, as well as the safety and health of employees. The management should also advise employees, customers, and the public of significant pressure vessel–related safety, health, and environment hazards.

A survey was undertaken to determine all of the probable causes of unsafe conditions and to make safety assessments for preventing pressure vessel accidents. The survey questionnaire was sent to 100 pressure vessel manufacturers, repairers, and users. Table 3 indicates common causes for pressure vessel accidents according to survey responses:

In order to control pressure vessel hazards, the management should give serious consideration to the following safety practices.

Design Vessels According to Code

The design of a pressure vessel according to the code is the first step in ensuring that vessel structure is safe. The study shows that 9% of accidents occur as a result of faulty design.

TABLE 3

Causes of Pressure Valve Accidents

Causes	Percentage of Accidents
Operator error	29%
Poor maintenance	27%
Controls and safety devices failure	10%
Faulty design	9%
Faulty construction	8%
Improper installation	6%
Irregular inspection	5%
Improper repair	4%
Unknown	2%

The safest code for designing pressure vessel is ASME *Boiler and Pressure Vessel Code,* Section VIII—Pressure Vessels (ASME 2007a). It is the responsibility of the owner and user to provide sufficient information to design engineers. In addition to the ASME Code, design engineers must comply with the jurisdictional requirements of locations where vessels are going to be installed. The American Petroleum Institute's codes and standards should be used for designing pressure vessels for petrochemical industry uses.

Construct Vessels According to Code

The manufacturer is responsible for construction of pressure vessels. Construction includes fabrication, provision of materials, inspection and tests during fabrication, and stamping of pressure vessels. The study shows that 8% of the accidents occur because of faulty construction.

A pressure vessel is fabricated according to the design, drawings, and calculations done by a design engineer or professional engineer. In addition to the ASME Code, Section VIII (2007a), the manufacturer must use ASME Section IX (2007f) and V (2007g) codes. Again, the manufacturer must comply with the American Petroleum Institute's codes and standards for fabrication of pressure vessels to be used in petrochemical industries. The manufacturer must have a Certificate of Authority from the American Society of Mechanical Engineers to use code symbol stamps (U, U2, and U3). The shop inspection during construction should be inspected by an authorized inspector to ensure that the manufacturer complies with the code requirements.

Install Vessels Properly

Pressure vessels are installed according to manufacturers' recommendations and the requirements of the local jurisdiction. The proper Installation is key to controlling hazards from long-term use of pressure vessels. The survey shows that 6% of the pressure vessels fail because of improper installation.

Generally, a mechanical contractor installs pressure vessels in equipment rooms or at outdoor locations. The contractor is responsible for obtaining necessary permits from the jurisdictions involved prior to installation and startup. The owner must ensure that the contractor has applied proper codes, and manufacturer's recommendations for installation. When the job is completed, the owner or owner's representative must check and inspect the entire pressure vessel system. If a violation of local code is found, an authorized inspector must affirm that the violation has been corrected.

Take Care of Controls and Safety Devices

Controls and safety devices are used on pressure vessels for efficient and safe plant management. The survey shows that 10% of the pressure vessel accidents occur as a result of control and safety device failure.

These controls and safety devices are very sophisticated instruments specially designed by control engineers. The types of controls and safety devices depend on the application of pressure vessels, but, once installed, they are supposed to function smoothly without any problem. Owners and users must ensure their proper operation by checking and testing devices at regular intervals. It is advisable to follow manufacturers' recommendations and local jurisdictional requirements for installation, operation, maintenance, and repair of these devices.

Engage Qualified Operators

The survey shows that 29% of all the pressure vessel accidents occur because of operator error. It is essential that the management hire, train, and maintain qualified engineers and operators to operate pressure vessel systems.

Operators in process plants make crucial decisions at times. Because most jurisdictions don't have regulations for qualifying operators, the owner and user are responsible for giving proper training to the operating personnel. An operator must be familiar with all operating conditions including hazardous conditions related to pressure vessels. There should be written procedures including emergency for each pressure vessel and its associated system. The operators should have detailed knowledge of startup, shutdown, process upsets, or other unusual conditions.

Maintain Vessels Properly

The study shows that 27% of pressure vessel accidents occur because of poor maintenance. It is very important that management establish a maintenance program for all types of maintenance on pressurized systems.

In addition to preventing accidents, maintenance is performed on pressure vessels to increase efficiency, ensure safety, and prevent unscheduled shutdowns. The maintenance department should have a separate program for routine, preventive, shutdown, and overhaul maintenance. Every pressure vessel should have a maintenance log sheet. Any maintenance done on the equipment must be recorded in detail. The maintenance department is also responsible for readying the pressure vessel for inspection by an authorized inspector.

Inspect Vessels Regularly

Inspection is necessary to ensure that pressure vessels are fit for service. The study shows that 5% of failures occur because of noninspection or irregular inspection.

The objectives of inspection are the evaluation of the operational integrity of pressure vessels in service and the issuing of certificates in accordance with jurisdictional laws and rules. An authorized inspector, qualified by the jurisdictional authority, conducts in-service inspection. In large industrial complexes, the owner may establish an inspection department staffed with inspectors qualified by the jurisdictional authority. The owner's inspectors must inspect the equipment according to jurisdictional laws and rules. The management should implement the authorized inspectors' recommendations for controlling plant hazards.

Repair Vessels Properly

Repair work is done to restore a pressure vessel to safe and satisfactory operating conditions. The survey shows that 4% of the pressure vessel accidents occur because of improper repair.

The management should engage qualified repair firms to undertake all types of repairs. The various forms of repairs are major repair, minor repair, alteration, rerating, and modification. While undertaking any type of repair, the repair organization must follow the original code of construction. Approval from the authorized inspector is required before and after the repair. Vessels should be tested hydrostatically if required by the inspector. The repair organization must submit a repair report duly signed by the authorized inspector for repair work done. A pressure vessel repair report on Form R-1 is shown in Appendix C.

Investigate Unknown Factors

It may be difficult to determine the causes of accident in some cases. Every effort should be made to determine possible causes to avoid repetition of such accidents in future. Thorough investigation should be made to find out if the accident is related to terrorism, sabotage, or natural disaster. Operating personnel should be questioned to avoid missing any possible clues or links to outsiders.

Maintain Records

The management must ensure that proper records are maintained for each piece of equipment. These records include vessel identification, design and construction information, operation history, inspection history, and repair record.

In addition to the above records, the management should document training programs, qualification records, statistical analyses, and any reports by technical experts. The latest editions of all codes and standards related to pressure vessels should be available.

Comply with Jurisdictional Regulations

The management must comply with all local, state, and federal laws, rules, and regulations applicable to locations where pressure vessels are installed. Penalties or other consequences may occur when such regulations are not followed.

Many jurisdictional laws require that pressure vessels be offered for internal and external inspection at regular intervals. An inspector authorized by the jurisdiction performs such inspections and issues certificate of compliance. The law requires that the certificate of compliance be posted in the equipment room. The owner is also responsible for informing the jurisdiction of all accidents.

Conclusion

An unsafe act, an unsafe condition, a pressure vessel accident—all are symptoms of something wrong in the safety management system. Safety should be managed as any other company function. The management should direct the safety effect by setting achievable goals and by planning, organizing, and controlling in order to achieve them. An effective safety management program should be established to minimize pressure vessel hazards.

In order to achieve pressure vessel safety, management should establish an effective pressure vessel safety program. Typical components of a safety program are:

- Management's statement of policy
- Safety rules
- Authority and responsibility
- Training of employees
- Inspections
- Investigations
- Obtaining services from experts
- Recordkeeping

Each component of the safety program should be designed with the objective of reducing the number of pressure vessel accidents. The top management must support, and middle management participate, in such a program. The industrial safety survey shows that a safety manager's role is crucial to the success of a safety program. A safety manager must structure, program, investigate, and analyze safety functions to ensure proper functioning.

Pressure vessel safety programs determine and define causes that allow accidents to occur by (1) searching for root causes of accidents and (2) controlling the variables related to root causes. The highest risk is mostly associated with a small percentage of pressure vessels in a plant. Risk-based inspection (RBI) based on ranking of equipment, experience, and engineering judgment may be established to help carry out pressure vessel safety functions. ANSI/API RP-580, *Risk-Based Inspection* (2002) may be used for setting up a risk-based inspection program.

The use of risk-based inspection procedures and good engineering practices reduces the number of catastrophic pressure vessel failures. Management should include pressure vessel safety management as an element of production in order to control hazards. Only by safety management can the health and welfare of employees be ensured and the environment protected from the effects of dangerous pressure vessel explosions.

References

American National Standards Institute (ANSI)/American Petroleum Institute (API). 2007. Std 660, *Shell-and-tube Heat Exchangers.* Washington, D.C.: API.

______. 2006. Std 661, *Air-Cooled Heat Exchangers for General Refinery Service Petroleum and Natural Gas Industries.* Washington, D.C.: API.

American Petroleum Institute (API). 2001. RP 572, *Inspection of Pressure Vessels,* 2d ed. Washington, D.C.: API.

______. 2006. API 510, *Pressure Vessel Inspection Code Maintenance Inspection, Rating, Repair, and Alteration,* 8th ed. Washington, D.C.: API.

______. 2004. STD 620, *Recommended Rules for Design and Construction of Large, Welded, Low-pressure Storage Tanks,* 10th ed. Washington, D.C.: API.

______. 2002. RP 580, *Risk-Based Inspection.* Washington, D.C.: API.

______. 2007. API 650, *Welded Steel Tanks for Oil Storage,* 11th ed. Washington, D.C.: API.

American Society of Mechanical Engineers (ASME). 2007a. *Boiler and Pressure Vessel Code,* Section VIII. New York: ASME International.

______. 2007b. *Boiler and Pressure Vessel Code,* Section VIII, Division 1—Rules for Construction of Pressure Vessels. New York: ASME International.

______. 2007c. *Boiler and Pressure Vessel Code,* Section VIII, Division 2—Alternative Rules. New York: ASME International.

______. 2007d. *Boiler and Pressure Vessel Code,* Section VIII, Division 3—Rules for Construction of High Pressure Vessels. New York: ASME International.

______. 2007e. *Boiler and Pressure Vessel Code,* Section II, Materials—Parts A–D. New York: ASME International.

______. 2007f. *Boiler and Pressure Vessel Code,* Section IX—Welding and Brazing. New York: ASME International.

______. 2007g. *Boiler and Pressure Vessel Code*, Section V—Nondestructive Examination. New York: ASME International.

National Board of Boiler and Pressure Vessel Inspectors. 2002. "Pressure Vessel Safety Survey." www.nationalboard.org

Occupational Safety and Health Administration (OSHA). 1989. Publication 8-1.5, *Guidelines for Pressure Vessel Safety Assessment*. Washington, D.C.: OSHA.

______. *OSHA Technical Manual*. www.osha.gov/dts/osta/otm/otm-iv/otm-iv-3/html

Tubular Exchanger Manufacturing Association (TEMA). 2007. *TEMA Standards*, 9th ed. Tarrytown, NY: TEMA.

APPENDIX A

(*Courtesy:* ASME International.)

The American Society of Mechanical Engineers

CERTIFICATE OF AUTHORIZATION

This certificate accredits the named company as authorized to use the indicated symbol of the American Society of Mechanical Engineers (ASME) for the scope of activity shown below in accordance with the applicable rules of the ASME Boiler and Pressure Vessel Code. The use of the Code symbol and the authority granted by this Certificate of Authorization are subject to the provisions of the agreement set forth in the application. Any construction stamped with this symbol shall have been built strictly in accordance with the provisions of the ASME Boiler and Pressure Vessel Code.

COMPANY:

AEREX INDUSTRIES, INC.
3504 INDUSTRIAL 27TH STREET
FORT PIERCE, FLORIDA 34946

SCOPE:

MANUFACTURE OF PRESSURE VESSELS AT THE ABOVE LOCATION AND FIELD SITES CONTROLLED BY THE ABOVE LOCATION

AUTHORIZED: **SEPTEMBER 24, 2003**
EXPIRES: **DECEMBER 1, 2006**
CERTIFICATE NUMBER: **29,984**

Chairman of The Boiler
And Pressure Vessel Committee

Director, Accreditation and Certification

Appendix B

(*Courtesy:* ASME International.)

FORM U-1 MANUFACTURER'S DATA REPORT FOR PRESSURE VESSELS

As Required by the Provisions of the ASME Code Rules, Section VIII, Division 1

1. Manufactured and certified by: ______________________
(Name and address of Manufacturer)

2. Manufactured for: ______________________
(Name and address of Purchaser)

3. Location of installation ______________________
(Name and address)

4. Type: __________ (Horiz, vert., or sphere) __________ (Tank, separator, jkt. vessel, heat exh, etc.) __________ (Mfr's serial No.)

__________ (CRN) __________ (Drawing No.) __________ (Nat'l Bd No.) __________ (Year built)

5. ASME Code Section VIII Div 1 __________ [Edition and Addenda (date)] __________ (Code Case No.) __________ [Special Service per UG-120(d)]

Items 6-11 incl. to be completed for single wall vessels, jackets of jacketed vessels, shell of heat exchangers, or chamber of multi-chamber vessels.

6. Shell (a) No. of course (s): __________ (b) Overall Length (ft & in.): __________

Course(s)			Material	Thickness		Long. Joint (Cat. A)			Circum. Joint (Cat. A, B, & C)			Heat Treatment	
No.	Diameter	Length (ft & in.)	Spec./Grade or Type	Nom.	Corr.	Type	Full, Spot, None	Eff.	Type	Full, Spot, None	Eff	Temp.	Time

7. Heads: (a) __________ (Matl. Spec. No., Grade or Type) (H.T. – Time & Temp) (b) __________ (Matl. Spec. No., Grade or Type) (H.T. – Time & Temp)

	Location (Top, Bottom, Ends)	Thickness		Radius		Elliptical Ratio	Conical Apex Angle	Hemispherical Radius	Flat Diameter	Side to Pressure		Category A		
		Min.	Corr.	Crown	Knuckle					Convex	Concave	Type	Full, Spot, None	Eff.
(a)														
(b)														

If removable, bolts used (describe other fastenings) __________
(Mat'l Spec. No., Grade, Size, No.)

8. Type of jacket __________ Jacket closure __________
(Describe as ogee & weld, bar, etc.)

If bar, give dimensions __________ If bolted, describe or sketch

9. MAWP ______ (internal) ______ (external) psi at max temp. ______ (internal) ______ (external) °F Min. design metal temp. ______ °F at ______ psi

10. Impact Test __________ [Indicate yes or no and the component(s) impact tested] at test temperature of ______ °F

11. Hydro., Pneu., or comb. test press. __________ Proof Test __________

Items 12 and 13 to be completed for tube sections.

12. Tubesheet __________ [Stationary (Mat'l Spec. No.)] __________ [Dia., In. (subject to press.)] __________ (Nom. thk., in.) __________ (Corr. Allow., in.) __________ [Attachment (welded or bolted)]

__________ Floating (Mat'l Spec. No.)] __________ (Dia., in.) __________ (Nom. thk., in.) __________ (Corr. Allow., in.) __________ (Attachment)

13. Tubes __________ (Mat'l Spec. No., Grade or Type) __________ (O.D., in.) __________ (Nom. thk., in. or gauge) __________ (Number) __________ [Type (Straight or U)]

Items 14-18 incl. To be completed for inner chambers of jacketed vessels or channels of heat exchangers.

14. Shell (a) No. of course (s): __________ (b) Overall Length (ft & in.): __________

Courses			Material	Thickness		Long. Joint (Cat A)			Circum. Joint (Cat. A, B, & C)			Heat Treatment	
No.	Diameter	Length (ft & in.)	Spec./Grade or Type	Nom.	Corr.	Type	Full, Spot, None	Eff.	Type	Full, Spot, None	Eff	Temp.	Time

15. Heads: (a) __________ (Matl. Spec. No., Grade or Type) (H.T. – Time & Temp) (b) __________ (Matl. Spec. No., Grade or Type) (H.T. – Time & Temp)

	Location (Top, Bottom, Ends)	Thickness		Radius		Elliptical Ratio	Conical Apex Angle	Hemispherical Radius	Flat Diameter	Side to Pressure		Category A		
		Min.	Corr.	Crown	Knuckle					Convex	Concave	Type	Full, Spot, None	Eff.
(a)														
(b)														

If removable, bolts used (describe other fastenings) __________
(Mat'l Spec. No. Grade, size, No.)

(*Courtesy:* ASME International.)

Form U-1 (Back)

16. MAWP ________ (internal) ________ (external) psi at max temp. ________ (internal) ________ (external) °F Min. design metal temp. ________ °F at ________ psi

17. Impact Test ________________________ (Indicate yes or no and the component(s) impact tested) at test temperature of ____________ °F

18. Hydro., pneu., or comb. test press. ____________ Proof Test ____________

19. Nozzles, inspection, and safety valve openings:

Purpose (Inlet, Outlet, Drain, etc)	No.	Diameter or Size	Flange Type	Material		Nozzle Thickness		Reinforcement Material	How Attached		Location (Insp. Open.)
				Nozzle	Flange	Nom.	Corr.		Nozzle	Flange	

20. Supports: Skirt ________ (Yes or No) Lugs ________ (No.) Legs ________ (No.) Others ________ (Describe) Attached ________ (Where and How)

21. Manufacturer's Partial Data Reports properly identified and signed by Commissioned Inspectors have been furnished for the following items of the report: (List the name of part, item number, mfr's name and identifying number)

____

22. Remarks ____

CERTIFICATE OF SHOP COMPLIANCE

We certify that the statements in this report are correct and that all details of design, material, construction and workmanship of this vessel conform to the ASME Code for Pressure Vessels, Section VIII, Division 1.
U Certificate of Authorization No. ____________ Expires ____________

Date ________ Name ____________ (Manufacturer) Signed ____________ (Representative)

CERTIFICATE OF SHOP INSPECTION

I, the undersigned, holding a valid commission issued by the National Board of Boiler and Pressure Vessel Inspectors and/or the State or Province of ____________ and employed by ____________ of ____________ have inspected the pressure vessel described in this Manufacturer's Data Report on ____________ and state that, to the best of my knowledge and belief, the Manufacturer has constructed this pressure vessel in accordance with ASME Code, Section VIII, Division 1. By signing this certificate neither the Inspector nor his employer makes any warranty, expressed or implied, concerning the pressure vessel described in this Manufacturer's Data Report. Furthermore, neither the inspector nor his employer shall be liable in any manner for any personal injury or property damage or a loss of any kind arising from or connected with this inspection.

Date ________ Signed ____________ (Authorized Inspector) Commissions ____________ (Nat'l Board incl endorsements State, Province and No)

CERTIFICATE OF FIELD ASSEMBLY COMPLIANCE

We certify that the statements onthis report are correct and that the field assembly construction of all parts of this vessel conforms with the requirements of ASME Code Section VIII, Division 1. U Certificate of Authorization No. ________ Expires ________

Date ________ Name ____________ (Assembler) Signed ____________ (Representative)

CERTIFICATE OF FIELD ASSEMBLY INSPECTION

I, the undersigned, holding a valid commission issued by the National Board of Boiler and Pressure Vessel Inspectors and/or the State or Province of ____________ and employed by ____________ of ____________, have compared the statements in this Manufacturer's Data Report with the described pressure vessel and state that parts referred to as data items ____________, not included in the certificate of shop inspection, have been inspected by me and to the best of my knowledge and belief, the Manufacturer has constructed this pressure vessel in accordance with ASME Code, Section VIII, Division 1. The described vessel was inspected and subjected to a hydrostatic test of ____________ psi By signing this certificate neither the inspector nor his employer makes any warranty, expressed or implied, concerning the pressure vessel described in this Manufacturer's Data Report. Furthermore, neither the inspector nor his employer shall be liable in any manner for any personal injury or property damage or a loss of any kind arising from or connected with this inspection.

Date ________ Signed ____________ (Authorized Inspector) Commissions ____________ (Nat'l Board incl endorsements State, Provonce and No)

APPENDIX C

FORM R-1 REPORT OF REPAIR

in accordance with provisions of the *National Board Inspection Code*

1. Work performed by: ______________ (name of repair organization) ______________ (Form R No.)

______________ ______________ (PO No., Job No., etc.)

2. Owner: ______________ (name)

______________ (address)

3. Location of installation ______________ (name)

______________ (address)

4. Unit identification ______________ (boiler, pressure vessel) Name of original manufacturer ______________

5. Identifying nos.: ______________ (mfg serial No.) ______________ (National Board No.) ______________ (jurisdiction No.) ______________ (other.) ______________ (year built)

6. NBIC Edition / Addenda: ______________ (edition) ______________ (addenda)

Original Code of Construction for Item: ______________ (name / section / division) ______________ (edition / addenda)

Construction Code Used for Repair Performed: ______________ (name / section / division) ______________ (edition / addenda)

7. Repair Type: ☐ Welded ☐ Graphite Pressure Equipment ☐ FRP Pressure Equipment

8. Description of work: ______________ (use supplemental sheet R-4, if necessary)

____

____

____

______________ Pressure test, if applied ______________ psi MAWP ______________ psi

9. Replacement Parts. Attached are Manufacturer's Partial Data Reports or Form R-3's properly completed for the following items of this report.

____

10. Remarks: ______________

____

____

CERTIFICATE OF COMPLIANCE

I, ______________, certify that to the best of my knowledge and belief the statements in this report are correct and that all material, construction, and workmanship on this Repair conforms to the *National Board Inspection Code*. National Board "R" Certificate of Authorization No. ______________ Expires on ______________, 20

Date ______________ ______________ (name of repair organization) Signed ______________ (authorized representative)

CERTIFICATE OF INSPECTION

I, ______________, holding a valid commission issued by the National Board of Boiler and Pressure Vessel Inspectors and certificate of competency issued by the jurisdiction of ______________ and employed by ______________ of ______________ have inspected the work described in this report on ______________ and state that to the best of my knowledge and belief this work complies with the applicable requirements of the *National Board Inspection Code*.

By signing this certificate, neither the undersigned nor my employer makes any warranty, expressed or implied, concerning the work described in this report. Furthermore, neither the undersigned nor my employer shall be liable in any way for any personal injury, property damage or loss of any kind arising from or connected with this inspection.

Date ______________ Signed ______________ (inspector) Commissions ______________ (National Board and Jurisdiction No.)

SECTION 1

RISK ASSESSMENT AND HAZARD CONTROL

LEARNING OBJECTIVES

- Understand the rationale for cost analysis and budgeting from a safety management perspective.
- Be able to explain the mechanics of the budgeting process.
- Learn the definitions of key budgeting and financial analysis terms.
- Be able to identify methods of loss control.
- Outline the methods of simple cost determination.

Cost Analysis and Budgeting

Mark Friend

ACCORDING TO renowned economist Milton Friedman (Davis 2005), "The business of business is business." Management expert Peter Drucker, with years of experience working with top American corporations, gives us new perspective on safety when he says (Drucker 1986): "It is the first duty of business to survive. The guiding principle is not the maximization of profits but the minimization of loss." The bottom line in most business decisions is strictly business. Safety managers may have difficulty with this concept, because they tend to focus on compliance or some of the technical or human aspects of safety. These issues are what they know best, and sometimes it might be difficult for them to comprehend the importance of the financial aspects of safety.

The rising costs of health care (Ginsberg 2004, 1593), workers' compensation, equipment repair and replacement, insurance, and litigation arising from safety problems should make every safety professional aware that the job of the safety professional is increasingly mired in the business of business. There must be a marrying of the missions between management and the safety function (Pope 1985). The business of both is business.

Safety managers may not receive the support from management they believe they deserve. Executive management supports the areas management believes will provide the greatest returns. In spite of personal desires or the best of intentions, business decisions are often—and, in fact, must be—made based purely on financial considerations. How much profit or how much cost savings will result?

Safety managers and the whole management team face difficult decisions involving allocation of resources to various projects.

Their decisions must be pragmatic and cost-effective to best serve the owners of the business they represent. If the decision regarding making the workplace safer costs anything, that cost must be weighed against the cost benefit of the decision, as well as against the costs of other opportunities the business may forego as a result of making the workplace safer.

The safety function competes with every other function in the organization for resources. Funds spent on safety translate into equivalent amounts not being spent for new product development, employee raises, capital improvements, or dividends to investors or stockholders (Ferry 1985). In most organizations, a manager looks at the request from the safety department and compares that request to others he or she receives. The request perceived to provide the greatest benefit, typically in terms of financial gain, is funded, although sometimes that gain is expressed in terms of goodwill, positive employee attitude, or other assets not necessarily directly associated with company profit and loss statements. Generally, management is seeking a clear contribution to the profitability of the company.

Either way, managers will knowingly or intuitively build consideration of both approaches into their decision-making processes. Both ultimately involve money. When building a business case for any safety program, it is the responsibility of the safety manager to bring to the table the cost and the value of any item requiring funds beyond the normal allocated safety budget. When making plans within the budget, financial considerations must be paramount to ensure maximum use of resources by the firm. For example, the safety department has analyzed its records and found that during the previous three years there have been four injuries resulting from contact with rotating saw blades. A safety technician has researched the problem and determined it can be corrected by replacing current equipment with new equipment containing soft-tissue sensors that will stop a rotating saw blade on contact with hands, fingers, or other body parts. The cost of replacement is compared to the total cost of injuries over a specified period of time. The dollar savings provides a return comparable to the return on other opportunities presented to management for use of the funds. In addition to the monetary return, management may also consider the value of the reduction of human loss and suffering, providing psychic benefits to decision makers and those potentially affected by the decision. All is ultimately weighed against the dollar costs involved less the potential dollar return.

Ultimately, the responsibility for the safety function rests not with the safety professional, but with line management (Morris 1985), because line management is ultimately held accountable for the financial decisions and the profitability of the organization. The role of the safety professional is to monitor safety and advise management on the steps it must take to improve or maximize safety for the employees.

The language of business is that of accounting and finance. The successful safety professional, seeking the support of top management, will learn to speak in the terms used by executive management. These terms include those typically found on corporate financial statements and the standards by which those items are measured. "Return on investment," "tax benefit," and "capital" versus "budget expenditures" become relevant vocabulary terms. According to William Pope (Friend 1986, 51), the safety professional will realize that anything done regarding safety is done because management wants it to be done. Management only wants things done that have a positive financial impact on the organization (Pope 1983).

Budgeting

The task of budgeting should be familiar to any safety professional. Most businesses and government entities operate on budgets, as do their subunits. Corporate and government budgets are usually developed on an annual or more frequent basis in a number of ways.

Budgets tend to be based on expenditures from previous periods. The assumption is that performance will continue much as it has in the past, and that budgets will be similar this year to what they were in the past. Often a percentage growth factor is built in, depending on the profitability and trends within the organization. For example, if growth for the company tends to be around 5 percent per year, then each line for the next planned budget may be increased by 5 percent. Exceptions may be made for special projects

General Corporation
Safety Department Budget
For the Year 2006

Personnel wages and benefits		$216,000
Operating expenses		
Equipment purchases	113,221	
Materials	36,414	
Supplies	19,652	
Total operating expenses		169,287
Total estimated safety department expenses for 2006		$385,287

FIGURE 1. Sample budget
(*Source:* Horngren, Sundem, and Stratton 2002)

or for capital expenditures requiring additional funds beyond the normal budget.

The budget often resembles a income or profit and loss statement. When the amount to work with is known, it is shown in the first line; otherwise, only expenses required to operate are listed. Figure 1 is an example of a typical budget.

The budget is developed for a specific period of time, as noted in the figure's title. This particular budget simply lists items for which funds are expected to be needed. Normal protocol suggests that items are listed from more durable to less durable; thus, equipment is listed before materials, which is listed before supplies. Subcategories are indented, with totals appearing opposite the last item in the list or as a total of the list in the final column. Subtotal amounts are always indented one column. Subtotals of subtotals are indented two columns. Budgets may be developed for one year, six months, a quarter, or any period desired.

A budget may also be built for a particular project. It may include items as outlined above. There may be separate columns for maximum/minimum estimates, quotes, and final project amounts. Contingency amounts may also be included to cover unexpected costs.

Zero-based budgets are those requiring justification in each new period for each expenditure (Investopedia 2005). Instead of basing the budget for the coming year on the budget from last year, the budget manager is asked to deliver a bedget proposal for the forthcoming year, with justification for each line item. In essence, cost estimates are built from scratch or from zero. In reality, most managers usually ask for more than they want, hoping to get a percentage of it and thereby wind up with the actual amount desired.

Normal practice suggests that the budget manager prepares three budgets. The first represents the optimum amount with all the wants and desires built in, just in case there happens to be more than enough money to spread around. The second budget represents the very least amount the manager needs to survive during the coming year. Any negotiations may permit cutting to this point, but no lower. The last budget is an actual, realistic budget that accurately reflects the goals of the planning period. An appropriate approach in some situations is to present top management with this budget, with optimistic and pessimistic figures in mind for each line item.

Fixed and Variable Costs

Costs involved in any budget fall into two major categories: fixed and variable. *Fixed costs* are operating costs that exist regardless of activity or level of production (Kiger, Loeb, and May 1987, 888). For example, in a manufacturing operation, a fixed cost is the cost of the real estate and buildings in which the operation is run. *Variable costs* are those incremental costs that vary according to production levels (Kiger, Loeb, and May 1987, 889). Materials or component parts are examples of variable costs.

Consideration is often given to fixed and variable costs in the budget because management is particularly attuned to the additional costs involved in adding a new program. Ongoing costs, such as safety staff salaries, utility costs for the offices, and apportioned lease costs on the office space, may not be considered when new budgets are proposed. Management may make budgetary decisions based only on the variable costs introduced for new projects or programs.

Before a budget is assembled, a determination of expected costs to be included is made. Additional variable costs are always included, but some ongoing, fixed costs may also have to be added so that a fully informed decision can be made by management. Many companies simply apply a standard overhead rate to all budgets to account for the fixed costs. This saves time, in that they don't have to be calculated every time a budget or proposal is prepared.

Calculating how much a project will cost or how much a department will spend is the easy part of the task. The difficult part is justifying the expenditures.

Risk and Loss Exposure

Risk describes the expected value of potential: both the probability and the severity of a loss-event loss (Roland and Moriarty 1990, 303). All companies face risk, and often embrace it. Of course, risk implies an exposure to the potential for loss. There are two types of loss exposure: speculative and pure. *Speculative loss* offers the opportunity for gain as well as loss exposures (Williams et al. 1981, 5). The purchase of a lottery ticket, the investment in a new product, or the move to a new location could all provide positive cash flows or loss of profits. Most companies engaged in marketing, investment, or other business ventures face speculative exposures; they are not within the realm of risk management.

On the other hand, *pure* loss exposures are those which only allow for the possibility of loss (Williams et al. 1981, 6). The most that can be hoped for is a break-even. Fires, thefts, natural disasters, and other accidents can cause loss of profits but do not provide the business with opportunity for gain. These are within the purview of the risk manager. Risk management deals only with pure loss exposures by identifying, analyzing, and minimizing risk through cost-effective means. Risk management is a responsibility of the safety manager.

Organizations and their employees are subject to risk on a regular and ongoing basis. They are always exposed to potential losses, so the possibility of loss from various threats is always present. This loss exposure has three elements (Williams et al. 1981, 4–6):

1. the item that might be lost or suffer damage
2. the perils or agents that might cause the loss
3. the adverse reaction, which usually comes in the form of financial impact

Intervention measures can minimize loss from any of these elements, and all should be considered in any risk-management problem.

Organizations and their employees are always at risk from various perils. These may be controllable or completely uncontrollable. Any level of risk exposure over a long enough period of time will result in loss. Any item exposed to any peril enough times will suffer loss. Obviously, the more often an item is exposed to a peril, the more likely it is to suffer loss.

It is impossible to eliminate all accidents and resulting losses because it is impossible to eliminate all risk. Any exposure to any risk lends itself to the possibility of an accident, and thus a loss. For example, consider a hypothetical company engaging in one activity and having only one risk. The company is able to nearly eliminate the risk by reducing it to one-tenth of one percent (0.1 percent or 0.001). In other words, every time the company engages in the activity there is a chance of loss of only one in a thousand. In one exposure, the risk is very low, and the probability of an accident resulting is so miniscule as to be virtually nonexistent. It could happen, but it is highly improbable that it will happen. That is the ideal place for all probability of accidents to be.

What if the company repeated its procedure 1000 times per day? In one day, the probability of an accident is $1 - (1 - 0.001)^{1000}$, or 0.73. If the procedure is repeated 200 days per year, with 1000 exposures per day, an accident is almost a certainty. There is a 0.73 probability of having an accident on any given day, and the probability of having an accident over the course of a year is so high that multiple accidents are nearly a certainty: $1 - (1 - 0.001)^{260{,}000}$ (assuming 260 workdays per year). This certainty is the result of the law of large numbers at work. As the number of exposures is increased, the predictability of the accident becomes more certain. In this case, the company will likely have an accident at least seven days out of ten. Reducing the probability lower than it already is may be an impossibility, depending on the complexity of the system at work.

At one time, it was believed that if enough money and manpower were spent on a problematic safety situation, the situation could be resolved. The *Challenger* and *Columbia* disasters proved this was not true. Although NASA benefited from huge budgets to finance the shuttle program (AAAS 2006), each of the two shuttles failed, killing the occupants, destroying the craft, and delaying the space program for months. Even though the probability of failure of any one part may have been extremely low, the large number of parts and complex systems, combined with the political realities at NASA, made disaster inevitable. NASA was under tremendous pressure to succeed, and to keep the launches on schedule. Repeating risk eventually leads to loss. Repeating the space shuttle launches ultimately resulted in disastrous consequences. The probabilities of failure were not and could not have been reduced to zero. Unfortunately, they were not kept low enough to preclude disastrous failure twice in less than 100 exposures (NASA 2006).

Eliminating Loss

There are companies with goals of zero losses and zero accidents. The philosophy pervading these entities is that safety is of paramount importance, and therefore the organization should do whatever it takes to eliminate the accident problem. This is an impossible goal that cannot be sustained. Attempts to do so may lead to falsified reports, sandbagged claims, and hidden information.

A more likely goal is one of continuous improvement in terms of fewer accidents and incidents. Worthwhile targets might be expressed in terms of lower insurance or workers' compensation premiums. Emphasizing working more safely, improving processes, or eliminating hazards potentially leading to accidents may be more beneficial than trying to reach goals not reachable on a large-scale basis. Although zero accidents and the resultant zero losses are not likely, there are certain steps that can be taken to minimize or avoid exposure to loss. These techniques are collectively referred to as *managing risk* or *risk management*.

Risk Management

Managing risk consists of certain systematic steps that apply the resources of the organization to lower the probability of loss by ameliorating exposures, events, or results of incidents. The steps involved include:

- systematically identifying significant exposures and analyzing those exposures in terms of loss-producing potential and probability
- identifying and implementing the most cost-effective method of reducing or eliminating the probability of the loss or lessening the effects of the loss
- periodically and systematically analyzing the system to ensure that overall system loss reduction is maximized in terms of the item, the perils, and any adverse reactions that can be reasonably anticipated

When problems are found or loss exposures exist, an effort is made not only to correct the problem, but to correct any fault in the management system that permitted or even encouraged the problem to exist. For example, if there have been reports of injuries that could have or should have been avoided through the use of appropriate personal protective equipment (PPE), why did these problems exist? If the appropriate PPE program was not in place, or if a significant portion of the PPE program was overlooked, why did that occur? Was the original PPE analysis not thorough enough? Were changes made after the PPE program was put into place and, if so, why were they not incorporated? In other words, what is wrong with the management system that it did not respond to change? If the problem was with employees not wearing PPE because they chose not to do so, the same type of analysis must be made. Why did employees perceive they could ignore the program and proceed with their job without wearing appropriate PPE? Don't just fix the problem. Fix whatever deeper problem existed in the management system to allow or encourage the surface problem to exist. Always look for problems in the management system that can and should be corrected.

Identifying and Analyzing Exposure

There are many ways to identify personnel and property subject to loss. Most approaches involve determining hazards and the effects of unattended exposure of people or the property. Typical approaches to simple hazard determination are as follows:

- ***Record analysis.*** The first approach most companies take is to analyze past records. The OSHA 300 log, workers' compensation, first aid, and insurance records may all be used to point to problem areas. Multiple injuries of the same type or in the same area merit the highest levels of concern. Not only should the equipment, procedure, or person causing the problem be addressed, but also the deeper management system fault must be corrected.
- ***Inspection forms.*** Inspection forms can be obtained from OSHA, insurance companies, or safety textbooks. OSHA has checklists for specific types of businesses and operations, as well as generic checklists for general safety problems. With experience in an operation, a safety professional may prefer to develop an in-house form appropriate to the specific enterprise. The enterprise is simply evaluated against the checklist to aid in finding hazards. An example of a typical checklist appears in Figure 2.
- ***Production flow analysis.*** A common approach to identifying hazards is to follow a product through the production cycle, from receipt of raw materials through assembly or processing to shipment of finished goods. Evaluate each step in the production cycle and identify the hazards associated with it. Any hands or equipment touching the product must be considered as potential recipients and providers of hazards.
- ***Financial statement review.*** Line-by-line a review of each item on the financial statement, such as a balance sheet or income statement, can reveal areas where the company may be vulnerable financially. This goes beyond the typical inspection to review potential liability items in terms of litigation or legal obligation. This type of review tends to focus more scrutiny on business activities, as opposed to the tasks revealed in the production flow.
- ***Employee interviews.*** Time spent with employees to determine what hazards they note can be invaluable. Typical interviews include asking employees what near-misses they have observed, trouble spots they have noted or foresee, and loss-producing events they know about. Periodic discussions with production employees and their supervisors will sometimes reveal areas they are reluctant to write up.
- ***Incident and near-miss reports.*** Forms can be completed and turned in by employees any time they want to give insight into near-misses that might otherwise go unnoticed. These should only be used in conjunction with formal incident reports, similar to those required by the OSHA 301 form. The goal of any subsequent investigation is to determine fault in the management system so that corrections can be made. Investigations creating the impression that the intent is to point the blame at an individual or group have a diminished probability of having a valid conclusion.

All of the above methods are used to determine loss exposures or areas where loss might occur. Unfortunately, these methods may not be able to uncover all potential loss exposures. Even though a manager has access to previous OSHA 301 forms and the written description of the incident or accident, he or she may still not be aware of the real cause. Such phrases as "operator error" or "employee mistake" only serve to cover the true cause of the incident. Sometimes more sophisticated techniques will help reveal the actual cause or causes of an incident or accident.

FAULT TREE ANALYSIS (FTA)

Identifying the loss-producing hazard is not always a straightforward task. It may involve investigation into the cause of a particular incident or accident. Fault tree analysis may be useful in a number of situations. Widely used in system reliability studies, it offers the ability to focus on an event of importance, such as a highly critical safety issue. Fault tree analyses

Responsibility, Authority, and Accountability Checklist

In order to accomplish anything in the workplace, supervision must have the tools of the trade available. For the supervisor, the essential tools are the assignment of responsibility for a function or activity, the authority to do the job, and accountability to senior management to see that it is done. Using the following checklist, supervisors can determine if, in fact, they do have the tools necessary to do their job in safety. For self-directive work teams, the team must decide these issues.

Typically, these tasks are the responsibility of the supervisor for which he/she has complete authority and for which they are held accountable. They should also agree that these are of key importance to safety and health. Where check marks fall outside the "Yes" or "Complete authority" boxed, the recommended action is for the supervisor to discuss the situation with senior management and agree on steps necessary to assume appropriate responsibility, authority, and accountability. Note however, it is acceptable to delegate some of these items to assigned employees. This is part of the empowering process.

Is this your responsibility? ... *Yes!*	Is this your responsibility? ... *No!*	Do you have COMPLETE authority?	Do you have the authority to DECIDE; BUT TELL?	Is your authority limited to DECIDE; BUT CHECK FIRST?	Do you have NO authority?	Are you measured for accountability? ... *Yes!*	Are you measured for accountability? ... *No!*	Is this issue of key importance? ... *Yes!*	Is this issue of key importance? ... *No!*	
										Ensure equipment, materials, facilities, and conditions are safe.
										Provide for safety and health training
										Require employee compliance with safety requirements and rules.
										Recognize and reinforce safe behaviors.
										Make safety and health part of job standards and procedures.
										Request safety and health technical assistance.
										Obtain safe work permits
										Investigate accidents and take appropriate corrective action.
										Conduct inspections, audits and surveys.
										Establish emergency procedures for area.
										Hold safety meetings and workshops
										Correct unsafe conditions and behaviors.
										Stop production for safety reasons.
										Delegate authority for safety to others.

FIGURE 2. Sample responsibility, authority, and accountability checklist (*Source:* OSHA 2007)

are performed using a top-down approach. Begin by determining a top-level event, such as an accident, and then work down to evaluate all the contributing events that may ultimately lead to the occurrence of the top-level event. Probability can be used in fault tree analysis to determine the likelihood of the top-level event in any number of scenarios.

The fault tree diagram is a graphical representation of the chain of events in the system or process, built using events and logical gate configurations (Relex 2004). It can be used to:

- identify potential system reliability or safety problems during design phase of an operation
- assess reliability or safety of a system during operation
- identify components that may need testing or more rigorous quality assurance scrutiny

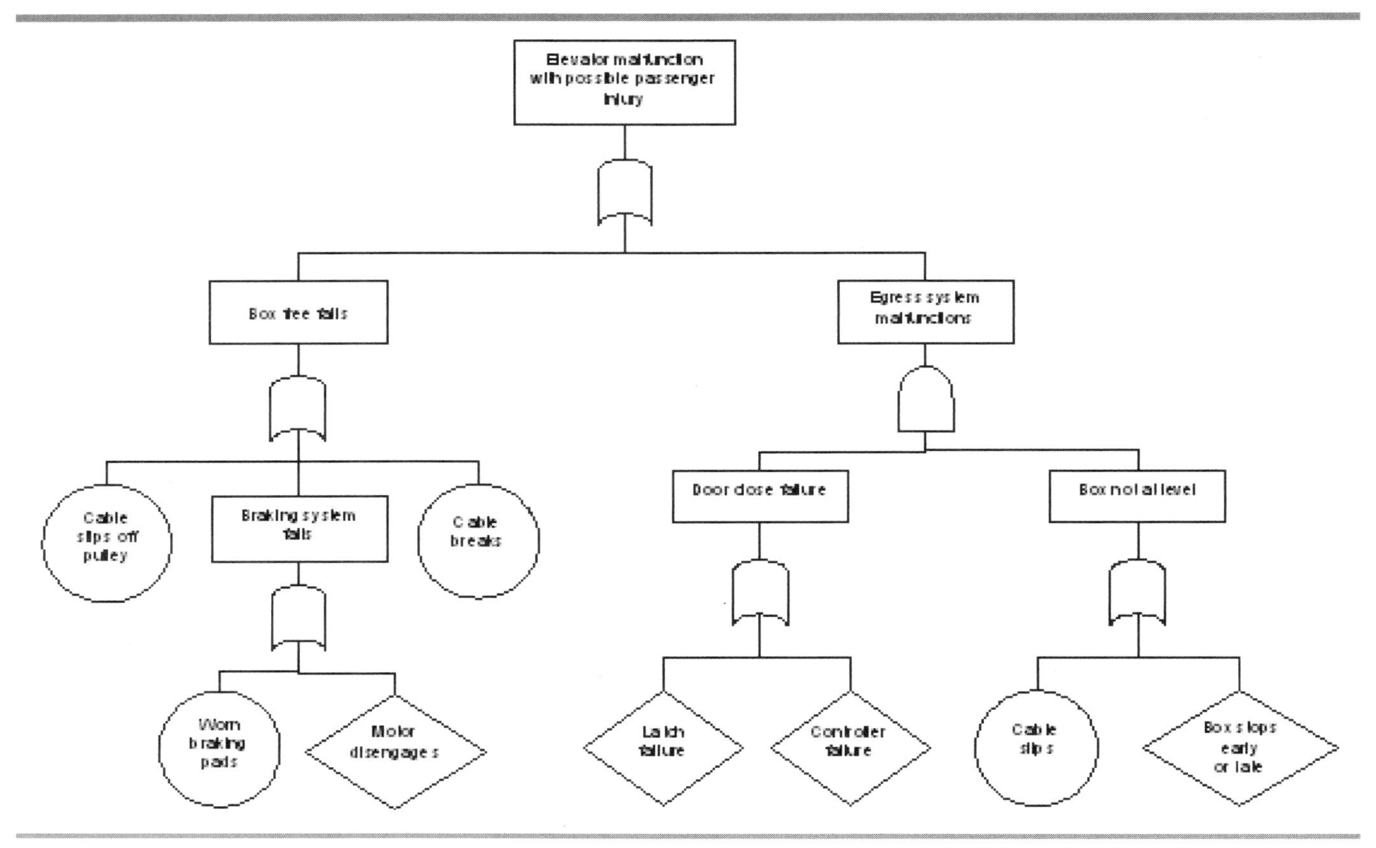

FIGURE 3. Sample fault tree analysis (Courtesy of Dr. Clarence Rodrigues, Embry-Riddle Aeronautical University)

- Identify root causes of equipment failures, accidents, or incidents

The analysis may also be combined with probabilities to determine mathematical risk (the fault tree shown in Figure 3 does not use the assignment of probabilities). The fault tree analysis is conducted using the following basic steps:

1. Determine the head event or the event to be analyzed.
2. Determine the next level event or events that caused the head event to occur.
3. Determine the relationship between or among causal events using the terms "and" and "or."
4. Continue the process through lower-level events until all possibilities are exhausted.
5. Assign probabilities to lowest-level events and systematically convert them to probabilities upward through the system.

Figure 4 defines the symbols that are used in fault tree analysis diagrams.

Determining Severity and Quantifying Risk

Once the potential loss exposures have been identified, the exposures must then be analyzed in terms of severity. MIL-STD-882, *System Safety Program Require-*

Symbol	Name	Description
	Rectangle	Any event in the system
	Circle	Primary fault with no further development
	Diamond	Secondary fault with no further development. These are usually expected to happen
	AND Gate	All outcomes connected with an AND gate must occur in order for the next higher-level event to happen
	OR Gate	Any outcome connected with an OR gate must occur in order for the next higher-level event to occur

FIGURE 4. Fault tree diagram symbols

ments, identifies severity categories as catastrophic, critical, marginal, and negligible. It lists frequency of occurrence as frequent, probable, occasional, remote, and improbable. The resulting hazard assessment matrix can guide the safety professional in combining the two categories to help determine overall risk.

The following matrix is derived from MIL-STD-882D, *Standard Practice for System Safety* (DoD 2000). It provides one systematic method for assigning a hazard level, based upon severity and frequency.

The hazard level consists of one number and one letter. The number represents the severity of the event. The numbers represent:

(1) Death, system loss, or irreversible environmental damage
(2) Severe injury, occupational illness, major system damage, or reversible severe environmental damage
(3) Injury requiring medical attention, illness, system damage, or mitigable environmental damage
(4) Possible minor injury, minor system damage, or minimal environmental damage

The letter assigned to the hazard level represents the frequency of occurrence. The letters represent:

(A) Expected to occur frequently
(B) Will occur several times in the life of an item
(C) Likely to occur sometime in the life of an item
(D) Unlikely, but possible to occur in the life of an item
(E) So unlikely, it can be assumed occurrence may not be experienced

As can be seen from Figure 4, each hazard level is associated with a risk category. Risk categories assist risk managers in differentiating credible high-hazard threats that may result in loss of life and property from less probable risks, therefore aiding management in risk versus cost decisions.

The risk assessment matrix is useful in determining overall risk and assigning resources to alleviate risk.

Frequency of Occurrence	Severity			
	(1) Catastrophic	(2) Critical	(3) Marginal	(4) Negligible
(A) Frequent	1A	2A	3A	4A
(B) Probable	1B	2B	3B	4B
(C) Occasional	1C	2C	3C	4C
(D) Remote	1D	2D	3D	4D
(E) Improbable	1E	2E	3E	4E

Risk Categories: High, Serious, Medium, Low

FIGURE 5. Risk assessment matrix (*Source:* Safety Management Services, Inc. 2002)

Figure 5 can be useful in determining overall risk, and thereby ensuring that resources are allocated to the most pressing need. Exposures are assigned to a position on the table by determining where they fall relative to frequency of occurrence and severity. Items falling in the upper left portion of the table deserve immediate attention and maximum application of resources. Those appearing in the lower right part of the table deserve little or no attention. Those in the middle category are addressed only after the more pressing needs are attended.

Identifying and Implementing Loss-Control Techniques

Reducing or controlling losses typically involves using engineering or administrative techniques or personal protective equipment (PPE). There are numerous control measures available for dealing with loss exposures or risk. These can be categorized as loss-control techniques and financing techniques (Williams et al. 1981, 53–100). Loss-control techniques attempt to reduce loss by intervening in the process. They include engineering techniques, administrative techniques, or personal protective equipment to deal with potential loss-producing exposures. For hazards potentially affecting personnel, engineering techniques are the first line of defense, followed by administrative techniques and PPE. Engineering approaches may not be cost-effective or feasible; thus, less-effective approaches may be considered. In certain operations, loss is inevitable, and a better approach may be to avoid it or to transfer it to another party.

Risk Avoidance

A large sporting goods manufacturer might have among its products helmets and pads for football players. Over a period of years, the company finds itself suffering from an increasing number of lawsuits alleging that the helmets are defective. Rather than continue producing helmets, the company may choose to withdraw altogether and purchase helmets from an outside vendor. Employers engage in this type of risk avoidance regularly when they refuse to participate in a certain type of activity due to the risk.

An alternate strategy may be to contract with a third party to pursue the activity and then possibly transfer the risk to them. As long as the company continues to manufacture the helmet or retain the risk, it should be aware of the risk and periodically evaluate it in terms of cost, severity, and the resulting potential losses.

Lease

Contracting or subcontracting is not the only method of transferring risk. Another common vehicle for transferring risk is through the use of a lease. A lease can be used to transfer risk from the occupier of property to another party. For example, Company A builds a building for its own use, sells it to Landlord B, and rents it back. Depending on the terms of the lease, certain responsibilities and risks may also transfer to the landlord. The landlord may be responsible for damages and upkeep. Certainly, market risk or the risk that the value of the real estate will decline transfers to Landlord B. The contract and common law will spell out many of the relationships.

Disclaimer

Disclaimers may also be used in an attempt to transfer risk. For example, a parking lot owner assumes no risk to vehicles parked there, as stated on the parking ticket and on signs. A renter of snow skis or jet skis may be requested to sign a statement releasing the owner of the equipment from liability in the event of injury.

The fact that a vehicle owner or an equipment renter accepts a statement refusing liability on the part of the premise or equipment owner does not necessarily release the owner from that liability. The owner may still be sued, and a judge or jury will determine whether liability on the owner's part exists. The decision may rest upon whether or not the owner was negligent, as well as on the reasonableness of the claim. Simply refusing to accept liability does not release the owner from it. For example, a passenger on a common carrier such as a bus, taxi, or airline expects to be delivered safely to his or her destination. If a taxi driver operates the taxi in a reckless manner and injures the passenger, the taxi driver and the owner of the taxi would reasonably expect to pay damages, regardless of any claims to the contrary on receipts or signs.

Attempting to transfer risk to another party without the benefit of insurance carries many potential pitfalls. A claim against any party assumes that the party has the wherewithal to cover the cost of losses and/or litigation. If not, the original transferor of the liability may ultimately be held responsible anyway. Any ambiguities in the contract may be construed against the originating party, so care must be taken in carefully outlining responsibilities. Insurance may be the best defense when all else fails.

Insurance

Insurance is purchased when all or part of the risk-producing process is retained. It provides a last-resort safety net when all else fails. The results of purchasing insurance are multifold:

- The insurer may assume all or part of the risk. The value of the insurer's assumption varies with the premium paid. Typically, it is less expensive for the insurer to cover the last dollar rather than the first dollar of the loss. In other words, it is less expensive to pay for a policy with high dollar limits on it and a high deductible than it is to pay for one with relatively low limits and little or no deductible.
- The insurance company may provide risk-management services designed to aid the insured in minimizing or controlling losses. The insurance company may employ loss control specialists whose job it is to help clients identify and rectify potential loss-producing

situations. These services may be included as part of the benefits package purchased with the premium.

- Due to the nature of insurance and insurance companies, as few as one loss may result in a policy cancellation. The insurance company may only be willing to cover the loss one time before refusing to renew. If the insurer continues coverage for more than one loss, the premium will reflect the loss experience.
- Insurance companies must not only cover the value of the loss, but also overhead and profits for those insurance companies with stockholders. Obviously, the stockholders, who actually own the company, expect a return on their investment from the stock insurance company, paid in the form of a dividend. A dividend is a payment of a portion of the profits to the company owners.

  Insurance companies owned by the policyholders return the profits or dividends to them. These companies are known as mutual companies. Although dividends may be higher when paid to a mutual company, standard wisdom holds that long-term premiums are actually lower, due to the fact that no profits are paid to outsiders.
- A major benefit of owning an insurance policy may be the legal representation provided by the insurer in the event of a major claim.The insurance company, as the entity paying the benefit, may find it advantageous to contest payment of benefits in court. If they do so, they pay for legal counsel and other related fees, thus transferring the cost of counsel from the insured to the insurer.

Self-Insurance

Many companies choose to maintain a certain level of risk themselves and not purchase insurance. They may even have a plan to cover the losses if they do occur. The primary result of not purchasing insurance or retaining the risk is that the company self-insures. Self-insurance is no insurance. The company assumes the cost of all losses and all legal fees associated with any litigation. This may be referred to as a calculated risk, but in many cases, it is simply risk.

An alternate plan is to purchase a very high-dollar deductible policy. In the event of a loss, the company will pay the first dollar amount and the insurance company picks up the loss after that. For example, a company may purchase a policy to insure it for losses up to $100 million or more, with a $1 million deductible. The deductible is high, but the insurance may be relatively inexpensive, due to the fact that losses rarely ever exceed $1 million. In the event of a catastrophic loss, even though the company won't like paying the first $1 million, the insurance policy may save the company from bankruptcy by paying any losses over $1 million, up to the $100 million of coverage.

Small insurance companies often do the same and hedge their own risks by purchasing insurance policies to cover huge losses that exceed a certain dollar amount. Although the risk is low, the potential severity is high, so they pay the premiums in order to ameliorate their risks.

Another way to help alleviate the pain of losses is to establish a sinking fund, whereby monies are set aside periodically to offset the cost of any losses due to certain events. For example, a large shipping company may decide against paying a $5 million dollar annual premium on one of its ships, but instead places that amount of money in the bank each year. In the event of a loss, that fund will be tapped to cover the loss. The risk, of course, is that enough funds will not be available at the time of the loss or that multiple losses could wipe out the fund in a short period of time. Purchasing an insurance policy in the early years, until the sinking fund is adequate to handle the losses, may offset this risk.

In addition, a high-dollar deductible insurance policy may be maintained.

The absolute worst approach to take with any risk or potential loss-producing situation is to ignore it and call it a "calculated risk."

Time Value of Money

Calculating the present value of future benefits is useful when considering alternative projects on a limited budget. Present-value calculations compare the cost of implementation against the dollar value of the benefit. For example, a safety manager is considering committing $10,000 to a project that will save approximately $17,000. However, the savings will not be realized for five years. Should the company commit? Can the safety manager convince management that this is a worthwhile investment? If financial considerations are a primary concern, it must be determined whether the benefit the company will receive later will be greater than the investment. This question is easily answered by using the following formula to determine present value (PV) of a dollar:

$$PV = 1/(1 + I)^n \quad (1)$$

In this formula, I equals the discount rate used and n equals the number of years that will elapse until funds are received. The discount rate is determined by considering alternative uses for the funds at similar risk. If management believes it can receive a return of 10 percent by investing in a venture of similar risk, the discount rate to be used is 10 percent.

Substitute these figures into the equation and solve for PV. The answer is for $1 and must be multiplied by the number of dollars saved. In this case, solve for present value by substituting as follows:

$$PV = 1/(1 + 0.10)^n \quad (2)$$

Alternatives to this formula include using a financial calculator or the present-value table (see Table 1). To use this table, determine the number of years until the return is received. For this example, it is 5 years. Find 5 in the "Period" column on the left, and go across the row to the figure listed under the designated discount rate. The factor for this problem is 0.621, which means that every dollar to be received 5 years from now, at a discount rate of 10 percent, is worth $0.621 today. Multiply $0.621 by $17,000 to determine the current value of the amount to be received in 5 years. The answer is $10,557.

At the 10 percent discount rate, this proposal will provide slightly better than a break-even return. The total dollar benefit is $557. If a 12 percent or higher discount rate were used, the proposal would no longer be cost-beneficial. If the applicable discount rate decreases to below 10 percent, the proposal becomes increasingly beneficial.

Table 1 is used to determine the value of $1 to be received at n periods in the future assuming a given interest or discount rate.

Calculating Project Savings

It is evident that potential dollar savings to be realized are attractive or unattractive based upon prevailing

TABLE 1

Present Value of $1

Present value interest factor of $1 per period at i% for n periods, PVIF(i,n).

Period	1%	2%	3%	4%	5%	6%	7%	8%	9%	10%	11%	12%	13%	14%	15%	16%	17%	18%	19%	20%
1	0.990	0.980	0.971	0.962	0.952	0.943	0.935	0.926	0.917	0.909	0.901	0.893	0.885	0.877	0.870	0.862	0.855	0.847	0.840	0.833
2	0.980	0.961	0.943	0.925	0.907	0.890	0.873	0.857	0.842	0.826	0.812	0.797	0.783	0.769	0.756	0.743	0.731	0.718	0.706	0.694
3	0.971	0.942	0.915	0.889	0.864	0.840	0.816	0.794	0.772	0.751	0.731	0.712	0.693	0.675	0.658	0.641	0.624	0.609	0.593	0.579
4	0.961	0.924	0.888	0.855	0.823	0.792	0.763	0.735	0.708	0.683	0.659	0.636	0.613	0.592	0.572	0.552	0.534	0.516	0.499	0.482
5	0.951	0.906	0.863	0.822	0.784	0.747	0.713	0.681	0.650	0.621	0.593	0.567	0.543	0.519	0.497	0.476	0.456	0.437	0.419	0.402
6	0.942	0.888	0.837	0.790	0.746	0.705	0.666	0.630	0.596	0.564	0.535	0.507	0.480	0.456	0.432	0.410	0.390	0.370	0.352	0.335
7	0.933	0.871	0.813	0.760	0.711	0.665	0.623	0.583	0.547	0.513	0.482	0.452	0.425	0.400	0.376	0.354	0.333	0.314	0.296	0.279
8	0.923	0.853	0.789	0.731	0.677	0.627	0.582	0.540	0.502	0.467	0.434	0.404	0.376	0.351	0.327	0.305	0.285	0.266	0.249	0.233
9	0.914	0.837	0.766	0.703	0.645	0.592	0.544	0.500	0.460	0.424	0.391	0.361	0.333	0.308	0.284	0.263	0.243	0.225	0.209	0.194
10	0.905	0.820	0.744	0.676	0.614	0.558	0.508	0.463	0.422	0.386	0.352	0.322	0.295	0.270	0.247	0.227	0.208	0.191	0.176	0.162
11	0.896	0.804	0.722	0.650	0.585	0.527	0.475	0.429	0.388	0.350	0.317	0.287	0.261	0.237	0.215	0.195	0.178	0.162	0.148	0.135
12	0.887	0.788	0.701	0.625	0.557	0.497	0.444	0.397	0.356	0.319	0.286	0.257	0.231	0.208	0.187	0.168	0.152	0.137	0.124	0.112
13	0.879	0.773	0.681	0.601	0.530	0.469	0.415	0.368	0.326	0.290	0.258	0.229	0.204	0.182	0.163	0.145	0.130	0.116	0.104	0.093
14	0.870	0.758	0.661	0.577	0.505	0.442	0.388	0.340	0.299	0.263	0.232	0.205	0.181	0.160	0.141	0.125	0.111	0.099	0.088	0.078
15	0.861	0.743	0.642	0.555	0.481	0.417	0.362	0.315	0.275	0.239	0.209	0.183	0.160	0.140	0.123	0.108	0.095	0.084	0.074	0.065
16	0.853	0.728	0.623	0.534	0.458	0.394	0.339	0.292	0.252	0.218	0.188	0.163	0.141	0.123	0.107	0.093	0.081	0.071	0.062	0.054
17	0.844	0.714	0.605	0.513	0.436	0.371	0.317	0.270	0.231	0.198	0.170	0.146	0.125	0.108	0.093	0.080	0.069	0.060	0.052	0.045
18	0.836	0.700	0.587	0.494	0.416	0.350	0.296	0.250	0.212	0.180	0.153	0.130	0.111	0.095	0.081	0.069	0.059	0.051	0.044	0.038
19	0.828	0.686	0.570	0.475	0.396	0.331	0.277	0.232	0.194	0.164	0.138	0.116	0.098	0.083	0.070	0.060	0.051	0.043	0.037	0.031
20	0.820	0.673	0.554	0.456	0.377	0.312	0.258	0.215	0.178	0.149	0.124	0.104	0.087	0.073	0.061	0.051	0.043	0.037	0.031	0.026
25	0.780	0.610	0.478	0.375	0.295	0.233	0.184	0.146	0.116	0.092	0.074	0.059	0.047	0.038	0.030	0.024	0.020	0.016	0.013	0.010
30	0.742	0.552	0.412	0.308	0.231	0.174	0.131	0.099	0.075	0.057	0.044	0.033	0.026	0.020	0.015	0.012	0.009	0.007	0.005	0.004
35	0.706	0.500	0.355	0.253	0.181	0.130	0.094	0.068	0.049	0.036	0.026	0.019	0.014	0.010	0.008	0.006	0.004	0.003	0.002	0.002
40	0.672	0.453	0.307	0.208	0.142	0.097	0.067	0.046	0.032	0.022	0.015	0.011	0.008	0.005	0.004	0.003	0.002	0.001	0.001	0.001
50	0.608	0.372	0.228	0.141	0.087	0.054	0.034	0.021	0.013	0.009	0.005	0.003	0.002	0.001	0.001	0.001	0.000	0.000	0.000	0.000

returns of other investments. However, the typical safety project provides a return over a period of years, as opposed to a one-time return. To calculate the savings gained from this project, the present value of an annuity formula is used:

$$PVA = \sum_{n=1}^{n} \frac{1}{(1+I)^n} \tag{3}$$

For example, if a project provides a return of $2000 per year for the next 7 years at the discount rate of 10 percent, present value is determined by substituting 7 for n and 0.10 for I.

The present value of an annuity table can also be used to determine project savings (see Table 2). To use this table, determine the number of years over which savings will be received—for this example, 7 years. Then, find the 7 in the "Period" column on the left. Go across the row to the figure listed under the designated discount rate. The factor for this problem is 4.868. If $1 is received each year over the next 7 years at a discount rate of 10 percent, the total amount received is worth $4.87 today.

Table 2 is used to determine the value of $1 to be received at the end of each of n periods in the future assuming a given interest or discount rate.

Multiply $4.87 by the $2000 that will be saved each year to determine current value of the total savings over 7 years. The answer is $9736. In this case, based strictly on dollar savings, this project would not be cost-beneficial. However, if the prevailing return or discount rate was only 9 percent, the return would be worth $10,066, making this a sound project in which to invest.

As is evidenced by the numbers, the higher the dollar saving potentially realized or the lower the potential return with alternatives, the greater the attractiveness of investing in any given safety project.

When considering the total value of potential savings of any project, it is tempting to factor in related costs, such as the associated time spent by the personnel department interviewing temporary replacement help, or the time spent investigating an accident. Management often views these costs as part of overhead and may refer to them as "utilization of idle time." To maintain a convincing argument, it is essential to point out costs that arise only as a result of not implementing the new program or developing the proposed project.

Expected Value Technique

The expected value technique, combined with these other methods, increases the ability to make sound decisions about where to spend budget dollars. It also is an additional tool for convincing management of the fiscal soundness of a proposal. When attempting

TABLE 2

Present Value of an Annuity

Present value interest factor of an (ordinary) annuity of $1 per period at i% for n periods, PVIFA(i,n).

Period	1%	2%	3%	4%	5%	6%	7%	8%	9%	10%	11%	12%	13%	14%	15%	16%	17%	18%	19%	20%
1	0.990	0.980	0.971	0.962	0.952	0.943	0.935	0.926	0.917	0.909	0.901	0.893	0.885	0.877	0.870	0.862	0.855	0.847	0.840	0.833
2	1.970	1.942	1.913	1.886	1.859	1.833	1.808	1.783	1.759	1.736	1.713	1.690	1.668	1.647	1.626	1.605	1.585	1.566	1.547	1.528
3	2.941	2.884	2.829	2.775	2.723	2.673	2.624	2.577	2.531	2.487	2.444	2.402	2.361	2.322	2.283	2.246	2.210	2.174	2.140	2.106
4	3.902	3.808	3.717	3.630	3.546	3.465	3.387	3.312	3.240	3.170	3.102	3.037	2.974	2.914	2.855	2.798	2.743	2.690	2.639	2.589
5	4.853	4.713	4.580	4.452	4.329	4.212	4.100	3.993	3.890	3.791	3.696	3.605	3.517	3.433	3.352	3.274	3.199	3.127	3.058	2.991
6	5.795	5.601	5.417	5.242	5.076	4.917	4.767	4.623	4.486	4.355	4.231	4.111	3.998	3.889	3.784	3.685	3.589	3.498	3.410	3.326
7	6.728	6.472	6.230	6.002	5.786	5.582	5.389	5.206	5.033	4.868	4.712	4.564	4.423	4.288	4.160	4.039	3.922	3.812	3.706	3.605
8	7.652	7.325	7.020	6.733	6.463	6.210	5.971	5.747	5.535	5.335	5.146	4.968	4.799	4.639	4.487	4.344	4.207	4.078	3.954	3.837
9	8.566	8.162	7.786	7.435	7.108	6.802	6.515	6.247	5.995	5.759	5.537	5.328	5.132	4.946	4.772	4.607	4.451	4.303	4.163	4.031
10	9.471	8.983	8.530	8.111	7.722	7.360	7.024	6.710	6.418	6.145	5.889	5.650	5.426	5.216	5.019	4.833	4.659	4.494	4.339	4.192
11	10.368	9.787	9.253	8.760	8.306	7.887	7.499	7.139	6.805	6.495	6.207	5.938	5.687	5.453	5.234	5.029	4.836	4.656	4.486	4.327
12	11.255	10.575	9.954	9.385	8.863	8.384	7.943	7.536	7.161	6.814	6.492	6.194	5.918	5.660	5.421	5.197	4.988	4.793	4.611	4.439
13	12.134	11.348	10.635	9.986	9.394	8.853	8.358	7.904	7.487	7.103	6.750	6.424	6.122	5.842	5.583	5.342	5.118	4.910	4.715	4.533
14	13.004	12.106	11.296	10.563	9.899	9.295	8.745	8.244	7.786	7.367	6.982	6.628	6.302	6.002	5.724	5.468	5.229	5.008	4.802	4.611
15	13.865	12.849	11.938	11.118	10.380	9.712	9.108	8.559	8.061	7.606	7.191	6.811	6.462	6.142	5.847	5.575	5.324	5.092	4.876	4.675
16	14.718	13.578	12.561	11.652	10.838	10.106	9.447	8.851	8.313	7.824	7.379	6.974	6.604	6.265	5.954	5.668	5.405	5.162	4.938	4.730
17	15.562	14.292	13.166	12.166	11.274	10.477	9.763	9.122	8.544	8.022	7.549	7.120	6.729	6.373	6.047	5.749	5.475	5.222	4.990	4.775
18	16.398	14.992	13.754	12.659	11.690	10.828	10.059	9.372	8.756	8.201	7.702	7.250	6.840	6.467	6.128	5.818	5.534	5.273	5.033	4.812
19	17.226	15.678	14.324	13.134	12.085	11.158	10.336	9.604	8.950	8.365	7.839	7.366	6.938	6.550	6.198	5.877	5.584	5.316	5.070	4.843
20	18.046	16.351	14.877	13.590	12.462	11.470	10.594	9.818	9.129	8.514	7.963	7.469	7.025	6.623	6.259	5.929	5.628	5.353	5.101	4.870
25	22.023	19.523	17.413	15.622	14.094	12.783	11.654	10.675	9.823	9.077	8.422	7.843	7.330	6.873	6.464	6.097	5.766	5.467	5.195	4.948
30	25.808	22.396	19.600	17.292	15.372	13.765	12.409	11.258	10.274	9.427	8.694	8.055	7.496	7.003	6.566	6.177	5.829	5.517	5.235	4.979
35	29.409	24.999	21.487	18.665	16.374	14.498	12.948	11.655	10.567	9.644	8.855	8.176	7.586	7.070	6.617	6.215	5.858	5.539	5.251	4.992
40	32.835	27.355	23.115	19.793	17.159	15.046	13.332	11.925	10.757	9.779	8.951	8.244	7.634	7.105	6.642	6.233	5.871	5.548	5.258	4.997
50	39.196	31.424	25.730	21.482	18.256	15.762	13.801	12.233	10.962	9.915	9.042	8.304	7.675	7.133	6.661	6.246	5.880	5.554	5.262	4.999

to prioritize the placement of resources, this technique combines educated assessment of the situation with simple arithmetic to help determine greatest need.

For example, management is considering undertaking three major safety programs; each will be a considerable investment. Which ones should be pursued, and in which order? The following step-by-step approach may be useful.

Step One: Determine the probability of loss for each problem area if the course of action is not pursued. Probabilities can be determined by examining past company records or records of similar organizations in similar situations. Many times, an insurance carrier can also provide useful data.

One example would be considering a program to counteract the effects of a flood. The probability of a flood occurring can be determined by reviewing experience from previous years and applying it to current years. If a company is located in an area that has experienced a damage-producing flood twice in the last 50 years (and no effective countermeasures have been taken), it appears the probability of such a flood occurring in any given year would be about 2 in 50, or 4 percent.

The same technique can be applied to fire, theft, or any other loss-producing event. Percentages can range from virtually zero for a remote probability (for example, damage from a volcanic eruption in the midwestern United States) to 100 percent for fire and theft. While the probability is only an estimate, it should be as accurate as practicable.

Step Two: Estimate expected dollar loss from all such events during an entire year. This can be accomplished by estimating from past records or by enlisting

CASE STUDY

Acme Widget Company

Acme Widget Co. has limited funds to spend on a major project during the coming year. All company departments, including the safety department, are vying for these funds. Management's decision will be based on its perception of where the highest rate of return can be earned without excessive risk.

The safety manager is considering submitting a proposal asking for funds to alleviate problems in one of three target areas.

Process A on the production line has experienced four injuries in the last 5 years. The average cost of each injury accident was calculated at $15,000. Probability of loss is 4/5 = 0.80.

Process B has experienced 22 injuries in the last 5 years. The average cost of each injury accident was calculated at $2500. Since the company is experiencing multiple losses per year, the likely number of losses in a given year is 22/5 = 4.40.

Process C has only been online for 3 years. Two injuries have occurred in that time. The average cost of each injury accident was calculated at $14,000. Probability of loss is 2/3 = 0.67.

From a cost standpoint, which problem is the highest priority? Calculate expected value for each as follows:

$$EV_A = \$15,000 \times 0.80 = \$12,000 \quad (6)$$
$$EV_B = \$2500 \times 4.40 = \$11,000$$
$$EV_C = \$14,000 \times 0.67 = \$9380$$

Process A has the highest expected value and would, therefore, receive highest priority. Expected savings from correcting these problems would be $12,000 annually. Expected savings from correcting Process B problems would be $11,000 and for Process C would be $9380.

Process A will be completely overhauled at the end of 6 years. How much would correcting the problems in this process be worth today? The company would save $12,000 annually for 6 years, a total of $72,000. The present value of that savings is based on the rate of return management expects on alternative investments of low risk–in this case, 12 percent. By using the present value of annuity table, the value of $1 to be received at the end of each year for 6 years at 12 percent is $4.111; therefore, multiply 4.111 by $12,000 to learn that total savings is equivalent to $49,332. If correcting Process A cost Acme Widget Co. $20,000, total savings in today's dollars will be $29,332.

The assumption in using this table is that all returns are gained at the end of each year. In application, however, it will not likely work this way. Therefore, a slight math error makes actual returns higher than they appear.

Savings realized on a permanent basis would typically be capped at no more than 10 years. Otherwise, it would be infinitely profitable to invest in any venture with a positive rate of return. With business uncertainties and the potential for new technology to eliminate any process that might be corrected, it may be impractical to project dollar savings beyond a 10-year period. The decision as to whether to limit the number of years to consider in present value is a judgment based on the individual business.

As any safety manager knows, safety budgets are always competing with other investment-returning projects within the organization. The safety manager must be able to use the same tools as other departments to formulate and present convincing arguments to management for increases in budgets and project funding.

the help of the company's insurance carrier. Contacts through professional associations may also be helpful.

Step Three: Multiply the expected annual loss in a category by the probability of a loss occurring. The resulting figure is the expected value, which can be compared to expected values determined for other categories.

For example, after reviewing company records, it is determined that fire losses have occurred four times in the last ten years. Employment records indicate that the workforce has been stable during that time, and that no major changes have occurred in the facility that would have significantly impacted the loss record. Thus, these figures should be representative of future figures. If it appears this is not a representative figure, adjust it accordingly. (The safety manager must use his/her best judgment in this area if no objective evidence is available.)

Obviously, basing the estimate on data from a longer time period enhances the probability that the figures are representative. In this case, the probability of a loss occurring in one year is the number of times a loss occurred over the 10-year period divided by the number of years in the period (Annual Probability of Loss = Number of Losses ÷ Number of Years) or

$$\text{Probability} = 4/10 = 40\ \% \tag{4}$$

Step Four: Calculate the total dollar cost of each loss by reviewing records. Losses from ten years ago will obviously not reflect current dollar costs. These amounts can be adjusted, however, by using an index (such as a consumer price index) to update them to current dollar figures.

Assume that the average fire loss is estimated to cost a company approximately $19,000. Multiply this average loss by the probability of a loss occurring in a given year to obtain the expected value:

$$\text{Expected Value} = \$19{,}000 \times 40\ \% = \$7{,}600 \tag{5}$$

This expected value represents the amount a company can expect to lose from fire losses in any one year. It can be compared to insurance coverage costs, other expected losses or the cost of investing in a fire prevention program. When comparing such figures for three different projects, the highest priority is that with the highest expected value. Always consider tax implications when making comparisons of this nature.

International Standards and Budgeting

As companies move toward international markets, questions arise concerning the role of international standards on safety and safety decisions. The economic and social constraints placed on companies vary from country to country, as do the relationships existing between companies and the local governing bodies. For example, the financial constraints on a decision may be completely different in Sweden than in Mexico.

An international safety and health standard was discussed at the International Organization for Standardization (ISO) meeting in Geneva, Switzerland in 1996, but it was generally agreed that it was not feasible. A majority of the participants failed to support any standard, as did the American National Standards Institute (ANSI). Much of the lack of support was due to cultural and labor law differences. The benefits and economic impacts also affected the decision (Swartz 2000, 290).

Conclusion

Effective handling of budgetary matters on the part of the safety manager requires a knowledge and use of fundamental accounting techniques. The safety practitioner should be able to identify costs associated with hazard exposure, as well as those associated with amelioration of the exposures. Safety program decisions are ultimately made by line management, so presentation of potential costs as well as those of remediation are critical to the decision-making process. In order to be effective, the safety manager must also be aware of alternate techniques or methods to solve the risk problem. These may range from risk avoidance to insurance. Long-term cost savings may also result in long-term benefits. The value of savings over time can be instrumental in persuading management of the soundness of a safety proposal. Consideration must also be given to the costs and benefits of operating in overseas environments for a company doing business on a multinational basis.

References

American Society for the Advancement of Science (AAAS). 2006. *NASA R&D Gains, But Steep Cuts Loom for Research* (retrieved November 1, 2006). www.aaas.org/spp/rd/nasa07p.pdf

Davis, Ian. 2005. *What is the Business of Business?* www.mckinseyquarterly.com/article_page.aspx?ar=1638&L2=21&L3=3

Department of Defense (DoD). 2000. MIL-STD-882D, *Standard Practice for System Safety*. www.safetycenter.navy.mil/instructions/osh/milstd882d.pdf

Drucker, Peter. 1986. *The Practice of Management*. New York: Harper Collins.

Ferry, Theodore S. Interview by Mark A. Friend, August 1985.

Friend, Mark A. "Safety Management Philosophies of Seven Major Contributors to Safety Management." PhD diss., West Virginia University, 1986.

Ginsberg, Paul B. 2004. "Controlling Health Care Costs." *New England Journal of Medicine* (October 14) 351:1591–1593.

Horngren, Charles T., Gary L. Sundem, and William O. Stratton. 2002. *Introduction to Management Accounting*. Upper Saddle River, NJ: Prentice Hall.

Investopedia. 2005. *Zero-based budgeting – ZBB* (retrieved December 21, 2005). www.investopedia.com/terms/z/zbb.asp

Kiger, Jack E., Stephen E. Loeb, and Gordon S. May. 1987. *Accounting Principles*. 2d ed. New York: Random House.

Morris, Julius. Interview by Mark A. Friend, August 1985.

National Aeronautics and Space Administration (NASA). 2006. *Volume 3 Space Shuttle Mission Chronology 2005–2006* (retrieved December 19, 2005). www.pao.ksc.nasa.gov/kscpao/nasafact/pdf/SSChronologyVolume3.pdf

Occupational Safety and Health Administration (OSHA). 2007. *Responsibility, Authority and Accountability Checklist*. www.osha.gov/SLTC/etools/safetyhealth/mod4_tools_checklist.html

Pope, William C. Interview by Mark A. Friend, August 1985.

______. 1983. *Principles of Organization and Management of Risk Control 2-A-3-4*. Alexandria, VA: Safety Management Systems, Inc.

Relex Software Corporation. 2004. *Fault Tree/Event Tree* (retrieved October 25, 2004). www.relexsoftware.com/products/ftaeta.asp

Roland, Harold E., and Brian Moriarty. 1990. *System Safety Engineering and Management*. New York: John Wiley & Sons.

Safety Management Services. 2002. MIL-STD-882B *Hazard Risk Assessment Matrix* (retrieved October 22, 2004). www.smsink.com/services_pha_matrix.html

Swartz, George. 2000. *Safety Culture and Effective Safety Management*. Chicago, IL: National Safety Council Press.

Williams, C. Arthur, George L. Head, Ronald C. Horn, and G. William Glendenning. 1981. *Principles of Risk Management and Insurance*. 2d ed. vol 1. Malvern, PA: American Institute for Property and Liability Insurance.

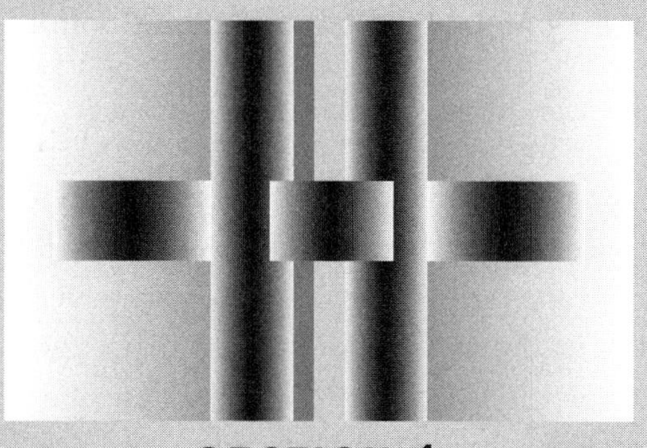

SECTION 1

RISK ASSESSMENT AND HAZARD CONTROL

LEARNING OBJECTIVES

- Understand the concepts of reliability and validity and how they are critical to the evaluation of any measurement process.
- Be able to apply these concepts to performance measurement.
- Conduct better performance measurements based on an understanding of research that evaluates the reliability and validity of incident-based measures, audits, and surveys.
- Understand the limitations of performance measurement and the hazards of incentives for performance that can lead to manipulating the results.
- Be able to identify instances of manipulation of results.
- Develop the ability to design, conduct, and evaluate a productive benchmarking study.

BENCHMARKING AND PERFORMANCE CRITERIA

Brooks Carder and Pat Ragan

PERFORMANCE MEASUREMENT is a fundamental step in risk assessment. In a stable system, performance will remain the same until the underlying process changes, so a measure of current performance constitutes an assessment of future risk (Deming 1982, 1993). Performance measurement and benchmarking are both methods that can assist in hazard control, by revealing opportunities for process improvement.

THE RELATIONSHIP BETWEEN PERFORMANCE MEASUREMENT AND BENCHMARKING

Performance measurement and benchmarking are obviously intertwined. Merriam Webster's online dictionary (www.m-w.com) defines *benchmarking* as "the study of a competitor's product or business practices in order to improve the performance of one's own company." However, the term derives from the noun *benchmark*. The definition of a benchmark includes "a point of reference from which measurements may be made" and "something that serves as a standard by which others may be measured or judged." Performance measurement is usually not very meaningful unless there is a benchmark for comparison. If you are asked how fast someone is going, and you get an answer of 100 miles per hour, you would think that was extremely fast on a bicycle, fast in a car, not very fast in a racing car, and extremely slow in a jet plane. To the extent that performance measurement is evaluative, there must be an explicit or implied benchmark.

On the other hand, to the extent that benchmarking represents an attempt to improve performance, it is necessary to find benchmarking partners that have excellent processes and excellent performance (Camp 1995). The objective is to identify and implement the processes that lead to superior performance in other companies. Thus, benchmarking cannot be done effectively in the absence of good performance measurement. Keeping these interrelationships in mind, the chapter first addresses performance measurement and then benchmarking.

Performance Appraisal

Defining Performance Appraisal

On the face of it, performance appraisal in safety should be very simple. One can simply count injuries, deaths, and property loss. In reality, however, the problem is very difficult. The following problems arise:

- Some industries and activities are inherently more hazardous than others.
- Over a short period of time or with a relatively small population, the inherent variability of these counts is high, making judgment based on the numbers very inaccurate.
- In an environment where a major disaster could occur, such as with an airline, a chemical plant, or a refinery, assessing the likelihood of a major event should be a top priority. These are so rare that, thankfully, in most sites, there is nothing to count, even though the danger may be high.

Dictionary.com defines *safety* as freedom from danger, risk, or injury. Conditions are easily conceived in which there is no history of injury but great risk of future injury. Of course this appears to be the case with shuttle flights up to the time of the Challenger and Columbia disasters. Although there was no history of injury, the engineers working on the flights estimated the probability of the loss of a vehicle in the range of 1 in 100 (Feynman 1999).

Ideally, a measure of performance would tell us the level of freedom from danger, risk, and injury. The measure would not be a picture in the rearview mirror, but rather an accurate forecast of future expectations, so long as the system is not changed. Many readers may believe that incident counts are indeed an accurate forecast of overall safety. But the available evidence indicates that this is not the case. Part of the problem lies in the lack of reliability of incident counts because the standards for OSHA-recordable events can vary from company to company, and even from day to day in the same company (Carder and Ragan 2004). An article in *Professional Safety* describes how recordable counts can be altered through "medical management of injuries" (Rosier 1997). This practice is widespread and is a significant limitation to the reliability of accident counts (Carder and Ragan 2004). Another problem is that accident counts have proven to be a poor predictor of catastrophic events (Manuele 2003, Petersen 2000, Wolf and Berniker 1999).

Objectives of Performance Appraisal

An important objective of performance appraisal is to provide information to guide improvement efforts. Another is to track the effectiveness of improvements that are implemented. This is the plan-do-study-act cycle described by Deming (1982). Closely related to this is the need to evaluate the performance of managers and to provide guidance for establishing reward systems.

Hazards of Performance Appraisal

The first question to be asked is whether an accurate meaningful assessment can really be made. This chapter suggests that one can indeed make a useful assessment of the safety performance of an organization or subunit. The second question is, whose performance is being appraised? A safety manager in a plant is part of a system. He or she usually has very limited control over the larger system. The system includes such practices as hiring policies, education and training, manpower decisions, budgets, capital expenditures, and much more. All of the things mentioned have an impact on safety. Although one can measure the performance of the system, it is much more difficult to measure the performance of individuals working in that system. Deming (1982) argues continuously and eloquently that attempting to evaluate the performance of individuals working in a complex system is a waste of time. Nevertheless, it is unlikely that business will move away from this anytime soon. However, the reader should be aware of the limitations of such evaluations when they are used.

Consider the following actual case study: Many years ago a marketing company had a young man in sales who was very bright and energetic. However, his performance in sales was continuously disappointing to his superiors. He was labeled an underachiever and, in private conversations, much worse. He constantly asked his managers to be allowed to sell in a different way and was constantly told that the company had a system of proven success and that he should sell exactly as he was told. The sales rep wanted to uncover marketing problems that confronted the customer and return to his office to prepare a solution. The solution would be presented to the customer on a subsequent visit. He was told that he needed to present a solution and close the order on the initial visit, like all of his successful colleagues. One day the management system changed, and his new manager told him to go out and sell in the way he wanted. Within a year he was the company's top salesperson and a leader in the industry. Up to this time, the company had considered a $10,000 order to be very large. After the system change, this rep wrote orders as large as $500,000, at higher margins. Changing the system dramatically changed his performance. At best one can measure only the interaction between an individual and the system in which he or she works (Deming 1982).

One of the worst risks of conducting a performance appraisal is that when rewards are based on that appraisal, it can provide an incentive to game the system. Levitt and Dubner's recent book (2005), *Freakonomics*, describes, in considerable detail, a number of cases of how reward systems lead to cheating. This is not an accusation that managers commit fraud in order to secure bonuses. Although this has happened, as evidenced by the accounting fraud convictions in the cases of Enron and World Com, it is hopefully rare. However, there is an inherent conflict of interest in basing the pay of a person who is measuring something on the result of that measurement process. An example of this kind of manipulation is seen in Figure 1 (Carder and Ragan 2004).

Figure 1 shows a control chart of recordable accidents for Group 2, one of several manufacturing units in a large plant. Each point on the x-axis represents one month. The y-axis is the rate of recordable accidents. There is an upward shift around months 23 to 28. This shift illustrates a process shift in the wrong direction with seven consecutive points above the previous mean. For rules of interpreting control charts, the reader can refer to Nelson (1984). Although the output of a stable process will vary, certain patterns in the variation indicate that the process has changed, indicating a special cause. Special causes need to be investigated. Some special causes indicate that there is something wrong with the measurement process. On closer examination, the next series of points is very close to the new mean. According to the rules of control charts (Nelson 1984), a special cause requires the finding of fifteen points within one standard deviation of the mean. In this case, this condition is not met because there is another process shift at month 30. However, the points between months 23 and 29 are much closer to the mean than one standard deviation, suggesting the presence of a special cause. Upon investigation, it was found that because of the upward shift in the incident rate, managers in Group 2 put extreme pressure on the group to hold down the accident rates. In their zeal to turn the trend, they did not realize they had gotten exactly what they asked for. People stopped reporting accidents. Accidents happened at the rate expected for the process; they simply

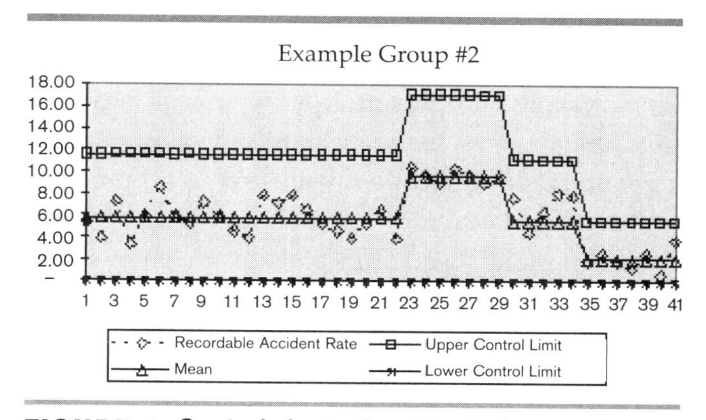

FIGURE 1. Control chart of accidents by month, showing process shifts

did not report those they could hide or classify as not being recordable. As the number of accidents in the month increased, pressure to not report also increased. Part of the reason the rates did not drop farther was that some accidents were just too serious to hide. This actual case is a clear illustration of manipulating the data to achieve an outcome and shows the value of using effective measurement tools to identify when this happens. Here the safety managers knew that the existence of nonrandom patterns indicated that something was wrong with the measurement process. Such patterns are frequently an indication that someone is manipulating the data (Deming 1982). The safety managers used that understanding to focus attention on the causes of the statistical anomaly and acted to correct it.

In order to guard against the risks involved in assessing performance, one must have an understanding of the principles of measurement. Application of these principles is critical to a meaningful assessment of performance. Incidentally, the application of these principles to Enron, World Com, and Health South would probably have revealed the problems early in the game. For example, Banstetter (2002) points out that several analysts who looked carefully at a variety of measures of Enron's performance were able to foresee serious problems. While revenues were growing, net margins were shrinking. Potential risks were obscured by keeping them off the balance sheet. Also, senior management was dumping a lot of stock.

Principles of Measurement

The quality of a measure is determined by its reliability and validity. No measure is perfect, and all measures have limitations on their reliability and validity (Deming 1982). Deming frequently illustrates this point by referring to measurements of the speed of light, which is an important constant in physics. He charted the variation of this measurement over time. Because the speed of light is assumed to be constant, the chart shows the variation in the measurement. Another illustration he used is variation in the census, depending on the method used. He noted that there is "no true value of anything. There is a measurement method and a result." If the method changes, it is likely the result will change.

Reliability of a Measurement

Essentially, *reliability* refers to the repeatability of a measure. For example, there is a particular golf course in central California at which the yardage markers appear to be unreliable. The markers that signify 150 yards to the green appear to be out of place on the eighth and ninth holes, reflecting a distance of closer to 160 yards. This may be explained by the assertion of one of the course employees that the persons who put in the markers used a 150-yard rope to make the measurements. The rope was nylon and got wet as they proceeded. The wet rope stretched, yielding longer distances on the later holes.

One could use many methods to measure the course, ranging from pacing off the distance to using a steel cable, laser, or global positioning system. Measuring a distance repeatedly with each method, would likely yield a spread of numbers around an average in each case. This is called *spread variation*. Variation is quantified by computing the standard deviation. The standard deviation is a statistical estimate that quantifies the variability of a measure. The greater the standard deviation, the greater the variability of the measure. The variability of measurements from the laser, presuming it was working properly, would be much smaller than the others. The wet nylon rope would very likely have the most variation. One would conclude that the laser was more reliable. There is no such thing as a perfectly reliable measure. All measures will show variation.

In some cases reliability is assessed by looking at the variation between observers. For example, audits are scored based on the judgment of an auditor or audit team. To judge the reliability, ask whether a different audit team, unaware of the initial team's evaluation, would give the same or a similar score. In practice, have two auditors conduct audits of a number of sites. Each auditor would be unaware of the scores given by the other. Then compute a correlation coefficient between the scores given by the two auditors. A correlation coefficient assesses the degree to which one measure predicts another. In this case one is assessing whether the score assigned by one team predicts the score of the other. A high value, meaning that the scores of one team are good predictors of the scores of the other, would indicate good reliability. A very low or even negative correlation would mean that the process has no reliability at all, in this particular test. It is still possible that one of the auditors is very accurate. However, there are two problems with this: (1) one can't know which auditor was correct, and (2) even if that could be determined, the process would be dependent on one person's judgment. In this case the measure is neither reliable nor useful. In order to develop a reliable and useful measure, one might attempt to clarify and better define the criteria and methods, retrain the auditors, and test again for reliability on a different set of plants.

No matter how reliable a measure might be, it will still have variation. If two auditors consistently report exactly the same score for the audit of a plant, management should question whether they are really operating independently. As mentioned previously, anomalies in variation often signal a problem with the measurement process. If a measure has no variation at all, one is not looking closely enough, the gauge is broken, or the numbers are being manipulated by the observer.

Reliability is not a property of the instruments, but of the entire measurement process. This includes the tools, the instruments, the procedures, and the people. Subjective judgment can be very reliable in some cases, and measurement with the finest instruments can be unreliable if the process that uses these instruments is flawed.

To repeat, any measure will vary. If the measure does not vary, or if the pattern of variation is not normal, an investigation is warranted. Assuming there are no anomalies in the variation, then the less the variation, the higher the reliability. For any particular purpose, there is a level of reliability that is acceptable for the task. When carpeting a room, one can use a tape measure, but not one that is elastic. When measuring length in order to construct a complex optical system, the laser might be required.

In the safety area of a business, it must be realized that all measures have reliability limitations and that all measures are subject to manipulation. It is important to understand the limitations of each measure and to take these limitations into account when making decisions based on the measurements.

It is important to realize also that a measure with high reliability may still be worthless if it does not convey anything useful about the what one is trying to measure. Just because a measure is reliable does not mean it will help management take effective action. This leads to the concept of validity.

Validity of a Measurement

Validity relates to whether one is measuring what he or she wants to measure. When measuring the width of a room, the question of validity usually does not arise. When measuring a complex process such as aptitude to perform well in college, or the ability of the safety management system in a plant to prevent future loss, validity becomes a serious question. Scientists (Cronbach and Meehl 1955) generally define three categories of validity: content-related validity, criterion-related validity, and construct-related validity.

CONTENT-RELATED VALIDITY

Often called *face validity*, content-related validity asks whether the content of the measurement process is, on its face, related to the purpose of the measurement. A good example is found in safety audits. If a question asks whether employees use personal protective equipment on the factory floor, that question has face validity. If one asks whether employees go out for beer together after work, that lacks face validity because nothing in the content appears to have anything to do with safety. However, the question might have criterion-related validity and construct-related validity. It could turn out that when employees have close personal relationships the plant is safer, and that going out for a beer (or bowling, or to church, etc.) after work is evidence of such relationships. Of course, this is not an assertion that this is actually the case.

CRITERION-RELATED VALIDITY

This is sometimes called *predictive validity*. It deals with whether one measure correlates with other measures that could be called criteria. For example, the SAT test is an attempt to measure the likelihood that a student will succeed in college. Obviously the criterion here is college performance. Yale University has used the SAT for many years. Although they admit it is not a very good predictor, they also say that it is the best they have. This is because different high schools have very different criteria for grading, so that an A in one might be equivalent to a C in another. Over the years, the Pearson correlation between SAT scores and Yale grades has run in the range of 0.2–.03. This is statistically significant, meaning that the SAT does indeed have criterion-related validity. However, this level of correlation means that, at best, what is measured by the SAT is accounting for no more than 9 percent of the variation in college grades. (Squaring the correlation coefficient of 0.3 gives us the percent of variation accounted for, 0.09.) The other 91 percent is presumably accounted for by other things, such as motivation, the quality of the student's secondary education, the difficulty of the courses chosen at Yale, luck, or any number of other variables each student must face and overcome to "make the grade" at the university.

When dealing with large populations, it may make good economic sense to use measures such as the SAT, which have relatively low criterion-related validity. However, individuals who are negatively affected by such measures will always have a pretty good argument that the measurement was unfair to them.

It is also important to realize that the criterion is arbitrary. After all, the success of a Yale career should not be measured by grades. Yale and most other universities are interested in producing good, productive citizens and leaders. Does the SAT predict that?

In the safety field, injuries and monetary losses are certainly useful criteria against which to test other measurements. However, they are not the only possible criteria, and they may not be the best criteria. Ultimately, one would like to know the ability of the safety management system to prevent future accidents and losses. Although a burned finger may be of some concern in a chemical plant, it is trivial in comparison with the release of a toxic chemical that could injure or kill thousands of people. Because catastrophes are fortunately infrequent, they are inconvenient to use as criteria in a validation study. An excellent safety measure would enable the prevention of catastrophic events. Minor incidents such as burned fingers are associated with cost and suffering, but they are neither equivalent nor directly related to catastrophic events (Manuele 2003, Peterson 2000).

Because criteria are usually somewhat arbitrary, and because there is often no single, ultimate criterion, it is best to use several criteria when attempting to establish criterion-based validity. This not only will make it more likely that validity will be established, but also will likely increase understanding of the measure being tested and the results attained. This leads to the concept of construct-related validity.

CONSTRUCT-RELATED VALIDITY

Construct-related validity goes to the understanding of what is being measured. In the measurement of safety performance, there has been little work on construct validity. Carder and Ragan (2004) used a reliable and valid safety survey to measure performance. They found that the critical constructs measured by the survey were (1) management's demonstration of commitment to safety, (2) education and knowledge of the workforce, (3) quality of the safety supervisory process, and (4) employee involvement and commitment. Coyle, Sleeman, and Adams (1995), working with a different survey, identified a similar set of constructs.

Safety professionals have been inclined to take the measures they are using for granted as a result of their content-based validity. If their intuitive feeling is that the measure is valid, they use it. They look to the content of their measurements to determine exactly what is being measured. There is a serious limitation in this practice. The most obvious case is the way incidents are counted. A minor cut or burn is recorded and investigated. On the other hand, many more important events, such as a chemical reaction going temporarily out of control, are often neither recorded nor investigated. The assumption, based on faith rather than evidence, is that the burn and the control of the chemical process are the same thing. Although they may be related, given they are both outputs of the overall management system, the evidence indicates that they are not the same thing (Wolf and Berniker 1999, Manuele 2003, Petersen 2000).

A better understanding what is being measured, leads to better judgment of what management actions are suggested by the measurement (Carder and Ragan 2004).

Managers prefer measures they believe to be highly reliable, in spite of the lack of any evidence regarding the validity of those measures. Many managers believe that surveys and interviews are of doubtful reliability and that the measurement of incident rates is quite reliable. In fact the data show that surveys and interviews can be very reliable, whereas the recording of incidents is frequently unreliable. In a recent book (Carder and Ragan 2004), we describe numerous examples from our own personal experience of how incident rates can be unreliable. More important is the limited ability of incident rates as a measure to enable the prevention of future catastrophic loss (Wolf and Berniker 1999, Manuele 2003, Petersen 2000). This is an important limitation of the construct-based validity of incident-rate measures.

A measure with moderate reliability and high construct-related validity is much preferred over a measure with high reliability and little or no evidence of construct-related validity. The latter measure may be very accurate, but it does not provide anything useful to guide management actions. If it is used to guide action, the effort will likely be wasted. Because it has little construct validity, it tells very little about the process one is attempting to improve. It would be like using a map of Ohio to drive in Pennsylvania.

Usability of a Measurement

Usability refers to the ease and cost of the following:

- collecting the data
- analyzing the data
- communicating measurement results
- using results to devise action plans

COLLECTING DATA

In 1994, a safety survey (Carder and Ragan 2003, 2004), which is discussed in detail later on, was developed. The survey was conducted in more than 50 plants of a major chemical manufacturer. In the previous year, the company had established a manufacturing strategy team (MST) to evaluate the quality of the management system in the same plants. When the plants' survey scores were compared with the MST scores, the correlation coefficient was –0.58. The correlation is negative because on the MST, lower scores indicated better performance, whereas with the safety survey, higher scores indicated better performance. This correlation indicates that the two processes were measuring many of the same things. However, the MST process required two to three highly trained and experienced staff members to conduct several days of interviews at each plant. The survey process required the employees to fill out a simple yes/no survey that took 20–30 minutes, usually during an already scheduled safety meeting. The scoring of the surveys was automated, and the analysis of the resulting data was relatively simple. Moreover, the survey had high reliability. The reliability of the MST process had not been evaluated. Thus, as a measure of performance, the survey was less expensive and simpler to implement than the MST process. It had much higher usability. Actually, the strong correlation between the survey and MST represented a validation of the MST process because the survey had been validated against other criteria as well.

An important lesson taken from this experience was that a good management system is likely to yield good safety performance. A tool that is used to measure the quality of the safety-management (accident-prevention) system also gives a good evaluation of the quality of the overall management system.[1] The survey was a simple, quick, and inexpensive tool that provided insight into how well a group was being managed. The alternative approaches to measuring the overall quality of the management system had a cost that was orders of magnitude higher, required much more effort, and produced less statistically supportable results. The survey tool is in the public domain, as are descriptions of how it can be used for measurement and for process improvement (Carder and Ragan 2003, 2004).

ANALYZING DATA

Managers often want to reduce performance evaluations to numerical data. This simplifies comparison between employees, plants, or other management units. It is relatively easy to accomplish with incident counts and surveys. It is more difficult with observational data and interviews. For example, developing a method to score an audit is a complex process because of the need to decide the proper weighting of various components. This is highlighted in the very extensive study of the audit process by Kuusisto (2000). To generate a numerical score, the investigation is broken into categories and subcategories. Each category must then be given a weight. Kuusisto shows that there is considerable variation in category weighting among different, widely used audit processes. The study also highlights the difficulty in establishing a reliable scoring process. This does not mean that interviews and observation should not be employed—because they can sometimes discover conditions that would not be revealed in a survey. However, it does mean that observation and interviews are at a disadvantage with respect to ease of data analysis.

COMMUNICATING MEASUREMENT RESULTS

Because performance evaluation is useful only to the extent that it can generate positive action or limit negative action, communicating measurement results to those who are impacted by it or who need to respond to it is critical. In the example being followed, processes such as the MST evaluation sometimes have a usability advantage. If a manager has a high degree of confidence in a staff member, then reports by the staff member are likely to generate action. Of course there is a problem with this, in that the usability of this kind of report is dependent on the relationship between the manager and the staff member. It may not always be possible to find the right staff member for a particular evaluation. With survey data, our personal experience is that when the reliability and validity of the survey have to be carefully explained to the manager, action is usually generated (Carder and Ragan 2004). Although we have rarely had much difficulty in accomplishing this, it should be done carefully and thoroughly by someone who has a good understanding of the survey process.

USING MEASUREMENT RESULTS

Measurement results should be used to devise effective action plans. This is the last consideration, but a critical one. For example, counting accidents tells one little about what to do. Often managers see an accident rate that is too high and tell their subordinates to try harder to be safe. This is not useful. Deming (1993) describes exactly this kind of event, and we have seen it ourselves many times. Deming recommends "looking into the process" that produced the accidents, rather than exhorting people to do better. A way of looking into the process is to find the true root causes of accidents using an effective investigation process. Then one can work to eliminate those causes, thereby reducing the potential for many future accidents.

If there are two methods with relatively equal reliability and validity, it makes sense to employ the one with the highest usability. However, although usability is important, remember that if a measure has very poor reliability or validity, the strong usability is worthless. It is like a man who is found one evening looking for his car keys on the street. A passerby questions him, finally asking if this is where the keys were lost. The man tells him he actually lost them in the middle of the block. The passerby asks why the man is looking for them on the corner. The searcher replies that he is searching at the corner because the light is better.

Effective Use of Available Measures for Performance Appraisal

The sections that follow examine some widely used types of measurement for reliability, validity, and usability, and ultimately for their value as measures of safety performance.

Incident-Based Measures

An almost universal measure of safety performance is based on incidents: the recording of accidents and the investigation of their causes. Since 1970 OSHA has required companies with eleven or more employees to maintain a record of accidents and injuries. Many other countries use incident-based measures with different incident definitions and different normalizing factors. In the United States the number is based on the number of incidents per 200,000 exposure hours. In Europe and many other countries the number of exposure hours used is 1,000,000.

The mere fact that accident counts are the fundamental method by which the government measures safety suggests that they can be measured very reliably. Although there are frequent and very articulate complaints that incident-based measures are not very helpful in process improvement (Petersen 1998), there is rarely a question about their reliability or validity. But, in fact, an examination of reliability and validity suggests that there are serious limitations to performance measures based on the counting of incidents. Many of these limitations are discussed in the following sections. For a more extensive treatment of this issue, see Carder and Ragan (2004).

Reliability of Incident Rates

There are two sources of limitation on the reliability of incident rates: variation in interpretation of the criteria for recording an incident and inherent variation in the statistic itself.

VARIATION IN RECORDING CRITERIA

Recording incidents is relatively complex. Studies of the reliability of classifying events as recordable or not recordable are lacking. However, there are many examples of stretching of the criteria:

1. Use of over-the-counter ibuprofen instead of prescription dosages. If the prescription is not used, then recording the incident is not required. Many doctors will cooperate with this approach. This cooperation is rationalized by the assumption that this is what the employee or employer wants, that it provides the same relief for the injured employee, and that it is less expensive than prescription medication.
2. First aid given for increasingly serious injuries that could have or, in some cases, should have warranted medical treatment. An article in *Professional Safety* (Rosier 1997) actually recommends setting up first-aid stations to "prevent the accident from falling into the OSHA recordable category."
3. Liberal interpretations of preexisting conditions are made to avoid recording an incident or to count a case as one case instead of two. Sometimes the first injury or illness case will have been in a different year, so even if the second is counted, the first incident is not included in the measures for the time being considered. It can be recorded with no negative effect because most companies do not factor in such historical changes.
4. Classification of more events as not being work-related. The pendulum swings from taking the employee's word for a case being work-related to requiring the employee to prove the case is work-related beyond any doubt.
5. Employees being offered full pay to work at home when their injuries prohibit their working at their normal workplace or traveling to and from their workplace. The requirement is that they go along with the story that the injury did not result in lost work.
6. Employee job definitions being used to define work relationship. In one case, an employee fell from a scaffold, breaking both wrists and suffering multiple other injuries. The injuries resulted in a hospital stay, but the case was initially not included on the injury log because it was argued that the employee was not doing his regular job. It seems this was an infrequent task that was not in his regular job description.
7. Manipulation of medical diagnosis. For example, an employee was cut and received fifteen stitches. After consulting with a physician, the employer argued that the stitches were "cosmetic," not "therapeutic," and therefore the case was not counted.

These are only a few of the many approaches used to avoid counting. It makes one feel there may be some truth in an old joke about an accident: An employee fell from a rooftop, and his scream drew the attention of his supervisor. It was lucky for the company that he screamed because it saved their twelve-year no-lost-day case record. It seems the supervisor was able to fire him for a safety violation before he hit the ground and was injured.

Although these might be extreme examples, they illustrate the difficulty in comparing the incident rates of two companies and being certain that the one with the lower rate truly has better safety performance. Petersen (1998) notes that accident rates do not discriminate between good and poor performers. Consider the plausible scenario where one company has a low accident rate because it underreports, and another company has a relatively high accident rate because it sees the reporting of accidents as an opportunity for investigation and improvement. The company with the higher reported record may actually have the better safety management system.

The problem noted here is not so much that employers are scofflaws and do everything they can to avoid the record-keeping rules out of contempt or malice. Frequently managers do not even know when "marginal" reporting is taking place. In almost every case of which we are aware, the companies involved went to great efforts to prevent unlawful practices and routinely disciplined those who knowingly committed these acts, including using termination in some cases.

A serious problem is that the events being measured are so complex that systems to define and count them must be elaborate and cumbersome. In addition, often strong incentives attached to the numbers can lead to manipulation of the count (Flanders and Lawrence 1999). The complexity provides more and more opportunity for errors or manipulation, intentional and unintended. Although the regulations may stand as an attempt to create a useful operational definition of a recordable accident, in practice, this has not been entirely successful. Petersen (1998) argues that accident data are not "useful for anything," although he concedes that some safety professionals disagree, and many executives object to the removal of these statistics. However, he argues, "it would be best for practitioners to wean their companies from dependency on such worthless figures."

VARIATION IN THE STATISTIC

Even if one determines that the recording process in a particular population could be reliable, the fact that the variability of incident rates is quite high for small groups or short periods of observation still remains. The appropriate statistic for computing the standard deviation of an accident rate is the U statistic, based on the Poisson distribution (Carder 1994): U = square root of (R/N), where R is the recordable rate, and N is the number of 200,000 work-hour exposure units. The 200,000 hours is the expected annual work of 100 workers.

When computing control limits of a process that has an average OSHA recordable rate of 3, the range of the control limits increases greatly as the population decreases. For a plant with 25 workers, the standard deviation is U = square root of (3/0.4), or 2.74. The upper control limit is three standard deviations from the mean, or $3 + (3 \times 2.74)$, or 11.22. This is based on one year of data. If the observation period shrinks to a month, then the units of exposure go from 0.4 to 0.033, and the upper control limit rises to 31.60. This looks high, but how many incidents can happen in a month for the measure to remain within control limits? The answer is that one accident in a month will yield a rate of 30.3 for that month, just inside the control limits. At this example site, two recordables in the month would exceed control limits. The point here is that comparing the recordable rates of small sites, or the rates of larger sites computed over short periods, is not very useful. If there are two plants with 40 employees, and if over a one-year period, one has a rate of 8 and the other of 0, there is not a sound statistical basis for concluding that they are different. This difference is within the normal random variation of measure being used. The incident rate of 8 required only three recordable events to reach.

Figure 2 shows how the control limits change as the plant population changes. The upper control limit (UCL) and lower control limit (LCL) are plotted on

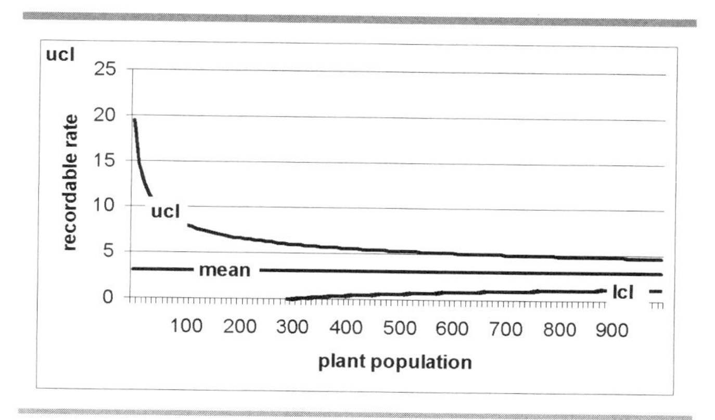

FIGURE 2. Control limits for accident rates as a function of plant size for a mean accident rate of 3

the x-axis, the group size on the y-axis. The mean recordable rate is 3. The control limits are for one year of data. Monthly control limits would be much wider.

The meaning of this is important. Safety managers would generally conclude that a recordable incident rate of 1 was good performance, an incident rate of 10 was not very good, and a rate of 30 was terrible. With a plant of 25 persons, an underlying rate of 3, and one year of observation, one might observe a rate of 1, or one might observe a rate of 10. In fact, one might see a monthly rate as high as 30! Any of these rates would be within the control limits and therefore would not signify that the process has changed. Remember that these observations are based on the same underlying safety process. Being aware of the statistical constraints, one will need several years of data to draw a definitive conclusion about the safety performance of this plant.

Validity of Incident Rates

Incidents are often used as a criterion for validation of other measures (see Carder and Ragan 2004). However, it is important to recognize that all measures have limitations on validity (Deming 1982). Moreover, a measure such as incident rates, which has low, or at best, moderate reliability, must, of necessity, have limited validity. It is important to discuss these limitations.

CONTENT-RELATED VALIDITY

Accident rates have obvious face validity as a measure of the safety process. After all, most recordable accidents are instances of something we are attempting to prevent. At the same time, a cut finger, on its face, is very different from the release of a large cloud of toxic gas. To the extent that one is concerned more about the release than about the cut finger, the cut finger may have little content-related validity. If the release does not result in an injury, in many current measurement schemes, the cut finger will receive higher importance and demand more resources be directed toward its prevention. It might be argued, as Heinrich essentially did, that the release and the cut finger are both instances of exactly the same thing and that "the severity of an incident is largely fortuitous." Therefore, cut fingers (incidents) are a measure that is valid in all respects. However, the available evidence indicates that this is not correct. Severe losses are associated with high energy, nonrepetitive work, and construction (which involves both high energy and nonrepetitive work) (Petersen 1998). In an environment in which resources are limited, the incidents that are related to potentially serious outcomes should be investigated first, whether they happen to be recordable or not.

Realize that the relationship between minor incidents and serious incidents is an empirical question, and no amount of argument or opinion can settle it in the absence of data. Figure 3 depicts an alternative look at Heinrich's (1959) assertion that minor incidents, moderate accidents, and major accidents are causally related (Carder and Ragan 2004). However, showing that the ratio of minor injuries to major injuries is 29:1 or something similar does not in any way demonstrate that the injuries derive from the same process or that a reduction in minor injuries will lead to a reduction in major injuries.

The success of treating all events the same, and assuming that by recording minor incidents one proportionally estimates the likelihood of more serious incidents, is not supported by available evidence. This frequently draws attention and resources away from finding and correcting the system hazards that could produce very serious outcomes, such as the chemical release that may be overlooked because the company was too busy investigating steam tracer burns.

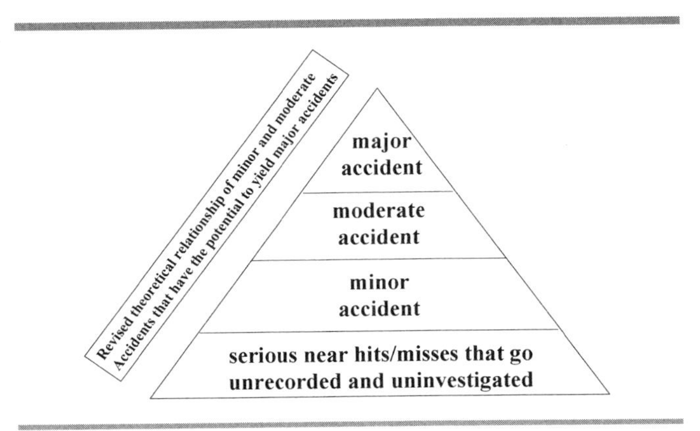

FIGURE 3. Revised version of Heinrich's accident triangle

CRITERION-RELATED VALIDITY

Accident rates themselves are frequently used as a criterion. However, safety is not necessarily the same as an absence of accidents. If safety is defined as freedom from danger, risk, and injury, then conditions are easily conceived in which there is no history of injury but great risk of future injury. Space shuttle flights, described previously, are an example of this. Although there was no history of injury prior to the Challenger flight, the engineers' estimates of the probability of loss of a vehicle indicated that shuttle flights were quite risky.

Perrow (1984), in his theory of normal accidents, argued that what he called normal accidents—disruptions of the flow of complex processes—were related to potential catastrophic events. Most of these disruptions in flow would not be classified as recordable accidents. In fact, plants that present the most significant hazard to workers and the public are typically high-capital, nonlabor-intensive plants that are likely to have very low rates of recordable accidents (Wolf and Berniker 1999).

In a test of Perrow's theory, Wolf and Berniker (1999) studied data on hazardous releases at 36 refineries from 1992 to 1997. Their assumption was that reportable releases are instances of normal accidents. They also provided data on the total case rate and catastrophic incidents. There were three fatal accidents in the period. They argued from the data that the incident rate was a poor measure for prediction of catastrophic loss. Better predictors were the complexity of the system and the number of toxic releases. More complex, tightly coupled (continuous-processing) systems have the highest danger, and the data showed that toxic releases were much better indicators of system problems than were recordable accidents. Although there was a small positive correlation between case rate and releases ($r = 0.23$, $n = 22$), it was far from statistical significance ($p = 0.30$, 2-tailed). The relationship between case rate and catastrophic accidents is not statistically significant, but the relationship between releases and catastrophic accidents is very significant ($p < 0.01$).

Incident measures certainly have some criterion-based validity. Because they represent injury and cost, they can be categorized as a criterion. They are not, however, the only important criterion, nor are they a substitute for measures of catastrophic loss or for measures that more accurately forecast the probability of such loss.

The suggestion from Wolf and Berniker's data is that releases may be a promising form of incident to use as the basis of safety-performance measurement. At the time of this writing, there is insufficient data to suggest anything beyond the exploration of this possibility, however.

CONSTRUCT-BASED VALIDITY

To consider construct-based validity, one must address the question of what it is that accident counts are measuring. This leads to the consideration of accident causation. The logic is that accidents are providing information about a set of underlying processes, which are labeled as causes.

Heinrich (1959) suggested that in 85 to 93 percent of cases, the primary cause is a human cause, so that incident rates may provide us with a measure of unsafe behavior. In fact, it is likely that 100 percent of all accidents involving anything manmade or involving people in any way reflect a human element.[2] In some way a human decision, action, or inaction was a contributing cause of the accident. Drawing on the logic of the cause-and-effect diagram that is widely used in quality improvement (Scherkenbach 1990), the

American Institute of Chemical Engineers has developed a system in which causes are divided into three major components: human, material, and organizational (AIChE 2002). The assumption of the process is that causes will exist in all three areas.

The human element consists of those actions or failures to act based on knowledge, beliefs, attitudes, physical abilities or capabilities, motivation, and so on. Usually the actions of operators are listed as a principal human cause. It is necessary to consider the entire community of humans who had influence on the event, including some who are no longer employees and some who might never have been employees. The roles of people can be categorized according to Figure 4.

The internal lines are dotted because the distinctions are not necessarily sharp. Individuals can have both managerial and technical responsibility, for example. The staff/technical category includes the designers of the system, who may have departed from the company a long time ago or who may have never been employees. Other similar circles can be added for peripheral or indirect human-cause groups, such as agencies, customers, suppliers, family, and so on.

The material element consists of physical aspects such as oil on the floor, a failed relief valve, a plugged line, and so on. It includes the physical work environment, working conditions, equipment, supplies, and materials. Based on the work of Wolf and Berniker, the complexity of the physical environment could also be included, with increased complexity creating higher risk. Of course, the physical environment is mostly the result of human activity. But like culture, it has a life that is somewhat independent of the persons currently involved in the system and is therefore worthy of identification as a causative factor. Because 100 percent of all accidents involve a human element, an organizational element, and a material element, systems theory tells us that one element cannot be changed without affecting the others. Nor can one element cause an accident independently of the rest of the system.

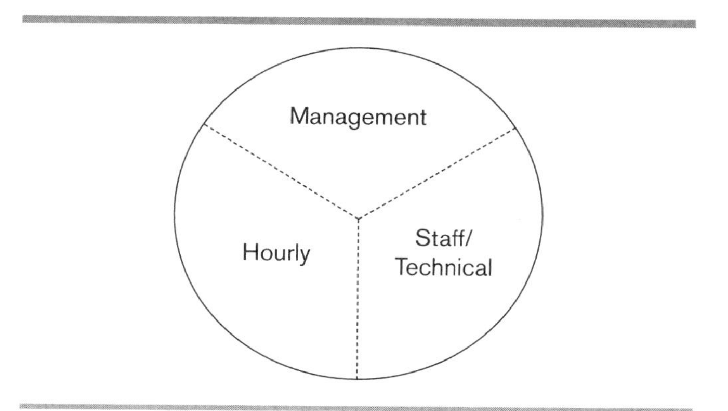

FIGURE 4. Division of sources of human causes

An ***organizational element*** relates to the structure, practices, procedures, policies, and so on, in place in the organization, as well as its history. This would include hiring policies, compensation policies, policies for promotion, reporting structures, training practices, work standards, regulations, and enforcement practices. These include both the stated policies and practices and the actual policies and practices, referring to what is actually done as opposed to what is written or said.

The human, material, and organizational factors come together to form the ***cultural system of the organization***. The cultural system represents the attitudes, beliefs, and expectations of the people in the organization. These are the product of human experience over time with the organizational factors and the material. All events are interpreted through the lens of this culture. In spite of changes in emphasis, rules, procedures, incentives, and personnel, the culture is resistant to change (Deal and Kennedy 1982, Senge et al. 1994). For example, in organizations that have a history of short-lived initiatives that are soon replaced by new initiatives, employees come to expect the "flavor of the month" (an initiative that is in line with the current trend but likely to change soon). Thus, any new initiative is greeted with resistance (McConnell 1997). Change is stalled, and the expectation of short-lived initiatives becomes a self-fulfilling prophecy. Changing culture requires changing the experience of the people over time. An organization in which senior management has paid little real attention to safety cannot expect to see an immediate change when some attention is given. However, if the attention and involvement of senior management

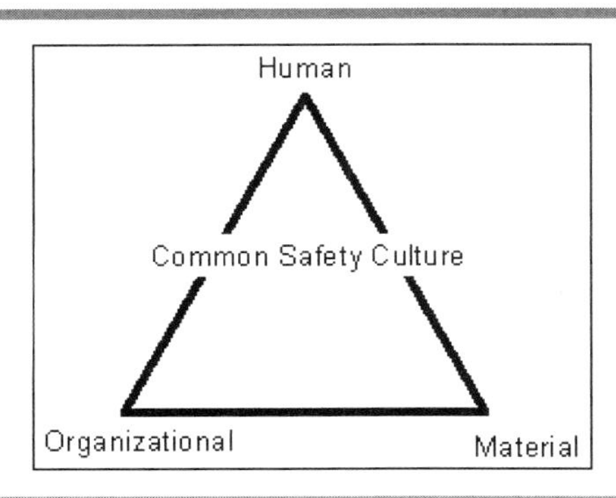

FIGURE 5. Three components create a common safety culture

is maintained over a period of time, on the order of years not months, the culture will change.

This system of accident causation is diagrammed in Figure 5, with the three areas combining to create an organizational culture. This is presented as an example of how this problem can be approached. Many ways of doing this can be productive, as long as the coverage of potential causal factors is comprehensive.

Attempting to summarize the issue of construct validity for incident-based measures, means measuring the performance of a very complex accident-producing process that includes the people, the physical environment, and the organizational system. Some incidents may reveal problems with these processes that could produce catastrophic outcomes. Other incidents may offer no insight at all into such problems. To the extent that recordable accidents are what is to be prevented, this limited validity is not an issue. However, if the primary objective is to prevent major or catastrophic loss, then management needs to be very conscious of finding systemic causes and directing its actions first at the most serious hazards.

Of course, a frequent assumption of performance measurement is that what is really being measured is the performance of the manager who controls a particular system of human, material, and organizational elements. With senior management, this may be appropriate. With the manager of a small plant or a department head, this is a stretch. These lower-level managers often have far less than complete control over much of the material and organizational systems in which they work. The incident rates that are produced by that system are certainly related to the performance of the manager, but they are also related to important elements beyond the manager's control, including the existence of an established culture that the manager has not had time to change.

Proper Use of Accident Statistics

Incident rates should be plotted on a control chart. The purpose of this is to see if the process is changing over time. This enables the user to detect system problems that have developed if the rate goes up or to evaluate the effects of improvement efforts that should drive the rate down.

The first step is to *establish a process mean*, which would be the mean recordable rate over the past twelve months or so. The next step is to *calculate control limits*. The upper control limit is equal to the mean plus three standard deviation units as calculated by the *U* statistic.[3] The lower control limit is the mean minus three standard deviation units. Of course it cannot go below zero. The mean and control limits are laid out on a chart, and each month the recordable rate is plotted. The following events indicate the process has changed, either improving or getting worse (Nelson 1984):

- A single point above or below the control limits
- Two out of three consecutive points in the upper third of the chart (more than two times *U* from the mean) or two of three consecutive points in the lower third
- Seven consecutive points above the mean
- Seven consecutive points below the mean
- Six points in either ascending or descending order.

Should any of these criteria be met, it is appropriate to try to find out what happened. If there is a run of points above or below the mean or several excursions outside of control limits, it is appropriate to compute a new mean and new control limits, going back to the time that the run started or that the first out-of-control point was observed.

As long as the process stays within limits, it is not useful to assume that the performance of the system has changed or to question what went wrong or what went right in a particular month. One is simply observing random variation. Trying to track down the cause would be like trying to find out why a roulette wheel came up on 17 on a particular spin or hit red four times in a row. On the other hand, if the wheel hit red eight times in a row, investigation might be worthwhile. That is the logic of the control chart. Look for the cause of a variation only when the process varies in a way that is extremely unlikely to have been a result of chance.

In fact, the control-chart interpretation rules are designed to identify nonrandom conditions in the data. No set of rules can cover all such conditions. If there is a clear and repeating pattern in the data, a special cause is likely present. For example, if points consistently alternate, one above the mean and one below the mean, there is likely a special cause. Special causes are indicated where the data are not random (Duncan 1986). A good example of a nonrandom pattern was presented by plotting a control chart of accidents for a particular pipeline company. A series of peaks, near but not beyond control limits, appeared to repeat on a 12-month cycle. Investigation revealed that these peaks occurred in summer. Each summer the company hired students to do landscaping and housekeeping chores. This group had a high rate of incidents. This special cause was dealt with by instituting safety training as part of the orientation of these students, and the peaks went away.

It is critical that normal fluctuations of the measure, within the control limits, not be treated as signs that the process has changed. Acting on these random fluctuations as though they represent process changes is called *tampering* (Deming 1982). Tampering not only wastes resources, but usually makes the performance of the system worse. An example of tampering would be to have a special safety meeting to figure out why so many accidents occurred in March when the number did not exceed control limits. There is nothing special about March. The focus should be on the causes of accidents, not on what happened in March.

Audits and Inspections

Audits and inspections are probably as widely used and accepted in safety as incident-based measures. Many companies use audit scores as a measure of performance.

The online *Merriam-Webster Dictionary* (2007) lists the following definitions for audit:

> **1 a :** a formal examination of an organization's or individual's accounts or financial situation **b:** the final report of an audit
> **2 :** a methodical examination and review.

Obviously, the second definition would apply to a safety audit. The word *audit* implies a standard. The purpose of the audit is to find out if the standard is being met. This could be a standard for the number and location of fire extinguishers, whether operators have met certain training requirements, whether the relief valves on pressure vessels have been tested in the time period required, or virtually anything else that management deems essential to safety. Audits can involve any or all of the following: physical inspections, interviews, and review of documents and records.

The basis of an audit is the assumption that the designer of the audit understands the process and has created an instrument to ensure that the process performs properly. Audits, then, derive their content from the theory of accidents held by the author of the audit. Depending on the author's point of view, the audit might emphasize unsafe conditions, unsafe acts, deficiencies in the safety management system, or any number of other personally held accident-causation philosophies.

Standardized Auditing Methods

There are a number of widely used and standardized audit systems. Kuusisto (2000) reviews some widely used systems, including a process developed by Diekemper and Spartz (D&S), the Complete Health and Safety Evaluation (CHASE I&II), the safety map, the OSHA VPP protocol, and the International Safety Rating System (ISRS), which appears to be the most widely used system.

All of the systems have a list of questions or areas of investigation, along with a method for scoring each question. Typically the audit involves interviews with site employees to obtain answers to the questions. The

TABLE 1

Portion of Diekemper and Spartz Audit Protocol

Activity	Level 1 (Poor)	Level 2 (Fair)	Level 3 (Good)	Level 4 (Excellent)
1. Statement of policy responsibilities assigned	No statement of safety policy policy. Responsibility and accountability not assigned.	A general understanding of safety, responsibilities, accountability, but not in written form.	Safety policy and responsibilities written and distributed to supervisors	In addition to the previous items, safety policy is reviewed annually. Responsibility and accountability is emphasized in supervisory performance evaluations.
2. Direct management involvement	No measurable activity	Follow up on accident problems	Active direction on safety measures. Management reviews all injury and property damage reports. and supervises the corrective measures.	Safety matters are treated the same way as other operational parameters (e.g., quality or production design). Management is personally involved in safety activities.

(*Source:* Kuusisto 2000)

audit protocol lists various specific areas of investigation and describes the criteria for scoring the company's performance in that area. Table 1 shows a very brief section of a Diekemper and Spartz protocol as used by Kuusisto.

It is apparent from this brief snapshot that this is not a simple process. For example, with regard to Activity 1, consider the problem of judging whether there is "a general understanding of safety." There is no operational definition here. An operational definition would very specifically outline how to make the judgment (Deming 1982). For example, the only way to really know if the understanding was *general* would be to interview a large number of employees. The operational definition might include "at least 63 percent of employees can describe the company's safety policy." Of course, that does not end the problem. What is the standard for deciding that their description is adequate? Lacking such definitions, there will be variation between auditors and diminished or nonexistent reliability.

A second issue is the fact that both questions have potentially equal weight in the final scoring. Are they equally important? That is an empirical question. To answer it one would want to see data relating the item score to some independent measure of the safety management system, such as injury rate, costs, and so on. We are not aware of any such published studies.

In fact, there is wide variation in the points of emphasis of the various standardized auditing systems. Kuusisto divides the areas of inquiry into four broad categories: (1) policy, organization, and administration; (2) hazard control and risk analysis; (3) motivation, leadership, and training; and (4) monitoring, statistics, and reporting. Which components are emphasized in the final score depends on the system. Kuusisto describes the differing emphases of three systems in Table 2: D&S, CHASE-II, and ISRS.

These differences are obviously quite substantial. Unfortunately, there is no published research which affords the most reliable and valid measurement of the safety-management system.

Reliability of Audits

There is very little published work on the reliability of audit systems. Kuusisto (2000) conducted an extensive study on the reliability of auditing methods. His dissertation reviews the literature on this topic and reports an original study of the inter-observer reliability of the D&S method and of an *improved* audit method

TABLE 2

Factor Emphasis of Three Audit Systems

Category	D&S	Chase II	ISRS
1. Policy, organization, and administration	20%	35%	33%
2. Hazard control and risk analysis	40%	48%	19%
3. Motivation, leadership, and training	20%	6%	19%
4. Monitoring, statistics, and reporting	20%	11%	29%
TOTAL	100%	100%	100%

(*Source:* Kuusisto 2000)

labeled Method for Industrial Safety and Health Activity Assessment (MISHA). In an initial test, Kuusisto audited six companies in the United States using the D&S method. He compared the audit scores that he recorded with the scores of internal audits conducted by a company employee who also used the D&S method. Statistical testing indicated that the reliability was "poor to moderate." In only one of the companies did reliability reach the moderate level. Two companies reached the *fair* level, two reached the slight level, and one was *poor*.

In a second test of the D&S method, Kuusisto examined three Finnish companies and compared the scores he recorded with scores independently recorded by his students. Here the reliability ranged from fair to almost perfect. The company employees in the first test were generally more strict in their interpretation than Kuusisto. In explaining the difference, he points out that the company auditors, because they worked in the company, had much more knowledge about the company than he did, whereas the students were working from the same information he was.

In the study of the MISHA method, Kuusisto compared his scores on a Finnish company with the scores recorded by the company's personnel manager, safety director, safety manager, and safety representative. He found *fair* agreement between his scores and the scores recorded by the safety manager and safety director and only *slight* agreement between his scores and those of the other two observers.

A reasonable conclusion from Kuusisto's studies is that audits can be made relatively reliable if they are conducted by individuals with similar training who are working from the same information. The highest reliability was achieved by comparing Kuusisto with his students, who were working from the same information. An intermediate level of reliability was obtained when comparing Kuusisto's scores with those of other safety professionals who were employees of the company. The safety professionals had specialized safety training and experience but different information because they knew much more about the company. If audits are conducted by persons with different training and different information about the company, then reliability is very low. It must be pointed out that this was not a determination of whether the finding of the audit was an accurate representation of the performance of the safety management system. That is a question of validity. The test of reliability means that the results of the audit are reproducible. The auditors with more experience, training, and personal knowledge (company employees) gave scores that differed more from Kuusisto's scores than did the scores produced by his students. Whose scores provided the most accurate or valid representation of the systems performance is left to question.

Overall, published literature suggests that under the best conditions, audits can be a reliable measure.

Content-Based Validity

Because safety audits are typically designed by safety professionals, they would be assumed to have content-based validity. Certainly the questions are those that are deemed by the designers to be the most important indicators of safety performance. The fact that, at least in some circumstances, reliable scores can be generated should satisfy us that the method has content-based validity. The presence of reliability indicates that several auditors are using similar definitions of the content, which provides a validation of that content.

Criterion-Based Validity

There are very few studies on criterion-based validity of audits. Bailey and Petersen (1989) and Polk (1987) describe an attempt to relate safety program characteristics with accident statistics and monetary losses in a very large study of railroads. They surveyed eighteen railroads and scored them on the following areas of safety programs:

1. safety program content
2. equipment and facilities resources
3. monetary resources
4. reviews, audits, and inspections
5. procedures development, review, and modifications
6. corrective actions
7. accident reporting and analysis
8. safety training
9. motivational procedures

10. hazard-control technology
11. safety authority
12. program documentation

Only two program areas had the appropriate correlation with the measures of loss: monetary resources and hazard-control technology. Companies that spent more and employed more advanced hazard-control technology had fewer and less costly incidents. Those companies report that two areas had counterintuitive correlations (meaning the higher the score, the higher the accident rate): equipment and facilities resources and reviews, audits, and inspections. Thus, the finding is that the more extensive the audits, reviews, and inspections, the less effective the safety program. This could mean that the audits and inspections make matters worse, or it could mean that railroads with frequent incidents and greater monetary losses increased their use of reviews, audits, and inspections, perhaps as a remedial measure. Although this is not a direct test of validity, the finding at least suggests that the audits in this study had no criterion-related validity and, more importantly, that the auditing may have actually impaired safety performance.

The strongest positive evidence for criterion-based validation comes from a study by Ray et al. (2000) of 25 manufacturing plants in Alabama. They developed an audit on the maintenance function of these plants. The audits were conducted by outside auditors from the University of Alabama. Audit scores were correlated with the recordable rate. Using a rank-order correlation, they found a statistically significant negative correlation between the audit scores and the incident rate, meaning that the better the audit score, the lower the incident rate.

Although this may be called a pilot study, it is a strenuous test of validity. Because the 25 plants were from different industries and clearly had different levels of risk, there was huge variation in the incident rates. This "noise" vastly decreases the likelihood of finding the expected correlation. A significant limitation of the study is that it is confined to the maintenance function. Another limitation was that the authors did not test reliability. However, low reliability would create even more noise to interfere with validation. It is important to note that this study was limited to an audit of the maintenance function and cannot be generalized at this point to different audits of different functions.

Construct-Based Validity

There appears to be nothing in the literature regarding what underlying constructs might be measured by audits. They generally measure compliance with standards that the authors of the audit believe to be essential to safe performance. It is apparent from Kuusisto's work on emphasis that the experienced safety professionals who write audits can have very different opinions about what is most important.

Usability of Audits

It is necessary to distinguish between the use of audits and inspection to maintain safety standards and their use to measure safety performance. In the former case, current practices may be sufficient. However, if audits are to be used as a performance measure, careful attention must be given to establishing a reliable and valid process. Ensuring reliability requires standardization of the audit protocol, training of the auditors, and testing of the reliability. Testing of the reliability requires that at least two auditors evaluate a number of sites independently, meaning that no auditor is aware of the scores assigned by another auditor. Then the scores are compared. The degree of correlation is a measure of the reliability. Although it is not likely to be perfect, there should be a strong positive correlation.

The second issue to address is validation. Sites with high audit scores should have other evidence of safety excellence, such as low incident rates and low monetary losses. Again, validation is established by correlating some other performance criteria with audit scores. The only systematic study that found in the literature that confirms a criterion-based validity, Ray et al. (2000), found a rank-order correlation of −0.336 between an injury-frequency index and audit scores on the maintenance function. This is statistically significant. Nevertheless, because the information is so limited, and the confirmation of validity is so narrow, anyone who wants to properly use audits as a performance measure has a long hill to climb in order to establish the reliability and validity of the process.

CASE STUDY

Safety Surveys

Some of the issues involving safety surveys can be illustrated by the following case study of a large chemical company. This company had grown in the previous several years through a set of acquisitions. It had over 6000 employees in the United States, distributed across more than 50 plants. Only two plants had more than 500 employees. Most had in the range of 25 to 50 employees.

This small plant size presented a problem for safety performance measurement. In 1993 the company was well along with the implementation of a total quality management (TQM) program. Based on the recent training in variation and the use of control charts, the safety managers had begun to use U-charts when looking at the incident-rate measures. The wide limits of variation encountered with small plant sizes convinced them that incident rates would not be particularly useful as a measure of performance in the company, at least for small sites and short time periods.

The safety director convened a high-level committee of safety professionals and plant managers to address the issue. Essentially, this was a process-improvement team focused on the problem of companywide safety measurement. The committee brought in consultants and industry experts for a series of seminars.

The most prominent survey at the time was the Minnesota Safety Perception Survey. This survey had been developed in the 1980s by a team that included safety professionals from the Association of American Railroads and scientists from the Army's Aberdeen Proving Ground (Bailey and Petersen 1989, Bailey 1997).

Bailey's team had designed and validated the Minnesota survey, which had subsequently been administered to over 100,000 employees in a number of companies. Bailey's database included survey results from six chemical companies, providing a good opportunity for benchmarking. Consequently, the company chose that survey. Chuck Bailey made a presentation to the team to explain how the survey had been constructed and how it could be used as a tool for improvement.

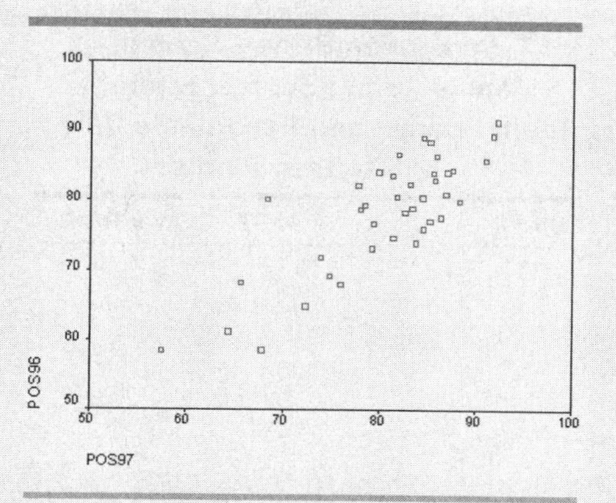

FIGURE 6. Scatterplot of scores for plants tested in two years
(*Source:* Carder and Ragan 2003)

To test the criterion-based validity of the survey, the group designed a pilot study. The basis of the study was the assumption that, if a question was valid, then a site with an effective safety program should have a higher proportion of positive answers on that question than would a site that had a weak safety program.

The survey that was tested consisted of the 74 questions from the Minnesota survey, along with 14 additional questions written by the team. The additional questions were written to cover some issues not covered in the original survey but that were important to a chemical company, including process safety and emergency response. In addition, the team added some questions covering incident reporting and the use of drugs and alcohol. The plan was to conduct a pilot study and then roll out the survey to the entire company if the pilot study demonstrated validity.

Reliability

The pilot studies tested only the validity of the survey, under the assumption that if it was valid, it would have to have at least moderate reliability. Subsequently, reliability was quantified in two ways (Carder and Ragan 2003). One method utilized the split-half technique, in which a person's total response (percent of favorable answers) to one-half of the survey's items (randomly selected) are correlated with his or her total response to the other half. If a particular respondent has 67 percent favorable answers on one-half of the questions, you would expect a similar percent favorable from that person on the other half of the questions. Application of this method yielded reliability coefficients in the range of 0.9, indicating a high degree of reliability.

The second method for testing reliability was the test-retest method. This took advantage of the fact that the survey was repeated in most of the plants in several consecutive years.

Figure 6 is a scatterplot of scores for the same plants when tested in 1996 (POS96) and 1997 (POS97) using the same survey instrument. The correlation between the two test scores yielded a Pearson *r* of 0.82, indicating again a high degree of reliability. If you were studying the reliability of an IQ test, you might want a somewhat higher number, under the assumption that IQ does not change materially over time. You would not want a perfect correlation with the safety survey. It is assumed that the performance of plants can either improve or decline, based at least in part on the actions of management. It should be expected that some sites will do more than others to implement changes based on the survey and that the scores of these sites will improve in comparison to the scores of sites that do not take any significant action.

In addition, subsequent administrations of the survey included questions that asked whether the site had taken action on the previous year's survey. There was a positive correlation between the proportion of respondents saying that action was taken and the change in the survey score of the site. In one study of 23 sites for which there were scores from 1996 and 1997, the Pearson r correlation between the questions on the site's response to the survey and the change in the site's overall score from 1996 to 1997 was 0.51. The probability of this being a result of chance is 0.02.

Safety Surveys (cont.)

Content-Based Validity

The original Minnesota survey and the fourteen questions that were added were all written by teams of safety professionals. Because the judges of content-based validity were safety professionals, the questions had de facto content-based validity.

Some might argue that the survey was measuring only the perceptions of workers and that such perceptions may have no relationship to reality, and therefore, one should not assume content-based validity. The response is that if workers have the *perception* that safety is not a high priority for management, then the workers are less likely to follow safety procedures. It does not matter what management actually thinks. The worker is guided by his or her own belief. He or she cannot read the mind of the manager.

Criterion-Based Validity

In spite of arguments that incident rates are not an ultimate criterion, they offered an obvious and readily available criterion against which to assess the validity of the survey. In addition, the team used the judgment of safety professionals regarding the quality of the safety program at the site. Here there were two levels: strong and weak. It must be realized that these criteria are not truly independent. When a safety professional is aware that a particular plant has a high incident rate, it is unlikely that he or she would judge it to have a strong safety program. It is possible that he or she might judge a site with a low rate as weak. He or she may be aware of serious hazards that the site has not effectively dealt with and conclude that they have been lucky so far.

The validity of each question was assessed using a statistical test to ensure that sites with strong safety programs and better accident records had better scores than sites with weak programs and high accident rates. The methodology is described extensively elsewhere (Carder and Ragan 2004)

Some of the original questions from the Minnesota survey were not valid in this test environment, even though they had been validated using a similar procedure when the Minnesota survey was developed (Bailey, personal communication). Validation studies were subsequently conducted in another company (Carder and Ragan 2003). This second company manufactured copy machines. Ten questions from the original Minnesota survey were not valid in either company. They were subsequently dropped from the survey that had been developed by the team.

Although Bailey (personal communication) has reported that questions would be valid only when the scores of hourly workers were used for the validation test, the studies described here showed that most questions would also be valid when only the scores of managers were used. Although managers generally had a higher proportion of favorable answers, these scores usually showed the same pattern of strengths and weaknesses. When there are severe problems at a site, the scores of managers may actually be lower than those of hourly workers (Carder and Ragan 2004).

An alternative method of assessing criterion-based validity is to look at the correlation between scores on the entire survey and some criterion, such as incident rates. Table 3 depicts the average recordable accident/incident rate for 1992–93 (Aver RAIR) and the survey score (total % favorable) for each site (Pcpt). In addition, there is a column titled MST. This is the rating given each of the sites by the company's manufacturing strategy team based on their analysis of the quality of the management system in operation at the site. This was not an attempt to rate safety management, but to rate the quality of the management system for manufacturing in general. The higher the number, the weaker the management system was judged to be. Because these scores were available, they offered an additional criterion against which to evaluate the survey. Although the accident record is not independent of the safety professional's rating of the site, this MST score should be independent from both.

TABLE 3

Perception Survey Scores, Manufacturing Scores, Strategy Team Scores, and Recordable Rates for Thirteen Plants

Site #	Pcpt	MST	Aver RAIR
15	0.73	8	4.95
17	0.70	8	5.15
19	0.83	2	0.00
27	0.83	4	1.80
35	0.82	8	6.00
38	0.83	2	3.70
44	0.78	5	2.55
56	0.69	8	3.30
58	0.76	4	2.55
72	0.62	6	4.90
75	0.78	5	2.65
78	0.55	8	8.25
80	0.71	4	1.80

(*Source:* Kuusisto 2000)

Table 4 is a matrix of the Pearson correlation coefficients (RAIR = Aver RAIR, PCPT SVY = Pcpt, and MST = MST).

There is a negative correlation between the RAIR and the survey score, as you would expect. The more positive the survey score, the lower the RAIR should be. A correlation this strong would occur in random data only 19 times in 1,000, so this correlation is statistically significant and provides further evidence for criterion-based validation.

It is very interesting to note that the MST rating is also correlated with both the survey score and the RAIR. The correlation between the survey score and MST is negative because a high survey score and a low MST score are both indicative of good performance.

TABLE 4

Matrix of Pearson Correlation Coefficients for Table 3

	RAIR	PCPT SVY	MST
RAIR	1.000		
PCPT SVY	−0.639 $p < 0.019$	1.000	
MST	0.756 $p < 0.003$	−0.577 $p < 0.039$	1.000

Safety Surveys (cont.)

The correlation between MST and RAIR is positive because the lower the MST and the lower the RAIR, the better the site. The difference in the strengths of these correlations may or may not be important. What is clearly important is that the existence of a correlation between all three measurements indicates that all three measures must involve some common factors. The fact that the MST rating, which is not based on safety, correlates strongly with two measures of safety suggests that both are dealing with some common characteristics of the plant.

One could argue from these data that quality and safety are fundamentally related and that the survey, in its most general sense, is a measure of the effectiveness of the management system and is not isolated to safety. The MST rating was an attempt to measure the quality of the management system. Although it is not surprising that this is a strong predictor of safety, it is the only statistical evidence that we are aware of that demonstrates the connection between safety and the quality of the overall management system. As discussed previously, the survey was a much less laborious process than the MST ratings. The survey may be a much more cost-effective method to evaluate the quality of the management system with regard to quality and productivity as well as safety. In fact, our personal experience indicates that when these questions are edited to remove the safety specificity, they can be an effective tool in helping understand the strengths and weaknesses of the overall management system.

To put the strength of these correlation coefficients into context, remember the example from another field–that the correlation between SAT scores and Yale grade-point average is in the range of 0.2 to 0.3. This means that the SAT, the best available predictor, can only account for about 10 percent of the variation in college GPA. In comparison with the SAT, the perception survey appears to be a much better predictor, accounting for 25 to 50 percent of the variation. It is also reasonable to suspect that the survey is a much more reliable measure than the MST rating, which represents the subjective judgments of a small staff group.

Costruct-Based Validity

The study of construct-based validity has been described extensively elsewhere (Carder and Ragan 2003). The original Minnesota survey purported to measure "20 factors influencing safety performance" based on an analysis by the survey's authors. An initial attempt to verify these twenty components measured the correlations between questions. More than one question is identified in each component. The logic of this is if there are three questions that measure "goals for safety," the answers to these three questions should have a relatively strong correlation. It turned out that questions within one of the twenty components often correlated more strongly with questions in other components than with other questions in their own component.

A factor analysis was conducted to better understand what was being measured by the survey. This is a statistical process that segregates the questions into groups of factors. Each factor represents a construct measured by the survey. An extensive series of focus groups was then conducted, with both hourly employees and managers, to determine what was being measured by each set of questions. Of course it is the respondents who know what they are telling the interviewers when they answer a question, not the designers of the questions. The basic design of the focus group was to present the group with a set of questions that represented one of the statistical factors. They were asked to explain what the set of questions was really getting at in a facilitated discussion, which ended with the group suggesting a name for the factor. Each focus group session lasted from three to six hours.

The groups named the following seven factors measured by the survey:

1. *Management's demonstration of commitment to safety.* Do management's actions convey the message that safety is very important?
2. *Education and knowledge of the workforce.* Are workers properly trained to do their jobs, and do they receive proper safety training? Do they understand their jobs and how to work safely?
3. *Quality of the safety supervisory process.* Does the company have standards for work, and are these standards enforced?
4. *Employee involvement and commitment.* Are employees involved in the planning process, and are they sufficiently committed to cautioning coworkers about unsafe practices?
5. *Drugs and alcohol (fitness for duty).* Is drug and alcohol use prevalent and tolerated?
6. *Off-the-job safety.* Does the company have an effective off-the-job safety program?
7. *Emergency preparedness.* Are employees at the site properly prepared to respond to an emergency?

The first four factors appear to be relatively universal, being similar to the results of other attempts to identify the critical components of the management system. Table 5 (from Carder and Ragan 2003) depicts the factors derived from several sources: (1) factor analysis of a database from the application of a safety-perception survey entitled the "Safety Barometer" developed by the National Safety Council (NSC, Carder and Ragan 2003); (2) a factor analysis conducted by Coyle et al. (1995) on data from a safety survey that he developed; and (3) the factors identified by a group of managers at Dow Chemical in the 1980s that formed the basis for the company's "self assessment process."

The first four factors are present in all of the surveys except for the NSC safety barometer, where we found no "education and knowledge" component.

In a similar vein, OSHA, through their VPP Star program, has identified "four major elements of an effective safety program" (1989). These are (1) management commitment and employee involvement, (2) work-site analysis, (3) hazard prevention and control, and (4) safety and health training. Again, these overlap with our factors. According to OSHA,

Safety Surveys (cont.)

hazard prevention and control must include a "clearly communicated disciplinary system." This is similar to our "quality of supervisory process" factor. Taking this into account, the OSHA elements cover all of ours.

Usability

The largest hurdle to clear is to explain the reliability and validity of the process to managers who are likely to hold a bias that incident counts are concrete and directly reflect reality and that survey scores are highly contrived and subject to many more sources of error.

Experience suggests that this hurdle can be successfully overcome. After the survey had been in use for two years in our target company, we conducted a survey of plant managers. Figure 7 (from Carder and Ragan 2003) shows their appraisal of the usefulness of the survey.

On the *x*-axis are the ratings, ranging from 7, very beneficial, to 0, not beneficial at all. Each bar represents the number of managers giving that particular rating. The majority (17 of 22) thought it was a useful process. None of the 22 respondents thought it was not useful at all (G-1).

TABLE 5

Management System Factors from Four Sources

Carder/Ragan Survey	Safety Barometer	Coyle Survey	Dow Self-Assessment
Management's demonstration of commitment	Management's demonstration of commitment	Maintenance and management issues	Line management leadership
Education and knowledge	not present	Training and management attitudes	Training
Quality of supervisory process	Quality of supervisory process	Accountability	Operating discipline
Employee involvement and commitment	Employee involvement and commitment	Personal authority	Total employee involvement
Off-the-job safety	not present	not present	Off-the-job safety
Emergency preparedness	not present	Included in company policy	not present
Drugs and alcohol	not present	not present	not present

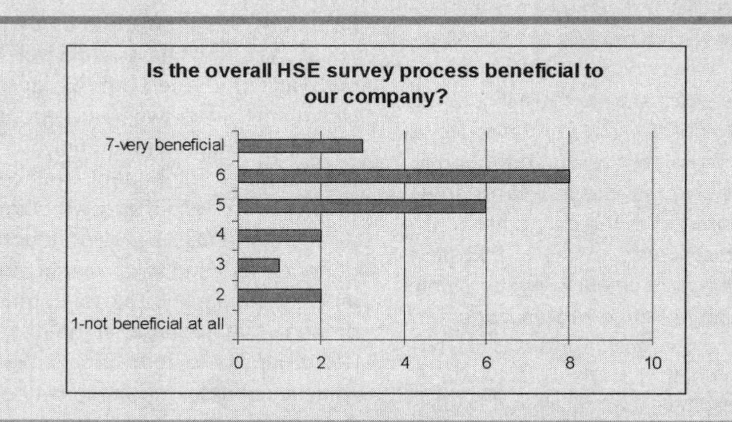

FIGURE 7. Plant managers rate the benefit of the survey process

Summary of Performance Measurement

This chapter has discussed the most commonly used systematic measures of safety performance, incident rates, and audits. Surveys, which appear to have certain advantages over the other methods, have also been discussed. Many other measures might have been included, such as costs and severity measures. They are not sufficiently developed and standardized to include in this chapter. However, it is not time to close the book on the development of better measures. The development of the survey has been a systematic search for a better measure by safety professionals, scientists, and managers. As the understanding of safety continues to improve, new and better measures will appear. The most important function of the preceding discussion is to lay out the ground rules for the evaluation of a performance measure: It must have reliability and at least content-related and criterion-related validity. Ideally it would also have construct-related validity so that it could serve as a basis for prescribing improvement actions.

BENCHMARKING

Benchmarking is properly viewed as one method of collecting information to enable the development of a plan for process improvement. Logically, benchmarking is a part of the plan phase for a plan-do-study-act cycle as described by Shewhart and Deming (Deming 1993). The proper way to engage in the planning phase is to develop information in order to guide the plan. Of course this must include information on internal processes. Benchmarking is another important source of data. One way to look at it is as a form of reality testing. If one wants to know what would happen if a certain practice were implemented, he or she may be able to find a benchmarking partner that has already implemented the practice and can study that partner's results.

As defined here, benchmarking is the study of a competitor's product or business practices in order to improve the performance of one's own company. This is a very complex process. First, it requires identification of superior performance. Second, it requires identification of the particular processes or activities that are responsible for that performance. Third, it requires the implementation of those processes or practices in one's own company.

Although none of these is simple, the implementation phase may be the most difficult. As an example, the Israeli Air Force has developed remarkable[4] safety performance over the last 25 years, driving their incidence of class 5 accidents (loss of pilot, aircraft, or damage over $1 million) from over 30 per year in the early 1980s to near zero by 2001 (Carder and Ragan 2004). According to Mr. Itzhak Raz, who was deeply involved in this improvement process, a critical change was the understanding in the culture that when pilots fly in a safe way, according to regulations, they are actually better fighters than pilots who cut corners and take risks. This runs counter to the macho culture that tends to pervade many air forces. Implementation of this cultural change required years of commitment from the most senior commanders to be effective. It is often easy to see what lessons are to be learned from the success of these measures. Implementation, however, is another matter entirely.

According to the American Society for Quality (Hill 2000), the reasons to engage in benchmarking are as follows:

- It avoids starting from scratch.
- It enables the transfer of tacit knowledge.
- It creates urgency and accelerates change.
- It identifies performance gaps.
- It helps develop performance goals.
- It helps develop realistic objectives.
- It encourages continuous innovation.
- It creates a better understanding of the industry.
- It establishes a process of continuous learning.

Against this backdrop of positive reasons, consider the following pitfalls.

- ***The Enron problem.*** As of February 2001, Enron had been ranked as the most innovative company by *Fortune* for six consecutive years and was ranked eighteenth on *Fortune*'s list of most admired companies. The measures of Enron's financial performance were, of course, quite flawed. One must be very careful about the performance measures that define outstanding companies. In safety, incident rates are not difficult to manipulate. Never select a benchmarking partner in safety merely on the basis of a low incident rate. It is critical that the benchmarking identify excellent processes rather than simply focusing on numerical results (DeToro 1995).
- ***The low-ceiling problem.*** Do not set ultimate targets based on the performance of others. The idea that the road to excellence is through imitation of superior performers already suggests that the company may be confined to second place. Even second place might be temporary. The company might discover processes and practices that will enhance its own performance, but the idea of trying to model one company after another is no guarantee of success. First, the company might set its sights too low and expend energy adopting processes that are soon to be obsolete. For many years GM was the target of Ford, in cost and product quality. Then along came Toyota and Honda, raising the bar by an order of magnitude. Table 6, published in 1990 (Womack, Jones, and Roos 1990), depicts the relative quality and cost parameters of GM and Toyota as of 1986. Fortunately for Ford, they had changed their target to Toyota by the early 1980s.

TABLE 6

Comparison of GM and Toyota Productivity and Profit

	A Good System	A Better System
	GM	Toyota
Assembly hours/car	31	16
Assembly space/car	8.1	4.8
Parts inventory	2 wk	2 hr
Defects/100 cars	130	45

(*Source:* Womack et al. 1990)

A second danger of copying is that the processes copied may not be compatible with the circumstances and culture of this particular company. If this is the case, it's possible these efforts do more harm than good (Stauffer 2003).

These cautions are not a reason to avoid benchmarking. However, anyone who develops a benchmarking project should keep them in mind. At any given time there will be strong performers and weak performers. Usually this changes over time. Do not look for "the player of the week"; look for superior processes that drive superior performance in the long term.

The American Society for Quality (ASQ) offers the following suggestions for successful benchmarking.

- Tie the benchmarking to the strategic initiatives of the organization. If this is not done, successful improvement efforts are unlikely, and those efforts will probably be wasted.
- Involve the process owners. First, their knowledge is critical to the company's benchmarking process. Second, their buy-in is necessary for any successful intervention.
- Understand one's own process. This is not as simple as it sounds. Discussions with managers in other companies in the course of benchmarking can assist in developing a better understanding of one's own process. Moreover, it is wrong to assume that the company will always find a better process than the one already in place. Even if it does, the company may determine that it would not be appropriate to implement in the firm (Stauffer 2003).
- Benchmark inside and outside. In any large company, there are likely to be units with excellent processes that may be as good as or better than anything one can find outside the company. An excellent process that already exists inside the company has one strong advantage: there is evidence that this process can succeed inside the culture and structure of this particular company.

Determining What Companies or Processes to Benchmark

This might appear on the surface to be a simple process. Find the company with the best performance, understand the processes that lead to this performance, and establish similar processes. Consider the application of this approach to baseball. The New York Yankees have had the one of the best performances over the past few years. One feature of their process has turned out to involve the expenditure of huge sums to hire star free-agent players away from other teams.

This process seems to have been adopted by a few teams. In 2003 the Yankees had a payroll of $180 million. Seven teams spent over $100 million: the Mets, the Dodgers, the Red Sox, the Braves, the Cardinals, the Rangers, and the Giants. The average payroll was $80 million, and the average team won 81 games, so that the average win cost about $1 million, but the Yankees, who won 101 games, were spending $1.8 million for each win. The most interesting anomaly in the data is the Oakland A's. They won 96 games with a payroll of $57 million, thus spending $0.59 million per win. Over the past several years, they have won nearly as many games as the Yankees with one-third the payroll.

Oakland's *process* has been eloquently described in the best-selling *Moneyball* by Michael Lewis (2003). Lewis described Oakland's use of statistical analysis of what player characteristics win baseball games and looks for players that will be undervalued in the marketplace, relative to their contribution to winning. For example, on-base percentage is more important than batting average. On-base percentage includes hits, walks, and being hit by a pitch. A player with a high batting average will be expensive, even if he rarely gets a walk. A player who has a relatively low batting average but who gets lots of walks and consequently has a higher on-base percentage will be much cheaper but also more valuable in terms of winning games.

The point of this is that benchmarking cannot be based simply on the identification of superior performance. Certainly in 2003 the Yankees had superior performance and an effective process. However, attempts to copy it have not been entirely successful. Moreover, most teams cannot even try because their revenue is not sufficient to spend even $100 million on payroll.

The following criteria can assist in the selection of effective benchmarking partners:

- ***Excellent performance.*** It does not have to be the best, but it has to be very good. Consider how the performance is measured and by whom. If the company is using accident records, be certain that the company is using the same recording criteria as its competitors? Does the company place such a strong emphasis on numerical goals, giving one reason to suspect the numbers they produce?
- ***Excellent processes.*** There are many sources of information about this. What do experts say? What do industry colleagues say? What has been written about them, excluding self-serving PR efforts? What kind of emphasis do they put on education? Education is a powerful driver of innovation. What is the quality of their education and training?[5] Excellent processes require excellent education and training.
- ***Interest in a cooperative effort.*** Look for partners who will share information candidly in a give-and-take relationship. Fortunately, in safety it is not likely competitors will want to protect the secrecy of their processes. Although safety may indeed be a competitive advantage, it is in everyone's interest to have excellence across an industry. A disaster at one chemical plant or refinery will put an entire industry on the defensive. However, look for more than a passive partner. The company will succeed most with a partner who will view the relationship as a win-win situation and who will put energy into the sharing process.

How to Proceed with Benchmarking[6]

The following list should be helpful in approaching the benchmarking process.

1. Clearly define the safety objective. Improving safety performance might seem to be a sufficient objective, but more thought is required. For example, is the company able to devote more resources to safety? Does the safety manager have the ability to implement actions based on the findings? When setting out to improve the processes of safety measurement, it is best to have the backing of senior management and a relatively large amount of control over the measurement process in the company. It is important to define an objective that is realistic, rather than setting out on a fishing expedition.
2. Understand one's own process, including the perception of its strengths and weaknesses. This actually goes hand in hand with the first objective. The study of one's own process will assist in defining the project. Most safety professionals have a pretty good intuitive idea of the strengths and weakness of their system.
3. Compare the company's performance with that of potential benchmarking partners. Use as many independent sources of information as possible. Examine the methods of measurement to understand reliability and validity. Include the subjective reports of professionals who have experience with the potential partners.
4. Select partners based on evidence of excellent performance, suitability for the company's strategic objectives, applicability to its processes, and the willingness of the potential partner to actively engage in the process.
5. Meet with the partners to develop a joint plan for the process. The plan should benefit both

parties. It should be specific, but also sufficiently flexible so that unexpected discoveries can be followed up.

6. Engage in a thorough process of data collection. Use as many sources as possible, including interviews, site visits, reviews of documents, and reviews of whatever numerical data exist. One may want to construct surveys, although it is a good idea to have surveys reviewed by a professional so that the results will be meaningful. Another option is to use TQM methods for data collection, such as flow charts and fishbone diagrams. Follow the findings as the process proceeds, so that promising leads are not missed because they were not discovered until the data-collection process was completed.
7. Integrate the data collected from the company's benchmarking process with the data that was collected in order to develop the action plan.

A final word (or two) on benchmarking:

Professionals in the fields of safety, incident prevention, risk management, and so on are likely to have an advantage over most others when benchmarking. They have the luxury of working in a field in which professionals are usually willing to share successes and failures. Managers are less likely to provide information on successful processes in manufacturing, sales, and marketing when these processes represent their competitive advantages.

Benchmarking, properly done, is a valuable assessment and improvement tool. One word of caution, however: benchmarking involves comparing outside practices to existing practices, and sometimes existing practices are not what one should be looking at. Often even *best practices* are based on current concepts. Sometimes a company is looking for a paradigm shift. Looking at what is being done and doing it as well as possible is essentially an example of continuous improvement. Taking a new look at how to do something unconstrained from existing practice is a valuable approach when a company seems to have reached the limits of improvement using existing systems and processes.

The field of safety experienced a paradigm shift in the early 1900s when people such as Heinrich suggested that accidents were the results of linear systems (Heinrich's domino theory) and that through work on minor accidents, the occurrence of serious accidents would be reduced because there was a ratio of minor to serious events.[7] Professionals started to look at accident trends and events that foreshadowed the occurrence of accidents so that they might prevent more accidents. At the time, many accidents were caused by inadequate guarding or exposure to unsafe work environments. The obvious answers were guarding and using personal protective equipment (PPE). Major improvements were made.

Another major paradigm shift occurred when guarding and PPE began to reach the limits of their ability to reduce accidents. Behavioral safety became an important topic of discussion in the 1980s. This included the work of Petersen and his colleagues (1991) and Krause and his colleagues (1997). Although the two groups took very different approaches to the problem, they remained focused on the leverage that could be developed by focusing on behavior.

Much of what has been discussed in this chapter is an example of yet another shift—viewing accidents as being caused by an interconnected system. In the authors' experience, dividing this system into human causes, organizational causes, and causes in the work environment is an effective approach.

ENDNOTES

[1]This is the ability of the management system to manage important outcomes such as quality, cost, productivity, innovation, safety, and profit.

[2]The only accidents that do not have a human element are those that in no way involve humans. For example, lightning striking a tree and causing a fire could be an accident with no human element, unless of course the tree was planted by a person or someone cut down all of the surrounding trees, leaving this one as the highest and most vulnerable to lightning strikes. Only pure acts of nature appear to have no human element. Even then the consequences are frequently affected by where humans build, remove, or otherwise alter the natural course of events. A tidal wave is an act of nature. The destruction of a coastal city has a human element be-

cause humans built there in the first place. Earthquakes are an act of nature, but the degree of damage is greatly dependent on where people built and the way they built their structures.

[3]The *U statistic* is based on the Poisson distribution. U = square root of (R/T) where R is the recordable rate and T is the number of 200,000-hour exposure units on which that rate calculation was based (Deming, 1993; Carder, 1994).

[4]In fiscal year 2002 the U.S. military had 97 class A aviation accidents, including 35 in the Air Force (see www.denix.osd.mil/denix/Public/ESPrograms/Force/Safety/Accidents/fy03_final.html).

[5]Deming has insisted on the differentiation of *education*, which is the teaching of theory, and *training*, which is the teaching of a skill. We agree with the distinction. The two are obviously related, and both are very important.

[6]This list is largely derived from Hill (2000).

[7]Heinrich suggested the accident triangle to illustrate this theory.

References

American Institute of Chemical Engineers (AIChE). 2002. *Guidelines for Investigating Chemical Process Incidents*. New York: AIChE.

Bailey, C. 1997. "Managerial Factors Related to Safety Program Effectiveness: An Update on the Minnesota PerceptionSurvey." .*Professional Safety* 8:33–35.

Bailey, C. W., and D. Petersen. 1989. "Using Safety Surveys to Assess Safety System Effectiveness." *Professional Safety* 2:22–26.

Banstetter, T. "Some Saw Through Enron's Ploys." *Fort Worth Star-Telegram*. February 10, 2002.

Camp, R. C. 1995. *Business Process Benchmarking: Finding and Implementing Best Practices*. Milwaukee: ASQC Quality Press.

Carder, B. 1994. "Quality Theory and the Measurement of Safety Systems." *Professional Safety* 39:23.

Carder, B., and P. T. Ragan. 2003. "A Survey-Based System for Safety Measurement and Improvement." *Journal of Safety Research* 34:157–165.

______. 2004. *Measurement Matters: How Effective Assessment Drives Business and Safety Performance*. Milwaukee: ASQ Quality Press.

Cronbach, L. J., and P .E. Meehl. 1955. "Construct Validity in Psychological Tests." *Psychological Bulletin 52*, pp. 281–302.

Coyle, I. R., S. D. Sleeman, and N. Adams. 1995. "Safety Climate." *Journal of Safety Research* 26:247–254.

Deal, T. E., and A. A. Kennedy. 1982. *Corporate Cultures: The Rites and Rituals of Corporate Life*. New York: Addison-Wesley.

Deming, W. E. 1982. *Out of the Crisis*. Cambridge, MA: MIT Center for Advanced Engineering Study.

______. 1993. *The New Economics for Industry, Government, Education*. Cambridge, MA: MIT Center for Advanced Engineering Study.

DeToro, I. 1995. "The 10 Pitfalls of Benchmarking." *Quality Progress* 28:61–63.

Duncan, A. J. 1986. *Quality Control and Industrial Statistics*. Homewood, IL: Irwin.

Feynman, R. P. 1999. *The Pleasure of Finding Things Out*. Cambridge, MA: Perseus.

Flanders, M. E., and T. W. Lawrence. 1999. "Warning! Safety Incentive Programs Are Under OSHA Scrutiny." *Professional Safety* 44:29–31.

Heinrich, H. 1959. *Industrial Accident Prevention*. 4th ed. New York: McGraw Hill.

Hill, C. 2000. *Benchmarking and Best Practices*. American Society for Quality, www.ASQ.org

Krause, T. R. 1997. *The Behavior-Based Safety Process: Managing Involvement for an Injury Free Culture*. 2d ed. New York: Van Nostrand Reinhold.

Kuusisto, A. 2000. "Safety Management Systems: Audit Tools and the Reliability of Auditing." Doctoral dissertation, Technical Research Center of Finland.

Levitt, S. J., and S. J. Dubner. 2005. *Freakonomics: A Rogue Economist Explores the Hidden Side of Everything*. New York: HarperCollins.

Lewis, M. 2003. *Moneyball*. New York: Norton.

Manuele, F. 2003. *On the Practice of Safety*. 3d ed. New York: Wiley.

McConnell, M. C. 1997. "Formula Management: In Search of Magic Solutions." *Health Care Supervisor* 16:65–78.

Merriam-Webster Online Dictionary. Retrieved March 28, 2007. www.m-w.com

Nelson, L. S. 1984. "Shewhart Control Chart—Tests for Special Causes." *Journal of Quality Technology* 16:237–239.

Occupational Health and Safety Administration (OSHA). 1989. Safety and Health Program Management Guidelines; Issuance of Voluntary Guidelines. Federal Register 54:3904–3916.

Perrow, C. 1984. *Normal Accidents: Living with High Risk Technologies*. New York: Basic.

Petersen, D. 1998. "What Measures Should We Use and Why?" *Professional Safety* 43:37–41.

______. 2000. "Safety Management 2000: Our Strengths and Weaknesses." *Professional Safety* 45:16–19.

Petersen, D. E., and J. Hillkirk. 1991. *A Better Idea: Redefining the Way American Companies Work*. Boston: Houghton Mifflin.

Polk, J. F. 1987. Statistical Analysis of Railroad Safety Performance, 1977–1982. Final Report of Contract DTFR 53-82-X-0076, Federal Railroad Administration.

Ray, P. S., R. G. Batson, W. H. Weems, Q. Wan, G. S. Sorock, S. Matz, and J. Coynam. 2000. "Impact of Maintenance Function on Plant Safety." *Professional Safety* 45:45–49.

Rosier, G. A. 1997. "A Case for Change." *Professional Safety* 42:32–35.

Scherkenbach, W. W. 1990. *The Deming Route to Quality and Productivity: Roadmaps and Roadblocks*. Milwaukee: ASQ Quality Press.

Senge, P. M., C. Roberts, R. B. Ross, B. J. Smith, and A. Kleiner. 1994. *The Fifth Discipline Field Book*. New York: Doubleday.

Stauffer, D. 2003. "Is Your Benchmarking Doing the Right Work?" Harvard Management Update 8.

Wolf, F., and E. Berniker. 1999. *Validating Normal Accident Theory: Chemical Accidents, Fires and Explosions in Petroleum Refineries*. Retrieved from www.plu.edu/~bernike/NormAcc/Validating%20NAT.doc

Womack, J. P., D. T. Jones, and D. Roos. 1990. *The Machine That Changed the World: Based on the Massachusetts Institute of Technology 5-Million-Dollar 5-Year Study on the Future of the Automobile*. New York: Rawson Associates.

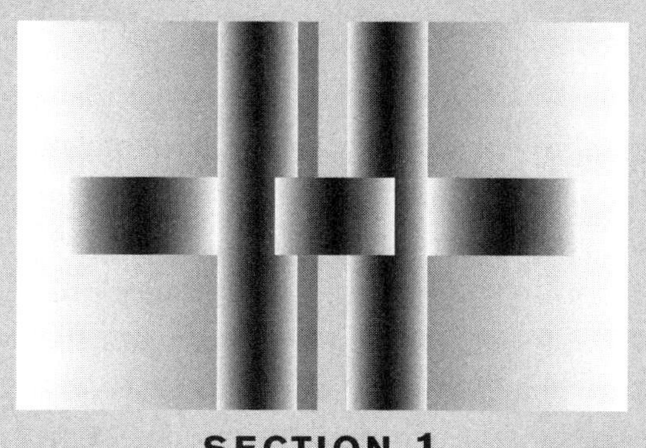

SECTION 1
RISK ASSESSMENT AND HAZARD CONTROL

BEST PRACTICES

Stephen Wallace

LEARNING OBJECTIVES

- Learn about the various codes, standards, and strategies used to identify risks and how to ensure that necessary safeguards are in place to prevent incidents.
- Be able to identify best practices in process design, including installing active and passive controls.
- Explain hazard reviews and the steps necessary to make processes inherently safer.
- Know what equipment is related to fire control and suppression.
- Recognize the steps necessary to ensure that building occupants are safe and that maintenance is being performed in an effective manner.
- Understand the steps that must be taken to ensure that facilities remain secure from internal and external threats.
- Be able to discuss special material hazards, such as reactivity and asphyxiation.

ASSESSING RISKS and controlling hazards in chemical and manufacturing plants often involves the use of sophisticated tools to evaluate detailed engineering designs. As noted in this chapter, the areas that must be considered are broad and include aspects such as design of pressure-relief valves, electrical safety, and materials of construction. It may be difficult for the practitioner to be well versed in all aspects. Various evaluation tools may be used to determine whether the risk involved with certain processes is acceptable and what provisions need to be made to lower it. Such tools can be qualitative or more complex, such as fault trees.

Regulatory compliance alone does not guarantee that an organization will assess and control risks in the most effective manner. For example, OSHA 29 CFR 1910.119(d)(3)(ii) states that organizations must document that equipment complies with *recognized and generally accepted good engineering practices*, sometimes referred to as RAGAGEPs (OSHA 1996a). Fortunately, there are several sources available to aid practitioners in understanding good practices (sometimes referred to as *best practices*) in this area. To reflect the jargon used in the chemical and manufacturing industries, the terms "good practices" and "best practices" will be used interchangeably in this chapter.

VOLUNTARY GUIDELINES, CODES, AND CONSENSUS STANDARDS

Best practices for assessing and controlling risks go beyond regulatory compliance and come from a variety of sources. These practices are developed to provide specific guidance in the design, operation, and maintenance of equipment processing hazardous materials. Good practices can be voluntary guidelines, recommended practices, codes, and consensus standards. For the purposes of this chapter, we will not differentiate between the legalities of these different types and instead will present the relevant information in the content of each.

Trade organizations with member companies often develop standards to assist their membership in assessing and controlling hazards. Some property and business insurance companies in the chemical and manufacturing industries develop guidelines for those they insure to follow to minimize the risk of loss. Even companies and individuals that are not members of these organizations can often obtain these guidelines, codes, and standards, generally for a fee.

Some good-practices organizations, such as the Center for Chemical Process Safety (CCPS), grew out of larger professional societies and were established to address the needs of members who desired more specific guidance on ways to ensure that their processes were designed, operated, and maintained in the safest manner. Individual practitioners or members of academia with expertise in a particular subject area may also publish and present their ideas and findings. Companies sometimes develop internal guidelines, although this information is often considered proprietary and is not always publicly available.

The last portion of this chapter provides a listing and description of some of the organizations that develop voluntary guidelines and consensus standards. Although this listing is not comprehensive, it includes organizations whose products are frequently referenced and cited throughout industry.

Best Practices in Risk Assessment and Hazard Control

As noted previously, the field of risk assessment and hazard control is very broad. Risks must be evaluated and controlled during all stages of process operation, including design, commissioning, operation, maintenance, and decommissioning. This section will address risk assessment during some of the most critical phases. A number of best practices resources will be referenced, and an explanation of these follows at the end of the chapter. It is important to note that the guidance provided in this chapter is not meant to replace any codes and standards but rather to make the practitioner aware of sources that exist and critical components of their contents. The guidance presented in this chapter is meant to distill some of the fundamental concepts and is not meant to be a comprehensive discussion of each area. Also, codes, standards, and good-practice guidelines are subject to revision, so users must ensure that they are using the most recent editions and the ones most applicable to their situations.

The information in this chapter has broad applicability to several industries. Although specific sectors such as refining and petrochemical processing are discussed, the lessons of hazard identification, inherently safer design, and adequate layers of protection are applicable to a variety of industries, including the defense and aerospace industries as well as manufacturers of pulp and paper, food, pharmaceuticals, and explosives. Also, many of the examples in this chapter are taken from the process-safety field, but the concepts are appropriate in a variety of settings, including evaluating occupational hazards. For example, adequate hazard identification and inherently safer design are useful concepts to apply whether an organizaton is trying to mitigate a chemical release or prevent injury to a worker using a complex machine with several moving parts.

Finally, requirements listed throughout this chapter are noted as good practices. Several factors determine whether a requirement is mandatory, such as determinations by the authority having jurisdiction (AHJ). Because of this, the chapters generally do not specify legal requirements, except those specifically noted in OSHA regulations. This chapter lists good practices and does not specifically make "shall"-type statements. The practitioner is encouraged to consider all of these practices based on their merits. Specific codes and the AHJ should be consulted to determine which practices are mandatory and which are suggested.

Process Design

The most critical phase in identifying risks is the design phase. It is during this phase that errors can most easily be corrected. Flaws that exist after the design phase will require a larger expenditure of capital to correct. In other words, it is easier to make corrections *on paper* before the equipment is actually in place.

To effectively control risk, both the likelihood and the consequences of accidents must be reduced. It is during process design that the technology, process monitoring and control instrumentation, materials of construction, and mitigation equipment are chosen.

In *Guidelines for Engineering Design for Process Safety*, the Center for Chemical Process Safety (CCPS) notes that redundant levels of safeguards, often referred to as "layers of protection," should be built into processes. The CCPS recommends that designers minimize risks by making processes inherently safer, installing design features that do not need to actively function to be effective, and installing additional equipment that may be relied upon to function if needed to avoid process deviations (CCPS 1993, 6). The last step in protecting the safety of the process is implementing procedural safeguards, often referred to as *administrative safeguards*. Inherent safety is the most effective design strategy and is addressed first in this section.

Inherent Safety

The concept of *inherent safety* can most effectively be implemented during process design. The CCPS notes that during the early stages of design, the process engineer has maximum degrees of freedom to consider alternatives in technology, chemistry, and location of the plant (CCPS 1993, 5). Applying the concept of inherent safety means that the process designers use strategies of safer chemistry, rather than control systems and operator intervention, to prevent injuries and environmental and property damage. The elements involved in inherent safety include

- ***intensification*** (using smaller quantities of hazardous materials)
- ***substitution*** (replacing materials with less-hazardous substances)
- ***attenuation*** (using less-hazardous conditions or a less-hazardous form of a material)
- ***limiting of effects*** (designing facilities that minimize the impact of releases)
- ***simplification/error tolerance*** (designing facilities that make operating errors less likely or less consequential).

The first three elements of inherent safety stated above involve using smaller quantities or less-hazardous materials. These issues are generally easily understood by practitioners, although at times it may be challenging to implement them. Designers should consider whether the operating unit can tolerate having smaller supplies of raw materials on hand. While a lower inventory of raw material may cause production concerns if there are upsets in the supply, it also means that there is less material to be released in the event of a failure. However, caution must still be exercised when these principles are applied. For example, by decreasing the size of chlorine cylinders kept on site, more change-out of cylinders may be needed, which is a risk that must be managed. Designers should also consider whether less-hazardous materials can be used in the process, such as aqueous rather than anhydrous ammonia.

Limiting of effects involves designing facilities and equipment in such a way as to minimize the impact of releases. Separation distances should be adequate between hazardous material unloading, storage, and processing. Facilities should be designed to minimize the need for transportation of highly hazardous materials. Distances to sensitive receptors, such as residential communities, should be considered in the design. Adequate buffer spaces should exist between sensitive receptors and hazardous installations. Facilities should be located in proximity to utilities, emergency response support, and adequate water supplies.

Finally, equipment in facilities should be designed to minimize errors or make them less consequential. Equipment can be designed to ensure that flow rates are within safe limits by choosing pumps and sizing lines appropriately. Vessels should be designed for a full vacuum [–14.7 pounds per square inch gauge (psig)] if the possibility exists for pumping the materials out with the vents closed or if hot vapor may become trapped and condense. Reactors should be designed to be robust enough to withstand the maximum allowable pressure and eliminate the need for a large emergency relief system. Piping systems may be designed to eliminate components susceptible to leakage by using sight glasses, flexible connectors, and so on.

Protective Equipment–Passive and Active Safety Systems

Once inherent safety is considered and implemented to the extent possible, engineers should install equipment and devices to provide passive safety protection for workers and equipment. *Passive* safety measures, as opposed to *active* safety measures, have the advantage of being able to function with no additional operation or moving parts necessary. For example, pressure vessels typically contain relief valves that open to relieve pressure in a vessel to prevent rupture. While safety relief valves are generally designed with high reliability, the valves still must operate to perform their function, but if a vessel is designed to withstand a higher pressure than can be generated in the process, the pressure rating of the vessel works in a *passive* manner to keep the vessel from rupturing.

Active devices include controls, interlocks, and shutdown systems that detect deviations from design intent and take action to correct the situation before a problem develops. As an example, Figure 1 shows a tank with redundant high-level alarms and interlocks to both trip the feed pump and close the block valve on the inlet line. Particularly hazardous systems are generally designed with multiple active devices to ensure that redundant layers of protection exist. For example, in a system that is likely to generate pressure, a pressure indicator is installed, which allows operations personnel to monitor the pressure. This device typically has high-pressure alarms (and high-high-pressure alarms) coded into it to warn of a pressure excursion (deviation from a normal path). A high-pressure shutdown will be designed to take action, such as opening a vent or closing a heating source. Ultimately, an emergency shutdown can be installed to take a number of actions, including stopping the flow of all feed material to the unit where the upset is detected. Such shutdown systems may be programmed to activate based on various types of inputs, including high pressure, temperature, and concentration of contamination.

When instrumentation is added to equipment during design, the designer must install sufficient levels of protection in the most economical manner. In ISA 84, *Safety Instrumented Systems for the Process Industry Sector*, the Instrumentation, Systems, and Automation Society (ISA) has guidelines for installing levels of protection based on how hazardous the material is and how critical the device should be (ISA 2004). A comprehensive discussion of the methods employed by the ISA is outside the scope of this chapter, but a device that is critical and operating in a hazardous service should have a high reliability rating, on the order of 99.99 percent reliable. (Stated another way, the safety system should have a high availability, on the order of 0.9999.) This level is generally not achieved by a single safety device but by including redundant devices.

Procedural Safeguards

The final safeguards at the facility level are often procedural or administrative. When active and passive safety devices fail to control an event, operations personnel must intervene and bring the system back to a stable state. This may involve correcting the situation, if possible, or instituting a controlled shutdown.

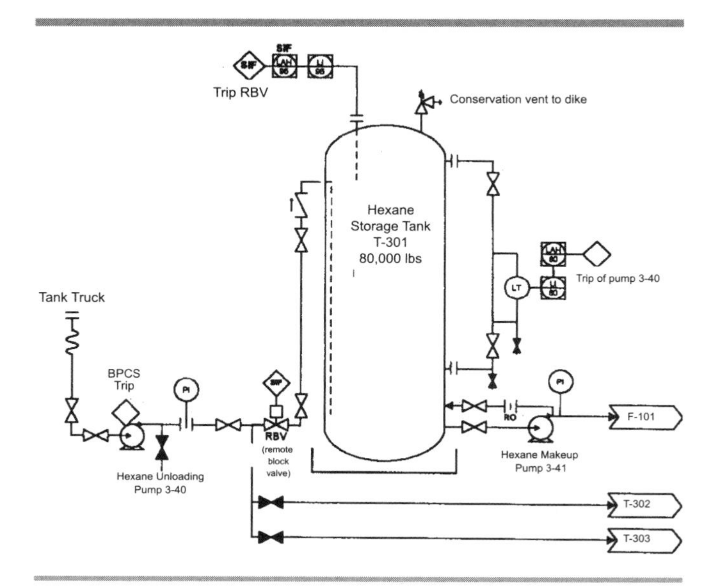

FIGURE 1. Example of a tank with redundant sensors, controls, and shutdowns (*Source:* CCPS 2001c)

While this level of safeguard is typically not considered a *design* issue, the plant can be designed to make the job of operations personnel easier or more difficult. The facility must be designed with inputs to help personnel quickly diagnose a situation. Newer plants are likely to have computer-based systems monitoring process control. Critical parameters should be alarmed, and alarms should be prioritized. A temptation for designers is to add numerous alarms to a process. This can be done for the sake of safety; however, if a process is overalarmed, it can be difficult for operations personnel to determine the critical parameters during an emergency.

Control of ignition sources is often achieved by using procedural safeguards. Preparation for hot work—such as welding and cutting—and placement of vehicles that could be ignition sources, such as vacuum trucks, are best controlled by administrative safeguards. However, control of other ignition sources, such as the placement of furnaces, deflagration vents, and electrical equipment, is best addressed during design.

Hazard Reviews during Design

As mentioned previously, the design phase is the best time to make changes to improve the safety of a process. In *Guidelines for Engineering Design for Process Safety* (CCPS 1993, 5), the CCPS notes that the opportunity for ensuring maximum inherent safety occurs during the early design stages. If the process to be designed is a well-established technology, there may already be templates for equipment (including safety equipment) to be installed. However, if the process conceived is a new technology or a complete reapplication of an existing technology, the design phase can be extensive, lasting several months or even years. New technologies are often tested in smaller pilot plants before designs are scaled up to full production capacity.

The design phase itself consists of a number of iterative steps, including conceptual design, design review, and revisions. Of particular interest to safety practitioners is the design review stage. At this point, the general concept of the process is on paper, but many of the details may not have been completed. Early in the design phase, a simple review technique such as what-if may be sufficient to identify items that need to be included. As the design concept becomes more advanced, techniques such as hazard and operability analyses (HAZOPs) are probably necessary to identify equipment needs.

Managing Changes during Design

Since the design process is by nature a dynamic one, it is easy for changes to be made that may not be widely communicated. If changes are not discussed, alterations may be made that could impact the safety of the process. Therefore, similar to the system used to evaluate changes once the process is running, it is important to have a system for managing changes during design. In *Loss Prevention in the Process Industries*, Lees notes that such a system should include declaration and explanation of the proposed change, authorization, and communication of changes (Lees 1996, 11.7). A thorough technical evaluation is also critical. Similar systems must be in place for construction, commissioning, and operation.

Equipment Design

Process design should be undertaken only by knowledgeable persons who are familiar with all applicable codes and standards. Control and instrumentation should be available for critical parameters, such as pressure, temperature, level, and flammable and toxic concentrations. Alarms should alert operators to deviations, then interlocks should take automatic action if the situation is not corrected.

Below are some general design guidelines for specific equipment types. Detailed standards should be consulted for more specific information. Note that design standards for some equipment, such as pressure vessels and control systems, are addressed elsewhere in this handbook and therefore are not discussed here. Furthermore, fire protection is covered later in this section. General information is presented on storage vessels, piping systems, and reactors.

Storage Vessels

Storage tanks contain a variety of toxic, flammable, and corrosive materials. Since large quantities of materials are found in storage tanks, preventing a loss of containment is particularly important. Tanks should be provided with adequate pressure relief for the controlling case, which is generally an external fire in the case of tanks storing flammable materials. Design should also consider vacuum relief, with conservation vents for atmospheric tanks to avoid collapse of the tank wall (API 2002a, 7-1). Fire protection, such as monitors, spray and foam applicators, along with flame arrestors, should be installed to prevent flames from entering the tank. Foundations and supporting structures should be capable of handling the weight of the vessel and the maximum amount of liquid that it will hold. Materials of construction and corrosion allowances should be appropriate for the material.

Low-pressure storage tanks should be designed to withstand a combination of gas pressure (or partial vacuum) and static liquid head (Figure 2). Tanks should be capable of handling the anticipated volume, and they should be provided with an overflow system that discharges to a safe location. Consideration should be given to providing an inerting blanket in the vapor space on tanks storing flammable material. A floating roof design can also be considered if the vapor pressure of the fluid is high or a cone roof when emissions should be minimized but do not need to be collected. As a guide, a fixed roof can be used for materials with vapor pressures below 1.5 pounds per square inch absolute (psia); for vapor pressures between 1.5 and 11 psia, a floating roof should be considered. For materials (such as ammonia) stored as pressurized liquid, pressure spheres are generally used. Horizontal tanks can be used for high-pressure storage. Walls of the tank should not have any pockets where gases may become trapped. The temperature of liquids in the tank should not exceed 250°F. Additional metal thickness must be provided as a corrosion allowance where corrosion is expected. API (American Petroleum Institute) 620, *Design and Construction of Large, Welded, Low-Pressure Storage Tanks*, should be consulted for additional information regarding minimum wall thickness, stress allowances, and other design issues (API 2002a).

FIGURE 2. Example of a low-pressure tank (Photo courtesy of Roy Sanders)

Piping Systems

Piping should be chosen based on the ranges of operating temperatures and pressures, properties of the fluids and materials, flow conditions such as two-phase flow, and any special valving needs. The American Society of Mechanical Engineers (ASME) B31 piping code lists a number of restrictions. Of note to the practitioner, cast iron shall not be used above ground within process unit boundaries in hydrocarbon or other flammable fluid service at temperatures above 300°F, and high silicon iron and lead, tin, and associated alloys shall not be used in flammable fluid services. Also, nonfireproof valves should be avoided. ASME B31 should be consulted for more details (ASME 1999).

Reactors

Chemical reactors are often the heart of a chemical plant, because it is in the reactors that reactants are turned into useful, salable product. The different types of reactors are *batch* and *continuous*; continuous reactors are further categorized as *continuous stirred tank reactors (CSTRs), tubular reactors, and fluidized bed reactors*.

The CCPS's *Guidelines for Engineering Design for Process Safety* provides good guidelines on loss prevention considerations for chemical reactors (CCPS 1993, 118–125). Reactors should have adequate pressure protection. Reactors should also have adequate cooling capabilities. The issue of adequate pressure relief and cooling is often dependent on the potential for a runaway reaction. This potential may mean that the relief protection needs to consider two-phase flow. Also, traditional pressure relief may not be adequate because of the need to relieve a runaway reaction. Multiple relief devices, placement lower on the vessel closer to the liquid level, and additional safeguards to prevent a runaway reaction may need to be considered. Also, because temperature and pressure can increase rapidly in a reactor, redundant instrumentation that monitors temperature and pressure—and actuates an alarm in advance of reaching critical setpoints—must be installed. Batch reactors must be designed to minimize the possibility of making sequencing mistakes (e.g., adding components in the wrong order or for the wrong amount of time).Where materials are added manually, operators need to have accurate measures of how much material has flowed into the vessel. Valve sequencing should be part of the control scheme to ensure that valves on reactants are not opened prematurely.

Fire-Protection and Fire-Mitigation Systems

Preventing the release of flammable and toxic materials is the best way to ensure that losses will not occur due to uncontrolled fires and explosions; therefore, the emphasis during design should be to avoid such situations. However, even though efforts are made to prevent releases and minimize ignition sources, fires and explosions can still occur in facilities processing flammable materials. Fire-protection equipment is designed to minimize losses during such events and is more sophisticated than simple sprinkler systems. Fire-protection equipment can be categorized much as process equipment is, with both active and passive systems.

Isolation Valves

Isolation valves should be placed in piping throughout the process to ensure that large inventories of flammables or toxics can be segregated and protected if there is a leak downstream in the process (see Figure 3). The locations of such valves depends on inventory, condition of the material, and ability to isolate the inventory by other means. Often such emergency isolation valves are located at the battery limits of the process to afford emergency access. Valves can be power operated or manual, and controls can be local or remote. Automatic valves should be considered when the configuration of piping will not ensure that operators can easily access valves during a fire. For example, in piping leading from a vessel to the suction of a pump when the pump is located directly under the vessel, a pump seal fire may hinder efforts to reach and close manual valves on the bottom of the tank. Isolation valves should still function when utilities are lost. Manual valves should be placed in an accessible position. If the valve is elevated, a chain wheel should be provided. However, the chain should be secured in such a way that it is not a hazard to personnel walking or riding bicycles through the process

FIGURE 3. Isolation valve in a chemical plant
(Photo courtesy of Roy Sanders)

area. Valves with fusible links are sometimes used to ensure that the valves will operate during a fire.

Approaches to Fighting Fires

For a fire to occur, three elements must be present: fuel, oxidant, and an ignition source. These three items collectively are known as the *fire triangle*. The CCPS notes that generally, fire protection systems involve removing one or more of the elements (CCPS 1993, 490). Fire protection often consists of a combination of systems that automatically activate based on some parameter being exceeded and systems that must be activated by responders. Responders may choose to extinguish a fire to prevent a large-scale loss. However, depending on the circumstances, it may be preferable to let the fire burn until all the fuel is exhausted, if extinguishing the fire causes unburned material to be released. The unburned material might be toxic, or it might contact an ignition source and result in an explosion. The NFPA's (National Fire Protection Association) *Fire Protection Handbook* is a good source for more information about the technology of fires and fire suppression/detection systems (Cote 2003).

If a fire is to be extinguished, this can be accomplished by exhausting the fuel source, manually fighting the fire, or by using fixed fire-fighting systems. Equipment to accomplish all three methods will be discussed below.

Detectors and Alarms

Effective fire protection starts with detection systems. A fire that is detected and extinguished in its incipient stage will not spread and result in a catastrophic incident. Therefore, fires or conditions that could result in fires need to be detected as soon as possible. Personnel cannot be available at all times to detect fires, so it is important to install automatic detectors in areas that process flammable and combustible materials. Detectors can trigger alarms, activate deluge systems, or both. Detectors that trigger alarms generally come in two categories: *combustible gas detectors* and *fire detectors*. The functioning of each type and some good practices are discussed below. Much of the information in the following sections on detectors can be found in *Guidelines for Engineering Design for Process Safety* (CCPS 2003, 491–515). The *Fire Protection Handbook* is also an excellent source to provide additional information on this subject (Cote 2003).

Gas Detectors

Combustible gas detectors can detect vapors from a liquid release before a fire occurs. The detector senses the presence of potentially flammable mixtures of vapor in air before they reach an explosive limit. Such detectors are used in facilities where a potential leak source exists. These systems often have a warning alarm to indicate to personnel in the area that there is a problem and an automatic action alarm that is set at a higher level. The actions taken automatically can include shutting down processes or activating deluge systems.

Gas detectors should be located in such a way as to consider ignition sources, wind direction, and gas density. They can be located throughout an area to provide coverage or strategically placed at potential flammable release points. Gas detectors should be maintained according to manufacturers' guidelines. It should also be noted that calibration of gas detectors can present a challenge, since calibration depends on the specific gas to be detected. Inaccuracies can result if a mixture of gases is released.

Some of the more common types of detectors are diffusion-head-type *catalytic oxidation detectors*, infrared sensors, metal oxide semiconductors, and thermal incineration flame cell sensors. Catalytic oxidation detectors work by oxidizing the gas and heating an internal element, which causes a measurable change in the resistance. *Infrared detectors* sense a change in radiation absorbed at specific wavelengths, with higher gas concentrations absorbing more infrared radiation. *Metal oxide semiconductors* respond to certain gases by experiencing a change in electrical current at a given voltage. *Thermal incineration flame cell sensors* operate as their name indicates. They pass the flammable gas through a flame (that is constantly burning) and measure the increase in heat.

Fire Detectors

Fire detectors differ from gas detectors because they identify changes in the atmosphere after a fire occurs by sensing either smoke, heat, or flame (radiant energy). *Radiant energy detectors* can be used if rapid response is critical.

The CCPS provides information on detectors. *Thermal detectors* are of two types: those that identify rises in heat by sensing the rate at which the temperature is increasing and those that sense when the temperature reaches a predetermined setpoint. A third type of thermal detector known as a *rate-compensated fixed-temperature detector* is a combination of the other two types. Because fixed-temperature detectors are typically set at high temperatures, a rate-compensated detector usually operates faster than a fixed-temperature detector. But rate-compensated detectors also have the advantage of being able to sense a slow-rising, gradual fire, which may not be detected by a rate-of-rise sensor. However, the distinction between these types disappears in modern *addressable heat detectors* because the heat sensor sends a continuous signal to a central microprocessor that can use a wide variety of signal-processing algorithms to monitor temperature variations from background ambient values (CCPS 1993, 494–496).

Fire detectors should be located in accordance with NFPA standards, particularly NFPA 72 (NFPA 2007d). Protection such as shielding should be installed on detectors that are outdoors so that they are not compromised by the elements. Detectors can be set up to sound an alarm locally and/or remotely. Heat detectors can also be used to close valves and stop a fuel source that may be feeding a fire. For example, plastic tubing, which can act as a crude heat detector, can shut air-operated valves if they are burned, causing a loss of air through the tube. More-sophisticated commercial spot and linear heat detectors (as well as other fire-protection equipment) are tested and certified by organizations such as Underwriters Laboratories (UL) and Factory Mutual (FM) and their counterparts in other countries.

Smoke Detectors

Smoke detectors work by detecting airborne combustion particulates. The two general types of smoke detectors are *ionization* and *photoelectric* detectors. The CCPS notes that ionization detectors have widespread use in industry (CCPS 1993, 495). Smoke detectors may give an alarm or can be set to automatically actuate fire-suppression systems. In chemical and manufacturing facilities, false alarms may occur (for example, due to dirty environments), so it is often prudent to have voting logic for detectors so that a false positive from one will not result in an inadvertent trip of a suppression system. Smoke detectors are often used in areas such as offices, control rooms, and rooms with electrical and computer equipment (CCPS 1993, 495–496).

Radiant Energy Detectors

The other property besides a rise in temperature and combustion particulates that is measured by fire detectors is radiant heat emitted by a flame. *Ultraviolet* (UV) and *infrared* (IR) *radiation* detectors are the most common detectors of this type. Often, redundant and multiple-wavelength UV/IR or IR/IR detectors are used to decrease the chances of a false trip being set off by sunlight or devices such as welding arcs and flashing lamps. Since these devices sense optical signals through a lens, it is imperative to keep the lens clean. Some detectors are self-cleaning. Those that are not should be located where they can be maintained as needed. Of course, a balance must be struck in that the detectors should be located where routine necessary maintenance can be performed but also must be situated so that they will sense radiant heat and afford the maximum detection benefit (CCPS 1993, 496).

Manual Alarms

Even in facilities where there are automatic alarms and shutdowns, manual alarms are installed so that operations personnel can activate them if they observe problems. There are two common types of pull stations: *pull-lever* and *break-glass* design. These alarms generally require two distinct operations to avoid a false trip. Alarms should be located in areas with normal means of egress. Buildings and process areas should have as many manual stations as deemed appropriate by code and design engineers. There should be at least one station that is in clear view at all times, accessible from any point in the building or area, and have a maximum distance to travel no more than 200 feet. The activation of a manual alarm should generally also activate the main alarm system, alerting personnel in other parts of the facility of the situation so that they can respond appropriately or avoid the unit with the emergency (CCPS 1993, 496–497).

Fire-Suppression and Mitigation Systems

In addition to detection systems, equipment must be in place to suppress and mitigate the effects of a large fire if one should occur. A comprehensive package of suppression and mitigation equipment consists not only of sprinkler or deluge systems, but also includes active and passive systems to aid in extinguishing the fire and protecting surrounding structures. Active systems must be activated to function, whereas passive systems provide protection by their mere existence (CCPS 1993, 497–515).

Portable Fire Extinguishers

The most basic suppression system is a portable fire extinguisher. Extinguishers should be used only on very small fires. If the fire grows and spreads beyond the incipient stage, the appropriate response for personnel using portable extinguishers is to withdraw and use fixed equipment that can be operated from a distance. The *Supervisors' Safety Manual* provides detailed information on fire extinguishers (NSC 1997, 318). Extinguishers come in four classes and are designed to be used on different types of fires:

- Class A: Used on ordinary combustible materials, such as wood, paper, cloth, rubber, and many plastics.
- Class B: Used on fires involving flammable liquids, greases, oils, tars, oil-base paints, lacquers, and similar materials.
- Class C: Used on fires in or near live electrical equipment, where the use of a nonconductive extinguishing agent is of first importance.
- Class D: Used on fires that occur in combustible metals, such as magnesium, lithium, and sodium.

Fixed Fire-Suppression Systems

Fixed fire-suppression systems generally consist of equipment in place to deliver water or some other agent to extinguish the fire. These systems are very important since mobile fire-fighting strategies result in delays (due to responders' need to collect and don protective equipment and execute a plan). Systems can be manual or automatic, and the number of units, size, capacity, and specifications of the system vary with the size of the building or facility and the amount of flammable material present (CCPS 1993, 497–499).

Fire-Water Supply

The CCPS notes that the two critical elements of fire protection equipment are adequacy and reliability (CCPS 1993, 497). The equipment needs to be of adequate capacity to extinguish the largest credible worst-case scenario fire a facility is likely to experience. The flow requirements for an anticipated fire typically range from two to four hours. The system must also be able to continue functioning after explosions, so redundancy in piping, pumps, and supply is important to ensure availability. Redundant equipment prevents single failures from impairing the entire system.

Redundant supply pumps should be installed with different sources of power. Diesel-engine pumps are generally more reliable, although electric pumps can be adequate if they have a reliable power source. Pumps should start automatically and have manual shutdown switches. A *makeup pump* (often called a *jockey pump*) is often installed and is triggered by the initial pressure drop due to the opening of sprinklers. An automatic switch can start the primary pump when the jockey pump no longer has the capacity to supply the equipment. Pumps should be located away from areas where hazards exist so that the pumps themselves will not be damaged, and as much fire-water piping as possible should be buried to protect it from explosions and direct fire exposures. Redundant pumping stations are also desirable in the event that shrapnel from an incident damages one. Installing piping in loops allows different sections to be used if some sections are damaged during an incident, and adequate valves must be installed to isolate damaged sections.

A good practice is to have the water dedicated to fire protection completely separate from the process water. Dedicated fire-water pumps and storage facilities should be installed. Combination systems should be avoided if possible because using fire-water for process needs may diminish the supply of fire-water; process-water equipment may not be designed for the pressures needed during a fire; and in combined process/fire-water systems, water can be redirected from process needs that are critical during an emergency, such as cooling water. If supply tanks or outdoor reservoirs are shared, passive safeguards should be in place to ensure that enough fire-water will be present during emergencies. Such safeguards include installing the takeoff from the process-water supply tank sufficiently high that if the tank is drained down to that level, there will still be an adequate amount of fire-water.

The supply system must deliver copious amounts of water to multiple areas simultaneously. For design purposes, consider the maximum demand, both instantaneous and continuous, and consider the maximum number of water sprays, deluge systems, and sprinklers that may have to operate at the same time. Also consider needs for manual fire-fighting. Total demand can be as high as 10,000 gallons per minute (gpm) for process or storage areas. In the *Guidelines for Engineering Design for Process Safety*, CCPS recommends that the on-site, dedicated fire-water supply should be capable of being replenished within 24 hours after being used (CCPS 1993, 499).

Automatic Water-Delivery Systems: Sprinklers, Water Sprays, and Deluge Systems

Automatic water-delivery systems are commonplace throughout industry in commercial buildings and in various manufacturing settings. These fixed delivery systems include sprinklers, water sprays, and deluge systems. Several codes from the NFPA govern the design specifications for these fixed water-delivery systems based on the application.

TYPES OF SPRINKLERS

There are various types of fixed automatic water-delivery systems. According to the *Accident Prevention Manual*, the wet-pipe system is the most common type of sprinkler installed. It is referred to as *wet-pipe* because piping in the system is filled with water that is under pressure. When one or more sprinkler heads actuate because of heat exposure, the water automatically sprays out of the open sprinkler heads (NSC 2001, 330).

According to NFPA 13, pressure gauges shall be installed in each system riser and above and below each alarm check valve (NFPA 2007a). Relief valves of at least one-quarter inch are to be installed on every gridded system unless a pressure-absorbing air reservoir is present. The relief valve should be set to operate at 175 pounds per square inch (psi) or ten psi above the maximum system pressure. One caution about wet-pipe systems—because water is present at all times in the pipe, there is a constant concern with freezing in climates where freezing temperatures occur. NFPA 13 provides guidance on adding antifreezing agents to water in such services (NFPA 2007a). Wet-pipe systems can also supply auxiliary dry-pipe, preaction, or deluge systems if the water supply is adequate (NSC 2001, 330).

A *dry-pipe* system is one in which the piping contains pressurized air that holds the water in the system until a sprinkler opens. When the sprinkler opens, the air is released, and a *dry-pipe valve* that is holding the water back opens to allow water to flow into the risers. The dry-pipe valve is a critical device in these systems, and NFPA 13 lists several requirements to protect it, including specifications for the valve enclosures, supply, and protection against accumulated water above the clapper. Except under circumstances where the system is quick acting, NFPA 13 recommends that each dry-pipe valve control no more than 750 gallons of water. Pressure gauges should be installed throughout the system, including on both sides of the dry-pipe valve (NFPA 2007a, 7.2).

Dry-pipe systems are often a better choice in areas where the piping is susceptible to freezing. The NSC notes that a good rule of thumb is to use dry-pipe systems when more than twenty sprinklers are involved. One caution regarding dry-pipe systems is that the action of depressuring the system and opening the valve to allow water flow results in a delay of actuation (NSC 2001, 330–331). The NFPA recommends that quick-opening devices be installed on dry-pipe valves if they contain more than 500 gallons of water (NFPA 2007a, 7.2.3.3).

The pressure in the system can be supplied by either air or nitrogen, and must be maintained throughout the year. The pressure should be in accordance with the manufacturer's recommendations for the dry-pipe valve or twenty psi higher than the calculated trip pressure of the valve.

A *preaction system* is similar to a dry-pipe system; however, it generally reacts faster. The preaction valve controls the water supply to the system's piping and is actuated either manually or by a fire-detection system. An alarm is generally installed with such systems that annunciates when the valve opens and starts to allow water to flow. In a preaction system, water is released into the piping, but there is no discharge unless a sprinkler head has actually opened. One advantage of preaction systems is that sensors can send a signal to the control panel at the same time they send the signal to the valve, so that the fire may be extinguished before the sprinkler heads open, thus limiting water damage. This approach may be effective where valuable merchandise is stored and water damage (along with fire damage) is regarded as especially problematic. Some sources note that one disadvantage of preaction systems is that they are more complex and are considered less reliable than standard water-spray sprinkler systems. Often a combination dry-pipe and preaction system may be used in areas that are larger than one dry-pipe valve can accommodate. NFPA notes that when this combination is used, it should be designed such that the failure of the detection system shall not prevent the entire system from functioning as a conventional automatic dry-pipe system (NFPA 2007a, 7.4.2).

Deluge systems spray water into an entire area by allowing water to flow to sprinkler heads that are open at all times. The entire area receives water spray from all sprinklers. These systems are used to protect areas where the risk of a widespread fire is significant and it is believed that regular sprinklers would not act quickly enough over a large enough area. Some areas deluge systems may be chosen to protect include buildings that contain large quantities of flammable materials, explosives plants, and airplane hangars. The deluge valve can be actuated either manually or by thermal

FIGURE 4. Water-spray system on a tank (Photo courtesy of Roy Sanders)

or flame detectors. Deluge systems may discharge 5000 gpm or more, and may require an on-site water supply in addition to municipal water systems.

Water-spray systems protect specific hazards, such as tanks that contain flammable material, whereas sprinkler systems protect broader areas including buildings and structures (see Figure 4). Water-spray systems are similar to deluge systems except that there are spray nozzles in the water-spray systems (instead of the open sprinklers in the deluge systems.) NFPA 15, *Standard for Water Spray Fixed Systems for Fire Protection*, provides in-depth guidance on water-spray systems (NFPA 2007b, ch. 7).

One additional type of water-supply sprinkler is a *limited water-supply system*. As the name indicates, this type of system is installed when facilities do not have access to a large supply of water. When this type of system is installed, great care must be taken to ensure that the fire-extinguishing needs of the facility are still being met adequately.

PURPOSE OF WATER-SPRAY SYSTEMS

Most people know that water-spray systems are used to extinguish fires and cool surrounding equipment, but spray systems are also used in some industries to prevent fires and explosions. The systems accomplish this through a variety of mechanisms. Water-spray systems activate when they detect a release of material that may act as a fuel and act to dilute, disperse, and quench it. The activation of the spray may also induce air into the fuel and reduce the concentration below flammable limits. The water may also absorb some of the material if it is soluble, thereby lowering the amount of fuel available.

NFPA 15, *Standard for Water Spray Fixed Systems for Fire Protection*, provides limited information to help practitioners design a water-spray system to prevent fires and explosions. Although the standard acknowledges that systems can be designed to accomplish this purpose, the guidance essentially states that application rates should be based on field experience or actual test data with the product (NFPA 2007b, 7.5). In *Guidelines for Fire Protection in Chemical, Petrochemical, and Hydrocarbon Processing Facilities*, the CCPS notes that the design model for water-spray systems includes choosing scenarios, calculating the area of concern, determining dilution and/or absorption required, and determining water-spray characteristics. Once these factors are known, the required flow rates can be determined. Practitioners who design spray systems should consider the limitations of the systems in preventing fires and explosions. Additional research is needed to determine under what circumstances these systems will actually prevent fires and explosions and how they should be designed to accomplish this task (CCPS 2003, 133–136, 251–255).

WHERE TO INSTALL SPRINKLERS

Automatic sprinklers are generally installed for protection in buildings of combustible construction or occupancy. Such areas typically include laboratories and warehouses. NFPA 13 governs the spacing, location, and position of sprinklers (NFPA 2007a, 8.1). NFPA 30 provides specific sprinkler protection requirements for flammable liquid warehouses and processing facilities (NFPA 2003 ch. 6 and 7).

SIZING OF SPRINKLERS

In years past, pipes in sprinkler systems were sized using the schedule system, which used tables and assumptions to calculate sizes. However, virtually all sprinkler, deluge, and water-spray systems are now sized based on hydraulic calculations. This method involves a number of steps, and the calculations are generally facilitated using computer programs. In order to use the hydraulic-calculation method to design sprinkler piping systems, several pieces of information must be known, including the minimum rate of water application, the area to be protected, information about the sprinkler system itself (such as coverage for each sprinkler and a discharge constant for each nozzle type), pipe sizes, friction losses, and required pressure at reference points. The hydraulic design procedure can be used to calculate pressure along branches, pressure at cross mains, and pressure in risers. Several reference books have application problems that can demonstrate this method (Lindeburg 1995, 38–39).

Maintenance and Inspection

The components of a fire-protection system must be maintained and inspected periodically to ensure that the components will function properly when necessary. NFPA 25 provides guidance on inspecting, testing, and maintaining fire-protection systems. It also addresses potential changes after initial design (NFPA 2002). Building owners must perform an evaluation before changing the occupancy or processing, use, and storage of materials. The evaluation must consider items such as changing office space into storage, process or material changes, and revisions to the building (such as relocating walls).

Equipment must be inspected and tested periodically to ensure that it will be able to function when needed. Where possible, automatic supervision should be installed so that a monitoring system is in place to display the status and indicate abnormal conditions with equipment. It is critical to verify the condition of the equipment in the water supply system. The water temperature and heating system should be inspected daily during cold weather, or weekly if the system is automatically supervised. If the water level is part of a system that is not automatically supervised, it should be inspected monthly. The tank exterior should be inspected quarterly, and the interior should be inspected every three to five years. Temperature alarms and high-temperature switches should be tested monthly during

TABLE 1

Inspection, Testing, and Maintenance Frequencies for Selected Equipment

Equipment	NFPA Reference	Inspection	Testing	Maintenance
Water supply tank	25 Table 9.1	Internal: 3–5 yrs; External: quarterly	--	--
Piping	25 Tabl e 11.1	Quarterly (for corrosion and damage)	Annually (water supply piping)	See individual components
Preaction/deluge valve	25 Chapter 12	Externally: monthly Internally: annually (every 5 years if the valve can be reset without removing faceplate.)	Annually (full flow)	Annually
Fire pump system	25 Table 8.1	Weekly	Weekly (no flow); Annually (flow)	Annually (motor)

(Adapted from NFPA 25, 2002)

cold weather. Water-level alarms should be tested semiannually and level and pressure indicators every five years (see Table 1).

Manual Water-Based Protection Systems

Manual fire-fighting protection is usually provided by hydrants, monitor nozzles, fire trucks, and hose lines. Monitor nozzles can supply large amounts of cooling water to equipment exposed to fire. Unlike hydrants, monitors can be placed into service quickly and then operate unattended. Critical areas should be covered by at least two monitor nozzles. Remotely operated nozzles decrease the risk to personnel even further.

Fire truck pumps can supply as much as 1000 gpm. *Fire hydrants* are generally chosen from standard designs based on operating characteristics. Hydrants should have at least two 2.5-inch capped outlets. A 4- or 5-inch pumper connection can be used to supply water to a monitor or fire truck if necessary. Spacing for hydrants depends on a number of factors including layout, hazards, and drainage. Due to limited hose lengths, hydrants should be located within 300 to 400 feet of the buildings they protect. Lindeburg notes that typical spacing is 300 feet for storage and distribution of petroleum oils and 400 feet for warehouses (Lindeburg 1995, 26). Spacing intervals between hydrants range from 250 to 400 feet. Hydrants should be installed close to paved roads for accessibility, and the pumper connections should face the street.

Chemical Extinguishing Systems

In addition to water-distribution systems for extinguishing fire, special circumstances may call for fire-fighting foams and chemical and gaseous agents to be used. These systems are often installed to supplement, rather than replace, traditional sprinkler systems. The agents include foam, carbon dioxide, steam, hydrofluorocarbons, and inert gas systems (CCPS 1993, 502–506).

Foams

The CCPS discusses the different types of foams used in industry (CCPS 1993, 502–504). Several types can be effective extinguishing agents. They work partly by excluding oxygen from the fire by smothering and partly by absorbing the radiant heat flux from the flame to the burning liquid pool. Two types of foam that are widely used in industry are *synthetic* and *protein* foams. A popular type of synthetic foam is *aqueous film-forming foam* (AFFF), which works to extinguish a fire by forming a vapor seal on the surface of hydrocarbon liquid. Protein foams have a high burnback resistance and can be used to cover vertical surfaces as an insulating blanket to aid in confining fire. These foams are especially useful when foam must adhere to the hot side of a vessel. However, because higher concentrations of protein foams are required, larger quantities of material must be on hand, and they are more costly. Synthetic foams, like AFFF, have the advantage that they can be applied through fog or water-spray nozzles as well as conventional foam nozzles. However, AFFF in particular has limited usefulness on petroleum fires since it cannot adhere to vertical surfaces.

Foam is rated based on the expansion rate when it is mixed with water and aerated. Low-expansion foams have an expansion ratio of 20 to 1. Medium-expansion foams can have expansion rates of up to 200 to 1, and high-expansion foams can have rates up to 1000 to 1. Medium- and high-expansion foams are often used to flood indoor areas and confined spaces. NFPA 16 provides guidance for the installation of foam-water sprays and sprinkler systems (NFPA 2007c, ch. 7).

NFPA 11 provides standards for foam systems. It notes, among other things, that the foam concentrates and equipment should be stored in an area where they are not exposed to the hazard they protect (NFPA 2005a, 4.3.2). Concentrate should be available in sufficient quantities to protect the largest single hazard or group of hazards. Different types of foam should not be mixed together and stored, but they can be applied to a fire either sequentially or simultaneously. Since foam works by being injected into water, the discharge pressure ratings of foam pumps must exceed the maximum water pressure available at the point of injection (NFPA 2005a, 4.3, 4.4, 4.6).

Piping in the hazard area should be constructed of steel or another alloy rated for the pressure and temperature involved. The NFPA specifically states that pipe carrying foam concentrate shall not be galvanized. All valves used for the water-and-foam solution should be of an indicator type, such as a post indicator valve (PIV). The tank requiring the largest foam solution flow can serve as the basis for the capacity of the foam system.

Foam can be mixed with water and applied to fires, or it can be injected directly into tanks containing flammable material. For example, methods for protecting fixed-roof tanks include foam monitors, surface application with fixed discharge outlets, subsurface application, and semisubsurface injection methods. The NFPA provides design criteria for tanks containing hydrocarbons. Fixed-roof (cone) tanks containing hydrocarbon are a common type of vessel in which nozzles supply foam. When fixed foam-discharge outlets are used in this type of tank, good practice requires that minimum discharge times and application rates should be determined in the design. For example, for crude petroleum, the minimum application rate is 0.10 gpm (per ft^2), and the minimum discharge time for a type-I discharge outlet is 30 minutes. (A type-I discharge outlet is an approved discharge outlet that conducts and delivers foam gently onto the liquid surface without submergence of the foam or agitation of the surface.) Similarly, subsurface foam injection systems can be used for liquid hydrocarbons in vertical fixed-roof atmospheric storage tanks. When these systems are used, the foam discharge outlets should be located at least one foot (0.3 m) above the highest water level to prevent destruction of the foam. Piping that is inside dikes can be buried to protect it against damage.

Special requirements exist for medium- and high-expansion foams. They can be used on ordinary combustibles and flammable and combustible liquids and should be discharged to cover the hazard to a depth of at least two feet (0.6 meters) in two minutes (NFPA 2005a, 6.13.3). High-expansion foams should be used on liquefied natural gas vapor and fire control. In large areas or those where egress is difficult, remote-control stations can be installed for manual actuation. Reserve supplies of foam must be on hand in order to put the system back into service after operation. Specifically, NFPA recommends maintaining enough high-expansion foam concentrate and water to permit continuous operation for 25 minutes or to generate four times the submergence volume (whichever is less) but at least enough for fifteen minutes of full operation (NFPA 2005a, 6.12.9, 6.13.3).

Other Extinguishing Media

The CCPS discusses other common extinguishing media, including carbon dioxide and halon systems. Carbon dioxide extinguishes fires by excluding oxygen and smothering the fire. Carbon dioxide can be delivered by manual and fixed applicators and is often delivered through portable extinguishers. Carbon dioxide may be chosen to extinguish fires involving flammable liquids and electrical equipment. However, care must be exercised when using this extinguishing medium, as the flooding application needed for effectiveness is between 30 percent and 40 percent, and it becomes an asphyxiation hazard at these levels (CCPS 1993, 505–506). NFPA 12 can be referenced for good practices involving carbon dioxide systems (NFPA 2005b, ch. 4).

Other common extinguishing media are halon and halon alternatives. Halon is used for both manual and fixed-system applications. Halon systems are often installed in control rooms and in sensitive areas such as switchgear and motor-control centers. Halon becomes toxic to humans at higher concentrations, both because it is an asphyxiation hazard and because it emits toxic decomposition products. The production of halon gases has now been phased out and alternatives are being produced. For halon systems that are still functional, NFPA 12A provides guidance on designing and maintaining them (NFPA 2004, ch. 5 and 6).

See Table 2 for a list of the major NFPA standards that offer guidance on fire-fighting equipment.

TABLE 2

Selected Listing of NFPA Standards Applicable to Fire-Fighting Equipment

NFPA Standard	Title
11	Standard for Low, Medium, and High-Expansion Foam
12	Carbon Dioxide Extinguishing Systems
13	Installation of Sprinkler Systems
14	Standard for the Installation of Standpipe and Hose Systems, 2003 Edition
15	Standard for Water Spray Fixed Systems for Fire Protection
16	Standard for the Installation of Foam-Water Sprinkler and Foam-Water Spray Systems, 2003 Edition
20	Standard for the Installation of Stationary Pumps for Fire Protection, 2003 Edition
17	Standard for Dry Chemical Extinguishing Systems, 2002 Edition
17A	Standard for Wet Chemical Extinguishing Systems, 2002 Edition
22	Standard for Water Tanks for Private Fire Protection, 2003 Edition
24	Standard for the Installation of Private Fire Service Mains and Their Appurtenances, 2002 Edition
25	Standard for the Inspection, Testing, and Maintenance of Water-Based Fire Protection Systems, 2002 Edition
30	Flammable and Combustible Liquids Code, 2003 Edition
58	Liquefied Petroleum Gas Code
69	Standard on Explosion Prevention Systems, 2002 Edition (addressing inerting and explosive venting)
86	Standard for Ovens and Furnaces, 2003 Edition (discusses steam used to extinguish fires in enclosures)
230	Standard for the Fire Protection of Storage, 2003 Edition
430	Liquid and Solid Oxidizers
484	Combustible Metals, Metal Powders, and Dusts
654	Fires and Dust Explosions from Combustible Particulate Solids
2000	Clean Agent Extinguishing Systems

Passive Protection

Similar to the philosophy applied to process equipment design, fire equipment can be active or passive. Previous sections discussed active equipment—equipment that requires an action (e.g., electrical, mechanical, or manual) in order to be effective. Such equipment includes deluge systems that actuate based on a signal from a flame detector. There is also equipment that, by its inherent nature, prevents incidents or mitigates consequences by simply existing—so-called *passive equipment*. Similar to passive process equipment, fire equipment of this nature is considered reliable because it does not need to have functioning components. However, this equipment should be inspected and maintained regularly in order to ensure that it does not deteriorate (CCPS 1993, 507).

Barriers

Barriers are put in place to minimize the consequences of fires by limiting their spread. Fire barriers are typically built from materials such as concrete or masonry. Such materials generally have a resistance rating that indicates how long the barrier can resist heat and flame. Ratings typically range from half an hour to four hours. Caution should be exercised, however, in considering ratings to be absolute. Ratings determined by simulating the heat from ordinary combustibles may not accurately represent the way the barrier will behave during fire fueled by a flammable liquid, and the actual resistance in process settings may be less than the specified value.

Two types of barriers include fire walls and fire partitions. Fire walls provide better protection than partitions. A standard fire wall has no openings, has a resistance rating of four hours or more, and is designed to remain standing even if the structure around it collapses. End walls should be provided if a fire can spread around fire walls. Openings in any barriers should be made with caution, and protections should be provided to maintain the barrier's integrity. If possible, penetrations should be avoided and conduit should be routed around fire barriers. Partitions provide less protection than fire walls (CCPS 1993, 507–509). The resistance rating is generally less for partitions, and they are not designed to continue standing independent of surrounding structures. Locations of barriers are based on several factors, including types of fire hazards and company and insurance requirements. In general, barriers should be considered between areas with high fire hazards and those that are occupied by personnel or that contain critical operations such as instrument rooms. CCPS provides additional examples of areas where barriers should be considered, such as between occupancy types (e.g., warehouse and production), and between separate unrelated processes (CCPS 1993, 508). Where barriers are used, they should be designed and constructed in accordance with local code requirements and engineering designs such as those published by Underwriters Laboratory.

Fireproofing

Fireproofing is used in office buildings and warehouses, to provide insulation for steel structures, as steel can fail if it is exposed to an intense fire for a prolonged period of time. According to Lees, steel members should not reach a temperature greater than 1000°F. In a facility that contains flammable material, fireproofing serves to prevent the failure of equipment supports. Such a failure could result in additional releases of flammable material that could feed the fire. Fireproofing in the form of thermal insulation is also installed on equipment that contains flammable material to provide protection to the vessels in the event of a fire. This can be done in conjunction with, or instead of, water-sprinkler systems. Care must be taken when choosing a method of protection, and experienced designers should be consulted.

Fireproofing comes in three basic types: *spray-on* or *coated systems*, *wrap systems*, and *box systems*. When using spray-on fireproofing material, the substrate surface must be properly prepared so that the material will adhere to it. The material should be clean and rust free before insulation is applied. Fireproofing must be sufficient to resist damage from normal plant activities and from water from fire hoses. Factors that need to be considered in the choice of fireproofing include the type and height of the structure and the degree of protection needed. Lees quoting a 1967 study by Waldman notes that the height above grade that fireproofing must be applied is variable, but but it must be applied at least 35 feet above grade (Lees 1996, 16.266). The CCPS adds that fireproofing must be applied not only above grade, but also above other objects where flammable liquids could pool. The CCPS further notes that consideration should be given to installing fireproofing within fifteen to twenty-five feet of potential fires, including drainage paths for liquids (CCPS 1993, 511). As noted previously, when deciding on the degree of protection, organizations must consider that a steel member should not be allowed to reach temperatures in excess of 1000°F (Lees 1996, 16.2.266). In deciding the degree of fireproofing, consideration should also be given to existing fire-protection systems, including water sprays.

Fireproofing should be periodically inspected for damage or deterioration. Insulation should be applied and maintained so that water cannot penetrate it and corrode the member underneath, because corrosion under insulation is difficult to detect. Insulation should also accommodate expansion and contraction of the underlying member.

Separation to Minimize Fire and Explosion Impacts and Allow Access

Using the principles of inherent safety, it is important to locate equipment in such a way as to minimize the impact of fires and explosions on surrounding vessels, piping, and structural supports (Wallace et al. 1994). One approach is to separate units and equipment within units so that the heat from a fire will not severely impact the surrounding area. Using this strategy can also minimize the impact of debris from explosions in a way that other fire-fighting strategies, such as sprinklers and fireproofing, cannot. Another consideration in the unit and equipment layout is to provide separation for access by emergency vehicles.

Separation distances can be determined in different ways. Combustible materials can be separated to decrease the intensity of a potential fire. Organizations can use separation charts developed by industry (including insurers) that provide guidelines for minimum distances between areas with hazardous materials and equipment. Another method is to apply a risk-ranking method and make decisions about spacing based on the results. A number of tools available commercially have traditionally been used to do this type of ranking.

The developers generally caution, however, that the tools are meant only to provide a relative ranking of risks and any separation distances yielded from the calculations should be considered in context.

A more sophisticated method of determining separation distances involves calculating the heat that may be received by objects and setting separation distances based on the results. While this method often yields smaller distances (and thus space can be used more efficiently), it requires a high level of expertise. Various computer programs can assist in this analysis (CCPS 1993, 512–513). For the purposes of this chapter, we will explore the traditional method of using spacing charts and general guidelines.

Insurance companies such as IRI (Industrial Risk Insurers) (now a division of GE Insurance Solutions) have developed charts listing different types of equipment and proposed separation distances between them. The charts are periodically revised and are available for purchase. Previous versions of charts have been published publicly and are available in a number of places, such as *Guidelines for Engineering Design for Process Safety* (CCPS 1993, 70). These charts offer generally conservative estimates of safe distances and discuss spacing between units, equipment, and particular types of storage tanks. To effectively use such charts, one needs to judge whether a processing unit represents a moderate hazard, an intermediate hazard, or a high hazard. The minimum distance between units deemed to be of moderate hazard is given as 50 feet, whereas the minimum distance between high hazard units is 200 feet. The minimum distance between control rooms and processes considered moderate, intermediate, and high hazard is given as 100, 200, and 300 feet, respectively. More information on designing control rooms is given in the section on building design.

Special consideration should be given to the spacing of storage tanks, since they often hold large inventories of hazardous material. Arranging storage tanks in groups allows for the common use of fire-fighting equipment. Although many sources advise practitioners to use common diking as well, caution must be exercised since a release from one tank directly affects other tanks in the same containment area. Storage tanks should be located remote from process areas to keep upsets in processes from endangering the inventories in storage. Lees advises that the distance between process and storage should not be less than fifteen meters, or approximately 50 feet (Lees 1996, 10.17–26). Regarding access for emergency vehicles, there should be access on all sides of a storage area, and access routes should be connected so that if one route is affected by a fire, others will be available.

The NFPA also has tables on separation distances between storage tanks and public ways or important buildings on a site. NFPA 30 should be referenced directly for guidance on distances, as they vary depending on the stability of the liquid stored, the type and capacity of tank, protection provided, and the operating pressure of the tank (NFPA 2003b, 4.3.2). For example, a tank with a capacity of 100,000 to 500,000 gallons should be separated from property lines by at least 80 feet or from important buildings by at least 25 feet, but these distances may be adjusted depending on the factors noted in the previous sentence.

The CCPS also provides guidance on spacing, which includes locating fired heaters away from units that could leak flammable materials. It also recommends avoiding placing rotating equipment, such as pumps and compressors, in a single area and around vessels and vulnerable equipment because rotating equipment can often be the source of leaks (CCPS 1993, 71).

CHALLENGES WITH APPLYING A SEPARATION STRATEGY

While the benefits of maintaining separation distances have been explained, there are challenges in this approach of which practitioners should be aware. Land on which facilities are built can be limited by factors outside practitioners' control, which may make it difficult to adhere to ideal separation distances. Also, as plant expansions occur, distances between units and equipment within units decreases. Plant expansions are changes that must be managed with the same diligence that accompanies equipment changes in facilities. Practitioners must clearly define and maintain distances as much as possible during plant expansions and advocate for extra separation distances during design if future expansions are expected.

SEPARATION BETWEEN CHEMICAL FACILITIES AND SENSITIVE RECEPTORS

One sometimes-controversial aspect of layout considerations is where to site a chemical or manufacturing facility in relation to the community. Buffer spaces between facilities and sensitive community receptors tend to be governed by local rather than national codes. NFPA 30 provides some guidance regarding locating storage tanks close to public ways (NFPA 2003b, 4.3.2). Sensitive receptors include residential homes, community services such as hospitals and schools, and environmental receptors, such as waterways. The risk management plan, as mandated by the Clean Air Act Amendments of 1990 as required by the Environmental Protection Agency (EPA 1990), attempts to have facilities that handle or process highly hazardous materials determine the potential effects on the surrounding community and communicate those effects. Other environmental legislation, such as the Emergency Planning and Community Right-to-Know Act (EPCRA) (EPA 1986a) and the Superfund Amendments and Reauthorization Act of 1986 (SARA), Title III (EPA 1986b), previously fostered this concept.

A challenge with this aspect of applying separation distances is that it often does not lend itself to the hard science of conducting a risk assessment by determining the impact based on the properties and quantities of materials handled, but is often governed by the climate of the relationship between the facility and its neighbors. Also, sometimes a facility was actually in place before the residential community was. However, these realities do not relieve the company owners/operators from using science to determine, as best they can, the potential impact on the community from a release. Factors to consider include credible worst-case scenarios, population densities, and the ability to notify and control the movement of people in an emergency. Other aspects that companies need to address are their relationship to the community, their role in ensuring that community notification equipment and procedures are state of the art, and their partnerships with other industrial neighbors and emergency responders that could provide mutual aid in the event of a release. Since communities often grow up around chemical or manufacturing plants, causing community expansion toward the plant if there are not local codes prohibiting it, communication between the facility and its neighbors must be ongoing to be most effective.

Drainage

Drainage serves the purposes of containing flammable and toxic liquid spills and discharging runoff from water used to fight a fire. Drains should be designed to take liquids away from the area of the spill so that they can be properly treated, so as not to exacerbate a situation. Potential soil and groundwater contamination should be considered in the design of sewers. One of the primary aspects of designing such systems is to consider the volume of material that is likely to be released during a spill. Systems must be able to handle the largest accidental and anticipated release of material as well as expected rainfall. Other factors that should be considered include type of surface, spacing of facilities, and possible interactions between chemicals that could lead to reactive chemical incidents. Special consideration must be given to providing drainage from storage areas. Flammable materials must not be allowed to collect at low points or flow to points in the facility where there are ignition sources.

Drainage systems must also be designed to be effective during a fire. The discharged water from a fire-fighting event must be controlled and contained; otherwise, the flammable materials may spread to other areas. The containment system can use a combination of curbing, grading, trenches, ditches, and diking. The system should safely handle burning liquid in areas that contain flammable liquids. Practitioners should consider whether drains can be left open or should be closed. Enclosed systems should contain traps or other means to prevent flames from entering the system. Systems should be designed to handle the combined flow of spray systems that will operate simultaneously, streams from hoses and monitors that may be used during a fire, the largest anticipated or accidental release of liquid, and rainwater. The containment system should be able to handle the flow of material for the expected duration of a fire unless authorities grant approval to handle the flow for a smaller period of time.

Inerting

Inerting prevents explosions by displacing oxygen and replacing it with inert gas. Nitrogen and carbon dioxide are common inerting agents. Inerting is generally used to pad the space above flammable liquid storage tanks. Inerting may also be used to sweep flammable materials out of equipment prior to maintenance.

Building Design

Of special concern to safety practitioners is the design and integrity of buildings in facilities. This is because personnel are particularly vulnerable in buildings when there is a fire, explosion, or toxic release. In some high-profile incidents, including Flixborough in Europe in the 1970s and Phillips in the United States in the late 1980s, the majority of the fatalities were personnel inside the control room. Unless buildings are located and/or designed in such a manner as to protect inhabitants, people inside can be impacted by the pressure wave and shrapnel if there is an explosion, toxic gases if there is a release, and the inability to escape if there is a fire.

Besides the requirement of protecting personnel inside buildings, another incentive to ensure that buildings are adequately protected is the ability to allow operators to continue to control the process in the early stages of an incident. By maintaining control, personnel and automatic control systems may decrease the chances that the incident will escalate into a catastrophic event. All on-site buildings can be of concern; however, of particular importance are control rooms. Large concentrations of employees may be gathered in these places that are located close to hazardous sources. Also, critical activities may need to be conducted from control rooms during emergencies, including shutting down processes. For these reasons, control rooms are given special attention in the following sections.

Location of Important Buildings and Service Functions

The location of important buildings, especially control rooms for process units, is critically important in keeping people safe. The desire from an operational standpoint to have the control room located close to the unit so that operations personnel have ready access to the equipment must be considered along with the desire to have people exposed to the smallest risk possible. Historical tables from IRI suggest the minimum distance from control rooms to process units with moderate, intermediate, and high hazards is 100, 200, and 300 feet, respectively (CCPS 1993, 70, referencing IRI 1991). Frank Lees recommends minimum distances from the plant to control rooms in the range of twenty to thirty meters, or around 65 to 100 feet, although he notes that the distance may need to be increased based on hazard studies (Lees 1996, 10.33).

Of equal importance to the location of the control room is the location of other important service functions. Control rooms often contain additional services, including laboratories, instrument shops, offices, meeting rooms, locker rooms, and break rooms. While it is convenient to have these functions located in the control room, placing them there means that more people may be unnecessarily exposed to hazards than if these service functions were place in a location remote from the control room, and therefore farther from the process unit. Several sources advise practitioners to strongly consider the merits of securing control rooms so that the only functions contained in them are those that are essential for the control of the facility. Other functions can be moved to be farther from the process areas.

Also regarding service functions, control rooms should not function as emergency control centers, since operations will continue from them, and operational activities should not interfere with emergency response. Competition between process control during the critical phases of an incident and emergency control is undesirable.

Designing Buildings to Protect Against Explosions

One of the hazards for personnel in buildings during an event is the risk of an explosion that could impact the building. If a process unit contains substances that can explode, such as flammable or chemically reactive materials, this factor must be considered in the design and location of its building. The American Petroleum Institute's (API) RP 752 governs the management of hazards to plant buildings (API 2003a). Although this organization primarily writes standards for the petrochemical industry, the guidelines are useful in the design of buildings in chemical and manufacturing facilities as well. Their suggested methodology in evaluating building design involves a three-stage approach.

Building and Hazard Identification

The first step is to identify both the hazards and the buildings that are on site. The hazard-identification step includes an analysis to determine whether there is a potential for an explosion. All general sources that can cause explosions from vapor clouds, chemical decomposition, dust, boiling liquid expanding vapors (BLEVEs), or mechanical failures of vessels must be considered.

Once it is determined that an explosion can occur, the buildings must be evaluated. Building evaluation involves taking an inventory of buildings that exist or are planned for the site and then determining the occupancy load for each building. Once occupancy loads are determined, buildings can be compared to company standards to see whether they merit further quantitative evaluation or whether a simple checklist (see "Checklists for Evaluating Buildings" on page 268 for an example) will be sufficient to determine risks.

There are several aspects involved with determining the occupancy and establishing company standards for screening purposes. One aspect is the normal occupancy load for the building each week. *Load* in this case refers to the number of hours that people are expected to be inside the building. This is determined by calculating the collective number of hours that personnel spend in the building each week. Two other aspects of occupancy are the individual and peak occupancies. The *individual occupancy* is the percentage of time an individual typically spends in the building. *Peak occupancy* is the number of people who may be exposed at a given time. For example, meeting rooms, kitchens, and maintenance shops may have high numbers of personnel at peak times. Practitioners should realize that peak occupancy should be seriously considered when determining the actual risk posed to personnel. Determining the hours of occupancy only on an annualized (or some other normalized) basis will underestimate the actual risk to personnel in situations in which people gather on a temporary basis. Episodic events, such as regular meetings, represent times when large numbers of personnel may gather in buildings and be at risk, even if it is for a short period of time.

Since the concept of screening buildings is based on risks, it should be noted that the criteria used by companies varies. API 752 notes that some organizations choose occupancy loads that range from 200 to 400 person-hours per week (API 2003a, 5), while others base individual occupancy on 25 to 75 percent of an individual's time. Numbers for peak occupancy ranging from five to 40 people are given as a typical standard. It should be noted, however, that these numbers are given as references only, and organizations must consider several factors in determining their own occupancy standards. For example, the occupancy load of 400 person-hours per week referenced above equates to ten people each working a full 40-hour work week in a building, and it may not be advisable for a company to accept this degree of risk.

Building Evaluation

After the hazards and buildings of interest are identified, the next step is to perform evaluations of the buildings. Options for evaluating buildings include comparing building design and spacing to industry standards, performing a consequence analysis, or conducting a screening risk analysis. The three methods will be briefly discussed below.

COMPARISON TO INDUSTRY STANDARDS

Buildings can be compared to industry standards, such as the spacing standards discussed earlier in this section. The effects of explosions on buildings decrease as the distance from the process increases, so spacing guidelines can be used to determine whether the building is remote enough from the process to be likely to escape the major effects of an explosion.

STANDARDS FOR BUILDING DESIGN

There are also standards for building design. Lees's *Loss Prevention* series has used several sources to collect good practices for the design of control rooms, which are generally the most vulnerable buildings on site due to proximity to the unit, occupancy, and role in controlling the process (Lees 1996, 10.33). As discussed previously, control rooms in hazardous areas should contain only necessary process-control functions, which limits occupancy in the event of an explosion. Control rooms should generally be located only one floor above the ground. The roof of a control room should not hold heavy machinery. There should be no tall structures (such as tall distillation columns) around the building that could fall on the control room during an explosion. Its windows should be reinforced—or not present at all—to eliminate the possibility of shattering glass affecting personnel inside. Subsequent to an explosion, broken windows and perforations in buildings can serve as points of entry for toxic materials. However, the desire to minimize the effects of broken windows must be balanced against the operational advantages of being able to view the process directly from inside the control room, which makes reinforced glass a possible alternative. The control building should be constructed of ductile materials such as steel and reinforced concrete rather than brittle materials that collapse easily during an explosion, such as brick and masonry.

If the building is in a vulnerable location in proximity to the process, it should be reinforced to withstand an overpressure that the walls and ceiling are likely to experience during an explosion. This is a complex and evolving subject, and practitioners should be aware of the newest codes and standards related to designing buildings to withstand explosions. Some guidance is derived from conducting an evaluation to determine the likely overpressure, and others present a minimum overpressure loading that the building should be designed to withstand. (These references are provided to allow comparison to historical standards. See the section on "Consequence Analysis for Building Evaluation" for additional information on determining overpressure.)

Some sources provide a combination distance and overpressure loading, indicating that buildings are best protected by both placement and design. Other guidance, not accounting for distance from the process, tends to be more conservative in the design requirements. Various guidelines list a pressure that the building should be able to withstand for a certain period of time, typically between 20 and 100 milliseconds (ms). According to the guidance available, buildings should be designed to withstand pressures ranging from three to ten psi at a distance of 100 feet. For high-hazard areas, buildings may be designed to withstand ten to fifteen psi for a duration of 30 to 100 ms. Roofs can generally be designed to withstand less overpressure than walls, but they should still be designed to withstand between three and ten psi in high-hazard areas.

CONSEQUENCE ANALYSIS

In addition to comparison with preestablished standards, another approach is to perform a *consequence analysis* to determine the potential overpressure the building could be exposed to and design it to withstand that load. In this method, release scenarios are chosen (generally from a process hazard-analysis-type of study), considering passive and active mitigation. Then, based on the quantity of material released and other factors related to the characterization of the facility, the potential overpressure that could be generated is calculated. A number of methods exist to perform these calculations, and opinions vary regarding which method is best, so each should be considered on its merits. Some widely used methods include the TNT-equivalency, multienergy, and Baker-Strehlow. These methods are complex and beyond the scope of this chapter; however, more detailed information can be found in these CCPS publications: *Guidelines for Evaluating the Characteristics of Vapor Cloud Explosions, Flash Fires, and BLEVEs* (1994) and *Guidelines for Evaluating Process Plant Buildings for External Explosions and Fires* (1996).

Buildings must be designed to withstand an overpressure in excess of what is calculated. If the study is evaluating existing buildings, the calculated overpressure should be compared to the design rating of the building. If the design is adequate to protect personnel, the evaluation is completed, and the only remaining step is to complete a simple checklist to determine whether the material inside the building is properly anchored, whether there is adequate fire protection in the building, and whether the ventilation system is adequate If the study finds that a building is not designed to withstand the calculated pressure, an additional evaluation must be conducted and remedial actions developed regarding the building's structure or occupancy.

It is also important to make a distinction between explosion-resistant and explosion-proof. *Explosion-resistant* means that a building is expected to withstand the effects of an explosion without collapsing and should therefore be able to protect personnel inside, although the building may suffer structural damage and need extensive repair prior to being placed back into service. In contrast, an *explosion-proof* building is one that is not expected to be damaged significantly in an explosion. Different levels of reinforcement affect the degree to which a building will be functional after an explosion. Explosion-proof construction is generally significantly more expensive than explosion-resistant construction, which often makes it a less attractive alternative.

RISK-SCREENING ANALYSIS

A third method, along with comparison to established standards and consequence assessment, is risk-screening analysis. This method generally requires more time and effort than the methods previously mentioned. It is similar to consequence analysis, except that the frequency of an explosion is considered along with its consequences. Information on explosion frequencies may be limited, but a frequency for the unit being studied should be used if available. API 752 provides a table of generic frequencies of major explosions for various types of petrochemical units, including a value of 4.3×10^{-4} as the frequency of explosions per year of operation as a value for all units (API 2003a, 25). (Stated another way, this represents the probability of one explosion in about 2300 years of operation.) Once the frequency is found, charts showing the *probability of fatality* are used to determine the risk to occupants based on the anticipated peak overpressure. An aggregate risk to all occupants is then determined and compared to the company's accepted criteria to determine whether additional evaluation is necessary.

Additional Evaluation for Buildings at Risk

Buildings that are shown to be of concern from earlier analyses should undergo a risk assessment, during which specific release events are identified and the overall potential risk to building occupants is determined. A ranking of the potential risks, based on frequency and severity, should be applied. The Center for Chemical Process Safety proposes that a four-by-four risk matrix plotting the consequence versus the frequency can be a useful tool to assist in qualitative risks decisions (CCPS 1996, 88) (see Figure 5). Scenarios that are of higher consequence and higher frequency will be ranked as a higher risk on the matrix. The risk for a particular building is then compared to a company's risk-acceptance criteria to determine whether

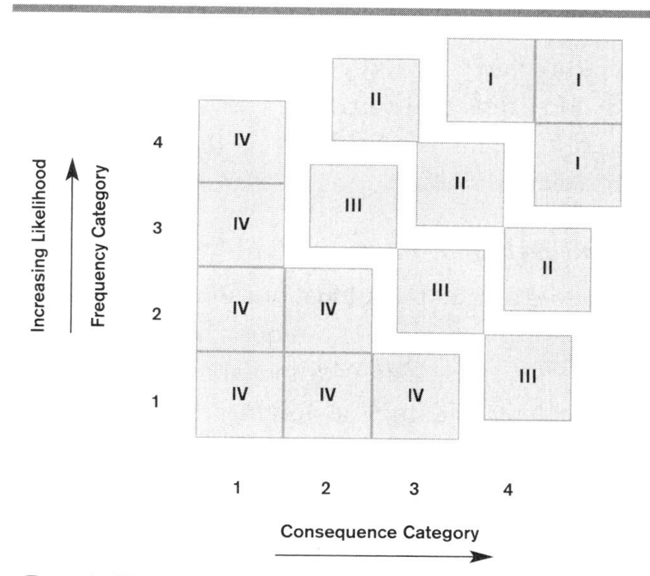

Example Risk Ranking Categories

Number	Category	Rank
I	Intolerable	Should be mitigated with engineering and/or administrative controls to a risk ranking of III or less within a specified time period, such as six months.
II	Undesirable	Should be mitigated with engineering and/or administrative controls to a risk ranking of III or less within a specified time period, such as 12 months.
III	Tolerable with controls	Should be verified that procedures or controls are in place.
IV	Tolerable as is	No mitigation required.

FIGURE 5. Example of a 4-by-4 matrix
(*Source:* CCPS 1996)

mitigation measures should be applied or the building is within acceptable limits. CCPS cautions that for events having major catastrophic potential, more quantitative methods may be appropriate.

The methods of conducting risk assessments of this type are complex, and the general outline is discussed here for exemplary purposes only. Detailed guidance can be found in CCPS *Guidelines for Evaluating the Characteristics of Vapor Cloud Explosions, Flash Fires, and BLEVEs* (1994), *Guidelines for Evaluating Process Plant Buildings for External Explosions and Fires* (1996), and others. Such methods should be applied by someone knowledgeable in the field. Also, the acceptance criteria chosen by a company are important variables in determining whether further mitigation should be applied to buildings. Several issues must be considered when companies develop such criteria, including the hazard, factors of importance to those generating and regulating the risks, and public attitudes. In general, the field of quantitative risk assessment is more advanced in the nuclear field than in the chemical and petrochemical fields.

Risk-Reduction Measures for Buildings

If the series of building evaluations shows that a building is at risk, a number of options exist for reducing the risk, including taking additional measures to prevent incidents, altering the design, and changing the occupancy loading. Options for preventing explosions include applying inherently safer principles, such as reducing inventories of hazardous materials and changing conditions in the process to reduce the potential for runaway chemical reactions or releases due to corrosion. Additional engineering controls can be added, such as redundant instrumentation. Also, procedural safeguards can be enhanced. Inspection frequencies for piping and equipment can be increased. Stringent permit systems for line opening, hot work, and maintenance can be developed, implemented, and enforced.

A number of mitigation actions can be considered for buildings that are at risk. As mentioned previously, windows can be reinforced or eliminated (although the risk-reduction potential of eliminating windows must be balanced against the operational advantage of being able to see the unit from the control room.) Also, doorways can be strengthened. Modifications can be made to buildings to reinforce them structurally, or a new building with more robust supports and a sufficient distance from the hazard can be constructed.

A final remedy to buildings at risk may be to study the occupancy and reduce the hours personnel work inside the building. Companies should consider opportunities to relocate personnel who are not critical to the function of the unit to another location. Several good practices for building design and spacing were discussed above.

Building Design to Protect Against Fires

Fires pose a hazard to buildings and their occupants. Fires can result in additional hazards, such as explosions and toxic releases—which are covered elsewhere in the chapter—but this section presents good practices only meant to protect against the direct hazards of a fire.

The degree of risk to which building occupants are exposed depends on a number of factors, including the materials the building is made of, the occupants' distance from likely fire hazards, and the effectiveness of fire-suppression systems. A key aspect of preparing occupants for a fire is to ensure that emergency plans have appropriate measures for occupants to take regarding safe shutdown of the process, escape, and sheltering-in-place as well as instructions to anyone who has responsibility for fighting incipient fires inside buildings. Escape routes should be clearly noted in emergency-response plans, and personnel should be trained in advance on where and how to evacuate properly and where to assemble once they are out of the building.

The same practices that apply to reducing the potential for explosive incidents that were discussed above apply to reducing the potential for fires as well. Mitigation measures to protect occupants include locating buildings away from areas with fire hazards. In addition, drainage outside of the building should direct spills of flammable and combustible materials away from the building. If buildings are in proximity to hazardous areas, they should be constructed with fire-resistant exteriors. Fire- and smoke-suppression systems should be installed, including water sprays inside the building and monitors outside the building. The ventilation system for the building should stop entry of outside air when products of combustion are detected.

Building Design to Protect against Toxic Releases

Toxic materials affect personnel inside buildings in different ways than flammable and explosive materials do. They impact personnel largely through inhalation, and even small amounts of some materials can affect people's ability to make critical judgments during an incident. Assessment of the potential risks to building occupants involves an analysis process of identifying whether any toxic materials of concern are handled on site. If toxics are handled, API 752 notes that buildings should be surveyed to see if any buildings exceed occupancy criteria (API 2003a, 16). The potential effect of toxics on building occupants is judged based on site conditions. If it is determined that a toxic material could enter a building (or buildings), the company must examine its mitigation and emergency response systems to determine whether they are adequate.

If it is deemed necessary to take remedial actions to protect occupants, the risk reduction measures described above to prevent explosions will also be helpful in preventing toxic-release incidents. If control rooms are susceptible to entry by toxic materials, they must be gas tight. Windows (if they are present) should be nonopening. Door and window frames should be designed to minimize gas entry. There should be no more than two entrances/exits, and they should contain air locks. Emergency response plans should include instructions on sheltering-in-place and evacuating. Appropriate PPE should be provided. Self-contained breathing apparatuses (SCBAs) should be considered for personnel who will normally occupy the building. The building functions should be examined to determine whether the occupancy can be reduced. Pressurized ventilation systems should be seriously considered, and the building must be reviewed thoroughly to determine whether there are areas of vulnerability that must be sealed (Lees 1996, 10.43).

Ventilation in Process Buildings

Of special consideration in industrial or office buildings that may be exposed to toxic releases and smoke is the ventilation system. Ventilation systems must have detectors that indicate whether a contaminant has infiltrated the system, and the detectors should shut down the intake system so that additional contaminated air will not continue to be drawn into the building. These detectors should also activate an alarm in the control room to let occupants know of the change in atmosphere. Intake stacks should be elevated so that contaminants that are heavier than air will not be drawn into them, and stacks should have sensing devices that measure contaminant levels and trip or sound alarms when the level is too high (CCPS 1993, 87–88).

The entire heating, ventilation, and air conditioning (HVAC) system should be designed of material that will not corrode. The system (including the intake) should be inspected and maintained frequently so that the integrity of the entire system will not be compromised.

The occupants should be provided with a supply of air for a period of time. If the ventilation system needs to be shut off, which will cause the building to lose pressurization, this is an important protection against ingress of contaminants. As noted previously, SCBAs should be available for occupants. Alternatively, it may be possible to supply air from a clean source remote from the control building. Lees suggests that, if possible, positive pressure in the building should be maintained between 0.5 and 1.0 inches of water (gauge), which requires two to fifteen air changes per hour depending on building size and integrity (Lees 1996, 10.43).

Trenches directly under or close to the control room should be tightly sealed or avoided if possible. Instrument sample lines for toxic materials should not be routed into the control room, as their leakage could expose occupants. The control room should have an indication of the wind direction and a water spray inside the building that can be activated automatically or manually. Such a system can provide additional protection against toxic ingress. A communication system in addition to the phone system should be installed in the control room to allow communication with a remote command center during an emergency.

Building Design to Allow Egress and Functioning during an Emergency

A critical aspect of building design is ensuring that personnel will be able to escape quickly if necessary. The design of buildings is governed by a number of state, local, and insurance codes. In general, these codes are influenced to a great extent by NFPA publications such as NFPA 1, *Uniform Fire Code* (NFPA 2006a) and NFPA 101 *Life Safety Code* (NFPA 2006b), which governs building construction, means of egress, and fire protection. A few general requirements are noted below, and the NFPA codes and the *Fire Protection Handbook* (Cote 2003) should be referenced directly for more information.

Buildings should have at least two escape routes that are clearly illuminated, even during a power failure. Emergency lighting should be provided and maintained to ensure functionality. Exits should be free from obstacles, and designers must consider that they need to be accessible to personnel who may be partially incapacitated or wearing bulky breathing equipment. Exits can include doors, walkways, stairs, ramps, fire escapes, and ladders. There must be a means to effectively notify occupants that an emergency has occurred and tell them the proper response to take. This can be accomplished by a combination of communication strategies (such as a public announcement system or alarm warning system) and training in advance of an incident. Locking or fastening devices on doors should not be permitted, as they could hinder the free escape of personnel inside the building. Fire alarms should be installed and maintained to function properly during an emergency. Initial construction and significant modifications must be reviewed by the AHJ to ensure that the intent of the codes is met. Equipment, devices, and systems intended to provide safety to occupants should be continually maintained to ensure their integrity during an emergency.

Exit doors and adjacent areas must be free of mirrors, which could confuse the direction of egress. Doors should be of the side-hinge or pivotal type and should swing out (in the direction of egress). Exits must discharge at a public way or at an exterior exit discharge. In order to facilitate emergency exiting, the means to communicate the direction of egress should be clearly understandable, noncontradictory, consistent with societal expectations, and in languages and symbols that can be clearly understood by the population occupying the building.

Checklists for Evaluating Buildings

Even if buildings in process areas are deemed to be of low occupancy or are remote enough from the process to be less at risk, it is a good idea to complete a checklist to determine whether there are additional opportunities for risk reduction. While it is difficult to design a checklist for all situations, the following is a list of typical questions, adapted from API 752 (API 2003a), that should be asked:

1. Is the building located upwind of the hazard?
2. Is the building included in an emergency response plan for fire and toxic release? Are the occupants trained in emergency response procedures? Are evacuation instructions posted?
3. Are large pieces of office equipment or stacks of materials within the building adequately secured?
4. Are the lighting fixtures, ceilings, or wall-mounted equipment well supported? Are process controls mounted on interior walls?
5. Is there heavy material stored on the ground floor only?
6. Have all the exterior windows been assessed for potential injury to occupants?
7. Are there doors on the sides of the building opposite from an expected explosion or fire source?
8. Is there exterior and interior fire-suppression equipment available to the building?

9. Is there a detection system within the building or in the fresh air intake to detect hydrocarbons, smoke, or toxic materials?
10. Is the air intake properly located?
11. Can the ventilation system prevent air ingress or air movement within the building? Are there hydrocarbon or toxic detectors that shut down the air intake? Does the building have a pressurization system?
12. Are there wind socks visible from all sides of the building?
13. Is there a building or facility alarm or communication system to warn building occupants of an emergency?
14. Is there sufficient bottled fresh air or fresh supplied air for the occupancy load?
15. Are all sewers connected to the building properly sealed to prevent ingress of vapors?

Maintenance

Along with proper design and operation of a chemical or manufacturing facility, a third critical component is maintenance. Appropriate maintenance keeps equipment functioning properly and corrects malfunctioning equipment. This section discusses the different types of maintenance and best practices associated with different maintenance activities. Maintenance activities may be performed on specific pieces of equipment or parts or a process while the rest of the process is operational, or it may be performed on several pieces of equipment simultaneously, such as in the case of an entire unit shutdown or turnaround. Large-scale turnarounds are a tremendous undertaking in a large manufacturing operation, generally requiring the addition of many temporary workers to accomplish all the tasks. The length of time between turnarounds varies from process to process, but it can run anywhere from every few months to three to five years. Once a schedule for unit shutdowns and maintenance is established, it should be adhered to. Even with production pressures to continue operating, a general shutdown, which allows maintenance to be performed on a broad scale throughout the unit, should not be compromised or delayed without serious consideration of the potential consequences. Publications from a number of organizations were used in developing this section, including CCPS, API, and OSHA.

Types of Maintenance

This section explores the different types of maintenance, including planned and unplanned. Although some of the examples cited are of equipment in processing plants, all machines must be maintained at some point, so the concepts are broadly applicable throughout a variety of facilities. A section is also devoted to reliability-centered maintenance (CCPS 2001b, 203).

Planned Maintenance

It is preferable to perform maintenance in advance of a breakdown so that problems can be prevented. This is accomplished by developing a list of equipment to maintain and specifying the appropriate activities to keep the equipment from breaking down (CCPS 2001b, 205). The frequency of such activities should also be thoughtfully established. When maintenance is planned according to a schedule, parts, materials, and trained personnel can be available. Conducting maintenance according to a routine plan can also ensure that equipment is properly emptied, purged, and isolated. In order for planned maintenance to be successful, management must be committed to the idea that shutting down a piece of equipment that is currently operating well is appropriate; however, planned maintenance can reduce equipment breakdowns and downtime and prove economically justified. Only equipment that is not critical to operations and safety, and thus can wait to be fixed if it breaks, does not need to be included in a planned maintenance program (CCPS 2001b, 205–207).

Preventive Maintenance

Preventive maintenance involves the process of establishing fixed schedules for routine inspection, replacement, and service of equipment. In broad terms, it encompasses all planned maintenance activities from replacing seal oil in pumps to verifying the functionality of instrumentation systems to calibrating instruments and emptying, cleaning, and inspecting tanks.

Preventive maintenance involves cleaning and servicing equipment while it is operating and dismantling and checking or inspecting equipment while it is shut down.

In *Guidelines for Safe Process Operations and Maintenance*, the CCPS notes that preventive maintenance should begin as soon as possible, even during the design phase of a project CCPS 2001b, 207). Equipment selected should be dependable and easy to maintain. For example, locating heat exchangers at the perimeter of a facility allows maintenance personnel to extract bundles without having to lift them over other equipment. Necessary activities and schedules should be established based on plant experience, codes and industry standards, and manufacturers' recommendations. First- and second-line maintenance and operations supervisors must be proactive in recognizing the need to establish preventive maintenance programs or in altering the activities or schedules of existing programs. The CCPS notes that they are also the ones who should recommend whether a piece of equipment should be placed into a predictive maintenance program, discussed in the next section (CCPS 2001b, 207).

Predictive Maintenance

Predictive maintenance is similar in concept to preventive maintenance. The CCPS notes that while preventive maintenance operates according to a predetermined schedule, predictive maintenance takes into account real-time conditional variables. Parameters such as temperature, vibration, flow rates, and motor current are analyzed to determine whether service requirements have changed and should be adjusted (CCPS 2001b, 207).

Critical equipment should be included in a predictive maintenance program. This type of program lends itself especially well to rotating equipment, pumps, and compressors, because of the types of information that can be collected on an ongoing basis (CCPS 2001b, 207). Also, schedules for predictive maintenance on electronic equipment can be based on statistical failure information.

Unplanned Maintenance

Although all maintenance, even breakdown maintenance, involves some degree of planning, the term *unplanned maintenance* refers to activities that take place after a failure has occurred. If the failure involves equipment that is not critical to safety or operations, the process may not need be to be shut down, and maintenance may be performed during normal operations. However, if the equipment is critical to safety or operations, maintenance activities often must take place in an expeditious manner to keep the process running. Because of these pressures, unplanned maintenance of critical equipment carries additional risks to personnel and the process. Qualified personnel, tools, and the parts necessary to conduct unplanned maintenance activities may not be readily available (CCPS 2001b, 212).An incident at a refinery in Martinez, California, killed four people when personnel cut into a line containing naphthalene before it was properly emptied.The maintenance was being performed while the unit was operating, and the hot surfaces of surrounding equipment may have acted as ignition sources (CSB 2001).

Even when unplanned maintenance must be performed on critical equipment, it is important that enough preparation be made so that the job can be performed safely. A good practice is to bring together a multidisciplinary team representing maintenance and operations to perform a hazard analysis in order to determine potential problems as well as steps to ensure that any problems encountered can be addressed (Lees 1996, 21–22). Also, consideration should be given to problems with equipment preparation (e.g., draining and purging) so that maintenance personnel are not put at additional risks. In the incident mentioned in the previous paragraph, personnel had trouble draining equipment due to lines being plugged with corrosion products. Some good practices for the different activities involved with preparing for maintenance are discussed later in this section.

Reliability-Centered Maintenance (RCM)

One approach that is widely used in chemical and petrochemical facilities is *reliability-centered maintenance*, or RCM. As the name suggests, RCM is a method that is largely based on the reliability of equipment, often determined through an analysis process similar to a failure modes and effects analysis (FMEA). RCM begins by establishing the primary and secondary functions of equipment. Ways in which the equipment can fail to function are then determined. The effects of failures are evaluated, including so-called "hidden consequences," such as failures of relief devices that may go unannounced but may impede their functioning at critical times later. Finally, maintenance tasks that can eliminate or reduce failures and are considered feasible and effective are identified for each failure mode. If a specific preventive maintenance task is not identified for a failure mode, certain default actions are taken, which may include redesign, inspection, or no action if the risk is deemed to be low. RCM is a formal process, and a great deal of literature exists to assist in its implementation.

Developing Critical Equipment and Instrument Lists

The first step in an effective maintenance plan is to identify what equipment and instrumentation must be included in the mechanical-integrity program (CCPS 2001b, 203). This list includes equipment such as pressure vessels, storage tanks, piping, pressure-relief systems, pumps, alarms and interlocks, fire protection, and emergency shutdown devices. A key step is to determine which specific equipment should be included in the list of *critical* equipment and instrumentation—equipment that is essential to the safety, operation, and environmental functioning of the unit. A facility must establish a list of criteria for critical equipment and revisit it often to ensure that it reflects changes in equipment and operational modes.

The critical-equipment list should include all equipment (including instrumentation) necessary for the continued safe operation of the plant, but caution should be exercised so that the list remains meaningful. This list can then be used to develop a preventive and predictive maintenance inventory so that the service schedule for that equipment can be developed more accurately. Having a critical-equipment and instrument list also allows supervisors to assign reasonable priorities when writing work orders.

Several considerations go into developing a list of critical equipment. Use of a modified hazard-analysis technique, augmented by discussions of operational reliability and criticality for each piece of equipment and instrument may be appropriate for determining which equipment requires greater scrutiny. When developing a list of critical-instruments, the severity of the consequences if the instrument system fails should be a major consideration. Safety alarms and interlocks, fire protection, and equipment whose failure would result in an automatic and sudden shutdown of the process are typical candidates for the list. Instruments that protect against unplanned releases of flammable or toxic material are typically considered critical, especially in cases where there is limited redundancy and both the process control and safety functions rely on the same equipment (e.g., when level control, alarm, and shutdown are all on the same circuit and rely on the same sensor). Along with the functioning of the equipment, other factors to consider include availability of spare parts (e.g., presence of redundant pumps) and whether failures are likely to be announced, especially in the case of instrumentation and shutdown systems. The availability of such devices must remain high for the plant to continue to function safely.

Causes of Eventual Failures in Piping and Vessels

Corrosion and erosion that are allowed to progress in piping and vessels can result in failures in equipment and a release of material (CCPS 2001b, 217–219). Vessels and piping designed to be used in corrosive services generally have additional thickness, referred to as a *corrosion allowance*. Monitoring strategies such as ultrasonic thickness (UT) testing allows maintenance personnel to determine whether areas of piping are thinning at an accelerated rate.

Corrosion can have the effect of thinning vessel and piping walls and also depositing products of corrosion that can plug equipment downstream. Corrosive systems, such as those containing sulfides and chlorides, must be monitored. Attention must be paid to temporary piping as well, since it may not be designed to withstand corrosion. Of particular concern to maintenance technicians is corrosion under insulation, in part because it is very difficult to detect visually. Such corrosion can remain hidden but eventually result in loss of containment. Intruding water can bring in corrosive elements and ultimately result in stress corrosion cracking. Representative sections of insulation should be stripped off periodically so that the piping can be inspected. Doing this is especially important if the temperature of carbon steel piping is between approximately 25°F and 250°F (or if austenitic stainless steel piping is between 150°F and 400°F), because at these temperatures water will not quickly be driven off. Areas of special concern include those where insulation is penetrated (such as at deadlegs) or damaged and at low points in the system.

Another type of corrosion, referred to as pitting corrosion, occurs when cavities are produced in the walls of piping and equipment. *Pitting* is a localized form of corrosion and is often more difficult to detect than uniform corrosion. It can be caused by localized damage to a protective coating or imperfections in the metal structure.

Erosion is another problem that can cause failures in equipment. Erosion can occur when liquid at high velocity wears down the walls of piping at injection points and elbows. Water hammer can occur when fluid flows through sharp turns. Particulates in fluid can also erode piping walls over time. Other potential causes of failure include wear (e.g., in pumps and valves) and intermittent operation. Intermittent operation can cause additional stress on equipment (including instrumentation) because the equipment is always in a state of flux.

Testing and Inspection

A common theme in maintenance programs for all types of equipment is to collect data—by testing or inspection—that may indicate a problem in advance. When discussing systems that act to maintain safety or operations, such as instrumentation loops or emergency shutdown systems, failures are generally detected by testing the system's functionality (CCPS 2001b, 221–223), whereas with piping systems and vessels in which the failure is generally a breach of a vessel wall, or a pump in which a failure may be a blown seal, problems are identified by conducting inspections to determine the integrity of equipment. Piping inspection can be quite challenging and is not dealt with in this section.

Testing of Interlocks

Safety interlocks and process control loops are critical to the safe operation of a unit. It is important that these systems be reliable and available when needed. Because of the nature of some interlocks, they may not be activated until an emergency occurs. It is important that instrument failures be detected. Some instruments have a self-checking capability. But for many interlocks, testing for functionality, or *proof-testing*, is important. Instrument loops contain several parts, including sensors, transmitters, transducers, and actuators, and it is important that all of the parts work properly. Instrument loops are often tested as entire circuits. Frequency of proof-testing depends on the instrument and the service, but a typical schedule would be every six to twelve months. Proof-testing is accomplished according to certain procedures (CCPS 2001b, 221–223). Two methods are an *actual test* and a *simulated test*. In the actual test, the process is brought to the trip point, and the functionality of the instrument circuit is observed. As the name suggests, a simulated test only simulates these conditions. A comprehensive functionality test of an interlock system involves inspecting the system from sensor to actuator to ensure it is not deteriorated at any point. This is more complex than simply ensuring that the interlock works.

Critical instruments are those that require greater scrutiny because their failure could result in serious safety consequences. These instruments should be maintained and calibrated according to an authorization system. The test frequency can be determined based on engineering judgment from maintenance studies. When establishing testing frequencies, facilities should consider the potential frequency of the failure. The test frequency must be at much smaller intervals than the expected failure frequency. If an emergency shutdown device is tested annually, for example, it could potentially be in an unannounced failed state for an entire year. Testing frequencies must consider the criticality of the system, which is why development of a meaningful critical-equipment and instrumentation list is important. When critical-instrument lists are developed, the severity of the consequences if the instrument system fails should be a major consideration.

Critical instruments should be tested more frequently than those that are not as critical.

Establishing Inspection Schedules

A number of sources can be consulted to establish inspection schedules. Various codes and standards provide guidance on external and internal inspections for different types of equipment. The API, with codes including API 570 (1998), 576 (2000), and 653 (2001), is an excellent source even for facilities that are not units in a refinery. Where a corrosion rate is not known, a short inspection interval is recommended. Risk-based inspections use inspection strategies based on the likelihood and consequences of failures. Intervals between inspections should be established based on corrosion rates, the remaining life of the equipment, piping service classifications, the judgment of subject-matter experts (such as inspectors), and applicable jurisdictional requirements.

API 570 provides maximum inspection frequencies for different classes of piping. Class 1 is piping that has a high potential for an emergency if a leak were to occur, and includes flammable services (such as liquid propane) that may auto-refrigerate, leading to brittle fractures and pressurized services that may vaporize rapidly, creating an explosive mixture. This class also includes pipes carrying material that has a high concentration of hydrogen sulfide, hydrogen chloride, or hydrofluoric acid. Class 2 involves much of the piping in petrochemical plants and includes materials such as hydrocarbons that do not vaporize rapidly, hydrogen, fuel and natural gas, and strong acids and bases. Class-3 piping involves materials that are flammable but vaporize slowly upon release. Distillate and product lines to and from storage are also included in this class (API 1998, 6-1). Table 3 shows recommended maximum inspection intervals for the various classes. Risk-based inspection is sometimes used to adjust schedules such as these, but caution should always be exercised when increasing the interval between inspections to ensure that the equipment will continue to be adequately protected.

Guidance on the frequency of inspections for buried piping depend on the soil resistivity. Table 4 provides some guidelines for piping without cathodic protection. (Cathodic protection applies a current and reduces the corrosion potential by bringing the metal to an immune state.)

The API does not prescribe an inspection interval for pressure vessels but notes that the interval should be based on operating experience, manufacturers' recommendations, and recommendations of applicable regulatory bodies. API 576 does, however, note that the maximum interval between inspections of pressure-relief devices is ten years (API 2000, 21.) This time frame, along with the other factors mentioned previously, may be used as one basis for equipment inspections as well. The inspection frequency for low-pressure storage tanks should be based on a number of factors, including the nature of the product stored, results of visual inspections, corrosion rates, changes in operating mode, and changes in service. External visual inspections should be made at least every five years. When the corrosion rate is not known, external thickness measurements should be taken ultrasonically at least every five years. Internal inspection frequencies should not exceed twenty years, and if the corrosion rate is not known, they should not exceed ten years.

TABLE 3

Recommended Maximum Inspection Intervals for Piping

Type of Circuit	Thickness Measurements	Visual External
Class 1	5 years	5 years
Class 2	10 years	10 years
Class 3	10 years	10 years

(*Source:* API 1998)

TABLE 4

Recommended Maximum Inspection Intervals for Buried Piping

Soil Resistivity (Ohm-cm)	Inspection Interval
< 2000	5 years
2000 to 10,000	10 years
> 10,000	15 years

(*Source:* API 1998)

Establishing Inspection Methods

Inspection and surveillance methods usually fall into one of the following categories: internal visual, external visual, vibrating piping, and thickness measurement. Supplemental measurements can also be conducted as necessary. *Internal visual inspections,* while common with vessels, are generally not performed on piping unless it has a large diameter. Internal inspections of vessels should include the vessel shell, linings, and internal devices such as trays. *External visual inspections* determine the condition of the outside of vessels and piping. Painting and coating need to be observed, as well as the condition of the insulation system. Signs of leakage, vibration, corrosion, movement, or the misalignment of piping can also be observed.

Thickness measurements are taken at various locations along the vessel or pipe circuit and are used to determine the presence of localized corrosion. An average, or more conservatively, the thinnest reading, is used to calculate the rate of corrosion and the remaining life, along with the time for the next inspection. The number and locations of measuring points should be determined based on risks. A higher number of readings should be taken in areas where the consequences of failure are higher, where materials are deemed to be more prone to corrosion, and where there is a high number of fittings, branches, deadlegs, and other such configurations. For piping and vessels with low potential for safety and environmental consequences and those in relatively noncorrosive systems, fewer thickness readings are necessary. If the consequences are extremely low or if the service is noncorrosive, thickness readings may be eliminated.

The minimum thickness can be located by ultrasonic scanning or radiography. Ultrasonic scanning is the most accurate means for measuring thickness on piping with a nominal size greater than one inch, whereas radiographic techniques are better for piping with diameters of one inch or smaller. API 570 notes that radiographic profile techniques can be used to locate areas to be measured in insulated systems or areas where local corrosion is expected. Then ultrasonics can be used to determine actual thicknesses. Inspectors must correct measurement readings for temperature, especially at temperatures above 150°F.

These methods can be used when equipment is in operation. When equipment is not in service, thickness measurements can be taken directly with calipers. If there is pitting, special pit-depth measuring devices may also be used (API 1998, 5–8).

Other nondestructive methods of measuring thickness discussed in literature include magnetic methods, dye-penetrant methods, stress-wave-emission monitoring, holography, and electrical measurements. One other method of determining equipment integrity should be mentioned: hydraulic or pneumatic pressure-testing of piping systems, although this method is not normally conducted as part of a routine inspection. API 570 has more detailed information about conducting pressure tests (API 1998, 5–9).

A unique challenge is inspection of underground piping. Because of inaccessibility and the fact that most of the piping is not visible, inspections are difficult. Techniques used include above-ground visual surveillance to detect changes that may indicate a leak, such as discoloration of soil, pool formation, and odors. The soil resistivity around piping may be tested to determine the likelihood of corrosion. Low resistivity indicates a more corrosive environment than high resistivity. A number of inspection methods exist, including intelligent pigging (passing a device through the line to determine the internal conditions), video cameras inserted into the piping, and excavation, which allows for an external inspection.

Finally, personnel performing inspections must be properly trained and certified. The API issues certification to personnel who have demonstrated competency at the craft of inspection and interpretation of results. These certifications require the completion of an exam to show that the requirements of API standards are well understood. There are also minimum education and experience requirements.

Hazards of Maintenance Work and Preparation for Maintenance

Maintenance includes several functions, such as testing, inspecting, cleaning, calibrating, repairing, and installation of replacement equipment. The nature of maintenance work brings with it inherent hazards.

Technicians must perform work on machinery with moving parts and on systems that contain hazardous materials. Hazards faced by maintenance workers include entanglement, crushing, burns, exposure to toxic materials, electrocution, asphyxiation, and injury by impact. Also, since maintenance work can affect operations, improper restoration of the equipment after maintenance can result in hazards when the system is restarted. Such problems can include release of toxic and flammable materials or hazards due to improper guarding of equipment.

To address these issues, an effective work-permit system should be implemented. Lees notes that the purposes of such a system are to ensure that hazards of conducting work are considered and precautions are properly thought out and specified and to make sure that all of these issues are understood by the persons involved (Lees 1996, 21.14–16). The system also facilitates communication, especially between operations personnel, who generally prepare equipment for maintenance and start it up after maintenance, and maintenance personnel, who conduct the work (see Figure 6). Permit systems allow the coordination of work, as maintenance jobs can be large in scope and involve several people. Lees also notes that a clear system of handover is a critical part of the permitting process. The permitting system should ensure that everyone involved with the job understands the hazards and how to comply with safe codes of practice. Permit forms vary from company to company, but in general permits include a description of the work, safety requirements, discussion of potential hazards, authority to start, acceptance of conditions, and acknowledgement of completion (Lees 1996, 21.14–21). Refer to the "Permit to Work" chapter of this handbook for additional details on the permitting system and the various hazards of maintenance.

Tasks that should be permitted prior to commencing work include those on equipment where toxic, flammable, or corrosive substances are present, especially when equipment needs to be opened for servicing and/or cleaning. Such tasks include:

- work on equipment that is pressurized or that operates at extreme temperatures
- maintenance requiring hot work or evacuation
- work that requires entry into confined spaces or areas where there is a potential for oxygen deficiency or enrichment
- work on machinery, especially machines with moving parts, such as conveyors, hoists, and lifts, and includes work requiring heavy equipment to be lifted, such as with a crane.

First- and second-line supervisors are a critical key to an effective permitting process. They are responsible for making sure that maintenance work is properly authorized and controlled. Checklists are often used in the process to prompt personnel about potential hazards of the process. Also, provisions must be made to stop work if the requirements of the safe preparation of equipment, such as draining and isolation, cannot be met (Wallace et al. 2003, 214–215). A reasonable approach if such a situation occurs is to stop the work and bring together a multidisciplinary team to plan a strategy for safely accomplishing the job. As mentioned previously, a serious incident occurred in Martinez, California, when personnel cut into a line containing flammable material before it had been properly evacuated (CSB 2001, 11)

Communication is another function of the permit process. Since equipment that may be dangerous or contain hazardous material is being handed off from one department to another, it is imperative that the departments communicate. Operations personnel must communicate with maintenance personnel and vice versa. For critical jobs, a prejob conference between operations and maintenance departments at the job site should be conducted. Supervisors, operators, and craftspersons who will be performing the work should all attend the conference. Safety precautions and potential problems must be discussed and action items must be developed to resolve any potential issues. The equipment that requires maintenance must be clearly identified on the permit, and operations personnel should consider having a discussion with maintenance personnel at the job site if maintenance personnel are unfamiliar with the equipment—or if the equipment to be maintained is in a confusing maze of piping and other equipment—to decrease

Hot Work / IGNITION SOURCE Permit

Issue Date:_________	Start Time: AM PM	End Time: AM PM	**Welding, Burning, Open Flame** **Ignition Source**
Unit / Location:			
Permit Issued To:			
Detailed Description of Job/Including Tools and equipment:			

Operations / Equipment Owner (Permit Issuer)

LEL/O_2	Required	Not Required	**Calibration Date:**_________
	Initial Result/Time	Revalidation Result/Time	**Chemical Name(s)**
Oxygen (19.5-23.5%)			
LEL (0%)			

	YES	**N/A**
1. Can material to be worked on be removed to a shop area?		
2. Has process equipment been cleared of all flammable material?		
3. Have all control points been identified, locked & or blinded? Spools removed?		
4. Is a water hose or a proper fire extinguisher on site and in working order?		
5. Is a fire watch required?		
6. Have Flammable/Combustible materials in the vicinity or on lower floor levels been removed or protected?		
7. Have all drains, sewers, cracks, and openings been covered or shielded to prevent sparks or hot slag from entering?		
8. Have combustible materials been removed from the hot work area or shielded with fire retardant material such as fire blankets?		
9. Have precautions been taken to prevent heat transfer from object of hot work to other combustible materials?		
10. For overhead work, has the work area(s) below been properly barricaded or Fire Watch posted?		
11. Have conveyors and ducts been shut down or protected to prevent sparks from traveling?		

Maintenance Foreman / Construction Foreman

1. Has job been reviewed with workers (Special Precautions & JSI)?		
2. Has the Fire Watch been advised of his/her duties?		
3. Has appropriate PPE been assigned for the task?		

I have verified the work area to be clear and approve this permit. (Signature and Phone/Pager)

WBOF - Equipment Owner/Supervisor:	
WBOF - Maintenance/Construction Foreman:	
Ignition Source - Operations Representative:	

Please Notify Permit Issuer Upon Completion or Changes in Conditions.

FIGURE 6. Example of a permit for hot work (*Source:* CCPS 1996)

the chances of the wrong equipment being opened. Personnel should be available to ensure communication throughout the job, similar to the concept of a fire watch when hot work is being performed. Also, it is vital that communication occur from shift to shift if the job extends beyond one shift. In some cases, if safety testing (such as testing for flammables) is not being done continuously, it should be redone when a new shift comes on.

Supervision and training are important elements in a maintenance preparation and execution system. While permit systems may establish authority, appropriate supervision must continue throughout the execution of critical work. Also, maintenance personnel should be adequately trained in their craft, in safe preparation tasks such as isolation, and in the permitting system. Refresher training and continuing education must be included for all crafts. All maintenance personnel, whether employees or contractors, must be trained in the general hazards at the facility and the hazards of the specific chemicals used there (Wallace et al. 2003, 213).

Line Breaking, Equipment Opening, and Isolation

Because maintenance often requires dismantling equipment that processes hazardous material, properly evacuating that material and purging the system should preface any job where equipment will be opened. As mentioned previously, identification of the correct equipment is vitally important. Use of a mini-conference at the job site should be considered if the job is particularly hazardous, if personnel performing the work are new or unfamiliar with the hazard, or if equipment is in a confusing maze or adjacent to other similar equipment. Tagging is also important, especially on flanges that must be opened.

After the correct equipment is identified, the general steps for opening equipment containing hazardous liquids and gases are depressuring, cooling, isolating, emptying liquids (and solids), purging, and cleaning. Liquids and gases may be conveyed to a different part of the unit or to a scrubber or flare system. Material should not be vented to the atmosphere unless it can be proven that doing so is safe and the material is benign. For processes that operate at elevated temperatures, the equipment should be cooled prior to opening. Caution should be exercised, however, because equipment that contains hot material that is cooled prior to maintenance may be at risk for vacuum. To protect against this hazard, equipment should be properly vented during cooldown. Inert gas may be injected to ensure that the vapor space continues to be occupied.

Prior to opening, the equipment must be isolated from process fluids and high pressure and temperature sources. Machinery and automatic valves must be isolated from sources of power and energy. Methods of isolating vessels and piping include, in order of effectiveness, closing block and isolation valves, double-blocking and bleeding (DBB) the piping, installing blinds, and disconnecting the equipment. Closing valves is the least effective method because valves leak. DBB is more effective because it involves closing two valves and opening a bleed between them, which keeps material from accumulating and pressure from building up between the valves. Installation of blinds and physical disconnection are the best methods for isolation; however, care must be taken to use blinds that are of appropriate thickness to withstand the highest pressure to which they may be exposed. Also, caution must be exercised when the piping system is separated to install a blind, as residual material may be present and maintenance technicians may be at risk. When lines are physically disconnected, a blank should be installed so that the end of the piping is not exposed to the atmosphere. In the event of a breech of isolation upstream, a blank will prevent material from being released. Physical disconnection is the preferred method if personnel are entering confined spaces and for relief and vent lines, due to the various connections that may be involved. The system of isolation should guarantee that blinds are uninstalled after work is completed and that physically disconnected piping is reconnected prior to restarting the process .

Machines and other devices that operate from an energy source must be isolated from their sources of power, including electrical, hydraulic, and pneumatic, to ensure that there is no unexpected energization. A lockout/tagout system consistent with the requirements in OSHA 1910.147 (1996b should address these concerns. *Lockout* refers to installing locks on circuit breakers, switches, and flanges or valves so those devices cannot be opened while equipment is being maintained. Everyone involved with the job, from operations to maintenance to electricians, will typically add a lock to the equipment so that it cannot be operated until all locks are removed. *Tagout* involves the use of tags instead of locks, and this method should be used only if it can be proven to be as effective as lockout. Although typically thought of in conjunction with electrical equipment, a lockout/tagout system should be used with equipment whenever premature use could result in injury or a release of material. Electrical sources may be isolated by removing the fuses or locking off the isolator. Refer to the "Permit to Work" section of this handbook for more information.

Once the equipment is isolated, the next step is to remove residual hazardous materials from the equipment to be serviced. If it is liquid, it can be pumped to a different location or drained if it is not hazardous. Then any vapor must be evacuated from the equipment by ventilation, flushing with water, purging, or steaming. Purging involves replacing a toxic or flammable gas with an inert gas and then with air. It is important that an inert gas, such as nitrogen, be used initially so that flammable material and air do not form a flammable mixture. Also, it is important that an air purge follow the inert gas purge before personnel enter the equipment; otherwise, there is a risk of asphyxiation. The sequence of shutdown steps is very important, especially during the purge phase. In the case of flammable gas, it is during the purge phase that the material in the vessel is likely to pass through the flammable range. Inert gas or steam is often used to control this hazard.

Under no circumstances should a vessel or piping be opened unless operations and maintenance personnel can be certain of the conditions inside the equipment. Personnel must be able to verify that the system is free of pressure, high temperature, and the presence of flammable, toxic, or reacting material (Wallace et al. 2003, 213). Gauges and analyzers should be used where available, and must be included in the design of the process. Indirect measures, such as analysis of the material on exit from the equipment and purge time and rate can be used if necessary, but they are not the preferred method. However, even when gauges are present and are used, personnel should be aware of conditions inside equipment that may render gauges useless. As an example, in Augusta, Georgia, three operators died during an incident in which they were opening a vessel that contained what they believed to be fully reacted polymer. A portion of the polymer in the core of the vessel was still unreacted and, due to the buildup of pressure inside the vessel, exploded when bolts on the vessel cover were loosened. The polymer had plugged the lone pressure gauge on the vent line, and no other effective means existed to ensure that there was no pressure inside the vessel. This case speaks to the need for redundancy of gauges and instrumentation to monitor conditions in process equipment (CSB 2002, 47).

The final step in preparing a vessel for entering is often cleaning. There are a number of methods of cleaning equipment prior to entering, including flushing with water, chemical cleaning, steaming, and manual cleaning. The method chosen will depend on the type of facility and nature of the material that is present in the equipment.

Welding, Cutting, and Other Hot Work

A number of maintenance jobs require some type of hot work to remove or repair equipment. Hot work brings a source of ignition into a potentially flammable or toxic atmosphere, so protections must be in place to minimize the hazard. The first step is to conduct a job analysis to determine whether hot work is required or there is a lower-risk way to accomplish the work. The analysis should be completed by competent personnel. If it is determined that hot work is required, then the hazards must be determined and appropriate precautions taken.

Many of the precautions that must be taken prior to hot work, such as isolation and purging, are standard steps that must be taken in conjunction with most permitted work (API 2002b, 13). Some of the specific requirements for hot work include ventilation, testing for hazards, and fire watches. Note that some references state that welding should be confined to manual electric-arc methods.

VENTILATION

Ventilation is required in a variety of circumstances, including those in which hot work is performed in areas with limited air movement (API 2002b, 14). The intent of ventilation is to prevent worker exposure to hazardous fumes by removing contaminated air and adding fresh air to the environment. *Air-movers* are mechanical devices that facilitate ventilation. They should direct fumes and vapors away from working personnel. Air-movers should be bonded so they are less likely to generate static electricity. (Bonding involves connecting different parts of equipment or containers that may have a different potential charge due to separation resulting from gaskets or caulking. A conductor, such as a copper wire, equalizes the potential charge.) Local ventilation may also be an alternative in some applications. With this approach, a high velocity of airflow (e.g., 100 cfm) is directed at, or close to, the hot work.

FLAMMABILITY TESTS

A competent person must perform flammability tests in the area before hot work commences, and the detector must be calibrated and maintained appropriately. Hydrocarbons that may be in the liquid phase (not detectable by gas monitoring) must be found and removed before hot work is started. Testing should be conducted after purging with steam and inert gases has occurred, but testing personnel must be aware that steam vapors can affect detector results. The surrounding area should be thoroughly tested as well to ensure that no flammable vapors are present. Low points and confined areas must be included in the testing. Tests must be conducted at multiple locations on runs of piping. If there is any detectable reading of flammable vapor, work should be stopped or not allowed to begin until the source is found and eliminated.

Consideration should be given to the frequency of testing. One-time testing conducted only before the work begins has limitations. Conditions can change during the course of work, and continuous monitoring rather than initial testing is often necessary to ensure that any change in flammable conditions is detected. During an incident in Delaware City, Delaware a flammables test was conducted at the beginning of a maintenance shift prior to the commencement of welding on a walkway around tanks containing flammable material. During the course of the day, conditions changed, and flammable vapor that had leaked out of a hole in a tank was apparently blown in the direction of the welding operation, which resulted in an explosion, causing a fatality and resulting in a large release of acid to a river.

FIRE WATCHES

According to OSHA 1910.252, *Welding, Cutting, and Brazing* (1998b), fire-watch personnel are required any time there is a permit for performing hot work. The purpose of the fire watch is to provide constant surveillance in the area. Fire-watch personnel look for sparking and potential fires. They extinguish fires only if they are able to do so with a fire extinguisher or a water hose. If the fire watch cannot extinguish a fire, he or she should sound a fire alarm. OSHA 1910.252 also notes that the fire watch should keep vigil for at least 30 minutes after completion of hot work (OSHA 1998b). As a precaution, fire blankets should be used to protect other equipment in the area from sparks. The fire watch should make sure these precautions are in place and that general safety is maintained in the area.

Hot Tapping

Hot tapping is the procedure of fitting a branch onto a pipe or vessel that is in service and then cutting a hole in the attached fitting. Hot taps allow modifications to be made without shutting the plant down, but there are several potential hazards associated with the procedure. In addition to the usual hazards of welding, there other hazards: a leak may occur during the operation, the fluid in the pipe may decompose explosively due to the heat applied during welding, and the modified equipment may fail at some point later. Because of these hazards, hot tapping should be avoided whenever possible.

If hot tapping is necessary, it requires careful preparation and consideration of the process fluid, operating conditions, materials of construction and their dimensions, and requirements of the specific job. API 2201, *Procedures for Welding or Hot Tapping on Equipment in Service*, provides good guidance on performing such activities (API 2003b):

- The pressure should be reduced as much as possible so the hazard is reduced.
- The material of construction and the thickness of the pipe or vessel wall should be thoroughly checked.
- Welding must not take place if the shell has deteriorated to the point that it is unsafe.
- To prevent burn-through (a condition in which the unmelted area beneath the weld cannot support the pressure), API 2201 recommends using a 3/32-inch (or smaller) welding electrode for piping less than 1/4 inch thick. For pipe walls greater than 1/2 inch, larger diameter electrodes can be used (API 2003b, 2).
- Flow through pipes during hot tapping generally helps to dissipate heat and prevent burn-through; however, a balance must be maintained because high flow rates can increase the cooling rate and increase the risk of cracking.

Therefore, a minimum flow should be maintained but high flow should be avoided.

- A minimum wall thickness of 3/16 inch is recommended for most hot tapping applications.
- Hot tapping close to connections and seams must be avoided, and it should not be conducted within eighteen inches of a flanged or threaded connection or within three inches of a welded seam.
- If a hot tap is performed on a tank, it should be performed at least three feet below the top of the liquid level at the point of the cut to ensure that the cut is not being made into the vapor space.
- Hot taps should be avoided on vapor and oxygen mixtures that are close to their flammable range. They should also be avoided on hydrogen systems (unless a special engineering review approves it) and on systems containing chemicals that are likely to decompose on heating, including unsaturated hydrocarbons such as ethylene that may experience decomposition upon heating (API 2003b, 4).

Confined Space Entry

Entry into a confined space poses unique hazards, as personnel are particularly vulnerable to the atmosphere inside and have limited mobility. The hazards personnel are likely to face include flammables, toxic substances, oxygen deficiency, and oxygen enrichment. Entry into a confined space should be avoided if at all possible.

OSHA 1910.146 (1998a) lays out regulations for confined spaces:

- Vessels that have contained flammable or toxic material must be thoroughly cleaned before entry.
- Personnel must ensure that all sources of flammables are positively isolated from the vessel.
- The use of only a single valve for isolation is not recommended. Physical disconnection is the recommended isolation strategy.
- If an oxygen-deficient environment is expected, personnel must use a respirator, either self-contained or with an air line and a reliable source of oxygen.
- The atmosphere should be tested for flammables, toxics, and oxygen concentration. If possible the tests should be conducted without entering the vessel. The tests should continue as necessary to verify that the space is still safe for entrants.
- Ventilation must be provided to ensure that the space remains nonhazardous.
- Lighting should be sufficient to allow personnel inside the vessel to see to work and escape if necessary.
- A plan of rescue should be developed prior to personnel entering a vessel. The plan should consider the configuration inside the vessel, as many reactors and columns contain trays and supports that can hinder rescue. Personnel should wear a safety line to facilitate rescue if necessary.
- An attendant must stand watch to monitor workers in confined spaces.
- A system of communication must be set up so that personnel in a confined space can quickly alert personnel outside of any problems.

Quality Control and Documentation Program

Even the best maintenance organizations can have serious problems unless a comprehensive quality-assurance program is in place. Such a program will ensure that proper construction materials are used in design and replacement, that installation and inspection procedures are effective, and that the system for spare parts and service products (such as gaskets, packing, and lubrication) are of the correct type. The system should also ensure that inspections and preventive maintenance activities occur on schedule, and that the action items resulting from those activities are addressed promptly. Data on equipment failure, repair, and availability should be readily at hand. Records of maintenance on safety-critical and protective devices are particularly important and should be accurately kept. Protective devices include relief valves, vents, safety interlocks and shutdowns, and fire-protection equipment. In such a record-keeping system, as-built drawings, certifications of compliance with codes and standards for equipment, and verification of materials of construction are documented in such a way that they can be easily referenced. Audits of equipment suppliers that have occurred (or are planned), along with documentation, should also be part of the system.

Replacement parts should be inventoried and stored in a warehouse. Keeping critical parts in inventory is a significant purpose of a quality-control and documentation system. The inventory may be used to determine when parts are likely to run out and need to be reordered. A proactive system considers usage, criticality, lead time, and costs in making decisions on inventories. The system controls equipment that is checked out to ensure that the correct parts are selected for installation in the field. Many organizations have adopted the standards set forth by the International Organization for Standardization (ISO), such as ISO 9000 or ISO 14000. While these programs may still have errors and should be audited occasionally by the end user, participation in ISO generally means that documentation is kept and the programs are easier to audit. Finally, particular attention should be paid when equipment is reused. The integrity of the equipment should be checked and the intended use should be verified. Mixing reconditioned parts with new parts is not advised. The fitness of each piece of used equipment must be considered before putting it into service.

Contractor Safety

Recent trends in industry have been to use contractors for more work, not only in construction activities, but also on more routine tasks (Lees 1996, 21.5). OSHA 1910.119 requires facilities that fall under the jurisdiction of the Process Safety Management Program (PSM) standard to evaluate contractors' safety performance, inform contractors of hazards, and explain emergency procedures to them. The owner/operator of the facility is also required to develop procedures to control contractor entrance and exit into hazardous areas and maintain an injury/illness log for contractors (OSHA 1992).

The obligation of evaluating contractors' safety performance begins before a contracting organization is hired. A contractor's recent injury and illness logs as well as incident reports from significant injuries can be reviewed. Employers seeking contractors should ensure that the contractors have the appropriate skills, certifications, and knowledge to perform the jobs they will be assigned. Past experience with the particular process or chemicals can prove to be useful. The owner or operator must train contractors on any known material or process hazards. This training must be in a language and manner that is understandable to contract employees. The safety of contract workers should also be monitored and evaluated throughout a job. Close interaction between facility employees and contractors is essential. Contractor companies must ensure that their employees are trained in the work practices necessary to perform the tasks they are likely to be assigned. They should ensure that their personnel have received training on hazards and must document that it was understood and that contract employees follow the safety rules of the facility. Contractors must advise owners/operators of hazards created by their work and any hazards found during their work.

Material Handling

There are a number of safety aspects related to the handling of materials, much of which includes lifting and moving material either manually or with mechanical equipment. These activities can result in numerous types of occupational injuries, such as back strains. This section addresses only those aspects of material handling that are related to process-safety issues, such as knowing the inherent properties of the materials and reducing sources of ignition. Readers should refer to the "Ergonomics" section in this handbook for additional information on the ergonomics issues of material handling.

In *Guidelines for Engineering Design for Process Safety*, the CCPS states that to safely handle materials, facilities must thoroughly understand their physical and chemical properties. CCPS notes that the general properties, such as boiling point, vapor pressure, and critical pressure and temperature, should be determined and their application thoroughly understood. Other important properties include reactivity, flammability, toxicity, and stability. Once these properties are determined, safe-handling procedures should be developed (CCPS 1993, 56–61).

Of special concern in material handling is avoiding ignition sources. Such sources should be avoided during the design of the facility as much as possible and controlled through the permitting system. However, additional protections are needed to prevent the buildup of static electricity. It can occur due to product flow in piping, particulates passing through conveyors, filling operations in containers, and personnel wearing nonconductive shoes. Personnel grounding and the use of antistatic footwear should be considered where appropriate. Equipment should be grounded, and facilities should be aware that static can accumulate in various media, such as plastic-lined pipes. Another strategy for avoiding static discharge is allowing relaxation time for charge bleed-off. A study should be conducted to determine whether static electricity is a hazard in a particular process. Brush discharges should be avoided, as should splash filling into barrels and tanks. Caution should be exercised during handling of solids, as small particles may ignite.

Facility Security

Since the terrorist attacks of September 11, 2001, federal and state governments have been concerned about installations that could potentially become targets for those wishing to harm the country's citizens. Chemical and manufacturing facilities qualify as such facilities because they often contain hazardous materials that can impact workers on site and the surrounding community if the materials are released through acts of sabotage (CCPS 2002, 1). This concern is prevalent in all sectors of the chemical, petrochemical, manufacturing, transportation, and storage industries. A study of vulnerabilities should be conducted at all chemical, paper production, weapons production, waste and water treatment, and pharmaceutical facilities, as well as at any other installation that contains material or equipment that could be used for malevolent purposes by an adversary. In 2006, the federal government established a requirement that certain chemical facilities must conduct vulnerability assessments.

Often, facility security is equated with the physical security at the perimeter of the building or unit which is designed to keep persons without a need from entering the processing units. However, a comprehensive security-vulnerability analysis involves a number of steps including assessing the threat, vulnerability, target attractiveness, likelihood of adversary success, countermeasures, and emergency response in the event that an event does occur. Good practices guidelines are available from a number of sources. Most of the guidance presented in this section comes from the American Chemistry Council (ACC) and the Center for Chemical Process Safety (CCPS) and deals largely with the threat of terrorism. However, the techniques that are used to analyze this threat can be used to address other security issues. Excellent guidance is also available from organizations such as the National Safety Council (NSC), which notes that security issues include prevention of drugs in the workplace, fraud, liability, theft, and violence. As this subject continues to be an evolving aspect of safety, practitioners should stay abreast of updates in the concept of plant security and in mandates from federal, state, and local governments.

Facility Characterization

Facility characterization involves determining assets, hazards, and potential consequences for a particular installation This information can be developed as a separate step in the analysis or in conjunction with the threat and vulnerability assessments (CCPS 2002, 43).

Critical Assets Identification

The team analyzing security vulnerability should identify assets for the installation (CCPS 2002, 49). Assets include both material and nonmaterial items that allow a facility to operate and could be used for malevolent purposes by adversaries. Examples of critical assets include

- chemicals that are processed, stored, generated, or transported

- storage tanks, process vessels, and interconnected piping
- the process-control system
- operating personnel
- sophisticated machinery that can be used for multiple purposes
- raw materials and finished product
- utilities and waste treatment facilities
- business information and management systems
- community and customer relations.

Hazard Identification

The security-analysis team must identify and understand the hazards of the assets at the facility (CCPS 2002, 50). Development of a potential target list should include the following items:

- *Highly hazardous chemicals:* These materials include all raw, intermediate, and finished materials. The team must document not only the presence of such materials, but also the locations, concentrations, and states of the chemicals. Also, any particular hazards associated with a chemical—such as the ability to be a chemical weapons precursor of an inhalation poison—should be noted. Lists of highly hazardous chemicals, such as those in OSHA 29 CFR 1910.119, EPA 40 CFR Part 68 (1992), and on the U.S. Chemical Weapons Convention Web site (www.cwc.gov) can be referenced to determine whether there are chemicals of concern at the site. Additionally, all flammables, corrosives, environmental damagers, carcinogens, and explosives must be included.
- *Safety information:* All information on the assets of the facility, such as the design basis for equipment, plot plans, flow diagrams, and hazard studies, should be collected for the evaluation. Information on the population and sensitive environmental receptors in the surrounding community should be included as well.

Consequence Analysis

Similar to the way a hazard evaluation for accidental events focuses on the potential consequences, a security vulnerability analysis must identify the potential consequences of a successful attack on the assets of an installation. Different scenarios must be considered to accurately identify consequences. Scenarios developed in conjunction with the EPA Risk Management Plan (EPA 2004) can be used for guidance on the release of material. For instance, in the case of the release of a toxic chemical to the atmosphere, the atmospheric concentrations of concern can be estimated at various distances downwind. Also the potential consequences to a neighboring population or the environment must be assessed. If the situation has to do with the release and subsequent ignition of an explosive material, potential effects of overpressure and radiant heat must be estimated. If the scenario of concern is the theft of a chemical or piece of sophisticated machinery, the consequences to others from future misuse must be considered. If the problem of a cyber-attack is to be evaluated, the consequences would include loss of production or the sudden shutdown of a process.

Attractiveness of the Target

One of the important considerations in assessing the threat to an installation is how attractive the target is to adversaries. Examples of considerations include the proximity of the installation to a symbolic target such as a national landmark, a high corporate profile among terrorists, and proximity to large populations (CCPS 2002, 54).

Layers of Protection

To develop reasonable recommendations for countermeasures, the analysis team must determine the safeguards already in place to prevent an incident. These include physical security, cyber-security, administrative controls, and other safeguards (CCPS 2002, 55).

The information collected during the facility characterization phase can be used during the subsequent steps to identify a list of specific potential targets and the vulnerability of each.

Threat Identification

An assessment of the threats to a facility includes identifying potential sources of threats, types of threats, and their likelihood. This step involves identifying adversaries and investigating their intentions and capabilities. The information collected during the facility characterization step will help to inform the analysis team during this phase (CCPS 2002, 55–60).

The list of potential threats to industrial facilities includes

- release of hazardous materials to cause a fire, explosion, or dispersion of toxins
- theft of hazardous chemicals or confidential information
- damage to the infrastructure of the installation
- contamination of products
- vandalism of equipment
- cyber-attack
- disabling of safety and security systems.

These acts may be carried out by internal or external adversaries who can be terrorists (foreign or domestic), criminals, violent activists, or disgruntled employees. All of these groups represent threats that must be evaluated and addressed. However, the major threat to consider is external adversaries, such as terrorists, who intend to inflict a large number of casualties. This threat should be considered first. An especially dangerous situation is one in which someone with inside knowledge of all the materials and processes on site is working in collusion with external terrorists.

The information collected during this phase, along with the information collected during the facility characterization phase, will allow a facility to assess the threat to its installation. Such information includes a list of all possible adversaries; assessment of their capabilities; information about the materials on site, including their quantities, toxicities, and locations; and information about existing security measures.

Vulnerability Analysis

There are two approaches to assessing the vulnerability of a facility—the *asset-based approach* and the *scenario-based approach*. One advantage of the asset-based approach is that it is generally less labor intensive than the scenario-based method.

In the asset-based approach, a list of critical assets, or potential targets, is developed. This may be accomplished by breaking the facility down into zones and considering the assets in each zone. The critical assets may include material that is released or systems that will affect the operation of a company, such as computer systems or irreplaceable equipment. Much of this information may already have been gathered during the threat-assessment and consequence-analysis steps of the process. Then the organization considers what the worst-case scenario is for the loss or damage of each asset, along with protective layers in place to prevent loss or damage of the asset. Such layers include security, lighting, and barriers. Where layers are not deemed sufficient to prevent an attack, the target may be considered vulnerable, and the team should develop a list of such targets. Targets can then be characterized as very high, moderate, or very low risk by using tools such as a three-by-three matrix that considers both severity and likelihood. Of particular concern are high-value targets that are vulnerable. These so-called high-value, high-payoff targets are listed on the priority scale as very high.

The other approach is the scenario-based method of vulnerability analysis. While this approach generally requires more time and effort, it has the advantage of producing recommendations for countermeasures that are more cost effective, since they are tailored to the scenarios that are developed. The analysis team collects information from on-site inspections and interviews to develop a list of potential targets. The analysis for this method is similar to the asset-based method in that the assets are identified and potential consequences are determined. However, in the scenario-based approach, the analysis is carried further to determine how a target might be attacked. For example, the team considers the possibility that material in a storage tank can be released, but also considers how an attacker may accomplish this task, such as driving a truck into the tank. Once the list of scenarios is developed, protective layers to prevent each scenario are considered and the team rates the likelihood of the success of an attack as high, medium, or low. The team reviews all scenarios and then determines which ones are representative and should be analyzed further (CCPS 2002, 60–68).

Once the scenarios from either method are developed, they should be ranked in order from highest consequence to lowest consequence. The scenarios can be ranked on a simple numerical scale (e.g., 1–5), or a risk matrix can be used to rank each scenario based on likelihood and consequences.

Identifying Countermeasures

The final step in the security vulnerability analysis is to consider the adequacy of existing countermeasures to manage risk and recommend additional countermeasures when there are shortcomings. Countermeasures can be proposed that deter an attack, detect an attack if it occurs, or delay an attack until an intervention can occur. The team should determine whether there are ways to reduce the profile of the facility or apply the principles of inherently safer design. As with recommendations to prevent accidents, recommendations to prevent terrorist actions should be feasible and written so that individuals who were not part of the analysis team are able to clearly understand the proposed action. Cost-benefit analysis can be conducted on recommendations based on the potential likelihood and consequences of scenarios.

In their paper "Site Security for Chemical Process Industries," Gupta and Bajpai note that countermeasures can be proposed that address vulnerabilities in a number of areas (Gupta and Bajpai, 2005, 301–309). Some examples are shown in Figure 7.

Special Hazards

Much of this chapter has dealt with addressing the general hazards presented by flammable and toxic materials. There are additional groups of materials that, due to the insidious nature of the hazards they present, require specific approaches to minimize risks. These approaches may include special considerations during hazard assessment and design or additional procedural safeguards. In this section we will address three of these classes of materials: reactive materials, asphyxiants, and combustible dusts.This section borrows heavily from work conducted by the United States Chemical Safety and Hazard Investigation Board. While this section is not intended to be a comprehensive list of special hazards, these materials have been responsible for a number of deaths and injuries over the years, and a body of knowledge has been accumulated regarding good practices.

Reactive Chemicals

Reactivity can be defined as the tendency of a material (or combination of materials) to undergo chemical changes under the right conditions (CCPS 2001d, 2). Materials can be reactive by themselves—for example, unstable materials that decompose when exposed to heat. Another example is reactive interactions that occur when stable materials are combined to produce products that react to give off heat, pressure, or a volume of vapor. Reactive chemicals have been responsible for some of the highest-profile accidents in the chemical industry, such as the reaction between methyl isocyanate and water in Bhopal, India, in 1984. The Chemical Safety and Hazard Investigation Board (CSB) conducted a study of reactive chemical incidents and

Vulnerable System	*Potential Countermeasures*
Information and cyber security	• Provide adequate physical security and control access to computer and server rooms • Protect networks with firewalls and password controls
Physical Security	• Improve perimeter fencing and ensure proper lighting • Restrict movement of vehicles within the plant
Policies and Procedures	• Establish strict procedures for visitor and contractor entry into the facility • Create a program to survey surrounding area • Conduct background investigation on all employees • Train employees and contractors on their duties in addressing emergencies, bomb threats, hostage situations, etc. • Encourage employees and contractors to report unknown personnel, unidentified vehicles and packages

FIGURE 7. Vulnerability systems and potential countermeasures (*Source:* Gupta and Bajpai 2005)

found that 165 such incidents between 1980 and 2002 resulted in an average of five deaths each year.

In the safety alert *Reactive Material Hazards: What You Need to Know*, the CCPS notes that the determination of what action needs to be taken to address reactive hazards at chemical and manufacturing facilities starts with an analysis that answers four questions:

1. How do we handle reactive materials?
2. Can we have reactive interactions?
3. What data do we need to control these hazards?
4. What safeguards do we need to control these hazards? (CCPS 2001d, 2)

The first step in this analysis is to determine whether a facility contains any materials that are intrinsically reactive, independent of any interactions with other materials. Material safety data sheets (MSDSs) are a good starting point; however, because the information on them varies, a good practice is to consult multiple MSDSs from a variety of sources. Descriptions such as *unstable, polymerizing, pyrophoric, water-reactive,* and *potential to be an oxidizing agent* should be listed on MSDSs to alert practitioners to such hazards. Other useful references include *Sax's Dangerous Properties of Industrial Materials* (Lewis 2004), NFPA 704 (2007e) and the U.S. Department of Transportation's *Emergency Response Guidebook* (DOT 2004).

The next step is to determine whether there is a potential for dangerous reactions due to the interaction between materials that may come into contact with each other. Consideration must be given to materials that may be intentionally present but in the wrong concentrations (such as charging too much catalyst) as well as those that may be present by accident. Also, varying conditions, such as temperature and pressure, should be considered, since these changes can alter the behavior of materials. This part of the process commences with determining what materials are on site and then determining which ones can react violently when mixed. The likelihood of such mixing can then be determined and necessary safeguards described. Once all the materials on site are inventoried, the consequences of their various interactions can be considered with tools such as compatibility charts. In a simple compatibility chart, a matrix is created in which all the materials on site are listed on the horizontal and vertical axes, and then their interactions are considered and recorded. While creating a compatibility chart in a matrix format is a relatively straightforward method, it does have limitations. The chart is only as valuable as the quality of the sources used to determine reactivity; multiple sources should be considered when possible. Practitioners should be aware that the chart considers only mixtures of two materials, whereas hazardous interactions may not occur until three or more chemicals combine. Air, water, oil, and other materials that may inadvertently be present (such as after maintenance) should be included. Other tools to assist with assessing chemical interactions include the Chemical Reactivity Worksheet, available from the U.S. National Oceanic and Atmospheric Administration (NOAA 2006) and the American Society for Testing and Materials (ASTM) *Standard Guide for Preparation of Binary Chemical Compatibility Chart and Materials* (ASTM 2000).

Once it is determined that there is an intrinsic or interaction-reactive hazard, the next step is to compile the information necessary to determine whether a facility is protected against a reactive incident. Such data includes compatibility with different materials of construction, contaminants to avoid, proper spill and emergency response procedures, and other special considerations. Some of this information may be available from MSDSs, technical bulletins from chemical suppliers, or other data sources. Some information may require chemical testing. Heat-of-mixing information can often be used to determine how much heat or gas will be given off from a chemical interaction.

When appropriate information has been gathered, the final step is to identify and implement safeguards. The principles of inherent safety, which were discussed earlier in this chapter, can decrease the risks associated with reactive materials. Eliminating such materials as feasible and reducing inventories can decrease the chances of a reactive chemical incident. Also, good practices from the material supplier and codes and standards should be referenced and implemented. Multiple safeguards should be in place to prevent catastrophic reactive incidents. Compatible materials of construction and adequate emergency venting must be part of the design strategy. Storage of such materials should be remote from operating areas if possible, and positive separation strategies for incompatible materials must be employed. Dedicated fittings can be used to ensure that incompatible materials are not inadvertently mixed. Reactive materials must be monitored constantly for temperature and pressure excursions. Storage and process areas must have appropriate fire protection, and facilities must consider the possibility of a reactive chemical incident in their emergency response plans.

Practitioners should be aware that handling practices can often play a role in these types of incidents, and that good practices must be followed during all phases of operation, including when a processing unit is shut down. The CSB investigated an incident in Pascagoula, Mississippi, that involved mononitrotoluene (MNT), a material that reacts when heated (CSB 2003a, 11). The material had been left in a distillation column while the unit was shut down, but only a single manual valve was closed to isolate the steam supply from the column. Steam leaked through the valve and the material inside the column reacted energetically, destroying the column and propelling shrapnel off site and onto the property of neighboring facilities. This incident speaks to several issues involved in managing reactive chemical hazards. If the MNT had been evacuated from the column while it was shut down, if the steam supply had been better isolated (perhaps isolated automatically based on the temperature in the column), or if pressure relief had been adequate, the incident could have been prevented.

Asphyxiants

Asphyxiants present the hazard of displacing oxygen in the lungs, thereby essentially smothering victims. Asphyxiants are not poisons in the sense that chemicals like hydrogen sulfide are, but in elevated levels, they can be just as deadly. A variety of materials used in chemical and manufacturing facilities, such as carbon dioxide and steam, can act as asphyxiants. However, one of the most common scenarios involves nitrogen asphyxiation, and it will be discussed in more detail in this section. The good practices that prevent nitrogen asphyxiation also can be applied to other asphyxiants.

Nitrogen asphyxiation is a particularly insidious hazard because nitrogen is present in the air around us, so its risks are not always fully appreciated. Nitrogen is also used in chemical and manufacturing plants to make the facilities safer. Nitrogen is used to keep equipment free of contaminants through inerting and to purge toxic materials from equipment prior to opening. However, when elevated levels of nitrogen enter the breathing zone, victims can quickly become incapacitated. The CSB conducted a study of the issue and determined that in the ten-year period between 1992 and 2002, nitrogen asphyxiation incidents accounted for an average of eight deaths each year. The CSB found that the majority of victims were working in and around confined spaces. Specific causes of nitrogen asphyxiation incidents included failure to detect an oxygen-deficient atmosphere, mistaking nitrogen for breathing air, and attempted rescue involving entering spaces that contained elevated levels of nitrogen without proper breathing protection (CSB 2003b, 3–5).

Good practices must be employed to prevent incidents of nitrogen asphyxiation. A continuous monitoring system must be used in and around confined spaces that personnel will enter to ensure that the atmosphere is fit for breathing upon entry and that it does not change over time. The entire space should be monitored to the extent practicable, not just the entry portal. Protective systems such as alarms and auto-locking entryways that prevent access can be used to warn workers of hazardous atmospheres. Personnel monitors can also measure oxygen concentration and activate audible or vibration alarms. Fresh-air ventilation should be maintained in spaces where personnel will enter, not just at the commencement of jobs, but throughout them as well. Ventilation systems must be properly designed and maintained.

Rescue methods should be planned before personnel enter confined areas. Harnesses and lifelines must be attached to personnel so that they can be retrieved quickly from such areas. This approach benefits the potential rescuers as well since they will probably not have to enter the confined area to retrieve personnel inside. Personnel should attempt rescue only if they are appropriately trained, have the correct rescue equipment, and have a dependable source of breathing air. The CSB study showed that approximately 10 percent of the fatalities in such incidents happened to personnel attempting rescue (CSB 2003b, 5).

When workers are using supplied air, the integrity of the air source must be preserved, and the air must be protected from interruption. Air compressors should have alternate sources of power, and the air supply should be continuously monitored. Air hoses must be routinely inspected and replaced, and vehicular traffic should be restricted in the area of supply hoses. Escape packs worn by workers allow them an additional five to ten minutes of breathing air in the event of failure of the primary source. Management systems must be in place to prevent the mix-up of nitrogen and breathing air. Special incompatible fittings must be used for cylinders containing nitrogen and breathing air. Personnel should understand that the fittings are in place for a reason and that adaptors defeating their purpose will not be tolerated. Cylinders should be clearly labeled, and color-coding helps to identify systems. An incident occurred at a nursing home when a nitrogen cylinder was mistakenly delivered with a batch of oxygen cylinders. Even though the nitrogen cylinder had nitrogen-compatible fittings, an employee reportedly removed a fitting from an empty oxygen cylinder and used it as an adaptor to connect the nitrogen cylinder to the oxygen systems. Four deaths occurred as a result of pure nitrogen being delivered to patients (CSB 2003b, 6).

Finally, a comprehensive training program must be implemented (CSB 2003b, 8–9). Good practices are effective only if personnel are adequately trained. Personnel must be trained on appropriate confined-space entry procedures. They should also be trained on atmospheric monitoring systems—both how to use them and how to tell if they are not working properly. Safe handling of air and nitrogen delivery systems should be stressed as well as precautions to take when working around equipment that may contain elevated levels of nitrogen. Personnel should also be trained on rescue/retrieval systems and understand the warning not to enter atmospheres that may be hazardous unless they are properly trained and equipped. Finally, training should cover new and revised procedures for entering confined spaces. Contractors as well as employees should be trained, and personnel should be trained in a language and using a method that they can comprehend.

Combustible Dusts

Dust in high concentrations can be explosive. This hazard is well known in some industries, such as agriculture and coal mining, but has not been widely recognized in many manufacturing sectors. A common scenario with dust explosions is that some initiating event lofts accumulated dust into the air, and it explodes upon contact with an ignition source. Facilities that use powdery raw materials or slurries containing such materials are at risk for dust explosions. Since the lofting of accumulated dust is a common element in dust explosions, the design, operation, and cleaning practices at facilities must act collectively to minimize such accumulations. The principles of inherent safety should be applied if possible to select the least-hazardous materials. Less-hazardous materials include those with larger particulate sizes, higher minimum ignition energies, and higher K_{st} values. (***Note:*** K_{st} is the maximum rate of pressure rise normalized to a 1.0 cubic meter volume and is a measure of explosion severity.) If possible, dust should be handled in closed systems that do not allow an accumulation of material. If this is not possible, surfaces where material could accumulate, such as I-beams, should be minimized. Fire walls, blast-resistant construction, and deflagration venting should be installed in production areas.

Production areas where dust can accumulate should be cleaned frequently, including areas above production lines. It is important to note that dust can accumulate in several areas not readily visible from floor level. Beams, conduit lines, and false ceilings are all areas that should receive frequent attention. Tools that disperse dust, such as compressed air tools, should be avoided. Workers must be trained in the hazards of combustible dusts so that they recognize hazardous situations.

Dust explosions have been the cause of a number of catastrophic incidents in the manufacturing sector. Design, operation, maintenance, and cleaning strategies should be employed to minimize the risks. NFPA 654, *Standard for the Prevention of Fire and Dust*

Explosions from the Manufacturing, Processing, and Handling of Combustible Particulate Solids, contains essential guidance on the subject (NPFA 2006d).

References

American Petroleum Institute (API). 1991. RP 2202, *Dismantling and Disposing of Steel from Above-ground Leaded Gasoline Storage Tanks*. 3d ed. Washington, DC: API.

______. 1998. API 570, *Piping Inspection Code: Inspection, Repair, Alteration, and Rerating of In-service Piping Systems*. 2d ed. Washington, DC: API.

______. 2000. RP 576, *Inspection of Pressure-Relieving Devices*. 2d ed. Washington, DC: API.

______. 2001. Std 653, *Tank Inspection, Repair, Alteration, and Reconstruction*. 3d ed. Washington, DC: API.

______. 2002a. API 620, *Design and Construction of Large, Welded, Low-Pressure Storage Tanks*. 10th ed. Washington, DC: API.

______. 2002b. RP 2009, *Safe Welding, Cutting, and Hot Work Practices in the Petroleum and Petrochemical Industries*. 7th ed. Washington, DC: API.

______. 2003a. RP 752, *Management of Hazards Associated With Location of Process Plant Buildings*. 2d ed. Washington, DC: API.

______. 2003b. RP 2201, *Procedures for Welding or Hot Tapping on Equipment in Service*. 5th ed. Washington, DC: API.

American Society of Mechanical Engineers (ASME). 1999. ASME B31.3, *Process Piping*. New York: ASME

American Society for Testing and Materials (ASTM). 2000. E2012-99, *Standard Guide for Preparation of Binary Chemical Compatibility Chart*. West Conshohocken, PA: ASTM

Center for Chemical Process Safety (CCPS). 1993. *Guidelines for Engineering Design for Process Safety*. New York: AIChE.

______. 1994. *Guidelines for Evaluating the Characteristics of Vapor Cloud Explosions, Flash Fires, and BLEVEs*. New York: AIChE.

______. 1996. *Guidelines for Evaluating Process Plant Buildings for External Explosions and Fires*. New York: AIChE.

______. 2001a. *Essential Practices for Managing Chemical Reactivity Hazards*. New York: American Institute of Chemical Engineers (AIChE).

______. 2001b. *Guidelines for Safe Process Operations and Maintenance*. New York: AIChE.

______. 2001c. *Layer of Protection Analysis: Simplified Process Risk Assessment*. New York: AIChe.

______. 2001d. *Reactive Material Hazards: What You Need to Know*. New York: AIChE.

______. 2002. *Guidelines for Analyzing and Managing the Security Vulnerabilities of Fixed Chemical Sites*. New York: AIChE.

______. 2003. *Guidelines for Fire Protection in Chemical, Petrochemical, and Hydrocarbon Processing Facilities*. New York: AIChE.

Chemical Safety and Hazard Investigation Board (CSB). 2001. Report No. 99-014-I-CA, *Refinery Fire Incident, Tosco Avon Refinery, Martinez, CA*. (March) Washington, DC: CSB.

______. 2002a. Report No. 2001-05-I-DE, *Refinery Incident, Motiva Enterprises LLC, Delaware City, DE*. (October) Washington, DC: CSB.

______. 2002b. Report No. 2001-03-I-GA, *Thermal Decomposition Incident, BP Amoco Polymers, Inc. Augusta, GA*. (June) Washington, DC: CSB.

______. 2003a. Report No. 2003-01-I-MS, *Explosion and Fire, First Chemical Corporation, Pascagoula, MS*. (October) Washington, DC: CSB.

______. 2003b. Report No. 2003-10-B, *Hazards of Nitrogen Asphyxiation*. (June) Washington, DC.: CSB.

Cote, Arthur E. ed. 2003. *Fire Protection Handbook*. vols I and II. Quincy, MA: National Fire Protection Association.

Department of Transportation (DOT). 2004. *Emergency Response Guidebook*. Washington, DC: DOT.

Environmental Protection Agency (EPA). 1986a. Emergency Planning and Community Right-to-Know Act. 42 USC 116. Washington, DC: EPA.

______. 1986b. Superfund Amendment and Reauthorization Act f 1986. 42 USC 103. Washington, DC: EPA.

______. 1990. Clean Air Act Amendments of 1990. P.L. 101–549. Washington, DC: EPA.

______. 2004. *Chemical Accident Prevention Program; Risk Management Plan*. 40 CFR Part 68, Subpart G. Washington, DC: EPA.

Gupta, J.P. and Bajpai, S. 2005. "Site Security for Chemical Process Industries." *Journal for Loss Prevention in the Process Industries*. (Jul–Nov) 18(4–6):301–309.

Industrial Risk Insurers (IRI). 1991. *IRInformation Manual. 2.5.2: Plant Layout and Spacing for Oil and Chemical Plants*. Hartford, CT: IRI.

Instrumentation, Systems, and Automation Society (ISA). 2004. ANSI/ISA-TR84.00.01-2004. *Standard—Safety Instrumented Systems for the Process Industry Sector Parts 1 and 2*. Research Triangle Park, NC: ISA.

Lees, Frank P. 1996. *Loss Prevention in the Process Industries*. 2d ed. Oxford, UK: Reed Educational and Professional Publishing Ltd.

Lewis, R. 2004. *Sax's Dangerous Properties of Industrial Materials*. 11th ed. New York: John Wiley and Sons, Inc.

Lindeburg, Michael R. 1995. *Fire and Explosion Protection Systems: A Design Professional's Introduction*. 2d ed. Belmont, CA: Professional Publications, Inc.

National Fire Protection Association (NFPA). 2002. NFPA 25, *Standard for the Inspection, Testing, and Maintenance of Water-Based Fire Protection Systems*. Quincy, MA: NFPA.

______. 2003. NFPA 30, *Flammable and Combustible Liquids Code*. Quincy, MA: NFPA.

______. 2004. NFPA 12A, *Standard on Halon 1301 Fire Extinguishing Systems*. Quincy, MA: NFPA.

______. 2005a. NFPA 11, *Standard for Low, Medium, and High-Expansion Foam*. Quincy, MA: NFPA.

______. 2006a. NFPA 1, *Uniform Fire Code*. Quincy, MA: NFPA.

______. 2005b. NFPA 12, *Standard on Carbon Dioxide Extinguishing Systems*. Quincy, MA: NFPA.

______. 2006b. NFPA 101, *Life Safety Code*. Quincy, MA: NFPA.

______. 2006c. NFPA 220, *Standard on Types of Building Construction*. Quincy, MA: NFPA.

______. 2006d. NFPA 654, *Standard for the Prevention of Fire and Dust Explosions from the Manufacturing, Processing, and Handling of Combustible Particulate Solids*. Quincy, MA: NFPA.

______. 2007a. NFPA 13, *Standard for the Installation of Sprinkler Systems*. Quincy, MA: NFPA.

______. 2007b. NFPA 15, *Standard for Water Spray Fixed Systems for Fire Protection*. Quincy, MA: NFPA.

______. 2007c. NFPA 16, *Standard for the Installation of Foam-Water Sprinkler and Foam-Water Spray Systems*. Quincy, MA: NFPA.

______. 2007d. NFPA 72, *National Fire Alarm Code*. Quincy, MA: NFPA.

______. 2007e. NFPA 704, *Standard System for the Identification of the Hazards of Materials for Emergency Response*. Quincy, MA: NFPA.

National Oceanic and Atmospheric Administration (NOAA), Office of Response and Restoration. 2006. *Chemical Reactivity Worksheet*. Version 1.7. (February) Washington, DC: NOAA.

National Safety Council (NSC). 1997. *Supervisors' Safety Manual*. 9th ed. Itasca, IL: NSC.

______. 2001. *Accident Prevention Manual: Engineering and Technology*. 12th ed. Itasca, IL: NSC.

Occupational Safety and Health Administration (OSHA). 1992. 29 CFR 1910, *Standards for General Industry*. Washington, D.C.: OSHA.

______. 1996a. 29 CFR 1910.119, *Process Safety Management of Highly Hazardous Chemicals*. Washington, DC: OSHA.

______. 1996b. 29 CFR 1910.147, *The Control of Hazardous Energy (Lockout/Tagout)*. Washington, DC: OSHA.

______. 1998a. 29 CFR 1910.146, *Permit-Required Confined Spaces*. Washington, DC: OSHA.

______. 1998b. 29 CFR 1910.252, *Welding, Cutting & Brazing; General Requirements*. Washington, DC: OSHA.

Wallace, S. 1994. "Optimize Facility-Siting Evaluations." *Hydrocarbon Processing: A Journal of Gulf Publishing* (May) 73(5):85–96.

______. 1999. "Take Action to Resolve Safety Recommendations." *Chemical Engineering Progress: A Monthly Journal of the American Institute of Chemical Engineers* (March) 99(3):67–71.

______. 2000. "Catching Near Hits." *Professional Safety*. 11:30–34.

______. 2001. "Using Quantitative Methods to Evaluate Process Risks and Verify the Effectiveness of PHA Recommendations." *Process Safety Progress: A Quarterly Journal of the American Institute of Chemical Engineers* (March) 20(1):57–62.

Wallace, S. et. al. 2003. "Know When to Say "When": A Review of Safety Incidents Involving Maintenance Issues." 2003. *Process Safety Progress: A Journal of the American Institute of Chemical Engineers* (December) 22(4):212–219.

Appendix: Best Practice Resources

Following is a list of some sources that may be beneficial to practitioners. This list is not meant to be comprehensive; however, it presents a number of organizations that develop important standards and guidance on a variety of subjects as well as useful books and pamphlets. Some Web-site addresses are provided, although organizational names and contact information are subject to change and it is not possible to guarantee the continuing accuracy of this list. Much of this information is taken from the organization's promotional literature and is not meant to endorse any particular organization but rather to raise awareness of the mission of each one. Practitioners are encouraged to ensure that the reference cited is the appropriate one for their purposes.

American Petroleum Institute (API), www.api.org

The American Petroleum Institute is a trade organization representing the oil and natural gas industry in the United States. It is based in Washington, DC. It develops a variety of standards for the industry (available for purchase) that can also be used for other industries. It has developed a number of standards regarding safe maintenance and design of equipment.

Center for Chemical Process Safety (CCPS), www.aiche.org/ccps

The Center for Chemical Process Safety has published several books to assist practitioners in the field of process safety, and a number of those publications were referred to throughout this chapter. One of the publications, *Guidelines for Engineering Design for Process Safety* (1993), focuses specifically on design for process safety. Other publications focus on analyzing appropriate layers of protection, building design and location, and maintenance.

American Society of Safety Engineers (ASSE), www.asse.org

The American Society of Safety Engineers is a professional organization dedicated to assisting safety professionals. Its members manage, supervise and consult on safety, health, and environmental issues in industry, insurance, government, and academia. ASSE has specialty divisions and a number of chapters around the country. ASSE is responsible for the creation of this reference handbook and numerous safety publications and is the ANSI Secretariat for many national standards. The society's headquarters are in Des Plaines, Illinois.

American Chemistry Council (ACC), www.americanchemistry.com

The American Chemistry Council, formerly the Chemical Manufacturers Association, is a major trade organization in the chemical industry. Based in Arlington, Virginia, the council is committed to improved environmental, health, and safety performance through Responsible Care, commonsense advocacy designed to address major public policy issues, and health and environmental research and product testing.

American Society for Testing and Materials (ASTM), www.astm.org

ASTM International was originally known as the American Society for Testing and Materials. It is one of the largest voluntary standards-development organizations in the world, and it has developed standards that provide guidance on design and manufacturing. ANSI standards developed at ASTM are the work of ASTM members, including technical experts representing producers, users, consumers, government, and academics from several countries.

National Fire Protection Association (NFPA), www.nfpa.org

The National Fire Protection Association is the premier organization in the United States for publishing standards regarding fire protection and emergency response. The association publishes various standards, including those regarding sprinkler systems, storage tank spacing, and the uniform fire code. The association is also responsible for the *Fire Protection Handbook* (Cote 2003). NFPA holds several seminars throughout the year. Its information can be purchased.

Instrumentation, Systems, and Automation Society (ISA), www.isa.org

The Instrumentation, Systems, and Automation Society is an educational organization that fosters advancement in the theory, design, manufacture, and use of sensors, instruments, computers, and systems for automation.The society hosts conferences to promote the subject of automation and publishes a number of books, magazines, and standards. Of particular interest to safety practitioners is ISA-84, which can be used to establish safety integrity levels for instruments in hazardous-material services.

National Safety Council (NSC), www.nsc.org

The National Safety Council is dedicated to influencing society to adopt safety, health, and environmental policies, practices, and procedures that prevent accidents. The council is located in Itasca, Illinois, and has several chapters throughout the United States. NSC is responsible for several safety publications, including the *Supervisors' Safety Manual* (NSC 1997).

American Welding Society (AWS), www.aws.org

The American Welding Society, or AWS, is a nonprofit organization whose goal is advancing the technology and application of welding and related disciplines. The society produces a number of standards related to welding, cutting, and brazing.

National Association of Corrosion Engineers (NACE), www.nace.org

NACE International, originally known as the National Association of Corrosion Engineers, is dedicated to the study of corrosion issues. NACE has developed several corrosion prevention and control standards and conducts several workshops, and classes each year. It also has a number of technical committees.

Occupational Safety and Health Administration (OSHA), www.osha.gov

The Occupational Safety and Health Administration is a regulatory body dedicated to protecting workers in the United States. OSHA produces regulations and technical assistance bulletins. Of particular interest to safety practitioners regarding process safety is OSHA 1910.119, *Process Safety Management of Highly Hazardous Chemicals* (1972).

Chemical Safety and Hazard Investigation Board (CSB), www.csb.gov

The Chemical Safety and Hazard Investigation Board is a nonregulatory federal agency dedicated to preventing accidents in the chemical industry by investigating incidents that occur and making recommendations to prevent recurrence. The board produces public reports of the incidents and studies they conduct that are available free of charge from their Web site.

American Society of Mechanical Engineers (ASME), www.asme.org

The American Society of Mechanical Engineers is a professional society that sets internationally recognized industrial and manufacturing codes and standards that enhance public safety, including the boiler and pressure vessel code and several codes on designing and maintaining piping.

The Chlorine Institute, www.chlorineinstitute.gov

The Chlorine Institute is dedicated to promoting safe practices in industries that produce and handle chlorine and various other chemicals. The institute produces a number of pamphlets recommending good safety practices. The institute is headquartered in Arlington, Virginia.

Insurance Companies

A variety of companies, such as Factory Mutual and Zurich, insure manufacturing and chemical facilities. Many of these companies develop good practices based on their research and observations, which are often available for purchase but are sometimes considered proprietary.

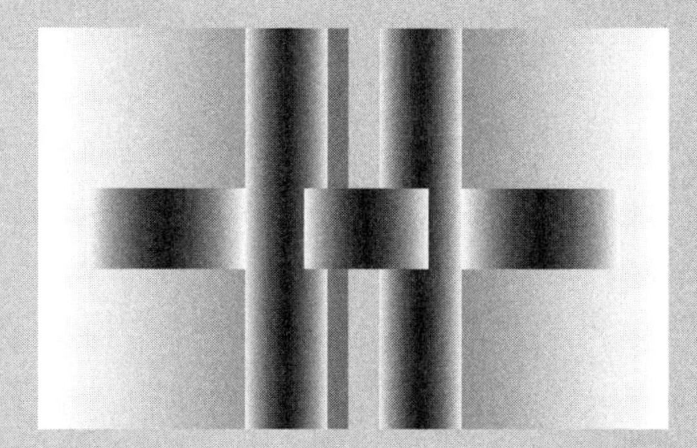

SECTION 2
EMERGENCY PREPAREDNESS

Regulatory Issues

Applied Science and Engineering

Cost Analysis and Budgeting

Benchmarking and Performance Criteria

Best Practices

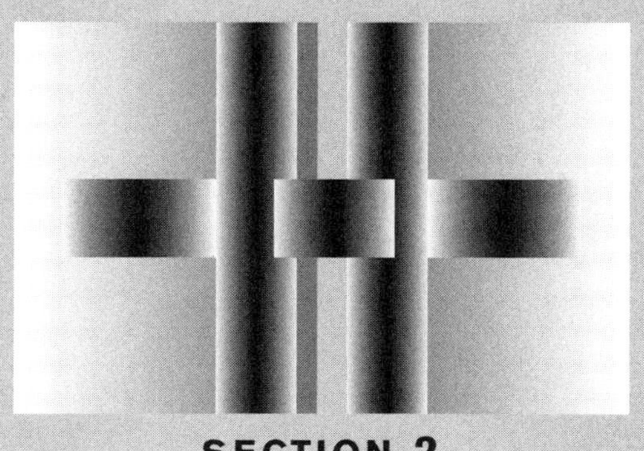

SECTION 2
EMERGENCY PREPAREDNESS

LEARNING OBJECTIVES

- Recognize the benefits of developing an emergency action plan (EAP).
- Be able to list the essential elements of an emergency action plan.
- Know what regulations and guidelines exist concerning emergency response preparedness and where to go to locate them.
- Become familiar with OSHA, EPA, DOT, NFPA, NRC, and FEMA, their roles, and their influence on emergency response preparedness.
- Understand how the community is affected by emergencies and the role of the Local Emergency Planning Committees (LEPC).
- Develop familiarity with the common acronyms used in emergency response regulatory issues.

REGULATORY ISSUES

Jon J. Pina

SINCE THE UNITED STATES suffered terrorist attacks to the World Trade Center and the Pentagon on September 11, 2001, there has been worldwide recognition that no one is immune to man-made or natural disasters. We know emergencies can and do happen, and that we need appropriate parties to respond. The goal of this chapter on preparedness planning is to explain how preparedness planning can lessen the impact of emergency situations on organizations, employees, the environment, and surrounding communities.

Safety professionals, in concert with management, play an important role in the development and implementation of emergency response plans. Many facilities, not just "mom-and-pop" businesses, depend on the local fire departments, police, and emergency medical services. Facilities such as chemical plants, refineries, tank farms, and electrical generating plants are primary targets for terrorists and pose severe risks to local communities that experience major natural disasters or catastrophic process failures.

The actions taken after an unplanned event occurs can affect the number of deaths, the extent of injuries, and the severity of economic losses, property damage, and/or damage to the environment. Therefore, developing a written emergency action plan (EAP) is important in preparing for an unplanned event. This plan could include purchasing the appropriate personal protective equipment (PPE) and maintaining it, selecting key personnel to act as emergency responders, conducting periodic emergency drills, auditing the management system, and updating the plan as needed. Safety, health, and environmental (SH&E) professionals play a major role in the planning of emergency response procedures in the workplace.

Types of Emergencies

Organizations should plan for a variety of emergency situations. Natural disasters include floods, hurricanes, tornados, earthquakes, winter storms, avalanches, forest fires, mudslides, and volcano eruptions. The effects of natural disasters, as such, can cause catastrophic results, such as ruptured pipelines; utility outages; highway accidents; and chemical, biological, or radiological releases. Man-made emergencies might stem from terrorist attacks; bomb scares; workplace violence; sabotage; equipment or process failures; operator errors; and vehicular, railway, or aircraft accidents.

Developing an Emergency Action Plan

It takes a lot of thought and resources to fashion an emergency action plan that covers all possible worst-case scenarios. OSHA's Process Safety Management Program (PSM) and the EPA's Risk Management Program (RMP) are examples of two such plans. They reflect how significant it is for management to emphasize the importance of emergency planning by being aware of emergency numbers to call and knowing who should be the designated responders. Management's support of such programs can increase productivity and improve employee morale while decreasing employee days off and lowering workers' compensation rates.

Even a well-developed EAP can turn out to be "not good enough" when a true emergency situation arises. Odds are high that without a well-developed EAP results can be catastrophic. Therefore, emergency preparation achieved through periodic drills is highly recommended. These drills need to replicate any present scenarios and should also include potential consequences of other exposures that are considered highly probable. The emergency plan can benefit greatly from conducting an appropriate risk assessment to identify exposure levels. The following is an outline of OSHA's emergency action plan (OSHA 2002a and 2002c).

OSHA

I. EAP

A. General Industry 29 CFR 1910.38

Emergencies that May Require a Plant to Be Evacuated:

1. Fires
2. Explosions
3. Toxic material releases
4. Chemical spills
5. Radiological or biological accidents
6. Natural disasters

Minimum Requirements:

1. Emergency escape procedures and emergency route assignments
   a. Clearly show escape routes and refuge areas
   b. Exits must be wide enough so that all personnel can evacuate in a timely manner
   c. Exits must be unobstructed
   d. Exits must not expose employees to additional hazards
   e. Designate a meeting location in a safe area
   f. Employees must be told what actions they are to take in the event of an emergency
2. Procedures followed by those who must stay behind and shut down critical plant operations before evacuating the building
   a. List in detail the duties of each employee who must stay behind to care for essential plant operations
      i. Monitor power supplies
      ii. Monitor plant processes that cannot be safely left unattended
3. Procedure to account for all employees after evacuation
4. Rescue and medical duties for employees who have been designated to perform them
5. Report all fires and other emergencies
6. Names of persons or departments to be contacted for further information or explanation of duties under the plan

Establishing a Chain of Command:

1. Appoint responsible individuals to coordinate the response team
   a. Must appoint a response team coordinator and a back-up coordinator
2. Duties of the response team coordinator
   a. Assess the danger posed by the situation and decide whether or not to activate the planned emergency procedure
   b. Direct all efforts to evacuate the plant in the safest way possible
   c. Ensure that medical aid and the local fire department are called in if necessary
   d. Direct the shutdown of plant operations when necessary

Communication:

1. Designate an alternative area where incoming and outgoing phone calls can be made and received
2. Equipment should be available in the alternate headquarters so that employees can be informed of the situation and the proper authorities can be contacted
3. Alarms must be able to be heard by all people in the building and must be run by a back-up power supply in addition to the regular power supply
4. All employees should be informed of the different ways to report an emergency in the plant
   a. Manual pull box alarms
   b. Public address systems
   c. Telephones with emergency phone numbers written nearby

Training of Emergency Response Teams:

1. Physical capabilities must first be tested to determine if an employee can handle the duties of the team; for example, can the rescuer climb steps to the top floor of a facility without use of the elevator?
2. Use of all the different fire extinguishers in the plant
3. First aid and cardiopulmonary resuscitation (CPR)
4. Procedure for shutting down operations
5. Evacuation procedure
6. Procedure for a chemical spill
7. Use of self-contained breathing apparatus (SCBA)
8. Incipient- and advanced-stage fire fighting
9. Trauma counseling

Training for All Employees:

1. Evacuation plans
2. Alarm systems
3. Reporting procedures for personnel
4. Shutdown procedures
5. Types of potential emergencies

When Training Programs Must Be Provided:

1. When plan is initially developed
2. For all new employees
3. When new equipment, materials, or processes are introduced
4. When procedures have been updated or revised
5. When employee performance must be improved
6. At least annually for portable fire extinguishers for those expected to use them

Hazardous Circumstances During an Emergency:

1. Chemical splashes and exposure to toxic materials
2. Falling and flying objects
3. Unknown atmospheres
   a. Toxic gases
   b. Vapors or mists
   c. Inadequate oxygen
4. Fires and electrical hazards
5. Violence from employees, visitors, or intruders

Safety Equipment/PPE Needed During an Emergency:

1. Goggles or face shields for eye protection
2. Hard hats and boots

3. Proper respirators
   a. Air purifier
   b. Supplied air
   c. SCBA
   d. Escape masks
4. Chemical suits, gloves, hoods

Medical Assistance:

1. In the absence of a medical facility within close proximity, there must be people on site capable of performing first aid
2. Eye-wash stations must be provided if eyes and skin may be exposed to corrosive materials

Security:

1. Off-limits area must be established by roping off the area
2. Notify law enforcement or hire private security to keep unauthorized persons out
3. Protect and secure important records and documents (OSHA 2002a)

OSHA Terrorist Incident Risk Categories:

1. Green Zone
   a. Workplaces not likely to be a target of terrorists
   b. Characterized by limited vulnerability, limited threat, and limited potential of significant impact
   c. If your plant is considered a Green Zone, but there are workplaces around you in a higher risk zone, you may want to adopt the preparations of the higher-risk workplace
2. Yellow Zone
   a. Workplaces that may be targets
   b. These places are characterized by only one of the following
      i. High vulnerability
      ii. High threat
      iii. Potentially significant impact
   c. If there are workplaces considered Red Zones in close proximity to your plant, you may want to take the same precautions that they take
3. Red Zone
   a. Workplaces most likely to be targeted
   b. Characterized by two of the following
      i. High vulnerability
      ii. High threat
      iii. Potentially catastrophic impact

Worksite Terrorist Risk Assessment List:

1. Uses, handles, stores, or transports hazardous materials
2. Provides essential services
   a. Sewer treatment
   b. Electricity
   c. Fuel
   d. Telephone
3. Has high volume of pedestrian traffic
4. Has limited means of egress
   a. High-rise building
   b. Underground operations
5. Has high volume of incoming materials
   a. Mail
   b. Imports/exports
   c. Raw materials
6. Considered a high-profile site
   a. Water dam
   b. Military site
   c. Classified site
7. Is part of the transportation system
   a. Shipyard
   b. Busline
   c. Trucking station
   d. Airline (OSHA 2002a)

B. Construction 29 CFR 1926.35

This section applies to all emergency action plans required by a particular OSHA standard. The emergency action plan shall be in writing and shall cover those designated actions that employers and employees must take to ensure employee safety from fire and other emergencies. It should be noted that an industry classified as "General Industry" may fall under "Construction" if certain activities are performed (example – employees painting facility) or the status of a facility is temporarily changed (example – power plant outage).

The following elements, at a minimum, shall be included in the plan:

1. Emergency escape procedures and emergency escape route assignments

2. Procedures to be followed by employees who remain to operate critical plant operations before they evacuate
3. Procedures to account for all employees after emergency evacuation has been completed
4. Rescue and medical duties for those employees who are to perform them
5. The preferred means of reporting fires and other emergencies
6. Names or regular job titles of persons or departments who can be contacted for further information or explanation of duties under the plan

The employer shall establish an employee alarm system:

If the employee alarm system is used for alerting fire brigade members, or for other purposes, a distinctive signal for each purpose shall be used.

The employer shall establish in the emergency action plan the types of evacuation to be used in emergency circumstances.

Before implementing the emergency action plan, the employer shall designate and train a sufficient number of persons to assist in the safe and orderly emergency evacuation of employees.

The employer shall review the plan with each employee covered by the plan at the following times:

1. Initially when the plan is developed
2. Whenever the employee's responsibilities or designated actions under the plan change
3. Whenever the plan is changed

The employer shall review with each employee, upon initial assignment, those parts of the plan which the employee must know to protect the employee in the event of an emergency. The written plan shall be kept at the workplace and made available for employee review. For those employers with ten or fewer employees, the plan may be communicated orally to employees and the employer need not maintain a written plan (OSHA 2002b).

Written Safety and Health Program for Employees Involved in Hazardous Waste Operations Shall Consist of the Following Components:

1. An organizational structure which includes the following:
   a. General supervisor in charge of all hazardous waste operations
   b. Safety and health supervisor with the responsibility to develop and implement a safety and health plan and verify compliance
   c. General duties and functions of all employees involved in hazardous waste operations
   d. Lines of authority, responsibility, and communication
   e. The organizational structure must be reviewed and altered as needed
2. A comprehensive work plan:
   a. Address the tasks and objectives of the site operations and resources required to reach those tasks and objectives
   b. Address anticipated clean-up activities as well as normal operating procedures
   c. Define work tasks and objectives and identify the methods for accomplishing those tasks and objectives
   d. Establish personnel requirements for implementing the plan
   e. Provide for the implementation of initial training
   f. Provide for the implementation of the required informational programs
   g. Provide for the implementation of the medical surveillance program
3. A site-specific safety and health plan that addresses the following:
   a. A safety and health risk analysis for each site, task, and operation listed in the work plan
   b. Employee training assignments that assure compliance with the requirements for training
   c. Personal protective equipment required by employees for each of the tasks and operations being conducted on the site

   d. Medical surveillance requirements
   e. Frequency and types of air-monitoring, personnel-monitoring, and environmental sampling techniques and instrumentation to be used, including methods of maintenance and calibration of monitoring and sampling equipment to be used
   f. Site-control measures which comply with the site-control program
   g. Decontamination procedures
   h. An emergency response plan for safe and effective responses to emergencies, which includes the proper PPE as well as other necessary equipment
   i. Confined space entry procedures
   j. A spill containment program
   k. Provide preentry briefings which are to be held before initiating any site activities and at other times as they are needed to ensure employees are well acquainted with the site safety and health plan
   l. Inspections shall be conducted to grade the effectiveness of the plan
4. A safety and health training program
5. A medical surveillance program
6. The employer's standard operating procedures for safety and health
7. Any necessary interface between general program and site-specific activities

The Program Shall Be Made Available to the Following People and Parties:

1. Contractors/subcontractors
2. Employees or employee representatives
3. OSHA personnel
4. Federal, state, and local agencies with regulatory authority over the site

Monitoring:

1. Monitoring shall be conducted if the site evaluation indicates the potential for the following conditions:
   a. Ionizing radiation or immediately dangerous to life and health (IDLH)
   b. Combustible atmospheres
   c. Oxygen deficiency
   d. Toxic substances
   e. If site evaluation produces sufficient evidence to rule these conditions out
2. Monitoring devices
   a. O_2/LEL (lower explosive level–combustible gas) meters
   b. Volatile organic compound (VOC) meters
   c. CO (carbon monoxide) and H_2S (hydrogen sulfide) meters
   d. Meters for common site-specific hazardous gases
   e. Colorimetric detector tubes
   f. Radioactive indicating meters
   g. Visual observations
3. Continuous air monitoring shall be performed
4. Monitoring of high-risk employees:
   a. After the clean up of hazardous operations is over the employer should begin monitoring the employees who were likely to have high exposure to any hazardous substances and other health hazards.

Risk Identification and Considerations:

1. Employees must be informed of any risks that have been identified
2. Risks to consider:
   a. Exposures above the permissible explosive level (PEL)
   b. IDLH concentrations
   c. Potential skin absorption and irritation sources
   d. Potential eye irritation sources
   e. Explosion sensitivity and flammability ranges
   f. Oxygen deficiency

Training:

1. Any employee exposed to hazardous substances, health hazards, or safety hazards along with their supervisors and management must receive proper training.
2. Elements of the training program:
   a. Names of employees and any alternates who are in charge of safety and health
   b. Any hazards present on site
   c. Personal protective equipment to be used

   d. Work practices that should be used to minimize the risk of hazards
   e. Engineering controls
   f. Equipment on site
   g. Medical surveillance, including recognition of any signs and symptoms that would indicate overexposure
3. Initial training
   a. General site workers
      i. A minimum of 40 hours of off-site training
      ii. A minimum of 3 days on site under the direct supervision of a trained supervisor
   b. Workers who will occasionally be on site but are unlikely to be exposed at levels which exceed the PEL or workers on site who work in areas where it has been determined that they will not be exposed at levels above the PEL
      i. 24 hours off-site training (minimum)
      ii. One day under supervision of experienced supervisor
   c. On-site supervisors and management shall receive an additional eight hours of specialized job training
   d. On-site supervisors, management, and workers shall receive eight hours of refresher training annually unless they can prove through documentation or certification that their work itself has provided this training.

<u>Engineering Controls, Work Practices, and PPE for Employee Protection:</u>

1. Engineering controls:
   a. Use of pressurized caps or control booths on equipment
   b. Use of remotely operated material-handling equipment
2. Work practices:
   a. Removing all nonessential employees while opening drums
   b. Wetting down dusty operations
   c. Locating employees upwind from possible hazards
3. Selecting PPE shall be based on the following criteria:
   a. The requirements and limitations of the site
   b. The conditions and time period of the task being performed
   c. The hazards and potential hazards on site
4. Types of PPE and when they should be used:
   a. SCBAs shall be used when chemical exposures present a serious risk of immediate death, illness, or injury
   b. Level-A suits shall be used when absorption of a hazardous substance creates the potential for immediate death, illness, or injury
5. PPE shall be upgraded when new information or new site conditions indicate that increased protection is needed to keep employees below the PEL.

<u>Handling Drums and Containers:</u>

1. Drums and containers shall be inspected and their contents must be determined before they are moved.
2. Unlabeled drums and containers shall be treated as hazardous substances until they can be properly identified.
3. Site operations shall be organized to minimize the movement of drums and containers.
4. All employees involved in the movement of drums must be notified of the risks associated with the substance contained in the drum.
5. A program must be implemented to contain and isolate hazardous substances when the possibility of a major spill exists.
6. Drums and containers that cannot be moved without spilling the contents must be transferred into another container.
7. A detection device shall be used to estimate the depth and location of buried drums.
8. Fire-extinguishing equipment shall be on hand to put out any incipient fires.

<u>The Following Procedures Shall Be Followed When Opening Drums or Containers:</u>

1. Employees not involved in opening drums or containers shall keep a safe distance when they are opened.

2. If employees must work within close distance to drums that are being opened, a protective shield shall be used that does not interfere with their work.
3. Controls used when opening drums or containers shall be located behind the explosion-proof barrier.
4. Drums and containers must be opened in a manner which allows excess pressure to be safely released.

Shipping and Transporting Drums and Containers:

1. Drums and containers must be identified and classified before packaging them for shipping.
2. Drum or container staging areas shall be kept to a minimum number to safely identify materials and prepare them for transport.
3. Staging areas shall be designed with adequate access and egress routes.
4. Bulking hazardous wastes together is permitted only after adequate characterization of the materials is completed.

Decontamination Procedures:

1. A decontamination procedure must be implemented and known by employees before any employees or equipment may enter areas where potential for exposures to hazardous substances may occur.
2. A standard operating procedure shall be developed to minimize employee exposures to hazardous substances or equipment that has come into contact with hazardous substances.
3. All employees leaving a contaminated area shall be properly decontaminated and must dispose of or decontaminate their clothing and equipment.
4. Decontamination procedures must be monitored and changed if found to be ineffective.
5. Decontamination must take place in an area which minimizes exposure of uncontaminated equipment and people to those who have been contaminated.
6. Protective clothing must be cleaned, decontaminated, or replaced as needed.
7. Employees wearing permeable clothing that becomes wetted with a hazardous substance must remove the clothing and get to the shower.
8. Protective clothing and equipment cannot be removed from the change rooms by unauthorized personnel.
9. Commercial laundries that decontaminate protective clothing must be informed of the risks posed by the hazardous substances.
10. Showers and change rooms must be provided where indicated in the decontamination procedure.

Emergency Response:

1. A written emergency response plan shall be developed and implemented to deal with anticipated emergencies.
2. The program must be available to employees, employee representatives, and OSHA personnel.
3. The emergency response plan must cover the following areas:
   a. Preemergency planning and coordination with outside parties
   b. Personal roles, lines of authority, communication, and training
   c. Emergency recognition and prevention
   d. Safe distances and places to seek refuge
   e. Site security and control
   f. Evacuation routes and procedures
   g. Decontamination
   h. Medical treatment and first aid
   i. Emergency alerting and response procedures
   j. Critique and follow up of response
   k. PPE and emergency equipment
   l. Emergency response organizations may use local or state plans, or parts from each
4. Procedures for handling an emergency response:
   a. A senior emergency response official shall be in charge of the Incident Command System (ICS) and all emergency responders and their communication will go through this official.

b. The first senior officer to arrive on site shall take charge and pass the duties along to higher-ranking officials as they arrive.
c. The senior official must identify any hazardous substances and conditions and address the use of proper engineering controls, maximum exposure limits, and hazardous substance handling.
d. The individual in charge of the ICS shall implement appropriate emergency operations based on the hazardous substances and conditions.
e. The individual in charge of the ICS must ensure that employees wear necessary PPE.
f. SCBAs must be worn by employees who face inhalation hazards while engaged in the emergency response.
g. The individual in charge of the ICS must limit the emergency personnel to those who are actually performing the emergency operations.
h. Back-up personnel shall be on site and prepared to provide assistance or rescue.
i. At minimum, basic life-support personnel must be on site with medical equipment and transportation.
j. The individual in charge of the ICS shall appoint a safety officer who knows the operations of the emergency response plan to identify and evaluate hazards.
k. When activities are determined to be an IDLH or to pose an imminent danger, he/she has the authority to alter, suspend, or terminate the operations.
l. If operations are terminated, the officer in charge of the ICS shall implement decontamination procedures.

Training:

1. Training shall be based on an individual's duties and the function of the emergency response organization.
2. Training must be completed before an individual is permitted to take part in actual emergency operations (OSHA 2005).

It should be noted that 1910.120/1926.65, Hazardous Waste Operations (HAZWOPER) standards cover general hazardous waste worker protection. Emergency hazardous material clean-up activities, usually conducted by hazardous material (HAZMAT) teams are referred to in (q) of 1910.120/1926.65. Facilities with the potential for hazardous material spills should have designated personnel who maintain current HAZWOPER certification to perform clean-up activities such as small-quantity spills.

Environmental Protection Agency

Preventing and Handling Chemical Emergencies

The potential danger of accidental releases of hazardous substances has increased over the years as accidents have occurred all around the world. When more than 40 tons of methyl isocyanate was released in Bhopal, India, killing well over 2,000 people, public concern rapidly intensified. Another release occurred in Institute, West Virginia, sending more than 100 people to the hospital. These two accidents increased the public's awareness, especially after they saw that such incidents can and do happen in the United States. In response to the public concern and existing hazards, the EPA created the Chemical Emergency Preparedness Program (CEPP) in 1985. The CEPP is a voluntary program that encourages both state and local authorities to identify hazards in their areas and plan for potential emergencies. This local planning actually complements emergency response planning carried out at the national and regional levels by the National Response Team and Regional Response Team (other EPA chemical emergency preparedness programs). These teams provide resources for and coordination on preparedness, planning, response, and recovery activities for emergencies involving hazardous substances, pollutants, contaminants, HAZMAT, oil, and weapons of mass destruction in a disaster (U.S. National Response Team 2005).

In 1986 Congress passed the Emergency Planning and Community Right-to-Know Act (EPCRA) in response to concerns regarding the environmental and safety hazards posed by the storage and handling of toxic chemicals. EPCRA is also known as Title III of the Superfund Amendments and Reauthorization Act of 1986 (SARA) and has four major provisions: emergency planning (Section 301–303), emergency release notification (Section 304), hazardous chemical storage reporting requirements (Sections 311–312), and toxic chemical release inventory (Section 313). The regulations that implement EPCRA (EPCRA Overview) are all contained in the Code of Federal Regulations (CFR). SARA Title III requires states to establish state emergency response commissions (SERCs) and local emergency planning committees (LEPCs) to put together emergency response plans for each community. Facilities must provide an emergency release notification and written follow-up notice to the LEPCs and SERCs if a hazardous substance is released into the environment that is equal to or exceeds the minimum reportable quantity set in the regulations. EPCRA (Section 304, 40 CFR 355) requires reporting of the following two types of chemicals: (1) extremely hazardous substances (EHS) and (2) Comprehensive Environmental Response, Compensation and Liability Act (CERCLA) hazardous substances (Emergency Planning 2006).

The following must be included in the emergency notification:

- the chemical name
- an indication of whether the substance is extremely hazardous
- an estimate of the quality released into the environment
- the time and duration of the release
- whether the release occurred in the air, water, and/or soil
- any known or anticipated acute or chronic health risks associated with the emergency, and where necessary advice may be obtained regarding medical attention for exposed individuals
- proper precautions, such as evacuation or sheltering
- name and telephone number of contact person

There are also several required elements of a community emergency response plan. Those elements addressed by LEPCs include:

- identify facilities and transportation routes of extremely hazardous substances
- describe emergency response procedures, both on and off site
- designate a community coordinator and facility coordinator to implement the plan
- outline emergency notification procedures
- describe how to determine the probable affected area and population
- describe local emergency equipment and facilities and the persons responsible for them
- outline evacuation plans
- provide a training program for emergency responders
- provide methods and schedules for exercising emergency response plans (Emergency Planning 2006)

The EPA also established the Chemical Accident Prevention Program in 1986 and integrated it with the CEPP. Then, under the Chemical Accident Prevention Program, the EPA developed the Accidental Release Information Program (ARIP) and the Chemical Safety Audit Program. The ARIP is used to collect data on the causes of accidents and steps that facilities should take to prevent reoccurrences. The Chemical Safety Audit Program was developed to gather information on practices at facilities to help prevent accidents.

Section 305(b) of SARA requires the Environmental Protection Agency to conduct a review of emergency systems to monitor, detect, and prevent chemical accidents at facilities across the country.

Oil-Spill Prevention and Preparedness

To address the potential environmental threat posed by petroleum and nonpetroleum oils, the U.S. Environmental Protection Agency has established a program designed to prevent, prepare for, and respond to any oil spill affecting the inland waters of the United States. The EPA's oil program has a long history of responding to oil spills, including several major oil spills such as the Exxon Valdez spill in 1989 (EPA 2004).

In July 2002, the EPA published the Spill Prevention, Control and Countermeasure (SPCC) rule (40

CFR 112) and the rule became effective on August 16, 2002. The EPA requires that certain facilities develop and implement oil-spill prevention, control, countermeasure, or SPCC plans. Unlike oil-spill contingency plans that typically address spill clean-up measures after a spill has occurred, SPCC plans ensure that facilities have set up contaminant areas to prevent oil spills from reaching navigable waters (EPA 2004a). Under the EPA's oil-pollution prevention regulations, facilities must also apply spill prevention by combining planning and enforcement measures in their SPCC plans. Also, the EPA enforces the oil-spill liability and penalty provisions under the Oil Pollution Act of 1990, which provides incentives to facility operators to take the necessary steps to prevent oil spills (EPA 2004b). A spill contingency plan is required as part of the SPCC plan if a facility is unable to provide secondary containment. Also, facilities are required to submit a variety of information after they have had a single discharge of more than 1,000 gallons or two or more discharges (over 42 gallons) in a 12-month period. To ensure that facilities comply with the spill-prevention regulations, the EPA occasionally conducts on-site facility inspections. The SPCC plan must be available to EPA for on-site review and inspection during normal working hours and a copy of the entire SPCC plan must be maintained at the facility if it is normally attended for at least four hours per day. Otherwise, it must be kept at the nearest field office (EPA 2004c).

Oil spills occur despite prevention efforts, and an emergency of any magnitude can happen anywhere, at any time, and in any weather conditions. Therefore, emergency preparedness requires significant planning and training. The two principal elements of the EPA's oil-spill preparedness program are the development and coordination of contingency plans and performance of oil-spill prevention and response training. Well-designed local, regional, and national contingency plans assist response personnel in their efforts to contain and clean up any size spill by providing information that the response teams will need before, during, and after an oil spill occurs. Training ensures that emergency responders, whether they are facility personnel, response contractors, or state and local government officials, know the what, when, and how of responding to an oil spill (EPA 2004a).

EPA requires owners and operators of facilities subject to the oil-pollution prevention regulations to conduct training on facility-specific oil-spill prevention and response measures. Under the regulations, EPA requires operators to instruct their personnel on the operation and maintenance of equipment to prevent discharges of oil and to have a designated person who is accountable for oil-spill prevention. The current regulations also force facility owners or operators to conduct spill-prevention explanations for their operating personnel as often as needed to ensure a sufficient understanding of the SPCC plan for that facility (EPA 2004).

In 1994, EPA added requirements for oil-spill response training for facilities that are required to prepare a facility response plan. Facility owners or operators are especially required to develop and implement a facility response training program if their facility is determined to pose substantial harm to the environment in the event of a spill. In addition to that, facilities are required to develop and implement an oil-spill drill/exercise program. This program should include tabletop and operation exercises that are both announced and unannounced and that include participation in larger area drills and exercises. To satisfy the drill/exercise program, facilities may participate in the federal government's Preparedness for Response Exercise Program (EPA 2004).

In 1991, the EPA proposed revisions to the SPCC regulations to clarify the nature of the spill-prevention training requirements. Proposed training requirements include:

- All employees who are involved in oil-handling activities would be required to receive 8 hours of facility-specific training within one year of the final regulations.
- In subsequent years, employees would be required to undergo 4 hours of refresher training.
- Employees hired after the training program has been initiated would be required to

receive 8 hours of facility-specific training within one week of starting work and 4 hours of training each subsequent year (EPA, 2005).

Additionally, the Office of Emergency Management (OEM) is working with the EPA, federal agencies, state and local response agencies, and industry to prevent accidents and maintain superior response capabilities to ensure the United States is better prepared for environmental emergencies. The goal is to provide national leadership to prevent, prepare for, and respond to health and environmental emergencies (EPA 2004d).

The Department of Transportation

The Department of Transportation (DOT) was established on October 15, 1966 when the DOT Act was signed by President Lyndon Johnson. However, the Department's first official day of operation was April 1, 1967. The stated mission of the DOT is to "Serve the United States by ensuring a fast, safe, efficient, accessible, and convenient transportation system that meets our vital national interests and enhances the quality of life of the American people, today and into the future."

The Department of Transportation is made up of the Office of the Secretary and eleven other individual operating administrations. These eleven administrations are the Federal Aviation Administration, the Federal Highway Administration, the Federal Motor Carrier Safety Administration, the Federal Railroad Administration, the National Highway Traffic Safety Administration, the Federal Transit Administration, the Maritime Administration, the Saint Lawrence Seaway Development Corporation, the Research and Special Programs Administration, the Bureau of Transportation Statistics, and the Surface Transportation Board. These individual departments make it easier for DOT to regulate and control all aspects of transportation.

The Office of the Secretary (OST) is known as the lead agency of DOT. The secretary provides leadership for the DOT and advises the president on all issues pertaining to federal transportation. The Bureau of Transportation Statistics (BTS) collects information on transportation and other necessary areas, which they compile and analyze to create statistics on the nation's transportation.

The Federal Aviation Administration (FAA) oversees the safety of civil aviation. It enforces the regulations and standards associated with the manufacture, operation, certification, and maintenance of aircraft. It also enforces regulations under the Hazardous Materials Transportation Act for shipments by air.

The Federal Highway Administration (FHWA) increases the country's safety, economic vitality, quality of life, and the environment by initiating highway transportation programs. The Federal Motor Carrier Safety Administration (FMCSA) works with the federal, state, and local enforcement agencies to prevent and control commercial motor-vehicle-related accidents and deaths.

The Federal Railroad Administration (FRA) ensures the safety of railway passengers and oversees safe environmental rail transportation. The Federal Transit Administration (FTA) helps in developing mass transportation systems for cities and communities.

The Maritime Administration, or the MARAD, encourages the development and maintenance of a well-balanced U.S. merchant marine, which carries the nation's domestic and foreign waterborne commerce. Another interesting duty of this administration is that it can act as a secondary or back-up navy and military in case of a national emergency.

The National Highway Traffic Administration (NHTSA) is in charge of reducing deaths from highway accidents. It also enforces safety regulations for motor vehicles. This administration puts up the signs telling you "DUI you can't afford it"—all as a way to reduce motor-vehicle accidents.

The Research and Special Programs Administration (RSPA), oversees rules governing the safe transportation and packaging of hazardous materials by every means of transportation except water transportation. Pipeline safety is controlled by RSPA. This administration deals with emergency preparedness and hazardous material emergencies, helping state and local authorities train for hazardous materials emergencies.

Regulations for the Department of Transportation can be found in 14 CFR, which addresses Aeronautics and Space. These regulations pertain to the Federal Aviation Administration. However, the majority of regulations for DOT are found in 49 CFR, which is titled Transportation; here the individual operating administrations are organized into parts.

Subtitle A–Office of the Secretary of Transportation, Parts 1–99, lists the regulations for the Office of the Secretary of Transportation. Under Subtitle B–Other Regulations Relating to Transportation, Chapter 1 deals with the Research and Special Programs Administration in Parts 100–185 and 186–199 and Chapter 2, Parts 200–299, lists regulations for the FRA. In Chapter 3, Parts 300–399, are all the regulations enforced by the Federal Motor Carrier Safety Administration. Parts 400–499 in Chapter 4 are regulations for the Coast Guard and Homeland Security. Chapter 5, Parts 500–599, covers the National Highway Traffic Safety Administration. The Federal Transit Administration falls under Chapter 6 in Parts 600–699. The National Railroad Passenger Corporation, also known as (AMTRAK), is dealt with in Chapter 7, Parts 700–799. Chapter 8, Parts 800–899, lists the regulations for the National Transportation Safety Board. The Surface Transportation Board is found in Chapter 10, Parts 1000–1999 and 1200–1399. Chapter 11, Part 1420, covers the Bureau of Transportation Statistics. Chapter 12, Parts 1500–1599, deals with the Transportation Security Administration. Any regulations pertaining to these administrations can be found in 49 CFR.

The administration that focuses mainly on emergency preparedness is the Research and Special Programs Administration. This administration develops safety programs and training. These regulations can be found in 49 CFR, Parts 107 and 110. The RSPA also offers plans for oil spills and hazard prevention. They are located in 49 CFR, Part 130. (U.S. DOT 2003 and 2004)

Most of the U.S. DOT's concerns about emergencies are centered on hazardous materials that can cause harm to life and/or the environment if not stored, labeled, or transported correctly. Regulations related to the transport of such substances are published in the Hazardous Materials Transportation Act of 1975 (HMTA), which is the major transportation-related statute affecting transportation of hazardous cargos. The objective of the HMTA, according to the policy stated by Congress, is ". . . to improve the regulatory and enforcement authority of the Secretary of Transportation to protect the Nation adequately against risks to life and property which are inherent in the transportation of hazardous materials in commerce."

These regulations apply to ". . . any person who transports, or causes to be transported or shipped, a hazardous material; or who manufactures, fabricates, marks, maintains, reconditions, repairs, or tests a package or container, which is represented, marked, certified, or sold by such person for use in the transportation in commerce of certain hazardous materials."

The Hazardous Materials Table found in 49 CFR, Part 172.101, designates specific materials as hazardous for the purpose of transportation. It also classifies each material and specifies requirements pertaining to its packaging, labeling, and transportation. Hazard communication consists of documentation and identification of packaging and vehicles. This information is communicated in the following formats: shipping papers, package marking, package labeling, and vehicle placarding. Upon determining the proper shipping name of the hazardous material, as shown in the Hazardous Materials Table, the table will specify its correct packaging. Packaging authorized for the transportation of hazardous materials is either manufactured to DOT standards or, if it does not meet DOT standards, is approved for shipment of less-hazardous materials and limited quantities. The shipper is responsible for determining the shipping name. The shipper must also ascertain the hazard class, United Nations identification number (if required), labels, packaging requirements, and quantity limitations. The Hazardous Materials Transportation Act preempts state and local governmental requirements, unless those requirements afford an equal or greater level of protection to the public than the HMTA requirement. DOT has broad authority to regulate any hazardous materials that are in transport, including the discretion to determine which materials will be classified as "hazardous." These materials are placed in one of nine categories, determined by their chemical and physical properties. Based on the classification of the material, the DOT is also responsible for determining the appropriate packaging materials for shipping or transport. Finally, also based on the material classification, strict guidelines are furnished for the proper labeling/marking of packages of hazardous materials offered for transport and for placarding of transport vehicles.

Materials Classification

Class 1: Explosives
- Division 1.1 Explosives with a mass explosion hazard
- Division 1.2 Explosives with a projection hazard
- Division 1.3 Explosives with predominantly a fire hazard
- Division 1.4 Explosives with no significant blast hazard
- Division 1.5 Very insensitive explosives
- Division 1.6 Extremely insensitive explosive articles

Class 2: Gases
- Division 2.1 Flammable gases
- Division 2.2 Nonflammable gases
- Division 2.3 Poison gas
- Division 2.4 Corrosive gases

Class 3: Flammable liquids.
- Division 3.1 Flash point below –18°C (0°F)
- Division 3.2 Flash point –18°C and above, but less than 23°C (73°F)
- Division 3.3 Flash point 23°C and up to 61°C (141°F)

Class 4: Flammable solids, spontaneously combustible materials, and materials that are dangerous when wet
- Division 4.1 Flammable solids
- Division 4.2 Spontaneously combustible materials
- Division 4.3 Materials that are dangerous when wet

Class 5: Oxidizers and organic peroxides
- Division 5.1 Oxidizers
- Division 5.2 Organic peroxides

Class 6: Poisons and etiologic materials
- Division 6.1 Poisonous materials
- Division 6.2 Etiologic (infectious) materials

Class 7: Radioactive materials
- Any material, or combination of materials, that spontaneously gives off ionizing radiation; it has a specific activity greater than 0.002 microcuries per gram

Class 8: Corrosives
- A material, liquid or solid that causes visible destruction or irreversible alteration to human skin or a liquid that has a severe corrosion rate on steel or aluminum

Class 9: Miscellaneous
- A material that presents a hazard during transport, but which is not included in any other hazard class (such as a hazardous substance or a hazardous waste)

ORM-D: Other regulated material
- A material which, although otherwise subjected to regulations, presents a limited hazard during transportation due to its form, quantity, and packaging (DOT 2004)

Nuclear Regulatory Commission

The Nuclear Regulatory Commission (NRC) was established by the Energy Reorganization Act of 1974 to enable the nation to safely use radioactive materials for beneficial civilian purposes while ensuring that people and the environment are protected. The NRC regulates commercial nuclear power plants and other uses of nuclear materials, such as in nuclear medicine, through licensing, inspection, and enforcement of its requirements.

Originally the NRC was called the Atomic Energy Commission. This agency was established to control the use of nuclear sources and waste. Their primary goal is to protect the public and environment from nuclear materials, waste, reactors, and radiation. The NRC's attitude is prepare for all emergencies beforehand. This will allow personnel to evaluate the situation very quickly and perform the appropriate actions. In the case of nuclear reactors, people do not have much time to think about what to do if an emergency happens. They must have a plan already written so that they can just act on it (U.S. NRC 2004).

The Three Mile Island incident in 1979 showed the NRC that there must be better coordination between nuclear power plants and the federal, state, and local levels of government. This led to the standard that nuclear power plant operators must submit an emergency preparedness plan to local and state governments. It must be submitted to any government agency within a 10-mile radius plume exposure pathway as well as a 50-mile ingestion pathway (Sylves 1984).

Federal responsibilities during a nuclear emergency include oversight of the nuclear plant's safety procedures. They do a thorough review of the emergency preparedness plans to make sure they are adequate. The NRC is also responsible for training the local and state officials to make sure there is accurate synchronization between the nuclear power plant and all federal response agencies. The NRC deals more with protecting the surrounding public in case of an emergency on a state and local level. They are to make sure that the public is informed of an emergency if one occurs. This may range from distributing potassium iodide pills to a full evacuation (NRC 2004a).

One of the components of the NRC's Emergency Response Plan is licensing of plant operators. The training provides the plant operator with knowledge on how to properly protect the public in case of a radiological emergency. These plant operators must have an Emergency Preparedness (EP) Plan as part of their licensing requirement. The NRC conducts many thorough inspections of the nuclear plants to ensure the EP is effective enough. The NRC has four regional offices throughout the United States (King of Prussia, PA; Atlanta, GA; Lisle, IL; and Arlington, TX) staffed with people who do the inspections. Aside from these regional offices, the NRC has *resident inspectors* at each Nuclear Plant to do the inspections on a day-to-day basis. Finally, the NRC requires the plant operators to do a full-scale exercise of the plan at least once every two years (NRC 2004b).

The NRC also helps nuclear plants prepare for an emergency by delineating four emergency classes. These classifications help workers know what to expect if a specific warning is given and what kind of protection will be needed or if a complete evacuation must occur. The classifications, in order of increasing severity, are as follows: Notification of Unusual Event, Alert, Site Area Emergency, and General Emergency. Notification of Unusual Event means that there are events and processes going on that indicate potential degradation in the level of safety in the plant. This is the lowest emergency classification. This level requires no outside assistance unless the problem worsens. The Alert classification means that there is a substantial problem with the safety of the facility. A Site Area Emergency means that there are processes in action or conditions are present that can severely affect the public. This poses a severe threat to the safety of the workers and surrounding community. In this category, any radioactive material released should not exceed the EPA's protection guideline, except near the facility. The final and most hazardous level of emergency is the General Emergency. This level means that there is sufficient melting of nuclear fuel or damage to the core that loosing containment of it is a possibility. The release of radioactive material to the atmosphere would be substantial (NRC 2004c).

The NRC plays a pivotal role not only in the safety of the nuclear power plant but in the surrounding community. Their philosophy of "preparing for emergencies before they happen to simplify the decision-making process" is a great idea (NRC 2004d). This gives all plant workers a heads up on what is going on so they can simply react and not have to think about what to do.

Federal Emergency Management Agency

The Federal Emergency Management Agency (FEMA) was designed for first response to a major emergency or disaster, such as a hurricane, flood, earthquake, terrorist activity, nuclear disaster, or any major spill. It is not a regulatory agency like OSHA or the EPA. Its original objectives focused on nuclear disasters, which were a major concern after the events of Three Mile Island and Chernobyl. In March of 2003, it became part of Homeland Security. It is now designed to protect against, plan for, and recover from major disasters. Most of FEMA's work recently has been with response to hurricanes. Prior to 1979, there were many individual associations that dealt with disasters and federal emergencies, but President Carter initiated an executive order that combined all of them into the Federal Emergency Management Agency. Some of the previous organizations were the National Fire Prevention and Control Administration, the Federal Preparedness Agency of General Services, and the National Weather Service Community Preparedness Program.

FEMA's role is to ensure that disaster response and recovery efforts comply with federal environmental laws and executive orders, protect people, and do not cause additional damage to the environment. For example, FEMA, in collaboration with local, state, and federal partners, provides guidance on environmental requirements related to the selection of temporary housing sites, debris removal and disposal, and the reconstruction of infrastructure. Federal environmental requirements are meant to assure public health and safety and protect the environment during recovery operations. Through technical assistance and mitigation grant programs, FEMA provides incentives for limiting construction in flood plains and protecting environmentally sensitive areas. In addition, communities participating in FEMA's National Flood Insurance Program agree to adopt and enforce floodplain management ordinances, particularly with respect to new construction (FEMA 2004).

FEMA and many other organizations have joined Homeland Security to improve preparedness for and prevention of terrorism, such as the events that took place in New York and Washington on September 11, 2001. Homeland Security, as defined by Wikipedia encyclopedia, refers to governmental actions designed to prevent, detect, respond to, and recover from acts of terrorism, as well as respond to natural disasters. The term became prominent in the United States following the September 11, 2001 attacks; it had been used only in limited policy circles prior to those attacks. Before this time, such action had been classified as civil defense (FEMA 2004).

As its name implies, anything that comes under the category of an emergency is affected by FEMA. Since various types of emergencies require differing responses, it is apparent FEMA must be very diverse in its handling of all crises. Therefore, it delegates many responsibilities to different levels of government. FEMA permits state and local governments to develop emergency plans applicable to potential problems that could arise in their jurisdiction. Some grants are available only when a presidential declaration of a major disaster or emergency is designated. To find out what grant programs have been activated, reference FEMA's Web site and view the Disaster Federal Register Notices (DFRNs) for a specific disaster (FEMA 2004).

FEMA has six major goals in prevention and response. The first goal is to reduce the loss of life and property damage. The second goal is to minimize suffering and major disruption caused by such disasters. Another goal is to prepare the nation for and address the consequences of terrorism. It is also to serve as the portal for emergency management information and expertise. FEMA's fifth goal is to stimulate motivation and prepare the environment for work. The last goal is to see the agency ranked among the foremost disaster response organizations in the world, setting an international standard of excellence.

The agency uses a three-tiered approach to plan and manage major disasters. The first tier is the annual performance plan. This plan sets the major financial goals for the following year. The second tier consists of the management plans. This tier sets the long-term goals, which are essentially the building blocks of the agency. The last tier is a strategic plan. This plan defines how the agency will cope with such disasters.

One division of FEMA that was designed to help with both disaster clean up and providing assistance is the mitigation division. One of its functions is to align disasters such as a hurricane into different regions for insurance agencies. Under this division are the National Flood Insurance Program, National Dam Safety Program, National Earthquake Hazards Reduction Plan, the National Hurricane Program, and several others. FEMA oversees all of these programs and many more (FEMA 1999).

The nation is broken down into ten different geographical regions under FEMA. The main office is located in Philadelphia (Division three). The major disasters in that region are from hurricanes, floods, and tornados. In its history, all division-three responses have been to natural disasters—except for the terrorist attacks in 2001 (FEMA 2004).

The Federal Response Plan is also monitored by FEMA. This plan has multiple facets in response to terrorism. It is used to reduce U.S. vulnerability to terrorist attacks and to deter and respond to terrorism. Weapons of mass destruction are a major concern since they can be very destructive to human life and the environment. Response to terrorism involves crisis management and addressing the consequences. The plan gives authority to state and local governments to act upon those who are a threat and also to respond to a threat.

The National Response Center

The National Response Center (NRC) is the sole federal point of contact for reporting oil and chemical spills. It should not be confused with the Nuclear Regulatory Commission which has the same acronym (NRC). Every health and safety plan and every contingency plan should list the National Response Center and its toll-free number. The National Response Center is staffed by Coast Guard personnel who maintain a 24-hour-per-day, 365-day-per-year telephone watch. The NRC watches standers enter telephone reports of pollution incidents into the Incident Reporting Information System (IRIS) and immediately relay each report to a predesignated Federal On-Scene Coordinator (FOSC). The NRC also provides emergency response support to the FOSCs. This includes extensive reference materials, state-of-the-art telecommunications, and the operation of automated chemical identification and dispersion information systems (National Response Center 2004).

The National Fire Protection Association

The National Fire Protection Association (NFPA) is an organization that has been dedicated to fire prevention since 1896. Due to the accidental nature of fires, emergency preparedness has been on the forefront of discussion ever since the creation of the NFPA. The NFPA also focuses heavily on fire prevention, but when the prevention techniques fail, it is necessary to have plans in place to deal with the resulting emergency.

The NFPA breaks up the topic of emergency preparedness into multiple sections. NFPA 1620 deals with pre-incident planning. It lays out a recommended practice that addresses the protection, construction, and operational features of specific occupancies to develop pre-incident plans for use by responders to manage fires and other emergencies using available resources. Pre-incident planning involves evaluating the protection systems, building construction, contents, and operating procedures that can impact emergency operations. Fire departments use NFPA 1620 to develop contingency plans for specific locations within their district. For example, if a new building is built, the local fire department will evaluate the fire protection systems of the building, such as the sprinkler systems and the fire alarms. They also will evaluate the building's construction, contents, and operating procedures. Evaluation of the building's construction is crucial because, in order to put out a fire as quickly as possible, you need to know what is burning. Knowing the contents of the building and the operating procedures is also essential because it will help identify any special hazards, including combustible chemicals or hazardous waste, a main objective of SARA III reporting requirements. NFPA also requires the person developing the pre-incident plan to get data on the number and type of occupants within the building, the availability of water, and other site-specific considerations as well. Some other site-specific considerations might include whether the building is next to a densely populated neighborhood or highly explosive weapons manufacturer (NFPA 2003b).

NFPA 1221 is the standard for the installation, maintenance, and use of emergency communication systems. This standard includes requirements for communication centers, methods of communication, and other communications-related areas of emergency response. Requirements for the methods of communication mainly deal with the procedures for reporting emergencies over various types of communication devices, such as two-way radios and phones. It also lists procedures for public reporting and dispatching (as a result of 911 calls). Regarding communication center requirements, this standard sets minimum specifications for things like security, building construction, power, location, and fire protection to make sure that, in the event of an emergency, the communication center will still be operational. NFPA 1221 also lists requirements for communication center operations (NFPA 2002).

NFPA 1600 is entitled Disaster and Emergency Management. This section deals with a broad spectrum of areas concerning emergency preparedness. The main goal of this section is to provide people with the criteria to assess current programs, or to develop, implement, and maintain a program. It also sets guidelines for how to mitigate, prepare for, respond to, and recover from a disaster or emergency. NFPA 1600 establishes a common set of criteria for disaster management, emergency management, and business continuity programs.

The National Fire Protection Association (NFPA) put together a committee whose purpose was to develop documents and procedures that could be used during a disaster and/or emergency. In 1995 the committee finished their development of NFPA 1600, which was then defined as the "Practice for Disaster Management." In 2000, the committee revised the document to incorporate the appropriate materials necessary to change NFPA 1600 from a recommended practice to a standard. With added insight from the Federal Emergency Management Agency (FEMA) and the International Association of Emergency Planners, sections on disaster/emergency management were added, and NFPA 1600 was renamed the "Standard on Disaster/Emergency Management and Business Continuity Programs."

The American National Standards Institute (ANSI) recommended to the 9/11 Commission on April 30, 2004, that the NFPA 1600, Standard on Disaster/Emergency Management and Business Continuity Programs, be recognized as the national preparedness standard (IAFF 2003). This revised standard is designed to be a description of the basic criteria needed for a comprehensive program that addresses disaster recovery, emergency management, and business continuity. Clearly a benchmark, and potentially a requirement, NFPA 1600 should be an important influence on any emergency program. This chapter discusses the standard and its implications for emergency, continuity, and disaster recovery planners (NFPA 2004).

NFPA standards are developed through a consensus standard's development process approved by the American National Standards Institute. The NFPA develops standards that are routinely adopted by state and local lawmakers for building, life safety, and electrical issues. The NFPA's mission is to "reduce the worldwide burden of fire and other hazards on the quality of life by providing and advocating scientifically based consensus codes and standards, research, training and education" (NFPA 2004).

Contained in NFPA 1600 are the tools and guidelines a company needs to build a disaster/emergency plan for their business. In the year 2000 version, these tools are confined to three chapters and one appendix entitled "Explanatory Material." The first chapter is a very short chapter dedicated to definitions that introduce the language of the material to the person(s) writing the program. Chapter two, named "Program Management," is even shorter than the first chapter. "Program Elements" is the title of the third chapter, which is the longest of the three chapters. This chapter covers many topics including hazard mitigation, resource management, training, and plan elements. The explanatory material is not required under the standard but is there for informational purposes and pertains to the material in the previous chapters. Each chapter and section has a corresponding section in the explanatory chapter which provides more depth. Soon after it was made into a standard, NFPA 1600 was put to the ultimate test when terror struck on September 11, 2001. According to the Sept./Oct. 2004 issue of the *NFPA Journal*, the 9/11 Commission adopted NFPA 1600 as a basis for a program they call the National Preparedness Standard, which can be useful to the private sector. Department of Homeland Security Secretary Tom Ridge explains in the NFPA journal article (Nicholson 2004) that the:

> Standards encourage mutual respect, cooperation, and open communication—essential elements for our national approach to readiness. Voluntary standards like these—and the process used to develop them—help make us smarter about how to perform our duties better, and give us direction and guidance in the areas we need them most. They are just one tool—but an important one—in our effort to make our country more secure.

After 9/11, audits and inspections revealed that many facilities in the private sector still did not have an emergency action plan. The article stated: "Preparedness in the private sector and public sector for rescue, restart, and recovery of operations should include 1. A plan for evacuation, 2. Adequate communications capabilities, and 3. A plan for continuity of operations." For this reason, the NFPA 1600 Standard on Disaster/Emergency Management and Business Continuity Programs was adopted as a guide.

In 2004, NFPA 1600 was revised to suit the new style of the National Fire Codes. The standards were not changed, but a chart that draws parallels between FEMA's "Capabilities Assessment for Readiness" (CAR), emergency management functions (EMF), NFPA 1600, and BCI and DRII professional practices were added, as well as a new chapter for definitions. More explanatory information was also added to the standard, and a date in the year 2007 was set for another revision of the most recent version of the standard (Nicholson 2004, NFPA 2003a, and NFPA 2004).

NFPA 1600 may become a mandatory requirement. Presently, OSHA compliance officers may reference NFPA 1600, as a guideline, as they cite related "recognized hazards" under the General Duty Clause (NFPA 2004).

The NFPA standards are based on the consensus of volunteer professionals in related fields and are widely accepted. Although these standards are not laws, they are often referenced in the Code of Federal Regulations and often hold up in courts. NFPA 471 and 472 are both standards used as requirements for hazardous material incidents. NFPA 471 recommends operating procedures for hazardous material incidents. NFPA 472 recommends competencies of first responders, trained at different levels, to respond to hazardous material incidents. The NFPA 471 and 472 standards should be followed in terrorist incidents involving hazardous materials, while keeping in mind the added threat (NFPA 1997).

NFPA 471: Responding to Hazardous Materials Incidents

The formation of NFPA 471 began with approval from the NFPA Standards Council during their July 1985 meeting. The council was aware of the increased seriousness and likelihood of hazardous material incidents, so a committee was formed to create an appropriate standard. The committee met a total of six times and included mostly people with backgrounds in working with hazardous materials and in the fire service. NFPA 471 specifically focuses on the recommended work done by emergency responders during a hazardous material incident. Although NFPA standards are not laws by themselves, NFPA 471 is contained in 29 CFR 1910.120 and 40 CFR 311.

The purpose of NFPA 471 is to lay out a detailed minimum requirement of the safety controls needed in a hazardous material incident. It also includes standard operating procedures that should be followed in all emergency responses to hazardous materials. NFPA 471 includes recommendations for the following six categories: personnel and training, site safety, personal protective equipment (PPE), incident mitigation, decontamination, and medical monitoring (NFPA 1997).

NFPA 472: Professional Competence of Responders to Hazardous Materials Incidents

NFPA 472 was written to give responders a guide to what minimum training requirements are needed to perform, safely, the tasks required at a hazardous materials incident. As established in NFPA 471, there are different levels of incidents based on the specific characteristics of an incident. Responders should only perform certain tasks at an incident that they have been appropriately trained to do according to NFPA

472. The standard has set qualifications for responders at the awareness, operations, and technician levels. NFPA 472 also has recommendations for the competencies of incident commanders and safety officers (NFPA 1997). There are other NFPA standards, such as NFPA 600, that have sections that may also apply to emergency response.

Summary

It is not really practical to attempt to memorize regulations. Aside from being difficult to remember, they often change without much notice. The safety, health, and environmental professional should know what resources are available and, most importantly, know how to identify and use selected resources. This chapter attempts to provide an overview of the most prevalent regulations and guidelines. However, the chapter does not cover all the resources, especially the state and local agencies.

Emergency response preparedness is achieved through extensive emergency risk assessment, development of an EAP, and following up with audits and drills. The objective of planning for an unexpected emergency is to minimize consequences of exposure, such as fatalities, severity of injuries, fires, threats to the community, and financial losses. The role of the safety, health, and environmental professional is to know what is required, what is recommended, and what the limitations are of emergency response for their organization. His or her most important function is to recommend the proper course of actions to management.

DISCLAIMER

This handbook provides a general overview. Since regulators make changes frequently, it will not be able to keep up with regulatory revisions, rendering the book out of date in certain areas. It is recommended that specific regulations and guidelines be checked for updates before referencing them.

References

Barbalace, K. L. 2007. *USDOT Hazardous Materials Transportation Placards.* www.environmentalchemistry.com/yogi/hazmat/placards

Emergency Planning for Chemical Spills (accessed 2006). *Other EPA Chemical Emergency Preparedness Programs.* www.chemicalspill.org/Right-To-Know/ceppo1.html

Environmental Protection Agency (EPA). Nov. 2004a. *Oil Spill Training.* www.epa.gov/oilspill/oiltrain.htm

______. Nov. 2004b. *Preparing for Oil Spills.* www.epa.gov/oilspill/prepare.htm

______. Nov. 2004c. *Preventing Oil Spills.* www.epa.gov/oilspill/prevent.htm

______. Nov. 2004d. *Spill Prevention, Control, and Countermeasure.* www.epa.gov/oilspill/spcc.htm

______. Nov. 2004e. *Office of Emergency Management.* www.epa.gov/ceppo/about.htm

______. Nov. 2004f. *Oil Spill Profiles.* www.epa.gov/oilspill/oilprofs.htm

______. 2005. *EPCRA Overview.* www.yosemite.epa.gov/oswer/ceppoweb.nsf/webprintview/epcraoverview.htm

Federal Emergency Management Agency (FEMA). 2004. *Disaster Federal Register Notices (DFRNs).* www.fema.gov

______. 1999. *Terrorism Incident Annex.* www.au.af.mil/au/awc/awcgate/frp/frpterr.htm

Gaade, Rem. 2002. "The New NFPA 472." *Hazardous Materials Management Magazine* (retrieved 2005). www.hazmatmag.com

International Association of Fire Fighters (IAFF). Copyright © 2003 (updated: 4/30/03). www.daily.iaff.org/education/learningmods/Intro1710/whnjs.htm

National Fire Protection Association (NFPA). 1997. NFPA 471, *Recommended Practice for Responding to Hazardous Materials Incidents.* www.safetynet.smis.doi.gov/nfpa471.pdf

______. 2002. NFPA 1221, *Installation, Maintenance, and Use of Emergency Services Communications Systems.* www.constructionbook.com/nfpa-1221-standard-for-installation-maintenance-use-of-emergency-services-communications-systems-2002-edition

______. 2003a. NFPA 1600. www.nfpa.org

______. 2003b. NFPA 1620, *Recommended Practice for Pre-Incident Planning.* www.constructionbook.com/nfpa-1620-recommended-practice-for-pre-incident-planning-2003-edition

______. 2004. NFPA 1600. www.davislogic.com/NFPA1600

National Response Center. Nov. 2004. www.nrc.uscg.mil/organizetxt.htm

National Response Team (accessed 2005). www.nrt.org
Nicholson, John. "The 9/11 commission and NFPA 1600." *NFPA Journal*, September/October 2004, pp. 58–59.
Occupational Safety and Health Administration (OSHA). 2002a. 29 CFR 1910.38, *Employee Emergency Plans* and 1910.39 *Fire Protection Plans*. www.osha.gov
______. 2002b. *Emergency Preparedness: Fire and Explosion Planning Matrix*. www.osha.gov/dep/fire-expmatrix/index.html
______. 2002c. 29 CFR 1926.35, *Employee Emergency Plans and Fire Protection Plans*. www.osha.gov/pls/oshaweb/owadisp.show_document
______. 2005. 29 CFR 1910.120/1926.65, *Hazardous Waste Operations and Emergency Response* (accessed 11/2005). www.osha.gov/pls/oshaweb
Sylves, Richard T. "Nuclear Power Plants and Emergency Planning: An Intergovernmental Nightmare." *Public Administration Review* (Sept.–Oct. 1984), 44(5):393–401.
U.S. Department of Transportation (DOT). Nov. 2004. www.dot.gov
______. 2003. 49 CFR Parts 100–180. Washington, D.C.: U.S. Department of Transportation.
U.S. Department of Transportation Special Programs Administration. 1990. *An Overview of the Federal Hazardous Materials Transportation Law*. Washington, D.C.: U.S. Department of Transportation.
U.S. Nuclear Regulatory Commission (NRC). Nov. 2004a. *Emergency Classifications*. www.nrc.gov/what-we-do/emerg-preparedness/emerg-classification.html
______. Nov. 2004b. *Emergency Preparedness and Response*. www.nrc.gov/what-we-do/emerg-preparedness.html
______. Nov. 2004c. *Evacuation and Sheltering*. www.nrc.gov/what-we-do/emerg-preparedness/evacuation-sheltering.html
______. Nov. 2004d. *Federal, State, and Local Responsibilities*. www.nrc.gov/what-we-do/emerg-preparedness/federal-state-local.html
______. Nov. 2004e. *How We Prepare to Protect the Public*. www.nrc.gov/what-we-do/emerg-preparedness/protect-public.html
______. Nov. 2004f. *Who We Are*. www.nrc.gov/who-we-are.html

SECTION 2
EMERGENCY PREPAREDNESS

LEARNING OBJECTIVES

- Categorize disasters by their main cause or hazard type.
- List and describe the four phases of emergency management.
- Be able to define the terms *emergency* and *disaster*.
- Describe the objectives of emergency planning.
- List the specific skills/knowledge a safety manager must possess for effective emergency management.
- List examples of the materials and information an on-site safety manager will be expected to provide during an emergency.

APPLIED SCIENCE AND ENGINEERING

Susan M. Smith and Kathy J. Council

EVERY DAY THE NEWS announces emergencies and disasters occurring in communities across the United States and around the world. These disasters range from floods and tornados to bombings and anthrax scares. The human toll and costs associated with disasters will continue to increase as coastal areas in the United States continue to become more densely populated (Smith, Council, and Rogerson 2006). Other types of disasters can occur anywhere and are not limited to coastal areas (e.g., tornados, severe winter weather, derailments, and acts of terror).

Safety managers need to understand how emergency plans for facilities are developed, what emergency plans for facilities include, and how employees of facilities can prepare themselves to follow the emergency plan in the event of a disaster or a high-risk situation. Knowledge of the different types of disasters and different stages in emergency management can help each employee be better prepared if a disaster or major emergency occurs at a facility.

IMPORTANCE OF THE SAFETY MANAGER IN COORDINATING SAFETY, SECURITY, AND EMERGENCY MANAGEMENT FUNCTIONS

The importance of maintaining a relationship between the operational levels of safety, security, and emergency management within a company or agency has greatly increased since the terrorist attacks on the World Trade Center in New York City and the Pentagon in Washington, D.C., on 9/11/01. Terrorist events, which have occurred both within this country and throughout the world, have proven the need for terrorism to be of greater priority in emergency planning at all levels (McEntire 2007). As the importance of preventing a disaster initiated by terrorists increases, the importance of collaboration between the manager, safety manager, the security office, and those responsible for emergency response in large companies has increased (Smith, Council, and Rogerson 2006). Terrorist threats may still be a low priority for some companies.

Although different businesses and industries may use a title other than safety manager (e.g., safety officer, safety professional, safety engineer, or some other title), for simplicity, this chapter uses the term *safety manager* throughout. Small- or medium-sized companies, or individual plant locations, may group all safety responsibilities, as well as risk management, under the safety office. Quite a few U.S. companies have also consolidated many of the functions of site-specific emergency response and security under the office of safety and risk management. This demonstrates how important it is for a safety manager to understand the role, responsibilities, and actions required of a company safety office in maintaining an environment that provides the day-to-day safety and security for employees. In addition, through mitigation and planning, the safety manager can help reduce the impact of environmental risks if an emergency response due to terrorism or a natural disaster is required.

An appendix at the end of the chapter provides a ten-point checklist for emergency preparedness to assist safety managers in hazard analysis and emergency preparedness (Smith 2001 and Smith and Rogerson 2002). Some of the key safety practices a safety manager and his or her staff must implement to mitigate (prevent or minimize) damage if an emergency situation occurs include: (1) the proper storage and use of hazardous materials, (2) effective policies to prevent workplace violence, (3) high-quality and well-maintained security systems, (4) a work site designed to minimize safety hazards, (5) entrances and exits to facilities that support the safe transport of goods and employees into the facility, and (6) adequate evacuation procedures (Smith, Council, and Rogerson 2006).

The presence on site, or at a facility, of an individual designated as the safety manager whose main responsibility is to facilitate actions and behaviors that reduce hazards, minimize risk, and support safe behaviors has been linked to a reduction in reported injuries by the construction industry (Findley, Smith, Kress, and Petty 2004). In many cases, the safety manager will be the closest person at an individual plant or company who the plant manager can rely on to coordinate immediate response in case of an emergency or disaster (Smith, Council, and Rogerson 2006).

Since the safety manager is the person who is typically the designated contact in the event of an on-site chemical spill resulting from process or transportation accidents that require hazardous material precautions, the safety manager is also the "natural person" a plant supervisor or manager is going to expect to take the lead in at least certain types of emergency situations. Fires and spills that are entirely contained in a plant or on facility property would, in most cases, be considered emergencies. The safety manager will certainly not be able to say this is "not my area." Rapid response and functional coordination with local community or county emergency response agencies will be expected.

In an emergency or disaster situation an on-site safety manger will be expected to have ready access to and provide:

- detailed building and site maps
- precise locations of all areas using or storing hazardous materials with information on each major chemical area, including ready access to material safety data sheets (MSDS)
- contact information and up-to-date knowledge of how to contact appropriate county or city emergency response personnel
- an operational plan focused on evacuating plant or company personnel and customers and providing directions that focus on an effective emergency response while minimizing the exposure of responders to further hazards
- information on the best route for emergency vehicles to approach all buildings on the affected site, and the ability to implement a plan to have all roads on and off the site secured and monitored

- support for the on-site plant manager in maintaining a critical communication link between the actual plant site and corporate headquarters during an emergency (Smith, Council, and Rogerson 2006).

In a nonemergency situation, the safety manager's day-to-day facilitation and implementation of programs and actions to maintain a safe plant, company, or institutional environment will be critical to the prevention or reduction of injury and property damage in the event that an emergency situation arises. A facility with an excellent housekeeping record, appropriate management of MSDS for all hazardous chemicals, ongoing training for employees in critical safety and emergency response areas, and periodic safety inspections and inventories with follow up to correct deficiencies to maintain all facilities and grounds in a safe condition can lead to more effective response in an emergency situation (Smith and Council 2006). Each of these functions is typically conducted under the ongoing supervision of an on-site safety manager. When a safety manager is focusing on the ability of personnel within and external to the facility to respond effectively during an emergency, many of the skills already needed for effective day-to-day safety operation of a facility or institution are in more critical demand (Smith and Council 2006).

If a safety manager has been designated or functions as the emergency response coordinator, he or she must know about the emergency, infrastructure, and governmental services provided within the community that can be called upon. The skills and knowledge a safety manager brings to a safety management position are also the skills a safety manager can bring to an emergency situation. They include:

- effective communication
- demonstrated leadership and ability to work as part of a team
- knowledge of risk assessment and hazard analysis
- prioritization and implementation skills for mitigation actions
- practice in engaging in and implementing ongoing planning
- the ability to perform critical record management and retrieval in order to obtain necessary information, emergency contacts, and outside contacts promptly during an emergency situation
- the ability to coordinate and work with other managers on site and at the corporate level
- knowledge of those managers within the community who focus on safety, security, or emergency response (Smith and Council 2006).

In addition to these skills and possessing site-specific cognitive knowledge, the safety manager also should anticipate needs and ensure that systems are in place to:

1. access critical response information rapidly
2. allow decisions that will contain a potential problem situation to be made quickly
3. access rapidly the names and specific capabilities and training of key plant personnel (e.g., those trained in HAZMAT response, safety, security, giving medical emergency aid, team leadership, equipment shut-down procedures, and communications)
4. define and itemize what the specific planned roles of off-site emergency response personnel will be prior to the actual occurrence of an emergency or disaster in case the most probable crisis situations do occur
5. allow for flexibility in working as part of a team to improvise and problem solve (Smith, Council, and Rogerson 2006).

During a complex emergency, the safety manager who has anticipated and worked with other key administrators at his or her facility rather than working alone in a bubble is much more effective (Smith, Council, and Rogerson 2006). Safety managers who provide support for managers or administrators in everyday operations can be effective in responding to the immediate on-site emergencies. Rapid and effective response can reduce injury and property damage during an emergency regardless of its cause. Examples could be response to spills on land and in water, fires, derailments, or weather-related events such as tornados or hurricanes. A decision either to evacuate or to shelter in place, whether it involves employees, clients, or visitors, must be made rapidly, and any evacuation, such as the evacuation of patients from a health care facility, must be carefully planned and drilled.

Emergency Management Plans

The objectives of all emergency planning and actions include:

- keeping an emergency or crisis event from expanding into a disaster
- getting back to a "normal" state of operation from an emergency quickly
- maintaining safety for all employees and customers during an emergency condition (Haddow and Bullock 2003; and Smith, Council, and Rogerson 2006).

Every facility should have a current emergency management plan and an ongoing process for preparing and training employees according to that plan. This emergency plan should outline procedures for responding to such situations as:

- fires
- bomb threats
- severe weather
- utility service failures
- floods
- earthquakes
- toxic fumes
- chemical spills

Figure 1 shows an example of a table of contents for an emergency plan from the Federal Emergency Management Agency's *Emergency Management Guide for Business and Industry*. According to FEMA, "This guide provides step-by-step advice on how to create and maintain a comprehensive emergency management program. It can be used by manufacturers, corporate offices, retailers, utilities or any organization where a sizable number of people work or gather."

Table of Contents

Section 1–4 Steps in the Planning Process

- Step 1 Establish a Planning Team
- Step 2 Analyze Capabilities and Hazards
- Step 3 Develop the Plan
- Step 4 Implement the Plan

Section 2 – Emergency Management Considerations

- Function: Direction and Control
- Function: Communications
- Function: Life Safety
- Function: Property Protection
- Function: Community Outreach
- Function: Recovery and Restoration
- Function: Administration and Logistics

Section 3 – Hazard-Specific Information

- Hazards: Fire
- Hazards: Hazardous Materials Incidents
- Hazards: Floods and Flash Floods
- Hazards: Hurricanes
- Hazards: Tornadoes
- Hazards: Severe Winter Storms
- Hazards: Earthquakes
- Hazards: Technological Emergencies

Section 4 – Information Sources

- Sources: Additional Readings from FEMA
- Sources: Ready-to-Print Brochures
- FEMA Regional Emergency Management Offices
- State Emergency Management Offices

FIGURE 1. Table of contents for an emergency plan (Federal Emergency Management Agency, *Emergency Management Guide for Business and Industry*, www.fema.gov, 2006)

Describing a Disaster Situation

When a situation requires major shifts in operation, manpower, or outside assistance to provide even limited services, the emergency event for that facility has the potential to become a disaster and would require activation of the emergency. If the resources are overwhelmed and a site cannot handle the emergency without a major disruption of its normal services, or the event causes substantial fatalities to occur or a severe loss of property or equipment, that event would also be classified as a disaster (Smith, Council, and Rogerson 2006). A hurricane may require an emergency management response as well as the definite activation of a facility plan to reduce the impact of this potential disaster, which may disrupt many company or plant operations.

An effective emergency plan should address both large and small issues and identify the capacity to deal with an emergency. *Capacity* is the ability of a facility or a community to deal or cope with a situation with the available resources. From the Centre for Research on the Epidemiology of Disasters (CRED), the United Nations provides a definition of a *disaster* as "a situation or event, which overwhelms local capacity, necessitating a request to the national or international level for external assistance . . . an unforeseen and often sudden event that causes great damage, destruction and human suffering" (www.unisdr.org).

Basically, a disaster can be described as a situation that occurs when the disruption to a facility or community due to a natural or man-made hazard results in many injuries, the loss of lives, or the destruction of a large amount of property (Smith, Council, and Rogerson 2006).

Classifying Disasters

To achieve emergency planning, it is important for safety managers to understand how disasters are classified based on their causes. Many of the emergency situations caused by weather or terrorism could potentially create disasters or major emergencies for a facility or community. Therefore, in developing emergency plans, emergency management officials classify disasters by four main causes, or hazard groups (Drabek and Hoetmer 1991):

- natural disasters
- technological disasters
- civil disruption or violence
- long-term or ecological changes

The type of hazard that causes a disaster may determine preparedness, planning, and recovery and response methods selected by a facility. Examples of such hazards include:

- natural hazards caused by weather, including floods, tornados, and severe storms
- technical hazards caused by man-made technology, as when a tanker accident releases a toxic gas or a radiation leak
- civil disasters caused by disruption or violence, such as bombings or the intentional release of harmful viruses
- ecological disasters that cause a long-term change in the Earth, such as rising levels of the ocean due to melting of the ice caps

Of the four, ecological disasters are the hardest to prepare for and respond to (Drabek and Hoetmer 1991).

Skill/Knowledge: A safety manager should be able to specify the four main categories for classifying disasters by hazard type in emergency planning and correctly state their characteristics.

Planning a General Response for Emergencies

Since an emergency can be defined as a situation resulting from a man-made or natural hazard that requires a rapid response to minimize damage, the facility emergency plan must provide guidelines for each department to use when responding to an emergency or a natural hazard (Smith, Council, and Rogerson 2006). For example, an emergency management plan must describe how a facility will provide backup electricity or natural gas during a power outage in order to maintain essential services when the normal power system is interrupted.

A safety manager must know the services provided by his or her facility that must be maintained during an emergency. Some of these will be services that will prevent damage to equipment while others will be vital to maintaining a healthy environment for employees or emergency workers. Facilities must have strategies to reduce damage in times of emergency.

These strategies might include purchasing and maintaining back-up generators to maintain power to essential services during power outages, and purchasing could control shipments coming in or out and any emergency equipment required.

Skill/Knowledge: A safety manager should know the safety department's role as it is described in the emergency plan and be able to implement the plan.

The Four Phases of Emergency Management

Emergency plans for managing emergency situations must include many necessary actions. When facilities prepare emergency plans, they should be required to include four major action areas that must be addressed in all types of emergencies, regardless of cause or hazard (Haddow and Bullock 2003). These four major action areas, or phases, are called emergency:

- preparedness
- response
- recovery
- mitigation

Emergency preparedness means, "to organize for emergency response before an event" (Drabek and Hoetmer 1991, p. 34). Examples of being prepared for emergencies could include:

- continuous training, drills, and review of exercises to ensure industries, businesses, or health care facilities are prepared if and when a disaster strikes
- maintaining a system of emergency generators to keep the electric power going during a power outage

An *emergency response* should include procedures for moving employees or customers out of a facility and evacuating from the facility grounds if a toxic chemical spill or an explosion makes a facility unsafe. Plans may also include directions for the safest locations to find shelter within the buildings of the facility (Smith, Council, and Rogerson 2006).

Recovery involves having a plan to get the facility back open and in full operation after any major disaster, including a utility outage or water supply shutdown. The safety manager at a facility would speak with the plant manager to determine what role the safety department would have in a recovery operation and what actions the safety staff might be required to take in support of the facility recovery.

In emergency planning, *mitigation* refers to any action designed to prevent or reduce the damage of an emergency. For example, making sure the facility is not built in an area that floods, is one method of preventing a future disaster. The creation and use of such rules to prevent problems is called a *mitigation action*. Mitigation can occur before a disaster situation and also can be incorporated into the recovery process.

Most facilities are required to have an emergency management plan, but all facilities should have effective emergency management and planning within the facility. To be effective, emergency management must cover all four of these main areas:

- Preparedness—training continuously to be ready to respond correctly in emergency situations. Planning, drilling, training, and education are key components of preparedness (Haddow and Bullock 2003).
- Response—having a plan to ensure that employees can respond rapidly and effectively to any type of an emergency, whether it requires sheltering in place or an evacuation, whichever best guarantees the safety of employees, customers, patients, or other personnel.
- Recovery—having the capacity to quickly return operations to normal after a disaster
- Mitigation—having an ongoing planning and operational system to reduce or prevent hazards before they can cause an emergency situation or disaster.

(Smith, Council, and Rogerson 2006)

Skill/knowledge: A safety manager should be able to name, define, develop, and implement a plan utilizing the four categories or phases of emergency management required in emergency planning.

How Facilities Plan for Disasters and Emergencies

A facility should maintain an ongoing emergency planning process. This process typically includes representatives of company management, as well as representatives from each major department. Some facilities provide opportunities for all employees to be involved in the overall planning process. Employee involvement can contribute to a more effective plan within the facility.

The emergency facility planning process includes:

1. identifying the hazards
2. determining which hazards are the greatest risks
3. outlining actions that prevent or reduce hazards
4. outlining actions to take in advance to ensure that the response and recovery from an emergency, if it occurs, will represent the best possible effort

Written plans and procedures must be in place. All plans must include descriptions and actions for relocation to shelter in place and/or for the evacuation of employees and other occupants, such as customers. However, each facility's plan is unique and is designed to meet that facility's special needs. It is important for a safety manager to be very familiar with and participate in the continuous process of improving a facility's emergency plan, even if he or she is not in charge of creating the plan at his or her specific facility (Smith, Council, and Rogerson 2006).

A facility plan should outline what to do if an emergency occurs and support preparedness through drills and other training programs. For example, a facility should prepare exercises in which employees will practice following emergency procedures for dealing with severe cold or hot weather conditions. Each facility's plan and drills should be designed to address the specific needs of that facility and should address the possible hazards that may surround the specific facility and are considered hazards of that geographic location, such as low-lying coastal areas or flood plains.

A facility emergency plan should address how the facility should interact with local community emergency response agencies as well as other state and federal agencies that will be involved if an emergency situation arises. Plans should identify hazards, determine which hazards are the greatest risks, outline actions that prevent or reduce hazards, and recommend actions to take in advance to mitigate or prevent a problem from becoming a major emergency. By prioritizing hazards based on risk assessment, potential damage can be reduced or eliminated, and appropriate response options evaluated.

Skill/knowledge: A safety manager should be able to describe why a facility should develop plans for dealing with disasters and emergencies and give recommendations of how this might be accomplished (Smith, Council, and Rogerson 2006).

A Facility Hazard Analysis

An effective hazard analysis must begin with thorough hazard identification. During the hazard identification process, the hazards faced by a community or industry need to be identified and grouped by types, potential impacts, and potential mitigation strategies. If hazards with common or similar characteristics can be grouped together, the safety manager can develop mitigation strategies that are appropriate for more than one hazard, such as evacuation methods and procedures (Smith, Council, and Rogerson 2006).

Often, natural hazards can be grouped by geographic location. Certain areas can be mapped and identification made; for example, fault lines for earthquake risk or coastal areas for hurricane risk. Some natural hazards give ample warning, such as hurricanes; others, such as tornados, give very little warning. Business and industry can work together with their local governments to take a lead in developing specific mitigation strategies to reduce or eliminate risk to their business or industry. For example, a business or industry should not be allowed to be rebuilt in an area that has been destroyed multiple times by hurricanes. The risk for repeat destruction would be obvious.

Technological hazards are increasing as the developed environment grows denser and more people are put at risk. Transportation accidents caused by train derailments and chemical spills have accentuated the need for pre-incident hazard identification and on-site emergency preparedness plans. Most technological hazards will occur rapidly and give little warning. They can also impact a large area and be very destructive like the nuclear incident at Chernobyl. On the other hand, technological hazards can sometimes be insidious and develop slowly over time without people being aware of the impact. The impact of the contamination at Love Canal due to improper disposal and burial of toxic waste was man-made, and the eventual need to relocate the entire town because of the high levels of toxic waste is an example of a technological hazard (Drabek and Hoetmer 1991).

Civil hazards include famine, war, and hostile acts of terrorism. Many civil hazards are difficult events to manage since they occur without obvious warning. The events of 9/11 in New York and at the Pentagon are examples of civil hazards since both were impacted by terrorist attacks. The events of 9/11 in New York City have been used by Kendra and Wachtendorf (2003) to illustrate the need to be flexible and to improvise when faced with a situation that requires the evacuation of a half-million people.

Once hazards have been identified, the actual and perceived risk and vulnerability of critical structures need to be made part of the hazard analysis. The analysis needs to include a dollar figure and engineering information (e.g., type of structure, square footage, and use) for facilities. The number of people occupying the structure and the types of materials stored in the structure should also be included in the hazard analysis (FEMA 2002.) The specific location of the hazards must be included in the analysis and any mapping that is available. Flood maps exist for most areas from the National Flood Insurance Program (NFIP), but many industries, cities, and counties do not have up-to-date flood maps, and the older maps may not reflect current hazards accurately due to recent building and expansion. However, in certain areas, geographic information systems (GIS) are becoming more available for use in the development of local maps.

Mitigation: A Safety Manager's Tool

Mitigation refers to the use of sustained actions that are designed to reduce or eliminate the damage and destruction caused by hazards (FEMA 2003 and Pine 2007). Mitigation includes structural and nonstructural methods. When a community or industry conducts a hazard analysis, specific mitigation strategies that need to be considered should also be identified.

The basic steps needed to determine effective mitigation strategies involve conducting a hazard assessment, a risk analysis, and a vulnerability assessment of facility assets (Haddow and Bullock 2003). The results of an effective hazard analysis can help identify the types of mitigation needed to address a particular hazard. The earlier a safety manager can identify and understand what hazards exist, the greater the potential to reduce risks and loss of life or property (Pine 2007). A vulnerability assessment can be conducted to address one or more perceived threats. A business or industry that promotes the implementation of effective mitigation strategies can realize many benefits. Preventing or reducing a hazard can protect critical infrastructure, potentially keeping the business from being destroyed in the event an emergency situation occurs. Practicing mitigation can save lives, property, assets, and money. Mitigation actions may include support for effective land-use planning or support for roads that promote both safe transport of hazardous materials and effective evacuation routing if an emergency situation occurs. Mitigation involves keeping people away from areas of potential danger through zoning, building codes, incentives, and other methods that might be utilized prior to a potential disaster occurring (Smith, Council, and Rogerson 2006).

Skill/knowledge: A safety manager must have a working knowledge and understanding of effective mitigation tools since "mitigation is the cornerstone of emergency management" (Haddow and Bullock 2003).

Safety Manager and Department Roles During an Emergency

A safety manager should identify all departments within a facility and assist them in practicing appropriate responses for their department and for employees designated as key response-team leaders during an emergency. The emergency plan should outline the responsibilities of each department during an emergency. For example, if a utility outage occurs, the actions that should be taken by the utility, maintenance, and equipment departments will be outlined. Actions that should be taken by other departments that are directly responsible for production or a service should also be outlined in the plan.

For response to a disaster caused by a specific hazard, the facility's emergency management plan should include an outline of actions by department, including:

- staff duties by department and assignment
- measures to take for equipment failure
- shelter-in-place locations
- evacuation procedures

(Smith, Council, and Rogerson 2006)

Each facility determines how it will communicate emergency roles and actions to its employees. If the safety manager is not directly involved in decisions related to emergency response, he/she must be informed and have the authority to implement the planned responses should the need arise. A safety manager needs to speak with the company manager to find out how a facility provides information on its emergency plan to employees.

All facilities should have an effective emergency management and planning process. Ongoing planning must include operational training and drills since training employees can enhance response capacity. The facility emergency plan should outline required training of all employees designated as responders (Perry and Lindell 2007). According to FEMA (2003), a drill is "a small and limited exercise to improve a single function in response and recovery operations." According to McEntire in *Disaster Response and Recovery* (2007), a drill should be composed of the following components:

- identification of the purpose of the drill
- who is involved in the drill
- the location of the drill
- the type of hazard to be simulated
- a narrative introduction to the drill
- potential problems or situations arising from the hazard
- actions to be taken
- an evaluation

Skill/Knowledge: It is important for the safety manager to assist in the identification and development of the roles that departments and employees will play in an emergency response situation, since the safety department will be expected to play a major role in assisting a facility to effectively respond to an emergency situation.

Emergency Management Responsibilities at the Local, State, and Federal Levels

The local county or city organizational structure for emergency management usually includes the mayor or county manager; a local emergency management board that includes community and business representatives; staff from local homeland security, emergency management, police, and fire departments; and representatives from other local departments. A comprehensive community-level emergency plan should be updated on a regular basis and should provide specific recommendations to address emergencies impacting local businesses or industrial facilities. The traditional role of local governments in emergency management planning has been to focus primarily on emergency preparedness and response (Drabek 1991).

Responsibility and decision-making authority at the state level for emergency management are given to the governor or a person designated by the governor. Examples of the types of authority vested in the governor are: suspending state statutes; directing evacuation efforts; providing for the procurement of needed personnel, materials, and facilities without regard to existing limitations; authorizing the release of emergency funds; controlling access to the disaster area; issuing emergency declarations; and applying for federal disaster assistance and emergency response. Daily emergency management functions are conducted by a lead state agency designated to perform that function. The state emergency management office may fall under one of several departments or agencies. It may be under the governor's office, a state council, or the state police, or, as it does in 22 states, come from a division under a civilian department in the adjutant general's office (Drabek and Hoetmer 1991). The state emergency management office administers federal funds and is in charge of coordinating resources within the state. Often, the safety manager is the liaison with governmental officials (Smith, Council, and Rogerson 2006).

The federal government gets involved rapidly if a disaster is multistate and crosses state lines. The basic role of the federal government is to protect life and property in case of disaster. The first action of the federal government is to issue a Presidential Declaration of Disaster. The federal government also assists state and federal governments in recovery after a disaster. To facilitate this, Congress can appropriate additional funding for a disaster area. The Federal Emergency Management Agency (FEMA) can issue loans to communities, businesses, and homeowners affected by a disaster. Created in 1978, FEMA holds the lead role in the coordination of national functions in case of natural, technological, and civil disasters. FEMA was designed to deal with disasters using an all-hazards approach. However, it is currently located under the Department of Homeland Security. Budget reductions over the past five years within FEMA have shifted funds away from federal planning for natural disasters as a result of response and recovery operations to the events of 9/11 (Haddow and Bullock 2003). These multiyear budget reductions have reduced the operational capacity of FEMA. The events related to the response of FEMA after Hurricane Katrina and the flooding of Louisiana and Mississippi in August and September 2005 demonstrated clearly the ongoing need for improvement in the ability of the federal, state, and local governments to respond to natural disasters in the United States.

A Safety Manager's Role in Responding to Disaster

A safety manager should know the general needs of emergency preparedness and response prior to taking on a professional position; however, he or she will learn more about specific emergency roles for safety management at a specific facility through information provided by the plant manager and by direct participation of the safety department in plan development, implementation of drills, and sponsoring special emergency management training conferences. Specific exercises with different components will be organized by a facility to improve its emergency preparedness, response, recovery, and mitigation in the face of various disasters. For example, a safety manager might lead or participate in training exercises that demonstrate the differences in actions required among responses to hurricanes, chemical spills, and a flu pandemic.

Rapid growth in the need for effective safety managers has led to a change in emphasis from safety engineering to safety management. As Planek and Fearn (1993) noted, "Data suggest that the shift in expert opinion today has been toward a higher valuation of the safety manager's role, motivation, and communication, with less emphasis on engineering, record keeping, and related activities." This has brought about the need for clarification of the role of the safety manager as greater skills are required to function in this role. The need for continuing education and continual review of plans is also necessary in these changing times. The specific role of the safety manager in emergency management will continue to evolve as hazards are identified and deficiencies in education and training are noted. The safety manager plays a vital role in the economy and in the ability of an organization to survive (Smith, Council, and Rogerson 2006).

As the safety manager learns more about the role of safety management during an emergency through education and training, and through direct participation in developing plans, the safety department can improve its response to critical situations. In an emergency, every department will have a role in maintaining a safe environment, but the role of the safety department will be key to an effective response. Knowing the role a safety department is expected to carry out can help the safety manager support companywide efforts to maintain the safety of employees and customers at a facility during a disaster. Emergencies require careful planning and practice in special areas that will be the sole responsibility of the safety department. It may be appropriate for safety managers to complete specific emergency management courses, which could include case studies and examples of emergency plans applicable to a specific industrial sector.

Points to Remember

Characterizing Disasters by Their Causes

In developing emergency plans, emergency management officials classify disasters by four main causes, or hazard groups:

- natural disasters
- technological disasters
- civil disruption or violence
- long-term or ecological changes.

The Four Emergency Management Action Areas

The four major action areas, or phases, used for emergency management are:

- emergency preparedness
- emergency response
- emergency recovery
- emergency mitigation.

Steps to Use in Preparing Emergency Plans

All plans must include descriptions and actions for relocation and evacuation of employees and customers. The process of developing an emergency plan for a facility includes:

1. identifying hazards
2. determining which hazards are the biggest risks
3. outlining actions that prevent or reduce hazards
4. outlining actions to take in advance to ensure that the response and recovery from an emergency, if it occurs, represents the best possible effort.

Important Terms

Disaster: A situation that occurs when the disruption to a facility or community due to a natural or man-made hazard results in many injuries, the loss of lives or a large amount of property is destroyed.

Emergency: A situation resulting from a man-made or natural hazard requiring a rapid response to minimize damage

Emergency management: The process of addressing the preparedness, response, recovery, and mitigation actions required to allow an area to respond to a major emergency or disaster and return to normal operations as soon as possible with minimum damage.

Hazard: Any event that can cause a major emergency, disrupting service and creating unsafe conditions.

References

Centre for Research on the Epidemiology of Disasters (CRED). School of Public Health, Université Catholique de Louvain. *Disaster Statistics* (retrieved December 28, 2005). www.unisdr.org/disaster-statistics/introduction.htm

Cox, Robert D. April 13, 2006. *Hazmat: Hospital and Community Planning for Hazmat Incidents* (retrieved January 3, 2007). www.emedicine.com/emerg/topic228.htm

Drabek, T., and G. Hoetmer, eds. 1991. *Emergency Management: Principles and Practice for Local Government*. Washington, D.C.: The International City Management Association.

Federal Emergency Management Agency (FEMA). 2003. *State and Local Mitigation Planning How to Guide*. FEMA 386-1, FEMA 386-2, FEMA 386-3, FEMA 386-4, FEMA 386-5.Washington, D.C.: FEMA.

______. *Emergency Management Guide for Business & Industry* (retrieved April 30, 2007). www.FEMA.gov.

Findley, M., S. Smith, T. Kress, and G. Petty. "Construction Safety Program Elements that Lead to Decreases in On-the-Job Injuries and Reduction in Workers Compensation Related Costs." *Professional Safety* (February 2004) 49(2):14–18.

Haddow, G., and J. Bullock. 2003. *Introduction to Emergency Management*. Burlington, MA: Butterworth Heinemann, an imprint of Elsevier Science.

Hall, H. I., G. S. Haugh, P. A. Price-Green, et al. "Risk Factors for Hazardous Substance Releases that Result in Injuries and Evacuations: Data from 9 States." *American Journal of Public Health* (June 1996) 86(6):855–57.

Kendra, J., and T. Wachtendorf. 2003. "Creativity in Emergency Response to the World Trade Center Disaster" in Monday, Jacqueline L. (ed.), *Beyond September 11th: An Account of Post-Disaster Research*. Program on Environment and Behavior Special Publication #39. Boulder: University of Colorado, Institute of Behavioral Science: Natural Hazards Research and Applications Information Center.

Krieger, G., and J. Montgomery, eds. 1997. *Accident Prevention Manual For Business and Industry*. Itasca, IL: National Safety Council.

McEntire, D. A. 2007. *Disaster Response and Recovery*. Hoboken, NJ: John Wiley and Sons.

Occupational Safety and Health Administration (OSHA). *Job Hazard Analysis*. Revised 2002. OSHA 3071 (retrieved January 3, 2006). www.osha.gov

Perry, R., and M. Lindell. 2007. *Emergency Planning*. Hoboken, NJ: John Wiley and Sons.

Pine, J. C. 2007. *Technology in Emergency Management*. Hoboken, NJ: John Wiley and Sons.

Planek, T. W., and K. T. Fearn. 1993. "Reevaluating Occupational Safety Priorities: 1967 to 1992." *Professional Safety* 38(10):16–21.

Smith, S. 2001. *The Ten-Point Checklist for Emergency Preparedness*. Developed at The University of Tennessee, Safety Education Department. Reviewed and distributed by the Educational Resources Division of the National Safety Council. www.nsc.org/issues/prepare.htm

Smith, S., K. Council, and W. Rogerson. "Facility Emergency Planning for Safety Managers." Presentation at the National Safety Council Annual Meeting, November 6, 2006, San Diego, CA.

Smith, S., L. Peoples, and K. Council. 2005. "Effective Evacuation Planning: An Assessment of Evacuation Planning Efforts Implemented During Hurricane Season in the United States." Proceedings of the International Emergency Management Society, Faroe Islands.

Smith, S. M., and W. Rogerson. "The 10-Point Checklist." Proceedings of the International Emergency Management Society, May 2002, Waterloo, Canada.

Appendix: Ten-Point Checklist for Emergency Preparedness

This checklist was developed at the University of Tennessee Safety Education Department by Dr. Susan M. Smith, Assistant Professor of Safety and Public Health, and students participating in the UT Safety Workshop. The workshop was a graduate-level course that focused on the emergency-preparedness needs of small businesses and nonprofit organizations.

NOTE: Evaluators should indicate the status of facilities inspected (see below for a suggested rating key) to help ensure that repairs, upgrades, and replacements are made as needed.

Rating Key:
S/A = substandard or adequate
W/B = works or broken
U = unknown

	STATUS				
	YES	W/B	S/A	NO	U
WARNING SIGNALS					
VISUAL ALARMS					
Are visual alarms present and detectable from work areas?					
Comments:					
Are visual alarms activated automatically by a sensor system?					
Comments:					
Are visual alarms activated manually?					
Comments:					
Are visual alarms placed 80 inches above the highest floor level or 6 inches below the ceiling, whichever is lower?					
Comments:					
In large rooms and spaces exceeding 100 feet across, are the visual alarms spaced around the perimeter no more than 100 feet apart or are they suspended from the ceiling?					
Comments:					

	STATUS				
	YES	W/B	S/A	NO	U
AUDIBLE ALARMS					
Are audible alarms present?					
Comments:					
Are audible alarms operational?					
Comments:					
Are audible alarms:					
Activated automatically by a sensor system?					
Comments:					
Electric powered?					
Comments:					
Electric powered with battery backup?					
Comments:					
Are smoke/heat detection systems present?					
Comments:					
Are smoke/heat detection systems operational?					
Comments:					
Are there separate alarms or signals for different kinds of emergencies?					
Comments:					
Are units installed with tamper-proof screws to prevent tampering and injury?					
Comments:					
Is there a regular maintenance schedule for alarms and detectors?					
Comments:					
Is there a regular maintenance schedule for alarms and detectors?					
Comments:					

	STATUS				
	YES	W/B	S/A	NO	U
AUDIBLE ALARMS (cont.)					
Does the fire alarm automatically notify security?					
Comments:					
Does the fire alarm automatically notify the fire department?					
Comments:					
Does the fire alarm have an "ALL CLEAR" signal?					
Comments:					
Is the alarm control panel accessible when the building is occupied?					
Comments:					
Does the control panel show where the fire is located?					
Comments:					
COMMUNICATION					
Is there a phone accessible on every floor?					
Comments:					
Is the phone equipped with a TDD (telecommunication device for the deaf)?					
Comments:					
Are emergency numbers attached to or posted near the phones?					
Comments:					
Is there an alternate means of communication in case of a power outage?					
Comments:					
Is there access to two-way radio communication?					
Comments:					
Is there access to Doppler Radar?					
Comments:					

	STATUS				
	YES	W/B	S/A	NO	U
EVACUATION					
Are evacuation routes posted?					
Comments:					
Are evacuation routes properly marked?					
Comments:					
Are secondary routes of exit identified?					
Comments:					
Are the evacuation routes the proper width (at least 22" wide)?					
Comments:					
Are exits marked with an exit sign?					
Comments:					
Are exits signs Illuminated?					
Comments:					
Are exits signs provided wit the word "EXIT" in lettering at least 5" high and stroke lettering of at least 1/2" wide?					
Comments:					
Are emergency lights present?					
Comments:					
Are emergency lights working properly?					
Comments:					
Are exits supported with emergency lighting?					
Comments:					
Is there an emergency light in each hallway?					
Comments:					

	STATUS				
	YES	W/B	S/A	NO	U
Is there an emergency light in each lobby?					
Comments:					
Are exit routes well lighted?					
Comments:					
Are exit routes unobstructed?					
Comments:					
Are there at least two exits in all occupied rooms?					
Comments:					
Are exits accessible to people with disabilities?					
Comments:					
If an exit door is not accessible, are there signs to indicate the nearest accessible exit?					
Comments:					
Are there sufficient exits to permit prompt escape in case of an emergency?					
Comments:					
Are "NOT AN EXIT" routes clearly marked?					
Comments:					
Are exit doors easily opened (without a key) from the direction of exit traveled?					
Comments:					
Do glass exit doors meet safety requirements for human impact?					
Comments:					
Do exit doors function properly?					
Comments:					

	STATUS				
	YES	W/B	S/A	NO	U
Are exit doors of the type that allow emergency escape from inside?					
Comments:					
Is panic hardware installed on exit doors?					
Comments:					
Are there any architectural problems (such as narrow aisles or evacuation stairs) in the building?					
Comments:					
Do self-closing fire doors work properly?					
Comments:					
Are stairwells at least 22" wide?					
Comments:					
Are guardrails sturdy?					
Comments:					
Are guardrails free from rough edges?					
Comments:					
Are handrails sturdy?					
Comments:					
Are handrails free from rough edges?					
Comments:					
Are standard guardrails provided wherever aisles or walkway surfaces are elevated more than 30" above any adjacent floor or the ground?					
Comments:					
Are the building's access roads for EMS (emergency medical services) kept free of obstructions?					
Comments:					

	STATUS				
	YES	W/B	S/A	NO	U
Are access walkways to the building for the EMS personnel kept free of obstructions?					
Comments:					
Is there a designated assembly point?					
Comments:					
Are each building's evacuation points set at a safe distance?					
Comments:					
UTILITIES/ELECTRICAL CONTROL					
Is the building equipped with gas shut-off valves?					
Comments:					
Are the gas valves marked?					
Comments:					
Are the gas valves accessible?					
Comments:					
Is the building equipped with shut-off switches for electricity?					
Comments:					
Are the electrical switches marked?					
Comments:					
Are the electrical switches accessible?					
Comments:					
Do junction boxes close properly?					
Comments:					
Are electrical cords in good condition, not frayed, etc.?					
Comments:					

	STATUS				
	YES	W/B	S/A	NO	U
Are extension cords located so they do not present a trip hazard?					
Comments:					
Are electrical outlets overloaded?					
Comments:					
Are electrical enclosures such as switches, receptacles,, etc., provided with tight-fitting covers or plates?					
Comments:					
Is storage around electrical equipment safely arranged?					
Comments:					
Is all defective equipment taken out of service?					
Comments:					
Is lighting adequate?					
Comments:					
Are light fixtures operable?					
Comments:					
Is the building equipped with steam shut-off valves?					
Comments:					
Are the steam shut-off valves accessible?					
Comments:					
Does the building have a back-up energy source?					
Comments:					
Are rooms equipped with back-up lighting?					
Comments:					
Are evacuation routes equipped with back-up lighting?					
Comments:					

	STATUS				
	YES	W/B	S/A	NO	U
FIRE SUPPRESSION					
Are the appropriate type of fire extinguishers:					
Available in the correct number?					
Located in all appropriate places?					
Charged?					
Inspected annually?					
Free from obstruction?					
Visible?					
Comments:					
Are fire alarm pull stations:					
In place?					
In a good state of repair?					
Free from obstruction?					
Visible?					
Comments:					
Is there a sprinkler system?					
Comments:					
Are the sprinkler heads free from obstruction?					
Comments:					
Is there a fire hydrant located in the vicinity of the building?					
Comments:					
SEVERE STORM/TORNADO SHELTER					
Is there a plan to provide shelter in an alternative facility in case of severe weather?					
Comments:					
Will this shelter accommodate the number and type of individuals designated or assigned to it?					
Comments:					
Does the shelter provide adequate protection from severe storms/tornados/earthquakes?					
Comments:					

	STATUS				
	YES	W/B	S/A	NO	U
Are emergency evacuation routes communicated?					
Comments:					
Are the appropriate shelter areas properly marked?					
Comments:					
Is there a means of communication available in the shelter?					
Comments:					
Is an alternate source of power available for the entire building or designated shelter area?					
Comments:					
Is survival equipment (food, water, first-aid equipment, blankets, flashlights) available in the shelter?					
Comments:					
MANAGEMENT ISSUES					
Does the facility have an emergency preparedness and evacuation plan?					
Comments:					
Are evacuation drills performed regularly?					
Comments:					
HOUSEKEEPING					
Are floor finishes in a good state of repair?					
Comments:					
Is the building interior clean and orderly?					
Comments:					
Are storage areas clean and orderly?					
Comments:					

	STATUS				
	YES	W/B	S/A	NO	U
Is equipment properly stored?					
Comments:					
Does furniture restrict egress from the building?					
Comments:					
Is the outside of the building clearly marked with a name and number?					
Comments:					
Are all rooms numbered?					
Comments:					
Are cleaning materials stored in a secure cabinet or room?					
Comments:					
Are elevated surfaces more than 30 inches above the floor or ground provided with a standard guardrail?					
Comments:					
Are stairwells being used for the storage of materials?					
Comments:					
Are flammable liquids appropriately stored?					
Comments:					
Is the area free of an accumulation of combustible materials?					
Comments:					
BOMB THREATS					
Are telephones equipped with recording devices?					
Comments:					
Are call-in Bomb Threat Checklists available at telephones?					
Comments:					

	STATUS				
	YES	W/B	S/A	NO	U
Is a bomb threat procedure in effect?					
Comments:					
Is an emergency contact list available?					
Comments:					
Have prearranged signals been established to announce bomb threats?					
Comments:					
Does the Facility Crisis Coordinator conduct threat assessments?					
Comments:					
Are all rooms numbered?					
Comments:					
Have bomb search teams been trained?					
Comments:					
Are employees trained in letter/parcel bomb recognition?					
Comments:					
Are withdrawal distances to evacuation assembly areas "safe" from bomb blast and fragmentation?					
Comments:					
Has a Crisis Command Center been identified?					
Comments:					
SECURITY ISSUES					
Does the organization have a written critical incident and business continuity plan?					
Comments:					
Has an emergency procedures and evacuation plan been written and provided to employees with training?					
Comments:					

	STATUS				
	YES	W/B	S/A	NO	U
Have back-up plans for securing office space, equipment, etc., been developed?					
Comments:					
Have integrated protocols with emergency services been developed?					
Comments:					
Does every employee have a photo ID?					
Comments:					
Does building security permit anyone on site after hours without a current ID?					
Comments:					
Are individuals who do not have a photo ID directed to security?					
Comments:					
Are employees empowered/trained to challenge anyone on site to present their badge or call security/management?					
Comments:					
Are employee calendars maintained as well as client/guest sign-in logs?					
Comments:					
Is mandatory security and response training in place for all receptionists?					
Comments:					
Are panic alarms installed on each floor?					
Comments:					
Have call-in procedures been established for after emergencies?					
Comments:					
Are procedures in place for client contact after an emergency?					
Comments:					

	STATUS				
	YES	W/B	S/A	NO	U
Are security escort procedures in place for employees working during "off" hours?					
Comments:					
Have all mailroom procedures been evaluated?					
Comments:					
Are procedures/protocol in place for family notification and grief counseling?					
Comments:					

Building Director/Representative ______________________ Date ______________

Evaluator ______________________ Date ______________

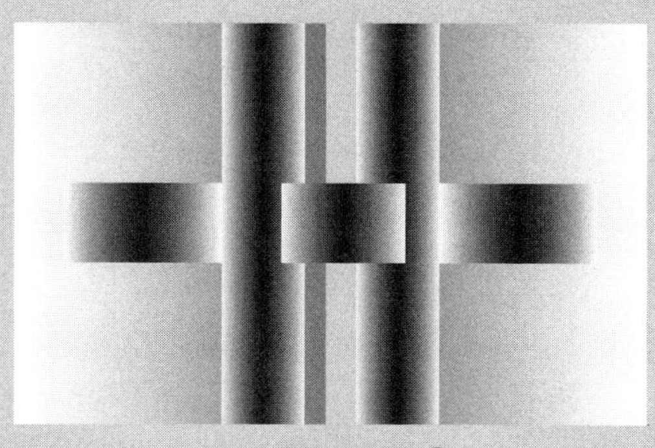

SECTION 2
EMERGENCY PREPAREDNESS

COST ANALYSIS AND BUDGETING

Pam Ferrante

LEARNING OBJECTIVES

- Gain a working knowledge of the development and implementation of an employer-based Emergency Response Team (ERT), specifically the types of activities they participate in and how they are strategically managed by an in-house safety professional.
- Understand the factors and their associated methods that should be considered when determining the best method of providing for emergency response: a fully staffed in-house ERT, a fully outsourced ERT, or some combination of both.
- Be able to use the above knowledge and methods to analyze the preferred method for providing emergency response within the safety professional's organization and present the analysis in a narrative format for review and consideration by an employer.

THIS CHAPTER WILL FOCUS ON the development of employer-based emergency response teams (ERTs) from a cost analysis and budgeting perspective. In part, it will provide a format to assist the safety professional in assessing the need for an employer-based team based upon the actual hazards present in the workplace; it will also provide a method for determining whether contracting out emergency response services is a more economically practical alternative. A significant portion of this chapter will be devoted to a detailed discussion of the various factors that must be fully evaluated from a cost perspective, from initial outlays for employee training and equipment to reviewing and planning for the ongoing costs of maintaining an in-house ERT. Both issues are significant to any proper review of this issue.

ERTs are developed to address a variety of emergency situations that may occur, and typically include responses at varying levels to fires, hazardous materials incidents, any natural disasters, and confined space rescues. (For purposes of brevity, this chapter will confine its discussion to those four.) Each employer can choose which, if any, of those four situations will be handled internally and to what degree. The safety professional is often called upon to provide a cost/benefit analysis to be used in the decision-making process. Moreover, if an employer chooses to create an internal ERT, the safety professional will often be tasked with implementing the activities under the team's purview as well as managing the budget for its operation.

Regulatory Review

One of the first tasks to be undertaken by the safety professional is a determination of exactly what the employer is required to do by regulation; this default position typically represents the minimum an employer can undertake. At the opposite end of the spectrum is what the employer may choose to do based upon its own preferences. Organizations may choose to create an ERT that exceeds the minimum requirements in an infinite number of varying degrees.

The regulatory requirements for each of the major types of activities that fall under ERTs vary. Below is a brief summary of the various federal regulations that apply to the four types of emergencies being discussed in this chapter. It should be noted that the safety professional must also undertake a proper review of any additional state and local statutes that may apply with regard to reporting spills and/or releases, but the content of that review is beyond the purview of this chapter.

Hazardous Materials Incidents

There are two key federal statutes that require some type of emergency response by an employer for hazardous materials incidents: Title III of the 1986 Superfund Amendments and Reauthorization Act (SARA Title III), also called the Emergency Planning and Community Right-to-Know Act (EPRCA); and the Comprehensive Environmental Response, Compensation and Liability Act of 1980 (CERCLA). These two pieces of legislation require that the owner of a facility notify the appropriate authorities as soon as the release of a reportable quantity (RQ) of a hazardous substance, as defined by SARA and CERCLA, is known. This notification is not required if the spill or release is contained solely within the facility boundaries or does not have the potential to enter the environment. These regulations are concerned with the potential impact if the spill enters the soil, air, or water and allow the employer substantial discretion in how the spill or release is handled within the facility's boundaries. For the safety professional, this flexibility has a substantial impact on how ERTs are developed and maintained and how to determine what level of hazards the facility is capable of responding to versus when the response is turned over to a local public hazardous materials team.

In addition to the above environmental regulations that define when a response is required for a hazardous materials incident, the Occupational Safety and Health Administration (OSHA) Hazardous Waste Operations and Emergency Response (HAZWOPER) regulations, found at 29 Code of Federal Regulations (CFR) 1910.120, also detail the requirements for the protection of workers that may be involved in an internal response to a hazardous materials incident as well as when an Emergency Response Plan (ERP) must be developed by an employer. The ERP addresses numerous requirements of an emergency response that are part of the process of analyzing what an employer is required to do versus what an employer may choose to do [29 CFR 1910.120(q)(1)].

Fire Response

An employer's responsibility with regard to workplace fires is essentially limited to developing work practices that prevent them, such as the prohibition of smoking areas where flammable substances are stored or used and frequent removal of combustible materials from the workplace. OSHA further requires the placement and maintenance of fire extinguishers and suppression systems appropriate for the types of fires that may occur, as well as compliance with the National Fire Protection Association's (NFPA) Life Safety Code with regard to exit routes and evacuation plans. The primary goal of these is the safe and orderly evacuation of employees and visitors, rather than an attempt to fight the fire (NFPA-101 2000).

The OSHA regulations for fire prevention and the use of fire protection systems for the construction industry are found at 29 CFR 1926.24 and for general industry at 29 CFR 1910, Subparts E and L. OSHA does address internal fire-fighting capabilities by employers at 29 CFR 1910.156 (Fire Brigades). However, this section addresses the optional use of fire brigades, and does mandate their development.

Confined Spaces

Confined spaces are addressed by OSHA in the general industry regulations (29 CFR 1910.146), which are focused substantially on developing a written program for safe entries through the use of a permitting system. Nonentry rescues can be performed by employees trained at the basic level, but require no specialized equipment beyond a full body harness on any entrant, which is attached to a lifeline and some mechanical means for extricating the entrant from the space, such as a tripod. The regulations allow for an internal rescue team to be developed at the employer's discretion. This team may be trained and equipped to perform entry-type rescues as well as initial medical response, including first aid and cardiopulmonary resuscitation. An exception to the voluntary development of a confined space rescue team is in situations where the employer's location is so remote that access to emergency medical response and rescue is not in close proximity. In such cases, an internal rescue team is no longer an option, but a requirement, if the employer chooses to perform confined space entries.

Natural Disasters

OSHA addresses emergency response for other types of situations, such as weather-related emergencies and natural disasters, in the general industry regulations at 29 CFR 1910.38 and requires the development of an Emergency Action Plan. This plan is only required to address safe evacuation of the workplace and a means of accounting for personnel. Similar to fire response, employees are not required to take any other actions.

Reviewing Other Emergency Plans

In addition to the regulatory review for the four major types of emergencies noted above, the safety professional will also need to be aware of any other emergency-type plans that are in place within his/her organization, all of which are required by various regulatory bodies. These include a Spill Prevention Control and Countermeasures Plan, required by the Oil Pollution and Prevention Act (40 CFR Part 112), to provide measures for preventing an oil spill into navigable waters and a Hazardous Materials Security Plan, required by the Department of Transportation (49 CFR Part 172), for organizations that ship or transport various quantities of hazardous materials. Prior to the development of an ERT, these existing plans will need to be reviewed to determine what requirements the organization may already have put into place with regard to emergency response and to evaluate how to integrate existing requirements into any new response capabilities. Finally, any new emergency response capabilities may require that existing plans be revised and adjusted accordingly.

Once the regulatory requirements for what the ERT must be prepared to handle are ascertained, the process of calculating and analyzing cost factors can begin with what is required. Optional activities can be added into the proposal submitted by the safety professional based upon his/her assessment of the need/value of undertaking a greater level of response or based upon what the employer has previously indicated a willingness to undertake.

Assessing the Economic Perspective

In later portions of this chapter specific details related to the individual aspects of managing an ERT will be reviewed. These include training requirements, record-keeping issues, and equipment maintenance. In this introductory section, however, a few words about performing an economic analysis with regard to the overall functioning of the ERT are essential. This economic analysis should be composed of four key elements: reviewing the data on actual costs of developing and implementing an ERT, performing a comprehensive economic analysis of the use of an ERT, reviewing the organization's economic performance to include how the costs of managing an ERT will (or will not) contribute, and comparing the use of funds to develop and maintain an ERT within the overall investment strategy of the company. (For a general overview of fundamental economic analysis for the safety professional, the reader is referred back to the "Basic Economic Analysis and Engineering Economics" chapter at the beginning of this text.

Data Compilation and Review

Assuring that sufficient, accurate data is reviewed and considered is the first and most crucial step. Considerable information will need to be made available to those in the organization who participate in making decisions about investment allocation. The list below provides several resources for each of the four types of emergency responses that can be utilized to assist in the compilation of development and implementation costs. These resources are listed because they provide the safety professional with needed details that assist in the process of defining what the organization will do, and what it needs to accomplish those goals.

- Hazardous Materials Incident Response Teams
  1. *Hazmat Team Planning Guidance* (U.S. EPA 1990). The reader is cautioned that the cost estimates listed in this document reflect 1990 prices and should not be used while performing a cost analysis; however, this document does provide the safety professional with a clear method of walking through the steps necessary to gather all of the essential data needed for a cost review. (The reader is referred to the subsequent section on equipment selection.) Calculating current costs will require consultation with vendors and others who offer the equipment for sale. This document also provides additional reference materials that can be consulted.
  2. *Occupational Safety and Health Guidance Manual for Hazardous Waste Site Activities* (NIOSH 1985).
  3. *Hazardous Materials Emergency Planning Guide* (National Response Team 1987)
  4. National Fire Protection Association (NFPA), *Recommended Practice for Responding to Hazardous Materials Incidents* (NFPA-471 2002) This document and the subsequent one provide comprehensive listings of what an ERT must be prepared to do in the event of an emergency.
  5. *Standard for Professional Competence of Responders to Hazardous Material Incidents* (NFPA-472 2002).
  6. Occupational Safety and Health Administration (OSHA), Title 29, Code of Federal Regulations (CFR), Part 1910.120 (*Hazardous Waste Operations and Emergency Response*).

The NFPA is an organization that has published consensus codes and standards relating to fire safety and emergency response since 1896. It should also be noted that, as a voluntary organization whose prime purpose is the development and publication of consensus standards and codes, NFPA updates its documents on a more frequent basis than any of the governmental entities named above. This allows for inclusion of newer technologies and best practices that are essential in any data analysis process, and the safety professional is encouraged to utilize sources from both groups.

- Fire Response Teams
  1. *Standard on Industrial Fire Brigades* (NFPA-600 2000). This document provides much of the essential information needed by the safety professional to enable a comprehensive calculation of the costs of establishing and implementing an internal fire-fighting capability at any of three levels: incipient response, advanced exterior response, and interior structural response. In addition to specific equipment lists for each level, it also includes management responsibilities, training requirements, and medical surveillance requirements.
  2. *Standard on Fire Department Occupational Safety and Health Program* (NFPA-1500 2003). This document is specific to the medical surveillance requirements and prescreening criteria used for fire-fighting response teams.
  3. *Emergency Response Team Handbook* (Cocciardi 2004) This book contains several chapters

that address the issue of developing and implementing internal fire-fighting capabilities, again at all of the three standard levels noted above.

- Confined Space Entry Rescue Teams
  1. *Standard for Rescue Technician Professional Qualifications* (NFPA-1006 2003). Although this standard addresses specific rescue situations that are beyond the discussions of this chapter, such as surface water rescues and subterranean rescues, one section does provide excellent coverage of the development of a confined space entry rescue team, including preplanning, monitoring of atmospheric hazards, and entry techniques.
  2. *Emergency Response Team Handbook* (Cocciardi 2004). Coverage of confined space rescues is somewhat brief in this book, but several helpful points are made and additional information on the requirements of a confined space rescue team is presented.
- Natural Disaster Preparation
  1. *Standard on Disaster/Emergency Management and Business Continuity Programs* (NFPA 1600 2004). This standard provides a comprehensive look at emergency response and business continuity planning and includes several important pieces of information that will assist in developing an internal response to such emergencies as weather, fires, and terrorism attacks.
  2. The Federal Emergency Management Agency has prepared a number of documents that can be used to evaluate and prepare a workplace for response to various natural disasters. These documents are available online and can be reviewed and downloaded at http://www.ready.gov/business/index.html (FEMA 2006).

Once the safety professional has obtained and reviewed a sufficient variety of resources for this step of the process, a review document should be generated that summarizes essential elements that must be taken into consideration. Different methods can be used for this process, including narrative text or spreadsheets. Combinations of styles can be effective in providing differing perspectives and appealing to differing management styles. The safety professional should endeavor to determine which method is required by internal procedures or which is traditionally utilized within the organization for this type of analysis and replicate it in order to assure key decision makers attend to the content. The safety professional should also be conscious of the language used in this document. Business and accounting terms are more likely to be read and responded to than safety language and terms.

Fundamental in this step is a focus on general issues regarding ERTs and the cost of developing and maintaining them. As noted, later sections of this chapter will provide details for specific aspects of ERTs; however, the safety professional's job at this juncture is to provide necessary information in a concise document dedicated to global cost-analysis issues that address the organization's regulatory requirements for emergency response as well as some beginning discussion of how an ERT would function within the organization's structure. This should include a discussion of the types of activities in which it must and should participate. One way to look at this initial cost analysis is to consider this section an executive summary.

Economic Analysis

In order to compete in the organization's overall investment-allocation process, the safety professional must provide the decision makers with two key pieces of information. The first addresses the amount of investment that must be made in order to achieve regulatory compliance. This can be regarded as the least amount that can be spent for the organization to remain in compliance and is important as the safety professional attempts to provide justification for additional expenditures beyond compliance requirements.

This portion of the analysis must also lay out the clear implications should the organization choose not to comply with regulatory requirements. This review should be two-pronged. First, the fines and citations that would be levied by various governmental authorities following such actions/inactions as the lack of report filings (e.g., toxic chemical release reports mandated by SARA Title III) or the failure to properly train employees in the use of fire extinguishers. It should also include the citations and penalties that would be levied following an on-site inspection by authorities from OSHA, EPA, or the organization's state-based authorities, resulting from the lack of internal emergency response preparedness. This would likely be due to the absence of training for employees; lack of equipment to protect employees in the event of a hazardous materials release, fire, or natural disaster; lack of a confined space rescue team that was not in close enough proximity; or insufficient preventative and/or warning systems. The costs of these penalties are widely available to the safety professional through researching the various regulations promulgated by OSHA, EPA, and other state authorities. Each regulatory publication will have a fine and penalty section that details the specific costs that can be calculated based on their applicability to a specific organization. Tables 1 and 2 delineate the current penalty structure for violations of OSHA regulations (OSH Act of 1970, Section 17) and for violations of the Pennsylvania Act 1990-165, promulgated by the State of Pennsylvania to comply with Title III of the Federal Superfund Amendments and Reauthorization Act.

TABLE 1

Penalty Structure for Violations of Occupational Safety and Health Regulations

Definition of Violation	Penalty
Willful or repeat violations of the requirements of section 5 of the OSH Act, any standard, rule, or order promulgated pursuant to section 6 of the OSH Act, or regulations prescribed pursuant to the OSH Act	A civil penalty of not more than $70,000 for each violation, but not less than $5000 for each willful violation
A serious violation of the requirements of section 5 of the OSH Act, of any standard, rule, or order promulgated pursuant to section 6 of the OSH Act, or of any regulations prescribed pursuant to the OSH Act	A civil penalty of up to $7000 for each such violation.
A violation of the requirements of section 5 of the OSH Act, of any standard, rule, or order promulgated pursuant to section 6 of the OSH Act, or of regulations prescribed pursuant to this Act, and such violation is specifically determined not to be of a serious nature	A civil penalty of up to $7000 for each violation.
Failure to correct a violation for which a citation has been issued under section 9(a) within the period permitted for its correction ("failure to abate")	A civil penalty of not more than $7000 for each day during which such failure or violation continues.
Willful violation of any standard, rule, or order promulgated pursuant to section 6 of the OSH Act, or of any regulations prescribed pursuant to the OSH Act, and that violation caused death to any employee.	Upon conviction, punishment of a fine of not more than $10,000 or imprisonment for not more than six months, or both; except that if the conviction is for a violation committed after a first conviction, punishment shall be a fine of not more than $20,000 or imprisonment for not more than one year, or both.

(Partial listing OSH Act of 1970, Section 17 "Penalties")

TABLE 2

Fee for Violations of Pennsylvania Hazardous Materials Emergency Planning and Response Act

Definition of Violation	Penalty
Failure to pay local Hazardous Chemicals fees as established by County ordinance, or file Local Emergency Planning Council (LEPC) in-kind service approval to the county Treasurer by March 1 of each year for each chemical listed on the facility Tier II (Act 165, Section 207 (b)(2)	$35.00–$75.00
Failure to pay Hazardous Chemical fee to the Department of Labor and Industry by March 1 of each year for each chemical listed on the facility Tier II (Act 165, Section 207(c)	$10.00
Failure to pay Toxic Chemical Release Fee for each chemical reported on the Toxic Chemical Inventory Form as required by SARA Title III, Section 313 by July 1 of each year to Department of Labor and Industry (Act 165, Section 207(e)	$250.00 per chemical not to exceed $5000.00
Failure to pay Emergency Planning fee as established by county ordinance or file LEPC in-kind service approval to the county treasurer by March 1 of each year by each facility with threshold-planning quantities (TPQ) of extremely hazardous substances (Act 165, Section 207(f)	$100.00

(PA Act 1990-165)

The second prong to be analyzed is the resultant cost to the organization if there is insufficient internal preparedness or no internal preparedness. The implications are significant due to the likelihood of damage to the organization's property and equipment as well as the additional costs of clean up following the mitigation of the emergency. In addition to the costs, the organization is likely to be facing business interruption of some duration. Calculating these costs will involve some assistance and information from the various financial departments within the organization and should include both personnel costs as well as lost income due to the inability to produce a product to sell to customers. If possible, the chance that customers will go elsewhere to purchase what they need should be noted and quantified. Various annual financial reports to shareholders can be used to gather data for this part of the analysis.

Of tremendous benefit to the safety professional is the ability to calculate the long-term financial impact of regulatory citations as well as the costs of incidents related to a lack of emergency planning. Though more difficult to quantify numerically, the public relations cost needs to be addressed in some manner. This type of information and cost analysis lends substantial strength to a presentation to investment-allocation decision makers; the importance of investing time in developing this aspect of the cost analysis by the safety professional cannot be understated.

Economic Performance Review

In order to provide a rationale for investing resources in developing and maintaining an ERT, the safety professional will need to communicate the value of the ERT in the organization's overall economic performance. Some of these issues were touched upon in the previous section on economic analysis; however, the safety professional needs to look further into how investments in an ERT contribute to the company's economic performance, not simply how they prevent future cost expenditures. Undoubtedly this is a difficult case to make since it is unlikely that the investments provide tangible increased services or sales to customers. However, it is this author's opinion that the argument that intangible benefits, such as the reputation of the organization, can be made. In addition, well-trained employees may be more productive due to increased morale and confidence in their abilities and, as a result, may gain a more positive perception of their employer. In this area, the safety professional will need to rely on less direct conclusions based upon expenditures and will need to provide more indirect inferences regarding how the ERT improves the overall functioning of the organization.

One aspect to emphasize would be the organization's overall status as an industry leader and to demonstrate how, in that leadership position, the organization also understands that it must be fully prepared for all manner of situations. An inference can be developed that the organization can utilize its internal emergency preparedness in marketing and public relations documents and activities.

In addition to the above, the external activities of the organization's employees that relate to overall community emergency preparedness and response should be highlighted and promoted in this section of the analysis. Many large organizations usually employ a number of employees who participate actively in their community's emergency response capabilities, whether on the local volunteer fire department, hazmat team, or medical emergency response units. As these employees take the message of the preparedness of their employer into the community and their activities become associated with their employer, the organization's public image may be enhanced.

As noted above, the line from ERT expenditure to direct organizational economic benefit is less clear in both of these scenarios. However, the safety professional must include both direct and indirect calculations in an economic analysis and must endeavor to quantify these less direct benefits in some credible manner.

Critical Issues Affecting Selection of Emergency Response Capability

Previous discussion has centered on a review of regulatory requirements and guidelines as a means of first defining the comprehensive level of emergency response capability that can performed by an employer and then using that information to begin to flush out which elements the employer *must* perform along with which elements the employer *chooses* to perform. The following section helps narrow the selection process in two different ways, first by taking a brief look at the different levels of response by defining both ends of the continuum and then taking a closer look at the variations available to an employer. Once this framework has been established, the specific details inherent in developing and implementing an ERT can be discussed.

Levels of Emergency Response

After regulatory requirements are satisfied, employers often have the ability to decide how much control they wish to have over how the incident is managed and for how long they wish to retain direct control of the incident. As the various levels are discussed, the reader should take note of the overlapping nature of the various ERTs that may reasonably be involved in any emergency response (Cashman 1983, 184). It should be noted that, whichever model is best suited for a specific organization, the public response to any emergency must also be considered.

- Fully Internal Capability: This level assumes that the organization will manage all aspects of the emergency response for all four of the major types of emergencies that are being addressed in this chapter. This includes:
  1. The ability to contain and confine a hazardous materials release as well as complete much of the clean up.
  2. The ability to contain a fire of varying proportions through the implementation of the varying fire response levels. These include incipient fire response, advanced exterior response, and interior structural response. (It should be noted that these levels are consequential in that an employer may choose to do incipient only; incipient and advanced exterior; or incipient, advanced exterior, and interior structural. The employer may not choose the first and third level and skip the second.)
  3. The ability to perform an entry rescue of an employee during a confined space entry.
  4. The ability to deal with the planning and preparation for natural disasters that includes the ability to shelter employees in place for brief periods of time if it is not safe to evacuate and move about the community.

  This level also often includes the ability to treat injuries that do not require hospital-type treatment. Because of the economies of scale required to operate at this capacity, this model is traditionally limited to larger organizations.

- Fully External Capability: This level assumes that nearly all of the activities outlined above are contracted out to a secondary organization. In reality, in order to assure that the incident does not rapidly progress into devastating consequences, most organizations will still need to be able to manage the initial moments or even hours of the incident. The point at which responsibility is turned over is determined largely by the time it would take for the secondary organization to arrive at the facility. For this reason, the use of this model is less widespread and contained to those organizations where quick response is physically located nearby.
- Combination of Internal and External Capability: This level allows for an organization to contract out its internal emergency response to a secondary organization in varying degrees; therefore, there is significantly more variation than in either of the two previously discussed levels. In some cases, such as hazardous materials incidents, the organization manages the incident through the majority of activities. In others, such as fire response, the organization limits itself to simple incipient response, which requires the use of basic fire-extinguishing media by a specially trained cadre of employees, but turns over all fires that are not extinguished in this manner to the local volunteer or paid fire departments. In this scenario, organizations are able to utilize local fire response capabilities, although this is much more difficult to manage when the local fire department is not paid and response times may vary. (It should be noted that, while the level of service provided is not necessarily a result of whether or not the service is volunteer or paid, the response times do vary and should be thoroughly evaluated by the safety professional if the intention is to utilize volunteer services.)

  It should also be noted that in some communities there may be limited public access onto a private facility's site for emergencies, whether they are related to hazardous materials incidents, fires, or confined space rescues. In these situations, the employer must be able to provide at least a minimum level of response and may not have the option of contracting anything to an external provider. For example, in the case of most emergencies an organization must be able minimally to move its employees to a safe location. It must also be able to summon necessary assistance for response organizations and have procedures in place for dealing with situations when communications are disrupted.

Specific Cost Analysis Factors

The discussion to this point has centered on the process a safety professional should undertake to determine the manner and type of emergency response and organization. The process used a regulatory framework to determine the minimum level of response required and then considered all possible emergencies as a means of deciding the full range of response both required and desired by the organization. What remains at this juncture is the core of emergency response capabilities that need to be developed by the organization based upon the regulatory requirements and the safety professional's judgment of which response capabilities are needed.

The remainder of the chapter will focus on evaluating the specific costs associated with an ERT and providing the safety professional with the tools and techniques to perform such an analysis for their own organization. One of the fundamental goals of the process is a prudent use of funds. Chief Don Eversole, the Hazmat Coordinator for the Chicago Fire Department believes that the major difficulty in developing and implementing ERTs is the need to "be ready and right" all the time (Cashman 2000, 203). In light of the amount of time an ERT actually spends responding to emergencies, the calculated per-hour or per-unit cost can be quite high.

In addition to the initial cost outlays required to create an ERT, the safety professional must also be prepared to analyze the ongoing cost of an ERT. Initial expenditures will likely include mostly capital assets (equipment and protective clothing); however, the organization must be able to commit to making the ERT an ongoing part of the budget with due consideration given to ongoing operating costs, including projections for annual increases in the cost of maintaining the readiness of the ERT over time.

Given this background, what are the major line-item costs that will need to be considered in the safety professional's analysis? Those line items are listed below and discussion on them follows.

1. Equipment and supplies
2. Training, drills, and other preparatory activities
3. Medical surveillance
4. Miscellaneous costs—staff time, insurance, and record keeping

Equipment and Supplies

It is helpful at this point to provide a definition of the terms equipment and supplies. *Equipment* typically refers to items that are considered capital assets that have an expected useful life, which will vary depending upon the item. For accounting purposes, equipment purchases are generally expensed where the cost is taken in the year it occurred or depreciated via an accounting method that attributes the cost of an asset across the useful life of the asset, whether it is one, five, ten, or more years. *Supplies* are items that are consumable over a shorter period of time or at once. These items need to be replaced frequently and they are expensed in total in the year in which they are purchased.

At this juncture, the safety professional needs to start with the development of an asset and supplies list that will vary based upon the type of emergencies to which the ERT team will respond. Most emergencies will include the need for some manner of personal protective equipment (PPE) as well as supplies and other necessary items. Other types of emergencies, particularly hazardous materials incidents and fires, will require air-monitoring instrumentation and decontamination equipment. In situations where the hazards are well understood and where the organization does not have any exceptionally hazardous substances or unique physical situations that would make an emergency response more complicated, the asset and supplies list will contain many standardized items that are fairly easy to obtain from a wide variety of manufacturers or suppliers. Fortunately, this situation is more often the norm (Cashman 2000, 3–6).

In situations where the organization utilizes some exotic or extremely dangerous type of hazardous substances or where the organization decides that the ERT will perform either advanced exterior fire or interior structural firefighting and/or emergency medical response, the amount and type of equipment will be more complicated and therefore more expensive and complicated to obtain.

For hazardous materials response, one resource that should be consulted is the *Hazmat Team Planning Guidance* (U.S. EPA 1990, 12–13). As noted previously in this chapter, this document contains a list of personal protective equipment as well as containment equipment and tools generally required to perform hazardous materials emergency response. It also provides general information on the type of monitoring equipment that is typically required on an incident site. The prices in this list, however, reflect costs when the document was written in 1990, but the list provides an excellent base from which to begin. Tables 3, 4, and 5 show how the specific personal protective equipment list, the containment equipment and tools list, and the air-monitoring equipment list from this document can be updated with current prices to develop a preliminary budget (EPA 1990, 13).

For fire response, the prime resource to be consulted for developing a list of necessary equipment for each of the fire response levels is the *Standard on Industrial Fire Brigades* (NFPA-600 2000). In addition the *Emergency Response Team Handbook* (Cocciardi 2004, 59–61) also provides additional supportive information useful in developing an equipment and supplies list.

For the development of an internal rescue team for confined space entries, the OSHA regulations should be consulted. They provide details of some of the requirements, particularly with regard to preparation for a rescue. Specifically, OSHA requires that a Confined Space Rescue Team perform at least an annual drill utilizing a mock victim or manikin [29 CFR 1910.146(k)(2)(4)]. The organization will need to

TABLE 3

Sample Hazardous Materials Response Personal Protective Equipment List

Budget Line Item	Quantity	Price	Cost
Tychem® Level-A encapsulated suits (Responder)	4	$850.00	$3400.00
Tychem® Level-B suits (CPF 4)	4	$165.00	$660.00
Tychem® hooded coveralls (CPF 2)	8	$20.00	$160.00
Nomex® coveralls	4	$300.00	$1200.00
Hazmat chemical boots	8	$58.00	$464.00
Latex boot covers	8	$3.75	$30.00
Nitrile gloves	8	$1.60	$12.80
PVC gloves	8	$2.50	$20.00
Butyl gloves	8	$18.50	$148.00
Box latex gloves (inner gloves)	1	$20.00	$20.00
Chemtape®	4	$20.00	$80.00
Hard hats	8	$10.50	$84.00
TOTAL			**$6278.80**

TABLE 4

Sample Hazardous Materials Response Containment Equipment and Tools List

Budget Line Item	Quantity	Price	Cost
Nonsparking bung wrench	2	$16.50	$33.00
Nonsparking drum box wrench	2	$13.50	$27.00
Nonsparking tool kit	1	$931.00	$931.00
Polypropylene shovels	2	$36.00	$72.00
Rolls caution tape	2	$22.00	$44.00
Pushbrooms	2	$32.00	$64.00
Traffic cones	12	$16.00	$192.00
85-gallon epoxy lined drums	2	$123.00	$246.00
30-gallon epoxy lined drums	2	$82.00	$164.00
8-gallon drums	2	$47.00	$94.00
25-gallon poly pails	2	$37.00	$74.00
Hazmat "A" kit	1	$1895.00	$1895.00
3" x 4' oil sorbent boom (by case)	3	$46.00	$138.00
Cases (100 each) sorbent pads	4	$37.00	$148.00
Bags of inert sorbent material	6	$15.00	$90.00
Cases (150 each) 55-gallon PE bags	1	$150.00	$150.00
TOTAL			**$4362.00**

TABLE 5

Sample Hazardous Materials Response Recommended Minimum Air-Monitoring Equipment

Instrument	Approximate Cost
Combustible gas indicator	$565.00
Oxygen meter	$250.00
Colormetric pump	$337.00
Colormetric tubes	$20.00–$95.00 per box of 10
Photoionization detector (ToxiRAE®)	$2500.00–$3500.00
Flame ionization detector (Foxboro®)	$7995.00
Mulitgas monitor (oxygen, combustible gas, up to two additional gases)	$1500.00–$4600.00

have on hand all of the equipment for the actual rescues in addition to completion of these drills.

Finally, preparations for natural disasters and other emergencies do not typically require a substantial amount of specialized equipment and supplies. However, items that may need to be purchased include water, food, and medical supplies to be used while sheltering-in-place; methods of communication for employees and others in the event of a power failure; and possibly some specific signage to be posted throughout the facility, notifying employees of evacuation routes, assembly locations, and general procedures to be followed. Drills can generally be conducted without any specialized equipment. Again, a review of the federal government documents referenced previously to help an organization develop an emergency response plan should assist in the development of a list of items to be purchased.

Once the initial written list has been generated, the safety professional should also consider consulting with additional resources to refine and revise the list prior to actual cost calculation. Opportunities to talk with other professionals in the field allow the safety professional to benefit from their experiences. It also provides a means to view products that will need to be purchased first hand. Some suggested resources to consult include:

- The local Emergency Management Coordinator (typically based at the county level) and local Fire Services Commander, both of whom are likely to be knowledgeable resources, having been involved in the development and ongoing implementation of ERTs at the public level.
- Local hazmat teams and local fire clubs and organizations will likely share their specific inventory lists as well as provide a consultative review and critique of the list.
- Suppliers and distributors of the items likely to be purchased often have technical services departments and representatives that can be consulted. The safety professional is alerted to the supplier's bias in promoting their products for sale; however, suppliers and distributors do have a level of technical experience and access to source documents that may be very beneficial.
- Local emergency management conferences and expositions offer an opportunity to view products that are used by ERTs and allow the safety professional to speak first hand with a number of suppliers of the same items. This situation provides definite advantages when performing comparisons between pricing and specifications is needed.
- Trade groups where similar-sized organizations may have already been through the process of developing an ERT can share positive and negative experiences as well as best practices.

Once the safety professional is assured that the equipment list is complete and accurate, an additional level of analysis needs to be completed before costs can be calculated—determining the duration of the expected response and the number of responders expected to participate. These two factors are necessary to determine how much of each of the items will need to be purchased, not only initially, but also to be maintained on an ongoing basis.

Information contained in the following sidebar comes extensively from OSHA regulations that deal with hazardous materials emergency response and the use of respiratory protection. Additional sources include published texts that assist in the determination of the proper level and type of chemical protective clothing

SIDEBAR

Equipping a Hazmat Team

There are several factors that need to be considered while developing equipment and supply lists for an internal hazmat team, but the primary issue concerns how much of the response the internal ERT will handle and how much will be turned over voluntarily to external providers, which has been addressed previously. The level and duration of response provides the safety professional with crucial information regarding the number of responders who will likely participate, which is based on how many hazmat teams will be dressed out and how many employees will be in each team.

For hazmat incidents, the absolute minimum number of responders in a team is two so that the "buddy system" can be practiced to assure that each responder can be seen by at least one additional employee. The use of the "buddy system" is a requirement of the OSHA HAZWOPER Regulations [29 CFR 1910.120(q)(3)(v)]. As the dangerousness of the response escalates, the 2-member buddy system must increase to four or five responders in each team.

Also of note are OSHA regulatory requirements for the provision of standby teams in addition to the response team. Entrance by a response team into an environment considered immediately dangerous to life and health (IDLH) requires a back-up team standing by to initiate rapid emergency rescue should the response team encounter problems and need to be removed from the environment with assistance [29 CFR 1910.120(q)(3)(vi) and 29 CFR 1910.134(q)(3)].

Equally important to the determination of the number of responders and teams to be prepared for is the selection of the chemical protective clothing that will match the specific hazardous substances expected to be encountered. Fortunately, safety professionals who work at specific facilities know what hazardous substances are likely to be encountered, thus allowing the selection of the clothing to be based on known hazards instead of having to prepare for a wider variety of hazards as is required for public hazmat teams that may encounter any number of hazardous substances in an emergency. In situations where the potential hazards are known, several excellent databases exist that detail the clothing choices available, both by material and manufacturer's copyrighted suit name. These include the *Quick Selection Guide to Chemical Protective Clothing* (Forsberg and Mansdorf 2002) and *Personal Protective Equipment for Hazardous Materials Incidents: A Selection Guide* (Ronk, White, and Linn 1984).

The selection of the specific chemical protective clothing must also work jointly with the level of personal protective clothing chosen for the hazmat team. The OSHA HAZWOPER regulations outline the four levels of chemical protective clothing, define situations when each level should be used and establish the general requirements of the ensemble for each level (Appendix B to the HAZWOPER regulations). For purposes of brevity, this Appendix is summarized here. (The reader is referred back to Table 3 for additional information on the cost of obtaining the listed personal protective equipment for levels A and B.)

Level A – to be selected when the greatest level of skin, respiratory, and eye protection is required.

- Positive-pressure, full-facepiece, self-contained breathing apparatus (SCBA), or positive-pressure-supplied air respirator with escape SCBA, approved by the National Institute for Occupational Safety and Health (NIOSH)
- Totally encapsulating chemical-protective suit
- Coveralls
- Long underwear
- Gloves, outer, chemical-resistant
- Gloves, inner, chemical-resistant
- Boots, chemical-resistant, steel toe and shank
- Hard hat (under suit)
- Disposable protective suit, gloves and boots (depending on suit construction, may be worn over totally encapsulating suit)

Level B – to be selected when the highest level of respiratory protection is necessary but a lesser level of skin protection is needed.

- Positive-pressure, full-facepiece, self-contained breathing apparatus (SCBA), or positive-pressure-supplied air respirator with escape SCBA (NIOSH approved)
- Hooded chemical-resistant clothing (overalls and long-sleeved jacket; coveralls; one- or two-piece chemical splash suit; disposable chemical-resistant overalls)
- Coveralls
- Gloves, outer, chemical-resistant
- Gloves, inner, chemical-resistant
- Boots, outer, chemical-resistant steel toe and shank
- Boot covers, outer, chemical-resistant (disposable)
- Hard hat
- Face shield

Level C – to be selected when the concentration(s) and type(s) of airborne substance(s) is known and the criteria for using air-purifying respirators are met.

- Full-face or half-mask air-purifying respirators (NIOSH approved)
- Hooded chemical-resistant clothing (overalls; two-piece chemical splash suit; disposable chemical-resistant overalls)
- Coveralls
- Gloves, outer, chemical-resistant
- Gloves, inner, chemical-resistant
- Boots (outer), chemical-resistant, steel toe and shank
- Boot covers, outer, chemical-resistant (disposable)
- Hard hat
- Escape mask
- Face shield

Level D – to be selected as a work uniform affording minimal protection: used for nuisance contamination only.

- Coveralls
- Gloves
- Boots/shoes, chemical-resistant, steel toe and shank
- Boots, outer, chemical-resistant (disposable)

- Safety glasses or chemical splash goggles
- Hard hat
- Escape mask
- Face shield

This part of the process also requires an evaluation of such issues as the use of disposable, limited-use, or reusable protective clothing along with estimations of the sizes needed. Since emergencies are by their nature unplanned, the safety professional can never be certain which members of the ERT might be working during an emergency, requiring that a larger stock of protective clothing and gear is available at all times. In recent years, the development of high-quality and affordable, disposable, and limited-use clothing has made it easier to purchase large amounts of clothing in varying sizes, helping to resolve this issue.

Additionally, the safety professional will need to analyze whether decontamination workers will be needed to assist responders and to what level they need to be dressed. Decontamination workers are commonly used in the first stages of decontamination when the outer protective clothing as well as the self-contained breathing apparatus is being removed. As seen in the photo in Figure 1, a Level-A hazmat ensemble includes a chemical protective suit that has attached gloves, making unzipping the suit by the individual responder impossible. (Some responders carry sheathed knives inside their suit, but these are for use in an emergency only.) It would be similarly difficult for another member of the team wearing a Level-A suit to provide that assistance, although it should be noted that there are some situations where team members in Level A assist each other.

In this part of the analysis, the safety professional will also have to assess the ability to share equipment. In most cases, about the only items that can be interchanged among responders are the air-supply cylinders. Therefore, the safety professional's calculation in the development of the initial equipment list will have to include a determination of how many cylinders will be needed. Assuring a steady supply of additional filled cylinders during any response is crucial and the initial equipment cost will also involve either the purchase of the necessary equipment to refill cylinders or making arrangements with an outside party to provide this service, as long as the outside party is able to provide this assistance rapidly. In some cases, the latter form of assistance may not exist or may not be available quickly enough, thus limiting the options available as it would be an extremely disastrous situation if the emergency response had to be suspended because the supply of air cylinders was depleted (Strong 1996, 123).

FIGURE 1. DuPont™ Tychem® TK Commander Level-A Suit (*Source:* DuPont Tychem TK; Courtesy of DuPont.)

Another facet of sharing equipment that must be considered concerns what level of response the ERT is being prepared to perform. If the response will include hazardous materials incident response and fire fighting, separate gear will need to be purchased as the gear is typically not interchangeable. In addition, not only is PPE for fire fighting different than that for emergency response, but neither ensemble typically includes what would be worn by employees of the organization during normal operations (Rand 2002, 16–18).

The type of respiratory protection (air-purifying or atmosphere-supplying) required during a hazardous materials incident is often one of the most costly line items in this part of the budget. Rarely will the less expensive air-purifying respirators provide sufficient protection, and self-contained breathing apparatus that are operated in a positive pressure mode will nearly always be required for all initial entries until atmospheric conditions can be evaluated and there is assurance that the situation is stable and not likely to deteriorate. As noted above, maintaining a sufficient supply of cylinders as well as a mechanism to refill them is an expensive part of this process. The cost of atmosphere-supplying protection will vary greatly by manufacturer; however, a recent review of two commonly used manufacturers reveal prices ranging from $1700.00–$3700.00 per ensemble, not including the costs of additional cylinders (Scott C100 – $1800.00; Scott Air Pak – $2800.00; MSA Air Hawk – $1700.00, MSA MMR Extreme – $3700.00).

The final type of equipment required involves various communication devices that will be used during the response. These devices serve as a method of communicating internally with other members of the team and externally with outside resources and organizations. Traditionally, land-line or cellular telephones are the most common means of communicating externally; however, the safety professional should be aware of problems with these devices in the event of a power failure or service failure as would be typical in a natural disaster covering a large area. In such cases, ham radios or two-way radios may serve as a good back-up system. The safety professional should also consider contacting local businesses to discuss sharing communication devices if needed. Finally, consultation with the local emergency planning unit (typically at the county level) will help identify the overall community's plan for communicating in emergencies and help develop a proper back-up system for the organization.

Internal communication devices used among emergency response team members are typically two-way radios. Newer versions of such communication equipment provide for headsets that allow hand-free communication and work well inside of chemical protective clothing, such as a Level-A suit. It is crucial to understand that all communication devices used on any emergency site must meet certain guidelines to assure that they do not provide the source of ignition in an explosive environment.

The *National Electrical Code*® (NEC®) published by the National Fire Protection Association provides important information on the classification of hazardous locations where emergency response might need to occur. The NFPA system defines hazardous locations based on two factors: the type of flammable or combustible material present and the conditions under which the hazard will be present (see Figure 2). These two factors are combined to provide a hazardous location classification and all communications equipment that may potentially be the source of ignition must be properly rated for the location in which it may be used (OSHA 1996).

Summary of Class I, II, III Hazardous Locations

Classes	Groups	Divisions	
		1	2
I Gases, vapors, and liquids (Art. 501)	A: Acetylene B: Hydrogen, etc. C: Ether, etc. D: Hydrocarbons, fuels, solvents, etc.	Normally explosive and hazardous	Not normally present in an explosive concentration (but may accidentally exist)
II Dusts (Art. 502)	E: Metal dusts (conductive and explosive) F: Carbon dusts (some are conductive, and all are explosive) G: Flour, starch, grain, combustible plastic or chemical dust explosive)	Ignitable quantities of dust normally are or may be in suspension, or conductive dust may be present	Dust not normally suspended in an ignitable concentration (but may accidentally exist). Dust layers are present.
III Fibers and flyings (Art. 503)	Textiles, wood working, etc. (easily ignitable, but not likely to be explosive)	Handled or used in manufacturing	Stored or handled in storage (exclusive of manufacturing)

FIGURE 2. Classification of hazardous locations (*Source: National Electrical Code*®)

used for hazardous materials emergencies and a NFPA standard for the use of air-monitoring equipment in potential hazardous atmospheres. Where appropriate, the specific source is cited in the text of the sidebar.

Once the safety professional is assured that the equipment list is complete, several decisions will need to be made if the safety professional has some leverage as to where the equipment and supplies are purchased. In some cases an organization has purchasing policies that make the use of anything but existing vendors difficult; some do not permit the practice at all.

Fundamental to this process is whether or not the purchase will be made from a sole source contract with one or two suppliers or whether the purchase will be put out for competitive bid. As seen in Figure 3, there are pros and cons to both scenarios.

In addition to the calculation of the initial purchase costs of the needed equipment and supplies, the safety professional will also need to take into account ongoing costs. Each major piece of equipment will have its own requirements for inspections, maintenance, and calibrations, some of which may need to be done by external sources. Even if these tasks can be completed in-house, sufficient staff time will have to be allocated; whether it is the time of the safety professional or some other member of the organization's workforce. Determining ongoing costs requires specific knowledge and information that is typically available from the purchasing source. Moreover, most disposable equipment, particularly chemical protective clothing, will have a shelf life established by the manufacturer. Knowing when items will need to be disposed of or how to properly rotate stock will be an important task of the staff person performing the regular equipment inspections.

The safety professional will also need to build in expected increases in costs due to the normal price increases that occur over time. Assistance in determining the percentage that should be used and how to include this cost in the budgetary calculations can typically be obtained through co-workers in the financial department of the organization. In some cases, sole-source contracts will have built-in price guarantees for specific periods of time. An extensive number of purchases allows manufacturers to offer a longer period of price guarantees as well as providing the safety professional with more leverage for bargaining.

There are two final costs of equipment purchasing that need to be calculated: (1) any potential salvage cost should the equipment be able to be resold when the organization no longer is able to use it or when the organization updates its equipment and (2) disposal costs that may be incurred at the end of the equipment's useful life.

Safety professionals are urged to use due consideration and care in the early stages of evaluating prices for equipment. The price range for any single piece of equipment can be fairly wide and requesting approval to buy a less expensive item in order to get approval for a budget can be tempting; however, poor quality of equipment over time may end up costing more than an initial purchase. Carefully reviewing the specifications, talking with colleagues who may be users, and perhaps negotiations for a trial usage period are all effective strategies in this process.

Sole-Source versus Competitive Bid Process

Sole-Source Contract	Competitive Bid
• Leverage of purchase power – vendor is willing to meet needs in order to gain the business • Lack of competition may reduce amount of price value obtained from one distributor • Some limits on the variety of selections – can only purchase what distributor makes available • Can be a quicker process to go from bid request to contract signature	• In smaller companies, minimal amount of cost reduction realized due to limited quantity of product being purchased • Competition between bidders helps to realize the largest price reductions; allows smaller companies the ability to negotiate for lower prices • Can obtain bid from supplier for specific or unique item that is not readily available through the general distribution network • Can be a slower process to contact distributors, obtain and review bids, and sign contract

Note: Regardless of the relative merits of each option, organizations may already have processes in place that require either option, eliminating choices available to the safety professional.

FIGURE 3. Comparison of sole-source and competitive bid processes (*Source:* Ferrante, empirical experience)

In addition to reviewing the specifications of the equipment, there are numerous references that can be consulted, including the American National Standards Institute (ANSI), NFPA, Underwriter's Laboratory (UL), and the U.S. Fire Administration to assist in this task. As discussed earlier in the sidebar, the *National Electrical Code®* published by the NFPA provides important information on the classification of hazardous locations where emergency response might need to occur. In addition to communications equipment, all monitoring equipment that may potentially be the source of ignition must be properly rated for the location in which it might be used (OSHA 1996).

The final issue to be considered by the safety professional with regard to equipment and supplies is their storage. Central to this decision-making process is the need for the equipment to be easily available near the specific area where an emergency will occur. This could mean developing several small satellite locations where initial supplies are maintained and a central warehouse where additional supplies can be accessed as needed from any location throughout the facility. It could also mean purchasing carts, trailers, or some other mobile mechanisms to allow the equipment and supplies to be quickly taken to any location throughout the facility (Strong 1996, 123). General storage considerations such as keeping all equipment clean, dry, and away from dust and moisture are fairly universal, but the specifications of each piece of equipment should be consulted for any unique requirements that need to be met.

Once all of the issues have been resolved and the safety professional is assured that the list is accurate and complete, the process of getting approval for purchase needs to be addressed. As noted above, some safety professionals will use a sole-source contractor and some will put the list out for competitive bid. Figure 3 noted some of the pros and cons of each scenario; this discussion will end with an example of how the safety professional can compare and contrast the actual bidded amounts that he/she may receive. One of the simplest means to achieve this goal is to look at what is commonly called the *time value of money*. The formulas used to calculate the time value of money are discussed in the "Cost Analysis and Budgeting" chapter in Section 1 (Vol. 1) of this handbook. The reader is referred to this chapter for an in-depth discussion. However, a useful formula from this section can be used to understand some of the variations available to the safety professional as he/she attempts to determine how best to secure bids for the cost of equipment.

$$P = F(1 + i)^{-n} \quad (1)$$

where:

P = Present Cost

F = Future Cost

i = rate of return (expressed as a percentage)

n = number of years

As mentioned in Figure 3, one of the values of having a sole-source contract with a distributor is the ability to leverage price guarantees in exchange for a long-term contract. This leverage may not be available in a situation where competitive bids are solicited and where a built-in price increase may occur.

Table 6 shows two different potential contractual situations where a safety professional has solicited a 5-year sole-source contract with a price guarantee. In exchange for this guarantee, the yearly cost is higher than what is seen for the competitive bid contract, which has a built-in 5 percent per-year price increase. However, the yearly price for this contract starts at a lower amount.

In this example, although the sole source contract provides for a price guarantee for quite a lengthy period of time, the overall cost at present value of the money actually exceeds the competitive bid contract where there is a built-in price increase.

The safety professional will need to spend some time carefully evaluating the various options for purchase of any equipment or supplies used for the development and implementation of an internal ERT.

Training, Drills and Other Preparatory Activities

Next to the cost of purchasing and maintaining equipment and supplies, the largest cost of an ERT is the training of employees to provide a specific emergency response. [The reader is referred to the chapter "Cost Analysis and Budgeting" in Section 4 (Vol. 1) by Brent Altemose for an in-depth review and discussion of evaluating the costs of the various types of training programs.] Decisions previously made by the safety professional regarding what level of response should be provided and for how long will need to be reviewed at this stage. The specific source documents utilized to answer these questions should be consulted again,

TABLE 6

Example Analysis of Two Equipment Contract Alternatives Using Present Cost Based on the Time Value of Money

	Conversion of Yearly Costs to Present Cost at Year 1			
	Sole-Source Contract		Competitive Bid	
Year 1		P = $75,000		P = $65,000
Year 2	F = $75,000	P = $51,124	F = $68,250	P = $46,524
Year 3	F = $75,000	P = $51,124	F = $71,663	P = $48,850
Year 4	F = $75,000	P = $51,124	F = $75,246	P = $51,292
Year 5	F = $75,000	P = $51,124	F = $79,008	P = $53,857
Present Cost		P = $279,496		P = $265,523

A = Yearly Cost; P = Present Cost = $F(1 + i)^{-n}$; n = number of years; i = rate of return = 8% or 0.08 for this example.

as many of them also contain requirements for employee training. In addition, the OSHA HAZWOPER and confined space entry regulations also provide detailed requirements as to the topics and information that must be provided during training. As noted previously, the confined space regulations require that an annual rescue drill be performed by members of the team and, in addition to the topics required for HAZWOPER training, there are mandatory minimum training hours established for the various training categories.

In order to assure that all of the pertinent training needs and capacities are addressed, the safety professional needs to create a training plan or matrix that details what information or courses need to be provided and how many workers need each level of training. This matrix can be created by hand in a format such as Word or Excel, or the safety professional can look into the purchase of any number of software programs that will assist in the development and management of the organization's training needs. (It should be noted that these software programs, while helpful in managing employee training, cannot provide assistance in calculating the cost of training.) Included in the matrix will be an evaluation of the need for drills during the year and what type of drills are mandated or are recommended. (More information about drills is discussed later in this section.) When developing the training matrix, the safety professional will need to assess and evaluate which portions of the training can be provided by in-house staff, including the safety professional him/herself, and which portion must be contracted out. Although the safety professional may be qualified to provide the training, consideration must be given to the time that it takes to prepare and deliver this type of training. The cost of the safety professional's time should be compared to the cost of bringing in outside contracted instructors. Figure 4 shows the type of training required by the OSHA HAZWOPER regulations [29 CFR 1910.120(e)] as well as recommended drills for an internal 25-member hazmat ERT.

The safety professional is further advised to review the potential for cross-training the ERT. If the organization is planning to provide some or all of the emergency response, fire fighting, and initial medical treatment, it may be that employees can cover more than one type of task (Cashman 2000, 203). In addition, the safety professional should seek out those employees in the workforce who may already be trained at some level because they participate in their community's hazmat teams, volunteer fire service, or EMS. These employees may already have completed all or some of the training needed, at least on a generic level. Perhaps the content and time requirements of the courses can be reduced to deal only with the hazards specific to the organization.

The training plan and budget must include calculations on the cost for providing initial training for employees as well as the required and recommended refresher training, which typically needs to occur at least annually. The logistics of training new employees must also be considered and calculated. If the organization is large enough, a monthly class (or one held at some regular frequency) may suffice. But if hiring is slower or more sporadic, employees may not be able to participate in any ERT activities until the training has been completed.

Finally, comprehensive calculation of the costs of training requires an assessment of how the organization's production capabilities will be maintained while the ERT members are in training classes and not performing their typical duties. The American Society for Training and Development has estimated that these costs can run up to 80 percent of the total cost of the training program (Head 1994, 4). This might involve scheduling additional staff if the employees' work tasks must be done despite their absence. This may lead to overtime costs since the organization may not be able to reduce production to allow for training. There are also situations in which employees might need to attend training at a time when they are not scheduled to work, also causing additional overtime costs. And while some organizations utilize employees who volunteer to be a member of the ERT, these employees would require some type of compensation for their time when responding to an emergency event.

Training Courses and Drills Matrix for a Hazardous Materials Response Team

(25-member team: 10 members First Responder Operations; 10 members Hazardous Materials Technician, 4 members Hazardous Materials Specialist; and 1 member Incident Commander)

	TRAINING COURSE			
Drill Type[1]	**First Responder Operations (8 Hours)**	**Hazardous Materials Technician (24 Hours)**	**Hazardous Materials Specialist[2]**	**Incident Commander[2]**
	10	10	4	1
Facility/Equipment Familiarization (2x per year)	X	X	X	X
Tabletop Exercise (once per year)	X	X	X	X
Tabletop Simulation (2x per year)	once	once	both drills	both drills
Tabletop Gaming (2x per year)			X	X
Field Exercises/Simulations (every 18–24 months)	X	X	X	X

[1]The types of drills are described as follows (McCall 2005):

- Facility/Equipment Familiarization – Allows for responders (on site and off site) to visit the facility to orient to physical plant and layout and understand where the hazardous situations may occur. Employees and visitors have a chance to hear from facility management about the plan for emergency response.
- Tabletop Exercise – These are first-level drills, primarily based upon discussions and take place in a conference room or similar environment. Key players and senior staff attend and "talk through" responses to various emergencies. One of the main goals is role familiarization.
- Tabletop Simulation – Similar to above but with a more dynamic process and with a more specific scenario. The moderator of the exercise can alter or modify the original scenario to test out players' responses to changing conditions.
- Tabletop Gaming – This drill provides a more interactive progress as decisions made at earlier stages of the drill affect the consequent decisions; the script is much more fluid in this type of drill. Its primary goal is to explore decision-making processes and the consequences of those decisions.
- Field Exercises/Simulations (Full-Scale Exercises) – These are drills in which the scenario is played at the location of the incident. These exercises test all of the operational systems of the incident response and are complicated to plan and implement.

[2]Per 29 CFR 1910.120(q) – The minimum number of hours of initial training for First Responder Operations and Hazardous Materials Technicians is specified. Hazardous Materials Specialists and Incident Commanders must receive training at least at the Hazardous Materials Technician level. The additional number of hours of training is not specified. The required competencies are determined by the employer based upon an assessment of the types of responses that may be required. Internal ERT members need only be trained for the types of hazards the organization may face, while a public ERT requires more extensive training due to the wider variety of hazards that may be encountered.

FIGURE 4. Training matrix for a hazardous materials response team

Figure 5 takes the training matrix presented in Figure 4 and begins the process of developing a working budget for providing a 25-member hazmat team with training over the course of five years. This budget includes the use of both internally and externally provided courses.

At this point in the process, the safety professional will likely need to be able to justify the cost of training an internal ERT. There are several methods that can be used: calculation of the return on investment (ROI) and the beak-even point help to provide a basis for the use of funds.

Training Costs Matrix for a Hazardous Materials Response Team

(25-member team: 10 members First Responder Operations; 10 members Hazardous Materials Technician, 4 members Hazardous Materials Specialist; and 1 member Incident Commander)

Type of Cost[1]	TRAINING COSTS			
	Estimated Unit Cost	**Unit Multipliers**	**Total Cost[2]**	**Timing of Expense[2]**
		INTERNAL COURSES		
Curriculum Development[1] – Initial Class				
Internal salaries – loaded cost[2]	$70/hour	96 hours (1 24-hr class, 1 8-hr class)	$6720.00	Year 1
Written materials (copying, paper, etc.) (once per year)	$75.00	2 classes	$150.00	Year 1
Video purchases	$99.00	8 videos	$792.00	Year 1
Equipment for dress-out activities (reusable for life of project)	NA	NA	$2500.00	Year 1
Supplies for dress-out activities (replaced for each class)	$75.00	2 classes	$150.00	Year 1
Curriculum Development[1] – Refresher Class				
Internal salaries – loaded cost[2]	$70/hour	24 hours (2 4-hr classes)	$1680.00	Year 2
Supplies for dress-out activities (replaced for each class)	$25.00	8 classes	$200.00	Years 2–5
Implementation Costs: Instructor Costs				
Internal salaries – loaded cost	$70/hour	32 hours (1 8-hr class, 1 24-hr class)	$2240.00	Year 1
Internal salaries – loaded cost[2]	$70/hour	32 hours (8 4-hr refresher classes)	$2240.00	Years 2–5
Attendee Cost – First Responder Operations				
Internal salaries – loaded cost	$30/hour	8 hours × 10 attendees	$2400.00	Year 1
Written materials	$25.00/attendee	10	$250.00	Year 1
Meals (catered in)	$7.50/person	10	$75.00	Year 1
Other business cost (coverage for attendees on plant floor)	$24/hour	8 hours × 10 attendees	$1920.00	Year 1
Internal salaries – loaded cost	$30/hour	4 hours × 10 attendees	$1200.00	Years 2–5
Written materials	$5.00/attendee	10	$50.00	Years 2–5
Meals (catered in)	$7.50/person	10	$75.00	Years 2–5
Other business cost (coverage for attendees on plant floor)	$24/hour	8 hours × 10 attendees	$1920.00	Years 2–5

[1]Based on 3 hours of curriculum development per 1 contract hour

[2]Includes salary and benefits

[3]Cost of attendee's time

FIGURE 5. Training expenditure matrix for a hazardous materials response team

Training Costs Matrix for a Hazardous Materials Response Team (cont.)

(25-member team: 10 members First Responder Operations; 10 members Hazardous Materials Technician, 4 members Hazardous Materials Specialist; and 1 member Incident Commander)

	TRAINING COSTS			
Type of Cost[1]	**Estimated Unit Cost**	**Unit Multipliers**	**Total Cost**[2]	**Timing of Expense**[2]
	INTERNAL COURSES			
Attendee Cost – Hazardous Materials Technician				
Internal salaries – loaded cost	$30/hour	24 hours × 15 attendees	$10,800.00	Year 1
Written materials	$40.00/attendee	15	$600.00	Year 1
Meals (catered in)	$7.50/person	15	$112.50	Year 1
Other business cost (coverage for attendees on plant floor)	$24/hour	24 hours × 15 attendees	$8640.00	Year 1
Internal salaries – loaded cost	$30/hour	4 hours × 15 attendees	$1800.00	Years 2–5
Written materials	$15.00/attendee	15	$225.00	Years 2–5
Meals (catered in)	$7.50/person	15	$112.50	Years 2–5
Other business cost (coverage for attendees on plant floor)	$24/hour	4 hours × 15 attendees	$1440.00	Years 2–5
	EXTERNAL COURSES			
Course Cost				
Hazardous Materials Specialist (Initial)	$225.00	4	$900.00	Year 1
Meals	$10.00/day	2 days × 4 attendees	$80.00	Year 1
Other business costs[3]	$50/hour	16 hours × 4 attendees	$3200.00	Year 1
Hazardous Materials Specialist (Refresher)	$150.00	4	$600.00	Years 2–5
Meals	$10.00/day	1 day × 4 attendees	$40.00	Years 2–5
Other business costs[3]	$50/hour	8 hours × 4 attendees	$1600.00	Years 2–5
Incident Commander (Initial)	$225.00	1	$225.00	Year 1
Meals	$10.00/day	2 days	$20.00	Year 1
Other business costs[3]	$50/hour	16 hours × 1 attendee	$800.00	Year 1
Incident Commander (Refresher)	$150.00	1	$150.00	Years 2–5
Meals	$10.00/day	1 day	$10.00	Years 2–5
Other business costs[3]	$50/hour	8 hours × 1 attendee	$400.00	Years 2–5
Year 1 Cost			**$42,574.50**	
Yearly Cost, Year 2			**$1680.00**	
Yearly Cost, Years 2 to 5			**$48,250.00**	
Total Program Cost, 5 Years			**$92,504.50**	

[1]Based on 3 hours of curriculum development per 1 contract hour

[2]Includes salary and benefits

[3]Cost of attendee's time

FIGURE 5. Training expenditure matrix for a hazardous materials response team

ROI is equal to the net benefits of the expenditure and includes the savings that are accumulated from the expenditure or profits that are recognized from the expenditure after the costs are subtracted and is expressed as a percentage.

$$\text{ROI} = \frac{\text{savings/profits (in \$)}}{\text{expenditure (in \$)}} \times 100 = X\% \qquad (2)$$

The break-even point is how long it takes an organization to recoup its investment and is calculated by dividing the program cost by the yearly benefits. This calculation is typically expressed in years or portions thereof.

$$\text{Break-even} = \frac{\text{expenditure (in \$)}}{\text{yearly benefits (in \$)}} = X \text{ years} \qquad (3)$$

Using the budget prepared in Figure 5, the following are the calculated ROI and break-even point for the first year's costs, assuming a savings of $18,000.

$$\text{ROI} = \frac{\$18{,}000.00}{\$42{,}574.50} = .423 \times 100 = 42\% \qquad (4)$$

$$\text{Break-even} = \frac{\$42{,}574.50}{\$18{,}000.00} = 2.4 \text{ years} \qquad (5)$$

The safety professional will need to investigate the expected ROI and break-even point for their individual organization, as this amount varies greatly. However, in general, it is helpful not to isolate one calculation from the other in the analysis of the reasonableness of expenditures. In the example above, the ROI of 42 percent is favorably high and the organization can expect to recoup its investment in two years, another favorable number. This lends credence to the safety professional's argument that the cost of the training provides sufficient benefits to justify it. However, if one or the other of these numbers is "off," the situation looks different. For example, if the ROI was 10 percent and the break-even point was 1.1 years, the expenditure would look less favorable since the return was small, even though the expenditure is recouped rather quickly. But a wise safety professional would further argue that the short period of time it would take to recoup the investment would lessen the problems associated with the low ROI. On the other hand, if the ROI was a higher number, such as 60 percent, and the break-even point was in excess of 10 years, convincing the organization to spend the money to implement the training program might be harder. There are risks to an organization in waiting such a long period to recover expenditures, particularly one such as training, which is an ongoing expense since it traditionally is not a one-shot expense and requires some frequency of refresher classes.

Formal classroom training provides only a small portion of what is required to assure that an ERT is "right and ready" all the time. Even though the training curriculum for many emergency response situations typically involves a variety of dress-out activities, it is the formal drills that occur on an ongoing basis outside of scheduled training classes that truly prepare an ERT to manage the wide variety of emergencies that may occur (see Figure 6). Drills are an essential part of preparing an ERT because they (1) provide a controlled environment in which mistakes can be safely made and corrected, (2) test response capabilities and equipment, and (3) provide opportunities to develop and refine the relationships between supervisory-level staff of different departments who may not typically work together on a day-to-day basis within the organization. In addition to the internal relationships that are required during an emergency response, managers will often have to work with outside agencies that will assist in an

FIGURE 6. Evacuee checked with radiation detector while drill evaluator looks on.
(*Source: Pittsburgh Post-Gazette*, 2006. All rights reserved. Reprinted with permission.)

emergency. These relationships need to be developed as well and represent another purpose of drills and other preparatory activities.

The critical value of drills was reinforced shortly after the 9/11 attacks in this country when a conference was convened by the Rand Corporation in New York City. Various panel discussions repeatedly reflected the belief that the emergency response at the Pentagon site was more successful than at the World Trade Center. Panelists and participants generally agreed that it was due to the more extensive drills that the Pentagon ERT responders had participated in prior to the attack:

> . . . unless practices are ingrained before a major incident and the use of equipment and procedures is part of preparedness, responders are unlikely to absorb training fully in the heat of the battle to save lives or to be predisposed to wear personal protective equipment as prescribed. (Rand 2002, 41)

and

> The need for pre-disaster planning came up repeatedly in many panels. Extensive training, particularly among military units, was cited in the discussions as a reason for the comparatively smooth handling of the disaster response and use of PPE at the Pentagon site.. . . . Such exercise, it was argued, can play a very important role in building relationships that enable managers to deal more expeditiously with the problems that arise in the heat of a real emergency." (Rand 2002, 61)

A wide variety of drills exist, each with associated costs and recommended frequencies. These include facility/equipment familiarization drills, tabletop exercises, tabletop simulations, tabletop gaming, and field exercises/simulations (McCall 2005). All but field exercises are relatively inexpensive to conduct, save for the cost of staff time to participate in the drill—time when they are not performing their typical work tasks. (See Figure 4 for a training matrix delineating the drills by type and frequency for a 25-member hazmat team.) The safety professional must also account for the time he/she must utilize to plan and prepare for the drill as well as the time of others who assist him/her in the effort. Once the drill is completed, additional time will be needed to review the evaluations of the drill and prepare a report. Finally, recommended changes in procedures, types of equipment used, and various other parts of the emergency response will need to revised, taking additional staff time, primarily that of the safety professional and command-level ERT members.

While the various tabletop exercises provide excellent opportunities for relationship building and procedure testing, they generally involve only supervisory and command-level employees, with a limited number of front-line responders. This obviously limits the ability of responders to successfully prepare for an actual emergency. For example, the ability to quickly and correctly don and doff a Level-A ensemble or fire-fighting gear is not something that can be practiced once a year in a training class. The need to develop this competency requires that some level of more formal drills occur, preferably at varying points during the year. Formal drills can be full scale; however, these are extremely time-consuming to prepare for and expensive to conduct. The safety professional will have to devise a plan that meets the skill-building needs of the members of the ERT while also remaining cost effective.

When developing a drill plan as part of the overall safety training matrix, the safety professional must balance the cost issues as well as assure that all members of the ERT have sufficient opportunities to practice and refine their skills. One method of holding costs down while still providing plenty of drill opportunities is to assure that the formal training time includes a number of hands-on activities that serve as drills. It is also a sound practice to assure that the shorter refresher training classes that occur annually use the bulk of their available time for drills and other practical activities rather than using the lecture and video format, increasing the cost effectiveness of the training time.

One final note before concluding this section regards the judicious use of expired or slightly used, but thoroughly decontaminated, PPE and other types of equipment for training purposes, both during classroom activities and drills. This is a cost-effective way for the safety professional to create a realistic scenario for practice purposes. While many manufacturers of PPE offer training suits that are specifically designed to mimic the regular suits at a substantially lower cost, the use of these types of suits still adds to the budget. Whatever option is utilized, the safety professional must be careful not to intermingle regular PPE and training PPE; some method of marking the training PPE must be used to avoid the dangerous and possibly deadly situation in which a responder will don a training suit for an actual incident.

Medical Surveillance

Various regulations provide a framework for the mandatory requirements of medical surveillance for ERTs, depending on the type of emergency they are prepared to address:

- OSHA HAZWOPER regulations [29 CFR 1910.120(f)], the OSHA Respiratory Protection regulations [29 CFR 1910.134(e)], and NFPA 471 (NFPA-471 2002) address the need to provide medical surveillance for members of an ERT responding to hazardous materials emergencies.
- NFPA 1500, Standard on Fire Department Occupational Safety and Health Programs (NFPA-1500, 2002) and the OSHA regulations for Fire Brigades [29 CFR 1910.156(c)] address medical surveillance requirements for fire response.
- OSHA Confined Space Entry regulations (29 CFR 1910.146) do not identify any medical surveillance requirements specific to participation on a rescue team, and there are also no specific medical surveillance requirements mentioned under the various documents used to plan for natural disasters. However, the regulations do note the need to have rescue team members trained in first aid and cardiopulmonary resuscitation.

The NFPA defines *medical surveillance* as:

> . . . the ongoing, systematic evaluation of response personnel who are at risk of suffering adverse effects of heat/cold exposure, stress, or hazardous materials exposure. . . . The monitoring is done for the purpose of achieving early recognition and prevention of these effects in order to maintain the optimal health and safety of on-scene personnel. (NFPA-471 2002, 14–15)

All of the sources listed provide descriptions of who should participate in the medical surveillance program as well as what types of testing they should receive and how often. They also detail the situations in which additional surveillance should occur, such as when exposure to particular hazardous substances should be monitored in urine or blood or when an ERT member reports symptoms that may be the result of overexposure to a hazardous substance. It should be noted that the OSHA regulations provide for mandatory requirements that the safety professional must be assured are completed, while the NFPA consensus standards provide "best practice" recommendations. Figure 7 details the medical surveillance requirements for a fire ERT with some likely associated costs for one year (NFPA-1500 2002).

The costs of a medical surveillance program include both the cost of the actual medical examination and procedures as well the cost of staff time while participating in the various medical examinations and tests, time when the members of the ERT are not performing their typical work tasks. It should be noted that all of the OSHA standards require that medical surveillance of any kind take place during normal work hours.

If the organization is large enough, the ability to leverage healthcare providers for decreased costs should be investigated; however, even smaller organizations should take the time to investigate what reductions may be available. Perhaps a local consortium of smaller organizations already exists or could be developed by the safety professional for the purpose of achieving the same type of cost savings enjoyed by larger organizations. In some cases where an organization maintains some sort of on-site medical facilities, healthcare providers can be required to come directly to the organization, a practice that represents both a significant time and cost savings. In other cases, larger healthcare providers often have mobile screening vans that can come directly to the organization and perform much of the testing and evaluations. Finally, for larger organizations that utilize

Medical Surveillance Requirements for a Fire Emergency Response Team

(25-member team: 10 members Incipient Level only, 10 members Advanced Exterior Level, and 5 members Interior Structural Level)

	Tests Required[1]						
Level	**Physical Examination[2] ($125)**	**Work and Exposure History ($45)**	**EKG ($250)**	**Pulmonary Function Test ($125)**	**Chest X-ray ($450)**	**Blood Work ($95)**	**Fitness Test ($75)**
Incipient	X						
Advanced Exterior	X	X	X	X	X	X	X
Interior Structural	X	X	X	X	X	X	X

[1]All tests are recommended at preplacement and annually.

[2]Rule out those employees with known heart disease, epilepsy, or emphysema.

Projected Medical Surveillance Costs per Year

Level	**Total Price All Tests Above**	**Total Cost of Tests**	**Cost of Staff Time[1]**	**Total Medical Surveillance Cost**
Incipient	$125.00 × 10	$1250.00	$30/hr × 1 hr × 10 attendees	$1550.00
Advanced Exterior	$1165.00 × 10	$11,650.00	$30/hr × 4 hrs × 10 attendees	$12,850.00
Interior Structural	$1165.00 × 5	$5825.00	$30/hr × 4 hrs × 5 attendees	$6425.00
			Total Cost – One Year	**$19,425.00**

FIGURE 7. Medical surveillance matrix for a fire emergency response team

their own in-house medical services, this resource, whether it is registered nurses or physicians or some other type of health care personnel, can often double as a member of the medical surveillance team, thus minimizing the additional cost to the ERT budget.

A final cost for medical surveillance is incurred in assuring that during a response sufficient medical back up is available for the responders. The medical team serves two purposes. The first is to triage and provide initial medical treatment to responders or victims who may be either injured or exposed during the incident. The second is to monitor the health status of the responders as they enter and exit the decontamination unit. The most critical issue is usually heat stress, but other types of stress-related complications need to be carefully monitored, including problems with heart and blood pressure. As noted previously, it is cost effective for the members of this part of the medical surveillance team to be employees of the organization, preferably those who already possess the necessary certifications through their work in the community as paramedics or emergency medical technicians. This will avoid the need to provide for additional training. Another option would be to train members of the team to perform the monitoring.

Miscellaneous Expenses

The ability to capture the actual costs associated with the time of the safety professional and other members of the organization who participate in the development and implementation of the ERT requires deliberate calculation of the number of hours per week, per month, or whatever time increments work best for the internal accounting structure. As direct costs, these are fairly easy to capture. However, the indirect costs of staff time, benefits, and other expenses, as well as the time spent not performing required work tasks, must also be included. Depending on the organization, benefits are typically in the neighborhood of 25–35 percent of the base salary (Head 1994, 36) and are termed the "loaded" cost, but the safety professional will need to access this specific percentage from either the Human Resource or Accounting Department of his/her organization.

Another miscellaneous cost of developing and implementing an ERT involves record-keeping requirements for an ERT. The costs associated with this issue include staff time to check and maintain training records, equipment calibrations and inspections, and medical surveillance, among others. In addition to accounting for staff time for record-keeping activities, there is also the need to plan for storage of records, both those that are current and those that are kept for archival purposes. The OSHA regulations pertaining to the maintenance and retention of medical records requires that they be kept for 30 years beyond the employment of the individual [29 CFR 1910.1020(d)(1)(i)]. This will require a storage area that is safe from exposure to damaging elements, such as water or vermin, and that also provides for security due to the confidential nature of the records. For larger organizations, this may require a separate building or even an off-site facility because of the large number of records that accrue given the length of time the records must be kept and the relative infrequency with which they must be accessed.

The final issue that merits discussion in this chapter involves insurance coverage. The use of an internal ERT by an organization will necessarily need to be reflected on the organization's various insurance coverages, although having an ERT should not make a significant difference in the cost. In fact, it is more likely to decrease the premiums because having a fully functioning ERT is an excellent risk reduction strategy. Workers' compensation and general liability will most likely be affected; however, the increases should be at least partially offset by the ability of the organization to handle internal emergencies in a way that minimizes damages to property and equipment and also reduces the interruption of business activities.

Conclusions

Understanding and planning for the costs associated with developing and implementing an in-house ERT at any organization are complicated issues for any safety professional. This chapter has discussed the predominant factors that must be considered as well as the types of costs that will be incurred for the four major emergencies that an ERT may be expected to respond to: fires, hazardous materials incidents, confined space entry rescues, and natural disasters. The safety professional needs to recognize that each organization's situation is unique and the decision of whether or not to maintain an internal ERT and what response level it will be capable of performing must be carefully considered. In order to obtain support for the effort, both financial and otherwise, the safety professional needs to be deliberate in his/her evaluations of the costs and be certain that all other relevant factors have been fully and completely evaluated.

Most assuredly the safety professional needs to be able to understand the type of asset-allocation decision making that is performed in their individual organization and mimic it when preparing any materials for review by management as this will increase the likelihood of approval. The here-and-now costs must be carefully calculated as well as future costs, including any associated cost-of-living increases. As has been noted, some of the costs are more difficult to capture as they involve calculating the time of various staff members to perform the necessary duties. It is also important to attempt to capture the various indirect costs such as business interruption, replacement of damaged property and equipment, and possible fines and penalties from regulatory agencies.

This chapter has provided the safety professional with a myriad of resources that should be consulted during this process. These resources include traditional voluntary consensus standards from organizations such as NFPA as well as publications that have been prepared by regulatory agencies like EPA,

OSHA, and NIOSH. All of these sources provide excellent information, but the safety professional is encouraged not to limit research to written materials. There are numerous personal resources in any community that can provide information and assistance in all facets of developing and implementing an ERT. These persons often participate in the LEPC or other community response capabilities for hazardous materials emergencies or local fire-response organizations, both paid and volunteer. The safety professional should also not overlook the capabilities found within his/her organization among employees who are volunteer members of community-based ERTs. In addition, manufacturers and suppliers of the products and equipment that will need to be purchased and maintained also have technical assistance staff who can assist, and while these persons may have a sales goal in mind, the information they can provide is usually helpful as long as the safety professional is mindful of the potential conflict of interest. Finally, the safety professional may want to consider seeking the services of an outside consultant in those areas where specialized knowledge and experience is required.

References

Cashman, John R. 2000. *Emergency Response to Chemical and Biological Agents*. Boca Raton: Lewis.

______. 1983. *Hazardous Materials Emergencies: The Professional Response Team*. 3rd ed. Lancaster, Pennsylvania: Technomic Publishing Company.

Cocciardi, Joseph. 2004. *Emergency Response Team Handbook*. Massachusetts: National Fire Protection Association.

Forsberg, Krister, and S. Z. Mansdorf. 2002. *Quick Selection Guide to Chemical Protective Clothing*. 4th ed. New Jersey: John A. Wiley and Sons.

Head, G. 1964. *Training Cost Analysis*. Alexandria, VA: American Society for Training and Development.

McCall, Harry, A., M.A., CSP. 2005. "Recipe for a Successful Emergency Response Drill." Proceedings Paper from the American Society of Safety Engineers Professional Development Conference.

National Fire Protection Association (NFPA). 2002. NFPA Standard 471-2002, *Recommended Practice for Responding to Hazardous Materials Incidents*. Quincy, MA: NFPA.

______. 2002. NFPA Standard 472-2002, *Standard for Professional Competence of Responders to Hazardous Materials Incidents*. Quincy, MA: NFPA.

______. 2000. NFPA Standard 600-2002, *Standard for Industrial Fire Brigades*. Quincy, MA: NFPA.

______. 2003. NFPA Standard 1006-2003, *Standard for Rescue Technician Professional Qualifications*. Quincy, MA: NFPA.

______. 2003. NFPA Standard 1500-2002, *Standard on Fire Department Occupational Safety and Health Program*. Quincy, MA: NFPA.

National Institute for Occupational Safety and Health (NIOSH). 1985. *Occupational Safety and Health Guidance Manual for Hazardous Waste Site Activities*. Washington, D.C.: National Institute for Occupational Safety and Health/Occupational Safety and Health Administration/U.S. Coast Guard/EPA.

National Response Team (NRT). 1987. *Hazardous Material Planning Guide (NRT-1)*. Washington, D.C.: NRT.

Rand Corporation. 2002. "Protecting Emergency Responders: Lessons Learned from Terrorist Attacks" (conference proceedings). California: Rand Corporation.

Ronk, Richard, Mary Kay White, and Herbert Linn. 1984. *Personal Protective Equipment for Hazardous Materials Incidents: a Selection Guide*. Morgantown, West Virginia: National Institute for Occupational Safety and Health, Division of Safety Research.

Strong, Clyde B. 1996. *Emergency Response and Hazardous Chemical Management: Principles and Practices*. Delray Beach: St. Lucie Press.

U.S. Department of Labor. 1996. *OSHA Construction Safety and Health Outreach Program – Hazardous (Classified Locations)* [online]. U.S. Department of Labor, Occupational Safety and Health Administration, Office of Training and Education [cited May 1996]. Available from World Wide Web (http://www.osha.gov/doc/outreachtraining/htmlfiles/hazloc.html)

U.S.Environmental Protection Agency. 1990. *Hazmat Team Planning Guidance*. Washington, D.C.: EPA Office of Solid Waste and Emergency Response.

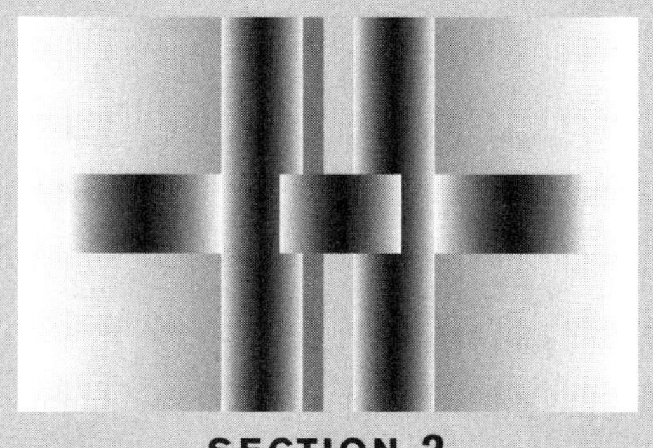

SECTION 2

EMERGENCY PREPAREDNESS

LEARNING OBJECTIVES

- Be able to evaluate hazards or risks.
- Learn to develop an emergency preparedness plan.
- Recognize what is required to put the plan into practice.
- Know what drills to conduct for testing an organization's ability to resolve an emergency event.
- Understand the importance of continually reviewing and revising an emergency preparedness plan as needed.

BENCHMARKING AND PERFORMANCE CRITERIA

Bruce J. Rottner

CHILDREN IN PUBLIC SCHOOL once were required to practice fire drills twice a month. The hazard (fire) was identified as the major threat to life in a school. An evacuation plan was developed, and it was practiced routinely. This lesson in emergency preparedness provides the first three elements of establishing an effective emergency preparedness program: *evaluate*, *develop*, and *put into practice*. Although the plan was practiced, it was never tested, reviewed, or reevaluated.

If an emergency response plan has been sitting on a shelf, it should be taken down and read again. The company should decide if it is still appropriate, figure out how to practice without waiting for a real emergency, and revise it as needed. Then the process should start all over again.

Benchmarking in emergency response is difficult because each plan is different, depending on the values, resources, and specific hazards of an organization. Instead of a process of sharing and measuring against other plans, it is an internal process of self-evaluation of plans and policies with a goal of continuous improvement.

Within the first five years of the twenty-first century, the world's ability to respond to unexpected events has been severely tested. Starting with the 9/11 terrorist attacks that brought about the collapse of the World Trade Center in New York in 2001, countries have been challenged to address both man-made and natural disasters of unusual proportions. Terrorism, by its nature, has highlighted the need for emergency preparedness and has served to consolidate many diverse efforts into a unified approach to preserve life and protect property and the environment. Companies that routinely handle biological, radiological, or chemical hazards have steadily continued to diminish the probability of events related to these hazards by complying with rigid regulatory requirements and self-imposed safety programs. For example, the nuclear power industry has been responsible for developing large-scale emergency plans, even though the probability of nuclear events was thought to be low, because of the amount of work that went into the safety design of the facilities and because the proper handling of nuclear material was well understood. On the other hand, terrorist acts that threaten to initiate biological, radiological, or chemical emergencies have increased the probability of these types of emergency events, even though the location of such events is unknown. Although terrorism might have highlighted the uncertainty of the location for a chemical or radiological event, the possibility of a transportation mishap involving these materials has been present for a longer period of time. In the planning considerations discussion in the Hazardous Materials section of FEMA's *Emergency Management Guide for Business and Industry* (FEMA 1993), it is suggested that a facility identify any highway, railroad, or waterway where hazardous materials are transported; and then determine how a hazardous-material event would impact the facility.

Although terrorism receives the most attention, there are other areas where emergency preparedness is being tested to its limits. For example, what started out as a localized problem regarding electrical transmission lines known as the Lake Erie loop resulted in an electrical fault that was rapidly transmitted through 21 major power stations within three minutes. The outcome was a blackout lasting as long as 30 hours, affecting approximately 50 million people in areas extending west to Detroit, Michigan; east to New York; and north into Canada. Although it is suggested that industry deregulation and an aging infrastructure contributed to this event, the massive impact to such a large area is disturbing and cannot be easily dismissed (CNN 2003). Not to be outdone by man, nature imposed its wrath in the Indian Ocean and surrounding lands. On December 26, 2004, an earthquake with an estimated magnitude of 9.0 occurred on the floor of the Indian Ocean. This earthquake had enough energy to alter the earth's rotation. At the fault line, there was lateral movement of the earth's surface of about ten yards and vertical movements of some several yards. These shifts in the earth's surface displaced a large volume of water and created a tsunami. In this case, the tsunami claimed approximately 300,000 lives. (National Geographic News 2005).

Then there was Hurricane Katrina to teach people a strong lesson about emergency preparedness. This most destructive natural disaster in U.S. history impacted 93,000 square miles stretching across Florida, Mississippi, Louisiana, and Alabama; it resulted in 1300 fatalities, and flooded the city of New Orleans. According to the report, "The Federal Response to Hurricane Katrina: Lessons Learned" (The White House 2006), "individual, local and state plans, as well as relatively new plans created by the Federal government since the terrorist attacks on September 11, 2001, failed to adequately account for widespread or simultaneous catastrophes." The report goes on to identify the federal government's role as "coordinator" for public and private entities under the National Incident Management System (NIMS) and the National Response Plan (NRP); however, the federal government cannot coordinate if the initial responding entities are not functioning as a result of the emergency event. The report states that "this framework does not address the conditions of a catastrophic event with large-scale competing needs, insufficient resources, and the absence of functioning local government" (The White House 2006).

Emergency preparedness has become highly specialized. Today a fire is not often equated to an emergency event. But studying this more traditional hazard and the use of building codes to address deficiencies identified in fires that cause large losses gives us a clear roadmap to a continuous improvement process.

Building codes address a number of life-safety and property-conservation issues, and they are monitored through the design and construction process. Unfortunately, this process is still retrospective rather than proactive. To avoid experiencing a large loss of life, institutions develop and regularly practice evacuation plans, including alternate exit routes. Thus, children learn the value of participating in fire drills. In terms of property damage, the value of a sprinkler system is almost universally accepted. One such fire incident that changed workplace safety occurred on March 25, 1911, at the Triangle shirtwaist factory; it resulted in the deaths of 146 workers, 123 of whom were women. The workers on the ninth floor, encountering locked doors and heavy fire in the stairwell from the eighth floor, took the only available option—to jump from the windows. Fire Chief Edward Worth, who arrived on the scene two minutes after the alarm sounded, found the windows along Washington Place "full of people." In 1911, the tallest ladder in New York City was approximately 30 feet too short to reach the ninth floor (Von Drehle 2003). Today, manufacturers of aerial ladders such as Seagrave (2007) and American LaFrance (2007) indicate that the tallest ladder trucks available reach maximum heights of about 100 to 110 feet. This means that the currently available equipment can reach only the eleventh floor of a high-rise building. Occupants on higher floors must rely on interior stairwells and other features of the fire-safety building design for survival. Despite the success of available equipment, loss of life and property continue to the present day as a result of fires.

The ability to prepare, respond, mitigate, and recover from an emergency event has become routine in safety planning and brought to issue to the forefront of public- and private-sector organizations of all sizes. In the past, the Occupational Safety and Health Organization (OSHA) and the Environmental Protection Agency (EPA) mandated programs requiring companies such as those in the chemical industry with obvious risks to develop emergency response initiatives. Private organizations without major obvious internal hazards were exempt from these regulations. Now, no one is exempt from developing strategies for emergency response.

Emergency preparedness is the last step in a hazard-evaluation process that attempts to reduce the potential occurrence of an emergency event or, at a minimum, to reduce the consequences if the event were to occur. Once a facility develops protective measures against an adverse event, emergency response measures are put in place to contain, notify, and evacuate as necessary. Even though the focus of private-sector programs is inside the fence line, they also protect the surrounding communities. Terrorist acts designed to produce the most damage and destruction and to impact the most vulnerable areas within society, on the other hand, are initiated from outside the fence line, and a facility has no reason to anticipate such an event. Moreover, the private sector "controls 85 percent of the critical infrastructure in the nation. Indeed, unless a terrorist's target is a military or other secure government facility, the 'first' first responders will almost certainly be civilians. Homeland security and national preparedness therefore often begins with the private sector" (National Commission on Terrorist Attacks upon the United States 2004).

As we begin to deal with the more abstract hazards and their consequences, it is difficult, if not impossible, to measure, by traditional standards, the success of an effective emergency preparedness program. As was the case with early biologists and chemists, organization management today is being asked to believe in something that cannot be counted, measured, or seen. In fact, resources spent on *intangible* emergency preparedness programs are more evident than the gains realized through this commitment of resources. Even when an emergency event occurs, how is the minimization of loss to be measured? Is there a tangible way to measure the success of countless hours and resources dedicated to this effort? The best measure of success is to evaluate the ability to respond. This is not hazard-specific but identifies how well the organization can respond to an extraordinary event of any type:

- Does the plan protect life, property, and the environment?
- Have sufficient resources been stockpiled to overcome the majority of incidents identified?
- Is the organization trained and ready for its role in emergency response?

In the author's opinion, these are the types of questions that must be answered when assessing an organization's emergency preparedness.

Compliance or Benchmark

Complying with regulations involving hazardous materials is, in itself, a difficult process. OSHA's hazardous waste operations (HAZWOPER) standard and process safety management (PSM) require a multitude of programs that are expensive to initiate. Add to the mix EPA's Risk Management Program (RMP) under the Clean Air Act, and the cost could rise even more.

Beyond compliance, programs such as Responsible Care® have proven to be successful. Responsible Care® is a global performance-based program that is implemented throughout the chemical industry; it is administered in the United States by the American Chemistry Council. Their Web site (www.responsiblecare-us.com), reports that Responsible Care® has:

- reduced environmental releases by 78 percent.
- achieved an employee safety record four-and-a-half times better than the U.S. manufacturing sector.
- completed 100 percent of security vulnerability assessments.
- reduced process safety incidents by 45 percent over the past decade as the industry continued to grow.

Additionally, the threat of terrorism has widened the playing field to include biological and radiological hazards, which may not be part of an organization's portfolio of hazards. Based on the experience of the author, a facility must now acquire expertise to prepare for hazards it is normally unfamiliar with.

Setting benchmarks and establishing intermediate milestones is a more effective measure of one's success in emergency preparedness. This process is not about meeting a goal (e.g., compliance), but it is a process of continuous improvement. Success in emergency preparedness within an organization will be realized by continuously repeating the steps outlined below:

- Adequately assess potential hazards and realistically evaluate the ability to mitigate an emergency event resulting from those hazards.
- Develop a written plan to address hazards and emergency events.
- Conduct drills and exercises.
- Critically evaluate response capabilities and identify key areas for improvement.

Now that a process for obtaining success has been established, effective measurement of that success is necessary to maintain management's commitment throughout, hopefully, long, safe, and uneventful periods. Measuring the success of emergency preparedness through response to an actual emergency event is not practical and may be disastrous. In the meantime, consider measuring success in other practical ways. For instance, has the organization done the following:

- conducted risk assessments and identified all types and magnitude of emergencies
- implemented preventive measures and process improvements
- developed a plan that has identified resources and lines of authority to react immediately to the situation (must include flexibility in filling key personnel positions throughout the emergency management team, including that of the incident commander (IC)
- established an effective mechanism to move nonemergency personnel to a safe location
- identified mechanisms to notify and seek assistance and resources from the surrounding community
- promoted interest in emergency preparedness by facility employees at all levels

The most important measure of success is the ability of an organization to return to normalcy after an emergency. This scenario allows the organization to focus on its main mission with the least amount of interruption of services and loss of market share.

Hazard and Risk Assessment

What places an organization at risk? On the surface this seems like a simple question. The answer requires events to be defined, events that would prohibit an organization from functioning and achieving stated business goals. It may be important within a manufacturing facility to list those events that would halt production or cause the organization to lose its market share of a particular product.

To assist in the efforts of identifying risks and types of hazards, the following components of the FEMA *Emergency Management Guide for Business and Industry* (1993) are included:

1. Critical Products, Services, and Operations:
   - company products and services and equipment needed to produce them
   - supplies and source vendors needed
   - utilities
   - human resources
2. Potential Emergencies:
   - historical
   - geographic (e.g., proximity to flood plains, nuclear plants, and major transportation routes and airports)
   - technological (e.g., process or system failure)
   - human error

Emergency Response Plans that Work

No matter what level of response it chooses, every private-sector organization now has the responsibility to be adequately prepared for an emergency. If an organization is of sufficient size and its plan is to just call 911 and let community response groups deal with the problem, then maybe it is time the organization rethink its emergency response plan. At a minimum, an organization must account for its employees during emergencies, but in today's environment this effort may not suffice.

For extremely small companies with limited resources, the only response feasible may be to await assistance from community-based emergency response agencies. This is sometimes a practical solution if the inherent hazards are minimal. In any case, an organization must decide on its level of participation during an emergency event.

If an organization can only evacuate and account for its personnel, then that should be the limit of its response; however, the organization should still develop a plan and practice its role in emergency management. Successfully accounting for plant personnel (and visitors) allows outside response groups to focus on mitigation sooner rather than spending initial efforts on rescue operations. In its endorsement of the American National Standards Institute (ANSI) recommended standard for private preparedness, the 9/11 Commission (2004) affirms that "Private sector preparedness is not a luxury; it is a cost of doing business in the post 9/11 world. It is ignored at a tremendous potential cost in lives, money and national security." This message is restated in the lessons learned from Hurricane Katrina. In the Scope and Methodology section of the report, it is clear that to "pursue a real and lasting vision of preparedness," it will involve "every level of government, private sector, individual citizens, and communities (The White House 2006).

When developing an emergency response plan, an organization must identify what constitutes an emergency. According to the *Merriam-Webster Online Dictionary*, an emergency is defined as "an unforeseen combination of circumstances or the resulting state that calls for immediate action" (2007). It could be suggested, however, that the dictionary definition is somewhat in error. An emergency will continue to its conclusion if it is not interrupted by *immediate action*. For example, the December 26, 2004, tsunami ran its course without immediate action. If intervention is undertaken during an emergency event, the outcome will be altered, and not necessarily in a positive way. By definition, the sense of urgency during emergencies is exaggerated, but for certain types of emergencies (e.g., catastrophic release of a gas), prompt response might not be possible. Does an emergency require or demand an action? The answer is no. It is perfectly acceptable to decide that the safest response is no response at all.

To assist in developing an emergency response plan, FEMA's *Emergency Management Guide for Business and Industry* (1993) lists the following components as those a plan should address:

- direction and control
- communications
- life safety
- property protection
- community outreach
- recovery and restoration

Incident Command System

On February 28, 2003, President George W. Bush issued Homeland Security Presidential Directive (HSPD)-5, which required the development of NIMS. The adoption of a single national system should add consistency to emergency response. It should also bring familiarity and understanding to the terminology used in such situations. It is not unusual for a frequent rider of the New York City subway system to hear several languages spoken at the same time; however, because one hears those languages does not mean that one understands them. By adopting a single system, the terms discussed will have exactly the same meaning to everyone participating in the emergency response; this situation would be like everybody in that subway speaking and understanding one common language.

HSPD-5 has set a high mark for cooperation. According to the directive, "The system (NIMS) will provide a consistent nationwide approach for federal, state, and local governments to work effectively and efficiently to prepare for, respond to, and recover from domestic incidents regardless of cause, size, or complexity" (Department of Homeland Security 2003). Those who have experience with multiagency emergency response know that such objectives are not always accomplished despite the need for an immediate and cohesive plan of action. One of the cornerstones of NIMS is the Incident Command System (ICS), which, additionally, the 9/11 Commission recommends be adopted by all emergency response agencies nationwide.

HSPD-5 is mostly directed to the government or the public sector. What about the private sector? The private sector has the responsibility to interact with public response agencies. Everyone on the "playing field" should play by the same rules. Imagine two teams on a field playing a game in which one team follows the rules of soccer and the other, football.

ICS provides the framework to manage the incidents. Although the structure may be the same, an ICS structure will look substantially different throughout the various phases of an emergency event. The emergency and rescue phase will more than likely have a high-ranking officer, known as the incident commander, within emergency response groups. This may not necessarily be true as the event progresses from rescue to recovery and ultimately to the final phases that become heavily weighted toward the financial side of the emergency event. It could be argued that this phase is not really part of emergency preparedness, but if one accepts the concept that emergency preparedness brings an organization back to *normalcy*, then the event lasts longer than is commonly thought. There are several versions of the ICS. One source of information on ICS is the National Fire Protection Association's *NFPA 1561, Standard on Emergency Services Incident Management Systems* (2005).

ICS is an emergency management structure that can be expanded to address all types of emergencies. ICS works because there is one person in charge, the IC. During the phases of an emergency response, the IC may change; however, at any given time, there is only one IC. The remaining functional staff within ICS is divided into four main areas: operations, planning, logistics, and finance/administration. Additionally, the command staff could have as many as three positions that report directly to the IC. These three positions are safety, liaison (with outside emergency response agencies), and public information (see Figure 1).

The most critical position to fill is that of incident commander. It is essential for the IC to have the authority to commit resources and also the responsibility to accept consequences of his or her decisions. Equally important, the IC must be well trained in emergency response. This second requirement is most often overlooked, but it could be the difference between success and failure of an emergency response. In the Federal Emergency Management Agency's (FEMA's) *Emergency Management Guide for Business & Industry* (1993), the responsibilities of the IC are as follows:

- Assume command.
- Assess the situation.
- Implement the emergency management plan.
- Determine response strategies.
- Activate resources.
- Order an evacuation.
- Oversee all incident response activities.
- Declare the incident is over.

Because the ICS is an expandable system, it is possible, in very small events, for the incident commander to serve as the command staff as well as the functional area staff personnel.

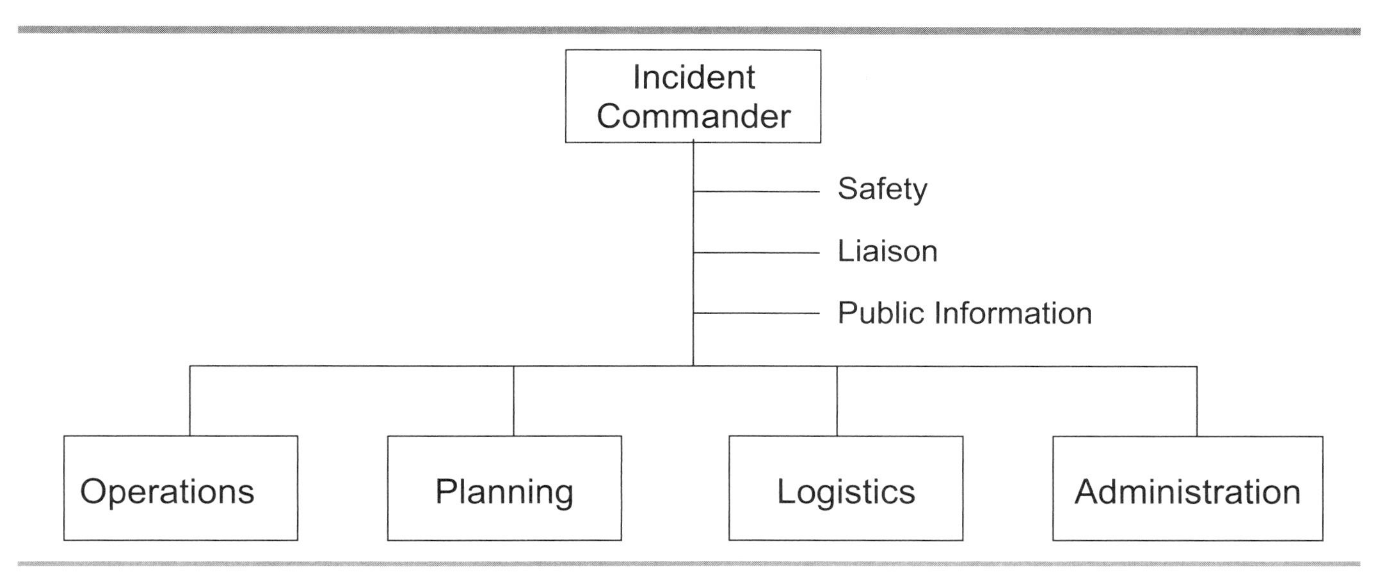

FIGURE 1. Incident command structure (Adapted from NFPA 2005)

Command Staff

The safety position monitors the emergency response and provides recommendations to the IC on matters of safety and health. Preservation of life and health is universally accepted as one of the critical incident goals, one that includes the health and safety of emergency responders. Besides the IC, the safety position carries the most responsibility and should be filled with great care and consideration. Safety has the right and authority to overrule the IC whenever an imminent hazard arises. The person filling this safety position, therefore, must be equally skilled in emergency management and have an understanding of safety principles.

The liaison position maintains contact with all government representatives and private/public organizations. During emergency events managed by public emergency response organizations, the private sector, except for the specific organizations involved in the emergency, is usually excluded from immediate contact with the IC and the command staff. The liaison position is essential for maintaining an open line of communication with all the nonemergency response agencies involved.

The public information position provides information and establishes communication with the public and the media. Keeping the *outside world* informed is better than having the public speculate as to what is going on. In the absence of information, speculation becomes truth. In the private sector, this position is often filled by corporate-level company representatives. Larger companies usually have an existing public relations department or corporate legal counsel that is responsible for disseminating information under normal circumstances; these departments assume the public information role during emergencies.

Functional Structure and Staff

The *operations* section is responsible for the hands-on functioning of the emergency event and the direct implementation of tactical incident goals. This group is usually made up of the fire department, the police department, the hazardous materials units, and other public- and private-sector emergency response groups. The person in charge of the operations staff reports directly to the incident commander and manages all tactical response activities.

The *planning* section is responsible for collecting and evaluating incident information, and for providing summaries of this information to the IC. The planning section is also responsible for assisting the IC in identifying incident objectives. This group includes technical specialists from the private sector. IC's in the private sector usually support the public emergency response as an adjunct to the safety position or as a technical specialist within this section. Planning also has a very important responsibility during long-lasting events for identifying staffing and resources needed to ensure consistency over the entire event.

The *logistics* section provides those resources identified by the IC and the planning section to effectively manage the incident. This section is also responsible for resources such as food, medical services, housing, and personal supplies for the emergency responders. For example, the planning section may identify the need for a large quantity of a special extinguishing agent, and the logistics section would need to find a source to purchase those products from and establish delivery to the operations section.

The *finance/administration* section is responsible for maintaining detailed records of time, cost, and purchases related to an emergency event. This unit is also responsible for receiving billing and authorizing payment to outside vendors. This section is usually only needed during large emergency events that span a relatively significant amount of time. Most often the work of this section occurs after the emergency is over. For organizations seeking reimbursement for various costs from the government, this section plays a very significant role in emergency response. The 9/11 Commission Report recommended that Congress endorse a voluntary preparedness program for the private sector. *NFPA 1600, Disaster/Emergency Management and Business Continuity Programs* (NFPA 2007) establishes a common basis for all emergency response plans and programs. Appendix E of this standard identifies several sources on emergency management, including ICS. One such source is *NFPA 1561, Standard on Emergency Services Incident Management System* (2005).

Components of a Written Response Plan

There are some common elements to all successful emergency response plans. First and foremost, the plan should incorporate the ICS. As previously discussed, an ICS by definition is a flexible management plan for handling all types of emergencies (NFPA 2005).

Additionally, a plan should include:

- the definition of an emergency
- available resources
- clearly defined roles and levels of authority
- memorandum of understanding (MOUs) with outside response agencies
- MOUs with outside vendors
- lists of internal resources
- provisions for effective risk communication
- training requirements for all response levels, including management

As simple as it sounds, the definition of emergency is a key component of a response plan. A plan cannot be developed if one does not understand when it is to be used and what it is intended to protect against. A common understanding of an emergency is required to determine when to activate the plan. It certainly would be a waste of time to develop a plan but never use it because no one knew when to implement it.

Within the emergency response plan, an organization must identify clearly defined roles and levels of authority. Key personnel must be identified for each critical position in the ICS.

Once the emergency management plan has been written, there must be a mechanism for evaluating it annually and after each use. There is no better learning tool than an actual emergency. Event debriefing or critiquing is essential to identify problematic areas and ways to improve the plan. Learn from mistakes and improve the plan for the next time.

For example, if specific equipment were needed to resolve the emergency, some questions to ask regarding it might include the following:

- Was the specialized equipment available to the emergency responders?
- Was there sufficient quantity to complete the emergency response?
- Is additional equipment, not previously identified, more helpful or needed in addition to that which was used?

Another example might include evaluating the personal protective equipment (PPE) requirements of the emergency response team:

- Was PPE available in sufficient quantity and sizes for the emergency responders?
- Did the emergency responders know how to use the equipment without hesitation?

- Was the PPE the right type for the emergency that occurred?
- Is any other PPE, not previously identified, required to handle emergencies at this facility?

As illustrated above, a drill or actual emergency event must be evaluated in order to ensure that the existing emergency management plan is still adequate. Shortly after the drill or emergency event, all participants should gather to discuss and self-evaluate their performance. The IC should lead the discussion and start by listing his event goals and objectives, followed by his critique of the entire operation. All other participants should be given the opportunity to state both positive and negative comments about their participation, the availability of equipment, communication between the command staff and field staff, and so on. The discussion should be documented and forwarded to those responsible for maintaining the plan. All discussion points should be evaluated and either accepted by modifying the plan or denied with written details of why the particular issue did not prompt a change in the plan. All evaluations of a drill or actual emergency should be maintained on file with the master copy of the file.

Finally, in the context of today's wide-ranging list of potential emergency events, the plan must be a living document that has a mechanism for change during an ongoing emergency event. Maybe the situation has never been encountered before and improvisation is necessary. Incorporation of this new material must be possible. Without this flexibility, the author, based on experience, finds it conceivable that a plan could be obsolete before the event is concluded.

Coordination with Outside Emergency Response Agencies

Successful management or the ability to achieve a positive outcome with the least loss of life, damage of property, or damage to the environment during an emergency event between a private organization and public emergency services is based on the development of a working relationship; it is also based on the development of an open and trusted communication, where the emphasis is on the words *trusted* and *relationship*. Both organizations must feel comfortable assisting each other. Neither a private organization nor a public emergency response service can handle an emergency event by themselves once it has become significantly complex.

This process can start with a plant tour. Public emergency responders must be familiar with the plant layout; a tour of the buildings and an explanation of activities will further that familiarity. Emergency responders should be comfortable with entering a facility even if it is three o'clock in the morning. Because public emergency responders often arrive well before their private-sector counterparts, and because they must be able to respond safely, it is important that they understand the hazards associated with the organization's operation.

Essential to this relationship between the private and the public sector is *disclosure*. All of the hazards identified in a risk or hazard assessment must be part of the dialogue between the public and the private emergency responders. This disclosure process also applies to the physical plant. Changes in the utilization of space and the configuration of equipment may occur over time to increase efficiency or incorporate new technology. From an emergency preparedness perspective, these changes may cause confusion for public emergency responders who are not aware of them.

In the past, private-sector operations cloaked their operations in secrecy. Exposing hazardous operations to regulatory organizations was frowned upon because it brought with it the threat of scrutiny, restrictions, and registration (e.g., permits). Change has been ongoing, and there is more cooperation between the private and the public sectors. The brick wall has been replaced by a chain-link fence. The amount of openness and the development of rapport between private-sector management and public officials will determine success.

It is also important to remember that the emergency responders may be victims of the emergency event as well. Relying solely on their response to resolve the organization's emergency response needs might not be practical. This is mostly true during natural disasters and transportation emergencies involving hazardous materials.

A good written plan is one that provides an organized response to an emergency event. It is an extension of the overall safety program of an organization. Additionally, it must be flexible to allow for proper decision making in the event that identified resources might not be available. The only benchmark of a good plan is that it provides resolution to all drills and emergencies that occur in a specific organization. A plan of one organization cannot be compared to that of another. Organizations with multiple facilities may want to share plans between facilities if the operations and the safety culture are similar.

Training and Personnel Competency

Now that the written response plan is complete, it is time to gather human resources within an organization to fill the various roles identified in the plan. Whatever predefined roles are established for the personnel, training is necessary to ensure competence. This is especially true for the IC. Of all the training that goes into emergency response, the IC often receives the least amount of it. Because the IC is a prominent high-level person within an organization, it is often assumed that he can manage an emergency response without a basic understanding of the organization, the terminology, and the principles of emergency management.

Anyone that participates in emergency response must be trained for the responsibilities they are to assume. Front-line emergency response teams must receive a level of training commensurate with the tasks they are asked to perform. Training cannot be limited to once in a career or even once a year. Emergency response training is an ongoing process, but there is absolutely no room for on-the-job training. The proper application of emergency response techniques and equipment must be practiced repeatedly. Additionally, the training participants should practice in the required PPE (whatever level necessary for the protection of the responders). This experience can be illustrated by wearing self-contained breathing apparatus (SCBA) under a Level-A encapsulating suit while practicing to stop a leak from a compressed gas cylinder. If this level of protective equipment is required, it is a good idea, if participants are not in compliance, to practice interrupting the process and allowing different responders to don the necessary PPE. Having limited air supply could prevent the assigned personnel from completing of an emergency task, thereby requiring the back-up team to complete it for them. For facilities located in areas with extreme weather patterns due to seasonal changes, emergency responders should practice their technique in extremely cold or hot conditions. Such extreme conditions will negatively impact performance.

The practice of wearing PPE requires a consistent training program. Unless an emergency responder uses the same level of PPE for his regular job responsibilities and emergency response, continual training is necessary. It is important to remember that PPE is at the lower end of the hierarchy of controls available to safety and health professionals. It works as a protective barrier for the response personnel, and if used incorrectly, could lead to an exposure incident.

Whatever level of response an organization chooses, training is necessary. Organizations selecting regular evacuation drills should also be training personnel in the practice of sheltering in place. In some types of emergencies, the best response may be to stay inside an employee's place of work, isolated from the outside environment. This type of strategy is counterintuitive to the almost instinctual desire to flee the area of harm. Telling people that they cannot leave a building for their own protection will be challenging, given the availability of rapid communication. Computer online capabilities and cellular telephones will quickly alert occupants of a building of the events that are taking place outside.

Those organizations that choose to develop a fire brigade to a structural level or a spill response team to a technician level should examine their decision carefully. Emergency response organizations assigned these tasks are deliberately put in harm's way. At these levels, the responders will adopt offensive action and will need specialized PPE. (*Note:* Defensive action entails protecting personnel from exposure and arresting the growth of the event, whereas offensive action entails direct effort to curtail the emergency event by either reducing or putting out a large fire or stopping a spill.) Proficiencies in response technique and the proper use of PPE at this level will require a commitment of time and money. This level of commitment will also impact the training requirements of the IC. An IC must always receive equal or more training than the responders under his command.

The true benchmark for training is a successful response to an emergency event, but this is a considerably risky method for ensuring personnel competence. Emergency management and response training requires both classroom and hands-on proficiency. No emergency response organization should perform actions above the level of their training and ability. The *Emergency Management Guide for Business and Industry* (FEMA 1993) recommends general training for all members of an organization in the following areas:

- individual roles and responsibilities
- information about threats, hazards, and protective action
- notifications, warning, and communication procedures
- means for locating family members in an emergency
- emergency response procedures
- evacuation, shelter, and accountability procedures
- location and use of common emergency response equipment
- emergency shutdown procedures

Conducting Drills

Practice for Timing

Every emergency preparedness program should conduct one exercise to understand the amount of time necessary to identify a hazardous situation and respond appropriately. One should not forget, however, that such an exercise will not adequately reflect the actual amount of time spent to identify, mobilize, and respond to an emergency. The critical lesson learned from the above exercise will be crucial in developing an emergency response program. Some emergency events are so rapid (e.g., chemical vapor releases, dust explosions, or vapor explosions) that the response is to the effects of the event rather than the event itself. Private-sector emergency programs that rely heavily on first-shift plant human resources should conduct a drill at rush hour or on a holiday weekend. The critical lessons learned through those drills will come in handy in a situation where two primary responders are expected, but two back-up responders, fully prepared to address the emergency, arrive before them. The public as well as the private sector will benefit from this type of exercise. Based on the author's experience, the public sector must also realize that private-sector responders may not be available or may be just blocks away, held back at a public safety perimeter.

Promoting Cooperation, Mutual Respect, and Understanding

Emergency exercises and drills should promote cooperation between public and private sectors. The public sector may depend on the private entity for its economic contribution to the community. If the private sector fails to resume normal operations or there are extensive delays, the public sector will be adversely impacted. When an event is initiated on private property, the private entity often needs the resources and assistance of public emergency responders. At the same time, the public sector must respond to facilities once thought to be safe that have obvious hazards. Both the private and public sectors have an interest in resuming normal activities as quickly as possible. This situation demands that both sides realize that cooperative efforts will bring about positive results in a more timely and cost-efficient manner.

Based on the author's experience, the private sector must also recognize that it can be adversely affected by a public-sector emergency. Disruption of services or disruption of access to a facility will negatively impact business. Raw materials that cannot make it into a facility or finished product that cannot be sent out will definitely have a negative impact on the bottom line.

When planning emergency response drills, it is essential to avoid having them for the wrong reasons. The top three wrong reasons, based on the author's many years of participating in these activities, are discussed below.

Success Is the Only Option

Facility personnel responsible for emergency response have spent a considerable amount of time and resources to develop a program that not only meets the requirements of the regulations, but one that also guides the organization through any emergency. After convincing management to fund the development of the program and obtaining essential equipment, the same emergency response leadership is now tasked with conducting an emergency response drill. Should the drill be designed to stress all aspects of the plan and possibly identify weaknesses, or should it be orchestrated to achieve a successful result? Unfortunately, more times than not, the drill will be predetermined to be an overwhelming success. There is no other option.

Drills by the Numbers

Sometimes drills and tabletop exercises are conducted for all the wrong reasons. In some cases drills are just done by the numbers. Regulations may specify a specific number of drills annually. Content and quality are not defined. For example, a company decides to hold an off-hours event when its site occupancy is minimal. The plant manager stays behind to initiate the alarm and then waits for the guards to dial 911. The community emergency response groups descend on the site and are greeted by the plant manager. Everything goes according to plan and another successful drill is documented for the record.

Impending Doom

There are other drills that are designed to result in a self-fulfilling prophecy of doom. The emergency response committee has developed a plan to the best of their abilities; however, they are aware that there are gaps and areas that need improvement. Understanding the weaknesses, the committee can steer the drill away from those areas, but, most of the time, they will either drive head on to make upper management listen to their concerns, or they will not purchase a piece of equipment that they believe to be essential. Thus, the drill is designed to highlight the inherent weakness. After the exercise or drill, the ceremonious debriefing occurs. In this case, the drill highlights weaknesses in the emergency response.

Now that the wrong reasons for having drills have been explored, here are some drill scenarios that will effectively test an organization:

- resource management in a multiday event
- staffing for an extended event
- community notification
- implementing MOUs with outside responders or contractors
- risk communication to an organization during an emergency event

Based on the author's experience, a great mechanism to benchmark a plan and resources is to see what the neighboring facilities are doing. Although they may be reluctant to share, valuable information can be gained by cosponsoring a drill or exercise with them.

HOMELAND SECURITY VERSUS THE "RIGHT TO KNOW"

When the EPA first published its Risk Management Program under the Clean Air Act, a facility with the potential for a chemical release to exceed the perimeter of its property had to define, evaluate, and publicize its worst-case scenario. This practice was predicated on the idea that an informed community could preplan and ultimately respond successfully to an emergency event. It was not unusual for these worst-case scenarios to be published in the press, discussed at town meetings, or otherwise disseminated to anyone interested in listening. In fact, the information was so well known that it was difficult not to know what potential harm a facility could cause to the surrounding community.

Unfortunately, terrorism has mandated that we reevaluate this policy. With the involvement of the Department of Justice (DOJ), a community's right to know must be balanced against society's right to protect itself from larger harm. The availability of information describing the extent of damage caused by a chemical release could be exploited by those with harmful intentions to cause an intentional release.

In August of 2000, the EPA and DOJ issued EPA 550-FOO-012, the Final Rule on the Chemical Safety Information, Site Security and Fuels Regulatory Relief Act: Public Distribution of Off-Site Consequence Analysis Information. This program does not discourage or prevent the development of data from an off-site consequence analysis (OCA); instead, it limits the access to the information by storing it in a more controlled environment.

As an alternative to publishing OCA, the EPA has implemented a vulnerable zone indicator system (VZIS) for accessing *vulnerability* information about a particular address. Within this system, one could find out one's proximity to a potentially harmful (if released) chemical or industrial source.

Homeland Security programs have infused a tremendous amount of money into equipping, protecting, and training emergency responders. As these programs continue to grow and become more complex, the involvement of private-sector experts cannot be minimized. Knowledge is the only commodity that this money cannot buy—consider the importance of the Material Safety Data Sheet (MSDS) to underscore this point. This MSDS is prepared by experts to provide written support to end-users, including emergency responders. During an emergency, it would be extremely beneficial to tap into the resources of the facility expert who created such a document.

As an area becomes more densely populated, residential, mercantile, and building occupancies begin to encroach on industrial facilities. What once was distant from populated areas becomes surrounded by housing, parks, stores, and the like. This scenario can be seen in many northern sections of New Jersey. Conducting hazard assessments and determining the impact on the community is traditionally the responsibility of the more hazardous operations and should remain so if they want to build a new facility or add a process directive to an existing plant. But what if the hazardous occupancy exists first? Should they be responsible for protecting those who impinge on a hazardous operation, thereby eliminating any buffer zone? In this scenario, less hazardous or nonhazardous occupancies must assume the responsibility for evaluating risk in a reverse risk assessment (RRA). RRA is a concept developed by the author to make hazard assessments the responsibility of all entities to address their hazards as well as those in the surrounding areas. This is often the result of community growth and the need for these communities to coexist with recognized hazards or existing high-hazard operations. Whether it is the ice cream store that wants to locate next to the dry cleaner that utilizes perchloroethylene (PERC) in its cleaning process, or the day care or senior center that builds downwind to a chemical company, the responsibility to evaluate and provide mitigation to the hazards should fall upon the new occupants. Beyond chemical hazards, issues such as potential terrorism targets, radiological exposure, and other threats to life must be identified and evaluated before the specific occupancy may exist. Location, location, location is now the responsibility of all.

Impact of 9/11 on Emergency Response

Human beings have achieved what once was thought to be impossible. The National Aeronautics and Space Administration (NASA) has landed men on the moon not once but twice. Although there has been no return to the lunar surface recently, those landings stand as a high point of human achievement in space travel. The space program has embarked on many diverse and complex programs such as the space shuttle. In a sense, the World Trade Center disaster was the lunar landing of emergency response. This event was the unthinkable, the mark that could never be achieved. If the event had not unfolded before our eyes, it would have been difficult to believe. If the analogy holds, how do we then begin to evaluate what may occur next or to think about the next potentially bigger and more horrific scenarios? By considering 9/11 as the new standard of disaster, emergency response has been catapulted to a new level. Unfortunately, the resulting response and planning has also created levels upon levels of detailed planning and organization that will be difficult to initiate in events of smaller magnitude. That the expandability and flexibility of ICS is desirable in an emergency management structure has been forgotten. There must be a return to placing emphasis on the more probable events rather than the worst-case scenarios. Response to terrorist acts is part of the process, but it cannot be the only focus of our efforts.

After spending ten months at the World Trade Center site ("Ground Zero"), it is easy for the author to see that enumerating all the lessons learned in response management would probably require a book. Painfully, it is obvious that an emergency response to a worst-case scenario will tax every available resource, including human resources. That type of event will only enter the realm of believability once it happens; therefore, the ability to tailor a response to such an event is almost impossible.

In the beginning of the event, long-term planning is conducted in measures of minutes instead of hours and days. Before the ink is dry on a set of incident objectives, they probably will have changed. At the early stages of an event, it will be difficult to remain within the working limits of the emergency plan.

Ground Zero was a unique working environment. Emergency responders and construction workers were working side by side. Notice that the words are *side by side* and not *together*. Each person was comfortable within the realm of his or her own expertise, yet uniformed emergency responders found it difficult to understand the important role that many civilians played in the response. Once these two worlds collided in a single emergency response, there was extreme difficulty in the merging of their cultures. Balancing the needs of both sides should work in theory; however, adding the emotional component to the mix makes reasonable and prudent decisions often sound unacceptable. Consider the scenario where rescue workers are looking for victims in the debris as it is hoisted away by heavy construction equipment. The need for the rescue force to be close to the digging is contrary to rules allowing work in and around heavy equipment. On the other side of the argument is the premise that operating at a safe distance from the heavy equipment would hamper locating victims. Heavy-equipment operators at Ground Zero had to learn to work within a crowd despite efforts to establish a safe zone of operations.

The response to the World Trade Center disaster also proved that unified command does not work as well as incident command. Daily meetings at an uptown pier were as far removed from the activities at Ground Zero as was the physical distance. The consolidation of control under the New York City Department of Design and Construction was a pivotal point in uniformly applying safety procedures throughout the emergency site.

Perimeter control is essential during of an emergency event. Chemical response teams are trained to pause and evaluate the situation before creating various perimeters and points of entry. When faced with the enormity of lives in the balance, it is doubtful that controlled access can be established immediately in a catastrophic event. On the other hand, a large emergency event may mask a biological or radiological component that could prove disastrous.

Besides being an emergency operation, Ground Zero was also a federal crime scene. Terrorist attacks are the primary source of emergency events and will continue to create this mixed environment; however, any emergency event that results in a fatality will include similar restrictions. Law enforcement is an exclusionary process. The detailed processing of evidence and the maintenance of a chain of custody are necessary to processing a crime scene. Contrast this with a multitude of personnel descending on an emergency event to perform rescue-and-recovery operations. Multiagency emergency events often do not have a clear set of incident objectives, especially early in the event. At the World Trade Center Emergency Project, the needs of law enforcement impacted the site in the following areas:

- establishment of perimeter security
- credentialing of emergency responders
- processing of criminal evidence
- response of volunteers and accountability

If these systems were in place, the initial response would have been severely delayed; however, a true identification of all of the initial responders would have been a useful resource for follow-up medical surveillance.

The success of the safety program at Ground Zero to facilitate the emergency preparedness response required numerous dedicated safety professionals working countless hours and keeping constant vigilance over activities at the site. The first requirement in any emergency response is to protect the health and life of people, including emergency responders.

(*NOTE:* This section on 9/11 is based on the author's own observations, experience, and knowledge as the Assistant Director of Environmental, Health, and Safety for the New York City Department of Design and Construction at the World Trade Center Emergency Project.)

Summary

Once the smoke has cleared, the cloud has dispersed, and the clean up is all but complete, the last benchmark is whether policies and programs that return the organization from emergency to normalcy have been introduced. Employee confidence after an event is usually at an all-time low. The reality of the hazards of an employee's occupation or existence has been made very apparent. Regulatory agencies and the community want reassurance that the process has been made safe. All eyes are on the decision maker. Is it safe? Is it time to attempt restarting a process that has caused some sort of unanticipated result? Do we now have the ability, in hindsight, to realize when the process is malfunctioning? Managing emergency preparedness as a continuous process of improvement and assessing and measuring success on a multitude of levels should make some of these questions easier to answer.

References

American Chemistry Council. 2007. *Tracking Performance, Sharing Results*. www.responsiblecare-us.com

American LaFrance. 2007. *Aerials*. www.amercanlafeance.com

CNN News. 2003. *New Yorkers Face New Emergency* (August 8, 2003). www.CNN.com/2003/US/Northeast/08/14/NYC.powerout/index.html

Cornell University ILR School. *The Triangle Factory Fire: Investigation, Trial and Reform*. www.ilr.cornell.edu/trianglefire/narrative6.html

Department of Homeland Security. 2003. *Homeland Security Presidential Directive (HSPD-5)*. www.dhs.gov/xnews/realeases/press-release_0105.shtm

Environmental Protection Agency (EPA), Chemical Emergency Preparedness and Prevention Office, and U.S. Department of Justice (DOJ). 2000. *EPA 550-FOO-012: Chemical Information, Site Security and Fuels Regulatory Act: Public Distribution of Off-Site Consequence Analysis Information (Final Rule)*. www.epa.gov/ceppo.web

Federal Emergency Management Administration (FEMA). 1993. *Emergency Management Guide for Business and Industry (FEMA 141)*. www.fema.gov/pdf/business/guide/bizindst.pdf

Linder, Douglas. 2002. *Famous Trials: The Triangle Shirtwaist Fire Trial (1911)*. www.law.umkc.edu/faculty/projects/ftrials/triangle/trianglefire.html)

Merriam-Webster Online. 2007. "Emergency." www.m-w.com/dictionary/emergency

National Commission on Terrorist Attacks Upon the United States. 2004. *The 9/11 Commission Report: Final Report of the National Commission on Terrorist Attacks Upon the United States*. New York: Norton.

National Fire Protection Association (NFPA). 2005. *NFPA 1561, Standard on Emergency Services Incident Management System*. Quincy, MA: NFPA.

_______. 2007. *NFPA 1600: Standard on Disaster/Emergency Management and Business Continuity*. Quincy, MA: NFPA.

National Geographic News. 2005. *The Deadliest Tsunami in History?* www.news.nationalgeographic.com/news/2004/12/1227_041226_tsunami.html

Occupational Safety and Health Administration (OSHA). *29 CRF 1910.120*. www.osha.gov/pls/oshweb/owadips.show_document?p_table+STANDARD&p_id=9765

______. *29 CRF 1910.138*. www.osha.gov/pls/oshweb/owadips.show_document?p_table+STANDARD&p_id=9788

Seagrave Fire Apparatus. 2007 (retrieved October 31, 2007). www.seagrave.com

White House. 2006. *The Federal Response to Hurricane Katrina: Lessons Learned*. www.whitehouse.gov/reports/katrina-lessons-learned

Von Drehle, David. 2003. *Triangle: The Fire That Changed America*. New York: Atlantic Monthly Press.

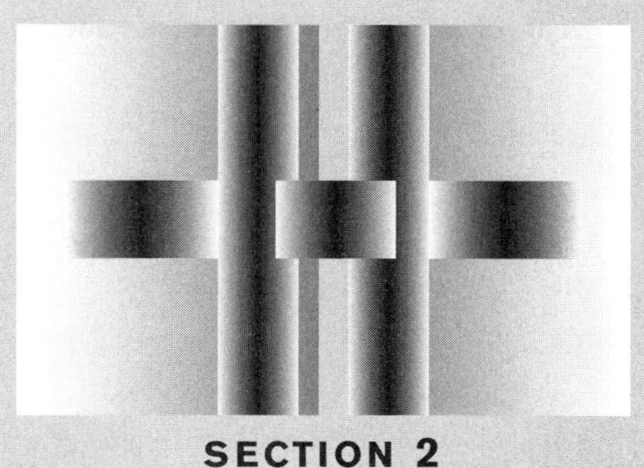

SECTION 2

EMERGENCY PREPAREDNESS

LEARNING OBJECTIVES

- Be able to incorporate the life cycle of emergencies and its role in developing a comprehensive emergency management program.
- Be able to apply the concepts of scalability, standardization, and common terminology to emergency management.
- Distinguish between hazard-driven and event-driven models in emergency management plans.
- Be able to apply the risk matrix to prioritize and select emergency management program elements.
- Be able to utilize the elements of NFPA 1600, the DRII/DRJ Generally Accepted Practices, and the National Incident Management System/National Response Plan to develop a comprehensive emergency management plan.

BEST PRACTICES

Philip E. Goldsmith

AN *EMERGENCY* IS any unplanned event that can cause death or significant injuries to employees or the public, or that can disrupt a business or its operation, cause physical or environmental damage, or threaten an organization's financial standing or public image. The process of dealing effectively with the impact of such events is known by many names, some of which imply a narrow or specific focus, such as

- ***Disaster recovery plans:*** Often used by the information technology (IT) community, these plans address an *interruption in IT service* by implementing a plan to restore the organization's critical business functions.
- ***Business continuity plans:*** Often used by the private sector, these programs aim to ensure the *long-term survival* of the entity following an event.
- ***Continuity of operations plans (COOPs):*** Often used by government entities, these plans provide for *continuity of the essential functions* of the organization, emphasizing alternate sites to substitute for a temporary inability to access or operate the primary site.
- ***Emergency response plans:*** Evolving from the old Civil Defense program and the Fire Service, these programs address *immediate response actions* that are taken to prevent further personal injury or property damage after an event.
- ***Occupant emergency programs:*** Often used by the federal government, these programs focus on *safeguarding building occupants* during an emergency, historically focusing on evacuation, but more recently including shelter in place.

- ***Comprehensive emergency management programs:*** Among the more recent alternatives, but rapidly becoming the umbrella term for a *comprehensive program* that targets all of the more probable hazards, considers all of the events that those hazards may trigger, and does so within the context of a program with multiple elements addressing the four phases of an emergency: mitigation, preparedness, response, and recovery.

While the broadest term might be *comprehensive emergency management*, this chapter uses the simpler term *emergency management* to include all aspects of the above programs. A comprehensive emergency management program may have more than one "plan" document; it may have a series of interrelated plans addressing specific phases or actions to be taken. For example, a university may have a strategic plan for emergency management, an IT disaster recovery plan, a campus emergency response plan, individual building occupant emergency plans, as well as a business continuity plan.

Phases of Emergency Management

Emergencies don't just pop up and go away. They have a life cycle of occurrence (FEMA 2006 Sec. 1, 11-13) that can be matched by four phases of management planning.

- ***Mitigation:*** actions taken to *eliminate or reduce the chance of occurrence* of the event or reduce its severity or consequences, either prior to or following a disaster/emergency.

  Example: Reinforcing the roof structure of a building won't prevent the next hurricane from occurring, but it can reduce the effect of the wind. On the other hand, moving the operation out of a hurricane zone is also a way of practicing mitigation.
- ***Preparedness:*** activities, programs, and systems developed and implemented *prior to* an emergency that are used to support and enhance response to and recovery from the emergency. Preparedness can help the organization protect people and property during an emergency by having a well-defined response and recovery plan, by training various levels of employees, and by providing resources that might be needed.

  Example: An organization that is prepared for hurricanes has documented the actions that must be taken as the storm approaches, during the storm, and after the storm (at various levels of impact). It has also trained employees in their roles and stocked supplies that might be needed to protect the property from further damage after the storm.
- ***Response:*** activities designed to *address the immediate effects* of the disaster/emergency. Response activities need to be as well thought out as possible, be specific, and yet be easily understood without ambiguity.

  Example: An organization ready to respond to a hurricane has identified the steps to be taken to prevent further damage to the building and its contents, including shutting off gas and water lines in the building to prevent fire, explosion, and flood should those lines be damaged by the collapse of part of the building.
- ***Recovery:*** activities and programs designed to *return conditions to normal or to a level that is acceptable* to the organization. Short-term recovery returns critical systems to acceptable levels. Long-term recovery, which could take months or even years, attempts to return to pre-emergency or better levels.

  Example: An organization recovering from a hurricane hopes to have the ability to conduct at least interim operations, either on site or at an alternate facility. Long-term recovery may mean rebuilding a customer base, resupplying inventories, and so on.

Best Practices in Emergency Management

Emergency plans don't come off the shelf. Although various checklists, templates, and computer programs can assist in the development and execution of plans, each organization and each facility needs to craft its own emergency management program to suit its own needs.

Having made the case for specificity, a seemingly contradictory approach must also apply: emergency management is often planned individually but executed within a much larger environment. Although plans must be specific and unique, they must be scalable and standardized and use common terminology so they can be used in that larger environment.

This chapter looks at best practices in emergency management through an examination of three industry standard documents. None of these documents is an outline for an emergency management program or plan, but each introduces best-practices concepts against which a given program or plan may be measured.

- NFPA 1600: Standard on Disaster/Emergency Management and Business Continuity Programs (NFPA 2007)
- Generally Accepted Practices for Business Continuity Practitioners (Disaster Recovery Journal and DRI International 2006)
- The National Incident Management System (U.S. Department of Homeland Security 2006a)

Scalability

Emergencies come in all sizes. Within an organization, emergencies can

- be specific to an individual, such as an employee injury
- be internal to the facility, such as a fire
- affect the entire organization, such as a computer network failure
- affect the immediate area, such as a power outage
- affect a large geographic area, such as a severe weather or terrorist event.

Emergencies can also be either direct, impacting the organization itself in some way, or contingent, impacting a business affiliate such as supplier or vendor, producing a contingent effect on the organization.

Emergency programs and plans must provide a consistent, flexible, and adjustable framework within which all levels of the organization can work together to manage incidents, regardless of their cause, size, location, or complexity. This flexibility applies across all phases of incident management: prevention, preparedness, response, recovery, and mitigation.

Standardization

The organization's emergency plan must include a set of standardized processes, procedures, and systems to facilitate emergency management at various functions or geographic areas, as well as standardized organizational structures, such as emergency response teams, communications teams, and so on, to carry out those activities.

Common Terminology

An emergency response organization is made up of individuals who may not work together as a team except during the response operation. When they come together, the use of common terminology is essential to efficient and effective communications, both internally and with other organizations responding to the incident.

Emergency Management Models

Emergency management is both a management and a decision-making process. Like any other management process, it has four common elements:

Plan → Organize → Lead → Control

As a decision-making process, it can be described in six decision-making steps (Baranoff et al. 2005 Sec. 2, 1–28):

Identify exposures → Analyze exposures →
Examine feasibility of techniques →
Select techniques → Implement techniques →
Monitor and revise

Emergency management allows personnel to make a critical set of decisions: to prioritize events on the basis of their impact on the organization and to apply controls on the basis of their effectiveness in minimizing the impact of the events.

Two common emergency management models may be considered, one focusing on the events themselves, the other focusing on the impact of the events on the organization. Both are effective and can find applications in any type of organization, public or private. Both use the above decision-making concept.

The *hazard-driven model* is based on the perils and their resulting hazards that the organization might face. Its steps are:

1. Identify the *hazards* an organization may face:
   - naturally occurring hazards:
     - geological
     - meteorological
   - human-caused events:
     - accidental
     - intentional
2. Analyze the impact of those events.
3. Select alternative controls to minimize the impact.
4. Implement controls.
5. Monitor/follow up.

Hazard-Driven Model: NFPA 1600

National Fire Protection Association standards are familiar to most safety professionals, especially in the design of *things*, such as fire suppression systems, buildings, and so on. A relatively recent addition to NFPA's standards is NFPA 1600: *Standard on Disaster/Emergency Management and Business Continuity Programs* (NFPA 2007, 5–8).

NFPA 1600 applies a hazard-driven model. This standard provides basic criteria to assess current programs or to develop and maintain a program. The standard addresses the four common emergency management program elements of preparedness, response, recovery, and mitigation.

First issued in 1995 as a recommended practice, NFPA 1600 closely parallels the Federal Emergency Management Agency's (FEMA) Capabilities Assessment for Readiness (CAR).

The standard calls for an emergency management program to include

- laws and authorities
- risk assessment
- incident prevention
- mitigation
- resource management and logistics
- mutual aid/assistance
- planning
- incident management
- communications and warning
- operational procedures
- facilities
- training
- exercises, evaluations, and corrective actions
- crisis communication and public information
- finance and administration.

Impact-Driven Model

The *impact-driven* model focuses on the *effects* of the event on people, property, or the organization. Its steps are:

1. Identify critical business elements in the organization
2. Assess the impact of disruption on each of those elements
3. Select alternative controls for the exposures
4. Implement controls for the exposures
5. Monitor/follow up.

This model is exemplified by the DRII/DRJ Generally Accepted Practices (Disaster Recovery Journal and DRI International 2006).

Some factors in assessing business impact are:

- ***Extent of the emergency:*** In order to balance its approach to emergency management, the organization should begin with an *all-hazards* approach, then adjust its plans and programs based on the probable impact.
- ***Frequency/severity:*** Risk management analyses commonly examine both the probability and severity of incidents and the resulting

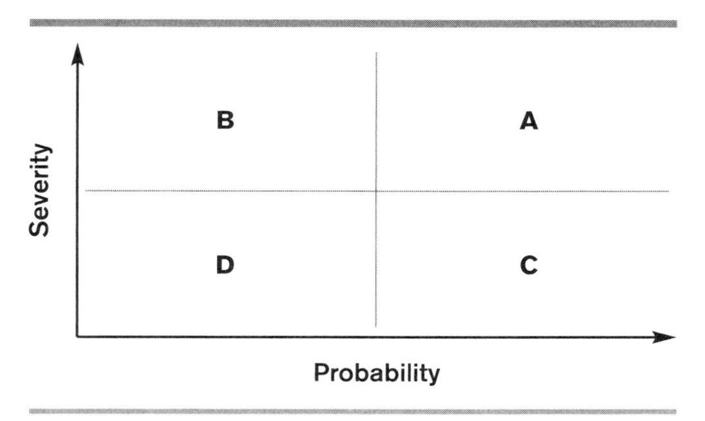

FIGURE 1. Risk matrix

loss or disruption. A qualitative view of this concept is shown in Figure 1, representing probability and frequency as simply high or low (in fact, they may each be represented as a continuum). Using this simplistic high/low approach, exposures in area D (low probability, low severity) might commonly be accepted as routine, and the controls for them are part of routine operations. Exposures in areas B or C (high probability or high severity) should be identified and specific controls should be implemented to deal with frequent occurrences (e.g., power outages) and catastrophic but rare occurrences (e.g., a terrorist attack). Disruptions that are both probable and severe (area A) are best dealt with by a major shift in operation or mission.

- ***Length of interruption:*** The length of the disruption caused by the incident must also be considered. Three relative periods of impact should be examined:

  1. Emergency response: the first few hours
  2. Incident management: the first 1 to 5 days
  3. Recovery: 5 to 30 days

  These time periods may differ depending on the type of emergency. The impact of the length of interruption also varies widely. The loss of a system for a day might be an annoyance; for a week, a substantial inconvenience; for a month, a significant financial loss.

Consensus Standards

While not as regulated in its entirety as other areas in which the safety professional often practices, there are many regulations at the federal and state level. Chapter 1 of this section ("Regulatory Issues") covers the OSHA regulations that impact emergency management programs.

Several codes and standards promulgated by the National Fire Protection Association may also impact the emergency management program, including

- NFPA 101, Life Safety Code
- NFPA 730, Guide for Premises Security
- NFPA 1620, Pre-incident Planning.

In addition to being consensus standards, some NFPA codes and standards, such as the Life Safety Code, have been commonly adopted into local and state building codes.

In larger organizations, there may be hierarchical policies, procedures, and so on. Individual facility programs should reflect corporate requirements as well as local needs.

As emergency programs and plans are written and updated, such programs should include strategies to review the most current applicable codes and standards to assure the continuation of compliance as applicable.

Risk Assessment

As shown in the model concept, the organization should begin its emergency management program development with a hazard assessment. This assessment should address both the likelihood of occurrence (probability) and the vulnerability of people, property, the environment, and the organization itself (severity).

Impact analysis is a term often used in the emergency management area to describe the potential for negative impact on the organization. The impact analysis should address potential harm to

- the health and safety of persons in the affected area at the time of the incident
- the health and safety of personnel responding to the incident

- continuity of operations
- property, facilities, and infrastructure
- delivery of services
- the environment
- the economic and financial condition of the organization
- regulatory and contractual obligations
- reputation of or confidence in the organization
- regional, national, and international considerations.

Based on factors such as location, operations, size, and so on, each organization should identify the specific events that may impact it, considering both the probability and the severity of the event if it occurs. While this list is not all-inclusive, the analysis process should address such events as:

1. naturally occurring events:
   - geological events, such as earthquakes, landslides, or subsidence
   - meteorological events, such as hurricanes, tornadoes, floods, tidal surges, extreme heat or cold, or lightning
   - biological events, such as plague, smallpox, anthrax, West Nile virus, or pandemic flu
2. human-caused events:
   - accidental events, such as spills or releases of hazardous material (chemical, radiological, biological), explosions or fires, building collapses, or pollution
   - intentional events, such as terrorism (conventional, chemical, radiological, biological, cyber-), sabotage, civil disturbance, or arson
   - technology-caused events, such as the failure of communications equipment or utility supplies.

Incident Prevention

Historically, many emergency management programs have addressed naturally occurring events, such as hurricanes, which cannot be "prevented." The 2007 edition of NFPA 1600 introduced a concept familiar to safety professionals: that many incidents can, in fact, be prevented, including many human-caused events.

Incident prevention provides a strategy to prevent incidents identified through the risk assessment process. For those that cannot be prevented, the remainder of the program elements will apply.

Mitigation

Mitigation is a process by which management provides the resources to limit or control the consequences, extent, or severity of an incident on the organization. Every emergency management plan should include a mitigation component, considering the following alternatives:

- eliminating the hazard
- limiting the amount or size of the hazard
- segregating the hazard from that which is to be protected
- providing protective systems or equipment
- establishing hazard warning and communication procedures
- duplicating essential personnel, critical systems, equipment, information, operations, or materials.

Resource Management and Logistics

The organization needs to know what resources will be available, especially at the critical moment of response. External resources can be found commercially, through government agencies, and by sharing with others (see Mutual Aid/Assistance below). Internal resources can be found at the corporate level, but also at other facilities with similar hazards and operations. Resources include not only easily measurable ones such as personnel, equipment, facilities, and funding, but also harder-to-define resources such as expert knowledge and training. Regular inventory and reassessment of all resources is needed to overcome shortfalls. Considering the quantity, response time, capability, limitations, cost, and liability connected with using them is important; just because they are available doesn't mean they are appropriate.

Logistics is the ability to locate, acquire, store, distribute, maintain, test, and account for the program's services, personnel, resources, materials, and facilities. The emergency management program should document the logistics process and describe the ability to deploy the listed resources.

Mutual Aid/Assistance

Mutual aid is an important concept in emergency management. Sharing of resources not only expands the ability to apply response and recovery actions, but reduces the cost to all who share. Partnering with similar industries, local organizations and government agencies, and parent organizations can make emergency management budgets stretch farther. The program documents should describe agreements in detail. The benefits of mutual aid, however, may not be available to a given organization in the event of a communitywide emergency.

Planning

One of the most common mistakes made by emergency program planners is the concept of *the book*: assuming that the entire plan should be housed in one document, no matter how big. This will doom all but the simplest plans to failure. Maintenance becomes unrelenting, and information is hard to find because it is meant to serve everyone.

Another approach is to consider the emergency management program is made up of several interrelated documents or plans:

- ***The strategic plan:*** Contains the program's vision, mission, goals, and objectives
- ***The emergency operations/response plan:*** Assigns responsibilities to organizations and individuals for specific actions at projected times and places in an emergency or disaster
- ***The mitigation plan:*** Establishes interim and long-term actions to eliminate hazards or to reduce the impact of those hazards that cannot be eliminated
- ***The recovery plan:*** Provides restoration of services, facilities, programs, and infrastructure
- ***The continuity plan:*** Maintains critical and time-sensitive applications, vital records, processes, and functions, as well as the personnel and procedures necessary to do so, while the damaged entity is being recovered.

Since the emergency management program and its various plans or elements involve both internal and external contacts, resources, and organizations, it is important that the functions, responsibilities, and authority of all parties involved be defined.

Incident Management

Incident management emphasizes both response and recovery—what the organization needs to do after the event.

Since the emergency management program and its various plans or elements involve both internal and external contacts, resources, and organizations, it is important that the functions, responsibilities and authority of all parties involved be defined.

The organization should establish a logical control and coordination function using an incident management system. It should include the ability to coordinate response, continuity, and recovery activities with public authorities.

Communication and Warning

In many larger emergencies even basic communications systems can be compromised. The emergency management program should address the ability to communicate information at all levels, internally and externally, throughout the time frame of the emergency.

Operational Procedures

In order to capture the critical procedures to be used in the event of an emergency, they should be documented in a clear and consistent manner. In addition to the operational procedures, the critical procedures should address the safety, health, and welfare of people and the protection of property and the environment.

Facilities

Many emergencies may render the organization's primary facility inoperable for a short or extended period.

The program should identify alternate facilities to support not only the response/recovery effort, but to continue the organization's business operations. Regular testing and maintenance of these facilities is appropriate to assure readiness when needed.

Training

Training is especially helpful in emergency management, because many of the players in an emergency response or recovery operation have other duties as their regular job. The emergency management program should assess training needs and implement appropriate training to create awareness and enhance the skills required to develop, implement, maintain, and execute the program.

Exercises, Evaluations, and Corrective Actions

We all learn from experience, but emergencies rarely occur, and it is difficult to learn from something that has not happened. Program plans, procedures, and capabilities should be assessed through postincident elements such as incident reports and lessons learned. The program should include exercises designed to test both the plan itself and the individuals involved in it. Exercises serve to ensure that the plan is practical and workable and to train those designated to implement it. Corrective action should be taken and documented for identified deficiencies.

Crisis Communication and Public Information

Emergencies are news. The media are likely to want more information than the organization has or is willing to release. Releasing information prematurely can cause irreparable harm to individuals and the organization. Procedures to disseminate and respond to requests for predisaster, disaster, and postdisaster information are needed. The targets include internal audiences such as employees and families, as well as external audiences, such as the media and their viewers, customers, suppliers, and the public at large. The program should address:

- a central contact facility for the media
- a disaster/emergency information-handling system
- prescripted information outlines or templates
- a method to coordinate and clear information for release
- the capability of communicating with special needs populations
- protective action guidelines/recommendations (e.g., shelter in place or evacuation).

No amount of prescripting will address every conceivable situation. Advance preparation of outlines or templates can ensure the ability to cover the expected questions ("When will you reopen?" "How many people were injured?") with answers that are factual while controlling speculation.

Finance and Administration

The moment of an emergency is not the time to look for funding, especially for resources that could reasonably have been foreseen. The emergency management program should include procedures for funding as well as other administrative matters. For example, having the money to pay employees is one thing; having the ability to get it to them is another.

Procedures must be established to ensure that fiscal decisions can be expedited in accordance with established authority levels and accounting principles.

These financial and administrative procedures should address:

- establishing and defining responsibilities for the program finance authority
- program procurement procedures
- payroll
- accounting systems to track and document costs.

Generally Accepted Practices for Business Continuity Practitioners

As FEMA was developing its Capabilities Assessment for Readiness model (CAR), the business continuity community was developing a similar model, focusing primarily on the response and recovery phases.

DRI International, a certification organization for business continuity planners, in conjunction with the Business Continuity Institute (BCI), its counterpart in the United Kingdom, has developed a set of professional practices to address the following ten areas:

- project initiation and management
- risk evaluation and control
- business impact analysis
- developing business continuity strategies
- emergency response and operations
- developing business continuity plans
- awareness and training programs
- maintaining and exercising business continuity plans
- public relations and crisis communications
- coordination with public authorities.

Out of these professional practices (the "what") has evolved a set of generally accepted practices (the "how") (Disaster Recovery Journal and DR International 2006, 3). The generally accepted practices outline alternative methodologies to use in the development of a business continuity plan.

Project Initiation and Management

As with any major program, a business continuity plan requires support from the organization's senior management and the ongoing coordination of the program itself. The need must be defined, a steering committee or other governance body may be appropriate, budget requirements must be presented, and ongoing project document needs should be established.

Risk Evaluation and Control

Planning for emergencies implies an understanding of the events and external issues that can impact the organization and its operations. The types of events, the loss that could be caused by those events, and potential control methods must be determined.

Business Impact Analysis

Based on the risk assessment, the organization can now identify the impact of each defined event on the organization. This allows the organization to prioritize the development of plans based on a *worst first* order. The business impact analysis should consider both frequency and severity of incidents.

Developing Business Continuity Strategies

These strategies are high-level alternatives to be evaluated based on their control of the potential impact on the organization. Strategies often evolve from and should mesh with the organization's mission and broad operating strategies.

Emergency Response and Operations

Each identified event must have at least a framework of a response plan. The more complex the event, the more detailed the plan should be.

Developing Business Continuity Plans

Here, *business continuity* refers to long-term implications—in a sense, the *recovery* phase. After the immediate response, preplanned recovery actions should evolve.

Training and Awareness

Recognizing that an emergency management plan requires participation across the organization, the organization should take steps to ensure that employees, vendors, customers, and others are aware of their roles and trained in their expected actions, no matter how small.

Maintaining and Exercising Business Continuity Plans

All plans need testing and exercising. Preplanning for these exercises should be an integral part of the process. Exercises should be constructed in such a way that they test specific plan goals and objectives, and after-action processes should be used to improve plans.

Public Relations and Crisis Communications

The organization must communicate its status during an emergency with internal stakeholders, such as employees and managers, as well as with external stakeholders, such as shareholders, the media, suppliers, and customers. Detailed contact information and a prepared communications outline will assist in providing a balanced view of the situation.

Coordination with Public Authorities

Recognizing that no organization exists in a vacuum, the organization must identify public authorities and other organizations, such as the Red Cross, that can assist or be assisted during an emergency.

The National Incident Management System (NIMS) and the Incident Command System (ICS)

Homeland Security Presidential Directive 5 (HSPD-5) calls for a national incident management system to coordinate response and recovery activities of federal, state, and local governments and others in the response and recovery community. While intended for these government agencies, there is much that can be applied by any organization.

The result of HSPD-5 was the implementation of the National Incident Management System (NIMS) (US Department of Homeland Security 2006a). It provides a consistent, flexible, and adjustable framework within which public and private entities can work to manage emergency response and recovery operations. A major component of NIMS is an updated application of the Incident Command System (ICS).

As a management system, NIMS has six major components:

- command and management
- preparedness
- resource management
- communications and information management
- supporting technologies
- ongoing management and maintenance

These components work together to provide a framework for managing incidents, regardless of their cause, size, or complexity.

Command and Management

In order to effectively respond to an emergency, the responders need an organizational system in which to work. This system and its staffing and hierarchies may be different from the *normal* structures. NIMS is based on three key organizational systems: the incident command system, multiorganizational coordination systems, and public information systems.

The Incident Command System

One of the most important best practices incorporated into NIMS is the Incident Command System (ICS). The concept of ICS was developed in the 1980s by the U.S. Forest Service and other agencies to improve response to wildfires, situations in which many emergency responders are thrown together without an otherwise defined management structure. Incident command essentially determines who's in charge in fast-moving and stressful situations. The Incident Command System calls for an organization with five major functions: command, operations, planning, logistics, and finance and administration—with an optional sixth functional area, information and intelligence.

Multiorganizational Coordination Systems

At the federal, state, and local government level, this means the coordination of response activities by these entities. However, even in the private sector, there should be an understanding of the interactive management components through mutual aid agreements and other assistance arrangements. For example, tenants in a multitenant building must work together and with the property manager to coordinate response and recovery actions.

Public Information Systems

For incidents involving the public, there must be a process and procedures to disseminate timely and accurate information during emergency situations.

Preparedness

Despite the implication of its name, NIMS does include elements of preparedness. It recognizes that when response and recovery work are well planned, preparedness is what makes them work. It requires that activities be conducted before the potential incident, including planning, training, exercises, personnel qualification and certification standards, equipment acquisition and certification standards, and publication management processes and activities.

Planning

Plans are the elements of a program that describe how personnel, equipment, and other resources are used to support emergency response actions. Plans allow priorities to be set in advance and internal and external resources to be identified.

Training

As with any program or system, training is needed to ensure that the people responding will know what to do. It is even more important in emergency management, since the responder may never have experienced "the real thing."

Exercises

Organizations and personnel must participate in realistic exercises to ensure that individuals' actions are effectively integrated with others's actions.

Personnel Qualification and Certification

In addition to being trained, personnel involved in some aspects of emergency management should be specifically qualified or certified in order to establish minimum performance capabilities. These can range from certification programs for emergency managers to certificates of competency in first aid/CPR, fire fighting, hazmat response, and so on.

Equipment Acquisition and Certification

Incident management organizations and emergency responders at all levels rely on various types of equipment to perform their mission-essential tasks. A critical component of operational preparedness is the acquisition of equipment that will perform to certain standards, including the capability of being interoperable with similar equipment used by other jurisdictions.

Mutual Aid

Mutual aid agreements are the means for one jurisdiction to provide resources, facilities, services, and other required support to another jurisdiction during an incident. Each jurisdiction should be party to a mutual aid agreement with appropriate jurisdictions from which they expect to receive or to which they expect to provide assistance during an incident.

Publications Management

Publications management refers to forms and forms standardization; developing publication materials; administering publications, including establishing naming and numbering conventions; managing the publication and promulgation of documents; and exercising control over sensitive documents; as well as revising publications when necessary.

Resource Management

NIMS defines standardized mechanisms and establishes requirements for processes to describe, inventory, mobilize, dispatch, track, and recover resources over the life cycle of an incident.

Communications and Information Management

NIMS identifies the need for a standardized framework for communications, information management (collection, analysis, and dissemination), and information sharing at all levels of management.

Incident Management Communications

Incident management organizations must be able to communicate across agencies and jurisdictions. Interoperability of equipment and systems allows organizations that might otherwise use different network protocols, radio frequencies, and so on, to standardize equipment and software.

Information Management

In addition to hardware and software interoperability, the architecture of information management should be standardized.

Supporting Technologies

Supporting technologies include technology and technological systems such as voice and data communications systems and information management systems (e.g., record keeping and resource tracking). Also included are specialized technologies that facilitate ongoing operations and incident management activities in situations that call for unique technology-based capabilities.

Ongoing Management and Maintenance

As with any system, even a management system, ongoing review and maintenance is needed to ensure compatibility with new parts of the organization and outside resources.

Additional Comment about Resource Planning

Some managers may have trouble planning for the unknown in terms of what they might need. One model to follow to comprehensively plan the resources and support that might be needed is found in the National Response Plan support function annex. These sections attempt to plan for any resource or support needed. They are listed below, each with an example that may be useful in the industrial setting:

- *Transportation:* Will we need to transport personnel to another location? If we need to conduct business at an alternate site, will we need to move equipment or files?
- *Communication (phones and network):* Should we have spare cell phone batteries?
- *Engineering and public work:* Will I need contract engineering support to power down or restart my operation? For example, will I need structural engineers to evaluate building integrity after an event?
- *Firefighting (a term used in the National Response Plan):* This section includes mobilization centers for distribution of supplies. Where will the shelter-in-place kits be kept? How will hazmat response materials arrive if needed?
- *Emergency management:* Who will be in charge of the facility and response? If an in-house fire brigade is in place, the chief will be in charge. How will the emergency response be coordinated with management?
- *Mass care:* What will be needed for employees (e.g., blankets)? Is there is an evacuation place to go in the winter? Are there shelter-in-place kits?
- *Resource management:* Who would I call to get more resources or a temporary facility for operation?
- *Public health:* What kind of medical response is available for the employees? What types of medical information and screening are available for events like pandemic flu?
- *Rescue:* Do the local responders know their way around my building?
- *Hazmat:* Are there procedures for hazmat response, supplies, safe distances, and locations?
- *Agriculture:* This is probably not as applicable in the industrial setting.
- *Energy:* What if power goes off? What are the generator and emergency lighting capabilities? How can we charge cell phone batteries?
- *Public safety:* What types of interaction with local officials is necessary?
- *Long-term recovery:* Do we have a business plan for recovery?

- ***External affairs:*** What legal and public media issues are involved and how do we deal with them?

References

Baranoff, Etti G., Scott E. Harrington, and Gregory R. Niehaus. 2005. *Risk Assessment*. Malvern, PA: Insurance Institute of America.

Disaster Recovery Journal and DRI International. 2006. *Generally Accepted Practices for Business Continuity Practitioner.* www.drj.com/GAP/gap.pdf

Federal Emergency Management Agency (FEMA). 2006. *FEMA Independent Study Program: IS-1 Emergency Manager*. www.training.fema.gov/EMIWeb/IS/is1.asp

National Fire Protection Association (NFPA). 2007. *NFPA 1600: Standard on Disaster/Emergency Management and Business Continuity Programs*. www.nfpa.org/assets/files/PDF/NFPA1600.pdf)

U.S. Department of Homeland Security. 2006a. *The National Incident Management System*. Washington, DC: US Department of Homeland Security.

______. 2006b. *National Response Plan*. Washington, DC: US Department of Homeland Security.

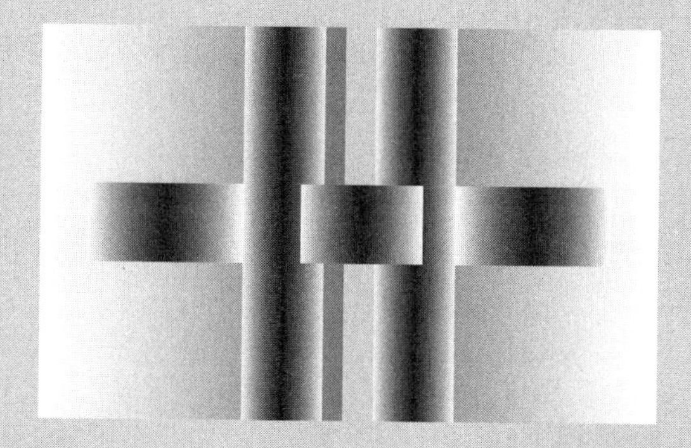

SECTION 3
FIRE PREVENTION AND PROTECTION

Regulatory Issues

Applied Science and Engineering

Fire Dynamics

Fire Prevention and Control

Fire Suppression and Detection

Cost Analysis and Budgeting

Benchmarking and Performance Criteria

Best Practices

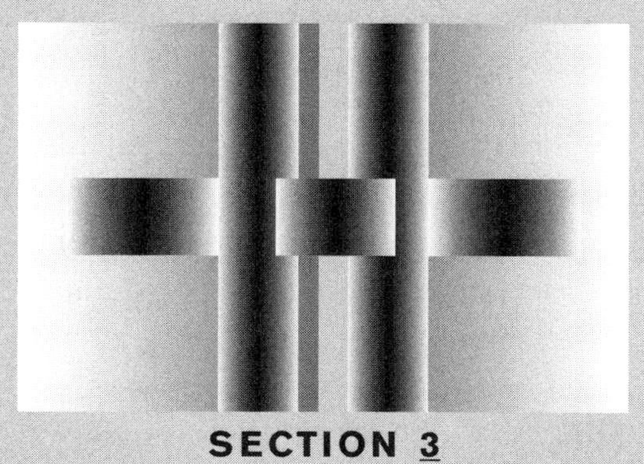

SECTION 3

FIRE PREVENTION AND PROTECTION

LEARNING OBJECTIVES

- List the rules and responsibilities of the state fire marshal under the *Uniform Fire Code*.
- Understand how insurance requirements are determined by insurance carriers.
- Discuss how OSHA and the NFPA standards vary as illustrated by the discussion of portable fire extinguishers.
- List the three separate and distinct parts of a means of egress.
- Understand the primary types of fixed extinguishing systems and know which NFPA Codes pertain to each.

REGULATORY ISSUES

James H. Olds

TO UNDERSTAND the regulations, codes, and standards that affect fire prevention and protection, we first need to differentiate between *codes* and *standards* as well as between *prevention* and *protection*.

A *code* is a standard that has been adopted by one or more governmental agencies and by virtue of its adoption, has the force of law. A *standard*, on the other hand, is a set of definitions and guidelines designed for voluntary use. This difference may cause difficulty for the reader, and therefore, throughout this section, the terms "code" and "standard" may be used interchangeably.

Prevention is generally considered to be eliminating the potential of a fire starting; *protection* is considered to be actions taken to limit the effects of a fire once it has started. *Prevention* is accomplished by disrupting the fire tetrahedron by eliminating one or more of its four components: heat, fuel, oxygen, and chemical reaction. Thus, *prevention* requires education more than *protection* does.

Protection includes intentional actions designed to reduce fire hazards through inspections, facility design, or process design, as well as the installation of fire suppression or detection systems (Ferguson and Janicak 2005, 3). *Protection* is the focus of various codes and standards from 24 agencies or organizations. Of these organizations, the National Fire Protection Association (NFPA) is the most notable for fire-protection and related systems.

The need for protection was identified early in history. In 200 B.C., Ctesibius of Alexandria invented an early form of fire protection—a hand pump capable of delivering water to a fire. The first modern form of extinguisher did not appear, however, until 1819, when British inventor Captain George William Manby, of Denver, England, best known for his shipboard rescue device called the Manby Mortar, used a copper container to dispense a potassium carbonate solution expelled by compressed air. Another early extinguisher system existed from the late 1800s into the early 1900s—a glass ball or vial known as a glass fire-extinguisher bomb. These devices were normally hung on or near doorways and were designed to be thrown at the base of a fire in an effort to control the incipient stage. Sprinkler technology first emerged in England in 1806. It made its way to the United States, and in 1852 James B. Francis installed the first perforated-pipe system in Lowell, Massachusetts. This early design was soon improved upon by the attachment of a spray mechanism to increase the water flow (Grant 1996).

On August 11, 1874, U.S. Patent No.154, 076 was obtained by Henry S. Parmelee of New Hampshire for a sprinkler head (Grant 1996); by 1884 this technology had grown to include fifteen different types of heads from various manufacturers. Due to the various types of sprinkler heads and designs, insurance companies began to question the reliability and functioning characteristics of each. To satisfy this need, C. J. H. Woodbury, of Factory Insurance Companies, was the first to conduct evaluations of the various sprinkler heads. This evaluation included response times; how long it took for each system to release water after a gradual temperature build-up.

In 1895 a group of men representing sprinkler and insurance interests gathered in Boston to discuss the inconsistencies in piping standards and sprinkler spacing requirements. During this meeting, as a way of highlighting the problem, it was noted that within 100 miles of Boston there were nine different standards for piping and sprinkler spacing. The meeting took place in the office of Everett Crosby, from the Underwriters Bureau of New England. Also in attendance were John Freeman, of the Factory Mutual Fire Insurance Companies (known today as Factory Mutual (FM)); Uberto C. Crosby, chairman of the Factory Improvement Committee of the New England Fire Insurance Exchange (and father of the host, Everett Crosby); W. H. Stratton, of the Factory Insurance Association (renamed in later years as Industrial Risk Insurers); Fredrick Grinnell of the Providence Steam and Gas Pipe Company (known today as Grinnell Fire Protection); and F. Eliot Cabot, of the Boston Board of Fire Underwriters (Grant 1996). After this initial meeting, subsequent meetings were held, and by March 19, 1895, the group had released the *Report of Committee on Automatic Sprinkler Protection* (Grant 1996). It is from this initial report that NFPA 13 on automatic sprinklers and the National Fire Protection Association (NFPA) itself can trace their heritage.

NFPA Codes and Standards

Today, with a quick call to the NFPA customer service line, one can learn that the NFPA has published approximately 300 safety-related codes and standards. The number is approximate because standards are constantly being revised, rescinded, or created. Each code focuses on a particular area of interest. The two most commonly cited are NFPA 1, *Uniform Fire Code* (NFPA 2006a) and NFPA 101, *Life Safety Code* (NFPA 2006c). Although they are called "codes," they are actually standards and are not regulatory. Each code refers to an entity called the *authority having jurisdiction* (AHJ). This could be a local fire marshal or the individual who sets policies and develops ordinances for a particular area. NFPA defines it as ". . . an organization, office, or individual responsible for enforcing the requirements of a code or standard, or for approving equipment, materials, an installation or a procedure" (NFPA 2006a). In order to determine whether a code applies to a specific area of interest, one must contact the local AHJ.

The bylaws of the NFPA as posted on its main Web page state in part: "The purpose of the Association shall be to promote the science and improve the methods of fire protection and prevention, electrical safety and other related safety goals; to obtain and circulate information and promote education and research on these subjects . . ." (NFPA 1971). As such, the NFPA is considered unique in that its standards process is a full, open, consensus-based process, meaning that anyone can participate with fair and equal treatment.

NFPA Standards: The Development Process

First, proposals for new or revised safety-related projects, submitted by anyone with concern, are forwarded to the Standards Council for a preliminary review. The Standards Council is made up of representatives from various areas—consumers, insurance companies (often in association with Factory Mutual), organized labor, manufacturers, enforcement agencies or AHJs, installation or maintainance organizations, and applied research or testing laboratories such as Underwriters Laboratory (UL) or Factory Mutual, as well as various *subject matter experts* (commonly referred to as SMEs).

If the Standards Council deems it appropriate, a *call for proposals* is published in public notices such as *NFPA News*, the U.S. *Federal Register*, and appropriate trade journals. Interested parties have 24 weeks to respond with specific proposals.

Once proposals are received, a technical committee convenes and reviews them. Approval of submitted proposals requires a two-thirds vote from the technical committee. The *Report on Proposals* (ROP) is published twice a year and is sent automatically to everyone who presented proposals. Other interested parties may request copies. After release of the ROP, there is a 60-day comment period during which anyone may submit a comment.

The committee reconvenes at the end of the 60-day period and reviews all submitted comments. Interested parties are encouraged to address the committee in an open forum. As before, approval of submitted comments requires a two-thirds vote, thus forming a *consensus* on the standard. These standards differ from other agencies' standards in that other agencies (e.g., American Society of Mechanical Engineers (ASME), American Petroleum Institute (API), and the American Society for Testing and Materials (ASTM) rely more on technical experts and engineers to develop standards, whereas the NFPA relies on a committee of interested parties to form a consensus regarding the development of each standard.

Once the comments that are received are agreed upon, the proposed standard is released, as before, twice a year as the Report on Comments (ROC). It is made available to anyone for a review period of seven weeks.

Twice a year (in May and November), during the NFPA Association Meeting, the proposed standard or change to an existing standard is open for public review and debate. Anyone who has been a member of record for at least 180 days may vote on the final recommendation.

If approved, the final recommendation is then brought back to the Standards Council that started the process for issuance. The entire process takes approximately two years.

NFPA codes or standards are not enforceable; they are recommendations that may or may not be adopted as industry standards. For example, the insurance industry might adopt a standard and provide it to the insured during a policy-renewal process or a periodic survey as a "recommendation." It is also possible for the NFPA codes to be the basis for insurance recommendations that are stricter than the codes themselves. In either case, the NFPA standards—or those espoused by Factory Mutual—are enforceable as law only if they are adopted by the AHJ.

Uniform Fire Code

The *Uniform Fire Code* (NFPA 2006a) establishes the office of the state fire marshal and details the rules and responsibilities of that office. The *Uniform Fire Code*, adopted in 1971, is unique in that it applies to all new and existing structures. It sets forth the following tenets of fire protection:

- Establishes the office of the fire marshal
- Establishes the right of inspection, plans review, and permitting procedures for special events
- Establishes the duties and powers of the incident commander
- Regulates and controls special events and interior decorating based on fire load
- Establishes acceptable conditions for fire-fighter safety
- Establishes parameters for acceptable storage, handling, and transportation of flammable or hazardous materials
- Establishes a *baseline* for fire protection by defining topics such as general fire safety and means of egress, defined in the *Uniform Fire Code* as "a continuous and unobstructed way of travel from any point in a building or

> structure to a public way consisting of three separate and distinct parts: (1) exit access, (2) the exit, and (3) the exit discharge" (NFPA 2006a).

The *Uniform Fire Code* also provides definitions for occupancy load, building water-flow requirements (referring back to NFPA 13), and fire hydrant locations and distribution, which may also be governed by local ordinance.

As its title indicates, the *Uniform Fire Code* is the main source on which all other fire-related codes are based, and it is from this code that the other NFPA codes are derived. Therefore, certain aspects of this code may not apply to every situation or location. It is always advisable for safety professionals to meet and get to know the local AHJ (NFPA 2006a).

Code Enforcement

The state fire marshal's office sets the particular codes that are to be followed within each state. Once this basic level of enforcement is established within the state, individual counties or municipalities that have an AHJ may adopt specific codes or standards that apply to their specific needs or interests. Once these codes are adopted, they become law where incorporated in local ordinances. The AHJ may vary from state to state. In one state the AHJ at the local level could be the fire marshal; in another it could be zoning officials or AHJ status could rest with code-enforcement activities. Regardless of how codes are administered, it is only due to this type of mechanism that NFPA codes or any other standards become enforceable.

If a standard is presented as a *recommendation* by an insurance carrier, it may become a matter of negotiation between the insured and the carrier when rates are being set. It is not enforceable as a law but is used in conducting business. Insurance requirements are applied and selected by insurance companies. Often their recommendations are chosen based on research or past loss history. Factory Mutual (FM) standards are promoted by Factory Mutual Research and Engineering as a service to its member insurance companies. FM provides good guidance for property protection and building codes, while the NFPA looks, comparatively, more at life safety. FM sets international standards as part of the *International Organization for Standardization* (ISO) system. The FM system offers approval and certification of fire equipment, detection systems, and ISO 9000 registration throughout Canada, Europe, and the United States (Factory Mutual Global 2008).

OSHA Enforcement

OSHA is a federal enforcement agency, and its standards are enforceable as law. However, OSHA does not incorporate by reference every NFPA standard. When advising clients, safety professionals must remember to check to see if the referenced standard is incorporated by reference into the OSHA standards. If not, advise the clients accordingly. Sometimes OSHA may incorporate an NFPA standard at the time of its promulgation, but the standard may become out of date. Or an OSHA standard may require a specific action to be performed but not detail how it is to be performed. In these circumstances, OSHA has the ability to enforce the requirement for the action specified. OSHA's enforcement is therefore limited to checking to see that an action was performed as described in the wording of the standard but not how it was performed. As an example, OSHA 1910.157(e)(3) states, "The employer shall assure that portable fire extinguishers are subjected to an annual maintenance check" (OSHA 2002c). Subjecting the extinguishers to this check is enforceable by OSHA. How the check is accomplished is not enforceable, but it is described in NFPA 10, which was revised and made more user friendly in 2007 (NFPA 2007a). This NFPA standard, if adopted by an AHJ, becomes law. In most cases an NFPA standard is adopted at state level and therefore is regulated by the state, which in this case would require the maintenance check to be conducted by a licensed individual, generally an outside contractor who holds a specific license from the state.

In 29 CFR 1910.157, OSHA also requires extinguishers to be placed a specific distance apart (OSHA 2002c). However, NFPA 10 bases the selection of the size and type of fire extinguisher on the occupancy (NFPA 2007a). The earliest NFPA standard for portable fire extinguishers was published in 1921, and it has been revised frequently ever since. At one point there were two separate standards, one for installation and one for maintenance. Within the OSHA Construction Standard, under Subpart F, the NFPA standard is not incorporated by reference but is referred to as NFPA 10A-1970 for servicing and maintenance requirements. This reference by OSHA relates back to the 1970s, when the NFPA split NFPA 10 into two standards. Since 1974, however, the two standards have been recombined. In 1998, NFPA 10R was introduced, covering recommendations for portable fire extinguishers in family dwelling units and living areas. NFPA 10R has since gone out of print. Although *not* incorporated by reference in 29 CFR 1910.157 (2002c), NFPA 10 provides the guidance to be adhered to when selecting, installing, and maintaining portable extinguishers (NFPA 2007a). The OSHA standards outlined in Subpart L are not applicable to the maritime, construction, or agriculture industries (OSHA 2002c). However, all three—NFPA 10 (2007a), 1910.157 (OSHA 2002c), and the construction standard found in OSHA 1926.150 (OSHA 1996b), discuss selection of portable extinguishers based on the hazard categories Class A, B, and C. The NFPA takes this discussion further and defines these hazard areas as:

Low Hazard (Class A combustibles): offices, classrooms, churches, assembly halls, and guest room areas of hotels/motels.

Ordinary Hazards (Class A combustibles and Class B flammables): dining areas, mercantile shops and allied storage, light manufacturing, auto showrooms, parking garages, workshops, and warehouses.

Extra High Hazard (Class A combustibles and Class B flammables are present in quantities that exceed those expected in areas of ordinary hazards): woodworking shops; auto repair, aircraft, and boat servicing areas; cooking areas; manufacturing processes that include painting, dipping, coating, or flammable liquid handling (NFPA 2007a).

NFPA 10 also lists obsolete portable fire extinguisher mediums that are to be removed from service such as soda acid, chemical foam (except film-forming foam), vaporizing liquid (i.e., carbon tetrachloride), cartridge-operated water, cartridge-operated loaded stream, copper or brass shell joined by soft solder or rivets, carbon dioxide with metal horns, and solid charge-type aqueous film forming foam (AFFF) (NFPA 2007a).

The NFPA standard also requires extinguishers to have a label, tag, or stencil detailing the product name of the medium as it appears on the manufacturer's material safety data sheet (MSDS) relating the extinguisher back to requirements under 29 CFR 1910.1200 (OSHA 1996a).

Placement recommendations for the extinguishers vary among the three standards. In some cases the NFPA standards are more stringent; in other cases the OSHA requirements are. In this case the most stringent requirement should be followed, with the cardinal rule being "more is better." As an example, the 1910.157 standard states that extinguishers for Class B hazards must be within 50 feet or less. The construction standard (1926.150) states that an extinguisher having a rating of not less than 10B shall be placed no more than 50 feet from 5 gallons or more of a flammable liquid. The NFPA standard provides a chart, and although the chart is compatible with the 1926 standard, it also lists an extinguisher with a 5B rating that can be placed within 30 feet of a Class B hazard.

Both the 1926 and 1910 standards discuss periodic maintenance and servicing of extinguishers. However, the NFPA standard describes how, when, and by whom this is to be done. From this reference individual states and AHJs (NFPA 2006) provide licensing criteria for people who perform such work.

NFPA 101

Life safety is in the title of NFPA 101: *Life Safety Code* (2006c). This code is the result of numerous years of research and study. Its origins can be traced to a 1912 pamphlet entitled *Exit Drills in Factories, Schools, Department Stores and Theaters*. Updated and revised almost continuously since its inception, this code was placed on a three-year revision cycle in 1967. It remains today one of the most important reference documents and is incorporated by reference in the OSHA standards. The 42 chapters of the code focus on issues of fire safety within structures. Each chapter addresses specific topics; commencing with chapter eleven, focus is on either existing or new facilities based on their occupancy classification. One full chapter is dedicated to special structures and high-rise buildings (NFPA 2006).

Occupancy Classifications

In defining occupancy, certain conditions are defined. Paragraph 4.6.10.1 reads:

> "No new construction or existing building shall be occupied in whole or in part in violation of the provision of this Code, unless the following conditions exist: a plan of correction has been approved; the occupancy classification remains the same; no serious life safety hazard exists as judged by the authority having jurisdiction (NFPA 2006c)."

One should note that the last reference refers back to those discussions found in NFPA 1, *Uniform Fire Code* (NFPA 2006a). This one paragraph opens many doors for interpretation. First, who has to approve the plan? Once again, approval must be given by the authority having jurisdiction (AHJ) in the local community. The "classification of occupancy" is also referred to. To understand this, one must move two chapters further into the code to discover that there are thirteen classifications of occupancy, some with several subsets: special structures, assembly, educational, day care, health care, ambulatory health care, detention or correctional; residential, residential board and care; mercantile; business; industrial; and storage. This becomes much more complex than the levels of occupancy outlined for fire extinguishers. Each of the thirteen occupancy classifications is addressed in a separate chapter in the *Life Safety Code* that details occupant load, hazard classifications, specific means of egress, minimum construction requirements, approval of layouts, and requirements for fire detection or suppression systems (NFPA 2006c).

Special structures is a classification that includes specific structures such as high-rise buildings, permanent membrane structures, tension-membrane structures, electric and fired heaters, and standby power systems.

Assembly occupancy is for building that have gatherings of 50 or more people. A subset provides for small assembly occupancies—less than 50 people. Assembly occupancies of 300 or more must be protected by an automatic sprinkler system, except for gymnasiums, places of worship, or a single multipurpose assembly of less than 12,000 square feet. Further, assembly occupancies of 300 or more people and all theaters must have an approved fire alarm system that can be initiated manually, equipped with an emergency power source, and monitored by a constantly attended receiving station.

An entire section of the assembly chapter addresses fire protection for theatrical facilities. These facilities are unique in that they require roof vents over the stage area and a drop-down curtain that acts as a barrier between the performers and the audience. This barrier, at one time made from asbestos and extremely friable, is now made from high temperature coated fiberglass material and is referred to as a *Proscenium Wall* in the code. Other unique features in regards to these facilities is that all stages that are over 1000 square feet require not only a standpipe but also an attached 32-foot $1^1/_2$-inch hose.

Assembly occupancy offers unique challenges in that the occupants may not be familiar with the exits, protection systems, or established procedures. Therefore it is incumbent upon the staff that works at the assembly structure to be knowledgeable about such systems and procedures and to act as guides in the event of an emergency (NFPA 2006c).

Educational occupancy is for educational facilities used by six or more students through the twelfth grade for four or more hours per day or more than twelve hours per week. These facilities have the same alarm and sprinkler requirements as those for assembly occupancy.

For new educational facilities, there are provisions for the location of rooms occupied by pre-school, kindergarten, or first-grade students on the level of exit discharge. Second-grade students cannot be located in classrooms more than one floor above an exit discharge. The travel distance to the exits cannot exceed 150 feet unless the facility is equipped with an automatic sprinkler system, and then it is extended to 200 feet.

The difficulty with educational facilities lies with the occupants. In most educational facilities the students greatly out number the staff, but the staff is responsible for ensuring proper evacuation procedures and control of the students (NFPA 2006c).

Day-care occupancy is for facilities with twelve or more clients receiving care, maintenance, or supervision by someone other than a relative or legal guardian for less than 24 hours per day. Adult day-care facilities have a similar definition except that occupancy is for three or more clients. Adults in this category must not exhibit behavior considered harmful to themselves or others.

Under day-care occupancy, alarm initiation is manual and through operation of smoke detectors or sprinkler systems. However, clients may not be capable of understanding what the alarms mean, so day-care staff must escort them to a place of refuge.

A large concern with day-care occupancy is how to provide care for the clients, whether they are infants or adults, during an evacuation. In educational facilities, students are generally taken outside to a consolidation point to await instruction. For infantile or adult clients, this is not always possible due to the elevated levels of care necessary. Some may require additional medical support during evacuation or may not be capable of self-evacuation (NFPA 2006c).

Health-care occupancy is for facilities where four or more individuals receive medical care or other treatment and are incapable of self-preservation due to age or physical or mental disabilities. Although standards are similar to those of day-care facilities, unlike day-care facilities, health-care occupancies include sleeping accommodations for the clients.

Due to the potential fragility of the clients, health-care occupancies must be designed and maintained to minimize the possibility of evacuation for the clients. Staff members may be expected to perform limited fire-safety functions to limit the spread of fire to the room or area of origin (NFPA 2006c).

Ambulatory health-care occupancy is for facilities where services are provided to four or more persons receiving emergency care, urgent care, or anesthesia, or to persons who are incapable of self-sustaining care.

At this level of occupancy, the client's health care may not be adequately maintained in the event of an evacuation. Further, because of the type of care provided, some windows or doors may be locked for security purposes. Therefore this standard is written with the assumption that staff is available to perform certain fire-safety functions intended to limit the spread of fire to the area of origin. If an emergency required the evacuation of clients, suitable transportation would be necessary as well as a suitable receiving facility to ensure life safety (NFPA 2006c).

Detention or correctional occupancy is for facilities that house four or more persons under some degree of restraint. Due to security requirements, this occupancy is subdivided into five groups:

Condition I: Free Egress. Free movement is allowed from the sleeping area to the exterior.

Condition II: Zoned Egress. Free movement is authorized from the sleeping area to a smoke compartment. Paragraph 3.3.35.2 of the *Life Safety Code* (NFPA 2006a) defines a *smoke compartment* as "space within a building enclosed by smoke barriers on all sides, including the top and bottom."

Condition III: Zoned Impeded Egress. Free movement is allowed within individual smoke compartments (such as sleeping quarters) to a group activity space. Egress is impeded by remote-controlled release.

Condition IV: Impeded Egress. Free movement from occupied spaces is restricted. Remote-controlled release is provided to allow movement from all sleeping quarters and areas of activity within a smoke compartment to another smoke compartment.

Condition V: Contained. Free movement from an occupied space is restricted. Staff-controlled release mechanisms are required to permit movement from one occupied space to another.

Residential occupancy defines more than single-family home structures. Under this category of occupancy there are five subsets, all classified as ordinary hazards:

- *One and two family dwellings*
- *Lodging or rooming house*, which is defined as containing sixteen or fewer individuals
- *Hotel*, which is defined as containing more than sixteen individuals and used primarily for transient lodging with or without meals
- *Dormitory*, defined as containing more than sixteen individuals who are not related and without meals
- *Apartment building*, which is defined as three or more units with independent cooking and bathing facilities.

In these structures the standard stipulates that, if they are equipped with a sole means of egress, it shall not pass through a nonresidential occupancy within the same building. For lodging or rooming houses, hotels, and dormitories, there is are added requirements: every story that is larger than 2000 square feet must have automatic sprinkler protection, and every area used for sleeping must have both primary and secondary means of escape.

As with other occupancy classifications, residential occupancies may include individuals who live permanently within the structure and thus could be considered knowledgeable about its layout and methods of escape. Unfortunately, residential occupancies present challenges because they are subjected to little oversight by inspection officials. On the other hand, this classification of occupancy may also include individuals who are more transient in nature (rooming house or hotel tenants) and who therefore should not be considered knowledgeable about the layout or means of escape. In such structures there are requirements for emergency lighting and marking of exits unlike what is required for single-family homes or for structures in which the guest rooms are equipped with an exit to the outside at street or ground level (NFPA 2006c).

Residential board and care occupancy is for facilities handling four or more residents, not related by blood or marriage, for the purpose of providing personal services. The requirements for means of egress are similar to those of residential rooming houses and dormitories in that each sleeping area must be equipped with a primary and secondary means of egress. Further, these facilities must be fitted with smoke detectors on every floor, including the basement and the sleeping quarters. These facilities must be equipped with a manual fire alarm providing an alert to the occupants without delay in the event of a fire (NFPA 2006c).

Mercantile occupancy standards are for facilities used for the display and sale of merchandise. *Business* occupancy standards are for facilities used for account and record keeping or transaction of business other than mercantile business. *Industrial* occupancy standards are for places where items are produced, repaired, manufactured, mixed, processed, packaged, finished, decorated or assembled.

The *mercantile* occupancy is broken down into three subclassifications:

Class A, structures with a gross area of more than 30,000 square feet or occupying more than three stories for sales purposes.
Class B, structures of more than 3000 square feet but less than 30,000 square feet that occupy no more than three stories.
Class C, structures not more than 3000 square feet with no more than one story.

Means of egress are important in these structures due to the number of individuals who may be inside who are not considered to be knowledgeable about the contents or means of egress (primarily because they may be in the structure for the first time). For this reason, staff members must be trained and take charge to ensure an orderly exit if the need arises. In Class A mercantile occupancies there must be at least one aisle that is at least 60 inches wide and leads directly to an exit. In all classes of mercantile occupancy, if the exits are protected by automatic sprinkler systems, the occupants are permitted to pass through storerooms provided that (1) not more than 50 percent of the exits are through storerooms, (2) the storerooms cannot be locked, (3) the main aisle through the storeroom is not less than 44 inches wide, and (4) the path is continuous, unobstructed, and defined with fixed barriers.

Mercantile occupancies must also abide by the provisions of Subpart E (*Exit Routes, Emergency Action Plans and Fire Prevention*) of 29 CFR 1910 (OSHA 2002a). This subpart requires written emergency planning and training as well as employee training on portable fire extinguishers.

Storage occupancy standards are for places that are used primarily for the storage or sheltering of goods, merchandise, products, vehicles, or animals. Contents for storage occupancies are based on the degree of hazard they represent: low, ordinary, and high. This is the same classification system that is used in NFPA 10 (NFPA 2007a). For low- and ordinary-hazard storage facilities, only one means of egress is required from any story or section as long as it does not exceed the distance for a *common path of travel,* defined as 100 feet for storage occupancies that are equipped with an automatic sprinkler system and 50 feet for those that are not. For high-hazard storage, every section or story is considered separate, and therefore each must have two means of egress.

Multiple occupancy standards are considered separate from the aforementioned thirteen. They provide guidance for situations in which two or more classes of occupancy exist, possibly presenting special hazards.

Means of Egress

Chapter 7 of the *Life Safety Code* contains standards related to *means of egress*, and goes into more detail than the basic definition in the *Uniform Fire Code* (NFPA 2006c). Included in this chapter are general requirements for corridors. For occupant loads exceeding 30, corridors shall have a separation from the other parts of the building consisting of walls which are no less than 1-hour fire resistant rated (NFPA 2006c).

In addition to the corridor provisions, the *means of egress* are further subdivided into other components involved in departing or gaining access to buildings, such as doors (which are described in terms of width, clearance width, swing force, and capability of swinging open from any direction), locks, latches, headroom, measurement, number of doors required, clearance, floor level, controlled access, and so on.

This chapter also contains requirements regarding the source and performance of illumination, including periodic (monthly and annual) testing requirements. Internally illuminated signs must comply with the requirements detailed in UL 924 (UL 2006), which is referenced within the *Life Safety Code*. It requires each device to be marked according to its specifications. Note that it is the *Life Safety Code* (NFPA 2006c) that establishes the requirement to illuminate exits, whereas it is Underwriter's Laboratory that establishes the requirements for the illuminating devices.

Wattage requirements for such illuminating devices as well as all other illuminated areas both inside and outside of structures are different. Wattage requirements are established in the Illuminating Engineering Society of North America's LEM-1-05, *Recommended Procedure for Determining Interior and Exterior Power Allowances* (IESNA 2000). According to LEM-1–05, exit signage mounted indoors shall not exceed 5 watts at the face. This document is actually an excerpt from ANSI 90.1–04, with which UL correlates its test data. On the other hand, the *Life Safety Code* requires an illumination level of 54 lux (5 foot-candles) at the surface for externally illuminated signs. NFPA 5000, *Building Construction and Safety Code,* sets the illumination level for emergency egress as being not less than 1 foot-candle (10 lux) at any one point at the surface, and not less than 0.1 foot-candle (1 lux) at the floor (NFPA 2005b).

Means of egress is also addressed in building codes. According to the International Building Code as well as NFPA 5000 (NFPA 2005b), *means of egress* has a slightly different definition than what is found in the *Life Safety Code*. The International Building Code defines it as "a continuous and unobstructed path of vertical or horizontal egress travel from any point in a building or structure to a public way. A means of egress consists of three separate and distinct parts, the exit access, the exit, and the exit discharge (ICC 2006)." NFPA 5000 defines it as "comprising the vertical and horizontal travel and includes intervening room spaces, doorways, hallways, corridors, passageways, balconies, ramps, stairs, elevators, enclosures, lobbies, escalators, horizontal exits, courts and yards (NFPA 2005b)." This differentiation becomes important when one is addressing those aspects of Subpart E, 29 CFR 1910 (OSHA 2002a), because it is incorporated by reference and required compliance in 29 CFR 1910.35 (OSHA 2002b). The International Building Code and OSHA require a means of egress, but it is the *Life Safety Code* that provides clear requirements in a general manner. For specific occupancy classifications within the *Life Safety Code*, additional egress requirements are tailored to and dependent upon the occupancy classification (NFPA 2006c).

Drills are also addressed in the *Life Safety Code* and are referenced to specific occupancy levels. In general, drills must be conducted frequently enough to familiarize the intended occupants with the proper procedures to follow in the event of an emergency. Emphasis is on orderly evacuation, conducting drills at both expected and unexpected times, choosing a location where evacuees consolidate, and creating a written record of each drill. Each occupancy classification requires periodic fire/evacuation drills. Most classifications require drills to be conducted only with the employees who work within that classification or state that drills are to be conducted so as to familiarize the occupants within a particular classification. The exceptions are drill requirements for day-care occupancies and educational occupancies, both of which require monthly drills. Before designing emergency evacuation plans, safety professionals should refer to the appropriate occupancy level chapter to establish the required drill frequency (NFPA 2006c).

Two chapters of the *Life Safety Code* discuss the fundamental requirements of general fire safety required for the structure and link the code to other NFPA publications. These fundamental requirements call for structures to be constructed, designed, maintained, and equipped to provide a reasonable level of life safety and property protection from actual or possible hazards (NFPA 2006c).

Both the *Life Safety Code* (NFPA 2006c) and the *Uniform Fire Code* (NFPA 2006a) provide guidance on basic fire protection. One important subject is the *means of egress*. The *Uniform Fire Code* (Section 4.4.3) states that the means of egress from all parts of the building shall be unobstructed, with no locking devices that prevent free escape. In addition, each means of egress shall be marked in a clear manner and its illumination shall be included in the lighting design for the structure (NFPA 2006a). Occupancy load is based upon the *Life Safety Code* (Section 7.3.1.1). It is dependent upon the type of structure and is not less than the number of persons determined by dividing the floor area by a load factor provided in Table 7.3.1.2–Occupant Load Factor, in Chapter 7 of the *Life Safety Code* (see Table 1 of this chapter for excerpts) (NFPA 2006c). The load factor ranges from 1.5 square feet to 200 square feet depending on the occupancy classification. The number of occupants any structure may have is determined by this load factor. This factor also determines the minimum number of means of egress. From any story with occupancy loads of more than 500 but less than 1000, there must be at least three means of egress. For occupancies over 1000, there must be at least four means of egress from every story. For occupancies less than 500 there may not be less than two means of egress unless exceptions are specifically addressed in the appropriate chapter of the *Life Safety Code* that discusses the occupancy in question (NFPA 2006c).

Once the occupancy load is determined, the egress capacity—how large each egress must be to accommodate the stated number of individuals—and the minimum travel distance to each egress must be determined. As with the occupancy load factors, this information is contained in the *Life Safety Code* (NFPA 2006). Taken from Table 7.3.3.1 of the *Life Safety Code*, Table 2 depicts egress capacity. The occupancy load divided by the number of exits must not exceed the number contained in this table.

The *Life Safety Code* does not provide a definition of the term *reasonable level* but allows it to be defined as the condition obtained by the application of the various standards or codes (such as OSHA, ANSI, or others) in conjunction with insurance requirements and what is accepted by the AHJ (NFPA 2006c). NFPA 5000 gives more details regarding fundamental fire and life safety by providing twelve basic guidelines:

TABLE 1

Occupancy Load Factor and Minimum Travel Distance

Occupancy Classification	Square Footage/Person	Minimum Travel Distance in Feet for Sprinklered Buildings
Assembly		250
Concentrated	7	
Bench-type	18 linear inches per person	
Fixed seating	Per number of seats	
Kitchens	100	
Stages	15	
Educational Use		200
Classrooms	15	
Day-Care	35	150
Health-Care		200
Inpatient	240	
Detention and Correction	120	200
Residential		
Hotel and dormitory	200	125
Apartment	200	125
Boardinghouse	200	75
Industrial	100	200
Business	100	300
Mercantile		250
Sales on street floor	30	
Sales on two or more street floors	40	
Sales below street level	30	
Sales above street level	60	

(*Source:* Excerpted from NFPA *Life Safety Code*, NFPA 2006c)

TABLE 2

Egress Capacity

Occupancy	Stairway Width/Person (in.)
Board and Care	0.4
Health-Care	
Sprinklered	0.3
Non-sprinklered	0.6
All Other Occupancies	0.3

(*Source: Life Safety Code*, NFPA 2006c)

1. Provide adequate levels of safety without dependence on a single safeguard.
2. Provide for appropriate levels of life safety after considering the size and constitution of the occupancy.
3. Have a back-up or secondary means of egress.
4. Ensure that egress paths are unobstructed and clear.
5. Make sure the egress path is clearly marked
6. Make sure the egress is illuminated. This is where the code establishes the minimum acceptable levels illumination.
7. Make provisions for early warning in the event of a fire. This goes back to OSHA standards 29 CFR 1910.164 and 1910.165 (1980b).
8. Ensure that vertical openings are covered.
9. Ensure proper installation of equipment and components.
10. Take measures to limit fire spread.
11. Maintain structural integrity during a fire to prevent premature collapse.
12. Maintain all features in proper working order (NFPA 2006).

A subset of the *Life Safety Code*, NFPA 101A (2007f), *Guide on Alternative Approaches to Life Safety*, provides other approaches. These alternative approaches are not mandatory, but if one is applied, then the mandated procedures must be followed. The guide includes methods for establishing evaluation systems for health-care facilities, detention centers, and roominghouses. The evaluations are conducted using worksheets provided for each listed occupancy classification. The evaluator uses a worksheet to assess compliance with the established code for the particular structure. If it is determined that the existing conditions or structure is not in compliance, safety factors are calculated for the occupancy load, interior finish, means of egress, smoke control, detection and alarm systems, and many other factors depending on the type and construction of the structure. If, after calculating these factors, the evaluator determines that the existing structure is as good as or better than what the applicable code requires for each and every fire zone, the the structure or occupancy is considered to be in compliance. This evaluation system is necessary when the more conventional system described in NFPA 101 does not fit given conditions (NFPA 2006c).

BUILDING CODES

No discussion on classifications of structures would be complete without mentioning the various building codes that affect fire protection. The first building codes or ordinances in the United States focused on the potential for fire to spread from building to building. However, today there are two general types of building codes. *Specification* or *prescriptive* codes describe what materials can be used, the building size, and how to assemble it *Performance* codes describe the objectives to be met by the code and scoring criteria to determine whether the objectives have been met (Cote and Grant 2000, 1–42).

Building codes in general address four areas of fire protection:

1. enclosures of vertical openings
2. evacuation exits, requiring a minimum of two for every building or structure (they do not consider occupancy classifications as the *Life Safety Code* does)
3. flame spread of interior finishes
4. automatic fire suppression systems (Cote and Grant 2000, 1–44).

One difference between building codes and fire codes is that the requirements for exits and fire extinguishing equipment are found in building codes, whereas the servicing and maintenance of such systems are found in fire codes.

NFPA 5000

One building code is NFPA 5000, *Building Construction and Safety Code* (2005b). The basic philosophy of NFPA 5000 is to provide the *umbrella*—the primary coordinating document—for NFPA's *Consensus Code Set* (C3) (NFPA 2005b). The C3 codes comprise a set of documents developed under the criteria of and accredited by the American National Standards Institute (ANSI) (Cote and Grant 2000, 1–45). The ANSI code development process is considered to be more accessible to the general public than the process used by other code organizations. The committee that eventually produced this code was formed in 1913 after the devastating Triangle Shirtwaist Factory fire in New York City. They first produced a series of pamphlets on how to exit buildings, and by 1927 had published the first version of the NFPA *Building Exits Code*. The core requirements are presented in the first fourteen chapters, the purpose of which is to provide the minimum design regulations to safeguard life, health, property, and *public welfare*. (Note the reference to "public.") The Code does not address features that solely affect economic loss to private property (Paragraph. A.1.1).

The *Building Exits Code* applies to construction, alteration, repair, equipment, use and *occupancy* (NFPA 2005b), maintenance, relocation, and demolition of every building or structure. It also applies to existing buildings when they experience change of use or occupancy classification; repair, renovation, modification, reconstruction, or addition; relocation; or are considered unsafe or a fire hazard, generally when the costs of said activities exceed 50% of the appraised taxed value of the structure or building in question.

NFPA 5000 provides for the establishment of a *department of building and safety* under the control of a *director* that is considered to be the AHJ. In some cases the AHJ may be delegated to a subordinate unit within the local government and may be split into departments such as building inspection and code enforcement—or it may remain within the local fire department. However the delegation is arranged, three basic conditions must exist: the entity creating or adopting the law must be empowered to enforce it; the law must be in writing and available to all who are subject to it; and an entity authorized to enforce it must be specified (NFPA 2005b).

NFPA 5000, *Building Construction and Safety Code* provides

- specific governance of the *right of entry* for the AHJ and other code enforcement agencies;
- the *right of stop-work orders,* so that unsafe activities may be shut down by the code enforcement agencies;
- the right to order occupancy to be discontinued (or evacuated);
- the right of permitting;
- the right to have concealed work uncovered for purposes of inspection;
- the right—either on its own initiative or after report is filed—to examine any building appearing unsafe;
- for the creation of a Board of Appeals consisting of five members and empowered to adjudicate decisions of the AHJ (NFPA 2005b).

This code does not necessarily provide a description of what an unsafe building is or what constitutes a fire hazard, but it defines a safe building, in paragraph 4.1.3.3, as one that would protect people against injury in the performance of the tasks they would typically perform within the building. Unsafe conditions would include unsanitary conditions, insufficient means of egress, improper stair dimensions or guardrail height, or failure to address features that could result in falls. The code also states that buildings should be constructed in such a manner as to provide for safe crowd movement and be designed and constructed to provide reasonable safety for occupants and workers during construction (NFPA 2005b).

International Code Council

Notwithstanding NFPA 5000, in December 1994 three code agencies were active in the United States. To eliminate technical disparities among the three codes, to provide a single code reference on which to base future construction, and to facilitate competitiveness within the building industry, an alliance was formed. It was originally grouped with the NFPA but later broke away due to technical inconsistencies. Effective January 1, 2003, the International Code Council (ICC) became the official code organization within the United States and several other countries. The former code groups—the Building Officials and Code Administrators International (BOCA), serving primarily in the western half of the country; the International Conference of Building Officials (ICBO), serving primarily the northeastern portion of the United States and extending into the Midwest; and the Southern Building Code Congress International (SBCCI), serving the southeastern portion of the United States—combined efforts to produce a single set of codes that professionals work with throughout the United States. (ICC n.d.).

The ICC has already established fourteen international standards including standards for fire (similar to the NFPA's *Life Safety Code* and *Uniform Fire Code*), plumbing, and mechanical installations. Additionally, it has incorporated NFPA 70 (2008f), the *National Electrical Code* and NFPA 70E, *Standard for Electrical Safety in the Workplace,* regarding arc-flash protection during electrical work (NFPA 2004a).

These building codes are developed in a different manner than those in the NFPA system. The codes of the three agencies (BOCA, ICBO, and SBCCI) are developed by a governmental consensus process. (Cote and Grant 2000, 1–46). This process draws a consensus from governmental member code-enforcement entities and not from the general public. Only building officials participate in the final vote.

Adoption of the code is left to the each state. In most cases states require adoption of the code by their political subdivisions. Depending upon which lobbying groups are strongest in the various state capitals, the state may adopt either the NFPA or the ICC code.

Aside from the various building codes and inclusive of the discussions on means of egress; regardless of what is required by NFPA 5000 (2005b), the *Life Safety Code* (NFPA 2006c), or the *International Building Code*, the United States government requires any structure, facility, or building that is accessible to individuals with disabilities to have a minimum clearance width for walking surfaces of at least 36 inches (ADA/ABA 2004). For areas where the accessible route makes a 180-degree turn around an element that is less than the 48 inches wide, the approach must be a minimum of 42 inches wide, 48 inches at the turn, and again 42 inches when leaving the turn to allow for wheelchair access. This applies to the design or construction of facilities or buildings accessible to individuals with disabilities. This stipulation impacts on the requirements for emergency evacuation planning contained in Subpart E to 29 CFR 1910 (OSHA 2002a). For such structures, facilities, and buildings accessible to individuals with disabilities, special care must be taken to ensure life safety during an evacuation. This may mean providing for an escort, assisting individuals into protected stairwells, and then communicating their location to the fire incident commander on the scene or some other method deemed appropriate. More assistance may be obtained by contacting the local servicing fire department or AHJ.

Additionally, recognizing that boilers and similar closed vessels add significantly to rated fire hazards, all states (except South Carolina) and a few counties and cities have adopted a boiler code, sometimes referred to as a *Boiler Safety Act*. This act is enforceable under the state agency that governs the installation, examination, periodic inspection, and licensing requirements for these vessels. In many cases the applicable state law will incorporate aspects of ASME's *Boiler and Pressure Vessel Code*, including the potential for periodically exercising pressure relief valves and replacing them every three years (ASME 2007).

Although the NFPA is not the only agency that prepares codes or standards that affect fire systems, it is a comprehensive location at which to begin a search for applicable regulations. Other agencies, such as the American Petroleum Institute (API), the National Fire Sprinkler Association (NFSA), the National Association of Fire Equipment Distributors (NAFED), the National Fire Academy (NFA), the American Welding Society (AWS), the Compressed Gas Association (CGA), the Society of Fire Protection Engineers (SFPE), and Underwriters Laboratories (UL), publish codes or standards specific to their industry. (See Appendix B at the end of this chapter for a more complete listing.) The next several paragraphs are intended to provide the reader with a quick overview of specific NFPA standards that affect selected industries or applications. This overview is not intended to provide detailed explanations or discussion for any particular area but rather to serve as a guide for further study.

There are numerous standards for fixed systems based on the type of extinguishing agent or hazard present. For a complete listing of all the NFPA standards on fixed systems, refer to either the NFPA Web site (NFPA.org) or the *NFPA Journal*. Beyond basic fire-extinguisher guidelines as previously mentioned, the most common types of fixed systems are addressed in the following section.

NFPA Standards Concerning Extinguishing Systems

NFPA 12, Carbon Dioxide Extinguishing Systems (NFPA 2008a)

These systems are used wherever flammable liquids or electrical components are stored, used, or maintained (e.g., fuel islands or stations, flammable storage areas or vaults, computer rooms, or electrical vaults).

These systems are also frequently installed as a part of restaurant hood systems. In such systems the nozzles must be situated directly over the hazard area to provide adequate coverage and replaced periodically (relates to NFPA 25 (2008c) for servicing requirements). Further, these systems do not provide for or support periodic cleaning of hood systems. Grease and other trapped products are often the cause of restaurant fires. Therefore, contracting with a local restaurant hood-cleaning contractor is essential.

Another hazard associated with carbon dioxide systems is that this gas is heavier than air. When this system is used in subfloorings, such as in computer rooms, it presents a potential hazard to those who must make repairs after system discharge. To eliminate such hazards, purging is required.

NFPA 13, Installation of Sprinkler Systems (NFPA 2007b)

This standard originates from the NFPA Committee on Automatic Sprinkler Systems, and was originally entitled *Rules and Regulations of the National Board of Fire Underwriters for Sprinkler Equipment, Automatic and Open Systems*. This work was found in the 1896 *NFPA Proceedings* and has been updated in successive editions ever since.

Although this standard is not incorporated by reference in the OSHA standard, 1910.159 (Automatic Sprinkler Systems) (a)(2) states: "For automatic sprinkler systems used to meet OSHA requirements and installed prior to the effective date of this standard (1980a), compliance with the National Fire Protection Association (NFPA) or the National Board of Fire Underwriters (NBFU) standard in effect at the time of the system's installation is acceptable as compliance with this section."

This standard provides the classifications of occupancies and stored commodities to be used in determining the installation, design, and water supply requirements of sprinkler systems. These classifications are somewhat different from those outlined in NFPA 101 (2006c) or NFPA 5000 (2005b) and are presented below:

Light Hazard: Occupancies where the quantity of combustibles and/or the combustibility is considered low and a fire would produce a low amount of heat.

Ordinary Hazard

Group I: Occupancies where the combustibility is low but the quantity of combustibles is moderate but do not exceed 8 feet.

Group II: Occupancies with quantities of combustibles that are considered to be moderate to high but not exceeding 12 feet.

Extra Hazard

Group I: Occupancies where the quantity and combustibility of the contents is very high and dust, lint, or other such material is present.

Group II: Occupancies that contain moderate to substantial amounts of flammable or combustible liquids.

The standard also provides for four classifications of commodities, the storage of which will affect the placement, installation, design, or style of the sprinkler system.

In short, this NFPA standard provides detailed information on the installation of sprinkler systems that are normally designed by architectural firms and reviewed by the AHJ, with acceptance of the installation reviewed and approved by those individuals empowered for *Code Enforcement*. So unless one is constructing a new building or is remodeling an existing building by more than 50 % of their taxed values, once the system is installed, refer to NFPA 25 (NFPA 2008c) for maintenance an inspection criteria.

NFPA 14, Installation of Standpipe and Hose Systems (NFPA 2007c)

On the inside front cover of NFPA 14 (2007c), an inscription states that this standard dates from 1912 from the initial report made by the Committee on Standpipes and Hose Systems. It was revised in 1914 and adopted by the NFPA in 1915. This standard is incorporated by reference in OSHA 1910.158 (Standpipe and Hose Systems) (OSHA 1998c). The standard details five types of standpipe systems and then divides those types into three classes defined by the size of hose that can be connected to the standpipe.

Class I Systems: Provide for a single 2½-inch hose connection to supply water for use by fire departments and those trained in handling heavy fire streams.

Class II Systems: Provide for a single 1½-inch hose to supply water primarily for building occupants or for members of the fire department upon initial response.

Class III Systems: Provides for two hose connections, one 1½-inch to be used primarily by building occupants and one $2^{1}/_{2}$-inch to be used primarily by responding fire department personnel and others who are trained in handling heavy fire streams.

OSHA Regulation 1910.158 (OSHA 1998c) addresses both Class II and Class III systems but makes an exception for Class I systems. Under this OSHA regulation, employers are required to ensure that each standpipe installed after January 1, 1981, is equipped with a lined hose, and each hose must have a shut-off-type nozzle.

Employers must designate trained individuals to perform required inspections. These inspections are detailed further in NFPA 25 and must be carried out by individuals authorized by the AHJ (NFPA 2008c).

OSHA 1910.156 (1998b) provides training requirements and requirements for personal protective clothing for individuals who perform interior firefighting—individuals who could reasonably be expected to use a standpipe system with hose attachment (OSHA 1998b). (This OSHA regulation does not, however, apply to individuals engaged in airport crash rescue or forest fire fighting.)

As described in Appendix C of Subpart L (Fire Protection) to the OSHA regulation (OSHA 1999), the requirements listed for training and personal protective equipment (PPE) selection follow the guidelines established in NFPA 197 (NFPA 2008g), NFPA 1001 (NFPA 2008h), and NFPA 1971 (Protective Clothing for Structural Fire Fighting) (NFPA 2007g), as well as numerous others. Interestingly, however, the OSHA regulation does not mention NFPA 600 (2005a), the standards for industrial fire brigades.

The NFPA also provides information on where standpipes are to be located based on the class. In general, for Class I systems, they must be installed in stairwells, and in Class II and III systems they must be installed so that all portions of each floor level can be reached with a 130-foot hose.

NFPA 15, Water Spray Fixed Systems (NFPA 2007d)

Water spray fixed systems are used to supplement other forms of systems and are not meant to replace sprinkler systems. They are considered to be adequate fire protection for the following hazards: gaseous and liquid flammable materials; electrical hazards such as transformers, oil switches, motors, cable trays, and cable runs; ordinary combustibles such as paper, wood and textiles; and certain hazardous solids such as propellants and pyrotechnics.

Water runoff must be controlled, especially when these systems are used for fire protection of petroleum-based products where water is often used for purposes of; surface cooling, smothering by the production of steam, emulsification, or dilution. To control the water runoff, water flow must be channeled into a suitable storage area capable of containing the full amount of water. Once contained, the water must be tested to determine whether it may be disposed of in the sanitary sewer system or must be disposed of as a contaminated waste.

NFPA 17, Dry Chemical Extinguishing Systems (NFPA 2002)

Dry chemical extinguishing systems are used where the following hazards exist: flammable or combustible liquids, gases or plastics, electrical hazards such as oil-filled transformers, and ordinary combustibles, as well as in textile operations subjected to flash surface fires and in restaurant hood systems.

These systems are not considered suitable for chemicals containing their own oxygen supply; metals such as sodium, potassium, magnesium, titanium, and zirconium; or deep-seated or burrowing fires in ordinary combustibles.

Dry chemicals extinguish oil and grease fires by creating a foam barrier separating the fire from the oxygen source. The foam, which is created by a hydrolysis process known as *saponification,* allows the oil or grease to cool below the auto-ignition point. The chemicals that create this condition are potassium bicarbonate and sodium bicarbonate. They should not be confused with the dry chemicals used in standard ABC extinguisher; they use monoammonium phosphate, which has a detrimental effect on the saponification process and therefore should never be used on kitchen fires. The proper extinguisher to use for kitchen fires is a "K" UL-rated extinguisher.

NFPA 17A, Wet Chemical Extinguishing Systems (NFPA 2008b)

Wet chemical extinguishing systems are used in the following hazard areas: restaurant, commercial, and institutional hood systems; plenums, ducts, and filter systems associated with cooking appliances; grease-removal devices; odor-control devices; and areas designed for energy-recovery systems installed in exhaust systems.

These systems use either potassium carbonate or potassium acetate with water content between 40 and 60 percent. As with dry systems [NFPA 17 (2002)], these wet chemicals also extinguish oil and grease fires by creating a foam barrier separating the fire from the oxygen source, and the foam is created by saponification, which allows the oil or grease to cool below the auto-ignition point.

Since 1969 the insurance industry has required restaurant owners to install wet or dry systems in order to obtain fire insurance. If an installed fire system has been in service since before 1994, organizations should check to see whether the system is compliant with standards detailed in UL 300, which contains a classification of hood systems specifically designed for restaurant use. Restaurants are shifting away from the traditional animal fat used in deep fryers and are switching to vegetable oils. These oils burn at a much hotter temperature than animal fat, and therefore foam barriers may be inefficient. The UL 300 system corrects this deficiency. Organizations should check with either their insurance carrier or the local AHJ to determine whether their system should be upgraded.

NFPA 25, Inspection, Testing and Maintenance of Water-Based Fire Protection Systems (NFPA 2008c)

Of all of the fire protection standards, this one is of particular importance to safety professionals because it details the scheduled checks and services that should be done on each type of system. Impairments are of utmost importance and related to the Factory Mutual standards. Understanding this standard may be critical in accident or incident investigations as well as for the establishment of a loss-control program. It provides guidance on impairments, sample inspection forms, and information about impairment tagging.

For specific information on scheduled maintenance and repair of water-based fire protection systems by a state-certified inspection firm, organizations should refer to the appropriate NFPA standard. In general, however, the following inspection guidelines should be followed:

Sprinkler systems must be checked quarterly, and once every five years the valves themselves must be inspected and the inspection dates must be stenciled onto the valve body.

Hose cabinets and standpipes must be checked semiannually. New hoses must have the installment date stenciled on the hose no less than three feet from the nozzle. New hoses must be hydrotested in five years and in three-year increments thereafter. (In most cases hoses are for internal fire brigade use only and not for fire department use. The AHJ may provide guidance regarding such hoses.) Most of the time it is more cost-effective to replace fire hoses every five years than to continue hydrotesting periodically. Once a hose is hydrotested, it must be dried before storage. The most efficient way to dry hoses is in a hose dryer, normally located at the local fire department.

Engineered and fixed systems must be checked semiannually. These systems may be found in various applications, but one of the more prominent is commercial restaurant hoods. Each system, regardless of application, has a discharge nozzle. This nozzle must be replaced semiannually when the check is performed.

Alarm systems must be checked annually, with a minimum of 10 percent of all attachments such as pull stations, strobes, smoke-evacuation systems, and detectors checked.

Other NFPA Codes and Standards

NFPA 30, Flammable and Combustible Liquids Code (NFPA 2008d)

This code applies to activities necessary for the storage, handling, and use of flammable and combustible liquids, but does not apply to the transport of such liquids; fuel oil storage; liquids with a flash point of 100°F or greater; or cryogenic, aerosol, mist, or spray products. The code lists requirements for storage in belowground or aboveground fixed or portable tanks including safety cans. (For aboveground or belowground storage tanks, organizations should check with their state governing agency for state-specific requirements.) The requirements outlined in NFPA 30 are essential when addressing similar issues under 29 CFR 1910.106, which is incorporated by reference the 1969 edition of NFPA 30.

This code also addresses the management of fire hazards and the inspection criteria for storage vessels and provides definitions for both flammable and combustible liquids that follow the guidelines from 1910.106 (OSHA 2005a). However, the OSHA standard goes further; it provides information on Class 1A, 1B, and 1C flammable liquids and Class II and Class IIIa and IIIb combustible liquids. The 1926 construction standard does not provide definitions for these two types of liquids but incorporates the provisions of the 1969 code.

The OSHA construction standards 1926.152, *Flammable and Combustible Liquids* (1998d) and 1926.153, *Liquefied Petroleum Gas* (1993) state the following:

- No more than 25 gallons of liquid may be stored outside of an approved storage cabinet.
- No more than 60 gallons of flammable liquid or 120 gallons of combustible liquid may be stored in any one cabinet, and no more than three such cabinets may be stored in any one storage area. These criteria are the same as those listed in NFPA 30, but the NFPA standard provides additional guidance on what constitutes a suitable storage cabinet.
- Inside storage of liquefied petroleum gas (LPG) is prohibited.

The OSHA regulations provide guidelines on the venting and placement of tanks containing flammable or combustible liquids. However, the NFPA provides more definitive data and is considered to be more stringent; therefore, organizations should consult the NFPA code.

The NFPA code also provides guidance on removal and temporary or permanent closure of storage tanks. Organizations that consult their AHJ regarding tank closure or removal may find that the AHJ merely provides the information contained in the NFPA standard. Most states, however, have specific environmental rules governing such activities.

NFPA 51B, Fire Prevention During Welding, Cutting, and Other Hot Work (NFPA 2003)

NFPA 51B (2003) is implied under the general requirements for welding in both the 1910 and 1926 OSHA standards. It is applicable whenever the following processes are involved: welding and allied processes; heat treating; grinding; thawing pipe; powder-driven fasteners; or hot riveting. It does not include use of candles, cooking operations, use of pyrotechnics, use of electric soldering irons, design and installation of gas cutting equipment, lockout/tagout during hot work, or additional requirements that may be necessary for confined-space operations. OSHA 1910 is also written in accordance with ANSI Standard Z49.1, *Safety in Welding, Cutting, and Allied Processes* (ANSI 2005).

The standard establishes a permit system detailing the responsibilities of the permit authorizing individual (PAI), the hot work operator, and the fire watch. It also provides information about permit-required areas and nonpermissible areas as well as the permit itself.

Nonpermissible areas are divided into three types:

- Areas not authorized by management.
- Areas in buildings where the sprinkler system is impaired. (This relates directly to those standards expressed in Factory Mutual standards and in NFPA 25 (2008c).

- Areas that have an explosive atmosphere; uncleaned areas containing combustible, flammable, or explosive contents; and areas that contain combustible, flammable or explosive dusts.

This standard states that the fire watch must clear a 35-foot area around the area where hot work is to take place and must remain in place for at least half an hour after the hot work is completed.

NFPA 54, National Fuel Gas Code (NFPA 2006b/ANSI Z223.1)

This code covers piping systems that extend from the point of delivery to the point of connection for fuel-gas piping systems, fuel-gas utilization equipment, and related accessories. It also sets the maximum operating pressure at 125 psi, with exceptions for gas-air mixtures within flammable ranges (10 psi) and liquefied petroleum gas (LPG) piping systems (20 psi). However, there are nineteen areas where this code does not apply, including LPG installations at utility plants; oxygen-fuel-gas cutting and welding operations; fuel-gas systems using hydrogen as fuel; portable LPG equipment of all types that are not connected to fixed fuel-piping systems; and liquefied natural gas (LNG) installations. These areas, particularly those associated with petroleum and the petrochemical industries, use guidelines published by the American Petroleum Institute (API), such as RP 2001, *Fire Protection in Refineries* (API 2001). The API publishes twelve standards that specifically address fire protection in refineries and on piping systems and the control of ignition sources. For a complete listing, visit the API Web site at www.api.org.

Two other organizations should be mentioned here. As with the API, which focuses on the petroleum and petrochemical industries, the American Society of Testing and Materials (ASTM) and the American Society of Mechanical Engineers (ASME) provide guidance on other aspects of systems frequently located in petrochemical plants, refineries, and electrical generation plants. The ASTM focuses on how items are tested to ensure performance quality, whereas the ASME focuses on how items should be engineered, maintained, or repaired.

As an example, the metals used to fabricate a pressure vessel or pipeline must meet specific standards set by the ASTM (ASTM Fire Safety, 2008). If a pressure vessel or pipeline were to be exposed to excessive heat, as it would during a fire, this metal could become compromised, leading to an unintentional release of the vessel's contents. Therefore, it is imperative that if pressure vessels, pipelines, tanks, or other such structures are exposed to excessive heat, testing must be conducted as soon as reasonably possible and before the structure is placed back in service to ensure functional integrity.

If equipment is to be repaired or maintained, the ASME provides guidance on how to perform the work. In conjunction with the American Welding Society (AWS), ASME provides guidance on which welding procedures to use and how to guard against the incipient stages of fire while welding on pressure vessels, pipelines, and other structures.

Installation of systems that use fuel gas must be performed by a qualified agency—an individual, firm, or corporation that is familiar with the proper equipment, installation, servicing, and connections required as well as knowledgeable in the precautions prescribed by the AHJ.

If systems are interrupted, it is the duty of the qualified agency to notify all affected users. Likewise, potential ignition sources must be controlled. Additional measures the agency must take include providing for electrical continuity before alterations are made in a metallic piping system and prohibition of smoking, open flame, lanterns, welding, cutting torches, or other sources of ignition.

NFPA 58, Liquefied Petroleum Gas Code (NFPA 2008e)

This standard covers containers used for the transport, storage, and delivery of LPG, including pipelines and all associated equipment for the proper dispensing of LPG. In accordance with Department of Transportation (DOT) requirements found in both 49 CFR 173 (DOT 2004) and 49 CFR 178 (DOT 2003), LPG tanks must be constructed of steel or aluminum. Recently, however, advances have been made with composite materials so that tanks can be fabricated from composites similar to those used for SCBA tanks. At present, two manufacturers have been approved by the DOT to produce such tanks for conventional use outdoors. The requirements described in this standard are imperative when addressing issues under 29 CFR 1910.110 (OSHA 1998a), which was incorporated by reference the 1969 version of NFPA 58 (NFPA 2008e). Within the OSHA requirements (29 CFR 1910.110) are general requirements for the storage and handling of LPG, such as requiring the gas to be odorized and to be stored in containers that meet either API (2001), ASME (2007), or DOT standards (49 CFR Part 178) (2003). The OSHA regulation also provides for minimum storage distances that the LPG can be stored from buildings as well as safeguards that apply when handling and working with LPG. However, it does not cover the use of LPG at power plants, chemical plants, refineries, or underground storage facilities; transport by air; or use in recreational vehicles. To clarify, OSHA governs the workplace, whereas the DOT governs transportation. Hence, if LPG is to be stored and/or handled within a workplace, OSHA standard 1910.110 applies (OSHA 1998a). If, however, LPG is to be transported to another location, 49 CFR Part 178 applies (DOT 2003).

Requalification of cylinders is addressed within NFPA 58 (2008e): cylinders must not be continued in service with expired requalification dates. They must be requalified within twelve years after manufacture and then every five years. On cylinders used for industrial truck-fueling applications or service, the pressure-relief valve must be replaced with a new or unused valve twelve years after manufacture and then every ten years. The replacement date must be stamped on the cylinder neck along with fill instructions. Some cylinders are designed to be filled only in the horizontal position, and these cylinders must be stamped accordingly. These points are part of the preoperational checks required by 29 CFR 1910.178 (OSHA 2006).

The standard sets the number of cylinders that may be placed inside structures, which is dependent upon the occupancy classification as stated in NFPA 101 (2006c), proximity to a source of ignition, and security from vehicle incursions.

The standard requires training for anyone involved in fueling or filling containers or tanks. This becomes an issue for industrial truck operators if they are expected to refuel tanks and should, therefore, be included in the training requirements of 29 CFR 1910.178 (OSHA 2006).

NFPA 70, National Electric Code (NFPA 2008f)

This code, when used in conjunction with NFPA 70E, *Standard for Electrical Safety in the Workplace* (NFPA 2004a) and the *National Electric Safety Code* (C2-2007) from the Institute of Electrical and Electronics Engineers, Inc. (IEEE), form the baseline from which all other electrical safety issues stem. These are essential documents of reference and referral for compliance with Subpart S (Electrical) of 29 CFR 1910 (General Industry Standard), (OSHA 1999b), and Subpart K (Electrical) to 29 CFR 1926 (Construction Standard) (OSHA 2003a).

NFPA 70 provides for the standards for wiring, wiring protection, wiring methods, motors, and circuits and provides guidance based on occupancy classifications (NFPA 101) (NFPA 2008f).

This standard also describes levels of personal protection required for workers who are engaged in high-voltage (over 600 volts) as well as low-voltage (under 600 volts) work.

NFPA 70E (2004a), which incorporated technical data from the IEEE (IEEE-1584), was first drafted in 1976 at the request of OSHA. It defines electrical hazard areas in the workplace as hazard/risk-category classifications: –1, 0, 1, 2, 3, and 4. These classifications allow organizations to determine, for each level, personal protection requirements, approach distances, and who is qualified to work on particular equipment. It also provides flow charts based on management oversight risk tree (MORT) analysis to determine hazard levels; a sample lockout/tagout policy; job briefing and planning checklists; energized-electrical-work permits; and in Annex H, a simplified two-category, flame-resistant clothing system, which is essential for people working at certain hazard levels such as racking or unracking breakers. IEEE-1584 goes beyond this safer approach and provides equations and calculations to determine a more exact method of predicting the potential energy release (IEEE 2007).

Employers must identify hazard/risk categories in a written electrical-safety program. An evaluation process, conducted by or for the employer, must be accomplished prior to the commencement of work on or near live electrical parts operating at 50 volts or more. This evaluation is used to determine the actual hazard/risk categories. As previously stated, conducting such an evaluation may require calculating the potential stored electrical energy. This is where the calculations provided in IEEE-1584 come into play (IEEE 2007).

- *Class –1* refers to electrical components that are less than 10 kiloamps (kA) available short circuits and allows the evaluator to reduce protection levels down by 1 (from Class 0) i.e., if the components are rated at either less than 25 kA or 10 kA (depending upon other factors), the established protection level of 0 can be reduced to –1.
- *Class 0* refers to the hazard/risk category in which normally energized parts can be worked on with doors or other closure devices installed so that no live or energized parts are exposed to the worker.
- *Class 1* refers to conditions in which protective covers are off and workers are performing such tasks as voltage testing or removal of bolted covers on components rated at 240 volts or less; working on fused-switch operations on components rate at 240 to 600 volts; opening hinged covers to expose energized parts on 600-volt motor control centers (MCCs); and miscellaneous equipment cover removal. In these cases the electrical hazard, although present, is considered low.
- *Class 2* refers to conditions in which work is done on energized parts, including voltage testing on switchboards or panel boards rated between 240 and 600 volts; work on control circuits with energized parts of 120 volts or less, exposed; removal of bolted covers to expose bare, energized parts on MCCs rated at 600 volts or application of safety grounds after voltage testing on same; racking or unracking circuit breakers from cubicles containing 600-volt-class switch gear; racking or unracking starters of components rated between 23 kilovolts and 7.2 kV; and working on control boards with voltage rated at 120 or less.
- In *Class 3* the hazards are more severe. Operations include working on 600-volt switch gear to remove bolted covers, thus exposing energized parts; working on control circuits in the voltage range of 2.3 to 7.2 kV, exposing energized parts; insertion or removal of starters on components in the same voltage range (2.3 to 7.2 kV); and opening of hinged covers on components rated at 1 kV and above.
- *Class 4* work, the highest class, presents the greatest hazard. Activities include removing bolted covers to expose energized parts on components rated between 2.3 kV and 7.2 kV; circuit-breaker or fused-switch operations with enclosure doors open on metal-clad switch gear rated at 1 kV or more or working on energized parts of metal-clad switch gear rated at 1 kV or more, including circuit-breaker racking and unracking; and working on energized parts, including voltage testing and removal of bolted covers to expose energized parts on equipment rated at 1 kV or more.

These hazard/risk categories drive the personnel protective equipment (PPE) requirements for electrical workers. For hazard classes –1 and 0, workers must wear a non-melting or untreated natural-fiber long-sleeve shirt with work pants that meet the protective standards described in ASTM F 1506-00 (ASTM 2002). Fire resistant (FR) clothing is required when electrical workers work with hazards in Class 1 or above. The range of protective clothing goes from a long-sleeve shirt with FR pants or coveralls to an arc-flash-rated flash suit and face shield.

The *National Electrical Safety Code* (C2-2007), by the IEEE: This Code applies mostly to the transmission and distribution industries as well as substation operations, telecommunications, and the general operation of electrical lines including switching and tagging procedures, fall protection, and general rules for employees. It is essential to refer to this code when issues arise under OSHA 1910.269 (OSHA 1994).

NFPA 72, National Fire Alarm Code (2007e)

Dating back to 1898, this code establishes the minimum level of installation, application, location, performance, and maintenance requirements for fire-alarm systems and their related components.

The code requires that fire alarm systems have the following characteristics installed by a trained and certified individual: at least two independent power sources, each monitored to ensure reliability; indicating lights to alert individuals of the function status of the unit; a method of silencing and deactivating audible alarms; and distinctive signals. In short, the alarm must have the ability to alert responsible parties about a dangerous condition.

The provisions in this code do not necessarily apply to systems established under 29 CFR 1910.165 (OSHA 1980c), and NFPA 72 is not incorporated by reference into 1910.165. Under 1910.165, employee alarms that are capable of being supervised should be; and those that are nonsupervised must be tested once every two months. Alarms should be capable of being perceived above normal noise and light levels. Employers must explain to each employee the preferred means of reporting emergencies. As with many OSHA regulations, this code provides special provisions for employers with ten or fewer employees (OSHA 1980c).

NFPA 409, Standard on Aircraft Hangars (NFPA 2004b)

The NFPA provides a standard just for aircraft hangars as a special-case style of construction. The original guidance was contained in a publication from the National Board of Fire Underwriters (NBFU), which is now the American Insurance Association. In 1954 the NFPA published its first version of the standard, which was adopted in its entirety by the NBFU in the same year.

This document specifies fire protection based on a group classification system for hangars. It lists five basic types of hangars: Group I, Group II, Group III, Group IV, and Paint Hangars. These groupings are cross-referenced with the style of construction provided in NFPA 220 (2006d).

In NFPA 220 there are five types of building construction—Type I through Type V, each type having two to three subsets represented by a numerical identifier. Types I through IV are constructed from the listed materials and are rated as noncombustible or limited combustibility. Type V construction is made from materials considered to be combustible (NFPA 2006d).

A *Group I* hangar has at least one of the following characteristics: an access door height over 28 feet, a single fire area in excess of 40,000 square feet, or the ability to house an aircraft with a tail height over 28 feet. Group I hangars must have either a foam-water deluge system, a combination of automatic sprinkler protection and a low-level, low-expansion foam system, or a combination of automatic sprinklers and a low-level, high-expansion foam system. If the foam-water deluge system is desired, it must be installed with a supplementary system, which could be one of three types of low-expansion foam systems.

If unfueled aircraft are housed in a Group I-style hangar, the classification drops to *Group II*. Group II hangars have either access doors that are no more than 28 feet tall or a single fire area between 12,000 and 40,000 square feet, dependent upon the type of construction. Group II hangars must have fire-protection systems that are similar to those for Group I—a combination of automatic sprinkler protection and a low-level, low-expansion foam system, a combination of automatic sprinklers and a low-level, high-expansion foam system, or a closed-head foam-water sprinkler system that can be either wet pipe or pre-action.

With both Groups I and II, the water supply must sufficient to provide at least 30 minutes of protection unless unfueled aircraft are stored; then the minimum is 60 minutes.

Group III hangars may be of any of the construction types listed in NFPA 220 (2006d), and are limited to one story. If there is a potential for a flammable liquid spill, the floor must be made of noncombustible material and designed to prevent the flammable liquid from entering the service area. Within Group III hangars, if hazardous operations are conducted, such as fueling, defueling, cutting, welding, and so on, fire protection must be the same as for Group II hangars. Otherwise, Group III hangars are only required to have fire extinguishers.

Group IV hangars are those that are constructed from a membrane-covered rigid-steel-frame structures. These structures must provide one form of fire protection listed for Group II hangars, including hand hose systems that deploy water or foam. Total fire floor area cannot exceed a single-hangar fire area, and the protection system chosen must achieve control of the fire within 30 seconds and extinguishment within 60 seconds. Total The total water supply shall be capable of supplying water for a minimum of 45 minutes.

The last group is *Paint Hangars*, for areas where there are ignitable concentrations of flammable gas or vapors. These hangars must comply with the fire-protection requirements for Group II hangars. In addition, special consideration is given for their ventilation system and electrical equipment, which must meet requirements listed in NFPA 70 (*National Electric Code*) for Class I, Division I, Electrical (NFPA 2008f). Electrical components must be rated as Class 1, because the release of flammable gases or vapors could cause failure in regular (non-Class 1) electrical components and thus cause the electrical equipment to act as a source of ignition.

OSHA standards are considered to be the minimum acceptable, essential standards in establishing safe work practices and procedures. Even though there may be special provisions for employers with ten or fewer employees, employers are still responsible for providing a safe workplace. In most cases OSHA standards require meeting certain requirements or actions but do not elaborate on how this is to be accomplished. In other cases, such as in fire protection, the OSHA standards do not include all of the required periodic checks and services required by state-licensed individuals. The standards or codes discussed above must be referenced if not already referenced in the OSHA standard.

When OSHA standards incorporate other standards by reference, the referred-to publications are listed within the standard either in a separate section, at the end of the subpart, or within the text of the standard itself. In numerous cases, those standards, which are referred to by year, are not current. Organizations must always apply the most recent and/or strictest standard.

Factory Mutual Standards (FM Global 2008) offer worldwide certification and testing services of industrial and commercial loss-prevention products (ISO 9000). These standards are espoused by the general insurance industry and may be incorporated as *company-specific standards* that are presented as recommendations by insurance carriers. In many cases these standards may be stricter than the required level of protection but are considered necessary for the maintenance/continuance of property insurance or are required to maintain a specific insurance rate. The insurance carrier is often an active partner with the property holder and should be used in a consultative mode. In other cases where an economic impact study can be presented to a carrier who is requiring retro-fits, other acceptable options may be pursued.

Most property-insurance carriers have standardized forms and programs that assist the insured in following the standards, whether they are NFPA-related or Factory Mutual-related. One such program is that for fire-system impairments referenced under NFPA 13, Annex A (NFPA 2007b) and NFPA 51B (NFPA 2003). This program, which is company specific, normally includes forms, tags, and procedures to follow in notifying the local fire department as well as a representative within the insurance company whenever there is impairment to the fire systems.

Other available programs from insurance carriers associated with the Factory Mutual system include periodic checks on fire systems, including the fire-extinguisher inspection logs; periodic inspections of boiler systems, air-conditioning systems, fire pumps, and security surveillance systems (these checks are mandatory under many states' boiler safety acts); smoking policies; guidance on new construction; and emergency management guidance on bomb threats and fires as well as on hurricanes, earthquakes, and other natural disasters. Insurance companies can also provide guidance for proper housekeeping and hazard analysis. For specific services that may be offered, organizations should check with their property insurance carrier.

Conclusion

Fire safety and prevention has been and shall continue to be a concern for safety professionals. It is only through due diligence and cross-referencing that safety professionals can gain a true understanding of what standards apply to a given situation and how the various standards interact. Numerous resources are available, and many agencies provide standards. Only through the development of a careful understanding of the various sources can a well-balanced fire safety and prevention program be established.

References

American National Standards Institute (ANSI). 2005. Standard Z49.1-2005, *Safety in Welding, Cutting, and Allied Processes.* Washington, SC: ANSI.

American Petroleum Institute (API). 2001. Standard RP 2001, *Fire Protection in Refineries.* Washington, D.C: API.

______. 2006. *Safety and Fire Protection.* Washington, D.C: API. www.api.org/

American Society of Mechanical Engineers (ASME). 2007. *Boiler and Pressure Vessel Code.* New York: ASME.

American Society for Testing and Materials (ASTM). 2002. F 1506–00 (Revised), *A Standard Performance Specification for Flame Resistant Textile Materials for Wearing Apparel for Use by Electrical Workers Exposed to Momentary Electric Arc and Related Thermal Hazards.* (May 10) West Conshohocken, PA: ASTM.

______. Fire Safety Standards, 2008. www.astm.org

Cote, Arthur E., and Casey C. Grant. 2000. "Building and Fire Code Standards." In NFPA, *Fire Protection Handbook.* 18th ed. 1–42 through 1–54. Quincy, MA: NFPA.

Department of Transportation (DOT). 2003. 49 CFR 178, *Specifications for Packaging* (accessed March 2006). www.ecfr.gpoaccess.gov

______. 2004. 49 CFR 173. Shippers; General Requirements for Shipping and Packages (accessed June 1, 2007). www.ecfr.gpoaccess.gov

Ferguson, Lon and Christopher A. Janicak. 2005. "Introduction to Industrial Fire Protection." In *Fundamentals of Fire Protection for the Safety Professional,* 1–28. Lanham, Maryland: Scarecrow Press Incorporated.

Fire Safety Advice Centre. n.d. "History of the Portable Fire Extinguisher" (accessed March, 14 2007). www.firesafe.org.uk/html/fsequip/exting2.htm

Factory Mutual (FM) Global. 2008. "FM Approvals: Setting the Standard in Third-Party Certification" (accessed January 23, 2008). www.fmglobal.com/page.aspx?id=50000000

Grant, Casey C. 1996. "The Birth of the NFPA." Quincy, MA: NFPA (accessed March 14, 2007). www.nfpa.org/ItemDetail.asp?categoryID=500&ItemID=18020+URL=About+Us+History

Illuminating Engineering Society of North America (IESNA), Lighting Engineering Management. 2000. IESNA 2000, LEM-1–05, *Recommended Procedure for Determining Interior and Exterior Power Allowances.* New York: IESNA.

Institute of Electrical and Electronics Engineers (IEEE). 2007. *National Electric Safety Code®.* New York: IEEE Publishing.

International Code Council (ICC). n.d. *Introduction to the ICC* (accessed on January 23, 2008). www.iccsafe.org/news/about

______. 2006. *2006 International Building Code®.* Washington, D.C.: International Code Council.

International Organization for Standardization (ISO). 2005. (ISO) 9000, *Quality Management Systems.* Geneva, Switzerland: ISO.

National Fire Protection Association (NFPA). 1971. NFPA Bylaws. www.nfpa.org/assets/files/PDF/CodesStandards/Directory/NFPABylaws2007.pdf

______. 2002. NFPA 17, *Dry Chemical Extinguishing Systems.* Quincy, MA: NFPA.

______. 2003. NFPA 51B, Fire *Prevention During Welding, Cutting, and Other Hot Work.* Quincy, MA: NFPA.

______. 2004a. NFPA 70E, *Standard for Electrical Safety in the Workplace.* Quincy, MA: NFPA.

______. 2004b. NFPA 409, *Standard on Aircraft Hangars.* Quincy, MA: NFPA.

______. 2005a. NFPA 600, *Standard for Industrial Fire Brigades.* Quincy, MA: NFPA.

______. 2005b. NFPA 5000, *Building Construction and Safety Code Handbook*. Quincy, MA: NFPA.

______. 2006a. NFPA 1, *Uniform Fire Code*. Quincy, MA: NFPA.

______. 2006b. NFPA 54/ANSI Z223.1, *National Fuel Gas Code*. Quincy, MA: NFPA.

______. 2006c. NFPA 101, *Life Safety Code*. Quincy, MA: NFPA.

______. 2006d. NFPA 220, *Standard on Type of Building Construction*. Quincy, MA: NFPA.

______. 2007a. NFPA 10, *Portable Fire Extinguishers*. Quincy, MA: NFPA.

______. 2007b. NFPA 13, *Installation of Sprinkler Systems*. Quincy, MA: NFPA

______. 2007c. NFPA 14, *Installation of Standpipe and Hoses*. Quincy, MA: NFPA.

______. 2007d. NFPA 15, *Water Spray Fixed Systems*. Quincy, MA: NFPA.

______. 2007e. NFPA 72, *National Fire Alarm Code*. Quincy, MA: NFPA.

______. 2007f. NFPA 101A, *Guide on Alternative Approaches to Life Safety*. Quincy, MA: NFPA.

______. 2007g. NFPA 1971, *Protective Clothing for Structural Fire Fighting*. Quincy, MA: NFPA.

______. 2008a. NFPA 12, *Carbon Dioxide Extinguishing Systems*. Quincy, MA: NFPA.

______. 2008b. NFPA 17A, *Wet Chemical Extinguishing Systems*. Quincy, MA: NFPA.

______. 2008c. NFPA 25, *Inspection, Testing and Maintenance of Water-Based Fire Protection Systems*. Quincy, MA: NFPA.

______. 2008d. NFPA 30, *Flammable and Combustible Liquids*. Quincy, MA: NFPA.

______. 2008e. NFPA 58, *Liquefied Petroleum Gas Code*. Quincy, MA: NFPA.

______. 2008f. NFPA 70, *National Electric Code*. Quincy, MA: NFPA.

______. 2008g. NFPA 197, *Initial Fire Attack*. Quincy, MA: NFPA.

______. 2008h. NFPA 1001, *Fire Fighter Professional Qualifications*. Quincy, MA: NFPA.

Occupational Safety and Health Administration (OSHA). 1980a. 29 CFR 1910.159, *Automatic Sprinkler Systems*.

______. 1980b. 29 CFR 1910.164, *Fire Detection Systems*.

______. 1980c. 29 CFR 1910.165, *Employee Alarm Systems*.

______. 1993. 29 CFR 1926.153, *Liquefied Petroleum Gas (LP-Gas)*.

______. 1994. 29 CFR 1910.269, *Electric Power Generation, Transmission and Distribution*.

______. 1996a. 29 CFR 1200, *Hazard Communication*.

______. 1996b. 29 CFR 1926.150, *Fire Protection*.

______. 1998a. 29 CFR 1910.110, *Storage and Handling of Liquefied Petroleum Gas*.

______. 1998b. 29 CFR 1910.156, *Fire Brigades*.

______. 1998c. 29 CFR 1910.158, *Standpipe and Hose Systems*.

______. 1998d. 29 CFR 1926.152, *Flammable and Combustible Liquids*.

______. 1999a. 29 CFR 1910 Subpart L, *Fire Protection*.

______. 1999b. 29 CFR 1910 Subpart S, *Electrical*.

______. 2001. 29 CFR 1926, Subpart F, *Fire Protection and Prevention*.

______. 2002a. OSHA 29 CFR 1910, Subpart E, *Exit Routes, Emergency Action Plans, and Fire Protection*.

______. 2002b. 29 CFR 1910.35, Compliance *with NFPA 101–2000, Life Safety Code*.

______. 2002c. 29 CFR 1910.157, *Portable Fire Extinguishers*.

______. 2003a. 29 CFR 1926, *Safety and Health Regulations for Construction*.

______. 2003b. 29 CFR 1926.15, *Relationship to the Service Contract Act; Walsh-Healey Public Contracts Act*.

______. 2005a. 29 CFR 1910.106, *Flammable and Combustible Liquids*.

______. 2005b. 29 CFR 1926, Subpart K, *Electrical*.

______. 2006. 29 CFR 1910.178, *Powered Industrial Trucks*.

Underwriters Laboratory (UL). 2006. UL 924, *Standard for Emergency Lighting and Power Equipment*. Northbrook, IL: UL.

U.S. Access Board. "Americans with Disabilities Act Comparisons" (accessed June, 21 2007). www.access-board.gov/ada-aba/comparison/index.htm

Appendix A: Recommended Reading Sources for Standards and Codes

ADA-ABA Guidelines. 2007. "New ADA Accessibility Guidelines Side-by-Side Comparison." www.access-board.gov/ada-aba/comparison/index.htm
Alliance of American Insurers. www.allianceai.org
Americans with Disabilities Act Comparisons. www.access-board.gov
American National Standards Institute (ANSI). www.ansi.org
American Petroleum Institute (API). www.api.org
American Society for Testing and Materials (ASTM). www.astm.org
American Welding Society (AWS). www.aws.org
Compressed Gas Association (CGA). www.cganet.com
Conroy, Mark. 2006. "User Friendly NFPA 10." *NFPA Journal,* November/December.
Department of Transportation (DOT) 2004. 49 CFR 173. Shippers; General Requirements for Shipping and Packages (accessed June 1, 2007). www.ecfr.gpo access.gov
Factory Mutual (FM). www.fmglobal.com
Friedman, Raymond. 1998. *Principles of Fire Protection Chemistry and Physics.* Quincy, MA: NFPA.
Moore, Wayne D. 2007. "Waiting for the Fire." *NFPA Journal,* 101:30.
National Association of Fire Equipment Distributors (NAFED). www.nafed.org
National Fire Academy (NFA). www.usfa.fema.gov/fire-service/nfa.shtm
National Fire Protection Association (NFPA). www.nfpa.org
National Fire Sprinkler Association (NFSA). www.nfsa.org
Nunez, Ron. March 2007. "Fire in the Workplace: Fundamental Elements of Prevention and Protection." *Professional Safety,* pp. 47–48.
Society of Fire Protection Engineers (SFPE). www.sfpe.org
Underwriters Laboratories (UL). www.ul.com
U.S. Fire Administration (USFA). www.usfa.fema.gov

Appendix B: Sources of Standards and Codes

In general, code organizations provide a wealth of knowledge on the particular subject on which they focus. Each has customer service hotlines and information available on the Internet, including the ability to join as a member or order applicable publications and standards. The following is a partial listing of code organizations:

Alliance of American Insurers. www.allianceai.org
American National Standards Institute (ANSI). www.ansi.org
American Petroleum Institute (API). www.api.org
American Society for Testing and Materials (ASTM). www.astm.org
American Welding Society (AWS). www.aws.org
Compressed Gas Association (CGA). www.cganet.com
Factory Mutual (FM). www.fmglobal.com
National Association of Fire Equipment Distributors (NAFED). www.nafed.org
National Fire Academy (NFA). www.usfa.fema.gov/fire-service/nfa.shtm
National Fire Protection Association (NFPA.) www.nfpa.org
National Fire Sprinkler Association (NFSA). www.nfsa.org
Society of Fire Protection Engineers (SFPE). www.sfpe.org
Underwriters Laboratories (UL). www.ul.com
U.S. Fire Administration (USFA). www.usfa.fema.gov

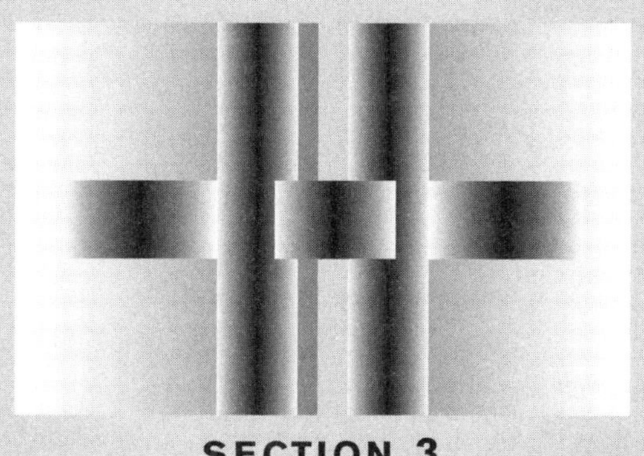

SECTION 3

FIRE PREVENTION AND PROTECTION

Applied Science and Engineering: Fire Dynamics

David G. Lilley

LEARNING OBJECTIVES

- Develop an understanding of the phenomena that encompass fire dynamics–from fuel release, dispersion, and ignition through fire spread, growth, and possible flashover–culminating in the devastation produced and the subsequent investigation.
- Gain an appreciation of safety and loss-prevention issues involved in the occurrence of fires and of factors that can mitigate the extent of subsequent damage.
- Develop the ability to calculate useful information from empirical equations that describe observed fire phenomena, permitting deductions for other situations that are interpolated or extrapolated from a limited amount of experimental data.
- Learn to use mathematical modeling concepts related to computer-based calculation procedures known as the zone method and the computational fluid dynamic (CFD) method for simulating fire growth, smoke production, and travel from fires.

THE ULTIMATE GOAL of this chapter is to improve scientific understanding of fire behavior leading to flashover in structural fires. Topic areas are release, spread, and ignition of flammable materials; burning rates; radiant ignition of other items; fire spread rates; ventilation limit imposed by sizes of openings; flashover criteria; and fire modeling. Within each topic area is information dealing with background and theory, followed by comments. The references given provide a good cross-section of experimental and theoretical studies related to the understanding of real-life fires. These provide useful test data for development of the theory of fire growth and its application to real-world fire situations.

The development of structural fires can be understood more readily when technical knowledge puts the phenomena on a firm scientific footing. The theory can assist in understanding and applying scientific information to real-world fire situations. Recent extensive study, research, experimentation, and field observations and measurements have led to the need for critical evaluation of phenomena associated with fire dynamics—the term chosen to represent topics associated with fire behavior, including ignition, fire development, and fully developed fires. The scientific topics of chemistry, physics, aerodynamics, and heat transfer all play their part. Technical information about fuels, burning rates, fire spread, and flashover and backdraft phenomena is also relevant. These topics are addressed in the light of the relevant equations embodied in such computer programs as FPETool, CFAST, HAZARD, and FDS. The references provide further details.

The purpose of this chapter is to review a few of the ideas involved in appreciating how technical understanding or engineering analysis can assist in determining "responsibility," that is, to determine where the fault lies as to why the incident occurred.

Investigators are aided by knowledge of failure modes of appliances, fire causes and ignition, dynamics of fuel release and dispersion, and dynamics of fire behavior. The extensive reference list provides additional information. Various aspects of safety related to fire protection are addressed in handbooks from the Society of Fire Protection Engineers (SFPE) (2002) and the National Fire Protection Association (NFPA) (Cote 2003). Emmons (1985), Thomas (1974), Zukoski (1985), and other references address technical engineering aspects of fire phenomena. Appliance operations are discussed in American Gas Association (AGA) (1997 and 1988), Berry (1989), Carroll (1979), Patton (1994), and elsewhere. This chapter builds on the author's previous fire-related papers, including his short course (Lilley 1995a and 2004) and his general review (Lilley 1995b). Specific theories, equations, and data about fuel burning properties are given in Lilley (1997a and 1997b). Graphic illustrations are now given to exemplify the theories and their application to real-world situations. More extensive treatment of the discussion here is given in Lilley (1998a and 1998b), with theory and calculation methodology included. The present brief contribution concentrates on the phenomena, specific problems, and graphical illustration of the sample calculations.

Fire modeling plays an important role in fire protection engineering and building design engineering. Two different modeling techniques are commonly used in fire modeling: field modeling (computational fluid dynamics technique) and zone modeling. In field modeling, computational demands are very large, and correct simulation ultimately depends on the empirical specification of things such as ignition, burning rates, fire spread, ventilation limitation, and so on. At this time, field models are being developed and applied for the simulation of structural fires. But the zone method represents a more mature field for fire and smoke transport; the methods have been developed and applied for many years.

Fire Investigation

Major structural fires require investigation. The analysis of failure modes is important in determining why the fire occurred—what went wrong—and whether it was incendiary or otherwise. If incendiary, was it an act of arson? If otherwise, was it a design defect, manufacturing defect, installation error, operating error, repair error, or something else? Engineering analysis is often needed to determine the cause of the fire—what happened, why the incident occurred, and what can be done to prevent future incidents—in short, to assist in having a safer situation in the future.

Origin and cause investigations deal with the *where* and *how* of the fire. *Why* the fire began is the issue in question when *responsibility* for the fire is addressed. Of particular concern is the source of the fuel for the fire, and the ignition source, as either or both of these might well be associated with responsibility. For example, it may be that a welder, plumber, or other workman using an acetylene torch might be responsible for causing a fire by supplying the ignition through negligent actions. On the other hand, the workman might not have been negligent, having taken prudent precautions, but was working in an environment where gaseous vapors were (without his knowledge) present. Then responsibility may be on the imperfection that caused the release of the gaseous vapors. Affixing responsibility for a fire is not always a clear-cut situation, and engineering analysis is often needed.

When research, experiments, theory, or calculations are needed to confirm or deny a suggested fire scenario, a complete and thorough investigation is undertaken. Beyond origin and cause of fires, the technical engineering investigation focuses on the fire scenario and the real cause of the fire. Deficiencies are identified with regard to adherence to codes and standards, whether installation and maintenance procedures were adequate, and whether design or manufacturing defects contributed. Product-liability issues are addressed, with questions of whether an item was defective and unreasonably dangerous. Safety engineers are particularly concerned with all these issues.

Fire Severity

Assessment of fire-hazard potential involves more than just enclosure surface area and total fuel load in terms of mass of equivalent ordinary combustibles per unit area (in kg/m^2 or lbm/ft^2). The arrangement of combustibles, their chemical composition, physical state, ease of ignition, rates of fire growth, and so on are all factors to be evaluated, in addition to room geometry and size, ventilation capability, and fire protection facilities. Generally, the faster a fire develops, the greater the threat will be. Typical fire loads are given by the NFPA (Cote 2003) for different occupancies.

Interior finishes of walls, ceilings, and floors can significantly affect the initial fire growth. Fire spread over interior finishes therefore deserves special attention, and it may be noted that fire spread rates vary enormously. The relative hazard of an interior finish is usually determined by the so-called Steiner tunnel test in accordance with ASTM E 84-07(b) (2007), NFPA 255 (2006b) and UL 723 (2003). Classification of interior finish materials is A, B, or C according to flame spread rating 0–25, 26–75, or 76–200. Smoke development and fuel-contribution ratings are also found in the test via temperature and smoke density recordings. NFPA 101, *Life Safety Code* (2006a) limits the use of interior wall, ceiling, and floor finish. These issues are important because combustible interior finishes, such as low-density fiberboard ceilings and plastic floor coverings, offer significant factors in fire development permitting a fire to spread to objects far from the origin of the fire.

Fires that reach flashover or full-room involvement produce acutely lethal products of combustion, including smoke, carbon monoxide, hydrogen cyanide, acrolein, hydrogen chloride, nitrogen oxide(s), and so on, in addition to heat and insufficient oxygen. These gases are produced in large quantities and driven to remote areas, causing life-threatening situations—inhalation of fire gases (toxic products of combustion) being the major cause of death in fires. For these reasons, any combination of finishes, combustible building materials, or contents and furnishings that could result in flashover (full-room involvement) in a few minutes represents a severe fire hazard in many types of occupancies. Protection by automatic sprinklers and fire-rated construction separations is often needed and is sometimes mandated by relevant codes. Computer fire models include the mathematical characterization of experimental information related to whether flashover will occur. It turns out that a particular fire size is required (in terms of heat release rate in kW or Btu/hr), this value depending on the extent of ventilation, room size, insulation rating of the walls, and so on. One can then assess whether an especially hazardous situation exists—whether a given fire scenario has the potential to develop flashover or full-room involvement.

Combustion

The Basic Elements

In the release and dispersion of flammable material, a vapor cloud is formed that spreads and disperses as the released material mixes and is carried downstream in the atmosphere. It is important to know if a potential ignition source is located within the flammable or explosive limits, these being the percentage (usually by volume) of an in-air substance that will burn once ignited. Most substances have both a lower (lean) and an upper (rich) flammable (explosive) limit, called the LFL and UFL, respectively (or LEL and UEL, respectively). Some typical values for several fuels are given in Table 1, for fuel-in-air mixtures on a volume percent basis. Either too much or too little fuel in the vapor-air mixture can prevent burning. But, at the edge of a fuel-rich vapor cloud will be a region with other more flammable fuel-air ratios, so that, as such a cloud approaches, ignition is certainly possible from an igniting source. There is a wide range of fuels with different flammable limits. Higher temperatures and/or higher pressures generally increase the range over

TABLE 1

Examples of Flammable Ranges

Fuel	Lower Limit (%)	Upper Limit (%)
Gasoline vapor	1.4	7.6
Methane (natural gas)	5.0	15.0
Propane	2.2	9.6
Butane	1.9	8.5
Hydrogen	4.0	75.0
Acetylene	2.5	81.0
Carbon monoxide	12.5	74.0

which a given fuel-air mixture is capable of being ignited and burned. Beyler (2002) gives an extensive table of values for many gaseous fuels, and discusses temperature and pressure effects on the LFL and UFL. Drysdale (1998) discusses the chemistry and physics of fire. Slye (2002) concentrates on flammable and combustible liquids, and Babrauskas (2003) provides a wealth of information about the ignition problem. Crowl and Louvar (1990) address chemical safety issues of, among other things, limits of flammability of gaseous mixtures, temperature and pressure effects, and explosions.

Most gases and vapors burn with a chemical reaction that uses the oxygen readily available in the surrounding air. Gases that do not burn include nitrogen, helium, carbon dioxide, steam, and carbon tetrachloride. Some chemical reactions occur without oxygen, using, for example, chlorine or nitrous oxide that interact with the fuel. Mixtures of gaseous or vapor fuels with air can be too lean or too rich in the relative amount of fuel to permit burning.

Very rapid burning will cause explosions. Explosions are produced when oxygen-fuel mixtures are within flammable limits and an ignition source is present. In explosions, great volumes of gases are produced. These burning gases expand so rapidly and violently they produce heat, light, sound, and kinetic energy, which is associated with motion. For a fire to be produced, the following conditions must exist: there must be a fuel, that fuel must be heated to its ignition temperature, and there must be sufficient oxygen. Two concepts are used: the fire triangle and the fire tetrahedron.

The Fire Triangle

The fire triangle illustrates the interdependency of the three ingredients necessary to sustain fire. If one side of the triangle is missing, the figure is incomplete. In like manner, if one of the three ingredients (oxygen, heat, or fuel) of fire is missing, there can be no fire.

While the fire triangle is useful in teaching the basic elements of fire, the observed phenomena and the research performed on the theory of fire suggests a more complicated reaction taking place within the flame than can be accounted for with simply the fire triangle. These observations suggest there is another basic ingredient necessary to sustain fire. A need for a *fourth side of the triangle* became evident. Research identified this fourth ingredient as the *chemical or uninhibited chain reaction*, resulting in the production of free radicals. Certain combinations of elements, which are unstable and chemically highly reactive, are produced in all flames and are a necessary part of the burning reaction. The model chosen to illustrate this fourth side is the fire tetrahedron.

The Fire Tetrahedron

The tetrahedron is a three-dimensional, four-sided pyramid with each side being a triangle. Each side of the tetrahedron is labeled with one of the four ingredients of fire. Similar to the edges of the fire triangle, the fourth side is labeled "chemical chain reaction," or "uninhibited chain reaction."

Note that uninhibited chain reaction occurs when excess heat from the exothermic reaction radiates back to the fuel to produce vapors and causes ignition in the absence of the original ignition source. From this tetrahedron theory has come the halon extinguishing system, which functions by interrupting the chain reaction, thus breaking the tetrahedron and extinguishing the fire without removing either heat, fuel, or oxygen. Halon is a coined and generic name for several halogenated hydrocarbon compounds that have found significant use as fire-extinguishing agents. The fire-extinguishing mechanism is that heat breaks down the halogenated compound into reactive hydride radicals that interrupt the chain reaction of the combustion process.

Chemical Reactions

This discussion is restricted to two typical types of fuel: hydrocarbon fuels of the form C_xH_y and alcohol and other oxygen-containing fuels of the form $C_xH_yO_z$. The general burn involves the *air-bracket* multiplier (m) and how this compares with the *correct* or *stoichiometric* multiplier (m_s). The reaction is then classified as

1. Stoichiometric $m = m_s$
2. Fuel rich $m < m_s$ and $\phi = m_s/m > 1$

3. Fuel lean $m > m_s$ and $\phi = m_s/m < 1$
4. Minimum oxygen $m = m_{min}$

Here m is the multiplier of the air bracket in the chemical balance equation. Also m_s is the correct or stoichiometric multiplier of the air bracket, for which there is just enough oxygen to complete the carbon and hydrogen reactions to carbon dioxide and water vapor, and there is no excess oxygen in the products. With this correct multiplier, there is enough oxygen that no carbon monoxide is in the products unless there is inadequate mixing of the fuel and oxygen. The symbol $\phi = m_s/m$ is the equivalence ratio, which is a nondimensional number representing the degree of fuel richness that the reactant mixture has with respect to fuel and oxygen.

Focusing on 1 kg-mole of fuel (an amount of fuel equal to its molecular weight written in kg), the equations that follow apply.

Hydrocarbon fuels:

$$C_xH_y + m(O_2 + fN_2) \rightarrow n_1CO + n_2CO_2 + n_3H_2O + n_4N_2 + n_5O_2 + n_6H_2 \quad (1)$$

Alcohol and oxygen-containing fuels:

$$C_xH_yO_z + m(O_2 + fN_2) \rightarrow n_1CO + n_2CO_2 + n_3H_2O + n_4N_2 + n_5O_2 + n_6H_2 \quad (2)$$

Here m determines the amount of oxidant supplied relative to the amount of fuel supplied. The value of f determines the amount of nitrogen relative to the amount of oxygen in the air. The value of $f = 3.76$ gives the familiar combination that represents standard air, with 21 percent oxygen and 79 percent nitrogen, on a volume basis. The n_1, n_2, n_3, n_4, n_5, n_6 values are the molar (kg-mole) amounts of product species from the burning of 1 kg-mole of fuel. Neglecting dissociation (the partial break up at higher temperatures of CO_2 to CO and O_2, and the partial break up of H_2O to H_2 and O_2), $n_6 = 0$ for all cases considered.

Hydrocarbon fuels have the form C_xH_y, where x and y are known for the fuel of interest. The most familiar paraffins have $y = 2x + 2$. Examples are methane CH_4, propane C_3H_8, and butane C_4H_{10}. Alcohol fuels have the form $C_xH_yO_z$. Examples are methanol CH_3OH and ethanol C_2H_5OH. Other well-known fuels, such as cellulose ($C_6H_{10}O_5$), are of the form $C_xH_yO_z$ and can be handled in the same fashion.

If stoichiometric burning occurs, the equation becomes

$$C_xH_yO_z + (x + y/4 - z/2)(O_2 + fN_2) \rightarrow n_2CO_2 + n_3H_2O + n_4N_2 \quad (3)$$

and it is clear that this amount of oxidant produces entirely CO_2 and H_2O in the product stream, their amounts are easily determined, and there is no CO or O_2 or H_2 in the products with

$$m = m_s = x + y/4 - z/2$$
$$n_2 = x$$
$$n_3 = y/2$$
$$n_4 = mf$$
$$n_1 = n_5 = n_6 = 0$$

If fuel rich burning occurs, the equation becomes

$$C_xH_yO_z + m(O_2 + fN_2) \rightarrow n_1CO + n_2CO_2 + n_3H_2O + n_4N_2 \quad (4)$$

and it is deducible that there is no O_2 or H_2 in the products. Hence

$$n_1 = 2(m_s - m)$$
$$n_2 = 2(m - m_{min})$$
$$n_3 = y/2$$
$$n_4 = mf$$
$$n_5 = n_6 = 0$$

with the following formulas for stoichiometric and minimum air-bracket multipliers:

$$m_s = x + y/4 - z/2$$
$$m_{min} = x/2 + y/4 - z/2$$

If fuel lean burning occurs, then

$$C_xH_yO_z + m(O_2 + fN_2) \rightarrow n_2CO_2 + n_3H_2O + n_4N_2 + n_5O_2 \quad (5)$$

Now there is no CO or H_z in the products and the product stream amounts are

$$n_1 = 0$$
$$n_2 = x$$
$$n_3 = y/2$$
$$n_4 = mf$$
$$n_5 = m - m_s$$
$$n_6 = 0$$

If minimum oxidant burning occurs, then

$$C_xH_yO_z + (x/2 + y/4 - z/2)(O_2 + fN_2) \rightarrow n_1CO + n_3H_2O + n_4N_2 \quad (6)$$

It is seen that this amount of oxidant is just enough to oxidize all the fuel to CO and H_2O and there is no CO_2 or O_2 in the products (minimum amount of air required) with

$$m = m_{min} = x/2 + y/4 - z/2$$
$$n_1 = x$$
$$n_3 = y/2$$
$$n_4 = mf$$
$$n_2 = n_5 = n_6 = 0$$

Note that $n_3 = y/2$ and $n_6 = 0$ occurs in all the cases, the no-dissociation assumption prohibiting hydrogen from occurring in the products.

Temperature of the Combustion Products

The mean specific heat of the products C_{Pm}(J/kg K) can be calculated by taking the molar weighted average of the molar specific heats of the products. Approximating this mean specific heat at 1000 K requires individual species values of specific heat C_P at 1000 Kelvin. Table 2 provides these values along with the molecular weight (MW) of the product species of interest. The mean specific heat can be expressed as

$$C_{Pm} = [n_1 MW_{CO}\, C_{P,CO} + \text{etc}]/M \quad (7)$$

where

MW_{CO} = Molecular weight of CO, etc

M, as the total mass of the product per kg-mole of fuel (kg), is determined by

$$M = [n_1 MW_{CO} + \text{etc}]$$

The specific gas constant for the product mixture R is a constant and can be calculated from the relation

$$R = \bar{R}/MW \quad (8)$$

where

$\bar{R}$ is the universal gas constant, the value of which is 8314 J/kg-mole K

MW is the mean mixture molecular weight of the products in kg/kg-mole K

$$MW = [n_1 MW_{CO} + \text{etc}]/n_T, \text{ where}$$
$$n_T = n_1 + n_2 + n_3 + n_4 + n_5 + n_6$$

The adiabatic flame temperature is defined as the temperature of the combustion products when all the energy released by the fuel on burning goes to heat up the products. The adiabatic flame temperature (AFT) of the products is

$$AFT = 298 + \Delta T \quad (9)$$

where

ΔT = temperature rise = (heat liberated)/[(total mass) (C_{Pm})]

$$\Delta T = h_{rp}/[(M)(C_{Pm})]$$

Here M is the total mass (in kg) of the product per kg-mole of fuel, C_{Pm} is the mean specific heat of the products (in J/kg K), and h_{rp} is the heat liberated (from reactants to products) per kg-mole of fuel (in J/kg-mole of the fuel). That is, h_{rp} for the fuel is equal to $H_c \times MW_{fuel}$, where H_c is the heat of combustion of the fuel in mass units (that is, in J/kg) and MW_{fuel} is the molecular weight of the fuel (in kg/kg-mole). Lilley (2007) gives the details.

Flammability Limits

As introduced above, most substances have both a lower (lean) and an upper (rich) flammable (explosive) limit, called the LFL and UFL, respectively. Some typical values are given in Table 1, for fuel-in-air mixtures on a volume percent basis. Either too much or too little fuel in the vapor-air mixture can prevent burning. Higher temperatures and/or higher pressures

TABLE 2

Molecular Weight and Specific Heats C_P at 1000K (J/kg K)

Species	MW	C_p (J/kg K)
CO	28	1185
CO_2	44	1234
H_2O	18	2290
N_2	28	1167
O_2	32	1090
H_2	2	1510

generally increase the range over which a given fuel-air mixture is capable of being ignited and burned [see discussion and figures in Slye (2003), for example].

For some situations, it may be necessary to estimate the flammability limits without experimental data. Then the approximate expressions are recommended by Crowl and Louvar (1990):

$$\text{LFL} = 0.55\ C_{st} \tag{10}$$

$$\text{UFL} = 3.50\ C_{st} \tag{11}$$

where C_{st} is the stoichiometric volume percent of fuel in fuel plus air. In fact, C_{st} may be found directly from the stoichiometric chemical balance equation, (Equation 3) via

$$C_{st} = \left[\frac{1}{1 + m_s/0.21)}\right] \tag{12}$$

and then estimates of the lower and upper flammability limits follow from Equations 10 and 11. For many fuels, the estimates are very good compared with accepted data. However, for some alkanes, although the estimate of the LFL is reasonable, the estimate for the UFL is higher than found in practice. For example, with propane C_3H_8 the estimates compute as 2.2 to 14.1 percent, whereas the familiar values are 2.2 to 9.6 percent, these being the values given by NFPA (Cote 2003) and elsewhere.

Evolution of Flammable Material

The Basics

The release of flammable material falls within the science of chemical process safety, health and loss prevention, topics addressed in SFPE (2002), NFPA (Cote 2003), Lees (1980), Crowl and Louvar (1990) and AIChE (1990). These and other authors discuss the many principles, guidelines, and calculations that are necessary for safe design and operation, and analysis of failures, including fires and explosions, vessel overpressure protection, hazards identification and risk assessment, source models, and dispersion modeling. For example, material may be released from holes and cracks in tanks and pipes, from leaks in flanges, pumps, and valves, and from a large variety of other sources.

Source models represent the material release process. They provide useful information for determining the consequences of an accident, including the rate of material release, the total quantity released, and the physical state of the material. Source models are constructed from fundamental or empirical equations, which represent the physicochemical processes occurring during the release of materials. Several basic source models are available, each applicable to its particular release scenario:

1. flow of liquids through a hole
2. flow of liquids through a hole in a tank
3. flow of liquids through pipes
4. flow of vapor through holes
5. flow of vapor through pipes
6. flashing liquids
7. liquid pool evaporation or boiling

Problem parameters come into play in determining the amount of release or rate of release. The purpose of the source model is to determine the form of material released (solid, liquid, or vapor), the total quantity of material released, and the rate at which it is released.

Energy Balance Equation

Source releases are driven by pressure differences, and resulting velocities depend on the density of the fluid (liquid or gas), whether or not the fluid is compressible, and whether or not choking (maximum flow) conditions occur. The volume flow rate (m^3/s or ft^3/s) and mass flow rate (kg/s or lbm/s) may be determined via introduction of the flow passage cross-sectional area and density of the flowing material. A correction factor is introduced to cater for the flow being less than ideal, via multiplication by C_0 (called the discharge coefficient, a number less than one). A mechanical energy balance describes the various energy forms associated with flowing fluids,

$$\int \frac{dP}{\rho} + \Delta\left(\frac{V^2}{2\alpha g_c}\right) + \frac{g}{g_c}\Delta z + F = 0 \tag{13}$$

where

P = the pressure (N/m^2 = Pa or lbf/ft^2)
ρ = the fluid density (kg/m^3 or lbm/ft^3)

V = the average instantaneous velocity of the fluid (m/s or ft/s)
g_c = the conversion factor constant (= 1 kg-m/N-s^2 or 32.2 lbm · ft/lbf-s^2)
α = the unitless velocity profile correction factor with the following values,
= 0.5 for laminar flow
= 1.0 for turbulent flow with almost a flat velocity profile
g = the acceleration due to gravity (= 9.81 m/s^2 or 32.2 ft/s^2)
z = the height above some datum (m or ft)
F = the net frictional loss term (J/kg or ft·lbf/lbm)

For incompressible liquids the density is constant and the first term in Equation 13 is replaced by $\Delta P/P$.

Flow of Liquid Through a Hole

Because no altitude change occurs, only the pressure and kinetic energy (by virtue of the square of the velocity) terms remain. There is a trade-off between pressure energy and velocity produced. That is, high pressure and zero velocity inside the container, with the resulting liquid escaping into zero gage pressure at a high velocity. In practice the driving pressure gives a lower exit velocity than the idealized equation gives, because of losses associated with the flow going through the area reduction. This is taken into account by multiplying the ideal answer by a number less than unity, this number being called the discharge coefficient (C_0). Some advice is given after Equation 15 about the value to use for this coefficient. The resulting equation for the velocity (m/s or ft/s) of fluid exiting the leak is

$$V = C_0\sqrt{\frac{2g_cP_g}{\rho}} \tag{14}$$

and the mass flow rate Q_m (in kg/s or lbm/s) due to a hole of area A (m^2 or ft^2) is given by

$$Q_m = \rho VA = AC_0\sqrt{2\rho g_cP_g} \tag{15}$$

The total mass of liquid spilled is dependent on the total time the leak is active. In these equations g_c is the familiar conversion factor (= 32.2 lbm-ft/lbf-s^2 in U.S. customary units and = 1 kg-m/N-s^2 in SI units) and P_g is the gage pressure pushing the fluid through the hole to an ambient zero gage pressure.

The discharge coefficient is a complicated function of the Reynolds number of the fluid escaping through the leak, the contour of approach to the exit, and the diameter of the hole. The following guidelines are suggested.

1. For sharp-edged orifices and for Reynolds numbers greater than 30,000, C_0 approaches the value 0.61. For these conditions, the exit velocity of the fluid is independent of the size of the hole.
2. For a well-rounded nozzle the discharge coefficient approaches unity.
3. For short sections of pipe attached to a vessel (with a length-diameter ratio not less than 3), the discharge coefficient is approximately 0.81.
4. For cases where the discharge coefficient is unknown or uncertain, use a value of 1.0 to maximize the computed flows.

Flow of Liquid Through a Hole in a Tank

When the hole in a tank is below the liquid surface level, there is an additional hydrostatic pressure due to depth acting as well as the pressure in the region just above the liquid's surface level. The resulting equation for the instantaneous velocity (m/s or ft/s) of fluid exiting the leak is

$$V = C_0\sqrt{2\left(\frac{g_cP_g}{\rho} + gh_L\right)} \tag{16}$$

where h_L (m or ft) is the liquid surface height above the leak. The instantaneous mass flow rate Q_m (in kg/s or lbm/s) due to a hole of area A (m^2 or ft^2) is given by

$$Q_m = \rho VA = \rho AC_0\sqrt{2\left(\frac{g_cP_g}{\rho} + gh_L\right)} \tag{17}$$

As the tank empties, the liquid height decreases and the velocity and mass flow rate will decrease.

Flow of Vapor through Holes

Free expansion leaks for gas (vapor) through holes differs from that of liquid flow because of the much lower density of the fluid and because it is compressible. The gas obeys the ideal gas law

$$P = \rho \bar{R} T / M \qquad (18)$$

where

P = the absolute pressure (N/m^2 = Pa or lbf/ft^2)
ρ = the fluid density (kg/m^3 or lbm/ft^3)
$\bar{R}$ is the universal gas constant
(= 8314 J/kg-mole K or 1545 ft-lbf/lbm-mole °R)
T = the absolute temperature (degrees K or °R)
M = the molecular weight of the gas (kg/kg-mole or lbm/lbm-mole)

For the isentropic (no heat transfer and no mechanical losses) expansion of the gas P/ρ^{γ} = constant where γ is the ratio of the heat capacities, $\gamma = C_p/C_v$. In general γ is the specific heat ratio of the gas emerging, which is equal to 1.67, 1.40, and 1.32 depending on whether the gas is monotomic, diatomic (or air), or triatomic, respectively. Introducing the discharge coefficient and keeping absolute pressures, yields the equation for mass flow rate (in kg/s or lbm/s) as

$$Q_m = C_0 A P_0 \sqrt{\frac{2 g_c M}{R_g T_0} \frac{\gamma}{\gamma - 1}\left[\left(\frac{P}{P_0}\right)^{2/\gamma} - \left(\frac{P}{P_0}\right)^{(\gamma+1)/\gamma}\right]} \qquad (19)$$

where P is the absolute pressure of the external surroundings, P_0 is the absolute pressure of the source (both in N/m^2 or lbf/ft^2), and $R_g = \bar{R}$ stands for the universal gas constant. The above equation holds only for flows that are not choked, that is flows with upstream absolute pressure less than about twice the downstream absolute pressure (which is usually the atmosphere). For high-pressure ratios (with upstream absolute pressure greater than about twice the downstream absolute pressure), the flow becomes sonic at the hole. Then the flow rate only depends on the upstream, driving, absolute pressure (P_0) and the mass flow rate (in kg/s or lbm/s), which is given by

$$(Q_m)_{\text{choked}} = C_0 A P_0 \sqrt{\frac{\gamma g_c M}{R_g T_0}\left(\frac{2}{\gamma + 1}\right)^{(\gamma+1)/(\gamma-1)}} \qquad (20)$$

where M is the molecular weight of the escaping vapor or gas, T_0 the absolute temperature of the source, and R_g the universal gas constant.

For sharp-edged orifices with Reynolds numbers greater than 30,000 (and not choked), a constant discharge coefficient C_0 of 0.61 is indicated. However, for choked flows, the discharge coefficient increases as the downstream pressure decreases. For these flows and for situations where C_0 is uncertain, a conservative value of 1.0 is recommended when the maximum (worst) possible scenario is being estimated.

Dispersion of Released Material

The Basics

The information about source release is required for any quantitative dispersion-model study. Dispersion models describe the airborne transport of materials away from the accident site. After a release, the airborne material is carried away by the wind in a characteristic plume or a puff. The maximum concentration occurs at the release point, which may or may not be at ground level. Concentrations downwind are less, due to turbulent mixing and dispersion of the substance with air. A wide variety of parameters affect atmospheric dispersion of toxic materials:

1. wind speed
2. atmospheric stability
3. ground conditions, buildings, water, trees
4. height of the release above ground level
5. momentum and buoyancy of the initial material released

Two types of vapor cloud dispersion models are commonly used: the plume and puff models, collectively known as the Pasquill-Gifford model, using dispersion coefficients that empirically specify the rates of spread of the dispersing material. The plume model describes the steady-state concentration of material released from a continuous source. The puff model describes the temporal concentration of material from a single release of a fixed amount of material. Models are available that permit concentration (kg/m^3) and volumetric concentration percent (%) to be calculated at location (x, y, z) as a function of time (t) after initial release. It is important to determine if the mixture is within the flammability limits at a nearby ignition source, in the case of fuel release. Useful references include AIChE (1990), ASME (1973), Crowl and Louvar (1990), Lees (1980), Hanna and Drivas (1987), and Lilley (1996). Sprays of liquid droplets and particles may be handled via computation of particle trajectories, including air resistance and wind effects, as outlined by Chow (1979) and Lilley (2008).

In both plume and puff models, it is common to utilize the dispersion coefficients, σ_x, σ_y, and σ_z. These represent the standard deviations of the concentration in the downwind, crosswind, and vertical

TABLE 3

Atmospheric Stability Classes for Use with the Pasquill-Gifford Dispersion Model

Wind Speed (m/s)	Day Radiation Intensity			Night Cloud Cover	
	Strong	Medium	Slight	Cloudy	Calm and Clear
<2	A	A–B	B		
2–3	A–B	B	C	E	F
3–5	B	B–C	C	D	E
5–6	C	C–D	D	D	D
>6	C	D	D	D	D

Stability classes for puff model are A and B, unstable; C and D, neutral; E and F, stable.

(Adapted from well-known sources, see Lilley 2004)

TABLE 4

Equations and Data for Pasquill-Gifford Dispersion Coefficients

Equations for Continuous Plumes

Stability Class	σ_y from fitted equation (m)
A: Unstable	$\sigma_y = 0.493x^{0.88}$
B: Unstable	$\sigma_y = 0.337x^{0.88}$
C: Neutral	$\sigma_y = 0.195x^{0.90}$
D: Neutral	$\sigma_y = 0.128x^{0.90}$
E: Stable	$\sigma_y = 0.091x^{0.91}$
F: Stable	$\sigma_y = 0.067x^{0.90}$

Stability Class	x (m)	σ_y from fitted equation (m)
A: Unstable	100–300	$\sigma_z = 0.087x^{1.10}$
	300–3000	$\log_{10}\sigma_z = 1.67 + 0.902 \log_{10}x + 0.181(\log_{10}x)^2$
B: Unstable	100–500	$\sigma_z = 0.135x^{0.95}$
	500–2×10^4	$\log_{10}\sigma z = -1.25 + 1.09 \log_{10}x + 0.0018(\log_{10}x)^2$
C: Neutral	100–10^5	$\sigma_z = 0.112x^{0.91}$
D: Neutral	100–500	$\sigma_z = 0.093x^{0.85}$
	500–10^5	$\log_{10}\sigma_z = -1.22 + 1.08 \log_{10}x - 0.06(\log_{10}x)^2$
E: Stable	100–500	$\sigma_z = 0.082x^{0.82}$
	500–10^5	$\log_{10}\sigma_z = -1.19 + 1.04 \log_{10}x - 0.070(\log_{10}x)^2$
F: Stable	100–500	$\sigma_z = 0.057x^{0.80}$
	500–10^4	$\log_{10}\sigma_z = -1.91 + 1.37 \log_{10}x - 0.119(\log_{10}x)^2$

Equations for Puff Releases

Stability Class	σ_y from fitted equation (m)	σ_z from fitted equation (m)
A and B: Unstable	$\sigma_y = 0.14x^{0.92}$	$\sigma_z = 0.53x^{0.73}$
C and D: Neutral	$\sigma_y = 0.06x^{0.92}$	$\sigma_z = 0.15x^{0.70}$
E and F: Stable	$\sigma_y = 0.02x^{0.89}$	$\sigma_z = 0.05x^{0.61}$

(Data and equations adapted from well-known sources, see Lilley 2004.)

(x, y, and z) directions, respectively. The dispersion coefficients are a function of atmospheric conditions and the distance downwind from the release. The atmospheric conditions are classified according to six different stability classes shown in Table 3. The stability classes depend on wind speed and quantity of sunlight. During the day, increased wind speed results in greater atmospheric stability, while at night the reverse is true. This is due to a change in vertical temperature profiles from day to night. The Pasquill-Gifford dispersion coefficients ($\sigma_x = \sigma_y$ for horizontal dispersion and σ_z for vertical dispersion) for a continuous source are given in Table 4, showing fitted curves simplified from data in Lees (1980) and Crowl and Louvar (1990), and described further with examples of their use in Lilley (2004).

Plume Model

Consider only release at ground level. This simplifies immensely the resulting equations for ground-level concentration further downstream and sideways from the source. For a plume with continuous steady state source (at ground level) of constant mass release rate Q_m (kg/s) and a wind velocity u(m/s) in the x-direction, the ground concentration is given at $z = 0$ by

$$C(x,y,0) = \frac{Q_m}{\pi\sigma_y\sigma_z u}\exp\left[-\frac{1}{2}\left(\frac{y}{\sigma_y}\right)^2\right] \quad (21)$$

where C represents mass concentration (kg/m^3) of the released material in air. This is at the (x, y, z) location. Here x is in the wind-driven direction, y is sideways, and $z = 0$ is the vertical direction. Because the values of the dispersion coefficients σ_x, σ_y, and σ_z are functions of the downstream distance x, it is clear that the concentration given by Equation 21 reduces with downstream distance x and sideways distance y. This concentration can be transferred to mass concentration (comparing with the density of the carrying medium air being approximately 1.2 kg/m^3), and to parts per million (ppm) by mass. The concentration can also be transferred to volumetric concentration, via introduction of the molecular weights of the dispersed material and the air (molecular weight of air being about 29 kg/kg-mole K), and hence to parts per million (ppm) by volume. The concentration along the centerline of the plume directly downwind is given at $y = z = 0$ and reduces Equation 21 to

$$C(x,0,0) = \frac{Q_m}{\pi\sigma_y\sigma_z u} \quad (22)$$

Observe that for continuous ground level release the maximum concentration occurs at the release point.

Puff Model

As with the plume model, dispersion coefficients σ_x, σ_y, and σ_z are also utilized for puff release of material, with the values given in Table 4 and figures in Lilley (2004). Again, only ground-level release is considered for simplicity. The downstream concentrations on the ground ($z = 0$) for a puff of mass release Q^*_m (kg) with a wind velocity u(m/s) in the x-direction is given by

$$C(x,y,0,t) = \frac{Q^*_m}{\sqrt{2}\pi^{3/2}\sigma_x\sigma_y\sigma_z}\exp\left[-\frac{1}{2}\left(\frac{x-ut}{\sigma_x}\right)^2 + \frac{y^2}{\sigma_y^2}\right] \quad (23)$$

This equation gives the mass concentration (kg/m^3) of the released material at a distance x downstream in the direction of the wind, and a crosswise distance y to the side (both x and y being distances in meters). The ground-level concentration along the x-axis is given at additionally imposing that $y = 0$ and Equation 23 reduces to

$$C(x,0,0,t) = \frac{Q^*_m}{\sqrt{2}\pi^{3/2}\sigma_x\sigma_y\sigma_z}\exp\left[-\frac{1}{2}\left(\frac{x-ut}{\sigma_x}\right)^2\right] \quad (24)$$

The center of the cloud is found at coordinates (ut, 0, 0) because, at time t, the distance $x = ut$. Thus, the concentration at the center of this moving cloud is given by

$$C(ut,0,0,t) = \frac{Q^*_m}{\sqrt{2}\pi^{3/2}\sigma_x\sigma_y\sigma_z} \quad (25)$$

Burning Rates

Background

The *burning rate* is usually expressed either as a mass loss rate (kg/s) or as a heat release rate (kW) with the latter being more commonly used. Calculation procedures for fire effects in enclosures require knowledge of the energy release rate of the burning fuel. The term *energy release rate* is frequently used interchangeably with *heat release rate*, and it is usually expressed in units of kilowatts (kW) and symbolized by $\dot{Q}$. Major references are the *Fire Protection Handbook* (NFPA 2003) and the *Handbook of Fire Protection Engineering* (SFPE 2002). The energy release rate from burning fuels cannot be predicted from basic measurements of material properties. It depends on the fire environment, the manner in which the fuel is volatilized, and the efficiency of the combustion. Therefore, one must rely on available laboratory test data. In addition, knowledge of the complete history of the energy release rate may be required for many situations. This is particularly desirable where the fuel package exhibits unsteady burning. For those cases where only limiting conditions or worst-case analysis is required, it may be reasonable to assume that the fuel is burning at a constant rate, which simplifies the calculation considerably.

Budnick, et al. (2003) discuss simplified calculations for enclosure fires in their chapter in the NFPA handbook (NFPA 2003). They show the simple calculations for evaluating fire conditions in enclosures during the preflashover fire-growth period. They include discussion about the maximum mass loss rate at ventilation-limited burning conditions. The rigorous treatment of energy release rates is available for selected material types such as wood cribs, wood and plastic slabs, and liquid pool fires where experimental correlations have been established. Section 3, Chapter 15, in the SFPE *Handbook of Fire Protection Engineering* (SFPE 2002), provides a detailed discussion of the prediction of burning rates for liquid pool fires. Data on mass loss rates for selected fuel packages are available in several publications: Alpert and Ward (1983), Babrauskas and Krasny (1985), Babrauskas, et al. (1982), and Lawson, et al. (1984). Also, detailed discussions of energy release rates for specific fuels are available in publications by Babrauskas (1985), and Lawson and Quintiere (1985).

Babrauskas, in his chapter in the SFPE handbook (SFPE 2002), also discusses heat release rates and heat flux rates in full-scale, intermediate-scale, and bench-scale experimental measurements. He discusses modeling of these, and the prediction of full-scale heat release rates from bench-scale data. He usefully includes heat release rates (in kW) versus time (in seconds) for a large variety of real products, including categorizing a great deal of previous research information into sections dealing with pools (with liquid or plastic fuel), cribs (regular arrays of sticks), wood pallets, upholstered furniture, mattresses, pillows, wardrobes, television sets, Christmas trees, curtains, electric cable trays, trash bags and containers, and wall and ceiling lining materials. Thus, a wealth of information is available with detailed referencing to the original data source.

Also, data are available for the heat release rate versus time for many items in Babrauskas and Grayson (1992). Babrauskas and Williamson (1979), and Pettersson, et al. (1976) introduce postflashover fires in a worst-case approach and a schematized approach respectively. Where an exact burning rate is not required, the worst-case approach can be applied. Also, the schematized approach can be used where all of the burning-rate information is expressed solely as a fuel loading. Janessens and Parker (1992) developed the principle of oxygen consumption calorimetry as a measurement technique. More researchers have studied burning rate, and Babrauskas (2002) summarizes the available data in the SFPE handbook (SFPE 2002).

Theory

The energy generated by the fire is the primary influence on the temperature in a compartment fire, and much research has been conducted in characterizing the energy release rate of many fuels under a variety of conditions. The rate of energy release is equal to the mass loss rate of the fuel times the heat of combustion of the fuel (see Budnick, et al. 2003).

$$\dot{Q} = \dot{m}_f \times \Delta H_c \tag{26}$$

where

$\dot{Q}$ = energy release rate of the fire (kW)
$\dot{m}_f$ = mass burning rate of the fuel (kg/s)

ΔH_c = effective heat of combustion of the fuel (kJ/kg)

The effective heat of combustion is the heat of combustion that would be expected in a fire where incomplete combustion takes place. This is less than the theoretical heat of combustion as measured in the oxygen bomb calorimeter. The effective heat of combustion is often described as a fraction of the theoretical heat of combustion.

The most common method for measuring heat release rate is known as oxygen consumption calorimetry. The basis of this method is that for most gases, liquids, and solids, a more or less constant amount of energy (heat) is released per unit mass of oxygen consumed. This constant has been found to be 13,100 kJ per kilogram of oxygen consumed and is considered to be accurate with very few exceptions to about ±5% for many hydrocarbon materials (see Huggett 1980). This corresponds to about 3,000 kJ per kilogram of air, called the *heat of combustion of air*. See Babrauskas (2002) for the heat of combustion of various burning items.

In fuel-controlled fires, there is sufficient air to react with all the fuel within the compartment. In ventilation-controlled fires, there is insufficient air within the compartment, and some of the pyrolysis products will leave the compartment, possibly to react outside the compartment. For calculating the temperatures produced in compartment fires, the primary interest is in the energy released within the compartment.

The pyrolysis rate of the fuel depends on the fuel type, on its geometry, and on the fire-induced environment. The energy generated in the compartment by the burning pyrolysis products then depends on the conditions (temperature, oxygen concentration, etc.) within the compartment. Although the processes involved are complex, and some are not well understood, there are two cases where some simplifying assumptions can lead to useful methods for the approximation of energy released by the fire:

1. ventilation-limited fires
2. nonventilation-limited fires

The following equation for stoichiometric fuel pyrolysis can be used to estimate the mass loss rate of the fuel at which ventilation-limited burning will occur.

$$\dot{m}_{st} = \frac{1}{r_s}(0.5)A_0\sqrt{H_0} \qquad (27)$$

where

$\dot{m}_{st}$ = stoichiometric mass loss rate of the fuel, burning rate (kg/s)
r_s = stoichiometric air/fuel mass ratio
A_0 = area of ventilation opening (m^2)
H_0 = height of ventilation opening (m)

This equation actually follows from the maximum air mass flow rate (kg/s of air) associated with the ventilation limit through an opening of area A_0 and height H_0, as discussed later in the section "Ventilation Limit" and Equation 50. Equation 27 simply divides the air mass flow rate by the stoichiometric air-fuel mass ratio to obtain the mass flow burning rate (kg/s) of the fuel. For wood fuel, $r_s = 5.7$. For values of r_s for other materials, see NFPA (Cote 2003) and Budnick, et al. (2003). For most hydrocarbon fuels, $r_s = 15$ is a good approximation for air-fuel mass ratio.

Comments

The ventilation-limited burning rate equation may be applied to a compartment containing a single typical rectangular ventilation opening. This equation is primarily intended to be used in evaluating the fully developed fire, which mostly happens in the post-flashover fire. The fully developed fire can be simplified with a condition of constant energy release rate. The approximate method produces estimates generally accurate to within ±5 percent and typically within ±3 percent. However, this method is not recommended for accurate prediction of the initial start-up phase of fires in which plenty of room air is available.

Babrauskas (2002) also discusses the ability to predict burning rates of full-scale combustibles. This rate varies greatly with the combustible. In no case is it yet possible to predict the burning rates of practical combustibles simply on the basis of thermophysical and thermochemical data. Even in the *ideal* cases of a liquid pool, the relationships require actual test data.

Burning Rates of Typical Items

Emphasis is often placed on the growth phase of the fire. Slow, medium, fast, and ultrafast fire growths may be specified by the t^2-fire growth model, where, after an initial incubation period,

$$\dot{Q} = \alpha_f (t - t_0)^2 \tag{28}$$

where α_f is a fire-growth coefficient (kW/s^2) and t_0 is the length of the incubation period (s). The coefficient α_f appears to lie in the range from 10^{-3} kW/s^2 for very slowly developing fires to 1 kW/s^2 for very fast fire growth. The incubation period (t_0) will depend on the nature of the ignition source and its location, but data are now becoming available (see Babrauskas 1984) on fire growth rates on single items of furniture (an upholstered chair, a bed, etc.) that may be quantified in these terms. Suggested values for the coefficient α_f are also given in the formula section of Makefire—a subset of the FPETool and FASTLite computer programs. The specification there for the fire-growth coefficient α_f (kW/s^2) is

Slow	0.002778 kW/s^2
Medium	0.011111 kW/s^2
Fast	0.044444 kW/s^2
Ultrafast	0.177778 kW/s^2

and these correspond to growth times of the fire from zero size to 1 MW total heat output in

Slow	600 seconds
Medium	300 seconds
Fast	150 seconds
Ultrafast	75 seconds

Experimental furniture calorimeter data are available for a variety of items, giving heat release rate $\dot{Q}$ (kW) versus time (seconds). The modeled data for 34 items are given directly in numerical form in the extensive Table 5. Each of these sets of data is in conformity with several parameters that completely characterize the situation:

t_0 time to the onset of ignition
t_{1MW} time to reach 1 MW
t_{lo} level-off time

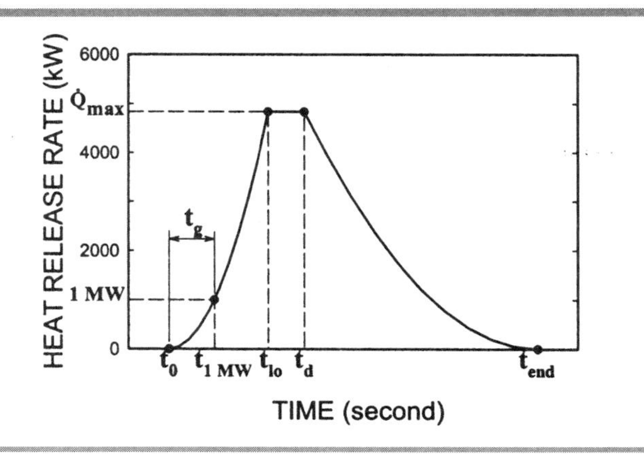

FIGURE 1. Heat release rate versus time in t^2-fire characterization (Adapted from Lilley 2004)

t_d time at which $\dot{Q}$ decay begins
t_{end} time at which $\dot{Q}$ equals zero

Notice that both the ascent and descent are characterized by t^2-fire activity:

$$\dot{Q} = \alpha_g t^2 \text{ where } t = t - t_0$$

$$\dot{Q} = \alpha_d t^2 \text{ where } t = t_{end} - t$$

where α_g and α_d are the fire-growth and fire-decay coefficients (kW/s^2), respectively. Figure 1 illustrates the situation.

These heat release rates $\dot{Q}$ (in kW) versus time t (in seconds) are active only in the growth ($t_0 \le t \le t_{lo}$) and decay ($t_d \le t \le t_{end}$), respectively. The maximum heat release rate $\dot{Q}_{max}$ (kW) occurs when $t_{lo} \le t \le t_d$. The growth time to reach 1 MW = 1000 kW of heat release rate $\dot{Q}$ is $t_{1MW} - t_0$ seconds, and this is related to the fire-growth parameter α_g (kW/s^2) via

$$\alpha_g = 1000/(t_{1MW} - t_0)^2 \tag{29}$$

Similarly the fire-decay parameter α_d (kW/s^2) is found via

$$\alpha_d = \dot{Q}_{max}/(t_{end} - t_d)^2 \tag{30}$$

Also note that the maximum heat release rate $\dot{Q}_{max}$ (kW) is related to other parameters via

$$\dot{Q}_{max} = 1000[(t_{lo} - t_0)/(t_{1MW} - t_0)]^2 \tag{31}$$

Careful perusal and interpretation of the information will enable the discerning reader to deduce what

TABLE 5

Heat Release Rate Versus Time for Various Items in t^2-Fire Characterization

Code	t_0	t_{1MW}	t_{lo}	t_d	t_{end}	Q_{max}	a_g	a_d
Wardrobe 1	0	35	60	90	500	2938.8	0.816327	0.017482
Wardrobe 2	0	40	100	110	140	6250.0	0.625000	6.944444
Wardrobe 3	0	30	70	80	400	5444.4	1.111111	0.053168
Wardrobe 4	0	90	150	160	450	2777.8	0.123457	0.033029
Wardrobe 5	0	150	170	670	2000	1284.4	0.044444	0.000726
Chair 1	0	1000	650	660	1900	422.5	0.001000	0.000275
Chair 2	0	100	140	160	500	1960.0	0.100000	0.016955
Chair 3	0	200	175	176	900	765.6	0.025000	0.001461
Chair 4	0	60	60	210	430	1000.0	0.277778	0.020661
Chair 5	0	350	275	475	1000	617.3	0.008163	0.002240
Chair 6	0	50	70	90	315	1960.0	0.400000	0.038716
Chair 7	0	2000	210	310	1000	11.0	0.000250	0.000023
Chair 8	0	400	275	475	1000	472.7	0.006250	0.001715
Chair 9	0	200	90	310	550	202.5	0.025000	0.003516
Chair 10	0	75	50	250	1250	444.4	0.177778	0.000444
Chair 11	0	425	347	367	1000	666.6	0.005536	0.001664
Chair 12	0	80	160	170	420	4000.0	0.156250	0.064000
Chair 13	0	200	187	200	500	874.2	0.025000	0.009714
Bed	0	1100	680	1080	1300	382.1	0.000826	0.007896
Lounge chair 1	0	350	170	220	350	235.9	0.008163	0.013960
Lounge chair 2	0	120	20	21	150	27.8	0.069444	0.001669
Lounge chair 3	0	275	230	430	900	699.5	0.013223	0.003167
Lounge chair 4	0	500	130	140	300	67.6	0.004000	0.002641
Loveseat 1	0	400	350	400	2000	765.6	0.006250	0.000299
Loveseat 2	0	80	130	160	400	2640.6	0.156250	0.045844
Loveseat 3	0	350	330	430	1500	889.0	0.008163	0.000776
Metal wardrobe 1	0	250	125	150	500	250.0	0.016000	0.002041
Metal wardrobe 2	0	50	40	47	200	640.0	0.400000	0.027340
Patient lounge chair	0	170	80	90	150	221.5	0.034602	0.061515
Sofa 1	0	500	260	460	800	270.4	0.004000	0.002339
Sofa 2	0	100	170	250	430	2890.0	0.100000	0.089198
F21 Chair	140	215	250	250	360	2151.1	0.177778	0.177778
F31 Loveseat	90	165	215	265	390	2777.8	0.177778	0.177778
F32 Sofa	75	150	205	270	400	3004.4	0.177778	0.177778

(Adapted from Lilley 2004)

the values of the defining parameters are. Finally, $\dot{Q}$ versus t is given by

$$
\begin{aligned}
\dot{Q} &= 0 && 0 \le t \le t_0 \\
\dot{Q} &= \alpha_g(t - t_0)2 && t_0 \le t \le t_{lo} \\
\dot{Q} &= \alpha_g(t_{lo} - t_0)2 && t_{lo} \le t \le t_d \\
\dot{Q} &= \alpha_d(t_{end} - t)^2 && t_d \le t \le t_{end} \\
\dot{Q} &= 0 && t_{end} \le t \le \text{Infinity}
\end{aligned}
$$

with the parameters taken directly from Table 5 for the particular item under consideration.

Simplified Calculations

Flame Heights

Estimates of flame height L can be important in determining exposure hazards associated with a burning fuel. Experimentally determined *mean* flame heights have been correlated by several researchers. A simple correlation for flame heights for pool or horizontal burning fuels has been developed by Heskestad (1983):

$$\frac{L}{D} = -1.02 + 15.6N^{1/5} \quad (32)$$

where

L = mean flame height (m)
D = diameter of fire source (m)

The nondimensional parameter N is given by

$$N = \left[\frac{C_p T_\infty}{g\rho_\infty^2(\Delta H_c / r_s)^3}\right]\frac{\dot{Q}^2}{D^5} \quad (33)$$

where

C_p = specific heat of air at constant pressure (kJ/kg K)
T_∞ = ambient temperature (K)
g = acceleration of gravity (9.81 m/s^2)
ρ_∞ = ambient air density (kg/m^3)
ΔH_c = heat of combustion (kJ/kg)
r_s = stoichiometric air/fuel mass ratio
$\dot{Q}$ = total heat release rate of the fire (kJ/s) or (kW)
D = diameter of the pool fire (m or ft)

For noncircular fuel packages, an effective D can be estimated by an equal area circle of diameter D using

$$D = 2(A_f/\pi)^{1/2} \quad (34)$$

where

D = effective diameter (m)
A_f = area of fire (m^2)

Under usual standard atmospheric conditions, this equation often can be simplified to the most often used formula

$$L = -1.02D + 0.23\dot{Q}^{2/5} \quad (35)$$

Because flames are unstable, the mean flame height (L) is generally taken to be the height above the fire source where the flame tip is observed to be at or above this point 50 percent of the time. The above correlation is considered suitable for pool fires or for horizontal surface burning.

Another expression for flame height is the NFPA 921 (2008) equation

$$H_f = 0.174(k\dot{Q})^{0.4} \quad (36)$$

where

H_f = flame length (m)
$\dot{Q}$ = heat release rate of fire (kW)

In this equation, the proximity of a wall or corner may be taken into account via the value of k. That is, $k = 1$ (no nearby walls), $k = 2$ (adjacent to a wall), and $k = 4$ (adjacent to a corner).

Plume Centerline Temperature and Velocity

The plume centerline excess temperature and velocity at elevations above the mean flame height can be estimated from the following equations:

$$\Delta T_0 = 9.1\left[\frac{T_\infty}{gC_p^2\rho_\infty^2}\right]^{1/3}\dot{Q}_c^{2/3}(Z - Z_0)^{-5/3} \quad (37)$$

$$U_0 = 3.4\left[\frac{g}{C_p\rho_\infty T_\infty}\right]^{1/3}\dot{Q}_c^{1/3}(Z - Z_0)^{-1/3} \quad (38)$$

where

ΔT_0 = excess centerline temperature $(T_g - T_\infty)$(K)
T_g = gas temperature (K)
T_∞ = ambient temperature (K)
g = acceleration of gravity (9.81 m/s^2)
C_p = specific heat of air at constant pressure [(kJ/kg)/K]
ρ_∞ = ambient air density (kg/m^3)
$\dot{Q}_c$ = convective heat release rate of the fire (kJ/s) or (kW)
Z = elevation above burning fuel fire source (m)
Z_0 = location of virtual fire source (m)
U_0 = centerline mean velocity (m/s)

In many cases of standard atmospheric conditions, these equations simplify to

$$\Delta T_0 = A\dot{Q}_c^{2/3}(Z - Z_0)^{-5/3} \quad (39)$$

$$U_0 = B\dot{Q}_c^{1/3}(Z - Z_0)^{-1/3} \quad (40)$$

where

A = 25.0 K m$^{5/3}$ kW$^{-2/3}$
B = 1.03 m$^{4/3}$ s^{-1} kW$^{-1/3}$

These two equations permit plume centerline excess temperature and velocity to be estimated at elevations above the flame. While methods exist to calculate excess temperatures and velocities at locations other than along the plume centerline, the highest confidence is placed on centerline estimates.

Modeling Correlations for Other Phenomena

Simplified empirical equations for the calculation of other events during the evolution of an enclosure fire are available in NFPA (Cote 2003) and SFPE (2002) handbooks. The algebraic equations are based principally on experimental correlations, and permit the user to make estimates of the results of a fire burning inside a given structure. The equations correlate experimental data and results versus other parameters for several cases of special interest for the fire dynamicist:

1. Room model for smoke layer depth and temperature
2. Atrium smoke temperature
3. Buoyant gas head pressure
4. Ceiling jet temperature
5. Ceiling plume temperature
6. Egress time
7. Fire/wind/stack forces on a door
8. Mass flow through a vent
9. Lateral flame spread
10. Law's severity correlation
11. Plume filling rate
12. Radiant ignition of a near fuel
13. Smoke flow through an opening
14. Sprinkler/detector response
15. Thomas' flashover correlation
16. Upper-layer temperature
17. Ventilation limit

The relevant equations are embodied in computer programs like FPETool, FASTLite and HAZARD, so that making calculations of fire behavior becomes straightforward, provided one appreciates correctly the physics involved. Further details appear in Bukowski, et al. (1989), Peacock, et al. (1994), and Portier, et al. (1996).

Radiation Ignition

Background

It is of interest to estimate the radiation transmitted from a burning fuel array to a target fuel positioned some distance from the fire to determine if secondary ignitions are likely. Will the second item ignite? This very important question is discussed at length in the detailed handbook on ignition, see Babrauskas (2003). The formulas for estimating the energy level required of a free burning fire to ignite a nearby item have been developed, on the basis that the exposed fuel item is not close enough to be in contact with the flame of the exposure fire. The relationships expressed by this procedure were developed empirically from tests and reported by Babrauskas (1981). Budnick, et al. (2003) discuss the simple expression in their chapter in the NFPA handbook (NFPA 2003). Tewarson (2002) explains fire initiation (ignition) and shows experimental data in the view of concepts governing generation of heat and chemical compounds in fires in the SFPE handbook (SFPE 2002). Flame spread tests are probably the best known fire-performance tests. The most widely used fire-performance test is the Steiner Tunnel Test. This test attempts to simulate the spread of fire across a plane surface and may include the imposition of a known external heat radiant flux (see Clarke 1991, Belles 2003). (For details about the test method as an ASTM standard, see ASTM 1969).

Theory

The radiant heat flux received from a flame depends on a number of factors, including flame temperature and thickness, concentration of emitting species, and the geometric relationship between the flame and the *receiver*. Although considerable progress is being made toward developing a reliable method for calculating flame radiation, a high degree of accuracy is seldom required in *real-world* fire engineering problems, such as estimating what level of radiant flux an item inside a plant might receive from a nearby fire in order to design a water spray system to keep the item cool. Two approximate methods will now be considered.

Considerations of inverse square distance lead to

$$\dot{q}''_0 = \frac{P}{4\pi d^2} = \frac{x_r \dot{Q}}{4\pi d^2} \quad (41)$$

where

$\dot{q}''_0$ = incident radiation on the target (kW/m²)
d = distance (radius) to target fuel (m)
P = total radiative power of the flame (kW)
x_r = radiative fraction (typically between 0.2 for alcohol fuel and up to 0.6 for petroleum product fuels)
$\dot{Q}$ = total heat release rate (kJ/s) or (kW)

In a second approximate method, the flame is approximated as a vertical rectangle and the radiant flux is calculated using *view factor* information. This takes into account the large size of the flame and its estimated (constant) temperature, with angles and orientations being accounted for appropriately between the flame and the target. Flux levels close to the flame are then more accurately handled (see Drysdale 1998 and the FASTLite computer program, see Portier, et al. 1996). Radiant heat can also be bounced off from other surfaces en route to the target. Experimental work led to simple equations related to the *nonpoint-source* assumption. Babrauskas (1981) reported his experimental results as three curves relating the energy level of a fire to the distance at which the incident radiation on an exposed target would be 10, 20, or 40 kW/m². These incident radiations are considered the approximate levels of incident radiation necessary for ignition of materials that are easily ignited, within the normal range of ignitability, or difficult to ignite, respectively.

- ***Easily ignited materials*** are those that respond rapidly to incident energy: ignite when they receive a radiant flux of 10 kW/m² or greater. Typical examples are thin curtains, loose newsprint, or draperies.

$$\dot{Q} = 30.0 \times 10^{\frac{d+0.08}{0.89}} \quad (42)$$

- ***Normally ignited materials*** include those combustibles that have increased resistance to heating: ignite when they receive a radiant flux of 20 kW/m² or greater. These are typified by upholstered furniture and other materials with significant mass but small thermal inertia.

$$\dot{Q} = 30.0\left(\frac{d = 0.05}{0.019}\right) \quad (43)$$

- ***Difficult-to-ignite materials*** include those combustibles that have greater resistance to heating: ignite when they receive a radiant flux of 40 kW/m² or greater. Typical examples are 1/2-inch (0.013 m) or thicker materials with substantial thermal inertia such as wood and thermoset plastics.

$$\dot{Q} = 30.0\left(\frac{d = 0.02}{0.0092}\right) \quad (44)$$

Values of distance at which ignition occurs for a given fire heat release rate used by Equations 42 to 44 are smaller than those from the point-source approach of Equation 41. This shows that, in reality, one needs to get closer to a given flame for ignition than the point-source approximation would suggest. This is expected because, as you get close to a flame, the point-source approach will be less and less valid. In fact, a fraction of the heat released from the flame will be some distance away and sideways from the target, whereas the point-source approach assumes that all the heat released is directly ahead of the target. Hence, there is a recommendation (see "Comments" section) that the point source should be used only when $d/R > 4$ where R is the flame radius and d is the distance away from the target.

Comments

This procedure applies to conditions where the exposed item is too close to the exposing fire for that source to be considered a point source. Rather, the exposed item views the broad cross-section of a fire typical of that produced by a free-burning upholstered chair or couch. Usually x_r ranges from 20 to 60 percent, depending upon the fuel type. The distance d should be measured from the center of the flame. Experimental measurements indicate that the theoretical equation of inverse square distance has good accuracy for $d/R > 4$ where R is the flame radius, and the point-source nature of the heat from the flame is then a reasonable assumption. For radiation at $0.5 < d/R < 4$ refer to the SFPE handbook (SFPE 2002) for a more exact analysis. Babrauskas' simple empirical equations are deduced from 11 data points, ranging from 0.05 m to 1.4 m of distance d, with about ±10 percent error for difficult to ignite materials, and much greater accuracy for easily and normally ignitable materials.

Minimum Safe Distance from Flames

Earlier in the chapter information was provided about the burning rate (heat release rate versus time) of a single specified item in the burn room. What happens next? Either the item burns out without further damage to the surroundings, or one or more nearby items ignite and add fuel to the fire. This can occur by direct flame contact (if the second item is judged to be sufficiently close) or, more frequently, by radiant heat energy becoming sufficiently large on the surface of the second item. Direct-flame contact requires time to pyrolyze the fuel and time to heat the gases produced to their ignition temperature. The radiant-flux ignition problem is a very complicated issue, and depends on many factors. The radiant energy comes from the flame above the first item, the upper layer and room surfaces, but simplifying assumptions are sometimes used. As the radiant energy flux rate increases from the first item to the second, often a simple criterion for ignition of the latter is used. A good approximation is that the radiant heat flux (arriving on the surface of the second item) necessary to ignite the second item is

- 10 kW/m^2 for easily ignitable items, such as thin curtains or loose newsprint
- 20 kW/m^2 for normal items, such as upholstered furniture
- 40 kW/m^2 for difficult-to-ignite items, such as wood of 0.5-inch or greater thickness

For many enclosure fires, it is of interest to estimate the radiation transmitted from a burning fuel array to a target fuel positioned some distance from the fire to determine if secondary ignitions are likely.

Useful related calculations are exhibited in Table 6. These give the minimum safe distance from pool fires of different diameters, for the eight fuels being considered. Notice that the more volatile, quicker-burning gasoline has a much greater propensity to ignite remote objects than do the slower-burning and less radiant heat-releasing fuels. The data given show the heat release rate ($\dot{Q}$ in kW) and minimum safe distance (R_0 in meters) from pool fires of various sizes (square of side-length from 10 cm to 5 m) and fuels (methanol, ethanol, and so on). The minimum safe distances are associated with easy-, normal-, and difficult-to-ignite items requiring radiant heat-flux values of 10, 20, and 40 kW/m^2 arriving on their surface. Notice in Table 6 that

1. A radiative function x_r of 0.2 is used with the alcohols (methanol and ethanol) and 0.4 is used with the other fuels.
2. The burning rate $\dot{m}''$ (kg/m^2s) is reduced for pools of diameter less than 1 m, except in the case of alcohol fuels (which do not exhibit this reduction of burning rate phenomenon) and polypropylene and polystyrene (for which data are not available).
3. The equivalent diameter D is approximately 13 percent larger than the side length of a square.

In actuality, ignition is not immediate when the particular level of incident radiant heat flux reaches 10, 20, or 40 kW/m^2, respectively, for easy-, normal-, and difficult-to-ignite items. These values are used as simple rules of thumb in applied calculations (see Lilley 2004). Fundamental ignition principles, outlined for example in the SFPE handbook (SFPE 2002), suggest that, for fire initiation, a material must be heated above its critical heat flux (CHF) value (CHF value is related to the fire point). It was found that as the surface is exposed to heat flux, initially most of the heat is transferred to the interior of the material. The ignition principles suggest that the rate at which heat is transferred depends on the ignition temperature T_{ig}, ambient temperature T_a, material thermal conductivity k, material specific heat c_p, and the material density ρ. The combined effects are expressed by the thermal response parameter (TRP) of the material.

TABLE 6

Heat Release Rate and Minimum Safe Distance from Pool Fires of Various Sizes and Fuels

Area (m^2)	Pool Size	Fuel Type	Heating Value H in MJ/kg	Burning Rate $\dot{m}''_\infty$ in kg/m²s	Radiative Fraction x_r	Parameter $k\beta$	Burning Rate $\dot{m}''$ in kg/m²s	Heat Release Rate $\dot{Q}$ in kW	Flame Height (m)	Minimum Safe Distance (m)		
										Easy	Normal	Difficult
0.01	0.1 m	Methanol	20.0	0.017	0.2		0.017	3.4	0.26	0.07	0.05	0.04
	*0.1 m	Ethanol	26.8	0.015	0.2		0.015	4.0	0.29	0.08	0.06	0.04
		Gasoline	43.7	0.055	0.4	2.1	0.012	5.1	0.33	0.13	0.09	0.06
		Kerosene	43.2	0.039	0.4	3.5	0.013	5.5	0.34	0.13	0.09	0.07
		Jet A fuel	43.0	0.054	0.4	1.6	0.009	3.8	0.28	0.11	0.08	0.06
		PMMA	24.9	0.020	0.4	3.3	0.006	1.5	0.16	0.07	0.05	0.04
		Polypropylene	43.2	0.018	0.4		0.018	7.8	0.41	0.16	0.11	0.08
		Polystyrene	39.7	0.034	0.4		0.034	13.5	0.54	0.21	0.15	0.10
0.04	0.2 m	Methanol	20.0	0.017	0.2		0.017	13.6	0.42	0.15	0.10	0.07
	*0.2 m	Ethanol	26.8	0.015	0.2		0.015	16.1	0.47	0.16	0.11	0.08
		Gasoline	43.7	0.055	0.4	2.1	0.021	36.3	0.74	0.34	0.24	0.17
		Kerosene	43.2	0.039	0.4	3.5	0.021	36.8	0.74	0.34	0.24	0.17
		Jet A fuel	43.0	0.054	0.4	1.6	0.016	28.1	0.64	0.30	0.21	0.15
		PMMA	24.9	0.020	0.4	3.3	0.011	10.5	0.36	0.18	0.13	0.09
		Polypropylene	43.2	0.018	0.4		0.018	31.1	0.68	0.31	0.22	0.16
		Polystyrene	39.7	0.034	0.4		0.034	54.0	0.90	0.41	0.29	0.21
0.25	0.5 m	Methanol	20.0	0.017	0.2		0.017	85.0	0.78	0.37	0.26	0.18
	*0.5 m	Ethanol	26.8	0.015	0.2		0.015	100.5	0.88	0.40	0.28	0.20
		Gasoline	43.7	0.055	0.4	2.1	0.038	417.1	1.99	1.15	0.81	0.58
		Kerosene	43.2	0.039	0.4	3.5	0.034	362.7	1.85	1.07	0.76	0.54
		Jet A fuel	43.0	0.054	0.4	1.6	0.032	345.1	1.81	1.05	0.74	0.52
		PMMA	24.9	0.020	0.4	3.3	0.017	105.2	0.91	0.58	0.41	0.29
		Polypropylene	43.2	0.018	0.4		0.018	194.4	1.32	0.79	0.56	0.39
		Polystyrene	39.7	0.034	0.4		0.034	337.5	1.79	1.04	0.73	0.52
1	1 m	Methanol	20.0	0.017	0.2		0.017	340.0	1.22	0.74	0.52	0.37
	*1 m	Ethanol	26.8	0.015	0.2		0.015	402.0	1.38	0.80	0.57	0.40
		Gasoline	43.7	0.055	0.4	2.1	0.050	2178.7	3.83	2.63	1.86	1.32
		Kerosene	43.2	0.039	0.4	3.5	0.038	1652.3	3.31	2.29	1.62	1.15
		Jet A fuel	43.0	0.054	0.4	1.6	0.045	1940.2	3.60	2.49	1.76	1.24
		PMMA	24.9	0.020	0.4	3.3	0.020	486.0	1.58	1.24	0.88	0.62
		Polypropylene	43.2	0.018	0.4		0.018	777.6	2.15	1.57	1.11	0.79
		Polystyrene	39.7	0.034	0.4		0.034	1349.8	2.96	2.07	1.47	1.04
4	2 m	Methanol	20.0	0.017	0.2		0.017	1360.0	1.82	1.47	1.04	0.74
	*2 m	Ethanol	26.8	0.015	0.2		0.015	1608.0	2.11	1.60	1.13	0.80
		Gasoline	43.7	0.055	0.4		0.055	9614.0	6.71	5.53	3.91	2.77
		Kerosene	43.2	0.039	0.4		0.039	6739.2	5.52	4.63	3.28	2.32
		Jet A fuel	43.0	0.054	0.4		0.054	9288.0	6.59	5.44	3.84	2.72
		PMMA	24.9	0.020	0.4		0.020	1992.0	2.50	2.52	1.78	1.26
		Polypropylene	43.2	0.018	0.4		0.018	3110.4	3.44	3.15	2.22	1.57
		Polystyrene	39.7	0.034	0.4		0.034	5399.2	4.85	4.15	2.93	2.07
25	5 m	Methanol	20.0	0.017	0.2		0.017	8500.0	2.83	3.68	2.60	1.84
	*5 m	Ethanol	26.8	0.015	0.2		0.015	10050.0	3.42	4.00	2.83	2.00
		Gasoline	43.7	0.055	0.4		0.055	60087.5	13.01	13.83	9.78	6.91
		Kerosene	43.2	0.039	0.4		0.039	42120.0	10.52	11.58	8.19	5.79
		Jet A fuel	43.0	0.054	0.4		0.054	58050.0	12.75	13.59	9.61	6.80
		PMMA	24.9	0.020	0.4		0.020	12450.0	4.24	6.30	4.45	3.15
		Polypropylene	43.2	0.018	0.4		0.018	19440.0	6.19	7.87	5.56	3.93
		Polystyrene	39.7	0.034	0.4		0.034	33745.0	9.14	10.36	7.33	5.18

The minimum safe distances are associated with easy-, normal-, and difficult-to-ignite items requiring radiant heat flux values of 10, 20, and 40 kW/m² arriving on their surface.

(Adapted from Lilley 2004)

$$\text{TRP} = \Delta T_{ig} \sqrt{k \rho c_p} \qquad (45)$$

where ΔT_{ig} ($T_{ig} - T_a$) is the ignition temperature above ambient in degrees K, k is in kW/m-K, ρ is in kg/m^3, c_p is in kJ/kg-K, and TRP is in kW-s$^{1/2}$/m^2. The TRP is a very useful parameter for the engineering calculations to assess resistance of ignition and fire propagation in as-yet uninvolved items. The ignition principles suggest that, for thermally thick materials, the inverse of the square root of time to ignition is expected to be a linear function of the difference between the external heat flux and the CHF value.

$$\sqrt{\frac{1}{t_{ig}}} = \frac{\sqrt{4/\pi}\,(\dot{q}''_c - \text{CHF})}{\text{TRP}} \qquad (46)$$

where t_{ig} is time to ignition in seconds, $\dot{q}''_c$ is the external heat flux kW/m^2, and CHF is in kW/m^2. Most commonly used materials behave as thermally thick materials and satisfy this equation. This equation then rearranges to the following form, with the ignition temperature highlighted on the left-hand side:

$$t_{ig} = \frac{\pi}{4}\left[\frac{\text{TRP}}{\dot{q}'' - \text{CHF}}\right]^2 \qquad (47)$$

The CHF and the TRP values for materials derived from the ignition data measured in the Flammability Apparatus and the Cone Calorimeter, by Scudamore, et al. (1991), are given in Lilley (2000). He also shows in tables and figures how the ignition time (t_{ig}) may be determined from the heat flux ($\dot{q}''$) and the CHF and TRP. Readers are directed to that study to see fully how the size and material of a pool fire determines the total heat release rate ($\dot{Q}$), the heat flux ($\dot{q}''$) on a target fuel, and the time required for ignition to occur. Typical values of CHF are in the range of 10 to 40 kW/m^2 and typical values of the TRP are in the range 100 to 800 kW-s$^{1/2}$/m^2. The smaller values of these parameters give easier and faster ignition. Calculations from the ignition time required equation give results as shown in Tables 7, 8, and 9 for the cases of CHF equal to 10, 20, and 40 kW/m^2, respectively.

Fire Spread Rates

Background

Fire spread applies to the growth of the combustion process and includes surface flame spread, smoldering

TABLE 7

Ignition Time t_{ig} in Seconds for Critical Heat Flux CHF = 10 kW/m^2

External Heat Flux $\dot{q}''$ kW/m^2	Thermal Response Parameter (TRP) kW-s$^{1/2}$/m^2			
	100	200	400	800
15	314.2	1256.6	5026.5	20106.2
20	78.5	314.2	1256.6	5026.5
30	19.6	78.5	314.2	1256.6
40	8.7	34.9	139.6	558.5
50	4.9	19.6	78.5	314.2
100	1.0	3.9	15.5	62.1
150	0.4	1.6	6.4	25.6

(Adapted from Lilley 2004)

TABLE 8

Ignition Time t_{ig} in Seconds for Critical Heat Flux CHF = 20 kW/m^2

External Heat Flux $\dot{q}''$ kW/m^2	Thermal Response Parameter (TRP) kW-s$^{1/2}$/m^2			
	100	200	400	800
25	314.2	1256.6	5026.5	20106.2
30	78.5	314.2	1256.6	5026.5
40	19.6	78.5	314.2	1256.6
50	8.7	34.9	139.6	558.5
60	4.9	19.6	78.5	314.2
100	1.2	4.9	19.6	78.5
150	0.5	1.9	7.4	29.7

(Adapted from Lilley 2004)

TABLE 9

Ignition Time t_{ig} in Seconds for Critical Heat Flux CHF = 40 kW/m^2

External Heat Flux $\dot{q}''$ kW/m^2	Thermal Response Parameter (TRP) kW-s$^{1/2}$/m^2			
	100	200	400	800
45	314.2	1256.6	5026.5	20106.2
50	78.5	314.2	1256.6	5026.5
60	19.6	78.5	314.2	1256.6
70	8.7	34.9	139.6	558.5
80	4.9	19.6	78.5	314.2
100	2.2	8.7	34.9	139.6
150	0.6	2.6	10.4	41.5

(Adapted from Lilley 2004)

growth, and the fireball in premixed flame propagation. In flame spread—and in fire growth generally—the rate of spread is highly dependent on the temperatures imposed by any hot smoke layer heating the unburned surface, as well as gravitational and wind effects. The flows resulting from the fire's buoyancy or the natural wind of the atmosphere can assist (wind-aided) or oppose (opposed-flow) flame spread. Quintiere and Harkleroad (1984) derive formulas and describe material properties involved in the lateral speed of a fire spreading in a direction other than that impinged by flame from the burning material; generally this means lateral or downward spread from a vertical flame. Quintiere (1998) summarizes surface-steady flame spread:

1. Spread on solid surface
   a. Downward or lateral wall spread
   b. Upward or wind-aided spread
2. Spread though porous solid arrays
3. Spread on liquids

Some typical fire spread rates are

- smoldering: 0.001 to 0.01 cm/s
- lateral or downward on thick solids: 0.1 cm/s
- wind-driven spread through forest debris or bush: 1 to 30 cm/s
- upward spread on thick solids: 1.0 to 100 cm/s
- horizontal spread on liquids: 1.0 to 100 cm/s
- premixed flames: 10 to 100 cm/s (laminar) and up to 10×10^5 cm/s (detonations)

Magee and McAlevy (1971) studied the rate of flame spread over strips of filter paper of different orientations (inclined angles); see Drysdale (1998) for discussion. According to Hicks (2003), fire spread rarely occurs by heat transfer through, or structural failure of, wall and floor-and-ceiling assemblies. The common mode of fire spread in a compartmented building is through open doors, unenclosed stairways and shafts, unprotected penetrations of fire barriers, and non–fire-stopped combustible concealed spaces. Even in buildings of combustible construction, the common gypsum board or lath-on-plaster protecting wood stud walls or wood joist floors provides 15 to 30 minutes' resistance to a fully developed fire, as discovered in a standard fire test (see NBFU 1956). When such barriers are properly constructed and maintained and have protected openings, they will normally contain fires of maximum expected severity in light-hazard occupancies. However, no fire barrier will reliably protect against fire spread if it is not properly constructed and maintained and if openings in the barrier are not protected.

Fire can spread horizontally and vertically beyond the room or area of origin and through compartments or spaces that do not contain combustibles. Heated, unburned pyrolysis products from the fire will mix with fresh air and burn as they flow outward. This results in extended flame movement under noncombustible ceilings, up exterior walls, and through noncombustible vertical openings. This is a common way that fire spreads down corridors and up open stairways and shafts. Hicks (2003) explains fire spread and suggests fire protection in concealed spaces, vertical openings, room or suite compartments, corridors, building separations, roofs, and fire walls. Ramachandran (2002) shows stochastic models of fire growth to be governed by physical and chemical processes. Quintiere (2002a) discusses flame spread applied to the phenomenon of a moving flame in close proximity to the source of its fuel, originating from a condensed phase—that is, solid or liquid. Details are in their chapters of the SFPE handbook (SFPE 2002).

Theory

The concept of ignition temperature is key to understanding flame spread in simple, correct terms. Pilot ignition temperature must be taken into account, because a pilot (the flame itself) is always present. The following defines the position x_p at the ignition temperature T_{ig}. The temperature ahead of the flame, not affected by direct heating from the flame, is taken as T_s. The distance along the surface affected by the flame's heat transfer is defined as δ_f. In general, the flame can heat the region ahead of it in many ways. These depend on the mode of spread (orientation, wind) and on the nature of the solid or liquid fuel. An observer riding on the flame front, at position x_p, sees the new fuel coming toward him at the flame spread velocity V or the velocity the flame has in spreading along the fuel (which is at rest). The flame spread velocity formula states that the rate of energy supplied to this newly heated fuel, arriving at temperature T_{ig}, is equal to the net heat transfer rate from the burning region $\dot{q}$. Using a balance for spread, the steady flame spread velocity is

$$V = \frac{\dot{q}}{\rho c A(T_{ig} - T_s)} \qquad (48)$$

where

V = flame spread velocity (m/s)
ρ = the fuel density (kg/m3)
c = the fuel-specific heat (kJ/kg K)
A = the cross-sectional area of the burning material perpendicular to the flame travel direction (m^2)
$\dot{q}$ = rate of heat supplied (kW)
T_s = the fuel temperature beyond the range of the flame's heat (K)
T_{ig} = the fuel's ignition temperature (K)

Comment

These formulas should be considered as methods for use in making general estimates. Again, orientation, wind, and the nature of the fuel all make a difference. Most specific cases of flame spread can be derived from this formula by more carefully describing $\dot{q}$ and A (see Quintiere 1998 and 2002a).

Ventilation Limit

Background

One of the enclosure effects is the availability of oxygen for combustion. If the air in the space, as well as that drawn in through openings and that blown into the space by HVAC systems or other means, is insufficient to burn all the combustible products driven from the fuel package, only the amount of combustion supportable by the available oxygen can take place. This situation is referred to as ventilation-limited burning and, when it occurs, the combustible products driven from the fuel package and not burned in the room often burn when they combine with air outside the room, appearing as flame extensions from the room. Furthermore, ventilation-limited burning changes the mass loss rate.

Kawagoe and Sekine (1963) originally presented the idea that the fire heat release rate within a compartment could be limited by ventilation geometry. This idea has been followed, and many subsequent postflashover experiments have been performed—see, for example, Babrauskas (1979) and Fang and Breese (1980). Heskestad (2003) covers the application of venting practices to nonsprinklered buildings and Campbell (2003) covers fire modeling in ventilation-limited fires in the NFPA handbook (NFPA 2003). Emmons (2002) shows equations for the measurement of velocity, volume flow, and mass flow for vent flows relating to an orifice or nozzle, and vent flows for buoyant and nonbuoyant flows. Quintiere (1995) reviews vent-limited effects on zone models. Walton and Thomas (2002) show the simplified mass flow rate equation using the ventilation factor in the SFPE handbook (SFPE 2002).

Theory

Fires nearing flashover and postflashover fires have an interface between the upper and lower layers located near the floor; here flow rates reach their maximum for a given upper gas temperature. Rockett (1976) has shown that temperature dependence on the flow becomes small above 150 °C, and the flow into the compartment can be approximated as a constant times the so-called ventilation factor

$$A_0\sqrt{H_0} \qquad (49)$$

where

A_0 = area of opening (m^2)
H_0 = height of opening (m)

Rockett calculated values for this constant of 0.40 to 0.61 $kg/s \cdot m^{5/2}$, depending on the discharge coefficient of the opening. The value most commonly found in literature is 0.5 $kg/s \cdot m^{5/2}$ (Thomas and Heselden 1972). The very simple, useful, and well-known relationship for the air mass flow into an enclosure through an opening (with smoke flowing out of the upper part of the opening and air flowing in through the lower part of the opening) can thus be arrived at:

$$\dot{m}_a = 0.5 A_0 \sqrt{H_0} \tag{50}$$

where, additionally,

$\dot{m}_a$ = mass flow rate of air (kg/s)

The corresponding maximum heat release rate (in kW) in a ventilation-limited fire then evaluates to

$$\dot{Q}_{\max} = 3000 \times \dot{m}_a = 1500 \times A_0 \sqrt{H_0} \tag{51}$$

using the typical heat of combustion of air as 3,000 kJ/kg of air. This value follows from the air-fuel ratio by mass being about 15:1 and the heating value of a typical fuel being about 45,000 kJ/kg of fuel. The development is similar for other fuels burning in air, with different stoichiometric air-fuel ratios and different heating values similarly giving the same result for the heating value of air participating in the burn (approximately 3000 kJ/kg). The first use of this type of opening flow analysis for evaluating postflashover fire-test data is attributed to Kawagoe (1958). See Emmons (2002) for other calculations, including velocity and volume flow.

Comments

The equations are deduced from more than 30 data points for $0.02 \le A_0 \sqrt{H_0} \le 15$ in metric units of $m^{5/2}$ with ±20% error; see Babrauskas (1980) and Drysdale (1998). One can now estimate the air mass flow rate in through an opening by knowing only the area and the height, from bottom to top, of the vent. Equation 50 gives an estimate of the mass flow rate for fires where the gas temperature is at least twice that of the ambient temperature (measured in absolute degrees Kelvin) and where the enclosure temperature can be assumed to be uniformly distributed over the entire volume. In practice, this means that the gas temperature should be higher than 300 °C (≈573 K, slightly less than twice the ambient temperature of 293 K). The two conditions are most often met in postflashover fires, where temperatures are well in excess of 800 K and the enclosure is more or less filled with smoke of roughly uniform temperature. The equation has therefore been very useful in the analysis of postflashover fires. According to Quintiere (1994), large vents and temperatures well below 800 °C have lower flow rates than the theoretical maximum from the equation just given. Thus, Equations 50 and 51 should be considered as methods for making estimates of air-flow rates and associated fuel-burning rates.

Flashover

Background

Flashover is characterized by the rapid transition in fire behavior from localized burning of fuel to the involvement of all combustibles in the enclosure—see Walton and Thomas (2002). High-radiation heat transfer levels from the original burning item, the flame and plume directly above it, and the hot smoke layer spreading across the ceiling are all considered to be responsible for the heating of the other items in the room, leading to their ignition. Warning signs are heat build-up and *rollover* (small, sporadic flashes of flame that appear near ceiling level or at the top of open doorways or windows of smoke-filled rooms). Factors affecting flashover include room size, ceiling and wall conductivity and flammability, and the heat- and smoke-producing qualities of room contents. Water cooling and venting of heat and smoke are considered to be ways of delaying or preventing flashover. Three methods for estimating flashover have been developed. They will be described in the next section.

Theory

It is a well-established fact that furnishings are frequently major contributors to fire growth. Recent developments make it possible to determine whether furnishings in a given environment are capable of producing sufficient energy to cause full room involvement. In this section, three simplified equations are given for estimating the rate of heat release required for flashover to occur in a room. These contain a mathematical formula for estimating the amount of energy that must be present in a room, or similar confined space, to raise the upper-level temperatures to a point likely to produce flashover.

1. ***Method of Babrauskas***

According to Babrauskas (1980), at flashover the minimum fire heat release rate in kW is

$$\dot{Q}_{f0} = 750 A_0 \sqrt{H_0} \quad (52)$$

where

A_0 = window or door ventilation area (m^2)
H_0 = height of opening (m)

The amount of heat output suggested by Babrauskas is exactly *half* the maximum burning permitted by way of a single ventilation opening of size A_0 and height H_0. This correlation is derived from 33 test fires having energy release rates from 11 to 3820 kW and ventilation factors from 0.03 to 7.51 $m^{5/2}$. A gas temperature for flashover of 600 °C (873 K), a specific heat of air 1.0 kJ/kgK, and emissivity of 0.5 are used to simplify the development.

2. ***Method of Thomas***

From experimental data, Thomas (1981) developed an average for net radiative and convective heat transfer from the upper gas layer (kW) $\dot{q}_{loss}$ of $7.8A_w$ (see Walton and Thomas 2002). The combination of an upper-layer temperature of 600 °C for flashover criterion and C_p = 1.26 kJ/kgK led to the following expression for the minimum fire heat release rate in kW for the occurrence of flashover:

$$\dot{Q}_{f0} = 378 A_0 \sqrt{H_0} + 7.8 A_w \quad (53)$$

where

A_w = wall area (m^2)

The wall area A_w is the total for the surrounding walls, ceiling, and floor of the enclosure. Sometimes, the floor area is omitted, because relatively little heat is lost through floors. These two models will henceforth be referred to as "Thomas 1" and "Thomas 2," respectively.

3. ***Method of McCaffrey, Quintiere, and Harkleroad***

The method of McCaffrey, et al. (1981) for predicting compartment-fire temperatures may be extended to predict the energy release rate of the fire required to result in flashover in the compartment. See Lawson and Quintiere (1985) and Walton and Thomas (2002) for further information. Specializing the equation for an upper-gas temperature of 522 °C and ambient temperature of 22 °C or ΔT_g = 500 °C for flashover, and substituting values for the gravitational constant (g = 9.81 m/s^2), the specific heat of air (C_p = 1.0 kJ/kg · K), and the density of air (ρ_∞ = 1.18 kg/m^3), then rounding 607.8 to 610, yields

$$\dot{Q}_{f0} = 610 \left[h_k A_T A_0 \sqrt{H_0} \right]^{1/2} \quad (54)$$

where

h_k = effective heat transfer coefficient (kW/m^{2}K)
= $(k\rho C/t)^{1/2}$ for $t \le t_p$
= k/δ for $t > t_p$
A_T = total area of the compartment surfaces (m^2)
= $A_{walls} + A_{floor} + A_{ceiling} - A_{openings}$
A_0 = area of opening (m^2)
H_0 = height of opening from bottom to top (m)
k = thermal conductivity of the compartment surface material (kW/m K)
δ = thickness of compartment surface (m)
ρ = density of the compartment surface (kg/m^3)
C = specific heat of the compartment surface (kJ/kg K)
t = exposure time (s)
t_p = thermal penetration time (s): $(\rho C/k)(\delta/2)^2$

Comments

These three correlations for flashover determination are deduced from laboratory experimental data about fires ranging in size from small-scale to full-scale, and their estimated accuracy is ±30%—typically ±10%. It is necessary, however, to assess how they compare to each other, and which of the methods for predicting flashover is most *conservative*—in the sense that minimal fire size for flashover corresponds to minimal time from the onset of the fire to flashover. The same is true of fire growth rate. The use of conservative estimates can increase the safety of those making estimates of the time it is safe to spend in the vicinity of a growing fire, whether for those trapped or those fighting the fire. Babrauskas (1980) has compared the effect of room wall area and type of wall structure on the energy release required for flashover, finding that the energy required for flashover depends intimately on both, normalized by the ventilation factor. It may be observed that over the range of compartment sizes of most interest, all of the methods produce similar results, and all are conservative representations of the data.

Flashover is characterized by

1. Temperatures reaching approximately 500 °C (932 °F) to 600 °C (1112 °F) in the upper portions of the room
2. Heat flux of from 20 to 25 kW/m^2 (6340 to 7925 Btu/hrft^2) occurring at floor level, with near-simultaneous ignition of combustibles not previously ignited
3. The filling of almost the entire room volume with smoke and flames

Generally, very high heat release rates occur after flashover, and (subject to oxygen availability) most ignitable items in the room burn. The heat increases, and the windows break and melt, making oxygen more readily available, thereby increasing the severity of the fire.

Single-Room Flashover Calculations

It has been observed that in practice flashover occurs when the upper room temperature of the smoke layer reaches between 300 and 700 °C (572 and 1292 °F). Flashover depends on many factors, but lower temperatures should be used to obtain conservative safe estimates of the time left before its occurrence. According to the calculations shown here, the time to reach flashover is characterized in the FAST and FASTLite computer programs' single-room simulation by the upper layer temperature reaching 600 °C (1112 °F). Ten parameters of interest affect the time required to reach flashover conditions. Their standard (default) values are

1. Floor area = 4 m × 4 m = 16 m^2
2. Vent width = 2 m
3. Vent height = 1 m (distance from bottom to top of the vent)
4. Vent height above floor = 1.5 m (distance from floor to midpoint of the vent)
5. Ceiling height = 2.4 m
6. Fire specification = medium fire
7. Fire location = fire in center of floor
8. Wall and ceiling material = 0.016 m (5/8") thick gypsum board
9. Fire radiation fraction = 0.3 (radiation heat loss fraction from the flame)
10. Fire maximum heat release rate = 3 MW

Previous studies using the FAST computer program (Kim and Lilley 2002a and 2002b) have shown that the major parameters affecting flashover are fire growth rate, ventilation opening area, and room area. Hence, the focus of the present calculations, and on the differences between the various equations, will be on these parameters. During fire growth, conduction heat loss is most pronounced through the ceiling and walls; little heat is lost through the floor. In the Thomas flashover criterion, the total enclosure surface area is used, but the contribution of floor area to the total surface area is sometimes omitted. The calculations shown here clarify this effect. Use of the Thomas equation that includes the floor area in the total enclosure surface area is indicated by "Thomas 1," and omission of the floor area by "Thomas 2."

Figure 2 shows the calculation of the time to reach flashover conditions versus fire growth specification and ventilation factor for a room of floor area 3 × 4 m^2. Increased room size increases the time required to reach flashover, according to the Thomas and FAST equations (see Kim and Lilley 2002a and b). It has no effect in Babrauskas's equation because enclosure surface area does not play a part in the Babrauskas theory. It may also be noted that, as room size increases, the FAST computer code gives very large flashover times, which conflict with the other equations considered.

These data are provided in Table 10, where fire growth specification, room size, ventilation factor, and flashover theory are all considered. These four parameters strongly affect the calculated time required to progress from inception of the fire to flashover conditions. Notice that Xs in some locations in the table indicate that flashover conditions were never reached in the computer calculations in the cases of the largest room sizes and the two largest ventilation factors considered.

TABLE 10

Calculating Time (s) to Reach Flashover According to Ventilation Factor, Rapidity of Fire Growth, Room Area, and Four Different Theories

		6 $m^{5/2}$ (2 m high x 2.12 m wide)				4 $m^{5/2}$ (2 m high x 1.41 m wide)				2 $m^{5/2}$ (2 m high x 0.71 m wide)			
Vent Factor		T1	T2	B	FL	T1	T2	B	FL	T1	T2	B	FL
Slow	3*4*2.4	957.1	940.3	1239.3	933.5	813.5	793.7	1011.9	785.6	638.4	612.9	715.5	602.4
	4*6*2.4	1009.1	977.0	1239.3	949.7	874.1	836.8	1011.9	857.7	714.0	667.8	715.5	747.1
	6*8*2.4	1094.2	1034.1	1239.3	1088.3	971.1	902.9	1011.9	968.9	829.9	749.0	715.5	874.4
	8*12*2.4	1236.7	1128.7	1239.3	2041.1	1129.3	1009.8	1011.9	1282.7	1010.4	874.9	715.5	1076.6
	12*16*2.4	1463.8	1277.4	1239.3	X	1374.2	1173.6	1011.9	X	1278.4	1059.8	715.5	2039.9
Medium	3*4*2.4	478.6	470.1	619.6	466.7	406.8	396.8	505.9	392.8	319.2	306.5	357.8	301.2
	4*6*2.4	504.6	488.5	619.6	506.0	437.1	418.4	505.9	461.9	357.0	333.9	357.8	410.7
	6*8*2.4	547.1	517.1	619.6	705.0	485.5	451.4	505.9	545.1	414.9	374.5	357.8	486.2
	8*12*2.4	618.4	564.3	619.6	1677.4	564.6	504.9	505.9	915.5	505.2	437.4	357.8	715.9
	12*16*2.4	731.9	638.7	619.6	X	687.1	586.8	505.9	X	639.2	529.9	357.8	1689.0
Fast	3*4*2.4	239.2	235.0	239.2	233.3	203.3	198.4	252.9	196.4	159.6	153.2	178.8	150.6
	4*6*2.4	252.2	244.2	239.2	300.3	218.5	209.2	252.9	250.0	178.5	166.9	178.8	225.5
	6*8*2.4	273.5	258.5	239.2	524.2	242.7	225.7	252.9	354.2	207.4	187.2	178.8	283.6
	8*12*2.4	309.1	282.1	239.2	1500.0	282.3	252.4	252.9	738.5	252.6	218.7	178.8	542.1
	12*16*2.4	365.9	319.3	239.2	X	343.5	293.3	252.9	X	319.5	264.9	178.8	1517.7
Ultrafast	3*4*2.4	119.6	117.5	119.6	116.7	101.7	99.2	126.5	98.2	79.8	76.6	89.4	75.3
	4*6*2.4	126.1	122.1	119.6	207.6	109.2	104.6	126.5	145.4	89.2	83.5	89.4	123.3
	6*8*2.4	136.7	129.2	119.6	437.4	121.4	112.8	126.5	263.5	103.7	93.6	89.4	193.1
	8*12*2.4	154.6	141.1	119.6	1417.6	141.1	126.2	126.5	652.9	126.3	109.3	89.4	457.1
	12*16*2.4	183.9	159.6	119.6	X	171.7	146.7	126.5	X	159.8	132.4	89.4	1432.4

T1 = Thomas 1 theory, including floor area in total surface area; T2 = Thomas 2 theory, excluding floor area in total surface area; B = Babrauskas theory; FL = FASTLite computer calculations to reach 600 °C.

(Adapted from Lilley 2004)

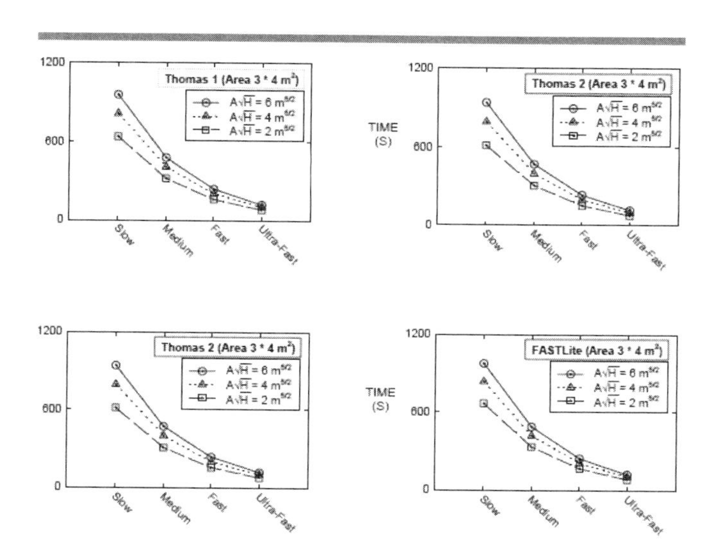

FIGURE 2. Time to reach flashover conditions versus fire growth specification and ventilation factor for a room of area 3 m x 4 m (Adapted from Lilley 2004)

Computer Modeling

Background

According to Friedman's (1991) international survey of computer models for fire and smoke, 36 actively supported models are identified around the world. Among those, 12 models are zone-type models currently in use. A later updated survey by Olenick and Carpenter (2003) shows a significant increase in available fire models over the last 10 years, including field models. Zone models solve the conservation equations for distinct regions. A number of zone models exist, varying to some degree in the detail of their treatment of fire phenomena. The dominant characteristic of this class of model is that it divides rooms into hot upper layers and cold lower layers, assuming that properties can be approximated throughout zones by some uniform function. Temperature, smoke, and gas concentrations are assumed to be uniform at every point in a zone. Zone modeling has proved a practical method for providing estimates of fire processes in enclosures. Experimental observations show that the assumption of uniform properties within zones yields good results (see Jones, et al. 2000).

Potentially greater accuracy in simulating enclosure fires is possible using field models and computational fluid dynamic (CFD) models, in which structures are divided into many subvolumes and in which the basic laws of conservation (representing mass, momentum, and energy) are written as partial differential equations representing the variation of the dependent variables (velocity components, density, temperature, and species) as a function of position and time throughout the 3-D domain of interest (perhaps one, or many, rooms, corridors, and stairways within a building). This CFD approach has been used extensively for nonfire applications over the last 40 years in many areas of fluid flow, turbulence, heat transfer, and combustion with many other sophistications included in the simulations; see, for example, Gupta and Lilley (1984). Only recently, however—in the last ten years—has this fundamental differential equation method been applied to fire situations (see Karlsson and Quintiere 2000, SFPA 2002, NFPA 2003).

Zone Models

The zone-modeling equations are embodied in computer codes, allowing users to readily calculate the effects of a specified fire (and subsequent fire spread, other fires, etc) on temperatures, smoke-layer heights, combustion product species, and so on. This computerized zone-modeling approach emerged in the mid-1970s, when the effort to study fires developing in compartments intensified.

The CFAST (Consolidated Model of Fire Growth and Smoke Transport) is a zone model that predicts the effect of a specified fire on temperatures, various gas concentrations, and smoke-layer heights in a multicompartment structure. CFAST is based on solving a set of equations that predicts changes in enthalpy and mass over time. The equations are derived from the conservation equations for energy, mass, and momentum, as well as an equation of state—in this case, the ideal gas law. The conservation equations are fundamental to physical systems and must hold in all cases. These equations are rearranged to form a set of predictive equations for the sensible variables in each compartment. CFAST is formulated as a set of ordinary differential equations and was the first model of fire growth and smoke spread to cast the entire model in this form, utilized because of the efficiency of solving the conservation equations in this way. The FAST (Fire Growth and Smoke Transport) computer code is one of several noteworthy multiroom zone-type computer programs. CFAST is embodied in the recent FAST code as the program's actual calculation routine. It is a multicompartment (up to 30 rooms for version 3.1.6) zone-type computer model that predicts the temperature, heat transfer, and smoke-hazard development in each compartment, as well as the type and location of the fire. See Cooper (1980), Cooper and Stroup (1982), Jones (1985), Jones and Peacock (1989), and Walton, et al. (1985) for details.

FASTLite is a collection of procedures that builds on the core routines of the earlier FIREFORM and a simplified version of the computer model CFAST to provide engineering calculations of fire phenomena for building designers, code enforcers, fire protection engineers, and fire safety–related practitioners. This program provides quantitative estimates of some of the likely consequences of fires, and the underlying models have been subjected to a range of verification tests to assess the accuracy of their calculations. FASTLite is a small version of FAST (and CFAST) that can simulate up to 3 rooms through a graphical user interface. Kim and Lilley (1997 and 1998) have used FASTLite to study the time required to reach flashover conditions, as well as various parameters' effects on this time, but FASTLite is no longer supported by NIST, which favors FAST/CFAST.

HAZARD I (see Bukowski, et al. 1989) presents an advanced mathematical modeling approach to simulating fire development in a multiroom building that complements the experimental approach and postfire on-site investigations. HAZARD I is the first fully integrated modeling tool in the world. The HAZARD program initially used FAST as its solver, but the latest version uses CFAST (see Peacock, et al. 1994). The latest microcomputer version of the code can handle buildings with up to 15 areas, as well as multiple fires, HVAC connections, and ceiling, floor, and wall vents. This code is a generalized version of the earlier program allowing for user-friendly input and output and using a helpful database of experimental properties in its calculations. The latest version (1.2) is described by Peacock, et al. (1994).

The CFAST Computer Code

Fire-analysis models have been developing from the late 1960s. CFAST is a member of a class of models referred to as zone models, and it is the most popular one in use today (see Karlsson and Quintiere 2000). In a zone-element model, each room is divided into a small number of volumes (called zones), each of which is assumed to be internally uniform, having unvarying temperature and smoke and gas concentrations. In CFAST, each room is divided into two layers. Because these layers represent the upper and lower parts of the room, conditions within a simulated room can only vary from floor to ceiling, not horizontally. This assumption is based on experimental observations that burning rooms stratify into two distinct layers. We can measure variations in conditions within a layer, but these are generally negligible in comparison to differences between layers. This assumption places some limitations on the predictive capability of such models, but as modeling evolves, many of these assumptions are receding.

CFAST solves a set of combined equations simulating changes in enthalpy and mass over time. The starting point is the set of conservation equations (for energy, mass, and momentum) that are fundamental to physical systems. Subsidiary equations are the ideal gas law and definitions of density and internal energy. The resulting equations are rearranged to form a set of predictive equations used to calculate sensible variables in each compartment. In CFAST the simulation is formulated as a set of ordinary differential equations. CFAST and FASTLite are distributed by the Building and Fire Research Laboratory (BFRL) of the National Institute of Standards and Technology (NIST). CFAST is the kernel of the zone fire models (FAST, FASTLite, FireCAD, and FireWalk) supported by the BFRL. An amalgamation of FAST 18.5 and CCFM was developed as CFAST 1.0 in 1990 that was functionally equivalent to FAST 18.5 but included more modules added from CCFM. Multiple fires, multiwall radiation, distributed mechanical ventilation ducts, ceiling jets, and 3-D positioning of fires, and a more robust ODE solver (DASSL; see Brenan, et al. 1989) were added later, and the maximum number of compartments was increased to 15. In 1993, CFAST was updated to version 2.0 with the addition of a new supporting file (THERMAL.DF), a new conduction routine, and a new convection routine. The problems of flow through horizontal openings, very large fires in small rooms, interactions between fire size and plume entrainment, and optional ceiling-jet calculations were fixed later. The comparison method and usage of different versions (1.4, 1.6.4, and 2.0.1) of CFAST can be found in Alvord (1995).

In 1996, CFAST 3.0 was released with a new user interface for CEdit and the ability to simulate a new phenomenon, ceiling/floor heat transfer, in its approximation of intercompartmental heat transfer. In November of 1999, a full graphical user interface version (3.1.6) was released after some minor corrections and the addition of the simulation of vertical heat flow. Later, version 4.0.1 was released and added the phenomenon of horizontal heat conduction, permitting up to 30 rooms and 30 ventilation sources. Version 6 is available as of 2008.

CFD Models

This approach has been used extensively for nonfire applications over the last 40 years in many areas of fluid flow, including many other sophistications in the simulations (see, for example, Gupta and Lilley 1984). Only recently, however—over the last ten years—has this fundamental differential equation method been applied to fire situations. In CFD models, the structure is divided into a very large number of subvolumes, and the basic laws of conservation representing mass, momentum, and energy are written as partial differential equations. These represent the variation of the dependent variables (velocity components, density, temperature, and species) as a function of position and time throughout the 3-D domain of interest—whether one or many rooms, corridors, or stairways within a building.

The fundamental balance equations are supplemented by phenomenological equations that model to some degree or another such associated subprocesses as combustion, turbulence, radiation, and soot modeling. Some CFD fire-modeling codes include

1. SOFIE (Simulation of Fires in Enclosures), which contains a multitude of submodels especially developed for fire application
2. SMARTFIRE, a user-friendly program that accepts several add-on packages
3. JASMINE, which includes many submodels important to fire simulation
4. FDS, a multiroom computational fluid dynamics (CFD) fire-dynamics simulator model

Fire Dynamics Simulator (FDS) is a computational fluid dynamics program that calculates the flow field of a fire-driven regime. FDS was developed by the National Institute of Standards and Technology (NIST) for use in the field of structural fires. It can be used with more reliability when the fire size is specified and the building size is large relative to the fire; see McGrattan (2004 and 2005). FDS calculates the temperature, velocity, species concentration, pressure, heat release, and density of the gaseous field, as well as the material temperature, heat flux, and burning rate at solid boundaries. An associated study by Forney and McGrattan (2004) explained how the companion Smokeview code could be used to visually reveal the results of the computations, a major problem in 3-D studies. Version 5 is available in 2008.

FDS is written in FORTRAN 90 and uses an explicit predictor-corrector finite differences algorithm to solve the Navier-Stokes equations on a rectilinear grid. Reflecting reality, FDS is sensitive to its boundary conditions. All materials in the computation require accurate thermal material properties specification. As in the case of fire-driven flow, FDS approximates low-speed, thermally driven fluid, emphasizing smoke transport and heat transfer. Smoke and sprinkler regimes are simulated by Lagrangian particles. Two options exist for incorporating turbulence. By default, the Smagorinsky form of the Large Eddy Simulations (LES) is used, but FDS is capable of performing Direct Numerical Simulation (DNS). All modes of heat transfer are included in FDS. Conduction and convection are handled in traditional ways within the confines of the numerical grid and the simplified Navier-Stokes equations. Radiation heat transfer is handled through a *finite volume method* analogous to that of convective transport. A nonscattering, gray gas is used in which the solution domain is divided into a series of discrete angles. From there, the view factors from any grid point to any other grid point are calculated. FDS uses a mixture fraction combustion model that tracks the fraction of the product gases that entered the calculation domain as fuel. This leads to diffusion-limited (rather than chemical-limited), infinitely fast combustion. Product composition is then calculated from the mixture fraction through a series of empirical expressions via well-known techniques (see Gupta and Lilley 1984).

Kameel and Khalil (2004) have advanced their 3-D modeling to that of transient effects of smoke coming from a specified developing fire by using an updated general code with a k-ε turbulence model and modeling the fire as a heat and volume source. They provide useful code by which to look at the effects of a targeted fire on the environment, specifically dealing with the volume, temperature, and location of smoke in the enclosure as it varies from point to point in the 3-D space.

Conclusion

The development of fire analytical modeling has accelerated over the last 30 years. Zone-type and field-type CFD approaches to multiroom structural fire modeling were reviewed in this chapter. Their background, methodology, and applicability were discussed. In particular, emphasis was placed on the theory and methodology of the CFAST model (Consolidated Model of Fire Growth and Smoke Transport) and its simpler variant, the FASTLite model, which are zone-type approaches in wide use. More recent CFD studies were also identified, such as FDS, and their applications described. Studies of this type assist in the understanding of structural fires and the development and assessment of the predictive capability of computer modeling studies.

Sample Calculations

Flow of Liquid Through a Hole in a Tank

Jet A fuel is stored in a large above-ground tank to a depth of 50 ft. The exit piping from the bottom of the tank develops a leak equal in area to that of a round hole of diameter 0.25 inches.

(a) Calculate the exit velocity of leaking fuel.
(b) Calculate the volume flow rate.
(c) Calculate the heat release rate in kW if the fuel burns.

Solution

(a) Torricelli equation (equating the pressure due to depth to the exit velocity produced) gives the ideal velocity as

$$V = \sqrt{2gh} = [2(32.2)(50)]^{1/2} = 56.75 \text{ ft/s}$$

Using a discharge coefficient of 0.61 (estimate for turbulent flow, sharp-edge orifice)

$$V_{real} = 0.61 \quad V_{ideal} = 34.61 \text{ ft/s}$$

(b) $Q = \text{Velocity} \times \text{Area}$

$$= (34.61)\left[\frac{\pi}{4}\left(\frac{0.3}{12}\right)^2\right] = 0.0170 \text{ ft}^3/\text{s}$$

$= 1.02$ CFM [because 1 min = 60 sec]
$= 7.62$ gpm [because 1 ft^3 = 7.48 gallons]

(c) Heat Release Rate (HRR) is

$$\dot{Q} = \dot{m}\Delta H_c = \rho Q \Delta H_c$$

$$= (0.82 \times 1000 \text{ kg/m}^3)(0.0170 \text{ ft}^3/\text{s})\left(\frac{1 \text{ m}}{3.2808 \text{ ft}}\right)^3(43{,}300 \text{ kJ/kg})$$

$$= 17{,}000 \text{ kW} = 17 \text{ MW}$$

Graphic Example

Figure 3 shows the volume flow rate in gallons per minute of jet A liquid fuel emerging from a sharp-edged round hole at the bottom of a tank. Three different hole sizes are illustrated in the three lines shown. Tanks with depths of fluid from zero to 80 ft are considered as the scale on the abscissa. Notice that the greater the depth and the larger the hole, the greater the volume flow rate. Figure 4 gives corresponding heat release rates in MW for the cases considered in Figure 3 on the assumption that the emerging fuel is ignited.

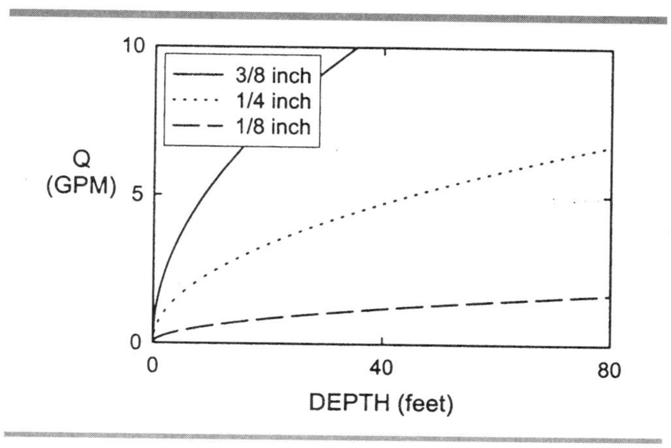

FIGURE 3. Liquid volume flow rate versus tank depth for jet A liquid fuel release (Adapted from Lilley 2004)

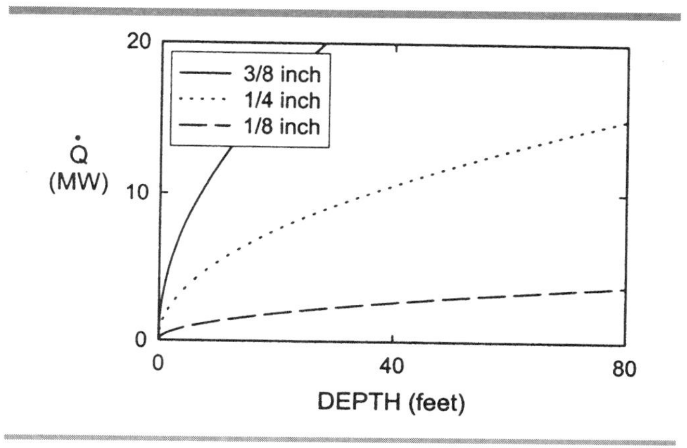

FIGURE 4. Heat release rate versus tank depth for jet A liquid fuel release (Adapted from Lilley 2004)

Flow of Liquid Through a Hole

A hole of diameter 1/32 inch develops in a liquid line from a propane tank at 70 °F. Calculate

(a) exit velocity
(b) volume flow rate
(c) heat release rate if ignited

Solution

(a) At 70 °F, the gage pressure inside a propane tank is approximately 132 psig. The basic equation reduces to $p_1 - p_2 = 1/2\ \rho V^2/g_c$ so that

$$\text{Ideal } V = \left[\frac{2g_c(p_1 - p_2)}{\rho}\right]^{1/2}$$

$$= \left[\frac{2(32.2)(132 - 0)(144)}{(0.509)(62.4)}\right]^{1/2} = 196 \text{ ft/s}$$

where the specific gravity of propane is 0.509. Hence the actual $V = (0.61)(196) = 120$ ft/s using a discharge coefficient of 0.61, a suitable value for turbulent flow through exits.

(b) Q = Volume flow rate = Area × Velocity

$$= \left(\frac{\pi}{4} D^2\right)(V) = \frac{\pi}{4}\left[\frac{1}{(32)(12)}\right]^2 [120 \text{ ft/s}]$$

$$= 6.392 \times 10^{-4} \text{ ft}^3/\text{s} = (6.392 \times 10^{-4})(60)$$

$$= 0.0383 \text{ ft}^3/\text{min [CFM] of liquid propane}$$

$$= (0.0377)\frac{270 \text{ ft}^3 \text{ vapor}}{1 \text{ ft}^3 \text{ liq}}$$

$$= 10.35 \text{ CFM of vapor propane}$$

(c) Heat release rate

$$\dot{Q} = (10.35 \text{ ft}^3/\text{min})(2500 \text{ Btu/ft}^3)\left(\frac{60 \text{ min}}{1 \text{ hr}}\right)$$

$$= 1{,}553{,}000 \text{ Btu/hr}$$

$$= (1{,}553{,}000 \text{ Btu/hr})\left(\frac{29.31 \text{ kW}}{100{,}000 \text{ Btu/hr}}\right)$$

$$= 455 \text{ kW} \approx 0.5 \text{ MW}$$

Graphic Example

Computations of this type are illustrated graphically in Figures 5 and 6. The volume flow rate of propane is given in ft³/min CFM of propane vapor as a function of hole diameter (the three lines corresponding to 1/8-, 1/16-, and 1/32-inch-diameter holes) and driving pressure [from zero to 300 pounds per square inch gage pressure (psig) as the scale on the abscissa]. This gaseous volume has been calculated on the basis of a 270:1 expansion ratio in transition from liquid to vapor, appropriate for vapor at standard pressure and temperature (14.7 psia and 60 °F). It may be noted that pressure-driving forces in propane liquid-vapor mixtures are correlated with temperature, and that temperatures of 60, 100, and 130 °F correspond roughly to commercial propane gage pressures of 100, 200, and 300 psig, respectively. Heat release rates in MW are given in Figure 6.

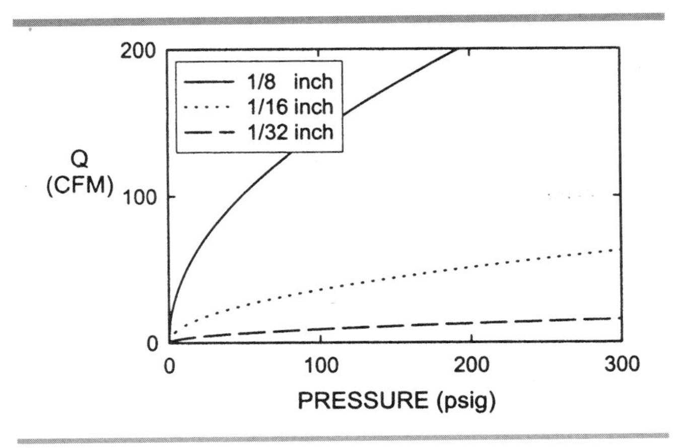

FIGURE 5. Gaseous volume flow rate versus driving pressure for liquid propane release (Adapted from Lilley 2004)

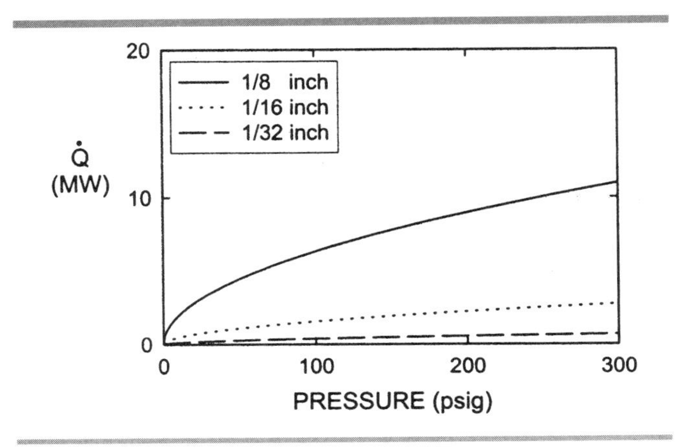

FIGURE 6. Heat release rate versus driving pressure for liquid propane release (Adapted from Lilley 2004)

Flow of Vapor Through a Hole

A leak develops in a natural gas pipeline equal in area to that of a round hole of 1/4-inch diameter. The gas pressure is 25 psig in the large-diameter pipe. Assume a discharge coefficient of 0.84.

(a) Calculate the mass flow rate.
(b) Calculate the volume flow rate.

(c) Calculate the energy release rate if the leaking gas is ignited.

Solution

(a) The mass flow rate for choked conditions is given by

$$(Q_m)_{\text{choked}} = (0.84)\frac{\pi}{4}\left(\frac{1}{48}\right)^2[(25 + 14.7)(144)] \times \left[\frac{(1.4)(32.2)(16)}{(1545)(520)}(0.335)\right]^{1/2}$$

using $\left(\frac{2}{\gamma + 1}\right)^{(\gamma+1)/(\gamma-1)} = 0.335$ when $\gamma = 1.4$

so that the calculation yields

$$(Q_m)_{\text{choked}} = 0.0284 \text{ lbm/s} = 102 \text{ lbm/hr}$$

(b) Volume flow rate = mass flow rate ÷ density

$$= (102 \text{ lbm/hr})\left(\frac{1 \text{ ft}^3}{0.075x\ 16/29 \text{ lbm}}\right)$$

using the molecular weights of natural gas (taken as methane) and air as 16 and 29, respectively

$$= 2470 \text{ ft}^3\text{/hr} = 41 \text{ CFM}$$

(c) $$\dot{Q} = (2470 \text{ ft}^3\text{/hr})\left(\frac{1000 \text{ Btu}}{\text{ft}^3}\right)$$

$$= 2{,}470{,}000 \text{ Btu/hr}$$

$$= (2{,}470{,}000 \text{ Btu/hr})\left(\frac{29.31 \text{ kW}}{100{,}000 \text{ Btu/hr}}\right)$$

$$= 724 \text{ kW} = 0.72 \text{ MW}$$

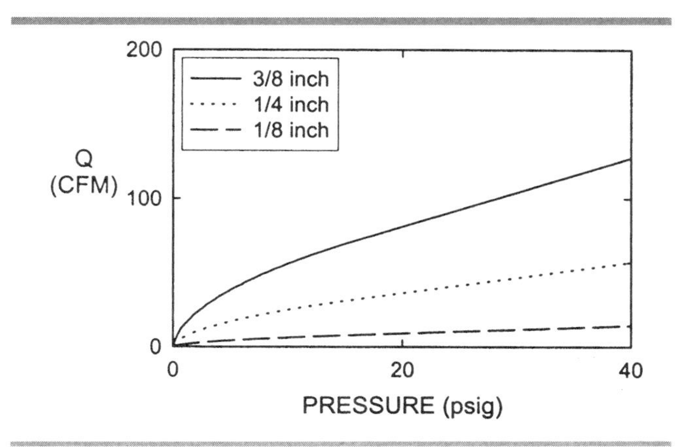

FIGURE 7. Gaseous volume flow rate versus driving pressure for natural gas release (Adapted from Lilley 2004)

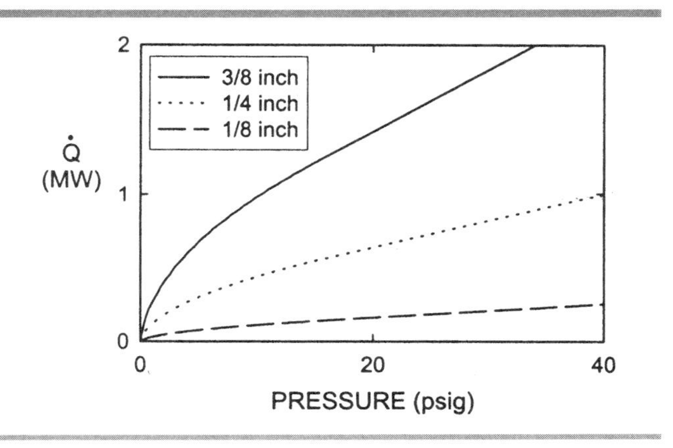

FIGURE 8. Heat release rate versus driving pressure for natural gas release (Adapted from Lilley 2004)

Graphic Example

Natural gas emerging through holes with absolute driving pressures greater than about twice the external downstream absolute pressure (14.7 psia) gives rise to what is called *choked* flow with sonic conditions at the hole. The compressible flow situation then has volume flow rates (usually quoted in standard ft^3/min—as if the gas were at 60 °F and 14.7 psia) that do not depend on downstream pressure; the absolute driving pressure is paramount in determining efflux rates. Shown in Figure 7 is the volume flow rate (in standard ft^3/min) of natural gas as a function of driving pressure quoted in psig. Three lines are given representing three different hole diameters. Figure 8 gives corresponding heat release rates if the emerging gas is ignited. Notice that both sets of lines in these two figures are straight lines, illustrating the linear dependence of flow rate upon driving pressure under choked conditions.

Also shown in Figures 7 and 8 are computed values when the driving pressure is less than 14.7 psig, thereby creating nonchoked flow and the corresponding equation. In those nonchoked cases, the flow rate depends on both the upstream and downstream pressures.

Dispersion of Continuous Fuel Release

On an overcast day, an emission of 10 kg/s (22 lbm/s) of butane occurs continuously from a ground-level release point. The wind speed is 3 m/s (10 ft/s). Calculate the mean concentration (in volume percent) of butane at 20 m (66 ft) downwind and 4 m (13 ft) to the side.

Assume the gas is neutrally buoyant and refrain from considering the dense gas situation. If a small pilot flame is located at that point, is ignition likely to occur or not? Discuss.

Solution

Take stability class C, a neutrally stable atmospheric day. Data in Lilley (2004) gives fitted curves for the dispersion coefficients for the plume in terms of the downstream distance x as

$$\sigma_x = \sigma_y = 0.195x^{0.90} = 2.89 \text{ m}$$
$$\sigma_z = 0.112x^{0.91} = 1.71 \text{ m}$$

$$C(x,y,0) = \frac{Q_m}{\pi\sigma_y\sigma_z u}\exp\left[-\frac{1}{2}\left(\frac{y}{\sigma_y}\right)^2\right]$$

(a) At $x = 20$ m, $y = 0$ m, $u = 3$ m/s, exp [ ] = 1

$$C(20,0,0) = \frac{10}{\pi(2.89)(1.71)(3)} = 0.215 \text{ kg/m}^3$$

Air has 1.2 kg/m^3, so volume concentration percent

$$= \frac{0.215/58}{1.2/29} \times 100 = 9.0\%$$

(b) At $x = 20$ m, $y = 4$ m, $u = 3$ m/s

$$C(20,4,0) = C(20,0,0)\exp\left[-\frac{1}{2}\left(\frac{4}{2.89}\right)^2\right]$$

$$= (0.215)(0.384) = 0.0825 \text{ kg/m}^3$$

Volume concentration percent

$$= \frac{0.0825/58}{1.2/29} \times 100 = 3.4\%$$

Ignition is likely to occur, because the fuel-air mixture is within flammability limits at this point. Flammability limits for butane are approximately 1.9 to 8.5 percent.

Graphic Example

When the vapor of natural gas or a liquefied petroleum gas (LPG) is released into the atmosphere continuously, a continuous plume cloud results. The volume concentration percent reduces both downstream and sideways. For the situation just described, a variety of computations are displayed in Figures 9 through 12 that show downstream decay of volume concentration percent and sideways decay at 10, 20, and 30 m (33, 66, and 99 ft) downstream, respectively. The three lines in each case represent the effect of wind speed—3, 6, and 9 m/s (10, 20, and 30 ft/s). All the calculations shown assume the atmosphere to be the same stability class, C. Also, in these four figures (as well as in the sample calculation), it may be noted that butane (C_4H_{10}), with molecular weight 58, is the material under consideration. The graphs clearly show which points

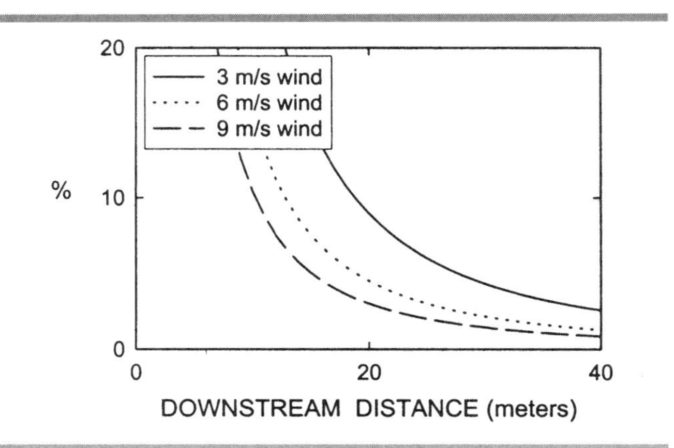

FIGURE 9. Downstream volume concentration percent decay for three wind speeds with 10 kg/s (22 lbm/s) continuous butane release (Adapted from Lilley 2004)

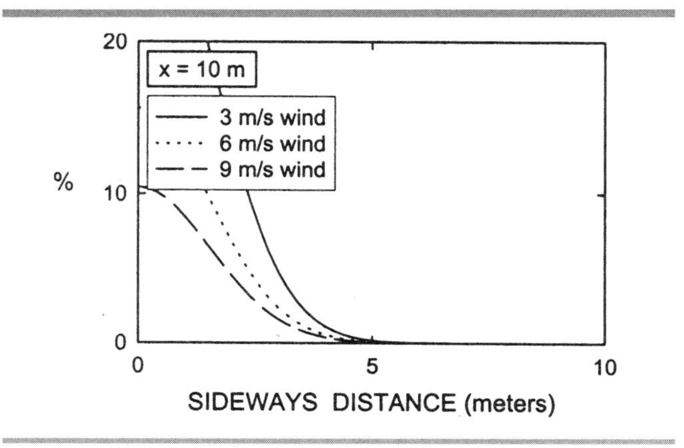

FIGURE 10. Sideways volume concentration percent decay at 10 meters (33 feet) downstream for three wind speeds with 10 kg/s (22 lbm/s) continuous butane release (Adapted from Lilley 2004)

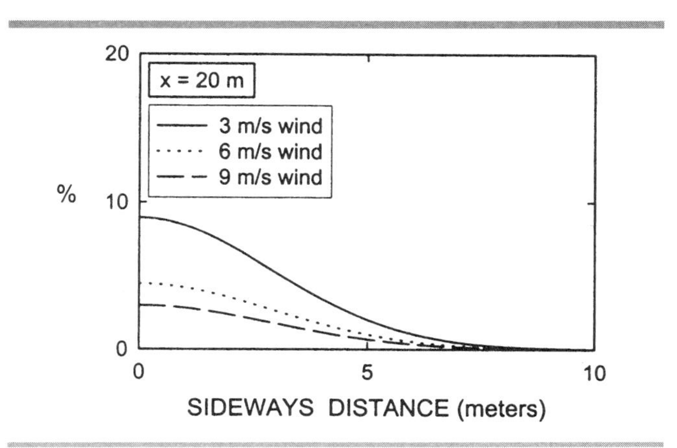

FIGURE 11. Sideways volume concentration percent decay at 20 meters (33 feet) downstream for three wind speeds with 10 kg/s (22 lbm/s) continuous butane release (Adapted from Lilley 2004)

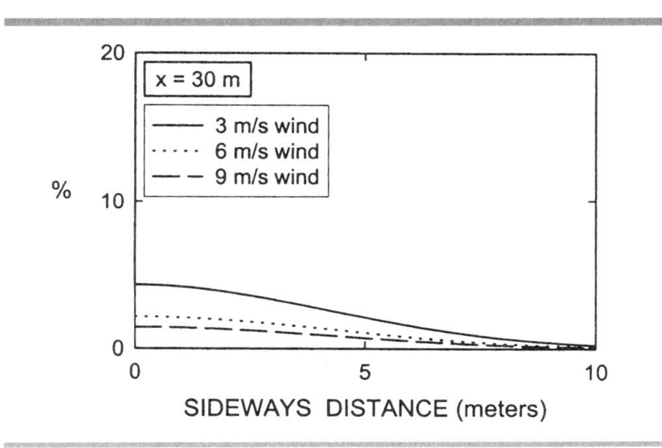

FIGURE 12. Sideways volume concentration percent decay at 30 meters (33 feet) downstream for three wind speeds with 10 kg/s (22 lbm/s) continuous butane release (Adapted from Lilley 2004)

are located within flammability limits; for butane these are 1.9 to 8.5 percent. Also, it may be noted that wind speed and direction are rarely uniform, and that there will be an oscillating, meandering time effect on the actual ground footprint of the cloud because of wind gusts and turbulence. This effect is such that ignition sources too far away from the cloud, on average, to allow ignition to take place might very well provide an ignition source for the gas cloud because of this transient location of the cloud.

When commercial propane is being considered instead of butane, its lower molecular weight (44 instead of 58) comes into play. Its volumetric concentration percent is increased from that of butane by the ratio 58:44 provided that emerging efflux rate remains the same (10 kg/s for the case being considered here). Figures 13 through 16 correspond to Figures 9 through 12 but take commercial propane as the material released.

Notice that 10 kg/s flow rate is approximately equivalent to 5.22 gallons of liquid propane per second, 0.70 ft^3 of liquid propane per second, and 189 ft^3 of gaseous vapor per second. This flow rate is approximately the same as that produced by 70 °F (132 psig) liquid propane emerging through a one-square-inch hole—about 1000 times greater than the amount emerging through a 1/32-inch-diameter hole. The corresponding heat

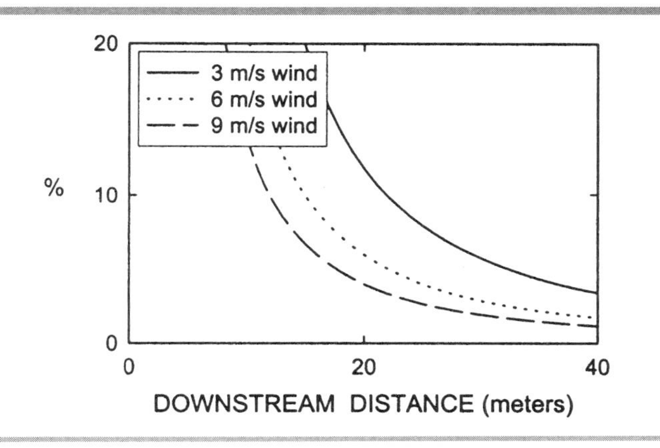

FIGURE 13. Downstream volume concentration percent decay for three wind speeds with 10 kg/s (22 lbm/s) continuous propane release (Adapted from Lilley 2004)

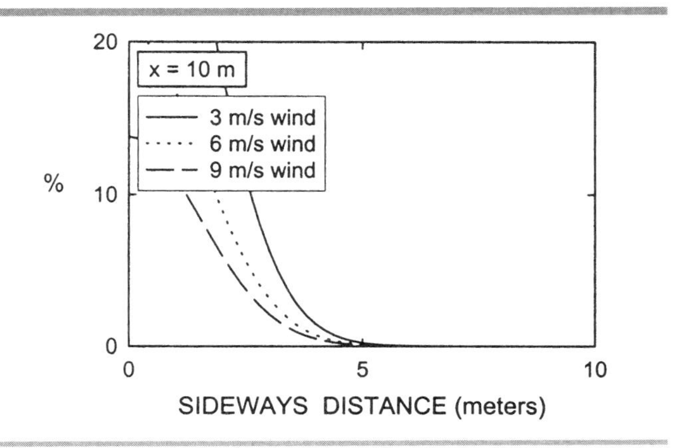

FIGURE 14. Sideways volume concentration percent decay at 10 meters (33 feet) downstream for three wind speeds with 10 kg/s (22 lbm/s) continuous propane release (Adapted from Lilley 2004)

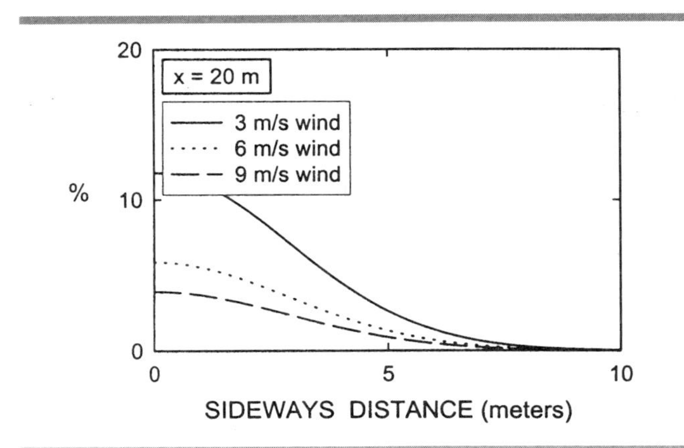

FIGURE 15. Sideways volume concentration percent decay at 20 meters (33 feet) downstream for three wind speeds with 10 kg/s (22 lbm/s) continuous propane release (Adapted from Lilley 2004)

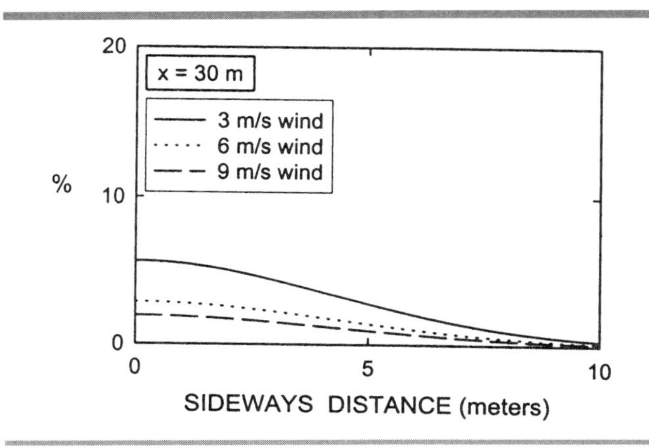

FIGURE 16. Sideways volume concentration percent decay at 30 meters (33 feet) downstream for three wind speeds with 10 kg/s (22 lbm/s) continuous propane release (Adapted from Lilley 2004)

release rate when ignition occurs is approximately 500 MW—extremely high. This helps to explain why pressure-regulator failures (permitting high downstream pressures) are particularly dangerous, and why pressure-driven liquefied petroleum gas (LPG) releases are extremely hazardous when their vapor clouds are ignited.

Heat Release Rate of Pool Fires

Calculate the heat release rate in kW for a kerosene pool fire of diameter

(a) 0.5 m
(b) 1 m
(c) 1.5 m

using data for large-pool burning-rate estimates.

Solution

Take

$$\Delta H_c = 43.2 \text{ MJ/kg}$$
$$\dot{m}''_\infty = 0.039 \text{ kg/m}^2\text{s}$$

so that

$$\dot{m}'' = 0.039 \text{ kg/m}^2\text{s}$$

for pools of a diameter greater than 1 m, and

$$\dot{m}'' = \dot{m}''_\infty[1 - \exp(-k\beta D)]$$

for smaller pools, where the product $k\beta$ is given by $k\beta = 3.5 \text{ m}^{-1}$ for kerosene. Values of this for other common fuels are given in Lilley (2004). Thus

(a) $\dot{m}'' = 0.039[1 - \exp(-(3.5)(0.5))] = 0.032 \text{ kg/m}^2\text{s}$
$\dot{Q} = \dot{m}'' .A. \Delta H_c = 271 \text{ kW}$

(b) $\dot{m}'' = 0.039[1 - \exp(-(3.5)(1))] = 0.038 \text{ kg/m}^2\text{s}$
$\dot{Q} = 1290 \text{ kW}$

(c) $\dot{m}'' = 0.039 \text{ kg/m}^2\text{s}$
$\dot{Q} = 2980 \text{ kW}$

Graphic Example

Figure 17 shows the calculated heat release rate for pool fires of various diameters. Three different fuels are illustrated by the three lines shown. Notice that gasoline burns very much more readily than alcohol—it

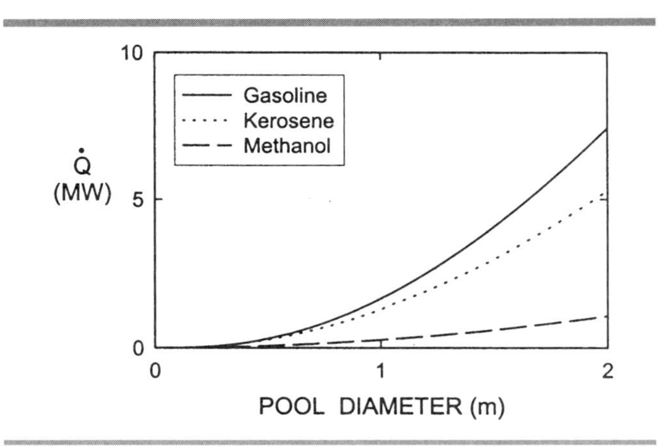

FIGURE 17. Flame length versus pool diameter for different fuels (Adapted from Lilley 2004)

is much more volatile and has a lower flash point. Not only does it burn quicker in terms of regression rate (meters/second), but its energy per unit liquid volume vaporized is also higher, leading to the calculations shown.

Flame Heights of Pool Fires

Estimate the flame height for each of the three pool fires specified in the last section.

Solution

Use the equation

$$L = -1.02D + 0.23\dot{Q}^{2/5}$$

so that

(a) $D = 0.5$ m, $\dot{Q} = 271$ kW, and $L = 1.65$ m
(b) $D = 1.0$ m, $\dot{Q} = 1290$ kW, and $L = 3.02$ m
(c) $D = 1.5$ m, $\dot{Q} = 2980$ kW, and $L = 4.11$ m

Using the equation given in NFPA 921 (2008) of

$$H_f = 0.174(k\dot{Q})^{0.4}$$

with $k = 1$ (no nearby walls) gives similar results:

(a) $H_f = 1.64$ m
(b) $H_f = 3.05$ m
(c) $H_f = 4.27$ m

Graphic Example

Figure 18 gives computed flame heights [according to the simplified Heskestad (1983) formula] for pool fires of various diameters for the three different fuels under consideration. Notice that the more volatile gasoline burns faster and produces a higher flame than the other fuels.

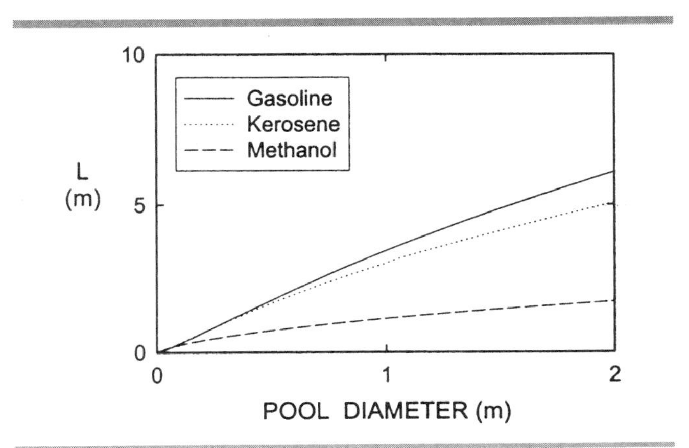

FIGURE 18. Flame length versus pool diameter for different fuels (Adapted from Lilley 2004)

Maximum Upward Velocity in a Fire Plume

Estimate the maximum temperature and maximum upward velocity on the centerline in the plume above the three pool fires specified in the last section.

Solution

Assume the room temperature to be 20 °C and take maximum temperature from

$$\Delta T_0 = 650 \text{ K}$$

in each case so that maximum temperature is 670 °C. Maximum velocity is found from

$$u_{0,\max} = 1.97\dot{Q}_c^{1/5}$$

where $\dot{Q}_c$ is the convective heat release rate carried in the plume. That is,

$$\dot{Q}_c = (1 - x_r)\dot{Q}$$

where x_r is the radiative fraction of the total heat released $\dot{Q}$. Although larger fires radiate a lower fraction than smaller fires, and taking kerosene to give a sooty luminous flame with a relatively high radiative fraction of $x_r = 0.4$, maximum upward velocities calculate as

(a) $D = 0.5$ m, $\dot{Q}_c = 0.6(271)$ kW, and $u_{0,\max} = 5.45$ m/s
(b) $D = 1.0$ m, $\dot{Q}_c = 0.6(1290)$ kW, and $u_{0,\max} = 7.45$ m/s
(c) $D = 1.5$ m, $\dot{Q}_c = 0.6(2980)$ kW, and $u_{0,\max} = 8.81$ m/s

Graphic Example

Maximum upward velocities in the plume above the fires of different diameters are illustrated graphically in Figure 19. Again, the same three fuels are considered. The more volatile and more quickly burning gasoline generates a faster maximum upward velocity in the fire plume than do the other two fuels considered.

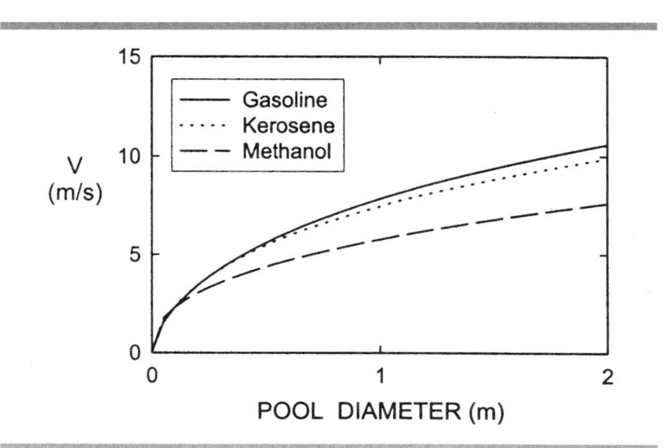

FIGURE 19. Maximum upward velocity in flame plume versus pool diameter for different fuels (Adapted from Lilley 2004)

Minimum Safe Distance from Flames

Using an estimated value of the radiative fraction of the pool fire heat that is radiated, and using a simple inverse square distance law, determine the minimum distance away that

(a) easily ignitable items
(b) normal items
(c) difficult-to-ignite items

would have to be for safety in the case of the three pool fires specified in the last section.

Solution

Taking the radiative fraction to be $x_r = 0.4$ and using

$$\dot{q}'' = \frac{x_r \dot{Q}}{4\pi R_0^2}$$

with easy-, normal-, and difficult-to-ignite items needing 10, 20, and 40 kW/m^2, yields

(a) Easily ignitable items:

$$R_0 = \left[\frac{(0.4)(271)}{4\pi(10)}\right]^{1/2} = 0.93 \text{ m } [D = 0.5 \text{ m}]$$

$$R_0 = \left[\frac{(0.4)(1290)}{4\pi(10)}\right]^{1/2} = 2.03 \text{ m } [D = 1.0 \text{ m}]$$

$$R_0 = \left[\frac{(0.4)(2980)}{4\pi(10)}\right]^{1/2} = 3.08 \text{ m } [D = 1.5 \text{ m}]$$

(b) Normally ignitable items:

$R_0 = 0.65$ m [$D = 0.5$ m]
$R_0 = 1.43$ m [$D = 1.0$ m]
$R_0 = 2.18$ m [$D = 1.5$ m]

(c) Difficult-to-ignite items:

$R_0 = 0.46$ m [$D = 0.5$ m]
$R_0 = 1.01$ m [$D = 1.0$ m]
$R_0 = 1.54$ m [$D = 1.5$ m]

Graphic Example

Computations of this type lead to Figures 20, 21, and 22. These illustrate the minimum safe distance from pool fires of different diameters for the three fuels being

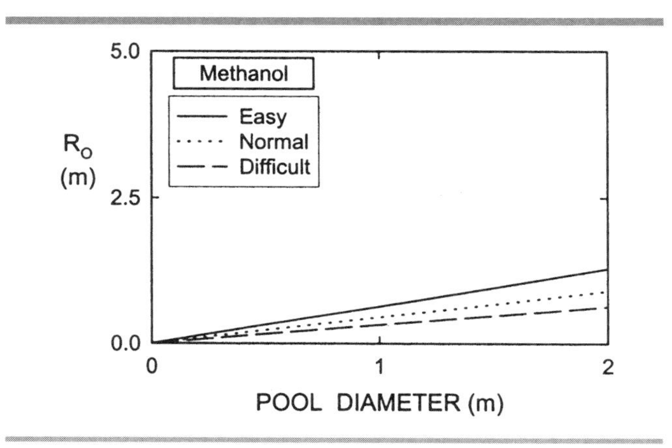

FIGURE 20. Minimum safe distance from pool fires with methanol fuel (Adapted from Lilley 2004)

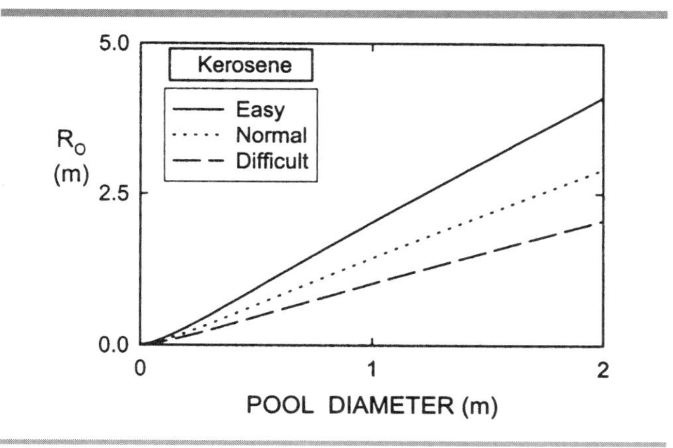

FIGURE 21. Minimum safe distance from pool fires with kerosene fuel (Adapted from Lilley 2004)

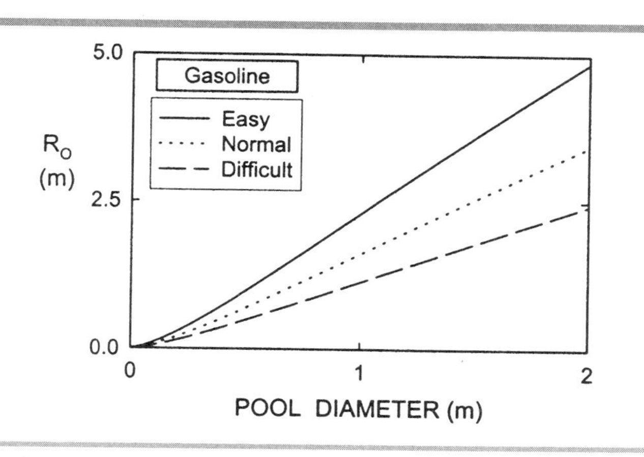

FIGURE 22. Minimum safe distance from pool fires with gasoline fuel (Adapted from Lilley 2004)

considered—methanol, kerosene, and gasoline, respectively. Notice that the more volatile, quicker burning gasoline has a much greater propensity to ignite remote objects than has the slower burning methanol.

Typical Room Fire Loads and Corresponding Ventilation Factor for 20-Minute Burn

Take a typical fire load in a

(a) living room
(b) family room
(c) bedroom
(d) dining room
(e) kitchen

of floor size 12 m², and assume it burns to completion steadily in 20 minutes. Calculate the heat release rate and the minimum value of the ventilation parameter for this to occur.

Solution

Typical fire loads in various rooms of a house are given in Hicks (2003) with values of 3.9 lb/ft², 2.7 lb/ft², 4.3 lb/ft², 3.6 lb/ft², and 3.2 lb/ft², respectively, for the five rooms specified here. These convert to 18.25 kg/m², 12.64 kg/m², 20.12 kg/m², 16.85 kg/m², and 14.98 kg/m², respectively. For a constant heat release rate, take

$$\dot{Q} = m\Delta H_c/t$$

where

$\dot{Q}$ = heat release rate (kW)
m = mass (kg)
ΔH_c = heating value of ordinary combustibles (kJ/kg)
t = time (seconds)

so that for the five cases of interest

(a) Living room = fire load 18.25 kg/m²

$$\dot{Q} = \frac{(12\text{ m}^2)(18.25\text{ kg/m}^2)(18{,}608\text{ kJ/kg})}{(20\text{ min})\ (60\text{ s/min})}$$

$$= 3396\text{ kJ/s} = 3.40\text{ MW}$$

(b) Family room = fire load 12.64 kg/m²

$$\dot{Q} = 2.35\text{ MW}$$

(c) Bedroom = fire load 20.12 kg/m²

$$\dot{Q} = 3.74\text{ MW}$$

(d) Dining room = fire load 16.85 kg/m²

$$\dot{Q} = 3.14\text{ MW}$$

(e) Kitchen = fire load 14.98 kg/m²

$$\dot{Q} = 2.79\text{ MW}$$

Corresponding ventilation factors, $A\sqrt{H}$, can be calculated from the ventilation limit

$$\dot{Q} = \dot{m}_{\text{fuel}}\,\Delta H_{c,\text{fuel}} = \dot{m}_{\text{air}}\,\Delta H_{c,\text{air}}$$

where

$\dot{Q}$ = heat release rate kW
$\dot{m}_{\text{fuel}}$ = burning rate of fuel (kg/s)
$\Delta H_{c,\text{fuel}}$ = heating value of fuel (kJ/kg)
$\dot{m}_{\text{air}}$ = consumption rate of air participating in combustion (kg/s)
$\Delta H_{c,\text{air}}$ = heat release value of air participating in combustion (kJ/kg) [≈ 3000 kJ/kg]

and

$$\dot{m}_{\text{air}} = 0.5\,A\sqrt{H}$$

where

A = area of opening (m²)
H = height of opening (m)

Hence

$$\dot{Q} = 0.5\,A\sqrt{H}(3000) = (1500)\,A\sqrt{H}$$

and rearranging to isolate the ventilation factor gives

$$A\sqrt{H} = \dot{Q}/1500$$

Finally, the ventilation opening values follow for the five cases of $\dot{Q}$ just determined as

(a) $A\sqrt{H} = 3400/1500 = 2.27\ m^{5/2}$
(b) $A\sqrt{H} = 2300/1500 = 1.57\ m^{5/2}$
(c) $A\sqrt{H} = 3740/1500 = 2.49\ m^{5/2}$
(d) $A\sqrt{H} = 3140/1500 = 2.09\ m^{5/2}$
(e) $A\sqrt{H} = 2790/1500 = 1.86\ m^{5/2}$

Graphic Example

Corresponding to the typical fire loads just described in rooms of floor size 3 m × 4 m in area, the above calculations are shown graphically in Figure 23 for total burns in a 20-minute period. Figure 24 shows corresponding values of the ventilation factor. This value represents the ventilation opening size requirement in order to permit sufficient air for combustion to enter the enclosure and the products of combustion to leave.

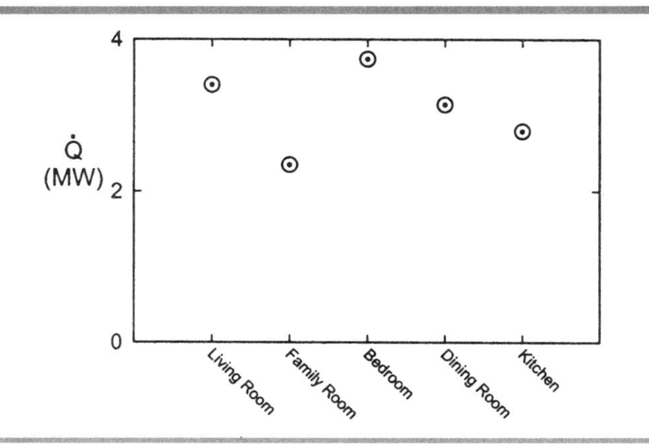

FIGURE 23. Heat release rate for 20-minute burn (Adapted from Lilley 2004)

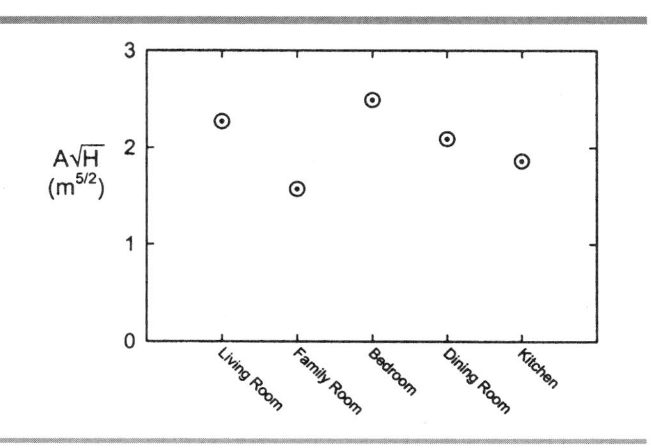

FIGURE 24. Minimum ventilation factor needed for 20-minute burn (Adapted from Lilley 2004)

Maximum Upward Velocity for Various Pool Fires

Calculate the maximum temperature and maximum velocity in the plume above a 1 m² pool of burning

(a) methanol
(b) ethanol
(c) gasoline
(d) kerosene
(e) jet A fuel
(f) polymethylmethacrylate
(g) polypropylene
(h) polystyrene

Solution

Assume room temperature to be 20 °C and take the maximum temperature from

$$\Delta T_0 = 650\ K$$

in each case so that maximum temperature is 670 °C. Maximum velocity is found from

$$u_{0,\max} = 1.97\dot{Q}_c^{\,1/5}$$

where $\dot{Q}_c = (1 - x_r)\dot{Q}$ and where x_r is the radiative fraction of the total heat released $\dot{Q}$. Radiative fraction estimates are taken from SFPE (2002). Thus for 1 m² pool fires, the velocities are computed and given in Table 11.

Graphic Example

Maximum upward velocities in the fire plume for a 1-meter-square pool fire are shown in Figure 25 for eight different fuels. Notice that the more volatile fuels (gasoline, kerosene, and jet A) exhibit higher velocities.

TABLE 11

Maximum Plume Velocity

Fuel	Radiative Fraction x_r	Burning Rate $\dot{m}''$ kg/m²s	Heating Value H_c MJ/kg	Maximum Plume Velocity (m/s)
methanol	0.20	0.017	20.0	6.0
ethanol	0.20	0.015	26.8	6.3
gasoline	0.40	0.055	43.7	8.4
kerosene	0.40	0.039	43.2	7.9
jet A	0.40	0.054	43.0	8.4
PMMA	0.40	0.020	24.9	6.2
polypropylene	0.40	0.018	43.2	6.7
polystyrene	0.40	0.034	39.7	7.5

(Adapted from Lilley 2004)

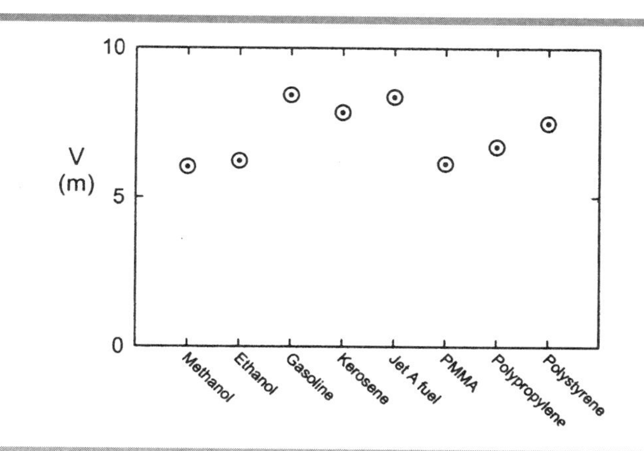

FIGURE 25. Maximum upward velocity in flame plume of a 1-meter-square pool fire for different fuels (Adapted from Lilley 2004)

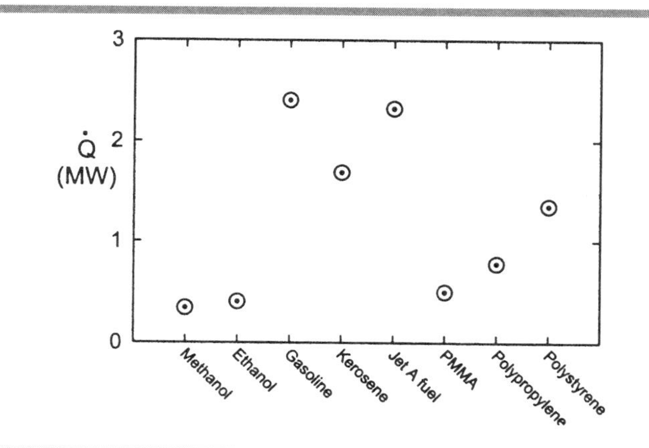

FIGURE 26. Heat release of a 1-meter-square pool fire for different fuels (Adapted from Lilley 2004)

Corresponding heat release rates $\dot{Q}$ in MW are given in Figure 26 (for eight different fuels, each in a 1-meter-square pool fire). Notice again that gasoline, kerosene, and jet A fuels exhibit greater heat release rates for the same base area of burning.

Flame Height for Various Pool Fires

Estimate the flame height and fraction of heat radiated for the eight pool fires specified in the last section.

Solution

Flame heights are found using the Heskestad (1983) equation

$$L = -1.02D + 0.23\dot{Q}^{2/5}$$

or the NFPA 921 (2008) equation

$$H_f = 0.174(k\dot{Q})^{0.4}$$

with $k = 1$ (no nearby walls), where

L = flame length (m)
D = pool diameter (m)
$\dot{Q}$ = heat release rate of fire (kW)
H_f = flame length (m)

Thus, when $D = (4A/\pi)^{1/2} = 1.13$ m, $A = 1$ m², and radiative fractions as given before (alcohols with 0.2 and the other fuels with 0.4),

(a)	methanol	$L = 1.22$ m	$H_f = 1.79$ m
(b)	ethanol	$L = 1.38$ m	$H_f = 1.92$ m
(c)	gasoline	$L = 4.02$ m	$H_f = 3.92$ m
(d)	kerosene	$L = 3.34$ m	$H_f = 3.40$ m
(e)	jet A	$L = 3.95$ m	$H_f = 3.86$ m
(f)	PMMA	$L = 1.61$ m	$H_f = 2.09$ m
(g)	polypropylene	$L = 2.14$ m	$H_f = 2.49$ m
(h)	polystyrene	$L = 2.96$ m	$H_f = 3.11$ m

Graphic Example

Computed flame heights are portrayed in Figure 27 for the range of eight different fuels considered, each pool of size 1 m². Both the Heskested equation (Method 1) and the NFPA 921 (2008) equation (Method 2) are used, with slightly different values given for each case, with the latter equation giving, if anything, larger values.

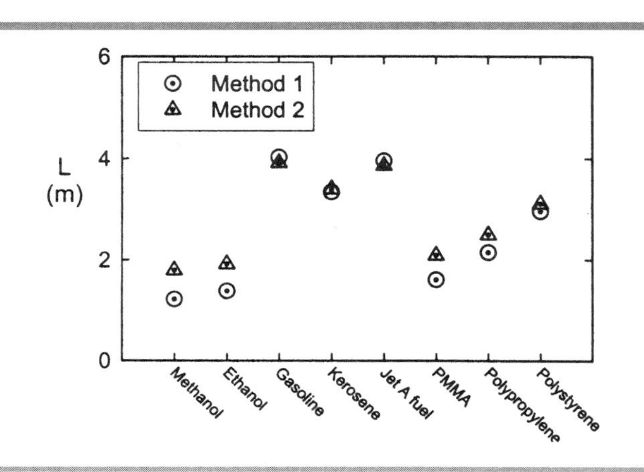

FIGURE 27. Flame length of a 1-meter-square pool fire for different fuels (Adapted from Lilley 2004)

Notice that the more volatile fuels considered (gasoline, kerosene, and jet A) have greater flame lengths.

Minimum Safe Distance from Various Pool Fires

Using an estimated value of the radiative fraction of the heat radiated from the pool fire, and using a simple inverse square distance law, determine the minimum distance away that

(a) easily ignitable items
(b) normal items
(c) difficult-to-ignite items

would have to be for safety in each of the eight pool fires specified in the last section.

Solution

Using the equation

$$\dot{q}'' = \frac{x_r \dot{Q}}{4\pi R_0^2} \text{ or } R_0 = \left[\frac{x_r \dot{Q}}{4\pi \dot{q}''}\right]^{1/2}$$

with easy-, normal-, and difficult-to-ignite items needing 10, 20, and 40 kW/m² yields

(a) Easily ignitable items:

methanol	$R_0 = 0.74$ m
ethanol	$R_0 = 0.80$ m
gasoline	$R_0 = 2.77$ m
kerosene	$R_0 = 2.32$ m
jet A	$R_0 = 2.72$ m
PMMA	$R_0 = 1.26$ m
polypropylene	$R_0 = 1.57$ m
polystyrene	$R_0 = 2.07$ m

(b) Normally ignitable items:
Similar calculations

(c) Difficult to ignite items:
Similar calculations

Note that the calculations given here have used the radiative fraction x_r as 0.2 for the alcohols (methanol and ethanol) and 0.4 for the other fuels considered. This is consistent with data in SFPE (2002).

Graphic Example

For the eight 1 m² pool fires considered, calculations of this kind generate values as seen in Figure 28 for the minimum safe distance for easy-, normal-, and difficult-to-ignite items. Clearly the volatile fuels (gasoline, kerosene, and jet A) have a greater propensity to ignite items at some distance away.

Useful, related calculations are exhibited in Figure 29. Here the heat flux in kW/m² landing on a target is given as a function of total heat release rate $\dot{Q}$ in MW and distance away in meters. A radiative fraction x_r of 0.4 is used in these calculations so that the data may be compared readily with the total heat release rate of pool fires (for example) given earlier. Table 12 also provides the data.

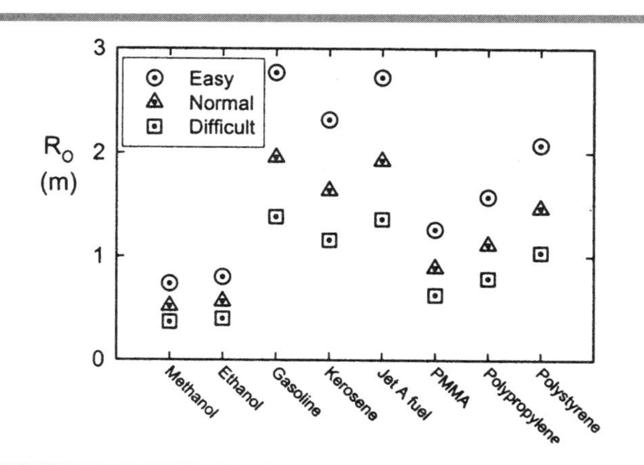

FIGURE 28. Minimum safe distance from a 1-meter-square pool fire for different fuels (Adapted from Lilley 2004)

Flashover

After how many seconds will flashover occur in a typical room 3 m wide, 4 m long, and 2.4 m high that is subjected to a t^2-fire with an incubation period of 75 seconds and

(a) slow
(b) medium
(c) fast
(d) ultrafast

fires with a single ventilation opening of size 2 meters wide and 2 meters high? Take growth parameters as

Slow	0.00293 kW/s^2
Medium	0.01172 kW/s^2
Fast	0.04690 kW/s^2
Ultrafast	0.18760 kW/s^2

TABLE 12

Heat Flux $\dot{q}''$ in kW/m^2 on Target

Source Total Heat $\dot{Q}$ MW	Distance of Target from Source			
	1 m	2 m	5 m	10 m
1	31.8	7.96	1.27	0.32
5	159.0	39.80	6.37	1.59
10	318.0	79.60	12.70	3.18

(Adapted from Lilley 2004)

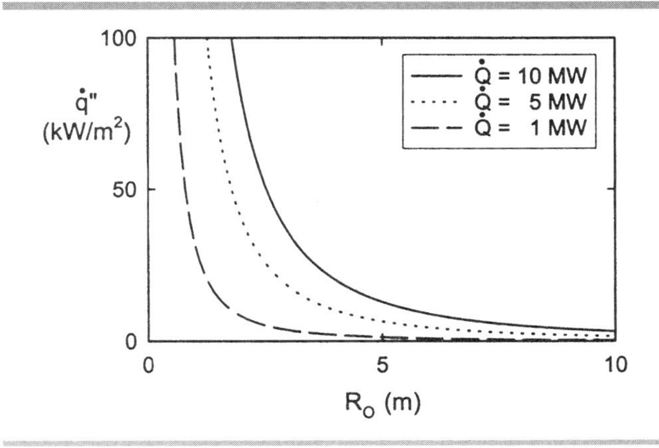

FIGURE 29. Heat flux on target versus source heat release rate and distance away (Adapted from Lilley 2004)

Solution

Using the flashover criterion of Thomas, the required heat release rate is

$$\dot{Q} = 378\, A_0\sqrt{H_0} + 7.8A_T$$

Use of the given data with a single opening of size 2 meters wide and 2 meters high calculates that

$$A_0\sqrt{H_0} = 5.66 \text{ m}^{5/2}$$
$$A_T = 53.60 \text{ m}^2$$

so that the heat release rate needed for flashover evaluates as

$$\dot{Q} = 2558 \text{ kW}$$

The t^2-fire is characterized (after the initial incubation period) by

$$\dot{Q} = \alpha t^2$$
$$= (1000/t_g^2)t^2$$

where α is the fire growth parameter in kW/s^2 and t_g is the growth time in seconds to reach a heat release rate of 1000 kW. Remembering to include in the calculations the incubation period of 75 seconds, the time to reach $\dot{Q}$ = 2558 kW calculates that

(a) Slow
$t = 75 + [2558/0.00293]^{1/2} = 1009$ seconds
(b) Medium
$t = 75 + [2558/0.01172]^{1/2} = 542$ seconds
(c) Fast
$t = 75 + [2558/0.0469\ ]^{1/2} = 309$ seconds
(d) Ultrafast
$t = 75 + [2558/0.1876\ ]^{1/2} = 192$ seconds

Graphic Example

Figure 30 shows heat release rate versus time from the start of ignition for four different fire growth specifications. It should be noted that the medium fire growth is typical of many room fires and, even for this fire specification, the rate of heating is such that rates in excess of 2 MW occur in as little as 400 seconds. The level of heat release rate necessary for room flashover is a function of several parameters, the most important of which is ventilation. Illustrated in Figure 31 is the heat release rate needed for flashover

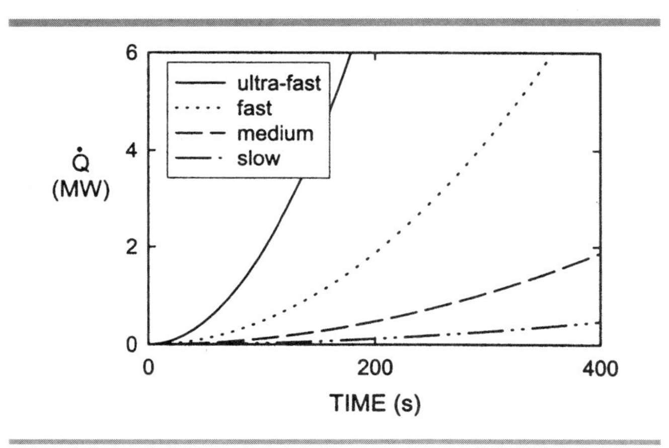

FIGURE 30. Heat release rate versus time for different fire growth specifications (Adapted from Lilley 2004)

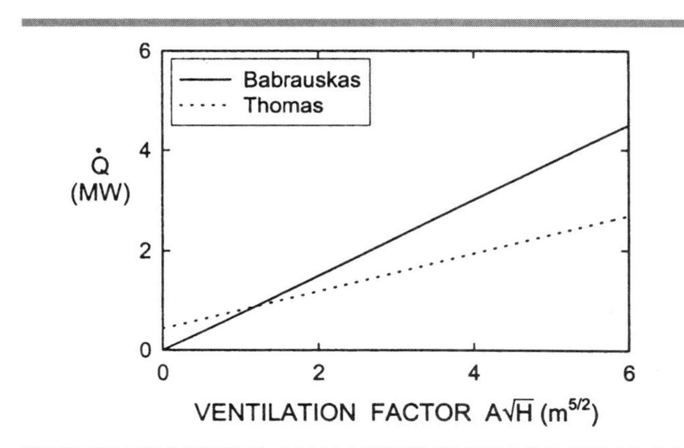

FIGURE 31. Heat release rate required for flashover versus ventilation factor according to two different theories (Adapted from Lilley 2004)

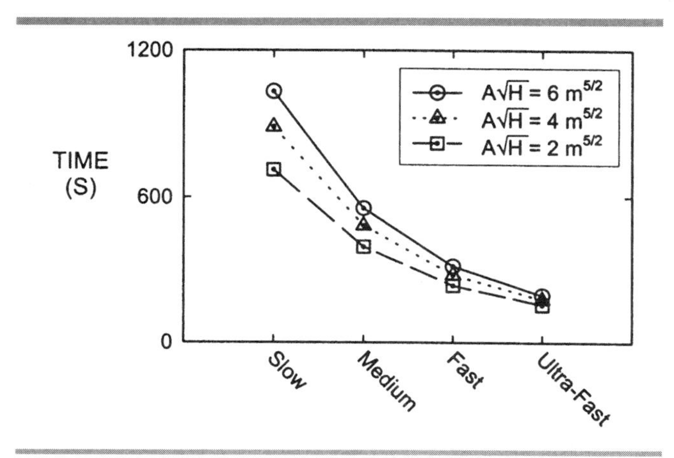

FIGURE 32. Time to reach flashover conditions versus fire growth specification and ventilation factor (Adapted from Lilley 2004)

conditions to exist versus ventilation factor, according to the two different theories of Babrauskas and Thomas. This calculation is specific to a room having a floor size of 3 meters by 4 meters, with a 2.4-meter-high ceiling. Another way of discussing the calculations is to consider the time needed from ignition to reach flashover conditions with the different types of fire specifications. The details are shown in Figure 32 (using Thomas's equation) for three different values of the ventilation factor. Notice that the 75-second incubation period is included in these values. The trends are clearly evident: the rapidity of fire growth is extremely important in determining flashover conditions, and ventilation is of only lesser importance. Notice that medium-to-fast fire growths exhibit the onset of flashover in just a few minutes.

Chapter Conclusion

This chapter discussed important information related to fire dynamics: evolution, spread, and ignition of flammable material from leaks in appliances and process equipment; combustion; burning rates; fire and flame size; radiant ignition of other items; fire spread rates; ventilation limit imposed on the fire growth because of the size of openings that allow the entry of fresh air; and prospects for the occurrence of flashover. These are the main components related to fire dynamics' issues in real-world fires. These data have been used to enhance understanding of the physics of fires and increase the capacity to more accurately predict their occurrence and simulate their growth and hazards. Information about experimental data and analysis helps in developing and extending the range of knowledge, as well as modeling capabilities. Information, analysis, and calculations of the phenomena involved in fire dynamics can assist in digesting and applying scientific principles to real-world fire situations. Safety engineers can develop an appreciation for the technical and scientific aspects of these phenomena and their applicability to actual situations. Sample calculations were exhibited to illustrate a number of aspects of fire dynamics.

REFERENCES

Alpert, R. L., and E. J. Ward. 1983. SFPE TR 83-2, *Evaluating Unsprinklered Fire Hazards*. Boston: SPFE.

Alvord, D. M. 1995. *CFAST Output Comparison Method and Its Use in Comparing Different CFAST Versions*. NISTIR 5705. Springfield, VA: National Technical Information Services.

American Gas Association (AGA). 1988. *Fundamentals of Gas Appliances*. Arlington, VA: AGA.

______. 1997. *Directory of Certified Appliances and Accessories*. Cleveland, OH: AGA.

American Institute of Chemical Engineers (AIChE). 1990. *Safety, Health and Loss Prevention in Chemical Processes*. New York: AIChE.

American Society of Mechanical Engineers (ASME). 1973. *Prediction of the Dispersion of Airborne Effluents*. New York: ASME.

American Society for Testing and Materials International (ASTM). 1969. ASTM 286-89, *Standard Method of Test for Surface Flammability Using an 8-ft (2.44 m) Tunnel Furnace*. Philadelphia, PA: ASTM.

______. 2007. ASTM E84-07b, *Standard Test Method for Surface Burning Characteristics of Building Materials*. West Conshohocken, PA: ASTM.

Babrauskas, V. 1979. Technical Note 991, *COMPF2—A Program for Calculating Post-Flashover Fire Temperatures*. Springfield, VA: National Bureau of Standards.

______. 1980. "Estimating Room Flashover Potential." *Fire Technology* 16:94–104.

______. 1981. NBSIR Report 81-2271, *Will the Second Item Ignite?* Gaithersburg, MD: National Bureau of Standards.

______. 1984. "Upholstered Furniture Room Fires—Measurements, Comparison with Furniture Calorimeter Data, and Flashover Predictions." *Journal of Fire Sciences* 2:5–19.

______. 1985. "Free-Burning Fires." Proceedings of the Society of Fire Protection Engineers Symposium: Quantitative methods for fire hazard analysis. College Park, MD: University of Maryland.

______. 2002. "Heat Release Rates." From *Fire Protection Engineering*, 3d ed. Boston, MA: NFPA and SFPE.

______. 2003. *Ignition Handbook*. Issaquah, WA: Fire Science Publishers and SPFE.

Babrauskas, V., and J. F. Krasny. 1985. *Fire Behavior of Upholstered Furniture*. NBS monograph. Gaithersburg, MD: National Bureau of Standards.

Babrauskas, V., and R. B. Williamson. 1979. "Post-flashover compartment fires—application of a theoretical model." *Fire and Materials* 3(1):1–7.

Babrauskas, V., and S. J. Grayson, eds. 1992. *Heat Release in Fires*. New York: Elsevier Applied Science.

Babrauskas, V., R. L. Lawson, W. D. Walton, and W. H. Twilley. 1982. *Upholstered Furniture Heat Release Rates Measured with a Furniture Calorimeter*. NBSIR 82-2604. Gaithersburg, MD: National Bureau of Standards.

Belles, D. W. 2003. "Interior Finish." From *Fire Protection Handbook*, 19th ed. A. E. Cote, ed. Quincy, MA: NFPA.

Berry, D. J. 1989. *Fire Litigation Handbook*. Quincy, MA: NFPA.

Beyler, C. 2002. "Flammability limits of premixed and diffusion flames." From *Handbook of Fire Protection Engineering*, 3d ed. Boston, MA: NFPA and SFPE.

Budnick, E. K., D. D. Evans, and H. E. Nelson. 2003. "Simplified Calculations for Enclosure Fires." From *Fire Protection Handbook*, 19th ed. A. E. Cote, ed. Quincy, MA: NFPA.

Bukowski, R., et al. 1989. *The HAZARD-1 Computer Code Package for Fire Hazard Assessment*. Gaithersburg, MD: National Bureau of Standards (National Institute of Standards and Technology).

Carroll, J. R. 1979. Physical and Technical Aspects of Fire and Arson Investigation. Springfield, IL: Thomas Publishers.

Chow, C-Y. 1979. *Computational Fluid Mechanics*. New York: Wiley.

Clarke, F. B. 1991. "Fire Hazards of Materials: An Overview." From *Fire Protection Handbook*, 17th ed. Quincy, MA: NFPA.

Cooper, L. Y. 1980. "A Mathematical Model for Estimating Available Safe Egress Time in Fires." NBSIR Report 80-2172. Gaithersburg, MD: National Bureau of Standards.

Cooper, L. Y., and D. W. Stroup. 1982. "A Concept for Estimating Available Safe Egress Time in Fires." NBSIR Report 82-2578. Gaithersburg, MD: National Bureau of Standards.

Cote, A. E., ed. 2003. *Fire Protection Handbook*, 19th ed. Quincy, MA: NFPA.

______. 2003. "Chemistry and Physics of Fire." From *Fire Protection Handbook*, 19th ed. A. E. Cote, ed. Quincy, MA: NFPA.

Crowl, D. A., and J. F. Louvar. 1990. *Chemical Process Safety: Fundamentals with Applications*. Englewood Cliffs, NJ: Prentice Hall.

Drysdale, D. 1998. *An Introduction to Fire Dynamics*. 2d ed. Chichester, England: Wiley.

Emmons, H. W. 1985. "The Needed Fire Science." From *Fire Safety Science*. C. E. Grant and P. J. Pagni, eds. New York: Hemisphere.

______. 2002. "Vent Flows." From *Handbook of Fire Protection Engineering*, 3d ed. Boston: NFPA and SFPE.

Fang, J. B., and J. N. Breese. 1980. NBSIR 80-2120. *Fire Development in Residential Basement Rooms, Interim Report*. Gaithersburg, MD: National Bureau of Standards.

Forney, G. P., and K. B. McGrattan. 2004. NIST Special Publication 1017. *User's Guide for Smokeview Version 4*. Gaithersburg, MD: NIST.

Friedman, R. 1991. *Survey of Computer Models for Fire and Smoke*. 2d ed. Norwood, MA: Factory Mutual Research Corp.

Gupta, A. K., and D. G. Lilley. 1984. *Flowfield Modeling and Diagnostics*. Tunbridge Wells, England: Abacus Press.

Hanna, S. R., and P. J. Drivas. 1987. *Vapor Cloud Dispersion Models*. New York: American Institute of Chemical Engineers.

Heskestad, G. 1983. "Virtual Origins of Fire Plumes." *Fire Safety Journal* 5:103–108.

______. 2003. "Venting Practices." From *Fire Protection Handbook*, 19th ed. A. E. Cote, ed. Quincy, MA: NFPA.

Hicks, H. D. 2003. "Confinement of Fire in Buildings." From *Fire Protection Handbook*, 19th ed. A. E. Cote, ed. Quincy, MA: NFPA.

Huggett, C. 1980. "Estimation of Rate of Heat Release by Means of Oxygen Consumption Measurement." *Fire and Materials* 4:61–65.

Janessens, M., and W. J. Parker. 1992. "Oxygen Consumption Calorimetery." From *Heat Release in Fires*. London: Elsevier Applied Science.

Jones, W. W. 1985. "A Multicompartment Model for the Spread of Fire, Smoke and Toxic Gases." *Fire Safety Journal* 7:55–79.

Jones, W. W., and R. D. Peacock. 1989. NIST Technical Note 1262, *Technical Reference Guide for FAST Version 18*. Gaithersburg, MD: NIST.

Jones, W. W., G. P. Forney, R. D. Peacock, and P. A. Reneke. 2000. NIST Technical Note 1431, *Technical Reference for CFAST: An Engineering Tool for Estimating Fire and Smoke Transport*. Gaithersburg, MD: NIST.

Kameel, R., and E. E. Khalil. "Transient Numerical Simulation of Air Characteristics in Ventilated Compartment Fire." Paper AIAA-2004-0478. 42nd Aerospace Sciences Meeting, Reno, NV, Jan. 5–8, 2004.

Karlsson, B., and J. G. Quintiere. 2000. *Enclosure Fire Dynamics*. Boca Raton, FL: CRC Press.

Kawagoe, K. 1958. "Fire Behaviour in Rooms." Report of the Building Research Institute, No. 27, Japanese Government.

Kawagoe, K., and T. Sekine. 1963. "Estimation of Fire Temperature–Time Curve for Rooms." BRO Occasional Report, No. 11. Building Research Institute, Ministry of Construction, Japanese Government.

Kim, H.-J., and D. G. Lilley. "Flashover: A Study of Parametric Effects on the Time to Reach Flashover Conditions." Paper DETC 97/CIE-4427. Proceedings of the ASME 17th International Computers in Engineering Conference/Design Conference, Sacramento, CA, Sept. 14–17, 1997.

______. "A Comparison of Flashover Theories." Proceedings of the ASME International Computers in Engineering Conference, Atlanta, GA, Sept. 13–16, 1998.

______. 2002a. "Flashover: A Study of Parametric Effects on the Time to Reach Flashover Conditions." *Journal of Propulsion and Power* 18(3):669–673.

______. 2002b. "Comparison of Criteria for Room Flashover." AIAA Paper 99-0343. *Journal of Propulsion and Power* 18(3):674–678.

Lawson, J. R., and J. G. Quintiere. 1985. NBSIR 85-3196, *Slide-Rule Estimates of Fire Growth*. Gaithersburg, MD: National Bureau of Standards.

Lawson, J. R., W. D. Walton, and W. H. Twilley. 1984. NBSIR 83-2787, *Fire Performance of Furnishings as Measured in the NBS Furniture Calorimeter, Part 1*. Gaithersburg, MD: National Bureau of Standards.

Lees, F. P. 1980. *Loss Prevention in the Process Industries*. London: Butterworths.

Lilley, D. G. 1995a. *Fire Dynamics*. Short course. Presented initially in Tulsa, OK, March 1995.

______. 1995b. "Fire Dynamics." Paper AIAA-95-0894. 33rd Aerospace Sciences Meeting. Reno, NV, Jan. 9–12, 1995.

______. "Fire Hazards of Flammable Fuel Release." Paper AIAA-96-0405. 34th Aerospace Sciences Meeting. Reno, NV, Jan. 15–18, 1996.

______. 1997a. "Structural Fire Development." Paper AIAA-97-0265. 35th Aerospace Sciences Meeting. Reno, NV, Jan. 6–10, 1997.

______. 1997b. "Estimating Fire Growth and Temperatures in Structural Fires." Paper AIAA-97-0266. 35th Aerospace Sciences Meeting. Reno, NV, Jan. 6–10, 1997.

______. 1998a. "Fuel Release and Ignition Calculations." Paper AIAA-98-0262. 36th Aerospace Sciences Meeting. Reno, NV, Jan. 12–15, 1998.

______. 1998b. "Fire Development Calculations." Paper AIAA-98-0267. 36th Aerospace Sciences Meeting. Reno, NV, Jan. 12–15, 1998.

______. 2000. "Minimum Safe Distance from Pool Fires." *Journal of Propulsion and Power* 16(4):649–652.

______. 2004. *Fire Dynamics*. 2d ed. Stillwater, OK: Lilley & Associates Course.

______. 2007. "Volume and Temperature of Smoke in Fires." Paper AIAA 07-0600. 45th Aerospace Sciences Meeting. Reno, NV, Jan 8–11, 2007.

______. 2008. "Particle Trajectories in 3-D Space with an Excel/VBA Code." Paper AIAAA-08-1168. 46th Aerospace Sciences Meeting. Reno, NV, Jan 7–10, 2008.

Magee, R. S., and R. F. McAlevy III. 1971. "The Mechanism of Flame Spread." *Journal of Fire and Flammability* 2:271–297.

McCaffrey, B. J., J. G. Quintiere, and M. F. Harkleroad. 1981. "Estimating Room Temperatures and the Likelihood of Flashover Using Fire Data Correlations." *Fire Technology* 17(2):98–119.

McGrattan, K. B., ed. 2004. NIST Special Publication 1018, *Fire Dynamics Simulator (Version 4), Technical Reference Guide*. Gaithersburg, MD: NIST.

______. 2005. NIST Special Publication 1019, *Fire Dynamics Simulator (Version 4), User's Guide*. Gaithersburg, MD: NIST.

National Board of Fire Underwriters (NBFU). 1956. *Fire Resistance Ratings of Less than One Hour*. New York: NBFU.

National Fire Protection Association (NFPA). 2008. NFPA 921, *Guide for Fire and Explosion Investigations*. Quincy, MA: NFPA.

______. 2006a. NFPA 101, *Life Safety Code*. Quincy, MA: NFPA.

______. 2006b. NFPA 255, *Standard Method of Test of Surface Burn Characteristics of Building Materials*. Quincy, MA: NFPA.

Olenick, S., and D. Carpenter. 2003. "An Updated International Survey of Computer Models of Fire and Smoke." *Journal of Fire Protection Engineering* 13:87–110.

Patton, A. J. 1994. *Fire Litigation Sourcebook*. 2d ed. New York: Wiley.

Peacock, R. D., et al. 1994. NISTIR 5410, *An Update Guide for HAZARD 1 Version 1.2*. Gaithersburg, MD: National Institute of Standards and Technology.

Pettersson, O., S. E. Magnusson, and J. Thor. 1976. *Fire Engineering Design of Steel Structures*. Stockholm, Sweden: Stalbyggnadsinstitutet.

Portier, R. W., R. D. Peacock, and P. A. Reneke. 1996. NIST Special Publication 899, *FASTLite: Engineering Tools for Estimating Fire Growth and Smoke Transport*. Gaithersburg, MD: NIST.

Quintiere, J. G. 1998. *Principles of Fire Behavior*. Albany, NY: Delmar Publishers.

______. 2002a. "Surface Flame Spread." *Handbook of Fire Protection Engineering*. 3d ed. Boston, MA: NFPA and SFPE.

______. 2002b. "Compartment Fire Modeling." From *Handbook of Fire Protection Engineering*, 3d ed. Boston, MA: NFPA and SFPE.

Quintiere, J. G., and M. Harkleroad. 1984. "New Concepts for Measuring Flame Spread Properties." Symposium on Application of Fire Science to Fire Engineering, ASTM and SFPE, Denver, CO, June 27, 1984.

Ramachandran G. 2002. "Stochastic Models of Fire Growth." From *Handbook of Fire Protection Engineering*, 2d ed. Boston, MA: NFPA and SFPE.

Rockett, J. A. 1976. "Fire Induced Gas Flow in an Enclosure." *Combustion Science and Technology* 12:165–175.

Scudamore, M. J., P. J. Briggs, and F. H. Prager. 1991. "Cone Calorimetry—A Review of Tests Carried Out on Plastics for the Association of Plastics Manufacturers in Europe." *Fire and Materials* 15:65–84.

Slye, O. M. 2003. "Flammable and Combustible Liquids." From *Fire Protection Handbook*, 19th ed.A. E. Cote, ed. Quincy, MA: NFPA.

Society of Fire Protection Engineers (SFPE). 2002. *Handbook of Fire Protection Engineering*. 3d ed. Boston, MA: NFPA and SFPE.

Tewarson, A. 2002. "Generation of Heat and Chemical Compounds in Fires." From *Handbook of Fire Protection Engineering*, 2d ed. Boston, MA: NFPA and SFPE.

Thomas, P. H. 1974. "Fires in Enclosures." From *Heat Transfer in Fire*, P. L. Blackshear, ed. New York: Halsted-Wiley.

______. 1981. "Testing Products and Materials for Their Contribution to Flashover in Rooms." *Fire and Materials* 5(3):103–111.

Thomas P. H., and A. J. M. Heselden. 1972. Fire Research Note No. 923, *Fully Developed Fires in Single Compartments*. Borehamwood, UK: Fire Research Station.

Underwriters' Laboratories (UL). 2003. UL 723, *Standard for Test Burning Characteristics of Building Materials* (Aug 29, 2003). Northbrook, IL: UL.

Walton W. D., and P. H. Thomas. 2002. "Estimating Temperatures in Compartment Fires." From *Handbook of Fire Protection Engineering*, 2d ed. Boston: NFPA and SFPE.

Walton, W. D., S. R. Baer, and W. W. Jones. 1985. NBSIR 85-3284, *User's Guide for FAST*. Gaithersburg, MD: National Bureau of Standards.

Zukoski, E. E. 1985. "Fluid Dynamic Aspects of Room Fires." From *Fire Safety Science*, C. E. Grant and P. J. Pagni, eds. New York: Hemisphere.

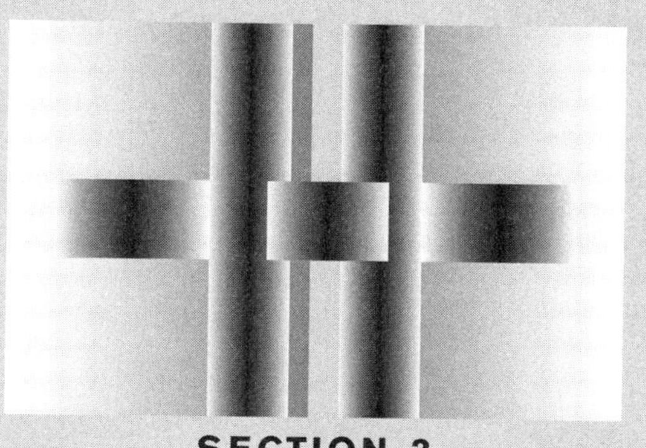

SECTION 3
FIRE PREVENTION AND PROTECTION

APPLIED SCIENCE AND ENGINEERING: FIRE PREVENTION AND CONTROL

Craig Schroll

LEARNING OBJECTIVES

- Identify key elements associated with life safety
- Define the three components of the means of egress
- Explain the purpose of compartmentation
- Identify essential elements of fire-loss prevention and control programs
- Define a fire hazard
- Identify at least three common sources of ignition
- Explain the key items concerning conducting hot work safely
- Explain the role of housekeeping in fire-loss prevention
- Explain the difference between flammable and combustible liquids
- Identify at least three key concepts concerning flammable liquids safety
- Identify at least one type of portable fire extinguisher rated for each of the five classes of fire
- List three safety criteria that must be considered prior to using a portable fire extinguisher

THIS CHAPTER focuses on reducing the hazards that may permit a fire to start and mitigating the impacts of fires that do occur. Managing fire risks is a critical aspect of any effective safety program.

Much of our experience with fire relates to using it in a controlled way—for example, for heating and cooking. Controlled fires are essential to our way of life. Uncontrolled fires can and do destroy life and property.

In a book of this type, it would be inaccurate to describe the coverage on any single topic as comprehensive, and this chapter is no exception. The intent here is to provide an overview of several essential aspects of fire-loss prevention and control. Additional resource materials listed at the end of this chapter offer additional depth on topics that may be of particular interest to readers.

Significant regulatory and code-compliance issues are involved with fire prevention and control. This chapter will not delve deeply into citing specific regulation- and code-based requirements, as they are covered in the regulatory chapter of the fire unit. Regulatory requirements relative to fire vary considerably from one jurisdiction to another. Safety professionals must refer to local requirements when dealing strictly with compliance concerns.

LIFE SAFETY

Safety of the occupants in a facility is the highest priority of fire prevention and control. There is a long, sad history of failing to fulfill this responsibility (Table 1).

From a strict compliance perspective, the National Fire Protection Association's (NFPA) *Life Safety Code* (NFPA 101 2006b)

TABLE 1

Selected Historical Large-Loss-of-Life Fires

Facility	Date	Deaths
The Station Nightclub, West Warwick, RI	February 20, 2003	100
Happy Land Social club, Bronx, NY	March 25, 1990	87
Beverly Hills Supper Club, Southgate, KY	May 28, 1977	165
Our Lady of the Angels School, Chicago, IL	December 1, 1958	95
Winecoff Hotel, Atlanta, GA	December 7, 1946	119
Ringling Brothers Barnum & Bailey Circus, Hartford, CT	July 6, 1944	168
Cocoanut Grove nightclub, Boston, MA	November, 28, 1942	492
Rhythm Club, Natchez, MS	April 23, 1940	207
Consolidated School, New London, TX	March 18, 1937	294
Ohio State Penitentiary, Columbus, OH	April, 21, 1930	320
Cleveland Clinic Hospital, Cleveland, OH	May 15, 1929	125
Eddystone Ammunition Company, Eddystone, PA	April 10, 1917	133
Triangle Shirtwaist Company, New York, NY	March 25, 1911	146
Iroquois Theater, Chicago, IL	December, 30, 1903	602

is one of the best references to use as a guide for providing adequate life safety. This code may or may not carry the force of law in your area depending upon whether it has been adopted by the local authority having jurisdiction (AHJ).

When using the *Life Safety Code*, it is best to start by determining the occupancy type of the facility. The occupancy chapters (11–42) of the *Code* drive the requirements. Most occupancy types have separate chapters for new and existing facilities. This organizational concept allows the introduction of more restrictive requirements over time without forcing retrofitting of existing structures. If an issue is determined important enough by the *Code*'s technical committee, requirements may also be added to the existing occupancy chapter. For example, after the Station Club fire in Rhode Island, the existing assembly-occupancy chapter was modified to require sprinklers in assembly occupancies such as nightclubs, bars with live entertainment, and dance halls with a capacity of 100 persons or more. Detailed information on specific requirements is provided in the topic chapters—primarily chapters seven through ten. For ease of use, similar sections of the occupancy chapters are numbered in the same way. For example, in chapter twelve, on new assembly occupancies, the requirements for occupant load are in section 12.1.7, and in chapter 39, on existing business occupancies, the requirements for occupant load are in section 39.1.7. With a few exceptions, this standardization of the section numbering is consistent in all of the occupancy chapters.

Life safety includes several fundamental issues that must all be handled effectively in order to improve the opportunity for people to be protected from fire and escape while conditions within the structure are still survivable.

Elements of Life Safety

Escape Time

Providing time for escape is one element of life safety. It is accomplished through a combination of early warning (detection and alarm systems) and slowing the growth and spread of a fire and fire products through construction, control of fuel, and suppression and other systems. These areas are covered by codes and regulations that must be reviewed for compliance. Relevant systems are covered elsewhere in this book.

Means of Egress

Another fundamental issue with life safety is maintaining a sufficient number of exits and a clear, unobstructed, well-identified, protected, and properly illuminated path of travel from all occupied areas of the structure to the outside. These paths are the *means of egress*. They include the exit access, the exit, and the exit discharge. The *exit access* includes all of the areas of the structure that an occupant might have to travel through to reach an exit. The *exit* (Figure 1) is the transition from inside the building to outside the building. In its simplest form, it is a door through an exterior wall at the level of exit discharge. It can also include a door into a protected stair tower, the stair tower itself, and the door from the stair tower to the outside or a protected corridor configuration that travels horizontally. The *exit discharge* (Figure 2) begins immediately outside the building and continues to the street.

Three major measures are used to determine the number, size, and location of exits within a facility (NFPA 2006b)—occupant load, egress capacity, and travel distance. The *occupant load* is the number of people permitted to be in a particular area of a building. It is based upon the occupancy type and the square footage of the area being considered. It limits the number of people permitted in the area at one time to a number that can safely exit the area in a reasonable time. The *egress capacity*—the number and size of exits—is based on code requirements and the number of people that can occupy the space. *Travel distance* is the maximum distance an occupant must walk before reaching an exit. Travel distance limits help to ensure that exits are positioned effectively. Details on specific requirements for number, size, and location of exits may be found in NFPA, *Life Safety Code* (NFPA 2006b).

Exits and access to them must be marked with signs (Figure 3). With a few exceptions, such as individual offices and sleeping rooms in hotels or dormitories, most areas of facilities must be provided

FIGURE 1. Exit (*Source:* © FIRECON 2007)

FIGURE 2. Exit discharge (*Source:* © FIRECON 2007)

FIGURE 3. Exit sign (*Source:* © FIRECON 2002)

FIGURE 4. Self-contained emergency light (*Source:* © FIRECON 2002)

with exit signs. These signs must be positioned to allow occupants to identify the path of egress from the building. Doors that could easily be mistaken for exits should be marked "Not an Exit." Exit signs must be illuminated internally, externally, or with photoluminescence. The illuminating method must provide for backup power if normal electric power is lost. Signs should be checked monthly.

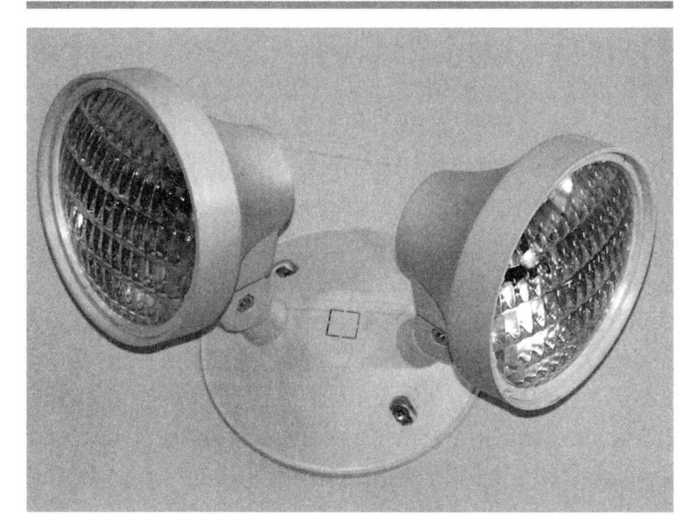

FIGURE 5: Generator-supplied emergency light (*Source:* © FIRECON 2005)

Adequate normal and emergency lighting must be provided so that people can see to safely travel the paths to the exits. Emergency lights may be self-contained units (Figure 4) with battery packs or may be connected to an emergency generator (Figure 5). Emergency lighting should be tested monthly for basic function and annually for the ability to provide illumination for at least 90 minutes. The annual test should be conducted by interrupting power to the normal lighting circuit so that the automatic activation feature of the emergency lighting is also tested.

One of the key challenges for safety professionals is maintaining the readiness of life-safety components and systems. Regular inspections are required to ensure that the means of egress are continuously in proper condition. Proper condition should be assessed against the fundamental criteria described above and detailed requirements from the code in force at a facility's site.

Effective emergency planning, covered elsewhere in this book, is also a critical component of providing for the life safety of occupants.

COMPARTMENTATION

Compartmentation—dividing structures into separate areas—is an effective method for reducing the spread of a fire. Restricting the fire to the compartment of origin is one of the primary goals of compartmentation.

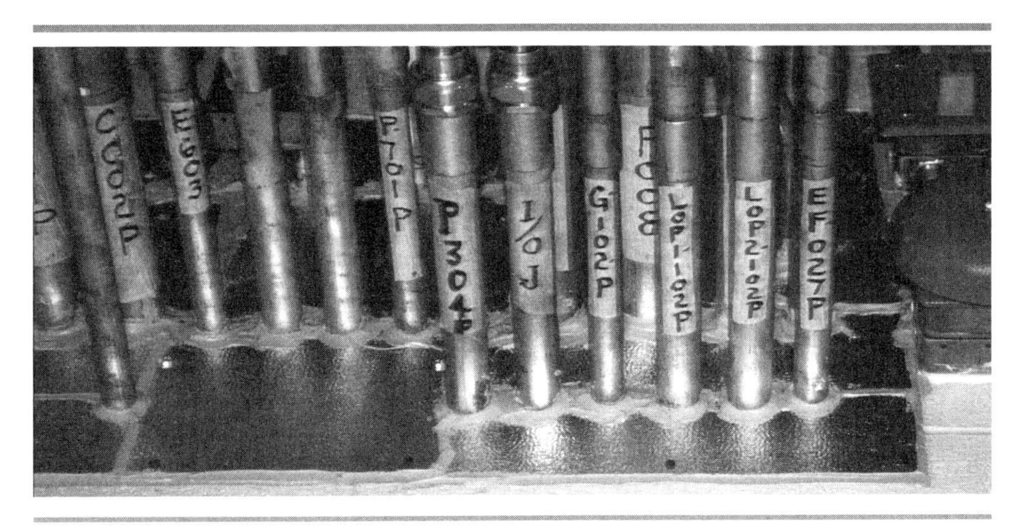

FIGURE 6. Fire barrier around conduit (*Source:* © FIRECON 1999)

FIGURE 7. Pass-through without putty (*Source:* © FIRECON 2005)

Another use of this technique is to protect the means of egress from the structure. The two phases of this process are initial design and construction and maintaining the integrity of the features over the life of the structure.

Initial design and construction is usually the easiest portion of this effort. Numerous codes clearly describe construction standards and methods for achieving effective separation within a structure. NFPA Standard 5000, *Building Construction and Safety Code* (NFPA 2006c) is an example of a code in which detailed requirements may be found. NFPA Standard 80, *Standard for Fire Doors and Other Opening Protectives* (NFPA 2007b), provides detailed requirements for fire doors and other devices designed to protect openings in fire-rated construction.

While codes are also clear about *maintaining the integrity of compartmentation,* this phase is more difficult to achieve during the life of a structure. Any breach to fire-rated wall, ceiling, or floor assemblies must be properly sealed to maintain compartmentation effectiveness. Unfortunately, this is not always done correctly. No penetration may be permitted through a fire wall or other rated enclosure that would allow the passage of heat, fire gases, or smoke. Figures 6 and 7 show both good (Figure 6) and bad (Figure 7) examples of handling a penetration through a fire wall.

Loss Prevention and Control Programs

Fire-loss control programs are specifically focused efforts aimed at reducing the risk of losses, the magnitude of potential losses, and the impact of those losses on continuing operations. Programs must address fire risk before it occurs (prevention), during a fire (prevention of further loss), and recovery following a fire. They are based on the assumption that it is more effective and ultimately less costly to manage the loss potential of an operation prior to an incident rather than deal with the impact after a loss has occurred.

No matter what size an organization is, what it produces, what services it provides, or where it is located, a loss prevention and control program is essential to its success. The size, scope, and complexity of the program will vary considerably from one organization to another, but the need for a program is always present. There are many approaches to loss prevention and control. No single approach is likely to be a perfect fit for every organization. The general approach described in this section comes largely from the author's experience with a wide variety of organizations.

The first step in the loss prevention and control process involves an analysis of the organization's loss potential. Two primary areas are evaluated initially: what can be lost and what may cause a loss. When examining possible causes, evaluations are made of the probability that an event will occur and the severity of the impact if it does. What can be lost may also be divided into several categories—for example, people, property, environment, continuity of operations, and heritage (Schroll 2002b).

Elements of an Effective Fire-Loss Prevention Program

Figure 8 shows a fire-loss control flowchart.

Based on my experience, effective loss prevention and control programs have several common elements that must be addressed (Figure 9).

- ***Obtain management commitment:*** The first element that must be present in an effective a loss-control program of any type is commitment on the part of management to having a program that is aimed at getting results and not just looking good on paper.

This commitment must come in several forms, the most important of which is also usually the most difficult to obtain. The organization's top management must recognize that loss control is an essential business priority. The fact that effective loss control is necessary not only from a human perspective but also in terms of costs and benefits seems fairly obvious to safety professionals, but it is not obvious to many people. Top management personnel often view loss prevention and control as a cost, not a benefit, because the money spent on loss control activities is easy to identify and count, but the money saved is more difficult to effectively document. One of the keys to having an effective program is convincing management that loss control must be an integral part of the organization's operation. Many managers have the impression that loss control is an add-on activity. The same individual who would not dream of running his or her business without insurance can't seem to grasp why money needs to be spent on training the workforce in fire prevention. This gap in understanding must be bridged with education. One of the more important roles of safety professionals is to educate top management.

Once people in top management understand the essential nature of loss prevention and control, they will accept the need for its components: an effective written policy and management support.

An *effective overall written loss prevention and control policy* must originate from the top management personnel of an organization. For a policy to be effective, it must be important to those who write it and be understood and acted upon by those who read it. The individuals responsible for safety fall in the middle. Top management will need advice on what policies to write and how they should be written, and employees will

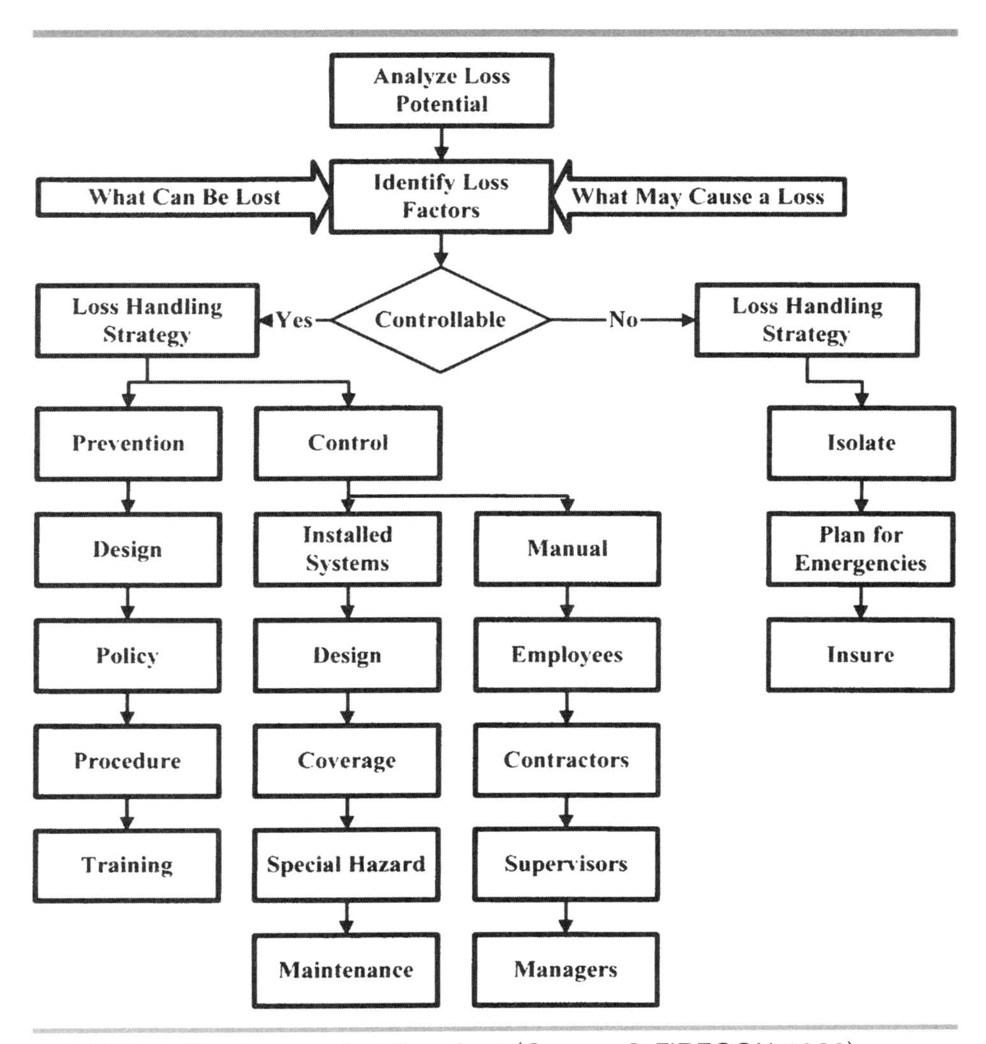

FIGURE 8. Fire-loss control flowchart (*Source:* © FIRECON 1989)

need training and assistance to use the policies and procedures effectively.

Support from management personnel usually takes two basic forms: time and money. Through their time, managers demonstrate visibly that the loss-control program is important to the organization and to them personally. By allocating money, they authorize safety professionals to use the organization's resources to have an impact on losses. Expenditures of time and money must be justified on a case-by-case basis. This is as it should be, but the general management philosophy concerning whether loss control efforts are worth an investment of time or money must be positive for the overall program to be successful. This does not mean that management will approve every loss control and prevention project or idea, but it must be willing to be receptive.

Setting an example is a powerful communication tool. Members of top management will demonstrate their true commitment to loss control by the example they set. What they *do* will have a far greater impact on how employees view their commitment to loss control than what they *say*.

Management personnel must also set a personal example by always following safety policies and using safe practices. Small things, like not wearing safety glasses during plant

FIGURE 9. Common elements of effective loss prevention and control programs. (*Source:* © FIRECON 1989)

visits, are noticed by employees and can have a major negative impact on a loss-control program.

- *Define and assign responsibility:* Management personnel must define who is responsible for each aspect of the loss-control program. Due to the nature of the loss-control process, doing this can at times present difficulties. Often, the responsibility for loss control must be given to a number of individuals, each having specific areas of concern. This works well as long as the individuals know which areas are their responsibility and understand what is involved with each area of responsibility. Too often, this assignment is left to chance or given as an add-on assignment to someone who already has too much to do, creating the impression that loss control is not really a priority issue.
- *Delegate authority:* Delegating authority is also critical. An individual should not be held accountable for a task if sufficient authority to complete the task has not been provided. Sometimes a person is made responsible for loss control as an add-on task to his or her main job but has not been given authority to accomplish anything. This frustrates the individual and prevents significant loss-control efforts from being established.
- *Establish program objectives:* The principle of establishing effective goals and objectives applies to any endeavor, including loss control. Safety professionals need the focus provided by objectives to gain maximum benefits. Poorly planned loss-control efforts will result in haphazard expenditure of resources and failure to realize benefits that could have been obtained with better planning. It is vital that top management be involved in the process of establishing the objectives of the overall loss-control program. It is much easier to obtain support from managers who have been involved in setting specific targets for a program. The process of involving top management also forces safety professionals to look more objectively at the goals they have set for the loss-control program. This self-analysis is essential to keeping their thinking on track.

The objective-setting process also provides a way of measuring how successful loss-control programs are. Without specific objectives, it is impossible to determine whether the desired results have been accomplished.

Objectives should be as specific as possible. For example, if one objective is to reduce the number of fires in the plant by 10 percent, a completion date should be set so that this objective can be evaluated. As part of the objective-setting process, the methods that can be used and the resources that will be needed to accomplish the objective should be evaluated. Once this is done, the objective can become an action plan. When safety professionals explain in detail how objectives will be accomplished and what resources will be needed, they can win support from management personnel, who can see what results are intended from the expenditure of resources.

- ***Involve supervisors and employees:*** For a loss-control program to have long-term success, it is crucial that the program have the support not only of top management, but also of intermediate-level supervisors and all other employees. Without it, some progress may be made, but it will be more difficult to achieve and usually will not be long lasting.
- ***Emphasize loss-control prevention in all areas:*** Loss control must be a consistent effort that is reflected in all phases of an operation, not just the most convenient ones. Maximum positive results can be achieved only if loss-control efforts have an impact on all phases of the operation. This seems like a small point, but loss control is not a part-time activity or one that can be applied successfully to only a few operations within an organization. If a commitment to a loss-control program is not made throughout the organization, the results of the program will be less than the best possible. Loss prevention and control must become an integral part of the way the organization operates.
- ***Provide adequate training:*** Training is an essential element in the overall success of a loss-control program. Personnel cannot be expected to perform their tasks safely if they have not been taught how to do so. An organization's first training challenge is to educate all personnel concerning the importance of loss prevention and control and their roles and responsibilities in the overall loss-control program.

  Employees must be provided with the opportunity to obtain the knowledge and develop the skills that allow for safe performance of their jobs. Training on safe ways to conduct operations should be integrated into the overall employee training process. Within the area of fire protection, two distinct types of training must be provided. The first is training concerning prevention of fires and reduction of their impact. The second is how to respond when a fire does occur. Even if your personnel are not engaged in manual fire-control efforts, they must at least be trained on safe evacuation procedures (Schroll 2002c). Training is also covered in the "Safety and Health Training" section of this handbook.
- ***Perform regular audits and inspections:*** The physical features of a loss prevention and control program must be thoroughly and regularly inspected. Procedures and their application must be reviewed frequently. The only way to ensure that things are the way they should be is to look. Inspections are advantageous because they make the safety professional a highly visible figure in plant operations and indicate to all personnel that loss control is important.

  To ensure that facilities remain safe from uncontrolled fires, an audit-and-inspection process is essential. *Audits* are evaluations of existing conditions compared to a target that has been set *internally*. For example, if an organization's policy is that each employee will receive training on using portable fire extinguishers annually, an audit may be conducted to confirm that this has occurred or to identify individuals who have not attended training. In essence, audits are used to confirm that an organization is doing what it said it would do. Another use of audits is to check processes. For example, a safety professional could use inspection records to reexamine a sampling of extinguishers to confirm that the information on the inspection report is accurate (Schroll 2002b). *Inspections* are observations and evaluations that compare conditions to a *code or standard requirement.* If a safety professional checks all of the plant's fire extinguishers to ensure that they meet the requirements outlined in NFPA Standard 10, *Portable Fire Extinguishers* (NFPA 2007a), that is an inspection.

- ***Investigate all losses:*** To have an impact on the future, one must understand the past. All losses are caused by something, and determining the cause of a loss is important in preventing future losses. An investigation may also identify a need for improvement in an organization's loss prevention and control policies or the enforcement of those policies. Even small losses should be investigated completely. A loss may have been small this time, but the cause of the loss, if not corrected, may create a major loss next time. An investigation must take place as soon as possible after a loss so that valuable evidence is not lost or damaged. A thorough investigation also provides information needed for insurance and other legal considerations.
- ***Maintain adequate records:*** Effective records are essential to tracking progress toward established objectives. In addition, maintening adequate records permits organizations to gain valuable information from past losses that can be applied to prevent or more effectively control future ones. Trends can also be seen over time, and good records provide the necessary information to make adjustments to future loss-control efforts.

  Legal requirements often specify which documents are to be retained. Injury reports, training records, and many other records must be available to satisfy the requirements of regulatory agencies. Insurance companies also frequently stipulate that certain records must be maintained. Records required by an insurance company will typically at a minimum include fire-protection system inspection, testing, and maintenance records. Records must also be maintained that track the progress of loss-prevention efforts.
- ***Enforce policies consistently:*** The policies developed for loss-control programs are useless if they are not enforced properly. Policies quickly lose their impact if employees perceive that they will not be executed. Even worse than not enforcing them is to enforce them inconsistently. Consistency in distribution of rewards and punishments is also critical. If employees perceive that a policy has been applied unequally based on who failed to adhere to the policy or where the infraction occurred, the policy will lose most of its effectiveness.
- ***Plan for emergencies:*** Planning for emergencies is essential. No matter how effective a loss prevention and control program may be, an emergency will occur eventually. How much is lost during that emergency will depend largely on the extent and quality of prior emergency planning. Planning is covered in detail elsewhere in this book.

Fire Loss-Control Program Conclusions

The items that have just been discussed are the major components of a successful loss-control program. To summarize, they include:

- Management commitment
- Definition and assignment of responsibility
- Delegation of authority
- Establishment of program objectives

- Involvement of supervisors and employees
- Emphasis on loss prevention and control in all areas
- Provision of adequate training
- Performance of regular audits and inspections
- Investigation of all losses
- Maintenance of adequate records
- Consistent enforcement of policies
- Planning for emergencies

Although some success can be attained without addressing all of these elements, lasting success requires that they all be addressed. The amount of commitment to any individual component will vary considerably from one organization to another (Schroll 2002b).

Hazard Evaluation

An adequate understanding of the hazards present in a facility is an essential element in controlling or eliminating those hazards. A *hazard* is an inherent property of a material or process (ANSI/AIHA 2005). In the context of this chapter, we will focus on fire-related hazards. One of the most fundamental concepts of fire prevention is the separation of ignition sources and fuels.

Sources of Ignition

Items that may serve as sources of ignition will be considered first. There may be many potential ignition sources in a facility, some of which may be specific to a particular type of operation. In this section we will focus upon several that are common to most facilities.

Hot-Work Activities as a Source of Ignition

Hot work includes welding, torch-cutting, grinding, and other activities that create ignition sources that might easily ignite a fire or initiate an explosion. Hot work has historically been a significant source of fire ignition. Hot work can be done safely, but too often it is not. One of the best references for handling hot work appropriately is NFPA 51B, *Standard for Fire Prevention During Welding, Cutting, and Other Hot Work* (NFPA 2003). OSHA 29 CFR 1910.252 also establishes some basic requirements that must be addressed.

We will focus in this section exclusively on the fire-safety aspects of hot work. Many other potential health and safety issues may be involved with welding and similar operations, but they are beyond the scope of this chapter. Fire safety during hot work covers two primary concerns—preventing setting workers on fire or involving them in explosions and preventing igniting materials in the area.

During hot work, the ignition source must be able to accomplish its task, so special attention must be paid to fuels in the area. The ideal way to avoid doing hot work near fuels is to take all hot work to the controlled environment of a shop, but this is not always possible and often is not practical. To prevent fires when hot work must be performed away from a shop, special precautions are essential:

- ***Removal of combustible materials:*** Combustible materials should be removed from the area where the hot work will be performed. Ideally a clear area of at least 35 feet in diameter around the area being worked on should be secured. Combustibles that cannot be moved should be covered with fire-retardant covers. Wall and floor openings within 35 feet should also be protected so that hot particles cannot pass through the openings into other areas. Special caution must be exercised when hot work is done above an area containing combustibles. Slag and other hot materials may travel a larger distance when the original source is at the ceiling of an area.
- ***Hot-work permit system:*** A hot-work permit system is a useful way to manage hot work conducted outside of a welding shop. This system requires written authorization for hot work and provides an approval process that should include a checklist of the major safety issues that need to be addressed. An individual or a small group of people within

HOT WORK PERMIT

Prevent Fires

For welding (gas or electric), burning or any work requiring or producing heat, flame, sparks, or electric current.

Date Issued: ______________ **Time** ______________ A.M. / P.M.
Expires: ______________ **Time** ______________ A.M. / P.M.
Location: ______________
Work To Be Performed: ______________

NECESSARY PRECAUTIONS

All items must be checked Yes or Not Applicable (N/A)

General

Yes N/A

- ☐ ☐ The work will be confined to the area or equipment specified on the permit.
- ☐ ☐ Sprinklers (water) are in commission and will not be taken out of service.
- ☐ ☐ Areas provided with fire suppression systems capable of causing immediately dangerous to life conditions (e.g. Cardox) have been evaluated. (check one):
  - ☐ Fire suppression systems are to remain in service during the work.
  - ☐ Fire suppression systems have been taken out of service and alternate protection provided as prescribed by local fire safety officer.
- ☐ ☐ Ample portable extinguishing equipment such as a charged hand hose or suitable portable fire extinguishers are provided at the work site.
- ☐ ☐ The area has been assessed for potential explosion hazards from gases, vapors, combustible dusts and explosive atmospheres eliminated.

Precautions within 35 ft. of work

- ☐ ☐ Flammable liquids removed; other combustibles, if not removed, protected with fire-resistive tarpaulins or metal shields.

Yes N/A

- ☐ ☐ All wall and floor openings covered.
- ☐ ☐ Fire resistive tarpaulin suspended beneath work to catch sparks and hot slag.

Work on walls or ceilings

- ☐ ☐ Construction is noncombustible and without combustible covering or insulation.
- ☐ ☐ Combustibles moved away from other side of wall or floor/ceiling.

Work on enclosed equipment

- ☐ ☐ Tanks, containers, or other enclosed equipment cleaned and purged of all combustibles/flammables. The absence of an explosive atmosphere has been verified by a combustible gas detection instrument.
- ☐ ☐ Equipment has been isolated and locked out according to established written procedures.

Fire watch

- ☐ ☐ A person is assigned to watch for dangerous sparks during the work and for at least 30 minutes after work, and during any coffee or lunch breaks.
- ☐ ☐ Fire watch is trained on use of extinguishing equipment.

THE SIGNATURE BELOW SIGNIFIES THAT:

- The precautions checked above have been taken to prevent fires.
- All requirements outlined in facility guidelines for fire protection/prevention on repair jobs requiring burning or welding have been met.
- All changes in conditions that may affect the potential for fire to occur will be re-evaluated and a new permit issued.

______________ ______________ ______________
Signature *Symbol Number* *Date*

TO REPORT A FIRE

☐ Contact ______________ by Phone Dial ______________
☐ Contact ______________ by Radio and Request Assistance.

Original to Supervisor
Duplicate to Job Site

FIGURE 10. Hot-work permit

a facility should be authorized to issue these permits. These individuals examine the area where hot work is to be performed and confirm that appropriate precautions are in place before approving the permit. The person approving the permit should not also be doing the work, so that a double check is built into the system. Figure 10 is an example of a hot-work permit.

- ***Assigning fire watches:*** Assigning a *fire watch* is a critical part of performing hot work safely. The fire watch is a person who observes the work and is prepared to immediately control fires that may start. Fire watches must have a fire extinguisher or hose available for instant use and must position themselves where they have a clear view of the work area. In situations where the area being affected by the work is too large or where it covers multiple floors, more than one fire watch may be needed. After the work is completed, fire watches remain in the area where the work was performed for at least 30 minutes to ensure that no fires begin from smoldering sources of ignition.
- ***Ensuring that installed fire-protection systems are operational:*** Hot work should never be performed in properties with installed fire-protection systems if those systems are out of service.
- ***Wearing appropriate clothing:*** Welders should wear clean, fire-retardant clothing. Although the "clean" part is frequently overlooked, it is important, because fire-retardant garments that have been saturated with oil or grease will not perform effectively. Shirts or jackets should have long sleeves and no pockets and should extend over the top of the trousers. Trousers should not have cuffs. Welders should wear leather gloves with gauntlets that extend beyond the end of the sleeve.
- ***Observing closed-container cautions:*** Hot work conducted on closed containers or closed spaces that are part of larger pieces of equipment constitutes a special hazard. Drums that may previously have held flammable liquids are a common example. The best plan is to dehead drums with a mechanical device instead of a torch. If a torch is used, the worker must ensure that no flammable residue remains in the drum, and this can be difficult to assess. If heavy sludge remains in the bottom of a container, it may not release ignitable vapors until heating has begun.
- ***Maintaining equipment:*** Welding and cutting equipment must be properly maintained and regularly inspected. Gas cylinders must be stored and secured properly. Fuel gases must be separated from oxygen cylinders.

Although the precautions necessary to do hot work safely are not complex, improperly conducted hot work has been responsible for many deaths and injuries among those performing the work and for many fires.

Electrical Sources of Ignition

Electrical fires are also relatively common. A proper initial installation is the beginning point to ensure that electrical hazards are properly handled. NFPA 70, *National Electrical Code* (NFPA 2008) covers the requirements for electrical installations in great detail and is used by many jurisdictions as their legally required code. Inspection and maintenance are required over each electrical system's life to confirm that effective protections remain in place.

Overloading of electrical circuits is a relatively common problem in many facilities. It is more prevalent in older facilities that were not designed for the number of electrical devices currently in use. Temporary wiring and inappropriate use of extension cords is another common problem.

Special electrical installations are required for locations that normally have flammable vapors or gases, combustible dusts, or combustible fibers present. These areas are referred to as hazardous (classified) locations. Detailed coverage of this specialized topic is beyond the scope of this chapter, but NFPA 497, *Recommended Practice for the Classification of Flammable Liquids, Gases, or Vapors and of Hazardous (Classified) Locations for Electrical Installations in Chemical Process Areas* (NFPA 2004) contains information that should be helpful for obtaining additional information. Article 500 of NFPA 70, *National Electric Code* (NFPA 2008) contains requirements for electrical installations in hazardous locations.

Hazardous (classified) locations are divided into three major classes. Class I locations contain flammable vapors or gases, Class II locations contain combustible dust, and Class III locations contain ignitable fibers or flyings. Within each class, divisions are used to separate those areas where the hazard is routinely present from areas where it may be present occasionally.

Heat-Producing Appliances as Sources of Ignition

Heat-producing appliances can be ignition sources. This category may include building heating devices, devices used for cooking, and numerous types of processing equipment. As with the electrical hazards discussed above, proper initial installation and continuing maintenance are important.

Clearance from potential fuels is of key importance for fire prevention. Combustible materials must be kept a safe distance from heaters and similar devices.

Smoking as a Source of Ignition

Smoking is a common ignition source. Controlling this hazard is best accomplished by providing a safe and controlled environment for smokers. Banning smoking throughout a facility may sound like an effective option, but often it leads to smokers finding places to smoke that are not the best choices from a fire-prevention perspective.

Housekeeping

Maintenance of a clean and orderly environment within a facility is one of the most basic steps necessary to improve safety in general and fire prevention in particular. Housekeeping is also a good general indicator of the status of loss prevention and control within a facility. If an organization maintains good housekeeping, generally the other portions of their loss-prevention efforts are also effective. An environment that demonstrates poor housekeeping usually has deficiencies in other areas of the loss-prevention process.

Fire-Related Aspects of a Housekeeping Program

Housekeeping is simple but not always easy. The specifics will vary somewhat among different facilities, but a few common elements can be applied to all types of operations.

- **Control of combustibles:** Control of combustibles is necessary in all facilities. Trash, cardboard boxes, scrap, and similar items should not be allowed to accumulate in operational areas of the facility. These items should be collected in appropriate containers that are emptied frequently enough to avoid excessive accumulation.

- **Maintenance of unobstructed aisles:** Aisles should be laid out to allow access to all areas and must be kept unobstructed. This is of much greater importance than it may at first appear. Obstructed aisles can be a concern for several reasons. The most important is that obstructed aisles may compromise life safety by blocking an egress path. Storage of materials in aisles may also compromise the design of installed fire protection systems by adding fuel load to the facility in excess of what was intended. Blocked aisles may also interfere with access for manual fire-control operations.

Storage, Transport, Dispensing, and Use of Flammable and Combustible Liquids

Flammable and combustible liquids are used as fuels, solvents, lubricants, and for many other purposes (Schroll 2002a). Certain types of operations require more of these substances than others, but few operations are without at least some. NFPA 30, *Flammable and Combustible Liquids Code* (NFPA 2000) is one of the best references for detailed requirements concerning this issue.

To help in determining necessary hazard-control measures, flammable and combustible liquids are classified according to their degree of hazard. The degree of hazard is based upon the flash point and boiling point of each liquid. Hazard-control measures must be more stringent for more-hazardous liquids. Table 2 illustrates these classifications. This classification system is used primarily to determine appropriate code requirements for storage, transportation, use, and protection. For example, smaller quantities of class IA liquids would be permitted in a storage area than class IC liquids due to the higher hazard classification of class IA liquids.

The major distinction between flammable and combustible liquids is ease of ignition, not fire severity. As Table 2 demonstrates, flammable liquids have flash points at or below normal room temperature. Under normal ambient conditions, *flammable* liquids produce enough vapors that even small sources of ignition may be enough to start a fire. Vapors produced by flammable liquids ignite and burn. Anything that enhances the production of vapors improves the chances that a fire will start. *Combustible* liquids require either a stronger ignition source, preheating of the liquid, or both in order to be ignited. For example, if kerosene were heated to 120ºF, it would become much easier to ignite than it is at lower temperatures because at that temperature sufficient vapors are released from the liquid to form, with air, an ignitable mixture.

TABLE 2

Flammable and Combustible Liquids Classification

FLAMMABLE LIQUIDS		
A flammable liquid has a flash point below 100°F and a vapor pressure not exceeding 40 pounds/square inch (absolute) at 100°F.		
CLASS IA	Flammable liquids with a flash point below 73°F and boiling point below 100°F.	Examples: Acetaldehyde, Ethylene Oxide, Ethyl Ether, Methyl Chloride, Methyl Ethyl Ketone.
CLASS IB	Flammable liquids with a flash point below 73°F and boiling point at or above 100°F.	Examples: Acetone, Ethyl Alcohol, Gasoline, Hexane, Methanol, Toluene.
CLASS IC	Flammable liquids with a flash point at or above 73°F and below 100°F.	Examples: Butyl Acetate, Butyl Alcohol, Propyl Alcohol, Xylene.
COMBUSTIBLE LIQUIDS		
A combustible liquid has a flash point at or above 100°F.		
CLASS II	Combustible liquids with a flash point at or above 100°F and below 140°F.	
CLASS IIIA	Combustible liquids with a flash point at or above 140°F and below 200°F.	
CLASS IIIB	Combustible liquids with a flash point at or above 200°F.	

(*Source:* NFPA 2000)

The form of the liquid can also have an impact on ease of ignition. If kerosene were sprayed over a candle flame as a mist, it would easily ignite because the mist is closer to the vapor state than liquid kerosene is. The cautionary lesson one should take from this is that the presence of combustible liquids does not necessarily mean there is no fire risk, just that under most circumstances there is less risk than with flammable liquids.

When selecting materials for use in a process, the product with the highest flash point that will obtain the desired results should be chosen when other considerations are equal. Other characteristics that influence both fire and personnel exposure issues are vapor pressure, boiling point, and evaporation rate. These all have an impact on how readily a material will become a vapor. Low vapor pressure, low evaporation rate, and high boiling point are preferred.

FIGURE 11. Type I safety can
(*Source:* © FIRECON 2007)

Storing Flammable and Combustible Liquids

Flammable and combustible liquid storage should be designed to prevent ignitable vapors from reaching an ignition source. Loss prevention should address both controlling flammable liquid and its vapors and controlling ignition sources. Control must be maintained throughout the storage, dispensing, use, and disposal phases of using flammable liquids.

Two factors must be considered in controlling stored flammable liquids: the containers the liquids are stored in and the area the containers are stored in.

- **Containers:** Containers made to hold five gallons or less are called *safety cans*. Safety cans are designed to control flammable vapors and contain liquid. They provide a means for safe transporting and storage of flammable liquids. Safety cans must be approved for use by a recognized testing organization such as Underwriters Laboratories or FM Global (NFPA 2000). They must be leak-proof, automatically vent excessive pressures to avoid container rupture, prevent external flames from coming into contact with the flammable liquid inside, and have filling and dispensing openings that close automatically. Their contents must be clearly marked. Safety cans are available in two styles: type I and type II. Type I safety cans (Figure 11) have a single large opening for filling and dispensing and are primarily designed for dispensing liquids into large containers. Funnel attachments are available for type I cans that allow dispensing into narrow openings. Type II safety cans (Figure 12) have separate filling and dispensing openings as well as an attached, flexible dispensing spout. Safety cans constructed from polymer materials are also available.
- **Storage cabinets:** Safety cans and other small containers should be stored in flammable liquid cabinets (Figure 13) when not in use. These cabinets are generally

FIGURE 12. **Type II safety can** (*Source:* © FIRECON 2007)

yellow and must be labeled, usually in red letters, "Flammable Keep Fire Away." The amount of a liquid that can be stored in a cabinet is based on the degree of hazard the liquid presents. The number of cabinets permitted in any single area is also restricted by fire codes. NFPA 1, *Uniform Fire Code* (NFPA 2006a) limits the number of cabinets to three in any single fire area.

Cabinets are typically constructed of metal with an air space between the exterior and interior sheeting. They are equipped with a catch basin in the bottom to contain a limited amount of spilled material. They are equipped with a bung opening (a threaded cap like that in the lid of a 55-gallon drum) that may be used for venting the cabinet or may be capped. If the cabinet is vented, the vent must be to the outside of the building. If the vent is not used, the opening must be capped.

Drum cabinets (Figure 14) should be used to enclose drums of flammable liquid that must be stored inside a structure for dispensing. Drums for waste accumulation should also be in cabinets.

- **Storage facilities:** Large quantities of flammable and combustible liquids should be stored in remote areas away from process operations. This may be done in several ways:
  - ***First choice:*** The ideal storage arrangement is to store flammable and combustible liquids in a separate shed or building. This is safest because the separation of the building prevents vapors or spilled material in the storage area from coming into contact with ignition sources in the facility's main building. If a fire occurs in the separate flammable-liquids storage area, it is not likely to spread to the main building areas.
  - ***Second choice:*** Second choice is an attached room with an exterior entrance and no opening directly into the main building.
  - ***Third choice:*** Third choice is an attached building connected by an interior door.
  - ***Fourth choice:*** Fourth choice is a separate room within the building that has exterior

FIGURE 13. Flammable liquid cabinet (*Source:* © FIRECON 2001)

FIGURE 14. Flammable liquid cabinet for drum (*Source:* © FIRECON 1998)

access. This arrangement allows the interior walls to be constructed without the penetration of a doorway and improves vapor and fire containment.

- ***Fifth choice:*** The fifth-choice arrangement, which is common, is an interior room with interior access. The requirements of an operation may make an arrangement with interior access the only practical choice. If this is the case, vigilance will be required to ensure that the openings into the room are kept closed. These storage rooms may often be equipped with forklift-size openings in addition to personnel doors to allow for the transport of pallets of material.

All of these storage areas should have several basic features:

- ***Doors:*** Every door should be a rated self-closing fire door, and all doorway openings should be provided with curbing to prevent the escape of liquid.
- ***Drainage:*** The areas should be equipped with suitable drainage to trap liquids in a containment sump or underground tank. The drains, however, should not be connected to regular sewer lines.
- ***Ventilation:*** Floor-level ventilation (Figure 15) should be provided to remove flammable vapors.
- ***Electrical equipment:*** Explosion-proof electrical equipment should be used. Explosion-proof electrical equipment is designed to avoid electrical ignition of the flammable vapors and to contain vapor explosions that might occur inside the electrical device.
- ***Fire detection and suppression systems:*** Fire detection and suppression systems of the appropriate type for the specific hazard should be installed. For example, an inside storage room for flammable liquids may be protected by a total flooding carbon dioxide fire-suppression system.

FIGURE 15. Floor-level ventilation (*Source:* © FIRECON 2002)

- **Transferring flammable and combustible liquids:** Storage within operating areas of a facility should be limited to the quantity required for one day's use. Proper methods must be used to transfer liquids from storage areas to operating areas.
  - ***Transferring flammable liquids from drums into smaller containers:*** Drums should be fitted with self-closing faucets and pressure-relief vents. A drip pan should be used below the faucet to prevent spills incidental to dispensing. The drip pan is designed to contain small amounts of flammable liquids that may drip from storage containers.
  - ***Controlling static electricity:*** Static electricity is a significant ignition risk when transferring flammable liquids. During dispensing and any other product transfer operation, containers must be grounded and bonded to prevent the development of a difference in electrical potential that might result in an electric arc during dissipation of static charges. *Grounding* (Figure 16) is accomplished by establishing an electrically conductive path between a container and an electrical ground. *Bonding* is achieved by making an electrically conductive connection between two or more containers. Both the dispensing and receiving containers should be grounded, and the two containers should be bonded together.
  - ***Using a solvent tank:*** When many small containers of flammable liquids are used throughout the work area, a portable solvent tank may be used to store, transport, and dispense the liquids. A portable solvent tank is a large tank equipped with a pump. The tank allows flammable liquid to be safely transported to the area of use, and it is more efficient than taking each small container to a central storage area.

FIGURE 16: Example of grounding
(*Source:* © FIRECON 1998)

- *Maintaining operator attention:* Operator attention during dispensing is critical. Numerous spills and fires have resulted from transfer operations that were left unattended. Spring-loaded valve handles help to prevent incidents caused by operator inattention, but operators may figure out ways to block the handles open, a practice that must not be permitted.

The use of flammable liquids always involves a certain degree of risk. However, fires can be prevented through the use of procedures, safeguards, and vigilance. All personnel involved in the use of flammable and combustible liquids should be thoroughly trained in safe operating procedures and emergency actions.

Generally, the methods for controlling the hazards of flammable liquids are containment of the liquids and their vapors, restrictions on the quantities of liquids in the building, safe work practices, adequate personnel training, and installed fire-protection systems. Each specific risk should be evaluated using these basic criteria.

Preventing the Buildup of Hazardous Vapors

Liquid and vapor containment is most easily accomplished by keeping liquids in appropriate containers and keeping those containers closed when not in use. In operating areas where vapors cannot be contained due to the nature of the process, the vapors must be controlled. This is most effectively done with ventilation systems. The design of these systems is critical but beyond the scope of this chapter; however, safety professionals must have a basic understanding of the need for adequate ventilation and the systems that can be used to control vapor accumulation. Competent professional assistance must be used when initially designing these systems or when planning system modifications.

Ventilation-system designs must often include process-control features linked to vapor concentrations within the system. Ventilation systems often include flammable-vapor monitors that sound an alarm and/or automatically shut down processes if high vapor levels are detected—usually 25 percent or more of the lower explosive limit. Maintenance of ventilation systems is an essential part of ensuring their continued effectiveness. Controls, fans, interlocks, and other critical components must be regularly inspected and tested, and problems discovered must be corrected promptly. Processes protected by essential ventilation systems should never be run when the systems are not functioning properly.

Safe Work Practices

Of the many safe-work practices, one of the most important is the appropriate use of the correct material. Incidents frequently occur because people use materials inappropriately—for example, they use gasoline to clean something. Gasoline is a poor choice as a cleaning solvent. Many cleaning materials are more appropriate and less hazardous than gasoline. Unfortunately, the ready availability of gasoline leads to its selection, sometimes with disastrous results. Organizations should develop written procedures to aid personnel in implementing safe work practices. To be effective, the use of safe practices must be enforced consistently.

Training is another important element of flammable liquid safety that ties in closely with work practices. Personnel should thoroughly understand the hazards and behavior of the materials they are expected to use on the job. People cannot be expected to perform effectively without adequate training. At a minimum, training should cover the hazards of the materials and processes used; proper storage, dispensing and use practices; and emergency procedures.

Portable Fire Extinguishers

Portable fire extinguishers are an important first line of defense against small fires if people are properly trained in their safe use (Schroll 2002c). NFPA 10, *Standard for Portable Fire Extinguishers* (NFPA 2007a) is one of the best references for requirements related to extinguishers.

Many types of fire extinguishers are currently available, and each has certain advantages and disadvantages. For that reason, it is important to understand the capabilities of each type of extinguisher when selecting one to provide protection for a specific hazard or area.

Water Extinguishers

Water extinguishers are available in two types. The pump-type extinguisher discharges the water through a pump operated by the user, and the pressurized extinguisher uses pressure to discharge the water.

Pump-type extinguishers are not commonly used. Although some operations may still have and use them, most often pressurized water extinguishers are employed in situations where water is the appropriate extinguishing agent.

Pressurized water extinguishers are pressurized with 100 psi of compressed air or nitrogen (Figure 17).

FIGURE 17. Pressurized-water extinguisher (*Source:* © FIRECON 2002)

Capacity: two and one-half gallons.
Type of fire: Class A fires only.
Discharge time: About one minute. They
Discharge range: 30 feet when initially discharged.
Advantages: In many situations, water is easy to clean up and causes little or no additional damage. The extinguisher is inexpensive to recharge and can easily be returned to service by in-house personnel with some basic training. In addition, the soaking ability of water is a major advantage in fighting class A fires.
Disadvantages: They are effective only on class A fires and can be dangerous if used on any other class of fire. They are also relatively heavy. In any area that is not continuously heated, they are subject to freezing. Antifreeze can be added to prevent freezing, but that makes maintenance more difficult.

Portable Foam Extinguishers

Portable foam extinguishers (Figure 18) are modified pressurized-water extinguishers and are also available in two types. The *premix foam extinguisher*, which is the most common, contains a mixture of water and foam concentrate within the extinguisher. An aerating nozzle is used to make foam from the foam

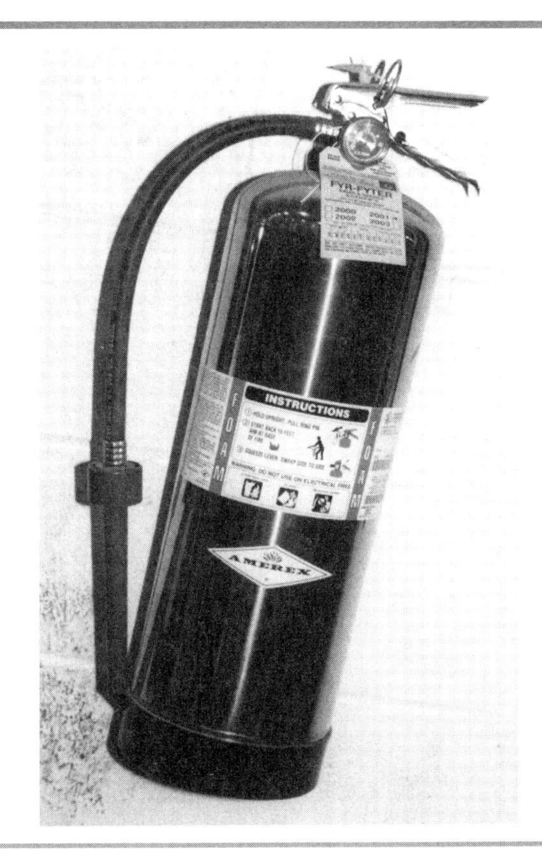

FIGURE 18. Pressurized foam extinguisher (*Source:* © FIRECON 2001)

solution as it is discharged. They are pressurized to 100 psi with compressed air or nitrogen.

The *cartridge-type foam extinguisher,* which is no longer manufactured, contains water only in the body of the extinguisher. At the time of discharge, the water flows through a cartridge containing pellets that make foam when mixed with the water. The other construction features of the extinguisher are similar to a water extinguisher.

Type of fire: Class A and B fires.
Capacity: Two and one-half gallons.
Discharge time: About one minute.
Discharge range: Somewhat less than the 30 feet typical of water extinguishers.
Advantages: In many situations, foam is easy to clean up and causes little or no additional damage. The extinguisher is relatively inexpensive to recharge and can be returned to service by in-house personnel with proper training. Class B fires can be controlled with foam extinguishers. Unlike any other type of portable fire extinguisher, foam extinguishers can cover Class B spills with foam to help prevent ignition. The soaking ability of water is a major advantage in fighting class A fires.
Disadvantages: Similar to those of water extinguishers.

Foam extinguishers are also available in wheeled units, and their most common size is 33 gallons. These units have a range of approximately 30 feet, and a discharge time of one minute.

Portable Dry-chemical Extinguishers

Portable dry-chemical extinguishers are available in two types. Both types are available with either major class of dry-chemical agent: regular dry chemical or multipurpose dry chemical.

A *stored-pressure dry-chemical extinguisher* (Figure 19) contains a dry-chemical agent and pressurizing gas, usually nitrogen, within the extinguisher.

Type of fire: Regular dry chemical, class B and C fires; multipurpose dry chemical, class A, B, and C; Purple K, flammable liquids fires.
Capacity: Two and one-half to 30 pounds.
Discharge time: Eight to 20 seconds.
Discharge range: Five to 30 feet.
Advantages: Rapid fire control, particularly for class B fires. The units are not subject to freezing, so they can be placed outside or in unheated areas.
Disadvantages: The extinguishers cannot be recharged in-house without costly, specialized equipment and more-extensive personnel training. Regular dry chemical is not effective on class A fires. Multipurpose dry chemical, although effective and approved for use on class A fires, is not as effective as water. Also, both types can make a mess that is difficult and costly to clean up. In situations involving delicate electronic equipment, the cleanup costs can be higher than cost of the initial fire damage.

FIGURE 19. Stored-pressure dry-chemical extinguisher (*Source:* © FIRECON 2001)

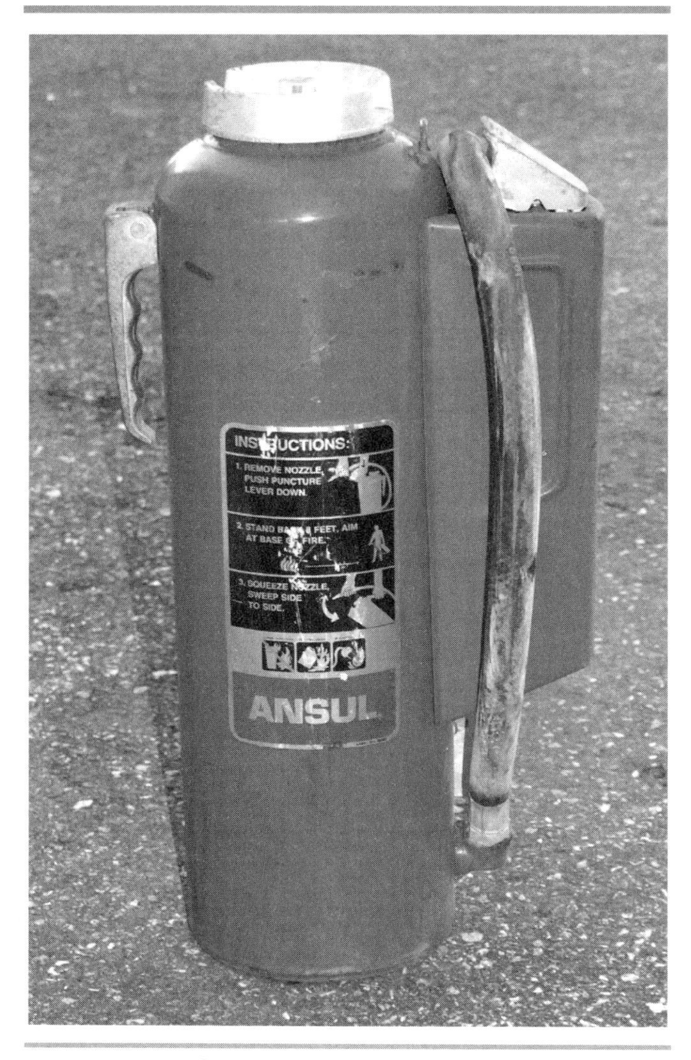

FIGURE 20. Cartridge-operated dry-chemical extinguisher (*Source:* © FIRECON 2007)

Cartridge-operated dry-chemical extinguisher: (Figure 20) This type contains a dry-chemical agent in the body of the extinguisher and a separate cartridge of pressurizing gas. When the extinguisher is needed, its main body is pressurized from this cartridge.

Type of fire: Regular dry chemical, class B and C fires; multipurpose dry chemical, class A, B, and C; Purple K, flammable liquids fires.
Capacity: Two and one-half to 30 pounds.
Discharge time: Eight to 20 seconds.
Discharge range: Five to 30 feet.
Advantages: Rapid fire control, particularly for class B fires. The units are not subject to freezing, so they can be placed outside or in unheated areas. They are relatively inexpensive to recharge and can be returned to service by in-house personnel with proper training.
Disadvantages: Regular dry chemical is not effective on class A fires. Multipurpose dry chemical, although effective and approved for use on class A fires, is not as effective as water. Also, both types can make a mess that is difficult and costly to clean up. In situations involving delicate electronic equipment, the cleanup costs can be higher than cost of the initial fire damage.

FIGURE 21. **Wheeled dry-chemical extinguisher**
(*Source:* © FIRECON 2005)

Dry-chemical extinguishers are also available in wheeled units (Figure 21) in sizes from 75 to 350 pounds. Their discharge range varies from fifteen to 45 feet, and their discharge time is from 30 to 150 seconds.

Carbon Dioxide Extinguishers

Carbon dioxide (CO_2) extinguishers contain carbon dioxide forced into its liquid state by pressure. On small carbon dioxide units (Figure 22) the discharge horn is attached by tubing, whereas on larger units (Figure 23) the discharge horn is at the end of a flexible hose.

Type of fire: Class B or C.
Capacity: Two and one-half to 20 pounds.
Discharge time: Eight to 30 seconds.
Discharge range: Five to eight feet.
Advantages: Since carbon dioxide does not leave a residue, clean-up is not a problem. The units are not subject to freezing, thus they can be placed outside or in unheated areas.
Disadvantages: Carbon dioxide extinguishes fires by depleting oxygen in the immediate area. If the extinguisher is used in a relatively confined environment, this could create a serious risk to personnel. The units are not suitable for class A fires. Static electricity can build up during discharge of the extinguisher and could possibly ignite an explosive atmosphere or damage sensitive

FIGURE 22. Small carbon dioxide extinguisher (*Source:* © FIRECON 2002)

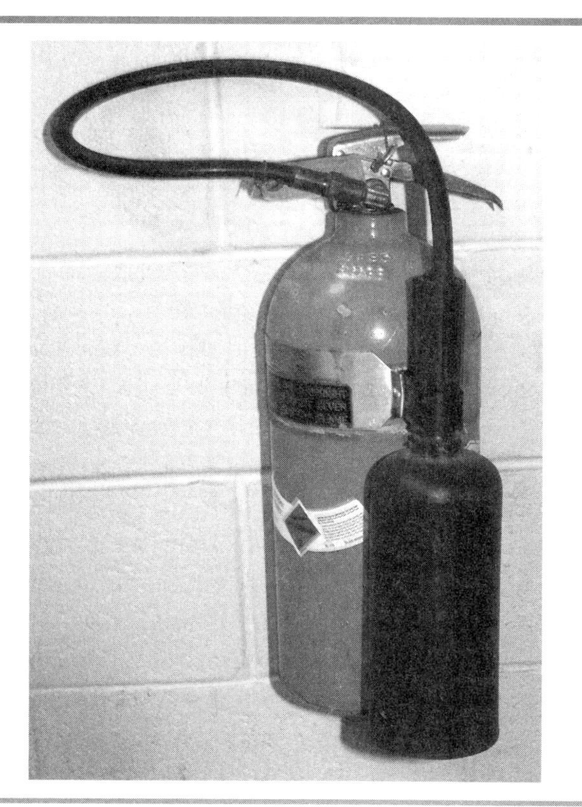

FIGURE 23. Large carbon dioxide extinguisher (*Source:* © FIRECON 2002)

FIGURE 24. Wheeled carbon dioxide extinguisher (*Source:* © FIRECON 2001)

electronic circuits. These extinguishers cannot be recharged in-house without specialized equipment and personnel training.

Carbon dioxide extinguishers are also available in wheeled units (see Figure 24) in sizes ranging from 50 to 100 pounds. Their discharge range is from three to ten feet, and their discharge time is from ten to 30 seconds.

Halon Extinguishers

Halon extinguishers (Figure 25) contain a halogenated hydrocarbon agent forced into a liquid state by pressure. The most common halons for portable extinguishers are Halon 1211 (bromochlorodifluoromethane) and Halon 1301 (bromotrifluoromethane). Due to the environmental concerns about halon, numerous other clean agents have been introduced, and research continues to develop new agents. Many of these agents, including Halotron and several of the FE series from DuPont, are used in portable fire extinguishers.

Type of fire: Class B or C; larger units are approved for class A.
Capacity: One and one-half to 22 pounds.
Discharge time: Eight to 30 seconds.
Discharge range: Nine to fifteen feet.
Advantages: Halon leaves no residue, so cleanup is not a problem. The units are not subject to freezing, so they can be placed outside or in unheated areas. Generally, they offer

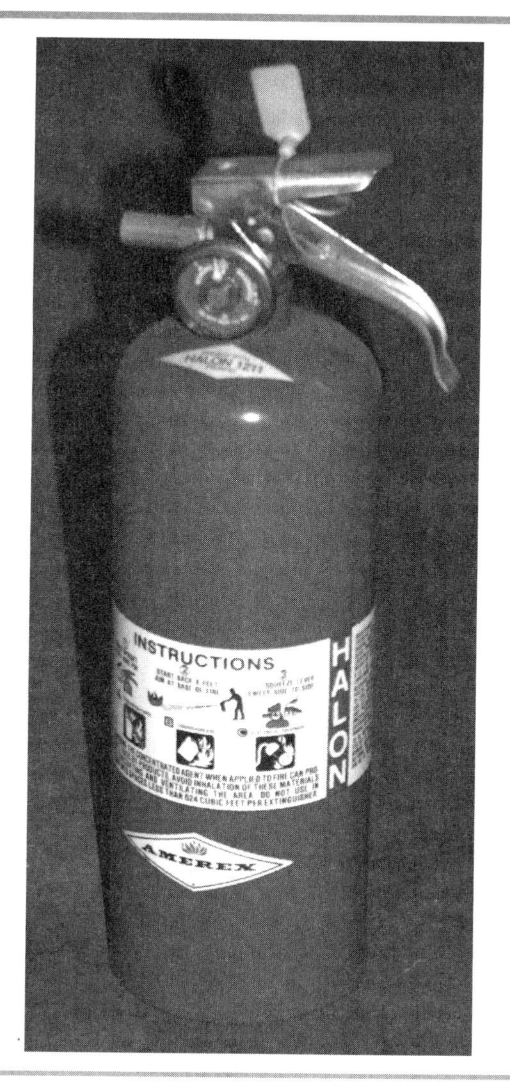

FIGURE 25. Halon extinguisher
(*Source:* © FIRECON 2007)

substantially more-effective fire control than carbon dioxide extinguishers do.

Disadvantages: The smaller units are not suitable for class A fires, and halon is expensive. Recharge costs can approach or exceed the initial purchase price of the extinguisher. Halons are a type of chlorofluorocarbon (CFC) and have been identified as contributing to the depletion of the ozone layer in the atmosphere. The Montreal Protocol prohibits manufacture of additional halon agents. However, the agents are still available and the extinguishers may continue in service.

FIGURE 26. Wheeled halon extinguisher
(*Source:* © FIRECON 2005)

These extinguishers are also available in wheeled units up to 150 pounds (Figure 26) with discharge times of 30 to 35 seconds and ranges of ten to eighteen feet.

Dry-powder extinguishers

Dry-powder extinguishers (Figure 27) contain one of several dry-powder agents.

Type of fire: Class D.

Capacity: Thirty-pound cartridge-operated units. The agents are also obtainable for manual application with a scoop.

Discharge time: Thirty to 60 seconds.

Discharge range: Six to eight feet.

Advantages: Can be recharged in-house.

Disadvantages: Dry powder extinguishers are suitable only for class D fires and are the only effective fire-control agents for this type of fire. Dry-powder agents must be evaluated in relation to the specific combustible metal that is being protected.

Dry-powder extinguishers are also available in 150- and 350-pound wheeled units.

FIGURE 27. Dry-powder extinguisher
(*Source:* © FIRECON 2001)

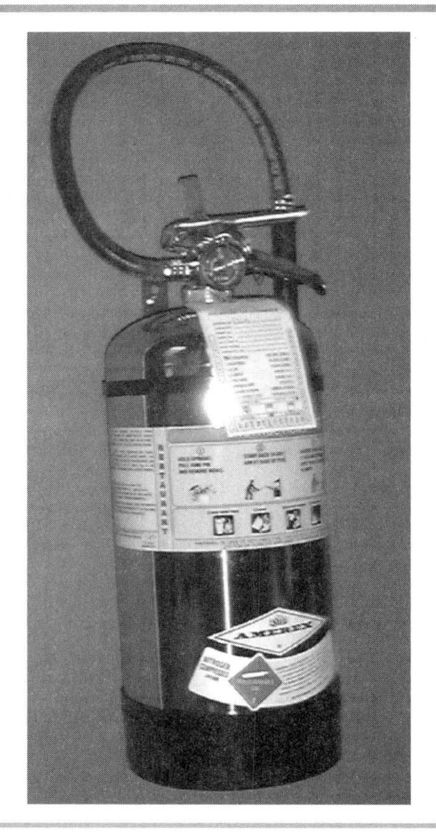

FIGURE 28. Wet-chemical extinguisher
(*Source:* © FIRECON 2001)

Wet-chemical Extinguishers (Figure 28)

Type of fire: Class K fires.
Capacity: Six liters and two-and-one-half-gallons.
Discharge range: Ten to twelve feet.
Discharge time: From slightly less than one minute to about one and one-half minutes.

Water-mist Extinguishers

This is a relatively new type of extinguisher with which deionized water is discharged in a fine spray.

Type of fire: Class A and C fires.
Capacity: Two and one-half gallons (there are also other sizes).
Discharge time: 80 seconds
Discharge range: Ten to twelve feet.

Extinguisher Ratings

Extinguisher ratings are used to determine which classes of fire can be controlled effectively and safely with the extinguisher being tested. Ratings also provide a guide to the size of fire an extinguisher will control.

All extinguishers receive a class rating. Extinguishers for class A fires are given a numerical rating from 1A to 40A using a relative scale. A 4A extinguisher, for example, will control a fire approximately twice as large as a 2A extinguisher. Class B extinguishers are rated numerically from 1B to 640B based on the approximate square footage of a spill fire they can control. For example, a 10B extinguisher will extinguish a spill fire of approximately ten square feet.

Fire tests are conducted using wood and excelsior for class A extinguishers. Class B extinguishers are tested on n-heptane fires in square pans using a liquid depth of two inches. Extinguishers for class C fires do not undergo fire tests but are tested for nonconductivity of electricity. Tests using specific combustible metals are employed for class D extinguishers.

The two largest testing laboratories for portable fire extinguishers are Underwriters Laboratories (UL) and Factory Mutual (FM). Extinguishers should have been approved by one and preferably both of these organizations. Extinguishers with these approvals will have symbols on their labels.

The Occupational Safety and Health Administration (OSHA) does not require portable fire extinguishers in facilities because OSHA regulations are aimed at the protection of people, not property. If extinguishers are present, OSHA regulations (29 CFR 1910.157) address the issues of testing, inspection, and maintenance. If the policy of the organization allows for employee use of extinguishers, the regulations require employee training.

The insurance carrier for a facility may also impose requirements relative to the number, type, and placement of portable fire extinguishers. In addition, local and state authorities may have regulations regarding extinguishers.

Selecting Extinguishers

The selection of an appropriate extinguisher for a particular area depends upon several factors:

- ***Class of fire expected:*** The extinguisher must be rated for the class or classes of fires that could be expected in the area being protected.
- ***Materials being protected:*** In a warehouse storing paper goods, for example, a water-type extinguisher would be more suitable than a multipurpose dry-chemical extinguisher. Although the dry-chemical extinguisher is rated for class A fires, it does not offer the penetration ability that water does for deep-seated fires in class A materials.
- ***Type of equipment and property being protected:*** Some extinguishing agents offer faster fire control than others but may have disadvantages such as the potential to cause additional damage to equipment. For example, dry-chemical agents generally offer the most rapid control of flammable liquid fires, but if electronic equipment is present in the same area, the residue left by these agents may cause more damage to it than the fire does. In this example, halon would be a more effective agent. The opposite situation can also occur. If an area outside a structure—a loading dock, for example—is being protected, dry-chemical agents are much less affected by wind than halon is. Every extinguishing agent used for portable fire extinguishers has advantages and disadvantages that must be considered.
- ***Agent compatibility:*** For example, if the area being protected contains methyl ethyl ketone, a polar solvent, a regular foam agent will not be effective; alcohol-resistant foam must be used.
- ***The potential severity of the fire:*** The potential severity must be considered in order to determine the size of extinguisher required. If, for example, an extinguisher must protect a quality-control lab area that uses flammable liquids in quantities up to a pint, it will need only limited fire-control capability. If the area to be protected is a 4-by-8-foot dip tank, much greater fire-control capability is necessary.

  NFPA 10 (NFPA 2007a) divides fire-severity potential into three classifications: light (low), ordinary (moderate), and extra (high). These classifications, within the context of the standard, are used to determine specific requirements for items such as the distribution of extinguishers throughout a facility. Used in general terms during the evaluation of extinguisher protection, the hazard classifications help to assess the severity of a potential fire in a specific area.
  - **Low hazard** includes areas such as offices, classrooms, and other general-use areas. The quantity of class A materials in these areas is relatively insignificant. Small amounts of class B materials may be present as well.
  - **Moderate hazard** includes light manufacturing, stores, workshops, and some warehouses. The quantity of class A and B materials is greater than in low hazard areas.
  - **High hazard** includes manufacturing, vehicle repair, and warehouse facilities, where the quantity of class A and B

materials is greater than in moderate hazard areas.

- ***User capabilities:*** The training individuals have in using extinguishers is the primary consideration in choosing an extinguisher. If training is limited or nonexistent, extinguishers should be simple and small. The user's physical ability to operate an extinguisher may be affected by physical or mental handicaps, and strength limitations may affect the size of the extinguisher that can be handled safely by an individual.

  Although it is true that larger extinguishers can be used on larger fires, the size and capability of the extinguisher can provide a false sense of security to an untrained operator. Fighting large fires requires special training, and occupants without sufficient training should not fight large fires.
- ***The extinguisher's location:*** Where an extinguisher is placed must be considered, because the use environment can have a major impact on which extinguisher will be most appropriate. For example, temperature affects water and foam extinguishers. They are subject to freezing and must either not be used in unheated areas or be modified with antifreeze solutions. Wind and draft have an impact on the effectiveness of any gaseous agent. Carbon dioxide in particular is difficult to use effectively in windy conditions.
- ***Agent reactivity and contamination:*** Some extinguishing agents may react with certain materials and chemicals. Contamination can also occur. For example, the use of a dry chemical extinguishing agent in a food processing area would require the disposal of all exposed food items. On the other hand, a carbon dioxide extinguisher used in this same area would not contaminate the food.
- ***Use in a confined area:*** The use of certain extinguishing agents in a confined area may present hazards to the individual using the extinguisher. Carbon dioxide, for example, will exclude the oxygen in an area if a sufficient concentration is allowed to accumulate.
- ***Initial cost:*** The initial cost of fire-extinguisher protection can vary significantly depending on what kinds of extinguishers are used to achieve adequate protection. One way to improve the selection of extinguishers in terms of cost is to evaluate the facility as a whole rather than as individual areas. A decision must be made regarding whether obtaining the most effective protection or meeting minimum requirements is the goal.

  Of course, the quality of the extinguishers used will have a significant impact on the cost, but money spent on high-quality extinguishers is often well worth the initial investment. The cost of the extinguisher should be evaluated over the life of the unit.

  Agent prices also vary, and the agent type selected will affect the initial cost. Clean agents are much more expensive than carbon dioxide, but the added fire-fighting capability may provide more-effective overall protection.

  The size selected will also affect initial costs. Purchasing many small units costs more than purchasing fewer large units of the same type of extinguisher. However, placement considerations may make purchasing a greater number of smaller units a better overall choice.

  When making the initial purchase of extinguishers, quotes should be obtained from several sources, because final purchase prices can vary considerably from dealer to dealer. Extinguishers should be purchased from a reputable source, preferably one that will also provide long-term maintenance and recharging services.
- ***Ongoing costs:*** For example, regular inspections must be performed. The time it takes to perform these routine inspections will depend on the number of extinguishers and the complexity of the inspection. The

maintenance cost of extinguishers is influenced by the same factors as those affecting inspection costs. Hydrostatic testing must be performed periodically, and the cost of this maintenance and testing depends on the type and quantity of extinguishers.

Some varieties of extinguishers are easier to maintain in-house than others. If in-house maintenance is used, the selection of extinguishers can affect the cost involved and the difficulty of ongoing maintenance. The cost of recharging extinguishers depends on the type of agent and on whether it is done in-house or by an outside contractor. The recharging of the extinguishers should be part of the initial selection criteria.

- ***Standardization:*** Standardization of extinguishers throughout a facility offers some advantages with respect to inspection, maintenance, and training. If extinguishers are already in place, continuing to use the same type may be the most effective choice.
- ***Placement:*** Extinguishers must be placed where they will provide effective fire protection and meet code requirements. Two primary criteria affect placement: travel distance and coverage. The *travel distance* is the distance an individual must traverse to get to an extinguisher. Travel distances to class A extinguishers may not exceed 75 feet. Travel distances to class B extinguishers are typically 30 or 50 feet depending on the rating of the extinguisher and the hazard classification of the area. *Coverage* refers to the number of extinguishers needed to provide adequate protection for an area based on its square footage. In addition to placing fire extinguishers where they will meet code requirements, they should be set up in areas where they will provide the most effective use. Locating extinguishers near exits encourages users to maintain an escape path while using an extinguisher. Extinguishers should also be placed in close proximity to high-hazard processes and equipment but not within a likely fire area.

Extinguishers should be placed in such a way that they are readily accessible and easily visible. Signs can be used to improve the visibility of an extinguisher location.

Extinguishers should be securely installed on a wall or in an approved extinguisher cabinet. Specific code requirements may apply to the installation of extinguishers and should be checked prior to setting them up in a facility.

Inspection and Record Keeping

Extinguishers must be inspected regularly to ensure that they are ready in the event of a fire. Inspections should generally be conducted monthly, although specific local requirements may vary. The monthly inspection should be thorough and cover several basic items. A checklist of the following items helps to avoid missing items that should be examined.

- ***Numbering:*** When purchased, each extinguisher should be assigned a unique number that will be used to identify it in an internal record system.
- ***Location:*** The location of each extinguisher should be checked to ensure that it is located where it is supposed to be and that the location is still satisfactory. Modifications to the structure, equipment position, or storage of materials may require that an extinguisher be moved to a different location.
- ***Accessibility:*** Extinguishers should be easily accessible and clearly visible. Obstructions around extinguishers should be removed immediately. Items that cannot be removed during an inspection should be noted so that the obstruction can be removed later. A note should be made even when items blocking access to an extinguisher are removed. This allows inspection records to reflect patterns that may be corrected by relocating an extinguisher.
- ***Signage:*** When the extinguisher cannot be seen due to visual obstructions, a sign should be installed to indicate its location.

The inspector should check the placement and condition of such signs.

- ***Charge:*** Extinguishers should be fully charged. An extinguishers equipped with a pressure gauge can be checked visually. Extinguishers without gauges, such as carbon dioxide and cartridge-operated dry-chemical extinguishers, must be weighed to check the charge. Carbon dioxide extinguisher heads are stamped to indicate full and empty weights, as are cartridges for dry-chemical extinguishers of that type. The inspector should also ensure that extinguishers are not overcharged.
- ***Seal:*** Extinguishers should be sealed with a breakable seal to guard against tampering and accidental discharge. If this seal is broken and the extinguisher checks out as charged, the inspector should replace the seal and make a note of the replacement.
- ***Hose:*** The hose should be checked for signs of damage or wear. The tightness of the connection to the valve head of the extinguisher should also be checked. Any extinguishers with damaged hoses should be taken out of service for maintenance.
- ***Nozzle and discharge valve:*** The nozzle should be examined for damage and obstructions. The extinguisher discharge valve should also be inspected.
- ***Condition of dry chemicals:*** Dry-chemical extinguishers are subject to caking and packing of the agent, which could prevent an extinguisher from operating properly. *Caking* occurs when moisture in the extinguisher causes the agent to form clumps or a solid mass. *Packing* is the settling of the agent in the extinguisher to the extent that the agent forms a single hard mass. These conditions can be determined by raising and lowering the extinguisher rapidly. The inspector should be able to hear and feel a slight delay in the shifting of the agent, which indicates that it is loose.
- ***Overall condition:*** The overall condition of the exterior of the extinguisher shell should be checked. It should be free of rust and corrosion, and the label should be in place and legible. There should be no dents or other damage to the cylinder.

One individual should be assigned the responsibility for the inspection of all extinguishers. This person may, but does not have to be, the person who actually performs the inspections. In smaller facilities one person is usually enough to perform all inspections. Often, an individual from the maintenance department is assigned the task of inspection. If security guards are used at the facility, inspections can be made part of their routine rounds.

Outside contractors can also provide inspection services. The costs of in-house versus contractor inspections must be compared to determine the most effective alternative.

Fire-Extinguisher Maintenance

Fire-extinguisher maintenance is essential to ensure that extinguishers are in usable condition in the event of fire and to comply with regulations.

Maintenance involves a more comprehensive check than the monthly inspections and is generally done annually. Maintenance is also necessary any time the extinguisher has been used or is found to need repair during the monthly inspection. Extinguishers also require periodic internal examinations and hydrostatic tests. Time intervals for these tests vary for different extinguishers. Generally for water, foam, and carbon dioxide extinguishers, the interval is five years. For dry-chemical and clean-agent extinguishers, the internal exam must be conducted every six years and the hydrostatic test every twelve years.

Manual Fire Control

Manual fire control is the third and final option of the three major methods of handling fire protection. *Fire prevention and hazard reduction* is the first and most important. *Engineering solutions that limit fire growth and development, including automatic suppression and extinguishing systems,* are the second major approach.

Fighting fires should be the last option considered as part of fire-protection preparations. Even though it is a last resort, it is an important part of being prepared for fires.

The most common manual fire-control equipment used by employees is a portable fire extinguisher. The following general guidelines apply to extinguisher use in all situations and with all types of extinguishers.

When an individual discovers a fire, the first priority is to notify other occupants. Life safety is the primary concern. The second priority is to call the fire department, since a fire is more easily controlled in the early stages of development. Delay in summoning the fire department can substantially increase the loss caused by a fire. From the fire department's perspective, it is much better to arrive and find that a fire has been extinguished with a fire extinguisher than to arrive at a fire that has grown and spread because the fire was not reported until after an occupant had unsuccessfully attempted to extinguish it. Even if a fire has been extinguished, it is a good practice to have fire department personnel check to ensure that it is, in fact, completely out.

Only after completing these two actions should an individual consider using an extinguisher.

The individual must evaluate the fire situation based on several factors and decide whether or not to fight the fire. Personal safety is the first factor. Next is the availability of an operable fire extinguisher. If no fire extinguisher is accessible or one that is present cannot be operated, the rest of the considerations are academic. The next consideration is the magnitude and intensity of the fire and the rate at which it is growing. An individual should not risk personal injury by attempting to control a fire. If the fire has grown large or is developing rapidly, it is not safe to fight it. The individual must also be concerned with the products being created by the fire. Depending on its nature, the smoke, heat, and gases generated can make the area surrounding the fire extremely dangerous in a relatively short period of time.

If the individual determines that it is safe to attempt to control the fire, the next step is to find a fire extinguisher suitable for the class of fire to be controlled and bring it to the area. When bringing the extinguisher to the fire area, safety is more important than speed. Personnel should be trained to walk—not run.

For a fire inside a structure, the person using the extinguisher should stand so that an exit is at his or her back so that if the attempt to control the fire fails, the exit will allow a safe escape. If the fire is outside, wind should be coming from behind the individual to prevent smoke and heat from being blown toward the person and to aid in controlling the fire by carrying the extinguishing agent over the fire.

The person with the extinguisher must make preparations to fight the fire from a minimum of 20 feet from the fire. Preparations include checking the condition of the fire extinguisher, pulling the safety pin, and test-discharging the unit to ensure that it operates. Only after all of these actions have been completed should the individual move in to control the fire.

After the fire is extinguished, the individual should back away from it so that if reignition occurs, he or she will be aware of it immediately.

Conclusion

This chapter covers several of the issues that safety professionals need to consider to effectively handle fire protection. As stated at the beginning, it is essential that local codes and standards be consulted to confirm the specific requirements that will be applied to a facility.

References

ANSI/AIHA (American National Standards Institute/American Industrial Hygiene Association). 2005. Standard Z10-2005, *Occupational Safety and Health Management Systems*. Fairfax, VA: AIHA.

NFPA (National Fire Protection Association). 2000. NFPA 30, *Flammable and Combustible Liquids Code.* Quincy, MA: NFPA.

______. 2003. NFPA 51B, *Standard for Fire Prevention During Welding, Cutting, and Other Hot Work.* Quincy, MA: NFPA.

______. 2004. NFPA 497, *Classification of Flammable Liquids, Gases, or Vapors and of Hazardous (Classified) Locations for Electrical Installations in Chemical Process Areas.*. Quincy, MA: NFPA.

______. 2006a. NFPA 1, *Uniform Fire Code.* Quincy, MA: NFPA.

______. 2006b. NFPA 101, *Life Safety Code.* Quincy, MA: NFPA.

______. 2006c. NFPA 5000, *Building Construction and Safety Code.* Quincy, MA: NFPA.

______. 2007a. NFPA 10, *Standard forPortable Fire extinguishers.* Quincy, MA: NFPA.

______. 2007b. NFPA 80, *Standard for Fire Doors and Other Opening Protectives.* Quincy, MA: NFPA.

______. 2008. NFPA 70, *National Electrical Code.* Quincy, MA: NFPA.

Occupational Safety and Health Administration (OSHA). www.osha.gov

Schroll, R. Craig. 2002a. "Flammable Liquid Safety." *Occupational Health & Safety*: March 2002.

Schroll, R. Craig. 2002b. *Industrial Fire Protection Handbook.* 2d ed. Boca Raton FL: CRC Press.

Schroll, R. Craig. 2002c. "Manual Fire Control." *Occupational Health & Safety*: February 2002.

APPENDIX: RECOMMENDED READING

National Fire Protection Association (NFPA). 1990. *Industrial Fire Hazards Handbook.* 3d ed. Quincy, MA: NFPA.

______. 2003. *Fire Protection Handbook.* 19th ed. Quincy, MA: NFPA.

______. 2005. Standard 600, *Standard on Industrial Fire Brigades.* Quincy, MA: NFPA.

______. 2007. Standard 1081, *Standard for Industrial Fire Brigade Member Professional Qualifications.* Quincy, MA: NFPA.

Society of Fire Protection Engineers (SPFE). 2002. *SFPE Handbook of Fire Protection Engineering.* 3d ed. Quincy, MA: NFPA.

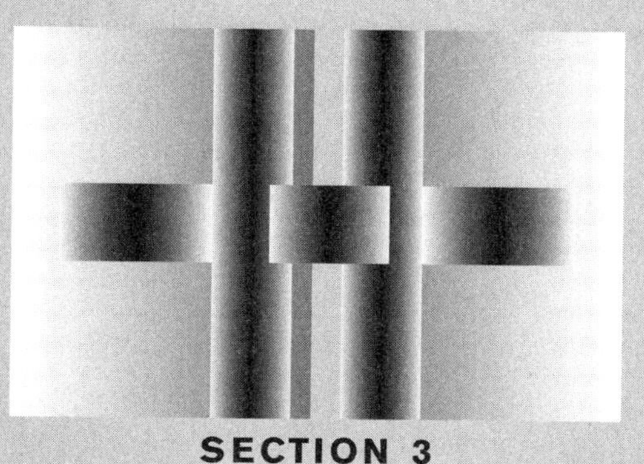

SECTION 3

FIRE PREVENTION AND PROTECTION

LEARNING OBJECTIVES

- Understand fire development stages and how they affect the speed and accuracy of fire detection.
- Become familiar with detection methods and design basics, including spacing and environmental issues.
- Learn how to use fire detection methods to actuate, supervise, and control extinguishing systems.
- Understand fire-extinguishing phenomena.
- Know which extinguishing agents to select, weighing pros and cons for extinguishing systems by fuel type.
- Understand water as an extinguishing agent, including use of water to limit structural damage and extinguish fire, and what additives can be used to enhance water's use as an extinguishing agent.
- Understand special hazard extinguishing agents (including carbon dioxide, dry chemicals, Halon, and clean agents) and their use.
- Become familiar with types of sprinkling and water-spraying systems, including their basic system components and use.

APPLIED SCIENCE AND ENGINEERING: FIRE SUPPRESSION AND DETECTION

Dick Decker

IT HAS OFTEN been said by firefighters that detection of a fire within the first few minutes is more important than the first few hours of firefighting. Firefighting can be thought of as occurring in three steps

- Fire commencement until alarm
- Alarm until use of extinguishers
- Use of extinguishers until extinguishing

Fires can be started in many ways that depend upon the type and configuration of fuel, as well as upon the oxygen concentration, ignition source, and flame chemical reaction characteristics. A fire in an upholstered chair could develop slowly from a smoldering cigarette, but that same cigarette could cause rapid ignition and a violent explosion when discarded near a gasoline spill. The multitude of variables associated with fire ignition and development require total evaluation of the fuel, ignition sources, oxygen concentration, and burning characteristics when selecting fire detection and extinguishing systems for use in safeguarding life and property.

FIRE DETECTION METHODS

This section will discuss fire-detection and alarm-system basics as well as the relationship of the detection process to alarm and extinguishing.

Stages of Fire

Most fires develop through stages or steps. Progression through the stages varies with the decomposition rate of the fuel. Some chemicals and oils progress rapidly to the final stage, but others—such as insulating materials—may take considerable time. The stages are normally identified as (1) the incipient stage, (2) the smoke stage, (3) the flame stage, and (4) the intense heat stage. During the incipient stage, no smoke is visible. No flame exists, nor any appreciable heat release. During this stage the products of combustion are invisible. Particles released are minute solids and liquids—mostly carbon and sulfur and some water vapor. These particles are not visible to the human eye and are usually less than 5 microns in size. In the second, or smoke stage, the particles released have aggregated into visible masses over 5 microns in size. These larger smoke particles present a threat to life upon inhalation. As the fire continues to burn, increased combustion products are released, and the fire gases become incandescent and luminous. When the last stage is reached, the highly exothermic chemical reactions occurring in the fire cause significant heat release. In other words, the chemical reaction goes from a low-heat, small-particle release to a high-heat, high-combustion product release including radiant energy release (Scherer 1972).

Speed of development is a very significant issue when selecting a detection method. Detectors require combustion by-products or radiation to travel from the fire location to the detector sensor. Particle detectors, normally located under ceilings or roofs, sense rising and spreading combustion gases, burn rate of combustibles (particle generation rate), coagulation rate (aggregation rate of the particles), and air movement. Hot combustion by-products are lighter than their surrounding air, causing them to rise until they strike a barrier (such as a ceiling or some sort of capture device or canopy). Upon doing so, the gases spread until encountering a barrier to horizontal spread, after which they build downward until a break in the barrier to horizontal movement again permits their upward or lateral spread.

Burn rate, or the weight of combustibles oxidized in a predetermined time period, is another very important consideration during fire detection. A single smoker, for example, releases very few combustion by-products in a room during 10 minutes of smoking, but 20 men could release over an ounce during the same period, adequate for detection. An ounce of newspaper burning in a trash can 20 feet from a detector in the same room could also release enough combustion by-products to activate an alarm (Scherer 1972). Smoldering fires typically create low concentrations of combustion products because of their low rate of burn. These types of fires also normally have a low spread rate because of their deep-seated slow burning characteristics. Quicker detector response times often indicate a fire closer to the detection device.

The rate at which combustion particles gather into bunches—often referred to as particle coagulation—influences detection. As particles aggregate, they enter the air. Because of the natural attraction of such particles, this occurs continuously in a space where a combustion reaction is occurring. At low burn rates, the coagulation rate could be high because of the presence of large particles of incompletely reacted (burned) fuel, causing the particles to settle out before detection. These types of fires are typically oxygen deficient at the reaction zone, often causing incomplete combustion accompanied by heavy, unreacted fuel particles. These incompletely reacted fuel particles can settle out on floors and walls after extinguishing. At low rates of burn, particles may coagulate and settle out at a rate faster than that of their generation, and an alarm may fail to occur.

Fire Detection Devices

Numerous devices exist for detecting fire or environmental changes indicative of fire. They must be appropriate for the location and the environment to be protected. These devices incorporate varied detection speeds and reliability that have to do with fire characteristics, detection methods, and the type of space in which the detectors are located. Ionization detectors provide for quick detection of slow-developing fires by detecting incipient-stage combustion by-products. Smoke detectors that depend on light obscuration for actuation depend upon larger particles of combustion. Heat detectors work best in cases of high heat release, rapidly developing fires and often monitor temperature change rate. Infrared and ultraviolet detectors need radiant energy release from a flame to cause actuation and work best in cases of organic-liquid and gas fires that have luminescent flames (Cote 2003; Scherer 1972). The following discussions will review the characteristics of (and variations upon) different types of detectors, explaining their proper applications.

Smoke Detectors

Ionization or photoelectric detectors, or a combination of the two, are frequently used to detect early-stage fires. Smoke detectors meet the needs of most areas containing primarily ordinary combustibles. Both ionization and photoelectric detectors can activate during the smoke stage of fire development, and ionization detectors can even detect invisible particles (less than 5 microns) in the incipient stage. As the particles aggregate, they are more easily detected by light obscuration methods. Smoke detectors are suitable for

- Indoor areas having low ceilings, such as offices, closets, electrical rooms, and small storage rooms for ordinary combustibles.
- Areas that are relatively clean, having minimal amounts of dust and dirt.
- Areas containing solid fuels such as wood, paper, fabric, and plastics.

Ionization Detectors

The basic ionization technique uses a small radioactive source to create an electrical current (ionization current) within a defined space or detection chamber similar to the one shown in Figure 1.

FIGURE 1. Ionization smoke detector (*Source:* Notifier Company)

Detectors of this type are exposed to air drawn from the space to be protected. (It should be noted that this ionization current has nothing to do with ions produced within the flame during combustion.) Combustion products, 1000 times the size of air molecules, hamper production of ions by the detector radiation source because of increased absorption and slower movement within the electron field. Combustion products are less mobile than air molecules and less likely to be neutralized before hitting the plate (chamber walls). This condition causes the field resistance to increase. The air within the chamber is a simple variable resistor. Normal air allows sufficient current to flow to maintain the detector in its normal state, but airborne combustion products increase the resistance, decreasing the current flow and causing the alarm to activate. Detectors vary widely in capability and operation and must be able to compensate for atmospheric and environmental changes from smoke and fire. Because ionization detectors are designed to use air as the conducting medium, the presence of foreign gases in the detector chamber could affect sensitivity. Most ionization detectors, for example, will be activated by high concentrations of carbon dioxide. Areas safe for habitation, however, should not see levels of foreign gases high enough to affect detector sensitivity. Vapors or gases from manufacturing processes, however, could cause false alarms. Some companies manufacture detectors that have a dual chamber design reducing the effects of air contaminates. Dust that settles on elements inside the detection chamber can become an insulating layer, increasing the possibility of false alarms and requiring frequent maintenance for prevention.

Photoelectric Detectors

Photoelectric smoke detectors such as the one shown in Figure 2 usually use a light-reflection photocell to sense fires. Smoke passes through a grill and enters a black-coated chamber containing a photocell and a light source. Smoke particles entering the chamber reflect light, causing the photocell to generate voltage that, when amplified, triggers a transistorized relay controlling the signal circuit. Smoke must be of sufficient concentration (2–4% obscurity) and duration (5–10 seconds) to actuate the detector. Light sources should be checked and cleaned periodically.

Projected-beam smoke detectors operate on principles similar to those of spot-type photo detectors but are more sensitive to real fires and less prone to sounding nuisance alarms. The projected beams generally have adjustable response thresholds (typically between 20% and 70% total obscuration), and some systems also have adjustable sensitivity settings. These alarms project a beam of light at a receiver calibrated to respond when the amount of light reaching the receiver falls below a certain level. Obstructions, such as solid objects, can cause trouble alarms. In cases such as high-smoke, aviation-fuel fires, the total block feature on some systems can itself be blocked. Projected-beam detectors work well where a string of spot-type detectors would be difficult to access for individual detector testing.

FIGURE 2. Photoelectric smoke detector (*Source:* Notifier Company)

Decomposition Product Sensing

During a fire, the gas composition of the atmosphere continuously changes. Levels of decomposition products can be detected rapidly by analyzing a sample of these fire decomposition products. Several system designs draw samples of the atmosphere into the protected area and analyze them for smoke particles. These include cloud-chamber systems, continuous-air-sampling systems, and spot-type aspirating detectors. Cloud-chamber systems use piping or tubing to draw samples from one or more locations in the area being protected. The samples are filtered to remove dust and humidified, after which air pressure is reduced by means of a vacuum system, forming a cloud inside the sensing chamber. By measuring the density of the cloud, combustion can be detected. Continuous-air-supply systems pull samples in a similar way, passing them across a very sensitive photoelectric detector cell. Spot-type devices use a small fan inside an individual detector to pull samples through a filter into the photoelectric cell. These work very well in dusty or dirty air in which fibers are present, such as in textile mills.

Heat Detectors

Historically, most fire detectors have monitored ambient temperature or temperature increase rate to sense fires. The oldest detector may well have been the automatic sprinkler that first appeared in the 1860s, which not only detected fires but also applied extinguishing agents. (Sprinklers are discussed in detail later in this section.) Heat detectors connected to electrical circuitry appeared some time afterward. These devices are very reliable and rarely give false alarms. They work best in confined spaces where rapidly developing, high-heat fires are expected. Heat detectors are usually located near ceilings, depending upon convective heat transfer for actuation. Detection can depend upon a predetermined temperature or rate of temperature increase. Rate-of-rise detectors are particularly used in cases in which fire development is expected to be very quick

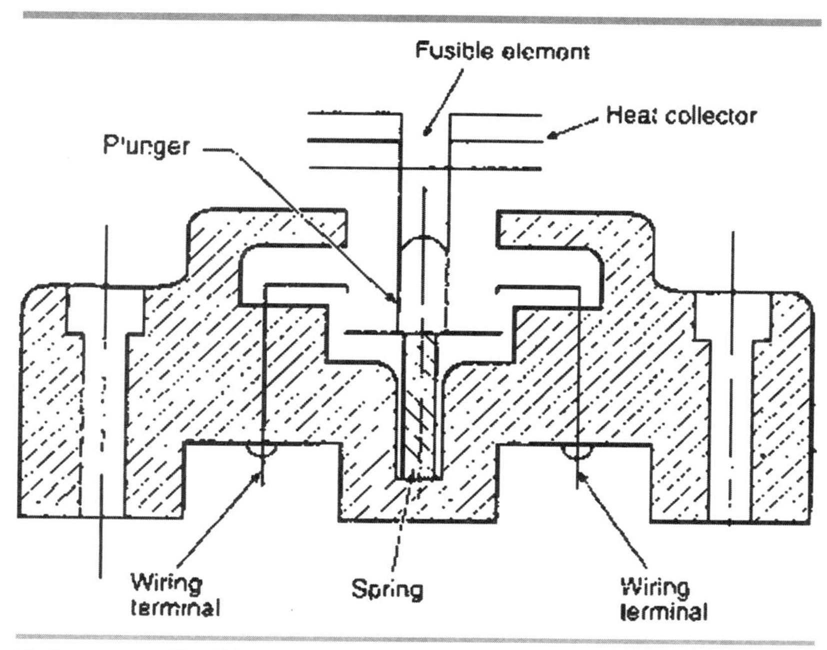

FIGURE 3. Fusible element detector (*Source:* Cote 2003)

(because the potential fuel source in the area being protected has a rapid burn and heat release rate after ignition, such as might be found in a flammable liquid storage room).

Fixed-Temperature Detectors

Fixed-temperature detectors include eutectic-element, lineal, and bimetallic detectors. The target temperature is selected based on the normal ambient temperature in the protected area. Typical temperatures settings available are 135°F, 165°F, 212°F, and 312°F, but the surrounding temperature can exceed the target level because of the "thermal lag" caused by the time it takes to heat the detector's operating element to the set level.

Eutectic-Element Detectors

Fusible-element heat detectors similar to the one in Figure 3 use eutectic alloys of bismuth, tin, lead, and cadmium to hold contacts in a predetermined position until the melting point of the alloy is exceeded, activating an alarm. Such detectors must be replaced after activation.

Lineal

Several types of continuous-line detectors are used for fire detection. One is a pair of insulated wires (shown in Figure 4) connected to a (normally open) circuit; another features a coaxial center conductor inside a steel capillary tube separated by temperature-sensitive glass semiconductor material (Figure 5).

In the first case, the wire insulation melts at the set point and the circuit is closed, causing an alarm.

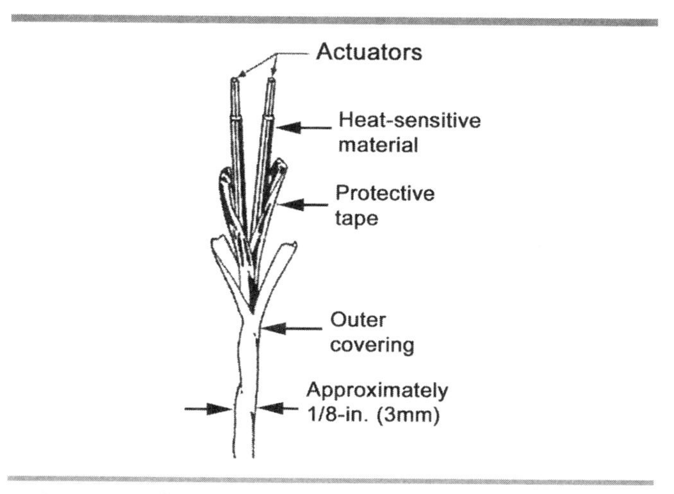

FIGURE 4. Lineal wire detector (*Source:* Cote 2003)

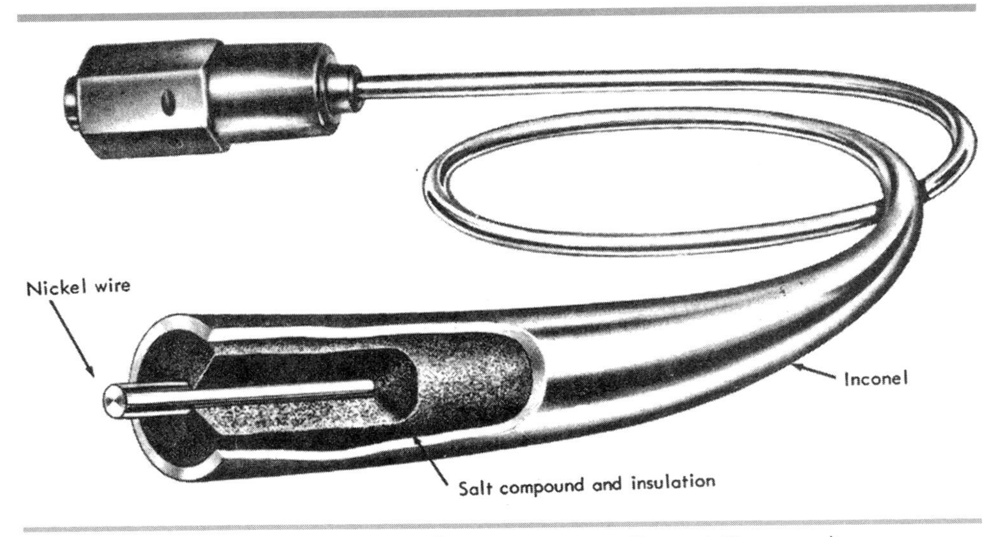

FIGURE 5. Lineal heat detector (*Source:* Allison Control Company)

The wires are under tension in a braided sheath. After alarm activation, the fused section of the conductors must be replaced before the system can be returned to service. In the second case, a small current flows between conductors through the semiconductor material at normal temperatures. At elevated (fire exposure) temperatures, the resistance of the media drops, allowing increased current flow and causing an alarm via the control circuitry. The sensing element activates the alarm when any portion of it exceeds a predetermined temperature. The lineal detector can be installed at ambient temperatures very close to the alarm temperature, because the set temperature must be exceeded prior to alarm activation. The elements are strings, coils, or form-fit to the contours of the areas they protect; they are also moisture-resistant. Available in lengths over 100 feet, they are rugged and reusable even when exposed to temperatures of 2000°F. They operate in gas, liquid, and solid environments.

FIGURE 6. Bimetallic detector (*Source:* Standard Steel LLC)

Bimetallic

Bimetallic heat detectors (as in Figure 6) use two metals of different coefficients of thermal expansion bonded together. When the detector is heated, the metal with the higher expansion rate bends or flexes toward the other element, causing the normally open circuit to close. The low-expansion metal is Invar' (36% nickel and 64% iron). Other alloys, such as manganese, copper, and nickel or nickel, chrome, and iron or stainless steel, are used for the high-expansion component.

Bimetal components are used for the operating elements of a variety of fixed-temperature heat detectors. They take the form of a bimetal strip or snap disc. As they are heated, the bimetal strip deforms in the direction of the contact point. The gap between the contacts varies with the predetermined temperature—a wider gap means a higher activation temperature. Disc-based detectors use a disc to help collect heat. As the disc heats, internal stress causes concave movement, closing the contact. When the disc cools, the detector resets (Cote 2003).

Rate-compensated: Rate-compensated heat detectors respond when temperatures about the detector exceed the set point (see Figure 7).

The rate of temperature change does not influence the response. These detectors have low thermal lag and are automatically self-restoring.

Rate-of-rise: Rate-of-rise detectors monitor the rate of air temperature increase (as occurs in flaming fires) (Figure 8).

Rate-of-rise alarms typically sound alarms upon 12–15°F temperature rise per minute. They compensate for rates of temperature change less than the low end of the set point. Features incorporated into these units (Figure 8) include a diaphragm containing a small vent orifice positioned inside the heat collecting semisphere. As the temperature increases outside the semisphere, heat transfer into the sealed chamber causes the air to expand. When the air expands at a rate faster than can be vented through the small orifice, the diaphragm is pushed outward, causing an electrical contact to close and initiating an alarm signal. If the temperature change is slow, the vent orifice relieves the air at the rate at which it expands and the contacts are not closed. Some of these units also have a fixed-temperature feature where a contact closure occurs at a predetermined temperature regardless of rate of temperature rise. They should not be used in areas subject to surges in temperature (such as in front of space heaters).

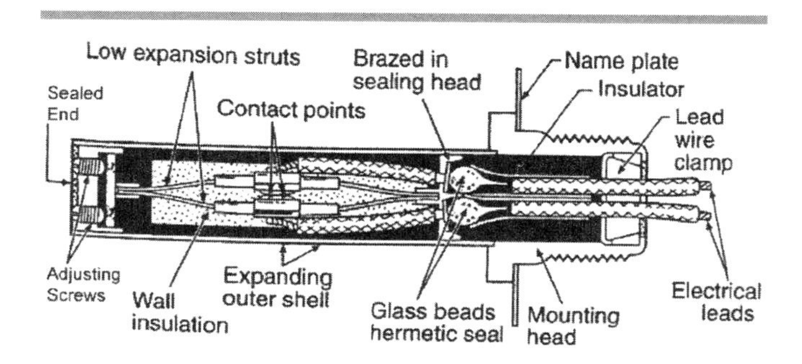

FIGURE 7. Rate-compensated detector (*Source:* Cote 2003)

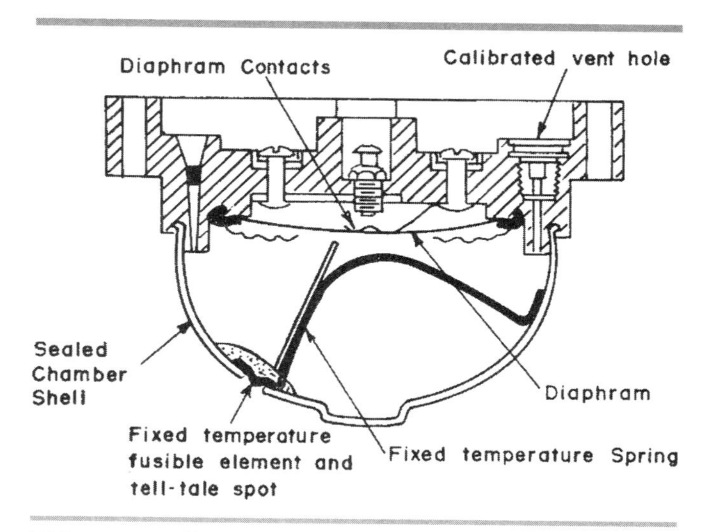

FIGURE 8. Rate-of-rise detector (*Source:* Notifier Company)

Radiant Energy

Detectors that respond to radiant energy include ultraviolet and infrared devices, line-of-site devices that monitor a specific light range that includes flames emitted during combustion. This includes radiation in ultraviolet, visual, and infrared spectra. When the detector recognizes any light in the range, it initiates an alarm signal. Radiant energy detectors are best at protecting areas with high ceilings and open spaces, such as warehouses and auditoriums. They are also used in outdoor areas where wind can inhibit other detection methods, including areas where fires can rapidly reach the flame stage (such as flammable liquid processing and storage areas). The sensitivity of radiation detectors is inversely proportional to the square of the distance from the source; doubling the fire distance requires a fire of quadruple size to produce the same detection speed.

Ultraviolet Detectors

Ultraviolet flame detectors typically use vacuum photodiode tubes to detect UV radiation given off in the 1000 to 4000 angstroms wavelength range by flames. A control unit monitors UV photons striking the detector and initiates an alarm after they reach a certain level. Such detectors are sensitive to most fires involving hydrocarbons, hydrogen, and combustible metals such as magnesium. If heavy smoke exists, however, it can absorb ultraviolet rays and inhibit operation. Nuisance alarms can also be caused by lighting, X-rays and arc welding (Scherer 1972).

Infrared Detectors

Infrared fire detectors respond directly to the presence of a flame, sensing infrared radiation emanating in the 8000–10,000 angstroms range. Normally these detectors are designed to wait for flamelike flickering and sustained radiation. They are intended for use in cases in which fires may develop rapidly with little or no incipient combustion. They are effective in detecting flammable liquid fires and should be mounted high, able to monitor a wide area. These detectors react to flickering light having a frequency of 5–30 cycles per second (cps). It is not uncommon to mount these detectors at heights over 50 feet above hazard areas. At greater heights, the protection area can be larger, but larger fires are needed to activate the alarm. These detectors can be used to protect areas up to 10,000 ft^2, depending on ceiling height and equipment arrangement. A 20-foot ceiling height might reduce the coverage below 3000 ft^2 (Scherer 1972).

Selection Considerations

Fire detectors must be selected based on the burning characteristics of the fire load present in the area, the arrangement of the protected space, and the physical atmosphere normally present. Smoke detectors are most suitable for indoor areas with low ceilings, such as offices and residential areas. Heat detectors are ideal for areas where flammable gases and liquids may be present, or areas where fires will quickly cause marked changes in temperature. They are suitable for smoky, dirty areas normally free of drafts or wind that could prevent heat from reaching and enveloping a detector. They work well in manufacturing areas where significant amounts of fumes, gases, or vapors are present and are adaptable to areas that could have products of combustion present under normal circumstances, such as kitchens, boiler rooms, and areas near vehicle exhausts. Radiation detectors are most useful in protecting open areas where visible flames are expected.

Location and Spacing Issues

Air movement is important to consider when locating and spacing fire detectors. Combustion by-products tend to rise, a tendency influenced by drafts, air diffusers, fans, and other such sources of air currents that may hinder travel to the detection devices. Detectors should never be placed within 4 feet of a fan, supply diffuser, or other strong draft source. It may be advantageous, however, to place detection devices closer to air-return connections. Combustion products rise to ceilings and other obstruction to vertical movement before spreading; when such obstructions include eight- to ten-foot ceilings, the spacing of detectors is not so much of a factor as it is in areas with higher ceilings.

When considering spacing and location of fire detection devices, the capabilities of individual detectors must also be considered. Here are some general guidelines for spacing detectors:

- Put at least one detector in each room, storage area, and hallway, following the manufacturer's guidelines.
- Place detectors as close as possible to the center of the spaces they are to protect.
- Place detectors at the tops of elevator towers, stairwells, and other vertical shafts.
- Never locate detectors within 4 inches of a corner formed by the intersection of vertical and horizontal barriers (Cote 2003).

Fire Alarm and Annunciation Systems

Fire alarm and annunciation systems gather fire detection signals, evaluate their meaning, sound alarms, and initiate corrective actions. Systems may also be used to monitor the condition of fire water supply system components such as valves and pumps, as well as the status of fire-extinguishing system components. Systems must be designed and installed in accordance with NFPA 72, *National Fire Alarm Code* (NFPA 2007e) and incorporate redundancy adequate to ensure acceptable levels of reliability. System components include primary and backup power supplies, data-gathering circuits, alarm-annunciation circuits, corrective-action control circuits, memory- and data-recording devices, alarm-signaling devices, and remote communication features. Figure 9 depicts a typical system.

Power Supplies

If a fire alarm and annunciation system is to function reliably, it must be provided with a reliable main or primary supply of power as well as a backup or secondary supply. The power-supply reliability must be maintained at the protected premises and at any remote supervisory station. Public-utility-generated power from a dedicated circuit generally serves as the primary source. The circuit disconnect or breaker

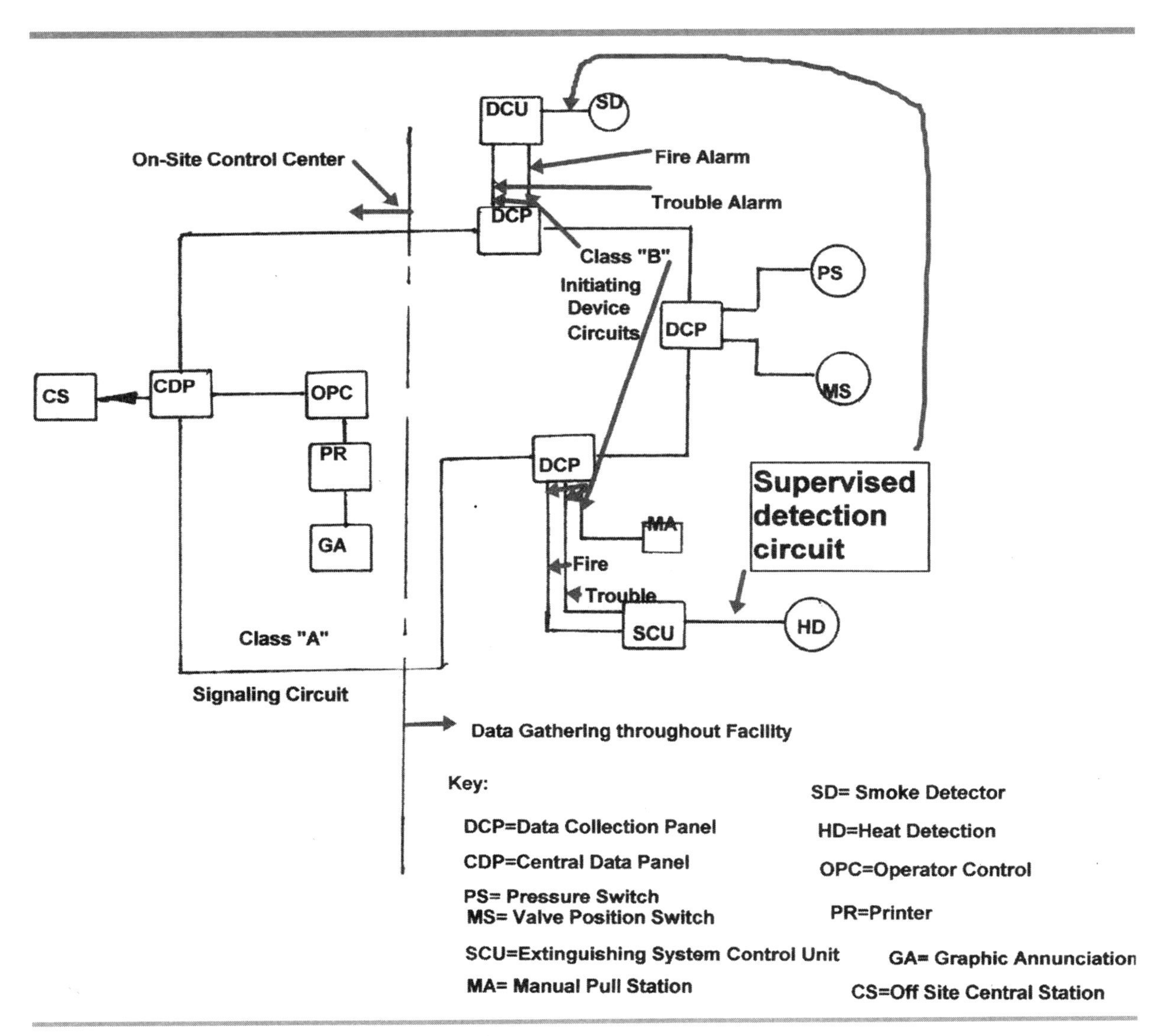

FIGURE 9. Fire alarm system schematic

must be "red" and the circuit properly protected from mechanical damage. The circuit controls must be protected against unauthorized access, and the location of the power supply control devices must be identified in the fire-alarm control cabinet.

Secondary power must be supplied within 30 seconds of the failure of the primary supply. Generally, backup power is supplied by storage batteries but can also be provided by a combination of emergency generators and batteries depending on the system power requirements and the extent of the alarm system. The size of the supply is based on the length of time it is expected to operate the system. NFPA 72 requires a five-minute actual alarm period and a 24-hour standby backup power supply for local, central-station, remote-station, proprietary, and auxiliary systems. Where emergency voice alarms are provided, the power supply must be adequate for an emergency operation period of two hours. The NFPA defines the two hours of emergency operation as equivalent to 15 minutes of operation of all input and output devices under full load. The backup power supplies must be dedicated to the fire-alarm system or facility fire safety features. Under some very stringent requirements, an emergency power system designed to ensure operational continuity of a facility can be used for supplying the fire alarm system power by connecting it to the emergency bus system in lieu of a backup supply (NFPA 2007e).

Fire Alarm Circuits

Fire alarm systems incorporate various electrical circuits to gather signals, sound alarms, and initiate corrective action. NFPA 70 and 72 define installation requirements (NFPA 2008b; NFPA 2007e). The initiating device circuits are information-gathering circuits that connect detection devices, manual fire-alarm pull boxes, fixed fire-extinguishing supervisory and system actuation devices, and building-monitoring and process-alarm devices to a central control processor unit. Alarm annunciation circuits that transmit an alarm signal to audible or visual annunciation devices—sometimes called notification appliances circuits—cause an audible device to sound an alarm in a particular area or cause some type of visual signal such as a strobe light or annunciation screen to communicate the alarm. The third type of circuit, corrective action control circuits, permits communication with an actuation device and (in some cases) allows two-way communication. When two-way communication can take place on the circuit, the circuit is often called a signaling-line circuit. The communication can be between a fixed fire-extinguishing system or an off-site alarm-supervising central station.

A typical circuit used in fire alarm systems consists of a two-wire circuit with an end-of-line resistor. Initiating devices, normally having open contacts, are connected in parallel. A small amount of electrical current flows through the wire to monitor the wiring integrity. A break in the wire causes a trouble condition at the system control unit. Everything electrically beyond the break in the wiring is out of service until repairs are made to the circuit. Operation of an initiating device shunts the resistor, thus increasing the current, causing the system control panel to respond in an alarm condition.

A circuit configuration that operates up to the single fault (open-circuit or ground-fault condition) is a Class B circuit. A circuit configuration that requires the wiring to be returned to the system control unit, which is configured to allow a device or appliance to operate on either side of a single open-circuit or ground-fault condition, is a Class A circuit. Both of these designations can be used to initiate device, notification appliance, and signaling-line circuits.

CENTRAL CONTROL PROVISIONS

Many fire alarm system control units have more than one initiating device circuit, allowing fire location to be indicated on an annunciation panel by floor, wing, subsection, or even room. The annunciation panel can be built into the control unit or located in a lobby, maintenance area, telephone switchboard room, or other location where it is accessible to building and fire-service personnel. Annunciation is principally designed to serve those responsible for managing the facility and occupant safety, as well as to assist the fire department, with whom coordination is important if the annunciation of alarm is to be effective.

New annunciation panels and addressable technology make it possible to annunciate not only fire alarms but also supervisory alarms giving the status of fixed fire-extinguishing systems, water supplies, ventilation systems, and exits, assisting the overall management and control of the facility. Such control is very useful to a fire department responding to an emergency. If their response has been preplanned and they are familiar with the facility, they can quickly evaluate the status of all supervised elements of the fire defense system. Some systems' central control provisions allow voice instructions to be given to occupants—technology is constantly changing to keep pace with our high-tech society.

Signal Processing

Fire-alarm and supervisory signals are collected throughout a facility by a data-gathering circuit connected to a central processing unit (CPU). Such gathering circuits often are looped design so that a single break will not prevent the transmission of alarm signals back to the CPU. The system supervises the circuit, identifying faults and giving trouble signals that describe problems. These systems should be set up to receive signals whenever a water flow signal occurs (to report system operation), a system valve is shut off (to indicate that the extinguishing agent supply is shut off), and any electrical fire detection device activates (to signal fire presence). CPUs can also be used to collect signals from manual fire-alarm pull stations. A typical schematic is shown in Figure 9.

Alarm Record Retention

Older systems printed a list of fire or trouble alarms, often in code, on a tape; modern systems, however, store these events in solid-state memory and can print complete reports on demand.

On-site Alarm Annunciation

The primary purpose of a fire alarm system is the automatic indication of the existence of fire and the communication of the need for evacuation to an open or otherwise safe area. Both audible and visual notification are desirable. Combination units (similar to that depicted in Figure 10) provide both.

It is also important to announce changes in the status of fixed fire-extinguishing systems and in agent supply to these systems.

Visual

Modern systems provide visual annunciation at the CPU on a monitor screen. The alarm locations are often shown on schematic diagrams of the facility so that operators know exactly where the initiating device is physically located. Strobe or other emergency lights can also be placed in key areas, including exits, to communicate the alarm visually to facility occupants.

Audible

Alarms are sounded throughout facilities using various bells, horns, buzzers, and similar annunciation devices. Some systems sound a general alarm, but others actually sound coded signals. Systems can be configured to provide voice alarms that include evacuation instructions.

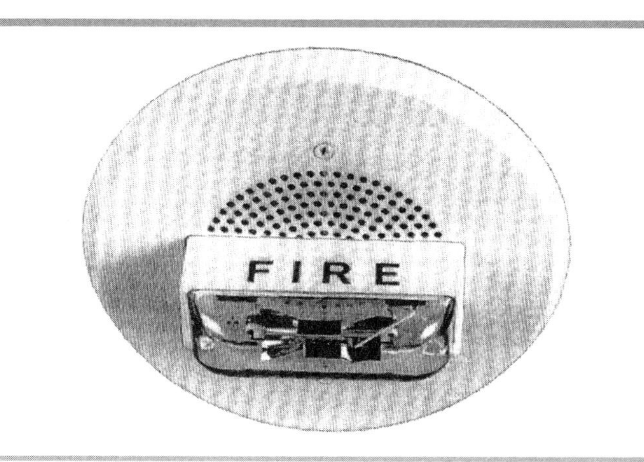

FIGURE 10. **Combined audible buzzer and strobe light** (*Source:* Notifier Company)

Remote Station Notification

The secondary purpose of a fire alarm system is to notify local emergency services. In order to do this, central control units for facilities that are not manned continuously often send any fire or trouble signals to a constantly manned remote site over phone lines or via radio transmission. Sometimes such stations are public emergency communication systems or fire department dispatch centers, but they may also be a commercially operated central station services.

Fire-Extinguishing Agents

Selection of agents to be used in extinguishing a fire begins by discovering the types of fuel involved, the size of the fire, and which agents are available in sufficient quantities to control the anticipated fire.

Classification of Fires

Fires are classified as Class A, Class B, Class C, Class D, or Class K based on their fuels and extinguishing requirements. Class A fires involve ordinary combustible materials such as wood, paper, cloth, rubber, and many plastics. Class B fires involve flammable liquids, oils, greases, tars, oil-based paints, lacquers, and flammable gases. Class C fires involve energized electrical equipment in cases when electrical nonconductivity of extinguishers is important. Powered-down electrical equipment may be safely controlled with Class A and B extinguishing agents. Class D fires involve combustible metals such as magnesium, titanium, zirconium, sodium, lithium, and potassium. Class K fires are fires in cooking appliances that involve combustible cooking media such as vegetable oil, animal oils, or fats (Cote 2003; NFPA 2007).

Extinguishing agents must be selected to match the class of fire expected. If multiple classes are possible, the selection may require a certain level of compromise.

Extinguishing Phenomena

The fire triangle (Figure 11) often comes to mind when discussing extinguishing phenomena. Its three sides represent the fuel, oxygen, and heat required for combustion. Removal of any of these elements causes extinguishing. A fourth mechanism—an uninhibited chemical reaction—is also present in all fires. Thus, fires can be extinguished by physically withholding fuel, oxygen, or heat from the flame and by introducing chemicals that inhibit or modify the combustion's chemistry. "Combustion" refers to an exothermic (heat-producing) chemical reaction between fuel and oxygen. Most extinguishing agents (water, foam, dry-chemical) work by a combination of mechanisms, and their value in any circumstance depends upon the fuel involved, as well as upon any other peculiarities of the situation.

Temperature Issues

Because a fuel must be in a gaseous state to participate in flaming combustion, temperatures must be sufficient to deliver the fuel to the flame as a gas. Some fuels, such as natural gas, butane, or propane, either are gas or will rapidly become gas at atmospheric conditions. Other fuels, such as flammable and combustible liquids, must vaporize before they can participate in combustion. The temperature at which a liquid fuel emits enough vapors above the liquid surface to be ignited by a pilot ignition source (such as a spark) is called the "flash point" of the

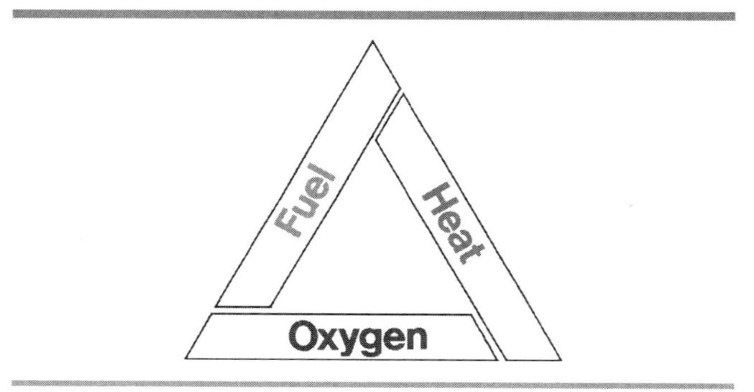

FIGURE 11. Fire triangle (*Source:* Safety Services of Texas)

liquid and provides a good measure of the degree of hazard presented by the fuel. Another temperature that is important when evaluating degrees of fire hazard is the autoignition temperature. The autoignition temperature is the temperature at which a liquid fuel will ignite without the presence of a spark. Detailed laboratory test procedures exist by which to determine flash points and autoignition temperatures.

Oxygen Issues

Most fires occur at ambient conditions. The combustion process usually gets its oxygen from air, which contains approximately 20.9% oxygen, 79% nitrogen, and lesser amounts of minor gases. When oxygen levels increase above levels in ambient air, the atmosphere is called an "oxygen-enriched" atmosphere. When combustion occurs in an enriched atmosphere, the process is more rapid and heat release is greater. Acetylene cutting torches, for example, make use of compressed oxygen supplies. In contrast, if oxygen is reduced the combustion process is slowed, as might occur in a deep-seated fire in baled paper (Cote 2003).

Fuel Issues

Fire is a combustion (oxidation) process in which a fuel is oxidized and heat released (an exothermic chemical reaction). A chemical analysis of the combustion products involved identifies the presence of certain molecules, as well as the combination of oxygen atoms with other atoms. Typical products include carbon monoxide, nitrogen oxide, and so on. In flaming combustion (fire), heat is generated faster than it is dissipated, causing temperatures to rise hundreds or thousands of degrees. All fuels must be gaseous in order for combustion to occur. Flame is actually a gaseous oxidation reaction occurring in a space at a much higher temperature than the surrounding air, emitting light. A portion of the radiant heat from the flame is fed back into the fuel, causing liquids and solids to vaporize. During smoldering combustion, no visible flame is present. Smoldering occurs in porous materials that form a char when heated. As oxygen slowly diffuses into the pores, a glowing reaction zone occurs without visual flame. The high temperatures in the glowing reaction zone vaporize the solid fuel.

Chemical Inhibition

High temperatures in excess of a few thousand degrees Fahrenheit are required for combustion, and the chemical reaction must occur quickly enough to prevent reaction-zone cool-down. At the adiabatic flame temperature, heat generation is only adequate to keep the combustion reaction going without releasing energy. If anything upsets the heat balance—such as water-cooling—combustion probably will stop. Extinguishing can occur by either cooling the gaseous combustion zone or by cooling the actual fuel. If the fuel is cooled, flammable vapors to fuel the fire will not be generated. Another way to remove heat from the combustion zone is to modify the air supply by injecting gas (often inert) that has adequate heat capacity to take heat from the reaction. A decrease in temperature of a few hundred degrees can cause a very significant slowdown of the reaction. Combustion consists of rapid chain reactions involving hydrogen (H) atoms and other active species such as hydroxyl free radicals (OH) and free oxygen (O) atoms. In addition to the use of cooling to slow the chemical chain reactions in a flame, an actual chemical inhibitor can be introduced into the flame to interfere with the process. Dry-chemical extinguishing agents, such as sodium bicarbonate and potassium bicarbonate, and various halogenated agents, such as Halon 1301 or 1211, provide free radicals that interfere with the reactions occurring in the combustion zone, causing the reactions to stop (Cote 2003).

Fire-Extinguishing Agents

Agents used for extinguishing fires include water, foams, carbon dioxide, dry chemicals, Halons, and clean agents. Each of these agents has features that make it the agent most desirable for a particular fire-hazard situation.

Water

Water is the most commonly used fire-extinguishing agent, largely because it is widely available and inexpensive. Because it is liquid under conditions of normal use, it can be easily delivered and applied to fires.

Physical characteristics: A single gallon of water can absorb 9280 Btu of heat as it increases from a 70°F room temperature to become steam at 212°F. Thus, even in cases in which water from sprinklers will not suppress fire, the cooling ability of water spray can protect structural elements of a building, containing the fire until it can be extinguished by other means (Cote 2003).

Fire-extinguishing characteristics: In the case of a fire involving a solid combustible in air, several extinguishing mechanisms are involved simultaneously. The solid fuel is cooled by the water, which causes the rate of its conversion to gas (referred to as "pyrolysis" or "gasification") to slow. The chemical reactions in the flame gases are cooled, reducing both radiant heat feedback to the combustible solid and the endothermic reaction, causing the solid to become gaseous. In cases where conditions are confined, application of water can result in steam generation, which may prevent oxygen from reaching the flame. Radiant heat transfer can be blocked by the water if it is applied as a fog. Water mixed with foam solutions can form foam extinguishing agents.

Additives: Additives can be injected into water to reduce friction losses and form foams that will provide an effective cover for burning hydrocarbons or mix with them, diluting them until combustion cannot be sustained. Normally, hydrocarbons will float on top of the water, continuing to burn (and possibly spreading). To combat such fires, foam solutions can be introduced into the water to provide an effective cover, smothering the fire.

Advantages and disadvantages: The primary advantage of water as an extinguishing agent is its abundance and the inexpensiveness of large-scale use. Its disadvantages lie in its inappropriateness for use in controlling certain water-reactive materials. In some cases, the use of water can *produce* heat, flammable or toxic gases, or even explosions. The quantities of such products must be considered, however, because application of sufficient amounts of water can overcome such reactions. However, another drawback of water lies in its density, which is greater than that of most hydrocarbon fuels, and in its immiscibility. Liquid hydrocarbons of low specific gravity may float on the surface of water and be carried off-site, causing environmental pollution. In some cases water can be absorbed by building contents, rendering them unusable.

Foam

Application of an aqueous foam blanket is another way to apply water—for example, to flammable liquid-fuel fires. Several extinguishing mechanisms could be present in such a case. Foam acts as an insulating blanket, keeping radiant heat from the flame from vaporizing the fuel. The pressure of the fuel vapor could be reduced and the liquid cooled if the fire point of the flammable liquid is higher than the temperature of the foam. Where the flammable liquid is water-soluble, as is alcohol, it could be diluted by the water in the foam and the vapor pressure of the flammable liquid reduced.

Physical characteristics: Firefighting foam is an aggregate of gas-filled bubbles formed from aqueous solutions of specially formulated, concentrated liquid-foaming agents. In most cases, the gas is air, but inert gases can also be used. Because foam is lighter than the aqueous solution from which it is formed, it floats on flammable and combustible liquids, acting as a blanket separating the surface from the air above it. Foam is produced by mixing water with a concentrate and then aerating and agitating the solution to form bubbles. Foams are produced in various consistencies; some are thick and viscous and form strong, heat-resistant blankets over liquid surfaces and vertical structures, but others are structured to spread rapidly over surfaces because of their thin construction. Some produce a vapor-sealing, surface-active film over the flammable or combustible liquid. Each foam is categorized according to its expansion ratio: that of the final foam volume to the volume of the original solution before aeration. Foams tend to be divided into three ranges: low expansion (up to 20:1), medium expansion (20:1 to 200:1), and high expansion (200:1 to 1000:1). Low-expansion foams are typically used to blanket flammable liquid-surface fire exposures. Medium-expansion foams are used to control fires that may be more deep-seated in nature. High-expansion foams are best suited to hazards that are enclosed and which may involve vertical surfaces. Foams are also classified according to their effectiveness in controlling hydrocarbon fuels, water-miscible fuels, or both (Cote 2003).

Fire-extinguishing characteristics: Foam's primary extinguishing mechanisms are oxygen removal, heat shielding, and fuel-vapor control. The foam bubbles form a blanket separating the surface from the air above, cooling the surface and preventing vapor release and combustion.

Advantages and disadvantages: Some of the advantages of foam include its capability of making water light enough to float on flammable and combustible liquids, as well as its increase of extinguishing volume during the foaming process. Its chief disadvantage is the inconvenience of obtaining and using special equipment to create foam from the foam solution, forming foam bubbles. Furthermore, bubbles are not compatible with some other extinguishing agents and may break down sooner than is desired.

Carbon Dioxide

As a gas at ambient conditions, carbon dioxide has been used effectively for years to extinguish flammable-liquid and electrical fires as well as, to a lesser extent, fires in ordinary combustibles. It can be stored as a liquid in high-pressure cylinders or in low-pressure, refrigerated containers near 0°F. High-pressure systems operate properly at temperatures of 32°F to 120°F. Discharging carbon dioxide has a white cloudy appearance because of the ice particles that form when the liquid flashes to vapor.

Physical characteristics: Carbon dioxide is noncombustible and (in most cases) nonreactive and has a sufficiently high vapor pressure to allow it to discharge from the storage container. At atmospheric pressure, it is a gas capable of spreading throughout a protected enclosure and penetrating a fire. As a gas or a solid (dry ice), it does not conduct electricity and can thus be safely use to control fires proximate to live electrical parts.

Fire-extinguishing characteristics: The primary extinguishing mechanism of carbon dioxide is smothering, or oxygen removal. The cooling effect is quite small compared to an equal weight of water but may be of some help, particularly if applied directly to burning material.

Advantages and disadvantages: An advantage of carbon dioxide is that it leaves no trace, thus requiring no additional cleanup. This makes it a good choice around food-preparation areas. Some disadvantages of carbon dioxide include its harmful effects on humans at levels at or above 6% concentration; it causes loss of consciousness at 9%. Solidified gas (dry ice) can produce frostbite. Carbon dioxide will extinguish or suppress fire in most combustible materials, but it will not extinguish fires in a few active metals, metal hydrides, and materials with available oxygen (such as cellulose nitrate). In the case of reactive metals such as sodium, magnesium, titanium, and zirconium, as well as metal hydrides, carbon dioxide actually decomposes. Another concern involves extreme climactic conditions, which require special design considerations. Carbon dioxide produces a static charge (from ice particles) at discharge that can serve as an ignition source if not grounded (Cote 2003).

Dry Chemicals

Dry chemicals are powdered mixtures of compounds that are normally effective at extinguishing A, B, and C fires. There are also special dry-powder compounds for use on specific class D (combustible metal) fires and class K fires involving food-grade oils. Most dry chemicals are used as the extinguishing agent in portable fire extinguishers but may also be used to protect specific hazards in fixed systems.

Chemicals: The most common dry chemicals are sodium bicarbonate, potassium bicarbonate (Purple K), and monoammonium phosphate (multipurpose). Two other chemicals used as extinguishing agents are borox and urea-potassium bicarbonate.

Physical characteristics: Monoammonium phosphate, unlike most dry chemicals, is acidic. Inadvertent mixture of alkaline and acidic dry chemicals can initiate undesirable chemical reactions, generating carbon dioxide gas and even damaging equipment. Various additives are mixed with the base materials to improve flow, storage life, and water repellency. The additives coat the product particles to make them free-flowing and resistant to caking. These chemicals should not be stored above 120°F lest some of the additives melt and cause sticking problems. Particles range from 10 to 75 microns in size, which has a definite effect on the extinguishing efficiency. Each dry chemical has its own unique size characteristics that affect vaporization and decomposition of chemicals as well as flow in hoses or piping.

Fire-extinguishing characteristics: Use of dry chemicals to extinguish fires incorporates four of the basic extinguishing mechanisms: chemical inhibition, fuel coating, flame cooling, and radiant energy transfer blocking. Dry chemicals are exceptionally efficient in extinguishing flammable- and combustible-liquid fires. They are also very useful in controlling electrical fires. With the exception of multipurpose chemicals, dry chemicals are of limited help in fighting Class A fires, especially deep-seated ones. Dry chemicals cause almost instantaneous extinguishing when injected into the flame near the base of the fire. Potassium bicarbonate and monoammonium phosphate dry chemicals are faster acting than sodium bicarbonate, because the particles of these chemicals interfere greatly with the chemical chain reactions in the flame. The concentration of "free" radicals present within the flame is lower with these chemicals (Cote 2003).

Advantages and disadvantages: The primary advantage of dry chemicals is their ability to produce rapid extinguishing. Dry-chemical extinguishing agents are stable at low and normal temperatures and are both nontoxic and noncarcinogenic. Exposure can, however, cause irritation of mucus membranes and possible chemical burns of the skin, eyes, and respiratory track. The primary disadvantage is that accompanying cleanup of the residue after use. For the most part, the applications of dry-chemical extinguishing agents are limited to specific hazards and are not suited to total facility protection (Cote 2003).

Halogen Agents

Halogenated extinguishing agents are hydrocarbon chemicals that have had one or more hydrogen atoms replaced with fluorine, chlorine, bromine, or iodine atoms. The resulting chemicals are nonflammable and have flame-extinguishing properties. The flame-extinguishing phenomenon of halogen agents is not totally clear but appears to involve a chemical reaction interfering with the actual combustion process. Halogen agents are believed to block the chemical chain reactions present in the flame. A more complete discussion of halogen agents can be found in the *NFPA Handbook*, Section 11, Chapter 1, as well as in references at the end of Chapter 1 (Cote 2003).

These chemicals have been identified as ozone-depleting chemicals. Consequently, many of them, including Halon 1301 and Halon 1211, are no longer produced and are now replaced by clean agent systems. At present, recycled supplies of these chemicals are all that remain available for extinguishing-system recharging.

Chemicals: The most-used halogen agents were Halon 1301 (bromotrifluoromethane), Halon 1211 (bromochlorodifluoromethane), and Halon 2402 (dibromotetrafluoroethane). Whereas Halon 1301 was commonly used proximate to vital electronics equipment and computer rooms, the higher boiling point of Halon 1211 adapted it to hand-held fire extinguishers.

Physical characteristics: These agents are high vapor pressure compounds that must be stored under pressure to keep them liquefied. At low concentrations, these chemicals have little effect on personal health. Halon 1301 is relatively safe to 7% by air volume and

Halon 1211 to 3%. Decomposition products, however, are of higher toxicity.

Fire-extinguishing characteristics: The extinguishing mechanism appears mostly related to chemical reaction interference with the combustion process. The agents act by moving the active chemical species involved in the chemical chain reaction that occurs in the flame. This process is known as chain breaking and is rapid.

Advantages and disadvantages: Halogenated extinguishing agents have several advantages over carbon dioxide. Some of the halogenated compounds are effective at very low volumetric concentrations, leaving sufficient oxygen in the air for safe breathing after application or flooding of an enclosure. Because of the low vapor pressure of some of the halogenated compounds, some can stay liquid while being discharged via nozzle, making it possible to project them farther.

Drawbacks of halogenated agent use have to do with the corrosiveness and toxicity of decomposition products and with the chemicals' detrimental effects on the ozone layer (Cote 2003).

Clean Agents

Clean agents are fire-extinguishing agents that vaporize readily and leave no residue. They have been developed to replace halogen agents. These agents include halocarbon compounds and inert gases and mixtures. The compounds known as halocarbons include compounds containing iodine, fluorine, bromine, and chlorine. The five groups of these compounds exhibit a wide variety of characteristics but share several common attributes

- Nonconductive
- Gaseous after discharge
- Residueless vapor
- High vapor pressure chemicals
- Can be discharged via existing Halon 1301 extinguishing systems (usually require nitrogen pressurization for discharge purposes)
- Require more agent (both by weight and volume) than Halon 1301
- Produce greater amounts of hydrogen fluoride (HF) and other toxic decomposition products than does Halon 1301
- More expensive than Halon 1301

Chemicals: The chemicals found in clean agents include hydrobromofluorocarbons, hydrofluorocarbons, hydrochlorofluorocarbons, perfluorocarbons, and fluoroiodocarbons.

Fire-extinguishing characteristics: Fire-extinguishing using halohydrocarbon clean agents is caused by a combination of chemical flame interaction and physical mechanisms. Some extinguish by flame radical scavenging, thereby interrupting the flame chemical reaction. Some extract heat from the flame zone and reduce flame temperature below the adiabatic flame temperature, the temperature needed for reaction to continue. Flame temperature reduction may be caused by a combination of things, including removal of heat needed for vaporization of the agent, heat capacity of the material, and energy absorbed by the decomposition of the agent. The literature, including the *NFPA Handbook,* discusses the extinguishing process in considerable detail (Cote 2003).

Advantages and disadvantages: The primary advantage of these agents is their capability of extinguishing fires of complex geometry with relatively small agent concentrations that leave no residue. They also can be substituted as the agent in Halon 1301 systems with relative ease. The primary disadvantage relates to ozone depletion. Other disadvantages include the need for the hazard to be enclosed, thus containing the low extinguishing concentration. These agents must be held in the combustion zone for a period adequate to achieve the desired interaction of the agent with flame, referred to as a soaking period (Cote 2003).

Fire Protection Hydraulics

This section discusses hydraulics as applied to supply, distribution, and application of water as a fire-extinguishing agent. The discussions will review principles applied to the flow of water through pipes, hydrant outlets, nozzles, sprinklers, and the valves and accessories associated with the supply and distribution of water to a fire. It does not review the theory and derivation of the mathematical relationships associated with the formulas and empirical data included but does explain how to apply the information to the solution of fire protection problems. Refer to the "Fire Dynamics" chapter in this section of the text for more detail on fluid mechanics.

Water's Properties

Water generally used for fire-extinguishing is normally freshwater but could be saltwater from a natural body of water. Although the density of water, like all liquids, varies with its temperature, normal practice takes 62.4 lbm/ft^3 (pound mass per cubic foot) as the density of water. Sea water has a density of approximately 64.1 lbm/ft^3. One cubic foot of water equals 7.48 US gallons, so a gallon of water weighs about 8.34 lb. Viscosity is a measure of a fluid's resistance to flow and is an important factor, although the formulae used in fire protection hydraulics do not account for changes in viscosity.

Pressure: When discussing pressures in fire water distribution systems, it is essential to indicate whether the water in question is still or in motion (inside a pipe network or flowing from an orifice). Static pressure is the pressure of a column of still water, such as a storage tank. Because water is always moving in public distribution grids, the static pressure measured at a hydrant or standpipe gauge with no water actually discharging from it or any other nearby outlet of consequence can be assumed to be the "system" static pressure. Static pressure is also taken as the elevation difference between two points in an interconnected piping system. A residual pressure is the amount of pressure at a particular location in a piping system with a volume of water flowing. Pressure drop therefore, is the difference between the measured static pressure and the residual pressure. Total pressure in a system is equal to the normal pressure plus the velocity pressure. Total pressure at an orifice (hydrant outlet) is measured with a pitot tube.

Normal pressure: Normal pressure is pressure measured perpendicular to the side of a container or pipe full of water, whether or not flow is occurring. Without flow, it is equal to the static pressure; with flow, the residual.

Velocity pressure: Velocity pressure (Vp) is a measure of the energy required to put water in motion in a pipe. It can be calculated from the following equations:

$$\text{Velocity Head Equation: } h = V^2/2g \tag{1}$$

or

$$Vp = 0.00673\ V^2 \tag{2}$$

where V is the velocity of water flowing in a pipeline.

Velocity pressures for flows in different sizes of pipes can be obtained from the graph in Figure 12.

Bernoulli's Theorem: The physical law of conservation of energy applied to problems of incompressible flow of fluids is expressed by Bernoulli's theorem.

The theorem is explained in the *NFPA Handbook* as follows: "In steady flow without friction, the sum of the velocity head, pressure head and elevation head is constant for any incompressible fluid particle throughout its course" (Cote 2003). This means that in any system, the total pressure is the same at all positions. All of the heads, such as total head, velocity head, pressure head, and elevation head are expressed in feet. Therefore, all units must be converted to feet, or all terms expressed as pressures must be expressed as pounds per square inch (psi).

Bernoulli's theorem can be expressed mathematically when applied to two locations of different elevation, as follows:

Bernoulli's equation:

$$V_1^2/2g + P_1/w + Z_1 = V_2^2/2g + P_2/w + Z_2 \tag{3}$$

where V = velocity in fps, g = acceleration because of gravity (32.2 fps), P = pressure (psi), Z = elevation head in feet, and w = weight of the water (64.4 pcf).

Orifice Flow and Discharge Calculations

As a liquid leaves a pipe, conduit, or container through an orifice and discharges into the atmosphere, the normal pressure is converted to velocity pressure.

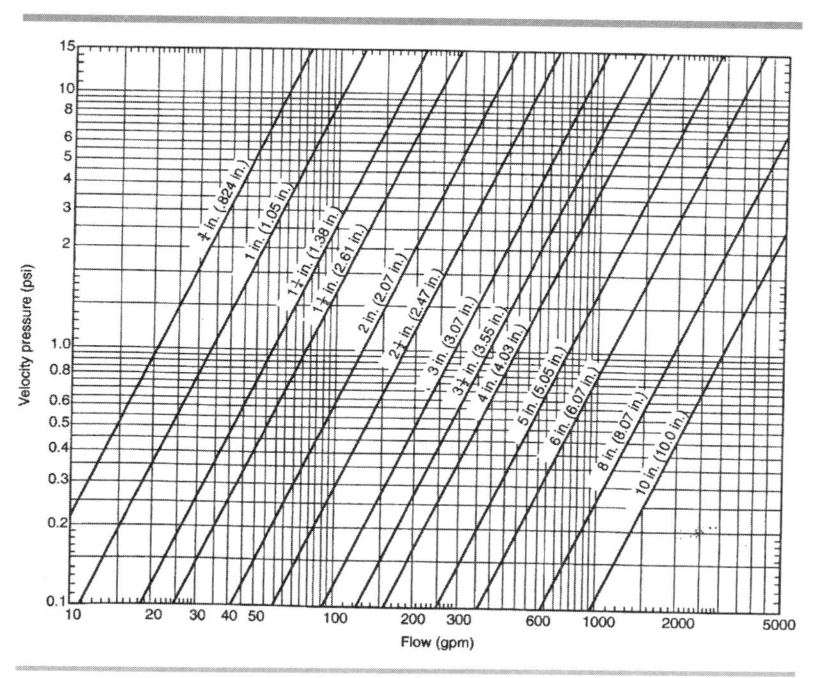

FIGURE 12. **Velocity pressure for water flow in pipes**
(*Source:* Cote 2003)

The rate of flow through an orifice can be expressed in terms of velocity and cross-sectional area of the stream, the basic relations being

Orifice discharge equation: $Q = av$ (4)

where Q = rate of flow in ft^3/s, a = area of cross section in ft^2, and v = velocity at the cross section in ft/s.

The velocity produced in a mass of water by the pressure head acting upon it is equal to $\sqrt{2gh}$ or

Velocity equation: $v = \sqrt{2gh}$ (5)

where g is the acceleration due to gravity.

The basic flow formula in Equation 4 can be rearranged to provide the velocity head, as follows:

Orifice flow using head (ft): $Q = a\sqrt{2gh}$ (6)

Hydraulic head (h), expressed in feet, can be converted to pounds per square inch (psi) using Equation 7:

Head to PSI formula: $h = 2.307p$ (psi) (7)

It follows that, with the orifice diameter in inches, Q in gal/min, and h in psi,

Orifice flow based on velocity pressure:

$Q = 29.84d^2(\sqrt{Pv})$ (8)

The above equations assume that

- the jet is a solid stream the full size of the discharge orifice.
- 100 percent of the available total head is converted to velocity head, which is uniform over the cross section (Cote 2003).

In actual flow from nozzles or orifices, the velocity, considered to be average velocity across the entire cross section of the stream, is somewhat less than the velocity calculated from the head. The reduction is because of the friction of the water against the nozzle or orifice and turbulence within the nozzle and is accounted for by a coefficient of velocity, designated C_v. Values of C_v are determined by laboratory tests. With well-designed nozzles, the coefficient of velocity is nearly constant and approximately equals 0.98.

Some nozzles are designed so that the actual cross-sectional area of the stream is somewhat less than the cross-sectional area of the orifice. This difference is accounted for by the coefficient of contraction, designated C_c. Coefficients of contraction vary greatly with the design and quality of the orifice or nozzle. For a sharp-edged orifice, the value of C_c is about 0.62.

The coefficients of velocity and contraction are usually combined as a single coefficient of discharge, designated C_d

Combined coefficient of discharge

$$C_d = C_v C_c \quad (9)$$

The basic flow equation can now be written as

Orifice flow equation with coefficient of discharge

$$Q = 29.8 C_d^4 (d^2 \sqrt{P_v}) \quad (10)$$

The coefficient of discharge, C_d, is defined as the ratio of the actual discharge to the theoretical discharge. For any specific orifice or nozzle, values of C_d are determined by standard test and found in various references such as the *NFPA Handbook*, Section 10, Chapter 5 (Cote 2003). The coefficient is applied to the formula to get a more accurate result.

Pipeline Water Flow

Water flow results when there is a pressure difference in an interconnected system or when a space containing it is opened to the atmosphere below the liquid surface. Water flows from a point of higher pressure to one of lower pressure. The pressure is normally reported in pounds per square inch gauge (psig), and is that amount which is more than atmospheric pressure (which is approximately 14.7 psig). The atmospheric pressure can be ignored in most cases, because it is applied to everything and applied equally in all directions. The term "head" is frequently used when discussing pressure, including velocity head, pressure head, elevation head, friction head, and head loss. A one-foot-high column of water causes one foot of head at the base, or 0.434 psig pressure. Pressure or head loss may be caused by a change in velocity or elevation or by friction loss. Hydraulic head losses also occur at lateral entrances and are caused by hydraulic components such as valves, bends, control points and orifices (Lee 2000).

Pressure Loss Formulas

Although various formulas are available in the literature for calculating water flow and pressure loss, the one most often used by water system designers is the Hazen–Williams formula, the relationship that will be used in this discussion of fire protection hydraulics.

Hazen–Williams Formula: The Hazen–Williams formula (the friction–flow formula commonly used in fire protection hydraulics) was developed by experiment and experience but contains no terms allowing for the effects of changes in temperature, density, or viscosity. It assumes that no additives are present and that water temperatures are near 60°F. Other equations used to determine pressure loss, such as the Darcy–Weisbach equation, do consider such factors (Cote 2003; Lee 2000; Wood 1961).

Darcy–Weisbach equation:

$$h = f \times l/d \times v^2/2g \quad (11)$$

where h = friction head, f = friction factor, l = length of the pipe, d = pipe diameter, v = velocity, and g = the acceleration caused by the force of gravity.

Hydraulic formulae used in fire protection are usually exponential in form and are variations of the Chezy formula

Chezy formula: $$v = Cr^x s^y \quad (12)$$

where v is velocity, C is the coefficient of friction, r is the hydraulic radius (area divided by circumference), and s is the hydraulic slope (loss of head divided by length).

The most popular exponential formula is the Hazen–Williams, which in its basic form is

Hazen–Williams formula:

$$v = 1.31 C r^{0.63} s^{0.54} \quad (13)$$

where v = velocity, C = coefficient of friction, r = the hydraulic radius, and s = the hydraulic slope.

The friction coefficients in formulae of this type are constant for a specific roughness of pipe and are independent of velocity; thus the accuracy of these formulas is variable. However, the fixed values generally assumed for viscosity and density are considered adequate for most fire-protection hydraulic work.

In order to make the formula more usable, it can be rearranged to express pressure in PSI instead of velocity, in terms of pipe diameter instead of hydraulic radius, and in gallons per minute (gpm). The hydraulic slope (s) is simply the pressure loss divided by the length. Because the Hazen–Williams formula is usually used to determine the pressure loss per foot of pipe, the length (L) can be replaced by the pressure loss (P), again in feet.

Because it is desirable to use flow rather than velocity, it is known from the discussion on velocity pressure that

Hazen–Williams formula:

$$v = Q/a = 0.4085\ Q/d^2 \tag{14}$$

where v = velocity, Q = flow in gpm, a = pipe cross-sectional area in square inches, and d = diameter in inches.

Hazen–Williams pressure loss formula:

$$P = 4.52Q^{1.85}/c^{1.85}d^{4.87} \tag{15}$$

where P = pressure loss per foot of pipe (psig), Q = flow in gpm, C = coefficient of friction, and d = diameter in inches.

This is the formula most often used for fire protection purposes.

The solutions of many fire protection problems involving pipe flow and friction do not require direct calculation using formulas, because tables and charts are readily available. However, in using the simplifying charts and tables, great care must be taken to identify the C value (coefficient of friction) upon which the chart or table is based. Where the type or condition of a pipe necessitates the use of a different value of C, the friction loss from the table must be multiplied by a conversion factor to obtain the correct results for the desired C value.

Figure 13 is a graphical representation of friction loss in $C = 120$ pipe taken from the *NFPA Handbook* (Cote 2003). It gives reliable results for pipe up to 10 inches diameter but because of scale is less accurate than tables found elsewhere in related literature. Other charts and tables are available in NFPA, and other, literature (Center 1959; Clark 1996; DiNenno 1998; Lee 2000; Wood 1961). Because the figure is for pipe with $C = 120$, it must be modified per Table 1 to provide pressure loss for C values other than $C = 120$.

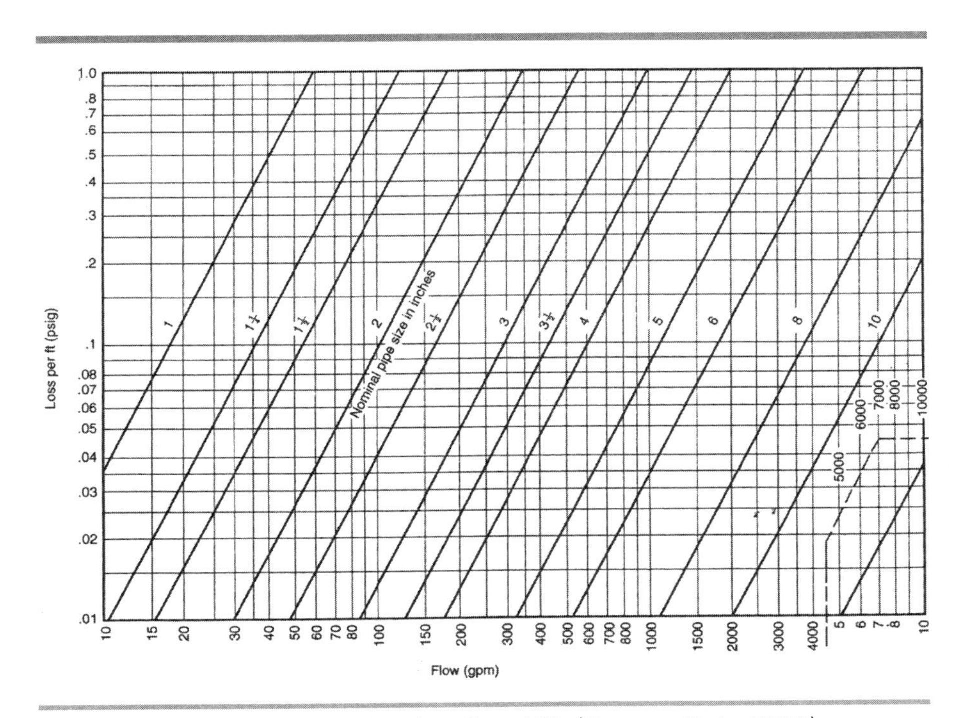

FIGURE 13. Friction loss in pipe C = 120 (*Source:* Cote 2003)

Hazen–Williams *C* values for pipes of different kinds and service periods have been measured and can be used to estimate *C* values for particular pipes. Table 2, which gives data from the *NFPA Handbook*, Section 10, Chapter 5 (Cote 2003), can serve as a guide for *C* value estimation.

Manufacturers of pipe can provide *C* factors for the particular pipe they manufacture when the pipe is new. The literature, including the *NFPA Handbook* (Cote 2003), Wood (1961), and Lee (2000), has tables of *C* factors for pipe in service and new pipe.

Experience shows that *C* values are most reliable when the flow velocity is close to that at which the *C* was determined. Many were determined near 3 ft/sec velocity. Sprinkler system flows range from 10 to 20 ft/sec.

TABLE 1

Hazen–Williams *C* Factor Conversion Chart

Value of C	Multiplying Factor
80	2.12
100	1.40
120	1.00
130	0.86
140	0.75
150	0.66

(*Source:* Cote 2003)

TABLE 2

Estimating Guide for Hazen–Williams *C* Value

K-Kind of Pipe	Value of C
Unlined cast iron	
10 years old	90
10 years old	65
50 years old	50
Cast iron, new	120
Cast iron, cement-lined	140
Cast iron, enamel-lined	140
Average steel	140
Asbestos-cement	140
Reinforced concrete	140
Plastic	150

(*Sources:* Cote 2003; Wood 1961; Lee 2000)
Assumes moderate corrosive water (charts available for other conditions)

Equivalent Lengths

The concept of equivalent lengths refers to the practice of substituting a pipe of equal friction pressure loss for another pipe or fitting (possibly of a different diameter or material) having identical friction loss with the same amount of water flow.

Minor Losses

Friction loss within a pipe causes most of the head loss, but head losses also occur when pipe size changes occur and when fittings or valves are inserted into the piping. Although these losses are usually considered minor, the configuration of some valves and devices installed in piping systems can cause significant losses. Fitting and valve losses are usually expressed as equivalent length or as a coefficient. Tables similar to Table 3 exist elsewhere in the associated literature, as in the *NFPA Handbook* (Cote 2003) and Wood's *Automatic Sprinkler Hydraulic Data* (1961), which express friction loss of fittings as an equivalent pipe length. By multiplying the number of fittings in a pipe run by the equivalent length of a single fitting, the total equivalent can be obtained.

TABLE 3

Fittings and Valves Expressed in Equivalent Feet of Pipe

Pipe Fitting	1"	1 1/4"	1 1/2"	2"	6"	8"	12"
45 Elbow	1	1	2	2	7	9	13
90 Elbow	2	3	4	5	14	18	27
Gate Valve				1	3	4	6
Swing Check Valve	5	7	9	11	32	45	65
Tee	5	6	7	9	17	24	37

(*Sources:* Cote 2003; Wood 1961)

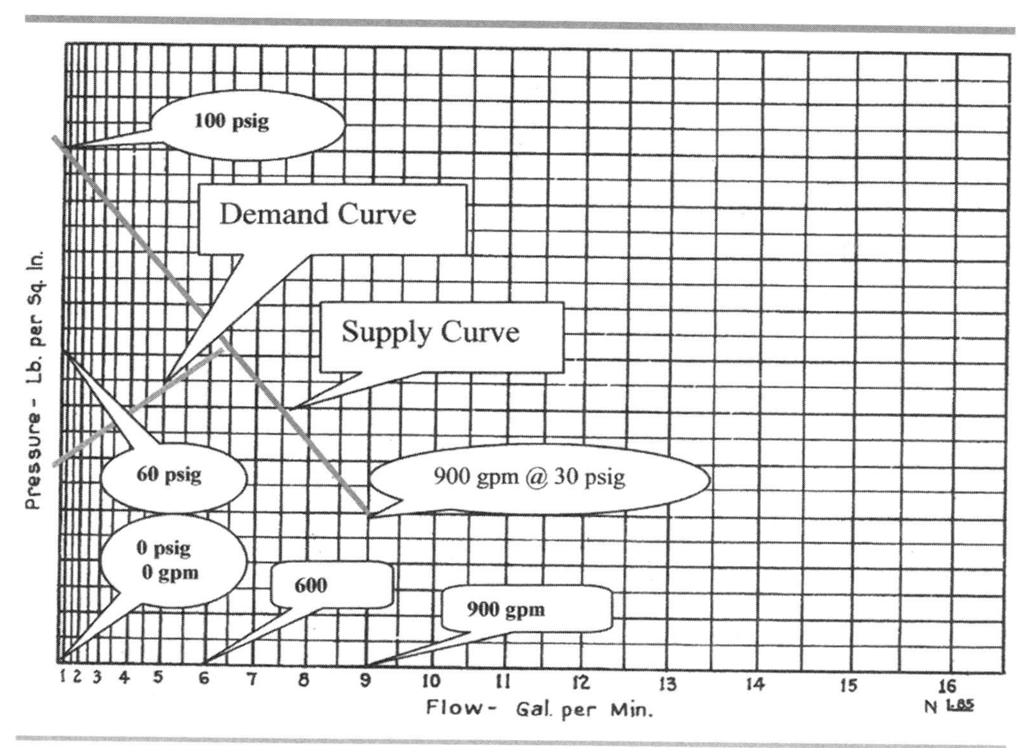

FIGURE 14. Example plot of pressure: gpm on $\text{Log}^{1.85}$ paper

Water Flow

For fire protection purposes, water always flows through piping from a point of supply to a point of demand. This can take the form of a single pipe or a compound piping system or network. There are three major types of compound piping systems: pipes in series, pipes in parallel, and branching pipes. Pipes in parallel occur in a network when two or more paths are available between two points. Pipe branching occurs when water can flow to or from a junction of three or more individual pipes from independent outlets or sources. Sometimes two or more pipe sizes can be involved. In order to evaluate the pressure losses between supply points and fire protection demand points, the possible flow paths must be determined so that estimations can be made of the flow distribution and associated pressure drop.

Hydraulic Flow Curves

It is often necessary to evaluate a water supply to determine whether it is adequate for a fixed-system or hose-stream demand. A quick and reliable way to do the evaluation is to plot flow test results and system demand on a graph using semi-exponential paper called $N^{1.85}$ paper or hydraulic paper. The design of $N^{1.85}$ paper is based on the Hazen–Williams pressure drop formula. In that relationship, head loss, static pressure less residual pressure is proportional to the 1.85th power of the flow. (An example flow curve is shown in Figure 16.) Flow tests plot as a straight line on $N^{1.85}$ paper. In the example shown in Figure 14, the demand is 600 gpm at 60 psig, and the supply curve shows in excess of 700 gpm available at 60 psig (Wood 1961).

Loop Flow

Estimation of flow characteristics and friction loss in loop and parallel pipe systems is often necessary when discovering whether the water supply can adequately supply a fixed extinguishing system. In loop systems, the pressure drop is equal in each path between the same nodes on the loop. This is true regardless of the size, condition, and length of the pipe. The flow in the leg having the highest losses will be

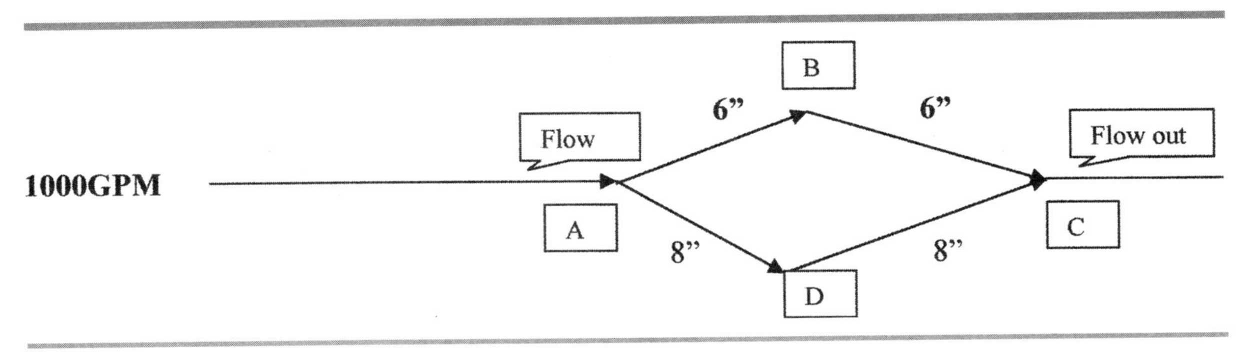

FIGURE 15. Simple water system piping loop

less than the flow in the leg with less loss, making the pressure drop by both routes the same. As systems become more complex, it becomes more difficult to estimate flows in the legs and balance the pressure at each node. Simple loops and parallel pipe systems can be analyzed graphically or using equivalent pipe methods, as explained in the literature; but this is beyond the scope of this chapter. Figure 15 illustrates a simple water system piping loop with water flow from point A to point B to point C and from point A to point D to point C. The flow will balance through the two routes to result in equal pressure loss between point A and C regardless of route.

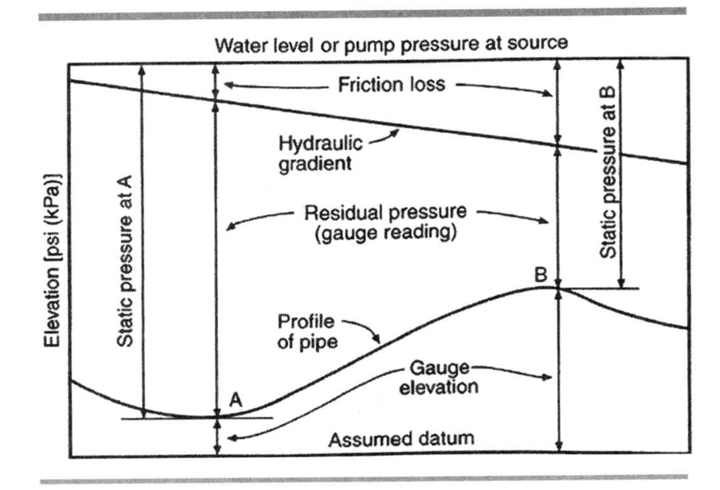

FIGURE 16. Hydraulic gradient diagram (*Source:* Cote 2003)

Hydraulic Gradients

Presentation of the flow characteristics of a pipeline in graphical form as a profile of residual pressure is referred to as a hydraulic gradient. Hydraulic gradients are useful in design of water supply mains and for evaluating the condition of the supply.

Following are some basic axioms that apply to hydraulic gradients

- The higher the elevation of the pipe, the lower the static pressure.
- Static pressure readings measure distance below the source.
- Static pressure plus gauge elevation (PSI) is constant for all points along the pipe.
- Static pressure minus residual pressure equals total friction loss from the source to the point of measurement.
- Residual pressure plus gauge pressure equals hydraulic-gradient elevation.
- Friction loss is independent of elevation (Cote 2003).

See Figure 16 for a graphical explanation.

A hydraulic gradient diagram illustrates pressure loss in a piping network with changes in static pressure and flow losses. Pressure at any point in the network is equal to the initial supply pressure less friction loss plus or minus elevation pressure differences. The hydraulic gradient diagram helps visualize the phenomenon.

Hardy Cross Analysis

Most water distribution networks as well as sprinkler systems and water spray systems are complexes of loops and branch pipes. The solution methods used for single and parallel pipes and single loops are not suitable for more complex networks. A trial-and-error procedure needs to be used with a view to causing the pressures to balance to a certain tolerance at each common node. The Hardy Cross is the most widely used loop method. To analyze flows in a network of pipes, certain conditions must be satisfied

- The algebraic sum of the pressure drops around each circuit (loop) must zero.
- Continuity must be satisfied at all nodes or junctions, causing inflow to equal outflow at the junction.
- Energy loss must be the same for all paths.

One technique, in which small loops within a complex pipe network are replaced by single hydraulically equivalent pipes, can simplify the process. The following example illustrates how to apply the equivalent pipe technique to a simple four-pipe loop (see Figure 17).

Example Equivalent Pipe Problem: Assume all pipes are at the same elevation.

Pipes 1 and 2 have 1000 feet of 6-inch pipe (C = 120), and pipes 3 and 4 have 1000 feet of 6-inch pipe (C = 120). Pressure loss from Figure 12 for 500 gpm through 6-inch C = 120 is 0.009 psig/ft or about 9 psig for the total run and is equivalent to flowing 1000 gpm through a single 1000-foot 8-inch C = 120 pipe.

A detailed review of Hardy Cross procedures is beyond the scope of this chapter, but these procedures are explained elsewhere in the literature (Cote 2003; Lee 2000; DiNenno 1988; Wood 1961). Numerous computer software, including some in the public domain (such as the EPA's EPANET2), exists to help complete this tedious task.

Fire Water Supply Methods, Demands, and Systems

Water is the most commonly used extinguishing agent, being easily obtained in sufficient quantities to extinguish fires. It can be supplied from a variety of sources both public and private that include municipal and public water supply systems, private on-site elevated storage tanks, pumps drawing water from grade-level storage tanks or private water reservoirs, natural water bodies, and pressurized storage tanks. When water is supplied to fixed fire-extinguishing systems, such as building sprinkler systems, it is common (and also required by most regulatory authorities) to incorporate not only a main or "primary" supply but also a backup, or "secondary," supply. In municipal areas and other areas having fire hydrants suitable for fire department pumper suction, it is common to provide a connection that the fire department can use to boost the water pressure in the fixed system. In this case, the fire department connection is accepted as the secondary supply. Where this is not possible, secondary supplies require some combination of supply methods to achieve adequate redundancy.

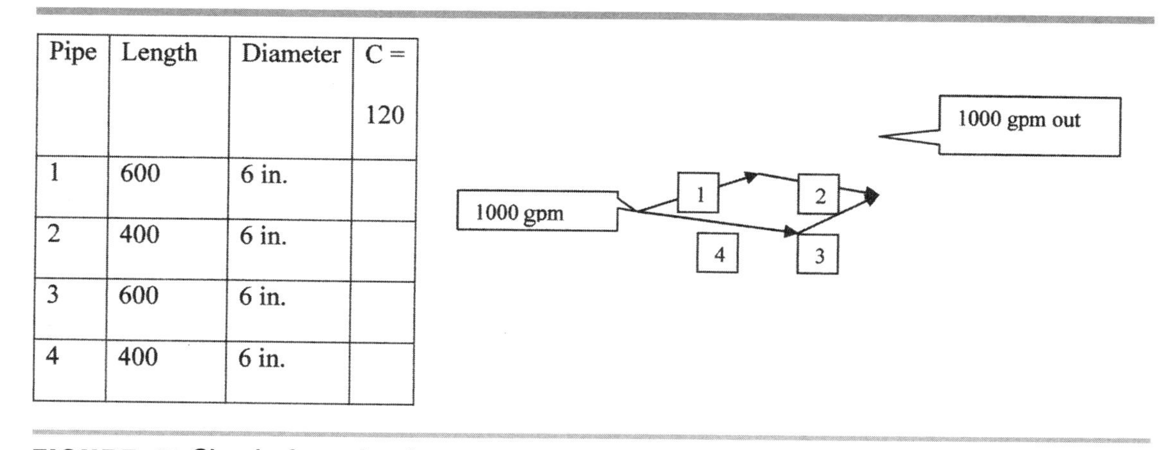

Pipe	Length	Diameter	C = 120
1	600	6 in.	
2	400	6 in.	
3	600	6 in.	
4	400	6 in.	

FIGURE 17. Simple four-pipe loop

Municipal/Public Supplies

Municipal and public water systems distribute water for domestic, commercial, and industrial consumption and serve as a supply of water for fire emergencies. Such systems use an interconnected piping grid to deliver water to demand points. When evaluating the adequacy of the municipal supply, consideration must be given to peak consumption for nonfirefighting purposes. For the supply to be acceptable, it must provide the required fire flow for fixed systems and hose streams in addition to meeting other demands.

Normally municipal or public systems supply water from large natural water bodies such as rivers, lakes, or aquifers. In cases where the source is at a higher elevation than the point of demand, gravity can provide the energy to deliver the water. However, the impetus is quite often provided by pumps in combination with elevated standpipes and storage tanks. These tanks float on the system to stabilize pressure (Cote 2003).

Elevated Water Tanks

On-site, private, elevated storage tanks can provide a reliable source of water for fire protection. After they are filled with water, they stand by waiting for fire system operations. Such tanks need to be high enough and of adequate capacity to supply the demand at adequate pressure for the required duration. Each foot an elevated tank is raised above the point of demand adds 0.434 psi to the head. It is not uncommon to have tanks over 150 feet high (including tanks placed on top of tall buildings or other structures). Some elevated tanks serve both domestic and process needs as well as providing fire protection reserves. In such cases, piping must be arranged to prohibit drawdown below the fire protection reserve. In cold climates, such tanks must be heated—especially those tanks used only for fire protection. Detailed design requirements are found in NFPA 22, *Standard for Water Tanks for Private Fire Protection* (NFPA 2003).

Stationary Fire Pumps

Pumps that draw from a reliable water source connected to a reliable energy source provide a very good way to supply or boost the pressure of a fire water system. Pumps can be used to develop the total supply by taking suction from a storage tank or water body and developing the total flow at a certain discharge head, or they can take suction from a low-pressure piping network, boosting it to a higher pressure. (The latter case may be necessary in a high-rise building.) Such pumps can be arranged for automatic operation or (as is often the case in pumps that are secondary supplies) may require manual starting. Pumps must be dedicated to fire protection use and be approved by a listing agency such as UL or FM for fire protection service. They must be installed in accordance with NFPA 20, *Standard for the Installation of Stationary Fire Pumps for Fire Protection* (NFPA 2007d).

Suction Sources

Suction sources for dedicated fire water pumps include above- and below-grade storage tanks, natural water bodies such as rivers and lakes, and other large, reliable water sources such as cooling-tower basins.

Lift

Lift is the distance a pump can draw or lift water. The pressure exerted on a surface at sea level by the atmosphere is 14.7 psi. Each increase of one foot in water elevation causes a pressure increase of 0.434 psi, so atmospheric pressure at sea level is equal to the pressure exerted by a 34-foot-high column of water. The reason a pump can lift water is that it creates a negative pressure in the suction pipe, the pressure of the atmosphere pushing the water up the suction pipe until it reaches the pump. Because of friction and entrance losses in the real world, atmospheric pressure can only push water about 15 feet up a pipe. Thus, NFPA 20 (2007d) permits a maximum of 15 feet of lift. Because of this, pumps arranged to operate under a negative suction head must be arranged so that the total suction lift, including entrance losses through the foot valve and friction losses, do not exceed 15 feet. (This applies to centrifugal fire water pumps lifting water from a lower elevation, not to vertical turbine pumps, because the impellers of vertical turbine are below the water level and therefore have a positive suction head) (Cote 2003).

Drivers

Just like the actual water supply, the dependability of the energy source driving a pump is critical to the supply of fire water to a point of demand. Various drivers have been used for fire pumps over the years, including steam turbines, electric motors, and gasoline and diesel engines. The key concern surrounding drivers is unit reliability; is the energy supply dependable, and is it adequate to supply the driver for the duration of the demand? Although steam turbines were often used to drive pumps in the past, the increased reliability of electrical supplies and the improved starting and control features of stationary engines have seen them become the driver of choice for modern fire pumps.

Engines

Engines for driving stationary fire pumps can be either diesel- or gasoline-fueled. In most cases, unless a pump serves entirely as a backup supply, they are set to automatically start upon system pressure drop. The engine and the control unit (which monitors the batery and control circuits and actually starts the engine upon pressure drop) must be approved for fire-protection use by a recognized testing laboratory such as UL. They must be provided with a fuel supply meeting the requirements of NFPA 20 (NFPA 2007d).

Electric Motors

Electric motors used to power stationary fire pumps must be set to automatically operate upon distribution system pressure drop or water flow and must be supplied with reliable electrical power in accordance with NFPA 20 (usually a direct, dedicated circuit to the fire pump controller from the public utility service) (NFPA 2007d).

Piping Arrangement

The piping arrangement depends on the type of pump and the configuration of the supply. Horizontal centrifugal pumps are the most common and can be arranged to draw from a storage tank under a positive suction head or from a source lower than the pump under a negative head. If taking suction under lift, a foot valve is usually placed at the base of the suction line to keep the suction line charged. Pumps require isolation valves, check valves, and relief valves in the piping. A typical arrangement is shown in Figure 18.

Some pumps are of a vertical turbine design in which the pump impeller is submerged in the water supply (see Figure 19).

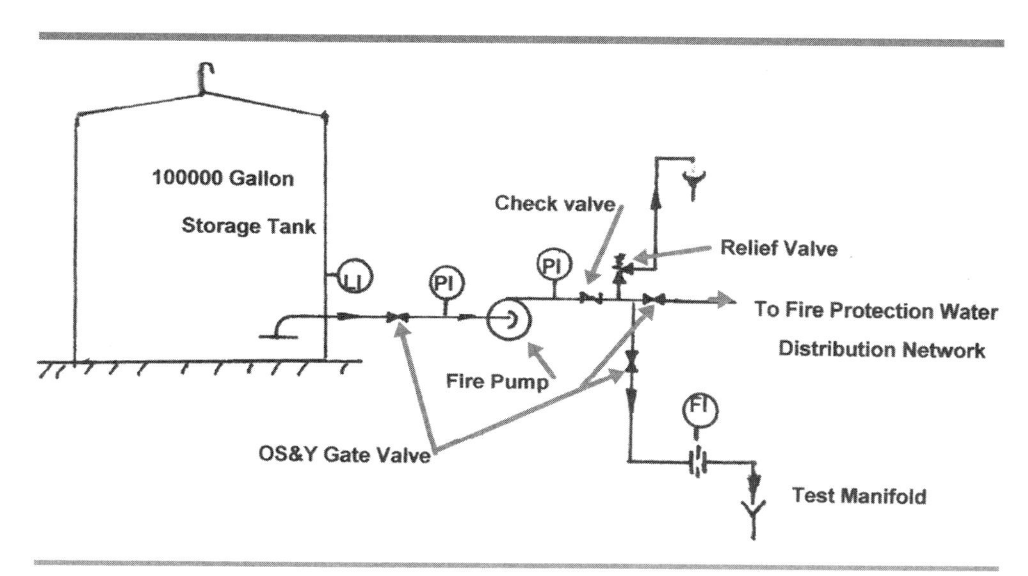

FIGURE 18. Typical horizontal fire pump piping schematic

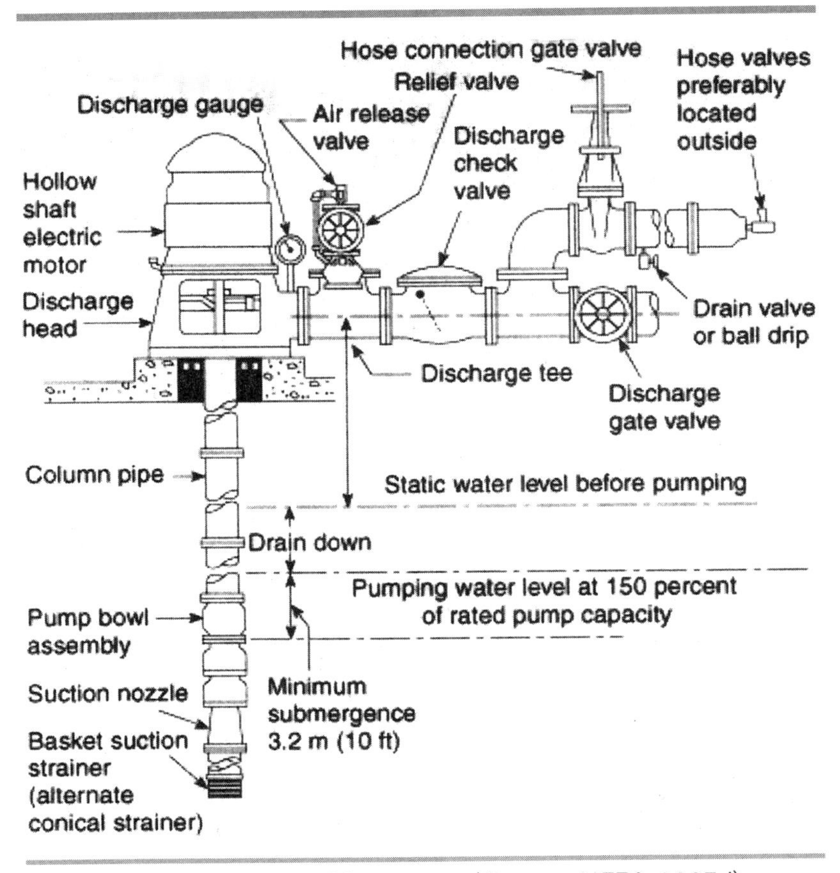

FIGURE 19. Vertical turbine pump (*Source:* NFPA 2007d)

Detailed piping requirements and physical arrangement recommendations are given in NFPA 20 (NFPA 2007d), Factory Mutual Insurance Requirements (Center 1959), and in other sources. Pump suppliers often provide a complete skid-mounted fire pump unit incorporating pumps, drivers, controllers, accessory valves, and devices such as the Aurora pump unit (shown in Figure 20).

Pressure Tanks

Pressure tanks provide a good fire protection water supply in cases in which the total demand is relatively small. In the case of a pressure tank, water is stored in a pressure-rated tank and the space above the liquid surface is pressurized with air or nitrogen. When the fire-extinguishing system requires water, the air pressure pushes the water from the tank to supply the system. Such tanks have a maximum drawdown that is reached when pressure in the tank has

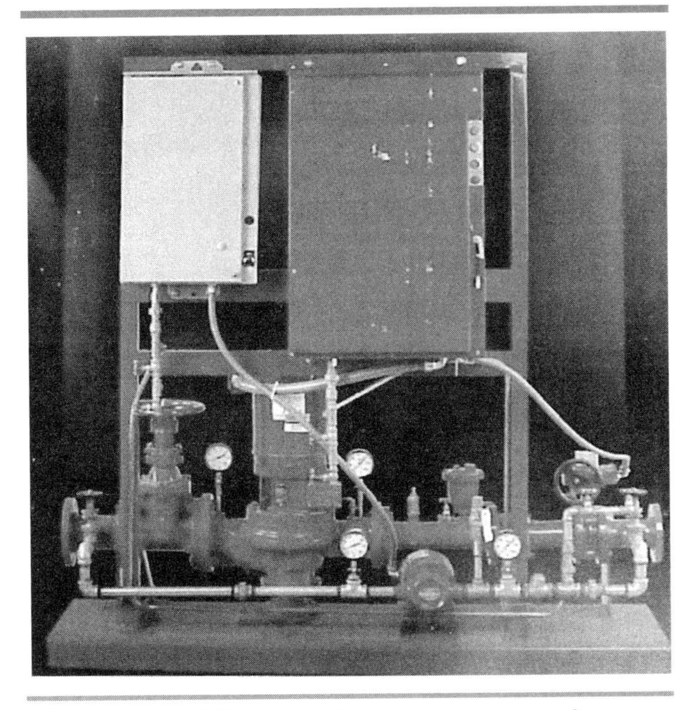

FIGURE 20. Inline centrifugal fire pump unit (*Source:* Aurora Pump Company/Pentair Water.)

been reduced to the minimum extinguishing-system supply pressure. These tanks require supervision of water levels and air or nitrogen supplies to ensure that they are available when needed (Center 1959).

Fixed Fire-Extinguishing Systems

Fixed fire-extinguishing systems are systems in which fire-extinguishing agent supply is delivered through a distribution piping network to the point of application. The supply can be piped to the system as-is (as is the water supplying a sprinkler system) or stored near the hazard in a pressure vessel (as are carbon dioxide or dry chemicals). All fixed systems have some sort of a distribution system and discharge nozzles or sprinklers and are a permanent part of the facility to be protected. They include sprinkler systems, carbon dioxide systems, water spray systems, Halon systems, dry chemical systems, foam systems, and other less applied systems such as fog systems and inert gas systems. This section will describe some of the more common fixed systems (Cote 2003).

Sprinkler Systems

Automatic sprinkler systems are considered to be the most effective and economical way to apply water for fire suppression. These systems use a rigid piping network to deliver and distribute water to the fire location. The water actually discharges through an orifice into a deflector that causes the water to spread in a pattern that extinguishes the fire and cools building components and contents (Cote 2003).

History and Principles of Operation

Sprinklers had their origin in the nineteenth century and have improved in reliability and performance since that time. The sprinklers used into the early 1950s are known as "old-style" sprinklers. At that time, the spray sprinkler was developed and gained wide use because of its more efficient "umbrella" spray pattern. (The old-style sprinkler directed approximately 40% of its discharge upward, spreading the remainder downward over the fire.) New sprinkler designs having varied discharge characteristics have continued to be developed since the 1980s. Sprinkler system designers must take into account discharge patterns, flow rates, and fire area characteristics to achieve effective coverage of the hazards.

Under normal operation, a cap, plug, or valve keeps the orifice shut, holding back the water or air inside the piping network. The orifice-closing mechanism is held in place by liquid-filled glass bulbs, thermosensitive devices, levers, and soldered elements (or, in the case of on/off sprinklers, by a heat-actuated flow control valve).

One of the most common types of sprinklers operates when a link or other device separates, at a predetermined melting point, the solder holding the devices or links together. The solder is a eutectic alloy composed of metals such as lead, tin, bismuth, and cadmium. Another style uses a glass bulb nearly full of liquid. The liquid expands as it heats, rupturing the glass bulb. Other thermosensitive elements used to achieve sprinkler actuation at predetermined temperatures include fusible alloys, chemical pellets, and bimetallic discs similar to those used in fire detection devices. All of these mechanisms incorporate mechanical forces many times that exerted by the water pressure in the piping to which they are connected, thereby withstanding water hammer surges without leakage (Cote 2003).

Sprinkler Types, Temperature Ratings, and Capabilities

Automatic sprinklers are designed to operate at fixed temperatures in sufficient time to control fires and prevent fire spread. Their speed of operation depends upon the shape, size, and mass of the thermosensitive mechanism, as well as upon the surrounding air temperature and the movement of fire gases past the thermosensitive element. Most sprinklers put into use today are "standard spray sprinklers" similar to that shown in Figure 21.

Standard spray sprinklers are configured for use in either the "upright" or "pendent" position. Upright

FIGURE 21. Standard "upright" spray sprinkler

sprinklers are attached to the top side of the distribution piping, and water is discharged upward from the sprinkler orifice when it is activated, striking a deflector that redirects the water downward, spreading it over the protected area. Pendent sprinklers are connected to the bottom side of the distribution piping and in similar fashion spread the water over the protected area when the water strikes a deflector of slightly different configuration. Many adaptations of the basic design have been developed over the years. These include flush, concealed, dry pendent, large-drop, on–off, residential, and side wall sprinklers. A complete discussion of different types of sprinklers can be found in the *NFPA Handbook* (Cote 2003).

Water discharge and distribution of a sprinkler depends upon orifice size, flow characteristics, and deflector configuration. When the water is discharged, the stream hits the deflector attached to the sprinkler frame and is redirected into a spray pattern covering a certain area. Actual spray patters are influenced not only by deflector design but also by the shape, operating pressure, and size of sprinkler structural elements. Testing laboratories establish requirements for water distribution at minimum operating pressures for sprinklers.

In addition to their different physical design configurations, sprinklers are manufactured with a variety of orifice sizes. A half-inch diameter is most common and is found in the vast majority of systems in use today. Different orifice sizes can produce varying flows at different pressures. As orifice size decreases (assuming the same operating pressure), flow decreases; similarly, as orifice size increases, flow increases. Flow from a sprinkler is equal to a flow coefficient *K* factor multiplied by the square root of the operating pressure

Sprinkler discharge formula:

$$Q\ (\text{gpm}) = K\sqrt{p} \qquad (16)$$

Most half-inch-orifice standard spray sprinklers' $K = 5.6$. Each manufacturer's literature lists the *K* factor for each sprinkler it manufactures. *K* is discovered by flow-testing each different sprinkler head. The effect of the *K* factor on discharge is given as percentage in Table 4.

Sprinkler operating temperatures are established in the laboratory by immersing the sprinkler in an oil bath and increasing the temperature until the release mechanism operates. Ratings are stamped on the fusible links or sprinkler housing, or color-coding is used to identify the temperature rating. Typical temperature ratings, classifications, and color codings are shown in the Table 5.

TABLE 4

Effect of the *K* Factor on Discharge as a Percentage

K-Factor	% of Discharge @ Same Temperature Compared to K = 5.6
1.4	25
2.8	50
4.2	75
5.6	100
8.0	140
14.0	250
25	450

(*Source:* Cote 2003)

TABLE 5

Sprinkler Temperature Ratings

Temperature Rating	Temperature Classification Degrees F	Maximum Ceiling Temperature °F	Color Code	Glass Bulb Colors
135–170	Ordinary	100	Uncolored	Orange or Red
175–225	Intermediate	150	White	Yellow or Green
250–300	High	225	Blue	Blue
325–375	Extra High	300	Red	Purple
400–475	Very Extra High	375	Green	Black
500–575	Ultra High	475	Orange	Black
650	Ultra High	625	Orange	Black

(*Source:* Cote 2003)

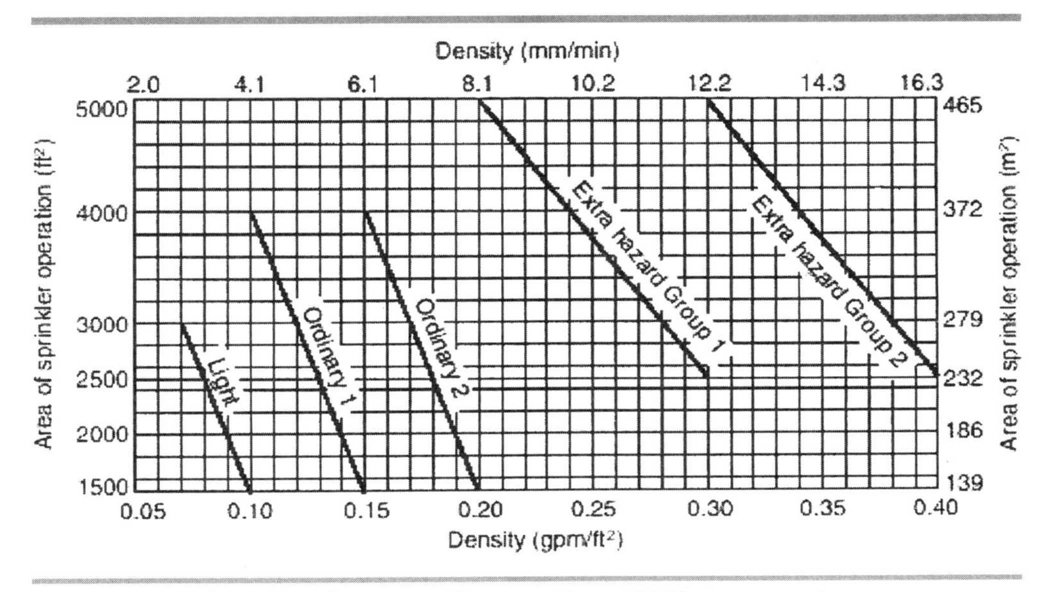

FIGURE 22. Area/density curve (*Source:* Cote 2003)

Orifice Sizes and Discharge Characteristics

Sprinkler discharge changes with flow characteristics and operating pressure. A detailed discussion of this issue can be found in the *NFPA Handbook* (Cote 2003). Nominal *K* factors are listed against *K* factor ranges for common sprinklers in Table 10.10.6 of the *NFPA Handbook* for easy use by system designers.

Area/Density Considerations

Historical experience forms the basis for standard recommendations on water density and area of operation for hydraulically designed sprinkler systems. Figure 22 from the *NFPA Handbook* gives area-density curves used by system designers (Cote 2003).

System Types

There are four basic types of sprinkler systems, including the wet-pipe system, the dry-pipe system, the deluge system, and the preaction system.

Wet-pipe system: A wet-pipe system is by far the most common type of sprinkler system. It consists of a network of piping containing water under pressure. Automatic sprinklers are connected to the piping in such a way that each sprinkler protects an assigned area of the building. The application of heat to a sprinkler will cause that single sprinkler

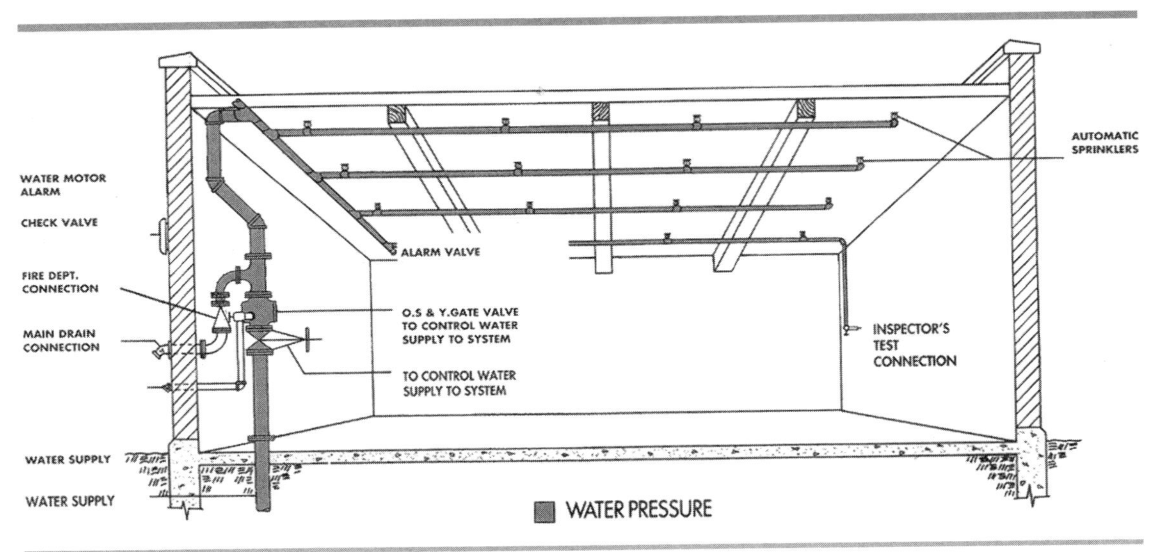

FIGURE 23. Typical wet-pipe sprinkler arrangement (*Source:* Tyco GEM)

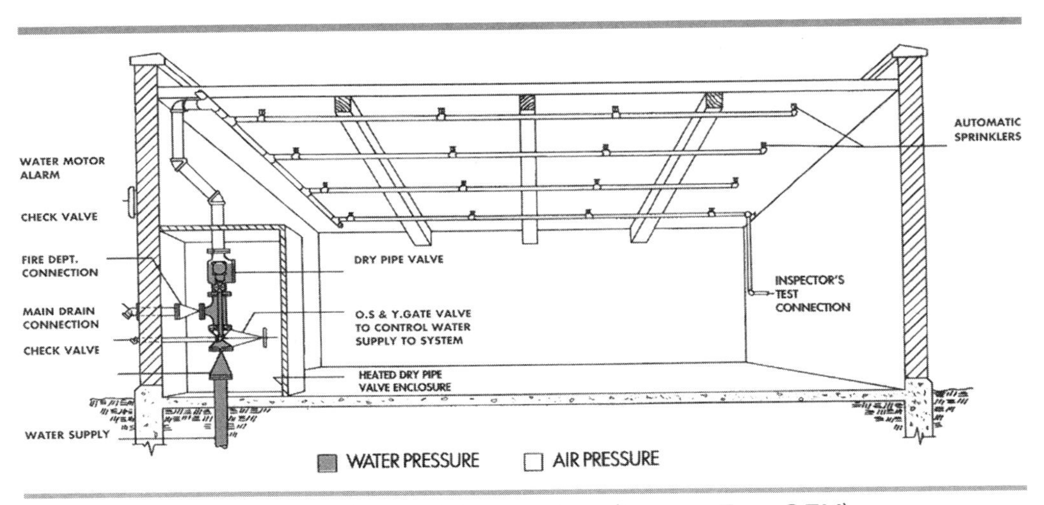

FIGURE 24. Typical dry-pipe sprinkler system (*Source:* Tyco GEM)

to operate, discharging water over its area of protection. A typical piping arrangement is shown in Figure 23.

Dry-pipe system: A dry-pipe system is similar to a wet-pipe system, but water is held back from the piping network by a special dry-pipe valve. The valve is kept closed by air or nitrogen pressure maintained in the piping. The operation of one or more sprinklers will decrease the air pressure, operating the dry valve and permitting water to flow into the piping, suppressing the fire. Dry systems are used in situations in which water in the piping would be subject to freezing. Figure 24 shows typical dry-pipe sprinkler system piping.

Deluge system: A deluge system is one that uses open instead of automatic sprinklers. A special deluge valve holds back the water from the piping and is activated by a separate fire-detection system. When activated, the deluge valve admits water to the piping network, and water flows simultaneously from all of the open sprinklers. Deluge systems are used to control rapidly spreading, high-hazard fires.

Preaction system: A preaction system is similar to a deluge system, but closed automatic sprinklers are used, and a small air pressure is usually maintained in the piping network to ensure that the system is airtight. As with a deluge system, a separate detection system is used to activate a deluge valve, admitting water to the piping. Because automatic sprinklers are used, however, the water is usually stopped from flowing unless heat from the fire has activated one or more sprinklers. Some special arrangements of preaction systems permit variations on detection-system interaction with sprinkler operation. Preaction systems are generally used where there is particular reason to avoid accidental water discharge, as in areas containing valuable computer equipment.

These four basic types of systems differ in how they discharge water into the area of the fire. Although many types of sprink1er systems exist that are classified according to the hazard they protect (including such types as residential, in-rack, or exposure sprinklers), the additives they incorporate (such as antifreeze or foam), and the special system connections they allow (such as multipurpose piping), sprinkler systems can still be categorized as one of the basic four types (Cote 2003).

Design Considerations

Design of a sprinkler system starts with evaluation of the occupancy of the area to be protected and the degree of fire hazard presented by the operations and commodities contained in the area. Similar (but not identical) classifications of occupancies and commodity hazards are available in insurance guides, NFPA standards, and building codes. Each attempts to establish minimum design requirements for similar risk. In an effort to achieve some uniformity, these organizations usually distribute occupancies of similar risk levels into groupings such as *light hazard*, *ordinary hazard*, and *extra hazard*. An office, for example, might be a light hazard, a grocery store an ordinary hazard, and a manufacturing plant using plastics an extra hazard. These groupings are then tailored according to sprinkler spacing and water density specifications, construction characteristics, and system performance needs.

Occupancy

Exact definitions of occupancies vary between regulations and standards but are typically similar to those defined in the following paragraphs.

Light hazard: These occupancies, with a low amount of combustible contents, are expected to have a relatively low heat-release rate during a fire.

Ordinary hazard: These occupancies, with low-combustibility contents in moderate quantities at heights less than 12 feet, exhibit moderate heat-release rates.

Extra hazard: These occupancies (or portions of occupancies) have a great quantity of highly combustible contents in which fire is expected to rapidly spread (possibly because of the presence of dust or lint), moderate to substantial amounts of flammable or combustible liquids, or shielding. They have contents with high rates of heat release.

In addition to these occupancy considerations, where storage is present, the actual commodities being stored must be considered when selecting fire protection provisions. The protection requirements should be based on individual storage units (as pallet loads) and the variation of commodities between units. Normally, the protection should be based on the highest classification of commodities and storage arrangement. NFPA 13 breaks commodities into classes and establishes minimum sprinkler location and spacing guidelines and water discharge densities for four commodity classes (Classes A–D) by the actual commodities involved (for example, metal parts on wood pallets on single-layer corrugated cardboard boxes). These classifications are further defined by the amount of plastics, elastomers, and rubber present in the commodity. The further classification has three groups of such materials identified as groups A, B, and C (NFPA 2007b).

Construction

Construction issues needing consideration when designing sprinkler systems have to do with ceiling arrangement insofar as depth, spacing, and openness of structural members affect sprinkler performance. These structural features can affect activation time and water distribution patterns. These issues influence allowable coverage areas and sprinkler positioning. NFPA 13 discusses two types of ceiling construction: obstructed and unobstructed. Ceiling construction that inhibits heat flow and water distribution

TABLE 6

Pipe Schedule System Water Supply Requirements

Occupancy	Minimum Residual Pressure (psi)	Flow Required @ Pressure (gpm)	Duration (Minutes)
Light Hazard	15	500–750	30–60
Ordinary Hazard	20	850–1500	60–90

(*Source:* NFPA 2007b)

is considered obstructed. Unobstructed ceiling construction evinces open structural member arrangement and design and does not impair sprinkler performance (NFPA 2007b).

Size of Protected Area

Automatic sprinkler systems are either hydraulically designed or designed using pipe schedules from NFPA 13, which lists the number of sprinklers permitted to be supplied by a particular size of pipe (NFPA 2007b). Pipe schedule design methods in the past were used for all systems, but now pipe schedule methods can only be used for light-hazard and ordinary-hazard applications. Required water supplies for pipe schedule systems are shown in Table 6.

A much better and more accurate way to determine sprinkler system water demand is through hydraulic calculations incorporating design density and area of estimated operation. Hydraulic calculation methods provide a more cost-effective way to determine the needed water supply. The following example illustrates the use of hydraulic calculations for designing a sprinkler system.

Sprinkler System Hydraulic Design Example

Problem: A sprinkler system is needed for a large warehouse operation of an Ordinary 2 NFPA 13 hazard class (NFPA 2007b). It is a wood frame construction with obstructions. (See *NFPA Handbook*, Section 10, Chapter 11, Figure 10.11.15 for more information on obstructed construction.) At the farthest end of the building from the offices is a 1500 ft^2 rectangular storage area needing protection. It is depends upon a separate water network from the rest of the warehouse and is furthest from the fire water supply pumps. Use half-inch-orifice upright standard spray sprinklers to provide the protection. Discover the required number of sprinklers, as well as their appropriate spacing and layout; make a sketch of the suggested layout. Calculate the pressure and flow for the three most remote sprinklers on the system.

1. Choose a 15 × 30 ft area and a density of 0.20 gpm/ft^2 from NFPA 13 Figure 6-10R; assume that no additional hazards are present.
2. For Ordinary 2, Hazard Level: Combustible Construction with Obstructions, the coverage required is <130 ft^2 per sprinkler at a 15-foot maximum spacing. By dividing coverage area by protected area per sprinkler, the minimum number of sprinklers is determined. Therefore, the number of sprinklers required is equal to 1500 (ft^2)/130 (ft^2 per sprinkler), or 11.5 (so 12), sprinklers.
3. Because water must be distributed from 12 sprinklers to control the fire, the approximate number of sprinklers per branch line can be estimated by multiplying 1.2 by the square root of the area and dividing by the maximum spacing (15 feet). $(1.2\sqrt{1500})/15$, or 3.1 (so 4), sprinklers per branch line. Assuming the remote area to be a 30 × 50 ft rectangle, three branch lines with sprinklers spaced 10 feet apart would be needed. Spacing between the branch lines would equal 30 feet divided by 3, or also 10 feet. See Figure 25 for a sketch of the sprinklers.
4. The flow from the three most remote sprinklers is 1500 (ft)/12 (sprinklers), or 125, ft^2 of protected area per sprinkler. At a density of 0.2 gpm/ft^2, each sprinkler needs to flow a minimum of 125 × 0.2, or 25, gpm. The K factor for a half-inch orifice sprinkler is about 5.6. Using the sprinkler discharge Equation 16,

$$Q\ (\text{gpm}) = K\sqrt{p} \qquad (17)$$

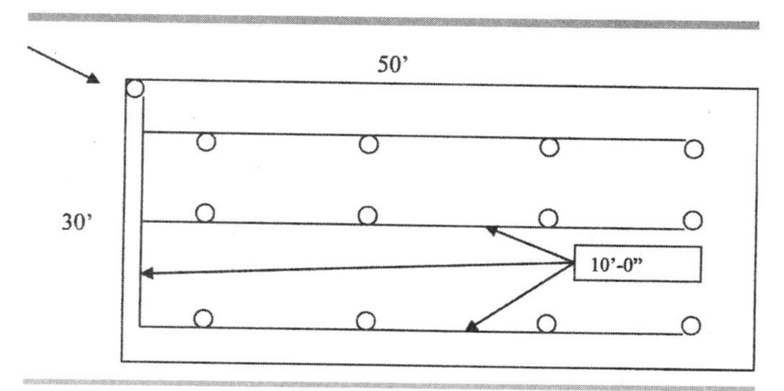

FIGURE 25. Sketch of sprinklers

and the required pressure p at the end sprinkler to achieve the desired flow and water density becomes

$$p = (Q\ (\text{gpm})/K)^2 = (25/5.6)^2 = 19.92 \text{ psig} \quad (18)$$

5. The flow from the last sprinkler back to the next-to-last sprinkler is approximately 10 feet, assuming that a one-inch $C = 120$ pipe results in friction loss (according to Figure 12) of 0.12 psig/ft, or a total of 1.2 psig for the 10 feet, in addition to flow loss through a 90° elbow with an equivalent length of 2 feet, or a total of 0.24 psig, for a total pressure at the second-to-last sprinkler of 1.2 + 0.24 + 19.92, or 21.36, psig.
6. Flow from the second-to-last head:

$$Q\ (\text{gpm}) = K\sqrt{P} = 5.6\sqrt{21.36} = 25.88 \text{ gpm} \quad (19)$$

7. Flow into T at second-to-last sprinkler:

$$25 + 25.88 = 50.88 \text{ at } 21.36 \text{ psig} \quad (20)$$

8. During flow from the third-to-last sprinkler to the second-to-last sprinkler, assuming a one-and-a-quarter-inch $C = 120$ pipe having a side-outlet T having an equivalent length of 6 feet, 50.88 gpm must flow through a total equivalent length of 10 + 6, or 16, feet with a loss, according to Figure 13, of 0.18 psig/ft or 2.88 psig. Pressure at the third-to-last sprinkler:

$$21.36 + 2.88 = 24.24 \text{ psig} \quad (21)$$

9. Flow from the third-to-last head:

$$Q\ (\text{gpm}) = K\sqrt{P} = 5.6\sqrt{24.24} = 27.57 \text{ gpm} \quad (22)$$

10. Total demand at the inlet to the T for the third-to-last head:

$$25 + 25.88 + 27.57 = 78.45 \text{ gpm at } 24.24 \text{ psig} \quad (23)$$

System Components

System components include water control and isolation valves in supply piping, a piping network, sprinklers, an auxiliary drain, and testing valves. Isolation valves can be of a post indicator, indicating butterfly or outside stem and yoke design. Control valves for wet-pipe systems are called alarm check valves; dry-pipe systems have dry-pipe valves, and deluge or preaction systems have deluge valves. Because these components are designed for a 175 psig maximum allowable working pressure, systems in high structures may need to be separated to avoid overpressurization. The only other pressure issue in need of consideration is "water hammer," the force developed by the sudden reduction or elimination of water velocity in a pipe (often caused by rapid valve closing). The energy from the moving water is transferred to the pipe wall, compressing the water slightly. As the shock of this energy change occurs, pulsations or shock waves move back and forth, generally accompanied by a series of rapidly succeeding noises like hammering on the pipe. This problem generally does not occur when outside stem and gate valves are closed, because they take a longer time to close; but it can occur if a quick-closing plug-style valve is closed. Pressures applied to the piping are many times the normal operating pressure but last for such a short time that they normally do no damage.

Listing Requirements

All valves and devices used in sprinkler systems must be tested and approved for fire-protection use by a recognized testing laboratory such as UL or FM.

Pipe and Fittings

Steel, copper, and nonmetallic (CPVC) pipes and fittings meeting the requirements of NFPA 13, *Standard for the Installation of Sprinkler Systems,* as well as associated materials and dimensional standards of ASTM and ANSI, may be used for sprinkler systems. Systems with dry piping may not use pipe and fittings subjected to heat damage during pre-actuation fire exposures. Requirements in the NFPA standards should be consulted when selecting, and during specification of, pipe and fittings (NFPA 2007b).

Valves

Control valves used for control of water to sprinkler systems must be indicating types. By looking at the valve it can be readily determined whether the valve is open or closed. The most familiar type of valve with this designation is the *outside stem and yoke* (OS&Y) valve. When the stem (screw) is visible, the valve is open. A second type of indicating valve used in fixed fire-extinguishing supply mains is the *post indicator valve* (PIV). These can be used through building walls or with underground valves. PIVs have operating mechanisms displaying open or closed signs to indicate the position of the valve, as shown in Figure 26.

A third type of indicating valve is the *indicating butterfly valve,* which uses an arrow mechanism to indicate whether it is open or closed. All these valves are capable of being fitted with electrical sensors that electrically supervise valve positions, and sending signals to fire alarms or annunciation systems (Cote 2003).

Wet-pipe sprinkler systems have *alarm valves* or water flow switches on the system side of the shutoff valve to detect water flow in the system. The alarm valve is a check valve that passes a portion of the flow when it opens to a pressure- or flow-sensing device. In many cases, the flow-sensing device is a water motor operated gong; in others it is a pressure or a flow switch (see Figure 27).

Dry-pipe systems use a *dry-pipe valve* instead of an alarm valve (see Figure 28).

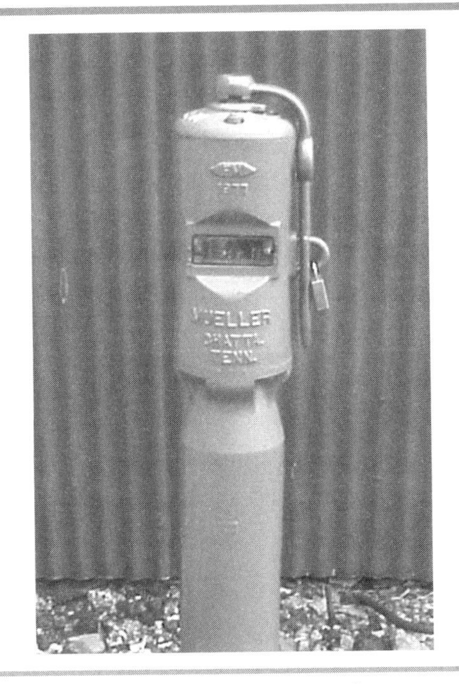

FIGURE 26. Post indicator gate valve (*Source:* Standard Steel LLC)

FIGURE 27. Alarm valve (*Source:* Standard Steel LLC)

FIGURE 28. Dry-pipe valve (*Source:* Standard Steel LLC)

FIGURE 29. Deluge valve
(*Source:* Standard Steel LLC.)

Likewise, both the preaction and deluge systems use a deluge valve instead of an alarm valve (see Figure 29).

In addition, a two-inch drain test valve is installed on the system side of the shut-off valve to provide indication of water supply to the system and for use in draining the piping network. Fire department connections are also connected to the supply main via a check valve. Normally the fire department connection is downstream of the alarm valve, but not always. If it connects before the alarm valve (and in the case of dry-pipe, preaction, and deluge systems), a separate check valve is needed in the supply main to prevent backflow of water into the supply main.

Coverage Area and Extent of Coverage

It is desirable to install sprinklers throughout the premises to ensure an adequate level of property protection and life safety. Because locations of fires are not wholly predictable, building codes and standards such as NFPA 13 (2007b) require sprinklers to be installed throughout a facility with minor exceptions such as residential bathrooms and certain electrical equipment rooms having two-hour-fire-rated separation. Whenever concealed spaces (such as above-dropped ceilings) are of noncombustible construction, sprinklers are not normally required. The area of protection for sprinklers must consider the hazard being protected, the type of construction, the structural member spacing, and the sprinkler type (Cote 2003).

Table 7 gives coverage areas for standard upright and pendent sprinklers. Similar information for other types of sprinklers is given in NFPA 13 (NFPA 2007b).

Foam Extinguishing Systems

Delivery of foam solution to a hazard being protected, as well as initial foam generation, can be achieved in several ways. Normally, the foam concentrate is stored in a building protected from freezing. The concentrate is pumped or drawn from the storage container or tank and transferred to the

TABLE 7

Maximum Spacing and Protection Area for Sprinklers

Hazard Type	Construction Type	System Type	Protection Area (Sq Ft)	Spacing (Ft)
Light	Noncombustible			
	(Obstructed & Non-Obstructed)	Pipe Schedule	200	15
	Combustible (Non-Obstructed)	Hydraulic Calculated	225	15
	Combustible (Obstructed)	All	168	15
	Combustible (W/Member < 3 Ft)	All	130	15
Ordinary	All	All	130	15
Extra	All	Pipe Schedule	90	12
		Hydraulic > .25gpm/sq ft	100	12
		Hydraulic < .25 gpm/sq ft	130	15
High Piled Storage	All	Hydraulic > .25gpm/sq ft	100	12
	All	Hydraulic < .25 gpm/sq ft	130	15

(*Source:* Cote 2003)

injection point, where it is mixed with water. The resulting solution travels from the injection point to the foam-generation device, where it is aerated and becomes foam. After generation, it is applied to the protected surface. This process varies with the hazard being protected and the expansion ratio of the foam in use. As a general rule, the more gently the foam is applied (in other words, the less the bubbles break down), the more rapidly the fire is extinguished and the lower the amount of required concentrate. Foams must be applied at adequate rates to establish and maintain their protective blanket. Application rates are described in terms of solution per square foot of protected surface. For example, foam with a 10:1 expansion ratio and an application rate of 0.15 gpm/ft^2 would create 1.5 gallons of finished foam per square foot every minute. Increasing the rate within suggested limits normally speeds extinguishing. NFPA 11, *Standard for the Installation of Low-, Medium- and High-Expansion Foam* (2005), provides the recognized criteria for foam extinguishing systems. Generally, air foams are more stable if water used for generation is at or near ambient temperatures. Air supplies for foam-making are best drawn from atmosphere containing little or no combustion gases. It is important to abide by the operating pressure requirements of foam-making devices to ensure that foam proportioning and generation occurs as designed. Not all vaporizing liquid and dry chemical extinguishing agents are compatible with all foams. These chemicals can at times create foam-generation problems and detract from blanket durability (Cote 2003).

Applications

Foam extinguishing systems are generally used to protect flammable and combustible liquid storage tanks, as well as processing and handling areas for such materials. Foam solution can be injected at the base of a tank (in what is called subsurface application) or be introduced by foam-makers mounted at the top of the tank. The foam concentrate–water solution is supplied via rigid piping. Dip tanks and other open-top quenching tanks typically have similar foam-making devices mounted around the top edge of the tank. Process or handling areas where leaks or spills can occur may use foam-making sprinklers, monitor nozzles, or portable hose lines. Aircraft hangars and similar hazards that could be subject to flammable liquid spills are often protected with high-expansion foam generators mounted along the walls. The exact provisions and arrangement are engineered to protect the hazard based on the physical arrangement of, and activity in, the area.

Design Specification

All systems are designed to meet requirements of NFPA 11 (NFPA 2005). Discharge rates and foam-generation methods are selected based on the size and location of the hazard. A small dip tank may use a venturi inductor proportioner, and a large flammable liquid tank farm may have a *foam house,* incorporating a foam concentrate storage tank and a pump-style proportioner.

System Components

Each system must include a water supply, provision for foam concentrate storage, a proportioning method to inject the concentrate into the water (thereby making the foam solution), and a foam maker to introduce air or gas into the solution, forming bubbles. Systems include piping, valves, and accessories necessary to supply the foam to the hazard to be protected. Figure 30 shows a typical tank farm system.

Flammable Liquid Storage Tank Foam System Design Example

A storage tank 60 feet in diameter and 40 feet high, with a conical roof, is being used to store heptane with a specific gravity of 0.68. The tank is assumed to be filled to a maximum depth of 37 feet. Application of foam solution will be through a single subsurface connection (see Figure 31).

$$\text{Surface area: } \pi D^2/4 = \pi 60^2/4 = 2830 \text{ ft}^2 \quad (24)$$

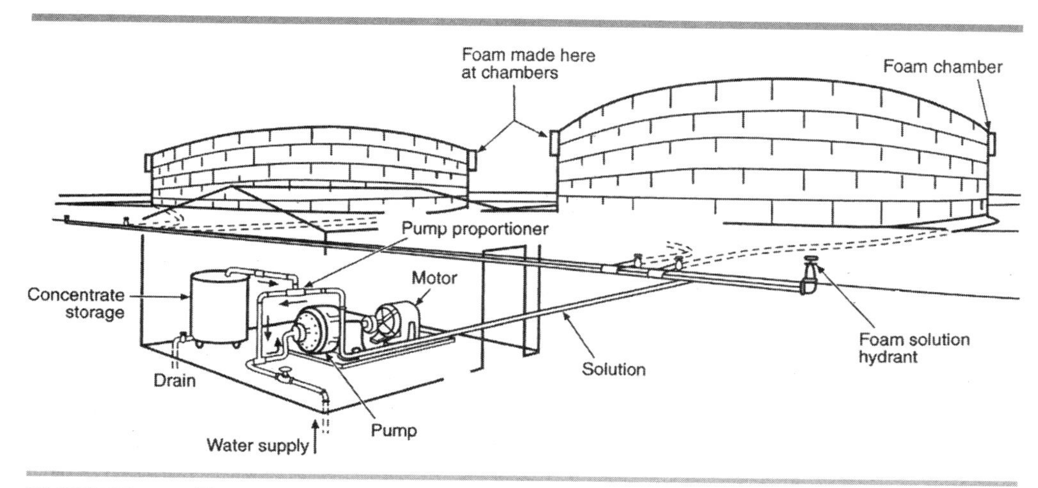

FIGURE 30. Foam extinguishing system for a tank farm (*Source:* Cote 2003)

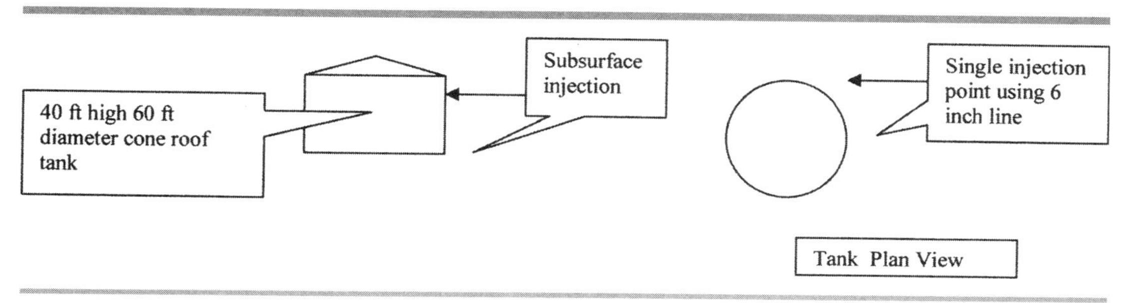

FIGURE 31. Subsurface foam injection example

Select a 0.1 gpm/ft^2 application rate with an 8:1 expansion rate resulting in a foam generation rate of 0.8 gpm/ft^2.

The total foam solution rate needed to protect the tank can be obtained by multiplying the surface area by the application rate: in this case, 0.1 gpm/ft^2 × 2830 ft^2 = 283 gpm.

A 3% foam concentrate would require 8.5 gpm of foam liquid concentrate and 274.5 gpm of water at a press where injected into the tank above the liquid head of 38 feet times the specific gravity of heptane, or 25.84 ft ≈ 11.21 psig. In addition, the designer must consider pressure loss incident to getting the solution to the base of the tank.

Water Spray Systems and Exposure Protection Systems

The primary difference between a sprinkler system and a water spray system is that the water spray system provides protection of a specific hazard or piece of equipment instead of a general area. Water spray systems direct water toward hazards with a goal of achieving a protective film of water over structures, walls, tanks, or other equipment being protected. These systems are designed in accordance with NFPA 15, *Standard for Water Spray Fixed Systems for Fire Protection* (2007c).

Applications

Water spray can be used to control and extinguish fires, to protect exposures, and to prevent fire. Water spray extinguishes a fire by cooling the combustion zone, producing steam that restricts oxygen, and diluting and emulsifying flammable liquids—or by some combination of these mechanisms. Water spray is ideal exposure protection, because it forms a heat-absorbing and heat-removing film on the surface of protected equipment and structures. Water spray can also be used to cool equipment such as reactive chemical tanks, preventing its failure and reducing the potential for fire.

Materials and equipment that can be protected with water spray include flammable gas and liquid processing and storage equipment, oil-filled electrical power transformers, electrical cable trays, and ordinary combustibles (Cote 2003).

Design Specifications

The design of water spray systems must be in accordance with NFPA 15 (NFPA 2007c). These systems are engineered to provide over the protected surface a density of water normally between 0.10 gpm/ft^2 and 0.50 gpm/ft^2. The protected surface is the actual surface area of the piece of equipment being protected. Location and placement of nozzles is based on surface characteristics and the pattern of the spray nozzle. Some consideration is also given to water rundown (Cote 2003).

System Components

Water spray systems use open directional-spray nozzles connected to a piping network to deliver water to protected hazards. Water to the system is controlled by a sprinkler system deluge valve actuated by hydraulic, pneumatic, or electrical fire detectors. All systems can also be activated manually. Upon detection, the water is released to the piping system and delivered to the nozzles, whence it is distributed over the protected hazard. A typical system is illustrated in Figure 32.

Piping Arrangement Issues

Piping must be arranged to position nozzles to achieve the desired coverage, and it must be designed to drain after operation. In cases in which protection is of high-voltage power transformers, adequate electrical clearance must be maintained of live electrical parts. Nozzles are generally positioned several

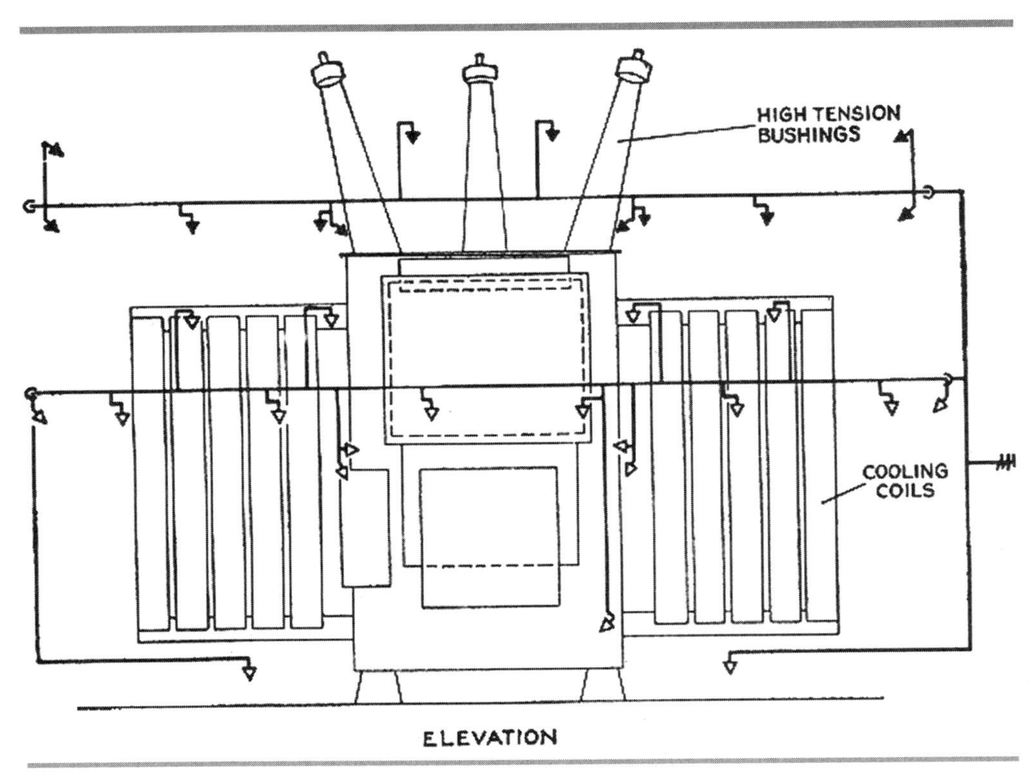

FIGURE 32. Schematic of water spray system on electrical transformer (*Source:* Center 1959)

feet from the protected surface and arranged with fittings that allow spray patterns to be adjusted.

CARBON DIOXIDE EXTINGUISHING SYSTEMS

Because of the danger presented by carbon dioxide to personnel, audible and visible predischarge alarms are needed on all systems. Odorizers can be added to carbon dioxide stores to help personnel detect leaks. Systems should be locked out when workers need to enter a confined space protected with carbon dioxide, or self-contained breathing apparata should be available for rescue and escape.

Supply System

Carbon dioxide is stored near the point of demand in high-pressure cylinders or low-pressure refrigerated tanks.

Low-Pressure Systems

Low-pressure carbon dioxide systems store the extinguishing agent as a refrigerated liquid at 0°F. The storage units are pressure vessels with a design working pressure of at least 325 psig. They are insulated and refrigerated at 0°F. The vapor pressure of carbon dioxide is 300 psig at 0°F. A pressure switch controls the refrigeration compressor. A pressure relief device on the unit protects it upon refrigeration failure. As some of the liquid is vaporized during rise in temperature, it is vented and more evaporates. This process causes a self-refrigeration effect that reduces tank pressure. Heaters are installed inside any tank where ambient temperatures can be expected to drop below −10°F to be sure adequate pressure is present to supply the demand. Multiple systems can be supplied from one unit or tank by using selector valves to direct the carbon dioxide to the proper extinguishing system. Storage units can be located up to several hundred feet from the protected hazard (see Figure 33).

High-Pressure Systems

High-pressure carbon dioxide systems use cylinders to store the gas as a liquid at ambient temperatures. The temperature of the air outside the cylinder determines the pressure inside. The cylinders are designed to withstand maximum expected temperatures. They have a maximum filling density of 68% of the weight of the water capacity at 60°F. They are designed with a dip tube so that liquid is discharged. Low storage temperatures greatly affect the discharge rate. These systems have a pressure of about 760 psig at ambient conditions (Cote 2003).

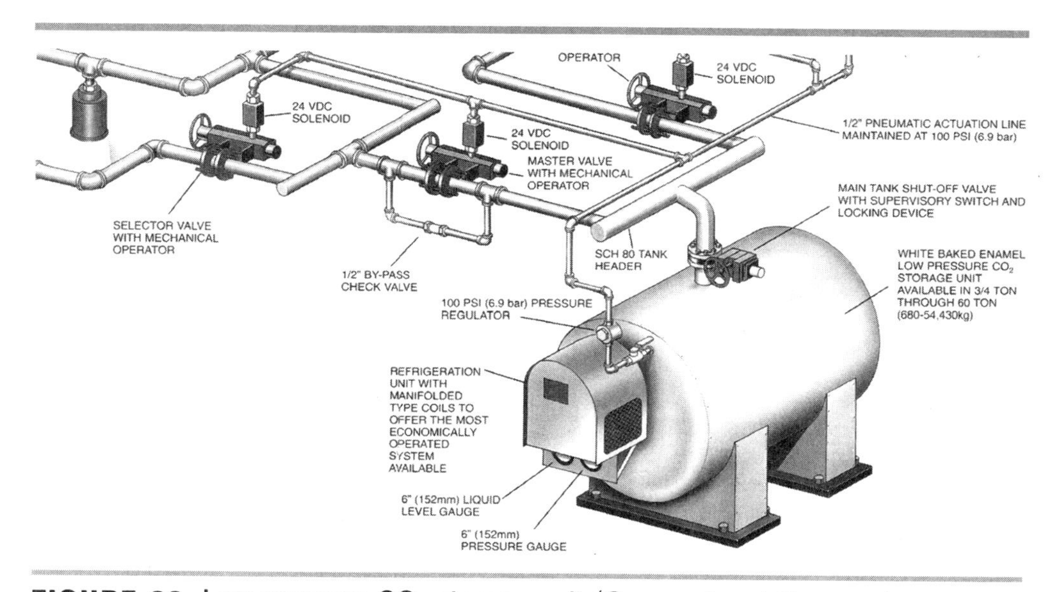

FIGURE 33. Low-pressure CO_2 storage unit (*Source:* Ansul Company)

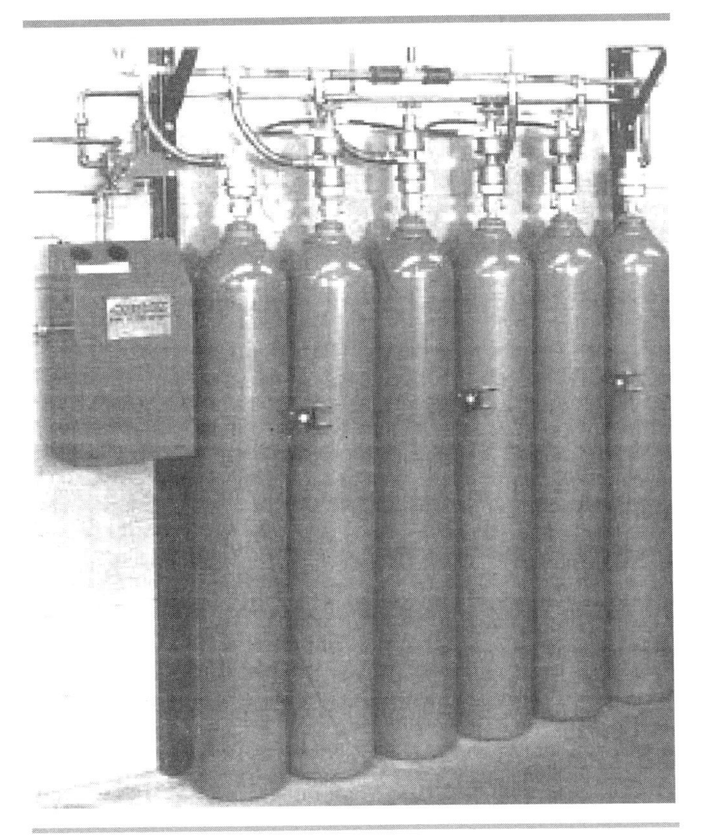

FIGURE 34. High-pressure CO_2 supply manifold (*Source:* Chemetron Corporation)

A typical high-pressure supply manifold is shown in Figure 34.

Application Method

Carbon dioxide can be applied to a fire hazard via a fixed system by filling an enclosed space or volume with an extinguishing concentration (total flood) or by applying the gas to the actual hazard or fire (local application). Concentrations are determined by the fuel source and combustible loading per requirements in NFPA 12, *Standard on Carbon Dioxide Extinguishing Systems* (NFPA 2008a).

Total Flood

Application of carbon dioxide via a total flooding system uses nozzles designed and located to achieve a uniform distribution throughout the protected enclosure. The volume needed is calculated by enclosure volume and the amount of combustibles present. Minimum design concentrations used in total flooding range from a minimum of 34% up to 75%, based on the hazard present in the enclosure. Enclosure continuity is very important to avoid leaking and loss of extinguishing concentration, especially where a soaking period is needed to extinguish deep-seated fire. Leakage must be kept to a minimum, especially in the lower parts of the enclosure (because carbon dioxide is heavier than air). Vents to prevent overpressurizing during system discharge should be placed high in the enclosure near the ceiling; high vents have little effect on leakage.

Local Application

Carbon dioxide is applied directly to the surface of the fire using directional nozzles arranged to provide a pattern that covers the surface of the hazard. Areas adjacent to the actual fire exposure are also protected if there is a potential for fire spread. Nozzles are specially designed for low-velocity discharge to avoid splashing and air entrapment. Systems should be automatically actuated to ensure fast response and to minimize heat buildup. Some shielding or enclosure is helpful to hold the carbon dioxide concentration at the hazard.

System Components

All carbon dioxide extinguishing systems require a storage system, as well as piping to transfer the agent to the point of demand, nozzles to dispense the agent, and a detection and release system. Storage can be in high-pressure cylinders manifolded together or a low-pressure refrigerated storage vessel. Most carbon dioxide systems have piping that is empty and open to the atmosphere but which becomes charged after the agent is released from the storage vessel. Flow rate is a critical element of the fire-extinguishing process. The minimum pressure in the pipe line must be kept above the triple point of 75 psia. If pressure falls below the triple point, dry ice can form and block nozzle orifices. Design limits in NFPA 12 require minimum nozzle design pressures of 300 psia for high-pressure systems and 150 psia for low-pressure systems (NFPA 2008a).

Carbon dioxide enters piping as a liquid, and friction loss during flow causes loss of pressure. As

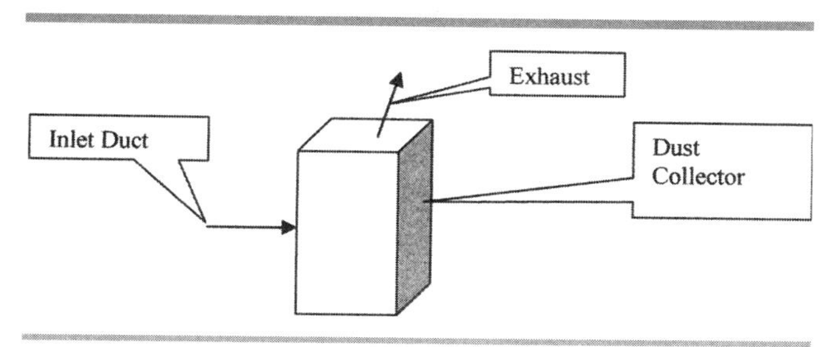

FIGURE 35. Simplified dust collector carbon dioxide protection system

pressure drops, the carbon dioxide boils and vaporizes, resulting in a two-phase (liquid and gas) flow, something that calculations must consider. These calculations are tedious if done manually but can be performed relatively easily with the aid of computers.

Valves and other components must be able to withstand maximum operating pressures, must seal tightly, and must be capable of manual and automatic operation. All components must be approved by a recognized testing laboratory for carbon dioxide service (Cote 2003).

Carbon Dioxide System Example: A small dust collector, 4 × 4 × 6 ft, is to be protected using a carbon dioxide fire-extinguishing system. NFPA 12 requires an extinguishing concentration of 75% for dust collectors (NFPA 2008a). For this example, assume that automatic dampers, actuated by heat detectors, close dampers in inlets and exhaust openings to isolate the enclosure (see Figure 35).

Volume: $4 \times 4 \times 6 \text{ ft} = 96 \text{ ft}^3$

NFPA 12 requires that 4.375 ft^3 of this hazard be protected per pound of carbon dioxide.

Quantity needed to provide protection:
$96 \div 4.375 = 21.94 \text{ lb}$

Clean Agent Extinguishing Systems

Clean agent systems are similar to Halon and high-pressure carbon dioxide systems. Lower vapor pressure agents require superpressurization with nitrogen to achieve liquid discharge from the storage unit. Clean agents are stored in pressurized containers and are released to the piping network upon control valve actuation.

Application Method

Like Halon systems, clean agent systems can be of total-flood or local-application design.

System Components

System components include a storage vessel, control valves, a piping network, and nozzles.

Dry-Chemical Fire-Extinguishing Systems

Fixed dry-chemical extinguishing systems are used to protect specific hazards. Systems are arranged to provide total flooding around a hazard and as a supply for hand hose lines.

Agent Type

Dry-chemical systems can use any of the dry chemical agents, including sodium bicarbonate, potassium bicarbonate, and monoammonium phosphate.

Application Methods

In total flooding systems, a predetermined amount of dry chemical agent is discharged through fixed piping and nozzles into an enclosed space or enclosure around a hazard. Upon actuation of the system, nitrogen gas is released to pressurize the agent storage vessel, and the agent is expelled through the fixed piping to the nozzles. Hand hose line systems are supplied in a similar fashion and are used for manual attack of surface fires. The hose lines are primarily used for knockdown, and water is use for extinguishing the fire, especially if it is deep-seated.

System Components

System components include detection and control features with a control unit similar to Figure 36, including dry-chemical and nitrogen storage vessels, piping and nozzles for fixed systems, and hose lines and reels for manual hose line systems.

FIGURE 36. Dry-chemical system control unit (*Source:* Standard Steel LLC)

Halogenated Hydrocarbon Systems

These systems are similar to high-pressure carbon dioxide systems; the agents are stored in high-pressure cylinders and distributed through a rigid piping network with open-discharge nozzles.

Supply System

The supply system consists of a supply of agent stored singly or in a manifolded group of joined high-pressure cylinders connected to an automatically actuated control valve or valves and to a piping network with distribution nozzles.

Application Method

Much as carbon dioxide systems, these systems can be of total-flooding or local-application design. Higher vapor pressure agents are more easily applied by total flooding, but lower vapor pressure agents can be more effectively applied locally.

Total Flooding

In total-flood systems, nozzles are arranged to distribute agents within an enclosure or room with the goal of achieving a uniform extinguishing agent concentration throughout the protected volume. The volume needed to produce such a concentration is calculated according to enclosure volume and the amount and nature of combustible present.

Local Application

Like local-application carbon dioxide systems, these are arranged to apply the agent to the actual hazard.

System Components

System components include storage vessels or cylinders, control valves, and piping and distribution nozzles. All components must be UL- or FM-approved for the application and must be constructed of materials compatible with the agent. Details of construction requirements and system arrangements can be found in NFPA 12A, *Standard on Installation of Halon 1301 Fire Extinguishing Systems* (NFPA 2004). A typical system arrangement is shown in Figure 37.

Portable Fire Extinguisher Requirements

Portable extinguishers are provided as a first line of defense to handle incipient fires. They are provided in most buildings, whether unprotected or protected by sprinklers or other fixed systems. Hand fire extinguishers are for use in quick occupant response. Extinguishers are rated for use on Class A (ordinary combustibles), B (flammable liquid), C (electrical), D (combustible metal), and K (cooking oil) fires. Occupants should check the rating prior to attempting to extinguish a fire with a manual fire extinguisher.

Design Characteristics

Portable extinguishers are devices that contain an extinguishing agent that can be expelled under pressure for the purpose of manually extinguishing a small fire. Most extinguishers have the agent and the expellant gas in a single container under pressure.

FIGURE 37. Typical Halon 1301 supply manifold
(*Source:* Standard Steel LLC)

Some have the expellant gas stored in a separate cartridge that pressurizes the agent storage vessel after actuation, discharging the agent. Extinguishing agents used in portable extinguishers include water, carbon dioxide, dry chemicals, and higher vapor pressure Halon agents. Small units can be handled manually, but larger units are often wheel-mounted.

Operating Features

All extinguishers have discharge nozzles, many of which are attached to a short, flexible hose. They also have release valves, storage containers, and (if pressurized) pressure gauges.

Ratings

Portable fire extinguishers are classified for use on certain types of fires and are rated by testing laboratories according to their relative extinguishing effectiveness at 70°F. Class A ratings are based on wood and excelsior fires, Class B upon two-inches of normal heptane in square pans, Class C upon live electrical parts (confirming nonconductivity), and Class D upon tests specific to particular combustible metals. These ratings are identified on a permanent name plate on the extinguisher. Ratings are based on the agent and extinguisher as configured when tested; agents can not be changed without voiding the rating.

Location, Access, and Travel Distance

The number of fire extinguishers needed to protect a property is dependent on the construction and occupancy hazards. Construction hazards usually require Class A capability, and occupancy hazards usually need Class B, C, or D capability. NFPA 10, *Standard for Portable Fire Extinguishers* (NFPA 2007a), gives guidance on size and placement of extinguishers, similar to that given in Table 8.

TABLE 8

Hand Fire Extinguisher Distribution

	Light (Low) Hazard	Ordinary (Moderate) Hazard	Extra (High) Hazard
Minimum Rating Single Extinguisher	2-A	2-A	4-A
Minimum Floor Area Per Unit of A	3000 Sq Ft	1500 Sq Ft	1500 Sq Ft
Max Flam Lqd Area Per Extinguisher	11250 Sq Ft	11250 Sq Ft	11250 Sq Ft
Maximum Travel Dist. to an Extinguisher	75 Ft	75 Ft	75 Ft

(*Source:* Cote 2003)

Fire extinguisher size and placement for Class B fires other than fires of appreciable liquid depth should be as required in NFPA 10 (NFPA 2007a), much as in Table 9.

Extinguishers with Class C rating are needed where energized equipment could be involved in a fire, being nonconductors.

Class D extinguishers should be provided within 75 feet of travel distance of any combustible metal fire hazard.

TABLE 9

Class B Extinguisher Placement

Type Hazard	Minimum Extinguisher Rating	Maximum Travel Distance (ft)
Light (Low)	5-B	30
	10-B	50
Ordinary (Moderate)	10-B	30
	20-B	50
Extra (High)	40-B	30
	80-B	50

(*Source:* Cote 2003)

References

Center, Charles C., ed. 1959. *Handbook of Industrial Loss Prevention.* New York: McGraw-Hill.

Clark, John W., and Warren Viessman, Jr. 1996. *Water Supply and Pollution Control.* Scranton, PA: International Textbook Company.

Cote, Arthur E., ed. 2003. *Fire Protection Handbook.* Vols. I and II. Quincy, MA: National Fire Protection Association.

DiNenno, Philip J., ed. 1988. *SFPE Handbook of Fire Protection Engineering.* Quincy, MA: National Fire Protection Association and Society of Fire Protection Engineers.

Lee, C. C., ed. 2000. *Handbook of Environmental Engineering Calculations.* NY: McGraw-Hill.

National Fire Protection Association. 2003. NFPA 22, *Standard for Water Tanks for Private Fire Protection.* Quincy, MA: NFPA.

______. 2004. NFPA 12A, *Standard on Installation of Halon 1301 Fire Extinguishing Systems.* Quincy, MA: NFPA.

______. 2005. NFPA 11, *Standard for Low-, Medium-, and High-Expansion Foam.* Quincy, MA: NFPA.

______. 2006. NFPA 1, *Uniform Fire Codes.* Quincy, MA: NFPA.

______. 2007a. NFPA 10, *Standard for Portable Fire Extinguishers.* Quincy, MA: NFPA.

______. 2007b. NFPA 13, *Standard for the Installation of Sprinkler Systems.* Quincy, MA: NFPA.

______. 2007c. NFPA 15, *Standard for Water Spray Fixed Systems for Fire Protection.* Quincy, MA: NFPA.

______. 2007d. NFPA 20, *Standard for the Installation of Stationary Pumps for Fire Protection.* Quincy, MA: NFPA.

______. 2007e. NFPA 72, *National Fire Alarm Code.* Quincy, MA: NFPA.

______. 2008a. NFPA 12, *Standard on Carbon Dioxide Extinguishing Systems.* Quincy, MA: NFPA.

______. 2008b. NFPA 70, *National Electrical Code.* Quincy, MA: NFPA.

Scherer, Gerald M. 1972. *Designing an Early Warning Fire Detection System.* Northford, CT: The Notifier Company.

Wood, Clyde M. 1961. *Automatic Sprinkler Hydraulic Data.* Youngstown, OH: Automatic Sprinkler Corporation of America.

Appendix: Recommended Reading

International Code Council (ICC). 2003. *International Building Code.* Country Club Hills, IL: ICC.

McKinnon, Gordon P., ed. 1990. *Industrial Fire Hazards Handbook.* 3d ed. Quincy, MA: National Fire Protection Association.

Nam, S. 2005. "Case Study: Determination of Maximum Spacing of Heat Detectors." *Journal of Fire Protection Engineering* (February).

Solomon, Robert E., ed. 1994. *Automatic Sprinkler Handbook.* 6th ed. Quincy, MA: National Fire Protection Association.

Truve, Richard L. 1976. *Principles of Fire Protection Chemistry.* Boston, MA: NFPA.Viking Data Book. 2004. Hastings, MI: Viking Group.

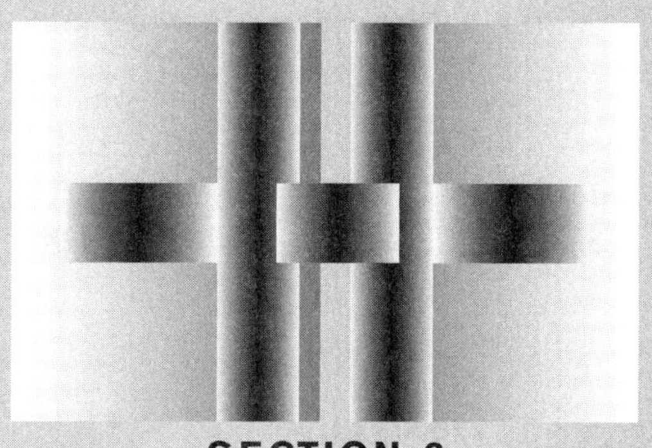

SECTION 3
FIRE PREVENTION AND PROTECTION

Cost Analysis and Budgeting

James G. Gallup and Ken Lewis

LEARNING OBJECTIVES

- Understand the magnitude of the cost of fire protection in the United States
- List the resources available to assist with fire-protection decision making
- Describe the basic function of risk management
- Understand the basic structure of risk assessment
- Explain how the safety professional can interface with fire-protection cost-benefit issues
- Determine whether specific fire-protection features are cost effective
- Understand both the technical and practical concepts involved in performing cost-benefit analyses on fire-protection options

SAFETY PROFESSIONALS in industrial and commercial facilities are asked from time to time to address cost issues associated with fire-protection systems or fire-protection programs. These requests could come from interests internal to the facility or could originate because insurance interests, municipal requirements, or the Occupational Safety and Health Administration (OSHA) wish the facility to consider fire-protection improvements.

This chapter provides a basis for addressing fire-protection benefits and costs from a monetary standpoint. Sufficient information is provided to prepare a simplified analysis, to be done by the safety professional or the safety staff, or to develop a request for proposal (RFP) to hire outside firms to provide a more rigorous analysis. Although highly theoretical approaches are available, this chapter also provides a practical approach.

Cost-benefit analysis is a tool that may be used when conducting risk analysis as a portion of the property risk-management function. This chapter introduces general risk-management and risk-assessment information as an entrée into cost-benefit analysis.

The Magnitude of the Cost of Fire Protection

The total cost of fire protection includes not only the cost of fire-protection systems and fire-protection features in buildings but also the losses from actual fires and the cost of the fire service. John Hall, of the National Fire Protection Association (NFPA), has estimated the total yearly cost of fire protection in 2004 to be between $187 and $251 billion (Frazier 2005). Writing in *Fire*

Protection Engineering, Patricia Frazier estimates the cost breakdown for fire protection is approximately as follows (Frazier 2005):

- Cost of fire services 30%–45%
- Fire protection in buildings 25%–35%
- Estimates of human loss 10%–15%
- Indirect economic losses 5%–15%
- Direct economic losses 5%–6%
- Net fire insurance 5%

Indirect costs of fire protection beyond system costs are difficult to measure. These costs include such considerations as training, special transportation for hazardous chemicals, fire-insurance overhead, and other hidden costs.

Losses do to fires in the United States are estimated to be about 2 percent of the gross domestic product (Frazier 2005, 32). During 2006, there were 524,000 structure fires resulting in direct fire damage losses of $9.6 billion, 3245 civilian fire deaths, and 16,400 injuries (Karter 2007). But a limited number of small fires account for a high percentage of the total dollars lost each year (Watts and Ramachandran 2005).

Beyond direct fire damage and business interruption losses, fires cause significant indirect losses that are not easily measured. These indirect losses include loss of market share, human suffering, cost of temporary lodging, effects on the local economy, and tax losses.

Safety Professional Interface

The safety professionals who are most likely to need to perform fire-protection cost-benefit analysis are those employed by manufacturing firms, commercial corporations, government agencies, and institutional corporations. Many times the analysis will be needed in order to develop budgets for correction of deficiencies identified by fire-insurance carriers and municipal fire-prevention inspections.

At times a safety professional may be asked to provide a cost-benefit analysis for fire protection for new construction or for modernization of existing facilities. Sometimes the analysis is needed for upgrades to systems due to improvements in fire-protection technologies. Often an analysis will be necessary to compare benefits and costs of projects that have more than one acceptable fire-protection solution.

Approaches to Analysis

Overview

Safety professionals may be asked to prepare cost-benefit analyses for fire-protection upgrades. This section provides an overview of a rigorous means of providing analyses using outside firms as well as simplifications that will permit safety professionals to perform analyses themselves. The type of analysis used will depend on the needs of the organization.

Both the rigorous and the simplified analysis require some knowledge of the risk-management function.

Risk Management

Based on the experience of this author, the decision to spend money on fire protection depends on many factors. Items to consider include building codes, fire codes, other municipal requirements, and the consequences of noncompliance. Property-insurance recommendations and the consequence of noncompliance also require exploration.

Another factor is the level of risk an organization is willing to accept. The acceptable level of risk is either explicitly defined by an organization or is intuitively defined by the decisions the organization makes. Most organizations assign property risk management to a corporate officer. In large corporations, the corporate officer is the *risk manager*. The field of study is *risk management*.

Risk has two components: the probability that an event (a fire) will occur under a specific set of circumstances and the severity or expected value of the loss if the event should occur. *Risk analysis* is the systematic examination of how fire risk changes as assumptions change.

Probability is defined as the likelihood that a fire will occur under a certain set of circumstances per unit of time. This calculation is based on historical information, educated estimates, and experienced judgment. In general, calculations of the probability of occurrence are beyond what is expected of a safety professional.

Calculating the severity or consequences of a fire may also be beyond what is expected of a safety professional. The severity is generally quantitative and is based on worst-case scenarios for a specific set of circumstances.

Risk management provides value judgments in establishing levels of risk. Risk can be broadly managed by the following three risk management techniques (NFPA 1990, 5):

1. Avoid the risk.
2. Transfer the risk.
3. Control the risk.

Risk can be avoided by not performing the tasks generating the risk. Risk can be transferred using fire insurance. Risk can be controlled by managing the frequency of the event occurrence and/or by managing the size of the event once the event has occurred.

Fire protection has to do with controlling the risk. Therefore, the safety professional should review the other risk-management technologies with the risk manager prior to exploring fire-protection control measures.

Fire hazards can be controlled in three general ways. When developing cost estimates all three should be considered. Fire-control options include (NFPA 1990, 24):

1. Hazard reduction
2. Fire control
3. Loss limitations

Hazard reduction is one fire-prevention option. Preventing fires is generally the least expensive and the most cost-effective option. Fire prevention can be achieved through education, engineering, and enforcement.

Education for fire protection includes developing safe task procedures regarding fire exposures and the education of employees on those safe procedures. Engineering involves eliminating fire hazards by design, limiting the ignition sources and the size of the fire by design, or isolating the sources by design.

Fire control is accomplished by using passive and active fire-protection systems. Passive systems include fire walls, floors, roofs, and associated protective opening devices such as fire doors. Detection and suppression systems are active fire-protection systems.

Loss limitation refers to increasing the likelihood that fire will not propagate beyond designated levels. Loss limitations can be accomplished by spacial separation and fire-resistive construction.

When deciding on an analysis technique to be used for the cost-benefit analysis, two viewpoints are available. The first is an altruistic (societal) cost-benefit view. Using this view, broad items can be included such as the value of firefighter lives saved, the wages lost to the community due to the fire, the ripple effect of those lost wages, the tax cost of a major employer moving to a different community, the effects on the environment, and perhaps the reduced municipal fire-fighting requirements due to improved fixed fire-protection in a major building.

The second viewpoint is an analysis of the benefits and costs directly related to the corporation. A comprehensive cost-benefit analysis will include, among other items, a prediction of the direct and indirect costs of a fire using the various options, as well as tax impacts, human life exposure, reliability of the system, insurance costs, impact of inflation, salvage value, resale value of the facility with various options, cost of temporary or partial production and storage facilities, and the exposure to fire-department-caused water damage. Other items that can be considered are the impact on corporate image and reputation, the impact on suppliers, the impact on customers, the loss of market share, the loss of skilled employees, and the damage to trademarks.

Regarding the cost of the loss, *direct cost* usually refers to the cost of the building and equipment burned as well as business-interruption costs. *Indirect costs* include damage to the organization through loss of employees or damaged customer relations. Indirect costs are difficult to measure and thus difficult to predict. Experienced judgments are required.

A cost-benefit analysis is based on the accepted level of risk defined by the organization and describes the least expensive means to fulfill the organization's risk-reduction requirements. The analysis compares the cost of improved fire protection to the predicted losses due to a fire without added fire protection.

The analysis involves the following steps (NFPA 2003, 405):

- Define the objectives.
- Assess the available data.
- Determine the alternatives.
- Set the basic criteria.
- Predict the outcome of the alternatives (both benefits and costs).
- Choose the alternative with the best benefit-to-cost ratio.
- Analyze the levels of uncertainty.

Benefits and costs are stated in dollar equivalents. A decrease in risk or construction cost is viewed as a benefit. An increase in risk or construction cost is viewed as a cost.

The cost of fire-protection systems or building features must be based on a life-cycle cost. Life-cycle costs are often larger at the beginning of the time period when systems are installed. The life-cycle costs include the cost of design and engineering; the cost of the equipment itself; installation costs; start-up costs; inspection, test, and maintenance costs; operational costs; training costs; and salvage costs.

Benefits and life-cycle costs of various options can then be plotted on a timeline for comparison. Sometimes this will be viewed as a cash-flow diagram. As stated in the Engineering Economics chapter, cash-flow diagrams can be analyzed by a number of different techniques based on the time value of money. Available techniques include annual cash flows, present worth, payback, return on investment, and break even.

Highly Technical and Theoretical Approach

The size, complexity, or visibility of an analysis could lead an organization to hire consultants or university staffs to complete the study and analysis. If this occurs, the safety professional is valuable in assisting the development of a well-defined scope of work for the outside groups to use in proposals. Additionally, the safety professional can assist in locating firms to provide the study. The information from the risk management portion of this chapter should be used to help structure the scope of work.

The organization will first want to determine whether a societal viewpoint or an organizational viewpoint is most appropriate. Industrial and commercial facilities are most likely to want the cost-benefit analysis to relate directly to the organization.

If safety professionals are assigned to coordinate proposals from outside firms, they should involve the risk manager, the fire-insurance purchaser, the insurance company, the engineering department, the corporate fire-protection engineer, and the contracting department prior to preparing the requirements for the proposal. If the organization is small, these functions may be assigned to individuals with other duties, but the coordinating functions of the safety professional would be the same.

The level of detail and the cost range should be defined in the request for proposals. The most sophisticated analysis techniques will include detailed computer models. Event tree analysis and probability analysis models may be included. Fire modeling may be used to develop estimates for the timing of fire growth, smoke development, and damage assessments (SFPE 2002, 3–162) (NFPA 2003, 3–69). Outside firms will probably need to assist in determining whether the loss of human life should be considered in the analysis.

Practical Simplifications

Safety professionals can develop rough, yet meaningful, cost-benefit estimates by collecting information from various sources, simplifying the analysis, and using the Engineering Economics chapter of this volume to analyze the costs and benefits. If help with engineering economics is needed, safety professionals should contact the facility's accounting department or engineering department to assist with that portion of the analysis.

As with the more rigorous analyses done by outside firms, safety professionals should first contact the individuals responsible for risk management, the engineering department, the purchaser of property insurance, the corporate fire-protection engineer, and the contracting department to collect scope and administrative requirements.

In smaller corporations with limited staff, safety professionals may have to perform several of the functions listed above themselves. The process can still be performed effectively by utilizing the economic data and expertise of the facility's insurance and fire-protection professionals.

Municipal building code and fire prevention code requirements will require research; the safety professional must review the NFPA standards for possible options in meeting the code requirements, and be sure that the written requirements of the property insurance carrier are defined.

The subjects that will not be included are human-life loss costs, municipal fire protection impacts, community lost wages, and loss of market share. The analysis will also neglect inflation and tax impacts, salvage value, resale value, and the impact on customers and suppliers as well as computer modeling of fires, probabilistic analyses, and sensitivity analysis of the results.

The goal will be to reduce benefits and costs to dollar values, plot the dollar values on a timeline, and use the Engineering Economics chapter to compare the costs.

The first step is to list all options. The property insurance carrier, the fire department, the risk manager, and the corporate fire-protection engineer will have substantial information on options. The safety professional will then be able to reduce the analysis to the following for each option, which will include benefits and costs:

- The cost of the fire protection systems
- Similar costs for optional systems
- The cost of installation and testing for each option
- The yearly maintenance costs for each option
- The insurance-premium savings for each option
- Qualitative analysis of the reliability of each option
- The time period to be analyzed
- The cost of the fire at the end of the time period for each option

The cost of the fire-protection features for each option will have several sources. Safety professionals can prepare a rough scope of work that can be formally or informally priced by vendors. In general, vendors will be willing to provide this service free of charge in order to be included on the ultimate bidders' list. The corporate fire-protection engineer, the insurance carrier, and the fire department may have knowledge of current costs of fire protection features that will often be in a usable form, such as dollars per square foot. These same individuals can also provide estimates of the cost of installation, testing, and yearly maintenance.

At times, safety professionals may need to use published cost-estimating publications. One of the recognized sources is *Means CostWorks* (R. S. Means 2005), which contains cost data for building construction, electrical work, mechanical work, plumbing work, work per square foot, and for repair and remodeling.

Although seemingly simple, firm figures for the costs and benefits for property-insurance-related issues can be difficult to obtain. These numbers will also be related to the cost of the loss (size of the fire) at the end of the time period. Although the cost of the loss from the facility viewpoint could be estimated as the property insurance deductible, this is not necessarily accurate, because the loss may also result in increased premiums. Situations in which an organization uses partial or total self-insurance will also be more complicated to estimate.

One way to deal with insurance costs is to use only changes in premium levels. The yearly reduction in the fire-insurance premium can be estimated

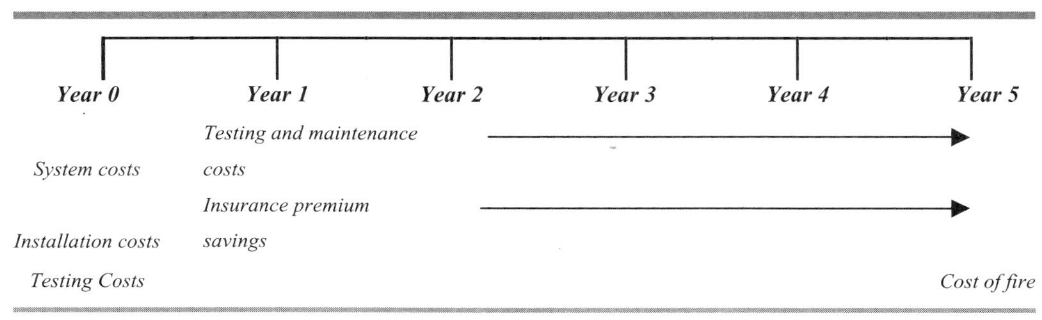

Figure 1. Timeline

by the insurance company through the risk manager. The deductible can be neglected if the total cost of the loss with each option is used.

The cost of the loss with each option can be obtained by discussion with the insurance carrier, the fire department, the corporate fire-protection engineer, and the facility controller. The insurance carrier will have published maximum foreseeable loss (MFL) estimates that can be used directly or modified to fit the options. For example, the MFL without sprinklers might be the entire area of the facility between fire walls. The MFL with the sprinkler option might be 2000 square feet plus smoke damage.

The time period used will generally be five to ten years. Periods shorter than five years will distort the analysis, and periods longer than ten years are not practical for business planning.

The dollar figures for benefits and costs will then be plotted on time lines. The time lines may look like Figure 1 for each option.

Although several economic analysis options may be available, the present value (present worth) option from the Engineering Economics chapter may be easiest to understand and present. Using this option, each cost and benefit is converted to a present value.

The controller for the organization can assist with choosing the proper interest rate. Terminologies that safety professionals might use when speaking with the controller are "discount rate," "cost of capital," or "internal rate of return." The goal is to use a single interest rate to calculate present values. The rate is likely to be in the 10 percent range.

The result of a present value analysis will be a present value cost for each option. The qualitative estimate of the reliability can then be used with the hard numbers to make final decisions about the best option to choose.

Other Information

Fire Prevention

Fire-prevention activities tend to be the most cost-effective fire-protection measures because equipment seldom needs to be purchased and tasks can be completed by existing personnel. Inspections conducted by the property-insurance carrier are generally funded through the property premium. If additional inspections are required by the facility, the costs can be obtained through the insurance company. The benefits of fewer and less-severe fires will be difficult to estimate.

Fire-prevention inspections are generally part of the housekeeping tours provided by first-line supervisors, the safety department, and mid- and upper-level managers. Their cost includes the preparation of inspection forms regarding reducing ignition sources or limiting combustible materials. The cost also includes the portion of the inspections relating to fire protection.

The education, engineering, and enforcement portions of fire prevention can also generally be conducted with existing personnel. The incremental costs should be negligible.

Estimates of the cost of inspection, testing, and maintenance by outside vendors can be obtained from the vendors.

If the facility decides to conduct inspections, perform testing, and maintain of fire-protection systems using in-house personnel, several costs may be incurred. The initial training and refresher training sessions will generally be completed by outside vendors. Training will also be needed when employees performing the tasks leave and new ones are hired. Some systems will also require equipment for testing as well as replacement parts for equipment that is tested in a destructive manner.

Fire History

In addition to the cost-benefit analysis, several other factors are worth considering. The NFPA (www.nfpa.org) is one easily referenced source of fire-history data. The NFPA studies most fire fatalities and looks in detail at the root causes of the deaths. This data can be used in addition to the cost-benefit analysis to justify expenditures. For example, the cost-benefit analysis may show that it would cost $3.00 per square foot to retrofit a building with an automatic fire-sprinkler system. This may not be required by code, but if a recent fire in a similar building caused extensive damage and the deaths of two employees and one firefighter, the fire history along with the cost-benefit analysis may be the information organization executives need to justify the expenditure.

Here are a few recent examples:

1. On January 29, 2003, in Kingston, North Carolina, a dust explosion and subsequent fire at the West Pharmaceutical Services manufacturing plant killed six workers and injured thirty-eight others, including two firefighters. Direct property costs were reported at $150 million.
2. On February 20, 2003, in Warwick, Rhode Island, 100 occupants of the Station nightclub perished in a fire ignited by indoor pyrotechnics during a rock concert.
3. On October 17, 2003, in Chicago, a fire in the 36-story Cook County Administration Building killed six occupants and injured twelve.

Fire Codes

Another factor to consider in planning is that following certain fire codes may allow easing of some regulations. For example, the NFPA 101 *Life Safety Code* allows greater travel distances to fire exits and greater allowances for dead-end corridors when a building is provided with an automatic fire sprinkler system. Organizations using these allowances may be able to make design changes that enhance utilization and efficiency. Installing these systems during initial construction is always more cost effective than retrofitting.

Fire Protection Systems

At times building and fire codes permit automatic sprinklers to reduce the fire-resistive levels of a building. Safety professionals may need to research the building code when pricing this type of alternative. Publications such as *Means Building Construction Cost Data* can assist with cost estimates for fire-resistive construction.

CASE STUDY

An aluminum-rolling mill intends to install a new cold-rolling mill in a fully sprinklered building. The risk manager and the engineering manager have asked the safety professional to conduct a cost-benefit analysis for a local suppression system on the mill. The options are a carbon dioxide system and a closed-head sprinkler system.

John, the safety professional, reviews the building code and fire prevention code and finds them silent regarding the suppression-system options. He reviews the NFPA code requirements for the two types of systems and the written requirements of the property-insurance carrier. The insurance company prefers a carbon dioxide system but will not raise the premium rate if sprinklers are used.

Corporate history and insurance records indicate that two major fires have occurred on a cold-rolling mill each year. Historical fire damage when carbon dioxide systems are used is $8000, including rearming the system.

The corporate fire-protection engineer expects the heat detectors for the carbon dioxide system to operate at approximately the same time as the first sprinkler activates. However, the carbon dioxide system will completely discharge before additional sprinklers open. The damage from a fire extinguished with the carbon dioxide system will be approximately one-eighth the damage of a fire extinguished with sprinklers.

The sprinkler vendor estimates that a sprinkler system inside the cold mill will cost $27,000. A carbon dioxide system will cost $82,000.

The controller estimates the cost of capital at 10 percent. Sprinklers are inspected by the insurance company and plant maintenance personnel quarterly. There are no incremental costs.

An outside vendor will inspect the carbon dioxide system quarterly. The cost will be $1500 per quarter.

The sprinklers are substantially more reliable because no separate detection system is needed and and there are fewer parts.

	Sprinklers	*Carbon Dioxide*
Initial Cost	$27,000	$82,000
Maintenance (yearly)	-	$ 6,000
(2) Fires (yearly)	$128,000	$16,000

Figure 2. Costs
Use the compound interest tables at the back of this book to establish present value factors for annuities at 10 percent for 5 years.

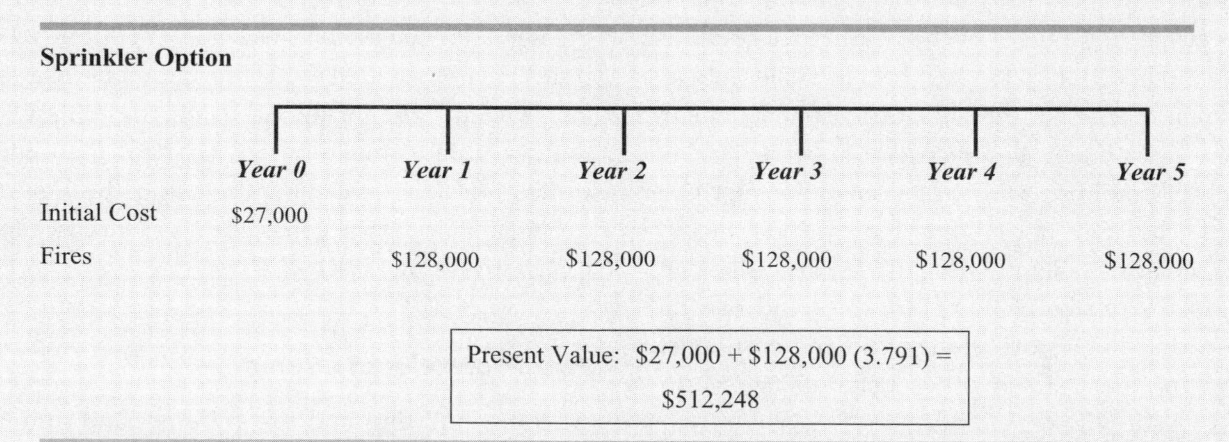

Figure 3. Sprinkler option

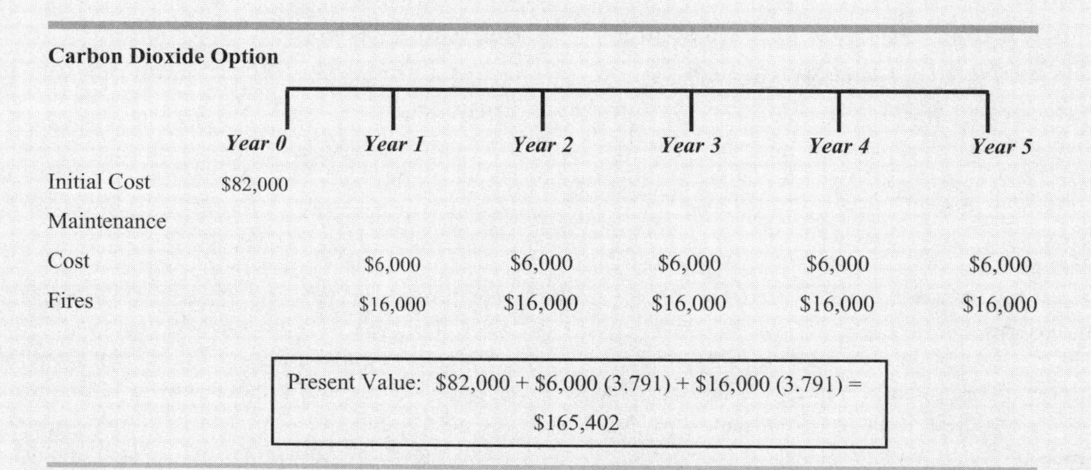

Figure 4. Carbon dioxide option

Conclusion

Although carbon dioxide systems are less reliable, the present value of the yearly cost of fires favors that type of system. The insurance company and the safety professional recommend the carbon dioxide system over sprinklers.

References

Bukowski, Richard W. and Wayne D. Moore. 2003. *Fire Alarm Signaling Systems*. Quincy, MA: , National Fire Protection Association.

Frazier, Patricia. Spring 2005. "Total Cost of Fire in the United States." *Fire Protection Engineering* 26:32–36.

Karter, Michael J., Jr. September 2007. *Fire Loss in the United States During 2006*. Quincy, MA: NFPA.

Means, R. S. 2005. Means CostWorks, release 9.0 (CD-ROM). Kingston, MA.

NFPA (National Fire Protection Association).1990. *Industrial Fire Hazards Handbook*. 3d ed. Quincy, MA: NFPA.

______. 2003. *Fire Protection Handbook*. Quincy, MA: NFPA.

______. 2006. Life Safety Code, NFPA 101. Quincy, MA: NFPA.

Society of Fire Protection Engineers. 2002. *SFPE Handbook of Fire Protection Engineering*. 3d ed. Bethesda, MD: SFPE.

Watts, John M., Jr., and G. Ramachandra. Spring 2005. "Fire Protection Engineering Economics." *Fire Protection Engineering* 26:10–17.

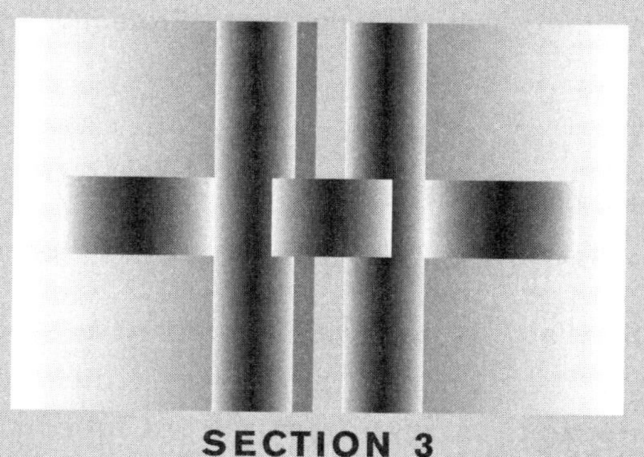

SECTION 3

FIRE PREVENTION AND PROTECTION

Benchmarking and Performance Criteria

Wayne Onyx

LEARNING OBJECTIVES

- Discuss the different fire inspection schedules and the guidelines for each.
- Explain the procedure for testing a dry-pipe sprinkler system.
- List the safeguards for storing flammable gas and liquids.
- Explain the procedure for a 2-inch drain test of sprinkler systems.
- Understand the procedures for testing fire pumps.
- List the three components of a public water supply test and discuss the test procedures for each.

INSPECTION, TESTING, and maintenance of fire protection equipment are critical components of a well-maintained facility. It is essential that these activities are performed properly and timely by qualified personnel.

Plant inspectors should be

- Completely familiar with the facility—hazards, layout, operations, and protection.
- Knowledgeable of how the fire protection systems function along with the location of the control valves, fire extinguishers, fire hoses, and detection systems. Also, they must be able to recognize conditions that would make sprinklers inoperative.
- Capable of identifying and eliminating ignition sources.

Training for plant inspectors can be provided through courses offered by NFPA, property insurance companies, fire-protection contractors, and "on-the-job" experience.

Contractors (outside vendors) should be certified in accordance with state and local requirements.

Fire Protection Inspection Schedules

In-House Inspection Frequencies

If an in-house inspection is carried out by the company employees or staff as opposed to an outside contractor, the following is a general guideline. More descriptive guidelines and forms are contained in the following sections.

Weekly

If the sprinkler control valves are not locked or electronically supervised, they should also be inspected weekly. If dry-pipe, pre-action, or deluge systems exist, they need to be checked for normal air and water pressure. Fire pumps should be churn-tested to verify they are starting properly and operating without any obvious problems.

Monthly

Monthly inspections include visually recorded checks of fire extinguishers, housekeeping, flammable and combustible liquids, and sprinkler-system gauge pressures.

Quarterly

Quarterly inspections involve testing of fire protection equipment alarms, verifying that sprinkler control valves are fully open and turn freely, and testing the alarm devices via inspectors' test connections. These tests are sometimes done monthly.

Semiannually

These inspections generally cover the fire alarm systems and are typically carried out by the alarm-service provider or alarm contactor.

Annually

Annual inspections involve the physical examination and maintenance of all fire protection equipment. The purpose of the tests is to ensure that systems are functioning properly and operating at their expected output. The tests include performing 2-inch drain tests and trip tests of dry-pipe, pre-action, and deluge valves, and functional tests of fire pumps, water supplies, and alarm monitoring panels.

Other

There are certain aspects of fire protection equipment which have two-, three-, five-, and ten-year intervals on required maintenance and testing.

Detailed Information on Fire Protection Inspections

Weekly Fire Protection Inspection

Notice: To avoid false alarms to a supervisory service or authority, the receiver of any alarms should be notified prior to the beginning of the test procedures and following the conclusion of the testing procedures.

Control Valves

If sprinkler control valves are not secured by locks or electronically supervised, inspect them weekly, verifying that they are in the wide-open position.

Dry-Pipe Sprinkler System

Inspect air-pressure and water-pressure gauges. Be sure that air and water pressures are within the normal range for each particular system. Record pressure readings. It is recommended that the normal pressure be noted on the gauge or a tag attached to it. A loss of air pressure of more than 10% should be investigated.

Fire Pump

- Perform churn-testing for electric and/or diesel-driven fire pumps. Start the tests via system pressure drop to confirm the automatic start feature of the pump. This is done by operating the test valve on the water-pressure test line to the fire-pump controller. Follow the procedures below and record the results on an inspection form:

  1. Pump starting pressure.
  2. System suction and discharge pressure.

- Observe and record these driver-specific items:

  1. Electric Motor: Run pump for a minimum of 10 minutes.
  2. Diesel Engine: Run pump for a minimum of 30 minutes.

3. Record the engine oil-pressure gauge, speed indicator (RPMs), and the water and oil temperatures periodically while the engine is running. Record any abnormalities.

- Pump House/Pump Room

1. Maintain temperature at no less than 40°F, or 70°F if diesel pump is without an engine heater.
2. Check to ensure that ventilation louvers are clear and free to operate for proper ventilation.

- Pump Systems: Check to ensure that

1. Pump suction, discharge, and bypass valves are fully open.
2. All piping is free of leaks.
3. Suction line pressure gauge reading is normal.
4. System line pressure gauge reading is normal.
5. Suction reservoir is full.
6. Pump packing glands show a slight discharge and there is no overheating. Adjust gland nuts if necessary.

- Electrical Systems: Check to ensure that

1. Controller pilot light (power on) is illuminated.
2. Transfer switch normal pilot light is illuminated.
3. Isolating switch is closed—standby (emergency) source.
4. Reverse phase alarm pilot light is off.
5. Oil level in vertical motor sight glass is normal.

- Diesel Engine Conditions: Check to ensure that

1. Fuel tank is two-thirds full.
2. Controller selector switch is in auto position.
3. Batteries' voltage readings are normal.
4. Batteries' charging current readings are normal.
5. Batteries' pilot lights are on.
6. All alarm pilot lights are off.
7. Engine running time meter is operational.
8. Oil level in right-angle gear drive is normal.
9. Crankcase oil level is normal.
10. Cooling water level is normal.
11. Electrolyte level in batteries is normal.
12. Battery terminals are free from corrosion.
13. Water-jacket heater is operating.

Water Storage Tank

- When the water temperature is electronically supervised, during cold weather (i.e., if the tank is subject to freezing), the following items should be performed, inspected, and recorded weekly

1. Verify and record the temperature of the water.
2. Inspect the heating equipment to ensure the system is operational and effective. *Note:* If the water temperature is not supervised, these items should be addressed daily.

- Under normal conditions, check to make sure that the tank is full and that the gauge is functioning.

Fire Alarms

Check for trouble signal.

Monthly Fire Protection Inspections

Notice: To avoid false alarms to a supervisory service or authority, the receiver of any alarms should be notified prior to the beginning of the test procedures and following the conclusion of the testing procedures.

The following inspections and tests should be performed monthly. The results should be recorded on a monthly fire protection equipment report form. Repairs and corrective action should be completed promptly.

Fire Extinguishers

Check all to be certain they are accessible, not damaged, hung properly, charged with the seal intact, and with the current annual inspection tag in place. Initial this tag monthly.

Sprinkler Control Valves

Check and test all sprinkler riser control valves, yard main divisional valves, and water supply pit valves to make sure they are accessible, open and sealed, electronically supervised, or locked. Sprinkler control valves accessible to the public should preferably be locked in the open position. All other sprinkler control valves should be sealed open or have electronic supervision.

Fire Hoses

Check each fire hose, if present, to ensure that the hose is ready for use and mounted on its rack (or reel) with nozzle attached. Visually inspect the hose as to condition (creases are not worn or split and there is no water in the hose). The use of mounted fire hoses has been discontinued at many facilities.

Fire Hydrants

Visually inspect to make sure they are unobstructed, undamaged, and ready for use.

Fire Department Connections

Check to see that they are unobstructed, caps in place and couplings free to rotate.

Fire Alarms

Check lead-acid and dry cell batteries.

General Fire Safety Conditions

- Limit smoking to designated areas.
- Free stairwells and corridors of obstructions and storage.
- Ensure emergency lights and exit lamps are in working order.
- Post evacuation drawings.
- Flammable Gas and Liquid Safeguards

  1. Store all flammable and combustible liquids and aerosol cans in approved flammable liquid cabinets with the doors kept closed.
  2. Store liquids outside of cabinets. Do not exceed one day's supply or 25 gallons of Class IA liquids (flash point less than 73°F) or 120 gallons of Class IB, IC, II, or III liquids (flash point less than 200°F).
  3. Use safety cans with self-closing lids for the handling and storage of small amounts of gasoline and other flammable liquids.
  4. Chain oxygen and acetylene cylinders in the upright position, separated by 20 feet or a noncombustible barrier, and with protective caps in place when not in use.

- In sprinklered buildings, do not store combustibles in areas shielded from automatic sprinklers. Likely shielded areas are

  1. under conveyors
  2. in the pit
  3. beneath stairwells

- General Housekeeping

  1. Do not store anything in electrical rooms or on electrical equipment.
  2. Keep waste combustible materials in proper containers.
  3. Store excess pallets outside at least 50 feet from buildings.
  4. Store packing materials, such as cardboard boxes, Styrofoam filler, and plastic bags in a clean area away from possible ignition sources. Keep materials in packing areas to a minimum and keep the area clean and free of clutter.
  5. Store used rags contaminated with grease, oil, or solvents in an approved fire can with a self-closing lid while awaiting vendor pickup.
  6. Keep materials clear of light fixtures and bulbs. Provide wire mesh guards to protect bulbs and fixtures from damage where needed.

Quarterly Inspections

Notice: To avoid false alarms to a supervisory service or authority, the receiver of any alarms should be notified prior to the beginning of the test procedures and following the conclusion of the testing procedures.

Sprinkler Water Flow

Test each sprinkler system water flow alarm by operating the inspector's test connection on wet-pipe systems or the alarm bypass valves on dry-pipe systems. The alarm bypass valve may be used on wet-pipe systems. However, it is preferred that the inspector's test connection be used since this simulates the operation of only one sprinkler. By flowing the inspector's test connection for 60 seconds, the water motor gong and any electronic water flow monitors should operate.

Valve Tamper Devices

Test each valve tamper device by turning each control valve three complete turns in the closed direction and then fully reopen. This should trigger the alarm.

Dry-Pipe Sprinkler Systems

Determine priming water level by slowly opening the priming water level test valve. If only air escapes, close the test valve and add priming water. This is done by closing the lower priming valve, opening the upper priming valve, and adding approximately 1 quart of water through the priming funnel. The upper priming valve is then closed and the lower priming valve opened, which allows the water to run into the dry-pipe valve. Again, check the test valve. If water does not run out, repeat the procedure. When sufficient water has been added so that water drains from the test valve, allow it to drain until air begins to escape; then close the valve securely. Also be sure that upper and lower priming valves are closed securely.

Semiannual Inspections

Fire-Alarm Batteries

Check nickel cadmium and sealed lead–acid batteries.

Fire-Alarm Control Unit Trouble Signals

INITIATING DEVICES

Check the following (typically done by alarm-company contractors)

- air sampling
- duct detectors
- electromechanical releasing devices
- suppression system switches and fire-alarm boxes
- heat detectors
- smoke detectors
- CO detectors

Annual Tests

Fire Extinguishers

Annually, have all fire extinguishers serviced by a fire-extinguisher service contractor. Also, depending on the type of extinguisher, discharge and hydrostatic tests are required at specific intervals. NFPA 10 (2007b) specifies the required intervals for each type of extinguisher. The extinguisher contractor should be able to show the test intervals for the extinguishers.

All Sprinkler Systems

- Conduct a 2-inch drain test. This involves noting and recording the pressure on the gauge on the lower side of the sprinkler valve (static water supply pressure). Open the 2-inch main drain fully; after the flow has stabilized, note and record the pressure on the gauge again (residual water supply pressure). After recording the residual pressure, slowly close the main drain valve and note the time it takes for the water supply pressure to return to the original static pressure.
- Lubricate all valve stems. Apply graphite or graphite in light oil to the valve stem. Fully close the valve and then reopen to test its operation and distribute the lubricant on the valve stem.
- Clean strainers if provided. This will generally involve shutting off the water supply

and removing the strainer to clean it. Some strainers are self-cleaning and simply require rotation of the operating wheel.

Dry-Pipe Valves

- Trip-test the dry-pipe valve. Before the trip test, fully open the main drain valve and flush out the water supply until the water flows clean. If a hydrant is located on the system supply, flush it before flushing the main drain. This flushing will help to reduce the amount of debris getting into the dry-pipe system. Trip-test each dry-pipe valve, including quick opening devices, if provided. This test should be done in the spring, after freezing weather is over, with the water supply control valve only partially open. Once the valve trips, quickly close the water control valve so that the system is not filled with water. (Caution: some dry-pipe valves will not operate properly without a flow of water adequate to fully lift the clapper valve.) Trip the valve by opening the inspectors test valve, which releases air pressure within the system. After the test, open the 2-inch main drain valve to drain the system. Remove the valve cover and thoroughly clean the valve interior. Replace worn or damaged parts as required, reset the valve, and replace the cover. Add priming water and open the air supply to fill the system with air. When the air pressure has reached its proper level, open the 2-inch drain to reduce the chance of a water hammer tripping the system, and then slowly open the water supply valve. When the water supply valve is fully open, slowly close the 2-inch main drain.
- Every three years, fully trip-test the dry-pipe valve with the water supply valve fully open. Terminate the test when clean water flows from the inspector's test connection. Also, conduct a full trip test whenever the sprinkler system undergoes a major alteration or extension. Water should reach the inspector's test connection within 60 seconds. If greater than 60 seconds, then an accelerator may need to be provided.
- In addition, at 10-year intervals (5-year intervals if a full flushing is ever required) have a flushing investigation performed by a qualified sprinkler contractor in accordance with NFPA 25 (2008). Notify the insurance carrier's local fire protection office in advance so they may have a representative present during the testing. Dry-pipe systems of non-coated ferrous piping should be inspected after being in service for 15 years, 25 years, and then every 5 years thereafter.

Notice: To avoid false alarms to a supervisory service or authority, the receiver of any alarms should be notified prior to the beginning of the test procedures and following the conclusion of the testing procedures.

Fire Pumps

- Full water flow tests are performed annually to evaluate performance of the fire pumps. A maximum load is induced on the pump and driver to make sure the unit is ready and able to perform under an actual fire-load condition. Pump capacity tests are performed by discharging water through an approved flowmeter, through hoses attached to a hose header, or through yard hydrants. Most installations have a test header.
- The test is typically performed by a qualified fire-pump service contractor sometimes in conjunction with a property insurance company representative. However, it can be conducted by other qualified individuals who have access to the necessary testing equipment. It should be performed when a high volume of water can be discharged into the yard.
- A copy of the manufacturer's test curve and previous test results, including the field acceptance tests, should be available for comparison of results. Should the annual

results not be within 10% of previous test results, an investigation should be made to determine the problem. Corrective action should be implemented as soon as possible.

- Equipment needed to perform this test may include

  1. A 50-foot length of 2½-inch lined hose with Underwriter's playpipe nozzle. A length of hose and play pipe is needed for each test header outlet. Also, a method to secure the hose and pipe is needed (not required when using an approved flowmeter)
  2. Pitot tube with gauge (or listed flow meter) and test pressure gauges (not required when using an approved flowmeter)
  3. Previous test data, field acceptance data sheet, and manufacturer pump curve
  4. RPM meter and amp meter
  5. Hose wrench and spanner

- During the pump performance test all of the controller functions and alarms should be tested and verified and the strainer should be dismantled and cleaned. For additional maintenance needs, consult the manufacturer's guidelines.

Hydrants

- Inspect dry barrel yard hydrants to be sure the barrel is drained.
- Inspect wall hydrants to be sure the pipe to the outside outlets is free of water.
- Flow test all hydrants every three years to make sure they are working properly and to keep the threads and working parts free of foreign material.

Fire Hose

Hydrostatic test by a service contractor should be performed five and eight years after installation, and every two years thereafter.

Fire Alarm

The fire-alarm contractor should check the panel.

Water Supply Testing and Evaluation

There are several reasons for flow testing water supplies:

1. Fire flow available to a given area can be compared with specific demands.
2. Closed valves and other obstructions become apparent.
3. Water supply deterioration can be detected and monitored over time.
4. Weakened underground pipe sometimes ruptures under fire-pump or fire-department pressures. It is far better that this condition be discovered during a test than during a fire.
5. Problems with the fire pump can be detected and repaired.

Public Water Supply Testing

- Public water supplies should be tested periodically to ensure that the flow and pressures remain adequate for fire protection. This can normally be done in cooperation with the public water department. Hydrants on private property may be tested independently or in cooperation with a property insurance carrier if the proper testing equipment (Pitot tube and pressure gauges) is available.
- The weather and surrounding conditions should be considered prior to the beginning of any tests. If the temperature of the air and or ground is near freezing, precautions may be necessary. Also, traffic patterns and eventual flow of water should be considered.
- The components of a water test are static pressure, water flow, and residual pressure. These three components provide the majority of the information required to analyze the water supply.

Static Pressure

The static pressure is the normal pressure at the sprinkler riser base (or near ground level) with normal domestic usage and water flow. This can typically

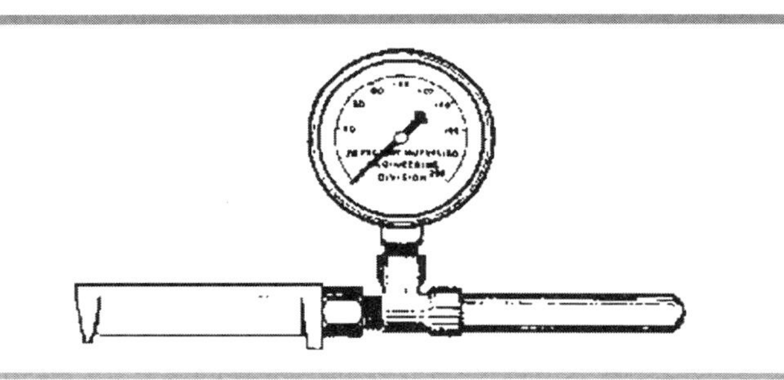

FIGURE 1. Pitot tube with gauge and air chamber (*Source:* FM Global 2006. Copyright 2006 FM Global. Reprinted with permission. All rights reserved.)

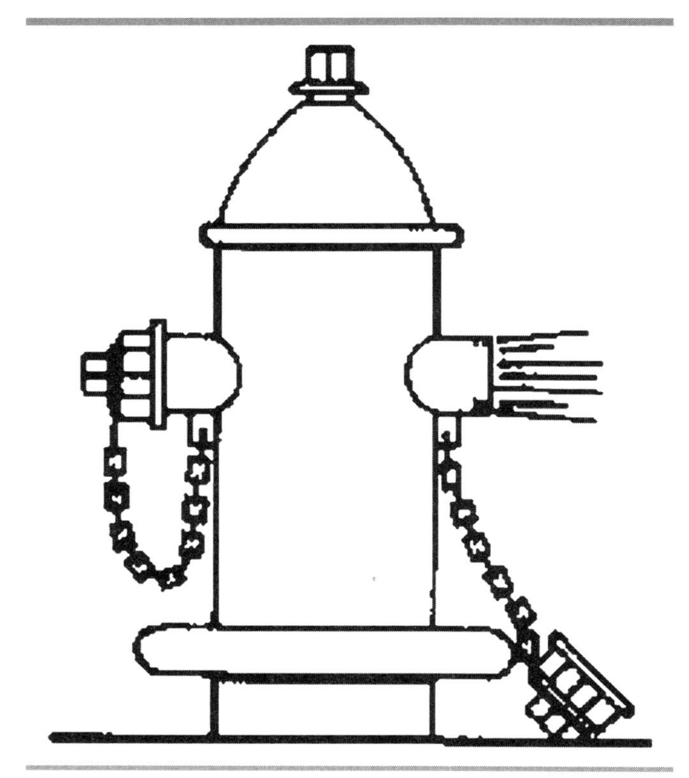

FIGURE 2. Hydrant butt (*Source:* FM Global 2006. Copyright 2006 FM Global. Reprinted with permission. All rights reserved.)

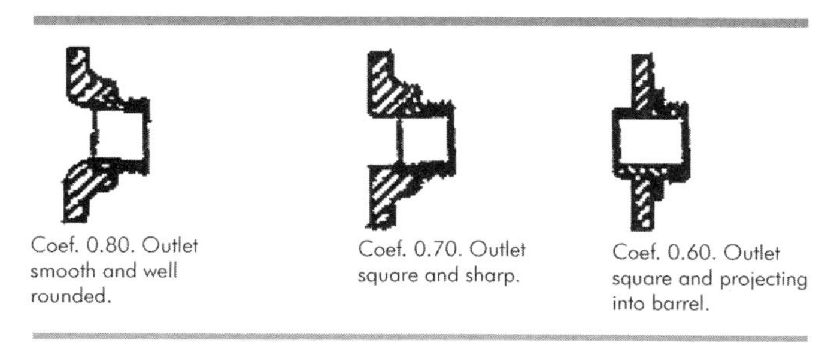

FIGURE 3. Typical hydrant butts (*Source:* FM Global Insurance, Data Sheet 3–0. 2006. Copyright 2006 FM Global. Reprinted with permission. All rights reserved.)

be read directly off the pressure gauge located below the sprinkler check valve. The upper gauge sometimes will show a higher pressure that is "trapped" above the check valve.

Water Flow

A typical flow test involves opening one or more hydrant outlets and measuring the flow through the use of a Pitot tube (Figure 1). The Pitot tube measures velocity pressure of the water discharging from an orifice. The diameter of the opening and the corresponding discharge coefficient of the discharge orifice along with velocity pressure must be known to estimate flow. Most modern hydrants have a discharge coefficient of 0.8, but a smooth playpipe would have a coefficient of 0.97, which provides a less turbulent flow (Table 1).

"Discharge coefficients used in calculations of flow from hydrants depend upon the character of the hydrant outlet (Figure 2). Figure 3 shows three general types of hydrants." (FM Global 2006).

The Pitot tube orifice is held firmly in the center of the water-flow opening with the blade pointing back toward the flow of water (Figure 4).

The scientific equation for converting the pressure created by the water velocity to flow is Equation 1.

$$Q = ac(d)^2(P_v)^{1/2} \tag{1}$$

TABLE 1

Discharge Coefficients

	Typical Discharge Coefficients
Hydrant butt, smooth, well-rounded outlet	0.80
Hydrant butt, square outlet	0.70
Hydrant butt, inset outlet	0.60
Smooth Underwriter nozzles	0.97
Deluge nozzles	0.99
Open pipe, smooth and well-rounded, at least 10 diameters long	0.90
Open pipe, burred opening or less than 10 diameters long	0.80

(*Source:* FM Global Insurance 2006. Copyright 2006 FM Global. Reprinted with permission. All rights reserved.)

TABLE 2

Conversion Table, English (U.S.) to Metric

Measurement	Unit (abbreviation/symbol) U.S.	Metric	Conversion factor (formula)
Volume			
gallons	gal		3.7854×10^{-3}
			gal × 0.003785 = m^3
cubic meters		m^3	m^3 × 264.2 = gal
			m^3 × 35.32 = ft^3
			ft^3 × 0.02832 = m^3
Flow			
gallons per minute	gpm		gpm × 0.06309 = l/s
liters per second		l/s	l/s × 15.85032 = gpm
			l/s × 0.001 = m^3/s
cubic meters per second		m^3/s	gpm × 6.3090×10^{-5} = m^3/s
Discharge density			
gallons per minute per square foot	gpm/ft^2		gpm/ft^2 × 40.75 = mm/min
millimeters per minute		mm/min	mm/min × 0.0245 = gpm/ft^2
Temperature			
Fahrenheit (degrees)	°F		(°F − 32) × 5/9 = °C
Celsius (degrees)		°C	°C × 5/9 + 32 = °F
Pressure			
pounds per square inch	psi		bar × 14.5 = psi
pounds-force/square inch		bar	psi × 0.0689 = bar

where

Q is the rate of flow, gpm (dm^3/min)
a = constant = 29.8 (0.666) (0.0666) (Use 0.666 for P_v in bars and 0.0666 for P_v in kPa.)
c = discharge coefficient
d = orifice diameter, inches (mm)
P_v = velocity pressure or Pitot pressure, psi (b) (kPa)

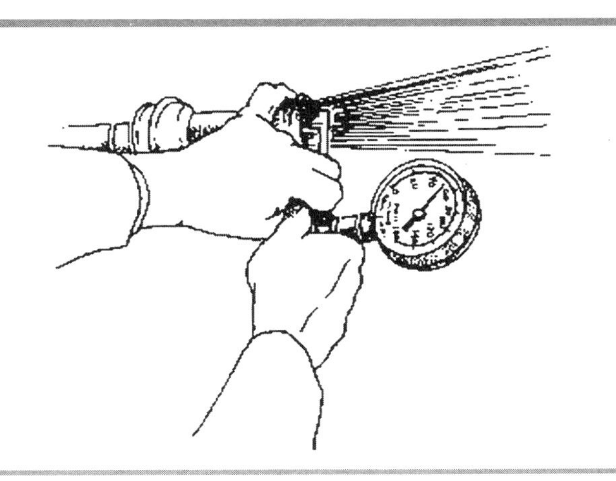

FIGURE 4. Taking Pitot readings at hose nozzle (*Source:* FM Global 2006.)

Table 2 provides a conversion chart.

Manufacturers of most equipment do not produce exact metric–Imperial equivalents, but rather a range of sizes that are accepted as nominal equivalents. Listed in Table 3 are some commonly encountered equivalents.

TABLE 3

Metric Equivalents for Pipe Sizes

Imperial (in.)	Metric (mm)
1	25
1.25	32
2	50
2.5	65
3	80
3.5	90
4	100
5	125
6	150
8	200
10	250
12	300

(*Source:* FM Global Insurance 1999.)

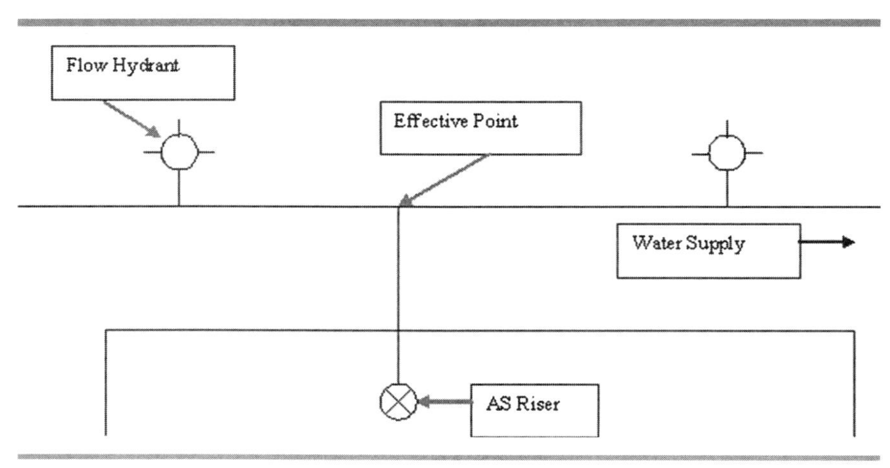

FIGURE 5. Residual pressure effective point

Residual Pressure

While the water is being flowed, the pressure drops due to friction loss, resulting in what is called the *residual pressure*. The residual pressure should be read from the same spot that the static pressure is read, ideally the closest sprinkler riser to the hydrant. The static and residual pressures can also be read at an upstream hydrant, in which case the *effective point* of the test is at that hydrant. The effective point of the test is where the flowing water meets the nonflowing water (Figure 5).

If this is a great distance from the plant or sprinkler system, some calculations are required to determine the additional friction loss accumulated over this distance to determine the actual supply available to the base of the riser (BOR).

Fire-Pump Testing

- Full water flow-pump tests are performed annually to evaluate performance of the fire pumps. A maximum load is induced on the pump and driver to make sure the unit is ready and able to perform under an actual fire load condition. Pump capacity tests are performed by discharging water through an approved flow meter, through hoses attached to a hose header, or through yard hydrants. Most installations have a test header.
- The test should be performed by a qualified fire-pump service contractor or a property loss-control representative from an insurance carrier. It should be performed when a high volume of water can be discharged into the yard. Extreme care should be taken to properly secure the hoses as they can create severe injury or death if they become loose during high flows. Also, care should be taken so that the water discharges into a safe area as the water streams will exert a large force.
- A copy of the manufacturer's test curve and previous test results, including the field acceptance tests, should be available for comparison of results. Should the annual results not be within 10% of previous test results, an investigation should be made to determine the problem. Corrective action should be implemented as soon as possible.
- Equipment needed to perform this test may include:

  1. A 50-foot length of 2½-inch lined hose with underwriter's playpipe nozzle. A length of hose and play pipe is needed for each test header outlet. Also, a method to secure the hose and pipe is needed.
  2. Pitot tube with gauge and test pressure gauges.

3. Previous test data, field acceptance data sheet, and manufacturer's pump curve.
4. RPM meter and amp meter.
5. Hose wrench and spanner.

- During the pump performance test, all controller functions and alarms should be tested and verified, and the strainer should be dismantled and cleaned. For additional maintenance needs, consult the manufacturer's guidelines.
- Other comments:

1. When accuracy is critical, plan to use hose streams rather than hydrant butts.
2. Check for local regulations which may affect planning. For example, it may be necessary to notify the public water department before flow testing private systems.
3. Flow an amount of water equal to or greater than the demand, near the demand area.
4. An approved flowmeter can be used in lieu of a full flow test.

Testing to Reduce Fire Losses

Many of the above testing procedures will not prevent fire losses, but are designed to minimize the extent of the loss and damage when a fire occurs. Testing increases the chances that fire protection systems and water supplies will function when needed. Fire losses can be prevented through good housekeeping, safe flammable liquid handling, and adequate hot work permits and smoking procedures.

The importance of an adequate sprinkler valve inspection program can be demonstrated by records kept by FM Global. FM Global Insurance is one of the larger property insurance companies that provide coverage for major manufacturing and industrial locations. (FM Global 2002a).

Since 1958, there have been 940 fires at FM Global-insured locations where fire protection valves were closed, resulting in more than $1.3 billion in losses. Each year, FM Global engineers typically discover close to 1000 improperly closed valves (ICVs) and impaired fire protection systems. In a survey conducted between 1995 and 1999, FM Global Engineers discovered approximately 3800 ICVs. A majority of these ICVs (64%) were neither provided with a non-breakable lock nor equipped with a supervisory monitoring device. In addition, a majority (60%) consisted of valves that typically are found indoors or within pits. (FM Global 2002a).

Surprisingly, however, the survey also revealed that, of the ICVs discovered, 19% were locked at the time of discovery, 13% were equipped with tamper switches, and 4% were both locked and equipped with a tamper switch. This tends to indicate that simply providing a lock and/or tamper switch alone is no guarantee that a fire protection system control valve won't be improperly closed or impaired.

Some of the common reasons for the improperly shut valves are (FM Global 2002a)

- Sprinkler system repair
- Building alterations
- Maintenance
- Cold weather
- Error (not realizing the valve is part of the sprinkler system)
- Malice (including arsonous intent)

In addition, periodic testing of the fire protection systems is designed to detect potential failures prior to the emergency where they are needed. For example, a 2-inch drain test confirms that there are no significant blockages in the water supply line and helps verify that the supply has not deteriorated over time. An inspector's test flow confirms that the water flow alarms will operate in the appropriate amount of time. A fire pump can not be relied on to operate in an emergency if it has not been periodically run and exercised. All of the suggested testing is important to verify that the systems are operational and in good order. Almost all major fire losses are a partial result of inadequate fire protection testing and maintenance.

Cause	Number
Electricity	1,368
Hot surfaces/radiant heating	473
Hot work	387
Arson/incendiarism	684(est.)
Exposure	222
Friction	297
Open flames (excluding hot work)	268
Smoking	236
Molten substances	68
Overheating	398

FIGURE 6. Top ten ignition sources (*Source:* FM Global 2002b. Copyright 2002 FM Global. Reprinted with permission. All rights reserved.)

Fire Losses

FM Global conducted a study of their clients' most costly losses from 1992–2001. Figure 6 shows the results of that study.

Fire Incident Data and Effectiveness of Extinguishing Systems

It is difficult to draw accurate conclusions from fire-loss data, because most cases where fire protection is effective in controlling or extinguishing a fire never get reported. However, if a fire protection system should fail to control a fire, the resultant loss and failure are almost always reported. NFPA stopped tracking sprinkler performance in 1970, mostly because of this concern. Nevertheless, NFPA data available from 1925–1969 indicate the different reasons for unsatisfactory sprinkler performance. The following sections summarize that data.

Failure to Maintain Operational Status of Systems–53%

The failures that fall into this category include shut sprinkler valves, inadequate maintenance, sprinkler distribution obstructions, and frozen piping. Performing the recommended periodic inspection testing and maintenance would likely have prevented the majority of these failures.

Inadequate Sprinkler System–22%

This includes both design inadequate to protect the current hazard and inadequate coverage of protection. As a result of the number of hydraulically designed systems installed since 1970, this percentage has grown. Many of the hydraulic designed systems are installed with little foresight or future planning. Many of these systems are designed for the immediate planned use, with little thought given to potential future changes in occupancy or increased hazards as a result of changes in processes or addition of storage areas. Most AHJs and insurance consultants who review plans for adequacy are looking at only the current planned use, with little regard to potential future changes. If the hazard increases for any of the above reasons, it can be difficult and costly to upgrade these systems to make them adequate. Also, there is sometimes little incentive for them to do so, because most local codes do not require existing systems to be modified to meet current NFPA 13 design criteria (NFPA 2007a).

Defects not Involving Sprinkler Systems–16%

This includes inadequate water supplies and faulty building construction. Some of these failures could also be considered to be due to inadequate sprinkler systems as noted above.

Inadequate Performance of Sprinkler System–6%

This includes problems associated with antiquated systems, slow operation of sprinklers, and defective dry-pipe valves. Some of these failures could also fit into some of the categories above.

Other–4%

This category includes exposure fires and unknown causes.

Vulnerability Analysis

A vulnerability analysis can be carried out by a variety of fire protection consultants, or by the property insurance carrier. FM Global, Allianz, and Zurich are some of the insurance companies that provide these services. Typically, these are reports that identify areas of exposure and offer recommendations to reduce the risk of loss. A major part of this analysis deals with the adequacy of inspection, testing, and maintenance of fire protection systems. Recommendations usually cite FM Global or NFPA standards.

References

FM Global Insurance Company (FM Global). 1999. Data Sheet 9–12/17–14. "Metric System Terms and Conversions." Johnston, RI.

______. 2002a. Publication P7133, "Controlling the Shut Valve Hazard." Johnston, RI.

______. 2002b. Publication P8610, "Ignition Sources: Recognizing the Causes of Fire." Johnston, RI.

______. 2006. Data Sheet 3–0, "Hydraulics of Fire Protection Systems." Johnston, RI.

National Fire Protection Association (NFPA). 2007a. NFPA 13, "Installation of Sprinkler Systems." Quincy, MA.

______. 2007b. NFPA 10, "Standard for Portable Fire Extinguishers." Quincy, MA.

Appendix: Recommended Reading

FM Global Resource Collection

FM Global Publications

National Fire Protection Association (NFPA). 2007. NFPA 20, "Standard for the Installation of Stationary Pumps for Fire Protection." Quincy, MA.

______. 2007. NFPA 24, "Standard for the Installation of Private Fire Service Mains and Their Appurtenances." Quincy, MA.

______. 2007. NFPA 72, "National Fire Alarm Code®." Quincy, MA.

______. 2008. NFPA 25, "Standard for the Inspection, Testing, and Maintenance of Water-Based Fire Protection Systems." Quincy, MA.

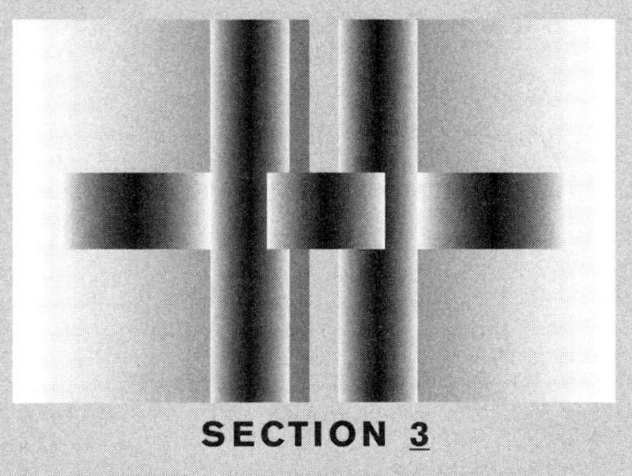

SECTION 3
FIRE PREVENTION AND PROTECTION

LEARNING OBJECTIVES

- Increase understanding of the importance of fire prevention and fire protection in the industrial and commercial setting, particularly the oil and gas industry.
- Discover best practices in fire prevention and protection.
- Understand roles in performing the services of a fire protection engineer or professional charged with fire prevention and protection responsibilities.
- Discover reference resources for additional information and in-depth study.

BEST PRACTICES

Craig A. Brown

IN DEFINING best practices safety professionals must apply fundamental principles of personal and professional integrity and act in terms of fundamental canons in fulfilling their duties. To do less is to fail in their duty to protect and serve people, the environment, and property under their stewardship. Best practices have been developed and are available to professionals in the form of legislation, regulations, specifications, codes, standards, recommended practices, and guidelines. This section will touch on some of those, but it is ultimately the duty of the safety professional to search out the most current and authoritative information on the specific subject in whatever forms available (whether academic textbooks, professional society coursework, networks of professionals, or Internet media), apply the information thus collected to whatever situations they encounter.

The American Society of Safety Engineers has published a Code of Professional Conduct (ASSE 2002). This Code should guide safety professionals' conduct regardless of the state of their affiliation with ASSE; see Sidebar 1.

BEST PRACTICES

Best practices can be summarized by the following:

- Prevention first
- Protection based on evaluated risk second

Prevention First

The concept is simple: eliminating potential (likelihood) or consequence, thus eliminating the risk. But how can this be done consistently?

SIDEBAR 1

AMERICAN SOCIETY OF SAFETY ENGINEERS CODE OF PROFESSIONAL CONDUCT
(ASSE 2002)

Membership in the American Society of Safety Engineers evokes a duty to serve and protect people, property, and the environment. This duty is to be exercised with integrity, honor, and dignity. Members are accountable for following the Code of Professional Conduct.

Fundamental Principles

1. Protect people, property, and the environment through the application of state-of-the-art knowledge.
2. Serve the public, employees, employers, clients, and the Society with fidelity, honesty, and impartiality.
3. Achieve and maintain competency in the practice of the profession.
4. Avoid conflicts of interest and compromise of professional conduct.
5. Maintain confidentiality of privileged information.

Fundamental Canons

In the fulfillment of my duties as a safety professional and as a member of the Society, I shall:

1. Inform the public, employers, employees, clients, and appropriate authorities when professional judgment indicates that there is an unacceptable level of risk.
2. Improve knowledge and skills through training, education, and networking.
3. Perform professional services only in the area of competence.
4. Issue public statements in a truthful manner, and only within the parameters of authority granted.
5. Serve as an agent and trustee, avoiding any appearance of conflict of interest.
6. Assure equal opportunity to all.

Approved by House of Delegates June 9, 2002
Source: http://www.asse.org

Safety professionals can prevent danger by applying best practices and conducting quantitative or qualitative risk assessments. Keep in mind that it is not always possible or even prudent to attempt to eliminate all risk. Protection must be considered, reviewed, and, as appropriate, included in even the best risk-based design.

Protection

Protection takes two forms:

- Constructed protection facilities (including fixed detection and fixed automatic protection, such as sprinklers, water supplies, and automated process control systems) involving automated systems that require minimal human intervention to initiate protective response
- Personnel-operated systems (fire brigades, fire departments and their associated mechanical equipment, and operator/tenant-managed systems such as fire extinguishers and operator-controlled process control systems) that make use of human intervention (nonautomated actions) as part of the initiation of protective response.

Constructed protection facilities should be viewed as mitigating infrastructure losses and not as replacements for sound preventive measures. Life safety codes require minimum exit dimensions and require that exits be operational whenever people are inside a facility. Therefore, enhanced automatic detection and protection systems should not be considered as justification for allowing exits to be blocked or locked.

Likewise, personnel operated systems mitigate unplanned, unforeseeable events, or crises, which to date are best handled by human intervention. Fire, medical, and police departments (and now the Department of Homeland Security) require highly skilled, competent responders. Those events that are foreseeable in nature (wildfires, property damage, medical emergencies, accidents) find responders with clear operating procedures. But what about unforeseen, unplanned events (Bhopal, Chernobyl, September 11)—crises?

When such events occurs, best practice demands that designers and responders focus on a guiding principle.

Guiding Principle

As a principle, *always* provide emergency responders (be they facility operators, tenants, first responders/brigades, or professional emergency responders) with the time, tools, and training needed to mitigate emergencies with the lowest probability of sustaining personal injury. The injury of an emergency responder while responding to a fire emergency is not acceptable.

Even in an unforeseeable event, time can be provided through a prudent and risk-assessed design that considers response activities that address difficult questions such as how to respond to a September 11–type event, criminal acts, or acts of war. As will be seen below, the design phase provides the opportunity for safety professionals to consider design aspects that give responders time to consider risks and response tactics that can mitigate them.

Once time is provided, tools can be identified to protect responders and help them mitigate the emergency. Training that focuses responders on the need to first protect themselves and properly use the tools at their disposal provides a level of assurance that they will resolve emergencies instead of becoming casualties of them.

As a professional, keep in mind that the only operating manual responders may have to rely upon is the time they have to make a judgment to engage, contain, or back away from an emergency. The only resources they may have are the tools at their immediate disposal and the training they have received in the use of those tools. Do not put responders at undue risk in by requiring them to rush response or take chances in order to accomplish their response goal: saving lives and protecting the environment and property. Urgency must be balanced by a focus on response, safety, and incident mitigation.

Facility Layout and Construction Considerations for Fire Safety

When involved with facility layout and construction considerations, the safety professional has numerous codes and standards to inform design considerations. Those codes and standards include those of the National Fire Protection Association, the American Petroleum Institute, and hospitals and hospices, as well as building codes and fire codes.

Various publications also exist to which practicing professionals may wish to subscribe (these may be available online):

- *Building Safety Journal*
- *NFPA Journal*
- *Fire Protection Engineering*

Layout and Spacing

One of the most inherently safe methods of preventing a fire or explosion from spreading to adjacent property or equipment is proper layout and spacing. Layout and spacing can also prevent a fire or explosion by separating potential sources of fuel from potential sources of ignition. Many factors must be considered when arranging and spacing process facilities and equipment, including

- Public safety (public exposure, proximity to highly populated areas).
- Environmental (proximity to sensitive environmental habitats and natural drainage routes—where does firewater runoff go?).
- Economics (brownfields, or developments within existing developed zones, and greenfields, new developments in zones not currently used for industrial or commercial purposes, may have high development costs because of space limitations or the availability of cost-effective land; designers should balance costs against spacing requirements through provisions of spacing between equipment, provision of buffer zones with other facilities/activities, and, in

the case of highly technical and congested facilities, such as offshore oil production platforms, distilleries, and electronics plants, using separation barriers such as fire- and blast-proof walls to mitigate or isolate specific hazards).

- Regulatory requirements (spacing from property lines and easements; see local fire and building codes).
- Process risks (high-pressure plants handling highly volatile liquids or gases will require larger spacing than low-pressure plants or plants not handling hydrocarbons).
- Emergency response capability (facilities lacking strong internal or municipal emergency response support should consider increased spacing and drainage control to minimize losses in the event of a fire).
- Atypical hazardous facilities (unique facilities, including new technology, should include vapor cloud and radiant heat–modeling calculations).
- Within structures (codes typically limit maximum allowable floor area between fire-rated separation walls, separations between floor levels, travel distances to exits, and separation distances between exits; from a design perspective, this is critical in locations where large groups of people can congregate, such as nightclubs or theaters, or where uses of the structure limit ease of mobility through the structure, as in high-rack storage).
- Tall structures (the World Trade Center attack demonstrated the need for designers to consider design and response capabilities to tall structures; keep in mind that responders may need to use the same pathways for access that the occupants are using for egress. FEMA, the Federal Emergency Management Administration, commissioned a report, available on its Web site, providing valuable information into the response of a structure under catastrophic conditions [FEMA n.d.]).

Layout and spacing must be used in conjunction with other risk mitigation practices, including selection of construction materials, fireproofing, and occupancy isolation.

Equipment Spacing, Fireproofing, and Electrical Classification

Equipment spacing is intended to minimize the chance that a vapor cloud from one piece of equipment may come into contact with a potential source of ignition. It is also intended to

- Reduce the potential that a fire on one piece of equipment may damage adjacent equipment.
- Reduce the potential that a blast overpressure may be magnified, resulting in detonation.

Distances are based on fire hazardous zones and historical fire spread. Many engineering design companies maintain their own spacing guidelines. Safety professionals should benchmark their spacing guidelines with others and balance economics and risk.

Fireproofing dimensions are intended to describe when a piece of equipment or structural steel is within a fire hazardous zone and should be fireproofed. Although these distances will often resemble the equipment spacing distances, they will not always be the same. The same guidance is available as is described above.

Electrical area classifications (*e.g.*, NFPA, the International Electrical Code, or IEC) contain distances describing potential areas in which a flammable vapor cloud may form. Fixed electrical equipment within these areas must meet the described electrical classification criteria. These distances do not address fire hazardous zones and hence may differ from both the fireproofing and equipment spacing distances.

Design engineers should be aware of and apply all three considerations when laying out equipment and buildings.

Legal Requirements

Most countries, states, municipalities, and other administrative districts have legislation and regulations addressing layout and spacing for hazardous industries, commercial, institutional, and residential facilities. The NFPA 30 requirement is the spacing of hazardous process facilities from public right of ways and property lines that could be built upon (NFPA 2005b). Verify the regulatory requirements that apply in local building and fire codes, as well as hazardous facility regulations; check local land title offices to identify any existing easements that must be maintained. Regulatory requirements must be met unless variances or other specific approvals are obtained from the enforcing authority. In the United States, the Department of Homeland Security (DHS) Web site provides general requirements for ensuring layout conditions.

Design for Local Conditions

New facilities and modifications should be designed to accommodate local conditions that may affect the safety of personnel and equipment. If the historical record indicates a reasonable probability of extreme conditions, the designer should make appropriate provisions in laying out facilities. Historical records may be obtained by researching local newspaper weather data, county extension agency offices (farm bureaus), and insurance bureau reports. Extreme conditions include hurricanes, tornadoes, violent hailstorms, heavy snow loads, severe and prolonged freezing temperatures, airborne saltwater spray, cooling tower fog (fog developed by a cooling tower when operating during atmospheric conditions that allow air in the vicinity of the tower to reach its dew point), lightning storms, floods, tides, landslides, rock slides, and earthquakes.

Layout Objectives

It is important to protect the public from exposure to spills, fires, explosions, heat, smoke, odors, hazardous releases, and noise. Consider both present and future activities on adjacent property; open fields today might become small communities, shopping malls, or hospitals next year. According to NFPA 30, facilities with fire hazardous exposures (for example, grassland, adjacent processing plants, and flammable construction) should have a 20-foot perimeter firebreak consisting of road, gravel, or similar noncombustible construction (NFPA 2005b).

It is important to protect people and equipment inside facilities. Open space should be provided between process units as a buffer zone so that a fire or release from one unit cannot impact adjacent units. Where possible, space should be left adjacent to roadways to prevent mechanical damage to vehicle traffic. Open areas are purposefully left for maintenance "lay down" areas during maintenance turnarounds to avoid congestion and other hazards. Offices, maintenance shops, and utilities are typically separated from process units by open areas or nonhazardous facilities. It is critical to refrain from increasing risks by building new (or expanding existing) process units within such buffer zones.

The drainage system design must also be considered when laying out process facilities. It can carry burning hydrocarbons or other flammable or combustible liquids closer to adjacent property lines, evacuation routes, and critical equipment within the plant, thus exposing them to fire, heat, and smoke hazards. Designers should also consider the effect these drains may have on the local environment. Consider how the drain will handle significant fire water flows; will it contain the volume? Will the receiving facility of the drain (for example a sump, treatment facility, or lagoon) hold the expected maximum surge flow? If not, the responders should be aware, and mitigation plans should address emergency response impacts.

Protection of Critical Equipment

Critical equipment should be located and designed for maximum protection. For the purposes of this section, critical equipment is defined as that equipment which is necessary for safe, normal commercial facility/industrial plant operation and control, as well as equipment necessary for safe shutdown during plant upsets, fires, and other emergencies. Critical equipment may include instrument air supplies, process control systems, electrical power, substations, main process block valves, certain pumps and compressors, emergency shutdown (ESD) and relief/depressuring systems, and fire water systems.

High-value equipment may warrant extra separation. A risk assessment with cost–benefit analysis should be conducted if loss of the equipment has the potential to significantly impact ongoing operations (resulting in more than one month of business interruption or the like).

Consideration also should be given to commercial operations nearby the public. At an equipment rental commercial operation, the filling of LPG bottles for homeowner use is often done with the public (clients) nearby. A professional reviewing this type of operation should consider the level of training and experience both the operator of the filling operation has and the access the owner of the LPG bottle has to the operation while it is in progress. The designer may consider providing both separation barriers and specific training information.

Protection for Utilities

Power-generating plants, boiler plants, and substations generally serve several process plants or facilities. They should remain operable when one or more of the process plants is in distress. It is prudent to have greater spacing between boiler plants and a high-pressure process plant than between two process plants, because a boiler plant draws significant air flow and is a source of ignition; greater spacing provides distance, reducing the risk of the boiler acting as an ignition source.

Plant Equipment Access

Equipment access and layout is also critical to safe normal operations, routine maintenance and shutdowns, and emergency response.

Spacing and access areas should be checked against the size and turning radii of various cranes and other equipment required for maintenance and shutdowns. In some situations it is prudent to conduct a "dropped object study" to determine if the risk of a dropped piece of equipment is adequately mitigated by the layout and spacing. The concept of the study is to qualitatively review the lifting path that the load will take and ascertain the impact of a dropped load on the area in question.

Security

Plant location and degree of public access may indicate that plant borders and entrances should be supervised and protected with various passive and active systems (such as cameras or motion sensors). Evaluate the risk of fire or other hazards entering the facility through the fence line. The Department of Homeland Security (DHS) provides additional information with which designers and responders can learn about critical facilities security.

Evacuation

Evacuation routes should take into account the location of potential releases, prevailing winds, and drainage patterns. Best practice is to provide a minimum of two evacuation routes remote from each other. In the event that one route is blocked by the incident, people and equipment can make use of the second route. Give consideration to the needs for emergency equipment entering the site and focus efforts on avoiding conflicts between those evacuating and those responding (Cote 2003).

Block Layout and Roads

A large plant composed of several major units should be laid out in a rectangular or block pattern with adequate roadways giving access to major elements. Streets that separate blocks are excellent firebreaks that also facilitate movement and use of equipment.

Roads, pipeway and electrical overpasses, bridges, pipe tunnels, and curves must be capable of handling the largest equipment needed in the area—

cranes, fire apparatus, drilling rigs, delivery trucks, and so on. Ensure that roads and curves are adequately designed to allow adequate turning radii and load support for emergency vehicles, and that additional fencing has not limited access of equipment with overhangs during vehicle movement (turning). Remember that fire trucks are heavy; speed bumps should have approach and departure angles that limit the impact loads transferred between the vehicle and road surface as it passes over the speed bump.

For purposes of fire protection and the economy of operations and maintenance, main access roadways are needed. Within operating areas, every unit and facility should have roadways that allow for ease of access of firefighting equipment. Within enclosed facilities, space considerations may lead to value engineering approaches that reduce access pathways to the minimum dimensions for lift trucks, small vehicles, and normal operations access. These approaches need to be balanced against considerations for response during an emergency. Remember that it is not enough to gain access; there are times when emergency equipment needs to exit the area quickly as conditions change. Do not create traps for firefighters or their equipment. Likewise, operational plans need to ensure that emergency responders consider evacuation for their protection; remember that the best practice for responders is to leave themselves a way out.

During an emergency, it may be necessary to block certain roadways; therefore, each unit should have two or more approaches. Block layout will generally provide two-direction access. Within a given area, consideration should be given to positioning of mobile apparatus; it may be necessary for other equipment to pass this piece of equipment during the emergency. Consider three pieces of equipment that are positioned; what if the middle apparatus needs relocated? Can it pass the other equipment without shutting down their operation?

Likewise, consider facility access. Given security concerns, the desire may be to have limited (and thus easily controlled) access. However, provide two-direction, independent, all-weather access. Remember that snow removal and icing conditions may need to be considered if the secondary access is not normally used.

Restricted and Unrestricted Roads

Motor vehicles may be an ignition source. According to NFPA 30, to minimize the risk of a vehicle igniting flammable material in a facility, any road within 25 feet of potential release sources of flammable or toxic material is designated as a restricted road. Restricted roads should not be used for routine plant travel of operations and maintenance vehicles (NFPA 2005b).

Traffic on restricted roads can be closely controlled by installing barriers and signs at the entrance to the facility and by requiring procedures such as hot work permits for vehicles that must enter the facilities. This usually requires that operators test for flammable atmospheres in the area with appropriate gas detection equipment prior to vehicular entry.

Unrestricted roads are those roads located more than 25 feet from release sources; they do not require traffic control.

Protection of the Public through Spacing

Where practical, use offices, warehouses, and other low-risk buildings as buffers between the process plants and the public. Greenbelts and planted areas are often used to this end. These areas give the industrial facility a friendly, modern image that can make them more acceptable to the public.

Plant-to-Plant Spacing

These are the major issues to consider in plant-to-plant spacing:

- Personnel safety and operator access: operation, maintenance, and emergency response to a facility are all enhanced when adequate space is provided to allow people to do their job without feeling cramped or enclosed.

- Explosion damage prevention: economic considerations dictate that land be utilized efficiently, but consider the effects of an explosion causing damage to other surrounding facilities and possibly increasing damage and lost production.
- Spill spread control: effective spacing can allow spill control to be channeled more effectively.
- Vapor releases and vapor cloud travel: spacing allows vapors to disperse before reaching other potential ignition sources.
- Fire and fire spread: as with explosions, potential fires can and have been known to spread across open areas and ignite other facilities.
- Flood and fire water control: spacing can allow for better flood control if the facility is in a flood-prone area. Spacing also can be used to channel fire water runoff away from equipment and response paths.
- Maintenance access: if equipment has to be pulled, spacing can make the job easy and reduce production downtime.
- Firefighter access: reducing conflict between those evacuating and those entering a site by greater spacing.
- Radiant heat from fire: radiant heat from a fire is the square of the distance; increasing the distance reduces the heat flux and thus the potential for other combustibles becoming secondary fires.
- Economical construction: land prices in most locales are at a premium. Balancing the cost for more land and the risk of closer spacing is the role of the designer. Qualitative and quantitative methods are available to help designers propose alternatives to owner. Some methods available include quantitative analysis of blast over pressure and determining the maximum overpressure a process may generate. Through use of isopleths, the designer can show that overpressure structures may be exposed in a credible scenario. Structural engineers can then design structures that can withstand those forces, providing guidance on impact to those involved.

Temporary Buildings

Temporary buildings should always be of concern for the safety professional. They often are overlooked and defeat the concept of minimizing the employee population around process plants or other hazardous locations. Do not underestimate the impact a temporary office can have at a sewage treatment plant where digesters create methane gas or at small LPG storage facilities. Right-to-know laws and local emergency response committees (LERC) are beginning to recognize these locations as at risk locations for terrorist or other subversive activity. Therefore, temporary buildings should abide by the following guidance whenever they are located adjacent to hazardous (chemical, biological, or physical) facilities. For further information on this topic, refer to the Environmental Protection Agency (EPA) Chemical Emergency Preparedness and Prevention Office (CEPPO) Web site (EPA n.d. a).

Because of their mobile design, temporary buildings can be less resistant to damage from fire or vapor cloud explosions. Blast overpressure or toxic release scenarios usually represent the most stringent spacing criteria for temporary buildings. Electrical components within a temporary building may also provide a source of ignition to a flammable vapor cloud.

Temporary buildings and trailers should not block access and evacuation or egress routes, and vehicle traffic patterns should be evaluated during temporary building siting. Does the temporary building introduce increased private vehicle traffic or result in additional deliveries to the inner areas of the process plant (a security concern)? Temporary buildings and trailers should not be located where personnel frequently have to, or will choose to, traverse high hazard areas in order to access the structures. The addition of temporary buildings may require the installation of barricades or barrier tape in plant areas to minimize the likelihood of increased employee population in high-risk areas of the facility. Emergency routes should not be impeded or blocked by the location of the temporary facilities, and emergency response plans must be updated to include mustering and emergency response procedures whenever temporary buildings are added to a facility.

Prompted by recent high-profile incidents in which fatalities occurred, many jurisdictions are now looking at strengthening the regulations controlling temporary buildings. Designers can again rely on best practice principles—prevention first. Locate the temporary building out of the plant environment. It is worth requiring a longer walk or drive to keep people away from risk; in the event of incident, workers can stay in their work locations to help support the response rather than evacuating the site.

Selection of Building Materials

When selecting building materials, consider the economic life of the project and the use (and potential uses) of the facility, selecting materials that provide

- Fire and blast overpressure resistance
- Sustainable resource

Life Safety in Building Design

The first question to ask in building design is, "What is the building's purpose?" Is the building

- Industrial?
- Commercial?
- Institutional?
- Residential?
- Mixed-use?

Once this is known, life safety considerations can be found systematically documented in NFPA 101, *Life Safety Code*, as well as in local building and fire codes (NFPA 2006). Best practice is to follow these standards and consider the following, based on a cost–benefit analysis:

- Upgrade to noncombustible materials of construction, gaining in decreased insurance premiums, potentially use of recycled materials, and increased life of the facility.
- Install fire water sprinkler systems, gaining in reduced insurance premiums and increased safety for the building occupants—especially residential or institutional buildings where people sleep or may be infirm.
- Use "green" or renewable materials; sustainability is a best practice. Using materials that have been or are recyclable contributes to community well-being.

Design Drawing Review and Approval

Each professional will have his or her own style in conducting drawing review. Individual licensure requirements in different states will define the approval process that professionals use. It is imperative that professionals know the approval process and follow it. Doing so reduces rework, demonstrates increases professional integrity, and solidifies the reputation of professionals with local authorities.

A good reputation can make the difference when trying to obtain approval from local authorities to use technology that has not previously been employed locally. Local authorities often have budget constraints, lack current technology knowledge, and have varying views on the value of new technological development to the community. They may well be caught in the middle of two choices. Integrity and due diligence, coupled with willingness to educate and support local authorities in dispute resolution between various organizations will enhance a safety professional's reputation and may prove the difference between client gaining approval or not.

Professionals should understand the objective of the project. By understanding the objective and goals of the project, they are then in positions to review the details (the plans, drawings, and calculations) that become the final product. Therefore,

SIDEBAR 2

GUIDELINES FOR PEER REVIEW IN THE FIRE PROTECTION DESIGN PROCESS

(Source: SFPE 2002)

1.0 GENERAL

These guidelines address the initiation, scope, conduct, and report of a peer review of a fire protection engineering design. In these guidelines, peer review is defined as the evaluation of the conceptual and technical soundness of a design by individuals qualified by their education, training, and experience in the same discipline, or a closely related field of science, to judge the worthiness of a design or to assess a design for its likelihood of achieving the intended objectives and the anticipated outcomes. A peer review could be conducted on any or all components of a design, such as the fire protection engineering design brief, conceptual approaches or recommendations, application or interpretation of code requirements, or supporting analyses and calculations.

2.0 SCOPE OF A PEER REVIEW

2.1 Overview

The scope of the peer review might be a complete review of the entire documentation, including compliance with applicable codes and standards and the appropriateness of the assumptions, engineering methods, and input data used to support the design. Alternatively, the scope of the peer review might be limited to specific aspects of the design documentation, such as specific models or methods and their associated input data and conclusions drawn from the output data.

Agreement on the scope of the peer review should be achieved between the contracting stakeholder and the peer reviewer. The scope should be explicitly identified at the time of execution of the agreement to undertake the peer review. Any changes to the scope must be agreed to by both the contracting stakeholder and the peer reviewer.

The peer review should be limited to only the technical aspects of the design documentation. The peer review should not evaluate the education, experience, or other personal aspects of the person or company that prepared the design.

The peer review should examine both the internal and external appropriateness of the design. External appropriateness considers whether the correct problems are being solved. Internal appropriateness considers whether the problems are being solved correctly.

2.2 Third Party Inspection vs. Third Party Review

Some stakeholders may also utilize third parties to undertake inspections of completed installations. As the scope of these inspections is typically related to compliance of the completed installation with the previously reviewed design documents, such inspections are outside the scope of a peer review as covered by these guidelines.

2.3 Details of a Peer Review

Whether the scope of the peer review is the complete documentation of a project or some specific aspect of it, the peer reviewer should consider the following details, as appropriate to the design being reviewed:

- Applicable codes, standards and guides
- Design objectives
- Assumptions made by the designer (*e.g.*, performance criteria, design fire scenarios, material properties used in correlations or models)
- Technical approach used by the designer
- Models and methods used to solve the design problem (see Appendix F of the *SFPE Engineering Guide to Fire Protection Analysis and Design of Buildings*)
- Input data to the design problem and to the models and methods used
- Appropriateness of recommendations or conclusions with respect to the results of design calculations
- Correctness of the execution of the design approach (*e.g.*, no mathematical errors or errors in interpretation of input or output data)

3.0 INITIATION OF A PEER REVIEW

3.1 Overview

The decision to initiate a peer review is typically made by a project stakeholder, whose interest might be safety, financial, environmental, or cultural. A peer review is often commissioned by an enforcement official; however, other stakeholders may also commission such a review. This decision usually follows the design development of a project and is occasionally a prescribed part of the design review and approval. A determination to initiate a peer review might be made by a stakeholder during a preliminary project meeting, when presented with a project design brief, or when presented with a complete set of design documents.

3.2 When to Conduct a Peer Review

The decision as to whether or not to conduct a peer review is up to individual stakeholders. The motivation might be a desire to have a better understanding of the quality, completeness, or scientific bases of the design. The decision to conduct a peer review might also be made by a stakeholder who has resource limitations and wishes to bring in outside assistance to evaluate the fire safety features of the design. Another possible reason to initiate a peer review might be to provide additional quality assurance for the design.

4.0 CONDUCT OF A PEER REVIEW

4.1 Standard of Care for a Peer Review

Peer reviews should be conducted in accordance with the SFPE Canons of Ethics. Within the agreed-to scope, a peer review should be performed to the same standard of care that would be expected of a responsible designer during the evaluation of trial designs. Section 2.3 of these guidelines identifies the attributes of a performance-based design that should be evaluated during a peer review. However, if a peer reviewer discovers deficiencies that fall outside of the scope of the review, those deficiencies should be brought to the attention of the contracting stakeholder. A peer review is often intended to ensure that the public's safety goals or the fire -protection goals of other stakeholders are met. Generally, improvement of the design or value engineering is not the purpose of a peer review. The design team will typically accomplish improvement of the design.

4.3 Standard of Reasonableness

Peer reviewers should not be influenced by matters of their own design preference, since there will frequently be more than one acceptable solution to a design problem. Technical issues that would not be expected to have a significant effect on the performance of the design should be identified as observations or findings rather than as deficiencies.

4.5 Confidentiality

Normally, the results of a peer review should be communicated only to the contracting stakeholder. At the discretion of the contracting stakeholder, the results may be communicated to the design engineer of record. In some instances, when dictated by professional ethics, communication of the results to the appropriate enforcement officials may be necessary.

4.6 Intellectual Property Rights

During the peer review process, the peer reviewer should treat the information and materials as confidential and with privilege and should not extract, copy, or reproduce through mechanical, electronic, or other means any or all of the concepts or approaches developed by the design engineer.

5.0 REPORT OF A PEER REVIEW

5.1 Documentation

At the conclusion of a review, the peer reviewer should prepare a written record that identifies the scope of the review and the findings. The report should identify whether, in the peer reviewer's opinion, the design meets the design objectives. The items shown in Section 2.3 of these guidelines should be addressed in the report. Peer reviewers should substantiate any comments on appropriateness by references to published technical documentation.

5.2 Supplemental Information

Resolution of differences in the conclusions between the design team and the peer reviewer may require supplemental technical documentation to resolve the differences. It is important for the designer and the peer reviewer to realize that peer review is only a tool to make an informed decision.

6.0 ADDITIONAL INFORMATION

More information on fire protection engineering, performance-based fire protection design, and peer review in the fire protection design process can be found on the Society of Fire Protection Engineers Web page at http://www.sfpe.org.

they should review design drawings, design calculations, fixed protection (detection and extinguishing) system drawings, and construction drawings and may also be called to review designs when local codes or private insurance demands that owners have a professional review the plans.

Although professionals may wish to review projects independently, peer review is very effective in bringing together people knowledgeable in many facets of a project, improving the final product with the sum of their collective learning.

Sidebar 2 is an extract from the Society of Fire Protection Engineers (SFPE) providing guidelines for a design review process.

Fire Extinguishing System Selection (Sprinkler and Special Hazard)

Overview

This section provides fire extinguishing system selection details and those considerations widely accepted as best practice.

Single Fire Concept

The fire extinguishing system selection for a facility is designed to handle firefighting efforts associated with one major fire at a time. The design capacity of major firefighting facilities is determined by the largest single fire contingency.

However, some system components are sized to handle less significant contingencies. For instance, foam concentrate requirements for special hazard systems are usually determined by a tank fire rather than by the worst contingency, which may be a fire in the process area.

Firefighting Methods

Fire extinguishing consists of one or more of the following methods:

- Quenching (cooling)
- Smothering (blanketing)
- Flame suppression (heat absorption)
- Flame propagation interruption (free radical–chain breaking)

Use of Water as an Extinguishing Agent

Water continues to be the most widely used and universally accepted fire extinguishing material, being both economical and effective. Used properly, it has excellent quenching capabilities, cooling effectiveness, and (for some materials) vapor dispersion characteristics. A gallon of water applied at 50°F and entirely vaporized into steam at 212°F removes over 9000 BTUs of heat. A general understanding of the latent heat of vaporization, and a comparison between water and another common fire extinguishing agent, carbon dioxide, is available online (Wikipedia 2007a).

When sprayed lightly, water cools the surface of combustible or flammable liquids and combustible materials such as paper or cardboard. It may form froth on viscous oils, which can further cool the fuel to below its flash point, resulting in "extinguishing by frothing"—a special case of quenching. Sprayed water is also a flame suppressor. It reduces the size and intensity of the flame and cools and protects materials exposed to flames—both products and structures. Even as a spray, however, water is not usually capable of extinguishing burning gases or the vapors of volatile oils.

Water can also be used as a smothering agent, particularly in fighting fires involving liquids heavier than water (*e.g.*, carbon disulfide).

The steam generated as fire vaporizes water can displace or exclude air, extinguishing the fire by smothering. Smothering is aided by confining the steam generated to the combustion zone.

Flammable materials that are soluble or miscible in water (*e.g.*, methyl alcohol) may, in some instances, be extinguished by dilution.

Layout and Size

In climates where freezing does not occur, aboveground installation of steel fire water distribution lines has the advantages of low initial cost and ease of inspection and repair. Pipelines should be routed to minimize fire or mechanical damage. In cold climates, distribution lines should be buried below the frost line. Recommended depth of cover in feet for fire water systems in the United States is given in Figure A-8-1.1 of NFPA 24 (2007d).

When possible, fire water mains should be arranged in loops around process facilities and industrial or commercial facilities. Shutoff valves should be located to allow isolation of system segments for maintenance purposes or in case of line failures, while still providing water for all facilities. The minimum water rate of a section of pipe out of service should be at least 60 percent of the design rate at design pressure for that area.

At least a 4-inch fire water header should be provided in each facility area to serve incipient stage hose stations. Branch lines to hose stations should be at least 2 inches in size. Fire water mains and headers looping the facilities should not be less than 8 inches in diameter. Laterals supplying single hydrants or monitors should not be less than 6 inches in diameter. In fire water systems using saltwater or nonpotable sources, the pipe diameter should be increased one size to allow for deposits and scale buildup.

Best Practice Considerations for Water Sprinkler Systems

The following are some considerations designers may need when deciding which type of system to use:

1. Wet pipe water sprinkler—Use in locations where combustible construction materials or occupancy warrant rapid response for small, preflashover fires. Typical examples are hotel rooms, hospitals, retail occupancies, and even computer centers. Often designers are concerned about the potential for water to damage electronic equipment, but the standard of reliability of wet pipe systems, combined with the danger posed to a room by fire, make it clear that the sooner an extinguishing agent is applied the less total damage occurs, making it evident that wet pipe systems can be effective in many occupancies.
2. Dry pipe sprinkler system—Use in locations where freeze potential exists or where the nature of the occupancy can tolerate no water exposure potential unless the area is involved in an active fire.
3. Deluge system—Use where massive water flows are needed to provide additional cooling to limit the potential for greater damage. An example would be a deluge system on liquefied petroleum gas (LPG) storage tanks at a bulk storage facility.
4. Pre-action systems—Consider a pre-action system in conjunction with either a dry pipe or deluge system. A pre-action dry pipe system may be used in a warehouse environment where freeze potential exists, but in the event of incipient stage fire detection, the system would be charged with water; should an individual sprinkler head activate, the protected area would have immediate protection. In the case of a deluge system the pre-action would be part of the automatic response. Although this is seldom needed in staffed facilities, unstaffed facilities may warrant the pre-action system's beginning water flow at the incipient stage detection of fire.

Foam Systems

Fixed Systems

Fixed foam systems should be justified based on a risk evaluation. However, semifixed systems are frequently warranted for large storage tanks—especially for low-flashpoint products. The basic design requirements given in NFPA 11 should be followed (NFPA 2005a).

Testing

The quality of foam samples should be tested annually. Reliable testing of foam ensures its effectiveness during an emergency situation.

Testing frequency varies depending on how and where the foam is stored. The manufacturer may, as a free service, test the foam sample for pH, specific gravity, sedimentation, and quality. If the sample fails one of these tests, the manufacturer can perform a fire test at a nominal fee. Contact the specific foam manufacturer for details on sending samples.

Best Practice Considerations for Foam Fixed Systems

The following provides some considerations when contemplating the use of a fixed foam system:

1. Low-expansion foam is used primarily for extinguishing flammable or combustible liquids.
2. Medium- or high-expansion foam may be used to fill basements or ship hulls. These types of foams can also be used effectively to control liquefied natural gas (LNG) spill fires and to help disperse the flammable vapor cloud that accompanies such spills.
3. Foam is conductive and should not be used on electrical fires.

Fixed Water Spray Systems

Refer to NFPA 15, "Water Spray Fixed Systems for Fire Protection," for additional information (NFPA 2007c).

Water in spray form is more effective than straight streams, especially on burning surfaces and on surfaces to be cooled.

In most places, water spray streams can be applied with hand-directed nozzles on hoses or monitors after a fire starts. However, fixed sprays are justified in some facilities. Conditions that may justify fixed sprays include

- Process vessels containing 2500 gallons or more of flammable liquid under pressure where monitor streams cannot reach all exposed surfaces above the normal liquid level.
- Mechanical equipment containing liquids above their auto-ignition temperature or that are volatile and located under other high-value equipment.
- Where pumps are handling hydrocarbons above 600°F or above their auto-ignition temperature.
- Where high-value, long-delivery, critical pumps are located under other high-value equipment, such as air coolers.
- Such critical surfaces such as valves, manifolds, and headers where large volumes of high temperature hydrocarbons are processed and where effective cooling is required.
- Where critical equipment is located on offshore production platforms, such as wellhead production and compression equipment areas.
- Where critical equipment resides in unattended facilities or where firefighting personnel may not be immediately available.
- Where sprays are used as an alternative to fireproofing for structural members or critical instrument cables.
- Where goods are stored in high-rack storage, a typical warehousing practice in some distribution centers.

Water spray systems should be tested at least quarterly to verify that the system is working properly, that nozzles are not plugged, and that coverage is adequate. If the system protects high-value goods or equipment that should not be wetted, air or inert gas may be considered as an alternative—but ensure that no pressure is recorded on the system, making sure that pneumatic pressure does not rupture the pipe or send projectiles from the system.

Fixed Water Spray Requirements

The possible variables encountered during fires of flammable liquids or gases in petroleum handling facilities make precise calculations difficult. Volume, pressure, and temperature of the materials being handled—as well as weather conditions and the structural configuration involved—are all factors that influence water application rates. Other factors to consider include available water supply, drainage capacity, and dispersion of flammable or possibly toxic materials.

The following sections give recommendations for minimum water application rates (densities) for fixed water spray systems.

Spray Systems for Pumps

Pumps and other devices that handle flammable liquids or gases should have shafts, packing glands, connections, and other critical parts enveloped in directed water spray at a density of not less than 0.5 gpm per square foot (gpm/ft^2) of area covered. For a given nozzle, the "area covered" equals the area of the nozzle's circle of coverage at the pump centerline, assuming a horizontal circular pattern of spray coverage at pump centerline.

Interference from piping may require that one or more spray nozzles be located higher or lower than the normal 4 or 5 feet above pump centerline. Narrow-angle nozzles have a long reach and a narrow coverage area. Wide-angle nozzles have a short reach and a wide coverage area.

Lateral lines coming from the top of the header minimize most nozzle plugging problems. Other recommended features are main lines sloped to drain and a flush valve at the end of each main line.

Spray Systems for Vertical Vessels

Water should be applied to vessels at a rate of not less than 0.25 gpm/ft^2 of exposed uninsulated surface. To ensure adequate coverage, the horizontal distance between nozzles must be close enough to permit the meeting of spray patterns. The vertical distance between nozzles may be as much as 12 feet, provided rundown is expected. Nozzles should be no more than 4 to 6 feet from the vessel.

Spray Systems for Spheres or Vessels

Water sprays on spheres or horizontal cylindrical vessels should be capable of discharging 0.25 gpm/ft^2 surface area of the upper half of the vessel.

Surfaces of the lower half are not always wetted by water rundown from above; additional coverage may be required by handheld hoses or monitors if the vessel is likely to be less than half full of liquid. Grading and drainage from under vessels are important factors to minimize heat input to the lower vessel surface.

Water sprays are not effective in providing cooling for high-velocity, jet-impinging fires. The velocity of jetting gases blows the water spray droplets away from the vessel shell. For LPG storage vessels, water monitors are required in addition to sprays. Refer to API 2510 (2001) for additional information.

Deluge Systems for Spheres

Deluge systems, or high-capacity water spray systems, are preferred on LPG storage spheres (tanks). A density coverage of 0.25 gpm/ft^2 of surface area above the equator is recommended. For example, about 1600 gpm of fire water would be required to adequately protect one 65-foot diameter sphere.

The main components of a water deluge system are

- An adequate water supply line to the top of the sphere, terminating in an open-ended pipe that spills the water onto the top of the sphere.
- Weir box for even distribution of water over the top of the sphere, or two or three water distributor rings spaced above 2 feet apart to further distribute flow over the sphere surface. Provide drain holes to prevent retention of rainwater.
- A valve and drain line in the water line located at least 50 feet from the sphere. This is normally a quick-opening (quarter-turn) manual valve but could be operated by a fire detector in unattended locations. The valve must be located away from the drainage path from the sphere.

Structures and Miscellaneous Equipment

Where projections such as vessel access flanges, pipe flanges, support brackets, or vessel legs obstruct water spray coverage (including rundown on vertical surfaces), additional deflectors or nozzles may be needed to maintain the wetting pattern of pressure-containing surfaces.

Nonmetallic Electrical Cable and Tubing Runs

Open cable trays/conduit banks (unfireproofed) may be protected by fixed sprays when there is potential for fire exposure, such as above hot-oil pumps or near furnaces. The preferred protection is to route critical control and power wiring away from fire risk areas. Where routing outside a fire risk area is not feasible, use properly installed sprays instead. An application rate of 0.30 gpm/ft^2 of projected area is recommended.

System Components

Components of a fixed water spray installation should be standardized to provide an interchangeable system. Systems may be operated automatically or manually, depending on the anticipated degree of hazard.

According to NFPA 15, equipment exposed to corrosive atmospheres should be constructed of corrosion-resistant materials or covered with protective coatings to minimize corrosion (NFPA 2007c).

Pipe, tubing, and fittings should be designed to withstand a working pressure of not less than 175 pounds per square inch, gauge (psig). Include a strainer and full-flow bypass in the system.

Nozzle Selection

Nozzles producing a solid cone spray pattern are effective for most fire control and surface cooling applications. However, flat spray or other patterns may be more suitable for certain applications.

Select a nozzle with an angle of discharge and capacity at the pressure available that gives the needed density on the surface, considering the distance to the nozzle mounting location.

Spray nozzles are manufactured in a variety of configurations. Take care to ensure proper application of the nozzle type. Distance of "throw" or location of the nozzle from the surface is limited by the nozzle discharge characteristics (BETE Nozzle Company n.d.).

Select nozzles that are not easily obstructed by debris, sediment, sand, and rust deposits in the water. The nozzle orifice size should be at least 3/8 inch. Use the largest practical nozzle size. Installing a few large nozzles is preferable to installing a greater number of smaller nozzles. Nozzles with no internal parts are less likely to plug. Include approved strainers with full capacity bypass and flush-out connections in cases when debris may cause plugging problems.

Stainless steel nozzles are recommended, but brass and other materials are available.

Water Supplies

The type of water used is important. Freshwater has the advantage of causing less plugging and corrosion than salt or nonpotable water. If salt or nonpotable water is used, a freshwater flush is recommended after each use of the system, especially the sections downstream of the main control valve that are exposed to the atmosphere. Left unflushed, these sections can suffer scale build-up that can reduce the spray system performance.

The water supply flow rate and pressure should be able to maintain water discharge at the design rate and duration for all systems designed to operate simultaneously. Allow for the flow rate of hose streams and other fire protection water requirements when determining the maximum water demand for fixed sprays.

Manual control valves or remote actuation points should be located at least 50 feet from hazards and identified to ensure accessibility during emergencies. When system actuation is automatic, provide manual overrides.

The water supply should be from reliable sources, such as connections to city water systems, fire pumps, or to fire-department connections for mobile fire pumpers.

Size of System

Protect separate fire areas with separate spray systems. Keep single systems as small as is feasible—generally consider independent spray systems for each piece of equipment (for example, one spray system per vessel, pump, or similar piece of equipment). The reason is that one large system, if damaged, may be unable to adequately protect the facility, let alone the individual piece of equipment. By keeping systems small, and providing individual systems for individual tanks and separations in high rack storage, as well as other such protections, the possibility that an incident may cause the overall protection to fail is severely limited.

Separation of Fire Areas

Typical fire areas are

- Operating sections that can be shut down independently of other sections.
- Offshore platform modules.
- Process sections such as distillation, exchanger banks, manifolds, or reactor sections.
- Natural firebreaks (such as pipeways and aisleways).
- Rack storage units.

Drainage

It is important to make provisions for drainage of water or foam solution that is likely to be discharged into a fire area. Drainage capacity should allow for the expected amount of spilled oil plus fire water flows.

As an example, drainage for the 65-feet diameter vessels mentioned above would require containment of the liquefied petroleum gas spill, along with the 1600 gpm of cooling water.

Automatic Sprinkler Systems

In permanently staffed industrial facilities, sprinkler systems are generally not automatic. However, in offices, laboratories, and warehouses, automatic heat-actuated systems are commonly used. Sprinkler system design should follow NFPA 13 (2007b).

Multistory living quarters on offshore facilities should be sprinklered. Such systems are normally freshwater with provision for saltwater makeup if the system is activated.

The need for actuation of systems to transmit an alarm to a fire station is based on local code requirements and depends on whether the facility is manned continuously. The more usual method is to notify the local fire department by telephone.

The following material does not instruct professionals how to design automatic systems; but this knowledge is available via a number of online courses.

Fixed Foam Systems

From a risk management perspective, fixed foam equipment is seldom recommended for use except on large floating roof hydrocarbon storage tanks (over 120 feet in diameter) and in special situations such as manifold pits or onboard tank vessels. NFPA 11 provides further information on this subject (NFPA 2005a). The role of foam as an extinguishing agent is covered later.

Fixed Halon Systems

Halons are vaporizing liquids that chemically inhibit combustion by interrupting flame propagation much as dry chemicals do. The two most widely used Halons are Halon 1301 and Halon 1211. Their installation and use is discussed in NFPA 12A (NFPA 2004).

Because Halon can harm the ozone in the atmosphere, as of January 1, 1994, Halon can no longer be legally produced in the United States, among many other developed countries. No new Halon systems should be installed (Cote 2003).

Halon Alternatives

Some Halon substitutes have received EPA approval as part of the Significant New Alternatives Program (SNAP). These products require significant redesign of existing fixed suppression equipment. Approved substitutes require storing and dispensing from 1.7 to 10 times the volume of Halon 1301. A substance that allows simple exchange of gas in existing storage cylinders does not today exist. However, the list is changing, and information quickly becomes obsolete. The EPA's Significant New Alternatives Policy Web site provides current information about accepted and approved alternatives. The National Fire Protection Association (NFPA) also provides guidance in NFPA 2001 (2004).

Existing Fixed Halon Systems

Removal of existing Halon systems is not required. Maintaining an existing system, however, could be expensive. If unnecessary releases occur, lost Halon may be sourced but will be expensive. The fire protection professional involved in a review of existing Halon systems should undertake a risk assessment and apply risk management principles to discover whether the existing system can be removed, alternatives can be installed, the system can be converted (if the need is discovered to be great) to manual release, or a limited surplus supply of Halon can be sourced and stockpiled from known vendors.

Fixed Dry Chemical Systems

Fixed dry chemical systems may be installed to protect an area of unusual hazard where the powder would not cause additional damage or where other media would be substantially less effective. Systems can be installed either inside or outside and should be designed in accordance with NFPA 17 (2002a). Consider the effects of wind for outdoor systems. A disadvantage of fixed dry chemical systems is that they must extinguish a fire with one discharge, else it will continue unabated.

Fixed Carbon Dioxide Systems

Carbon dioxide (CO_2) extinguishes almost entirely by smothering, although it does have a negligible cooling effect of about 100 BTU per pound (refer to the latent heat of vaporization mentioned above). Liquid carbon dioxide is stored under pressure in steel cylinders. When the valve on the cylinder opens, the rapid expansion of the liquid into gas produces a refrigerating effect that solidifies part of the carbon dioxide to a "snow" that soon sublimes into gas, absorbing heat from the burning material or surrounding atmosphere. The gas extinguishes fire by reducing the oxygen content of surrounding air below the flammable limit of the fuel.

Unless this concentration of gas is maintained for an extended period, carbon dioxide does not normally extinguish fires in materials that smolder or produce glowing embers, such as paper and wood. Its greatest effectiveness is on flammable liquid fires that do not involve material that might cause a reflash after the CO_2 has dissipated. It is especially suitable for laboratories and has wide application in the protection of delicate electrical and electronic equipment, where cleanup after extinguishing is an important consideration.

Carbon dioxide is clean and leaves no residue to damage equipment. It is recommended that all entry points to a protected space have signage warning people entering of the risk in event of discharge. A sample warning may read as follows:

> *Warning:* Air into which CO_2 has been discharged will not sustain life. CO_2 cannot be used safely in closed staffed facilities unless warning alarms are first sounded and personnel are either evacuated before release or don self-contained breathing apparatuses.

Carbon dioxide systems should be designed in accordance with NFPA 12 (2008a).

Design considerations include a number of factors, including quantity of stored CO_2, method of actuation (manual, automatic), the location of predischarge alarms (to allow adequate egress time), ventilation shutdown, pressure venting (so as not to overpressurize a room), and things that will help ensure personnel safety and effective extinguishing (Cote 2003).

Fire Detection Systems

Fire or smoke detection systems are desirable in installations where fire might go undetected for considerable time or in industrial and commercial facilities with significant public exposure or potential environmental impact. In some areas, detection systems may be required by the authority having jurisdiction (*e.g.*, local building department or fire department).

Fire detection systems should be considered for the following types of industrial or commercial facilities:

- Petrochemical plants—Consider both flame and fire detection. Typically need exists to identify fires at the incipient stage and eliminate fuel sources while applying extinguishing agents.
- Electronics plants—Consider fire and smoke detection. Typically concerns are of electrical and electronic failures that could lead to fire but which are generally best detected from the products of combustion generated, by means of smoke detectors.
- Warehousing—Consider smoke detection and possibly fire detection. The nature of each warehouse (*e.g.*, high-rack storage or private individual unit storage) will need to be considered.

When selecting a detector for a specific application, consider the location of probable fire, as well as whether immediate flame or smoldering is likely and with what precision the location of fire can be pinpointed. Detectors can be made to sound an alarm locally or at a remote location, as well as to shut down and depressurize equipment (*e.g.*, pumps and compressors), close valves, shut down ventilating systems, discharge extinguishing agents, and perform still other operations.

All fire detection and alarm systems except those detectors having parts that destruct on exposure (as by melting), should be tested periodically by causing them to actuate. Develop a suitable test program for each unit that will ensure that detectors, alarms, and other intended functions will operate in case of fire. Test detectors at least every six months—or more often, depending upon the location and the environment to which the device is exposed. Maintain the test records and correct any deficiencies immediately.

Explosion Suppression

Suppression of explosions is possible under certain conditions, because a short but significant period of time elapses before destructive pressures develop. If conditions are right, it is possible to use the time available to operate a suppression system.

Effective use of the rate of pressure rise to suppress an explosion requires three major considerations in the design of suppression systems (Cote 1990):

1. The explosion must be detected in its incipient stage to allow time for operation of suppression equipment. Because of the short time available, detection and suppression must be automatic, equipped with provisions to discriminate between an explosion and ambient variables that normally exist.
2. The mechanism for dispersing the extinguishing agent must operate at extremely high speed to fill the enclosure completely within milliseconds after detection of the explosion. The detection must automatically actuate to prevent time lag. The extinguishing agent must be dispersed in a very fine mist form at rapid speed—normally through the use of an explosive release mechanism.
3. The extinguishing agent is normally a liquid that is compatible with the combustion process to be encountered. Methods integral to the suppression mechanism are the same as those use to extinguish fires—cooling, inerting, blanketing, and combustion inhibiting.

Explosion suppression systems are not in general use in the petroleum industry, but they may be considered for the protection of high-hazard, high-value operations where an explosion would have very serious consequences and where normal methods of fire protection are not adequate. Explosion suppression systems are more commonly encountered in dust-handling processes and airplane maintenance facilities. NFPA 69 (2008b) provides further information on this subject.

Fixed Fire Water Systems

All new installations should be flow-tested with water to ensure that nozzle layout, discharge pattern, and overall performance is adequate. When practicable, the maximum number of systems that may be expected to operate in a fire should be tested simultaneously to ensure adequate water supply. Open the main supply header flush valves at the start of testing.

Test water spray systems at least quarterly to ensure reliability. Note that no leakage or misalignment problems have been experienced as a result of testing or using water sprays over hot pumps. The rain-like drops of water do not quench localized areas and thus are less risk than hose streams.

Where sprays are to be tested on painted surfaces, discoloration from rust and pipe deposits can be minimized by testing during local rainfall or by wetting the surface with clean fire water before testing. A deluge system can be pickled to remove rust and silica from piping and storage vessels. Pickling solution is generally a low concentration of passivated hydrochloric acid (and sometimes hydrofluoric acid). Remember to flush piping well after treatment is finished and dispose of testing liquid in an environmentally friendly manner.

Inspector's Test Drain

Fixed systems have a test loop that allows periodic inspection and maintenance of the system. The inspector's test drain provides the mechanism to ensure that the system is functional (see Figure 1). Monthly, the system should be checked to ensure that valves are in their correct position (either open or closed). Refer to NFPA 25 (2002b) for more details.

With recent increase in home sprinkler systems, the responsibility to inspect the system is placed upon the home owner. Local and state governments have numerous programs by which they educate the public. As a professional it is appropriate to support these efforts and, when communicating with clients, to provide them with information.

MATERIAL STORAGE

Combustibles

Because many different materials and business needs exist for storage of combustible materials, it is beyond the scope of this section to provide information that will allow designers or fire protection professionals to consider every area. However, the following principles should apply to prevention and protection

- Eliminate the potential (likelihood) or eliminate the consequence, thus eliminating the risk. Consider storage volumes, storage heights, and fire separation areas.
- Do not put responders at undue risk by forcing them to rush their response. Urgency tempered by a focus on response, safety, and mitigation of the incident is the goal. High-rack storage presents its own set of issues. Consider responders battling a fire and the impact of the fire on supporting structures.

The solutions available to professionals include (Zalosh 2003)

- Test methods to provide a better measure of commodity protection requirements.
- New, less flammable (and potentially less environmentally damaging) packaging materials.
- Effective sprinkler systems for warehouse storage.

Storage best practices include (Zalosh 2003; Cote 2003)

- Understanding how commodity characteristics and storage height and configuration influence fire hazard and protection requirements.
- Documenting the heat release rate characteristics and sprinkler system requirements for representative ordinary combustible and plastic commodities and typical storage configurations.

- Identifying and working with local fire officials to understand what the storage facility handles and how the facility-inherent fire safe design (one should exist that is based on the prevention and protection principles noted above) can operate in unison with the responders.

Flammable Liquids and Gases

Storage of flammable liquids and gases is well covered by NFPA 30 (2005b) and the *Fire Protection Handbook* (Cote 2003). These are available from NFPA and other online sources (IHS n.d.).

Inspection of Building for Fire Safety

Table 1 informs fire protection professionals of the types of risks potentially (and commonly) found in buildings.

Testing of Fire Protection Equipment

Fire Extinguisher and Equipment Inspection and Maintenance

General

All fire extinguishers shall be inspected at least monthly or at more frequent intervals where conditions warrant, and they shall be given more detailed maintenance servicing annually or whenever the monthly inspection indicates need. This is a legal requirement (Cote 1990).

Table 2 provides an overview of the inspections that should, at minimum, be done.

Inspection

The inspection is a "quick check" to ensure that an extinguisher is in its designated place, is accessible, has not been actuated or tampered with, and has no obvious physical damage, corrosion, or condition that will prevent operation. The frequency of inspections

TABLE 1

Sample Fire Protection Checklist

Fire Protection Checklist			
Checklist completed by:			**Date:**
Buildings:			
Item	**Yes**	**No**	**Corrective action or observations**
Are emergency response plans available?			
Are egress routes clearly identified?			
Is firefighting equipment clearly identified?			
Is egress for persons with disabilities considered, including for those persons who have temporary disabilities caused by accident, illness, or treatments (*e.g.*, broken legs, colds, surgeries)?			
Are infrastructure facilities (*e.g.*, boiler rooms and electrical and communications closets) in good repair and housekeeping in good order?			
Are exits unlocked and open in the direction of travel?			
Are audible and visual alarms functional?			
Are fire protection systems and portable firefighting equipment in good order?			
Are personnel identified to provide evacuation directions and subsequently account for people?			
Are assembly locations clearly marked and equipment and resources readily available for conducting triage or communicating actions to be taken?			

TABLE 2

Fire Extinguisher Inspections

	Water				Dry Chemical Units		
List of Extinguisher Inspections	**Stored pressure**	**Pump tank**	**Carbon dioxide**	**Halon 1211**	**Disposable shell**	**Cartridge operated**	**Stored pressure**
1. Ensure extinguisher is in designated place, clearly visible, and accessible	×	×	×	×	×	×	×
2. Ensure visual seal(s) are intact	×	×	×	×	×	×	×
3. Check pressure indicator or gauge	×	×	×[1]	×	×		
4. Take from bracket, heft for proper weight	×	×[2]	×	×	×	×	
5. Examine for damage, corrosion, etc.	×	×	×	×	×	×	×
6. Check nameplate and instructions for legibility	×	×	×	×	×	×	×
7. Examine hose for cuts, weather cracks, etc.	×	×	×	×	×	×	
8. Check nozzle and hose for plugging and operation	×	×	×	×	×	×	×
9. Check mounting bracket	×	×	×	×	×	×	×
10. Return extinguisher to proper location	×	×	×	×	×	×	×
11. Record inspection on tag and Check Sheet	×	×	×	×	×	×	×
12. Initial and date Monthly Inspection Check Sheet	×	×	×	×	×	×	×

[1]If equipped with pressure indicator.

[2]CO_2 extinguishers should be weighed every six months.

(*Source:* Cote 2003)

will vary based on the needs of the situation; they should normally be conducted at regular intervals not to exceed one month. Inspections should be completed by facility operators (as opposed to outside contractors). This will familiarize operators with the location and operation of extinguishers. The value of an inspection lies in the frequency, regularity, and thoroughness with which it is conducted. The inspector should include inspection items appropriate for the type of extinguisher in use (refer to the above list).

In locations where caking of the powder in cartridge-operated dry chemical extinguishers is not proven to be a problem, do not open the extinguisher to dump and screen the powder at monthly inspections. This procedure itself can cause caking by allowing moisture to enter.

Maintenance (Cote 1990)

The maintenance check is a "thorough check" of the extinguisher that is performed annually. Its purpose is to give maximum assurance that an extinguisher will operate effectively and safely. A complete maintenance check should also be performed whenever the need is indicated by the monthly inspection.

Maintenance checks should be performed in accordance with the instructions on the manufacturer's label and the requirements of NFPA 10 (2007a). Following are some specific comments on the maintenance of selected types of extinguishers.

Stored-Pressure Water or Antifreeze

These extinguishers need not be recharged annually. If the seal is broken or missing, the liquid level should be checked by weight or observation, recharged as necessary, and resealed.

Stored-pressure water extinguishers should also be serviced when

1. Partially or completely discharged.
2. Pressure is below the operating range, or weight is less than required.
3. Showing signs of physical damage or corrosion.

Pump Tank Water Extinguishers

The water charge should be dumped and replaced every 12 months. These extinguishers should also be serviced when

1. Partially or completely discharged.
2. Pressure is below the operating range, or weight is less than required.

3. Showing signs of physical damage or corrosion.

The pump should be tested periodically by placing the nozzle in the tank fill opening and operating the pump for a few strokes. Hydrostatic pressure testing is not required for these extinguishers.

Carbon Dioxide (CO_2) Extinguishers

CO_2 extinguishers need not be recharged annually. The weight of the full charge and the gross weight are normally indicated on the extinguisher. CO_2 extinguishers should be weighed every six months and should be serviced when

1. Partially or completely discharged.
2. The weight is less than required (90% of the full charge indicated on the extinguisher).
3. Showing signs of physical damage or corrosion.

If a seal is broken or missing, the extinguisher should be leak-tested and the weight checked. The unit should be recharged if a loss of 10% or more of the weight of the full charge has been sustained. The extinguisher should then be resealed.

New or recently recharged extinguishers should be inspected quarterly for the first six months to detect any leakage; if none occurs, semiannual weighing should be adequate. Special equipment is required to recharge this type of extinguisher; all recharging must be done by a trained individual.

Halon 1211 (bromochlorodifluoromethane) Extinguishers (Cote 2003)

These units need not be depressurized and recharged annually. They should receive a thorough maintenance servicing and be recharged if there has been a weight loss of 5% or more of the Halon charge. Halon 1211 extinguishers should also be serviced when

1. Partially or completely discharged.
2. The weight is less than required (95% of the charge weight indicated on the nameplate).
3. Showing signs of physical damage or corrosion.

As a reminder, new Halon 1211 extinguishers should not be sourced. If a Halon 1211 extinguisher is used, the operator should review protection needs while planning to replace the extinguishing agent with a suitable alternative for the location being protected.

Dry Chemical Extinguishers (NFPA 17 (2002a))

Factory-sealed disposable extinguishers should be inspected and maintained in accordance with manufacturers' nameplate instructions. The annual maintenance check normally consists of removing the head from the cylinder to verify that the extinguisher has not been discharged. If the factory seal assembly has not been perforated, the extinguisher should be reassembled, resealed, and returned to service.

Weighing the cylinder is not recommended (unless required by law), because

1. Without special scales, loss of pressure on a weight basis is difficult to determine because of the small amount of pressuring gas used.
2. The shells are factory-pressurized and tested for leakage.
3. The probability of leakage through a factory installed and tested cartridge seal assembly is considerably less than through valve assemblies such as those used on rechargeable extinguishers.

If an extinguisher is equipped with a pressure gauge or indicator, visually inspect the gauge. Those shells indicating less than rated pressure should be depressurized by operating the extinguishers and then discarded.

Stored-pressure extinguishers equipped with pressure indicators or gauges are not normally required to be depressurized each year. However, in some jurisdictions, it legally required that all dry chemical extinguishers—except factory-sealed disposable types—be depressurized, the chemicals removed, and the extinguisher given a thorough examination at each annual maintenance check.

Each extinguisher should be thoroughly examined whenever the monthly inspection indicates a need or whenever the pressure indicator or gauge shows that extinguisher pressure is not within the operable range. Stored-pressure dry chemical extinguishers on a 12-year hydrostatic test schedule should be emptied every six years and subjected to the maintenance procedures listed in NFPA 10 (2007a).

Stored-pressure dry chemical extinguishers should be serviced annually or when

1. Partially or completely discharged.
2. The weight is less than required.
3. Showing signs of physical damage or corrosion.
4. A seal is broken or missing, in which case the unit should be leak tested and weighed, refilled to correct weight (as required), and resealed.

To recharge most models of the stored-pressure type, invert the unit from its normal operating position and open the valve to release any pressure remaining in the extinguisher. With the extinguisher in an upright position, remove the valve mechanism and dump out any remaining chemical into a clean container. After verifying that the chemical agent is all free-flowing, return it to the extinguisher and make up any deficiency with new chemical to the level designated by the manufacturer. Disassemble the valve to clean chemical and all other material off all gaskets and other sealing surfaces before reassembling and replacing valve assembly.

Repressurize the extinguisher to the pressure designated on the unit, following the detailed procedure printed on the label. Most types of extinguishers are pressurized with nitrogen, but some makes may be approved for air. Where air is obtained from an ordinary compressor, some type of moisture trap should be used in the air line to prevent moisture from entering the chemical chamber.

The operating mechanism should be sealed, the date of recharge recorded on the tag, and the tag signed. The pressure gauge should be checked about 24 hours after pressurizing to detect any leakage.

Cartridge-operated dry chemical extinguishers should be given a thorough maintenance check in accordance with the manufacturer's nameplate instructions and the requirements of NFPA 10—annually or whenever a monthly inspection indicates the need (NFPA 2007a). This maintenance check should include removing the cartridge, checking that the seal has not been punctured (some state or local regulations may be more stringent, requiring weighing of the cartridge and replacement under specific conditions), looking for and noting signs of corrosion, especially around moving parts and the bottom seam. The moving parts should be operated to make certain they are free.

Do not apply oil or grease to any moving part of the extinguisher. If these parts cannot be restored to good working condition by cleaning and buffing with emery cloth, steel wool, or such methods, they must be replaced.

Check condition of the hose. The hose and internal piping should be checked annually for possible obstruction with packed chemical by blowing air through the nozzle, hose, and piping.

Cartridge-operated dry chemical extinguishers should be serviced annually or when

1. Partially or completely discharged.
2. Showing signs of physical damage or corrosion.

If the seal is broken or missing, examine the seal disk on the expellant cartridge to be sure it has not been punctured; check the chemical's condition and level.

To recharge most models of the cartridge-operated type, replace the cartridge with a full cartridge of the correct size and with the correct inert gas for the service, filling the dry chemical container to the prescribed level or weight with recharge material.

Before removing the fill cap, generally release any residual gas and blow out the hose. However, absence of a release of gas may indicate a plugged hose, and pressure may still remain in the unit; the fill cap should be unscrewed cautiously to allow any gas to escape through the vent holes in the cap.

Dump all remaining chemical into a clean container and make certain it is all free-flowing. In all cases, the nozzle, hose and piping should be checked for freedom from obstructions by blowing air through them as described earlier prior to recharging. Refill with the required amount of chemical.

When installing the replacement gas cartridge, the hose and nozzle must be placed in the normal storage position to prevent the plunger from rupturing the cartridge seal. Detailed recharging instructions are given in manufacturer instructions on extinguishers.

The date of recharge should be recorded on the tag, and the tag should be signed. The operating mechanism should be sealed to make inspections easier.

TABLE 3

Hydrostatic Test Interval for Extinguishers

Extinguisher Type	Test Interval (Years)
Stored-pressure water and/or antifreeze	5
Carbon dioxide (CO_2)	5
Dry chemical with stainless steel shells	5
Dry chemical, stored-pressure, with mild steel, brazed brass, or aluminum shells	12
Dry chemical, cartridge or cylinder, operated, with mild steel shells	12
Halon 1211 (bromochlorodifluoromethane)	12

(From NFPA 10. 2007a. National Fire Protection Association, Quincy, MA. Used with permission.)

Hydrostatic Pressure Test

Fire extinguishers other than disposable shells and pump tank extinguishers should be hydrostatically tested at intervals not exceeding those specified in Table 3. Hydrostatic tests should also be performed any time an extinguisher shows evidence of corrosion or mechanical damage, or the unit should be replaced entirely. Extinguishers under the jurisdiction of the U.S. Department of Transportation (DOT) are subject to its test requirements. Most jurisdictions make reference to or follow the testing requirements specified in NFPA 10 (2007a). However, some jurisdictions may differ. Determine what requirements are applicable to particular facilities.

Record Keeping

According to NFPA 10 (2007a), each extinguisher should have a securely attached tag or label indicating the month and year the annual maintenance check was performed and identifying the person performing the service. The same tag or label should indicate whether recharging was performed. Refer to NFPA 10, "Portable Fire Extinguishers" (2007a), as well as to local (city, county, and fire marshal) codes. Discover what requirements are applicable to particular facilities.

A file should also be kept for each extinguisher that should include the date of annual maintenance check and the name of the person or agency performing the work, the date last recharged, the date of the hydrostatic test, and a description of its physical condition. Consider also including the monthly "quick check" inspections on this file record (see Table 3).

Dry Chemical Extinguisher Clogging and Packing

Nozzle clogging has been attributed to failure to completely clear hoses of residual chemical after use, as well as to inverting extinguishers or handling them roughly so as to permit chemical to work up into the outlet hose.

Because dry chemical extinguishers (except stored-pressure types) are not hermetically sealed, a chance always exists that rain or excessive atmospheric moisture may enter the case and be absorbed in the powder, causing it to cake. Reports of some difficulty have been received, principally from regions of high humidity or extreme temperatures.

There have also been some reports of packed dry chemical plugging the outlet elbow and rendering the unit inoperative in extinguishers carried on automotive equipment.

Various procedures have been used to detect these conditions. The presence of lumps of caked powder can be detected by screening; one extinguisher user has adopted an inspection routine that involves dumping the chemical out into a clean container through a funnel incorporating a 16-mesh screen. Lumps that do not break up are discarded. Before refilling, the hose is shaken to dislodge any residual chemical. This user has not experienced clogging of the outlet elbow as a result of vibration and does not feel it necessary to blow through the hose to make sure that it is free. Extinguishers on automotive equipment are checked once a month and all others each six months.

Another user, however, has concluded that packing caused by vibration is particularly important. This conclusion was based on failure of several extinguishers to discharge properly after they had been carried for several months on automotive equipment. To guard against a repetition of such failure, this user instituted an inspection routine involving blowing back through the outlet nozzle with a tire pump.

Some users have concluded that the formation of hard lumps in the powder is not a significant problem. Their testing routine employs the tire pump as described above. The top of the extinguisher is removed before the test so that the operator can look in and verify that the chemical fluffs up as the air is pumped through it. Extinguishers on automotive equipment are carried upright and are inspected monthly. Extinguishers in stationary installations are checked every three months.

Operability and Reliability Experience

The petrochemical industry has experienced high reliability of cartridge-type fire extinguishers in many different indoor and outdoor environments. However, there have been some problems—most related to inspection and maintenance. The solutions discussed below can help gain knowledge from these experienced problems and avoid similar problems in the future.

Maintenance difficulties with cartridge-operated dry chemical extinguishers have involved loss of pressure in actuating cartridges, rusting or sticking of operating parts, and packing or caking of chemical powder caused by moisture or other causes. Inspection routines require the following five basic steps:

1. Check the CO_2 cartridge.

   Checking the CO_2 cartridge involves weighing to the nearest 1/4 ounce, something that requires an accurate scale. In general, factory-filled cartridges should be used in the maintenance of dry chemical extinguishers. Where this is done, periodic weighing of the cartridge is not considered necessary because of the improbability of leakage. A visual inspection of the cartridge seal disk assembly, discovering whether it has been perforated or damaged, is adequate.

   Cartridges recharged by company employees or extinguisher maintenance organizations should have their weights checked annually.
2. Check the condition of the puncture pin mechanism.

   The puncture pin mechanism on early models of Ansul extinguishers gave considerable trouble from corrosion in cases in which the units were unprotected and the climate severe. This trouble has been partially corrected in later models by the use of corrosion-resistant materials and rubber seals. (In cold climates an extinguisher may also be rendered temporarily inoperative if snow or moisture freezes on parts of the nozzle or on the puncture pin mechanism.)

   To check, break the tamper seal and remove the CO_2 cartridge. Remove the safety pin (if the extinguisher is so equipped) and work the mechanism several times to verify that it moves freely. Check the cartridge gasket and inspect the seal disk to verify that it is intact. Replace the tamper seal and the safety pin, and replace the cartridge.
3. Check the functionality of the hose nozzle valve.

The hose nozzle valve should be worked to see that it operates freely. Manufacturer instructions caution against the use of oil and suggest that hoses and nozzles be shaken to dislodge any loose powder that may have worked into the hose.

4. Check the hose and outlet tube for clogging.
5. Check the condition of the dry chemical itself. It is here that differences in service experience and testing routines are most marked.

Investigating and Reporting on Fire Losses

Overview

This section discusses the general procedure for investigating a fire and preparing the fire report. Note that the process described here can be used in any incident or accident investigation.

When a fire occurs it is important to learn why it happened in order to prevent similar fires in the future. In all but the most obvious and lowest-risk cases, investigation is warranted.

Fire Investigation Flow Chart

Figure 1, below, is an example of a fire investigative flow chart.

Data Collection (BoM 1991)

It is important to set aside a few hours to spend concentrating on the fire scene. Taking certain basic steps and concentrating on the evidence at hand often leads to ascertaining the primary cause of the fire. The following steps are recommended:

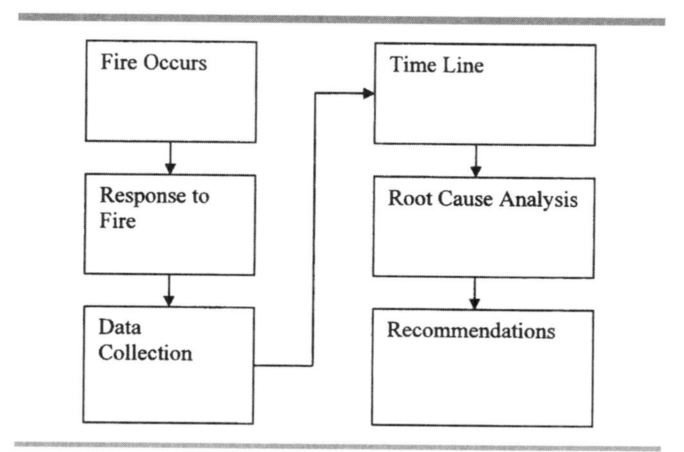

FIGURE 1. Fire investigative flow chart

1. Begin the investigation as soon as the fire area is safe to enter, before any clean-up or repair work is begun.
2. Obtain eyewitness reports before personnel leave the facility. Interview operators on shift, firefighters, and any other eyewitnesses. Try to note as much detail as possible, including the sequence of events, estimates of the event's duration, the color of fire and smoke, and noises and smells accompanying the event. If possible, tape-record discussions with eyewitnesses.
3. Review records and logs. Immediately following a serious fire, all logs and records should be dated, time-marked, and retained for future reference. Other useful records include material samples, inspection records, drawings, weather records, and any other information that can help in the reconstruction of the conditions and events leading up to the fire. Note that some computerized systems overwrite data every 24 hours.
4. Inspect in detail the physical evidence and the damage. If a fire has burned in a confined area, sketch a plot plan to show the outer limits of the damage and the locations that flame actually touched.

   Flame contact with any object heats the object rapidly to 1000°F or above. If flame has not reached an object, its temperature will seldom exceed 400°F. Table 4 provides guidance for estimating fire temperatures in an area during a fire. Use this guidance to draw "temperature circles" on the plot plan.

   Insulation fibers, grass, and leaves will be singed and charred by the ignition of a vapor cloud. These clues can help define the extent of the vapor cloud prior to ignition.

   The source of the fuel is usually located near the center of the highest temperature circle. An unexpected cool zone inside the

TABLE 4

Material Response to Temperature

Material	Characteristics	Temperature, °F
Polyvinyl chloride (PVC)	Distorts	185
Polystyrene	Distorts	210
Paint	Scorches	250
Plastic	Melts	250
Plastic	Chars	250–400
Paint	Chars	250–400
Nylon	Distorts	300–360
Wood	Ignites	380–510
Dry coke	Formed	400–1000
Steel	Strength lost	1000–1200
Silver solder	Melts	1165–1450
Aluminum	Melts	1220
Glass	Softens	1400–1600
Brass	Melts	1600
Steel	Excessive scaling and grain coarsening	1600
Concrete	Spalling	1800–2000
Copper	Melts	1980
Stainless steel	Melts	2600

(*Source:* Cote 2003)

otherwise hottest area might point to a fuel source that continued to flow throughout the fire, providing localized cooling. This could be the initial leak or failure.

Photographs taken as a permanent record can be helpful during the review stage of the investigation.

5. Identify the fuel source. After sketching the "circles of temperature," look carefully at the area inside the hottest circle to identify damage or leaks that occurred during the fire. Rule out these items as initial sources of fuel. For example, piping that splits during a fire will have knife-thin edges along the tear. These thin edges were created by flame that weakened the steel enough for it to bulge, become thin, and tear. On the other hand, failed piping that shows no thinning at the edge of the failure except for possible corrosion should be noted as a likely initial source.

   A sample of the suspected fuel is helpful when calculating its vapor pressure or flash point to support conclusions based upon examination of the fire site.

6. Identify the ignition source. After deciding upon the probable fuel source and type, inspect the fire area again and review eyewitness reports to try to identify the source of ignition. Common sources of ignition have to do with
   - Welding, cutting, drilling, or burning.
   - Open flame, such as that from fired boilers or heaters.
   - Oil-soaked insulation on piping above 350°F.
   - Hot bearings on pumps.
   - Engines.
   - Pyrophoric compounds.
   - Spontaneous ignition.
   - Static electricity.
   - Electrical sparks or arcs.

   Do not be intimidated by what looks like total destruction, an impossible mess, and a shapeless pile of junk. Look past rubble to visualize the original conditions. Search out detailed clues. Identify the pattern of temperature and flame contact in order to surround the fuel source.

7. Document details of the emergency response to the fire, including

- Use of detection, alarm, and shutdown systems.
- Effectiveness of prefire plan.
- Number of firefighters, including mutual aid, and their response time.
- Amount (and type) of firefighting equipment used.
- Level of effectiveness of firefighting equipment and tactics, including fixed equipment, mobile equipment, and the facility fire water system.

In cases of large fires, it is helpful to map locations of fire trucks, monitors, and hose lays, as well as the outline of the fire area.

Investigation Team

Employee (first line supervisor, operator, contractor employee) participation on the investigation team is critical. Involving employees brings valuable knowledge to the investigation and avails safety professionals of a valuable resource through which to communicate findings back to the impacted group. Include team members who have expertise in process, materials, machinery, or instrumentation, as necessary. A team leader who has investigative experience can be a great help to the group effort.

Supervisory and management personnel need to understand that their presence on the investigation team can sometimes cause intimidation, limiting the team's effectiveness. It may also be intimidating to have someone viewed as contributing to or somehow responsible for the incident as part of the team. However, the goal of the investigation should be to identify the cause of the incident, not to blame an individual. Insurance industry experience indicates that less than 10% of losses can be attributed to human action, and less than 1% to malicious action. By involving persons (or representatives of persons) involved directly in the incident, the investigation team can demonstrate commitment to finding the cause rather than assigning blame. If there is indication that blame should be placed, legal counsel should be contacted and the matter handled in accordance with applicable laws.

Timeline

The first step in analyzing the data collected is establishing a detailed timeline of events leading up to the fire—especially any changes in plant operations during the few days prior to the event. The investigation team should discover the approximate times of fuel release, ignition, and fire discovery. Subsequent equipment failures that added fuel to the fire, as well as operator and firefighter response to the incident, should be documented on the timeline until the point of control and extinguishing.

Root Cause Analysis

Root cause analysis is a method used to identify the underlying causes of incidents. Using root cause analysis, the investigative team tries to answer such questions as

- What was the source of ignition, and why was it in proximity to a source of fuel?
- Why did the fire progress beyond the incipient stage?
- What systems contributed to (and detracted from) the size of the incident?

Each incident includes both physical causes (such as power surges) and human causes (such as inappropriate maintenance or operator response). Management system causes (such as inadequate training or procedures) often allow human causes to occur. Root cause analysis is a method designed to identify each type of cause and to help prevent its recurrence. The "root" causes of an incident are the most basic causes that can reasonably be identified and can be controlled. There are numerous commercial products of this type on the market, and an Internet search can easily identify various vendors. For more general information on root cause analysis, readers should review Wikipedia's article on the topic (Wikipedia 2007b). Although this is not a peer-reviewed site, it offers general information that can be helpful in understanding the common aspects of root cause analysis.

Other causes can be physical, human or behavioral, and systems-related.

Physical causes can lie in the failure or changing of equipment or devices, or in physical conditions having some effect that leads to undesired consequences. Some examples are broken piping, vibrating pumps, failed temperature indicators, furnace tubes developed coking, tanks struck by lightning, or shorted electrical circuits.

Human or behavioral causes lie in human actions (or lack of action) that caused undesirable physical conditions or events. Examples might include failure to open a valve, to perform inspections for corrosion, to light a furnace only after sufficient purging, to read gauges incorrectly, to use correct design codes, or to observe all steps in a shutdown procedure. Yet these things may or may not be classified as human error in a particular instance—sometimes the human titularly at fault may have merely followed an incorrect procedure or acted in the absence of proper training or instruction.

System causes occur when a procedure or management system failed to support humans in taking appropriate actions. The systems may be absent, unenforced, incorrect, or otherwise unusable. Usually these are synonymous with root causes.

Effective incident investigation systems look beyond human and behavioral causes to determine root causes in management systems. In this sense, a management system is the mix of equipment, procedures, training, and culture that supports people's performance of their duties.

For an example of the three types of basic causes, consider an incident in which a pump bearing failed (see Table 5).

TABLE 5

Example of a Cause for Pump Bearing Failure

Cause	Basic Types of Causes	Fix
Bearing overheated	Physical	Replace bearing
Operator failed to oil bearing	Human/behavior	Remind operator to keep bearing oiled
Operator not required to oil bearing	Management system	Require operator to oil bearing

Fire Reporting

Importance

Reports of fires are of great importance in a fire protection program. They provide

- Reasons for fire occurrence that can help in taking corrective measures.
- Basis for changes in the design of facilities, in expenditures for fire prevention and firefighting equipment, underwriting insurance, and training, especially when stored at governmental and commercial facilities, augmenting fire loss experience that has been accumulated over many years.
- Valuable data for discussions with regulatory bodies regarding legislation affecting industrial and commercial operations.
- A source of material for fire training programs.

What Fires Should Be Reported?

All fires should be reported and documented. Fires are defined as any occurrence of fire, combustion explosion, or spread of fire involving properties, products, operations, or employees that is not intentionally ignited for a useful purpose, irrespective of resulting deaths, injuries, or damages.

Definitions

Reportable fire: All fires, regardless of size, dollar loss, or injury, should be reported.

Every fire provides a lesson in fire prevention, and every report broadens communal knowledge of fire protection techniques, making it important to report all fires, regardless of levels (or absence) of loss. Two objectives of a report are describing the effectiveness of fire control measures and revealing the circumstances that permit a fire to start in unexpected places or in an unexpected manner. The question should be asked: "Was the fire an expected result or a part of the work being done?" Even a small fire may reveal a situation that, under slightly different circumstances, could have caused serious loss. This type of information, combined with experience in other areas, can provide the basis for important steps in a fire protection program.

Recordable fire: In order to standardize across the industry, it is important that the fire reporting program used has a defined recordability statistic. Often low-value, low-risk potential fires are reported but not recorded. A fire reporting program should define loss levels in excess of which to record fires. Insurers, government agencies, and other parties may define these limits. If not, use damage loss value, loss potential, and lost product value as guidance when deciding whether or not to record a fire. The important thing is to report fires, identifying root causes so that actions can be taken to reverse negative trends in fire loss.

Preparation of a Report

The following makes use of the National Fire Incident Reporting System (NFIRS), which has been widely used since 1980.

Following the investigation, a report is prepared. A sample NFIRS report form may be found at the NFIRS website (USFA n.d.). Note that all questions apply to every fire; answers to some questions may require more space than that provided on the form.

It is important that the information be accurate and complete, and that statements based on opinion rather than fact be so designated. Reports should contain as many facts as possible.

Sketches or photographs showing the area affected by fire, and indicating significant dimensions or distances, are desirable.

Because the prevailing legal system is adversarial in tone and sometimes appears to oppose the no-blame culture needed to effectively investigate and explain fires in such ways as to improve fire prevention programs, it is important to try to prepare fire reports in such ways that possible legal concerns are not compromised. The following points give some guidance

- Report only facts; avoid opinions.
- Avoid such words as "detailed," "full," or "complete." If a final report unintentionally falls short of that goal, use of such words might be difficult to justify.
- Assume that whatever is written will become public information.
- If the possibility of legal action exists, it may be advisable to seek legal counsel to review the report. Attorney-client privilege concerns can be discussed and are outside the scope of this chapter. Suffice it to say that a belief that the fire may lead to litigation is cause to consult an attorney with experience in the appropriate industry and in handling such investigations.

However, safety professionals must not let possibility of legal action deter them from investigating and documenting the fire investigation. Legal reasons, and in today's environment Homeland Security reasons may prevent them from sharing all details, but they should share and communicate what they can. Sharing learning with others is a best practice.

Report Form

The NFIRS report form asks for specific information on the fire being reported—where it was, what was involved, what was damaged, who was injured, when it occurred, and when it was controlled and extinguished (USFA n.d.). Section E1 includes a simple timeline to help organize the description of the incident. Describe what happened, including

- The events leading up to the fire. The most valuable lessons from incidents often come from a thorough and objective examination of the events leading up to fire.
- The discovery of the fire and initial response. What happened to either increase or limit losses between the time the fire was

discovered and the time it was controlled? Were fuel sources isolated? Was an emergency shutdown system used? Was anyone hurt at the start of the incident or during firefighting?

- Control and extinguishing of the fire. Describe the procedures used to fight the fire. Who responded—the fire brigade? the municipal fire department? Were firefighting tactics effective? What were they? How long did it take to extinguish the fire? Was anything unusual about the control methods that should be passed on to others in hopes of developing or improving their fire protection programs?

Root cause analysis of the fire provides information that is important to the investigation and a useful tool for gaining experience to prevent future similar fires. A description of each section follows.

Sources of Fuel and Oxygen

Indicate the source of fuel (*e.g.*, a tank, pump, piping). Air is the usual source of oxygen. However, check the box marked "other" if another oxygen source was involved (*e.g.*, hydrogen peroxide, a leak). Describe what happened to allow the fuel to mix with oxygen (*e.g.*, the tank was overfilled, the pump seal failed, the tank inerting system failed).

Source of Ignition

Why was the source of ignition in proximity to the flammable mixture? Why did the fire start?

EXAMPLE 1: A lightning strike in the vicinity of the Buffalo Wallow tank setting caused vapors from the pressure/vacuum vent to ignite. It is thought that the pressure pallet inside the vent valve stuck in the open position.

EXAMPLE 2: A pump bearing failure caused internal friction (heating) and vibration that led to failure of the mechanical seal. After a release of flammable material occurred when the mechanical seal failed, the release was ignited by hot pump surfaces.

What facility systems contributed to the severity of this fire (or reduced its impact)? Several choices are provided. Check all those responsible for allowing a small easy-to-control fire to escalate into a large fire or those that effectively limited losses or the extent of the fire. This information is helpful because

- It can be used to increase the effectiveness of a fire prevention program.
- It may be useful at other locations similar to the one in question.

Explain how the systems checked above affected the fire (*e.g.*, how they failed to limit losses or why they were successful in reducing fire losses).

Elaborate on the advantages or shortcomings of the systems that affected fire losses. Feel free to add marginal notes or attach a second sheet (for example, equipment could not be isolated from fuel sources because brass block valves melted).

What incident-prevention systems could have warned of the fire or prevented its occurrence? Check all the process safety management (PSM) systems that were in place but which were not used or which were needed to reduce risk of fire. Process safety management is an OSHA requirement for the petrochemical industry; more information about it can be found at the OSHA website (OSHA n.d.).

Is action necessary to prevent recurrence? List any proposed actions or attach root cause analysis and lessons learned. Section X is the final step in analyzing the data contained in the fire report. Compare the PSM systems listed in Section IX with the data contained in Sections IV–VIII and ask the following questions:

Should existing facility or PSM systems or other management controls be modified to

- Prevent a flammable mixture from forming?
- Eliminate ignition sources?
- Reduce the impact of a fire?
- Improve emergency response?

How to Estimate Loss

In most cases, reported loss includes the cost of in-kind replacement of damaged structures or equipment, including removal and clean-up of damaged equipment and the value of the product or material consumed in the fire. It does not include any losses associated with future production unless insurance coverage includes such clauses. Report the replacement cost or value of any nonfacilities (*i.e.*, third-party) losses. If the damaged equipment is abandoned, report book value. The loss reported should be that resulting from the fire or explosion; do not include the value of substantial improvements made when the structure or equipment is replaced.

Only losses resulting from fire should be reported as fire losses.

Fire Prevention and Inspection Training for Employees

Fire Protection Checklists

Tables 6 and 7 contain examples of checklists related to fire prevention and suppression that are pertinent to industrial facilities. Such checklists can be used during fire prevention and protection reviews to determine the level of fire safety in a facility, and can also use these checklists during facility inspections, risk assessments, Pre-Startup Safety Reviews (PSSR), and Management of Change (MOC) evaluations.

Managers

Managers oversee an organization's implementation of fire prevention and protection strategy. Training of managers is related to communicating and auditing performance of the staff and operations they manage. Training programs that stress communication both internally and externally are critical.

Supervisors

Supervisors organize their staff to accomplish specific roles within functions (whether operations, maintenance, emergency response, or staff).

Training for supervisors is focused on plans and procedures to complete necessary actions. Training should also focus on development in areas of communication and feedback on performance. The goal of the training should be to ensure that all employees, contractors, and visitors clearly understand and can plainly articulate the expectations for fire safety.

TABLE 6

Fire Protection Checklist Example I

Fire Protection Checklist			
Reviewed by:			**Date:**
Emergency Response Plans:			
Item	**Yes**	**No**	**Corrective Action**
1. Is emergency response plan available and up-to-date?			
2. Have all facilities established emergency procedures that are tested periodically?			
3. Are public agencies and mutual assistance organizations included in practice sessions?			
4. Is incident command system (ICS) being used? Is incident commander positively identified?			
5. Are emergency supplies and equipment readily available?			
6. Are phone lists for company personnel callout, mutual aid, local fire and police departments, emergency supplies, etc., up-to-date and reviewed periodically?			
7. Is a radio communication channel available for dedicated emergency use?			
8. Are hypothetical drills performed monthly? Are written copies of the hypothetical drill report available for review?			

TABLE 7

Fire Protection Checklist Example II

Fire Protection Checklist			
Reviewed by:			**Date:**
Housekeeping			
Item	**Yes**	**No**	**Corrective Action**
1. Is maintenance cleanup complete?			
2. Are drainage channels clear and free-flowing?			
3. Are combustibles stored in safe areas of the plant, control room or shop?			
4. Are process areas weed-free, preventing fire from approaching the plant or equipment?			
5. Is firefighting equipment freely accessible?			
6. Are materials stored neatly and orderly, and are they current?			
7. Are spills promptly cleaned up?			

TABLE 8

Hot Work Permit Checklist

The hot work permit procedure contains a list of safety items to be examined before starting work. The following should be considered for inclusion in the hot work permit procedure:

1. Is hot work justified in the planned area, or can the work be done in a less hazardous area?
2. Can the job be done by cold work?
3. Has the site been inspected by a responsible individual?
4. Have pipelines or equipment that may release flammable or combustible materials been blinded off or disconnected? Are there any leaks that cannot be stopped?
5. Are pipe lines or equipment to be worked on vapor-free or vented?
6. Has the hydrocarbon been drained from all low points?
7. Are bleeders open and unplugged using proper rodding-out procedure?
8. Have openings to sewers and underground drains within 35 feet of the work been sealed with water and plugs or sand bags? (Be sure to remove seals and/or sand bags when the job is finished.)
9. When it is necessary to chip, gouge, saw, or drill metal where flammable vapors may be present, use hand tools only and apply a cooling liquid to prevent heat buildup. The hot work permit should document these special precautions.
10. Is there a potential for vapor emissions from equipment, piping, valves, and vents in the vicinity? Are flammable vapor tests necessary? If so, how often? At what reading is it safe for work to begin? (See Section 374)
11. Are adjacent areas safe? Are any water-reactive or other chemicals present in the vicinity that could be a hazard to firefighters? Is it necessary to redirect venting from relief systems away from the hot work area?
12. Are workers in adjacent areas and affected plants aware that the work is going on?
13. Will working conditions remain safe for the duration of the job? Is it necessary to inspect or monitor valves, flanges, or equipment frequently to ensure that leaks do not develop or become serious?
14. Is the fire watch qualified, having adequate instruction in what to look for and what to do if hazardous conditions arise? Are radio communications between the fire watch and the control room or supervisor needed? Do workers know how to report a fire?
14. Is the fire watch qualified, having adequate instruction in what to look for and what to do if hazardous conditions arise? Are radio communications between the fire watch and the control room or supervisor needed? Do workers know how to report a fire?
15. Are combustible materials (*e.g.,* wood, paper, cloth) present that could be ignited? Is it likely that a hydrocarbon spill elsewhere will run into the hot work area?
16. Are fire screens and blankets required to contain sparks and slag? Are charged fire hoses and portable extinguishers available? Does the fire watch know how to use the equipment properly?
17. Are operating activities (*e.g.,* venting, sampling) likely to release flammable vapors or liquid to the atmosphere within 50 feet of the hot work area? If so, have hot work activities been coordinated with operations?
18. Are high-velocity, high-volume fans required to direct potentially flammable vapors away from the area? If the fans are electric-motor driven and located inside an electrically classified area, do they meet NFPA 70 Class I Division 1 requirements?
19. Should the hot work area be inspected when workers leave the area (*e.g.,* at shift change, breaks, and work completion)? (The fire watch should remain on-site for at least 30 minutes per OSHA requirements.)
20. If the hot work will occur above grade, what is the chance that windborne hot particles could drift outside the hot work area or hot slag and drop from one level to another (as when working on an elevated platform constructed of an expanded metal deck)?

Employees

Employee roles are basic: they should be expected, to follow plans, procedures, and rules related to their job functions.

Training should focus on the skills needed to safely perform jobs and to identify and communicate to other employees or supervisors when their work may create, or be exposed to, a hazard.

Hot Work Permits

A hot work permit program is critical to the safety of a facility and is required under OSHA regulations. Table 8 provides an example for the safety professional, including the types of questions that a hot work permit program should address.

Figure 2 represents an example of a hot work permit.

HOT WORK PERMIT

BEFORE INITIATING HOT WORK, ENSURE PRECAUTIONS ARE IN PLACE!
MAKE SURE AN APPROPRIATE FIRE EXTINGUISHER IS READILY AVAILABLE!

This Hot Work Permit is required for any operation involving open flames or producing heat and/ or sparks. This includes, but is not limited to: Brazing, Cutting, Grinding, Soldering, Thawing Pipe, Torch-Applied Roofing, and Cadwelding

INSTRUCTIONS

1. Verification below is to be completed by a qualified employee.
2. The completed original is to be dropped off at the central control room and copies sent to the safety office.
3. Must be submitted 24 hours before work is started.

HOT WORK BEING DONE BY:

❑ Employee
❑ Contractor____________________________

Date: ____________ W.O. # ____________
Start Time: ________

Location / Building / Floor

Nature of Job / Object

Name of Person Doing Hot Work

I verify the above location has been examined, the precautions checked on the Required Precautions Checklist have been taken to prevent fire, and permission is authorized for work.

Signed:

Permit Expires | Date | Time AM PM

Fire Detection Disabled | Reactivated

Date / Time: __________ __________

Initial: __________ __________

THIS PERMIT IS GOOD FOR ONE DAY ONLY

REQUIRED PRECAUTIONS CHECKLIST

❑ Automatic Fire Detection Disabled?
❑ Available sprinklers, hose streams, and extinguishers are in service/operable?
❑ Hot work equipment is in good repair?

Requirements within 10 m (35 feet) of work:

❑ Flammable liquids, dust, lint, and oil deposits removed?
❑ Explosive atmosphere in area eliminated?
❑ Floors swept clean?
❑ Combustible floors wet down, covered with damp sand or fire-resistant sheets?
❑ Remove other combustibles where possible. Otherwise protect with fire-resistant tarpaulins or metal sheets?
❑ All wall and floor openings covered?
❑ Fire-resistant tarpaulins suspended beneath work?

Work on walls or ceilings/enclosed equipment:

❑ Construction is non-combustible and without combustible covering or insulation?
❑ Combustibles on other side of walls moved away?
❑ Danger exists by condition of heat into another area?
❑ Enclosed equipment cleaned of all combustibles?
❑ Containers purged of flammable liquids/vapors?

Fire Watch / Hot Work area monitoring:

❑ Fire watch will be provided during and for 30 minutes after work, including any coffee or lunch breaks?
❑ Fire watch is supplied with suitable extinguishers?
❑ Fire watch is trained in use of this equipment and in sounding alarm?
❑ Fire watch may be required for adjoining areas, above and below?
❑ Monitor hot work area 30 minutes after job is completed.

Other precautions taken:

❑ Confined space entry permit required?
❑ Area protected with smoke or heat detection?
❑ Ample ventilation to remove smoke/vapor from work area?
❑ Lockout/tagout required?

FIGURE 2. Sample hot work permit

Conclusion

Professionals responsible for fire prevention and protection must have a broad understanding of the practical aspects of designing prevention into the development and operation of a facility. This responsibility is first of all to ensure life, environment, and facility safety. Professionals must balance prudent prevention and protection measures against mitigating or eliminating hazards, risk-managing their work product.

Professionals have many avenues of information available them through technical resources that include the NFPA codes and standards. It is recommended that professionals involved in fire prevention or protection activities make periodic reference to the resources presented in this chapter.

References

American Petroleum Institute (API). 2001. API Std 2510, *Design and Construction of LP Gas Installations at Marine and Pipeline Terminals, Natural Gas Processing Plants, Refineries, Petrochemical Plants and Tank Farms.* 8th ed. New York: API.

______. 2003. RP 752, *Recommended Practice for the Management of Hazards Associated with Location of Process Plant Buildings.* 2d ed. New York: API.

American Society of Safety Engineers (ASSE). 2002. "Code of Professional Conduct." Des Plaines, IL: ASSE (retrieved 25 May 2007). www.asse.org/search.php?varSearch=code+of+professional+conduct

Benedetti, R. P., ed. 1996. *NFPA Flammable and Combustible Liquids Code Handbook,* 3d ed. Quincy, MA: NFPA.

BETE Nozzle Company. n.d. . "BETE Nozzle Company, Products." www.bete.com

Bureau of Mines (BoM). 1991. AD 771 191, *Fire and Explosion Manual for Aircraft Accident Investigation.* Washington, D.C.: Bureau of Mines.

Cote, A. E., ed. 1990. *Industrial Fire Hazards Handbook,* 3d ed. Quincy, MA: National Fire Protection Association.

Cote, A. E., ed. 2003. *Fire Protection Handbook,* 19th ed. Quincy, MA: National Fire Protection Association.

Department of Homeland Security (DHS). n.d. "Hazard Prevention and Protection." www.dhs.gov

Environmental Protection Agency (EPA). n.d. a. "Chemical Emergency Preparedness."www.epa.gov/ceppo

______. n.d. b. "Emergency Planning Community Right-to-Know." www.epa.gov/emergencies/context/epcra.index.htm

______. n.d. c. "Ozone Layer Depletion—Alternatives—SNAP" (retrieved May 25, 2007). www.epa.gov/ozone/snap/fire/index.html

Federal Emergency Management Agency (FEMA). n.d. "Resource Record Details: World Trade Center Performance Study" (retrieved June 12, 2007). www.fema.gov/library/viewRecord.do?id=1728

International Handling Systems. "IHS/Global: Your Source for Standards and Specifications." www.global.ihs.com

National Fire Protection Association (NFPA). 2002a. NFPA 17, *Standard for Dry Chemical Extinguishing Systems.* Quincy, MA: NFPA.

______. 2002b. NFPA 25, *Standard for the Inspection, Testing, and Maintenance of Water-Based Fire Protection Systems.* Quincy, MA: NFPA.

______. 2004. NFPA 12A, *Standard on Halon 1301 Fire Extinguishing Systems.* Quincy, MA: NFPA.

______. 2005a. NFPA 11, *Standard for Low-, Medium-, and High-Expansion Foam and Combined Agent Systems Quincy,* MA: NFPA.

______. 2005b. NFPA 30, *Flammable and Combustible Liquids Code.* Quincy, MA: NFPA.

______. 2006. NFPA 101, *Life Safety Code.* Quincy, MA: NFPA.

______. 2007a. NFPA 10, *Portable Fire Extinguishers.* Quincy, MA: NFPA.

______. 2007b. NFPA 13, *Standard for the Installation of Sprinkler Systems.* Quincy, MA: NFPA.

______. 2007c. NFPA 15, *Standard for Water Spray Fixed Systems for Fire Protection.* Quincy, MA: NFPA.

______. 2007d, NFPA 24, *Installation of Private Fire Service Mains and Their Appurtenances.* Quincy, MA: NFPA.

______. 2008a. NFPA 12, *Standard on Carbon Dioxide Extinguishing Systems.* Quincy, MA: NFPA.

______. 2008b. NFPA 69, *Standard on Explosion Prevention Systems.* Quincy, MA: NFPA.

Occupational Safety and Health Administration (OSHA). n.d. "Safety and Health Topics: Process Safety Management." www.osha.gov

Society of Fire Protection Engineers (SFPE). 2002. "Guidelines for Peer Review in the Fire Protection Process" (retrieved March 6, 2007). www.spfe.org

US Fire Administration (USFA). n.d. "National Fire Incident Reporting System (NFIRS)." www.usfa.dhs.gov/fireservice/nfirs/index.shtm

Wikipedia, 2007a. "Latent heat" (retrieved June 3, 2007). en.wikipedia.org/wiki/Latent_heat

Wikipedia, 2007b. "5 Whys" (retrieved June 3, 2007). en.wikipedia.org/wiki/5_Whys

Zalosh, R. G. 2003. *Industrial Fire Protection Engineering.* New York: John Wiley and Sons.

APPENDIX: RECOMMENDED READING

American Petroleum Institute (API). www.api.org

______. 1997. API 500, *Classification of Locations for Electrical Installations in Petroleum Facilities*, 2d ed. New York: API.

National Fire Sprinkler Association (NFSA). n.d. "National Fire Sprinkler Association." www.nfsa.org

National Fire Protection Association (NFPA). 1999. NFPA 50A, *Gaseous Hydrogen Systems at Consumer Sites*. Quincy, MA: NFPA.

______. 2004. NFPA 59, *Utility LP—Gas Plant Code*. Quincy, MA: NFPA.

______. 2004. NFPA 2201, *Clean Agent Fire Extinguishing Systems*. Quincy, MA: NFPA.

______. 2008. NFPA 58, *Liquefied Petroleum Gas Code*. Quincy, MA: NFPA.

______. 2008. NFPA 70, *National Electrical Code*. Quincy, MA: NFPA.

National Institute of Health (NIH), Office of Research Facilities (ORF). *NIH Design Policy and Guidelines*. Washington, D.C.: NIH/ORF. orf.od.nih.gov/PoliciesAndGuidelines/DesignPolicy/HTMLVer/Voume4/FireProtection.htm

SECTION 4 INDUSTRIAL HYGIENE

Regulatory Issues

Applied Science and Engineering

General Principles

Chemical Hazards

Physical Hazards

Biological Hazards

Cost Analysis and Budgeting

Benchmarking and Performance Criteria

Best Practices

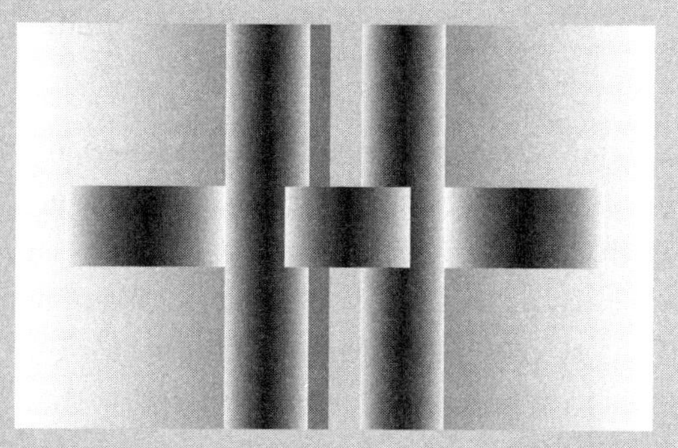

SECTION 4

INDUSTRIAL HYGIENE

LEARNING OBJECTIVES

- Obtain a historical overview of the federal regulation of occupational safety and health.
- Learn about the various federal agencies responsible for workplace safety and health, with a focus on OSHA.
- Learn about the formation of OSHA and the health standards it has promulgated to date.

REGULATORY ISSUES

Gayla McCluskey

THE U.S. GOVERNMENT was slow to respond to the changing occupational environment in the late nineteenth and early twentieth centuries, when the country developed into a leading industrial nation. Industrialization introduced new chemicals, dusts, and dangerous machines into factories, mines, and cities. Workers began experiencing diseases such as silicosis in workplaces that had inadequate ventilation and unsanitary conditions (MacLaury 1981).

The regulation of health and safety initially fell to state labor agencies. The Massachusetts Bureau of Statistics urged the development of legislation that dealt with "the peril to health from lack of ventilation." Massachusetts passed the nation's first factory inspection law in 1877, which required equipment guarding and emergency fire exits. By 1890, nine states had factory inspection laws, thirteen required machine guarding, and 21 had narrow provisions for health hazards.

Prince Otto von Bismarck initiated the first Workers' Compensation system in Germany in 1884 for the purpose of compensating injured workers. In the United States, organized labor opposed the system because it was not a preventive measure, until the Pittsburgh Survey was released in 1907–08. This survey detailed lives and working conditions in Allegheny County, Pennsylvania. It found that injured workers and the survivors of those killed bore the economic brunt of accidents, even though most accidents were thought to be the fault of the employer. The study's authors suggested that employers should bear a substantial share of the economic burden, thus giving them an incentive to reduce the causes of injury.

The Pittsburgh Survey was supported by both labor and business, and as a result, a limited Workers' Compensation law was passed in 1908 for federal employees. Wisconsin established the first state workers' compensation program in 1911 and, by 1921, most states had established programs. However, there was little incentive for a company to invest in safety improvement because insurance premiums were so low. In addition, very few states included compensation for disease, despite what was known at the time about occupational illness.

Federal Regulation of Health and Safety

The federal government began investigating industry-related diseases early in the twentieth century. The Bureau of Labor published the first studies of death and disease in the dusty trades in 1903. In 1910, it published a study of the horrors of phosphorus necrosis, a disfiguring and sometimes fatal disease of workers in the white phosphorus match industry (MacLaury 1981).

The U.S. Commissioner of Labor, Charles Neill, met Dr. Alice Hamilton[1] at a 1910 European conference on occupational accidents and disease. As Director of the Illinois Commission on Occupational Diseases, Dr. Hamilton was conducting a study of lead exposure. Neill invited Hamilton to work for the Bureau of Labor and, in that capacity, she traveled around the country until 1921, visiting lead smelters, storage battery plants, and other hazardous workplaces (MacLaury 1981). Her findings were so scientifically persuasive that they resulted in sweeping reforms, both voluntary and regulatory, leading to the reduction of occupational exposure to lead. Other investigations for which she is best known include studies of carbon monoxide poisoning in steelworkers, mercury poisoning in hatters, and "dead fingers" syndrome among laborers who used jackhammers (NIOSH 2006).

Congress created the Department of Labor in 1913, appointing William B. Wilson, a former coal miner and union official, as Secretary. The Bureau of Labor Statistics started to compile accident statistics, initially in the iron and steel industry. In 1914, federal responsibility for industrial safety and health was placed in the Office of Industrial Hygiene and Sanitation under the Public Health Service (MacLaury 1981).

The United States' entry into World War I worsened the health and safety conditions found in the factories that were trying to meet the wartime needs of the military. In response, Congress created a service to inspect war production sites, advise companies on reducing hazards, and help the states to develop and enforce safety and health standards.

In 1933, President Franklin D. Roosevelt selected Frances Perkins[2] as Secretary of Labor, the first woman appointed to the Cabinet. Perkins was appointed to the New York Industrial Commission in 1918 and later named its Commissioner by Roosevelt when he was governor of the state. Perkins created the Bureau of Labor Standards in 1934, the first permanent federal agency to promote health and safety for all workers. However, the Bureau's efforts continued to focus on helping state governments improve their administration of state health and safety laws.

Industrial Hygiene evolved significantly after the passage of the Social Security Act of 1935, which allocated funds to the Public Health Service for research and grants to states for industrial hygiene programs. Before 1935, only five state health departments contained industrial hygiene units; by 1936, seventeen units existed in state and local health departments (Corn 1992).

The Walsh-Healy Act of 1936 enabled the federal government to establish health and safety standards for federal contractors. In spite of this Act, enactment of safety and health laws were largely left to the states until 1958, when Congress gave the authority to set safety and health standards for longshoremen to the Department of Labor (MacLaury 1981).

The Department of Labor issued mandatory health and safety standards under the Walsh-Healy Act in December 1960. These were the first federal regulations that applied to the whole range of industry. Until that time, most of these standards had previously been used as informal guidelines to aid federal and state inspectors. The business community was caught by surprise, since the regulations were enacted without hearings or prior announcement. Business community criticism caused the Department to announce revisions to the standards, and a hearing on the new federal guidelines was held in March 1964. The Department of Labor decided to examine all of its safety programs in order to develop a more coordinated health and safety policy. An outside consultant's (identity not given) review recommended all of the safety programs be consolidated under a single agency.

In 1965, the U.S. Public Health Service issued a report entitled, "Protecting the Health of Eighty Million Americans." The report noted that a new chemical enters the workplace every twenty minutes, cited evidence showed a strong link between cancer and the workplace, and concluded that old problems were far from being solved. The report called for a national occupational health effort centered in the Public Health Service.

OSHA's Beginning

In 1968, President Lyndon B. Johnson's administration submitted a bill to the 90th Congress to create the first federal comprehensive safety and health program. The Occupational Safety and Health Bill was introduced in the House by James O'Hara (D-Michigan) and in the Senate by Ralph Yarbrough (D-Texas). The bill contained provisions to conduct research to determine a basis for standard-setting, granted authority to the Secretary of Labor to set and enforce standards and impose sanctions for violation of the standards, and provided resources to the states to develop and strengthen their programs (Ashford 1976).

Congressional hearings were held during which Secretary Wirtz cited two casualty lists facing the country at that time: the military toll in Vietnam and the industrial toll at home. Wirtz asked "whether Congress is going to act to stop a carnage" which, he said, continued because people "can't see the blood on the food that they eat, on the things that they buy, and on the services they get." Labor supported the bill, but the business community argued that it vested the Secretary of Labor with too broad a power and undermined the role of the states (MacLaury 1981).

The Johnson proposal did not reach the floor of either body of Congress. After President Richard Nixon was elected and confirmed, he asked Secretary of Labor James C. Hodgson to prepare a health and safety bill for introduction in 1969.

The Republicans introduced bills sponsored by Congressman William Ayres (R-Ohio) and Senator Jacob Javits (R-New York). The Democrats introduced Occupational Health and Safety Bills HR 3809, sponsored by James O'Hara (D-Michigan) and S 2193, sponsored by Harrison A. Williams (D-New Jersey) that included similar provisions to the legislation previously introduced (Ashford 1976).

The Republican-sponsored bills differed from those sponsored by the Democrats by vesting the authority to set and enforce standards to a new safety board, consisting of members appointed by the president, rather than vesting the authority to the Department of Labor. In addition, the Nixon administration's proposal limited the Labor Department's role to workplace inspection.

Hearings were held on the Nixon proposal, during which Secretary of Labor George Shultz asked Congress to "work out our differences and get something done." Organized labor opposed the new safety board and argued that the program should be in the Labor Department. The U.S. Chamber of Commerce, the National Association of Manufacturers, and other industry groups supported the Nixon proposal (MacLaury 1981).

The Senate approved a compromise version of S 2193 in November 1970, in which an independent review commission charged with final review of enforcement activities would be created, while standards-setting and initial enforcement authority were given to the Secretary of Labor. The bill also included a proposal to elevate the status of the former Bureau of Occupational Safety and Health and give it the authority to perform research and make recommendations to the Department of Labor (Ashford 1976).

The House approved HR 19200, a revised administration bill sponsored by William A. Steiger (R-Wisconsin) and Robert Sikes (D-Florida). The House bill proposed an independent board to set standards, with enforcement to be provided by an independent appeals commission. The Labor Department would then share enforcement responsibilities and conduct inspections.

The two bills were sent to a Congressional conference committee and the resulting compromise version became the Occupational Safety and Health Act of 1970. The legislation was signed into law by President Nixon on December 29, 1970, and became effective 120 days later.

The Occupational Safety and Health Administration (OSHA) was created within the Department of Labor to implement the requirements of the Act. George C. Guenther, at the time Director of the Labor Standards Bureau, was named as OSHA's first Assistant Secretary (MacLaury undated). The agency set up ten regional offices and 49 area offices to cover the 50 states, the District of Columbia, Puerto Rico, the U.S. Virgin Islands, Guam, and American Samoa (Showlater 1972).

Initial Standards Setting

Section 6(a) of the Act required that OSHA promptly set up a regulatory framework by adopting any national consensus standard and any established federal standard as an occupational safety or health standard, unless OSHA determined a standard would not result in improved health and safety. This authority was allowed for a period from the effective date of the Act until two years after the effective date (U.S. Congress 1970) .

On April 28, 1971, OSHA issued the *Compliance Operations Manual*, which set policies. In it, OSHA set priorities for inspections in the following order:

1. Inspection of fatalities and catastrophes
2. Response to employee complaints
3. Inspection of target industries and facilities with targeted health hazards
4. Random inspections

The Target Health Hazards Program included asbestos, cotton dust, silica, inorganic lead, and carbon dioxide (Ashford 1976).

On May 29, 1971, OSHA adopted a body of voluntary consensus standards. These took effect on August 27, 1971, and in some cases, February 15, 1972. The consensus standards included over 400 pages of standards from the American National Standards Institute (ANSI) and the National Fire Protection Association (NFPA), both national consensus organizations.

On August 13, 1971, OSHA made effective immediately those standards having an effective date initially set for February 1972. These included standards for noise, personal protective equipment, and nearly 500 permissible exposure limits (PELs) for approximately 400 chemical substances (Ashford 1976)[3] .

These PELs were largely based upon the 1968 American Conference of Governmental Industrial Hygienists (ACGIH)[4] threshold limit values (TLVs) that had been included as part of the Walsh-Healy Public Contracts Act of 1969. ACGIH had separated out the chemicals it considered to be carcinogenic and included them as an appendix to the TLV list, with zero as a recommended exposure limit. The Department of Labor concluded the appendix was not part of the Walsh-Healy Act and was, therefore, not a federally established standard that OSHA could adopt.

After the two-year start-up period, OSHA was required to follow Section 6 (b) of the Act to promulgate any new standard. This section requires OSHA to publish the standard in the Federal Register as a proposed rule and allows interested persons a period of 30 days after publication to submit written data or comments. Within the comment period, any interested person may file written objections to the proposed rule and request a public hearing to be held to hear these objections. Within 60 days after the comment period expires, or within 60 days after a hearing is held, OSHA is required to decide whether or not to issue the rule. The rule may contain a provision delaying the effective date, not to exceed ninety days, as determined by the Secretary, to allow employers and employees to become familiar with the provisions of the standard (U.S. Congress 1970).

Standards that apply to toxic materials or harmful physical agents, also must fulfill these requirements:

> The Secretary, in promulgating standards dealing with toxic materials or harmful physical agents under this subsection, shall set the standard which most adequately assures, to the extent feasible, on the basis of the best available evidence, that no employee will suffer material impairment of health or functional capacity even if such employee has regular exposure to the hazard dealt with by such standard for the period of his working life. Development of standards under this subsection shall be based upon research, demonstrations, experiments, and such other information as may be appropriate. In addition to the

CASE STUDY

George Guenther Administration, 1971–1973: A Closely Watched Start-up (USDOL, undated)

The Occupational Safety and Health Act of 1970 heralded a new era in the history of public efforts to protect workers from harm on the job. This Act established for the first time a nationwide, federal program to protect almost the entire work force from job-related death, injury and illness. Secretary of Labor James Hodgson, who had helped shape the law, termed it "the most significant legislative achievement" for workers in a decade.[1] Hodgson's first step was to establish within the Labor Department, effective April 28, 1971, a special agency, the Occupational Safety and Health Administration (OSHA) to administer the Act. Building on the Bureau of Labor Standards as a nucleus, the new agency took on the difficult task of creating from scratch a program that would meet the legislative intent of the Act.

Secretary Hodgson selected George Guenther, an experienced public official who was then director of the Labor Standards Bureau, to head OSHA as Assistant Secretary of Labor for Occupational Safety and Health. Guenther had come to the Labor Department from the post of deputy secretary of the Pennsylvania Department of Labor and Industry. Before that he headed a hosiery manufacturing firm in Pennsylvania.

At Senate hearings on April 2, 1971, on his nomination to head OSHA, Guenther stressed that under him the agency would strive to be "responsive and reasonable" in its administration of the Act. Senator Jacob Javits, a New York Republican who had played a major role in the passage of the Act, told Guenther:

> I hope you will remember one thing. . . . We have not finished with the occupational safety law and . . . (Y)ou have the lives of millions of workers in your hands. . . . Don't be afraid of anybody who tries to intimidate you, whether it is business or labor or politics (sic). You have got friends here and advocates. Come to us.[2]

During OSHA's start-up phase some of its actions and policies were reasonably successful, others were less so. The organizing and establishment of the agency within the Labor Department went smoothly. A decision to seek voluntary compliance and avoid a punitive approach to enforcement was well-received by the business community. Because of limitations on resources to protect workers in five million workplaces nationwide, OSHA loosely targeted its enforcement in a worst-case-first approach which emphasized investigation of catastrophic accidents and employers' compliance in the most dangerous and unhealthy workplaces. Partly at the urging of organized labor, OSHA tried to emphasize the "H" (for Health) in its name. The first standard that it set was for asbestos fibers.

On the other hand, a move that brought OSHA long-lasting notoriety as a "nitpicker" was the verbatim adoption and sometimes unreasonable enforcement of a body of voluntary consensus standards developed by industry associations. While adoption was specifically mandated by the Act, OSHA chose to promulgate the rules en masse and immediately, having them take effect in August 1971 instead of using the full two-year phase-in period which the law allowed. Another unsuccessful early move was the decision to develop state programs as the primary means of realizing the goals of the Act. Counting on the bulk of the states to participate, OSHA limited the development of its own staff of enforcement officers. It quickly became apparent that the states were not going to participate as extensively as OSHA had hoped. As a result, the agency soon found itself inadequately prepared to directly enforce the law on a nationwide basis. A damaging legacy from the start-up period came to light during the 1974 Watergate investigations in the form of an internal memorandum which sought ideas on ways to tailor OSHA's program that would increase business' support of President Nixon's reelection campaign in 1972. There is no evidence that this "responsiveness" program affected OSHA significantly at the time, but its revelation in 1974 did considerable damage to the agency's reputation.[3]

OSHA's principal client groups–organized labor and the business community–played active roles during the start-up phase and throughout the agency's history. Organized labor had been instrumental in the passage of the OSH Act. Labor leaders and workers' advocates, from AFL-CIO president George Meany to Ralph Nader, spoke out frequently on OSHA's conduct and policies. In June 1971 Meany told a special union meeting on workers' safety and health that when the OSH Act passed, "the AFL-CIO served notice that we were going to be watching over the government's shoulder. . . .We warned that if this law isn't fully and effectively implemented, we would scream bloody murder."[4] In April 1972 Nader announced the completion of "Occupational Epidemic," a study he sponsored condemning government, business and labor for failing to deal with illness and injury on the job. Published later under the title *Bitter Wages*, it was prepared at his Center for the Study of Responsive Law by a task force of lawyers, law students, and graduate students in medicine and engineering. Nader and his report had harsh words for the Labor Department, which he called a "hostile environment" for enforcement of the OSH Act.

While organized labor and its supporters agreed that OSHA's initial effort to enforce the Act was not effective, the business community did not have a uniform position on OSHA. Many in the small business community strongly opposed the agency. On the other hand, the National Association of Manufacturers (NAM), *Fortune* magazine, and other entities associated with large enterprises were relatively tolerant of it. Their equanimity resulted to a large extent from the fact that, unlike most small businesses, large corporations were used to dealing with federal regulation and usually had longstanding safety programs of their own.

This is not to say that big business had no concerns about the Act or its implementation. Leo Teplow, a safety expert from the steel industry, called the

OSH Act "the most extensive federal intervention into the day-to-day operation of American business in history."[5] Employers were concerned about the competence of OSHA inspectors, the costs of compliance with federal standards, and the burden on business of meeting the Act's requirements to report and record injuries, illnesses and deaths on the job. The consensus standards that OSHA adopted were familiar to large employers, but it was a different matter to employers now that these rules were to be enforced as law.

OSHA took several steps to meet business objections. In January 1972, after receiving numerous petitions from employers for exceptions from recordkeeping and reporting rules, OSHA eased the requirements somewhat. The previous September the agency granted the Boeing Co. the first interim variance under the OSH Act, which allows temporary exemption from a standard if an employer can convince OSHA that he can provide equivalent protection. OSHA also started to revise its consensus rules. This was part of a three-phase standards development program consisting of corrections of minor mistakes in the consensus standard, changes in these standards, and adoption of new safety and health rules.[6]

Despite such accommodations to business, however, for many smaller employers OSHA loomed as a threat to their economic well-being and even survival. In the spring of 1972 a movement centered in Wyoming developed among small businessmen who sought to curtail enforcement of the OSH Act. They flooded their congressmen with letters about alleged harsh tactics by OSHA inspectors, such as forcing businessmen to close because of safety violations and threatening employers with jail sentences.

This campaign had a tremendous effect on Congress in an election year. More than 80 bills were introduced to limit OSHA's powers, principally by exempting small businesses from the Act. Committees held hearings on OSHA. Congressmen spoke out against it. Senator Carl Curtis of Nebraska said that attempts to limit OSHA were a reply to those who would allow the "arrogance of power" to run over small businessmen.[7]

Organized labor strongly opposed exemptions, but was sympathetic to the problems of small businesses. Jacob Clayman of the AFL-CIO, testifying before a House committee in June 1972, said that stories of harassment were blown out of proportion, but charged that inept administration was causing real problems for small businesses. He said that OSHA had made special efforts to inform large corporations about the Act but failed to include small businesses. George Taylor, also from the AFL-CIO, charged before later hearings that OSHA had been "unpardonably" remiss in its failure to tell small businessmen about their rights and duties under the OSH Act.[8]

The Labor Department worked hard to defeat the growing movement to exempt small employers. On Capitol Hill, OSHA spokesmen denied charges that the agency fined ranchers and farmers in Wyoming, ordered the closing of a public school, or drove a company out of business because of violations. These claims, they said, were based on nothing but rumors. Guenther told a House committee that he strongly opposed across-the-board exemptions of small businesses, but indicated that OSHA would support a move to give it the authority to omit inspections of very small employers at its discretion. He also told the committee that OSHA would soon exempt employers with eight or fewer employees from certain reporting requirements. OSHA was also willing to provide consultation visits, if given the legal authority to do so.[9]

Despite vigorous efforts by the Labor Department and labor unions, exemptions of small employers twice came within a presidential veto of enactment in 1972. In June the House passed an amendment to the Labor-HEW appropriations bill exempting employers with 25 or fewer employees from the Act. The Senate passed a narrower exemption for places with 15 or fewer workers. The conference committee adopted the Senate version of the amendment and sent the appropriations bill to the White House for President Nixon's signature. Nixon, concerned about the size of the appropriations for the departments, vetoed the bill. In the fall, Congress again included an exemption amendment in the new appropriations bill that it sent to the President. However, the Congress and the President were still at odds over the budget and Nixon once again vetoed the Labor-HEW appropriations bill, ending efforts for exemption for the time being.[10]

1. Safety Standards, March 1971.
2. U.S. Senate, Committee on Labor and Public Welfare, Hearings on the Nomination of George Guenther, Washington, 1971, p.11.
3. See discussion of "Watergate memo" below in text.
4. *National Journal* (*NatJ*), Dec. 4, 1971.
5. *Wall Street Journal* (*WSJ*), Dec. 1, 1971.
6. *Daily Labor Report* (*DLR*), Sept. 23, 1971, Jan. 22, March 3, 1972.
7. *DLR*, June 21, 1972.
8. *New York Times* (*NYT*), June 29, Sept. 28, 1972.
9. *DLR*, June 14, 21, 1972; *NYT*, Oct. 3, 1972; *WSJ*, June 32, 1972.
10. *NatJ*, July 1, 1972; *NYT*, June 16, 28, 29, Oct. 8, 1972; *DLR*, June 23, Aug. 18, 23, 1972; *Occupational Safety and Health Reporter* (*OSHR*), Nov. 2, 1972.

attainment of the highest degree of health and safety protection for the employee, other considerations shall be the latest available scientific data in the field, the feasibility of the standards, and experience gained under this and other health and safety laws. Whenever practicable, the standard promulgated shall be expressed in terms of objective criteria and of the performance desired.

The Act also gave OSHA the authority to issue emergency temporary standards (ETS) if the Secretary determines that:

- employees are exposed to grave danger from exposure to substances of agents determined to be toxic or physically harmful or from new hazards

- the emergency standard is necessary to protect employees from such danger.

An ETS is immediately effective on the date of its publication in the Federal Register.

OSHA's first ETS was issued for asbestos in 1971. It was not challenged in court (Martonik, Nash, and Grossman 2001). Subsequent ETSs for vinyl chloride and DBCP (1,2,-dibromo-3-chloropropane) were also not challenged. However, when the second ETS for asbestos was issued in 1983, four other ETSs had been rejected. The second asbestos ETS was also rejected, and OSHA has not issued an ETS for any hazard since that time.

General Duty Requirements

The OSH Act contains a broad-based provision to be used when a standard has not been promulgated to address a particular serious hazard. That provision, known as the General Duty Clause, is found in Section 5(a) of the Act, which states: "Each employer shall furnish to each of his employees employment and a place of employment which are free from recognized hazards that are causing or are likely to cause death or serious physical harm to his employees." (U.S. Congress 1970)

An early use of the General Duty Clause involved a citation dated July 7, 1971, issued to the American Smelting and Refining Company's (ASARCO) plant in Omaha, Nebraska. The 1971 citation was issued in July before the adoption of the TLVs in August of that year (see earlier discussion); otherwise, the citation would have listed a violation of 29 CFR 1910.1001. The citation read:

> Airborne concentrations of lead significantly exceeding levels generally accepted to be safe working levels, have been allowed to exist in the breathing zones of employees working in the lead-melting area, the retort area, and other work places. Employees have been, and are being exposed to such concentrations. This condition constitutes a recognized hazard that is causing or likely to cause death or serious physical harm to employees." (NIOSH 1973)

ASARCO contested the citation and the case was heard by the Occupational Safety and Health Review Commission. ASARCO argued that the lead levels in excess of the threshold limit value of 0.2 milligrams per cubic meter (mg/m^3) in its plant did not constitute a recognized hazard causing, or likely to cause, death or serious physical harm, because its employees used respirators and were part of a biological sampling program. The company said there was no evidence that any of its employees had been injured by the airborne concentrations of lead.

The Hearing Examiner found that ASARCO's first responsibility was to reduce the airborne levels to 0.2 mg/m^3, or as close to that figure as possible. The Hearing Examiner determined that proof of violating Section 5(a)(1) of the Act does not depend upon proof that injury has occurred. The original citation was upheld and the proposed penalty of $600 was affirmed.

In previous years, OSHA's position was that exposures above other exposure limits could be cited under Section 5(a)(1). In 1983, then-Assistant Secretary Thorne Auchter wrote in a letter of interpretation that pesticide exposures above the TLV could be cited as 5(a)(1) (OSHA 1983).

After the 1989 Air Contaminants Standard was vacated in 1992 (see further discussion in section entitled "PEL Update"), OSHA issued an interpretation to the compliance staff that Section 5(a)(1) could be used to cite exposures to the 164 substances that were not previously regulated by OSHA. This section could also be used to cite exposures between the 1989 proposed PELs and the original limits adopted in 1971 (OSHA 1993).

The guidance in 2005 from OSHA's *Field Inspection Reference Manual* is:

> Section 5 (a)(1) shall not normally be used to impose a stricter requirement than that required by the standard. For example, if the standard provides for a permissible exposure limit (PEL) of 5 ppm, even if the data establishes that a 3 ppm level is a recognized hazard, Section 5 (a)(1) shall not be cited to require that the 3 ppm level be achieved unless the limits are based on different health effects. If the standard has only a time-weighted average [P]ermissible[E]xposure [L]evel and the hazard involves exposure about the recognized [C]eiling[L]evel, the Area Director shall consult with the Regional Solicitor. (OSHA 1994a)

Hierarchy of Control

Long before OSHA was formed, health and safety professionals ranked controls by their reliability and efficiency in controlling hazards. This ranking was referred to as the hierarchy of control, and includes (in order of preference), engineering controls, work practices, and personal protection equipment.

National consensus organizations have also long supported the concept of the hierarchy of controls. In its consensus standard of 1959, the American National Standards Institute wrote:

> Respirators are used to supplement other methods of control of airborne contaminants rather than to substitute for them. Every effort should be made to prevent the dissemination of contaminants into the breathing zone of workers. In some instances, it is necessary to use respirators only until these control measures have been taken; in others, such measures are impracticable, and the continued use of respirators is necessary. (ANSI 1959)

In 1963, the American Conference of Government Industrial Hygienists (ACGIH) and the American Industrial Hygiene Association (AIHA) jointly published a guide to respiratory protection that made clear the preferred methods of controlling occupational health hazards:

> In the control of those occupational diseases caused by breathing air contaminated with harmful dusts, fumes, mists, gases, or vapors, the primary objective should be to prevent the air from becoming contaminated. This is accomplished as far as possible by accepted engineering control measures. . . . (ACGIH and AIHA 1963)

When OSHA was formed, the initial standards from consensus organizations adopted under Section 6(a) included three provisions establishing the hierarchy of control:

- The respiratory protection standard, 29 CFR 1910.134, included language from the 1969 ANSI consensus standard that stated the primary objective was to prevent atmospheric contamination. This was to be accomplished as far as feasible[5] by engineering controls. When engineering controls were implemented or where they were not feasible, respirators could be used per the requirements of this regulation. The current standard contains the same language.
- The noise standard, 29 CFR 1910.95, requires employers to comply with the permissible exposure limit through the use of feasible administrative or engineering controls.
- The air contaminants standard, 29 CFR 1910.1000, indicates that, in order to achieve compliance with the permissible exposure limits, administrative and engineering controls must first be determined and implemented whenever feasible (U.S. Congress, Office of Technical Assessment 1985).

All substance-specific standards issued by OSHA have required the use of engineering controls and administrative work practices. These are to be used to reduce exposures to or below the permissible exposure limit, or as low as possible, before personal protective equipment can be used.

The use of respirators is generally allowed in the following circumstances:

- during the time it takes to install engineering controls or implement work practices
- when engineering and administrative controls are not feasible, such as during equipment repair
- when engineering and administrative controls are not sufficient in themselves to reduce exposures below the permissible exposure limits
- for some specific substance standards, for cases of intermittent or short-term exposure
- where they are worn to further reduce exposures even when exposures are below the permissible exposure limits
- in emergencies

Health Standards

The Early Years (Start-up to 1984)

As previously mentioned, any new standards issued after the two-year start-up period must follow the multistep process described in Section 6(b) of the Act. Through 1984, OSHA issued eighteen separate health standards. The average time it took OSHA to finalize the rule after initial notification was about two years. These standards are listed in Figure 1.

Regulation	Final Standard Issued
Asbestos	1972
Fourteen carcinogens	1974
Vinyl chloride	1974
Coke oven emissions	1974
Benzene	1978
1,2–dibromo-3-chloropropane (DBCP)	1978
Inorganic arsenic	1978
Cotton dust/cotton gins	1978
Acrylonitrile	1978
Lead	1978
Cancer policy	1980
Access to employee exposure and medical records	1980
–Modified	1988
Occupational noise exposure/hearing conservation	1981
Lead – reconsideration of respirator fit-testing requirements	1982
Coal tar pitch volatiles – modification of interpretation	1983
Hearing conservations reconsideration	1983
Hazard communication	1983
Ethylene oxide	1984

FIGURE 1. OSHA health standards issued from 1972 until 1984 (*Source:* Office of Technical Assistance, 1985)

Ten of these regulations established new permissible exposure limits (PELs) and included requirements for monitoring and medical surveillance (for asbestos, vinyl chloride, coke oven emissions, benzene, DBCP, arsenic, cotton dust, acrylonitrile, lead, and ethylene oxide). The carcinogen (fourteen carcinogens) standard contained work practices and medical surveillance requirements. The hearing conservation standard included monitoring, audiometric testing, training, and record-keeping requirements.

The access to medical records standard provides employees and their authorized representatives access to medical and industrial hygiene monitoring records. The hazard communication standard, initially only applying to the manufacturing sector, is a very broad-based standard requiring containers of hazardous chemicals to be labeled. Chemical manufacturers must prepare material safety data sheets (MSDSs), provide the MSDSs to each purchaser of the material, and the purchaser must make them available to its employees. The cancer policy addressed future regulation of carcinogenic chemicals.

These standards resulted in new or revised requirements for 24 chemicals and for noise. The time frame for the development of these regulations (discussed above) did not include the time required to resolve legal challenges. The hearing conservation amendment, the benzene standard, and the requirements for one of the fourteen carcinogens were set aside by the courts (U.S. Congress, Office of Technical Assessment 1985).

Feasibility

From its very beginning, OSHA recognized the need to collect information concerning the technical feasibility of and costs associated with promulgating its standards in order to counter the arguments made by opponents of the regulations (U.S. Congress Office of Technical Assistance 1985).

In its first full standard, OSHA proposed to lower the PEL for asbestos from 12 fibers per cubic centimeter (f/cc) to 5 f/cc in 1972, with a further lowering to 2 f/cc in 1976. The delayed effective date was "to allow employers to make the needed changes for coming into compliance." The Industrial Union Department of the American Federation of Labor--Congress of Industrial Organizations (AFL-CIO) brought suit challenging, among other issues, the use of economic factors in setting the limit. The Industrial Union Department argued that the phrase "to the extent feasible" in section 6(b)(5)[6] of the Act should be interpreted to only mean whether or not the technology to control exposures was available. The District of Columbia Court of Appeals ruled that OSHA could take into account the costs of compliance with a new standard. The court said that a standard would be considered economically feasible if compliance with it would not threaten the viability of an industry as a whole, even if individual businesses might close.

Two subsequent decisions refined the two-pronged definition of feasibility. The AFL-CIO then challenged

OSHA's decision to relax the requirements of the mechanical power press guarding standard because OSHA determined that it was infeasible to meet the requirements with current technology. The Third Circuit Court of Appeals upheld the decision, saying that while the OSH Act was a "technology-forcing piece of legislation," OSHA's determination that the standard was technologically infeasible was supported. In an industry challenge to the vinyl chloride standard, the Second Circuit Court of Appeals upheld a "technology-forcing" standard even if the technology necessary for compliance was not readily available. A standard would be considered infeasible only if meeting the standard was shown to be "clearly impossible."

President Gerald Ford issued the first Executive Order requiring inflationary impact statements during the same time frame as these decisions. OSHA came under pressure from the Council of Wage and Price Stability (CWPS) to base its decisions on cost-benefit analysis. The CWPS argued that a new standard on coke oven emissions was not worthwhile based on its economic analysis. CWPS also recommended OSHA consider allowing the use of respirators to comply with the standard.

Before the standard was finalized, Morton Corn[7] was appointed Assistant Secretary of Labor for OSHA (Figure 2 lists the persons who have served in this position since the Agency's beginning). Regarding cost-benefit analysis, he engaged in an in-depth review and concluded that the methodology for disease and death effects is very preliminary, and one can derive any desired answer. When OSHA issued the coke oven emissions standard, it rejected the use of cost-benefit analysis (MacLaury undated).

OSHA issued final standards for benzene, DBCP, arsenic, cotton dust, acrylonitrile, and lead in 1978. Four were challenged in the courts. The cases concerning benzene and cotton dust are particularly relevant to OSHA's use of economic analysis (U.S. Congress, Office of Technical Assessment 1985).

Asst. Secretary	Term
George Guenther	April 1971–January 1973
M. Chain Robbins (Acting)	January 1973–April 1973
John Stender	April 1973–July 1975
Bert Concklin & Marshall Miller (Acting)	July 1975–December 1975
Morton Corn	December 1975–January 1977
Bert Concklin (Acting)	January 1977–April 1977
Eula Bingham	April 1977–January 1981
David Zeigler (Acting)	January 1981–March 1981
Thorne G. Auchter	March 1981–April 1984
Patrick Tyson (Acting)	April 1984–July 1984
Robert A. Rowland (Recess appointment; never confirmed)	July 1984–July 1985
Patrick Tyson (Acting)	July 1985–May 1986
John A. Pendergrass	May 1986–March 1989
Alan C. McMillan (Acting)	April 1989–October 1989
Gerard F. Scannell	October 1989–January 1992
Dorothy L. Strunk (Acting)	January 1992–January 1993
David Zeigler (Acting)	January 1993–November 1993
Joseph A. Dear	November 1993–January 1997
Gregory R. Watchman (Acting)	January 1997–November 1997
Charles N. Jeffress	November 1997–January 2001
R. Davis Layne (Acting)	January 2001–August 2001
John L. Henshaw	August 2001–December 2004
Jonathan L. Snare (Acting)	January 2005–April 2006
Edwin G. Foulke, Jr.	April 2006–Present

FIGURE 2. Occupational Safety & Health Administration Assistant Secretaries (Courtesy of the United States Department of Labor)

Benzene – Significant Risk

The benzene PEL was initially set in 1971 as 10 ppm. OSHA issued a standard in 1978, which included the 1 ppm PEL and a 5 ppm ceiling limit. OSHA adopted 1 ppm based upon its "lowest feasible level" policy. Epidemiological studies indicated that human beings contracted leukemia at concentrations significantly below the 10 ppm PEL. OSHA said that higher exposures carried a greater risk and that the determination of a level that presents no hazard could not be determined. Thus, a safe level could not be determined. Therefore, OSHA felt prudent policy required that the lowest feasible level, 1 ppm, be selected in favor of employee protection (Corn 1992).

The petroleum industry challenged OSHA's more stringent standard for benzene exposure. The Court of Appeals for the Fifth Circuit ruled that the phrase "reasonably necessary or appropriate" contained in Section 3(8)[8] of the Act meant OSHA could issue a more stringent regulation only if it estimated the risks addressed by the standard and determined that the benefits bore a "reasonable relationship" to the costs. The court invalidated the standard because it decided there was insufficient evidence that the new requirements would have any "discernible benefits" (U.S. Congress, Office of Technical Assessment 1985).

The decision was appealed to the Supreme Court, which voted 5 to 4 in 1980 to uphold the lower court's ruling, although it did not follow the same line of reasoning. Five separate opinions were issued by the court, but no single opinion had the support of more than four Justices. Justice Stevens presented the view of four of the Justices who voted to strike down the standard. In his opinion, the issue of whether the Act required OSHA to follow a cost-benefit rule was not addressed and OSHA had not made a "threshold finding" that the risk presented by benzene exposure was "significant:"

> By empowering the Secretary to promulgate standards that are "reasonably necessary or appropriate to provide safe or healthful employment and places of employment," the Act implies that, before promulgating any standards, the Secretary must make a finding that the workplaces in question are not safe. But "safe" is not the equivalent of "risk-free" . . . a workplace can hardly be considered "unsafe" unless it threatens the workers with significant risk of harm.
>
> Therefore before he can promulgate any permanent health or safety standard, the Secretary is required to make a threshold finding that a place of employment is unsafe—in the sense that significant risks are present and can be eliminated or lessened by a change in practices.

The court ruled that the record before OSHA did not contain "substantial evidence" to support such a finding and that OSHA exceeded its authority in issuing a more stringent standard. The court did not rule on whether cost-benefit analysis was required in addition to this requirement to demonstrate "significant risk." It provided only limited guidance as to what was meant by "significant risk," and said in a footnote that OSHA's decisions concerning the acceptable level of risk "must necessarily be based on considerations of policy as well as empirically-verifiable facts".

The term "significant risk" does not appear in the language of the Act. Justice Marshall wrote the dissenting opinion for the four judges who voted to uphold the standard. He argued that the plurality had improperly made its own findings about factual issues and had unfairly described OSHA's analysis. He said the requirement to demonstrate "significant risk" is a "fabrication bearing no connection with the acts or intentions of Congress and is based only on the plurality's solicitude for the welfare of regulated industries."

Cotton Dust – Significant Risk and Cost-Benefit Analysis

Cost-benefit and significant risk were taken up again in a case involving a more stringent standard for cotton dust exposure. OSHA issued a cotton dust standard in 1978 establishing various PELs for different textile operations. OSHA argued that the standard was designed to provide the highest degree of health and safety for workers, but it did not base exposure levels on a cost-benefit analysis. OSHA stated that:

> In setting an exposure limit for a substance, such as cotton dust, OSHA has concluded that it is inappropriate to substitute cost-benefit criteria for the legislatively determined directive of protection of all exposed employees against material impairment of health and bodily function (Federal Register 1978).

The textile industry representatives argued that OSHA had exceeded its statutory authority, because it had not conducted a cost-benefit analysis or explicitly determined that the benefits of the standard justified the compliance costs. OSHA argued that the Act did not require a cost-benefit analysis and it was required to issue the most protective standard where compliance was technologically and economically feasible once it determined that exposure to cotton dust presented a "significant risk" (U.S. Congress, Office of Technical Assessment 1985).

The United States Court of Appeals for the District of Columbia issued an opinion in 1979 upholding the standard, finding it to be economically and technically feasible. It said a formal cost-benefit analysis was not required. In 1981, the United States Supreme Court upheld the standard in a five to three decision. Justice Brennan, in the majority, said:

> In effect then, as the Court of Appeals held, Congress itself defined the basic relationship between costs and benefits by placing the "benefit" of worker health

above all other considerations save that of making "attainment" of the "benefit" unachievable. Any standard based on balancing the costs and benefits by the Secretary that strikes a different balance than that struck by Congress would be inconsistent with the command set forth in Section 6(b)(5). Thus, cost-benefit analysis by OSHA is not required by the state because feasibility is (Corn 1992).

Standards Post-1984

From 1984 until the present, the time frame from start to completion of health standards has significantly increased due to administrative requirements imposed on OSHA and the complexity of risk assessment and cost-benefit analysis. Many of OSHA's health standards have resulted from petitions from labor organizations or public interest groups and from decisions mandated by the courts. Martonik commented on how quickly OSHA's first health standard, Asbestos, was developed, saying, "This timetable seems remarkable when compared with more recent OSHA rulemakings, which have taken from 4 years (e.g., benzene, arsenic) to more than 10 years (e.g., 1,3–butadiene, methylene chloride) from proposal to completion" (Martonik, Nash, and Grossman 2001).

The health standards that have been issued since 1984 appear in Figure 3.

1,3–Butadiene

The history of the 1, 3–butadiene standard is important because, as of the date of this writing, it is the only OSHA standard that was developed as a consensus standard. Interested stakeholders representing employers and employees worked together to outline a voluntary agreement that eventually led to the standard. The final standard was not challenged in court (OSHA 1997b).

The interest in 1,3–butadiene began when the National Toxicology Program released results on an animal study in 1983 indicating 1,3–butadiene (BD) caused cancer in rodents. OSHA and EPA jointly published a request for information in January 1984. Several labor unions[9] petitioned OSHA to issue an ETS of 1 ppm or less. OSHA denied the petitions on the grounds that it was evaluating health data to determine whether regulatory action was appropriate.

The EPA published an Advanced Notice of Public Rulemaking (ANPR) in May 1984 to announce the initiation of a regulatory action by the EPA to determine and implement the most effective means of controlling exposures to the chemical BD under the Toxic Substances Control Act (TSCA). EPA was working with OSHA because available evidence indicated that exposure to BD occurs primarily within the workplace. As a result of the information received, EPA designated BD as a probable human carcinogen and asked OSHA to consider regulating exposure in the workplace. In October 1986, OSHA published an ANPR and then published its proposed rule in August 1990. Hearings were held in 1991 and the comment period closed in 1992.

In the only mediated rule to the date of this writing (mediation had been tried, but failed, on the development of the coke oven and benzene standards), representatives[10] of the industry and labor groups involved in the production and use of BD developed and submitted a voluntary agreement that included provisions the industry/labor group wanted to have included in the standard. The agreement proposed a lower permissible exposure limit, additional provisions for exposure monitoring, and a program designed to reduce exposures below an action level.

Regulation	Final Standard Issued
Asbestos standard revisions	1986
Formaldehyde	1987
Field sanitation	1987
Benzene	1987
Air contaminants update	1989
Chemical exposure in laboratories	1990
Cadmium	1992
MDA	1992
Asbestos standard revisions	1994
1,3–butadiene	1996
Methylene chloride	1997

FIGURE 3. OSHA health standards issued after 1984 (*Source:* USDOL-OSHA, Various and USDOL-OSHA, 1991)

In response to these recommendations, OSHA reopened the rule for further comment. The industry/labor group submitted examples of regulatory text that put their recommendations into specific requirements. OSHA adopted these recommendations and the standard became effective in 1997, thirteen years after the EPA-initiated rulemaking, despite cooperation from the regulated community and its employee representatives (OSHA 1997b).

PEL Update

Another novel approach was an attempt to update the PELs en masse when John Pendergrass[11] served as Assistant Secretary of Labor–Director of OSHA. As previously discussed, OSHA adopted approximately 400 PELs in 1971 for various substances that were largely based upon the 1968 ACGIH threshold limit values. Over the years, OSHA relied upon these values, although some of the guidelines from which the PELS were derived had been lowered.

To resolve this dilemma, OSHA undertook a major revision effort; in 1989, it published a final rule revising 212 existing exposure limits and establishing 164 new PELs. The new limits went into effect in September 1989, although engineering controls to comply with the limits were not required until January 1993. The use of respirators was allowed in the interim (de la Cruz and Sarvadi 1994).

Industry and labor groups challenged the proposed rule. The American Iron and Steel Institute argued that the agency did not demonstrate that the new limits would reduce or eliminate a significant risk or demonstrate the standard's economic and technological feasibility. There were also arguments made that the 7-month time frame in which the standard was developed did not allow for adequate review and comment. Other companies and trade associations were concerned about specific substances being regulated.

The AFL-CIO said that OSHA's evidentiary and factual findings were sound and lowered PELs should have been issued based upon these findings. The union objected to the short time frame the rulemaking took, the lack of specific monitoring or medical-surveillance requirements in the rule, and that the use of respirators was allowed for several years to comply with the reduced PELs instead of engineering controls.

In 1992, the 11th Circuit Court of Appeals vacated the entire air contaminants standard. The court found that OSHA could address multiple substances in one rulemaking but that the agency did not sufficiently demonstrate the proposed changes were necessary or that they were economically or technologically feasible. This decision to vacate both the new limits and the revised, more protective limits forced OSHA to return to the use of the original 1971 limits, still in effect today.

Methylene Chloride

The background of the methylene chloride (MC) standard is illustrative of competing objectives between environmental protection and occupational health and how regulatory priorities can change.

OSHA established a PEL of 500 ppm, a ceiling limit of 1000 ppm, and a maximum peak of 2000 ppm for methylene chloride in 1971. In 1976, NIOSH recommended a reduction of the PEL to 75 ppm. The EPA classified methylene chloride as a probable human carcinogen in 1985. The Consumer Product Safety Commission (CPSC) and the Food and Drug Administration (FDA) took actions to ban the use of methylene chloride in consumer products and cosmetics (OSHA 1997a).

The International Union, United Automobile, Aerospace, and Agricultural Implement Workers (in short, the United Automobile Workers or UAW) petitioned OSHA in 1985 to publish a hazard alert[12], issue an emergency temporary standard, and begin development of a permanent standard. In response to the scientific knowledge and the petition, OSHA issued guidelines in 1986 to control exposure. It denied the petition for an ETS, but began work on a permanent standard. OSHA issued an ANPR and began conducting site visits to assess how the standard would affect small business in response to comments received.

ACGIH lowered the TLV for methylene chloride to 50 ppm in 1988. The FDA banned the use of MC in cosmetic products in 1989, and in 1990 the CPSC required labels on products containing 1 percent or more of methylene chloride.

OSHA published a Notice of Public Rulemaking (NPRM) to address the risks of exposure to methylene chloride in late 1991, and a correction notice was published in January 1992. The proposal was turned over to an advisory committee and hearings were held.

In late 1990, the Clean Air Act Amendments (CAAA) were signed into law. They required the phase-out by 2000 of ozone-depleting chemicals, certain chlorofluorocarbons, and halons. The EPA was tasked to identify substitutes for these chemicals, and methylene chloride was one chemical being considered by the Agency. The CAAA also required EPA to address the residual risks of methylene chloride and other specified chemicals by establishing maximum achievable control technology (MACT) standards. In 1993, EPA issued a notice of proposed rulemaking on MACT rules for methylene chloride and other chemicals. In 1994, the EPA issued a final rule identifying methylene chloride as a substitute for ozone-depleting chemicals being phased out under the CAAA of 1990. The EPA found the use of methylene chloride acceptable in the production of flexible polyurethane foam, polyurethane integral skin foams, metal cleaning, electronics cleaning, precision cleaning, and adhesives, coatings, and inks.

OSHA reopened the rulemaking several times as a result of EPA decisions and to receive comments on engineering controls and other matters. The standard was eventually issued on January 10, 1997.

Shortly thereafter, the UAW, the Halogenated Solvents Industry Alliance, Inc. (HSIA), and others filed a motion asking OSHA to reconsider three aspects of the standard. These three aspects were the decision not to include medical removal benefit protection, the start-up dates for engineering controls, and the decision allowing the use of respirators in some specific uses of methylene chloride

In response, OSHA decided to add some limited medical removal benefits[13]. OSHA also extended the final engineering control start-up date, which had been limited to employers with fewer than twenty employees. The extension now applied to employers in specific application groups who had 20 to 49 employees and to foam fabricators who had 20 to 149 employees (OSHA 1998a).

Asbestos

OSHA has issued two ETSs, three major notices of proposed rulemaking (NPRMs), three final standards, and 31 Federal Register notices related to asbestos regulation. These modifications introduced a framework for the content of toxic substance standards that followed. Medical removal protection was introduced in the 1972 standard. In the1986 standard, the type of work performed, rather than type of business or product made, determined which of the two standards applied.

One innovation of the 1986 standard was the new approach for initial monitoring of employee asbestos exposure. Previously, every employer had to monitor each employee's asbestos exposure to make a determination for that employee. The 1986 standard allowed the use of "objective data"[14] for initial exposure determination. The use of objective data has been adopted in subsequent standards for cadmium, formaldehyde, 1,3–butadiene, and methylene chloride.

The 1994 standards differed from previous standards in many ways. Communication of hazards was required between employers, employees, and contractors. Mandatory work practices are specified for specific jobs and operations, regardless of exposure level (Martonik, Nash, and Grossman 2001).

OSHA initially regulated asbestos at a PEL of 12 f/cc when it adopted the ACGIH TLVs in 1971. In November 1971, shortly after the first asbestos PEL became effective, the Industrial Union Department of the AFL-CIO petitioned the Agency for an ETS based upon a report (date not given) from Dr. Irving Selikoff and colleagues at Mt. Sinai Medical School (Martonik, Nash, and Grossman 2001). Dr. Selikoff recommended the enactment of a work practice standard to control exposure rather than a standard that simply set a PEL (OSHA 1994b).

On December 7, 1971, OSHA issued an ETS that lowered the PEL to 5 f/cc instead of the 2 f/cc the union requested. The ETS did not include work practices or engineering controls, but did contain several specifications for respirator use. OSHA's first ETS was not challenged in court.

OSHA promulgated its first asbestos standard in only 183 days, issuing the new standard on June 7,

1972. It reduced the PEL to 5 f/cc immediately, with a PEL of 2 f/cc to become effective on July 1, 1976. The standard included additional provisions that were to serve as models for other health standards. These included methods of compliance, monitoring intervals, monitoring techniques, labels and warnings, and medical examinations.

The Industrial Union Department of the AFL-CIO petitioned for the court to review the standard, challenging the delay in requiring the lower PEL and objecting that the additional provisions were not sufficiently rigorous. The court upheld the standard, but asked OSHA for an explanation of the uniform application of 2 f/cc PEL in 1976 and the retention period for medical records.

OSHA published an NPRM in October 1975 to revise the asbestos standard and lower the PEL to 0.5 f/cc. Many comments were received in response to the notice and OSHA never adopted a standard based upon the 1975 proposal.

In June 1983, the International Association of Machinists and Aerospace Workers petitioned OSHA for a second ETS, asking that the PEL be reduced from 2 f/cc to 0.1 f/cc, work practices be modified, and other protective provisions be required in spite of exposure level. Sixteen other unions joined with this request.

In November 1983, OSHA granted the request and published an ETS that lowered the PEL to 0.5 f/cc. Included within the standard were numerical risk assessments as a result of the 1980 benzene decision discussed above. An industry trade association petitioned for judicial review and the ETS was vacated on March 7, 1984.

The next month, OSHA published an NPRM that included additional information. OSHA held hearings and in June 1986, issued two final standards—one applying to general industry and the second to construction operations. The approach OSHA took was unique in that the determination of which standard applied was based upon the work activities being conducted. Most activities involving previously installed materials were covered under the construction standard, including maintenance and repair. The construction standard also included specific work-practice requirements based upon the type of work being performed. Both standards included a PEL of 0.2 f/cc (Federal Register 1986).

The 1986 standards included a new approach for the initial monitoring of exposure. In previous health standards, employers were required to assess exposure through initial monitoring provisions. The asbestos standards allowed the use of "objective data"[15] in place of initial monitoring. This concept has been adopted in subsequent health standards.

Industry and labor petitioned the Court of Appeals for review of the standards. Labor argued that a lower PEL was both economically and technologically feasible. An industry trade association argued that the proposed PEL was neither.

The court upheld most of the provisions of the standards, but remanded some provisions back to OSHA for reconsideration. OSHA was asked to explain why a PEL of 0.1 f/cc was found to be infeasible and why certain provisions proposed by labor were not adopted.

OSHA next proposed revising the asbestos standard in June 1990 in response to the issues raised by the court. OSHA published two final standards on August 19, 1994, for construction and for general industry. OSHA lowered the PEL to 0.1 f/cc and required specific work practices for work activities regardless of exposure levels. The standard included provisions for hazard communications between building owners, employees, and contractors.

Industry and labor petitioned the court for judicial review. OSHA issued corrections to the 1994 rules that modified the provisions for working with roofing materials and exempted asbestos-containing asphalt roof coatings, cements, and mastics from the requirements of the standard.

Other Health Standards

Permissible Exposure Limits

The PELs adopted in 1971 appear in subpart Z of OSHA regulations, and are known as the Z-tables. These are expressed in terms of 8-hour time-weighted averages (OSHA 1971).

Benzene	Formaldehyde
Beryllium and Beryllium compounds	Hydrogen Fluoride
Cadmium–dust and fumes	Hydrogen Sulfide
Carbon Disulfide	Methylene Chloride
Carbon Tetrachloride	Mercury
Chromic Acid and Chromates	Organo(alkyl)mercury
Ethylene Dibromide	Styrene
Ethylene Dichloride	Tetrachloroethylene
Fluoride as dust	Toluene
	Trichloroethylene

FIGURE 4. Substances regulated in Table Z-2 (*Source:* OSHA)

Most PELs are found in Table Z-1. Table Z-2 contains PELs for the substances listed in Figure 4.

Table Z-3 contains PELs for exposure to mineral dusts, including silica (crystalline and amorphous), silicates, graphite, coal dusts, and inert or nuisance dusts.

PELs for substance-specific standards are listed in Table 1. The more recent standards also include an action level (AL), which is generally half of the PEL. All PELs and ALs are calculated as 8-hour time-weighted averages. Certain requirements are triggered at exposure above the AL. Some standards also include excursion limits (EL) or ceiling limits (C)[16], which are higher than the PELs. The EL or C values are restricted by the time allowed for exposure and are not-to-exceed limits.

Ventilation Standard

The OSHA ventilation standard, 29 CFR 1910.94, applies to abrasive blasting; grinding, buffing and polishing; and spray finishing operations. It was initially published in 1979 and was amended in 1993, 1996, and 1998.

Section A of the standard applies to operations where an abrasive is forcibly applied to a surface by pneumatic or hydraulic pressure or by centrifugal force. It is not applicable for steam blasting or steam cleaning, or other hydraulic cleaning methods that do not use abrasives.

Section B of the standard requires the use of enclosures and systems to remove dust, dirt, fumes, and gases generated through the grinding, polishing, or buffing of ferrous and nonferrous metals.

Section C of the standard applies to spray-finishing operations where organic or inorganic materials are used in dispersed form for deposit on surfaces to be coated, treated, or cleaned. Such methods of deposit may involve automatic, manual, or electrostatic deposition. The section does not apply to metal spraying or metallizing, dipping, flow coating, roller coating, tumbling, centrifuging, or spray washing and degreasing as conducted in self-contained washing and degreasing machines or systems (OSHA 1974a).

Respirator Standard

OSHA's original respirator standard, 29 CFR 1910.134, contained seven sections, the first six from ANSI Z88.2-1969, Practices for Respiratory Protection, and the seventh section from ANSI Standard K13.1-1969 (U.S. Congress, Office of Technical Assessment 1985). In late 1994, OSHA published a final rule revising the standard, which became effective in 1998.

The standard requires a Respiratory Protection Program that is managed by a qualified program administrator. Respirators must be certified by NIOSH.

TABLE 1

Substances Regulated by Specific Standards

Material	Standard	PEL
Asbestos	1910.1001 1926.1001	0.1 f/cc
Vinyl chloride	1910.1017	1 ppm
Arsenic, inorganic	1910.1018	10 μg/M³
Lead	1910.1025	50 μg/M³
Cadmium	1910.1027	5 μg/M³
Benzene	1910.1028	1 ppm
Coke oven emissions	1910.1029	150 μg/M³
Cotton dust	1910.1043	200 μg/M³
1,2-dibromo-3-chloropropane	1910.1044	1 ppm
Acrylonitrile	1910.1045	2 ppm
Ethylene oxide	1910.1047	1 ppm
Formaldehyde	1910.1048	0.75 ppm
Methylenedianaline	1910.1049	10 ppb
1,3–butadiene	1910.1051	1 ppm
Methylene chloride	1910.1051	25 ppm

(Source: OSHA, 1991)

Pressure-demand supplied-air respirators must be used in atmospheres deemed to be immediately dangerous to life or health (IDLH) and respirators for non-IDLH atmospheres shall either be atmosphere-supplying or air-purifying with an end-of-service-life indicator.

Employees must undergo a medical examination before being assigned a respirator and fit-testing must be performed prior to initial use and annually thereafter. The medical evaluation includes a mandatory questionnaire. The physician provides a written recommendation regarding the employee's ability to use the respirator. The recommendation should include any limitations on respirator use related to a medical condition the employee may have, or relating to the workplace conditions in which the respirator will be used, including whether or not the employee is medically able to use the respirator and any follow-up medical testing required.

The standard specifies the OSHA-accepted quantitative and qualitative fit-testing protocols. All respirators that rely on a mask-to-face seal need to be checked annually using either qualitative or quantitative methods to determine whether the mask provides an acceptable fit to a wearer. The qualitative fit-test procedures rely on a subjective sensation (taste, irritation, smell) of the respirator wearer to a particular test agent, while the quantitative test uses instruments to measure face-seal leakage. The relative workplace exposure level determines what constitutes an acceptable fit and which fit-test procedure is required. For negative-pressure air-purifying respirators, users may rely on either a qualitative or a quantitative fit-test procedure for exposure levels less than ten times the occupational exposure limit. Exposure levels greater than ten times the occupational exposure limit must use a quantitative fit-test procedure for these respirators. Fit-testing of tight-fitting, atmosphere-supplying respirators and tight-fitting, powered air-purifying respirators shall be accomplished by performing quantitative or qualitative fit-testing in the negative pressure mode.

Respirators must be maintained, cleaned, and disinfected. Respirator users shall be trained in specific provisions prior to initial use and annually thereafter (OSHA 1998b).

Noise and Hearing Conservation

The noise and hearing conservation standard, 29 CFR 1910.95, requires that engineering controls and work practices be used to reduce noise exposure below the levels published in its Table G-16, which appears as Table 2 in this chapter. If the controls and practices are insufficient to reduce levels to those in the table, personal protective equipment shall be used to do so (OSHA 1974c).

Monitoring shall be conducted when exposure levels are expected to meet or exceed an 85 dbA 8-hour, time-weighted average. A hearing conservation program is required for employees exposed at or above noise levels of 85 dbA, which includes audiometric testing, the use of hearing protectors, and employee training.

Ionizing Radiation

Both OSHA and the Nuclear Regulatory Commission (NRC) have jurisdiction of U.S. facilities that manufacture or use nuclear materials. The two agencies have signed a memorandum of understanding that

TABLE 2

Permissible Noise Exposures*

Duration Per Day (in hours)	Sound level (dBA slow response)
8	90
6	92
4	95
3	97
2	100
1 1/2	102
1	105
1/2	10
1/4 or less	15

* When the daily noise exposure is composed of two or more periods of noise exposure of different levels, their combined effect should be considered, rather than the individual effect of each. If the sum of the following fractions: C(1)/T(1) + C(2)/T(2) C(n)/T(n) exceeds unity, then, the mixed exposure should be considered to exceed the limit value. Cn indicates the total time of exposure at a specified noise level, and Tn indicates the total time of exposure permitted at that level. Exposure to impulsive or impact noise should not exceed 140 dB peak sound-pressure level.

(Source: OSHA)

TABLE 3

Allowable Quarterly Exposure

	Rems per Calendar Quarter
Whole Body: Head and trunk, active blood-forming organs; lens of eyes, or gonads	1 1/4
Hands and forearms; feet and ankles	18 3/4
Skin of whole body	7 1/2

(Source: OSHA, 1974b)

delineates the general areas of responsibility of the agencies (OSHA 1989).

The NRC regulates radiation exposure under 10 CFR Part 20. The allowable exposure limits are found in Section 10.1201, Occupational Dose Limits. The annual dose limit for adults is 5 rem total effective dose (NRC 1991).

The OSHA Ionizing Radiation standard, 29 CFR 1910.1096, sets forth the following limits of exposure that are for one calendar quarter of the year in its Table G-18, which appears as Table 3 in this chapter. A rem is a measure of the dose of any ionizing radiation to body tissue in terms of its estimated biological effect relative to a dose of 1 roentgen (r) of X-rays. The relation of the rem to other dose units depends upon the biological effect under consideration and upon the conditions for irradiation.

There are two exceptions to these limits. An employee may receive a dose to the whole body up to 3 rem in a calendar quarter or if his or her accumulated dose to the whole body does not exceed 5 $(N - 18)$ rem, where N is the employee's age in years at his or her last birthday. The employer must maintain records that indicate the dose will not cause the individual to exceed the amount authorized.

The standard includes requirements for monitoring, provision of individual exposure monitoring devices, warning signs and labels, and an immediate evaluation warning signal.

The standard also requires that OSHA be immediately notified, for employees not under NRC jurisdiction, when an employee is exposed to a whole body does of 25 rem (and other high doses) or when a release to the environment of a specified amount of radioactivity occurs. OSHA must also be notified within 24 hours when an employee is exposed to a whole body dose of 5 rem (and other doses).

OSHA Industrial Hygiene Inspections

Health inspections follow the same procedures as safety inspections, which are outlined in OSHA's *Field Inspection Reference Manual* (OSHA 1994a). Industrial hygiene sampling procedures are found with the *OSHA Technical Manual* (OSHA 1995) and sampling and analytical procedures are located in a searchable database, maintained by the Salt Lake Technical Center, on the OSHA Web site (OSHA 2005). Various manufacturers, such as SKC, Inc. (2007) or Mine Safety Appliances Company (2007), provide a wealth of information on their web sites and in the *NIOSH Pocket Guide* (NIOSH 2005).

OSHA compliance officers will generally perform air sampling to assess compliance with appropriate PELs. Once laboratory results are received, the compliance officer must consider sampling and analytical error in determining if a violation exists. The formulas used calculate the 95 percent confidence limits around the sample result and indicate the lower and upper confidence limits. These confidence limits are termed one-sided since the only concern is in being confident that the true exposure is on one side of the PEL (OSHA 2005).

Sampling and analytical errors (SAEs) have been developed for most materials. The current guidance is to contact the OSHA Salt Lake City Technical Center if no SAE is provided with sample results. OSHA personnel are to use the SAEs provided by an equipment manufacturer for detector tubes or direct-reading instruments.

Various examples are provided within the *OSHA Technical Manual* for single and multiple samples as well as for mixtures. The method used for one full-shift, continuous single sample is as follows:

1. Determine exposure severity (Y) by dividing the sampling result (X) by the PEL:

$$Y = X/\text{PEL} \tag{1}$$

2. Calculate the upper confidence limit:

$$\text{UCL}_{95\%} = Y + \text{SAE} \tag{2}$$

3. Calculate the lower confidence limit:

$$LCL_{95\%} = Y - SAE \quad (3)$$

These one-sided confidence limits are used to classify the sample result into one of three categories:

- in compliance—the result and the UCL both do not exceed the PEL
- in violation—the result and the LCL both exceed the PEL
- possible overexposure—the result does not exceed the PEL but the UCL does exceed the PEL, or the result exceeds the PEL but the LCL is below the PEL

In cases of possible overexposure, the compliance officer is directed to take additional samples. If further sampling is not conducted or the additional results fall into the same category, the compliance officer is directed to discuss the possible overexposure and encourage exposure reduction or additional sampling to determine if the exposure is below the PEL.

State Plans

The first state agency to deal with workplace safety was founded in 1867 in Massachusetts. By 1900, most heavily industrialized states had some form of legislation requiring employers to reduce or eliminate certain workplace hazards. An often separate effort was being made by state health agencies to control disease (Ashford 1976).

The language of the OSH Act in Section 18 encouraged states to administer their own health and safety laws:

> (a) Nothing in this Act shall prevent any State agency or court from asserting jurisdiction under State law over any occupational safety or health issue with respect to which no standard is in effect under section 6.
>
> (b) Any State which, at any time, desires to assume responsibility for development and enforcement therein of occupational safety and health standards relating to any occupational safety or health issue with respect to which a Federal standard has been promulgated under section 6 shall submit a State plan for the development of such standards and their enforcement." (U.S. Congress 1970)

The OSH Act goes on to say that the Secretary must approve a state plan if the plan:

> [P]rovides for the development and enforcement of safety and health standards relating to one or more safety or health issues, which standards (and the enforcement of which standards) are or will be at least as effective in providing safe and healthful employment and places of employment as the standards promulgated under Section 6 which relate to the same issues." (U.S. Congress 1970)

Currently, the states listed in Figure 5 have approved state plans.

Mine Safety and Health Administration

The Mine Safety and Health Administration (MSHA) is a separate federal agency within the Department of Labor, which develops standards and provides enforcement in mining operations. Prior to its formation, the Metal and Nonmetallic Mine Safety Act of 1966 provided for inspections and health and safety standards. In 1969, the Coal Mine Safety Act was enacted, which set mandatory health and safety standards, provided aid to states to improve compensation programs, and

Alaska	New Mexico
Arizona	New York
California	North Carolina
Connecticut	Oregon
Hawaii	Puerto Rico
Indiana	South Carolina
Iowa	Tennessee
Kentucky	Utah
Maryland	Vermont
Michigan	Virgin Islands
Minnesota	Virginia
Nevada	Washington
New Jersey	Wyoming

FIGURE 5. Approved state occupational safety and health plans (*Source:* OSHA) *Note:* The plans for Connecticut, New Jersey, New York, and the Virgin Islands cover public-sector (state and local government) employment only.

provided for research into coal workers' pneumoconiosis (black lung). The Act also provided compensation to miners disabled by black lung or to their widows. In 1973, the responsibility for enforcing this Act and the other Metal and Nonmetallic Mine Safety Act was transferred to a new Mining Enforcement and Safety Administration (Ashford 1976).

The Mine Safety and Health Administration was created by Public Law 95-164, the Federal Mine Safety and Health Act of 1977. Under the Act, the Secretary may "develop, promulgate, and revise as may be appropriate, improved mandatory health or safety standards for the protection of life and prevention of injuries in coal or other mines."

The Agency has standards for exposure to respirable dust. The following MSHA standards adopted the 1973 TLVs by reference:

- 30 CFR, Part 56: Safety and health standards—surface metal and nonmetal mines
- 30 CFR, Part 57: Safety and health standards—underground metal and nonmetal mines

MSHA adopted a Hazard Communication Standard in 2000, which includes provisions similar to those contained in the OSHA Standard. It amended its Noise Standard in 1999 to include hearing conservation provisions.

Environmental Protection Agency

The Environmental Protection Agency (EPA) was founded in 1970 during President Nixon's administration. Throughout the years, the EPA and OSHA have had many successes in working together while, in some cases, issues of jurisdiction have been troublesome.

The first jurisdictional battle occurred in the early 1970s concerning the regulation of pesticide exposure. OSHA issued a temporary emergency standard for 21 pesticides on May 1, 1973. The effective date of the standard was stayed by the court and the standard itself was vacated in 1974. Before the standard was vacated, two organizations and an individual filed suit against OSHA seeking a permanent standard. EPA issued its pesticide standard in 1974. The court decided that EPA had the authority for setting standards for pesticide exposure.

The EPA develops regulations under a number of statutes, including the Toxic Substances Control Act (TSCA) and recently began setting OELs using authority vested by the Act (15 USC s/s 2601). Under Section 5 of the Act, any person who plans to manufacture or import a new chemical substance must first submit notice to the EPA. The EPA conducts a risk assessment and, if the agency determines that the chemical may present an unreasonable risk of injury via inhalation exposure, it will set a New Chemical Exposure Limit (NCEL). These limits are for informational purposes only and are not legally enforceable limits. However, chemical manufacturers are bound by a TSCA Section 5 (e) Consent Order to follow the agency's recommendations. At the time of this writing, there are 35 limits posted on the EPA Web site with the title, Non-Confidential List of NCELs (EPA 2002).

OSHA in the Twenty-First Century

In 1991, The United Food and Commercial Workers Union (UCFW), along with the AFL-CIO and 29 other labor organizations, petitioned OSHA to publish an ETS on ergonomic hazards. OSHA concluded that there was not a sufficient basis to support issuance of an ETS, but agreed there was a need to initiate rulemaking to address ergonomic hazards. In 1992, OSHA published the ANPR requesting information for consideration in the development of an ergonomics standard. OSHA issued its first ergonomics guidelines in 1990 for meatpacking plants. In 1996, OSHA developed a strategy to address ergonomics through a four-pronged program including:

- training, education, and outreach activities
- study and analysis of the work-related hazards that lead to muscular-skeletal disorders
- enforcement
- rulemaking

OSHA published the ergonomics program standard on November 14, 2000. The standard became effective on January 16, 2001 (OSHA 2000).

CASE STUDY

General and Health-Related Milestones from OSHA's Thirty-Five Year Milestones

(USDOL, undated)

OSHA's mission is to send every worker home whole and healthy every day. Since the agency was created in 1971, workplace fatalities have been cut in half and occupational injury and illness rates have declined 40 percent. At the same time, U.S. employment has nearly doubled from 56 million workers at 3.5 million work sites to 105 million workers at nearly 6.9 million sites. The following milestones mark the agency's progress over the past 30 years in improving working environments for America's workforce.

December 29, 1970: President Richard M. Nixon signed the Occupational Safety and Health Act of 1970.

May 29, 1971: First standards adopted to provide baseline for safety and health protection in American workplaces.

January 17, 1972: OSHA Training Institute established to instruct OSHA inspectors and the public.

November–December, 1972: First states approved (South Carolina, Montana, and Oregon) to run their own OSHA programs.

May 20, 1975: Free consultation program created-nearly 400,000 businesses participated in past 25 years.

June 23, 1978: Cotton dust standard promulgated to protect 600,000 workers from byssinosis; cases of "brown lung" have declined from 12,000 to 700 in last 22 years.

January 20, 1978: Supreme Court decision setting staffing benchmarks for state plans to be "at least as effective" as federal OSHA.

April 12, 1978: New Directions grants program created to foster development of occupational safety and health training and education for employers and workers. (Nearly one million trained over 22 years.)

November 14, 1978: Lead standard published to reduce permissible exposures by three-quarters to protect 835,000 workers from damage to nervous, urinary, and reproductive systems. (Construction standard adopted in 1995.)

February 26, 1980: Supreme Court decision on Whirlpool affirming workers' rights to engage in safety and health-related activities.

May 23, 1980: Medical and exposure records standard finalized to permit worker and OSHA access to employer-maintained medical and toxic exposure records.

July 2, 1980: Supreme Court decision vacates OSHA's benzene standard, establishing the principle that OSHA standards must address and reduce "significant risks" to workers.

January 16, 1981: Electrical standards updated to simplify compliance and adopt a performance approach.

July 2, 1982: Voluntary Protection Programs created to recognize work sites with outstanding safety and health programs (nearly 700 sites currently participating).

November 25, 1983: Hazard communication standard promulgated to provide information and training and labeling of toxic materials for manufacturing employers and employees (Other industries added August 24, 1987).

November–December, 1984: First "final approvals" granted to state plans (Virgin Islands, Hawaii and Alaska) giving them authority to operate with minimal oversight from OSHA.

April 1, 1986: First instance-by-instance penalties proposed against Union Carbide's plant in Institute, West Virginia, for egregious violations involving respiratory protection and injury and illness record keeping.

January 26, 1989: "Safety and Health Program Management Guidelines," voluntary guidelines for effective safety and health programs based on VPP experience, published.

March 6, 1989: Hazardous waste operations and emergency response standard promulgated to protect 1.75 million public and private sector workers exposed to toxic wastes from spills or at hazardous waste sites.

December 6, 1991: Occupational exposure to bloodborne pathogens standard published to prevent more than 9,000 infections and 200 deaths per year, protecting 5.6 million workers against AIDS, hepatitis B and other diseases.

October 1, 1992: Education Centers created to make OSHA training courses more available to employers, workers, and the public (12 centers have trained over 50,000 workers and employers to date.)

January 14, 1993: Permit-required confined spaces standard promulgated to prevent more than 50 deaths and more than 5,000 serious injuries annually for 1.6 million workers who enter confined spaces at 240,000 workplaces each year.

February1, 1993: Maine 200 program created to promote development of safety and health programs at companies with high numbers of injuries and illnesses.

June 27, 1994: First expert advisor software, GoCad, issued to assist employers in complying with OSHA's cadmium standard.

August 10, 1994: Asbestos standard updated to cut permissible exposures in half for nearly 4 million workers, preventing 42 cancer deaths annually.

September 4, 1995: Formal launch of OSHA's expanded web page to provide OSHA standards and compliance assistance via the Internet.

June 6, 1996: Phone-fax complaint handling policy adopted to speed resolution of complaints of unsafe or unhealthful working conditions.

November 9, 1998: Strategic Partnership Program launched to improve workplace safety and health through national and local cooperative, voluntary agreements.

April 19, 1999: Site-Specific Targeting Program established to focus OSHA resources where most needed–on individual work sites with the highest injury and illness rates.

November 14, 2000: Ergonomics program standard promulgated to prevent 460,000 musculoskeletal disorders among more than 102 million workers at 6.1 million general industry work sites.

On March 6, 2001, the United States Senate passed a resolution of disapproval of the Ergonomics Program Standard under the Congressional Review Act. The House of Representatives then passed a similar resolution the next day. President George W. Bush signed the resolution into law, forcing OSHA to remove the standard from the Code of Federal Regulations (OSHA 2001).

John Henshaw[17] was appointed as Assistant Secretary of OSHA in August 2001. OSHA's focus has shifted further toward guidelines and compliance assistance in this century. OSHA has expanded the number of strategic partnerships with employers, associations, and labor unions, building a program that began in 1998.

Another focus is to expand the number of companies participating in Voluntary Protection Programs (VPP). VPP is an initiative to promote comprehensive safety and health programs through cooperative arrangements among management, labor, and OSHA (OSHA 2004).

Summary

In its 35-year history, OSHA has recognized the need to develop occupational health standards, with most of those developed under the 6(b) process directed to prevent occupational disease. However, OSHA has long exhibited a safety bias when it comes to enforcement. In 1973, the first full year of enforcement, OSHA conducted 48,409 inspections of which only 6.6 percent were health-related (Ashford 1976).

Not much is different in 2006, as OSHA planned to conduct 38,579 inspections, with the majority targeted to high-hazard industries (OSHA 2006). This was understandable in the beginning, when most of the initial compliance officers were predominantly safety-trained. OSHA employed 933 compliance officers in 1975, with a ratio of one industrial hygienist to five safety professionals (Ashford 1976). Today, federal OSHA employs approximately 1000 compliance officers, of whom approximately half are industrial hygienists.

The preamble to the 1972 asbestos standard, OSHA's first health standard, was only two pages long, while the preamble to the asbestos standard published in 1986 totaled 144 pages and the 1994 standard was 93 pages. It took only six months from the publication of the NPRM to publication of the final standard in 1972, but it was over four years from notice to final for the current asbestos standard.

The complexity of today's health standards, coupled with additional requirements imposed by the courts and various administrations, have led many to conclude that OSHA needs a more streamlined, standard-setting process. OSHA's own advisory committee, the National Committee on Occupational Safety and Health (NACOSH 2000) said in a 2000 report: ". . . the standards setting process is not working as intended in the Occupational Safety and Health Act of 1970."

NACOSH recommended a better system for managing the standards-setting process, more use of advisory committees and negotiated rulemaking, better partnerships with standards-setting bodies and professional associations, and specific actions by Congress and the Executive Branch. NACOSH suggested that Congress should give the Secretary authority to update the PELs without requiring Section 6 rulemaking and making it possible to issue and maintain Emergency Temporary Standards. NACOSH concluded that the Office of Management and Budget's role in reviewing OSHA standards should be limited to economic impact and paperwork burdens, not risk assessment, health effects, or means of control.

OSHA, in spite of its shortcomings and the roadblocks placed before it, has been instrumental in improving working conditions in this country and educating the public on the importance of health and safety. Since the agency was established in 1971, workplace fatalities have been cut by 62 percent and occupational injury and illness rates have declined 40 percent. At the same time, U.S. employment has nearly doubled from 56 million workers at 3.5 million work sites to 115 million workers at nearly 7 million sites.

Endnotes

[1]Alice Hamilton is widely considered to be the founder of industrial medicine and industrial hygiene in the United States. She would later write her autobiography, which she began by saying, "Thirty-two years ago, in 1910, I went as a

pioneer into a new, unexplored field of American medicine, the field of industrial disease" (Hamilton, 1995).

[2]Perkins served for 12 years and 3 months (longer than any other Secretary of Labor) and went on to serve as a member of the Civil Service Commission. The Department of Labor headquarters was named for her in 1980 and she was inducted into the Labor Hall of Fame in 1988 (USDOL 2007a).

[3]OSHA PELs are found in the Z-tables (Z1, Z2, and Z3) within 29 CFR 1910.1000. In addition to an 8-hour PEL, some materials also have a short term exposure limit or ceiling limit.

[4]ACGIH® established the Threshold Limit Values for Chemical Substances (TLV®-CS) Committee in 1941. This group was charged with investigating, recommending, and annually reviewing exposure limits for chemical substances. It became a standing committee in 1944. Two years later, the organization adopted its first list of 148 exposure limits, then referred to as Maximum Allowable Concentrations. The term "Threshold Limit Values (TLVs®)" was introduced in 1956 (ACGIH 2007).

[5]The definition of "feasible" to a professional generally means engineering controls are available to reduce exposures below the PEL. However, its legal definition has been argued before the Occupational Safety and Health Review Commission (OSHRC) and the courts. There are two components of feasibility – technological and economic. It is OSHA's burden to establish that a control is both technologically and economically feasible. Once OSHA has shown a control is technologically feasible, the burden of producing evidence shifts to the employer, who may raise issues of economic feasibility. It is then up to OSHA to establish that the benefit of the engineering controls justifies their cost relative to other methods (OSHRC 1989).

[6]"The Secretary, in promulgating standards dealing with toxic materials or harmful physical agents under this subsection, shall set the standard which most adequately assures, to the extent feasible, on the basis of the best available evidence, that no employee will suffer material impairment of health or functional capacity even if such employee has regular exposure to the hazard dealt with by such standard for the period of his working life. Development of standards under this subsection shall be based upon research, demonstrations, experiments, and such other information as may be appropriate. In addition to the attainment of the highest degree of health and safety protection for the employee, other considerations shall be the latest available scientific data in the field, the feasibility of the standards, and experience gained under this and other health and safety laws. Whenever practicable, the standard promulgated shall be expressed in terms of objective criteria and of the performance desired."

[7]Morton Corn was the first industrial hygienist to head OSHA. He was professor of occupational health and chemical engineering at the University of Pittsburgh at the time of his appointment. Dr. Corn was appointed in 1975 by President Gerald Ford and served until 1977.

[8]"The term 'occupational safety and health standard' means a standard which requires conditions, or the adoption or use of one or more practices, means, methods, operations, or processes, reasonably necessary or appropriate to provide safe or healthful employment and places of employment."

[9]United Rubber, Cork, Linoleum and Plastic Workers of America (URW), the Oil, Chemical and Atomic Workers (OCAW), the International Chemical Workers Union (ICWU), and the American Federation of Labor and Congress of Industrial Organizations (AFL-CIO).

[10]The letter transmitting the agreement was signed by J.L. McGraw for the International Institute of Synthetic Rubber Producers (IISRP), Michael J. Wright for the United Steelworkers of America (USWA), and Michael Sprinker (CWU). The committee that worked on the issues also included Joseph Holtshouser of the Goodyear Tire and Rubber Company, Carolyn Phillips of the Shell Chemical Company, representing the Chemical Manufacturers Association, Robert Richmond of the Firestone Synthetic Rubber and Latex Company, and Louis Beliczky (formerly of the URW) and James L. Frederick of the SWA.

[11]John Pendergrass was the second industrial hygienist to serve as Assistant Secretary of Labor for OSHA. He was nominated by Ronald Reagan and served from 1986 to 1989. Mr. Pendergrass also served as President of the American Industrial Hygiene Association from 1974 to 1975. Prior to being appointed to OSHA, Mr. Pendergrass was employed by 3M Company.

[12]OSHA has issued Hazard Alerts in the form of compliance directives and/or as guidance to employers. For an example, see http://www.osha.gov/pls/oshaweb/owadisp.show_document?p_table=DIRECTIVES&p_id=1799

[13]The benefits under this standard are, for each removal, an employer must maintain for up to six months the earnings, seniority, and other employment rights and benefits of the employee as though the employee had not been removed from MC exposure or transferred to a comparable job.

[14]The "objective data" must demonstrate that, under "the work conditions having the greatest potential for releasing asbestos," an activity coupled with a specific material, simply cannot result in excessive concentrations (OSHA, 1994b).

[15]Objective data is information which clearly demonstrates that employees cannot be exposed to asbestos at levels above the asbestos PELs. The employer is responsible for determining if and how this type of information can be developed for the particular conditions and work performed (OSHA 1999).

[16]Excursion limits and ceiling limits are short-term exposure limits which shall not be exceeded. They are averaged over a 15 to 30-minute time frame.

[17]John Henshaw is the third industrial hygienist to serve as Assistant Secretary of OSHA. He was appointed by President George W. Bush in 2001 and served until 2004. Prior to his appointment he had been employed at Monsanto and Solutia. Henshaw was president of the American Industrial Hygiene Association in 1990-1991.

References

American Conference of Governmental Industrial Hygienists and American Industrial Hygiene Association. 1963. *Respiratory Protection Devices Manual*. Michigan: Ann Arbor Publishers.

American Conference of Governmental Industrial Hygienists. 2007. *History* (retrieved April 27, 2007). www.acgih.org/about/history.htm

American National Standards Institute. 1959. *American Standard Safety Code for Head, Eye, and Respiratory Protection*, Z88.2. Washington, D.C: ANSI.

Ashford, Nicholas. 1976. *Crisis in the Workplace*. Cambridge, Massachusetts: The MIT Press.

Corn, Jacqueline. 1992. *Response to Occupational Health Hazards, A Historical Perspective*. New York: John Wiley & Sons, Inc.

De la Cruz, Peter, and David Sarvadi. 1994. "OSHA PELs: Where Do We Go From Here?" *American Industrial Hygiene Association Journal* (October) 55:894–900.

Environmental Protection Agency. 2002. *Nonconfidential List of TSCA Section 5(e) New Chemical Exposure Limits* (retrieved April 27, 2007). www.epa.gov/opptintr/newchems/pubs/nceltbl.htm

Federal Register. 1978. "Occupational Exposure to Cotton Dust." 27379.

Federal Register. 1986. "Occupational Exposure to Asbestos, Tremolite, Anthophyllite, and Actinolite." 22612.

Hamilton, Alice. 1995. *Exploring the Dangerous Trades*. Beverly, Massachusetts: OEM Press.

MacLaury, Judson. 1981. *The Job Safety Law of 1970: Its Passage Was Perilous* (retrieved April 5, 2007). www.dol.gov/asp/programs/history/osha.htm

______. Undated. *Dunlop/Corn Administration, 1975–1977: Reform and Professionalization* (retrieved April 27, 2007). www.dol.gov/oasam/programs/history/osha13corn.htm

Martonik, John, Edith Nash, and Elizabeth Grossman. 2001. "The History of OSHA's Asbestos Rulemakings and Some Distinctive Approaches that They Introduced for Regulating Occupational Exposure to Toxic Substances." *American Industrial Hygiene Association Journal* (62):208–217.

Mine Safety and Health Administration. 2007. *Title 30 Code of Federal Regulations* (retrieved April 27, 2007). www.msha.gov/30cfr/0.0.htm

Mine Safety Appliances Company. 2007. *Catalog* (retrieved April 27, 2007). www.msanorthamerica.com/catalog/

National Advisory Committee on Occupational Safety and Health. 2000. *Report and Recommendations Related to OSHA's Standards Development Process*. www.osha.gov/dop/nacosh/nreport.html#ISSUES

National Institute of Occupational Safety and Health (NIOSH). 1973. *The Industrial Environment—Its Evaluation and Control*. Washington, D.C.: U.S. Government Printing Office.

______. 2005. *NIOSH Pocket Guide to Chemical Hazards* (Publication No. 97-140) (retrieved April 27, 2007). www.cdc.gov/niosh/npg/

______. 2006. *Alice Hamilton History* (retrieved April 5, 2007). www.cdc.gov/niosh/hamilton/HamHist.html

Nuclear Regulatory Commission. 1991. Occupational *Dose Limits for Adults*. Part 20.1201 (retrieved April 27, 2007). www.nrc.gov/reading-rm/doc-collections/cfr/part020/part020-1201.html

Occupational Safety and Health Administration (OSHA). 1971. *Air Contaminants* (retrieved April 5, 2007). www.osha.gov/pls/oshaweb/owadisp.show_document?p_table=STANDARDS&p_id=9991

______. 1974a. *Ventilation* (retrieved April 14, 2007). www.osha.gov/pls/oshaweb/owadisp.show_document?p_table=STANDARDS&p_id=9734

______. 1974b. 29 CFR 1910.1096, *Ionizing Radiation* (retrieved April 27, 2007). www.osha.gov/pls/oshaweb/owadisp.show_document?p_table=STANDARDS&p_id=10098

______. 1974c. 29 CFR 1910.95, *Occupational Noise Exposure* (retrieved July 31, 2007). www.osha.gov/pls/oshaweb/owadisp.show_document?p_table=STANDARDS&p_id=9735

______. 1983. *Hazardous Exposures to Pesticides May Be Cited Under the General Duty Clause* (retrieved April 27, 2007). www.osha.gov/pls/oshaweb/owadisp.show_document?p_table=INTERPRETATIONS&p_id=19104

______. 1989. *CPL 02-00-086 - CPL 2.86 - Memorandum of Understanding Between the OSHA and the U.S. Nuclear Regulatory Commission* (retrieved April 27, 2007). www.osha.gov/pls/oshaweb/owadisp.show_document?p_table=DIRECTIVES&p_id=1658

______. 1991. *Hearing Report on OSHA Reform Senate Labor and Human Resources Committee, October 29, 1991* (retrieved April 5, 2007). www.osha.gov/pls/oshaweb/owadisp.show_document?p_table=INTERPRETATIONS&p_id=20443

______. 1993. *Compliance and Enforcement Activities Affected by the PELs Decision* (retrieved April 27, 2007). www.osha.gov/pls/oshaweb/owadisp.show_document?p_table=INTERPRETATIONS&p_id=21220

______. 1994a. *OSHA Field Inspection Reference Manual CPL 2.103* (retrieved April 5, 2007). www.osha.gov/Firm_osha_data/100007.html

______. 1994b. *Occupational Exposure to Asbestos* (retrieved April 27, 2007). www.osha.gov/pls/oshaweb/owasrch.search_form?p_doc_type=PREAMBLES&p_toc_level=1&p_keyvalue=Asbestos~(1994~-~Amended)

______. 1995. *OSHA Technical Manual* (retrieved April 27, 2007). www.osha.gov/dts/osta/otm/otm_ii/otm_ii_1.html

______. 1997a. *Occupational Exposure to Methylene Chloride* (retrieved April 14, 2007). www.osha.gov/pls/oshaweb/owasrch.search_form?p_doc_type=PREAMBLES&p_toc_level=1&p_keyvalue=Methylene~Chloride

______. 1997b. *1,3–Butadiene* (retrieved April 14, 2007). www.osha.gov/pls/oshaweb/owadisp.show_document?p_table=STANDARDS&p_id=10087

______. 1998a. *Methylene Chloride; Final Rule* (retrieved April 14, 2007). www.osha.gov/pls/oshaweb/owadisp.show_document?p_table=FEDERAL_REGISTER&p_id=13852

______. 1998b. *Respiratory Protection* (retrieved April 27, 2007). www.osha.gov/pls/oshaweb/owadisp.show_document?p_table=STANDARDS&p_id=12716

______. 1999. *Asbestos: Notification Requirements and Exposure Monitoring* (retrieved July 25, 2007). www.osha.gov/pls/oshaweb/owadisp.show_document?p_table=INTERPRETATIONS&p_id=22740

______. 2000. *Ergonomics Program* (retrieved April 15, 2007). www.osha.gov/pls/oshaweb/owadisp.show_document?p_table=FEDERAL_REGISTER&p_id=16305

______. 2001. *Ergonomics Program* (retrieved April 15, 2007). www.osha.gov/pls/oshaweb/owadisp.show_document?p_table=FEDERAL_REGISTER&p_id=16515

______. 2004. *OSHA's Budget Request for Fiscal Year 2005* (retrieved April 15, 2007). www.osha.gov/pls/oshaweb/owadisp.show_document?p_table=TESTIMONIES&p_id=348

______. 2005. *Chemical Sampling Information* (retrieved April 27, 2007). www.osha.gov/dts/chemicalsampling/toc/toc_chemsamp.html

______. 2006. *OSHA Enforcement: Effective, Focused, and Consistent* (retrieved April 27, 2007). www.osha.gov/

______. 2007. *Health Standards* (retrieved April 27, 2007). www.osha.gov/comp-links.html

______. 2007. *OSHA—A History of its First Thirteen Years, 1971-1984, U.S. Department of Labor Office of the Assistant Secretary for Policy* (retrieved April 5, 2007). www.dol.gov/asp/programs/history/mono-osha13introtoc.htm

Showlater, David. 1972. *How to Make the OSHA–1970 Work for You*. Ann Arbor, Michigan: Ann Arbor Science Publishers.

SKC, Inc. 2007. *Help with Air Sampling* (retrieved April 27, 2007). www.skcinc.com/help.asp

U.S. Congress, Office of Technical Assessment. 1985. *Preventing Illness and Injury in the Workplace*, OTA–H-256. Washington, D.C.: U.S. Government Printing Office.

U.S. Congress. 1970. Occupational Safety and Health Act of 1970, Public Law 91-596.

______. 1977. Federal Mine Safety & Health Act of 1977, Public Law 91-173.

U.S. Department of Labor. 2007a. *Frances Perkins* (retrieved April 27, 2007). www.dol.gov/oasam/programs/history/perkins.htm

______. 2007b. *George Guenther Administration, 1971–1973: A Closely Watched Start-up* (retrieved April 15, 2007). www.dol.gov/oasam/programs/history/osha13guenther.htm

______. 2007c. *OSHA 35-Year Milestones* (retrieved April 15, 2007). www.osha.gov/as/opa/osha30yearmilestones.html

______. 2007d. *State Occupational Safety and Health Plans* (retrieved July 25, 2007). www.osha.gov/dcsp/osp/index.html

SECTION 4
INDUSTRIAL HYGIENE

LEARNING OBJECTIVES

- Define and contrast risk, risk assessment, risk management, and risk communication.
- Be able to describe in general terms how a risk assessment is conducted.
- Learn how to determine the need for risk management.
- Compare the risk assessment/risk management model with the "anticipation, recognition, evaluation, and control" model for industrial hygiene.
- Define exposure assessment, and describe its role in the occupational and environmental health and safety (OEHS) program.
- Know the difference between compliance exposure assessment and comprehensive exposure assessment, and explain why comprehensive exposure assessment is preferred.
- Learn basic methodologies for determining concentrations of airborne contaminants.

APPLIED SCIENCE AND ENGINEERING: GENERAL PRINCIPLES

Deborah Imel Nelson, Shery Milz, and Susan Arnold

THE DEFINING ROLES of workplace safety and health professionals can be described in terms of risk: assessing the risks posed by workplace hazards, evaluating the acceptability of the risks, and managing the risks, while engaging in constant risk communication with affected parties (see Figure 1). Hazards—whether chemical, physical, biological, or mechanical—are the sources of risk. *Risk* can be defined as the probability and magnitude of harm resulting from exposure to hazards. *Risk assessment* is a matter of characterizing—qualitatively, semiquantitatively, or quantitatively[1]—the probability and magnitude of harm. We often speak of the risk of exposure to a chemical hazard as a function of the magnitude, duration, and frequency of exposure, and of the toxicity of the chemical. If there is no exposure to the chemical, then technically there is no risk (although one must always keep the *potential* of exposure in mind).

Risk assessment is followed by a judgment of the acceptability of the risk. Many scientific and nonscientific factors, influence the perceived acceptability of a risk and resulting risk-management decisions, including

- toxicological profile of a chemical (Are the effects acute and minimal? Is it an irritant or are the effects dreaded and insidious? Does it cause a reproductive health effect?)
- magnitude of estimated risk, dimensions of the risk (see Table 1), and potential public reactions
- availability of human, technical, and financial resources to conduct risk assessment/management (e.g., it might be less expensive to initiate risk management than to conduct a detailed risk assessment)
- estimated uncertainty/error of risk-assessment methods (a high degree of uncertainty would argue for more conservative risk-management methods)

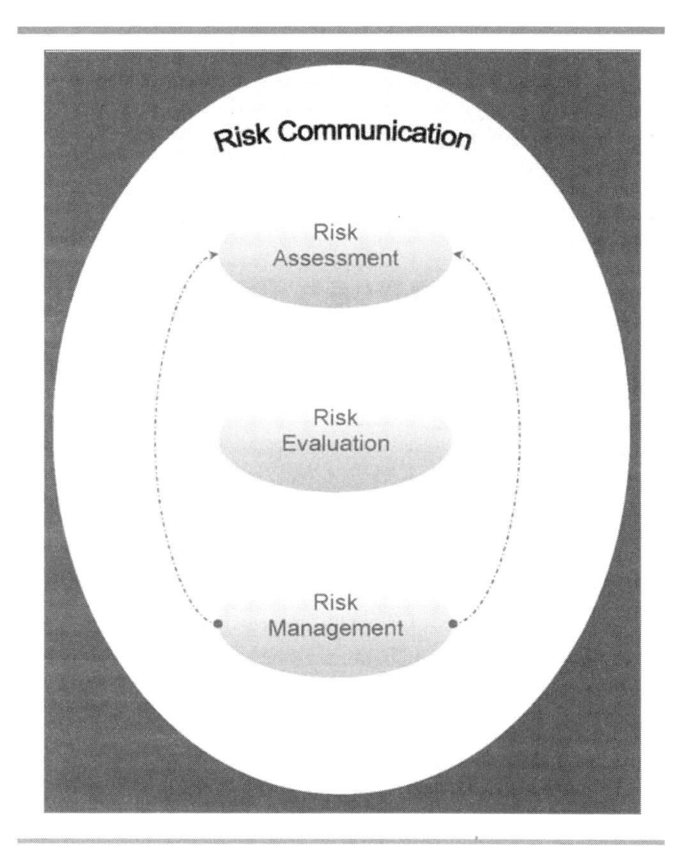

FIGURE 1. Relationship of risk functions

- social/cultural/moral/ethical/political factors (e.g., acceptable risk levels may be lower for children than for adults)
- risk perception of stakeholders/special interest groups
- time available for analysis (if time is limited, analysts may recommend proceeding directly to risk management)
- efficacy of risk-management methods (e.g., if a simple, inexpensive, and effective control method is available, it might be instituted immediately)
- policy decisions (e.g., acceptance of the Precautionary Principle in Europe may lead to more risk-management efforts) (Bales 2006)
- applicable laws or regulations (if an exposure level is below regulated levels, it might be deemed to be acceptable)

Determining how much risk is acceptable is a very difficult question to answer, because it cannot be determined scientifically. It has been stated that the "best level of risk is zero, if it is free, but it never is" (Caplan and Lynch 1996). Every stakeholder may have a different answer, and nobody is wrong. For example, consider a hazardous chemical used in a manufacturing plant upwind from an elementary school. As the exposure to this chemical (which is a function of both concentration and length of exposure) increases, the risk of experiencing adverse health effects goes up. A small amount of the chemical is routinely released from the facility; however, the downwind ambient concentration is orders of magnitude lower than the concentration inside the facility. What is the acceptable concentration in the plant, and in the elementary school? The dimensions of risk shown in Table 1 suggest that the acceptable concentration in the workplace is higher, because of the following:

- voluntariness (theoretically, the workers could quit their jobs)
- controllability (workers could use respirators)
- benefits and equity (workers are eligible to receive compensation)
- institutional trust (workers know and have a higher level of trust in the employer)
- understanding (workers routinely handle this chemical)
- characteristics of the potential victims (workers are adults)

TABLE 1

Selected Factors Affecting Risk Perception in the Workplace

Factor	Conditions Associated with Lower Risk Perception	Conditions Associated with Higher Risk Perception
Voluntariness	Voluntary	Involuntary
Controllability	Controllable	Uncontrollable
Benefits	Clear benefits	Little or no benefit
Equity	Fair distribution of risks	Unfair distribution of risks
Origin	Natural	Man-made
Institutional trust	Trusted source	Untrusted source
Catastrophic potential	Limited	Widespread
Understanding	Familiar	Not understood
Victims	Adults	Children

(Adapted from Fischoff et al. 1981; Covello et al. 1988)

The plant workers are thus more likely to tolerate higher concentrations of the chemical than would be considered acceptable in the school. The children's exposure is not voluntary or controllable; they receive no benefits; they are children rather than adults; they (or their parents) may not trust the manufacturer or understand the chemical; and they (or their parents) may have learned of catastrophic chemical releases through the media. The worker has a higher level of understanding and is generally more aware of when the exposure is taking place and thus is able to take appropriate protection measures. A member of the community is less likely to know when exposure occurs and is, therefore, less likely to find the exposure acceptable.

As illustrated by this example, the level of acceptable risk varies in environmental and occupational settings. Because of such factors as voluntariness and equity, workers tend to accept higher levels of risk in the workplace, ranging from 1 incident per 10,000 workers exposed up to 1 incident per 1000 workers exposed (often expressed as 10^{-4} to 10^{-3}). In contrast, in environmental settings, acceptable risks range from 1 incident per 1,000,000 people exposed up to 1 incident per 100,000 people exposed (often expressed as 10^{-6} to 10^{-5}). In other words, the acceptable risk in an occupational setting ranges from 10 to 1000 times higher than for the general public.

These levels of acceptable risk may be used by the Environmental Protection Agency (EPA) and the Occupational Safety and Health Administration (OSHA) in establishing acceptable concentrations—for residual soil concentrations of contaminants at hazardous waste site clean ups, for example. National Ambient Air Quality Standards (NAAQS) are established by the EPA to protect the public health, including sensitive populations such as asthmatics, children, and the elderly (EPA 2006). The OSHA law was written "to assure so far as possible every working man and woman in the Nation safe and healthful working conditions" (OSH Act 1970). The permissible exposure limits (PELs) were adopted from several sources, including the 1968 American Congress of Governmental Industrial Hygienists (ACGIH) Threshold Limit Values® (OSHA 1989). TLVs® represent "conditions under which it is believed that nearly all workers may be repeatedly exposed day after day without adverse health effects" (ACGIH 2005). Contrasting selected NAAQS with OSHA PELs indicates this trend of establishing acceptable occupational concentrations at higher values than acceptable environmental values (see Table 2).

TABLE 2

Comparison of OSHA PELs and EPA NAAQS

Contaminant	OSHA PEL*	EPA NAAQS
CO (carbon monoxide)	50 ppm, 8-hour time-weighted average	9 ppm, 8-hour time-weighted average
O_3 (ozone)	0.1 ppm, 8-hour time-weighted average	0.08 pm, 8-hour time-weighted average
PM_{10} (particulate matter less than 10 μg in diameter)	5 mg/m^3 (respirable nuisance dust)	150 μg/mg^3, 24-hour time-weighted average
SO_x (sulfur oxides)	SO_2: 5 ppm, 8-hour time-weighted average	0.03 ppm, annual arithmetic average
Pb (lead)	50 μg/m^3, 8-hour time-weighted average	1.5 μg/m^3, quarterly average

*ppm = parts per million parts; 1 ppm = 0.0001% concentration by volume. mg/m^3 = 1 milligram contaminant per cubic meter of air. 1000 μg/m^3 = 1000 micrograms per cubic meter = 1 mg/m^3.

(*Sources:* EPA 2006; OSHA 1971)

Based on the acceptability of a risk, a decision may be made to do nothing, to conduct further risk assessment, or to implement risk-management measures. *Risk management* is prevention or reduction of risk in a scientifically sound, cost-effective, acceptable manner.

Risk communication completes the risk-assessment and risk-management functions. At its simplest, *risk communication* is a two-way dialogue between risk managers and those who are potentially affected by a risk. Other stakeholders, such as the general public, workers' families, risk assessors, and other experts may also be involved. Risk communication is a necessary and powerful component of the risk function, because people are much more likely to *buy in* to a decision in which they have participated.

Making Risk Decisions

These four risk functions—assessment, evaluation, management, and communication—can be neatly separated on paper. The risk assessor, guided primarily by scientific factors and influenced by values, identifies and defines the risk in qualitative or quantitative terms; the risk evaluator, strongly influenced by values, determines how acceptable the risk is; the risk manager, guided primarily by values and influenced by science, does what is necessary to eliminate or reduce that risk to the acceptable level; and the risk communicator maintains a constant dialogue among all stakeholders. However, in reality, the OEHS professional may be wearing all four hats at once. How does he/she balance the potentially conflicting values guiding the different aspects of occupational health risk decision making?[2] Fortunately, there are several risk decision-making models available to OEHS professionals.

The traditional approach in the United States for controlling worker exposure to hazardous airborne chemicals has been to compare the results of an exposure assessment to the OELs.[3] The U.S. Department of Labor PELs are mandatory concentration limits, mostly in the form of 8-hour time-weighted average concentrations (OSHA 1971). The ACGIH® annually publishes lists of recommended OELs, including the Threshold Limit Values (TLVs®) (ACGIH® 2005). Other sources of OELs include the American Industrial Hygiene Association (AIHA) Emergency Response Planning Guidelines Series (ERPGs) and the Workplace Environmental Exposure Level Guide Series (WEELS) (AIHA 2002), the National Institute for Occupational Safety and Health (NIOSH) Recommended Exposure Limits (RELs) (NIOSH 2005), EPA Acute Exposure Guideline Levels (AEGLs) [EPA undated(a)], and OELs established by corporations and organizations for their own internal use.

Risk decision makers usually do not have the data, time, or resources to conduct a full quantitative or even semiquantitative risk assessment for all scenarios, and they often rely on experience and professional judgment. For example, he or she might observe a well-ventilated workroom where a small quantity of a nonvolatile, low toxicity chemical is occasionally used, and conclude that the risk level is very low. Simple matrices can be used to rank dissimilar risks, and to document the decision-making process. These are qualitative risk-assessment processes that do not necessarily use numbers, but nonetheless can provide valuable information to risk decision makers.

Rating	People	Assets	Environment	Image
0	None	None	None	None
1	Slight	Slight	Slight	Slight
2	Minor	Minor	Minor	Minor
3	Major	Local	Localized	Considerable
4	1-3 deaths	Major	Major	Major national
5	Many deaths	Extensive	Massive	Major international

Decision Models

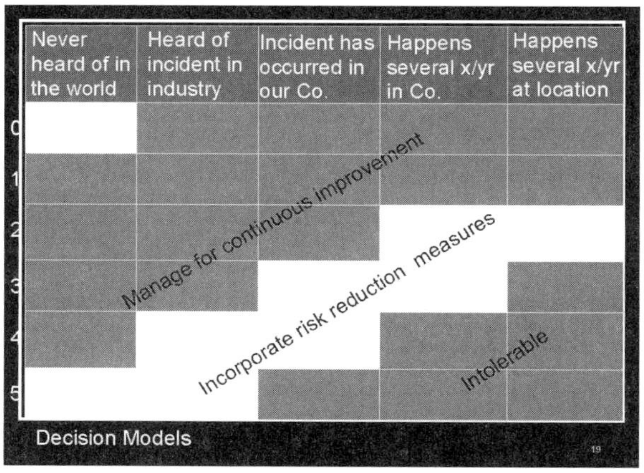

Never heard of in the world
Heard of incident in industry
Incident has occurred in our Co.
Happens several x/yr in Co.
Happens several x/yr at location
Manage for continuous improvement
Incorporate risk reduction measures
Intolerable
Decision Models

FIGURE 2. A simple risk decision-making matrix (Adapted from Nelson et al. 2003)

Figure 2 is based on a matrix developed by a major multinational corporation. The matrix is based on severity of impact and probability of event occurrence. The matrix is easy to use and communicate to stakeholders, and it has the advantage of allowing different types of impacts to be considered. To illustrate the use of this matrix, consider the removal of leaking underground storage tanks from service stations, a common occurrence in the 1990s. These tanks were often too large to remove from the site in one piece, and were cut with welding torches into several pieces prior to removal. In several instances, the welding torches sparked explosions of residual fuel in the tanks. From mid-1984 to mid-1990, at least nine deaths and seven injuries requiring hospitalization directly resulting from underground tank-removal operations were reported to OSHA. An additional thirty-eight deaths and thirty-two injuries requiring hospitalization resulted from activities around tanks not specified to be underground were also reported (Nelson 1991). How can the hypothetical tank-removal service, Yankee Tank Yank, Inc. (YTY), use this information to determine the appropriate level of risk management for explosion of fuel tanks during the removal process?

The rating on severity of impact will be examined first, followed by the rating of probability. The impact ratings of an explosion with respect to people, assets, the environment, and image range from 0 to 5. An explosion could kill one or more people, resulting in a rating of 4. The impact on assets and the environment would be local, for a rating of 3. YTY's image would potentially suffer at the national level, for a rating of 4. The highest impact rating is 4, resulting from impact on people and on image. To consider probability, move across the matrix to the right. Deaths resulting from tank explosions have occurred in the industry, but not at YTY. Crossing the row and the column leads to a white box labeled "Incorporate risk reduction measures." YTY should therefore make timely and significant efforts to protect their people, their assets, the environment, and their image, but not on the urgent basis implied by an "Intolerable" rating.

Control banding is a semiquantitative, matrix-based approach. The best developed and tested version was developed by the UK Health and Safety Executive (HSE) to help small and medium enterprises (SME) comply with the Control of Substances Hazardous to Health (COSHH) regulations, which require that a risk assessment be conducted in all workplaces (see Health and Safety Executive undated).

COSHH Essentials has roots in earlier efforts of the pharmaceutical and chemical industries to improve workplace health and safety, particularly when workers were handling little-known chemicals. These earlier approaches based control strategies on the toxicity of a chemical. The HSE added the dimension of exposure assessment, using the quantity of the chemical in use, along with "dustiness" or volatility, as surrogates of employee exposure. Please note that COSHH Essentials is not synonymous with *control banding*; other versions of control banding have been designed by other countries and by private companies.

The first step of the analysis is to obtain the information that will be used to determine the necessary level of control for a specific activity with a chemical:

a. a measure of its toxicity
b. an indication of its volatility (for liquids) or dustiness (for solids)
c. the amount of the chemical in use

In the COSHH Essentials system, the European risk phrases, or R-phrases, for human health effects are used as a surrogate for toxicity. These phrases, such as *harmful by inhalation* or *toxic by inhalation*, have been grouped in Categories A (irritants) up to Category D (very toxic). (Category E consists of carcinogens, mutagens, and reproductive hazards, and is reserved for use only by occupational health experts.) The volatility of liquids, or the dustiness of solids, indicate the potential of the compound to become airborne. The amount of the material in use (e.g., being transferred from one container to another) is related to the quantity that could become airborne.

This information is entered into a matrix that yields the appropriate level of control (see Figure 3). Each of the four matrices is keyed to a category, ranging from A–D. Columns in each table relate to volatility or dustiness, and rows refer to the amount in use. The numbers in each cell of the matrix refer the analyst to the following control levels:

1. good industrial hygiene practice
2. engineering control (e.g., ventilation)
3. containment, with small breaches only4
4. refer to expert advice

After obtaining the recommended control level, the user is then directed to the control guidance sheets (CGS). Available for many, though not all, workplace tasks, CGS follow a standard template: control level 1 is green, 2 is blue, 3 is lavender, and 4 is hot pink. Each of these sheets contains information on access, design of control equipment (with illustration), maintenance, examination and testing, cleaning, personal protective equipment (PPE), training, supervision, and an

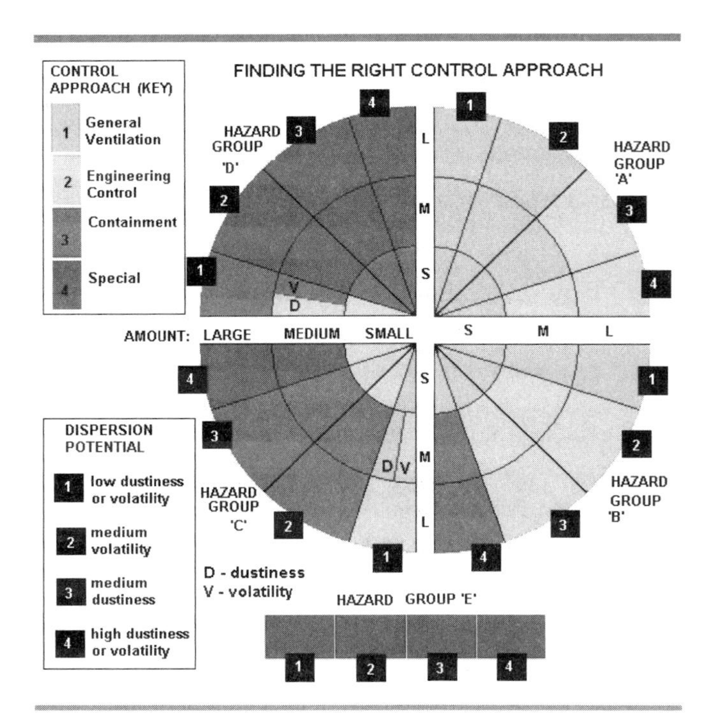

FIGURE 3. **COSHH Essentials wheel** (Crown Copyright: UK HSE)

employee checklist. This information is always located in the same place on every CGS. Thus, by providing minimum information, employees and employers can receive direct, appropriate guidance for reducing any exposure to chemical hazards.

Control banding is receiving much attention worldwide, for example, by the World Health Organization (WHO) and the International Labor Organization (ILO). An electronic workbook is available on the HSE Web site.

The Risk-Assessment Paradigm

The EPA has published the most extensively on risk assessment mainly because of its role in Superfund hazardous waste site management.[5] The EPA paradigm was based on the National Research Council "Red Book" (NRC 1983) and includes the following:

- hazard identification (which EPA implements as data collection and evaluation)
- dose-response (toxicity) assessment
- exposure assessment
- risk characterization

These four components have been memorably rephrased as the following (Barnes 1994):

- Is this stuff toxic?
- How toxic is it?
- Who is exposed to this stuff, for how long, and how often?
- So what?

Using Risk Assessment in Industrial Hygiene Practice

Industrial hygiene has been defined as the science and art of "anticipation, recognition, evaluation, and control of environmental factors arising in or from the workplace that may result in injury, illness, impairment, or affect the well-being of workers and members of the community" (Olishifski and Plog 1988). Anticipation and recognition can be difficult to distinguish, but anticipation can be thought of as an activity that occurs prior to an actual site visit, while recognition, evaluation, and control all occur at the workplace. Risk assessment and management can be roughly compared to anticipation, recognition, evaluation, and control, as seen in Table 3.

The remainder of this chapter will focus on the exposure-assessment aspects of industrial hygiene. While risk assessment requires both exposure assessment and toxicity assessment, the OEHS professional is far more likely to be involved in exposure assessment than in toxicity assessment. Exposure assessment is highly site-specific, while toxicity values are usually obtained from standard reference sources, such as

TABLE 3

Comparison of Risk Assessment/Management to Anticipation/Recognition/Evaluation/Control

Risk Assessment/Management	Anticipation, Recognition, Evaluation, and Control
Hazard identification	Anticipation, recognition
Exposure assessment, toxicity assessment, risk characterization	Evaluation
Risk management	Control
Risk communication	Hazard communication

(Adapted from Jayjock, Lynch, and Nelson 2000)

AIHA WEELs, ACGIH® TLVs®, OSHA PELs, and EPA AEGLs.

Defining Exposure Assessment

Exposure assessment answers the questions posed by Barnes (1994): "Who is exposed to this stuff, for how long, and how often?" The trend in conducting exposure assessments has been shifting from collecting samples of airborne contaminants for the purpose of determining compliance with OSHA regulations (compliance strategy) to a more comprehensive approach to conducting exposure assessments. According to Mulhausen and Damiano (2003), this comprehensive approach can be described as the "systematic review of the processes, practices, materials, and division of labor present in a workplace that is used to define and judge all exposures for all workers on all days." Other reasons for performing exposure assessments include addressing employee complaints, for medical and epidemiological studies, and for determining the effectiveness of engineering and administrative controls. It is important to understand that the results of the exposure-assessment process must account for today's as well as tomorrow's regulations (Mulhausen and Damiano 2003).

Exposure-Assessment Strategies

Comprehensive exposure assessment is an iterative or cyclical process. The early efforts can be aimed at the "low-hanging fruit," that is, information that is fairly easy to obtain. As information is collected, it can be used to focus further efforts to assess and control exposures. Mulhausen and Damiano (2003) have outlined seven steps in an exposure-assessment strategy:

1. ***Start:*** Establish an exposure-assessment strategy that includes determining the goals of the strategy (i.e., compliance, comprehensive, etc.) and developing a written program.
2. ***Basic characterization:*** Collection of the information needed to characterize the workplace, the work force, and the environmental agents.
3. ***Exposure assessment:*** Analysis of the information collected during the basic characterization to establish similar exposure groups (SEGs), to define the exposure profile of each SEG, and to determine exposure acceptability within each SEG.
4. ***Further information gathering:*** When exposures cannot be determined to be acceptable or unacceptable (i.e., undetermined), additional information such as additional workplace monitoring, modeling of exposures, biological monitoring, toxicological data, or epidemiological data can be collected to aid in the decision-making process.
5. ***Health hazard control:*** When exposures are determined to be unacceptable, prioritized control strategies are implemented. Priority depends on the number of employees exposed, the risk of those exposures, frequency, and uncertainty.
6. ***Reassessment:*** The iterative process of the exposure-assessment strategy requires that the assessment be performed on a regular basis. The frequency of reassessment generally depends upon the severity of the hazard. Common reassessment frequencies are annual for highly hazardous operations or operations with many workers, and up to once every four years for administrative spaces.
7. ***Communication and documentation:*** All results and decisions, both qualitative and quantitative, need to be recorded and reported to the employees.

Quantitative Exposure-Assessment Methodologies

When selecting appropriate quantitative exposure-assessment methodologies, the OEHS professional has to consider many factors. These can be summarized as the "who, what, when, where, why, and how" of exposure assessment. Of these, "why are we sampling" and "what are we sampling for" have the greatest impact on selection of monitoring methodologies. Workplace monitoring may be conducted for a variety of reasons, for example, to determine compliance with regulations, to locate contaminant source(s), to measure the effectiveness of controls, and so on, each of which calls for a slightly different approach. Determining regulatory compliance, for instance, usually requires calculation of 8-hour time-weighted average exposures; evaluating the effectiveness of engineering controls may be best accomplished either by sampling before and after a control has been instituted or by sampling with and without the control operating (Mulhausen and Damiano 2003).

The next most important factor to consider in selecting a quantitative exposure-assessment methodology is the nature of the hazard being evaluated, that is, "what are we sampling for." If the concern is for airborne contaminants, it is important to consider all chemicals, including raw materials, intermediates, by-products, waste products, and final products. However, it is usually too expensive and time-consuming to collect samples of all airborne contaminants, so OEHS professionals often prioritize their sampling and analytical approach based on the quantity of the chemical in use, its volatility/dustiness, relationship to OEL, and so on. The list of hazards (i.e., the hazard inventory) to determine the sampling priority is generally developed during the basic-characterization step. The hazard inventory should include the chemical hazards, often found on the material safety data sheets (MSDSs), the physical hazards (noise, radiation, etc.) as well as the biological hazards. Evaluating the potential for exposure in comparison to the toxicity of the contaminant can be thought of as qualitative or perhaps even semiquantitative exposure assessment (Bisesi 2004).

Together, the *why* and the *what* determine the how, who, when, and where of exposure assessment.

How? Quantitative exposure assessment can be performed in many ways. Workplace monitoring may be done at a screening level or a comprehensive level. Samples may be collected in the workers' breathing zone or in that general area. Semiquantitative assessment can be performed using mathematical models to estimate exposures. Each type of monitoring can be performed with a variety of equipment, which can include direct-reading instruments, colorimetric detector tubes, passive monitors, sampling pumps and media, and so on (see Figures 4 and 5, and Table 4). Unfortunately, no universal sampler exists. Therefore, several factors may need to be considered in the selection of instrumentation, such as the efficiency of the instrument, the reliability of the instrument in real-world situations, as well as the preference of the investigator (Bisesi, 2004).

Quantitative exposure assessment is just one part of a comprehensive exposure-assessment strategy. The quantitative approach encompasses workplace monitoring and may be performed to answer many questions, such as the level of personal employee exposures and the effectiveness of engineering controls. The methods utilized for workplace monitoring are varied;

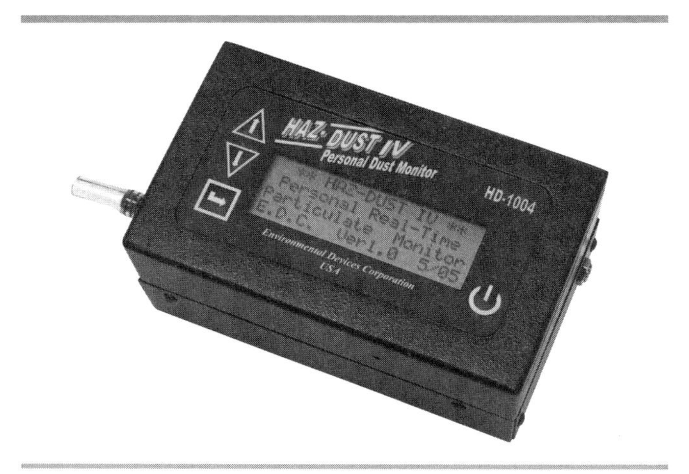

FIGURE 4. Personal dust monitor (Courtesy of SKC, Inc.)

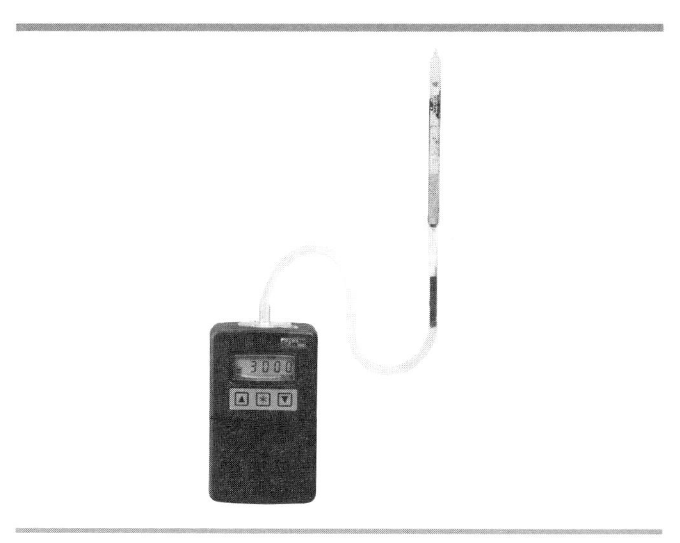

FIGURE 5. AirChek2000 low-flow pump (Courtesy of SKC, Inc.)

TABLE 4

Comparison of Air-Monitoring Methods

Method	Mechanism	Airborne Contaminant	Analytical Technique
Direct-reading instruments	Varies	Gases, vapors, particulates, noise, radiation	N/A
Colorimetric detector tubes	Chemical reaction on fixed media	Gases, vapors	N/A
Passive dosimeters	Adsorption	Gases, vapors	GC/FID, etc.
Air sampling pumps and media (e.g., filters, charcoal tubes, impinger solutions)			
Filters	Filtration	Particulates	Gravimetric, chemical analysis
Charcoal tubes	Adsorption	Vapors	GC/FID, etc.
Impinger solutions	Absorption, chemical reaction	Gases	Chemical analysis

(Adapted from Jayjock, Lynch, and Nelson 2000)

they depend on the reason the monitoring is being conducted. The summary below discusses a few of the considerations when selecting a sampling strategy. A more comprehensive discussion of sampling equipment and methods can be found in Bisesi (2004), Plog and Quinlan (2002), and DiNardi (2003). Advice on specific sampling and analytical techniques is well beyond the scope of this chapter, but it is available from OSHA (undated), NIOSH (1994), ASTM (2003), and ACGIH (2001, 2004). Various manufacturers, such as SKC, Inc. (undated) or Mine Safety Appliances (undated), provide a wealth of information on their Web sites.

Instantaneous, or Real-Time, Monitoring

Instantaneous, or real-time, monitoring is generally used for short-term sampling lasting from a few seconds up to fifteen minutes. However, instantaneous monitoring can also be performed over longer periods of time, such as eight hours or longer, in the data-logging mode. Instantaneous monitors provide results immediately but are best used as screening tools to determine what concentrations of contaminants are present in order to plan future assessment needs. Instantaneous monitoring is also useful during short-term operations or in determining peak concentrations (Bisesi 2004). Instantaneous monitors include direct-reading instruments and colorimetric detector tubes.

Integrated Monitoring

Integrated monitoring generally occurs over a period of several hours. Work shifts often last eight hours, and therefore integrated monitoring lasts eight hours or a full shift. OELs used for comparison purposes are usually based on an eight-hour time period. Integrated monitoring for a full shift can be performed by collecting one sample for the entire eight-hour shift. Additionally, multiple consecutive samples can be collected over the shift. A common approach is to collect two samples—one before the lunch break and one after. The individual samples are then combined to determine the eight-hour time-weighted average (8HR TWA) concentration, in which C is the measured concentration of each and T is the sample time of that sample in hours.

$$\text{8HR TWA} = \frac{C_1T_1 + C_2T_2 + \cdots + C_nT_n}{\text{8 hours}} \quad (1)$$

With integrated sampling, samples must be sent to a laboratory for analysis, resulting in a delay between the sampling event and the reporting of results (Bisesi 2004).

Nontraditional work shifts, such as 10-hour or 12-hour shifts, are becoming more common in today's workplaces. Sampling strategies should be designed not only for comparison to applicable OELs, but also to assess the workers' full-shift exposures. OSHA compliance officers are directed to either sample during the estimated "worst continuous 8-hour work period of the entire extended work shift," or to collect multiple samples over the entire work shift (OSHA 1999). In this way, time-weighted average exposures can be calculated for comparison to any 8-hour OELs, as well as for the entire work shift. Alternatively, OELs can be modified for application to unusual work shifts. For more information on the evaluation of these extended exposures, consult Paustenbach (2000), Brief and Scala (1975), the ACGIH *TLV Guide* (2007), Klonne (2003), or OSHA (1999).

Personal Monitoring

Personal monitoring can be performed with both instantaneous samplers and integrated samplers. Personal samples are collected on the worker, in the breathing zone for an inhalation hazard or in the hearing zone for a noise hazard. Breathing-zone samples are generally collected by integrated monitoring. Hearing-zone samples are generally collected by instantaneous sampling. The breathing zone refers to the area within a 1-foot radius of a worker's nose and mouth. Breathing-zone samples are often attached to a worker's collar near the collar bone. The hearing zone refers to the area within a 1-foot radius of the ear (Bisesi 2004).

Area Monitoring

Area monitoring takes place at a location, often at a source of contamination, and not on a person. Workers are often quite mobile and are usually at some varying distance from the source; thus, area-monitoring results are not representative of personal exposures. For this reason, area monitoring should not be compared to OELs. Area monitoring is, however, a good tool for evaluating background concentrations or the effectiveness of engineering controls, or for checking for leaks.

Both instantaneous monitors and integrated monitors can be used during area monitoring. Samples can be collected at multiple locations throughout a facility (Bisesi 2004).

Mathematical Modeling

Exposure monitoring cannot be performed for every person, to every contaminant, or on every day. Exposure modeling can therefore be used to estimate exposures when monitoring cannot be performed. In addition, exposure modeling can also be used to estimate exposures for processes that are no longer operational and even for processes that are still in the planning stages. However, it is important to remember that exposure models are just generalizations of reality and are actually estimates of the air concentration that a worker may breathe in rather than an estimate of worker exposure (Jayjock 2003).

Models range from the very simple to the complex and are generally applied in a tiered approach, beginning with the simple models and progressing through more complicated models. The simple models often overestimate exposures. However, if the exposure overestimate can be assessed as an acceptable exposure, then no further action needs to be taken. Generally, more complex models are needed to determine exposure acceptability. The more complex the model, the more resources are needed in terms of time, effort, and money, in order to reach an exposure estimate that is less of an overestimate. The tiered approach may begin with the saturation vapor pressure model, proceed to a general ventilation model, and end with a two-phase model (Jayjock 2003).

Exposure-modeling methodologies have been described in books (Keil 2000) and chapters (Jayjock 2003; Nicas 2006). The saturation vapor pressure model is often used as a first-tier estimate. All that is needed for this estimate is the vapor pressure of the contaminant of interest and the atmospheric pressure using the following equation:

$$C_{sat} = \frac{(10^6)(\text{vapor pressure})}{(\text{atmospheric pressure})} \qquad (2)$$

For example, if a vat of toluene is left uncovered in a workspace, the saturation vapor pressure model determines the maximum concentration that could exist in the workspace. If this maximum concentration is less than the OEL of the contaminant, then the exposure can be deemed acceptable. For toluene, the vapor pressure is 21 mm Hg (NIOSH 2005). Therefore, the saturation concentration when atmospheric pressure is 760 mm Hg (1 atm) is 27,631 ppm. Because this concentration is greater than both the PEL (200 ppm) and the TLV (50 ppm), the exposure cannot be determined to be acceptable, so another tier of modeling may be necessary.

As a further illustration of the need for mathematical modeling in exposure assessment, the Canadian Environmental Protection Act (CEPA) (Environment Canada 1999) and the European Registration, Evaluation and Authorization of Chemicals (REACH) (European Commission 2006 are causing a paradigm shift in the world of practical risk assessment. These regulations are based on the precautionary principle, under which scientific uncertainty can no longer be used to justify inactivity on the part of decision makers. Rather, it calls for a reordering of events such that the risk profile of the chemical must be established before it can be widely used in commerce.

Both regulations require the responsible party to assess the human health and environmental risks associated with the chemicals. For chemicals identified as having significant risks, additional requirements apply. In both cases, the regulations have provisions for banning chemicals deemed to present excessive risks.

Entities with products entering these countries at any point in their life cycle are subject to these regulations. Thus, the global supply-chain network is impacted and companies in the United States will be under increasing pressure to proactively assess their chemicals and products if they wish to do business in the EU or Canada. Further, the U.S. Government Accounting Office has recently reported that a reauthorization of the EPA Toxic Substances Control Act (TSCA 1976) could reflect both the Canadian and European frameworks (GAO 2006).

The need to quickly and efficiently assess thousands of chemicals necessitates the use of modeling tools. Unlike monitoring data, models permit the user to assess past, present, and future scenarios. Conservative, default assumptions are used in place of real-world data to inform the models. Thus, in concordance with the precautionary approach, conservatism is traded for data, and more sophisticated models can be used until the necessary level of refinement is achieved. These models purposefully overestimate exposures, but the degree to which they exaggerate is decreased as more real-world data are provided. Conversely, the simpler models require minimal expertise while the more sophisticated models have a steeper learning curve.

Physical-chemical models such as the Saturation Vapor Concentration and Box models and expert models, including EASE and Risk of Derm, can be employed for assessing single-chemical scenarios. Other models are available for assessing and screening hundreds or thousands of chemicals, such as the (freeware) Complex Exposure Tool (CEPST) (Lifeline Group 2007).

In all cases, the models are premised on a set of assumptions with which the user must be comfortable. These assumptions do not necessarily need to be correct, but the user must be able to accept them to have confidence in the output. The models operate from a single mathematical algorithm or a set of algorithms. The inputs to the algorithms represent the determinants of exposure and the output represents the predicted exposure. Models can be deterministic, providing a single-point estimate of exposure based on point estimates of the determinants; or they can be probabilistic, providing a distribution of exposure estimates based on ranges or distributions of input parameters. The latter enable the user to quantitate his or her level of confidence in the exposure estimate.

Who? As discussed above, sampling in the breathing zone of workers is the most common type of occupational air monitoring. Selection of the individual is strongly dependent on the reasons for the sampling and the nature of the contaminant being sampled. For a comprehensive strategy, a random sample of several employees within each SEG should be monitored.[6] For a compliance strategy, where compliance with OELs is the objective, only the employee with the highest potential exposure needs to be sampled (Ignacio and Bullock 2006). For certain contaminants, the sampling-and-analysis method may limit sample duration to no longer than 30 minutes, which would then require sixteen different samples (and therefore sixteen different analytical tests) per person to sample an entire 8-hour shift. Therefore, fewer people may be sampled because of cost restrictions than if only two samples per person need to be collected for another contaminant.

When and how often and how long? Workplace monitoring can be performed to establish baseline conditions for new processes and for existing processes that have not previously been sampled. Workplace monitoring can also be performed when equipment is changed, when chemicals are changed, when controls are implemented, or even on a regular schedule. Workplace monitoring needs to be performed on all shifts that the work is done, as well as during multiple seasons, particularly winter and summer (Ignacio and Bullock 2006).

The length of time for each sample is dependent upon the averaging time of the OEL being used and on the limit of detection of the analytical method. The analytical limit of detection is the minimum amount of material that the analytical method is able to determine. Sampling needs to be done long enough to collect enough material for the analytical method to detect. Samples should be collected in such a way that the averaging time is the same as for the OEL (8-hour time-weighted average, 15-minute short-term exposure limit, etc.) and that enough of the contaminant is collected to exceed the analytical limit of detection. A common approach is to collect multiple samples (2–4) consecutively throughout the workday in order to develop a full-shift, 8-hour time-weighted average exposure estimate (Ignacio and Bullock 2006).

Where? Samples may be collected in the breathing zone of a worker (personal), in the workroom (area), or near a specific operation (source). The location is determined by what information is needed from the workplace monitoring. Only personal samples (collected in the breathing zone) can be used to evaluate worker exposure, but area or source samples may be necessary to determine the effectiveness of an engineering control. (This is a subjective judgment that will depend on factors such as the actual reduction in source emissions and whether this reduction results in worker exposures that are within acceptable limits.)

In addition to the foregoing, the OEHS professional must consider a number of practical factors in designing and implementing a sampling strategy:

- cost of equipment, sampling media, sample analysis, and sampling technician's time
- availability of validated sampling and analytical techniques
- sampling considerations: need for bulk sample, or minimum detectable limit
- calibration of equipment
- personal safety
- chain of custody (physical control of the samples themselves)
- proper handling of samples, including labeling, shipping requirements, and maintenance of temperature

Each company will determine its own response to these factors and devise a sampling strategy that meets the needs of the company.

Use of Exposure Assessment Results

Sampling and analytical results are interpreted/evaluated/judged for acceptability, using accepted statistical techniques (the science) and experience/knowledge/expertise (the art) of the OEHS professional. If results are below the OEL by an adequate margin, then exposures are considered acceptable, and the only follow-up activity might be to document and communicate the results. An adequate margin is generally user defined at a company level and may or may not be statistically based. An adequate margin could be defined so that the sample average with its confidence interval does not overlap the confidence interval of the appropriate PEL. If the results are above the OEL, then exposures are considered unacceptable, and appropriate control measures (e.g., engineering controls, administrative controls, or PPE) should be selected, implemented, and maintained. Exposures between acceptable and unacceptable are the undetermined exposures. For these SEGs, additional information needs to be collected to further refine the assessment in order to better determine acceptability. To make statistically based decisions, six to ten samples need to be collected in each SEG, in order to minimize uncertainty without the cost of each sample outweighing the amount of uncertainty improvement (diminishing returns). These six to ten samples do not all need to be collected during one sampling event; they can be collected over time as long as the SEG remains stable. These samples should be randomly spaced and cover all shifts and seasons in which the work is done. If a truly statistically random sampling strategy is utilized, then each and every person in each and every shift of the SEG will have an equal chance of being monitored. This could be accomplished by assigning numbers to each person in every shift every day during one

CASE STUDY

Quantitative Exposure Assessment

Quantitative exposure assessment can be demonstrated in the following hypothetical evaluation of the staff supporting a firing range. The range is staffed daily by two range safety officers and one range master. The range staff can be exposed to lead up to eight hours per day. To assess exposures to this group, the staff can be divided into two separate SEGs based on their different exposure potential from the different tasks performed. One SEG would include the range master, and the second SEG would include the range safety officers. Therefore, monitoring of both the range master and the range safety officers is warranted. Because only three persons work the range, all three should be monitored, if possible. At a minimum, the range master and one range safety officer need to be monitored.

The selection of the sampling day should be as random as possible. In other words, each day of operation should have an equal opportunity of being selected for sampling. For this hypothetical operation, lead exposure needs to be assessed at least twice each year. Each day the range is operating during each six-month period could be assigned a number in order from 1 to 130 (5 days per week for 26 weeks). A random number table could then be used to determine which day to sample. Suppose Day 73 is selected. If the numbering began at Monday, January 1, Day 73 would be Wednesday, April 11. Sampling would then be performed on Wednesday, April 11.

Now that the date has been picked, sampling preparation can begin. First the sampling method must be determined. The *NIOSH Manual of Analytical Methods* (NIOSH 1994) is a good source of sampling methods. Analytical laboratories are also good sources for determining the method. Both the laboratories and the *Manual of Analytical Methods* can then provide the details necessary to complete the sampling. For lead, NIOSH Method 7082 is used to determine the amount of lead by flame atomic absorption spectrophotometry. The method calls for the sample to be collected on a 0.8-micrometer cellulose ester membrane filter. The sample must be collected at a flow rate of 1–4 liters per minute (lpm) for a total collected volume between 200 and 1500 liters.

The OEL for lead is reported as an 8-hour time-weighted average. Therefore, the sampling should be performed for a full 8-hour shift. As stated above, collecting multiple samples is better than collecting just one 8-hour sample. For this hypothetical firing range, the air is not very dusty and therefore two 4-hour samples can be collected without overloading the filters. At a flow rate of 2 lpm, each 4-hour sample would have had a sampling volume of 480 liters pulled through the filters. This sampling volume meets the requirements of NIOSH Method 7082, and the samples can therefore be collected at a flow rate of 2 lpm.

Prior to the sampling event, the sampling media (in this case the filters preloaded into cassettes) must be ordered from the analytical laboratory. To ensure accuracy of the results, it is best to make sure the laboratory is accredited by the AIHA.[7] The cost of the media is generally included in the cost of the analytical method. Before the media arrive, the integrated sampling pumps and calibrator must be fully charged.

On the day of sampling, Wednesday, April 11, the fully charged integrated sampling pumps must be pre-calibrated with a filter in place in the sampling train to set the pumps at an operating flow rate of 2 lpm. This pre-calibration can be done either in the office or on site, but must be done on the day of sampling. Then upon arriving at the site, the pumps should be placed on the two range safety officers and the range master before the workers enter the firing range for the morning session. The pumps need to be turned on to begin collecting the sample, and the flow rate should be checked to ensure the pump is still operating at 2 lpm. The pump is generally attached to the belt and worn on the side or the back of the person so that it does not interfere with work procedures. The filter is attached to the collar in the breathing zone, generally within 12 inches of the nose and mouth (Bisesi 2004). After sampling has begun, the flow rate should be checked multiple times during the 4-hour sampling event; suggested times could be at 30 minutes, 1 hour, 2 hours, and 3 hours into the sampling event. At the end of the morning session, the sampling pump is removed from the worker, the flow rate checked one final time, and the sample stopped. After the workers return from lunch, the same procedure is used for the 4-hour afternoon session. Post-calibration can then be performed in the field or back in the office. With a calibrated rotameter, the post-calibration can be done as the final flow rate check in the field prior to the pump's being turned off. The stopped sampling pumps can also be returned to the office with the sampling media still in place for the post-calibration.

After the sampling event has been completed, all six filters (two for each of the three workers) along with two field blanks[8] are sent back to the AIHA-accredited laboratory for analysis.[9] The field blanks are required per NIOSH Method 7082 and are taken to the site and treated in the same manner as the samples except that no air is pulled through them (Jordan 2003).

year. A random-number generator can then be used to determine which person on which day working which shift will be sampled. In the real world, a more *representative* sampling approach is generally used. This representative approach may entail selecting a convenient sampling date, but without regard to process operations, then rolling a die to determine which shift on that date, and finally drawing numbers to determine who will be sampled (Ignacio and Bullock 2006).

The data collected can then be statistically analyzed to aid in the decision-making process. A first step is to determine whether the data are log-normally distributed, as are most sampling data. This determination can be done with such tests as the Shapiro and Wilk W test or with probability plotting. After determining the data is log-normally distributed, other calculations can be made, including the arithmetic mean, geometric mean, 95th percentile, exceedance fraction, and upper tolerance limit. Depending on the policy of each company, any one or all of these statistics can be used to determine acceptability. Examples of acceptability cutpoints include an exceedance fraction less than 5 percent or an upper tolerance limit less than the OEL. The exceedance fraction is the proportion of all possible exposures in a SEG that could exceed the OEL and the upper tolerance limit is the 95 percent upper confidence limit of the 95th percentile. (Typical benchmarks include an exceedance fraction of less than 5–10 percent or when the 95 percent upper confidence limit of exposures is less than the OEL.) An extensive discussion on these statistics, along with step-by-step instructions and a spreadsheet program, is available in Ignacio and Bullock (2006).

Impacts of Uncertainty and Data Quality

Two additional concepts—uncertainty and data quality—need to be addressed relative to acceptability of occupational exposures and subsequent decisions to conduct risk management. In the occupational environment, there are two sources of uncertainty: natural variability and ignorance (lack of knowledge). The latter tends to be the largest source of uncertainty in occupational risk assessment. This lack of knowledge impacts the risk assessment-management-communication chain because an assessment cannot be conducted without any information on the agent or scenario, and it is very difficult to manage risks that have not been properly assessed. Likewise, it is difficult to communicate an issue and strategy without any facts. Because of this difficulty, risk assessment tends to be iterative, with data traded for conservatism until satisfactory answers and actions are defined.

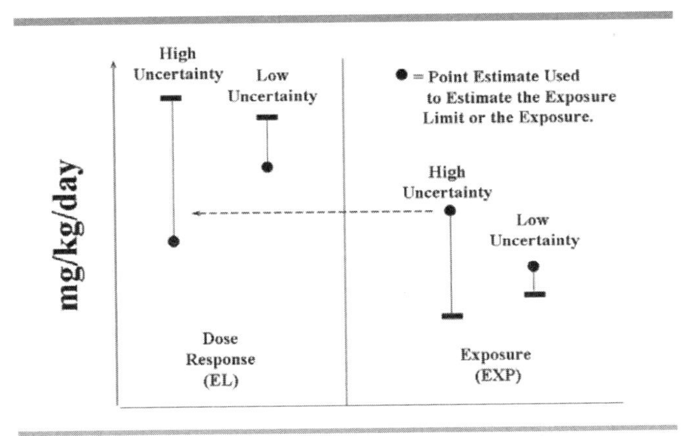

FIGURE 6. Levels of confidence in exposure and dose-response assessments (*Source:* Jayjock, Lynch, and Nelson 2000)

Conclusions regarding the acceptability of risk are impacted by the quality of data for both the exposure limit (EL) and the exposure (EXP). If EXP is below the EL, any combination of high-quality (low-uncertainty) estimate for either the EL or EXP will result in the conclusion of acceptability, because the upper-bound estimate of exposure will be less than the lower-bound estimate of the EL (Jayjock, Lynch, and Nelson 2000).

Figure 6 illustrates this concept, showing the point estimate and range of values for both the EL and EXP. In this illustration, the upper-bound exposure estimate with high uncertainty overlaps with the lower-bound estimate of the EL with high uncertainty; the exposure and risk would be deemed unacceptable. As either or both estimates are refined and the uncertainty is reduced, the two do not overlap and one would conclude that the exposure and risk are acceptable (Jayjock, et al. 2000).

Role Within OSH Programs

Exposure assessment has matured from the compliance mode (e.g., is the highest exposure less than the OEL?) to comprehensive assessment (e.g., characterization of all exposures on all days to all workers). The comprehensive approach allows the OEHS professional to focus resources on areas of high priority and therefore operate in a more efficient manner than the compliance approach. Additionally, the iterative nature of the comprehensive approach forces the OEHS professional to update information for each SEG on a regular cycle. Therefore, the OEHS professional is better able to manage risks to protect workers at the present time, and to anticipate and avoid future risks.

Conclusion

OEHS professionals have increasingly incorporated the concepts of risk assessment and management into their practice. Conducting a risk assessment requires knowledge of the toxicity or hazard of environmental agents, and of worker exposure to these environmental agents. Toxicity measures are typically obtained from standard reference sources, but exposure assessment is a site-specific activity. Risk is characterized by comparing the results from the exposure assessment with the toxicity measures. Risk-management activities are initiated by a finding of unacceptable risk resulting from exposure to occupational hazards. Risk communication is a continual, two-way dialogue involving all relevant stakeholders. A comprehensive understanding of the toxicity and other hazardous properties of an environmental agent, exposures to that agent, and the resulting risk leads to better management of OEHS programs.

Endnotes

[1]A qualitative risk characterization would utilize verbal descriptors, such as low, medium, or high risk. A quantitative process might express the risk of exposure to a toxicant under a particular scenario as 1 case per 1000 persons exposed, or 1×10^{-3}. A semiquantitative risk description would have both verbal and numeric descriptors—for example, a high probability of 15–20 percent of a work force absent with pandemic flu during a 2–3-week period.

[2]Unlike "pure" risk assessment, in which the assessment is conducted entirely by scientists with little or no field experience, occupational risk assessment and management tend to be more integrated. The familiarity of the risk profile and more realistic expectations of control measure options promote a more grounded approach to risk management. That is, the OEHS professional tends to better appreciate the practical limitations of engineering controls, PPE, and administrative controls and can make risk management recommendations accordingly.

[3]In some situations, more in-depth methods—including wipe samples, biomonitoring, dermal absorption, and/or evaluation of environmental exposure—are needed to manage exposures to occupational airborne hazards.

[4]A breach is a rupture or break in containment—for example, for sampling (intentional) or a leak (unintentional).

[5]OSHA's introduction to risk assessment was the July 1980 U.S. Supreme Court decision striking down the benzene standard. The court found that OSHA failed to show that the benzene standard would prevent a "significant risk in the workplace," suggesting that 10–3 is a significant risk and 10–7 is not. OSHA has thus incorporated risk assessment into the standard-setting process (e.g., ethylene oxide, formaldehyde, and methylene chloride).

[6]In random sampling, each worker has an equal probability of being selected for personal monitoring. This is a basic premise of common statistical tests used in exposure evaluation and should be the goal of any sample selection method. Much sampling intended to be "random" is actually "representative." For a comprehensive discussion of the designation of similar exposure groups, and the selection, number, and duration of samples, refer to Ignacio and Bullock (2006).

[7]For more than 30 years, the AIHA's Industrial Hygiene Laboratory Accreditation Program (IHLAP) has been accrediting labs that analyze samples collected in the workplace. Labs that have been accredited through IHLAP must meet defined standards for performance based on a variety of criteria, resulting in a high level of professional performance. Criteria evaluated include personnel qualifications, results from participation in the Proficiency Analytical Testing (PAT) program, facilities, quality control procedures, laboratory records, methods of analysis, and the results of site visits to the laboratories. The AIHA IHLAP is the largest program of its kind in the world and has been in operation since 1974.

[8]Blanks are the same sampling media used to collect the personal samples. The blanks are generally opened in the area where samples are taken and immediately resealed. They are analyzed along with the personal samples to ensure that the sampling media are not contaminated.

[8]The limit of detection (LoD) must be distinguished from the limit of quantitation (LoQ). LoD is the smallest quantity or concentration required by a method to reliably determine presence or absence of a compound. LoQ is the smallest quantity or concentration required by a method to reliably determine the amount present. The LoD is lower than the LoQ.

References

American Conference of Governmental Industrial Hygienists (ACGIH). 2001. *Air Sampling Instrumentation.* 9th ed. Cincinnati: ACGIH.

______. 2004. *Air Monitoring for Toxic Exposures.* 2d ed. Cincinnati: ACGIH.

______. 2007. *TLVs® and BEIs®. Threshold Limit Values for Chemical Substances and Physical Agents: Biological*

Exposure Indices. Updated annually. Cincinnati: ACGIH.

American Industrial Hygiene Association (AIHA). 2002. *Emergency Response Planning Guidelines and Workplace Environmental Exposure Level Guides Handbook*. Fairfax, VA: AIHA Press.

American Society for Testing and Materials (ASTM). 2003 (May 10). ASTM D6196-03. *Standard Practice for Selection of Sorbents, Sampling, and Thermal Desorption Analysis Procedures for Volatile Organic Compounds in Air* (retrieved July 21, 2007). West Conshohocken, PA: ASTM. www.techstreet.com/cgi-bin/detail?product_id=1094393

Bales, J. 2006. "Health Fears Lead Schools to Dismantle Wireless Networks: Radiation Levels Blamed for Illnesses." *Times Online* (retrieved July 21, 2007). www.timesonline.co.uk/tol/life_and_style/education/article642575.ece

Barnes, D. G. 1994. "Times Are Tough—Brother, Can You Paradigm?" *Risk Analysis* 14:219–223.

Bisesi, M. S. 2004. *Bisesi and Kohn's Industrial Hygiene Evaluation Methods*. 2d ed. Boca Raton, FL: Lewis Publishers.

Brief, R. S., and R. A. Scala. 1975. "Occupational Exposure Limits for Novel Work Schedules." *American Industrial Hygiene Association Journal* 36:467–469.

Caplan, K. J., and J. Lynch. 1996. "A Need and an Opportunity. AIHA Should Assume a Leadership Role in Reforming Risk Assessment." *American Industrial Hygiene Association Journal* 57:231–237.

Covello, V. T., P. M. Sandman, and P. Slovic. 1988. *Risk Communication, Risk Statistics, and Risk Comparisons: A Manual for Plant Managers*. Appendix C, p. 54 (retrieved January 15, 2007). Washington, DC: Chemical Manufacturers Association. www.psandman.com/articles/cma-appc.htm

DiNardi, S. R., ed. 2003. *The Occupational Environment: Its Evaluation, Control, and Management*. Fairfax, VA: AIHA Press.

Environment Canada. 1999. "Canadian Environmental Protection Act, Environmental Registry." Gatineau, Quebec, Canada (retrieved March 18, 2007). www.ec.gc.ca/CEPARegistry/the_act

Environmental Protection Agency. (EPA) Undated(a). *The Development of Acute Exposure Guideline Levels (AEGLs)* (retrieved July 21, 2007). www.epa.gov/opptintr/aegl/

______. Undated(b). *Human Health* (retrieved July 21, 2007). www.epa.gov/ebtpages/humanhealth.html

______. 2006. *National Ambient Air Quality Standards (NAAQS)* (retrieved February 25, 2007). www.epa.gov/air/criteria.html

European Commission. 2006. "Registration, Evaluation, Authorisation, and Restriction of Chemicals" (retrieved March 18, 2007). Brussels, Belgium: European Commission, Joint Research Centre. ec.europa.eu/enterprise/reach/index_en.htm

Fischhoff, B., S. Lichtenstein, P. Slovic, and D. Keeney. 1981. *Acceptable Risk*. Cambridge: Cambridge University Press.

Government Accounting Office (GAO). 2006. GAO-06-1032T. *Chemical Regulation: Actions Are Needed to Improve the Effectiveness of EPA's Chemical Review Program*. Washington, DC: U.S. Government Printing Office.

Health and Safety Executive (HSE) (Great Britain). Undated. "COSHH Essentials Wheel." www.coshh-essentials.org.uk/

Ignacio, J. S., and W. H. Bullock. 2006. *A Strategy for Assessing and Managing Occupational Exposures*. 3d ed. Fairfax, VA: AIHA Press.

Jayjock, M. A. 2003. Chapter 8, "Modeling Inhalation Exposure." In S. R. DiNardi, ed. *The Occupational Environment: Its Evaluation, Control, and Management*. 2d ed. Fairfax, VA: AIHA Press.

Jayjock, M. A., J. R. Lynch, and D. I. Nelson. 2000. *Fundamentals of Risk Assessment for the Industrial Hygienist*. Fairfax, VA: AIHA Press.

Jordan, R. C. 2003. Chapter 42, "Quality Control for Sampling and Laboratory Analysis." In S. R. DiNardi, ed. *The Occupational Environment: Its Evaluation, Control, and Management*. 2d ed. Fairfax, VA: AIHA Press.

Keil, C. B., ed. 2000. *Mathematical Models for Estimating Occupational Exposure to Chemicals*. Fairfax, VA: AIHA Press.

Klonne, D. R. 2003. Chapter 3, "Occupational Exposure Limits. " In S. R. DiNardi, ed. *The Occupational Environment: Its Evaluation, Control, and Management*. 2d ed. Fairfax, VA: AIHA Press.

The Lifeline Group. 2007. *Complex Exposure Tool (CEPST)* (retrieved March 18, 2007). www.thelifelinegroup.org

Mine Safety Appliances (MSA). Undated. Catalog. Pittsburgh, PA (retrieved July 21, 2007). www.msanorthamerica.com/catalog/

Mulhausen, J. R., and J. Damiano. 2003. Chapter 6, "Comprehensive Exposure Assessment." In S. R. DiNardi, ed. *The Occupational Environment: Its Evaluation, Control, and Management*. 2d ed. Fairfax, VA: AIHA Press.

National Institute for Occupational Safety and Health (NIOSH). 1994. Schlecht, P. C., and P. F. O'Connor, eds. DHHS-NIOSH Publication 94-113, *NIOSH Manual of Analytical Methods (NMAM®)*. 4th ed. (August, 1994). 1st Supplement Publication 96-135 (1996). 2d Supplement Publication 98-119 (1998). 3d Supplement 2003-154 (2003) (retrieved July 21, 2007). www.cdc.gov/niosh/nmam/

______. 2005. DHHS-NIOSH Publication No. 2005-149, *NIOSH Pocket Guide to Chemical Hazards* (retrieved July 21, 2007). www.cdc.gov/niosh/npg/default.html

National Research Council (NRC), Committee on the Institutional Means for Assessment of Risks to Public Health, Commission on Life Sciences. 1983. *Risk Assessment in the Federal Government: Managing the*

Process. Washington, DC: National Academy Press. (Also known as the "Red Book.")

Nelson, D. I. 1991. "Health and Safety Training for Removal of Underground Storage Tanks." *Applied Occupational and Environmental Hygiene* 6(12):1015–1019.

Nelson, D. I., F. Mirer, G. Bratt, and D. O. Anderson. 2003. Chapter 9, "Risk Assessment in the Workplace." In S. R. DiNardi, ed. *The Occupational Environment: Its Evaluation, Control, and Management*. 2d ed. Fairfax, VA: AIHA Press.

Nicas, M. 2006. Appendix I, "Estimating Airborne Exposure by Mathematical Modeling." In J. L. Ignacio and W. H. Bullock, eds. *A Strategy for Assessing and Managing Occupational Exposures*. 3d ed. Fairfax, VA: AIHA Press.

Occupational Safety and Health Act (OSH Act). 1970. Public Law 91-596, 84 STAT. 1590, 91st Congress, S.2193, December 29, 1970, as amended through January 1, 2004 (retrieved February 25, 2007). www.osha.gov/pls/oshaweb/owadisp.show_document?p_table=OSHACT&p_id=3356

Occupational Safety and Health Administration (OSHA). 1971. *Limits for Air Contaminants*. Code of Federal Regulations, Title 29, Section 1910.1000.

______. 1989. *Air Contaminants (Amended Final Rule, January 1989). Preamble to Final Rule. Section 6—VI. Health Effects Discussion and Determination of Final PEL* (retrieved July 21, 2007). www.osha.gov/pls/oshaweb/owadisp.show_document?p_table=PREAMBLES&p_id=770

______. 1999. OSHA Policy Regarding PEL Adjustments for Extended Work Shifts, 11/10/1999 (retrieved September 3, 2006). www.osha.gov/pls/oshaweb/owadisp.show_document?p_table=INTERPRETATIONS&p_id=22818

______. Undated. *Safety and Health Topics: Sampling and Analysis* (retrieved July 21, 2007). www.osha.gov/SLTC/samplinganalysis/index.html

Olishifski, J. B., and B. A. Plog. 1988. Chapter 1, "Overview of Industrial Hygiene." In B. A. Plog, G. S. Benjamin, and M. A. Kerwin, eds. *Fundamentals of Industrial Hygiene*. 3d ed. Itasca, IL: National Safety Council.

Paustenbach, D. J. 2000. "Pharmacokinetics and Unusual Work Schedules." In R. L. Harris, ed. *Patty's Industrial Hygiene*. 5th ed., vol. 3, pp 1787–1901. New York: John Wiley & Sons.

Plog, B. A., and P. J. Quinlan. 2002. *Fundamentals of Industrial Hygiene*. 5th ed. Itasca, IL: National Safety Council.

SKC, Inc. Undated. "Help with Air Sampling" (retrieved July 21, 2007). www.skcinc.com/help.asp

Toxic Substances Control Act (TSCA). 1976. Title 15, Chapter 53, 15 USC s/s 2601 et seq. www.access.gpo.gov/uscode/title15/chapter53_.html

Appendix: Recommended Reading

Milz, S. A., R. G. Conrad, and R. D. Soule. 2003. Chapter 7, "Principles of Evaluating Worker Exposure." In S. R. DiNardi, ed. *The Occupational Environment: Its Evaluation, Control, and Management*. 2d ed. Fairfax, VA: AIHA Press.

Ramachandran, G. 2005. *Occupational Exposure Assessment for Air Contaminants*. Boca Raton, FL: Taylor & Francis Group.

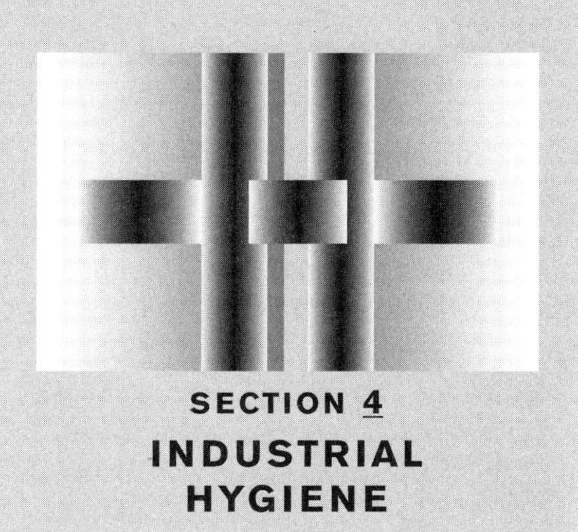

SECTION 4
INDUSTRIAL HYGIENE

Applied Science and Engineering: Chemical Hazards

William Piispanen

LEARNING OBJECTIVES

- Understand chemical terminology as it applies to workplace safety.
- Possess a basic understanding of industrial hygiene chemical-sampling techniques and the limitations of the methods.
- Be familiar with safe workplace practices for working with chemicals.

CHEMISTRY IS one of the oldest and most basic of the sciences. Early chemists were concerned with the basic essential elements—the air we breathe, the water we drink, and fire as the source of energy. The quest for many scientists was identifying the basic elements that constituted our world. Further experimentation turned to the transforming of materials from lesser value to richer value through alchemy. Early experiments by the English physicist and chemist John Dalton led to the concept that basic elemental materials were constituents of everything that made up the living world. Finally in 1869, the idea of a periodic table of the elements was developed to explain similarities in material character (Pauling 1970, 144–145). Additional research based on this ordering of chemical substances led to the theory of elements combining to form new chemical compounds. With this growing understanding of chemical properties came the use of new chemical compounds and materials in industrial applications. The world was changing as new products were introduced into the marketplace and new jobs were developed to create those products.

We have come a long way from these early discoveries, but new elements are still being added to the periodic table as complex experiments continue to probe the very nature of matter. From a safety and health perspective, we know that some combinations of elements are not only safe to use but also essential to life, whereas the same elements combined in a different formula can be poisonous or can introduce hazardous chemicals to the workplace. Sodium chloride, or table salt, is a good example. Considered by many people to be absolutely necessary for the enjoyment of some foods, this compound can be disassociated to the elements of sodium and chlorine. The element sodium (Na) is an extremely hazardous material in the workplace because of its ability to generate explosive hydrogen gas when it contacts water. Chlorine (Cl) gas is poisonous to any life form and is also corrosive and reactive in nature. But where would we be without common table salt, NaCl? This chapter discusses the role of chemistry in the practice of safety, identifies sources for chemical safety information, and provides a basis for understanding how the professional practice of industrial hygiene protects worker health.

Chemicals in the Workplace

Although the understanding that some workplace materials can contribute to illnesses and disease is hardly a new revelation—the published studies of Ramazzini in 1713 (Ramazzini 1964) showed correlations between disease and certain occupations—the study of occupational diseases associated with worker exposure to specific chemicals, either as raw materials or as unwanted waste products, is relatively recent. Early correlations of workplace exposure to hazardous chemicals is usually traced to the late eighteenth and early nineteenth century when studies showed the correlation between London chimney sweeps and incidences of scrotal cancer. In the twentieth century, studies of mortality from lung and bladder cancer in steel workers identified coal tar pitch as a hazardous workplace chemical. Later studies by the U.S. Public Health Services looked at incidence rates of lung cancer in asbestos workers and other lung disease in granite quarry workers exposed to crystalline silica dust.

The Occupational Safety and Health Act was passed in 1970, and with it was created the Occupational Safety and Health Administration (OSHA). This agency of the federal government was charged with setting workplace exposure limits for select hazardous chemicals and materials. The initial permissible exposure limits (PELS) were based on existing occupational exposure limits published by the American Conference of Industrial Hygienists (ACGIH) or the American National Standards Institute (ANSI). In the United States there are published occupational exposure limits for well over 400 substances considered hazardous to human health.

Occupational exposure limits are defined for acceptable limits of worker exposure to potentially hazardous materials. It is important to understand that these limits are defined for healthy workers with an assumption of lifetime exposure based on 8 hours per day and 40 hours per week exposure. Some chemicals may pose a potential risk of cancer in either animals or humans, and these chemicals, mixture of chemicals, or exposure circumstances associated with technological processes are identified in technical literature; see U.S. Department of Health and Human Services (HHS), *Report on Carcinogens*, published as a biennial report per Section 262 of Public Law 95-622 (HHS 2004). In the case of substances that are known to be or reasonably anticipated to be human carcinogens, worker exposure should be minimized to the lowest feasible concentration (LFC) unless there are quantitative occupational exposure-level limits based on analytical methods of detection or technological feasibility (NIOSH 1994a, Appendix A).

The following definitions include those used in regulations and technical literature to describe chemical hazards and the methods used to measure or monitor these materials in the workplace (Washington Group International 2005, 4.2–4.4). Other terminology and units of measure may be found in technical literature concerning community heath regulations (such as Title X of Housing Urban Development, which describes lead levels in public and family housing) or environmental regulations (such as leachable levels of lead from lead-contaminated soil, as described in Environmental Protection Agency (EPA) regulations, or the transportation of a hazardous material such as lead on public means of transit, as described in Department of Transportation regulations 49 CFR 172).

Recognizing Common Occupational Health Hazards

An occupational illness is any abnormal condition or disorder, other than an occupational injury, that is caused by exposure to environmental factors associated with the work environment.

There are six major categories of occupational illnesses (ANSI 1977, A2.8):

1. ***Occupational skin diseases or disorders.*** Examples: dermatitis (the most common occupational illness), eczema, or rash caused by irritants such as cement or poisonous plants; chemical burns from contact with strong acids or bases; chloracne, which may be a symptom of skin contact with chlorinated chemicals such as PCBs; or generalized inflammation or allergic response from exposure to chemicals such as isocyanates in paint formulations.
2. ***Dust diseases of the lungs.*** Examples: silicosis, asbestosis, coal workers' pneumoconiosis, and other pneumoconiosis.
3. ***Respiratory conditions caused by toxic agents.*** Examples: pneumonitis, pharyngitis, rhinitis, or acute congestion. This may be the result of allergic reactions in some cases.
4. ***Poisoning.*** Examples: poisoning by lead, mercury, arsenic, or other metals; carbon monoxide or other gases; organic solvents; insecticide sprays; and chemicals such as resins.
5. ***Disorders caused by physical agents.*** Examples: heatstroke, sunstroke, heat exhaustion, freezing, frostbite, caisson disease, hearing loss, effects of radiation and X-rays, sunburn, and flash burn. Although not directly attributable to chemical exposure, some physical injuries do result from contact with chemical products, such as direct skin contact with liquefied nitrogen causing freezing of the skin.
6. ***Disorders caused by repeated trauma.*** Examples: hearing loss, swelling of the joints, and other conditions due to repeated motion, vibration, or pressure. These disorders are not generally considered related to chemical exposure.

When considering possible occupational diseases that may be related to chemical exposures in the workplace, it is necessary to examine the chemicals in their physical forms that would exist in typical work environments. In many cases, it is possible for chemicals to exist in more than one physical form, so the following discussion is not to be considered exclusive, and the safety professional should review the MSDS and other technical information regarding the safety of any industrial chemical in use on the job site or in the work environment.

GASES, VAPORS, MISTS, AND FUMES

Industrial Sources

Gases, vapors, mists, and fumes in the workplace originate from many sources. These sources are normally controlled to limit the release of the materials into the workplace. In the case of flammable storage areas, volatile vapors are captured by air-handling systems, and containment of the solvents in containers or storage units limits the release of vapors. In chemical processes, generated vapors and gases may be captured by air-handling systems such as covers and air shrouds over the process, as in the case of dip or pickling tanks. In operations such as spray-painting, the over-spray and excess paint spray is captured in spray booths or with water curtains to limit the spread of paint-spray materials. Fumes that can be generated in hot torch-cutting or welding operations are more difficult to contain, but point-of-use air capture is available to limit worker exposure to the airborne metal fumes. An unplanned or uncontrolled release, such as spills of solvents or breaks in process lines, have the potential for high exposure levels in the immediate work area and for dissipation to other work areas and the environment. In all cases, it is critical to understand both the degree of toxicity to the worker and the physical parameters that affect the degree of dispersal.

Potential Routes of Occupational Exposure

Toxic or irritant gases, vapors, fumes, or mists can enter the human body in four ways (Schaper and Bisesi 1998, 67):

1. Inhalation—breathing in through the nose or mouth.
2. Ingestion—generally considered hand-to-mouth transfer of material.
3. Absorption—typically considered direct skin contact and transfer through the pore structure.

4. Injection—mechanical penetration of material through the outer skin layer.

The most common form of exposure is through inhalation, which is the focus of this section. The degree of hazard associated with mist and fume inhalation depends on the toxicity, concentration, size, and solubility of the aerosol particles. Gases and vapors are subject to molecular diffusion characteristics, and the degree of hazard is more dependent on the chemical toxicity of the compound or the degree of irritation or chemical reactive damage caused by chemical reaction with throat and lung tissues.

- Particles larger than 10 microns in size are usually trapped in the nasal passages, throat, larynx, trachea (windpipe), and bronchi (tubes from windpipe leading into lungs), causing local irritation or more permanent tissue damage.
- Particles smaller than 10 microns in size may be deposited deeply within the lungs. These particles may then affect large areas of the body, producing systemic effects that are typical of heavy metals such as lead, or they may damage the lung tissue locally as with silica or asbestos.
- Certain inert gases and vapors have the capability of displacing oxygen and causing simple asphyxiation. Some typical examples are nitrogen, propane, helium, methane, argon, carbon monoxide, and hydrogen sulfide. Simple asphyxiants (e.g., argon and nitrogen) displace oxygen in the surrounding air, and chemical asphyxiants (e.g., carbon monoxide and hydrogen sulfide) displace oxygen in the body.
- The acceptable standards for the maximum exposure to regulated hazardous substances are listed in 29 CFR, in Part 1910.1000, Subpart Z, of the OSHA general industry standards and in Part 1926.55 of the construction standards.
- The acceptable standards for the maximum allowable exposure to unregulated hazardous substances are NIOSH's recommended exposure limits (RELs), ACGIH threshold limit values, and/or AIHA workplace exposure limits (see definitions for ACGIH, AIHA, and NIOSH in the definition section for assistance in sourcing these publications).

The principal route of occupational exposure to gases, vapors, mists, and fumes is inhalation (Stokinger 1977, 11). Ingestion could be a route if the material were to condense or accumulate on a surface or on a worker's hand and then be transferred to the mouth. Skin contact also could occur through a condensation or deposition mode. Injection would be unlikely except for an incident such as high-pressure gas-stream impact on exposed skin. Gases, vapors, mist, and fumes are typically associated with industrial and workplace exposures where these materials are commonly in use and the air distribution of the material is uncontrolled.

Particulates and Fibers

Industrial Sources

Industrial sources of particulates and fibers are widespread. Fibrous materials such as asbestos and fiberglass are used in insulation, gaskets, valve packings, siding, roofing materials, and other applications where both thermal and acoustic insulation properties are important.

Particulates are found both as raw materials, such as cement, coal dust, and blasting grit, and as raw chemical products. Particulates result from mechanical operations on solid materials, such as grinding on painted or metal surfaces, cutting or drilling operations, milling and machining of solid materials, and even windblown dust or direct application of air pressure to particulate material. Particulate material, therefore, represents distinct chemical properties of the original material, but the size distribution is a function of the mechanical process that distributes the particulate into the air.

Potential Routes of Occupational Exposure

Fibers are principally an inhalation hazard. Some fibers, such as fiberglass, can cause skin irritation but without significant health consequences. Fibers in the

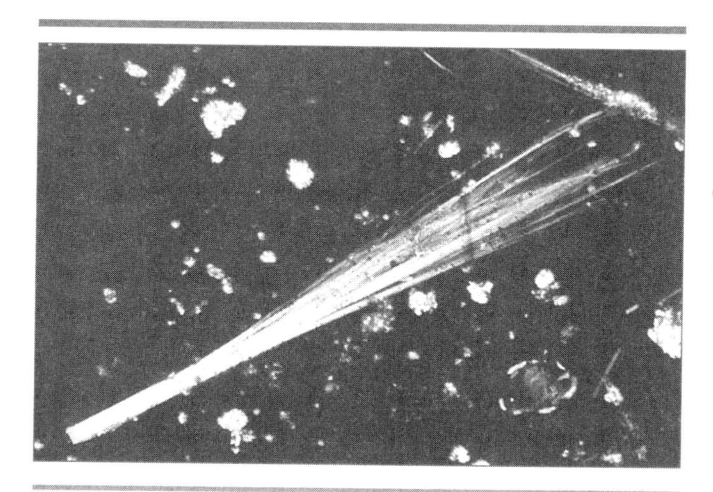

FIGURE 1. Asbestos fiber viewed using scanning electron microscope (Photo courtesy Industrial Hygiene Resources, Boise, Idaho.)

workplace can represent chronic or acute health hazards. But in most cases, the concern is chronic exposure, and the fiber of major concern is asbestos. A photomicrograph of an asbestos fiber is shown in Figure 1.

Asbestos is a mineral material composed primarily of silicates. A number of silicate minerals exist as fibers, but according to the U.S. Bureau of Mines, to be asbestos, the material must have crystal growth along two planes and must have sufficient fiber growth so that the fibers can be identified, separated, and processed. For a comprehensive definition of asbestos, see 29 CFR 1910.1001, Appendix J, 3.5 (Campbell et al. 1977). The use of asbestos in the workplace has been widespread, and worker exposure is found in many industries and trades. The effects of occupational exposure to asbestos fibers are not dependent on the chemical reaction of the fiber; in fact, the material is chemically very inert. Instead, the mode of action is primarily mechanical injury. Asbestos is considered a human carcinogen. The U.S. Department of Health and Human Services' *Report on Carcinogens* first listed asbestos in its 1980 edition (HHS 1980). It is characterized by a *marker* type of cancer usually associated with pleural and peritoneal mesotheliomas (see Agency for Toxic Substance and Disease Registry publication ATSDR-HE-CS-2002-005 for information on asbestos diseases). The primary route of occupational exposure to asbestos is inhalation of fibers and deposition of these fibers within the lung.

Particulate material is also inhaled into the lung and can be deposited in the lung tissue, can be absorbed or digested, or can leach or otherwise release adsorbed chemicals into the lung tissues and fluids. Thus, the toxicology of inhaled particulate material is complex and depends on particle size, aerodynamic characteristics, chemical composition, and sometimes crystallography, as in the case of some silica minerals.

In many industrial and work conditions, the primary concern of worker exposure is particulate from heavy metals. Of particular concern are lead (found in paints and coatings), cadmium (also found in paints and coatings), and chromium (particularly the hexavalent form, which is listed as a known human carcinogen). In 2004 the CDC reported data on workplace injuries and illnesses that showed the 4-year mean prevalence rate for adult workers with blood lead levels (BLLs) above 40 micrograms of lead per deciliter of blood was 2.9 per 100,000 workers. This value is a decline from the value of 3.9 reported for the years 1994–1997 (CDC 2004, 55), which indicates that worker uptake of lead is declining because of better controls and a reduction of lead materials in the workplace.

Lead and similar chemicals can be found in a particulate form when surface coatings are cut or abraded. These chemicals can also form aerosol sprays if applied in a paint formulation and can form a fume if heated above the boiling point of the metal or compound. In the case of heavy metals, the form is often task-dependent on the source and type of worker exposure. The most important step for eliminating worker exposure is reviewing the material safety data sheet for potential sources of heavy metal constituents and then limiting worker exposure to these materials through appropriate engineering controls and respiratory protection.

Solvents, Acids, and Bases

Industrial Sources

Solvents, acids, and bases are commonly encountered in the work environment, and not all exposures represent occupational health risks. Solvents are frequently used in cleaning operations for paints, grease, oils, and other organic-based materials. Acids such as muriatic acid (HCl) may be used to clean concrete, and phosphoric acid (H_3PO_4) and nitric acid (HNO_3) are used in metal surface cleaning for paint prep. Caustic-based materials are found in soaps and cleaning agents used to remove dirt and other surface contamination. Acids and bases are also found in many raw materials used in industrial processes.

Some of these materials can represent potential hazards. For example, benzene as a solvent is recognized as a known human carcinogen, and yet in 1987 OSHA estimated that about 237,000 workers in the United States were potentially exposed to benzene (OSHA 1987). It is not known if this number has changed since that time (ACS 2006). Benzene can constitute, by volume, between 1 and 5 percent of petroleum products (commercial gasoline usually contains between 1 and 2 percent benzene) and is considered both an inhalation health hazard and a possible skin hazard. The NIOSH short-term exposure limit (15-minute maximum level per workday) is 1 ppm (NIOSH 2005, 26). Repeated skin exposure to benzene can cause sensitization and skin lesions and should be avoided.

Acids and bases represent less of a chronic or carcinogenic risk, but in exposures resulting from spills and splashes, there can be severe skin, eye, and throat irritation due to the corrosive nature of these chemicals on body tissues. Hydrofluoric acid (HF) is a particularly dangerous industrial material in that skin exposure to even dilute solutions may not cause immediate pain but can develop later into painful ulcers. HF in contact with skin results in the fluoride ion easily passing through the skin pores, after which it rapidly binds with calcium and magnesium in the body tissues and bone material. The result is a significant alteration of calcium and magnesium levels in the body tissues (Gosselin, Smith, and Hodge 1984, 188–190). Treatment for HF exposure requires immediate flushing with water and medical treatment of the exposed and underlying subcutaneous layers. Salts of acids are also potential hazards, and many are considered poisonous, both chronically and acutely.

Potential Routes of Occupational Exposure

Organic solvents, such as naphtha, mineral spirits, gasoline, turpentine, and alcohol, can be hazardous for two reasons:

1. They affect the central nervous system to some extent, acting as depressants and anesthetics. Effects can range from mild unnoticed irritation to narcosis and death from respiratory arrest.
2. All solvents that contact and wet the skin can cause dermatitis (inflammation, rashes) and allergic reactions. Solvents contacting the eyes can cause slight irritation and, in some cases, permanent blindness. Many chlorinated solvents, for example carbon tetrachloride, can also penetrate the intact skin and cause systemic effects in other organ systems such as the liver. Certain solvents are also carcinogenic.

Protective measures for controlling solvent exposure include the following:

- Substitution of a less toxic alternative. An example is the use of Stoddard solvent, a petroleum spirit, for parts-degreasing operations instead of methylene chloride solvent, which is a possible human carcinogen (HHS 2004, 91).
- Engineering controls installed at the point of vapor generation. An example is installation of an air-moving ventilation system over a parts-cleaning operation to remove solvent vapors from the workers' breathing zone.
- Adequate ventilation of the work area. An example is paint-spray containment and ventilation system to remove excess paint aerosol from the painting process.
- Proper work practices. An example is using tongs to remove parts from an acid cleaning bath to prevent direct skin contact with the acid solution.
- Personal protective clothing. PPE would include suitable eye protection, face splash shields, chemically impervious aprons, chemical gloves,

and impervious boots specified to reduce the skin exposure to hazardous chemicals.

- Barrier skin creams. These provide limited protection, however. Some skin barriers contain silicone and other suspended metals in oil- or lanolin-based materials and may not be effective for all chemical exposures. In most applications, barrier creams are limited to certain water-based solutions (i.e., coolants), alcohol-based solvents, and some mild alkaline solutions. The intent of barrier creams is the prevention of mild chemical dermatitis, not permeation of the skin by a toxic chemical.

Acids are liquids or solids with a pH value below 7. A strong acid has a pH below 4. The EPA defines a corrosive acid as less than pH 2.0 (EPA 2005, 34561). Table 1 is a tabular depiction of the pH scale of relative acid and base strengths. Note that pH measurements are actually logarithmic powers of hydrogen ion activity in solution and are used as representation of relative strength of activity. However, as in the case of the EPA, pH measurements may be used to classify liquids as potentially corrosive with the ability to cause serious skin and tissue injury.

Acids cause localized injury at the area of contact, causing burns. Acid mist can irritate the eyes and the lungs if inhaled. Pulmonary edema or fluid buildup in the lungs can also occur. Examples of common acids include sulfuric acid, hydrochloric acid, and nitric acid. Potential routes of exposure include contact with liquids in dip-cleaning tanks, entry into storage vessels containing acids, spraying or applying acids during surface cleaning, transfer operations of bulk liquids, and line breaks or leaks of industrial sources. While performing any of these operations, it is important to provide proper skin, eye, and respiratory protection against acid liquids and mists.

Bases are liquids or solids with a pH above 7. A strong base has a pH above 10. The EPA defines a corrosive base as one with pH greater than 12.5 (EPA 2005, 34561). Bases are irritants to the skin, eyes, and lungs if inhaled. Common bases such as ammonia, sodium hydroxide, cement, and lye can cause tissue damage, which may initially go unnoticed. Preventing eye exposure is critical when handling bases. Skin contact may produce delayed effects, so it is essential that any potential areas of skin or eye contact be washed with water and treated as a potential chemical burn or irritation.

Acceptable Risks and Occupational Limits

The three modes of action by toxicological substances in the body are physical, chemical, and physiological. Of these, the physiological is the most common. The physiological role of toxic substances in the body may be simply classified as either Reaction I or Reaction II (Stokinger 1977, 25). With Reaction I the toxic agent acts on the body—for example, carbon monoxide can rapidly bind with the iron (hemoglobin) in the blood, thus reducing the ability of the red blood to take up oxygen in the respiration process. With Reaction II, the body acts on the toxic agent, as with liver detoxification of pesticides and chlorinated solvents such as methylene chloride, which is metabolized to carbon monoxide in the body (Torkelson and Rowe 1981, 3450). In both cases, carbon monoxide is involved in the physiological effects, but the rate of reaction is significantly different for the two chemicals, and the acute and chronic effects

TABLE 1

Scale Representation of pH based on 40 CFR 261.22

14	12.5	10	8	7	6	4	2.0	0
Very strong base	Corrosive base	Strong base	Weak base	Neutral solution	Weak acid	Strong acid	Corrosive acid	Very strong acid

(Source: EPA 2005)

of exposure can be significantly different for these two chemicals.

Toxins may also be classified according to toxic mode of actions in the following categories: (1) irritants, (2) asphyxiants, (3) anesthetics, (4) systemic poisons, (5) sensitizers, and (6) particulate matter other than systemic poisons (Dinman 1978, 137–140). Chemicals may act in multi-toxic modes, and the mode may be dependent on concentration, exposure time, and method of entry into the body. Thus, occupational exposure limits consider acceptable levels of exposure based on chemical species and physical characteristics, as well as physiological responses of the body to the chemical toxin.

In some toxic modes, the toxin is metabolized in the body and excreted as waste product. In other cases, the toxin may be absorbed into body tissues as the toxin, a metabolite of the toxin, or a conjugate of the toxin. In many cases the metabolites or conjugates of the toxin are less toxic than the original material and, therefore, allow the body to cope with the potential injury. This ability to cope with the toxin provides the baseline for acceptance of a "no toxic effect" dose limit. This no-toxic-effect limit provides the basis for occupational exposure limits that set the workplace limits for exposure to chemicals where the body has the ability to tolerate and repair damage or eliminate the toxin through physiological responses.

Occupational exposure levels (OELs) have been established for workers that reflect the body's natural capacity to protect against injury and illness from exposure to hazardous chemicals and materials. Often, these exposure levels are based on a daily exposure for 8 hours a day, 40 hours a week, and a lifetime of 40 years of work activity. In many cases these exposure models assume a recovery period between exposures for detoxification and elimination of the toxin and also assume a standard 5-day workweek. For nonstandard work schedules, the OEL may need to be adjusted.

In some chemicals and products the effects of exposure are much more immediate (Reaction I type), and the occupational exposure levels reflect short-term effects and damage of the chemical. These levels are typically set to reflect a 15-minute exposure, although other time limits are also established.

Additionally, there are short-term exposures that should never be exceeded in the workplace at any time because of the potential for immediate injury or health effects. Ceiling limits are established for these materials.

There are other OELs that apply to skin exposure and potential carcinogenic or teratogenic materials. In these instances, the goal is to minimize or eliminate to the extent possible any worker exposure.

Finally, it should be recognized that there are additional OELs that recognize the toxin metabolites and conjugates as markers of human exposures. These biological exposure indices (published by ACGIH) establish levels of indicators in the blood, urine, and fecal material that would be indicative of a potential chemical exposure or other physiological condition. The use of BEIs can assist the occupational health professional to assess body burden of a contaminant, reconstruct past exposures in the absence of other exposure measurements, detect any nonoccupational exposures, test the efficacy of control methods, and monitor all work practices (ACGIH 2006, 91). Other studies have also examined exhaled breath as an indicator of exposure. Even hair samples that indicate arsenic exposure and saliva tests showing ethyl alcohol consumption are recognized biological indicators of exposure.

In all exposure-level models there is an element of acceptable risk. There is a level that is presumed safe for humans for most chemicals (excluding carcinogenic and other chemicals that cause irreparable or very rapid damage). Returning to our earlier example of the chemical sodium chloride—table salt—there is a presumption that most levels of exposure are safe (this is excluding synergistic effects on other system functions such as blood pressure, heart function, and kidney). But actually, salt water can be quite irritating to the eyes—try opening your eyes while swimming in the ocean—and even toxic in large quantities: toxicity is estimated as between 0.5 and 1.0 grams of NaCl per kilogram of body weight (Medtext 2004).

Methods for Determining Exposure Levels

Personal Sampling

Monitoring vapors, gases, particulates, and fibers in the workplace is generally conducted using two methods:

1. By using extractive sampling techniques where the material of interest is trapped on a filter media or adsorbent
2. By a direct-reading instrumental method in which the concentration of material of concern is correlated to a concentration through a detection media and signal amplification.

Both exhibit advantages and limitations in their approach to assessing workplace exposure to chemical hazards.

Direct-reading instruments generally are used to monitor organic vapors, inorganic gases, aerosols, and particulates and fibers suspended in the air. There are a variety of detection methods. The detection system for gases and vapors often uses infrared, ultraviolet, photo ionization, thermal conductivity, electron capture, colorimetric, or electrochemical means of discriminating the physical characteristics of a target analyte. Aerosols, particles, and fibers generally rely on light-scattering properties or mass attenuation of the material of interest and may be detected by photometric detectors, beta attenuation monitors, and condensation nuclei counting devices. There are also direct reading instruments for surface contamination, such as X-ray fluorescence for lead in paint. Levels of detection for portable direct-reading gas and vapor analyzers range from the subparts per million up to percent levels (see Table 9.1 in *The Occupational Environment—Its Evaluation and Control*, AIHA 1998). Accuracy of the detection method varies with the instrumental method, and some methods for vapors and gases are subject to interferences (a quenching or masking of the signal) from other chemicals as well as false positive data from similarly responding chemicals (leading to erroneously high readings).

Portable aerosol and particle monitors generally can measure particle concentrations in the range of 0.01 to 200 or more milligrams per cubic meter. The problem with aerosol and particulate monitoring instruments is that they provide only a quantitative measure of particle concentration—no information on composition is provided. Also, many particle monitors do not provide discrimination of the particle size, which is often critical in assessing the potential health hazard of particulate material in the breathing zone (though some of the new monitoring systems use size-range channels to provide size-distribution frequency data). For a complete discussion of aerosol photometers, see Section G of the *NIOSH Manual of Analytical Methods*, "Aerosol Photometers for Respirable Dust Measurements" (NIOSH 1994a).

As a result of potential inaccuracies, integrated sampling may be needed to provide samples that can be analyzed outside of the monitoring system and under more controlled laboratory systems. This integrated sampling may include media such as charcoal, selective polymer resins (XAD and others), silica gel, and polyurethane foam as a collection media for organic vapors and gases. Inorganic gases may be collected on chemically treated sorbents; sometimes liquid adsorbent media are used for acid gases. In all integrated sampling it is critical that the flow rate and time of sampling be noted accurately because this is included in the calculation. Therefore, sampling systems normally include some type of air mover and calibrated flow control or measuring system.

When collecting particulate and fibers, it is important to determine if the material being sampled is representative of the material that is being inhaled into the lung (i.e., respirable size with potential to travel into the airway passages or total particulate that the worker is exposed to in the breathing zone without regard to particle size). Generally, this is accomplished using a sampling cyclone in conjunction with the filter, but other methods include impactors, elutriators, and even direct-reading instruments that use light-scattering methods. An example of a personal sampler impactor is shown in Figure 2.

In most cases, the sizing of the particle or fiber is dependent on the shape, size, and density of the particle, which are generally unknown parameters, and

FIGURE 2. Parallel particle impactor (Photo courtesy of SKC, Inc.)

as a result, particle size is generally correlated to the aerodynamic equivalent diameter (a.e.d.) which is the diameter of a unit-density sphere that would exhibit the same aerodynamic-characteristic settling velocity (Johnson and Swift 1998, 248). Other factors, such as particle shape and density, together with slip factors and sampling turbulence, may impact the sizing of the particles. In most cases, particle-sizing instrument flow rates correlate well to particle aerodynamic diameters for the purposes of determining the percentage or proportion of particles that could potentially travel into the lungs and the lower air exchange areas of the lungs.

An example of this application is the method for determining the respirable fraction of crystalline silica in dust (NIOSH Method 7500). This method recommends sampling using a cyclone separator operated at a flow rate of 2.5 liters per minute (l/min) with a polyvinyl chloride (PVC) filter media. At a flow rate of approximately 2.5 l/min, using an aluminum (SKC) cyclone, the "cut point" of the cyclone will produce a sample of approximately 50 percent of 4-micron a.e.d. or less. In this case the 50 percent cut point (d50) represents the particle size captured with 50 percent efficiency.

For crystalline silica, the PVC filter sample is analyzed using X-ray diffraction microscopy. In asbestos sampling, the fibers are trapped on a mixed cellulose ester (MCE) filter, but the analysis of the fibers (including size information) is determined directly by microscopic methods. Figure 3 shows an elemental analysis of a typical dust particle from a filter media, but this information does not identify the material as crystalline silica. Therefore, additional analysis would be required to assess the health risk.

There are also sampling methods that do not involve either direct reading instruments or integrated sampling using air-collection techniques. These methods can be referred to as passive sampling—sampling by simple grab or wipes or by diffusive-type air samplers. Passive sampling offers a simple and nonintrusive method to provide a quantitative assessment of potential chemical-exposure hazards. Passive sampling methods have been developed and validated for a number of chemicals. See "Chlorinated and Organonitrogen Pesticides," NIOSH Method 9201, for an example of dermal patch, passive sampling method, or "Nitrogen Dioxide," NIOSH Method 6700, for a treated sampling badge technique (NIOSH 1994a). These types of sampling systems may be subject to error when used in lower atmospheric pressures (high altitudes) or in high-dust work locations, but their ease of use makes them ideal for controlled workplace exposure monitoring. The ANSI/ISEA 104-1998 Standard (ANSI/ISEA 1998) provides guidance to manufacturers and users on the methods to evaluate passive sampling devices for a variety of chemicals and conditions.

Passive samplers are also useful in fixed locations to monitor general work-area concentrations, for example, diffusive charcoal tubes suspended in the work area to monitor potential benzene levels. By adjusting the length of time the samplers are exposed, very low

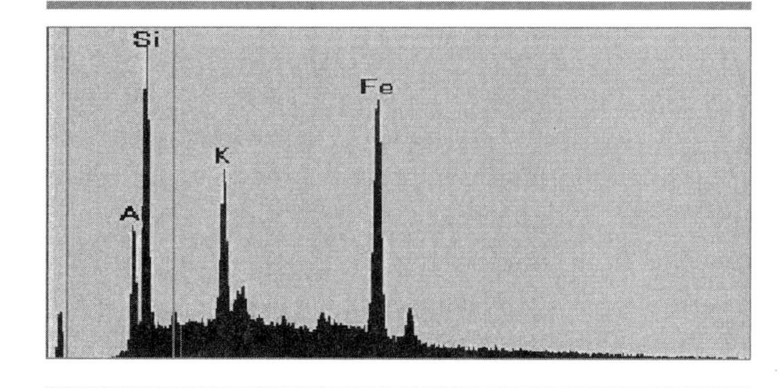

FIGURE 3. Elemental analysis of a typical dust particle (*Source:* Piispanen 2001)

levels can be detected. Results are used to approximate potential area concentrations of contaminants in the work area.

Area Sampling

Area sampling is often used in conjunction with or in support of personal sampling in assessing chemical hazards in the workplace. Area sampling may include direct reading instruments used in a survey mode, where the work area is surveyed for potential contamination level. Handheld photoionization detectors and chemical cell devices are readily available for this application. The direct reading data can be collected by an instrument data logger and later downloaded or managed to provide maximum, minimum, and average concentrations. In data-logging mode, multiple instantaneous readings over time are recorded on a built-in data system. Data summary may be displayed directly on the instrument meter or downloaded to a computer data-management program.

This data is useful when assessing a work zone for potential chemical hazards from organic solvent vapors and specific inorganic species such as ammonia, chlorine, carbon monoxide, hydrogen sulfide, nitric oxides, sulfur dioxide, hydrogen cyanide, phosphine, and other compounds and gases. In most cases where direct-reading instruments are used in a survey mode, there is usually some *a priori* knowledge of the potential contaminant so that the appropriate instrument is selected. The error of the method is that additional materials may be present or that other materials present may mask the response signal or result in a false positive signal.

Therefore, area sampling also includes samples that are collected and analyzed at an off-site laboratory using instrumental methods as shown in Figure 4. In this application, samples are collected in a manner that preserves the sample, protecting it from degradation or chemical alteration. A common collection media is the SUMMA® canister or equivalent product shown in Figure 5. These are preevacuated, specially polished, spherical stainless steel canisters that are used to collect volatile organic gases. The frequently referenced methods for the analysis of these SUMMA® samples are EPA TO-14A and EPA TO-15 (EPA 1999). These EPA analytical methods provide quantitative analysis of many volatile species at concentrations ranging from 1 ppb to 10,000 ppm, depending on the size of the container and the analytical detection limits.

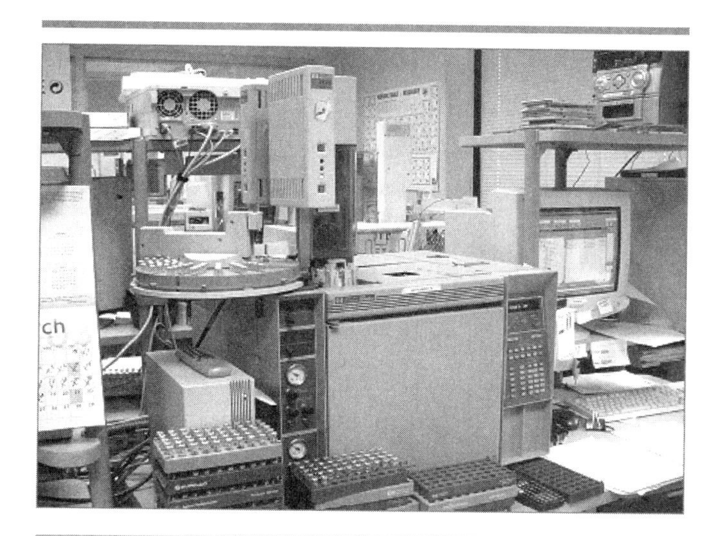

FIGURE 4. Analytical instruments for semivolatile organic analysis (Courtesy Washington Group International)

FIGURE 5. SUMMA® evacuated canister sampler (Courtesy Washington Group International)

A list of organic compounds known to have been analyzed by the canister method can be found in ASTM Method D-5466-95, Table 1 (ASTM 1999, 89). There are also new canisters with specially coated interior surfaces for active polar and low-sulfur-containing organic species. (See ASTM Standard Guide D-6345-98 for discussion of whole air sampling with canisters). These coated canisters are identical in appearance to standard preevacuated cylinders except for the addition of the interior surface coating, which may include silica or polymer materials. Silica-treated canisters may be referred to as SilcoCans.

The evacuated canisters are set out in preselected work locations, and then, through a control bleed orifice, air samples are collected in a controlled flow rate. The advantages of using evacuated canisters are that there is no requirement for power or a pump in the sampling location, and the results provide very detailed analysis of potential air contaminants. However, the results are limited in that they do not necessarily represent the direct worker exposure and can be influenced by environmental conditions, changes in operations, and local heating and ventilation systems.

It is critical, therefore, that air movement and environmental conditions be monitored and recorded during passive area sampling. Frequently, carbon dioxide (CO_2) is monitored both as an indicator of building occupancy and as a measure of ventilation dilution of the air. ASTM Method D-6245-98 provides guidelines on using carbon dioxide concentrations to evaluate indoor air ventilation rates. In this method, the percent of outdoor air (OA) is calculated based on CO_2 concentrations as follows:

Calculation of Outdoor Air Dilution (ASTM 1999, 193)

$$\% \text{ OA} = 100 \times (C_r - C_s) \;/\; (C_r - C_{out}) \tag{1}$$

where

C_r is the concentration of CO_2 (ppmv) in the recirculation air stream.

C_s is the concentration of CO_2 (ppmv) in the ventilation supply stream

C_{out} is the concentration of CO_2 (ppmv) in the outside ambient air.

Many direct-reading indoor air-quality monitors are programmed to perform this calculation directly. The results can be used to determine the impact of outside air either as a dilution factor in a workspace or as source of contamination to the workspace.

Other factors that should be monitored include temperature and relative humidity because these are often reported as indicators of worker comfort levels. An inventory and use pattern of chemicals in the work area should be maintained during area sampling. Samplers should not be placed near known sources of chemicals unless the intent is to identify specific components of the source as a contributor to the overall area contamination.

Area sampling data provides a picture of the potential air contaminants and constituents that could potentially impact worker health and comfort levels. Of course the task now is to determine how to use the data collected and to identify and correct potential contamination or discomfort conditions. (*Note:* This discussion does not address biological contaminants or odors that also may contribute to worker health and comfort levels. Additional sampling and analysis may be required to assess these contributions as well.)

Modeling of Exposure

Modeling of exposures is an approach to determine the temporal and spatial variations and impacts of sources of air contaminants in the workplace (ASTM 1999, 97). Models are also available to estimate potential impact of water or soil sources of contaminants in the workplace (see Marlowe 1994 for a developed model). Modeling is used to understand the factors in the work environment that may expose workers to hazardous chemicals. With this understanding, controls can be designed to reduce or eliminate the exposure potential.

Most models are based on the principle of mass conservation (i.e., the mass of materials in the system that is maintained, including factors such as the workspace and any ancillary air-handling system). Chemicals may undergo chemical or physical transformation due to thermodynamic reactions or mechanical displacement mechanisms. Uncertainty factors or safety factors (usually between 4 and 100) are often incorporated to provide an additional measure of protection to workers.

Models generally focus on a single material or a similar class of materials. Models of particulate material-based chemicals must account for physical factors that influence suspension, transport, removal, coagulation, and deposition. Gaseous and vapor phase materials are subject to volatilization, sorption, and condensation factors. Most models provide some average flux measurement or emission rate for the material of interest based on input values for the source of the material. Ultimately, the value and accuracy of the model is dependent on the precision of measurement of the input values and the applicability of the physical laws that drive the model.

The true test of any model is its ability to predict the desired results with accuracy—for example, the average daily airborne exposure level to lead dust from a contaminated soil–removal operation or the short-term concentration to carbon monoxide from a combustion source. Additional data is usually collected during the operations using personal or area sampling techniques to validate the model results and to determine the amount of error in the prediction. Additionally, sensitivity analysis methods may be applied to the model to evaluate the relative impact of input parameters. With this information, the input values may be reexamined to determine the accuracy of the data, and if necessary, the sample size of the input data may be increased to provide a better statistical chance of accuracy. For a discussion of methods for evaluating air-quality models, refer to ASTM Method D-5157-97 (ASTM 1999).

General Methods of Controlling Potential Exposures

When chemical hazards in the workplace are predicted or discovered, steps must be taken to eliminate or control them. General methods of control might include the following (based on "General Methods of Controlling Environmental Factors," Procedure 1, of Washington Group International 2005, 11):

- Substituting less harmful materials for materials that are dangerous to health. For example, using a citrus-based degreaser product instead of a petroleum solvent product.
- Changing or altering the process or work operation to reduce the number of persons exposed. This might include scheduling dip tank change-out or cleaning operations to night shift operations when fewer workers are in the area.
- Isolating or enclosing the process or work operation to reduce the number of persons exposed, such as containing the operation in a glove box or enclosure.
- Using wet methods, such as water spray or mist, to reduce dust in excavation and other construction activities. Water trucks are used on construction and demolition sites, and water mist sprays and water curtains are used to control chemical product use areas.
- Incorporating engineering control measures, such as point-of-use air suction/blowers, to remove or disperse contaminants at the source before they reach the worker, such as the slot hood dust capture shown in Figure 6.
- Ventilating work area and chemical storage areas with clean makeup air, thus reducing the concentration of the material below recommended workplace limits.

FIGURE 6. Local exhaust ventilation, slot hood (Courtesy of Gayla McCluskey, 2007)

- Employing good hygiene work practices including hand- and face-washing; prohibiting eating, smoking, or chewing in contaminated areas; and controlling spread of contamination on workers' clothing by maintaining decontamination procedures.
- Wearing personal protective equipment and devices such as chemical gloves and boots, spill aprons or chemical-resistant clothing, respirators, and eye and face protection, as exemplified by a worker wearing a powered air-purifying respirator in Figure 7.
- Maintaining good work-area housekeeping, including cleanliness in the work area, proper waste disposal, adequate washing facilities, clean lavatories and eating facilities, safe drinking water sources, and control of insects and rodents.
- Continuous or frequent personal and area sampling with monitoring devices, complemented by worker medical-surveillance programs when required to detect potential intake of toxic materials. This may also include installing continuous chemical monitors on process lines to warn of chemical leaks in the process.
- Maintaining adequate training and educational programs to supplement engineering controls. Training should include Hazard Communication training as specified in 29 CFR 1910.1200, as well as task-specific hazard analysis of potential chemical hazards that may be encountered in the work area or from other sources of chemical hazards.

FIGURE 7. Powered air-purifying respirator (PAPR) (Courtesy of Gayla McCluskey, 2007)

Safe Work Practices Summary

Chemicals in use in the workplace exhibit defined properties such as physical state, chemical composition, solubility, and flammability, as well as the potential ability to cause negative health effects and irreparable harm to humans and the environment if used incorrectly. Yet the ability to use chemicals in the creation of new materials, drugs, and other beneficial materials is what creates and sustains work in the world. The safe use of chemicals in the workplace is not in need of a "Philosopher's Stone" or an "Elixir of Life" as sought by the early alchemists but instead relies on prudent safety practices based on chemical principles and good practices. With an understanding of these principles and an appreciation for safe practices, the risk of health impact and irreparable injury from exposure to workplace chemicals can be managed.

Safe work practices for working with chemicals in laboratories are based on OSHA regulation 29 CFR 1910.1450 and in the general workplace under 29 CFR 1910.1200 but in addition should include the following as a minimum:

- Review and understand the chemical material's composition and physical and chemical properties. This would include understanding physical and chemical transformation that may occur in any use, either planned or unplanned, such as an emergency condition, spill, or release to the environment.
- Within the planned use of a chemical material (including high-potential, unplanned releases), the points of potential exposure and the time frame of potential exposure must be identified. This provides information needed to design sampling programs as well as a basis for identifying protective measures when required to limit occupational exposures.

- Determine the potential exposure levels using both worst-case scenarios as well as average exposure scenarios. Sometimes a hazard-ranking model can be used to assess potential risk factors. It is important to consider additive and synergistic effects of certain chemicals. Consider including an assessment of the physical conditions in the workplace that could contribute to additional safety problems, such as open flames when flammable solvents are in use, incompatibility in material storage or containment facilities, and thermal and ventilations systems that could disperse or distribute the materials outside the immediate work area.
- Establish controls that reduce the potential exposures to below acceptable occupational exposure levels. Controls may include engineering controls such as ventilation, administrative controls such as establishing work-entry boundaries to hazardous areas, or the use of personal protective equipment such as respirators and protective clothing.
- Monitor the work area to assure that the controls are effective. It is critical to assure that changes to the materials in use and the process are not made without reassessing the potential hazards. It is also important that workers report any unusual conditions or problems immediately because these could be indicative of changing or unexpected conditions.
- Finally, it may be necessary or desirable to include medical monitoring or other surveillance programs of worker health to ensure that there are no detectable health impacts. Medical monitoring may require blood samples to determine potential lead exposure or urine samples to determine potential cadmium exposure. Exposure to a number of hazardous chemicals and tasks requires medical monitoring as specified in the OSHA standards. It is important to use the information gathered in the previously outlined steps to support the medical monitoring program.
- Worker training is a critical element to any chemical safety program. Information is presented to workers in a variety of media, but it is important that there be a thorough understanding of the potential for exposure and the need to follow safe work practices to minimize any potential risk.

Chemistry is not a subject that is within the technical discipline of many workers, including many safety professionals. A wealth of technical information is available, and there are a number of professional industrial hygienists who can assist in the understanding of the material. With this available information and technical assistance, a safety program that incorporates good chemical safety practices is readily available. The history of chemistry in the workplace is a long history, often focused on the negative impact of chemicals on workers. Chemical substances exist in all facets of life both on and off the job. It is the responsibility of the safety professional to identify safe work practices for those substances with the potential to create any negative impacts.

Important Terms

American Conference of Governmental Industrial Hygienists (ACGIH): Technical Information Office, 6500 Glenway Avenue, Building D-7, Cincinnati, OH, 45211-4438. Source for threshold limit value (TLV) handbook and other technical information.

Acid: An inorganic or organic compound that (1) reacts with metals to yield hydrogen, (2) has a pH of less than 7.0, (3) dissociates in water to yield hydrogen or hydronium (H_3O^+) ions, and (4) neutralizes bases or alkaline media. Many acids are corrosive to human tissue and must be handled with care. Acids are used in industrial pickling processes and also as raw material in chemical processes. Acids also constitute industrial waste streams.

American Industrial Hygiene Association (AIHA): 2700 Prosperity Avenue, Suite 250, Fairfax, VA, 22031, (703) 849-8888. Source for technical publications and laboratory accreditation information.

American National Standards Institute (ANSI): 23 West 43rd Street, New York, NY, 10036, (212)642-4900.

Source for many safety and engineering consensus standards.

Asbestos: A commercial term applied to a family of magnesium-silicate minerals in fibrous form. They have electrical and thermal insulating properties and can be woven into fabrics. These mineral forms have been used extensively in the construction industry and for other uses such as brake and clutch linings.

American Society for Testing and Materials (ASTM): 100 Barr Harbor Drive, West Conshohocken, PA, 19428, (610) 832-9693.

Base (caustic): An alkaline substance that (1) dissociates in water to liberate hydroxide (OH^-) anions, (2) has a pH greater than 7.0, and (3) can sometimes be used to neutralize acids. Bases can be corrosive to living tissue and must be handled with care. Caustics are often found in strong industrial cleaning products as well in chemical processes as a raw material.

Biological exposure indices (BEIs): Guidance values, determined by the ACGIH, for assessing biological monitoring results based on determinants that are likely to be observed in biological samples collected from healthy workers who may be exposed to chemicals to the same extent as inhalation exposure.

Calibrate: To check, adjust, or systematically standardize a measuring instrument (e.g., an air-sampling pump or combustible gas indicator per the manufacturer's instructions and specifications).

Cassette: The filter holder used when sampling for particulates.

Centers for Disease Control and Prevention (CDC): An agency of the U.S. Department of Health and Human Services. Its main office is located in Atlanta, Georgia. The CDC also includes the National Institute for Occupational Safety and Health (NIOSH) and the Agency for Toxic Substance and Disease Registry (ATSDR). (See www.cdc.gov for information and access to CDC information.)

Contaminant: An undesirable substance or constituent in an unwanted location.

Cyclone: An air-sampling device that is used for sampling respirable particulates. A cyclone *swirls* the incoming air stream and, through centrifugal action, causes the larger nonrespirable particles (less than 10 microns) to drop out. The respirable particles then travel with the air stream and are captured on a filter cassette. The nonrespirable particles accumulate in the bottom of the cyclone and are usually discarded.

Dust: Solid particulate matter suspended in air by mechanical means such as handling, crushing, grinding, abrading, or blasting. Dusts can become an inhalation, fire, or explosive hazard.

Flammable: Any solid, liquid, or gas that ignites easily and burns rapidly. A flammable liquid is typically defined by OSHA as one with a flashpoint below 100 degrees F (37.8°C). A vapor or gas is flammable when its concentration in air is higher than its lower explosive limit (LEL) or lower than its upper explosive limit (UEL). The flammable range for gas or vapors is between these values. The LEL and UEL are expressed in percent of gas or vapor in air by volume.

Flow rate: The volume of air or other fluid that flows within a given period of time.

Fume: Minute, solid particles (generally less than one micron in diameter) dispersed in air from the heating of a solid. The heating is often accompanied by a chemical reaction when the particles react with oxygen to form an oxide. An example is metal fume production during welding or torch-cutting operations.

Gas: A formless fluid that occupies the space of its enclosure. Gases change to a liquid or solid state with increased pressure and decreased temperature. Gases have low densities and viscosities and can expand or contract greatly.

Hazard Communication Standard: An OSHA standard that requires chemical manufacturers and importers to assess the hazards associated with the materials that they produce. Material safety data sheets, which are used to communicate hazard information to the user, are a result of this regulation (see OSHA 29 CFR 1910.1200 for general industry or 1926.59 for construction).

Heavy metals: Metallic elements with high atomic weights such as iron, lead, arsenic, cadmium, vanadium, and uranium. Many heavy metals represent potential health hazards when inhaled or ingested in high concentrations.

International Agency for Research on Cancer (IARC): An agency of the World Health Organization (WHO) located in Lyons, France, IARC publishes the *IARC Monographs on the Evaluation of Carcinogenic Risks of Chemicals to Man*. (See www.iarc.fr for access to information.)

Immediately dangerous to life and health (IDLH): A level of exposure determined to ensure that a worker could escape without injury or irreversible health effects in the event of a failure of a respiratory protection system. These values are documented in Documentation for Immediately Dangerous to Life and Health Concentrations (NIOSH 1994b).

Indicator tube: A glass tube containing a solid material that changes color when brought into contact with a specific chemical or class of chemicals. When a known volume of air is pulled through the tube, the length of the resulting stain is proportional to the concentration of the contaminant. These tubes are used primarily as screening tools to approximate the concentration of a contaminant in the workplace.

Inorganic: Designates compounds that generally do not contain carbon. Exceptions are carbon monoxide, carbon dioxide, carbides, and carbonates.

Mists: Suspended liquid droplets formed by condensation from the gaseous to the liquid state or by mechanically breaking up a liquid into a dispersed form by splashing, foaming, or atomizing.

Material safety data sheets (MSDS): Describe the safety precautions and components of a hazardous material. MSDSs are prepared by the product manufacturer, and the hazard information required on the MSDS is defined in the Hazard Communication Standard (29 CFR 1910.1200g).

National Institute for Occupational Safety and Health (NIOSH): 4676 Columbia Parkway, Cincinnati, OH, 45226 (513) 533-8287. Source for criteria documents and technical guidance.

Organic: Compounds or chemicals that contain carbon, hydrogen, and other elements in chain or ring structures. These may be solid, liquid, or gaseous.

Occupational Safety and Health Administration (OSHA): 200 Constitution Avenue NW, Washington, D.C. An agency of the U.S. Department of Labor established by the Williams-Steiger Occupational Safety and Health Act of 1970. OSHA provides sources of health and safety regulations as well as enforcement of published regulations. Regulations enacted by Congress under OSHA are codified under Title 29 of the Code of Federal Regulations (CFR). Part 1910 of 29 CFR applies to general industry rules, and Part 1926 applies to construction industry rules. Under Subpart A, states may also administer occupational safety and health regulations under state plan programs. (See www.osha.gov for information and access to material published by OSHA or check state government Web sites for state programs.)

Polychlorinated biphenyls (PCBs): Very stable materials of low flammability that contain 12 to 68 percent chlorine. They have been used as insulating materials in electrical capacitors and transformers, as plasticizers in waxes, in paper manufacturing, and for other industrial purposes and are still encountered in the workplace today.

Permissible exposure limit (PEL): The allowable personal exposure level averaged over an 8-hour work shift. This is an OSHA legislative value (see 29 CFR 1910.1000 through 1910.1050 for general industry values).

Personal protective equipment (PPE): Includes clothing and equipment worn to shield or isolate individuals from chemical, physical, and biological hazards that may be encountered in the workplace. PPE is used to protect the respiratory system, skin, eyes, hands, face, head, body, and hearing.

Parts per million (ppm): A unit of measure, equal to 0.000001. When applied to a gas concentration in air, it is equal to 1 liter of gas or vapor per 1000 cubic meters of air and is frequently referred to as ppmv to denote that the units refer to volumes. The term is also used as a mass-based unit, such as for lead concentration in soil; in mass-based measurements, only ppm is used. In this case 1 ppm is equal to 1 milligram per kilogram.

Respirable: The respirable portion of the total particulate collected. Respirable particulates are usually those having an aerodynamic diameter of ten microns or less.

Silica: Silica is a mineral that, with prolonged exposure, can cause lung diseases. The OSHA standard for "crystalline silica" is currently based on the percentage of quartz in the respirable fraction.

Threshold limit value–Ceiling (TLV-C): The concentration of a contaminant that should not be exceeded during any part of the working exposure. For some substances such as irritant gases, only one category—the TLV-C—may be relevant. For other substances, the TLV-STEL or TLV-TWA may be relevant depending on the contaminant's physiologic action. If any one of these three TLVs is exceeded, a potential health hazard is presumed to exist. (*Note:* TLVs® are prudent industrial guidelines established by the American Conference of Governmental Industrial Hygienists and are not regulatory limits. The use of the term TLV in any of the forms in this chapter implies the use of the ACGIH registered name and disclaimer of liability with respect to the use of the guideline.)

Threshold limit value–Short-term exposure limit (TLV-STEL): The concentration to which workers can be exposed continuously to a substance for a short period of time without suffering from (1) irritation, (2) chronic or irreversible tissue damage, or (3) narcosis of sufficient degree to increase the likelihood of accidental injury, impair self-rescue, or materially reduce work efficiency, and provided that the daily TLV-TWA (see next definition) is not exceeded. STEL is not a separate independent exposure limit; rather, it supplements the time-weighted average (TWA) limit where there are recognized acute effects from a substance whose toxic effects have been reported from high short-term exposures in either humans or animals. A STEL is defined as a 15-minute time-weighted average exposure, which should not be exceeded any time during a workday even if the 8-hour time-weighted average is within the TLV. Exposures at the STEL should not be longer than 15 minutes and should not be repeated more than four times per day. There should be at least 60 minutes between successive exposures at the STEL. An average period other than 15 minutes may be recommended if observed biological effects warrant.

Threshold limit value–Time-weighted average (TLV-TWA): The time-weighted average concentration for normal 8-hour workday and a 40-hour workweek, to which nearly all workers may be repeatedly exposed—day after day—without an adverse effect.

Total dust: The collection of all particulates, regardless of size, onto a filter or other measuring device.

Vapors: The gaseous state of a material suspended in air that would be a liquid under ordinary conditions.

References

Agency for Toxic Substances and Disease Registry (ATSDR). 1997 (rev. 2000). "Case Studies in Environmental Medicine, Asbestos Toxicity." Report ATSDR-HE-CS-2002-0005. Washington, D.C.: U.S. Department of Health and Human Resources.

American Cancer Society (ACS). 2006 (rev. 02/02/2006). "Benzene." www.cancer.org

American Conference of Governmental Industrial Hygienists (ACGIH). 2006. *Threshold Limit Values for Chemical Substances and Physical Agents and Biological Exposure Indices*. Cincinnati, OH: ACGIH.

American Industrial Hygiene Association (AIHA). 1998. *The Occupational Environment—Its Evaluation and Control*. Salvatore R. DiNard, ed. Fairfax, VA: AIHA Press.

American National Standards Institute (ANSI). 1977. *American National Standard for Uniform Recordkeeping for Occupational Injuries and Illnesses*. ANSI Z 16.4-1977. New York: ANSI.

American National Standards Institute and International Safety Equipment Association. 1998. *American National Standard for Air Sampling Devices-Diffusive Type for Gases and Vapors in Working Environments*. ANSI/ISEA 104-1998 (R2003). Arlington, VA: International Safety Equipment Association.

American Society for Testing Materials (ASTM). 1999. *ASTM Standards on Indoor Air Quality*. ASTM No. IAQ00. West Conshohocken, PA: ASTM.

Campbell, W. J. et al. 1977. "Selected Silicate Minerals and Their Asbestiform Varieties." U.S. Department of

Interior, Information Circular 8751. Washington, D.C.: U.S. Government Printing Office.

Centers for Disease Control. 2004. *Worker Health Handbook*. U.S. Department of Health and Human Services publication DHHS 2004-146. Washington, D.C.: U.S. Government Printing Office.

Dinman, B. D. 1978. "The Mode of Entry and Action of Toxic Materials." In G. D. Clayton and F. E. Clayton, eds., *Patty's Industrial Hygiene and Toxicology*, vol. 1. New York: Wiley.

Gosselin, R. E., R. P. Smith, and H. C. Hodge. 1984. *Clinical Toxicology of Commercial Products*. Baltimore, MD: Williams and Wilkins.

Johnson, D. L., and D. L. Swift. 1998. "Sampling and Sizing Particles." In S. R. DiNardi, ed., *The Occupational Environment—Its Evaluation and Control*. Fairfax, VA: AIHA Press.

Marlowe, C. 1994. "Action Levels for Hazardous Waste Site Work." AIHA Seminar Publication. Fairfax, VA: AIHA Press.

MEDTEXT. 2004.Thompson Micromedex [online searchable database]. www.csi.micromedex.com

Occupational Safety and Health Administration. 1987. Occupational exposure to benzene. Final Rule. Federal Register 52:34460-34578.

Pauling, L. 1970. *General Chemistry*. New York: Dover.

Piispanen, W. H. 2001. Unpublished data, collected in New York City.

Ramazzini, B. 1964. *Diseases of Workers*. Translated from *De Morbis Artificum Diatriba*. New York: Hafner Publishing.

Schaper, M. M., and M. S. Bisesi. 1998. "Environmental and Occupational Toxicology." In S. R. DiNardi, ed., *The Occupational Environment—Its Evaluation and Control*. Fairfax, VA: AIHA Press.

Stokinger, H. E. 1977. "Routes of Entry and Modes of Action." In *Occupational Diseases: A Guide to their Recognition*. U.S. Department of Health Education and Welfare Publication DHEW 77-181. Washington, D.C.: U.S. Government Printing Office.

Torkelson, T. R., and V. K. Rowe. 1981. "Halogenated Aliphatic Hydrocarbons Containing Chlorine, Bromine, and Iodine." In G. D. Clayton and F. E. Clayton, eds., *Patty's Industrial Hygiene and Toxicology*, vol. 2B. New York: Wiley.

U.S. Department of Health and Human Services. 1994a. *NIOSH Manual of Analytical Methods*. 4th ed. DHHS Pub 94-113. Washington, D.C.: U.S. Government Printing Office.

______. 1994b. *Documentation for Immediately Dangerous to Life and Health Concentrations*. PB-94-195047. Springfield, VA: National Technical Information Service.

______. 2005. *NIOSH Pocket Guide to Chemical Hazards*. DHHS Publication 2005-149. Pittsburgh: U.S. Government Printing Office.

U.S. Department of Health and Human Services, Public Health Service. 1980. *1st Annual Report on Carcinogens*. Research Triangle Park, NC: National Institute of Environmental Health Sciences.

______. 2004. *11th Report on Carcinogens*. Research Triangle Park, NC: National Institute of Environmental Health Sciences.

U.S. Department of Labor, Occupational Safety and Health Administration (OSHA). 2006. 29 CFR 1926, *Safety and Health Regulations for Construction*, and 29 CFR 1910, *Occupational Safety and Health Standards for General Industry*. Washington, D.C.: U.S. Government Printing Office.

U.S. Environmental Protection Agency (EPA). 1999. *EPA Compendium of Methods for the Determination of Toxic Organic Compounds in Ambient Air*. 2d ed. EPA Pub. 625/R-96-010. Washington, D.C.: U.S. Government Printing Office.

______. 2005. "Part 261—Identification and Listing of Hazardous Waste." In 70FR, June 14, 2005. Washington, D.C.: U.S. Government Printing Office.

Washington Group International. 2005. *Industrial Hygiene Procedures Manual*. W. H. Piispanen, ed. Boise, ID: Standard Register Publishers.

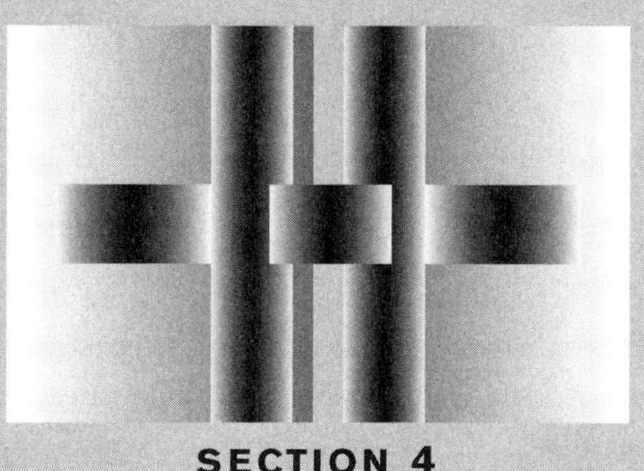

SECTION 4
INDUSTRIAL HYGIENE

Applied Science and Engineering: Physical Hazards

James C. Rock

LEARNING OBJECTIVES

- Describe the characteristics that distinguish chemical from physical agents in workplaces.
- Summarize the bases for Permissible Exposure Limits (OSHA PEL) and Threshold Limit Values (ACGIH TLV).
- Be able to list the four types of electromagnetic radiation.
- Summarize the key principles of wave propagation and their application when selecting commercially available engineering controls.
- Describe how the principle of conservation of energy allows the evaluation of the performance of engineering controls
- Recognize how time, shielding, and distance are the keys to the safe use of physical agents.
- Learn the criteria for the proper selection of welding goggles and the dangers of short-wavelength light sources to the eye.

THE POTENCIES OF physical agents in the workplace arise from the rate and quantity of energy delivered to tissue by mechanical, electrical, and magnetic forces.[1] In contrast, the potencies of chemical hazards are based on chemical reactivity with molecules of tissues. This distinction is clouded, in practice, because many modalities of physical agents deliver energy in ways that either create reactive chemicals in tissue or facilitate, by thermal heating or molecular rotation, bending and vibration, which are metabolic chemical reactions that damage tissue. This chapter emphasizes engineering design that prevents absorption of damaging quantities of energy at damaging power levels from physical agents.

Physical Agents

A Chemical Reaction Is Not a Physical Agent

A chemical reaction is said to occur when one or more valence electrons from one molecule are spontaneously donated to another molecule (ionic bond) or are shared with another molecule (covalent bond). Chemical reactions are either exothermic, releasing energy as the electron transfer occurs, or endothermic, absorbing energy from the immediate environs to complete the electron transfer. When energy is taken from or deposited in the wrong place, a toxic event occurs, sometimes in the form of a cascade of chemical reactions. At low doses, most of this damage is quickly repaired by homeostatic processes. When the repair mechanism is overwhelmed, damage accumulates, and a poisoning, or toxic, event is said to have occurred.

Physical Agents and Their Interactions

In a physical reaction, energy is delivered to molecules by means other than electron sharing between atoms in molecules. Several such phenomena are discussed in this chapter: noise, vibration, thermal stress, static or quasistatic electromagnetic fields, nonionizing radiation, and ionizing radiation. These occupational stressors, known collectively as physical agents, have potencies characterized by intensity and frequency. The typical physical agent standard involves specifications for free-field intensity, frequency, permissible duration, and a tissue absorption coefficient.

Workplace monitoring and control requires careful attention to these parameters. For example, at the same intensity, noise at frequencies between 20 and 40 Hz is better tolerated by human ears than is noise between 2 and 4 kHz. Likewise, vertically polarized electromagnetic radiation at frequencies between 100 and 300 MHz delivers more energy to tissue of a standing adult human than either the same intensity electromagnetic (EM) radiation with horizontal polarization at the same frequency or any polarization at frequencies between 30 and 100 kHz.

Carrying this EM insight further, X-rays at a frequency higher than 30 PHz (30×10^{15} Hz) ionize molecules in tissue and body fluids, producing free radicals and peroxides that disrupt RNA and DNA molecules.

Hazard, Risk, Causation, and Threshold

Diseases associated with physical agents are also associated with other factors. It is important that the safety professional understand core concepts when discussing physical agent exposures. The first is the concept of dose-response introduced by Paracelsus in 1567 (Ottoboni 1991):[2] "What is it that is not a poison? All things are poison and nothing is without poison. It is the dose only that makes a thing not a poison."

At doses below a sufficient dose, hazardous forms of energy are beneficial.[3] Hazard is a property of a stressor that causes harm at high doses. It is distinct from risk, which means conditions of use that produce a high dose (suprathreshold) with a high probability of harm.

Some argue that carcinogens, including ionizing radiation, have no threshold. Alice Ottoboni points out that there are practical threshold doses, even for carcinogens (Ottoboni 1991). As a population dose increases, members of the population show increasing cancer incidence rates and decreasing induction periods. A practical threshold exists when the exposure to a carcinogen is sufficiently small to reduce its cancer incidence rate to less than one case in the entire population, or to increase its induction period to more than the population life expectancy.

Cause is demonstrated when an exposure produces the same pathological symptoms in all exposed persons. For example, because an IR LASER beam burns a hole in the skin of every person who is exposed, the LASER is said to cause the burn. Likewise, when every hot dog in a microwave oven is heated, the oven is said to cause the heating. In contrast, 10+ years of cellphone use has not been shown to produce observable pathological effects in humans (Schüz 2006).

High ionizing radiation doses are associated with calculable risk of radiation sickness (acute effect) and of some cancers (chronic effect). Low doses are not. Even in the high-dose cohorts, not everyone develops cancer. Herman Cember discusses this situation quite cogently (Cember 1996, 233–237).

> Deterministic effects are characterized by three qualities:
>
> 1. A certain minimum dose must be exceeded before the effect is observed.
> 2. The magnitude of the effect increases with the size of the dose.
> 3. There is a clear, unambiguous causal relationship between exposure to the noxious agent and the observed effect.
>
> The stochastic [random] effects are those that occur by chance, and they occur in exposed as well as in unexposed individuals. Stochastic effects are therefore not unequivocally related to exposure to a specific noxious agent. . . . No pathologist can say with certainty that the cancer would not have occurred if the person had not been exposed to the carcinogen.

Review of Scientific Principles for Physical Agents

Energy is the ability to do work, measured in Joules (J). Power is energy per unit of time, or the rate at which work is done, in Watts (W). Intensity is power per unit area and has dimensions of W/m^2 or J/m^2 s. Frequency is measured in cycles per second, called Hz in the SI system of units. In an interaction between an electromagnetic field or an acoustic field and tissue, a portion of the intensity is absorbed by the tissue. The absorption efficiency is a function of the frequency (wavelength in tissue) and tends to peak when the tissue has a long dimension that is about 40% of the incident wavelength.

Primer on Physical and Physiological Systems of Units

The exposure to physical agents is controlled by the intensity of exposure. Intensity here is used as a precisely defined concept in physics that indicates the power transmitted by an electromagnetic wave or an elastic wave (acoustic wave) across a unit area perpendicular to the direction of propagation. To calculate intensity, it is sometimes convenient to multiply the energy density (J/m^3) of a traveling wave by its propagation velocity (m/s) to obtain intensity ($J/m^2s = W/m^2$).

Although all physical agent standards could, in principle, be expressed simply in units of J, W, and W/m^2, without reference to the time history or waveform of the exposure, the resulting standards would be overly restrictive. When a physical agent interacts with tissue, only that portion of the energy having appropriate frequency characteristics is absorbed. Thus, exposure standards are frequency dependent and based on the proportion of incident energy absorbed by critical organs.

Vibration standards are set with specific purposes. Along the spinal axis, the goal is to avoid destructive resonance that causes severe bending of the spine. Transverse acceleration, across the spinal axis, is set to avoid resonance in internal organs that might be bruised while bouncing off vertebral bodies or the rib cage. The hand–arm vibration endpoint is soft tissue damage in the forearm–wrist area and has different values in different directions relative to the bones.

Noise standards use units of power (W) to represent the radiated power from noise sources and units of intensity (W/m^2) to represent the rate at which acoustic power is transmitted, and they quantify intensity at a given point by using measurements taken with microphones that report root-mean-squared (*rms*) sound pressure (Pa). Exposure guidelines limit intensity and duration of exposure in order to limit absorbed energy, in J. In hearing conservation practice, standard frequency-weighting curves are applied to the pressure signal before these acoustic quantities are measured. The A-weighting approximates transmission of the acoustic power from the air to the inner ear at normal listening levels of 40 to 70 dBA. The C-weighting approximates the auditory system's transmission of acoustic power at higher occupational exposure levels of 90 to 115 dBC. The outer ear canal acts as an organ-pipe resonator to emphasize frequencies in the 1 to 4 kHz range. The middle-ear bones interact with the elasticity of the ear canal and oval window with more resonant enhancement in the same region, which is detuned by the stapedial reflex in the presence of loud sound, leading to the flatter pass-band of the C-weighting filter. Other weightings, called B- and D-weighting, are used for environmental and community noise standards.

Hearing conservation program guidelines are typically given in terms of A-weighting. Protection factors of hearing defenders (earplugs and earmuffs) are given in terms of C-weighting.[4] Infrasound standards have a weighting based on emotional effects and on soft-tissue resonances in the abdomen. Ultrasonic standards are based on a combination of emotional and subharmonic effects occurring as a result of nonlinear transmission of high-intensity signals from the environment to the inner ear and other tissues.

Electromagnetic (EM) standards for extremely low frequencies from microwave frequencies to infrared frequencies are based on limiting the direct heating of tissue and avoiding both localized and total-body thermal effects. As frequencies increase above those of microwaves, the depth of heating in tissue decreases and then increases again.[5] Radiated power is measured in W, intensity of a propagating wave in W/m^2, and average doses in J deposited during specified averaging periods—typically 6 minutes to account for the protection afforded as circulating blood carries localized thermal energy safely away (ACGIH Worldwide 2003).

In the spectral region near and including the *visible light* frequencies, the primary organs of concern are the eye and skin. The photon energy at these frequencies is sufficient to cause chemical reactions in rods and cones. Two complementary models apply: one measures intensity of radiation (W/m^2) and the other measures intensity in terms of the photon flux in [photons/(s m^2)]. The photon flux, measured in narrow frequency bands, can be summed to estimate total photon energy in J or intensity in J/m^2.

In the visible light region of the EM spectrum, two sets of units are in general use.

Radiometry is the science of quantifying the phenomena of electromagnetic radiation. Radiometry criteria are expressed in thermodynamic quantities: W, W/m^2, and J.

Photometry is the science of quantifying electromagnetic radiation in the visible light region of the spectrum[6] in units weighted by frequency according to the response of the human eye. Photometry is used to design lighting and projection systems for all kinds of human activities: entertainment, living, learning, driving, working, and so on. The photometric units have special names, recognized in the SI system, used to differentiate them from the radiometric units. The following table (Ohno 1997, 4–6) summarizes these differences. The problem set for this chapter includes exercises to illustrate the details more clearly.

Under an international standard created by the International Commission on Illumination (CIE), the dark-adapted standard eye (scotopic eye) has its best sensitivity at 505 nm wavelength (green light) and the light-adapted eye (photopic eye) has its best sensitivity at 555 nm (yellow light).[7] Light meters used for photometry may have both scotopic and photopic filters used to verify scene-lighting levels for both dark-adapted and light-adapted eyes, respectively.

Note that high intensity infrared ($\lambda > 780$ nm) or ultraviolet light ($\lambda < 380$ nm) registers no intensity on the photometric scale. The retina in a normal human eye does not respond to these frequencies because they are absorbed in the cornea, lens, and fluids of the eye. The radiometric standards for these frequencies exist because they heat skin and eye tissue (infrared) and high doses of UV energy are associated with sunburn, corneal burn, cataracts, and various cancers.

Table 1 summarizes the units of radiometry and photometry used by safety professionals and industrial hygienists.

The TLV uses spectral radiance of a light source (in Watts per steradian per square meter per nanometer of wavelength, $W\ sr^{-1}\ m^{-2}\ nm^{-1}$) or spectral irradiance incident to the eye (in Watts per square meter per nanometer of wavelength, $W\ m^{-2}\ nm^{-1}$). The coefficients describing this are called *spectral weighting coefficients* by ACGIH.

Although English units are discouraged in the United States, they continue to exist in safety and hygiene literature and in exposure limits promulgated by regulatory agencies. Illuminance in footcandles (fc) is defined as a lumen per square foot. Luminance in footlamberts (fL) is defined as (candela/π) per

TABLE 1

Quantities and Units for Photometry and Radiometry

Photometric Quantity	Unit	Radiometric Quantity	Unit
Luminous flux	lm (lumen)	Radiant flux	W (watt)
Luminous intensity	$lm\ sr^{-1} = cd$ (candela)	Radiant intensity	$W\ sr^{-1}$
Illuminance	$lm\ m^{-2} = lx$ (lux)	Irradiance	$W\ m^{-2}$
		Spectral irradiance factor	$W\ m^{-2}\ nm^{-1}$
Luminance	$lm\ sr^{-1}\ m^{-2} = cd\ m^{-2}$	Radiance	$W\ sr^{-1}\ m^{-2}$
	(per sr per m^2 of source)	Spectral radiance factor	$W\ sr^{-1}\ m^{-2}\ nm^{-1}$
Luminous exitance	$lm\ m^{-2}$	Radiant exitance	$W\ m^{-2}$
Luminous exposure	$lm\ m^{-2}\ s = lx\ s$	Radiant exposure	$W\ m^{-2}\ s = J\ m^{-2}$
Luminous energy	lm s	Radiant energy	W s = J (joule)
Color temperature	K (kelvin)	Radiant temperature	K

TABLE 2

Ionizing Radiation Measurement Units

Units	Radioactivity	Absorbed Dose	Dose Equivalent	Exposure
Common	curie (Ci)	rad	rem	roentgen (R)
	3.7×10^{10} per s	10 mJ/kg	QF × AbsDos	
SI	becquerel (Bq)	gray (Gy)	sievert (Sv)	
	1/s	1 J/kg	QF × AbsDos	C/kg
Conversion	1 mCi = 37 MBq	1 rad = 0.01 Gy	1 rem = 0.01 Sv	1 R = 258 μC/kg
Conversion	1 Mbq = 0.027 mCi	1 Gy = 100 rad	1 Sv = 100 rem	1 C/kg = 3880 R

square foot, the reflected luminance that results when a perfect diffuser is illuminated with 1 fc.[8] The luminance of a scene depends on its illumination, the orientation of its surfaces, and the proportion of incident light that is reflected specularly or diffusely, refracted, absorbed, or transmitted at each wavelength.

The *ultraviolet* frequencies increase from UV-A through UV-B to UV-C. There is a progressive increase in photon energy and in the proportion of molecules that can be ionized by such photons. The UV-C band ends at a wavelength of 0.1 nm, and all higher frequencies (shorter wavelengths) are considered strongly ionizing.

X-rays and *gamma rays* are photons with energies higher than UV light, the X-rays arising from orbital electrons and their interactions and the gamma rays arising from nuclear decay. These two types of photons are indistinguishable when they pass through tissue. The X-ray, γ-ray, and high-energy particle radiations are collectively known as *ionizing radiation*. Health hazards associated with ionizing radiation are the domain of the health physics profession and are expressed in specialized units, summarized in Table 2. As in the case of light, one set of SI units expresses transport of energy, and a parallel set expresses biological effectiveness in depositing energy in susceptible tissues.

The term radioactivity denotes the number of disintegrations per unit time. An atom may be unstable (*radioactive*) because its nucleus has too much mass, charge, or energy. To become stable, it emits mass, charge, and/or energy in a disintegration event. The rate at which this occurs in a given quantity of atoms is measured per second by the SI unit becquerel (Bq) or in the older unit, curie (3.7×10^{10}). The intensity of the resulting radiation field depends on the energy per emitted particle, whether alpha, beta, gamma, neutron, or other subatomic particle.

The SI unit of *absorbed dose* is the gray, the name for a quantity of absorbed energy totaling 1 J per kg of material. The older unit rad represents 0.01 J per kg of material. This is a physical unit and is not a good measure of the risk of adverse biological effect. For example, 1 Gy of internal alpha radiation poses far more risk than 1 Gy of external beta radiation. Rapidly dividing cell types are more susceptible than slowly dividing cell lines.

The *dose equivalent* is the product of absorbed dose in tissue multiplied by a quality factor—and then sometimes multiplied by other necessary modifying factors at the location of interest. It is expressed numerically in SI units called sieverts, or the older unit rem.

In the context of health physics, the term *exposure* represents the level of ionization realized in exposed material. Ionization is one means by which energy is deposited in tissue. It produces large quantities of peroxides and free radicals as well as a small amount of direct damage to life-critical molecules. Much, if not most, radiation damage is caused by secondary chemical reactions with peroxides and free radicals created in body fluids by ionization.

Primer on Conservation Laws

Four quantities associated with physical agents are conserved in isolated systems: energy, linear momentum, angular momentum, and electric charge.[9] *Linear momentum* is a vector quantity formed by the product of particle mass and the velocity vector, with relativistic

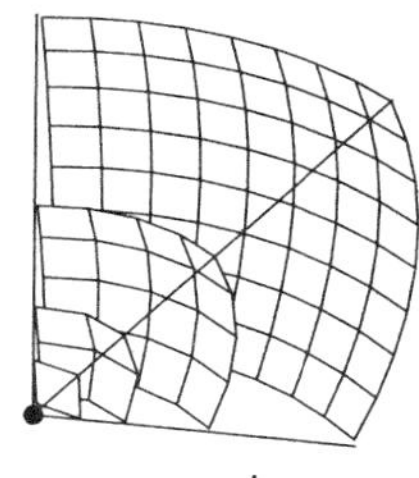

Point Source: $I[r] = \frac{I_o}{4\pi r^2}$

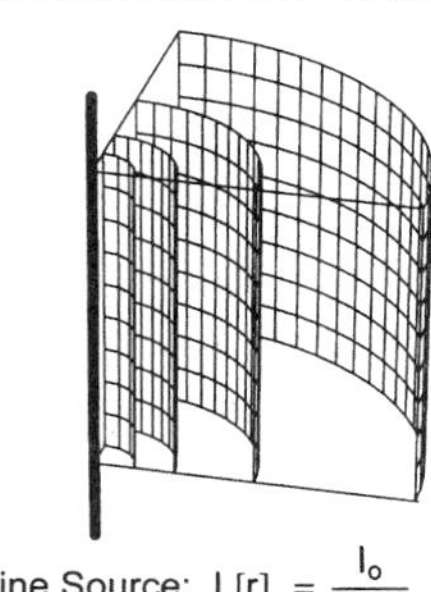

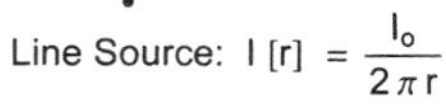

Line Source: $I[r] = \frac{I_o}{2\pi r}$

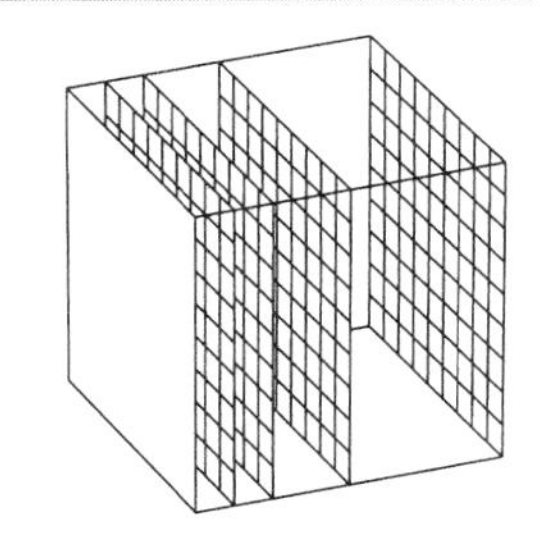

Area Source: $I[r] = I_o$ for r^2

FIGURE 1. Radiation geometry

corrections for sufficiently high velocities. *Angular momentum* is a vector quantity formed by the product of the moment of inertia and the angular velocity vector, or, equivalently, by the cross-product of linear velocity and radius to the reference point times the mass. *Electric charge* is a property of subatomic particles that is positive, negative, or zero and exists only in multiples of the proton charge.

Shielding for physical agents abides by these conservation principles, and its efficacy can often be checked by verifying conservation laws. Consider an incident wave with an intensity of 100 W/m^2. Measurements show that after it passes through the body of a worker, its intensity falls to 95 W/m^2. The principle of conservation of energy says that energy is neither created nor destroyed but may be changed in form; thus it is seen that 5 W/m^2 was either deposited in the tissue by the passing wave or reflected in another direction. If deposited, it may have caused physical motion, molecular ionization, molecular excitation, or simply thermal heating. If a portion is reflected, bystander exposures are possible.

Now consider the interaction of a photon with orbital electrons of a molecule, using the hydrogen atom to illustrate the principles involved. In this collision, charge, energy, linear momentum, and angular momentum are all conserved. Three energy regions of interest exist during a collision between a photon and orbital electron: absorption, scattering, and ionization. A photon with energy equal to the transition energy between electron orbits[10] is absorbed, adding kinetic energy to the electron, and the atom is said to be in an excited state. Later, when the excited electron relaxes to a lower energy orbit, it emits a photon of the same or lower frequency (energy). If an excited state is metastable, as is the case in a carbon dioxide molecule, many electrons can be pumped to the excited state and then triggered to emit their radiation simultaneously in phase, a process called Light Amplification by Simulated Emission of Radiation (LASER). A photon with energy between absorption bands may be scattered or partially absorbed and its excess energy scattered as lower-energy photons. A photon with sufficient energy (> 13.6 eV for hydrogen) ejects an electron from the atom, leaving behind a charged ion. The energy of the ejected electron is the difference between the energy of the incident photon and the ionization energy of the atom.

Primer on Radiation Geometry

Consider four radiation source geometries: point source, line source, surface source, and dipole source.[11] Understanding these allows intuitive professional judgment about numerous additional geometries that can be sketched as a mixture of these and analyzed by linear superposition (see Figure 1).

A point source radiates its power, P_{ps}, uniformly in all directions through a 4π steradian solid angle. At a radius, r_{ps}, the surface area is the product of the solid angle of a wave front and the square of its distance from the source: $4\pi r_{ps}^2$. The intensity of a spherical wave front is the ratio of power to its surface area, because energy is conserved. This is the inverse square law. I_{pso}

is defined as the intensity measured at $r_{ps} = 1$ m from the point source. Sources having a maximum dimension smaller than 0.1 wavelength can be approximated by a point source.

Equation: Point source intensity versus distance

$$I_{ps}(r_{ps}) = \frac{P_{ps}}{4\pi r_{ps}^2} \sim \frac{I_{pso}}{r_{ps}^2} \qquad (1)$$

A line source radiates its power, P_{ls}, uniformly in all directions through a coaxial cylindrical surface. At a radius, r_{ls}, and length, L_{ls}, the surface area of a wave front is $2\pi r_{ls}L_{ls}$. The intensity is the ratio of power to surface area and is inversely proportional to r_{ls}.

Equation: Line source intensity versus distance

$$I_{ls}(r_{ls}) = \frac{P_{ls}}{2\pi r_{ls}L_{ls}} \sim \frac{I_{lso}}{r_{ls}} \qquad (2)$$

A plane surface, or area source, radiates its power, P_{as}, uniformly in all directions away from a plane. At a distance, r_{as}, from a plane with dimensions length and width, the surface area of a wave front remains the product of its length and width, $A_{as} = L_{as}W_{as}$. In the near field, its intensity is the ratio of power to surface area, a constant independent of r_{as}.

Equation: Area source intensity versus distance

$$I_{as}(r_{as}) = \frac{P_{as}}{A_{as}} \sim I_{aso} \qquad (3)$$

In the extreme near field, called the *reactive near field*, nonradiating energy is stored in both potential and kinetic modes. The intensity of the field includes all such energy, whether stored or radiating. For example, dipole sources follow an inverse cube law in the reactive near field, and exposures close to the source are much stronger than those associated with the radiating fields.[12]

Equation: Dipole source intensity versus distance in reactive near field

$$I_{dps}(r_{dps}) = \frac{I_{dps}}{r_{dps}^3} \qquad (4)$$

A dipole source is two point sources close to one another. In this geometry, both constructive and destructive interference occur, and the two sources tend to fully or partially cancel one another in some directions while reinforcing in others. In the far field, intensity from a dipole source falls off as the inverse square of distance, just like a point source.

Primer on Waves: Elastic and Electromagnetic

A wave is a disturbance or variation that transfers energy progressively from point to point in a medium. It may take the form of elastic deformation, variation of pressure, magnetic intensity, electric or gravitational potential, or temperature (*Merriam-Webster's Online Dictionary* 2006). A traveling wave is described by its parameters: wavelength, frequency, period, amplitude, impedance, and intensity (as illustrated in Figure 2).

Wavelength, λ, is the ratio of wave speed, c, to its frequency, f. The speed of a traveling wave depends on the properties of the media through which it propagates; its intensity is proportional to its frequency.

Equation: Speed, wavelength, period, and frequency

$$c = \lambda f = \frac{\lambda}{T} = \frac{\lambda\omega}{2\pi} \qquad (5)$$

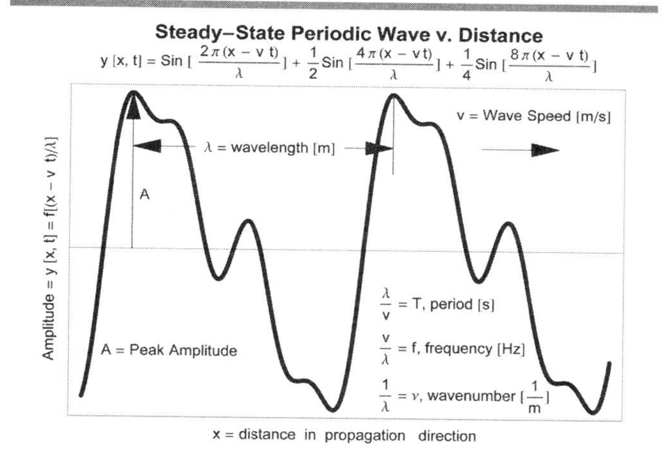

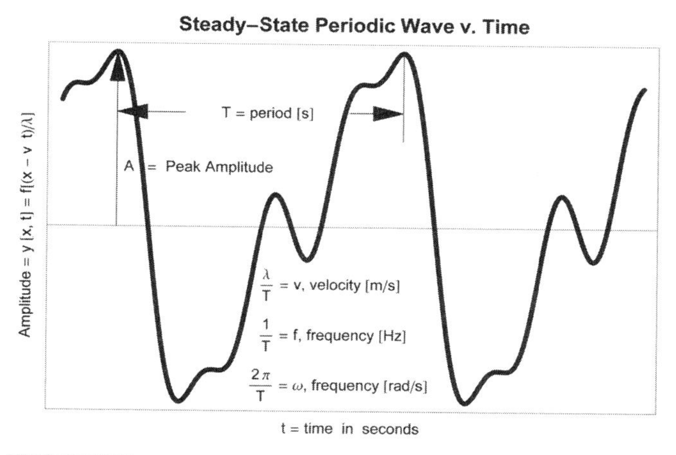

FIGURE 2. Traveling wave parameters

TABLE 3

Speed, Impedance, and Intensity of Traveling Waves

Wave Type	Wave Speed	Impedance	Intensity
(Eq. 6) Compression wave in ideal gas	$c_{gas} = \sqrt{\frac{\gamma_{gas} T}{M_{gas}}} \sqrt{T}$ [m/s]	$Z_{gas} = \rho_{gas} c_{gas}$ [rayl]	$\frac{P_{rms}^2}{Z_{gas}}$ [W/m²]
	M in g/mol, T in K	ρ in kg/m³, c in m/s	P in Pa, Z in rayl
(Eq. 7a) Sound in air	$c_{air} = \sqrt{\frac{\gamma_{air} RT}{M_{air}}} = 20.05\sqrt{T}$ [m/s]	$Z_{air} = \rho_{air} c_{air} = \frac{69.83\, P_{air}}{\sqrt{T_{air}}}$ [rayl]	$\frac{P_{rms}^2}{Z_{air}}$ [W/m²]
	M in g/mol, T in K	ρ in kPa, T in K	P_{rms} in Pa, Z in rayl
(Eq. 7b) Ref values at 293 K = 20°C	$c_{0\,air} = 343$ [m/s]	$Z_{0\,air} = 400$ [rayl]	$\rho_{0\,air} = 1.2$ [kg/m³]
(Eq. 8) Compression wave in liquid	$c_{liq} = \sqrt{\frac{\gamma_{liq} B_{liq}}{\rho_{liq}}}$ [m/s]	$Z_{liq} = \rho_{liq} c_{liq}$ [rayl]	$\frac{P_{rms}^2}{Z_{liq}}$ [W/m²]
(Eq. 9a) Acoustic wave in pure water at P = 1 atm	$c_{water} \approx 1483 + 2(T - 293.15)$ [m/s]		For T = 293.15 K $c_0 = 1481.4$ m/s $Z_0 = 1{,}480{,}000$ rayl $\rho_0 = 998.2$ kg/m³
(Eq. 9b) Acoustic wave in sea water	$c_{sea\ water} \approx 1449 + 4.6T_c - 0.055T_c^2 + 0.0003T_c^3 + (1.39 - 0.012T_c)(S - 35) + (0.017d)$ [m/s] $T_c = (T - 273.15)$, with T in K; S = salinity in ppt_{wt}; d = depth in m		$c_0 = 1500$ m/s $Z_0 = 1.54\ 10^6$ rayl $S_0 = 35$ parts/1000 $d_0 = 1$ m $T_0 = 286.15$ K = 13°C $\rho_0 = 1026.4$ kg/m³
(Eq. 10) EM in homogeneous isotropic material	$c_m = \frac{1}{\sqrt{\mu_m \varepsilon_m}}$	$Z_m = \mu_m c_m = \sqrt{\frac{\mu_m}{\varepsilon_m}}$ in [Ω = V/A]	$\lvert \vec{S} \rvert = \frac{\lvert(\vec{E} \times \vec{B})\rvert}{\mu_m} \approx$ $\frac{\vec{E} \cdot \vec{E}}{Z_m} \approx$ $\left(\frac{\vec{B}}{\mu_m} \cdot \frac{\vec{B}}{\mu_m}\right) Z_m$ [W/m²]
(Eq. 11) EM in vacuum	$c_0 = \frac{1}{\sqrt{\mu_0 \varepsilon_0}}$ [m/s] = 299,792,458 [m/s]	$Z_0 = \sqrt{\frac{\mu_0}{\varepsilon_0}} = 376.73\ \Omega$	$\lvert \vec{I}_{EM} \rvert = \left\lvert \vec{E} \times \frac{\vec{B}}{\mu_0} \right\rvert$ [W/m²] $\lvert \vec{I}_{EM\,free\ field} \rvert \approx Z_0 \left(\frac{B_{rms}}{\mu_0}\right)^2$ $\approx \frac{(E_{rms})^2}{Z_0}$ [W/m²]

The speed of an elastic compression or transverse wave equals the square root of the ratio of the bulk modulus (Pa = N/m²) or string tension (in N) to its mass density (in kg/m³ or kg/m). Similar concepts apply to electromagnetic waves. Permeability and permittivity of materials are analogous to elasticity and mass density. EM speed equals one over the square root of the product of permeability and permittivity. EM impedance is the square root of the ratio of permeability to permittivity.

As shown in Table 3, wave propagation speed is determined by the ratio of force to inertia. Impedance is the ratio of force to velocity. Intensity is the ratio of force squared to impedance, or, equivalently, the product of force and velocity. At a boundary between two materials (say air and human tissue), the ratio of wave impedances determines the proportions of energy reflected from, and penetrating, the boundary.

The speed of light in a vacuum, c_0, is a constant of nature found in Maxwell's equations that are identified in Einstein's theory of general relativity as the speed limit for all radiation, mass, and energy. When an electromagnetic wave passes through substances, it travels more slowly than c_0, because permittivity and permeability in materials are no less than those in a vacuum.

The impedance reveals important insights about behavior of a propagating wave. In real media $c_m < c_0$, and Z_m may differ from Z_I. When an electromagnetic wave enters or leaves such media, its frequency remains constant (conservation of energy) while its wavelength and its speed change in proportion (see Equation 5).[13]

The impedance of a traveling wave (the ratio of its potential to kinetic parameter) is approximately constant in the far field. There, average intensity is proportional to one of the *rms* quantities squared: P^2_{rms}, V^2_{rms}, F^2_{rms}, E^2_{rms}, B^2_{rms}, $dvol^2_{rms}$, I^2_{rms}, or Vel^2_{rms} and the constant of proportionality is the impedance or its reciprocal, the conductance. Most EM survey instruments measure E or H and calculate intensity as $E^2_{rms}/377$ or $H^2_{rms} \times 377$, and most noise survey instruments measure P_{rms}, calculating intensity as $P^2_{rms}/400$. These instruments are designed to measure the far-field intensity of propagating waves and may underestimate intensity in the near and transition fields, where the intensity of nonpropagating induction fields may exceed the intensity of the propagating fields by a large factor and the ratio of potential to kinetic quantities is neither constant nor equal to the impedance of the free field. Table 3 summarizes key features of acoustic and electromagnetic traveling waves that must be addressed when designing engineering controls to limit emission and absorbent and reflecting shielding to limit propagation of these occupational stressors.

Primer on Near and Far Fields

In general, the spatial distribution of intensity in the near field of an extended source is difficult or impossible to measure with common survey instruments. It is in the far field that the radiation field is well approximated by impedance between the potential and kinetic quantities. Further out, the intensity follows the simple inverse square law.

There is no single distance from a source beyond which near-field behavior suddenly becomes far-field behavior. OSHA has adequately summarized the situation for its field compliance officers (OSHA). Table 4 summarizes key points that apply equally to acoustic and EM radiation as a function of its wavelength, lambda (λ), originating from a source whose maximum transverse dimension is D, located a distance r away from the measurement point.

In the reactive near field, much more stored energy exists than does radiating energy. This energy can damage the tissue of any worker present in the field, even though it does not travel from its source. In the near and transition fields, the impedance is an adequate model neither for the ratio of the electric field to the magnetic field nor for the ratio of acoustic pressure to acoustic velocity. A satisfactory survey at distances closer than the far-field boundary requires direct measurement of all three vector quantities of interest for safety purposes: electric field strength, magnetic field strength and EM intensity for electromagnetic fields; or acoustic pressure, acoustic velocity, and acoustic intensity for airborne or waterborne sound fields. In the far field, intensity measurement is straightforward. The impedance at the point of measurement allows for the measurement of either the potential field (E_{rms} for EM, or P_{rms} for acoustic) or the kinetic field (H_{rms} for EM or Vel_{rms} for acoustic) and allows for the calculation of the intensity from the measured *rms* values.

TABLE 4

Near-Transition Far-Field Regions as a Function of $\lambda = c/f$

Region	Computatiion
Reactive near field: very high proportion of stored energy	(Eq. 12a) $r_{reactive\ near\ field} < \lambda/2\pi < 0.16\lambda$
Near field: more stored than propagating energy	(Eq. 12b) $r_{reactive\ near\ field} < r_{near} < \text{Max}\ [\lambda, D^2/\lambda]$
Transition region: more propagating than stored energy	(Eq. 12c) $r_{near} < r_{transition} < r_{far}$
Far field: for free-field impedance	(Eq. 12d) $r_{far} \leq \text{Max}\ [2\lambda, 2D^2/\lambda]$
Far field: for inverse square law behavior	(Eq. 12e) $r_{far} > \text{Max}\ [5\lambda, 2D^2/\lambda]$

Primer on Parameter Dynamic Range

Many physical agents interact with human organs over wide dynamic ranges. It is common practice to describe their risk criteria, their perception, and the protection factors of personal protective equipment on a logarithmic scale. The two most important types of logarithmic scales for industrial hygiene and safety are the optical density and intensity of acoustic and EM fields.

Optical density is defined as the base 10 logarithm of the reciprocal of the transmittance, T_r. For a protective lens it is expressed as a dimensionless fraction.[14] The transmission loss is the optical density expressed in dB.

Equation: Optical density (OD), transmittance (T_r), and transmission loss (TL)

$$D = \log_{10}(1/T_r);\text{ and } TL = 10\ OD \text{ in dB} \qquad (13)$$

Level, used in acoustics and in EM surveys, is the base 10 logarithm of the dimensionless ratio of two intensities, powers, or energies. A *level* in decibels (dB) is 10 times the base 10 logarithm of the ratio between the two quantities. The dB can express a ratio or be written as "dB re Q_{ref}" where Q_{ref} identifies the size and units of the reference quantity. Reference quantities commonly encountered in safety guidelines include: 1 pW, 1 mW, 1 mV, 1 μPa, 20 μPa, 1 pW/m^2, and 1 mW/cm^2. The equation pair below converts between physical units and dB.

Equation: Level in decibel versus Q_{ref}

$$L_Q = 10 \log_{10}(Q/Q_{ref}) \text{ [dB re } Q_{ref}\text{]} \qquad (14)$$

$$Q = Q_{ref}\ 10^{L_Q/10}$$

Because power is energy per unit time, a level in decibels may be used to describe the ratio between two power quantities when the length of observation for both is identical, as well as to describe the ratio between two intensities when the illuminated area and length of observation are the same. Average power is the product of an *rms* potential quantity (pressure, voltage, force, electric field, and so on) and a companion *rms* kinetic quantity (volume change, electrical current, velocity, magnetic field, and so on). It is important to know the inner workings of every survey instrument. Some measure intensity directly, but most measure either a potential or kinetic quantity, square it, average it, and multiply or divide by an assumed impedance[15] to report power levels. These common types may misrepresent risk in near and transition fields.

Primer on Thermal Effects

Small changes in tissue temperatures lead to dramatic changes in metabolic rates. Thermal energy is conveniently understood as independent, incoherent vibrations of molecules in material substances. In humans, a useful rule of thumb is that metabolic rates double with a temperature rise of 3 K (5°F) and are halved with a temperature fall of –3 K. To prevent acute physiological damage, including inability to complete complex mental tasks, like self-rescue, thermal stress standards are set to avoid more than ±1 K (2°F) changes in human body core temperature caused by occupational exposure factors. EM guidelines for specific organs are designed to prevent unacceptable temperature rises in those organs, considering blood as a coolant, and so typically have averaging times measured in minutes and seconds.

To prevent skin burns, limit the infrared intensity for radiated thermal energy, and limit the temperature for contact with gases, liquids, and solids. In a 1991 study, Suzuki, et al. (1991, 443–451) demonstrated that rat skin suffers superficial, deep, and full-thickness skin burns from extended contact with solids at surprisingly low temperatures: 37.8°C (100°F), 41.9°C (108°F), and 47°C (119°F). This study shows the wisdom of the longstanding engineering design guideline that equipment with exposed surfaces situated such that personnel contact is likely should be insulated to operate with surface temperatures below 57°C (135°F).

Primer on Physical Acoustics

In physical acoustics, sound pressure (P_{rms}) is a potential quantity easily measured by a microphone. Acoustic intensity is calculated from P_{rms} with acoustic impedance, which can be estimated by use of the ideal gas law:

Equation: Sound intensity in air

$$\vec{I}_{air} = \text{Mean}\left[P_{air}\overrightarrow{Vel}_{air}\right]$$

so that

$$|I_{air}| = \frac{P^2_{rms}}{Z_{air}} = \frac{P^2_{rms}}{Z_0} \cdot \frac{Z_0}{Z_{air}} \qquad (15)$$

with P in kPa and Z in rayl.

Equation: Acoustical impedance of air

$$Z_{air} = \rho_{air}c_{air} = \frac{M_{air}P_{air}c_{air}}{RT_{air}} = \frac{69.846\ P_{air}}{\sqrt{T_{air}}} \qquad (16)$$

Hearing-damage risk is proportional to average intensity. Sound-level meters are designed to average P^2, take its square root and reporting

$$P_{rms} = \sqrt{\text{Mean}[P^2(t)]}$$

in terms of L_P in dB re 20 μPa. This makes L_P numerically equal to L_I when $Z_{air} \sim 400$. It is a happenstance of history that hearing conservation standards are presented in terms of sound pressure levels.[16] It is better practice to use sound intensity levels for comparison with hearing damage risk criteria. In any occupational setting involving wide departures from barometric pressure near 100 kPa or temperatures near 293 K (20°C), it is good practice to estimate the acoustic impedance and calculate the sound intensity level associated with the measured sound pressure level. First, look at the definitions of sound pressure level and of sound intensity:

Equation: Sound pressure level versus *rms* sound pressure

$$L_P = 10\log_{10}\left(\frac{P^2_{rms}}{P^2_0}\right) = 20\log_{10}\left(\frac{P^2_{rms}}{P_0}\right) \qquad (17)$$

in [dB re 20 μPa]

$$P_{rms} = P_0\,10^{L_P/20}\ [\mu\text{Pa}]$$

Equation: Sound intensity level versus *rms* sound intensity

$$L_I = 10\log_{10}\frac{I}{I_0}\ [\text{dB re } I_0 = 1\ \text{pW/m}^2] \qquad (18)$$

$$I = I_0\,10^{L_I/10}\ [\text{pW/m}^2]$$

The corrected sound intensity level, L_I in [dB re 1 pW/m^2] is a function of L_P measured with a microphone in [dB re 20 μPa], the air temperature during measurement [T_{air} in K] and the absolute atmospheric pressure at the location of the measurement [P_{air} in kPa].[17]

Equation: Sound intensity level pressure and temperature correction

$$L_I = L_P + 10\log_{10}\left(\frac{400}{\rho_{air}c_{air}}\right) \qquad (19)$$

$$L_I(L_P, T_{air}, P_{air}) =$$

$$L_P + 10\log_{10}\left(\frac{5.72693\sqrt{T_{air}}}{P_{air}}\right)[\text{dB re 1 pW/m}^2]$$

The correction factor to convert L_P to L_I, evident above, should be applied whenever the temperature is outside the range 200 K to 500 K (–100 to 440°F) or the atmospheric pressure is outside the range 80 kPa to 125 kPa (0.8 to 1.25 atm). People at work are seldom exposed to such temperatures, so the correction factor is needed primarily in workplaces at altitudes greater than ~5000 ft and in pressurized workplaces such as those connected with tunneling, commercial diving, and hyperbaric medicine.[18]

Primer on Photons

Electromagnetic phenomena exhibit two properties: waves and particles. During scattering events with molecules, the conservation of energy and momentum are explained by behavior of a particle called a photon. During propagation of energy through empty space, through materials, and in electromechanical equipment, the collective behavior of photons is well explained by Maxwell's wave equations with the Lorentz force equations.

A photon is a massless particle that travels at the speed of light, has a finite extent, behaves as a wave, and carries linear momentum, angular momentum (spin), and energy between molecules. It is the quantum of the electromagnetic field and, according to quantum electrodynamic theory, mediates the forces between electromagnetic fields and molecules. Photon energy and momentum can be transferred to a molecule or carried away by other subatomic particles during a collision. The principles of conservation of energy and conservation of momentum apply. The photon collision either raises the internal energy of the molecule or knocks an electron from the molecule in an ionizing event. When an ionized molecule captures an electron or an excited molecule relaxes from an excited state back into its state of equilibrium, photons may be emitted to carry away the excess energy and momentum. Both the energy and the momentum of a photon are proportional to its frequency, f, and the constant of proportionality is Plank's Constant, h. The speed of light equals the ratio of photon energy to photon momentum.[19]

Equation: Photon energy, photon momentum, and speed of light

$$U_{photon} = \frac{hc}{\lambda} = hf = hbar\ \nu \text{ [J]} \tag{20}$$

$$P_{photon} = \frac{h}{\lambda} = \frac{hf}{c} = \frac{hbar\ \nu}{c} \text{ [kg m/s]}$$

$$c = \frac{E_{photon}}{P_{photon}} \text{ [m/s]}$$

Primer on the Electromagnetic Spectrum

The EM spectrum displays the energy density of radiation as a function of frequency or vacuum wavelength.[20] Portions of the spectrum are assigned specific uses by national laws and international treaties. The dose of electromagnetic energy in tissue depends on tissue properties and radiation frequency. This primer lays the foundation for understanding the basis for regulations and guidelines discussed later in the chapter.

For nonionizing electromagnetic radiation (lower frequencies), models of collective effects are sufficient to specify health-related guidelines. Forces are proportional to an electric field, E in volts/meter, and a magnetic field, B in Tesla or $H = B/\mu$ in amperes/m. The force on a stationary charged particle with net charge q in coulombs is a vector having the same sense as the electric field vector when q is positive and the opposite sense when q is negative. A magnetic field applies forces to moving charged particles that are at right angles to the magnetic field and to the velocity of the particle. The total Lorentz forces are the sum of the electric forces on all charged particles and the magnetic forces on moving charged particles.

Equation: EM forces experienced by a charged particle

$$\vec{F}_{elec} = q\vec{E} \text{ [N]} \tag{21}$$

$$\vec{F}_{mag} = q\overrightarrow{Vel} \times \vec{B} \text{ [N]}$$

$$\vec{F}_{Lorentz} = \vec{F}_{elec} + \vec{F}_{mag} = q\vec{E} + q\overrightarrow{Vel} \times \vec{B} \text{ [N]}$$

The (relativistic) relationships between electrical charge and electrical current distribution and their associated electric and magnetic fields are nearly[21] completely described by four equations known as Maxwell's equations. These equations predicted the speed of light as a constant of nature and form the basis for modern telecommunications, as well as the basis for safety guidelines for the use of electromagnetic devices in occupational settings. The differential form of these important equations is offered in Table 5, along with an intuitive description.[22]

In the far field, or free field, there are no sources of electric or magnetic fields, neither charges nor currents. Here, the symmetry between Faraday's Law and Ampere's Law predicts that an EM wave will propagate long distances from its source through free space (charge-free, current-free space). Under these conditions, a simple relation exists between the radiating time-varying electric and magnetic field vectors. In space, they are orthogonal to each other and to the direction of energy propagation. In time, they are in a fixed phase with each other. Further, their amplitudes are in proportion to the speed of light. With E in V/m, H in A/m (so that $B = \mu_0 H$ is in T), and c in m/s,

Equation: Free-field ratio E to B (no charge or current sources)

$$\frac{E_{rms}}{B_{rms}} = c_0 \text{ with } c_0 = \frac{1}{\sqrt{\mu_0 \varepsilon_0}} \text{ in vacuum} \tag{23}$$

$$\frac{E_{rms}}{B_{rms}} = c_m \text{ with } c_m = \frac{1}{\sqrt{\mu_m \varepsilon_m}} \text{ in isotropic material}$$

The Poynting Vector, S, provides the intensity and direction of power flow in free space for a traveling EM wave. The Poynting vector allows the expression of the power density of a traveling EM wave in its far field in terms of the magnetic permeability of the medium and its time varying electric and magnetic fields. In the equation below, E is in V/m, H in A/m, B in T, and power density is in W/m^2.

Equation: Free-field Poynting Vector

$$\vec{S}_{EM} = \vec{E} \times \vec{H} = \vec{E} \times \frac{\vec{B}}{\mu_m} \text{ [W/m}^2\text{]} \tag{24}$$

TABLE 5

Maxwell's EM Equations in Free Space

Differential Form (Nave, 2007)	Sketch (Rock, 2007)	Comment (Rock, 2007)
(Eq 22 a) $Div(\bar{E}) = \frac{\rho}{\varepsilon_0}$ and $\int_{All A} \bar{E} \times d\bar{A} = \frac{\rho}{\varepsilon_0}$ A is a surface around charge ρ. For a sphere of radius r, with a point charge ρ at its center, $A = 4 \pi r^2$.; $\lvert \bar{E} \rvert = \frac{\rho}{4 \pi r^2 \varepsilon_o}$	+ −	(Eq 22a) Gauss' Law$_E$: Electric field strength through a closed surface is proportional to quantity of net charge enclosed by that surface. The Electric field lines diverge from the positive point charge (on left) and converge to negative charge (on right). Solid lines are lines of constant potential in volts. Arrows show direction and intensity of the E field. To imagine this image in its proper 3-D form, visualize two dandelion pods: field lines converge or diverge uniformly from each core.
(Eq 22 b) $Div(\bar{B}) = 0$ and $\int_{All A} \bar{B} \times d\bar{A} = 0$. A is any closed surface.	S N	(Eq 22b) Gauss' Law$_B$: Magnetic field lines do not diverge; they are closed on themselves (think rubber bands). There is no magnetic monopole.
(Eq 22 c) $Curl(\bar{E}) = -\frac{\partial \bar{B}}{\partial t}$ and $\int_{All S} \bar{E} \times d\bar{S} = -\frac{\partial \Phi_B}{\partial t}$ with $\Phi_B = \int_{All A} \bar{B} \times d\bar{A}$; S is the boundary of an open surface named A.	$\vec{E}$ $\frac{d\vec{B}}{dt}$	(Eq 22c) Faraday's Law: A time-varying magnetic field passing through an open surface creates an encircling electric field that curls around that surface. The minus sign agrees with Lenz' Law: **E** is directed to induce current flow in conductors that resist the change in the magnetic field. When dB/dt is negligible, the electric field resulting from a long charged wire is dominated by Eq 22 a.
(Eq 22 d) $Curl(\bar{B}) = \frac{1}{c^2}\left(\frac{\bar{J}}{\varepsilon_o} + \frac{\partial \bar{E}}{\partial t}\right)$ and $\int_{All S} \bar{B} \times d\bar{S} = \mu_0\, i + \frac{1}{c_o^2}\frac{\partial \Phi_E}{\partial t}$ with $\Phi_E = \int_{All A} \bar{E} \times d\bar{A}$ and $i = \int_{All A} \bar{J} \times d\bar{A}$ S is the boundary of an open surface named A.	$\vec{B}$ $\vec{J}$ $\frac{d\vec{E}}{dt}$	(Eq 22d) Maxwell-Ampere Law: The net current through an open surface creates an encircling magnetic field that curls around that surface. Net current, in amperes, is the sum of ionic current and displacement current: $(\mu_o \bar{J})$ and $\frac{1}{c^2}\frac{\partial \bar{E}}{\partial t}$. When dE/dt is negligible, B at a distance r from a long wire carrying current I, is $\frac{\mu_o I}{2 \pi r}$

The intensity at any point is defined as the time average of the Poynting vector at that point. This definition allows use of simple survey instruments measuring the *rms* values of E or B to measure the intensity. In the equation that follows, note that $Z_{EM} = Z_{tissue} \approx Z_{ohms} \approx 377$ ohms.

Equation: Free-field EM intensity from measured E_{rms} and B_{rms} (valid in far field)

$$I_{EM} = E_{rms} H_{rms} = \frac{E_{rms}^2}{Z_{EM}} = Z_{EM} H_{rms}^2 = \frac{c_m B_{rms}^2}{\mu_m} \; [\text{W/m}^2] \tag{25}$$

with $Z_{tissue} \approx Z_0 \approx 377$ [ohms]

Radiating EM waves exert pressure on the surfaces they illuminate that arises from the momentum of photons. According to Newton's third law, an object that reflects radiation must recoil with enough momentum to stop the incoming wave and send it out again—twice the momentum of the wave. If the energy is absorbed, however, the object must only stop it, recoiling with the momentum of the wave. When a photon is absorbed, its energy and momentum are transferred to the molecule that absorbs it. When a photon is totally reflected, its momentum change is twice its own momentum and the molecule that reflects it recoils twice as much as it would have from an absorption event (this is the principle underlying solar sails for space craft).[23] Radiation pressure is quantified by the equations below. The intensity of EM radiation is the time average of the Poynting vector,

$$I = \langle \vec{S} \rangle$$

The radiation pressure, or P_{EM} (force per unit area), exerted by electromagnetic radiation is proportional to its intensity.

Equation: EM radiation pressure from intensity I

$$P_{EM_{absorption}} = \frac{I}{c_0} \text{ [Pa/m}^2\text{]} \quad (26a)$$

$$P_{EM_{reflection}} = \frac{2I}{c_0} \text{ [Pa/m}^2\text{]} \quad (26b)$$

Intense LASER pulses deliver sufficient energy that retinal damage (and repair) can occur from both localized heating and mechanical impulses caused by momentum change.[24] Photon and radiation momentum are easily estimated from the ratio between energy and the speed of light.

Equation: Momentum change from a photon with energy, U_{photon} [J]

$$\Delta p_{photon\,aabsorbed} = \frac{U_{photon}}{c_0} \text{ [kg m/s]} \quad (27a)$$

$$\Delta p_{photon\,reflected} = \frac{2U_{photon}}{c_0} \text{ [kg m/s]} \quad (27b)$$

$$F_{photon} = \frac{\Delta p_{photon}}{\Delta t} \text{ [N]} \quad (27c)$$

A wide range of EM frequencies (and therefore of photon energies) may be encountered in occupational settings. A table or graph showing properties of electromagnetic waves across the full spectrum of interest is too large to publish in this text,[25] but an abridged illustration is offered in Figure 3 to illustrate some of the regions of interest for industrial hygiene and health physics.

According to classical electromagnetic theory, maximum energy transfer between radiated EM waves and an electrically conducting target occurs when the target size is about 0.4 wavelengths (if electrically isolated) or 0.2 wavelengths (if electrically bonded to a ground plane). That means for humans, energy from radio waves with frequencies between 30 and 300 MHz (wavelengths between 10 and 1 meter) are most strongly absorbed. Wavelengths shorter or longer than these tend to pass through the body while depositing only a small fraction of their propagating energy.[26] See Tables 6, 7, and 8 for wavelength and frequency summaries.

At low frequencies (wavelengths that are long in comparison to molecule and organ dimensions), exposure is controlled by controlling the electric field's strength in tissue. At low frequencies, high electric field strength can induce AC currents, but it is primarily contact with voltage sources that causes damaging current flow. Ohmic heating is the primary stressor for frequencies between DC and VLF. However, caution must be exercised; neuromuscular disturbance—such as cardiac fibrillation by contacting current with a frequency near 50 to 60 Hz—can be fatal.

TABLE 6

DC to Very Low Frequency EM Spectral Bands

Name	Symbol	Frequency	Wavelength/km
Approximate direct current	~DC	0 Hz to1 Hz	> 300,106
Extremely low frequencies	ELF	1 Hz to 300 Hz	300,000 to 1000
Voice frequencies	VF	300 Hz to 3 kHz	1000 to 100
Very low frequencies	VLF	3 kHz to 30 kHz	100 to 10

TABLE 7

Radio-Frequency Spectral Bands

RF Band Name	Symbol	Frequency/MHZ	Wavelength
Low frequency	LF	0.03 to 0.3	10 km to 1 km
Medium frequency	MF	0.3 to 3	1 km to 100 m
High frequency	HF	3 to 30	100 m to 10 m
Very high frequency	VHF	30 to 300	10 m to 1 m

TABLE 8

Microwave Spectral Bands

Band Name	Symbol	Frequency/GHZ	Wavelength
Ultra high frequency	UHF	0.3 to 3	1 m to 10 cm
Super high frequency	SHF	3 to 30	10 cm to 1 cm
Extremely high frequency or millimeter band	EHF or mm band	30 to 300	10 mm to 1 mm
Micrometer band	μm band	300 to 3 000	1 mm to 100 μm

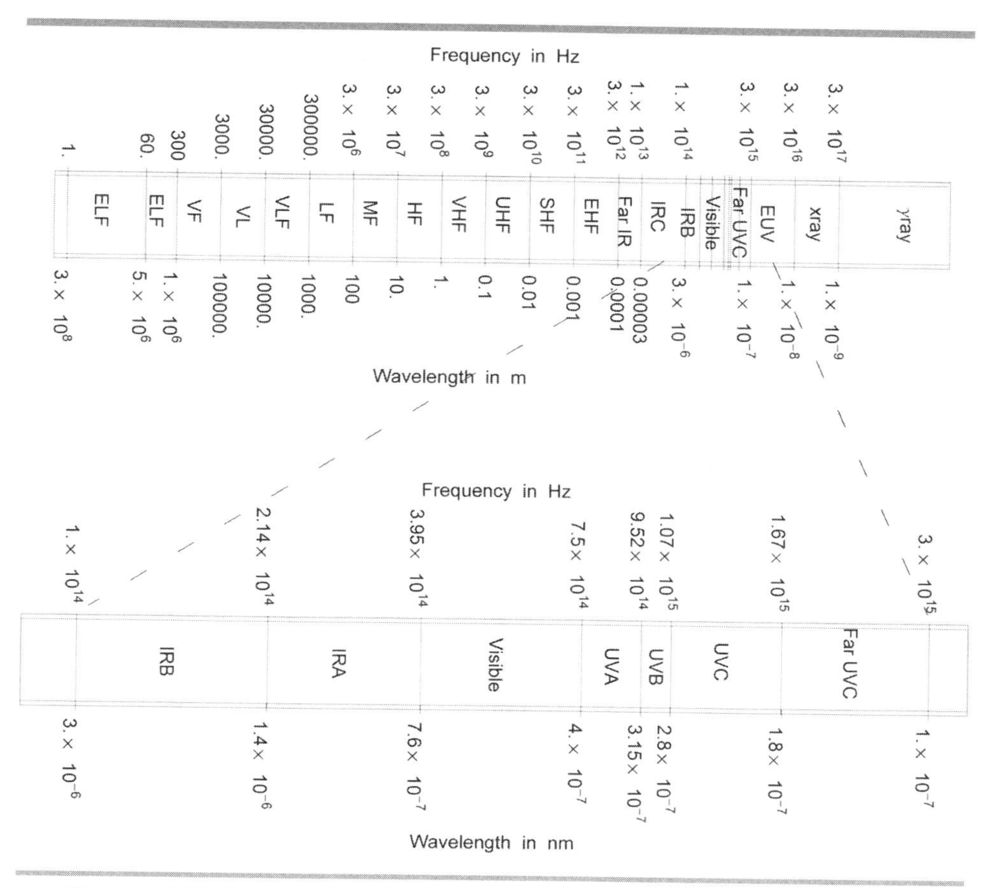

FIGURE 3. EM: Spectrum allocation

Electronic and communications engineers describe various regions of the spectrum with names that may be unfamiliar to safety professionals. Common names for specific regions of the radio-frequency spectrum are summarized in Tables 6–9.

Radio-frequency bands pose primarily an ohmic heating risk. Microwave frequencies pose skin-effect risks where induced currents are concentrated near the surface and of microwave resonance absorption by some molecules common in tissues.

In far infrared frequencies, the penetration depth is small, causing primarily surface heating. Symptoms include welder's flash and skin erythema that looks like sunburn. At frequencies of visible light, very little penetration occurs except in specialized tissues of the cornea, lens, and aqueous humor of the eye. At ultraviolet frequencies, ionization occurs as photons interact with molecules and are almost completely absorbed near the surface of the human body (see Table 9).

At frequencies above extreme UV, a high proportion of X-rays and gamma rays penetrate the body completely, allowing medical imaging of the skeleton and of some organ systems. A small proportion of X-ray and γ-ray photons collide with tissue molecules and produce copious quantities of ionization.

Ionization events are widely associated with changes in RNA and DNA. If these are not repaired in a timely fashion, they are associated with increased risk of cancer in some tissue types. Fortunately, humans are equipped with robust repair mechanisms; peer-reviewed epidemiological studies have shown that low levels of radiation exposure actually reduce the risk of cancers. An undated critical report available on the Internet that criticizes these studies is referenced here for the convenience of readers who wish to evaluate minority arguments (Nussbaum, et al., undated). However, subthreshold levels seem to represent exposures low enough that the rate of damage remains below the

TABLE 9

Summary Effects of IR, Vis, and UV Photons

Band	Name	Photon Energy/eV	λ/nm	Eye	Skin
Far IR	IR-C	0.00124–0.0413	1×10^6–30×10^3	Corneal burn	Skin burn
Thermal IR	IR-C	0.0413–0.413	30×10^3–3×10^3	Corneal burn Aqueous flare IR cataract	Skin burn
Near IR	IR-B	0.413–0.886	3000–1400	Corneal burn Aqueous flare IR cataract	Skin burn
Near IR	IR-A	0.866–1.63	14000–760	IR cataract Retinal burns	Skin burn
Visible	Vis	1.63–3.1	760–400	Photochemical Retinal burns	Skin burn Photosensitive reactions
Near UV	UV-A	3.1–3.94	400–315	Photochemical UV cataract	Pigmentation Skin burn DNA changes
Medium UV	UV-B	3.94–4.43	315–280	Photokeratitis	Sun tan Pigmentation Skin aging
Far UV	UV-C	4.43–6.89	280–180	Photokeratitis Germicidal	Skin cancer Skin burn Germicidal
Far UV	FUV	6.89–12.4	180–100	Photokeratitis Germicidal	Skin cancer Skin burn Germicidal
Extreme UV	EUV	12.4–124	100–10	Photokeratitis Germicidal	Skin cancer Skin burn Germicidal

Photobiological system: UV-A1 (400–340 nm) tans skin, a protective mechanism; UV-A2 (340–315) changes DNA; UV-B provides energy for producing Vitamin D.[1] UV-C, FUV, and EUV are germicidal and carcinogenic.

Physics system: far UV (FUV) is 280–100 nm, and extreme UV (EUV) is 100–10 nm. Vacuum UV extends from 200–10 nm, is absorbed in the atmosphere and propagates only in a vacuum. Near IR is 0.76–3 μm , thermal IR is 3–30 μm , and far IR is 30–100 μm.

[1] The ozone layer partially filters UV-B and nearly completely filters UV-C, FUV, and EUV from the sun. Arguably, the ozone layer makes life on earth possible.

rate of repair, keeping damage from accumulating. (See Figure 4 on the biological effects of EM radiation.)

Primer on Ionization

Ionizing radiation interacts with tissue (or any material) when high-speed subatomic particles (including photons) collide with electrons or nucleons of atoms in molecules. These collisions conserve mass, charge, and momentum. An ionizing event occurs when an electron or other charged particle is added to, or removed from, a molecule. Particle energy for ionizing weakly bound orbital electrons is as low as ~7 eV for amines; most molecules are ionized by energies above 12.4 eV.[27] Ionization in modern occupational settings is typically caused by common high-energy subatomic particles, including helium nuclei (alpha particles), electrons (beta$^-$ particles), positrons (beta$^+$ particles), photons (UV, X-, and γ rays), protons, neutrons, and cosmic rays (subatomic and subnucleonic particles).

Ionization in material, including in human tissue, produces ion pairs, conserving charge in the universe. The first ionization occurs when an ionizing particle knocks one or more electrons out of their orbital position, giving the parent molecule a net positive charge. A second ionizing event occurs quickly when the freed electrons are captured by another molecule, which becomes a negatively charged ion. When a particle has sufficiently high energy, an ion track may be created as the particle bounces between molecules, successively ionizing a series of molecules along its path. Very high-energy charged particles and uncharged particles can disrupt the nucleus of an atom by emitting secondary ionizing radiation in the form of gamma rays,

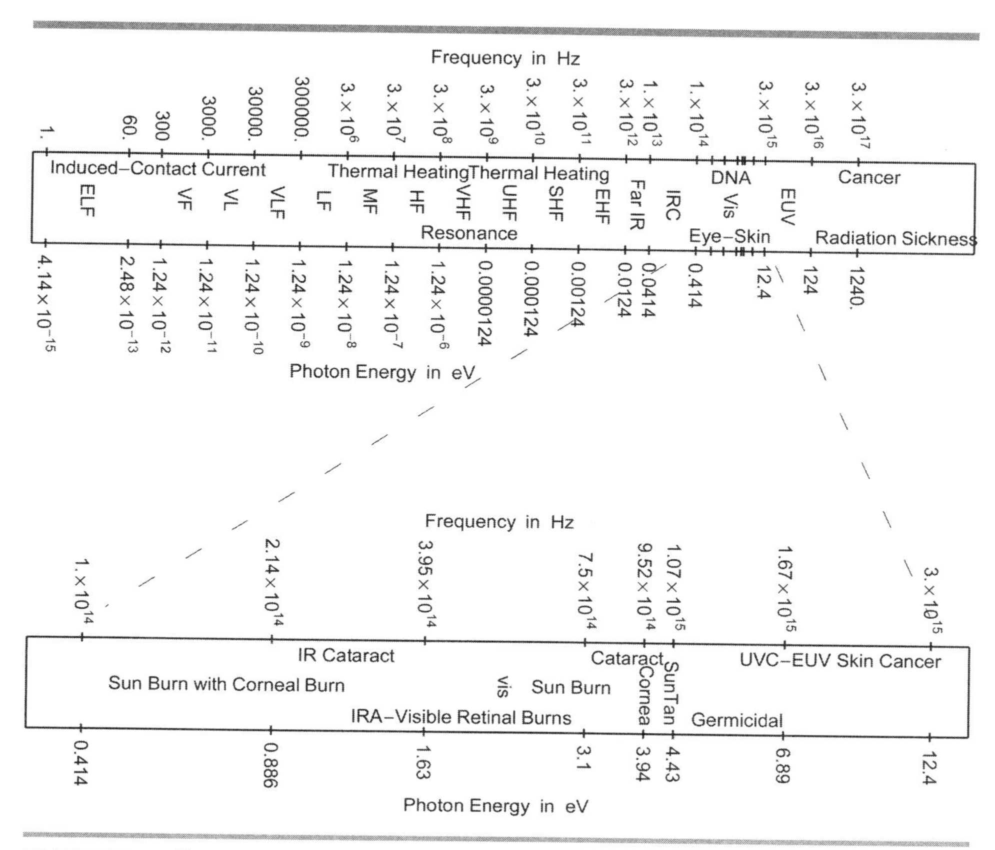

FIGURE 4. Summary of biological effects of EM radiation

neutrons, protons, beta particles, neutrinos, muons, and so on. This principle is used in neutron activation analysis, in which neutrons excite nuclei so that characteristic gamma rays are emitted, allowing a gamma spectrometer to identify the elements in the unknown substance.

Primer on Ionizing Radiation

It will be helpful to understand the nature of several of the various forms of energy, which range from the mildly ionizing ultraviolet light to the highly ionizing X-, alpha, beta, and γ rays and high-energy neutrons.

The dominant tissue-damage mechanism of ionizing radiation is the oxidation of life-critical molecules by free radicals and peroxides created along the aqueous ion track left in tissue by a passing ray of highly ionizing radiation. Excess consumption of oxidant-rich foods produces the same kinds of damage as low-level exposure to ionizing radiation. No matter the source of damage, the same repair mechanisms work to repair damaged tissue. If the damage rate is lower than the repair rate, neither acute nor chronic injuries are observed clinically. We live in a radiation-filled environment that includes cosmic background radiation and radiation from naturally occurring environmental radio nuclides, and our bodies experience internal radiation from radio nuclides in tissue that enter in chemicals from air, water, and food. This has been the case throughout human history.

Environmental Background Radiation Levels

At sea level, the average radiation level is approximately 0.03 microsieverts per hour. As altitude increases, radiation exposure increases exponentially. Mexico City, which is 2240 m above sea level, is exposed to about 0.09 microsieverts per hour; La Paz, Bolivia, at ~3640 m the highest city in the world, endures cosmic and solar radiation of about 0.23 microsieverts hourly, more than seven times the tropical sea-level value (*The World of Nuclear Science*, undated).

Naturally occurring background radiation is the main source of exposure for most people. Levels typically range from about 1.5 to 3.5 millisieverts per year. In Brazil and Sudan many people have average exposures up to 40 mSv yearly. Several places are known in Iran, India, and Europe where natural background radiation, including radon gas, gives an annual dose exceeding 50 mSv and even up to 260 mSv (at Ramsar in Iran). Lifetime doses from natural radiation range from 120 to 20,000 mSv and there is no evidence of increased cancers or other health problems arising from these high natural levels.[28] Clearly, the 50 mSv/yr (5 rem/yr) PEL for annual occupational exposure is well below the threshold for identifiable disease in humans.

There are many common sources of radiation in our lives. Table 10 summarizes some of these, and an EPA Web site (EPA 2007) allows users to estimate their own annual radiation dose by answering some simple multiple-choice questions.

Ultraviolet Radiation (UV Rays)

UV radiation is divided into three bands, called UV-A (400–315 nm, or 3.1–3.94 eV/photon), UV-B (315–280 nm, or 3.94–4.43 eV/photon) and UV-C (280–100 nm, or 4.43–12.4 eV/photon). Some UV-B and most UV-C photons are able to ionize some biochemical molecules. In most occupational settings vacuum UV wavelengths (180–100 nm, or 6.89–12.4 eV/photon) are of no consequence, because these photons are so completely absorbed that they travel no further than 4 cm (1.5 inch) in air at sea level. Commercial IH survey instruments, called photo ionization detectors, measure the airborne concentration of some vapors and gases based on UV ionization. Typically, photo ionization detectors have interchangeable UV sources that produce 8.4 eV, 9.5 eV, 10.0 eV, 10.6 eV, or 11.7 eV photons (International Sensor Technology undated).

TABLE 10

Common Sources of Everyday Radiation

Source	Amount
1 kg of coffee	1000 Bq
1 kg super-phosphate fertilizer	5000 Bq
The air in a 100 m^2 Australian home (radon)	3000 Bq
1 adult human weighing 70 kg (100 Bq/kg)	7000 Bq
The air in many 100 m^2 European homes (radon)	30,000 Bq
1 household smoke detector (with americium)	30,000 Bq
Radioisotope for medical diagnosis	70,000,000 Bq
Radioisotope source for medical therapy	100,000,000,000,000 Bq

Alpha Particles or Alpha Rays

Alpha radiation is the name for fast-moving, positively charged helium nuclei emitted from a nucleus during a disintegration event. Alpha particles have a mass number $A = 4$ and atomic number $Z = 2$, indicating a double positive charge. They have high energy—typically in the MeV range—but because of their large mass are stopped by just a few inches of air, the human epidermis (the outer layer of skin, made up of dead cells), or a piece of paper. Alpha particles are emitted having discrete energies, and alpha spectra are used to identify the parent radio nuclide. Because alpha radiation does not penetrate the skin, external exposure poses little hazard. When alpha-emitting radio nuclides are inhaled or ingested, their radiation causes extensive localized internal damage.

Beta Particles or Beta Rays

Beta particles are fast-moving, charged particles emitted from a nucleus during a disintegration event that changes its net charge, and thereby its position in the periodic table. They carry either a single negative charge (electron) or single positive charge (positron) and typically have energies in the range of a few hundred keV to several MeV. Radio nuclides responsible for spontaneous beta emissions exhibit continuous energy distributions. The maximum energy of beta emission is characteristic of the nuclide present. Beta rays are more penetrative than alpha particles but less so than gamma rays. Beta particles penetrate several feet of air, several millimeters of plastic, and lesser amounts of very light metals.

Gamma-Ray and X-Ray Photons

X-rays and gamma rays are photons having higher energy (higher frequency and shorter wavelengths) than UV-C. Industrial hygienists consider energies greater than 12.4 eV to be ionizing for soft X-rays. Health physicists note that those of significant health concern typically range from several keV to several MeV in occupational settings. X-rays and gamma rays differ only in their sources; X-rays originate from the sudden acceleration of a charged particle (bremsstrahlung radiation) outside the nucleus of an atom, and gamma rays originate within the nucleus of the atom when nucleons relax to lower energy states from an excited state. Gamma rays are emitted at characteristic frequencies and gamma spectroscopy is useful for identifying the elemental content of unknown substances. Like all photons, X-rays and gamma rays carry no charge, have zero rest mass, and deliver energy and momentum to an atom during a collision. X- and gamma-ray photons damage living cells by energy transfer when they scatter from molecules in living tissue. A collision of this sort most frequently leaves a free radical or ion that can produce a cascade of damaging chemical reactions. High-Z materials are the most effective shield for high-energy photons, which interact primarily with their numerous orbital electrons. To a first approximation, more electrons per atom mean better shielding from a solid made up from that atom. Soft X-rays can be stopped by a thin piece of aluminum foil, but hard X-rays can penetrate several inches of lead.

Transitions that result in gamma emission leave A (the mass number) and Z (the atomic number) of the element unchanged and are called *isomeric transitions*. Gamma emissions carry energy and momentum away from an unstable nucleus as it adjusts to a lower energy state and are emitted as photons whose frequencies are characteristic for each atom. Because gamma photons are usually emitted following radioactive decay that also includes alpha, beta, or neutron emissions, a full health physics assessment is required to discover the limits on time, shielding, and distance that will protect workers.

Neutrons

A neutron is uncharged and has slightly more mass than a proton. Free neutrons are unstable and decay into protons by beta emission, having a half life of about 11 minutes (Christenson, et al. 1972, 1628–1640). Neutrons are stable only when in a nucleus and bound to protons. Neutrons are emitted as a by-product of fission or by nuclear reactions involving other particles. Neutrons can cause much damage in tissue and cells by creating ions, peroxides, and free radicals in all tissue types. Secondary reactions by the free radicals and ions created by neutrons are the primary cause of biochemical damage—especially that occurring to life-critical RNA and DNA molecules.

Primer on Human Vision

The human eye is a remarkable organ that has automatic control of focal length, of the retinal illumination level, and of retinal photochemical gain. The cornea and lens are made of specialized cells organized as highly parallel bundles of proteins wrapped in bilipid layers. These are transparent to photons in the visible spectrum, between infrared and ultraviolet light (700–400 nm). The retina is a complex, multilayered structure having photosensitive elements (called *rods* and *cones*) at the back; initial image processing occurs in layers of neurons toward the front of the eye. The neural network acts like a screen door, casting shadows on the image plane containing the rods and cones. However, the short distances between neurons in the front layers allow real-time image preprocessing with minimal time delay. The processed image is collected and sent to the visual cortex through the fovea, a hole in the retina through which the optic nerve passes. It forms a blind spot in the visual field of each eye. The overlapping images received in the visual cortex have

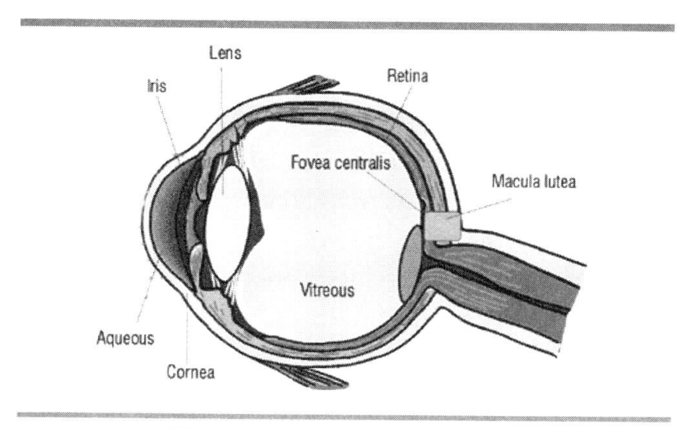

FIGURE 5. Wavelength sensitivity of tissues in the eye

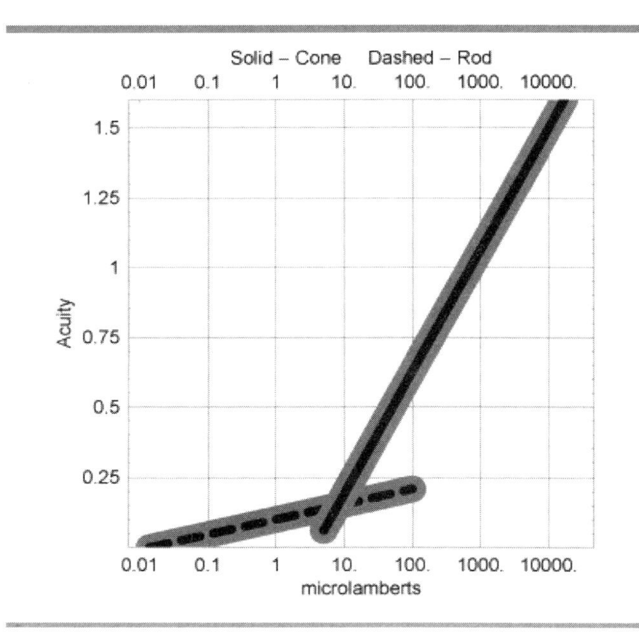

FIGURE 6. Human visual acuity

nonoverlapping blind spots, making the perceived image with binocular vision complete.

Occupational injuries to the eyes include physical trauma, corneal burns, lens burns, and retinal burns. Figure 5 shows the nature of injury caused by high-intensity radiation of various wavelengths.

As shown in Figure 6, the idealized human eye responds to light intensity over nearly nine orders of magnitude, the brightest (6) by the response of the cones and the dimmest (3) by the rods (Graham 1965). Visual acuity is monotonic with intensity and is best near about 10 millilamberts for the cones and about 100 microlamberts for the rods. The best visual acuity of rods is much poorer than the best of cones. To maximize performance in critical work areas, consider the level of illumination and the quality of contrast, and limit reflections to maximize visual performance.

Recent evidence supports tetrachromic response of light-sensitive molecules in rods and cones, as illustrated in Figure 7 (Fulton 2006). The human retina responds well to ultraviolet photons. In a normal eye, this does not happen, because UV photons are absorbed in the cornea and lens. In an aphakic eye, the lens has been removed and the UV sensitivity allows color perception that is more varied than that normally experienced. Additionally, normally tolerated UV photon flux can cause retinal damage in an aphakic eye. Therefore, separate exposure limits apply to individuals who have no lens (aphakic individuals).

In terms of eye damage, one must consider both the frequency (a measure of energy per photon) and intensity (number of photons per unit area). Damage occurs when too much energy is absorbed (primarily by the cornea, the lens, and the retina, or when an intense LASER pulse is reflected). The wavelengths responsible for such damage are detailed in the exposure guidelines sections that follow.

Primer on Human Hearing

Noise is a wave that propagates through material.[29] It is convenient to think of noise as collective coherent vibrations moving as elastic waves through a gas, liquid, or solid. The elastic waves have alternating compression and rarefaction peaks along their direction of travel. Occupational noise usually originates from vibrating solids or turbulent fluid flow. When the noisy medium (air, liquid, or solid) is in contact with some portion of the body, it can transfer mechanical energy from its alternating pressurized and rarified peaks. This energy creates vibrating waves in that portion of the body. The induced motion, if it exceeds the elastic limit of the moving parts, may cause direct mechanical damage. If it merely moves some parts—including molecules—relative to one another, frictional heating increases temperature and the resulting increased molecular motion increases reaction rates. At ultrasonic

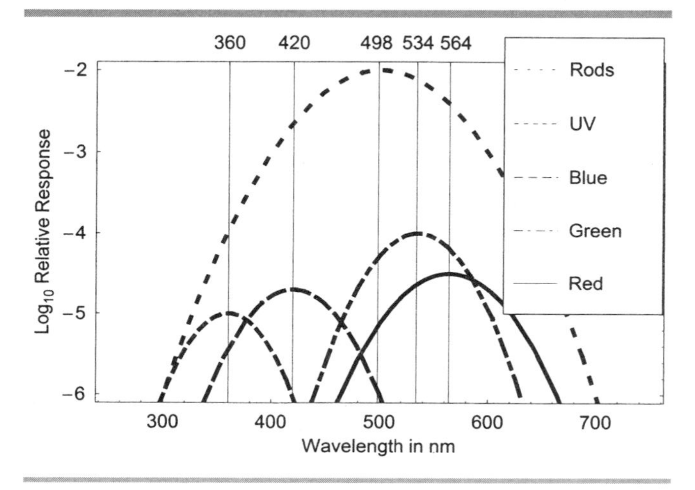

FIGURE 7. Tetrachromic vision

frequencies there may be a direct effect on the rate of chemical reactions in tissue as individual molecules are rapidly rotated, flexed, and bounced into each other, producing reaction rates similar to those expected at higher temperature. Focusing of acoustic energy in fluids is known to cause cavitation bubbles that experience transient pressures sufficient to melt most metals by adiabatic heating when they collapse.

Noise is here defined as small pressure oscillations of less than a few Pa around an average ambient air pressure of ~50,000–100,000 Pa. These alternating pressurization and rarefaction peaks induce vibrations in the tympanic membrane (eardrum). In turn, the eardrum vibrates the middle-ear bones (hammer, anvil, and stirrup). The footplate of the stapes transfers the vibrations through the oval window into the fluid in the inner ear. The fluid motion bends the inner and outer hair cells of the cochlea. The cochlea acts as a harmonic analyzer and directs different frequencies to different neurons in the auditory nerve. Signals from the left and right auditory nerves combine and spread out over the surface of the auditory cortex. The complex features appear spatially separated so that humans are able to identify speakers, identify threats, localize the source of sound in their 3-D world, to communicate by means of spoken language.

Remarkably, because noise and speakers' phonemes tend to be localized in differing regions of the auditory cortex, verbal communication remains possible in the presence of interfering noise.

Sound that vibrates the eardrum and stimulates the cochlea lies in the frequency range of 20 Hz to 20 kHz. The outer ear canal acts as a tuned organ pipe with one end open and the other closed by the eardrum. Its resonant behavior emphasizes sound with frequencies from 1–4 kHz while attenuating sound at much higher or much lower frequencies. The middle-ear bones also emphasize sound in this preferred speech range. When sound intensity is dangerously high, small muscles contract to pull the middle-ear bones into a protective geometry that reduces signal strength delivered to the cochlea, a protective mechanism called the *stapedial reflex*.[30]

In the cochlea, the organ of Corti is a nonlinear frequency-tuned transmission line that moves with high-frequency sound near its base and with low-frequency sound around the 2.75 turns of its helix all the way to its distal end. As it bends, those of its hair cells located near the base of the cochlea respond to all frequencies, and those near the apex respond preferentially to low frequencies. The hair cells stimulate neurons in the auditory nerve; these signals are sent to the auditory cortex for processing. The net result is that the human sensitivity to hearing behaves with maximum sensitivity in or near the range 2 to 4 kHz for most adults and at higher frequencies for young children, who have shorter external ear canals.

The ear adjusts to intensities over 12 orders of magnitude, from 10^{-12} to 1 W/m^2. The root-mean-square sound pressures associated with these intensities range from 20 μPa (197×10^{-12} atm) to 20 Pa (197×10^{-6} atm) in normal occupational settings.

Damage to the ear is mediated by sound power and total energy received. Incident sound power is a quantity that equals the product of sound intensity and the area of the tympanic membrane (eardrum). Under normal occupational conditions of sea-level barometric pressure (P_{air} ~100 kPa) and comfortable temperatures (T_{air} ~293 K), the reference quantities for sound pressure level and sound intensity level are such that the two are approximately numerically equal. It is best practice to compare the sound intensity level in dB re 1 pW/m^2 with the appropriate hearing-damage risk criteria expressed in dB.

Exposure Guidelines for Physical Agents

Thermal Stress

Thermal stress occurs when the tissue temperature rises (heat stress) or falls (cold stress) too far from its normal temperature. Occupational thermal stress standards are designed to keep the core temperature within ±1.0°C (2°F) of the normal human body temperature of 37°C (98.6°F) and to keep the surface temperature of skin within the approximate range of 16–35°C (60–95°F). Hand and finger temperatures need to be kept warmer than 16°C (60°F) to preserve manual dexterity. There are ear-canal thermal monitors that can be used for real-time monitoring of a core-temperature surrogate—and surface-temperature monitors can be used as well.

Heat Stress

Heat stress[31] depends upon environmental conditions (temperature, humidity, air velocity, radiant-heat load) and on the worker (metabolic rate, type of clothing worn, acclimatization, sweat rate, evaporation rate, cardiovascular health). When in doubt of stress levels, or when conditions are extreme, encourage use of continuous ear-canal temperature monitoring. When conditions are within a normal range, a description of the work environment using the wet-bulb globe temperature (WBGT) is sufficient to establish the baseline work schedule.

The WBGT is calculated by one of two formulas: the first is used in cases not involving a solar load and the second is used when outdoors working with a solar load.

Equation: Wet-bulb globe temperature (WBGT)

$$WBGT_{no\ sun} = 0.7\ T_{NWB} + 0.3\ T_{GT} \quad (28a)$$

where

T_{NWB} = natural wet bulb; T_{GT} = globe temperature.

$$WBGT_{in\ sun} = 0.7\ T_{NWB} + 0.2\ T_{GT} + 0.1\ T_{DB} \quad (28b)$$

where T_{DB} = dry bulb.

The work–rest schedule outlined by OSHA has been modernized by the TLV committee since OSHA adopted its older version into law. The revised version, reproduced as Table 11, will provide reasonable assurance that assigned tasks will not result in heat exhaustion, heat stroke, or worse.

These TLVs are based on the assumption that nearly all acclimatized, fully clothed[32] workers with adequate water and salt intake should be able to function effectively under the given working conditions without exceeding a deep body temperature of 38°C (100.4°F).[33] They are also based on the assumption that the WBGT of the resting place is identical or near to that of the workplace. When the WBGT of the work area is different from that of the rest area, a time-weighted average may be used to estimate the effective WBGT each hour.

In addition to the work–rest schedule, other control measures for heat stress include access to cool nonalcoholic drinks (10–16°C, or 50–60°F), wearing cooling vests (or vortex coolers with air-supplied hoods when mobility is not required), air-conditioned rest areas, and scheduling hot work during cool parts of the day. Although fans can help, their use must be judicious. In any environment where the WBGT of the air streaming from the fan is higher than 35°C (95°F), the convective heat transfer from the air will cause a temperature increase in the exposed worker. Fans are primarily useful when the WBGT of the air currents directed on workers is less than the tabulated value for 100% work under the appropriate metabolic rate. This means fans are truly useful when they produce drafts with measured WBGT below 26–29.5°C (79–85°F).

All workers should be trained to recognize the symptoms of heat stress in themselves and others so heat exhaustion and heat stroke can be prevented. The OSHA technical manual points out that the signs and symptoms of *heat exhaustion* are headache, nausea, vertigo, weakness, thirst, and giddiness. Fortunately, this condition responds readily to prompt treatment. *Heat cramps* are usually caused by performing hard physical labor in a hot environment and have been attributed to an electrolyte imbalance caused by sweating. Cramps can be caused by either too much or too little salt. Because sweat is a hypotonic solution (±0.3% NaCl), excess salt can build up in the body if the water lost through sweating is not replaced. Thirst cannot be relied on as a guide to the need for water;

TABLE 11

Work–Rest TLV for Controlling Heat Stress

	Workload			
Work-Rest Regimen	**Light**	**Moderate**	**Heavy**	**Very Heavy**
Continuous work	29.5°C	27.5°C	26.0°C	
75% work–25% rest each hour	30.5°C	28.5°C	27.5°C	
50% work–50% rest each hour	31.5°C	29.5°C	28.5°C	27.5°C
25% work–75% rest each hour	32.5°C	31.0°C	30.0°C	29.5°C

Light: Sit with moderate hand–arm motion; use table saw; some walking.
Moderate: Walk at 6 km/hr carrying 3-kg load on level terrain; scrubbing while standing.
Heavy: Carpenter sawing by hand; intermittent shoveling dry sand; pick and shovel use.
Very heavy: Continuously shoveling wet sand; emergency rescue wearing SCBA.

instead, water must be taken every 15 to 20 minutes when in hot environments. In *heat collapse*, the brain does not receive enough oxygen because blood pools in the extremities. An exposed individual may lose consciousness suddenly and with little warning. To prevent heat collapse, the worker should gradually become acclimatized to the hot environment. *Heat rashes*, however, are the most common problem in hot work environments. *Prickly heat* manifests as red papules and usually appears in areas covered by restrictive clothing. As sweating increases, these papules give rise to a prickling sensation in that skin which is persistently wetted by unevaporated sweat. Heat-rash papules may become infected if they are not treated. In most cases, heat rashes will disappear when the affected individual returns to a cool environment.

Cold Stress

On cold, windy days, the rate of cooling is a function of both windspeed and air temperature. Effective November 11, 2001, NOAA published the modern windchill formula, intended for worldwide use (National Weather Service). When windspeed is less than 3 mph, the windchill formula is unnecessary; use the dry-bulb temperature instead. For higher windspeeds, the windchill temperature is computed from the dry-bulb temperature (T) and windspeed (V):

$$T_{\text{windchill }°F} = 35.74 + 0.6215\,T_{°F} - (35.75 - 0.4275\,T_{°F})\,Vel_{mph}^{0.16};\ \text{for } V_{mph} > 3 \quad (29a)$$

$$T_{\text{windchill }°C} = 13.13 + 0.6215\,T_{°C} - (13.95 - 0.4863\,T_{°C})\,Vel_{mps}^{0.16};\ \text{for } V_{mps} > 1 \quad (29b)$$

Exposed skin will suffer frostbite in a few minutes at low temperatures. Table 12 summarizes the approximate effect of windchill temperature on exposed skin and also contains recommendations for work–rest schedules for use by properly clothed personnel. Together, these two aspects of the human response to cold can be used to conservatively manage occupational cold exposure in a program that includes specific worker training about symptoms requiring self-rescue and use of a buddy system—allowing every worker to observe and be observed by another.

Noise and Vibration

Noise and vibration standards are designed to prevent direct tissue damage and to limit discomfort from nondamaging sound pressure levels. Because damage and control mechanisms differ, separate noise exposure guidelines exist for infrasonic sound (< 20 Hz), acoustic sound (20 Hz to 20 kHz), and ultrasonic sound (> 20 kHz). Similarly, vibration damage mechanisms differ depending on the included angle between the vibration vector and key body axes. The vibration standards are separately specified for vibration directed along or transverse to the spine for standing, sitting, and prone positions, and along or transverse to the forearm.

The tissues most sensitive to acoustic energy are those found in the outer, middle, and inner ears. Control measures have been well summarized by OSHA and are available via the Internet.

Noise

All noise control, whether using the PEL[34] (legally required) or more protective TLV[35] (ISO standard), involves only three equations: $D(C,T)$, $T(L)$, and $L(C,D)$.

TABLE 12

Guidelines for Cold Stress

$T_{windchill}$ (°F)	Time for Frostbite of Exposed Skin[1]	Maximum Continuous Work	Number of 10-min Warm-Up Breaks in 4-hr Shift
Warmer than −18	> 30 minutes	115 minutes	1
−20 to −24	~30 minutes	115 minutes	1
−25 to −29	~30 minutes	75 minutes	2
−30 to −34	10 to 30 minutes	50 minutes	3
−35 to −39	< 10 minutes	40 minutes	4
−40 to −44	5 to 10 minutes	30 minutes	5
Colder than −45	< 5 minutes	Emergency work only	

Notes:

Be more cautious and shorten these times in the cases of young children and senior citizens.

Do not administer hot liquids to hypothermia victims; tepid liquids are better.

Do not heat the extremities of a hypothermia victim, which drives cold blood to the heart.

Seek immediate medical attention for any hypothermia victim.

[1]The complete NOAA windchill chart indicates slightly longer times to frostbite at lower temperatures than these at some wind speeds. Full details are on the Internet (www.nws.noaa.gov/om/windchill/).

TABLE 13a

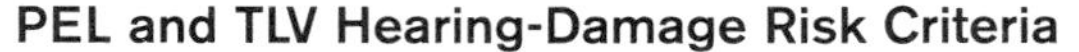

PEL and TLV Hearing-Damage Risk Criteria

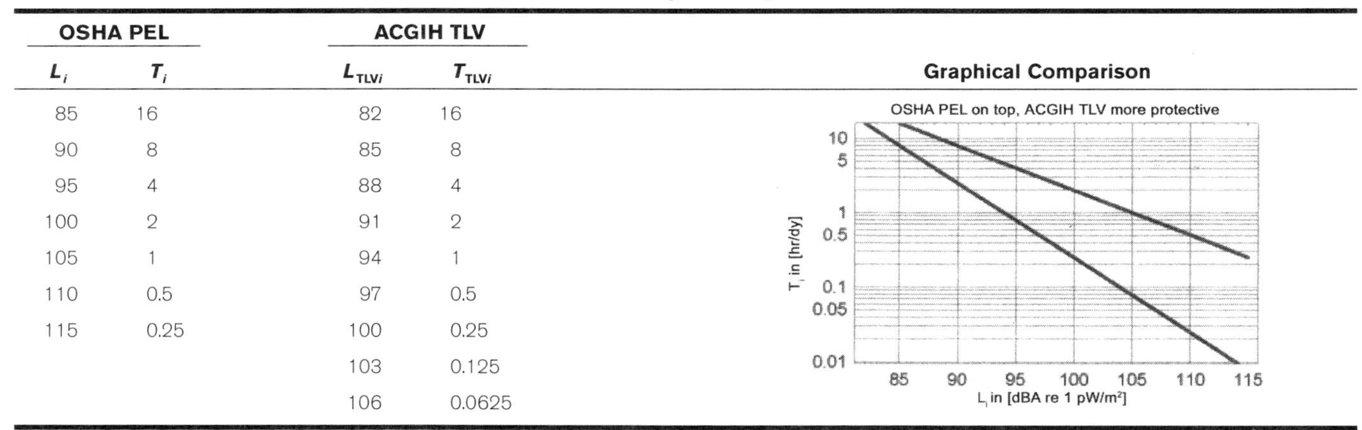

OSHA PEL		ACGIH TLV	
L_i	T_i	L_{TLVi}	T_{TLVi}
85	16	82	16
90	8	85	8
95	4	88	4
100	2	91	2
105	1	94	1
110	0.5	97	0.5
115	0.25	100	0.25
		103	0.125
		106	0.0625

Graphical Comparison

TABLE 13b

Equations for PEL and TLV Hearing-Damage Risk Criteria

(Eq. 30) OSHA Noise Dose (29CFR1910.95)

$$T_{PEL_i} = 2^{\frac{105 - L_i}{5}} \text{ [hr]}$$

$$D_{OSHA} = \sum_{i=1}^{N} \frac{C_i}{T_{PEL_i}} \text{ [unitless]}$$

$$C_{total} = \sum_{i=1}^{N} C_i \text{ [hr]}$$

$$L_{OSHA} = 105 - 5 \log_2\left(\frac{C_{total}}{D_{OSHA}}\right) = 105 - 7.2135 \log_{10}\left(\frac{C_{total}}{D_{OSHA}}\right)$$

in dB re 1 pW/m^2

(Eq. 31) ACGIH Noise Dose (ACGIH TLV Booklet 2007)

$$T_{TLV_i} = 2^{\frac{94 - L_i}{3}} \text{ [hr]}$$

$$D_{TLV} = \sum_{i=1}^{N} \frac{C_i}{T_{TLV_i}} \text{ [unitless]}$$

$$C_{total} = \sum_{i=1}^{N} C_i \text{ [hr]}$$

$$L_{TLV} = 94 - 3 \log_2\left(\frac{C_{task}}{D_{TLV}}\right) = 94 - 4.3281 \log_{10}\left(\frac{C_{total}}{D_{TLV}}\right)$$

in dB re 1 pW/m^2

D is the noise dose, portrayed here as a fraction between 0 and 1. Doses > 1 daily or 5 weekly should be avoided by appropriate use of personal protective equipment (earplugs and earmuffs). *L* is the sound pressure level and is numerically equal to the sound intensity level in most workplaces having ambient pressures near 100 kPa (1 atm). Good hearing conservation programs reduce *L* by (1) substituting quieter machinery for older, noisy machinery to the maximum extent practical; (2) adding engineering controls such as sound barriers, which contain noise and reduce transmitted waves; and (3) installing sound-absorbing materials to minimize noise reflection.

Infra sound at frequencies between 1 and 80 Hz is to be kept below 145 dB. Compliance may be inferred when an unweighted sound-level meter reads less than 150 dB re 1 pW/m^2. Lower levels may be required if workers complain of resonance in their lungs (50–60 Hz).

The OSHA hearing-damage risk criteria, expressed as T_I, is the maximum allowed duration of exposure at sound intensity level L_I. The TLV criteria, expressed as T_{TLVi}, is the recommended maximum duration of exposure at sound intensity level L_I. (See Tables 13a and b.) Noise dose in either system is based on measured sound intensity levels, L_I, each lasting for a period C_I hours. Note the difference between these two criteria in Table 13a. In other words, the OSHA PEL uses a 5-dB doubling rate, and the TLV uses a 3-dB doubling rate. OSHA allows 8 hours of exposure daily to 90 [dBA re 1 pW/m^2],[36] and the TLV allows exposures of 85 [dBA re 1 pW/m^2]. Protect hearing with nonoccupational exposures below 80 dBA and sleeping areas below 70 dBA to allow the ears to rest.

When noise emissions have not been controlled at sources, hearing protection must be used for exposed workers. In the United States, earplugs and earmuffs are labeled with a noise reduction ratio (NRR) that averages 22 dB. Extensive field testing has demonstrated remarkably lower performance in practice (Berger 2000). It is recommended that the effective NRR be computed as half the labeled NRR. When C-weighted sound intensity levels are available, the A-weighted ear-canal level is estimated to be NRR/2 smaller than the free-field C-weighted level. When only A-weighted sound intensity levels are available, the estimated ear-canal level is reduced by about (NRR – 7)/2. For example, when NRR = 22 dB, the practical A-weighted ear-canal level is approximately 11 dB smaller than the free-field C-weighted level and approximately 7.5 dB smaller than the free-field A-weighted level. With proper training and care, an individual can achieve the advertised NRR = 22 dB. With sloppy use, however, hearing loss is likely.

$$L_{A,Ear\ Canal} = (L_C - \mathrm{NRR}/2) \quad [\mathrm{dBA\ re\ 1\ pW/m^2}] \tag{32a}$$

$$L_{A,Ear\ Canal} = (L_A - (\mathrm{NRR} - 7)/2) \quad [\mathrm{dBA\ re\ 1\ pW/m^2}] \tag{32b}$$

The analysis above makes it clear that earplugs alone are insufficient in workplaces with OSHA equivalent levels above 95 dBA. Elliott Berger has confirmed this by measuring threshold shifts in workplaces. When the noise is more intense, use both earplugs and earmuffs. NRR values are not additive; when used together there is an additional 5-dBA augmentation of the NRR of the better-performing of the two devices. Earplugs with an NRR of 22 dB and circumaural muffs with an NRR of 28 dB have a combined NRR of 33 dB.

The TLV for ultrasound at 1/3 octave center frequencies between 10 kHz and 100 kHz is separately specified depending upon whether the worker's head is underwater or not. In water, the ceiling values are 167 dB between 10 kHz and 20 kHz and 177 dB between 20 kHz and 100 kHz. In air, the ceiling values are 105 dB from 10 to 20 kHz, 110 dB at 25 kHz and 115 dB from 31.5 to 100 kHz. Furthermore, the 8-hour-TWA TLV is 88 dB at 10 kHz, 92 dB at 16 kHz, and 94 dB at 20 kHz, although subjective annoyance and discomfort may occur in some individuals when exposed to tonal sound above 75 dB in this frequency range.

Occupational hearing protection with the TLV considers those chemicals and drugs that increase susceptibility to noise-induced hearing loss. Chemicals already of concern include toluene, lead, manganese, and n-butyl alcohol, and chemicals proposed for addition to this list include arsenic, carbon disulfide, carbon monoxide, mercury, styrene, toluene, trichloroethylene, and xylene. When these are present, periodic audiograms should be administered and carefully reviewed.

Vibration

Vibration damage mechanisms depend upon whether the vibrating object is in direct contact with a small spot on the body, with the whole body, or with only the hand or arm. Within the variety of possible whole-body vibration exposures, the potency of the vibration depends on the included angle between the vibration vector and critical body axes, as well as on the frequency and the amplitude of vibration. Vibrations or motion at frequencies below 1 Hz cause motion sickness in some individuals and are not part of the vibration TLV. Vibrations above 4 kHz are unimportant from a whole-body or hand–arm standpoint, but localized injury can occur when ultrasonic transducers in the range of 4 kHz to 100 kHz or more are applied directly to the skin.[37]

Vibration survey meters have one or more internal frequency-weighting networks.[38] For hand–arm vibrations, the frequencies between 4 and 8 Hz are measured without attenuation and higher frequencies are attenuated progressively at –20 dB per decade so that the weighting at 160 Hz is –20 dB (a factor of 1/10) and at 1600 Hz is –40 dB (a factor of 1/100). The weighting network for longitudinal acceleration (in the direction of the spine) is flat between 4 and 8 Hz, rolls off at –20 db per decade between 8 and 80 Hz, and rolls off at –3 dB per octave between 4 Hz and 1 Hz. The weighting network for transverse whole-body acceleration is flat between 1 and 2 Hz with a gain of 1.414 to represent an observed resonance of the human body in these vibration directions. The response rolls off at –20 dB per decade between 1 and 80 Hz; these features are illustrated in Figure 8.

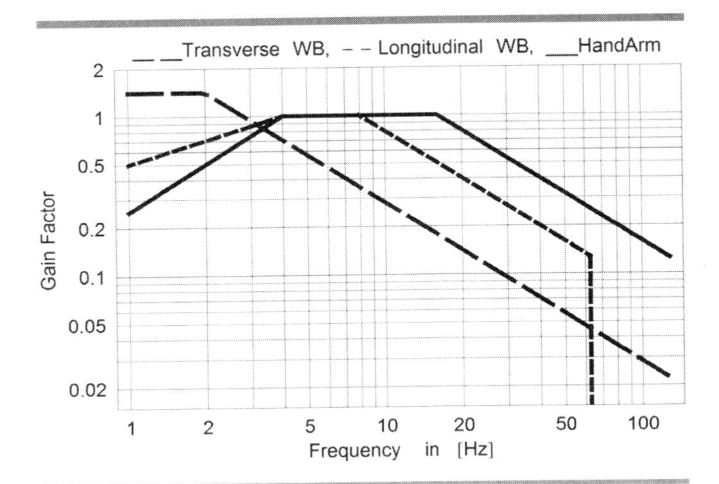

FIGURE 8. TLV vibration guidelines, frequency weighting, $W(f)$

Vibration is properly measured with 3-axis accelerometers. Along each axis, either adjust the raw data to reflect body-frequency sensitivity with weighting from Figure 8, or make the measurements with an instrument equipped with equivalent internal filter networks.[39] Each measurement period is characterized by three numbers, the *rms* value of acceleration along each axis in m/s².

Equation: Frequency-weighted acceleration for *x*-axis

$$A_{wxn} = \sqrt{\sum_{f_{min}}^{f_{max}} [W(f_i) \cdot A_{xn}(f_i)]^2} \ [\text{m/s}^2] \qquad (33)$$

Vibration intensity may vary during a work shift. If so, a time-weighted average (TWA) acceleration is computed from the frequency-weighted acceleration for each axis for each period. Use *rms* addition for the time-weighted average calculation, as shown in Equation 34.

Equation: TWA acceleration along one axis, illustrated by *x*-axis

$$A_{wx} = \sqrt{\frac{\sum_{n=1}^{N}(C_n \cdot A_{wxn}^2)}{\sum_{n=1}^{N} C_n}} \ [\text{m/s}^2] \qquad (34)$$

The *rms* resultant vibration is computed from the TWA values for individual axes.

Equation: Resultant TWA frequency-weighted acceleration

$$A_{wt} = \sqrt{A_{wx}^2 + A_{wy}^2 + A_{wz}^2} \ [\text{m/s}^2] \qquad (35)$$

The TLV committee and the European Commission both use an action level of 0.5 m/s² for 8 hours of exposure to the resultant whole-body acceleration, as computed above. The implied OEL is 1 m/s². The TLV committee recommends 4 m/s² as the TLV for 4- to 8-hour periods of hand–arm vibration exposure. Periods other than 8 hours have a nonlinear relationship between acceleration level and allowed duration of exposure. The data are tabulated in summary in the TLV booklet and in their entirety in the ANSI standards. However, there is a shortcut for those who have computed the resultant acceleration from Equation 35. The allowed duration of exposure depends on whether the area of concern is hand–arm or whole-body vibration.[40] For hand–arm vibration, the allowed duration in hours is calculated from the resultant *rms* hand–arm acceleration vector in m/s² with a valid range of 10 minutes to 16 hours.

Equation: Hand–arm vibration criteria

$$T_{HA} = \frac{98}{A_{wt}^2} \ [\text{hr}] \qquad (36)$$

For whole-body vibration, the allowed duration in hours is calculated from the resultant *rms* whole-body acceleration vector in m/s² with a valid range of 1 minute to 24 hours.

Equation: Whole-body vibration criteria

$$T_{WB} = \frac{0.7994}{A_{wt}^2} - \frac{0.006532}{A_{wt}^4} \ [\text{hr}] \qquad (37)$$

Let C_i be the observed duration of resultant *rms* acceleration A_{wt} that has an allowed daily duration of T_{TLV} hours. The daily dose is C_i/T_i and should be maintained less than 1 daily and less than 5 weekly.

A summary of steps for vibration analysis during a workday includes:

- Measuring the vibration frequency spectrum along three orthogonal axes. (Units for *rms* acceleration are [m/s²] along each of three orthogonal axes.)
- Applying frequency weighting reflecting human sensitivity.
- Recording the duration for each period of this vibration in hours, called C_i.
- Estimating the TWA *rms* acceleration along each axis.

- Calculating the resultant vibration in *rms* acceleration units of m/s^2.
- Calculating the allowable daily exposure duration.
- Calculating the daily exposure dose.
- Verifying that vibration dose is < 1 daily and < 5 weekly.

Electromagnetic Radiation

Electromagnetic radiation is energy propagating in the form of oscillating electric and magnetic fields that spans an infinite range of frequencies. One of the most complete graphical summaries of the electromagnetic spectrum is that copyrighted by Unihedron in 2006.[41]

This section is organized and moves progressively through the electromagnetic spectrum from the lowest frequencies to the highest frequencies. At the low frequencies, up to radio frequencies and microwaves, the standards limit electrical current flow in tissue to minimize tissue heating rates below normal homeostatic thermal regulatory-system capabilities. As frequencies increase from infrared up through extreme ultraviolet, photon energies increase from a fraction of an electron volt (eV) to more than 100 eV. The genetic damage threshold due to ionization starts near 3.7 eV and continues beyond 100 eV, sufficient for ionization damage to all tissues.

Photons with vacuum wavelengths shorter than 100 nm have photon energies greater than 12.4 eV and frequencies greater than 3 PHz and are considered to be ionizing. When these are scattered by a molecule, free electrons (ionization event) and lower-energy photons typically result.

Nonionizing Radiation

Nonionizing radiation is freely propagating electromagnetic radiation. Orthogonal electric and magnetic fields propagate at the local speed of light in all media. The photons are nonionizing for frequencies ranging from very low frequency of seconds per cycle through radio frequencies, microwave frequencies, infrared frequencies, and visible light frequencies. The EM waves discussed in this section do not have enough energy to knock electrons free from the parent atoms or molecules; these low-energy photons do not cause ionization.

The primary mechanisms of action for nonionizing EM radiation are the vibration, flexing, and twisting of polyatomic molecules. Such vibrations may directly increase reaction rates by bending reactant molecules enough that they fit one another, or indirectly—by causing a chemical chain reaction resulting in increased temperature. People have experience with nonionizing radiation when they use microwave or stone ovens to heat food or when they warm themselves by a campfire on a cool evening or bask in the warm sun on a beach.

Microwaves and infrared light are used in industry for accelerating two-part glues, drying coatings, curing some polymeric materials, and heat-sealing some thermal plastics. Control of exposures to nonionizing radiation is based on the twin concepts of limiting local temperature rise in susceptible tissue and limiting induced electrical current flows to levels that do not interfere with normal neuromuscular events such as fine motor control, heart rate, blood pressure, and respiratory rate.

Engineering control recommendations have evolved separately and are discussed separately for quasistatic electric and magnetic fields, extremely low-frequency (ELF) electromagnetic fields, radio-frequency (RF) electromagnetic fields, infrared (IR) radiation, visible light, ultraviolet (UV) light, and the coherent monoenergetic light from a LASER (primarily in the infrared to ultraviolet frequencies). Work is underway to extend the frequency range of LASERs, and exposure guidelines are under consideration.

Static Electric and Magnetic Fields

Between 0 and 100 Hz, the electric field strength should be kept smaller than 25 kV/m.

Between 0 Hz and 1 Hz, the magnetic flux density should not exceed a ceiling of 2 T, with an 8-hour TWA of 60 mT for whole-body exposures or 600 mT for extremity exposures.[42]

Sub-Radio-Frequency Electric and Magnetic Fields

The ACGIH TLV book is a convenient source of guidance for these frequencies. Table 14 has been constructed by estimating values for H and D, which are equivalent to the B and E criteria given in the TLV guidelines. Measure the free field without the worker. Figure 9 shows sub-radio frequency for E and H fields.

TABLE 14

TLV Guidelines for Electric and Magnetic Fields

Frequency/Hz	B = Magnetic Flux Density/(mT) 8-hr TWA	H = Magnetic Field Strength/(A/m) 8-hr TWA	D = Electric Field Displacement/(nC/m²) Ceiling	E = Electric Field Strength/(V/m) Ceiling
< 1	60*	47,700	221	25,000
$1 < f < 100$	$60/f$	$47{,}700/f$	221	25,000
$100 < f < 300$	$60/f$	$47{,}700/f$	$221/f$	$2{,}500{,}000/f$
$300 < f < 4{,}000$	0.2	15.9	$221/f$	$2{,}500{,}000/f$
$4000 < f < 30{,}000$	0.2	15.9	5.53	625
	$B = \mu_0 H$		$D = \varepsilon_0 E$	

*60 mT 8-hr TWA is for whole body with 2 T ceiling and 600 mT 8-hr TWA is for extremities.
Keep ceiling exposures below 0.1 mT for pacemaker and medical device patients.
In this table, f is in hertz (Hz). $\mu_0 = 4\pi\ 10^{-7}$ [V s/(A m)], $\varepsilon_0 = 1/(\mu_0 c_0^2) \approx 10^{-9}/(36\pi)$ [A s/(V m) = C/(V m)].

(*Source:* ACGIH 2007)

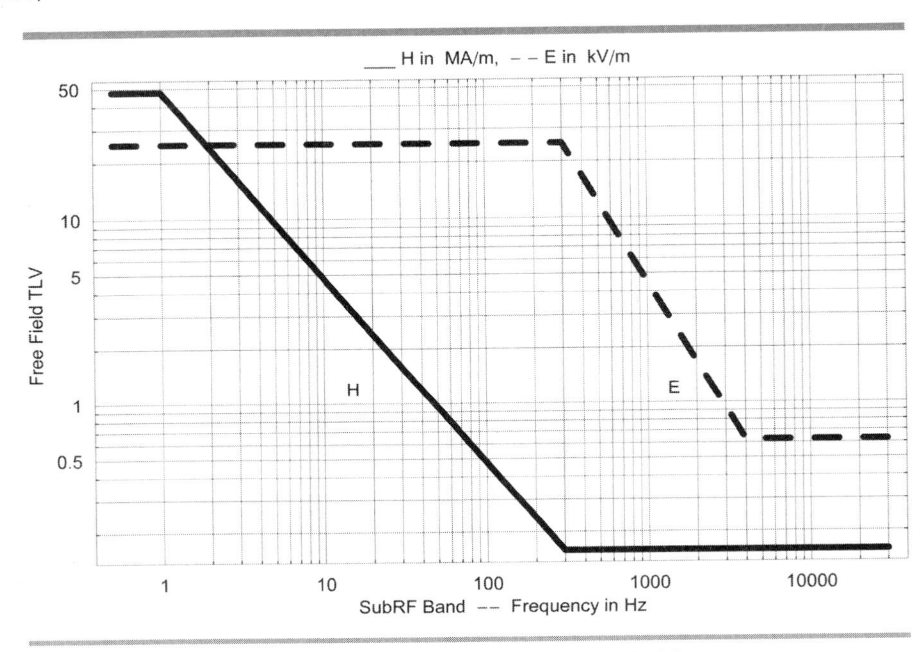

FIGURE 9. TLV for sub-radio-frequency E and H fields
(*Source:* ACGIH 2007)

Radio Frequency and Microwave Electromagnetic Radiation

Radio frequency and microwave radiation limits are set to limit the *rms* body currents induced by the electromagnetic field over the frequencies ranging from 30 kHz to 300 GHz. The TLV guidelines limit the potential for electrostimulation, *shock* (for frequencies < 0.1 MHz) or perceptible heating (for frequencies > 0.1 MHz).

In all cases, the whole-body specific absorption rate (SAR) should be kept below 0.4 W/kg. The exposure limits listed assume the worker has no physical contact with electrically conductive devices (including metallic jewelry) and is insulated from earth ground. When these conditions are not met, consult more detailed references.[43]

TLV guidelines are conservatively tabulated in Table 15. These values are deemed tolerable. Under special conditions identified in the documentation for TLVs, some of these values can be relaxed.

The TLV guidelines for the radio-frequency and microwave portions of the electromagnetic spectra are summarized in Table 15 and Figure 10. Note that for frequencies above 100 MHz it is often easier to measure power density than to measure either the magnetic or the electric field strength. Additionally, the averaging time is 6 minutes for frequencies between 30 KHz and

15 GHz.[44] The TLVs apply to a spatial average of the electromagnetic field strength over the cross-sectional area of the worker. When surveying, measure the field intensity at the worker's location without the presence of the worker.

Ultra Wideband Radiation (UWB)

This is radiation that typically takes the form of short pulses, $T_{pw} << 1\ \mu s$ with rise times < 1 ns. UWB pulses have bandwidth greater than their center frequency and may have measurable energy in frequencies between DC and 30 GHz. UWB signals may interfere with safety-related electronic equipment, including fire alarms, automatic process controls, avionics, and communications and computing equipment. Emerging evidence shows that UWB signals can be used to preferentially destroy cancerous tissue while producing minimal damage in nearby healthy tissue with pulse durations shorter than 1 μs and rise times of a few ns (Schoenbach 2006, 20–26).

TABLE 15

EM Guidelines for RF Through Far-IR

Frequency*	**Wavelength**	**$S = E \times H$** ** *rms*/(mW/cm²)	**E = Electric Field** *rms*/(V/m)	**H = Magnetic Field** *rms*/(A/m)	**Averaging Time† for E^2, H^2, or S**
30–100 kHz	10–3 km	10^5–10^5	614	163	360 s
0.1–3 MHz	3000–100 m	10^5–3333	614	$16.3/f$	360 s
3–30 MHz	100–10 m	3333–3.33	$1842/f$	$16.3/f$	360 s
30–100 MHz	10–3 m	3.33–1	61.4	$16.3/f$	360 s
10–300 MHz	3–1 m	1	61.4	0.163	360 s
0.3–3 GHz	100–10 cm	$f/300$			360 s
3–15 GHz	100–20 mm	10			360 s
15–300 GHz	20–0.1 mm	10			$(36.95 \times 10^6)/(f^{1.2})$
300 GHz	0.1 mm	10			10 s

*In this table, when the frequency appears in a formula, its units are in MHz.

**The Poynting Vector, S, is the cross-product of E and $H = B/\mu$. S gives the *rms* power density and propagation direction of a radiating electromagnetic field. $S/(\text{W/m}^2) = E/(\text{V/m}) \times H(\text{A/m})$. There is additional energy in the nonpropagating reactive near field, close to a source of radiation.

†The impedance of free space is $z = (\mu_0/\varepsilon_0)^{1/2} = 376.7$ ohms. Because $z = E/H$, it follows that the Poynting Vector amplitude can be estimated by $S = E^2/z = zH^2$. In units listed in this table, $S = EH/10$, $S = E^2/3767$, and $S = 37.76\ H^2$. These are plotted in Figure 10.

(*Source:* ACGIH TLV 2006)

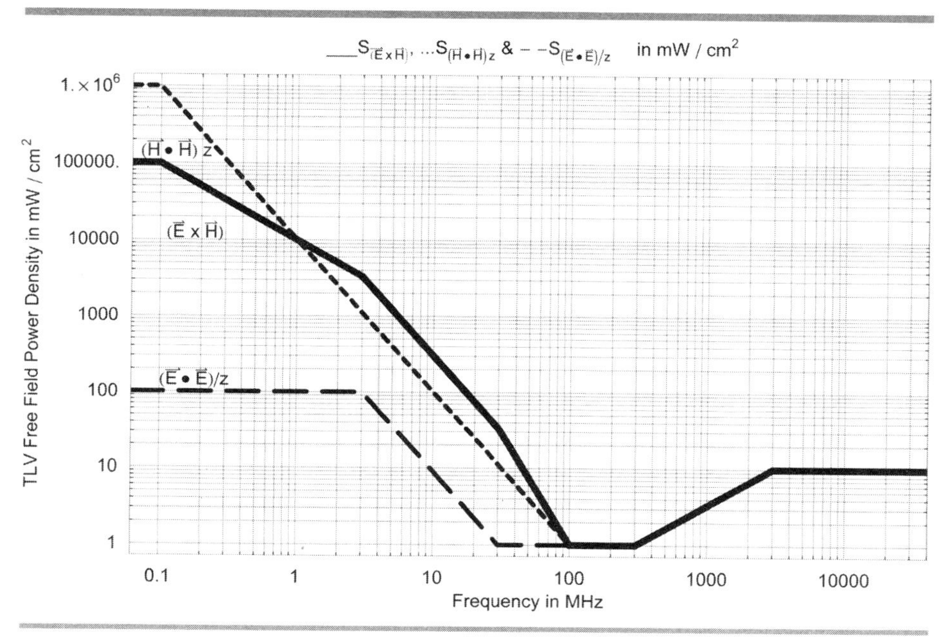

FIGURE 10. TLV for RF and microwave exposures

The TLV (ACGIH 2006) for this exposure category is designed to keep specific absorption below W_{al} = 0.4 (W/kg) for periods longer than 6 minutes and below a specific energy absorption limit of, E_{al} = 144 (J/kg), in any 6-minute period. With incident radiation power density, S (mW/cm^2), conservative SAR coefficient, S_c = 0.25 (W/kg)/(mW/cm^2), and pulse duration (including ring-down time), T_{pw} (s), the specific absorption per pulse is SA_{pp} (J/kg). The specific absorption rate is the product of the specific absorption per pulse and the pulse repetition rate in pulses/s.

Equation: Specific absorption per UWB pulse and specific absorption rate

$$SA_{pp} = S \bullet T_{pw} \bullet S_c \text{ [J/kg]} \tag{38}$$

The permitted pulse repetition rate is a function of the *SA* per pulse, the 6-minute energy absorption limit (E_{al} = 144 J/kg), and the 6-minute thermal averaging time (T_{at} = 360 s).

Equation: Allowed *PRF* for UWB radiation

$$PRF = \frac{E_{al}}{T_{at} \bullet SA_{pp}} \text{ [pulses/s]} \tag{39}$$

If the actual average pulse repetition rate, *prf*, is higher than the allowed *PRF* estimated above, exposure duration must be shortened to less than 6 minutes. The permissible exposure duration, T_{max} < 360 s, is inversely proportional to the square of the specific absorption rate. The numerator is the product of tissue energy and tissue power absorption limits: ($E_{al} \times W_{al}$).

Equation: Allowed duration of UWB radiation exposure at *PRF* [pulse/s]

$$T_{max_{UWB}} = \frac{E_{al} W_{al}}{SA_{pp}\, prf} \text{ [s]} \tag{40}$$

Infrared (IR) and Visible Radiation

These frequencies form a continuum near the portion of the spectrum called the visible light. Infrared light is of lower frequency and longer wavelengths than the red end of the visible light spectrum. Visible light (760–400 nm) passes through the cornea, the lens, and the aqueous humor before it excites rods and cones in the retina through photosensitive chemical reactions that make vision possible. Infrared radiation primarily serves to warm the skin, even while air temperature is cool.[45] Photochemical damage is possible if the intensity of short-wavelength visible (blue to violet) light is not controlled. Such damage is called the blue light hazard.

To avoid retina thermal injury, use the retinal weighting function, $R\lambda$, to adjust measured spectral radiance according to its thermal damage potential. Keep the total effective radiance within the limits in the following equation, where t is in s, and α is the mean angle[46] (radian) subtended by the lamp.

Equation: Retina thermal TLV; for $T_{max} \leq 10$ [s]

$$L_{retina} = \sum_{\lambda=380}^{1400} (L\lambda R\lambda \Delta\lambda) \leq \frac{5}{\alpha (T_{max})^{0.25}} \text{ [W/cm}^2\text{ sr]} \tag{41a}$$

$$T_{max} \leq \left(\frac{5}{\alpha L_{retina}}\right)^4 \text{ [s]}$$

Equation: Retina thermal TLV; for $T_{max} > 10$ [s]

$$L_{retina\,No\,Vis} = \sum_{\lambda=380}^{1400} (L\lambda R\lambda \Delta\lambda) \leq \frac{0.6}{\alpha} \text{ [W/cm}^2\text{ sr]} \tag{41b}$$

$$T_{max} \leq \left(\frac{5}{\alpha L_{retina\,No\,Vis}}\right)^4 \text{ [s]}$$

To avoid photochemical injury to rods and cones in the retina, limit *blue* light exposure. Use the blue weighting function, $B\lambda$, to adjust measured blue spectral radiance according to its photochemical damage potential. This TLV is not protective for LASER light at these frequencies because of the especially potent effects of monochromatic and coherent light. The TLV in the following equation can be relaxed for very small light sources, using procedures in the TLV booklet (ACGIH 2007).

Equation: Blue light TLV (photochemical retina damage), for $T_{max} \leq 10^4$ [s]

$$L_{blue} = \sum_{\lambda=305}^{700} (L\lambda B\lambda \Delta\lambda) \leq \frac{100}{T_{max}} \text{ [W/cm}^2\text{ sr]} \tag{42a}$$

$$T_{max} \leq \frac{100}{L_{blue}} \text{ [s]}$$

Equation: Blue light TLV (photochemical retina damage), for $T_{max} > 10^4$ [s]

$$L_{blue} = \sum_{\lambda=305}^{700} (L\lambda B\lambda \Delta\lambda) \leq 0.01 \text{ [W/cm}^2\text{ sr]} \tag{42b}$$

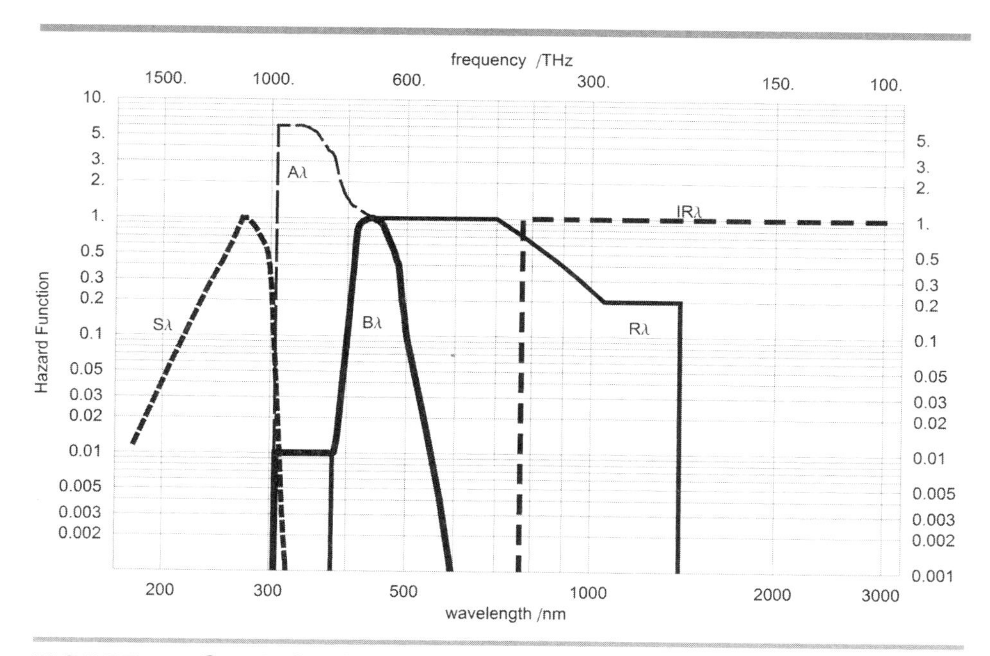

FIGURE 11. Spectral weighting for IR, visible, and UV TLV guidelines

To protect from photochemical damage in an aphakic eye (no lens and no blue light filter), use the aphakic weighting function, $A\lambda$.

Equation: Aphakic TLV (photochemical damage, no lens), for $T_{max} \leq 10^4$ [s]

$$L_{aphakic} = \sum_{\lambda=305}^{700} (L_\lambda A\lambda \Delta\lambda) \leq \frac{100}{T_{max}} \text{ [W/cm}^2\text{ sr]} \quad (43a)$$

$$T_{max} \leq \frac{100}{L_{aphakic}} \text{ [s]}$$

Equation: Aphakic TLV (photochemical damage, no lens), for $T_{max} > 10^4$ [s]

$$L_{aphakic} = \sum_{\lambda=305}^{700} (L_\lambda A\lambda \Delta\lambda) \leq 0.01 \text{ [W/cm}^2\text{ sr]} \quad (43b)$$

To protect from thermal heating damage to the lens and cornea, it is sufficient to use the weighting function called $IR\lambda$ in the following equation. $IR\lambda$ is a function that equals unity between 770 and 3000 nm. The TLV booklet distinguishes between IR sources that include visible light and those that are invisible to the human eye. The following equations are at least as protective as those in the TLV booklet under these circumstances.

Equation: Cornea and lens TLV (thermal heating); for $T_{max} \leq 10^3$ [s]

$$E_{IR} = \sum_{\lambda=770}^{3000} (E_\lambda T_\lambda \Delta\lambda) \leq \frac{1.8}{(T_{max})^{0.75}} \text{ [W/cm}^2\text{]} \quad (44a)$$

$$T_{max} \leq \left(\frac{1.8}{E_{IR}}\right)^{4/3} \text{ [s]}$$

Equation: Cornea and lens TLV (thermal heating); for $T_{max} > 10^3$ [s]

$$E_{IR} = \sum_{\lambda=770}^{3000} (E_\lambda T_\lambda \Delta\lambda) \leq 0.01 \text{ [W/cm}^2\text{]} \quad (44b)$$

Figure 11 shows five defined spectral weighting functions: $A\lambda$, $B\lambda$, $R\lambda$, $IR\lambda$, and $S\lambda$. The aphakic weighting (absent a lens to absorb UV and blue light), $A\lambda$, is defined over the wavelength range of 305 nm to 700 nm and shows relative potencies of photon energies for photochemical damage in the retina. The blue weighting function, $B\lambda$, is defined over the same range and shows the photochemical potency of photon energies to damage the retina, taking into account the filtration of the lens. The retinal weighting function, $R\lambda$, defined over the wavelengths between 380 nm and 1400 nm, shows relative potency of these wavelengths for overheating the retina, also accounting for cornea, lens, and optic media filtration. The actinic UV weighting function, $S\lambda$, defined over the wavelengths between 180 nm and 400 nm, shows relative potency of these wavelengths for skin and cornea damage.

Ultraviolet Radiation

OSHA does not have a PEL for UV radiation, instead referring employers to the ACGIH TLV (OSHA, 2003). The UV TLV represents conditions to which it is believed nearly all healthy workers may be repeatedly exposed without acute adverse health effects, such as erythema of the skin and photokeratitis of the cornea and conjunctiva. The TLVs apply to UV radiation from arcs, gas and vapor discharges, fluorescent sources, incandescent sources, and solar radiation, but not from UV lasers. The UV spectrum lies above the high-frequency (violet) end of the visible spectrum and has photon energies sufficient to ionize electrons held in relatively weak chemical bonds.

The TLVs may not be protective for aphakes (missing lens) or for individuals who are photosensitized by chemicals, drugs, genetic makeup, or diseases. Many chemicals and some liver diseases cause skin or corneal photosensitization reactions. Examples include some antibiotics (such as tetracycline and suphathiazole), some antidepressants (such as imipramine and sinequan), extracts from some agricultural plants (such as celery, lime oil, and blue-green algae) and many other chemicals used in cosmetics, dyes, and antipsychotic drugs.

The TLV for radiant exposure to UV-A is limited to energy of 1 J/cm^2 for periods < 1000 s, and the irradiance is limited to 1 mW/cm^2 for periods longer than 1000 s. Because the lens absorbs UV radiation, aphakes (no lens) are at risk of UV-induced retinal injury at lower levels than these.

Because ozone is produced in air by photons with $\lambda < 250$ nm (extreme UV-B plus UV-C), good hygiene practice requires that the ozone level be controlled below its TLV, which ranges from 0.05 ppm for heavy work to 0.1 ppm for light work.

The UV-weighting function shows that humans are most susceptible at wavelengths of 270 nm and have better tolerance for longer or shorter wavelengths. Because sodium vapor lamps have strong emission lines near 270 nm, they are normally encased in a UV filtering envelope. When that envelope is broken, exposed workers are at risk.

A UV survey is easy with a UV radiometer having a spectral response that implements the relative spectral effectiveness values provided in the TLV book and reports effective irradiance, E_{UV} in mW/cm^2. To protect eyes and skin from adverse consequences of UV light, use the actinic spectral effectiveness function, $S(\lambda)$, to adjust measured UV spectral irradiance according to its skin- and eye-damage potential.

Equation: UV TLV (skin and eyes genetic damage)

$$E_{UV} = \sum_{\lambda=180}^{400} (E_\lambda S_\lambda \Delta\lambda) \leq \frac{0.003\ [\text{J/cm}^2]}{T_{max}}\ [\text{W/cm}^2] \quad (45)$$

$$T_{max} < \left(\frac{0.003\ [\text{J/cm}^2]}{E_{UV}\ [\text{W/cm}^2]}\right)[\text{s}]$$

Note that the TLV for UV radiation can be exceeded by a few minutes in the noonday sun in the tropics or near the time of the summer solstice in the mountains of subtropical and temperate zones. For example, in a location where the effective irradiance is $E_{eff} = 0.001$ W/cm^2, the allowable duration is $T_{max} = 3$ s. Note also that unfiltered ophthalmologic slit lamps and some microscope illumination can exceed both the blue light and the UV TLV in a few minutes (Sliney 1997, 981–996).

LASER Radiation

OSHA notes the distinguishing features of LASER light sources:

1. a nearly single-frequency output (i.e., an almost pure monochromatic light beam)
2. a beam with a Gaussian intensity profile
3. a beam of small divergence
4. a beam of enormous intensity
5. a beam which maintains a high degree of temporal and spatial coherence
6. a beam that is, in many devices, highly plane polarized[47]
7. a beam with enormous electromagnetic field strengths (Liu 2001)[48]

It takes an entire book to properly describe all LASER safety considerations (ACGIH, 1990).[49] In the interest of space, the principles are described here and the reader is encouraged to use more extensive references as needed (see Table 16). OSHA citations for LASER violations are issued by invoking the general duty clause or, in some cases, Subpart I. In such cases, the employers are required to revise their reportedly unsafe workplace using the recommendations and requirements of such industry consensus standards as the ANSI Z136.1 standard and the ACGIH TLV and LASER monograph.

To implement a LASER safety program, a qualified LASER safety officer (LSO) will be appointed. The LSO will implement appropriate controls, not less than those required by ANSI, ACGIH, and the OSHA construction standard. In addition to labels, Class IIIB and Class IV lasers may require other precautions. A LASER-controlled area is required during periods of operation of any LASER installation that inadequately encloses the beam. The controlled area will encompass the entire no-hazard zone (NHZ) determined by the LSO using guidance from the OSHA technical manual and ANSI standards. Personnel inside the NHZ will wear LASER safety goggles with optical density OD = $\log_{10}(H_0/\text{MPE})$, where H_0 is the anticipated worst-case exposure in J/cm^2 for a pulsed LASER or W/cm^2 for a continuous LASER, as appropriate. MPE is the maximum level of laser radiation to which a person may be exposed without hazardous effects or biological changes in the eye or skin. The MPE is determined by the wavelength of the laser, the energy involved, and the duration of the exposure. The ANSI 136.1 standard tables 5, 6, and 7 summarize the MPE for particular wavelengths and exposure durations (Princeton 2007).

The following facts are useful for training laser personnel about symptoms of possible overexposure (Bader 2006):

- Exposure to the invisible carbon dioxide laser beam (10,600 nm) can be detected by a burning pain at the site of exposure on the cornea or sclera.
- Exposure to a visible laser beam can be detected by a bright color flash of the emitted wavelength and an after-image of its complementary color (e.g., a green 532-nm laser light would produce a green flash followed by a red after-image).
- When the retina is affected, there may be difficulty in detecting blue or green colors secondary to cone damage, and pigmentation of the retina may be detected.

TABLE 16

Hazards by LASER Classification

Class	UV	Vis	NIR	IR	Hazards: Direct Ocular	Hazards: Diffuse Ocular	Hazards: Fire
I	x	x	x	x	No	No	No
IA		x			Yes, < 1000 s	No	No
II		x			Yes, < 0.25 s	No	No
IIIA	x	x	x	x	Yes	No	No
IIIB	x	x	x	x	Yes	Yes, near 500 mW	No
IV	x	x	x	x	Yes	Yes	Yes

x indicates that a LASER in this wavelength range can be classified in this class by ANSI guidelines:

I: No emissions at known hazard levels (~0.4 μW continuous wave visible). Can mean enclosed LASER of higher power if enclosure is interlocked to prevent operation while open.

IA: LASERs "not intended for viewing" that are safe for direct exposures up to 1000 s. Power limit is 4 mW. Supermarket scanners are an example.

II: Low-power visible LASER < 1 mW. Human aversion to bright light will limit exposure duration. Minimal controls. Use ANSI "Caution Label."

IIIA: Intermediate power LASER (1–5 mW). Only hazardous for direct intrabeam viewing. Use "Caution Label" if < 2.5 mW, "Danger Label" if 2.5–5 mW.

IIIB: Moderate power lasers (5–500 mW continuous or the lesser of 10 J/cm^2 pulsed or the diffuse reflection limit). Not expected to ignite ordinary flammable substances. Controls are strongly recommended. Use ANSI "Danger Label."

IV: High-power LASER > 500 mW continuous or >10 J/cm^2 per pulse. Viewing can be hazardous in direct beam, or by specular or diffuse reflection/scattering. Type IV LASERs generally pose both eye and skin hazards. Controls and ANSI "Danger Labels" are required.

(*Source:* ACGIH TLV 2006)

TABLE 17

Regulatory Limits for Ionizing Radiation

	NRC		OSHA	
Whole body	50 mSv/yr	5 rem/yr	12.5 mSv/qtr	1.25 rem/qtr
Lens of eye	150 mSv/yr	15 rem/yr	12.5 mSv/qtr	1.25 rem/qtr r
Skin	500 mSv/yr	50 rem/yr	75 mSv/qtr	7.5 rem/qtr
Extremities	500 mSv/yr	50 rem/yr	187 mSv/qtr	18.75 rem/qtr
Embryo/fetus	5 mSv/yr	0.5 rem/ gestation	--	--
Minor	10% of adult		10% of adult	
Cumulative occupational from birth to age n yrs			< 50 (n−18) mSv	< 5(n−18) rem[1]
General public	1 mSv/yr	0.1 rem/yr	--	--
General public	0.01 mSv/hr	0.002 rem/hr	--	--

[1]Found in 29 CFR 1910.1096(b)(2)(ii)

- Exposure to the Q-switched Nd:YAG laser beam (1064 nm) is especially hazardous and may initially go undetected because the beam is invisible, and because the retina lacks pain sensory nerves. Photoacoustic retinal damage may be associated with an audible *pop* at the time of exposure. Visual disorientation caused by retinal damage may not be apparent to the operator until considerable thermal damage has occurred.

Ionizing Radiation

Many recommend that ionizing radiation exposures be kept as low as reasonably achievable (ALARA). This cannot mean zero exposure; practically speaking, humans are continuously exposed to ionizing radiation that includes solar radiation, cosmic radiation, external radiation from environmental radio nuclides, and internal radiation from radio nuclides in the air, water, and food that are incorporated into body tissues.

Most radiation sources in the United States are regulated by the Department of Energy and the Nuclear Regulatory Commission; OSHA and the EPA play only a minor role.[50] Key regulatory limits are summarized in Table 17.

To determine acceptable health protection, use radiation-weighted measures of dose, the sievert and the rem. These equivalent doses are derived from physical dose measurements reported in grays and rads by use of a quality factor specified by the Nuclear Regulatory Commission. The quality factors in Table 18 depend both on the type of radiation and on its energy.

X-Ray and Gamma Radiation

X-rays and gamma rays are best shielded with electron-rich materials such as the high-Z elements. Lead is commonly used. Because this type of radiation does not activate any elements, people exposed do not become radioactive. The quality factor is unity, so the effective exposure in sieverts is numerically equal to the measured dose in grays.

Alpha Particulate Radiation

Alpha radiation, which does not penetrate the outer layers of dead skin or even a thick sheet of fine paper,

TABLE 18

Quality Factors for Ionizing Radiation

Energy	Quality Factor	Neutron Fluence per Sievert	Neutron Fluence per Rem
Unknown γ	1	--	--
Unknown β	1	--	--
< 0.01 MeV n	2	(81–98) × 10^9/cm^2/s	810–980 million/cm^2/s
0.01 MeV n	2.5	101 × 10^9/cm^2/s	1010 million/cm^2/s
0.1 MeV n	7.5	17 × 10^9/cm^2/s	170 million/cm^2/s
0.5–1 MeV n	11	(3.9–2.7) × 10^9/cm^2/s	39–27 million/cm^2/s
Unknown n	10	(1.4–2.7) × 10^9/cm^2/s	14–27 million/cm^2/s
Unknown p	10	--	--
Unknown α	20	--	--

poses no external radiation hazard. If ingested and metabolized, however, alpha-emitting radio nuclides do pose a health threat. The best protection is to prevent inhalation and ingestion by engineering design. The quality factor for alpha particles emitted inside the body is 20. The effective dose in sieverts, to be compared with exposure guidelines, is 20 times the measured dose in grays.

Beta Radiation

Electrons and positrons penetrate several millimeters of tissue and deposit their energy in potentially damaging ways. Low-energy beta radiation is adequately shielded by electrical conductors—such as a few millimeters of aluminum—but higher-energy particles need more complete design, often employing a low-Z substance thick enough to stop all beta rays followed by a high-Z substance thick enough to absorb all bremsstrahlung photons generated in the first layer (Cember 1996). Beta particles have a quality factor of unity, so the effective exposure in sieverts is numerically equal to the measured dose in grays.

Neutron and Proton Radiation

Neutrons readily penetrate long distances in most materials and are best shielded by thick barriers of hydrogen-rich materials. Neutron shields are commonly made from pools of water, polyethylene, and concrete. The quality factor for neutrons depends on their energy, but a default value of 10 is authorized by the NRC when the energy is unknown.[51] Design of shielding systems for protons is a task proper to a fully qualified health physicist who specializes in such designs.

ENDNOTES

[1]Modern physics textbooks and two useful Web sites summarize concepts central to this chapter: www.site.uottawa.ca:4321/astronomy/index.html and www.hyperphysics.phy-astr.gsu.edu/hbase/hframe.html.

[2]Paracelsus' 1567 book was published in Switzerland 27 years after his death.

[3]Henry F. Smyth, Jr. introduced sufficient dose to toxicology in *Food and Cosmetics Toxicology* 5:51 (1967). It applied dose-response and is consistent with hormesis, the beneficial effects of low doses of hazardous stressors. The modern basis for radiation hormesis is found in a pivotal paper by T. D. Luckey, "Physiological benefits from low levels of ionizing radiation," *Health Physics* 43:771–789 (1982). For further details, see www.angelfire.com/mo/radioadaptive/inthorm.html and www.belleonline.com.

[4]See Review Questions 4–6 for a NIOSH summary of attenuation factors for A- and C-weighting filters.

[5]At specialized microwave frequencies, energy is deposited in the skin at the depth of pain endings of neurons and equipment exists for crowd control using this principle. At the higher frequencies of IR B and IR C some of the incident intensity passes completely through the human body—medical imaging devices are being produced that take advantage of this.

[6]As discussed in NIST SP 250-37 (July 1997), the sensitivity curve for human vision is given by proportional response coefficients called V_λ. The CIE visual response curve shows that $0.0001 < V_\lambda < 0.0001$ for wavelengths between 380 to 780 nm and essentially zero for all wavelengths outside this range.

[7]This easily demonstrated shift in wavelength of best sensitivity is called the Purkinje shift. Consider a yellow sunflower in a field of green grass. At high noon, light-adjusted eyes see the yellow appear brighter. As the sun sets, the eye's sensitivity to yellow falls and its sensitivity to green rises. At dusk, green appears brighter than the yellow to dark-adapted eyes. When color perception is important to safety, it is important to control illumination.

[8]In SI units, the luminance of a perfect diffuser is $(1/\pi)$ (cd/m^2) when it is illuminated with 1 lx = 1 lm/m^2.

[9]The equations for these terms are found in the glossary.

[10]For the hydrogen atom, these energies are characterized by the visible light Balmer Series, the Lyman Series in UV, the Paschen Series in near-IR, and the Brackett, Pfund, and Humphreys Series at longer wavelengths. Every atom and molecule has characteristic energy gaps that are revealed by characteristic spectra.

[11]A very nice animation of these geometries by Dr. Dan Russell of Kettering University, Applied Physics Department, was available, as of August 2006, at www.kettering.edu/~drussell/Demos/rad2/mdq.html.

[12]Well described in textbooks describing the physics of acoustics and of electromagnetic radiation. See also Straus, I. "Near and Far Fields—From Statics to Radiation." *Conformity* 2001. Curtis–Straus, LLC. Online at www.conformity.com/0102reflections.html as of Aug 2006.

[13]The energy of a photon is proportional to its frequency. Conservation of energy means constant frequency when a photon moves between materials. The wavelength and speed of propagation change as photon radiation passes from one material to another.

[14]For example, a welder's lens that transmits 0.0001 of the incident intensity has an optical density of 4. A welder's

lens with an optical density of 7 transmits 1.0×10^{-7} of the incident intensity.

[15]Impedance is usually taken to be 377 ohms for EM measurements and 400 rayls for acoustic measurements.

[16]From the 1940s through the 1960s, many thought that hearing damage was proportional to sound pressure with its 6-dB doubling rate (force per unit area). Physicists thought it was proportional to sound intensity (work per unit area) with its 3-dB doubling rate. In the early 1970s OSHA chose a 5-dB doubling rate as a conservative compromise between these two camps. It is now generally agreed that sound intensity is the appropriate metric, and NIOSH, ACGIH, and ISO all recommend the 3-dB doubling rate with an 85 dBA 8-hour TWA goal.

[17]Pressure reported by NOAA and the U.S. Weather Bureau is published after being corrected to sea level and is not useful for calculations during a sound survey taken at altitudes not at sea level. Atmospheric pressure at other altitudes is estimated from the NOAA standard atmosphere tables published as *U.S. Standard Atmosphere*, 1976, U.S. Government Printing Office, Washington, D.C., 1976. These have been republished in the *CRC Handbook of Chemistry and Physics*, which may be more generally accessible. An FAA rule of thumb says that pressure decreases by ~1 inch Hg for each 1000 feet above mean sea level up to about 12,000 ft MSL.

[18]It is good practice to carefully study the manual for your sound-level meter to discover if it is internally corrected for nonstandard T and P.

[19]A photon has linear momentum because of its speed plus angular momentum, called spin by modern physics.

[20]When an EM wave propagates through material, its wavelength is generally shorter than in a vacuum, a phenomenon expressed by means of the refractive index of the material.

[21]See www.distinti.com for evidence that an expanded set of equations more completely describes electromagnetic phenomena than Maxwell's Equations. For purposes of this chapter, however, the classical equations are sufficient. Also see www.infohost.nmt.edu/~pharis/ for evidence of a deeper connection between mass, gravitational forces, and electromagnetic forces than is yet incorporated into the standard model of physics. (Web sites verified August 2006.)

[22]Many textbooks discuss the use of these equations to solve practical problems. When charges are in the region of propagation, or if polarizable or magnetic materials are in region of propagation, a more complex set is required. For more detail, see, for example, www.hyperphysics.phy-astr-gsu.edu/hbase/electric/maxeq.html.

[23]See www.en.wikipedia.org/wiki/Solar_sail for more discussion of this device.

[24]The impulse momentum theorem calculates force as the ratio of momentum change to the time needed for the change. In its simple form, $F = \Delta p / \Delta t$.

[25]See a remarkably detailed EM spectrum which shows use, wavelength, frequency, photon energy, and other useful characteristics of EM radiation at www.unihedron.com/projects/spectrum/index.php.

[26]These guidelines are violated by resonant structures. The 9-cm ferrite antenna in an AM radio is too small to absorb meaningful energy from a 1 MHz AM radio wave with λ = 300 m unless it is tuned with a capacitor to resonate at 1 MHz. A hydrogen atom with a diameter of 0.5 nm absorbs and emits EM radiation at the H_{alpha}, a red line with wavelength = 656.3 nm. This transition is between quantum state $n = 2$ and $n = 3$ in the hydrogen atom. These states are reasonably viewed as resonant nonradiating states for the orbital electron.

[27]The ACGIH uses 12.4 eV as the boundary between ionizing and nonionizing radiation.

[28]See http://www.uic.com.au/ral.htm for a summary of radiation effects and sources in everyday life.

[29]See OSHA's online *Technical Manual*, Chapter 5, "Noise and Hearing Conservation." For additional information: www.osha.gov/dts/osta/otm/noise/index.html.

[30]Impulse noise, like a hammer hitting an anvil, is so short that it has ended before these reflexes can protect the cochlea. This is why impulse noise is more hazardous than steady sound intensity levels between 85 to 115 dBA.

[31]OSHA's quick card for heat stress is available online at www.osha.gov/Publications/osha3154.pdf. Detailed discussions for safety professionals are in the OSHA *Technical Manual*, Section III, Chapter 4, available at: www.osha.gov/dts/osta/otm/otm_iii/otm_iii_4.html.

[32]Assumes a light, loose-woven cotton summer work uniform allowing free air circulation to evaporate perspiration and maintain the evaporative cooling needed by the human thermoregulatory system. Add 3.5°C (6.3°F) to measured WBGT if close-woven coveralls are worn, and add 5°C (9°F) if two layers of woven or knit coveralls are worn. Compare the corrected WBGT with the guidelines in the table.

[33]See the current version of the booklet "TLVs and BEIs" from ACGIH for guidance on unacclimatized workers. OSHA recommends that a new worker be given five days to reach acclimatization with 20% exposure on day 1, increasing by 20% each day to 100% of the work schedule on day 5.

[34]OSHA PEL for Noise & Vibration: 29 CFR 1910.95 and 1926.10.

[35]The ACGIH TLV is consistent with the ISO standard and most other national standards while being more protective than the OSHA standard. Recommend use of TLV.

[36]90 dBA for 8 hours is the maximum OSHA dose. However, an OSHA noise dose of 50%, representing an OSHA equivalent 85 dBA for 8 hours, is the dose above which OSHA requires a hearing conservation program.

[37]The TLV committee recommends avoiding direct contact with transducers producing intensities greater than 15 dB above 1 g rms. This means keep direct contact below 5.6 g (55 m/s^2).

[38]These attenuation plots are based on data in the *ACGIH TLV Booklet for 2005*, Physical Agent section, under Hand Arm Vibration and Whole Body Vibration.

[39]The next three equations are discussed in the ACGIH TLV booklet, some textbooks, and ANSI standards.

[40]These equations for allowed duration of vibration as a function of *rms* acceleration were fitted by linear regression to tabulated values in the 2005 TLV book. The correlation coefficient for the fit was $R^2 = 0.995$ with p-value < 0.000000005 for the whole-body fit and was $R^2 = 0.997$ with p-value < 0.00004 for the hand–arm fit.

[41]This remarkable summary of the electromagnetic radiation spectrum remained available as of August 5, 2006 at www.unihedron.com/projects/spectrum/downloads/spectrum_20041024.pdf.

[42]$B = (\mu_0 \times H)$, with B in T, H in (A/m) and $\mu_0 = 4\pi\ 10^{-8}$ (V s)/(A m)

[43]ACGIH Worldwide 2003 and IEEE Std C 95.1-1999 or newer.

[44]Localized heating is the problem of concern, and blood flow through body tissues can control localized heating in an averaging time of 6 minutes.

[45]This is the principle of radiant heating, by which wall, floor, and ceiling panels are kept at a higher temperature than the air, conserving energy by minimizing the air-temperature change required in the tempered spaces.

[46]The solid angle of a lamp with measurements $l = 1$ m, $w = 0.03$ m at a distance of 2 m is the ratio of its average transverse dimension, $(l + w)/2 = 0.515$ m, to its distance from the observer, $\alpha = 0.515/2 = 0.257$ radian.

[47]A perfectly polarized beam would have its electric field vector continuously oscillating in one direction. A nonpolarized beam has an E field oscillating in random directions or in multiple simultaneous directions.

[48]For example, beam power in a 300 Hz, 21 fs pulsed laser focused to a 1.6 μm spot is reported to exceed 4×10^{18} W/m^2, corresponding to an electric field strength of nearly 4×10^{10} V/m. This has produced 100 keV X-rays from a 7 μm spot in solid targets.

[49]ANSI has six monographs dealing with LASER safety, numbered Z136.1 through Z136.6. All ANSI LASER standards are available from www.z136.org/sections.php?op = viewarticle&artid = 4. The overarching standard is: Z136.1-2000, "Safe Use of Lasers."

The OSHA *Technical Manual*, Chapter 6 is a very well-written summary of LASER safety considerations and is available at www.osha.gov/dts/osta/otm/otm_iii/otm_iii_6.html as of August 2006.

[50]Department of Energy (DOE): 10 CFR 835, "Occupational Radiation Protection."

Nuclear Regulatory Commission (NRC): 10 CFR 20, "Standards for Protection Against Radiation."

OSHA: 1910.1096, "Ionizing radiation (general industry)"; 1926.53, "Ionizing radiation (construction)"; 1915.57, "Uses of fissionable material in ship repairing and shipbuilding and ionizing radiation requirements for ship repairing and shipbuilding activities"; 1910.120 and 1926.65, "Hazardous waste operations and emergency response."

The two HAZWOPER standards regulate ionizing radiation at hazardous waste sites.

[51]There is more detail provided in NRC Table 1004(b).2 from 10 CFR 20.1004 and OSHA Table G-17—Neutron Flux Dose Equivalents from 29 CFR 1910.1096(a)(7)(v). See also the National Institutes of Health radiation safety Web site at www.nih.gov/od/ors/ds/rsb/exposure.html.

References

American Conference of Governmental Industrial Hygienists (ACGIH) Worldwide. 1990. *A Guide for Control of Laser Hazards*. www.acgih.org

______. 2003. *Documentation for the TLVs and BEIs*. www.acgih.org

______. 2006. *2006 TLVs and BEIs*. www.acgih.org

Armstrong T. J., T. E. Bernard, and M. S. Lopez. 2007. "Hand–Arm Vibration Exposure Guidelines in the United States with Special Reference to the ACIGH TLV." Proceedings of the 11th International Conference on Hand Arm Vibration, Bologna, Italy, pp. 674–680.

Bader, O., and H. Lui. 2006. *Laser Safety and the Eye: Hidden Hazards and Practical Pearls*. www.derm.ubc.ca/laser/eyesafety.html#What

Berger E. 2000. *The Naked Truth about NRRs*. Earlog Technical Monograph 20. www.e-a-r.com/pdf/hearingcons/earlog20.pdf

Cember, H. 1996. *Introduction to Health Physics*. New York: McGraw-Hill.

Christensen, C. J., A. Nielsen, A. Bahnsen, W. K. Brown, and B. M. Rustad. 1972. "Free-Neutron Beta-Decay Half-Life." *Phys*. 7:1628–1640.

Environmental Protection Agency (EPA). 2007. *Calculate Your Radiation Dose*. www.epa.gov/radiation/students/calculate.html

Fulton, J. T. 2006. *Processes in Biological Vision Including Electrochemistry of the Neuron*. www.4colorvision.com/pdf/17Performance1a.pdf

Graham, C. H., ed. 1965. *L. A. Visual Acuity*. New York: John Wiley & Sons. www.webvision.med.utah.edu/imageswv/KallSpat18.jpg

Griffen, M. G., et al. 2006. "Guide to Good Practice on Hand–Arm Vibration (V7.7 English)—Implementation of EU Directive 2002/44/EC" (December 6) (retrieved November 18, 2007). www.humanvibration.com/EU/VIBGUIDE/HAV%20Good%20practice%20Guide%20V7.7%20English%20260506.pdf

Liu, J., H. Wang, et al. 2001. "An Ultra High Intensity Laser at High Repetition Rate." www.sunysb.edu/icfa2001/Papers/th2-2.pdf#search = 'laser%20beam%20intensity

Merriam-Webster's Online Dictionary. www.m-w.com/cgi-bin/dictionary

National Weather Service (NWS). Undated. *NWS Windchill Chart* (retrieved July 6, 2005). www.nws.noaa.gov/om/windchill/

Nussbaum, R., and W. Kohnlein. Undated. *Radiation Hormesis & Zero-Risk Threshold Dose: Two Scientifically Refuted, but Stubborn Myths* (retrieved April 2007). www.gfstrahlenschutz.de/docs/hormeng2.pdf

National Institutes for Occupational Safety and Health (NIOSH). 1998. NIOSH Publication No. 98-126, *Criteria for a Recommended Standard: Occupational Noise Exposure*. www.cdc.gov/niosh/docs/98-126/chap4.html

Occupational Safety and Health Administration (OSHA). 1990. *OSHA RF Radiation Electromagnetic Field Memo*. www.osha.gov/SLTC/radiofrequencyradiation/electromagnetic_fieldmemo/electromagnetic.html

______. 2003. *OSHA Standard Interpretation: Workplace Exposure Limits for UV Radiation*. www.osha.gov/pls/oshaweb/owadisp.show_document?p_table = INTERPRETATIONS&p_id = 24755

______. STD 01-05-001, PUB 8-1.7, *Guidelines for Laser Safety and Hazard Assessment*. www.osha.gov/pls/oshaweb/owadisp.show_document?p_table = DIRECTIVES&p_id = 1705

______. Undated. *Noise Control. A Guide for Workers and Employers*. www.nonoise.org/hearing/noisecon/noisecon.htm

Ohno, Y. 1997. "Photometric Calibrations." United States Department of Commerce, National Institute of Standards and Technology.

Ottoboni, M. A. 1991. *The Dose Makes the Poison*. New York: Van Nostrand Reinhold.

International Sensor Technology. *Photoionization Detectors*. Undated. (retrieved August 2006). www.intlsensor.com/pdf/photoionization.pdf

Princeton University. *2007 Laser Safety Guide*. Section 4: "Laser Control Measures" (modified 10/3/2007). www.princeton.edu/sites/ehs/laserguidesec4.htm

Schoenbach, K. H., R. Nuccitelli, and S. J. Beebe. 2006. "Extreme Voltage Could Be a Surprisingly Delicate Tool in the Fight Against Cancer." *IEEE Spectrum*. www.spectrum.ieee.org/aug06/4257

Schüz, J., R. Jacobsen, J. H. Olsen, J. D. Boice, Jr., Joseph K. McLaughlin, and C. Johansen. 2006. "Cellular Telephone Use and Cancer Risk: Update of a Nationwide Danish Cohort." *Journal of the National Cancer Institute* (Dec.) 98(23):1707–1713.

Sliney, David H. 1997. "Optical Radiation Safety of Medical Light Sources." *Phys. Med. Biol.* 42:981–996.

Suzuki, T., T. Hirayama, K. Aihara, and Y. Hirohata. "Experimental Studies of Moderate Temperature Burns." *Burns* (Dec.) 17(6):443–451.

Taylor, Barry N. 1995. NIST Special Publication 811, *Guide for the Use of the International System of Units (SI)*. Washington, D.C.: Department of Commerce, National Institute of Standards and Technology. www.physics.nist.gov/cuu/Units/bibliography.html

Tominaga, Y. 2005. "New Frequency Weighting of Hand–Arm Vibration." *Industrial Health* 43:509–515.

The World of Nuclear Science. "Measuring Radiation." Undated. (retrieved April 2007). www.library.thinkquest.org/C004606/applications/measuringrad.shtml

Appendix: Glossary of Mathematical Symbols and Physical Constants

SI units are in square brackets [ ].

A Ampere, SI base unit of electric current

$\vec{A}$ Area element with direction orthogonal to surface area

A_{xxx} Acceleration [m/s²]

wxn—weighted along x-axis during nth time interval

wx—time weighted average acceleration along x-axis

wt—total resultant TWA acceleration, a composite of three-axis measurement data

$A\lambda$ Wavelength weighting function to prevent photochemical retina damage in lensless eye

Amu Atomic mass unit = Dalton—1/12 mass of ^{12}C: 1 amu = $1.6605402 \times 10^{-27}$ kg

$\vec{B}$ Magnetic field = magnetic flux density = $\mu\vec{H}$ [Tesla, T = V s/m²]

B_{med} Bulk modulus of a compressible medium [Pascal, Pa = N/m² = kg/ms²]

$B\lambda$ Blue light wavelength weighting function to prevent photochemical retina damage in normal eye

C Symbol for capacitance, measured in Farads [F = C/V = A s/V]

C Coulomb, quantity of charge or quantity of electricity [C = A s]

cd Candela, base unit for luminous intensity [cd]

c_{med} Speed of a wave in a compressible medium [m/s]

c_{air} Speed of sound in air [m/s]

c_m Speed of light in medium m, $c_m = \dfrac{1}{\sqrt{\mu_m \varepsilon_m}}$ [m/s]

c_0 Vacuum speed of light or electromagnetic wave; $c_0 = 299792458 \approx 3 \times 10^8$ [m/s]

cm Centimeter [m/100]

dB Decibel, base 10 logarithm of a ratio of two energy quantities [dB], or a ratio of one energy, Q, to a reference energy, Q_0 [dB re Q_0].

dBA A-weighted sound intensity level, used for hearing conservation [dBA re 1 pW/m²]

$\vec{D}$ Electric displacement field [A s/m² = C/m² = F V/m²]

D_{xxx} Noise dose where xxx = PEL for OSHA and xxx = TLV for ACGIH [unitless]

$\vec{E}$ Electric flux density [J/m C] = electric field = [V/m]

e Unit charge on electron and proton, 1.60218×10^{-19} [C]

EM Abbreviation for electromagnetic

E_{al} Allowed limit of specific energy deposited in tissue in a specified time, E_{al} = 144 [J/kg]

E_{IR} Effective irradiance incident to skin or eye in 770 nm $< \lambda <$ 1400 nm, [W/cm²]

E_{UV} Effective irradiance incident to skin or eye weighted by $S\lambda$, [W/cm²]

$E\lambda$ (λ) Spectral irradiance at wavelength = λ, incident to skin or cornea, [W/cm² sr nm]

eV Electron volt, energy of an electron accelerated by 1 V (1.60218×10^{-19}) [J]

f Frequency in cycles per second [Hz]

F Farad, quantity of capacitance [F = C/V = A s/V]

F Force [N]

G Gauss, cgs measure of magnetic flux density [T = 10^4 G]

h Planck constant = 6.6207×10^{-34} [J s = kg m²/s]

$hbar$ Dirac constant = Planck constant/2π; $h/2\pi$ = 1.0547×10^{-34} [J s]

H Henry, measure of inductance [H = s V/A]

$\vec{H}$ Magnetic field strength, $\vec{H} = \beta/\mu_0$ [A/m]

I Electric current [A = W/V = C/s]

I_{xxx} Intensity for radiation source geometries defined by xxx [W/m²]

xxx ps means point source

xxx ls means line source

xxx as means area source

xxx dps means dipole source

$\vec{I}_{air}$ Sound intensity [W/m²]

$\vec{I}_{EM}$ Electromagnetic intensity [W/m²]

$\vec{I}_{ff}$ Electromagnetic intensity in radiating far field = $|\vec{S}|$ [W/m²]

I_0 1 pW/m², reference sound intensity

J Joule, unit of energy J = Nm = kg m²/s²

$\vec{J}$ Electric current density vector [A/m²]

L Symbol for inductance, measured in Henrys, H = s V/A

L_I Sound intensity level, 10 $\log_{10}$ (I/I_0) [dB re I_0 = 1 pW/m²]

$$L_I = L_P + 10 \log_{10}\left(\frac{400}{\rho_{air} c_{air}}\right)$$

L_P Sound pressure level, 20 $\log_{10}$ (P_{rms}/P_0) [dB re P_0 = 20 μPa rms]

$L_{aphakic}$ Effective radiance of light source weighted by $A\lambda$ for lenseless eye [W/cm² sr]

L_{blue} Effective radiance of light source weighted by $B\lambda$ [W/cm² sr]

L_{retina} Effective radiance of light source weighted by $R\lambda$ [W/cm² sr]

L_λ Spectral radiance of light source at λ [W/cm² sr nm]

lm Luminous flux, lumen [lm = cd sr]

lux Illuminance [lux = lm/m² = cd sr/m²]

M Molar mass [g/mol]

M_{air} Molar mass of NOAA standard air, 28.964 [g/mol]

m Meter, SI base unit of length

m_α the alpha particle mass = 6.644656×10^{-27} [kg]

m_d the deuteron mass = $3.34358309 \times 10^{-27}$ [kg]

m_e the electron mass = 9.109382×10^{-31} [kg]

m_n the neutron mass = $1.6749272 \times 10^{-27}$ [kg]

m_p the proton mass = $1.6726216 \times 10^{-27}$ [kg]

m_{length} mass per unit length of a string or rod [kg/m]

mW milliWatt [10^{-3} watt]

n A neutron, an uncharged nucleon with $m = m_p$ and $q = 0$

N Newton, unit of force to accelerate 1 kg at 1 [ms^{-2}]; N = kg m/s²

OD Optical density, $\log_{10}(1/T_r)$ [unitless]

p A proton, a positively charged nucleon with $m = m_p$ and $q = e$

$\vec{p}$ Momentum vector of a particle with mass.

Linear momentum vector = $\dfrac{m\vec{v}}{\sqrt{1-(v/c)^2}}$;

Angular momentum (AM) vector = **L**, of a particle of position vector (**r**) and linear momentum (**p**) is defined as $\mathbf{L} = \mathbf{r} \times \mathbf{p}$.

p_{photon} Momentum of a photon [J S/m = kg m/s]

P_{rms} Pressure or *rms* sound pressure [Pa]

P_{EM} Radiation pressure of EMS wave [Pa]

P_{air} Barometric pressure with standard atmosphere, reference value = 101.325 [kPa]

Pa Pascal, the SI pressure exerted by a force of 1 Newton over 1 square meter [N/m²]

P_0 20 μPa, reference *rms* sound pressure

prf Observed pulse repetition frequency [pulses/s]

PRF Allowed pulse repetition frequency [pulses/s]

pW picoWatt = [10^{-12} Watt]

q Charge [Coulomb = C]

Q An energy or power quantity used in converting to and from dB

R Molar gas constant, $R = 8.31447$ J/KMole

rms Root mean square = square root of the average value of the square of a time-varying signal over a defined interval or over one period of a periodic signal.

r_{xxx} Distance between observer and source, [m]. See I_{xxx} for meaning of *xxx*.

$R\lambda$ Wavelength weighting function to prevent IR thermal damage to retina

s Second, SI base unit of time

$\vec{s}$ A vector along a line or path to be integrated

$\vec{S}$ Poynting vector indicating power density and direction of traveling EM wave $\vec{S} = \vec{E} \times \vec{H}$ in [W/m²]

$|\vec{S}|\,|\vec{E}| \times |\vec{H}|$ in a free-field plane wave [W/m² = A V/m²]

SA_{pp} Per pulse specific absorption of energy in tissue [J/kg]

$S_c(f)$ Specific absorption coefficient for tissue bathed by EM waves with frequency = f in Hz [$S_c(f)$ is unitless]

SI base units: meter: m; kilogram: kg; second: s; ampere: A; Kelvin: K; mole: mol; candela: cd

Partial SI prefixes

atto	a = 10^{-18}	Exa	E = 10^{18}
femto	f = 10^{-15}	Peta	P = 10^{15}
pico	p = 10^{-12}	Tera	T = 10^{12}
nano	n = 10^{-9}	Giga	G = 10^{9}
micro	μ = 10^{-6}	Mega	M = 10^{6}
milli	m = 10^{-3}	Kilo	k = 10^{3}
centi	c = 10^{-2}	Hecto	h = 10^{2}
deci	d = 10^{-1}	Deka	da = 10

S_λ Wavelength weighting function to prevent UV damage to skin and cornea

T Tesla, magnetic flux density, T = s V/m²

T Tension in a vibrating string [N]

T Period of a traveling wave, $T = 1/f = 2\pi/\omega$ [s]

T_{air} Absolute temperature of air in Kelvin [K]

T_{at} Averaging time for reversible tissue thermal effects, $T_{at} = 360$ [s]

$T_{°F}$ Temperature in degrees Fahrenheit or in degrees Celsius [°C] or Kelvin [K]

T_i OSHA allowed duration of exposure to sound intensity level, L_i [hr]

T_{pw} Pulse width including rise, fall, and ring-down time [s]

T_r Transmission through a medium, $0 \le T_r \le 1$

T_{max} Maximum duration of exposure to radiation [s]

T_λ Wavelength weighting function to prevent IR thermal damage to cornea and lens

U Energy [Joule = J]; $U_{photon} = hf$; hf is energy of a photon

V Volt, electrostatic potential difference, energy per unit charge, [V = W/A = J/C]

V_m Molar volume, ideal gas law estimate = RT/P [V in L when T in K and P in kPa]

Vel Velocity [m/s] or Vel_{mph} in [mi/hr] when subscript is shown

$\vec{Vel}_{air}$ Vector acoustic (oscillating) velocity in air [m/s]

W Watt, unit of work at rate 1 Joule per second [W = J/s = V A]

W_{al} Allowed limit of specific power deposition in tissue [W/cm^2]

Z_{air} Acoustic impedance of air with P_{air} in [kPa] and T_{air} in [K]

$$Z_{air} = \rho_{air} c_{air} = \frac{M_{air} P_{air}}{RT_{air}} (20.05 \sqrt{T_{air}}) = \frac{69.7758\ P_{air}}{\sqrt{T_{air}}}$$

in [Pa s/m]

Z_m EM impedance of medium *m* characterized by μ_m and ε_m: $Z_m = \sqrt{\frac{\mu_m}{\varepsilon_m}}$ [V/A = Ohm]

Z_0 EM impedance of a mass-free vacuum:

$$Z_0 = \sqrt{\frac{\mu_0}{\varepsilon_0}} \text{ [V/A = Ohm]}$$

Z_{0air} Reference acoustic impedance of air, $Z_{0air} \equiv 400$ Rayls = 400 [Pa s/m] = 400 [N s/m^3]

α Alpha particle, a fully ionized helium nucleus, He^{++} with m ~4 amu and $q = +2e$

β Beta particle, a free electron (β^-) or a free positron (β^+) with $m = m_e$ and $q = e$

γ Gamma particle, a photon with $m = 0$, $q = 0$, and finite momentum

γ_{air} Adiabatic constant for air and diatomic gases ~1.4

$\Delta\lambda$ A short wavelength interval on a spectrum [m or nm]

ε_m Permittivity in a material [A s/(V m) = F/m]

ε_r Relative permittivity, $\varepsilon_r = \frac{\varepsilon_m}{\varepsilon_0} \geq 1.0$ [unitless]

ε_0 Vacuum permittivity or dielectric constant, $\varepsilon_0 = 8.854187817 \geq 10^{-12} \approx 10^{-9}/36\pi$ [A s/V m = F/m]

μ_m Permeability in a material, $\mu_m \geq \mu_0$ [V s/A m = H/m]

ν (nu) = $2\pi f$, with frequency, f, in radians/s

μ_0 Vacuum permeability or magnetic constant, $\mu \equiv \pi/2{,}500{,}000$ [V s/A m = H/m]

μ_r Relative permeability, $\mu_r = \frac{\mu_m}{\mu_0} \geq 1.0$ [unitless]

Φ_E Electric flux $\int_A \vec{E} \cdot d\vec{A} \approx \vec{E} \cdot \vec{A}$ [N m^2/C] → [J m/C] → [W m/A] → [V m]

Φ_M Magnetic flux $\int_A \vec{B} \cdot d\vec{A} \approx \vec{B} \cdot \vec{A}$ [*T* m^2] →[N m/A] → [J/A] → [W s/A] → [V s]

λ Wavelength in [m], used herein for both elastic and electromagnetic waves

π (Circumference of circle)/diameter ≈ 3.14159, π radians = 180 degrees

ω Frequency in radians per second ($\omega = 2\pi f$) [rad/sec]

Ω Ohm 1 $|\Omega|$ = 1[V/A]

ρ_{air} Density of air, $\frac{M_{air}}{V_{molar\,air}} \approx \frac{M_{air} P_{air}}{RT_{air}}$ [ρ in kg/m^3 when *M* in kg/kmol, *P* in kPa, *T* in K]

ρ Mass density [kg/m^3] or charge density [C/m^3]

$\nabla \cdot \vec{A}$ = Div ($\vec{A}$) = Divergence of a vector field, $\vec{A}$, quantifying the flux away from or toward the point of analysis

$\nabla \times \vec{A}$ = Curl ($\vec{A}$) = Curl of vector field $\vec{A}$ is a vector operator quantifying its vorticity at the point of analysis

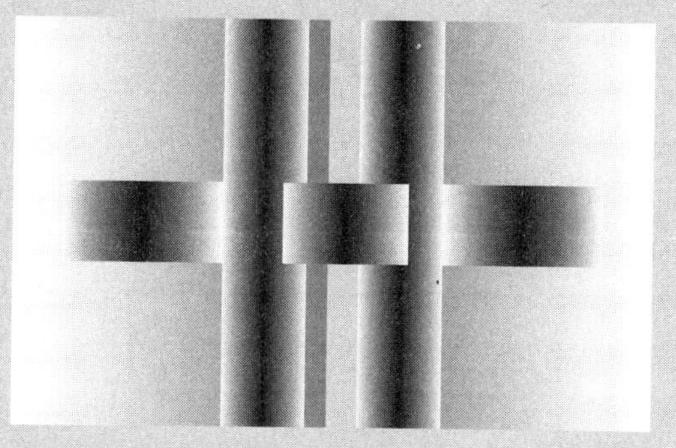

SECTION 4
INDUSTRIAL HYGIENE

LEARNING OBJECTIVES

- Become well-versed in the anticipation, recognition, evaluation, and control of biological hazards in an occupational setting.
- Learn where to find reference information on biological hazards in an occupational setting.
- Become familiar with the bloodborne pathogens and recombinant DNA regulatory and safety requirements.
- Outline the operation and uses of primary biological safety containment devices (e.g., biological safety cabinets).

Applied Science and Engineering: Biological Hazards

Michael A. Charlton

A RELATIVELY NEW paradigm in the last 50 years is the anticipation, recognition, evaluation, and control of biological hazards in the workplace. Exponential growth in basic sciences, microbiology, and medicine has led to unparalleled advances in human morbidity and mortality. Improving human health pushes researchers to study the fundamentals of disease genesis and transfer in our population. This chapter studies biological safety aspects of anticipation, recognition, evaluation, and control, especially in the occupational environment.

Human medicine in developed countries has evolved over the past 50 years. The ability to diagnose, treat, and ultimately cure infectious diseases has significantly increased human life span over the past century. The prevalence of infectious disease cases in developed countries has given way to other chronic diseases, such as coronary heart disease and solid tumor cancers.

The practicing safety professional should be aware of the techniques for anticipating, recognizing, evaluating, and controlling biological hazards in the workplace. Naturally occurring biological hazards exist in almost any workplace that involves plant and animal contaminants, meat processing, farming, laboratories, or health care (Heubner 1947; Sulkin and Pike 1951; Oliphant, Gordon, Meis, and Parker 1949). In particular, laboratory-acquired infections have become very significant safety concerns. More than 3500 laboratory infections and 160 deaths have been reported, worldwide (Wedum 1997).

Biological safety programs must be founded on fundamental safety principles. Pike (1979) once concluded that "common sense, good work practices and using the appropriate equipment should protect workers from the risks related to the use of hazardous biological agents." These guiding principles should be emphasized throughout acceptable biological safety programs.

ANTICIPATION AND RECOGNITION

The primary focus of biological safety in the United States today occurs in the highly specialized realm of clinical testing and microbiology laboratories. Through the years, numerous case-prevalence studies have reported biological hazards in these occupational settings. The most well known are a series of laboratory surveys published by Sulkin and Pike from 1949 to 1979. By 1976, Pike had reported a stunning 3921 cases of laboratory-acquired illness. The most common diseases reported were brucellosis, typhoid, tularemia, tuberculosis, and hepatitis. Most noteworthy, approximately 70 percent of Pike's reported cases were not associated with a known accident or incident (Pike 1976). Table 1 summarizes the reported causes of 3497 laboratory infections (Wedum 1997). It is presumed that these illnesses may have occurred via outdated laboratory techniques (e.g., mouth pipetting), a lack of engineering controls, or a loss of containment.

Routes of Transmission

These exposure data illustrate the most common exposure routes in an occupational setting. The National Research Council (NRC 1989) reported five primary routes of exposure in the biomedical research laboratory. The oral ingestion and respiratory routes are frequent safety risks for many hazardous materials. However, biological agents may also be transmitted through direct contact (e.g., droplets contacting unprotected skin), ocular contact, and self-inoculation (e.g., percutaneous or needlestick injury). These routes are especially important in industries that routinely utilize aerosol-generating equipment and sharp implements (e.g., syringes and scalpels).

Clinical testing laboratories were found to have biological safety hazards, particularly posed by blood-borne pathogens. Skinholj (1974) reported a retrospective case study in which Danish clinical testing laboratories reported an incidence of hepatitis that was seven times higher than the incidence in the general Danish population. Anticipation of biological safety hazards associated with blood-borne pathogens represents a critical program for safety professionals.

TABLE 1

Reported Cause of 3497 Laboratory-Acquired Illnesses

Reported Cause	Count	Percentage
Aerosol exposure	466	13.3
Accident (reported)	566	16.2
Unknown cause	2465	70.5
TOTAL	3497	100.0

(*Source:* Wedum 1997)

A clear record of occupational biological hazard is available in the literature. Table 2 shows an interesting distribution of infections as a function of self-reported occupation (Wedum 1997). These hazards are pronounced in the occupational setting but less established as a community health risk. Clearly, food- borne pathogens have resulted in community health risks, especially in sensitive populations, such as the elderly and oncology and pediatric patients. The record of community health risk from occupations handling biological agents and human specimens is not well established. Richardson (1973) observed no secondary cases (to the community) from 109 laboratory infections reported to the Centers for Disease Control and Prevention (CDC) from 1947 to 1973. A limited number of secondary cases are available in the literature, ranging from sexual transmittal of Marburg disease (Martini and Schmidt 1968) to Q-fever among commercial laundry workers handling contaminated linens (Oliphant et al. 1949). Q-fever transmission from close household contact (presumably as a result of infectious aerosol generation) and percutaneous transmission from a B-virus spousal exposure were also reported by Holmes, Hilliard, and Klontz. (1990). The highly infectious nature of Q-fever supports the notion that enclosed and improperly ventilated operations have the potential for creating secondary community infections. The sporadic nature of these secondary community infections makes it difficult for the safety professional to anticipate biological hazards. However, plans for post-exposure decontamination, contaminated laundry procedures, and proper hand hygiene appear to be warranted.

The two most common early biological safety hazards include aerosol transmission via centrifuging

TABLE 2

Percent Occurrence of Laboratory-Acquired Infection as a Function of Self-Reported Occupation

Self-Reported Occupation	Ft. Detrick Data (%) ($n = 369$)	Pike and Sulkin Data (%) ($n = 1286$)
Trained scientific personnel	58.5	78.1
Laboratory technical assistants	21.7	
Animal attendants	2.1	10.3
Contaminated glassware washer	3.8	
Housekeeping	0	
Administrative and clerical	3.7	
Maintenance employees	7.8	6.7
Personal contacts	2.4	
Students not in research	0	4.9
TOTAL	100.0	100.0

(*Source:* Wedum 1997)

and ingestion transmission via oral pipetting of infectious materials. Oral pipetting was reported to account for 6 to 18 percent of known accidental inoculations from 1893 to 1968 (Wedum 1997). Wedum noted various causes including a loose cotton plug, incorrect selection of the type of pipette, excessive suction, removal of the pipette from the liquid, and even presence of a nose cold in the laboratory technician. The onset of mechanical, calibrated pipettes has virtually eliminated the oral exposure pathway from most occupations in the past fifteen years.

Due to the ability of the pathogens to rapidly entrain in air, microbial aerosols containing pathogens can be released during centrifugation. In particular, aerosols can be created by infectious fluid remaining on the lip of the tube, leakage of a tube when the fluid head angle is less than 45 degrees, distortion of a nonrigid tube by centrifugal forces, or even residual fluid in the cap screws (Wedum 1997). Anticipation of these biological safety hazards is critical because two of the reported cases resulted in fatalities. A risk-assessment procedure that incorporates sealed rotors, routine decontamination, tube inspection, vigorous training, and emergency spill procedures is warranted from these data.

The data from Table 2 indicate that approximately 10 percent of laboratory-associated illnesses occur via percutaneous injuries. Although these data may be dated, safety professionals should heed the general trend of occupational biological hazards.

Types of Biological Agents

It is critical for the safety professional to understand the subtle differences in biological agents. This knowledge should yield improved control measures and emergency response procedures in the occupational environment. However, knowledge of biological interactions and human disease onset is not fully defined and is rapidly developing. Visit the Centers for Disease Control Web site for the current guidance on specific biological agents. Another excellent regulatory reference is from the Public Health Agency of Canada (PHAC 2006). These MSDS are produced by Canadian health authorities with the stated purpose of proving guidance to employees and employers regarding biological agents in the workplace. They contain succinct hazard information organized in alphabetical order and include excellent safety information for rapid reference.

A common source of health information is the *Merck Manual of Diagnosis and Therapy* (Beers and Berkow 1999), which conveys basic understanding of infectious diseases especially from the patient perspective. The following summary is drawn from the *Merck Manual.*

Bacteria

These common and familiar biological agents may be further subdivided into gram-positive cocci, gram-negative cocci, gram-positive bacilli, gram-negative bacilli, spirochetes, mycobacteria, and anaerobics. Each one has unique traits, especially in relation to the effect on humans and potential therapies.

Staphylococcal infections are extremely common in humans, especially in hospitals or related environments. These infections appear in circumstances involving newborns, postoperative surgeries, internal catheters, certain pulmonary disorders, and even some bone maladies.

Streptococcal infections are another group of common bacterial agents and are generally evaluated by growth on sheep-blood agar media. Another common way to classify streptococci is by the amount of carbohydrates in the cell wall. The most familiar of these gram-positive cocci infections are rheumatic fever, scarlet fever, and familiar tonsillitis/pharyngitis diseases. The most common disease among these agents is streptococcal pharyngitis characterized by sore throat, fever, and tonsillar exudate. Most streptococcal infections are generally considered virulent, and some subgroups have known resistance to drug therapies. However, many safety professionals will be familiar with the symptoms of these diseases because of their commonality. With proper diagnosis and pharmacological therapy, most healthy workers will have an excellent long-term prognosis from gram-positive cocci.

The general term *gram-positive bacteria* indicates a dark-colored reaction to a Gram-staining procedure. This purple-colored reaction is caused by incorporation of the Gram stain throughout the cell via the cell wall. This clinical laboratory procedure is extremely common in order to highlight bacterial cells against other cells. A counterstain involves the addition of safranin to the biological sample. This counterstain allows gram-negative bacteria to turn a light red or pink color to easily differentiate the dark purple gram-positive bacteria.

Pneumococcal infections are another type of gram-positive lancet-shaped encapsulated diplococcus, with more than 85 different types. These bacteria are readily spread person to person via aerosolized droplets (e.g., coughing or sneezing). These infections are so common that nearly half of a healthy workforce may harbor these pathogens during the winter months. Further, extreme care should be observed for special populations in the workplace (or patients) with immunological deficiencies, for oncology patients, and for those with sickle cell diseases.

Gram-negative cocci include the genus Neisseria. These bacteria differ from the gram-positive stain because their cell wall structure does not readily take up the primary Gram stain. Generally, gram-negative cells appear red or pink because of a secondary counterstain added during the staining procedure. The most familiar gram-negative cocci bacterial agent is *N. gonorrhoeae*, which is the causative agent for the sexually transmitted disease gonorrhea. Another significant human pathogen is *N. meningitides*, which causes a clinically significant bacterial meningitis. Normally, these gram-negative cocci are not strongly associated with occupational exposures. Gram-negative rod or bacilli include significantly more species than the cocci groups. Thousands of subspecies have been observed, including many clinically significant species. Of particular importance to a safety professional, in the gram-negative bacilli category, are the respiratory hazards caused by *Legionella pneumophila* and the ingestion/food-borne hazard posed by *Salmonella enteritidis* or *Salmonella typhi*.

GRAM-POSITIVE BACILLI

The most important occupational gram-positive bacilli is *Erysipelothrix rhusiopathiae*, which causes a common skin irritation. Bacterial infection with *E. rhusiopathiae* occurs via dermal abrasions and cuts resulting in local infection. The most commonly associated occupational-exposure settings include animal husbandry, handling animal carcasses or excreta, and extremity contact with animal derivatives.

GRAM-NEGATIVE BACILLI

This broad group of organisms is gram-negative, oxidase-positive, and readily cultured. These organisms also ferment glucose. One of the most common bacteria is *Escherichia coli*, which commonly inhabits the gastrointestinal tract of healthy humans. *E. coli* is extremely invasive and toxic when the host system is disrupted via trauma, disease, or pharmaceutical induction. External introduction of some strains into a human may occur through undercooked animal products, especially beef. *E. coli* is extremely common in research laboratories. Laboratory-acquired cases of *E. coli* enterohemorrhagic disease have been reported in the literature (Rao, Saunders, and Masterton 1996).

Many farm animals in the United States may host the toxic strains of *E. coli*. Occupational exposure to farm animal excreta, especially bovines, is generally considered a strong correlation with *E. coli*.

Gram-negative bacilli are extremely widespread and include common hospital-acquired infections (*Klebsiella, Enterobacter*, and *Serratia*), *Salmonella, Shigella,*

Haemophilus, *Brucella*, and *Cholera*. Of these, *Brucella* has been responsible for numerous occupational exposures. The infectious dose of *Brucella* spp is extremely low, and hence it was the first biological agent (*Brucella suis*) weaponized by the United States in 1954 (Gilligan and York 2004). Prohibitions on ingestion of unpasteurized milk (especially goat) have led to case-incidence reductions, and approximately 100 cases are reported in the United States each year (Gilligan and York). Occupational exposures to unvaccinated animals account for half of the cases in the United States. Brucellosis is also the most frequently acquired laboratory infection (Pike 1978). Table 3 outlines the ten most frequent laboratory-acquired illnesses based on Pike's work (Collins 1983).

Spirochetes are generally spread via nonvenereal body contact and include endemic syphilis, lyme disease (*Borrelia burgdorferi*), and rat-bite fever (*Streptobacillus moniliformis*). Tuberculosis, or TB, is associated with three mycobacteria (*M. tuberculosis*, *M. bovis*, and *M. africanum*). TB occurs in developed countries especially via aerosolized droplet nuclei. The case rate of TB in the United States is approximately 8 cases per 100,000 people. The droplet nuclei are generally considered dangerous because they may stay suspended in air for several hours post-generation.

Viruses

Viruses are extremely small (0.02 to 0.3 μm) protein or lipid covers for an RNA or DNA core. Viruses are dependent on host cells for replication. They depend on healthy cells for propagation and reproduction. Typical microbiological analysis via microscopy cannot detect the small virus particles. However, special chemical or biophysical marker tests can indicate the presence of a virus in a host body. The most prevalent viruses encountered by safety professionals involve blood-borne pathogen viruses (e.g., human immunodeficiency virus, hepatitis B virus, and hepatitis C virus). Vaccines for many viruses are available and should be emphasized via local public-health officials or occupational health programs. These common viruses include influenza, varicella, measles, mumps, poliomyelitis, rabies, rubella, and hepatitis A and B.

TABLE 3

The Top Ten Laboratory-Acquired Infections

Agent/Disease	Cases Reported	Mortalities Reported
Brucellosis	426	5
Q-Fever	280	1
Hepatitis	268	3
Typhoid fever	258	20
Tularemia	225	2
Tuberculosis	194	4
Dermatomycosis	162	0
Venezuelan equine encephalitis	146	1
Psittacosis	116	10
Coccidioidomycosis	93	2
TOTAL	2168	48

(*Source:* Collins 1983)

Fungi

In humans, fungi are generally seen as co-infections resulting from other underlying health conditions (e.g., immune-compromised patients). Fungi may occur in isolation of other conditions but are normally associated with a particular geography. For example, *Coccidioides immitis* is predominantly found in the southwest region of the United States, whereas *Blastomyces dermatitidis* may be found along the eastern seaboard of the United States.

EVALUATION

Routine evaluation of the occupational environment has been demonstrated to prevent injuries in many different workplaces. The safety professional should be prepared to conduct episodic facility evaluations that may include biological hazards. The anticipation of potential biological hazards should focus resources on engineering and administrative control measures for preventive purposes. Because of the less-than-uniform application of all control measures, the hazard evaluation phase buttresses biological safety programs in prevention and control phases.

The observation and feedback of existing workplace conditions is an important addition to any biological safety program. In common practice, these evaluation plans include visual review of posting, signage, immunization requirements, ventilation, decontamination processes, physical layout, and emergency procedures. Verification of certain biological safety records includes review of training documents, independent facility certifications, biosafety cabinet certifications, inventory records, autoclave or inactivation logs, and written handling and decontamination procedures. The use of some biological agents may require immunizations, medical testing, or routine health examinations. Occupational health records may also provide an understanding of the workplace. Using this type of employee record also poses special privacy challenges that must be thoughtfully considered prior to the evaluation process.

Many facilities employ a *team-based* evaluation approach. The evaluation team should have enough experience to provide diverse feedback on common biological safety principles and workplace operations. A model team might include a laboratorian, safety professional, medical health professional, veterinary health professional, employee union, and management representative. The contribution and viewpoint of each member of this ideal evaluation team assists in improving the overall evaluation process.

Sources for Conducting Biosafety Evaluations

An inclusive list of all biological safety considerations is not available because of the diverse levels of protection necessary for various biological agents. The primary references in the United States are issued by the National Institutes of Health (NIH), the CDC, and the Occupational Safety and Health Administration (OSHA). Each source contains a biological safety aspect that does not apply to all occupational settings. Therefore, while developing an evaluation protocol, refer to each applicable standard or guideline.

The NIH has published guidelines for the safe use of recombinant DNA molecules in research. The definition of recombinant DNA is given in section I of the NIH guidelines (NIH 2002). Recombinant DNA is defined in that document as either molecules that are constructed outside living cells by joining natural or synthetic DNA segments to DNA molecules that can replicate in a living cell or molecules resulting from the replication of these. This document is commonly referred to as simply "the NIH guidelines" among biosafety professionals. The NIH guidelines apply to all recombinant DNA research in the United States and its territories for any institution receiving NIH funding. Recombinant DNA research may take the form of benchtop science, human testing derived from recombinant DNA, or human clinical trials using recombinant DNA technology. According to Section I-D-2, any institution receiving NIH funding for recombinant DNA research must apply the NIH guidelines to *all* recombinant DNA research irrespective of the funding source (NIH 2002). Because of this fiscal requirement, the NIH guidelines should be interpreted as mandatory in nature if not in substance.

The NIH guidelines require initial procedure and containment assessment by an independent committee called the Institutional Biosafety Committee (or IBC). The IBC is also required to periodically review recombinant DNA research at the institution for compliance with the NIH guidelines. Further, many larger institutions are required to appoint a biological safety officer (or BSO). The general duties of the BSO may be thought of as the daily administration for the IBC. Under Section IV-B-3-c of the NIH guidelines, the BSO must conduct "periodic inspections" to ensure laboratory standards are followed (NIH 2002). No specific guidance is given to the BSO, but the implication is that the periodic laboratory safety inspections are intended to assess physical containment, standard microbiological procedures, special procedures, and specific biological barriers, all of which should be evaluated by the IBC prior to commencing biological agent research.

Conducting Facility Safety Assessments for Biological Hazards

The absence of mandatory regulatory language on biological hazards forces most safety professional to use alternative resources for developing and implementing a periodic facility safety evaluation program. The two primary regulatory references in the United States are OSHA's "Bloodborne Pathogens" regulations (OSHA 2001) and the CDC guidance document "Biosafety in Microbiological and Biomedical Laboratories" (BMBL) (CDC 1999). The design and operation of a facility must at a minimum strictly adhere to these minimum parameters (see Sidebars 1, 2, and 3).

The development of a facility safety assessment should include evaluation of notifications to employees. Notifications are of critical importance with biological agents because of variability of the infectious dose between various agents and the intra-agent susceptibility of human population segments. The notification and posting of mandatory protections yields written guidance before anyone enters a higher-hazard area. The notification generally consists of the *biohazard* symbol in contrasting colors and a written description of the biological agent. The standard biohazard symbol is referenced in 29 CFR Part 1910 (OSHA 1974). Information regarding specialized protective equipment is generally required to be posted at the entrance of biological agent containment facilities. Most protective notifications also include necessary immunizations and emergency response procedures.

The widespread availability of immunizations has reduced the incidence of many occupational biological agents, especially viral agents. The public-health promotion of immunizations is widely recognized as a key life-span factor in developed countries. Immunization public-health promotions are generally aimed at community populations; however, occupational usage is a critical facility safety evaluation parameter that should be included during safety evaluations. Immunization of personnel is a complicated and ethical dilemma. The risks of immunization created by negative side effects must be balanced against the protective benefits against the infectious disease. This analysis should involve trained and experienced medical professionals, infectious disease workers, safety professionals, and occasionally local leaders in the public-health area. An important tenet is that not all immunizations are indicated for all employees because of underlying health conditions. Any inadequate immunization processes are critical factors that should be included in the risk assessment conducted by the safety professional.

Laboratory facility design, work-practice controls, and safety equipment combine to form the primary factors. Consolidated tables for evaluating proper measures are available from a variety of sources (Herman et al. 2004). Table 4 outlines general guidance that can be used to evaluate laboratory facility design features at Biosafety Level 2 (BSL-2).

SIDEBAR 1: HISTORY OF BLOOD-BORNE PATHOGENS (Adapted from the CDC MMWR 1988)

1974—Skinholj reported an incidence of hepatitis seven times the general population incidence for Danish clinical chemistry laboratories.

1983—In 1983 the U.S. Centers for Disease Control (CDC) published an MMWR document titled "Guideline for Isolation Precautions in Hospitals," which contained a section on blood and body fluid precautions. Healthcare worker precautions were intended to be implemented when a patient was known or suspected to be infected with a blood-borne pathogen.

The purpose of this critical 1983 CDC publication was the prevention of nosocomial (or hospital-acquired) infections from patients with a variety of infectious conditions. The principal shortcoming of these recommendations was the application of blood and body fluid protections only to those patients likely to be infected.

1987—In August 1987, the CDC published an MMWR document titled "Recommendations for Prevention of HIV Transmission in Health-Care Settings.." The severity of this disease was widely recognized in the 1980s, hence the stringent recommendations outlined by this 1987 CDC publication.

These recommendations extended the 1983 blood and body fluid precautions to all patients. Because these safety precautions were being extended universally, the term Universal Blood and Body Fluid Precautions or Universal Precautions was applied.

In October of 1987, the CDC and OSHA distributed a joint federal advisory notice to all state public-health directors, state epidemiologists, and hospital infection-control committees. This notice outlining recommended protections against blood-borne pathogens was the first-ever joint federal notice addressing safety programs for healthcare workers.

1988—The CDC immediately recognized the inherent value of "Universal Precautions" and published an update extending these protections to HIV, Hepatitis B, and other blood-borne pathogens in healthcare settings. The updated pathogen-control plan reported,

Blood is the single most important source of HIV, HBV, and other blood-borne pathogens in the occupational setting. Infection control efforts for HIV, HBV, and other blood-borne pathogens must focus on preventing exposure to blood as well as on delivery of HBV immunization.

1992—The U.S. Occupational Safety and Health Administration (OSHA) published the first-ever regulatory standard specifically addressing the occupational hazards of the healthcare worker. The "Bloodborne Pathogens" regulations were developed over a five-year period and were based on physician, scientist, administration, and union consolidated testimony. These regulations were effective in the United States starting on March 5, 1992.

SIDEBAR 2: SYNOPSIS OF BLOOD-BORNE PATHOGENS RISK AND EPIDEMIOLOGY

(Adapted from the CDC 2003)

Many employees with clinical or research job responsibilities are at risk for occupational exposure to blood-borne pathogens, including hepatitis B virus (HBV), hepatitis C virus (HCV), and human immunodeficiency virus (HIV). Exposures occur through needlesticks or cuts from other sharp instruments contaminated with infected blood or through contact of the eye, nose, mouth, or skin with a patient's blood. Important factors that influence the overall risk for occupational exposures to blood-borne pathogens include the number of infected individuals in the patient population and the type and number of blood contacts. Most exposures do not result in infection. Following a specific exposure, the risk of infection may vary with factors such as these:

1. Pathogen
2. Type of exposure
3. Route of exposure
4. Amount of patient blood involved in the exposure
5. Health status of patient
6. Health status of the clinician
7. Use of post-exposure

Most institutions have a system for reporting blood-borne pathogen exposures in order to rapidly evaluate the risk of infection, inform employees about treatments available to help prevent infection, monitor the employee for side effects of treatments, and determine if infection occurs. This may involve testing the employee's blood and that of the source patient and offering appropriate post-exposure treatment. An injury-reporting system is described in blood-borne pathogens training and outlined in a facility's exposure control plan.

How Can Occupational Exposures Be Prevented?

Many needlesticks and other cuts can be prevented by using safer techniques (for example, not recapping needles by hand), disposing of used needles in appropriate sharps-disposal containers, and using clinical or laboratory devices with safety features designed to prevent injuries. Using appropriate barriers such as gloves, eye and face protection, or gowns when contact with blood is expected can prevent many exposures to the eyes, nose, mouth, or skin.

The type of biological agent containment laboratory must be initially defined by the researcher or clinician. For example, clinical testing laboratories and production laboratories differ significantly in terms of hazard potential for the same biological agent. Therefore, the risk-assessment process must include a determination of whether the facility is a clinical, research, animal, or production laboratory. Each facility uses or amplifies biological agents differently, and hence, each has specific exposure risks. The safety professional must conduct a risk-assessment process in order to assess the potential hazards.

A risk-assessment process is outlined in the fourth edition of the BMBL (CDC 1999). Stated factors of interest to a safety professional conducting a risk assessment include pathogenicity of the agent, route of transmission, agent stability, infectious dose, concentration (in units of infectious organism per unit volume), source of the agent, availability of historical human or animal exposure data, medical prophylaxis/surveillance, personnel training, and proposed alterations to the recombinant gene. This complex process may require the assistance of a team of professionals including the safety professional. In many instances, a risk-assessment team is formed by the Institutional Biosafety Committee.

TABLE 4

Common Laboratory Facility Design Features to Be Reviewed during the Evaluation Phase

Number	Design Feature
1	Physical separation between the laboratory and remainder of the building.
2	Physical security and restricted access must be maintained between the laboratory and other areas of the building.
3	Doors into BSL-2 laboratories are self-closing and self-latching in places where they intersect a public area.
4	Laboratory furniture is conducive to facilitating agent-specific decontamination.
5	An active vector-control program is established for the facility.
6	Hand-washing facilities are readily available in the laboratory.
7	Protective clothing is worn in lieu of street clothing. Laboratory personnel must have access to a change room or area.
8	Laboratory working surfaces (e.g., benchtops) are easy to clean, impervious to water, and resistant to laboratory chemicals and disinfectants.

Note: These data were intended to be used for Biosafety Level 2 laboratories. Additional facility-design characteristics and work-practice controls are required for Biosafety Level 3 and animal laboratories.

(*Source:* Herman et al. 2004)

SIDEBAR 3: SUMMARY OF OSHA BLOOD-BORNE PATHOGEN REQUIREMENTS

(*Source:* OSHA 2001)

Information and Training

1910.1030(g)(2)(i)

Employers shall ensure that all employees with occupational exposure participate in a training program which must be provided at no cost to the employee and during working hours.

1910.1030(g)(2)(ii)

Training shall be provided as follows:

At the time of initial assignment to tasks where occupational exposure may take place; Within 90 days after the effective date of the standard; and At least annually thereafter.

1910.1030(g)(2)(iii)

For employees who have received training on bloodborne pathogens in the year preceding the effective date of the standard, only training with respect to the provisions of the standard which were not included need be provided.

1910.1030(g)(2)(iv)

Annual training for all employees shall be provided within one year of their previous training.

1910.1030(g)(2)(v)

Employers shall provide additional training when changes such as modification of tasks or procedures or institution of new tasks or procedures affect the employee's occupational exposure. The additional training may be limited to addressing the new exposures created.

1910.1030(g)(2)(vi)

Material appropriate in content and vocabulary to educational level, literacy, and language of employees shall be used.

1910.1030(g)(2)(vii)

[These are the training content requirements.]

The training program shall contain at a minimum the following elements:

An accessible copy of the regulatory text of this standard and an explanation of its contents;

A general explanation of the epidemiology and symptoms of bloodborne diseases;

An explanation of the modes of transmission of bloodborne pathogens;

An explanation of the employer's exposure control plan and the means by which the employee can obtain a copy of the written plan;

An explanation of the appropriate methods for recognizing tasks and other activities that may involve exposure to blood and other potentially infectious materials;

An explanation of the use and limitations of methods that will prevent or reduce exposure including appropriate engineering controls, work practices, and personal protective equipment;

Information on the types, proper use, location, removal, handling, decontamination and disposal of personal protective equipment;

An explanation of the basis for selection of personal protective equipment;

Information on the hepatitis B vaccine, including information on its efficacy, safety, method of administration, the benefits of being vaccinated, and that the vaccine and vaccination will be offered free of charge;

Information on the appropriate actions to take and persons to contact in an emergency involving blood or other potentially infectious materials;

An explanation of the procedure to follow if an exposure incident occurs, including the method of reporting the incident and the medical follow-up that will be made available;

Information on the post-exposure evaluation and follow-up that the employer is required to provide for the employee following an exposure incident;

An explanation of the signs and labels and/or color coding required by paragraph (g)(1); and

An opportunity for interactive questions and answers with the person conducting the training session.

Training Records

1910.1030(h)(2)(i)

Training records shall include the following information:

1910.1030(h)(2)(i)(A)

The dates of the training sessions;

1910.1030(h)(2)(i)(B)

The contents or a summary of the training sessions;

1910.1030(h)(2)(i)(C)

The names and qualifications of persons conducting the training; and

1910.1030(h)(2)(i)(D)

The names and job titles of all persons attending the training sessions.

1910.1030(h)(2)(ii)

Training records shall be maintained for 3 years from the date on which the training occurred.

1910.1030(h)(3)

Availability

1910.1030(h)(3)(i)

The employer shall ensure that all records required to be maintained by this section shall be made available upon request to the Assistant Secretary and the Director for examination and copying.

1910.1030(h)(3)(ii)

Employee training records required by this paragraph shall be provided upon request for examination and copying to employees, to employee representatives, to the Director, and to the Assistant Secretary.

CONTROL

The control of biological hazards in the workplace begins with the risk-assessment process. Fundamentally, each process or manipulation involving a patient or focused agent must be evaluated prior to employee exposures (Caucheteux and Mathot 2005). Several U.S. regulatory agencies mandate *a priori* review; however, the general practice of biological safety has historically matched administrative and engineered controls against hazard potential.

A system of biological containment is commonly employed to determine necessary laboratory work practices, safety equipment, and facility design. The mandatory controls are designed to be dynamic based on the hazard posed by the agent. As the hazard of the biological agent increases, additional control measures are required (Richmond and McKinney 1999).

Biological Safety Cabinets

Biological safety (biosafety) cabinets or BSCs are a class of safety equipment dedicated to protecting the environment, the worker, and even the product contained inside the device from the biohazard. A properly functioning and certified BSC is a central consideration in establishing appropriate physical containment of a biohazardous agent. However, three primary types of biological safety cabinets are available for containment purposes.

Selecting the right biological safety cabinet is inherent to the risk-assessment process for pathogenic agents. Therefore, the safety professional should answer the following questions in order to assess the type of BSC required:

1. Do I need to protect the outside environment or community from the agent?
2. Do I need to protect the worker handling the biological agent?
3. Do I need to protect the cultures or "products" inside the BSC from the outside unfiltered air?
4. Are there other hazardous chemicals, vapors, or radioactive materials present that must be considered in addition to biological agents?

By answering these four questions, the safety professional can easily select an appropriate BSC. Each type of BSC contains inherent features that meet one or more of the containment requirements outlined in these guidelines.

The BMBL compares and describes BSCs. Table 5 compares different types of BSCs (CDC 1999). A Class I BSC is shown in Figure 1. A Class I BSC is under negative pressure with respect to the outside air and has an inward flow of approximately 75 feet per minute (fpm). The Class I BSC does not protect the product (e.g., cultures) contained inside the BSC. The exhaust air leaving the Class I BSC is high-efficiency particulate air (HEPA) filtered. Therefore, environmental protection from the biohazard is offered by this BSC.

The National Institute for Occupational Safety and Health (NIOSH) recently issued guidance on protecting building environmental systems from chemical, biological, and radiological agents (NIOSH 2003). This guidance includes an excellent fundamental review of HEPA air filtration, filter mechanisms, and air purification. The primary difficulty in air filtration involves the particle size and chemical composition knowledge of the air contaminants. These parameters may

TABLE 5

Comparison of Biological Safety Cabinets

Type	Face Velocity (lfpm)	Airflow Pattern	Radionuclides/ Toxic Chemicals	Biosafety Level(s)	Product Protection
Class I open front	75	In at front; rear and top through HEPA filter	No	2, 3	No
Class II Type A1	75	70% recirculated through HEPA; exhaust through HEPA	No	2, 3	Yes
Class II Type B1	100	30% recirculated through HEPA; exhaust via HEPA and hard-ducted	Yes (Low levels/volatility)	2, 3	Yes
Class II Type B2	100	No recirculation; total exhaust via HEPA and hard-ducted	Yes	2, 3	Yes
Class III	NA	Supply air inlets and exhaust through 2 HEPA filters	Yes	3, 4	Yes

(*Source:* CDC 1999)

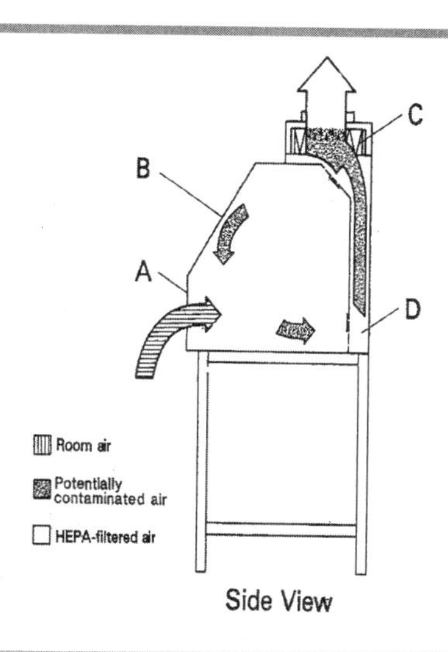

FIGURE 1. Class I biological safety cabinet
Legend: (A) front opening, (B) sash, (C) exhaust HEPA filter, (D) exhaust plenum. (*Source:* CDC 1999)

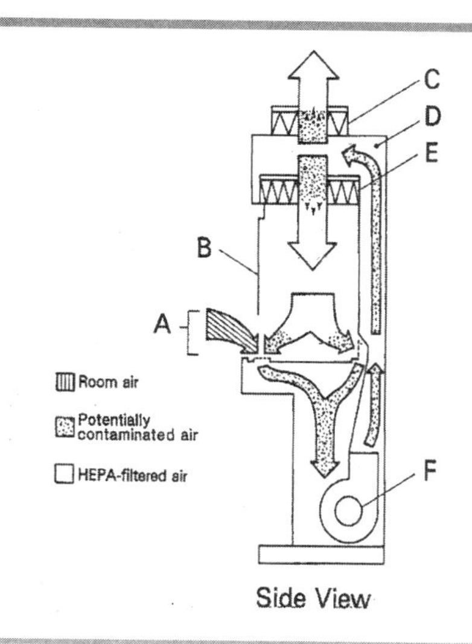

FIGURE 2. Class II, type A biological safety cabinet
Legend: (A) front opening, (B) sash, (C) exhaust HEPA filter, (D) rear plenum, (E) supply HEPA filter, (F) blower, (G) filter screen. (*Source:* CDC 1999)

not be fully known or quantified before engineering controls such as biosafety cabinets are installed. Particle sizes can range from 0.0001 μm (molecular sizes) up to 100 μm (human hair sizes). HEPA filters are designed to filter the size of particles that is most difficult to remove occurring from 0.3 μm to 1.0 μm. The NIOSH guidance document on protecting HVAC systems includes a table that compares filtration efficiency, particle sizes, and applications. Applications involving biological agents require the highest filtration efficiency in this document. One important caveat is that air filtration does not alter the oxygen level or significantly change the vapor phase of airborne chemicals. A thorough review of other hazardous materials present in the workplace is highly advisable prior to commencing operations in a HEPA-filtered BSC.

The Class II BSC is the most commonly encountered type of primary containment device. The Class II, Type A1 BSC is shown in Figure 2. The Type A1 BSC sends exhaust back into the laboratory room after the air is filtered via internal HEPA filters. Therefore, the Type A1 cabinet is not appropriate for any chemicals or radioactive materials that would not be readily filtered by a HEPA filter because of the possible recirculation of the contaminant into the room air. The use of natural gas inside a type A1 BSC is a prime example of how recirculation could significantly impact the workplace. Recirculating natural gas could conceivably build up an explosive atmosphere inside the BSC or the room because natural gas is not filtered by a HEPA filter. The Class II, Type B2 BSC is shown in Figure 3. This containment device offers the usual protection necessary for BSL-2 and BSL-3 laboratories. Room air enters the Class B2 cabinet and is immediately HEPA-filtered. The filtered air blows straight down, creating an air curtain in front of the product, thereby protecting the working surface from outside air. The Type B2 cabinet exhaust is also HEPA-filtered. Therefore, the Class B2 BSC HEPA-filters the incoming room air (for product protection), generates a straight-down air curtain (for product and personnel protection), and finally HEPA-filters the BSC exhaust (for personnel and environmental protection). The exhaust from a Class B2 BSC may be directly connected to the building exhaust system via a *hard* or *thimble* connection to the ductwork. As the name implies, a hard ductwork connection employs mechanical ducting to the exterior of the building. A thimble connection employs a 1-inch air gap between the cabinet exhaust and the building exhaust ductwork. The thimble connection is commonly used to reduce the effect of building exhaust-system fluctuations on the B2 BSC.

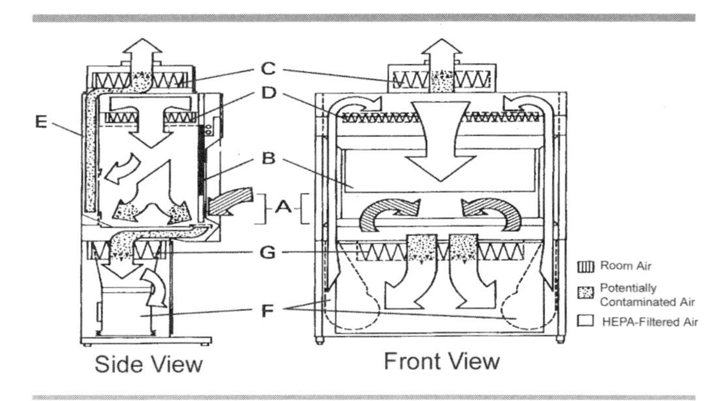

FIGURE 3. Class II, type B1 biological safety cabinet (classic design)
Legend: (A) front opening, (B) sash, (C) exhaust HEPA filter, (D) supply HEPA filter, (E) negative pressure exhaust plenum, (F) supply blower, (G) filter screen.
Note: The carbon filter in the building exhaust system is not shown. The cabinet exhaust needs to be connected to the building exhaust system. (*Source:* CDC 1999)

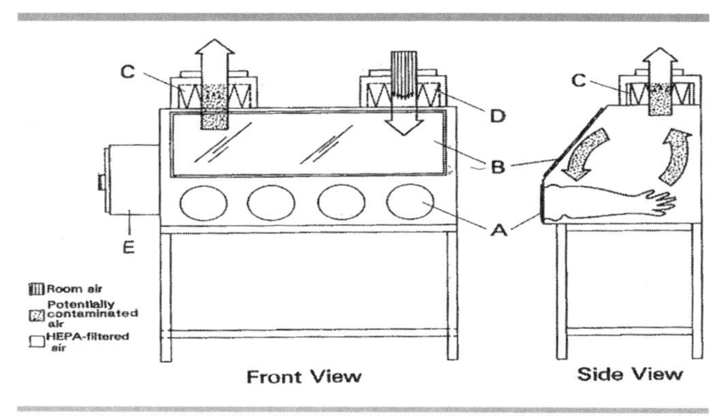

FIGURE 4. Class III biological safety cabinet
Legend: (A) glove ports with O-ring for attaching arm-length gloves to cabinet, (B) sash, (C) exhaust HEPA filter, (D) supply HEPA filter, (E) double-ended autoclave or pass-through box.
Note: A chemical dunk tank may be installed, which would be located beneath the work surface of the BSC with access from above. The cabinet exhaust needs to be connected to an independent building exhaust system. (*Source:* CDC 1999)

The Class III biological safety cabinet offers the maximum amount of protection for the product, worker, and outside environment. The inside of a Class III BSC may be thought of as totally isolated from the outside environment. Many safety professionals would be familiar with a similar protective device termed a *glove box*. Incoming air is HEPA-filtered, and the contained product has no direct connection outside of the BSC. Figure 4 depicts the Class III BSC. The Class III cabinet is normally found only in the highest level of biological agent containment (e.g., BSL-4). The exhausted air is HEPA-filtered when leaving the unit. Generally, the air is also HEPA-filtered a second time prior to entering the external exhaust fan.

The prudent safety professional also encourages routine testing of primary safety devices such as BSCs. The primary standard for testing BSCs is the National Sanitation Foundations Standard 49 (NSF 2002). This standard is commonly used to evaluate and test the operation of *biohazard cabinetry*. It also specifies training necessary for individuals to conduct assessment or certifications of BSCs.

Training

Employee training encompasses a primary focus of the safety professional for biological hazards. Specific biological safety training is *risk-based* in accordance with the BMBL (CDC 1999). A general rule of thumb is that higher levels of containment dictate additional biological safety training. The frequency of refresher training is also based on the selected level of containment. Training for laboratory staff is based on an assumed knowledge and competency of safe microbiological practices.

The following are abridged topics for biological safety training courses (NRC 1989):

1. ***Aseptic techniques:*** Fundamental microbiological techniques require laboratory techniques to prevent cross-contamination. The use of aseptic (or contamination-free) techniques is crucial when propagating infectious substances.
2. ***Laboratory and personnel hygiene:*** A review of historical accidents and exposures shows that laboratory practices and proper hand hygiene played an important role in laboratory-acquired diseases. Further, OSHA regulations also require proper hand hygiene when contacting blood-borne pathogens for exposure control purposes (OSHA 2001).
3. ***Laboratory safety practices:*** Laboratory safety practices generally involve the application of

primary containment devices, engineered safety equipment (e.g., sealed centrifuges or eyewashes), and risk assessment. These practices generally become more restrictive with the higher levels of containment.

4. ***Personal protective equipment (PPE):*** The availability of PPE varies as a function of the hazard posed by the pathogen. For example, respiratory protection will be required for any pathogen that could become airborne and infect a healthy adult worker. Additional protective equipment is generally associated with higher levels of biological containment.
5. ***Containment levels:*** The primary levels of biological containment are reviewed in the BMBL (CDC 1999). Higher containment of human pathogens also corresponds with higher hazard pathogens. Infectious diseases with no cures generally require rigorous containment facilities and procedures.
6. ***Decontamination, disinfection, and sterilization:*** The differences between decontamination, disinfection, and sterilization are reviewed in this chapter. Common procedures to monitor the efficacy of autoclave sterilization are also included.
7. ***Signage, notices, entrance criteria, and postings:*** For hazard awareness and protective equipment–donning purposes, areas that use biological agents must include appropriate notices and postings. Employees must refer to entrance criteria prior to entering a containment facility. These criteria may include protective equipment or immunization requirements.
8. ***Prudent management of biological/infectious wastes:*** Procedures for safe disposal of medical waste and sharps waste apply to many employers in the United States. Improper disposal of contaminated waste materials and laundry have been associated with secondary infections in the recent past.
9. ***Operation and use of biological safety cabinetry:*** The primary containment devices such as BSCs represent a fundamental part of biological containment. These devices provide protection to employees, their research specimen, and the surrounding public. An overview with figures from the BMBL (CDC 1999) is included in this chapter.
10. ***Emergency action, spill response, needlesticks and self-inoculation:*** The CDC gives health-care providers guidance on emergency response and needlestick treatment protocols in an effort to reduce the probably of transmitting a human pathogen. OSHA requires needlestick procedures under its blood-borne pathogens regulations (OSHA 2001).

Institutional Biosafety Committees

The NIH recommends that all individuals, corporations, or institutions not strictly covered by the NIH guidelines establish a properly constituted IBC. Even if an entity is not using recombinant DNA, a properly constituted safety committee should review the presence of biological hazards in the workplace. The IBC is responsible for the following on behalf of the institution:

1. Reviewing procedures, tasks, and protocols involving biohazardous agents, toxins originating from biohazardous materials, and blood and tissues from humans and nonhuman primates
2. Reviewing recombinant DNA research and approving those proposed projects that conform to the NIH guidelines. Specific IBC review items are detailed within the NIH guidelines (2002).
3. Notifying a principal investigator of the results of the committee protocol review
4. Lowering containment levels for certain agents
5. Setting acceptable containment levels for experiments involving whole plants or animals
6. Periodically reviewing recombinant DNA or other applicable research at the institution
7. Adopting emergency response plans
8. Reporting significant problems or violations of the NIH guidelines (2002) to the NIH

9. Not approving certain types of research unless prior approval is given by the NIH
10. Completing certain other duties as necessary

These complex duties require a diverse committee structure in order to fulfill the broad scientific and ethical reviews necessary under the NIH guidelines (2002). The IBC must have at least five members who have the collective expertise to review recombinant DNA research and assess the potential hazards presented by the proposed research. Additional experts (e.g., a plant pathogen agronomist) may be a necessary and temporary addition to an IBC depending on the type of recombinant DNA research. A conflict-of-interest resolution is required for IBC members involved in any protocols submitted by a member to the committee. The NIH also encourages institutions to make their IBC meetings open to the public and to release minutes of IBC activities upon request to members of the public (NIH 2002).

Personal and Environmental Sampling/Interpretation

The safety professional may elect to conduct environmental samples for biological hazards in an occupational setting. The difficulty of personal or environmental sampling is widely recognized in industrial hygiene. Occupational exposures to airborne infectious aerosols may yield disease transmission, hence the motivation to conduct *a priori* sampling of environmental air. Precautions aimed at eliminating or minimizing the spread of infectious diseases are generally termed airborne precautions (Garner 1996). However, airborne precautions are not the solitary protective measure; other types of prevention may also be effective (e.g., hand-washing or prohibitions on eating). The primary risks due to infectious aerosols are associated with speaking, sneezing, and coughing by contagious patients (Lenhart, Seitz, Trout, and Bollinger 2004). Research activities outlined in the laboratory-acquired illness section include centrifugation, shaking, and mechanical pipetting. Clinical activities would include aerosolizing medication, airway suctioning, intubation, and bronchoscopy (Lenhart et al. 2004).

The primary methods of controlling airborne hazards generally include engineering controls (e.g., BSCs), standard microbiological practices, proper hygiene, respiratory protection, host immunization programs, training, and antibiotic prophylaxis. Therefore, air sampling is generally not considered a feasible safety practice because of the difficulty in selecting personal air-sampling equipment that are portable and the necessary media to subsequently culture the airborne biological agent. In order to conduct many types of biological air sampling, the safety professional may need to know the suspected agent in order to select an appropriate growth media. Further, long-term time-weighted air sampling is not recommended because of the confounding air contaminants. Air sampling longer than even the shortest duration will result in a large confounding-agent population on the growth media.

Analysis of Air Samples

The analysis or counting of air samples is also a critical consideration when evaluating the occupational environment. False positive readings in sampling are routinely encountered and must be considered by the safety professional. For example, an air sample containing bacillus may incorrectly be interpreted as *Bacillus anthracis* because they are of the same genus. This particular false positive is routinely reported in the news because of the heightened public sensitivity resulting from *B. anthracis* attacks in 2001. Further, surrogate species may produce false positives because of similar toxin productions. An example of this is *B. thuringiensis*, which is used worldwide as an organic pesticide to protect crops from insects. *B. thuringiensis*, which is safe for humans but toxic to insects, belongs to the family of *B. cerus*. If *B cerus* and *B. thuringiensis* were misidentified in an environmental sample, the safety precautions and decontamination efforts would differ significantly because *B. cerus* is known to cause gastroenteritis in humans.

Interpretations

The complications presented by air and environmental sampling dictate strict proficiency in analyzing these samples and interpreting the results. Employing a trained and proficient microbiologist or pathologist is critical when interpreting biological samples. Routine proficiency testing, written test procedures, and previous sampling experience can assist the safety professional in obtaining an accurate hazard-assessment picture and excluding false positive events. The American Industrial Hygiene Association (AIHA) provides accreditation procedures for laboratories reporting and interpreting many air sample and mold sample results (AIHA 2007).

Air sampling is normally conducted to provide negative tests for certain high-risk procedures. For example, if an air hose is designed to carry an infectious dose of a biological agent, sentinel plates may be distributed throughout the work area to monitor airborne concentrations. The aerosols gravimetrically settle onto the sentinel plates, which are subsequently cultured in a controlled environment. The primary concern with sentinel plates is the proper placement of the plates and removal of local air eddy currents that could bypass the sentinel plate, yielding a false negative event.

Environmental wipe tests may also be used to control biological hazards in an occupational setting. The most common example would include a patient diagnosed with a complex respiratory disease that could survive on environmental surfaces (e.g., *M. tuberculosis*). In this situation, the safety professional would oversee the decontamination of a potentially contaminated building space, furniture, toilet fixtures, and so on. However, in order to reassure employees about safe working conditions, environmental wipe tests using sterile conditions would be prudent. The wipe-test media is then specifically cultured for the known infectious agent in order to control possible additional environmental exposures. This procedure is most common for those biological agents with a hardy, long-lasting spore (e.g., *B. anthracis* or *C. immitus*).

Clearance and Decontamination

A primary reason cited for conducting biological environmental sampling is the validation of decontamination or sterilization practices. For example, formaldehyde gas is commonly employed to decontaminate BSCs prior to maintenance work on the HEPA filters. This work would normally be considered highly hazardous because the pathogenic agents are diffused or impacted on the surface of the HEPA filter. In order to assess the positive results of the decontamination process, a biological indicator is commonly employed.

The term *sterilization* means destroying biological life, especially hardy bacterial spores. Sterilization is restrictive and difficult to achieve because it involves the thorough killing of biological life. A related term is *decontamination*, which means reducing or abating the number of biological organisms below the threshold necessary to cause disease. Disinfection is generally associated with decontamination because it means reducing all biological organisms below a certain level on an environmental surface. The safety professional must determine which of these goals he or she would like to achieve so that an appropriate methodology and subsequent biological indicator can be selected.

The most common biological indicator involves a 1.25-inch strip of filter paper containing *Geobacillus stearothermophilus* (for validating steam autoclaves and vapor phase hydrogen peroxide) or *Bacillus atrophaeus* (for validating ethylene oxide or dry heat sterilizers). The safety professional places a biological indicator strip inside the equipment to be sterilized prior to initializing the cycle. After completion, the spore strip is sent to a testing laboratory for subsequent culturing. If no growth appears after incubating for 24–48 hours, it is a positive indication that the pathogens in that sterilization cycle were effectively killed. If the biological indicators grow following sterilization, it is an indication that certain parameters were not met during sterilization. Common problems with steam autoclaves include overloading and low temperature. If the effective kill temperature is not achieved, harmful pathogens may potentially remain within the load.

Summary

The complex discipline of biosafety combines a diverse background of microbiology, engineering, industrial hygiene, chemistry, and education. The melding of these talents enables scientists, clinicians, veterinaries, and agronomists to safely engage the furtherance of their patient care or research goals. The safety professional can better anticipate, recognize, evaluate, and control biological hazards using the topics reviewed in this chapter. However, this information should be combined with collaboration with other competent medical or laboratory professionals to generate a team-based approach to biological safety. Biohazardous agents lead to complex diseases with multifactorial effects on the human population. A team-based approach with knowledgeable professionals will clearly complement the safety professional and lead to a safer work environment.

References

American Industrial Hygiene Association (AIHA) Environmental Microbiology Laboratory Accreditation Program (EMLAP). 2007. *ISO/IEC 17025 Accreditation* (retrieved January 7, 2008). www.aiha.org/Content/LQAP/accred/EMLAP.htm

Beers, M. H., and R. Berkow, eds. 1999. *Merck Manual of Diagnosis and Therapy*. 17th ed. Whitehouse Station, NJ: Merck Research Laboratories.

Caucheteux, D., and P. Mathot. 2005. "Biological Risk Assessment: An Explanation Meant for Safety Advisors in Belgium." *Applied Biosafety* 10:10–29.

Centers for Disease Control and Prevention (CDC). 1988. "Perspectives in Disease Prevention and Health Promotion Update: Universal Precautions for Prevention of Transmission of Human Immunodeficiency Virus, Hepatitis B Virus, and Other Bloodborne Pathogens in Health-Care Settings." *Mortality and Morbidity Weekly Report* 37(24):377–388.

______. 1999. *Biosafety in Microbiological and Biomedical Laboratories (BMBL)*. 4th ed. Washington, D.C.: CDC.

______. 2003. *Exposure to Blood: What Healthcare Personnel Need to Know* (retrieved December 14, 2007). www.cdc.gov/ncdid/dhqpl/bbp/Exp_to_Blood.pdf

Collins, C. H. 1983. *Laboratory Acquired Infections: History, Incidence, Causes, and Prevention*. Boston: Butterworth.

Garner, J. S. 1996. "Guideline for Isolation Precautions in Hospitals: Parts I and II." *Infection Control & Hospital Epidemiology* 17(1):53–80.

Gilligan, P. H., and M. K. York. 2004. "Sentinel Laboratory Guidelines for Suspected Agents of Bioterrorism: Brucella Species." *American Society of Microbiology*, pp. 3–4.

Herman, P., et al. 2004. "Biological Risk Assessment of the Severe Acute Respiratory Syndrome (SARS) Coronavirus and Containment Measures for the Diagnostic and Research Laboratories." *Applied Biosafety* 9:128–142.

Heubner, R. J. 1947. "Report of an Outbreak of Q-fever at the National Institutes of Health." *American Journal of Public Health* 37:431–440.

Holmes, G. P., J. K. Hilliard, and K. C. Klontz. 1990. "B-Virus Infection in Humans: Epidemiologic Investigations of a Cluster." *Annals of Internal Medicine* 112:833–839.

Lenhart, S. W., T. Seitz, D. Trout, and N. Bollinger. 2004. "Issues Affecting Respirator Selection for Workers Exposed to Infectious Aerosols: Emphasis on Healthcare Settings." *Applied Biosafety* 9:20–36.

Martini, G. A., and H. Z. Schmidt. 1968. "Spermatogenic transmission of the 'Marburg virus.' (Causes of 'Marburg simian disease')." *Klinische Wochenschrift* 46(7):398–400.

National Institutes of Health (NIH). 2002. *Guidelines for Research Involving Recombinant DNA Molecules* (retrieved November 15, 2006). www4.od.nih.gov/oba/rac/guidelines_02/NIH_Guidelines_Apr_02.htm

National Institute for Occupational Safety and Health (NIOSH). 2003. *Guidance for Filtration and Air Cleaning Systems to Protect Building Environments from Airborne Chemical, Biological, and Radiological Attacks*. DHHS Publication Number 2003-136. Washington, D.C.: NIOSH.

National Research Council (NRC), Committee on Hazardous Biological Substances in the Laboratory. 1989. *Biosafety in the Laboratory: Prudent Practices for the Handling and Disposal of Infectious Materials*. Washington, D.C.: National Academy Press.

National Sanitation Foundation (NSF) and American National Standards Institute (ANSI). 2002. *Class II Laminar Flow Biohazard Cabinetry. Standard 49*. Ann Arbor, MI: Academic Press.

Occupational Health and Safety Administration (OSHA). 1974. 29 CFR 1910, "General Industry Standards." Washington, D.C.: OSHA.

______. 2001. *Occupational Exposure to Bloodborne Pathogens; Needlestick and Other Sharps Injuries; Final Rule*. 66:5317–5325. Washington, D.C.: OSHA.

Oliphant, J. W., D. A. Gordon, A. Meis, and R. R. Parker. 1949. "Q-fever in Laundry Workers, Presumably Transmitted from Contaminated Clothing." *American Journal of Hygiene* 49:76–82.

Pike, R. M. 1976. "Laboratory-Associated Infections: Summary and Analysis of 3,921 Cases." *Health Laboratory Science* 13:105–114.

______. 1978. "Past and Present Hazards of Working with Infectious Hazards." *Archives of Pathology & Laboratory Medicine* 102(7):333–336.

______. 1979. "Laboratory-Associated Infections: Incidence, Fatalities, Causes, and Prevention." *Annual Review of Microbiology* 33:41–66.

Public Health Agency of Canada (PHAC). 2006. *Infectious Diseases* (retrieved November 15, 2006). www.phac-aspc.gc.ca/id-mi/index.html

Rao, G. G., B. P. Saunders, and R. G. Masterton. 1996. "Laboratory-Acquired Verotoxin Producing Escherichia coli (VTEC) Infection." *Journal of Hospital Infection* 33(3):228–230.

Richardson, J. H. 1973. "Provisional Summary of 109 Laboratory-Associated Infections at the Centers for Disease Control, 1947–1973." 16th Annual Biosafety Conference, Ames, Iowa.

Richmond, J. Y., and R. W. McKinney, eds. 1999. *Biosafety in Microbiological and Biomedical Laboratories*. 4th ed. Washington, D.C.: U.S. Government Printing Office.

Skinholj, P. 1974. "Occupational Risks in Danish Clinical Chemistry Laboratories." *Scandinavian Journal of Clinical and Laboratory Investigation* 33:27–29.

Sulkin, S. E., and R. M. Pike. 1949. "Viral Infections Contracted in the Laboratory." *New England Journal of Medicine* 241:205–213.

______. 1951. "Survey of Laboratory-Acquired Infections." *American Journal of Public Health Nations Health* 41(7):769–781.

Wedum, A. G. 1997. "History and Epidemiology of Laboratory-Acquired Infections (in Relation to the Cancer Research Program)." *Journal of the American Biological Safety Association* 2:12–29.

Appendix: Recommended Reading

Budavari, S., M. O'Neil, A. Smith, P. Heckelman, and P. Merck. 1996. *Merck Index: An Encyclopedia of Chemicals, Drugs, and Biologicals*. 12th ed. Whitehouse Station, NJ: Merck Research Laboratories.

Centers for Disease Control and Prevention (CDC). 2006. *Diseases and Conditions* (retrieved November 15, 2006). www.cdc.gov/node.do/id/0900f3ec8000e035

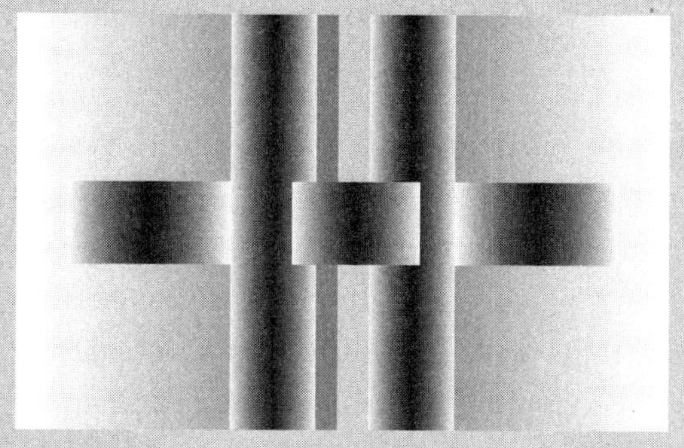

SECTION 4
INDUSTRIAL HYGIENE

Cost Analysis and Budgeting

David Eherts

LEARNING OBJECTIVES

- Learn to utilize strategic plans and tactical objectives in the budget-setting process.
- Be able to plan for and estimate annual expenses.
- Be able to plan for and estimate capital needs.
- Know how to establish budget priorities.
- Learn to assess the financial impact of safety and industrial hygiene (IH) programs.
- Examine the principles of cost-benefit analysis.
- Master the skills to communicate and sell the annual budget request to upper management.
- Be able to manage the traditional (or zero-based) budget for an IH or an environmental health and safety (EHS) department once it is successfully approved.

To be effective, industrial hygiene managers must understand the context of the business in which they work and be able to operate fully as business managers. That includes the ability to understand and build workable budgets and assess the financial and business impacts of safety and health programs and projects. This chapter provides a framework for addressing these issues and shows how it can be done. Some of the principles are illustrated through the application of a software tool, Return on Health, Safety, and Environmental Investment (ROHSEI), that streamlines the process of financial cost-benefit analysis. However, the same type of analysis can be done without software if one understands the principles of cost accounting or by using other tools available in the marketplace.

The concept of analyzing costs and benefits of proposed business projects is not new. What is relatively new is the use of traditional cost-benefit analysis to demonstrate the impacts that health, safety, and environmental decisions have on a business. This task is no longer necessarily solely within the purview of financial managers. Increasingly the skill to assess and articulate these business impacts is becoming a necessity for industrial hygienists and safety professionals. This chapter provides insights into the safety, health, and environmental budgeting process and offers examples of cost-benefit analyses that have been used successfully to demonstrate the business value of various safety and health initiatives.

The stress involved in proposing a sound EHS budget every year is something every safety and health professional will experience at one level or another, but some key activities can make the entire process not just easier, but more effective. This chapter will explore these key activities and propose some best practices that can be used to (1) accurately forecast the company's needs and (2) evaluate and justify a priority list of current and new activities requiring, and deserving, funding, while preparing a budget that aligns with and supports the company's overall strategic direction.

Planning

Everything starts with a plan. Reference the classic Deming Quality Cycle: Plan, Do, Check, Act (Deming 1986). Or consider Lewis Carroll's *Alice's Adventures in Wonderland* (Carroll 1862), in which one day Alice comes to a fork in the road and sees a Cheshire cat in a tree. "Which road do I take?" she asks. The cat's response is in the form of a question: "Where do you want to go?" "I don't know," Alice answers. "Then," says the cat, "it doesn't matter."

Most big companies start their annual budget process in August or September, but they do not start the objective-setting process until November or sometimes even January of the following year. How can people establish budgets when they do not know where they are going or what they are going to be doing?

The budget process ideally starts with a strategic plan that should establish general direction and priorities. Annual objectives should follow from the strategic plan and include any emerging issues, opportunities gleaned from periodic benchmarking, progress on the EHS management system, and opportunities discovered through audit programs and/or risk assessments. This type of budgeting has been called *program budgeting*, a term first coined by the U.S. Department of Defense in the early 1960s (Garret et al. 1998). Holtshouser (1988) points out, "Essentially, program budgeting transforms the annual preparation of a budget into a conscious effort to appraise and formulate an organization's goals via the long-range view." Though these initiatives normally flow down from upper management, every opportunity should be afforded for proposed objectives to come up through the ranks. Buy-in from the field is very important, but so is buy-in from upper management, so educating company bosses on the strategic direction, benchmarking results, and emerging issues (as well as keeping them up to date on audit and inspection results, near-misses, and accident root causes) will help ensure their buy-in and, therefore, appropriate budgeting.

Once an initial list of objectives is proposed, priorities should be set based on relative risk, cost-benefit analyses, and management evaluations. This chapter is not meant to be a treatise on strategic planning and annual objective-setting, but it should be emphasized that a good budget cannot be built without a thorough understanding of where the program is going and what it will be doing (Lalli 2003).

Once there is a tactical list of possible objectives, they should be prioritized first by risk reduction. Projects that reduce risk generally save the company money. For example, if a particular hazard could result in one million dollars of property loss and the chances of this situation occurring in any given year is one in one hundred, then at a very simple level, the company should put away $10,000 annually to cover this risk. The time value of money and the potential for the event to occur in year one certainly complicate the calculation, but the point is that companies pay for their risk. So once the risk-reduction opportunities are prioritized, cost-benefit analysis will assist in deciding which solutions to pursue and how to convince management to release the necessary funds.

Analyzing the Business Case for Health, Safety, and Environmental Investment

To make a sound business case, one must understand cost-benefit analysis. A number of tools are available for conducting financial analyses, but one is particularly well suited for health, safety, and environmental professionals: Return on Health, Safety, and Environmental Investment ROHSEI (Eherts 2005). This software tool is used in this chapter to illustrate the issues that can be addressed in safety and health cost-benefit analyses and the power in assessing them correctly.

ROHSEI was developed circa 1997 by Arthur Andersen Consulting and a task force of fifteen Organization Resources Counselors (ORC) member companies. The tool includes a software program that assists in measuring, demonstrating, communicating, and understanding how EHS investments impact EHS and business performance.

Companies use ROHSEI when evaluating capital and expense investment decisions, such as material selections, capacity expansion, program development investment, resource allocation, training approaches, and risk-management strategies. They also use it when they need to improve their understanding of the business impact of health and safety investments on business performance, including productivity, product quality, and customer satisfaction.

Since its launch in 1997 ROHSEI analyses have helped companies to decide

- the best way to protect structures from fires
- whether to offer on-site primary (nonoccupational) healthcare
- whether to in-source short-term or long-term disability (STD/LTD) management
- whether to recycle disposable coveralls
- the best method for soil remediation
- whether creating a Web site for an electronic HSE manual, Web-based training, material safety data sheets (MSDSs), and accident reporting is cheaper than using old-fashioned hard-copy manuals
- whether behavior-based safety programs are more cost effective than traditional audit/inspection/management system-based programs
- whether it is cost effective to offer primary healthcare in Europe (within a socialized medicine system)
- whether it pays to invest in computerized occupational health patient file and industrial hygiene database programs to facilitate real-time epidemiology
- whether PPE (respirators) or engineering controls (local exhaust ventilation) is more cost effective
- whether ergonomics programs for the lower back are cost effective
- whether ergonomics programs for work-station improvements are cost effective
- whether the sales calls missed for training in fleet safety programs is offset by lower accident rates
- whether it pays to buy and implement information technology software (EHS modules)
- whether starting a consulting firm from within a company's EHS department is preferable to downsizing.

Further information concerning the details of these analyses is available at www.orc-dc.com.

The ROHSEI methodology helps answer questions such as "What are the direct business impacts of each alternative vis-à-vis incident reduction, waste reduction, and raw material usage?" and "How does each alternative impact hidden business issues, such as worker productivity, product quality, and customer satisfaction?" (Eherts 2005). It gives answers in terms of net present value, return on investment, payback period, internal rate of return, production equivalent units, and supporting calculations and assumptions. For health and safety decisions there is also an indirect impact analysis tool to examine the hidden impacts. The tool provides qualitative rankings of investment alternatives, evidence to support the ratings, and graphs that display how each investment alternative impacts worker productivity, product quality, and customer satisfaction.

The process improves decision making in four ways: (1) integrating health and safety into the business-investment evaluation process, (2) communicating health and safety impacts in traditional financial and business terms, (3) improving understanding of the *hidden* impacts of health and safety investments, and (4) improving overall health and safety performance by encouraging better business decisions on resource allocation.

Being able to conduct financial analyses of safety and health decisions is important to leadership. For example, after seeing ROHSEI at an executive committee meeting in 1997, one company rolled it out at their headquarters and then throughout their geographical regions. Not only did their management accept the idea, they embraced it by requesting that major projects involving EHS issues complete a detailed ROHSEI analysis prior to approval. At the same time, they trained their area EHS directors in the process so that, as their sites requested assistance with the analysis, they were immediately ready to step in (Eherts 2005).

Although the process offers significant benefits, there is also the potential for misuse, especially if it is manipulated by untrained users. When accident or illness prevention is involved, it is imperative to utilize the program only when comparing several alternatives that will provide the same desired result. For example, there may be several methods of focusing attention on a safe and healthy workplace that would lead to decreased accident and illness rates. The system works very well when used to decide which method may be most effective, thereby freeing up funds within the EHS budget for other progressive initiatives. But an untrained user could attempt to use the program to determine, for example, whether a method that is forecast to decrease accident and illness costs by 25 percent is more cost effective than a method that has the potential to decrease accident and illness costs by 50 percent. Such analyses are sure to fail, not only because they foster mistrust by employees and/or their representatives, but also because they violate the basic ethics of the profession—that the safest and most healthful workplace possible must be provided.

Even with the limitation that the process be used only for alternatives yielding the same outcome with regard to accident prevention, the benefits are many. Experience has shown that management agrees with most proposals for workplace safety improvements. However, with an accompanying ROHSEI analysis, they often also seem to feel a greater sense of urgency to get improvements underway, which is understandable since the analysis allows them to readily see the benefits, and they are interested in realizing these gains as soon as possible. By using ROHSEI, EHS has the potential to be truly seen as a business partner. A number of case studies involving ROHSEI analysis follow (Eherts 2005).

CASE STUDY

Cost-Benefit Analysis of Various Methods of Fire Protection

This analysis, described in Figure 1, compares the costs and benefits of various methods of fire protection for a chemical manufacturing facility. One alternative, favored by the most American insurers, is to install a sprinkler system to protect the building. Another, favored by many European EHS professionals, is to employ a fire brigade. A third alternative is to ensure that there is a good fire-detection system and an overabundance of clear exit routes, and rely on a nearby fire company to put out fires.

Using a team approach, EHS professionals carefully consider and quantify such costs and benefits as insurance premiums, capital investment, ongoing maintenance and testing of a sprinkler system, initial and ongoing costs of an on-site fire brigade, insurance deductible for rebuilding costs, lost sales, and, more importantly, lost market share in the unlikely event of a fire (Figure 2).

EHS professionals factor in the probability of a fire and its consequences for each alternative. The software calculates net present value (NPV), return on investment (ROI), internal rate of return (IRR), and discounted payback period (DPP), using the following formulas:

$$\text{NPV} = \sum_{i=1}^{t}\left[\frac{(B_i - C_i)(1+n)^i}{(1+r)^i}\right] = \sum_{i=1}^{t}\left[\frac{B_i\,(1+n)^i}{(1+r)^i}\right] - \sum_{i=1}^{t}\left[\frac{C_i\,(1+n)^i}{(1+r)^i}\right] \qquad (1)$$

Equation 1 takes the future cash flows from a project investment and discounts them to the present. Essentially the net present value (NPV) is the difference between the present value of the future cash flows and cost of the investment. NPV, today's value of a series of future costs and benefits, is calculated by subtracting the total of all discounted benefits. Equation 1 calculates NPV based on costs, benefits, and the discount rate that is entered by the user (Birkner 1998).

Note: In equations 1–5, r = the discount rate; i = the period (the number of years in the future that the future value will be received); n = the interest rate, and t = time. The discount rate r is the reciprocal of 1 + n. For example, if the interest rate is 5%, the discount rate is approximately 0.9524.

$$\text{ROI} = \frac{\sum_{i=1}^{t}\left[\frac{(B_i - C_i)(1+n)^i}{(1+r)^i}\right]}{\sum_{i=1}^{t}\left[\frac{C_i\,(1+n)^i}{(1+r)^i}\right]} = \frac{\sum_{i=1}^{t}\left[\frac{B_i\,(1+n)^i}{(1+r)^i}\right] - \sum_{i=1}^{t}\left[\frac{C_i\,(1+n)^i}{(1+r)^i}\right]}{\sum_{i=1}^{t}\left[\frac{C_i\,(1+n)^i}{(1+r)^i}\right]} \qquad (2)$$

The formula for return on investment (ROI) is calculated by dividing the NPV by the present value of the project costs. The NPV is calculated by subtracting the present value of project costs from the present value of project benefits. Equation 2 calculates ROI based on costs, benefits, and the discount rate that is entered by the user (Birkner 1998).

$$\text{IRR} = r \text{ when } = \sum_{i=1}^{t}\left[\frac{B_i - C_i}{(1+r)^i}\right] = 0 \qquad (3)$$

The formula for internal rate of return (IRR) for a project's costs and benefits is the interest rate when the NPV of the project is equal to zero. The IRR is the most often used alternative to NPV. This rate relies only on the cash flows of the project, not on external rates. Equation 3 calculates IRR based on costs and benefits that are entered by the user (Birkner 1998).

$$\sum_{i=1}^{y}\left[B_i\,\frac{(1+n)^i}{(1+r)^i}\right]-\sum_{i=1}^{y}\left[C_i\,\frac{(1+n)^i}{(1+r)^i}\right]\geq 0 \qquad (4)$$

The formula for the discounted payback period (DPP) refers to the amount of time it takes for a project to pay for itself, accounting for the time value of money of each cash-flow element. As such, a project is "paid back" in the year that the cumulative net cash flows exceed the capital outlay of the initial year (year 0). Although this metric is traditionally used for projects with capital investments, it can be used for programmatic analyses as well, such as when substantial investments are made in training in the early years of a project. In this case, it would be more accurate to say that a project is paid back when the cumulative cash inflows exceed cumulative cash outflows (see Equation 4) (Birkner 1998).

It is important, however, to consider costs, such as closure, that occur late in a project's life. Initial capital costs may be recovered somewhere during the midpoint of the analysis, but late-in-project costs could be incurred after this time, making simple payback analysis misleading. If this is the case, consider looking at total project costs as opposed to cumulative project costs (see Equation 5). Note, however, that it is not an established metric.

$$\sum_{i=1}^{y}\left[B_i\,\frac{(1+n)^i}{(1+r)^i}\right]-\sum_{i=1}^{t}\left[C_i\,\frac{(1+n)^i}{(1+r)^i}\right]\geq 0 \qquad (5)$$

Two other popular methods of cost-benefit analysis are the *initial investment rate of return* (IIRR) and the *accounting rate of return* (ARR) (Holtshouser 1988). The IIRR is calculated by dividing the net savings after taxes by the initial investment. This method disregards the time value of money and is biased toward proposals that yield a quick return. It is mostly used by cash-strapped companies that need quick returns. The ARR is calculated by dividing the average net income by the average investment. It measures capital expenditures over the life of the project and takes into account the effects of depreciation and taxes. It specifically ignores discounting of future cash flows and, therefore, is the opposite of IIRR; it tends toward bias in favor of proposals that result in cash flows toward the end of the project life.

- *Business Case Summary Overview* -
1. Focus of opportunity
2. *Scenarios (options)*
3. Primary benefits and drawbacks
4. Recommendations

Return to the Navigation Screen

Print the Business Case Summary

Use the scroll bar to scroll down this screen.

2. Describe Scenarios (options)

Name the Scenarios

Name your scenarios using the button at right. Briefly describe each Scenario (or option), noting which, if any, investment alternative will serve as the "current situation" (or base case) in the analysis. While the model facilitates a quantitative analysis of each alternative, it is important for all involved to acquire a thorough understanding of each alternative.

1. *Leave the facility unprotected.*

2. *Utilize a paid on-site fire brigade (assuming full detection of heat and smoke). Current situation at Dagenham and Vitry.*

3. *Commit capital investment(s) to install sprinkler (fire suppression) system(s).*

FIGURE 1. Describe scenarios (*Source:* Birkner 1998)

- *Business Case Summary Overview* -
1. Focus of opportunity
2. Scenarios (options)
3. *Primary benefits and drawbacks*
4. Recommendations

Return to the Navigation Screen

Print the Business Case Summary

Use the scroll bar to scroll down this screen.

3. Primary benefits and drawbacks

Throughout the analysis, you should use this page to document important issues and assumptions about Scenarios and their benefits and costs. In addition, summarize the key benefits and costs for each Scenario.

1. Leave facility unprotected:
Costs: Business Interruption Insurance costs of $.15/$100 for an unprotected building. Property insurance costs of $.15/100 based upon same assumptions and $15MM in property value. In the rare event of a fire (1:1000/year, expected loss would be deductible of $2MM.) $500 per year for annual fire company support.
Benefits: No capital costs, salary, or expenses for fire brigade.

2. Fire brigade:
Costs: 6 person team, 3 shifts a day. Paid $35,000/year, 29% benefits for total cost of $45,150/member/year. $300/year/member for continued training. Cost of structure to house equipment, personnel $50,000 and equipment (truck) cost of $250,000, $10,000 annual equipment maintenance/replacement costs. Business Interruption Insurance costs of $.15/$100 for an unprotected building. Property insurance costs of $.15/100 based upon same assumptions and $15MM in property value. In the rare event of a fire (1:1000/year, expected loss would be deductible of $2MM.)
Benefits: Double as hazmat team, save training costs of $6000/year. Savings of $500/year for local fire company support (no longer necessary). Only 50% of property will be destroyed in the rare event of a major fire (but deductible is same as if complete loss). No capital sprinkler system costs or maintenance.

FIGURE 2. Primary benefits and drawbacks (*Source:* Birkner 1998)

Other calculations that are sometimes used include:

- ***Cash flow:*** The flow of cash into and out of a company associated with the case being studied, including simple (undiscounted) cash flow (does not consider the value of time) and discounted cash flow (adjusted to consider the value of time).
- ***Net cash flow:*** The cash value of the benefits minus costs for each option under consideration.

- ***Break-even analysis:*** The amount of sales income needed to equal the costs of the option–the point at which the project breaks even.
- ***Profitability index:*** The present value of benefits divided by the present value of costs.
- ***Total cost of ownership:*** The sum of all costs over the time period of the option being evaluated.

Comparison of Options

Calculation	Leave Unprotected	Fire Brigade	Sprinkler
Net Present Value	($0)	($5,184,901)	$398,736
Internal Rate of Return	NA	NA	32%
Return on Investment	(68%)	(66%)	88%
Discounted Payback Period	Does not pay back	Does not pay back	3.1 years

(*Source:* Birkner 1998)

If one utilizes a software program such as ROHSEI, there is no need to perform these calculations independently. The software performs them automatically. The results in this case are detailed in the table above.

CASE STUDY

Cost-Benefit of Recycling Protective Outer Garments

This case study involved a visit by a salesman attempting to convince a company to use a service to recycle their disposable protective outer garments. The salesman stated that there would be a good ROI if the company initiated the program. In testing this claim, the EHS staff calculated the time and effort of collecting and shipping used disposable garments to the service company (crumpled Tyvek is bulky and difficult to package efficiently) versus the cost of incinerating the garments under their existing system. The ROHSEI analysis revealed that it would cost more in time (productivity losses) and money to use the proposed recycling system. EHS pointed out to management, which was familiar with the ROHSEI tool, that with recycling they would lose money, but for a good reason, and suggested that after studying the analysis, management might feel that the cost was reasonable–and, as they were now certain (because of the ROHSEI analysis) that there were no hidden costs, they might decide to recycle despite the additional cost. In this case, ROHSEI did not quantify the benefits of employee perception that management cares about the environment and recycles where possible. These benefits may be realized in employee loyalty and decreased attrition (recruiting and retraining costs), not to mention productivity and quality gains. After all, recycling is the right thing to do.

There may be other cases where the initial ROHSEI analysis indicates that none of the alternatives considered have a positive ROI or NPV, especially when changes are being proposed to comply with OSHA regulations. But if utilization of this program is continued at deeper levels, it may be possible to show every initiative in a positive financial light. It is also important, if not imperative, to better understand the cost of each program or initiative. Sometimes, the best choice is the alternative that gives full compliance at the least cost. Other times, management may find that full compliance comes with significant cost savings.

Knowing the exact cost saved or spent on any health and safety initiative certainly helps to make the best decisions, for both employees and the business.

Finally, even though certain initiatives offer a clear cost benefit to the company, emerging issues and a changing regulatory environment may require that priority be given to other areas. Though intuitively an initiative that most reduces risk and offers a safer and healthier work environment should take precedence, the reality is that compliance with a new regulation may take center stage. So it is always important to monitor upcoming regulations and stay involved with industry groups and professional organizations, such as the American Society of Safety Engineers (ASSE), the American Industrial Hygiene Association (AIHA), the National Safety Council (NSC), and the National Fire Protection Association (NFPA), and make frequent visits to regulatory Web sites such as osha.gov to stay current with emerging issues and the changing regulatory environment.

Annual tactical goals can be established if there is a common, well-understood, and well-communicated strategic direction and a good understanding of emerging issues and changing regulations. These goals should reflect regulatory requirements and risk reduction and cost-benefit calculations to stand the best chance of full endorsement by management. With management's full support of EHS annual objectives, a budget can then be established to ensure the best chance of successfully implementing a long-term strategic plan and accomplishing the EHS department's vision.

CASE STUDY

Cost-Benefit Approach of Offering On-site Health Care

This study involves the question of whether to increase head count and invest in a larger medical facility in order to offer on-site health care (primary care) to employees for nonoccupational illnesses or injuries. On-site care may already be available within a company facility for occupational injuries or illnesses. This case study considers whether the cost of additional medical personnel and a larger clinic might be offset by early detection and less time lost to private doctors' visits for personal medical issues. Many large companies are essentially self-insured for medical benefits (the company directly pays for the costs incurred and their insurance companies simply manage the paperwork), so they would save the costs of the actual doctors' visit bills. Detailed analysis is shown in Figures 3 through 9.

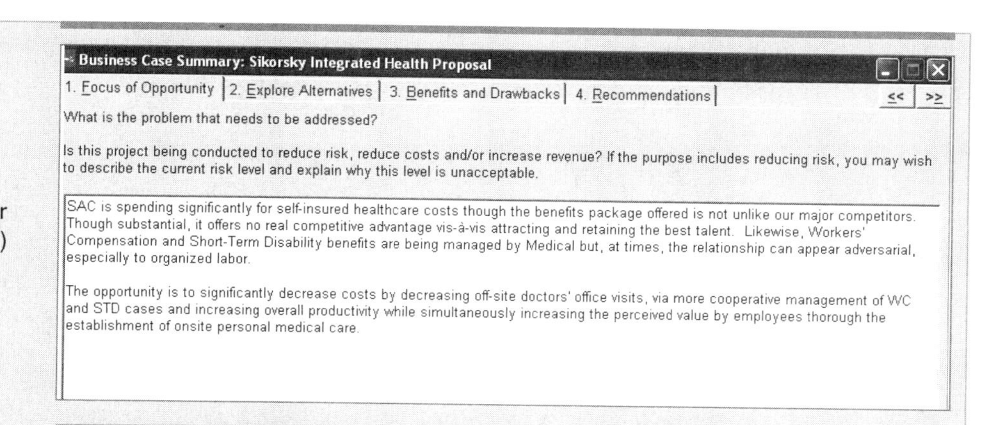

FIGURE 3. Business case summary (*Source:* Eherts 2005)

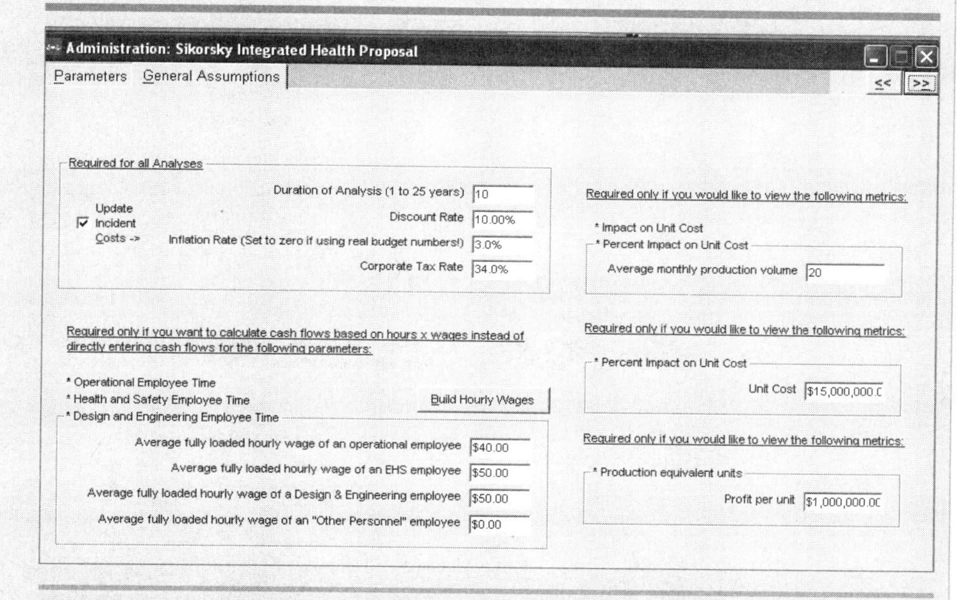

FIGURE 4. General assumptions (*Source:* Eherts 2005)

Business Case Summary: Sikorsky Integrated Health Proposal

1. Focus of Opportunity | 2. Explore Alternatives | 3. Benefits and Drawbacks | 4. Recommendations

Throughout the analysis, you should use this page to document important issues and assumptions about Scenarios and their benefits and costs.

In addition, summarize the key benefits and costs for each Scenario.

Current Situation: We operate an occupational health clinic to provide federal and state required first aid, medical surveillance and workers' compensation programs. The company sees regulatory compliance as a threshold expectation and the budget will continue at the current level.

Costs: Integrated Primary Care and Disability Management: Costs include two additional MDs and two additional RNs at $200,000 per MD and $75,000 per RN for a total of $550,000. Capital renovations necessary to accommodate the increased utilization are estimated at $250,000. An additional $50,000 of expense will be needed annually for additional medical supplies and pharmacy.

Benefits: If we conservatively estimate 0.5 personal visits per person, the clinic will save 5000 visits at 3 hours each at $40/hr or $600,000 in productivity and 5000 visits times self-insurance savings of $107 each or $535,000. Costs of Disability Case Management are estimated at 5000 lost days. Benefits associated with more diligent Case Management are estimated to be 500 days (x 8 hours/day x $40/hour) or approximately $200,000 in productivity savings, not to mention medical costs savings. Other savings include closure of legacy cases at significant discount to the accrued reserve and avoidance of unnecessary surgeries for an estimated combined savings of $300,000.

FIGURE 5. Benefits and costs (*Source:* Eherts 2005)

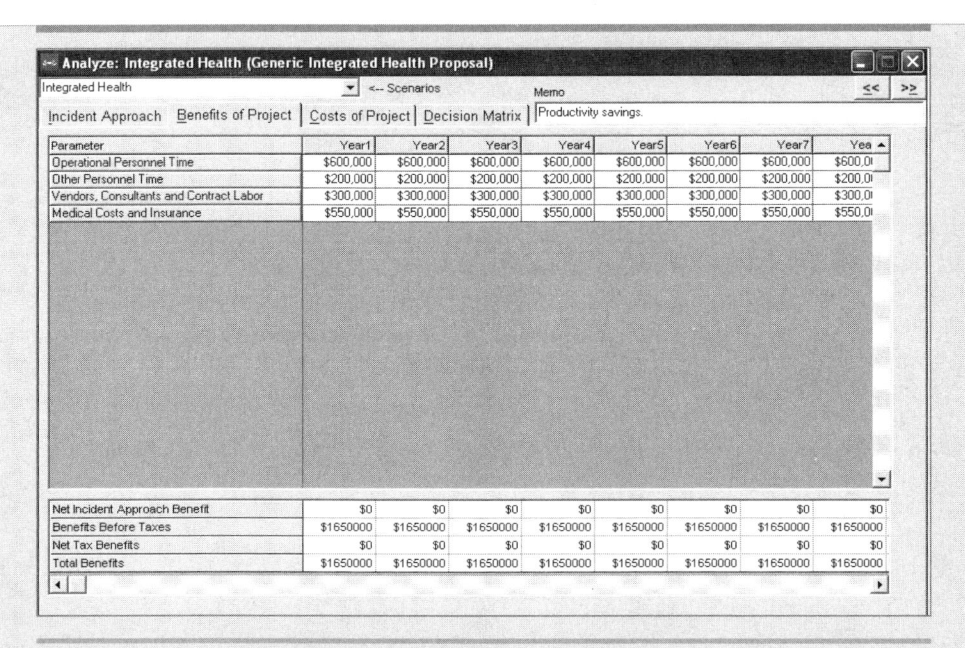

Parameter	Year1	Year2	Year3	Year4	Year5	Year6	Year7	Yea
Operational Personnel Time	$600,000	$600,000	$600,000	$600,000	$600,000	$600,000	$600,000	$600,0
Other Personnel Time	$200,000	$200,000	$200,000	$200,000	$200,000	$200,000	$200,000	$200,0
Vendors, Consultants and Contract Labor	$300,000	$300,000	$300,000	$300,000	$300,000	$300,000	$300,000	$300,0
Medical Costs and Insurance	$550,000	$550,000	$550,000	$550,000	$550,000	$550,000	$550,000	$550,0

Net Incident Approach Benefit	$0	$0	$0	$0	$0	$0	$0	$0
Benefits Before Taxes	$1650000	$1650000	$1650000	$1650000	$1650000	$1650000	$1650000	$1650000
Net Tax Benefits	$0	$0	$0	$0	$0	$0	$0	$0
Total Benefits	$1650000	$1650000	$1650000	$1650000	$1650000	$1650000	$1650000	$1650000

FIGURE 6. Benefits of project (*Source:* Eherts 2005)

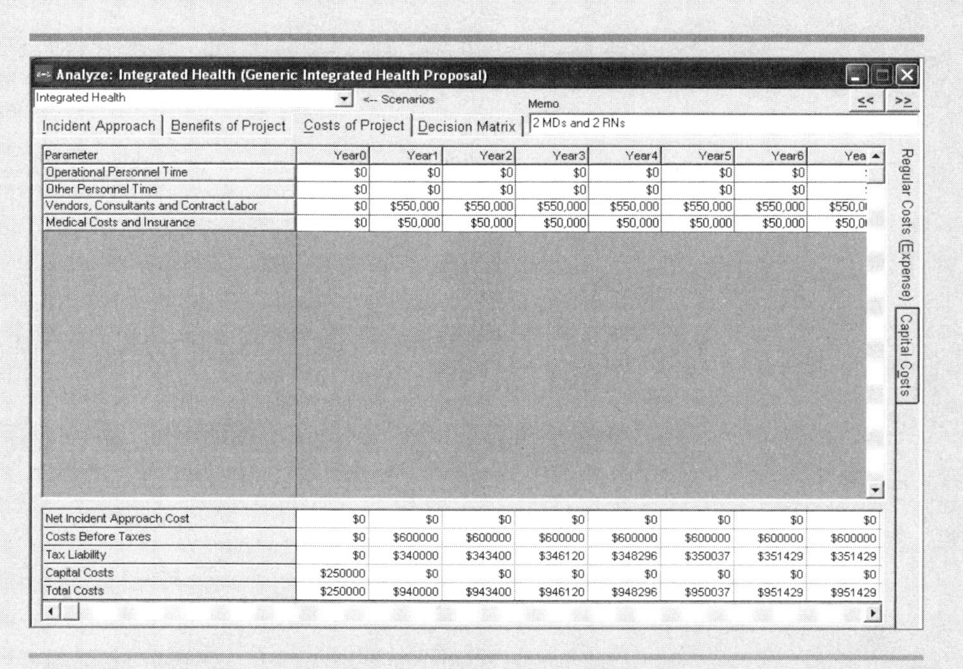

Parameter	Year0	Year1	Year2	Year3	Year4	Year5	Year6	Yea
Operational Personnel Time	$0	$0	$0	$0	$0	$0	$0	
Other Personnel Time	$0	$0	$0	$0	$0	$0	$0	
Vendors, Consultants and Contract Labor	$0	$550,000	$550,000	$550,000	$550,000	$550,000	$550,000	$550,0
Medical Costs and Insurance	$0	$50,000	$50,000	$50,000	$50,000	$50,000	$50,000	$50,0

Net Incident Approach Cost	$0	$0	$0	$0	$0	$0	$0	$0
Costs Before Taxes	$0	$600000	$600000	$600000	$600000	$600000	$600000	$600000
Tax Liability	$0	$340000	$343400	$346120	$348296	$350037	$351429	$351429
Capital Costs	$250000	$0	$0	$0	$0	$0	$0	$0
Total Costs	$250000	$940000	$943400	$946120	$948296	$950037	$951429	$951429

FIGURE 7. Costs of project (*Source:* Eherts 2005)

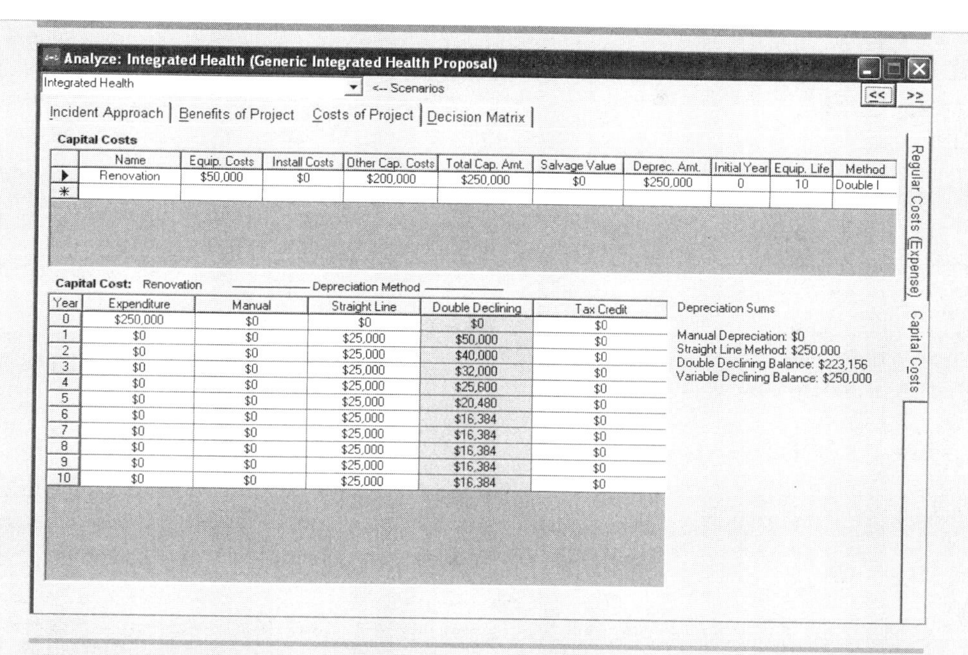

FIGURE 8. Project costs (*Source:* Eherts 2005)

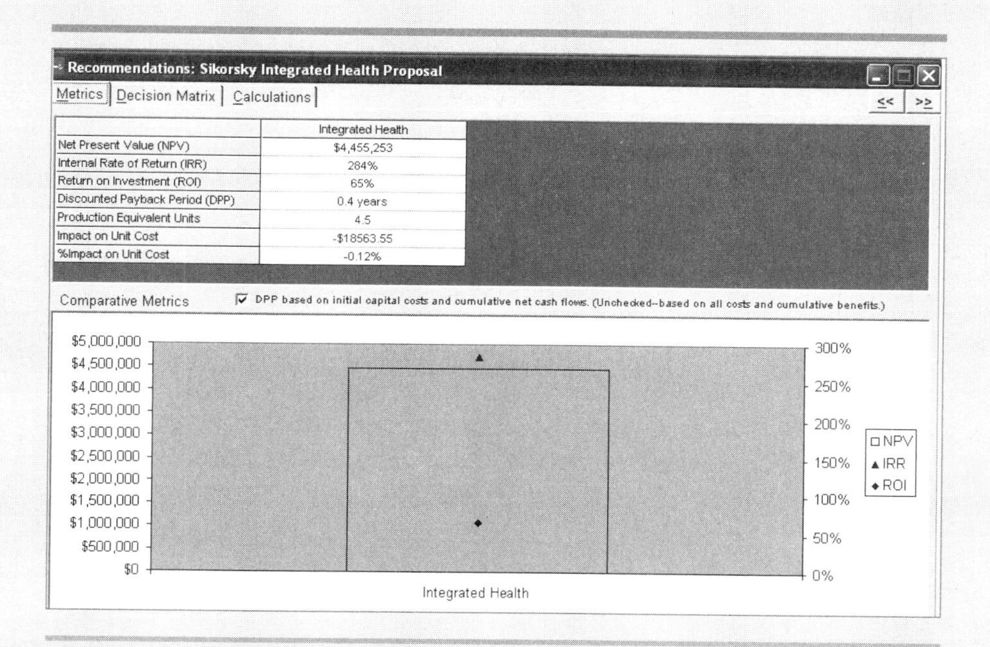

FIGURE 9. Metrics (*Source:* Eherts 2005)

CASE STUDY

Cost-Benefit Analysis of Hierarchy of Controls

This study involves the best way to control employee exposure while manufacturing potent pharmaceuticals or toxic chemicals and complying with regulatory agency rules for preventing cross-contamination. These are detailed in Figure 10. The costs and benefits are shown in Figure 11.

Entering these values into the ROHSEI programs yields results in favor of the engineering control investment, as shown in Figure 12.

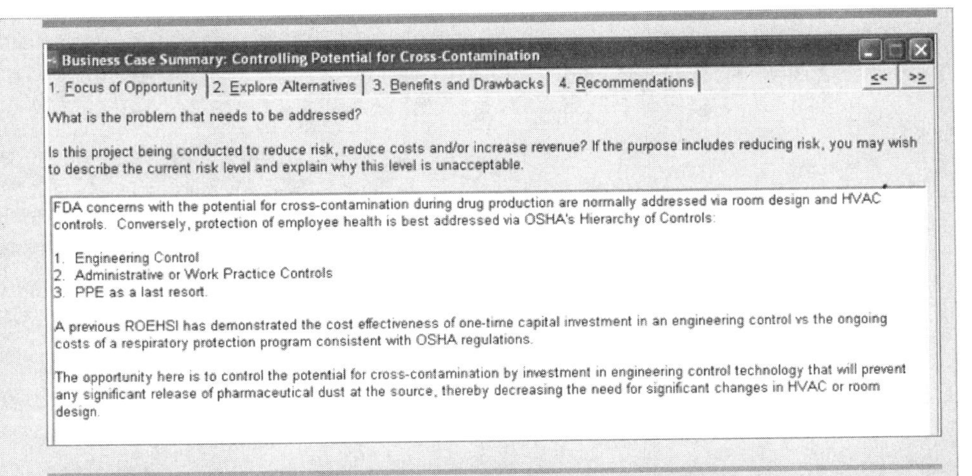

FIGURE 10. Focus on opportunity (*Source:* Eherts 2005)

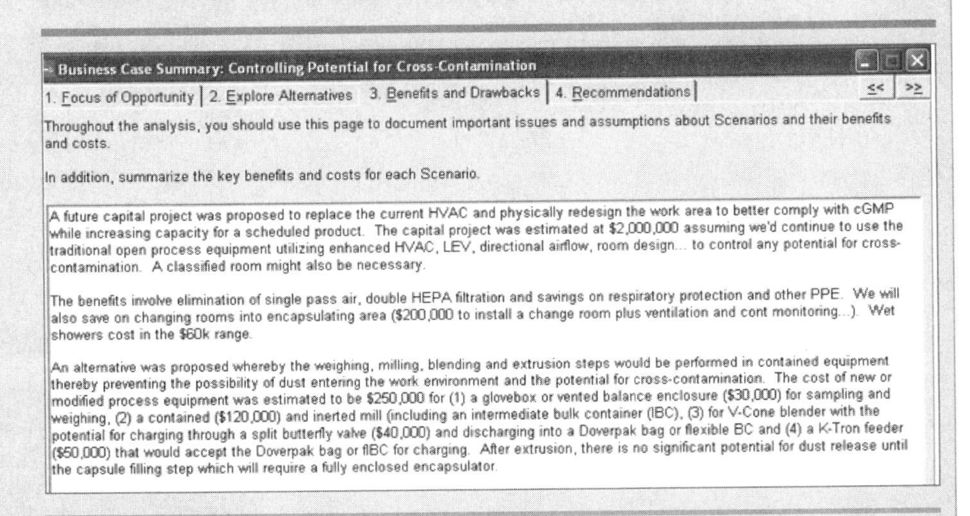

FIGURE 11. Benefits and drawbacks (*Source:* Eherts 2005)

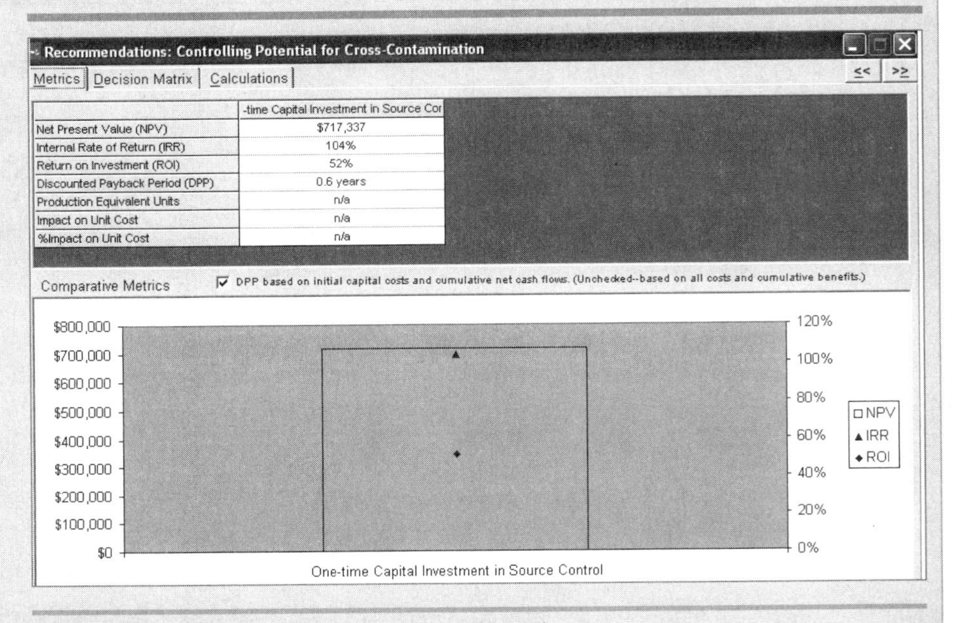

	-time Capital Investment in Source Cor
Net Present Value (NPV)	$717,337
Internal Rate of Return (IRR)	104%
Return on Investment (ROI)	52%
Discounted Payback Period (DPP)	0.6 years
Production Equivalent Units	n/a
Impact on Unit Cost	n/a
%Impact on Unit Cost	n/a

FIGURE 11. Metrics (*Source:* Eherts 2005)

Salaries and Compensation

Budgets should start with the goal of acquiring and maintaining the necessary human capital required to run a program. "Talent is a multiplier. The more energy and attention you invest in it, the greater the yield will be" (Labarre 2001).

As demonstrated in the section on ROHSEI, the costs of an exceptional team of professionals pays back handsomely in a more visibly committed management, significantly reduced risk, and complete employee buy-in.

Benchmarking Salary Data

Benchmarked salary data are available from many sources. ORC, for a membership fee, provides high-quality data on salary ranges based upon occupation, region of the country, experience, and education. ASSE, on its Web site, is another source, though the data are less specific; the same is true for AIHA. If one is seeking a job, these data may point to preferable industries. If an employer is searching for candidates, these data will tell what compensation is necessary to get the best talent.

Compensation Other Than Salary

Salary is only one piece of the total compensation equation. Potential for an annual bonus, good benefits and pension plans, paid continuing education and schooling, and the option to participate in a 401k plan and even a long-term incentive plan (LTIP) will also help attract and retain the most talented employees. Any working population can be divided into three categories (Buckingham 1999):

1. People who are engaged (loyal and productive): 26 percent
2. People who are not engaged (just putting in time): 55 percent
3. People who are actively disengaged (unhappy and spreading their discontent): 19 percent

The way to ensure an engaged, loyal, and productive workforce, in addition to giving proper compensation, is to demonstrate that employees are valued by the company.

If bonuses, pension plans, paid continuing education, and other expensive benefits are not available to a manager, the best candidates can be attracted and retained by adding benefits that do not cost as much but may be attractive based upon employees' sense of pride and the feeling that they are respected and valued by their employer. Examples of these benefits include company-leased vehicles, transit checks, flexible schedules or additional time off, combined business and vacation travel, and even annual events such as family days and planned social events at professional meetings and conferences.

Besides budgeting for salaries (as well as benefits, training, travel, and conferences), EHS professionals should also budget for such items as air sampling media, analyses, replacement of old equipment, consultants, subscriptions, software, safety supplies, equipment maintenance and calibrations, medical clinic and office supplies, and permits. An EHS department will usually be responsible for managing environmental waste disposal (Holtshouser 1988). Budget factors should include a mark-up consistent with inflation and the expected growth (or decline) of the amount of disposed waste, as well as costs associated with minimizing waste.

Zero-Based Budgeting

Some companies require the use of zero-based budgeting (Pyhrr 1973), in which the entire budget is built from scratch every year based on expected programs and demands regardless of what was budgeted and spent in prior years. In this model, every proposed expense needs to be justified again for the following year and the budget estimated accordingly. The significantly increased effort necessary to do zero-based budgeting every year is seen by many as a waste of money. On the other hand, fiscal watchdogs often encourage use of this process as a way to ensure that every expense is carefully considered and justified. Additional positive aspects of zero-based budgeting include increased restraint in developing budgets, a reduced entitlement mentality with respect to cost increases, and making budget discussions more meaningful during review sessions. Finally, zero-based budgeting can be useful on occasion for shaking up a process that may have grown stale and counterproductive over the years.

Capital Expense Planning

Business dictionaries generally define a *capital expenditure* as a payment for assets such as property, industrial buildings, or equipment (Harvey 2004), but not for day-to-day expenditures such as payroll, inventory, maintenance or marketing. Capital expenditures are expected to increase the value or quantity of company assets.

Companies may also define a capital expense monetarily (e.g., a $1100 piece of sampling equipment must be purchased with capital money, whereas a similar piece of equipment that costs $900 can be paid for with expense money). Capital budgeting is therefore also an extremely important responsibility of a safety manager, because the lack of a capital budget may prevent the purchase of equipment needed to successfully complete an important objective. Capital projects must be prioritized in the same way as expense budgets. ROHSEI is a good tool for determining and prioritizing the items on a capital budget.

An important difference between capital and expense money involves how they are treated from a tax perspective. Capital projects allow for depreciation. The reasons for treatment of capital depreciation in a certain way are beyond the scope of this chapter, but consistency with a company's finance department rules is a must. Communication is extremely important.

The Budget Cycle

Before the next year's budget process begins, it is extremely important to understand exactly what is required of an organization in the upcoming fiscal year. Many companies finalize the following year's budget by the end of third quarter, before they have completed their objective-setting process. If a company insists on finalizing next year's budget before the fourth quarter, it would be wise to ensure that strategic planning and the following year's objectives are finalized so that associated costs and any projected revenues can be included in the new budget.

Establishment of a process for tracking emerging issues and upcoming regulatory changes and a program for accurately determining the cost benefits of proposed initiatives are crucial to ensure that the budget is matched to priorities. Changes to state or federal EHS regulations, should be viewed as an opportunity to under take initiatives that would decrease risk and save the company money. These should be prioritized and their value well communicated to management.

Summary

The time to demonstrate one's value to the organization is long before business realities dictate that changes are necessary. One should be proving one's value on a daily basis and planning for next year's budget requests long before the process begins formally in the company.

Being seen as a true business partner will further health and safety in the organization beyond what can be achieved through old-fashioned programs based solely on trusting management to always "do the right thing." Decreasing risk and complying with emerging regulatory changes while saving the company money at every turn is the secret to long-term success.

References

Birkner, Lawrence. "Return on Health and Safety Investment" (a presentation at the American Industrial Hygiene Conference and Exhibition). Atlanta, Georgia. May 1988.

Buckingham, Markus and Curt Coffman. 1999. *First, Break All the Rules: What the World's Greatest Managers Do Differently*. New York: Simon & Schuster.

Carroll, Lewis. 1862. *Alice's Adventures in Wonderland*. New York: Bantam.

Deming, W. Edwards. 1986. *Out of the Crisis*. Cambridge, MA: MIT Press.

Eherts, David M. "An Insider's Take on Socially Responsible Companies—The Value of EHS" (a presentation as part of the SoundWaters Environmental Business Lecture Series). Stamford, Connecticut, May 17, 2005.

Garrett, Jack, Lewis Cralley, and Lester Cralley, eds. 1988. *Industrial Hygiene Management*. Hoboken, NJ: John Wiley and Sons.

Harvey, Campbell. 2004. *The Financial Glossary*. Huntingdon Valley, PA: Farlex, Inc.

Holtshouser, Joseph. 1988. "Budgeting." In Jack T. Garrett, Lewis J. Cralley, and Lester V. Cralley, eds. *Industrial Hygiene Management*, pp 353–371. New York: John Wiley and Sons.

Labarre, Polly. 2001. "Marcus Buckingham Thinks Your Boss Has an Attitude Problem." *Fast Company* (July) 49:88.

Lalli, William. 2003. *Handbook of Budgeting*, 5th ed. Hoboken, NJ: John Wiley and Sons, Inc.

Pyhrr, P. A. 1973. *Zero-Base Budgeting: A Practical Management Tool for Evaluating Expenses*. New York: John Wiley and Sons.

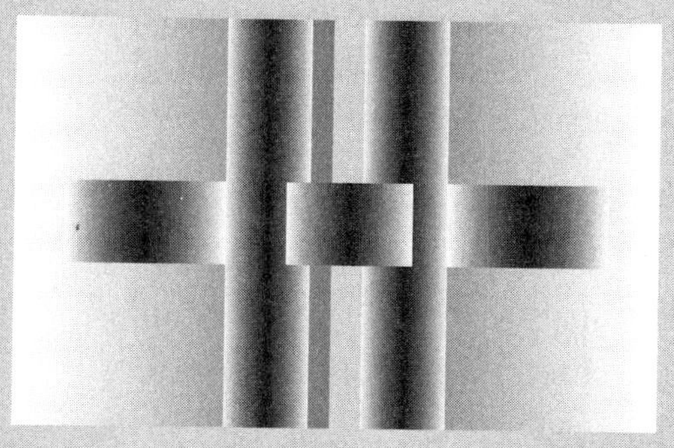

SECTION 4
INDUSTRIAL
HYGIENE

Benchmarking and Performance Criteria

Forrest Illing

LEARNING OBJECTIVES

- Discuss VPP selection criteria and the different VPP programs.
- Identify four of NIOSH's key areas of responsibility.
- Discuss the three categories used to classify the results of a workplace exposure assessment.
- Identify the four employer requirements of the OSHA HazCom Standard.
- Understand the statistical principles for evaluating whether worker exposure is below an acceptable occupational exposure limit (OEL) with 95 percent confidence.
- Discuss the key provisions of the OSHA Noise Standard, 1910.95, sec (i).
- Discuss the principles underlying the process of control banding in the COSHH Essentials model.
- Understand some of the factors that can affect the accuracy of inhalation exposure models.

BEST PRACTICES ARE "management practices and work processes that lead to world-class, superior performance" (Fletcher Challenge 1994). Their purpose is to achieve high performance with minimal risk. Organizations use the term *best practices* in a variety of ways. Best practices can describe a standard way of doing things within the organization, a process for improving performance, or a process of innovation. They have five components—best skills, best processes, best solutions, appropriate resources, and continuous improvement.

Benchmarking is a best-practices tool. Benchmarking is used to evaluate various aspects of an organization's processes or management systems with respect to a best practice. Then the organization can develop plans to adopt the best practice. *Gap analysis* is part of the benchmarking process. It is the analysis of the differences between the organization's current capabilities and the expected level of performance. For example, if a company discovers a group of workers with a noise exposure that puts them at risk for noise-induced hearing loss, it will want to include the workers in a hearing conservation program until the noise exposure can be reduced to an acceptable level. The initial benchmark will be compliance with government regulations. Additional benchmarks may include compliance with more protective guidelines such as the American Conference of Governmental Industrial Hygienists (ACGIH) Threshold Limit Value (TLV) for noise and investigating ways to reduce the noise exposure (ACGIH 2007).

Standard practices are contained in government regulations and nongovernmental standards and guidelines. Awareness of standard practices helps one avoid reinventing what has already been accomplished. Organizations that have not implemented standard practices can use best-practice management ideas to implement standard practices. Where standard practices have been implemented, continuous improvement will be needed as the standards and workplace evolve. Company audits of health and safety programs identify gaps between the company's performance and the standard.

An organization can seek to improve its performance. To minimize risk, the organization will identify another organization or management system that it recognizea as being best in class. This becomes the benchmark the organization wants to achieve, and managers will evaluate their own practices with respect to the benchmark. They must establish their baseline performance parameters so they can measure their progress toward the benchmark.

World-class organizations use best practices to manage health and safety concerns. The U.S. Occupational Safety and Health Administration's (OSHA's) Voluntary Protection Program (VPP) is the most recognized management system in this area. OSHA's Web site reports that VPP sites have a lost-workday rate 50 percent below the industry average (Fletcher Challenge 1994). Other standards include

- the American Chemistry Council's Responsible Care Program
- ANSI Z10 *Occupational Health and Safety Management Systems*
- the British Standards Institute's *Occupational Health and Safety Management Systems*
- the International Labor Organization's *Guidelines for Occupational Safety and Health Management System*
- the American Industrial Hygiene Association's *A Strategy for Assessing and Managing Occupational Exposures*

The following sections discuss government agencies with regulations that impact industrial hygiene and nongovernmental organizations with standards that can serve as best practices in the field of industrial hygiene.

Government Agencies

Occupational Safety and Health Administration (OSHA)

OSHA and the Mine Safety and Health Administration (MSHA) are the two principal agencies of the federal government that regulate safety and health in the workplace. OSHA regulates general industry, maritime operations, and construction. MSHA regulates mining and tunneling. OSHA 29 CFR 1910 contains the OSHA Standards for General Industry; 29 CFR 1915 contains the OSHA Standards for Shipyard Employment; 29 CFR 1917 contains the OSHA Standards for Marine Terminals; and 29 CFR 1918 contains the OSHA Standards for Longshoring.

The OSHA and MSHA Web sites (USDOL-OSHA undated; MSHA undated) provide access to up-to-date regulations, compliance information, hazard controls, and more. The National Institute for Occupational Safety and Health (NIOSH) provides access to information on workplace hazards and control of workplace hazards (NIOSH undated).

The following states and territories administer occupational safety and health regulations. Most administer the federal OSHA regulations, but a few have more stringent regulations. The federal OSHA Web site has links to each state and territory.

Alaska	Arizona	California	Connecticut
Hawaii	Indiana	Iowa	Kentucky
Maryland	Michigan	Minnesota	Nevada
New Mexico	New York	North Carolina	Oregon
Puerto Rico	South Carolina	Tennessee	Utah
Vermont	Virgin Islands	Virginia	Washington
Wyoming			

Achieving and maintaining a lost-workday incidence rate of 50 percent of the industry average is a benchmark to best practice. OSHA's Voluntary Protection Program (VPP) is for companies that have achieved this goal and want to improve. In 1982, OSHA created the VPP. Through its programs, OSHA works with companies that demonstrate excellence in occupational safety and health. The average VPP participant has a lost-workday incidence rate 50 percent below its industry average. To participate in a VPP, a work site must show a commitment to effective safety and health management systems and be an industry leader in occupational safety and health. OSHA exempts VPP sites from programmed inspections, but employers and employees retain their rights and responsibilities under the Occupational Safety and Health Act (USDOL-OSHA 2004).

VPP participants are selected on the basis of their written safety and health management system and their performance. OSHA conducts a thorough evaluation of the work-site's system and performance. Continuous improvement is expected. The Star Program is for work sites with injury and illness rates at or below the national average for their industry. The Merit Program is for work sites with good safety and management systems that have the potential and commitment to achieve Star status within three years. Star Demonstration is for work sites with Star-quality safety and health programs that want to test alternatives to Star eligibility and performance requirements. Star Demonstration sites are evaluated every 12 to 18 months, Star sites are evaluated every 3 to 5 years, and Merit sites are evaluated every 18 to 24 months (USDOL-OSHA 2004). In 2007, more than 1200 sites participated in VPP; 95 percent of the sites were Star participants; 35 percent of the sites had less than 100 employees (USDOL-OSHA 2007).

National Institute for Occupational Safety and Health (NIOSH)

The National Institute for Occupational Safety and Health was established by the Occupational Safety and Health Act of 1970. NIOSH is part of the Centers for Disease Control and Prevention (CDC) and is the only federal institute responsible for conducting research and making recommendations for the prevention of work-related illnesses and injuries. The institute's responsibilities include the following:

- Investigating hazardous working conditions as requested by employers or workers
- Evaluating hazards ranging from chemicals to machinery
- Creating and disseminating methods for preventing disease, injury, and disability
- Conducting research and providing scientifically valid recommendations for protecting workers
- Providing education and training to persons preparing for or actively working in the field of occupational safety and health (NIOSH 2004)

Nongovernmental Organizations

American Conference of Governmental Industrial Hygienists (ACGIH)

The ACGIH ". . . is a private, not-for-profit, nongovernmental corporation whose members are industrial hygienists or other occupational health and safety professionals dedicated to promoting health and safety within the workplace." It is an organization of industrial hygienists who are employed by government and educational institutions. Members do not act as representatives of their employers. They act in their professional capacity as industrial hygienists. The ACGIH is known for its threshold limit values (TLVs) and biological exposure indices (BEIs) (ACGIH 2007).

The ACGIH Web site provides access to its publications store where the organization's printed material and other industrial-hygiene-related publications may be obtained (ACGIH undated).

American Industrial Hygiene Association (AIHA)

"AIHA promotes healthy and safe environments by advancing the science, principles, practice, and value of industrial hygiene and occupational and environmental health and safety" (AIHA 2006).

The AIHA was founded in 1939. The association works with the American Board of Industrial Hygiene (ABIH) to promote the certification of industrial hygienists. The AIHA administers education programs and laboratory accreditation programs. It has 30 technical committees working on the challenges in industrial hygiene.

The AIHA Web site provides access to its publications store, where printed materials related to the field of industrial hygiene may be obtained (AIHA undated).

American National Standards Institute (ANSI)

"The American National Standards Institute (ANSI) is a private, nonprofit organization (501(c)3) that administers and coordinates the U.S. voluntary standardization and conformity assessment system. The Institute's mission is to enhance both the global competitiveness of U.S. business and the U.S. quality of life by promoting and facilitating voluntary consensus standards and conformity assessment systems, and safeguarding their integrity" (ANSI undated).

ASTM International

"ASTM International is one of the largest voluntary standards development organizations in the world—a trusted source for technical standards for materials, products, systems, and services. Known for their high technical quality and market relevancy, ASTM International standards have an important role in the information infrastructure that guides design, manufacturing, and trade in the global economy" (ATSM undated).

American Society of Heating, Refrigerating, and Air-Conditioning Engineers (ASHRAE)

"ASHRAE will advance the arts and sciences of heating, ventilation, air conditioning, refrigeration, and related human factors to serve the evolving needs of the public and ASHRAE members" (ASHRAE undated).

ASHRAE's Standard 62.1-2004, *Ventilation for Acceptable Indoor Air Quality*, provides guidelines for establishing comfortable indoor environments for offices. It relies on the EPA's Primary Ambient Air Quality Standards for the determination of air quality, and it references indoor air quality guidelines from other organizations such as the World Health Organization. ASHRAE Standard 55-2004, *Thermal Environmental Conditions for Human Occupancy*, provides guidelines for a comfortable thermal environment.

American Society of Safety Engineers (ASSE)

"ASSE is a global association providing professional development and representation for those engaged in the practice of safety, health, and the environment and those providing services to the private and public sectors to protect people, property, and the environment" (ASSE undated). ASSE was founded in 1911. It has thirteen practice specialties.

Corporate Environmental Health and Safety (EHS) Policy and Program

A corporation's EHS policy expresses its commitment to EHS and distributes responsibility to the functional areas (Leibowitz 1997). The safety professional should obtain the participation of all functional areas of the corporation, because the success of the EHS program will depend on their participation. The president and chief executive officer should sign the policy statement so that the corporation's commitment to safety and health is clear to everyone. The policy should be displayed in all of the company's facilities.

Management, from the top down to the line managers, supervisors, employees, and EHS staff, is responsible for the EHS program. EHS staff members have little authority to accomplish their tasks. They collect the information necessary to determine whether the company's processes require change to ensure the protection of the workers. Where change is necessary, the EHS staff must rely on management, supervisors, and workers to implement the change. Their cooperation is necessary for successful implementation of change.

Workplace Assessment

Workplace assessment is a component of the corporate safety and health program. Each workplace requires an assessment of the workers' exposures to health hazards. The assessment is necessary to anticipate, recognize, evaluate, and control health hazards arising in the workplace. A workplace assessment determines whether the company is in compliance with OSHA regulations and other applicable standards and guidelines.

A written workplace assessment program is a useful tool for organizing and guiding the assessment. The written workplace assessment program should include the following:.

- Purpose and goals of the workplace assessment
- Responsibilities of the industrial hygienists and others involved in the assessment
- Methods for characterizing the workplace, workforce, and environmental agents
- Methods for defining similarly exposed groups of workers
- Criteria for determining whether the exposure is acceptable, unacceptable, or uncertain
- Priorities for gathering more information about uncertain exposure assessments
- Criteria for conducting quantitative exposure assessment
- Methods of prioritizing and controlling unacceptable exposures
- Methods of communicating and documenting the assessment's findings
- Criteria for periodic reassessment of workplace exposures
- Program evaluation (DiNardi 1997)

The assessment should include the following:

- Location within the corporate structure and address
- Work area within the facility
- Exposure group
- Work assignments within the exposure group
- Job titles or job classifications within the exposure group
- Process description
- Materials used
- Exposure agents
- Exposure measurement statistics
- Assessment of the risk to the exposure groups' health from inhalation, skin contact, eye contact, ingestion, and injection routes of exposure
- Control measures used to keep the risks acceptable
- Periodic reassessment

The purposes of the assessment are to categorize employees into exposure groups with similar exposures and to evaluate the risk to each exposure group's health from the materials and physical agents to which they are exposed. The result of the assessment will be one of three categories: acceptable exposure, uncertain exposure, and unacceptable exposure. Unacceptable exposures must be controlled. Uncertain exposures require additional information to reduce uncertainty about the exposure. Acceptable exposures should be documented and reassessed periodically.

The workplace assessment is a way to recognize health hazards in the workplace, evaluate the severity of the hazards, and implement the necessary controls to minimize risks. It should be applied to existing operations and to new processes and new materials before they are introduced into the workplace in order to anticipate possible health hazards.

Risk Communication

Workers should be informed about the hazards of their work before they are exposed to the hazards. This information should include health hazards of the materials and agents to which the workers will be exposed. Otherwise, workers cannot make an informed decision to accept the work and its associated hazards. Many chemical and physical agents do not provide warning that they are injuring a worker's health. The health effects may take weeks, months, or years to manifest themselves, and often the injury is irreversible.

Some OSHA standards require informing workers about the hazards of materials to which they are exposed. These standards include the following:

1910.110, Liquefied Petroleum Gases
1910.111, Anhydrous Ammonia
1910.1001, Asbestos
1910.1003–1910.1016, 13 carcinogens:
 4-nitrobiphenyl, α-naphthylamine, methyl chloromethyl ether, 3,3′-dichlorbenzidine

and its salts, bis-chloromethyl ether, β-naphthylamine, benzidine, 4-aminodiphenyl, ethyleneimine, β-propiolactone, 2-acetylaminofluorene, 4-dimethylaminoazo benzene, and N-nitrosodimethylamine
1910.1017, Vinyl chloride
1910.1018, Inorganic arsenic
1910.1025, Lead
1910.1026, Hexavalent chromium
1910.1027, Cadmium
1910.1028, Benzene
1910.1029, Coke oven emissions
1910.1030, Blood-borne pathogens
1910.1043, Cotton dust
1910.1044, 1,2-Dibromo-3-chloropropane
1910.1045, Acrylonitrile
1910.1047, Ethylene oxide
1910.1048, Formaldehyde
1910.1050, Methylenedianiline
1910.1051, 1,3-Butadiene
1910.1052, Methylene chloride
1910.1200, All chemical hazards

The OSHA Hazard Communication Standard, 1910.1200, governs hazard communication for chemical substances in the workplace. The standard applies to chemical manufacturers, importers, distributors, and employers whose employees are exposed to chemical hazards. The standard has four requirements for employers whose employees are exposed to chemical hazards:

- A chemical inventory of all materials used and generated in the workplace
- Material safety data sheets (MSDSs) for all materials in the workplace
- Appropriate labels or the equivalent on all containers
- Informing employees of the hazards of all the materials to which they are exposed

The OSHA Occupational Exposure to Hazardous Chemicals in Laboratories Standard, 1910.1450, has similar requirements for informing laboratory workers about chemical and physical hazards in laboratories.

For physical agents,

1. OSHA Noise Standard, 1910.95, section (l) has requirements to inform workers about the hazards of noise and the location of harmful noise sources.
2. OSHA Ionizing Radiation Standard, 1910.1096, section (i) has requirements to inform workers about the hazards of working in or frequenting "radiation areas."

Employees in certain industries or engaged in special operations must receive training. They are:

1910.109, Transporting explosives
1910.177, Servicing of single- or multi-rim wheels
1910.254, Arc and resistance welding
1910.266, Logging operation
1910.268, Telecommunication
1910.269, Electric power generation
1910.272, Grain handling
1910.410, Commercial diving

Communication is an interactive process. In complying with OSHA's employee information and training requirements, the safety professional is presenting information to workers. Some of the OSHA standards require an opportunity for questions from the workers. A question-and-answer period provides workers with an opportunity to clarify information and to express their concerns with workplace safety and health. Safety professionals should have a means for follow-up to answer questions for which they do not have an answer.

Workers may assess hazards as more or less severe than the safety professional has assessed it. If a worker assesses the risk as less severe than the company's assessment, he or she may not participate adequately in the required control measures. The consequences of failure to participate should be clear to all—the company and the workers.

If a worker assesses the risk as greater than the company's assessment, he or she may want additional control measures. The safety professional should make clear what, if any, accommodations will be made. Often, companies make PPE available for voluntary use even though the company's assessment of the hazard does not indicate that PPE is necessary.

In addition to informing employees about workplace hazards, management must be informed of the hazards, and the public and the workplace's neighbors may also need to be informed. Management and the public may disagree with the assessment. The company should make clear to its employees that cooperation with management is required. With the public, the company should be prepared to discuss what additional control measures they are prepared to take if any.

Quality Control

Program Audits

Audits of safety and health programs maintain the effectiveness of those programs. An audit should be designed to determine whether the programs are achieving the company's internal and external goals (DiNardi 1997). The company's goals should include regulatory compliance and maintaining a healthy workforce.

OSHA and NIOSH have some program audits available on their Web sites. Many consulting companies and environmental, safety, and health publishing houses have safety and health program audits available for a fee.

An audit is a systematic method of examining the program elements necessary to protect the safety and health of the workforce. When an audit is completed, the audit report should include a list of deficiencies. The severity of the deficiencies should be ranked based on how seriously the deficiency affects the health of the workforce. Management can use the audit report to develop action plans to correct the deficiencies and improve the program's effectiveness. Each program should be audited by the facility and by teams from outside of the facility (DiNardi 1997).

Internal audits make use of the company's internal resources to perform the audit. They are less expensive than external audits. A team should be selected to perform the audit. Selection of the team members will involve considering experience and motivation. Also, a company may choose workers for participation on an audit team whom it wishes to involve in safety and health because the audit is an opportunity to educate the workers. If the company is subject to external audits, they should use the external audit format for their internal audits. This will help the company organize program records in a manner that will facilitate an external audit.

External audits are performed by corporate audit teams, consultants, or a combination of the two. Members of an internal audit team may be distracted by their jobs, whereas an external audit team is dedicated to the audit because it is their job. An internal audit team is more familiar with the facility than an external team. An external team will require more participation from management and workers to perform the audit. An external team is less likely to be influenced by social relationships with coworkers than an internal team. An external team is also more likely to have broad experience than an internal team.

Industrial Hygiene Consultant Selection

Companies and corporations retain consultants for a variety of reasons. Hiring consultants may be less expensive than developing and retaining an in-house resource. Consultants may have knowledge and expertise that is not readily available within the company. Even if an in-house resource is available, a consultant may be used because he or she can provide an objective evaluation or because of scheduling conflicts with the in-house resources.

The company or corporation should have objectives that it expects the consultant to meet. Well-defined objectives may not allow any modification by the consultant. Complex or less well-defined objectives may require consultation with a variety of consultants to select a protocol to meet the objectives.

When selecting a consultant, the company or corporation will consider the consultant's credentials, experience, reputation, availability, and cost. Often, the safety professional relies on the recommendations of friends and colleagues to find a consultant. Other sources of consultants are directories. The AIHA has a directory of members who provide consulting services that can be searched at the AIHA Web site

(AIHA undated). No matter how safety professionals find consultants, they should examine the consultant's credentials and references carefully.

Industrial Hygiene Laboratory Selection

Selecting an industrial hygiene laboratory or laboratories is similar to selecting an industrial hygiene consultant. Consider the laboratory's credentials, experience, reputation, availability, and cost. The laboratory's experience should include frequent analysis of the same types of samples you plan to submit.

The laboratory should be accredited by the AIHA. This indicates that the laboratory participates in the Proficiency Analytical Testing (PAT) program. Review the laboratory's performance in the PAT program for the past 24 months and examine the laboratory's references and quality assurance program carefully.

The AIHA Web site provides a list of accredited laboratories.

Evaluating Exposure-Monitoring Results

Evaluating exposure-monitoring results requires an understanding of the monitoring strategy and of the accuracy of the sampling and analytical methods used to obtain the results. Usually, industrial hygienists try to monitor workers and tasks likely to have the highest exposures so they can select control measures that are adequate to protect against the highest exposure.

Corporations have quality control procedures to maintain product quality. These procedures include the collection of a representative number of samples for product testing. Similarly, when evaluating workers' exposures, a sufficient number of samples should be collected to evaluate the exposure, and statistical analysis should be applied to the exposure data. However, instead of trying to keep the quality of the majority of the products above a certain minimum acceptable quality level, the industrial hygienist works toward keeping the majority of worker exposures below an acceptable occupational exposure limit (OEL).

When evaluating exposure-monitoring results, the industrial hygienist must do more than determine whether each exposure is below the appropriate OEL. He or she should also apply statistical analysis to determine whether the exposures are below the OEL with 95 percent confidence.

Industrial hygiene monitoring usually consists of full-shift sampling or short-term sampling (1 to 60 minutes). Obviously, one full-shift sample will not represent a worker's lifetime exposure to a substance. More than one sample will be necessary to determine the average exposure. NIOSH's *Occupational Exposure Sampling Strategy Manual* provides easy-to-use tables to determine the number of samples necessary to evaluate exposures and the statistical tools to interpret the sample results (NIOSH 1977).

The *Occupational Exposure Sampling Strategy Manual* recommends that sampling and analytical methods have an accuracy of ±50 percent at less than one-half of the OEL, ±35 percent at one-half the OEL to the OEL, and ±25 percent above the OEL.

The 95 percent upper confidence limit of each sample should be below the applicable OELs interval for a full-shift exposure

$$\text{UCL (95\%)} = x + 1.645 \times x \times cv_T \tag{1}$$

where

UCL (95%) is the 95 percent upper confidence limit
x is the full-shift exposure
cv_T is the total coefficient of variation for the sampling and analytical method

The employer should keep the 95 percent upper confidence limit of each exposure below the OEL. If the upper confidence limit is not below the OEL, then additional controls are needed.

When several full-shift exposure measurements are available from a similar exposure group, the 95 percent upper confidence limit of the distribution should be below the OEL. A *similar exposure group* is a group of workers who perform similar work. Their exposures will be distributed log-normally—that is, the logarithms of the exposures will be distributed normally. The normal distribution does not describe exposures because the normal distribution would predict negative exposure—exposures less than zero. The log-normal distribution does not predict exposure below zero.

The 95 percent upper confidence limit of the distribution should be kept below the OEL. The 95 percent upper confidence limit is

$$UCL\ (\leq 95\%) = gm \times gsd^{1.645} \quad (2)$$

where

UCL (≤95%) is the 95 percent upper confidence limit
gm = the geometric mean of the exposures
gsd = the geometric standard deviation of the exposures

The geometric mean is the antilogarithm of the mean of the logarithms of the exposures. The geometric standard deviation is the antilogarithm of the sample deviation of the logarithms of the exposures.

Personal Protective Equipment (PPE)

OSHA requires the use of personal protective equipment to reduce workers' exposure to hazards when engineering and administrative controls are not feasible or effective. Employers are required to determine all of the hazards in their workplace and determine whether PPE should be used (OSHA undated).

The OSHA Personal Protective Equipment Standard, 1910.132, governs PPE including PPE for eyes, face, hands, and extremities; protective clothing; respiratory devices; and protective shields and barriers. The standard requires a workplace assessment to determine the need for PPE, the selection of the appropriate PPE based on the hazard, and worker training.

The OSHA Eye and Face Protection Standard, 1910.133, governs the selection and use of eye and face protection.

The OSHA Respiratory Protection Standard, 1910.134, governs the selection and use of respiratory protection equipment.

The OSHA Head Protection Standard, 1910.135, governs the selection and use of head protection.

The OSHA Occupational Foot Protection Standard, 1910.136, governs the selection and use of foot protection.

The OSHA Electrical Protection Standard, 1910.137, governs the selection and use of electrical protective devices.

The OSHA Hand Protection Standard, 1910.138, governs the selection and use of hand protection.

The OSHA Occupational Noise Exposure Standard, 1910.95, governs the selection and use of hearing protection.

The OSHA Web site topic page on personal protective equipment provides links to useful information for complying with the OSHA requirements.

Each of the OSHA standards requires employers to select PPE appropriate to the hazard. Some of the standards, such as the Occupational Noise Standard, specify the selection process. Others do not.

Protection against chemical hazards is required by the OSHA Personal Protective Equipment Standard and the Hand Protection Standard. Skin and eye protection is necessary against corrosive substances and against substances that can be absorbed through the skin in toxic amounts. The ACGIH's *TLV Booklet* indicates substances that can be absorbed through the skin in toxic amounts with the "skin" notation for the substance. Similarly, Table Z-1 in OSHA regulation 1910.1000 contains a skin designation column to indicate substances that can be absorbed through the skin in toxic amounts.

The NIOSH *Pocket Guide to Chemical Hazards* lists recommended personal protection and sanitation for chemicals. Protective gloves and clothing must be made of material that can prevent the substance from penetrating the gloves or clothing (NIOSH 2007). The NIOSH Web site has a database, *Recommendation for Chemical Protective Clothing*, that lists chemical hazards and the materials suitable to resist the hazard. Also, the manufacturers of gloves and coveralls provide information on their products' resistance to chemical hazards (NIOSH undated).

Hearing Protection

The OSHA Noise Standard, 1910.95, section (i) governs the selection and use of hearing protection.

1. Employers must provide hearing protectors to all workers whose noise exposures are 85

dBA 8-hour time-weighted average (TWA) or more.

2. Employers must ensure that hearing protection is worn by all workers whose noise exposures are above 90 dBA and all workers with a standard threshold shift (STS) or without a baseline audiogram whose noise exposures are 85 dBA 8-hour TWA or more.
3. Workers must have the opportunity to select their hearing protectors from a variety of suitable protectors provided by the employer.
4. Employers must provide training in the use and care of all hearing protectors.
5. Employers must ensure the proper fitting and supervise the correct use of all hearing protectors.

Appendix B of the standard (Methods for Estimating the Adequacy of Hearing Protector Attenuation) prescribes the methods that must be used to ensure that the hearing protectors provide an adequate level of protection. The methods use the noise reduction rating (NRR). The most commonly used method is the following:

For workers without STS,

$$NRR = (TWA + 7) - 90 \quad (3)$$

For workers with STS,

$$NRR = (TWA + 7) - 85 \quad (4)$$

where

NRR = the noise reduction rating
TWA = the 8-hour time-weighted average noise exposure in decibels, A-weighted scale (dBA)

The NRRs are measured in laboratories. They represent the attenuation level that the majority of users can obtain with laboratory supervision. In practice, the attenuation provided by hearing protection varies much more than in the laboratory. The insertion of ear plugs is seldom supervised in the workplace. If ear plugs are not inserted properly, they provide much less protection than if they are inserted properly.

The *OSHA Technical Manual* advises employers to select hearing protection with higher NRRs than the OSHA Noise Standard requires (*OSHA Technical Manual*, section III, Health Hazards, chapter 5, Noise Measurement, section IX, New Developments in Hearing Protection Labeling). The manual advises employers to determine the NRR by using the following formulas:

For workers without STS,

$$NRR \geq 2 \times (TWA - 90) + 7 \quad (5)$$

For workers with STS,

$$NRR \geq 2 \times (TWA - 85) + 7 \quad (6)$$

NIOSH's experience is that hearing protection does not provide the attenuation indicated by the NRR. NIOSH recommends using 75 percent of the NRR for earmuffs, 50 percent for formable earplugs (foam earplugs), and 30 percent for all other earplugs (NIOSH 1996).

OSHA's and NIOSH's advice to provide better hearing protection than equations 3 and 4 require is one way to address the issue. In the author's opinion this is an area that requires improvement. I have observed many workers in noisy environments who have not inserted their ear plugs correctly. When asked why, many workers admit they know better and will insert their ear plugs correctly in the future. Workers and line supervisors need more commitment to inserting ear plugs correctly. This issue is also common to other forms of PPE.

Respiratory Protection

The OSHA Respiratory Protection Standard, 1910.134, governs the use of respiratory protection in the workplace. Section (a)(1) says, "In the control of those occupational diseases caused by breathing air contaminated with harmful dusts, fogs, fumes, mists, gases, smokes, sprays, or vapors, the primary objective shall be to prevent atmospheric contamination. This shall be accomplished as far as feasible by accepted engineering control measures (for example, enclosure or confinement of the operation, general and local ventilation, and substitution of less toxic materials). When effective engineering controls are not feasible, or while they are being instituted, appropriate respirators shall be used pursuant to this section." The following are the general requirements of the standard:

1. A written respiratory protection program
2. Procedures for selecting respirators
3. Medical evaluations of respirator users
4. Fit-testing procedures for tight-fitting respirators
5. Procedures for the use of respirators during routine and emergency situations
6. Procedures and schedules for cleaning, disinfecting, storing, inspecting, repairing, discarding, and otherwise maintaining respirators
7. Procedures to ensure adequate air quality, quantity, and flow of breathing air for atmosphere-supplying respirators
8. Training employees in regard to the respiratory hazard to which they are potentially exposed during routine and emergency situations
9. Training employees in the proper use of respirators
10. Procedures for regularly evaluating the effectiveness of the program

The OSHA Web site contains the entire respiratory protection standard and useful information on the interpretation of the standard. A sample program is contained in OSHA's *Small Entity Compliance Guide for Respiratory Protection Standard* (OSHA 1997).

Other OSHA regulations contain additional requirements for respiratory protection. These regulations include the following:

1910.94, Ventilation
1910.111, Anhydrous ammonia
1910.156, Fire brigades
1910.252, Welding, cutting, and brazing
1910.261, Pulp, paper, and paperboard Mills
1910.1001, Asbestos
1910.1003, 13 Carcinogens
1910.1017, Vinyl chloride
1910.1018, Inorganic arsenic
1910.1025, Lead
1910.1027, Cadmium
1910.1028, Benzene
1910.1029, Coke oven emissions
1910.1030, Blood-borne pathogens
1910.1043, Cotton dust
1910.1044, 1,2-Dibromo-3-chloropropane
1910.1045, Acrylonitrile
1910.1047, Ethylene oxide
1910.1048, Formaldehyde
1910.1050, Methylenedianiline
1910.1051, 1,3-Butadiene
1910.1052, Methylene chloride

Occupational Exposure Limits

Best practice is to keep employees' exposures to airborne contaminants as low as possible. The industrial hygienist evaluates workers' exposures to airborne substances against occupational exposure limits (OELs). OELs include the OSHA PELs (permissible exposure limits), ACGIH TLVs, NIOSH RELs (recommended exposure limits), AIHA WEELs (workplace environmental exposure limits), and other sources.

The OSHA Air Contaminants Standard, 1910.1000, Tables Z-1, Z-2, and Z-3, contains PELs for 501 substances or groups of substances. The PELs in Tables Z-1, Z-2, and Z-3 were established between 1966 and 1971. The PELs for 29 substances or groups of substances have been established since 1971. They are contained in standards 1910.1001 through 1910.1052.

NIOSH RELs are available in the NIOSH *Pocket Guide to Chemical Hazards* (NIOSH 2007).

For chemical substances, "Threshold limit values (TLVs) refer to airborne concentrations of chemical substances and represent conditions under which it is believed that nearly all workers may be repeatedly exposed, day after day, over a working lifetime, without adverse health effects. TLVs are developed to protect workers who are normal healthy adults." For physical agents, "As with other TLVs, those for physical agents provide guidance on the levels of exposure and conditions under which it is believed that nearly all healthy workers may be repeatedly exposed, day after day, without adverse health effects" (ACGIH 2007).

The TLVs and BEIs are updated annually. OSHA adopted the 1968 TLVs for the substances listed in 29 CFR 1910 subpart Z, Tables Z-1 and Z-3. MSHA adopted the 1973 TLVs. The 2007 TLVs contain over 100 new substances and over 100 revisions to the 1968 and 1973 TLVs.

The AIHA has WEELs for 102 substances or groups of substances. Many of the WEELs are substances for which OSHA and the ACGIH do not have exposure limits.

Control Banding

Control banding is a process that may be used to manage exposure to chemicals whether the chemical has an OEL or not. Seventy-five thousand chemical substances are used in or imported into the United States (EPA 2006). Most of these substances do not have PELs, TLVs, RELs, or WEELs. Control banding assigns the chemical to a hazard band based on its toxicity. The toxicity information from the chemical's material safety data sheet (MSDS) is all that is needed to assign the chemical to a hazard band. Each band has a control technology (see Table 1). NIOSH considers control banding a potentially useful tool for small businesses (NIOSH undated).

The Health and Safety Executive (HSE) of the United Kingdom has the most developed model, called COSHH Essentials, for control banding (HSE undated). The tool will assign a chemical or mixture to a hazard band and recommend exposure controls. "The German authority (Bundesanstalt für Arbeitsschutz und Arbeitsmedizin—BAuA) evaluated the system based on about 1000 personal measurements from field studies in eighteen industrial applications. They found that for solids (dusts and powders) and medium-scale use (liter quantities) of liquids, exposures were within the range predicted by COSHH Essentials or lower. For the use of small quantities (milliliters) of solvent-based products (such as paint or adhesive), exposures sometimes exceeded the range" (NIOSH undated).

Confined Space Entry

The OSHA Permit-Required Confined Spaces Standard, 1910.146, governs entry into permit-required confined spaces. OSHA defines a confined space as a space that has all three of the following characteristics:

1. Is large enough and so configured that an employee can bodily enter and perform assigned work.
2. Has limited or restricted means for entry or exit (for example, tanks, vessels, silos, storage bins, hopper, vaults, and pits are spaces that may have limited means of entry).
3. Is not designed for continuous employee occupancy.

OSHA defines a permit-required confined space as a confined space that contains one or more of the following characteristics:

TABLE 1

Control Bands for Exposure to Chemicals by Inhalation

Band Number	Target Range of Exposure Concentration	Hazard Group	Control
1	>1 to 10 mg/m³, dust >50 to 500 ppm, vapor	Skin and eye irritants	Use good industrial hygiene practice and general ventilation
2	>0.1 to 1 mg/m³, dust >5 to 50 ppm, vapor	Harmful on single exposure	Use local exhaust ventilation
3	>0.01 to 0.1 mg/m³, dust >0.5 to 5 ppm, vapor	Severely irritating and corrosive	Enclose the process
4	<0.01 mg/m³, dust <0.5 ppm, vapor	Very toxic on single exposure, reproductive hazard, sensitizer	Seek expert advice

(Source: NIOSH)

- Contains or has a potential to contain a hazardous atmosphere.
- Contains a material that has the potential for engulfing an entrant.
- Has an internal configuration such that an entrant could be trapped or asphyxiated by inwardly converging walls or by a floor that slopes downward and tapers to a smaller cross-section.
- Contains any other recognized serious safety or health hazard.

The standard requires employers to do the following:

1. Determine whether the workplace contains any permit-required confined spaces (PRCSs).
2. Post warning signs on PRCSs or otherwise inform exposed employees of the existence, location, and dangers posed by the PRCSs.
3. Take effective measures to prevent its employees from entering the PRCSs (if the employer has decided that its employees will not enter PRCSs).
4. Develop and implement a written PRCS program (if the employer has decided that its employees will enter PRCSs).

OSHA specifies three ways that an employer may allow its employees to enter a PRCS:

1. If the only hazard is a potential or actual atmospheric hazard that can be controlled by continuous forced-air ventilation, the employer may use the "alternate" procedures specified in 1910.146(c)(5).
2. If the PRCS poses no actual or potential atmospheric hazards and if all hazards within the space are eliminated without entry into the space, the PRCS may be reclassified as a nonpermit confined space for as long as the nonatmospheric hazards remain eliminated. 1910.146(c)(7).
3. All other PRCS entries must be governed by a written PRCS program that meets the requirements of 1910.146(c)(4) and 1910.146(d).

If an employer (host) arranges for another employer's (contractor's) employees to perform work that involves PRCS entry, the employer has a duty to do the following:

1. Inform the contractor that the workplace contains PRCSs.
2. Inform the contractor that PRCS entry is allowed only with a program that meets the requirements of 1910.146.
3. Inform the contractor of the hazards identified by the host that make the space a PRCS.
4. Inform the contractor of the host's experience with the PRCS.
5. Inform the contractor of any precautions or procedures that the host has implemented for the protection of employees in or near the PRCS.
6. Coordinate entry operations with the contractor when both the host's employees and the contractor's employees will be working in or near PRCSs.
7. Debrief the contractor at the conclusion of the entry operations regarding any hazards confronted or created in PRCSs during entry operations.

The OSHA Permit-Required Confined Spaces Standard, 1910.146, contains requirements for the training of entrants, entry supervisors, attendants, and rescue teams.

Ventilation System Program

Ventilation systems are used to distribute heating and cooling for human comfort. They are also used to capture and remove airborne contaminants from the workplace and to replace the air removed for contaminant control. If ventilation systems fail to provide adequate heating or cooling, the occupants will let you know they are uncomfortable. If the ventilation system fails to control airborne contaminants adequately, it may be minutes, days, weeks, months, or years before the effects on the workers' health become apparent.

Best practices for the commissioning, design, selection, installation, operation, maintenance, and testing of local exhaust ventilation systems may be found in ANSI/AIHA Z9.2-2001, Fundamentals Governing the Design and Operation of Local Exhaust Systems, and ACGIH's *Industrial Ventilation—A Manual of Recommended Practice*. The ACGIH manual is in its 25th edition. It contains more than one hundred ventilation sketches for specific industrial operations in addition to information on the commissioning, design, selection, installation, operation, maintenance, and testing of local exhaust ventilation systems.

The OSHA Ventilation Standard, 1910.94, covers the design and operation of ventilation systems for abrasive blasting; grinding, polishing, and buffing; and spray finishing.

Welding, Cutting, and Brazing Program

OSHA 1910 subpart Q governs welding, cutting, and brazing operations. Subpart Q contains the following general requirements:

1. Fire prevention and protection
2. Fire watch
3. Welding or cutting containers
4. Confined spaces
5. Protection of personnel; personal protective equipment
6. Health protection and ventilation
7. Fluorine compounds
8. Zinc
9. Lead
10. Beryllium
11. Cadmium
12. Mercury
13. Cleaning compounds
14. Industrial applications

OSHA 1910.253 governs oxygen-fuel gas welding and cutting. OSHA 1910.254 governs arc welding. OSHA 1910.255 governs resistance welding.

The American Welding Society (AWS) is a source of best practices for safety and health guidelines for welding, cutting, and brazing. The following guides are representative examples:

- AWS F1.3-1999, *A Sample Strategy Guide for Evaluating Contaminants in the Welding Environment*. Available at www.aws.org.
- AWS F3.2M/F3.2-2001, *Ventilation Guide for Weld Fume*. Available at www.aws.org.
- ANSI Z49.1-1999, *Safety in Welding, Cutting and Allied Processes*. Available at www.ansi.org.

Carcinogens Program

Carcinogens are handled more cautiously than many other hazardous materials because of the lack of treatment options for cancers. Best practice is to identify carcinogens in the workplace, to eliminate or restrict their use, to assess the risk to worker health, to keep exposure as low as reasonably achievable, and to require appropriate PPE to protect workers from contact with carcinogens.

The OSHA Hazard Communication regulation, 1910.1200(d)(4), recognizes carcinogens or potential carcinogens identified by the National Toxicology Program, the International Agency for Research on Cancer, and OSHA 1910 Subpart Z for the purposes of hazard communication. Carcinogenic substances must be listed on MSDSs if present in the material at concentrations of 0.1 percent or more unless the substance is a trade secret. Hazardous chemicals must be listed on MSDSs if present at concentrations of 1 percent or more.

OSHA regulates carcinogens in the same manner that it regulates other hazardous substances. The OSHA standards for carcinogens such as benzene (1910.1028) and cadmium (1910.1027) follow the same outline as the OSHA standard for lead (1910.1025).

The ACGIH TLV booklet's table of adopted values notes which substances are carcinogens. It has five classifications:

- A1—Confirmed human carcinogen
- A2—Suspected human carcinogen
- A3—Confirmed animal carcinogen
- A4—Not classifiable as a human carcinogen
- A5—Not suspected as a human carcinogen

Medical Surveillance Program

Medical surveillance is an exposure-assessment tool. Usually, industrial hygienists evaluate workers' exposures by measuring airborne contaminant concentrations or noise and radiation doses outside of the body. Biological monitoring is more intrusive, but it often is a better measure of the effects of contaminants on individual workers.

Some OSHA regulations require medical surveillance when exposure is above the action level. For example, OSHA's Occupational Noise Standard requires annual audiograms if workers have noise exposures at or above 85 dBA as 8-hour time-weighted averages (TWAs). Noise exposures above 85 dBA 8-hour TWA indicate an increased risk of noise-induced hearing loss. The audiograms determine whether an individual's hearing acuity has been affected by noise exposure.

It is best practice to establish a medical surveillance program any time workers are exposed to an amount above 50 percent of the OSHA PEL, ACGIH TLV, or other acceptable exposure limit. Sources of medical surveillance guidelines may be found in the following documents:

- OSHA's regulations (UDSOL-OSHA undated)
- NIOSH's Occupational Safety and Health Guidelines for Chemical Substances (NIOSH undated)
- NIOSH's Specific Medical Tests or Examinations Published in the Literature for OSHA-Regulated Substances (NIOSH undated)
- ACGIH's TLVs and BEIs (ACGIH 2007)

Modeling Inhalation Exposure

A model attempts to predict what will occur in a particular situation. We all use intuitive models based on our experience. We know what time to leave in order to arrive at work on time. When driving to new destinations, we may predict (model) how long the trip will take based on distance and speed.

Exposure modeling is a tool for assessing inhalation exposure for a particular task. In a large factory, the majority of the tasks may never be monitored because the industrial hygienist intuitively assesses the exposure potential based on his or her experience—similar exposures in the past measured below or above the applicable exposure limits. Best practice is to overestimate the exposure (err on the high side) so that workers are adequately protected from the hazard. When the industrial hygienist expects the exposure to be above applicable exposure limits, best practice is to monitor periodically to have an accurate measure of the exposure.

Sometimes industrial hygienists do not measure exposures that are above applicable exposure limits because the necessary exposure controls are part of an industry's standard operating procedures. For example, industrial hygienists usually do not measure an arc welder's exposure to ultraviolet light because this is a well-known hazard of arc welding. Welders are trained on the precautions necessary to protect their eyes and skin from excessive exposure to ultraviolet light during arc welding operations. They are trained to select the appropriate shade of eye protection based on the type of arc welding and the strength of the arc.

If industrial hygienists do not have experience with a particular task, they may measure a worker's exposure. This is a form of modeling. The results of the measurement will become part of their experience.

Industrial hygienists may try to predict the exposure using a model when exposure monitoring is not possible. Models may be used to predict exposures during accidents such as spills or when new materials or processes are anticipated.

The exposure model is the industrial hygienist's hypothesis of the exposure. The hygienist will choose a model that will overestimate the exposure in order to err on the side of safety.

A simple model of a solvent spill predicts that the solvent vapor concentration in the area of the spill will reach the equilibrium vapor pressure of the solvent at the temperature in the room. This is a worst-case assumption because it assumes no dilution of the vapors by ventilation. The equilibrium vapor pressure is the vapor pressure immediately above the liquid. This model can underestimate the exposure if a mist of the liquid solvent is generated.

For example, toluene has an equilibrium vapor pressure of 21 mm of Hg at room temperature. This is equivalent to 27,600 ppm, which is above the OSHA PEL of 200 ppm, the ACGIH TLV of 50 ppm, the IDLH (immediately dangerous to life or health) concentration of 500 ppm, and the lower explosive limit (LEL) of 1.1 percent or 11,000 ppm. This model predicts an explosion hazard and exposure 550 times the TLV concentration.

$$C_{Sat} = \frac{21 \text{ mm Hg}_{toluene}}{760 \text{ mm Hg}_{air}} \times 10^6 \text{ ppm} = 27{,}600 \text{ ppm} \quad (7)$$

More complicated models will include estimates of the effects of room size, ventilation, and contaminant generation rates. Whenever possible, best practice is to verify a model's predictions with exposure measurements.

References

American Conference for Governmental Industrial Hygienists (ACGIH). 2007. *2007 TLVs and BEIs*. Cincinnati, OH: ACGIH Worldwide Signature Publications.

American Industrial Hygiene Association (AIHA). Undated. (Accessed May 2, 2007.) www.aiha.org

______. 2006. (Accessed May 2, 2006.) www.aiha.org/content/aboutaiha/strategicplan.htm

American National Standards Institute (ANSI). Undated. (Accessed May 2, 2006.) www.ansi.org

American Society of Heating, Refrigerating and Air-Conditioning Engineers (ASHRAE). Undated. (Accessed May 2, 2006.) www.ashrae.org

American Society of Safety Engineers (ASSE). Undated. (Accessed May 2, 2007.) www.asse.org

American Society for Testing Materials (ASTM). Undated. (Accessed May 2, 2007.) www.astm.org

DiNardi, S. R., ed. 1997. *The Occupational Environment—Its Evaluation and Control*. Farifax, VA: AIHA Press.

Environmental Protection Agency (EPA). 2006. www.epa.gov/opptintr/newchems/pubs/invntory.htm

Fletcher Challenge, Ltd. 1994. (Accessed May 2, 2007.) www.walden3d.com/best_practices/bp_FCP_definition.html

Health and Safety Executive (HSE). Undated. (Accessed May 2, 2007.) www.hse.gov.uk/coshh/index.htm

Leibowitz, Alan. 1997. Ch. 37. "Program Management." In *The Occupational Environment—Its Evaluation and Control*. Salvatore R. DiNardi, ed. Fairfax, VA: AIHA Press.

Mine Safety and Health Administration (MSHA). Undated. (Accessed May 2, 2007.) www.msha.gov

National Institute of Occupational Safety and Health (NIOSH). Undated. (Accessed May 2, 2007.) www.cdc.gov/niosh/homepage.html

______. Undated. *NIOSH Safety and Health Topic: Control Banding*. (Accessed May 2, 2007.) www.cdc.gov/niosh/topics/ctrlbanding

______. Undated. *Occupational Safety and Health Guidelines for Chemical Substances* . (Accessed May 2, 2007.) www.cdc.gov/niosh/81-123.html

______. Undated. *Specific Medical Tests or Examinations Published in the Literature for OSHA-Regulated Substances*. (Accessed May 2, 2007.) www.cdc.gov/niosh/nmed/medstar.html

______. 1996. *Occupational Noise and Hearing Conservation Selected Issues*. (Accessed May 2, 2007.) www.cdc.gov/niosh/noise2a.html

______. 2005. *Pocket Guide to Chemical Hazards*. (Accessed April 27, 2007.) www.cdc.gov/niosh/npg

U.S. Department of Labor-Occupational Safety and Health Administration (USDOL-OSHA). Undated. (Accessed May 2, 2007.) www.osha.gov

______. Undated. *Safety and Health Topics: Personal Protective Equipment*. (Accessed May 2, 2007.) www.osha.gov/SLTC/personalprotectiveequipment/index.html

______. 1997. *Small Entity Compliance Guide for Respiratory Protection Standard*. (Accessed May 2, 2007.) www.osha.gov/Publications/Abate/abate.html

______. 2004. *OSHA Fact Sheet: Voluntary Protection Program*. (Accessed May 2, 2007.) www.osha.gov/OshDoc/data_General_Facts/factsheet-vpp.pdf

______. 2007. *Current VPP Statistics*. (Accessed May 2, 2007.) www.osha.gov/dcsp/vpp/charts.html

Appendix: Additional Resources

Resources for Corporate Policies and Programs

Garrett, J. T., L. J. Cralley, and L. V. Cralley, eds. 1988. *Industrial Hygiene Management*. New York: John Wiley & Sons.

Leibowitz, Alan. 1997. Ch. 37. "Program Management." In *The Occupational Environment—Its Evaluation and Control*. Salvatore R. DiNardi, ed. Fairfax, VA: AIHA Press.

OSHA. *Safety and Health Program Management Guidelines*. Federal Register, January 26, 1989 (54 FR 3908-3916). www.osha.gov/SLTC/etools/safetyhealth

______. 1996. *Occupational Safety and Health System: A Guidance Document*. Fairfax, VA: AIHA Press.

______. 1999. *Occupational Health and Safety Management System Performance Measurement: A Universal Assessment Instrument*. Fairfax, VA: AIHA Press.

Resources for Workplace Assessment

Armstrong, Thomas W. and Bernard D. Silverstein, eds. *User's Guide to a Strategy for Assessing and Managing Occupational Exposures*. 2d ed. 2000. Fairfax, VA: AIHA Press.

Bullock, William H., and Joselito S. Ignacio, eds. 2006. *A Strategy for Assessing and Managing Occupational Exposures*. 3d ed. Fairfax, VA: AIHA Press.

Nelson, Deborah Imel. Ch. 37. "Risk Assessment in the Workplace." In *The Occupational Environment—Its Evaluation and Control*. Salvatore R. DiNardi, ed. Fairfax, VA: AIHA Press, 1997.

NIOSH. *Occupational Exposure Sampling Strategy Manual*. NIOSH publication number 77-173. Available at www.cdc.gov/niosh/publistb.html#1977.

Resources for Risk Communication

Lee, S. L., D. R. Lillquist, and F. J. Sullivan. 1997. Ch. 41. "Risk Communication in the Workplace." In *The Occupational Environment—Its Evaluation and Control*. Salvatore R. DiNardi, ed. Fairfax, VA: AIHA Press.

National Research Council. 1989. *Improving Risk Communication*. Washington, DC: National Academy Press.

Sandman, P. M. 1991. *Risk = Hazard + Outrage: A Formula for Effective Risk Communication*. Fairfax, VA: AIHA Press (videotape).

Waldo, A. B., and R. Hinds. 1988. *Chemical Hazard Communication Guidebook: OSHA, EPA, DOT Requirements*. New York: John Wiley & Sons, Inc.

Resources for Program Audits

Labato, Frank J. 1997. Ch. 38. "Survey and Audits." In *The Occupational Environment—Its Evaluation and Control*. Salvatore R. DiNardi, ed. Fairfax, VA: AIHA Press.

Lynch, Jeremiah R., and James T. Sanderson. 1998. Ch. 14. "Program Evaluation and Audit." In *A Strategy for Assessing and Managing Occupational Exposures*. 2d ed. John R. Mulhousen and Joseph Damiano, eds. Fairfax, VA: AIHA Press.

Resources for Consultant Selection

McFee, Donald R., and Gary N. Crawford. 1998. Ch. 17. "Consultants and Their Use." In *A Strategy for Assessing and Managing Occupational Exposures*. 2d ed. John R. Mulhousen and Joseph Damiano, eds. Fairfax, VA: AIHA Press.

NIOSH. "Guidelines for Selecting and Using an Industrial Hygiene Consultant." In *Occupational Exposure Sampling Strategy Manual*. Publication number 77-173. Available at www.cdc.gov/niosh/publistb.html#1977.

Resources for Quality Control

Bullock, William H., and Joselito S. Ignacio, eds. 2006. *A Strategy for Assessing and Managing Occupational Exposures*, 3d ed. Fairfax, VA: AIHA Press.

NIOSH. *Occupational Exposure Sampling Strategy Manual*. NIOSH publication number 77-173. Available at www.cdc.gov/niosh/publistb.html#1977.

Resources for Personal Protective Equipment (PPE)

Anna, D. H., ed. *Chemical Protective Clothing*. 2d ed. Available at www.aiha.org.

ASTM. 1998. *Standard Practice for Chemical Protective Clothing Program*. F 1461–93 (11.03).Ordering information available at www.astm.org.

Forsberg, K., and S. Z. Mansdorf. *The Quick Selection Guide to Chemical Protective Clothing*. 4th ed. Available at www.aiha.org.

Genium Publishing Corporation. *Personal Protective Equipment Pocket Guide*. Available at www.acgih.org.

NIOSH. *Recommendation for Chemical Protective Clothing*. NIOSH publication number 2004-103. Available at www.cdc.gov/niosh/homepage.html.

______. *Pocket Guide to Chemical Hazards.* NIOSH publication number 2004-103. Available at www.cdc.gov/niosh/homepage.html.
OSHA. Web site: www.osha.gov.
______. *Assessing the Need for Personal Protective Equipment: A Guide for Small Business Employers.* OSHA 3151 (part of Small Business Management Series). Available at www.osha.gov.
Stull, Jeffery O. *PPE Made Easy.* Available from ACGIH at www.acgih.org.

Resources for Hearing Protection

NIOSH. *Compendium of Hearing Protection Devices.* NIOSH publication number 95-105. Available at www.cdc.gov/niosh/homepage.html.
OSHA. Regulation 29 CFR 1910.95.
______. "Noise Measurement." *OSHA Technical Manual,* Sec. III, Ch. 5. Available at www.osha.gov.

Resources for Respiratory Protection

AIHA *Respiratory Protection: A Manual and Guideline.* 3d ed. Available at www.aiha.org.
ANSI. *Standard for Fire Service Respiratory Protection Training.* ANSI/NFPA 1404-2002. Available at www.ansi.org.
______. *Commodity Gas Specification for Air.* ANSI/CGA G-7.1-1989. Available at www.ansi.org.
______. *Respiratory Fit Testing Methods.* ANSI Z88.10-2001. Available at www.ansi.org.
NIOSH. Web site: www.cdc.gov/niosh/homepage.html.
______. *Guide to Industrial Respiratory Protection.* NIOSH publication number 87-116. Available at www.cdc.gov/niosh/87-116.html.
______. *Guide to the Selection and Use of Particulate Respirators Certified Under 42 CFR 84.* NIOSH Publication Number 96-101. Available at www.cdc.gov/niosh/homepage.html.
______. *Pocket Guide to Chemical Hazards.* NIOSH Publication Number 2004-103. Available at www.cdc.gov/niosh/homepage.html.
OSHA. Web site: www.osha.gov.
______. *Small Entity Compliance Guide for Respiratory Protection Standard* (1910.134). Available at www.osha.gov/pls/publications/pubindex.list#118.
Respirator selection guides from respirator manufacturers.

Resources for Occupational Exposure Limits

ACGIH. *Documentation of Threshold Limit Values for Chemical Substances and Physical Agents and Biological Exposure Indices.* Current edition available at www.acgih.org.
______. *2007 TLVs and Biological Exposure Indices (BEIs).* Current booklet available at www.acgih.org.
AIHA. *Workplace Environmental Exposure Level Guides.* Available at www.aiha.org.
EPA. *New Chemical Exposure Limits.* Available at www.epa.gov/oppt/newchems/nceltbl.htm.
NIOSH. *Pocket Guide to Chemical Hazards.* Available at www.cdc.gov/niosh/npg/npg.html.
OSHA. Regulations 29 CFR 1910.1000 through 1910.1052.

Resources for Control Banding

NIOSH. *Safety and Health Topic: Control Banding.* Available at www.cdc.gov/niosh/topics/ctrlbanding/#9.
Silverstein, Bernard D. 2006. Ch. 24."Control Banding/COSHH Essentials." In *A Strategy for Assessing and Managing Occupational Exposures.* 2d ed. Joselito S. Ignacio and William H. Bullock, eds. Fairfax, VA: AIHA Press.
United Kingdom Health and Safety Executive (HSE). 2005. *COSHH Essentials.* Available at www.coshhessentials.org.uk.

Resources for Confined Space Entry

American Petroleum Institute. *Guidelines for Confined Space Entry on Board Tank Ships in the Petroleum Industry.* API RP 1141. Available at www.api.org.
______. *Guidelines for Work in Inert Confined Spaces in the Petroleum Industry.* ed ed. API Publication 2217A Available at www.api.org.
______. *Safe Entry and Cleaning of Above Ground Storage Tanks.* API Publication 2015 (1994). Available at www.api.org.
ANSI. *Safety Requirements for Confined Spaces* (provides the minimum safety requirements to be followed while entering, exiting, and working in confined spaces). ANSI Z117.1-2003. Available at www.ansi.org.
Compressed Gas Commodity Association. *Accident Prevention in Oxygen-Rich and Oxygen-Deficient Atmospheres.* Publication CGAP–14. Available at www.cganet.com.
National Fire Protection Association. *Recommended Practice on Materials, Equipment, and Systems Used in Oxygen-Enriched Atmospheres.* NFPA 53, 1999. Available at www.nfpa.org.
______. *Recommended Practice on Static Electricity.* NFPA 77, 1993. Order at www.nfpa.org.
______. Standard for the Control of Gas Hazards on Vessels. NFPA 306, 1997. Order at www.nfpa.org.
______. *Standard for the Safeguarding of Tanks and Containers for Entry, Cleaning or Repair.* NFPA 326, 1999. Order at www.nfpa.org.
______. *Standard Procedure for Cleaning or Safeguarding Small Tanks and Containers Without Entry.* NFPA 327, 1993. Order at www.nfpa.org.

NIOSH. Web site: www.cdc.gov/niosh/homepage.html.
OSHA. Web site: http://www.osha.gov.

Resources for Ventilation System Programs

ACGIH. *Industrial Ventilation—A Manual of Recommended Practice*. The current edition is available at www.acgih.org.

ANSI/AIHA. Z9.2-2001. *Fundamentals Governing the Design and Operation of Local Exhaust Systems*. Available at www.ansi.org.

______. Z9.3-1994. *Spray Finishing Operations—Safety Code for Design, Construction, and Ventilation*. Available at www.ansi.org.

______. Z9.4-1997. *Exhaust Systems – Abrasive Blasting Operations – Ventilation and Safe Practices for Fixed Location Enclosures*. Available at www.ansi.org.

______. Z9.5-2002. *Laboratory Ventilation*. Available at www.ansi.org.

______. Z9.6-1999. *Exhaust Systems for Grinding, Polishing, and Buffing*. Available at www.ansi.org.

AWS. F3.2M/F3.2-2001. *Ventilation Guide for Weld Fume*. Available at www.aws.org.

Resources for Modeling Inhalation Exposure

AIHA. 2000. *Exposure Assessment Strategies Committee. Mathematical Models for Estimating Occupational Exposure Limits to Chemicals*. Fairfax VA: AIHA Press.

Jayjock, Michael A 1997. Ch. 16.. "Modeling Inhalation Exposure." In *The Occupational Environment—Its Evaluation and Control*. Salvatore R. DiNardi, ed. Fairfax, VA: AIHA Press.

Nicas, Mark. 2006. Appendix I. "Estimating Airborne Exposure by Mathematical Modeling." In *A Strategy for Assessing and Managing Occupational Exposures*. 3d ed. Joselito S. Ignacio and William H. Bullock, eds. Fairfax VA: AIHA Press.

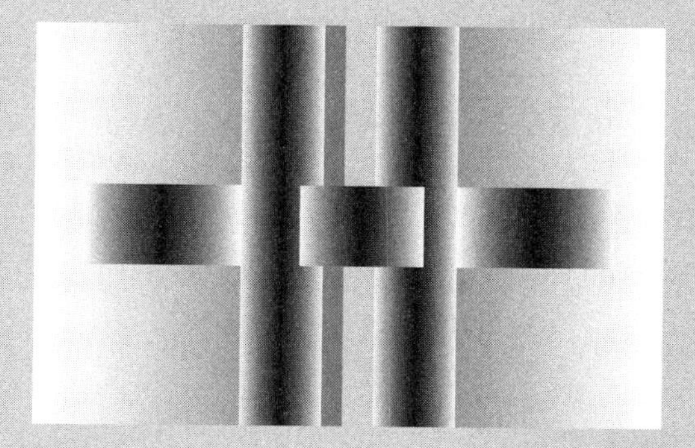

SECTION 5 PERSONAL PROTECTIVE EQUIPMENT

Regulatory Issues

Applied Science and Engineering

Cost Analysis and Budgeting

Benchmarking and Performance Criteria

Best Practices

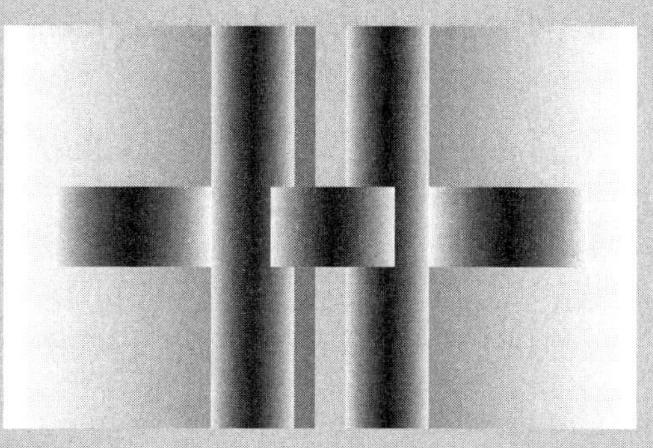

SECTION 5
PERSONAL PROTECTIVE EQUIPMENT

LEARNING OBJECTIVES

- Be able to explain the role that consensus standards play in the selection and use of personal protective equipment (PPE).
- Learn about the prominent role PPE plays in OSHA's General Industry and Construction Standards.
- Be able to discuss why relying on OSHA requirements alone may not provide up-to-date guidance in the selection and use of PPE.

REGULATORY ISSUES

Robert Edgar

PERSONAL PROTECTIVE EQUIPMENT (PPE) is one of the three essential elements in an effective occupational health and safety program. PPE, along with engineering and administrative controls, represent the fundamentals of any worker protection strategy (OSHA 1998, 29). Whenever practical, engineering controls are the first line of worker defense. When engineering controls are not feasible, are impractical, unaffordable, or simply not effective, then PPE in conjunction with work practices (administrative controls) are essential in worker protection (Plog and Quinlan 2002, 586; Harris 2000, 2386). Properly selected PPE can help protect workers against injury and exposure to a wide array of general and specific workplace hazards(Camplina 2003, 48–51).

The hierarchy of prevention of occupational injuries and illnesses requires four integrated elements. First, hazard identification and risk assessment. Second, hazard elimination and risk control. Third, material or process substitution and process design or redesign (including engineering and administrative controls). And lastly, the routine use of PPE in combination with training and supervisory oversight (Harris 2000, 2386; Heinrich 1941, 448; National Safety Council 1997a, 711 and 1997b, 387).

Around the world, government regulation has some influence on worker protection, perhaps to the extent that prevailing social, political, and economic conditions allow. Increased globalization and communication, to the extent it promotes greater occupational and environmental awareness, has made access to the various regulatory schemes used around the world much easier.

Safety and health professionals must remain abreast of current rules, regulations, standards, recommendations, and best practices in the selection and use of PPE. This systematic process is the foundation of PPE selection and, according to OSHA, required in the form of a PPE written certification as described in the PPE standard (29 CFR 1910.132).

The lack of PPE can lead to morbidity and mortality. During the Great Depression in the 1930s, workers in Gauley Bridge, West Virginia, carved a tunnel through rock that contained a high percentage of free silica without using wet methods and with little or no respiratory protection (Cherniak 1989, 204). Of the two thousand employed on the project, several hundred died and many more were permanently disabled by silicosis. Although silicosis usually takes several years to develop, some workers developed silicosis in only a few months.

PPE and OSHA

An electronic search of OSHA's General Industry Standard (29 CFR1910) found 1,634 PPE references in 2007. An identical search of the Construction Industry Standard (29 CFR 1926) mentioned PPE 556 times (BNA 2007). Additionally, the General Industry Standard referred to employee training 567 times, and in the Construction Standard, employee training appeared 286 times. While OSHA's authority in workplace safety dates to the early 1970s, employers today still receive citations for the failure to use PPE (Mendeloff and Gray 2005, 219–237).

PPE and Consensus Standards

The Occupational Safety and Health Act of 1970 provided the legal authority for the promulgation and enforcement of federal safety and health standards through a newly formed federal agency, the Occupational Safety and Health Administration (OSHA 1970). Congress gave this newly established federal agency the authority to create a set of start-up health and safety standards (without the formality of rulemaking) from existing consensus standards, most of which came from the National Fire Protection Association and the American National Standards Institute (Plog and Quinlan 2002, 586). An electronic search of 29 CFR 1910 for the term "consensus standards" results in 112 references, and 15 references in 29 CFR 1926 (BNA 2007).

Key organizations in the United States that have consensus standards relevant to PPE include the following:

- American Congress of Governmental Industrial Hygienists (ACGIH)
- American National Standards Institute (ANSI)
- American Society for Testing and Materials International (ASTM)
- The National Fire Protection Association (NFPA)
- The National Institute of Occupational Safety and Health (NIOSH)

PPE: Quality and Performance

In the United States, the NIOSH certification of respirators and respirator parts is a well-known requirement (NIOSH 2004, 32). The certification ensures that respirators used in the United States meet minimum performance criteria. Across the Atlantic, safety and health professionals in Europe rely on the certification standards developed by the European Committee for Standardization (CEN). CEN and its allied organizations are responsible for the development of European norms to ensure standardization of performance criteria for a variety of products, including PPE (Jung 1995, 108–117; Mayer and Korhonen 1999, 347–360). Increasingly, global harmonization of PPE certification standards means that products around the world will be more alike tomorrow than they are different today (Cunningham and Johnstone 1999, 423; Rothstein, Hood, and Baldwin 2001, 232).

Why PPE Must Exceed OSHA Standards

The process OSHA must use to change or update a regulatory standard to reflect a current consensus standard is not as simple as might be imagined. In fact, it may lag by several years. The consensus standards available in 1970 represented the best source of information on PPE at the time. As new information becomes available today, updated consensus standards for PPE appear when the convergence of technology and application reach a tipping point. OSHA, unlike organizations that develop consensus standards, has not kept up with best safety practices in many cases because of the strict, mandated updating requirements they must meet (Dinardi 1997, 1365).

OSHA's current eye and face protection standard is an example of a PPE standard that is not up to date. OSHA, in creating its original regulations in 1970, incorporated the 1968 version of ANSI Z87.1. Although ANSI updated this standard in 1989, OSHA did not adopt it until 1994 (OSHA 1994, 1). ANSI again updated this standard in 2003. It is prudent to consider OSHA requirements to be the minimum requirements in worker protection, not necessarily the best industry practices (ISEA 2002, 2).

References

Bureau of National Affairs (BNA). 2007. Environmental & Safety Library. www.bna.com/products/ens/eslw.htm

Camplin, Jeffery C. 2003. "Assessing the Need for PPE: Hazards Present Determine Level of Protection." *Professional Safety* 48(12):48–51.

Cherniak, Martin. 1989. *The Hawk's Nest Incident: America's Worst Industrial Disaster*. New Haven, CT: Yale University Press.

Cunningham, Neil, and Richard Johnstone. 1999. *Regulating Workplace Safety, Oxford Socio-Legal Studies*. Oxford, England: Oxford University Press.

Dinardi, S. R. 1997. *The Occupational Environment: Its Evaluation and Control*. Fairfax, VA: American Industrial Hygiene Association.

Harris, Robert L, ed. 2000. *Patty's Industrial Hygiene*. 5th ed. (4 vols.). New York: John Wiley & Sons.

Heinrich, Herbert William. 1941. *Industrial Accident Prevention: A Scientific Approach*. New York: McGraw-Hill.

International Safety Equipment Association (ISEA). 2002. "Personal Protective Equipment: An Investment in Your Workers' and Company's Future." In *Partnership for Worker Protection*. Arlington, VA: ISEA.

Jung, K. 1995. "CEN Standards for Testing and Certifying Personal Protective Equipment: Status Quo and Deficiencies: Examples." *International Journal of Occupational Safety and Ergonomics* 1(2):108–117.

National Safety Council (NSC). 1997a. *Accident Prevention Manual for Business and Industry: Administration & Programs*. 11th ed. (3 vols.). Gary Krieger and John Montgomery, eds. Itasca, IL: NSC.

______. 1997b. *Supervisors' Safety Manual*. 9th ed. Patricia Laing, ed. Itasca, IL: NSC.

Mayer, A,. and E. Korhonen. 1999. "Assessment of the Protection Efficiency and Comfort of Personal Protective Equipment in Real Conditions of Use." *International Journal of Occupational Safety and Ergonomics* 5(3):347–360.

Mendeloff, John, and Wayne B. Gray. 2005. "Inside the Black Box: How Do OSHA Inspections Lead to Reductions in Workplace Injuries." *Law & Policy* 27(2):219–237.

National Institute for Occupational Safety and Health (NIOSH). 2004. *NIOSH respirator selection logic 2004*. Edited by NIOSH: DHHS. Cincinnati, Ohio: NIOSH.

Occupational Safety and Health Administration (OSHA). 1994. *Non-Mandatory Compliance Guidelines for Hazard Assessment and Personal Protective Equipment Selection*. www.osha.gov/pls/oshaweb/owadisp.show_document?p_table=STANDARDS&p_id=10120

______. 1998. *Personal Protective Equipment* (OSHA 3077). www.ehsservices.com/Library/Personal%20Protective%20Equipment.pdf

Plog, Barbara, and Patricia Quinlan. 2002. *Fundamentals of Industrial Hygiene*. 5th ed. Itasca, IL: NSC.

Rothstein, Henry, Christopher Hood, and Robert Baldwin. 2001. *The Government of Risk: Understanding Risk Regulation Regimes*. Cambridge, England: Oxford University Press.

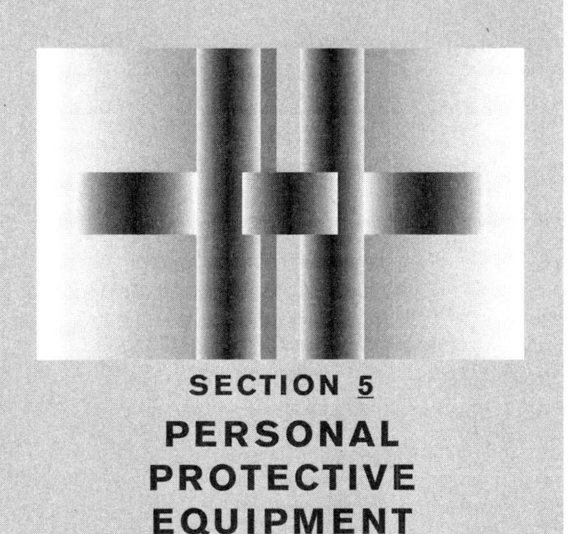

SECTION 5
PERSONAL PROTECTIVE EQUIPMENT

LEARNING OBJECTIVES

- Identify three levels of workplace control hierarchy and the role played by personal protective equipment (PPE).
- Identify head protection selection criteria and explain the difference between: Type I and II helmets, hard helmets and bump hats.
- Identify two methods for enhancing visibility of materials and garments and the utility of both.
- Discuss gloves and clothing performance standards for abrasion, cutting, and puncture resistance.
- Discuss PPE's role in reducing heat stress and creating it.
- Discuss the two types of hazards presented to electrical workers and how to choose PPE to maximize protection.
- Name three strategies for using fall protection and the three required components of PPE.
- Describe the differences between permeation, penetration, and degradation caused by agents on chemical protective clothing.
- Explain the difference between qualitative and quantitative respirator fit testing.

APPLIED SCIENCE AND ENGINEERING

David May

PERSONAL PROTECTIVE EQUIPMENT (PPE) plays a unique role in protecting workers from injury or occupational disease. PPE is often the last line of defense, providing a physical barrier between the worker and the hazard. The application of PPE in enhancing occupational safety and health is enormous, covering approximately 19.6 million workers in the United States alone, according to OSHA estimates (OSHA 1999c, 15418).

There are a multitude of risks workers face. These risks are grouped by section in this chapter by hazard type as follows:

- Impact and abrasion
- Thermal stressors (heat and cold)
- Nonionizing exposures
- Noise and vibration hazards
- Electrical-related hazards (shock and arc burns)
- Fall risks
- Chemical hazards (contact and respirable hazards)
- Biological hazards
- Ergonomic stressors

Additional hazards associated with the use of PPE follow these sections.

HIERARCHY OF CONTROLS

Deciding which method of controlling any occupational hazard to use follows recognition of the hazard in a hazard assessment. Hazard evaluation, discussed in detail elsewhere in this book, is an important element in the overall safety program. Understanding the basis of the hazard and the appropriate

TABLE 1

Hierarchy of Controls

Risk Prevention Level	Prevention Target	Example Controls
Primary	Elimination or reduction of exposure at source	Substitution Engineering controls Isolation, enclosure, guards Ventilation Zero energy
Secondary	Administrative reduction of exposure by time limits, procedure, or distance	Worker rotation Work practice changes Training
Tertiary	Control at the worker	PPE Biological monitoring

(*Source:* NIOSH 2006b)

application of PPE is necessary to ensure that the use of PPE will be effective in preventing injury or illness. PPE is the choice of last resort because of the limitations of PPE or the way that it is used. PPE often is a temporary measure used during the installation of engineering or administrative controls. Table 1 outlines the hierarchical methods of protection. Engineering controls are preferred more than administrative or work practice methods.

There are numerous reasons why PPE is the least-preferred control (both in principle and by regulation). They include limits on the ability to provide consistent protection, lack of use by workers, misuse of PPE rendering its protection moot, and complexity and difficulty of administering PPE programs. The use of PPE as the sole method of control is often not cost effective owing to PPE replacement and program oversight.

PPE does not provide unlimited protection. Residual risk may continue to be unacceptably high, even when PPE is used effectively. There are many references to situations where PPE failed to meet assumed risk reduction. For example, in a study of 1300 heat burns, most victims were wearing some PPE. The majority of these workers reported that their burns resulted from the hot liquid or flame going around the gear, usually shoes or gloves (Personick 1990). In a study of latex and vinyl gloves commonly found in healthcare facilities, 25 percent of the gloves leaked water (n = 4838) after use (Korniewicz et al. 1993). An assessment of the effectiveness of respirators used by coke oven workers (measuring benzene soluble particulates inside versus outside the respirator) found program protection factors of less than 2.0 for nearly half the workers (Wu 2002). While respiratory protection reduced the incidence of immunologic respiratory disease (from 10 percent to 2 percent per year) in a cohort of epoxy workers, it did not eliminate it (Grammer, Harris, and Yarnold 2002). Since all PPE has limited protective capability, it is important to determine how those factors contribute to the risk assessment. Manufacturer's data and performance criteria found in consensus or regulatory standards often indicate the limits of PPE effectiveness.

PPE fails to protect when it is not worn. For example, there have been many eye injuries from lasers when eyewear was available but not worn. The most commonly cited reason was that laser protective eyewear is dark, uncomfortable, and limits vision (OSHA 1991b). Surveys and interviews suggest comfort plays a large role. Nearly 800 workers who used PPE in either a metal refining or an automobile glass plant reported that less than 25 percent of them found their hard hats, cooling vests, respirators, and/or safety harnesses were comfortable. In the same study, only a third to one-half rated safety glasses, gloves, hearing protectors, coveralls, and safety shoes somewhat more comfortable. More specifically, workers reported that PPE was difficult to wear. Common complaints included being too heavy, interfered with communication, irritated the skin, caused uncomfortable pressure on the body, interfered with work, or was thought unnecessary (Akbar-Khanzadeh, Biseli, and Rivas 1995; Akbar-Khanzadeh 1998). A survey by the International Safety Equipment Association found that "equipment not being available or provided" or "employers do not require/enforce usage" were the two most common reasons why PPE use in heavy construction lags behind other industries (ISEA 2004).

To be truly effective, employers must provide and require PPE use. Often, workers bring their own PPE to work. The appropriateness and effectiveness of PPE remains the responsibility of the employer, regardless of how it came to work. Successful PPE programs also require solicitation of workers' opinions about the comfort and utility of PPE. Regulatory requirements occasionally reinforce this concept, such as requiring the availability of multiple choices of hearing protectors. In addition, many manufacturers are marketing more fashionable forms of PPE to increase their acceptability (ISEA 2004; Koppes and Pearson 2002; Sokol 2002).

The modification of PPE by the user can seriously compromise effectiveness. Often, workers alter PPE to make it more comfortable, accommodate smoking (for example, dust masks) or identify ownership. These changes can render the device less effective or, in some cases, useless. For example, some users of earmuffs have reduced the headband pressure to increase comfort or to drill holes in the ear cup to promote ventilation or drain perspiration, with obvious deleterious results (Berger 1986, 333, 352).

Regulations often reinforce the risk reduction hierarchy. For example, OSHA requires that administrative or engineering controls must be first determined and implemented whenever feasible to ensure that air contaminant exposures are below permissible exposure limits (PELs). Only when such controls are not feasible can employers use personal protective devices (such as respirators) to meet the PEL (OSHA 1974). Likewise, NFPA 70E and OSHA require that electrical systems must be de-energized prior to beginning work. Personnel can work on live circuits with appropriate PPE only if the employer can demonstrate that de-energizing (a primary control) introduces additional or increased hazards or is infeasible (Mastrullo 2004).

Simply stated, when engineering, work practice, or administrative controls are not feasible or fail to reduce the overall risk, PPE plays a critical role in worker protection as a complementary means of protection with guards, engineering, and administrative controls.

PHYSICAL HAZARDS AND PPE

PPE can protect workers from physical hazards they encounter in the workplace. As described in the following section, these hazards include impact, thermal, non-ionizing radiation, noise, vibration, electrical, and protection from falls.

Impact Hazards

Workers face many impact hazards where PPE can reduce the risk of incident or injury. These hazards include exposure of tissue, head, feet, and eyes to flying particles and falling objects, and body contact with equipment or vehicles. Both protection and visibility in PPE design are essential in these circumstances.

Head Protection

According to data from the United States Bureau of Labor Statistics, head injuries constitute approximately 3 percent of workplace injuries that result in at least a day or more away from work. The impact from falling, flying, swinging, or slipping objects cause 25 percent of all occupational head injuries. Typical head impact injuries consist of perforation of the skull due to a localized force, fracture of the skull from an excessive blunt force, or brain injury due to sudden displacement of the brain within the skull (Balty and Mayer 1998; BLS 2007a).

Helmets (hard hats) protect in areas where there is a potential injury to the head caused by falling objects and bumps against moving or fixed objects. Hard-hat manufacturers often construct helmets using polyethylene or polycarbonate plastics. While the shell is the most obvious part of head protection, of equal importance is the helmet's suspension, which plays a major role by stretching and absorbing a blow by spreading the impact force over a large area. It is important to maintain the suspension's ability to perform this function, including its periodic replacement whenever the harness shows signs of cracking, tearing, or fraying.

Helmets are designed to reduce the danger of impact, penetration, and force of the blow (and electrical shock and burn; see the section on electrical hazards PPE) through the performance requirements established by ANSI Z 89.1. For example, this standard requires an impact test of dropping a 3.6-kilogram weight such that it reaches a velocity of 5.5 meters per second before impact and must not transmit a force that exceeds 4450 Newtons. While helmets meeting these criteria reduce the amount of force transmitted to the head from an impact blow, it does not protect against extreme high-force impact. According to NIOSH, most blows to the head come from the back, front, or side of the helmet (OSHA 1994b; Gilchrist and Mills 1996). In the mid-1990s, ANSI added performance requirements for protection from blows to the side. Head protection classified as Type I reduces the force of impact only to the top of the head, while Type II helmets protect against top and sides. Impact, force transmission, and penetration tests are performed over a range of 0°F to 120°F. Helmets meeting the ANSI Z89.1 criteria are marked with type and class, size, name of manufacturer, and date of manufacture (ISEA 2003b, 36). See Table 19 for additional performance criteria.

Manufacturers continue to make hard hats more acceptable, versatile, and comfortable. This includes such additions as fleece liners for cold weather, brow pads, and evaporative cooling and ventilation for hot environments. ANSI-approved designer hard hats are also available that feature an assortment of colors, sport, and NASCAR logos. Some manufacturers now certify their helmets for forward or backward wear. Hard hats today integrate with other safety devices, including noise muffs, face protection, dust/chip screen visors, and illumination attachments (see Figure 1). Since attachments may affect the performance of the helmet, users should investigate compatibility with the accessory or helmet manufacturer and should not drill, cut, or glue the helmet unless so directed by the helmet manufacturer.

FIGURE 1. **A hard hat with an ear muff attachment** (Photo courtesy of OSHA)

Bump caps protect against minor hazards such as low-impact bumps into low-hanging objects. A European standard, EN 812, gives some performance criteria for the protection attributes of bump caps (European Committee for Standardization 2001). Carefully evaluate the use of bump caps and limit them to specific applications that do not involve the potential of impact hazards. In most cases, a hard hat is a better choice.

Finally, falling or misadjusted hard hats must not become a hazard. Chinstraps or ratchet suspension systems can prevent a helmet from falling or failing to protect when needed.

Eye and Face Protection

Eye and face protection equipment includes such devices as spectacles, goggles, face shields, welding helmets, and respirators (that cover the face). Eye and face protection can be used to protect workers against hazards of flying particles, dusts, molten metal, liquid droplets, chemical vapors, laser light, and ultraviolet and infrared radiation.

Unfortunately, workplace eye injuries are still common and account for a large portion of ocular injuries, despite the fact that they are largely preventable. In a one study, 81 percent of eye injuries that came to an emergency department occurred at the workplace (Henderson 1991). The Bureau of Labor Statistics estimates there were over 42,000 workplace eye injuries in the private sector that were serious enough to require at least one day away from work in 2002. Over half of all eye injuries occurred in manufacturing or trade (wholesale and retail). Welders, cutters, and nonconstruction laborers were the occupations with the most eye injuries (Harris 2004). Eye injury accounts for 11 to 12 percent of all workers' compensation injuries in construction, with the most common injuries involving overhead work, use of power tools, or both (Lipscomb et al. 1999; Welch, Hunting, and Mawadeku 2001). Many workers are injured because they were not wearing eye or face protection or, if they were, it was not appropriate for the task (Yu, Liu, H., and Liu, K. 2004; Welch, Hunting, and Mawadeku 2001). In a study of an automobile manufacturer, only 25 percent of workers had been using some form of eye protection at the time of injury (Wong et al. 1998). In another study, eye protection was not used despite it being provided in over half the cases needing treatment (Henderson 1991). An earlier BLS study found that 40 percent of the injured workers were wearing some form of eye protection when the accident occurred, most likely wearing eyeglasses without side shields (OSHA 1993b, 4). Protective side shields provide additional protection against lateral impact (OSHA 1994b; ANSI 2003, 15).

Choosing the correct device involves matching the anticipated hazards while ensuring comfort and clear vision. Eye protection is available in a number of sizes and styles. For workers needing corrective eyewear, ANSI-approved frames and lenses eliminate the need for additional eye protection. Among the barriers to wearing eye protection, workers cite the propensity of eye protection to fog in warm and moist conditions as a reason not to wear occupational eye protection, even when mandated. The use of anti-fog treatment can help in some situations. In one study, individuals performed a target detection task in environments that typically cause eyewear to fog. The study examined eye protection treated with anti-fog coating, eyewear not treated, and no eyewear situations. While the study concluded that the use of anti-fog coating is relatively effective, it noted that in environments where prolonged fogging occurs, water droplets form on the anti-fog coated lenses and disrupt visual performance in a manner similar to the fog it is trying to prevent (Crebolder and Sloan 2004). Similarly, miners considered conventional anti-fog treatments to be completely inadequate in the coal mining industry (Dain et al. 1999).

ANSI Standard Z87.1 outlines the requirements for occupational eye protection. The ANSI Z87 mark identifies eye protection that meets this standard. The most well-known test for these products is the impact and penetration test that involves dropping or otherwise propelling projectiles at the product. Spectacles, goggles, and face shields all have different testing criteria. For example, face shields are tested for high-mass impact by dropping a pointed projectile weighing 500 grams from a height of 50 inches. Additional face shield tests include high velocity impact (0.25-inch ball traveling 300 feet per second) and an impact test using a 1-inch steel ball dropped from 50 inches. Protectors marked with "Z87" comply with the basic impact testing, while those marked with "Z87+" comply with high-impact test requirements. Both frames and lenses of ANSI-approved spectacles are subject to impact and penetration tests. ANSI provides additional testing criteria of eye and face protectors (including the eye protection aspects of respirators). These criteria included optical properties, flammability, corrosion resistance, ease of cleaning, and transmittance of luminous, ultraviolet, and infrared radiation (ANSI 2003, 1–67).

Personnel should select eye and face protectors for any given task from a hazard assessment that includes a clear understanding of the work tasks and the anticipated physical and chemical hazards. Consider simultaneous hazards as well in the hazard assessment. See Table 2 for selection of eye/face protection summarized from various sources.

Foot Protection

The Bureau of Labor Statistics estimated that there were over 57,000 occupational injuries involving the foot that resulted in at least one lost day from work in 2006 in the private sector in the United States.

TABLE 2

Typical Selection of Eye/Face Protectors

	Hazard								
Protector	*Flying particles*	*Radiant heat or splash from molten metals*	*Chemical dust and splash hazards*	*Chemical mist or vapor*	*Aerosol pathogens*	*Arc viewing, welding, and cutting*	*Gas welding, cutting and viewing gas-fired furnaces*	*Torch brazing*	*Torch soldering*
Spectacles and side shields	LE	C1	—	—	P	C3	C3	—	U
Face shield	C5	C1	C2	—	P	—	C4	T	U
Goggles (cover or cup)	LE	C1	S1	S2	P	C3	C3	—	U
Welding helmet	—	—	—	—	—	U	U	U	—
Welding hand shield	—	—	—	—	—	C3	C3	—	—
Full-face respirator	LE	C1	U	U	P	—	—	—	—
Loose-fitting respirator	C6	C1	C6	U	—	—	U	—	—
Welding full-face respirator	—	—	—	—	—	U	U	U	U

Key:

U Utilized
LE Low energy impact particles
C1 Combination of metal wire screen (wire mesh face-shields can reduce infrared radiation by 30–50%) or a reflective face shield over spectacles, goggles or respirator
C2 Face shield can be added over goggles for severe hazards, face shields can not be used alone
C3 Welding helmet or hand shield combination with spectacles or goggles
C4 Welding face shield over spectacles
C5 Face shields can be added over spectacles or goggles for high energy impact particle hazards
C6 Spectacles worn under respirator where high energy impact particles or chemical splash hazards are anticipated
S1 Splash goggles (i.e., no direct air vents) for chemical splash hazards
S2 Nonvented goggles
P Protectors with mask; indirectly vented or non-vented goggles are preferable to spectacles with side shields; face shield should be chin-length
T Welding face shield

(*Source:* Adapted from ANSI 2003, NIOSH 2004a)

Being struck by an object (especially a falling object) or stepping on an object caused over 50 percent of these foot injuries (BLS 2007a).

The type and degree of protection offered by footwear should follow the personal protection risk assessment. Most common safety shoes minimize the severity of impacts to the foot and toe, using protective parts built into the shoe. For example, a toecap made of a strong, yet light material encases the toes, helping to protect against falling objects and crushing injuries.

Until recently, ANSI Z41 provided the minimum requirements for safety shoes (ANSI 1999). In 2004, the committee responsible for ANSI Z41 merged with the ASTM International Committee F13 on Safety and Traction for Footwear. The resulting F2413-05 standard specifies tests to determine the minimum performance of footwear protection of the toes, metatarsal area (between the foe and the ankle), and the foot bottom. The standard designates two classes of protection based on compression and impact protection for the toe, with the option of metatarsal protection (indicated by the marking "mt"), according to Table 3. Impact tests involved dropping an object from various heights to attain certain velocities and proscribed impact forces.

TABLE 3

Compression and Impact Requirements

Class	Compression Test (pounds-force)	Impact Test (foot-pounds)
50	1750	50
75	2500	75

(*Source:* ASTM 2005b, 1–7.)

The use of protective footwear should not create additional hazards from static shock hazards, insulation hazards caused by induced voltage, or as ignition sources. ASTM F2413 provides performance tests for electrical shock resistance, conductive, static dissipative, and dielectric protective properties. Table 4 lists the various markings used for these criteria.

In addition, footwear used around liquids with caustic, acid, or high temperature properties, including molten metal, should be designed or worn so that spilled liquid will not enter the shoe or boot (or alternatively used with spats, chaps, gaiters, or leggings).

Slip resistance is also a consideration for footwear worn in the workplace, since falls account for such a large portion of injuries. Recently, there has been considerable discussion on an acceptable approach to rating slip resistance for footwear, including the appropriate measuring technique or device (tribometer) that will give reliable results over a variety of conditions (DiPilla and Vidal 2002).

Typically, a slip rating of between zero and one reflects a range from low to high slip resistance. It has been recommended that dry walking surfaces have a slip-resistance rating of 0.5. Note that slip resistance is often confused with the static coefficient of friction, which is the force required to slide one object over another under dry conditions. Likewise, the dynamic coefficient of friction, the measure of the force necessary to keep an object in motion, often results in values less than its static counterpart. In contrast, slip resistance is a measure that includes these factors as well as others, such as presence of water or other contamination and likely nonconstant velocity of sliding objects (for example, shoe surface on a walking surface) (ASSE 2006, 9–21; Grieser, Rhoades, and Shah 2002).

When surfaces are wet or contaminated with oil, powders, or other materials that lower the slip resistance, control measures, such as slip-resistant footwear, may be appropriate in addition to surface treatment. Footwear traction is a function of tread pattern and hardness, sole material, wear, and shape. Softer materials and tread patterns that are random or perpendicular to the direction of travel and without enclosed spaces are generally more slip resistant (DiPilla 2006). Unfortunately, although some evaluation of footwear traction occurs for ranking purposes,

TABLE 4

Additional Footwear Approval Criteria

Test	Mark	Description
Conductive	Cd	To provide protection against static electricity buildup, typically to prevent the initiative of explosive atmospheres when used with grounded floors (defined as resistance ≤ 500,000 ohms)
Electrical hazard	EH	To provide secondary protection against accidental contact of open circuits of 600 volts or less under dry conditions (tested at 14,000 volts)
Puncture resistant	PR	To protect against penetration of the sole by sharp objects (tested at 270 lbf)
Static dissipative	SD	To protect against the accumulation of static electricity while protecting against electrical hazard due to live electrical circuits (defined as resistance of 1–100 megohms)

(*Source:* ASTM 2005b, 1–7)

FIGURE 2. Example of a whole-shoe tester protoype. (Copyright 2006 by William English Inc. Reprinted by permission)

there is no standard that is comprehensive and assists buyers who are searching for specific footwear traction performance (English 2003, 111–124) (see Figure 2).

Specialized standards for protective footwear also exist to provide cut resistance protection for power chain saw operators (ASTM 2004a, 1–5). These devices typically add a cover or incorporate cut-resistant material, such as ballistic nylon, into the footwear to prevent a chain saw from cutting into the shin, ankle, foot, and toes.

Visibility-Enhancing Clothing/Vests

Fatalities in the highway construction industry often occur in the work zone, that area between the first warning sign and the last traffic control device. Motorists passing through the work zone accounted for 318 deaths between the years 1992 and 1998 (Pratt, Fosbroke, and Marsh. 2001, 5). In addition to the use of administrative controls and engineering controls, such as barriers and separation to reduce risk of contact, workers need appropriate PPE to be visible to traffic that passes through work zones (for example, highway construction) or in their work area (such as parking lots and warehouses).

Visibility-enhancing garments are comprised of high-visibility colored cloth and retro-reflector trim. High-visibility materials serve to increase the ability of motorists or equipment operators to see workers in normal light contrasted against the work background. The usual recommendation is fluorescent colors. Retro-reflective materials are used in reduced light such as night, dawn, dusk, and inclement weather, and are designed to reflect light back toward a light source, such as vehicle headlights. Retro-reflective materials have the ability to reflect light back to the same point as the source of the light without significant scattering, regardless of its angle of incidence. The material is typically composed of thousands of micro-reflectors (glass beads or micro-prisms) in each square inch of material. Retro-reflective characteristics are measured by a coefficient of retro-reflection ($cd/lx/m^2$); the higher the number, the brighter the reflective properties of the material. The coefficient of retro-reflection (R_A) is dependent on the angle of entrance and observation. Minimum coefficient values have been developed for various combinations of these two angles to define Photometric Performance Levels 1 and 2 (lower and higher coefficient values respectively) (ANSI/ISEA 2004, 9).

These protective garments generally contain both highly visible and retro-reflective materials, since construction and similar work involves a high degree of environmental and illumination variability. Workers in highly visible, fluorescent-colored garments in the day can nearly be invisible at night if not marked with reflective material. It is important that the reflective material provide all-around (360-degree) visibility. The use of retro-reflective trim on garment sleeves (as opposed to only the torso) significantly improves personnel conspicuity at night (Sayer and Mefford 2004). Users should inspect high-visibility clothing regularly to ensure that these beneficial properties have not been lost over time due to use and washing (Pratt, Fosbroke, and Marsh 2001, 12).

The authoritative guide on the design of high-visibility and reflective apparel is ANSI/ISEA 107-2004 (ANSI/ISEA 2004). This standard defines three classes of conspicuity that specify coloration, the total area of the background material, and the photometric performance level of reflective materials.

TABLE 5

Conspicuity Classes

Class No.	Risk Criteria and Examples
1	Workers' full attention is to approaching traffic and vehicle speeds are less than 25 mph; includes parking lot attendees or warehouse workers in proximity to vehicle traffic
2	Workers' attention is to perform tasks that divert their attention from approaching traffic or higher vehicle speed, closer proximity to traffic; includes school crossing guards, roadway construction workers, accident site investigators
3	Workers who have high task loads that require attention away from traffic, significant vehicle speed; includes roadway construction personnel, flaggers, survey crews, accident site investigators, utility workers, emergency response personnel

(*Source:* ANSI/ISEA 2004, A-3)

Both the performance class and the photometric performance level are marked on each piece of apparel meeting ANSI/ISEA 107-2004. A work-site or work-task hazard analysis determines the necessary conspicuity class. Factors include the task of the employee, the proximity to traffic, vehicle speed, and the visual complexity of the background (see Table 5).

The *Manual of Uniform Traffic Control Devices* (MUTCD) maintains requirements for high-visibility clothing for workers on highway construction projects (FHWA 2003, 6E-1). OSHA has adopted, by reference, recommendations from several iterations of the MUTDC that specifically apply to roadway personnel (often called "flaggers") (OSHA 2002a). Flaggers should wear a Class 2 vest, shirt, or jacket in orange, yellow, or yellow-green (or fluorescent versions of these colors). Nighttime work requires that the outer garment be retro-reflective orange, yellow, white, silver, or strong yellow-green (or fluorescent versions of these colors).

Abrasion, Cutting, and Puncture (Mechanical Risks)

Personal protection equipment that falls into this category includes gloves and aprons for protection for the hands, arms, and torso. Many types of material provide protection against cuts, abrasions, and puncture. For example, leather gloves provide abrasion and puncture resistance. Specially designed leather gloves, worn over electrically-insulating rubber gloves, protect the insulation integrity of the inner glove. Some heavy-coated fabrics, thick rubber, or plastic materials are abrasion-resistant. Synthetic materials made from aramid (aromatic polyamides), thermoplastic fibers (extended chain polyethylene), or composites of thermoplastics and carbon fibers also are used with increasing frequency, since they can combine cut-resistant properties with other useful characteristics such as flexibility and heat resistance. Synthetic materials used in gloves for hypodermic needle protection, chaps for snakebite risk or chain saw protection, and ballistic protective vests help provide puncture and penetration resistance. Gloves using metal mesh protect against accidental cuts during the use of hand knives by food workers.

The *American National Standard for Hand Protection Selection Criteria* recommends three performance tests for mechanical protection for gloves (ISEA 2000, 1–23). To determine the ability of a material to protect against cut resistance, the amount of weight needed to cut through the material with a blade is used (ASTM 2005c, 1–10). Testing abrasion resistance is according to ASTM D3389, using an abrasion wheel that counts the number of cycles before abrading a hole through the film or coating (ASTM 1999a, 1–4). Puncture resistance is measured by the amount of force required to pierce the sample with a standard-sized point. All three of these tests result in a level of performance of zero to five or six, with the higher number being the more protective.

Other performance criteria include those by the National Institute of Justice (NIJ), which has developed criteria for cut, puncture, and pathogenic resistance for gloves used by law enforcement personnel (NIJ 1999, 1–9). Gloves that meet certain testing criteria

TABLE 6

Standards Used for Gloves for Law Enforcement

Standard Number	Title
NFPA 1999	Standard on Protective Clothing for Emergency Medical Operations
ASTM D 5151	Standard Test Method for Detection of Holes in Medical Gloves
ASTM F 1671	Standard Test Method for Resistance of Materials Used in Protective Clothing to Penetration by Blood-Borne Pathogens Using Phi-X174 Bacteriophage Penetration as a Test System
ASTM F 1342	Standard Test Method for Protective Clothing Material Resistance to Puncture
ASTM F 1790	Standard Test Method for Measuring Cut Resistance of Materials Used in Protective Clothing
ASTM D 5712	Standard Test Method for Analysis of Protein in Natural Rubber and its Products
ASTM D 2582	Standard Test Method for Puncture - Propagation Tear Resistance of Plastic Film and Thin Sheeting
BS EN 420:1994	General Requirements for Gloves
BS EN 388:1994	Protective Gloves Against Mechanical Risks

(*Source:* NIJ 1999, 1–9)

TABLE 7

Additional ASTM and ISO Standards on PPE and Mechanical Risks

Standard Number	Title
ASTM F1414-04	Standard Test Method for Measurement of Cut Resistance to Chain Saw in Lower Body (Legs) Protective Clothing
ASTM F1897-04	Standard Specification for Leg Protection for Chain Saw Users
ASTM F1818-04	Standard Specification for Foot Protection for Chain Saw Users
ISO 13998:2003	Protective clothing–Aprons, trousers and vests protecting against cuts and stabs by hand knives
ISO 13999-1:1999	Protective clothing–Gloves and arm guards protecting against cuts and stabs by hand knives–Part 1: Chain-mail gloves and arm guards
ISO 13999-2:2003	Protective clothing–Gloves and arm guards protecting against cuts and stabs by hand knives–Part 2: Gloves and arm guards made of material other than chain mail
ISO 13999-3:2002	Protective clothing–Gloves and arm guards protecting against cuts and stabs by hand knives–Part 3: Impact cut test for fabric, leather and other materials
ISO 14877:2002	Protective clothing for abrasive blasting operations using granular abrasives

(Adapted from ASTM and ISO)

for protection against biohazards, blades, and needles have the highest ratings (see Table 6).

There are additional standards that cover other specific occupational applications of PPE and the hazards of cuts, punctures, and abrasion (see Table 7).

Thermal Stress and PPE

Contact Thermal Stress

One of the most common uses of personal protective equipment is to protect human tissue against contact with hot and cold surfaces. Over 17,000 serious occupational heat burns or scalds occurred in the United States in 2006 (BLS 2007a); see Table 8.

There are several performance tests for protective clothing subjected to heat (firefighter gear is not addressed here). The first test determines whether the material will ignite when exposed to a flame. ASTM F1358 prescribes a two-point test where the material is subjected to a flame for 3 seconds and 12 seconds, and classified from 0 to 4 based on ignition of the material and how long it burned (level 4 is no ignition) (ASTM 2005a).

TABLE 8

Effect of Elevated Temperature and Duration on Skin

Skin Temperature	Time Duration	Effect on Skin
110°F	6.0 hrs	Cell breakdown begins
158°F	1.0 sec	Complete cell destruction
176°F	0.1 sec	Curable burn
205°F	0.1 sec	Third degree burn

(*Source:* Cooper Bussmann, Inc. 2004, 18. Reprinted by permission.)

The second test determines whether the material will degrade when heated. The heat resistance test is ISO/DIS 17493, which assesses material degradation (damage and shrinkage) from 176 to 608°F (ISO 2000).

The heat test ASTM F1060 determines the ability of the material to protect against a second-degree burn from contact with a hot surface. This assessment measures the highest contact temperature that provides at least 15 seconds to a second-degree burn and at least 4 seconds between onset of pain and the second-degree burn. This test measures rate of heat transfer with a calorimeter (ASTM 2001b, 1–6).

ASTM F1939 considers properties of clothing materials exposed to radiant heat. Materials exposed to 0.5 or 2.0 calories per square centimeter heat source are used to calculate a radiant protective performance value (ASTM 1999d, 1–6, 1–9; ISEA 2000, 1–23).

Heat transfer through a given material ($m^2°C/W$) determines its contact resistance to cold (that is, conductive thermal resistance). This is the same procedure as the European standard, EN 511, for gloves (European Committee for Standardization 2006). As with most heat indexes, the results are divided into 4 to 5 levels, with the higher number indicating better performance in that characteristic (ISO 1989, 1–7; ISEA 2000, 1–23).

Personal Protective Clothing/Equipment—Heat Stress

Hot environments can lead to several heat-related disorders. Heat stroke is the most serious of the disorders. The body's failure to regulate its own core temperature (loss of sweating) can lead to mental confusion, loss of consciousness, and death. Heat exhaustion presents itself as extreme fatigue, nausea, and/or headache. Insufficient fluid or salt intake (or both) can lead to heat exhaustion. Failure to replace lost salt causes heat cramps, which can occur during or after work. Heat rash is a condition of the skin when sweat does not easily evaporate from the body, generally found in humid environments. The presence of an infection can further complicate heat rash.

There are a number of heat stress indexes available to determine the degree of hazard, although no single one is universally used. For predicting weather-related heat stress, the National Weather Service uses a heat index that calculates an apparent temperature based on temperature and humidity. This widely used measure does not account for solar or other radiation loads, but can serve as a practical indicator in certain occupational settings (see Table 9).

A more comprehensive index, which factors in radiation, evaporative cooling, type of clothing, and

TABLE 9

Heat Index: Temperature versus Relative Humidity

	Relative Humidity					
°F	40%	50%	60%	70%	80%	90%
110°	135					
105°	121	133	148			
100°	109	118	129	142		
95°	98	106	113	122	133	
90°	90	94	99	105	113	121
85°	84	86	90	92	96	101
80°	79	80	81	82	84	85

Key:

Heat Index (HI)	Possible Heat Disorder
130° or higher	Heatstroke/sunstroke highly likely with continued exposure
105°–130°	Sunstroke, heat cramps, or heat exhaustion likely, and heat stroke possible
90°–105°	Sunstroke, heat cramps, and heat exhaustion
80°–90°	Fatigue possible with prolonged exposure and physical activity

(*Source:* NOAA 2005)

work demands on acclimatized or un-acclimatized workers, is based on a measurement of the wet-bulb globe temperature (ACGIH 2006, 182–199).

There are a number of heat stress controls, including climate control, use of work-rest regiments, providing plenty of drinking water, replacement of electrolytes, and acclimatization. Personal protective clothing/equipment also can play a role in reducing heat stress. Strategies to reduce thermal heat stress using PPE depend on the major factors in the heat balance equation. When air temperature is below skin temperature and there is no significant thermal radiation input, heat will be lost through both convection and evaporative cooling most efficiently through skin alone. If air temperature is higher than skin temperature, clothing will lessen the heating effects by convection. This advantage may be lost or reduced, however, if the material used prevents sweat evaporation. If thermal radiation is significant, it is appropriate to minimize exposed skin (clothing can absorb some radiant heat). Reflective clothing or equipment will further absorb and/or reflect infrared radiation. Aluminized reflective suits (see Figure 3) are used by workers exposed to high radiant sources such as molten metal, kilns, ovens, and high-temperature fire rescue (OSHA 1999a, 14–16; ACGIH 2000, 1–34).

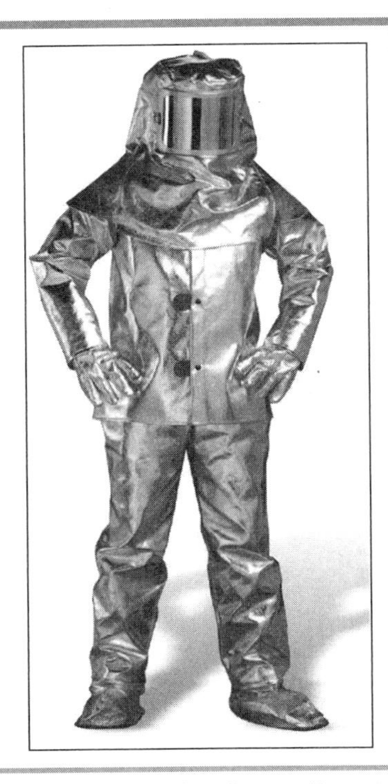

FIGURE 3. Aluminized reflective suit
(Reprinted by permission of Chicago Protective Apparel, Inc.)

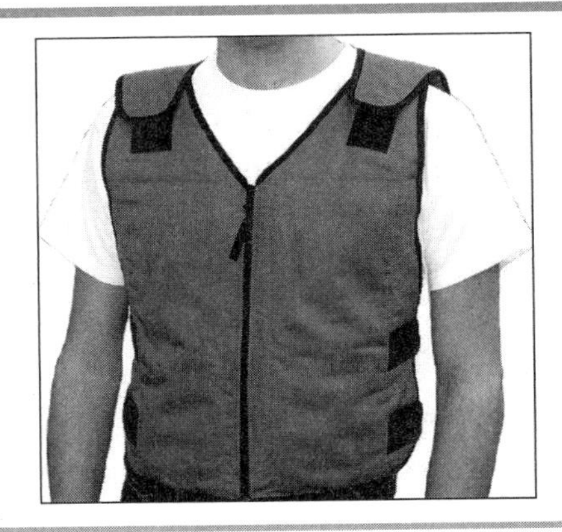

FIGURE 4. Phase-changing cooling vest
(Reprinted by permission of Polar Products, Inc.)

Cooling by conduction is possible using a vest that incorporates a phase-changing material in the lining that is pre-frozen and that provides a constant cool temperature next to the skin (see Figure 4). Other vest designs accept reusable cold gel or ice packs that fit into the panels or pockets. For special applications that do not require a high degree of mobility during cooling, suits are available that pump cold water via serpentine capillary tubes attached to a T-shirt. Increasing evaporative cooling is another approach, where vests lined with water-absorbing gel crystals evaporate water over the course of a shift. Soaking the vest in cool water recharges its cooling properties (OSHA 1999a, 14–16).

Still other cooling options include garments or vests that cool by convection (and evaporation) using compressed air blown through the garment. A vortex tube attached to an air outlet produces cool air due to expansion, allowing the worker some temperature control (OSHA 1999a, 14–16).

Personal Protective Clothing—Cold Stress

Personal protective clothing is an important defense in protecting workers in cold environments, such as those found in construction, agriculture, commercial fishing, and refrigerated storage areas. Health risks from cold environments include frostbite, hypothermia, and trench foot. Frostbite occurs when skin tissue freezes and ice crystals form between cells, leading to cellular dehydration. If exposure continues, frostbite may progress to deeper tissues, including muscles and tendons. Hypothermia begins to occur when the body temperature falls to approximately 95°F. Shivers, lethargy, and mild confusion accompany this stage of hypothermia. Further reduction of body temperature to 90°F affects cerebral and muscle function, with possible loss of consciousness and heart failure. A third condition, trench foot, occurs from exposure to a prolonged, cool (60°F or less) and wet environment and results in itching, pain, swelling, and tissue death (OSHA 1998a, 1–4).

For workers involved in outside work, wind chill values are a useful tool in determining the need for protective clothing by accounting for the combination of both wind and temperature in a single index. The National Weather Service (NWS) updated the Wind Chill Temperature Index in 2001, based on improved accuracy in determining frostbite threshold values (see Table 10).

Use the following equation to determine intermediate wind chill values:

$$\text{Wind Chill (°F)} = 35.74 + 0.6215T - 35.75(V^{0.16}) + 4275T\,(V^{0.16}) \quad (1)$$

where

T = Air Temperature (°F)

V = Wind Speed (mph)

Example 1: What is the wind chill if the temperature is 5°F and the wind speed is 5 mph?

Answer: Using Equation 1:

$$\text{Wind Chill} = 35.74 = (0.6215)(5) - 35.75(5^{0.16}) + 0.4275(5)(5^{0.16})$$

$$\text{Wind Chill} = 35.74 + 3.1075 - 46.24 + 2.76 = -4.6 \text{ or } -5°\text{F}$$

Thermal insulation resists heat conduction through clothing. While heat transfer though clothing involves conduction, convection and radiation, for simplicity, these terms combine to a single dry heat transfer term (K_{CL}). The heat exchanged is a function of the difference of skin temperature and outer environment and the thermal insulation of the clothing, as expressed in Equation 2 (Olesen 1982; Holmér et al. 1999).

$$K_{CL} = \frac{(t_s - t_{cl})}{0.155 I_{cl}} \quad (2)$$

where

K_{CL} = Heat exchanged though clothing (W/m²)

t_s = Mean skin temperature (°C), typically 33° or 34°C at 25°C ambient temperature

t_{cl} = Clothing surface temperature (°C)

I_{cl} = Thermal insulation of the clothing (clo)

This insulation calculation does not consider the effect of air movement (such as wind) and the move-

TABLE 10

National Weather Service Wind Chill Chart

Wind (mph)	Temperature (°F)								
0	40	30	20	10	0	–10	–20	–30	–40
10	34	21	9	–4	–16	–28	–41	–53	–66
20	30	17	4	–9	–22	–35	–48	–61	–74
30	28	15	1	–12	–26	–39	–53	–67	–80
40	27	13	–1	–15	–29	–43	–57	–71	–84
50	26	12	–3	–17	–31	–45	–60	–74	–88
60	25	10	–4	–19	–33	–48	–62	–76	–91

Frostbite times: 30 min | 10 min | 5 min

(*Source:* NOAA 2001, 3)

ment of the wearer, both of which tend to reduce insulation values (Holmér et al. 1999). The standard unit for measuring the thermal insulation of clothing (I_{cl}) is the clo. The clo is the insulation needed for comfort of an average person in a room for 8 hours with no activity at 20°C (1 clo = 0.155 m^{2}°C/W). Each piece of clothing has a clo value that, when added together, creates an ensemble value (Takahashi-Nishimura, Tanabe, and Hasebe 1997; Olesen 1982) (see Table 11 and Figure 6).

Example 2: Sedentary workers are engaged in light work (assume air velocity is negligible). The clothing surface temperature is 27°C and the clothing insulation is 0.5 clo. If the temperature of the skin is maintained at 33°C, how much should the insulation of the clothing be increased to maintain the same heat loss if the clothing surface temperature is dropped to 22°C?

Answer: Using Equation 2, the heat loss through the clothing is:

$$K_{CL} = \frac{(33^\circ\ C - 27^\circ\ C)}{0.155 m^{2\circ}\ C/W(0.5)} = 77.4\, W/m^2$$

Dropping the temperature to 22°C and solving for I_{cl}:

$$77.4\, W/m^2 = \frac{(33^\circ\ C - 22^\circ\ C)}{0.155 m^{2\circ}\ C/W(I_{cl})}$$

$I_{cl} = 0.9$ (a difference of $0.9 - 0.5 = 0.4$ clo)

Cold weather clothing is tested and marked with clo ratings that can help match the anticipated cold environment with the appropriate insulated clothing.

TABLE 11

Thermal Insulation of Clothing

Clothing Examples	clo
Shorts	0.1
Light summer clothing (ensemble)	0.5
Typical indoor winter clothing (ensemble)	1.0
Insulated coveralls	1.1
Business suit, shirt, socks, shoes, underwear (ensemble)	1.5
Long underwear, trousers, shirt, fleece jacket, heavy insulated parka, insulated over trousers, socks, boots, gloves, and hat (ensemble)	3.0

(*Source:* Olesen 1982; Holmér and Rintamäki 2004)

For example, Figure 5 gives maximum recommended exposure times for medium cold strain at different insulation values during light work and wind at 0.2 m/s.

Strategies for protecting workers in cold environments involve reducing heat loss to maintain body heat balance. The insulating properties of protective clothing depend on the ability of material to provide still air between the external environment and the

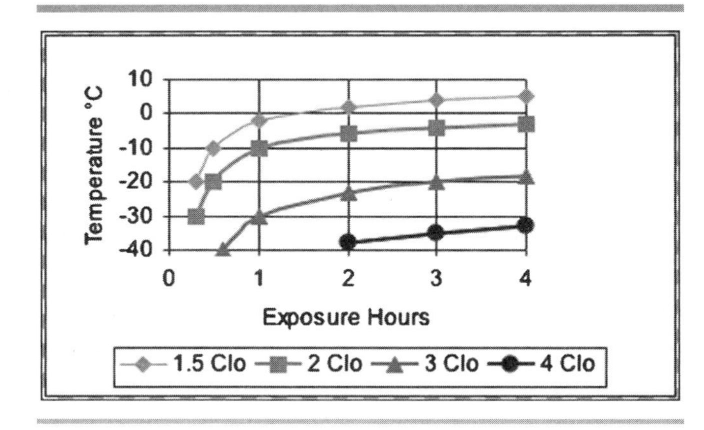

FIGURE 5. Maximal exposure times in a 0.2 m/s wind during light work for four different clothing insulation values (*Source:* Holmér and Rintamäki 2004; ISO 2006. Reproduced by permission from Ingvar Holmér and the Finnish Institute of Occupational Health.)

FIGURE 6. Polyester padded work vest with an indicated insulation value of 1.65 clo. (Reproduced by permission from Helly Hansen ASA.)

body. Both thick clothing and multiple thin layers provide insulation. The quiescent air barrier that attaches to the outside of the material is an important insulator. Thus, as the number of clothing layers increase, the number of air barriers also increase. This has the effect of providing greater insulation than the same thickness of a single layer of the same material, as indicated in Figure 7 (Havenith 1999).

Multiple, loose-fitting layers are generally preferred over bulky single layers, as this approach provides flexibility in adding or subtracting layers to match insulation with changing work demands and wind chill conditions. It is of similar benefit to be able to ventilate insulated clothing to release excess body heat and water vapor as needed. An outer layer should reduce the wind effect (intrusion of cold air) and provide protection from water, sleet, and snow. The middle layer should trap warm air and retain the insulation effects using wool, synthetic fibers, or down (provided it remains dry). An inner layer of synthetic weave (for example, polyester or polypropylene) should allow ventilation (AHRE 2004, 1–15).

If exposed, the human head can be a major source of heat loss (up to 40 percent). A hat or insulated form of head protection is vital in cold environments. Other extremities (feet and hands) are also vulnerable from exposure to cold air and direct contact with cold surfaces (AHRE 2004, 1–15). The presence of a steel toecap does not influence the thermal response of insulated footwear (Kuklane 2004). As with outer garments, the layered approach works well for gloves, using removable glove liners or windproof mitts. A synthetic sock liner under a synthetic blend or wool sock tends to wick moisture away from the foot. It is important to keep the clothing and footwear as dry as possible (including dampness from sweat), since moisture increases conduction of heat away from the body, and wet clothing loses insulating properties, leading to excessive cooling (OSHA 1998a; AHRE 2004, 1–15).

Hand, finger, and body warmers are also available as supplemental sources of heat. These devices are air-activated and heat through the catalytic oxidation of iron powder. These devices insert into gloves, pockets, between layers of clothing, or in specially-made vest panels. If not used properly, they can cause thermal or chemical burns.

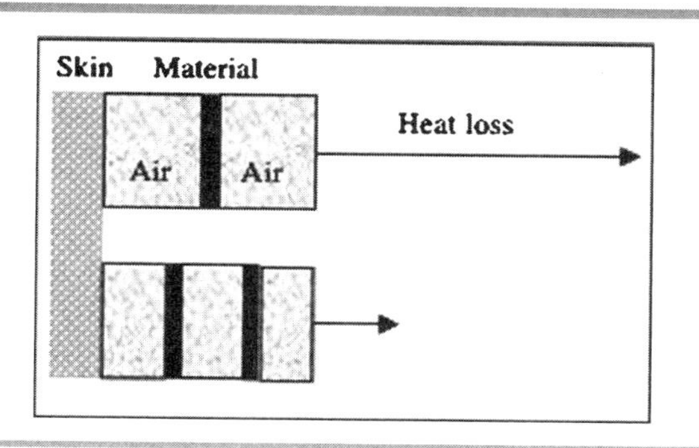

FIGURE 7. **Effect of multiple fabric and air layers on heat resistance** (*Source:* Adapted from Havenith 1999)

Nonionizing Radiation PPE (Welding, Lasers, and Solar Radiation)

Welding

Gas and arc welding/cutting can generate a multitude of hazards to the face and eyes. In addition to flying sparks and molten metal, welding/cutting produces an intense bright light comprised of visible, ultraviolet and infrared radiation. "Welder's flash" (keratoconjunctivitis) is a common affliction of the cornea caused by ultraviolet radiation. Infrared radiation produces heat that also may affect the eye.

The shade number of the filter lens in the welding helmet determines the amount of attenuation of the optical radiation. There are several criteria that have been provided to determine the appropriate shade number (ISO 1979; OSHA 1990; ANSI 2003, 1-67). They include the source of the radiation (for example, the type of welding) and, where applicable, the size of the electrodes, the amount of current in amps, and flow rates of acetylene or oxygen. Table 12 provides the typical ranges for the operation indicated. Note: always select the darkest shade that the work allows.

In addition to fixed-shade number welding helmets, there are auto-darkening helmets that increase their protective shade as soon as an arc forms. These are popular because the worker does not need to lift and lower the helmet to stop and start the weld. An

TABLE 12

Shade Number for Optical Radiation

Operation	Shade No. (Minimum)	Shade No. (Suggested)
Shielded metal arc welding (SMAW)	7–11	10–14
Gas metal arc welding (GMAW or MIG)	7–10	11–14
Flux covered arc welding (FCAW)	7–10	11–14
Gas tungsten arc welding (GTAW or TIG)	8–10	10–14
Air carbon arc cutting	10–11	12–14
Plasma arc welding (PAW)	6–11	6–14
Plasma arc cutting (PAC)	8–10	9–14
Carbon arc welding (CAW)	14	14
Torch brazing	3–4	3–4
Torch soldering	2	2–3
Oxy-fuel gas welding and cutting	4–6	3–8

(*Source:* OSHA 1994a; ANSI 2003, 1–67; AWS 2005, 7)

electronic device senses the presence of an arc and then darkens to a selected shade number in less than a millisecond (see Figure 8). When the arc is no longer present, it immediately lightens. ANSI Z87.1 specifies the criteria for switching times and other auto-darkening performance in sensing lenses (ANSI 2003, 49-58).

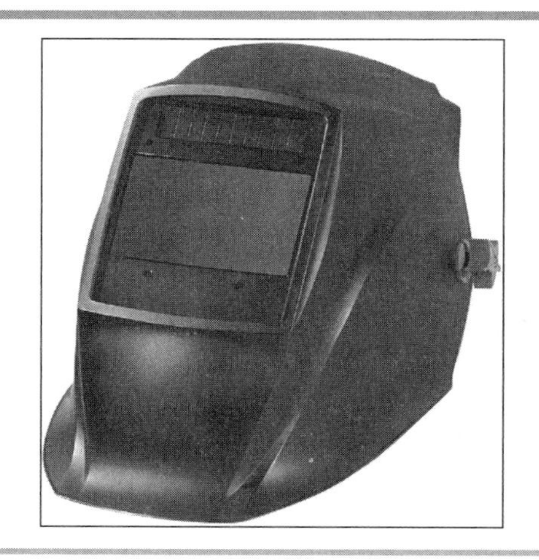

FIGURE 8. Auto-darkening welding helmet (Copyright 2006, by Miller Electric Mfg Co. Reproduced by permission from Miller Electric Mfg Co.)

Laser PPE

Lasers are classified based on their ability to cause biological damage, either through direct exposure to the beam or through specular reflection. In general, Class 1, 2, or 3A lasers pose little or no risk to personnel, while class 3B can pose ocular and skin risk. Class 4 lasers represent the highest risk for both skin and eyes (OSHA 1993a).

There are two approaches to laser eyewear protection: selective spectral absorption by colored glass or plastic, or selective reflection from the dielectric coating on the lens. Protective eyewear intended for use with laser beams must bear a label that includes the laser wavelength they protect against, the optical density for those wavelengths, and the percent visible light transmission. The optical density (OD) is a measure specifying the attenuation of the transmitting medium at a particular wavelength. It is in logarithmic units; thus, an OD of six attenuates a beam by 1,000,000. Eye protection needs to reduce the laser beam to a safe level (the maximum permissible exposure) that is dependent on the wavelength and intensity of the irradiance or radiant exposure. For example, a worst-case hazard evaluation for the necessary optical density of eye protection—assuming a wavelength of 0.4 to 1.4 micrometers (intra-beam viewing and a pupil size of 0.7 centimeters)—can be determined using the formula (OSHA 1993a, 15–17):

$$OD = \log_{10}\left[\frac{\dfrac{4\Phi}{\pi(d)^2}}{MPE}\right] \quad (3)$$

where

OD = Optical density

Φ = Laser radiant power output in watts for continuous wave lasers

D = Aperture in cm (in the case of visible light or near-infrared wave length, use the pupil radius of 0.35 cm)

π = 3.141

MPE = Maximum permissible exposure limit from ANSI Z-136.1 in W/cm^2.

The MPE varies by both wavelength and exposure time. For a pulsed laser, replace Φ with the radiant energy output in joules and use the MPE in J/cm^2. See ANSI Z-136.1 for additional methodology for choosing eye protection (ANSI 2000, 184).

Example 3: Determine the optical density for eye protection for a 0.514 μm wavelength, 1 Watt, argon laser (continuous wave). Assume an accidental exposure to a visible beam may be possible, resulting in an aversion response (0.25 sec).

Answer: Using Equation 3: Φ is 1 W; d^2 is 0.7^2 = 0.49 cm^2; and MPE from ANSI Z-136.1 for 0.25 sec is 2.5×10^{-3} W/cm^2

$$OD = \log_{10}\left[\frac{\frac{4(1)}{3.141(0.49)}}{2.5 \times 10^{-3}}\right]$$

$$OD = \log_{10}(1039.5) = 3.0$$

Solar Ultraviolet Radiation

In addition to the well-known risk of sunburn and photoaging, concerns about solar ultraviolet radiation (UVR) have increased as the incidence of malignant melanoma (a form of skin cancer) has doubled in the 20 years prior to the 1990s. Malignant melanoma is responsible for approximately 80 percent of skin cancer deaths (NRBP 2002, 5–9). Today, there is a greater awareness of the damaging effects of the sun and the wider use of protective creams and gels. The World Health Organization (WHO) developed a useful measure of potential sun exposure called the Global UV Index. The National Weather Service (NWS) and the Environmental Protection Agency (EPA) adopted this index in 2004 to forecast expected exposure to solar UV. A projected daily value is available for each zip code in the United States, using the UV intensity at the solar noon hour, based on season, altitude, and weather (see Table 13).

Use of tightly-woven clothing and a wide-brim hat may be the most effective approach to preventing skin damage. Clothing performance labeling standards are available that provide an ultraviolet protection factor (UPF) (ASTM 2000, 1–5). UPF values range from 15 to 50+ and correspond respectively to 93 to over 98 percent blocked ultraviolet radiation. The amount of ultraviolet radiation transmitted through clothing is a function of the type of material and treatment, percent moisture, tightness of the weave, and fabric color (Stone 2002, 1–2). A typical, light-colored T-shirt has a UPF of only 5 to 7, while one that is specially treated may be as high as 50 UPF.

Protective creams and gels provide sun protection, although some caution is necessary in interpreting the protection derived from its use. The index used to rate sunscreen is the sun protection factor (SPF), which ranges from 2 to 35+ SPF. Protection increases with increasing SPF value. An SPF value of 15 or greater is typical for occupational settings (OSHA 2003a, 1–2). Ironically, the use of sunscreen has been associated with an increased risk of malignant melanoma. Several theories may explain this association. For example, the early use of sunscreen with low SPF values permitted longer exposure to the sun, and thus a higher risk (Westerdahl et al. 2000). Secondly, the ability of a product to prevent sunburn defines the SPF value. Sunburn results

TABLE 13

Global Ultraviolet Index

Ultraviolet Index	Exposure Level
1–2	Low
3–5	Moderate
6–7	High
8–10	Very high
11+	Extreme

(*Source:* EPA 2004)

from exposure to ultraviolet radiation in the 280–320 nanometer (nm) (UVB) wavelength. Originally, many thought that UVB was responsible for melanoma, basal, and squamous cell carcinoma. The thinking was that preventing sunburn would prevent these skin cancers. In fact, most (95 percent) of solar ultraviolet radiation at ground level has a longer wavelength (UVA, 320–400 nm). Although not a major cause of sunburn, UVA is linked to photoaging and skin cancer (NRBP 2002, 5–9). Thus, SPF values can be misleading in protecting against skin cancer, since they are calculated from only a UVB endpoint—namely a sunburn (Bernerd et al. 2003). To protect against multiple biological endpoints (skin cancer, photoaging, and sunburn) sunscreen needs to reduce broad-spectrum UVA and UVB radiation.

Another concern about sunscreen includes inadequate application by users, resulting in one-half to one-quarter of the laboratory SPF value (Wolf et al. 2003; Poon and Barnetson 2002). This suggests using an SPF value of 30 or greater to achieve an effective protection of 15. Finally, there is a suggestion that sunscreen inhibits formation of nitric oxide in the skin, resulting in the reduction of cell-mediated tumor immune response which, in turn, may promote melanoma (Chiang et al. 2005).

Excessive sun exposure is associated with an increase in the incidence of cataracts and corneal sunburn, so the use of sunglasses is appropriate for outdoor workers. ANSI Z80.3–2001 outlines the performance criteria for sunglasses (ANSI 2001). Sunglasses that allow less than 1 percent transmittance of both UVA and UVB protect against ultraviolet radiation (Sheedy and Edlich 2004). Today, many safety glasses provide effective sun protection with impact protection and style in mind.

Noise and Hearing Protection Devices (HPDs)

Noise-induced hearing loss is an insidious occupational disease. Although it is entirely preventable, it remains one of the most common occupational diseases. An estimated 30 million workers were exposed to hazardous noise in 2001 (NIOSH 2001, 1–2). OSHA has estimated the number of recordable cases of occupational noise-induced hearing loss in the United States at 145,000 (OSHA 2002b). An effective hearing conservation program is essential in any workplace with hazardous noise levels. Personal hearing protection is required whenever noise reduction is not possible through some other more reliable means, such as engineering controls.

Evaluating Noise Hazards for PPE

A sound level meter or a noise dosimeter that integrates the noise over time measures potentially hazardous noise levels. Occupational noise is measured in decibels (dB) using a reference sound pressure of 20 micronewtons per square meter ($\mu N/m^2$)(20 micropascals [μPa]) that approximates the threshold of human hearing at 1000 hertz (Hz). Sound pressure levels are then measured using the equation:

$$L_p = 20 \log_{10} \frac{p}{p_{ref}} \tag{4}$$

where

L_P = Sound pressure level

p = Root-mean-square (rms) sound pressure

p_{ref} = Reference rms sound pressure (Bell 1982)

Decibels are logarithmic and added according to Equation 5:

$$L_{p\ total} = 10Log\left(10^{\frac{L_{p1}}{10}} + 10^{\frac{L_{p2}}{10}} + 10^{\frac{L_{pn}}{10}}\right) \tag{5}$$

where

L_{total} = Total sound level in decibels

L_{p1}, L_{p2} and L_{pn} = Noise sources or octave bands (Bell 1982)

(The center frequency names the octave band. The lower frequency limit is half the upper limit, for example, the 1000 Hz band has lower and upper limits of 707 and 1,414 Hz respectively.)

Example 4: What is the sound pressure level of 1000 Hz and 2000 Hz octave bands, each at 99 dB?

Answer: Using Equation 5:

$$L_{p\ total} = 10Log\left(10^{\frac{99}{10}} + 10^{\frac{99}{10}}\right)$$

$$L_{p\ total} = 10Log(7{,}943{,}282{,}347 + 7{,}943{,}282{,}437)$$
$$= 102\ dB$$

Sound-weighting networks, designed to model the ear's response, are the basis for noise measurements. The A and C scales are most common. The notation dBA refers to decibels using the A-weighting network on a sound level meter. The A-weighted scale is most sensitive between 2 to 4 kilohertz (kHz) and less sensitive at the low and high frequencies. The C-weighted network is essentially flat in response from 30 to 8000 Hz (Bell 1982, 37–63).

The structure of the ear consists of three main sections. The outer ear is comprised of the pinna and the auditory canal. Because of its shape and size, the pinna and the auditory canal both collect and amplify the sound. The eardrum (tympanic membrane) is located at the end of the canal and serves to transmit vibrations to the bones of the middle ear. The middle ear consists of three ossicular bones that transmit and amplify vibrations on the oval window which, in turn, transmits vibrations to the inner ear. Hair cells attached to nerve cells in the inner ear are sensitive to displacement caused by vibration that transmits electrical impulses to the brain via the auditory nerve (Clemis 1975). Noise-induced hearing loss results in permanent pathological changes in the inner ear. While there are other bone and tissue conduction paths whereby noise (albeit at less intensity) reaches the inner ear, they are unaffected by hearing protective devices. The role of the hearing protector is to limit the sound intensity reaching the eardrum (and thus the inner ear). The most common method is by attenuating the noise at one of several locations, such as outside the pinna, at the entrance of ear canal, or within the ear canal (Berger 1986).

Noise Exposure Criteria

Hearing protection must be able to reduce noise to acceptable levels. The generally accepted performance target for protecting workers with HPDs is 75 to 82 dBA as a time-weighted average over an 8-hour work shift. This includes measurement error and a safety factor below regulatory and recommended occupational exposure limits (OELs). See Table 14 for a summary of time-weighted averages (TWA), occupational noise exposure limits, and action limits from several organizations. The exchange rate refers to the number of decibels that are added to the allowable limit if the exposure time is cut in half. The threshold is the noise level at which the noise dose begins to accumulate.

Noise Reduction Ratings (NRR) for HPDs

There have been several approaches developed over the last three decades toward the accurate prediction of noise attenuation under real-world conditions. Unfortunately, regulations covering the testing and use of HPDs in the United States have not kept up with evolving studies, leaving the situation in need of Noise Reduction Rating derating schemes.

Since 1979, the EPA has required manufacturers of HPDs to label the attenuation of their products using a single-number Noise Reduction Rating (NRR) (EPA 1979). The NRR requirement was promulgated by EPA's Office of Noise Abatement and Control and remains in effect today. Manufacturers of HPDs are required to calculate NRRs, following ANSI Standard S3.19-1974 (ANSI 1974). This test protocol uses a group of subjects and determines hearing thresholds with and without the hearing protector of interest. The protocol adopted by the EPA allows the experimenter to fit the hearing protector on each

TABLE 14

Occupational Exposure Limits (OEL) for Noise

Organization	8 Hour Limits (OEL/Action Limit)	Exchange Rate	Threshold
OSHA (OSHA 1981, 4078–4179)	90/85 dBA	5 dB	90/80 dBA
MSHA (MSHA 1999, 49548–49637)	90/85 dBA	5 dB	90/80 dBA
NIOSH (NIOSH 1988, 1–122)	85 dBA	3 dB	80 dBA
ACGIH (ACGIH 2006, 113–117)	85 dBA	3 dB	80 dBA

of the test subjects and check the best fit, using a preliminary test noise. The mean attenuation at octave bands from 125 Hz to 8 kHz compares a subject's binaural unprotected hearing threshold with the occluded threshold (using the HPD). The procedure computes the NRR, assuming noise with equal energy across all six octaves of 100 dB for each octave band, for a combined level of 107.9 dBC (remember the decibel scale is logarithmic, not linear, so adding two equivalent dBs results in an increase of 3 dB, see Equation 5). Each octave is corrected to the dBA scale. The method then computes two standard deviations from the test data, which is subtracted from the attenuation for each octave band. This attempts to assure that it protects a large percent of the exposed population, at least to the extent of the published value (98 percent of the tested/exposed population) to account for individual variability. The combined attenuated octave bands result in a protected dBA value (Berger 1986).

The Noise Reduction Rating value includes a 3-dB safety factor to account for differences in assumed versus actual frequency spectrum. It is calculated as shown in Equation 6:

$$\text{NRR} = 107.9\ \text{dBC} - \text{protected dBA} - 3\ \text{dB} \qquad (6)$$

Calculating the worker's potential sound level exposure resulting from use of the hearing protector depends on the type of instrument used and whether or not one uses the dBA or dBC scale. Once a worker's 8-hour time-weighted average is determined (using dosimetry, octave band analysis, or representative sound level readings), the anticipated protected sound level can be calculated from the following equation:

$$PL_A = L_C - \text{NRR} \qquad (7)$$

or

$$PL_A = L_A - (\text{NRR} - 7\ \text{dB}) \qquad (8)$$

where

L_C or L_A = Noise exposure level, using the C or A network in dBC or dBA

PL_A = Estimated protected exposure in dBA

OSHA adopted the use of NRR values and Equations 7 and 8 for determining the suitability of HPDs in its Hearing Conservation Amendment (OSHA 1981). Following the inclusion of this methodology into regulatory standards, the testing method has met with substantial criticism. A number of studies have shown that NRR values derived by this method are often less than those obtained under field conditions (Royster 1980; Royster, L,. Royster, J., and Cecich 1984; Edwards et al. 1978, 1983; Paddilla 1976). The causes include inter-laboratory variability, testing sample size, use of expert versus novice HPD users, poorly-written instructions, and the use of HPDs that did not fit properly (Berger et al. 1982, 1998; ; Murphy et al. 2004; Royster et al. 1996). Task-related or environmental constraints can also affect the difference between laboratory and field values (see Figure 9).

The International Standard ISO 4869 (promulgated in the early 1990s and used in Europe) reflects certain changes in the HPD testing protocol (ISO 1990, 1994). That method calculates a single number rating (SNR) for a given percentage of the population. For example, an SNR_{84}, using one standard deviation, would protect 84 percent of the population, since we are concerned with only the lower end of the NRR distribution. In this test protocol, test subjects fit HPDs under the supervision of the experimenter.

It is also possible to use frequency information of the noise reduction capability of HPDs, using the multiple high, medium, and low (HML) ratings. This method uses the exposure measurement difference in the C and A networks to determine if the noise is primarily in the upper, mid, or lower frequencies.

If the measured difference is *less than or equal to* 2 dB, then the predicted noise reduction level is (NIOSH 2000):

$$PNRx = Mx - \left(\frac{Hx - Mx}{4} * (L_C - L_A - 2dB)\right) \qquad (9)$$

If the difference is greater than or equal to 2 dB, then the predicted noise reduction level is:

$$PNRx = Mx - \left(\frac{Mx - Lx}{8} * (L_C - L_A - 2dB)\right) \qquad (10)$$

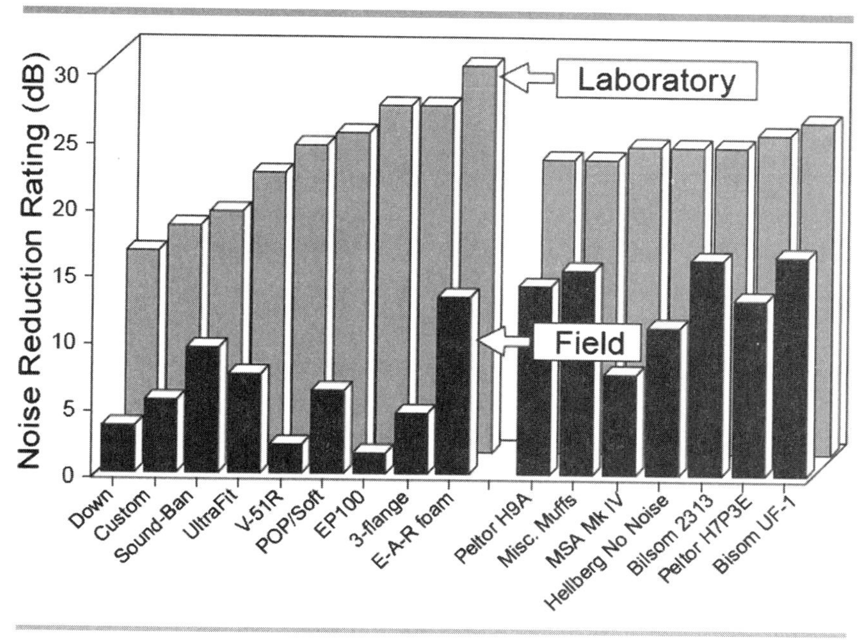

FIGURE 9 **Comparison of NRRs published in North America**
(*Source:* Berger 2000, 1. Copyright 2000 by Aearo Company. Reproduced by permission from Aearo Company.)

where

PNR = Predicted Noise Reduction level

x = Percent population protected (for example, 84 percent for one standard deviation)

H, M, L = HPD manufacturer's ratings (high, medium, low)

L_C, L_A = Noise exposure level on the C or A network in dBC or dBA

The noise under the hearing protector is:

$$PL_A = L_A - PNRx \tag{11}$$

Example 5: What would the noise under the hearing protector be if noise was measured as 101 dBA and 103.5 dBC (one standard deviation)? The manufacturer rated the hearing protector as SNR 27 dB, H = 31 dB, M = 24 dB, L = 16 dB.

Answer: Since the difference between the A scale and C scale is 2.5 dB, use Equation 10:

$$PNRx = Mx - \left(\frac{Mx - Lx}{8} * (L_C - L_A - 2dB)\right)$$

$$= 24 - \left(\frac{24 - 16}{8} * (103.5 - 101 - 2)\right)$$

$$= 23.5 \text{ dB}$$

Then, using Equation 11:

$$PL_A = L_A - \text{PNR} = 101 - 23.5 = 77.5 \text{ dBA}$$

ANSI S12.6-1997 (ANSI 1997) is also a more recent replacement for the NRR procedure. This method uses either subject-fit (NRR_{SF}) protocol, where experimenters provide subjects the manufacturer's instructions without intervention of the experimenter, or experimenter-assisted fit, where the experimenter is allowed to coach the subject (Murphy et al. 2004). As would be expected, the subject-fit protocol has the lowest mean attenuation, highest subject-to-subject variation, highest trial-to-trial variation, and the best agreement with real-world data (Murphy, Franks, and Krieg 2002; Royster et al. 1996).

Equation 12 demonstrates the use of the NRR_{SF} data:

$$PL_A = L_A - NRR_{SF} \tag{12}$$

Or, if using dBC data:

$$PL_A = (L_C - 5 \text{ dB}) - NRR_{SF} \tag{13}$$

where NRR_{SF} = Subject-fit noise reduction ratio.

TABLE 15

Difference in Methods for Calculating Single Number HPD Ratings

Method	Year	Index	Frequencies (Hz)	HPD Fitting Conducted By	Performance Level (Population protected)
ANSI-S3.19	1974	NRR	125 – 8K	Experimenter (EPA method)	2 SD (98%)
ISO-4869	1992	SNR HML	63 – 8K	Test subject	Variable (84% typical)
ANSI-S12.6	1997	NRR	63 – 8K	"A" Informed user "B" Test subject	1 SD (84%)

Table 15 compares the three test protocols.

Since manufacturers of HPDs continue to label their products with NRR developed by ANSI S3.19-1974 (ANSI 1974), various organizations have recommended that the NRR value be derated. Although not a regulatory requirement, OSHA recommends users derate the NRR attenuation by 50 percent as indicated in Equation 14 (OSHA 2005):

$$PL_A = L_A - 0.5(\text{NRR} - 7) \quad (14)$$

Likewise, NIOSH recommends that HPDs be derated to the percentages indicated in Table 16. However, NRR ratings calculated using subject-fit data are not derated (NIOSH 2003).

The resulting calculation would then be:

$$PL_A = L_A - (\text{NRR} \bullet d_r - 7 \text{ dB}) \quad (15)$$

Where d_r is the derated percent (or fraction) from Table 16 (NIOSH 2000).

NIOSH has collected NRR data and published a compendium. The compendium contains data on nearly 300 HPDs from over 20 manufacturers nationwide (NIOSH 1995a, 2003). More recently, NIOSH developed an accessible and searchable database for users at their website. In addition to the NRR attenuation data, the search mechanism provides the ability to enter worker exposure data in dBA or by octave-band levels to obtain a list of protectors that should result in noise levels between 75 and 85 dBA. In that calculation, NIOSH derates the NRR to account for real-world performance of the HPDs, as indicated in Table 16. When comparing HPD NRR values, a difference of less than 3 dB is inconsequential due to large variation seen in their use. Other factors, such as comfort and environmental suitability, are more important when the difference in NRR values is slight (Berger 1986, 332; Suter 2003, 24).

TABLE 16

NIOSH Recommendation for Derated Nonsubject-Fit NRR Values

HPD Type	Percentage of NRR (dr)
Earmuff	75%
Slow-recovery formable earplugs or custom molded plugs	50%
All other earplugs	30%

(*Source:* NIOSH 2000, 2003)

Example 6:

(1) What hearing protector $\text{NRR}_{(SF)}$ value is needed to achieve an 85 dBA protected noise level for slow recovery foam earplugs when used in a noise environment of 93 dBA, using the $\text{NRR}_{(SF)}$ rating (method B)?

(2) What would be the equivalent NRR value to have equal protection, using NIOSH derating if the NRR value was based on non-subject-fit data?

Answer:

(1) From *Equation 12*:

$$PL_A = L_A - \text{NRR}_{(SF)}$$

so

$$\text{NRR}_{(SF)} = L_A - PL_A = 93 - 85 = 8 \text{ dB}$$

(2) From *Equation 15*:

$$PL_A = L_A - (\text{NRR} \bullet d_r - 7 \text{ dB})$$

Solving for NRR:

$$NRR = \frac{(L_A + 7 - PL_A)}{d_r} = \frac{93 + 7 - 85}{0.5} = 30 \text{ dB}$$

In situations involving excessively-high noise (defined as greater than 100 dBA according to NIOSH), workers should wear both an earmuff and an earplug (NIOSH 1988, 1–61). To evaluate the use of double HPDs, determine the higher of the two field-adjusted NRR values and add 5 dB (OSHA 2005). Simply adding the individual protection factors together is not sufficient.

The NRR method is not recommended for impulse noise. According to a National Hearing Conservation Association Task Force, however, the attenuation of impulse noise is as least as great as indicated by the NRR value (Suter 2003, 25).

Types of HPDs

There are many different types of HPDs, each with its advantages and disadvantages (see Table 17). In an effective hearing conservation program, at least two different HPD types should be available for workers to ensure comfort and appropriate application in given environmental conditions.

Earmuffs consist of ear cups that seal around the ear, using a cushion that is usually foam or fluid-filled. A headband holds the ear cups in place, using tension over or behind the head or under the chin. Different wearing positions often have different NRR values. For special applications, such as construction work, hard hats can be adapted to include hearing protection. Noise-reducing acoustic material usually fills the ear cups. While earmuffs have a long service life, periodically check muffs for deterioration, such as hardening, puncturing or splitting of the cushion. Never modify the earmuff. A distorted headband results in reduced band force around the head and will inevitably reduce the effectiveness of this HPD.

Advantages of the earmuffs are numerous. They are easily visible, have a long service life, and are easy to fit and maintain. Additionally, they may provide some thermal comfort in cold environments. The disadvantages may include discomfort when worn for long periods or in hot environments, reduced effectiveness when glasses or hair interferes with the cushion seal, limited compatibility with certain PPE such as respirators, hard hats, or welding helmets, and use where headspace may be restricted.

Semiaural devices (canal caps) are similar to earmuffs because they use a headband to hold them in place. Unlike earmuffs, they are lighter and provide protection by using a plug or flexible tip that caps the ear canal. They are more compatible for use with glasses and hard hats than earmuffs, but they tend to be less comfortable than earplugs. Some semiaural devices are easy to dislodge and are best suited for intermittent use. Like the earmuffs, semiaural devices require periodic inspection for tightness of the headband and condition of the tips (see Figure 10).

Earplugs fit directly into the ear canal and come in a wide variety of styles and types. Premolded earplugs rely on a uniform shape but come in many different sizes to fit the ear canal. Custom-molded

TABLE 17

Comparison of HPD Types

HPD	Fit	Life	Use	PPE	Other Issues
Earmuff	U	Y	S	X	Hot environments, monitoring ease
Semi-aural devices	U	Y	I	X	
Formable earplugs	U	D	L		Fit training, dirty environments
Custom molded earplugs	P	Y	L		Initial expense
Premolded earplugs	M	M	L		Fit training
Active		Y			Initial expense

Key:

Fitting:	Fits most workers (U), requires multiple sizes (M), requires professional fitter (P)
Life of Device:	Expect (with care) up to several years (Y), months (M), days (D)
Use:	Typical use of the device during the day; short (S), intermittent (I), up to all day (L)
PPE:	May pose compatibility issues with other PPE (X)

(*Source:* Adapted from Michael and Byrne 2000)

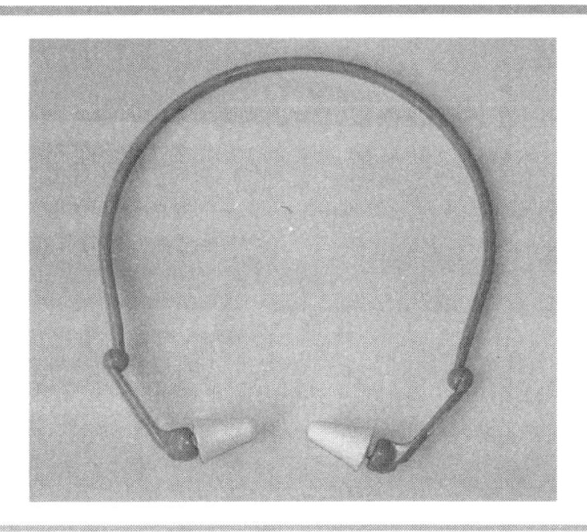

FIGURE 10. Example of a semiaural hearing protector

earplugs use an impression from each ear. Formable earplugs expand to fit inside the ear canal. Earplugs can come with or without an attached cord to hang around the neck when not in use. Earplug designs can be single use and disposable or multiple use and washable.

In general, earplugs have the distinct advantage of greater compatibility with other PPE and greater comfort for continuous use and in hot environments. The higher variability in fitting and fit effectiveness is a potential disadvantage. Proper insertion of earplugs into the ear canal is critical to achieve appropriate noise attenuation. Differences of 6 to 10 dB in attenuation between untrained and trained workers in fitting procedures has been documented (Toivonen et al. 2002). Since most earplugs require some manual dexterity to insert into the ear canal, not all workers can use them. In addition, the manual handling precludes the use of some of these HPDs in dirty environments (particularly those without insertion stems). In these situations, frequent removal and insertion can lead to the introduction of dirt and debris into the ear canal.

Formable earplugs expand to the shape of the ear canal using slow-recovery foam, spun fiberglass (Swedish wool) and silicon putty. These HPDs are more comfortable but require good insertion technique to achieve appropriate noise attenuation. Although simple cotton plugs have been used in the past, cotton alone has very little noise attenuation and should not be used today. Slow-recovery foam earplugs are inserted about halfway in the ear canal after being tightly compressed into a small cylinder. Noise attenuation depends on depth into the ear canal and snugness of fit (see Figure 11).

Custom-molded earplugs are custom made for each worker, using silicone putty or vinyl. These earplugs usually have a small handle to assist with insertion. These devices have proven to be no more effective than other well-fit earplugs (Berger 1986, 348).

Premolded earplugs are made of vinyl, silicones, and other elastomers, and often have one to five flanges that fit into the ear canal. They generally come in different sizes that determine both the comfort and the attenuation for the worker, and are reusable.

Active hearing protectors attempt to allow low levels of sound through the HPD (for example, where critical communications or signal detection are needed) while blocking high noise. These devices use circuitry built into the HPD (both earplugs and earmuffs). An *active level dependent* HPD design allows the transmission of harmless noise. When the device detects a loud impulse or harmful level of sound (greater than 85dB) the circuit shuts the processor down so that the HPD realizes its full protection. *Active noise canceling* HPDs reduce noise by inverting the wave phase. This device is usually effective for low frequencies (Berger 1986, 355). A third approach selectively attenuates noise at certain frequencies. These HPDs can improve hearing

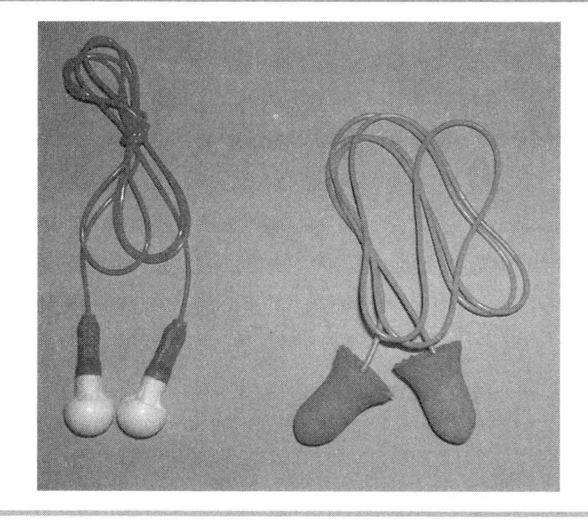

FIGURE 11. Example of preshaped and formable earplugs with strap

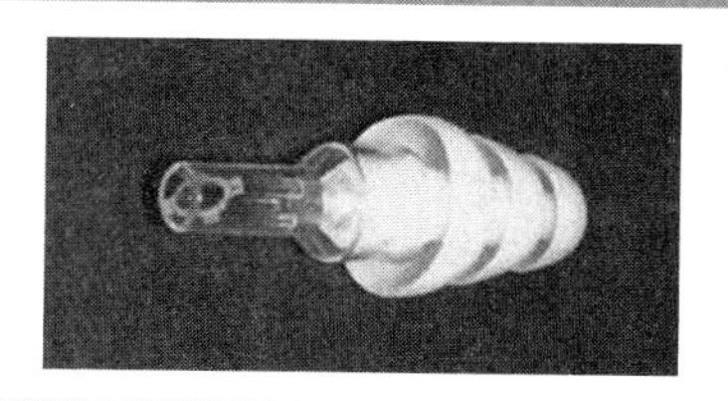

FIGURE 12. Example of a premolded flat attenuation earplug

acuity in noisy environments if properly matched to the application (Ong, Choo, and Low 2004).

Flat attenuation hearing protectors correct the nonuniform frequency response effect of most hearing protectors (most of which perform best above 3 kHz) Reducing noise equally across all frequencies results in a more natural perception (albeit at the expense of a higher NRR value). These devices allow high-fidelity sound to reach the ear and sometimes are called musician's HPD (see Figure 12).

HPD Selection

There are a number of factors on which to base HPD selection. The NRR comparison is just one factor in the decision-making process. Of equal importance is the wearer's comfort, since wearing hearing protectors throughout the noise exposure is critical. In one study, HPDs with lower NRRs, but perceived by workers as being more comfortable, tended to be worn more, and thus provided a higher degree of protection than HPDs with higher NRR but with lower comfort index (Arezes and Miguel 2002). If a worker does not wear the PPE for just 30 minutes out of an 8-hour noise exposure (assuming a constant noise level), the loss of protection would be equivalent to a drop in 7 dB from an HPD with a NRR value of 25 dB (Michael and Byrne 2000). Table 18 summarizes selection criteria.

Vibration PPE

Hand-arm vibration syndrome (HAVS) relates to vascular, neurological, and musculoskeletal disease and Raynaud's phenomenon (vibration-induced white finger), and occurs in occupations with significant exposure to hand-arm vibration. Vibration originates at the point of operation of a hand tool (blade, bit, or tooth) and transmits through the tool itself directly to the hand and wrist of the user. The Bureau of Labor Statistics (BLS) estimates that over 1 million workers in the United States are exposed to hand-arm vibration (Dong et al. 2006, 1). Occupational limits for hand-arm (segmental) vibration are available from the American Conference of Governmental Industrial Hygienists (ACGIH). The daily duration exposure limit is frequency-weighted (peaking at approximately 5 to 16 Hz) and measured as acceleration in the dominant vibration direction in meters per second squared (m/s^2) (ACGIH 2006, 128–131).

In addition to anti-vibration tools, anti-vibration gloves can help reduce vibration coupling of hands to tools. Constructed with some combination of impact-resistant materials, padding, or an air bladder, these gloves help protect where the hand makes contact with the tool (see Figure 13).

Anti-vibration gloves are generally more effective at high vibration frequencies (greater than 150 Hz). A potential limitation of anti-vibration gloves is the inevitable trade-off between vibration protection and greater grip force needed to hold a tool. Performance standards for anti-vibration gloves have been approved by ISO and adopted by ANSI (ANSI 2002). Gloves meeting this standard (and so labeled) must have a weighted transmissibility value (TR_M) less than 1 at 16–400 Hz (that is, no amplification at midrange frequencies) and a TR_H less than 0.6 at 100–1600 Hz (high frequencies). The transmissibility

TABLE 18

HPD Selection Criteria

	Hearing Protector Selection Criteria
1	Noise reduction based on wearer's daily equivalent noise exposure
2	Comfort (including climate aspects)
3	Variations in noise level (high variability could suggest multiple HPDs)
4	User preference
5	Communication needs
6	Hearing ability
7	Compatibility with other safety equipment
8	Physical limitations of the worker
9	Other working conditions (including need for highly visible HPDs)
10	Replacement and care issues

(*Source:* Suter 2003, 22)

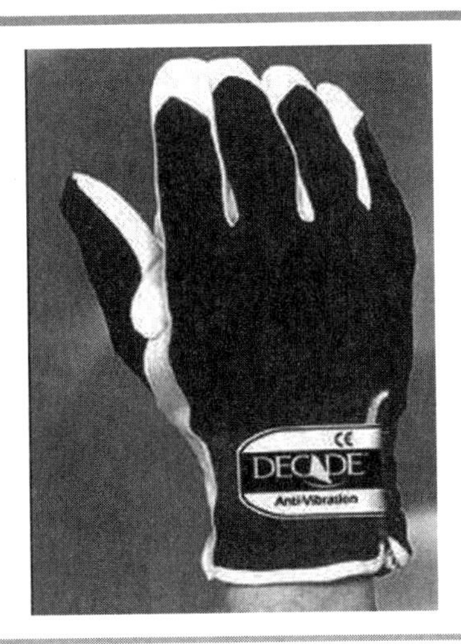

FIGURE 13. Anti-vibration glove
(Copyright 2005, by Chase Ergonomics, Inc. Reproduced with permission from Chase Ergonomics, Inc.)

value is the ratio of acceleration at the surface of the hand and at a reference point. The lower the transmissibility value, the greater the effectiveness in reducing vibration. It is important to note that measurement of vibration occurs in the palm of the hand and does not include the fingers.

However, some caution is required in choosing a glove with the intent of reducing vibration. ANSI warns that the transmissibility value derived from application of their standard is not yet sufficient to assess the health risk due to vibration (ANSI 2002). Unfortunately, many gloves offer reduction at only higher frequencies, and many tools have a frequency spectrum centered at much lower values (Griffin 1997). Gloves that keep hands warm and dry, regardless of their vibration reduction capacity, should serve to limit some synergistic effects of vibration and cold. Intervention studies showing a reduction in vibration or HAVS have generally used multiple controls, making it difficult to assess the impact of gloves alone (Jetzer, Hayden, and Reynolds 2003; Clarke et al. 1986).

Electrical PPE

In 2006, there were nearly 250 fatal occupational electrocutions in the United States, accounting for 12 percent and 5 percent of all fatal injuries in construction and manufacturing respectively (BLS 2007b). The appropriate use of electrical PPE mitigates two primary types of hazards: shock and arc flash. When working on energized circuits or conductors, workers must wear PPE that is appropriate for both the voltage involved and the energy from a potential arc flash. To the greatest extent possible, electrical systems must be de-energized first. There are only two circumstances where OSHA regulations allow work on energized equipment: (1) when continuity of service is required, and (2) when de-energizing equipment would create additional hazards.

Electrical Shock

Electrical shock occurs from touching or getting too close to current-carrying exposed live parts. It does not take a large value of current flowing through the human body to cause serious harm. A flow of current greater than 30 milliamps (mA) will potentially cause respiratory paralysis.

Helmets (hard hats) with insulating properties are necessary where there is potential contact exposure to electrical conductors with the head. Performance standards for helmets are available in ANSI Z89.1 (see the earlier discussion on impact hazards for more information on protective helmets). Helmets approved for Class G and Class E applications must provide insulation against 2200 and 20,000 volts (phase to ground) respectively. Class E helmets must withstand 30,000 volts to determine burn-through resistance (see Table 19). ANSI Z89.1 cautions users

TABLE 19

Classification of Protective Helmets

Classification	Criteria for Helmets
Type I	Intended to reduce force of impact only to the top of the head
Type II	Intended to reduce the force of impact from top or sides
Class G	Intended to reduce danger from contact with low voltage conductors
Class E	Intended to reduce the danger of contact with higher voltage conductors
Class C	Not intended to provide protection against contact with electrical hazards

(*Source:* ISEA 2003b, 2–3)

that, while their procedure tests helmets at these high voltages, they are not intended to protect workers at such extremes.

Rubber insulating gloves must provide a high dielectric strength in addition to strength, flexibility, and comfort (some gloves are used with liners to provide comfort, warmth, and/or to absorb sweat). Rubber insulating sleeves extend protection from the glove cuff to the shoulder. Gloves and sleeves need to meet specifications of ASTM D120 and D1051 respectively (OSHA 1994c; ASTM 2002b, 1–9; ASTM 2002c, 1–9). See Table 20 and Figure 14. In addition, it is necessary for workers to wear leather gloves over the rubber gloves to protect the insulation against damage (cut and puncture).

Electrical PPE should be inspected before each use to ensure its protective integrity. Gloves must be air-tested for leaks at least daily and electrically at least every six months. Workers must examine sleeves daily by rolling them up and looking for breaks and cracks, inside and out. Sleeves must be tested electrically every 12 months. See ASTM F1236 and ASTM 496 for additional visual inspection criteria and in-service care for electrical protection rubber products (ASTM 2001a, 1–11; ASTM, 2002a, 1–8).

Electrically-protective *footwear* bears the "EH" label (non-conductive), according to dielectric tests properties to protect against accidental contact of 600 volts (ANSI 1999). EH footwear has electrically-rated shock-resistant heels and soles. As with other PPE, the electrical protection properties will deteriorate with excessive wear. See Table 4 for additional approval criteria.

TABLE 20

Rubber Insulating Gloves and Sleeves[1]; Maximum Use, Proof Testing and Retesting by Class and Type

Class	AC (rms) Max. Use (volts)	AC (rms) Proof Test (volts)	DC Proof Test (volts)
00	500	2,500	10,000
0	1,000	5,000	20,000
1	7,500	10,000	40,000
2	17,500	20,000	50,000
3	26,500	30,000	60,000
4	36,000	40,000	70,000
Type			
I	Not ozone resistant		
II	Ozone resistant		

[1]Sleeves have classes 0–4 at the voltages indicated.

(*Source:* ASTM 2002b; ASTM 2002c)

FIGURE 14 Glove testing machine (Reproduced with permission from the VON Corporation)

Electric Arc Flash

Electrical arc flash is a serious hazard, resulting in an intense pressure wave, radiant energy, sound, molten metal, and shrapnel. Although of brief duration, the heat is sufficient to cause tissue burns even without direct contact and to ignite typical work clothing.

An *arc flash* occurs when current flows between separated energized conductors. Magnetic forces from the fault current cause the conductors to separate, resulting in an arc flash. Temperature as high as 35,000°F vaporize the conductors and causes copper vapor to expand to 67,000 times the volume of solid copper. The resulting ionized conducting vapors allow the fault current to flow until the current-limiting device opens. This event is life-threatening, particularly if the blast is directional out of an enclosure towards nearby workers. Simulated tests have measured blast pressures greater than 2000 pounds per squart foot (lbs/ft^2)(Crnko and Dyrnes 2000).

The guiding concept in hazard control of arc flash is the use of risk boundaries. PPE must protect any part of the worker's body that is within the arc flash zone. The intensity of the arc flash potential determines the appropriate level of PPE needed. The potential incident energy of the arc flash can be calculated or obtained from a table of common work tasks found in NFPA 70E (NFPA 2004). For any specific task, there are protocols for calculating an arc flash potential (IEEE 2002).

The potential energy (or hazard/risk category) corresponds to the need for voltage-rated gloves, flame resistant (FR) clothing, eye and face, head, and hearing protection. Textile materials are rated on protection by the amount of incident energy on material that is:

- Sufficient to cause a second-degree burn through the material (the arc thermal performance value or ATPV), or
- The point above which causes the material to break open (the E_{BT} value) and the ATPV can not be obtained.

The calorie per square centimeter (cal/cm^2) of a potential arc flash determines the rating needed for suitable clothing. In addition to the ATPV, some clothing manufacturers will also provide the percentage of heat reduction by the fabric from an arc flash exposure (heat attenuation factor or HAF). ASTM F1959 outlines the determination of arc flash ratings for heat flux rates of 2 to 600 cal/cm^2 (ASTM 2006a). Generally, burns are unexpected when the exposure is less than 1.2 cal/cm^2. NFPA does not address exposures above 40 cal/cm^2 because, regardless of the PPE protection against the heat, the pressure wave is likely to cause serious trauma (see Table 21) (NFPA 2004, 1–126; Hoagland, Shinn, and Reed 2004; Doughty et al. 2000, 2782–2789).

ASTM F1506 describes the requirements for fire-resistant clothing (ASTM 2002d). Outerwear, such as jackets and rainwear, must also be fire-resistant per ASTM F1891 (ASTM 2006b). It is also important to minimize the potential of burns from melting fibers by avoiding the use of synthetic materials (nylon or polyester) that melt below 600°F (Lovasic 2005, NFPA 2004).

In addition to clothing, protection from arc flash potential requires PPE that will protect the head, eyes and face, hearing, and feet. The selection of this equipment is also according to a hazard/risk category rating. Safety glasses are required for all hazard/risk categories of 1 or less. Risk Category 2 requires safety glasses or goggles combined with a full-face shield (minimum ASTM F2178 arc rating of 8) (ASTM 2006c). Category 3 requires the Category 2 ensemble plus a flash hood. Category 2 is also the threshold for hand protection(leather or FR gloves for arc flash and insulating rubber gloves worn underneath for shock protection), hearing protection, and heavy-duty, leather work shoes (NFPA 2004).

Personal Fall Protection

Falls are responsible for a large portion of this country's occupational mortality. More than 800 workers died from falls in 2006, which was approximately 14 percent of all occupational fatalities. Over one-half of all fall deaths are concentrated in the construction industry. The typical fall distance for a fatality is not great, with more than 90 percent of all them occurring from heights of 11 to 30 feet (Derr et al. 2001; Webster 2000; BLS 2007b).

TABLE 21

Protective Clothing and Arc Ratings

Hazard/Risk Category (HRC)	Typical Clothing Description (Number of clothing layers is given in parentheses)	Minimum Arc Rating of PPE, ATPV, or E^{BT} (cal/cm^2)
0	Non-melting, flammable materials (untreated cotton, wool, rayon, silk, or blends of these) with fabric weight of at least 4.5oz/yd^2 (1)	N/A
1	Fire-resistant shirt and pants or fire-resistant coverall (1)	4
2	Cotton underwear—conventional short sleeve shirt and brief/shorts, plus fire-resistant shirt and fire resistant pants (1 or 2)	8
3	Cotton underwear plus fire-resistant shirt and pants plus fire-resistant coverall or cotton underwear plus two fire-resistant coveralls (2 or 3)	25
4	Cotton underwear plus fire-resistant shirt and pants plus multi-layer flash suit (3 or more)	25

(*Source:* NFPA 2004)

Personal fall protection equipment plays an important role in reducing fall-related injuries and fatalities. A 25-year-old OSHA study that examined 99 fall-related construction fatalities suggested that nearly all of the fatalities were preventable (OSHA 1979). In the United States, there are two applicable consensus performance standards: ANSI A10.32 (ASSE 2004) for construction, and ANSI Z359 for non-construction applications (ASSE 2007a, 2007b).

There are three major components of personal fall protection equipment: the anchorage, the personal body harness or positioning belt, and the connection that links the body harness/belt to the anchorage (see Figure 15). Although discussed separately below, compatibility of the entire fall arrest system is critical, since components produced by different manufacturers may not be interchangeable. The identity of the manufacturer, part numbers, and maximum user weight (including tools), along with certain warning labels, are mandatory on fall protection equipment. Unless otherwise specified, fall protection components are not designed for more than 310 pounds, including the worker and tools. When the capacity exceeds 310 pounds, special customization of the fall arrest system is necessary (ASSE 2007a, 8).

There are three basic strategies for using personal fall protection. They depend on whether the worker is subject to a fall (*fall arrest*), is restrained by the device to prevent the worker from reaching the edge so no fall is possible (*fall prevention*), or is being positioned (*fall positioning*) to limit a fall to two feet or less. A positioning device is a body belt or body harness system rigged to support an employee on an elevated vertical surface, such as a wall, and work with both hands free (see Figure 15).

Anchorage

An *anchorage* is a secure point of attachment for lifelines, lanyards, or deceleration devices. The anchorage is often the most overlooked component of personal fall protection equipment. It must be able to take the impact force from a fall arrest without failing. The structure to which an anchorage is attached for a fall arrest system must be capable of supporting 5000 pounds per person attached, 3600 pounds if certified by a person or entity, or be designed and installed to maintain a safety factor of at least two. Unless designed for the purpose, guardrails do not have sufficient strength to be employed as anchorage points. Structures capable of a minimum load of 3000 pounds (or twice the anticipated impact force, whichever is greater) are necessary for connecting restraint and positioning systems. Since tying off a rope, lanyard, or lifeline around an I-beam can reduce its strength by 70 percent due to the cutting force of the beam edge, a webbing lanyard or wire-core lifeline or other attachment device should be used (ASSE 2007a, 10–59; OSHA 1994d).

A number of anchoring devices are unique to specific situations. Beam anchors, used in the steel erection industry, attach to the bottom flange of an overhead, horizontal I-beam or the top flange of a horizontal I-beam when at a worker's feet. Fixed

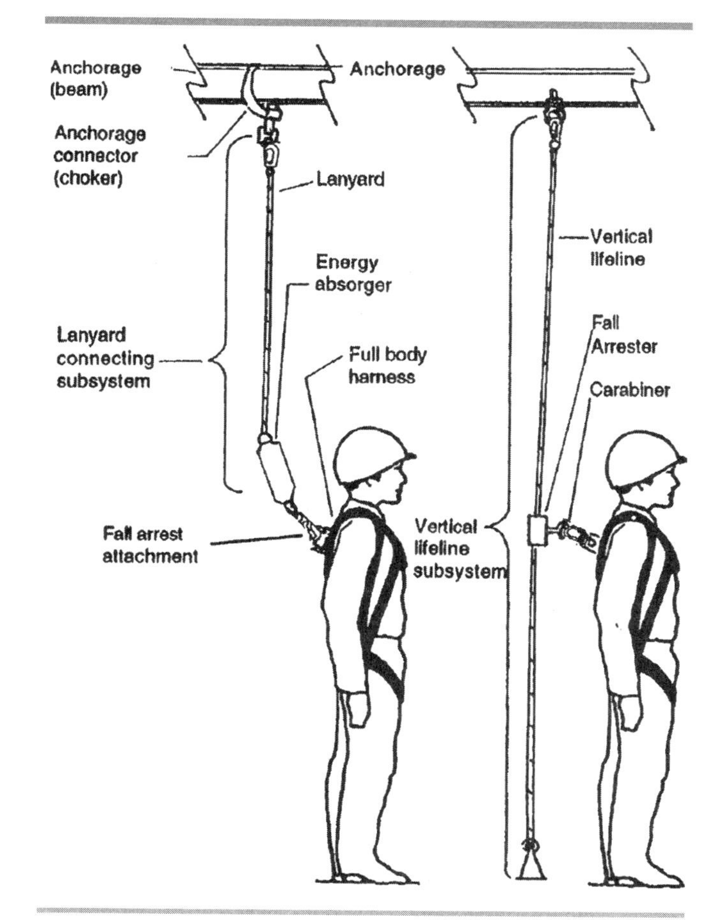

FIGURE 15. Examples of anchorage, body harness, and connection equipment
(*Source:* ANSI/ASSE Z359.1-2007, 83. Copyright 2007, ASSE, reproduced with permission from ASSE)

beam anchors attach to a horizontal beam at single locations, while sliding beam anchors are available to follow the worker across a beam during steel construction. Roofing anchors are available for metal and wood, flat and pitched surfaces (see Figure 16). Beam and trolley anchors on access platforms (for example, truck and railcar platforms) and mobile vacuum anchors are used in some applications where anchor penetration is not possible.

The device attaches by nailing it through the sheathing and into the rafters.

Harnesses and Belts

The body belt (safety belt) is a strap with means both for securing it about the waist and for attaching it to a lanyard, lifeline, or deceleration device. Since 1998, body belts can no longer be used as the fall arrest system in the construction industry in the United States, since there is the potential for serious impact injury from being suspended or from slipping out of the belt (OSHA 1994d). Although a harness is more protective, body belts still have use as positioning or restraining devices on construction sites (and they are still allowed as fall arrest devices in some general industry OSHA standards).

The body harness consists of straps that secure the worker and distribute the fall arrest forces more evenly over the thighs, pelvis, waist, chest, and shoulders. Despite the reduced likelihood of severe impact trauma by use of a harness rather than a belt, quick rescue of the worker is critical. Suspension in a harness for just a few minutes can allow

FIGURE 16. Example of a commercially available, reusable roofing anchor

blood to pool in the legs. The resulting reduction of blood flow to the heart and the brain (orthostatic intolerance) can lead to shock and death. Preplanning prompt rescue is critical in any fall protection program (Weems and Bishop 2003). Some harnesses allow the worker to relieve leg arterial and venous pressure by standing on a special attached strap or provide a sub-pelvic strap to provide extra support.

The human body sustains significant forces when dropped otherwise short distances during a fall event. Research indicates that the threshold for significant body injury occurs at approximately 2700 pounds when wearing a harness (ASSE 2007a, 9). ANSI and OSHA stipulate the maximum arresting impact forces on an individual wearing a body harness to 1800 pounds (ASSE 2007a, 9; ASSE 2004, 9; OSHA 1994d). Dropping a rigid, 220-pound test dummy, both head-first and feet-first twice the distance of the lanyard, is the standard test of a fall arrest system. For positioning or retrieval systems, the test fall distance is the length of the lanyard (ASSE 2004, 24–26).

Connection Equipment, Lifelines, and Lanyards

The connection equipment links the fall arrest system or positioning device systems to the anchorage. It is imperative that connective parts have sufficient strength to maintain integrity when subjected to impact forces caused by a fall. D-rings, O-rings, and carabiners are required to have a minimal tensile strength of 5000 pounds, and each device is proof-tested to 3600 pounds by the manufacturer (OSHA 1994d).

A lanyard is a flexible line of rope, wire rope, or strap that has a connector at each end for attaching the body belt or body harness to a deceleration device, lifeline, or anchorage. A tensile strength of 5000 pounds is required for all lanyards (3000 pounds for self-retracting lanyards) (ASSE 2007a, 15, 42).

Lifelines can serve as a means for connecting a personal fall arrest system to the anchorage in situations where the worker's task involves substantial movement. Lifelines consist of a flexible line made of synthetic materials or wire rope, and are hung in either a vertical or a horizontal position. There are performance tradeoffs among the choice of materials (maximum arrest impact forces on the body and anchorage points versus clearance distances below the worker). When used vertically, workers must have their own lifelines to prevent them from falling into one another. Although more complex to design, lifelines can also run horizontally. The degree of sag and tension in a horizontal lifeline influences the maximum impact force. A qualified person or engineer should carefully design horizontal lifelines with one or multiple worker attachments.

Deceleration (shock-absorbing) devices reduce impact forces on the body during a fall event. They can dissipate up to 65 to 80 percent of the impact force during a fall arrest and limit arresting impact forces to less than 650 pounds (Ellis 1993, 40–48). Such devices consist of a rip-stitch lanyard, tearing or deforming lanyard, a mechanism such as a rope grab, or an automatic self-retracting lifeline/lanyard. A rope grab is a deceleration device that travels on a lifeline and automatically, by friction, engages the lifeline and locks to arrest the fall of a worker. A self-retracting lifeline/lanyard contains a drum-wound line that slowly extracts or retracts under slight tension. After the onset of a fall, self-retracting lines automatically lock the drum and arrest the fall within a distance of no more than 4.5 feet (ASSE 2007a, 31).

Safe Use Considerations

Planning for the safe setup and use of personal fall protection is essential to ensuring the device will perform as expected. An effective fall protection program requires a qualified or competent person to oversee its administration. Safe use considerations include avoidance of swing falls, unintentional disengagement of connectors, improper side-loading of connectors and anchors, and planning of sufficient fall distance to allow the arrest device to function prior to hitting an obstruction.

When the anchorage is not directly above the point were a fall occurs, a *swing fall* is likely. A swing fall is dangerous since the worker may strike objects during the swing and lengthen the fall distance when a self-locking deceleration device is used. Elevating the anchorage point, moving it closer to a point overhead, or using a horizontal lifeline helps lessen this potential danger.

Another dangerous situation is roll out. *Roll out* occurs where a connector device unintentionally disengages from a snap hook or carabiner. The use of a non-locking or an incompatible component is the usual cause of this dangerous situation. OSHA and ANSI both require two separate forces to open the connector gate as part of a fall arrest or positioning system.

A similar condition (*burst out*) can occur when the impact load transfers to the gate or lock of a snap hook. Failure to use the anchor appropriately, such as allowing excessive side loading, has lead to anchor failure and fatal falls (OSHA 2004, 1–3).

There must be sufficient clearance from any obstruction below the worker to accommodate a potential fall. Some manufacturers sell indicators (like a plumb bob) that users can test if the distance between the anchorage and the lower level is sufficient. The following formula can be used to calculate the minimum clearance distance needed below the worker's feet (working elevation). The manufacture of the equipment should be able to provide most of the values.

$$C = L + (At - An) + A + D + E + H + Sf \tag{16}$$

where

C = Clearance required below worker's feet

L = Lanyard length

(At – An) = Elevation of the attachment point (At) on the worker (D-ring on the harness) minus the elevation of anchorage (An) This is a negative number when the anchorage is above the worker

A = Maximum activation distance needed (if any) to activate the fall arrest device, for example, the self-reacting lanyard free-fall distance or rope grab locking distance

D = Deceleration distance needed for the shock absorption device and the harness stretch effects. The deceleration distance is the measured distance between the location of a body harness attachment point at the moment of activation of the deceleration device at the onset of fall arrest forces and the location of that attachment point after the worker comes to a full stop. Generally, 3.5 feet or less is the design criteria for most deceleration devices.

E = Elongation of the lifeline due to the static and dynamic impact forces. This distance is a function of the type of lifeline material and is proportional to the length of the lifeline.

H = Harness effect (approximately 1 foot)

Sf = Safety factor. A distance of at least 1 foot is recommended (ISEA 2002). See the manufacturer's literature for specific recommendations.

Note: (At – An) = 0 when the lanyard is attached at the same elevation as the harness D-ring.

L + (At – An) = 0 for a self-retracting lanyard (that is, when there is no vertical slack in the lanyard)

L + (At – An) + A = Free fall distance, and should be less than or equal to 6 feet

Example 7: A harness will protect a worker with a deceleration (shock-absorbing) device and a 6-foot lanyard that is connected to a D-plate bolted to a structural I-beam above the worker. His safety harness D-ring is located 2 feet below the anchor. The manufacturer's literature states that the deceleration distance has a maximum distance of 3.5 feet and there is no elongation of the harness or lanyard. How much clearance below the working surface is necessary?

Answer: Using the Equation 16 above and assuming 1 foot for the harness effect, no activation distance, no elongation distance, and choosing a safety factor of 1 foot:

$$C = L + (At - An) + A + D + E + H + Sf$$
$$C = 6 + (-2) + 0 + 3.5 + 0 + 1 + 1 = 9.5 \text{ feet}$$

PPE for Chemical Hazards

Chemical Protective Clothing

Personal protective equipment plays a vital role in chemical safety. Chemically resistant PPE is the last line of defense against chemical exposure. Chemical protective clothing (CPC) includes protective suits, coveralls, hoods, gloves, sleeves, boots, and aprons. Unfortunately, no material protects against all chemicals and, for each risk assessment done, the analyst must examine the particular properties of the hazardous chemical against the potential protective material. In some cases, the lack of chemical-specific information for PPE makes informed choices difficult. For example, the NIOSH *Pocket Guide to Chemical Hazards* lists a skin protection requirement for approximately 450 organic compounds. Unfortunately, only 39 percent of the compounds have specific glove recommendations (Boeniger 2001; NIOSH 1997b).

Skin exposure can occur from multiple sources, such as contact with contaminated surfaces involving any single or combination of solid, semisolid (dust or paste), or liquid chemicals. Skin immersion (typically involving the hands) into a liquid is a common exposure scenario in the workplace. Exposures can also occur from aerosol deposition on the skin as a dust or liquid, or involve skin contact with a gas or vapor.

When controls fail to prevent direct contact with toxic chemicals, workers risk local effects on skin; skin absorption into the body across the tissue barrier, entry of chemicals through cuts, abrasions, or nonintact skin, or by ingestion via transfer from contaminated hands to food or tobacco products. Both local and systemic effects can lead to serious occupational disease. Direct local effects include dermatitis, irritation, corrosion, and skin cancer. Systemic effects result from the chemical entering the body and acting on a target organ or system, according to the toxicity of the substance. A single exposure to the skin of toxic substances, such as hydrofluoric acid, dimethylmercury, paraquat, and cholinesterase-inhibiting pesticides, have resulted in death (Nierenberg et al. 1998; Blodgett, Suruda, and Insley Church 2001; Wesseling et al. 1997; Fuortes, Ayebo, and Kross 1993; Takase et al. 2004). Repeated exposures to other chemicals can cause chronic effects and result in occupational disease.

Concern over potential transdermal exposure is common for many chemicals. Of the nearly 700 chemicals for which ACGIH has established occupational exposure limits, 27 percent have a skin notation. The skin notation applies to substances where the cutaneous route of entry can cause systemic effects

(ACGIH 2006, 3–78). Although the respiratory tract is often the primary route of entry for workplace chemicals, there are some cases where skin absorption is the greatest concern. Exposure to pesticides is a well-known example of how skin absorption is a primary route of entry into the body. The potential of dermal exposure is the basis of some occupational exposure levels. For example, OSHA concluded, in their rulemaking on methylenedianiline, that induction of liver disease in workers was because of dermal absorption and, therefore, the airborne limit includes an amount based on dermal deposition (OSHA 1992).

The amount of transdermal absorption depends on the properties of the chemical, the surface areas of the exposed skin, and the duration of the substance on the skin. Chemicals penetrate the stratum corneum (epidermis) and diffuse into the dermis. In the dermis they contact blood cells and enter into the circulatory system and/or move into the hypodermis, which consists of connective tissue or fat (adipose tissue). Hydrophobic properties and molecular weight are also important in potential exposure. Substances that are hydrophobic and have a low molecular weight generally have a higher dermal permeability than those that are hydrophilic and have a high molecular weight (Klaassen and Watkins 1999).

Exposure to chemical mixtures complicates the hazard assessment, particularly when the components involve widely-differing characteristics. Chemical mixtures can also behave in a synergistic manner, where the combination is more aggressive on the material (permeation or degradation) than either is alone. A *carrier effect* occurs when the substance having a greater permeation rate carries the slower substance along with it.

PPE materials with the broadest range of chemical protection may be the best initial choice to test for protective properties for the specific application. In cases where the material with the best chemical barrier property has poor puncture and tear qualities, additional measures (for example, double gloving) is necessary.

There are numerous materials used in the manufacture of chemical protective clothing, boots, and gloves. They include a variety of natural and synthetic elastomers that can be blended and/or layered as a laminate. Dipped or treated gloves use woven or nonwoven materials for shape and strength. Each material has characteristics that render it more or less suitable for a particular application. Three useful aspects to review in a risk assessment are the permeation, penetration, and degradation of the material when exposed to the chemicals in question.

Permeation

Permeation is the movement of a chemical through the barrier material at the molecular level. It involves a three-step process of absorption onto the material after initial contact, diffusion of the chemical through the molecules of the material, and desorption of the chemical at the far side of the material. Diffusion is the slowest step of the three and thus controls the rate of migration of the chemical. High permeation rates have been associated with substantial swelling of the material, up to 30 percent in some studies (Mikatawaga, Que Hee, and Ayer 1984; Purdham et al. 2001; NIOSH 1978, 14–15).

Migration of the chemical through the material results in eventual breakthrough to the opposite side of the material. The time it takes to break through is important, as this indicates the first point of failure of the PPE. *Breakthrough time* is the period from initial application of the chemical to the outside of material and its subsequent presence on the other side. A longer breakthrough time reflects a higher chemical permeation resistance. In addition to the specific material-chemical system, breakthrough time is a function of material thickness and analytic sensitivity of the detector and the chemical collection system on the far side of the material. Higher temperatures will also increase the permeation rates of low molecular solvents. For example, increasing the temperature from 73°F (as generally provided in manufacturer's data) to 95°F caused ethyl acetate breakthrough time to drop 40 percent in nitrile material (Evans, McAlinden, and Griffin 2001). As shown in Figure 17, after initial breakthrough, there is a point where some material-chemical systems (but not all) reach a steady permeation rate through the material, generally given in terms of micrograms per

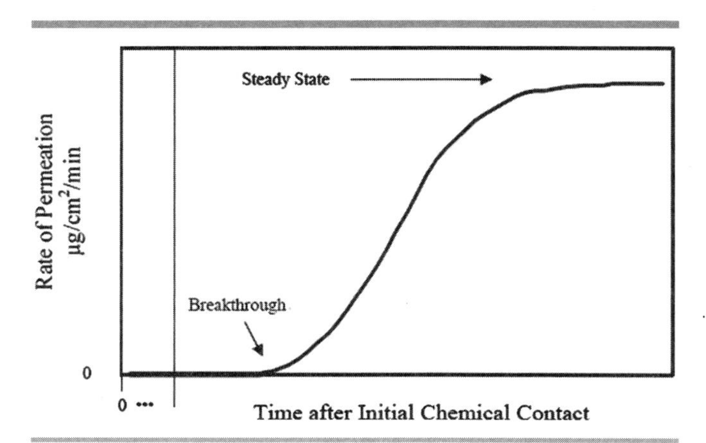

FIGURE 17 Permeation rate through CPC material (NIOSH 1978, 14–15)

square centimeter per minute ($\mu g/cm^2/min$). Knowledge of the breakthrough time and the rate of permeation of a particular material-chemical combination allows the assessment of the potential protection of the CPC material.

Changing to new PPE prior to the breakthrough point of the chemical (particularly for highly toxic substances) is the object of developing a service use time (for example, for gloves) where permeation occurs. Often, the worker cannot easily detect breakthrough of nonirritating or noncorrosive chemicals by other means, since perspiration will mask the sensation of the presence of the chemical. Whenever there is an obvious failure (irritation, burning, discoloration, or odor) or degradation of material, the worker must use another material.

There are specific permeation tests for CPC materials, often using a permeation test cell. ASTM F739 simulates CPC immersion for eight hours with a particular chemical. ASTM F739 tests chemicals that are normally a gas or a liquid that is volatile or soluble in water. A normalized permeation rate allows the comparison of different CPC materials. A permeation rate of 0.1 $\mu g/cm^2/min$ is the normalized breakthrough value. This allows comparison of materials independent of analytical detection limits. The method classifies the results into six levels according to breakthrough times, the longest (and safest) of which is 480 minutes (ASTM 1999b). An alternative method, ASTM F1383, uses intermittent contact with the liquid or vapor. The results from this method will usually result in longer breakthrough times and lower permeation rates (ASTM, 1999c, 1–9).

Since some chemical-material systems reach a steady-state permeation less than the normalized rate for reporting breakthrough, caution must be observed where there is potential for sensitization to small (but continuous) exposure of the offending chemical.

Although not covered by the above ASTM testing methods, solids can permeate polymeric materials in a manner similar to liquids (including breakthrough and a steady-state permeation rate). For example, a number of organic solids have measured breakthrough times with five different types of glove materials in as little as 2.4 to 9.5 minutes (Fricker and Hardy 1992).

Penetration

Penetration (particularly by liquids) results from structural problems in the protective barrier. These problems include the use of porous materials, small holes, punctures, tears, or other imperfections, or by the presence of faulty closures, zippers, or seams in the garment or device. Penetration concerns are particularly important where workers are subject to splash. ASTM D5151 covers the leak testing of gloves using water (ASTM 2006d).

Degradation

Degradation occurs when the chemical alters properties of the material that are important in providing protection. Specific chemicals may weaken the material structure, allowing mechanical punctures through the material more readily. For example, polyvinyl alcohol (PVA) polymer works well for certain organic solvents but degrades in the presence of water. Less noticeable, microscopic effects such as cavitations, convexities, (that is, loss of surface integrity), and cracking of material can also occur (Canning, McQuillan, and Jablonski 1998). Poor decontamination procedures can degrade protective materials (Gao, El-Ayouby, and Wassell 2005). ANSI 105 examines the degradation effect of chemicals on glove materials by measuring puncture resistance. In this procedure, the force required to puncture glove material is compared to the force required after the material has been soaked in a challenge chemical for one hour (ISEA 2000, 1–23).

Choosing and Using CPC Materials

As with many other forms of PPE, test conditions often differ from the work environment. Performance varies with differences in the product, material flexing, stretching, temperature variation, physical damage, age, and exposure to other substances (Perkins and Pool 1997; Mickelsen and Hall 1987). For example, an aggressive permeation testing protocol for 19 gloves that included elevated inside glove temperature (95°F) and prior stretching found breakthrough times as little as 33 percent of those obtained under standard testing conditions (Oppl 2002, 1–4).

Breakthrough times should be regarded only as a first step in choosing appropriate chemical-resistant PPE. Additional thought must be given to its actual use, user acceptance, reuse, and in-use testing (Boeniger and Klingner 2002).

Field studies of gloves under working conditions suggest there can be a significant exposure of contaminants inside protective gloves. Donning and doffing dirty gloves provides a common mode of contamination, since it is difficult to perform without touching the outside of the glove (Garrod, Phillips, and Pendleton 2001). Once the inside of a glove is contaminated, percutaneous absorption is an almost inevitable result. Moisture from perspiration inside the glove can hydrate the skin surface, causing a reduced effectiveness of the barrier properties of the stratum corneum for substances that are hydrophilic.

The potential consequences of reusing PPE is an essential ingredient in an overall risk assessment. Some chemicals will continue to migrate through the protective material after the initial exposure, resulting in significantly reduced breakthrough times, or in some cases, result in detectable contamination on the inside of the material at the beginning of the next day's shift (Zellers et al. 1992).

Several generic references list breakthrough times, permeation rates, and degradation data by chemical (Mansdorf 1998; Forsberg and Keith 1999; Forsberg and Mansdorf 2002). Glove manufacturers also provide data on the specific performance of the products they sell. Table 22 indicates criteria for selecting CPC materials.

Choosing CPC Ensembles

To protect emergency service personnel in hazardous material emergency response incidents, it may be necessary to enclose the wearer to prevent contact with highly toxic vapors or gases. Such vapor-protective ensembles use self-contained breathing apparatus

TABLE 22

A Protocol for Choosing CPC Material

Information Gathering and Testing	Consideration Factors
1. Chemical identification	Identity and concentration from MSDS, container, and/or product literature.
2. Chemical state, utilization parameters, and toxicity	Determine if gas, liquid, solid, or combination under use and temperature; likely route(s) of entry; toxicity; caustic properties; and possible breakdown products.
3. Chemical exposure	Will there be continuous, intermittent, or accidental contact?
4. CPC use/reuse	Potential exposures during donning and doffing contaminated CPC, effect of decontamination or cleaning on material properties.
5. Resistance of CPC	Potential for rips, tears, and cuts.
6. Task requirements	Interference that CPC has on job task dexterity; its contribution to heat stress.
7. Test material or match application with manufactures test data	Obtain samples of candidate CPC; test or obtain data according to ANSI F739, ANSI 105, ASTM F1383, or NIOSH recommendations (Roder 1990); establish trial use.
8. Monitor use of CPC and working conditions	CPC used during potential exposures; under glove sampling; wipe samples for breakthrough; biological monitoring.

(*Source:* Adapted from Roder 1990, 5–18)

(SCBA) under the protective suit and are equipped with a one-way exhaust valve that allows air inside to escape but prevents the intrusion of environmental contaminants back into the suit. Performance standards exist in NFPA 1991 and include the provisions of testing suit materials to provide at least one hour of permeation resistance for 21 specified chemicals (acids, solvents, and gases) that are representative of the classes of chemicals often found at hazmat emergencies (NFPA 2000a, 6–49). The suits and ensemble components are subject to nearly 30 different performance and resistance tests. Users (buyers) of this equipment can test prospective suits for comfort, function, fit, and integrity by referring to ASTM F1154, which includes wearing the suit under a variety of conditions (ASTM 2004b). For less hazardous situations, NFPA 1992 defines the requirements for liquid splash-protective clothing. NFPA 1992 specifically excludes known or suspect carcinogens, hazardous material vapor atmospheres, and chemicals with skin toxicity or with a skin notation in the ACGIH list of threshold limit values (TLVs) (NFPA 2000b, 6-9; ACGIH 2006, 3–80). See additional guidelines from OSHA in Table 23.

Respirators

The function of the lung is to obtain oxygen and expel carbon dioxide. The enormous surface area required for this gas exchange process provides an ideal route of entry for air contaminants found in the work environment. Entry of particles, fibers, mist, vapors, and gases can lead to systemic illness as well as disease at the site of deposition. Respirators play an important role in preventing contaminants from entering this vulnerable organ.

NIOSH estimates that 10 percent of all U.S. employers either require respirator use or allow voluntary use, which is about 600,000 workplaces (Doney et al. 2005). When engineering or administrative controls fail to reduce air contaminants below an occupational exposure limit (OEL) where workers may be exposed to harmful levels of an air contaminant, or where there is insufficient oxygen, respirators are required. There are several governmental and nongovernmental sources of OELs, and many substances have several limits (see Table 24).

However, most contaminants do not have any exposure limits. There are a number of approaches to the limited risk information. A recent concept in chemical risk assessment and control involves the use of chemical *control bands*. Typically, the methodology classifies chemical hazards based on toxicity, type of task, amount of material used, and process attributes. Hazards are grouped into bands with a control scheme specified for each band (NIOSH 2004b). For example, a vapor degreasing operation that uses one drum or less of solvent that toxicologically is irritating to the respiratory system, with a boiling point of 150°C, and used with an operating temperature of 50°C, falls into the containment control band, according to the ILO Chemical Control Toolkit (one of several control banding methodologies). The control specifies enclosure design criteria and local exhaust. It also recommends when respiratory protective equipment use is likely, such as for equipment cleaning, maintenance activities, and when dealing with spills (ILO 2005).

Choosing Respirators (Protection Factors and Fit Testing)

The selection, issuance, use, and storage of respirators is a complex practice and must be performed in accordance with a respirator program, such as that described by OSHA in 29 CFR 1910.134 (OSHA 2007a). Two key concepts underlie the use of respiratory protection: matching the appropriate respirator with the hazards and proper fit. The extent of infiltration (leakage) around the face piece largely determines the degree of protection. Infiltration sources include leakage through the air-purifying element, the exhalation valve, and the respirator-to-face seal (which is typically the greatest source).

There are a number of factors to consider when choosing an appropriate respirator for a given task. These include:

- Determining the airborne hazard (dust, mist, gas, vapor, and so on)
- The toxicological properties of the

TABLE 23

Hazardous Waste Emergency Response Guidelines

Description	Respirator	Protective suit	Gloves	Boots
Level A Greatest level of respiratory, eye and skin protection	Positive pressure, full face, SCBA, or air supplied with escape SCBA	Totally encapsulating (disposable protective suit may be worn over encapsulating suit)	Inner and outer, chemically resistant	Chemically resistant, steel toe and shank
Level B Same respiratory protection but lesser level of skin protection needed	Positive pressure, full face SCBA, or air supplied with escape SCBA	Hooded chemical resistant clothing (overalls and a long-sleeved jacket; coveralls; one or two-piece chemical-splash suit; disposable chemical-resistant coveralls)	Inner and outer, chemically resistant	Boots, (outer), chemically resistant, steel toe and shank Boot-covers, chemical-resistant (disposable) if needed
Level C Concentration and type of airborne substance is known (not IDHL)	Air purifying, full-face or half-mask Escape mask (if needed)	Hooded chemical resistant clothing (overalls; two-piece chemical-splash suit; disposable chemical-resistant jacket)	Inner and outer, chemically resistant	Boots (outer), chemical-resistant steel toe and shank (if needed) Boot-covers, chemical-resistant (disposable) if needed
Level D Atmosphere contains no known or potential hazard, work precludes contact with hazardous levels of chemical	Escape mask (if needed)	Coveralls	Gloves (if needed)	Boots/shoes chemical-resistant steel toe and shank Boots outer, chemical-resistant (disposable) if needed

Note: Levels A, B, C, and D: Add coveralls, hard hat, and face shield if needed. Safety glasses or chemical splash goggles for Level D if needed. Respirators need to be NIOSH approved.

(*Source:* OSHA 1989)

TABLE 24

Occupational Exposure Limits (OELs), United States

OEL	Source	Stated Characteristics
Permissible Exposure Limit (PEL)	(OSHA 1974)	Regulatory requirements
Threshold Limit Value (TLV)	(ACGIH 2006, 3–87)	Nearly all workers may be repeatedly exposed without adverse health effects
Workplace Environmental Exposure Limits (WEEL)	(AIHA 2000, 31–61)	Will provide protection for most workers
Recommended Exposure Level (REL)	(NIOSH 1997)	To reduce or eliminate adverse effects and accidental injuries
Emergency Response Planning Guidelines (ERPG 1 to 3)	(AIHA 2000, 1–28)	One hour exposure, range of outcome severity: nearly all individuals would not experience more than mild, transient adverse health effects (ERPG-1); nearly all individuals would not experience or develop life threatening health effects (ERPG-3)
Immediately Dangerous to Life or Health (IDLH)	(NIOSH 1994)	Up to 30 min, poses a threat of exposure to airborne contaminants when that exposure is likely to cause death or immediate or delayed permanent adverse health effects or prevent escape from such an environment

contaminant (including the potential for eye irritation and routes of entry)
- The air concentration
- The potential for an environment that is immediately dangerous to life and health (IDLH), including oxygen deficiency
- The applicable OEL (or PEL), if any
- An appropriate protection factor

In a limited number of instances, a specific OSHA standard prescribes the necessary respiratory protection. A useful hazard ratio using the OEL is:

$$HR = \frac{WE}{OEL} \tag{17}$$

where

HR = Hazard ratio

WE = Contaminant concentration of the worker's exposure

OEL = the PEL, REL (recommended exposure limit), TLV, WEEL (workplace environmental exposure limit), and so on, and may be in the form of a time-weighted average (TWA), short-term excursion limit (STEL) or ceiling (C)

The HR is unitless, and temporal considerations of the OEL and the worker's measured exposure must match. The HR determines the amount of protection a respirator must provide by comparing it to a respirator's assigned protection factor (APF) and the maximum use concentration (MUC) (Bollinger 2004, 7–8).

Assigned protection factors (APFs) are determined for classes of respirators and incorporate the quantitative effectiveness of a respirator in the workplace for a percentage of properly-trained users. An *assigned protection factor* typically uses a fifth percentile of the lognormal distribution of protection values. This value provides sufficient protection for 95 percent of wearers (assuming adequate user training). See Table 25 for assigned protection factors.

The *maximum use concentration* (MUC) is the upper limit of exposure to a particular contaminant for a particular respirator. Unless a lower MUC is recommended or required, the MUC is based on the assigned protection factor and the OEL as shown in Equation 18 (Bollinger 2004, 7–8):

$$MUC = APF \times OEL \tag{18}$$

The MUC is limited to concentrations below those immediately dangerous to life or health (IDLH). The MUC may also be limited by the performance limits of a cartridge or canister, or limits established by NIOSH certification. The lower MUC takes precedence over the calculated MUC (OSHA 2006).

Example 8: The highest result of the measurement of a worker's occupational exposure of a particular solvent was 64 parts per million (ppm) (as a short peak measurement). OSHA, NIOSH, and ACGIH all specify an occupational exposure limit of 5 ppm as a ceiling limit. NIOSH considers this solvent to have good warning properties and an IDLH concentration of 300 ppm.

(1) What is the hazard ratio?

(2) Recommend an appropriate respirator. Can it be based on APFs (see Table 25)?

(3) What would be the MUC if a half-mask, air-purifying respirator (not a powered air-purifying respirator) was desired for this job?

Answer:

(1) Using the ceiling OEL:

$$HR = \frac{WE}{OEL} = \frac{64}{5} = 12.8$$

(2) The IDLH is higher than the worker's exposure so APF methodology can be used. Since the HR is 12.8, the worker would require a respirator with an APF of greater value to assure protection. A full-face, air-purifying respirator (APF of 50) would be an example satisfying the parameters in this example.

(3) A half-mask, air-purifying respirator has an APF of 10. This would indicate a MUC of 10 × 5 ppm = 50 ppm. This type of respirator would not be a suitable recommendation for this scenario.

Once the class of respirator can be determined, it is necessary to ensure that the respirator will fit the worker. The anthropometrical factors used by manufacturers to size respirators do not work for all

TABLE 25

Selected Assigned Protection Factors (APF)

Type of respirator	Half Mask	Full Face Piece	Helmet/Hood	Loose-Fitting Face Piece
1. Air-purifying	10	50		
2. Powered air-purifying (PAPR)	50	1,000	25*	25
3. Supplied-air (SAR) or airline				
A. Demand mode	10	50		
B. Continuous flow mode	50	1,000	25*	25
C. Pressure-demand or other positive-pressure mode	50	1,000		
4. Self-contained breathing apparatus (SCBA)				
A. Demand mode	10	50	50	
B. Pressure-demand or other positive-pressure mode		10,000	1,000	

*An APF of 1,000 can be utilized if supported by certain test data; see 29 CFR 1910.134(d)(3)(i)(a)

(*Source:* OSHA 2006)

individuals. Therefore, respirators of different sizes and from different manufacturers should be available for fitting workers' unique faces.

Respirators that rely on a face-to-respirator seal must be quantitatively and/or qualitatively fit-tested for each individual. The *qualitative fit test* (QLFT) is a pass/fail test that checks the adequacy of the respirator seal against the face. The QLFT has limitations and may only be used when a fit factor less than 100 is the maximum that is needed. This includes a safety factor of 10, and therefore it can only be used when HR is less than 10. Qualitative fit testing involves exposing the wearer to a challenge atmosphere (isoamyl acetate, saccharin solution, or Bitrex™) and determining if the individual can detect the smell or taste of the test substance in the respirator. Although OSHA also allows the use of irritant smoke (stannic chloride smoke tubes) for simple qualitative respirator fit test, NIOSH does not recommend it, owing to the possibility of exposure to hydrogen chloride (Bollinger 2004, 29–30; OSHA 1998b).

A *quantitative fit test* (QNFT) numerically measures leakage for each respirator-wearer combination. Several procedures are available. A challenge aerosol of a nontoxic product, such as corn oil, measured inside (C_i) and outside (C_o) of the respirator is used to calculate the fit factor (C_o/C_i). Alternatively, measuring ambient aerosol condensation nuclei avoids the need to generate a challenge test atmosphere. It is also possible to measure the volume of exhaust, measure in milliliters per minute (ml/min), that is required to hold a predetermined negative pressure inside a respirator.

The measure of the actual protection provided in the workplace is termed the *workplace protection factor* (WPF). There are wide variations in WPFs, both within and between workers (Nicas and Neuhaus 2004). There also appears to be little correlation between WPF and quantitative fit tests such that protection factors obtained in the laboratory are generally higher than measurements in the field (Myers et al. 1984; Myers 2000).

Quantitative fit tests are useful for the confirmation of a good fit. Passing a quantitative fit test typically results in greater individual protection (Coffey et al. 2004). In an acceptable quantitative fit test, the results should be greater than the APF (generally by at least a factor of 10, for example, a fit factor of 100 for air-purifying half-masks and 500 for air-purifying full-face piece respirators) (Bollinger 2004, 2).

Many work environments contain mixtures of contaminants. Airborne contaminates require additional thought and consideration in planning protection strategies. Respirators are available with multiple-use NIOSH certifications (for example, dust and vapor).

Some respirators are solely for escape. Escape respirators allow a worker a limited time to self-evacuate from a sudden respiratory hazard. They are not assigned protection factors, but are selected based on application. Escape respirators could include an air-purifying respirator with filters or cartridges (which assumes sufficient oxygen and no IDLH atmosphere) or a gas mask (which assumes sufficient oxygen but can otherwise be used in an IDLH environment up to its maximum use concentration). A self-contained breathing apparatus (SCBA) or a self-contained self-rescuer (SCSR) allows escape from oxygen deficient or IDLH atmospheres. With some materials, the risk of obscured vision or eye irritation is an important consideration requiring a full-face respirator (Bollinger 2004, 17–21).

Air-Purifying Respirators

In general, there are two types of respiratory devices: air purifying and air supply. The *air-purifying respirator* takes air from the worker's environment and cleans it of noxious particles, vapors, or gases prior to its entry into the face piece, but does not provide oxygen. Normal oxygen levels are 21.3 percent volume per volume (v/v). Levels of oxygen below 16 to 19.5 percent, depending on altitude, are oxygen-deficient atmospheres. *Air-supply respirators* (SCBA or SAR) are the only devices designed for and approved for entry into an oxygen-deficient atmosphere or any other IDLH environment (OSHA 1998b).

With most air-purifying respirators, air contaminants are removed as the worker inhales outside air through a cleaning matrix (filter or sorbent bed) (see Figure 18). This design is termed a negative pressure respirator, since the pressure inside the face piece is temporarily reduced when the wearer inhales.

Another type of atmosphere-purifying respirator is the powered air-purifying respirator (PAPR). These use a battery-powered air pump that blows air through an air-purifying element into a tight- or loose-fitting facemask, helmet, or hood (see Figure 19).

Particulate-Filtering Respirators

The particulate-filtering respirator is perhaps the simplest of the air-purifying respirators. This respirator uses a filtering mechanism to clean the air of dusts, fumes (solid particles condensed in the air from molten metal), mists, and fibers. Particulate respirators must capture aerosols over a large-diameter size range, such as fumes as small as 0.001 micrometer (μm) to dusts, mist, and pollen as large as 100 μm. Respirators or dust masks balance filter efficiency in removing the particles against the discomfort of increased breathing resistance. As the filter becomes loaded with particles, capture efficacy of the filter material increases, as does breathing resistance. Some respirator manufacturers increase the cross-sectional area of the filter by unique folding and packing so as to increase the particle-loading capacity and decrease the breathing resistance (Bollinger and Schutz 1987, 13–27).

Filters clean contaminated air using a number of physical processes that depend on the size and

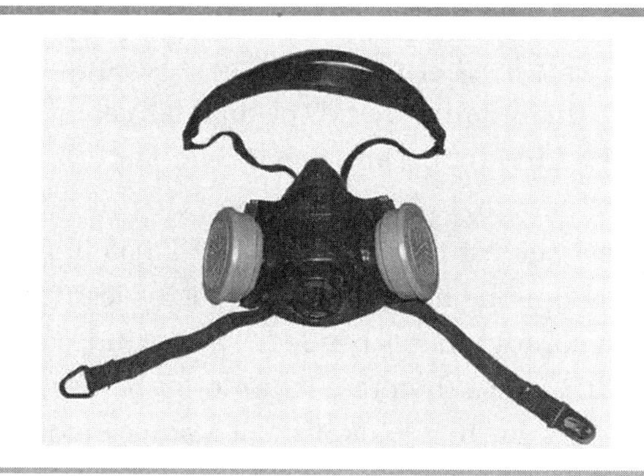

FIGURE 18. Air-purifying, half-face and full-face cartridge respirators

FIGURE 19. Powered air-purifying respirator (PAPR) (Figure courtesy of 3M Corporation 2006)

velocity of the particle. *Interception* occurs when particles in the airstream are caught in the fiber matrix. *Impaction* of particles occurs when the particle cannot follow the gas stream around the fiber because the particle has too much inertia to change direction. *Sedimentation* occurs when larger, heavier particles settle out of the air stream by gravity. The flow rate through the filter must be relatively slow for this mechanism to be important. *Diffusion* is another capture mechanism, caused by the random movement (caused by Brownian motion) of particles in the air stream. In this mechanism, particles randomly collide with filter fibers. Low flow rates (which cause the particle to remain near fibers longer) increase diffusion capture. *Electrostatic* capture occurs when the particle and the fiber are oppositely charged. The positive- and negative-charged surfaces attract, and capture takes place. This mechanism works together with interception and diffusion for increased filtration efficiency. A static charge given to some respirator filters can be lost when subjected to high humidity and certain types of particles (Hinds 1982, 164–186; Bollinger and Schutz 1987, 13–27).

Air-purifying filter respirators are further classified by their efficiency (percentage of particles removed) and resistance to oil particulates. Filters labeled *N* are not resistant to oil, *R* is resistant to oil, and *P* is oilproof (see Table 26). High-efficiency particulate air (HEPA) filters are equivalent to N100, R100, or P100 filters. Filter testing involves the use of sodium chloride particulate or highly-degrading oil and dioctyl phthalate, with a diameter of 0.3 μm. This diameter is used because it tends to be the most difficult to capture by a filter.

Gas- and Vapor-Removing Air-Purifying Respirators

Gas- and vapor-removing, air-purifying respirators function by sorption of the contaminant gas or vapor via an interaction of its molecules with a granular, sorbent material (such as activated charcoal) or by use of a catalyst that changes the gas or vapor to a less-toxic chemical. Several mechanisms remove the gases or vapors from the air stream. *Adsorption* captures the contaminant molecule on the surface of the media by physical attraction. Granules with large surface area are typically used, activated carbon being the most common. *Absorption* occurs when chemical bonding of the contaminant occurs within the media (rather than on the surface of the media). Since this process involves transport of the contaminant to the media interior, it is not as fast as adsorption. The third type of capture is by the use of a *catalyst*, which drives a chemical reaction to a less-toxic chemical. For example, hopcalite causes the conversion of carbon monoxide to carbon dioxide (Bollinger and Schutz 1987, 27–46; Myers 2000).

Unlike filters, these respirators do not become more efficient with exposure. The effectiveness of these devices to adsorb a gas or vapor diminishes with use.

TABLE 26

NIOSH Disposable Particulate Respirators

		HR Limits	
Type	**Description**	**Half Mask**	**Full Face**
N95	Filters ≥ 95% of airborne particles. Not resistant to oil.	<10	<20
N99	Filters ≥ 99% of airborne particles. Not resistant to oil.	<20	<100
N100	Filters ≥ 99.97% of airborne particles. Not resistant to oil.	<100	<100
R95	Filters ≥ 95% of airborne particles. Somewhat resistant to oil.	<10	<20
R99	Filters ≥ 99% of airborne particles. Somewhat resistant to oil.	<20	<100
R100	Filters ≥ 99.97% of airborne particles. Somewhat resistant to oil	<100	<100
P95	Filters ≥ 95% of airborne particles. Strongly resistant to oil.	<10	<20
P99	Filters ≥ 99% of airborne particles. Strongly resistant to oil.	<20	<100
P100	Filters ≥ 99.97% of airborne particles. Strongly resistant to oil.	<100	<100

(*Source:* NIOSH National Personal Protective Technology Laboratory)

Eventually, the exhausted media allows contaminants to break through the media into the facemask. Thus, the more sorbent that is available in the respirator, the longer the time to breakthrough, given a constant exposure level. Higher concentrations of the contaminant will reduce the service time of the respirator. By design, canisters contain larger volumes of sorbent and thus last longer than a cartridge (Bollinger and Schutz 1987, 27–6).

Canister or cartridge end-of-service-life indicators (ESLI) for determining change-out schedules are available for some chemicals. In absence of an ESLI, service time for canisters or cartridges should be determined for the conditions of use. These change-out times can be obtained either by testing the respirator against the maximum concentration gases or vapors encountered in the workplace in question, or by calculating the service life, using objective data with a sufficient safety factor (respirator manufacturers and OSHA supply helpful software). When calculating service times, the efficiency of chemical adsorption onto media depends on sorbent contact time (essentially the residence time of the contaminant in the media bed). This residency time is, in turn, dependent on the flow rate caused by the worker's inspiratory air velocity. Peak inspiratory flow and minute ventilation for workers executing rigorous physical work can significantly exceed manufacturers' design criteria, and potentially result in quicker breakthrough times (Kaufman and Hastings 2005). High humidity may also adversely affect service life.

It is important to match the contaminant with the sorbent capture properties. Canister and cartridge approval depend on the contaminates and the application. NIOSH testing involves a relatively small numbers of chemicals, with some results generalized to a class of contaminants, for example, organic vapors. Users of cartridge or canister respirators against other contaminants should consult published or manufacturers' literature for applicability. NIOSH does not certify cartridge respirators for substances that lack good warning properties below the OEL (odor, irritation, and so on), unless the respirator is equipped with an appropriate ESLI.

Atmosphere-Supplying Respirator

The atmosphere-supplying respirators provide clean breathing air to the worker without concern for the contaminate concentration in ambient air. The two major types are the supplied-air respirator (SAR) and the self-contained breathing apparatus (SCBA). These devices can provide air to the face piece using several methods, depending on their class. They include continuous flow, demand, and a pressure-demand mode. In the continuous flow mode, air flows at all times, and is controlled by a valve that provides at least 115 liters per minute (L/min) to a tight-fitting respirator or 170 L/min to a hood or helmet. The demand mode releases air into the face piece during inhalation. This results in the face piece having a slight negative pressure during inhalation and positive pressure during exhalation relative to the environment outside the mask.

The pressure-demand mode maintains a slight pressure inside the respirator face piece. Although air from the regulator still enters the respirator during inhalation, pressure-demand creates a positive pressure during both exhalation and inhalation. This mode normally limits outside contaminated air from entering the face piece, and thus is preferred over the demand mode, which is less protective. There is little difference between air consumption rates between demand and pressure-demand modes. Typically, workers use the demand mode during donning and adjusting the respirator and then switch to pressure-demand mode during normal use (Bollinger and Schutz 1987, 55–65).

Supplied-Air Respirators (SAR)

Supplied-air respirators are useful for situations where movement around the work area is less critical. SARs are operated in any of the three modes (continuous flow, demand, or pressure-demand). A special application airline respirator is available for abrasive-blasting, designated type CE by NIOSH. It contains additional protection from rebounding abrasive material. To use in an IDLH environment, an airline respirator must have an escape source of air (a 3- to

TABLE 27

Additional NIOSH/OSHA Requirements for Airline Respirators

Item	Criteria
Airline pressure	≤ 125 psi
Hose length	25–300 feet
Air flow range (helmet or hood)	170–425 l/m
Air flow range (tight fitting respirator)	115–425 l/m
Air quality for compressed gas	Grade D[1]
Carbon monoxide	<10 ppm

[1]Compressed Gas Association (CGA) specification G-7.1 (Type 1), which specifies air quality of CO < 10 ppm, CO_2 < 1000 ppm, oil (condensed hydrocarbons) < 5 mg/m^3, and a lack of noticeable odor. The air should be tested to assure that it continuously meets these criteria.

(*Source:* NIOSH 1995b)

10-minute high-pressure air cylinder) in case the primary source of air fails (see Table 27).

While wearing a SAR, a hose (airline) connects the worker to a stationary air source. The breathing air consists of a dedicated air compressor with an air receiver or compressed air cylinder(s). Although spatially constrained, these respirators are lighter than SCBA and are not limited to the shorter service times found with SCBAs.

It is important that the airline uses unique fittings that prevent attachment to any gas other than breathable gas. During a 12-year period, there were 23 deaths involving workers using airline respirators. The most common problem was compatible hose fittings that allowed workers to plug into air supplies that were nonbreathable (Suruda et al. 2003). This appears to be an ongoing problem, as a recent survey indicates that 24 percent of employers requiring the use of airline respirators either used compatible fittings or were unaware if their fittings were compatible with nonbreathable gas sources (Doney et al. 2005).

Self-Contained Breathing Apparatus (SCBA)

The SCBA, like the SAR, provides a high degree of protection but is portable, unlike the SAR. The SCBA uses its own source of air, which the wearer must carry (up to 35 pounds for the total respirator weight). SCBA respirators come in two general forms, closed-circuit and open-circuit. The *closed-circuit* respirator reuses exhaled air after adding oxygen and removing carbon dioxide. The *open-circuit* SCBA dumps expired air into the environment without any reuse by the wearer. Each system has its own advantages.

The closed-circuit SCBA reuses a portion of the exhaled air from the wearer. For this reason, closed-circuit designs are sometimes called *re-breathers*. These units have a one- to four-hour service cycle. As expired air passes through a solid sorbent, carbon dioxide is removed. Oxygen is reintroduced into the system using either a compressed gas cylinder or an oxygen-generating device. The oxygen-generating device generally consists of a peroxide compound that releases oxygen when contacted by water and carbon dioxide from the exhaled breath. A similar, albeit smaller, device is available as a self-rescuer and finds its primary application in mines.

An open-circuit SCBA does not recirculate air from the wearer, and as a result has a shorter service life, generally from 30 minutes to one hour, although high breathing rates will diminish the stated capacity. Escape (only) SCBAs are also available. Their service life ranges from 3 to 10 minutes. SCBA uses a cylinder of high-pressure air (2200–4500 psi) carried by the wearer. A one- or two-stage regulator reduces air pressure before entering the face piece. A bypass valve is available in case of a regulator malfunction. Users normally operate SCBAs in a pressure-demand mode in order to conserve air and provide a high degree of protection (Bollinger and Schutz 1987, 55–65).

Biological Hazards PPE

The use of PPE to protect against biological hazards has drawn increasing attention with the arrival and growth of acquired immunodeficiency syndrome (AIDS), drug-resistant tuberculosis, severe acute respiratory syndrome (SARS), and the occurrence of anthrax terrorism. Pathogens can be transmitted by direct contact with blood or other infectious materials through indirect contact with contaminated objects or surfaces, through contact of mucosa (oral, nasal, or conjunctiva), with droplets containing pathogens, and by inhalation of airborne particulates containing microorganisms. PPE use to prevent nosocomial transmission of disease to healthcare workers is an important prevention strategy, along with engineering controls, universal precautions, and infection control procedures.

Protective Barriers

Many common personal protective devices seen in other workplace settings are now protecting workers against biological hazards by providing a barrier against viruses, bacteria, and fungi. Risk of HIV transmission after a percutaneous exposure to HIV-infected blood is approximately 0.3 percent (CDC 1998). Use of protective barriers minimizes a route of entry through which the pathogen can enter the host. For example, unprotected mucous membranes of the eyes, nose, and mouth may serve as a portal for bloodborne pathogens (Ippolito et al. 1998). Eye protection with shields (or goggles) and a surgical mask (or NIOSH-certified particulate respirator) and/or a chin-length face shield are necessary to protect workers when there is potential exposure to spatter or droplets containing bloodborne pathogens (CDC 1998; OSHA 1991a). Note that a surgical mask is not a respirator and does not provide adequate protection against aerosol exposure. A NIOSH-certified respirator must be worn in situations where aerosol protection is required.

Protective clothing (aprons, lab coats, gowns, and jackets) provides barrier protection for the skin. There are several standard tests for resistance to biological agents (see Table 28).

Biological pathogens can penetrate the skin through very small cuts and abrasions in the hand. Disposable gloves provide a highly-effective exposure barrier. See Table 29 for additional glove standards in healthcare settings.

Latex is often the glove material of choice in protecting against viral hazards due to its fit, comfort, some ability of latex to reseal itself on puncture, and generally a lower failure rate when compared to vinyl or copolymer glove materials (Korniewicz et al. 2002). Likewise, in a simulated use test, vinyl gloves failed 12–61 percent of the time, while latex and nitrile performed significantly better, with failure rates of only 0–4 percent and 1–3 percent, respectively (Rego and Roley 1999). However, one drawback to the use of latex gloves is the possibility of latex allergy (see discussion of this issue at the end of this chapter). Regardless of the type of glove chosen, gloves should be changed regularly (or double-gloving used) as workers are frequently unaware of minute tears in gloves that occur during use that can compromise their barrier protection (CDC 2003).

The National Institute of Justice (NIJ) has published pathogenic- and puncture-resistance criteria for gloves for law enforcement personnel, using existing performance tests from the United States and Europe (NIJ 1999, 1–9). Table 6 gives additional NIJ performance criteria.

Respirators for Biological Hazards

The use of respirators in certain industrial sectors (principally health care) for biological hazards is rel-

TABLE 28

Tests for PPE Material Resistance to Viral or Blood Penetration

Test	Title
ASTM F1671	Standard Test Method for Resistance of Materials Used in Protective Clothing to Penetration by Blood-Borne Pathogens Using Phi-X174 Bacteriophage Penetration as a Test System
ASTM F1670	Standard Test Method for Resistance of Materials Used in Protective Clothing to Penetration by Synthetic Blood
ASTM F1862	Standard Test Method for Resistance of Medical Face Masks to Penetration by Synthetic Blood (Horizontal Projection of Fixed Volume at a Known Velocity)
ASTM F1819	Standard Test Method for Resistance of Materials Used in Protective Clothing to Penetration by Synthetic Blood Using a Mechanical Pressure Technique

TABLE 29

Specifications for Gloves and Clothing Used in Healthcare

Test	Title
ASTM D3577	Standard Specification for Rubber Surgical Gloves
ASTM D3578	Standard Specification for Rubber Examination Gloves
ASTM D5250	Standard Specification for Poly (vinyl chloride) Gloves for Medical Applications
ASTM D6319	Standard Specification for Nitrile Examination Gloves for Medical Application
ASTM D5151	Standard Test Method for Detection of Holes in Medical Gloves
NFPA 1999	Standard on Protective Clothing for Emergency Medical Applications

atively new but important in the prevention of the transmission of airborne pathogens. Reducing the risk of airborne transmission of infectious agents distinguishes between particles that can be respired and those that are nonrespirable. When protection is needed for contact against droplets (typically larger than 5 μm in diameter) that may come in contact with the mucus membranes (eyes, nose, or mouth), a simple barrier is needed. These particles settle out of the air quickly and are unlikely to enter the respiratory tract. Particles less than 5 microns are capable of inhalation and require true respiratory protection. Protection against the smaller-diameter aerosols, which can remain airborne for long periods, requires the use of a NIOSH-certified (N95) respirator (Leighner 2001). Airborne pathogens with a low infectious dose (similar to exposure to highly-toxic chemicals) dictate the use of respiratory protection with a higher degree of efficacy. Examples of such applications include occupational exposures to anthrax spores, Coccidioides immitis spores (a soil fungus), and Hantavirus (Nicas, Neuhaus, and Spear 2000; Nicas and Hubbard 2002; Mills et al. 2002).

Recently, OSHA placed the use of respirators for the protection from tuberculosis (TB) aerosols under its respiratory standard (OSHA 2003b). Since a number of studies indicate that a standard surgical mask does not provide adequate protection for filtering TB-containing aerosols, OSHA and NIOSH do not recognize it as a respirator. The diameter of infectious TB droplet nuclei ranges from one to five μm. Workers can use N95 disposable respirators for care of TB patients. NIOSH recommends a still-higher standard of respiratory protection for high-risk procedures, such as bronchoscopy or autopsy (NIOSH 1999, 5–10).

A similar situation regarding surgical masks exists for other exposures as well. Nurses that were exposed to severe acute respiratory syndrome (SARS) through assisting in high-risk procedures on patients had a lower risk of infection when consistently protected with an N95 mask than if they consistently used a surgical mask (although there was some benefit from use of the latter, it was not significant) (Loeb et al. 2004).

ERGONOMIC PPE

Kneepads (Contact Stressors)

Kneepads are an accepted form of personal protective equipment in certain industries. These industries include shipyards, mining, and construction, and employ welders, finishers, tile setters, floor layers, chippers, and grinders. By design, kneepads provide cushioning and reduce the contract stress derived from a static kneeling position by expanding the contact area. In addition, kneepads can provide limited protection from puncture injury. Hinges on some designs facilitate standing and kneeling without removal. A wedge-shaped knee support attached to the back of the pad helps prevent hyperflexion on some designs (Hudock 2003, 66). EN 14404:2004 outlines the performance criteria for kneepads in the European Union (CEN 2004).

Back Belts (Lifting Stressors)

For a number of years, back belts have been an implied form of PPE to lessen the risk of back injury from manual lifting. However, there are lingering questions about their usefulness. In 1994, NIOSH reviewed the efficacy of using back belts among uninjured workers from the scientific literature. NIOSH concluded that the effectiveness of back belts as PPE is equivocal at best. This conclusion was based on the view that there was insufficient evidence that back belts reduced the risk of injury from studies of workers, or that belts significantly reduced the biomechanical loading of the human trunk (NIOSH 1994a, 1–25). Recent studies and reviews of the literature have not clarified the role of back belts in preventing injury.

Several studies have suggested changes in trunk flexion or biomechanical properties while wearing belts (Willey 2001; Jonai et al. 1997; McGorry and Hsiang 1999; Magnusson, Pope, and Hansson 1996; Giorcelli et al. 2001). One study found that home attendants wearing back belts had some reduction of low back injury, compared to workers only trained in lifting techniques or workers with no training or back belts (Kraus et al. 2002). However, a number of other recent studies and reviews conclude that back belts do not show promise in preventing injury or reducing lumbar muscle fatigue (Jellema et al. 2001; Perkins and Bloswick 1995; Majkowski et al. 1998; Thomas et al. 1999; Ciriello and Snook 1995; Wassell et al. 2000). There currently are no generally-accepted performance standards for the selection or use of back belts to prevent occupational injury.

Hazards Associated with the Use of PPE (Additional Considerations)

The use of PPE can itself create significant hazards for the worker. These include heat stress, allergic reaction from the PPE material, and ergonomic, physical, and psychological stress. A reduction in vision, communication, and mobility may actually compromise worker safety when wearing PPE.

Metabolic Cost and Heat Stress of PPE

PPE can increase the metabolic heat production from its inherent weight and from additional work required to overcome resistance of movement. Table 30 gives the metabolic cost of PPE, expressed as watts per square meter of body surface area. The values are additive when items are worn together.

In addition to increased metabolic heat gain, workers face -increased risks of heat stress when the PPE substantially limits the loss of heat by convection and evaporation. A resulting build-up of internal body heat can quickly lead to hyperthermia, injury, and death. The body can be quickly incapacitated, even though the work conducted is light, or of short duration and in a cool environment. The annual incidence of heat-related illness could be as high as 1 out of 100 workers who routinely wear personal protective clothing in some industries (Crockford 1999).

When workers wear protective clothing ensembles, a number of studies suggest that adding an adjustment to the wet-bulb globe temperature (WBGT) is appropriate for heat stress assessment. Different clothing ensembles have been evaluated by comparing the average point at which body thermal regulatory control was lost for a given work demand. The high evaporative resistance (low water vapor permeability) of some clothing combinations accounts for most of the major adjustments in WBGT. Mean adjustments of 5 to 6°C for double-cloth overalls and water-barrier, vapor-transmitting coveralls, and 11°C for vapor barrier ensembles can be applied. Fully-encapsulating, vapor-barrier suits create their own microenvironment, often saturated with water vapor. With the exception of the critical importance of solar load, the conditions affecting heat stress are largely independent of conditions outside a fully-encapsulating suit (O'Connor and Bernard 1999; ACGIH 2006, 182–199; Kenney, Hyde, and Bernard 1993; Bernard 1999; Coles 1997).

Allergic Reaction

Unfortunately, PPE can also be the source or cause of injuries or illness that its use was intended to prevent. Awareness and surveillance of the potential hazards of PPE can minimize occurrence of these events. Irritation or allergic reaction to latex gloves is a common example. Additionally, once chemicals

TABLE 30

Metabolic Cost of PPE

Cost in W/m^2	Personal Protection Equipment (Item)
10	Safety shoes (walking or hiking)
20	Safety boots (walking or hiking)
20	Low performance respirator
40	High performance respirator
60	Self-contained breathing apparatus
20	Light, disposable chemical coverall
50	PVC chemical protective ensemble with hood, gloves, and boots
75	Heavy insulated clothing ensemble, for example, fire fighters' turnout gear

(*Source:* Hanson 1999)

enter or otherwise permeate the interior of a glove, occlusion of the chemical against the skin can exacerbate potential effects. Materials used in the manufacturing of gloves can cause nonallergic contact dermatitis or a Type IV (delayed) allergic reaction. Often implicated as the offending agents are excess rubber accelerators. Carbamates, thiazoles, and thiurams cause 30 percent, 5 percent and 60 percent of rubber accelerator dermatitis respectively (Groce 2005, 3).

Latex proteins can cause a more serious allergic reaction. Latex proteins cause a Type I immediate hypersensitivity reaction. The reaction often begins within minutes of the exposure and can include skin rash, hives, flushing, itching, asthma, and, in rare cases, life-threatening shock. NIOSH estimates that 8 percent to 12 percent of all workers routinely exposed to latex proteins are sensitized. Powder added to latex gloves to improve donning and wearing characteristics can carry proteins to the skin, resulting in increased incidence of latex allergic reactions.

Blood tests that detect latex antibodies can assist with the diagnosis of latex allergies. Workers with latex allergy should avoid contact with latex products and avoid areas where respirable latex proteins may be present (NIOSH 1997a, 14).

CONCLUSION

Although personal protection equipment holds the tertiary position in the hierarchical ranking of occupational hazard controls, it remains commonly used to isolate workers from the hazard. Being the final opportunity to mitigate a hazard, the importance of correct PPE application and usage cannot be understated.

The effective use of PPE control relies on a firm understanding of the underlying scientific and engineering principles of personal protective equipment. This knowledge allows safety and health practitioners to evaluate and choose PPE that have been appropriately tested and/or recommended through the experience of consensus standards and practices. This information helps assure that users recognize the limitations that accompany any PPE usage.

Poorly-instituted PPE programs are not difficult to find. Of the 389 categories of violations cited by OSHA in fiscal year 2006, the duty to have fall protection, respiratory protection, and head protection ranks third, fourth and sixteenth in frequency, accounting for nearly 12 percent of all penalties assessed by OSHA in fiscal year 2006 (OSHA 2007b).

There is still much research to accomplish as well. Given its importance, it is not surprising that the National Institute for Occupational Safety and Health (NIOSH) has included personal protective equipment in one of its top 21 National Occupational Research Agenda (NORA) priority research areas (NIOSH 2006a). NIOSH has also established the National Personal Protective Technology Laboratory (NPPTL). NPPTL is responsible for advancing federal research on personal protective equipment technologies for workers.

Finally, we owe professional diligence in our understanding, application, and decision making to the millions of workers who rely on PPE to be effective in preventing injury or illness.

REFERENCES

Akbar-Khanzadeh, F. 1998. "Factors Contributing to Discomfort or Dissatisfaction as a Result of Wearing Personal Protective Equipment." *Journal of Human Ergology (Tokyo)* 27(1–2):70–75.

Akbar-Khanzadeh, F., Michael C. Bisesi, and Ruben D. Rivas. 1995. "Comfort of Personal Protective Equipment." *Applied Ergonomics* 26(3):195–198.

Alberta Human Resources and Employment (AHRE). 2004. "Working in the Cold." *Workplace Health and Safety Bulletin*. Edmonton: Alberta Human Resources and Employment.

American Conference of Governmental Industrial Hygienists (ACGIH). 2000. *Heat Stresses and Strain. Documentation of the Threshold Limit Values and Biological Exposure Indices*. Cincinnati, OH: ACGIH.

______. 2006. *2006 TLVs and BEIs Based on the Documentation of the Threshold Limit Values for Chemical Substances and Physical Agents and Biological Exposure Indices*. Cincinnati, OH: ACGIH.

American National Standards Institute (ANSI). 1974. *ANSI S3.19-1974, American National Standard for the Measurement of Real-Ear Hearing Protectors and Physical Attenuation of Earmuffs*. New York: ANSI.

______. 1997. *ANSI S12.6-1997, Methods for Measuring the Real-Ear Attenuation of Hearing Protectors*. New York: ANSI.

______. 1999. *ANSI Z41-1999, American National Standard for Personal Protection—Protective Footwear*. Itasca, IL: National Safety Council.

______. 2000. *ANSI Z136.1-2000, American National Standard for Safe Use of Lasers*. Orlando: Laser Institute of America.

______. 2001. *ANSI Z80.3-2001, Ophthalmics: Nonprescription Sunglasses and Fashion Eyewear—Requirements*. New York: ANSI.

______. 2002. *ANSI S3.40-2002–ISO 10819:1996, Mechanical Vibration and Shock—Hand-Arm Vibration—Method for the Measurement and Evaluation of the Vibration Transmissibility of Gloves at the Palm of the Hand*. New York: ANSI.

______. 2003. *ANSI Z87.1-2003, American National Standard Practice for Occupational and Educational Personal Eye and Face Protection Devices*. Arlington, VA: International Safety Equipment Association.

American National Standards Institute/International Safety Equipment Association (ANSI/ISEA). 2004. *ANSI/ISEA 107-2004, American National Standard for High-Visibility Apparel and Headwear*. Arlington, VA: ISEA.

American Society for Testing and Materials (ASTM). 1999a. *ASTM D3389-94 (1999), Standard Test Method for Coated Fabrics Abrasion Resistance (Rotary Platform, Double-Head Abrader)*. West Conshohocken, PA: ASTM.

______. 1999b. *ASTM F739-99a, Standard Test Method for Resistance of Protective Clothing Materials to Permeation by Liquids or Gases Under Conditions of Continuous Contact*. West Conshohocken, PA: ASTM.

______. 1999c. *ASTM F1383-99a, Standard Test Method for Resistance of Protective Clothing Materials to Permeation by Liquids and Gases Under Conditions of Intermittent Contact*. West Conshohocken, PA: ASTM.

______. 1999d. *ASTM F1939-99a, Standard Test Method for Radiant Protective Performance of Flame Resistant Clothing Materials*. West Conshohocken, PA: ASTM.

______. 2000. *ASTM D6603-00, Standard Guide for Labeling of UV-Protective Clothing*. West Conshohocken, PA: ASTM.

______. 2001a. *ASTM F1236-96, Standard Guide for Visual Inspection of Electrical Protective Rubber Products*. West Conshohocken, PA: ASTM.

______. 2001b. *ASTM F1060-01, Standard Method for Thermal Protective Performance of Materials for Protective Clothing for Hot Surface Contact*. West Conshohocken, PA: ASTM.

______. 2002a. *ASTM F496-02a Standard Specification for In-Service Care of Insulating Gloves and Sleeves*. West Conshohocken, PA: ASTM.

______. 2002b. *ASTM D120–02a, Standard Specification for Rubber Insulating Gloves*. West Conshohocken, PA: ASTM.

______. 2002c. *ASTM D1051, Standard Specification for Rubber Insulating Sleeves*. West Conshohocken, PA: ASTM.

______. 2002d. *ASTM F1506-02ae1 Standard Performance Specification for Flame Resistant Textile Materials for Wearing Apparel for Use by Electrical Workers Exposed to Momentary Electric Arc and Related Thermal Hazards*. West Conshohocken, PA: ASTM.

______. 2004a. *ASTM F1818-04, Standard Specification for Foot Protection for Chain Saw Users*. West Conshohocken, PA: ASTM.

______. 2004b. *ASTM F1154-99a (2004) Standard Practices for Qualitatively Evaluating the Comfort, Fit, Function, and Integrity of Chemical-Protective Suit Ensembles*. West Conshohocken, PA: ASTM.

______. 2005a. *ASTM F1358-05. Standard Test Method for Effects of Flame Impingement on Materials Used in Protective Clothing Not Designed for Flame Resistance*. West Conshohocken, PA: ASTM.

______. 2005b. *ASTM F2413-05, Standard Specification for Performance Requirements for Foot Protection*. West Conshohocken, PA: ASTM.

______. 2005c. *ASTM F1790-05, Standard Test Method for Measuring Cut Resistance of Materials Used in Protective Clothing*. West Conshohocken, PA: ASTM.

______. 2006a. *ASTM F1959/F1959M-06ae1, Standard Test Method for Determining the Arc Rating of Materials for Clothing*. Conshohocken, PA: ASTM.

______. 2006b. *ASTM F1891-06 Standard Specification for Arc and Flame Resistant Rainwear*. Conshohocken, PA: ASTM.

______. 2006c. *ASTM F2178-06, Standard Test Method for Determining the Arc Rating and Standard Specification for Face Protective Products*. West Conshohocken, PA: ASTM.

______. 2006d. *ASTM D5151-06 Standard Test Method for Detection of Holes in Medical Gloves*. West Conshohocken, PA: ASTM.

American Society of Safety Engineers (ASSE). 2004. *ANSI/ASSE A10.32-2004, Fall Protection Systems for Construction and Demolitions Operations*. Des Plaines, IL: ASSE.

______. 2006. *ANSI/ASSE A1264.2-2006, Provisions of Slip Resistance on Walking/Working Surfaces*. Des Plaines, IL: ASSE.

______. 2007a. *ANSI/ASSE Z359.1-2007, Safety Requirements for Personal Fall Arrest Systems, Subsystems and Components*. Des Plaines, IL: ASSE.

______. 2007b. *ANSI/ASSE Z359.3-2007, Safety Requirements for Positioning and Travel Restrain Systems*. Des Plaines, IL: ASSE.

American Welding Society (AWS). 2005. *ANSI Z49.1-2005, Safety in Welding, Cutting, and Allied Processes*. Miami, FL: AWS.

Arezes, P. M. and A. S. Miguel. 2002. "Hearing Protectors Acceptability in Noisy Environments." *The Annals of Occupational Hygiene* 46(6):531–536.

Balty, I.., and A. Mayer. 1998. Head Protection. In *Encylopaedia of Occupational Health and Safety*, 4th ed., J. M. Stellman, ed.. Washington, D. C.: Brookings Institute Press.

Bell, L. H. 1982. Chapter 2, "Levels and Spectra." In *Industrial Noise Control: Fundamentals and Applications*, pp. 37–63. New York: Marcel Dekker, Inc.

Berger, Elliott H. 1986. Hearing Protection Devices. In *Noise and Hearing Conservation Manual*, ed. E. H. Berger, et al., pp. 321–381. Akron, OH: American Industrial Hygiene Association.

______. 2000. *E•A•RLog 20: The Naked Truth About NRRs* (retrieved February 28, 2008). www.e-a-r.com/pdf/hearingcons/earlog20.pdf

Berger, Elliott H., J. R. Franks, A. Behar, J. G. Casali, C. Dixon-Ernst, R. W. Kieper, C. J. Merry, B. T. Mozo, C. W. Wilson, D. Olin, J. D. Royster, and L. H. Royster. 1982. "Inter-laboratory Variability in the Measurement of Hearing Protector Attenuation." *Sound and Vibration* 16(1):14–19.

______. 1998. "Development of a New Standard Laboratory Protocol for Estimating the Field Attenuation of Hearing Protection Devices, Part III: The Validity of Using Subject-fit Data." *The Journal of the Acoustical Society of America* 103(2):665–672.

Bernard, T. 1999. "Heat Stress and Protective Clothing: An Emerging Approach from the United States." *The Annals of Occupational Hygiene* 43(6):321–327.

Bernerd, F., C. Vioux, F. Lejeune, and D. Asselineau. 2003. "The Sun Protection Factor (SPF) Inadequately Defines Broad Spectrum Photoprotection: Demonstration Using Skin Reconstructed in Vitro Exposed to UVA, UVB or UV-Solar Simulated Radiation." *European Journal of Dermatology* 13(3):242–249.

Blodgett, D. W., A. J. Suruda, and B. Insley Church. 2001. "Fatal Unintentional Occupational Poisonings by Hydrofluoric Acid in the U.S." *American Journal of Industrial Medicine* 40(2):215–220.

Boeniger, M. F. 2001. *Recommendation for Chemical Protective Clothing, Disclaimer and Additional Information* (retrieved March 30, 2005). www.cdc.gov/niosh/prot-cloth/ncpc2.html

Boeniger, M. F. and T. D. Klingner. 2002. "In-use Testing and Interpretation of Chemical-Resistant Glove Performance." *Applied Occupational and Environmental Hygiene* 17(5):368–378.

Bollinger, N. J. 2004. *NIOSH Respirator Selection Logic*. Cincinnati, OH: NIOSH.

Bollinger, N. J. and R. H. Schutz. 1987. *NIOSH Guide to Industrial Respiratory Protection*. Cincinnati, OH: NIOSH.

Bureau of Labor Statistics (BLS). 2007a. *Case and Demographic Characteristics for Work-related Injuries and Illnesses Involving Days Away From Work 2006; Tables R-13, R-16, R-32* (retrieved December 18, 2007). www.bls.gov/iif/oshcdnew.htm#02c

______. 2007b. *2006 Census of Fatal Occupational Injuries Data; Tables A-1, A-9* (retrieved December 18, 2007). www.bls.gov/iif/oshcfoi1.htm

Canning, K. M., P. B. McQuillan, and W. Jablonski. 1998. "Laboratory Simulation of Splashes and Spills of Organophosphate Insecticides on Chemically Protective Gloves Used in Agriculture." *Annals of Agricultural and Environmental Medicine* 5(2):155–167.

Centers for Disease Control and Prevention (CDC). 1998. "Public Health Service Guidelines for the Management of Health-Care Worker Exposures to HIV and Recommendations for Post-exposure Prophylaxis." *Morbidity and Mortality Weekly Report* 47(RR-7):1–33.

______. 2003. "Guidelines for Infection Control in Dental Health-Care Settings—2003." *Morbidity and Mortality Weekly Report* 52(RR17):1–61.

Chiang, T. M., R. M. Sayre, J. C. Dowdy, N. K. Wilkin, and E. W. Rosenberg. 2005. "Sunscreen Ingredients

Inhibit Inducible Nitric Oxide Synthase (iNOS): A Possible Biochemical Explanation for the Sunscreen Melanoma Controversy." *Melanoma Research* 15(1):3–6.

Ciriello, V. M. and S. H. Snook. 1995. "The Effect of Back Belts on Lumbar Muscle Fatigue." *Spine* 20(11):1271–1278.

Clarke, J. B., et al. 1986. "Chipping Hammer Vibration." *Scandinavian Journal of Work, Environment & Health* 12(4 Spec No):351–354.

Clemis, J. D. 1975. Anatomy, Physiology, and Pathology of the Ear. In *Industrial Noise and Hearing Conservation*, P. E. Olishifski, et al., eds., pp. 204–223. Chicago: National Safety Council.

Coffey, C. C., R. B. Lawrence, D. L. Campbell, Z. Zhuang, C. A. Calvert, and P. A. Jensen. 2004. "Fitting Characteristics of Eighteen N95 Filtering-Facepiece Respirators." *Journal of Occupational and Environmental Hygiene* 1(4):262–271.

Coles, G. V. 1997. Letter to the Editor. *Applied Occupational and Environmental Hygiene* 12(3):155.

Cooper Bussmann, Inc. 2004. *Safety Basics: Handbook for Electrical Safety*. 2d ed. St. Louis: Cooper Bussmann, Inc.

Crebolder, J. M. and R. B. Sloan. 2004. "Determining the Effects of Eyewear Fogging on Visual Task Performance." *Applied Ergonomics* 35(4):371–381.

Crnko, T., and S. Dyrnes. 2000. "Arcing Flash/Blast Review with Safety Suggestions for Design and Maintenance." Paper presented at Pulp and Paper Industry Technical Conference. June 19–23, Atlanta, GA.

Crockford, G. W. 1999. "Protective Clothing and Heat Stress: Introduction." *The Annals of Occupational Hygiene* 43(5):287–288.

Dain, S. J., et al. 1999. "Assessment of Fogging Resistance of Anti-fog Personal Eye Protection." *Ophthalmic & Physiological Optics* 19(4):357–361.

Derr, J., L. Forst, H. Yu Chen, and L. Conroy. 2001. "Fatal Falls in the U.S. Construction Industry, 1990 to 1999." *Journal of Occupational and Environmental Medicine* 43(10):853–860.

DiPilla, S. 2006. "Slip-Resistant Footwear: The Program Doesn't End After the Purchase." *Restaurant Hospitality* 90(9):84–86.

DiPilla, S., and K. Vidal. 2002. "State of the Art in Slip-Resistance Measurement." *Professional Safety* 47(6):37–42.

Doney, B. C., D. W. Groce, D. L. Campbell, M. F. Greskevitch, W. A. Hoffman, P. J. Middendorf, G. Syamlal, and K. M. Bang. 2005. "A Survey of Private Sector Respirator Use in the United States: An Overview of Findings." *Journal of Occupational and Environmental Hygiene* 2(5):267–276.

Dong, R. G., K. Krajnak, O. Wirth, and J. Wu. 2006. *Proceedings of the First American Conference on Human Vibration*. Morgantown, WV: National Institute for Occupational Safety and Health.

Doughty, R. L., T. E. Neal, G. M. Laverty, and H. Hoagland. 2000. "Electric Arc Hazard Assessment and Personnel Protection." Paper presented at Industry Applications Conference, Rome, Italy.

Edwards, R. G., A. B. Broderson, W. W. Green, and B. L. Lempert. 1983. "A Second Study of Effectiveness of Earplugs as Worn in the Workplace." *Noise Control Engineering Journal* 20(1):6–15.

Edwards, R. G., et al. 1978. "Effectiveness of Earplugs as Worn in the Workplace." *Sound and Vibration* 12(1):12–22.

Ellis, J. N. 1993. *Introduction to Fall Protection*. 2d ed. Des Plaines, IL: American Society of Safety Engineers.

English, W. 2003. Traction Testing of Footwear. In *Pedestrian Slip Resistance: How to Measure It and How to Improve It*. 2d ed., pp. 111–124. Alva, FL: William English, Inc.

Environmental Protection Agency (EPA). 1979. 40 CFR Part 211 - Product Noise Labeling, Subpart B - Hearing Protective Devices. *Federal Register* 44(190):56139–56147.

———. 2004. *Sun Wise Program: What is the UV Index?* (retrieved February 28, 2005). www.epa.gov/sunwise/uvwhat.html

European Committee for Standardization (CEN). 2001. *EN 812:1997/A1:2001, Industrial Bump Caps*. Brussels: CEN.

______. 2004. *EN 14404:2004, Personal Protective Equipment - Knee Protectors for Work in the Kneeling Position*. Brussels: CEN.

______. 2006. *EN 511:2006, Protective Gloves Against Cold*. Brussels: CEN.

Evans, P. G., J. J. McAlinden, and P. Griffin. 2001. "Personal Protective Equipment and Dermal Exposure." *Applied Occupational and Environmental Hygiene* 16(2):334–337.

Federal Highway Administration (FHWA). 2003. *Part VI. Manual of Uniform Traffic Control Devices*. Millennium Edition. Washington, D.C.: U.S. Department of Transportation.

Forsberg, K., and L. H. Keith. 1999. *Chemical Protective Clothing Performance Index*. 2d ed. New York: John Wiley & Sons, Inc.

Forsberg, K., and S. Z. Mansdorf. 2002. *Quick Selection Guide to Chemical Protective Clothing*. 4th ed. New York: John Wiley & Sons, Inc.

Fricker, C., and J. K. Hardy. 1992. "Protective Glove Material Permeation by Organic Solids." *American Industrial Hygiene Association Journal*. 53(12): 745–750.

Fuortes, L. J., A. D. Ayebo, and B. C. Kross. 1993. "Cholinesterase-Inhibiting Insecticide Toxicity." *American Family Physician*. 47(7):1613–1620.

Gao, P., N. El-Ayouby, and J. T. Wassell. 2005. "Change in Permeation Parameters and the Decontamination Efficacy of Three Chemical Protective Gloves After Repeated Exposures to Solvents and Thermal Decontaminations." *American Journal of Industrial Medicine*. 47(2):131–143.

Garrod, A. N., A. M. Phillips, and J. A. Pendleton. 2001. "Potential Exposure of Hands Inside Protective Gloves—A Summary of Data from Non-Agricultural Pesticide Surveys." *The Annals of Occupational Hygiene*. 45(1):55–60.

Gilchrist, A. and N. J. Mills. 1996. "Protection of the Side of the Head." *Accident, Analysis and Prevention*. 28(4):525–535.

Giorcelli, R. J., R. E. Hughes, J. T. Wassell, and H. Hsaio. 2001. "The Effect of Wearing a Back Belt on Spine Kinematics During Asymmetric Lifting of Large and Small Boxes." *Spine*. 26(16):1794–1798.

Grammer, L. C., K. E. Harris, and P. R. Yarnold. 2002. "Effect of Respiratory Devices on Development of Antibody and Occupational Asthma to an Acid Anhydride." *Chest* 121(4):1317–1322.

Grieser, B. C., T. P. Rhoades, and R. J. Shah. 2002. "Slip Resistance. Field Measurements Using Two Modern Slipmeters." *Professional Safety*. 47(6):43.

Griffin, M. J. 1997. "Measurement, Evaluation, and Assessment of Occupational Exposures to Hand-Transmitted Vibration." *Occupational and Environmental Medicine*. 54(2):73–89.

Groce, D. 2005. *What Causes Allergic Reactions to Gloves* (retrieved May 6, 2005). www.chemrest.com/what%20causes%20allergic%20reactions%20to%20gloves.pdf

Hanson, M. A. 1999. "Development of a Draft British Standard: The Assessment of Heat Strain for Workers Wearing Personal Protective Equipment." *The Annals of Occupational Hygiene*. 43(5):309–319.

Harris, P. 2004. *Nonfatal Occupational Injuries Involving the Eyes* (retrieved January 7, 2005). www.bls.gov/opub/cwc/content/sh20040624ar01p1.stm

Havenith, G. 1999. "Heat Balance When Wearing Protective Clothing." *The Annals of Occupational Hygiene*. 43(5):289–296.

Henderson, D. 1991. "Ocular Trauma: One in the Eye for Safety Glasses." *Archives of Emergency Medicine* 8(3):201–204.

Hinds, W. C. 1982. *Aerosol Technology*. New York: John Wiley & Sons.

Hoagland, H., B. Shinn, and V. Reed. 2004. "A Far Better 70E." *Occupational Health and Safety* 73(8):34–38.

Holmér, I., and H. Rintamäki. 2004. "Risk Assessment for Cold Work." *Barents Newsletter on Occupational Health and Safety* 7(1):5–7.

Holmér, I., H. Nilsson, G. Havenith, and K. Parsons. 1999. "Clothing Convective Heat Exchange-Proposal for Improved Prediction in Standards and Models." *The Annals of Occupational Hygiene* 43(5):329–337.

Hudock, S. 2003. *Compendium of Ergonomic Analyses of Shipyard Work Processes*. Cincinnati, OH: NIOSH.

Institute of Electrical and Electronics Engineers (IEEE). 2002. *IEEE 1584-2002, Guide for Performing Arc Flash Hazard Calculations*. Piscataway, NJ: IEEE.

International Labour Office (ILO). 2005. *Safework: International Chemical Control Toolkit. Draft Guidelines*. Geneva, Switzerland: ILO.

International Safety Equipment Association (ISEA). 2000. *ANSI/ISEA 105-2000, American National Standard for Hand Protection Selection Criteria*. Arlington, VA: ISEA.

______. 2002. "Calculating Total Fall Distance." *Protection Update* (Winter):12.

______. 2003a. *Eye and Face Protection. Use and Selection Guide*. Arlington, VA: ISEA.

______. 2003b. *ANSI Z89.1, American National Standard for Industrial Head Protection*. Arlington, VA: ISEA.

______. 2004. *ISEA Survey Shows Safety Equipment Use Trends Upward in Heavy Construction* (retrieved June 13, 2005). www.safetyequipment.org/04researchreport.htm

International Standards Organization (ISO). 1979. *ISO 4850:1979, Personal Eye-protectors for Welding and Related Techniques—Filters—Utilisation and Transmittance Requirements*. Geneva, Switzerland: ISO.

______. 1989. *ISO 5085-1:1989, Textiles—Determination of thermal resistance—Part 1: Low thermal resistance*. Geneva, Switzerland: ISO.

______. 1990. *ISO 4869-1:1990, Acoustics—Hearing Protectors—Part 1: Subjective Method for the Measurement of Sound Attenuation*. Geneva, Switzerland: ISO.

______. 1994. *ISO 4869-2:1994, Acoustics—Hearing Protectors—Part 2: Estimation of Effective A-Weighted Sound Pressure Levels When Hearing Protectors Are Worn*. Geneva, Switzerland: ISO.

______. 2000. *ISO/DIS 17493, Clothing for Protection Against Heat and Flame—Test Method for Convection Heat Resistance Using a Hot Air Circulating Oven*. Geneva, Switzerland: ISO.

______. 2006. *ISO-FDIS-11079:2006, Ergonomics of the Thermal Environment. Determination and Interpretation of Cold Stress When Using Required Clothing Insulation (IREQ) and Local Cooling Effects*. Geneva, Switzerland: ISO.

Ippolito, G., V. Puro, N. Petrosillo, G. DeCarli, G. Micheloni, and E. Magliano. 1998. "Simultaneous Infection with HIV and Hepatitis C Virus Following Occupational Conjunctival Blood Exposure." *JAMA–The Journal of the American Medical Association* 280(1):28.

Jellema, P., M. W. van Tulder, M. van Poppel, A. N. Nachemson, and L. M. Bouter. 2001. "Lumbar

Supports for Prevention and Treatment of Low Back Pain: A Systematic Review within the Framework of the Cochrane Back Review Group." *Spine* 26(4):377–386.

Jetzer, T., P. Hayden, and D. Reynolds. 2003. "Effective Intervention with Ergonomics, Antivibration Gloves, and Medical Surveillance to Minimize Hand-Arm Vibration Hazards in the Workplace." *Journal of Occupational and Environmental Medicine* 45(12): 1312–1317.

Jonai, H., M. Villanueva, M. Sotoyama, N. Hisanaga, and S. Saito. 1997. "The Effect of a Back Belt on Torso Motion—Survey in an Express Package Delivery Company." *Industrial Health* 35(2):235–242.

Kaufman, J., and S. Hastings. 2005. "Respiratory Demand During Rigorous Physical Work in a Chemical Protective Ensemble." *Journal of Occupational and Environmental Hygiene.* 2(2):98–110.

Kenney, W. L., D. E. Hyde, and T. E. Bernard. 1993. "Physiological Evaluation of Liquid-Barrier, Vapor-Permeable Protective Clothing Ensembles for Work in Hot Environments." *American Industrial Hygiene Association Journal* 54(7):397–402.

Klaassen, C. D., and J. B. Watkins. 1999. Toxic Responses of the Skin. In *Casarett & Doull's Toxicology: The Basic Science of Poisons.* 5th ed., pp. 441–446. New York: McGraw-Hill.

Koppes, B., and G. Pearson. 2002. "Is Your High-Visibility Apparel Working for You?" *Protection Update* (4):6–7.

Korniewicz, D. M., M. El-Masri, J. M. Broyles, C. D. Martin, and K. P. O'Connell. 2002. "Performance of Latex and Nonlatex Medical Examination Gloves During Simulated Use." *American Journal of Infection Control* 30(2):133–138.

Korniewicz, D. M., M. Kirwin, K. Cresci, and E. Larson. 1993. "Leakage of Latex and Vinyl Exam Gloves in High and Low Risk Clinical Settings." *American Industrial Hygiene Association Journal* 54(1):22–26.

Kraus, J. F., et al. 2002. "A Field Trial of Back Belts to Reduce the Incidence of Acute Low Back Injuries in New York City Home Attendants." *International Journal of Occupational and Environmental Health* 8(2):97–104.

Kuklane, K. 2004. "The Use of Footwear Insulation Values Measured on a Thermal Foot Model." *International Journal of Occupational Safety and Ergonomics* 10(1):7986.

Leighner, L. A. 2001. "Don the Barriers." *Critical Care Nursing Quarterly* 24(2):30–38.

Lipscomb, H. J., J. M. Dement, V. McDougall, and J. Kalat. 1999. "Work-Related Eye Injuries among Union Carpenters." *Applied Occupational and Environmental Hygiene* 14(10):665–676.

Loeb, M., A. McGeer, B. Henry, M. Ofner, D. Rose, T. Hlywka, J. Levie, J. M. McQueen, S. Smith, L. Moss, A. Smith, K. Green, and S. D. Walter. 2004. "SARS among Critical Care Nurses, Toronto." *Emerging Infectious Disease* 10(2):251–255.

Lovasic, S. 2005. "Matching Protective Apparel to Electrical Arc Flash Hazards." *Professional Safety* 50(1):56–58.

Magnusson, M., M. H. Pope, and T. Hansson. 1996. "Does a Back Support Have a Positive Biomechanical Effect?" *Applied Ergonomics.* 27 (3):201–205.

Majkowski, G. R., B. W. Jovag, B. T. Taylor, M. S. Taylor, S. C. Allison, D. M. Stetts, and R. L. Clayton. 1998. "The Effect of Back Belt Use on Isometric Lifting Force and Fatigue of the Lumbar Paraspinal Muscles." *Spine* 23(19):2104–2109.

Mansdorf, S. Z. 1998. *Recommendations for Chemical Protective Clothing. A Companion to the NIOSH Pocket Guide to Chemical Hazards* (retrieved March 1, 2008). www.cdc.gov/niosh/ncpc/ncpc2.html

Mastrullo, K. 2004. "NFPA 70E-2004: What You Need to Know." *Plant Engineering* (5):44–47.

McGorry, R. W. and S. M. Hsiang. 1999. "The Effect of Industrial Back Belts and Breathing Technique on Trunk and Pelvic Coordination during a Lifting Task." *Spine* 24(11):1124–1130.

Michael, K. M., and D. C. Byrne. 2000. "Industrial Noise and Conservation of Hearing." In *Patty's Industrial Hygiene.* 5th ed., R. L. Harris, ed., pp. 789–803. New York: John Wiley & Sons, Inc.

Mickelsen, R. L., and R. C. Hall. 1987. "A Breakthrough Time Comparison of Nitrile and Neoprene Glove Materials Produced by Different Glove Manufacturers." *American Industrial Hygiene Association Journal* 48(11):941–947.

Mikatawaga, M., S. S. Que Hee, and H. E. Ayer. 1984. "Permeation of Chlorinated Aromatic Compounds Through Vitron and Nitrile Glove Materials." *American Industrial Hygiene Association Journal* 45(9):617–621.

Mills, J. N., A. Corneli, J. C. Young, L. E. Garrison, A. S. Khan, and T. G. Ksiazek. 2002. "Hantavirus Pulmonary Syndrome—United States: Updated Recommendations for Risk Reduction." *Morbidity and Mortality Weekly Report* 51(RR-9):1–12.

Mine Safety and Health Administration (MSHA). 1999. "30 CFR Part 62; Health Standards for Occupational Noise Exposure; Final Rule." *Federal Register* 64(176): 49548–49637.

Murphy, W. J., J. R. Franks, and E. F. Krieg. 2002. "Hearing Protector Attenuation: Models of Attenuation Distributions." *The Journal of the Acoustical Society of America.* 111 (5 Pt 1):2109–2116.

Murphy, W. J., J. R. Franks, E. H. Berger, A. Behar, J. G. Casali, C. Dixon-Frost, E. F. Krieg, B. Mozo, J. D. Royster, L. H. Royster, S. D. Simon, and C. Stephenson. 2004. "Development of a New Standard Laboratory Protocol for Estimation of the Field Attenuation of Hearing Protection Devices: Sample

Size Necessary to Provide Acceptable Reproducibility." *The Journal of the Acoustical Society of America*. 115(1): 311–323.

Myers, W. R. 2000. "Respiratory Protective Equipment." In *Patty's Industrial Hygiene*. 5th ed. R. L. Harris, ed., pp. 1489–1550. New York: John Wiley & Sons, Inc.

Myers, W. R., M. J. Peach, K. Cutright, and W. Iskander. 1984. "Workplace Protection Factor Measurements on Powered Air-Purifying Respirators at a Secondary Lead Smelter: Results and Discussion." *American Industrial Hygiene Association Journal* 45(10):681–688.

National Fire Protection Association (NFPA). 2000a. *NFPA 1991, Vapor-Protective Ensembles for Hazardous Materials Emergencies*. Quincy, MA: NFPA.

______. 2000b. *NFPA 1992, Liquid Splash-Protective Ensembles and Clothing for Hazardous Materials Emergencies*. Quincy, MA: NFPA.

______. 2004. *NFPA 70E, Standard for Electrical Safety Requirements for Employee Workplaces*. Quincy, MA: NFPA.

National Institute for Occupational Safety and Health (NIOSH). 1978. *Development of Performance Criteria for Protective Clothing Used Against Carcinogen Liquids*. Cincinnati, OH: NIOSH.

______. 1988. *Criteria for a Recommended Standard. Occupational Noise Exposure* (Revised). Cincinnati, OH: NIOSH.

______. 1994a. *Workplace Use of Back Belts. Review and Recommendations*. Cincinnati, OH: NIOSH.

______. 1994b. *Documentation for Immediately Dangerous to Life or Health Concentrations (IDLH): NIOSH Chemical Listing and Documentation of Revised IDLH Values* (retrieved June 15, 2005). www.cdc.gov/nish/idlh/intridl4.html

______. 1995a. *Publication No. 95-105, Hearing Protector Compendium*. Cincinnati, OH: NIOSH.

______. 1995b. Respiratory Protective Devices; Final Rules and Notice. *Federal Register* 60(110):30335–30398.

______. 1997a. *Preventing Allergic Reactions to Natural Rubber Latex in the Workplace*. Cincinnati, OH: NIOSH.

______. 1997b. *Publication No. 97-140, NIOSH Pocket Guide to Chemical Hazards*. Cincinnati, OH: NIOSH.

______. 1999. *Respiratory Protection Program in Health Care Facilities. Administrator's Guide*. Cincinnati, OH: NIOSH.

______. 2000. *Method for Calculating and Using Noise Reduction Rating—NRR* (retrieve December 20, 2004). www2a.cdc.gov/hp-devices/pdfs/calculation.pdf

______. 2001. *Publication No. 2001-103, Work-Related Hearing Loss*. Cincinnati, OH: NIOSH.

______. 2003. *Hearing Protector Device Compendium* (retrieved December 17, 2004). www.cdc.gov/niosh/topics/noise/hpcomp.html

______. 2004a. *Eye Protection for Infection Control* (retrieved January 10, 2005). www.cdc.gov/niosh/topics/eye/eye-infectious.html

______. 2004b. *NIOSH Safety and Health Topic: Control Banding* (retrieved June 27, 2005). www.cdc.gov/niosh/topics/ctrlbanding/#1

______. 2006a. *About NORA* (retrieved January 16, 2007). www2a.cdc.gov/nora/NORAabout.html

______. 2006b. *NIOSH Program Portfolio: Engineering Controls* (retrieved September 23, 2006). www.cdc.gov/niosh/programs/eng

National Institute of Justice (NIJ). 1999. *NIJ Test Protocol 99-114, Test Protocol for Comparative Evaluation of Protective Gloves for Law Enforcement and Corrections Applications*. Washington DC: NIJ.

National Oceanic and Atmospheric Administration (NOAA). 2001. *NWS Wind Chill Index* (retrieved March 1, 2008). www.weather.gov/os/windchill/index.shtml

———. 2005. *Heat Index* (retrieved December 16, 2007). www.crh.noaa.gov/pub/heat.php

National Radiological Protection Board (NRBP). 2002. *Health Effects from Ultraviolet Radiation. Report of an Advisory Group on Non-ionizing Radiation* (retrieved March 1, 2008). www.hpa.org.uk/radiation/publications/documents_of_nrpb/pdfs/doc_13_1.pdf

Nicas, M., and A. Hubbard. 2002. "A Risk Analysis for Airborne Pathogens with Low Infectious Doses: Application to Respirator Selection against *Coccidioides immitis* Spores." *Risk Analysis* 22(6):1153–1163.

Nicas, M., and J. Neuhaus. 2004. "Variability in Respiratory Protection and the Assigned Protection Factor." *Journal of Occupational and Environmental Hygiene* 1(2):99–109.

Nicas, M., J. Neuhaus, and R. C. Spear. 2000. "Risk-based Selection of Respirators against Infectious Aerosols: Application to Anthrax Spores." *Journal of Occupational and Environmental Medicine* 42(7):737–748.

Nierenberg, D. W., R. E. Nordgren, M. B. Chang, R. W. Siegler, M. B. Blayney, F. Hochberg, T. Y. Toribara, E. Cernichiari, and T. Clarkson. 1998. "Delayed Cerebellar Disease and Death after Accidental Exposure to Dimethylmercury." *New England Journal of Medicine*. 338(23):1672–1676.

O'Connor, D. J., and T. E. Bernard. 1999. "Continuing the Search for WBGT Clothing Adjustment Factors." *Applied Occupational and Environmental Hygiene* 14(2):119–125.

Occupational Safety and Health Administration (OSHA). 1974. *29 CFR 1910.1000: Air Contaminants* (retrieved March 1, 2008). www.osha.gov/pls/oshaweb/owadisp.show_document?p_id=9991&p_table=STANDARDS

______. 1979. *Occupational Fatalities Related to Roofs, Ceilings, and Floors as Found in Reports of OSHA Fatality/Catastrophe Investigations*. Washington, DC: OSHA.

______. 1981. "Occupational Noise Exposure; Hearing Conservation Amendment; Final Rule." *Federal Register* 46(11):4078–4179.

______. 1989. *29 CFR 1910.120: Hazardous Waste Operations and Emergency Response* (retrieved March 1, 2008). www.osha.gov/pls/oshaweb/owadisp.show_document?p_table=standards&p_id=9765

______. 1990. *29 CFR 1910.252 Subpart Q: Welding, Cutting and Brazing* (retrieved March 1, 2008). www.osha.gov/pls/oshaweb/owadisp.show_document?p_id=9853&p_table=STANDARDS

______. 1991a. *29 CFR 1910.1030: Bloodborne Pathogens* (retrieved March 1, 2008). www.osha.gov/pls/oshaweb/owadisp.show_document?p_table=STANDARDS&p_id=10051

______. 1991b. *STD 01-05-001—PUB 8-1.7: Guidelines for Laser Safety and Hazard Assessment* (retrieved March 1, 2008). www.osha.gov/pls/oshaweb/owadisp.show_document?p_id=1705&p_table=DIRECTIVES

______. 1992. "29 CFR Parts 1910 and 1926 Occupational Exposure to 4,4′ Methylenedianiline (MDA); Final Rule." *Federal Register* 57(154):35630–35696.

______. 1993a. Laser Hazards. In *OSHA Technical Manual*, pp. 15–17. Washington DC: OSHA.

______. 1993b. *Eye Protection in the Workplace*. Washington DC: OSHA.

______. 1994a. *29 CFR 1910.133: Eye and Face Protection* (retrieved March 1, 2008). www.osha.gov/pls/oshaweb/owadisp.show_document?p_table=STANDARDS&p_id=9778

______. 1994b. "Personal Protective Equipment for General Industry; Final Rule." *Federal Register* 59:16334–16364.

______. 1994c. Electric Power Generation, Transmission, and Distribution; Electrical Protective Equipment. *Federal Register* 59:4320–4476.

______. 1994d. "Safety Standards for Fall Protection in the Construction Industry; Final Rule." *Federal Register* 59(152):40672–40753.

______. 1998a. *Fact Sheet No. OSHA 98-55: Protecting Workers in Cold Environments* (retrieved March 1, 2008). www.osha.gov/pls/oshaweb/owadisp.show_document?p_table=FACT_SHEETS&p_id=186

______. 1998b. "Respiratory Protection." *Federal Register* 63(5):1151–1300.

______. 1999a. Heat Stress. In *OSHA Technical Manual*, pp. 14–16. Washington DC: OSHA.

______. 1999b. Respiratory Protection. In *OSHA Technical Manual*, p. 6. Washington DC: OSHA.

______. 1999c. "Employer Payment for Personal Protective Equipment, Proposed Rule." *Federal Register* 64(61):15418.

______. 2002a. "Safety Standards for Signs, Signals, and Barricades." *Federal Register* 67:57722–57736.

______. 2002b. "Occupational Injury and Illness Recording and Reporting; Final Rule." *Federal Register* 67(126):44037–44048.

______. 2003a. *OSHA 3166-06R: Protecting Yourself in the Sun* (retrieved March 1, 2008). www.osha.gov/Publications/osha3166.pdf

______. 2003b. "Occupational Exposure to Tuberculosis; Proposed Rule; Termination of Rulemaking Respiratory Protection for M. Tuberculosis; Final Rule; Revocation." *Federal Register* 68:75767–75775.

______. 2004. "Hazards of Misusing Wire Form Anchorage Connectors for Fall Protection." *Safety and Health Information Bulletin*. Washington DC: OSHA.

______. 2005. *Noise and Hearing Conservation Technical Manual, Appendix IV:C. Methods for Estimating HPD Attenuation* (retrieved January 7, 2007). www.osha.gov/dts/osta/otm/noise/hcp/attenuation_estimation.html

______. 2006. "Assigned Protection Factors; Final Rule." *Federal Register* 71(164):50188.

______. 2007a. *29 CFR 1910.134: Respiratory Protection* (retrieved March 1, 2008). www.osha.gov/pls/oshaweb/owadisp.show_document?p_id=12716&p_table=STANDARDS

______. 2007b. *Frequently Cited OSHA Standards* (retrieved January 16, 2007). www.osha.gov/pls/imis/citedstandard.htm

Olesen, B. W. 1982. "Thermal Comfort." *Brüel & Kjaer Technical Review*, pp. 13–14.

Ong, M., J. T. L. Choo, and E. Low. 2004. "A Self-Controlled Trial to Evaluate the Use of Active Hearing Defenders in the Engine Rooms of Operational Naval Vessels." *Singapore Medical Journal* 45(2):75–78.

Oppl, R. 2002. "Selection, Testing, and Effectiveness in the Field of PPE and Gloves." Paper presented at International Conference on Occupational & Environmental Exposures of the Skin to Chemicals: Science and Policy, September 8–11, Washington, D.C.

Paddilla, M. 1976. "Ear Plug Performance in Industrial Field Conditions." *Sound and Vibration* 10(5):33–36.

Perkins, J. L., and B. Pool. 1997. "Batch Lot Variability in Permeation Through Nitrile Gloves." *American Industrial Hygiene Association Journal* 58(7):474–479.

Perkins, M. S., and D. S. Bloswick. 1995. "The Use of Back Belts to Increase Intra-abdominal Pressure as a Means of Preventing Low Back Injuries: A Survey of the Literature." *International Journal of Occupational and Environmental Health* 1(4):326–335.

Personick, M. E. 1990. "Heat Burns Sustained in the Workplace." *Monthly Labor Review* (7):37–38.

Poon, T. S., and R. S. Barnetson. 2002. "The Importance of Using Broad Spectrum SPF 30+ Sunscreens in Tropical and Subtropical Climates." *Photodermatology Photoimmunology and Photomedicine* 18(4):175–178.

Pratt, S. G., D. F. Fosbroke, and S. M. Marsh. 2001. *Building Safer Highway Work Zone: Measures to*

Prevent Worker Injuries from Vehicles and Equipment. Cincinnati, OH: National Institute for Occupational Safety and Health.

Purdham, J. T., B. J. Menard, P. R. Bozek, and A. M. Sass-Kortsak. 2001. "MCPA Permeation Through Protective Gloves." *Applied Occupational and Environmental Hygiene* 16(10):961–966.

Rego, A., and L. Roley. 1999. "In-Use Barrier Integrity of Gloves: Latex and Nitrile Superior to Vinyl." *American Journal of Infectious Control* 27(5):405–410.

Roder, M. M. 1990. *A Guide for Evaluating the Performance of Chemical Protective Clothing (CPC).* Morgantown, WV: NIOSH.

Royster, L. H. 1980. "An Evaluation of Effectiveness of Two Different Insert Types of Ear Protection in Preventing TTS in an Industrial Environment." *American Industrial Hygiene Association Journal* 41(3):161–169.

Royster, L. H., J. D. Royster, and T. F. Cecich. 1984. "An Evaluation of the Effectiveness of Three Protection Devices at an Industrial Facility with a TWA of 107 dB." *The Journal of the Acoustical Society of America* 76(2):485–497.

Royster, J. D., E. H. Berger, C. J. Merry, C. W. Nixon, J. R. Franks, A. Behar, J. G. Casali, C. Dixon-Frost, R. W. Kieper, B. T. Mozo, D. Olin, and L. H. Royster. 1996. "Development of a New Standard Laboratory Protocol for Estimating the Field Attenuation of Hearing Protection Devices. Part 1: Research of Working Group 11, Accredited Standards Committee S12, Noise." *The Journal of the Acoustical Society of America.* 99(3):1506–1526.

Sayer, James R., and Mary Lynn Mefford. 2004. "High Visibility Safety Apparel and Nighttime Conspicuity of Pedestrians in Work Zones." *Journal of Safety Research* 35(5):537.

Sheedy, J. E., and R. F. Edlich. 2004. "Ultraviolet Eye Radiation: The Problem and Solutions." *Journal of Long Term Effects of Medical Implants* 14(1):67–71.

Sokol, B. 2002. "Hearing Protection For Constructors Requires More Than Sound Blocking." *Protection Update* (4):8–9.

Stone, J. 2002. *Shirts and Stuff for Sun Safety.* Ames, IA: Iowa State University, University Extension.

Suruda, A., W. Milliken, D. Stephenson, and R. Sesek. 2003. "Fatal Injuries in the United States Involving Respirators, 1984–1995." *Applied Occupational and Environmental Hygiene* 18(4):289–292.

Suter, A. H. 2003. "History and Use of EPA's Hearing Protector Labeling Regulation." EPA Workshop on Hearing Protector Devices; March 27–28. Washington, D.C.

Takahashi-Nishimura, M., S. Tanabe, and Y. Hasebe. 1997. "Effects of Skin Surface Temperature Distribution of Thermal Manikin on Clothing Thermal Insulation." *Applied Human Science* 16(5):181–189.

Takase, I., K. Korio, A. Tamura, H. Nishio, T. Dote, and G. B. Anderson. 2004. "Fatality Due to Acute Fluoride Poisoning in the Workplace." *Legal Medicine (Tokyo)* 6(3):197–200.

Thomas, J. S., S. A. Lavender, D. M. Corcos, and G. B. Andersson. 1999. "Effect of Lifting Belts on Trunk Muscle Activation During a Suddenly Applied Load." *Human Factors* 41(4):670–676.

Toivonen, M., R. Pääkkönen, S. Savolainen, and K. Lehtomäki. 2002. "Noise Attenuation and Proper Insertion of Earplugs into Ear Canals." *Annals of Occupational Hygiene* 46(6):527–530.

Wassell, J. T., I. Gardner, D. P. Landsittel, J. J. Johston, and J. M. Johnston. 2000. "A Prospective Study of Back Belts for Prevention of Back Pain and Injury." *JAMA – The Journal of the American Medical Association* 284(21):2727–2732.

Webster, T. 2000. "Workplace Falls." *Compensation and Working Conditions* 5(1):28–38.

Weems, B., and P. Bishop. 2003. "Will Your Safety Harness Kill You?" *Occupational Health and Safety* 72(3):86–90.

Welch, L. S., K. L. Hunting, and A. Mawadeku. 2001. "Injury Surveillance in Construction: Eye Injuries." *Applied Occupational and Environmental Hygiene* 16(7):755–762.

Wesseling, C., C. Hogstedt, A. Picado, and L. Johansson. 1997. "Unintentional Fatal Paraquat Poisonings among Agricultural Workers in Costa Rica: Report of 15 Cases." *American Journal of Industrial Medicine* 32(5):433–441.

Westerdahl, J., C. Ingvar, A. Måsback, and H. Olsson. 2000. "Sunscreen Use and Malignant Melanoma." *International Journal of Cancer* 87(1):145–150.

Willey, M. S. 2001. "The Effects of Back Belts and Load on Selected Lifting Kinematics during a Simulated Patient Transfer." *Work* 17(1):31–38.

Wolf, R., et al. 2003. "Sunscreens—The Ultimate Cosmetic." *Acta Dermatovenerologica Croatica* 11(3):158–162.

Wong, T. Y., et al. 1998. "The Epidemiology of Ocular Injury in a Major US Automobile Corporation." *Eye* 12(Pt 5):870–874.

Wu, M. T. 2002. "Assessment of the Effectiveness of Respirator Usage in Coke Oven Workers." *American Industrial Hygiene Association Journal* 63(1):72–75.

Yu, T. S., H. Liu, and K. Liu. 2004. "A Case-Control Study of Eye Injuries in the Workplace in Hong Kong." *Ophthalmology* 111(1):70–74.

Zellers, E. T., H. Ke, D. Smigiel, R. Sulewski, S. J. Patrash, M. Han, and G. Z. Zhang. 1992. "Glove Permeation by Semiconductor Processing Mixtures Containing Glycol-Ether Derivatives." *American Industrial Hygiene Association Journal* 53(2):105–116.

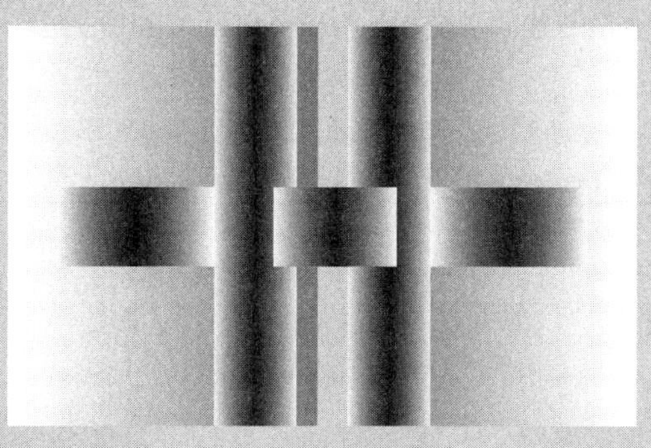

SECTION 5
PERSONAL PROTECTIVE EQUIPMENT

LEARNING OBJECTIVES

- Identify and list some of the costs associated with a personal protective equipment (PPE) program.
- Describe some of the advantages and disadvantages of using PPE versus using engineering controls in worker protection.
- Given some examples of PPE and engineering controls, identify recurring costs in each.
- List some instances in which the use of PPE is an acceptable alternative to engineering controls.

COST ANALYSIS AND BUDGETING

Kevin E. Stroup

THIS CHAPTER DISCUSSES the cost of administering a personal protective equipment (PPE) program, as well as its implications when planning a budget. It specifically compares a PPE-based approach to worker protection to the generally preferred approach of engineering controls, comparing the advantages and disadvantages of each. Understanding the advantages and disadvantages of each approach involves determining direct and indirect costs to the organization. Although engineering controls (including product substitution) is typically the most effective methods for protecting workers, PPE is appropriate and necessary in many circumstances when hazard elimination or administrative controls are not possible or are not yet in place (Krause and Weekley 2005, 34–40; Harris 2000, 2386).

To the greatest extent possible, incorporate personal protective equipment requirements into standard operating procedures. Handle the procedures for purchasing and dispensing personal protective equipment in the same way as purchasing and dispensing other supplies or production-related equipment. Failure to provide or properly use personal protective equipment increases the likelihood of injuries, illnesses, and their associated costs. Other possible consequences to the organization include poor worker morale, adverse publicity, fines, and government citations. To the greatest extent possible, integrate the purchase, availability, and use of personal protective equipment into the appropriate operation and maintenance (O&M) budgets. Give personal protective equipment the same consideration and importance as the tools, equipment, and other supplies required to do the job.

Various factors influence the decision to use personal protective equipment over other methods of control (National Safety Council 1997a, 711). These factors include regulatory requirements, type and location of work, the hazard(s) involved, and the duration of the task, operation, or project. Personal protective equipment should always be a last resort when other types of control measures (engineering, work practice, or administrative) are neither feasible nor practical. To some extent, technical and economic reasons may drive the reliance on PPE over other measures. In most cases, however, PPE augments and supplements other control measures. According to the Occupational Safety and Health Administration (OSHA), "PPE devices alone should not be relied on to provide protection against hazards, but should be used in conjunction with guards, engineering controls, and sound manufacturing practices" (OSHA 1998, 29). Good industrial hygiene practice dictates that the elimination of hazards is the highest priority.

When the option exists of using PPE instead of engineering controls, thoroughly consider the various advantages (pros) and disadvantages (cons) of this approach. A potentially valid economic reason for not choosing an engineering control measure is a situation in which a limited amount of hazardous material is involved. The limited production of the product may be important to the organization, but the procurement of costly engineering controls may not be economically feasible. This is especially true if the future production of this product is uncertain.

Figure 1 compares the various advantages and disadvantages of personal protective equipment and engineering controls. These factors can affect costs, both directly and indirectly, as well as other aspects of the business such as efficiency and quality.

Personal Protective Equipment	Engineering Controls
Pros:	**Pros:**
• Minimal capital costs • Adaptable to non-routine tasks or emergencies • Ease of use • Readily available	• Reliable and effective • Requires less supervision and enforcement • Eliminates or controls hazard at its source
Cons:	**Cons:**
• Requires proper selection, supervision and enforcement • Requires training • Requires proper fitting and use • Requires re-supply and replacement • Medical surveillance may apply • Does not eliminate or control the hazard at its source • "Last line of defense"	• Design and installation costs • Maintenance and inspection costs • May lessen productivity and/or efficiency • Space/layout restrictions • Retrofitting old equipment • May fail or be circumvented • Timing-design/installation time

FIGURE 1. Major advantages and disadvantages of PPE and engineering controls

PPE: Cost Considerations

Ideally, the decision to install engineering controls or to rely on personal protective equipment is part of the design process for new operations or facilities (Slote 1987, 744). This also applies to the redesign or renovation of existing processes or facilities. At this stage, the implementation of engineering controls is usually less expensive than retrofitting, because they are part of the planning process. Although elimination of the hazard through engineering controls is the preferred method for protecting workers, personal protective equipment often remains as a necessary part of the worker protection strategy. Among the factors that affect this decision include project cost limitations, regulatory requirements, the type of hazard, and long operational and maintenance (O&M) costs. Keep in mind that O&M costs are inherently recurrent and grow with time.

Situations can arise in project or process design in which where the reliance on personal protective equipment is a cost control measure—as when a process or activity is of limited duration or when the purchase and installation of new equipment or facilities is neither cost effective nor worth the investment. When PPE is the primary method of protection, carefully evaluate the alternatives before justifying this approach. Again, it is always prudent to install engineering controls or design the hazard out of the job if it is technically and economically feasible. In many cases, the need for personal protective equipment will be extended to other tasks. Activities such as hazardous material spill clean up, equipment cleaning, or confined space entry may be necessary. Hazard evaluation of a project in the design phase is the best time to review all aspects of a process or facility and to consider all available options for routine, nonroutine, and emergency tasks.

PPE in Existing Operations or Activities

The decision-making process for the implementation of personal protective equipment or engineering measures intended to control hazards for existing operations or activities is usually the result of a review triggered by some event or program requirement. Internal reviews, such as a job safety analysis (JSA), hazard and risk assessments, accident/incident investigations, and programmed audits or inspections are common examples of management activities that play an important role in assessing the effectiveness of the safety program (National Safety Council 1997b, 387).

In existing operations, cost may be an important factor when deciding to use personal protective equipment instead of engineering controls, because retrofitting and renovation may be costly. Costs for engineering controls are typically included in the project budget for new designs, whereas engineering retrofits and renovations would likely require additional funding for design and installation. Indirect costs are incurred when installing engineering controls, because operations and production can be slowed or even stopped while retrofitting or renovation takes place. To avoid the costs associated with renovation and engineering controls, management may choose to rely on personal protective equipment if it assures worker protection. The costs for maintaining and administering the personal protective equipment program may justify the expense of engineering controls methods over time.

The key cost considerations in a PPE program are:

- the initial cost in purchase
- PPE training (preparation, delivery, and employee wages during training)
- replacement, maintenance, and repair
- administration (recordkeeping, policies, procedures)
- medical clearance and surveillance

Brevity of the process or operation does not justify the capital investment into the purchase or modification of equipment or facilities

- short-term manufacturing process for small quantity of product

Activities that are dynamic, mobile, and, by the nature of the work, not performed in a single location within a building or structure

- construction work
- jobs performed outdoors
- landscaping work

Nonroutine operations where hazards may be inconstant or conditions changeable

- confined space entry
- plant/facility maintenance activities

Hazard elimination is neither possible, practical, nor complete

- live electrical work
- emergency response

When required by regulatory requirements

- entering IDLH environments
- asbestos removal
- aerial platforms

FIGURE 2. Situations in which PPE is cost effective and the most practical alternative

In certain situations, the use of personal protective equipment may be more cost-effective than engineering controls (see Figure 2).

Engineering Controls Versus PPE

A simple cost comparison can help indicate whether engineering controls or personal protective equipment makes greater financial sense. As previously stated, eliminating the hazard with engineering controls is always the preferred method. There are circumstances when the use of personal protective equipment can be an appropriate and economical alternative to engineering controls. For example, work in a specific area or location may be brief, making the installation or procurement of engineering controls too expensive. Some production operations may involve a limited product run for testing purposes, making investment in costly engineering controls or equipment modifications uneconomical. A cost comparison between engineering controls and the use of personal protective

Direct Costs of PPE	Direct Costs of Engineering Controls
• Purchasing PPE • Training • Replacement of worn PPE or PPE accessories • Cleaning PPE • Administrative oversight • IH support • Medical surveillance	• Engineering design • Software and programming • Purchasing equipment • Installation/modification • Maintenance • IH support

FIGURE 3. Direct costs associated with personal protective equipment and engineering controls

TABLE 1

Engineering Control Cost Calculator

Cost Category	Cost Type	Description/Calculation	Cost ($)
Materials	Direct	Hardware, parts, construction materials, safety devices, sensors, alarms, etc.	
Installation	Direct	(Hours) × (employee $/hour)	
Software	Direct	Base product and upgrades	
IT support	Direct	(Hours) × (technician $/hour)	
Preventive Maintenance–Labor	Direct	(Hours) × (employee $/hour)	
Preventive Maintenance–Parts & Supplies	Direct	Replacement drive belts, switches, sensors, etc.	
Training	Indirect	(Hours) × (employee $/hour)	
Lost manufacturing	Indirect	Lost profit	
Cost of labor while not in production	Indirect	(Hours) × (employee $/hour)	
		Total Cost for Engineering Control: $	

equipment may be appropriate in these situations. The previous table lists several situations where personal protective equipment would be more cost-effective than engineering controls.

Figure 3 compares the various direct costs associated with both options. The direct costs are the costs incurred from purchases and payments for equipment, services, or other requirements directly related to the control measure. Direct costs are "identifiable monetary costs (or costs that have monetary equivalence) involved in providing a specific service" (Biancardi 1987, 619–642).

In the case of PPE, *identifiable costs* are the costs associated with the purchase of equipment as well as other required purchases and functions that are part of the of the PPE program. The same definition holds true for engineering controls. Figure 3 lists the major direct costs for personal protective equipment and engineering controls.

In addition to the direct costs attributed to personal protective equipment and engineering controls, there are indirect costs too. According to one author, indirect costs "are any costs that cannot be directly linked with the program" (Biancardi 1987, 619–642). Indirect costs include downtime of production or processing equipment for renovation or upgrade, time spent by administrators or safety and health personnel to revise policies and procedures, and time away from production to train or outfit employees. Typically, this cost consists of employees' hours away from production multiplied by their hourly pay rates—as well as all profit not realized because production was interrupted. Determining costs usually requires estimating the various losses attributed to implementation. The total cost for engineering controls is outlined in Table 1.

A personal protective equipment program also incurs both direct and indirect costs. Accurate accounting

TABLE 2

PPE Program Cost Calculator

Program Element	Cost Type	Description/Calculation	Cost ($)
Program implementation	Indirect	(Hours) × (EHS member $/hour)	
Cost for training employees	Indirect	(Hours) × (employee $/hour)	
Trainer costs	Direct	(Hours) × (trainer $/hour)	
Overtime costs (if applicable)	Indirect	(OT Hours) × (employee OT $/hour)	
Training materials	Direct	• Rental/purchase • Printing • In-house production, etc.	
Miscellaneous costs	Direct	Storage, overhead costs, etc.	
PPE	Direct	• Purchase price • Replenishment • Replacements • Disposables, sundries	
Annual PPE maintenance	Direct	• Inspection • Cleaning • Service/repair	
Medical	Direct	• Physicals for respirators • Audiometric testing	
Industrial hygiene	Direct	Program costs, laboratory services	
Hazard communication	Indirect	Signs and labels	
		Total Cost for PPE Program: $	

of personal protective equipment purchases, consultant and service provider invoices, and employee time spent working on program-related activities reveals the total cost (see Table 2).

Consider long-term costs when determining the most cost-effective method of protecting workers from workplace hazards. Although implementation costs for personal protective equipment may be less in the early stages of an activity or operation, prolonged costs may make initially expensive engineering control measures more attractive. The following hypothetical scenario illustrates this point.

Engineering Controls Versus PPE: A Hypothetical Situation

Scenario

The weighing and dispensing of hazardous raw material ingredients in a batch production operation may put a group of workers at risk. These workers must open containers, scoop or pour a dry powder material into secondary containers, and provide the required weights and volumes for the production process. Although this activity is in a designated area—industrial hygiene monitoring reveals a significant amount of airborne dust. It poses a potential health hazard to the workers. Table 3 compares the costs over time for local exhaust ventilation and PPE.

Table 4 takes the same scenario described above and looks at the cost of a PPE-based approach.

TABLE 3

Capital, Operation, and Maintenance Costs of a Local Exhaust Ventilation System (values are hypothetical)

Ventilation System Costs ($)	Years of Operation						6-Yr Total ($)
	1	2	3	4	5	6	
Installation	10,000						10,000
Annual certification	500	500	500	500	500	500	3000
Annual maintenance	1000	1000	1000	1000	1000	1000	6000
Cost of energy	800	800	800	800	800	800	4800
Annual cost	12,300	2300	2300	2300	2300	2300	23,800

Local Exhaust Ventilation Versus Respirators

A key element of industrial hygiene is the prevention of employee exposure to hazardous airborne contaminants (Plog and Quinlan 2002, 586). Employers must control exposures to harmful dusts, vapors, gases, and fumes through substitution, engineering controls, or personal protective equipment. In industrial, production, or laboratory facilities, local exhaust systems may be the most practical and cost-effective method of controlling exposure. Laboratory hoods, canopy hoods, and local exhaust trunks are the most commonly used devices. There are situations where it is not practical or feasible to use ventilation systems to remove the contaminants from the work area, as in construction and some maintenance and emergency response situations. In these situations, some form of respiratory protection is required.

Figures 4 and 5 present key costs associated with respiratory protection and ventilation systems.

TABLE 4

PPE Cost Over Time (values are hypothetical)

Respiratory Protection Costs ($)	Years of Operation						6-Yr Total ($)
	1	2	3	4	5	6	
Initial purchase	250	250	250	250	250	250	1500
Training	6000	6000	6000	6000	6000	6000	36,000
Medical surveillance	5000	5000	5000	5000	5000	5000	30,000
Fit testing	600	600	600	600	600	600	3600
Replacements, accessories & cartridges	150	150	150	150	150	150	900
Annual cost	12,000	12,000	12,000	12,000	12,000	12,000	72,000

Purchase respirators
Required medical evaluation and clearance
Evaluation of atmospheric conditions
Fit testing
Training
Replacement respirators, accessories, and filter cartridges

FIGURE 4. Cost elements for respiratory protection

Design and engineering costs
Installation of general exhaust ventilation system(s)—capital costs
Installation of local exhaust ventilation system(s)—capital costs
Inspection, maintenance, and calibration of ventilation systems—O&M costs
Cost of energy costs for heating or cooling

FIGURE 5. Costs associated with engineering controls for airborne contaminants

Some activities require both local exhaust ventilation and respiratory protection. For example, certain spray painting operations where workers are required to move about the spray paint booth would require both types of protection. Permit-required confined space entry is another instance in which portable ventilation units and respiratory protection may go hand in hand.

Engineered Fall Prevention Versus Personal Fall Arrest Systems

Effective fall protection is possible using engineering methods or personal fall (fall arrest) protection systems (Ellis 2001, 514). The type of roof, activity, and work surface usually dictates both the safest and most cost-effective approach. In some cases, need for access is infrequent. In other cases, need for access is routine or constant.

Engineered solutions to fall protection include fixed platforms, scaffolds, guardrails or other fixed barriers. (Note that warning lines are not an engineering control.) Engineering controls for fall prevention also include relocating equipment or the process to ground level, the erection of portable scaffolds, permanent platforms or mezzanines with guardrails and toe boards, aerial lifts, safety nets, skylight covers, and fixed ladders. The primary benefits of an engineered solution is that it may eliminate the need for personal fall protection altogether and can be designed to meet regulatory requirements. As discussed earlier, planning for these systems during the initial design phase of a new project or during a renovation is more cost-effective than an "after-the-fact" installation.

Costs to consider when deciding whether to use personal fall protection include purchase of equipment, hardware, and anchor points, as well as replacement of worn or damaged components and training. Other costs may include storage, cleaning, and inspection. Also consider the need for increased supervision when using personal fall arrest systems—which have both direct and indirect cost implications. Under any circumstances, personal fall protection is necessary when performing construction or infrequent tasks and unexpected repairs at elevation.

Sometimes personal fall protection requires pre-engineering in the form of roof anchors or other points of attachment. It is not always possible to work at elevated heights safely by simply tying off to the nearest structural member or object. Figure 6 presents examples of situations where personal fall protection and engineering methods work together.

Noise Solutions

When practical, preventing worker exposure to noise (at or above 80–85 decibels) involves the use of engineering controls to limit vibration or sound generation at its source. Other engineering methods include enclosing noisy equipment or applying acoustical treatment to wall and ceiling surfaces that causes them to

Situation	Engineering Fall Protection Control	Personal Fall Protection Components
Working within 6 feet of roof edge	Fixed anchor points	Full body harness and lanyard
Climbing fixed ladders to high elevations	Fixed rail climbing system	Full body harness with a climbing-side D-ring
Working on utility lines or lighting	Aerial lifts or mobile work platforms	Full body harness and lanyard
Working on the topside of an aircraft	Horizontal lifeline with self-retracting lanyards	Full body harness

FIGURE 6. Personal fall arrest and engineering controls

absorb sound and reduce reverberation. Although these measures are preferred methods for permanent locations such as equipment machine rooms, production areas, and maintenance shops, the use of personal hearing protection is the only viable alternative in many other situations.

A cost analysis of the most economical method for preventing high noise levels must take into account several factors associated with the work environment. Is the work environment a permanent location? Does the nature of the job require workers to be at various locations for relatively short periods, as in cases of maintenance or construction? Does the sound come from existing machinery, or is it part of a process, or the result of equipment, in the design stage? Can personal protective equipment alone protect employees from overexposure to noise? How often are workers exposed to high noise levels? There are cost implications when comparing an engineered solution to personal hearing protection.

Eliminating noise through engineering controls is often the most cost-effective method if it occurs during the design, procurement, or installation phase of a process. Acoustical treatment of ceilings, walls, and floors, placement of noisy or vibrating machinery on vibration-isolation mountings, and installation of acoustic enclosures around noisy machinery are examples of effective noise-reduction methods. Purchasing inherently quiet equipment can reduce or eliminate the need for engineering controls or hearing protection.

Types of costs associated with the implementation and administration of a hearing protection program relying heavily on the use of hearing protection devices (earplugs or earmuffs) and the use of engineering controls are listed in Figure 7. These items represent direct costs resulting from the use of personal protective equipment to protect workers.

In addition to the direct costs of implementing a noise control strategy, excessive noise can adversely affect workers in other ways. Noise can be a contributing factor to other problems such as miscommunication and inability to concentrate. This can contribute to accidents or human error that detract from quality and productivity. Noise can lead to other social and physiological problems as well. People who work in noisy environments may more frequently fall prey to physical ailments such as high blood pressure, heart disease, ulcers, and circulatory problems (Bauer 1990, 337). Noise may also raise frustration levels in people whose difficulties in communicating increase irritability, anger, fatigue, and incidence of headaches. Aside from the obvious effects of excessive noise—namely, hearing loss—other adverse effects may be more difficult to quantify.

Engineering Considerations in Noise Abatement	Personal Hearing Protection
• Design and performance specification	• Training
• Space redesign	• Ear plugs/muffs
• Capital costs	• Audiograms/medical surveillance
• Equipment installation	• Recordkeeping
• Equipment maintenance	• Sound level surveys
• Noise-canceling enclosures	• Personal dosimeters
• Acoustic treatment of floors, ceilings, and walls	• Signs/other forms of warning
	• Administrative costs

FIGURE 7. Direct costs of engineering considerations in noise abatement and of personal protective equipment for hearing protection

Budgeting for PPE

Like any supply, managers and supervisors must plan for current, recurrent, and future PPE costs in their budgets, both project and annual. Forecasting future costs of administering a PPE program may simply require extrapolation of PPE costs from prior years' while estimating future costs of new equipment and services. Firsthand knowledge of the specific PPE requirements is an important factor in determining a realistic budget. Changes in process and employee staffing may also be a factor in organizational planning. The loss or breakage—and subsequent replacement—of PPE all add unexpected costs to any budget. Departments with specific responsibilities in the PPE program also need to consider current and future costs involved in meeting their responsibilities.

Strategically, PPE is a necessity for the protection of workers and for compliance with applicable safety and health regulations. In a tactical sense, PPE is a risk management tool because it may lessen, or even prevent, the adverse consequences of an accident, injury, or illness. For these reasons, the employer may fully fund some items, but others may be supplemented by a certain percentage up to a given dollar amount. Items subsidized by many employers include prescription safety glasses or safety footwear. Often worn outside of work, these items afford the employer the benefit of their protection off and on the job. Many employers have contractual arrangements with local vendors for purchase of prescription eyewear and footwear. Large or geographically remote employers (including many factories, refineries, and chemical plants) bring PPE vendors on site.

Conclusion

Analyzing the costs of personal protective equipment helps in making informed decisions about control measure implementation and resource allocation. The information derived from analyzing the costs associated with personal protective equipment is useful in comparing a PPE-only approach to the costs of engineering controls. This analysis method should consider both short- and long-term direct and indirect costs associated with personal protective equipment or engineering controls. Design-stage hazard evaluation reviews and safety reviews of existing operations are excellent opportunities to conduct cost analyses of proposed or existing controls. Safety practitioners are encouraged to calculate the costs associated with worker protection and to advocate for adequate resources in capital and O&M budgets. This investment yields both tangible and intangible benefits to employers and to employees.

References

Bauer, Roger. 1990. *Safety and Health for Engineers*. New York: Van Norstrand Reinhold.

Biancardi, Michael. 1987. "How to Consider Cost-Benefit Analysis in Occupational Safety Practice." In *Handbook of Occupational Safety and Health*, L. Slote, ed. New York: John Wiley & Sons.

Ellis, Nigel. 2001. *Introduction to Fall Protection*. 3rd ed. Des Plaines, IL: American Society of Safety Engineers.

Harris, Robert L., ed. 2000. *Patty's Industrial Hygiene*, 5th ed. 4 vols. New York: John Wiley & Sons.

Krause, Thomas, and Thomas Weekley. 2005. "Safety Leadership." *Professional Safety* 50(11):34–40.

National Safety Council (NSC). 1997a. *Accident Prevention Manual for Business and Industry: Administration & Programs*. 11th ed. 3 vols. Gary Krieger and John Montgomery, eds.Itasca, IL: NSC.

______. 1997b. *Supervisors' Safety Manual*. 9th ed. Patricia Laing, ed. Itasca, IL: NSC.

Occupational Safety and Health Administration (OSHA). 1998. *Personal Protective Equipment* (OSHA 3077). Washington, D.C.: U.S. Department of Labor.

Plog, Barbara, and Patricia Quinlan. 2002. *Fundamentals of Industrial Hygiene*. 5th ed. Itasca, IL: National Safety Council.

Slote, Lawrence, ed. 1987. *Handbook of Occupational Safety and Health*. New York: John Wiley & Sons.

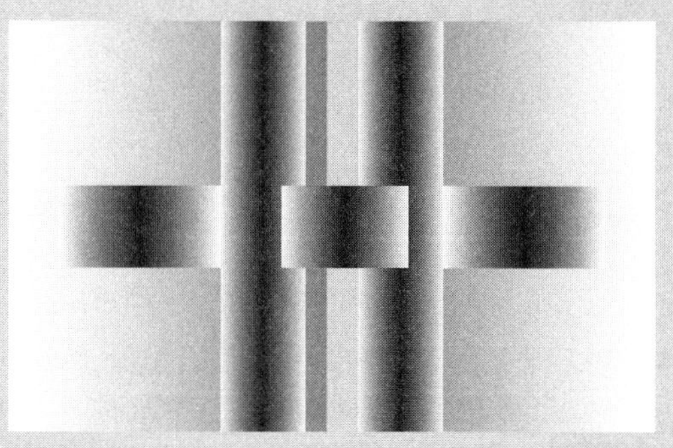

SECTION 5
PERSONAL PROTECTIVE EQUIPMENT

Benchmarking and Performance Criteria

Kevin E. Stroup

LEARNING OBJECTIVES

- List, then define, the basic concepts and techniques used in measuring the effectiveness of a PPE program.
- Describe some of the common benchmarks in the evaluation of a PPE program.
- Explain some of the common assessment methods used to test the effectiveness of a PPE program.

THIS CHAPTER DESCRIBES the process of evaluating a personal protective equipment (PPE) program using qualitative and quantitative criteria. Without periodic evaluation, even the best health and safety programs fade over time—becoming outdated, outmoded, and perhaps out of compliance. Other authors have discussed, in some depth, the various approaches to program evaluation and measurement in health and safety (Biancardi 1987, 619–642; Janicak 2003, 208; Weinstein 1997, 211). For a busy health and safety practitioner, simply finding the time to conduct a program evaluation is a challenging—but essential—responsibility. This chapter covers the basics of how to evaluate a PPE program to help keep it current and effective.

Across management disciplines, there are two important terms in the evaluation process—*benchmarks* (standards, objectives, and intended performance) and *appraisals* (measurement methods or "tools"). Benchmarks are the standards, objectives, and performance requirements used to measure program effectiveness (Porter and Tanner 2004, 468). As expected, compliance with regulatory requirements (specifically, OSHA's PPE Standard 29 CFR 1910.132) and accident, injury, and illness prevention are the most likely benchmarks of interest. Appraisals, in simple terms, are program measurements. Appraisals can be either formal or informal. Appraisals are the familiar audits, inspections, and managerial reviews that well-run organizations use to improve practice. Appraisals can include direct observation of employee and supervisor behavior. Taken together, appraisals compare practices to benchmarks (standards, objectives, and intended performance) to determine whether the program is meeting its stated goals and purpose.

Evaluation of a PPE program must include review of both the technical and performance aspects of the selection and the use of the PPE. The technical elements include an evaluation of the types and specifications of PPE used (Camplina 2003, 48–51). Essentially, a technical review encompasses the *protectiveness* of the PPE used in comparison to the hazards present and the risk (chance of occurrence) within in the context of applicable regulatory or consensus standards. Evaluation of the performance elements of the PPE program involves the roles and responsibilities of management along with the behavior of workers and first-line supervisors. Evaluating the technical and performance elements together provides a measure of overall program effectiveness, thereby allowing the organization to rely less on "after-the-fact" injury or cost experience. Increasingly, safety and health practitioners are interested in so-called *leading* measures of evaluation that include the safety *climate* and safety *culture* of the organization (Flin et al. 2000, 177–192; Reason 1998, 293–306).

Effective appraisals include a review of management practices, facilities, working conditions, staffing, and employee practices using established benchmarks. The intent of the appraisal process is twofold: first, to identify deficiencies or opportunities for improvement—essentially making recommendations on how to improve or eliminate deficiencies—and second, to identify positive contributions—those elements or practices that make the program successful. Periodic program evaluation is a *best practice* in any health and safety management system consistent with ANSI/AIHA Z-10-2005 (ANSI/AIHA 2005, 60).

In the development of safety policies and procedures, organizations must codify, to the extent possible, the roles and responsibilities for conducting audits, reviews, and inspections related to PPE. Figure 1 illustrates how various PPE-related activities are assigned to different levels in a hypothetical organization.

Selecting Appropriate PPE

OSHA's PPE standard (29 CFR 1910.132) requires a hazard and risk assessment as the first step in (1) determining the need for PPE and (2) selecting the proper PPE. The selection of PPE can range from simple and rather obvious to complex and nuanced, depending on the hazards present (OSHA 1998, 29). For obvious reasons, only certified PPE is acceptable. Data supplied by a reputable manufacturer is usually the first and best source of technical information on the applications and limits of a particular piece of PPE. In both the United States and Canada, the Safety Equipment Institute (SEI) serves as a definitive source of information on the standards used in PPE certification (see Figure 2).

Changing work conditions—which inevitably will occur—can affect the classification of hazards and the risks they represent. The more dynamic or complex the workplace, the more thoughtful and complete the technical review must be in order to protect employees. For this reason, some types of workplaces have adopted "you must look like this to enter" policies to account for the expected and perhaps unexpected hazards of a job or task (Dow 1995–2007). In these industries (chemical and energy industries, heavy industries,

Evaluation Activity	Who Is Responsible	Appraisal Done by
PPE hazard assessment	Supervisors/EHS	Management/third parties
PPE availability	Supervisors/EHS	Management/EHS
PPE use	End user–worker	Supervisors/EHS
PPE training	Supervisors/EHS	Management/EHS
PPE care and cleaning	End user–worker	Supervisors/EHS
PPE enforcement	Supervisors/management	Management/EHS
Program evaluation	EHS/management	Management/third parties
Program improvement	Supervisors/management	EHS/third parties

FIGURE 1. Suggested roles and responsibilities in the evaluation of a PPE program (Stroup/Blayney)

American National Standards Institute, Current Edition	
Protective Footwear	ANSI Z41-1999
Eye and Face Protection	ANSI Z87.1-2003
Protective Headwear for Industrial Workers	ANSI Z89.1-2003
Emergency Eyewash and Shower Equipment	ANSI Z358.1-1998
Safety Requirements for Personal Fall Arrest Systems, Subsystems & Components	ANSI 359.1-1992
American National Standards Institute/International Safety Eqiupment Association	
Disposable Coveralls	ANSI/ISEA 101-1996
High Visibility Safety Apparel	ANSI/ISEA 107-1999
National Fire Protection Association	
Powered Rescue Tools	NFPA 1936, 1999 Edition
Protective Ensemble for USAR Operations	NFPA 1951, 2001 Edition
Protective Ensemble for Structural Fire Fighting	NFPA 1971, 2000 Edition
Station\Work Uniforms for Fire Fighters	NFPA 1975, 1999 Edition
Protective Clothing for Proximity Fire Fighting	NFPA 1976, 2000 Edition
Protective Clothing and Equipment for Wildland Fire Fighting	NFPA 1977, 1998 Edition
Open-Circuit Self-Contained Breathing Apparatus for the Fire Service (SCBA)	NFPA 1981, 2002 Edition
Personal Alert Safety Systems (PASS) for Fire Fighters	NFPA 1982, 1998 Edition
Fire Service Life Safety Rope, Harness, and Hardware	NFPA 1983, 2001 Edition
Vapor-Protective Ensembles for Hazardous Chemical Emergencies	NFPA 1991, 2000 Edition
Liquid Splash-Protective Ensembles for Hazardous Chemical Emergencies	NFPA 1992, 2000 Edition
Protective Ensembles for Chemical or Biological Incidents	NFPA 1994, 2003 Edition
Protective Clothing For Emergency Medical Operations	NFPA 1999, 2003 Edition
Flame-Resistant Garments for Protection of Industrial Personnel Against Flash Fire	NFPA 2112, 2001 Edition
Canadian Standards Association	
Industrial Eye and Face Protectors	CAN Z94.3-02
Industrial Protective Headwear	CAN/CSA Z94.1-92
Protective Footwear	CAN/CSA Z195-1992
Full Body Harness	CAN/CSA Z259.10-M90
Shock Absorbers for Personal Fall Arrest Systems	CAN/CSA Z259.11-M92

FIGURE 2. Standards used in certification (adapted from the Safety Equipment Institute, 2007)

utilities, etc.), health and safety is simply the proverbial "cost of doing business" (Flin et al. 2000, 177–192; Flin and Yule 2004, ii45–ii51). Not surprisingly, companies in these industries have plenty of awareness or experience with potential mishaps (even if they rarely occur) or perhaps face a higher degree of regulatory scrutiny. The term *high reliability* nicely describes those industries where the cost of something going wrong is disproportionate to the cost of prevention.

The precautionary pendulum can swing in the other direction as well. Sometimes improvement in process design or operation—perhaps through new equipment (such as a process enclosure) and engineering controls (robotics, sophisticated ventilation) may eliminate the need for some PPE (Christensen and Manuele 1999, 279).

Performance Measures

Measuring performance in a PPE program includes direct observation of PPE use by workers and supervisors, audits, checklists, and examination of management effectiveness in the overall success of the program (Hess 1997, 573). Figure 3 provides some examples of areas in a PPE program that can be measured.

PPE: Other Sources of Useful Information in Evaluation

Supply and procurement system information is particularly useful in measuring performance against program requirements—for example, a review of chemical purchasing information against the selection and use

Requirement	Measured Behavior or Activity	Who is Responsible
PPE policy	PPE requirements incorporated into standard operating procedures	Management/EHS
PPE training	Attend training–training records Training documentation	Employees and supervisors EHS
Fall arrest systems	Inspection of harness and lanyard prior to usage Enforcement of PPE use	Employees/supervisors Supervisors/EHS
Eye and face protection	Distribution and supply –necessary types and preferred styles Frequency of use–documented observation	Supervisors/purchasing/EHS
Use of gloves	Wearing the proper glove for the hazards present Proper glove for the hazards	Supervisors/EHS Employees/supervisors
Storage of PPE	Proper storage	EHS through OSHA's "written certification" Employees/supervisors

FIGURE 3. Examples of measurable behaviors or activities in PPE program evaluation (Stroup/Blayney)

of chemical protective clothing or respiratory protection choices (this is a leading or predictive activity). A lagging or retrospective activity, such as reviewing PPE purchase, use, and reorder information, can provide helpful summary information, identify trends, and provide a basis for future decisions.

A PPE program evaluation is not complete unless it specifically addresses potential changes in the workplace since the last review or program inception and takes a close look at whether management helps or hinders the use, availability, and support for the program. Finally, sensitivity to employee preferences in fit, style, and comfort is important. This includes asking employees what they prefer and comparing one brand with another when equal choices exist. Asking employees and supervisors to suggest new products they come across and working with them to determine the suitability, cost, and usefulness for their needs builds credibility and helps to ensure the long-term success of the program.

Conclusion

The evaluation of a PPE program requires meaningful benchmarks and some useful methods of appraisal. Internal standards and policies, governmental regulations or consensus standards, and performance-based criteria related to PPE program activities are all good sources for developing benchmarks. Organizations should integrate appraisal into all levels of management in the form of periodic audits, inspections, and observational measures. All levels of management should be involved in the appraisal process of the PPE program—from setting expectations to modeling expected behavior to PPE selection, ordering, and distribution. To be useful, the appraisal process must provide insight into how effective the PPE program is in protecting individuals and complying with OSHA's PPE standard. With this information, organizations can highlight strengths and seek to improve areas of weakness in their program. While finding the time may be a challenge, program evaluation is an investment in the viability and effectiveness of any health and safety program.

References

American National Standards Institute/American Industrial Hygiene Association (ANSI/AIHA). 2005. *Occupational Health and Safety Management Systems.* Fairfax, VA: AIHA.

Biancardi, Michael. 1987. "How to Consider Cost-Benefit Analysis in Occupational Safety Practice." In *Handbook of Occupational Safety and Health*, edited by L. Slote. New York: John Wiley & Sons.

Camplina, Jeffery C. 2003. "Assessing the Need for PPE: Hazards Present Determine Level of Protection." *Professional Safety* 48(12):48–51.

Christensen, Wayne, and Fred Manuele. 1999. *Safety Through Design: Best Practices*. Des Plaines, IL: National Safety Council.

Dow, Chemical. 2007. *Dow Caustic Soda Solution 1995–2007* [cited 2007]. www.dow.com/causticsoda/safety/equip.htm

Flin, R., K. Mearns, P. O'Connor, and R. Bryden. 2000. "Measuring Safety Climate: Identifying the Common Features." *Safety Science* 34:177–192.

Flin, R., and S. Yule. 2004. "Leadership for Safety: Industrial Experience." *Quality & Safety in Health Care* 13 (Supplement II), pp. ii45–ii51.

Hess, Kathleen. 1997. *EH&S Auditing Made Easy: A Checklist Approach for Industry*. Rockville, MD: Government Institutes, Inc.

Janicak, Christopher. 2003. *Safety Metrics: Tools and Techniques for Measuring Safety Performance*. Rockville, MD: Government Institutes, Inc.

Occupational Safety and Health Administration (OSHA). 1998. *Personal Protective Equipment* (OSHA 3077). Washington, D.C.: U.S. Department of Labor.

Porter, L. J., and S. J. Tanner. 2004. *Assessing Business Excellence: A Guide to Business Excellence and Self-Assessment*. 2d ed. Oxford: Elsevier.

Reason, James. 1998. "Achieving a Safe Culture: Theory and Practice." *Work & Stress* 12(3):293–306.

Weinstein, Michael. 1997. *Total Quality Safety Management and Auditing*. Boca Raton: Lewis.

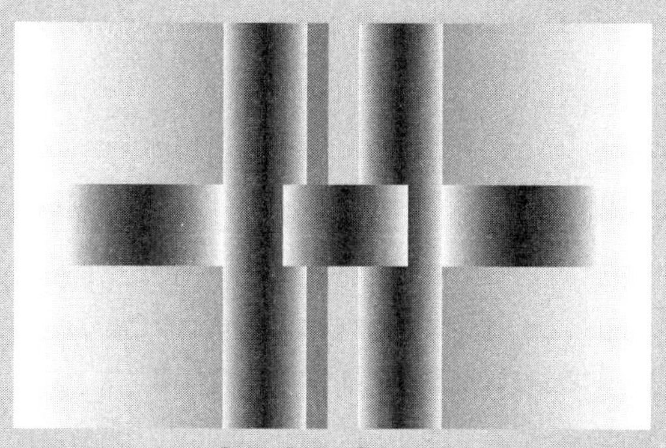

SECTION 5

PERSONAL PROTECTIVE EQUIPMENT

Best Practices

Michael B. Blayney

LEARNING OBJECTIVES

- List the four key elements of an effective PPE program. Briefly describe why each element is important in ensuring the success of the program.
- Describe how PPE fits into an overall health and safety strategy.
- Differentiate the role of training and education in a PPE program.
- Explain why evaluation is essential in the success of a PPE program.

THE DEVELOPMENT OF personal protective equipment (PPE) inevitably traces its origins to our need for warmth and environmental protection. With this basic assumption, what we think of as PPE today developed in response to the uniquely human traits of economics and technology. Warfare has also played an important role in the development of PPE (Oakeshott 1994, 358). As new technologies develop, new PPE will help reduce the potential for harm. To serve its intended purpose, PPE must match the potential hazard(s) and be effectively administered to ensure consistent use (CCOHS 2005). The purpose of this chapter is to describe best practices in the establishment and administration of a PPE program. This chapter also emphasizes the important role PPE has in numerous health and safety standards.

The history of medicine contains evidence of the need for PPE and its evolution. The 1713 edition of *De Morbis Artficum* by Bernardino Ramazzini described 54 occupations at risk of occupational illness (Ramazzini 1964, 549). Alice Hamilton, a physician in the early twentieth century and often considered the founder of modern occupational medicine, described how a worker employed a pie tin to protect his face when opening a metal furnace (Hamilton 1943, 433). In 1955, Donald Hunter reduced his clinical knowledge of work-related disease into an interdisciplinary book on occupational medicine (Baxter and Adams 2000, 1001). As the relationship between occupational injury and disease has become clearer and the methods and means to reduce harm more sophisticated, what we think of PPE today—hard hats, gloves, safety glasses, specialty fabrics—has become an otherwise integral aspect of on-the-job protection.

The classic hierarchy of methods and means in worker protection emphasizes that PPE is typically the last line of defense and is not an acceptable alternative when engineering controls are possible (Harris 2000, 2386; Herrick 1998, 30; Krieger and Montgomery 1997, 711). Depending on the industry, there are many occasions when PPE is the primary or even sole means of worker protection (Carpenter et al. 2002, 236–247; Schenker, Orenstein, and Samuels 2002, 455–464). A comprehensive review of prevention strategies in industrial hygiene found PPE discussed in one-third of nearly one hundred peer reviewed articles (Roelofs et al. 2003, 62–67).

While PPE may be among the first things a worker dons before beginning work, it may represent the last or only line of defense from exposure or harm (Jagt et al. 2004, 355–362; Sutter 2002, 768–789). Knowing the advantages and limitations of PPE in a given situation is just as important as having the appropriate items available (Melamed et al. 1996, 209–215; Kamal et al. 1988, 707–716). Training and education therefore play an essential role in an effective PPE program.

Today, the Occupational Safety and Health Administration (OSHA), along with similar safety and health agencies in other countries, requires PPE programs to be established or be part of an overall safety programs(Taylor 1990, 247–253). Consensus certification standards exist in both the United States and European Union to ensure that PPE meets both performance and specification requirements. In the United States, the American National Standards Institute (ANSI) in cooperation with the International Safety Equipment Association (ISEA) develops and publishes a variety consensus standards specific to PPE. The well-known ANSI certification in the United States tells us that the device is acceptable for its intended purpose. The Canadian Standards Association (CSA) has similar certification standards. In the European Union, the underlying intent of the Directive for Personal Protective Equipment (89/686/ECC) under the European Committee for Standardization has the same intent as ANSI—a harmonized approach to equipment certification. Well known to safety practitioners in Europe and around the world is the CE certification.

Elements of an Effective Personal Protective Equipment Program

Over the last one hundred years, safety and health programs in the United States developed for both practical and ethical reasons (Aldrich 1997, 415). Some of the first organized efforts to foster safety in the culture of American industries more than eighty years ago focused on the important role PPE played in protecting the individual worker. As the cost of worker injuries and illnesses became a greater issue for society, so did the need for prevention and mitigation (Heinrich 1941, 448; Anonymous 1926, 112). Organized labor, progressive corporations and relatively new government agencies shape what we know as the safety and health profession today (Robinson 1991, 223; Wilson 1985, 179). Today, employers and employees who participate in OSHA's Voluntary Protection Program include PPE as part of an overall safety and health management system and continually reassess the role of PPE in their operations through audit and review processes (OSHA 2004). The ISEA provides a range of useful "best practice" materials as part of their Partnership for Worker Protection Series (ISEA 2002, 2).

A PPE program is an essential part of the social contract safety and health plays in the employer/employee relationship (Swartz 2000, 450). Employers and employees share in all aspects of the PPE program process from hazard identification and classification to PPE selection and requirements for use (Lusk, Kerr, and Kauffman 1998, 466–470). Under OSHA's PPE standard (29 CFR 1910.133–138), employers are required to conduct a workplace hazard assessment, document the need for various types of PPE (via a "written certification") and ensure its proper use by employees.

An effective PPE program has at least three interrelated elements or steps that include: (1) hazard identification/risk inventory, (2) prescribing/issuance, and (3) training and education. A fourth element in an effective PPE program is evaluation. Evaluation can be part of the three interrelated elements or be a separate, distinct element covering all three. In all cases, evaluation helps to keep a health and safety program current and effective. Since human behavior predicates the use of protective equipment, PPE programs require periodic evaluation. Depending on the nature of the workplace and culture of the organization, evaluating PPE compliance could range from informal observation to formal written procedures. Regardless of what form it takes, some method of ensuring the proper use of PPE is essential.

The essential elements of an effective PPE program are

- *Hazard evaluation/risk inventory:* Identifying hazards and ranking them according to risk (frequency and severity).
- *Prescribing and issuance:* Selecting the appropriate items and determining distribution, maintenance, and replacement.
- *Training and education:* Incorporating PPE into overall training and educational efforts.
- *Evaluation and improvement:* Periodically checking to ensure that theprogram is meeting its intended purpose

Hazard Identification/Risk Inventory

The foundation of an effective PPE program is the identification of hazards in the workplace and an inventory of the risks (chances of occurrence) they pose. As mentioned earlier, appropriate PPE is required when engineering controls are not applicable, available, affordable, or sufficient for the hazards found in the workplace. Hazard identification has two distinct but closely related goals: to identify the obvious (or expected) and to identify the not-so-obvious (or unexpected) hazards. The effectiveness of the hazard identification/risk inventory process, in turn, rests with finding an appropriate balance between the risks posed by the hazards identified with the practical implications PPE imposes on the worker. For example, assigning a vapor-tight suit to a worker in a hazardous material clean-up operation may ensure unequivocal chemical protection, but if the hazards present do not require it, then you are merely substituting one set of hazards for others such as heat stress, immobility, and inefficiency. While under-protection rather than over-protection is more likely to occur, this hypothetical example does underscore some of the reasons that workers fail to wear PPE when needed. PPE can be uncomfortable to wear, bulky, awkward, or perhaps just unattractive (Akbar-Khanzadeh and Bisesi 1995, 195–198; Marr and Quine 1993, 73–77). Employers and employees are necessary partners in the hazard identification and PPE selection process.

The workplace hazard assessment underlies OSHA's PPE standard (Camplina 2003, 48–51). For both workers and safety practitioners, many workplace hazards are easy to point out. Obvious hazards are traceable to the environment, the required tasks, and the materials handled. For example, cutting, grinding, chipping, and abrasive blasting can cause eye injuries. For these tasks, a responsible employer would require ANSI-compliant eye protection. As required by OSHA, it is incumbent on the employer to work with employees to ensure the use of PPE when required and that consequences for noncompliance reflect the rationale for PPE in the first place—to prevent harm. Safety practice loses its meaning when it is reactionary and punitive. Discipline should always be the last recourse (Thomen 1991, 378).

Potential hazards in the workplace are not so obvious. These hazards may be some combination of factors that are specialized, complex, novel, episodic, or unexpected. The need for cut-resistant gloves when handling sheet metal is obvious and immediate—the hazard and potential consequences are readily apparent to most individuals. In simple terms, physical hazards are often easier to identify than the more insidious exposures posed by biological, chemical, or ionizing/non-ionizing radiation hazards. Additionally, trauma is immediate and understandable when compared to an undetected chronic exposure of some sort. As the hazards become less obvious, the role of worker education and training becomes much more important. In all situations, hazard recognition and awareness necessarily precedes choosing what PPE to wear. It is not surprising that many OSHA substance-specific standards such as the one regarding formaldehyde (29 CFR 1910.1048) have significant training requirements in addition to PPE selection and use.

Prescribing and Issuing

A well-thought-through method of prescribing and issuinge is the next element in a PPE program. The prescribing and issuing process must be part the written PPE program and thoroughly communicated to all involved. Depending on the circumstances, the prescribing and issuing process may be simple and direct, such as a supervisor distributing work gloves on a job site, or considerably more complex, involving several steps and individuals in a corporation-wide safety program. Regardless of the situation, an effective PPE program ensures that items are readily available, worn properly, maintained, and replaced or serviced as required.

Whenever possible, standardization of required PPE will go a long way in simplifying the prescribing and issuing process (Jung and Ziegenfub 1989, 87–89). Worker involvement in the selection of PPE for comfort, style, and ease of use will help to involve and invest everyone in the process. Additionally, occupational healthcare workers (physicians, nurses, physical therapists), purchasing personnel, human resources personnel (labor relations), and senior management all have a role to play in the selection, use, and support of a successful PPE program (Kellerman, Tokars, and Jarvis 1998, 629–634).

The promulgation of OSHA's PPE standard in the 1990s created a great deal of controversy over who pays for PPE—was it the employer or the employee? Was PPE a "tool of the trade" to be provided by the worker or was it the responsibility of the employer? Should some more personal items—such as safety shoes or prescription safety glasses—be paid for by workers and other items by employers? This controversy is not new. Not so long ago, workers often incurred the cost of PPE required by the employer (Aldrich 1997, 415). This issue aside, the costs of not having a PPE program include direct and indirect costs from injuries, regulatory fines, and intangibles such as harm to reputation and diminished quality of life. Simply stated, a good PPE program is money well spent—PPE use is a visible, positive, and indicative reflection of people-oriented values in any company or workplace.

Training and Education

> "Education is what survives when what has been learned has been forgotten."
>
> B. F. Skinner (*New Scientist* May 21, 1964)

Training and education is the third element in the PPE program process. Training and education are closely related concepts but are not one and the same. Training provides information toward some specific end or observable behavior, while education develops concepts and understandings that are useful in new situations. With PPE, both training and education are important for several reasons. Training provides specific information relevant to the situation or workplace—and is best described in behavioral terms. For example, "Wear these gloves when handling the following solvents." Often, statements such as "Select the appropriate chemical resistant glove when handling solvents" raise more questions than they answer (Blayney 2001, 233–236). The more direct and concise the message, the more likely it is to be retained and followed.

In occupational safety and health, the ability to apply knowledge in new or unique situations is evidence of education. For example, a safety program that discusses permeation and degradation when selecting chemical-resistant gloves helps increase the likelihood of an informed choice in a new situation. In training and education, the challenge is to strike a balance between the practical and the theoretical when determining worker needs and outcomes (Wallerstein and Rubenstein 1993, 166).

The following concepts are essential best practices in any PPE training and education program. First, emphasize that PPE is a last (or only) line of defense and that it is by no means fool proof. Second, emphasize the importance of proper fit or prior approval for use (regarding respirators, for example). Third, discuss the importance of personal hygiene in helping to minimize potential exposures and in preventing re-contamination on the job or taking contamination home. Finally, stress the proper use of the equipment, its useful life and replacement.

Putting It All Together

Taken alone, the three elements of the PPE program process discussed so far—hazard identification/risk inventory, prescribing and dispensing, and training/education—are important but lack context without a written program (or a simple policy statement in less complex situations) addressing both policy and practice. The written program should include the workplace hazard analysis; policies and procedures (or references) for dispensing; use, care, and maintenance; and consequences for noncompliance. The written program should be clear, concise, and easy-to-follow and be readily available to all affected employees. PPE hazard evaluations can be either task- or discipline-specific. Clearly define consequences for noncompliance after consultation with management (typically, the human resource or legal function). A willingness to define consequences without the willingness to carry it out thwarts any credibility the program may have in worker protection. Again, worker involvement is essential in the development and success of the program. Before the program is issued in final form, have affected workers provide feedback and suggestions. Program ownership helps ensure success.

Because PPE is an integral part of numerous other health and safety standards, some form of cross-referencing one standard with another is encouraged. For example, OSHA's Hazard Communication Standard (29 CFR 1910.1200) is the foundation for chemical safety in the workplace. It makes sense to reference that PPE standard in a hazard communication program. Numerous other occupational and environmental standards could be cross-referenced in this way as well.

FIGURE 1. PPE manufacturers provide an increasing array of product styles. (Photo courtesy of Gateway Safety)

Combining employee training and education with the evaluation and selection of PPE enhances both interest and compliance. Manufacturers and distributors of PPE have gone to great lengths to incorporate style and comfort into their products. Gone are the days of "one size fits all" and styles that hark back to a period where protection was the only consideration. Style is recognized as an important factor in a worker's propensity to use PPE (see Figure 1). PPE even plays a role in commercial advertising where workers are seen wearing safety equipment in ads for a variety of companies, services and industries.

Evaluation and Improvement

Evaluation is the fourth element in the PPE program process. Effective safety and health management is a dynamic process and does not occur unless there is some periodic evaluation. Regardless of how it is viewed, some method of program evaluation is needed to ensure that the program remains effective, current, and useful to those it is intended to serve (Richthofen 2002, 362). There are various methods of monitoring program effectiveness. These methods can be observational, anecdotal, or more formal, such a checklist audit. It is important to document the evaluation process and retain these records for both historical and regulatory compliance. Compliance depends on documentation.

Finally, the written PPE program or policy statement must include some method of assurance and enforcement. The human resources function of the organization must be part of this process. Line supervisors carry the burden of ensuring PPE use, but they cannot do this without the clear and visible involvement of management to set and maintain standards (Laing 1997, 387). The written program or policy statement must also define the consequences for noncompliance.

The Future of PPE

The future of PPE will mirror its history— shaped by the uniquely human traits of economics and technology. New technologies are likely to present new hazards. For example, nanotechnology involves the creation of tiny devices that are active in some way. While not alive, these new devices go beyond simple biology, chemistry, and physics and are capable of interacting with their environment. Discussion is already taking place in the health and safety profession as to the implications of nanotechnology for workers and the environment (Knowles 2006, 20–27; McShane 2006, 28–34).

PPE tomorrow will likely incorporate active sensing methods that detect and identify hazardous materials. Much of this technology will come from the defense industry (Ali et al. 1999, 504). PPE may also be able to adapt to the physical environment or the physiological demands of the wearer. Finally, PPE standards will inevitably join the tide of global harmonization.

Regardless of what the future brings, PPE programs must continue to protect workers from the potential hazard(s) in the workplace. Workers and employers both benefit and depend on PPE as an integral part of an overall health and safety program. Unlike many things in our profession, this will not change over time.

Summary: Best Practices in Administering a PPE Program

1. Involve those who will wear PPE and those who will support the program. *Let them know you need their input.*
2. Conduct a through hazard analysis and rank the risks. *Consider both obvious and insidious hazards.*
3. Match PPE to the hazards and specify its use based on the risks (prescribing). *Be practical by keeping instructions simple and easy to follow.*
4. Select and standardize PPE (issuing). *Compare choices with regard to worker preferences in terms of comfort and style.*
5. Develop a written program/standard operating procedures. *Again, keep it simple, concise, and written in behavioral terms.*
6. Train and educate. *Outline uses and limitations of PPE and develop knowledge applicable in new situations.*
7. Evaluate and improve. *An effective safety program evolves over time, but only if it is revisited and revised periodically.*

References

Akbar-Khanzadeh, Farhang, and Michael S. Bisesi. 1995. "Comfort of Personal Protective Equipment." *Applied Ergonomics* 26(3):195–198.

Aldrich, Mark. 1997. *Safety First: Technology, Labor and Business in the Building of American Work Safety 1870–1939*. Baltimore and London: The Johns Hopkins University Press.

Ali, Jave, Adrian Dwyer, John Eldridge, Frank Lewis, William Patrick, and Frederick Sidell. 1999. *Jane's Chemical-Biological Defense Guidebook*. Alexandria, Virginia: Jane's Information Group.

Anonymous. 1926. *Medical Care of Industrial Workers, Studies of Health Service in Industry*. New York: National Industrial Conference Board.

Baxter, Peter J., and Peter H. Adams, eds. 2000. *Hunter's Diseases of Occupations*. 9 ed. London: Hodder Arnold.

Blayney, Michael B. 2001. "The Need for Empirically Derived Permeation Data for Personal Protective Equipment: The Death of Dr. Karen E. Wetterhahn." *Applied Occupational and Environmental Hygiene* 16(2):233–236.

Camplin, Jeffery C. 2003. "Assessing the Need for PPE: Hazards Present Determine Level of Protection." *Professional Safety* 48(12):48–51.

Carpenter, W. Scott, Barbara C. Lee, Paul D. Gunderson, and Dean T. Stueland. 2002. "Assessment of Personal Protective Equipment Use among Midwestern Farmers." *American Journal of Industrial Medicine* 42:236–247.

CCOHS. 2005. "Designing an Effective PPE Program. Review of Reviewed Item." *OSH Answers*. www.ccohs.ca/oshanswers/prevention/ppe/designin.html

Hamilton, Alice. 1943. *Exploring the Dangerous Trades*. 2d ed. Fairfax, VA: American Industrial Hygiene Association.

Harris, Robert L, ed. 2000. *Patty's Industrial Hygiene*. 5th ed. (4 vols.) New York: John Wiley & Sons.

Heinrich, Herbert William. 1941. *Industrial Accident Prevention: A Scientific Approach*. New York: McGraw-Hill.

Herrick, Robert, ed. 1998. *Encyclopaedia of Occupational Health and Safety*. Edited by J. M. Stellman. 4th ed. (4 vols.) Vol. 1. Geneva: International Labour Office.

International Safety Equipment Association (ISEA). 2002. "Personal Protective Equipment: An Investment in Your Workers' and Company's Future." In *Partnership for Worker Protection*. Arlington, VA.

Jagt, Katinka van der, Erik Tielemans, Ingrid Links, Dick Brouwer, and Joop van Hemmen. 2004. "Effectiveness of Personal Protective Equipment: Relevance of Dermal and Inhalation Exposure to Chlorpyrifos Among Pest Control Operators." *Journal of Occupational and Environmental Hygiene* 1:355–362.

Jung, Kurt, and Bernd Ziegenfub. 1989. "Marking of Personal Protective Equipment." *Journal of Occupational Accidents* 11:87–89.

Kamal, Abdel Aziz M., Gaber M. Sayed, Mahmoud H. Hassan, and Aly. A. Massoud. 1988. "Usage of Personal Protective Devices Among Egyptian Industrial Workers." *American Journal of Industrial Medicine* 13:707–716.

Kellerman, Scott E., Jerome I. Tokars, and William R. Jarvis. 1998. "The Costs of Healthcare Worker Respiratory Protection and Fit Testing Programs." *Infection Control and Hospital Epidemiology* 19(9):629–634.

Knowles, Emory E. 2006. "Nanotechnology:Evolving Occupational Safety, Health & Environmental Issues." *Professional Safety* 51(3):20–27.

Krieger, Gary, and John Montgomery, eds. 1997. *Accident Prevention Manual for Business and Industry: Administration & Programs*. 11th ed. 3 vols. Itasca, Illinois: National Safety Council.

Laing, Patricia, ed. 1997. *Supervisors' Safety Manual*. 9th ed. Itasca, Illinois: National Safety Council.

Lusk, Sally, Madeline J. Kerr, and Sirkka A. Kauffman. 1998. "Use of Hearing Protection and Preceptions of Noise Exposure and Hearing Loss Among Construction Workers." *American Industrial Hygiene Association Journal* 59:466–470.

Marr, S.J., and S. Quine. 1993. "Shoe Concerns and Foot Problems of Wears of Safety Footwear." *Occupational Medicine* 43(2):73–77.

McShane, Brian. 2006. Nanotechnology:Is There Cause for Concern? *Professional Safety* 51(3):28–34.

Melamed, Samuel, Stanley Rabinowitz, Mabel Feiner, Esther Weisberg, and Joseph Ribak. 1996. "Usefulness of the Protection Motivation Theory in Explaining Hearing Protection Device Use Among Male Industrial Workers." *Health Psychology* 15(3):209–215.

Oakeshott, R. Ewart. 1994. *The Archaelogy of Weapons Arms and Armour from Prehistory to the Age of Chivalry*. 2d ed. Rochester, NY: Boydell Press.

Occupational Safety and Health Administration (OSHA). 2004. "Success with VPP: Management Leadership and Employee Involvement. Review of Reviewed Item. www.osha.gov/dcsp/success_stories/vpp/reg4_ss_milliken_elm.html

Ramazzini, Bernardino. 1964. *Diseases of Workers*. Translated by W. C. Wright. New York/London: Hafne. Original edition, 1940.

Richthofen, Wolfgang von. 2002. *Labour Inspection: A Guide for the Profession*. Geneva: International Labour Office.

Robinson, James C. 1991. *Toil and Toxics: Workplace Struggles and Political Strategies for Occupational Health*. Berkeley: University of California Press.

Roelofs, C.R. , E.M. Barbeau, M.J. Ellenbecker, and R. Moure-Eraso. 2003. "Prevention Strategies in Industrial Hygiene: A Critical Literature Review." *American Industrial Hygiene Association Journal* 64(1):62–67.

Schenker, Marc B., Marla R. Orenstein, and Steven J. Samuels. 2002. "Use of Protective Equipment Among California Farmers." *American Journal of Industrial Medicine* 42:455–464.

Sutter, Alice. 2002. "Construction Noise: Exposure. Effects, and the Potential for Remediation; A Review and Analysis." *American Industrial Hygiene Association Journal* 63:768–789.

Swartz, George, ed. 2000. *Safety Culture and Effective Safety Management*. Chicago: National Safety Council.

Taylor, C. B. 1990. "Personal Protective Equipment in 1992: The European Dimension." *Journal of Occupational Accidents* 11:247–253.

Thomen, James R. 1991. *Leadership in Safety Management*. New York: John Wiley & Sons, Inc.

Wallerstein, Nina, and Harriet Rubenstein. 1993. *Teaching about Job Hazards: A Guide for Workers and Their Health Providers*. Washington, D.C.: American Public Health Association.

Wilson, Graham K. 1985. *The Politics of Safety and Health: Occupational Safety and Health in the United States and Great Britian*. Oxford: Clarendon Press.

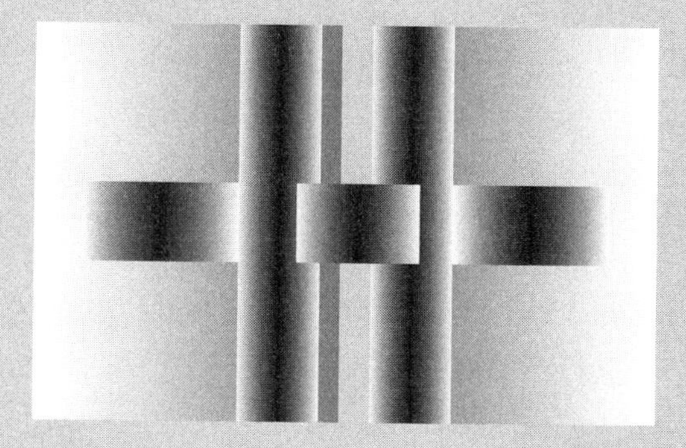

SECTION 6
ERGONOMICS AND HUMAN FACTORS ENGINEERING

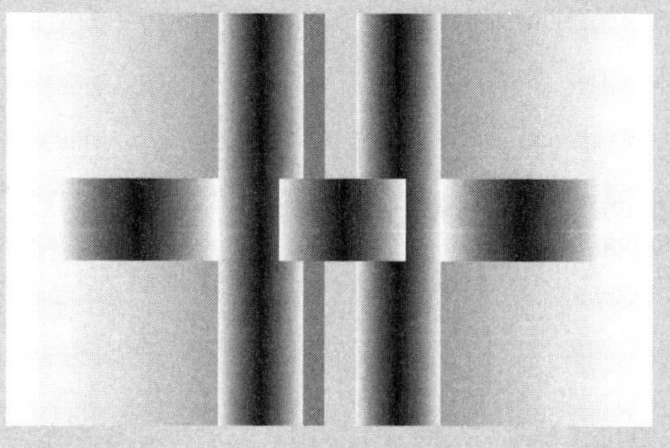

SECTION 6

ERGONOMICS AND HUMAN FACTORS ENGINEERING

LEARNING OBJECTIVES

- Understand how politics relates to standards-setting in the United States.
- Become familiar with the most important current standards and guidelines.
- Be aware of the current government policy and initiatives on ergonomics.
- Understand the influence of international standards.
- Recognize the relationship between workers' compensation and ergonomics.
- Know where to find standards.

REGULATORY ISSUES

Carol Stuart-Buttle

REGULATIONS HAVE ALWAYS BEEN entwined with politics. Early debates in the Congress of the United States raised fundamental differences about the relative control by federal and state governments over any aspect of residents' lives. Today there is debate on regulation versus no regulation, regardless of whether at the federal or state level. Ergonomics regulation has been at the heart of such contention.

Such regulation has drawn opposing views from groups representing business and those representing workers, including unions. The business community, harboring a philosophical resistance to intervention and oversight by the Occupational Safety and Health Administration (OSHA), opposes regulation. It argues that regulation increases costs (Karr 2002, Mugno 2002, and Greenhouse 2001). However, many companies that are members of associations arguing against regulation have established ergonomic initiatives and maintain them. Supporters of regulation state that workers are getting hurt in some industries when ergonomics is ignored, and the "bad apples" in business need regulation because it pushes them to implement measures that protect workers. Sometimes the message that ergonomics is good for business and, if done right, can be cost effective is lost among arguments by the most vocal groups on either side of the debate. Those in safety and health, as well as the human factors and ergonomics communities state that ergonomics is good economics, as repeatedly manifested by articles in professional and trade publications (Chapman 2006; McDermott, Lopez, and Weiss 2004; Hogrebe 2004; Dwyer and Lotz 2003; Rodrigues 2001; Maloney 2000; Dolan 1998; and Hendrick 1996).

In addition, the General Accounting Office (GAO) (renamed the Government Accountability Office) has reported positive results from ergonomics programs (GAO 1997).

One troubling aspect of politicizing ergonomics was exposed when, in the heat of fierce rhetoric over a federal standard, the field of ergonomics was portrayed as having no foundation in science. Nevertheless, Congress funded the National Research Council and Institute of Medicine (within the National Academy of Sciences) to conduct a scientific review to determine if work-related musculoskeletal disorders were founded in research. It concluded there is a scientific basis for both injury risk and interventions (NRC 2001).

But the political battles continue. The federal Ergonomics Standard of 2000 was rescinded a few months after its passage with the changing of a presidential administration.

The same groups that were involved in the federal debates also took on the Washington State Ergonomics Standard. Lawsuits were attempted on the basis that the regulation was invalid (BNA 2001). When that did not succeed, the opposition brought the standard to popular vote and advocated voting against it, resulting in its defeat.

Similarly, politics has affected voluntary standards. Groups opposing ASC Z-365, *Management of Work-Related Musculoskeletal Disorders*, introduced legal tactics that made the financial burden to the secretariat too great to carry (BNA 2003). The committee was discontinued. A California ergonomics rule has weathered petitions from both business communities opposing the rule and union groups stating the rule is too lenient. The most recent petition from the California Labor Federation has stalled and remains under review by an advisory committee (BNA 2004a).

On a summary note, as succinctly stated by R. D. Fulwiler, regulatory pressure is no longer the driver for safety and health as it once was, and it never should have been a key driver. However, many companies need the regulatory nudge to put mechanisms in place that support both business success and health and safety success (Fulwiler 1998).

Standards in the United States

OSHA Standards and Guidelines

The primary mandatory standard in the United States is the Occupational Safety and Health (OSH) Act of 1970 (OSHA 1970). Citations for ergonomic violations are based on the General Duty Clause of the Act [Section 5(a)(1)], which states: "Each employer shall furnish to each of his employees, employment and a place of employment which is free from recognized hazards that are causing or are likely to cause death or serious harm to his employees."

This high-level statement has led to contests over ergonomic citations, especially questioning what is a recognized hazard and whether some of the medical conditions that might arise from poor ergonomic design cause serious harm. There is no other federal standard specific to ergonomics. Despite contests against OSHA, some landmark cases (Pepperidge Farms and Beverly Enterprises) have been settled by courts that interpreted the general duty clause pertaining to ergonomics, particularly related to risk factors of repetitive-motion disorders and lifting (Abrams 2002, BNA 1997 and 2002). The outcome of these cases has made it possible for OSHA to continue issuing ergonomic citations with fewer contests arising and to reach settlement agreements with companies. From 2002 to 2004 OSHA conducted 1500 ergonomic inspections and issued thirteen ergonomic citations using the General Duty Clause (BNA 2004b).

Ergonomics Program Standard Repealed

OSHA released an Ergonomics Program rule in November 2000 that was repealed in March 2001 under a new federal administration. The standard was published in the Federal Register on November 14, 2000, Vol. 65, No. 220 (29 CFR 1910.900) and is still available in public records (Federal Register 2000a). The standard was developed over approximately ten years. The early drafts were prescriptive and detailed, while the final released version was more programmatic. Early drafts mandated companies establish an ergonomics program, whereas the final version provided a quick fix that, if effective, would not require a full ergonomics program. An extensive appendix provided guidance material on how to meet the standard.

The main elements of the standard were:

1. Provide basic information to employees on musculoskeletal disorders (MSDs).
2. Determine the work-relatedness of an employee's reported MSD.
3. Provide prompt medical management of work-related MSDs.
4. Implement a quick fix if two or less cases on the same job or go to a full program.
5. When more than two cases arise or a quick fix is ineffective, implement a full ergonomics program with the following elements:
   - management leadership
   - employee participation
   - job hazard analysis
   - hazard reduction and control
   - training
   - program evaluation.

Although the standard was rescinded, it might be useful to an industry as a guide for addressing its ergonomics issues. Very similar material is also available on the Washington state government Web site related to their state ergonomics standard, which was also repealed. (See the Washington State Ergonomics Standard later in this chapter.)

OSHA's Four-Pronged Approach to Ergonomics

After the repeal of the ergonomics standard in 2001, the administration charged the Department of Labor (DOL) with addressing ergonomics-related problems, usually termed occupational musculoskeletal disorders (MSDs). The Department of Labor, under Secretary Elaine Chao, took more than a year after the ergonomics standard was rescinded to delineate a comprehensive plan. OSHA conducted public hearings and stakeholder meetings, received comments, and studied alternative approaches. A DOL news release on April 5, 2002 announced the DOL plan (OSHA 2002a and 2002b). The four-pronged approach of the DOL plan outlines the steps OSHA will take to address MSDs. The plan is not a rule or standard. The four parts are:

a. *Guidelines*. Industry-specific, voluntary guidelines are being developed to provide effective and feasible solutions to ergonomics-related problems. The injury and illness incidence rates of an industry are determining factors for which industries need guidelines. (For more details, see "Guidelines Affecting Ergonomics.")

b. *Enforcement*. OSHA will continue to inspect facilities for ergonomics-related problems and issue citations or hazard alert letters based on the General Duty Clause of the OSH Act. In addition, OSHA conducts a Site Specific Targeting (SST) Inspection Program that identifies workplaces with high incident rates (annual rates based on 100 full-time workers). The list is generated annually based on the latest injury data. Identified facilities receive a notice that their incidence rates are high and they could be inspected. In 2004, facilities with injury or illness rates at or greater than fifteen, resulting in days away from work, restricted work activity, or job transfer for every 100 full-time workers [known as the Days Away, Restricted or Transfer (DART) rate] were sent letters. The primary SST list will also include sites that have a Days Away From Work Injury and Illness (DAFWII) rate of 10 or higher. The average national DART rate in 2002 for private industry was 2.8, while the national average DAFWII rate was 1.6. The SST list stems from OSHA's Data Initiative for the previous year, which surveys approximately 80,000 employers to attain the latest injury and illness data (OSHA 2004).

c. *Outreach and Assistance*. Many compliance tools and information are provided through the OSHA Web page (www.osha.gov), including "e-tools" on specific topics and industries, case studies, and training and education resources. The administration also has developed cooperative programs that include:

Alliances, Consultation, the Safety and Health Achievement Recognition Program (SHARP), Strategic Partnerships, and the Voluntary Protection Program (VPP). All of these programs existed before the four-pronged initiative, except for Alliances (formed in 2002), but the DOL plan has promoted cooperation. VPP and SHARP guide and encourage high standards of safety and health processes that include ergonomics. Alliances and Strategic Partnerships forge bonds between OSHA and many industries and associations. Alliances are formal agreements with groups and companies to cooperate on developing resources to address workplace safety and health. Currently, there are 70 national OSHA office alliances, but regional office alliances can also be formed. Of the national alliances, 26 are categorized as having an ergonomic emphasis. OSHA Strategic Partnerships (OSPs) have existed since 1998 but have grown under the four-pronged plan. OSPs are entered into with companies that want help in addressing specific safety and health issues. At the end of March 2005, OSHA had formed 373 partnerships, 228 of which remain open. Eighteen OSPs specifically emphasized ergonomics. However, ergonomics is frequently an important part of all these cooperative programs (OSHA 2005a).

d. *National Advisory Committee on Ergonomics (NACE).* This committee was formed at its initial meeting in January 2003 and was chartered until November 2004. The committee accomplished the following:

- provided information related to various industry or task-specific guidelines
- identified gaps in the existing research base related to applying ergonomic principles to the workplace
- identified current and projected research needs and efforts
- determined methods of providing outreach and assistance that will communicate the value of ergonomics to employers and employees
- provided ways to increase communication among stakeholders on the issue of ergonomics.

Guidelines Affecting Ergonomics

There are two guidelines that historically have influenced ergonomics: the Safety and Health Management Guidelines (Federal Register 1989) and the Ergonomics Program Management Guidelines for Meatpacking Plants (OSHA 1990). In addition, under the recent four-pronged initiative, OSHA issued three industry guidelines and another is under development. The following introduces each of these guidelines in more detail.

a. *Safety and Health Management Guidelines.* These guidelines were issued in 1989 as a result of successful Voluntary Protection Programs (VPP) (Federal Register 2000b). The guidelines provide criteria for organizing a managed safety and health program (Federal Register 1989). Despite the age of these guidelines, they are still referred to, and subsequent guidelines such as the Ergonomics Program Management Guidelines for Meatpacking Plants (OSHA 1990) have been based on them. The SHARP program also refers to these guidelines as the minimum standard from which to develop and maintain a safety and health management system (OSHA 2005b).

The guidelines are general and have four elements: management commitment and employee involvement, work-site analysis, hazard prevention and control, and safety and health training. The preamble emphasizes that an effective program is more important than a written one but that written guidance enhances communication.

b. *Ergonomics Program Management Guidelines for Meatpacking Plants.* OSHA issued these guidelines in 1990 in response to the high incidence of cumulative trauma disorders in the red meatpacking industry (OSHA 1990).

The guidelines emphasize management commitment and employee involvement and cover four main program elements: work-site analysis, hazard prevention and control, medical management, and training and education. Medical management was introduced as a program element because of the severity of the musculoskeletal disorders that were found in the meatpacking industry at the time. The guidelines also have detail on implementation and examples for red meatpacking. At around this time, the press reported a clamor by industry and some in Congress for guidance to control the "national epidemic" of repetitive-motion injuries, and a draft federal standard was then expected from OSHA fairly soon after the guidelines (BNA 1991, Occupational Hazards 1990). Other industries have often referred to the meatpackers' guidelines, especially since it was several years until there was a drafted federal ergonomics standard.

c. *Industry Guidelines*. Under the four-pronged OSHA initiative, three guidelines have been released in final versions and a fourth is being developed. NACE made recommendations of industries that might benefit from guidelines, and OSHA is expected to create more, but none has been announced to date. Those that have been developed are anticipated to be useful to industries other than the target audience. The guides are heavily illustrated and emphasize identifying and controlling hazards that are common in the industry. Copies are available at www.osha.gov. The guidelines in order of release are:

- Guidelines for nursing homes
- Guidelines for retail grocery stores
- Guidelines for poultry processing
- Guidelines for the shipyard industry (in process).

VPP and SHARP

The Voluntary Protection Program (VPP) and the Safety and Health Achievement Recognition Program (SHARP) are both successful national incentive programs for improving safety and health. To be recognized by either program a company is required to have injury and illness rates below the national average. The head of OSHA in 2002, John Henshaw, announced plans to increase company participation and stated that VPP work sites have 50 percent fewer workplace injuries than non-VPP work sites (BNA 2004c, Safety and Health 2002). VPP and SHARP are listed as part of the overall cooperative programs that are elements of the outreach and assistance prong of the OSHA ergonomics initiative. To demonstrate excellence in safety and health, a company needs to attain and maintain a low injury rate. To achieve that, many companies must address MSDs through ergonomics.

The VPP was established in 1982 and its requirements center on comprehensive management systems with active employee involvement to prevent or control the safety and health hazards at the work site (Federal Register 2000b). In 2000, these VPP guidelines were reorganized to be consistent with the 1989 Safety and Health Management Guidelines. As a result, there are now four rather than six elements to the VPP. This program can provide a safety and health incentive for a company as well as support mechanisms and mentoring from other VPP companies to help reach the high standard. In addition, sites are removed from programmed OSHA inspections during VPP participation. However, OSHA is still obliged to respond to any employee complaint or any fatality or catastrophe. There are three levels of VPP programs: Star, Demonstration, and Merit. Star recognizes workplaces that are self-sufficient in controlling hazards at the work site. The Demonstration level includes work sites with Star-quality safety and health programs but that require demonstration or testing of experimental approaches that are different from Star. The Merit level recognizes good safety and health programs, but additional steps are required to reach the Star level. The 2000 Federal Register notice announced more relaxed application criteria for VPPs to encourage companies to apply for the program and then strive to reach the Merit or Star status, which still requires meeting stringent criteria. Previously, the program posed tougher criteria for entry. In 2004,

VPPs were initiated to promote safety and health in the construction industry because that industry was not eligible for participation under the existing VPP (BNA 2004d).

The Safety and Health Achievement Recognition Program (SHARP), is a recognition program designed particularly for small employers. As with the VPP, SHARP exempts the facility from inspection during the time that the SHARP certification is valid, except when an employee complaint is filed or a fatality or catastrophe occurs. To participate in SHARP, a company needs to:

- request a consultation visit that involves a complete hazard identification survey
- involve employees in the consultation process
- correct all hazards identified by the consultant
- implement and maintain a safety and health management system that, at a minimum, addresses OSHA's 1989 Safety and Health Program Management Guidelines
- lower its Lost-Workday Injury and Illness rate (LWDII) and Total Recordable Case Rate (TRCR) below the national average
- agree to notify the state Consultation Project Office prior to making any changes in working conditions or introducing new hazards into the workplace.

If the requirements are satisfied, the work site will be recommended for approval and certification (OSHA 2005b).

American National Standards Institute

The American National Standards Institute (ANSI) is a voluntary consensus standards body that issues many standards in a year. The following are just a few of ANSI's main ergonomics standards that exist or are being developed. There are others that relate to safety and ergonomics in the workplace but pertain to more specific domains. Documentation may be purchased through ANSI, but often it is purchased directly from the group responsible for developing the standard in coordination with ANSI. Although voluntary, at times ANSI standards are cited in OSHA regulations.

Recently federal agencies and departments have moved toward adopting private-sector standards rather than military standards. This change of direction was provoked by laws such as the National Technology Transfer and Advancement Act of 1995 that encouraged all federal agencies and departments to use technical standards published by voluntary consensus standards bodies as much as possible (Public Law 2006).

The following are some of the general ergonomics standards that ANSI has released or has under development that are less industry specific.

a. *HFES 100, Human Factors Engineering of Computer Workstations (Canvass Draft 2005).* This standard, issued as a draft 3/31/02, is an update of the ANSI/HFS 100-1988 standard (HFES 2005a). It is in the canvass review stage that includes comments received in a first review in March 2005. Final approval as an ANSI standard is anticipated in 2007. There are four main sections: Installed Systems, Input Devices, Visual Displays, and Furniture. The draft provides basics on ergonomics of computer workstations, including workstation design specifications for the anthropometric range of 5th to 95th percentile and for four reference postures: reclined, upright, and declined sitting, and standing. Several input devices are covered, and there is a section on how to integrate all the workstation components for an effective system (Albin 2006).

b. *ANSI/AIHA Z10 Occupational Health Safety Management Systems.* The American Industrial Hygiene Association (AIHA) issued ANSI/AIHA Z10 in 2005 (ANSI/AIHA 2005). This document is applicable to organizations of all sizes and provides a concept and action outline to improve safety and health management systems. All safety and health management standards and guidelines are relevant to ergonomics because ergonomics is successful when conducted within a framework, whether that framework or process is an ergonomics process or integrated into the safety and health process. Other than introductory sections, the standard covers five main topics: management leadership and employee participation, planning, implementation and operation, evaluation and corrective action, and management review. The appendix includes some forms and audit materials as well as an example of a risk assessment matrix. The standard is a performance standard, not a specification standard, therefore it can be easily integrated into any other existing management systems. Z10 is intentionally compatible with the International Labor Organization's (ILO) Guidelines on Occupational Health and Safety Management Systems (ILO-OSH 2001) (Manuele 2006).

c. *HFES 200 Software User Interface Standard*. This is a four-part canvass-stage standard issued by the Human Factors and Ergonomics Society (HFES) under the auspices of ANSI (HFES 2005b). It closely mirrors the International Standards Organization, ISO 9241 standard on visual display terminals, except for original parts on accessibility and voice input/output. The four sections are: Introduction, Accessibility, Interaction Techniques, and Voice Input/Output and Telephony.

d. *ASC Z-365 Management of Work-Related Musculoskeletal Disorders—discontinued*. The ASC Z-365 was formed in 1991. The secretariat, National Safety Council (NSC), issued the most recent working draft that was near completion in October 2000. The draft contained similar elements to that of the repealed federal standard and the OSHA meatpackers' guidelines. The document was programmatic rather than specific in that it did not provide details on how to conduct analyses or implement interventions. In the fall of 2003, the secretariat withdrew from the process citing extraordinary costs involved with this particular standard. After twelve years and with an almost completed standard, the public review completed, and approval pending, the committee was disbanded as no alternative secretariat came forward (BNA 2003).

American Conference of Governmental Industrial Hygienists' TLV®

The American Conference of Governmental Industrial Hygienists (ACGIH) is well known for developing and publishing threshold limit values (TLV) for chemical substances and physical agents. These include TLV for hand-arm vibration, whole-body vibration, and thermal stress. In 2001, a new ergonomics section was published with TLV for hand activity level (HAL) as well as a draft of the lifting TLV (ACGIH 2001).

The HAL TLV is intended for monotask jobs performed for four hours or more. A rating system is used to assess the average hand activity level and related recovery time. The TLV is determined from a graph of HAL against normalized peak hand force. The TLV depicted by a solid line on the Action Limit graph is based on combinations of force and hand activity associated with a significantly elevated prevalence of musculoskeletal disorders. A lower alternative action limit indicates that some surveillance should be instituted and general controls implemented.

The lifting TLV provides weight limits based on frequency and duration of lift within 30 degrees of the sagittal plane. There are three lifting tables defined by the duration of the lifting and the rate of lift per hour. Each table portrays a grid of three horizontal distances from the low back and four height zones of the start of the lift. Weight limits are provided in the matrix for each circumstance. The TLV of weights under the defined conditions of each table are considered to be acceptable values for nearly all workers on a repeated daily basis without developing work-related lower back or shoulder disorders associated with repetitive lifting tasks. These tables are based on data that include more recent dynamic biomechanics (Marras and Hamrick 2006).

National Institute of Standards and Technology

The National Institute of Standards and Technology (NIST) helps to develop standards that address measurement accuracy, documentary methods, conformity assessment and accreditation, and information technology standards. A current initiative of NIST is to develop industry usability reporting guidelines that directly affect software ergonomics. NIST can also be a channel through which to link to military standards. Many of the standards are domain specific, such as automotive, chemical processing, construction, healthcare, and manufacturing, and are especially concerned with quality and system efficiency as well as equipment performance.

In 1988, NIST launched The Malcolm Baldridge National Quality Award that recognizes performance excellence and quality achievement by U.S. manufacturers, service companies, educational organizations, and healthcare providers. This award has driven many companies to address safety and health issues, in particular ergonomics, so they can achieve performance excellence and be a contender for the award.

Miscellaneous Standards-Setting Groups

There are many other sources of standards relating to ergonomics that may be important to certain domains or specialties. A few of the other organizations that develop standards are the American Society of Mechanical Engineers (ASME), the American Society for Testing and Materials (ASTM), the Institute of Electrical and Electronics Engineers (IEEE), and the Society of Automotive Engineers (SAE).

State Regulations

California Ergonomics Standard

This is a mandatory state law, made effective in 1997, that addresses formally diagnosed work-related repetitive-motion injuries that have occurred to more than one employee performing identical work activities (CAL/OSHA, 1997). The employer must implement a program to minimize the repetitive-motion injuries through work-site evaluation, control of the exposures, and training. The standard is nonprescriptive in that it does not state how to perform an evaluation, control the exposures, or conduct training. The brief two-page standard gained some notoriety as it was issued during the same period as early drafts of the federal standard, which were several pages long with extensive supporting material.

Maine's Video Display Terminal Law

Maine has a video display terminal operators law under the health and safety regulations that requires employers to provide an education and training program to all employees who use a computer for four or more hours a day when they are first employed, and annually thereafter (Safetyworks 1989). Although the term *ergonomics* is not used, the training required pertains to ergonomic principles of how to set up a computer workstation for user comfort and health.

Washington State Ergonomics Standard

In the year 2000, Washington state adopted a mandatory ergonomics rule (Labor and Industries 2006a). The rule stated that employers have to look at their jobs to determine if there are specific risk factors that make a job a "caution zone job" as defined by the standard. All caution zone jobs must be analyzed, employees of those jobs are to participate and be educated, and the identified hazards reduced. Although the rule was issued in 2000, there was a two-year grace period for compliance. However, the rule was repealed in November 2003 (effective December 2003) after state residents voted against the standard. Much of the information on the rule remains on the Washington State Labor and Industries Web site as a resource in ergonomics (Washington Department of Labor and Industries 2006b). The information includes resources such as case histories, checklists and calculators for determining the risk for MSDs, publications, and videos.

Michigan's Ergonomics Rule

In June of 2002 a steering committee was formed by two standing commissions of Michigan's OSHA, a General Industry Standards Commission and the Occupational Health Standards Commission. The steering committee was to develop a framework for addressing an ergonomics rule, and an Ergonomics Standard Advisory Committee was formed to address the feasibility of such a standard. The Ergonomics Standard Advisory Committee is in the process of writing the third draft of a possible ergonomics rule. The committee states that it plans to keep the rule as minimal as possible. Those who oppose an ergonomics rule are expected to introduce legislation to stop the rule's development. In 2004, the Governor of Michigan, Jennifer Granholm, vetoed a bill that banned funding an ergonomics standard (BNA 2005).

Workers' Compensation in the United States

Injury rates are one of the primary measures of the effectiveness of an ergonomics program. Injuries can indicate a design problem, and reduction in injuries through intervention is one measure of effectiveness. workers' compensation (WC) rates and costs provide a dollar amount related to the injuries in a company. They can be another benchmark along with the OSHA log rates. WC costs are also part of return-on-investment calculations. In ergonomic analyses, these WC measures are used along with other indicators, including performance measures, for figuring returns on investment (Oxenburgh and Marlow 2006, Stuart-Buttle 2006a and 2006b).

The WC laws were established in the early 1900s to provide a mechanism for processing work-related injury and illness claims. When an employer accepts the responsibility for an occupational injury, regardless of fault, and the injury is processed through the WC system, the employer is protected from common lawsuits alleging negligence. Although the worker forfeits the right to sue, he or she benefits from economic compensation of a percentage of wages lost while injured as well as medical treatment and rehabilitation. (Refer to Section 7 of this Handbook for an in-depth discussion of workers' compensation.)

There are specific controls within the WC system to prevent abuse (Maygar 1999). They include:

1. Injuries must occur during the course of normal employment.
2. Injured workers must submit to a medical evaluation, although they can refuse medical treatment.
3. Injuries must be reported within a prescribed time period.
4. Medical treatment and benefit limits are established.
5. Maximum medical improvement (MMI) exams are conducted and the extent of residual disability determined on more extensive cases.
6. Informal hearings or conferences are held to resolve disputes and come to mutual agreements with the Workers' Compensation Commission.

Careful recording and close monitoring of each case by the employer helps prevent confusion or disputes. Injury prevention and a prompt, effective, return to work reduces costs through lower WC premiums and reserves that are based on a company's claim history. By practicing injury prevention, a company also saves on indirect costs, such as lost productivity and temporary hiring costs. These cost-reduction strategies are often left to the employer to implement, but the company may need guidance on how to move a worker back to full employment without re-injury. This is best achieved through safety and ergonomic processes (Stuart-Buttle 2006b). (For more on Safety Management, see Section 3 of this handbook.)

The WC system is run at the state level and covers approximately 127 million workers nationally without any congressional role (Treaster 2003). The WC premiums vary by state, as do the related regulations pertaining to safety initiatives. Most of the safety regulations tied in with WC laws were adopted during WC reforms in the 1980s and 1990s in an effort to control occupational injuries and overall WC costs. Just over half the states (52 percent) have at least one type of safety requirement such as, safety committees, safety and health programs, insurance carrier loss-control services, and targeting initiatives (Smitha, Oestenstad, and Brown 2001). Twenty-four states (48 percent) have no safety requirement. Usually the employers are subject to penalties if they fail to comply with mandatory workplace safety requirements. The study by Smitha et al. (2001) of WC safety regulations focused on the four types of safety requirements that were mandatory in a state from 1992 to 1997. From one to three of these types of safety requirements may be present in a single state. Nine states had safety committee requirements, ten had safety and health program requirements, eleven had regulated insurance carrier loss-control services, and eighteen states had targeting initiatives in place, which means there were mandatory safety requirements for employers with above-average injury or WC loss rates. Between 1980 and 1990 claim costs increased an average of 11 percent per year, while between 1991 and 1995 the costs increased only 2 percent. Smitha et al. (2001) attribute this to prevention initiatives under the WC reforms. However, more recent figures reflect the national increase in medical costs that have increased WC costs as well. As reported in 2003, the average cost of WC insurance has risen 50 percent in the last 3 years (Treaster 2003).

Americans with Disabilities Act

The Americans with Disabilities (ADA) Act came into effect in 1990 (EEOC 1990). One part of the act addresses accessibility for the disabled and a second part pertains to employment. Two main points of the act under the employment section are:

- The ADA prohibits disability-based discrimination in hiring practices and working conditions.
- Employers are obligated to make reasonable accommodations for qualified disabled applicants and workers, unless doing so would impose undue hardship on the the employer. The accommodations should allow the employee to perform the essential functions of the job.

The act also has some bearing on ergonomics. Human Resource Departments need to be aware of the essential functions of a job, so many companies have modified their job descriptions to include them. However, in some industries, defining those functions and job demands is not always easy and can entail the assistance of safety and health personnel. Many different personnel in industry, including those working in ergonomics, need job descriptions that often provide information related to their area of responsibility. So those addressing ergonomics should be aware of what descriptions do exist and should be sure that essential functions are included (Stuart-Buttle 2006b).

If a disabled worker needs to be accommodated, the company may make workstation modifications. Often this can provide an opportunity to improve a job for all the workers. A job should be thoroughly assessed so that any changes would provide the necessary accommodation and also make an ergonomic improvement for others. Caution is advised in tailoring a workstation to one person to such an extent that no one else could work in that area if the person is absent. This implies the benefit of including adjustability that helps everyone. Designed appropriately, making the changes should provide a return on the investment rather than generating a cost to accommodate one person (Stuart-Buttle 2006b).

International Standards

International Standards Organization

The International Standards Organization (ISO) organizes its standards by general topic area into 40 main groups. There are three types: A and B standards that address principles and processes and C standards that are technically based and mostly of interest to specific types of industries. Technical Committees (TC) also have their own listings of standards. For example, TC 159 is the Ergonomics Technical Committee. Apart from specific ergonomics-related standards, some rules in other categories and TCs may be pertinent to a particular industry, such as the categories of electronics or material-handling equipment. Many of the industry-based or equipment-based C standards also pertain to manufacturing issues such as equipment dimensions or stability tests.

Technical committee activity can be found through accessing the ISO Web page (www.iso.org). Many ergonomics standards are issued under section 13.180 "Ergonomics" of category 13, "Environment, Health Protection, Safety." However, other sections of category 13 and elsewhere pertain to ergonomics as well. For example, section 13.140 is on noise and 13.160 is on vibration and shock with respect to human beings. The ergonomics section (13.180) has a number of standards dated from 1977 to 2001. To provide an idea of the types of ISO ergonomics standards available in this section, several of them are grouped by topic in Table 1.

British Standards Institute

The British Standards Institute (BSI) is the primary standards-setting body in the United Kingdom. There are two notable publications of BSI that have begun to affect business in the United States. They are the BS 8800 *Guide to Occupational Health and Safety Management Systems* (1996) and OHSAS 18001 *Occupational Health and Safety Management Systems-Specification* (1999) (BSI 1996 and 1999). The BS 8800 is based on the ISO 14001 environmental management systems model. The OHSAS 18001 is part of the Occupational Health and Safety Assessment Series and the 18001 document gives guidance to companies developing a management system that can be assessed and certified (Seabrook 2001). In 1996, there was an initiative by the International Standards Organization to propose ISO 18000 on occupational health and safety management systems (Markey et al 1996). At the time, during a large international workshop on the topic, the standard was rejected. ISO revisited the issue in 2000 and backed away from taking action. These management-process standards are relevant to ANSI/AIHA Z10 "Occupational Health and Safety Management Systems" that was recently issued, as noted earlier.

TABLE 1

Tabulation of Some Existing ISO Ergonomics Standards and Others in Development

Topic	ISO Document Number
Ergonomics Guiding Principles and Human-System Interaction	ISO 1503:1977, ISO 6385:1981, ISO 13407:1999, ISO/TR 18529:2000
Anthropometry Evaluation of Static Working Postures	ISO 15535:2003, ISO 7250:1996, ISO 11226:2000
Ergonomic Procedures for the Improvement of Local Muscular Workloads	ISO/AWI 20646 (Approved work item by TC)
Ergonomics Manual Handling Part 1: Lifting and Carrying (1.2) Part 2: Pushing, Pulling and Holding Part 3: Handling of Low Loads at High Frequency	ISO 11228-1:2003, ISO/CD 11228-2, ISO/CD 11228-3 Parts 2 & 3 are drafts.
Ergonomic Principles Related to Mental Workload (2 parts)	ISO 10075:1991, ISO 10075-2:1996 (Part 3 is a final draft of TC 159)
Signals and Controls and Displays	ISO 7731:1986, ISO 9355-1:1999, ISO 9355-2:1999, ISO 11428:1996, ISO 11429:1996
Ergonomic Design of Control Centres (4 parts available, total of 8 parts planned. Parts 4–8 will cover workstation layout, displays & controls, environment, evaluation of control rooms and specific applications.)	General principles and principles of arrangement and layout. ISO 11064-1:2000, ISO 11064-2:2000, ISO 11064-3:1999, ISO 11064-4:2004, ISO/DIS 11064-6, ISO/CD 11064-7
Principles of Visual Ergonomics – Indoor Lighting	ISO 8995:1989
Ergonomic Requirements for Office Work with Visual Display Terminals (17 parts - 8 software and 9 hardware). **Note:** There are recent (2000) amendments to some parts. Parts date from 1992–2000.	ISO 9241, Parts 1–9: General overview requirements of task, posture, and layout. Hardware requirements of visual display, colours, keyboard, and input devices. Parts 10-17: Software requirements usability, dialogue principles, dialogues of menu, command, form filling, and direct manipulation.
Ergonomic Requirements for Work with Visual Displays Based on Flat Panels	ISO 13406-1:1999 ISO 13406-2:2001
Ergonomic Design for the Safety of Machinery (2 parts)	ISO 15534-1:2000, ISO 15534-2:2000, ISO 15534-3:2000
Speech Communication	ISO 9921-2:1996
Thermal Environments	ISO 7243:1989, ISO 7726:1998, ISO 7730:1994, ISO 7933:1989, ISO 8996:1990, ISO 9886:1992, ISO 9920:1995, ISO 10551:1995, ISO/TR 11079:1993, ISO 11399:1995, ISO 12894:2001, ISO/TS 13732-2:2001

(Sherehiy, Rodrick, and Karwowski 2006)

International Labor Organization Guidelines

The ILO guidelines on occupational safety and health management systems (ILO-OSH 2001) were developed in cooperation with the International Occupational Hygiene Association (ILO 2001). These guidelines are designed for both governments and organizations. An ILO update in 2002 indicated the guidelines are being considered for adoption by 18 countries and the ILO is now trying to market them aggressively in competition with BSI (Johnson 2002).

The OHSAS 18001was issued before the ILO-OSH 2001, but they are now in direct competition to win over global corporations and be adopted as an international standard. Smaller companies in the United States and elsewhere will shortly follow any global lead. The underlying health and safety management system of a company is relevant to ergonomics, and the process of managing ergonomics-related problems is itself an essential component of effective control of such problems (NRC 2001).

Canadian Ergonomics Standards

British Columbia

In the fall of 1994, the Workers' Compensation Board of British Columbia (BC) issued a draft ergonomics regulation. The regulation failed to be adopted by the BC Legislature in 1995. However, since 1994, there has been a two-page section on ergonomics in "Part 4, General Conditions" of the Occupational Health and Safety Regulations of the Workers' Compensation Board of BC (British Columbia 2006). "Sections 4.46–4.53, Ergonomics (MSI) Requirements," charge employers with identifying factors that might expose workers to the risk of a musculoskeletal injury (MSI), to assess the identified risks, and to eliminate or minimize the risks. Employees are to receive education and training and be consulted by the employers. Evaluation of effectiveness is required. Since this is written into the regulations BC has useful materials on its Web site to support companies in addressing musculoskeletal disorders. Other provinces, such as Manitoba, have adapted these materials in their ergonomics guidelines to improve safety and health (Manitoba Labor and Immigration 1999).

Saskatchewan

The government of Saskatchewan has elements that address ergonomics in sections of The Occupational Health and Safety (OH&S) Regulations of 1996, although the term ergonomics is not used in the regulations. Sections 78–83 cover lifting and handling loads, standing for long periods of time, appropriate jobs for sitting and seating requirements, identifying and controlling musculoskeletal injuries, constant effort and exertion, and visually demanding tasks (Saskatchewan Labor 1996). A Code of Practice for Visual Display Terminals was issued in 2000 that provides guidance on complying with the OH&S regulations at computer workstations.

Ontario

The Ministry of Labour, Government of Ontario (ON) drafted legislation titled Physical Ergonomics Allowable Limits. The report was rescinded and shelved in 1995–96. Ontario's Occupational Health and Safety Act of 1979 was changed in 1990 with some significant additions. All employers have to have a health and safety policy and program and the officers of corporations have direct responsibility. In workplaces of greater than twenty workers there has to be a joint labor and management Health and Safety Committee that is responsible for health and safety in the workplace. The committee is to meet regularly to discuss health and safety concerns, review progress, and make recommendations. Workplaces of less than twenty must have a Health and Safety Representative. By 1995, employers had to certify that the members of their joint health and safety committees were properly trained (BNA 1994).

In February of 2005, the government of Ontario set a goal to reduce workplace injuries by 20 percent by 2008. In order to meet this goal the Ministry of Labour established an ergonomics working panel to make recommendations on how to reduce ergonomics-related injuries, coordinate fourteen health and safety associations to develop prevention strategies, provide research monies, allocate funds for bed lifts to reduce back injuries to nurses, and hire more inspectors. The government's approach does not include an ergonomics standard (Ontario Ministry of Labor 2005).

Canadian Standards Association

The Canadian Standards Association (CSA) produces voluntary standards pertaining to many areas. One guideline that is widely used is CSA-Z412 "Guideline on Office Ergonomics." The latest version (2000) is produced as an interactive CDROM, Adobe PDF file, or hardcopy. The guideline is comprehensive and practical, covering office systems, furniture and equipment, the environment, and conducting an ergonomic analysis (CSA 2000).

Other International Standards

The main ergonomics-related standards of the United Kingdom, Australia, and Japan are summarized in a chapter in the *Handbook of Standards and Guidelines in Ergonomics and Human Factors* (Stuart-Buttle 2006c). The handbook also has chapters providing detail on other international standards (Karwowski 2006).

References

Abrams, A. L. 2002. "OSHA & Ergonomics; Enforcement Under the General Duty Clause." *Professional Safety*, September 2002, pp. 50–52.

American Conference of Governmental Industrial Hygienists (ACGIH). 2001. *2001 TLVs and BEIs: Threshold Limit Values for Chemical Substances and Physical Agents and Biological Exposure Indices*. Cincinnati, Ohio: ACGIH.

Albin, T. J. 2006. Chapter 20, "Human Factors Engineering of Computer Workstations: HFES Draft Standard for Trial Use." In Karwowski, W., ed. *Handbook of Standards and Guidelines in Ergonomics and Human Factors*. New Jersey: Lawrence Erlbaum.

American National Standards Institute (ANSI) and American Industrial Hygiene Association (AIHA). 2005. *ANSI/AIHA Z10 Occupational Health and Safety Management Systems*. Fairfax, VA: AIHA.

Bureau of National Affairs (BNA). 1991. "Lantos Blasts OSHA for Standards Delay While Workers Suffer from 'National Epidemic'." *Occupational Safety and Health Reporter* (March 27, 1991), p. 1522.

______. 1995. "Ontario Ministry Moves Up Deadlines for Meeting Certification Requirements." *Occupational Safety and Health Reporter* (March 30, 1995), p. 1430.

______. 1997. "Ergonomic Hazards Properly Cited Under General Duty Clause, Commission Says." *Occupational Safety and Health Reporter*, April 30, 1997, pp. 1491–1492.

______. 2001. "State Businesses, National Associations Join Lawsuit on Washington State Rule." *Occupational Safety and Health Reporter* (October 25, 2001), pp. 965–966.

______. 2002. "Beverly Settlement Raises Complex Issue of Using Law, Not Rule, to Protect Workers." *Occupational Safety and Health Reporter* (January 24, 2002) 32(4):73–75.

______. 2003. "Safety Council Must Be 'Willing and Able' to Continue as Secretariat, ANSI Panel Says." *Occupational Safety and Health Reporter* (October 16, 2003) 33(41):1000–1001.

______. 2004a. "Ergonomics Process Falters in States; Defeat, Delay Plague 2003 Standards Efforts." *Occupational Safety and Health Reporter* (January 22, 2004) 34(1):78–79.

______. 2004b. "ABA Members Concerned Over Ergonomics, Repeat OSHA Violations, Targeted Inspections." *Occupational Safety and Health Reporter* (March 11, 2004) 34(11):267–268.

______. 2004c. "VPP Challenge, Corporate Pilots Started; OSHA Moving Forward with VPP Growth." *Occupational Safety and Health Reporter* (June 3, 2004) 34(23):567–568.

______. 2004d. "Construction Voluntary Protection Program Details Published; OSHA Requests Comments." *Occupational Safety and Health Reporter* (September 9, 2004) 34(36):901–902.

______. 2005. "Michigan Begins Drafting Possible Standard; State Bill Planned to Stop Ergonomic Rules." *Occupational Safety and Health Reporter* (March 3, 2005) 35(9):177.

British Columbia (BC) Workers' Compensation Board. 2006. *Occupational Health and Safety Regulations of the Workers' Compensation Board of BC*. British Columbia: WC Board.

British Standards Institute (BSI). 1996. BS 8800 *Guide to Occupational Health and Safety Management Systems*. Part 4, General Conditions. Ergonomics, Sections 4.46–4.53. www2.worksafebc.com/publications/OHSRegulation/Part4.asp

______. 1999. *OHSAS 18001 Occupational Health and Safety Management Systems Specification*. www.bsi-global.com

California Occupational Safety and Health Administration (CAL/OSHA). 1997. S5110. *Repetitive Motion Injuries*. California Code of Regulations, Title 8, Section 5110. www.dir.ca.gov/scripts/samples/search/query.id

Chapman, C. D. 2006. "Using Kaizen to Improve Safety and Ergonomics." *Occupational Hazards* (February 2006), pp. 27–29.

Canadian Standards Association (CSA). 2000. CSA – Z412. *Guideline on Office Ergonomics*. www.csa.ca

Dolan, S. 1998. "Health and Safety Initiatives Help Enhance Compaq's Profitability." *Safe Workplace*, Spring, pp. 23–25.

Dwyer, W., and C. Lotz. 2003. "An Ergonomic 'Win-Win' for Manual Material Handling." *Occupational Hazards* (September 2003), pp. 108–112.

Equal Employment Opportunity Commission (EEOC). 1990. *Americans with Disabilities Act of 1990*. Public Law 101-336, 101st Congress (July 26, 1990). Washington, D.C.: U.S. Government Printing Office.

Federal Register. 1989. *Safety and Health Program Management Guidelines*. Federal Register 54 (16) (January 26, 1989). Washington, D.C.: U.S. Government Printing Office.

______. 2000a. *Occupational Safety and Health Administration: Ergonomics Program; Final Rule*. Federal Register 65 (220) (November 14, 2000). Washington, D.C.: U.S. Government Printing Office.

______. 2000b. *Occupational Safety and Health Administration: Revisions to the Voluntary Protection Programs to Provide Safe and Healthful Working Conditions; Notice*. Federal Register. Notices. 65 (142) (July 24, 2000). Washington, D.C.: U.S. Government Printing Office.

Fulwiler, R. D. 1998. "People, Public Trust and Profit; Modeling Safety for Lasting Success." *Occupational Hazards* (April 1998), pp. 36–39.

"Gearing Up for OSHA's Ergonomics Guidelines." 1990. *Occupational Hazards* (April 1990), pp. 47–48.

General Accounting Office (GAO). August 1997. *Worker Protection: Private Sector Ergonomics Programs Yield Positive Results*. Washington, D.C.: U.S. Government Printing Office.

Greenhouse, S. 2001. "House Joins Senate in Repealing Rules Issued by Clinton on Work Injuries." *The New York Times*, March 8, 2001.

Hendrick, H. W. 1996. "Good Ergonomics Is Good Economics." Presidential address. Proceedings of the Human Factors and Ergonomics Society 40th Annual Meeting, Santa Monica, CA, pp. 1–15.

Human Factors and Ergonomics Society (HFES). 2005a. HFES100, *Human Factors Engineering of Computer Workstations*. Canvass Draft, 2005. Santa Monica, Calif.: HFES.

______. 2005b. HFES200, *Human Factors Engineering of Software User Interfaces*. Complete version. Santa Monica, Calif.: HFES.

Hogrebe, M. C. 2004. "Video Microscopes for our Video Generation." *ErgoSolutions Magazine* (February 29, 2004), pp. 22–24.

International Labor Organization (ILO). 2001. *Guidelines on Occupational Health and Safety Management Systems*. ILO/OSH 2001. Geneva, Switzerland: ILO. www.ilo.org

Johnson, D. 2002. "Are We Getting Closer to a Global Safety and Health Standard? The Race to Fill the Void." *The Synergist* (August 2002), pp. 24–27.

Karr, A. R. 2002. "Ergonomics 101: Business Resists Course." *Wall Street Journal*, January 26, 2002.

Karwowski, W., ed. 2006. *Handbook of Standards and Guidelines in Ergonomics and Human Factors*. New Jersey: Lawrence Erlbaum.

Maloney, D. 2000. "The Ergonomic Silver Lining." *Modern Materials Handling* (February 2000), pp. 53–58.

Manitoba Labor and Immigration. 1999. *Manitoba's Ergonomics Guideline*. Workplace Safety and Health Division, Manitoba, Canada. www.gov.mb.ca/labour/safety/ergoguide.html

Manuele, F. 2006. "ANSI/AIHA Z10-2005, The New Benchmark for Safety Management Systems." *Professional Safety* (February 2006), pp. 25–33.

Markey, D., S. Levine, and C. Redinger. 1996. "Conformance of ISO Occupational Safety and Health Management System Standards in Public-Sector Procurement Specifications to the GATT/WTO Requirements." *AIHA Journal* (October 1996) 57:936–944.

Marras, W. S., and C. Hamrick. 2006. The ACGIH TLV® for Low Back Risk. In W. S. Marras and W. Karwowski, eds. *Occupational Ergonomics Handbook*. 2d ed. "Fundamentals and Assessment Tools for Occupational Ergonomics," Ch 50. Boca Raton, FL: CRC Press.

Maygar, S. V. 1999. "Medical Claim Management; Controlling Workers' Compensation Losses." *Professional Safety* (March 1999), pp. 41–45.

McDermott, H., K. Lopez, and B. Weiss. 2004. "Computer Ergonomics Programs." *Professional Safety* (June 2004), pp. 34–39.

Minter, S. G. 2001. "Attack on TLV®: The American Conference of Governmental Industrial Hygienists Enters a High-Stakes Legal Battle Over Its Most Famous Product—TLVs. *Occupational Hazards* (May 2001), p. 6.

Mugno, S. A. 2002. "Voluntary Efforts Derail Ergonomic Injuries." *Safety + Health* (February 2002), pp. 25, 27.

National Research Council (NRC) and the Institute of Medicine. 2001. *Musculoskeletal Disorders and the Workplace: Low Back and Upper Extremities*. Panel on Musculoskeletal Disorders and the Workplace. Commission on Behavioral and Social Sciences and Education. Washington, D.C.: National Academy Press.

Occupational Safety and Health Administration (OSHA). 1970. *Occupational Safety and Health Act of 1970*. Public Law 91-596, 91st Congress, S. 2193, December 29, 1970. As amended by Public Law 101-552, 3101, November 5, 1990. Washington, D.C.: U.S. Government Printing Office.

______. 1990. OSHA 3123. *Ergonomics Program Management Guidelines for Meatpacking Plants*. Washington, D.C.: OSHA.

______. 2002a. "OSHA announces Comprehensive Plan to Reduce Ergonomic Injuries." OSHA National News Release. USDL 02-201, April 5, 2002. www.osha.gov/media/oshnews/apr02/national-20020405.html

______. 2005b. *A Four-Pronged, Comprehensive Approach*. April 8, 2005. www.osha.gov/ergonomics/ergo fact02.html

______. 2004. *Site-Specific Targeting 2004* (SST-04) Directive number: 04-02 (CPL 02). April 19, 2004. www.osha.gov/pls/oshaweb/owadisp.show_document?p_table=DIRECTIVES&p_id=3123

______. 2005a. *Partnership: An OSHA Cooperative Program*. April 30, 2005. www.osha.gov/dcsp/partnerships/index.html

______. 2005b. *OSHA's Safety and Health Achievement Recognition Program (SHARP)*. April 30, 2005. www.osha.gov/dcsp/smallbusiness/sharp.html

Ontario Ministry of Labour. 2005. *Repetitive Strain Injuries and Ergonomics*. Backgrounder Document d'information, February 28, 2005. www.gov.on.ca

"OSHA Seeks VPP Growth for Worker Protection." 2000. *Safety + Health* (November 2000), p. 18.

Oxenburgh, M., and P. Marlow. 2006. "Cost Justification for Implementing Ergonomics Intervention." In Marras, W. S., and W. Karwowski, eds. *Occupational Ergonomics Handbook*. 2d ed. "Interventions, Controls, and Applications in Occupational Ergonomics," Ch 4. Boca Raton, FL: CRC Press.

Public Law. 2006. *National Technology Transfer and Advancement Act of 1995*. Public Law 104-113, 104th Congress. 110 Stat.775. Washington, D.C.: U.S. Government Printing Office.

Rodrigues, C. 2001. "Ergonomics to the Rescue." *Professional Safety* (April 2001) pp. 32–34.

Safetyworks. 1989. *Title 26: Labor and Industry*. Chapter 5, "Health and Safety Regulations." Subchapter 2-A: Video Display Terminal Operators [Heading: PL 1989, c. 512 (new)] §251 and 252. Maine Department of Labor. janus.state.me.us/legis/statutes/26/title26 sec251.html

Saskatchewan Labour. 1996. *The Occupational Health and Safety Regulations, 1996*. Saskatchewan, Canada. www.labour.gov.sk.ca/safety/fast/ergonomics

Seabrook, K. A. 2001. "International Standards Update: Occupational Safety and Health Management Systems." Proceedings of American Society of Safety Engineers, 2001 Professional Development Conference, Anaheim, California, Session 507.

Sherehiy, B., D. Rodrick, and W. Karwowski. 2006. Chapter 1, "An Overview of International Standardization Efforts in Human Factors and Ergonomics." Karwowski, W., ed. *Handbook of Standards and Guidelines in Ergonomics and Human Factors*. New Jersey: Lawrence Erlbaum.

Smitha, M. T., K. R. Oestenstad, and K. C. Brown. 2001. "State Workers' Compensation, Reform and Workplace Safety Regulations." *Professional Safety* (December 2001), pp. 45–50.

Stuart-Buttle, C. 2006a. Injury Surveillance Database Systems. In Marras, W. S., and W. Karwowski, eds. *Occupational Ergonomics Handbook*. 2d ed. "Interventions, Controls, and Applications in Occupational Ergonomics," Ch 6. Boca Raton, FL: CRC Press.

______. 2006b. Ergonomics Process in Small Industry. In Marras, W. S., and W. Karwowski, eds. *Occupational Ergonomics Handbook*. 2d ed. "Interventions, Controls, and Applications in Occupational Ergonomics," Ch 8. Boca Raton, FL: CRC Press.

______. 2006c. Chapter 5, "Overview of National and International Standards and Guidelines." Karwowski, W., ed. *Handbook of Standards and Guidelines in Ergonomics and Human Factors*. New Jersey: Lawrence Erlbaum.

Treaster, J. B. 2003. "Cost of Insurance for Work Injuries Soars Across U.S." *The New York Times*, June 23, 2003.

Washington Department of Labor and Industries. 2006a. *Hazard Prevention; Ergonomics; Rule History*. www.lni.wa.gov/Safety/Topics/Ergonomics/History/default.asp

______. 2006b. *Hazard Prevention; Ergonomics*. www.lni.wa.gov/Safety/ Topics/Ergonomics/default.asp

Appendix: Internet Resources for Regulations

Federal and National Resources

www.acgih.org American Conference of Governmental Industrial Hygienists

www.aiha.org American Industrial Hygiene Association

www.ansi.org American National Standards Institute

www.asme.org American Society of Mechanical Engineers

www.asse.org American Society of Safety Engineers

www.astm.org American Society for Testing and Materials

www.access.gpo.gov General Printing Office

www.hfes.org Human Factors and Ergonomics Society

www.ieee.org Institute of Electrical and Electronics Engineers

http://standards.nasa.gov NASA Technical Standards Program

www.nist.gov National Institute of Standards and Technology

www.nsc.org National Safety Council

www.osha.gov Occupational Safety and Health Administration

www.sae.org Society of Automotive Engineers

State Resources

www.dir.ca.gov California Ergonomics Standard

www.safetyworksmaine.com Maine Department of Labor Outreach Program

www.lni.wa.gov/wisha Washington State, Department of Labor and Industries

International Resources

www.worksafebc.com British Columbia (BC) Government, Canada

www.bsi-global.com British Standards Institute, UK

www.csa.ca Canadian Standards Association, Canada

www.cenorm.be European Committee for Standardization (CEN)

www.europa.eu.int European Union (EU)

www.iea.cc International Ergonomics Association (IEA)

www.ilo.org International Labour Organization (ILO)

www.iso.org International Standard Organization (ISO)

www.useuosh.org Occupational Safety and Health Administration (USA) and European Union (EU)

www.perinorm.com Perinorm, private database of international, European and national standards

www.worksafebc.com Workers' Compensation Board of British Columbia, Canada

www.gksoft.com/govt/en/world Worldwide Governments on the WWW html

SECTION 6

ERGONOMICS AND HUMAN FACTORS ENGINEERING

LEARNING OBJECTIVES

- Gain an overview of the history of ergonomics.
- Understand the basics of ergonomics.
- Recognize the different risk factors that contribute to an ergonomic injury.
- Distinguish between cumulative trauma disorders (CTDs) and musculoskeletal disorders (MSDs).
- Be able to recognize the various types of ergonomic injuries.
- Learn about biomechanics.
- Grasp the basics of how to analyze and measure potential risk factors.
- Understand how the lifting formula (single task) designated by the National Institute for Occupational Safety & Health (NIOSH) is used.

APPLIED SCIENCE AND ENGINEERING: PRINCIPLES OF ERGONOMICS

Magdy Akladios

THE TERM ERGONOMICS stems from the Greek words *ergos*, meaning *work*, and *nomos*, meaning *laws*. Put together, it translates to the study of the *laws of work*. It is defined as the science that tries to adapt a task to conform to the capability of the human performing that task. Briefly put, ergonomics aims to fit the task to the human, rather than vice versa. Historically, ergonomics has been used synonymously with *human factors*. However, more and more they are becoming two distinct and unique sciences. While ergonomics deals with studying the physiological effects of work activities on people, human factors deals with the interaction between human beings and their work environment, including machines and other systems. Ergonomics aims at designing machines to reduce injuries to the humans who operate them, while human factors aims at reducing human error through the understanding of psychology, sensory input, error, and noise—in terms of outside interferences and factors (Eastman Kodak Company 1983).

HISTORY OF ERGONOMICS

Ergonomics is said to have started in ancient history. Imhotep (2667–2648 B.C.), the Egyptian chief architect, physician, and high priest of Heliopolis, was credited with the first report of documenting the treatment of back pain resulting from the Egyptians' work habits during the building of King Djoser's (2687–2668 B.C.) step pyramid at Saqqara.

The next known report came thousands of years later, when the physician Bernardino Ramazzini (1633–1714) wrote about the work-related complaints he saw during his medical practice in 1713. His publication was called *De Morbis Artificum*, or *The*

Diseases of Workers. However, up to that point, this science had no known or recognizable label.

It was in 1857 that Wojciech Jastrzebowski, a Polish scholar and professor of natural sciences at the Agronomical Institute in Warsaw-Marymount, coined the term ergonomics.

While the industrial revolution was taking hold in all manufacturing fields, many tasks were still performed by hand. In the early 1900s, a number of accommodations were developed to improve worker productivity and efficiency. At the time, ergonomics took a turn towards management, and was dubbed *Scientific Management*. In 1911, Frederick Winslow Taylor, a management consultant, developed the "One Best Way" method to enhance worker production. Taylor was ranked at the top, along with Darwin and Freud, as one of the daring and forward thinkers of modern times. Frank and Lillian Moller Gilbreth pioneered the science of time-motion studies and were able, through job evaluations and analysis, to reduce the time and effort by which a task could be conducted. After her husband's death, Lillian continued her work in the science of Scientific Management and was dubbed "The Mother of Management." The Gilbreths' efforts reduced the number of motions in bricklaying from 18 down to 4.5, hence enhancing productivity from 120 to 350 bricks per hour.

World War II (1939–1945) marked greater sophistication in military equipment and, subsequently, in nonmilitary applications. As a result, human-machine interaction captured greater interest, especially in airplane and cockpit design. This paved the way for such human factors designs to occupy center stage during the race for outer space supremacy during the late 1960s and early 1970s.

As the interest in ergonomics and human factors engineering grew, supporting societies and organizations started to sprout up around the world. The Ergonomics Research Society was first formed in Great Britain in 1949. This society started the *Ergonomics Journal* in 1957 (The Ergonomics Society 2007). In the United States, the Human Factors Society was founded in 1957, and has now grown to 22 technical groups and numerous local and student chapters. Known today as the Human Factors and Ergonomics Society (www.hfes.org), it publishes the journal *Human Factors*.

In 1961, the first International Ergonomics Association meeting was held in Stockholm, Sweden. This organization has members from the United States, the United Kingdom, Japan, Australia, Scandinavia, and other countries interested in that type of research.

In 1990, the Board of Certified Professional Ergonomists (BCPE) (www.bcpe.org) was established to certify human factors and ergonomics practitioners, researchers, and engineers in the United States.

Uses and Benefits of Ergonomics

Ergonomics recognizes that every human-machine system involves the interaction between humans and machines in an environment where information is being transferred back and forth between the two, with the human being in control, and the machine providing adequate feedback to the human. Hence, the science of ergonomics, typically, is a combination of a variety of sciences, including:

1. Physiology, which involves studying the effects of work on heart rate, oxygen consumption, perspiration, calories, extreme temperature stresses, and so on.
2. Anatomy, which involves studying the anatomical structure of the different body parts, their relationship to each other, how the human body works in general, and the effects of work on these body parts.
3. Psychology and human factors engineering, which involves the decision-making process of human beings, especially in critical situations, and recognition of human limitations in terms of visual, auditory, and other sensory capabilities.
4. Anthropometry, which is the study of physical dimensions in people, including the measurement of human body characteristics such as strength, size, breadth, girth, and distance between anatomical points. Anthropometry also includes segment masses, the centers of gravity

of body segments, and the ranges of joint motion, which are used in biomechanical analyses of work postures. These measurements are mostly useful for the design of things that are used by people, such as door heights, force needed to turn door knobs, and so on.

5. Biomechanics, which involves measuring forces, angles, loads, and the study of muscular activity (see a more detailed definition of biomechanics in this chapter).
6. Engineering, which considers the previous five sciences, uses engineering design concepts, and applies scholarly thinking to design better tools, machines, equipment, controls, and control rooms. Ergonomists typically find themselves working with a variety of engineers, including aeronautical, computer, information systems, nuclear, civil, and architecture (Wickens, Gordon, and Liu 1998).

As a result, the ultimate aim of ergonomics is to anticipate (and hence reduce and possibly eliminate) potential injuries, accommodate and enhance human performance, and provide an environment where humans and machines work seamlessly and in harmony with each other. This system ultimately enhances productivity, reduces claims costs, and pain and suffering, as well as provides workers with a life-long enjoyment at work, hence self achievement and worth (Sanders and McCormick 1993).

Recognizing Potential Ergonomic Problems Through the Anatomy and Physiology of Specific Disorders

Defining Cumulative Trauma Disorders (CTDs) and Associated Risk Factors

Ergonomic problems arise from minute injuries (microtrauma) to the musculoskeleton due to exposures to certain factors. When these exposures are repeated (repeated trauma) due to work habits and environment, they gradually develop into what is now known as cumulative trauma disorders (CTDs), also known as repetitive motion injuries (Putz-Anderson 1988).

CTDs were recognized in the early days of ergonomics and over the years by the trade with which they were associated. Names like bricklayer's shoulder, carpenter's elbow, stitcher's wrist, gamekeeper's thumb, housemaid's knee, telegraphers' cramp, and cotton twister's hand were recognized in the medical field as disorders that were caused by the type of work with which patients were involved. It was later that the factors that contribute to these disorders were recognized as what is known today as risk factors. While these factors may not be the sole cause of a CTD, the existence of one or more of them will enhance the probability that a CTD gradually develops.

These risk factors include:

- forceful muscle exertion
- repetitive motion
- awkward postures
- vibration
- contact stresses

Risk factors may also include exposure to extreme temperatures, and/or lack of rest to allow the body enough recovery time.

Types of Ergonomic-Related Injuries

Ergonomic-related injuries are either chronic (hence called cumulative trauma disorders), or acute, as a result of a blow, cut, or fall. These result in strains (injuring muscles and tendons) and/or sprains (injuring ligaments) (Putz-Anderson 1988).

The following paragraphs describe the different types of injuries and the different body structures that are affected by them.

Muscle Injuries: Muscles are made up of tiny fibers laid out in the same direction, and are filled with blood vessels to provide oxygen to the muscles and carry carbon dioxide and other waste products away from the muscles. Muscles can be injured in one of three ways: (1) muscle fiber can be strained or irritated; (2) tiny muscle fiber can be torn due to excessive use; and (3) muscle fiber can be crushed due to a severe blow (Putz-Anderson 1988).

Tendon Injuries: Tendons are ropelike structures that connect muscles to bones. Their job is to transfer the motion from the muscle to the bone. Like muscles, tendons can be torn like frayed rope or, because of use, can be inflamed, causing pain, such as tendonitis or inflammation of the tendons (*itis* is a Greek word meaning *inflammation of*). Tendon disorders of the upper limb can include tendonitis, tenosynovitis, stenosis tenosynovitis, trigger finger, De Quervain's disease, ganglionic cyst, unsheathed tendons, golfer's elbow, tennis elbow, and rotator cuff tendonitis (Putz-Anderson 1988).

Ligament Injuries: Ligaments are strong, ropelike fibers that connect bones to each other. Where there are joints, there are ligaments to bind bones together and to limit the range of motion. Injuries to ligaments can include torn fibers, or fibers that tear loose from the bones to which they are attached. Sometimes they can even be torn completely from the bone. This typically occurs due to a sudden strong blow to that area. Because ligaments have poor blood supply, it might take months for torn ligaments to heal (Putz-Anderson 1988).

Bursa Injuries: There are 160 bursae in the body. These are fluid-filled sacs that are located prevalently where there is potential for rubbing pressure due to motion of a tendon or a ligament on top of a bone. Their function is to act as a lubricant to facilitate and ease the motion of these ligaments, tendons, and bones against each other. If bursae did not exist where they are now, continuous rubbing action of tendons on bones will cause inflammation to the tendons, and will soon cause tears in these tendons/muscles/ligaments. Injuries to bursae can include torn sacs, leaky fluids, inflamed bursae, or hardened sacs. Injuries can also include sudden strong blows to the bursae, which can cause traumatic rupture of the sac (Putz-Anderson 1988).

Nerve disorders: As the name implies, nerve disorders are those that involve pressure on a nerve. Pressure can be applied by poorly designed tool handles, equipment, hard work surfaces, or nearby body parts that have been overworked or swollen, such as inflamed tendons, ligaments, or bones. Examples can include carpal tunnel syndrome and trigger finger. Pain and tingling sensations are similar to hitting the "funny bone" of the elbow (Putz-Anderson 1988).

Neurovascular disorders: These are disorders that involve some combination of nerves (*neuro*), and blood vessels (*vascular*). A prime example is when blood vessels between the neck and the shoulder are compressed, causing a limitation to the amount of blood supply to that area of the body. This compression can be caused by applying pressure directly to that area by a tool or equipment, or by continuously abducting and turning the shoulders, causing muscles of that area to apply the pressure. When blood supply (and the oxygen, nutrients, and other materials the blood carries) is limited, the recovery of damaged muscles and other parts is impeded or, at best, slowed down considerably. Examples can include thoracic outlet syndrome and a variety of vibration-related syndromes (Putz-Anderson 1988).

Some Common Disorders

As mentioned earlier, there are a large number of injuries that can occur as a result of exposure to risk factors. However, to limit our scope, we will only discuss some of the most prevalent injuries. These include:

- carpal tunnel syndrome
- trigger finger
- low-back pain
- thoracic outlet syndrome
- cubital outlet syndrome

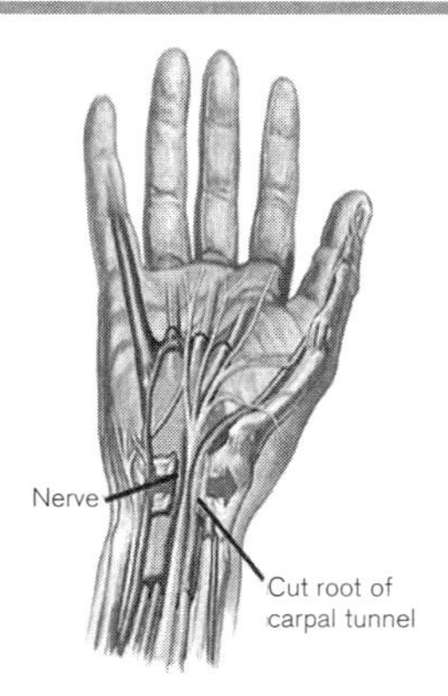

FIGURE 1 Carpal tunnel anatomy and surgery (*Source:* American Academy of Orthopedic Surgeons 2005)

Carpal Tunnel Syndrome

Carpal tunnel syndrome (CTS) is responsible for an average of 25 lost work days per case, resulting in an enormous amount of losses due to workers' compensation claims. Due to their smaller structure and hormones, females are at a higher risk of getting CTS than are males. Also, pregnancy causes a shift in female hormonal balance leading to inflammation as well.

Anatomy of CTS

CTS occurs when the median nerve, which passes through a tunnel that is located in the wrist area of the hand, is under pressure. The bottom of that tunnel is made out of tendons that are encased in tendon sheaths, and the transverse carpal ligament from the top. That ligament is as hard as bone and has no elasticity.

Causes of CTS

When an individual is exposed to awkward postures, excessive muscular motion of the hands and fingers, and other factors, the tendon sheaths on the bottom of the tunnel will get inflamed. This inflammation puts pressure on the median nerve which, in turn, is translated into symptoms of the CTS.

Symptoms of CTS

Symptoms can include numbness, tingling, atrophy, and general weakness in the opposition motion. These symptoms occur at the wrist area where the median nerve is located under the transverse carpal ligament, and radiates to where that nerve flows to from this point. This includes the thumb, the pointer finger, the middle finger, and the inside half of the ring finger (see Figure 1). Pain is mostly excruciating at night, as well as during exposure to the cause of the injury in the first place. Diagnosing CTS includes a medical examination, and in some cases, nerve conduction velocity testing, which rules out subjective diagnosis.

Treatment of CTS

Treatment of CTS includes rest and the wearing of braces to maintain the wrist in a neutral position, high doses of vitamin B6, aspirin or ibuprofen, which tend to reduce the inflammation of the tendon sheaths. Noninflammatory medicines are then administered. A cortisone injection can be administered to aid in the diagnosis process and to provide temporary relief of the symptoms. As a last resort, carpal tunnel release surgery or endoscopic carpal tunnel release surgery is conducted to cut open the transverse carpal ligament to release the pressure off of the median nerve. The body will then close that cut ligament by building scar tissue (Sechrest 1996a).

Trigger Finger

The tendons that control the motion of the fingers are held in place parallel to the finger via ligaments (pulleys). These pulleys are attached to the bones of the fingers and encase the tendons (see Figure 2).

Causes of Trigger Finger

As the name implies, trigger finger results from excessive and forceful use of the finger. Jobs such as spray painters and rivet gun operators are prime candidates for it. The excessive use of the finger results in inflammation of the sheath that encases the tendon. Since these ligaments are hard and have no give, the tendon sometimes gets too large to pass underneath the pulley. In worst-case scenarios, the tendon forms a bulge that prevents the tendon from gliding underneath the pulley. Similar to carpal tunnel syndrome, trigger finger is more common in women than men due to the smaller female structure. They occur most frequently in people who are between the ages of 40 to 60 (NIOSH 1997).

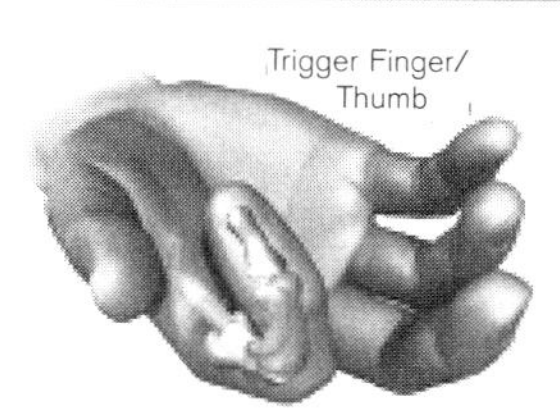

FIGURE 2 **Trigger finger/thumb** (*Source:* Medical Multimedia Group LLC 2001)

Symptoms of Trigger Finger

The symptoms include pain and numbness in that area, which in turn causes restricted motion of the finger. Forceful motion of the finger will result in excruciating pain. Diagnosis requires a physical exam. An occupational physician may also force the finger open. In doing so, the bulging tendon will glide underneath the pulley with an audible click and a lot of pain. The click is always a true sign of trigger finger.

Treatment of Trigger Finger

Like carpal tunnel syndrome, treatment might include braces, noninflammatory medicines, and in the worst-case scenario, surgery. During surgery, the ligament (pulley) is cut open to provide the inflamed tendon enough room to pass, thus eliminating the pressure and allowing the inflammation to subside. The body will then close that gap by building up scar tissue (Patel 1997c).

Back Problems

There are around 30,000 occupational back injuries per year, and 80 percent of the adult population in the United States complains of low-back pain. Back problems cost the highest as compared to other nonfatal occupational injuries. Back pain can be caused by: (1) irritation of the large nerve roots that go to the legs and arms; (2) irritation of the smaller nerves that innervate the spine; (3) strain of the large paired back muscles (*erector spinae*); (4) injury of the bones, ligaments or joints themselves; or (5) the disk space itself can be a source of pain due to an injury such as a herniated disk (Spine-Health.com 2005).

These injuries may be caused by trauma, trips and/or falls, aging-related, degenerative diseases of the bones, herniated disks, and/or lifting (either cumulative or abrupt). Lifting does not necessarily have to involve heavy objects to result in a back problem. In 1981, NIOSH developed a formula to be used as a guideline for jobs that require manual lifting as part of the day-to-day activities (NIOSH 1997). As specified by the NIOSH lifting formula (discussed in more detail in a later section of this chapter), factors that cause back problems due to lifting include the weight of the load, the angle of twist while lifting, the distance away from the body of the lifter, the frequency of lifting, the vertical distance of the lift, and the coupling of the load.

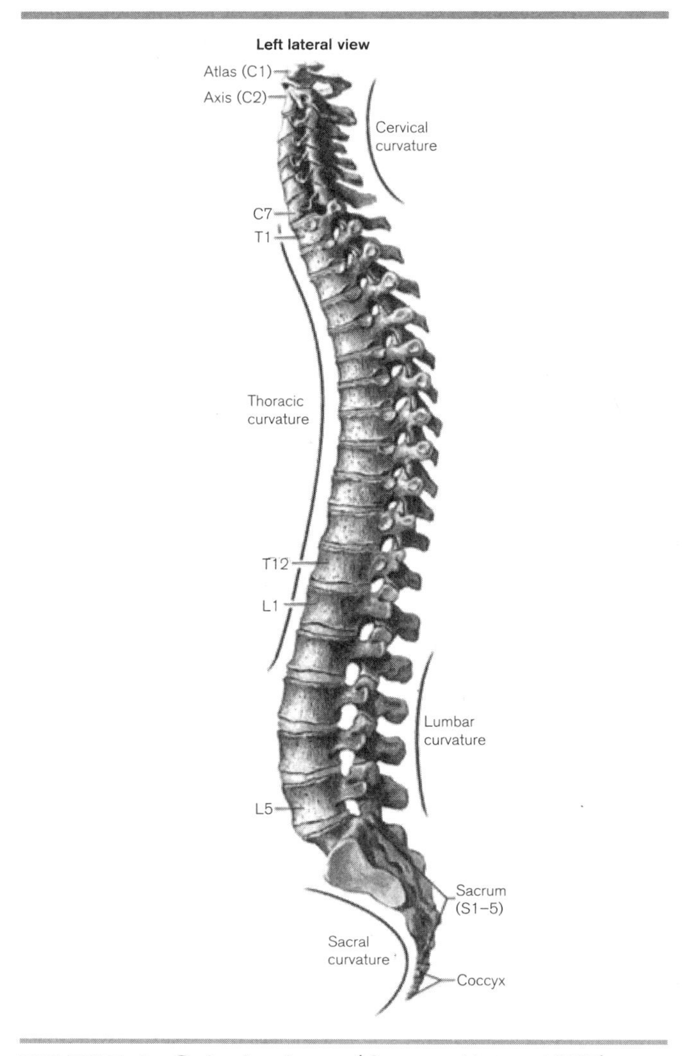

FIGURE 3 Spinal column (*Source:* Netter 1997)

Anatomy of the Back

The back consists of 24 vertebrae stacked on top of each other like a tower, with the thicker vertebrae on the bottom, and the finer, smaller vertebrae towards the top. When standing erect, a healthy spinal column should form a letter "S." The curves of the spine are to facilitate coupled motions of the back, including bending and squatting. These curvatures are called the *lordotic* and *kyphotic* curves (Meyer 2001).

There are four sections to the back: (1) the cervical section, which contains the seven vertebrae that make up the neck or upper back; (2) the thoracic vertebrae, which contain the twelve vertebrae that make up the middle back; (3) the lumbar vertebrae, which contains the five vertebrae that make up the lower back; and (4) the sacrum region. There are intervertebral disks separating the vertebrae (see Figure 3). Their function is to provide cushioning and separation between the vertebrae to prevent excessive rubbing action between vertebrae. Intervertebral disks are made out of an exterior annulus fibrosus, which is a cartridge-like shell, and an interior nucleus pulposus, which is a viscouslike paste.

Causes of Back Injury Due to Lifting

There are many scenarios by which a person can injure his or her back. One of the most common scenarios is as follows: A person who is standing up straight puts equal loads across the surface of the intervertebral disks. However, when this person bends at the back to lift an object, the load is shifted towards the front of the disk, and the rear of the disk is subjected to minimal forces. The resulting compression and sheer force can be enough to squeeze the interior pastelike material (the *nucleus pulposus*) out of the exterior shell (the *annulus fibrosus*) and push against the neighboring nerves. The pressure that these nerves are subjected to translates into pain. This breakage of the shell is commonly known as a *herniated disk*.

Treatment of Back Pain

Most back cases will heal with some physical therapy treatment sessions, if the pain resulted from causes other than a herniated disk. However, herniated disks are sometimes injected with a cortisone injection to help dissolve the nucleus pulposus that is pushing against the nerves. This is a series of two injections, six months apart. In a worst-case scenario, a procedure is conducted to surgically remove the cause of the pressure, sometimes removing the entire disk, and fuse the upper and lower vertebrae together. The success rate of this surgery is 75-80 percent. The procedure causes some range-of-motion disabilities and creates a fertile ground for further discomfort to the disks above and below the fused vertebrae.

A new treatment, developed and used in Europe and South Africa, calls for replacing the damaged disk(s) with an artificial material. This new procedure is now under extensive research in the United States in a number of large cities, such as New York, Los Angeles, and Chicago (Spinal Motion, Inc. 2005).

Thoracic Outlet Syndrome

The thoracic outlet is a space between the rib cage (*thorax*) and the collar bone (*clavicle*), through which the main blood vessels and nerves pass from the neck and thorax into the arm.

Causes of Thoracic Outlet Syndrome (TOS)

TOS is a neuro-vascular disorder that is caused by exposing the nerves and blood vessels that pass from the neck and thorax into the arm to excessive chronic pressure. The resulting symptoms are known as thoracic outlet syndrome. In some cases, the cause of compression is a reduction in the space through which these blood vessels and arteries pass (see Figure 4). This reduction in space and compression on the nerves and arteries can be due to repetitive activities that require the arms to be held overhead.

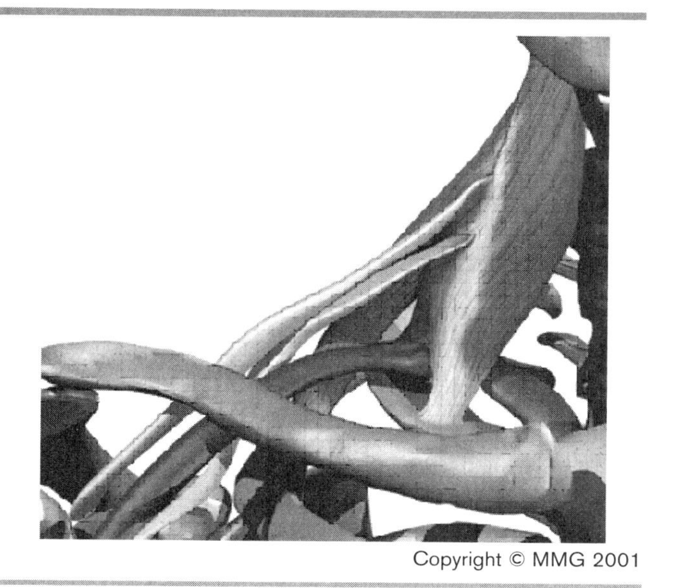

FIGURE 4 Thoracic outlet syndrome (*Source:* Medical Multimedia Group LLC 2001) 2001

Symptoms of TOS

Symptoms include a combination of pain, numbness, tingling (pressure on sensory nerves), weakness and fatigue (pressure on motor nerves), and/or swelling and coldness in the arm and hand (pressure on blood vessels). The symptoms can mimic many other conditions, such as a herniated disk in the neck, carpal tunnel syndrome, and even bursitis of the shoulder. As a result, TOS can be very difficult to diagnose.

Treatment of TOS

The length of time the arms are used in outstretched or overhead positions should be reduced and spaced out. Taking frequent breaks and changing positions are useful to prevent and eliminate TOS (Patel 1997b). If the symptoms persist, physical therapy and stretching may help alleviate the pressure. When all else fails, surgery for thoracic outlet syndrome is used to expand the tight space that caused the symptoms in the first place. This could be removing an extra rib, or removing the causes of the compression (Sechrest 1996b).

Cubital Tunnel Syndrome

The cubital tunnel is a tunnel that is located behind the funny bone (*medial epicondyle*) on the inside of the elbow. This tunnel is formed by the bone surrounded by muscles and ligaments. The ulnar nerve passes through the cubital tunnel on its way from the arm to the forearm and hand. The ulnar nerve runs to the little finger and the outside half of the ring finger (Eaton 1997).

Causes of Cubital Tunnel Syndrome

In normal subjects, bending the elbow causes the nerve to stretch several millimeters. When this is done repeatedly for activities that require repeated bending and straightening of the elbow in the workplace, the nerve becomes irritated and inflamed. In other patients, the nerve shifts and actually snaps over the prominence of the medial epicondyle. This snapping motion stretches and irritates the nerve. Leaning on the elbow, or resting the elbow on an elbow rest during a long-distance drive or while running machinery can cause repetitive pressure and irritation on the nerve. A direct hit on the tunnel can damage the ulnar nerve. When the ulnar nerve is exposed to pressure or is irritated at the elbow area, numbness and tingling is felt at the elbow area, and runs down the little finger and the ring finger of the hand.

Symptoms of Cubital Tunnel Syndrome

Early signs are numbness on the inside of the hand and in the ring and little fingers. Later there is weakness of the hand. There may be pain at the elbow. Tapping on the nerve as it passes through the cubital tunnel causes tingling or electric shock sensation down to the little finger.

Treatment of Cubital Tunnel Syndrome

The early symptoms of cubital tunnel syndrome usually respond to stopping the activity that is causing the symptoms. This is typically done by reducing the tasks that require repeated bending and straightening of the elbow. Frequent breaks from work for at least five minutes every half hour should be taken. In some patients, the symptoms might be worse at night because they sleep with their elbow bent. If the symptoms fail to respond to activity modifications, surgery might be required to stop progression of damage to the ulnar nerve. The operation moves the ulnar nerve from behind the medial epicondyle to the front of the medial epicondyle. This gives the nerve some slack and removes the stretching of the nerve (Patel 1997a).

Occupational Biomechanics

Biomechanics is a branch of the broader field of ergonomics. While ergonomics attempts to fit the task to the worker, biomechanics, according to Frankel and Nordin, ". . . uses laws of physics and engineering concepts to describe motion undergone by the various body segments and the forces acting on these body parts during normal daily activities." (Chaffin, Andersson, and Martin 1999)

Based on Chaffin, Andersson, and Martin (1999), occupational biomechanics is defined as "the study of the physical interaction of workers with their tools, machines, and materials so as to enhance worker performance while minimizing the risk of musculoskeletal disorders (MSDs)."

Simply put, biomechanics is the science that studies the effects of internal and external forces on the human body in movement and rest.

Chaffin defines two types of biomechanical injury mechanisms that are common to industry: (1) sudden force as a result of impact trauma, which has sudden outcomes such as amputations, fractures, contusions, and so on, and (2) volitional activity as a result of overexertion and CTD trauma, which has injuries that are cumulative over time, such as tendonitis, nerve entrapments, and other types of MSDs (Chaffin, Andersson, and Martin 1999). Occupational biomechanics is mostly interested in the latter type of activity.

History of Biomechanics

Biomechanics had early beginnings. Many early scientists developed systems to measure and count heart rate and other functions of the body. Leonardo da Vinci developed a great understanding of the functions of muscles and bones, and the musculoskeletal system in general. In the late 1500s, the physicist Galileo Galilei used a system of pendulum oscillations to simulate heart rates. William Harvey, in 1615, demonstrated the importance of capillaries in connecting veins and arteries. In the 1700s, Stephen Hales discovered the elastic functions of the aorta and its importance in producing a smooth flow from the heart to the body.

During the Industrial Revolution in the early 1900s, it was rationalized that the cost of a laborer was cheap, and that it was easier to replace a laborer if injured or killed than it was to fix a workstation. The twentieth century saw great progress in the field of biomechanics in terms of new discoveries, the development of new systems, programs, and standards in developed countries to reduce the pain and suffering experienced by workers. Today, the field of biomechanics uses many other fields of study to minimize the risk of musculoskeletal disorders (Chaffin, Andersson, and Martin 1999).

Methodologies in Biomechanics

As a result of the need for biomechanical analysis and studies, a variety of technologies and sciences started to develop. These include kinesiology, which includes kinematics and kinetics, biomechanical modeling, anthropometry, bioinstrumentation, motion classification and time prediction systems, and mechanical work capacities.

Kinesiology is the scientific study of human movement and the movements of implements or equipment that a person might use in a variety of forms of physical activity. Kinesiology includes kinematics and kinetics. Kinematics is the branch of mechanics dealing with the description of the motion of bodies, whole body, body segments, or fluids without reference to the forces producing the motion; whereas the science of kinetics studies these forces and other variables related to the motion of a body segment, such as acceleration, motion, or rate of change.

Biomechanical modeling is concerned with developing quantitative techniques to model and study the forces and moments associated with common manual tasks, such as lifting, on the human body. These developments started in the nineteenth century. However, the digital and computer age created excellent opportunities for great advancements in that field.

Anthropometry, as described earlier, is concerned with the collection, tabulation, and statistical analysis of different body segment measures of populations. New noninvasive techniques are now being adapted to accurately capture and log these measures for a variety of populations.

Bioinstrumentation is the development of new instruments, such as electromyography (EMG), that accurately collect data such as the forces produced by muscles during physical activities, force transducers and force plates, a variety of goniometers to measure angles of motion, and other meters and analysis instruments to measure a variety of forces and a variety of body segment angles as a result of motion during physical activities (Chaffin, Andersson, and Martin. 1999).

Motion classification and time prediction systems and mechanical work capacities are important to match the worker to the task. For example, NASA conducts aggressive studies on the cardiopulmonary functions of astronauts before, during, and after returning from space missions (Sawin et al. 2002). Similarly, some manual lifting tasks and extensive physical activity tasks require extensive screening to match the worker's work capacity to the demands of specific tasks. Again, the main objective is to increase worker productivity while reducing injuries associated with manual work.

Biomechanical Modeling

According to Chaffin, Andersson, and Martin (1999), biomechanical models have been used for various reasons. These include:

- enhancing the academic knowledge to obtain further insight as to how components of a system function and are coordinated to achieve desired outcomes
- providing the medical community with better understanding of how to estimate forces acting on different component body structures used for the prediction of the maximum allowable magnitude for a load held in various postures
- providing engineers with the necessary data to design tools, seats, and workplaces to provide workers with the least stressful configurations
- providing managers with early data to help consider a variety of alternative job conditions, work methods, and personnel stereotypes
- predicting potentially hazardous loading conditions on certain musculoskeletal tissues.

As a result, a number of biomechanical models have been developed and used over the past fifty years. These include planar static biomechanical models, three-dimensional modeling of static strength, dynamic biomechanical modeling, and special-purpose biomechanical models for occupational tasks.

Planar static biomechanical models include single-body-segment static models, such as those used to measure the forces acting on the one hand as a result of lifting an object. The single body segment in this case would be the forearm (see Figure 5).

Planar static biomechanical models also include two-body-segment static models, such as those used to measure the forces acting on the shoulder as a result of a load being applied at the hand. The two body segments in this case would be the forearm (hand to elbow) and the upper arm (elbow to shoulder).

The static planar model of nonparallel forces, such as those produced by an external force acting on the hand, is included and is at angle (α) to the vertical plane.

Planar static analysis of internal forces, such as those produced by the muscles at a distance of the joint to produce a state of static equilibrium, is also included. In this model, the forces in the single-body-segment static model are used to find the internal forces (F_M) exerted by the muscle at a distance (m) from the point of fulcrum of the elbow to maintain a state of static equilibrium (see Figure 5).

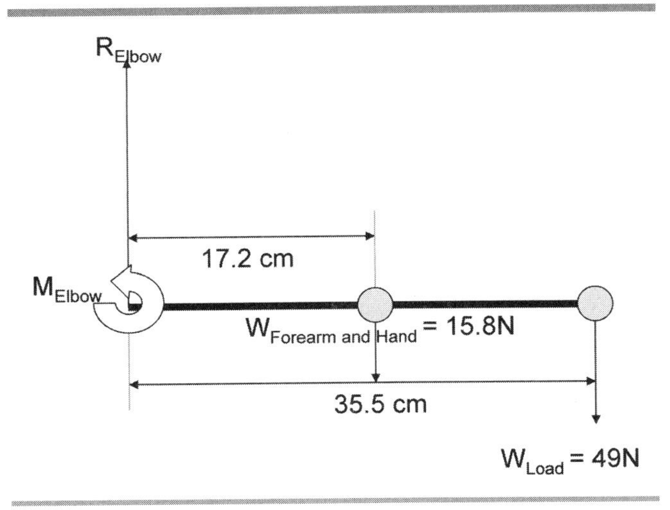

FIGURE 5 Single-body-segment static model of the forearm (Adapted from Chaffin, Andersson, and Martin 1999)

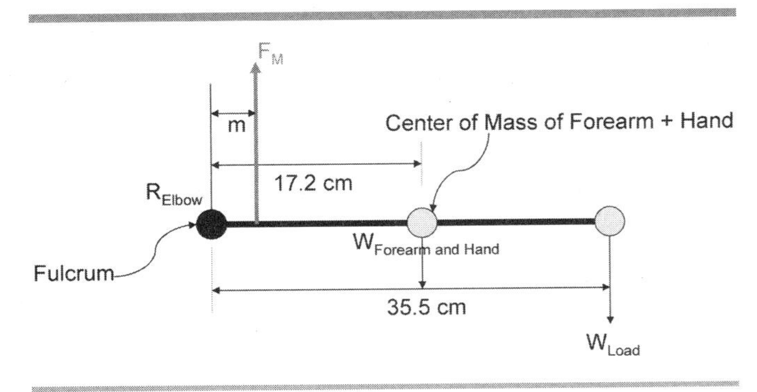

FIGURE 6 Planar static analysis of internal forces (Adapted from Chaffin, Andersson, and Martin 1999)

QUESTION FOR STUDY

Find the reactive forces and moment acting on the elbow as a result of a 5-kg weight lifted by the hand of an average-sized male, as shown in Figure 5.

According to Newton's second law,

$$F = ma \quad (1)$$

where

F = Force due to the load, or weight of the load (W_{Load})

m = Mass of the load (5 kg)

a = Acceleration due to gravity (9.8 m/sec²)

Therefore, $W_{Load} = 5 \text{ kg} \times 9.8 \text{ m/sec}^2 = 49 \text{ N}$ acting downward.

From anthropometric tables, the weight of the forearm and hand of an average-sized male is 15.8 N downward.

Finding R_{Elbow}:

Since this assembly is in static equilibrium, then the Σ_{Force} in the y-axis direction is 0. Hence,

$$\Sigma_{Force} = 0 \quad (2)$$

where

Σ_{Force} = the summation of all forces acting in the direction of the y-axis.

Therefore, $R_{Elbow} - (W_{Load} + W_{Forearm\ and\ Hand}) = 0$.

Therefore, $R_{Elbow} = W_{Load} + W_{Forearm\ and\ Hand} = 49 \text{ N} + 15.8 \text{ N} = 64.8 \text{ N}$ acting upward (in the opposing direction to the other two forces).

Finding M_{Elbow}:

Since this assembly is in static equilibrium, then the Σ_{Moment} about the elbow is 0. Hence,

$$\Sigma M_{Elbow} = 0 \quad (3)$$

where

Σ_{Moment} = the summation of all moments acting in the clockwise direction.

Then, $M_{Elbow} - [(W_{Load} \times 35.5 \text{ cm}) + (W_{Forearm\ and\ Hand} \times 17.2 \text{ cm})] = 0$.

Therefore, $M_{Elbow} = (W_{Load} \times 0.355 \text{ m}) + (W_{Forearm\ and\ Hand} \times 0.172 \text{ m}) = (49 \text{ N} \times 0.355 \text{ m}) + (15.8 \text{ N} \times 0.172 \text{ m}) = 20.11 \text{ Nm}$ counterclockwise (in the opposite rotation of the other two moments).

For the previous example, using a load of 5 kg, to find the forces exerted by the muscle (F_M) to maintain a state of static equilibrium, the Σ_{Moment} about the fulcrum of the elbow should be 0 (see Figure 6). Then

$$W_{Load} \times 35.5 \text{ cm} + W_{Forearm\ and\ Hand} \times 17.2 \text{ cm} = F_M \times m \quad (4)$$

where

$W_{Load} = 49 \text{ N}$

$W_{Forearm\ and\ Hand} = 15.8 \text{ N}$

$m = 5 \text{ cm (or 0.05 m)}$

Therefore, $F_M = (49 \text{ N} \times 0.355 \text{ m} + 15.8 \text{ N} \times 0.172 \text{ m}) \div 0.05 \text{ m} = 402.25 \text{ N}$ positive (upward).

The multiple-link coplanar static model is very similar to the above-mentioned two-body-segment static models, with additional multiple links. In this case, the body forces and moments are analyzed through the analysis of six links connecting six segments together (see Figure 7). These segments are: (1) the forearm, (2) the upper arm, (3) the upper torso, (4) the upper leg, (5) the lower leg, and (6) the foot. External loads are then added into the calculations based on the different angles and postures of the body. The links would naturally be: (1) the elbow, connecting the forearm to the upper arm; (2) the shoulder, connecting the forearm to the upper torso; (3) the hip, connecting the upper torso to the upper leg; (4) the knee, connecting the upper leg to the lower leg; (5) the ankle,

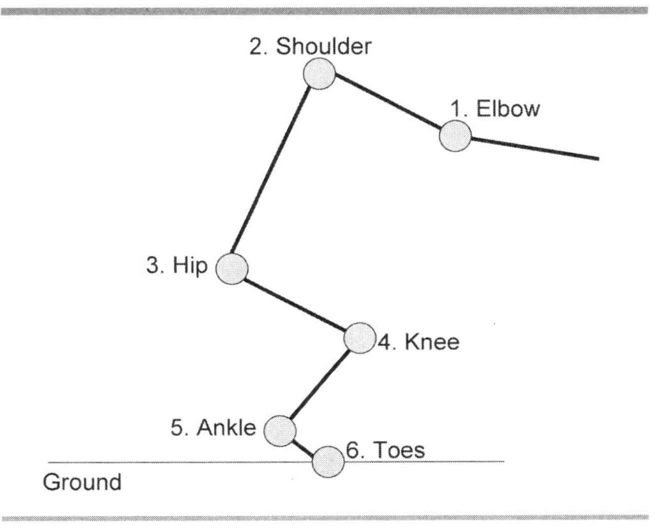

FIGURE 7 Six-segment body connected by six links (Adapted from Chaffin, Andersson, and Martin 1999)

connecting the lower leg to the knee; and (6) the toes, connecting the foot to the ground (see Figure 7).

Three-Dimensional Modeling of Static Strength

While two-dimensional modeling seemed simple and effective in solving and analyzing many occupational biomechanical settings, it proved to be inadequate where a worker has motions in the three axes *x*, *y*, and *z*. Examples include a worker pushing a cart with one hand while stabilizing the motion with the other. In cases such as these, the formulas may include the need to solve for six variables. These are the total forces (Σ_F) acting in the direction of each axis, and the total moment (Σ_M) around each axis. According to Chaffin, Andersson, and Martin (1999), Garg and Chaffin (1975) developed a linkage system model (see Figure 8) to represent specific anthropometric data, body postures, and loads of interest.

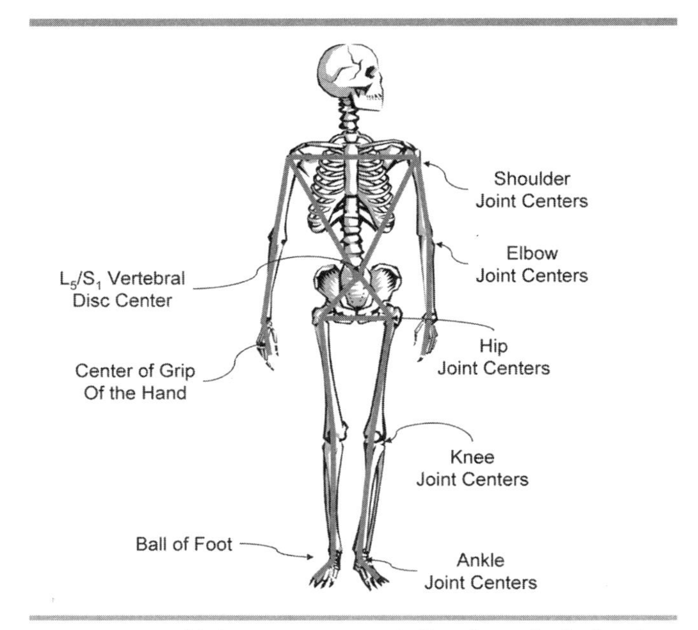

FIGURE 8 Linkage representation for whole-body biomedical modeling (Adapted from Chaffin, Andersson, and Martin 1999)

Outputs from this model include reactive forces and moments at each of the joints of the linkage. This model also computes the forces required by the muscles to perform a particular task, and thus determine the work load and physical strength required of workers performing the task. These complex vector algebraic calculations are now being performed more accurately and much faster by using computer systems, software, and programs such as the 3-D Static Strength Predicting Program (3D SSPP) developed by the University of Michigan's Department of Industrial Engineering (Chaffin, Andersson, and Martin 1999).

Dynamic Biomechanical Modeling

Dynamic biomechanical modeling presents different complexities from those presented by static modeling, such as kinematic measurements for the direction, velocity, and acceleration of body segments, as well as kinetic measurements for the forces and moments acting on these body segments.

Single-segment dynamic biomechanical models: Similar to single-segment static models, dynamic models present the same set of variables in addition to those introduced to the motion of a body segment. These new variables include inertia, radius of gyration, and centrifugal forces; hence, centripetal forces, velocity and acceleration, and the effects of that rotation on the joint centers adjoining that body segment. An example of the utilization of this model is modeling the forces and moments on the forearm resulting from lifting an object.

Multiple-segment biodynamic model of load lifting: Similar to multi-segment static modeling, biodynamic modeling calculates the forces and reactive moments resulting from a single segment's motion, then works its way backwards to calculate the resulting forces on the adjoining segments and joint centers, with the addition of other kinematic and kinetic considerations similar to those obtained in single-segment dynamic biomechanical calculations. An example of the utilization of this model is modeling the forces, moments, velocities, acceleration, inertia, and other variables from hand to the different joint centers, such as elbow, shoulder, hips, knees, and ankle joints via the different segments, such as the forearm, the upper arm, the upper torso, the upper leg, the lower leg, and the foot, as a result of lifting an object from floor to a table.

Coplanar biomechanical models of foot slip potential while pushing a cart: In this model, new external variables are introduced. These are the reacting forces of the cart on the worker's hand as a result of the push force, shear forces on the foot due to friction with the floor, and friction forces due to the worker's feet on the floor.

Special-Purpose Biomechanical Models for Occupational Tasks

These models use the previously discussed models to determine the required worker capability for manual tasks that are found to be taxing on the body. Special-purpose models have been developed for the wrist area, the back area, and other postures known to produce heavy muscle exertion on certain muscle groups. These models include modeling muscle strength, biomechanical models of the wrist and the hand, and lower-back biomechanical models.

Modeling muscle strength: The main purpose of modeling muscle strength is to predict human strength, based on age and gender, during a variety of tasks, lifting being one of them. While static muscle strength has been tested and examined, dynamic models that predict muscle strength are still under development.

Static models, as described earlier, state that, to maintain a state of equilibrium at a joint, the moment around that joint produced by the muscle should be equal to or higher than the moment produced by the external force exhibited by the load. Using this concept, a variety of data has been produced to predict the strength required by an individual (based on the average capability of each gender) using a variety of postures (Chaffin, Andersson, and Martin 1999).

For example, according to Chaffin, Andersson, and Martin, based on research done by Schanne and then corrected for population strength by Stobbe, the predicted shoulder flexion mean strength (S_S) in Newton meters (Nm) is given by this formula (see Figure 9):

$$S_S = [227.338 + 0.525\,\alpha_E - 0.296\,\alpha_S][G] \qquad (5)$$

where

α_E = Angle of the elbow joint
α_S = Angle of the shoulder joint
G = Gender adjustment = 0.1495 (for females), 0.2848 (for males)

Calculating the shoulder flexion mean strength (S_S) in Nm for a female when the angle of the elbow joint is 45° and the angle of the shoulder joint is 45°, since,

$$S_S = [227.338 + 0.525\,\alpha_E - 0.296\,\alpha_S][G]$$

then $S_S = [227.338 + 0.525 \times 45° - 0.296 \times 45°][0.1495] = 35.53$ Nm.

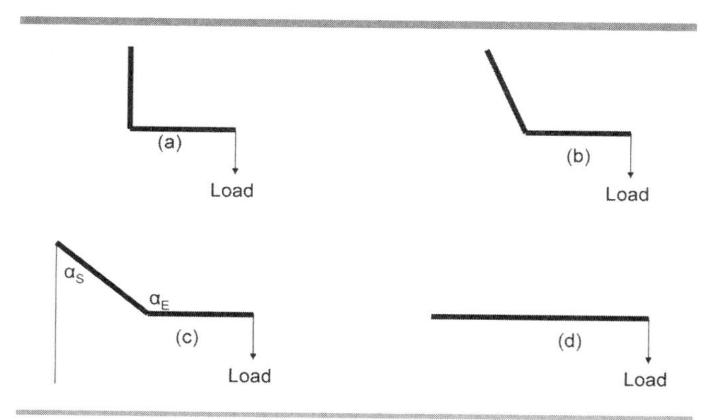

FIGURE 9 Postures: shoulder angles and elbow angles (Adapted from Chaffin, Andersson, and Martin 1999)

Table 1 shows different predicted shoulder mean strengths at different postures. As observed in the table, when substituting the different angles into the formula, the predicted shoulder mean strength for females is relatively constant at an average of around 42.5 Nm. Therefore, with the elbow extended (posture d in Figure 9), one may determine the maximum load capacity of a person by using the values for posture d in Table 1. Calculations show that this load should not exceed 69 N. Similarly, with the elbow close to the body, (posture a in Figure 9), the load should not exceed 155 N.

In addition to the average shoulder strength of females, figures such as these can be calculated for

TABLE 1

Predicted Shoulder Mean Strength at Different Postures

Posture	S_S (Nm)	α_E (degrees)	α_S (degrees)
a	41.050906	90	0
b	42.077971	120	30
c	43.105036	150	60
d	44.132101	180	90

(Adapted from Chaffin, Andersson, and Martin 1999)

other joints in the body, including the elbow, lower back, hip, knee, and ankle for both males and females. These predictions are very important in determining the amount of strength needed to perform a particular task, and the maximum loads that an individual may be able to handle. This may prove to be important for writing job descriptions, worker screening, and for manual-labor job placement to avoid potential injuries.

Biomechanical Models of the Wrist and Hand

Disorders of the wrist and hand have been followed and studied for the past 40 years. Expenses and lost work days as a result of these disorders are known to be costly. The most prevalent of these disorders include carpal tunnel syndrome, ganglionic cysts, and wrist tenosynovitis, which is the main cause of trigger finger (all discussed in a previous section of this chapter).

Biomechanical modeling was developed for the finger to measure the amount of force required to grasp a small object by the hand, to push a button with the finger, or to use the finger in trigger action motions, such as those used by spray painters. Data from finger anatomy, anthropometric data, tendons, and moment arms, and distances between centers of rotation were analyzed in terms of loads acting at the tip. Models similar to multi-segment systems were developed to measure the loads involved at each knuckle. The force required by the tendon to overcome the load applied at the finger tip can be calculated for a variety of hand sizes (see Table 2), where, F_t = force exerted by the tendon, and F_L = reactive force exerted by the load.

It is important to note that when forces are exerted by the tendon, it stretches by about 1–2 percent of its original length. Given time, the tendon does come back to its original length. According to Chaffin, Andersson, and Martin, this phenomenon was defined by Abrahams as *residual strain*. These exposures may cause ganglionic cysts and other forms of tendonitis (an inflammation of the tendons).

Regarding the wrist, the risk of developing wrist-related CTDs such as carpal tunnel syndrome, and the severity of such syndromes, was found to be related to the angle of flexion/extension, load, gender, wrist thickness, repetitive grip exertions, particularly with deviated wrists (flexed or extended), and the amount of stress the flexor tendons are exposed to (Silverstein, Fine, and Armstrong 1986). Furthermore, according to Rogers (1987), grip strength was found to be inversely related to the wrist angle of flexion or extension (see Table 3).

Lower-Back Biomechanical Models

Due to its functional importance, the frequency and high cost related to its injury, the back received a lot of attention from most ergonomists, ergonomics researchers, and the government. Furthermore, injuries associated with the back are most crippling and agonizing to their victims.

In addition to the NIOSH lifting formula described in the following section of this chapter, many other models were developed. Most researchers focused on the lower back because statistics showed this is where the majority of injuries occur to workers performing material-handling tasks. Also, biomechanical analysis showed that this is where most of the load is being handled. Therefore, it was determined that the lumbo-

TABLE 2

Tendon Grip Force (F_t) by Hand Size Based on Object Size

	Tendon Grip Force (F_t) by Hand Size	
Object Size	**Small Hand**	**Large Hand**
Grasping a small object	$F_t = 2.8\ F$	$F_t = 3.7\ F$
Pressing down, or grasping large object	$F_t = 3.1\ F$	$F_t = 4.3\ F$

(Adapted from Chaffin, Andersson, and Martin 1999)

TABLE 3

Grip Strength as It Relates to Wrist Deviation

Deviation	Angle of Deviation	Strength
Neutral	0°	100%
Ulnar	45°	75%
Radial	25°	80%
Flexion	45°	75%
Extreme Flexion	65°	45%
Extension	45°	80%

(Adapted from Chaffin, Andersson, and Martin 1999)

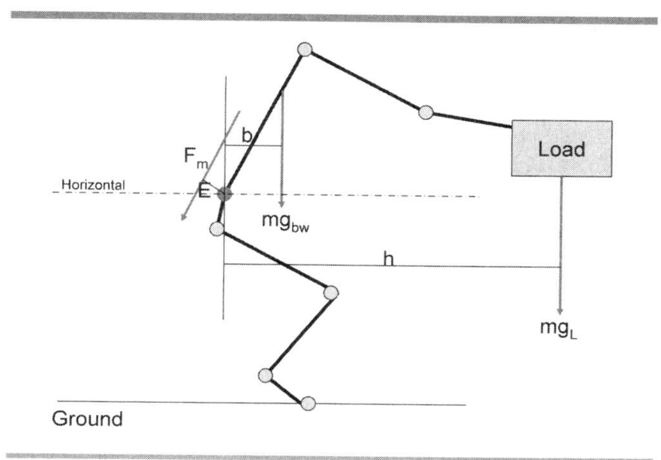

FIGURE 10 Simple lifting model (Adapted from Chaffin, Andersson, and Martin 1999)

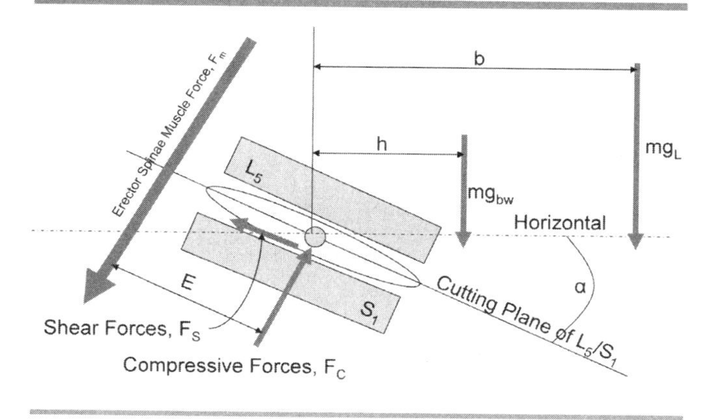

FIGURE 11 Compression and shear forces and angles acting on the L_5/S_1 disk (Adapted from Chaffin, Andersson, and Martin 1999)

sacral region of the back (L_5/S_1) should be used as the basis for setting limits for lifting and carrying tasks to avoid fatigue of the erector spinae muscle group, according to Chaffin, Andersson, and Martin (1999), based on research done by Tichauer.

Other research studies done by Krusen et al. and Armstrong and Smith et al. showed that 85 to 95 percent of all disk herniations occur equally at either L_4/L_5 or L_5/S_1 (Chaffin, Andersson, and Martin 1999).

Morris, Lucas, and Bresler determined there are two internal forces that act to overcome the external load moment. These are: (1) the force exerted by the erector spinae muscles located about 5 cm from the spine (this distance was later revised by other researchers to accommodate different anthropometric data), and (2) the internal abdominal pressure acting at the front of the spine (Chaffin, Andersson, and Martin 1999).

These findings were confirmed by other researchers as well. Chaffin (cited in Chaffin, Andersson, and Martin 1999), adopted the simple cantilever lower-back model of lifting as explained in Figure 8 where there is an imaginary link between the hip joint and the sacrum, and another between the sacrum and the shoulder, to represent the upper torso. Moments around the L_5/S_1 disk were calculated to obtain an idea of the resulting forces by the erector spinae muscle, and hence, compressive and shear forces acting on the L_5/S_1 disk. Figure 10 illustrates how the relationships are governed.

According to Figure 10, for equilibrium, there are no moments acting around the L_5/S_1 disk:

$$\Sigma M_{L_5/S_1} = b(mg_{bw}) + h(mg_L) - E(F_m) = 0 \qquad (6)$$

where

mg_{bw} = Upper body weight and mg_L = Load

F_m = Force acted by the erector spinae muscle group to maintain the body in equilibrium

b = Distance of the vertical component of the force representing the mg_{bw} from the L_5/S_1 disk

h = Distance of the vertical component of the force representing the load from the L_5/S_1 disk

E = Shortest distance of force acted by the erector spinae muscle group and the L_5/S_1 disk

Consider the following data: mg_{bw} = 350 N (upper body weight above L_5/S_1 disk) (Chaffin, Andersson, and Martin 1999):

mg_L = 450 N
F_m = Required
b = 20 cm
h = 30 cm
E = 6.5 cm

Hence, F_m = –3154 N (acting downward, or, according to Figure 10, counterclockwise).

To find the compressive forces and shear forces that the L_5/S_1 disk is exposed to (see Figure 11), one must first determine the angles at which all these forces are acting. Let α be the cutting plane of the L_5/S_1 disk and the horizontal.

According to Andersson (Chaffin, Andersson, and Martin 1999), that angle is equal to

$$40° + \beta$$

where

$\beta = -17.5 - 0.12T + 0.12e^{-2TK} + 0.5e^{-2T^2} - 0.75e^{-3K^2}$
T = Torso angle
K = Knee angle

Therefore, for the above example, if $T = 60°$ and $K = 120°$, then $\beta = 15°$ and $\alpha = 55°$.

For compressive forces:

$$\Sigma F_{\text{Comp}} = 0 \quad (7)$$

$$\Sigma F_{\text{Comp}} = \cos\alpha(\text{mg})_{\text{bw}} + \cos\alpha(\text{mg})_{\text{L}} + F_{\text{m}} - F_{\text{C}} = 0$$
$$F_{\text{C}} = \cos\alpha(\text{mg})_{\text{bw}} + \cos\alpha(\text{mg})_{\text{L}} + F_{\text{m}}$$
$$F_{\text{C}} = \cos 55(350) + \cos 55(450) + 3154 = 3612 \text{ N (downward)}.$$

For shear forces:

$$\Sigma F_{\text{Shear}} = 0 \quad (8)$$

$$\Sigma F_{\text{Shear}} = \text{sine}\,\alpha(\text{mg})_{\text{bw}} + \text{sine}\,\alpha(\text{mg})_{\text{L}} - F_{\text{S}} = 0$$
$$F_{\text{S}} = \text{sine}\,\alpha(\text{mg})_{\text{bw}} + \text{sine}\,\alpha(\text{mg})_{\text{L}}$$
$$F_{\text{C}} = \text{sine}\,55(350) + \text{sine}\,55(450) = 656 \text{ N (forward)}.$$

Note that, for the above example, the obtained values fall within the limits of a healthy individual. However, it is noteworthy to mention that these values change according to an individual's age, health, race, gender, and several other factors.

In addition to the previous back models, other models were also developed, including the *finite-element model*, which treats the disk and surrounding structures as a mesh of tensile elements, and calculates the stresses on each element as a result of lifting. According to Chaffin, Andersson, and Martin, other models included 3-D models to account for all other muscle groups in the cutting plane of the lumbar section. These included a ten-muscle 3-D model by Schultz and Anderson; a muscle geometry model where muscles were treated as pointwise connections from origin to insertion by Nussbaum and Chaffin; a vectors model where muscles were treated as 3-D vectors with directions, lengths, and velocities by Marras and Granata; and others as well. Other models involved dynamic lifting models of the back, including dynamic lifting, and dynamic push/pull models.

The National Institute of Occupational Safety & Health (NIOSH) Lifting Formulae

History and Development of the NIOSH Lifting Equation

The NIOSH lifting formula came about as a natural result of the fast developments in back modeling and analysis. It became a prime example, and one of the first special-purpose biomechanical models, for occupational tasks since manual lifting became a major concern and one of the biggest expenses incurred by employers, workers' compensation agencies, or anyone conducting business in this country in general.

The NIOSH lifting formula was initially published in 1981. However, new research showed that other components such as asymmetrical lifting, quality of hand-container couplings, work durations, and frequency of lifting played important roles in the task of lifting. Hence, a new NIOSH lifting equation was developed in 1991. The main purpose of the NIOSH efforts was to assist safety and health practitioners and ergonomists in assessing lifting tasks in an attempt to reduce material-handling-related back stresses and to stimulate further research and debate on the topic of lower-back pain.

Uses and Limitations of the NIOSH Formulae

NIOSH considered the research results of compression and shear strength of lower-back experiments conducted on cadavers by Genaidy et al. (Chaffin, Andersson, and Martin 1999). As a result, the current NIOSH lifting equation was built for a limit of 3400 N compressive force. This was determined to be the limit that 99 percent of male workers and 75 percent of female workers can withstand for occasional lifting tasks (fewer than 3 lifts/min). The NIOSH lifting equation assumes that:

- nonlifting activities are minimal
- other factors, such as unexpectedly heavy loads, slips, trips, and falls, are not accounted

for, and that environmental temperature and humidity are at comfortable levels (66–79° F and 35–50%, respectively)

- lifting is not conducted by one hand, while seated, kneeling, or constrained, or if the container has a shifting weight, such as those containing liquids
- the foot/floor surface provides at least 0.4–0.5 coefficient of static friction
- lowering or lifting pose the same hazard to the back.

According to NIOSH (1997), the NIOSH lifting equation does not apply to one-handed lifting/lowering when

- lifting/lowering for over 8 hours
- lifting/lowering while seated or kneeling
- lifting/lowering in a restricted work space
- lifting/lowering unstable objects
- lifting/lowering while carrying, pushing, or pulling
- lifting/lowering with wheelbarrows or shovels
- lifting/lowering with high-speed motion (more than 30 inches/second)
- lifting/lowering with foot/floor having least 0.4–0.5 coefficient of static friction
- lifting/lowering in an unfavorable environment, as indicated in item 2 above.

The equations basically state the following:

$$LI = L/RWL \qquad (9)$$

where

$$RWL = LC \times HM \times VM \times DM \times AM \times FM \times CM \qquad (10)$$

The following tables (Tables 4, 5, and 6) provide a list of definitions for the acronyms in the equation, as well as the values to use.

Note that tasks with more activities require the use of the more complicated, multitask analysis.

TABLE 4

NIOSH Lifting Equation: Acronym Definitions and Values

Acronym	Description	Definition	Value to Use in Equation	
			English	Metric
L	Load	The actual load being lifted (lbs in English, kg in Metric)	Actual load (in lbs)	Actual load (in kg)
LI	Load Index	L/RWL LI should be ≤ 1	Calculated	Calculated
RL	Recommended Weight Limit	The highest safe limit that nearly all healthy workers could lift for a substantial period of time (8-hrs) without adverse effects to their health	Calculated (in lbs)	Calculated (in kg)
LC	Load Constant	The highest safe limit that nearly all healthy workers could lift for a substantial period of time (8-hrs) without adverse effects to their health, provided that all other factors and conditions are not affecting the load negatively	51 lbs	23 kg
HM	Horizontal Multiplier	The factor affected by the horizontal distance of the load and the midpoint between the inner ankle bones of the lifter (H)	$10/H$ (H measured in inches)	$25/H$ (H measured in centimeters)
VM	Vertical Multiplier	The factor affected by the vertical distance of travel of the load (V)	$1 - (0.003\ \|V - 75\|)$ (V is in inches)	$1 - (0.0075\ \|V - 75\|)$ (V is in cm)
DM	Distance Multiplier	The factor affected by the horizontal distance of travel of the lifter carrying the load (D)	$0.82 + (4.5/D)$ (D is in inches)	$0.82 + (1.8/D)$ (D is in cm)
AM	Asymmetric Multiplier	The factor affected by the angle of twist of the lifter (A)	$1 - (0.0032A)$ (A is in degrees)	$1 - (0.0032A)$ (A is in degrees)
FM	Frequency Multiplier	The factor affected by the frequency of lifting (F).	From FM table (Table 2)	From FM table (Table 2)
CM	Coupling Multiplier	The factor affected by the coupling provided to the lifter in terms of cutouts, handles, or other means of providing control to the lifter.	From CM table (Table 3)	From CM table (Table 3)

Source: NIOSH 1997

TABLE 5

Frequency Multipliers (FM) Table

	Work Duration					
	< 1 hr		> 1 hr but < 2hrs		> 2hrs but < 8hrs	
Frequency, F (lifts/min)	$V < 30$	$V > 30$	$V < 30$	$V > 30$	$V < 30$	$V > 30$
< 0.2	1.00	1.00	0.95	0.95	0.85	0.85
0.5	0.97	0.97	0.92	0.92	0.81	0.81
1	0.94	0.94	0.88	0.88	0.75	0.75
2	0.91	0.91	0.84	0.84	0.65	0.65
3	0.88	0.88	0.79	0.79	0.55	0.55
4	0.84	0.84	0.72	0.72	0.45	0.45
5	0.80	0.80	0.60	0.60	0.35	0.35
6	0.75	0.75	0.50	0.50	0.27	0.27
7	0.70	0.70	0.42	0.42	0.22	0.22
8	0.60	0.60	0.35	0.35	0.18	0.18
9	0.52	0.52	0.30	0.30	0.00	0.15
10	0.45	0.45	0.26	0.26	0.00	0.13
11	0.41	0.41	0.00	0.23	0.00	0.00
12	0.37	0.37	0.00	0.21	0.00	0.00
13	0.00	0.34	0.00	0.00	0.00	0.00
14	0.00	0.31	0.00	0.00	0.00	0.00
15	0.00	0.28	0.00	0.00	0.00	0.00
> 15	0.00	0.00	0.00	0.00	0.00	0.00

Source: NIOSH 1997

TABLE 6

Coupling Factors (CM Table)

	Coupling Multiplier	
Coupling Type	$V < 75$cm (30")	$V > 75$cm (30")
Good: optimal handles	1.00	1.00
Fair: sub-optimal handles	0.95	1.00
Poor: no handles (e.g., bags)	0.90	0.90

Source: NIOSH 1997

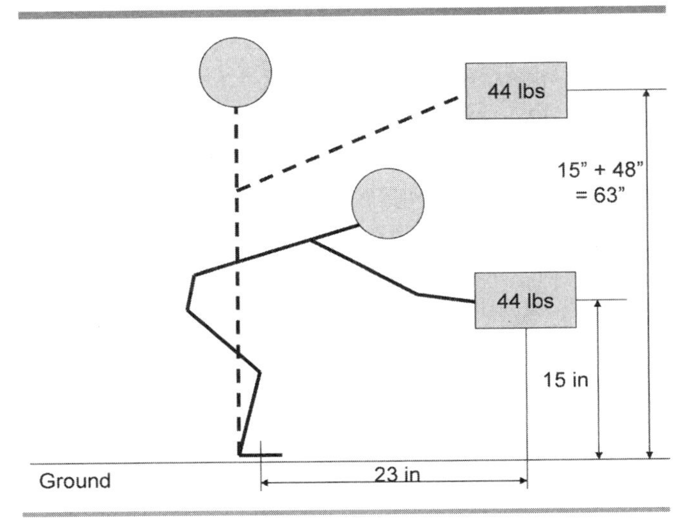

FIGURE 12 Lift "from–to" to illustrate the NIOSH lifting formula

QUESTION FOR STUDY

Given the following, refer to Figure 12 and apply the NIOSH lifting formula.

Load = 44 lbs
Hand location away from point connecting ankle bones at start of lift = 23" horizontally, and 15" vertically
Vertical distance of travel = 48"
Angle of twist = 0°
Duration = < 1
Frequency of lift = < 0.2 lifts/min
Coupling = Fair

Calculations:
It is important to calculate the *RWL* at both the origin and the destination of the lift to determine the most stressful situation.

Since LI = L/RWL

where

$RWL = LC \times HM \times VM \times DM \times AM \times FM \times CM$
and $LC = 51$ lbs

At the origin:

$HM = 10/H = 10/23 = 0.43$
$VM = 1 - (0.003 \mid V - 75 \mid) = 1 - (0.003 \mid 15 - 75 \mid) = 0.82$
$AM = 1$
$DM = 0.82 + (4.5/D) = 0.82 + (4.5/48) = 0.91$
$FM = 1$ (from Table 5)
$CM = 0.95$ (from Table 6)

Therefore, $RWL = 51 \times 0.43 \times 0.82 \times 0.91 \times 1 \times 1 \times 0.95 = 15.6$ lbs.

LI = $L/RWL = 44/15.6 =$ **2.82**

At the destination:

$HM = 10/H = 10/23 = 0.43$

$VM = 1 - (0.003\ |V - 75|) = 1 - (0.003\ |63 - 75|) = 0.964$

$AM = 1$

$DM = 0.82 + (4.5/D) = 0.82 + (4.5/48) = 0.91$

$FM = 1$ (from Table 5)

$CM = 0.95$ (from Table 6)

Therefore, $RWL = 51 \times 0.43 \times 0.964 \times 0.91 \times 1 \times 1 \times 0.95 = 18.28$ lbs.

LI = L/RWL = 44/18.28 = **2.4**

To improve this situation, one must consider increasing the smallest multipliers that are lowering the value of *RWL*, and in turn, lowering the value of LI.

From the example, the following multipliers were significantly low. Recommendations should be geared toward increasing their values: *HM* (0.43), *VM* (0.82), *DM* (0.91), and *CM* (0.95), respectively. To increase these values, the following suggestions could be applied:

- Bring the object closer to the body of the worker to reduce *H*, and therefore, increase *HM*
- Lower the destination of the lift to increase *VM* and *DM*
- Provide for better coupling.

References

American Academy of Orthopedic Surgeons (AAOS). 2005. *Your Orthopaedic Connection*. orthoinfo.aaos.org/main.cfm

British Broadcasting Corporation. 2007. *Historic Figures: Imhotep*. www.bbc.co.uk/history/historic_figures/imhotep.shtml

Chaffin, Don B., Gunnar B. J. Andersson, and Bernard J. Martin. 1999. *Occupational Biomechanics*. 3d ed. New York: John Wiley & Sons.

Eastman Kodak Company. 1983. *Kodak's Ergonomic Design for People at Work*. vol 1. New York: Van Nostrand and Reinhold.

Eastman Kodak Company. 1986. *Kodak's Ergonomic Design for People at Work*. vol 2. New York: Wiley & Sons, Inc.

Eaton, C. 1997. *E-hand.com—The Electronic Textbook of Hand Surgery*. www.e-hand.com

Medical Multimedia Group, LLC. 2001. eOrthopod.com. www.eorthopod.com/public

Meyer, Donald W. 2001. "Correction of Lordotic/Kyphotic S-Curves Without Extension Traction." *American Journal of Clinical Chiropractic*, April 2001. www.idealspine.com/pages/AJCC/AJCC_new/April2001/traction.htm

National Institute for Occupational Safety and Health (NIOSH). 1997. Publication 97-141: *Musculoskeletal Disorders and Workplace Factors—A Critical Review of Epidemiologic Evidence for Work-Related Musculoskeletal Disorders of the Neck, Upper Extremity, and Low Back*. www.cdc.gov/niosh/docs/97-141/

Netter, Frank H. 1997. *Atlas of Human Anatomy*. 2d ed. Summit, N.J.: Novartis Pharmaceuticals Corp.

Patel, M. R. 1997a. *Cubital Tunnel Syndrome*. www.handsurgeon.com/cubital.html

Patel, M. R. 1997b. *Thoracic Outlet Syndrome*. www.handsurgeon.com/thoracic_outlet

Patel, M. R. 1997c. *Trigger Finger*. www.handsurgeon.com/trigger.html

Putz-Anderson, Vern, ed. 1988. *Cumulative Trauma Disorders: A Manual for Musculoskeletal Diseases of the Upper Limbs*. New York: Taylor & Francis.

Ramazzini, Bernardino. 1713. *De Morbis Artificum* or *The Diseases of Workers*. Translated from the Latin by Wilmer Cove Wright (1964). New York: Hafner Publishing Company.

Rogers, S. H. 1987. "Recovery Time Needs for Repetitive Work." *Seminars in Occupational Medicine* 2(1):19–24.

Sanders, Mark S., and Ernest J. McCormick. 1993. *Human Factors in Engineering and Design*. 7th ed. New York: McGraw Hill.

Sawin, Charles F., Arnauld E. Nicogossian, A. Paul Schachter, et al. 2002. *Pulmonary Function Evaluation During and Following Skylab Space Flights*. www.lsda.jsc.nasa.gov/books/skylab/Ch37.htm

Sechrest, R. 1996a. *CTD: Carpal Tunnel Syndrome*. www.healthpages.org/AHP/LIBRARY/HLTHTOP/CTD/cts.htm

Sechrest, R. 1996b. *Thoracic Outlet Syndrome*. www.eorthopod.com/eorthopodV2/index.php/fuseaction/topics.detail/ID/79791a8f7dd9f446b38653cbeab9a955/TopicID/44a14668f9e0b1eafc5748bb059b5c60/area/6

Silverstein, B. A., L. J. Fine, and T. J. Armstrong. 1986. "Hand Wrist Cumulative Trauma Disorders in Industry." *British Journal of Industrial Medicine* 43:779–784.

Spinal Motion, Inc. 2005. *Kineflex*. www.SpinalMotion.com

Spine-Health.com. 2005. *Spinal Anatomy and Back Pain*. www.spine-health.com/topics/anat/a01.html

Wickens, Christopher D., Sallie E. Gordon, and Yili Liu. 1998. *An Introduction to Human Factors Engineering*. New York: Longman.

Appendix: Additional Resources

Abrahams, M. 1967. "Mechanical Behavior of Tendon in Vitro: A Preliminary Report." *Med. Bio. Eng.* 5:433.

Anderson, C. K., and D. B. Chaffin. 1986. "A Biomechanical Evaluation of Five Lifting Techniques." *Applied Ergonomics* 17(1):2–8.

Armstrong, J. R. 1965. *Lumbar Disc Lesions*. Baltimore, MD: Williams and Wilkins.

Board of Certified Professional Ergonomists. BCPE website. www.bcpe.org

Chaffin, D. B. 1987. "The Role of Biomechanics in Preventing Occupational Injury." Special Section, Conference on Injury in America. *Public Health Rep.* 102(6):599–602.

Chehalem Physical Therapy, Inc. 2000. *A Patient's Guide to Rehabilitation for Cubital Tunnel Syndrome*. Newberg, OR: Chehalem Physical Therapy, Inc.

Ergoweb, Inc. 2005. *History of Ergonomics*. www.ergoweb.com/resources/faq/history.cfm

Frankel, V. H., and M. Nordin. 1980. *Basic Biomechanics of the Skeletal System*. Philadelphia: Lea and Febiger.

Garg, A., D. B. Chaffin, and A. Freivalds. "A Biomechanical Computerized Simulation of Human Strength." AIIE Trans. March 1–15, 1975.

Genaidy, A. M. , S. M. Waly, T. M. Khalil, and J. Hidalgo. 1993. "Spinal Compression Tolerance Limits for the Design of Manual Manufacturing Handling Operations in the Workplace." *Ergonomics* 36(4):415–434.

Goldberg, Robert L. 2001. *The Medical Management of Upper Extremity Injury in the Occupational Setting*. www.neurometrix.com/papers_monos%20pdf/GoldbergMonograph.pdf

Human Factors and Ergonomics Society. HFES Web site. www.hfes.org

Kroemer, K. H. E. 1989. "Cumulative Trauma Disorders: Their Recognition and Ergonomics Measures to Avoid Them." *Journal of Applied Ergonomics* 20(4):274–280.

Kroemer, K. H. E., and E. Grandjean. 1999. *Fitting the Task to the Human*. 5th ed. Philadelphia, PA: Taylor & Francis.

Krusen, F., C. M. Ellwood, and F. J. Kottle. 1965. *Handbook of Physical Medicine and Rehabilitation*. Philadelphia: Sanders.

Maisel, M., and L. Smart. 1997. *Women in Science: A Selection of 16 Significant Contributors*. www.sdsc.edu/ScienceWomen/GWIS.pdf

Marras, W. S., and K. P. Granata. 1995. "A Biomechanical Assessment and Model of Axial Twisting in the Thoracolumbar Spine." *Spine* 20(3):1440–1451.

Martin, M. P. 2005. "Holistic Ergonomics: A Case Study From Chevron Texaco." *Journal of Professional Safety* (February) pp. 18–25.

Morris, J. M., D. B. Lucas, and B. Bresler. 1961. "Role of the Trunk in Stability of the Spine." *J. Bone Joint Surg.* 43A:327.

Nussbaum, M. A., and D. B. Chaffin. 1996. "Development and Evaluation of a Scalable and Deformable Geometric Model of the Human Torso." *Clinical Biomechanics* 11(1):25–34.

Roth, C. L. 2005. "How to Protect the Aging Work Force." *Occupational Hazards* (January) pp. 38–42.

Schanne, F. A. 1972. "A Three-Dimensional Hand Force Capability Model for the Seated Operator." Unpublished doctoral dissertation. Ann Arbor, MI: University of Michigan.

Schultz, A. B., and B. J. G. Andersson. 1981. "Analysis of Loads on the Lumbar Spine." *Spine* 6(1):76–82.

Schultz, A. B., B. J. G. Andersson, R. Ortengren, K. Haderspeck, and A. Nacherson. 1982. "Loads on the Lumbar Spine." *J. Bone Joint Surg.* 64-A:713–720.

Smith, A., E. M. Deery, and G. L. Hagman. 1944. "Herniations of the Nucleus Pulposus: A Study of 100 Cases Treated by Operation." *J. Bone Joint Surg.* 26:821–833.

Stobbe, T. 1982. "The Development of a Practical Strength Testing Program for Industry." Unpublished doctoral dissertation. Ann Arbor, MI: University of Michigan.

Tichauer, E. R. 1971. "A Pilot Study of the Biomechanics of Lifting in Simulated Industrial Work Situations." *J. Safety Res.* 3(3):98–115.

Waters, Thomas R., Vern Putz-Anderson, and Arun Garg. 1994. *Applications Manual for the Revised NIOSH Lifting Equation* (NIOSH Publication 94-110). Cincinnati, OH: National Institute for Occupational Safety and Health, Division of Biomedical and Behavioral Science.

Waterson, P., and R. Sell. 2007. *Chronology of the Society*. The Ergonomics Society. About the Society. www.ergonomics.org.uk/page.php?s=3&p=15

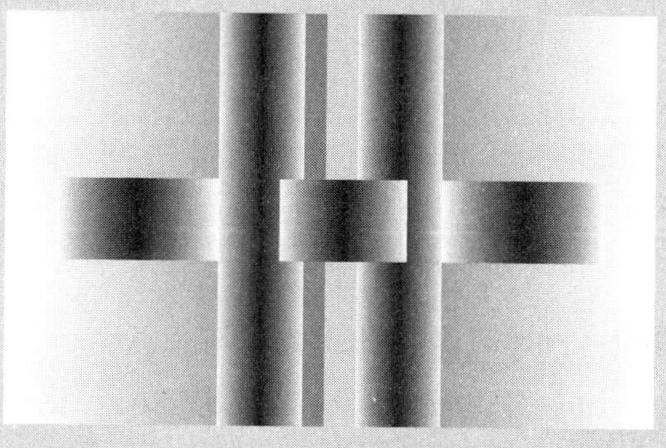

SECTION 6

ERGONOMICS AND HUMAN FACTORS ENGINEERING

APPLIED SCIENCE AND ENGINEERING: WORK PHYSIOLOGY

Carter J. Kerk

LEARNING OBJECTIVES

- Be able to define key terms.
- Apply introductory anthropometric design principles in work design using appropriate data and allowances.
- Understand the fundamentals of various systems in the body, including the skeletal, skeletal muscular, neuromuscular, respiratory, circulatory, and metabolic systems.
- Estimate energy requirements for walking and lifting.
- Understand the importance of aerobic work design.
- Be able to estimate oxygen consumption or uptake.
- Understand localized muscle fatigue, general physiologic and mental fatigue, and be able to identify possible control measures.
- Apply introductory work design principles to minimize fatigue and enhance performance.
- Understand work schedules and circadian rhythms.

WORK PHYSIOLOGY IS the study of physiological information about humans and how to apply that information in the evaluation and design of work. The term physiology is defined as the study of the processes and functions of an organism—in this case, the human organism. The realm of work physiology (including anthropometry for this chapter) includes: body systems (skeletal, skeletal muscular, neuromuscular, respiratory, circulatory, and metabolic), thermal stress, evaluation of cardiovascular capacity, fatigue, and work design. Work physiology and anthropometry can help the practicing safety professional to minimize occupational injuries while providing a safer workplace and improving productivity.

ANTHROPOMETRY

Definition and Use

The term *anthropometry* literally means *the measure of humans*. From a practical standpoint, the field of anthropometry is the science of measurement and the art of application that establishes the physical geometry, mass properties, and strength capabilities of the human body (Roebuck 1995). Anthropometric data are fundamental in the fields of work physiology (Åstrand and Rodahl 1986), occupational biomechanics (Chaffin, Andersson, and Martin 1999), and ergonomics/work design (Konz and Johnson 2004). Anthropometric data are used in the evaluation and design of workstations, equipment, tools, clothing, personal protective equipment, and products, as well as in biomechanical models and bioengineering applications.

Anthropometric Data

It is a fundamental concept of nature that humans come in a variety of sizes and proportions. Because there is a reasonable amount of useful anthropometric data available, it is usually not necessary to collect measurements on a specific workforce. The most common application involves design for a general occupational population. Some selected anthropometric data (body dimensions) are shown in Table 1. These data were collected on seminude subjects in rigid, erect postures; therefore, certain allowances must be applied for most practical uses. For shoe height, add 1 inch; for shoe weight, add 2 pounds; and for clothing weight, add 1 pound (Marras and Kim 1993). People rarely work in rigid, erect postures, so allowance for slumping may be appropriate for standing (subtract 1 inch) and sitting positions (subtract up to 1.5 inches) (Pheasant and Haslegrave 2006). For a comprehensive source of anthropometric data, Pheasant and Haslegrave (2006) and Roebuck (1995) provide extensive information about statistical aspects, data collection methods, gender differences, ethnic differences, and aging trends.

Anthropometric Design Principles

There are three general anthropometric design principles useful in the design of workspaces. Each design principle is described with its advantages and disadvantages.

1. **Design for Average.** With the *design for average* principle, you design a workspace for the average-sized person, or a one-size-fits-all approach. This is a commonly used approach by designers without knowledge of population variability and is generally not recommended. Another reason it is commonly used is it is normally the least-cost method. For example, a designer might design a standing workstation for light assembly work at the average standing elbow height for the 50th percentile male and female:

TABLE 1

Body Dimensions (Inches) of Seminude U.S. Adult Civilians

	Percentile							
	5th Fem	5th Male	50th Fem	50th Male	95th Fem	95th Male	Standard Develop Fem	Standard Develop Male
Heights (above floor)								
Height (standing)	60.1	64.8	64.1	69.1	68.4	73.5	2.50	2.63
Eye height	55.7	60.2	59.7	64.3	63.8	68.6	2.46	2.59
Shoulder height	48.9	52.8	52.5	56.8	56.4	60.9	2.28	2.44
Elbow height	36.5	39.2	39.3	42.2	42.3	45.4	1.76	1.89
Knuckle height[1]	26.4	27.6	28.7	30.1	31.1	32.7	1.46	1.61
Sitting Heights (above seat)								
Height (sitting)	31.3	33.6	33.5	36.0	35.8	38.3	1.37	1.40
Eye height	27.0	28.9	29.1	31.2	31.3	33.4	1.31	1.35
Shoulder height	20.0	21.6	21.9	23.5	23.8	25.4	1.13	1.17
Elbow height	6.9	7.2	8.7	9.1	10.4	10.8	1.06	1.07
Thigh height	5.5	5.9	6.3	6.6	7.1	7.5	0.48	0.50
Popliteal height (above floor)	13.8	15.5	15.3	17.1	16.9	18.8	0.93	0.98
Depths								
Forward reach (thumbtip)	26.6	29.1	28.9	31.5	31.4	34.1	1.43	1.54
Buttock-knee distance (sitting)	21.3	22.4	23.2	24.3	25.2	26.3	1.17	1.18
Buttock-popliteal distance (sitting)	17.3	18.0	19.0	19.7	20.8	21.5	1.05	1.05
Weight (lb)[2]			139.2	182.3				

[1]From Abraham, Johnson, and Najjar (1979)
[2]From Marras and Kim (1993)

(Unless otherwise noted, adapted from Kroemer, Kroemer, and Kroemer-Elbert 2001, p. 27)

[(39.3 + 42.2) ÷ 2] + 1 − 1 = 40.75 inches above floor level. (Remember to add 1 inch for shoe sole allowance and to subtract 1 inch for standing slump.)

The problem with the one-size-fits-all approach is that it fails to accommodate people at both ends of the population, specifically the shortest females and the tallest males. The shortest females forced to work at this assembly workstation will find the surface too high and may develop shoulder discomfort. The tallest males will find the surface too low and may develop low back or neck discomfort. If these discomforts lead to injuries and workers' compensation claims, then this will not be the least-cost method from a systems viewpoint.

2. **Design for Extreme.** The *design for extreme* principle is very useful in specific circumstances when it make sense to design a dimension at an extreme end of the distribution and, because of its function, the entire distribution is accommodated. Here are a few examples:

   - A doorway is designed so that extremely tall males and extremely broad people can fit through it. For example, an interior doorway may be designed to be 36 inches wide and 80 inches tall. Both these dimensions exceed the 99th percentile for height and body breadth. A doorway designed for the 50th percentile person would present problems, especially since there are life safety code and emergency egress concerns at play here.
   - If reach distances are designed for the shortest female to reach, then all will be accommodated. Do not design for fingertip reach, but for thumbtip reach, as this is more functional.
   - This principle should also be applied in general for strength requirements, with some precautions. If the weaker person has the strength capability for the task, then all will be accommodated. In some cases, this may set the strength requirement lower than practical and perhaps economically infeasible, so use reasonable judgment. Normally the people in the weakest tail of the strength distribution will self-select out of jobs with moderate to high strength requirements.

3. **Design for Range.** *Designing for the range* normally means designing an adjustable workspace. Returning to the standing workstation for light assembly, an adjustable-height workstation might be designed to accommodate elbow heights ranging from the 5th percentile female to the 95th percentile male, or 37.5 to 46.4 inches. Now the potential shoulder, neck, and low back discomfort discussed earlier may have been eliminated. Adjustability is one of the keys to effective ergonomic design. The adjustability function and features will come at a greater initial investment, but the potential increases in productivity, worker comfort, and reduced risk of workers' compensation claims will make this a favorable investment.

Most companies are not likely to initially practice the design for range principle in their workplaces because they fail to look at the economic advantage. Also, OSHA is not likely to include designing for range in regulations, so most companies will ignore this principle. However, it is important to note that this principle makes good economic sense and will improve productivity and worker satisfaction.

Practical Application of Anthropometric Data

As stated earlier, anthropometric data are useful in work physiology, occupational biomechanics, and ergonomic/work design applications. Of these, the most common practical use is in the ergonomic design of workspaces and tools. Here are some examples:

- The optimal *power zone* for lifting is approximately between standing knuckle height and elbow height, as close to the body as possible. Always use this zone for strategic lifts and releases of loads, as well as for carrying loads.

But minimize the need to carry loads—use carts, conveyors, and workspace redesign.

- Strive to design work that is lower than shoulder height (preferably elbow height), whether standing or sitting. (Special requirements for vision, dexterity, frequency, and weight must also be considered.)
- The upper border of the viewable portion of computer monitors should be placed at or below eye height, whether standing or sitting (Konz and Johnson 2004).
- Computer input devices (keyboard and mouse) should be slightly below elbow height, whether standing or sitting (Konz and Johnson 2004). Use split keyboards to promote neutral wrist posture (Konz and Johnson 2004). Learn keyboard shortcuts to minimize excessive mouse use. Use voice commands—speech recognition software is increasingly effective for many users and applications.
- For seated computer workspace, the lower edge of the desk or table should leave some space for thigh clearance (Konz and Johnson 2004).
- For seating, the height of the chair seat pan should be adjusted so the shoe soles can rest flat on the floor (or on a foot rest), while the thighs are comfortably supported by the length of the seat pan (Konz and Johnson 2004). Use knowledge of the popliteal (rear surface of the knee) height, including the shoe sole allowance from Table 1.
- The chair seat pan should support most of the thigh length (while the lower back is well supported by the seat back), while leaving some popliteal clearance (Konz and Johnson 2004). In other words, the forward portion of the seat pan should not press against the calf muscles or back side of the knees. Refer to the seated buttock-popliteal distance in Table 1.
- For horizontal reach distances, keep controls, tools, and materials within the forward reach (thumbtip) distance. Use the anthropometric principle of designing for the extreme by designing the reach distances for the 5th percentile female, thus accommodating 95 percent of females and virtually 100 percent of males.

Body Systems

In applying work physiology in the evaluation and design of work, it is essential to have fundamental knowledge of several relevant body systems, including the skeletal, skeletal muscular, neuromuscular, respiratory, circulatory, and metabolic systems. A brief introduction to each of these systems is presented to provide an elementary foundation. The practitioner is advised to learn these systems in more detail than is presented in this chapter. Recommendations on authoritative references include *Grant's Atlas of Anatomy* (Dalley and Agur 2004), *The Anatomy Coloring Book* (Kapit and Elson 2002), *The Physiology Coloring Book* (Kapit, Macey, and Meisami 2000), *Engineering Physiology* (Kroemer, Kroemer, and Kroemer-Elbert 1997), *Hollinshead's Functional Anatomy of the Limbs and Back* (Jenkins 2002), and *The Extremities: Muscles and Motor Points* (Warfel 1985).

The Skeletal System and Connective Tissue

A complete discussion of the skeletal system should include all connective tissue. Connective tissue includes bone, tendons, ligaments, fascia, and cartilage. Connective tissue provides support for the body and structural integrity of body parts and transmits forces (Chaffin, Andersson, and Martin 1999).

Ligaments, tendons, and fascia are dense connective tissue. Ligaments connect bone to bone and are quite significant in stabilizing joints. Tendons connect bone to muscle. The most significant tendon in the body is the tendocalcaneus (or Achilles' Tendon). Fascia covers muscle tissue, some internal organs, and holds a significant role in the makeup of the skin layers. Ligaments and tendons consist of dense, parallel fibers—both collagen (inelastic) and elastic—that are capable of powerful axial loads with a slight ability to stretch. The fibers in fascia are irregularly arranged and can resist loading in many directions about a plane, but not parallel to a plane, much like a fish net. There is virtually no vascularization (blood flow) to ligaments, tendons, and fascia, thus its healing ability is limited (Jenkins 2002).

There are two types of cartilage pertinent to our topic: hyaline and fibrocartilage. Hyaline cartilage covers the end of long bones at synovial joints. Synovial joints are lubricated, highly mobile joints throughout the body, including the knuckle, wrist, elbow, shoulder, hip, knee, and ankle joints. (Think of the white cartilage at the end of a chicken leg bone.) Hyaline cartilage is very thin and smooth, protecting bone wear and enhancing smooth joint mobility with the assistance of synovial fluid, while transmitting significant forces between bones.

The major location of fibrocartilage is found in intervertebral discs between each pair of vertebrae in the spinal column. It provides a cushion between the vertebrae and is extremely resistant to compressive forces (up to a point). Fibrocartilage in the intervertebral discs is not well designed to resist tension, shear, and torsion, a fact that provides some insight into some spinal injuries and ideas for proper work design. There is no vascularization (blood flow) in cartilage, but it does receive its nutrition by diffusion. Thus it heals much more slowly or less completely as compared to vascularized tissue (Jenkins 2002).

The final category of connective tissue is bone. Bone consists of collagen fibers in a mineralized (calcium) matrix. Bone is well vascularized (has a good blood supply). Bone functions as a support structure, a site of attachment for skeletal muscle, ligaments, tendons, and joint capsules, a source of calcium, and a significant site of blood cell development for the entire body.

There are five classifications of bones: *long* (clavicle, humerus, radius, ulna, phalanges, femur, tibia, and fibula), *short* (really cube-shaped): carpals (wrist) and tarsals (ankle), *flat* (cranial bones, ribs), *irregular* (scapula, pelvis, vertebrae, sternum), and *sesamoid* (patella) (Jenkins 2002). There are approximately 200 bones in the body. Many of these bones intersect at important joints. Synovial joints were discussed earlier. These joints provide for varying types of mobility, as well as allow for the passage of nerves and blood vessels, and consequently are sites for constriction or pressure.

TABLE 2

Selected Joint Motions or Functions

Joint Name	Degrees of Freedom	Motion or Function
Interphalangeal (finger and toe joints)	1	flexion, extension
Metacarpo-phalangeal, Metatarsal-phalangeal (hand to finger and foot to toe)	2	flexion, extension abduction, adduction
Wrist	2	flexion, extension radial & ulnar deviation
Elbow	2	flexion, extension forearm supination & pronation
Shoulder	3	flexion, extension abduction, adduction internal & external rotation
Hip	3	flexion, extension abduction, adduction internal & external rotation
Knee	2	flexion, extension internal & external rotation
Ankle	2	plantar flexion, dorsiflexion inversion, eversion

A summary of some of the major mobility joints of the body are presented in a simplified form in Table 2. A single degree of freedom consists of a single paired motion (for example, flexion and extension and abduction and adduction each represent one degree of freedom). Adduction means to *bring together*. Abduction means to *move apart*. If we move beyond the simplified presentation of joint movements, we will find that the shoulder, for example, also has movements of elevation, depression, protraction, and retraction because of the nature of the scapula floating on the posterior rib cage (Jenkins 2002).

A significant component of the skeletal system is the *vertebral column* or *spine*. It consists of four regions from superior (top) to inferior (bottom): *cervical*, *thoracic*, *lumbar*, and *sacral*. The cervical, thoracic, and lumbar regions provide varying (but somewhat limited) amounts of motion in three degrees of freedom (flexion/extension, side bending, and twisting). The spine provides a pathway and protection for the spinal cord, with pairs of spinal nerves emanating between intervertebral joints.

In addition to serving as a pathway and protector for the spinal cord, the vertebral column functions as a support structure for most of the body and is a site of attachment for a multitude of muscles and ligaments. The cervical region (neck) consists of seven vertebrae which support the neck and head [weighing approximately 6–8 pounds (2.7–3.6 kilograms)]. They are the most mobile vertebrae and provide the spinal nerves (brachial plexus) serving the upper extremities (the shoulder, upper arm, elbow, lower arm, wrist, hand, and fingers). The *thoracic* region (trunk) consists of twelve vertebrae (progressively larger moving inferiorly) which support the thorax, neck, and head, as well as articulation with twelve sets of ribs and twelve sets of spinal nerves. The thoracic region is much less mobile than the cervical region, largely because of articulation with the ribs. The *lumbar* region (lower back) consists of five vertebrae that are relatively quite massive and provide support for the entire upper body and torso. The lumbar region is capable of more mobility than the thoracic region. In part because of this mobility and considerable support demand, the lumbar region is the site of most clinical attention for injuries and surgical repair of herniated discs. When a disc ruptures (herniates) or bulges, there is great danger that it can compress against a spinal nerve, causing numbness, tingling, and sharp pain in parts of the body effected by that spinal nerve (sensory dermatomes). The *sacrum* is the final region of the spinal column and is actually a fused set of vertebrae, although the coccyx (tail bone) does float at the terminal end. The lateral sides of the sacrum form strong joints with the pelvis in the sacroiliac (SI) joint. The disc that sits atop the sacrum and below the fifth lumbar vertebra (the L5/S1 disc) supports the most weight of all the discs and is an important point of biomechanical and clinical interest (Jenkins 2002).

The Skeletal Muscular System

The human body possesses three types of muscle: skeletal, smooth (found in blood vessels and internal organs), and cardiac (found in the heart). The purpose of skeletal muscle is to move or stabilize body segments. There are 232 distinct skeletal muscles associated with the extremities, with at least another 42 distinct skeletal muscles devoted to the back. Skeletal muscles possess the ability to contract. Muscles are composed primarily of voluntary muscle fibers, but also contain small quantities of connective tissue and considerable blood vessels and nerves. Skeletal muscles are generally long and slender, and traverse a skeletal joint. Each end of a muscle is attached to a bone by one or more tendons. The body of the muscle consists of generally parallel muscle fibers (Jenkins 2002).

When a muscle contracts, we normally think of the muscle as shortening in length, exhibiting strength and providing force for work tasks. However, a contracting muscle may also retain its length (a so-called isometric exertion) or even lengthen (a so-called eccentric contraction). Take the example of holding a bucket with your elbow at a 90-degree angle. If you lift that bucket by flexing your elbow joint, your bicep muscle performs a shortening contraction, while the opposing tricep muscle performs a lengthening contraction. To lower the bucket by extending your elbow joint, the tricep muscle performs a shortening contraction, while the opposing bicep muscle performs a lengthening contraction (Kroemer, Kroemer, and Kroemer-Elbert 1997).

Each muscle consists of hundreds to thousands of muscle fibers, which are controlled in small groups by nerve cells. If you require a mild or controlling exertion, your body has learned to activate only a small group of nerve cells, which recruit a relatively small group of muscle fibers. If you require a strenuous, maximal exertion, your body has learned to activate as many nerve cells as possible, thus recruiting most, if not all, muscle fibers available in that muscle. In either case, these muscle fibers are activated on an *all-or-none principle* (Jenkins 2002). It is not possible to *partially* activate individual muscle fibers. But by the miraculous ability of the body to coordinate motor nerves, it is possible to smoothly control muscles on a continuum from a light, controlling contraction to an all-out maximal contraction. These amazing motor control programs are learned from the time of infancy through adulthood and are practiced and refined thousands of times. Consider that your body flawlessly executes complex motor control programs while climbing stairs or lifting a box.

A stroke may interrupt or erase such motor programs, but fortunately they can in some cases be relearned.

From an occupational safety and health viewpoint, it is important to focus on four points with regards to skeletal muscles: avoiding extreme exertions, avoiding overly excessive repetitive motions, avoiding awkward postures, and avoiding localized muscle fatigue.

Extreme Exertions: Overexertion from extreme use of force can produce an acute type of injury that can traumatically damage muscle, the muscle-tendon interface, the tendon-bone interface, and possibly rupture adjacent fibrocartilage (spinal discs). A good example of this case is forcefully attempting to lift an extremely heavy box whose weight you drastically underestimate. Also, force exertions at less than extreme levels that are repeated can lead to a chronic type of injury that develops over long periods of time.

Excessive Repetition: Overexertion from excessive repetition of joint motion can produce a chronic type of injury in the muscles and tendons. It may take weeks, months, or even years for these types of injuries to develop. In general, a job is considered repetitive if the basic cycle time is less than 30 seconds (Konz and Johnson 2004). The combination of repetition with excessive force may produce a detrimental effect that is worse than the sum of the parts.

Awkward Posture: Overexertion of skeletal muscles in awkward postures should be avoided, primarily because of poor mechanical advantage. In posture extremes, muscles may not be able to produce the forces required by the task, may place extreme stress on tendons, and may put pressure on nerve tissue and blood vessels. Awkward postures are usually at the ends of the range of motion for joints. The combination of awkward posture with excessive repetition and with excessive force may produce a detrimental effect that is worse than the sum of the parts.

Localized Muscle Fatigue: Prolonged isometric muscle exertions should be avoided. In these cases, the muscle motor units are over-used and the circulatory system is unable to provide oxygen and nutrients to the muscle cells, or remove carbon dioxide and lactic acid (Chaffin, Andersson, and Martin 1999). For example, avoid using the hand as a clamp or vise for extended periods of time. Further discussion of fatigue will follow later in the chapter.

The Neuromuscular System

The nervous system is an important control and regulation system in the human body. It gathers input from various sensors throughout the body, both internal and external. It processes information both in the brain and spinal cord to provide regulation of various body functions and control of motor activities. Our interests in work physiology focus primarily on thermal regulation and motor control. Thermal regulation will be addressed later in this chapter. This section will take an introductory look at motor control. Realize there is much more to the nervous system, but is beyond the scope of this chapter.

Anatomically, there are three major subdivisions of the nervous system: the *central nervous system*, the *peripheral nervous system*, and the *autonomic nervous system*. The central nervous system (CNS) includes the brain and spinal cord, and maintains primary controls. There are specific portions of the brain that regulate systems and aspects that are critical to the workplace, including respiration, cardiac function, digestion, attention, thermoregulation, learning, speech, vision, hearing, memory, emotions, and, most pertinent to our discussion, motor control. The CNS receives information from a multitude of sensors in the peripheral nervous system (Kroemer, Kroemer, and Kroemer-Elbert 1997).

The *peripheral nervous system* (PNS) includes the cranial and spinal nerves, transmitting signals to and from the brain along networks of nerve cells, or neurons. The PNS possesses sensors which respond to light, sound, touch, temperature, chemicals, pressure, and pain. Some of the most important sensors for motor control include *proprioceptors*, which provide information about the degree of stretch in muscles, the amount of tension in muscle tendons, the relative location of body joints, and even information about velocity and acceleration of joints. The vestibular system, which regulates equilibrium (or balance), is located in the inner ear and consists of three bony semicircular canals oriented at 90 degrees to one another. The PNS is a two-way system. So just as there are sensors bringing complex information to the CNS, the PNS is also delivering action signals from the CNS for motor control (Kapit and Elson 2002).

The *autonomic nervous system* (ANS) generates the *fright*, *flight*, or *fight response* and regulates involuntary functions such as cardiac muscle, internal organs, blood vessels, and digestion, all of which are critical to the workplace (Kapit and Elson 2002).

So, motor control is controlled from the CNS through the PNS to the muscular system. Some of this was discussed previously in the section title "Skeletal Muscle System." The CNS sends signals to specific motor neurons. Each motor neuron controls a set of muscle fibers and, when called upon, will stimulate those muscle fibers to contract with the help of a chemical neurotransmitter (acetylcholine) following the all-or-none principle discussed previously (Kapit and Elson 2002). This is important to our topic because of our earlier discussion of localized muscle fatigue and the more general discussion of fatigue that follows later in the chapter.

The Respiratory System

The *respiratory system* moves air to and from the lungs from the atmosphere. Within the lungs, part of the oxygen contained in the air is absorbed into the bloodstream. While delivering oxygen to the bloodstream, the air in the lungs simultaneously collects carbon dioxide, water, and heat from the bloodstream and then exhales them to the atmosphere. The respiratory path begins with the mouth and nose, passes through the throat (pharynx), voice box (larynx), and windpipe (trachea) and into the bronchial tree. The bronchial tree divides in 23 steps ending in microscopic spherical-shaped alveoli. There are hundreds of millions of alveoli, which provide as much as 80 square meters of gas exchange surface to the blood circulatory system in an adult (Kroemer, Kroemer, and Kroemer-Elbert 1997). The physiology of the bronchial tree and alveoli are of particular interest in industrial hygiene in the study of airborne contaminants (Nims 1999). Inspiration and expiration of air by the lungs is powered primarily by the contraction and relaxation of the thoracic diaphragm muscle and, to a lesser degree, by the intercostal muscles (between the ribs) (Kapit and Elson 2002).

Our primary interest in the respiratory system with regards to work physiology is both the quantity and quality of air moved in and out of the lungs. The quantity of air processed by the lungs is a function of body size, gender, age, conditioning, and work demand. Highly trained, large males have a lung capacity of 7–8 liters, of which about 6 liters is usable (vital capacity). Women have lung volumes about 10 percent smaller than male peers. Untrained persons have volumes of 60–80 percent of their athletic peers. At rest we breathe 10–20 times per minute, increasing to about 45 times per minute for heavy work or exercise (Kroemer, Kroemer, and Kroemer-Elbert 1997).

The quality of air processed by the lungs begins with the concentration of oxygen in the atmosphere. Normal atmospheric air consists of 20.9 percent oxygen. Concentrations less than 19.5 percent are considered oxygen-deficient and are a concern in confined spaces as well as other industrial environments. Without adequate oxygen, workers can become dizzy, uncoordinated, and pass out. Unless this situation is rectified quickly, brain damage or death can occur. Concentrations more than 23 percent are considered oxygen-enriched and can pose fire or explosion hazards (Nims 1999).

When at rest or during relatively sedentary activity, the body has little demand for oxygen and the exhaled concentration may be 19–20 percent oxygen. During more physically demanding work, the body demands and takes more oxygen. During extremely demanding work, the exhaled concentration may be as low as 14 percent oxygen. Athletes and trained persons also have the ability to process more oxygen than their less-trained peers at the same workload. This points to the importance of conditioning for workers, especially for physically demanding jobs, and for work-hardening when workers return from extended vacations or injury.

The Circulatory System

The *circulatory system*, using the heart as a pump and blood as the transport medium, plays an important role in transporting materials throughout the body. The blood carries oxygen, nutrients, carbon dioxide, heat, lactic acid, and water to and from various points about the body. The circulatory and respiratory systems are closely related at the alveolar interface in the lungs. The circulatory and metabolic systems are closely related through internal organs such as the stomach, intestines, and liver. The circulatory and musculoskeletal systems are closely related at the capillary interfaces in muscle cells (Kapit, Macey, and Meisami 2000).

The adult body contains 4–6 liters of blood, depending on gender and size. About two-thirds of this volume resides in the venous blood vessels (returning to the heart) and one-third in the arterial vessels (moving away from the heart). An important component of blood with regards to work physiology is hemoglobin, the iron-containing protein molecule of the red blood cells. Hemoglobin is a willing receptor for oxygen molecules, carbon dioxide, and carbon monoxide (Kapit and Elson 2002).

Working muscles and body organs place demands on the circulatory system for transport of oxygen, carbon dioxide, nutrients, lactic acid, and heat. The heart is the pump for this system. The output (liters per minute) of the heart pump has two primary input factors: the frequency of contraction [or heart rate (HR)] and the stroke volume. At rest, the heart pulses at 50–70 beats per minutes (bpm). This rate increases to as much as 200 bpm at extreme workload levels. Maximum heart rate can be estimated by the Karvonen Formula (Equation 1), which estimates that you lose about one beat per minute for each year of age after age twenty. This formula is only an estimate, and individual values can vary by plus or minus 15 bpm (Burke 1998).

$$\text{Heart Rate Maximum} = 220 - \text{Age} \quad (1)$$

A suggested estimate for older, fit individuals is shown in Equation 2 (Burke 1998).

$$\text{Heart Rate Maximum} = 205 - 0.5(\text{Age}) \quad (2)$$

In some cases, the heart rate can be used as an indicator of workload, but this is influenced by age, conditioning, and, to some extent, mental stress. Heart rate is a reasonable indicator of workload for light and moderate work demand. At rest, heart rate is influenced greatly by other factors, such as emotions and mental stress and cannot be relied upon as a physical indicator. The cardiac output at rest is about 5 liters per minute (l/min). At strenuous levels of exercise or work, this output can increase by fivefold to about 25 l/min. There are many factors that affect and control the cardiac output, starting with the workload demand, but also the conditioning and health of the heart, the flow resistance in the arteries and veins, and the capillary resistance in the muscle cells and organs (Kroemer, Kroemer, and Kroemer-Elbert 1997).

The four-chambered heart powers two closed-loop circulation systems. The right atrium receives deoxygenated blood from the muscles and organs through the venous return system. The blood is then pumped through the atrioventricular (AV) tricuspid valve to the right ventricle, which pumps the deoxygenated blood through the pulmonary semilunar valve via the pulmonary arteries to the alveolar interface in the lungs, where the hemoglobin exchanges carbon dioxide for oxygen. The oxygenated blood returns through pulmonary veins to the left atrium. The blood is pumped through the AV bicuspid (mitral) valve to the left ventricle, which in turn pumps the oxygenated blood through the aortic semilunar valve via the aorta and into the arterial system and on to various organs and muscle beds.

In the gastrointestinal tract, the blood gathers and transports nutrients from the digestive system for distribution to other organs and muscle sites. At the muscle sites, the blood is essentially trading oxygen and nutrients in the capillary beds for carbon dioxide, water, heat, and lactic acid. Finally, the blood flows back toward the heart through the venous return system.

This description is simplified and neglects other aspects, such as waste removal and filtration through the kidneys and other organs like the skin, bone, and brain, as well as fatty tissues and vessels supplying the heart muscle, and also the lymphatic system (Kapit and Elson 2002).

The Metabolic System

Now that we have covered a basic introduction to several body systems, including the skeletal, skeletal muscular, neuromuscular, respiratory, and circulatory, we can finally address the *metabolic system*, which is critical to our study of work physiology. The metabolic system is the process by which the body consumes and produces energy for the purposes of existence and work output. What follows is a simplified version of metabolism. References for this section include Åstrand et al. (2003); Kapit and Elson (2002); Kapit, Macey, and Meisami (2000); Konz and Johnson (2004); and Kroemer, Kroemer, and Kroemer-Elbert (1997).

The Human Engine and Energy Balance

Over time, the body maintains a balance between energy input and output. Most physiology texts describe the metabolic system as the engine in the human body. The metabolic system includes elements of several systems, including the digestive, respiratory, circulatory, and muscle systems. The inputs to the metabolic system are primarily oxygen and food. The digestive system receives and processes the food. The intestines and liver pass nutrients into the blood stream and they are delivered to various muscle sites about the body for metabolism of energy, usually in the presence of oxygen (aerobic), but sometimes without the presence of oxygen (anaerobic). The products from this metabolism are external energy (in the form of work) and internal energy (consumed to maintain body temperature and fuel internal organs). By-products of the metabolism are carbon dioxide, water, and heat. Some of the heat and water is transported close to the skin and lost through the skin by perspiration and convective heat loss. The remainder of the carbon dioxide, water, and heat is transported back to the alveolar interface in the lungs and expired by the respiratory system (Kapit and Elson 2002; Kapit, Macey, and Meisami 2000).

The energy balance equation, which is also used in the discussion of the thermoregulatory system, is as follows (Kroemer, Kroemer, and Kroemer-Elbert 1997; Åstrand, et al. 2003):

$$M \pm S \pm R \pm C \pm K - W - E = 0 \qquad (3)$$

where,

M = metabolic rate
S = heat storage rate
R = radiant heat exchange rate
C = convective heat exchange rate
K = conductive heat exchange rate
W = mechanical work rate
E = evaporative cooling exchange rate.

Units can be in watts (W) or in joules/second (J/s) (1 W = 1 J/s).

Equation 3 assumes that the person consumes exactly the same number of calories as burned through metabolism, otherwise the person would stand to lose or gain weight, or the equation would need a term for energy storage rate (lost or gained), primarily in the form of fat lost or added. In either case, the body is constantly striving for balance, or homeostasis. The metabolic rate is always positive as the body is constantly producing the energy needed for basic existence (basal metabolism: body temperature, base body functions, and blood circulation), plus what is required for current activity, and current digestive metabolism.

W is the mechanical work rate, or the external work produced. Note that W is always a loss (or zero in the resting case) from this equilibrium equation, as it is not normal to experience a gain of external work back to the system. Note also that evaporative cooling is always a loss (or zero) to the system.

S is positive if the body heat content increases, and negative for loss of body heat. Normally this number should remain close to zero, or at least be nonzero for very short periods of time, or else the body is risking hypothermia (negative) or heat stroke (positive). Radiant heat exchange would be positive due to exposure to the sun or a radiant heat source, such as a lamp or blast furnace.

R can also be negative if the body is radiating heat, thus the net R can be positive or negative. Convective heat gain occurs when air warmer than the skin temperature is encountered, and loss occurs for the opposite situation. Conductive heat gain occurs if the body is in physical contact with a warmer body, such as sitting on a heated surface, and heat loss occurs if seated on a cold surface.

Thermal Stress

Thermal stress is of particular interest in work physiology, as many work tasks must be carried out in extremes of heat and cold. The body's core temperature must be maintained close to 37°C (98.6°F). The key organs of the core are the brain, heart, lungs, and abdominal organs. Changes in core temperature of ±2°C affect body functions and task performance severely. Variations of ±6°C are usually lethal (Kroemer, Kroemer, and Kroemer-Elbert 1997).

Heat energy is circulated through the body by the blood. The blood flow can be controlled by vasomotor actions in the blood vessels. Vasomotor actions include constriction (narrowing of blood vessels), dilation (expanding of blood vessels), and shunting (shifting flow from superficial to deep blood vessels or vice versa). In a cold environment, heat must be conserved, and in a hot environment, heat must be dissipated, while gain from the environment must be prevented. The most efficient way to dissipate heat from the body to the environment is through the skin and, to a lesser extent, through the lungs. Involuntary shivering in the muscles results in internal heat generation. Of course, clothing and shelter can provide a significant effect on thermal control. Evaporative heat loss through the skin can take place when air moves against skin covered with perspiration. To promote evaporative loss, increase the amount of exposed skin by wearing less clothing and using fans to increase air flow. Lower humidity improves evaporative performance, while high humidity can make it much less effective.

In heat stress situations, it is desirable to limit radiant heat gain from sources such as the sun, a blast furnace, or open flames. Providing shade or a barrier can limit the radiant exchange. Clothing can be an effective barrier, but will interfere with evaporative cooling. In extreme heat stress situations, ice packs can be worn near the skin, the work rate may need to be limited, or the work time may need to be restricted. In extreme cold stress situations, it is important to keep skin covered and protected. Insulated clothing layers are necessary. Extra care must be taken to protect the fingers, feet, face and neck.

For a more in-depth coverage of this topic, consult Kroemer, Kroemer, and Kroemer-Elbert (1997) and Kapit, Macey, and Meisami (2000).

Energy Requirements for Work

Some examples of the energy required for activities (in watts per kilogram, W/kg) include sitting quietly or writing (0.4), standing office work (0.7), driving a car (1.0), washing floors (1.2), sweeping floors (1.6), heavy carpentry (2.7), cleaning windows (3.0), and sawing wood by hand (6.6). For total energy cost, add basal metabolism of 1.28 W/kg for males and 1.16 W/kg for females (Konz and Johnson 2004).

Watts are the SI unit of power. Power is the rate at which work is done, or (equivalently) the rate at which energy is expended (Hibbeler 2001). One watt is equal to a power rate of one joule of work per second of time. In these examples, the units are given in watts per unit of body weight in kilograms.

The metabolic cost of walking can be calculated as follows (Pandolf, Haisman, and Goldman 1976):

$$\text{Walking Metabolism} = C[2.7 + 3.2(v - 0.7)1.65] \quad (4)$$

in W/kg of body weight, where C, the terrain coefficient,

= 1.0 blacktop road, treadmill
= 1.1 dirt road
= 1.2 light brush
= 1.3 hard-packed snow
= 1.5 heavy brush
= 1.8 swamp
= 2.1 sand

and v = velocity in meters/second (m/s), where $v >$ 0.7 m/s.

The terrain coefficient represents the degree of difficulty presented by the walking. For example, it is much more difficult to walk in sand than to walk on a flat, hard surface. A 200-pound (90.7-kilogram) person, walking at 3 miles per hour (mph) (1.34 m/s), which is considered a normal walking speed, on a flat, hard surface ($C = 1.0$), would yield 4.23 W/kg of their body weight, or 384 W. A 150-pound (68.0-kilogram) person walking at 2 mph (0.89 m/s) in sand ($C = 2.1$) would yield 6.1 W/kg, or 415 W.

The metabolic cost of carrying a load in one's hands is 1.4 to 1.9 times the energy of carrying your own body weight (Soule and Goldman 1980).

The metabolic cost for lifting can be calculated as follows (adapted from Garg, Chaffin, and Herrin 1978):

$$\text{Lifting Metabolism} = [0.024(\text{BW}) + F(\text{Load Factor})] \div 0.014314 \quad (5)$$

in watts, where

BW = body weight (kg)
F = frequency of lifts (lifts/min).

$$\text{Load Factor} = \text{LFBW}(\text{BW}) + \text{LFL}(\text{W}) + \text{GF}(\text{W}) \quad (6)$$

where LFBW, the lift factor body weight,

= 0.00044, for arm lift
= 0.00265, for stoop lift
= 0.00419, for squat lift

LFL, the lift factor load,

= 0.02271, for arm lift
= 0.01147, for stoop lift
= 0.01786 for squat lift

GF, the gender factor,

= −0.00375(G), for arm lift
= −0.00617(G), for stoop lift
= −0.00507(G), for squat lift

G, the gender,

= 0, female
= 1, male

and W is the object's weight in kilograms.

For example, a 200-pound (90.7-kilogram) male performing arm lifts of boxes weighing 25 pounds (11.3 kilograms) at a rate of 4 lifts per minute would yield 223 W. A 150-pound (68.0-kilogram) female performing squat lifts of boxes weighing 15 pounds (6.8 kilograms) at a rate of 4 lifts per minute would yield 228 W.

In everyday activities only about 5 percent or less of the energy input is converted into work. The remainder is mostly converted into heat. In meeting work demands, the body is called to increase energy production up to 50 times that of resting state. In addition to the importance of temperature control in this equation, the body has a tremendous ability to meet the demand, largely depending on the circulatory and respiratory systems to serve the involved muscles in order to meet such a 50-fold requirement (Kroemer, Kroemer, and Kroemer-Elbert 1997).

TABLE 3

Interaction Between Aerobic and Anaerobic Processes to Meet Maximal Efforts

	Exercise Time, Maximal Effort		
Process	**10 seconds**	**10 minutes**	**2 hours**
Anaerobic			
kJ	100	150	65
kcal	25	35	15
Percent	85	10–15	1
Aerobic			
kJ	20	1000	10,000
kcal	5	250	2400
Percent	15	85–90	99
Total			
kJ	120	1150	10,065
kcal	30	285	2415

(Adapted from Åstrand and Rodah 1986, p. 325)

Aerobic and Anaerobic Processes

Under normal working conditions, it is extremely desirable for the body to be producing energy almost exclusively by aerobic (requiring oxygen) processes. In an aerobic task, a worker can easily take in all the oxygen required for the task with very little or no production of lactic acid in the muscle capillary beds. A good rule of thumb is that the worker should be able to talk easily while breathing during the work task.

When the body is asked to meet a demand of the greatest possible effort over short periods of time, it has the ability to meet that demand with anaerobic (without oxygen) processes, but at a very high cost. There will be a lactic acid buildup which will shut down the ability to continue to perform the effort. Buildup of lactic acid in the tissues quickly leads to muscular fatigue and a subsequent slowdown or shutdown of work. Significant buildup of lactic acid in the tissues takes considerable time to remove—one hour or more (Åstrand et al. 2003). Thus there is a heavy penalty for performing anaerobic work, and it should be avoided for normal working conditions. The relative interaction between aerobic and anaerobic energy processes is shown in Table 3. Note that a 2-hour maximal effort uses less anaerobic energy [65 kilojoules (kJ)] than a 10-second maximal effort (100 kJ). And a 10-second maximal effort uses less aerobic energy (20 kJ) than a 2-hour maximal effort (10,000 kJ) (Kroemer, Kroemer, and Kroemer-Elbert 1997, Åstrand et al. 2003).

The practical application of this information is to normally design jobs that are almost entirely aerobic to avoid the buildup of lactic acid. Four cases ranging from light work to extremely heavy work are given. An example of the physiologic response to work is shown in Figure 1 and illustrates the cases of light work and moderate-intensity work. Prior to onset of work, the body is in a state of equilibrium at resting level. When the work begins, the demand to perform the work is a step function. Even if the workload is relatively light, the aerobic response cannot satisfy the step function. But the body can meet the step demand with anaerobic processes. Soon the aerobic processes meet the demand and reach a state of aerobic equilibrium. Eventually the utilized anaerobic energy must be returned (or paid for), much like recharging batteries for their next use.

- During *light work*, the oxygen stored in the muscle plus the oxygen supplied from respiration and circulation will completely cover the oxygen need (repaying the oxygen deficit) (see Figure 1).
- During *work of moderate intensity*, anaerobic processes contribute to the energy output at the beginning of the task until aerobic processes can take over and completely cover the energy demand. The lactic acid produced diffuses into

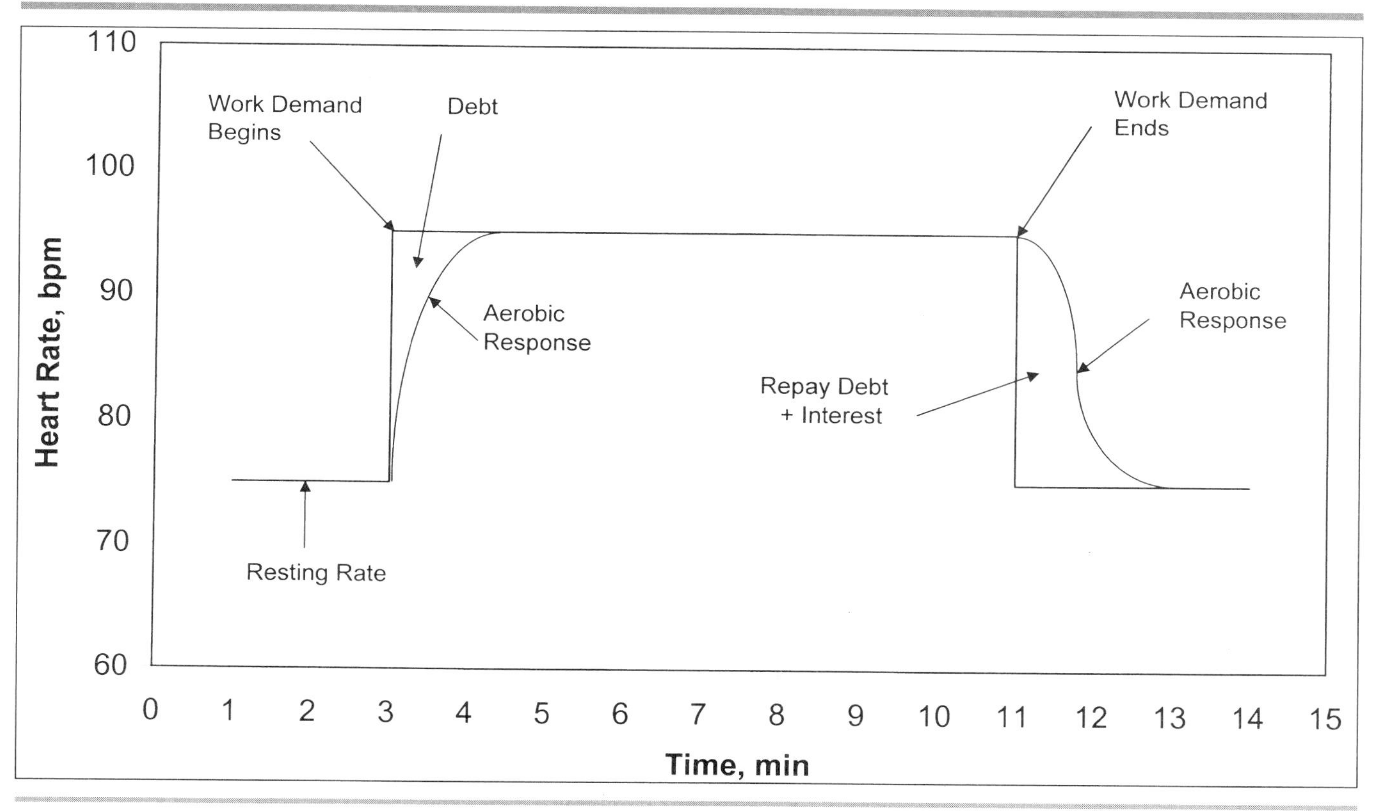

FIGURE 1. Physiologic response to work
(Adapted from Kroemer, Kroemer, and Kroemer-Elbert 1997, p. 217)

the blood, but the concentration returns to resting level and the work task can continue for hours (see Figure 1).

- During *heavier work*, the lactic acid production and concentration in the blood rises and remains high. The length of time the work rate can continue will be due largely to the motivation level of the worker.
- During *extremely heavy work*, the lactic acid concentration grows significantly and the oxygen deficit cannot be recovered. The work task cannot continue for more than a few minutes and it may take up to 60 minutes for the lactic acid concentration to recover to resting levels (Åstrand et al. 2003). When work tasks of this magnitude are required, it is necessary to implement work teams and generous work-rest ratios.

Part of this discussion centers on *demand versus capability*. As long as the aerobic capability of the worker exceeds the aerobic demand of the work task, you can design tasks suitable for continuous, full eight-hour shifts. A practical approach is to apply the rule of thumb suggesting workers should be able to talk easily while breathing. Otherwise you have the difficult task of estimating or measuring task requirements and estimating the aerobic capacity of the range of workers. This issue will be discussed in a later section.

The cardiovascular system has five different responses to work demand or exercise: changes in heart rate, heart stroke volume, artery-vein differential, blood distribution, and going into debt (as was just discussed). In response to increased workload or work demands, the body begins to accommodate and adjust by increasing the heart rate and stroke volume, thus delivering more oxygen to the muscle sites for metabolism. With the artery-vein differential response, the body takes more oxygen out of each unit of blood. Under resting conditions, the body only uses a portion of the available oxygen, therefore there is potential to use more when necessary. With the blood distribution response, the body has the ability to divert blood from areas of less critical immediate need to areas of greater need (see Table 4). Although the digestive system, kidneys, bone, and brain receive a smaller percentage of the blood volume during heavy work, the net volume of blood to these systems remains very similar to that supplied at rest because the cardiac output is up to five times greater during heavy work (Kroemer, Kroemer, and Kroemer-Elbert 1997; Åstrand et al. 2003; Kapit, Macey, and Meisami 2000).

Evaluation of Cardiovascular Capacity

Oxygen Uptake or Consumption

It would seem logical to evaluate the cardiovascular capacity of employees to determine safe and effective job placement. (Note that screening must be done carefully to maintain compliance with Federal Equal Opportunity Employment Laws and the Americans with Disabilities Act.) An individual's cardiovascular capacity can be determined by measuring maximum oxygen uptake, $\dot{V}O_2max$—or the rate at which one can process oxygen at maximal exertion (Åstrand, et al. 2003). But this is not a simple measurement to collect. On an absolute basis, the units for maximum oxygen uptake, $\dot{V}O_2max$; are in liters per minute (l/min); on a relative basis (with respect to body weight), the units for maximum oxygen uptake are in milliliters per kilograms per minute [ml/(kg/min)]. There are instruments that measure the concentration of oxygen exhaled as well as the volume of breathing air. This is normally done in an exercise physiology laboratory,

TABLE 4

Distribution of Blood at Rest and During Heavy Work

Organ/System	At Rest (%)	During Heavy Work (%)
Lungs/heart	100	100
Digestive system	20–25	3–5
Cardiac arteries	4–5	4–5
Kidneys	20	2–4
Bone	3–5	0.5–1
Brain	15	3–4
Fatty tissue	5–10	1
Skin	5	80–85
Muscle	15–20	

Note: Numbers are estimated percentages. Due to increased heart rate and stroke volume, the cardiac output can be five times greater when changing from rest to heavy work.

(Adapted from Åstrand and Rodah 1986, p. 152)

but there are portable instruments available for field studies (for example, Oxylog: PK Morgan, Kent, UK). Even the portable instruments need calibration and present a bit of an obstacle in the workplace. Fortunately, there is a relatively linear relationship between oxygen uptake (consumption) and heart rate during work, so the exercise rate can be used to estimate $\dot{V}O_2max$. (Kroemer, Kroemer, and Kroemer-Elbert 1997, Åstrand et al. 2003) (see Table 5).

The formulas for calculating maximum heart rate were previously given. The formula for calculating the percent of maximum heart rate (Equation 7 shows an example at 50 percent), requires an estimate of resting heart rate, which can be easily obtained (Burke 1998).

$$50\% \text{ of Max HR} = [(\text{Max HR} - \text{Resting HR})(0.5)] + \text{Resting HR} \quad (7)$$

For example, a 35-year-old worker has a resting heart rate of 65 bpm. Using Equation 1, we can estimate the maximum heart rate (Max HR) at 185bpm. Thus, using Equation 7, 25 percent of Max HR is approximately 95 bpm, 50 percent is approximately 125 bpm, and 75 percent is approximately 155 bpm.

There are maximal and submaximal stress tests that can be used to estimate $\dot{V}O_2max$. They each have their advantages and disadvantages. The treadmill tests have an advantage in that it is more of a whole-body test, and of course the individual must carry their entire body weight throughout the test. The speed and incline are dictated by the protocol.

The cycle ergometer tests have some advantages in that the instrument is more portable and does not require power. The ergometer test is more independent of body weight if that is desirable. The cadence and resistance is governed by the protocol. Instrumentation such as heart rate monitors and face masks are more stable with the ergometer.

The step tests are the most portable and may be best suited for field studies. The step height is dictated by the protocol. The step rate is governed by a metronome in an up-up-down-down sequence. The individual must carry their entire body weight. Note that with a fast cadence and fatigue setting in, the step test presents a trip hazard (Åstrand et al. 2003).

TABLE 5

Relationship Between Relative Heart Rate and Percent $\dot{V}O_2max$

Percent Max HR	Percent $\dot{V}O_2max$
35	30
60	50
80	75
90	84

Note: Regression equation for this data: Percent $\dot{V}O_2max$ = 1.07(Percent Max HR) − 10.2 (8)

(*Source:* E. R. Burke (ed.), *Precision Heart Rate Training*, 1998, p. 19)

The maximal tests are the most accurate, but require a trained staff (including emergency-trained medical personnel), more sophisticated laboratory equipment, and are more time consuming. The test is performed on a treadmill or cycle ergometer and the individual follows a test protocol that takes them to the point of complete exhaustion or collapse. For example, the maximal treadmill test calls for an increase in speed and slope every three minutes until even a world-class athlete reaches a collapse point. The maximum heart rate and $\dot{V}O_2max$ are measured directly just prior to collapse.

A set of submaximal stress tests are available for the treadmill, cycle ergometer, or raised steps. These tests are designed as submaximal and, by following the prescribed protocols, are inherently safer, especially for older or less-fit individuals. Heart rate data are sufficient, thus oxygen concentration and lung volume instrumentation is not needed. Laypersons can administer the test without medical personnel present. The test is essentially searching for the equilibrium heart rate required by the individual to perform at a fixed submaximal level. That equilibrium rate alone can be used to monitor the fitness of an individual performing the same test longitudinally. The equilibrium rate can also be used to extrapolate to the $\dot{V}O_2max$. It is not as accurate as the maximal test.

Two final field tests are presented that are quite simple in terms of equipment. All that is required is a timing device and a track or measured distance. Because the pace of these tests is not controlled, emergency-trained medical personnel should be present and a physician's consent for older or at-risk individuals should be obtained.

- Measure the distance a person can walk or run in 12 minutes (Konz and Johnson 2004).

$$\dot{V}O_2max = -10.3 + 35.3\,(DIST) \quad (9)$$

  - In ml/(kg/min)
  - DIST, miles covered in 720 s (12 min)

  For example, a person walks 1.0 mile in 12 minutes. Thus, their estimated $\dot{V}O_2max$ using Equation 9 is approximately 25 ml/(kg/min). If a person covers 1.5 miles in 12 minutes, their $\dot{V}O_2max$ is approximately 43 ml/(kg/min).

- Record the time required for a person to walk or run 2 kilometers (km) (Bunc 1994).
  - In ml/(kg/min)
  - $\dot{V}O_2max = 85.7 - 251.3(T)$ [for males] (10)
  - $\dot{V}O_2max = 61.9 - 124.2(T)$ [for females] (11)
  - T = time to run 2 km (hr)

  For example, a man and a woman each cover the 2-kilometer distance in 10 minutes (0.17 hour). Their approximate $\dot{V}O_2max$ using Equations 10 and 11 respectively is 43 and 41 ml/(kg/min).

In order to help evaluate the maximum oxygen uptake, some data for U.S. males are presented in Table 6. Females typically have a $\dot{V}O_2max$ 15–30 percent below males (Konz and Johnson 2004). If you compute absolute $\dot{V}O_2max$, in l/min and want relative $\dot{V}O_2max$ in ml/(kg/min), divide the absolute by the individual's body weight in kilograms.

The numbers in Table 6 represent a cross-section of U.S. males and are estimated for U.S. females at 75 percent of the male data.

Conversion of Oxygen Consumption to Energy and Energy Conversion

For *normal* adults on a *normal* diet doing *normal* work, we can calculate the energy conversion occurring in the body from the volume of oxygen consumed. Considering a *normal* nutritional diet of carbohydrates, fats, and proteins, the overall *average* caloric value of oxygen is 5 kilocalories per liter of oxygen (kcal/l O_2) or 21 kilojoules per liter of oxygen (kJ/l O_2). The energy conversion is fairly complex, but here are the necessary conversion factors:

$$1\ l\ O_2 = 5\ kcal = 21\ kJ \quad (12)$$

$$1\ kcal = 1\ cal = 1000\ cal \quad (13)$$

$$1\ W = 1\ J/s \quad (14)$$

$$1\ J = 1\ \text{Newton-meter (Nm)} = 0.239\ cal = 0.000239\ kcal = 10^7\ ergs = 0.938 \times 10^{-3}\ Btu = 0.7376\ ft\text{-}lb \quad (15)$$

$$1\ W = 0.85885\ kcal/hr = 0.014314\ kcal/min \quad (16)$$

Example

In an earlier discussion we learned that heavy carpentry required 2.7 W/kg + 1.28 W/kg for basal metabolism for males. Let us convert that to ml/(kg/min) of relative oxygen consumption: 3.98 W/kg (0.85885 kcal/hr per W) = 3.418 kcal/(kg/hr) = 0.057 kcal/(kg/min). Then multiply that by 1 l O_2/5 kcal = 0.01139 l O_2/(kg/min). = 11.39 ml O_2/(kg/min).

TABLE 6

Maximum Oxygen Consumption $\dot{V}O_2max$, in mL/(kg/min)

Cardiovascular Fitness	Male Age (Yeas)				Female Age (Years)			
	< 30	30–39	40–49	≥ 50	< 30	30–39	40–49	≥ 50
Very Poor	< 25	< 25	< 25	--	< 19	< 19	< 19	--
Poor	25–34	25–30	25–26	< 25	19–26	19–23	19–20	< 19
Fair	34–43	30–39	26–35	25–34	26–32	23–29	20–26	19–26
Good	43–52	39–48	35–46	34–43	32–39	29–36	26–35	26–32
Excellent	> 52	> 48	> 46	> 43	> 39	> 36	> 35	> 32

(Adapted from Cooper 1970, p. 28)

Let us assume we expect a person to work at that pace for an 8-hour shift. We are soon to learn that in the design of work, it is recommended an 8-hour shift not exceed 33 percent of $\dot{V}O_2max$. If we set 11.39 ml/(kg/min) as 33 percent of maximum, then the maximum would be 34.5 ml/(kg/min). Referring to Table 6, it would appear that males less than 30 years old would need to have *fair* or better cardiovascular fitness and those older than 30 would need to have, at least, *good* cardiovascular fitness to perform heavy carpentry work over an 8-hour shift and be able to avoid anaerobic metabolism and buildup of lactic acid. For females, accounting for a slightly lower basal metabolism, the maximum would be 33.5 ml/(kg/min). It would appear that females less than 30 years old would need to have *good* cardiovascular fitness or better and those older than 30 would need to have, at least, *excellent* cardiovascular fitness to avoid anaerobic metabolism for 8-hour shifts in heavy carpentry.

Classification of Work

Energy requirements for a job allow for judgment to be used to classify whether a job is easy or hard. Although these judgments rely on various factors, one such classification is given by Kroemer, Kroemer, and Kroemer-Elbert (1997) in Table 7. These measurements represent work performed over a whole work shift according to energy expenditure and heart rate.

TABLE 7

Classification of Light to Heavy Work

Classification	Total Energy Expenditure in kJ/min	in kcal/min	W	W/kg[1]	HR (bpm)
Light work	10	2.5	174	2.6	≤ 90
Medium work	20	5.0	349	5.1	100
Heavy work	30	7.5	523	7.7	120
Very heavy work	40	10.0	697	10.3	140
Extremely heavy work	50	12.5	872	12.8	≥ 160

[1]For a 150-lb (68-kg) person.

(Adapted from K. H. E. Kroemer, H. J. Kroemer, and K. E. Kroemer-Elbert 1997, p. 222)

Fatigue

A discussion of *fatigue* with regard to work physiology is essential. Some discussion of fatigue occurred earlier in this chapter. The following sections discuss localized muscle fatigue, general physiologic fatigue, and mental fatigue. Everyone experiences fatigue at some point during the day. It can be mental, physical, or both. It is usually accompanied by a loss of efficiency.

Localized Muscle Fatigue

Localized muscle fatigue was discussed in some detail earlier in the coverage of the skeletal muscular system. Prolonged isometric (static) muscle exertions should be avoided. In these cases, the muscle motor units are over-used and the circulatory system is unable to provide oxygen and nutrients to the muscle cells, as well as unable to remove carbon dioxide and lactic acid (Chaffin, Andersson, and Martin 1999). While prolonged static work should be avoided, it cannot be avoided altogether. All work tasks contain at least some elements of static work.

For example, while standing, several muscle groups in the legs, hips, back, and neck are tensed to hold that position. Standing tasks that include walking are much more comfortable because walking relieves the tension and ensures blood flow through the muscle tissues. When sitting down, the muscle tension in the legs, hips, and back are greatly reduced. However, prolonged sitting with little motion does not promote a healthy flow of blood through the muscle tissues. When lying down, almost all muscle tension is avoided. A recumbent position is the most restful and, as we know, is the best position for sleeping.

In summary, a standing/walking task is probably the healthiest and least likely to produce localized muscle fatigue. People who walk for a living need to be given a chance to sit occasionally, and be encouraged to wear comfortable, supportive footwear. People who stand for a living must be encouraged to move very frequently, sit occasionally, and wear comfortable, supportive footwear. People who sit for a living should use supportive, adjustable chairs, and should be encouraged to walk about frequently.

Static effort should be addressed if the following circumstances exist (Kroemer and Grandjean 1997):

- a high level of effort is maintained for 10 seconds or more
- a moderate effort persists for 1 minute or more
- a slight effort (about one-third of maximum force) lasts for 5 minutes or more.

General Physiologic and Mental Fatigue

In addition to localized muscle fatigue, there are other types of general physiologic fatigue and mental fatigue that can reduce or impair performance and lead to a general sensation of weariness. A feeling of weariness is not always unpleasant, especially if it is time to sleep or relax. It can be a protective mechanism to discourage one from overstraining and encourage rest and recuperation. In fact, rest and recuperation is the great balancing factor that acts to offset stress and fatigue. Recuperation takes place mainly during night-time sleep, but free periods during the day and all kinds of pauses during work also make their contributions. Some other types of somewhat distinguishable physiologic fatigue are (Kroemer and Grandjean 1997):

- Eye fatigue caused by straining the visual system
- General body fatigue, physical overloading of the entire organism
- Mental fatigue induced by mental or intellectual work
- Nervous fatigue caused by overstressing one part of the psychomotor system, as in skilled, often repetitive, work
- Chronic fatigue, an accumulation of long-term effects
- Circadian fatigue, part of the day-night rhythm and initiating a period of sleep

Some control measures for general physiologic fatigue that can be employed are:

- First, recognize the symptoms of fatigue. This is a responsibility for managers as well as workers.
- Realize that extended periods of overtime may help boost production, but possibly at a cost to the well-being of the workforce and to the quality of production.
- Work with ergonomists, safety engineers, industrial engineers, and industrial hygienists to ensure a sound design of work practices, work stations, tools, and the overall environment. Provide variety in the design of work. Use principles of job enlargement and job enrichment.
- Encourage worker participation. Use employee suggestion systems. Collect work feedback. Solicit workers for solutions to problems. Workers who participate in solving problems are more likely to accept changes in the workplace.
- Pay attention to lighting design. Fluorescents are generally over-used and can be harsh. Make use of indirect lighting where possible. Be vigilant for unnecessary glare on computer monitors. Tilt monitors down a few degrees to eliminate most glare. Align monitor surfaces perpendicular to external windows—never parallel.
- Provide comfortable break rooms that are separated from the work area. Maintain a comfortable temperature and humidity. Use indirect lighting. Provide an ample supply of fresh cool water and clean, easily accessible restrooms. Provide healthy food or snack alternatives, including protein.
- Train for and promote the concept of microbreaks. These are short pauses of less than a minute that are integrated into the workday, but do not significantly affect productivity. During these microbreaks, workers perform strategic stretching and strengthening exercises that promote blood flow and disrupt localized muscle fatigue.
- Provide regular employee training programs that discuss symptoms and causes of stress and fatigue. Educate employees on pertinent risk factors, with the realization that some of these risk factors come not only from the

workplace, but from hobbies, home life, and personal interrelationships.
- Provide access to health and fitness activities, or at least encourage participation.
- Provide employee assistance programs for those coping with financial, emotional, and substance abuse difficulties.

Work Design

What is the best way to design work? How much should you challenge workers physically? Should work be designed for gender or age differences? How should work schedules be designed? What effects do special work environments have (thermal stress, noise, and so on)? How can companies continue to meet the challenges of increasing productivity and improving quality?

There are ample opportunities in work design to address these issues. One of the major reasons for optimism is that most of these issues are misunderstood or ignored. Another reason for optimism is that addressing these issues will often lead to significant cost savings.

Design for Work Capacity

What portion of a person's overall physical work capacity is reasonable from a design standpoint? It was pointed out earlier that it is desirable to avoid anaerobic metabolism as much as possible. Here are some suggested guidelines. Note that the authors do not all agree, but together they do give a general sense of a reasonable work-design level. And just because workers can be pushed to these levels, does not mean they should be.

- 50 percent of maximum capacity for trained workers, 33 percent for untrained workers, reduce the level by 30 percent if the task is primarily upper-body work (Jorgensen 1985)
- Set lifting limits at 21–23 percent of uphill treadmill aerobic capacity or 28–29 percent of bicycle aerobic capacity, reduce to 23–24 percent of bicycle aerobic capacity for 12-hour shifts (Mital, Nicholson, and Ayoub 1993)
- 33 percent of maximum capacity for an 8-hour shift, 35.5 percent for a 10-hour shift, 28 percent for a 12-hour shift (Eastman Kodak 1986)
- 43–50 percent of maximum capacity for package handling for a 2-hour shift (Mital, Hamid, and Brown 1994)
- Assuming you want to exclude only a small percentage of the population, set your limits at about 350 W, 5 kcal/min, and 100–120 bpm (Konz and Johnson 2004)
- Do not exceed 110 bpm generally for a working day, do not exceed 130 bpm for intensive work periods (Wisner 1989)
- For extended periods of work, do not exceed a maximum of 35 bpm over resting level (Kroemer and Grandjean 1997)

Reduce cardiovascular stress first with engineering solutions, then administrative solutions. High metabolic rate jobs are prime candidates for mechanization. For material handling, consider use of conveyors, hoists, and forklifts. Workers should slide or lower objects rather than lift them. Use wheeled carts rather than carrying loads. Use powered handtools. Balancers, manipulators, and jigs can reduce static loads. Administrative solutions include job rotation and part-time work. When setting work standards, apply reasonable fatigue allowances (Konz and Johnson 2004).

Design for Gender and Age Differences

There are physiological differences between males and females. Females have a $\dot{V}O_2max$ 15–30 percent lower than males on average, largely because of a higher percent of body fat and a lower hemoglobin level. Females also have lower blood volumes and lung volumes on average. Rather than designing jobs differently for each gender, design jobs at a reasonable percentage of maximum for females and males will be accommodated also (Konz and Johnson 2004).

The physiological peak performance age is approximately between ages 25 and 30. After age 30, there is a steady decline in physiological performance. A decline in $\dot{V}O_2max$ is estimated at 1–2 percent per year after age 25, but there are large individual variations (Illmarinen 1992). Jackson, Beard, Weir, and Stuteville (1992) estimate the aging effect at 0.27 ml/(kg/min) per year, but emphasize that most of the decline is due to physical activity level and body fat and not actually aging. This implies that companies are wise to promote fitness and wellness programs. From a work design standpoint, light to moderate physical work is not sensitive to aging up to about age 65. But hard, exhausting work is strongly age-dependent, with a maximum capacity between ages 20 and 25 (Konz and Johnson 2004).

Work Schedules and Circadian Rhythm

Shift work is popular because companies want to maximize the use of machines and production facilities, as well as provide customer service 24 hours a day, 7 days a week. Many shift alternatives have developed in addition to the traditional work week of 8 hours a day, 5 days a week. Alternative compressed work weeks have become commonplace:

- 4 days of 10 hours/day
- 4 days of 9 hours/day, plus 4 hours on Friday
- 4 days of 9 hours/day in week 1, then 5 days of 9 hours/day in week 2
- 3 days of 12 hours/day in week 1, then 4 days of 12 hours/day in week 2

The advantages to these schedules are longer weekends and fewer commutes. A disadvantage is shorter overnight recovery times. As discussed earlier, the body needs adequate muscle recovery times rest to reduce overall fatigue.

Overtime is frequently used to extend the work week to meet increased production demand. Many hourly employees are motivated to work the overtime to increase their income. But overtime cuts into recovery time, with possible poor consequences for muscle recovery and fatigue. Short periods of overtime are acceptable, but prolonged overtime will begin to lose its advantages to the potential increases in fatigue, injuries, and workers' compensation claims. Part-time workers, temporary workers, and job-sharing can provide some relief for this issue.

Circadian rhythm is the body's internal cycle that lasts about 24 hours (between 22 to 25 hours) and has natural variations for numerous body functions, including body temperature, heart rate, blood pressure, respiratory volume, adrenalin production, mental abilities, release of hormones, and melatonin production. Cortisol (the "wake-up" hormone) peaks around 9 AM, and melatonin (the "go-to-sleep" hormone) peaks around 2 PM.

The most important function geared to circadian rhythm is sleep and, for most people, a normal pattern is to sleep at some stretch during the night and generally be awake and active during the day. Challenges arise when workers are asked to work at times that disrupt the circadian rhythm, such as work shifts during the evening hours or during the night-time hours. In general, the human organism is performance-oriented during the daytime and ready for rest at night (Kroemer and Grandjean 1997). Here are some of the challenges:

- Workers on permanent evening or night shifts struggle in attempting to adjust to a different schedule during the weekend so that they may function and socialize with family and friends.
- Evening or night shift workers may have difficulties finding regular eating times and finding quiet, dark places to sleep during the daytime.
- Some companies rotate shift workers so that (for example) every three weeks the evening shift rotates to days, the night shift rotates to evenings, and the day shift rotates to nights.
- Some workers (for example, military personnel, doctors, and security guards) are on-call for 24-hour shifts.
- Some workers fly across several time zones and land in a greatly altered daily routine.

It is easier to describe the challenges surrounding shift work and circadian rhythms than it is to solve the challenges and problems. Naps can be helpful, but the length and timing of the naps is controversial. One long night's sleep usually restores performance to a normal level, even after extensive sleep deprivation. During prolonged work, periods of vigorous exercise and fresh air may help maintain alertness. If shifts are necessary, either work only one evening or night shift per cycle, then return to day work and keep weekends free; or stay permanently on the same shift, whatever that is (Kroemer, Kroemer, and Kroemer-Elbert 1997).

Summary

This chapter provides the reader with an introduction to work physiology and anthropometry. Since this work is not intended to be a comprehensive treatment of these topics, readers should consult the references and other comprehensive works in these fields. Safety professionals are encouraged to learn more about how the human body is constructed and how it functions in the work environment, in order to be better able to evaluate existing work environments and to properly design new work environments. The safety professional's dual goals are to provide a safe and productive workplace.

References

Abraham, S., C. L. Johnson, and M. F. Najjar. 1979. "Weight and Height of Adults 18–74 Years of Age, United States, 1971–1974." *Vital and Health Statistics.* Series 11, No. 211, PHS 79-1659. MD: U.S. Department of Health, Education and Welfare.

Åstrand, P.-O., and Kaare Rodahl. 1986. *Textbook of Work Physiology: Physiological Bases of Exercise.* 3d ed. New York: McGraw-Hill.

Åstrand, P.-O., Kaare Rodahl, Hans A. Dahl, and Sigmund Stromme. 2003. *Textbook of Work Physiology.* 4th ed. Champaign, IL: Human Kinetics Publishers, Inc.

Bunc, V. "A Simple Method for Estimating Aerobic Fitness." 1994. *Ergonomics* 37(1):159–65.

Burke, E. R., ed. 1998. *Precision Heart Rate Training.* Champaign, IL: Human Kinetics Publishers, Inc.

Chaffin, D. B., G. B. J. Andersson, and B. J. Martin. 1999. *Occupational Biomechanics.* 3d ed. New York: John Wiley & Sons, Inc.

Cooper, K. 1970. *The New Aerobics.* New York: Bantam Books.

Dalley, A. F., and A. M. R. Agur. 2004. *Grant's Atlas of Anatomy.* 11th ed. Philadelphia, PA: Lippincott Williams & Wilkins.

Eastman Kodak. 1986. *Ergonomic Design for People at Work: Volume 2.* New York: Van Nostrand-Reinhold.

Garg, A., D. B. Chaffin, and G. Herrin. 1978. "Prediction of Metabolic Rates for Manual Material Handling Jobs." *American Industrial Hygiene Association Journal* 39:661–74.

Hibbeler, R. C. 2001. *Engineering Mechanics: Statics & Dynamics.* 9th ed. Upper Saddle River, NJ: Prentice Hall.

Illmarinen, J. 1992. "Design for the Aged With Regard to Decline in Their Maximal Aerobic Capacity: Part II The Scientific Basis for the Guide." *International Journal of Industrial Ergonomics* 10:65–77.

Jackson, A., A. Beard, L. Wier, and A. Stuteville. 1992. "Multivariate Model for Defining Changes in Maximal Physical Working Capacity of Men, Ages 25 to 70 Years." In proceedings of the Human Factors Society, pp. 171–74.

Jenkins, D. B. 2002. *Hollinshead's Functional Anatomy of the Limbs and Back.* 8th ed. Philadelphia, PA: W. B. Saunders Company.

Jorgensen, K. 1985. "Permissible Loads Based on Energy Expenditure Measurements." *Ergonomics* 28(1):365–69.

Kapit, W., and L. M. Elson. 2002. *The Anatomy Coloring Book.* 3d ed. San Francisco, CA: Benjamin Cummings.

Kapit, W., R. I. Macey, and E. Meisami. 2000. *The Physiology Coloring Book.* 2d ed. San Francisco, CA: Addison Wesley Longman.

Konz, S., and S. Johnson. 2004. *Work Design: Occupational Ergonomics.* 6th ed. Scottsdale, AZ: Holcomb Hathaway.

Kroemer, K. H. E., and E. Grandjean. 1997. *Fitting the Task to the Human: A Textbook of Occupational Ergonomics.* 5th ed. Bristol, PA: Taylor & Francis.

Kroemer, K. H. E., H. J. Kroemer, and K. E. Kroemer-Elbert. 1997. *Engineering Physiology: Bases of Human Factors/Ergonomics.* 3d ed. New York: Van Nostrand Reinhold.

______. 2001. *Ergonomics: How to Design for Ease and Efficiency.* 2d ed. Upper Saddle River, NJ: Prentice Hall.

Marras, W., and J. Kim. 1993. "Anthropometry of Industrial Populations." *Ergonomics* 36(4):371–78.

Mital, A., F. Hamid, and M. Brown. 1994. "Physical Fatigue in High and Very High Frequency Manual Material Handling: Perceived Exertion and Physiological Factors." *Human Factors* 36(2):219–31.

Mital, A., A. Nicholson, and M. M. Ayoub. 1993. *A Guide to Manual Material Handling.* London: Taylor and Francis.

Nims, D. K. 1999. *Basics of Industrial Hygiene.* New York: John Wiley & Sons, Inc.

Pandolf, K., M. Haisman, and R. Goldman. 1976. "Metabolic Energy Expenditure and Terrain Coefficients for Walking on Snow." *Ergonomics* 19:683–90.

Pheasant, S., and C. M. Haslegrave. 2006. *Bodyspace: Anthropometry, Ergonomics, and the Design of Work.* 3d ed. Boca Raton, FL: Taylor & Francis.

Roebuck, Jr., J. A, 1995. *Anthropometric Methods: Designing to Fit the Human Body*. Santa Monica, CA: Human Factors and Ergonomics Society.

Soule, R., and R. Goldman. 1980. "Energy Cost of Loads Carried on the Head, Hands, or Feet." *Journal of Applied Physiology* 27:687–90.

Tayyari, F., and J. L. Smith. 1997. *Occupational Ergonomics: Principles and Applications*. New York: Chapman & Hall.

Warfel, J. H. 1985. *The Extremities: Muscles and Motor Points*. 5th ed. Philadelphia, PA: Lea & Febiger.

Wisner, A. 1989. "Variety of Physical Characteristics in Industrially Developing Countries—Ergonomic Consequences." *International Journal of Industrial Ergonomics* 4:117–38.

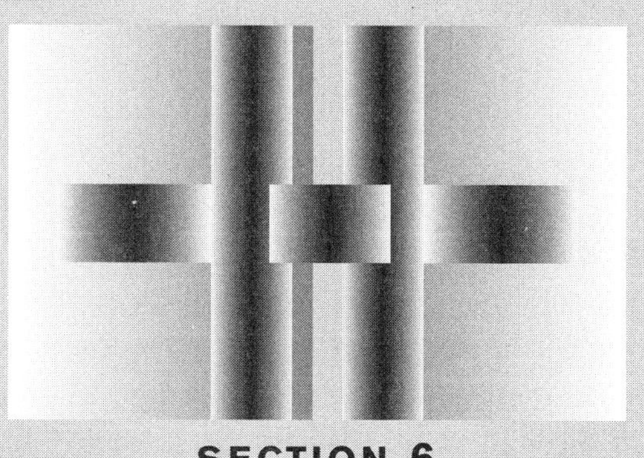

SECTION 6

ERGONOMICS AND HUMAN FACTORS ENGINEERING

LEARNING OBJECTIVES

- Gain a workable knowledge of the field of Human Factors Engineering (HFE).
- Understand the bases for design-induced accidents that are often mistakenly assigned to human error.
- Understand the scope and usefulness of HFE methods in preventing or reducing safety problems through improved design.
- Be able to describe a general process for avoiding perceptual, cognitive, and motor-related design flaws.
- Learn basic computations and models for use in HFE design and design review.
- Be able to demonstrate the need for HFE guidance in all stages of design.
- Learn how to access sources of information for HFE design guidance and further learning.

APPLIED SCIENCE AND ENGINEERING: PRINCIPLES OF HUMAN FACTORS

Steven F. Wiker

HUMAN FACTORS ENGINEERS (HFEs) practice in an interdisciplinary field of natural, physical, social sciences, and engineering that studies the performance capabilities and limitations of humans and applies that knowledge to the design of environments, machines, equipment, tools, and tasks to enhance human performance, safety, and health. Outside the United States, no formal differentiation is made between HFE and ergonomics, making this presentation a subset of ergonomics. There are a number of variants of HFE that include engineering psychology, human engineering, and so forth. Human factors practitioners are typically trained in experimental psychology; however, there are a number of academic fields that produce HFEs today.

Human factors engineers typically work on a variety of design issues and problems that focus on human–machine–task–environment system safety. One of the exciting aspects of human factors engineering is that development and application of such principles is continually challenging. The interplay of management demands, operating environments, task design, equipment design, user characteristics, and objectives influences the scope and approach for development and application of human factors engineering design principles. That said, one can systematically apply general human factors engineering principles and strategies to improve design performance and safety or to recognize when greater levels of expertise are required to handle the problem at hand.

Nearly all definitions of the safety process address the need to recognize the existence of hazards, or states of danger, in order to best use available information and resources either to eliminate or to mitigate such hazards. The overarching goal of elimination or mitigation is to reach a level of safety that avoids occurrences of injury or damage, or loss of life or of property, diminishing levels of occurrence to operational or societal levels of acceptability. This chapter addresses the importance of considering human factors engineering design principles and processes that must be addressed if interactions among humans, tasks, jobs, equipment, tools, products, and environments are to be both safe and functional.

Origins and Goals of HFE

There is a mistaken notion that HFE is a comparatively new field of endeavor in need of time for experimentation and maturation and acceptance of design principles and guides. It is true that the professional or organizational origins of Human Factors or Human Engineering only date to the end of World War II, but the majority of methods used by HFEs have roots in fundamental natural and physical sciences such as psychology, mathematics, engineering, and biological sciences. Many of the methods used can trace their origins to the period of the Industrial Revolution, or even earlier. As has the field of safety, HFE has continued to develop both empirically and heuristically throughout the development of humankind. Significant strides continue to be made in HFE with the advent of new technologies.

Historically, careful and systematic study of accidents revealed etiology in design errors found in many loci of the human–machine–environment–task (HMET) system design. Errors directly or indirectly responsible for accidents involved poor human factors engineering. Many of the problems found were attributable to a lack of understanding on the part of the designer or design team of basic human operator constraints, or to failure to consider how those demands changed when system failures occurred or when tasks, operating environments, or personnel changes occurred. Many good initial designs were subsequently compromised by *add-on* components or tasks or by expansion of the scope of operation without adequate HFE evaluation or testing before implementation of accompanying design changes.

For these reasons, HFE has continued to broaden its area of focus and activity to include transportation, architecture, environment design, consumer products, electronics/computers, energy systems, medical devices, manufacturing, office automation, organizational design and management, aging, farming, health, sports and recreation, oil-field operations, mining, forensics, education, speech synthesis, and many, many other arenas. Today, nearly all large corporate and military–industrial entities pair design engineers with HFEs in their searches for optimally usable and safe designs.

Why Do HFE Design Problems Affect Human Safety?

In 2006, 5703 people died on the job in the United States. The Bureau of Labor Statistics (BLS) of the United States Department of Labor catalogued death rates into occupations and published rankings of jobs that presented the greatest risk of death in the United States in 2006 (BLS 2006) (see Figure 1).

Closely associated with death rates are rates of equipment damage, facility damage, personnel selection and training costs, and other insidious costs. Because many of these accidents are attributed to human error, it is natural to ask why such problems exist. In-depth study of such accidents usually demonstrates a mismatch of the performance demands of equipment, environments, products, jobs, tasks, or system process designs with human capabilities and limitations. Most problems are originally *designed in* or created by ad hoc changes made to designs without careful design review and approval.

Five engineering design fallacies have been associated with use of anthropometric data (Pheasant 1988). Designers often suffer from one or more of the following fallacies:

- Because this design is satisfactory for me, it will be satisfactory for everybody else.
- Because this design is satisfactory for the average person, it will be satisfactory for everybody else.
- Human beings vary so greatly from each other that their differences cannot possibly be accommodated in any one design—but because people are so wonderfully adaptable, it doesn't matter anyway.

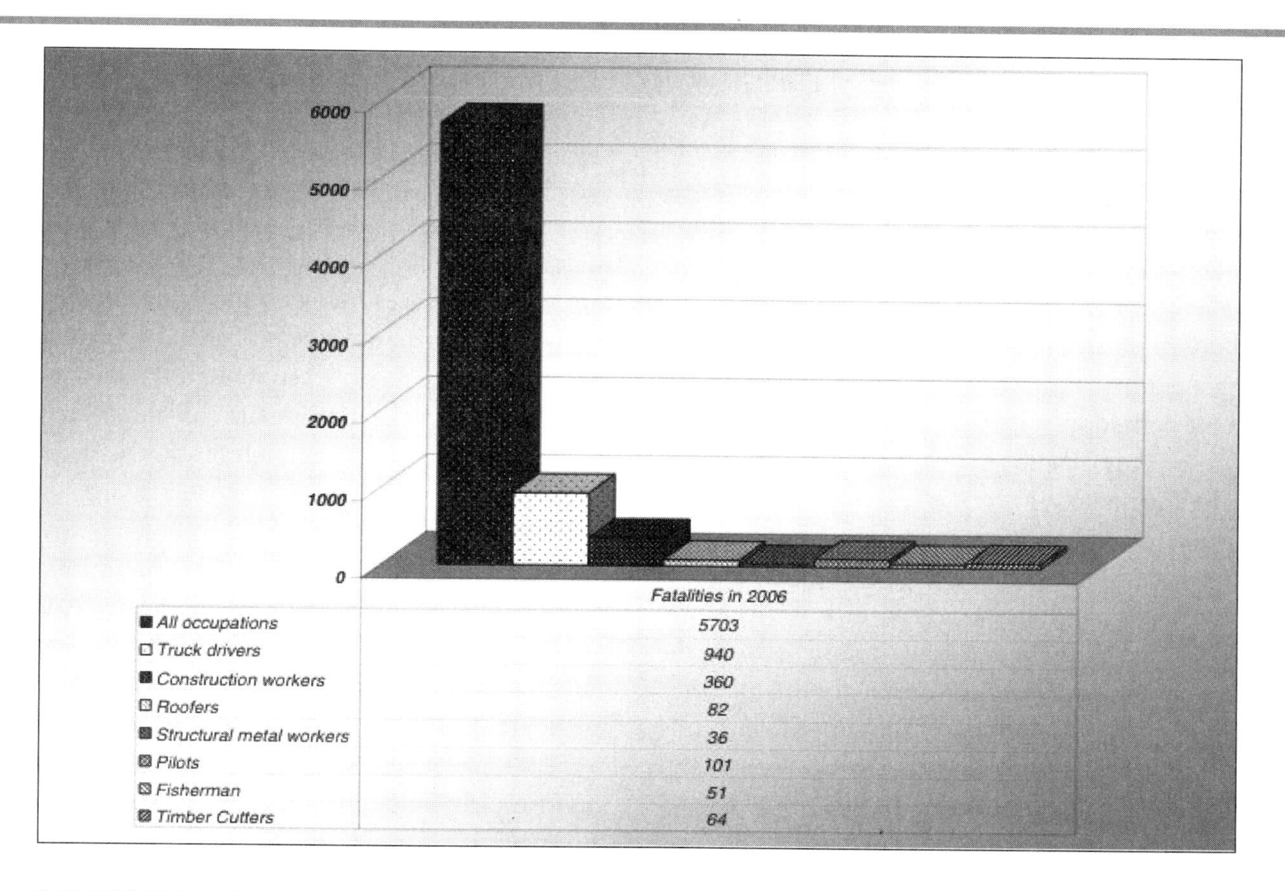

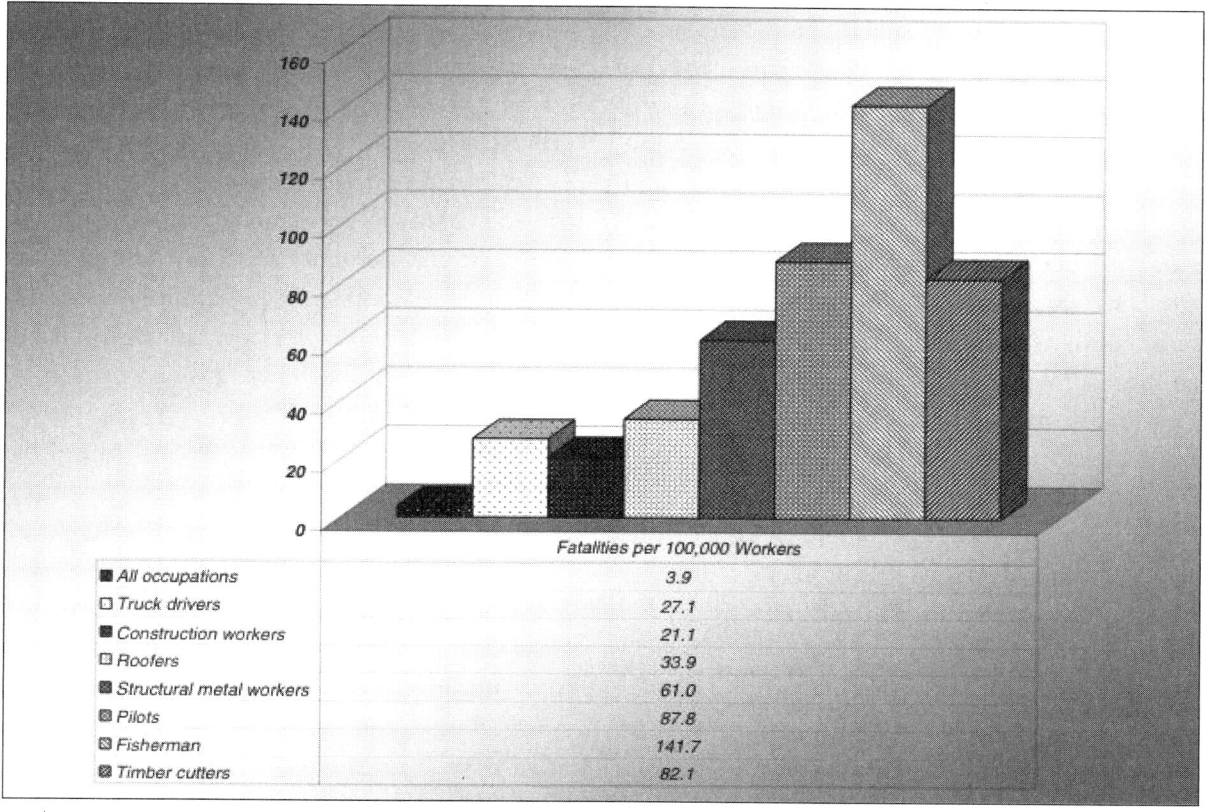

FIGURE 1. **Fatalities and fatality rates for selected occupations in 2006** (*Source:* BLS 2007)

- Because ergonomics is expensive, and because products are purchased for their appearance and styling, ergonomic considerations may be ignored.
- Ergonomics is an excellent idea. It is good to design with ergonomics in mind, but do it intuitively and rely on common sense rather than tables of data.

Many designers believe that human factors engineering is simply the process of using one's *common sense*. One definition of common sense is "sound, ordinary sense (or good judgment) shared by a group at large." Unfortunately, history is replete with examples of highly educated and otherwise very successful individuals or groups who in retrospect appear to have failed to exercise or understand common sense. Moreover, a design rationale that consistently satisfies the *majority* may be inadequate or unsafe for the minority.

A number of studies have shown that multiple population stereotypes exist for any given design and that the minority perspective can be held by large segments of the human population. Thus, one cannot presume that one's personal view of common sense matches that of the majority of individuals who are intended users of the design. In fact, it can be argued that designers and engineers who create new designs or modify existing ones are bright individuals who often "think outside of the box" to achieve breakthroughs by means of approaches, analyses, or resources used in unusual manners. These professionals are typically individuals having unique education, training, experience, and skills who by their nature are unlikely to work in the realm of common sense.

If product designs rely on individual designers' intuition or *common sense*, accidents and related problems will likely arise from the mismatch of designers' and users' mental models, expectations, and performance capabilities. To overcome this problem, it must be realized that individuals vary in their knowledge, belief systems, training, experience, skills, and physical and mental capacities to recognize hazards, make accurate or relevant judgments, and take appropriate corrective protective action without error.

Many people are frustrated by designs of television remote controls, by cellular phones that rely on the memorized use of multimodal button sequences, by electronic calculators, and by equipment and processes believed by their designers to follow common sense. A classic example of this problem is demonstrated by a calculator manufacturer that believed key entry should reflect the operational sequences that the computer actually used—but the key-entry sequence was not consistent with sequences followed by algebraic operations. Many customers found the number and operational keying sequence confusing, even though the designers felt the customer population would eventually recognize the benefits of using nonalgebraic key sequences.

Other designers, recognizing that a large segment of the customer base preferred to enter information into calculators using standard algebraic sequence, captured a large segment of the market by designing calculators that met users' expectations. Today, many calculators offer both modes of data entry, allowing users to select the mode they understand or prefer.

Engineers or designers who expect users to understand their designs or share their mental model of the application and scope of use of their designs are increasingly surprised by the lack of user agreement. Matters do not seem likely to improve in light of the general decline in educational levels and technical literacy of the general user population in the face of rapidly accelerating technology. As penetration of Third World markets increases, designers must also consider potential mismatches caused by cultural and educational differences as well. The mismatch problem has increased so much that it often serves as grist for comedians, who interview citizens on the street to demonstrate how little they know about the information that is part of daily life.

A national survey of Americans found that deficits in technological literacy were comparable to the decline in general literacy (Sum 1999). In surveys, investigators determined that

- 50% of American industrial workers have math skills below the eighth-grade level.
- 28% of workers cannot make correct change at cash registers.
- 33% of people do not know how a telephone works.

Too much knowledge or experience can be problematic as well. A phenomenon of negative transfer can occur when attributes of equipment, environs, and so on, suggest an action that is ultimately inappropriate, basing the suggestion upon past training and experience. Manual gearshift configurations, for example, are not universal in layout. Many times the configuration is placed on the gearshift knob, but the symbols are often worn off with use, and individuals experienced with one configuration will act as if using the configuration to which they are accustomed when introduced to different vehicles with unlabeled gear shifts.

Integrating human factors engineering requirements for any design should be determined at the outset—indeed, throughout the design development process. Integrating good human engineering design in the early stages of design reduces costs and the likelihood that post hoc modifications will have to be made, as well as safely increasing or restricting the breadth of user populations or product market shares. When early involvement of HFEs in the design process does not occur, usability, functionality, and safety are easily compromised.

Other problems develop when *quick fixes* are made to designs for any of a variety of reasons. Typically neither the original designers nor HFEs are involved in such design modifications, causing unintended problems to arise. It may be less expensive, for instance, to use a different-colored dye in polymers used to construct a product—so the color of the product is changed. But the color of the labels on the product may no longer offer the contrast sensitivity needed, causing operators to confuse controls, reducing functionality and potentially introducing safety hazards.

Sometimes designers forget to consider the capabilities and limitations of their intended and unintended user populations when making decisions about design parameters. The directions in which users will read symbols or icons can vary if a product will be used internationally. Some cultures read from right to left, others from top to bottom, still others from left to right, and so on. Failure to understand differences in the user population can make apparently safe designs suddenly unsafe.

Often designers forget that they are not merely designing a single, isolated product, piece of equipment, task method, or working or living environment. Rather, they are designing a component that will function within a larger system. Failing to consider this often leads to poor design, suboptimal applications, or creation of hazards that can all go unconsidered. Often personnel requirements—the numbers and kinds of personnel that will be needed to staff the system, or those who can are able to use the design—depend on the extent to which the systems approach is followed.

There are a variety of explanations for the development of HFE design flaws and safety hazards. Good human factors engineering design can only be achieved through rigorous analysis and comprehensive application of good design principles in the initial and the modification design phases. Project managers often misjudge the amount of effort required to engineer equipment, products, processes, and environments that must be functional and safe when they require interaction with humans. Nearly all engineering curricula do not require engineers or other professions to take coursework that integrates HFE design contributions into the design process.

Simply providing textbooks, design handbooks, and focused HFE standards for use by design teams is not always helpful. Design guidance is often too general and must be selected and tailored to meet projects' requirements. Instead, HFEs participate with designers to help them tailor their designs to meet users' needs, improve performance, and reduce the risk of causing safety hazards. Often HFEs design test and evaluation studies to confirm design objectives have been met.

The Scope of HFE Practices

The intent of this work is to illustrate the scope and usefulness of HFE methods in preventing or reducing safety problems. A general approach is recommended to help tackle major problems that will likely be encountered. One cannot rely solely on recommendations, guidelines, checklists, or standards to eliminate HFE problems or hazards. Guidelines are often too broad or too specific within the context of a particular HMET system to provide material support. A systems approach to design and evaluation is proposed to manage the risk of producing unsafe designs.

No short chapter can fully address all the theoretical foundations, methodologies, and tactics used by HFE professionals to reduce safety hazards. No attempt has been made to comprehensively summarize a wealth of knowledge, design methodologies, or approaches to safe design using HFE, but many excellent sources of such information do exist in which readers can find detailed information concerning such methods—including many of the chapters within this book.

This chapter is intended to provide an overview of the field of HFE and its role in promoting safety in the workplace, at home, and in recreational environments. It focuses on human sensory, cognition, and motor-performance limits and their roles in designing and maintaining safe environments, providing a methodological approach that can be followed to achieve these objectives.

The General HFE Process

Human factors engineering professionals prefer to follow a standard sequence of operations when assisting in the design, development, and implementation of HMET systems. Human factors inputs are required throughout system development and generally follow the process below:

- Understand the objectives and goals of the system and their impact upon human roles and performance requirements.
- Provide information about human performance's capabilities of shaping of the system's design, about functional allocation between humans and machines, and about potential risks for system failure and safety and health problems.
- Evaluate HMET system performance.
- Discover whether human–machine system performance meets design criteria.
- Discover whether safety standards involving humans are met.

Designs in complex systems are fluid; the steps above can be followed iteratively and heuristically, capitalizing on previous designs, documentation, and test results. At each step of the process, judgment depends greatly on the skill and expertise of those who are conducting the analysis.

Although it is to be hoped that the above work could be performed by means of handbooks, standards, and design guidelines, it is likely that new ground will have to be broken and that such activities will require experimentation, tests, and evaluations of behaviors before final design recommendations are made or specifications solidified.

General Steps

Although specific expertise and distribution of HFE activities will vary from design to design, HFEs tend to follow a systematic process for design development, review, and validation. The general process followed is outlined below:

Step 1. Define specific system performance objectives and constraints

Without a clear understanding of the system's objectives and design constraints, it is difficult to discover which designs or design options are appropriate and which risks of faults, failures, or other problems are tolerable. Typically, outcome metrics from the test results judged are provided, as is other HFE information describing the intended user population, the desired staffing levels, the training and skills of the user population, the performance envelopes involved, and any other information useful to the design team.

If the performance objectives and constraints are not clear at the outset of the project, this usually spells trouble in meeting time and cost milestones. Such a project is typically not well thought out, and corners may be cut, increasing the risks of errors in design accordingly.

Performance specifications delineate the goals and performance requirements for a design or system. Specifications are typically prepared by a team involved in the development of a system. They itemize functions, define parameters, and spell out design constraints. The design team creates or responds to a "design to" specification that states system performance capabilities and user requirements.

The specification typically translates the user's operational needs into system functions and requirements, allocates those requirements to subsystems, and allocates general functions to operators and service personnel. Other than expected or required functional allocations, the specifications do not address specific requirements in terms of human performance beyond gross statements about personnel requirements (e.g., will be used by airline pilots, firefighters, the general public, and so on).

The objective at this stage of design is to determine how the system design specifications map onto human performance requirements or demands. Essentially, the HFE must determine what the performance envelopes are and feed this back to the design team—particularly if the requirements are unreasonable.

To meet this objective, system requirements must be project-specific and written in a verifiable form. If the requirements are too broad, the HFE and other team members will face great difficulty attempting to prove that they have met their goals. Establishing requirements in operationally defined, verifiable manners provides a common perspective and basis of understanding among the design team and its sponsors or customers.

Step 2. Allocate roles, responsibilities and performance requirements

It is important to be apprised of the roles of humans in meeting the system design objectives and requirements. It must be understood how performance demands are to be functionally allocated among humans, machines, and software. HFEs typically document the basis for task allocation, performance demands, and safety responsibilities among humans, firmware, and software.

Step 3. Perform actual or simulated task or activity analysis

Activity analyses may be performed using a variety of approaches. However, functional, decision, and action flow analyses are typically used, along with simulations and mock-up analyses. Examination of similar systems and focus groups are used as resources for understanding human performance and the need for allocated responsibilities. Job safety analyses can also be performed during task analysis to discover sources of failures and accidents, as well as and failure modes and effects.

Step 4. Define design questions or problems encountered

It is inevitable that design questions or problems will develop during Step 3. Questions will arise of which method is best to employ, of the safety of performing a particular task or of human interaction with machines or the working environment, or of the possibility of using new technology in system design.

It is important to develop concise definitions of design questions or problems to define the scope of the effort of the HFE. Bounding the question or solution space enables the development of responsive answers in a timely, cost-efficient manner. Knowing when a functional answer to a designer's question has been produced allows efforts to be redirected to answering succeeding questions—with confidence. It also makes validation of the recommendations more straightforward and efficient.

Step 5. Understand the design questions

Interrelating design questions with system performance requirements is critical. If these steps are decoupled, a design can be recommended that optimally addresses the design question within a particular subsystem; but it may not yield an optimal outcome for the system at large. Design recommendations for controls, displays, and seating design in a race car intended to travel at the vehicle's maximum capacity may be entirely inappropriate and overly costly, as well as promote inappropriate operation, on highly trafficked downtown city streets.

Step 6. Select candidate design concepts for evaluation and testing

Often design options are provided by the designers. HFEs endeavor to discover which design candidate is best from a human–machine performance standpoint within the context of the overarching system performance objectives. The HFE may also find design modifications that can improve any given candidate's design value.

Experimentation in laboratory or field conditions is usually required in order to test design options.

Analysis, presentation, and interpretation of results are considered by the design team. Testing must be timely and cost efficient, but experimental findings should have adequate statistical force. Depending on the nature of the questions or problems addressed, testing and evaluation may iterate until a satisfactory design state is achieved by means of performance selection or utility metrics.

If multiple performance factors must be addressed, HFEs usually work with design teams to determine the weights to assign to each performance metric. The sum of the product of weighted or valued performance metrics is often used to rank design concepts by multiattribute utility (or by means of some form of decision support analysis).

HFEs should list all documents that will be or have been consulted in the development of the system. Any tailoring of information derived from standards should be documented. Ambiguously worded requirements in standards should be reworded or operationally defined so that HFEs are not later in disagreement with customers about whether a requirement has been met.

Step 7. Evaluate personnel selection and training requirements

If personnel selection criteria are set as part of the system design or performance requirements, analyses should confirm that the recommended design characteristics or candidate design options have not exceeded the intended user population capacities. If that is not feasible, it must be determined what population selection criteria or training requirements are now requisite to meet the system's mission or performance requirements. Failure to carefully evaluate personnel selection and training needs, documenting them for future use, inevitably creates problems that can lead to accidents.

Step 8. Document and justify recommended design features or options

Documentation of standards relied upon, rationale for tailoring standards to address design questions, testing methods, data, analyses, findings and interpretations of findings, and rationale for recommendations must be documented thoroughly.

Computer programs, drawings, mockups, and detailed reports should be delivered for use and archiving. The number and structure of reports, as well as the detail required, can vary from project to project. However, most HFE reports include

- human factors system function and operator task analysis
- human factors design approach
- human factors test or simulation plans
- procedural data documentation and analysis reports
- personnel and training plans

Other Considerations

Training Team Design Members

Short courses addressing the general utility of HFE are useful, but an intense course over a couple of days or hours cannot be expected to provide designers with the capacity to apply HFE principles in the design of complex human–machine systems. Although designers will become sensitive to general concepts, they will not be able to evaluate engineering tradeoffs or address specific questions.

That said, sensitivity training for team members across disciplines is always beneficial, because it serves to enhance collaborative interaction and appreciation for problems in their colleagues' domains of expertise. Of course, HFEs must be prepared to provide design recommendations that are well-defended and tutorial in nature. Designers are not happy about altering their preliminary, intended, or already realized designs without solid justification. Constructing well-documented, thorough justifications for improvements in designs helps HFEs both to gain acceptance and to provide on-the-job training for design teams. Helping design teams learn why design principles are applicable often helps integrate such principles into future preliminary designs, expediting HFE review and (it is to be hoped) reducing the need for future design changes.

Understanding the Level of Effort

Documentation is requisite in HFE design and justification, particularly when systems under development are large and may require years to produce, and when the need for future modifications or updates is anticipated in response to expected changes in system requirements, available technology, or user populations. All analyses, preliminary studies, tests and evaluations, and tailoring or shaping of standards in their application for the project must be documented. It is not unusual for design and engineering team members to move in and out of the project, thus requiring documentation review by onboard team members or project managers.

Documentation is also requisite for investigators of accidents, whether in defense of torts or product liability disputes or in cases of future designs that may need to build upon existing ones. If project performance objectives change, previous documentation can aid in understanding the implications of making such changes, as well as areas in which additional testing and evaluation will be required.

However, methodical analysis, documentation, and communication with design team members consume vast amounts of time. Because of the initial cost of HFE contributions, it must be ensured that outcomes make the initial investment and subsequent modifications cost effective; starting over with an HFE evaluation of complex systems is untenable.

Determining Human Interface Requirements

Once the HFE and team understand the functional flow of the work performed within or by the system, an even greater understanding of the actual tasks performed by humans is required, as well as how they interact with tools, equipment, other humans, their environment, and so on. Such information is requisite to beginning the mapping of functional demands on human perceptual, cognitive, and motor-performance requirements.

Human activity analysis may include a variety of analytical tools that help define human performance requirements for an existing or future system design (Chapanis 1965):

- function allocation
- function and operational flow analysis
- decision action analysis
- action information analysis
- task analysis
- timeline analysis
- link analysis
- simulation
- controlled experimentation
- workload assessment

Functional Flow Analysis

Functional flow analysis is a procedure for decomposing a system design into functional elements and identifying the sequences of functions or actions that must be performed by a system (Chapanis 1965). Starting with system objectives, functions are identified and described iteratively; higher, top-level functions are progressively expanded into lower levels containing more and more detailed information. As demonstrated in Figure 2, functions or actions are individually and hierarchically numbered in a way that clarifies their relationship to one another, permitting the tracing of functions throughout the entire system. The diagram in Figure 2 shows activities, sequences, and flows of functions.

Higher-level functions are carried out to such additional levels of detail as are required. The spatial and numeric organization is structured so that anyone can easily trace the input's flow through functions to output. Some HFEs identify the function's output on the connecting arrow to the next function, adding additional inputs to the next function.

Connector blocks are used to link functional flows from one page to the next; arrows enter function blocks from the left and exit from the right. Thus, the flow is from top to bottom, left to right. Functions subordinate to the general function are placed below it to show the normal sequence of system functions. Whenever arrows join or split, their junctions are shown, with "and," "or," and "and/or" gates encircled. An "and" gate requires all following or preceding functions to be or to have been performed; an "or" gate allows for one or more of the following or preceding functions to be or to have been performed.

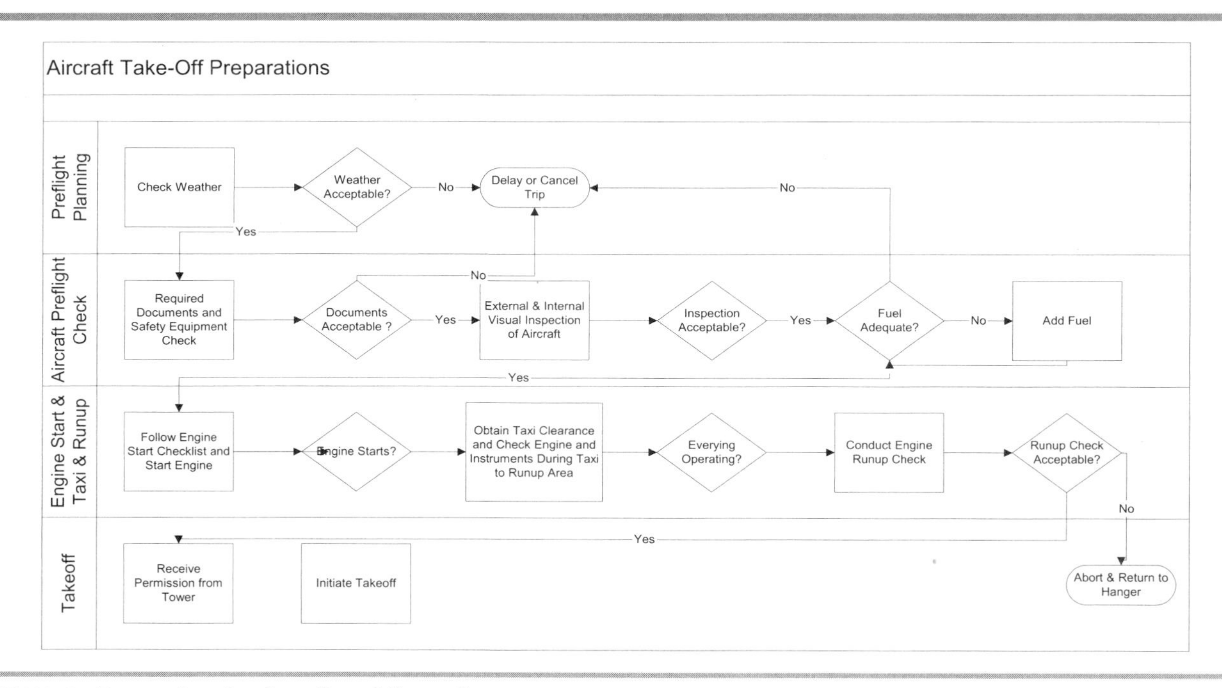

FIGURE 2. Example of a functional flow diagram

For each of the functions, the team can indicate the need for decisions by posing binary questions. Each function is a short verb–noun combination incorporating occasional adjectives or other modifiers. One decision might be "Projector status OK?" This can be decomposed into lower-level decisions that include "Power On?," "Projection Mode Properly Set?," and "Projector Connected to Computer?"

Each decision is placed in a diamond symbol and is phrased as a question that may be answered with a binary (yes or no) response. Function blocks and decision diamonds are given reference numbers.

Operational Analysis

An operational analysis is an analysis of forecast functions in an undeveloped system and is designed to obtain information about situations or events that will confront operators and maintainers (Kurke 1961, Niebel & Freivalds 2002). Scenarios and anticipated operations are created, and assumptions and operating environments are documented. Scenarios should be sufficiently detailed to convey an understanding of all anticipated operations and should cover all essential system functions, such as failures (both hard and soft) and emergencies.

Operational analyses are typically driven by the system performance objectives and requirements. Scenarios are usually developed by knowledgeable experts that include procedures, equipment use, environmental constraints, and potential rare or unanticipated outcomes. The scenarios are delineated in detail sufficient to allow designers and analysts to work together on design formulation and assessment.

From an HFE standpoint, the operational analysis provides an interactive description of human constraints and requirements (e.g., environmental conditions, skill

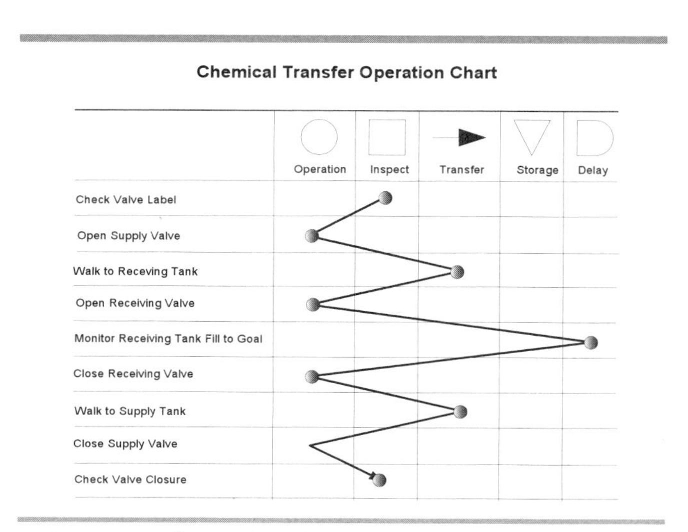

FIGURE 3. Operational analysis diagram symbology

and training requirements, and so forth), performance envelopes that can affect system performance, and failure modes and effects that must be considered when specifying HFE design requirements. It is therefore important the team producing the operational analysis is experienced with the technology and the operations addressed, and that the content experts (including HFEs) be participants. Typical operational and sequence diagrams and symbols used are shown in Figure 3.

Task Analysis

Task analysis is the first step in a systemic approach to design (Annett 1971; Annett & Stanton 2000; Desberg & Taylor 1986; Jonassen, Hannum, & Tessmer 1989; Jonassen, Tessmer, & Hannum 1999; Kirwan & Ainsworth 1992; School Library Manpower Project & NEA 1969; Shepherd 2001; Verdier 1960). Task analysis can focus on what actually happens in the workplace now but can also be used as a tool to specify the design of a new system. Task analysis provides a detailed description of how goals are accomplished, allowing designers to provide means for enhancing human performance. Analysis involves a systematic breakdown of a function into its underlying tasks, and then of tasks into subtasks or elements. The process creates a detailed task description of both activities, both manual and mental; durations of tasks and elements; frequencies, allocations, and complexities of tasks; environmental conditions; necessary clothing and equipment; and all other unique factors required for one or more humans to perform a given task. Estimates are made of the time and effort required to perform tasks, as well as of perceptual, cognitive, and motor-performance demands; training requirements; and all other information needed to support other system development activities.

To decide what type of information should be collected—and how it should be gathered—it is necessary to identify the focus of the analysis. It should be decided not only what system is to be the focused upon but also for what the results will be used (perhaps redesigning a new system, modifying an existing system, or developing training). Not all task information need be collected, but merely sufficient information to allow specific necessary objectives to be addressed (see Figures 4 and 5).

Just like functional diagrams do, task analyses use hierarchical number systems to help analysts understand the relationships of subtasks, tasks, and jobs. Once a task or subtask is numbered, it bears that number for the remainder of the design project.

Linking Tasks Together

Link analysis is a method for arranging the physical layout of instrument panels, control panels, workstations, or work areas to meet certain objectives (such as reducing the total amount of movement or increasing

Task Information	*Description*
Task and Subtask Structure	An organized, often hierarchical listing of the activities involved in a task with tasks and subtasks numbered accordingly.
Importance or priorities of subtasks	Assessment of the criticality of subtasks from a performance or safety perspective.
Frequency of subtasks	Frequency of occurrence of subtasks under different conditions.
Sequencing of subtasks	Order of occurrence of subtasks under different conditions.
Decisions required	Part of the sequencing may be based on a decision needed to choose the branch of activity and thus a given set of subtasks.
'Trigger' conditions for subtask execution	Execution of a subtask may depend upon the occurrence of a particular event or a decision made during a previous task or subtask.
Objectives	Performance objectives are provided including the importance to the overall mission.
Performance criteria	Human performance criteria are delineated such as perceptual requirements, psychomotor performance, accuracy constraints, etc.
Information required	Information that must be provided at each level of task or subtask that must be attended or used by the operator.
Outputs	Information, products or other results of human effort that result from the subtask.
Knowledge and skills required	Information and skills that the user utilizes in decision making and task performance.

FIGURE 4. Tasks analysis information typically gathered

Data Collection Method	*Description*
Direct and Videotape Observations	Use of time study or work sampling methods to develop job descriptions and psychomotor demands.
Interviews and Questionnaires	Interview workers or operators to determine the sequence of operations, tasks, subtasks, and hazards and difficulties they encounter in the performance of their job or task.
Focus group	Discussion with a group of typically 8 to 12 people, away from work site. A moderator is used to focus the discussion on a series of topics or issues. Useful for collecting exploratory or preliminary information that can be used to determine the questions needed for a subsequent structured survey or interview.
Interface Surveys	A group of methods used for task and interface design to identify specific human factors problems or deficiencies, such as labeling of controls and displays. These methods require an analyst to systematically conduct an evaluation of the operator–machine interface and record specific features. Examples of these methods include control/display analysis, labeling surveys, and coding consistency surveys.
Existing Documentation	Review any existing standard data sets from production engineering departments or other industries, operating manuals, training manuals, safety reports, and previous task analyses.
Link Analysis	A method for arranging the physical layout of instrument panels, control panels, workstations, or work areas to meet certain objectives, for example, to reduce total amount of movement or increase accessibility. The primary inputs required for a link analysis are data from activity analyses and task analyses and observations of functional or simulated systems.
Work Sampling Analysis	A method for measuring and quantifying how operators spend their time. Random or uniform sampling of activities are then aggregated over some appropriate time period (for example, a day), and activity-frequency tables or graphs are constructed, showing the percentage of time spent in various activities.
Checklists	Use a structured checklist to identify particular components or issues associated with the job. Available for a range of ergonomic issues, including workplace concerns, human–machine interfaces, environmental concerns.
Job Safety Analysis (JSA)	Using JSA, one can identify what behaviors in an operation are safe and correct. This analysis can be performed during task analysis. For each task or subtask, the analyst determines how the task should be performed and potential unsafe or hazardous methods. See chapters providing greater detail on JSA methods and benefits.
Critical Incident Analysis	Critical incidents can lead to accidents or system failures. This information is typically gained from human operators who supply first-hand accounts of critical incidents which are accidents, near-accidents, mistakes, and near-mistakes they have made when carrying out some operation. The critical incident technique only identifies problems—not solutions. The percentages of errors found in a critical incident study do not necessarily reflect their true proportions in operational situations, because the incidents are dependent on human memory and some incidents may be more impressive, or more likely to be remembered, than others. To be useful, the incidents must be detailed enough (a) to allow the investigator to make inferences and predictions about the behavior of the person involved, and (b) leave little doubt about the consequences of the behavior and the effects of the incident. The HFE then groups incidents into categories that have operational relevance: Mistakes and near-mistakes in reading an indicator, Mistakes and near-mistakes in using a control, Mistakes and near-mistakes in interpreting a label. The analyst then uses human-factors knowledge and experience to hypothesize sources of difficulty and how each one could be further studied, attacked, or redesigned to eliminate it. Studies are usually necessary to find ways of mitigating or eliminating those problems where elimination cannot be achieved based upon first principles alone.
Fault Tree Analysis	Some analysts use fault tree analysis in conjunction with critical-incident task analysis or job safety analysis or failure modes and effects analyses. Event trees, starting from a problem root such as equipment or other form of failure, follow through a path of subsequent system events to a series of final outcomes. This is also a probabalistic analysis that gives rise to likely failures and consequences that should be considered in the design of equipment, layouts, human capability demands, and so forth.
Failure Modes and Effects Analysis (FMEA)	FMEA is a method used to understand the consequences resulting from a failure within a system. Humans can be considered components that can fail for various reasons, and following FMEA methods, outlined in other chapters of this text, one can examine opportunities for eliminating or softening failure impacts. Typically, FMEA is used during functional flow, task, workload, linkage, and other HFE analyses. It helps in HFE analyses to determine what scenarios need to be interrupted by increasing the reliability of human–machine system performance.

FIGURE 5. Potential methods for data collection required for task analysis

accessibility) (Chapanis 1965, Sanders & McCormick 1993). A link is any useful connection between a person and a machine or part, between two persons, or between two machines or machine parts. If, for example, an operator walks to a supervisor to obtain permission or to consult about a problem, that activity represents a person–person link. If the operator returns to actuate the machine tool's cycle lever to press a metal part, that is a person–machine link. When the pressed part is automatically transferred to a transfer conveyor, that event represents a machine–machine link.

A number of steps are involved in link analysis:

- List all personnel and items to be linked.
- Determine frequencies of linkage among operators, tools, machines, and so forth.
- Classify the importance of each link.
- Compute frequency-importance values for each link.
- Sort link frequency-importance product values in order to focus on the links having the greatest cost.
- Fit the layout into the allocated space while minimizing linkage values.
- Evaluate the new layout in light of the original objectives.

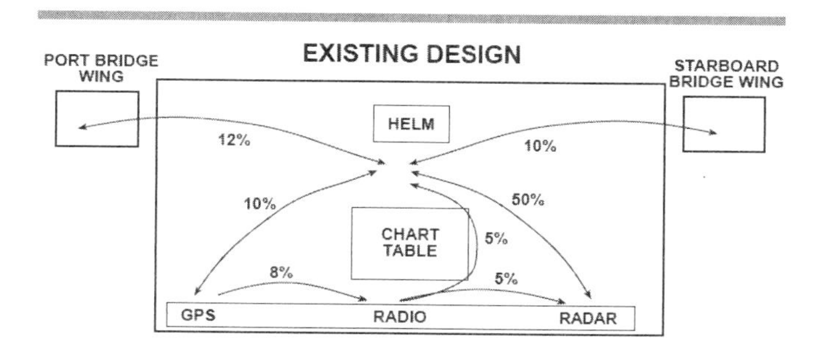

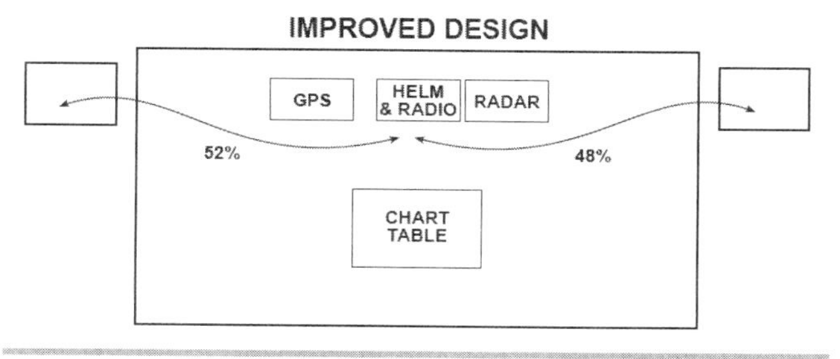

FIGURE 6. Linkage analysis showing relationships among ship bridge equipment placement and movement of an individual conning a vessel

Evaluation of various workplace or equipment arrangements can be made to assess move distances traversed during typical or unusual operations, as well as the crowding of activities and opportunities (clustering equipment to expedite operation) and the completion of multiple adjustments or areas that then require inspection. Linkage analysis also elucidates the focus and concentration of human–human and human–hardware interactions, information that is very useful in weighing the importance or directing focus on design problems. See Figure 6 for an example of linkage analysis showing relationships among ship bridge equipment placement and movement of an individual conning a vessel with a poorly laid-out bridge as opposed to a design that improves the operability of the vessel.

Workload Assessment

General workload assessment can also be performed at this point using timeline analysis (Chapanis 1965, Sanders & McCormick 1993). Timeline analysis follows naturally from task and linkage analyses and is concerned with the scheduling and loading of activities upon individual operators. Charts are produced that show sequences of operator actions, as well as the times required for each action and the times at which they should occur. The method produces plots of the temporal relationships among tasks, the durations of individual tasks, and the times at which each task should be performed.

Figure 7 shows that it is possible to integrate times from separate task analyses to show overlapping activities of two or more persons (Wickens & Hollands 2000). The greatest estimated workload occurs during the approach to land, and the least during landing.

Functional Allocation

Results of task analyses and workload timelines are often considered when making functional allocation decisions among workers, workers and machines, and among machines. Functional allocation is a procedure for assigning each system function, action, and decision to mixtures of hardware, software, and human

Activity	Sequence of Activities									
	Preflight	Taxi	Takeoff	Climb to Cruise	Cruise	Descend	Approach to Land	Land	Taxi	Postflight
Walk & Inspect	1									1
Radio Use	1	1	1	1	1	1	1	1	1	
GPS Programming	1			1	1	1	1			
Control Actuation		1	1	1		1	1	1	1	
Checklists	1		1	1			1			1
Instrumentation Checks	1	1	1		1		1			1
Monitor Gauges		1	1	1	1	1	1		1	1
Review Approach					1		1			
Workload	5	4	5	5	5	4	7	2	3	4

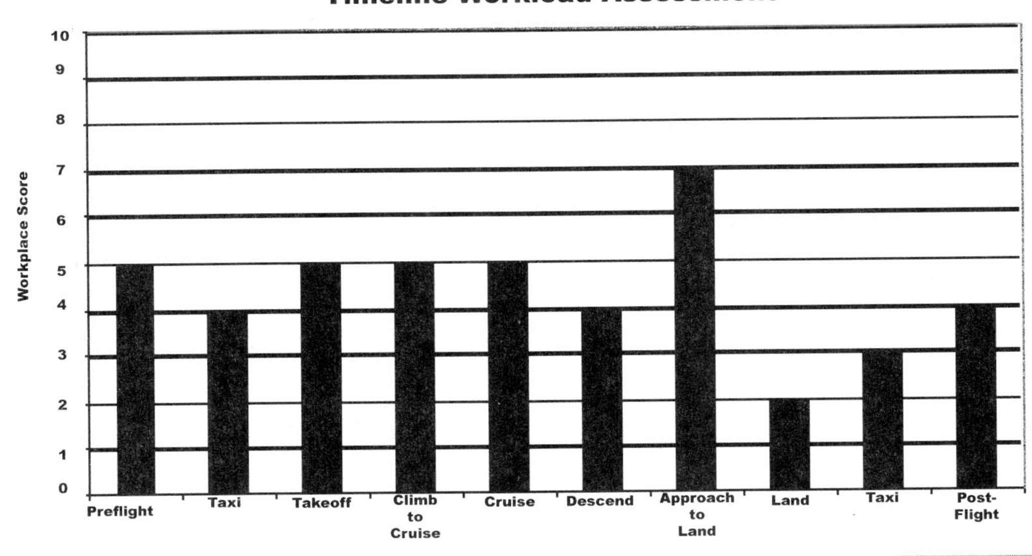

FIGURE 7. Workload timeline analysis (*Source:* Wickens & Hollands 2000)

operators. Allocation decisions can follow many approaches. They can range from *Fitts' Lists* of machine versus human performance strengths to *line-balancing* algorithms, in addition to results from expert, worker, and other inputs (Fitts 1951). For example, Fitts advocated the use of machines for massive counting or arithmetical tasks, pointing out that humans are better suited to handle novel events requiring complex assessment and decision making.

Professional judgment is involved in selecting criteria to be considered and in determining the weights to assign allocation criteria. In general, nearly all allocation decisions will have to address at least the following decision criteria (Price & Tabachnick 1968, Price 1985):

- implementation constraints
- maintainability and support logistics
- number of personnel required
- performance capability
- personnel selection and training costs
- political considerations
- power requirements
- predicted reliability
- safety
- technological feasibility.

Analysis of Similar Systems

If possible, consider designs and analyses of similar systems, something that can increase understanding of important design issues that have been considered in the past (and which may be quite applicable to current problems) (Chapanis 1965). If access to reports, tests, and evaluations is possible, as well as to interviews of users of such designs and records of accidents or accident investigation reports, the work of previous design teams can be capitalized upon.

But HFEs should be careful to not merely blindly accept previous designs without considering the differences between current design requirements and those of previous projects. Antecedent systems often provide better initial insights into operation and maintenance of proposed designs, into skills and training required, and into any history of usability or safety problems, but a design is intended for use in different environments, by expanded user populations, or under more stressful operating environments, previous designs may not be adequate, or may require modification.

Preliminary Focus Groups

Focus groups can help in assessing user needs, perspectives, and concerns early in the design process (Greenwood & Parsons 2000). A focus group's membership should be representative of the intended user group and small enough to encourage equal opportunity for expression of opinions and insights (e.g., five to ten members) about design issues and user needs. Preliminary focus groups can be much less formal than subsequent group meetings, in which greater moderation, and more focal or structured query will be needed to address specific issues.

Map Performance Requirements onto a Human User

For each system function assigned to a human, the HFE works bottom-up in specifying perceptual, cognitive, and motor demands that must be met to achieve required task performance using specific displays, controls, workplace configurations, and all other design variables affecting operator performance and safety. Often, stipulating design requirements requires design tradeoffs. HFEs must convey to design teams where tradeoffs are likely to occur, as well as the consequences of such tradeoffs (Chapanis 1965, Sanders & McCormick 1993).

Simply citing specific standards or sections of standards will not convince designers to read and fully understand HFE guidance. Dumping a stack of textbooks, design handbooks, or relevant standards on a design team's table while expecting its members to absorb and internalize the information in a fast-paced design process is very unrealistic and, furthermore, invites errors. Instead, the HFE must mete out information as needed, clearly describing tradeoffs and their consequences or conducting preliminary analyses to help direct designs along successful paths. HFEs should provide ranges of specifications, describing consequences clearly enough for design teams to understand and making them easy to take into account in design decisions.

It is also a mistake to allow designers to try to use disparate standards or design handbooks, cherry-picking what they believe important or relevant. When inexperienced designers are confronted with only general recommendations, such information can be interpreted in different ways, some of which may not result in easily usable systems.

Some demands or requirements placed upon human users do not have well-defined answers or specifications. This can be frustrating both to HFEs, who are under time constraints to arrive at recommendations for design teams, and to design teams, who are in holding patterns as they wait for input from HFEs. Sometimes design teams cannot wait for the development of specifications and must move forward. This can pose problems if no opportunity will arise later to correct decisions.

Are Acceptable Stimuli Presented to Users?

Is Stimulus Intensity Adequate?

A psychometric function describes the relationship between a parameter of a physical stimulus and the responses of a person who has to decide about a certain aspect of that stimulus (Stevens 1957). Change in perceived intensity with change in actual physical intensity obeys a power function:

$$\psi = kX^{p} \tag{1}$$

where

ψ = perceived or sensed intensity of the stimulus
k = empirically derived coefficient
X = physical intensity of the stimulus (e.g., sound pressure level)
p = exponent

Thus, different perceptual intensity changes are experienced with variations in actual stimulus intensity,

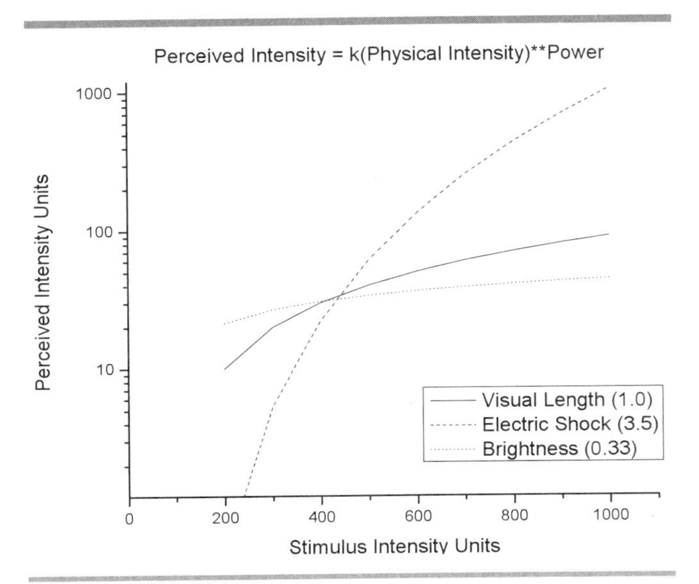

FIGURE 8. Impact of stimulus mode upon perceived stimulus intensity

and depending upon the stimulus type. See Figure 8 for the impact of stimulus mode upon perceived stimulus intensity with change in physical stimulus intensity change based upon exponent (noted in legend). Table 1 lists the stimulus power function exponents for several stimuli.

The sensory detection function usually resembles a sigmoid function with the percentage of correct responses (or a similar value) displayed on the ordinate and the physical parameter on the abscissa. The presence of the stimulus parameter very far toward one end of its possible range indicates that subjects will always be able to respond correctly. Conversely, presence at the other end of the range indicates that subjects never perceive the stimulus properly, making the likelihood of correct response merely a matter of chance. In between these two extremes exists a transition range where subjects have an above-random rate of correct response but do not always respond correctly.

The inflection point of the sigmoid function, or the point at which the function reaches midway between random occurrence of correct behavior and consistently correct response, is usually taken as sensory threshold. A stimulus that is less intense than a human's sensory threshold will not elicit any sensation. Psychophysicists who study the relationship between physical stimuli and their subjective correlates, or percepts, have established thresholds for various stimuli that are used by HFEs.

Several different sensory thresholds have been developed and are used by HFEs (Davis 2003; Fechner et al. 1966; Gescheider 1976; Gescheider 1984; Kaernbach, Schröger, & Müller 2004; Ljunggren et al. 1989; Manning & Rosenstock 1979; Psychonomic Society 1966; Stevens 1975; Yost, Popper, & Fay 1993).

Absolute threshold (AL) is the least stimulus that can be sensed half the time by a (typically young adult) human focused on the source of the stimulus in the absence of other stimuli. Absolute thresholds are generally determined by plotting detection rates against stimulus intensities or characteristics.

Recognition threshold (RL) is the level at which a stimulus can be not only detected but also recognized half the time.

Differential threshold (DL) is the level at which an increase in a detected stimulus can be perceived half the time. This threshold is often referred to as the Just Noticeable Difference (JND). Designers often want to use detection of change in stimulus intensity as an indication of the state of performance of a machine, tool, product, alarm, or other such device. These thresholds are referred to as difference thresholds or Difference Limens (DLs) and refer to the amount of increase that is required in a sensed stimulus before one can detect a Just Noticeable Difference. This magnitude depends upon the starting stimulus intensity.

Terminal threshold (TL) is the level beyond which a stimulus is no longer detected fifty percent of the

TABLE 1

Various Stimulus Power Function Exponents for Several Stimuli

Stimulus	Power Exponent	Description
Heaviness	1.45	Lifted weights
Lightness	1.20	Reflectance of gray papers
Loudness	0.67	Sound pressure of 3000 Hz tone
Muscle force	1.70	Static contractions
Pressure on palm	1.10	Static force on skin
Vibration	0.60	Amplitude of 250 Hz on finger
Vibration	0.95	Amplitude of 60 Hz on finger
Visual area	0.70	Projected square
Visual length	1.00	Projected line

(*Source:* Stevens 1957)

time. The upper end of the detectable stimulus range is set by the inability of sense organs to respond to increased stimulus intensity (or to respond when such exposures are injurious). Some upper thresholds are set by consensus because of the difficulty of studying such exposures without injuring humans.

Thresholds, or limens, are set at the midpoint of a cumulative detection probability threshold, because the inflection point of the function provides the greatest resolution of sensory response to changes in the stimulus.

Equal increments of physical energy do not produce equal increments of sensation between lower- and upper-stimulus thresholds. Often designers anticipate that increments in stimulus intensities of signals can be arbitrarily used between lower and upper bounds of stimulus intensities with equivalent results, but that is usually not the case. The DL, at a particular stimulus intensity of a given stimulus, complies with an approximating fraction known as Weber's Law (or Weber's Fraction). A stimulus having a Weber Fraction of 0.10 need only be increased by 10 percent if a JND is to be detected. A stimulus of intensity 10 has a JND of 1, but a stimulus of 100 must be increased by 10 units for to produce a JND.

For many sensory modalities, the JND is an increasing function of the base level of the current level of stimulus intensity. The ratio of the JND, or ΔI, and the current stimulus intensity is roughly constant. Measured in physical units,

$$\frac{\Delta I}{I} = k \qquad (2)$$

where I is the intensity of stimulation, ΔI the change in stimulus intensity from the level I, and k is the ratio, fraction, or constant for that particular type of stimulus (often referred to as the Weber Fraction or Weber Constant, this is discovered empirically).

Designers often overestimate detection percentage, forgetting that such values are only useful when humans are intensely attending to stimuli. A rule of thumb is that detection at 95 percent or 99 percent requires that the 50 percent detection threshold be multiplied by 2 or 3, respectively. However, if an operator is not intensely focused on a stimulus, or if the stimulus is changing or moving, threshold multiples of 10 or greater may be required to achieve intended design objectives. Standards, handbooks, and testing may be required to arrive at stimulus intensities that meet design requisite performance objectives.

TABLE 2

Relative Discrimination of Physical Intensities and Frequencies

Sensation	Number of JNDs
Brightness of white light	570 discriminable intensities of white light
Hues at medium intensities	120 discriminable wavelengths at medium intensities
Flicker frequencies between 1–45 Hz with on/off cycles of 0.5	375 discriminable interruption rates between 1 and 45 interruptions/sec at moderate intensities and an on/off cycle of 0.5
Loudness of 2000 Hz tones	325 discriminable intensities for pure tones of 2000 Hz
Pure 60 dBA tones between 20 and 20,000 Hz	1800
Interrupted white noise	460 discriminable interruption rates between 1 and 45 interruptions/sec at moderate intensities and an on/off cycle of 0.5
Vibration	15 discriminable amplitudes in chest region with broad contact vibrator within an amplitude range of 0.05–0.5 mm
Mechanical vibration	180 discriminable frequencies between 1 and 320 Hz

(*Source:* Van Cott & Warrick 1972, Mowbray & Gebhard 1958)

TABLE 3

Number of Absolutely Identifiable Sensations

Sensation	Number of Identifiable Levels
Brightness	3 to 5 with white light of 0.1-50 mL
Hues	12 or 13 wavelengths
Interrupted white light	5 or 6 rates
Loudness	3 to 5 with pure tones
Pure tones	4 or 5 tones
Vibration	3 to 5 amplitudes

(*Source:* Van Cott & Warrick 1972, Mowbray & Gebhard 1958)

Confusion in design can occur when designers wish users to detect changes in stimulus intensity level. Although a large number of JNDs exist for any given stimulus (see Tables 2 and 3), JNDs represent relative change detection capabilities (most stimuli are not amenable to relative comparison) and can only produce 7 ± 2 absolute detection levels (Miller 1956).

If, for example, the intensity of a pure tone is presented as a cue or alarm for the severity of oil temperature, perhaps only two levels of sound power level should be used to cue the vehicle operator. Attempting to use many levels of any given stimulus intensity with intentions of providing absolute cues may not be successful, confusing users, who may then respond inappropriately (see Tables 2 and 3).

Many investigators have contributed to our understanding of the specific types and ranges of physical energy (typically referred to as stimuli) that fall within human perceptual capabilities. Much early work has since been catalogued by others (Mowbray & Gebhard 1958, Van Cott & Warrick 1972). Today, many design guidelines and standards publish topically relevant thresholds for use by designers and engineers. Care must be taken, however, to control exposure to stimuli. If stimulus intensity is too great, users can experience sensory stimulus fatigue, causing psychometric function and threshold specifications to change dramatically; but designers and engineers often fail to understand this.

Visual Acuity Problems

Of all the human senses, vision is likely most important to system design. Many issues must be considered when specifying the nature of visual imagery, but visual acuity, or target size, contrast, and magnitude of illumination, are questions that must be addressed.

Visual acuity refers to the ability to see spatial detail and recognize images. Usually visual acuity is specified in minutes of arc, the angle subtended by the object viewed. For objects less than 10 degrees of arc, a small-angle tangent approximation can be used to estimate visual arc:

$$\text{Visual angle (minutes of arc)} = (57.3)(60)L/D \quad (3)$$

where

L = the size of the object measured perpendicularly to the line of sight

D = the distance from the front of the eye to the object

Generally speaking, acuities can be grouped into four categories (Safir et al. 1976, Sanders & McCormick 1993):

- detection of presence
- vernier, or detection of misalignment
- separation, or detection of space between dots, parallel lines, or squares
- form, or identification of shapes or forms

All these forms of acuity are affected by a number of physical and physiological factors. Among the former are illumination, contrast, time of exposure, and color. Engineering design handbooks provide significant guidance on the impact of various combinations of these factors (Booher & Knovel 2003; Galer 1987; Hancock 1999; Salvendy 1987; Stevens 1951; Tufts College & US Office of Naval Research 1951; Weimer 1993; Woodson 1981; Woodson, Tillman & Tillman 1992).

Designers should remember that visual thresholds are set at 50 percent detection rates. We normally increase the visual arc by a factor of 3 to obtain a 99 percent recognition threshold, but the size of the target may have to be increased if illumination, contrast, or movement will inhibit detection. Users may also have subnormal vision. If so, inverting the Snellen ratio and multiplying the required visual acuity for normal vision can encourage recognition by those who have, say, 20:100 vision (the visual acuity value is multiplied by 100/20, or 5) if it can be reasonably expected that users may not, for example, have the benefit of eyeglasses when attempting to exit a burning building, or when personal protective equipment makes impossible the use of glasses or contact lenses.

Operator age can materially affect detection of contrast and color variance. Designers should understand that visual detection and recognition is dependent upon targets' contrasts and illumination. If illumination decreases at dusk, inside buildings, or in vehicles used at night, the size of the target will have to be increased to provide an adequate level of contrast. Vibration and hypoxia (reduced oxygen content in the brain brought on by ascent to high altitudes or by the inhalation of certain kinds of gases) also reduce visual acuity. (Conversely, pilots who breathe pure oxygen at altitude at night frequently report dramatic increase in their color vision.) Figure 9 demonstrates the impact of target contrast and illumination upon spatial frequency and visual acuity. (Note the change in the visibility of words under different levels of illumination, clearly demonstrating the effects of the contrast and size of visual targets under different levels of illumination.)

Moreover, if the designer relies only upon color as a differentiating code for stimuli or information, reduced illumination shifts vision from color to black and white, making color cues difficult to discriminate. Color perception is also affected by the characteristics of the target (spectral content, luminance), by the environment in which it is viewed, and by the observer (Israelski 1978). Eight percent of men, and less than one percent of women, have some degree of color vision deficiency (Sanders & McCormick 1993).

The type of illumination used can materially affect color perception and discriminability. Two colors that appear identical in tungsten light may appear entirely different in daylight. High-pressure sodium lighting distorts almost all colors. Designers often fail to consider the extent to which variations in illumination characteristics can influence color detection across a wide range of operational environments.

Audition

Hearing is very important for human operators, because it allows communication by way of speech, as well as allowing users to attend to a variety of auditory cues such as bells, buzzers, beeps, horns, sirens, and so forth (Fisk & Rogers 1997, Salvendy 1987, Woodson 1981). Designers often rely upon auditory signals to alert multiple operators when visual stimuli cannot be relied upon to do so.

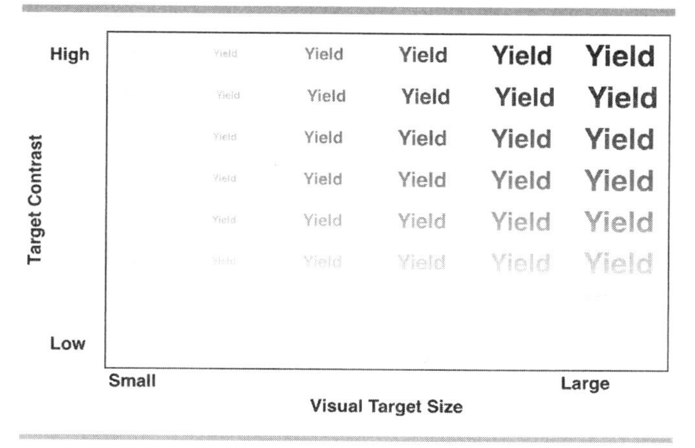

FIGURE 9. Impact of target contrast and illumination upon spatial frequency or visual acuity

Like vision, audition is a complex sensory system that defies comprehensive description in this document. But it is less difficult to describe a process that most HFEs go through in assessing the acceptability of speech or auditory display design. The transmitted information should match generally accepted selection criteria for auditory transmissions (Sanders & McCormick 1993):

- simple and short
- calls for immediate action
- does not refer to previous transmissions
- deals with events in time
- alleviates operators' overloaded visual systems
- supplements operation in areas of poor illumination
- provides instruction when absence from visual displays is required
- acoustical (within the range of operators' hearing)

Discrimination between frequencies and temporal sound patterns is much more reliable than is attempting to discriminate between simple increases in the sound intensity of a pure tone. To avoid ambiguity, systems should use no more than about six different auditory signals. Use of previously learned signals is preferred, and increases in repetition rate, rather than in amplitude, should be used to convey urgency (Woodson 1981; Woodson, Tillman, & Tillman 1992).

The chief challenge of the use of speech is in producing intelligible communication within a noise field. Noise in occupational environments, during emergencies, or during storms, among other situations, can mask speech that is not sufficiently loud. Normal speech averages about 65 dBA, and shouting can nearly reach 100 dBA. An extensive bibliography can be found elsewhere (Sanders & McCormick 1993).

By measuring the percentage of correct words transmitted from a speaker to a listener, a speech intelligibility score can be arrived at (e.g., percent correct). Studies have shown that speech intelligibility decays when the signal-to-noise ratio narrows across the voice spectrum.

When the one-third octave band frequency Sound Power Level (SPL) of a speaker's voice and that of the surrounding noise field is measured (the difference between the voice SPL and noise SPL, in dBA, is the signal-to-noise ratio of the speaker's voice) and the difference at each frequency band is weighed, the sum of those weighted differences is known as the Articulation Index (AI). The AI ranges between 0 to 1; 0 represents unintelligibility and 1 perfect speech intelligibility (Kryter 1985). See Table 4 for the frequency band weightings and an example of a solved AI computation of one-third octave band weights for voice-to-background noise signal-to-noise ratios.

The AI can then be used to look up operational consequences or capabilities for speech communication. The following intelligibility criteria for Articulation Index (AI) are used in the NASA 3000 Standard:

a. Very good to excellent intelligibility: AI = 0.7–1.0
b. Good intelligibility: AI = 0.5–0.7
c. Generally acceptable intelligibility: AI = 0.3–0.5
d. Unsatisfactory or only marginally satisfactory: AI = 0.0–0.3

TABLE 4

Articulation Index One-Third Octave Band Weights

1/3 Octave Band Center Frequency (Hz)	(Speech–Noise) Sound Power Level (dBA)	AI Weighting	dBa × Weighting
200	20	0.0004	0.0080
250	20	0.0010	0.0200
315	20	0.0010	0.0200
400	10	0.0014	0.0140
500	10	0.0014	0.0140
630	10	0.0020	0.0200
800	0	0.0020	--
1000	0	0.0024	--
1250	5	0.0030	0.0150
1600	5	0.0037	0.0185
2000	5	0.0038	0.0190
2500	10	0.0034	0.0340
3150	10	0.0034	0.0340
4000	10	0.0024	0.0240
5000	10	0.0020	0.0200
			AI = 0.2605

(Adapted from Sanders and McCormick 1993)

The average dBA readings for one-third octave bands can be taken at 500, 1000, and 2000 Hz and their average used as the Speech Interference Level (SIL) metric, querying the following figure to judge whether worker proximity and vocalization capacity are adequate for transmitting spoken messages in various noise fields (Sanders & McCormick 1993; Woodson 1981; Woodson, Tillman, & Tillman 1992). Figure 10 describes the effect of the SIL and the functional relationship between the proximity of speaker–listeners and the required speech sound power level.

Intelligibility can be enhanced by designing verbal messages that are simple and redundant, that use a reduced vocabulary set incorporating simple words,

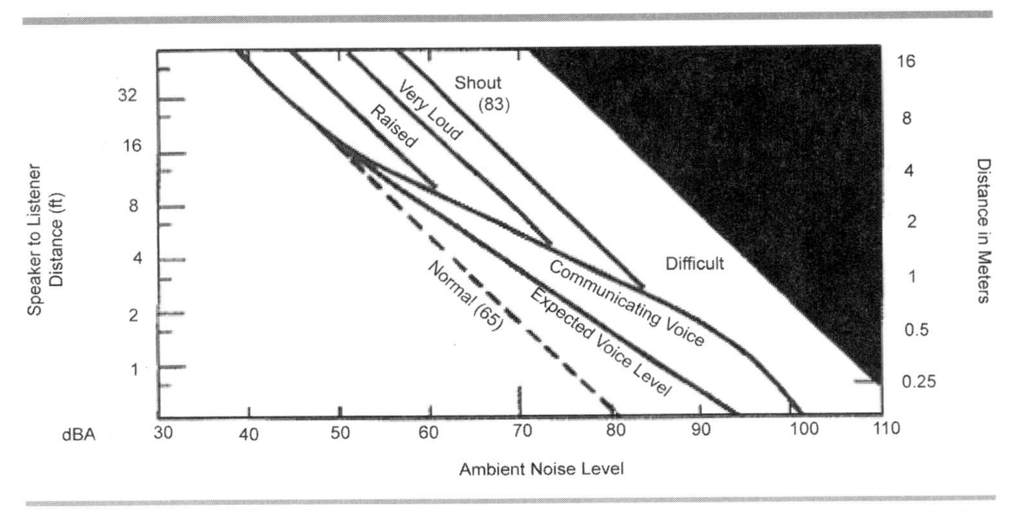

FIGURE 10. Effect of SIL and relationship between speaker–listener proximity and speech sound power level

and that use phrases having contextual meaning (e.g., pilots hear "cleared to land" on final approach).

Many designers assume that users of their systems will have adequate hearing, but hearing impairments can develop occupationally and with age. Unimpaired young operators can converse comfortably at 55 dBA. At the age of 50, speech sound power must average 67 dBA, and at 85, loud speech is required (about 85 dBA) (Coren 1994, Coren & Hakstian 1994). Because designers usually have no control over who will use their systems, most systems should be designed to allow for effective use by persons with hearing difficulties. Providing volume controls so listeners can pick their own listening levels is important.

Are Signals Detected as Expected?

Engineers or designers often believe that if a stimulus is suprathreshold, it will be dependably detectable, but that is not actually the case (Green & Swets 1966, Green & Swets 1974, Hancock & Wintz 1966, Helstrom 1960, Macmillan & Creelman 2005, McNicol 1972, Poor 1994, Schonhoff & Giordano 2006, Swets 1996, Wickens 2002). As shown in Figure 11, observers can detect the presence (hits) and absence of suprathreshold stimuli (correct rejections), but they can also miss signals (misses) and report the presence of nonexistent stimuli (false alarms). The magnitude of each of these outcomes depends upon the observer's expectations regarding the presence of signals, upon their sensorial or perceptual sensitivity (d'), and upon their decision criterion (β).

The β used by an observer is the magnitude of the physical stimulus intensity at which point he or she will uniformly report the presence of a stimulus, as well as absence if the intensity drops. The observer's sensitivity, or inherent ability to detect the signal, can change with age, fatigue, and other factors that can degrade sensation. The observer's response criterion (β) can also change based upon the expectation of the presence of the signal and upon values and costs associated with judgments. If β changes in the face of a constant d', material differences will manifest in operators' accurate reports of signal presence and in their false-alarm rates (see Figures 11 and 12).

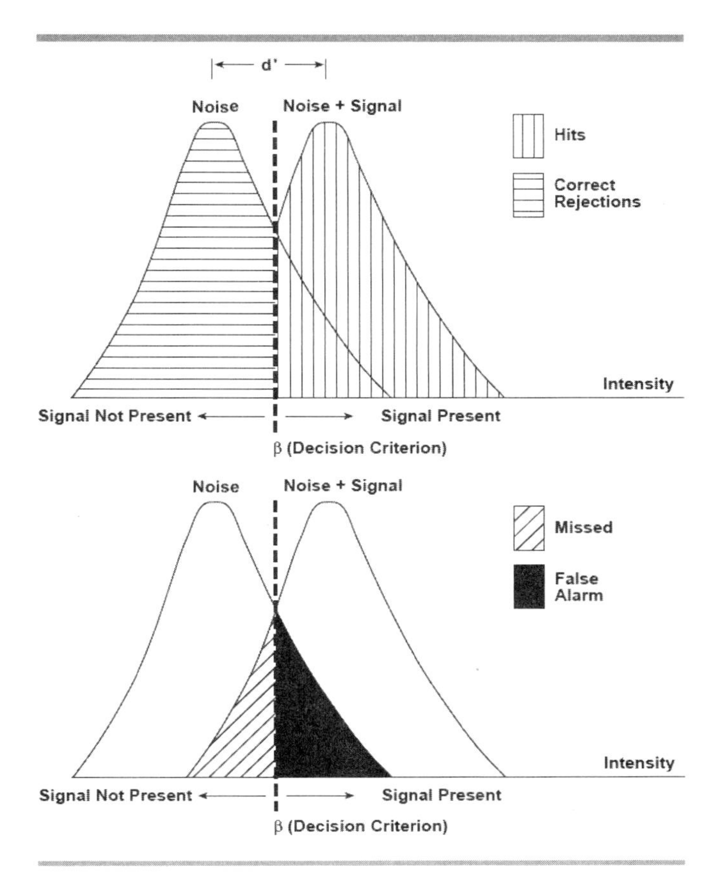

FIGURE 11. General stimulus signal detection behaviors

TABLE 5

Different Values for the Locus of Optimal Physical Intensity or Optimal β

Probability of Signal and Noise		Values		Costs		
Noise	Signal	Hit	Correct Rejection	Miss	False Alarm	Optimal β
0.99	0.01	100	100	100	100	99.00
0.99	0.01	100	100	100	100	99.00
0.99	0.01	100	100	100	100	99.00
0.99	0.01	100	100	100	100	99.00
0.99	0.01	1000	10	1000	10	0.99
0.99	0.01	1000	10	1000	10	0.99
0.99	0.01	10	1000	10	1000	9900.00
0.99	0.01	10	1000	10	1000	9900.00
0.01	0.99	100	100	100	100	0.01
0.01	0.99	100	100	100	100	0.01
0.01	0.99	100	100	100	100	0.01
0.01	0.99	100	100	100	100	0.01
0.01	0.99	1000	10	1000	10	0.00
0.01	0.99	1000	10	1000	10	0.00
0.01	0.99	10	1000	10	1000	1.01
0.01	0.99	10	1000	10	1000	1.01

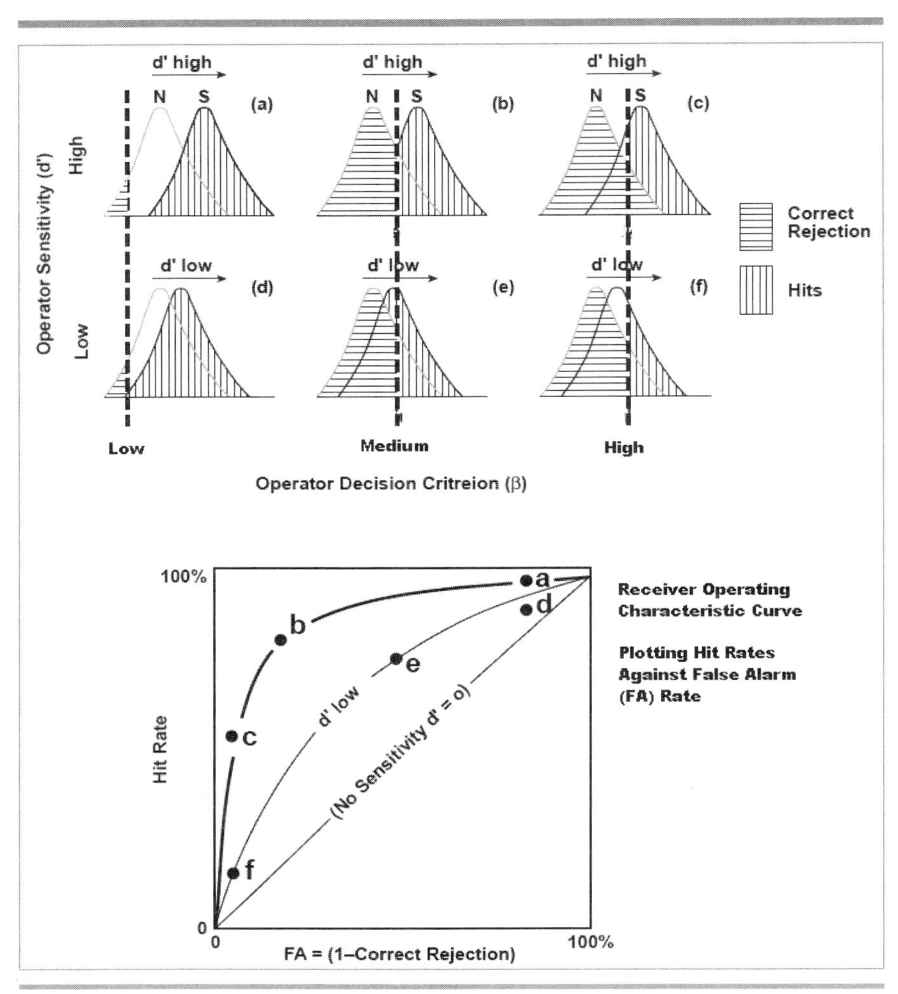

FIGURE 12. Impact of increasing β upon signal or hazard detection (Adapted from Wickens 2002)

One can attempt to improve the observer's β by providing accurate information about the probability of signal and noise, and the values of making hit, correct rejection and the positive magnitude of creating false alarms and misses. By adjusting the theoretical optimal β by a payoff matrix, observer SDT performance can be modulated.

$$\beta \text{ Optimal} = \frac{\text{PR (N)}}{\text{PR (S)}} \cdot \left(\frac{\text{Value (CR)} + \text{Cost (FA)}}{\text{Value (Hit)} + \text{Cost (Miss)}} \right) \quad (4)$$

As shown by Equation 4, and by Table 5, observers' response criterion can be easily biased by changing their perceptions of the prevalence of signals and of the values and costs of their decisions. The observed β for an observer is the ratio of the ordinal probabilities of signal and noise at the decision criterion. Thus, if the β is large, the criterion shifts to the right (i.e., physical stimuli must be substantial before observers will report the presence of signals). If the β is small, the criterion shifts to the left; small physical stimuli intensities will cause conclusions that signals are present.

Errors in signal detection can provoke unanticipated behaviors when cues to initiate behaviors are missed or inappropriate behaviors when cues are absent. Signal detection performance depends upon an observer's (1) expectation of a signal, (2) response criterion (β), and (3) observational sensitivity to the stimulus (d').

One can compute the observed operator's sensitivity, or d', by estimating the distance between signal plus noise distribution (SN) and the noise distribution (N) in units of z-scores, which are obtained by determining the miss and false-alarm rates and finding the z-scores from the means of each distribution to the response criterion.

Another method for evaluating an observer's d' and β involves plotting the hit rate against the false-alarm rate for various signal detection trials in which different frequencies of signals or payoff matrices are used to shift βs. The result is a receiver's or observer's operating characteristic curve. The greater the distance of the curve from the diagonal, the greater the signal-to-noise ratio and the greater the observer's capacity to detect the signal. The βs are determined by the tangent to the curve at the empirical plot point.

A receiver operating characteristic (ROC) curve allows the comparison of an observer's d' and β by plotting HIT rates against FA rates for various βs, which are changed by altering the probability of signals or the payoff matrix, or by allowing the observer to select different confidence levels for each decision regarding the presence of a signal or hazard. The ROC can be used to compare the d' values for different equipment designs among observers for personnel selection or to evaluate the impact of ambient and potentially masking phenomena (e.g., fog, rain, or darkness) upon the driver's capacity. In the ROC example given, observers A, B, and C have greater d', or sensitivities, than do F, E, and D, none of whom share the same β, or decision criterion. Some observers are very liberal, seeking to increase hit rates at the expense of increased false alarms (FA) (e.g., A and D), but others are very conservative (e.g., F and C), requiring greater signal intensities before calling a stimulus a signal. The conservative βs reduce both hits and false alarms.

The importance of understanding SDT is that it allows for the anticipation of significant differences in human detection of a signal or communication given differences that depend on the expectation of a signal and the values and costs associated with it. A designer should understand that stimulus being suprathreshold does not guarantee that humans will accurately respond to signals or correctly reject their presence.

Is the Information Confusing the Operator or User?

Even operators who reliably receive stimuli can confuse those that are very similar in nature or perception. Ambient noise can also mask attributions and produce miscommunications. An in-depth discussion of information theory approaches to the assessment of information sent and received is provided elsewhere (De Greene & Alluisi 1970).

If an operator's task requires the identification or classification of information, the sensory task has become perceptual task. The goal is to receive ($H_{received}$) all information that is sent (H_{sent}). If that occurs, the two circles in the Venn diagram shown will superimpose: all information is transmitted (H_t) without loss (H_{loss}) and no noise (H_{noise}). Loss of information includes sent information that is lost and noise that is erroneously classified as information.

To discover the amount of information that was transmitted, calculate the information that was sent and received and that which collectively represented noise and loss. Information is computed in terms of bits. In the case of a single event i, the information sent by that stimulus can be determined with the formula provided in Figure 13.

In the case of a large number of events, each possessing a different probability of occurrence, the average amount of information presented to the observer can be computed by the following formula for the average sent information.

$$H_{s\,average} = \sum_{i=1}^{N} P_i \log_2\left(\frac{1}{P_i}\right) \tag{5}$$

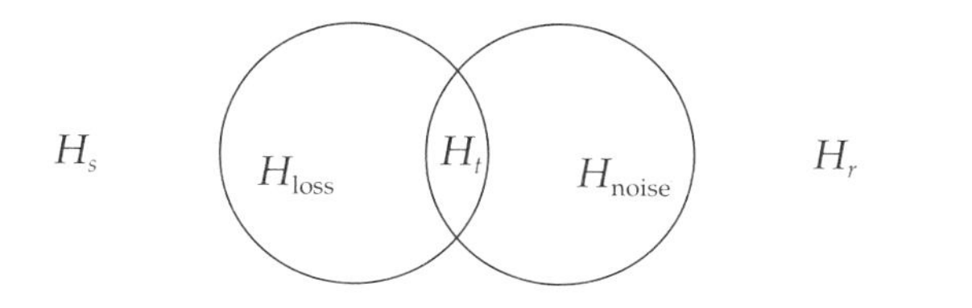

$$H_t = H_s + H_r - H_{sr}$$

FIGURE 13. General model for information transmittal (*Source:* De Greene & Alluisi 1970)

Response	Hazard Icons				
					Sums
Poison Hazard	900	0	0	0	900
Biohazard	100	0	300	300	700
Nuclear Hazard	0	500	400	300	1200
Alien Hazard	0	500	300	400	1200
Sums	1000	1000	1000	1000	**4000**

Response	Hazard Icons					
	A	B	C	D	Sums	Bits
A	0.23	0.00	0.00	0.00	0.23	0.48
B	0.03	0.00	0.08	0.08	0.18	0.44
C	0.00	0.13	0.10	0.08	0.30	0.52
D	0.00	0.13	0.08	0.10	0.30	0.52
Sums	0.25	0.25	0.25	0.25	1.00	
Bits	0.50	0.50	0.50	0.50		

Hs = 2.00 Bits
Hr = 1.97 Bits
Hsr = 3.15 Bits

Ht = 0.81 Bits

FIGURE 14. Confusion matrix demonstrating failure of warning symbols

Suppose that, when four hazard icons, as shown in Figure 14, are presented to 1000 people, who are asked the perceived meaning of each of the icons, their responses are as shown in the data set in Figure 14.

The distribution of responses provided by the subjects showed that the skull and crossbones produces 900/4000, or 23%, recognition as poison, and representing nearly half a bit of information:

$$0.49 \text{ bits} = (0.23)\log_2\left(\frac{1}{0.23}\right) \quad (6)$$

By computing the average information sent (H_s = 2.0 bits), the information received from the marginal probabilities of the rows (H_r = 1.97), and the information content inside the matrix (H_{sr} = 3.15 bits), the information that is transmitted can be determined ($H_t = H_s + H_r - H_{sr}$ = 0.81 bits). In the example above, one finds that the lost information ($H_s - H_t$ = 2.00 – 0.81 = 1.19 bits) caused by confusion and noise ($H_r - H_t$ = 1.97 – 0.81 = 1.16 bits) was equivalent and significant.

This analysis shows that there is hope for use of the skull and crossbones as an icon if one can discover why it was confused as a biohazard by 10 percent of the observers. Ideally, if one determines why the confusion between poisons and biohazards developed, the design can be corrected. However, if the desired response to poisons and biohazards was identical, the confusion might be acceptable.

Often, designers of icons, symbols, or other types of information that must be presented to users become too familiar with their designs and expect no confusion whatsoever. By conducting confusion matrix studies, as above, unexpected bases for confusion or equivocation among presented stimuli are often found.

Have the Affordances Been Adequately Considered?

One of the reasons that observers confuse the meanings of international symbols and other information, is that affordances differ among user populations. Gibson described an affordance as an objective property of an object, or a feature of the immediate environment, that indicates how to interface with that object or feature (Gibson 1966). Norman refined Gibson's definition to refer to a perceived affordance: one in which objective characteristics of the object are combined with the actor's physical capabilities, goals, plans, values, beliefs, and interests (Norman 1988).

An affordance is a form of communication that conveys an intended purpose and operation of a device or a message, as well as behaviors to be avoided. This is a powerful design element that can be useful if used wisely but punishing if it motivates inappropriate or undesired behaviors. Cognitive dissonance may also develop, leading to the use of objects in manners neither expected nor promoted by the affordance.

An example of a positive affordance is offered by the steering wheel of a new car. A driver who has never driven the car before expects, based on the design of the steering wheel and previous experiences, that rotating the wheel will cause the car to veer in the same direction.

An example of an undesired affordance is a colorful, lightweight plastic gun that resembles a squirtgun but fires high-velocity projectiles that present significant risk of injury.

In sum, affordances can be very powerful influences in the receipt and interpretation of sent stimuli that, if not evaluated through tests and evaluation procedures, can lead to dangerous behaviors.

Are the Cognitive Demands Acceptable?

Cognitive demands include handling information that is received from the sensory system, attended to, and combined with human memory to learn and to support decision making and selection of responses (Craik & Salthouse 2007, Durso & Nickerson 2007, Hancock 1999, Lamberts & Goldstone 2005).

HUMAN MEMORY LIMITS

Memory is an important tool for detection of problems, diagnostics, handling protocols, and learning new material or learning from mistakes (Bower 1977, Cermak & Craik 1979, Estes 1975, Gardiner 1976, Shanks 1997). Memory failures often directly or indirectly cause human performance failures. There are generally three types of memory: *sensory, short term, and long term.* Sensory memories act as limited buffers for sensory input. Visual iconic sensory memory is briefly present for visual stimuli and aural stimuli, producing *echoic sensory memory*. Other sensory inputs have brief sensory-memory periods as well. Stimuli captured by sensory memory must move rapidly into short-term memory through attention. If the stimuli are not attended to, the sensory memory essentially filters that information out, and it fades away.

Short-term memory serves as a workbench for reinforcement of selected sensory memory and for temporary recall of information under process. Short-term memory decays rapidly if not sustained by continuous stimulus or rehearsal and pulls information from long-term memory to help develop associations, feeding associative structures into long-term storage.

Interference often causes disturbance in short-term memory retention, and rehearsal and refreshment of the working memory is necessary to minimize loss. Interference can be sensory, distracting of attention, or memory overload.

Long-term storage may be classified as episodic memory (storage of events and experiences in serial form) or semantic memory (a record of associations, declarative information, mental models or concepts, and acquired motor skills). Information from short-term memory is moved to long-term memory by rehearsal. Rehearsal of information, concepts, and motor activities enhances transfer into long-term memory. Learning is most effective if it is distributed across time.

Often designers fail to (a) provide sufficient opportunity for users to focus on sensory memory, (b) allow entry and rehearsal in short-term memory, and (c) enable users to establish associations allowing them to recall information from long-term storage. Moreover, designers can rely too heavily upon operator learning and recall to prevent errors in sequences of operations or to support diagnostic and predictive decisions.

Designing down memory demands is always a good idea, but providing memory aids, such as checklists, increased display times, electronic to-do lists, attentional cues, large amounts of information *chunked* into smaller acronyms, and contextual cues can all promote accurate recall and sequencing of information needed to make good decisions or to operate machines and tools and perform tasks properly, among other activities (Sanders & McCormick 1993).

Does the Design Support Human Decision Making?

Despite effort, humans are often irrational decision makers. Many times decisions are based upon past experience using heuristics or rules of thumb. In other cases, decisions can be flawed because of innate limitations in human decision-making capacities (Wickens & Hollands 2000).

A decision occurs when one must choose from among options, predict or forecast outcomes, or diagnose a situation. The cognitive burden imposed by these activities is caused by having to recall or maintain a set of attributes or facts and their values while contemplating a decision. Metaphorically, one can view a complex decision as a long regression equation having many terms and coefficients. The greater the number of terms and coefficients, the more difficult the cognitive burden becomes, and likelier it is that the decision will not be ideal. Choices are typically easier to make than are predictions, because predictions often require additional mental algebra (Wickens & Hollands 2000).

Diagnosis typically produces the greatest burden, because the individual often has to mentally regress from a current state into an array of possible etiologies. In the early stages of diagnosis, hundreds of etiologies may exist. Further data gathering is required until the solution space can be adequately narrowed—but even when adequately narrowed, the potential solution space may be very large and can exceed human capacity to handle without high risk of error.

Facts of knowledge of attributes may take time to obtain and do not always arrive in an optimal or expected sequence or timeframe, which means decision makers may have to arrive at conclusions or make decisions when only partially informed. Initial attributes or weights associated with those attributes can decay with time, and decisions based upon an incomplete or erroneous attribute-weighting matrix can occur. Matching of weights to attributes may be out of phase, and early assessments based upon limited facts can lead to inappropriate initial hypotheses that, based upon prior experience or workload issues, are inappropriately adhered to. Initial hypotheses can serve as filters for new information as decision makers pay attention to only incoming facts that support initial guesses. Thus, from experience, decision makers attempt to arrive at decisions quickly to reduce the cognitive workload.

As the cognitive workload increases, as decision envelopes change, or as working memory is challenged, decision makers gravitate toward decision-making strategies that reduce their burden, an approach that often leads to error. Some of the types of errors encountered are summarized below.

When in doubt, correlate. When challenged, decision makers often seek causality by correlation. *Engine failures have a high correlation with exhaustion of fuel on long cross-country flights. Because an aircraft's engine stopped, it must have run out of fuel; an emergency landing should be attempted without further diagnosing the problem.* Leaping to such a conclusion may result in an unnecessary, dangerous emergency landing.

Rules of thumb or heuristic approaches to decision making simplify the process. Heuristics rely on a subset of attributes and exclude the need for further data gathering. Intellectual bigotry rules out many possibilities, reducing the need to attend to certain facts or attributes, thereby expediting the decision with less burden. *Men generally have greater upper-extremity strength. Because causing this accident required significant upper-extremity strength, it must have been caused by a man.* But large overlap exists in population strength capabilities across genders; insufficient evidence exists to exclude women from consideration.

Humans are not objective, statistical, or computational machines. Statistical assessment of data usually takes place by intuition rather than calculation and leads to errors in assigning weights to attributes. Humans tend to linearize curvilinear relationships and thus over- or underestimate future system behaviors. Humans also tend to overestimate variability when means or ranges in sampled data are greater and dislike arriving at estimates that are near extremes (i.e., they behave conservatively when estimating proportions). Estimates of means can be significantly influenced by modes, or the frequency of occurrence of a particular value (e.g., although the mean may be 20, there were six values of 4 and no other number was repeated, suggesting that the mean might be close to 4). Neither do humans use Bayes' Rule very well, which asserts that the probabilities of some events are changed by the existence or probabilities of prior events. Failure to correctly statistically characterize probabilities and magnitudes leads to a false perception of reality that can provoke inappropriate decisions.

When in doubt, choose conservative decision outcomes. This essentially regresses toward the mean. Humans bias their decisions based upon either primacy or recency factors. First impressions can take hold and bias all subsequent information gathering and information weighting. However, recent (perhaps negative) experience can shift the bias toward recent information, negating prior data or experiences. Either bias can provoke errors in assessment and decision. First impressions of workers are often incorrect, and a recent error that leads to a significant negative outcome (e.g., accident) does not necessarily reflect the worker's history of safe behavior.

Divide and conquer in the face of overwhelming data and choices. Throwing too much information at a human often leads to filtering. Workers seek to accept a small set of hypotheses (usually less than 3 or 4) and attend to information that principally supports one or more of their initial guesses. Ease of recall of an initially feasible hypothesis may be used to filter additional information, and the worker may rely upon heuristics, primacy bias, and other biasing behaviors to control the mental workload of decision making.

Reliable or highly diagnostic sources can be overweighted by decision makers, causing inappropriate selection of hypotheses and faulty data gathering. Experts used in accident investigations can provoke inaccurate conclusions because they have a demonstrated record of success and expertise. Second opinions and verification through testing, experimentation, simulation, or other objective measurements should be used to check expert opinions. One should also provide those opportunities when the expert is a machine, a gauge, or computer display. Humans often believe that computers do not make mistakes, but in reality they often do.

Overconfidence in one's ability to make decisions. Past good performance as a decision maker can cause overconfidence, biasing perspective and expectancy and producing poor decisions.

Negative consequences outweigh positive consequences. Decision makers are likely to select outcomes that are risk aversive, avoiding costs rather than pursuing gains.

Confirmation and Negative Information Bias

Once humans believe they have the answer, they tend to develop *cognitive tunnel vision* and resist attending to information that contradicts their belief. Designers often believe that humans are objective decision makers and that, if all the facts are presented, humans will objectively weigh them and arrive at a mathematically derived or optimal decision. That is not the case, particularly when decisions are complex or multifaceted.

To help decision makers avoid poor decisions and negative consequences (such as accidents, injuries or losses), designers should do the following (Wickens & Hollands 2000):

- Train decision makers so they understand what is causal and what is not.
- Reduce decision options as much as possible.
- Perform statistical computations and mathematical operations for decision makers and persistently present them for reinspection and reevaluation.
- Remind decision makers of all attributes that exist or that have been encountered, working to prevent their exclusion.
- Provide memory aids such as checklists and diagnostic fault trees to reduce the cognitive burden that makes cognitive shortcuts attractive.

Check Mental Workloads

In preliminary analyses, it may be acceptable to use timeline analysis to estimate the mental workload, or timesharing, that an operator may experience. Essentially, try to balance out workloads among workers and machines in the system. However, once the preliminary design is developed, evaluate the extant design using experimental workload assessment methods to determine whether workers or operators are likely to experience excessive mental demands. Workloads may be acceptable during normal operations but excessive when failures occur or when operation occurs under stressful conditions (Wickens & Hollands 2000).

Mental workload measurement has been classified into three categories: subjective (i.e., self-reporting) measures (Gopher & Braune 1984), performance measures, and physiological measures (O'Donnell & Eggemeier 1986). Performance measures can be made of the primary or actual work task, of secondary operational tasks, or of non–operationally relevant secondary tasks that tap the same resources that primary tasks do; reference tasks can also be used.

The underlying rationale for workload measurement is that humans have many (but nevertheless limited) resources with which to perform work. When one or many underlying resources are overallocated, performance suffers and psychological and physiological stress develops. Absolute workload assessments can be made when directly measuring primary and secondary tasks that burden the same resource pool. Relative resource demand assessments can be made when comparatively evaluating indirect measures, such as physiological strain. Which type of workload measurement should be used depends upon a number of factors.

Wickens proposed a multiple-resource theory for workload assessment and control in which different resources exist for different modalities (Wickens 1984). Auditory and visual processing resources, central processing resources, and different motor resources are required for the performance of psychomotor tasks. Resources are classified by type and have dedicated channels. If work is designed to pass through one particular type of channel, the people involved may experience resource limitation problems, degraded performance, physiological strain, and other negative outcomes. Redesigning a task to keep inputs from sharing a signal channel and to allow outputs to use multiple channels can increase workloads (to a point) without running into resource allocation constraints. Wickens is careful to point out that workload design is a little more complicated than this, but the general principles appear to work well.

The multiple-resource model is presented in elements. The processing stage addresses sensory encoding and perception, central processing, and response processing. The second element addresses input and response modes. The auditory, visual, and tactile modalities draw upon different resources, allowing cross-modal timesharing to perform better than intramodal timesharing. Listening to the radio while tracking the position of a car in a lane is much easier than listening to the radio while listening with eyes closed to instructions about when and how much to turn the steering wheel to maintain lane position. The third element is the processing mode, which can be verbal or spatial in nature. Overburdening a particular channel or switching between encoding, processing, and response modalities has negative consequences for mental workload.

There are a wealth of studies and supplementary discussions on the nuances of human workload that cannot be adequately addressed here. However, the general concept of unbalanced or excessive use of resources creates problems in mental workload and thereby increases performance decrements, strain, and the likelihood that unsafe shortcuts will be taken.

Mental Workload Assessment

A number of metrics have been developed for assessment of mental workload. No single metric is ideal because they all vary in their sensitivity, diagnosticity, primary task intrusion, implementation requirements (e.g., availability of equipment, expense, and other logistical constraints), and operator acceptance (e.g., certain operator populations are very resistant to admitting any problems with their ability to handle task demands, fearing they will be bypassed for promotion; other operators may not honestly respond, putting honest responders at a disadvantage if the information is available to decision makers). Sensitivity is an index of the responsiveness of the metric to changes in workload. Diagnosticity is the ability to discern the type or cause of workload, or the ability to attribute it to an aspect, or aspects, of the operator's task (Wierwille & Eggemeier 1993).

Primary task intrusion occurs when measurement of the workload metric interferes with task performance, thereby giving false indications of workload problems. Secondary tasks can be used to assess remaining resource capacity after primary task requirements are made and the primary and secondary task share the same resources. Sometimes subjects switch primary and secondary task priorities when they know that only secondary task performance is being measured. Self-report measures taken after completion of the task, as well as most physiological measures, seem to degrade primary task performance the least (Eggemeier & Wilson 1991; Eggemeier, Wilson, Kramer & Damos 1991).

The reader should understand that substantial warnings applying to the use of mental workload metrics are given in the literature; a book would be required to adequately address them. No reliable decision algorithm exists for selecting one or more metrics beyond the selection criteria noted previously, the nature of the tasks performed, the nature of the operator population, and the nature of questions that have to be addressed in the assessment.

Errors have occurred in past measurement of mental workload, putting operators at risk. Often errors occur because designers have used inappropriate metrics (based upon the selection criteria noted above) or have not corroborated outcomes of a single metric (e.g., failed to use a battery of metrics). For example, one may measure primary performance and find no decrement, then concluding that mental workloads are adequate. Yet such performance may have been achieved by extreme effort that produced psychological and physiological strain. With time, that situation will cause degradation of performance.

Another example is relying on a physiological metric merely because it is objective. Physiological metrics are integrated measurements that reflect totality of exposure and often have lag times, offering little diagnostic value. Personality structure can influence stress responses; resource overloads can be missed in individuals who may not share concern about performance decrements.

Self-report tools offer great value but are subject to operator cooperation, which can be inhibited by poorly described procedures, neophyte operators, or managerial factors (e.g., if I report problems with this new system and others don't, will I be at a future disadvantage for selection to use it?). Designers who are strong advocates of their design are not necessarily the individuals who should design, present, or assess the results of operators' self-reports via mental workload tools.

Dismissing weak indicators of mental workload problems or failing to consider the additional burdens that develop when components of systems are not operating as required or expected often produces situationally excessive workloads that go unaddressed during design. Often HFEs come in after accidents and must reassess mental workloads to rule that problem out during an accident investigation. Often mental workload is found to be high and inadequately considered in the design of the original system or after new modifications were made without adequate evaluation of the impact of the added equipment, tasks, or responsibilities.

Complex systems are typically designed concurrently as separate design teams work on components of a design. Often, however, they work too independently, failing to understand that HFE designs appropriate for their components are less so when all components are merged and integrated into an operational system. Often operators' visual, auditory, and motor resources are overtaxed during certain performance scenarios, particularly when handling system malfunctions or complex problems. Mental workload assessments in component design may be useful at those levels, but failing to replicate mental workload assessment at design completion or in the integration stages can be disastrous, particularly if systematic tests are not performed within an acceptable scope of failure modes.

Are Motor-Performance Demands Excessive?

Motor response, or performance, can be classified as single, sequential, discrete open-loop movements, or as closed-loop continuous movements. These activities can also be mediated by reaction-time components of manual performance. Textbooks recommended in this chapter's reference section have excellent, in-depth reviews of human reaction time and motor performance (e.g., Hancock 1999, Salvendy 1987, Weimer 1993). The following sections address sources of errors that designers have encountered when designing or evaluating the impact of their designs on motor-performance requirements and capabilities.

Are Anticipated Reaction Times Realistic?

The time taken by operators to detect and physically respond to some external event or input (such as a change in a traffic light, or a hazard signal) is referred to as reaction time. Reaction times are not static within operators; fatigue, attention or vigilance decrement, stress, aging, and yet other factors can degrade or enhance reaction time (Wiener 1964; Weiner 1987; Wiener, Curry & Faustina 1984; Wiener, Poock & Steele 1964). Reaction time is also influenced by the number of signals or events to which the operator has to attend. The Hick–Hyman Law demonstrates that increased numbers of events that must be attended to, increased information content, and increased information processing produce increased latencies in response times (Grossberg, AMA & SIAM 1981; Kamlet, Boisvert, & U.S. Army Human Engineering Laboratories 1969; Schulz & Lessing 1976; Welford & Brebner 1980).

$$\mathrm{RT} = \frac{150\ \mathrm{ms}}{\mathrm{bit}}\,[\log_2(n + 1)\ \mathrm{bits}] \tag{7}$$

where n = the number of choices or potential responses.

Reaction-time tasks are of three types: simple, disjunctive, or choice (Welford & Brebner 1980). Most reaction-time data have been collected under ideal laboratory conditions in which subjects are extremely focused on reaction-time tasks. As timesharing increases, reaction times increase—sometimes demonstrably (Hancock & Caird 1993; Hancock et al. 1995; Hancock et al. 1990; Kramer, Trejo, & Humphrey 1995).

Thus, if a nuclear power plant operator must attend to 10 signal lights that indicate a malfunction, then responding by pressing a single master alarm button, the minimum amount of time needed to react to the signal can be determined. This choice reaction time is computed as

$$\mathrm{RT} = \frac{150\ \mathrm{ms}}{\mathrm{bit}}\,[\log_2(10 + 1)\ \mathrm{bits}] = 519\ \mathrm{ms} \tag{8}$$

Designers often err in performing restricted laboratory or field tests of reaction times, presuming that such times will continue to be reflected when the operator is exposed to many timesharing activities; but that will probably not be the case. In such situations, designers should carefully consider the consequences of long response latencies. If such outcomes are problematic, design interventions will be needed (e.g., computer-assisted vigilance, an engineer in soft failures, increased operator training and awareness, and so on).

Are Open-Loop Performance or Discrete Movement Times Realistic?

Open-loop tasks in which feedback is not or cannot be used, such as a discrete rapid movement of a hand to a control location, can usually be performed more quickly than can closed-loop or tracking tasks, in which feedback is used to adjust movements. Fitts' Law provides an open-loop motor-performance prediction model based upon the concept that different movement amplitudes to different endpoint or target accuracy constraints produce different information-processing demands for humans (Fitts 1954). Human information processing rate capacity is limited, which means that one can anticipate that longer times will be required to perform longer movements or more precise movements, because more information must be processed within a channel capacity–constrained motor system.

Fitts and colleagues demonstrated a reliable relationship between movement times and task indexes of difficulty that were characterized in terms of bits of information (Fitts 1958, Fitts & Peterson 1964, Fitts et al. 1956)

$$\mathrm{MT} = a + b\log_2(2A/W) \tag{9}$$

where

a, b = regression model coefficients
A = amplitude or move distance
W = move endpoint accuracy requirement or target width

There are different degrees of molecularization of this model (Welford 1968; Wiker, Langolf, & Chaffin 1989). However, it serves as a warning to designers that if one designs movement tasks that approach the information processing constraints of the motor system, errors and mental workload levels will develop to unacceptable levels. If the cognitive demands are already high, adding additional challenges (such as large amplitude movements to very precise endpoints, or very precise movements in general) forces trade-offs on the part of the operator. Thus, while the motor task may be handled, the operator will have to reduce timesharing to achieve such performance.

Are Manual Tracking or Closed-Loop Performance Demands Excessive?

Closed-loop tracking tasks occur when driving a vehicle, aiming at a target, or performing any task in which motor behavior is adjusted to control error. Tracking can be pursuit (e.g., the operator sees both the target and the target tracking feature, as when driving along a roadway or acquiring a moving target with a cursor or gunsight) or compensatory (e.g., only the magnitude and direction of error is presented; the operator is asked to null the displayed error, as in an instrumental aircraft landing, when glide-slope indicators show only horizontal and vertical deviations from the desired glide slope). Pursuit tracking is an easier task, because the operator can see and forecast movement of the target (Sanders & McCormick 1993).

Tracking performance is materially affected by control order. Zero-order controllers produce displacement outcomes in response to control displacement. First-order controls increase response velocity in correspondence to the magnitude of the control's displacement. Second-order controls increase response acceleration in correspondence to the magnitude of control displacement. Third-order controls relate the magnitude of jerk response to control displacement. In general, zero- and first-order controls produce best results and reduce mental workload demands. When higher-order controls are used, they are difficult to master and to use to achieve desired results (in most, but not all cases) (Sanders & McCormick 1993).

In addition to control order, designers often introduce response lags into systems, causing overcontrolling, phase errors, control-system instability, and other problems. Often poorly designed controls create control–display response incapability (e.g., tillers on sailboats versus steering wheels; the operator moves the tiller to the left to make the vessel turn right) that cause tracking mistakes, lost time, and accidents (Jagacinski & Flach 2003, Woodson 1981).

Controls often need their gains set (gain being the magnitude of the response per unit change in control displacement). High-gain controls can be useful when large responses are needed to catch up with the target but are very problematic when the target is close and fine positioning (e.g., low-gain control) is required to acquire and maintain the target.

The operator's effective bandwidth in tracking tasks can be reduced by reducing control dead space (e.g., control slack or movement before underlying response is engaged) or control backlash (e.g., control kick at the end of movement within the dead space). If the required bandwidth exceeds the operator's capacity, phase and gain errors can occur, causing loss of stability or control.

If the nature of the control design (e.g., gain, bandwidth, dead space, kick, momentum or inertia, compensatory versus pursuit tracking) is not matched well with the assigned task, operator performance can be significantly challenged or fail. Matters only get worse when multiaxis or dual-limb tracking tasks are required.

Far too often, HFEs are brought in to resolve tracking task problems in which designers thought they were helping operators by introducing variable gain, variable order, and dual-tracking opportunities, often letting operators switch rapidly between modes. Whether or not such designs are helpful must be determined by realistic performance testing. Bode plots of phase gain errors in tracking behavior can give absolute and relative tracking performance capacity. As gains of any magnitude are coupled with large phase errors, the system is less controllable.

Designers often inherit control systems within the context of their new system design. Mental workloads and capacities to handle inherent tracking demands can differ materially from one system to another. Thus, looking at similar systems and planning on the control design that has been acceptable in the antecedent system does not guarantee that the design will be acceptable in a new application where perceptual, cognitive, and motor demands are quite different. Only by thorough testing in the context of system failures can one determine whether the control-system design is satisfactory. Excellent guidance in control theory applications to control design is provided elsewhere (Jagacinski & Flach 2003). An example of a tracking task error showing the relationship between phase and gain errors on tracking is provided in Figure 15.

Are Desired or Safe Behaviors Maintained?

Behavioral-based safety refers to an approach that seeks to reduce unsafe behaviors through programs of behavior modification (Geller 1992, Kohn 1988, McAfee & Winn 1989, Peters 1989, Roberts et al. 2005). HFEs always attempt to design systems that do not require behavior modification or reinforcement. When that is not possible, personnel selection, training, and behavior modification and reinforcement are necessary—endeavors typically referred to as administrative controls.

Managers often think that short-term cost savings afforded by an administrative control are a better option than *engineering out* hazards. However, administrative controls must be maintained daily, and behavior must be managed; the energy and costs of such efforts can persist for long periods, eventually costing more than the engineering control intervention. That said, HFEs often are faced with relying on administrative-hazard and unsafe-act control paradigms that are discussed at length in other chapters within this text. One example is the *NIOSH Work Practices Guide for Manual Lifting* (Waters et al. 1993).

Should Personnel Be Selected and Trained?

Nearly all systems require some level of personnel selection and training (Bullard 1994, Bureau of Business Practice 1979, Kroehnert 2000, Leigh & Leigh 2006). The nature of personnel selection is determined by the perceptual, cognitive, and motor-performance demands that have been imposed by allocation of tasks within

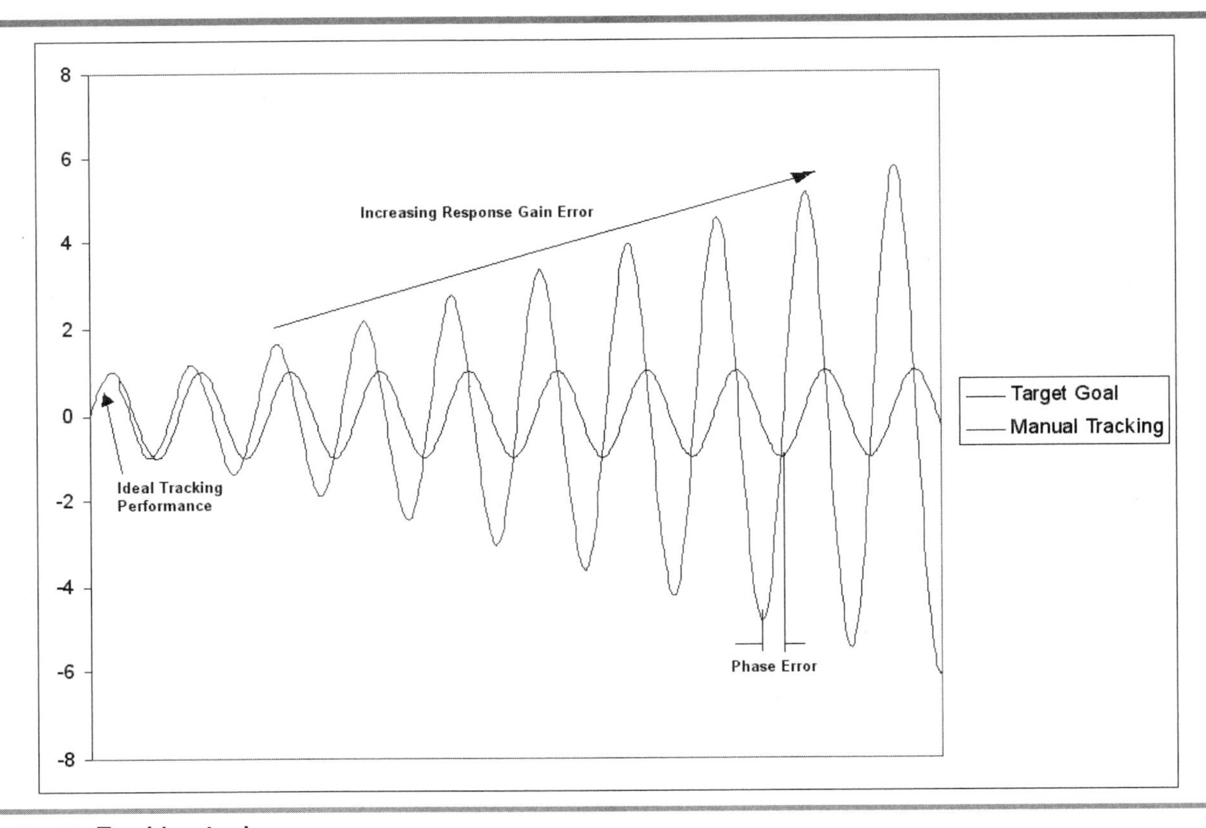

FIGURE 15. Tracking task error

the human–machine system. Because other chapters deal with these requirements, this chapter will merely reinforce the need for HFEs to ensure that personnel selection factors are considered and carefully matched with the system's design requirements and behaviors. This is particularly important when attempting to accommodate potential users who are impaired situationally or chronically.

Training is often used to relax personnel selection criteria and pressures. The more complex and difficult the system is to operate, the greater the likelihood that it will need personnel who have higher levels of education, training, and skill. Training programs are likely to be more extensive as well.

Training programs should include operator understanding and recognition of hazards, safe behaviors, and failure-response or emergency protocols. After recognizing a danger, individuals must act to protect themselves against accidents and possible harm. They must know what actions are correct and safe and complete all actions required. Knowing appropriate actions and performing them correctly requires training, practice, and reinforcement through design and use of appropriate warnings, labels, and other hazard-response communications.

Even the best selection and training programs typically do not select personnel with the precise skills needed to operate new systems or to produce consistently desired responses. Training helps to reduce the risk of accidents and the impact or severity of accidents that do occur. Far too often, however, HFEs are confronted with heavy designer reliance upon personnel selection and training to handle design problems. Moreover, many accident investigations conclude that personnel selection and training must be improved to reduce risk of future accidents—code language for poor design that requires humans of greater capacities or who have more training and practice if operation is to have desired margins of safety.

Design Using Appropriate HFE Standards, Codes, and Regulations

Standards, codes, and guidelines provide design information in a methodological manner [e.g., MIL-STD 1472D (1989), NASA-STD-3000 (1995), and ANSI/HFS 100 (1988)]. Most standards are classified as specification standards where specific guidance is provided. The guidance may not yield the best design outcome, but it should yield, if applied properly, an acceptable design.

Performance standards do not provide cookbook-style guidance. Instead, performance standards help designers understand what the requisite performance criteria are for an area of application. The Occupational Safety and Health Act's general duty clause does not provide specific details, simply stating that any design that provides a job or workplace free from "recognized hazards that are causing or are likely to cause death or serious physical harm to . . . employees" is acceptable. The designer is free to seek a variety of design options that meet that standard of care. This usually requires demonstrable proof that the design is safe (e.g., testing or citation of other studies of like design that have proven to be safe in comparable tasks, equipment, environments, and worker populations).

Standards result from consensus among content experts from industrial firms, trade associations, academia, technical societies, labor organizations, consumer organizations, and government agencies. Failure to pay heed to standards and codes may be disastrous, particularly if a product, machine, or system is ever involved in a product liability lawsuit. Anyone who fails to consider standards, codes, or guidelines in professional safety is often asking for accident, injury, death, loss of property, and even appearance in court.

One may deviate from standards or guidelines, but justification for such deviation should be strong and should produce a better and safer design. Standards are consensus based—in other words, reflective of lowest common denominators. Designers may need to improve performance or specifications to reflect a number of factors. However, although exceeding standards may need justification before the design team and those who authorize expenditures, substandard design is not acceptable.

It is often impossible for an evaluator or inspector, who is typically untrained in human factors, to decide whether a design does, or does not, meet a performance standard. A perfect world would see no opportunity for ambiguity or misinterpretation of HFE design standards, but that is simply not the case. Asking designers (who are not grounded in HFE engineering literature and principles) to interpret HFE standards without help is usually counterproductive.

Moreover, standards must be *tailored* or used with multifaceted design demands and constraints. Tradeoffs must be made to ensure that the intent of the standard is met. Sometimes standards present apparently contradictory or congruent information when, in fact, they do not. Moreover, not all requirements of a specification or performance standard must be *met* or designed to. Designers may spend significant amounts of time attempting to resolve apparently incongruent advice or requirements in a standard, a set of building codes, or a set of federal regulations. Terrible mistakes occur when designers latch onto specifications, use them, and remain unaware of an overriding specification in a separate standard to which they do not have access.

Standards, codes, and regulations cannot be used effectively by anyone who does not understand system performance objectives and requirements, as well as top-down design. Myopic use of the information provided in standards, design handbooks, building codes, and other consensus-based design guidance is an opportunity for misapplication of standards.

Design checklists do have their place, but far too often checklists are published without validation or assurance that, or even inquiry into whether, designs are effective and safe. Design checklists often act more as screening tools to grab the attention of those who have the education, skills, and expertise necessary to rule out problems by more thorough analysis.

Design checklists often query for binary responses on issues that are univariate in nature. Far too often, design problems represent interactions among a variety of design features. Additive responses may not provide any valid insight into the true magnitude of a problem or the acceptability of a design. Design checklists can be used to help detect the possibility of a problem, but they offer no protection when used blindly and improperly.

Conclusion

It is possible to design safe and effective systems that range from small- to large-scale, from simple to complex, and from development to testing. Human factors engineering principles work to improve the marriage between humans, machines, tasks, and operating environments. Good designs prevent accidents, mitigate the severity of the consequences of failures or accidents, and improve system performance. Those who are unfamiliar with HFE design principles should recognize the need to include HFE expertise in system design, test, and evaluation from the outset, as well as when modifications are to be made to existing system designs.

References

Annett, J. 1971. *Task Analysis*. London: H.M.S.O.

Annett, J., and N. Stanton. 2000. *Task Analysis*. London: Taylor & Francis.

Booher, H. R., and Knovel. 2003. *Handbook of Human Systems Integration*. Hoboken, NJ: Wiley-Interscience.

Bower, G. H. 1977. *Human Memory: Basic Processes: Selected Reprints with New Commentaries, from The Psychology of Learning and Motivation*. New York: Academic Press.

Bullard, R. 1994. *The Occasional Trainer's Handbook*. Englewood Cliffs, NJ: Educational Technology Publications.

Bureau of Business Practice. 1979. *Training Director's Handbook*. Waterford, CT: The Bureau of Business Practice.

Bureau of Labor Statistics (BLS). 2007. *Census of Fatal Occupational Injuries*. Washington, D.C.: U.S. Department of Labor, Bureau of Labor Statistics.

Cermak, L. S., and F. I. M. Craik. 1979. *Levels of Processing in Human Memory*. Hillsdale, NJ: Lawrence Erlbaum Associates.

Chapanis, A. 1965. *Research Techniques in Human Engineering*. Baltimore: The Johns Hopkins University Press.

Coren, S. 1994. "Most Comfortable Listening Level as a Function of Age." *Ergonomics*. 37(7):1269–1274.

Coren, S., and A. R. Hakstian. 1994. "Predicting Speech Recognition Thresholds from Pure Tone Hearing Thresholds." *Percept Mot Skills* 79(2):1003–1008.

Craik, F. I. M., and T. A. Salthouse. 2007. *The Handbook of Aging and Cognition*. 3d ed. New York: Psychology Press.

Davis, S. F. 2003. *Handbook of Research Methods in Experimental Psychology*. Malden, MA: Blackwell Publishers.

De Greene, K. B., and E. A. Alluisi. 1970. *Systems Psychology*. New York: McGraw-Hill.

Desberg, P., and J. H. Taylor. 1986. *Essentials of Task Analysis*. Lanham: University Press of America.

Durso, F. T., and R. S. Nickerson. 2007. *Handbook of Applied Cognition*. 2d ed. Chichester: John Wiley.

Eggemeier, F. T., and G. F. Wilson. 1991. "Performance-Based and Subjective Assessment of Workload in Multi-task Environments," in *Multiple-Task Performance*. London: Taylor & Francis, pp. 217–278.

Eggemeier, F. T., G. F. Wilson, A. F. Kramer, and D. L. Damos. 1991. "Workload Assessment in Multi-task Environments," in *Multiple-Task Performance*. London: Taylor & Francis, pp. 207–216.

Estes, W. K. 1975. *Handbook of Learning and Cognitive Processes*. Hillsdale, NJ: Lawrence Erlbaum Associates.

Fechner, G. T., H. E. Adler, D. H. Howes, and E. G. Boring. 1966. *Elements of Psychophysics*. New York: Holt Rinehart and Winston.

Fisk, A. D., and W. A. Rogers. 1997. *Handbook of Human Factors and the Older Adult*. San Diego: Academic Press.

Fitts, P. M. 1951. "Human Engineering for an Effective Air-Navigation and Traffic-Control System." Columbus: Ohio State University Research Foundation.

______. 1954. "The Information Capacity of the Human Motor System in Controlling the Amplitude of Movement." *J Exp Psychol* 47(6):381–391.

______. 1958. "Engineering Psychology." *Annual Rev Psychol* 9:267–294.

Fitts, P. M., and J. R. Peterson. 1964. "Information Capacity of Discrete Motor Responses." *J Exp Psychol* 67:103–112.

Fitts, P. M., M. Weinstein, M. Rappaport, N. Anderson, and J. A. Leonard. 1956. "Stimulus Correlates of Visual Pattern Recognition: A Probability Approach." *J Exp Psychol* 51(1):1–11.

Galer, I. A. R. 1987. *Applied Ergonomics Handbook*. 2d ed. London: Butterworths.

Gardiner, J. M. 1976. *Readings in Human Memory*. London: Methuen.

Geller, E. S. 1992. *Applications of Behavior Analysis to Prevent Injuries from Vehicle Crashes*. Cambridge, MA: Cambridge Center for Behavioral Studies.

Gescheider, G. A. 1976. *Psychophysics: Method and Theory*. New York: Lawrence Erlbaum Associates

______. 1984. *Psychophysics: Method, Theory, and Application*. 2d ed. Hillsdale, NJ: Lawrence Erlbaum Associates.

Gibson, J. J. 1966. *The Senses Considered as Perceptual Systems*. Boston: Houghton Mifflin.

Gopher, D., and R. Braune. 1984. "On the Psychophysics of Workload: Why Bother with Subjective Measures?" *Human Factors* 26:519–532.

Green, D. M., and J. A. Swets. 1966. *Signal Detection Theory and Psychophysics*. New York: John Wiley.

______. 1974. *Signal Detection Theory and Psychophysics*. Huntington, NY: R. E. Krieger Publishing Company.

Greenwood, J., and M. Parsons. 2000. "A Guide to the Use of Focus Groups in Health Care Research: Part 2." *Contemp Nurse* 9(2):181–191.

Grossberg, S., American Mathematical Society (AMS), and Society for Industrial and Applied Mathematics (SIAM). 1981. *Mathematical Psychology and Psychophysiology*. Providence, RI: AMS.

Hancock, J. C., and P. A. Wintz. 1966. *Signal Detection Theory*. New York: McGraw-Hill.

Hancock, P. A. 1999. *Human Performance and Ergonomics*. San Diego: Academic Press.

Hancock, P. A., and J. K. Caird. 1993. "Experimental Evaluation of a Model of Mental Workload." *Human Factors* 35(3):413–429.

Hancock, P. A., G. Williams, C. M. Manning, and S. Miyake. 1995. "Influence of Task Demand Characteristics on Workload and Performance." *Int J Aviat Psychol* 5(1):63–86.

Hancock, P. A., G. Wulf, D. Thom, and P. Fassnacht. 1990. "Driver Workload During Differing Driving Maneuvers." *Accid Anal Prev* 22(3):281–290.

Helstrom, C. W. 1960. *Statistical Theory of Signal Detection*. New York: Pergamon Press.

Israelski, E. W. 1978. "Commonplace Human Factors Problems Experienced by the Colorblind—A Pilot Questionnaire Survey." Paper presented at the Human Factors Society 22nd Annual Meeting, Santa Monica, CA.

Jagacinski, R. J., and J. Flach. 2003. *Control Theory for Humans: Quantitative Approaches to Modeling Performance*. Mahwah, NJ: Lawrence Erlbaum Associates.

Jonassen, D. H., W. H. Hannum, and M. Tessmer. 1989. *Handbook of Task Analysis Procedures*. New York: Praeger.

Jonassen, D. H., M. Tessmer, and W. H. Hannum. 1999. *Task Analysis Methods for Instructional Design*. Mahwah, NJ: Lawrence Erlbaum Associates.

Kaernbach, C., E. Schröger, and H. Müller. 2004. *Psychophysics Beyond Sensation: Laws and Invariants of Human Cognition*. Mahwah, NJ: Lawrence Erlbaum Associates.

Kamlet, A. S., Boisvert, L. J., and U.S. Army Human Engineering Laboratories. 1969. *Reaction Time: A Bibliography with Abstracts*. Aberdeen Proving Ground: U.S. Army Human Engineering Laboratories.

Kirwan, B., and L. K. Ainsworth. 1992. *A Guide to Task Analysis*. London: Taylor & Francis.

Kohn, J. P. 1988. *Behavioral Engineering Through Safety Training: The B.E.S.T. Approach*. Springfield, IL: C. C. Thomas.

Kramer, A. F., L. J. Trejo, and D. Humphrey. 1995. "Assessment of Mental Workload with Task-Irrelevant Auditory Probes." *Biol Psychol* 40(1–2):83–100.

Kroehnert, G. 2000. *Basic Training for Trainers: A Handbook for New Trainers*. 3d ed. Sydney: McGraw-Hill.

Kryter, K. D. 1985. *The Effects of Noise on Man*. 2d ed. New York: Academic Press.

Kurke, M. 1961. "Operational Sequence Diagrams in System Design." *Human Factors* 3:66–73.

Lamberts, K., and R. L. Goldstone. 2005. *Handbook of Cognition*. Thousand Oaks, CA: SAGE Publications.

Leigh, D., and D. Leigh. 2006. *The Group Trainer's Handbook: Designing and Delivering Training for Groups*. 3d ed. London: Kogan Page.

Ljunggren, G., S. Dornic, O. Bar-Or, and G. Borg. 1989. *Psychophysics in Action*. Berlin: Springer-Verlag.

Macmillan, N. A., and C. D. Creelman. 2005. *Detection Theory: A User's Guide*, 2d ed. Mahwah, NJ: Lawrence Erlbaum Associates.

Manning, S. A., and E. H. Rosenstock. 1979. *Classical Psychophysics and Scaling*. Huntington, NY: Krieger.

McAfee, R. B., and A. R. Winn. 1989. "The Use of Incentives/Feedback to Enhance Workplace Safety: A Critique of the Literature." *Journal of Safety Research* 20:7–19.

McNicol, D. 1972. *A Primer of Signal Detection Theory*. London: Allen and Unwin.

Miller, G. A. 1956. "The Magical Number Seven Plus or Minus Two: Some Limits on Our Capacity for Processing Information." *Psychol Rev* 63(2):81–97.

Mowbray, H. M., and J. W. Gebhard. 1958. *Man's Senses as Informational Channels* (No. CM-936). Silver Spring, MD: The Johns Hopkins University, Applied Physics Laboratory.

National Aeronautics and Space Administration (NASA). 1995. NASA-STD-3000, *Man- Systems Integration Standards*. Houston: Lyndon B. Johnson Space Center.

Niebel, B. W., and A. Freivalds. 2002. *Methods, Standards and Work Design*. New York: McGraw-Hill.

Norman, D. A. 1988. *The Psychology of Everyday Things*. New York: Basic Books.

Occupational Health and Safety Administration (OSHA). 1974. 29 CFR Part 1910, "Safety and Health Regulations for General Industry." Washington, D.C.: OSHA.

______. 1974. 29 CFR Part 1910.1030, "Bloodborne Pathogens." Washington, D.C.: OSHA.

O'Donnell, R. D., and F. T. Eggemeier. 1986. "Workload Assessment Methodology," in K. R. Boff, C. Kauffmann, and J. Thomas, eds., *Handbook of Perception and Human Performance*, Cognitive Processes and Performance vol 3. New York: Wiley & Sons, 42/41–42/49.

Peters, R. H. 1989. "Review of Recent Research on Organizational and Behavioral Factors Associated with Mine Safety." Washington, D.C.: U.S. Dept. of the Interior, Bureau of Mines.

Pheasant, S. 1988. *Bodyspace*. London: Taylor & Francis.

Poor, H. V. 1994. *An Introduction to Signal Detection and Estimation*. 2d ed. New York: Springer-Verlag.

Price, H. E., and B. J. Tabachnick. 1968. NASA CR-878, "A Descriptive Model for Determining Optimal Human Performance in Systems. vol 3. An approach for determining the optimal role of man and allocation of functions in an aerospace system (including technical data appendices)." Washington, D.C.: NASA.

Price, H. H. 1985. "The Allocation of Functions in Systems." *Human Factors* 27:33–45.

Psychonomic Society. 1966. *Perception & Psychophysics*. Austin: Psychonomic Society.

Roberts, K., J. F. Burton, M. M. Bodah, and T. Thomason. 2005. *Workplace Injuries and Diseases: Prevention and Compensation: Essays in Honor of Terry Thomason*. Kalamazoo: W.E. Upjohn Institute for Employment Research.

Safir, A., V. Smith, J. Pokorny, and J. L. Brown. 1976. "Optics and Vision Physiology." *Arch Ophthalmol* 94(5):852–862.

Salvendy, G. 1987. *Handbook of Human Factors*. New York: John Wiley.

Sanders, M. S., and E. J. McCormick. 1993. *Human Factors in Engineering and Design*. 7th ed. New York: McGraw-Hill.

Schonhoff, T. A., and A. A. Giordano. 2006. *Detection and Estimation Theory and Its Applications*. Upper Saddle River, NJ: Pearson Prentice Hall.

School Library Manpower Project, and the National Education Association of the United States (NEA). 1969. "Task Analysis Survey Instrument: Definitions of Terms, Checklist of Duties, Status Profile Sheet. A survey instrument of the School Library Manpower Project developed with the Research Division of the National Education Association. Phase I." Chicago: American Association of School Librarians.

Schulz, T., and E. Lessing. 1976. *Hick's Law in a Random Group Design: New Data for Cognitive Interpretations*. Bonn, Germany: Psychologisches Institut der Universität Bonn, Abt. Methodik.

Shanks, D. R. 1997. *Human Memory: A Reader*. New York: St. Martin's Press.

Shepherd, A. 2001. *Hierarchical Task Analysis*. New York: Taylor & Francis.

Stevens, S. S. 1951. *Handbook of Experimental Psychology*. New York: Wiley.

______. 1957. "On the Psychophysical Law." *Psychological Review* 64(3):153–181.

______. 1975. *Psychophysics: Introduction to Its Perceptual, Neural, and Social Prospects*. New York: Wiley.

Sum, A. 1999. NCES 1999-470, *Literacy in the Labor Force*. Washington, D.C.: Government Printing Office.

Swets, J. A. 1996. *Signal Detection Theory and ROC Analysis in Psychology and Diagnostics: Collected Papers*. Mahwah, NJ: Lawrence Erlbaum Associates.

Tufts College, Institute for Applied Experimental Psychology, and the United States Office of Naval Research. 1951. *Handbook of Human Engineering Data for Design Engineers (Report)*. Medford, MA: Tufts College.

US Air Force. *Air Force Manual*. Washington, D.C.: Government Printing Office.

US Navy. MIL-STD-1472D, *Human Engineering Design Criteria for Military Systems, Equipment and Facilities*. Philadelphia: Navy Publishing and Printing Service Office.

Van Cott, H. P., and M. J. Warrick. 1972. No. IP-39, *Man as a System Component*. Washington, D.C.: Government Printing Office.

Verdier, P. A. 1960. *Basic Human Factors for Engineers; The Task Analysis Approach to the Human Engineering of Men and Machines*. New York: Exposition Press.

Waters, T. R., V. Putz-Anderson, A. Garg, and L. J. Fine. 1993. "Revised NIOSH equation for the design and evaluation of manual lifting tasks." *Ergonomics* 36(7):749–776.

Webster, J. C. 1974. "Speech Interference by Noise." Paper presented at the Inter-Noise 74.

Weimer, J. 1993. *Handbook of Ergonomic and Human Factors Tables*. Englewood Cliffs, NJ: Prentice Hall.

Welford, A. T. 1968. *Fundamentals of Skill*. London: Methuen.

Welford, A. T., and J. M. T. Brebner. 1980. *Reaction Times*. New York: Academic Press.

Wickens, C. D. 1984. "Processing Resources in Attention," in *Varieties of Attention* (pp. 63–102). London: Academic Press.

Wickens, C. D., and J. G. Hollands. 2000. *Engineering Psychology and Human Performance*, 3d ed. Upper Saddle River, NJ: Prentice Hall.

Wickens, T. D. 2002. *Elementary Signal Detection Theory*. New York: Oxford University Press.

Wiener, E. L. 1964. "The Performance of Multi-Man Monitoring Teams." *Human Factors* 6:179–184.

______. 1987. "Application of Vigilance Research: Rare, Medium, or Well Done?" *Human Factors* 29(6):725–736.

Wiener, E. L., R. E. Curry, and M. L. Faustina. 1984. "Vigilance and Task Load: In Search of the Inverted U." *Human Factors* 26(2):215–222.

Wiener, E. L., G. K. Poock, and M. Steele. 1964. "Effect of Time-Sharing on Monitoring Performance: Simple Mental Arithmetic as a Loading Task." *Percept Motor Skills* 19:435–440.

Wierwille, W. W., and F. T. Eggemeier. 1993. "Recommendations for Mental Workload Measurement in a Test and Evaluation Environment." *Human Factors* 35(2):263–281.

Wiker, S. F., G. D. Langolf, and D. B. Chaffin. 1989. "Arm Posture and Human Movement Capability." *Human Factors* 31(4):421–441.

Woodson, W. E. 1981. *Human Factors Design Handbook: Information and Guidelines for the Design of Systems, Facilities, Equipment, and Products for Human Use*. New York: McGraw-Hill.

Woodson, W. E., B. Tillman, and P. Tillman. 1992. *Human Factors Design Handbook: Information and Guidelines for the Design of Systems, Facilities, Equipment, and Products for Human Use*. 2d ed. New York: McGraw-Hill.

Yost, W. A., A. N. Popper, and R. R. Fay. 1993. *Human Psychophysics*. New York: Springer-Verlag.

Appendix: Recommended Readings

Aasman, J., G. Mulder, and L. J. M. Mulder. 1987. "Operator Effort and the Measurement of Heart-Rate Variability." *Human Factors* 29:161–170.

Adams, J. A. 1982. "Issues in Human Reliability." *Human Factors* 24:1–10.

Bailey, R. W. 1982. *Human Performance Engineering: A Guide for System Designers*. Englewood Cliffs, NJ: Prentice-Hall.

Barnes. R. M. 1949. *Motion and Time Study*. 3d ed. New York: John Wiley & Sons.

Boff, K. R., L. Kaufman, and J. P. Thomas, eds. 1986. *Handbook of Perception and Human Performance, Volume I: Sensory Processes and Perception*. New York: John Wiley & Sons.

______. 1986. *Handbook of Perception and Human Performance, Volume II: Cognitive Processes and Performance*. New York: John Wiley & Sons.

Boff, K. R., and J. E. Lincoln, eds. 1999. *Engineering Data Compendium: Human Perception and Performance*. Wright-Patterson Air Force Base: Harry G. Armstrong Aerospace Medical Research Laboratory.

Blanchard, B. S., and W. J. Fabrycky. 1990. *Systems Engineering and Analysis*. Englewood Cliffs, NJ: Prentice-Hall.

Brenner, M., E. T. Doherthy, and T. Shipp. 1994. "Speech Measures Indicating Workload Demand." *Aviation, Space, and Environmental Medicine* 65:21–26.

Broadbent, D. E. 1958. *Perception and Communication*. London: Pergamon.

Chapanis, A. 1980. "The Error-Provocative Situation: A Central Measurement Problem in Human Factors Engineering," in W. Tarrants, ed., *The Measurement of Safety Performance*. New York: Garland STPM Press, pp. 99–28.

______. 1959. *Research Techniques in Human Engineering*. Baltimore: The Johns Hopkins University Press.

______. 1988. "Words, Words, Words Revisited." *International Review of Ergonomics*. 2:1–30.

Cooper, M. D., and R. A. Phillips. 1994. "Validation of a Safety Climate Measure." Proceedings of the British Psychological Society: 1994 Annual Occupational Psychology Conference. Birmingham, Jan 3–5, 1994.

Cooper, M. D., R. A. Phillips, V. J. Sutherland, and P. J. Makin. 1994. "Reducing Accidents Using Goal-Setting and Feedback: A Field Study." *Journal of Occupational & Organisational Psychology* 67:219–40.

Crossman, E. R. F. W. 1959. "A Theory of the Acquisition of Speed Skill." *Ergonomics* 2:153–166.

Cushman, W. H., and D. J. Rosenberg. 1991. *Human Factors in Product Design*. Amsterdam: Elsevier.

Department of the Air Force. 1988. DI-CMAN-K8008A, *Data Item Description: System Segment Specification*. Washington, D.C.: Department of the Air Force.

Department of the Army. 1987. Army Regulation 602-2, *Manpower and Personnel Integration (MANPRINT) in Material Acquisition Process*. Washington, D.C.: Department of the Army.

Department of Defense (DOD). 1974. MIL-STD-499A, *Military Standard: Engineering Management*. Washington, D.C.: DOD.

______. 1985a. MIL-STD-490A, *Specification Practices*. Washington, D.C.: DOD.

______. 1985b. MIL-STD-1521B, *Technical Reviews and Audits for Systems, Equipment, and Computer Software*. Washington, D.C.: DOD.

______. 1987. DOD-HDBK-763, *Human Engineering Procedures Guide*. Washington, D.C.: DOD.

______. 1994. DI-HFAC-80746A, *Human Engineering Design Approach Document—Operator*. Washington, D.C.: DOD.

______. 1994. DI-HFAC-80747A, *Human Engineering Design Approach Document—Maintainer*. Washington, D.C.: DOD.

______. 1994. DI-HFAC-80740A, *Human Engineering Program Plan*. Washington, D.C.: DOD.

______. 1994. DI-HFAC-80745A, *Human Engineering System Analysis Report*. Washington, D.C.: DOD.

______. 1994. DI-HFAC-80743A, *Human Engineering Test Plan*. Washington, D.C.: DOD.

______. 1994. DI-HFAC-80744A, *Human Engineering Test Report*. Washington, D.C.: DOD.

Geldard, F. A. 1953. *The Human Senses*. New York: John Wiley & Sons.

Green, A. E., ed. 1982. *High Risk Safety Technology*. New York: Wiley.

Grose, V. L. 1988. *Managing Risk: Systematic Loss Prevention for Executives*. Englewood Cliffs, NJ: Prentice-Hall.

Huey, B. M., and C. D. Wickens, eds. 1993. *Workload Transition*. Washington, D.C.: National Academy Press.

Hughes, P. K., and B. L. Cole. 1988. "The Effect of Attentional Demand on Eye Movement Behaviour When Driving," in *Vision in Vehicles II*. Amsterdam: North-Holland, pp. 221–230.

Itoh, Y, Y. Hayashi, I. Tsukui, and S. Saito. 1990. "The Ergonomic Evaluation of Eye Movement and Mental Workload in Aircraft Pilots." *Ergonomics* 33:719–733.

Johnson, A. K., and E. A. Anderson. 1990. "Stress and Arousal," in J. Cacioppo and L. Tassinary, eds., *Principles of Psychophysiology*. Cambridge: Cambridge University Press, pp. 216–252.

Jones, E. R., R. T. Hennessy, and S. Deutsch, eds. 1985. *Human Factors Aspects of Simulation*. Washington, D.C.: National Academy Press.

Jordan, P. W., and G. I. Johnson. 1993. "Exploring Mental Workload Via TLX: The Case of Operating a Car Stereo Whilst Driving," in A. Gale, ed., *Vision in Vehicles IV*. Amsterdam: North-Holland, pp. 255–262.

Jorna, P. G. A. M. 1992. "Spectral Analysis of Heart Rate and Psychological State: A Review of Its Validity as a Workload Index." *Biological Psychology* 34:237–257.

Kantowitz, B. H. 1987. "Mental Workload," in P. Hancock, ed., *Human Factors Psychology*. Amsterdam: North-Holland, pp. 81–121.

______. 1992. "Selecting Measures for Human Factors Research." *Human Factors* 34:387–398.

Kaufman, J. E., and H. Haynes, eds. 1981. *IES Lighting Handbook: Reference Volume*. New York: Illuminating Engineering Society of North America.

Kirwan, B. 1987. "Human Reliability Analysis of an Offshore Emergency Blowdown System." *Applied Ergonomics* 18:23–33.

Kryter, K. D. 1972. "Speech Communication," in H. Van Cott and R. Kinkade, eds., *Human Engineering Guide to Equipment Design*. Washington, D.C.: Government Printing Office, pp. 161–226.

Lowrance, W. W. 1976. *Of Acceptable Risk*. Los Altos: William Kaufmann.

Martin, D. K., and S. J. Dain. 1988. "Postural Modifications of VDU Operators Wearing Bifocal Spectacles." *Applied Ergonomics* 19:293–300.

Meister, D. 1989. *Conceptual Aspects of Human Factors*. Baltimore: The Johns Hopkins University Press.

______. 1971. *Human Factors: Theory and Practice*. New York: John Wiley & Sons.

Muckler, F. A., and S. A. Seven. 1992. "Selecting Performance Measures: 'Objective' Versus 'Subjective' Measurement." *Human Factors* 34:441–455.

National Research Council. 1983. *Risk Assessment in the Federal Government: Managing the Process*. Washington, D.C.: National Academy Press.

Norman, D. A., and D. G. Bobrow. 1975. "On Data-Limited and Resource-Limited Processes." *Cognitive Psychology* 7:44–64.

Parker, J. F. Jr., and V. R. West, eds. 1973. *Bioastronautics Data Book*. Washington, D.C.: Scientific and Technical Information Office, NASA.

Pelsma, K. H., ed. 1987. *Ergonomics Sourcebook: A Guide to Human Factors Information*. Lawrence, KS: The Report Store.

Phillips, C. 2000. *Human Factors Engineering*. New York: John Wiley & Sons.

Rouse, W. B., S. L. Edwards, and J. M. Hammer. 1993. "Modeling the Dynamics of Mental Workload and Human Performance in Complex Systems."

IEEE Transactions on Systems, Man, and Cybernetics 23:1662–1671.

Rowe, W. D. 1983. *Evaluation Methods for Environmental Standards*. Boca Raton, FL: CRC Press.

Salvendy, G., ed. 2006. *Handbook of Human Factors and Ergonomics*. 3d ed. New York: John Wiley & Sons.

Swain, A. D., and H. K. Guttmann. 1983. Report NUREG/CR-1278, *Handbook of Human Reliability Analysis with Emphasis on Nuclear Power Plant Applications*. Albuquerque: Sandia National Laboratories.

Teigen, K. H. 1994. "Yerkes-Dodson: A Law for All Seasons." *Theory & Psychology* 4:525–547.

Thomson, J. R. 1987. *Engineering Safety Assessment: An Introduction*. New York: Wiley.

Thorsvall, L., and T. Åkerstedt. 1987. "Sleepiness on the Job: Continuously Measured EEG Changes in Train Drivers." *Electroencephalography and Clinical Neurophysiology* 66:502–511.

Tversky, A., and D. Kahneman. 1974. "Judgment Under Uncertainty: Heuristics and Biases." *Science* 185:1124–1131.

US Army Missile Command. 1984. MIL-H-46855B, *Human Engineering Requirements for Military Systems, Equipment and Facilities*. Redstone Arsenal: US Army Missile Command.

______. 1989. DOD-HDBK-761A, *Human Engineering Guidelines for Management Information Systems*. Redstone Arsenal: US Army Missile Command.

Van Cott, H. P., and R. G. Kinkade, eds. 1972. *Human Engineering Guide to Equipment Design*, revised ed. Washington, D.C.: Government Printing Office.

Vivoli, G., M. Bergomi, S. Rovesti, G. Carrozzi, and A. Vezzosi. 1993. "Biochemical and Haemodynamic Indicators of Stress in Truck Drivers." *Ergonomics* 36:1089–1097.

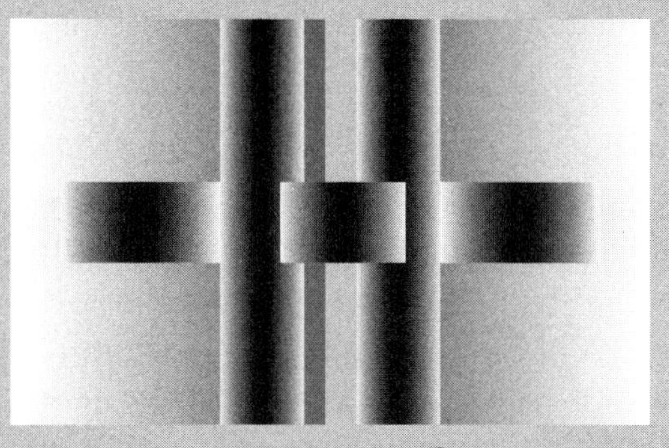

SECTION 6

ERGONOMICS AND HUMAN FACTORS ENGINEERING

LEARNING OBJECTIVES

- Define the role of ergonomists in achieving business goals.
- Describe some common cost and benefit concepts.
- Recognize the major cost-benefit categories associated with ergonomics projects.
- Illustrate, through a hypothetical case study, the basic elements of cost-benefit analysis for a sample ergonomics project.

COST ANALYSIS AND BUDGETING

Rani Muhdi and Jerry Davis

SUCCESSFUL ORGANIZATIONS continuously design, implement, and evaluate their business strategies to achieve certain goals. One of the primary goals of many organizations is to make a profit, which is basically accomplished when revenues exceed expenses over a period of time. This basic but critical concept of revenues versus expenses applies to all investment projects regardless of their size and level of complexity. Both revenues and expenses are usually summarized and controlled within the accounting system of an organization, but require the attention of safety professionals and engineers when attempting to compete for limited resources to fund their proposed projects. Therefore, it is paramount that potential ergonomics projects are competitive from a cost-benefit perspective in order to have the best chance of successfully obtaining funding. To do this, ergonomists and safety professionals must understand the fundamentals of cost justification and be able to concisely transmit this knowledge to senior management in their organization.

It is clearly understood that the role of safety, ergonomics, and health professionals is to establish, promote, and maintain a safe environment for all employees through the elimination and control of hazards. Safety professionals might be lulled into feeling that safety (or safety-related) projects will be funded because they are legislated and regulated by the OSH Act of 1970 and enforced by the Occupational Safety and Health Administration (OSHA). To routinely "drop the safety hammer" as a mechanism to obtain project funding is not a wise strategy for success. Having said that, it is fair to remind management that there are civil (and potentially criminal) penalties for violating safety and health regulations and standards. In reality, the consequences associated with noncompliance with the OSH Act do support, to some extent, safety professionals when proposing new projects (interventions) to management. However, it is still crucial that safety professionals communicate via business language in terms of revenues and costs and that they have the skills to build a business case for proposed safety/ergonomic interventions, other than simply preventing losses. Such communication helps the safety function continue to integrate into the business environment as a vital part of an organization by protecting workers' health and improving productivity and quality, while simultaneously adding to the bottom line.

Ergonomics is essentially based on the principle of *fitting the task to the worker* versus *fitting the worker to the task*. As such, one seeks to prevent physical harm and reduce injuries by making tangible changes in the design of the workplace (engineering controls), or implementing managerial procedures such as job rotation, lifting policies, and training (administrative controls). On the other hand, management might perceive the role of ergonomics and associated projects differently since it is not directly regulated and enforced in a manner similar to occupational safety, with the exception that employers are still obligated under the General Duty Clause of the OSH Act of 1970. Section 5(a)(1) of this act states that "each employer shall furnish to each of his employees employment and a place of employment which are free from recognized hazards that are causing or are likely to cause death or serious physical harm to his employees." In a manner identical to safety, OSHA compliance officers conduct inspections that place emphasis on ergonomic hazards, and some issue citations for failing to be in compliance with the General Duty Clause. For further reading on regulatory issues related to ergonomics, OSHA standards and guidelines, and other state and national standards, it is recommended the reader review the "Regulatory Issues" chapter by Carol Stuart-Buttle at the beginning of this section.

In order to impress the reader with the importance of organizations remaining compliant with OSHA guidelines for ergonomics under the General Duty Clause, a case study (company name withheld), summarized directly from a search of the General Duty Standard on the OSHA Web site, presents one of many inspections conducted by OSHA with an emphasis on ergonomic guidelines. In this instance, employees in a manufacturing company were required to perform tasks that had stressors commonly known to increase the risk of developing work-related musculoskeletal disorders (WRMSDs). Compliance safety and health officers (CSHO) who observed and evaluated work tasks associated with grinding and sanding suspected that certain employees were exposed to risk factors consistent with the development of back pain or other MSD-related injuries. Some of these risk factors included the lifting/lowering of frames to and from workbenches, flipping frames once on the workbench, manual material handling associated with pushing/pulling frames around the facility, and repeatedly bending the torso (awkward posture) to grind or sand frames close to the floor. CSHOs also expressed concern for those employees who were required to manually handle frames weighing in excess of 60 pounds. OSHA cited the plant for a total of ten violations (serious, willful, and repeated). The initial penalty assigned to these ergonomically related violations exceeded $100,000.

The term *compliance* is difficult for professionals in ergonomics to define. Unlike most matters in occupational safety where aspects of mandatory conformance with a standard are clear, no such guidance exists for issues related to ergonomics. Some might argue that to achieve compliance, jobs should be screened for risk factors and an analysis performed by qualified professionals using one of a number of tools, such as the Revised NIOSH Lifting Equation (NIOSH 1994, 1–47), the Rapid Upper Limb Assessment (RULA) (McAtamney & Corlett 1993, 91–99), or the Job Strain Index (JSI) (Moore & Garg 1990, 443–458), and changes subsequently made to workstations to reduce the exposure(s) to some lower levels identified by the tool. Others might argue that by following such procedures and eliminating most, if not all, ergonomically related injuries and illnesses, compliance is achieved. The point is that *compliance* is more of a relative term in ergonomics than it is in occupational safety. If an ergonomic strategy is solely based on this *compliance* aspect, management may eventually no longer sense a value in additional investment in ergonomics or related projects. Unfortunately, professionals in the field of ergonomics may not be trained in understanding the financial impact of ergonomics projects on the organization's primary goals. This lack of financial training is perceived to be one of the reasons that the contribution of ergonomics is not yet widely recognized by management (Hendrick 1996, 14).

One might be tempted to justify ergonomic solutions to management by simply assigning monetary values to the costs that could be avoided if injuries are reduced in the workplace. As previously alluded to, this approach employs only one of many facets of a comprehensive ergonomics cost-benefit justification and is rarely sufficient to justify a project (intervention). Rather, ergonomic interventions should be presented to management in the form of investment opportunities with specific economic measures such as present, annual, and future worth(s), return on investment (ROI)[1], and cost-to-benefit ratio. It is an added benefit if the employees also perceive the proposed ergonomic intervention as a personal opportunity to reduce injuries, alleviate some work-related stress, and contribute to a more favorable quality of life, both at and away from work.

The purpose of this chapter is to enhance the understanding of ergonomics practitioners (safety professionals) about cost-benefit analysis of ergonomic interventions. The chapter presents basic economic concepts that will assist professionals in health and safety/ergonomics to establish the awareness necessary to justify individual ergonomics projects. Further, it is suggested that practitioners become well versed in the project-management techniques associated with their specific place of employment and seek additional training in any areas related to these topics.

The chapter is divided into four primary sections. The first section briefly describes general cost and benefit concepts that are common to organizations along with some descriptive statistical data on the number and cost of injuries and illnesses in workplaces. The second and third sections, respectively, present the major cost and benefit categories associated with ergonomics projects. The chapter concludes with a case study that provides the reader with all the basic elements of cost-benefit analysis for a sample ergonomics project.

Cost Concepts

In order for ergonomics practitioners to promote successful interventions to management, they need to acquaint themselves with some basic understanding of costs. The word *cost* has many definitions and is dependent on the setting in which it is used. According to Canada, Sullivan, and White (1996, pp. 3–5), the cost of a *resource* is "the decrease in wealth that results from committing this resource to a particular alternative, or before any of the benefits of the alternative are calculated." Cost analysis is essential to all capital investments, including safety and ergonomics projects. On the personal side, people sometimes need to borrow money from a lending institution for any number of reasons, including the purchase of a home or automobile, sending a child to college, or even starting a business. Borrowers fill out an application, listing the revenues and obligations they currently have, and may be asked to secure the loan with some form of collateral. Generally, lending institutions have thoroughly evaluated the value of the potential purchase, the borrower's ability to repay this obligation, and any profit the lending institution intends to make off of the transaction. Given that people expect this level of financial scrutiny in their personal lives, why would they expect anything less when intending to "purchase" equipment and services at work? In summary, cost-benefit analysis is a technique available to safety, ergonomics, and health professionals for the purpose of comparing costs and benefits associated with proposed capital investments.

The accounting system of an organization maintains records of costs and revenues associated with all aspects of the business. These records include costs that are typically separated into three major categories: direct costs, indirect costs, and capital costs (Canada et al. 1996, 3–5).

1. Direct costs (tangible): Those labor and material costs that are directly charged to products or jobs. Some categories of direct

costs include personnel (salary, wages, and benefits), materials, supplies, equipment, and utilities. These costs can be calculated easily and charged directly against production. The labor cost of an operator on a production line and the cost of raw materials to produce a large number of a given product are examples of direct labor and direct material costs, respectively.

2. Indirect costs (intangible): Those costs that cannot be charged directly to a particular product or specific job, such as indirect labor (management and staff, including safety and health personnel), tools, inventory, floor space, and department supplies. The paycheck(s) of production manager(s), safety engineer, plant nurse, and security personnel are examples of indirect labor costs. Storage spaces and supplies for forklifts are other examples of indirect costs. These costs cannot be easily and conveniently traced to an individual unit of production since they are shared by several departments and personnel in an organization. Indirect labor and material costs are part of overhead costs, which comprise all production costs other than direct labor and material costs. Examples of overhead costs include power, maintenance, depreciation, and insurance.
3. Capital costs: Those costs associated with the acquisition of new assets. The construction of a new plant, expenditures for the purchase or acquisition of existing facilities, or even the total investment needed to start a business and bring it to a commercially operable status, are some examples of capital costs. If a potential ergonomics project is of sufficient magnitude, it may easily fall into this *capital cost* category. Generally, ergonomics projects incur direct and indirect costs, as discussed in the next two sections. However, the costs of large ergonomics projects, which may involve a significant piece of equipment being purchased and installed, might be classified as capital costs. This classification depends on an organization's financial investment policy for new projects. Ergonomists are sometimes given a limitation on the cost of projects they can undertake without going through the process of submitting funding requests to top management. The limit varies from one organization to another but typically is in the range of $500 to $1000 per project. Once the budget required for an ergonomics project (intervention) exceeds the threshold limit, the project might be considered significantly large and require a capital investment.

Cost of Ergonomic Hazards and Repetitive Strain Injuries

According to the 2004 Survey of Occupational Injuries and Illnesses by the Bureau of Labor Statistics (BLS 2005), a total of 4.3 million nonfatal injuries and illnesses were reported in private industry workplaces during 2004. For a list of ergonomics-related injuries and illnesses, please review the chapter titled "Applied Science and Engineering: Ergonomics" by Dr. Magdy Akladios earlier in this section. These cases occurred at a rate of 4.8 cases per 100 equivalent full-time workers. Approximately 2.2 million injuries and illnesses involved days away from work[2], job transfer, or restriction. Cases in these categories required recuperation away from work, transfer to another job, restricted duties at work, or a combination of these actions. The remaining 2.1 million injuries and illnesses were other recordable cases that did not result in time away from work.

Losses from injuries and illnesses in the workplace could also cause damage to property and equipment, with no firm estimate of the magnitude of these losses. When discussing the factors contributing to losses, one must also consider the loss of time, production, sales, investigations, legal and medical services, and rehabilitation, as they all constitute an actual dollar value to the bottom line of any company. Similar to the basic classification of costs, losses associated with workplace injuries and illnesses can be classified as direct and indirect costs. Direct costs typically include medical and emergency services, compensation benefits paid to an injured employee, insurance costs, legal and administrative costs, and expenses to repair or replace damaged materials, equipment, and the environment.

Indirect costs from injuries and illnesses are sometimes termed *hidden costs* because they are difficult to estimate for each case. Indirect costs are primarily related to time and production losses. When a worker is injured, not only his/her time is lost, but also the time of other employees, supervisors, and medical staff who assist an injured worker. Table 1, adapted from Brauer (2006, p. 23), based on work by Heinrich (1959), lists some categories of indirect costs associated with injuries and illnesses in the workplace. For a detailed list of indirect costs related to injuries and illnesses, the reader is advised to review the chapter by Dr. Michael Toole titled "Cost Analysis and Budgeting" in Section 3 of this handbook.

Although it is difficult to accurately estimate the direct and indirect costs of occupational injuries and illnesses in the United States, some researchers and insurance companies have made attempts to estimate such costs. The overall financial liability of work-related injuries in the United States was estimated at $116 billion in 1992, an increase from the 1989 estimate of $89 billion, and dramatically larger than the 1985 estimate of $34.6 billion (Leigh 1995, 107–133).

TABLE 1

Hidden Costs Associated with Injuries and Illnesses

Category	Description
A. Time losses	- Injured worker(s)
	- Other employees and supervisors to assist injured worker(s), investigate, and report to management
	- Medical and safety staff
	- Time to set up and restart the production process
	- Time to return to steady-state production with quality products
	- Recruitment and training time of replacement worker(s)
	- Time to fix or replace damaged property and equipment
B. Production losses	- Late or undelivered orders
	- Damage to tools, equipment, and/or environment
	- Reduction in productivity of the returned/injured worker(s)
	- Loss of reputation and public image[3]
	- Production losses because of low-quality products due to reduction in co-workers excitement or morale

(Adapted from Brauer 2006, p. 23)

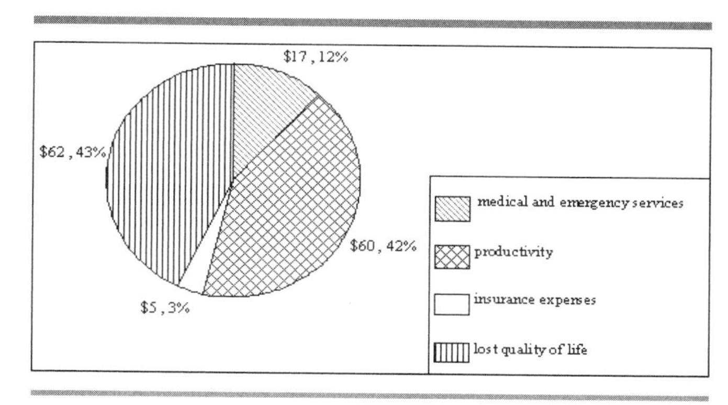

FIGURE 1. Costs and percentages of occupational injuries in the United States.
(*Source:* Miller & Galbraith 1995, pp. 741–747)

According to the *Liberty Mutual Workplace Safety Index* published in 2004, an estimate of $49.6 billion in wage and medical payments was made to workers injured on the job in 2002. In a previous study, Liberty Mutual found that businesses faced between $2 and $5 in indirect costs for every dollar spent on direct costs (Liberty Mutual 2002, 1). Although, the rate of growth in the cost of these injuries slowed significantly from 6.5 percent to 0.7 percent in 2003, the financial impact of workplace injuries remains. In fact, employers spent $50.8 billion in 2003 on wage payments and medical care for workers hurt on the job (Liberty Mutual 2005, 1). Furthermore, Miller and Galbraith (1995, pp. 741–747) estimated the annual workplace injury costs in the United States at about $140 billion. This includes $17 billion in medical and emergency services, $60 billion in productivity loss, $5 billion in insurance expenses, and $62 billion in lost quality of life. Figure 1 illustrates the costs (in billions) and the corresponding percentages of occupational injuries in the United States as estimated by Miller and Galbraith.

According to OSHA, each year occupational musculoskeletal disorders (MSDs) alone account for about $15–20 billion in workers' compensation costs, which is roughly $1 of every $3 spent on workers' compensation. MSD cases are higher than those for other minor injuries. The average per-case cost for carpal tunnel syndrome cases is $8070, which is more than double the $4,000 average per-case cost for all other minor injuries and illnesses in general industry. The average cost for lower back pain is ten times higher than the cost for carpal tunnel syndrome cases (Alexander 2006). According to the National Academy of Social Insurance, total workers' compensation benefits for injured workers rose by 2.3 percent to $56 billion, while insurance premiums rose by 7 percent to approximately $87 billion (National Safety Council 2006).

The Liberty Mutual Insurance Company concluded that lower back pain is the most prevalent and costly work-related MSD. Lower back pain MSDs account for 15 percent of all Liberty Mutual workers' compensation claims and 23 percent of the costs of these claims. For many occupations involving manufacturing or manual handling, MSD rates are high. In 1996, rates for lost work days (LWD) due to MSDs for occupations involving manufacturing and manual handling were as high as 30.4 and 42.4 per 1000 full-time employees (FTE), respectively.

Cost Analysis for Ergonomics Projects

The basic costing method for an ergonomics project is simply one of measuring cost avoidance. Ergonomics practitioners usually concern themselves with medical, lost time and production, legal, and administrative expenses associated with ergonomically preventable injuries. Avoided costs can be measured by calculating the frequency and severity of injuries and illnesses a particular ergonomic solution (preventative) is most likely to correct. Often, the most apparent category in which these costs are revealed and justified is in relation to workers' compensation. Eliminating the causes of these insurance and compensation claims allows ergonomists to attribute the associated savings as benefits of the ergonomics project. It is important to realize that an ergonomic solution (intervention) that prevents a cost from *occurring* will similarly prevent the cost from *recurring* in the future. However, cost avoidance never provides a complete economic justification for ergonomic interventions. Indeed, it may have the effect of forcing ergonomists into inactive roles in their organizations.

When attempting to perform a cost-justification analysis of an ergonomics project, it is important to fully account for the major cost categories associated with ergonomic injuries and illnesses. This approach helps to build a business case for management. Specifically, ergonomists should develop a robust cost-analysis methodology that will accurately reflect the costs associated with a company's values, including personnel, production, and even reputation. For the purposes of this chapter, these ergonomics-related costs are briefly categorized into personnel, equipment and material, productivity, and overhead. Table 2 lists some major cost categories.

External personnel costs. These costs are associated with outside personnel or consultants. They are typically easy to estimate since consultants usually have scheduled rates outlined in a contract and agreed upon by the organization before a project is undertaken. The task of an ergonomics practitioner then becomes to determine the benefits of this type of service to the ergonomics project compared to the costs of the service. External personnel costs include turnover and training costs that can arise with the addition of temporary, seasonal, and/or part-time personnel to the workforce.

Internal personnel costs. Personnel in an organization usually have specific job responsibilities. Generally, accounting systems do not track time spent on each responsibility. However, when in-house personnel are assigned to an ergonomics project, ergonomics practitioners should account for their time and associated costs with the help of the human resources department in the organization. When an assignment for internal personnel requires less than full-time responsibility (1 – FTE), costs are prorated according to the percentage of time committed to the project.

Turnover and training costs. These costs are related to replacing workers, either temporarily or permanently. Organizations allocate resources such as advertisements, personnel, travel, sign-on bonuses, and relocation, among other incentives, to recruit new employees for high-turnover-rate jobs. Once the recruitment process is finalized and a position is offered, organizations incur additional costs associated with training these new workers. High turnover rates could

TABLE 2

Major Cost Categories Associated with Ergonomics

Category	Description
A. Personnel	- External personnel - Internal personnel - Turnover and training costs - Employee(s) downtime
B. Equipment and material	- Purchase, storage, and upgrading - Installation and maintenance (preventative and corrective)
C. Productivity	- Losses in production due to undelivered orders - Losses in customers, direct, and indirect costs
D. Overhead	- Utilities - General administration

(Adapted from Oxenburgh 1997, pp. 150–156)

be an indication of a physically demanding, mentally tedious, or some other hazardous work environment. When planning a new ergonomics project to improve the safety of work conditions, practitioners need to consider turnover and training costs when justifying these projects.

Employee downtime costs. Unless management reassigns everyone whose workplace is affected by designing, installing, or maintaining a new ergonomic intervention, the organization will incur employee downtime costs. This can be avoided by developing a comprehensive plan during the design stage of the project. In addition, ergonomists should always ensure that temporary assignments are within a particular worker's physical and health capabilities. Employee downtime costs can be minimized by performing any major ergonomics projects during a planned shutdown phase.

Equipment and material costs. An ergonomics project may require the installation of new equipment or materials. The cost to purchase such equipment and materials should be added to the total cost of the project. Other costs associated with the purchase of equipment and materials might include insurance, shipping, and inspection. For large ergonomics projects that require significant design changes in the workplace, professional installation is needed. Depending on the size of the project, significant infrastructure/work-site preparation may be required in addition to providing utilities to the site. Scheduled and unscheduled maintenance may be necessary to maintain the functionality of the project. In some projects, upgrading or expanding an ergonomic system might be necessary due to changes in production demand, personnel characteristics, technology, or regulations and standards.

Productivity costs. When the installation of a new project interferes with the production schedule of an organization, losses in productivity and quality may be experienced. Usually this is overcome by installing new projects during scheduled shutdown periods. If the installation occurs at some other time, productivity losses can be estimated from the organization's production rate, cycle time, and time studies. Reduced productivity should be considered as a cost associated with the project. Additionally, if an ergonomic solution affects the productivity of a worker due to system limitations (permanent) or a worker's adjustment to the new system (temporary), the recurring costs should be added to the project's total cost.

Overhead costs. These costs are related to utilities and general administration. They are not directly related to production, but are still associated with the ergonomics project. Once the accounting system calculates them, overhead costs are typically charged as a percentage of direct costs. Therefore, a proportion of the unit production cost accounts for overhead costs. Though this may be difficult to measure, potential overhead savings are often noticeable after ergonomic interventions are implemented. As a result, the accounting department of an organization should recalculate the proportion of overhead costs in unit production costs.

Benefit Analysis for Ergonomics Projects

Understanding costs associated with ergonomic interventions is only a portion of the analysis process. If the *benefits* of these projects are not recognized, then ergonomics has the potential to be forced into a reactive and response-oriented role, rather than assuming a proactive one. If a project is proposed without fully capturing all associated savings, management could perceive ergonomics as a burden and a cost generator that potentially limits the progress toward an organization's primary goals. In this section, the benefits from a generic project are briefly categorized into personnel and productivity.

Reduced injuries and accidents. The most obvious and valuable personnel-related benefit of an ergonomic intervention is a reduction in injuries and accidents. This reduction is also coupled with a reduction in lost time from injuries and accidents. Lost time includes production time, staff and co-workers' time, time to set up and restart production, and time lost to reach steady-state as measured by the production of quality products. Another obvious benefit is a reduction in medical and workers' compensation premiums. According to Hendrick (1996, p. 8), an ergonomic intervention in a poultry plant reduced carpal tunnel syndrome, tendonitis, and tenosynovitis cases, and resulted in an annual saving in workers' compensation of approximately $100,000. This is simply one of several documented case studies that exist in virtually all areas of manufacturing and service-sector applications (Alexander 1999, 77–90; Kohn 1997, 2.5:1–15 & 2.8:1–33).

Increased worker productivity. A workstation, tool, or job redesign will often allow workers to perform their tasks more efficiently by reducing their physical and mental stress (load), or by inducing the operator to make fewer errors. After sufficient investigation, any sustainable gain in productivity as a result of the intervention should be directly attributed to the ergonomic improvement. Metrics indicative of increased productivity include reduced cycle time, the ability for operators to handle (load and unload) additional machines, more throughput per given unit of time (shift/day/week), and a reduction in defective product that needs to be scrapped or sent to be reworked. However, when the ergonomics project is part of a larger change in jobs or processes, these benefits could be more difficult to measure in isolation. Therefore, an increase in worker productivity due to an ergonomics project that is part of a larger change should be handled on a case-by-case basis. However, such an increase in productivity should not be ignored. A new design of five assembly workstations in an electronics plant resulted in a 15 percent increase in productivity, which translated into $2250–3000 per shift increase in productivity per worker (Hendrick 2003, 421).

Reduced training time. An ergonomic intervention that simplifies a job may also reduce the time associated with worker training. This reduction in training time could also translate into a reduction in training costs. Furthermore, simplifying a job through an ergonomic intervention may require less-qualified operators, thereby lowering the cost of direct labor. Thus, the savings in salary paid, as well as in deferred or avoided training costs, could be the direct benefits of an ergonomics project.

Reduced absenteeism and turnover. A common outcome of effective ergonomics projects is the reduction of absenteeism. Schneider (1985, pp. 26–29) reported a reduction in absenteeism from 4 percent to 1 percent when an ergonomic intervention was implemented in an office environment. The reduced absenteeism could also lead to a productivity increase due to less disruption of the work schedule and the need for fewer replacement personnel. This, in turn, could result in reduced training because fewer replacement personnel have to be trained (Hendrick 2003, 422). When ergonomics projects improve the workplace environment, it is common to experience a reduction in turnover rate, which could represent a significant financial benefit.

Reduced overtime salaries. Replacing an injured worker usually requires either that a new worker be trained (with associated costs incurred) or a trained co-worker becomes responsible for the tasks of the injured worker. This could mean overtime payments to workers before a replacement can be found and trained. Avoided overtime costs that could result from injury or missed work should be calculated and counted as benefits of the ergonomics project.

Improved quality of life. When an ergonomic intervention reduces the pain, injury, or cognitive stress associated with a job, the worker's quality of life might significantly improve. Such improvements may additionally increase worker productivity and enhance morale. A reduction in absenteeism and turnover rate could be related to metrics associated with improved quality of life. Ergonomists should work closely with management and human resources personnel to research and capture such data and incorporate this into their business case.

Increased quality. One area of potential improvement is in product quality. A properly designed workstation could lead to a reduction in errors and defective parts. The reduction translates into savings that can be calculated directly from the company's accounting system (how many units of a product were returned due to a defective part produced by the job in question, and at what cost), or from the production database system (how many units of a product were reworked due to error).

Reduced error, damage to equipment, and scrap. Operator errors may result in damage to tools or equipment. When an ergonomic intervention reduces such error, which in turn reduces damage to equipment, the replacement costs should be applied to the analysis of the ergonomic intervention. Additionally, an ergonomic design that considers all aspects of the production system may improve the time (efficiency) of maintenance performed or the amount and type of tools required to perform the maintenance. Once again, the savings, in terms of labor and tool costs should be applied to the analysis of the ergonomics project (intervention). Basic knowledge of time and motion studies and engineering work measurement are certain to enhance the skills of practicing ergonomists. Since ergonomic interventions tend to reduce the number of defective units produced, the amount of material wasted due to operator- or machine-induced error is also reduced. Table 3 is a summary of some of the major benefit categories associated with ergonomic interventions along with their descriptions.

TABLE 3

Major Benefit Categories Associated with Ergonomics Projects

Category	Description
A. Personnel	- Reduced injuries and illnesses
	- Increased worker productivity
	- Reduced training time
	- Reduced absenteeism and turnover
	- Reduced overtime expenditures
	- Improved quality of life
B. Productivity	- Increased quality
	- Reduced errors and damage to equipment
	- Reduced scrap

In the last few pages of this chapter, a brief understanding of cost and benefit categories is introduced. Clearly, the costs and benefits represented are among common categories professionals working in the field of ergonomics should consider in evaluating their projects. The categories, their monetary values, and their impact on an organization's primary goals directly depend on the nature of the proposed project and the specific job being analyzed[4].

The following case study is intended to provide the reader with a basic understanding of cost-benefit analysis for ergonomics projects. It is, however, not intended to take the place of valuable assistance from the accounting department of an organization. The numbers used in this case study are simply estimations, and do not reflect a specific industry or existing organization. For simplicity, the analysis is conducted without tax considerations. Practicing ergonomists can use the approach to learn how costs and benefits could be analyzed in similar settings. The case study provides a methodology for a simple cost-benefit analysis for ergonomic interventions. However, this approach is focused on an individual ergonomic solution, and is not applicable to analyzing ergonomics programs. For more detailed information on ergonomics programs, refer to the next chapter in this section. Written by Robert Coffey, it is titled, "Benchmarking and Performance Criteria."

For illustration purposes, the case study and its solution are presented in segments and in accordance with the material discussed in this chapter. The sequence followed in the solution reflects the concepts discussed in the cost and benefit analysis sections of this chapter. The case study is concluded with summary tables. Some of the basic economic concepts and formulas that appear in this case study are discussed and explained in other chapters within this handbook. It is suggested that the reader review Michael Toole's chapter, "Cost Analysis and Budgeting" in Section 3 of this handbook. The following is a brief description of the case study.

CASE STUDY

American Sunshine Electronics

Yazmin Ali, a recent graduate from an occupational safety & ergonomics program, has accepted a job as an ergonomist at American Sunshine Electronics (ASE). The company manufactures TVs, portable audio devices, and DVD players. This case study will focus on the TV production line. Two TV sizes are currently being produced, 20- and 24-inch flat screen TVs, with shipping weights of 56 and 68 pounds, respectively.

Yazmin truly believes that ergonomics plays a proactive role in an organization's progress, even in a small-sized company such as ASE. However, such a role is not achieved overnight, but rather through a continuous and collaborative effort between ergonomists and management. During the first three (3) months of her new position, Yazmin has reviewed and revised the organization's ergonomics program. One of the revisions she made, with management's consent, was to form a new ergonomics committee. The purpose of the committee is to enhance ergonomic conditions and increase awareness at the workplace.

The Safety and Health Division at ASE views the ergonomics committee as a collateral duty for those employees assigned to the team. However, the responsibility of implementing the recommendations provided by the committee falls on management. The objectives of the committee at ASE are to implement a comprehensive approach to assess ergonomic work-related risk factors and develop recommendations to alleviate these factors, communicate its assessment and recommendations to management, and finally to educate workers on ergonomic issues. The success of these objectives depends greatly on the commitment of its members, management, and, ultimately, the employees. The ergonomics committee consists of five (5) members: the ergonomist, an industrial engineer, the maintenance manager, a representative from the accounting department, and a senior employee from the production department. The average hourly wage/salary for the committee members is estimated to be $30/hour.

The ergonomics committee is currently preparing a cost-benefit analysis for a new ergonomic intervention on a manual assembly line. In order to develop the analysis, the committee has collected information on the current manual assembly line in terms of production capacity, scheduling, rework rate, number of workers, injury records, and workers' compensation expenses. The committee members estimate they will expend ten (10) working days, totaling eighty (80) hours (2 hrs/day/member), or 25 percent of their daily efforts on the project. The ergonomist is not contracting with outside personnel or consultants, therefore, only the internal direct personnel cost of the committee members involved in the ergonomics project will be prorated according to the percentage of time committed to the project (0.25 – FTE).

Internal direct personnel cost = 5 personnel × 8 hr/day × 10 days × 0.25 (2 hrs/day) × $30/hr = $3000.

The manual assembly line under analysis consists of seven (7) workstations operated by seven (7) workers and a supervisor. Each of the workstations requires a full-time operator. TV frames are brought to the assembly line in carts, which are lined up at the beginning of the assembly line according to a certain production schedule. Both TV models are assembled on the line. The sequence of the production units must be maintained since TV components, shipping packages, and distribution orders are synchronized with the production schedule. The operator on the first workstation manually unloads a TV frame from the cart to his/her workstation. Both manual and powered tools are used to assemble components into the TV frame.

Processing time at each workstation is set at three (3) minutes. Every three (3) minutes, the operator on the first workstation then loads the TV frame onto another cart, and pushes it to the next workstation. Due to the nature of the assembly process, size of the product, and workstation design, workers operate in awkward postures. It takes each TV about twenty-one (21) minutes to be assembled on the line. This is what is referred to in industry as "throughput time" or "lead time." When all workstations are busy, the assembly line produces one (1) TV every three (3) minutes (Takt time). The company runs on a single 8-hour shift, with an allowance rate of 15 percent (paid time) to provide for breaks, personal delay needs, and fatigue. The final product is inspected at the end of the assembly line.

Defective TVs are sent to the repair workstation. Previous production records estimate a rework rate of 0.3 percent. Reworked items (products) usually take about half the throughput time, or ten (10) minutes to repair. Operator error during assembly due to fatigue (especially toward the end of the work shift) and product mishandling due to the manual assembly process are the two primary causes of rework. The product (TV) is sold for $150. The direct material cost associated with the product is $50/item, while the indirect cost (indirect material, indirect labor, energy, depreciation, and maintenance) is $20/item. A worker on the assembly line makes approximately $20/hour including benefits. When overtime is needed after forty (40) hours per week, ASE pays one and a half (1.5) times the regular hourly wage. The operator on the rework workstation makes approximately 25 percent more than those on the assembly line ($25/hour). The difference in the hourly wage is justified by the nature of the work, which requires intensive manual handling and advanced experience. Figure 2 represents the manual production line.

The information presented in the previous segment of the case study should help the ergonomics committee calculate the production rate, direct labor cost, rework rate and cost, direct material cost, total cost, revenue, and profit.

Hourly production rate = 60 min/hr ÷ 3 min/item = 20 items/hr.

Therefore, the total annual production is calculated as:

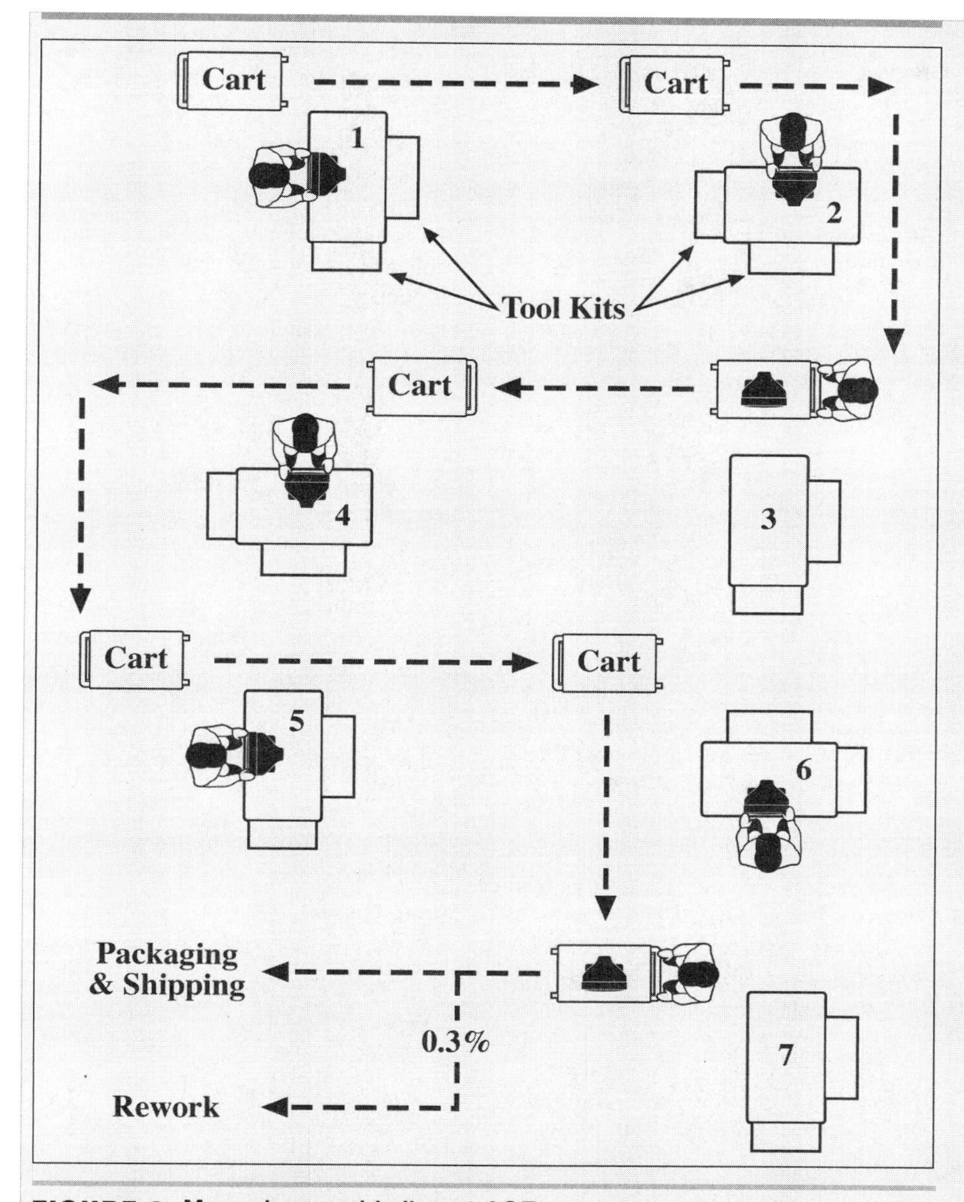

FIGURE 2. Manual assembly line at ASE

Total annual production = 20 items/hr × 8 hrs/day × 0.85 × 250 days/year = 34,000 items.

The annual rework rate is:

Annual rework rate = 34,000 items × 0.3% = 102 items.

Annual direct labor cost for rework = 102 items × (10 min/item ÷ 60 min/hr) × $25/hr = $425.

Annual direct material and indirect costs = 34,000 items × $(50 + 20)/item = $2,380,000.

We should think of direct labor cost as the wages of employees who directly worked on the product, including the rework station.

Annual direct labor cost = {7 workers × $20/hr × 8 hrs/day × 250 days/year} + $425/year = $280,425.

Annual total cost = $2,380,000 + $280,425 = $2,660,425.

Annual revenue = 34,000 items × $150/item = $5,100,000.

Annual operating profit = $5,100,000 − $2,660,425 = $2,439,575.

After reviewing the injury records for the workers on the manual assembly line, the ergonomist noticed a pattern of compensable injuries due to musculoskeletal disorders (MSDs). ASE is self-insured, paying all medical and compensation costs to sick and injured employees. The number of cumulative trauma disorder (CTD) injuries on the manual assembly line is historically determined to be about three (3) per year, with an average medical treatment cost of $2,500. One of the injuries is in an advanced stage (the worker is restricted from returning to the same job and must be assigned to a new job upon return, according to medical recommendations). Therefore, a new worker is recruited through a process that typically takes about two (2) weeks and includes training on a similar workstation and attending a short course on safety and ergonomic practices. For the purposes of this case study, the direct labor cost of the new worker for the first two (2) weeks will not be included in the direct labor cost of the production line, simply because the new worker is not productive on the line yet. During these two (2) weeks, the assembly line is run with six (6) workers.

An injured worker usually comes back to work after ten (10) business days. For simplicity, let us assume that all three (3) injuries happen at different time periods during the year (no overlap). Accordingly, the production line runs with six (6) workers for eight (8) weeks [two (2) weeks for training and six (6) weeks for three (3) injuries]. However, the assembly line runs on some Saturdays at full capacity to compensate for lost production due to training and injuries.

The processing time for each of the six employees is expected to increase.

Therefore, the time spent on each product at each workstation is calculated:

Processing time for each workstation = 21 min/6 workstations = 3.5 min/workstation.

In other words, one unit comes out of the assembly line every 3.5 min (Takt time). Furthermore, the hourly production rate is calculated as:

Hourly production rate = 60 min/hr ÷ 3.5 min/item = 17.14 ≈ 17 items/hr (rounded down)

The production loss is then calculated as:

Production loss due to injury in 8 weeks = (20–17) items/hr × 8 hrs/day × 0.85 × 8 weeks × 5 days/week = 816 items.

Therefore, potential profit loss due to production loss is (if not covered with overtime):

Operating profit loss = 816 items × \$72/item[5] = \$58,752.

The overtime cost to compensate for the production loss in 8 weeks is:

Overtime cost = 7 workers × 816 items × 3min/item ÷ 60 min/hr × \$30/hr ÷ 0.85 = \$10,080.

In addition,

Injury costs = 3 injuries × \$2500/injury = \$7500.

After an extensive search for the most appropriate material-handling equipment for the manual assembly line, the ergonomics committee has decided to install a powered conveyor and rearrange the workstations on the conveyor[6]. The cost for a 10-foot section of conveyor is \$2250. The production line is about 40-ft long. Currently, the conveyor is offered in two options. The first option requires purchasing each section for \$2650, including installation and training. This option provides annual maintenance at the rate of \$700 for the life of the equipment, which is expected to be ten (10) years before it has a salvage value of \$2000. The second option is to lease the new powered conveyor line at an annual rate of \$1500, including installation, training, and scheduled maintenance. Furthermore, rotational loading pallets are recommended to allow the workers to rotate the product without experiencing awkward postures. The pallets will be bought at \$50/pallet. A total of ten (10) pallets are needed for the production line [seven (7) pallets for the line, one (1) pallet for a new TV frame entering the production line, one (1) pallet for a finished product leaving the production line, and one (1) spare pallet]. Figure 3 illustrates the new production line. ASE's accounting system has established a minimum attractive rate of return (MARR) of 5 percent for capital expenditures.

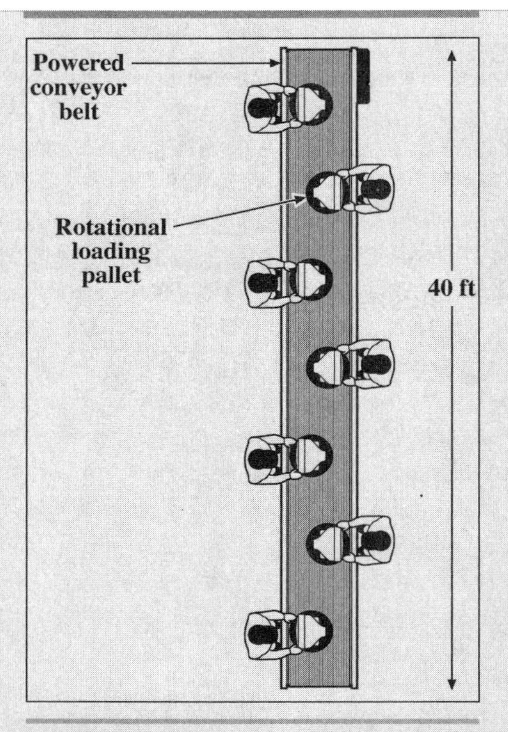

FIGURE 3. The new assembly line layout with a powered conveyor belt

Training in both options targets the workers on the assembly line and does not exceed four (4) hours. The training session, conducted by the ergonomist, includes safety and ergonomic issues related to conveyors and body postures. The ergonomics committee, with management consent, has decided not to redistribute all seven (7) employees to other departments in the plant during the equipment installation. Therefore, employees' downtime will be accounted for in this case[7]. However, training is conducted during the 2-day shutdown to eliminate additional training costs since employees are being paid during the shutdown. Furthermore, the trainer's cost, in this case the ASE ergonomist, is considered to be part of the time invested on the project, which has been categorized as an internal direct personnel cost.

After conducting time studies on the current production line, the industrial engineer on the ergonomics committee expects a decrease in throughput time of three (3) minutes to a new time of eighteen (18) minutes. In other words, Takt time will become (2.57) minutes (time it takes for a TV unit to leave the production line). The ergonomist, on the other hand, estimates a lower injury rate of one (1) CTD per year due to the ergonomics project (intervention). Finally, rework units are expected to drop to 0.1 percent. The new conveyor system takes about two (2) days to install and test (see Figure 3).

The proposed ergonomic intervention incurs new costs in terms of worker training, purchasing the new material-handling system, and temporary losses in productivity. If training is conducted prior to or after the 2-day shutdown, then training costs will be included in the total cost of the ergonomics project as:

Training cost = 7 workers × 4hrs × \$20/hr = \$560.

The cost for installing the new powered conveyor (capital expenditure) is analyzed for each option as:

1. First Option[8]:

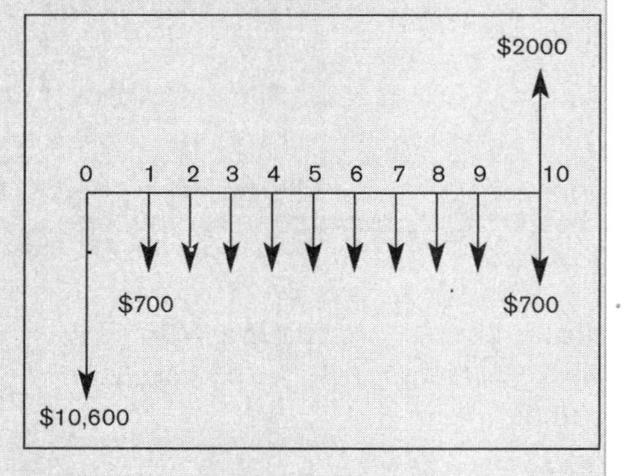

PW of a 40-ft conveyor = \$2650 × 4 = \$10,600

$NPW = -PW_{purchase} - AW_{maintenance} + FW_{salvage}$

$= -\$10,600 - \$700[(1 + 0.05)^{10} - 1] \div [0.05(1.05)^{10}] + \$2000(1.05)^{-10} = -\$14,778$

Total cost = \$14,778 + (10 pallets × \$50/pallet) = \$15,278

2. Second Option:

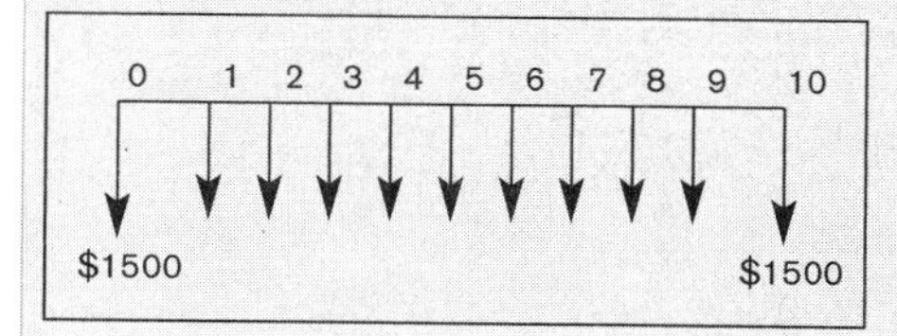

NPW = $-AW_{purchase}$ = $-\$1500[(1 + 0.05)^{10} - 1] \div [0.05(1.05)^{10}] = -\$11,583$

Total cost = $11,583 + (10 pallets × $50/pallet) = $12,083

Therefore, the committee will recommend the second option.

The new conveyor is expected to take 2 days to install. The productivity loss associated with this project is calculated as:

Productivity loss in 2 days = 20 items/hr × 8 hrs/day × 0.85 × 2 days = 272 items.

Therefore, potential profit loss due to production loss is (if not covered with overtime):

Operating profit loss = 272 items × $72/item = $19,584.

Since the workers on the assembly line are not distributed to other departments, additional downtime cost is incurred.

Downtime cost = 7 workers × $20/hr × 8 hr/day × 2 days = $2240.

Now, the overtime cost to compensate for the production lost in 2 days should be based on the new Takt time for the powered conveyor system:

Overtime cost = 7 workers × 272 items × (2.57 min/item ÷ 60 min/hr) × $30/hr ÷ 0.85 = $2879.

The following table summarizes the costs associated with the ergonomics project.

#	Cost	Value
1.	Internal personnel (ergo committee)	$ 3000
2.	Employees' downtime	$ 2240
3.	Equipment	$12,083
4.	Overtime	$ 2879
	Total cost due to ergonomics project[9]	$20,202

Postchange:

When the ergonomics project is introduced, the following rates and costs are impacted by such change.

Processing time for each workstation = 18 min/7 workstations = 2.57 min/workstation.

New hourly production rate = (60 min/hr ÷ 2.57 min/item) = 23.35 ≈ 23 items/hr. (rounded down)

Therefore, the new annual production is calculated as:

New annual production = 23 items/hr × 8 hrs/day × 0.85 × 250 days/year = 39,100 items.

Direct labor cost after introducing the conveyor system does not change for the workers on the assembly line. However, the rework rate is expected to change and so does the direct labor cost for rework.

New annual rework rate = 39,100 items × 0.001 = 39.1 ≈ 40 items. (rounded up)

New annual direct labor cost for rework = 40 items × (10 min/item ÷ 60 min/hr) × $25 = $167.

New annual direct material and indirect costs = 39,100 items × $(50 + 20)/item = $2,737,000.

Therefore, the annual cost is calculated as:

New annual direct labor cost = {7 workers × $20/hr × 8 hrs/day × 250 days/year} + $167 = $280,167.

New annual total cost = $2,737,000 + $280,167 = $3,017,167.

New annual revenue = 39,100 items × $150/item = $5,865,000.

New annual operating profit = $5,865,000 − $3,017,167 = $2,847,833.

Since one CTD injury is expected within one (1) year after the intervention is introduced, the injury cost associated with the case is estimated as follows. Let us think of this cost in terms of production rate, profit loss, and workers' compensation.

Since one (1) injury occurs, the processing time for each of the remaining six (6) employees postchange is expected to increase compared with prechange. Therefore, the time spent on each product at each workstation, known as processing time, is calculated as:

Processing time for each workstation = 18 min/6 workers = 3 min.

In other words, one unit comes out of the assembly line every 3 min (Takt time). Furthermore, the hourly production rate is calculated as:

Hourly production rate = (60 min/hr ÷ 3 min/item) = 20 items/hr.

The production loss is then calculated as:

Production loss due to injury in 2 weeks = (23−20) items/hr × 8 hrs/day × 0.85 × 2 weeks × 5 days/week = 204 items.

Therefore, potential profit loss due to production loss is:

Operating profit loss (if not covered with overtime) = 204 items × $72.83/item[10] = $14,859.36.

The overtime cost to compensate for the production loss in 2 weeks is:

Overtime cost = 7 workers × 204 items × (2.57 min/item ÷ 60 min/hr) × $30/hr ÷ 0.85 = $2158.80.

In addition,

Injury cost = 1 injury × $2500/injury = $2500.

The following table summarizes the impact of the ergonomic intervention on personnel and production annually.

#	Category	Prechange	Postchange	Δ	Up/Down
1.	Annual production	34,000 items	39,100 items	5100 items	↑
2.	Annual rework	102 items	40 items	62 items	↓
3.	Annual direct labor cost (rework)	$425	$167	$258	↓
4.	Annual operating profit	$2,439,575	$2,847,833	$408,258	↑
5.	Production loss due to injury	816 items	204 items	612 items	↓
6.	Overtime cost due to injury[11]	$10,080	$2158.80	$7921.20	↓
7.	Injury cost	$7500	$2500	$5000	↓

In order to calculate the benefit-to-cost ratio for the ergonomics project, the costs associated with the project and the benefits observed, after the intervention is introduced, should be compared. (Refer to the two tables on the right.)

#	Category	Δ (Benefits $)
1.	Annual direct labor cost for rework	258
2.	Annual operating profit	408,258
3.	Overtime cost due to injuries	7,921.20
4.	Injuries cost	5000
TOTAL		$421,437.20

#	Cost	Value
1.	Internal personnel (ergo committee)	$ 3000
2.	Employees' downtime	$ 2240
3.	Equipment	$12,083
4.	Overtime	$ 2879
TOTAL		$20,202

Benefit-to-cost ratio = $421,437.20/$20,202 = 20.86 (for every $1 spent on the ergonomic intervention, a monetary return value of approximately $21 is expected)

Budgeting for Ergonomics Projects

Safety professionals are generally familiar with the basic concepts of budgeting. Fundamentally, resources are allocated in anticipation of consumption during a future period. In the case of ergonomics, resources can be viewed as capital (dollars) and/or time (employees). Budgets for ergonomics would typically be contained within the larger safety and health budget submitted annually by the safety and health manager or director. Items covered in a general budget could include anticipated costs for consultants, any anticipated ergonomic training, professional books or periodicals, memberships, attendance at conferences, or a small amount of working capital for minor ad hoc fixes.

When determining how much funding to request in the pending budget cycle, or business plan, the safety practitioner is best served by looking back at historical financial data (previous budgets) as a starting point, and then modifying those estimates for use in projects anticipated during the funding period. It may also be prudent to try to benchmark against similar businesses, other entities in one's business unit, or even competitors if the opportunity arises.

Many manufacturers also require significant information about potential safety and health benefits relating to proposed capital projects, even those not specifically designed to be ergonomic in nature. The bottom line is that practitioners should be prepared to assist other departments with amplifying information to aid in project cost justification.

A Thought on Budgeting

Many times the financial and production data relating to cost justification are either difficult to glean or simply not available. A technique that some highly successful practitioners use is to couch the actual expenses associated with injuries in terms that business people (management) forthrightly understand. The reader should refer back to the case study for an example.

Suppose that there is an ergonomics project that one is trying to champion (obtain funding for) in next year's budgeting cycle, but it doesn't appear to be favorably received by management in preliminary discussions. A prudent ergonomics professional might state in a memo that the impacts of such a project are more far-reaching that it first appears. The expected profit for selling a TV was estimated at approximately $72 per unit[12]. Assume that the proposed project will eliminate one back injury per year. It is known from historical estimates (company documents) that back injuries experienced by the affected employees typically cost about $25,000, but let's not stop yet. After checking with the sales force, the success rate of salespeople is about 5 percent, meaning that they must establish contacts and write about twenty (20) proposals for each one that is successfully funded. Using the language of business, the anticipated elimination of one back injury per year, by selecting the project for funding, is approximately equivalent to having the sales force write proposals for $1,044,000 worth of TV sales! Needless to say, this simply grabs their attention.

$25,000 (cost per back injury) ÷ $72 (profit per unit) = 348 units

348 units × $150 per unit = $52,200

$52,200 × 20 (5% success rate with TV proposals) = $1,044,000

Share your successes in ergonomics. These words cannot be overstated. By spreading the word, formally (newsletters, bulletin boards, Web pages) and informally (word of mouth, email, etc.) after each successful project implementation, ergonomics and the funding it is receiving can gain momentum very quickly and eventually almost be self-sustaining. This can make ergonomists and safety practitioners very popular people.

In conclusion, ergonomists compete for limited company resources along with other departments. Unless the benefits of ergonomics projects are fully justified economically and transmitted clearly to top management, ergonomists will likely play an ineffective role in the technical operations of their organizations. There are several ways to justify ergonomics projects. Some are simple and straightforward, as illustrated in the case study. Additionally, some commercial software programs that can facilitate economic justification are also available. However, continuous improvement in ergonomics is the key to reducing costs and increasing benefits, as well as promoting collaboration between ergonomists and personnel in other departments.

ENDNOTES

[1]Spilling, Eitrheim, and Aarås (1985) reported a ROI of about nine for an ergonomic intervention by comparing a seven-year period prior to the ergonomic intervention to the seven years after intervention.

[2]According to 29 CFR 1904.4–Recording Criteria, each employer is required to keep records of each fatality, injury, and illness that is work-related, a new case, and meets one or more of the general recording criteria of 1904.7, or the application of specific cases of 1904.8 through 1904.12. Cases are work-related if:

(1) An event or exposure in the work environment either caused or contributed to the resulting condition, or
(2) An event or exposure in the work environment significantly aggravates a pre-existing injury or illness.

The term "lost work days" has been eliminated since the focus now is on days away, days restricted, or days transferred. This includes new rules for counting that rely on calendar days instead of work days. In this context, two terms have been used in industry to reflect such counting, DAFWII (Days Away From Work, Injury, or Illness) and DART (Days Away, Restricted, or Transferred). DAFWII counts the actual number of calendar days between first day missed due to injury and the returning day, without counting the day of injury. DAFWII only includes cases with days away from work and disregards those with restricted work activity and job transfer. It is calculated by dividing the number of cases involving days away from work by the number of hours worked by all employees during the calendar year. It is then multiplied by 200,000 to normalize the rate. Beginning with Site Specific Targeting (SST)-03, DAFWII was included based on the belief that an injury or illness, which requires a day away from work, is more serious than one that requires restricted activity. Therefore, incorporation of DAFWII as a targeting criterion would further identify establishments with the greatest number of serious hazards. DART, on the other hand, includes days of missed work as well as days of "light duty," and it is generally a larger number than DAFWII. It is calculated by dividing the number of cases involving days away from work, restricted work, or transfer by the number of hours worked by all employees during the calendar year. This quotient is then multiplied by 200,000 (100 employees working 2,000 hours per year) to normalize the rate. For further information, the reader is directed to "Your Establishment Received a Letter from OSHA: Now What?" by Davis, Ahuja, and Hollingsworth, published in 2005 in *Professional Safety*, 50(3):33–37.

[3]Though worth considering, it is difficult to quantify (by accepted means) a reasonable estimate. This concept is worthy of consideration for further research.

[4]"Determining the cost-benefits of ergonomics projects and factors that lead to their success" by Hal W. Hendrick, published in 2003 in *Applied Ergonomics*, 34(5):419–427, is an additional source on costs and benefits associated with ergonomics projects.

[5]Profit per unit = Annual operating profit ÷ Total annual production = $2,439,575 ÷ 34,000 = $72.

[6]Once again, the numbers in this case study do not represent an existing industry or organization. This is true for the suggested ergonomic intervention as well.

[7]The postintervention performance expectations are hypothetical in nature. The reader would have to make these estimations based on a number of factors, which may not be easy to forecast.

[8]NPW, AW, FW stand for net present worth, annual worth, and future worth, respectively. The reader is specifically advised to review Michael Toole's chapter titled "Cost Analysis and Budgeting" in Volume 1, Section 1 of this handbook for further explanation of such terms and associated formulas.

[9]For the purposes of the case study, the total cost due to the ergonomics project does not account for profit loss ($19,584) since it is assumed that overtime costs are included in the calculations. However, the authors ran the calculations for profit loss wherever applicable in the case study to provide additional calculations in case overtime is not accounted for and replaced with profit loss.

[10]Profit per unit = Annual operating profit ÷ Total annual production = $2,847,833 ÷ 39,100 = $72.83.

[11]If profit loss is accounted for instead of overtime due to injuries, then the benefit (saving) of profit loss is ($58,752 – $14,859.36) = $43,892.64, which will replace the overtime costs due to injury ($7,921.20). The benefit-to-cost ratio will be calculated accordingly.

[12]Profit per unit = Annual operating profit ÷ Total annual production = $2,439,575 ÷ 34,000 = $72.

ACKNOWLEDGEMENT

The authors would like to acknowledge and express gratitude to the following individuals for their guidance in assisting with this chapter.

- **David Alexander, PE, CPE.** David is the founder and president of Auburn Engineers, Inc., a leading ergonomics consulting firm. He has practiced ergonomics for three decades, and has produced seven texts on industrial ergonomics. He is the founder and former chair of the Applied Ergonomics Conference. David is both a Registered Professional Engineer and Certified Professional Ergonomist.
- **Eric Hollingsworth, BS.** Eric holds a bachelor's degree in Industrial Engineering from Columbia University. In the past, he served as a student research assistant in Occupational Safety and Ergonomics at Auburn University's Industrial and Systems Engineering Department. Eric is currently a research assistant working in medical response logistics in the Department of Public Heath at Weill Cornell Medical College.
- **Adam Piper, MS.** Adam earned his bachelor's (Mechanical Engineering) and master's (Industrial & Systems Engineering) degrees from Auburn University. He is currently a doctoral candidate in Industrial & Systems Engineering with a focus in Occupational Injury Prevention. Adam has participated in safety research and consulting in dozens of manufacturing facilities over the past ten years, and he recently completed a 10-month appointment as a guest researcher for NIOSH in Atlanta, GA.

REFERENCES

Alexander, D., ed., 1999. *Applied Ergonomics*. Volume 1. Norcross: Engineering & Management Press.

______. Personal communication, March 17, 2006.

Brauer, R. L. 2006. *Safety and Health for Engineers*. 2d ed. New York: John Wiley & Sons.

Bureau of Labor Statistics (BLS). 2005. Current injury, illness, and fatality data of 2004. Washington DC: Department of Labor. www.bls.gov/iif/home.htm

Canada, J., W. Sullivan, and J. White. 1996. *Capital Investment Analysis for Engineering and Management*. 2d ed. New Jersey: Prentice-Hall, Inc.

Davis, J., S. Ahuja, and E. Hollingsworth. 2005. "Your Establishment Received a Letter from OSHA: Now What?" *Professional Safety*, 50(3):33–37.

Heinrich, H. W. 1959. *Industrial Accident Prevention*. 4th ed. New York: McGraw-Hill.

Hendrick, H. W. 1996. "Good Ergonomics Is Good Economics." 1996 HFES presidential address. Proceedings of the Human Factors and Ergonomics Society, Santa Monica, CA, pp. 1–15.

______. 2003. "Determining the Cost-Benefits of Ergonomics Projects and Factors that Lead to their Success." *Applied Ergonomics*, 34(5):419–427.

Kohn, J. 1997. *The Ergonomic Casebook: Real World Solutions*. Boca Raton: CRC Press.

Leigh, J. 1995. *Causes of Death in the Workplace*. Westport, CT: Quorum Books.

Liberty Mutual. 2002. *Liberty Mutual Workplace Safety Index*. Boston, MA: Liberty Mutual. www.libertymutual.com/omapps/ContentServer?cid=1003349317278&year=2005&prid=1078447725811&pagename=CorporateInternet%2FPage%2FPressReleaseDarkBlue&c=Page

______. 2004. *Liberty Mutual Workplace Safety Index*. Boston, MA: Liberty Mutual. www.libertymutual.com/omapps/ContentServer?cid=1078439448036&pagename=ResearchCenter%2FDocument%2FShowDoc&c=Document

______. 2005. *Liberty Mutual Workplace Safety Index*. Boston, MA: Liberty Mutual. www.libertymutual.com/omapps/ContentServer?cid=1078448761767&pagename=ResearchCenter%2FDocument%2FShowDoc&c=Document

McAtamney, L., and E. N. Corlett. 1993. "RULA: A Survey Method for the Investigation of Work-Related Upper Limb Disorders." *Applied Ergonomics* 24(2):91–99.

Miller, T. R., and M. Galbraith. 1995. "Estimating the Costs of Occupational Injury in the United States." *Accident Analysis and Prevention* 27(6):741–747.

Moore, J. S., and A. Garg. 1990. "The Strain Index: A Proposed Method to Analyze Jobs for Risk of Distal Upper Extremity Disorders." *American Industrial Hygiene Association Journal* 56(5):443–458.

National Institute for Occupational Safety and Health (NIOSH). 1994. *Application Manual for the Revised NIOSH Lifting Equation* (Publication No. 94-110). Retrieved March 16, 2006 from NIOSH. www.cdc.gov/niosh/pdfs/94-110.pdf

National Safety Council (NSC). 2006. "In the news." *Safety + Health*, September 14–16, 2006.

Occupational Safety and Health Administration (OSHA). 1970. "General Duty Clause" of OSH Act of 1970. Washington, DC: Department of Labor. www.osha.gov/review_of_records.html

Oxenburgh, M. S. 1997. "Cost-Benefit Analysis of Ergonomics Programs." *American Industrial Hygiene Association Journal* 58:150–156.

Schneider, M. F. 1985. "Why Ergonomics Can No Longer Be Ignored." *Off. Admin. Autom.* 46(7):26–29.

Spilling, S., J. Eitrheim, and A. Aarås. 1985. "Cost-Benefit Analyses of Work Environment Investment at STK's Plant at Kongsvinger." In E. Corlett, J. Wilson, and I. Manencia, eds., *The Ergonomics of Working Postures: Models, Methods, and Cases*. London: Taylor & Francis, pp. 380–397.

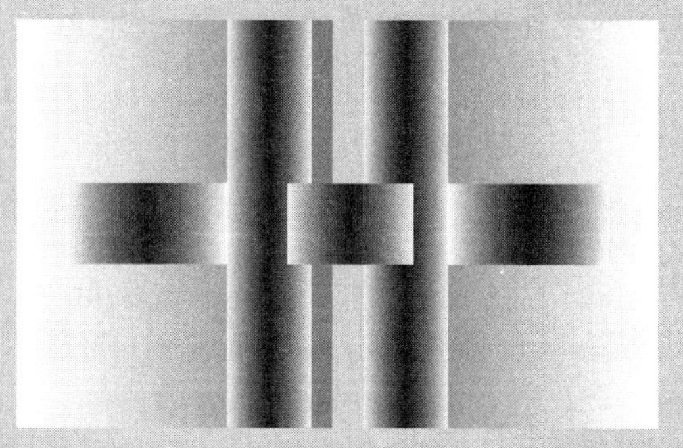

SECTION 6

ERGONOMICS AND HUMAN FACTORS ENGINEERING

LEARNING OBJECTIVES

- Outline the necessary elements of a system of metrics.
- Be able to explain the interaction between the formation of metrics and benchmarking.
- Recognize the advantages and disadvantages of leading and lagging metrics, and provide guideline for the formation of each.
- List the elements of an ergonomics program to be considered for metric formation.
- Be able to explain what a monitoring element is in metric formation and describe how to use such a function in metrics.
- Outline an ergonomic system of metrics based on activities being conducted in the learner's own facility.

BENCHMARKING AND PERFORMANCE CRITERIA

Robert Coffey

THIS CHAPTER was written to provide a basis for consideration as a safety and health professional contemplates the many aspects of measuring performance in the field of ergonomics. Although it does not provide specific instructions, it does outline many of the issues that must be considered when developing a performance-measurement system. Many of the metrics that can be considered when measuring the effectiveness of an ergonomics program or general safety program will be highlighted, and many complications that are associated with measuring the performance of systems that may not be purely quantitative will be presented. Although instruction is not provided on how to benchmark against other programs or other companies, some of the essentials for setting up an internal benchmarking effort are provided. Readers will be able to use the information in this chapter as a foundation upon which to begin their exploration of additional resources by means of which to develop and use sound metrics to measure internal ergonomics program performance, then benchmarking that performance against outside entities.

Before discussing how to benchmark and measure ergonomics programs, a moment should be taken to discuss benchmarking and metric formation generally. This discussion is of itself involved, but it is worthwhile—because the issues surrounding benchmarking ergonomics mirror the issue of benchmarking in general.

To benchmark is to provide "a point of reference from which measurements may be made" (Webster's 1983). After all, it is hard to change a situation without knowing what it is. Benchmarking is intended to establish a system that can track some feature (in the case of this chapter, safety in general—and ergonomics in particular). Benchmarking is a method for tracking data over time (perhaps at daily, weekly, monthly, or annual intervals).

So why benchmark? Benchmarking gauges (or substantiates) improvements (or deteriorations) in a system. The next questions, of course, are "How do I benchmark?" and "What do I benchmark?" These two questions are two sides of the same issue: developing a system of benchmarking. To continue further in this discussion, another term needs to be introduced, metric. According to *Webster's Dictionary* (1983), a metric is "a standard of measurement." (In this chapter, *benchmarking* indicates a system, and items being benchmarked are called *metrics*.) Certain foundational decisions must be made when developing a benchmarking system. *What metrics will be used?* A metric is not worthwhile unless it provides useful information about the system. Measuring monthly forklift cost fluctuations in an attempt to save money in a 1000–person operation that uses only one forklift will not be effective. *What will be used to benchmark against?* Benchmarking is often thought of as comparing data collected in different times and places about a particular metric (HSE 1999). This means that a unit (perhaps a department, or a facility, or an organization) can compare past data to present data. An excellent example is available by which to highlight the issues above and focus benchmarking and the formation of metrics on safety: the OSHA recordable.

Working Example: OSHA's Record-Keeping Standard

In 1971, the OSHA record-keeping standard, 29 CFR 1904, went into effect. This record-keeping standard applies to all industries that are under the Occupational Safety and Health Administration's (OSHA) purview and requires industrial businesses to track workplace injuries (OSHA 1971). It explains which injuries are to be recorded and categorizes injuries into types (such as recordable, restricted-duty, fatalities, and those causing lost work days). Injuries are to be tracked annually and the totals posted after the end of each year. The OSHA record-keeping standard created the largest U.S.–based benchmarking system ever used and made the OSHA recordable rate the most widely used metric in the United States (and the OSHA record-keeping standard the most comprehensive benchmarking system)—simply because the majority of industrial businesses, as well as civil employers, must comply with OSHA's record-keeping standard.[1] The system addresses all the issues raised above.

In theory, benchmarking is done to reduce employee injuries. In the case of the record-keeping standard, benchmarking defines the types of injuries that are *recordable* and tallies them yearly. Benchmarking is performed by all industry regulated by OSHA. In fact, the Bureau of Labor Statistics (BLS) tracks recordable rates by Standard Industry Codes (SIC) or by the newer North American Industry Classification System (NAICS)—United States, 2002. The data can be broken down to look at injuries by facility and company size, by general occupational type, by parts of body affected, or by types of event or exposure (BLS 2006). This gives companies an opportunity to benchmark against other companies in the NAICS, against similarly sized companies, against their industry in general, or just against their own data from previous years. By using a normalizing formula (the incident rate calculation), facilities of varying size and function can (theoretically) be compared (OSHA 2005). This data is scrutinized by governmental agencies, corporate offices, union committees, and all other parties who are interested in safety. Facilities even go so far as to track these metrics monthly, posting graphs or other visual aids to represent occurrences. OSHA has begun to use this data to decide which sites to inspect (OSHA 2006a).

Benchmarking and the formation of metrics are only tools—not ends in themselves. A system's quality is reflected in its success in meeting its objectives—in the case of safety, lowering injury rates. Benchmarking, as any tool, is not perfect. A benchmarking system that is not developed correctly will not meet goals. Poorly correlating comparison groups (such as a manufacturing site and an administrative office), creating too many metrics to track effectively, and developing inappropriate metrics can all detract from a benchmarking system's usefulness. OSHA's record-keeping system illustrates this point.

The appropriateness of 29 CFR 1904 has been debated. Although OSHA made efforts to improve the system, leading to the 2000 revision of the standard (OSHA 1971), detractors of the system point to its persisting deficiencies (Nash 2001). OSHA's recordability rules must be applied consistently in order to be effective, but interpretations of what is, and what is not, recordable are still debated. Some subjectivity exists in as regards attendant physicians, as well as some variability according to the natures of specific industries. An injury that does not affect one employee's ability to work at one employment may make another employee in another job unable to work (OSHA 2006b).

This chapter is neither an affirmation of nor an attack upon the OSHA record-keeping system. Rather, it is intended to aid the expression of general concepts of benchmarking and metric formulation. A comprehensive analysis of the system is well beyond the scope of this chapter and inconsistent with its purpose.

Metric Classification

Metrics, the gauges of systems, can be divided into two main types: pre-indication (leading) and post-indication (lagging). Pre-indication metrics are qualities that can be measured prior to an occurrence. Post-indication metrics are qualities that are measured after an occurrence—including the occurrence itself. The OSHA recordable illustrates this concept, being itself a post-indication metric. When an injury occurs, it is recorded on an OSHA 300 log according to the criteria it meets. Comparing amounts of workers' compensation dollars spent yearly is also a post-indication metric—something measured after it has occurred. In other words, to monitor injuries, an employer measures the number that occur, later remeasuring in the hope of finding that some reduction has occurred.

Because of the diversity of industry safety programs, however, pre-indication metrics are not nearly so clear-cut. They can, for example, track the completion of audit corrective actions (to ensure that unsafe conditions are corrected before injuries occur) or track the percentage of initial training that is conducted (to ensure that new employees are trained how to work safely before they are injured). Pre-indication metrics are used by analysts to discover which factors will lower the probabilities of injuries' occurrences, ensuring that these factors are then controlled.

Decisions about metrics—specifically about pre- and post-indication metrics—bring evaluators face to face with the confounding factors of time, resources, and complexity. Post-indication metrics are direct, observable results that are either outright goals or that correlate very closely with goals. Because no organization has infinite resources, much of the creation of post-indication metrics boils down to resource allocation. A new training program may show reductions in recordable rates and workers' compensation dollars for the year *after* it is implemented—but how can an evaluator know whether resources being applied *now* will affect future rates? No one wants to continue to expend resources in ways that will not help in meeting goals.

This is where developing pre-indication metrics may be helpful. Post-indication metrics are observable (they can be seen and measured in reference to goals), but pre-indication metrics are predictive (their measurements help predict whether certain goals will be met in the future). Pre-indication metrics take measurements while resources are being expended in order to predict the effectiveness of such expenditures. The chief difficulty associated with pre-indication metrics is deciding what to measure. Evaluators must decide what program elements will likely increase safety before implementing them. A perfect world would see the implementation of a safety-system "improvement," the tracking and measuring of some observable element of the program, and a later evaluation of whether the "improvement" decreased the occurrence of injuries. If so, the program should be continued. If not, the process should be repeated with a new safety-system "improvement." But the real world is not so forgiving. In reality, companies develop lists of possible improvements, prioritize them, and implement as well as possible those improvements most likely to decrease injury rates (a practice referred to as risk assessment). The amount and types of improvements implemented are governed by the resources available. One simple improvement can be implemented or multiple complex improvements can be implemented. An attribute of the program can be selected and tracked to ensure that the program is being implemented and maintained.

One of the largest difficulties when working with pre-indication metrics is choosing measurable quantities that are truly predictive. A company that has an extensive auditing program may find by monitoring and tracking the amount of audits conducted that the more audits a facility has, the lower its injury rates. The company would then set metrics for auditing: *the more audits the facility conducts, the fewer its injuries*. But the difficulty of pre-indication metrics is that they can be organizationally specific. Auditing and tracking audits is a worthwhile safety activity, but in some organizations it may not have the same significance as in others. Take, for example, Company B, which has noticed the association of injury reductions with the auditing metrics of Company A, described above. Company B decides to use the same metrics, even though it does not have the extensive auditing system of Company A. An auditing metric may not be the best predictor of goal attainment for Company B, which has no reason to expect a correlation between auditing and injury reduction.

Causality can be another confounding issue. Company B assumes that the auditing program itself decreases injury rates; but perhaps Company A's program includes aggressive corrective action, incorporating strict criteria for when corrective actions must be completed. This could have a great effect on the reduction of injuries—but these factors may not have been taken into account by Company B when it tracks the number of audits it conducts. Tracking audits may not have anything to do with injury reduction in Company B.

One difficulty associated with pre-indication metrics is the impossibility of proving the null hypothesis (Streiner 2003). In other words, "How can a negative be proven?" Relating pre- to post-indication metrics can be difficult. How can anyone quantify how many injuries *didn't* happen? Even when post-indication metrics are not within goals (as when recordable rates are high), how can it be truly ascertained that rates would have been higher if not for the efforts measured by the metrics?

Musculoskeletal Disorders (MSDs)

The discussion on benchmarking and metric formation in general extends to ergonomics in the specific. In fact, some of the issues discussed are compounded. Ergonomic injuries (referred to as musculoskeletal disorders, or MSDs) do not happen immediately but develop over time. In fact, some can take years to develop (NIOSH 1997). How can post-indication metrics (that is, injury rates) alone indicate whether improvements to the safety system are effective? The injuries recorded have developed over years and are only recognized when their symptoms interfere with everyday activities. Even if drastic efforts are taken after the diagnosis of the first few occurrences, further injuries will be diagnosed and recorded; such injuries develop over many years and the first have only now been recognized. On the other hand, poor pre-indication metrics may cause resources to be put into a safety program for years despite a complete lack of indication of the effectiveness of doing so. In ergonomics both post- and pre-indication metrics should be developed. Post-indication metrics are direct measurements that can include numbers of MSD recordables, numbers of repetitive motion injuries, and totals of workers' compensation dollars spent because of MSDs. But what are used as pre-indication metrics? Such items must have predictive value—and in the field of ergonomics, this is common. Risk factors, pain surveys, workplace design, and training are all possible fixes for ergonomic issues and can be used as pre-indication metrics. Suppose a company is implementing a new ergonomics program. The safety manager has discovered that the company is currently suffering five MSD injuries per 100 workers yearly. The company, which is paying about $175,000 annually because of such injuries, wants to put in place an ergonomics program to reduce these numbers and has set a goal of a 40 percent reduction within three years. The number of MSDs and the workers' compensation dollars are post-indication metrics and are used as a baseline against which to compare. As part of the ergonomics improvement program, the following elements are implemented: ergonomic task evaluation, new process ergonomic design review, worker training in lifting and repetitive motion, and the formation of an ergonomics committee. Pre-indication metrics are based on these elements. The metrics for the first one-year period may require that (1) all facilities will have formed an ergonomics committee, (2) all facilities will have conducted general training in ergonomics for their associates, (3) a new ergonomic process-review procedure will have been developed and implemented, and (4) a minimum of five work-task analyses will have been completed. These example metrics will depend greatly upon the organization's culture and the resources available. But even these examples emphasize a key factor: shifting metrics. These metrics may not have meaning in the future—something that can be true of both pre- and post-indication metrics. Recordable injuries may be reduced to nonexistence. Tracking recordables at this point may not be necessary. Measuring the amount of first-aid cases may be a more useful alternative—see the Hendricks pyramid (Heinrich 1931).

Metric Formation

When developing metrics, the following stipulations should be noted:

1. Pre-indication metrics, post-indication metrics, and goals must be decided upon and the differences between them understood. A goal is a desired outcome, for a metric, in a certain time frame—such as a 10 percent reduction in recordable injuries during a 2-year period.
2. Metrics should be quantifiable and measurable.
3. Pre-indication metrics should address interventions used to improve post-indication metrics.
4. As goals are met, new metrics and goals should be introduced that reflect the current situation.

Injury/Complaint Trending

Part of the intent of the OSHA record-keeping standard was to reduce the frequency of worker injuries. But how is writing down when people are hurt going to do this? The idea was that when injuries are recorded, problem areas and jobs will be evident (OSHA 1971). These areas and jobs are where the most attention to safety must be paid. This is trend analysis—the use of data about injuries to highlight priority areas where safety attention is needed. A set of data is analyzed to discover whether reoccurrences are so numerous that they constitute a pattern. Although OSHA's record-keeping standard is a good example, trend analysis can extend to any data set.

There are often unrecognized patterns in any system. Analysts try to find these patterns by means of trend analysis. But what is the point of finding a pattern of injuries? Two advantages are clearly evident: credibility and priority. Both of these advantages stem from predictabilities. A properly conducted trend analysis enables the prediction of yet-to-come occurrences (Wikipedia 2007a).

By understanding injuries, safety professionals can prioritize their safety efforts. Because they are often spread thin in the expenditure of their resources, money and time used on one project must often be removed to another project. From an economic standpoint, the cost of a project is not only the time and money used to complete it but also the loss of benefits from other projects that could have used that time and money. Accurate trend analysis allows the areas to be identified having the most severe injury issues, allowing for the development of programs to abate the associated hazards—but safety professionals must be cautious. The prioritization of trend-analysis data must occur case by case, and broad statements should be kept at a minimum. The frequency, severity, and possibility for fatality of injuries are all possible ways to prioritize resource expenditure. Arguments can be made for or against prioritizing any one of the above. The greater the frequency of injuries, the more people are getting hurt, and the higher is the probability of fatal injury. But what about numerous minor injuries, accompanied by a smaller percentage of very severe injuries? Shouldn't these severe injuries be addressed? All safety professionals (in fact, all business professionals) hold the prevention of fatalities as a high priority, but what happens when the data do not lend themselves to the prediction of fatalities? Most of the time, evident trends will themselves set the priorities. Consider the OSHA 300 log of a small, 50-employee manufacturer. The last three years of the OSHA log shows 18 recordable injuries—12 cuts to hands, 1 back injury, 1 case of heat exhaustion, 1 instance of hearing loss, 1 foot injury, 1 case of skin irritation, and 1 case of carpal tunnel syndrome. Hand and back injuries might be made priorities by some safety professionals: hand injuries because they constitute the bulk of this company's recordables for the last three years and back injuries because of their severity, being some of the most disabling and costly injuries. This data was analyzed by type of body part injured. Any of the data types on the OSHA 300 form (organized by employee, by date, by department, by job title, by body part, or by type) can be analyzed (OSHA 1971). Sorting the data by each category is a good place to begin a trend analysis. In the trend analysis above, the number of injuries (first criteria) is being sorted versus the body part affected (second criteria). Larger data sets may require the further expansion of these criteria. The first criteria could be the number of recordable injuries and the second could be organization by body part, and then the data could be re-sorted, in these subsets, by department. The first requirement of metric evaluation is the existence of a base of data broad enough to provide significant results.

Credibility is not an issue that is well defined or frequently used in trend analysis. The connection it has with prioritizing, however, is worth mentioning. To develop and implement programs to correct the hazards and decrease injuries requires resources (mostly of time and money). Convincing others of the effectiveness of expending these resources can be a daunting task. Trend analysis provides a foundation for justifying resource allocation. In fact, the trend analysis itself sets the stage for benchmarking and the formation of metrics. Trend analysis is simply self-benchmarking. When trend analysis is accurate and future outcomes can begin to be predicted, an organization can gain the ability to change outcomes by expending resources in an informed manner. If a trend can be show how, after resources are applied, a predicted change occurred, the credibility of safety professionals' recommendations for the expenditure of resources increases, making it easier to get approval for the allocation of resources.

As with other systems, the devil is in the details. Trend analyses can be informal, comparing two or three factors and representing them visually (charting or graphing), or formal (involving t-square analysis or normal distributions). The intended use of data, the mathematical skill of analysts, and the nature and extent of the data available will all affect the formality of trend analysis. Formal trend analysis that depends upon statistical principles will be affected by the interaction of sample sizes with the desired types of analysis and the desired level of uncertainty (Wikipedia 2007b). In fact, sample size plays a part even in informal trend analysis. When a frequency of occurrence is cited, and the sample size is small or the time period is short, it will be difficult—if not impossible—to draw predictable conclusions because of the marked increase in the effects of chance and variation (Wikipedia 2007b). Probabilities simply may not have had enough opportunity to manifest themselves. When a die is rolled six times, a high probability exists of it rolling a 1 at least once. But if a die is rolled only twice, that probability is much lower and (though present) may not manifest itself. But what happens when a probability is not known—when the equivalent of a four-, six-, ten-, twelve-, or twenty-sided die is rolled? Each of these approaches closer to the use of trend analysis to make predictions. If in twenty rolls five 1s are rolled, it is likely that a four-sided die is being rolled. If in eighteen rolls three 1s are rolled, a six-sided die seems likely—and so on. As in the case of trend analysis, observations (the numbers and results of die rolls) are used to evaluate probabilities. Such predictions make possible increasingly informed decisions about the efficient expenditure of resources. Remember, however, that as sample sizes increase, data should be more standardized. If only 20 recordable cases are compared, a fair amount of detail can be attached to each case; but if 2000 cases are compared, the information should be standardized to make comparison efficient and effective.

These methods are well-suited to the field of ergonomics. Because OSHA 300 injury logs, as the most common form of injury tracking in the United States, are also the most common tool for trend analysis, a definition of what injuries are ergonomic concerns is a good starting point. Ergonomics is the science of fitting each work area to the employees using it (thereby reducing stresses on them) (NSC 1997a). "Ergonomic injury," of course, is a misnomer—ergonomics is the cure, not the problem. Various names exist for the injuries that ergonomics can abate. Cumulative trauma disorders (CTDs) (NSC 1977b), repetitive strain injuries (RSIs) (NHS 2006), and musculoskeletal disorders (MSDs) (OR-OSHA 2005, 8) are all terms used to refer to disorders that affect muscles, tendons, ligaments, nerves, and blood vessels. General examples of these include strains, sprains, pulled muscles, back injuries. Specific conditions include tenosynovitis, tendonitis, epicondylitis, and carpal tunnel syndrome. For simplicity's sake, the acronym MSD will be used for the remainder of this chapter to refer to *ergonomic injuries*. OSHA included a MSD column on the OSHA 300 form in its original update to 29 CFR 1904 but subsequently removed it (OSHA 2001). The performance of this trend analysis will depend on the OSHA 300 log used; the number of occurrences will guide the trend analysis. The confounding factor here is the same for ergonomics as for performance measurement in general: the long onset time of injury. If, as Heinrich says (1931), minor injuries and near-misses lead up to major injuries (those considered *recordable*), will this not also be true for cases of ergonomic injuries? If these factors can be tracked, logic and Heinrich seem to dictate that they should be used to indicate the success or failure of efforts at safety. By tracking precursors, safety professionals can adjust or change programs as needed. But what precursors can be used? Two such items seem particularly applicable: complaints and risk factors.

Are there true *near-misses* for ergonomic injuries? Their nature would make them seem unlikely to point to an occurrence as "almost an injury." But ergonomics does have a near-miss analogy: complaints of pain and discomfort. In a chronic injury, an area is repeatedly injured beyond its ability to effectively repair itself. Infrequent occurrences are healed by the body, but consistent reinjury leads to chronic disease, and to pain or discomfort in the employee, including complaints of burning sensations, cramping, stiffness, and pain, among other symptoms. These complaints can be helpful indicators in benchmarking ergonomics. Unfortunately, these symptoms are often overlooked or ignored as *normal*. Just as not all employers track near-misses, so there may not be a system in place to track these complaints—an excellent opportunity entirely missed. One example of complaint tracking can be instituting a complaint form in an area with known ergonomic hazards. By counting the number and types of complaints after the implementation of various improvements, indications are given of the success or failure of these initiatives. The amount of complaints per department can be compared to provide some indication of where to set priorities. Note that personality types may strongly affect levels of complaints. One subset of the population may be reticent to complain (thinking, *I'm just getting old—this isn't anything big*), and another may be more vocal with its complaints, overshadowing those of others.

Risk (or contributing) factors are another area open to trend analysis (OR-OSHA 2005, 18). Risk factors are those conditions that may contribute to ergonomic injuries. The higher the number of risk factors associated with a particular task, the more likely is the occurrence of an ergonomic injury. By tracking the number of risk factors per job (or task, or department), the efforts of reduction programs can be evaluated and, if necessary, prioritized as the result of comparisons. Anything that is shown to be a precursor to an ergonomic injury and is able to be tracked can be used for trend analysis.

Workers' Compensation Loss

When benchmarking, one of the more useful things to consider can be workers' compensation data. The ethical and moral reasons for effective ergonomics programs are many, and they are supplemented by significant economic reasons. Workers' compensation data indicate the actual cost (to companies or insurance carriers) per injury and provide a way to gauge how injuries are directly (economically) affecting a company. Although workers' compensation data seem suitable for benchmarking, some cautions should be noted. The most basic workers' compensation data are a reflection of cost (money spent). Although workers' compensation data can be useful to a general safety program, the definition of *ergonomic injury* should be clear. Data must be compared around a certain subset of injuries, as opposed to the whole.

How a company's workers' compensation program is administered can also be a factor. Workers' compensation data can be affected more by the case-management skills of a facility than by the occurrence and severity of injuries. A company's ability to offer transitional work (light duty) can also play a part. Two companies can have identical incident and severity rates for ergonomic injuries, but one may have a very aggressive transitional work program, making the workers' compensation dollars spent by the company with the transitional work program lower because of reduction in lost-time indemnity.

Even workers' compensation data itself may be misleading, because the data do not reflect the indirect costs of incidents. Arguments have been made that the indirect costs of injuries (lost efficiency, employee morale, training time) can be 4 to 12 times greater than the direct costs, such as medical bills and indemnity (OR-OSHA 2005).

Despite these cautions, workers' compensation data can provide an array of benchmarking opportunities. Workers' compensation data move away from reliance on the OSHA recordable and are collected about any injury that generates expenses. Although first aid and first-aid supplies used on site are not factored into these data, including these costs is not difficult.

These data are also a bottom-line comparison and can offer an opportunity to show the company real cost savings from a successful program. These cost savings can take the form of lower premiums for a company insurance policy. Companies that insure themselves for Workers' Compensation may see even greater savings. Not only is the company saving on direct and indirect costs, it is also avoiding tying up its money in reserve (money set aside as the expected costs of workers' compensation cases). The workers' compensation administrator or the risk-management department can be asked to provide injury data broken down by types of injury, body parts injured, dates and times of injuries, work days lost because of injury, and more. These data are collected for any injury incurring financial expenses and can greatly aid metric formation and trend analysis.

Job/Task Analysis

Some good lagging metrics have been discussed, but leading metrics need the same sort of examination. Task analysis is a tool used in ergonomic assessments and improvements (NIOSH 1997). For the purpose of this chapter, an analyst performing a task analysis normally looks at a *job* (a job title, a specific department, or an individual responsibility) in its totality before breaking it down into specific tasks to be performed. These tasks are then analyzed for MSD hazards, and ergonomic corrective actions are advised. Consider a machine operator who has five responsibilities, or tasks, making up his or her job:

1. Ensuring that the machine continues running
2. Clearing part jams
3. Quality checking a certain amount of parts
4. Loading raw material
5. Housekeeping of the general area

An ergonomic analysis should identify MSD hazards and aid in developing work techniques—perhaps even indicate where to install equipment—for minimizing or eliminating any hazards (which are often prioritized).

If task analysis is a corrective measure, how can it be a metric? To answer this question, leading metrics must be discussed. Leading metrics, though not absolute, are predictive. Because task analysis is an ergonomics' tool, the corrective actions that follow it can be used as metrics (Sanders and McCormick 1993).

In its simplest form, a corporation could structure an ergonomic metric as follows:

All facilities in the corporation will:

1. Identify 100 percent of the work tasks performed in the facility and develop task analysis for 25 percent of the tasks in the first year.
2. Complete 75 percent of the task analysis for the facility in the second year.
3. Develop 100 percent of the task analysis by the third year.
4. Complete all corrective actions within 60 days of being generated.

If this were the company's only ergonomic metric, it would be lacking. Ergonomics, like safety in general, does not have a *one-size-fits-all* answer. Task analysis is a tool for ergonomic improvement, but it is not itself an ergonomics program. Other ergonomic elements would also need to have metrics developed. If, however, a company were developing a metric system for safety in general and wanted a single metric as a barometer for ergonomics, this metric might be acceptable. Again, it depends on the goal. It is possible to have too many metrics to follow; what is to be measured must be decided. If a combination of leading and lagging metrics is desired for indication of the progress of a safety program as a whole, and the entire scheme is such that a single metric on ergonomics is desired, the above may be adequate. If the desire is to develop a metric system specific to the ergonomics program, additional metrics should be developed that will center on the actions taken to improve ergonomics.

For some companies, the metric above may be too rigid. For a corporation made up of small manufacturing sites, it might be appropriate, but a company with facilities employing from 20 to 2000 employees, and having manufacturing and service sectors in global locations, may have trouble making this specific metric work.

The basic principles, however, can be adapted for other cases. In the metric above, an objective and a timeframe are set forth. When establishing a metric, the hazards involved must be identified. Because this is true of any leading metric, it is a good place to start. A fair amount of time should be devoted to this identification process for two reasons. The first has to do with thoroughness. An understanding at the beginning of the project of all the hazards involved will allow for knowledgeable prioritization. Any additional hazards that are discovered will require the original plan to be adapted, decreasing efficiency. The second reason is one of logistics. The facilities that have to comply with the new metrics will most likely need resources to aid them in doing so. By providing time during the identification phase of the metric, the facilities can develop schemes for obtaining the resources they will need, whether consultants, training, or internal corporate expertise—and all of these methods require funding. Consultants charge fees, training classes have costs (for both the training and possibly travel), and charge-back costs may accompany corporate aid. By allowing extra time during the identification phase, facilities will be able to develop budgets in support of ergonomic activities.

One other issue that must be decided during the identification phase is how to define a *task*. There is no clear-cut answer to this question. The *job* must be evaluated with an eye to recognizing its basic elements as distinct units. Think about the task list of the machine operator given a moment ago. Are the tasks listed adequate for the purpose of analysis? Perhaps—but further depth would be needed. The following may, in fact, be true:

1. Ensure that the machine continues running: strictly observe the machine.
2. Clear part jams: power down the machine, initiate lockout/tagout, physically (expending moderate effort) remove unfinished parts.
3. Perform quality checks n a certain amount of parts: nonstrenuously remove and measure five parts hourly.
4. Load raw material: raise a mechanical drum hoist.
5. Housekeep the general area: sweep and remove light debris.

But what if the job were changed slightly? What if the operator must retrieve drums of parts (with a forklift) and place them onto the drum hoist? What if housekeeping included machine maintenance? What if maintaining the machine includes lubricating it? All of these items will expand the task list.

A counterpoint to the above consideration is that a hazard identified for one *job* may be applicable to other jobs, limiting the necessity of repeated work for analysts. If all workers in a facility who are required to use a forklift do so in a similar manner, one task analysis for forklift use applies to those employees. Such common tasks can be prioritized for analysis, creating benefits for all employees very efficiently.

Once the identification phase is completed, further metrics can be developed around the task analysis. Amounts to be completed in set times (as in the example above) need not be percentages but may be specific numbers (such as ten task analyses completed in one year). Notice, however, that all the examples have a timeframe. Yearly repetition is convenient but is certainly adjustable to the specific needs and logistic issues of a company. If a company feels that the ability and resources are present to conduct task analyses for *all* jobs simultaneously, the use of percentages would be appropriate.

The task-analysis metric can be taken further. Once a facility has 100 percent of its tasks analyzed, what next? A metric requiring the facility to develop and implement a task-analysis review for all new projects and all modifications to work areas can be used to ensure continuation of ergonomic analysis to all work areas. Most importantly, when general hazards, as well as their causes, have been identified, the metric process can be continued as the prevention process.

Reduction Efforts

The task analysis above is actually an extended identification method that aims to identify ergonomic hazards, after which various reduction efforts come into play. What reduction methods will be used is an issue best decided by the specific approach chosen to address the problem. Whatever methods are chosen, metrics can be based on them.

Task analysis can be tied to reduction efforts, and simple metrics developed. The example used above can be extended even without a comprehensive plan to address ergonomics. A modified example could read as follows:

1. By the conclusion of the first year, have identified 100 percent of the ergonomic hazards, prioritized them, and conducted task analysis for 10 percent of them.
2. By the conclusion of the second year, have conducted task analysis for 35 percent of the hazards and developed corrective actions based on all completed task analysis.
3. By the conclusion of the third year, have completed task analysis for 55 percent of the hazards and developed corrective actions based on all completed task analysis.

This example can be continued. At root, some reduction effort is required to address whatever hazards are found; but it is up to each facility to decide how to correct them.

What if the facility above has decided that work area redesign should be used to eliminate the ergonomic hazard? A simple method, as above, can be used, or other factors can start affecting the metric. What if a metric is desired that takes frequency into account? The company in the above example wishes to use its resources to address ergonomic issues efficiently. The problem with the example above is that it assumes that all hazards can be addressed and that all corrective actions will be effective. Neither of these may be the case. Laying out the metric as above does not recognize that no effective fixes may exist for certain issues. Furthermore, when a corrective action "makes employees aware," is it truly effective? Can one say that a corrective action was implemented when no other course of action seemed appropriate? What if the severity is tied to the metric? Then the following might be effective:

- Using the facility OSHA 300 logs, prioritize the jobs having the highest ergonomic incident rates.
- Develop corrective actions (in this case process redesign) for the three tasks having the highest incident rates. Monitor these task symptoms during the first year.
- During the second year, develop redesigns for the tasks having the next four highest rates of incidence. Monitor these task symptoms, comparing them to the original three tasks from the year before and noting any reduction or increase.
- During the third year, develop redesigns for the tasks having the next four highest rates of incidence. Monitor these task symptoms, comparing their symptom rates to those of the task redesigns completed during previous years and noting any reduction or increase in symptoms.

The metric above is attempting to be efficient (by concentrating on the areas of most frequent injury), progressive (performing additional corrective actions yearly), and relevant (monitoring symptoms to evaluate the accuracy of the leading metric's predictions). Although the above scheme may be appropriate only for a limited number of organizations, its base concepts can provide guidance in forming metrics in general. The metric above tries to focus the company's efforts. No organization has unlimited resources.

In the first example above, the company decides which tasks to prioritize for corrective actions. Nevertheless, influences outside the safety realm may take precedence. Workload can cause a person to implement a less time-intensive and less effective corrective action in order to meet metric timelines. A leading metric can be met in such a way that it has no effect on a lagging metric. So long as multiple pressures exist in the workplace, the formation of leading metrics must take into account the final outcome desired (decreasing the number of injuries).

The second (more complex) example takes this into account, instructing the facilities to quantifiably correct certain hazards first (by using the OSHA 300 logs to determine the most hazardous tasks). If necessary, a facility can be called upon to justify its choice of corrective actions. The other advantage of the second example is its monitoring of symptoms. Leading metrics should be predictive; ergonomic injuries are years in the making. When symptoms are monitored, determining whether corrective actions are improving a situation can be accomplished to some extent. If symptoms associated with MSDs are not decreasing in frequency, the corrective action can be analyzed and, if determined to be ineffective, reworked. This monitoring also serves another purpose. Monitoring can assure that hazards are being addressed, but it can also indicate a systematic problem. It is likely that some corrective actions will have to be reworked, but what about a facility that reworks the majority of its corrective actions? Might this facility not have difficulty in developing effective corrective actions? The metric described above could not indicate whether such problems were related in particular to skills, management support, implementation, or any number of other factors, but it would emphasize the inefficient use of resources, identifying this area for analysis in order to discover the cause. One of the purposes of metrics is to ensure that limited resources are used meaningfully.

Both of the examples above are concrete and quantifiable. But what about something less quantifiable, such as a voluntary stretching program? How can metrics be developed around something like this? It can be assumed that an analysis was conducted and that it was determined that such a program would be a benefit. An oversimplified metric could read as follows:

> A voluntary stretching program will be developed that will have 85 percent employee participation.

This metric is rather vague and somewhat contradictory. How can 85 percent participation be ensured for a voluntary activity? A more quantifiable metric might read as follows:

- A voluntary, preshift stretching program will be developed. Employees will be encouraged to attend the stretching sessions five minutes before their shifts start.
- Symptoms will be monitored throughout the facility and those of employees who stretch will be compared to those of employees who do not.
- If after two years, employee participation is not at 50 percent and no mitigation of the symptoms of employees participating in the stretching program is evident, the program will be reevaluated to decide whether to continue it.

This metric could be used even if other reduction efforts are being used by changing the last criteria to read:

- If after two years, employee participation is not at 50 percent and a greater mitigation of symptom reduction is not noted in employees participating in the stretching program, the program will be reevaluated to decide whether to continue it.

This metric has some advantages. By means of monitoring, the effectiveness of the metric is recognizable, and a criterion is built in to decide on its continuation (this is especially important, because this program requires ongoing resources during its continuance). If the program is ineffective, the resources being expended on its behalf can be reassigned.

Diagnostic Tools

So far, the discussion on metrics has centered on either direct effects (such as injuries or dollars spent) or corrective plans (such as hazard identification or corrective actions). Are there any additional quantifiable ways to measure progress in ergonomics? Yes—when delving into occupational medicine. Because ergonomics centers on MSDs, methods that measure stress on the musculoskeletal system may act as beneficial metrics. Some of the diagnostic tools used in occupational medicine may fill the bill for metrics. However, several issues exist when using these tools as metrics. All of the methods mentioned require some level of training for effective use. Some of the methods are complicated and require input from licensed medical personnel. The time, money, or infringement of privacy required by some methods makes them prohibitive for use in the general employee population. With these things understood, however, applications do exist for these methods as metrics.

Job descriptions can be effective tools when they include not just verbal descriptions of the jobs themselves but also of the physical requirements for performing this job in a satisfactory manner. Such job descriptions include requirements to "lift 75 pounds," "stand for approximately 6 hours per shift," or "repeatedly lift over the head boxes weighing up to 50 pounds." These descriptions are quantifiable, stating the physical exertion required. An occupational medicine professional can help with the formation of a facility's job descriptions; but, with training, anyone can develop job descriptions (OSHA 2002). These job descriptions help return injured employees to work and serve as aids in hazard identification (the job descriptions having the highest weight limits can be flagged as hazardous). A metric can be developed around job descriptions requiring that, say, within two years the facility will have job descriptions for 100 percent of its jobs. Job descriptions can be used even more aggressively. Per NIOSH, employees should not be lifting loads over 51 pounds (NIOSH 1994, 13). A metric can be constructed that reads as follows:

- In the first year, job descriptions will be developed for all jobs in a facility and any job requiring employees to lift more than 51 pounds will be flagged.
- In the second year, the facility will prioritize all jobs requiring the lifting of more than 51 pounds, developing corrective actions for reducing the lifted weight or lift frequency for 20 percent of these jobs.
- In the third year, the facility will have reduced the lifted weight or lift frequency for 50 percent of all jobs requiring employees to lift more than 51 pounds.

This metric is quantifiable, and a monitoring function can be developed to determine whether its corrective actions are effective.

Preemployment screening (such as the Matheson System) is another tool which may serve a metric function (Matheson Discussion Group). Preemployment screenings are an extension of job descriptions. After a job description is developed, an occupational medicine professional can develop a questionnaire of past medical history and a series of functional *tests* (evaluating the employee's ability to reach, stretch, and lift) based on the job requirements. These screenings are fairly basic, and past injuries and illnesses play a large factor in them. Metrics formed from preemployment screenings will be similar to those developed from job descriptions.

Functional capacity evaluation (FCE), an extension of preemployment screening, normally is performed in connection with workers' compensation (Matheson). It measures impairment. Employees are put through a series of lifting, carrying, and moving tests that mimic job tasks. The testing is performed in an array of positions to simulate the work environment and to isolate specific muscles and muscle groups. Limb and back position determine the muscles used to perform work. If an injured muscle is isolated, it may not be able to handle the workload of a task as if it were one of a number used in the task. The FCE then outlines the capacities of the injured employee for use in the return-to-work program (RTW) and in the assignment of impairment ratings (California Department of Industrial Relations). An impairment rating is used to determine how *disabled* an employee is and is often the basis of a cash award for an injured employee. FCEs are normally performed a limited number of times; multiple performance reflects the possibility of some change in the level of impairment. Specific muscle-group testing can be conducted in addition to FCEs, discovering the capabilities of specific muscles or groups of muscles (unlike FCEs, which look at a worker's total abilities).

How can these tests be incorporated into metrics? In the way they are classically conducted, they are not effective; but with some modification, they may be of value. One of the objectives of the formation of metrics is the building of quantifiable metrics. When using a task analysis, this may at times be subjective. The above tests enable one the quantification of information from a task analysis. The hazards found during the hazard analysis can be measured for the work required to be performed. By incorporating the data on force into the task analysis, the tasks can be quantified.

These tests can also serve as monitoring functions. The force required to do a job can be measured initially and then again, after a reduction effort has been implemented. By implementing a reduction measure, some of the force required to perform the task should be reduced. A good example of this can be seen in measuring the force required to use certain hand tools. In one study, the force required to use an *ergonomically* designed pair of pliers was greater than the force required to use an ordinary pair (UC Berkeley ME). The ergonomic value of such a tool is questionable. Furthermore, the *ergonomically* designed tool cost more than an ordinary tool, making the *ergonomic* choice an inefficient use of resources.

Diagnostic tools help in the quantification of metrics, giving the ability to quantify, measure, and prioritize the use of resources.

Administrative Functions

Engineering controls are the preferred methods of hazard control (NSC 1997, 124). As long as engineering controls are in place, hazards are all but eliminated. Sometimes, however, engineering controls are not possible and must be replaced by the second level of the hierarchy—administrative control. Administrative control mitigates hazards by controlling how tasks are done, making hazards more dependent on human behavior. Can effective metrics be developed for behavior-driven items? Yes, and three common administrative ergonomic controls can be given as examples: training in ergonomics (NIOSH 1997, 34), ergonomics committees (NIOSH 1997, 8) and task rotation (NIOSH 1997, 34).

This discussion focuses on the systems being developed and not on the behavior of individuals. Safe versus unsafe ergonomic behavior is an entirely separate discussion from this section's discussion of recognizing a system's efficiency and efficacy by means of metrics.

Ergonomics training can take various forms but normally includes recognition of MSD hazards as well as discussion of ways to eliminate these hazards (NIOSH 1997, 13). The training can be given to various groups—all employees, only safety and ergonomics committee members, or members of management only. Can metrics be formed around ergonomics training? Yes—if the purpose of training is kept in mind. All training is conducted to impart knowledge (developing skills or competencies) or change behavior (NIOSH 1999, 5). It is easy to set simple metrics for training. One metric might require that 100 percent of a facility's personnel will have received ergonomics training within a one-year period.

Is this a fair measure of effectiveness? It is a leading metric, but its predictive value is limited and years are needed to recognize whether the training has affected the occurrence of injuries. Although training may have also been provided, no system was established to monitor the implementation of new knowledge or skills. Training must tie into other efforts. Once the purpose of training is clear, it can be tied into the information or skill being taught. If training is being conducted for the purpose of task analysis, make the training requirement part of the task-analysis metric. If the training is part of a reduction effort, tie it into that; make ergonomic improvement suggestions. A metric could read, for example, as follows:

- The facility will train 100 percent of its associates in ergonomic principles and the ergonomic suggestion process during the course of one year, collecting, tracking, and analyzing all suggestions.
- By the second year, the facility will have addressed within 30 days each ergonomic suggestion received and will have responded with its findings. If the facility has not received at least 25 viable suggestions by the end of the second year, the program will be reevaluated for continuation.

A metric developed for training requires an understanding of what the training should accomplish. As above, monitors can be built directly into the metric.

How can metrics be formed for other administrative items, such as forming ergonomics committees? A number of options are available; the simplest is to set a time limit, as in the following metric:

- During the first three months of the year, the facility will select the committee's members.
- During the second three months, the facility will provide the committee with training in the principles of ergonomics.
- By the year's end, the committee will have begun to generate corrective actions.

This is a very simple example, and its effectiveness is questionable. Not only is no monitoring component included, but no means exists for discovering whether the committee's efforts are affecting the rate of ergonomic incidents. The formation of the committee is the ultimate measurable outcome, but nothing is noted about how the incident rate might be lowered. What actions will the committee take? What actions will it particularly emphasize? Ergonomic hazards must be identified, corrective actions implemented, and the effectiveness of corrective actions monitored. The committee can assist in the accomplishment of all these functions. The following example incorporates them all:

- During the first quarter of the year, members of the ergonomics committee will be chosen and its first meeting will be held. During this meeting, the mission of the committee will be discussed and the members will draft suggestions for the committee's method of functioning.
- During the second quarter, the committee members will be trained how to recognize ergonomic hazards and how to develop ergonomic corrective actions. The committee will then decide upon a mission statement and brainstorm appropriate ways for the committee to:
  - Track symptoms
  - Recognize hazards
  - Suggest corrective actions
  - Follow up on implemented corrective actions
- During the third quarter, the committee will have communicated to management their finalized mission statement and finalized recommendations on the actions listed above. Management will respond with approval or alternative suggestions within the stated time frame.
- During the last quarter of the year, the committee will have met and will be operating according to its mission statement and stated procedures.

This metric for forming the committee has more structure than the previous example, but it still needs to measure and monitor effectiveness. This can be fixed by adding the following:

- During the third quarter, the committee members will determine goals for themselves for the next years regarding all of the committee elements, also developing a monitoring system for symptoms that will be regularly updated. For every corrective action implemented, the committee will provide a write-up describing the former procedure or work area setup, what was changed, and what the effects of implementation were. Because of the close correlation of ergonomics with efficiency, these effects may include piece rate, quality levels, comfort rating, or other such things.

This provides the committee with an opportunity to develop its own metric while giving the company an avenue by which to monitor the committee's improvements to the overall ergonomic processes (by monitoring symptoms and related items).

The final administrative function to look at is task rotation. Task rotation decreases the duration of exposure to hazards, increasing the time away (*healing time*). Employees are rotated from tasks having ergonomic hazards to those having no (or less, or different) ergonomic hazards. This procedure, similar to engineering principles, is highly structured. The confounding factors are task complexities and the skill sets required. In facilities having very structured job descriptions, task rotation may be a viable solution. In facilities having high levels of job differentiation—especially in tasks that rely on employee experience and knowledge—task rotation may not be a good solution to ergonomic issues. Thus, one of the first issues in metric formation is deciding if task rotation is an ergonomic answer for a specific facility. Another factor that complicates task-rotation metric formation is training. How can employees rotate to jobs they are not trained in? Often training will be part of an initial metric. An example of a task-rotation metric might read as follows:

- During the first six months of the year, the facility will conduct a hazards analysis for ergonomic hazards and identify tasks posing a high number of risk factors.
- During the second six months of the year, the facility will identify jobs appropriate for job rotation, cross-train employees to be able to perform such tasks, and create a rotation schedule.
- The second-year job rotation will be monitored, each department reporting the percentage of compliance with the rotation schedule.
- After the third-year job rotation, the program will be reevaluated and a recommendation made about its continuation.

As in other cases, monitoring symptoms provides enough data to evaluate the effectiveness of such a program. The metric above demonstrates that even program elements of more subjective natures can serve as the bases for effective metrics.

Audit Monitoring

The audit can use a standardized corporate audit format or even be a simple *walk-through* audit.

If a company already has a structured, defined audit protocol, ergonomic elements can be incorporated into it, from hazard identification to corrective measures—the gamut of ergonomic processes. If the company sets goals and metrics around audit scores for a facility, the ergonomic elements could be incorporated in that fashion. A company could develop an ergonomic audit protocol and then set metrics for scores on such an audit.

Even walk-throughs can have metrics attached to them. Metrics can deal with everything from repeat findings (requiring that a department will not have a repeat finding within a one-year period) to corrective actions (requiring that any ergonomic hazards noted must have a corrective action plan developed within 30 days). As a method for monitoring the safety process, auditing is an easy basis for metrics, including ergonomic metrics.

Conclusion

Many analytical approaches can be used to evaluate ergonomic hazards. These analytical approaches can (and should) be treated as any other tool. Once the reasoning behind goal and metric formulation is understood, safety professionals can decide what to measure and decide what criteria will be established. Ergonomics can be measured in the same way as any other process. Although the examples used here will likely not work for most facilities, the principles they exhibit can be the basis for a facility's, company's, or corporation's development of metrics that will work it. When the program and leading metrics are harmonized, the outcome will focus most on lagging metrics, or fewer injuries—the ultimate goal.

Summary

Forming metrics for ergonomics programs is a worthwhile endeavor. It allows safety professionals to evaluate the relevance of various components of such a program, assessing whether resources are being applied efficiently and outlining both immediate and (likely) future benefits. By following the general outlines offered in this chapter (that is, by measuring an essential element of a program's success over time), safety professionals can benefit from the formation of metrics and quantifiably assess their programs.

Endnote

[1]Prior to the mandatory OSHA record-keeping standard, voluntary reporting was conducted by organizations such as the National Safety Council. The membership of the voluntary reporting organizations did not include industrywide representation.

References

Bureau of Labor Statistics (BLS). 2006. *Injuries, Illnesses and Fatalities* (retrieved February 6, 2008). www.bls.gov/iif

California Department of Industrial Relations. *DWC Glossary of Workers' Compensation Terms for Injured Workers* (retrieved April 15, 2007). www.dir.ca.gov/dwc/WCGlossary.htm

Heath and Safety Executive (HSE). 1999. *Health and Safety Benchmarking: Improving Together*. Sudbury, Suffolk, UK: HSE.

Heinrich, W. H. 1931. *Industrial Accident Prevention*. New York; London: McGraw Hill Book Company.

Matheson Discussion Group, The. Undated. *The Matheson System: The Functional Evaluation System* (retrieved April 15, 2007). www.roymatheson.com/index.html

Nash, James. 2001. "Recordkeeping: OSHA Tries to Make It Simple." *Occupational Hazards Magazine* (October) 62(10).

National Institute for Occupational Safety and Health (NIOSH). 1994. *Application Manual for the Revised NIOSH Lifting Equation* (Publication 94-110). Cincinnati.: NIOSH

_______. 1997. *Elements of Ergonomic Programs* (Publication 97-117). Cincinnati: NIOSH.

_______. 1999. *A Model for Research on Training Effectiveness* (Publication 99-142). Cincinnati: NIOSH.

National Safety Council (NSC). 1997a. *Accident Prevention Manual for Business and Industry: Administrative & Programs*. 11th ed. Itasca, IL: NSC.

_______. 1997b. *Accident Prevention Manual for Business and Industry: Engineering & Technology*. 11th ed. Itasca, IL: NSC.

_______. 1997c. *Supervisors' Safety Manual*. 9th ed. Itasca, IL: NSC.

NHS Direct Online Health Encyclopedia. *Repetitive Strain Injury* (retrieved December 2006). www.nhsdirect.nhs.uk/en.asp?TopicID=389

Occupational Health and Safety Administration (OSHA). 1971. 29 CFR 1904, *Reporting of Fatality or Multiple Hospitalization Incidents*. Washington, DC: OSHA.

_______. 2001 (January 1). *OSHA Revises Recordkeeping Regulations*. Washington, DC: OSHA.

_______. 2002. *Job Hazard Analysis*. Washington, DC: OSHA.

_______. 2005. *OSHA Forms for Recording Work-Related Injuries and Illness*. Washington, DC: OSHA.

_______. 2006a. OSHA Compliance Directive 06-01 (CPL 02), *Site-Specific Targeting 2006* (SST-06). Washington, DC: OSHA.

_______. 2006b. *Standard Interpretations: Recording an injury when physician recommends restriction but no restricted work is available* (retrieved April 30, 2006). www.osha.gov/pls/oshaweb/owadisp.show_document?p_table=INTERPRETATIONS&p_id=25435

Oregon OSHA (OR-OSHA). 2005 (November). *Easy Ergonomics: A Practical Approach to Improving the Workplace*. Salem, OR: OR-OSHA.

_______. Undated. *Online Training Module Chapter 1—Management Commitment*. www.cbs.state.or.us/external/osha/educate/training/pages/100xml.html

Sanders, M. S., and E. J. McCormick. 1993. *Human Factors in Engineering and Design*. 7th ed. New York: McGraw Hill.

Stewart, J. M. 2002. *Managing for World Class Safety*. New York: John Wiley & Sons.

Streiner, David L. 2003. "Unicorns Do Exist: A Tutorial on "Proving" the Null Hypothesis." *Canadian Journal of Psychiatry* (December), pp. 756–761.

University of California at Berkeley, Mechanical Engineering Department. Undated. *Manual Crimping*. www.me/Berkeley.edu/ergo/casestudies/crimper.html

Webster's Ninth New Collegiate Dictionary. 1983. Springfield, MA: Merriam-Webster, Inc.

Wikipedia. 2007a. *Trend Analysis* (retrieved April 15, 2007). www.en.wikipedia.org/wiki/Trend_analysis

_______. 2007b. *Statistics* (retrieved April 15, 2007). www.en.wikipedia.org/wiki/Statistics

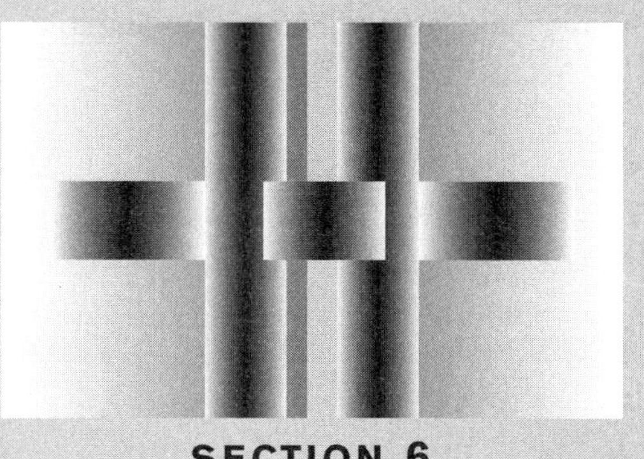

SECTION 6

ERGONOMICS AND HUMAN FACTORS ENGINEERING

LEARNING OBJECTIVES

- Understand the basic terminology of ergonomic hazards and repetitive strains (EHRS) best practices.
- Be able to apply a systematic approach to identify EHRS.
- Recognize which ergonomic tools to use for eliminating or reducing EHRS.
- Know which ergonomic tools to employ in the design of products and processes.

BEST PRACTICES

Farhad Booeshaghi

ACCORDING TO the Liberty Mutual Research Institute for Safety, disabling injuries due to repetitive motion were the sixth leading cause of on-the-job injuries in 2002, costing employers an estimated $2.83 billion (Liberty Mutual 2003). In this chapter, an attempt is made to study and outline the fundamental concepts of EHRS as they relate to the practices of various industries in eliminating or reducing repetitive motion injuries.

In the mid-1800s, Polish biologist Wojciech Jastrzebowski coined the term *ergonomics*, which means the study or science of work. Since World War II, many pioneers in the ergonomics field have refined this study and have referred to it as "fitting the task to the person." In fitting the task to the person, there are two types of workplace injuries to consider: acute injuries, resulting from an accident; and cumulative injuries, which develop over time. The science of ergonomics focuses mainly on cumulative injuries (Jastrzebowski 1857).

ERGONOMIC HAZARDS AND REPETITIVE STRAINS

A common workplace hazard is *repetitive strain*. Caused by repetitive motion, it is defined as "performing a task that requires the application or experience of force at a consistent frequency while holding a posture over time." Repetitive motions may become repetitive strains if the worker is exposed to undue physical stress, strain, or overexertion. These conditions include vibration, noise, heat, awkward postures, forceful exertions, contact stresses, and heavy manual material handling. Repetitive strains result in cumulative injuries known as *musculoskeletal disorders* (MSDs).

MSDs are injuries to the soft tissues—muscles, ligaments, tendons, joints, and cartilages—and the nervous system. Carpal tunnel syndrome, thoracic outlet syndrome, and tendonitis are examples of MSDs (Humantech 1996).

Conditions that may lead to repetitive strains are:

- vibration
- noise
- heat
- awkward postures
- forceful exertions
- heavy manual material handling

In the 1970s these disorders began to appear in increasing numbers on companies' injury and illness logs. OSHA cited, and continues to cite, companies for hazardous workplace conditions that cause problems such as tendonitis, carpal tunnel syndrome, and back injuries. The Bureau of Labor Statistics, an agency of the U.S. Department of Labor, recognizes MSDs as a serious workplace health hazard. In 2001 these injuries accounted for more than one-third of all lost-workday cases (BLS 2003).

Conditions that expose workers to repetitive strains are known as ergonomic hazards. Today's science of ergonomics looks at the cognitive or decision-making performance of humans as it relates to their daily interaction with their work environment, as well as the physical limitations of the interface, and attempts to eliminate or reduce ergonomic hazards. In best practices, engineers rely on the science of human-factors engineering and biomechanical engineering to identify, evaluate, and reduce or eliminate many workplace ergonomic hazards. Human factors that affect the mental, physical, and social behavior of workers and form their individual characteristics are functions of a plethora of variables including but not limited to age, gender, size, strength, habits, literacy, experience, training, and physical and mental conditioning. A worker's perception-reaction, decision making, physical performance, and visual and auditory responses are examples of such human factors and are directly related to the potential of the worker being exposed to EHRS while performing his or her daily job. Elements of a job to consider include:

Product
- usage
- usability
- controls

Process
- operation
- installation
- janitorial
- service and maintenance

User
- age
- experience
- gender
- stature
- language/culture

Human Factors Engineering

Although there are many opinions, it is for the aforementioned reason that many practicing engineers combine the term *ergonomics* with the term *human factors engineering* and refer to the combination as the *science of ergonomics and human factors engineering*. Biomechanical engineers study the kinetics (forces) and kinematics (associated path of motion) a person's body experiences in order to eliminate or reduce ergonomic hazards. Biomechanical engineering relies on bone mechanics, tissue engineering, and human anatomy and physiology to evaluate the exposure of workers to ergonomic hazards and repetitive strains. Issues associated to these hazards are ergonomic *risk factors*. Over time, ergonomic risk factors affect the human body and can cause permanent injuries. The following list outlines some common workplace ergonomic risk factors (Humantech 1996).

Hand and fingers
- pinch grip > 2 pounds
- power grip > 10 pounds
- finger press

Wrist
- radial deviation
- ulnar deviation
- flexion > 45 degrees
- extension > 45 degrees

Elbow
- forearm rotation
- natural position 15 degrees outward
- inward rotation
- outward rotation
- full extension ≥ 135 degrees

Shoulder
- raised > 45 degrees
- elbow behind body

Neck
- forward rotation > 20 degrees
- sideways rotation
- backward rotation
- twist

Back
- forward rotation > 20 degrees
- sideway rotations
- backward rotation
- twist

Leg
- straight stand
- one-leg stand
- squat
- kneel

Force
- magnitude
- direction
- surface area
- stress

Frequency (whether discrete or continuous)
- motion
- noise
- vibration
- duration

Repetitive strains begin with discomfort, evolve into pain, and eventually become a cumulative set of disorders. Occupational safety and health professionals have called these disorders by a variety of names, including cumulative trauma disorders, cumulative trauma injuries, repeated trauma, repetitive stress disorders, repetitive stress injuries, and occupational overexertion syndrome. The injuries are painful and often disabling, and they develop gradually—over weeks, months, and years (Karwowski and Marras 1997; Adler, Goldoftas, and Levine 1997; Yassi 2000).

EHRS are routinely observed when humans and machine interface. Hand tools, workstations (seated or standing), manual material handling, platforms, ladders and stairs, controls and displays, the visual work environment, and machine clearances and maintenance are examples of human-machine interfaces (BLS 2004).

EHRS are also observed in connection with these types of industries: construction (industrial/residential), medical and healthcare, nursing home, assisted care, child care, food service (restaurants/grocery stores), food processing and packaging, textile and fabric, mass storage and retrieval, manufacturing, retail and wholesale, transportation, postal service, service and repair, automotive and moving vehicle, household item, heating and air-conditioning, and cosmetology. They affect occupations such as custodians (janitors or cleaners), stock handlers and baggers, cashiers, carpenters, painters, electricians, roofers, cement masons, sheet metal workers, pavers, mail handlers, registered nurses, nurse's aides, orderlies and attendants, truck drivers, laborers, assemblers, typists, personal and office assistants, court reporters, barbers and hair stylists, teachers and lecturers, meat packers, ship builders, and more (Ernst, Koningsveld, and Van Der Molen 1997; Sitzman and Bloswick 2002; Melhorn 1998). The guidelines listed in this chapter may be applied to any industry or occupation that is not listed above and has similar occupational characteristics.

EHRS–Best Practices

Best practices for industries to reduce or eliminate EHRS rely on the application of scientific principles, methods, and data drawn from various disciplines, including biomechanics, industrial engineering, mechanical engineering, machine design, applied physical anthropometry, cognitive sciences, psychology, and physiology. The objective of best practices is to identify ergonomic risk factors associated with an industry, occupation, or task; to develop a best-practice ergonomic approach to eliminate or reduce those risk factors; and to outline specifications and criteria for the design of equipment, products, systems, and processes that are free from those factors.

Today, several industries use established control tools to eliminate or reduce EHRS effects on the human body. Examples of control tools include engineering controls that incorporate design, redesign, job-hazard analysis, work-site evaluation, human reliability analysis, strain index for reliability, and work-practices controls to manage ergonomic risk factors; administrative controls that include MSD management programs; training and education; risk-based programs; and personal protective equipment to guard against hazards.

The objective of this chapter is to outline some of the best practices industries use to eliminate or reduce the effects of EHRS on the human body. These guidelines are intended to provide the reader with the tools necessary to ensure that EHRS are addressed through each step in the product/process evaluation/design process. Since the fundamentals of EHRS best practices are the same among all industries, to accomplish this objective we begin by presenting the fundamentals of EHRS best practices followed by their applications in selected industries and occupations.

Fundamentals of Best Practices

As a part of the elimination or reduction of EHRS in the workplace we begin by trying to understand human-machine interface requirements. When a product or a process is being designed or modified, the first step in eliminating EHRS risks is identifying the risk factors associated with its worker interface. Examples of products or processes with human-machine interfaces include workstations (seated, standing, or seated/standing), hand tools; controls and displays; manual material handling; machine clearances and maintenance; and platforms, ladders, and steps. The next step is to design or redesign the product or the process to eliminate or reduce EHRS. To accomplish this, human-machine task allocation (HMTA) must be studied.

HMTA focuses on human strengths, such as decision making, thought processing, and teamwork. It also takes into account the machine's characteristics—such as repetition, force, and operation—and the work environment, which includes thermal, auditory, visual, chemical, and physical effects. To design or modify the workers' interface with the process or product, it is also important to consider the workers' gender, body size, age, training for proper tool use, and posture as well as the direction of tool and trigger travel (e.g., pull versus push).

Hand Tools

The need for greater productivity and less-expensive products and processes has mandated a more efficient workplace, resulting in safer hand tools. In the last decade, in order to reduce the problems associated with EHRS and simultaneously increase tool efficiency, improved hand tools have been designed and developed. These tools are often called *ergonomic tools*. Ergonomic hand tools such as wrenches, ratchets and sockets, screwdrivers, hammers, pliers, scissors, plumbing tools, punches, chisels, pry bars, gardening tools, cutters, files, and knives are now available. Most industries have adopted the concept of choosing the right tool for the job and using it correctly to avoid EHRS and potentially permanent and serious injuries.

It is impractical to identify one proper hand tool for every job or for every user. For example, a hammer with a large-diameter handle may decrease grip strength for a worker with a large hand, whereas it is unusable for a worker with a small hand. Each person has his or her own optimum grip diameter. The average person's optimum grip diameter, for a cylindrically shaped handle, is approximately 1.5 inches (Radwin and Haney 1996). The fit of a tool's handle, in terms of comfort and ease of use, is particularly important for tools used in long-term occupational activities such as a carpenter's hammer or a drill. A fitted tool handle reduces the magnitude of concentrated forces experienced by the worker's hand during the tool's normal usage. It also reduces the amount of gripping force needed to cradle and oppose the reactive forces generated when the tool is used.

OSHA Guidelines

According to the Occupational Safety and Health Administration (OSHA), the following pointers should be kept in mind when selecting tools for use. These pointers apply to frequently used tools for kitchen work, gardening, housekeeping, laundry, and other maintenance areas, as well as those used daily at an operator's job, such as carpentry or plumbing tools (OSHA 3080 2002):

- fits the job you are doing
- fits the work space available
- reduces the force you need to apply
- fits your hand
- can be used in a comfortable work position

Other guidelines for tool usage include using bent-handled tools to avoid wrist injury, using tools of an appropriate weight, selecting tools that have minimal vibration or vibration-damping devices, maintaining tools regularly, and wearing appropriate personal protective equipment.

Hand-Tool Design and Selection Guidelines

Variations in hand-tool type, usage, and performance, coupled with variations in human anthropometry, make hand-tool design and selection challenging. Hence, understanding the general ergonomic risk factors associated with hand tools and adapting them to specific tools become essential parts of the hand-tool design or selection process (Kroemer, et al. 2002). In general, the ergonomic risk factors for hand tools include

1. awkward or forceful gripping of a tool
2. awkward body posture
3. holding/supporting a tool's weight
4. repetitive motion
5. vibration transmitted to a person's body, including the hand and wrist
6. high noise-level exposure
7. poor lighting conditions
8. heat/cold exposure
9. the operator's characteristics, including gender, size, knowledge, skill, and training

To create a hand tool with ergonomic features, a designer must focus on eliminating or reducing the aforementioned risk factors. For example, tools should be designed, modified, or used in a manner that allows the forearm to be in a near-neutral position, fifteen degrees from pronation (the angle the forearm makes with the horizon when the palm is facing down). Heavy tools should be suspended from above so the bulk of the weight is not supported by the worker. Tool handles should extend the full length of the palm and be soft, shock-resistant, and large enough to grip easily. Trigger-activated tools should be modified to allow for multi-finger operation, which prevents the required activation force from being applied by only one finger (Humantech 1996).

Several common materials can be used to modify tools, especially the tool handle area, to improve hand grip and reduce vibration. Some hand tools are also designed for ease of one-handed use.

Awkward or Forceful Gripping

The ability to apply grip strength to a tool depends upon the tool-handle type, size, and length. In general, tools with longer handles provide more contact surface and require less strength, allowing the user to generate more leverage by applying a smaller force at a greater distance. A thick tool handle provides a greater surface for grasping and less stress on the hand. The stress is directly related to the amount of concentrated force applied over a small area on the hand. As either the applied concentrated force increases or the area where the force is applied decreases, the stress experienced by the tissues, muscles, tendons, ligaments, and bones in the hand increases. Repetitive exposure to high stresses may result in the loss of mechanical load-bearing capabilities for the human elements. For example, if a man routinely uses a pair of pliers with a 2.5-inch handle to remove rusted nails from boards, and the span of his hand is 3.5 inches, the bottom of the tool's handle section will exert a repetitive concentrated point force on the center of his palm. Over time, the stresses experienced may result in the loss of physical and mechanical characteristics of his hand. The main idea of redesigning the tool is to avoid concentrated pressure on small parts of the fingers or on the palm of the hand. For instance, plumbers add extensions to their pipe wrenches when loosening rusted pipes and use screwdrivers with thicker handles to generate higher torque, thereby reducing the overall required force. However, bigger and thicker handles do not always indicate an ergonomically safe tool. The user's ability, the tool's frequency of use and its application, and the allowable working space also affect the sizing of the handle. Changing the handle can reduce the force or grip strength necessary to use a tool. Even though larger and thicker tool handles facilitate application of grip force and reduce stresses on the palm, consideration should be given to the upper limit of the grip-handle design. An overextended tool handle may in itself become a problem and create new hazards and discomforts, such as catching on surrounding items or being too big for the available work space. Also, fine work may not be performed as efficiently with larger tools.

An optimum cylindrical gripping diameter is approximately 1.5 inches. The "OK" method is one way to determine the optimum grip diameter for an individual. In this method, the perimeter made by the thumb and index finger when making an "OK" sign indicates the optimum gripping diameter (Radwin and Haney 1996).

The best ergonomic designs for tools, based on the repetitive motions involved in their use, are those that avoid form-fitting contoured handles that fit only one hand size (Smith 2004). Today's ergonomically designed tools are made with slightly wider handles, which distribute the grip force over a large surface and thus decrease the necessary grip strength and resultant contact stress. Ergonomically designed tools that require opening and closing permit use by people wearing gloves, people with both small and large hands, and people who are either left- or right-handed. Hand tools with cushion grips provide comfort, slip resistance, and reduced grip force. A handle flange or handle taper is also used to reduce required grip strength.

Both variation in human anatomy and the way a tool will be used necessitate customization for a person and/or a job. A variety of materials can be used to customize or modify the handles of most hand tools to enhance their ability to be gripped. Commercial materials include Monoprene® thermoplastic elastomers, thermoset rubber, thermoplastic urethanes, Magic Wrap®; Plasti Dip®; GripStrip®; tool wraps, and pipe insulation. Several of these materials are readily available in most hardware stores. Gloves with slip-resistant material on the palm and fingers are also available. In addition to making tool handles thicker, materials such as thermoset rubber or epoxy putty can be applied to create custom molded finger grips. A wraparound handle allows the tool to stay on the hand with minimal effort, and a handle guard may be added to certain tools to prevent the hand from slipping forward onto the blade.

HANDLE TYPES AND SIZES

Hand tools that require two handles, such as pliers, generally have a 4.5-inch minimum handle length and a 3.5-inch maximum handle span. These tools are typically spring loaded to allow for easy separation of the handles. Spring-loaded handles eliminate the thumb force required to open the blades for sequential cuts and minimize hand fatigue (Kroemer, et al. 2002).

Hand tools with one handle, such as drills, are equipped with either an in-line handle or a pistol-grip handle. In tools with in-line handles, such as screwdrivers, the tool action is in line with the grip handle. In tools with pistol-grip handles, the tool action is at an angle to the handle, typically 100 degrees. In-line handles should be designed with a surface material such as rubber or closed-cell foam that has a high coefficient of friction, is compressible, and absorbs energy or vibration. However, materials that absorb moisture should not be used in a handle as they attract bacteria, can make the tool heavier, and can cause slippage by absorbing oils (Humantech 1996).

Pistol-grip handles provide greater torque while requiring less grip force than in-line handles. These handles should have a flange to prevent the fingers from slipping off, but the tool handle should not have finger grooves, since they cannot be designed to fit users' hands universally. Grooves also separate the fingers, causing localized pressure and stress concentration in the knuckles, joints, and finger ligaments, and they decrease grip force. Most pistol-grip tool handles are 1.5 inches to 2.5 inches wide (front to back) and the angle between the handle and the tool bit is around 100 degrees (Putz-Anderson 1998).

A majority of triggers on pistol-grip tools are at least 1.5 inches long to allow for two-fingered triggering using the index and middle fingers. Most three- or four-finger triggers are limited to use on suspended tooling systems. A good number of tools require four pounds or less static force to activate their triggers. Trigger strips or pressure-activated triggers are used on balanced or suspended tools. Thumb triggers are used in repetitive operations since our thumbs have greater capacity for limited action than our index or middle fingers, which are more prone to overexertion and injury. Lever triggers, which are activated by the palm or the inside of the thumb, are a minimum of 4 inches long and require four pounds or less to activate. They also have a safety catch to prevent unintentional activation. Pistol-grip power tools with a torque output exceeding 24 inch-pounds are equipped with a torque limiter to eliminate reactive forces.

Circular handles, averaging 1.5-inches in diameter, are used for power-grip applications. For tools used in precision operations, handles average 0.3 to 0.6 inches in diameter. In-line tool handles have oval grips with cross-sectional measurements of 1.25 to 1.75 inches. Straight or in-line power tools with a torque exceeding 14 inch-pounds are equipped with a torque limiter to reduce reactive forces. Slip-clutch, torque shut-off, hydraulic pulse, and torque reaction bars on balancers are examples of torque-limiting devices. Table 1 illustrates the sizing requirements for various kinds of hand tools (Helander 1995).

TABLE 1

Handle Types and Sizing Requirements

Handle Type	Required Handle Length	Required Handle Diameter (Span)
Two handles (pliers, etc.)	4.5 in.–5.5 in.	0.5 in.–1.0 in. (2.5 in.–3.5 in.)
One handle in-line	4.0 in.–5.0 in.	1.25 in.–1.75 in.
One handle pistol-grip	5.5 in.–6.5 in.	1.5 in.–2.5 in.
Circular handle power grip	4.0 in. minimum	1.5 in.
Circular handle precision	2.5 in. minimum	0.3 in.–0.6 in.

(*Source:* Humantech, 1996)

TOOL MAINTENANCE

Regularly scheduled maintenance of tools, including lubrication and replacing parts, keeps them at optimum performance and greatly reduces the forces exerted on the operator during use. For example, a worn drill bit requires more force to use than a sharp one; rusted pliers require more force to open and close than those with lubrication; and a broken saw blade requires a tighter grip to prevent the saw from kickback than a complete one. Saw blades that are Teflon coated or coated with other nonstick materials improve tool efficiency and reduce the forces the worker must apply. Sharp tools, maintained according to the manufacturer's specifications, also reduce the forces the worker must apply and thus reduce the possibility of injury.

Awkward Body Postures

Completing the task at hand, even given the proper tool for the job, sometimes requires an awkward body position. Posture greatly affects the forces experienced by body parts during a tool's usage. Affected areas include the hands, wrists, elbows, shoulders, neck, back, hips, knees, ankles, and feet. Whenever possible, tools should be designed so that the user can occupy a more neutral posture. Tools can also be fabricated so that the user is forced to employ a less injurious position during their use.

Posture risk factors associated specifically with the hands and wrists include pinch grip, finger press, radial deviation, ulnar deviation, flexion, and extension. Poor wrist positioning can also diminish grip strength. For instance, a study has shown that grip strength is decreased by 27 percent when a wrist is held in flexion, 23 percent in extension, 17 percent in radial deviation, and 14 percent in ulnar deviation (Terrel and Purswell 1976). Poor wrist positioning can also lead to repetitive strain injuries. It is best to use hand tools that minimize flexion, extension, and deviation. Several hammers and pliers are designed with a bent or curved handle to maintain a more neutral wrist position, and some tools, such as gardening tools or paintbrushes, can be modified with an add-on pistol grip that allows for a more neutral wrist position.

Arm extension and forearm rotation are risk factors that affect the elbows. Elbows should be kept in front of and as close to the body as possible during a hand tool's usage in order to eliminate risk factors affecting the shoulders. For instance, ergonomically designed computer keyboards are set at 15 degrees to eliminate forearm rotation. In addition, the posture required when using a hand tool should minimize or eliminate bending or rotating the neck forward, backward, or sideways. Hand-tool operations that require the operator to squat, stand on one leg, or kneel should be modified to eliminate the forces generated on the hips, knees, ankles, and feet.

In studies of workplace EHRS, *best work zone* and *preferred work zone* are commonly identified for operators performing a given task. These zones help to eliminate ergonomic risk factors and are defined in the following ways: the *best work zone* for an operator using a tool is between shoulder width, in front of the body, above the hip, below shoulder height, with the elbow next to the body; the *preferred work zone* is the region created by the rotation of the arms about the shoulder, in front of the body, and below shoulder height (Pulat 1997).

Obviously, it is best to stay within the best work zone or the preferred work zone when using a hand tool; however, this may not always be possible. Hence, pistol-grip and angled-handle tools should be used on vertical surfaces that are at elbow height and on horizontal surfaces that are below waist height. Straight or in-line tools should be used on horizontal surfaces that are elbow height and on vertical surfaces that are below waist height. For example, knives with angled and pistol-grip handles are designed for cuts made with a downward stroke. These knives may not be widely used; however, they keep the wrist in a neutral position while allowing for sufficient downward force to make a smooth cut.

Holding or Supporting a Tool's Weight

Hand tools that require substantial strength to support are enhanced by external means. Heavy tools are equipped with two handles and are suspended or counterbalanced. Even if a user can comfortably support the tool's weight for a short period of time, over a longer period, the weight will statically load the user's muscles and diminish ability to support the tool. To compensate for the muscle's diminishing ability to carry the tool's weight, forces are transmitted to the joints, tissues, and ligaments, which exposes the body to ergonomic risk factors that result in over-straining. Tool enhancements such as spring balancers, retractors, line reel balancers, Ergo-Arms® (articulated joints used in manufacturing lines to support the weight of a tool), and air hose support systems are used to eliminate or reduce static muscle loading associated with tool weight.

Repetitive Motion

Highly repetitive tool use, which includes short-cycle motions that repeat continually over a long period of time, may eventually result in EHRS injuries. Often, elimination or reduction of the repetitive motions necessary to complete a task requires a redesign of the entire process (Smith 2004). For example, if a repetitive assembly task has sufficient clearance, changing to tools with a ratcheting mechanism or gears can help reduce repetition. Maintaining hand tools properly (e.g., keeping saw blades and drill bits sharpened) and using proper operating methods (e.g., making pilot holes for drilling) can also reduce the required grip force and therefore reduce repetition. If the work environment allows for it, changing to a power tool might also reduce repetitive motions. If possible, switching to hand tools that have adjustable spring-loaded returns, such as pliers and scissors that open automatically, can reduce repetition. And finally, some innovative hand tools can also reduce repetitive motions because of their design. For example, the blade of the Stanley SharpTooth Tool Box Saw® which reportedly cuts 50 percent faster than a standard hand saw due to a unique tooth design that cuts in both directions (DoItYourself.com 2007).

If it is not possible to reduce the repetitive motions necessary to use a hand tool due to the nature of the job/task, it may be beneficial to plan or redesign the task itself. Sometimes a few hours of employee brainstorming and problem solving may not only increase morale, but may result in solutions that save time and costly medical bills later.

Vibration

Vibration in tools is generally associated with power hand tools that use sources of power such as air, electricity, and gas. Typically power hand tools are used when greater applied force is required, repetitive tasks are being performed, or time can be saved. With the advantages of power tools also may come some disadvantages, including vibration, repetitive strains such as trigger finger, and increased operator demands and requirements to handle and react to the forces generated by the power tool.

Among these disadvantages, vibration may be the greatest concern. Exposure to large amounts of vibration in a localized area, such as the hand, over a prolonged period of time, can increase the risk of chronic disorders of the muscles, nerves, and tendons. Some studies have shown vibration to cause temporary sensory impairments (Streeter 1970, Radwin et al. 1990).

Although vibration is sometimes a desired effect, as with sanders and grinders, most often it is an undesirable by-product of power tool use. Existing hand-tool guidelines tend to focus on areas where the effects on humans are measurable—for example, the amount of vibration transmitted (ISO 5349-1986, Acoustical Society of America ANSI S.270-2006). The amount of vibration transmitted by a power tool can be influenced by a tool's weight, design, and attachments (such as a power line). Proper maintenance of power tools is a top priority in order to prevent added vibration due to a failing bearing or worn, out-of-balance parts. Tools that produce low-frequency/high-amplitude vibration, i.e., 8–1000 Hz, are considered high risk.

Power tools designed with anti-vibration materials or anti-vibration mounts or handles have had limited success in reducing the amount of transmitted vibration. If the job or work environment requires the operator to use power tools for prolonged periods, it is best to consider redesigning the process, redistributing the work, or using some kind of external support for the power tool. Gloves made with materials that dampen vibrations transmitted to the hands, wrists, and arms may be worn, but their effectiveness may vary.

High Noise-Level Exposure

Excessive noise in the workplace results in operator exposure to ergonomic risk factors. Hearing loss, communication interference, annoying distractions, tuning out, and decrease in productivity are examples of such risk factors. In addition to the ergonomic challenges they present, these risk factors may lead to a decrease in productivity. The intensity, frequency, and duration of noise determine the level of risk. Exposure to a high noise level affects people gradually, beginning with reduction in quality and clarity of high-frequency sounds. Noise-level exposure in excess of 90 dBA across an 8-hour period is considered an ergonomic hazard. Where noise levels are excessive, administrative or engineering control tools should be integrated into the work process to minimize the noise associated with ergonomic risk factors. Requiring operators to wear hearing protection, shortening operators' exposure time to high noise levels, and educating and training the operators are examples of administrative controls. Using mufflers to reduce the noise from the release of air or exhaust from tools, isolating sound, and automating high noise-level tasks are examples of engineering controls. Eliminating the source of noise is far more beneficial than guarding against it.

Examples of ways to eliminate noise include

- replacing old machinery with newer more robust equipment
- lubricating moving machinery parts
- fitting the pneumatic tool's intake and exhaust system with mufflers
- modifying the material-handling process to eliminate the impact or sliding of metal on metal
- controlling resonance

Poor Lighting Conditions

Poor lighting conditions that result from illumination, glare, and color can affect the operator's vision, performance, and safety. Additionally, inadequate illumination can lead to eye strain and headache. The amount of luminance needed is a function of the task at hand, the duration of the task, the reflectance of the environment, and the operator's visual capability. Typically light levels are measured using a photometer, a device that is used to determine light levels in different planes. Light can be quantified in many ways—lux, lumens, footcandles (fc), candlepower, candelas, and so on. The two most popular scales are lux, the European measure, and footcandles, the U.S. measure. Lux equals the illumination of one square meter one meter away from a uniform light source. It is also defined as one lumen per square meter. One candela is equal to one lux. A footcandle is the illumination of one square foot one foot away from a uniform light source or one lumen per square foot. One footcandle is approximately ten lux (1 fc = 10.76 lx). Typically, a common area with dark surroundings is around 50 lux when illuminated. A workplace where fine work is performed over a long period should be about 5000 lux.

Quality of illumination can often be improved by reducing glare. Eliminate direct glare, where light directly affects the vision—such as sun in the eye. Surfaces that diffuse light, such as flat paint, matte paper, and textured finishes, are preferred. If the direct glare cannot be eliminated, use light shields, hoods, and visors to guard against it. If possible, change the orientation of the workplace, task, viewing angle, or viewing direction until maximum visibility is achieved. Use of color determines a person's perception of the work environment. Areas covered with long-wavelength colors, such as red, appear smaller than areas covered with shorter-wavelength colors, such as blue or green. Color brightness affects an operator's perception. Dark colors appear closer than the lighter ones. Also, cool colors, such as blue, green, and violet, are soothing, whereas warm colors, such as red, orange, yellow, and brown, are stimulating.

Heat or Cold Exposure

Exposure to extreme temperatures can affect workers' job performance and expose them to ergonomic risk factors. For example, air-powered tools can decrease exposure to temperature extremes by cooling the tool handle or blowing out exhaust air. Metallic tool handles should be insulated with a rubber, plastic, or cloth covering that shields the hand from extreme temperatures. Skin exposure to direct air exhaust can be eliminated by using exhaust mufflers and by directing the exhaust away from the operator. OSHA recommends that the operator's skin should not be exposed to temperatures less than 65 degrees Fahrenheit or greater than 95 degrees for a long period of time. Tools used for joining—such as a soldering iron, a welding rod, a hot glue gun, or a sealant tube—produce heat energy. Face shields, goggles, and gloves eliminate the probability of exposure to such heat energies. Electric motors, friction surfaces, and exhaust mufflers are also sources of heat energy. Face shields, goggles, and gloves help reduce exposure to heat.

Operator Characteristics

The human characteristics of age, gender, ethnicity, stature, physical ability, health, personal habits, knowledge, training, education, and past experience are directly related to the potential ergonomic risk factors workers may experience while performing their jobs. A task that might expose one operator to an ergonomic risk factor may not have any effect on another who has slightly different physical or emotional characteristics. Determining operators' characteristics using examination and testing allows for the selection and design of a work environment that minimizes or eliminates exposure to potential ergonomic risk factors. A positive work environment, the promotion of physical health, and incentive programs that encourage workers to break bad habits can also reduce risk factors associated with an operator's characteristics.

Workstations

Employers need to decide whether it is appropriate for particular employees to be seated or standing while performing their jobs based on the tasks they perform regularly. Seated workstations are best when the job requires one or more of the following: precise foot control, fine assembly, writing or typing tasks, body stability or equilibrium, long work periods, or work less than 5.9 inches (15 centimeters) above the work surface. An effective and ergonomic seated workstation will do the following: give the operator the correct eye position for viewing the task; allow the seat's height, depth, back angle, and foot rest positions to be adjusted for comfortable working conditions; allow for clearance of the legs and knees; and allow adjustment of the work surface's height and depth. Guidelines for designing effective seated workstations are available in many textbooks and on specific industrial Web sites. OSHA provides a good Web site on this topic at www.rmis.rmfamily.com/new/ergon60.htm.

Standing workstations are best in the following situations: no leg or knee clearance is available; heavy objects are regularly handled; high, low, and extreme reach are frequently required to complete the task; and mobility is necessary. An effective standing workstation will allow proper eye position with respect to viewing requirements, permit proper reach distances, and allow the work surface height and depth to be adjusted (Lang 1999).

Manual Material Handling

Workplace injuries are most commonly associated with material handling. Most of these injuries are caused by heavy lifting. To guard against material-handling injuries, companies dealing with this type of work should advise their employees to follow these simple procedures:

- Always bend at the knees and not at the waist. Bending at the waist puts excessive strain on the lower back, which can result in serious back injury.
- When lifting a heavy object from a crouched position, do not jerk the body, particularly the back. Keep the back straight and lift with your legs.

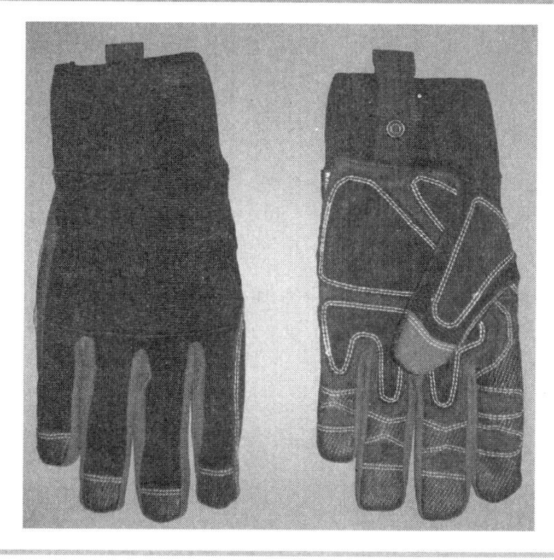

FIGURE 1. "Cargo Bull" by Ironclad® Performance Wear (*Source:* Ironclad Performance Wear, online at www.ironclad.com)

Platforms, Ladders, and Stairs

Very few best practices exist for climbing stairs on a repetitive basis. Generally, it is recommended that workers maintain proper posture and not bounce. Proper posture is generally defined as keeping the back straight and lifting the weight of the body with the legs. To prevent the concentration of weight in one area of the body, a hand should be placed on the handrail at all times to brace the body and to allow for more leverage when climbing. Bouncing is common when running up stairs. This places extra stress on the joints and may lead to injury, so it should be avoided.

Examples of Best Practices by Industry

Gardening

Today's garden gloves feature high-tech materials and advanced designs that reduce hand fatigue, eliminate friction, and dampen vibrations as well as protect the hand. Ironclad Performance Wear, which produces high-tech work gloves, suggests that the same performance offered in work gloves can be useful in gardening gloves (see Figure 1). Features such as a protected knuckle area; reinforced fingernail guards; washability; reinforced thumbs, saddles, and fingertips; breathability; adhesive grips; and seamless fingertips that increase the sense of touch make gardening easier and safer for the hands. Also, proper posture should be practiced when gardening in order to reduce common arm, hand, wrist, and back injuries. Maintaining a straight wrist can reduce some of these injuries.

Bent-handle cutters and pliers help the user maintain a straight wrist.

Pharmaceuticals

Pharmaceutical technicians hand-tighten dozens of vaccine jug caps daily. If not adequately tightened, the jugs could leak and spoil products worth thousands of dollars. However, most operators are poor judges of cap torque and are marginally capable of using the proper torque required to tighten caps adequately. Caps that are not properly tightened may result in significant unwarranted hand and wrist stress. One pharmaceutical company uses a dial torque wrench made with a special cap torque attachment and trains its technicians to use it. The cost is about $8 per worker.

Meatpacking

Meatpacking is one of the most hazardous industries in the United States because in assembly line processes such as boning meats, workers can make several thousand repetitive motions per day with no variation. The motions place physical stress and strain on the wrists and hands, resulting in carpal tunnel syndrome. One company has automated material-handling and packaging processes to reduce or eliminate operators' exposure to ergonomic risk factors.

Garment Manufacturing

Garment makers, who often perform fast-paced piecework operations involving excessive repetitive tasks, increase their risk of developing carpal tunnel syndrome. Garment-industry jobs often require workers to push large amounts of materials through machinery while sitting on nonadjustable metal stools. Workers doing these jobs can sustain disabling wrist, back, and leg injuries. Automation as well as engineering and administrative controls are used to alleviate some of the risk factors. A garment-manufacturing workstation is shown in Figure 2.

Ceramic Cooktops

At a glass ceramic cooktop plant, workers manually lift uncut plates of glass onto a waist-high conveyor belt, where they are stacked vertically on a nearby L-shaped holder. A forklift moves the strapped holder carrying the glass. The holder, however, presented the glass at knee height, making workers bend each time to pick up the glass until they devised a stand made from a wooden shipping crate and placed it beneath the holder to raise the glass to waist height.

Professional Offices

Employees in many offices experience pain from performing their daily tasks. Workers should be trained in the proper use of adjustments already provided in their chairs, computer monitors, and furniture systems. Changes in the placement of telephones, printers, and in-boxes can lead to better working postures. In addition, training and encouraging employees to take micro-breaks helps overused parts of the body to rest and recuperate. A professional office workstation is shown in Figure 3.

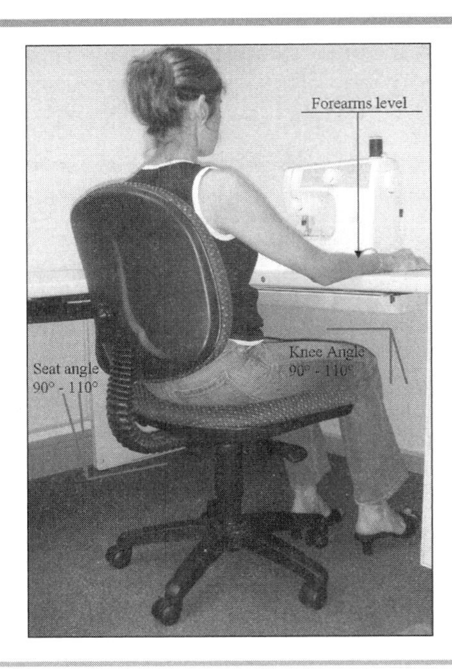

FIGURE 2. Garment-manufacturing workstation (*Source:* OSHA, www.osha.gov)

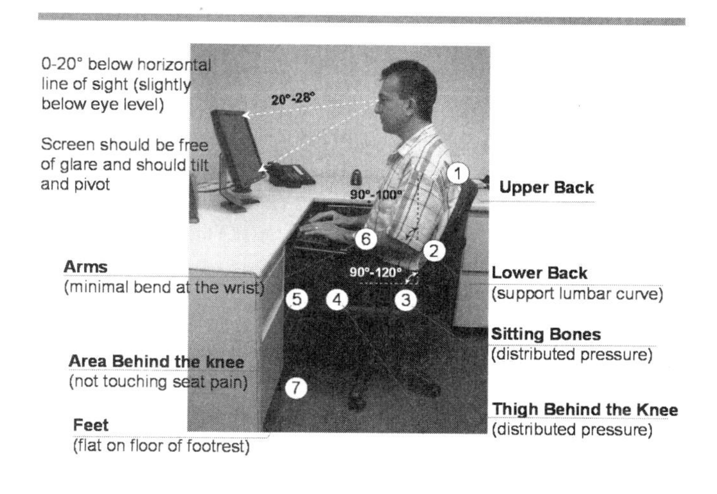

FIGURE 3. Professional office workstation
(*Source:* Nicholas Institute of Sports Medicine and Athletic Trauma, www.nismat.org)

Packaging

Workers used to pack items into rectangular boxes positioned so they had to reach repeatedly across the long axis of the boxes, exposing their backs, shoulders, and arms to physical stress. Rotating the boxes allowed them to reach across the shorter axis of the boxes, reducing the length of reach and the risk of injury. Manual lifting is often associated with the packaging industry. Excessively heavy boxes can place great stress on employees' muscles, leading to back strain or disc injury. By improving access to heavy boxes, allowing employees to pick them up without having to bend at the waist, and providing boxes with handles or handhold cutouts, the risks of injury can definitely be reduced (access the OSHA Baggage Handling eTool).

Poultry Processing

Poultry processing involves many risk factors, including repetition, force, awkward and static postures, and vibration, all of which may contribute to an increased risk of injury. The type of work performed, in conjunction with cold temperatures that are associated with poultry processing, may also be a contributing factor for MSDs. Poultry processors have created a number of solutions to reduce the duration, frequency, and degree of exposure to risk factors. Developing a job-rotation system helps to reduce the fatigue and stress of particular sets of muscles and tendons. Cross-training and providing floating employees ensures that staff support is always available to cover worker breaks and provide rotation alternatives. The employees in one poultry-processing plant complained that ill-fitting protective gloves did not provide adequate protection, so the company purchased gloves from several manufacturers to provide a wider range of sizes for better fit. Also, operators' stationary seats were replaced by adjustable seats. The Occupational Safety and Health Administration provides recommendations for the use of workstations that are appropriate for duties being performed (OSHA 2004).

Visual Display Terminals/Visual Work Environment/Computer Environment

Employees who spend much of their time in an office regularly stare at a computer screen or LCD display for large amounts of time. One common repetitive stress injury is damage to the eyes, often resulting in the need for glasses or contact lenses. It should be noted that flat-panel computer and workstation monitors emit much less radiation than some of the older model screens and are much better for the eyes.

Split keyboards are used in order to place less stress on wrists, hands, fingers, joints, and tendons when employees type for long periods. Much research has been done regarding split-keyboard technology over the last few years. For example, Timothy Muss, a graduate student in the Human Factors and Ergonomics Laboratory at Cornell University, along with Alan Hedge, professor of Design and Environmental Analysis at Cornell, developed a vertical keyboard that reduces many of the stress-related injuries associated with long periods of typing. Using gloves with special sensors, the stresses placed on various areas of the wrists and hands when using the new keyboard were carefully measured and compared to the stresses generated when using a traditional keyboard. The vertical split

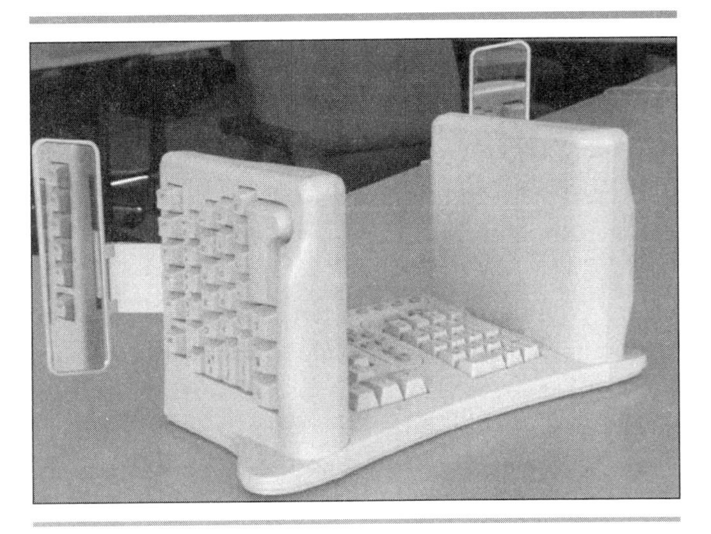

FIGURE 4. The VK, an ergonomically designed vertical keyboard (*Source:* www.news.cornell.edu)

keyboard, also known as the VK, was shown to be far superior to traditional keyboards in reducing repeated stress–related injuries (see Figure 4).

Machine Clearances and Maintenance

In the area of machine maintenance, best practices are generally specific to the job. Each machine has a different operating and maintenance procedure and its own best practices. Best practices common to all machines include easy access for maintenance and lubrication, clearances that allow for easy access to areas needing regularly scheduled maintenance, and engineering and administrative controls to reduce or eliminate workers' exposure to ergonomic risk factors while performing maintenance.

Construction–Industrial and Residential

In the construction industry, it is generally understood that workers use their own best practices to get jobs done fast. Most of the construction industry works on a strict timeline, and emphasis is on completing the job on time. An example of an ergonomic hammer, developed to reduce the continuous vibration that construction workers often endure, is the fiberglass California Framer®, pictured in Figure 5. This hammer, with its fiberglass handle and oversized striking face, reduces the stress that the wrist, hand, and arm often experience when hammering a nail.

Medical and Healthcare

There are many guidelines for proper procedures in all aspects of the medical and healthcare industry, most of which are controlled by OSHA. They include the proper ergonomic practices for repetitive motion that help prevent carpal tunnel syndrome. For example, an ergonomic scalpel handle for surgeons was developed by LM Instruments OY. Known as the ErgoHold 3 (see Figure 6), it is an improvement on the traditional scalpel because it has a firm grip and good balance, and it allows for precise blade control as well as controlled rotation.

FIGURE 5. Fiberglass California Framer® (*Source:* www.hammernet.com/news.htm)

FIGURE 6. ErgoHold 3 scalpel handle
(*Source:* LM Instruments, www.lminstruments.com)

Nursing Homes, Assisted Care, and Child Care

Throughout the nursing-home, assisted-care, and child-care industries, there are no clear-cut ergonomics best practices. In 2003, OSHA published some basic guidelines intended to reduce ergonomic risk factors associated with these industries (OSHA 2003). The main focus of those guidelines was on lifting and repositioning patients. OSHA has recommended a number of solutions, some of which include the use of powered sit-to-stand devices (see Figures 7 a and b), portable lift devices, ceiling-mounted lift devices, variable-position geriatric and cardiac chairs, built-in or fixed bath lifts, and toilet seat risers. This area could benefit from further research on reducing repetitive strain injuries in the workplace.

Food Processing and Packaging

There doesn't appear to ever have been a publicly released study on the best ergonomic box-cutter design in the industry. Several companies claim that their box cutter is the best and most ergonomic; such cutters are readily available in hardware stores. Ensuring that blades are sharp helps to reduce stress on the user. For an example, see Figure 8.

Textile and Fabric Manufacturing

In the textile and fabric industries, the use of precision cordless rotary shears reduces repetitive motion injuries due to cutting for long periods. One company instituted a program of monthly rotation of employees

FIGURE 7a. Sit-to-stand devices
(*Source:* OSHA, www.osha.gov)

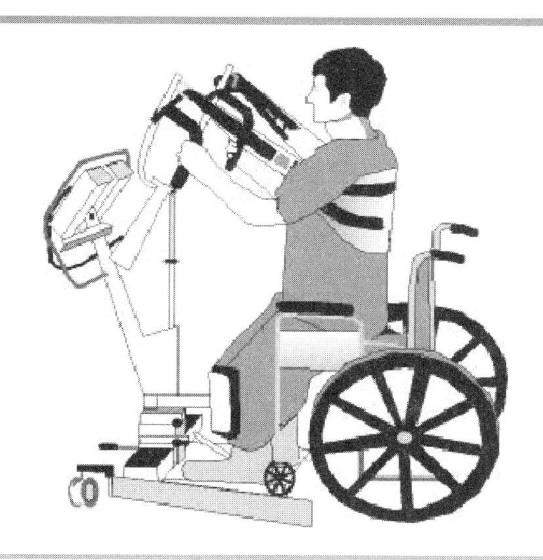

FIGURE 7b. Sit-to-stand devices
(*Source:* OSHA, www.osha.gov)

FIGURE 8. Box cutters (*Source:* Safety Knife Company, www.safetyknife.net)

with repetitive-motion jobs. They also designed their equipment with the operator and task in mind, redefined processes to ensure the safety of their employees, and educated employees about proper job procedures (BMP Center of Excellence for Best Manufacturing Processes 1994).

Service and Repair

Many products are available in the service and repair industry that incorporate ergonomics best practices in their design. Two examples are fatigue-reducing needle-nose pliers by Klein Tools and MaxGrip™ self-adjusting pliers by Stanley. Comfort-grip pliers have spatulated handles molded from flexible, non-slip material for softness, comfort, and a secure grip. Spatulated handles expand contact with the hand, reducing pressure and strain and improving power transfer. Two types of comfort-grip pliers are shown in Figure 9.

The author acknowledges the editorial contributions of Sharla R. Benedict.

Important Terms

Anthropometry: The study of human body measurements, used in developing design standards and requirements for manufactured products to ensure that they are suitable for their intended audience.

Biomechanics: A scientific and engineering field that explains the characteristics of the human body in mechanical terms.

Carpal Tunnel Syndrome: The compression and entrapment of the median nerve in the carpal tunnel, where it passes through the wrist into the hand. The median nerve is the main nerve that extends down the arm to the hand and provides the sense of touch in the thumb, index finger, middle finger, and half of the fourth, or ring, finger.

De Quervain's Disease: Inflammation of the tendon sheath of the thumb, attributed to excessive friction between two thumb tendons and their common

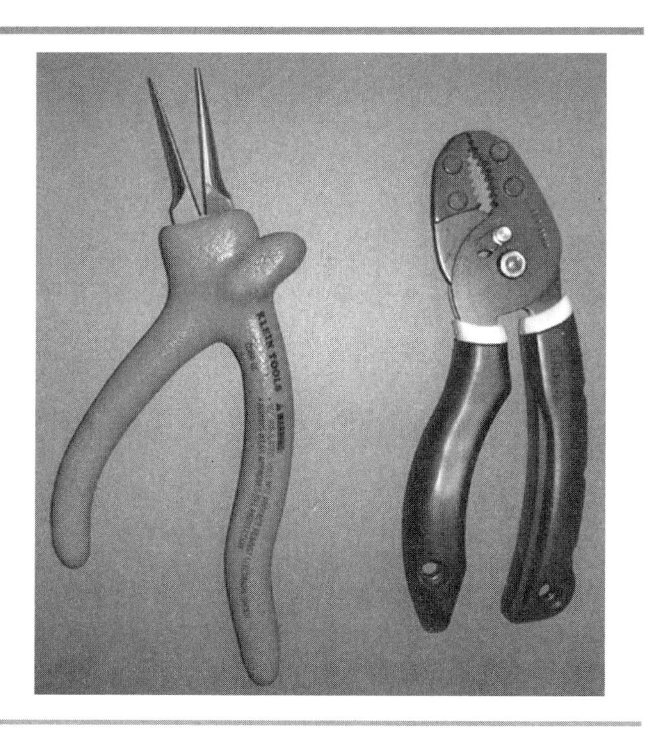

FIGURE 9. Klein fatigue-reducing pliers and Stanley MaxGrip™ self-adjusting pliers (*Source:* Grainger, www. grainger.com)

sheath. It is usually caused by twisting and forceful gripping motions with the hands.

Extension: A position in which a joint is straightened so that the angle between the adjacent bones is increased.

Finger Press: The motion of the tip of a finger or fingers pressing down on an object, such as a button or key.

Flexion: A position in which a joint is bent so that the angle between the adjacent bones is decreased.

Industrial Hygiene: The science of anticipating, recognizing, evaluating, and controlling workplace conditions that may cause worker injuries and illnesses.

Kinesiology: The study of the principles of mechanics and anatomy in relation to human movement.

Musculoskeletal Disorders: Injuries and disorders of the soft tissues (muscles, tendons, ligaments, joints, and cartilage) and nervous system.

Pinch Grip: A posture in which the hand does not fully encircle the handled object.

Radial Deviation: A posture in which the wrist is bent toward the thumb side of the hand.

Raynaud's Syndrome or White Finger: A condition in which blood vessels of the hand are damaged from repeated exposure to vibration for long periods of time. The skin and muscles do not get necessary oxygen from the blood and eventually die. Symptoms include: intermittent numbness and tingling in the fingers; pale, ashen, and cold skin; and the eventual loss of sensation and control in the hands and fingers.

Tendonitis: A tendon inflammation that occurs when a muscle or tendon is repeatedly tensed from overuse or unaccustomed use of the wrist and shoulder.

Tenosynovitis: An inflammation of or injury to the synovial sheath surrounding a tendon, usually due to excessive repetitive motion.

Trigger Finger: A tendon disorder that occurs when a groove is worn into the sheath of a flexing tendon of a finger, which is usually associated with using tools that have handles with hard or sharp edges. If the tendon becomes locked in the sheath, attempts to move the finger cause snapping and jerking movements.

Ulnar Deviation: A posture in which the wrist is bent toward the little finger side of the hand.

References

Adler, Paul S., Barbara Goldoftas, and David I. Levine. 1997. "Ergonomics, Employee Involvement, and the Toyota Production System: A Case Study of NUMMI's 1993 Model Introduction." *Industrial and Labor Relations Review*, (April) 50(3):416–437.

Acoustical Society of America. 2006. ANSI S.270-2006, *Hand Arm Vibration Standard*. Melville, NY: Acoustical Society of America.

BMP Center of Excellence for Best Manufacturing Processes. 1994. "Report of survey conducted at Mason & Hanger-Silas Mason Co., Inc." *Best Manufacturing Practices* (July). www.p2pays.org/ref%5C05/04934.pdf

Bureau of Labor Statistics, U.S. Department of Labor. News Release, 2003. *Number of Nonfatal Occupational Injuries and Illnesses with Days Away from Work Involving Musculoskeletal Disorders by Selected Worker and Case Characteristics, 2001* (March). www.bls.gov.

______. 2004. "Lost-Worktime Injuries and Illnesses: Characteristics and Resulting Time Away from Work, 2002." www.bls.gov/iif/oshwc/osh/case/osnr0019.pdf

DoItYourself.com. www.doityourself.com/invt/1493865

Ernst, A., P., Koningsveld, and Henk F. Van Der Molen. 1997. "History and Future of Ergonomics in Building and Construction." *Ergonomics* (October 1) 40(10).

Helander, M. G. 1995. *A Guide to the Ergonomics of Manufacturing*. Bristol, PA: Taylor & Francis.

Humantech, Inc. 1996. *Ergonomic Design Guidelines for Engineers*. Ann Arbor, MI: Humantech, Inc..

International Labour Organization (ILO). 1986. ISO 5349-1986, *Mechanical Vibration – Guidelines for the Measurement and the Assessment of Human Exposure to Hand-Transmitted Vibration*. Geneva, Switzerland: ILO.

Jastrzebowski,W. 1857. "Rys historyczny czyli nauka o pracy, opartej na prawach poczerpni´tych z Nauki Przyrody" ("An Outline of Ergonomics or The Science of Work Based Upon the Truths Drawn from the Science of Nature") *Przyroda i Przemys* (*Nature and Industry*) pp. 29–32. Reprinted 1997. Warsaw: Central Institute for Labour Protection.

Karwowski, Waldemar, and William S. Marras. 1997. "Design and Management of Work Systems, Principles and Applications in Engineering," *Occupational Ergonomics*, p. 15.

Kroemer, Karl, Henrike Kroemer, and Katrin Kreomer-Elbert. 2002. *Ergonomics – How to Design for Ease and Efficiency*. 2d ed. Upper Saddle River, NJ: Prentice Hall International.

Lang, Susan S. 1999. "Vertical Split Keyboard Offers Lower Injury Risk for Typists." *Cornell News*. (retrieved August 2, 2005) www.news.cornell.edu/releases/Nov99/vertical.keyboard.ssl.html

Liberty Mutual Research Institute for Safety. 2003. *Workplace Safety Index*. Boston: Liberty Mutual Group.

Melhorn, J. Mark. 1998. "Cumulative Trauma Disorders and Repetitive Strain Injuries—The Future." *Clinical Orthopedics & Related Research*. (June) 351:107–126.

Occupational Safety & Health Administration (OSHA). *OSHA Ergonomics eTools*. "Baggage Handling." www.osha.gov/SLTC/etools/baggagehandling/index.html

______. 2002.U.S. Government Printing Office.

______. 2003. *Ergonomics for the Prevention of Musculoskeletal Disorders—Guidelines for Nursing Homes*. www.osha.gov/ergonomics/guidelines/nursinghome/final_nh_guidelines.html/

______. 2004. *Ergonomics for the Prevention of Musculoskeletal Disorders Guidelines for Poultry Processing*. www.osha.gov/SLTC/ergonomics/index.html/

Pulat, B. Mustafa. 1997. *Fundamentals of Industrial Ergonomics*. Long Grove, IL: Waveland Press.

Putz-Anderson, Vern, ed. 1988. *Cumulative Trauma Disorders: A Manual for Musculoskeletal Disease of the Upper Limbs*. Bristol, PA: Taylor & Francis.

Radwin, R. G., T. J. Armstrong, D. B. Chaffin, G. D. Langolf, and J. W. Albers. 1990. "Hand-Arm Frequency-Weighted Vibration Effects on Tactility." *International Journal of Industrial Ergonomics* 6:75–82.

Radwin, R.G., and J.T. Haney. 1996. *An Ergonomics Guide to Hand Tools*. Fairfax, VA: AIHA Ergonomics Committee.

Smith, Sandy. 2004. "Off-the-Job Safety: Yardwork Leads to Blisters, Hand Injuries." *Occupational Hazards*. Accessed October 13, 2004. www.occupationalhazards.com/articles/12492

Sitzman, Kathy, and Donald Bloswick. 2002. "Creative Use of Ergonomic Principles in Home Care—Clinical Concerns." *Home Healthcare Nurse*. (February) 20(2):98–103.

Streeter, H. 1970. "Effects of Localized Vibration on the Human Tactile Sense." *American Industrial Hygiene Association Journal*. 31(1).

Terrel, R., and J. Purswell. 1976. "The Influence of Forearm and Wrist Orientation on Static Grip Strength as a Design Criterion for Hand Tools." *Proceedings of the Human Factors Society* 20:28–32.

Yassi, Annalee. 2000. "Work-Related Musculoskeletal Disorders., Nonarticular Rheumatism, Sports-Related Injuries, and Related Conditions." *Current Opinion in Rheumatology* (March) 12(2):124–130.

Appendix

Ben Cranor and Montral L. Walker

English and Metric Units of Measure

Abbreviations

a	ampere, amperes
atm	atmosphere
b	bar
Btu	British thermal unit
C	Centigrade
c	candle
cal	calorie, calories
cd	candela
cfm	cubic feet per minute
cm	centimeter, centimeters
cu	cubic
dB	decibels
E	volt, volts
EE	base 10
erg	erg
F	Fahrenheit
fpm	feet per minute
fps	feet per second
ft	foot, feet
g	gravitational constant
gal	gallon, gallons
gpm	gallons per minute
h	height
Hg	mercury
hp	horsepower
hr	hour, hours
in	inch, inches
J	joule, joules
k	kilo
K	Kelvin
kg	kilogram, kilograms
kgf	kilogram of force
km	kilometer, kilometers
kPa	kilopascal
kw	kilowatt, kilowatts
L	liter, liters, litre, litres
lb	pound, pounds
lbf	pound of force
Ln	linear
M	million
m	meter, meters, metre, metres
NTP	normal temperature and pressure
μ	micron
min	minute, minutes
mm	millimeter, millimeters
mi	mile, miles
mph	miles per hour
MW	molecular weight
N	Newton, Newtons
N m	newton meter
oz	ounce, ounces
Pa	Pascal, Pascals
pcf	pounds per cubic foot
psi	pounds per square inch
psig	pounds per square inch gauge
psf	pounds per square foot

pt	pint, pints
qt	quart, quarts
R	Rankine
rev	revolution, revolutions
rpm	revolutions per minute
rps	revolutions per second
s, sec	second, seconds
sq	square
thm	therm
vol	volume
W	Watt, Watts
WC	water column
wg	water gage (syn. gauge)
yd	yard, yards

Equivalents and Conversion Factors

Lengths and Distance

Units: m, cm, mm—all dimensions

Metric Conversion Factors

m = 2.54 EE–2 × in
= 0.3048 × ft
= 1609.34 × mi
= 0.9144 × yd
= EE–6 × μ

cm = 2.54 × in
= 30.48 × ft

mm = 25.4 × in

Metric Equivalents

m = 3.28 ft
= 39.37 in

cm = 0.3937 in

mm = 0.03937 in

English Conversion Factors

mi = 1.894 EE–4 × ft
= 0.6214 (.62) × km

ft = 3.28 × m
= 328 × km
= 3.28 EE–2 × cm

in = 3.937 EE–2 × mm
= 0.3937 × cm

English Equivalents

mi = 5280 ft
= 1.609 km

yd = 0.914 m

ft = 0.3048 (0.3) m
= 30.48 cm

in = 25.4 mm
= 2.54 (2.5) cm

Areas

Units: m^2, cm^2, mm^2—all areas

Metric Conversion Factors

m^2 = 0.0929 × sq ft

cm^2 = 6.452 × sq ft

mm^2 = 9.290 EE4 × sq ft

m^2 = 0.0929 × sq ft
= 6.452 EE–4 × sq in

cm^2 = 6.452 × sq in

mm^2 = 645.2 × sq in

Metric Equivalents

m^2 = 10.76 sq ft
= 1.55 EE3 sq in

cm^2 = 0.155 sq in

mm^2 = 0.00155 sq in

English Conversion Factors

sq ft = 10.76 × m^2
= 1.076 EE–3 × cm^2

sq in = 0.00155 × mm^2
= 0.155 × cm^2

English Equivalents

sq ft = 0.0929 m^2

sq in = 645.2 mm^2

Volume and Capacity

Units: m^3, cm^3, L—volumes

Metric Conversion Factors

m^3 = 3.785 EE–3 × gal
= 0.001 × L
= 0.02832 × cu ft
= 1.639 EE–5 × cu in

cm^3 = 1000 × L
= 473.18 × pt
= 16.378 × cu in
= EE6 × m^3
L = 3.785 × gal
= 0.94635 × qt
= 28.32 × cu ft
= 1.639 EE–2 cu in
= 1000 × m^3
= 0.001 × cm^3

Metric Equivalents

m^3 = 35.32 cu ft
= 1000 L
L (cubic decimeters)
= 0.26417 gal
= 1.057 qt
= 0.03532 cu ft
= 61 cu in
= 2.113 pt

English Conversion Factors

gal = 0.2642 × L
= 0.125 × pt
= 7.481 × cu in
= 264.17 × m^3
pt = cm^3 × 2.113 EE–3

English Equivalents

cu ft = 0.028 m^3
= 7.84 gal
= 28.32 L
= 1728 cu in
gal = 3.785 (3.8) L
= 3.785 EE–3 m^3
= 0.1337 cu ft
= 231 cu in
= 8.33 lb (water)
= 3.78 kg (water)
pt = 16 oz
= 0.473 L
= 28.875 cu in
= 0.5 qt
= 1.671 EE–2 cu ft
= 0.028 m^3
= 0.125 gal
= 3.463 cu in

Weight and Mass

Units: kg

Metric Conversion Factors

kg = 0.4536 × lb
MW (molecular weight) = kg/kg mole

Metric Equivalents

kg = 0.06854 slugs
= 2.2 lb

English Conversion Factors

lbm = 2.205 × kg
= 32.17 × slugs

English Equivalents

lbm = 0.4536 (0.45) kg
1 slug = 32.17 pound-mass

Density

Units: kg/m^3

Metric Conversion Factors

kg/m^3 = 16.02 × lb/cu ft
kg/m^3 = 16.02 × pcf
= 2.768 EE4 × lb per cu in
= 119.8 × lbm/gal

Metric Equivalents

kg/m^3 = 0.0624 pcf
1 g mole = 22.41 L

English Equivalents

62.4 lbs per cu ft
= 1000 kg/m^3
= 0.0361 lb per cu in
= 8.33 lbm/gal
lb per cu ft
= 16.02 kg/m^3
= 0.1337 lbm/gal
1 lbm mole = 359 cu ft

Weight and Force

Units: N, kgf—force

Metric Conversion Factors

Newton (J/m)
= 4.448 × lbf
= 9.807 × kgf
= EE-5 × dynes
kgf = 0.4536 × lbf

Metric Equivalents

Newton (J/m)
= 0.1020 kgf
= 0.2248 lbf
kgf = 2.2 lbf
g = 9.81 m/sec/sec

English Conversion Factors

lbf = 0.2248 × N
dyne = N × EE5

English Equivalents

lbf = 4.448 N
= 1.36 N m

Pressure

Units: Pa, kgf/cm²—pressure

Metric Conversion Factors

Pa (N/m²) = 2.491 EE2 × in H_2O
= 6894.76 × psi
= EE5 × bar
=1.013 EE × atm
= 47.88 × psf
= 6894.76 × psi
= 2.491 EE2 × in H_2O
= 0.981 EE3 × m (H_2O)
= 3 EE3 × ft (H_2O)
kPa = 0.0479 × lbf
kgf/cm² = 0.0703 × psi

Metric Equivalents

Pa (N/m²) = 20.89 EE–3 psf
= 0.145 EE–3 psi
= 1.02 EE–5 kgf/cm²
= 0.0075 mm Hg
= 2.953 EE–4 in Hg
kgf/cm² = 9.807 EE4 Pa
= 735.6 mm Hg
= 393.7 in H_2O
= 32.81 ft H_2O
= 28.96 in Hg
= 14.22 psi
= 0.9678 atm
H (m) (water) = 0.102 kPa

English Conversion Factors

psi = 6.944 EE–3 × psf
= 0.145 × kPa
= 14.5 × bar
= 0.4335 × h (ft)
= 0.0193 × mm Hg
psig = in Hg × 0.4912
in wc = 4.015 EE–3 × Pa

English Equivalents

psi = 0.0689 bar
= 6.895 kPa
= 0.0703 kg/cm²
= 27.67 in H_2O
bar = EE5 Pa
= EE2 kPa
= Mdynes/cm²
= 14.50 psi
= 0.9869 atm
atm = 14.7 psia
= 29.9 in Hg
= 1.013 EE5 N/m (Pa)
= 2116.22 psf
= 1.01325 bars
= 101.3 kPa
= 406.8 in H_2O
= 33.90 ft (water)
= 10.35 m (water)
= 760.5 mm Hg
= 0.08072 lbm/cu ft (air)
= 1.033 kgf/cm
Torr = 133.3 Pa
= 1 mm Hg
= 1.316 EE–3 atm

= 0.00136 Kgf/cm^2
= 0.5353 in H_2O
= 0.01934 psi
in w.c. = 249.1 Pa
= 0.25 kPa
= 1.868 mm Hg
= 0.00246 atm
= 0.0361 psi
= 0.07355 in Hg
= 0.0833 ft (water)
H (ft) (water)
= 0.4335 psi
= 3 kPa = 0.88 in Hg
= 0.0295 atm
= 12 in H_2O
= 0.03048 kgf/cm^2
= 22.24 mm Hg

Temperature

Units: K, C—temperature

Metric Conversion Factors

K = °C + 273.15 (273)

English Conversion Factors

C = 5/9 (°F – 32)
R = °F + 459.69 (460)
F = 32 + (9/5) °C

Power

Units: Watt

Metric Conversion Factors

W (n × m/sec)
= 2.26 EE–2 × ft-lbf/min
= 0.2931 × Btu/hr
= 745.7 × hp
= EE–7 × ergs/sec
= E × I
= 0.2931 × Btu/hr
= 745.7 × hp
kw = 1.758 EE–2 × Btu/min
= 0.7457 × hp
=1.055 × Btu/sec

Metric Equivalents

W (n × m/sec) = 1 J/sec
kw = 0.948 Btu/min
= 1.341 hp

English Conversion Factors

Btu/hr = Btu/min × 60
= 2.546 EE3 × hp
Btu/min = 42.41 × hp
= 56.88 × kw
= MBtu/hr × 1.667 EE4
Btu/sec = Btu/min × 1.667 EE–2
$Btu/ft^2/hr$ = 0.317 × W/m^2
Btu/ft/sec = 2.889 EE–4 × W/m
ft-lbf/min = 33,000 × hp
ft-lbf/sec = 550 × hp
= 737.46 × kw
hp = 1.341 × kw
= 3.989 EE–4 × Btu/hr
= 1.341 EE–3 × W
metric hp = 0.001843 × ft-lbf/sec = 1.0139

English Equivalents

Btu/hr = 0.393 W
Btu/min = 17.58 W
= 1.758 EE–2 kw
Btu/sec = 1.055 kw
Btu/ft^2 = 11.35 kw/m^2 (heat flux)
= 1.135 W/m^2
hp = 550 ft-lbf/sec
= 746 W

Energy

Units: J

Metric Conversion Factors

joule (Watt-sec, N m)
= 1.356 × ft-lbf
= 1055.1 × Btu
= 3.6 EE6 × kw-hr
= 2.684 EE6 × hp-hr
= 4.184 × cal
kw-hr = 2.931 EE–4 × Btu
= 2.778 EE–8 × erg
= 2.778 × J

kgf-m = 0.13826 × ft-lbf
erg = EE7 × J
kJ/kg (heat) = 2.33 × Btu/lb Δh (MJ/kg) + MJ/mole

Metric Equivalents

joule (watt-sec, N m)
= 0.7376 ft-lbf
= 0.9478 EE–3 Btu
= 0.2778 EE–6 kw-hr

kJ/kg (heat)
= 0.43 Btu/lb
= 5.1 MJ/L

English Conversion Factors

Btu = 1.285 EE–3 × ft-lbf
= 9.478 EE–4 × J
= 3412 × kw-hr
ft-lbf = 778 × Btu
Btu/cu ft = 26.8 × MJ/m^3
ft-lbf = 2.655 EE6 × kw-hr
= 0.7376 × J
hp-hr = 3.725 EE–7 × J
= 3.930 EE4 × Btu
therm = EE–5 × Btu

English Equivalents

Btu = 778 ft-lbf
= 252 cal
= 1055 J
= 1.055 EE–3 MJ
Btu/ft^2 = 11.356 kJ/m^2
Btu/cu ft = 0.0373 MJ/m^3

Velocity

Units: m/sec

Metric Conversion Factors

m/sec = 0.44704 × mph
= 5.1 EE–3 × fpm
= 0.3048 × fps
= EE–2 × cm/sec
cm/sec = 0.508 × fpm

Metric Equivalents

m/sec = 2.237 mph
= 196.85 fpm
= 3.281 fps
= 3.6 km/hr

English Conversion Factors

mph = 0.68182 × fps
fpm = 196.85 × m/sec
= 88 × mph
fps = 3.28 × m/sec

English Equivalents

mph = 1.467 fps
= 0.4470 m/s
fpm = 0.011364 mph
= 0.0051 m/sec
fps = 0.305 m/sec
= 18.29 m/min

Flow Rates

Units: m^3/sec, L/sec, and L/min

Metric Conversion Factors

Water

m^3/sec = 4.7 EE–4 × cfm
= 6.31 EE–5 × gpm
m^3/min = 0.0283 × cfm
= EE–3 × L/min
= 3.785 EE–3 × gpm
m^2/min = diffusivity
= 9.29 EE–2 × sq ft per sec
L/min = 3.785 × gpm
= 1000 × m^3/min
L/sec = 15.85 (16) gpm
= 2.118 cu ft/min
= 0.0631 × gpm
= 0.472 × cfm
= EE3 × m^3/sec
kg/sec = mass flow/mass loss rate
= lbs/min × 7.56 EE–6

Air

m³/sec = 4.7 EE–4 × cfm
m³/min = 0.0283 × cfm
L/sec = 0.47 × cfm

Metric Equivalents

m³/sec = 2118.8 cfm = vol flow, viscosity
= 1000 L/sec
= 1.6 EE4 gpm
= 35.31 cu ft
L/min = 26.42 EE–2 gpm
L/sec = 15.85 (16) gpm
= 2.118 cu ft/min

English Conversion Factors

cu ft/hr = 8.0208 × gpm
= 35.3 × m³/hr
cu ft/min = 35.34 × m³/min
cu ft/sec = 0.002228 × gpm
lb (water)/min = 62.43 × cu ft/min
gpm = 448.83 × cu ft/sec
= 0.2642 × L/min

English Equivalents

cu ft/min (cfm)
= 2.83 EE–2 m³/min
= 0.472 L/sec
= 7.48 gpm
= 4.72 EE– 4 m³/sec
cu ft/sec = 28.32 L/sec
= 0.0283 m³/sec
dynamic viscosity = slugs/ft-sec
gpm = 0.0631 L/sec
= 3.785 L/min
= 6.3 EE–5 m/sec
gpm/Ln ft = 12.3 L/min/m

Flow Densities

Units: m³/min, L/min, L/sec

Water Flow Densities

Metric

L/sec/m² = 0.679 × gpm/sq ft
L/min/m²
= 0.0245 gpm/sq ft
= 40.75 × gpm/sq ft

English

gpm/sq ft
= 40.75 L/min/m²
= 0.0245 × L/min/m²
= 1.473 × L/sec/m²

Other Units of Measure

sec = min × 60
1/sec = 1/min × 1.667 EE–2
min = hrs × 60
cd = c = lumens/steradians
radians = degrees × 1.74 EE–2
= min × 2.909 EE–4
= rpm × 6.283
coulombs = ampere-hrs × 3600
= faradays × 9.649 EE–4
= 1 ampere-second
1 ft-c = 10.76 lux
= 1 lumen/sq ft
= 10.76 lumen/m²
1 lux = 0.093 ft-c
5 ft-c = 54 lux
1 footlambert
= 3.426 cd/m²
= 3.1416 × cd/sq ft
= 0.2919 × cd/m²
rpm = °/sec × 0.1667
= 6°/sec
= 0.01667 rps
degrees = radians × 57.3
minutes = radians × 3437.7
seconds = 2.06 EE5 × radians
rev = 0.159 × radian

STATISTICS

Sample Mean

If the observations in a sample mean of size n are $X_1, X_2, \ldots X_n$, then the sample mean is:

$$\bar{X} = \frac{X_1 + X_2 + \ldots X_n}{n} = \frac{\sum_{i=1}^{n} X_i}{n}$$

Sample Median

Let $X_1, X_2, \ldots X_n$ denote a sample arranged in increasing order of magnitude; that is, X_1 denotes the smallest observation, X_2 denotes the second smallest, and X_n is the largest observation.

Then the median $\hat{X}$ is defined as the middle or $[(\frac{n+1}{2})]$th observation if n is odd and halfway between the two middle observations, if n is even $[(\frac{n}{2})]$th observation and the $[(\frac{n}{2})+1]$th observation

$$\hat{X} = \begin{cases} \left[X\left(\frac{n+1}{2}\right)\right]; \text{ if } n \text{ is odd} \\ \dfrac{X_{n/2} + X[(n/2)+1]; \text{ if } n \text{ is even}}{2} \end{cases}$$

Sample Mode

The mode is the observation that occurs most frequently in the sample.

Sample Variance

If $X_1, X_2, \ldots X_n$ is a sample of n observations, then the *sample variance* is

$$s^2 = \sum_{i=1}^{n} \frac{(X_i - \bar{X})^2}{n-1}$$

Sample Standard Deviation

The sample standard deviation s, is the positive square root of the sample variance. A normal random variable with $\mu = 0$ and $\sigma^2 = 1$ is called the standard normal variable. It is denoted as Z. Where σ^2 is the variance of the population and μ is the mean of the population.

Confidence Interval – on the mean with a known variance

If $\bar{X}$ is a sample mean of a random sample of size n from a population with a known variance $\sigma^2 100(1-\alpha)$, percent confidence interval on μ is given by

$$\bar{X} = Z_{\alpha/2}\left(\frac{\sigma}{\sqrt{n}}\right) \le \mu \le \bar{X} + Z_{\alpha/2}\left(\frac{\sigma}{\sqrt{n}}\right)$$

See the standard normal distribution table in this appendix for values.

Confidence Interval – on the mean of a normal distribution with an unknown variance

$$\bar{X} - t_{\alpha/2,\, n-1}\left(\frac{s}{\sqrt{n}}\right) \le \mu \le \bar{X} + t_{\alpha/2,\, n-1}\left(\frac{s}{\sqrt{n}}\right)$$

Where $t_{\alpha/2,\, n-1}$ is the upper $\alpha/2$ percentage of the t distribution with $n-1$ degrees of freedom

See *T*-Distribution Table for values of t.

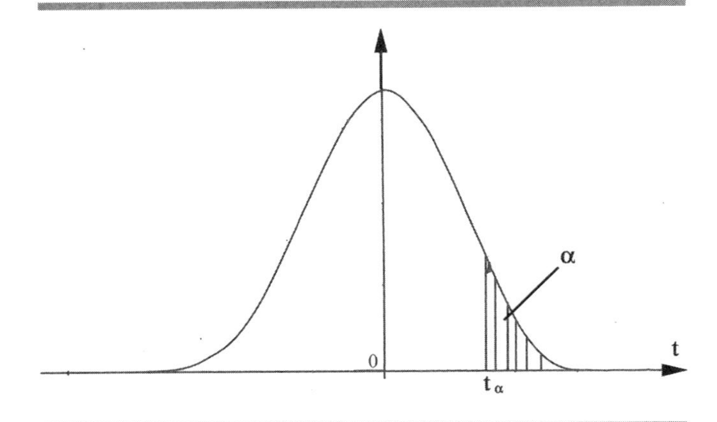

FIGURE 1.

TABLE 1

***T*-Distribution Table**

df	$\alpha = 0.1$	0.05	0.025	0.01	0.005	0.001	0.0005
∞	t_α=1.282	1.645	1.960	2.326	2.576	3.091	3.291
1	3.078	6.314	12.706	31.821	63.656	318.289	636.578
2	1.886	2.920	4.303	6.965	9.925	22.328	31.600
3	1.638	2.353	3.182	4.541	5.841	10.214	12.924
4	1.533	2.132	2.776	3.747	4.604	7.173	8.610
5	1.476	2.015	2.571	3.365	4.032	5.894	6.869
6	1.440	1.943	2.447	3.143	3.707	5.208	5.959
7	1.415	1.895	2.365	2.998	3.499	4.785	5.408
8	1.397	1.860	2.306	2.896	3.355	4.501	5.041
9	1.383	1.833	2.262	2.821	3.250	4.297	4.781
10	1.372	1.812	2.228	2.764	3.169	4.144	4.587
11	1.363	1.796	2.201	2.718	3.106	4.025	4.437
12	1.356	1.782	2.179	2.681	3.055	3.930	4.318
13	1.350	1.771	2.160	2.650	3.012	3.852	4.221
14	1.345	1.761	2.145	2.624	2.977	3.787	4.140
15	1.341	1.753	2.131	2.602	2.947	3.733	4.073

TABLE 2

Normal Distribution Table

Z	0.00	0.01	0.02	0.03	0.04	0.05	0.06	0.07	0.08	0.09
0.0	0.0000	0.0040	0.0080	0.0120	0.0160	0.0199	0.0239	0.0279	0.0319	0.0359
0.1	0.0398	0.0438	0.0478	0.0517	0.0557	0.0596	0.0636	0.0675	0.0714	0.0753
0.2	0.0793	0.0832	0.0871	0.0910	0.0948	0.0987	0.1026	0.1064	0.1103	0.1141
0.3	0.1179	0.1217	0.1255	0.1293	0.1331	0.1368	0.1406	0.1443	0.1480	0.1517
0.4	0.1554	0.1591	0.1628	0.1664	0.1700	0.1736	0.1772	0.1808	0.1844	0.1879
0.5	0.1915	0.1950	0.1985	0.2019	0.2054	0.2088	0.2123	0.2157	0.2190	0.2224
0.6	0.2257	0.2291	0.2324	0.2357	0.2389	0.2422	0.2454	0.2486	0.2517	0.2549
0.7	0.2580	0.2611	0.2642	0.2673	0.2704	0.2734	0.2764	0.2794	0.2823	0.2852
0.8	0.2881	0.2910	0.2939	0.2967	0.2995	0.3023	0.3051	0.3078	0.3106	0.3133
0.9	0.3159	0.3186	0.3212	0.3238	0.3264	0.3289	0.3315	0.3340	0.3365	0.3389
1.0	0.3413	0.3438	0.3461	0.3485	0.3508	0.3531	0.3554	0.3577	0.3599	0.3621
1.1	0.3643	0.3665	0.3686	0.3708	0.3729	0.3749	0.3770	0.3790	0.3810	0.3830
1.2	0.3849	0.3869	0.3888	0.3907	0.3925	0.3944	0.3962	0.3980	0.3997	0.4015
1.3	0.4032	0.4049	0.4066	0.4082	0.4099	0.4115	0.4131	0.4147	0.4162	0.4177
1.4	0.4192	0.4207	0.4222	0.4236	0.4251	0.4265	0.4279	0.4292	0.4306	0.4319
1.5	0.4332	0.4345	0.4357	0.4370	0.4382	0.4394	0.4406	0.4418	0.4429	0.4441
1.6	0.4452	0.4463	0.4474	0.4484	0.4495	0.4505	0.4515	0.4525	0.4535	0.4545
1.7	0.4554	0.4564	0.4573	0.4582	0.4591	0.4599	0.4608	0.4616	0.4625	0.4633
1.8	0.4641	0.4649	0.4656	0.4664	0.4671	0.4678	0.4686	0.4693	0.4699	0.4706
1.9	0.4713	0.4719	0.4726	0.4732	0.4738	0.4744	0.4750	0.4756	0.4761	0.4767
2.0	0.4772	0.4778	0.4783	0.4788	0.4793	0.4798	0.4803	0.4808	0.4812	0.4817
2.1	0.4821	0.4826	0.4830	0.4834	0.4838	0.4842	0.4846	0.4850	0.4854	0.4857
2.2	0.4861	0.4864	0.4868	0.4871	0.4875	0.4878	0.4881	0.4884	0.4887	0.4890
2.3	0.4893	0.4896	0.4898	0.4901	0.4904	0.4906	0.4909	0.4911	0.4913	0.4916
2.4	0.4918	0.4920	0.4922	0.4925	0.4927	0.4929	0.4931	0.4932	0.4934	0.4936
2.5	0.4938	0.4940	0.4941	0.4943	0.4945	0.4946	0.4948	0.4949	0.4951	0.4952
2.6	0.4953	0.4955	0.4956	0.4957	0.4959	0.4960	0.4961	0.4962	0.4963	0.4964
2.7	0.4965	0.4966	0.4967	0.4968	0.4969	0.4970	0.4971	0.4972	0.4973	0.4974
2.8	0.4974	0.4975	0.4976	0.4977	0.4977	0.4978	0.4979	0.4979	0.4980	0.4981
2.9	0.4981	0.4982	0.4982	0.4983	0.4984	0.4984	0.4985	0.4985	0.4986	0.4986
3.0	0.4987	0.4987	0.4987	0.4988	0.4988	0.4989	0.4989	0.4989	0.4990	0.4990

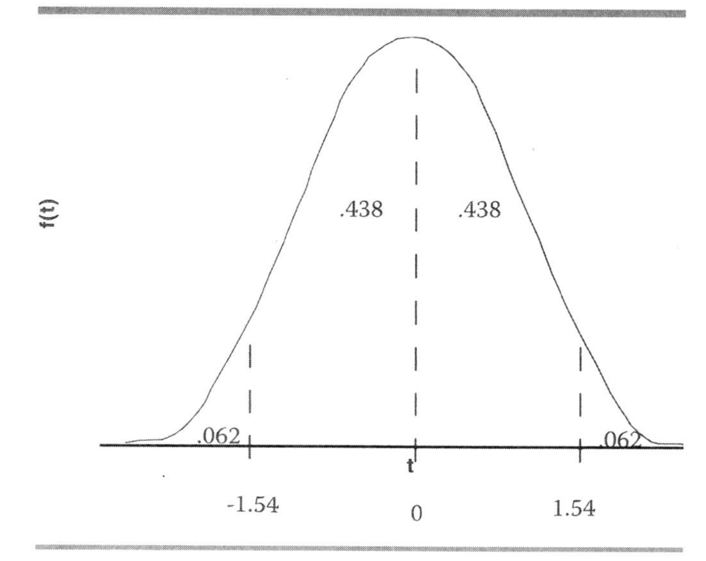

FIGURE 2.

Hypothesis Testing

Rejecting the null hypothesis when it is true is defined as a type-I error.

Failing to reject the null hypothesis when it is false is defined as a type-II error.

$H_0;\ \mu = \mu_0$ if σ^2 is known

Test statistic $Z_0 = \dfrac{(\bar{x} - \mu_0)}{\dfrac{\sigma}{\sqrt{n}}}$

Alternative hypotheses: $H_1;\ \mu \neq \mu_0$

$H_1;\ \mu > \mu_0$

$H_1;\ \mu < \mu_0$

Reject criteria: $|Z_0| > Z_{\alpha/2}$

$Z_0 > Z_{\alpha/2}$

$Z_0 < -Z_{\alpha/2}$

$H_0;\ \mu = \mu_0$ if σ^2 is unknown

Test statistic $t_0 = \dfrac{(\bar{x} - \mu_0)}{\dfrac{\sigma}{\sqrt{n}}}$

Alternative hypotheses: $H_1;\ \mu \neq \mu_0$

$H_1;\ \mu > \mu_0$

$H_1;\ \mu < \mu_0$

Reject criteria: $|t_0| > t_{\alpha/2,\, n-1}$

$t_0 > t_{\alpha/2,\, n-1}$

$t_0 < -t_{\alpha,\, n-1}$

Mathematics

A safety professional applies math knowledge and skills on a routine basis in analyzing and evaluating data, in preparing studies of quantifiable subjects, and in normal administrative duties. Thus, a basic competence in mathematics is expected for safety professionals and is required for applicants for the BCSP Safety Fundamentals examination.

In every mathematical operation, the unit terms (pounds, feet, miles, seconds, and so on) must be carried along with the numbers and must undergo the same mathematical operations as the numbers. Quantities cannot be added or subtracted directly unless they have the same units. However, any number of quantities can be combined in multiplication or division when the units as well as the numbers obey the algebraic laws of squaring, cancellation, and so on. For example, in converting 75 feet per second to miles per hour, notice that the minutes, seconds, and feet units cancel out, leaving miles and hours:

$$\frac{75 \text{ ft}}{1 \text{ sec}} \cdot \frac{1 \text{ mile}}{5280 \text{ ft}} \cdot \frac{60 \text{ sec}}{1 \text{ min}} \cdot \frac{60 \text{ min}}{1 \text{ hour}} = 51.14 \text{ miles/hr}$$

Scientific Notation

Very large and very small numbers can be expressed and calculated efficiently by means of scientific notation, a method that depends primarily on the use of exponents.

A number is said to be expressed in scientific notation when it is written as the product of an integral power of 10 and a rational number between 1 and 10. For example:

$$253 = 2.53 \times 10^2$$

The procedure for expressing a number using scientific notation is:

- Place the decimal point to the right of the first nonzero digit to obtain a number between 1 and 10.
- Multiply this number by a power of 10 whose exponent is equal to the number of places the decimal point was moved. The exponent is positive if the decimal point was moved to the left and negative if it was moved to the right.

By use of the laws of exponents and scientific notation, computations involving very large or very small numbers are simplified. For example:

$$\frac{378{,}000{,}000 \times 0.000004}{2000} = \frac{(3.78 \cdot 10^8)(4 \cdot 10^{-6})}{2 \cdot 10^3} = 7.56 \cdot 10^{-1} = 0.756$$

Engineering Notation

Engineering notation is a form of scientific notation in which the exponent of 10 is positive or negative 3 or a multiple of 3 expressed as follows:

10^9	giga-
10^6	mega-
10^3	kilo-
10^{-3}	milli-
10^{-6}	micro-
10^{-9}	nano-

For example:

10^3 watts = 1 kilowatt

10^{-6} watts = 1 microfarad

Significant Digits

Measurements, in contrast to discrete counts, often result in approximate numbers. As an indication of the accuracy of numbers, the scientific community has adopted the convention of significant digits.

The significant digits in a number are:

- Nonzero digits.
  In 241: 2, 4, and 1 are significant.
- A zero between 2 significant digits.
  In 1087: 1, 0, 8, and 7 are significant.
- A zero that is terminal on the right side of a decimal.
  In 1.0: 1 and 0 are significant.
- A zero that is known to be reliable or in some way significant.
  In 530: 5 and 3 are significant, 0 may not be significant unless proven so.

Computational Accuracy and Precision

In any computation involving sums and differences of approximations, the number with the smallest number of digits following the decimal point determines the number of decimal places to be used in the answer.

For example, for this sum:

$$\begin{array}{r} 2883.00 \\ 43.46 \\ +\ 0.1376 \\ \hline 2926.5976 \end{array}$$

the answer should be reported as 2927. Since 2883 has been rounded to the nearest integer, the answer should also be rounded to the nearest integer.

With products or quotients, the *accuracy* of the result depends on the number of significant digits in the component measurements. Thus, the number of significant digits to be retained in the result is the smallest number of significant digits in any of the components.

When comparing the *precision* of two or more approximate numbers, the number with the most significant digits to the right of the decimal point is the most precise.

Rounding

In performing computations, it is recommended that rounding be done after the calculation has been performed. This practice incurs the least possible error. The rules of rounding are:

- Add 1 if the succeeding digit is more than 5.
- Leave it unchanged if the succeeding digit is less than 5.
- If the succeeding digit is exactly 5, round off the number so that the final digit is even.

Exponents and Logarithms

Laws of Exponents

Exponents provide a shorthand method of writing the product of several like factors. If b is any number and n is a positive integer, the product of b multiplied by itself n times is denoted symbolically:

$$\underbrace{b \cdot b \ldots b}_{n \text{ factors}} = b^n$$

where n is the exponent and b is the base. It is read "b to the nth power." This definition can be extended to include exponents other than positive integers.

LAW I (Multiplication): When multiplying two or more factors with the same base, add the exponents and place the sum as an exponent on the same base.

$$A^m \cdot A^n = A^{m+n}$$

LAW II (Division): When dividing two quantities with the same base, subtract the exponent of the divisor from the exponent of the dividend and place the difference as an exponent on the same base.

$$A^m \div A^n = A^{m-n}$$

LAW III (Raising powers to powers): When raising a number with an exponent to a power, multiply the exponents and place the product as an exponent on the same base.

$$(A^m)^n = A^{mn}$$

ZERO EXPONENT: Any value (except 0) raised to the zero power is equal to 1.

$$A^0 = 1$$

NEGATIVE EXPONENT: A quantity raised to a negative power is equal to the reciprocal of that quantity with the corresponding positive exponent.

$$A^{-4} = \frac{1}{A^4}$$

SQUARE ROOT:

$$\sqrt{A} = A^{\frac{1}{2}} = A^{0.5}$$

Logarithms

Logarithms are exponents and behave accordingly. Hence, multiplication problems become addition of logarithms, division becomes subtraction, raising to a power becomes multiplication, and computing a root becomes division.

To perform calculations by means of logarithms, only a table of logarithms and a basic knowledge of algebra is needed.

DEFINITION: The logarithm of a number N to the base a is the exponent x to which the base must be raised to equal the number N.

$\log_a N$ = the logarithm of N to the base a

Then the definition states:

Properties of Logarithms

PROPERTY I: The logarithm of a product is equal to the sum of the logarithms of its factors.

PROPERTY II: The logarithm of a quotient is equal to the logarithm of the numerator minus the logarithm of the denominator.

PROPERTY III: The logarithm of the kth power of a number equals k times the logarithm of the number.

PROPERTY IV: The logarithm of the qth root of a number is equal to the logarithm of the number divided by q.

Common Logarithms

Logarithms that use 10 as a base are called *common logarithms*. Usually, these are written simply as log x, without any base indicated.

Note: Log 1 = 0 and log 10 = 1, since $10^0 = 1$ and $10^1 = 10$.

Thus, the integer portion of the logarithm of any number between 1 and 10 will be 1. Similarly, the integer portion of the logarithm of any number greater than 10 and less than 100 will be 2, and so on.

The integer portion of a logarithm is called the *characteristic*. The decimal portion is called the *mantissa*.

When finding the logarithm of a number using logarithmic tables, the characteristic is the exponent of 10, and the mantissa corresponding to the first part is taken from the table of logarithms. For example:

$$\begin{aligned}\text{Log } 52 &= \log(5.2 \times 10)\\ &= \log 5.2 + \log 10\\ &= 0.7160 + 1\\ &= 1.7160\end{aligned}$$

This same rule applies to numbers less than 1, although a characteristic with a negative value can be written as a digit with a line over it instead of as a negative number. The characteristic is normally written after the mantissa with a negative sign between them.

Prior to the popular use of calculators, lengthy tables were required to compute logarithms, especially when working backward to solve the original value of a logarithm (called an *antilog*). If a table is used, the number must be restated as a decimal minus a whole number. To use the previous problem as an example:

$$\begin{aligned}&\text{Find the antilog of } -2.2725:\\ &= \text{antilog } (0.7275 - 3)\\ &= \text{antilog } 0.7275 \times \text{antilog } (-3)\\ &= 5.34 \times 10^{-3}\\ &= 0.00534\end{aligned}$$

Natural Logarithms

The *natural*, or *Napierian*, system of logarithms has as its base the irrational number e, whose decimal expansion is 2.71828. Natural logarithms are denoted by the symbols *ln x* or *logex*. Natural logs occur in nearly all natural mathematical relationships.

Algebra

Algebra is a generalized form of arithmetic. It uses representative symbols and a few fundamental principles for reducing, or transforming, equations in order to determine a specific number in a specific problem.

Algebra uses signed numbers (+ or –) to identify positive or negative numbers.

Equations are the primary unit of expression of mathematical thought in algebra. Letters are used to represent numbers. By naming letters and mathematical symbols, short algebraic statements replace lengthy verbal statements.

The four steps of algebraic problem solving are:

1. *Representation* of unknowns by letters.
2. *Translation* of relationships about unknowns into equations.
3. *Solution* of equations to find the values of unknowns.
4. *Verification* of the value found to see if it satisfies the original problem.

Laws and Assumptions

Commutative Law: The result of addition or multiplication is the same in whatever order the terms are added or multiplied.

Associative Law: The sum of three or more terms, or the product of three or more factors, is the same in whatever manner they are grouped.

Distributive Law: The product of an expression of two or more terms by a single factor is equal to the sum of the products of each term of the expression by the single factor.

Algebra assumes:

- The sign preceding the product of two quantities is plus if the quantities have like signs, minus if unlike signs.
- Quantities equal to the same thing or equal things are equal.
- If equals are added to equals, the sums are equal.
- If equals are subtracted from equals, the remainders are equal.
- If equals are multiplied by equals, the products are equal.
- If equals are divided by equals, the quotients are equal. Division by zero is not permissible.
- If equals are raised to the same power, the results are equal.
- Like roots of both members of an equation are equal.
- A quantity may be substituted for its equal in any expression.

Order of Operations in Algebra

Order of Operations

The following is the order in which the operations in an algebraic expression are to be performed:

First, perform all operations within grouping symbols.
Second, perform all multiplications and divisions.
Third, perform all additions and subtractions.

Equations

Linear (one unknown, X):

$$AX + B = 0$$

Representation = straight line

Quadratic (one variable, X):

$$AX^2 + BX + C = 0$$

Representation = parabola

Quadratic (two variables, X and Y):

$$AX^2 + BXY + XY^2 + DX + EY + F = 0$$

Representation = curve

TRIGONOMETRY

Laws of Cosines and Sines

These two laws give the relations between the sides and angles of any plane triangle. In any plane triangle with vertices A, B, and C, and sides opposite a, b, and c, respectively, the following relations apply:

Law of Cosines

$$a^2 = b^2 + c^2 - \frac{2bc}{\cos A}$$

$$b^2 = a^2 + c^2 - \frac{2ac}{\cos B}$$

$$c^2 = a^2 + b^2 - \frac{2ab}{\cos C}$$

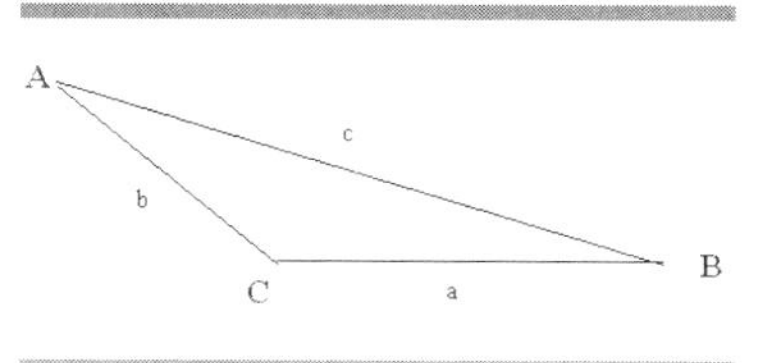

FIGURE 3.

Law of Sines

$$\frac{\sin A}{a} = \frac{\sin B}{b} = \frac{\sin C}{c}$$

or

$$\frac{\sin A}{\sin B} = \frac{a}{b} \quad \frac{\sin A}{\sin C} = \frac{a}{c} \quad \frac{\sin B}{\sin C} = \frac{b}{c}$$

These laws apply to any triangle, right-angled or not. In case the angle is between 90 degrees and 180 degrees, as in the case of angle C in the diagram above, then these rules can be applied:

$$\sin 0 = \sin (180 - 0) = 0.00$$
$$\sin 120 = \sin (180 - 120) = \sin 60 = 0.866$$
$$\cos 120 = -\cos (180 - 120) = -\cos 60 = -0.500$$
$$\cos 0 = -\cos(180 - 0) = 1.000$$

Solutions of Triangle Problems

Procedure in solving a triangle:

1. Construct a figure.
2. Label the figure, introducing single letters to represent unknown angles or lengths.
3. Outline the solution and clearly indicate the formulas that are used.
4. Arrange a good computation schematic and perform the arithmetic.

Engineering Economy

Annual Payment (from a Future Value)

$$A = F\left(\frac{i}{(1 + i)^n - 1}\right)$$

Terms and Units

A = annual investment or payment ($)
F = future value ($)
i = interest rate (%)
n = number of years

Example

How much must be invested annually over a period of 8 years at 5% to be worth $10,000 at the end of the 8-year period?

$$A = F\left(\frac{i}{(1 + i)^n - 1}\right)$$

$$A = \$10{,}000\left(\frac{0.05}{(1 + 0.05)^8 - 1}\right)$$

$$A = \$10{,}000\left(\frac{0.05}{1.477 - 1}\right)$$

$$A = \$10{,}000(0.105)$$

$$A = \$1050$$

An annual investment of $1050 for each of the next 8 years will be worth $10,000 at the end of the 8-year period.

Present Value of Future Dollar Amount

$$P = F(1 + i)^{-n}$$

Terms and Units

P = present value ($)
F = future value ($)
i = interest rate (%)
n = number of years

Example

What is the present value of $2000 to be received in 6 years, if it is invested at 4%?

$$P = F(1 + i)^{-n}$$

$$P = \$2000(1 + 0.04)^{-6}$$

$$P = \$2000\left(\frac{1}{1.04}\right)^6$$

$$P = \$2000(0.790)$$

$$P = \$1580$$

The present value of $2000 to be received in 6 years, if it is invested at 4%, is $1580.

Future Value of a Present Amount

$$F = P(1 + i)^n$$

Terms and Units

F = future value ($)
P = present value ($)
i = interest rate (%)
n = number of years

Example

If $6000 is invested for 8 years at 10% interest per year, what will the investment be worth at the end of the 8 years?

$$F = P(1 + i)^n$$

$$F = \$6000(1 + 0.10)^8$$

$$F = \$6000(1.10)^8$$

$$F = \$6000(2.144)$$

$$F = \$12{,}864$$

The future value of $10,000 if invested for 8 years at 10% interest will be $12,864.

Future Value

$$F = A\left(\frac{(1+i)^n - 1}{i}\right)$$

Terms and Units

A = annual investment or payment ($)
F = future value ($)
i = interest rate (%)
n = number of years

Example

If annual investments of $400 are made for 6 years and invested at 8%, what will the value of the investment be at the end of the 6-year period?

$$F = A\left(\frac{(1+i)^n - 1}{i}\right)$$

$$F = \$400\left(\frac{(1+0.08)^6 - 1}{0.08}\right)$$

$$F = \$400\left(\frac{1.587 - 1}{0.08}\right)$$

$$F = \$400(7.336)$$

$$F = \$2934.40$$

The value of annual investments of $400 made for 6 years at 8% will be $2934.40 at the end of the 6-year period.

Present Value (of Annual Amounts)

$$P = A\left(\frac{(1+i)^n - 1}{i(1+i)^n}\right)$$

Terms and Units

P = present value ($)
A = annual investment or payment ($)
i = interest rate (%)
n = number of years

Example

How much must be invested today at 5% so that annual returns of $1000 will be made from the investment for 10 years?

$$P = A\left(\frac{(1+i)^n - 1}{i(1+i)^n}\right)$$

$$P = \$1000\left(\frac{(1+0.05)^{10} - 1}{0.05(1+0.05)^{10}}\right)$$

$$P = \$1000\left(\frac{1.629 - 1}{0.05(1.629)}\right)$$

$$P = \$1000\left(\frac{0.629}{0.081}\right)$$

$$P = \$7765$$

$7765 invested today at 5% will provide annual returns of $1000 in each of the next 10 years.

Annual Payment (from a Present Value)

$$A = P\left(\frac{i(1+i)^n}{(1+i)^n - 1}\right)$$

Terms and Units

A = annual investment or payment ($)
P = present value ($)
i = interest rate (%)
n = number of years

Example

How much will an investment of $10,000 yield annually over 5 years at an interest rate of 15%?

$$A = P\left(\frac{i(1+i)^n}{(1+i)^n - 1}\right)$$

$$A = \$10{,}000\left(\frac{0.15(1+0.15)^5}{(1+0.15)^5 - 1}\right)$$

$$A = \$10{,}000\left(\frac{0.15(2.011)}{2.011 - 1}\right)$$

$$A = \$10{,}000\left(\frac{0.302}{1.011}\right)$$

$$A = \$10{,}000(0.299)$$

$$A = \$2990$$

$10,000 invested at 15% for five years will yield an annual payment of $2990.

Ergonomics

Lifting Index

$$LI = \frac{LW}{RWL}$$

Terms and Units

LI = lifting index
LW = loading weight
RWL = recommended weight limit

Example

Calculate the lifting index for a task with the following factors:

LW = 65 lb
RWL = 35 lb

$$LI = \frac{LW}{RWL}$$

$$LI = \frac{65}{35}$$

$$LI = 1.857$$

NIOSH Lifting Equation (English Units)

$$RWL = LC \times HM \times VM \times DM \times AM \times FM \times CM$$

$$RWL\ (\text{lb}) = (51)(10/H)[1 - (0.0075\ |V - 30|)][0.82 + (1.8/D)](1 - 0.0032A)(FM)(CM)$$

Terms and Units

RWL = recommended weight limit
LC = load constant
HM = horizontal multiplier
VM = vertical multiplier
DM = distance multiplier
AM = asymmetric multiplier
FM = frequency multiplier
CM = coupling multiplier

Example

Calculate the recommended weight limit (RWL) in pounds for a given task with the following factors:

LC = 60 lb
H = 15 in
V = 25 in
D = 45 in
A = 30°

frequency = 6 lifts/min
work duration = 2 hr
coupling type = fair

$$RWL = LC \times HM \times VM \times DM \times AM \times FM \times CM$$
FM = 0.50 (from table)
CM = 0.95 (from table)

$$60 \times \frac{10}{H} \times (1 - (0.0075|V - 30|)) \times \left(0.82 + \frac{1.8}{D}\right) \times (1 - 0.0032A) \times FM \times CM$$

$$60 \times \frac{10}{H} \times (1 - (0.0075|25 - 30|)) \times \left(0.82 + \frac{1.8}{45}\right) \times (1 - (0.0032 \times 30)) \times 0.50 \times 0.95$$

$$60 \times 0.667 \times 0.962 \times 0.860 \times 0.904 \times 0.50 \times 0.95$$

$$RWL = 14.217\ \text{lb}$$

NIOSH Lifting Equation (Metric Units)

$$RWL = LC \times HM \times VM \times DM \times AM \times FM \times CM$$

$$RWL\ (\text{kg}) = (23)(25/H)[1 - (0.003\ |V - 75|)][0.82 + (4.5/D)](1 - 0.0032A)(FM)(CM)$$

Terms and Units

RWL = recommended weight limit
LC = load constant
HM = horizontal multiplier
VM = vertical multiplier
DM = distance multiplier
AM = asymmetric multiplier
FM = frequency multiplier
CM = coupling multiplier

Example

Calculate the recommended weight limit (RWL) in kilograms for a given task with the following factors:

LC = 30 kg
H = 28 cm
V = 60 cm
D = 86 cm
A = 105°

frequency = 3 lifts/min
work duration = 4 hr
coupling type = poor

$$RWL = LC \times HM \times VM \times DM \times AM \times FM \times CM$$
$$FM = 0.55 \text{ (from table)}$$
$$CM = 0.90 \text{ (from table)}$$

$$30 \times \frac{25}{H} \times (1 - (0.003|V - 75|)) \times \left(0.82 + \frac{4.5}{D}\right) \times$$
$$(1 - 0.0032A) \times FM \times CM$$

$$30 \times \frac{25}{28} \times (1 - (0.003|60 - 75|)) \times \left(0.82 + \frac{4.5}{86}\right) \times$$
$$(1 - (0.0032 \times 105)) \times 0.55 \times 0.90$$

$$30 \times 0.892 \times 0.955 \times 0.872 \times 0.664 \times 0.55 \times 0.90$$

$$RWL = 7.324 \text{ kg}$$

General Science

Lower Flammability Limit for Mixtures

$$LFL_m = \frac{1}{\frac{f_1}{LFL_1} + \frac{f_2}{LFL_2} + \ldots \frac{f_n}{LFL_n}}$$

Terms and Units

LFL_m = lower flammability limit of a mixture (%)
$F_{1\ldots n}$ = the decimal fraction of the substance present for materials 1 . . . n (unitless)
$LFL_{1\ldots n}$ = the lower flammability limits for materials 1 . . . n (%)

Example

Determine the lower flammability limit of a mixture containing 25% toluene, 30% ethyl ether, and 45% acetone. The *LFL*s are: toluene 1.1%, ethyl ether 1.9%, and acetone 2.5%.

$$LFL_m = \frac{1}{\frac{f_1}{LFL_1} + \frac{f_2}{LFL_2} + \ldots \frac{f_n}{LFL_n}}$$

$$LFL_m = \frac{1}{\frac{0.25}{1.1\%} + \frac{0.30}{1.9\%} + \ldots \frac{0.45}{2.5\%}}$$

$$LFL_m = \frac{1}{0.565\%}$$

$$LFL_m = 1.769\%$$

Parts per Million: Pressure/Pressure

$$ppm = \frac{P_v}{P_{atm}} \times 10^6$$

Terms and Units

P_{atm} = atmospheric pressure (mm Hg)
ppm = parts per million (*ppm*)
10^6 = conversion factor to express answer in *ppm*
P_v = vapor pressure of agent or contaminant at a specified temperature (mm Hg)

Example

Calculate the equilibrium (saturation) concentration for xylene at 68°F and normal atmospheric pressure. The vapor pressure of xylene at 68°F is 9 mm Hg.

$$ppm = \frac{P_v}{P_{atm}} \times 10^6$$

$$ppm = \frac{9 \text{ mm Hg}}{760 \text{ mm Hg}} \times 10^6$$

$$ppm = 11{,}842\ ppm$$

Parts Per Million to mg/m³ Conversion

$$ppm = \frac{\text{mg/m}^3 \times 24.45}{\text{MW}}$$

Terms and Units

ppm = parts per million (*ppm*)
MW = molecular weight of contaminant
24.25 = molar volume of any gas or vapor at NTP
mg/m^3 = milligrams of contaminant per cubic meter of air (mg/m^3)

Example

Convert 2570 mg/m^3 of acetone to *ppm*. The MW of acetone is 58.1.

$$ppm = \frac{\text{mg/m}^3 \times 24.45}{\text{MW}}$$

$$ppm = \frac{2570 \text{ mg/m}^3 \times 24.45}{58.1 \text{ g}}$$

$$ppm = 1081 \; ppm$$

General Gas Law

$$\frac{P_1V_1}{nRT_1} = \frac{P_2V_2}{nRT_2}$$

Terms and Units

n = number of moles of gas or vapor (unitless)
T = absolute temperature Rankine (°R); Kelvin (K)
V_1; V_2 = gas or vapor volume under conditions 1 and 2 (1)
R = gas constant for air (0.082 liters atmosphere/ moles K)
P_1; P_2 = gas or vapor pressure under conditions 1 and 2 (mm Hg)

Example

An acetone/air mixture is collected in an impervious 5-L gas-sampling bag at 80°F and 710 mm Hg. Calculate the new volume for the bag when it is at NTP.

$$\frac{P_1V_1}{nRT_1} = \frac{P_2V_2}{nRT_2}$$

$$\frac{720 \text{ mm Hg} \times 5 \text{ liters}}{(80°\text{F} + 460)} = \frac{760 \text{ mm Hg} \times V_2}{(77°\text{F} + 460)}$$

$$\frac{3600}{540} = \frac{760\ V_2}{537}$$

$$V_2 = 4.7\ L$$

TLV of Mixture (Liquid Composition)

$$TLV_{mix} = \frac{1}{\frac{F_1}{TLV_1} + \frac{F_2}{TLV_2} + \ldots \frac{F_n}{TLV_n}}$$

Terms and Units

$F_{1 \ldots n}$ = weight % of chemical in liquid (unitless decimal)
$TLV_{1 \ldots n}$ = threshold limit value of the contaminant (mg/m³)
TLV_{mix} = threshold limit value of a mixture of chemical with additive effects (mg/m³)

Example

Determine the threshold limit value of a mixture containing 25% toluene, 40% xylene, and 15% hexane. The TLVs are: toluene 188 mg/m³, xylene 434 mg/m³, and hexane 176 mg/m³.

$$TLV_{mix} = \frac{1}{\frac{F_1}{TLV_1} + \frac{F_2}{TLV_2} + \ldots \frac{F_n}{TLV_n}}$$

$$TLV_{mix} = \frac{1}{\frac{0.25}{188 \text{ mg/m}^3} + \frac{0.40}{434 \text{ mg/m}^3} + \frac{0.15}{176 \text{ mg/m}^3}}$$

$$TLV_{mix} = \frac{1}{0.00310 \text{ mg/m}^3}$$

$$TLV_{mix} = 323 \text{ mg/m}^3$$

Heat Stress

Wet-Bulb Globe Temperature Index (No Solar Load)

WBGT = 0.7WB + 0.3GT

Terms and Units

WBGT = wet-bulb globe temperature index, °C
WB = wet-bulb temperature, °C
GT = globe temperature, °C

Example

Two temperature measurements are taken of the air surrounding an indoor task:

wet-bulb temperature = 30°C
globe temperature = 15°C

Calculate the wet-bulb globe temperature index.

WBGT = 0.7WB + 0.3GT

WBGT = 0.7(30) + 0.3(15)

WBGT = 21.0 + 4.5

WBGT = 25.5°C

The wet-bulb globe temperature is 25.5°C.

Wet-Bulb Globe Temperature Index (Solar Load)

WBGT = 0.7WB + 0.2GT + 0.1DB

Terms and Units

WBGT = wet-bulb globe temperature index, °C
WB = wet-bulb temperature, °C
GT = globe temperature, °C
DB = dry-bulb temperature, °C

Example

Three temperature measurements are taken of the air surrounding an outdoor task on a sunny day:

wet-bulb temperature = 30°C
globe temperature = 15°C
dry-bulb temperature = 45°C

Calculate the wet-bulb globe temperature index.

WBGT = 0.7WB + 0.3GT + 0.1DB

WBGT = 0.7(30) + 0.3(15) + 0.1(45)

WBGT = 21.0 + 4.5 + 4.5

WBGT = 30.0 °C

The wet-bulb globe temperature is 30.0°C.

Noise

Distance and Sound Pressure Level

$$dB_1 = dB_0 + 20 \log \frac{d_0}{d_1}$$

Terms and Units

dB_0 = noise level at distance d_0 (dB)
dB_1 = noise level at distance d_1 (dB)
d_0, d_1 = distance (any consistent units, e.g., m)

Example

A noise survey at a work station 6 m from a noise shows a reading of 124 dB. What will be the reading if the work station is moved further away, to 12 m?

$$dB_1 = dB_0 + 20 \log \frac{d_0}{d_1}$$

$$dB_1 = 124 \text{ dB} + 20 \log\left(\frac{6}{12}\right)$$

$$dB_1 = 124 + (-6)$$

$$dB_1 = 118$$

Sound Intensity Level

$$L_1 = 10 \log \frac{I}{I_0} \text{ dB}$$

Terms and Units

L_1 = sound pressure level (dB)
I = sound intensity (W/m^2)
I_0 = reference sound intensity (W/m^2)

Example

What is the sound power level for a measured intensity of 5^{-10} W/m^2 where the reference intensity is 14^{-8} W/m^2?

$$L_1 = 10 \log \frac{I}{I_0} \text{ dB}$$

$$L_1 = 10 \log\left(\frac{5^{-10}}{14^{-8}}\right)$$

$$L_1 = 22 \text{ dB}$$

Sound Pressure Level

$$L_P = 20\left(\log \frac{P}{P_0}\right)$$

Terms and Units

L_P = sound pressure level (dB)
P = measured sound pressure (Pa)
P_0 = reference sound pressure (Pa)

Example

What is the sound pressure level in dB when the measured sound pressure is 0.2 Pa?

$$L_P = 20\left(\log \frac{P}{P_0}\right)$$

$$L_P = 20\left(\log \frac{0.15\text{Pa}}{0.00002 \text{ Pa}}\right)$$

$$L_P = 78 \text{ dB}$$

Total Sound Pressure Level

$$L_{Pt} = 10 \log\left(\sum_{i=1}^{N} 10^{\frac{L_{Pi}}{10}}\right)$$

Terms and Units

L_{Pt} = total sound pressure level generated by N sources (dB)
L_{Pi} = individual sound pressure level of ith source (dB)
N = number of sound pressure levels

Example

Three machines will be situated in close proximity. Given their individual sound pressure levels, L_P, of 42, 54, and 78 dB, what is the approximate total sound pressure level, L_{Pt}?

$$L_{Pt} = 10 \log\left(\sum_{i=1}^{N} 10^{\frac{L_{Pi}}{10}}\right)$$

$$L_{Pt} = 10 \log\left(10^{\frac{42}{10}} + 10^{\frac{54}{10}} + 10^{\frac{78}{10}}\right)$$

$$L_{Pt} = 78 \text{ dB}$$

TWA Calculated from % Dose

$$\text{TWA} = 16.61 \log\left(\frac{D}{100}\right) + 90$$

Terms and Units

D = dose for noise exposure in percent
TWA = equivalent time-weighted average exposure in dB based on percent dose (D) for a shift

Example

The percent dose for a shift is determined to be 50%. What is the TWA exposure?

$$\text{TWA} = 16.61 \log\left(\frac{D}{100}\right) + 90$$

$$\text{TWA} = 16.61 \log\left(\frac{50}{100}\right) + 90 \text{ dBa}$$

$$\text{TWA} = 85 \text{ dBa}$$

Mechanics

Bending Moment (Concentrated Load at Center)

$$M = \frac{Pl}{4}$$

Terms and Units

M = maximum bending moment (ft-lb)
P = concentrated load applied at center of beam (lb)
l = length of beam (ft)

Example

A 5-ton hoist is suspended at the midpoint of a 12-ft beam that is supported at each end. What is the maximum bending moment in this beam? (*Note:* Neglect the weight of the beam.)

P = 5 tons = 10,000 lb
l = 12 ft

$$M = \frac{Pl}{4}$$

$$M = \frac{(10{,}000 \text{ lb})(12 \text{ ft})}{4}$$

$$M = 30{,}000 \text{ ft-lb}$$

Bending Moment (Concentrated Load Off-Center)

$$M = \frac{Pab}{l}$$

Terms and Units

M = the maximum bending moment (ft-lb)
P = concentrated load on beam (lb)
$a \neq b$ = respective distances from left and right supports of beam (ft)
l = length of beam (ft)

Example

A 10-ft beam supported at each end is loaded with a weight of 1400 lb at a point 3 feet left of its center.

What is the maximum bending moment in this beam? (*Note:* Neglect the weight of the beam.)

$P = 1400$ lb
$a = 3$ ft
$b = 7$ ft
$l = 10$ ft

$$M = \frac{Pab}{l}$$

$$M = \frac{(1400 \text{ lb})(3 \text{ ft})(7 \text{ ft})}{10 \text{ ft}}$$

$$M = 29{,}400 \text{ ft-lb}$$

Bending Moment (Uniform Loading)

$$M = \frac{wl^2}{8}$$

Terms and Units

M = maximum bending moment (ft-lb)
w = uniform weight loading per foot of beam (lb/ft)
l = length of beam (ft)

Example

A 12-ft beam supported at two points, one at each end, is uniformly loaded at a rate of 400 lb/ft. What is the maximum bending moment in this beam?

$w = 400$ lb/ft
$l = 12$ ft

$$M = \frac{wl^2}{8}$$

$$M = \frac{(400 \text{ lb/ft})(12 \text{ ft})^2}{8}$$

$$M = 7200 \text{ ft-lb}$$

Capacitance, (Electrical in Parallel Circuits)

$$C_{parallel} = C_1 + C_2 + \ldots C_n$$

Terms and Units

$C_{parallel}$ = equivalent capacitance of n capacitors wired in parallel in a circuit (farads or microfarads)
$C_{1,2\ldots n}$ = equivalent capacitance of individual capacitors in a circuit (farads or microfarads)

Example

Find the equivalent capacitance of a circuit having 3 capacitors wired in parallel if

$C_1 = 10\ \mu F$
$C_2 = 75\ \mu F$
$C_3 = 54\ \mu F$

$$C_{parallel} = C_1 + C_2 + C_3$$

$$C_{parallel} = (10 + 75 + 54)\ \mu F$$

$$C_{parallel} = 139\ \mu F$$

Capacitance, (Electrical in Series Circuits)

$$\frac{1}{C_{series}} = \frac{1}{C_1} + \frac{1}{C_2} + \ldots \frac{1}{C_n}$$

Terms and Units

C_{series} = equivalent capacitance in farads of n capacitors wired in series (farads or microfarads)
$C_{1,2\ldots n}$ = individual capacitance (farads or microfarads)

Example

What is the equivalent capacitance of a circuit having 4 capacitors wired in series if

$C_1 = 80\ \mu F$
$C_2 = 35\ \mu F$
$C_3 = 42\ \mu F$
$C_4 = 72\ \mu F$

$$\frac{1}{C_{series}} = \frac{1}{C_1} + \frac{1}{C_2} + \frac{1}{C_3} + \frac{1}{C_4}$$

$$\frac{1}{C_{series}} = \frac{1}{80\ \mu F} + \frac{1}{35\ \mu F} + \frac{1}{42\ \mu F} + \frac{1}{72\ \mu F}$$

$$C_{series} = 0.079\ \mu F$$

Electrical Resistance Length and Area of Conductor

$$R = \rho \frac{L}{A}$$

Terms and Units

R = resistance (ohms)
ρ = resistivity* (ohm-ft)
L = length of conductor (ft)
A = conductor's cross-sectional area (ft)

**Note:* ρ/A is resistivity in ohm/ft

Example

What is the resistance of 50 ft of #14 gauge copper wire? (*Note:* From table, we find the resistivity, ρ, for #14 gauge copper wire to be 2.525 ohms/1000 ft.)

$\rho/A = 2.525/1000 \text{ ft} = 0.002525 \text{ ohms/ft}$
$L = 50 \text{ ft}$

$$R = \rho \frac{L}{A}$$

$$R = (0.002525 \text{ ohm/ft})(50 \text{ ft})$$

$$R = 0.126 \text{ ohms}$$

Flow Calculation for Water Supplies

$$Q_2 = Q_1 \left[\frac{(S - R_2)^{0.54}}{(S - R_1)^{0.54}} \right]$$

Terms and Units

R = residual pressure for predicted flow rate (psi)
Q_1 = known flow rate from a group of open hydrants with a measured residual pressure, R_1 (gpm)
Q_2 = predicted flow rate at a different residual pressure, R_2 usually 20 psi (gpm)
S = static pressure in a hydrant system with no flow from any hydrants (psi)
R_1 = the residual pressure measured at a nonflowing hydrant during a flow test (psi)

Example

The static pressure of a hydrant system is 75 psi. When the flow from several hydrants combined reaches 3500 gpm, the pressure at the nonflowing hydrant reduces to 30 psi. (a) What is the expected flow for a residual pressure of 40 psi? (b) What would the static pressure need to be to obtain a flow of 5000 gpm at 30 psi?

Question a

$S = 75 \text{ psi}$
$Q_1 = 3500 \text{ gpm}$
$R_1 = 30 \text{ psi}$
$R_2 = 40 \text{ psi}$

$$Q_2 = Q_1 \left[\frac{(S - R_2)^{0.54}}{(S - R_1)^{0.54}} \right]$$

$$Q_2 = 3500 \left[\frac{(75 \text{ psi} - 40 \text{ psi})^{0.54}}{(75 \text{ psi} - 30 \text{ psi})^{0.54}} \right]$$

$$Q_2 = 3500 \left(\frac{(35)^{0.54}}{(45)^{0.54}} \right)$$

$$Q_2 = 3500 \left(\frac{38.18}{7.81} \right)$$

$$Q_2 = 3500(4.89)$$

$$Q_2 = 17{,}115 \text{ gpm}$$

Question b

It is known from question *a* that

$$\frac{Q_1}{(S - R_1)^{0.54}} = \frac{Q_2}{(S - R_2)^{0.54}} = \text{constant}$$

The constant for this system is 550.
The equation can be rewritten as

$$(S - R_2)^{0.54} = \frac{Q_2}{\text{constant}}$$

$$S = \left(\frac{Q_2}{\text{constant}} \right)^{1.85} + R_2$$

$$S = \left(\frac{5000}{550} \right)^{1.85} + 30$$

$$S = 89.4 \text{ psi}$$

Hydraulic Flow-Pressure Relationship

$$\frac{Q_1}{Q_2} = \frac{\sqrt{P_1}}{\sqrt{P_2}}$$

Terms and Units

P = pressure in psi
Q = flow in gpm

Example

A sprinkler system was hydraulically designed so that 1450 gpm at 30 psi was needed at the top of the riser to meet a certain storage arrangement demand. The warehouse manager would like to rearrange the storage in such a way that 2000 gpm will be required over the same area to satisfy the increase in the density of the demand. What will the pressure requirement be at the top of the riser to deliver the new flow requirement?

P_1 = 30 psi
Q_1 = 1450 gpm
Q_2 = 2000 gpm

$$P_2 = P_1(Q_2/Q_1)^2$$

$$P_2 = (30 \text{ psi})(2000 \text{ gpm}/1450 \text{ gpm})^2$$

$$P_2 = 57.1 \text{ psi}$$

Hydraulic Friction Loss Formula

$$P_d = \frac{4.52Q^{1.85}}{C^{1.85}d^{4.87}}$$

Terms and Units

Q = flow rate (gpm)
C = coefficient of friction
d = inside diameter of pipe (in)
P_d = pressure in lb per sq in per ft of pipe (psi)

Example

What is the friction loss in psi per ft for a 5-in pipe having a C of 85 with a 250 gpm flow?

Q = 250 gpm
C = 85
d = 5 in

$$p = 4.52Q^{1.85}/C^{1.85}d^{4.87}$$

$$p = (4.52)(250)^{1.85}/(85)^{1.85}(5)^{4.87}$$

$$p = 0.131 \text{ psi per foot}$$

Ideal Gas Law

$PV = nRT$

Terms and Units

P = absolute pressure (psia)
V = volume (ft^3)
n = the number of moles of gas (lb-moles)
R = the Universal Gas Constant (10.73 psi-ft^3/lb mole-°R)
T = absolute temperature (°R)

Example

A 40-lb cylinder of propane (C_3H_8 MW = 44) is mounted on a forklift truck that is garaged in a room having dimensions of 15 ft × 30 ft × 45 ft. If the cylinder is full (20 lb of gas in the tank) and the room temperature is 80°F at the time a leak develops, will a dangerous situation develop? Assume there is excellent air movement in the room to cause complete mixing of the gases with the air-handling system in a 100% recirculation mode with no leakage to the outside. The LEL for propane is 2.15%.

Or

Is the volume of 40 lb of propane at 80°F greater than 15 ft × 30 ft × 45 ft × 0.0215 = 435.4 ft^3?

V = ft^3 of 40 lb of propane at 80°F, 14.7 psia
n = 40 lb/44 lb/lb-mole = 0.9090 lb-moles
R = 10.73
T = 80°F + 460 = 540°R
P = 1 atm = 14.7 psia

$PV = nRT$

$V = (0.9090)(10.73)(540)/14.7$

$V = 358.3\ \text{ft}^3$

A dangerous situation.

Kinetic Energy

$$\text{K.E.} = \frac{mv^2}{2}$$

Terms and Units

K.E. = the kinetic energy (ft-lbf)
m = the mass of an object (slugs = W/g*)
v = the velocity of an object (ft/s)

**Note:* In order for the units of K.E. to come out correctly, if the weight (W) of an object is known [commonly stated as pounds (lbf)] it needs to be divided by g which has the value and units of 32.2 ft/s^2 to get the equivalent "slugs."

Example

How much work can be done by a 200-lb weight moving at a velocity of 60 mph?

$m = 200\ \text{lb} \div 32.2\ \text{ft/s}^2 = 6.2\ \text{slugs}$
$v = 60\ \text{mi/hr} \times 5280\ \text{ft/mi} \times \text{hr}/3600\ \text{s} = 88\ \text{ft/s}$

$\text{K.E.} = 1/2\ mv^2$

$\text{K.E.} = 1/2(3.1\ \text{slugs})(88\ \text{ft/s})^2$

$\text{K.E.} = 12{,}003.2\ \text{ft-lb}$

OHM's Law

$V = IR$

Terms and Units

V = voltage (v)
I = current (amps)
R = resistance (ohms)

Example

What is the current in an 18-gauge wire plugged into a 120-volt outlet that supplies a saw that has a resistance of 75 ohms?

V = 120 volts
R = 75 ohms

$I = V/R$

$I = 120/75$

$I = 1.6$ amps

Pipe Head Pressure

$$h_p = \frac{p}{w}$$

Terms and Units

h_p = pressure head (ft)
p = pressure (lbs/ft^2 or psf)
w = density (lbs/ft^3 or pcf)

Example

A pressure gauge at the base of a gravity tank reads 80 psi. How high is the level of the water in the tank?

$p = 80\ \text{psi} \times 144\ \text{si/sf} = 11{,}520\ \text{psf}$
$w = 62.4\ \text{pcf}$

$h_p = p/w$

$h_p = 11{,}520\ \text{psf}/62.4\ \text{pcf}$

$h_p = 184.6\ \text{ft}$

Potential Energy

$\text{P.E.} = mgh$

Terms and Units

P.E. = potential energy (ft-lb)
m = mass (slugs = $W\ (\text{lb})/g^{(\text{ft/s}^2)}$)
g = gravitational constant (32.2 ft/s^2)
h = height or elevation (ft)

Note: mg can be replaced by W, which simplifies the formula to P.E. = Wh.

Example

How much potential energy is stored in a 90,000-gal water gravity tank whose average elevation above ground level is 125 ft? (*Note:* Density of water = 62.4 lb/ft^3; 7.48 gal = 1 ft^3)

W = (62.4 lb/ft^3)(90,000 gal) ÷ 7.48 gal/ft^3 = 750,802 lb
h = 125 ft

P.E. = mgh

P.E. = mgh = Wh

P.E. = (750,802)(125)

P.E. = 6006.4 ft-lb

Friction

$F = \mu N$

Terms and Units

F = force (lb)
μ = the coefficient of friction, which is dimensionless
N = the force acting normal (i.e., perpendicular) to the surface in pounds (lbf)

Example

How much force is required to move a 500-lb box if the static coefficient of friction between it and the horizontal surface upon which it is resting is 0.90?

F = (0.90)(500 lb) = 450 lb

Or

A 300-lb box is placed on a plane incline at a 30° angle from horizontal. The plane has a static coefficient of friction of 0.80. What is the minimum push or pull force required to move the box down the plane?

Force required to overcome friction:

F = (0.80)(300 × cos 30)

= (0.80)(300 lb × 0.866)

= (0.80)(259.8 lb) = 207.84 lb

Note: Due to gravity, there is a preexisting force acting in the downward direction of the incline, which is equal to W sin 0 or 300 × 0.80 = 240 lb.

Therefore, the push/pull force required to begin the box's movement is 240 lb – 207 lb = 33 lb.

Force (Newton's Second Law)

$F = ma$

Terms and Units

F = force (lb)
m = mass (slugs; 1 slug = 32.2 lb) = W/g_c (lb ÷ 32.2 ft/s^2)
a = acceleration (ft/s^2)

Example

How much force must a seat belt be capable of withstanding to safely restrain a 130-lb woman when her car comes to a sudden (i.e., 1.0-second) stop if initially traveling at 60 mph? [Assume a safety factor (S.F.) of 3.]

m = 130 lb ÷ 32.2 ft/s^2 = 4.03 slugs
$a = v/t$ = (60 m/hr)(5280 ft/m)(hr/3600 s) ÷ 1.0 s = 88 ft/s^2

$F = ma$ × S.F. of 3

F = (4.03)(88) lb × 3 = 354.64 lb

Weight

$W = mg$

Terms and Units

W = weight in pounds—force (lb)
m = mass in slugs (1 slug = 32.2 lb)
g = acceleration of gravity (lb ÷ 32.2 ft/s^2)

Example

An object weighs 100 lb when measured at sea level. What is its mass?

W = 100 lb
g = 32.2 ft/s^2

$W = mg$

$m = W/g$

m = 100 lbf ÷ 32.2 ft/s^2 = 3.01 slugs

Resistance (Electrical in Series Circuits)

$R_{series} = R_1 + R_1 + \ldots R_n$

Terms and Units

R_{series} = equivalent resistance of n resistances wired in series (ohms)
$R_{1,2\ldots n}$ = resistance in ohms (ohms)

Example

There are three resistors in a series circuit having a 50-volt battery as a power source. Resistor A is a resistance of 30 ohms, resistor B is 25 ohms, and resistor C is 15 ohms. What is the voltage drop across resistor B?

$\Delta V_B = R_b I$

$R_B = 25$ ohm

$I = V/R_{series} = V/(R_A + R_B + R_C)$

$I = 50/23 = 2.17$ amps

$\Delta V_B = (25 \text{ ohms})(2.17 \text{ amps})$

$\Delta V_B = 54.24$ volts

Power (Electric)

$P = VI$

Terms and Units

P = power (watts)
V = voltage (volts)
I = current (amps)

Example

An electric heater is rated as 10.8 kw wired at 220 volts. What current draw will dictate minimum wire sizing and fuse protection for this appliance?

$P = 10.8$ kw = 10,800 watts
$V = 220$ volts

$P = VI$

$I = P/V = 10{,}800 \text{ watts}/220 \text{ volts}$

$I = 49.1$ amps

Ventilation

Effective Ventilation

$$Q' = \frac{Q}{K}$$

Terms and Units

Q' = effective ventilation rate (cfm)
Q = actual ventilation rate (cfm)
K = safety factor, ranging from 1 to 10

Example

What would be the actual ventilation rate if it were determined that 4000 cfm would control a contaminated concentration to the permissible exposure limit (PEL) under ideal conditions? Assume a safety factor of 4.

$$Q' = \frac{Q}{K}$$

$Q = Q'K$

$Q = 4000 \text{ cfm} \times 4$

$Q = 16{,}000$ cfm

Total Pressure in a Duct

$TP = VP + SP$

Terms and Units

TP = total pressure within a duct (″wg)
VP = velocity pressure within a duct (″wg)
SP = static pressure within a duct (″wg)

Example

Calculate the static pressure of a duct if the velocity pressure is 5 ″wg and the total pressure on the inlet side of the fan measures 0.9 ″wg.

$TP = SP + VP$

$SP = TP - VP$

$SP = -0.9 - 5.0$

$SP = -5.9$ ″wg

Note: The static and total pressures on the inlet/upstream/intake side of a fan are always negatively signed; therefore, the negative designation indicates the duct pressures are below atmospheric pressure.

Hood Flow Rate and Static Pressure

$$Q = 4005 C_e A \sqrt{SP}$$

Terms and Units

Q = volumetric flow rate (cfm)
A = area of the duct (ft^2)
4005 = constant
SP = hood static pressure (″wg)
C_e = hood entry coefficient (unitless)

Example

A static pressure tap was installed on an 8-in-diameter duct (A = 0.35 ft^2) with a plain opening. The manometer read 4.0 ″wg. What was the flow rate of the hood?

$$Q = 4005 C_e A \sqrt{SP}$$

$$Q = 4005(0.72)\ 0.35\ \text{ft}^2 \sqrt{4}$$

$$Q = 2018.52\ \text{cfm}$$

Note: C_e for a flanged hood is 0.82 or 82% efficient, and for a plain opening it is 0.72 or 72% efficient.

Coefficient of Entry

$$C_e = \sqrt{\frac{VP}{|SP_h|}}$$

Terms and Units

C_e = hood entry coefficient (unitless)
VP = velocity pressure (″wg)
$|SP_h|$ = absolute value of the hood static pressure (″wg)

Example

What is the hood entry coefficient for a flanged hood opening if the average velocity pressure for the duct is 0.50 ″wg and the hood static suction measures 2.15 ″wg?

$$C_e = \sqrt{\frac{VP}{|SP_h|}}$$

$$C_e = \sqrt{\frac{0.50\ \text{″wg}}{2.15\ \text{″wg}}}$$

$$C_e = \sqrt{0.23}$$

$$C_e = 0.48$$

INDEX

B

C

F

J

K

O

P

Q

R